CW00507760

The Birds of Durham

Edited by

Keith Bowey and Mark Newsome

Species texts by

Paul Anderson, Tony Armstrong, Chris Bell, Peter Bell, Keith Bowey,
Ian Mills, Fred Milton, Mark Newsome, David Raw and John Strowger

With the editorial assistance of

Tony Armstrong, Chris Bell, Peter Bell, Ian Mills and David Raw

DURHAM
BIRD
CLUB

Durham Bird Club

2012

© **Durham Bird Club 2012**

ISBN 978-1-874701-03-3

Front cover painting and other cover artwork by Geoff Watson.

All rights reserved. No part of this publication may be reproduced, stored in a retrieval system, or transmitted, in any form or by any means, electronic, mechanical, photocopying, recording or otherwise, without prior permission of the publishers.

The paper used for this book has been independently certified as coming from well-managed forests and other controlled sources according to the rules of the Forest Stewardship Council.

This book was printed and bound by Butler, Tanner & Dennis, an FSC certified company for printing books on FSC mixed paper in compliance with the chain of custody and on-products labelling standards.

Published with the support of the

Heritage Lottery Fund

Dedication

The year 2011 marked the sixtieth anniversary of the publication of George Temperley's *A History of the Birds of Durham* by the Natural History Society of Northumberland, Durham and Newcastle upon Tyne. It was the centenary of the birth of Fred Grey and sadly, it saw the untimely death of Brian Unwin. Together, they provided the main impetus for the founding of the Durham Bird Club. This book is dedicated to their memory and the work of those who went before, from Marmaduke Tunstall through Thomas Bewick, to John Hancock. This book is dedicated in recognition of their sterling work, the resonance of their achievements over the generations and the inspiration they have provided to those who are still drawn along in their wake.

Contents

		Page
Foreword		1
Introduction		1
The production process		2
About the book		4
Acknowledgements		7
List of abbreviations used in the text		10
Chapter 1	Durham Geography: Landscape, Habitats, Climate and Birdlife	11
Chapter 2	A Short Environmental History of County Durham	31
Chapter 3	The Study of Birds in County Durham	41
Chapter 4	A Short History of Durham Ornithology	71
Chapter 5	Birds of Durham - the Durham List	77
The Systematic List		89
Appendices		
Appendix 1	Category D and E species	937
Appendix 2	List of Authors	954
Appendix 2a	Researchers and Research Topics	955
Appendix 3	List of photographers and Illustrators	957
Appendix 4	Species sponsors and species supporters	959
Appendix 5	Map and gazetteer of bird watching sites	962
References		969
Index to bird species		1006

Foreword

Writing to the editors of *British Birds* in 1947, of his intention to produce a book about the birds of the county, George W. Temperley said, "*The County has been much neglected by ornithologists in the past. The only published list of its birds of any note was that compiled by Canon H. B. Tristram for the Victoria County History of Durham in 1905, now long out of date...I propose to show the changes which have taken place in the course of a century, as well as the present status of each species*". He duly and with great verve fulfilled his promise, with *A History of the Birds of Durham* published through the Natural History Society in 1951.

The publication of Temperley prompted a greater interest in the birds, not just of the north east, but of the land between the Rivers Tyne and Tees, the Land of the Prince Bishops, County Durham. However, it has taken over half a century to re-visit the subject in a comprehensive fashion; perhaps this is a measure of the enormity of the task. The years since Temperley was published saw a revolution in both ornithology and what is now called bird watching or, in modern parlance, 'birding'. County Durham has benefited from a slew of initiatives in this field. Amongst them: the birth of modern ecology (including the work of the University of Durham); the establishment of bird clubs covering the north and south east of the area, during the period 1959 to 1961 (the Tyneside, later the Northumberland and Tyneside Bird Club and the Teesmouth Bird Club); the establishment of its own organisation (the Durham Bird Club) in the mid-1970s; and, the creation of dedicated study groups (the Durham Upland Bird Study Group). In addition, those intervening years have seen the work and dedication of some truly great ornithologists and field naturalists. Nonetheless, until now, Durham has to a certain extent found itself in the equivalent modern position that George Temperley found it in 1947. Today, it is arguably, in ornithological terms, the most significant county in Britain not to possess a modern county avifauna. This is somewhat ironic as the work of the late George Temperley set something of a benchmark for such publications when it was published.

The creation of a new county avifauna has been a long-held aspiration of the Durham Bird Club. It was only with the support of the Heritage Lottery Fund for the development of *The Birds of Durham Heritage Project* (2010-2012) however, that a mechanism was developed by which this aspiration might be realised. *The Birds of Durham Heritage Project* was a multi-faceted project that set out to "*bring the wildlife of the past to the people of tomorrow*". A £30,000 Heritage Lottery Fund grant helped the project partners to: formulate a programme of public events, in the form of lectures and guided walks; create nine Bird Heritage Interpretation Panels; and, produce a suite of on-line resources, all of which told the tale of County Durham's bird heritage. Central to all of the above activities, the content of a new county avifauna, *The Birds of Durham*, was created. This authoritative book celebrates the sixty years of ornithological endeavour since the publication of the last complete overview of the county's birdlife. The Project was led by the Durham Bird Club with support from its Partners: the Durham Upland Bird Study Group, the Durham Wildlife Trust, the Natural History Society of Northumbria and the Teesmouth Bird Club.

Introduction

In the early 1930s, George Temperley was urged by George Bolam, author of The Birds of Northumberland published in 1932, to write up the history of the birds of Durham. On Bolam's death in 1934, his sister gave to Temperley volumes of Bolam's manuscript notes on his north country records. In the mid-1930s Temperley took up the task and in 1947 he published his 'call to arms' in *British Birds* asking for information and support for the work from ornithologists in the area. Temperley compiled the annual Ornithological Report for Durham and Northumberland for the years 1935 to 1956. In 1953, he was awarded the British Trust for Ornithology's Bernard Tucker medal for services to ornithology. In 1951, George Temperley published his ground- breaking book A History of the Birds of Durham. This effectively covered the period between Tristram's 1905 Durham list and 1950. Although published 60 years ago, it remained until now the only comprehensive review of all of the county's birdlife, the complete record of Durham's avifauna; the work is often known simply as Temperley.

The year 2011 was a significant year in many ways for the study of ornithology in the county of Durham. It marked the sixtieth anniversary of the publication of George Temperley's 'A History of the Birds of Durham'. It was one hundred years since the birth of Fred Grey, an inspirational figure who contributed in many ways to the development of bird-watching in County Durham and to the history of the Durham Bird Club. Sadly, it also marked the passing of one of the club's key founding members, the man perhaps most responsible for the formation of the club, Brian Unwin. The publication of Birds of Durham is intended to mark all of these events and to serve as a celebration of the achievements of all of these important ornithologists, and the birds of the county.

At the mid-point of the first decade of the 21st century, Durham, alongside perhaps Cumbria, was one of the few 'significant' English counties, in an ornithological sense, which had not had a full, modern account of its bird life produced. In the time since Temperley, some counties or areas have had one or even two such books published. To produce a high quality 'Temperley for the 21st century', and publish this in 2011 in celebration of sixty years since the publication of Temperley's work, was a long-term ambition of the Durham Bird Club. This book attempts to document the history and status of every bird species ever recorded in the 'county' of Durham (i.e. between Tyne and Tees).

It is hoped that this publication will serve as a benchmark statement on the status of birds in County Durham for the foreseeable future. Considerable effort has been made to ensure that the information it contains is correct, but as with any work of this magnitude, some mistakes and omissions will inevitably have been made. The editors apologise and accept responsibility for these. In some instances information may have been excluded or carefully edited due to its sensitive nature, or to agreements regarding confidentiality. The editors hope that this work, despite any deficiencies, will be considered the most complete account of all species documented in the Durham recording area, but they are more than happy for the text to be made out of date by new studies, dedicated application and improved knowledge of the birds of Durham.

The production process

Countless hours have gone into the production of this publication. Work commenced in some way, the day after George Temperley published 'A History of the Birds of Durham'. It began in earnest around 2006, with a small 'false dawn' in the period 1988 to 1990, before that initiative fell by the wayside.

Ostensibly, the task was a simple one. Pull together all of the known information on all species known to have been recorded within the stated geographical area, and put this into a form that is readable and makes sense for the next generation of ornithologists. As ever, the idea sounded better in the pub.

Work on this publication was started informally around 2006, by Keith Bowey, who began the process of creating outline drafts for all species, though many of these, particularly on rare and scarce species, were little more than outline sketches. Chris Bell's hugely authoritative database and analysis of county rarities and scarce migrants was invaluable and forms a crucial element of this work. The county's breeding atlas, A Summer Atlas of the Breeding Birds of County Durham, was used to provide basic details and background information on the breeding birds of the county. The book's systematic list of species texts were then further developed using a range of 'standard' sources. These were extensively 'polished', refined, improved, or in many cases re-written by the relevant local experts on that species. The overall text and balance of the book was overseen by an editorial panel, which included experienced members of the Durham and Teesmouth Bird Clubs, under the steer of Keith Bowey (Chairman, Durham Bird Club 1995-2006). This panel included standing members, Dave Raw (Chair of the Durham Bird Club 1988-1995 and Chair of the Durham Upland Bird Study Group 1992-2011) and Tony Armstrong (County Ornithological Recorder 1988–2006). This group also had contributions from Paul Anderson (Chairman, Durham Bird Club 2006 to present) and Mark

Newsome (County Ornithological Recorder (2006 to present). Chris Bell took on the lead role for the production of the texts for rare and scarce species, with assistance from Mark Newsome. Much of this was based upon the database that he had developed over a number of years. Chris also led on the production of the detailed chapter on the occurrence patterns of escaped and feral species in the county.

The panel of authors, all of them chosen because of their experience of the county's birdlife, was put together in the autumn of 2010, and species texts were allocated amongst them according to their special interest or knowledge. This panel of authors comprised: Paul Anderson, Tony Armstrong, Chris Bell, Peter Bell, Keith Bowey, Ian Mills, Fred Milton, Mark Newsome, Dave Raw and John Strowger, with additional inputs to some species texts by Derek Charlton, Mike Harbinson and David Sowerbutts (British Trust for Ornithology, Regional Representative for Durham, 1974- 2011). The species texts developed by each author are listed in the Appendices)

To support the work, the authors were provided with access to a range of key reference materials, including an electronic version of *The History of the Birds of Durham*, and guidance notes on how to develop the species texts. These authors took the outline draft texts and added, re-wrote or complemented them with additional research and personal knowledge. The final texts were then sent for sub-editing and proof-reading. The editorial oversight was provided by Keith Bowey, supported by a panel of sub-editors which included: Tony Armstrong, Chris Bell, Peter Bell, Ian Mills, Mark Newsome and David Raw, with some additional inputs from Paul Anderson. Wherever possible, the individual species accounts were scrutinized by observers with a special interest, knowledge or experience of that species in the county or region. In some instances, the observers may have been involved in an academic study of the species, have worked upon its conservation, or in some instances spent a lifetime studying its local activities in an amateur or professional capacity. Most of the completed texts were read over by Tony Armstrong, County Recorder from 1988–2006, and Mark Newsome, County Recorder from 2006 to present, to ensure content accuracy and consistency.

In addition, a vast amount of extra information came from many reference sources. Thanks go to all of those many authors that provided information via papers and publications on a whole range of bird-related topics over the years, and the specific scientific papers that grew from work within the county. These are acknowledged as appropriate in the texts for species and in the reference section, with apologies for any unintended oversights or omissions.

Information from Temperley to the 1970s is largely drawn from the annual Ornithological Reports for Northumberland and Durham, published through the Natural History Society in the Transactions of the Society from 1951 onwards. The most important body of data for this work were the *Durham County Bird Report* from 1970 to 1975, the annual reports of the Durham Bird Club, *Birds in Durham*, between 1976 and 2010, and the reports of the Teesmouth Bird Club, the *Cleveland Bird Report* over the same period. The information contained in all species accounts is based on a complete review of these reports and, in relation to the county's breeding birds, a review of the detail contained in *A Summer Atlas of the Breeding Birds of County Durham (2000)* and the *Breeding Birds of Cleveland (2008)*.

In the species texts, references to the *Atlas* indicate *A Summer Atlas of the Breeding Birds of County Durham* (Westerberg & Bowey 2000). As the texts are based upon a review of all published material relating to the county, each annual report is not referenced in every instance, specific reports are only referenced in relation to a key record or a particular piece of information that requires highlighting, such as a change in trend, a change in status or a feature of that species' story, for example an unusual record or a highest count. This means that a long sequence of repetitive references that would break up the text's flow is avoided.

Ringing data was gleaned from many sources, firstly and foremost the ringers themselves, the published reports of the British Ringing Scheme, from both early days when published in *British Birds*, and then by the British Trust for Ornithology. Further data was obtained from work over the years by Durham University at Teesmouth, the ringing databases of the Durham Ringing Group and the Durham Dales Ringing Group, and from the ringing reports appearing in Birds in Durham between 1979 and 2000. In most instances, these sources are not individually referenced, excepting where the detailed collation of ringing records for the Teesmouth area, as published in the *Birds of Cleveland* by Martin Blick (2009), was used as a source. The reference section was checked and cross-referenced by Peter Bell.

Very careful attention has been paid to citing reference materials and a standard presentation format has been adopted. However, for the sake of brevity and clarity, an abbreviated form of reference to records in the annual British Birds rarities reports has been used within the species texts where individual reports and page numbers have been listed for convenience.

In some instances species texts were supplemented by information contained within the monthly bulletins of the club, which from 2002 was known as *The Lek*; these were published as a seasonal magazine, rather than a monthly bulletin, from 2006.

The county map is adapted from that produced for the Durham Bird Club's publication *A Summer Atlas of the Breeding Birds of County Durham* from an original by Stephen Westerberg.

About the book

The Birds of Durham is intended to fulfil a number of functions. For the general reader it is hoped that it provides an insight into the history and current status of all bird species recorded in the County of Durham. It is hoped that it helps to show how the status and distribution of these species relates, at least in part, to the geography and habitats of the area. For the ornithologist it is hoped that the book provides a first serious attempt since 1951 to document the totality of the knowledge of the county's birdlife. The book attempts to place these records into the modern context or understanding, based on knowledge accrued both locally and nationally, in recent decades.

The book attempts to bridge the gap between publications of the past and the present day, so it embraces the timeless quality provided by black and white illustrations, it incorporates older photographic images of key species and records, regardless of quality, as objects of record, and it also includes some of the best of the modern generation's outstanding, digitally captured images of birds. Hence the book is illustrated in a number of ways. The sections of colour photographs, particularly of modern records, include typical, significant and rare species, and also images of some of the county's key habitats for birds.

Readers should bear in mind when reading species texts, that a book such as this can only ever be a synthesis of existing work and knowledge; it is only part of the story. As such, many birds, both individual and species will have gone unrecorded, and in some instances the species texts will have missed information whether published or personally accrued. In any of this no slight is intended and the authors and editors have endeavoured to give a true picture for each species both through time and in the space that we know as County Durham. This book attempts to provide a balanced overview so that future generations can better understand change, in which ever direction that takes place. In this sense, it is hoped that the book sets down a standard for the future recording and documentation of birds in the county.

The Birds of Durham sets out to:

- Provide an overview of the status, and where appropriate, documents all or the pattern of records, of every species recorded in the County of Durham, i.e. the recording area of the Durham Bird Club
- Summarise and analyse in an objective, but readable fashion, all of the known information about all species recorded in the county

The book is sequenced into three main parts:

- Introductory chapters - which provide background to the birds of the county, the ornithologists that have studied them and the history and geography of the area as it pertains to the county's birdlife
- The systematic list - which provides accounts for every species known to have been recorded in Durham, providing a short historical review and a status context that relates to all aspects of this species' year-round life-cycle whilst it is present in Durham
- Appendices and endpapers – which provide a range of additional information, including information about those species of birds that have been recorded in Durham but do not feature as part of the official record, either because of their unknown origin or their unconfirmed status as truly wild birds

The cut-off point for the data to inform the species accounts was the end of 2010, although references and information that informed the record up to 2010, from 2011 and early 2012, when available, were used. New species recorded in the county in 2011, plus additional significant records from 2011, have also been included, to make the book as up to date as possible at the time of publishing. The vernacular names for birds in the text are those used in *Birds in Durham*. These follow *The British List*, as published by the BOU (British Ornithologists' Union) at 15 December 2011, as does the systematic order used. The agreed IOC (International Ornithologists' Congress) species name is offered in brackets where this differs from the BOU name. Plant names follow Clapham, Tutin & Warburg's *Excursion Flora of the British Isles 3rd Edition* (1985).

Species texts vary in length according to the amount of information available and the perceived importance of the species in the context of the ornithology of the county. Nonetheless, the following elements are, to a greater or lesser degree, featured in most of the species accounts. Each species section within the Systematic List includes as detailed an account of the species as was possible according to current knowledge, which includes a review of all of the accepted records of the species to the end of 2011. Each text contains the following information about the species:

- Common name – i.e. those names used commonly by birdwatchers in the County of Durham area, in the order featured in the latest British Ornithologists' Union list, alongside the 'accepted BOU common name', shown in brackets
- Scientific binomial - binomials are those that appear in the latest publication of the BOU British list, or where recently amended as appropriate
- Summary sentence on the status of each species
- A detailed species account – this summarises and interprets the details of the species' occurrence and distribution in the county, its status, documenting the trends and changes in its occurrence over time, including all published records for rarer species, i.e. those recorded less than 20 times in the county; including the use of graphs and tables to illustrate trends or patterns
- Comments on winter and breeding distribution as appropriate
- Early/late dates of migrants and trends in changes of seasonal occurrence.

Where pertinent, reference to the historical work in adjacent counties of North Yorkshire and Northumberland has been made, where it is believed it sheds light upon the situation in Durham.

To provide structure for the preparation of the texts, and a certain amount of standardisation for the user of the book, the editors developed a format for the species accounts that was heavily based upon *The Birds of Shetland* (Pennington *et al.* 2002). Consequently, each species account was structured using the following headed sections:

- Historical review (i.e. in the main, referring to the period prior to 1970)
- Recent status (for non-breeding species, passage migrants, scarce and rare species)
- Recent breeding status (for breeding species or those species that have attempted to breed)
- Recent non-breeding status (for breeding species outside of the breeding season, or primarily wintering species)
- Distribution & movements – documenting information about the species' wider distribution outside of the area, its movements into, through and out of the county, as understood at the present time, with particular reference to relevant ringing data.

It is acknowledged that this does not always work to the best advantage of readability for every species account, but it was felt necessary to create a clear framework around which authors could work and by which the reader could more easily understand and structure their use of the book along the lines of a coherent story. The format works well for many species, although in some instances, the section categories blur and for some species, such as Little Auk, and other passage seabirds, more or less the whole of the species status is a commentary on its movements to and from, or along, the county's coastline.

All records published here have been accepted by the British Birds Rarities Committee (where relevant), the Durham Bird Club and/or the Teesmouth Bird Club records sub-committees, or have been previously published in a reputable source document for the county, where the relevant record had achieved the criteria for acceptance at that time. On occasions, new information not previously available for reasons of confidentiality, or other considerations, is included here in species accounts, especially where it adds to the understanding of the occurrence in the county of the species in question. In these instances, the reference sources are often ones of personal communication from local experts or personal observation of the species authors.

What the book did not set out to do was to review all records on the county list from the perspective of modern knowledge or understanding. It started from the assumption that what was previously published and accepted by the standards of the day remained acceptable today. Unless there was new or compelling evidence to say otherwise (i.e. the withdrawal of a record, a taxonomic review, or evidence of fraud or a deliberate attempt to mislead), the record stood as 'proven'. Old records were only reviewed out, or reviewed in, when they had been inadvertently overlooked in the past or where there was new information/evidence available. Such instances are few. Should modern knowledge or study methods have developed to such a degree that previously published information is now seen in a radically different light, then the relevant text attempts to contextualise and provide perspective on the previously published record, though the record remains 'as published'. In many instances, particularly from a historical perspective, research revealed new information, or by combining and reviewing existing knowledge, created new insights in relation to a species' occurrence in the county. In some cases, previously unavailable evidence or information has been included where it throws light on important records.

Many local ornithologists supplied information, much of it very detailed, and they deserve a special mention. It should go without saying that the authors accept the responsibility for any omissions or mistakes within the text, and that they would be grateful to anybody alerting them to such errors.

Acknowledgements

This work could not have been completed without the hundreds of thousands of observations of local birdwatchers upon which the whole text is based. In addition, many observers made available their own research and survey results from activities such as Common Bird Census, Wetlands Birds Survey, wildfowl counts, bird ringing and a range of other surveys and work. Incalculable hours of fieldwork, tens of thousands of hours of computing time and innumerable days poring over reference works and texts during the writing and editing processes, have been distilled down and incorporated into the production of this book. Numerous surveys, both national and local, hundreds of birdwatchers and many organisations have fed in information that led to the creation of its contents. Due acknowledgment is made to all of the many unnamed members of the Durham and Teesmouth Bird Clubs, The Natural History Society of Northumbria, Durham University research groups, local BTO members and many ornithologists and birdwatchers that have in so many ways, sometimes unbeknown to themselves, contributed to the process.

Wherever possible, the species accounts have been written or commented upon by a local specialist author with particular knowledge or experience of the species in the context of the county. The details of these are listed in Appendix 1. In some instances the authors may have been involved in an academic study of the species; they may have worked upon the conservation of the bird or spent a lifetime studying its local activities in an amateur capacity. The Durham Bird Club would like to offer its most sincere thanks to all contributors, whether as authors or specialist consultants, for their immeasurably important contributions.

The Durham Bird Club would like to offer its thanks to the following organisations which allowed use of data and offered support during the production process: the Durham Upland Bird Study Group; the British Trust for Ornithology; the Teesmouth Bird Club; and to Martin Blick and the Tees Valley Wildlife Trust for access to texts and information contained in *The Birds of Cleveland*, which proved hugely helpful, particularly in assessing some of the ringing data from the south east of the county.

A huge debt is owed to the many individual birdwatchers, researchers and experts who allowed their personal expertise and data to be used in order to assist the species' authors in preparing the texts. These include: Richard Barnes (for collating ringing information and details on a range of species from Durham Ringing Group reports); John Barrett; Dr. John Coulson (for extensive and helpful comments on gull species texts); Paul Danielson (for help in drawing together historical profiles); Martin Davison; Dr. Chris Gibbins (for his input to the texts on large gulls); Ian Findlay; Ian Fisher; John Fletcher (for access to his excellent work on the history of Teesmouth bird watching); Dr. Mike Harbinson (for translation of old Latin texts from the Durham Cellarer's rolls); John Hawes (Secretary of the Durham Dales Ringing Group, for provision of much ringing data); Dr. Andrew Hoodless (for helpful comments on the snipe and woodcock texts); Ian Kerr (for access to Northumberland texts and proof-reading some of introductory chapters); Mike Leakey (for comments on a range of wildfowl and wader texts, with particular reference to WeBS work and counts at Teesmouth); Dr. Fred Milton (for his historical research of newspapers and other publications and inputs to introductory chapters); Les Milton (for his ringing data and insights); Gordon Simpson; Allan Snape (for provision of Teesmouth tern ringing data, and helpful comments on tern texts); David Sowerbutts BTO Representative for Durham (who provided invaluable inputs to a number of species texts, created a number of maps and gathered much information about the early history of the Durham Bird Club); Robin Ward (for his insightful comments on species texts and his boundless knowledge on the waders of Teesmouth); Brian Walker; Dr. Phil Warren (Game Conservancy Trust for inputs to gamebirds texts); Stephen Westerberg (for comments on woodland birds, Kingfisher and Dipper texts and advice on bird ringing and a range of other matters); and, John Wood.

Others who helped with various stages of the production, including a wide range of tasks such as typing and proof-reading, include: Brian Pollinger (for extracting and collating some of the ringing data); Dougie Holden, Steve Egglestone and Hilary Chambers (for their assistance in extracting and collating data from

Birds in Durham); Richard Cowen (for extracting data from The Vasculum); and Dr. Mike Harbinson, Dr. Mike Smith, Peter and Janette Bell for proof-reading texts.

Previous recorders and editors of annual reports, Ken Baldridge, Graham Bell, John Coulson and the late Brian Unwin, made many and various inputs to aspects of the texts and they are thanked for these, their memories and for putting up with a plethora of questions from 'whipper-snappers'. Jim Edwardson, Colin Freeman, Tom Palmer and David Simpson contributed a range of comments and reminiscences about the history of bird watching in the county, and Joan and John Proudlock contributed reminisces of George Temperley.

Acknowledgements for their co-operation and advice are due to various members of the Teesmouth Bird Club, in particular Martin Blick, Chris Brown, Derek Clayton, Tom Francis, Geoff Iceton, Graeme Joynt, Alistair McLee, Ted Parker, John Regan, and Richard Taylor. Their help, knowledge and insights into the birds of the Teesmouth area proved invaluable.

The editors would like to offer our most sincere thanks to all contributing authors for their outstanding efforts.

Special thanks go to Ken Baldridge for his many wonderful illustrations, which greatly enhance the book, and his ever knowledgeable and friendly contributions on a number of species, especially Common Gull and Dunlin. Geoff Watson's magnificent artwork for the front cover admirably captures the essence of the western uplands of Durham and hints at what so many would love to see quartering the county's moors, one day.

Thanks go to Dan Gordon of the Tyne & Wear Museums and Archives for assistance with studies of the collections of the Natural History Society, and from the Natural History Society of Northumbria, Jim Edwardson, June Holmes, Les Jessop, Ian Moorhouse and James Littlewood all made significant contributions. June and Les in particular made important contributions regarding the NHSN collections and various aspects of the history of ornithologists in the county, and made possible the photographing of important historical specimens.

In addition, data were supplied by the Wetland Bird Survey (WeBS), a joint scheme of the British Trust for Ornithology, The Royal Society for the Protection of Birds and the Joint Nature Conservation Committee (the latter on behalf of the Council for Nature Conservation and the Countryside, the Countryside Council for Wales, Natural England and Scottish Natural Heritage), in association with the Wildfowl and Wetlands Trust. Although WeBS data are presented within this report, in some cases the figures may not have been fully checked and validated. Therefore, for any detailed analyses of WeBS data, enquiries should be directed to the WeBS team at the British Trust for Ornithology, The Nunnery, Thetford, IP24 2PU (webs@bto.org). For their assistance in this matter, thanks go to Peter Lack, Neil Calbrade and Greg Conway at the BTO.

Photographs were researched and collated by Chris Bell and Mark Newsome, and were drawn from a wide range of photographers, both local and from elsewhere in the country. A full list is detailed in Appendix 2. Sincere thanks go to all of these photographers, as the generous use of these photographs, old and new, enhance this publication immensely.

The task of formatting and preparing the book for the final production process was undertaken by Mark Newsome. Thanks also go to Digital Print and Sign Ltd. (Seaham) who gave much valuable advice and acted as a shock absorber, softening a potentially bumpy ride during this process.

Financial support for the production of *The Birds of Durham* was provided by the partners of *The Birds of Durham Heritage Project,* the Heritage Lottery Fund and a number of businesses and organisations that sponsored species texts (see the list in Appendix 3 and acknowledgements on the relevant species text pages). Furthermore, a number of individuals made donations to support the production of chosen species texts that had a particular resonance for them. These are listed in Appendix 3a. Without the generous support of these organisations and individuals, the production of this publication would not have been possible.

The book has been inspired by a wide range of the excellent county avifaunas that have been produced over the last ten to fifteen years. Not least amongst these were *Birds of Shetland*, *Birds of Essex* and *Birds of Wiltshire*. All of these publications are different in form and specific context to *Birds of Durham*, but the authors would like to acknowledge a debt of gratitude to the respective production teams of those volumes, and others, from which we have drawn inspiration.

A huge vote of thanks goes to all of the partners and families of all the authors and those who worked on the book in any capacity for their tolerance and support – so often the unheralded heroes of works such as this. Finally, to Julie and Robert Bowey, for their immeasurable patience and fortitude over many years when husband and dad always seemed to be 'busy', "Thanks!"

List of Abbreviations Used in the Text

Atlas	A Summer Atlas of the Breeding Birds of County Durham
'Atlas'	The Atlas of Breeding Birds in Britain and Ireland
a.s.l.	above sea level
BBRC	British Bird Rarities Committee
BBS	Breeding Bird Survey
BOU	British Ornithologists' Union
BTO	British Trust for Ornithology
c.	*circa* - 'about'
contra	Against or in contradiction of
CBC	Common Birds Census
CES	Constant Effort Site
CP	Country Park
DBC	Durham Bird Club
DRG	Durham Ringing Group
e.g.	for example
ESA	Environmentally Sensitive Area
ha	hectare
km	kilometre(s)
km^2	square kilometre
hr	hours
m	metre
'New Atlas'	The New Atlas of the Breeding Birds of Britain and Ireland
NR	Nature Reserve
LNR	Local Nature Reserve
NE	Natural England
NHSN	Natural History Society of Northumbria
NNR	National Nature Reserve
OD	Ordnance Datum
pers. comm.	personal communication
pers. obs.	personal observation
RAMSAR site	Internationally important wetland site
RBBP	Rare Breeding Birds Panel
Res.	Reservoir
RSPB	Royal Society for the Protection of Birds
sp.	Species (singular)
spp.	Species (plural)
SoCC	Species of Conservation Concern
SSSI	Site of Special Scientific Interest
TBC	Teesmouth Bird Club
tetrad	2km by 2km square of the British National Grid
vc	Watsonian vice-county
WBS	Waterways Bird Survey
WeBS	Wetlands Bird Survey
WWT, Washington	Wildfowl and Wetlands Trust, Washington
2nd CY	Second calendar year
10km square	10km x 10km square of the British National Grid

Chapter 1

Durham Geography: Landscape, Habitats, Climate and Birdlife

Introduction

The area covered in this publication corresponds to the area recorded by the Durham Bird Club in its annual publication *Birds in Durham*. This is based upon the Watsonian Vice-County system, proposed in 1873 and widely adopted during the late Victorian period for the purpose of botanical recording. In this instance, the county's core recording area is Watsonian vice-county 66. Today this can be visualised as the modern county of Durham plus that part of Tyne & Wear south of the River Tyne and that part of the old county of Cleveland to the north of the River Tees (this area is also recorded by the Teesmouth Bird Club). In addition, the rural district of Startforth (which is part of vc65) was acquired by Durham from North Yorkshire in 1974 as part of the extensive local authority boundary review of the time; it has been recorded as part of Durham since then. The boundaries of this recording area are formed by the coast in the east, the Rivers Tyne and Derwent in the north, the watershed ridge between Durham and Cumbria in the west, and stretches of the Tees, Lune and the Balder in the south. When the phrase 'County Durham' is used in the text it should be interpreted as the recording area outlined above.

Durham is located at the narrowest point of mainland England: it stretches from its eastern boundary, the north east of England coast, westwards to the Pennine uplands and its border with Cumbria. At its greatest extent, from South Shields on the north east coast, to the southernmost limit of Sockburn Parish in a meander of the Tees south of Darlington, it is 60km from north to south. At its widest point, from the junction of Crookburn Foot in upper Teesdale east to Hartlepool Headland (The Heugh), it is just over 75km from west to east. Durham has something like half the land area of its larger neighbour Northumberland and is considerably smaller than Yorkshire to the south; it covers an area of just over 1,000 square miles or almost 2,600 square kilometres. Despite this relatively small size, County Durham has a varied geography and geological form, and it manages to fit a large amount of birdlife into its small size and array of habitats.

The image that best encapsulates the ornithological landscape of County Durham is probably a vision of the windswept slopes of the North Pennines. Springtime Lapwings plunging over the valley bottoms, Skylarks carolling in the skies above Teesdale's white-washed farmhouses and Curlews bubbling over the sheep-dotted, grassy slopes of the valley and onward, up to the brooding moorlands above. Whilst this image is emblematic of the jewel in the county's avifaunal crown, it is not alone amongst its riches. Though small it is varied, with a wide diversity of landscape. From the cliffs and low shoreline of its North Sea coast to some of the highest land in England, Durham possesses an attendant wealth of ornithological interest. To date the county has remained relatively unappreciated and, in some respects, undervalued by birdwatchers living beyond its boundaries. Sandwiched between the larger counties of Northumberland and North Yorkshire to its north and south, with Cumbria to the west, Durham sits on the North Sea shore and its birdlife has tended to be somewhat over-shadowed by that of its better-known neighbours.

Landscape and climate

Describing the geography and landscape of a county as varied, or as being one of contrasts, is customary in introductory accounts such as this but in the case of Durham it is an accurate one; the county's associated complement of habitats is consequently varied. The physical aspects of the county are complex, with some broad plains along the coast and, a little way inland, the rolling hills of the magnesian limestone plateau, the Pennines high hills in the west and, in between, a patchwork of agriculture on a rolling lowland landscape. The county's topography and ultimately the flora and fauna that sit atop this, are expressive of its underlying geology; this in turn dictates its geography, both physical and human, although climate and hydrology have had a profound effect on the fine detail of such arrangements. The distribution of habitats and the birds reliant upon them are dependent upon soil types, which are influenced by the solid and drift geology that underlies them. Over the last two millennia, the influence of man has also had a significant impact upon the extent and nature of these.

In geomorphologic terms, Durham is largely an upland county, with over half of its total land surface area rising to above the 150m contour. Durham exhibits one of the greatest variations in altitudinal range of any English county, rising from sea level at the east coast to over 700m elevation at its highest point amongst the Pennine tops. As a consequence, its bird-life is correspondingly rich and varied. Of particular note are the birds associated with its uplands; foremost amongst these are perhaps Red and Black Grouse, the large populations of breeding waders and the largest population of breeding Wigeon in England.

The geography of the county's climate is dictated principally by altitude and the proximity of the North Sea. The monthly average temperatures across Durham reveal the tempered nature of the coastal climate and the more extreme nature of inland locations. In terms of climate the county of Durham is, in many respects, a relatively dry, if cool, place sitting in the rain-shadow of the Pennines. Relative to the climate of Cumbria for example, on similar latitude, Durham is colder but also appreciably drier and enjoys brighter weather; a very good summer month in Durham can have over 250 hours of sunshine, with May usually the brightest of all. The climate is dominated by the mid-latitude, westerly wind circulation and for close to two-thirds of an average year, winds are from points of the compass between the south west and north west. This same wind provides much of Britain's rainfall, but less so in County Durham, as air moving from the west has passed over the Pennines and is descending from higher altitudes, therefore becoming warmer and drier. The county's climate is strongly influenced by altitude. It is clear to even the most casual observer that to go from the coast, or even the shelter of the western river valleys, up on to the Durham moors is to enter an environment that is considerably cooler, where rainfall is a more common accompaniment to even summer days. As the land rises from east to west one finds that rainfall, wind speed and exposure all increase, whilst temperature and sunshine amounts move in the opposite direction. Likewise, low-lying inland areas furthest away from the moderating influence of the North Sea tend to have warmer summers, but colder winters, than the coastal margins.

In broad terms, the county's climate is not one of wide extremes and there is a relatively small difference between average winter and summer temperatures. For example, winters in Durham are by no means as cold as on the nearby European land mass but neither is summer as warm. The mean annual temperatures of the county are relatively low, (mean daily temperatures of 2-4°C in January, which is the coldest month, and around 15°C in July or August) though the range varies considerably between the upland areas and the coastal strip. Temperatures fall off at a rate of approximately 0.65°C for each 100m increase in altitude.

The general pattern of precipitation (both rain and snow) corresponds closely with relief. The driest areas are the lowest in elevation and the high ground in the far west, approaching Cross Fell just over the border into Cumbria, is amongst the wettest locations in northern England. The relationship between rainfall and altitude is a close one. As a 'rule of thumb', mean annual precipitation increases by nearly 2mm for each metre of altitude gained. Sunshine levels tend to increase the further east one moves in the county. The south east of the county has one of the driest climates in the UK, with an average of just 625mm of rain per annum. In Durham, whilst westerly airflows predominate, during the winter months in particular when these turn to the east or north, temperatures are significantly reduced. In winter anticyclones bring cold conditions and in summer fair, dry periods.

It is perhaps useful to have a base level or yardstick from which climatic data can very roughly be estimated depending upon where one is located within the county. Data from the University of Durham over the period 1961-1990 showed that mean temperatures there ranged from a low of 3.0°C in January to a high of 14.9°C in July. Rainfall totals in Durham City averaged 649mm per annum, with the driest months being February and then April, whilst August is the wettest month. Sunshine levels there total slightly less than 1,330 hours per annum, and are not surprisingly at their lowest in January and December when day length is at its shortest, with the sunniest month on average being May. The county often experiences late springs, which is significant in that the growing season in Durham is somewhat later overall than in other northerly parts of the country. This comes about from a combination of the effects of the North Sea at the coast and the large mass of high altitude land in the inner portions of the county. Such a contraction of the growing season has obvious impacts upon breeding bird species.

By virtue of its location on the north east coast, the county attracts a variety of rare migrants and vagrants, particularly in the autumn. Spring however, can be a notoriously variable time of year in Durham, and late frosts, cold spells and even snow on the uplands can all have profound effects on its birdlife. Returning migrants can be particularly affected by this unpredictable weather and they invariably arrive later than in the west, for example on the Cumbrian coast. This time-lag in migration is evident even on the east coast, where species' arrival dates in Durham can regularly be a full week later than those of birds in Yorkshire, just to the south.

As elsewhere, the impacts of extreme weather on birds can be evident in Durham, especially when winter weather systems come in from the east, such as in the severe winter weather of 1962/1963. In this instance freezing fog was especially frequent in early December and again in late December. This covered trees in parts of the country with layers of ice; in Durham a thin twig was found surrounded by a shell of ice four centimetres in diameter and every tree within miles was icebound. The impact upon small insect-eating species such as titmice and Long-tailed Tit can be imagined.

The geology of Durham is dominated in the east by magnesia limestone, very significantly in terms of economy, cultural and historical perspectives, by the coal measures over much of the centre and east of the county, with millstone grits and mudstone in the west, and a variety of other smaller but important components. Perhaps most notable amongst these is the 'sugar limestone' of upper Teesdale. The dominant rocks of the county, sandstones, limestone and shales were laid down during the Carboniferous geological period some 345 to 280 million years ago. Much of the county's surface landscape features, including many of the glacial deposits of till, sands and gravel, were fashioned by processes generated by the extreme cold that was visited upon the north east of England during the most recent cold phase of geological history, the Devensian. This commenced some 75,000 years before the present with the advance of ice across the British landscape. Subsequent retreat of ice found all of Britain and Durham ice-free by about 13,000 years before the present. The post-glacial drift geology of Durham has been largely determined by the huge quantities of sands, gravels, clays and boulder clays, of up to 100m thick in some places, which were left or moved around the county by vast ice sheets. This drift geology comprises largely boulder clays in the east and south, with peat in the west and north west and glacial sands and gravel in the south east. Excavation of some of the county's glacially deposited gravels (some of them supplemented by later alluvial action) along the main river valleys, such as the Wear, have created important wetlands such as at Low Barns near Witton-le-Wear. In the north of the county, along the Barlow and Blaydon Burns which flow into the Tyne, glacial sand deposits left a landscape of moraines the excavation of which has created a string of quarries that for many years have held some of the county's largest Sand Martin colonies.

In many respects, County Durham is defined and delineated by its three principal rivers. In the north these are the Tyne and its major tributary the Derwent, which form the border with historical Northumberland; at the southern frontier lies the Tees, abutting North Yorkshire, and the only one of the three major rivers to be wholly contained within the county, the Wear. These rise high up on the extensive, flat-topped summits of the Pennine Hills and in many ways they dominate the area's avifaunal components. The landscape in the west of the county is an area of national importance for its upland birds and even when its habitats and open landscapes fall quiet in the winter, its influences flow, courtesy of the rivers that they spawn, down through the lowlands and out to the North Sea. The valleys of these rivers, still cloaked in remnant woodlands, hold important populations of birds and a suite of classic, north country riparian species along their water courses.

From its upland origin, the Wear dips gently eastwards, falling into the broad basin of the middle Wear, once the centre of the Durham coalfield. To the east, the coal measures are capped by the arc of the East Durham hills, limestone over coal, and these in their turn dip down to form the cliffs of the Durham coastline. The lowland stretches of the county's rivers have been much influenced, in both ecological and economic terms, by their industrial past. Through time, densely populated areas and urban sprawl developed around the lower stretches of the county's principal rivers; at Tyneside, Wearside and Teesside. These lowland stretches of the county's rivers have been much influenced, in both ecological and economic terms, by their industrial past. Ship building was of major importance on both the Tyne and the Wear – Sunderland was once one of the biggest ship-building towns in the world - whilst the petrochemical industries have proved influential in the recent life of Teesmouth. These areas are strongly associated with the industries that were the economic mainstay of the area, from before the Industrial Revolution until the latter part of the 20th century. These included coal mining, ship building and heavy engineering and, more recently, the petrochemical industries. Even in such seemingly unfavourable areas, birds often utilise artificial habitats; most graphic in this respect and amply illustrating birds' capacity to adapt are the Kittiwakes that have famously nested in and around the Tyne at Gateshead, on the Baltic Flour Mill, now the Baltic Centre for Contemporary Art, for almost fifty years. These conurbations aside, a large proportion of the rest of the county, though having some urban areas such as Durham City, is in relative terms quite sparsely populated, especially so in the west. Today, perhaps just 12,000 people live in Durham's upland areas, probably less than 50% of the number that did so 150 years ago at the height of lead mining and other extractive industries for which the west of the county was renowned.

In all of its guises mining, whether it be deep coal extraction, drift mining or upland mineral extraction, has left an indelible imprint on the county's landscape. In both the uplands and the lowlands this is pertinent not just in terms of the physical characteristics of the place, and its places, but in the pattern of local settlements, local names, culture and traditions – the people themselves. The working of the Pennine ore field encouraged the woodland clearances in the west Durham dales and the washing out of ore by the practice of hydraulic mining left great gorges, termed hushes, down the lines of the mineral veins in some of the county's valley sides. The released flood-waters and the resultant out-wash of debris from these works re-worked and further in-filled the deposits that cover the upper dale's valley floors. In some cases, this so altered the chemistry and the faunal complement of some of Durham's upland watercourses that some bird species, such as Dipper, remain excluded from some of these locations even to this day.

There has been no greater impact upon the Durham landscape than that which came along from the long-term dumping of coal waste and spoil along the length of coast from Crimdon north to Seaham. The disposal of tens of thousands of tonnes of waste along the county's coastline in this fashion did immeasurable ecological and cultural damage and it blighted the pretty 'little' coast for generations. It resulted in the permanent scarring both literal and metaphorical of everything associated with Durham's coast and its infamous 'black beaches'; thankfully they are all now gone.

Despite this industrial imprint, in terms of land use, Durham is, and always has been, predominantly an agricultural county, with such activity making up over 70% of the county's total land area. This comprises a range of mixed arable and pastoral systems; much of the latter is in the uplands and is extensive in nature. Across much of the county, the nature of agriculture is determined by the quality of land upon which it is practised. The most productive soils are largely concentrated in the south east, around the Tees Lowlands, where Agricultural Land Classification grades much of this in categories two and three. Elsewhere, although there are pockets of more productive arable land such as over a large proportion of the East Durham Plateau, much of the county's land is relatively poor in quality and many areas across large parts of the landscape are of a low grade. As a consequence of the greater productivity of the soils, the agriculture of the south east of Durham has, in the period 1970-2010, tended towards greater intensification. There is a consequent pattern of larger fields and fewer hedgerows compared to the central, eastern and northern central portions of the county where many hedgerows have persisted, albeit in a poorly maintained state and in these areas, fields are to a large extent smaller. The adoption of agri-environmental schemes and similar initiatives has been important in some parts of the county in aiding the recovery, or putting a brake on the decline, of some farmland bird species over the period 2000-2010.

The rural nature of Durham is evidenced by the fact that County Durham, excluding the most heavily built-up areas, held fewer than 600,000 people at the 2001 census. By contrast, the Borough of Gateshead alone, covering less than 10% of the land area, had almost 200,000 inhabitants. The major land uses in the county (disregarding the Tyneside, Wearside and Teesside conurbations, which are largely built-up) comprise arable and improved grassland in one form or another which makes up over 50% of the total land area; semi-natural habitats comprising 34%; built-up land just under 7% and plantation 4.65% (Clifton & Hedley 1995). The largest and most important semi-natural habitat in the county is 'mire and heathland' in the form of blanket bog, peat lands and dry heather moor. County Durham is relatively poorly endowed with woodland, only 6% of its area (excluding south Tyne & Wear and north Teesside) is forested – the largest example of this is at Hamsterley in the Wear's upper watershed.

Whilst each particular land type or habitat has its own suite of typical bird species, the county's built-up areas are by no means devoid of birds; they host many of the county's breeding House Sparrow and Starling and there are good numbers of House Martin. Suburban areas are the principal habitat for Collared Dove with increasing numbers of Wood Pigeon and Magpie and some of the commoner finches. Chaffinch, Greenfinch and Goldfinch can all be found in suburban areas, especially in those areas of housing constructed during the building boom of the late sixties and early seventies, when developers tended to be generous with landscaping on the edge of estates and along road corridors. These areas, with their mature trees, now provide good nesting opportunities for these species. Perhaps less-expected in the urban context are nesting gulls. Herring Gulls have been nesting on the rooftops of buildings close to the Rivers Tyne, Wear and Tees in the east of the county for over four decades and, since the early 1980s, this species has been joined by Lesser Black-backed Gulls in Sunderland, Hartlepool and on Tyneside, with the largest breeding population of the latter now found inland at Darlington.

One of the key features from a wildlife point of view of post-industrial sites is change, and much wildlife has adapted to accommodate this. Brownfield sites in Durham have attracted many birds of interest including nesting Common Tern in Sunderland and wading birds, such as Ringed Plover, on some sites in Gateshead and very extensively around the Tees estuary, where some reclaimed land, left undeveloped, has become very important for several species of birds. In some instances the habitat that develops on such locations is herb-rich grassland, which in itself attracts abundant invertebrates and developing scrub. These can be important for Grey Partridge, Linnet and Reed Bunting in the context of Teesmouth. Linnet and Goldfinch often feed in such areas late in the summer when abundantly flowering ruderal weeds such as dock and knapweed provide them with foraging opportunities.

Residential areas have become more important for the winter survival strategies of many common 'garden' birds, particularly as the public has started to consider their gardens as places to encourage wildlife, increasingly feeding birds during the winter. As a result, some birds that would not have been considered garden birds just two human generations ago are now frequent visitors to nut-bags and scattered grain in the gardens of Durham. Such species include finches like Siskin and Goldfinch, insectivores like Long-tailed Tit, and what were once thought of as spectacular woodland birds like Great Spotted Woodpecker and, anywhere that mature woodlands are near at hand, Nuthatch. Other garden visitors including Wood Pigeon, Jay, Magpie, Pheasant are all now expected species at many urban-fringe and even truly urban garden feeding stations. The extensive areas of planted woodland and scrub around locations such as the 'new town' of Washington, as it matures to secondary woodland, provide an important habitat for birds such as Greenfinch.

No habitats in Durham can be considered entirely free from threat. Upland areas are often particularly at risk and potential threats to these delicate upland communities, such as over-grazing of heather moor, aforestation, reclamation of rough pasture and the agricultural improvement of hay meadows and other traditionally managed habitats are all pertinent. Despite the demise of the deep coal-mining industry in Durham, opencast coal excavations have the potential to damage many habitats. Within the urban conurbations and on the urban fringes there are important habitats that might be severely impacted by further urban expansion, in particular the uncontrolled development of green-field sites, the building of bypasses and other further expansion of the road infrastructure. All these activities may affect both the complement and the distribution of the county's birdlife at both a major and minor level. Underlying these local threats are the natural fluctuations to which bird populations are subject, weather variations and the as yet unknown risk of climate change.

The Uplands

The Durham uplands are of national importance and the greater part of the county's moorlands and hinterland habitats are included within the North Pennines Area of Outstanding Natural Beauty. On Durham's western moors, the hills are broadly conical though relatively smooth-topped, and they are high, with at least 15 of them rising to more than 600m above sea level. Although in some areas there are more rugged cliffs and rocky exposures, Durham does not possess the high mountain slopes or rugged screes that are so widespread in the Lake District of Cumbria to the west. The county's high ground is much cooler than that of the coast and it can experience quite severe weather conditions at almost any time of the year. Birds breeding in these areas need to have the ability to survive the extremes or to quickly replace lost clutches and broods once conditions have improved. Through the uplands there are numerous small burns winding their way down the slopes through steep-sided gills and draining into the wider river-valley bottoms of the Tees, Wear and Derwent.

The Pennine uplands cover the western portion of County Durham, from Langdon Common in the south to Muggleswick in the north. The area holds around 30% of the UK's remaining upland hay meadows, over 20% of England's upland heathland and over 60% of England's Black Grouse, as well as rare arctic alpine plants and over 20,000 pairs of breeding wading birds. This area includes extensive tracts of heather moorland, the highest point of which is to be found at 790m above sea level on Mickle Fell, high above Teesdale. Considerable areas of land lie above 600m ordnance datum and much of this is covered with blanket peat, the formation of which took place more than three thousand years ago, a habitat typical of areas with high rainfall where the soil has low fertility. The landscape of the North Pennines comprises an intricate mix of semi-natural habitats, many of which support their own special birds. The highest, most exposed parts of the Pennines endure some of the most inclement conditions in Britain; they are wet and cold, experiencing over 240 rain days per annum and, on some of the most exposed tops such as Widdybank Fell in upper Teesdale at an altitude of over 530m above sea level, more than 1,500mm of rain falls annually. On this high ground, summer is very much the driest season. For their elevations, Durham's upland sites have some of the lowest average temperatures anywhere in Britain which reflect their cool, easterly-facing aspect. For much of the year Durham's high moors and the summit plateaux of the North Pennines can be bleak and windswept, where few birds other than the ubiquitous Red Grouse are to be seen. On high ground, snowfall may account for 20% of annual precipitation totals and such weather, if late in the season, can be disastrous for ground-nesting species. In the more hospitable times of spring and summer, this landscape comes alive with the songs and calls of hosts of displaying waders, with trilling Skylarks and dashing Merlins, the latter harrying and chasing down the myriad Meadow Pipits that haunt this area between March and September.

These uplands owe their existence to a geological feature known as the Alston Block and, until the intervention of our Neolithic ancestors and their descendants, Durham's moorlands would have once had an extensive woodland cover of birches *Betula spp.*, alder *Alnus spp.*, hazel *Corylus avellana* and pines *Pinus spp.* but this was reduced by the encroachment of heath and bog vegetation as well as the effects of man's activities. It is not that many hundreds of years since the county's Pennine and then still-forested uplands would have been the home of Capercaillie and Goshawk, until the woodland clearances by the county's early inhabitants put paid to the once-widespread birch and pine woodlands. One consequence of such woodland clearance was that the faster run-off of flood waters from these less absorbent uplands exacerbated drainage damage and river incision at lower altitudes.

Today, much of the vegetation in this area is upland heath, a hugely important habitat for the birds of the county though a large proportion of this area is permanent grassland of one kind or another. The moors, which developed over the hard upland strata and poorly drained acid soils deficient in nutrients and depleted by centuries of deforestation, and their adjacent enclosed pastures are characterised by a distinctive suite of breeding wading birds. These include Lapwing, Curlew, Redshank and Snipe, which flourish in the hay meadows and pastures of upper Teesdale and Weardale. Dunlin can be found in small numbers on the exposed, wet tops of the most barren fells, and Golden Plover nest on sparse short turf and the recent heather burns that result from management for Red Grouse. Oystercatcher and Common Sandpiper frequent the stream sides and the areas around the county's many upland reservoirs. Drier heaths are covered by heather *Calluna vulgaris* and bilberry *Vaccinium myrtilus* and acid-loving grasses, much of this being purple-moor grass *Molinia caerulea* but also by wetter habitats sometimes dotted with the white flowing heads of cotton-grasses *Eriophorum spp.* indicating that active peat formation is continuing in such areas.

This complement of upland waders can be found at internationally significant densities on the moorland fringes and extensively managed grasslands of the upper dales. In some of these areas surveys have revealed breeding densities of over 90 pairs of waders per square kilometre, some of the highest densities in Britain (Clifton & Hedley 1995). There are few areas in Britain, and probably none in England, that hold such a variety and density of wading birds during the spring and early summer. On the fringing habitats between heather moor and managed farmland, in the damp rush-dominated pastures, known locally as white moor can be found the species that is the adopted symbol of the Durham Bird Club, the Black Grouse. Durham probably holds around 60% of England's remaining population of this once much more widespread species and one of the largest display sites, or leks, in the country can be seen in Upper Teesdale. A number of these upland birds range over several of these upland habitats and are dependent upon their presence in juxtaposition for maximum benefit.

The western uplands are principally sheep walk, often dominated by heather and criss-crossed on the hills' lower slopes by a network of dry-stone walls – a testament to the muscle and rigour of generations past. Above the walls, the land supports large populations of Red Grouse, a species that is of considerable importance in the economy of our uplands and perhaps more than any other contributes to how the uplands appear. Much of the county's upland heather moors are actively managed to encourage the Red Grouse; the principal management method is by the sequential burning of plots of heather to create a mosaic of micro-habitats which favours the species year-round. Juniper 'woodland' is an important and relict habitat in Durham which is largely restricted to the upland areas of the western dales, the largest area of this being found in upper Teesdale.

One of the least appreciated habitats of Durham's uplands are the hundreds of kilometres of dry-stone walls which, in terms of construction date from the early enclosures and before, have provided an important feeding and nesting resource for upland birds such as Wheatear for hundreds of years. They trace their journeys from the valley floors up to the edge of the unenclosed moors, capturing the 'in-bye' land and providing structure in a landscape that is poorly appointed with trees and shrubs.

Other breeding species that are of considerable importance in this area include nationally important numbers of breeding Wigeon and a healthy and long-studied population of Merlin. Twite and Yellow Wagtail were breeders, the former perhaps now lost and the latter rapidly declining. Among other passerines Meadow Pipit are seemingly ubiquitous during the breeding season the commonest bird in these habitats. Ring Ouzel and Wheatear are relatively common in areas with rocky outcrops and Skylark remains abundant on Durham's heather moors and upland permanent pasture, despite its well-documented decline across Britain in the latter part of the 20th century. As well as its birdlife, upper Teesdale is recognised as being one of the most important botanical sites in the British Isles and parts of it are protected as a National Nature Reserve (Moor House and Upper Teesdale NNR). The area around Widdybank Fell is internationally renowned for its beautiful botanical specialities (the Teesdale Assemblage) which include bird's-eye primrose *Primula farinosa*, spring gentian *Gentiana verna* and the Teesdale violet *Viola rupestris*.

The uplands drain into the fast-flowing upper reaches of the principal rivers: the Tees, the Wear and, in the north west, the Tyne via the Derwent. These have cut beautiful dales which support a mixed but largely still extensively practised agriculture and a variety of wildlife and birds. Typical birds of fast-flowing stretches of river are Dipper, Common Sandpiper and Grey Wagtail, with Goosander where woodlands provide suitable nest sites. Some of the upper dales, such as Weardale, whilst much less populated than in times past, retain a large scatter of upland settlements and this is something of a surprise as one tracks west up the Wear valley, encountering settlements every few kilometres of the way almost to the border with Cumbria. In the more fertile parts of the dales, in the valley bottoms, farms dot the landscape though these become sparser as altitude increases. Most characteristic of these are the white-washed farmsteads of the Raby Estate in Teesdale that speck the landscape for as far as the eye can see.

The higher pasturelands are fringed by dry-stone walls and the landscape is punctuated by farm buildings and stone byres. Pied Wagtail and Swallow are common with pairs on almost every farm or old building and the open fields provide good feeding places for Mistle Thrush which requires trees in which to breed but open areas for foraging. During the winter months large flocks of Fieldfare and Starling utilise the open pastures and many hundreds of Common Gull do likewise, many of the latter staying inland and finding safe haven overnight in their roosts on the larger reservoirs. The county's western uplands are dissected by deeply-cut river valleys, which even today retain much of their dales woodlands. As the rivers flow eastwards they enter the lowlands, leaving wider valleys and more gentle hills. Along these more meandering sections can be found Kingfisher, Mallard and Moorhen and, where eroded bank-sides occur and the substrate is suitable, small Sand Martin colonies can occasionally be found, even in the west.

County Durham's river systems

The three principal river systems of the county, the Tees, the Wear and the Derwent, which flows into the Tyne on the county's northern edge, all rise from the peat lands and flushed wetlands of the Durham uplands. Down their respective valleys they flow fast and free, north in some instances, but predominantly eastwards to the North Sea, softening and slowing into meanders through the county's lowland landscape as they descend. As they tumble from their dizzying source heights (the Tees rises at an altitude of over 890m in the Pennine fells) they reveal other sides to their characters. The river valleys, particularly in their upper reaches, provide a much more sheltered environment in comparison to the high fells that lie just above the westerly dales.

The upper Tees is a wild river, rising over the border in Cumbria on the eastern slopes of Cross Fell. It commences as a Durham watercourse at the Crookburn Beck above Cow Green Reservoir before it comes crashing over two of England's most spectacular waterfalls, the long cascade of Cauldron Snout and England's single highest waterfall, High Force. In between these its takes water from two important upland tributaries, the Harwood and Langdon Becks. In its early course, before assuming a somewhat more benign character as it tracks east past the dales settlements and market towns such as Barnard Castle, the river is joined by the Ettersgill, Bowless Burn, Hudeshope Burn and the Eggleston Burn. Into the Tees in the far south west of the county flow the Lune and the Balder, both upland rivers, but smaller and less dominant than the Tees. Both of these river courses have been extensively modified by the construction of five great dams to create reservoirs along their relatively short lengths. Nonetheless, both of these rivers rise on some of the wildest landscape in the county and it is a hardy soul who is not impressed with the wild open sweep of Lunehead.

Before the construction of the Cow Green Reservoir in the late 1960s, the upper reaches of the Tees above Cauldron Snout spilled into a large tarn-like arrangement with attendant wet grasslands and damp upland pastures known as the Weel. This was a hugely important area for both plants and breeding birds and is the site where the first known breeding of Wigeon in Durham took place.

Further downstream, the river accepts the Deepdale Beck, which drains land once considered part of Yorkshire but now in Durham, just upstream of Barnard Castle. Further east along the Tees, at the Meeting of the Waters, the Greta, also once part of Yorkshire, connects to the Tees. It rises many miles to the south west, on the county's fringes with Cumbria and North Yorkshire before it scuttles across the rough landscape inhabited by waders, breeding gulls and specialities of this part of the county, Teal and Wigeon. In its lower course, it retains some charming woodland cover along its valley close to Barningham, Brignall and down to Greta Bridge before sneaking through Rokeby Park and into the Tees. Downstream the Tees cuts through steep limestone cliffs, in the shadow of Tunstall's memory, at Wycliffe-on-Tees, before wandering on.

Below Whorlton, the river decelerates and adopts a more meandering course in its lower journey, connecting with the Dyance Burn, home of one of the county's longest-established heronries near Piercebridge, upstream of Darlington. Downstream, at Croft, the easternmost of the Tees' tributaries, the

Skerne joins it. This is something of a sad and impoverished river, draining a huge area of the southern Durham lowlands, from its stream headwaters at Hetton in the north to Trimdon in the south. It once followed a wayward and hugely meandering course through some of the county's greatest winter wetlands, but it has now been extensively canalised to drain the more fertile landscapes north of Darlington, through the area that was once home to Durham's extensive inland wetlands, the Carrs. At its mouth during the 1970s, the Tees was one of the most polluted in the region, but water quality has improved markedly in this area since then, as it has elsewhere in the county.

The River Wear meanders over the greatest distance through the county, almost 115km. It forms on some of the highest tops of the county by the joining of the Killhope and Wellhope Burns to the east of Lanehead in upper Weardale and, joined by the Burnhope Burn at Wearhead, it becomes the Wear. A series of burns and streams hitch along the river's ride as it flows down the valley: Ireshopeburn, Middlehope Burn, Swinhope Burn, Western Burn and, near Eastgate on the northern bank of the river, one of the river's largest tributaries, the Rookhope Burn which forms a significant upland valley in itself. Below Frosterley, the river is joined by the Bollihope Burn, which is drawn from a broad moorland catchment important for its bird life. The Bedburn Beck, collecting the waters from much of Hamsterley Forest, joins the Wear at Hamsterley village and from there many lesser steams connect to it on its way to Durham City.

A major tributary of the Wear is the River Gaunless. This river's upper reaches, near Eggleston, trace a winding 30km journey eastward, through countryside in part wooded but much affected by various aspects of the extractive industries. In its mid-section towards West Auckland it is of a lowland character, with an extensive flood-plain that brings attendant flood risk to nearby settlements. In some of these areas, such as near Spring Gardens, major wetland management schemes have been designed to accommodate flood overspill and provide wetland habitats. The Gaunless joins the Wear at Bishop Auckland.

In its mid-county wanderings the Wear wraps itself around the heart of Durham, the magnificent Cathedral and Castle sitting atop the rivers' incised meander around which the City of Durham is built. The river's major lowland tributaries include the Deerness and the Browney, which drain the county's rolling northern hills in the central portion of the county. The River Browney, rises some 22km west of Durham City near Waskerley, flows through an attractive valley between the Derwent and the Wear flowing past Lanchester and Witton Gilbert, before it is joined by the Deerness at Langley Bridge and accesses the Wear below Sunderland Bridge, close to Croxdale to the south of Durham City. From there the Wear ambles towards the sea, bending lazily towards Chester-le-Street, where it picks up flow from the Cong Burn, before heading towards Sunderland, coastward past riverside cliffs and through a narrow limestone gorge at Wearmouth, before releasing its long-travelled waters to the sea.

In the north of the county, the Derwent starts with the merging of the Beldon and Nookton Burns, which rise on fells in Northumberland and Durham to north and south respectively, at Gibraltar below the old lead mining village of Hunstanworth, and from that point forming the county's northern boundary for over 30km as far as the Milkwellburn near Blackhall Mill. A little further north, it reaches the Tyne at Derwenthaugh not far from Swalwell.

The Tyne is a Northumbrian river in origin, its North and South branches forming in that county, though the latter has a short section of its upper reaches in Cumbria. The Tyne meets Durham at the Stanley Burn to the west of Gateshead, forming the northern boundary of County Durham from about three kilometres west of Ryton. The Tyne abuts the county over only a relatively short distance on its journey to the North Sea, some 34.5km, making it in many ways the least important of the big three rivers in terms of the influence it has had on the birdlife of the county, though in some ways it has had a disproportionally large effect, courtesy of its role in the industrial development of the whole of the north east region.

The River Team, formed by the merging of small tributaries that rise around Dipton and Annfield Plain, drains down through the Beamish Burn and, as the Team (from the Norse for water), flows through Gateshead as an extensively modified and canalised ribbon, damaged in the 1940s to facilitate the development of the Team Valley Trading Estate, and connects to the Tyne at Dunston.

The last of the Tyne's southern tributaries is the Don, which appears to the east of Eighton Banks, as little more than a series of converging agricultural land drainage systems in the flat open lands between Sunderland and South Shields. It reaches the Tyne at Jarrow, close to the important ecclesiastical church of St. Paul's, Jarrow, at a location that was once a major part of Jarrow Slake. This was a large complex of inter-tidal mudflat, in an unusual inland location, that was once a magnet for wildlife, in particular wildfowl and waders. It was in-filled, largely with the products of slum clearance in neighbouring settlements of Jarrow and South Shields. Some elements of this area remained until the early 1990s, but most of this was lost to development before the turn of the Millennium, and only a tiny fraction now remains. It is interesting to speculate on what birds might have been utilising Jarrow Slake at the time of the Venerable Bede's residency of St. Paul's.

The water quality of all of Durham's main rivers, especially in their lower reaches, improved dramatically in the period from the early 1980s to 2010. This took place courtesy of major capital investment improvements that led to the reduction of point-source pollution in all of the main rivers. This came about because of the closure of sewage outfalls and the improved treatment of discharges from sewage treatment works across the county. In some places, such as at Birtley Sewage Treatment Works, large reedbeds were created during the early 2000s; these provided very important wildlife habitats and hugely improved the quality of water discharges into the River Team, which at that time remained the most polluted of the county's lowland rivers. Most of these improvements have had considerable benefits for locally breeding and wintering birds, although most environmental change tends to create winners and losers. Whilst riparian species such as Kingfisher, Dipper and Grey Wagtail benefited from the clean-up, one major loss was the county's largest flock of wintering Goldeneye at Ryton after the termination of sewage inputs to the River Tyne during the 1990s.

Away from the rivers, the county is networked by a filigree of smaller watercourses, creating an intimate drainage pattern across the county. Many of them are only locally named, if at all, and most are rural in character, but where such features penetrate urban fringes and abut people's activities, they inevitably allow dependent species to likewise penetrate along the habitat corridors that they create. In such situations they provide local people with some kind of contact with wildlife, birds in particular, and in that sense are important.

Two of the county's big three rivers, the Tyne and the Wear, reach the sea less than 11km apart along the northern portion of the county's coastline (at South Shields and Sunderland), whilst the Tees joins the sea at the most southerly point on the coast of Durham. Along the longest section of the county's coastline, no rivers flow. This arrangement pertains courtesy of the eastern limestone plateau, which forms a block to the passage of the main rivers, these watercourses having navigated around this block through the millennia since the last Ice Age. The rivers leave the East Durham Plateau to the work of small streams that have carved out the Durham denes, percolating their way to the sea though the porous rocks in this part of the county.

Woodlands and forest

Overall, Durham is not well endowed with significant areas of woodlands, either semi-natural or plantations, especially in comparison with continental Europe. It has a forested area that is lower than the national average of around 12%, although this varies hugely across the county footprint. Some parts such as the Derwent valley and parts of the Wear watershed, which holds Hamsterley Forest, have the greatest densities of woodland in the area. Exceptions are to be found principally in the main river valleys and in some of the coastal denes. Small parts of the upland areas and moorland fringes have been lost to coniferous forest in Durham, with smaller areas of commercial timber at some lowland sites although, in the main, the county has escaped the worst elements of upland aforestation. Woodlands in the wider Durham landscape are patchily distributed, largely on a pattern that developed around the 18th and early 19th centuries, fired by the need for timber production or game preservation, though in many instances the location of such features may have been based upon the presence of older features or traditions. The county's broadleaved woodland canopy is varied and mixed in terms of species composition, though ash *Fraxinus excelsior*, tends to predominate on base-rich soils such as over the limestone, with sessile oak *Quercus petraea* in more acid areas, although a mix of these and other species can be found from east to west. Prior to the Dutch elm disease outbreak of the 1970s and 1980s, wych elm *Ulmus glabra* was a significant timber tree but, like many other areas, Durham lost a significant number of mature trees during this episode. Bird species that lost out as a consequence included Rooks for many established rookeries in elms were lost at that time. The tree deaths however also brought new opportunities for hole-nesting species including woodpeckers and Nuthatch and the disease may have been in part responsible for local increases in a number of such species through the 1980s and early 1990s.

The south east and north east quadrants of the county are particularly poorly forested. In the south east, less than 5% of the land area to the north of the river Tees up to the old borough boundaries of Hartlepool and Stockton is covered by woodlands and the most extreme north eastern quadrant of the county from the River Wear north to the Tyne and South Tyneside in particular, is even less so. The largest areas of mature trees in this area comprise the mature plantations of the public parks in South Shields which were laid out during Victorian times.

The valley bottoms of the upper to middle reaches of the main rivers are dotted with pretty villages and market towns and it is these sections which often retain their attractive cloaks of woodland. A good example of this can be seen at Barnard Castle, where Flatts Wood and Deepdale cover the sloping sides of their respective waterways. The sheltered valleys of the middle Tees and the Wear provide another local climatic regime with a welcome refuge from the exposed moors that dominate their skylines and feature well-established oak and birch *Betula spp.*, woodlands of restricted under-storey and low ground flora. The steep-sided, wooded valleys in the dales often hold a range of typical, sessile oak woodlands and dependant bird species. Most of the county's remaining broad-leaved woodland habitat is restricted to these sites and the major coastal denes of Castle Eden and Hawthorn. Particularly in the west of the county this habitat features thin acidic soils and the associated ground flora are distinctive, often with wavy-hair grass *Deschampsia flexuosa*, bilberry and woodrushes *Luzula spp*. As one progresses westwards and upwards, the diversity of breeding birds decreases but the make-up of the fauna becomes perhaps more interesting and insectivorous summer visitors abound. The species represented in these areas include the rapidly declining Wood Warbler and the more widespread Redstart, Pied Flycatcher and Tree Pipit, often with breeding Woodcock and Tawny Owl, and in more recent years, Buzzard have become well-established.

Almost the entire native broadleaved woodland habitat of the county is now restricted to such riverine sites. The Derwent valley and the National Nature Reserve at Castle Eden Dene now retain the county's largest tracts of natural woodland and are consequently very important for specialist woodland species. The largest network of woodlands in northern Durham comprises those that follow the river Derwent downstream in an almost contiguous ribbon from Nookton Farm at the head of one of the streams that give birth to the river. These woodlands follow the river down to where it meets the Tyne, via Ruffside, the Derwent Gorge & Muggleswick Woods NNR and the Consett woodlands, Pont Burn, Milkwellburn, Chopwell Woods, the Gibside Estate, Spen Banks and the Derwent Walk Country Park. The majority of this vast woodland mosaic is deciduous, some of it secondary but much of it is ancient. There are large conifer blocks, for example Chopwell Woods, where previous ancient woodland was over-planted with softwoods. The breeding birdlife of the Derwent woodlands is as rich as anywhere in the county, probably the county's greatest haven for woodland species.

At lower altitudes, the low-lying areas of land to the east of the hills are more protected. In such places the Nuthatch is present in mature deciduous woodland and this species has significantly expanded its range in the county over the last forty years or so, spreading along and out of these wooded valleys into most available suitable habitat. Castle Eden Dene on the east coast is of national importance because of its ancient yew *Taxus baccata* woodland but for birds it is also the best example of Durham's coastal dene woodlands (Eden is derived from the Saxon word for yew, *yoden*). These are characterised by a diverse complement of common woodland species, most notably a healthy population of Marsh Tit.

Where plantations are primarily coniferous or mixed they can also be good for birds, particularly if the woodlands are of a varied age structure and species mix. Some of the best examples of these can be found at Hamsterley Forest and Chopwell Woods. Hamsterley, with an area of nearly 1,500ha, sprawls up the sides of the Euden and Bedburn Becks onto Woodland Fell and Eggleston Common. The main tree species here is Sitka spruce *Picea sitchensis* with a mix of Douglas fir *Pseudotsuga menziesii*, Norway spruce *Picea abies*, European and Japanese larch *Larix decidua* and *L. kaempferi* as well as Scots and lodgepole pine *Pinus sylvestris* and *P. contorta*. Nonetheless, there remain large areas of deciduous woodlands, not to mention neutral grassland and hay meadows, within the Forest boundary. This area can be good for breeding birds such as Goldcrest, Coal Tit and Siskin with Crossbill in some numbers during irruption years. Such habitats also hold numerous breeding Sparrowhawk and, in very much smaller numbers, Goshawk. The combination of felled forests and open country also makes this area attractive to Nightjar, a species which is largely restricted to this part of the county. Treecreeper is fairly widespread, but Nuthatch is very much more evident in the more mature woodlands of lower elevations.

Elsewhere around the county, coniferous blocks of woodland, especially in the western fringes, provide woodland habitat where, away from the valleys, little exists. In that sense the farm shelter-belts and copses are important as an additional habitat that complements others in these areas, though *per se* they are not of major importance to woodland birds due to their largely dense and species-poor nature. Many of the smaller woodlands were planted after the Second World War.

The lowlands

The Durham uplands are able to squeeze a considerable measure of precipitation from the prevailing westerly winds. By contrast the county's coastal strip is one of the driest parts of the British Isles. As the county's rivers flow eastwards, the level of the land falls away and they journey through wider valleys and skirt gentler hills than in the west.

Some might consider a large proportion of the Durham lowlands rather featureless in comparison with the scenic riches of the western dales, but these areas have their attractions. A short distance inland from the coast, between the Wear and the sea, there is a relatively rapid increase in altitude to the East Durham Plateau. This is high limestone escarpment is one of the principal landscape features of the eastern part of the county and is formed from an outcrop of Permian magnesian limestone. It runs north to south down Durham's eastern edge it dips toward the coast from a crest that averages 180m above ordnance datum, though it rises to a peak of 218m above sea level in the south west of the plateau area. This limestone ridge has profoundly influenced the botany of the area. It outcrops at Hartlepool Headland and the whole of the area is pock-marked with a series of both disused and working quarries from the north of the county at Sunderland and thence to Houghton and Hetton, across the main parts of the plateau at Cassop and Thrislington, all the way south east almost to Hartlepool. In the east of the county, over the limestone and along some of the coastal fringes, small numbers of Corn Bunting can still be found in isolated pockets, although they have disappeared over much of the rest of Durham.

From the high limestone reaches, the strata dip gently eastward, topped by glacial drift. At the coast, these strata form, by and large, low limestone cliffs that have been weathered and eroded by the sea. These are cut by the deep, easterly draining, heavily wooded coastal denes which are an almost unique feature of the Durham coast, and run from Ryhope in the north in a punctuated progression, via Seaham, Dalton, Horden and Hesledon, south to Crimdon, where the Crimdon Beck meets the sea. Each of the denes has its own particular qualities, but the largest ones, at Castle Eden and Hawthorn, dwarf the others in terms of area and ecological interest. They formed as a result of water eroding through the relatively soft limestone strata. The steep sides of the denes have subsequently protected the woodlands from the negative impact of agricultural and housing development. The vegetation and habitats present range from magnesian limestone grassland, ameliorated by the overlying drift, through scrub to mature valley woodlands. Although they are not strictly speaking ancient woodlands in the ecological sense, it is true that they have probably been wooded since the land recovered from the last Ice Age, though both of the major denes have been extensively managed and have therefore changed through time. Castle Eden Dene for instance has large numbers of introduced trees, many dating back to the 18th century, when the Burdon family managed the Dene for timber production. Whilst the great number of yew trees, which provide a major source of food for a variety of woodland birds, appear to be native, the impressive beech trees *Fagus sylvatica* are introduced. As well as containing many rare species of flowering plants and bryophytes, the latter resulting from the damp microclimates within the Dene, it has long been one of the county strongholds for the Marsh Tit. The variety of habitats contained within the Dene means that it has one of the most extensive lists of breeding birds of any site in the east of the county. Hawthorn Dene has a mixture of old woodland combined with secondary plantations and also supports a wide variety of woodland birds. Many of the smaller denes are important refuges for passage migrants, as well as for scrub-nesting species such as Wren which enjoy the protection of virtually impenetrable vegetation at locations such as Fox Holes Dene.

Much of this eastern lowland part of the county area is overlain with a variety of soil types that make it varied in vegetation; only in restricted areas do limestone outcrops materially influence the vegetation to such a degree that magnesian limestone grassland, one of the county's most important botanical habitats, develops. In these areas the ornithologist will be more interested in the scrubby fringes and steep woodlands and copses found throughout the plateau. These are probably best represented at Penshaw, Herrington, Houghton and Cassop Vale. Gorse *Ulex europaeus* is almost ubiquitous as a colonising shrub in such areas, as is bramble *Rubus spp.* and dog rose *Rosa canina agg.* When such scrub matures, hawthorn *Crataegus monogyna* and elder *Sambuca nigra* often become dominant and all of these plants supply a rich source of berries and seeds for feeding birds in the late summer and early autumn, and in spring these areas are some of the most important for nesting Linnet. In summer, herb-rich grassland means a good crop of invertebrates and these provide sustenance for a variety of birds, including Little Owl, which favour the limestone quarries, both old and new. The open ground of the limestone grassland slopes provides good hunting for Kestrel, whilst the adjoining scrub affords ideal prospecting perches for Sparrowhawk.

The sites of greatest ornithological interest in the lowlands are often those subject to a degree of human intervention, namely the disused quarries where fringing established grassland, colonising limestone flora and invading scrub, often blackthorn *Prunus spinosa*, provide a mosaic of habitats ideal for birds such as Whitethroat and Grasshopper Warbler. The classic sites for this type of habitat fall within an area delineated by Cassop, Thrislington and Wingate, but some of the north eastern sites, in Sunderland and South Tyneside, are also valuable. Lowland breeding Peregrine are increasingly prominent in this part of Durham.

In the south east of the county, the area around the broad estuary of the River Tees, the Tees Lowlands, lies below 120m above sea level over lacustrine deposits. It is drained, principally, by the river Skerne into the Tees. Long ago post-glacial lakes lay here becoming filled in as bogs, and they grew deep deposits of peat, up to 12m at Bradbury Carrs. At one time this area, 'the Carrs', made up of Mordon, Bradbury, Preston and Swan Carrs would have provided winter floods for masses of wildfowl and the damp and marshy conditions, now extensively drained, would have swarmed with bird life through the summer. The landscape remains punctuated by the cultural references to its watery past in the names of its hamlets and dwellings such as Great Isle and Little Isle Farms. In general terms the land in the Tees lowlands is dominated by arable farming as it comprises some of the most productive soils in the county.

Lowland temperatures in the summer are not infrequently below 15.5°C or 5.5°C in the winter though the coastal strip is, in relative terms, somewhat cooler in summer and respectively warmer in the winter. In parts of the wider more open areas of the central lowlands and the south east of the county, there are some areas that provide frost hollows, holding on to cold air during winter days and, in some situations, lowland fogs. This is especially so where these have a north-facing aspect to their orientation.

In the centre of the county and fanning out northwards from Bishop Auckland, the River Wear lowlands comprise a gently rolling topography that has been heavily moulded by glaciations. They extend eastward from below 120m in the Pennine foothills to the limestone scarp of the East Durham Plateau. The Wear lowlands are much narrower than their Tees counterparts because of the topography of the river valley itself and the proximity of the raised plateaux to the east and west. Despite this, the ecological interest is probably greater than that of the Tees lowlands. This is mainly on account of the more extensive valley woodlands which stretch from the Valley Burn at Spennymoor almost all the way into the urban area of Sunderland at South Hylton with only brief punctuations at Durham and Chester-le-Street. This central portion of Durham, which roughly corresponds to the exposed coalfield, in some respects is of relatively limited ornithological value in terms of specialist suites of species, although it has abundant common species like Chaffinch and Linnet and other species associated with mixed and open farmland. Many of the common hedgerow species, both resident and summer visitors alike, are also present, along with hole-nesting species like Stock Dove and predatory species such as Kestrel and Little Owl. Man-made or influenced wetlands, such as mining subsidence pools, today have breeding Mallard, Coot and Moorhen, Tufted Duck, Little Grebe and the occasional pair of Great Crested Grebe. The river valleys with their adjacent woodlands form some of the most interesting bird habitats of this area. Many of the woodlands are semi-natural and their flora and fauna reflect this. Ash, wych elm, and sycamore *Acer pseudoplatanus* often predominate in the canopy but pedunculate oak *Quercus robur* is also common. Whilst these Wear lowland woodlands are not of the highest quality, they do serve as home to a wide variety of woodland birds, such as Blackcap, Chiffchaff and Willow Warbler, woodpeckers and Tawny Owl. A large proportion of this part of the county, if not farmed or reclaimed after mining, is built up. Over much of this part of Durham the small areas of copse and plantation woodland are dispersed through the landscape.

River alluvium, silts and clays are probably some of the most widespread of the county's post-glacial deposits and many of these can be found at lowland locations in the central, north and south eastern portions of the county, often in association with the river floodplains. The county's agriculture is largely a mixed one, in the lowland areas at least, with arable areas scattered across much of the north, central, eastern and south eastern areas (Clifton & Hedley 1995). In relation to the depredations of this habitat elsewhere in England, the county still has a relatively large number of hedgerows, although there has been much degradation of these over the last six or seven decades, largely as a result of poor management rather than removal. The majority of hedgerows in Durham date from successive periods of enclosure between the 16th and 19th centuries. Most of the county's lowland hedges date from the enclosure of village town fields by private agreement in the 16th and 17th centuries. Those in the pastoral uplands and upland fringes were largely created later, under the 18th and 19th century Parliament Enclosure Acts. More ancient features such as medieval parish and township boundaries are scattered only thinly across the county. Along these hedgerows many typical species such as Blackbird, Song Thrush, Robin, Dunnock and Wren can be found in good numbers, whilst in agricultural habitats Grey Partridge remain relatively widespread in the field margins and hedge banks and are still reasonably common in many areas, often occurring up to the fringes of the uplands. Where hedgerow trees occur, species like Greenfinch, Goldfinch and Chaffinch are found and during summer Willow Warblers and Whitethroat are common. In some areas Tree Sparrow can still be found, despite some local declines.

In the south of the county, which has never been heavily exploited for coal, the landscape is pleasantly rural and criss-crossed with hedgerows and dotted with copses. To the south and east of Bishop Auckland the countryside becomes flatter and the scenery is characterised by scattered farmsteads and spread with a patchwork of agriculture and uncultivated land. Large numbers of Wood Pigeon together with other species associated with agricultural landscapes such as corvids and Starling are typical of such areas.

Wetlands

If Durham is poorly endowed with any lowland habitats, aside from semi-natural ancient woodlands, it is natural wetlands. Many of the county's wetland systems, such as those once present at the Skerne Carrs, were concentrated in the low-lying ground in the south east of the county and have long since disappeared in any recognisable form as a result of the extensive drainage of land for agriculture.

Ponds of all descriptions, mainly small in size, are widely distributed across the county but there is nothing remotely resembling a natural lake in the Durham area. Such small wetlands are important for wildlife although, unless of a more extensive size, they are unlikely to be important for birdlife beyond an occasional pair of Moorhen or Mallard. Many such sites are found on low-lying ground or on land where local drainage conditions mean that water accrues more rapidly than it can drain away, such as on clay-rich soils. The few lakes that are present in the county have been created for landscape purposes, often in connection with the development of the grander estates often built with mining wealth from the late 18th century onwards. Such locations would include Axwell Park Lake in the north, Brancepeth and Whitworth Parks in the mid-county, Hardwick Hall in the south and Wynyard in the south east.

Most of the significant extant wetlands, at least away from the Tees estuary, are either man-made or much influenced by his activities; many came about as a result of mining subsidence. One of the most telling wetland features of the county is its suite of reservoirs; these are principally located in the higher dales though there are some lowland examples. Some of these, especially the larger ones where a range of habitats exist around their edges, are very good for birds.

Some of the most important habitats for birds are the fringing edges of wetlands, reed swamp and marsh. These occur around ponds and in water-filled ditches with common reed *Phragmites australis* and other fringing vegetation growing next to areas of wet grassland and open water. They are one of the most productive of lowland habitats in the county for birds, including Water Rail and Reed Warbler, but they are relatively restricted in the Durham context. The largest concentration of these habitats is situated around the Tees estuary, though there are other important sites such as at Brasside Pond, Barmston Pond, Shibdon Pond and WWT Washington.

The county's largest water bodies are all man-made, constructed to provide water for the urban areas and industrialised conurbations. The largest of the upland reservoirs is Derwent Reservoir in the north-west, in the Derwent valley, followed by Cow Green Reservoir in the very highest reaches of upper Teesdale, whilst there is a series of others that dot the upper dales in all of the principal river valleys. These include Selset and Grassholme in Lunedale, Balderhead, Hury and Blackton in Baldersdale, Burnhope and Tunstall in the Wear valley, and north towards the Derwent valley lie Waskerley, draining into the Wear, and Smiddyshaw and Hisehope draining down to the Derwent. Some of these, especially those in Lunedale and Baldersdale, off the Tees valley, and Derwent Reservoir are particularly important in ornithological terms (Westerberg *et al.* 1994). The northern part of Derwent Reservoir lies in Northumberland. This site has seen collaboration between Northumbrian Water and the Durham Wildlife Trust leading to a large part of the shallower western end of the reservoir being developed into a nature reserve. Some of the upland reservoirs hold Black-headed Gull colonies but this species is as likely to be found around small moorland tarns or boggy moorland wetlands scattered in the west and south west. Some of the reservoirs and moorland wetlands support small but significant breeding populations of Wigeon. Scattered across the western uplands are many smaller reservoirs such as Burnhead Dam, Sikehead Dam and Burnhope Dam, at the head waters of the Burnhope Burn which, 11km to the north east, drains into the River Derwent. Many others, their function in feeding the mineral hushes long since gone, sit silent in the landscape, unnamed but often loved by a pair or two of Teal, an occasional pair of Ringed Plover, and almost always Oystercatcher and Common Sandpiper. These larger reservoirs were mainly constructed in the early part of the 20th century, though Derwent Reservoir was built during the 1960s and opened in 1967.

In the lowlands there is a cluster of reservoirs that service the Teesside area, most notably Hurworth Burn Reservoir, long one of the best wetland sites for birds in the county, and nearby Crookfoot Reservoir - two sites that for many years have shared populations of birds, such as the flocks of geese that commute between them. Towards the North Sea coast is the county's most easterly such water body, Hart Reservoir; within easy flying distance of the sea it attracts an occasional sea duck or diver.

The county's few lowland wetlands of any significant size have largely been formed as a result of mining subsidence, such as at Shibdon Pond and the wetlands in the Bishop Middleham area, or by mineral extraction, for example McNeil Bottoms and Low Barns, up and downstream respectively of Witton-le-Wear. Along a 30km stretch of the Wear valley, between McNeil Bottoms and Croxdale, there are a number of wetland sites, Low Barns, Butterby Oxbow, Holywell Hall Lake, Brancepeth Beck, Page Bank, Byers Green Hall, Cobey Carr, Escomb Lakes and Beechburn, which are important for birds. These sites provide wintering grounds for a variety of wildfowl, whilst the spring and summer brings Kingfisher and breeding birds such as Common Sandpiper, with scarcer waders, grebes and the odd rarity occasional visitors during migration periods. Man-made wetlands include the important complex of habitats at the Wildfowl and Wetlands Trust at Washington, in the lower Wear valley, where the county's largest breeding colony of Grey Heron is located. This site's development was catalysed by the impending loss of one of the north east's best lowland wetlands in the early 1970s, Barmston Pond at Washington. This site amassed an enviable list of passage wading birds, favourably comparable to any site in the area, and many years later it still remains one of, if not the best single wetland in terms of the range of waders recorded (a total of 34 different species), illustrating the attraction a wetland site can have when such habitat is in short supply. The remaining pond, whilst important in the local context, is just a remnant of what was once present.

The north east of Durham is particularly impoverished in terms of freshwater habitat so the winter floods at Boldon Flats and the small water body at SAFC Academy are both attractive to wetland birds. Boldon Flats is drained for grazing from early spring until late autumn, but it provides winter bathing and roost sites for large numbers of gulls and some waders. Being low-lying ground, the area is prone to flash floods during late summer storms, when large numbers of species such as Ruff, Green and Wood Sandpiper can occur before the floodwaters subside.

Primarily since the turn of the century, a set of wetlands in the centre of the county at Bishop Middleham has both expanded, courtesy of active mining subsidence, and developed to form one of the most important such complexes in the area. The main locations of Castle Lake, Stoneybeck Lake, Alan's Pools and the A1 Flashes are all located within 3km of the village of Bishop Middleham. Castle Lake is a subsidence pond fed by runoff from surrounding farmland and is actively managed by the Durham Bird Club. It boasts an impressive array of passage species, mainly waders but also scarcer wildfowl, and it holds a significant breeding population of Gadwall. Not only is the area rich in wetlands but the surrounding farmland holds the only sizable remnant Corn Bunting population in the county, along with large populations of other threatened farmland species.

Urban wetlands, particularly in the shape of formal ponds and park areas, occasionally assume quite a high level of importance for birds, one example being the formal hard-edged lake at South Shields Marine Park, which is close to the North Sea, and was one of the most important sites for Mute Swan in the county over a long period through the 1990s and early 2000s.

Coastline and cliffs

At the coast, temperature variations are buffered by the proximity of the North Sea. This ameliorating effect means that the average summer maximum temperature usually occurs in July or August, the winter minimum in January or February. The North Sea waters between Flamborough Head and Berwick are the coldest around the British Isles so the summer temperatures at the county's coastal sites are limited compared to elsewhere in Britain (Manley 1935). Thus, Sunderland's summer maxima do not reach those of inland Durham City, but neither do the winter minima descend so low. The maritime effect of the North Sea in Durham thus helps create warmer winters and cooler summers, with one obvious effect of this reduction in summer temperature and increased moisture being the predominance of 'haars' or 'sea frets' as locally called. These occur most often in the warmer months of the year during calm anticyclonic weather and are influenced by the steep temperature gradient between the cool sea and the warmer air radiated from the land. The effect of this is that a thin band of fog covers the coastal strip, making it considerably cooler than just a few kilometres inland.

Along the coastal strip, the sheltered denes aside, the predominant vegetation characteristics are of scrub or open grasslands. Where scrub is present it tends to be of low and, on the exposed coastal frontage, wind-sculpted, comprising species such as hawthorn and blackthorn with much gorse on free-draining soils. Where the larger Durham denes meet the sea, mature scrub and woodland finds its nearest approach in the county to the North Sea.

In the north east of Durham, where the magnesian limestone escarpment meets the sea, beautiful, buff-coloured cliffs, deeply fissured and wonderfully weathered, harbour myriad caves and crevices and host significant seabird colonies. These coastal limestone cliffs, overlaid with boulder clay and herb-rich grasslands, reach their highest, over 30m, at Marsden and Lizard Point. At Marsden Bay the cliffs form, with some impressive sea stacks, the basis of the largest seabird colony between the Farne Isles in Northumberland and Bempton Cliffs in North Yorkshire. The sights, sounds and smell of a 'seabird city' in summer are a memorable experience. Over 3,000 pairs of Kittiwake, up to 150 pairs of Cormorant, over 150 pairs of Herring Gull, almost 200 pairs of Fulmar and about 30 pairs of Razorbill nest. George Temperley (1951) noted that only from about 1930 had the cliffs "been colonised successively by Fulmar, then Kittiwake, the Herring Gull and the Lesser Black-backed Gull...". Temperley cited Hutchinson (1840) as proof that none of these species was breeding in Durham in the early part of the 19th century and there is no authoritative evidence for the previous large-scale breeding of seabirds at Marsden. Although, writing of Marsden Rock in his "View of the County Palatinate of Durham" (Newcastle 1834), Mackenzie affirmed that "vast numbers of seafowl used to build their nests upon its crest", there is, other than hearsay, little evidence to base the statement upon. In the mid-1920s, several pairs of Fulmars prospected these cliffs in the course of that species' colonisation of the British Isles. This was added to when Kittiwakes commenced their tenancy of Marsden Rock, in the early 1930s. The closure of the coast to public access during the 1939-1945 war gave Herring Gulls the opportunity to increase considerably, leading to their colonisation of roof tops in South Shields and Sunderland during the 1960s. Today, smaller colonies of Kittiwake and Fulmar can be found scattered south along the coast - a particularly unusual feature of the county being the inland nesting of Kittiwake on buildings along the River Tyne and at the mouth of the River Wear and at Hartlepool.

Where headlands jut out into the sea, such as at Whitburn and Hartlepool, the movements of seabirds can be observed from land and, in the right weather conditions, these promontories provide thousands of autumn migrants with their first opportunity for landfall after an arduous North Sea crossing. Though the coastal margin of Durham is relatively dry in terms of precipitation it 'enjoys' the bracing proximity of the North Sea, being exposed to the chilling effects of northerly and easterly wind and weather (Wheeler 1992). Such winds, during migration, are responsible for bringing 'falls' of migrants to the area. These invariably make for the first available cover which, in the north, is often best afforded by man-made features such as quarries, parks and cemeteries, whilst further south, the scrubby mouths of the coastal denes are important, as are the suburban gardens of the houses on the Heugh, site of the ancient mediaeval settlements, on Hartlepool Headland.

There are around 48km of coastline between North Gare at Teesmouth and the mouth of the River Tyne, just north of Littlehaven Beach at South Shields. This area has been hugely and very famously disrupted, damaged and defaced by man's industrial activities from the mid-19th century until the mid-1990s. The east Durham coalfields were first exploited, at an industrial level, in the 1840s. The disposal of spoil from these rapidly became a problem and the expedient way of dealing with this problem, of both liquid and solid coal waste from six collieries along the coast, was to tip it directly onto the beaches of Durham, for the sea and elements to deal with. This process ceased in 1993 with the closure of the last of these collieries at Easington. From that date, time and tide, the natural processes of weathering, have ostensibly returned the coast to a more normal profile and state. The wildlife of such areas is at long last returning and the offshore marine fauna has staged a remarkable recovery, with benefits for locally wintering seabirds.

Coastal sand dunes and related grasslands are a rare habitat in Durham, small amounts being found at Sunderland and South Shields but some of the best examples are located to the north of Hartlepool at Hart Warren and further extensive examples, over 300ha, at Seaton Dunes and Common to the north of the Tees estuary. Hart Warren is a particularly unusual example of such a habitat as, courtesy of the underlying geology, it is influenced by the magnesian limestone, making it a unique example in Britain. Such areas, often dynamic in terms of substrates, are inhospitable to birdlife but they usually hold good populations of breeding Skylark and Meadow Pipit and, where sufficient scrub develops, a few pairs of Stonechat, whilst in the winter they may attract Snow Bunting. Most prominently, the sandy beaches in front of the dunes at Crimdon, and occasionally Seaton, host the county's only colony of Little Terns.

Estuaries

In the south east corner of the county, the broad tidal reaches of the Tees and its attendant lowland plain contrasts markedly with the east Durham Plateau to the north and, in the far-distant west, the Pennine massif. Much of the geology of the Tees Plain or lowlands is composed mainly of Bunter Sandstone and Keuper Marl, rocks deposited in the Triassic period; these sweep in a broad arc from Seaton Carew west to Darlington and are overlain by a thick layer of glacial and post-glacial boulder clays, sands, gravels and river deposits; alluvial deposits that the Tees has worked upon for millennia.

Where the Tees reaches the sea, a large estuary exists. Four hundred years ago the teasing Tees meandered to its mouth in a natural state, flowing out to meet the North Sea over a sand bar that delineated county from coast. A correspondent of Sir Thomas Chaloner (1561-1615) put it more quaintly *"Neere unto Dobhoome (the port in the mouth of Tease for named) the shore lyes flatt where a shelfe of sand raised above the highe water marke interteines an infynite number of sea fowle which laye their egges heere and there scatteringlie in such sorte that in tyme of breedinge one can hardly sett his foote soe warelye that he spoyle not many of their nestes"* (Fowler 2010). It is clear that Teesmouth was once bounded by huge areas of wetland, marshes and commons, and was one of Britain's most important estuarine systems for birds. At the sand bar, it was said a dense cloud of seabirds came each spring to make their colonies at the mouth of the river. From the 18th century onwards, this area was increasingly brought under man's control, the engineers, enclosure and embankment, held ever greater sway and a wild way was lost. Sadly, much of the original estuary has been claimed from the sea and only a fraction of the splendour for wildlife that was the Teesmouth is left as the North Tees Marshes. According to one account (Heslop-Harrison 1928), by the early part of the 20th century, some 26,000 acres (10,500 ha) of salt marsh and other estuarine habitats had been lost. By 1850, the remaining estuary had been reduced to around 2,400ha, largely as a result of the embankment of land to exclude the sea and to create additional summer grazing for livestock. The claiming of this land from the once extensive marshes and estuarine habitats and the subsequent development of the petro-chemical and other heavy industries seriously depleted the area's ornithological interest. Nonetheless, what remains is still of great importance for a wide variety of birds, although they are somewhat overshadowed by the industrial backdrop. Teesmouth remains of such significance, both nationally and internationally, that it has been designated a Special Protection Area, has a National Nature Reserve (Teesmouth NNR) at its heart and is an internationally important RAMSAR site. It is particularly important for its wintering waders and wildfowl, its passage waders and some of its breeding waterfowl, as well as marshland birds such as Reed Warbler and Water Rail, and in recent times Cetti's Warbler has bred.

Abutting the Tees estuary are the remnant salt marshes and mud-flats of Seal Sands. These represent one of the most established and productive systems in the coastal region of the area. The only other recognisable areas of salt marsh in Durham are on the River Wear at South Hylton (Sunderland), where the River Don meets the Tyne at Jarrow Slake, and a small section on the River Team at Dunston. These are of little significance compared with the many acres flanking the Tees and Greatham Creek. Colonising salt marsh species, such as glasswort *Salicornia spp.* are generally poorly represented in Durham but they occur here. More abundant, where it occurs, is the sea blight *Suaeda maritima*; a source of sustenance for plant-feeding wildfowl and waders, as well as some passerines such as Twite during winter.

Today Teesmouth remains nationally important for a range of wildfowl and wader species. In recent years an increasing concern for birds and the wider environment has developed at Teesmouth, with growing co-operation between conservationists and industrialists on Teesside. As a result, some areas of land owned by the chemical industries are now being managed specifically for their wildlife interest. Through the early part of the 2000s the RSPB, working with the Teesside Environmental Trust, created a 380ha complex of wetlands and visitors' facilities at RSPB Saltholme (opened in 2009), complementing the existing wetlands and the groundwork undertaken in this area by a number of agencies over decades. The impact of human intervention on Teesside cannot be over-stated.

The future

Whilst many elements of the county's geography referred to are fixed in historical timescales, there will be changes in coming decades and many of these will affect the area's birdlife; none is more potentially impactful than the effects of global climate change. The underlying rocks will remain, the soils will remain but man's influences to come, for good or bad, will have the greatest influence upon the 'way the wind blows' and what comes on that wind.

Chapter 2

A Short Environmental History of County Durham

The modern landscape, flora and fauna of Durham is the product of its history; to better understand these, it is important to understand a little of the area's environmental past. The underlying rocks were laid down scores of millions of years ago. During the sequence of glaciations experienced by Durham through more recent millennia, the Pennine peaks were rounded, the valleys of the Tees, Wear and Tyne were carved out and their courses scoured and re-shaped as the kilometres-thick ice sheets scraped the rocks and then melted at the end of the last Ice Age. The glaciers and their melt-waters gouged out the surface deposits and wore away the underlying rocks under the vast hydro-static pressure built up from these – and then they re-distributed vast quantities of sediments around the lowland landscape. It was these powerful, landscape shaping forces that blasted a way to the sea for the River Wear through the limestone of the East Durham Plateau and left us with the topography we are familiar with.

The modern fauna and flora of Durham County developed only after the end of the last glaciation, some 10,000-13,000 years ago. This was a cold, bare landscape that remained after the ice's retreat. It would have been first colonised by low-growing plants that, in time, would have resembled the highest of today's northern uplands or the tundra of Fennoscandia. Sedges, grasses and herbs, mosses lichens and dwarf shrubs would have been prominent, and ultimately plants of the heather and willow families would have come to dominate; providing a living for the species now associated primarily with north European habitats and climes: Snow Buntings, Dotterel and perhaps Lapland Bunting. Geography and geology, supported by altitude, combined to create a landscape and flora in upper Teesdale that more resembled the high tops of northern Scotland, than the lowlands of east Durham. Through the centuries, this post-glacial landscape persisted in Durham and in few other places in England. It is likely that such a landscape would have been home to a typical array of northern tundra species, skuas, Snowy Owl, Ptarmigan, and numbers of breeding waders and wildfowl, many of which are today scarce passage visitors to the lowland areas of the county, as they shuttle between their now more northerly breeding grounds and the locations they inhabit during our winter.

Through subsequent millennia, birch scrub and colonising woodlands climbed the hillsides from the river valley bottoms, creeping up the valley sides, and taking their attendant bird populations with them. These would have comprised, at least initially, Brambling, Willow Warbler and Black Grouse, but also more open-country species such as Wheatear, Red Grouse, Twite and what would now be considered exotic breeders, such as Redwing and Fieldfare. No doubt some of Durham's familiar breeding species would have also had a presence, such as Merlin, Redpoll and Ring Ouzel. At the time of the ice melt sea levels rose by scores of metres, and the shoreline of modern Durham, with its large cliff barriers to the sea, barring subsequent erosion and degradation, would have taken on some of their modern profile. The areas of the Tees lowlands however, must have then seemed very different, with vast coastal marshes and lowland woodlands, sweeping out into the area now covered by the North Sea, only to be inundated through the years, as the sea rose to be replaced by the shallow seas of today's Hartlepool and Tees Bays.

Around 9,000 years ago much of Durham would have been covered by woodland. This would have comprised a wide array of tree and shrub species, including lime, elm and alder along the river valleys, with ash on the limestone areas, Scots pine on more exposed high ground, oak and abundant hazel on the free-draining acid soils. These longer-lived tree species would have followed in the wake of the by then existing birch woodlands. The primeval woodlands would have been rich in birdlife, much more so than modern woodlands and plantations that are species poor and lack the abundance of critical features, such as standing dead-timber, compared to these earlier, more extensive habitats.

Shortly after the end of the last glaciation, humans entered the Middle Stone Age and at this time, hunters in Yorkshire were documented as preying upon Cormorant, Mallard, Goldeneye, swan and Jay. It seems reasonable to assume that many of these species would have been present in Durham. The remains of Red

Kite and White-tailed Eagle have been found at such sites; presumably as trophies, rather than food. In terms of the first significant human impacts upon Durham's landscapes and habitats, this is not definitely known but is believed to have occurred around 7,000-8,000 years ago in the Mesolithic period. Some of the earliest evidence of human presence is of worked flints discovered near to the mouth of Castle Eden Dene; these were presumably imported by early hunter-gatherers as the stone of which they are made is not found locally.

The Mesolithic hunter-gatherers and those of the later Neolithic, left middens, including bird remains in cave sites around Britain, including Durham. They would have survived on a mixed diet gleaned from the landscape seasonal fruit, fish and birds. Many of the sites from this time contain the remains of Great Auk, indicating that the species was quite widespread around the British coastline some 2,000-3,000 years ago. The evidence of man's presence in Durham, in the shape of Neolithic stone implements, arrow-heads, hand axes and the like, is quite widespread.

Perhaps the first observers of Durham's bird life were roving bands, searching for Black Grouse, Hazelhen and Capercaillie amongst the birch and pine woodlands before foraying on to the higher ground in search of wader and gull eggs to plunder? Such species would have been breeding on the extensive bogs and peatlands of the west of the county. These people, mobile but small in number, would have had relatively little impact upon the world around them. There is no certain information about the Durham avifauna at these times, though excavations of Mesolithic settlements in neighbouring counties have revealed the presence of species such as, White Stork and Crane, Great Crested Grebe, Red-breasted Merganser, Lapwing and Buzzard.

During the later Neolithic period, some 5,500 years before the present, agriculture probably came to Durham. By 2500 BC, in most of lowland Britain, agriculture had become the dominant way of life, even in the uplands, with developing forms of wheat and barley the main crops. A change in the vegetational composition of the major flora in the county through this period in history is indicated by analyses of the pollen record. This shows that there was an increase in the relative proportion of grass to tree pollen and a concomitant reduction in the amount of tree pollen making up the total proportion of pollen, indicating that many trees were at this time disappearing from Durham. Such woodland clearance took place alongside the development of early farming systems. It was from this time that man began to become a major shaping force upon both landscape and fauna. The upland's woodland cover, which may have been sparse in the extreme west of Durham, would have been driven 'down the hill' to be replaced in the long-term by dwarf heath. As well as clearing upland woodlands, the early Durham inhabitants would have begun to cultivate the lowlands. The more extensive keeping of animals: pigs, sheep, and cattle, the clearance of land for year-round forage and grazing would have created more open habitats, new habitat for open country species. The late Stone Age saw the appearance of ritualised burials sites around the county, such as that at Muggleswick Park, and many of these contained evidence of an increasing reliance on agriculture; implying an ever greater impact upon the landscape of the area.

By 2000 BC, features such as Quern Stones, believed to have been used to grind flour, appear at sites such as Butsfield near Consett. The tools of these early people improved through the Bronze and subsequent Iron Age (from about 2,100 to 700 BC). The effect of these developments was to speed up woodland clearances and such de-forestation supported a growth in human populations; leading to even greater losses of woodland particularly in the uplands and a resultant degradation of soils in such areas. It is conceivable that some of the first permanent colonisations of the county may have taken place in the uplands, with movement to lower elevations as upland soils became less productive. There would have then been a wider influence upon lowland soils as a result of woodland clearance and tillage. By the end of the Iron Age there was a fully developed farming economy in Durham, with a range of livestock being reared locally and increasing levels of cereal production, as indicated by the excavation of Iron Age remains at Thorpe Thewles and Hartlepool Headland. As a result of the appearance within the Durham landscape, of grass-related crops (oats, wheat and barley, farmland birds would have begun to prosper, although some woodland species would have probably lost out.

Climate deterioration from around 700 BC would have contributed to the shaping of Durham's extensive westerly uplands, on acid soils, leached of nutrients as a result of deforestation. These uplands became dominated by low-growing shrubs, largely of the heather family, though juniper would have been widespread in some areas. In terms of its inhabitants, at this time, Durham would have been within the territory of the great northern tribe of ancient Briton, the Brigantes. These people, skilled in the use of iron and with a highly developed artistic culture, were a loose federation of tribal families led through a fierce, royal dynasty. Because of their Celtic roots, they would have venerated the oak and the extensive oak woodlands, particularly of north and west Durham would have been important to them. For instance, the name Derwent comes from the Celtic word, *Derw*, meaning 'valley of the oaks' and at that time the oak woodland of that valley would have extended from the banks of the Tyne to the edges of the high moorlands, around Nookton and Hunstanworth. In those uplands there was by this time, perhaps as little as 5% woodland cover around the Pennine massif.

Agriculture changed in a dramatic way when the wheeled plough came north with the Romans, leading to a large increase in the amount of ploughed land and a growth in the volume of grain produced on the most productive of the county's lowlands. The Roman conquest led to significant Roman settlements and camps at South Shields, at the mouth of the Tyne in the north, and at various locations close to the Roman Road of Watling Street (also known as *Deor* or Dere Street), that ran north south through the county. Settlements grew at Piercebridge, Binchester, Lanchester and Ebchester. Excavations of Roman sites has shown that Mallard, Crane, Red Kite, a number of corvids and goose species, including Greylag, Pink-footed and Barnacle, were all present; as were either Mute or Whooper Swans.

Following the decline of the Western Roman Empire, our understanding of the history of the early Middle Ages (the 'Dark Ages') in Durham is largely drawn from ecclesiastical records, including those of Cuthbert and the Venerable Bede. Through this period of successive invasions by Saxons, Danes and Vikings, which commenced around 450AD, monasteries were a key feature of the area, both as seats of learning and focal points for social gathering. One of the earliest documented settlers of this time, who offered his name to the landscape, was *Ceoppa* who in died in 685 and was buried near Horsegate above the location of his farmstead, that had been cut into the woodland named after him, Chopwell. Important monasteries were located at Gateshead as early as 653 and on Hartlepool Headland a little later according to Bede. After further forest depredations though the Saxon, Viking and Mediaeval periods, Durham's woodland cover was but a tiny proportion of its original area and largely located, as it is today, in ribbons along the main river valleys.

The north east of England today, incorporating County Durham, is in many ways analogous to the ancient kingdom of Northumbria. At the height of its powers, the kingdom of Northumbria stretched, as implied by the name, between the rivers of the Humber, north to the river Forth. This area became prominent at the beginning of the 7th century, as a kingdom of Angles, Danes and the offspring of marauding Norwegians. This bold, independent, distinctive and some might say, culturally rebellious, nature persists in the north east to the present.

Through time, climate conditions in the north east of England have varied hugely. This is illustrated by the fact that within the period of recorded history appreciably warmer periods have occurred, most notably during the Anglo-Saxon to Norman period (750 to 1250). Such warmer interludes alternated with colder times, such as the 'Little Ice Age', which lasted from the 12th to the mid-19th century, reaching its coolest period during the 17th century when, famously, the rivers froze in southern England, allowing frost fairs to take place on the Thames. Less extreme warmer periods occurred, for instance between the middle of the 19th to the mid-20th century. Such fluctuations of temperature and attendant climate factors will have had profound effects on the wider flora and fauna of Durham. The status and distribution of species, both resident and migratory ones, would have changed through time. In a county of relative altitudinal extremes, such as Durham, these effects would have been more pronounced than in other areas. During the colder episodes of history the county's highest tops would have seemed even more extreme and been more attractive to what are now considered northern bird species. By contrast, the coastal lowland margins in the warmer periods would have been balmier still, presumably allowing the penetration north into the county of some continental or southern species.

In some respects, the Durham area remained somewhat impervious to the Norman invasion of the 11th century; the national assay that was the Domesday Book not reaching into what was a wild uncultured land. Instead, an analogous document, known as the Boldon Book, did a similar job in the region, ordered by the Bishopric of Durham in 1183. Newcastle was founded in 1080, a bid by the Norman governors to exert control over the area, using the key crossing of the Tyne, once utilised by the Romans, as a strategic check to more rebellious instincts. Prior to this the region was known as centre of religious art and study which perhaps found its zenith of expression in the production of the Lindisfarne Gospels. It is littered with ornate illustrations of birds, no doubt many of them indicative of the species seen by the monks and scholars working on the texts and illustrations, sadly, not many are recognisable as species (barring what might have been a White-tailed Eagle).

Partly in response to the area's less than whole-hearted acceptance of the Norman Conquest, the area was colonised by a number of Norman fortifications. As a result, the north east is dotted with castles and fortified manors, a memory of its troubled and bloody past, from the southern access point to County Durham, at Barnard Castle, to the northern fortifications of Berwick upon Tweed. One of the most prominent is that of Durham, constructed after William I's visit to the north from around 1072, on the site of what had been an older fortification utilised by Cuthbert's followers after their remove from Chester-le-Street; prior to the establishment of the Cathedral. This castle eventually came to be the adopted home of Durham's Prince Bishops, until the early 19th century.

Durham was along with Cheshire and Lancashire, one of three English 'counties palatine'. This might be described as a territory, in this case a county, in which the chief lord, the Prince Bishop, exercised certain royal gifts and privileges (the *jura regalia*), including the rights to 'own' courts, mint coinage, run militia, appoint sheriffs and grant pardons. Essentially the Bishops acted, in many respects, as a regional sovereign in absence of the monarch. The origin of these privileges is not clear, but date to the Anglo-Saxon period. The power of the Prince Bishops was exercised in varying degrees by generations of Bishops until they ceased to be used on the death of the last of these, Van Mildert, in 1836. They wielded huge power, deciding how land was managed and what was to be hunted, where and by whom.

Coal mining has been a major economic driver and landscape shaper in Durham for centuries and the dawn of such mining activities in the area, may date from the Roman occupation, though deeper deposits would not have been worked for fuel until much later than this. Some twelfth century references to coal may in fact relate to charcoal, which would have been extensively produced from Durham's oak woodlands. The county's even depleted woodland resource was greatly prized. The heavily wooded Derwent valley was known to have been cropped to provide timber for the construction of Durham Cathedral in the 10th and 11th centuries. In 1294, Edward I requisitioned wood from the woodlands in Durham for warships; the shipwrights of the Tyne obtaining oak from Chopwell, a few miles up the Derwent. Centuries afterwards, Charles I repeated the compliment; extraction of timber for ship-building commencing in the Crown estates of 'Choople Woods' in June 1635. The cumulative impact on the area's woodland bird life must have been considerable through the centuries.

Lead mining and smelting has been conducted in the Durham uplands from a remote period of the past, possibly even prior to the Roman's location in the north east. Lead is known to have been exported to Rome 200 years prior to the invasion of the Legions, and much of this would have been sourced from the Pennine ore fields, including those of west Durham. During the Mediaeval period, the Bishop of Durham, Hugh Pudsey was granted permission to conduct lead mining in Weardale by King Stephen (1135-1154). However, it was coal, above all other minerals however, that both shaped and defined the north east, economically and culturally for nearly 800 years. By the latter half of the 13th century coal was being shipped out of the river Tyne. Sea-coal was collected along the coast of Durham by the early years of the 14th century. By the early modern period, the amount of coal shipped annually was measured in hundreds of thousands of tons, but this figure was dwarfed in the centuries to follow.

Through the Middle Ages, some of the relatively little available evidence about Durham's local birds came from the evidence of species gathered for food and documented by religious institutions. One of the most spectacular examples was the 50 unspecified swans provided by the Bishop of Durham for King Richard II's visit in 1387. During the 1530s, the Bursar of Durham Monastery, according to the Cellarer's rolls of that institution, over the years purchased: Brent Geese, Mallard, Oystercatcher, Lapwings, Ruff, Snipe, Woodcock, Whimper, Curlew and Redshanks for the table. Most were probably sourced locally, though Puffins were transported south from Northumberland for similar purposes.

In the north east of the county, salt manufacture was once a major industry at South Shields, this was documented as early as the 13th century and continued on through the reigns of Elizabeth I and on to Charles I. By 1667 there were 121 salt-pans at 'Shields' and this had risen to 143 by the height of the trade in 1696; subsequent decline meant the trade had ceased by the end of the 19th century. One might speculate on what wetlands bird species might have been attracted to such features through time.

In the early days of human habitation in Durham, local foraging would have been important in determining the location of settlements. Often these would be sited close to rivers that provided fresh water and hunting opportunities, such as fishing, at a time when local migratory fish stocks would have been prodigious in comparison with modern periods. Such fisheries have been well documented along the lower Tyne and other rivers, where tidal influences came up against shallow rocky shoals, offering easier fishing, such as at Blaydon and Ryton. The Tyne has long been renowned for its salmon fisheries. A Bishops Charter to the prior and monks of Durham in 1103 names twenty eight fisheries between Tynemouth and Wallsend, and during an inquisition by Henry I a further 48 fisheries were listed upstream from Jarrow to the Hedwin Stream – doubtless fish-eating birds and other animals would have associated with such productive habitats.

In a number of respects, there was relatively little change in the farming footprint of Durham, from the end of the Iron Age to the major land-use changes associated with the Agricultural Revolution of the late 18th century. It was at this time that the effects of a series of enclosure of land acts began to manifest themselves in Durham. These led to a large increase in the length of hedgerows and associated copses in the county, and all of the benefits both to nesting and distribution for a wide range of common woodland edge and 'hedgerow' species, which prior to this would have been very much less common.

Game-rearing commenced in earnest during the main period of enclosure in Durham, around the late 18th century, and over this period the blueprint of much of Durham's central and lowlands agricultural landscape would have been laid out. This is one that is characterised by scattered small woodlands and copses many of which were probably first planted as game coverts. These would have been connected by 'quick-thorn' hedgerows surrounding moderately-sized fields, with larger scale woodlands traversing the landscape, in a principally east west direction, with some local deviation, along the main river corridors. Bird and game preservation probably intensified through the 19th century as muzzle-loading guns gave way to more efficient breech loaded shotguns. By the mid-Victorian period 'walking-up shooting' was overtaken by 'driven' shooting that led to vastly increased game bags and a rapid surge in popularity for game shooting. To maximise game bags, the range of bird species classed as vermin rapidly increased leading to many birds of prey being ruthlessly persecuted and some species, such as Red Kite, being lost to the area.

The lowland landscape map set at this time largely remains recognisable in Durham to the modern day. From the 18th century, larger amounts of information about the county's birds began to become available. Much of this was courtesy of the increasing number of gentleman naturalists and collectors; the most significant in the early Durham era was probably Marmaduke Tunstall of Wycliffe (1743-1790), though he was followed by many important names in the following two centuries. Across the 18th and 19th centuries, as wealth derived from mining and other industrial developments accrued, there were established a number of major estates with the planting of many non-native and decorative trees. Examples include those at: Gibside, Ravensworth, Lumley, Brancepeth, Witton Park, Wynyard and Raby. These were often established by wealthy industrialists and land-owners such as Charles William Vane (Marquis of Londonderry), Lord Lambton and Lord Ravensworth; and doubled as shooting estates.

From the late 17th century, with the commencement of the Industrial Revolution, man's activities began to have a major affect on the look and quality of landscape, and its carrying capacity for wildlife and dependant birds. One of the big changes was the development of engineered features at a landscape level, largely from around the middle of the 18th century, such as the development of extensive wagon ways, used for the movement of coal around the county. The first of these appeared in the north of Durham around 1620 and at a later phase, it could be said that these laid the foundation for the rail network that was to grow from its 1822 base, when the first rail was laid along the Stockton and Darlington railway. This commenced a period of rampant railway development fuelled by the north's great engineers and innovators, which led to many new features on the Durham landscape, cuttings, embankments and viaducts across deep valleys all distributing the wealth of the area.

The major engineering of land for agricultural end uses took on a new facet in Durham during the 18th century, which brought the beginnings of encroachment upon the Tees estuary, with the embankment of land at Saltholme in 1740. As engineering expertise grew in the early 19th century, a large meander of the river, close to the Tees estuary was bypassed, with the opening of a canal that cut across this feature. From the mid-19th century, industrialisation of the county's main river mouths was well underway and such activities around Teesmouth ate more and more into the area of habitats, with the developments there of Victorian iron and steel works and a growth of shipbuilding.

There was a similar remodelling of the Tyne to serve industrial growth, driven by the Tyne Improvement Act of 1861. For much of its history, the Tyne had been a relatively shallow river but by 1866, 5.2 million tons of material had been removed to transform the river's geography. Up-river at Dunston, the 30-acre island, Kings Meadows, the largest in a group of small islands, called the Clarence Islands, was dredged away in 1885 to accommodate the growing industry. In just under fifty years, the Tyne had been completely remodelled. Through this time, what flowed down the county's rivers became ever more polluted. Water pollution caused by human effluent is often regarded as a 19th century impact and in many cases this is true, but it is significant that it was not until 1920, that Tyne estuary readings of oxygen fell to zero. The water quality of the river for much of the 20th century remained extremely poor. As late as 1959, it was reported that the water was lethal to fish and dangerous to humans, the resulting effect of 270 sewers pumping 35 million gallons of untreated sewage into the river. It was not until the 1970s, that these issues began to be seriously addressed.

The late 19th and early 20th centuries brought an increasing pace of change to Durham's countryside and intensification in the management of agriculture in the county, with no doubt consequent impacts on farmland and the birds of the wider countryside. Around this time, came the genesis of the chemical industry, which would have such profound landscape impact upon the life of Teesside. In Billingham, for instance, beginning with nitrate production during the 1914-1918 war, chemical production rapidly expanded from 1926 as ammonia, chemical fertiliser and methanol were added to the products list. The village of Billingham, with just 354 inhabitants in 1801, is now subsumed by the greater Teesside conurbation. The growth of industry not only consumed land, but led to despoilment and pollution. As early as the mid-17th century coal ships entering the Tyne were dumping their ballast onto the 300 acre expanse of mudflats at the outflow of the River Don, described by one contemporary as 'a parcel of Land or Waste on the River Tyne, called Jarrow Slike'. These expansive mudflats would have undoubtedly been a rich area for wildlife.

By 1882, the rapid industrialisation of the Tyne riverbanks was having a serious detrimental effect, with one individual observing that 'few would care to linger by the banks of the Lower Tyne' and that it required 'imagination to find the faintest suggestion of a meadow' and the 'grass and herbage' were 'black and shiny'. This was probably typical of many areas along the lower stretches of the county's rive systems, in the pre-regulatory days. This was having a demonstrable effect upon wildlife. John Hancock was of the opinion that Jarrow Slake had been 'in great measure destroyed as a resort of wildfowl'. He observed that the 'straight lines of the engineer' had replaced the 'beautiful sweeping reaches, and projecting headlands, that diversified both shores'. However, despite the obvious degradation of the river, the Tyne was still a key habitat for birds, with one observer recording that 'the Kittiwake, the Lesser Black-backed Gull, the Black-headed and other

Gulls' were still present on the river at Felling in 1884. Water quality was also markedly affected, with an 1882 description of the Don at Jarrow noting the *"the odours arising from this slimy ooze… were enough to make enthusiasm faint"*. Pollution and degradation of land got worse as the 20th century approached and the north east region specialised ever more in heavy industry.

Through the 19th century and in the early part of the 20th century, the spread of urban areas, particularly around the mouths of the major rivers, led to the merging of smaller settlements into ever greater urbanised zones (the population of Britain grew from 16 million in 1850 to 42 million in 1950). The creation of conurbations meant fewer and fewer species of bird as green spaces and habitats within them were lost to development and industry. For example, in the north end of the Team valley an area of wet floodplain holding substantial populations of breeding waders and wildfowl, from 1936 onwards was developed as an industrial trading estate to alleviate high rates of local unemployment. The River Team was hidden in a culvert and grazing meadows lost to roads, industrial units and manicured lawns.

In the wider landscape, the early part of the 20th century brought increased land use change across Durham, the rate of such change in particular in relation to agriculture, accelerated, after the Second World War. The intensification of agriculture brought new challenges to the birdlife of lowland Durham, and some changes in the uplands. The mechanisation of farm work led to the loss of many important open-county features, ponds, hedges, previously rough and uncultivated corners, and the loss of the horse from the wider countryside economy, after many centuries, meant impacts for many bird and wildlife species. A measure of the changes that occurred locally in agricultural landscapes is shown, in microcosm by the changes to the landscape of the county's coastal strip in the north east of the county in less than fifty years. In the early part of the 20th century, the shore and coastal fields stretched, uninterrupted, from South Shields to Marsden Village and beyond, to Sunderland. The coast in those days was a mixture of pasture where cattle grazed and barley, kale and turnip fields where Lapwings, Skylarks and Grey Partridge nested. Corn Buntings, Yellowhammers and Linnets were plentiful in this area and sang from the straggling hedgerows and limestone walls which marked out the patchwork of fields. Spring and autumn migrants were attracted by the variety of shelter and the winter stubble fields held large flocks of buntings and finches. The key factor in all of this is a loss of diversity, at both a small and grander scale; inevitably birds lose out.

Away from the industrial areas, agriculture was the biggest influence on the County Durham landscape, the intensification of this through the late 19th and 20th centuries took a heavy toll on Durham's breeding birds. The changes included a growth in field sizes, the loss of field boundaries, the felling of hedgerow trees drainage of damp margins, a move away from the over-wintering of stubbles, to autumn ploughing. The development, and widespread application, of persistent pesticides and herbicides, some of them manufactured at Teesside, to agriculture through the 1950s and early 1960s, had a hugely adverse affect on some of the county's birdlife. By way of example, it was at this time that one of the county's most common predators, the Sparrowhawk, disappeared for almost a generation from many parts of the lowlands.

The Great War resulted in the felling of 0.5million acres of woodland in Britain. With the cessation of hostilities, there was an imperative governmental plan to expand forest area and increase timber production. The Forestry Act of 1919 established the Forestry Commission which planned an ambitious planting programme. In Durham this resulted in the so-called 'State Forest' at Hamsterley which was first planted in the 1930s and covered some of the valley slopes of the Bedburn Beck upstream of Hamsterley Village, on to the slopes of Eggleston Common. This is now the largest forest in County Durham and covers more than 2,000 hectares and supports a wide range of breeding birds, including Nightjar, Goshawk and Buzzard. In the year 2000, the area of woodlands and forests in the UK was twice that in 1900, although with a cover of just 11% of the total land area it was one of the lowest in Europe.

As the region's population grew and industry developed, demand for water became more intense. To satisfy this, at the head of the major rivers, reservoirs of various sizes were built. Derwent Reservoir was completed in 1967, the largest such water body in the county. Damming rivers and flooding some sensitive upland habitats proved controversial and was much contested. In order to service a planned ammonia works for ICI on Teesside, Parliament agreed on the creation of a new reservoir at Cow Green in upper Teesdale in

1967. This despite much opposition from a range of conservation organisations, the resultant reservoir flooded land of the highest botanical importance, but the poor consultation prior to this controversial and much opposed development led the rejection of a number of other planned reservoirs elsewhere in the country on environmental grounds.

As the industrial fortunes of the region waned, many industrial processes ceased and the sites that had produced pollution and proved a problem for birdlife became an opportunity to exploit. Through time, derelict industrial sites became plentiful in Durham in particular following the collapse of manufacturing and industry. Some of these were unsightly and sometimes polluted, and in terms of the numerous coal spoil heaps of the Durham coalfield, they covered a large area of land. Nonetheless such areas, once out of production quickly became havens for wildlife. For example, one of the area's new economic hubs, the Metrocentre at Gateshead, stands on part of the site of Dunston Power Station, which closed in 1981 leaving spoil heaps and deep lagoons. Despite the contaminated ground, this environment proved attractive to Stonechat and Little Ringed Plover. Interestingly, during the various phases of the shopping centre's development in the 1990s shallow lakes in areas of the undeveloped site were left relatively undisturbed and they supported breeding Little Ringed and Ringed Plover, small numbers of breeding wildfowl, and were attractive to passage waders, including Wood Sandpiper and large number of roosting gulls. When left to revert to nature, such areas can, in time, develop into first-class habitats. As exemplified by Shibdon Pond, once an armaments dump, then a colliery, bounded by railways, and finally a site earmarked as a local authority refuse tip, but saved at the last minute and afforded protection as a nature reserve.

Another legacy of the county's industrial past is the numerous disused mineral railway lines that criss-cross the county, some beginning in the uplands. These once vital arteries of industrial and human carriage still function as popular transport networks, with many being converted to long-distance footpaths, bridleways or cycle paths. They form linear habitats and have the effect of providing flyways for small bird species, often through open, less hospitable farmland. In addition, because of the poor soil quality of railways embankments and cuttings (which were often formed of coal waste), some of the habitat that has developed is of a dry heathland type, a scarce environment in the county. Here, Stonechat, Whitethroat and Willow Warbler, among other species, have benefitted from this marginal habitat.

In Britain, the number of cars grew from 5.5 million in 1960 to 20 million in 1990, and this has had many negative environmental effects as the growing road network absorbed land and vehicle emissions accounted for new forms of pollution. The road network in County Durham is not as extensive as other parts of the country. Began in 1963, the A1 dual carriageway (now the A1M) runs along a north-south spinal route through the county and its construction revealed much about the region's past, as it navigated the peat deposits of a post-glacial silt lake in the River Skerne carrs and extensive outcrops of dolomitic limestone near Bowburn. This and other roads through Durham account for countless bird deaths every year, as is evident to car travellers on the Chester-le-Street section of the A1, which is usually littered with the corpses of Pheasants, released by adjacent shooting estates. Whilst road construction has eaten away at habitats, created noise and disturbance, the road verges that have resulted have proved important undisturbed habitat for some species.

Even in the late 1960s and early 1970s, the deprivations of land claim at Teesmouth had not abated and more land was claimed from the estuary and the sea, for the benefit of development over this period. With some degree of irony, the majority of this was never utilised, and areas of this became, with the passage of a few decades, increasingly important for wildlife. By the 1970s, there were only some 200ha of remaining estuarine habitats from an area that was in pre-historic times calculated to have measured well over 10,000ha. Man's landscape impacts did not cease with the last of Teesmouth's 1970 land claims, for a barrage across the river Tees was constructed in 1994, meaning that the areas upstream of Stockton on Tees were no longer tidal.

A significant landscape development during the second half of the 20th century was the creation of wildlife reserves and National Nature Reserves by statutory and non-statutory organisations. These proved important in protecting some of the county's best habitats for wildlife, including birds. Key to this process was the founding of the Northumberland and Durham Naturalists' Trust in the early 1960s, and its subsequent division in 1971 into two separate wildlife trusts for Durham and Northumberland. This led to the formation of a number of nature reserves which were managed primarily for their wildlife interests. The Durham Wildlife Trust manages a number of important wildlife sites across the county today, including some that are important for their birdlife; ironically some of these are on post-industrial sites, such as Low Barns near Witton-le-Wear and Rainton Meadows near Houghton-le-Spring.

Today, County Durham remains relatively replete with what seems unspoilt countryside. Along the Durham coastline there are beautiful beaches, many of which are recovering from generations of industrial despoilment and in this attractive landscape live a proud, passionate people. In this sense Durham can be seen as a sub-set of the wider north east of England region, whose distinct dialects in some ways define the identity of its folk and represent the finer details of the landscapes they inhabit. To the north the Northumbrian dialects are famous for their distinctive and readily recognisable nature; the soft rolling 'Rs' of the Northumbrian twang. Moving south there is the harsher, world-famous 'Geordie' accent - which swaggers in its self-deprecating toughness and in the sweat and genius of the people that speak it, knowing they powered much of the Industrial Revolution. South again, come the softer tones of the central part of rural Durham and the western dales. In the far south of Durham, along the Tees these accents begin to be influenced by and, in the west reaches of that river, to merge with North Yorkshire's 'Tyke' tones. These voices, recognisably and inherently 'northern' have, in the main, grown out of the early Germanic languages of the Angles and Vikings and the Celto Romano-British tribes who peopled the north east of England, modified its landscapes and now, study its birdlife.

Acknowledgments
This chapter was compiled using a wide range of both formal and informal background sources by Dr. Fred Milton and Keith Bowey. Prime sources include those highlighted below:

Bibliography & References
Benett R. 1844 *The Durham household Book (liber bursarii ecclesiae Dunelmensis); or the accounts of the Bursar of the Monastery of Durham from Pentecost 1530 to Pentecost 1534* London Surtees Society 18

Burton J.F. 1995 *Birds and Climate Change* Christopher Helm London

Boyle J. R. 1892 *Comprehensive Guide to the County of Durham* The Walter Scott Publishing Co. Ltd. Durham

Clark J.G. D. C. 1954 *Excavations at Star Carr* Cambridge

Embleton D. 1884 Note on the Birds seen at Nest House, Felling Shore, in May and June 1884 Newcastle, Bell, 1884, Medical Tracts Vol. 54.

Fisher J. 1966 *The Shell Bird Book (3rd Ed. 1973)* Ebury and Michael Joseph, London

Fletcher J. 2010 *Birdwatchers of Teesmouth 1600–1960, The Events leading up to the formation of the Teesmouth Bird Club* The Teesmouth Bird Club

Gardner-Medwin D. 1985 *Early Bird Records for Northumberland and Durham Birds: Ancient and Modern* Transactions of the Natural History Society of Northumbria Volume 54

Graham G.G. 1988 *The Flora and Vegetation of County Durham.* The Durham Flora Committee and the Durham County Conservation Trust, Durham

Hancock J 1874 *A Catalogue of the Birds of Northumberland and Durham* Transactions of the Natural History Society of Northumberland and Durham & Newcastle upon Tyne, Vol. VIV

Harrison C. 1988 *The History of the Birds of Britain* William Collins London

Kerr I. 2001 *Northumbrian Birds: Their history and status up to the 21st Century* Northumberland & Tyneside Bird Club Newcastle

Miles J. 1992 *Hadrian's Birds* Miles and Miles of Countryside, Castle Camrock

Palmer W. J. 1822 *The Tyne and its Tributaries* George Bell London

Spray C., Fraser M. and Coleman J. 1996 *The Swans of Berwick upon Tweed* Northumbrian Water

Temperley G.W. 1951 *A History of the Birds of Durham* Transactions of the Natural History Society of Northumberland, Durham & Newcastle-upon-Tyne, Vol. IX

Chapter 3

The Study of Birds in County Durham

Any work on the wildlife of an area is also a work on the people who have studied and documented it, as acknowledged by George Temperley, who included information on the history of ornithology and the ornithologists of County Durham in *A History of the Birds of Durham*. This chapter provides a summary of some of the developments in the observation and documentation of birds and short biographies of some of those who have contributed to the study of birds in Durham.

Over more than 200 years, the county has been studied by many naturalists, especially botanists and ornithologists. The principal areas of interest have always been the upper dales (in particular Teesdale), Teesmouth and the Derwent valley, although many other areas have been studied. The Marsden and Whitburn areas have attracted much attention, particularly from the early 20th century when seabirds began to be established as breeders and in the latter part of the century when interest in sea passage and rare migrants blossomed, Hartlepool Headland became a prime destination.

Early Days

David Gardner-Medwin (1985) and Nick Rossiter (1999) discussed the historic ornithology of the region, the first covered both Durham and Northumberland, the latter was primarily about Northumberland but both had some relevance to the history of Durham's birds.

The earliest 'ornithologist' relevant to the county that can be named is Reginald of Durham, who between 1156 and 1172 documented an account of the legends associated with St. Cuthbert, which related various stories about birds, including those of Eiders on the Farne Islands. A few centuries later, the Cellarer's rolls of the Abbey of Durham during the 14th century included information about a surprisingly wide a range of species purchased for the functioning of the Abbey as investigated by Canon J.T. Fowler and published in 1898 as *Extracts from the Account Rolls of the Abbey of Durham*, through the Surtees Society. This listed species, the purpose of their purchase and prices, under their medieval names, such as '*morecok*' or '*murkoke*' (Red Grouse), or in Latin, and they provided a fascinating insight to some of the species that were present locally at the time, usually by way of their presence on the monks' table.

Another early source of information, at least of edible species, in County Durham is *The Durham household book (Liber bursarii ecclesiae Dunelmensis)*. Written by Robert Benett the Bursar of the Monastery of Durham, it covered the period from Pentecost in 1530 until Pentecost 1534 and detailed expenditure on food and other items for the monastic community. Not surprisingly, many items related to edible birds, the majority of which would have been sourced locally. Some listed were clearly domesticated species, for example, *capones, pacoks*, aucea *(geese) and pulli columbarum* (pigeon poults); many however, referred to wild birds. The catalogue contained records of at least 13 species, some bought from local huntsmen. Clearly, it is impossible to ascertain that these were all killed within what is now County Durham, but in many instances, most probably were local. Gardner-Medwin highlighted the Tees and other marshland areas (probably Bradbury/Preston and Mordon Carrs) as being likely sources. Amongst those listed in the Bursar's records were: Brent Goose (*rutgoys*), Oystercatcher (*seapye*), Mallard, (*mawlert*), Plover, *Wype* (Lapwing), *Styntt, Dunling, Ree* (Reeve), *Snype, Wodcok, Whympernell* (Whimbrel), Curlew and *Reydshank*.

An early ornithological publication that has some relevance to Durham was that of John Ray, who produced *The Ornithology of Francis Willughby,* in 1676 in Latin, an English version being printed a couple of years later. Ray (1627-1705), son of an Essex blacksmith, was primarily a botanist; who with Francis Willughby (after they had met at Trinity College, Cambridge) toured Europe, collecting specimens and making natural history notes. In England they made a series of journeys both together and separately, including tours to the north of England by Ray, in 1661 and 1671, where they liaised with local collectors and specialists (Raven 1942). On these he met Ralph Johnson, setting up a correspondence, which led to the documentation of a number of first records of species for the county and one or two for Britain. Willughby's book was a general ornithological text and not devoted to Durham. It covered around 230 species, a number of which related to early records from their travels into Yorkshire, Durham and Northumberland. There was a particular Teesmouth connection to their observations as they had been in correspondence with Ralph Johnson of Brignall, a small hamlet upstream from Greta Bridge. Johnson wrote to Ray with observations of a number of intriguing early records, particularly of wildfowl and waders, many of which were collected by him in the Teesmouth area where he regularly hunted. For example, Johnson reported that Whimbrel "*is found upon the sands of Teezmouth*". In addition, via Ray and Willughby, Johnson was responsible for the first documented British record of Long-tailed Duck, which he found on the Tees.

John Wallis (1714-1793)

Some of the early documentation of the region's birdlife was undertaken by the Reverend John Wallis (1714-1793), though most of his work was of greater relevance to Northumberland. He moved from Northumberland to become a curate in the Darlington area in 1775 where he lived until he retired to Norton in summer 1793 but there appear to be no writings by him on the bird life of these areas.

Marmaduke Tunstall (1743-1790)

Marmaduke Tunstall of Wycliffe Hall was a keen naturalist and perhaps the most important figure in the early ornithological history of Durham. He was a descendant of Sir Thomas Tunstall, of Thurland Castle, Lancashire and the Wycliffes of Wycliffe-on-Tees. This family estate located in a small village on the south bank of the Tees was originally in North Yorkshire but is now in Durham. The family could trace its ancestry to the reign of Edward I.

Born at Burton Constable, Yorkshire, Tunstall was originally called Marmaduke Constable, his father Cuthbert having changed his name on succeeding to the estate of Burton Constable. After his father's death when Marmaduke was four he was raised by his mother who later sent him to a college at Douai in France. After finishing his education, he lived in London and amassed his natural history collection which included live specimens as well as preserved items which later formed the foundation of his museum.

Marmaduke Tunstall
Courtesy of the Natural History Society of Northumbria

His uncle, Marmaduke Tunstall died a bachelor, aged 89, in 1760, leaving the young Marmaduke, then just 17, the family estates of Scargill, Hutton, Long Villers and Wycliffe. In order to inherit, he had to take the family name of Tunstall. At this time he cultivated the friendship of Thomas Pennant, George Latham and other well-known naturalists and also corresponded with Linnaeus. He was elected a Fellow of the Society of Antiquaries of London in 1764 and in 1771 privately published his list of British birds *Ornithologia Britannica*, probably the first British work to use binomial nomenclature.

He began to rebuild Wycliffe Hall in about 1773 and included in this a handsome, large, airy room to re-house his museum. On his marriage in 1776 to Miss Markham of Hoxley, Lincolnshire. In 1777, he was elected a Fellow of the Royal Society and his collection was transferred from London in about 1780. He enjoyed a quiet studious life at Wycliffe and by then his museum was the third largest private collection outside London and one of the finest in England. It was especially rich in mounted birds. Many well-known naturalists, including Thomas Bewick (although not until after Tunstall's death) used the collection as models for natural history books. Tunstall was credited by T. H. Nelson in his *Birds of Yorkshire* (1907) with the earliest references to that county for several species, including Song Thrush, Blackbird and Redstart, all being found at Wycliffe.

Tunstall, a Roman Catholic, was known as 'a man of gentle manners and high principle'. He was broadminded and had the greatest respect for the opinions of others even when they differed from his own. He was also a philanthropist and was described as *'the poor man's friend'*. He died suddenly in October 1790, aged 48, and was buried in the chancel of his own church at Wycliffe. Whilst Tunstall added hugely to our knowledge of local birds, he could not be considered an ornithologist in any modern sense. There is no evidence that he took an interest in living birds. His main activity was in collecting a systematic series of bird specimens.

John Goundry (dates unknown)

John Goundry is known from George Townshend Fox's *Synopsis of the Newcastle Museum* and Thomas Bewick's autobiographical *Memoir*. He lived at Wycliffe-on-Tees and did taxidermy work for Marmaduke Tunstall (1743-1790). Fox referred once to 'Old John Goundry', Mr Tunstall's *"joiner and bird-stuffer"*, who was still living at Wycliffe in 1827. Goundry was a tenant on the Wycliffe estate, his house being shown on a plan of 1789 beside the church and river, with a garden of 20 perches. Thomas Bewick on visiting Wycliffe to draw specimens from Tunstall's collection, lodged in John Goundry's house and ate at his father's house, George Goundry who was the local miller. John Goundry was preparing specimens in 1790, when he stuffed a Stoat for William Salvin (as documented in a letter from Tunstall to William Salvin, on 24 May 1790). Goundry probably also prepared a male and female Scaup in December 1788, the first specimens of that species Marmaduke Tunstall had seen.

George Allan (1736-1800)

George Allan spent much of his youth in Wakefield, Yorkshire, where he was educated at the Queen Elizabeth Grammar School. In adult life he became a solicitor who lived at Blackwell Grange near Darlington but his passion was as an antiquarian. He produced a number of works on aspects of County Durham's history and assisted William Hutchinson in the preparation of the *History and Antiquities of the County Palatine of Durham*. He possessed a printing press, which he used to produce several works including a reprint of the *Legend of St. Cuthbert*. For much of his life he was an enthusiastic collector of specimens. There is little evidence of him taking an active interest in ornithology, as a field study, but in 1791 he purchased the contents of the Wycliffe Museum from Edward Sheldon who had inherited the Tunstall estates when Marmaduke Tunstall died in1790. This purchase was made for £700 and along with his collection, became known as the Allan Museum. The collections occupied two large rooms at the Grange and Allan opened his museum to the public in June 1792, receiving 7,327 visitors in three and a half years. He added considerably to the Wycliffe collections. When Allan died the collection passed to his son.

George Allan
by J Collyer, Courtesy of the Natural History Society of Northumbria

The Newcastle upon Tyne Literary & Philosophical Society was founded in 1793 and in its early years members and friends donated a number of natural history items. George Townsend Fox (1782-1848) was a businessman and botanist and a prominent member of the society when the Allan Museum came up for sale in 1822. Fox loaned £400 to the society for the purchase of this, which eventually passed on to the Hancock Museum.

Thomas Bewick (1753-1828)

Bewick was probably the greatest of the early natural history illustrators. A renowned wood engraver and naturalist, he was born at Cherryburn House, Mickley in south Northumberland. He was apprenticed to Ralph

Bielby, a Newcastle engraver, eventually going into partnership with him later in life. In the early 1770s he illustrated a number of children's books and famously producing wood engravings for Thomas Saint's books of *Fables* in 1779 and 1784.

Throughout his career, he was one of the foremost and influential artists and wood-engravers to work in the north of England. His books, *A General History of Quadrupeds* and *History of British Birds*, set new standards for illustrations with lifelike depictions of birds and mammals and these helped popularise the study of natural history. Bewick had an intimate knowledge of the habits of animals and birds, which he acquired during regular excursions into the countryside and his love for walking.

Thomas Bewick
by John Burnet, 1817 Courtesy of the Natural History Society of Northumbria

In 1789 Marmaduke Tunstall commissioned one of Bewick's best known engravings, the Chillingham Bull, yet they were never to meet. After Tunstall's death in 1790, Bewick spent two months at Wycliffe during 1791 using the extensive collection of stuffed birds as models. A number of drawings he made at the time are now in the archives of the Natural History Society of Northumbria although he preferred to draw from life whenever possible. A *General History of Quadrupeds* was published in 1790, and Bewick's great achievement, the *History of British Birds*, was published from 1797 and 1804. There were two volumes, *Land Birds* and *Water Birds*, with a supplement in 1821.

The *Quadrupeds* dealt with mammals of the world and is particularly thorough on some domestic animals. It includes bats and seals but does not include whales or dolphins; *Birds* is specifically British. A number of first documented records of species for Durham, such as Little Auk, come from this work.

In 1812 he and his family moved to Back Lane (later re-named West Street) in Gateshead where he lived for the rest of his life. He had retired from the workshop in 1825, handing the business over to his son Robert, and worked on his engravings from home by the time he was visited by the famous American naturalist John James Audubon in 1827.

Audubon lavished praise on the veteran artist, later describing him as "*a complete Englishman, full of life and energy although now 74, very witty and clever, better acquainted with America than most of his countrymen and an honour to England.*"

At the Natural History Society's first lecture on 20 October 1829 at the still occupied premises of the Newcastle Literary & Philosophical Society, the subject was the identification of a new species of swan, discovered by Richard Wingate, a Newcastle taxidermist, from a specimen taken at Haydon Bridge, Northumberland. Two years after Bewick's death, the new species was named Bewick's Swan *Cygnus bewickii* in his honour. A stone plaque was erected on the side of his family home in Gateshead in 1928 on the 100th anniversary of his death.

Into the Nineteenth Century

The late 18th century and the early 19th century saw a more systematic approach to the recording of Durham's birds. The late 18th century also saw the birth of people like the Rev. John Graves and Rev. John Brewster who later included lists of birds in the Cleveland area in larger works. Brewster's work would later include the bird list of John Hogg. In 1829, The Natural History Society of Northumberland, Durham and Newcastle its first volume of Natural History Transactions being published in 1831; one of the first such societies of its kind in the country.

Prideaux John Selby (1788–1867)

Selby of Twizell, Northumberland, was, according to George Temperley, the first ornithologist to pay significant attention to the birds of Durham. His first written work was *A Catalogue of the Birds hitherto met within the Counties of Northumberland and Durham* and was published in 1831. It provided the first relatively full bird list for both counties. He is best known for his *Illustrations of British Ornithology* (1821–1834), the first set of life-sized illustrations of British birds. He also wrote *Illustrations of Ornithology* with William Jardine and *A History of British Forest-trees* (1842). Many illustrations in his works were drawn from specimens in his own collection. In addition to the above works he contributed to Jardine's *Naturalist's Library* volumes on Pigeons (1835) and Parrots (1836). Selby concentrated mainly on the birds in his native north Northumberland but also studied those of Durham.

Prideaux John Selby
by T R Goddard Courtesy of the Natural History Society of Northumbria

John Hutchinson (1797-1855)

Hutchinson was born in Durham, the son of John Hutchinson, an alderman and mayor of the City of Durham during the late 18th and early 19th century. Though wishing to enter the church, he became a solicitor in Durham, subsequently becoming town clerk. In 1831 he moved Lanchester, where he worked as clerk to the local magistrates. In 1840, he produced a catalogue called *Birds of Durham*. It was produced in two volumes along with another on *Durham fishes, Reptiles and Quadrupeds.* Although never published, Temperley regarded this document as valuable in giving the status and distribution of birds in Durham at that time. Although he was a member of the Linnaean Society and corresponded with the renowned ornithologist William Yarrell, other than with William Proctor of the Durham University Museum, he does not seem to have consorted much with other local naturalists. The manuscript document of Hutchinson's list was discovered in a second-hand bookshop in the late 1940s by H. M. S. Blair and donated to the Natural History Society.

From the late 18th century through to the early 20th century, four generations of the Backhouse family, bankers with business interests in Darlington and, in particular Sunderland, formed something of a natural history dynasty in the north east. Though their primary interest was botanical these ranged across many subjects and they can be rightly considered all round natural historians, who excelled in their fields.

James Backhouse (1794-1869)

James Backhouse was the fourth son of James Backhouse (1757-1804) and Mary Backhouse. Born into a prominent Quaker family of Darlington, with considerable banking interests, his father died when he was a child and his mother brought him up in a religious atmosphere. He was a sickly child and in consequence he spent many childhood summers in the fresh air of upper Teesdale. This led to him having a lifelong love of that part of County Durham. Poor health and a need for outdoor pursuits led to him and his brother Thomas purchasing, in 1815, a nursery and gardening business in York. In 1822 he married, had two children and in 1824 was admitted as a minister in the Religious Society of Friends. In 1831, after his wife's death in 1827, he travelled to Australia, and then South Africa, on Quaker missionary work, returning in February 1841. On his return he recommenced work in the nursery business, and when his brother died in 1845, James his son was brought in. He then recommenced his visits to Teesdale alongside his son - one room of the High Force Hotel being kept as Mr. Backhouse's room - they largely studied botany but also documented some of the birdlife of the dale.

John Hogg (1800-1869)

John Hogg was born in Norton and educated at St. Peter's College Durham; he travelled widely in the region as a barrister. Hogg was largely a botanist - a fellow of the Linnaean Society – but he was also active in the wider natural history scene and a very capable ornithologist. He made extensive studies into many of the rarer bird species at a time when relatively little was known. During his life he published articles on fish, molluscs, barnacles, echinoderms and of most interest in relation to birds a catalogue called *Birds frequenting the country near Stockton* that was published in 1824. This contained a list of 126 species and is notable for naming Garganey for the first time in Durham. He was described by Temperley in 1951 as *"another very competent ornithologist"*. In 1844 he produced another paper entitled *Catalogue of birds observed in South Eastern Durham and in North Western Cleveland,* in the *Zoologist*. He was a Fellow of the Royal Society and a member of the Linnaean Society. Most of John Hogg's observations were made without optical aids and it is therefore understandable that he and others of the time used guns to bring birds down for study in-the-hand and for identification.

Edward Backhouse (1808-1879)

Edward Backhouse was born in Darlington on 8 May 1808, the son of Edward Backhouse of Darlington. In 1816, the family moved to Sunderland where the first Backhouse Bank was established by Edward's father, Edward Backhouse Senior (1781-1860). Edward remained in Sunderland for much of his life thereafter. He became a partner in the family banking firm of Backhouse & Co. but he did not take an active role in the business. His unpublished manuscript *A Catalogue of the Birds of the County of Durham*, written in 1834, was not widely available but was probably the earliest attempt at a complete listing of County Durham's birds, listing the 203 species that he had collated as having been recorded in Durham. This catalogue gave a status for each species but full details, such as dates and locations, were only listed for the scarce species. Away from wildlife, he worked as a minister for the Society of Friends from 1854 and was also active in establishing or supporting a range of charitable endeavours in the Sunderland area. Edward Backhouse frequently visited Seaton Carew and it was here that he saw County Durham's first documented Osprey in 1828. His natural history interests ranged widely and included: Lepidoptera, molluscs, geology and botany. He donated much of his extensive collection of specimens to the Sunderland Museum.

William Proctor (1798-1877)

William Proctor was a carpenter's apprentice who turned to natural history and specialised in taxidermy. Proctor was a taxidermist and the first sub-curator of the Durham University Museum. His best-known exploit was a collecting trip to Iceland. Durham University's museum was founded in 1833, the year after the founding of the University, the second university museum in England to be opened to the public. Proctor was appointed *"to the charge of the Birds in the Museum"* in 1834 at a stipend of £25. He was responsible for publishing the first list of exclusively County Durham birds called, not unsurprisingly, a *List of Birds found in the County of Durham*. It comprised 204 species, with limited additional information about each species and was appended to the Rev. George Ornsby's *Sketches of Durham* in 1846.

John Hancock (1808-1890)

Among the many famous naturalists who have been members of the Natural History Society, Hancock, a taxidermist of national renown is the best known. He was instrumental in helping to raise money to build a new museum for its collection. This 'New Museum of Natural History' was opened in 1884 and re-named The Hancock Museum in John and his brother Albany's, honour in 1891. He was a noted ornithologist, producing his *Catalogue of the Birds of Northumberland and Durham* in 1874, which documented 266 species. He donated his magnificent collection of mounted British birds, some of which can still be seen in the Hancock at the Great North Museum. Hancock spent time in Durham collecting specimens, for example, the first Tree Sparrow for Tyneside was a specimen he shot in Gateshead in 1834. He was a prodigious collector and during migration seasons would often set off from home in Newcastle at 3am to walk to the coast in the days before local railways were built. His skill as a taxidermist seemed incredible. One contemporary once suggested to him that if a bird was completely plucked and the feathers put in a bag and shaken up he could put it back together. Hancock is said to have thought for a moment and replied: "*Yes, I believe I could.*" John Hancock died aged 83 years. John and Albany Hancock (1806-1873) lived, with their sister Mary, at 4 St. Mary's Terrace, a house opposite the site on which the Hancock Museum was built by the Society to which they dedicated their lives.

John Hancock
by F. H. Michael, 1891, courtesy of the Natural History Society of Northumbria

Thomas Caverhill Jerdon (1811-1872)

Jerdon was a physician, zoologist and botanist and famous ornithologist. He was born at Biddick House in the north east of County Durham. He studied at Edinburgh University and went on to become assistant-surgeon in the British East India Company, in India. He is best known for his pioneering work on the ornithology of India and he made little contribution to the study of birds in his native county. His most important publication was *The Birds of India* (1862–1864), which documented over a thousand species. While living in Assam he suffered a severe attack of fever necessitating his return to England in June 1870; he died at Norwood on 12 June 1872. He is commemorated in the names of several species of birds, including Jerdon's Courser (Elliot 1876).

Canon Henry Baker Tristram (1822–1906)

Tristram was born at Eglington, Northumberland, into a staunchly Anglican family. Initially, he studied wildlife around his home village and later drew up the first comprehensive list of the birds of County Durham. He was educated at Durham School and Oxford and was ordained a priest in 1846. He later carried on his work in the Middle East where he had moved for heath reasons. He was a great traveller visiting North America, North Africa, various locations in Europe and Japan. Through his life he held various curacies around County Durham, including that at Greatham, near Hartlepool. Ultimately, from 1873, he served as Canon of Durham Cathedral. He published one of the first articles in support of Charles Darwin. At over 80-years of age in 1905, he produced the first fully authoritative list of the birds of Durham, as *Birds*, in the *Victoria County History of Durham Vol. 1*. It contained 247 species seen exclusively in Durham, only 19 less than in Hancock's 1874 catalogue for both Durham and Northumberland. He was a founder member of the British Ornithologists' Union and he contributed too many early issues of the Union's journal, *Ibis*. His travel and main researches took him repeatedly to the near and Middle East and led to his name being commemorated in the names of a number of species, such as: Tristram's Grackle, Tristram's Warbler and Tristram's Serin.

James Backhouse (1825-1890)

The son of James Backhouse, James was born on 22 October 1825 in York. He was a botanist, archaeologist and geologist. Through his life he worked in Norway, Ireland, and Scotland, and was particularly known for his work on the flora of upper Teesdale, alongside his father. Like his father he was an ardent and very gifted botanist and he shared with him many expeditions to upper Teesdale. Both of them were something of natural history Renaissance men, being skilled in geology and archaeology – exploring the caves of upper Teesdale – and finding some significant ornithological remains in the process. After his father's death he explored Teesdale with his son, also named James (1861-1945).

Abel Chapman (1851-1929)

Chapman was born at Silksworth Hall, Sunderland, where he spent much of his early life. Although born in Durham, he made Northumberland his own by adoption. He came from a privileged background in a brewing family and spent much of his life pursuing his gentlemanly interests of wildfowling, hunting, angling and specimen collecting, alongside his business activities. During his early years in the Sunderland area he and his brother, Alfred, began their interest and made their first collections of local birds and eggs. Although a big game hunter, Chapman was an early pioneer of wildlife conservation. He was a prodigious traveller, ranging from the Arctic to the Equator, in pursuit of sport with rod and gun. He was a field-naturalist whose delight lay as much in studying quarry in life as in bagging it. In Spain, Chapman and a colleague became lessees of a large area of marshland and coast, the Coto de Doñana, one of the richest wildlife areas in Europe. He is credited with helping to save the Spanish Ibex from extinction, discovering the main breeding-place of Flamingos in Europe and documenting the plumage changes of migrating waders. In South Africa, Chapman and friends set up a protected game reserve that later became the famous Kruger National Park. He published extensively, including his first book *Bird Life of the Borders* in 1889, which recounted some of his observations and thoughts on hunting techniques in the Durham and Northumberland uplands. Chapman was vocal in his opinions on shooting, realising that if the random killing of some species continued they might become extinct. He campaigned against the needless slaughter of Black Grouse in late summer when hundreds of young birds were shot. His fears for this species were borne out as the species became ever rarer in the region. Many of Chapman's collection of local birds and eggs found their way to the Hancock at the Great North Museum.

From the late 18th century, the development of large country estates, with their wildlife often documented by both landowners and gamekeepers, provided a source of much historical information on the birds of the county. There were many such estates and some owners took an active interest in wildlife. These included the Ravensworth Estate on the west bank of the River Team at Gateshead. During the mid-19th century Lord Ravensworth *"gave sanctuary to birds on his estate."*

The modern systematic and co-ordinated collection of data – of birds seen 'what, when, where and by whom' – can be seen, in part, as a development of the acquisition and documentation of ornithological specimens. This occurred in Durham mainly in the 19th and early 20th centuries, and the collection of bird specimens has been important in building the knowledge of the county's ornithology. The shift away from displays of mounted birds has been so complete that it is worth reminding ourselves of what formerly existed.

The Newcastle Museum was largely founded with the acquisition of George Allan's (1736-1800) collection in 1822; the collection had been assembled by Marmaduke Tunstall (1743-1790). A few specimens survive to the present. The collection grew through the 1800s and was augmented in 1883 by the gift of John Hancock's collection: a centrepiece of the newly developed Hancock Museum. Until developed in the 1980s the birds were displayed in systematic order and when the gallery was refurbished with a display about bird biology the Hancock collection was retained on the balcony as a 'British Series'.

In Durham, the intellectual capital of the region, the University's museum was founded in 1833, the year after the founding of the University of Durham, and it was the second university museum in England to be opened to the public. The Museum was first housed in the Fulling Mill was subsequently moved to South Bailey and later to Bishop Cosin's Almshouses, on Palace Green. During these moves and refurbishments much of the information associated with many of the specimens was mislaid or lost. This museum was important for a number of years though ultimately many of the specimens were lost when in 1917, the University decided to disperse much of the natural history collection and presumably most of the specimens went to the departments for teaching purposes.

The industrial towns had their museums, the largest being in Sunderland. Here, through the generous support of Edward Backhouse and others, a large collection was built up in the late 1800s. Typical of most Town Museums, until the 1960s the lion's share of the display space in the museum was devoted to Natural History, of which birds played a large part. In the 1960s most of the non-British birds were removed into store, but many specimens were in poor condition and were discarded.

In Gateshead the Shipley Art Gallery (and later, Saltwell Towers) had a display of very fine quality taxidermy, the collection of the Earl of Ravensworth. These were the inspiration for a young James Alder, whose school-teacher sent him to the museum to draw the birds. The birds, still in their cases but now in store, are now cared for by Tyne and Wear Archives and Museums.

The strength of the Natural History display at South Shields was the mammal collection of William Yellowley, but people still remember the White-tailed Eagle with spread wings (which was also his work). The bird is now on display in Sunderland. Darlington, with its proud tradition of Natural History societies, had a charming little museum until it was closed in 1997 and the collections dispersed.

Of all the displays of birds, once so important in presenting the public face of ornithology to people at large, the only one remaining in the region in its original scope is in the Dorman Museum in Middlesbrough, where the 'Nelson Room' displays the collection of T. H. Nelson, author of *The Birds of Yorkshire*.

The roots of some of the most important collections relating to the county can be traced back from the existing Natural History Society collection, now held in the Hancock at the Great North Museum. Much of the usefulness of this collection can be traced back to John Hancock's care in cataloguing specimens and ensuring that provenances and processes were all correctly detailed. Edward Backhouse donated many of his extensive collection of specimens to the Sunderland Museum on his death and from there some of the specimens went to the 'Hancock'. This NHSN collection has specimens from Edward Backhouse, via the Sunderland Museum, John Hancock himself, the Earl of Ravensworth, via the Gateshead Saltwell Towers Museum. Just to the south of the county, in Middlesbrough, the Dorman Museum forms a repository for some of Nelson's old Teesmouth specimens.

Many of the people who secured specimens for these collections were anonymous hunters but some became quite well known in their own right. Some, such as Joseph Duff of Bishop Auckland, were correspondents with some of the region's best ornithologist and the great collectors of the day. Some of these collectors of specimens were in themselves important, going as far back as Ralph Johnson of Brignall, who was a correspondent of Ray's. Over the last two hundred and fifty years, these collections and their location in museums, has been important from the perspective of the academic work that has been able to be done with them but also in the way they have been utilised to inspire people about birds. How many children over the years have studied the exhibits 'in the Bird Gallery of the Hancock and other local museums' and consequently had their imagination and enthusiasm for natural history stimulated? Today, the specimens themselves remain a resource for future study.

Into the Twentieth Century

Through the early 20th century there was a move from the Victorian activity of field collection to ever greater in-the-field study; away from the gun and towards field glasses and telescopes. Observing ornithologists made ever greater contributions to the knowledge of the bird life of the county, some analysed and published their notes and knowledge; others contributed one-off sightings to journals. Many of these worked in isolation or with only limited contact with each other, excepting through organisations such as the Natural History Society, geography was the enemy of communication and it was difficult to work at opposite ends of the county on anything like a regular basis. Straddling the turn of the of the century and the county, the Rev. G. F. Courtenay published his experiences of over 20 years birdwatching experiences at Hurworth Burn, *Birds of Hurworth Burn* in 1931. This was followed by his *Birds of Teesmouth* in 1933 and in May 1934, he also wrote of his time in the north of Durham "*some notes on birds in Sunderland*" listing the birds he saw in that area back to 1906.

Thomas Hudson Nelson (1856-1916)

Nelson was the second of three sons of Ralph Nelson, a leading citizen and magistrate in Bishop Auckland. His mother died when he was only ten years old and his boyhood was greatly influenced by the headmaster of King James's Grammar School who was the *Marquis de Keruen de Limoclan* (who fled France during the war of 1870). Nelson became head boy at this school, and he took up law as a profession. Before adulthood he was struck down with a serious heart condition and had to abandon his career and live for much of the rest of his life as an invalid. He moved to Redcar and regularly studied the birds of Teesmouth. Through some 40 subsequent years, he sailed the Tees estuary, walked the mud flats and links, and went inland to the moors, valleys, meres, tarns, streams and woods of Yorkshire. He wrote papers for *The Zoologist* and *The Naturalist*, the journal of the Yorkshire Naturalists' Union, and was a member of the British Ornithologists Union (BOU) from 1882. In August 1876, he watched three Pallas' Sandgrouse at Teesmouth through a telescope and in 1892 he spotted two Long-tailed Ducks about a mile offshore through the Redcar lifeboat`s telescope. He established such a reputation that the Yorkshire Naturalists' Union asked him in 1888 to take on the task of preparing a book on the county's avifauna. *Birds of Yorkshire* was published in two volumes in 1907. An old acquaintance told of how Nelson slept with a string tied to his foot, the other end hanging out of the window so that at dawn fishermen might wake him when there was "*anything particular about*".

George Bolam (1859-1934)

Bolam was mainly active in Northumberland, publishing extensive lists in 1912 and 1932, which made some contributions to the history of Durham's ornithology. According to Temperley he was "*for many years the leading light on the bird life of the Northern Counties*". He deplored the fact that Durham had received less documentation of its birdlife than other areas in the north and urged Temperley to remedy the situation, leaving him his various papers to help him.

Through the late 19th and the early part of the 20th century, the geography of Durham was covered by a number of field clubs and societies, many affiliated to the Natural History Society. These often published their own newsletters and bulletins, and included the: Tyneside Naturalist's Field Club; Cleveland Naturalist's Field Club; Darlington and Teesdale Naturalists' Field Club; Vale of Derwent Naturalist's Field Club (founded 1887); Weardale Naturalists' Club (founded 1896); and, the Sunderland Naturalists' Association (founded 1911).

George Bolam
Courtesy of the Natural History Society of Northumbria

James Backhouse (1861-1945)

In 1885, James Backhouse produced *Notes on the Avifauna of Upper Teesdale*, which mainly dealt with the birds of the valley above Middleton-in- Teesdale, listing 117 species. A broader piece of work *Upper Teesdale Past and Present* was published in 1898 and this included an updated chapter on the bird life of the area. He was also the author of *A Handbook of European birds for the use of field naturalists and collectors* (1890) and a number of other publications. He was a member of the British Ornithologists' Union and became the Honorary Curator of ornithology in the Museum of the Yorkshire Philosophical Society, though most of his natural activities away from Teesdale in his later life were outside of Durham.

Charles Milburn (1880-1933)

Born at Guisborough, south of the Tees, he lived in Middlesbrough. Records indicate that Millburn's abilities became evident at a young age, with the publication of his paper *Ornithological Notes from Cleveland and Teesmouth, 1899.* He reported that he had found the nests of 65 species in this area and this paper was the first based on extensive observations of the birds of Teesmouth. It is clear from his documentation that he did not have binoculars. Milburn`s abilities in finding nests was illustrated by an exhibit of eggs he collected in 1901, included those of Nightjar, Green Woodpecker, Hawfinch and Goldcrest. The following year, 1902, he and C. Braithwaite of Seaton Carew, discovered the county's first Little Bunting. In June 1915 the first copy of *The Vasculum* contained one of his papers. and it was in Volume II of this journal that his extraordinary observation, in October 1915, of a Rough-legged Buzzard being driven off a slag-heap close to Grangetown, by two White-tailed Eagles, was published. He was close friends with Joseph Bishop of Stockton, their friendship dating from 1901 when they watched Brent Geese together on a reservoir near Wynyard. A year later they found a Kentish Plover at Teesmouth and they were together again twenty-two years later when they found two more. George Temperley made numerous references to his records in *A History of the Birds of Durham.* He died relatively young, in 1933 of cancer of the pancreas.

Joseph Bishop (1882-1939)

Bishop was born in Stockton to George, an iron moulder, and Margaret (née Wiley). In the early 20th century he was a close birdwatching friend of Charles Milburn. In 1931 his knowledge and enthusiasm led to his being enrolled as an 'Official Watcher' for the RSPB, on Temperley's recommendation, who was then honorary secretary of the Natural History Society. That he was a dedicated birdwatcher was shown by his spending a night on the North Tees Marshes in spring 1932, hoping to hear a Spotted Crake. Temperley made many references to his records from Teesmouth. Bishop was tireless in his work to ensure the safety of breeding birds around the Tees estuary and also watched the small water-logged brickfield, Charlton's Pond, near Billingham. Apart from publishing a number of notes on local birds, he sent regular reports to the RSPB, up to at least July 1939. He died at his home in Stockton in November 1939 from chronic bronchitis. Appropriately, his stated occupation on the death certificate was 'Bird Watcher.' There is little doubt that his observations enormously increased the knowledge of the birds of Teesmouth.

Thomas Robson (birth date not known–1944)

Robson of Winlaton published the *Birds of the Derwent Valley* in 1896. He was active in the valley over the latter half of the 19th century and early 20th century. Robson documented his and other ornithologists' sightings, collected specimens and carefully examined the reports of bird occurrences in the district. Although Robson's book is concerned with the Derwent valley, the majority of the text refers to the lower portion. He also provided much information from the current Gateshead stretch of the Tyne valley from Ryton east to Redheugh. In 1906, he published an extensive note on the breeding of Honey Buzzard on the Gibside Estate in the publication of the Vale of Derwent Naturalists' Field Club. Confusingly, there were two active ornithologists of this name in the lower Derwent valley in the mid- to late Victorian period. One was the above author, the second contributed records to this publication. This second Thomas Robson lived at Swalwell and for some time near Istanbul, where he discovered a new sub-species of Long-tailed Tit. He also found Britain's first Great Reed Warbler in the Swalwell Mill Race in 1847.

George William Temperley (1875–1967)

Temperley was born in Newcastle in October 1875 to Nicholas–and Alice Marian (née Cocking), who then lived in Low Fell, Gateshead. His father a businessman, who served on Gateshead Council, was actively engaged in various social reform activities, found time to take his children on excursions to the seaside and country, establishing George's lifelong love of wildlife. He was a 'delicate' child and did not attend school until he was 12 he then attended Gateshead High School for Boys, where he became head boy. His association with the Hancock Museum commenced when he was just nine years old, in 1884. He joined the Literary & Philosophical Society aged 17 and the Natural History Society in 1906, remaining a member for the rest of his life, contributing through the various posts he held and in the work he did at the Hancock Museum, where he was eventually an honorary curator. During his young adult life he was actively involved in a raft of organisations and in 1906; he won the prestigious Hancock Prize, for the best essay submitted by a member on a natural history topic. Initially he worked in the family company as a provisions importer. He was married in 1901 and later had a son. His father retired in 1908 and George followed a career in social work. From 1909 to 1913 he lived in Sunderland and during that period became a founder member of the Sunderland Naturalists' Association. In 1913, he moved to Scarborough to be Secretary to the Council of Social Welfare. He returned to Newcastle in 1918 as secretary of the Citizens' Service Society where he remained until his retirement in 1928. He then devoted

George Temperley with 'Robert of Restharrow', 1937
Image courtesy of the 'Natural History Society of Northumbria (NHSN) Archive Collection

himself to natural history. He began publishing nationally, in *British Birds*, from 1919, in the north east he had active links to both the Vale of Derwent Naturalist's Field Club (founded in 1887) and Weardale Naturalists' Club (founded 1896). Later, he was a founding member then president of the Wallis Club, named after one of the north east's first great natural historians, the Reverend John Wallis and he became a member of the Council of the Natural History Society in 1922. In February 1929, in *The Vasculum*, he pleaded for a survey of the bird life of Northumberland and Durham, stating that it was over half a century since Hancock had written his "*Catalogue*". He was concerned that changes were taking place that could affect the bird life of both counties and as a result the present status of birds should be more faithfully documented. Clearly, he was putting down a marker for future work.

In the winter of 1890, when he was 15, he found two Shore Lark between Gateshead and Low Fell and he spent considerable time bird watching in the Team valley. Temperley's interest in the Team valley would last for at least another 40 years, when in 1934 he documented the first known breeding of Willow Tit in the county, publishing his account in 1935 in *British Birds*. Over the period 1920 to 1954, he was a regular correspondent to that publication. He contributed to a number of a national species reviews and surveys (such as Grey Heron, Great Crested Grebe and Corncrake) and published numerous notes on subjects as widely ranging as breeding Montagu's Harriers in Durham, influxes of Waxwing to the north east, the song of Italian Chaffinches and the first occurrence of Ferruginous Duck in Durham, at Hebburn Ponds in 1947.

In 1931, he became the Joint Honorary Secretary of the Natural History Society, a post he held until 1951. He long harboured an ambition to work as a museum curator and from 1933 to 1939 he was one of the Honorary Curators of the Hancock Museum. In 1935 he became Recorder of the Ornithological Section of the museum and compiled and edited the ornithological reports for both Northumberland and Durham for 21 years before he handed over to Fred Grey. On his retirement from the Management Committee of the Hancock Museum he was elected Vice-president of the Natural History Society and in 1955 he was made an Honorary Member.

He was urged by George Bolam to write up the history of the birds of Durham and on Bolam's death was given many of his papers by Bolam's sister. In 1947, he published a request in *British Birds*, "*I am collecting material for a History of the Birds of the County of Durham and shall be most grateful for any notes or records that any of your readers may be able to send me which contribute information with regard to the past or present status, distribution or habits of any species therein*". Temperley's greatest ornithological legacy *A History of the Birds of Durham* was published in 1951 (as Vol. IX of the Transactions of the Natural History

GWT (looking through telescope), 23 March 1951,
from right to left, Roger Clissold, N. Southern and T. Hird
*Image courtesy of the 'Natural History Society of Northumbria (NHSN)
Archive Collection*

Society), the first truly authoritative avifauna of County Durham. It was compiled with, as Fred Grey wrote in 1988, *"the detailed and painstaking research typical of the man whose informed love of the natural world his friends were privileged to share"*. Through the period of developing this he was in contact with people who could kept him informed of ornithological events in Durham, such as Dr Hugh Blair in South Shields, who had a wide circle of correspondents.

Temperley was also a very capable botanist, worked on the mammals of the region and he was appointed regional representative of the British Trust for Ornithology in 1948, a position he retained until 1956, being awarded the BTO's Bernard Tucker Medal in 1953, for services to ornithology. A greater recognition was the honorary degree of Master of Science conferred upon him by Durham University in 1952 for his contributions to natural sciences.

He died in November 1967 at his home, Restharrow, Stocksfield, two years after his 90th birthday. Fred Grey wrote: *"One's abiding memory of George William Temperley is that of a kindly and courteous gentleman infinitely helpful and patient with beginners on field outings and delighting in the Hawfinches that annually visited the single cherry tree in the garden of his lovely house at Stocksfield, fittingly named Restharrow"*. It is clear that Temperley's Durham avifauna and his efforts in encouraging others provided the spur for the development of subsequent individual county recognition and ornithological recording in the north east. Those inspired by Temperley, such as James Alder, Fred Grey and Brian Little, a subsequent recipient of the Bernard Tucker Medal, in their turn inspired the people who were influential in setting up the region's ornithological infrastructure between, 1960 and 1975.

Maurice Guy Robinson (1910-1981)

Robinson attended Darlington Grammar School gaining a scholarship to Oxford, before becoming Head of English at The Royal Grammar School, Newcastle. As a boy he had a prodigious list for the Darlington area, recording Nuthatch with young on the way to school in 1926 and Crossbills over his school field. He was an intuitive watcher but meticulous observation was his hallmark. Throughout his life he kept detailed notebooks, recording Lesser Spotted Woodpecker near Sockburn in the 1920s, Corncrakes at Neasham Lane in 1926 and Lesser Whitethroat nesting in gorse in 1928. In 1939, with close friend Bill Almond and Bertie Nicholson, they published *'The Birds of the Tees Valley'*, which focused on Teesmouth but also recognised the importance of Darlington Sewage Farm. The publication was described by Temperley as 'a model for catalogue of its kind'. In 1938, he visited the Camargue with G. K. Yeates finding the nests of two pairs of Black Kites, which was not previously known to breed in the area. These experiences with raptors were to prove crucial, for, when travelling home in August 1947, a large bird flew across the Edmundbyers to Stanhope Road. He recognised it as a male Montagu's Harrier. He spoke to the local gamekeeper and later informed the Natural History Society. Thus began the saga of the breeding Montagu's Harriers on the

Durham moors. He remained active in the field and contributed records for Teesmouth up to 1974. Whether introducing George Temperley to his first Wryneck or pointing out a Grey Phalarope to the young Graham Bell, Robinson was always, patient, modest and self-effacing.

Hugh Moray Sutherland (H.M.S.) Blair (1902-1986)
Dr. Blair, of South Shields, was a close associate of Temperley and a later friend of Fred Grey. He established a national reputation in ornithological circles through expeditions to Norway in collaboration with several eminent people in that country over the period 1924-1934. Between 1924 and 1934 he made five trips to Finnmark, where the northern waders particularly attracted him. His detailed paper *On the Birds of East Finnmark* was published in *Ibis* in 1936. He also made expeditions in 1936, 1938, 1949 and 1950 to Hardanger Vidda in west Norway. In 1965 he visited the Oslo fjord and his ability to read and write Norwegian led to a wide circle of correspondents in Fennoscandia.

In the 1940s, Blair made extensive contributions to David Bannerman's 12 volumes of *Birds of the British Isles,* contributing more original material to these than any other ornithologist. Desmond Nethersole-Thompson commented that for the later volumes, particularly those on the waders that Blair's name could justifiably have appeared on the title page.

He visited Spain six times between 1953 and 1963 and he watched birds in France in 1959, Belgium in 1961 and Holland in 1964 and 1970. He was elected to the British Ornithologists' Union in 1933 and served on the Council from 1944 to 1947. He also gave great help to those who were studying the birds of Fennoscandia, including major assistance with translations in the early volumes of the *Birds* of *the Western Palaearctic.*

As a young man, he had considered a career in ornithology but he qualified in medicine at Durham University in 1922 and his father persuaded him to join the family practice in South Shields. When he retired, in 1974, Blairs had doctored in South Shields for 120 years. Snippets of news from patients supplied him with interesting information, most notably the finding of a Manx Shearwater's egg on Marsden Rock. One of the families he looked after were the Harrisons of Newton Garths Farm, Whiteleas, South Shields. This was part of the tract of low-lying damp pastures of which Boldon Flats is the most easterly remnant. When Garganey nested there, it was Blair who invited George Temperley to view the growing family.

Hugh Blair died in June 1986, aged 84. Over the last 40 years of his life his acquaintance with Fred G. Grey deepened into a close friendship. Desmond Nethersole-Thompson wrote in *Ibis* in Blair's obituary described him as "*Our greatest boreal ornithologist and scholar*".

The first copy of *The Vasculum* was independently published in June 1915, as The North Country Quarterly of Science and Local History. This later became the regularly published journal of The Northern Naturalists' Union which was founded in 1924 and carried numerous important reports of birds and ornithological developments through subsequent decades. The ornithological reports for Northumberland and Durham, coordinated by the Ornithological Section of the Natural History Society of Northumberland, Durham and Newcastle upon Tyne, were compiled by George Temperley, and published in *The Vasculum* from 1936 to 1939, then in *The Naturalist* from 1941 to 1950, from then subsequently, in the Transactions of the Natural History Society until 1969. From their inception these were compiled by George Temperley until he handed over the role to Fred Grey in 1958, for the 1957 report. In 1961 John Coulson (a former pupil of Fred Grey) took over the compilation of the report for 1960. Then the following compilers took up the task over the next decade or so: Graham Bell for the 1961 to 1966 reports; Graham Bell (Durham) & Jim Parrack (Northumberland) from 1967 to 1968; and, Jim Parrack, Russell McAndrew, Hazel Johnson and Grace Hickling for the 1969 report.

The Tyneside Bird Club, today the Northumberland & Tyneside Bird Club to more accurately reflect its recording area, was founded in 1958. Its first bulletin included records from Jarrow Slake, South Shields, Marsden, Whitburn and Durham City. In November 1960 a number of people assembled at the Dorman Memorial Museum, Middlesbrough, and decided to form the Teesmouth Bird Club.

A significant contributor of observations to *The Birds of the Tees Valley* was W. Kenneth Richmond, who published *A short account of the present state of birdlife in the Teesmouth*, in *British Birds* in 1931. This summarised the status of birds at Teesmouth since the publication of *Birds of Yorkshire* in 1907, often concentrating upon the changes resulting from increased industrialisation. He later went on to publish, in 1934, *Quest for Birds*, which included a chapter on the Tees estuary and one on Darlington Sewage Farm.

After the War and Post-Temperley

With the general mobilisation on the outbreak of the Second World War, there was a dearth of observers across the county over the period of the conflict although Temperley noted that some coastal species profited from the closure of beaches and installation of coastal defences, such as those around Teesmouth.

Post-Second World War, the interest in, and study of, birds began to grow dramatically and the systematic approach to such studies spread away from those wealthy amateurs and collectors of former times, to include ever more observers from different social strata; for a start, the hobby was prevalent amongst a new generation of returning servicemen. By the late 1940s, there were already indications of an impending influx of young birdwatchers onto the Durham birdwatching scene; this was particularly evident around Teesmouth where names that would make a long-term contribution to the ornithological life of the county began to rise to prominence. Names such as Alan Baldridge, a friend of Philip Stead, and his younger brother Ken, who later in life would become the Ornithological Recorder for Durham and Chair of the Durham Bird Club, were prominent at this time.

Birdwatchers were still rare at the time and somebody with a pair of binoculars was pretty much guaranteed to be watching birds. In the early part of the decade many still conformed to the 'well dressed country gentlemen' description. Overall, things were quieter and the pace of life less hectic than the present. There were fewer birdwatchers, less traffic on the roads and less disturbance, but there was also a paucity of birdwatching essentials, such as optical equipment and bird reference books. By the second half of the fifties things were changing and a rapid democratisation of the hobby took place. By then, a new generation of birdwatchers, the majority in their teens or early twenties, were growing up to replace the many lost during the War. At this time, a second wave of local bird clubs began to emerge as the country recovered from the conflict. Amongst these in England were: Kent in 1952; Liverpool in 1953; Surrey in 1955; Derbyshire in 1954 and, in the northeast region, the Tyneside Bird Club (1958) and the Teesmouth Bird Club (1960). In many respects the study of birds increased in academic competence from then towards the modern day.

At this time there was a lack of decent bird books. Things changed in 1952 with the publication of the *Collins Pocket Guide to British Birds* by Fitter & Richardson and two years later came the revolution, with Peterson's *Field Guide to the Birds of Britain and Europe*, which became the model for most bird identification guides since.

Peter Laurence Hogg (1920-2009)

Born on Teesside, Peter's father had taken his son bird-nesting as a boy and Peter began birdwatching at Hartlepool as an eleven year-old. It was at Repton School that he discovered that members of the lapsed Field Club were allowed to have bicycles, so he and a friend resurrected the Field Club, in order to get bikes for birdwatching. After the War, Peter returned from serving in the RAF to his career in the shipping industry. In the early 1950s he collated the ineffective local bye-laws for bird protection for George Temperley, in order to assist the RSPB to get The Protection of Birds Act 1954 on to the statute book. With a friendly RSPCA Inspector he became involved in the Inspector's activities against bird-catchers, particularly a youth at Horden who was catching Bullfinches and Chaffinches to sell to the Coal Board. On another occasion he helped to have two men prosecuted, using the *Piers, Docks and Harbours Clauses Act 1861* which forbade the carrying of a loaded firearm on or near any harbour works, after they shot a swan near Graythorp. Peter was a grandson, son, brother and father of Magistrates: a strong moral code permeating the generations.

It was in the early 1950s perhaps soon after the discovery of the county's first Terek Sandpiper in September 1952, that David Graham Bell started visiting Teesmouth. Graham was studying French at Durham University and spent his weekends, birdwatching - Saturdays on Tyneside and Sundays on Teesside. He was born in 1934 at Didsbury, Manchester, developing his interest in birds from his father. Completing his studies, he taught at Grangefield Grammar School for boys in 1956; he would go on to be the compiler of the ornithological reports for Northumberland and Durham during most of the 1960s and a key player in the early days of the Teesmouth Bird Club.

In the north of the county, Fred Grey's activities around South Shields were stimulating the activities of young observers in that area, and would continue to do so over the next three decades. John Coulson was an active birdwatcher and bird ringer in that area, who in a short space of time, discovered two new species for the county, just in time for them to be squeezed into a short appendix of Temperley's *A History of the Birds of Durham*. He would go on to have a career as an academic ornithologist, becoming the world authority on the Kittiwake, a species he started to study as a boy at Marsden Bay.

Fred Grey (1911-1997)

Fred Grey was born in South Shields on 6 January 1911, to Richard Gustaf and Gertride Aline Grey. Educated at Westoe Secondary School, 1922-1929, he flourished as a scholar and an athlete. The South Shields coast was where his passion for wildlife was sparked and grew. He studied English at Armstrong College, Durham University with Latin, French and philosophy as subsidiary subjects, and was awarded a BA, which he converted into an M.A. in 1943. Fred became a teacher at Stanhope Road Senior Boys School, 1933-1938, moving to South Shields Grammar Technical School for Boys where he taught 1976, being Head of English for many years. He married Marty in 1940, and had two daughters, Anne and Joan.

Fred Grey (centre left)
image courtesy of the Natural History Society of Northumbria

Fred's field skills and his ability as an educator were legendary. His experience and wider gifts were recognised when, in 1957, he succeeded Temperley as the Honorary Secretary of the Ornithological Section of the Natural History Society and by his later appointment in 1975 as first chairman of Durham Bird Club. His most enduring legacy was the way he touched and even changed the lives of hundreds of people through his inspirational passion for nature, and above all, for birds. In the late 1950s he asked Derek Watson, one of his pupils who would go on to become an acclaimed wildlife artist, if he would illustrate an article on the spring migration of Wheatears for the *Shields Gazette*. This was the first in a series called *Birds to see now* which ran for over 20 years. Temperley drew on Fred's systematic records, particularly on the growth of sea-bird colonies around Marsden. Fred's was elected as a member of the British Ornithologists' Union in October 1944. He was also a bird ringer and the first time many people saw a wild bird in the hand was when assisting him with mist-netting at Marsden Hall during migration. In the late

1940s, he was involved with a pioneering scheme with James Alder, Temperley and others to put up nest boxes for Pied Flycatchers in Hamsterley Forest.

As a teacher, Fred was an awe-inspiring figure to most boys, keeping a stuffed Kittiwake in his cupboard, which he produced for dramatic effect during class. In the mid-1930s he played for the successful Westoe Rugby Club and afterwards served as a coach for Durham Schools. He had been a keen boxer and was known by the boys as "Basher", an epithet that could not have been further from his real nature. Involvement with Fred and the school Bird Club changed the lives of many boys and he introduced many to wildlife and the Natural History Society. Through such contact, it dawned on some that biology and wildlife could be a career as well as an exciting hobby, and many took that route. Fred was equally influential with the adults who enrolled for his WEA classes in ornithology. In an obituary on Fred in the Durham Bird Club Bulletin in 1997, Brian Unwin wrote: *"His Workers Educational Association classes were the stuff of legend. Many of the adults who attended his sessions caught his passion for birds and were permanently hooked."*

Fred wrote for his own enjoyment, published few scientific papers, and such was his modesty that even his Shields Gazette column, published appeared anonymously. After retirement he took part in a BTO Breeding Bird Survey in the Derwent valley for several years until his health began to deteriorate. The Durham Bird Club owes much to Fred Grey; the campaign to save Barmston Ponds, the setting up the first Whitburn Bird Observatory, the production of the Durham County Bird Reports from 1970, were all heavily influenced by him. In 1975, he became the Club's first Chairman and its first honorary member on his retirement from that post. His influence on bird-watching in the county and beyond was considerable and achieved mainly through his ability to enthuse people about the things which meant most to him. A great deal is owed to this 'scholar, ornithologist, athlete, inspirational teacher' and a truly remarkable man.

An important record in the north east of the county, occurred on 4 December 1955, when Derek Watson, one of Fred Grey's pupils at South Shields, visited Jarrow Slake and found an unfamiliar wheatear frequenting bare ground between the south shore of the 'Slake' and Pyman Bell's timber yard. It was seen on 5th and again on 6th when Fred Grey concluded that it was a male Desert Wheatear. During the next twelve days the bird was seen by a number of observers, including some of the region's best known ornithologists: George W. Temperley, Dr. H. M. S. Blair, James Alder, Alan Baldridge, Brian Little and Philip Stead, some of the key individuals in the subsequent formation of the region's three principal bird clubs, and one of the soon to be founder members of the Tyneside Bird Club, Jim Edwardson, another pupil of Fred's.

During this period, Teesmouth became a hot bed for the meeting of young birdwatchers and returning servicemen, establishing connections and friendships that would later crystallise into the Teesmouth Bird Club. The year of the Coronation, 1953, saw the arrival on Teesside of James Denis Summers-Smith, who by this time had already commenced his famous studies of House Sparrows.

The RAF provided another important, albeit temporary, observer to the area in the shape of Cliff J. Henty, who was at RAF Seaton Carew 1953-1955. Ken Baldridge recalls Cliff seeing 30 Pomarine Skuas and phoning Philip Stead; the first ones in the area since Nelson's days. He followed this with the discovery of Teesmouth's first significant influx of Lapland Buntings, in autumn 1953. Cliff went on to become Recorder for Upper Forth, Scotland but these sightings showed the way forward for local birdwatchers; scarce migrants and sea passage.

Further north into Durham around this time, David Simpson was commencing his life-time study of the Shotton Colliery area, which by 2010 had lasted over 50 years. Such 'local patch' work was to become more of a feature of birdwatchers' activities in the county over the next 30 years or so, before it subsided somewhat in the light of modern telecommunications developments and greater mobility. David's earliest birdwatching memory was of his father taking him to see some local Waxwings in 1948. A highlight of his early birdwatching years was meeting George Temperley, whom he described as an 'absolute gentleman'. In 1954, on his 'patch' he found the county's first Little Ringed Plover and a Golden Oriole at Edderacres in 1956, but he did trek south to see the 1959 Dusky Thrush at Hartlepool. Like most birdwatchers of the time, his was essentially a lone activity, though he occasionally connected with observers from Teesmouth such as Peter Hogg, with whom he watched Black-necked Grebes on Thornley Flooded Fields in 1954.

The clearance of mines and other defences from local beaches allowed for greater access by birdwatchers around Teesmouth and other coastal locations. At this time, the mid-1950s, most of the young birdwatchers walked, travelled by public transport or rode bicycles, which meant long rides for many to get to places such as the Headland at Hartlepool or Jarrow Slake from South Shields. As a result the majority of birdwatching was confined to areas close to the centres of population, such as Tyneside and Teesside. Large and bulky ex-service binoculars were the norm as optical aids until in the late fifties new specification binoculars began to appear in the shops. Initially, they limited in both magnification and optical brightness, and expensive. Telescopes were becoming more prevalent, though not common; invariably they were of the draw-tube type. Nonetheless, serious sea-watching got under way in both the north and the south of the county, but without the benefit of appropriate shelters, such activity was testing in the extreme.

In the late fifties birdwatchers across the region, as elsewhere, were becoming increasingly frustrated by the lack of information on recent bird sightings. It was during this period, that a small group of birdwatchers around Teesmouth, including Dennis Seaward, Philip Stead, Graham Bell, Peter Evans and Denis Summers-Smith, began occasional informal get-togethers. Around this time Philip Stead and Graham Bell determined to more systematically collate their records, and publish these as a monthly Teesmouth report. Records were collected informally by Graham Bell and he used the services of his school secretary to type up the results, the resultant copies being distributed in an *ad-hoc* way. The first single-page issue, for September 1958, was entitled the *Teesmouth Ornithologists' Bulletin* and it listed thirteen observers: Alan Baldridge, Graham Bell, Fred Grey, Jimmy Henderson, John Nicholson, Geoff Proctor, Pete Reid, Dennis Seaward, Ken Smith, Philip Stead, Ian Stewart, Denis Summers-Smith and Alan Vittery; a list that gave a glimpse into birdwatching's future in the area.

Philip Stead (1930-2005)
It is said that Philip John Stead started birdwatching from his pram, as a two year-old. At that time his family lived near the Dorman Museum, Middlesbrough, and after his first visit there as a child he often urged his mother to take him back to study the exhibits. As a youngster, his father ferried him about by car until he acquired his red 350cc BSA motorbike, on which Graham Bell was once given a lift to see his first Waxwing. Given this mobility, his brilliant eyesight, and friendly, helpful nature he was regarded by the up-and-coming generation as the guru of the local birdwatching community. His career in the steel industry started in 1946 as an apprentice draughtsman in the Bridge and Construction Works of Dorman Long, eventually qualifying as a Chartered Structural Engineer. National Service in the army (1951-1953) got in the way of birdwatching and reduced his time at Teesmouth but he still managed to provide some records during that period. Philip Stead, alongside Denis Summers-Smith, was the driving force in the setting up the Teesmouth Bird Club and was its first Chairman, from 1961-1962. He was regarded by many as the 'father' of the club. He wrote the *Birds of Tees-side* which was published through the Natural History Society in 1964, the aim being to update the work done by some 25 years previously by Maurice Guy Robinson, Bill Almond and Bertie Nicholson. This publication was followed by the *Birds of Tees-side 1962-67*, which was published by the Teesmouth Bird Club. He moved out of the area in 1972 and was made an honorary member of the Teesmouth Bird Club in 1973. Occasional trips back to Teesside meant that he made ad-hoc contributions to local ornithology over the years until 2004. He died in 2005.

To the north, this publication was noted on Tyneside, and there Dave Howey and Mike Bell began to canvass opinion on the production of a comparable bulletin for that area. In 1958, the Tyneside Bird Club was founded, with its first bulletin published in November of that year; this included records from Jarrow Slake, South Shields, Marsden, Whitburn and Durham City. This was compiled by Dave Howey and had contributions from eleven observers, eight of them were teenagers; Graham Bell being the only 'fully-fledged' adult observer. These eleven became the nucleus of the Tyneside Bird Club; seven lived north of the Tyne, three in South Shields and one on Teesside. The bulletin became the *Monthly Bulletin of the Tyneside Bird Club* in September 1959, though the initial intention was not to form a club but merely to produce a bulletin for active birdwatchers.

As the early Teesmouth Bird Report became more widely known more observers were drawn in. For example, Horden-born Brian Unwin (1945-2011) recalled meeting up with friends on Saturdays in the late 1950s and early 1960s, at Greatham Creek Bridge to watch birds along the Long Drag.

Seawatching grew in popularity in the 1950s, largely undertaken from the Headland at Hartlepool. It was discovered during this period that the old observation post in the Heugh Battery was the best position in rough weather. Eventually, for a reasonable rent, this was secured for observing sea passage. One of the pioneers of this activity in the north of the county was J. R. Crawford, one of the first local ornithologists to fully appreciate and document the significant movements of seabirds along the Durham coast, during the late 1940s and early 1950s. He watched mainly from Whitburn and, as recounted by John Coulson, was an ardent cyclist and often used a monocular in the field. Furthermore, he discovered the county's' first breeding Black–necked Grebes in 1946.

What was an undeniably momentous event in birdwatching in the county was the discovery of Britain's second Dusky Thrush by 14 year-old Russell McAndrew and Geoff Proctor at Hartlepool Headland on 12 December 1959. For a number of reasons, this might be seen as the point at which the 'modern birdwatching' era began locally, though many of the elements and infrastructures had been starting to develop prior to this. This record served to catalyse subsequent events because of the way it made observers reappraise the area's worth for migrant birds. The bird remained through the winter, to 24 February 1960, and was caught and ringed on 10 January 1960. What this record did was help focus wider attention on Hartlepool Headland, the observatory there, and the whole of the north east of England in terms of its significance as a birdwatching location.

By November 1960, the Cleveland Naturalists' Field Club was felt to be placing too little emphasis on birds. A number of people who would assemble after its meetings expressed this thought and it was it was at one of these sessions that Philip Stead and Denis Summers-Smith decided to form a bird club to cover Teesmouth. It was Philip Stead who drove this process onwards. News spread that there was to be a meeting at the Dorman Museum in November to discuss the possibilities. The Teesmouth Bird Club was set up *"principally to record the birds around the coastline from Hartlepool to Redcar, on the Tees marshes and around Teesside"* (Blick 2009). From the start the area of coverage included part of the then County Durham. The Teesmouth Bird Club celebrated its 50th anniversary in 2010.

It was during the 1960s that birdwatching began to shake off its old image and evolved into the shape we know today. This was in part thanks to some inspirational television wildlife programmes and ever-improving reference and portable books. For various reasons, County Durham was a little slower than neighbouring areas to move in this direction. By the end of the decade, Teesside and Tyneside both had thriving bird clubs but there had been no attempt to set up a similar organisation covering the territory in-between. One reason for this was that the county's comparatively few bird-watchers were geographically polarised, being concentrated mainly in its north-east and south-east corners and as a consequence they gravitated to whichever of the two existing bodies that they were closer to.

The Modern Period

From 1933 the Natural History Society had been responsible for compiling the *Ornithological Report for Northumberland and Durham* and in 1956 it appointed a special committee, made up of experts from Durham, Teesmouth, Northumberland and Tyneside, to scrutinise sight-records of less common and more difficult species. In 1967 this changed to two panels working independently, one for Northumberland and Tyneside, the other for Durham and Teesmouth. In October 1971, the Council of the Natural History Society decided that it was no longer able to continue publication of bird reports for the two counties and prompted by Dr. John Coulson of Durham University's Zoology Department, a former NHS ornithological report editor, an emergency meeting of 'interested parties' took place in November 1971, ultimately this would lay the foundations of the Durham Bird Club, some four years later.

From the late 1960s through to the early 2000s, the Zoology and then the Biological Sciences department at the Durham University became one of the leading ornithological research institutes in the country. Over this period, the main research group leaders in relation to birds were Professor Peter Evans (largely supervising work on wading birds at Teesmouth) and Dr. John Coulson (working mainly on seabirds and upland wading birds), with other, more latterly, inputs from Dr. Chris Thomas and Professor Brian Huntley, of the School of Biological and Biomedical Sciences; the latter specialising in issues relating to climate change. These research groups were engaged in a wide range of studies both within and outside of the county. As with most academic bodies, the research was not restricted to Durham's county boundaries. A large proportion of the work on the breeding waders was undertaken in upper Teesdale and west on to Chapel Fell; some of the westerly most portions of County Durham over to Cross Fell in Cumbria. For a selected list of some of the researchers involved and the topic researched over this period, see the Appendices.

Peter Evans (1937-2001)

Born on July 20th 1937 in Thornbury, Gloucestershire, he was educated in Bristol and in Yorkshire. He had an early interest in birds and bird migration, which was to lead him to undertake pioneering work using airfield radar to examine the movement of flocks of birds along the coast of north east England. Initially he studied chemistry at St. Catharine's College, Cambridge, and, in 1961, he completed a doctorate, later teaching chemistry at Ampleforth College before his ornithological interests led him to study bird migration and navigation, at the Edward Grey Institute, Oxford. He worked here for three years before moving north and joined Durham's Zoology Department as a lecturer in Ecology in 1968. He spent the rest of his career studying the physiological, ecological and behavioural strategies of his main interest, wading birds. He personally, and through his research group, conducted a long-term study of wading bird activity on the Tees estuary, in a multi-disciplinary approach that over decades, revealed much about the complex ecological relationships of that estuarine system.

Outside the life of the University, he quite frequently appeared on natural history programmes on both television and radio and gave of his time by providing lectures to local natural history societies and clubs. He was a founder member of the International Wader Study Group and in 1996 was awarded the prestigious Godman-Salvin Medal of the British Ornithologists' Union for his role in a range of ornithological studies and the training of students to further such work, at Durham University and more recently at Stockton College.

His search for applied outcomes of science led to much collaboration with industry and government and he was an early advocate of the statutory designation of land at Teesmouth for its wildlife importance. This ground-breaking work led to many important conservation outcomes in that area and laid the foundations for work that would be latterly realised by organisations such as Industry and Nature Conservation Association (INCA) and the RSPB.

He enthused generations of Durham University students with a passion for studying animals in their natural environment. He was directing his extensive research group and planning future projects to the end of his life when he died of cancer, aged 64, on 28 September 2001.

Brian Unwin (1945-2011)

Born the son of a miner in Horden, east Durham, in May 1945, he developed an early interest in the natural world and acquired his early birdwatching skills in the coastal denes near his home and on family visits to the Durham dales. Professionally, he was a journalist on the region's two morning newspapers, *The Journal* and *The Northern Echo*, before becoming regional reporter for the national news agency, the Press Association. He covered the Sunderland and Durham patches for The Northern Echo, from 1965 to 1972, and later he spent 14 years as the regional staff reporter for the Press Association, writing weekly Birdwatch notes in the Northern Echo. During a stint working on The Journal in the early 1980s, he took a sabbatical to join HMS

Brian Unwin (Mark Newsome)

Endurance, visiting the Falklands shortly before the conflict with Argentina, enabling him to become something of a regional expert in his coverage of the war. For the last 20 years of his career he was a freelance writer becoming 'Freelance Journalist of 1986' in recognition of his regular 'Wildlife Watch' column in the Newcastle Journal. Around this time, he undertook a study of Little Gulls along the Durham coast, which led to a re-evaluation of their local status and distribution.

In the early 1970s, he campaigned to save the old Barmston Ponds, and when that proved unsuccessful he was instrumental in persuading the Washington Development Corporation to encourage the Wildfowl Trust to establish what would become the Wildfowl and Wetlands Trust Washington at nearby Low Barmston. Brian was one of the founding inspirations behind the formation of the Durham Bird Club in the mid-1970s, coordinating the January 1975 meeting that led to the formation of the Club. He was the Club's first Secretary and Recorder, and the editor of the annual ornithological reports between 1973 and 1979. His special interest was migration and he chose to live in Whitburn to facilitate his enjoyment of migrants at the coast. In the mid-1980s, he persuaded the developer of the ex-Whitburn Colliery site to name the housing estate they were building 'Shearwater'. He and his family duly moved into a house there and he set about organising his garden to make it attractive to birds. The result, several rare species in the garden including: Bluethroat, Red-breasted Flycatcher, Common Rosefinch and Pallas's Warbler. Brian became ill in 2007 and fought a long battle with cancer of the oesophagus. At the time of his death, he was working on a north east edition of the *Best Birdwatching Sites* series for the Buckingham Press. He died on 29 December 2011, aged 66, leaving his widow, Jennifer, son Barry and daughter Beverley, and seven grand children, not to mention a significant legacy for all birdwatchers in the Durham area.

The early 1970s – the Evolution of the Durham Bird Club

The formation of the Northumberland and Durham Naturalists' Trust in the early 1960s was an important step not just in the study of birds and wildlife but also for the management of land for wildlife. The single trust divided in 1971 along county lines, becoming two separate wildlife trusts for Durham and Northumberland. The Durham County Conservation Trust paid for the construction of the first Whitburn Observatory building, which provided a focus for sea-watching efforts in that area. The Durham County Conservation Trust became Durham Wildlife Trust in June 1988.

On the back of the campaign to save old Barmston Ponds, confirmation that a wetlands centre would be constructed within the James Steel Park at Washington, not far from the site of those ponds, came in 1972; the centre was opened in 1975. This would later become the Wildfowl and Wetlands Trust Washington and a focus for much ornithological activity.

In the early 1970s, the Sunderland Natural History Society ran several birdwatching trips, most led by Brian Unwin and these grew into the programme of field trips and outings organised by the Durham Bird Club on its inception. Over the early part of the decade local RSPB members' groups catered for some birdwatcher involvement and a number of Young Ornithologists' Clubs, such as that run by Hazel Johnson in Durham, helped a number of young people to enjoy and develop their interests. A marker of the level of observational effort in 1973 was that 245 species were recorded in the Durham that year a figure not that dissimilar to those of 35 years later (265 were recorded in 2008).

From November 1971, a committee was established to publish a separate *Durham County Bird Report*. Initially, Russell McAndrew took on the task of compiling, with impressive rapidity, the 1970 and 1971 editions. In March 1972, the first of this new series of ornithological reports, covering only Durham during 1970 was published. From 1972, compilation was by a team, consisting of Fred Grey, Ken Smith, Ian Stewart and Brian Unwin. When in 1973, Ian Stewart retired from the Records Committee Brian Unwin became Secretary and effectively the County Recorder.

The original DBC symbol

By this stage, there was a clear impetus towards the foundation of a new club. A number of local ornithologists worked together to assess records and produce the annual ornithological report, there were enough contributors to this to indicate that there was a potentially viable membership base and that indoor meetings and field trips were likely to be supported. A public meeting in late 1974 at Sunderland Museum and Art Gallery to discuss bird watching and recording was well-attended. This was followed on 8 January 1975 by another meeting and enrolment for membership of the Durham Bird Club began shortly afterwards. The first Club committee comprised: Fred Grey (Chairman), Brian Unwin (Secretary and Recorder), Ian Hogg (Treasurer), David Sowerbutts (Indoor Meetings Secretary), Dan Mold (Field Meetings Secretary), along with Brian Armstrong, John Coulson, Tom Palmer and Morris Skilleter. The new Durham Bird Club was then responsible for producing the *Durham County Bird Reports* for 1974 and 1975. From April 1974, after local government reorganisation, Startforth Rural District became a part of County Durham and the 1974 ornithological report was the first to fully cover this area. Brian Unwin oversaw the compilation of the annual reports from 1973 through to 1980, including the first independent Durham Bird Club report, *Birds in Durham 1976*. By the end of 1975 Club membership had reached 103, rising to 130 in 1976. The choice of a symbol for the Club proved straightforward, given the importance of the population of Black Grouse in the county. From the formation of the Club, the advantages of Durham Bird Club membership were listed as: a free copy of the Annual Report, *Birds in Durham*; a monthly bulletin (this became a quarterly magazine in 2006); use of the Whitburn Observatory; field trips to places of ornithological interest; a programme of illustrated talks from a variety of expert guest speakers; and, participation in local and national projects and surveys. The Club's programme of indoor talks began on 12 March 1975, when appropriately John Coulson spoke about the Kittiwake; in 2011, that speaker's monograph of the species, distilling 50-yeares of study, was published.

The Secretary's report for 1976 recorded that Club members had participated in the national Wildfowl Counts and the BTO's Rookery Survey and it looked ahead to a time when Club activities would include *"more active participation in conservation, both of species and of habitat"*. This would come to pass in the 1980s and, in particular, through the early 1990s. At the beginning of 1977, Fred Grey, who had contributed so much to the study of birds in county prior to the foundation of the Club, retired as Chairman. Morris Skilleter took over until the end of the year and Fred accepted an invitation to become the Club's first Honorary Life Member.

The 1970s and 1980s saw an ever greater mobility of birdwatchers, a more rapid and abundant exchange of news and information and faster response to the presence of rare bird across the county. The demand for up-to-date information led to the establishment during the mid-1980s of the Durham Bird Club telephone grapevine to facilitate exchange of information about local birds. The increased interest in rarities and migration led to increasing reports of scarce birds and the adoption of a more rigorous approach to records from sometimes unknown sources and observers. Consequently, from 1982 the Club adopted a policy of asking for field descriptions to accompany the submission of records of a range of listed species, in addition to those considered rarities by the national British Birds Rarities Committee. These descriptions and the maintenance of the county list was in effect overseen by the Records Sub-committee, which since that time has been made up of up to six experienced observers drawn from the ranks of the Club. This works under the chairmanship of the county Ornithological Recorder and is charged with assessing submitted records and the associated descriptions of these, the objective is to maintain a high standard of accuracy in terms of the determination of records of unusual species in the county.

Through the 1980s, there was an increased involvement of birdwatchers in projects and surveys. These included the on-going national surveys such as wildfowl and wader counts, the Birds of Estuaries Enquiry and the Winter Atlas project. Alongside these were a number of BTO surveys (for example: Nightjar 1981; Wood Warbler 1984; Lapwing 1987) and the Club's own locally organised surveys, particularly Black Grouse.

During the late 1970s the content of the annual report *Birds in Durham* expanded somewhat, with a ringing report being included from 1977 to 1980, but there was always a tension between the content and the production values, as dictated by budgetary constraints. At this time the report also adopted the accepted Voous order, for the systematic list, replacing the previously used Wetmore listing. A further development in 1978 saw the inclusion of a paper on the distribution of Little Gulls along the Durham coast, by Brian Unwin. This was the first of many such papers although the trend for such inclusions did not develop fully until the later part of the 1980s and early 1990s. The observer coverage of the county has never been even but the editorial comments in the reports through the 1970s and early 1980s drew attention to the skewed nature of the observer pattern and the implications for the species and numbers recorded. In the 1970s, the pattern of active observers was said to be patchy at best.

From the late 1960s, through the 1970s and continuing into the early 1990s, the designation of various Sites of Species Scientific Interest, a number of National Nature Reserves, for example those at: Moor House and Upper Teesdale; Castle Eden Dene; Teesmouth and Derwent Gorge & Muggleswick Woods, provided protection for some of the county's most important ornithological sites. A slew of Local Nature Reserves were designated, particularly in the early 2000s, and an increasing area of land in public ownership, and by the Durham Wildlife Trust, was ever more effectively managed for its wildlife interest. These developments meant that as time wore on, many more sites that held important species or varied populations of birds were better protected than ever before.

Club logo adopted in 1984, designed by Sandra Armstrong

The Durham Ringing Group was formed in 1979 around Anthony Roberts as key trainer. It remained active until the early 1980s, but after Anthony moved less systematic ringing was undertaken. This changed in the late 1980s with the rejuvenation of the Group, driven by the activities of what had been the Derwent Valley Ringing Group, which had formed in 1987 and developed around ringing trainer Les Milton and his group of active bird-ringers in the Gateshead area. The subsuming of this into the Durham Ringing Group led to the inclusion of a more coordinated set of ringing initiatives, including the re-commencement of an upland nestbox scheme, originally started in 1979, focussed on ringing pulli of Pied Flycatchers and Redstart extensive study of titmice boxes, the running Constant Effort Site ringing studies, particularly in the Derwent valley, migrant ringing in Marsden Quarry where ringing had taken place intermittently since the mid-1960s and the start of nocturnal Storm Petrel ringing. Data from these studies was integrated in a more systematic fashion into the work of the Durham Bird Club than had been the case for a number of years. The activities and findings of the Durham Ringing Group was laid out in *Birds in Durham* in Ringing Reports from 1988 until 2000, the production of these was coordinated by Stephen Westerberg, although not all ringers in the county contributed to this. Some ringing led to more academic studies, on Kittiwakes and Sedge Warblers, in association with Durham University, and detailed local studies on colour-ringed Dipper. The group effectively folded in 2006 as key personnel moved away or joined the Northumbria Ringing Group.

The first Durham Bird Club publication, other than the bulletin and the annual ornithological reports, was produced in 1984; a 32-page, A4 booklet called *Where to Watch Birds in County Durham*.

The occurrence of an amazingly confiding Baillon's Crake in Mowbray Park, Sunderland in May 1986 attracted widespread national interest and an associated bucket collection among visiting birdwatchers raised almost £200 and this formed the beginning of the Club's Conservation Fund. An interest in such rare birds and the activities of rarity-watchers had commenced, in earnest, during the 1980s. From this time, as well as many other rare and scarce species the county experienced a run of extreme rare birds this interest was fuelled, and sustained, by the greater knowledge and activities of a growing band of young enthusiastic birdwatchers. This list of major rarities included: the second Long-toed Stint for Britain in August 1982; Britain's third White-tailed Plover in May 1984; Britain and Europe's first Double-crested Cormorant in 1989; Britain's second Great Knot in 1996; England's first Short-billed Dowitcher in 1999; in 2002, the first successful breeding of Bee-eaters in England in 47 years; the first record of 'Amur' Wagtail in the Western Palaearctic in 2005; England's second Siberian Rubythroat in an urban back garden in Sunderland in autumn 2006; in 2009, Britain and Ireland's first Eastern Crowned Warbler; Britain's fifth, Brown-headed Cowbird singing in a Seaburn garden in May 2010; and, the north of England's first Common Nighthawk in 2010. These were followed in 2011, by Britain's third and fourth records respectively of White throated Robin and Sandhill Crane.

From a conservation standpoint, the Club became far more active through the late 1980s and 1990s as the increasing pressure of land development gained momentum. In truth consultation about such matters had started early in the Club's existence, when in 1976 the Club was approached for advice about, for example, the Silksworth Colliery redevelopment. One campaign saw the Club involved in efforts to improve the management for birds of Boldon Flats. South Tyneside Council was involved in managing the 'winter flood' at this site from 1986 and by that time had designated the site as a Local Nature Reserve. By 1988, the Club, at the invite of the council, became a member of the Boldon Flats Steering Group. On an increasing range of conservation issues it was involved in discussions with the Nature Conservancy Council, English Nature and Natural England, as that agency went through re-organisation and name changes, about threatened SSSIs and it embarked on the compilation of bird lists for sites that were endangered or under the threat of development. Discussions with South Tyneside Council about a replacement Bird Observatory at Whitburn began in August 1988, arising from a consultation on Local Nature Reserves in the borough.

At Whitburn, the original wooden seawatching hut, which had been relocated from upper Teesdale in 1971, was destroyed by fire to be replaced by a brick structure that lasted until 1984. By 1987, ideas for a much sturdier replacement were in development. Tony Armstrong, Ornithological Recorder for Durham, co-ordinated the comments of local observers and produced detailed designs of a new bespoke observation hide for South Tyneside Borough Council's consideration. The Council approved funding in 1989 and the new observatory was built in 1990. In August 1990, the Club received a lease for the Observatory for management purposes before ownership of the building and the coastal park was handed

Whitburn Observatory (Mark Newsome)

over to the National Trust. The formal opening of the Observatory was carried out by Jessica Holmes of the BBC Natural History Unit on 16 May 1991 and the first bird seen through the shutters thereafter was a Swallow. In subsequent years, the more systematic approach to the documentation of seabird passage, linking to such movement past Hartlepool Headland in the south of the county, as a consequence of the facilities provided to observers by the Observatory, greatly increased the understanding of the significance of such movements in the wider context.

Expansion through the 1990s

Around the turn of the decade, a team to support the survey work of the Club and the BTO's area representative was formed. The early 1990s period saw an unprecedented amount of survey and census activity by Club members, after the establishment of the Club's Project & Surveys Group in 1990, under the guidance of David Sowerbutts county BTO representative. this first The existence of this group led to an expansion of such work, not just the implementation of national surveys but on the development of bespoke survey activities in the county. In the early 1990s this included work on: Wood Warbler, Yellow Wagtail (1991), Barn Owl (1992), Golden Plover in winter, Tree Sparrows and the Coordinated Coastal Migrant counts (1992 to 1996). There was also the development of a site-based recording system to help identify important locations for birds in the county and to meet the growing need for conservation purposes, from planners and developers for site-based information. From the beginning of the Club, BTO coordination and representation in Durham came from David Sowerbutts. His unstinting efforts through the years was particularly important in administering and writing up survey reports and recording this aspect of the Club's activity. He was deservedly awarded the BTO long-service medal in 2011. Over the years the programme of major survey works included two breeding atlases, the winter atlas and more latterly the ongoing Breeding Birds Survey as well as many BTO single species surveys.

By the early 1990s, thanks to increasing membership and focused projects, the quality and quantity of data included in *Birds in Durham* improved annually. Yet, the skew in observer effort, concentrated towards the coast and the centres of population, left obvious gaps in knowledge in the important western uplands. The formation of the Durham Upland Bird Study Group in 1992 was a response to redress this anomaly. The Group brought together DBC members with an interest in upland species with previously independent, highly experienced field-workers. The group's aim was "*to help conserve birds and their habitats in the Durham Pennines*" by monitoring breeding populations of key indicator species. The groundwork for the Group's formation had been laid by DBC's seminal survey of Black Grouse leks initiated in the mid-1980s by Brian Bates. This model demonstrated how confidential information on sensitive species could be collated and properly archived. Also contributing to the formation of DUBSG's work was the assessment of the county's Merlin population, undertaken by Mike Nattrass for the RSPB in the mid-1980s.

The DUBSG aimed to bring together the results of upland fieldwork by tracking the fortunes of Schedule 1 listed species, such as Merlin, Peregrine, Hen Harrier, Goshawk and Short-eared Owl, and several other important indicator species, including: Wigeon, Buzzard, Golden Plover, Raven and Ring Ouzel. The group acted as local contact for the issue of Schedule 1 Species Disturbance Licences by Natural England and Ringing Permits by the BTO and summarised data on such species for *Birds in Durham*. Species coverage has evolved with time. The late Denis Luckhurst clarified the important status of Wigeon in the county after a three-year study in the early to mid-1990s whilst in some years sample surveys of Ring Ouzel have been undertaken. The comprehensive study of Black Grouse was relaxed as the *North of England Black Grouse Recovery Project* was initiated in the late 1990s by the Game Conservancy Trust, the RSPB and other partners. In contrast, studies of Peregrine, Merlin, Raven and Buzzard took on more of a 'constant-effort' approach with high levels of coverage each year. Group resources were used to secure county coverage of national surveys for Merlin, Peregrine and Hen Harrier.

In 2006, the DUBSG joined with other raptor and upland bird study groups across the north to form the Northern England Raptor Forum (NERF), which aims to represent the views of dedicated fieldworkers on matters relating to the conservation of upland birds at a national level. The data collected by DUBSG feeds into an Annual Review published by NERF, which contributes to the work of the national Rare Breeding Birds Panel; thereby informing conservation and species protection policies. The challenges for species and habitat protection in the uplands remained as great in the early 21st century as when the DUBSG was formed and the conservation status of some species such as Hen Harrier and upland Peregrines were more precarious, making their monitoring all the more important. The DUBSG has been chaired by David Raw since its inception.

In October 1992, the Durham Bird Club published *Where to Watch Birds in Durham* a detailed book about the best birdwatching sites in the county, edited by Keith Bowey; it was the first fully comprehensive guide to birdwatching sites in the Durham area, it covered over 100 sites with maps, photographs and illustrations. In the same year, the Club produced *A Complete Species Checklist for Durham*, coordinated by David Raw, which summarised the status of 371 species recorded in the county to that point – the first attempt to do this since the publication of Temperley. The following year, the first local area treatment of birds in the county since Stead's *Birds of Tees-side* in the 1960s was published. This was *The Birds of Gateshead*, which was produced by Gateshead Council and written by Stephen Westerberg, Stephen Rutherford and Keith Bowey; the first and last had become members of the Durham Bid Club committee around the early 1990s.

In 1996, Lindsay Rewcastle became the first female to serve on the Club's committee, filling the role of Club secretary from 1997-1999. In mid-1996, the Club appointed Mark O'Connell as its first dedicated Conservation Officer, the role being adopted by Chris Cox in early 1997, and this important role continued thereafter, with John Olley serving in this capacity for over a decade. Through the mid-1990s, the Club made efforts to grow its membership base and this increased steadily through that decade from less than 140 in the mid-1980s, to 179 by 1991 and over 249 by the end of the decade. The production of *Birds in Durham* over this period saw a step-change in the content and quality of information contained within the reports, with greater analysis, increasing use of graphics and tabulation to illustrate trends. Survey summaries and one-off studies of sites and species were now routinely included as end-papers in the report; adding depth and stimulating interest.

Sadly in 1997, Fred Grey, 'father' of the Durham Bird Club, died at the age of 86; the end of an era.

Gateshead Kittiwake Tower, Image © Gateshead Council

Collaborative work by the Club, with Durham Wildlife Trust, in advising Gateshead Council on issues around the displacement of Kittiwakes by the development of the Baltic Centre for Contemporarily Arts led to the provision of a revolutionary design and the erection by the Council, in 1998, of the purpose-built Kittiwake Tower next to the 'Baltic' on the south bank of the River Tyne. The tower was subsequently moved about half a mile downstream in 2001. This structure proved hugely successful in attracting nesting birds and illustrating the value of innovation and collaboration in such projects.

Into the Twenty-first Century

In 2000, the Durham Bird Club published A *Summer Atlas of the Breeding Birds of Durham*; the fieldwork for this was carried out, initially, as part of the survey work for the 'New Breeding Atlas', between 1988-1991, with supplementary fieldwork in 1992 and 1993. Despite such publishing success, delays on the production of *Birds in Durham* in the early 2000s, led to a compressing of the publishing schedules (with six publications between 2005 and 2007) and a commitment to improve such work through the subsequent years. The decade saw very considerable improvements of the production values of this report, consolidating the improvements in content that had been made in the 1990s. The ever-increasing shift to the digital era of photography led to stunning bird images becoming readily available, and in response, the overall appearance of the club's publications, such as *Birds in Durham* and *The Lek*, took huge leaps forward.

During the 2000s, the club was increasingly involved in efforts to protect, enhance and then interpret the wildlife value of the wetlands around Bishop Middleham. The initial approach to this, largely through the efforts of County Ornithological Recorder, Tony Armstrong, was to have the various wetlands recognized in the planning system via the Durham County Council's County Wildlife Liaison Group. In March 2002, the

County Council agreed the importance of Castle Lake, and by March 2004, all three wetlands in this area had been designated as County Wildlife Sites.

From 2005, in a move driven by new County Ornithological Recorder, Geoff Siggens, the club moved away from its paper-based recording methods to electronic recording, a process that was first experimented with during the mid-1990s. This led to the development of an easily interrogated database of site-specific knowledge, which could be used to react to potentially adverse development threats at important sites. The process of managing records and writing club literature improved considerably with the shift to '100% electronic', and the sharing of important bird data assembled by club members with conservation organisations such as the BTO and RSPB became common place. The electronic era brought increased opportunities to contribute at all levels, with club seawatch data from Whitburn Observatory being amalgamated into both the national and European record.

With a change in Club committee personnel in the mid-2000s came a change in members. The membership total broke the 300 barrier for the first time ever and the rise in bird watching as a pastime brought a new keen breed of observer, albeit increasingly driven by computer based communication. Modern developments in the hobby of birdwatching saw an ever more rapidly changing nature in the tenor of how birdwatching was conducted. This might be characterised as a change from birdwatching to birding? The pastime became more competitive for some observers and increasingly facilitated, or perhaps occasionally dominated, by technology. At some levels, there was an inference that the more species seen, then the more proficient the observer. Such developments did not always go hand-in-hand with a commitment to regular survey or local patch work, particularly during migration periods. Whilst the number of observers and their capacity to observe grew hugely over this period, the amount of observation undertaken, rather perversely, may even have declined on a *per capita* basis? Nonetheless, in the round, the knowledge of the county's birds at 2011 was incomparable to 60 years previously and George Temperley, and others, would have envied the potential that modern observers have to add to the sum of ornithological knowledge of the county.

The late 2000s saw a number of important and authoritative publications relevant to the study of birds in Durham, with a particular emphasis on the south east of the county. First came the publication by the Teesmouth Bird Club in 2008 of *The Breeding Birds of Cleveland* by Graham Joynt, Ted Parker and Vic Fairbrother; a breeding atlas which mapped and provided detailed breeding information, on 127 breeding birds of the old county of Cleveland (which was in existence form 1974 to 1996). In 2009, the Tees Valley Wildlife Trust published *The Birds of Cleveland* by Martin Blick. This documented all species recorded in that area. In celebration of its 50th anniversary, the Teesmouth Bird Club published *The Birdwatchers of Teesmouth 1600-1960*, written by John Fletcher it provided a detailed and fascinating insight to the people who have studied birds in that area.

Castle Lake hide (M. Newsome)

The Durham Bird Club's ongoing involvement in conservation issues in the Bishop Middleham area was emphasised by the formal opening of its hide at Castle Lake on 24 May 2009, by John Taylor of the County Durham Environmental Trust. Further habitat creation and management by a dedicated band of volunteers, *The Shovelers*, saw the area grow into one of the most prolific inland wetlands in Northern England. Club Conservation Officers, John Olley and Alan Jones, also spear-headed habitat improvement work at Hurworth Burn Reservoir, amongst other sites, and broadened the efforts of the club into an ever widening world of bird conservation, wind farms and land development.

The year 2009 also saw the official opening of RSPB Saltholme by Kate Humble, Vice President of the RSPB, on 6 March. This was the first RSPB reserve in the county, and featured a state of the art visitor centre and numerous public hides. Although Saltholme Pools had long been a favoured haunt for large numbers of birds, the development of the area saw habitat management and protection taken to a new level.

On 1 April 2010, the Durham Bird Club secured a £30,000 grant from the Heritage Lottery Fund and embarked upon the Birds of Durham Heritage Project. This, spanning the period 2010 to 2012, was a multi-faceted project that set out to 'bring the wildlife of the past to the people of tomorrow'. To deliver it, the project partners formulated a programme of: public events, lectures and guided walks; created and erected nine Bird Heritage Interpretation Panels; and, produced a suite of on-line resources, all of which told the tale of County Durham's bird heritage. In the process of researching the above activities, the contents of this county avifauna were developed.

Bibliography & Acknowledgments

This chapter was compiled using a wide range of both formal and informal sources. These include: *A History of the Birds of Durham* by George Temperley; articles by Ian Kerr on Abel Chapmen and other historical figures first published in the Northumbrian Magazine; and, work on the background of the Natural History Society by Ian D. Moorhouse. It is hugely indebted to the extensive use of John Fletcher's publication on the early years of Teesmouth's birdwatching. Les Jessop provided an important section on the history of collections and museums. The notes and research of David Sowerbutts formed a vital element on the contents for the modern period of this chapter. Some of the pre-DBC details and background owes a debt to the articles produced about the formation of the Tyneside Birds club, by John Day and the other historical reference work by David Howey. Certain sections also refer to notes by the late Fred Grey, and very important reminiscences, inputs and contributions came from: Jim Edwardson, Colin Freeman, Mike Harbinson, David Sowerbutts and the late Brian Unwin. Other historical information sources that were used include Nick Rossiter's paper on the work of Wallis and David Gardner-Medwin's article on early ornithological records, both of which were published by the Natural History Society of Northumbria. A number of people helped to improve the text with comments and various inputs; these include Paul Anderson, Tony Armstrong, Paul Danielson, June Holmes, Ian Kerr and Dave Raw.

Durham Bird Club Officials and Committee Members 1975-2011

Between 1975 and 2011, at least 67 different Club members served on the Club committee for varying lengths of time and in various roles. One person served continuously over this whole period, David Sowerbutts; an outstanding commitment to birds and their study over more than forty years.

Chairman		Secretary	
Fred Grey	1974-1977	Brian Unwin	1975-1977
Morris Skilletter	1977 (part of)	Peter Gill	1978-1984
Tim Bennett	1978	Malcolm Steele	1985-1995
Anthony Roberts	1979	Stephen Westerberg	1996
Ken Baldridge	1980-1984	Lindsay Rewcastle	1997-1999
Brian Bates	1985-1988	Kevin Spindloe	2000-2003
David Raw	1989-1994	John Todd	2004
Keith Bowey	1995-2006	Steve Evans	2005-2007
Paul Anderson	2006 to present	Paula Charlton	2007-2009
		Richard Cowen	2010 to present

Treasurer		County Recorder	
Ian Hogg	1975-1977	Brian Unwin	1975-1980
John White	1978-1979	Ken Baldridge	1981-1986
John Orton	1980-1981	Tony Armstrong	1987-2005
Tom Bell	1982-1994	Geoff Siggens	2006
Fred Milton	1995-2004	Mark Newsome	2006 to present
Steve Skelton	2005-2007		
Allan Rowell	2008 to present		

References

Benett, Robert (1844) *The Durham household Book (liber bursarii ecclesiae Dunelmensis); or the accounts of the Bursar of the Monastery of Durham from Pentecost 1530 to Pentecost 1534* London   Surtees Society 18

Day J.C. 2008 *The First Fifty Years: A Short History of the Northumberland and Tyneside Bird Club* Northumberland and Tyneside Bird Club Bulletin

Elliot, Sir W., 1876. *Memoir of Dr T.C. Jerdon. History of the Berwickshire Naturalists' Club* 7

Fletcher J. 2010 *Birdwatchers of Teesmouth 1600–1960, The Events leading up to the formation of the Teesmouth Bird Club* The Teesmouth Bird Club

Graham G.G. 1988. *The Flora and Vegetation of County Durham.* The Durham Flora Committee and the Durham County Conservation Trust, Durham.

Howey D.H. 2008 *Birding in the 1950s and the formation of the Northumberland and Tyneside Bird Club* Northumberland and Tyneside Bird Club Bulletin

Jessop L. 1999 *Bird Specimens Figured by Thomas Bewick Surviving in the Hancock Museum* Newcastle upon Tyne Transactions of the Natural history Society of Northumbria Vol. 59 part 3

Nethersole-Thompson D. & Nethersole-Thompson M. 1986 *Obituary of H. M. S. Blair* Ibis 129   British Ornithologists' Union

Raven J. 1942 *John Ray Naturalist.His Life and Works.* Cambridge University Press

Chapter 4

A Short History of Durham Ornithology

This list includes races through over three centuries of events that have shaped the study or added to the list of the county's birds. For more details about some of these see the systematic list or Chapter 2.

1661 and 1671	Visits of John Ray to the north east region, meeting with Ralph Johnson of Brignall, setting up subsequent correspondence, which led to the documentation of a number of first records of species for the county and one or two for Britain, such as Long-tailed Duck
1678	*The Ornithology of Francis Willughby* published by John Ray, with accounts of a range of species from the Durham area
1784	Marmaduke Tunstall (1743-1790) publishes his Manuscript record of birds
1791	Part of Tunstall's collection purchased by George Allan for £700
1816	A *History of Hartlepool* was published by Sir Cuthbert Sharp, although only two pages of this were dedicated to birds; it listed a number of sea and coastal birds recorded in the area with some 88 species were noted
1822	The Newcastle Society purchases Marmaduke Tunstall's extensive collection of museum specimens
1823	A young James Backhouse (then aged 15) provides the first documented evidence of Hen Harrier breeding in Durham, on Wemmergill Moors; unfortunately the eggs were taken for his collection
1824	Catalogue of Birds Frequenting the County near Stockton, produced by John Hogg, originally as part of *The History of Stockton* (1824)
1828	The first Durham and UK record of Sooty Shearwater was "obtained" in Tees Bay by Mr G.Marwood.
1829	The Natural History Society of Northumberland, Durham and Newcastle upon Tyne, later the Natural History Society of Northumbria, is founded
1831	The only record of Pine Grosbeak for Durham and the first acceptable record for Britain was shot at Bill Quay, Pelaw in east Gateshead, *"sometime before 1831"*
1831	The first volume of the Natural History Society's Transactions is published, and includes *A Catalogue of the Birds hitherto met with in the Counties of Northumberland and Durham* by PJSelby
1834	A *Catalogue of the Birds of County Durham* by Edward Backhouse was a manuscript document that was not published. It listed 203 species seen in Durham, the status of each was given but only the less common birds show any detail of dates and locations
1840	*Birds of Durham* is produced in two volumes as a Manuscript document by John Hutchinson of Lanchester, with an accompanying volume on *Durham Fishes, Reptiles and Quadrupeds*. It was discovered in a second-hand bookshop by H.M.S. Blair
1845	The first confirmed UK breeding record of Black Redstart took place in Durham, at Crook Hall in Durham City
1847	The first Great Reed Warbler for County Durham was shot by Thomas Robson close to Swalwell on 28 May, the first record for Britain & Ireland
1855	The first record of Arctic Redpoll for County Durham and Britain was killed at Whitburn on 24 April
1863	The first of a few hundred Pallas's Sandgrouse was seen at Port Clarence on 13 May
1865	George William Temperley is born on 29 October
1874	John Hancock publishes a *Catalogue of the Birds of Northumberland and Durham*
1884	Robert Calvert published a *Geology and Natural History of County Durham*, that included a list of 205 bird species, many of these copied from earlier authorities, but with some additions of his own

1885	*Notes on the Avi-fauna of Upper Teesdale* by John Backhouse Junior, who studied the plants and other wildlife of the upper Tees valley, is published in The Naturalist. This mainly dealt with birds in the valley above Middleton-in-Teesdale and it listed 117 species
1896	*The Birds of the Derwent Valley* produced by Thomas Robson of Winlaton, documenting 137 species from the valley
1898	*Upper Teesdale Past and Present* published by John Backhouse with an updated chapter on the bird life in the area
1902	The first Little Bunting for Durham, and 2nd for UK, was shot at Seaton Snook
1905	Canon Tristram publishes *"Birds" in Victorian County History of Durham*
1907	*The Birds of Yorkshire* by T.H.Nelson, and covering parts of Teesmouth, is published
1915	The first copy of *The Vasculum* published June, independently, as The North Country Quarterly of Science and Local History.
1924	The Northern Naturalists' Union was founded
1931	*Birds of Hurworth Burn* by the Rev. G. F. Courtenay was published in the *Transactions of the NHSND&N*
1931	John Bishop was appointed as the official RSPB watcher for Teesmouth
1933	*Birds of Teesmouth* by Rev. G. F. Courtenay was published in The Vasculum
1935	From 1935 the annual ornithological reports for Northumberland and Durham were published firstly in *The Vasculum* (1936 to 1939) then in *The Naturalist* (1940 to 1950) and from then on in the *Transactions* of the Natural History Society
1939	*Birds of the Tees Valley* by W.E. Almond, J.B. Nicholson and M.G. Robinson was published in the *Transactions of the Northern Naturalists` Union*
1940s	Early ringing schemes began in Durham county around Hamsterley Forest
1947	Montagu's Harrier bred on Muggleswick Common; one pair, the first confirmed breeding in Durham since 1835 and a forerunner of this species breeding annually in Durham for the next decade
1951	*A History of the Birds of Durham* by George William Temperley, the first fully authoritative document about the county's birds was published by the Natural History Society
1955	Durham's only Desert Wheatear was found on the edge of a timber yard at Jarrow Slake on 4 December; it remained until 18th being then, only the tenth British record
1957	Fred Grey takes over the editing of the annual ornithological reports for Northumberland and Durham commencing with the report for 1957
1958	The Tyneside Bird Club was founded, with its first bulletin including records from Jarrow Slake, South Shields, Marsden, Whitburn and Durham City
1959	The only Durham record of Dusky Thrush was discovered on 12 December 1959 at Hartlepool Headland; at the time this was only the 2nd record for Britain
1960	In November a number of people assembled in the Nelson Room of the Dorman Museum, Middlesbrough, to discuss the formation of a bird club; this came to be the inaugural meeting of the Teesmouth Bird Club
1961	John Coulson takes over the editing of the annual ornithological reports for Northumberland and Durham commencing with the report for 1960
1961	Collared Dove bred for first time in Durham in West Hartlepool
1962	The formation of the Northumberland and Durham Naturalists' Trust in the early 1960s was important not just for the study of birds and wildlife, but for the management of land for wildlife and wider conservation issues
1962	The first Pallas's Warbler for County Durham was seen at Hartlepool Headland on 12 and 13 October, only the seventh for Britain
1967	Death of George William Temperley at the age of 92
1968	Grey-cheeked Thrush, only the sixth British record and one of only a handful of records for mainland Britain, was found dead at Horden on 17 October 1968
1969	Paddyfield Warbler, the first for County Durham was found on Hartlepool Headland on 18 September 1969, at the time only the third for Britain

1970	An adult Ross's Gull and a 1st winter Ivory Gull were together in the Tyne Estuary from 24th to 31 December
1971	The north east's single wildlife trust divided in 1971 along county lines, becoming two separate wildlife trusts for the counties of Durham and Northumberland; the Durham County Conservation Trust, becoming the Durham Wildlife Trust in the early 1990s
1971	The Whitburn Observatory building was provided by the Durham County Conservation Trust, providing a focus for sea-watching in that area. The original wooden hut was destroyed by fire and was replaced by a brick structure which lasted until 1984
1971	In October, the Council of the Natural History Society found that it was no longer able to continue publication of bird reports for the two counties. A committee was established to publish a separate *Durham County Bird Report*
1972	In March, the first in a new series of ornithological reports covering only Durham (from 1970), was published, using facilities at the University of Durham for production
1974	From 1 April, Startforth Rural District was part of the new County Durham after local government reorganisation. As it was not covered in ornithological reports from other areas it was adopted by 'Durham' as part of the county recording area, the 1974 ornithological report being the first to fully cover this area. On the same date the two boroughs in the south east part of Durham, namely Hartlepool and Stockton, became the northern half of the county of Cleveland. The annual bird report for the new county was published by the Teesmouth Bird Club as it is today
1974	Founding meetings of interested parties coordinated and catalysed by Brian Unwin led to the establishment of a new bird club in the north east, the Durham Bird Club: Fred Grey became its first Chairman
1975	A public meeting on 8 January 1975 attracted an attendance estimated at over 100 and enrolment of Club members began shortly after this. The initial committee was made up of Fred Grey (Chairman), Brian Unwin (Secretary and Recorder), Ian Hogg (Treasurer), David Sowerbutts (Indoor Meetings Secretary) and Dan Mold (Field Meetings Secretary), along with Brian Armstrong, John Coulson, Tom Palmer and Morris Skilleter. David Sowerbutts has served continuously since in a variety of roles, a quite remarkable achievement
1975	Washington Wildfowl Park opens, later to become WWT Washington
1977	The first Surf Scoter for Durham flew north along the entire county coastline, being seen as it passed Hartlepool and then Whitburn on 17 September
1978	A pair of Brambling that was present in Thornley Woods, in the lower Derwent valley, from late April into June appeared to have successfully fledged young
1980	The only record of Cirl Bunting for County Durham was found singing to the north west of Langley Park on 24 May, holding territory into summer; it re-appeared in spring 1981
1981	The first White-billed Diver for Durham was present in Hartlepool Docks from 14th to 22 February
1982	The first Long-toed Stint for Durham, the second for Britain and the Western Palaearctic, was at Saltholme Pool from 28 August until 1 September
1983	Gateshead Council commence a ranger service in the lower Derwent valley and Shibdon Pond, that would lead to the involvement of a number of active individuals in the study of the county's birds in the late 20th century
1984	The first White-tailed Plover for Durham and only the third for Britain was discovered at Cleadon on 21 May
1984	The first Durham Bird Club publication, other than the bulletin and the annual ornithological reports, was produced; a 32-page, A4 booklet called *Where to Watch Birds in County Durham*
1988	The first Bridled Tern for Durham was at Hartlepool on 9 August
1989	The first Double-crested Cormorant for Britain and Europe was at Charlton's Pond, Billingham from at least 11 January until 29 April

1990	Red-eyed Vireo, the first for County Durham was found in Mere Knolls Cemetery, Seaburn on 27 October 1990; it was the first mainland record on the north east coast of England
1990	A major Parrot Crossbill invasion took place with flocks of birds at Castle Eden Dene, Chopwell Woods and Hamsterley Forest, some staying into 1991, and possibly breeding
1991	Formal opening of the "new" Whitburn Observatory building on 16 May; it's still going strong to this day
1992	Booted Warbler, the first for Durham, was at Hartlepool Headland on 7th and 8 June; the first spring record for Britain
1992	Formation of the Durham Upland Bird Study Group; this brought together DBC members having an interest in upland species with previously independent but highly experienced field-workers
1992	First Coordinated Coastal Migrant Count; these continued annually until 1996
1992	Durham Bird Club publishes *Where to Watch Birds in Durham*
1993	*Birds of Gateshead* published by Gateshead Council
1996	The first Great Knot for Durham and 2nd for Britain was at Teesmouth from 13 October until 5 November
1997	Death of Fred Grey at the age of 86
1998	Erection of the purpose built Kittiwake Tower on the south bank of the River Tyne in Gateshead; the tower was subsequently moved about half a mile downstream in 2001
1999	The first Short-billed Dowitcher for Durham was at Teesmouth (first seen in Scotland) from 29 September until 30 October
2000	*A Summer Atlas of the Breeding Birds of Durham* published by the Durham Bird Club
2002	Bee-eaters breed successfully at the Durham Wildlife Trust reserve in Bishop Middleham; the first UK breeding record for 47 years
2005	An 'Amur' Wagtail or White Wagtail of the Far Eastern subspecies *leucopsis* was present on 5-6 April on the old site of the Vane Tempest colliery at Seaham on the Durham coast; this was the first record of this race in the Western Palaearctic
2006	Red Kites breed in Durham, in the lower Derwent valley, for the first time in over 170 years
2006	Avocets breed in Durham, at WWT Washington, for the first time
2006	The county's first Siberian Rubythroat, a first-winter female, spent three days, 26-28 October, in an urban back garden at Roker, Sunderland; the sixth record for Britain
2008	The first Glaucous-winged Gull for Durham and only the second for Britain was present at Teesmouth from 31 December until 10 January 2009
2008	Publication of *The Breeding Birds of Cleveland* by the Teesmouth Bird Club
2009	A Northern Harrier is seen and photographed near Selset Reservoir in Lunedale on 22 February; the first for Durham and only the second for Britain
2009	Publication of *The Birds of Cleveland* by Martin Blick
2009	Formal opening of the new RSPB Saltholme reserve by RSPB Vice President, Kate Humble on 6 March
2009	Formal opening of the Durham Bird Club hide at Castle Lake, Bishop Middleham on 24 May
2009	Britain and Ireland's first Eastern Crowned Warbler was found in a small clump of sycamores at Trow Quarry, South Shields on 22 October, and stayed for three days
2010	Durham Bird Club secure a grant from the Heritage Lottery Fund and embark upon the Birds of Durham Heritage Project
2010	Cetti's Warbler bred for first time in northern England, at Dorman's Pool, Teesmouth
2010	The county's first, and Britain's fifth, Brown-headed Cowbird was a singing male photographed in a Seaburn garden on 10 May

2010	The first Durham record of Common Nighthawk was found on 11 October close to Warren House Gill, Horden
2011	The first White-throated Robin for Durham and only the third for Britain was trapped at Hartlepool Headland on 6 June, remaining for a further four days
2011	The first Sandhill Crane for Durham was tracked flying south along the coastline of Durham from South Shields to Hartlepool on 29 September
2011	The first Pallid Harrier for Durham was present on the North Tees Marshes for four days from 20 October; part of an unprecedented influx into the UK
2011	Death of Brian Unwin on 29 December, at the age of 66
2012	A new county avifauna *The Birds of Durham* is published by The Durham Bird Club; the first definitive account for over 60 years of all the birds ever recorded in County Durham

Chapter 5

Birds of Durham - the Durham List

Introduction

This chapter looks at the development of the list of birds recorded in County Durham through time, particularly in comparison with that as documented in *A History of the Birds of Durham* by George Temperley in 1951. From that point, it examines some of the trends and developments since then, why these have taken place in terms of the location and geography of Durham and what might happen in the future based upon the experience of recent decades.

The very nature of bird recording, regardless of how systematic and scientific the process is, is essentially about creating lists, sometimes with supplementary information, and data, but lists nonetheless. What is a list and why is it important to keep one? This is a simple question but the answers go straight to the *raison d'être* of this publication. A list:

- Provides a succinct summary of all the birds that have occurred within a recording area, this can be just a simple list of the species' names or, more valuably, it can include notes on their status
- Whilst specific to the time of its compilation, it allows for temporal comparisons to be made and provides a vehicle for analysis of the changes
- Provides a trustworthy record against which future research can be carried out – essential if we are to understand the dynamics of our avifauna
- Can include significant sub-species which subsequently might be upgraded to full species status, and without such records their history would be lost
- Provides a note of the species present, or recorded, in a particular area, it allows for comparisons to be made between that site or area and the county/regional or national context; and, when combined with respective status notes it becomes a key tool in informed conservation decision making
- Can highlight omissions and prompt questions about why these are so, why have they not occurred and suggest future admissions; and, finally
- For most, birdwatching is a hobby which they do for pleasure. Keeping county or area lists is something which can enhance the enjoyment of all observers across the spectrum from the casual to the frenetic and obsessive. By recording and comparing their monthly, annual or life lists against the established order, they can contribute to the scientific archive at the same time, and through time.

Past lists

When *The Ornithology of Francis Willughby* was published in the English version, in 1678, it covered around 230 species for the whole of the country. One of the first published lists relating to the birds of the Durham area was that of Sir Cuthbert Sharp in *A History of Hartlepool* published in 1816. This covered some of the birds in the south east of the county and listed 88 species.

According to George Temperley (1951), the first ornithologist to pay major attention to the birds of Durham was Prideaux John Selby, whose first written work was called *A catalogue of the Birds hitherto met within the counties of Northumberland and Durham* and was published in 1831. This listed 214 species for both counties but it is unknown how many of these were specifically from Durham.

The earliest known complete list purely for Durham County was probably that made by Edward Backhouse whose *Catalogue of the Birds of County Durham* was an unpublished manuscript document listing 203 species seen in Durham, and providing a status for each of these but detail of dates and locations were only given for the scarce species. Another early Durham only list was that of John Hutchinson of Lanchester in 1840, which was produced as a manuscript document, with an accompanying volume on fish and quadrupeds. It documented over 150 species that were known to have been recorded in County Durham but most importantly, Hutchinson's list commented upon each species' status. Shortly after this, William Proctor, correspondent of Hutchinson's, published the first list of birds exclusively recorded in County Durham, unsurprisingly called a *List of Birds found in the County of Durham*. Comprising 204 species, this was published as an appendix to the Rev. George Ornsby's *Sketches of Durham* in 1846.

Subsequently, John Hancock repeated Selby's exercise producing his *Catalogue of the Birds of Northumberland and Durham* in 1874, which documented 266 species for the two counties. In 1884, within a *Geology and Natural History of County Durham*, Robert Calvert documented a list of 205 bird species recorded in Durham, but many of these were merely copied form earlier authorities, although there were some additions of his own.

In 1905, Tristram produced the first fully annotated catalogue of the birds of Durham that did not make reference to Northumberland. This was published in the *Victoria County History of Durham Vol. 1* and contained reference to 247 species seen exclusively in Durham.

In 1951 in *A History of the Birds of County Durham,* Temperley listed 273 bird species for County Durham. He broke these down into a number of categories. Namely: *"resident that bred"*, 82 species; *"breeding summer visitors"* 33; and, species that occasionally breed 8. This gave a total breeding number of 123 species. He thought that there were 71 regular passage migrants and 77 were occasional visitors or rare vagrants with two documented extinct species.

The county

County Durham is, in an ornithological sense, anything but typical of Britain; it is a small county yet it contains much ornithological variety in its small area. It stretches from some of the highest parts of England, down to sea level in a relatively short distance, accommodating all of avifaunal elements in-between those altitudinal extremes.

In Britain and Ireland, most biological recording follows the vice-county system established by H.C. Watson and published in 1873/1874 in his Topographical botany. This splits England, Scotland and Wales into 112 vice-counties of roughly equal area and ecological importance and was also based on the political county boundaries of that time. The recording area for the county and the Durham Bird Club is based upon the traditionally used Watsonian vice-county system. In this instance, the county's core recording area is vice-county 66, which covers all of the area between the Tyne and the Tees and is identical to the nineteenth-century County Durham. It was this area that was referred to in the county's mid-20th century avifauna by George Temperley (1951). Then, following a local government boundary change in 1974, a large area of north west North Yorkshire i.e. Startforth Rural District, transferred into County Durham and it made sense to include this within the county recording area. Today, the recording area is best described as the modern county of Durham, Tyne & Wear south of the River Tyne and that part of the old county of Cleveland to the north of the River Tees, along with the area acquired from vice-county 65 in North Yorkshire, during the boundary re-organisation of 1974. This extension of geographical area might indicate a concomitant growth in the size of the county list and certainly when considering the number of species recorded, some allowance for the size of the Durham list must be made. However, the county's relatively northerly position in England, and Europe more generally with southerly climes accommodating a wide range and greater number of species, militates against the number of species attracted to it. By and large, the variety of species decreases with an increase in latitude and Durham being one of the three most northerly English countries, has a lesser number of breeding species in comparison with more southerly climes. Over the years, even relatively 'common rarities' in the UK context have been much less common in the north east of England. This was once the case for species such as Little Egret, which has undergone a revolutionary change in its status in the British Isles over the last three decades, but has more recently spread into Durham, with future breeding increasingly likely.

Diversity is dictated by the range of climate, habitats and landscape features that can be accommodated within a given area of land, and which is in turn responsible for attracting species. The breeding birds of a county are particularly determined by these factors. A greater variety of habitats, even in a small county, is likely to attract a wider range of species than a paucity of environmental diversity, although some species require large tracts of their favoured habitats to prosper or persist. In Durham, the principal habitat areas that attract both large numbers and a great variety of birds are: the uplands, river valley woodlands, wetlands and the coastal strip. In the higher areas of the county, there upland specialists, wading birds and some birds of prey, though in such areas, these birds are increasingly pressurised by human influences. These Durham uplands hold truly significant numbers of some resident species such as Red Grouse. The river valley woodlands attract a range of upland woodland species and the coastal strip is the principal resort for rarities and vagrants. However, true vagrants can appear at sites even far inland (for example, the Killdeer at Cronkley Scar in March 1990). Wetlands attract wildfowl and associated wetland-fringe birds such as warblers, and if such wetlands sites have shallow edges or seasonal draw-down zones then passage waders expand the range of species that such sites attract.

The estuarine habitats of the Tees estuary, and to a lesser extent that of the River Tyne, are important for birds though since Temperley's time the extensive mudflats that the Tyne once boasted, largely at Jarrow Slake, were lost in large part during the 1970s, but ever more during the 1980s and again in the 1990s, until almost nothing remains. The mix of coastally located habitats between South Shields and the Tees estuary attract a wide range of species. These include migrant passerines in spring and autumn, seabirds to cliffs and beaches during the breeding season, and waders on both passage and in the winter to inter-tidal habitats such as Whitburn Steel.

Geography is a key determinant in the ability of a location or area to attract migrants and scarce species. For example, river valleys act as flyways and migration routes. High quality habitats in these situations or at the intersection of such features are likely to attract a greater variety of species than would equivalent habitat in a less geographically significant location.

Durham's avifauna is comprised of resident species, summer visitors, winter visitors, passage migrants and a, somewhat surprisingly wide, array of rarer visitors from far flung locations. Though in reality, many populations of even apparently resident species are in a seasonal state of flux, as has been shown by ringing studies. The species making up the Durham list, originate from a wide compass; the east, the north, the south, the Far East, and, somewhat bizarrely for an English east coast county, across the Atlantic. Not least amongst these visitors are Double-crested Cormorant, 'Northern' Harrier, Sandhill Crane, White-tailed Plover, Great Knot, Long-toed Stint, Short-billed Dowitcher, Glaucous-winged Gull, Eastern Crowned Warbler, Dusky Thrush, Siberian Rubythroat, White-throated Robin, 'Amur' Wagtail, and Brown-headed Cowbird.There is little doubt that the sites that attract the majority of rare species are also those that attract the largest number of observers, and this pattern reinforces itself, though there are many locations along the Durham coast that presumably attract unseen migrants and rarities, as evidenced by the occurrence of the 2010 Common Nighthawk serendipitously found at a site rarely visited by birdwatchers.

The county's western uplands provides the county's most significant suite of breeding species and one of the most important areas for upland birds in Britain. During the 20th century, 152 species of bird were recorded as 'breeding' in the county, 138 of these bred, or probably bred, during the survey period for *The Summer Atlas of Breeding Birds of County Durham* (1988-1994). Not all of these could be considered regular breeders or have, in some instances, been proven to have fledged young. A number of rare species have attempted to breed on only one or two occasions, such as Brambling, Dotterel and Golden Oriole. *A History of the Birds of County Durham* documented 123 breeding species for the area, so there has been some considerable change through the second half of the 20th century, with further additions at the outset of the 21st century. Obvious additions to the list of breeding birds since George Temperley's time include: Collared Dove, Ruddy Duck and Canada Goose, whilst the losses, though fewer, numbered amongst them Corncrake, Ruff (although it has exhibited breeding behaviour on a number of occasions) and, with some degree of irony, Ruddy Duck, although this has been brought about by planned extermination. In the early 1990s, some thought was given to what might be breeding in the next twenty years in Durham and it was suggested at that time that Avocets might colonise Teesmouth and that Common Rosefinch might begin breeding at some sites in the county (Anderson 1994). The first of these predications came true, but the latter species has not only remained a scarce species in the county, it has not built on the tentative breeding foothold it established in Britain in the early 1990s. Over this same twenty-year period many common breeding birds have declined dramatically and some species have arrived, colonised and then gone again for example, Ruddy Duck.

From Temperley to today

The growth in the number of observers post-Second World War and the increasing sophistication of equipment, digital, optical and printed reference material, particularly through the decades 1960-2000, had much to do with the growing number of species recorded. The increase in popularity of bird watching through the 1960s and 1970s led to an ever increasing number of local records and an ever-greater number of species recorded within the county. The growth in the number of species on the county list is in some sense, ironic as over this period there took place a real decrease in the actual ornithological value of many habitats. Bird-rich areas at both a macro and micro level have been squeezed into ever smaller geographical areas of the county, and corners of larger sites. It is somewhat anecdotal, but the average bird richness of any sample one-kilometre square in Durham is probably now a fraction of what the same area would have been 50 years ago, and likewise 50 years before that. Bird hotspots are probably that much hotter in the modern sense for a number of reasons. There are more observers, these observers are more skilled and in some respects there are fewer places for the birds to prosper within the wider environment of the county. The consequence is that birds inevitably appear at such hotspots. There have been major changes in the composition of species in the county since Temperley's publication. Extinct species have obviously remained extinct, though some of his lost breeding species, for example Red Kite, have returned, courtesy of major conservation projects and new breeding species have been added. Species noted by Temperley as 'having bred within the last hundred years', but no longer breeding, included Reed Warbler, which today is much more widespread across Durham and at Teesmouth its population is now counted in hundreds of territories. Likewise, Peregrine is a regular, though restricted breeder, but has been largely forced out of the uplands. Marsh Harrier has bred again, though it is yet to establish itself as regular feature among the county's breeding birds, although this may be just a matter of time.

Resident species that had severely declined according to Temperley included Raven, Stonechat, Lapwing, Ringed Plover and Black Grouse. This situation remains the case for the first of these, as Raven has never recovered its former status locally. By contrast, Lapwing remains common, particularly in Durham's uplands. Stonechats almost disappeared as a breeding bird from Durham, only to return with renewed vigour during the 1990s and early 2000s. Ringed Plover is perhaps more common as a breeding bird than ever in Durham, but Black Grouse has continued on its long downward trajectory, though Durham has England's largest remaining population of this species, and land management initiatives have been introduced to halt its decline. Other breeding species in recent decades going the way of Temperley's negative list above include: Wood Warbler, Whinchat, Tree Pipit and perhaps Twite; though a lack of knowledge seemingly hinders an understanding of that species' true status. House Sparrow, once so familiar as an almost 'domestic' breeder, is now largely absent from many residential areas, but is responding positively to local nest box provision, as does its near cousin, the Tree Sparrow, which suffered similar, but not quite as evident, declines given its more rural habitat preference.

By way of contrast some species have prospered quite spectacularly, since Temperley's time. For example, Mute Swans are today much more widespread as a breeding bird, partly as a result of reduced persecution at nests and the curtailing of the use of lead as a weight for fishing purposes. Few permanent waters in the county do not now have a pair of breeding Mute Swans yet in times not too far past, successful breeding was a rarely recorded phenomenon. In similar fashion, some song birds have seen long term upward trends. Just thirty years ago, Chiffchaff was a much less common bird in Durham than was its close relative Willow Warbler, and it was a bird of woodland. Today, not only is it more common than Willow

Warbler in some areas, but it can now be heard singing from scrub and marginal land throughout the county, including built up areas. Additionally Magpies seem to be present in ever increasing numbers, with birds in ever more obvious locations and in greater numbers.

The position of some species has seen little change, as is evident with the plight of the county's upland Hen Harriers. Despite the county's large tracts of suitable moorland habitat, even the tenuous, sporadic attempts at breeding of the recent past, now seem unlikely to return without a major change in the attitude of the land managers.

Considerable declines have occurred in the occurrence of some wintering species, one of the most noticeable amongst these is Hooded Crow, whose absence from the county is only now being properly appreciated following its recognition as a separate species from Carrion Crow. Some wintering species that have increased include Waxwing, as was noted by Temperley, though its numbers vary wildly from year to year, and most gull species, for which there has been a huge increase in interest amongst observers. Some common wintering species have nonetheless declined in numbers since the 1960s; Starling is one of the most obvious. This species once occupied roosts that could be estimated in the hundreds of thousands. Whilst it remains common, nothing like these numbers now inhabit the winter countryside of Durham.

The knowledge of the occurrence pattern of some species has changed out of all recognition since Temperley's time, and in some instances this knowledge has accrued relatively recently. It is just a few decades ago that Storm Petrel was thought to be a rare passage migrant along the Durham coast, but the advent of nocturnal ringing efforts for this species, from the late 1980s onwards, demonstrated that it is a regularly present species along the coast, although rarely noted during the day by increasing amounts of seawatching.

Patterns and trends in the occurrence of new species since 1951

The publication of a new complete county checklist by the Durham Bird Club in 1992 (Raw 1992), set the Durham County list of species in categories A, B and C of the then British list at 343, a large increase from Temperley's 273. Around that time, the list was considered to be growing at a rate of one or two species per annum (Raw 1994) and the actual increase has continued at a steady rate averaging just under two species per annum over a 61 year period, from 1951 to 2011. In many instances, additions to the county list have been birds that were genuinely new to the county, but some have 'arrived' not on the back of favourable weather conditions but as a result of taxonomic changes, for example: Balearic Shearwater, Yellow-legged Gull, Water Pipit and Hooded Crow. In a number of instances, species formerly regarded as merely escapes from captivity have proven to be self-sustaining in the wild, the most obvious examples of these being Canada Goose and Ring-necked Parakeet.

Since 1951, 120 new species have been recorded, an average of 1.97 species per annum (between 1951 and 2011). The highest average addition of species to the county list occurred between 1960 and 1989, when the average number of new species per annum was 2.3; the best decade was the 1970s with a total of 28 new species. The respective mean totals for those six decade periods, where: nine in the 1950s; 21 in the 1960s; 28 in the 1970s; 21 in 1980s; 21 in the 1990s; 13 in the 2000s; and, seven (to date) in the 2010s. The best period for the addition of new species to Durham came between January 1977 and the end of 1985; in this nine-year period 32 species were added to the county list. One presumes that as time progresses, there is an ever smaller body of new potential species for any county; the number of new species waiting to be observed is in some ways limited. It is remarkable that since Temperley was published in 1951, there have been just eight years in this 61 year period in which a new species has not been recorded in the county.

To the end of 2010, 382 species had been recorded in the recording area in an apparently wild state. A further three potential additions from 2011, subject to BBRC ratification, are: White-throated Robin, Sandhill Crane and Pallid Harrier. Of these, around 255 (about 65% of the total) might be recorded in any one year, although the year total topped the 270 mark for the first time ever in 2011. The majority of these species do not and have not bred in the county.

With the passage of time, competing pressures drive the rate of addition of new species in opposing directions. Firstly there is the increased number and ability of observers which tends to be counteracted by the ongoing reduction in numbers of likely additions to the list. The surge of birdwatchers and progress in identification techniques and knowledge is evident in the number of additions to the county list from the mid-1960s to the late 1990s. As time progresses, the additions tend to be of those species that are undertaking a wholesale distributional shift (Little Ringed Plover through the 1950s), or comprise the more unexpected species, that might have once been thought to be beyond consideration as appearing in Durham, for example Glaucous-winged Gull. Through time, the expectation of what might be a 'first' for Durham changes. Sixty years ago, it is unlikely that anyone would have predicted: Long-toed Stint, Eastern Crowned Warbler or Double-crested Cormorant in Durham and all 'firsts' for Britain, at the time of their finding in the county, though the status of the first of these has been superseded by an earlier Cornish record.

Mechanisms that have driven the growth of the Durham list, particularly over the period 1980 to 2010, include a growth in the interest in seawatching and an increased interest in migration and rare species through the 1980s, 1990s and 2000s. The permanent establishment of the Observatory at Whitburn, officially opened in May 1991, has provided a base which is as weather-proof as can be expected in the often stormy conditions that bring pelagic wanderers close in-shore. The resultant long hours of observation have produced record counts of many species and a diverse range of additions. With most pelagic species, the number recorded is directly proportional to the amount of time undertaken in seawatch studies, and this level of occurrence can be expected to continue as long as the time investment remains constant. In the last three decades of the 20th century, the number of seabirds added to the Durham list bears comparison with almost any other group of birds; among these were: King Eider, Lesser Crested Tern, Bridled Tern, Fea's/Zino's Petrel and Wilson's Petrel.

Partly as a result of the interest in rarities, the number of 'rare migrants' as a proportion of the total Durham list has changed considerably since Temperley's time, for example Yellow-browed Warbler, which had not been recorded in Temperley's time now averages over 20 records annually. The improvement in optics, the ornithological literature and the advent of the digital age and telecommunications have all driven this process. The publication of records in reports and the kudos of having one's name associated with a rare species has also driven some observers to become rarity hunters. This was amplified as the recording of birds in the north east moved from a regional level in the 1930s to the late 1960s, to a county footprint in the early 1970s and onwards, with the development of the annual publication of *Birds in Durham* from 1976.

The attraction of Teesmouth to waders from east and west is long established, but perhaps it was in its heyday in the late 20th century when water levels, particularly on Dorman's and the Reclamation Pools were lower. Additions at that time included: Buff-breasted Sandpiper, American and Pacific Golden Plover, Semipalmated Sandpiper, Great Knot, Long-toed Stint, and Spotted Sandpiper. One slightly unusual quirk of Durham's ornithological record is that it documented its first Short-billed Dowitcher, before adding the much more common Long-billed to the county list. Both nonetheless, came courtesy of observers at Teesmouth, during the last two decades. It is to be hoped that the advent of the RSPB's Saltholme Reserve will allow for a more bird-friendly management of the hydrological systems and the area will continue to attract exciting birds.

Long-term trends in vagrancy patterns, the ebb and flow of species 'natural' distribution and an increased knowledge of identification criteria all influence occurrence. The Ring-billed Gull was not recorded in Britain until 1973, and then 1982 in Durham, and whilst it remains a scarce bird locally, the fact that this has now been recorded 17 times, shows how quickly the status of a species can change and that this is in part driven by a greater understanding and knowledge on behalf of those searching for it.

During the strong growth of the Durham list over the period from 1980 to 2011, an unprecedented number of rare and unexpected birds appeared in County Durham. Pre-eminent amongst these was a run of extreme British rarities that occurred during the first decade of the 21st century and continued into the second. Amongst these were: the Western Palaearctic's' first Amur Wagtail on transient pools at the edge of a new housing estate at Seaham; a Siberian Rubythroat in an urban back garden at Roker in 2006; and, a

Brown-headed Cowbird, just a little distance away from this and in similar circumstances, at Seaburn in May 2010. The latter two 'garden' birds had something in common despite being from far away to the east and west – they were both found by people with an interest in, but limited knowledge of, birds. Nonetheless both realised that they had seen something unusual and shared the information with more knowledgeable observers. If they had not, both records would have been lost from the county's history. As the hobby grows then these occurrences are also likely to increase.

Similarly, in 2009, Britain's second North American 'Northern' Harrier was found in Lunedale in February, Britain's first Eastern Crowned Warbler in Trow Quarry in October, followed by the county's first Common Nighthawk in October 2010. All shared something in common, courtesy of a developed trait; they were identified from photographs. In the case of the Harrier, some months later, but within a few hours for the warbler, leading to what was then probably Durham's largest mass birdwatch on the bird's second day in the county. The Nighthawk was photographed by a construction worker who was able to show it to a visitor the next day, though it was never seen by a seasoned birdwatcher.

Given the growth in the numbers of birdwatchers who carry increasingly powerful digital cameras in the field, it is likely that this trend of post-event identification will continue in a benign reflection of the Victorian collector's approach '*What's shot is history, what's not is mystery*'.

Comparisons with neighbouring counties

To the end of 2011, there were 40 species that had not been recorded in Durham but had occurred in either Northumberland or Yorkshire, or both. By contrast, there were 12 species (including the White-throated Robin of 2011) on the Durham list which had not been seen in Northumberland or Yorkshire. Foremost among these were Britain's first Eastern Crowned Warbler and Double-crested Cormorant, and other extreme rarities, including: White-tailed Plover, Great Knot, Long-toed Stint and Glaucous-winged Gull. Compared to those neighbouring counties, Durham is small and has a relatively short stretch of coastline and a limited number of prime freshwater wetlands, consequently the range, and sheer number of species recorded in Durham is inevitably fewer, but nonetheless considerable for its size.

What has not been recorded in Durham

There are seventeen species that have been recorded in both Northumberland and Yorkshire but not in Durham and it is from this list that the next new species for Durham seem most likely to come. Amongst these the prime candidates for future vagrancy to Durham are perhaps Western Bonelli's Warbler and Black-headed Bunting, both of which have in excess of four records in both Northumberland and Yorkshire. In respect of bird migration and vagrancy however, making predictions is always problematic.

During the 18th and early 19th centuries, there were at least 14 species recorded in either Northumberland or North Yorkshire that were not noted in Durham. Some of these, because of their declining international status, are most unlikely to occur again - this would include Greater Spotted Eagle and Macqueen's Bustard - but a number of them have become increasingly regularly noted as rarities in southern England and some of these may re-occur in the north east of England, more specifically in Durham. Perhaps these might include: Little Crake, Black-eared Wheatear and Alpine Accentor?

One somewhat surprising absence from the Durham list is Nutcracker. This species has been recorded in both Yorkshire and Northumberland, in 1819 and 1958 in the latter county although British records have been far few between since the last major influx of 1968, though such a future influx might bring a first record in Durham at some time. Going back in time, the fact that there are no documented records of Great Bustard in Durham is something of a puzzle. Never a common bird in Britain, it was none the less once a regularly breeding species in the Yorkshire Wolds (Mather 1986), and possibly in north Northumberland in the 15th century (Kerr 2001). In centuries past, it would probably have also been more common as an overshooting migrant from the then more buoyant eastern and southern populations. The most likely repository for such a Durham record, of what was an eminently edible species, would have been the documentary evidence provided by the Cellarer's rolls of the Durham Monastery. No such records are known from that source and given the decline across much of its former European range, perhaps the species will never be recorded in the county?

The future

What will happen with Durham's birds in the future? Whilst many identification barriers have fallen, others remain to be 'cracked open'. Twenty years ago, it was considered that an in-hand description was necessary to secure the identity of a potential Blyth's Reed Warbler, with great attention required on wing formulae. Today, it is almost a familiar bird to the pursuers of autumn migrants on the east coast, and was added to the county list in 2007. Sykes's Warbler, still to be recorded in Durham, is a new challenge to be overcome, *viz.* the recent decision of the BBRC to ascribe a warbler at Marsden Quarry in 2009 an identity of Sykes's or Booted Warbler; highlighting the complex identification of this recently split species. Whilst this bird was considered most likely a Booted Warbler, Sykes's Warbler was recorded in Northumberland for the first time in 2010 and may occur in County Durham in time.

To the end of 2011, 592 species had been recorded in Britain. At just over 65% of this total, there is still considerable potential for additions to the Durham list. Additions in coming years may come by the way of developments relating to the study of taxonomy, which may lead to changes in how we define species. Such additions to Durham's list may come as species that were formally separated from closely related races to which they are currently, taxonomically associated, an example of which may include the Northern Harrier. Habitat developments, such as those at Teesmouth around RSPB Saltholme, are likely to provide new opportunities for established species and new species, and such improvements might support an expansion of breeding birds as well as attracting rare vagrants.

Global warming and the associated increasingly unpredictable weather patterns are undoubtedly having an impact on the UK's birds and thus on Durham's too. The addition of the Glaucous-winged Gull in December 2008 was reasonably speculated to be linked to the opening up of Canada's North West Passage through the growing summer retreat of the pack ice northwards. Likewise, auks and other seabirds from the northern Pacific may find it easier to cross the High Arctic and move into our waters. Recently, from the east have come Red-flanked Bluetails, first recorded in 2002 but now assuming the status of a scarce autumn migrant and, in 2011, Pallid Harrier. Both are experiencing a major westward shift in their breeding ranges in which weather patterns are likely to be playing a part.

Species from the south were once a relatively small element of the county list, for example Red-rumped Swallow was not recorded in Durham until 1995, and has subsequently been recorded on eight further occasions to the end of 2010. Likewise, Melodious Warbler which was not added to the county avifauna until 2003. Should global warming effects impact in a more emphatic fashion in the future, then it seems likely that species of a more southerly origin may make it as far north as Durham on more regular basis and species such as Squacco Heron, Sardinian Warbler and Iberian Chiffchaff may become 'Durham birds'. Many species' occurrence patterns have changed not just in terms of presence or absence but in terms of their seasonal first and last dates in Durham, and it seems likely that many species will make ever greater shifts in the timing of when they are present, or absent, from Durham in the future.

Species new to Durham since 1951

To the end of 2011, a total of 120 new species of bird were recorded within the boundaries of County Durham since publication of *A History of the Birds of Durham*. These are:

1951	Yellow-browed Warbler	**1964**	---
1952	Terek Sandpiper	**1965**	---
1953	Water Pipit	**1966**	Bearded Tit
1954	Little Ringed Plover		Icterine Warbler
1955	Desert Wheatear	**1967**	Lesser Yellowlegs
	Mediterranean Gull	**1968**	Grey-cheeked Thrush
1956	---		Ortolan Bunting
1957	---		Tawny Pipit
1958	Rustic Bunting	**1969**	Paddyfield Warbler
1959	Dusky Thrush	**1970**	Ross's Gull
	Balearic Shearwater		Richard's Pipit
1960	Lesser Grey Shrike		Greenish Warbler
	Barred Warbler	**1971**	Woodchat Shrike
	Cory's Shearwater	**1972**	---
1961	Snow Goose	**1973**	Green-winged Teal
	Broad-billed Sandpiper		American Wigeon
	Collared Dove		Collared Pratincole
1962	Pallas's Warbler	**1974**	Egyptian Goose
1963	Wilson's Phalarope	**1975**	Purple Heron
	Sharp-tailed Sandpiper		Subalpine Warbler
	White-rumped Sandpiper	**1976**	Radde's Warbler
	Red-throated Pipit		Killdeer
	Marsh Sandpiper	**1977**	Surf Scoter
	Red-crested Pochard		Buff-breasted Sandpiper

	Night Heron		Marsh Warbler
	Common Rosefinch		Lesser Crested Tern
	Bonaparte's Gull		Macaronesian Shearwater
	Franklin's Gull	**1985**	Yellow-legged Gull
	Ring-necked Duck		Thrush Nightingale
	Ruddy Duck	**1986**	---
1978	Gull-billed Tern	**1987**	---
	Mandarin	**1988**	Bridled Tern
1979	Arctic Warbler	**1989**	Ring-necked Parakeet
	Bee-eater		Semi-palmated Sandpiper
	Little Egret		Double-crested Cormorant
	American Golden Plover	**1990**	Red-eyed Vireo
	Baird's Sandpiper	**1991**	Yellow-breasted Bunting
	Cattle Egret	**1992**	Penduline Tit
1980	Cirl Bunting		Booted Warbler
1981	Dusky Warbler	**1993**	Great White Egret
	Caspian Tern		Black-winged Stilt
	Laughing Gull	**1994**	Hume's Warbler
	Black Kite		Pied Wheatear
	Nightingale		Blue-winged Teal
	White-billed Diver		Citrine Wagtail
1982	Parrot Crossbill	**1995**	Red-rumped Swallow
	Long-toed Stint		Pacific Golden Plover
	Savi's Warbler		Great Spotted Cuckoo
	Ring-billed Gull	**1996**	Great Knot
1983	Short-toed Lark	**1997**	Spotted Sandpiper
1984	White-tailed Plover	**1998**	King Eider

	Pine Bunting	**2007**	Blyth's Reed Warbler
1999	Pallid Swift		Long-billed Dowitcher
	Isabelline Shrike	**2008**	Glaucous-winged Gull
	Short-billed Dowitcher	**2009**	Whiskered Tern
	Lesser Scaup		Eastern Crowned Warbler
2000	Olive-backed Pipit	**2010**	Black-throated Thrush
2001	Caspian Gull		Brown-headed Cowbird
2002	Red-flanked Bluetail		Pallas's Grasshopper Warbler
	Fea's/Zino's Petrel		Common Nighthawk
2003	Melodious Warbler	**2011**	White-throated Robin
2004	---		Sandhill Crane
2005	Cetti's Warbler		Pallid Harrier
	Wilson's Petrel		
2006	Siberian Rubythroat		

References:

Anderson P. 1994 *Birding the First Twenty Years* Birds in Durham 1993 Durham Bird Club Durham

Kerr I. 2001 *Northumbrian Birds* Northumberland and Tyneside Bird Club Durham

Raw D. 1984 *Predicted Additions to The County List* Birds in Durham 1993 Durham Bird Club Durham

Raw D. 1992 *Birds of Durham Complete Species Checklist* Durham Bird Club Durham

Newsome M. 2007 *Birds of Durham Complete Species Checklist (Revised)* Durham Bird Club Durham

Newsome M. 2010 *Birds New to County Durham – What Next?* The Lek The Durham Bird Club Durham

Temperley G.W. 1951 *A History of the Birds of Durham* Transactions of the Natural History Society of Northumberland, Durham & Newcastle-upon-Tyne, Vol. IX.

The Systematic List

Mute Swan

Cygnus olor

Although a widespread resident, the Mute Swan is scarce as a breeding species; it is also a limited passage migrant.

Historical review

This species is widespread but not common; the number of breeding pairs and cygnets reared has been slowly increasing since the mid-1980s. A number of researchers believe that the Mute Swan was introduced to Britain by the Romans (Yalden & Albarella 2009), although remains of the species have been found in peat deposits dating from the Pleistocene era (Harrison 1988), indicating that the species, at least at one time, occurred here in a truly wild state. Evidence for the presence of swans in the north east of England comes from a letter sent by a Roman soldier stationed on Hadrian's Wall, just to the north of the county requesting nets to catch swans. This may have referred to the Whooper Swan *Cygnus cygnus* (Kerr 2001), but nonetheless, it demonstrates the presence of one or the other in the region at that time.

In the Cellarers' rolls of the Monastery of Durham, 'swan' was frequently mentioned as a provisioning species and clearly it was kept locally by the Priors to provide them with food (Ticehurst 1923). The earliest specific mention of the Mute Swan in County Durham is in the Cellarers' rolls of 1329, which documents a bird bought for the Easter festivities between April 20 and 27 of that year. Such birds seem to have been the property of the Prior, for the Cellarers' rolls for 1388, state *In ij signis per Prior em, nif* and birds seem to have been distributed over the different manors of the Priory. Payments are recorded in 1330 and 1349/1350 for conveying birds to Ketton, in Aycliffe, and *ad diverse loca prioratus* ("*to other locations within the priory*"), while in 1383 half-a-crown was paid for repairing the mill-dam at Bewley for their use (Fowler 1898).

The Mute Swan was commonly kept as a domestic fowl until at least the middle of the 18th century and there are records of local purchases from as far away as York (Gurney 1921). In the latter part of the 19th century, Hancock (1874) considered it a domestic bird and did not catalogue its occurrence in the region.

Historical records suggest that the Mute Swan was a reasonably common bird at the beginning of the 20th century, with mention of 40 birds at Hebburn Ponds as far back as August 1932 (Temperley 1951). This was almost certainly a moult flock of local birds, which had gathered at this site. Temperley (1951) said that it was "*a common resident wherever there are suitable lakes, ponds or rivers, breeding freely where protection is afforded*" and noted that it was numerous enough to be found on almost any water, for example a pair "*haunting the polluted waters of the Tyne, between Gateshead and Scotswood Bridge*" in the winter of 1943. Today, birds routinely inhabit this section of the now much cleaner River Tyne. Temperley observed that the Mute Swan was increasing in numbers, noting that for many centuries, it had been semi-domesticated in England, though originally of wild stock. By the mid-20th century, he wrote that "*it has once more returned to its feral state and few birds are now kept in a pinioned condition even in the public parks*" *(*Temperley 1951*)*. The species bred twice at the Shotton Brickworks Ponds in the late 1940s and early 1950s (Simpson 2011).

During the 1960s, breeding records in County Durham were sparse. A party of 22 at Seaton Carew on 6 June 1964 was the largest group reported in Durham since the early 1930s. Its status changed somewhat in the early 1970s and two breeding records took place in 1970, at Brasside Ponds and Witton-le-Wear, whilst up to 15 were at Teesmouth in the winter of that year. The largest single site gatherings during the 1970s were 28 at Witton-le-Wear on 16 January 1977 and 32 at Washington on 30 December 1979. Nationally, the British population remained fairly stable from the late 1960s to the mid-1980s, with increases thereafter (Mead 2000).

Through the 1970s, birds bred on more than one occasion at widespread localities such as Brasside Pond, Charlton's Pond, Cowpen Marsh, Gateshead, Haverton Hill, Saltwell Park and Witton-le-Wear. However, these breeding attempts often failed to produce young. There were up to three pairs at Teesmouth at this time, but in most years nests failed because they were robbed of the eggs. In 1974, birds bred at Shibdon Pond and then not successfully there for almost a decade. The most frequently used site in the county during this period was Witton-le-Wear, with two pairs rearing young there in most years through the late 1970s, but only two out of every three nesting attempts were successful. Over the whole decade, there were probably fewer than ten locations where nesting was known to have been attempted, and success was only secured at a small number of these sites in any one year. There were probably never more than three to four successful pairs in any one year during the 1970s, a dark decade for the species. Direct human persecution and the extensive use of lead shot by fishermen probably accounted for the death of many cygnets and the failure of the majority of nests.

Recent breeding status

Breeding numbers seem to have remained relatively stable through the 1980s, though the number of successful breeding attempts increased considerably through this decade. Unfortunately both adult and cygnet mortality rates were high at this time, the primary cause being lead poisoning from the ingestion of fishing weights. This remained the case until the middle of the 1990s when the species underwent a dramatic increase in the county, which has continued to the present, though at a rather reduced rate. This is illustrated by the fact that a single bird on the Marine Park Lake, South Shields was considered noteworthy in 1978. This site regularly attracted in excess of 100 birds through the late 1990s and early 2000s and a gathering of 200 birds in January 2002 proved to be the record total for the county (Newsome 2007).

The number of nesting attempts within the county was more or less stable over the 1990s with 17 pairs being the best count in 1994, but only five pairs attempted to breed in 1991 and 15 in 1999. Human disturbance and cruelty towards this species were all too prevalent, with particularly unpleasant examples from Joe's Pond and Rainton Meadows in 1997, when eight birds were shot and strangled. In a similarly depressing fashion, a family of nine were killed at Fishburn in 1999.

As illustrated by the *Atlas* (Westerberg & Bowey 2000), Mute Swans nest along all of the county's three main river valley systems: the Tyne, the Wear and the Tees. The majority of sites that hold breeding birds are in the lowland east of the county, just a few sites west of the A68 being regularly occupied. The furthest west the species is usually recorded is around the mid-Wear valley, such as McNeil Bottoms and Morley, near Butterknowle, and birds are relatively much scarcer at most of the westerly reservoirs.

Over the period 1988-1994, the species was surveyed annually in north east England as part of a long term, detailed study. It was discovered that in County Durham, the species was found to nest predominantly on still waters close or adjacent to rivers. Of all logged breeding attempts over the period 1988-1994, almost 93% were on still waters, with less than 6% on running water. Over 65% of still water sites used were located within 2km of one of the three main rivers, emphasising the significance of these watercourses and their valleys to the birds. Further inland, many "*apparently suitable*" sites exist that are isolated from rivers, but most of these have no historical record of breeding (Coleman 1991). The geography of most of the rivers in County Durham means that they are unsuitable for nest building by Mute Swans as they have relatively few islands, tend to be shallow and are largely fast flowing with little weed growth as a food source (Westerberg & Bowey 2000).

For many years, the county's main breeding concentrations were centred in two principle areas, the borough of Gateshead and the wetland sites on the north side of the Tees Estuary. These areas offer a number of suitable territories in relatively close proximity. Natal site fidelity appears to have had a positive influence on the development of these concentrations of breeding birds, with a few sites producing large numbers of cygnets, which then return to the same area to breed (Bone *et al.* 1995). This has the effect of clustering pairs of breeding birds, into "*extended colonies*". The Gateshead cluster is largely due to the breeding success of birds at Shibdon Pond from the early 1980s onwards. At least eleven sites in that area have held successful breeding pairs from 1980-2007, including: Axwell Park, Acer (Dunston) Pond, Clockburn Lake, Far Pasture Wetland, the MetroCentre Pools, Pelaw Quarry Pond, Ryton Willows, Saltwell Park Lake, Stargate Quarry Ponds, and Watergate Forest Park Lake, with unsuccessful breeding attempts also occurring on two or three other waters. In 1994, pairs held territory at seven sites in this area, the highest number at that time and easily the largest concentration of breeding birds in the county. Birds successfully colonised wetland sites in Gateshead throughout the 1990s, for example: Far Pasture, 1994; the MetroCentre Pools, 1998; Watergate Forest Park Lake, in 1999 and into the new century, Clockburn Lake in 2001.

The increase in numbers in the north of the county, which has subsequently rippled out to other areas of Durham, is mirrored on the North Tees Marshes. Here, the numbers breeding in the early part of the 21st century were almost double what they were in the early 1990s. Around Teesmouth, the history of breeding swans is somewhat more chequered. Historically, breeding numbers around Teesmouth were low, mainly due to human interference. Prior to 1978, a pair or two reared cygnets, but there were almost no successfully breeding birds in the late 1980s and up to 1990, with the exception of a pair that reared broods of cygnets at Haverton Hole in 1985, 1986 and 1987. Since 1992 however, there has been a steady increase in breeding numbers in this area. In 1998, nine pairs built nests and four of these reared young. Since 1998, the number of breeding pairs has continued to increase and by 2006 most suitable waters were occupied. About 20-25 pairs have nested in this area since 2000, with about 50 cygnets reared annually (Blick 2009).

The *Atlas* recorded that over a six-year period in the early 1990s, 35 separate territories were located around the county, varying numbers of which were used in each season (Westerberg & Bowey 2000). In total, over this time, there were 44 nesting attempts. Of these, 18 failed and the outcome of a further four was unknown. Of the failures, 23% were due to human interference and 11% to flooding. People therefore represent the most significant threat to Mute Swans in County Durham. The influence of human interference on breeding Mute Swans is best shown at Shibdon Pond, Blaydon. Regular wardening was introduced at that site in the early 1980s and it has become one of the few places in Durham where cygnets are reared in most years in which breeding is attempted (Bone *et al.* 1995).

The known paired population in Durham more than doubled in the *Atlas* period, from eleven pairs in 1989 to twenty five pairs in 1994. This increase was in line with a rise in numbers throughout the UK (Delany *et al.* 1992) and across north east England (Coleman 1991). In Northumberland, the number of breeding pairs rose by 100% between 1978 and 1990 (Coleman 1991). Since the late 1980s, the British population of Mute Swan has increased considerably and was estimated to be 37,500 pairs in the early 2000s. The increase was attributed to a decrease in juvenile mortality believed to be largely due to milder winter weather. This was reflected in the county, where the population also continued to increase during the latter part of the 1990s, but seems to have reached a plateau as birds have now occupied the majority of suitable breeding sites.

Breeding was confirmed on at least 23 waters during the first decade of the 21st century. The county population now probably stands at around 35 to 40 breeding pairs, only marginally more than the 30 breeding pairs stated in the *Atlas*. Alongside milder winter weather, the banning of lead as a fishing weight during the late 1980s was probably one of the main factors in the dramatic recovery of breeding swans across Durham and there are probably now more birds breeding in the county than at any time in the last 200 years.

Unusual observations of breeding behaviour include a pair at Hardwick Hall in 2007, which chased off another nesting pair and claimed their four cygnets alongside their own brood of eight. An example of very early nest building was noted at Brasside Ponds on 16 January 2007; this activity normally begins in late February or March. Occasionally 'Polish' type juveniles have been noted in the county; for instance, at Charlton's Pond in 1991 where a pair reared ten cygnets and a 'Polish' youngster. Other such birds were reared on Ropner Park lake, Stockton in 2002 (Blick 2009) and at Herrington Country Park in 2007.

Recent non-breeding status

The main concentrations of moulting birds occur at South Marine Park Lake, South Shields, on the River Tees at Stockton and on the North Tees Marshes. During the early 1990s, significant winter gatherings occurred at two sites in Gateshead, with up to 15 birds at the MetroCentre Pools in 1993/1994 and up to 25 birds on Saltwell Park Lake in 1994/1995.

In the late 20th and early 21st century, South Marine Park Lake in South Shields attracted considerable winter gatherings. The largest number recorded was 165 birds in 1999, a very considerable increase since 1990, when a maximum count of six was noted. A survey using colour- ringed birds recorded 127 different birds at the site in 1995, increasing to 270 individuals in 1999, indicating a rotation of birds at this location. Numbers peaked in the early 2000s, with 192 counted in January 2001 and 200 in January 2002. However, regular organised feeding ceased here in the mid-2000s and coupled with the partial draining of the lake for cleaning purposes in 2009, numbers soon plummeted. Yearly peaks in the late 2000s were 80 in January 2008, 59 in January 2009 and 70 in January 2010. The decrease in numbers at South Marine Park Lake was followed by a significant increase of Mute

Swans at Chester-le-Street riverside. Peaks of 42 in April 2008 and 68 in March 2009 were followed by a high of 148 in December 2010, possibly pointing towards this locality becoming the prime site for the species in the county.

Post-breeding, there is generally an influx of birds to Teesmouth marshes in late summer, the highest WeBS count there being of 102 birds in August 2009. In late 1993, the newly formed ponds at Belasis Technology Park, Billingham, attracted a sizeable gathering of up to 58 birds, but these birds now moult around Saltholme Pools or on the River Tees. This pattern of activity has led to a reduction in what had become a regular moult migration of north east coastal birds to Berwick-upon-Tweed during the late 1980s and the 1990s. Today, many of the non-breeding birds in the south east of the county gather on the River Tees, upstream of the Tees Barrage where the maximum count at Stockton to date is 75 birds (Blick 2009).

Peak counts at other wetlands regularly holding high numbers of Mute Swans during the non-breeding period include: Barmston Pond (27 in December 1985), Brasside Ponds (51 in December 2010), Castle Lake (49 in May 2006), Herrington CP (52 in May 2009), Hetton Lyons CP (42 in Sept 2001), Rainton Meadows (33 in Oct 1999), Saltwell Park (30 in March 2007) and Shibdon Pond (22 in October 1988). Today, it is estimated that between 300 and 400 Mute Swans are present in County Durham in most winters.

The Mute Swan remains very scarce in the west of the county. Few wander to the moorland reservoirs and even at the more sheltered Derwent Reservoir, the species is seldom seen. More regular western localities include Barningham and Thorpe Farm, Knitsley (near Castleside), McNeil Bottoms and Morley (near Butterknowle), whilst breeding is limited to the occasional single pair at established wetlands such as Low Barns, McNeil Bottoms and Sled Lane Pond.

As a passage migrant, this species occurs principally on the way to and from its preferred moulting and wintering sites. The coastal strip forms the major north/south flyway in the region, yet on the coast itself, sightings are rarely recorded. Small numbers are occasionally noted at Whitburn, largely in spring, late summer or autumn, with an average of five sightings per year in the period 2006-2010. Group sizes vary from one to six, but most consist of one to three birds. Mid-winter movements generally only occur in periods of hard weather, when the freezing of wintering sites forces birds to move elsewhere and thus were relatively infrequent throughout the 1990s and 2000s. Exceptionally, birds also settle on the sea in sheltered bays, harbours and river mouths in harsh conditions and can remain for a day or two until the weather improves. Small parties are also occasionally seen passing other coastal localities such as Crimdon Dene, Hartlepool Headland and Seaham, though reports are relatively few in number.

The hard weather of 1979 stimulated a small influx of continental migrants to the county, in early January of that year. At this time, when the species was considered uncommon, 18 were at Teesmouth on 6 January and three were in the mouth of the Tyne on 7th, with 24 at Washington the following day. This influx included the only long distance recovery of a foreign ringed swan in County Durham, a second year bird that had been ringed in Germany in February 1977, 890km to the east (Blick 2009). The number of birds which reach the north east and Durham in such winters is unknown, but the national picture would indicate that this is a rare event (Wernham et al. 2002).

Being large, obvious birds, dead swans are easily found and the cause of death can often be determined. A study in the Gateshead area during the late 1980s and through the early 1990s, which looked at the known cause of death of 22 birds, found that the majority of fatalities resulted from collisions with overhead cables (at least 14 in a nine-year period, most of the birds involved being juveniles) (Bone et al. 1995). In addition, three birds died from lead poisoning, two were shot, and a single bird died as a result of swallowing fishing line, whilst another was apparently clubbed to death by egg thieves. Even more recently, this species has continued to suffer greatly from the actions of humans, and fishing line injuries and shootings were common throughout the 2000s. In 2007, four out of the 18 juvenile birds at Herrington CP had to be taken into care with injuries caused by the ingestion of fishing tackle, whilst at Hetton Lyons four cygnets were taken into care after the adults were shot. Unfortunately, direct human persecution still has a major impact upon this species. In 2009, an adult and four cygnets were shot by youths at Ryton Willows, with other birds killed at Silksworth Lake and further human interference causing nest failure at Hetton Lyons. Occasionally natural predation is implicated in breeding failure, such as in 2008, when all cygnets at Low Barns were taken by a Fox *Vulpes vulpes*.

Distribution & movements

The Mute Swan has the most southerly breeding range of Eurasian swans. Mute Swans breed across most of Europe and Asia and are widely found in wildfowl collections across the world. Many birds live in areas mild enough for them not to have to migrate during the non-breeding season, though northerly populations are, at least in part, migratory. In most autumns, there is a small influx of birds into County Durham. The Mute Swan's winter distribution in Britain is broadly similar to its breeding distribution, except for localised movements to coastal waters, especially in cold weather, and to freshwater marshes and, in some areas, agricultural fields.

Since 1989, the species has been subject to an intensive ringing programme in north east England, which has resulted in over 85% of the birds in Durham bearing rings. This has shown that there is a regular movement of local swans from Durham to Northumberland, Tyneside, Teesside and, to a lesser extent, to southern Scotland, Yorkshire and Humberside. More specifically, studies of the large flock at the South Marine Park Lake, South Shields, have shown that there is a high turnover of birds throughout the year. In 1996, 287 different birds were recorded at the site over the year, though the monthly maximum count was only 62. Ringing recoveries have also established that there is a strong link between County Durham and neighbouring Northumberland, as birds regularly migrate north to commence their summer moult at Berwick-upon-Tweed. Interestingly, this migration does not normally begin until the bird's second calendar year. There are also many other widely scattered recoveries of birds from as far as northern Scotland, Norfolk and Wales (Bell 2008).

Specific noteworthy ringing recoveries include a bird ringed at Portobello, Midlothian, on 22 July 1961 which was re-trapped on 21 September 1961 at Amble, Northumberland, 70 miles to the south east, and then found dead at Frosterley, Durham on 4 July 1963, 44 miles to the south west. Three birds ringed in Ward Jackson Park, Hartlepool, in 1961 and 1962 moved north to Loch Leven, Kinross, then to Druridge Bay in Northumberland and finally to Shibdon Pond, Gateshead. One survived until at least January 1974, thus being 13 years old. A one year old female ringed at the Loch of Strathbeg, Aberdeenshire, on 30 July 1983 relocated to Haverton Hole, where it bred and hatched five young in 1985 and three young in 1987 (Blick 2009). Also at Teesmouth, two were seen in May 1987 that had been ringed at Berwick-on-Tweed after a major oiling incident. After treatment, they had been released in March 1986 and were still at Teesmouth in February-March 1988. One of these was at Charlton's Pond early in 1989 and was found dead there on 7 April of the same year. Also, an adult ringed at Billingham in November 1993 was seen at Welney, Norfolk, in February 1994 (Blick 2009).

Bewick's Swan (Tundra Swan)
Cygnus columbianus

Sponsored by
The Bewick Society

A scarce passage migrant and rare winter visitor.

Historical review

This species is, by some margin, the rarest of the three swans recorded in Durham, with most sightings occurring between November and February. It was first recognised as a separate species from the Whooper Swan *Cygnus cygnus* in 1829. Famously, the original specimen was shot just to the north of Durham, at Haydon Bridge, Northumberland. Its distinctive features were first pointed out in a paper read to the Natural History Society by R. R. Wingate on 20 October 1829 and it was named in honour of Thomas Bewick (Kerr 2001), who was a son of Northumberland but a resident of Gateshead in County Durham for much of his illustrious life. The species name was later adopted by the Linnaean Society. The first County Durham record after the bird's formal recognition came in the winter of 1836/1837, when one was shot near Seaton (Temperley 1951).

Hutchinson (1840) considered, according to the specimens he had examined, that it was *"as common as the whooper"*, which, with the benefit of another century and a half's records, seems unlikely. Hancock (1874) also considered it to be as common a winter visitor as the Whooper Swan, but this was not borne out by Temperley's experience from the late Victorian period and through the first half of the 20th century (Temperley 1951).

Tristram (1905) wrote that it was *"by no means so rare as frequently supposed...it visits us irregularly in hard winters"*, and three were documented as being *"taken together"* at Blaydon in 1887. However, as Temperley (1951) pointed out in the mid-20th century, this species *"has for many years been only a rare casual visitor to County Durham"*. He was able to document the following records: three at Boldon Flats in November 1909 (two of these were shot); three at Teesmouth on 3 March 1925; and, on 29 October 1930, one was *"taken"* from a flock of 33 swans of undetermined species that were flying south over the Tees. This flock may have comprised Bewick's or a mix of Whoopers and Bewick's. On 1 January 1939, one was present for a short time on Darlington Park Lake; in November 1947, a singleton was at the same place; and, on 6 March 1948, one was on Hell's Kettles near Darlington. There were therefore just six definite records of ten birds over a forty year period, in Temperley's summary.

Post-Temperley, records of the species remained intermittent over the next few decades. At Thornley Flooded Fields, near Shotton Colliery, a single adult was present from 21st to 28 February 1954 and two adults were seen there on 14th to15 April of the same year. During the rest of the 1950s, there was just one report, from Boldon Flats on 15 March 1959.

In 1960, there was an exceptionally large flock present at Teesmouth, nine birds in January building up during the winter to 33 in late March and early April. This is still the largest feeding flock recorded in Durham. During the same year, two were also present at Brasside Ponds on 26 March, and three were at Smiddyshaw Reservoir on 30 December.

During the early 1960s, the species occurred with increasing regularity but only in small numbers at scattered sites such as Smiddyshaw Reservoir in December 1961, Shibdon Pond in December 1962 and Barmston Pond in November 1968. A group of seven birds was at Hurworth Burn and Crookfoot Reservoirs from 25 November through to 7 December 1961. In 1962, two were at Hurworth Burn Reservoir during February/March and eight were at the Reclamation Pond on 26 December. In this period, David Simpson, of Shotton Colliery, reported several records from the Thornley Brickworks area, but noted that they had been very scarce in the eastern part of County Durham ever since, the only documented record being of one from 31 October to 6 November 1966. The year 1968 was exceptional, starting with four birds at Teesmouth from 3 February to 3 March and nine at Hurworth Burn Reservoir from 28 January to 4 February. In the autumn, 13 were at Teesmouth on 27 October, rising to 23 by 9 November. Perhaps some of the same birds were involved in the flocks that were at Teesmouth in early 1969, with up to three there in January, whilst 20 were noted on the marshes on 1 November.

Recent status

More was seen of this species through the 1970s than in any previous decade in the county and increasingly, small numbers of birds were noted spending at least part of the winter in the Teesmouth area. There was also a scattering of records from elsewhere in the county in most years, at locations such as Barmston Pond, where, for example, seven were noted on 24 October 1973. The maximum number of wintering birds noted at Teesmouth was 20 in January 1972. The largest ever passage recorded in the county also occurred in this period, when a flock of 49 flew over Hartlepool Docks on 18 January 1972 (Blick 2009).

The longest documented stay of a flock of Bewick's Swans in Durham relates to five birds that were found in the Teesmouth area on 12 November 1974 and which remained around Saltholme and Dorman's Pools until 16 March 1975 (Blick 2009). One of the best years of the decade for the species was 1975, with up to 19 birds at Teesmouth and a small number being recorded at eight other sites during November and December. Up to ten birds were also noted at Teesmouth in November and December 1977. Less was seen of the species as the decade wore on and Teesmouth sightings declined between 1978 and 1979, with some shift of focus towards the mid-Wear, with birds being seen at sites such as Lambton Ponds, Witton-le-Wear, where up to 15 were present in late November and December 1978. Other more unusual westerly sightings at this time included one or two near Bishop Auckland in March/April 1978, Willington in January 1979 and Baldersdale in December 1979.

The wintering population of Bewick's Swan in north west Europe showed a marked increase between the mid-1970s and mid-1990s, culminating in a peak of approximately 29,000 birds in 1995 (Beekman 1997), but evidence points to a decline through the late 1990s and 2000s. The trend in County Durham through this period broadly follows this fluctuation, with widespread and regular sightings through the 1980s declining to a minimal presence in the late 1990s. Indeed, between 1994 and 1999, there was only a single record per year in the whole of the county. Prolonged stays by this species are not common and most recent records have been of short-staying birds or individuals on passage.

The majority of records in the county in the last 30 years have come from the south east, with the North Tees Marshes and Hurworth Burn and Crookfoot Reservoirs remaining prominent. Notable gatherings have included 14 on the North Tees Marshes in November 1981, 10 at Bishopton on 18 November 1983, 24 south over the Long Drag on 3 November 1984, 16 on the North Tees Marshes on 18 November 1984, 12 at Hurworth Burn Reservoir on 2 November 1991, 13 at Crookfoot Reservoir on 9 November 1992 and 20 at Greatham Creek on 1 January 2001. Most other years through the 1980s and early 1990s, saw groups of up to nine birds lingering for several days or weeks, with a strong concentration between November and January. Large parties noted away from the south east include a group of 25 to 30 birds flying north between Whitburn and Cleadon Village on 14 February 2001 and 24 at Castle Lake, Bishop Middleham on 3 January 2004.

More unusual sightings have included single birds at Jarrow Slake on 3 January 1981, Longnewton Reservoir on 12 January 1985 and Sadberge on 23 March 1985, along with 9 at Hart Reservoir on 20 January 1985. Boldon Flats has regularly attracted passage birds for short periods, with sightings in November 1975, January 1989, January, February and November 1993, January 1996, December 2002, November 2009, November 2010 and November 2011. The complex of wetlands around Houghton-le-Spring has occasionally attracted passage birds, such as at Joe's Pond in October 1983, October/November 1984 and November 1991, Rainton Meadows in November 1998, Seaton Pond in October 1989, October 1990 and December 2000, and Hetton Lyons CP in November 2011. Brasside Ponds featured in November 1973 and November 1980 before the locality lost some of its attraction to wildfowl, whilst Barmston Pond was also a traditionally productive wetland in the 1970s, but due to site degradation has not attracted this species since 1991. In the west of the county, sightings have always been rare. Small numbers have been seen at Derwent Reservoir on several occasions; four in December 1982, three in December 1987 and two in February 1990. Four at Hury Reservoir on 18 November 1989 were also notable for the south west.

The earliest autumn arrivals in the last 40 years were at Teesmouth on 12 October 2002, Whitburn on 19 October 2005 and Greatham Creek on 21 October 1999. The majority of arriving passage birds move through the county in late October and early November. As regards departing birds, there are very few records in the county after late March. Unusually, two were near Bishop Auckland up to 4 April 1978, but the latest staying individual involves one which arrived at Barmston Pond on 8 February 1976 and moved next day to WWT Washington. It paired with a captive bird here and stayed until 27 April.

In contrast to the Whooper Swan, there are few sightings of this species on sea watches. Autumn arrivals observed at Whitburn include nine on 3 November 1984, seven on 5 November 1984, seven on 19 October 2005 and seven on 18 November 2010, whilst spring passage has been noted on just one occasion; a single bird flew north in the company of Whooper Swans on 30 March 2008. At Hartlepool Headland, two passed on 14 November 1986, 10 flew south on 31 December 1998 and four were also on the sea at Hart Warren on 19 November 1982.

As with the other swan species, there is occasionally conflict with human activity. On 11 November 1984, four birds settled on a waste lagoon at a chemical works at Teesmouth, but despite attempts to flush them away from the area, they died soon afterwards. One at Seaton Pond in October 1990 was unfortunately found dead on 2 November, whilst singles were found dead under overhead wires at Dorman's Pool on 26 February 1996 and 18 February 1998. More disturbingly, one was shot at Teesmouth in November 1971 and two were shot in the same area in January 1972.

Distribution & movements

This species breeds on the Arctic tundra across the northern Russian Palaearctic from the Kanin Peninsula to Kolyuchin Bay in the Chukchi Sea. Two subspecies have been identified in the Western Palaearctic; the race C. c. bewickii breeds mainly in the Siberian coastal lowlands from the Kola Peninsula east to the Pacific and the population west of the Taimyr Peninsula migrates to overwinter in north west Europe. Its main wintering range in the UK is much further south than the north east of England, being centred on the Ouse and Nene Washes in Cambridgeshire, the Norfolk Broads, Martin Mere and the Ribble Estuary in Lancashire and Slimbridge in Gloucestershire. The nominate race (known as 'Whistling Swan') breeds in the coastal plains of Alaska and Canada and winters as far south as Florida and northern Mexico. It is known only as a vagrant in the Western Palaearctic and has not occurred in County Durham.

There has only been one ringing recovery of Bewick's Swan in County Durham. A second-year male ringed at Slimbridge on 3 November 1971 was found freshly dead near Romaldkirk on 15 March 1972.

Whooper Swan
Cygnus cygnus

A scarce, but regular, passage migrant and winter visitor in small numbers.

Historical review

Of the yellow-billed swans, this is the species that is most inclined to spend the winter in County Durham, albeit in very small numbers. The first documented occurrence of 'wild swans' in the county comes from a single reference in the Bursar's rolls of the Durham Monastery, where in 1338/1339 *Et in vj cignis campestr., vidz. Elkes, empt., ix s ("six wild swans were purchased for nine shillings")*. So clearly, the Whooper Swan has been occurring in County Durham for as long as people have been documenting the area's wildlife. This quotation is of interest in that the use of the word "*elk*" (from the Old English *elfetu*) for a wild swan suggests a date of occurrence for the species two hundred years earlier than was previously recognised (Yalden & Albarella 2009).

More recently, John Hogg (1827) recorded that Whoopers visited the Tees Marshes in large flocks in hard winters, and several were killed in the winters of 1823 and 1827. In 1845, Hogg documented that they arrived *"not much before Christmas, not just at Teesmouth but also at Morden Carrs"*. Hutchinson (1840) recorded that it had been killed on the Rivers Wear, Tees, Gaunless, Browney "*and other brooks of the County"*. In the winter of 1880, the highest ever known count for the county occurred when a vast herd of swans was seen flying north west, past Teesmouth. When the leading birds of this flock arrived at Greatham shore, the rearmost portion of the flock was said to still be in Yorkshire and it was "*computed*" that the flight must have contained at least 1,000 birds (Temperley 1951). Regardless of the accuracy of this calculation, it is obvious the flock was of an exceptional number. Tristram (1905) recorded that it was frequently taken, but that its visits, even at that time, were becoming fewer. The drainage of Mordon Carrs during the late 19th century, and the industrialisation of Teesmouth, made these areas less attractive to the species.

It was noted by Temperley (1951) that herds of Whoopers wintered around the loughs of neighbouring Northumberland and there, numbers had been increasing, "*in recent years"*. In relation to Durham in the 20th century, Temperley described its status as "*an irregular and uncommon winter visitor, less frequently recorded now than formerly"*. Following Temperley, the bird was reported intermittently at a number of wetlands around the county during the 1960s. One of the highest counts of the 20th century occurred when 47 were noted at Crookfoot Reservoir on 21 March 1961 (Blick 2009). Also from 2 November to 20 December 1961, 35 birds were on the Tan Hill Reservoir, which would in 1974 become the southern border of the county (Mather 1986). The next large count in Durham was several years in coming and concerned about 40 birds at Seaton Carew, on the beach and on the sea, on 29 December 1964. Unusually, one was at Witton-le-Wear in July 1964, presumably a bird that had been delayed in migration, perhaps through injury or illness. The Whooper Swan was mainly a scarce species in Durham for much of the 1960s with just a handful of sightings of small numbers of birds between 1965 and 1968 at Derwent Reservoir, Houghall and Teesmouth. On autumn passage, the largest number noted over this period was 24 over Washington on 20 October 1968. Nonetheless, this species was recorded by David Simpson nearly every year from 1950 to 1980 at Shotton Brickwork Ponds, largely as a passage bird in early spring or late autumn, and this is probably reflects its status in the county over this period.

During the 1970s, birds tended to be recorded from October to the end of the year, then in the late winter and again in a light early spring passage through the county. The species was recorded regularly around the North Tees Marshes through the decade but never in large numbers, but early 1971 brought an influx of 24 birds at Teesmouth on 9 January. Over the early part of the decade, Witton-le-Wear attracted occasional birds, usually only one or two at any one time. Hurworth Burn Reservoir and in particular, Derwent Reservoir also regularly attracted small numbers. There was a strong autumn passage in 1974 which involved small parties at a number of sites and four herds of between 10 and 22 birds were noted on 24 October alone.

Through the mid- and late 1970s, this species was considered an annual winter visitor in small numbers, usually single figures, to the reservoirs of upper Teesdale, and in particular those in Lunedale and Baldersdale (Bradshaw 1976). Lambton Ponds at Witton-le-Wear also attracted eight birds on 21 October 1978, rising to 16 by the 29th, and all remained to 26 November, whilst up to 25 were there in December 1979. There were also widespread reports of small numbers of birds on northbound spring passage in 1979, from late February through into early April.

Recent non-breeding status

Even though the overall pattern of occurrence since 1980 has remained similar, the numbers of Whooper Swans now seen in County Durham are far higher than in the 1960s and 1970s. The main autumn arrival takes place in the second half of October and early November, although occasional family parties have ventured south much earlier than this. The earliest sighting in Durham is of a single bird south past Whitburn on 19 September 1998, followed by five at Whitburn on 22 September 2004, seven at Crimdon Dene on 23 September 2001 and 12 at Dorman's Pool on 27 September 1995. Autumn passage consists mainly of family parties and small herds, rather than the mass exodus witnessed in spring, and is channelled down the coast, as opposed to the cross-country northbound movement. Peak autumn day counts have included 25 at Whitburn on 25 October 1986, 28 at Saltholme Pools on 11 October 2003 and 55 south over Cleadon on 8 November 2009, whilst at least 50 were judged to have passed through the North Tees area in the last 3 months of 2000 and 60 did likewise in late autumn 2006.

During the early 1980s, small numbers still wintered in the county, mainly on the Lunedale and Teesdale reservoirs. Peaks counts included 14 at Hury Reservoir on 4 January 1981, 15 at Selset Reservoir on 26 January 1986 and 17 at Selset Reservoir on 23 December 1990. A party of 10 at Longnewton Reservoir in early February 1987 may also have been associated with the Lunedale birds. However, by the early 1990s, this small wintering presence was no longer reliable and sightings in this area became erratic and usually involved short-staying individuals. By the 2000s, any sighting of a bird feeding on a wetland in County Durham was considered unusual and indeed, the first three months of 2003 produced just three sightings anywhere in the county. Prolonged stays are infrequent, with two in the Seaton Pond area from 6 January to 24 February 2004 being atypical.

Occasionally, adverse weather in mid-winter can produce a movement of Whooper Swans between wintering grounds within the UK, as happens with Pink-footed Geese *Anser brachyrhynchus*. Such movements are relatively rare, with recent examples including 25 flying west along the River Tyne at Ryton Willows on 13 December 1986, 24 north past Hartlepool on 28 February 2005 and six off South Shields on 15 January 2006. Six birds also settled in Hartlepool Fish Quay on 1 January 1982 during hard weather (Blick 2009).

The largest herds witnessed in the county in recent years have been during early spring passage as they head back to Iceland. During the 1980s, movements generally involved flocks of no more than 22 birds with more impressive sights including 30 over WWT Washington on 12 April 1987, 33 west over Ryton Willows on 27 March 1989 and 38 north over Billingham on 14 March 1988. This last flock constituted the largest herd seen in the North Tees area since December 1962.

A clear increase then took place through the 1990s and 2000s with far more birds moving through the county. Between 12th and 30 March 2009, a total of 389 were logged from widespread locations, whilst between 2 March and 10 April 2010, a total of 520 passage birds were reported. Impressive lowland herds have included: 70 north over Cleadon on 2 April 2006; 78 at Hurworth Burn Reservoir on 12 April 2006; 94 north past Whitburn on 29 March 2008; 59 north past Hartlepool on 17 March 2009; 82 at Herrington Country Park from 22nd to 24 March 2009; and, 56 north past Hartlepool on 7 March 2010.

As well as northbound flocks following the coast, there is clear evidence of large parties following the river valleys north west. The moorland reservoirs have regularly attracted large groups in recent years, such as 25 at Smiddyshaw Reservoir on 4 April 1997 and 66 there on 17 March 2009, plus 90 at Tunstall Reservoir on 1 April 2000 and 50 there on 25 March 2004. A party of 49 also flew north west over Romaldkirk on 27 March 2001. The largest recorded northbound herd involved 120 flying north west over Cow Green Reservoir on 26 March 2011. Small numbers have also occasionally dropped into Balderhead, Derwent, Selset and Waskerley Reservoirs, and parties could be encountered anywhere in the west of the county on days of passage.

Although the vast majority of birds have headed north by early April, occasional stragglers have been seen in early May. The latest spring record of an apparently wild bird is at Dorman's Pool on 25 May 1999, with others noted at WWT Washington on 22 May 1982, Saltholme Pools on 6 May 2002 and Whitburn on 12 May 2005. One or two have also occasionally been noted as late as the first five days of May at locations such as Barmston Pond, Crookfoot Reservoir, Hurworth Burn Reservoir, Saltholme Pools and Seaham.

There have been several instances of birds hitting overhead wires and illegal shooting over the years. More disturbing was the case of six birds which were seen at Fish Lake in Lunedale in late March 1980. All six were found dead in early April and tests found high levels of lead and copper in their livers and kidneys, possibly ingested from the bottom of lake and caused by run-off from the adjacent Barytes mine.

Recent breeding status

Birds have bred in the county, but only as feral pairs. A pair of pinioned adults, which were released on Ropner Park Lake, Stockton, in December 1990, went on to rear a youngster in 1991 (Blick 2009). Over the years, there have also been occasional summering birds noted. An adult summered around Greatham Creek and Seal Sands in 1983, being present from 24 April to 21 August. Some, or perhaps all, of these unseasonal birds have been feral or escaped birds from collections, such as the well-documented wandering bird that appeared at several locations in the county through the 1990s. This 'tame' adult had been reared at Studley Park, North Yorkshire, and was often present at Teesmouth from late 1992 until it was found dead there in May 1998. More recently, three birds of unknown origin were at Far Pasture on 20 August 2006 and another bird summered on Wynyard Lake, being there from about 3 June to 12 October 2009 (Blick 2009).

Distribution & movements

Whooper Swans breed across subarctic Eurasia and Siberia, with western populations having a fragmented winter range encompassing north west Europe, the Mediterranean basin and the Black and Caspian Seas. The British and Irish wintering population of approximately 16,000 birds consists mainly of birds from Iceland, with smaller numbers from north west European breeding grounds (Garðarsson 1991). There was a distinct shift in wintering grounds through the 1980s, with the population on the Ouse Washes, Cambridgeshire, doubling in ten years (Cranswick *et al.* 1997). This population shift is reflected in the trends of occurrence in County Durham.

The origin of the birds seen in the north east is indicated by the history of an adult noted on Saltholme Pools in late October 1991. This bird had been ringed as a cygnet in Iceland in July 1988. It was seen in Cork, Ireland, in February 1989 and in Cumbria during January 1990. It was then seen at Martin Mere, Lancashire in the winter of 1990/1991 and was back in Iceland in April 1991 (Blick 2009). In similar fashion, a colour ringed adult seen at Boldon Flats on 2 March 2010 was originally ringed in Iceland on 7 August 1999. It was seen again near Reykjavik, Iceland, on 24 October 1999, by the River Foyle, County Tyrone, on 17 December 1999 and at Booragh Bog, County Offaly on 25 February 2001 (Siggens 2005). Also, a bird ringed as a juvenile at Martin Mere, Lancashire, in February 2003 was on Saltholme Pools on 31 October 2005 (Blick 2009). Re-sightings of individually marked Whooper Swans in Ireland have shown that these birds are particularly mobile, frequently moving between Britain and Ireland, as well as using several sites within Ireland, in a single winter (McElwaine *et al.* 1995). Nevertheless, Whooper Swans show a high degree of winter site fidelity (Black & Rees 1984).

Of recoveries involving feral birds, a wandering immature, originally ringed in North Yorkshire, was noted at Far Pasture, Shibdon Pond and the MetroCentre Pools in early July 1993. Also, an adult female ringed at Cawood, North Yorkshire on 26 September 2010 was present at Chester-le-Street riverside from late September 2010 to April 2011, and again from 17 October 2011 into 2012.

Bean Goose
Anser fabalis

**A scarce migrant and winter visitor from northern Europe and Siberia,
recorded from October to June, though with most from November to March; two subspecies occur.**

Historical review

The first record for County Durham is effectively unknown, as the species complex still included the Pink-footed Goose *Anser brachyrhynchus* until 1839 when the two became widely regarded as distinct species in their own right. Both Selby (1831) and Hutchinson (1840) make reference to the *"Bean Goose"*, with Hutchinson writing *"The Bean Goose, like the other species of Geese which resort here usually frequents the marshes and adjacent lands in the eastern parts of the County and the vicinity of the coast. It is most frequently killed by wild-fowl shooters near the Tees, but sometimes at considerable distance inland"*. However, it is unknown to which species this account refers. Hancock (1874) was the first to attempt to separate records of the two species in the county stating that the Bean Goose was *"an autumn and winter visitant; usually seen in considerable flocks flying in lines"*, whilst he referred to the Pink-footed Goose as *"not an uncommon autumn and winter visitant"*. An increased knowledge of the pattern of occurrence of both species gained during the latter part of the 20th century has demonstrated that all of the *"flocks flying in lines"* were probably Pink-footed Geese, which, in effect, casts doubt on all of the earlier records, although the Bean Goose did undoubtedly occur in the Victorian era. For example, there is a specimen of a Bean Goose in the Hancock Museum, Newcastle-upon-Tyne, which is said to have been shot at Jarrow Slake some time during 1837.

Subsequent to Temperley's publication (1951), there was just a single record of the Bean Goose before the close of the 1960s. A single bird was shot from a party of 20 at Greatham Creek, Teesmouth in December 1954, with the remaining 19 birds present in the same general area until 27 February 1955 when they were last seen flying over Seal Sands.

Recent status

A total of over 350 Bean Geese were recorded in County Durham during the period 1970 to 2011, though the numbers present each winter are usually small, and there have been eighteen blank years. Periodic influxes in response to cold weather on the Continent have boosted the county total, with 33 birds in January and February 1979, and an exceptional total of 69 birds in January to March 1996, including a skein of 27 birds in off the sea at Whitburn on 25 February. Other large flocks noted during 1996 include 12 at Boldon Flats from 28 January until 25 March, and a different flock of 16 over WWT Washington on 25 March. The 69 birds recorded in 1996 represented the best year for this species in the county to 2010, and was more than double the high totals of 33 birds recorded in both 1979 and 2006. The pattern of occurrence however, was unusual in 2006, with a small influx of 14 Taiga Bean Geese *Anser f. fabalis* in January and February, and the annual total boosted by a flock of 19 birds of unspecified race flying in off the sea at Whitburn on 19 December.

The late autumn of 2011 produced a record influx of Tundra Bean Geese, associated with a very large arrival of European White-fronted Geese *Anser albifrons*. Following the first arrivals on 12 November, parties of up to 17 birds were seen at wetlands such as Boldon Flats, Crookfoot Reservoir, Hurworth Burn Reservoir and on the North Tees Marshes. The influx was noted at other sites on the east coast of England and was in response to strong easterly winds and cold temperatures across Northern Europe.

Most records of Bean Geese are from along the coastal strip, between Tyne and Tees, in particular from the south east coastal plain, though birds have penetrated as far inland as Derwent Reservoir, Spennymoor, and Brasside Ponds. The majority of birds do not linger. In recent years, the North Tees Marshes have been the best area for the species, with small numbers occurring on an annual basis. Further inland, Crookfoot and Hurworth Burn Reservoirs and the Bishop Middleham area have also produced records.

The majority of Bean Goose records in County Durham fall in the winter period from November to March, though there are two October records, the earliest of which was of two birds flying north with Barnacle Geese *Branta leucopsis* at Whitburn on 3 October 2008. There are just two April records, a single bird of unspecified race at Crookfoot Reservoir from 9th until 17 April 1994, and a single Tundra Bean Goose *Anser fabalis rossicus* at Saltholme Pools, Teesmouth on 11 April 2009. May has also produced two records, one of which related to a long staying Tundra Bean Goose which lingered at Saltholme Pools until 9 May 2005 and the other concerned two birds of unspecified race at the MetroCentre Pools, Gateshead on 15 May 1996. There are also two June records, both of which refer to birds of suspect origin.

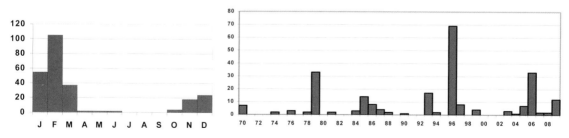

Records of Bean Geese in County Durham 1970-2009

There was a single of unspecified race on the Reclamation Pond, Teesmouth, from 18 June until 2 July 1993, and a similar bird at Derwent Reservoir on 16 June 2005.

Two forms of Bean Geese, currently treated as one species by the BOU, are known to occur in Britain, the nominate Taiga Bean Goose and the smaller Tundra Bean Goose. Both forms have been recorded in County Durham, though many older records were not separated as the identification criteria were not widely known until recently.

Taiga Bean Goose
Anser fabalis fabalis

All records specifically identified as this race are listed below:
1979	between Washington and Jarrow, sixteen, 16th to 24 February.
1979	Seaham Hall, 23rd to 26 February.
2003	Saltholme Pools, Teesmouth, three, 15th to 30 November.
2005	Haverton Hole, Teesmouth, three, 13th to 16 November.
2006	Port Clarence, Teesmouth, seven, 28th to 29 January.
2006	Stoneybeck Lake, Bishop Middleham, two, 29 January.
2010	Castle Lake, Bishop Middleham, 7th to 8 February.
2010	Saltholme Pools, Teesmouth, twenty two, 13 February.

The party of 22 at Saltholme Pools, Teesmouth, on 13 February 2010 was the largest flock of Taiga Bean Geese recorded in County Durham. Present for just over an hour, they flew off south and were later relocated in fields close to Coxwold, North Yorkshire, where they spent much of February (Newsome 2012).

Tundra Bean Goose
Anser fabalis rossicus

The first Bean Geese identified as belonging to this race were three at Saltholme Pools and Dorman's Pool, Teesmouth from 3 January to 10 March 1987. Sightings have come in a further nine years since then, and it is clear from sightings during in the first part of the 21st century that the Tundra Bean Goose is by far the most common form recorded in the county, with records between 2000 and 2011 totalling at least 116 birds, whereas there have been just six records of Taiga Bean Geese in the same period, relating to a total of 38 birds. The flock of 27 seen to arrive in off the sea at Whitburn on 25 February 1996 is the largest single flock of Tundra Bean Geese ever recorded in County Durham (Armstrong 1998). Both forms have occurred in the period November to February, though it is clear that most of the birds occurring during February and March relate to Tundra Bean Geese, particularly the larger flocks, which normally occur during periods of harsh weather on the Continent. This pattern of occurrence mirrors the status of the Tundra Bean Goose in the rest of the British Isles.

Distribution & movements
The Taiga Bean Goose nests in the taiga zone from Scandinavia east to the Urals, and winters mainly in the Low Countries and central Europe. Only two wintering flocks are regular in Britain; up to 250 frequent the Avon valley in Clyde, Scotland, from early October until late February, and around 100 winter in the Yare Valley, Norfolk, from late November until mid-February. Away from these two areas, they are genuinely rare birds in Britain.

The Tundra Bean Goose nests on the northern Russian tundra east to the Taimyr Peninsula, wintering primarily in the Low Countries, Sweden, Denmark, Germany and France, with small and declining numbers in Spain. There are no regular wintering flocks in Britain, though small numbers occur in most years, particularly along the North Norfolk Coast, and the east coast of England, especially during spells of harsh weather on the Continent.

Pink-footed Goose
Anser brachyrhynchus

A common passage migrant and scarce winter visitor.

Historical review

The status of the Pink-footed Goose in the county has changed considerably over the last 50 years. A scarce but regular passage and winter visitor in the early 20th century, it has become, by the beginning of the 21st century, the most commonly observed goose species on migration over County Durham. The earliest known record of this species in the north east region relates to remains that were excavated from Roman sites along Hadrian's Wall (Miles 1992). Historically, 'grey geese' were seen in their thousands over the Tees Marshes in the 1800s, with perhaps the majority of these being Pink-footed Geese, judging by the birds that were shot at that time (Blick 2009).

Robson (1896) documented the taking of a single Pink-footed Goose near Dunston in March 1893. However, there must be a degree of doubt over this identification, as the Bean Goose *Anser fabalis* and Pink-footed Goose were considered to be races of the same species at that time. Temperley considered the Pink-footed Goose *"an irregular autumn and winter visitor"* mid-20th century and quoted Tristam (1905) as stating that the Bean Goose was the commoner of the two, especially inland, when this specimen was taken. Confusion about the identity of the Bean and Pink-footed Goose means that many of the old records even up to the early 20th century are unreliable.

Up to the mid-1930s, Teesmouth was still a regular roosting area for geese and Bishop reported parties of grey geese there from August to the end of September 1933. He noted that flocks of 60 birds were present from 28 or 29 September, which may have been Pink-feet (Temperley 1951). Boldon Flats also used to attract grey geese, possibly this species, on a regular basis, but less so after the extensive drainage of that area which took place in the period after World War 2.

The mid-1940s produced a better run of sightings, particularly in the north of the county. On 7 November 1945, a flock of 51 Pink-footed Geese flew from the south east, across the beach at Whitburn and inland. On 16 November, 50 birds, possibly the same flock, were seen flying out to sea from Cleadon Hills, perhaps from Boldon Flats a little way inland. Birds were last seen in this area on 24 November, the flock still being 50-strong (Temperley 1951). On 26 November 1945, 36 birds were identified at Teesmouth. From 3rd to 5 November 1947, a flock of 80 was in the Whitburn area, and in 1948, a flock of up 17 birds was also there.

There is little documentation of the species in the county's ornithological records through the 1950s, with slightly more detail emerging during the 1960s. According to Blick (2009), the species was a regular passage visitor over the Teesmouth area during September and October, with flocks often in their hundreds being noted each autumn. For example, a skein of 165 flew north at Hartlepool on 28 November 1964. At that time, the peak passage period appeared to encompass the last four months of the year with the return spring passage being much less evident (Blick 2009). The information for the rest of the decade is sparse and reports for 1967 indicate only a handful of birds. For example, 11 were noted at Seaton Carew and Port Clarence in March and November, and a couple of skeins flew over Durham City in early October (Bell 1968).

Recent non-breeding status

Reports of this species through the 1970s and early 1980s were relatively few and far between. Although the species was encountered as a passage migrant, skeins numbering more than 20-30 birds were only rarely noted. For instance, in 1971, there was just a single record at Teesmouth on 6 February (Coulson 1972). In 1972, sightings were limited to just three birds passing Hartlepool on 31 December. Reports from the early part of this decade suggested that the occurrence of the species had undoubtedly declined (Unwin 1974).

The mid-1970s saw a slight upturn in fortunes, as occasional skeins of up to 80 passed over the county in autumn, whilst the 100 observed flying west over Darlington on 31 December 1974 was the largest flock encountered in Durham since 1965. Also notable at the end of this decade was a party of 99 seen at Teesmouth on 8 December 1979, which was likely to have been the same flock that had passed over Derwent Reservoir earlier that day. A similar theme was evident during the early 1980s, although there were signs of increased passage activity. In 1982, 144 flew south over Sunderland on 2 November and in 1983, 105 lingered at West Boldon on 26 January.

From the mid-1980s, the appearance of autumn skeins down the eastern third of the county became more expected. Notable sightings included 180 over Boldon Flats on 13 November 1984, 230 over Sunderland on 29 September 1987 and five skeins totalling 542 birds flying south past Hart Warren on 14 November 1993. New heights were achieved in 1994 when 974 moved over Teesmouth on 16 November, with 400 at Whitburn and 160 at Hawthorn the same day. Such numbers were not imaginable 15 years earlier.

Through the mid-1990s and 2000s, the number migrating through the county rose in direct correlation with the numbers utilising the North Norfolk coast as a wintering area. By way of confirmation of this trend, in the long-studied Gateshead area, no birds were positively identified until 1985, when 20 flew west over Sunniside on 12 October (Bowey et al. 1993). Through the late 1980s and 1990s, further flocks of up to 100 birds were infrequently noted, but increases through the 2000s resulted in more regular flocks witnessed on both southbound and northbound passage, including 150 over Chopwell Woods on 9 January 2008 and 150 over Lamesley on 19 February 2009. The species can now be considered a regular passage migrant.

The first returning autumn birds tend to be reported from the second week of September; for example, 37 past Sunderland Docks on 9 September 2009, 70 past Ryhope on 11 September 2005 and 47 past Whitburn on 13 September 2007. These birds often first appear at locations within a few miles of the coast, but birds moving cross-country along the river valleys, especially as movements increase, are a routinely noted phenomenon. More pronounced movement through the county usually takes place in late September or early October and continues into early November, although late autumn movements are possibly more related to local weather conditions with birds relocating between wintering areas, rather than migrants arriving directly from their breeding grounds.

The largest ever passage of Pink-footed Goose through the county occurred in 2008, when approximately 2,500 birds flew south over Hurworth Burn Reservoir during the morning of 28 September, with 1,310 logged moving through the Whitburn area later the same day and several skeins of 70-120 noted at localities further inland (Newsome 2009). It is likely that over 4,500 birds moved south through the county on this date. In 2007, the 2,040 birds counted flying south over Hartlepool on 15 September indicated another huge day of passage, whilst day counts of 500-1,000 birds are now routinely noted at migration watch points in the east of the county.

Late autumn passage is less prolific and more sporadic in appearance, although impressive numbers can still be noted in late October and early November. On 5 November 2005, 1,050 were noted moving south past Whitburn, 890 flew over the North Tees Marshes on 24 October 2003 and 810 passed over Whitburn on 25 November 2007. Up to 700 have been frequently noted at other coastal localities in this period.

It has been estimated that at least 3,000-5,000 Pink-footed Geese fly south along the Durham coast each autumn, but the true scale of the actual passage is hinted at by the observation of an estimated 8,000 which passed over the old county of Cleveland, between mid-September and late November 2004 (Blick 2009). However, only a small percentage of the birds that pass through County Durham get recorded, with many skeins no doubt moving over under cover of darkness, indicated by anecdotal evidence of large flocks of geese heard but not seen. Water Birds in the UK 2008/09 (Calbrade 2010) showed a minimum of 115,000 Pink-footed Geese in Norfolk in December 2008, and many of these are likely to have moved down the north east coast or over the county.

The northbound return passage from wintering grounds in Norfolk is often protracted and movements can become apparent as early as January, sporadically continuing through to early May. This is clearly an indication of staging movements between the various populations as birds would not be returning to their breeding grounds in January and February, but merely relocating to favoured areas in Lancashire and south west Scotland. The level of passage varies from year to year, as does the route taken, but recent examples include flocks of 100-320 at six localities between 8th and 12 February 2008, four skeins of 130-525 birds on 11 February 2006 and approximately 1,000 north past Whitburn on 30 January 2005. In contrast to autumn passage, flocks at this time of year often cut north west through the county following the river systems of the Tees, Wear and Tyne, with regular sightings at inland localities such as Derwent Reservoir, Langdon Beck and Tunstall Reservoir.

Conversely, cold weather after the New Year can result in further southbound movement of birds relocating from Scotland to Norfolk, although within the current climate trends, this is a rare occurrence. Such conditions in 2010 resulted in approximately 2,000 south over the Houghton-le-Spring area on 1 January, a further 2,000 southbound birds reported during the following week and more than 1,000 over South Shields between 1st and 7 February. By early April, northerly passage has decreased considerably and flock sizes have reduced. Occasionally, the last returning flocks from Norfolk are detected passing north up the Durham coast and recent examples observed at Whitburn include 180 on 5 May 2010, 42 on 6 May 2007, 200 on 23 April 2006 and 500 on 4 May 2004.

The number of birds actually resting on the ground and feeding, as opposed to birds moving through the county, is small in comparison. Any wetland or undisturbed farmland in the east of the county can attract a passing flock for a few days, but the larger skeins seldom touch down. The only area to attract birds on a regular basis is the North Tees Marshes and in the first decade of the 21st century, small parties of 20-30 birds have increasingly lingered here from early January to the third week of April and occasionally into early May. Recent peaks have included 82 on 1 April 2007 and up to 42 in the late winter of 2008 and indications are that this minimal wintering population will be maintained. Both Wynyard and Crookfoot Reservoir have also regularly attracted this species, although up to 119 at the latter site in February 2003 was exceptional (Blick 2009). Other sites such as Boldon Flats, Castle Lake (Bishop Middleham), Hurworth Burn Reservoir and the wetlands in the Houghton-le-Spring area have proved popular from time to time, but with little regularity and the parties of up to 30 birds soon move on. The appearance of up to three birds, often with Greylag Geese *Anser anser*, at localities such as Chester-le-Street, Derwent Reservoir, Herrington Country Park, Lamesley Pastures, Low Barns, Rainton Meadows and Saltwell Park is usually unrelated to movements of the species through the county and the origin of such birds is open to question.

Recent breeding status

Occasional singles, or a scattering of one to three birds, can be found in the county during the spring and summer months. These largely consist of feral or injured birds, which may associate with resident goose flocks (mainly Greylag Geese). Through the decade of the 2000s, such birds were seen at localities including Boldon Flats, Little Stainton, RSPB Saltholme, Saltwell Park and Wynyard. Despite being in areas where Greylags breed, there has never been any evidence of a nesting attempt.

During the *Atlas* survey period, a pair was noted over an upland area, close to Derwent Reservoir in the north of the county on 22 May 1989, with possibly the same pair being seen in a nearby area shortly afterwards (Westerberg & Bowey 2000). Although these birds were in a suitable breeding habitat, there was again no indication that breeding might have taken place.

Distribution & movements

The Pink-footed Goose breeds principally in Svalbard, Iceland and eastern Greenland. The populations in the latter two areas are mainly responsible for British wintering birds. The Greenland and Iceland populations, numbering over 220,000 individuals, breed primarily in central Iceland and in smaller numbers along the east coast of Greenland. The Icelandic population spends the non-breeding season entirely within Britain, in sites as widely spread as eastern Scotland, from the Moray Firth to the Borders, also the Solway Firth, down into Lancashire and increasingly in East Anglia.

The only ringing data from the county involve an adult ringed in Dumfriesshire on 23 November 1953, which was recovered at Sunniside on 26 April 1964 and one ringed at Dupplin, Perthshire on 11 October 1957 which was found near Hartlepool 25 years later, on 10 October 1982.

White-fronted Goose (Greater White-fronted Goose)
Anser albifrons

A scarce passage and winter visitor; two subspecies occur.

European White-fronted Goose
Anser albifrons albifrons

Historical review

Today, this species is amongst the rarest of the seven regularly occurring species of goose, both in County Durham and the whole of the north east. It may have been more frequent in the 19th century, but its overall status has probably changed little. Hutchinson (1840) said that it was *"by no means common"*, though he thought it more common than the Greylag Goose *Anser anser,* which he described as the *"rarest of its tribe"*. Hogg (1845) quoted John Grey of Stockton, who said it *"frequents our marshes in small flocks"*, obviously referring to Teesmouth, while Tristram (1905) said that it was *"not uncommon in hard weather at the coast, though not in large numbers"*. Temperley (1951) summarised the status as *"a rare winter visitor, only seen in severe winter weather and seldom observed far from the coast"*. Despite this, the first fully documented record of the species in the Teesmouth area in the 20th century was not until February 1954 (Blick 2009).

During the winter of 1941/1942, Temperley recorded the presence of a flock of *"grey geese"* roosting nightly on an area of land close to Dunston Staithes, adjacent to the Tyne. The birds were present from 3rd to 16 January 1942 and were reported by the observer to be White-fronted Geese. No specimens were taken and in light of this, Temperley felt it unwise to accept the identification as certain (Bowey *et al.* 1993). Through the 1950s and 1960s, the White-fronted Goose was infrequently recorded in Durham. In 1964, five birds, four adults and an immature were noted at Hurworth Burn Reservoir from 27 February to 5 March, with one at Teesmouth from 10th to 27 April in the following year. Another European race bird was at Teesmouth from 26th to 30 October 1968.

Recent status

Between 1970 and 2010, the species became almost an annual visitor to County Durham, though it was never been present in large numbers, with the vast majority of records relating to the nominate European form. Most records fell between mid-November and early March, with December and January probably being the peak months for sightings. Many of these observations related to small parties of between one and six birds, though occasionally, larger numbers were seen. Such examples include 60 west over Darlington on 7 October 1974; 24 over Cowpen Marsh on 2 March 1978; 19 at Cowpen Marsh on 29 January 1984; 16 at Crookfoot Reservoir on 30 December 1985; up to 55 on Saltholme Pools in early December 1987; 69 again on Cowpen Marsh from 16 February to 3 March 1993; up to 43 at Boldon Flats from 19 January to early February 1996; and 42 at Crookfoot Reservoir from 31 January to 5 February 1996.

Most sightings have involved birds in the eastern third of the county, with a high proportion of records relating to birds seen around the Teesmouth complex of wetlands, with the principle areas being Cowpen Marsh, Dorman's Pool and Saltholme Pools. However, almost any area which holds large numbers of geese could attract this species and White-fronted Geese have been seen at the following localities on at least one occasion: Bishop Middleham, Boldon Flats, Hetton Bogs, Hetton Lyons CP, Houghton Gate, Lambton, Low Butterby, Rainton Meadows, Shotton Pond, South Shields, Whitburn, WWT Washington and Wynyard. Occasional 'fly-overs' also contribute to the distribution of records, such as the bird flying south over Birtley with Pink-footed Geese *Anser brachyrhynchos* on 26 October 1996; the first modern record for Gateshead.

Birds are much less frequently seen in the west of the county. Derwent Reservoir has been the most productive western locality with birds seen in at least seven years between 1970 and 2010. More unusual westerly sightings have included two at Witton-le-Wear from 23rd to 31 October 1970, one at Hury Reservoir on 17 Mar 1986 and seven there on 15 February 1993, two at Adder Wood, near Hamsterley, on 9 March 1996, one at Low Barns NR on 17 April 1999, four at Black Bank Plantation on 22 January 2009 and four at Scargill from 19th to 22 February 2009.

Seawatch records are distinctly few, although birds clearly arrive onto the north east coast from continental Europe. At Hartlepool Headland, single birds flew past on 29 March 1989 and 4 November 2007, with nine south on 12 November 2011, while at Whitburn, the only recent records are of five flying south on 19 November 1997, four north on 27 October 2002 and one south on 24 November 2011.

As with many wintering wildfowl, there have been occasional unseasonal mid-summer records, such as the single birds at Dorman's Pool on 29 July to 1 August 1979 and at Saltholme Pools on 16 July 1989, but their origin is clearly open to question. Genuine autumn arrivals in September are exceptional, although there have been several sightings in the last ten days of the month. The earliest sighting was a single bird at Dorman's Pool on 24 September 1994. The latest sightings of departing birds in spring have been seven flying northwest over Dalton Piercy on 12 May 1982 and one at Hurworth Burn Reservoir on 17 May 2009.

The scarcity of this species in Durham during the first decade of the 21st century is in line with decreasing wintering numbers observed elsewhere in Britain (Hearn 2004). The formerly favoured haunts of the Severn Estuary, south east England and east Norfolk now hold just a fraction of the numbers of 40 years ago. The number of White-fronted Geese visiting the UK peaked at more than 10,000 in the late 1960s (Lack 1986), but numbers have declined steadily since then, with only 3,862 recorded in the winter of 2000/2001 (Hearn 2004).

Despite the steady decline in numbers through the early 21st century, the winter of 2011/12 saw a record influx into the county. Following the first arrivals on 10 November 2011, numerous parties were noted at many sites in both coastal and lowland areas, totalling perhaps some 350-400 birds. Larger flocks included up to 51 at Boldon Flats, up to 47 at Hurworth Burn Reservoir, 43 at Chourdon Point and 42 at Little Stainton, and a peak of 41 around the North Tees Marshes in December, whilst up to 31 were also recorded at Birtley, Bishop Middleham, Crookfoot Reservoir Hardwick Carrs, Rainton Meadows, Seaton Pond, and in the Whitburn area. In addition to these birds, evidence of southward displacement is provided by three large flocks over Saltholme Pools; 75 on 19 November; 27 on 3 December; and 66 on 27 December; whilst 37 also flew south over Wynyard on 17 December. Six at Balderhead Reservoir on 19 November were also notable for the west. This arrival was associated with a strong easterly airflow and cold temperatures in Northern Europe, and was witnessed in many parts of eastern England.

Greenland White-fronted Goose
Anser albifrons flavirostris

The first reference to the Greenland race in Durham concerns two birds at Derwent Reservoir on 3 March 1968. However, the true status of this subspecies is clouded due to many records of White-fronted Geese not being identified or published down to a sub-specific level. Small numbers have been identified in at least 19 years between 1980 and 2011, but visits have often been from single birds and are unpredictable in date and location. One of the more favoured sites has been Crookfoot Reservoir, although birds have also been seen at: Bishopton, Boldon Flats, Dalton Lodge, Derwent Reservoir, Hetton Lyons CP, Hurworth Burn Reservoir, Longnewton Reservoir, North Tees Marshes, Seaton Pond and Sadberge.

All records of this race fall between 22 September and 29 April, with the majority being between November and February. The earliest sighting, of two at Hetton Lyons on 22 September 1996, constitutes the earliest autumn date for any White-fronted Goose in the county. The largest groups noted include up to five at Seaburn and Boldon Flats over 24th to 29 January 1984, and up to five at Pitfield Farm, Sadberge, between 8 February and mid-March 1997.

This subspecies appears to be less prone to arrivals triggered by hard weather conditions and sightings presumably relate to wandering individuals or family parties. Extended stays appear to occur more regularly than with the nominate form, and there have been at least seven instances of birds staying at one site for longer than a month. This is in contrast to nominate race birds which usually pay only brief visits.

Distribution & movements

The White-fronted Goose is the most widespread and numerous goose species in the Western Palaearctic. There are five subspecies, which describe an almost circumpolar distribution. Two of these races have been recorded in the British Isles, and both occur in small numbers in County Durham. Nominate birds *A. a. albifrons* breed across northern Scandinavian and northern Russia, wintering largely in northern and eastern Europe, with the population estimated to be around 600,000 individuals, a ten-fold increase from the 1960s (Rose & Scott 1997).

The Greenland race, *A. a. flavirostris* breeds in western Greenland and winters in Ireland and the Western Isles of Scotland; largely on Islay and the Inner Hebrides. The world population of this subspecies is less than 27,000 individuals, having declined by 25% since the late 1980s (Fox *et al.* 2006), with the vast majority wintering in Ireland and western Scotland.

The general decline in numbers reaching Britain is thought to be due to birds remaining in wintering sites further east ('short-stopping'), especially in the Netherlands, possibly due to improved feeding opportunities, reduced hunting pressure, and milder winters in recent years (Owen *et al.* 1986). The species is still prone to occasional winter influxes from the Continent in response to adverse weather, but such arrivals seldom reach north east England. Distinct arrivals were noted in: January 1972; December 1987/January 1988; February 1993; December 1995/January 1996; January 2006; and November/December 2011, but such appearances may become less frequent.

Greylag Goose
Anser anser

A common feral resident and a winter and passage visitor in considerable numbers.

Historical review

The oldest references to the species in the north east comes from alleged Capercaillie *Tetrao urogallus* bones, collected in 1861 from Heathery Burn Cave in Weardale and dating from the early Bronze Age, dated at 1,000 years BC, which were held in the British Museum and later re-identified as being those of Greylag (Harrison 1980). Greylag bones have also been excavated from Roman sites along Hadrian's Wall (Miles 1992). As the Romans famously used domesticated geese for many purposes, it is conceivable that these may have been the remains of birds that were not of a wild origin (Kerr 2001). In terms of other historical records, there is only a single known reference to 'wild geese' in the Cellarer's rolls for the Durham Monastery and this is presumed to refer to Greylags, but it cannot be proven. It comes from the Cellarers' roll of about 1375, *In vij aucis indomitis, ij s. ix d.* ("*where seven were bought for two shillings and nine pence*"), and nine "*aucis fens*" were purchased in 1416. Geese were clearly a species of high cultural profile in the region during this period since Gosforth, in modern Newcastle, just to the north of Durham, derives its name from the Old English "*gos, ford*" meaning 'goose ford' (Yalden & Albarella 2009). Although there is still a wild breeding population in north western Scotland and the Hebrides, the species became extinct as a breeding bird in England around the middle of the 19th century. Native Greylags were almost certainly present in what was Cleveland until the medieval period, but it is not known when the local extinction occurred.

The Greylag Goose has long been a scarce winter visitor to County Durham, though numbers have increased dramatically in the last 40 years mainly due to the rapid expansion of the feral population in the UK. Temperley (1951) described the species as a "*very rare winter visitor*". He commented that indeterminate grey geese were fairly frequently noted in autumn and winter as visitors to the county, often in large skeins, but that as the birds could only be observed at a distance, it was not possible to be certain of their identification. Moreover, the specimens shot by wildfowlers were usually of little use unless they had been preserved. Furthermore, Temperley noted that over time the relative status of the different species may have been undergoing change, particularly in the late Victorian period. He said that the Greylag had always been a rare winter visitor to Durham, quoting Hutchinson (1840) as saying, "*the visits of the Grey Lag-goose to this County are by no means regular and it must be considered a rare bird*". Hutchinson (1840) stated that it was sometimes found around the fens at Mordon, Mainsforth and Bradbury Carrs. Tristram (1905), who was based at Greatham, thought that it was present in small numbers on an annual basis at Teesmouth: "*it occurs in the marshes near the Tees mouth in winter, but in very small numbers. The scarcest of all our familiar wild geese though......, seldom a season passes without a specimen at least being brought to me*". Nelson (1907) summarised its late 19th century status as being a regular winter visitor, although always heavily outnumbered by Pink-footed Geese *Anser brachyrhynchus*.

Since the early 20th century, Temperley noted that the visits of grey geese of all species had become "*fewer and briefer*", and observed that Greylags had become "*proportionally scarcer*". The only recent definite record at the time was of a flock of grey geese seen by W.K. Richmond and M. G. Robinson at Teesmouth on 27 December 1938, which contained at least three Greylags (Almond *et al.* 1939). Temperley concluded that as the species was now wintering in north Northumberland in greater numbers, visits to Durham might become more frequent.

The first modern introductions of the species into County Durham occurred in the 1930s and were based on the Wynyard Estate. During the 1960s, the Wildfowlers Association organised a progressive re-establishment of this species in England, with birds from Scotland (Owen & Salmon 1988). The species was still rather scarce in Durham during the early 1960s. For example, there were only two reports in 1964, these being of single birds at Cowpen Marsh in May and Hurworth Burn Reservoir in October, though there must have been records of passage skeins in late autumn. In 1963/1964, around 40 Greylag goslings were released in Wynyard Park. Birds then bred in that area over the next two decades, with up to six pairs breeding in Wynyard, and occasional pairs setting up territory at Crookfoot Reservoir (Joynt *et al.* 2008).

During 1967, there were no reports of breeding by feral birds outside Wynyard and indeed the Greylag was still relatively poorly reported as a passage migrant, most references being in single figures. In 1969, more free-flying birds were apparently released in Wynyard Park, with reports following at Hartlepool and Teesmouth perhaps being the precursor to the colonisation of Durham by feral birds. By 1970, records were increasing around Teesmouth, which probably referred largely to feral birds from Wynyard. Flocks of 'wild' birds however, included seven at Derwent Reservoir on 31 October.

By the middle of the 1970s, although there is almost no documentation of breeding there, it was clearly occurring on the Wynyard estate, as the flock had increased to around 100. In neighbouring Northumberland, breeding by feral birds was first documented at Sweethope Lough in 1974 (Kerr 2001). The Wynyard flock appears to have expanded rapidly and exported birds, which established another flock in the Crookfoot and Hurworth Burn Reservoir area towards the end of the 1970s. Breeding was first confirmed at Witton-le-Wear in 1975, at Crookfoot in 1978 and around this time, smaller groups of feral birds started to become established at Derwent Reservoir in the north west of the county, with other birds being noted at WWT Washington (Bell 2008).

Recent breeding status

In modern times, this feral breeding species is generally associated with lakes in large parks, gravel pits and reservoirs, all of these habitats being man-made environments where suitable short grass for grazing abuts a body of water. Favoured breeding sites usually have islands or are in areas where nesting can take place away from human disturbance. At WWT Washington, for example, nests were often hidden under dense bramble *Rubus fruticosa* (Westerberg & Bowey 2000). The availability of this habitat locally restricts the species' distribution in the county. As a breeding bird, it is scattered around the lowland wetland sites of the county, with a considerable population located around the reservoirs in some of the uplands, particularly around Selset Reservoir in the south west and Derwent Reservoir in the north west. Numbers of birds at both of these locations began to grow in the early part of the 1980s, after the initial establishment in the late 1970s.

Although there were large-scale increases in the population during the 1970s and 1980s, breeding was restricted to a relatively small number of pairs at a number of widespread sites. The rapid growth of the county's breeding populations through the 1990s and the early 2000s is not easily reconciled with the knowledge that many of these early breeding attempts were unsuccessful and that production of young was low for a long period of time. The obvious inference is that birds were either being released in the county or that there was recruitment from other parts of the country, possibly from across the Pennines, from the well-established population in the Lake District.

Nationally, the species underwent some expansion during the 1970s and 1980s (Marchant *et al.* 1990), and this process has been very much evident locally since 1990. Despite the invocation of control measures in some areas, the population is still growing and if this continues at the present rate the bird could soon be considered to be a pest (Bell 2008). The species' breeding range has increased considerably, and birds can now be found breeding in many new areas, including Gateshead and Rainton Meadows, where they were not present a decade and a half ago.

During the early 1990s, the four main local strongholds for the species, together with a few isolated breeding pairs away from these areas, gave an estimated county population of around 20 pairs (Westerberg & Bowey 2000). However, by the late 2000s, the breeding population had expanded to possibly 200-300 pairs, at least 80% of which were located in upland areas. Examples of populations noted in the second half of this decade, determined by counting family parties before the young were mobile, included 20 pairs at Balderhead Reservoir, 10 pairs at Edmundbyers, 10 pairs at Smiddyshaw Reservoir and 15 pairs at Waskerley Reservoir, while the Hamsterley Forest population consisted of perhaps 20-40 pairs, despite the lack of any large body of water. As the species can nest well away from water, the true number of breeding birds is difficult to assess. However, early summer increases to 600-800 birds at both Derwent and Selset Reservoirs, including many family parties, indicates the true size of our moorland and reservoir edge breeding population.

In the late 2000s, all the lowland populations (away from the North Tees Marshes) were no more than five pairs at any one locality. Favoured areas include Bishop Middleham, Brasside Ponds, Low Barns, Seaton Pond, Shibdon Pond and WWT Washington, plus several sites in the mid-Wear Valley. In the North Tees area, the expansion was described as an *"explosion"* in *The Breeding Birds of Cleveland* (Joynt *et al.* 2008) and by 2006, there were probably around 30-35 pairs breeding in this area. All local breeding birds are considered to originate from introduced stock.

Recent non-breeding status

The highest counts of Greylag have traditionally occurred during the autumn and winter months, though no doubt the vast majority of these birds consist of feral stock from elsewhere in the UK. Truly wild birds of genuine Icelandic origin continue to occur in Durham on a regular basis during the winter months, as has been proven by the recent sightings of birds bearing neck-collars fitted in Iceland (Bell 2008), but the differentiation between wild birds and those of feral stock is now almost impossible. Through the 1970s and early 1980s, when the county population of feral birds was not so high, flocks of wild birds tended to start arriving in, or passing through, the county from late September. The majority of sightings fell between September and December, with a noticeable peak in November. Many records of Greylag referred to birds heading in a westerly direction, perhaps across to the Solway Firth, and in most years, birds were also noted heading south along the coast. Coastal flocks noted in this period included 100 heading north west over Marsden on 2 November 1984, 98 south over Peterlee on 21 November 1988 and 168 south at Dawdon Blast Beach in November 1995.

During the 1990s and in particular through the 2000s, the pattern of occurrence changed dramatically. The appearance of continental immigrants is now nowhere near as obvious and coastal passage is on a very much reduced scale. The winter peak in numbers has been replaced by a late summer and autumn peak, as local breeding birds flock together in the uplands, then move to lowland wetlands. There is evidence of a south easterly shift in non-breeding populations, with birds leaving upland reservoirs, assembling at central Durham sites such as Castle Lake through August and early September, and then moving down to the North Tees Marshes later in the autumn.

The population of feral birds in County Durham has grown tremendously since the early 1980s; a pattern illustrated by the peak counts on the North Tees Marshes and at Derwent Reservoir. Between 1980 and 1993, there were no gatherings greater than 21 birds on the Tees Marshes, whereas the next six years produced a maximum of 184. An explosion in numbers occurred in 1999 when there was a new all-time high of 705 birds in October, and this was followed by further increases to 1,021 in September 2007 and 1,273 in September 2009. At Derwent Reservoir, up to 1985, the main presence of the Greylag Goose was as a wintering bird with a peak of 60 in the winter of 1984/1985. However, just four years later the peak count reached 517 birds, and in December 1991, a new county record of 844 birds was noted. The yearly peak varied between 300 and 800 birds through the remainder of the 1990s and 2000s, and the locality is now a well-established post-breeding gathering point.

The Wynyard area featured strongly through the 1970s and 1980s and formed an initial foothold for feral birds in the south east of the county. No doubt there is much interchange between this site, the Crookfoot and Hurworth Burn Reservoir area, and the North Tees Marshes. A dip in the Wynyard population occurred between 1989 and 1993 due to the temporary draining of Wynyard Lake and counts in several years did not exceed 41 birds, but subsequent years saw the population rise once more.

The five-yearly peaks in four of the main areas are shown below:

	Crookfoot Res./ Hurworth Burn Res.	Derwent Res.	North Tees	Wynyard
1980-84	400	46	11	290
1985-89	530	537	9	310
1990-94	410	844	146	530
1995-99	750	526	705	420
2000-04	700	800	730	631
2005-09	480	713	1,273	460

Other sites which have become increasingly prominent during autumn and winter include Bishop Middleham (900 in August 2008), Rainton Meadows (600 in October 2009), Low Barns to McNeil Bottoms (331 in March 2000), WWT Washington and Barmston Pond (170 in September 2001) and Boldon Flats (170 in September 2007). Less reliable localities which have occasionally attracted flocks of up to 420 birds include Brancepeth Beck, Brasside Ponds, Cobey Carr, Hetton Lyons, Houghton Gate, Lamesley Pastures, Longnewton Reservoir, Raby Castle, Seaton Pond and Thorpe Farm.

With so many Greylag Geese now commuting between various wetlands around the north east of England, it is inevitable that large skeins are occasionally witnessed moving overhead. An observer at any inland site may see large flocks from time to time, particularly between December and February, but also in late summer and early spring as birds head from or to breeding grounds, perhaps in Iceland but more likely on the Durham or Northumberland uplands.

Distribution & movements

The Greylag Goose has the widest distribution of any of the European goose species, and is found throughout the Old World, eastwards across Asia to China. Many different populations occur throughout Europe and it is highly migratory. A number of distinct populations of the nominate subspecies are recognised (Scott & Rose 1996). Birds from the Icelandic breeding population winter exclusively in Great Britain and Ireland (Hagemeijer & Blair 1997, Madsen *et al.* 1999), with most wintering in Scotland. Greylag Geese south of the line between the Isle of Man and Teesmouth are sedentary re-established birds, from geese translocated from the Hebrides since the 1930s, but there is also a large re-established flock in south west Scotland. In addition to feral birds, there are small populations of native breeding Greylag Geese in the Western Isles, and western and northern Scotland.

Over the last fifteen years or so a number of birds bearing engraved neck-rings have been noted in Durham. The first such occurrence was of a bird on the North Tees Marshes from December 1995 to February 1996. The same bird was present again from October 1996 to March 1997 and from October 1997 to March 1998. It had been ringed in a Swedish nest in June 1990, and was seen in Sweden in 1991, 1992 and 1993. It also made visits to Caithness in June 1994, appeared in North Yorkshire in November 1994 and August 1995, was in Orkney in March 1996 and dropped in at Loch Leven in July and August 1997 (Blick 2009). Over the period from December 2000 to March 2001, up to seven neck-collared birds were in the Crookfoot Reservoir area. All of these had been ringed at Loch Eye, Ross-shire, Scotland in October or November 2000, and they were considered to have originated in Iceland. Two of these birds were seen at Crookfoot on 16 March 2001 and were reported in Iceland on 29 April of the same year. Another collared bird, ringed in Iceland in July 1997, was seen in Scotland each year between 1998 and 2001, and was at Crookfoot Reservoir in January 2003. At least six other birds from this population have been recorded at Crookfoot since 2001, and three of the 14 have been seen in Wynyard (Blick 2009).

Of more local origin, a bird with an orange neck ring, 'BVJ', originally fitted at East Chevington, Northumberland, in May 2006, was at Saltwell Park intermittently from summer 2006 through to 2010, having also made excursions through the Scottish borders and elsewhere in Northumberland. Another bird ringed at Nosterfield, North Yorkshire, in February 2004 was in the Saltholme Pools area every winter from October 2006 to at least December 2009. It had also been seen at Ashington, Northumberland, in June and November 2008 (Blick 2009).

Snow Goose
Anser caerulescens

A rare vagrant from North America or from naturalised populations in Scotland and continental Europe, though most are probably escapes from captivity. There are 22 records, involving 65 individuals recorded in all months.

The first record for County Durham relates to a single white phase adult seen by Pete Reid flying south offshore at Hartlepool Headland on 11 September 1961. What was presumed to be the same bird was also seen at North Gare, Teesmouth on 27 December 1961 by Peter Harland and Robert Lightfoot, again heading south, though this time accompanied by two Canada Geese *Branta canadensis,* a rare bird in the county at the time (Stead 1964). Both of these records are thought to involve a Lesser Snow Goose *Anser c. caerulescens* which commuted between Lockwood Beck and Scaling Dam Reservoirs in North Yorkshire from 12 June 1961 until 9 January 1962. The occurrence date of this bird does not inspire confidence in its status as a wild vagrant, and it seems most likely to have been an escape from captivity. Nonetheless, the record was accepted by BBRC, with some caveats (*British Birds* 55: 570).

All records:
1961   Hartlepool Headland, white phase, flew south, 11 September, same North Gare, flew south, 27 December.

1975   South Shields, 30 April.

1971   Whitburn, eight flew north, 24th June

1977   Hurworth Burn Reservoir, Crookfoot Reservoir, & Wynyard, blue morph, 14 May until at least 18 March 1984.

1978   Hury Reservoir, blue phase, October to December.

1982   Derwent Valley, white phase, flew south west, 20 June.

1982   Bishopton, three immatures, 29th to 30 September, same Sunderland, flew north, 4 October.

1984   Hargreaves Quarry, Teesmouth, blue morph, 24th to 30 May.

1984   Hartlepool, ten, white phase, 21st to 22 September, same Dorman's Pool, Teesmouth 21 September.

1985   Shibdon Pond, Blaydon, two, white morph and blue morph, 18 May.

1986   Shibdon Pond, Blaydon, white morph, 11 June.

1988   Derwent Walk Country Park, Ryton Willows and Shibdon Pond, eleven, white morph 29th to 30 April. Presumed two of same Ryton Willows, 2 May.

1989   Derwent Reservoir, 16 April.

1991   Saltholme Pools, Teesmouth, four, one white morph and three blue, 22 June.

1996   Longnewton Reservoir, two, white morph, 16 October, same North Tees Marshes, 26 October, Hurworth Burn Reservoir, 29 January 1997, Pitfield Farm, Little Stainton, 9th to 15 February 1997, and Crookfoot Reservoir, 1 March 1998.

1996   Sedgeletch, Houghton-le-Spring, two, white morph, 22 December (possibly the same as above).

2001   Barmston Pond, Washington, white morph, 19 February.

2001   Houghton Gate, 22 October.

2003   Saltholme Pools and Cowpen Marsh, Teesmouth, white morph, 17th to 22 June.

2004   Whitburn, eight, four white morph and four blue morph, flew north, 16 May.

2009   Seaton Carew, North Gare and Seal Sands, white morph, flew south, 17 October.

2011   Bowesfield Marsh and Saltholme Pools, four white morph, 15th to 17 October.

The majority of Snow Geese recorded in County Durham are presumably either of feral origin or escapes from captivity, but several records are worthy of comment. It is interesting to note that most of the records of large flocks in the county, in particular ten in September 1984, eleven in April 1988, and eight in May 2004, coincide with the main migration period in North America. Snow Geese finish breeding in late August and reach their wintering grounds on the American prairies in mid- to late September, returning to the Arctic to breed from late May, involving a migration of over 3,000 miles. It is possible that these records could represent genuine vagrants, as vagrancy has been proved from ringing data. The record most likely to be a genuine vagrant in Durham is the adult that flew south past Seaton Carew on 17 October 2009. This bird was accompanying a migrating flock of Pink-footed Geese *Anser brachyrhynchus,* and was later relocated in Norfolk where it spent the winter.

Distribution & movements

The Snow Goose breeds in the Arctic, in Alaska, Canada, and north western Greenland, as well as on Wrangel Island in the far north east of Siberia. A long distance migrant, the entire population winters in the southern United States. The total population is estimated to be in excess of five million birds, having shown a 300% increase since the mid-1970s, thought to be linked to a shift in winter feeding to surplus grain on the agricultural lands of the American prairies. It is an annual vagrant to the British Isles in small numbers, most frequently amongst flocks of wintering Pink-footed Geese, Greenland White-fronted Geese *Anser albifrons flavirostris,* and to a lesser extent Icelandic Greylag Geese *Anser anser.* Most of these birds are believed to be true vagrants, though the situation is complicated by the presence of a large number of feral birds in Europe, many of which breed sporadically. The species was given dual categorisation by the BOU in 2005, when it was added to Category C of the British List on the basis of a self-sustaining population which has existed on the islands of Coll and Mull, in the Inner Hebrides, since the mid-1950s. Evidence of proven vagrancy to Europe comes from a ringed bird amongst a flock of 18 in

The Netherlands in April 1980. It had been ringed three years earlier near Churchill, Manitoba, Canada, suggesting the whole flock was of Nearctic origin (Brown & Grice 2005).

Greater Canada Goose
Branta canadensis

A common resident, as well as a small-scale passage migrant from population centres located elsewhere in Britain.

Historical review

This North American goose was initially introduced to King Charles I's waterfowl collection at St. James Park in London in 1665 (Kirby 1999), and was subsequently added as an ornamental bird to the parks and estates of many wealthy landowners through the 17th century. Since then, there have been numerous other introductions across the British Isles, and from these it has spread to areas throughout England, and, to a lesser extent, Scotland, Wales and Ireland. It was well-established as a breeding bird in Yorkshire before 1938, when it was accepted as a British species and included in *The Handbook of British Birds* (Mather 1986). Temperley (1951) made reference to birds living in a feral state in County Durham, though he also stated that most observations were of birds that were either hand-reared or direct escapes from captivity. The species has undergone a considerable change in status since the 1950s, when it was considered only a rare visitor to the area. By the late 1950s, birds were nesting on upland moors in Nidderdale, North Yorkshire, to the south of Durham, a habit which was later to be recorded in the south west part of Durham by the early 1970s (Mather 1986).

Two birds that were noted at the Reclamation Pond on 17 May 1957 appeared to be the first 'wild' birds at Teesmouth in the 20th century (Joynt *et al.* 2008). A mass release of birds for hunting purposes took place by The Wildfowlers Association of Great Britain and Ireland in the late 1950s and this undoubtedly contributed to the spread of the species in the north east. In this respect, 50 birds were known to have been released at Sweethope Lough, Northumberland around this time (Kerr 2001). Even during the 1960s, the bird was still relatively scarce. One of the county's first breeding records related to an introduced pair that were on Rossmere Park Lake, Hartlepool, in the late 1960s, but these birds were thought to have never reared any young (Blick 2009). By 1967, breeding was certainly occurring on the nearby Northumbrian loughs, for example Sweethope, and small numbers of birds were recorded on an annual basis moving through the county in spring and late summer. The well-documented Yorkshire to Inverness-shire moult migration flight was established by the early 1960s (Mather 1986), so birds were presumably over-flying Durham on an annual basis, in late spring and early summer, in this period. The cluster of records at the time tends to confirm such movements.

By 1964, the species was described as an *"irregular summer and winter visitor to the estuary and marshes of the Tees, at least in small numbers"* (Stead 1964). Between 1965 and 1973, however, the bird was still relatively scarce at Teesmouth, though the flock on the Wynyard Estate numbered 23 individuals after the 1975 breeding season. The first mention of a resident feral flock in Durham seems to be from Wynyard in 1973, when 28 were present and breeding was confirmed at that site in June. It is interesting to note that this flock included a bird that had been ringed at Malton in North Yorkshire, though it is unknown if the rest of the flock had been directly released in that area. Clearly the species was breeding on the Wynyard Estate around this time, perhaps as early as 1967, when 14 were present in April of that year (Blick 2009). It was initially slow to spread as a breeding species although nesting was attempted at Brasside Ponds in April 1974 and subsequently, with varying success. The birds were unsuccessful here until 1978, when young fledged at the fifth attempt.

By the early to mid-1970s, the species was already known to be breeding in the south west of the county, at Balderhead Reservoir in Baldersdale (Bradshaw 1976). It is possible that birds in this part of Durham may have come north along the Pennines from similar breeding locations in Yorkshire. There were increasing sightings from elsewhere in the south west of the county through the late 1970s, in particular in Baldersdale and Lunedale. Birds were at Blackton Reservoir from 12 January to April 1975, but there was, apparently, no evidence of breeding. Birds nested again at Balderhead Reservoir in 1975, but were disturbed by water skiers. However, they did fledge young here in 1979. Away from potential breeding sites, most records in the county through the 1970s occurred

111

between April and June, and often involved sizeable flocks, such as the 30 birds noted at Marsden on 20 June 1970. The volume of passage built up through the decade, with 63 flying north off Sunderland on 11 June 1977 being notable. Return passage was particularly obvious in late August 1979, with flocks of up to 50 birds seen moving over the Washington/Felling area on 22nd, Teesmouth and Washington on 23rd and over West Boldon and Washington on 24th.

Recent breeding status

The species has undergone a considerable change in breeding status in the county over the latter part of the 20th century. The origins of the feral population in County Durham would appear to stem from the large feral population in West Yorkshire, where the species was established from the 1940s and possibly to a lesser extent, from a similar population in the West Midlands (Bell 2008). In evidence, of 22 birds on the North Tees Marshes in August 1979, 17 had been colour-ringed at Gouthwaite Reservoir, North Yorkshire, and 2 at Cannock, Staffordshire.

The species became more widespread during the 1980s, with consolidation of the group of breeding birds established in the south west of the county. It seems likely however, that the population might have been supplemented by immigrants from elsewhere as, although there were four counts of birds that totalled over 100 birds in 1989, there were still, at that time, just ten known breeding pairs in the whole of the county (Bell 2008). The species was entering an expansionist phase locally through the 1980s and early 1990s, and thirteen pairs attempted to breed at Baldersdale in 1990, with 17 pairs there in 1994, of which nine nests of eggs were destroyed.

The *Atlas* (Westerberg & Bowey 2000) showed that the Canada Goose had a rather localised breeding distribution in the county in the period 1988-1991, being predominantly a lowland species, though there has been a considerable expansion in breeding range since then. Of the then 18 recorded breeding sites, ten were located at below 120m above sea level and 15 were in the east of the county. The three in the west referred to the well-established breeding birds on the Teesdale reservoirs, particularly in Lunedale and Baldersdale. All Durham records noted in the *Atlas* were from still water sites. Large areas of the county, some with suitable habitat, had no geese breeding and the distribution pattern in Durham followed that displayed in other areas of low Canada Goose population (Gibbons *et al.* 1993). This comprised isolated pockets of breeding concentrations, which formed a basis for expansion outwards as the population increases, as it did through the late 1990s and early 2000s.

Up to the early 1990s, the population in the UK had been increasing by up to 8% per annum (Owen *et al.* 1986) and stood at 61,000 in 1991 (Delany 1993). A rapid increase took place in the county during the 1990s, as reflected in the BTO's Breeding Bird Survey which indicated an increase of 161% in the UK population over the period 1994-2006 (Raven *et al.* 2007). In Durham, the increase in numbers appears to have been accompanied by an expansion in the range of sites occupied. Over this period, there was a considerable increase in the population based around the North Tees Marshes. In this area, pairs have attempted to breed since about 2001 (Blick 2009) and by around 2006 many still waters around the estuary held breeding birds, giving a population of between 25 and 28 breeding pairs (Joynt *et al.* 2008).

From the mid-1990s, additional localities were colonised, including the mid-Wear Valley wetlands, such as Beechburn, Cobey Carr, Low Barns NR, McNeil Bottoms and Page Bank, and also lowland sites such as Joe's Pond and Shibdon Pond. The pattern of spread and consolidation seems set to continue for some time and may occur in two phases, with birds coming into the county from outside the area, and through 'internal production', from the main breeding concentrations within the county itself. The county population is now estimated to be perhaps 60 to 80 pairs, although the exact number is difficult to establish. Countering this upward trend is the fact that Canada Geese have been labelled 'pests' due to their aggressive behaviour towards other wildlife and the damage they can cause to crops whilst grazing. In some areas, organised culls and egg pricking have taken place. Natural failures also occur frequently, such as at Shibdon Pond in 1996 when the resident male Mute Swan *Cygnus olor* killed all five goslings, whereupon the adult Canada Geese left the site. Destruction of young goslings by Mute Swans has since been noted at other sites.

As with many waterfowl, cross breeding with other species is frequently noted. At Longnewton Reservoir in 2008, a total of 13 hybrid Canada x Greylag Geese *Anser anser* were noted, along with a mixed pairing between a Canada Goose and a feral Barnacle Goose *Branta leucopsis*. Hybrid Canada x Greylags have also been recorded within goose flocks at Bishop Middleham and on the North Tees Marshes with increasing regularity. A female Canada Goose was paired with a domestic 'farmyard' Greylag at Shotton Pond in 1999. Such hybridisation is likely

112

to contribute to plumage variations such as leucism and four such birds were present at WWT Washington in August 1999.

Recent non-breeding status

Many non-breeding birds are noted at the main wetland localities throughout the year, but in some instances, post-breeding groups of birds join up to form larger flocks and move around the county. The North Tees Marshes is the main non-breeding stronghold for this species, with recent peaks of 526 birds present in February 2008, 684 in December 2009 and 716 in November 2010. It is clear from the yearly peaks that this species is continuing to increase in County Durham. Other large single site counts in recent years have included: 508 on Bowesfield Marsh on 14 September 2006; 500 at Selset Reservoir on 1 July 2005; 435 at Longnewton Reservoir in September 1999; 373 at Rainton Meadows in September 2010; 300 at Hury Reservoir in March 2010; 298 at Thorpe Farm on 22 February 2009; and, 253 at Grassholme Reservoir in December 2003. Other wetlands which have held between 150 and 250 birds include: Balderhead Reservoir, Bishop Middleham, Blackton Reservoir, Brasside Ponds, Derwent Reservoir, Herrington CP, Hurworth Burn Reservoir, Pitfield Farm, Raby Castle, Saltwell Park and WWT Washington, while numerous other sites have occasionally hosted flocks of 100-150.

Through the 1970s and 1980s, many records involved flocks moving through the county in May and June, and again in September and October. It seemed highly likely that this pattern of records was related to the well-documented moult migration of failed and non-breeding birds from the Yorkshire population (Walker 1970). These were known to migrate to the Beauly Firth in eastern Scotland, with many birds passing over County Durham in early June, but then returning in a more haphazard fashion throughout August, September and October. A study of the pattern of occurrences in Gateshead during the late 1980s and early 1990s supported this theory and showed that the species was, at that time, a passage migrant, with a tendency to appear in a seven-week period between mid-April and the first week of June, with a further peak of occurrences in early autumn, during August (Bowey *et al.* 1993). The pattern became more pronounced through the late 1990s and into the 21st century. As well as flocks moving north through locations in central Durham, coastal passage became a prominent feature. The well-watched Whitburn area, and to a lesser extent Hartlepool, produced regular sightings of flocks of 15-55 birds in late spring, while more recent examples of north-bound passage in the Whitburn area include 284 logged in June 1999, a total of 350 between 3rd and 6 June 2006 (including 184 on 4th and 138 on 6th), and 156 noted on eight dates between 28 May and 20 June 2008. Returning autumn flocks are more difficult to detect but 40 past Whitburn on 27 August 2008 and 150 over Old Quarrington on 28 September 2008 were likely to have involved such migrants.

Distribution & movements

The breeding range of the Canada Goose in North America is extensive, covering the whole of the northern part of the Continent. In Europe, the introduced population is restricted to Britain, the Netherlands, Belgium, Germany and Scandinavia. The British population is demonstrating a steady rate of growth of about 8% per year, with the population estimated to be around 82,000 birds in 1999 (Baker *et al.* 2006).

The American Ornithologists Union has recognised 12 different forms ranging from Canada Goose *Branta canadensis* to Cackling Goose *Branta hutchinsii* since 2004. The British Ornithologists' Union followed suit, but adopted the names Greater Canada Goose and Lesser Canada Goose (BOURC 2005). The species taxonomic situation is complex and understanding is developing. Several of the smaller forms reach the Britain isles infrequently as vagrants, particularly to Ireland and western Scotland. There are no acceptable records of any of these in Durham. A bird, apparently of the race *parvipes,* which breeds in Alaska and winters in Washington and Oregon was at Boldon Flats in March 2005 (Newsome 2006); though its companions were local Canada Geese and its racial identify not clear cut. There have been several other small race birds or hybrids seen at various wetlands around the county, but none have been in situations indicating a wild origin. The interbreeding of feral Canada Geese frequently results in birds showing features of other races, such as chin-straps and darker chests.

Two birds with engraved rings were at WWT Washington on 10 December 1976. They had been ringed on the Beauly Firth, Scotland, in July 1975 and remained at Washington to the end of the year. Another interesting series of recoveries involves four birds ringed on the Beauly Firth on 1 July 1995. Single birds were found dead at South Park, Darlington, on 21 September 1996 and on 24 November 2009, at Bishopton on 31 October 1996, and the final bird was also found dead near Darlington on 28 January 2000. Other ringing recoveries in County Durham have involved birds ringed in Derbyshire, Dumfries and Galloway, North Yorkshire and West Yorkshire.

113

Barnacle Goose
Branta leucopsis

An uncommon passage and winter visitor.

Historical review

There are records of bones of Barnacle Goose being excavated from the western settlements along Hadrian's Wall (Miles 1992), so the species had certainly reached the North of England at that time, but whether this included County Durham, is unknown.

Hogg (1845) stated that it was scarce near Hartlepool in the early 19th century, yet Nelson (1907) wrote that he had been informed that *"fifty years ago, it was by no means uncommon"* around Teesmouth, with *"14 killed at one shot"*, presumably with a punt gun, around 1857. This may suggest the intermittent presence of large flocks that gave wildfowlers considerable sport. George Temperley (1951) called the bird *"a very rare visitor in autumn and winter"*, but its status has certainly changed since then. In Temperley's time, it was said to frequent the coast, rarely being seen inland and then really only at Teesmouth. Temperley quoted two inland records: two or three birds shot on Stanhope Common *"about 1882 "*, and one shot at Middlehope in October 1897. Records decreased as Teesmouth was 'reclaimed', the last documented occurrence being in 1883. During the early part of the 20th century, the Barnacle Goose was a decidedly rare bird in Durham, with no observations of birds on the ground or of flocks moving through. This situation continued until the early 1960s, when more frequent reports began to appear.

The highest passage counts in the 1960s came from the Teesside area, when 28 were over Hartlepool on 18 October 1961. The year 1964 was significant with records in October, when passage birds were documented as moving through, though a 'tame' bird on the cliff top grasslands at South Shields on 23 September was the only specific Durham record. This was the first such positive observation for the species in the 20th century. A couple of years later, in 1967 a huge flock of 2,000 was seen flying west on 10 January at Haltwhistle, Northumberland, through the Tyne Gap, indicating the scale of westerly movement that might sometimes occur along the Tyne. Whether these birds had accessed the Tyne valley at the river's estuary, thereby travelling Durham's northern border, or had cut across from some point on the south Northumberland coast is not known.

Recent non-breeding status

The Barnacle Goose is today a regular winter visitor to County Durham in varying numbers each year, some of which are no doubt feral individuals, although wild birds occur annually, supported by ringing recoveries. The majority of the autumn passage involves the Svalbard population which is destined for the Solway, where the wintering population increased to between 24,000 and 27,000 during the late 2000s.

From the 1970s to the mid-1990s, the pattern of occurrence was similar to the present time, but on a generally reduced level. Passage flocks were noted at the coast in late September and early October in several years, but rarely numbered more than 90 birds. More exceptional were movements of 340 past Hartlepool Headland on 1 October 1979, 110 on the Long Drag on 28 September 1986, 320 past Whitburn on 29 September 1987, 150 at Whitburn on 26 September 1990, 476 on the Long Drag on 1 October 1991, and eight skeins totalling at least 420 south west over Teesmouth on 3 October 1992. Most movements were in this anticipated period, but occasionally delayed parties were seen arriving later in the autumn. Examples included 150 flying west over Gateshead on 25 November 1985, 70 over WWT Washington on 25 October 1997 and 23 north west over Shibdon Pond on November 1997.

Through the late 1990s and 2000s, the first autumn arrivals were sometimes difficult to determine due to the presence of an increasing number of feral birds in northern England. However, those clearly arriving from the Continent and bound for the Solway often appeared *en masse* in the last two weeks of September and sightings continued through to mid-October. The earliest skeins included five at Hartlepool on 17 Sep 2007, 70 at Teesmouth on 20 September 1996, 84 at Hartlepool on 22 September 2002 and 450 at Whitburn on 22 September 2004, but the bulk of the passage now occurs in the last week of September and the first week of October.

Although autumn passage is now witnessed in most years, numbers vary greatly and the volume of passage is very much weather dependant. Without the influence of a strong northerly wind, many birds cut across north Northumberland towards the Solway, to the north of County Durham. However, when strong northerly winds coincide with the arrival of Barnacle Geese in the north east, the results can be impressive. The highest ever passage occurred on 4 October 1999, when almost 3,500 were logged at Whitburn alone and at least 5,700 passed through Teesmouth to the south. No doubt there will be a high degree of duplication between these two areas, but some flocks in the south of the county will have re-orientated through central Durham towards the Solway (Bell 2008, Unwin 2003). As confirmation of the destination of these migrants, 5,000 Barnacle Geese arrived at Caerlaverock during the same day. Other high day-counts in recent years have included 1,200 west over Teesmouth on 28 September 2003, 1,035 at Hartlepool on 7 October 2004, 818 past Whitburn on 25 September 2010 and 711 north past Whitburn on 23 September 2003. Skeins amounting to several hundred birds are now routinely expected during late September and early October, and may move to the north west anywhere in the county, but particularly following the main river valleys of the Tees, Wear and Tyne. During the main period of autumn passage large numbers sometimes pass south down the Northumberland coast and then turn westward to head along the Tyne valley, sometimes cutting over Newcastle itself, and heading ribbon-like along the River Tyne, mostly unseen by the populace below (Kerr 2001).

The eastern third of the county is responsible for most sightings of Barnacle Geese, however, this is perhaps more a reflection on the distribution of observers. Occasional skeins are noted further to the west, but such occurrences are unusual. Several parties of wild birds have been seen at Derwent Reservoir, and were also noted over Langley Moor on 31 October 2004, Tunstall Reservoir and Harperley in late September 2009 and on Selset Reservoir on 7 October 1990.

Since the late 1990s, there has been evidence of a northbound spring passage over Durham, perhaps of late departing Dutch birds. The first such flock was 14 heading west over Crook on 27 April 1997, followed by skeins noted between 28 April and 18 May in seven years during the 2000s. Most flocks have been seen on seawatches at Hartlepool, Seaham and Whitburn and have consisted of between two and 28 birds, but 65 flew over Birtley and 8 passed Whitburn on 5 May 2009 and sightings on 10 May 2010 included 88 at Seaham and 109 at Whitburn.

Since the early part of the 21st century, an unusual feature has been the appearance of wintering birds on the North Tees Marshes. Although small numbers occasionally visited the North Tees Marshes during the 1980s and 1990s, their stay had been generally short-lived. After four birds overwintered in 2002/2003, up to 21 were present each winter from 2006/2007 onwards. Whether this presence will increase and whether it involves genuinely wild birds remains to be seen. It is often difficult to differentiate between genuinely wild birds and those from feral stock, but three present with Pink-footed Geese Anser brachyrhynchus and a Bean Goose Anser fabalis on farmland near Seaton Pond on 12 November 2001 were likely to have been displaced Dutch birds.

Undoubtedly, feral birds have appeared at many other wetlands in recent times and are usually immediately suspect due to the company they keep and their tameness. Over the years, up to five have been seen at sites such as Bishop Middleham, Brasside Ponds, Hurworth Burn Reservoir, Low Barns NR, McNeil Bottoms, Rainton Meadows, Seaton Pond, South Marine Park Lake and Wynyard, but any locality holding a goose flock may be visited. The origin of these birds is uncertain; they could be recent escapes, but equally, they could have originated from one of the many established feral populations around the UK. On 6 August 1979, eight feral birds were introduced on to Shibdon Pond and these were responsible for numerous local reports of the species here and at other local sites for a number of years. By late 1982, the numbers of feral birds at Shibdon had dwindled to four, with three present in 1983, and the final two birds were last seen flying north off the pond in April 1984. One or two feral birds have been seen intermittently there since this period.

Recent breeding status

Feral birds have attempted to breed at Shibdon Pond on a number of occasions. In the early part of the 21st century a pair nested and laid eggs though young were never successfully reared. The first successful breeding occurred in 2005 when a pair raised a single chick. The Barnacle Goose may well become established as a breeding species in the county in the future, as populations elsewhere in Britain and Ireland are increasing. In 2008, a Canada Goose Branta canadensis paired with a Barnacle Goose at Longnewton Reservoir; however, the outcome of the breeding attempt is unknown.

Distribution & movements

The Barnacle Goose has three distinct populations, located in Svalbard, northern Russia (Novaya Zemlya) and Greenland (Parkin & Knox 2010). These remain distinct both as wintering and breeding groups. The Svalbard population winters largely on the Solway Firth, and as many Durham sightings are of birds crossing the country during autumn, this is probably the source of the majority of the county's records. The population from northern Russia winters primarily in Holland and may be the origin of mid-winter birds seen in the county following hard weather on the Continent. The Greenland population winters exclusively in Ireland and western Scotland, and is unlikely to be involved in County Durham records.

An insight into the migratory speed of this species came in 1994, when a flock of nine Barnacle Geese was tracked along the Durham coast during the 1994 Coordinated Migrant Count on 2 October. At just before 08.20hrs they were seen flying south at Dawdon and were tracked along the Durham coast before heading west and inland at Hart Warren, at around 08.30hrs. They covered a distance of around 12km in some 12 to 15 minutes, against fresh south-south westerly winds, at a ground speed of over 48km per hour (Bowey 1999c).

Colour-ringed birds have been noted on a number of occasions in Durham. In January 1983, two birds from a flock of 13 on Seaton Common were ringed. One was ringed as a juvenile at Caerlaverock in October 1976, being seen there for most of the following winter and in Norway in May 1978, 1980 and 1981. This bird was also trapped on Spitzbergen in July 1977. The other individual had been ringed as an adult, also at Caerlaverock, in January 1980 and was observed there during the following winter and in Norway in May 1980, 1981 and 1982. Of a flock of 110 that were present by the Long Drag on 28 September 1986, at least 15 had colour rings that had been placed on the birds in Spitsbergen. Four of these birds were then reported at Caerlaverock within a week of being seen at Teesmouth. One ringed as an adult on Spitzbergen in August 1995 was at Caerlaverock in the winters of 1995/1996 and 1996/1997, before being seen on Seaton Common on 16 October 1997. Another that had been ringed as a pullus on Spitzbergen in the summer of 1997 stayed in the Teesmouth area from January to May 1998, becoming tame enough to come for bread when it accompanied Canada Geese in Stewart and Albert Parks, Middlesbrough. If this bird had not sported a Spitzbergen ring, it would have almost certainly been considered an escape from captivity (Blick 2009).

Brent Goose (Brant Goose)
Branta bernicla

An uncommon passage and winter visitor; two subspecies occur.

Historical review
Through the recorded period, most observations of this species have come from coastal areas of the county, mainly from the Teesside area, and more recently on passage, from Whitburn in the north of the county. The earliest reference to it in Durham comes from the birds listed in the Bursar's records of the Monastery of Durham, where Brent Goose was listed as *rutgoys* (Gardner-Medwin 1985). Nelson (1907) cited *Willughby's Ornithology* (1678), which recorded the occurrence of this species under the name of *'Rat or Rhode Goose'*. It documented the earliest known reference to the species in Yorkshire. However, the specimen referred to was taken from *"Mr Johnson, who showed us this bird at Brignall"* (Mather 1986). This related to a stuffed specimen of a Brent Goose shown to Ray and Willughby, by Ralph Johnson (of Brignall, near Greta Bridge, in the southern part of what is now County Durham) who regularly shot over the Tees Marshes. *Willughby's Ornithology* further recorded that it was a *"very heedless fowl"*, i.e. tame; *"if a pack of them come into the Tees, it is seldom one escapes away"*, as they *"suffer the Gunner to come openly upon them"*. It was clear that the south east coast of Durham was important for the Brent Goose at this time (Gardner-Medwin 1985).

The Brent Goose was more abundant in the region during the 19th century, the species being described as *"very common at Teesmouth"* in the early 1800s. Hogg (1827) said *"flocks of this species were very numerous near the Tees in the months of January and February, 1823"*, also saying that it was *"plentiful"* at Teesmouth in severe weather. In 1869, a wildfowler was reported to have shot 65 in the season from an immense flock that was then frequenting Teesmouth (Mather 1986). At this time, to the north of County Durham, truly vast numbers were occasionally noted around the traditional wintering area of Lindisfarne in Northumberland, such as the 20,000 that were estimated to be there in March 1886, after a heavy snowstorm (Kerr 2001).

Abel Chapman, referring to the occurrence of this species at Lindisfarne (Holy Island), Northumberland, said that on 3 March 1886, he saw *"not less than 30,000"* Brent Geese and the only other report of large numbers at this time came from Durham, presumably Teesmouth. This may well have referred to the same influx of birds (Atkinson & Matthews 1960). Other historical records include two at Teesmouth from August to October 1898 and 17 at Teesmouth during the first week of June 1900 (Blick 2009).

In the early part of the 20th century, Tristram (1905) said that it was a *"common autumn and winter visitor"*. Nelson (1907) also quoted the fact that it was *"formerly very common in the estuary of the Tees, where large quantities were killed by the professional fowlers"*, implying that it was, at that time, not so common. It would appear that numbers had declined somewhat by Nelson's time. Most unusually, Joseph Bishop recorded a flock of nine birds inland, near Wynyard, on 3 June 1901. Courtenay (1933), writing between 1907 and 1929, said he used to see them *"fairly frequently"* at Teesmouth *"in the first three months of the year. The largest flock I have ever come across numbered about forty"*.

By Temperley's time, most records were of passage birds along the coast, being seen from any time after October, with the greatest numbers recorded from December to March, especially if the Baltic was frozen. Temperley (1951) documented the fact that both dark-bellied *B. b. bernicla* and pale-bellied *B. b. hrota* forms were autumn and winter visitors of increasingly regular occurrence, mainly in severe weather. At his time of writing, Temperley said that the only place that it was to be found in Durham was at the mouth of the Tees. He was told by an informant that three-fifths of the birds seen and shot at Teesmouth *"in the last seventeen years"* were *"of the Pale-breasted form"*.

Post-Temperley through to the 1960s there is little information concerning this species. This might suggest that it was genuinely scarce in Durham. The first significant record was of five birds, three of which were dark-bellied individuals, noted at Teesmouth in late January 1954, and two of these remained into the second half of March (Temperley 1955). The only other records of the decade referred to birds passing along the coast. In 1958, there was a movement of four parties of birds, totalling 63 individuals, which passed Hartlepool Headland in an hour on 31 March, and on 18 November 1959 seven dark-bellied birds flew south close inshore at Hartlepool.

The early 1960s brought more records of this species, with a flock of 17 in the Tees Estuary on 29 January 1961. In the summer of 1962, most unusually, a strong-flying individual, i.e. not a 'pricked' bird, was at Greatham Creek on 2 June. The next few years saw small numbers in the Tees Estuary through the winter periods, with slightly larger flocks of 14 on 31 March 1964 and a maximum of 17 on Seal Sands on 17 February 1966. Occasional passage birds were noted, such as the three that flew south at Hartlepool on 8 April 1967, and the same number north there on 6 November 1968. Clearly, it was not a commonly recorded species in Durham over this period. Seven birds were back on Seal Sands in December 1968 and in early 1969, there were unusually high numbers recorded at Teesmouth; 28 dark-bellied birds were at Greatham Creek on 21st to 23 March, eleven staying until 27 March, and single birds were at Hartlepool and South Shields in February and March.

Into the early 1970s, both pale and dark-bellied birds were recorded at Teesmouth and Whitburn, but the period seems to have been something of a nadir for this species in the county with relatively few reports. In some years, there were only single observations, such as in 1973, when the only record concerned four off Hartlepool on 25 February. Even when the species was noted at Teesmouth, flocks were never great and no more than eight birds were reported in most years up to 1976. Very occasionally, sea passage resulted in larger parties passing our coastline. A flock of 35 pale-bellied birds flew north off Whitburn on 26 October 1971, while 22 dark-bellied birds flew into the estuary and then headed south on 25 December 1972.

In late 1978, there was an influx of birds from the Continent stimulated by hard weather, which were believed to be on their way to Lindisfarne, Northumberland. In December 1978 and January 1979, small numbers were present on Seal Sands, including nine on 26 December 1978. In January, there was a relatively heavy coastal movement of birds off Whitburn, which was described as one of the greatest movements of this species ever recorded in the county, though it clearly pales into insignificance compared to some of the nineteenth century records. Nonetheless, 50 past Whitburn on 2 January 1979 was the highest day-figure for passage birds of the decade. Birds were then noted regularly through the first two months of 1979, with passage continuing into February and small numbers being found at Teesmouth and elsewhere on the coastal strip.

Recent status

This species has, in recent decades been a relatively small scale visitor to the county. Both races are regularly seen in Durham, and in broad terms, the dark-bellied race is a regular winter visitor in small numbers, while the pale-bellied race is a regular passage migrant, largely through September and October and often in much larger numbers.

Dark-bellied Brent Goose
Branta bernicla bernicla

Dark-bellied Brent Goose *Branta bernicla bernicla* is not normally seen until late September and October, so the earliest autumn record of seven passing Whitburn on 28 August 2007 is clearly unusual. The main passage period is late autumn and winter, from the end of October though November, and occasionally into January. The highest ever day count for Durham occurred on 5 December 2002, when 249 flew south past Hartlepool Headland (Little 2003). Other days of heavy passage have included 65 at Whitburn on 6 November 2000, 54 there on 24 January 2005 and 39 at Hartlepool on 25 November 2007. Movements of 20-30 birds are much more the norm, even on days of heavy wildfowl passage.

Teesmouth remains the stronghold for *bemicla* feeding in the county today. It is not unusual for this area to dominate sightings, at least in the early part of the year. Seal Sands, the Brinefields and Seaton Common tend to be the most productive areas, with the short grass of the adjacent golf course providing attractive feeding. More rarely, the marshes at Cowpen, Greenabella and Dorman's Pool have attracted small numbers (Blick 2009). Through the whole review period, up to 20 birds were frequently present for several weeks, particularly between late November and March. Higher counts have included up to 21 in the winter of 1983/1984, 38 in March 1996, 60 in October 1998 and up to 42 on Seaton Common from late October to mid-November 2007. Flocks occasionally linger into early spring, such as up to 13 in early May 1993, but late spring and summer sightings are decidedly rare. A single dark-bellied bird was in the Greatham Creek area from 1st to 28 May 2008, and in 2009, single *bemicla* were at Seaton Common on 25 May, Saltholme Pools on 28 May and Seal Sands over 16-19 July.

Away from Teesmouth, birds have occasionally remained for several days in suitable short grass areas, such as on Hartlepool Headland, South Shields Leas and Whitburn Firing Range.

Dark-bellied birds have been seen inland on few occasions. Singles were at: Crookfoot Reservoir on 27 October 1985; Derwent Reservoir on 2 November 1995; Hury Reservoir on 21 December 1997; Rainton Meadows on 3 May 2000; Wynyard on 17 February 2006; and, near Sedgeletch on 22 November 2008, whilst two were at Portrack Marsh on 11 February 2006. More unusually, up to four dark-bellied birds fed at Offerton, on the outskirts of Sunderland, between 23 January and 3 February 1996 and regularly flew in to WWT Washington to bathe. There have also been a number of inland records of birds of unspecified race. On 14 February 1986, three birds flew over the Fellside area of Whickham and almost exactly three years later, in 1989, a party of six were noted over the same area. Singles were noted at Hury Reservoir on 4 March 1989 and at Joe's Pond on 10 December 1994, while a flock of 18 on Boldon Flats from 31 January to 8 February 1991 was most unusual.

Pale-bellied Brent Goose
Branta bernicla hrota

Pale-bellied Brent Goose *Branta bernicla hrota* are usually the first to arrive in autumn, well ahead of the first dark-bellied birds. The earliest recorded in County Durham are birds at Hartlepool on 22 August 1995 and Whitburn on 22 August 2002, but small flocks in the final few days of August are not unusual. The main passage period is through September and given suitable weather conditions, numbers passing can be impressive. On 9 September 2001, 310 were logged at Whitburn, contributing to a total count of 580 birds noted at coastal watch points on this day. Other high daily counts of northbound passage at Whitburn have included: 217 on 27 September 2007; 145 on 21 September 2004; 115 on 31 October 2008; and, 92 on 2 September 2006. A similar situation is evident at Hartlepool where the days of heavy passage have included: 230 on 13 September 1993; 172 on 9 September 2002; and, 80 on 31 October 2008. As a measure of how the levels of passage for this subspecies have risen in recent years, the highest day-count at Hartlepool in the 1980s was of just 11 birds on 24 September 1988, which was, at the time, the highest for at least 50 years.

Occasionally, adverse weather in the middle of the winter period can result in days of movement, although such instances are rare. On 24 January 2000, 48 pale-bellied birds flew north past Whitburn, whilst 10 passed the same locality on 26 January 2006. Most Brent Goose movements in mid-winter are of dark-bellied birds.

Prior to the year 2000, pale-bellied birds were distinctly in the minority, especially at Teesmouth where the dark-bellied race had always predominated. However, during the 2000s, small numbers of pale-bellied birds have lingered around Seal Sands, such as the flock of up to 23 between 25th and 29 October 2008. Up to six birds also made prolonged stays in other years, usually between late December and late February, and it is possible this trend may increase further.

Spring and summer sightings of pale-bellied birds are unusual, but not unprecedented. In the 2000s, one lingered at Saltholme Pools in the company of Greylag Geese *Anser anser* from 9 June to at least 5 July 2001, a total of seven passed Whitburn between 4th and 7 June 2009, and there were a further four sightings of one to two birds at Hartlepool, Lizard Point, Saltholme Pools, Seal Sands, Seaton Common and Whitburn between 31 May and 15 June.

Pale-bellied Brents have been seen inland on at least nine occasions. Single birds were at Hurworth Burn Reservoir on 17 December 1978 and Crookfoot Reservoir from 9th to 23 December 1978, one was at Shibdon Pond on 4th to 5 December 1985 with possibly the same bird later seen at Big Waters and Caistron in Northumberland. Further single birds were at: Shibdon Pond on 21 February 1994; Boldon Flats on 14 November 2001; WWT Washington in early January 2002; Brasside Pond on 6 March 2009; WWT Washington on 27 March 2009 (possibly the Brasside bird); Derwenthaugh and Shibdon Pond over 10th to 30 January 2010; and, Hurworth Burn Reservoir from 24 January to 3 March 2010.

Distribution & movements

The nominate form, Dark-bellied Brent Goose, breeds on the Arctic coasts of central and western Siberia, spending the non-breeding season in Western Europe. Over half the population winters in southern England, the remainder in northern Germany and northern France. There are four recognised populations of the Pale-bellied Brent Goose, two of which occur in Europe. The eastern Canadian breeding population winters almost entirely in Ireland with small numbers in Wales, the Channel Islands and the north French coast, whilst the whole population from Svalbard, east Greenland and Franz Josef Land winters in Denmark and at Lindisfarne, in Northumberland. A further subspecies occurs in North America and Asia, the 'Black Brant' *Branta bemicla nigricans*. This form has not been recorded in County Durham.

The numbers in all populations of Brent Geese crashed in the early 1930s, caused in part by a reduction in availability of eel-grass *Zostera* due to disease (Salomonsen 1958), together with probable high levels of shooting. Populations of dark-bellied and Canadian Brent Geese recovered considerably through the 1980s and 1990s, but the Svalbard population remains low and vulnerable, possibly due to predation from Polar Bears *Ursus maritimus* and Arctic Foxes *Alopex lagopus*, and competition from an expanding Barnacle Goose *Branta leucopsis* population (Owen *et al.* 1986, Madsen *et al.* 1992).

A single Pale-bellied Brent Goose on South Shields Leas in November 2001 had been colour-ringed at Lindisfarne in 1997. Further investigation of this bird's movements showed that it was present in Denmark in April 1997, again in Denmark in January 1998, in Greenland in the July of this year, back in Denmark in April 1999 and once again in England in November 1999 and January 2001 (Siggens 2005).

Egyptian Goose
Alopochen aegyptiaca

A rare vagrant from naturalised populations in the UK and continental Europe, there are 14 records, involving 16 individuals.

The first record for County Durham was of two birds at Barmston Ponds from 26 August until 10 September 1974. Accompanied by two Canada Geese *Branta canadensis*, these birds remained in the Washington area until at least mid-October, though they also visited nearby East Boldon on a number of occasions. From here they relocated to Hallington, Northumberland, until 14 December when one returned to the Washington area accompanied by what was presumably one of the same Canada Geese; the second remained in Northumberland (Kerr 2001). Present until 25 March 1975, this remaining bird, identifiable by an injured leg, was also seen at

Brasside Ponds on 5 April 1975 and at WWT Washington from 15 April 1975 until 9 May 1975, and again from 16 September 1975 until 23 January 1976.

All records:

1974	Washington and East Boldon, two, 26 August to mid-October, one of same, WWT Washington, 14 December to 25 March 1975, Brasside Ponds, Durham, 5 April 1975, WWT Washington, 15 April to 9 May 1975, and 16 September 1975 to 23 January 1976.
1989	Boldon Flats, flew south east, 11 June.
1990	Gainford, 20 March.
2002	Castle Lake, Bishop Middleham, 17 March and 6 July, same Saltholme Pools 22-23 June.
2007	Derwent Reservoir, two, 1st to 14 July, same Castle Lake, Bishop Middleham and Hurworth Burn Reservoir, intermittently 8 August to 25 October, Derwent Reservoir, 11 February 2008, and Thorpe Farm, Greta Bridge, 25 September 2008.
2007	Saltholme Pools, Teesmouth, 1st to 16 October.
2008	Thorpe Farm, Greta Bridge, 18 March.
2008	Castle Lake, Bishop Middleham, 10 May.
2009	Bowes Moor, 29 May, presumed same, WWT Washington, 7 June to 24 August, Rainton Meadows DWT, 15 August, Boldon Flats, 19th and 23rd to 27 September, and Barmston Pond, Washington, 20 September.
2010	Hury Reservoir, 7 March.
2010	Marine Park, South Shields, 18 March.
2010	WWT Washington, 3 June to 31 July and 17 August, same Boldon Flats 7th to 15 August, Brasside Pond 19 August and Rainton Meadows 24th to 29 August.
2010	Derwent Reservoir, 10 October.
2011	Rainton Meadows 23 March, presumed same Hurworth Burn Reservoir intermittently 23 March to 8 May, Boldon Flats 9th to 12 May, Houghton Gate 10th and 16 May, Rainton Meadows 28 May. Also presumed same Haverton Hole and North Tees Marshes intermittently, 9 April to 2 June.

The dramatic increase in the number of records since 2002 is undoubtedly linked to the expansion of the breeding population within the British Isles, though a large feral population also exists in the Netherlands. The exact number of individuals 'at large' within the county is difficult to assess as most are long-staying and have a tendency to wander around the north east accompanying the large population of feral geese, particularly Greylag Geese *Anser anser*.

Distribution & movements

Egyptian Goose is widely distributed across much of sub-Saharan Africa, with its natural breeding range extending into the Western Palaearctic at Lake Nasser in Egypt. This species was introduced to the UK from its native Africa in the late 17th century, and was added to Category C of the British List in 1971, on the basis of a naturalised population, which by the end of the 20th century numbered in excess of 1,000 birds. The nucleus of this feral population is centred on East Anglia, particularly in Norfolk, though there is evidence of a spread during the early part of the 21st century, with a smaller populations becoming established in the Home Counties, as well as at Rutland Water in Leicestershire and adjacent Nottinghamshire. The species bred in Yorkshire for the first time in 2009 and, given the upsurge in records since 2007, may breed in County Durham in the future.

Ruddy Shelduck
Tadorna ferruginea

A very rare vagrant from south eastern Europe and Asia, with one accepted record of an apparently wild bird. All post 1949 records are thought to be escapes from captivity or from feral stock elsewhere in Europe.

Historical review

The first and only truly acceptable record for County Durham is listed by Tristram who states: *"The only recorded occurrence is the appearance of a small flock in the interior of the County, one of which was shot and brought to Mr Cullingford for preservation on September 23rd 1892"* (Tristram 1905). This record does not appear to have been widely publicized as it was not listed by Vinnicombe and Harrop (1999). George Temperley however, thought it acceptable: *"It is stated in the Handbook of British Birds that 'Many occurred June to September,1892, various parts Ireland, in Cumberland, Berwick, Sutherland, Morayshire, Lincs., Norfolk and Suffolk; flocks of ten to fifteen and even twenty in some places.' In the light of this report on the 1892 influx, this Durham record does not appear so unique as at first sight. It is unfortunate that more details are not now forthcoming"* (Temperley 1951). Presumably, Temperley was unable to trace the specimen and the whereabouts of this remain unknown. There seems no reason to doubt the provenance of this record, the significance of which was probably not realised at the time. Forming part of the 1892 influx, this should be recognised as an acceptable record of the species in County Durham.

The second record was of an adult female on a flooded field in north east Durham from 5th until 22 May 1945, observed by the late Fred Grey. It seems likely that this was at Boldon Flats, an area he was known to observe at the time. Temperley noted that another bird was seen at Alnmouth, Northumberland during the same period and commented, *"It is probable that these birds had strayed from ornamental waters"*, and as such, it is best regarded as being of doubtful origin (Temperley 1951).

Recent status

All post 1949 records:

1988	Seal Sands, and adjacent North Tees Marshes, Teesmouth, 19 May to 21 June; same 23 March to 22 June 1989, 23 August to 15 September 1989, 10 February to 8 July 1990, 6 September 1990, 31 December 1990 to 20 January 1991, and 22 June to 3 July 1991.
1992	WWT Washington, three juveniles, 4 August to 4 September.
1994	Hurworth Burn Reservoir, four juveniles, 30 July, same Shibdon Pond, Blaydon, 31 July, with two remaining until 1 August.
1998	McNeill Bottoms, Witton-le-Wear, flew north west, 29 November.
2004	Whitworth Hall, Spennymoor, female, most of June.
2004	Saltholme Pools, Teesmouth, two adult females, 28 June.
2004	Bowesfield Marsh, Stockton-on-Tees, three, adult and two juveniles, 1st to 7 August, with two birds remaining until 8th.
2004	Whitburn, four flew north, 14 August.
2005	Shibdon Pond, Blaydon, 6 April, presumed same Saltwell Park, Gateshead, 10 April.
2005	Saltholme Pools, Teesmouth, male, 18 June.
2005	Whitworth Hall, Spennymoor, two, 21 November.
2008	Rainton Meadows DWT, two, 1 December.
2009	Derwenthaugh Meadows, 28 November.

All of the British post-1949 records are considered as being of feral origin and as such do not form part of the 'official' county record, although they are included here for completeness. The party of four juveniles at Hurworth Burn Reservoir on 30 July 1994, which were later seen at Shibdon Pond, coincided with a large influx into Fennoscandia during the late summer of that year, which involved upwards of 250 birds. These were thought by some observers to represent a genuine influx of wild birds from Asia, similar to that of 1892. A review of all records of Ruddy Shelduck (Harrop 2002) however, found no evidence for this or to upgrade the species to Category A of the British List.

Distribution & movements
The main breeding range of Ruddy Shelduck extends from south east Europe across into central Asia as far east as Lake Baikal, though a small population exists in North Africa, south into Ethiopia. Asian birds are migratory and winter to the south of their breeding range, from Afghanistan eastwards through the Indian subcontinent to south east China; large numbers winter in Iran. Outside of this period, birds are known to disperse widely in almost any direction, mainly in response to water shortage and drought (Parkin & Knox 2010). The species remains on the British List on the basis of an influx of at least 59 birds to Britain and Ireland in 1892, this being part of a wider influx into Europe, with birds being recorded as far west as Iceland and western Greenland. Most records subsequent to this are thought to represent birds of captive origin and the species is retained in Category B of the British List on the basis that it has not occurred in wild state since the end of 1949 (Harrop 2002).

Shelduck (Common Shelduck)
Tadorna tadorna

An uncommon and restricted breeding species, but a common passage and Winter visitor that was once present in nationally important numbers at Teesmouth.

Historical review
This duck is a common winter visitor to Teesmouth, with smaller numbers remaining during the summer, a minority of which breed around the estuary. It is also a regular passage migrant along the coastline, especially in autumn, and an occasional visitor to inland waters.

The earliest direct reference to this species in Durham comes from the Cellarers' rolls for the Monastery at Durham. There are three extracts from the Cellarers' rolls for 1326 and 1330 that refer to Shelduck and one from the Bursar's roll for 1338/1339 that mentions the purchase of certain birds under the name *"aucce rosettce"* or *"rosatce"*. As many as twelve Shelduck were bought for the Easter feast in April 1330 and another four, one week later; their value in 1338 was five pence halfpenny each. In his glossary on this work Gurney suggested that *russettce* should be understood to denote the Shelduck (Gurney 1921). Clearly the Shelduck has been a 'Durham bird' for a long time though it was not shown as breeding in Durham in the first part of the 19th century by Holloway (1996). Yet, Sir Cuthbert Sharp's *List of Birds observed at Hartlepool* (1816) reported it as breeding in rabbit burrows in sand hills near Hartlepool. Hogg (1827) said that they were common, though he may have been referring primarily to wintering birds. Tristram (1905) said *"the bird is now only an occasional straggler"*, but recalled that in the past (*c.*1840), it was a *"well-known breeding species in the sandhills and rabbit warrens by the coast, especially at Seaton and Teesmouth"*. He noted it nesting at Middleham, now in the heart of Hartlepool.

Temperley (1951) observed that when the birds were driven from their rabbit holes, they found safer sites in the raised banks of the Tees, which were reinforced with industrial waste slag, which left numerous deep holes and crevices. The birds first began to use this slag wall habitat in 1883 (Nelson 1907) and by 1899 four broods were noted to be nesting (Milburn 1901). Clearly, a steady increase occurred and 14 pairs were nesting on the Durham side of the estuary by 1939 (Temperley 1941).

Temperley (1951) also reported that the breeding population was small compared to the numbers wintering. Likewise, Nelson (1907) reported a spring passage with large flocks on the coast and in the Tees estuary, commenting that on 13 June 1882, a flock of 300 birds was observed from the Tees Light Vessel. Courtenay (1933) said flocks were common at Teesmouth between 1907 and 1929, particularly in the first three months of the year, and *"numerous"* into the early summer months.

Nationally, the species declined during the 19th century possibly as a result of persecution, but during the 20th century there was a sustained increase with birds spreading to inland sites from the 1940s.

On 7 May 1939, Bishop reported *"a goodly numbers of non-breeders are in the estuary"*. Peak numbers occurred in February and September, and around that time 270 were present on 16 February 1947. The non-breeding adults departed in July and immature birds arrived in the estuary in September, with the adults re-appearing in the winter months. Autumn birds were recorded on Darlington Sewage Farm and on 22 December 1938 on the Boating Lake at Darlington (Almond *et al.* 1939). Temperley (1951) recorded the Shelduck as *"a resident, breeding in small numbers at the Tees mouth"* and *"a spring and autumn passage migrant and a winter visitor".* He also noted that the bird was sporadically observed at Jarrow Slake and the nearby Hebburn Ponds, with other inland records being unusual, and possibly related to passage birds. There is a single historical reference to an inland record in Gateshead, when a pair was present on flood-water near Lamesley in the Team Valley, on 16 November 1941 (Bowey *et al.* 1993). Despite this, Holloway (1996) recorded that inland nesting was reported in good numbers as early as the 17th century in Norfolk.

During the 1960s, Teesmouth hosted one of the largest winter gatherings of the species in Britain, after birds had returned from their moult migration in the Helgoland Bight (Brown & Grice 2005). Over this period the mid-winter maximum consistently exceeded 1,800 birds and a series of record counts were set, commencing with the 2,700 that were there on 9 January 1964, rising to 2,935 on 14 January 1968. There were 3,300 present on 12 January 1969 and a peak of 4,443 birds on Seal Sands in January 1970 (Stewart 1970). The importance of Teesmouth for this species during the 1960s is emphasised by the fact that the count of 2,711 birds at Teesmouth in January 1967 constituted one twelfth of all of the Shelduck recorded in the UK during the international wildfowl census at that time. Wintering numbers have since declined with the subsequent infilling of the mudflats around the estuary for port-related industry (Joynt 2006). Between 1964 and 1970, between four and eight broods of ducklings were counted at Teesmouth on an annual basis, with occasional pairs noted elsewhere, such as the pair at Derwent Reservoir in early April 1967. In July 1964, 50 flew east down the Tyne at Hexham and 40 flew east at Blaydon, and similarly flocks of 50 and 70 over-flew South Shields during late July in 1969. No doubt these were all part of the moult migration.

Recent breeding status

In the early 1970s, breeding birds were confined to Teesmouth, with seven broods noted in 1970, four broods of ducklings in 1973 and around 100 young counted in 1974. Since the early 1970s, the county's breeding population has expanded to sites away from the Tees estuary itself. The first inland nesting was recorded at Crookfoot Reservoir in 1974, when two pairs nested in haystacks; one nest was destroyed by a fall of bales, but nine young hatched from the other. The following year, there were three broods at this site and birds also nested at Newton Bewley. Birds bred for the first time at Hurworth Burn Reservoir in 1976, then through the rest of the decade also at Hart Reservoir with increasing records further afield. However, nestboxes installed around Cowpen Marsh in the 1970s were largely unsuccessful. The spread slowed in the 1980s, though WWT Washington was colonised during this decade, as most other suitable sites seemed to have already attracted birds. The year, 1991 seems to have been one of the best for some considerable time for successful breeding, when eight broods were reported from Teesside and four from Hurworth Burn Reservoir.

Durham's breeding population of Shelduck represents only a tiny proportion of Britain's national population of 12,000 pairs (Owen *et al.* 1986). Although the species is large and conspicuous, their nesting sites are concealed; nonetheless, territorial birds with broods are unlikely to be overlooked. Within the county, three main centres of breeding are evident. They are: the estuarine fringe of the Tees lowlands, the reservoirs of the south east Durham plateau and around the WWT Washington on the River Wear. Birds have also bred along the tidal stretches of the River Tyne between Dunston and Stella since 1996, with a maximum of three pairs in 1999. Territories are found from sea level to elevations of approximately 120m, the latter around the reservoirs referred to previously.

Shelducks are traditionally thought of as estuarine birds and in County Durham the North Tees Marshes provide the classic breeding habitat, with nest sites in slag reclamation walls located conveniently close to productive foraging areas on inter-tidal mud flats and brackish lagoons. The *"New Atlas"* graphically documents the species' colonisation of inland breeding sites, and this national trend is, to an extent, reflected in Durham (Gibbons *et al.* 1993). As breeding birds, Shelduck remain relatively scarce away from the North Tees Marshes and confirmed breeding came from just nine sites across the county during the 2000s. The few inland sites that are used regularly, such as Houghton Gate, are characterised by the availability of suitable nest sites (often rabbit holes) and their proximity to shallow bodies of water, which provide security for the ducklings. In May 2001, pairs were present on wetlands at Kibblesworth, Lamesley and Far Pasture in various parts of Gateshead, indicating that occasional opportunistic birds will try to colonise unpopulated wetlands. In 2007, successful breeding was only confirmed at three sites away from the two main centres of Teesmouth and Washington. This comprised two pairs at Hurworth Burn Reservoir, two pairs in the Bishop Middleham area, and a pair, which raised five young at Houghton Gate, though breeding was also suspected at Crookfoot Reservoir.

In 2008, breeding was confirmed at just five locations away from Teesmouth, these being Bishop Middleham (Castle Lake), Houghton Gate, Hurworth Burn Reservoir, the River Tyne and WWT Washington. In 2010, successful breeding was reported at six sites around the county, with 37 juveniles gathering at Greatham Creek in late July. By this time, there were probably a minimum of eight pairs breeding along the section of the River Tyne from Dunston Staithes to Lemington Gut, with perhaps a total of 15 pairs along the river's tidal reach, from its mouth to Wylam in the west.

At Teesmouth, crevices in the estuary's numerous slag walls probably hold the majority of breeding pairs, particularly along the Long Drag and Greatham Creek. Up to three pairs have been known to breed on Crookfoot Reservoir since 1974, and single pairs have been seen with ducklings on Wynyard Lake, Hart Reservoir, Bowesfield Pond and on the sea off Hartlepool Headland on occasions. In 1999, up to ten broods were noted at Crookfoot Reservoir (Blick 2009). Artificial nest-holes have been created in two areas of the Tees Marshes.

The observation of broods of ducklings usually provides proof of breeding. The *Atlas* estimated the county's breeding population to lie in the region of 10-25 pairs, with some 50% of these, or more, being in the Teesmouth area (Westerberg & Bowey 2000). It seems likely that this figure may have recently risen. The Breeding Bird Survey summary for the period from 1994 to 2006 (Raven *et al.* 2007), indicated that there had been a 68% increase in England over that period. The estimate contained in *The Breeding Birds of Cleveland* stated that Teesmouth had possibly 50-55 pairs between 2006 and 2007 (Joynt *et al.* 2008) but many pairs will be breeding outside of Durham, on the south side of the estuary. So with some local increases, such as along the Tyne, perhaps the county population now stands in the region of 30-40 breeding pairs.

Recent non-breeding status

The county's breeding population is dwarfed by Teesmouth's non-breeding population (typically 300-400 in May and June). In contrast to the estuary's wintering population, spring numbers have remained remarkably stable over the last twenty years. The presence of these birds can confuse the determination of breeding data in that area.

Although numbers are much reduced on the North Tees Marshes compared to the 1970s and 1980s, it is still the primary location for this species within the county, with figures in the mid-hundreds in the final and first quarters of each year, such as the 652 noted there in January 2007. These figures pale into insignificance when compared to those of the 1970s, when a visit to Seal Sands between September and March would have revealed several thousand Shelduck on the mudflats. The count of over 4,440 birds on the whole estuary on 16 January 1970 was, at that time, the largest for any site in the British Isles (Stewart 1970); it has never been equalled at Teesmouth, although about 3,570 were noted on 21 November 1976. Durham University's monitoring of the land-claim area of Seal Sands, from 1973 onwards, showed that "Immediately following the land-claim of Seal Sands during the early 1970s, low water usage of Seal Sands by Shelduck increased to a peak count of 3,227 birds in winter 1976/1977. The population has since been in continual decline, the rate of decline reduced during the 1990s when numbers were already below those of the early 1970s" (Ward et al. 2003). The record counts were made when Seal Sands was about three times its present size. A considerable reduction in the number of wintering birds at Teesmouth followed, with a maximum count of only 1,400 in 1971, compared to over 4,000 the previous winter. This was the lowest count since 1963 and was probably related to disturbance associated with the commencement of works within the estuary.

In the 1980s, regular counts showed that Teesmouth was still one of the top ten or twelve most important estuaries in Britain for wintering Shelduck. In January 1990 this nationally important site recorded 1,637 birds, the best figure for the decade. In 1993 the number of birds in the same month had fallen to 981, though numbers increased again over the following years, when an average of about 1,000 to 1,300 birds formed the mid-winter peak. This fell again to 702 in January 1999 and 487 in February, the lowest known maximum counts. By the year 2000, Teesmouth experienced its lowest maximum mid-winter count for a generation, numbers declining by 5% even from the low level of 1999. Today, it no longer supports a nationally important wintering population of this species.

The reasons behind the initial increase and subsequent decline are complex. In the early 1970s, the land claim removed 60% of the intertidal area of Seal Sands. Unexpectedly the fine silts from the dredging used in the land claim filtered through the clinker walls to create an invertebrate-rich veneer across the remaining mudflats. This was to the benefit of predators such as Shelduck and Dunlin *Calidris alpina*, initially boosting the numbers using the area (Evans 1998). Thereafter, these fine silts consolidated there was a shift to coarser sediments on Seal Sands and the spread of *Enteromorpha* which *"may have modified prey availability on some mudflats"*; all factors implicated in the decline of Shelduck (Ward *et al.* 2003). In summary, the number of wintering Shelduck at Teesmouth underwent a significant reduction between 1975 and 2000, in contrast to the national situation, where there was no overall change (Ward *et al.* 2003).

Away from the two principal wintering areas of the North Tees Marshes and the River Wear at Washington, where 20-30 birds are frequently present, counts of birds tend to be very low. The River Tyne in the Dunston area sometimes attracts small flocks of between ten and twenty birds, and this area is growing in importance as the local breeding population becomes ever more established. Most other wintering records come from the coastal strip or occasionally other Durham wetlands, such as Boldon Flats.

Coastal movements in the county are logged on a regular basis at the main coastal watch points, such as Whitburn and Hartlepool Headland, but never in large numbers. On 17 July 1995, 90 birds flew east in two flocks passing Souter Point, presumably bound for the Waddenzee and the late summer moult. In 2007, at Whitburn Observatory passage of Shelducks was logged on just 49 dates in the year, which is indicative of the relatively small scale of observed movement. In 2008, coastal passage was logged on 68 dates, totalling 357 birds through the year. Because of these relatively low numbers, it has been suggested that the return from Germany to Seal Sands (and other estuaries in eastern Britain) must take place at least partly during the hours of darkness (Blick 2009). At Hartlepool, which is adjacent to the main wintering area, in some years very few are seen moving along the coast, the highest count being 171 on 20 December 1964, and it is rare for more than 20-30 birds to be recorded in any one day (Blick 2009).

The species is also a spring and autumn passage migrant of annual occurrence to inland sites, though only in small numbers. Spring sightings slightly outweigh those in autumn, whilst there are also occasional records of birds in winter and mid-summer.

Distribution & movements

Shelduck can be found breeding across much of coastal Europe and across to Asia; the largest numbers of the species occurring in the north of its range, where it can be found breeding as far north as southern Norway. In the post-breeding period, most adult Shelduck from Britain move to the German section of the Wadden Sea, where they moult alongside breeding birds from across north west Europe. Much smaller moult concentrations of the British breeding population utilise the Forth, Humber, Wash and in particular Bridgewater Bay (Wernham *et al.* 2002). Consequently, the majority of birds present in late summer at Teesmouth are largely 'birds of the year' together with a few adult 'baby-sitters' (Blick 2009).

A bird recovered at Seaton Carew on 20 January 1957 had been ringed on the estuary of the River Weser, Germany on 31 August 1952 (Blick 2009).

During the late 1970s and early 1980s, Durham University conducted an extensive colour-ringing and surveying exercise on the Shelduck of Teesmouth, which revealed that Seal Sands is an important 'staging post', and not just a wintering ground, for this species. In the winter of 1978/1979, dye-marking studies identified the passage through Seal Sands of at least 2400 Shelduck; more than twice the number recorded on any individual count (Evans 1984). At any one time, there may be 1,000 to 2,000 birds on Seal Sands between November and February, but between 3,000 and 4,000 individuals will probably have fed in the area over that period (Ward *et al.* 2003).

Birds breeding at Teesmouth, and those ringed as pulli during 1976-1980 on both the Firth of Forth and at Hauxley in Northumberland, have been recorded at Teesmouth in the following winter. Others recorded at Seal Sands include a bird ringed at Hambleton, Lancashire in 1974, which was on Seal Sands on 1 November 1979 (Bell 1980) and one from the Isle of Mull, Strathclyde, ringed in 1981 and seen on Seal Sands on 2 December 1983. Another bird ringed on Seal Sands was recorded at Frinton-on-Sea, Essex seven years later (Blick 2009).

Birds ringed as full-grown juveniles on Seal Sands have been reported as follows: one on 23 August 1977 was not seen at Teesmouth again until April and May 1980, it was then found at Bolton-on-Swale, North Yorkshire, in early May 1981. One ringed 1 November 1979 was found at Dornoch, Sutherland, in August 1984 and another ringed 20 November 1979 was recovered at East-Agder, Norway, on 3 April 1982. A bird ringed in January was seen on the Firth of Forth three months later. A juvenile ringed on the Somme Estuary, France on 16 December 1981 was at Seal Sands on 3 June 1982, whilst one ringed on Seal Sands 9 September 1994 was reported taking bread in Locke Park, Redcar, during the latter half of 1997. A bird ringed on Seal Sands in August 1997 was (illegally) shot at Hartlepool in November 1998 (Blick 2009).

Adults ringed at Teesmouth in November have been reported as follows: at Washington, Tyne and Wear, in the same winter; and in counties Antrim, Donegal and Down in Ireland in the following winter. Four birds ringed in November 1977 lived for at least eight years; one was found dead at Blakeney Point, Norfolk, in January 1985, another at Longtown, Cumbria, in April 1985, another in Germany in October 1985 and the fourth at Alnmouth, Northumberland, in March 1989. Three others, ringed in November 1977, were part of a small moulting flock at Grangemouth on the Firth of Forth in summer 1979, but they were back at Teesmouth in late autumn 1979 and again in subsequent autumns. One ringed on 4 November 1980 was found dead on Jersey in January 1983 (Blick 2009).

Adults ringed at Teesmouth in December have been seen at Widnes, Cheshire, in late autumn, Isle of Gigha, Strathclyde, in April and in Bangor, Gwynedd, in late March. Another bird, ringed in December 1983, was seen on Texel, Holland, in January 1987. Of some adults ringed in January 1979, one was seen in the Netherlands in February 1980 and May 1981 and another was at Port Glasgow, Strathclyde, on 4 July 1982. Others ringed in January have been noted on the Clyde Estuary in February and July and in Norfolk in February and April, one 11 years later. An adult ringed in February 1988 was in Schleswig-Holstein, Germany, on 1 October 1995, and one ringed on 3 June 1982 had apparently moved north to winter, being recorded at Budle Bay, Northumberland, on 28 January 1984 (Blick 2009). An adult ringed at Seal Sands on 10 February 1994 was found in the Netherlands on 21 April 1998, suggesting that it was breeding locally (Armstrong 1999).

Mandarin Duck
Aix galericulata

A scarce passage migrant and a resident in small numbers, as well as an occasional escape from wildfowl collections.

Recent status

Today, the Mandarin is a scarce bird in County Durham, with a handful of records per annum, its status changing quite dramatically between 1990 and 2011. The first record of Mandarin in Durham was of a pair of birds at Crimdon Dene on 24 April 1978 (Unwin 1979); earlier records may not have been documented, or dismissed as escapes from captivity. The next record did not occur until 13 April 1985, when a drake was at Shibdon Pond, followed two years later by a bird on Cleadon Hills on 10 April. A drake at Haverton Hole from 24th to 28 April 1990 was the first record of the species at Teesmouth and confirmed the emerging spring occurrence pattern. The species remained something of a rarity until the beginning of 1995, aside from the resident birds that were present at WWT Washington from 1991 to 1995 and beyond; there were just 12 records up to that point.

The species has been recorded almost annually in Durham since 1990. It was still a rare visitor at the outset of the decade and through much of the 1990s, this despite the fact that it bred in the county during this period and was in the process of establishing small colonies in Yorkshire and Northumberland. Shibdon Pond attracted birds on three occasions through the decade: a ringed drake on 29 December 1990 and further drakes on 3 April 1991 and 22 May 1996. Only four other sites attracted the species during the decade. The more regular pattern of occurrences over this period was presumably linked to the species breeding with greater regularity both north and south of Durham.

Through the 1990s, the pattern of occurrence became more dispersed, with Teesdale records assuming greater prominence later in the decade, some suggestion of a coastal bias and a scatter of Wear valley records, which may or may not have been related to the WWT Washington records. Through the 2000s, birds became more widespread and were recorded at wetlands and river locations in many parts of the county. Well over 30 sites documented the species through the decade and most of the county's main wetlands attracted birds on one occasion or another. In this period, many of the site records were probably accounted for by one or two wandering birds logging up many observations, though clearly a number of additional birds were involved, particularly in April 2009, when a total of around seven individuals were recorded in the county. In 2010, birds were recorded in greater numbers than ever before and from a suite of at least 12 different wetlands spread from the north to the south of the county. Birds were noted at a minimum of 13 sites during 2011, including at least three birds at Chester-le-Street in both winter periods and five (two males and three females) on the River Wear at Shincliffe on 13 October.

In the species' early pattern of occurrences in the north east, from 1971 to 1991, there was a distinct peak during April and May, and a slightly lower one in the September to October period. This suggested that it occurred in County Durham largely as a passage migrant from established populations elsewhere in the British Isles (Bowey 1992). This pattern persists, at least in part, to the present, as over 70% of all records, long-staying birds excepted, are of birds between March and May, most of them in April. There is a secondary, much less obvious, peak in September and October. The pattern of Mandarin occurrence in the south east of the county also indicates that there is a peak of passage in early spring (Joynt et al. 2008). Falling into this category were drakes at Ward Jackson Park, Hartlepool, on 6 April 2004 and on the River Tees near Yarm on 29 March 2007 (Joynt et al. 2008). The record of a pair flying north past Whitburn Observatory on 24 April 2004 supports the theory that the species behaves as a passage migrant through Durham, as does the occurrence of a male at South Shields Marine Park on 29 April 2001. This spring movement has been less obvious since around 2007, with the year-round presence of some birds in the county (although no birds were recorded in Durham during 2006). This is probably related to the fact that the species has been breeding to both the north and south of County Durham in recent decades, 'damping down' the effect of passage birds.

Three drakes at Crookfoot Reservoir on 26th to 27 September 2002 were apparently autumn passage birds. A number of other records fall into this category, such as the drake at Far Pasture in the lower Derwent valley on 4 October 2001 and one at Saltwell Park Lake, Gateshead on 6 November and 4 December 2005. A more surprising early autumn record concerned an eclipse drake on rock pools off Souter Point on 5 August 1998. This bird remained in the area, either on the nearby nature reserve pool or on the beach, until 4 September (Armstrong 1999b).

Recent breeding status

In 1991, a feral male bred with a captive female at WWT Washington and raised a single youngster, representing the first breeding record for the county. After this, a feral pair was at WWT Washington from 1994 to 1999 and breeding was recorded, but only once successfully, in 1995, when this pair laid a single egg in a Stock Dove *Columba oenas* nestbox. In the early 2000s a feral pair was again present at Washington, but there was no evidence of breeding. In 2001, two males and a female were there, unusually when one of the males died in June another male appeared the following day.

A breeding record, which perhaps deserves to be considered the first attempt by 'wild birds' in Durham, was of a pair of birds on the River Balder at Balder Grange in 1998; these birds raised a single youngster. Breeding had been suspected on the River Tees in 1997 at Cotherstone and another pair was present in this area, on the River Balder, in April 1999.

Records of birds from this part of the county during the early part of 2002 suggested that a small population of Mandarin had become established along the upper reaches of the River Tees, with pairs noted at Cotherstone in 2002 and at Romaldkirk and Middleton-in-Teesdale in 2003. Successful breeding was confirmed at this latter site in both 2004 and 2005; a female with two young being there on 22 June 2005. In 2008, birds were recorded at five locations and a female with two young was on the River Tees near Egglesburn on 16 June. At the outset of 2008, up to three birds (two drakes and a duck) were on the River Wear at Cox Green, with presumably the same birds visiting WWT Washington and Cox Green over the next couple of months. A pair, possibly the same birds, was at Butterby on 23 January and again on 30 April, raising the prospect of a breeding presence somewhere on the lowland part of the River Wear. Birds appeared to have bred successfully in 2010, when Mandarin was in breeding habitat in three areas of Teesdale during the breeding season. Possible breeding birds were on the Gill Beck, near Lartington, on the Balder, downstream of Hury Reservoir and, later in the season, on the Tees near Cotherstone, when two drakes were there with two females and a party of four juveniles in early October. A similar situation was evident in 2011, when pairs were detected in at least two areas of Teesdale in early spring.

There was an obvious population increase and expansion in the range of the Mandarin in some parts of northern England during the 1990s and early 2000s. During 1994, one or two pairs escaped from a wildfowl collection at Ridley Stokoe in west Northumberland, which culminated in a breeding attempt at Kielder in 1994 and successful breeding in that area in 1995 (Kerr 2001). It seems likely that this and the species' later spread into the eastern part of Northumberland in 1998 (Kerr 2001), was in part responsible for its more regular occurrence across the north east of England, and the subsequent colonisation of Durham during the first decade of the 21st century. The fact that Shibdon Pond attracted birds on a number of occasions through the mid-1990s and 2000s may also relate to the birds based in Northumberland, moving to and/or from their breeding locations along the Tyne valley. In North Yorkshire, gatherings of up to 60 birds were recorded on the River Wharfe (Newsome 2008), less than 30km to the south of County Durham during 2007. Further south in Yorkshire, to the north of Leeds, the species has been breeding in the Harewood Estate since at least 2002.

Distribution & movements

This ostentatiously colourful duck is a native of the Far East, being originally a resident of Japan, eastern China and the adjacent parts of eastern Russia (Parkin & Knox 2010). It was first recorded in an English wildfowl collection in 1747 (Long 1981), and the first captive breeding was recorded in 1834 (Lever 1977). It was not until the second half of the 20th century that it became established as a self-sustaining breeding bird in Britain; it was formally added to Category C of the British list in 1971. The British stronghold has long been around the Home Counties, but through the 1980s and 1990s birds spread through southern and central England, with small populations becoming established in northern England during the late 1990s and early 2000s; the national total being around 550 pairs by 2004 (Parkin & Knox 2010). British birds are non-migratory, though they are dispersive, particularly the first summer males (Bowey 1992).

Sponsored by

Wigeon (Eurasian Wigeon)
Anas penelope

A scarce resident and localised breeding bird, common as a passage and winter visitor.

Historical review

In the early 19th century, Hutchinson (1840) stated that in Durham *"the Wigeon is one of the most common ducks during the winter, being sent in considerable numbers to market from the coast. It is also found in small numbers on the rivers and brooks in the interior of the County. It is met with from the latter end of September till the end of March. It retires northward to breed".* Later that century, Robson (1896) described the shooting of one on Shibdon Flats during October 1891; it was said to be *"in the company of a few others"*. During the 19th century, a disproportionately large amount of bird recording was conducted with the assistance of the gun and Nelson (1907) wrote

that up to 23 Wigeon had been shot on one occasion on the coast (Blick 2009); his earliest and latest records of this species at Teesmouth were on 11 August 1883 and 15 May 1902.

Whilst John Hancock (1874) documented that this was a plentiful autumn and winter visitor to the north east, he mentioned no evidence of its breeding in Northumberland or Durham in the mid- to late 19th century. Just a few years later, Abel Chapmen (1924) noted that it was breeding at Broomlee Lough, Northumberland, in 1912, with other sites in that county attracting breeding birds in the following years, including Hallington and possibly Holy Island Lough (Bolam 1932). In the 19th century, barely any birds were shown as breeding in England and there were certainly none 'mapped' as breeding in the north east of England at that time (Holloway 1996).

In the mid-20th century, Temperley (1951) said that it was present in the county from September to March, noting that it was a *"common autumn and winter visitor to the coast"*, less frequently to inland waters, after which it *"retires northwards to breed"*. He noted that flocks on the sea during the day came inshore to feed at night around the Tees Marshes. Also it was noted on rivers and ponds where it was *"not molested"*. An expansion of the Scottish breeding population was said to have commenced in the first half of the 19th century, with some indication of 'regular breeding' being noted to the south of Durham, in Yorkshire, near Whitby, around 1888 (Mather 1986, Holloway 1996).

In 1929, Temperley reported that the Wigeon had become more common since Hancock's day and that it was 'now' breeding in Northumberland (*The Vasculum* Vol. XV 1929) and breeding occurred there in the 1920s, and again in 1949, and from 1978 (Kerr 2001). Temperley (1951) mentioned that it had been suspected of breeding in Durham, but was unable to offer any definite proof. It was reported as breeding, unsuccessfully, on a small moorland tarn in the North Riding of Yorkshire in 1939 (Mather 1986); a site which since 1974 has been included in the recording area of County Durham.

Ken Baldridge recalls his first view of breeding Wigeon in what is now County Durham, as being on 14 June 1953 when he and John Lumby found a pair on Fish Lake in Lunedale. The following year, on 3 June 1954, a female was found with a duckling. Breeding occurred here again in 1955, when a pair, an adult male and an immature bird, was seen on 7 May. On 7 June 1956, the breeding duck at this site was taken by a Peregrine *Falco peregrinus*. In spite of this, the breeding population of Wigeon spread in the south western portion of County Durham during the early 1960s. In 1965, three broods were found at a small tarn on Bowes North Moor. Birds may have bred at the site in 1958 and certainly did so in 1960. By 1966, birds were reported as breeding at Fish Lake, on the Bowes tarn site and also at Crag Pond, near Lartington, where it was presumed the species had been breeding for a number of years, as four broods were present in 1966. In 1971, there were 20 pairs at the Bowes tarn site, which produced at least 40 young. In one year between 1969 and 1970, numbers doubled here from nine to 18 pairs, and 76 young were raised. By 1981, birds were breeding in Lunedale, on Blackton Reservoir in Baldersdale and at the Bowes North Moor tarn, and no doubt at a number of other sites in the west of the county (Mather 1986). There were also occasional summering and even breeding birds, including ducks with broods, around the extreme upland stretches of the River Tees in the early to mid- 1960s, prior to the construction and flooding of the Cow Green Reservoir in 1967 (I. Findlay pers. comm.). A pair was seen on the 'Weel' in this area of the River Tees (K. Baldridge pers. comm.) and breeding was confirmed there in 1968 (I. Findlay pers. comm.). This location was swallowed up by Cow Green Reservoir. These may well, therefore, be the first records of breeding in vice-county 66, inside the 'old' County Durham boundary.

The building of Derwent Reservoir was a significant event for this species in Durham, for soon after its creation in 1967, birds began to use the site, both for wintering and ultimately, breeding. By 1968, up to 150 were noted there in winter, this count being exceeded the following year, when up to 400 were present; making the site more important than Teesmouth for the species at that time.

In 1970, up to 300 were at Teesmouth on 15 February and the species summered in this area, but there was but no proof of lowland breeding. The following year up to 310 were at Derwent Reservoir on 17 January and in 1971 breeding was proved at the same site.

Recent breeding status

In County Durham, the Wigeon is near the southern edge of its regular breeding range in the UK (Gibbons *et al.* 1993), and as with many species, it is unclear as to what future climatic change may have in store. The Wigeon is a rare breeding species nationally, with some 39-118 pairs in 2007 (Holling 2010), so the county's breeding stock is significant in the context not just of the English, but the UK population as a whole. In Durham, breeding Wigeon are largely restricted to the 200-400m altitude zone and are found mainly around the shores of the upland reservoirs as well as on the streams and high moorlands that make up the their catchment area. Nests are most usually found in rough tussock-strewn grassland, with attendant rushes *Juncus spp.* and/or bracken *Pteridium aquilinum*. The successful ducks move on to wetland pools and open water with their ducklings, shortly after they have hatched. In certain areas, predation by stoat *Mustela ermina* has been suggested as being a problem (Cramp *et al.* 1977).

It has been estimated that 20-25 pairs breed in Northumberland, largely in the south west of the county, with the most important breeding site being Derwent Reservoir, which lies across the Durham-Northumberland border (Kerr 2001). In 1992, eleven broods of ducklings were noted at this one location (Kerr 2001). Intensive fieldwork between 1994 and 1996, in particular by the late Denis Luckhurst, revealed that there were few breeding pairs other than those in the extreme south west and north west of the county. Population modelling and estimates based on figures derived by Denis Luckhurst (D. Luckhurst pers. comm.), suggested a county population of at least 63 breeding pairs. Most of these were found in the far south west of the county, in upper Teesdale, in Baldersdale, Lunedale and along the course of the River Greta (D. Luckhurst unpub. data). This estimate was produced by using counts of females with identifiable young or of females noted in suitable habitat during the breeding season. It should be noted that a population of up to 20 pairs has also been estimated to be present in the north west of the county (Day *et al.* 1995), largely around Derwent Reservoir and its catchment area.

By combining the detailed observations from the south west and counts made of males in April and May during the *Atlas* survey period (Westerberg *et al.* 1994), from the north west and scattered locations elsewhere in the county, Westerberg & Bowey (2000) suggested a potential breeding population of up to 90 pairs. At that time, this was a significant percentage (more than 20%) of the total estimated UK breeding population of less than 400 pairs (Gibbons *et al.* 1993). In Durham, breeding numbers increased from four pairs in 1990 to 36 pairs on 14 different sites in 1999, with successful breeding being recorded from at least six of these.

There has been little additional work undertaken on the species, since that of the early to mid-1990s, though casual sightings and individual fieldwork has been collated for the Rare Breeding Birds Panel of British Birds. In 1997, 31 pairs were seen and in 1999, fourteen broods of ducklings were recorded with a total of 36 pairs noted. In 2001, two broods of ducklings were observed at one upland reservoir in Teesdale during June, but there were no systematic estimates of numbers at the main breeding locations, though a count of 25 birds came from one site at this time. During 2002, 28 pairs were reported in the uplands and in 2003 at least 25 bred. One observer visited 11 potential sites in the Teesdale area and noted a minimum of 21 broods. This was considered a 'poor showing', possibly related to the unseasonable weather during June of that year. Two further pairs were noted at sites in Weardale although no confirmation of breeding was obtained, an additional pair with two young was seen at Derwent Reservoir, in the north west. In 2006, 21 potential breeding sites were checked, and 12 of these were occupied. In 2010, seven females were observed with broods in the western dales, although this was felt to grossly under-represent the county's breeding population.

Overall, there seems to be no evidence to indicate any major reduction in numbers or a contraction in range for this important breeding bird within the county, and the population estimates given by Westerberg & Bowey (2000) should be considered as still valid. Over the summer period, small numbers of birds are also sometimes noted at lowland sites in the county, but there is little evidence of breeding at any of these. For example, around Teesmouth, Wigeon can be found in every month of the year, albeit in very small numbers in June and July, yet there has been no indication of breeding to date (Blick 2009).

Recent non-breeding status

Over the second half of the 20th century, there was a gradual increase in the number of Wigeon wintering in the county. This seems to be due to a variety of factors, notably the reduction in shooting pressures and its consequent disturbance, and widespread habitat improvement.

Although the largest numbers were historically on the coast, particularly at Teesmouth, the inland reservoirs and lakes as far west as upper Teesdale and Lunedale have on occasions held several hundred birds. In the pre-Temperley era, records for Teesmouth were generally in the low hundreds but 600 were present on 28 February 1954, 800 in October 1963 and 643 in February 1977. However, inland reports from Derwent Reservoir referred to numbers comparable to those on the coast, with up to 400 being present in January 1969, October 1972 and January 1974. Hurworth Burn Reservoir, Crookfoot Reservoir and Witton-le-Wear also held winter flocks but the Lunedale and Baldersdale reservoirs frequently held significant numbers, with about 200 being present during January-March 1975. Selset Reservoir held 72 on 22 September 1981 which may have been a reflection of the heavy coastal passage at the time. Derwent Reservoir's record count came in 1992 with 434 in February of that year. In 1997, 200 were at Witton-le-Wear and 225 at Hury Reservoir.

The county's most important wintering area is still around Teesmouth, centred upon Seal Sands and the North Tees Marshes. During the 1980s numbers there were increasing with the high hundreds being frequently recorded, such as 973 in November 1984. Numbers continued to increase during the 1990s and a total of 2,922 were recorded on the whole of the Tees estuary in January 1995 and 2,480 in January 2008, although the usual winter maximum was between 1,200 and 1,500 (Blick 2009). In 2007, the North Tees WeBS count of 2,366 in December was the highest since 1996 and the wintering population levels demonstrate the importance of this site in the north east.

The improved fortunes of the Wigeon were also reflected inland with 299 in January 2004 at Balderhead Reservoir and 350 in December 2009. Selset Reservoir also attracted 175 in October 2004. Also in 2009, 136 were at Grassholme Reservoir and 72 at Selset Reservoir. Longnewton Reservoir attracted 519 in January 2009. Bishop Middleham is becoming increasingly important for the species with 520 in January 2008. Hurworth Burn

Reservoir hosted 358 in December 2004 and Boldon Flats attracted 300 in February-March 2005. The range of still waters attracting Wigeon is considerable, with many sites now individually attracting greater numbers than did Teesmouth fifty years ago.

Numbers of this species can vary quite considerably from one winter to another, depending on the weather conditions further north and east and also on other factors, such as the amount of winter forage.

Coastal passage is prominent from September to November, and to a lesser extent from late August to early December. During strong onshore autumn winds, large numbers can be seen passing coastal watch points such as Whitburn Observatory and Hartlepool Headland, several hundred frequently being recorded on good days. The phenomenon of coastal passage has been well recognised within the county with a large passage of birds off Hartlepool on 22 November 1969, when 1,449 were recorded moving north (Blick 2009). On 3 November 1984, at least 1,000 passed Hartlepool Headland and on occasion, during such movements, large flocks settle on the sea, where 500 were seen off Seaton Carew. In 2007, sea passage of Wigeon was noted off Whitburn Observatory on a total of 65 dates, the majority occurring between 8 August and 27 November. An August total of 288 included 140 on 21st, but September was the peak month with 3,392 logged on 21 dates. The best days of passage were 348 on 11th, 295 on 26th, 1,527 on 27th and 609 on 28th.

Distribution & movements

The species breeds from Iceland, through the Faeroe islands and Northern Britain, across the boreal zones and sub-Arctic zones from Scandinavian eastward across to Siberia. The British wintering population largely originates from Siberia and Scandinavia, with some birds from Iceland.

British breeding birds are believed to be largely resident, although there is little objective evidence (Cramp *et al.* 1977). Those birds breeding in Fennoscandia and the USSR migrate in autumn to wintering areas in Europe, including Britain. Post breeding moulting flocks are encountered in Estonia, southern Sweden, Denmark and the Netherlands (Cramp *et al.* 1977). Birds leaving these moulting areas may move west or north west to their wintering grounds. This would account for the significant coastal passage in September and October along the Durham coastline, often in a northerly direction.

Despite the bulk of sea passage being recorded in late September, numbers do not significantly increase on the county's wetlands until early November, suggesting that many of the birds that pass our coast are destined for other wintering areas. This also seems to be the case for some birds which have bred locally. A bird ringed as a pullus on Stainmoor, in the far south west of Durham, in July 1979 was shot at Southport, Lancashire on 11 October 1980 (Baldridge 1981).

American Wigeon
Anas americana

Rare vagrant from North America with 12 records, involving 14 individuals, between October and June.

The first record for County Durham is of an adult drake found by Richard Wakely, at the mouth of Greatham Creek, Teesmouth on 4 December 1973. Although present for much of the day, feeding amongst wintering Wigeon *Anas penelope* on Seal Sands, it was not seen subsequently (*British Birds* 67: 316).

All records:
1973	Seal Sands, Teesmouth, male, 4 December
1985	Dorman's Pool, Teesmouth, male, 2 June (*British Birds* 79: 534)
1986	Long Drag and Cowpen Marsh, Teesmouth, male and female, 26th to 29 May (*contra BB*) (*British Birds* 80: 526)
1988	Saltholme Pools, Long Drag, Seal Sands and Greatham Creek, Teesmouth, male, 5th to 7 May, and 29 August to 5 February 1989 (*British Birds* 82: 514)
1990	Greatham Tank Farm, Teesmouth, first-winter male, 11th to 16 March (*British Birds* 84: 459)

1994	Cowpen Marsh & Greatham Tank Farm, male & female, 30 April (*British Birds* 88: 501)
1997	McNeill Bottoms and Low Barns, Witton-le-Wear, male, 8 February to 25 March (*contra BB*) (*British Birds* 91: 465)
1997	Seal Sands, Teesmouth, male, 4th to 9 October (*British Birds* 91: 465)
1998	North Gare, male, 19 April (*British Birds* 92: 563)
2000	Cowpen Marsh and Saltholme Pools, Teesmouth, female, 6th to 7 May (*British Birds* 94: 461)
2006	Boldon Flats, East Boldon, eclipse male, 22nd to 29 October
2010	Saltholme Pools, Teesmouth, male, 19th to 25 April

Additionally, a drake showing some of the characteristics of a hybrid Wigeon x American Wigeon was at McNeil Bottoms on 20 February 2000. The majority of records in County Durham have been in spring, with two autumn and two winter occurrences. This is consistent with the northward dispersal of birds that have crossed the Atlantic previously and wintered further south in Britain or Europe. The two autumn records both relate to adult drakes in October amongst Wigeon and are perhaps most likely to be earlier vagrants heading south rather than transatlantic arrivals.

Distribution & movements
American Wigeon breed through much of the northern United States, Canada, and Alaska, wintering mainly on the Atlantic and Pacific coast of the United States, Central America and northern South America. An annual and increasingly regular vagrant to the British Isles, it was removed from the list of species considered by BBRC in 2001, by which time there had been over 350 records.

Gadwall
Anas strepera

A rare breeding bird and a scarce, but increasingly common, passage migrant and winter visitor.

Historical review
This rather subtly-plumaged yet dapper duck is today present throughout the year in Durham, though it was once a scarce passage migrant and winter visitor. The Gadwall was first recorded in County Durham on 18 February 1843 when a single was shot near Stockton-on-Tees (Hogg 1845). At this time, and for much of the 19th century, it was considered only a rare wintering species in Britain (Holloway 1996).

Hancock, in 1874, detailed no Durham records, whilst Tristram (1905) did not list it at all, indicating just how scarce it was at that time. Abel Chapman recorded seeing six on Boldon Flats on 26 March, but gave no year, although this was prior to 1889 (Chapman 1889). The species was not recorded again until October 1896, when three birds were shot at Saltholme Pools, Teesmouth, by a punt-gunner. In the 20th century one was at Saltholme in September 1929, but clearly Gadwall was a rare bird in the county over this period.

A drake was at Jarrow Slake on 20 May 1945 and in the same year a pair remained on a flooded field in north east Durham, which may have been Whiteleas Pond, from 27 April to 2 May. A drake was at Jarrow Slake in April 1946 and another at Hebburn Ponds in October 1947. On 13 April 1948 another pair was at Teesmouth. Inland, a flock of eight birds was seen at Waskerley Reservoir in April 1947, at that time, the largest number recorded for the county. Four were at the same site in April 1948, eight there again on 6 February 1949 and eight in 1950 (Temperley 1951), indicating either a small scale spring passage through the county or local releases of birds, perhaps by wildfowlers.

By the mid-20th century, Temperley (1951) was able to say that this was a *"very rare winter visitor"*, *"now showing a tendency to remain until the spring"*. Before 1948, there were only five records of this species at Teesmouth (Stead 1964).

Prior to 1950, this was a scarce bird in Durham, but between 1951 and 1973 many more were seen, which seemed to reflect the general increase in the number of Gadwall breeding in Britain at that time (Gibbons *et al.*

1993). On 24 August 1953, David Simpson found a party of eight birds at the Shotton Brickworks Pond and, on 31 October 1953, he found a dead female in Shotton Brickyards.

In 1962, it was reported that 900 Gadwall had been released by wildfowlers in the Lake District (Mather 1986) and birds may have spread from there across the Pennines and into Durham. Similarly, in 1964 birds were reported to have been released by wildfowlers on Cowpen Marsh (Blick 2009).

Despite such activity, throughout the 1960s, the Gadwall remained scarce in Durham, with the next documented record being of one at Crookfoot Reservoir on 3 February 1968 and three at Tanfield Ponds on 31 August of that year. In 1969, there were only a handful of records, with a pair at Teesmouth on 12 May and one there on 2 September. At Hartlepool, four flew by on 22 November, the first observation of coastal passage in the county, and one was present on 20 December, both rare examples of passage birds in the north east region.

During the 1970s, birds were increasingly noted, but were still rather scarce. A major shift in status occurred in 1970 when the Gadwall bred at Derwent Reservoir; a duck with ducklings being seen there in summer, the first breeding record for the county (Kerr 2001). The next few years saw an increase in the number of reports, principally around the passage period of April. At Teesmouth, regular spring and autumn birds were recorded from 1974, with counts in single figures until wintering numbers soared, reaching 31 in 1988 (Blick 2009).

Two birds were at Witton-le-Wear in the autumn of 1973, with birds still being present in December, culminating in the next successful breeding record for the county in 1974. Gadwall bred again at Low Barns (Witton-le-Wear) in 1975 (10 young seen on 4 June) and again in 1976 (a female seen with chicks), though these appear to have been isolated incidents and did not lead to the establishment of a regular breeding presence. In 1977, full-winged captive birds were known to be 'at large' around the WWT Washington area. On Teesside, Gadwall numbers continued to increase from 1974, and today there are birds on the North Tees Marshes throughout the year, and regular records from many other wetland sites around the county.

It is clear that there has been a very considerable change in the status of this species over the last thirty years. An example of how this developed can be taken from the well-watched Gateshead area during the decade of the 1980s (Bowey *et al.* 1993). Recent expansion is thought to be largely natural and not related to the release of birds (Joynt *et al.* 2008).

Recent breeding status

Gadwall first bred in Yorkshire in 1954 and had increased to at least six pairs by 1980 (Mather 1986) whilst breeding was first confirmed in Northumberland in 1965 with three pairs being present there in 1983 (Kerr 2001). Across the British Isles, the status of this species has changed dramatically during the late 20th and the early part of the 21st centuries. It very considerably expanded its range from its base in East Anglia into Lancashire, Cheshire and Anglesey during the 1980s (Marchant *et al.* 1990). The bird remains a lowland species in Britain, as in Durham, and it prefers vegetation-rich, eutrophic waters (Gibbons *et al.* 1993). The available breeding habitat for this species within the county is quite limited but there is still much room for expansion of the existing breeding population.

At the time of the production of the *Atlas*, the mapped distribution was believed to be a largely accurate illustration of the species' limited breeding distribution in the county, mostly relating to birds of feral origin in the Washington area. The first indication of breeding in that area came in 1987, when five pairs raised 32 young, no doubt their ancestors had originated at the WWT facility. Barmston Pond recorded its first breeding pair in 1991 and there were up to six pairs breeding at WWT Washington in 1993; the highest total there during the 1990s. Since then the species has spread to a number of new areas across the county and it is uncertain as to whether this is the result of the single, Washington-based, population, or expansion from other areas. Between 1987 and 1996 a total of 273 young were known to have been raised in the immediate Washington/Barmston area.

Breeding probably occurred on a pond in Cleveland in 1993 (Blick 2009), but was first confirmed on the North Tees Marshes in 1994, Gadwall rearing three broods there for the first time. This increased gradually to eleven pairs by 1999, with a corresponding increase in the number of locations occupied. Breeding was confirmed in every year of the first decade of the 21st century, after 2001. In the early 21st century, there was a concentration of breeding birds around the North Tees Marshes, at the Reclamation Pond, Dorman's and Saltholme Pools, with other groups at Haverton Hole and Cowpen Marsh (Joynt *et al.* 2008). The species has now spread as a breeding bird to much of the eastern part of the county, though breeding is mainly restricted to four core areas: Bishop Middleham, Hurworth Burn Reservoir, Seaton Pond and the North Tees Marshes. In 2007 and 2008, breeding was confirmed at Bishop Middleham, Hurworth Burn Reservoir and the North Tees Marshes. Prospective pairs were also present at Barmston Pond, Rainton Meadows and near Waldridge Fell, although there was no evidence of successful breeding at these sites. The total number of breeding pairs has continued to increase at Teesmouth, with between 15 and 25 present on the North Tees Marshes in the first part of the 21st century (Joynt *et al.* 2008). Breeding numbers are still relatively low, with less than 30 pairs recorded annually.

Clearly the suggested small breeding population estimate of "*up to eight pairs*", as stated by the *Atlas* (Westerberg & Bowey 2000) is now out of date. Since the end of the survey period for that publication the species' summer presence in Durham has increased dramatically in particular at Teesmouth, with other breeding birds established in the Hetton/Houghton area and also around Bishop Middleham. The most up to date national population estimate (Gibbons *et al.* 1993) was of almost 800 pairs but this is probably now also an underestimate. The Gadwall is known to be sensitive to autumn shooting pressures on its main sites (Gibbons *et al.* 1993). In County Durham the main breeding and wintering areas are protected or largely free from shooting although wildfowling still occurs at Cowpen Marsh, where small numbers of Gadwall are shot annually, but this is not likely to be of major concern.

Recent non-breeding status

Today, the species' principal wintering station is around the North Tees Marshes, from October through to March or April. Gadwall continued to be a relatively scarce passage migrant until the 1990s, but from that time onwards there has been quite a dramatic increase in the number of wintering birds.

In the early 1990s, the Washington and Barmston Pond area was the major stronghold in Durham, with a record count of 70 birds there in September 1991. However, by 1992 a major distributional shift had taken place and numbers dropped in the Washington area and increased at Teesside. By 1999, only seven birds were recorded at Barmston compared to 129 at Teesmouth, an increase of 72% from the previous year's maximum.

By 2007, birds were recorded at 26 different sites around the county and from the late 1990s, wintering numbers at Teesmouth rose steeply from maximum counts of 108 in 1998, to 332 around the North Tees Marshes in November 2003, with 422 in October 2006. Most of these birds were in the area of the Reclamation Pond, peaking at 511 in November 2005 (Joynt *et al.* 2008). By 2009, the nationally important site threshold was 171, a number exceeded each year, which qualified Teesmouth as the 10th most important site for this species in the UK (Blick 2009). Teesmouth's future wintering numbers might be adversely affected by the loss of favoured sites in that area. The Tees WeBS data for autumn 2009 peaked at 480, the highest recorded for that survey, but the Bishop Middleham wetlands were also increasingly important as a wintering site around this time.

In County Durham, birds are infrequently noted on spring passage, though this phenomenon is becoming slightly more regular. In 2007, four flying north past Whitburn Observatory on 22 December was the only record of coastal passage during that year. In autumn, a few birds pass coastal headlands, in particular Hartlepool, during the months of October and November. Despite the recent increases very few Gadwall are seen passing at sea and no more than nine per day have been recorded in recent years. Autumn arrival is mainly from September, as evidenced by an increase in numbers at the Teesmouth sites. Passage birds in the region are presumably from the Scottish or Icelandic populations.

Distribution & movements

The Gadwall breeds across North America and Eurasia in a Holarctic fashion, though probably not in the UK until around the middle of the 19th century. It can be found from Iceland and Scandinavia in the north, and the species' breeding range extends further south than almost any other dabbling duck. Wintering and breeding populations have increased dramatically since the 1960s and this has been, in part, supplemented by the release of captive bred birds. The origin of the Durham birds is unclear, though the early pattern of records is consistent with migrants of wild origin. However, many breeding birds may well be derived from captive stock.

Teal (Eurasian Teal)
Anas crecca

A common winter visitor, much less widely encountered as a breeding bird in summer, mainly in the western uplands of the county.

Historical review

In the records from the Cellarers of Durham Cathedral in the 14th century, there are just five mentions of this species, which is variously referred to as *"teill", "teles"* or *"tells"*. From this it has been deduced that *"it was evidently nothing like as numerous as the Mallard"*. Nevertheless, the species was clearly present in the county at this time and it was being caught for food, with specific reference to its purchase for eating at Christmas, which in 1348 included: 63 Mallard *Anas platyrhynchos*, Teal, Curlew *Numenius arquata* and Plover *Vanellus vanellus*. In the Cellarer's roll for Christmas week 1344, Teal were purchased alongside six dozen Skylark *Aluada arvensis* (Ticehurst 1923).

In the latter half of the 19th century, Thomas Robson (1896) reported the species as being *"commonly met with on Shibdon Flats, also on the river at Lintzford"* (Bowey *et al.* 1993). The species has probably always been present as a breeding bird in reasonable numbers in the county's western uplands, and it was documented as breeding in most counties of Britain and Ireland during the 19th century (Holloway 1996). Though it has never been that well recorded in Durham there are reports of it being known in this context by moorland keepers and the like (Chapman 1889). Before extensive land claim reduced the size of the Tees estuary, Teal did not appear to be that much more numerous than they are in the modern context. The highest recorded bag prior to the 20th century was of 23 shot in one day in September 1863 (Nelson 1907). This record, taken from the diary of an old punt gunner at Teesmouth (Mather 1986), shows a relatively conservative number compared to some shooting totals.

Temperley (1951) referred to it as a *"common resident; also a passage migrant and winter visitor"*. He thought this to be Durham's commonest species of duck, after the Mallard, breeding *"freely on the moorlands of the west of the County; wherever there are rushy pools, marshes and streamlets"*. He highlighted that in the east, *"it nests in a few watery places where it can be reasonably free from molestation, such as reservoirs, ponds and marshes along the banks of rivers"*. In the south east of the county, a nest with eggs was found at Cowpen Marsh on 28 April 1928. This was the first documentation of breeding at Teesmouth, on the Durham side of the river. Much later, Stead reported breeding on three occasions at Cowpen Marsh *"in recent years"* (Stead 1964). Breeding also occurred at the same location in 1966 (Joynt *et al.* 2008). Inland, two broods were at both Derwent Reservoir and Middleton-in-Teesdale in 1968, and five broods were at the former site in 1970, with two pairs breeding on Pikestone Fell in 1972 and up to eight broods at Derwent Reservoir in 1977. Through the 1970s breeding was suspected in the Derwent valley, Weardale and Teesdale; being confirmed in the latter area, when a female was seen with five chicks near Middleton Common in June 1978.

Temperley (1951) also documented flocks off the Durham coast during daylight, in the same manner as Mallard and Wigeon *Anas penelope*, passing down the coast in September and October and large winter influxes *"of birds from abroad"*. However, records for the 1960s and 1970s refer to counts in the low hundreds, e.g. 310 at Teesmouth on 22 February 1964. Teal became increasingly common from the late 1970s onwards at Teesmouth, with annual peaks including 857 in 1976.

Recent breeding status

Teal frequent areas of open moorland with small pools or marshy ground (Fox *et al.* 1989), breeding at low density and requiring high water quality with little pollution, as well as freedom from disturbance (Owen *et al.* 1986). The breeding distribution of the Teal in the county is imperfectly known, but it is largely concentrated in the west. Targeted effort during the development of the *Atlas* (Westerberg & Bowey 2000) confirmed this observation and demonstrated that breeding was largely concentrated along the upper catchments of our rivers and around the associated moorland tarns. The nest is usually in thick cover, well-hidden and often near water (Fox 1986). Small young are adept at hiding in the dense vegetation along the moorland streams, small water bodies and reservoirs that are so common in the Durham uplands. Nesting sites in Durham can be in quite remote areas, breeding birds are usually shy and secretive and, as a consequence, many potentially breeding birds will go unrecorded, even by experienced fieldworkers. In the local context, Teal nests are prone to predation from ground predators such as

135

mustelids (Westerberg & Bowey 2000), and some studies indicate that its habit of nesting close to water may help to reduce predation from such threats (Fox 1986).

Obtaining confirmation of breeding is very dependent on observer effort and as a result breeding has only been recorded in 25 of the past 40 years. These records have come mainly from upper Teesdale, Weardale and the upper Derwent valley. In 1980 birds were reported from nine moorland sites. In 1993, 15 pairs were noted, but only three broods were seen. Rarely are rivers specifically mentioned, but in the 1990s Dennis Luckhurst recorded breeding annually on the River Greta, until his untimely death in 1997. The highest annual count was of seven broods totalling 55 young in 1992. Breeding pairs were at Grassholme Reservoir in 1994 and 1996. A clutch of seven eggs was found on Cotherstone Moor in 1997 (Dennis Luckhurst pers. comm.). Breeding was again thought to have taken place on the River Greta in 2009. No doubt breeding took place in the intervening years, but this emphasises the importance of observer cover in establishing the true status of any species. Elsewhere, a brood of five was in Eastgate Quarry in 2005, three young were noted at Derwent Reservoir in mid-July of the same year, and a family party was at Smiddyshaw Reservoir in June 2007.

The Teal is generally thought to be undergoing some decline in Britain, with a considerable contraction of range documented between the two national atlas periods (Sharrock 1976, Gibbons *et al.* 1993). The national population estimate for breeding Teal, as made by (Gibbons *et al.* 1993) was between 1,500 and 2,600 pairs. Westerberg & Bowey (2000) determined that at three to five pairs per occupied ten-kilometre square in Durham the population estimate was between 33 and 55 pairs for the upland areas of the county. Consultation with fieldworkers familiar with the species indicated that this was likely to be an underestimate and 'rounding up' was used to state that the county held some 40-60 pairs.

Mostly, records of 'summering birds' on wetlands below the county's 300m contour probably refer to non-breeding birds. However, in lowland situations, small numbers of birds do occasionally breed, but probably at very low densities. This is exemplified in the south east of the county, in what was north Cleveland. In this area, with its extensive wetland habitats and high density of observers, there were just twelve confirmed breeding records between 1975 and 1995, with a number of these being on the North Tees Marshes (Bell 1976-1996). On 1 July 1989, a duck with one duckling was seen near the mouth of the Tees, and a duck with five young was at Teesmouth in early August 1991. A brood of ducklings was noted at Teesmouth in July 1995. There were no further documented successful breeding records between 1995 and 2005. Thereafter breeding was confirmed by the survey work for *The Breeding Birds of Cleveland*, a female with two ducklings being noted on 11 August 2006 at Haverton Hole (Joynt *et al.* 2008). Other lowland examples include the pair that successfully raised young at Saltholme Pools in 2007, and breeding records at Shibdon Pond and Haverton Hole in 2008 and 2009. The species had long been suspected of breeding at Shibdon Pond, based on the evidence of small numbers of birds summering in most years, from the late 1980s and through the 1990s.

Recent non-breeding status
The North Tees Marshes consistently hold the highest numbers in the county through the winter period, but other wetlands occasionally produce impressive numbers. At times in winter, the Teal is the most numerous duck around Teesmouth. This area saw a significant increase in winter numbers from the late 1970s. The mean maximum in the 1970s was 474, in the 1980s it was 1,216 and in the 1990s it was 1,047. Numbers have declined in the last decade with a mean maximum of 848. However, this figure was considerably boosted by two very large counts in the winters of 2003 and 2009. Peak counts at Teesmouth were 1,753 in November 1982, 1,829 in October 1984, 1,446 in January 1994, 1,436 in January 1995 and 1,495 in October 2003. Since 2000, the Bishop Middleham complex of wetlands has become increasingly significant, with numbers regularly peaking in the high hundreds, with 1,000 in October 2002, 800 in March 2004 and 780 in December 2009. This site is, however, prone to disturbance and the effects of fluctuating water levels.

Many other sites, which hold Teal on a regular basis, can also play host to significant numbers on occasions. Shibdon Pond held 397 in September 1996, a site record; 353 were at WWT Washington in October 1992; 386 were at Dunston in December 1996 and the Gateshead area, including the Tyne at Dunston, Elswick and Shibdon Pond, held up to 600 in January 2001. Boldon Flats which normally hosts a maximum of 150 birds, held 215 in November 2006 and in the same year Derwent Reservoir held 300 in October. Numbers at some sites vary over time. For instance, in 2000 it was observed that WWT Washington's wintering population had fallen by 62% in just four years. This site was once second only to Teesmouth in importance for wintering birds in the county, with up to 509 recorded there in November 1990.

The lower reaches of the rivers Tyne and Wear often held 20-30 through the winter periods, and small flocks can also turn up on urban streams and ponds, such as the River Don in Jarrow and River Team in Gateshead. Jarrow Slake, into which the Don flows, and which historically was a very important wetland site before reclamation, held 280 in December 1991. The upland reservoirs, little frequented by observers in winter, have held birds in recent years with 60 at Selset in December 2008 and 46 at Grassholme in January 2008.

Some small scale coastal passage often occurs in June and July, and at this time birds are often reported from lowland sites such as Far Pasture, Hetton Lyons, Rainton Meadows, Shibdon Pond, Seaton Pond and Barmston Pond, as well as from along the coastal strip. From mid-July onwards coastal passage is evident. The July passage is probably of failed breeders. The main passage through Europe occurs during October and November (Cramp *et al.* 1977). Recent records for County Durham however, indicate significant passage during August and September. At Teesmouth numbers peak in September then remain high, but slightly lower, during October to January.

Winter distribution and movements are heavily dependent on weather (Cramp *et al.* 1977). A decrease in numbers occurs from February to April which indicates return passage to the breeding grounds. Numbers are then very low, typically less than 30 at Teesmouth during May and June, until they start to increase significantly in August. Although numbers are smaller, a similar pattern occurs at Derwent Reservoir, albeit with numbers peaking in October. Numbers do tend to fall thereafter through the winter to about 60% of the October figure in December and January, before falling further to leave summering and breeding birds from April onwards.

Passage of this species along the Durham coastline, mainly between early August and mid-November, is predictable in terms of timing but variable in its scale from year to year. In 2007, the species was recorded as passing along the coast on 88 dates during the year, from the Whitburn Observatory; the total number of birds noted over the year amounted to 5,564. The equivalent figure for 2008 was 5,994 on 93 dates. The vast majority of this activity was concentrated in the period August to November, with just 92 birds of the 2007 total noted in the first six months of the year. In 2008, peak passage occurred in the second half of August when 3,165 were logged, with 1,350 of these on 30 August. In 2009, 5,994 were recorded on 93 dates, with 3,547 of these occurring in the first two weeks of September. The peak was 1,504 on 3 September. Subsequently numbers fell with 539 on 30 September, 228 on 10 October and 106 on 22 October. The difficulty in judging the significance of these numbers is reflected in the record of 1,120 flying north and 435 flying south between 3rd and 16 September at Hartlepool, suggesting that some of these movements may just be local. Some of the highest counts of birds passing Hartlepool Headland include 381 on 22 November 1969 and 720 on 25 August 2003 (Blick 2009). The last three months of the year tend to show a decline in movements of the species, even on days when there are significant movements of other wildfowl.

Distribution & movements

The Teal in its nominate form breeds across Eurasia as far east as Siberia, and beyond this is represented by a different race on the Pacific Aleutian Island. Its breeding range is not as far north as many wildfowl species and occurs more in a mid-latitudinal belt. European breeding Teal winter largely around the North Sea, with large numbers concentrated around the Baltic and in the Low Countries (Cramp *et al.* 1977).

Older ringing returns indicate that birds visit Durham from Sweden, Denmark and Holland. More recently, one ringed near Copenhagen, Denmark, on 10 March 1992 was found dead under wires at the Reclamation Pond on 2 March 1998. A female ringed at Seal Sands on 11 October 1994, was shot on 20 August 1995 at Turku-poro 1,627km to the north east of its ringing location, thus adding Finland to the list of source countries. Not all wintering birds in the county's wetlands come from such far flung destinations; a bird ringed at Shibdon Pond in 1988 was killed by a mink during the breeding season, at Kielder Burn in Northumberland in 1991. The earliest known ringing recovery in the county, a bird ringed at Longtown, Cumbria, on 3 March 1925 that was recovered on the River Wear, somewhere between Sunderland and Durham on 21 January 1928, indicates that at least some birds move east-west and west-east, presumably along winter temperature gradients, across the north of England. Some birds travel much further to the south west in winter, as was shown by a female bearing a nasal saddle that was recorded on the Long Drag in October 2008, which proved to have been ringed at San Jacinto Dunes Reserve in Portugal on 10 February 2008 (Blick 2009).

Green-winged Teal
Anas carolinensis

A rare vagrant from North America; 18 records, all between November and June.

The first record for County Durham is of a drake found by Peter Salmon at Hurworth Burn Reservoir on 23 December 1973, which remained until 28 January 1974. At the time of the occurrence this species was still considered to be a race of Teal *Anas crecca* (*British Birds* 67: 316).

All records:

1973	Hurworth Burn Reservoir, 23 December to 28 January 1974
1982	Crookfoot Reservoir, 6 November (*British Birds* 104: 627)
1985	Jarrow Slake, 21 February (*contra Birds in Durham 1985*) (*British Birds* 79: 535)
1985	Long Drag, Teesmouth, 31 March to 1 April *(contra BB)* (*British Birds* 79: 535)
1986	Greatham Tank Farm, Teesmouth, 20 April to 1 May *(contra BB)* (*British Birds* 80: 527)
1993	Dorman's Pool, Teesmouth, 18 April
1995	Greenabella Marsh, Teesmouth, 15th to 20 May
1996	Crookfoot Reservoir, 3rd to 5 April
1997	Crookfoot Reservoir, 7th to 13 April
2002	Long Drag, Teesmouth, 19th to 20 May
2003	Dorman's Pool, Cowpen Marsh and Saltholme Pools, Teesmouth, intermittently, 26 May to 23 June
2003	Greatham Creek, Teesmouth, 3rd to 20 June
2004	WWT Washington, 29 February and 3 March, same Barmston Pond, 4 March
2005	Shibdon Pond and Far Pasture Wetland, Gateshead, 27 February to 9 March, presumed same Shibdon Pond and Far Pasture Wetland, Gateshead, 29 January to 18 March 2006
2008	Saltholme Pools, Teesmouth, 6 March
2009	Bowesfield Marsh, 4th to 11 February
2009	Greenabella Marsh, Teesmouth, 7 March to 21 April
2011	Bowesfield Marsh, 7 February to 1 March

All records refer to males. The exact number of records is clouded by potentially returning individuals, though it is notable that like many other vagrant Nearctic wildfowl in the county there is a distinct peak of records in spring, with relatively few mid-winter sightings. These birds have presumably wintered further south and are migrating northwards amongst Teal with the onset of spring, having crossed the Atlantic some time previously.

Distribution & movements

The Green-winged Teal is a common breeding bird across Canada and much of the United States, wintering mainly in the southern USA, Central America, and the Caribbean. Once considered to represent the North American race of Teal, it has been considered a distinct species by the BOU since 2001. Hybrids between the two species have been recorded on both sides of the Atlantic. An increasingly common vagrant to Britain, it was removed from the list of species considered by the BBRC at the end of 1990 by which time there had been over 300 records. Some twenty years later this figure had more than doubled, with the species averaging around 20 records per annum.

Mallard
Anas platyrhynchos

A very common resident and numerous winter visitor.

Historical review

"The Mallard is the most widely distributed species of wildfowl in England, and is instantly recognisable to most people, thriving in all types of wetland, from the smallest farm pond to the largest of reservoirs" (Grice & Brown 2005). It was probably domesticated by the Iron Age (Harrison 1988) and in Durham it has a long and distinguished history, especially as a bird for the table. As far back as the 14th century, the Cellarers' Rolls for the Monastery of Durham give numerous entries of the bird being purchased for clerical consumption, indicating that the Mallard was common and easily available at that time. For Christmas 1348, no less than 63 Mallard were obtained, along with Teal *Anas crecca*, Curlew *Numenius arquata* and Plover *Pluvialis sp.* The records are careful to distinguish between the wild Mallard, *Maulard de Ryver* and the domestic duck, the latter simply being listed as *anales*. Benett writing later notes that a *"mawlert"* was bought in 1533 for 2d (Gardner-Medwin 1985).

As a breeding bird, the Mallard's occurrence in County Durham was noted as long ago as 1840, when Hutchinson stated that *"Common Wild Duck and Teal were the only species of duck that breed in the County of Durham"*, though clearly, as documented above, the species had been present for much longer (Bell 2008). According to Hancock (1873), the Mallard's *"nest is to be found wherever there is a suitable locality in the wilder districts of the two counties"*, and it was, no doubt, a widespread breeding species throughout the 18th and 19th centuries. There is however, little specific documentation for this period. Robson (1896) noted *"it nests in Axwell Park and Gibside"* and *"could be met with at Shibdon Flats"* and *"on the Derwent"*.

In the 20th century, Temperley (1951) thought the Mallard *"the commonest breeding duck in the county"* and also noted large winter gatherings along the coast. *"Large flocks spend the daylight hours afloat off shore, flying in to feed at dusk. These flocks may be seen even off the busiest ports, as in the mouth of the Tyne, Wear and Tees."* He documented a count of 1,700 birds at the mouth the Tees in December 1945, with a total of 4,000 counted along the Durham Coast in December 1947. On 28 February 1964 a new record count of 1,750 was set at Teesmouth, mainly on Seal Sands, a figure repeated on 17 November 1968 (Blick 1978). These large winter gatherings however, have now become a thing of the past (Joynt *et al.* 2008). There is still a noticeable winter influx, though not as remarkable as before. The Mallard has also been released for sporting and ornamental purposes in County Durham for some generations. Although there are few details of such events, mention is made of hand-reared birds being released on Billingham Pond in 1966, and also of 200 birds released for shooting at Seaton Pond each year during the 1990s (Bell 2008). Temperley noted that winter coastal flocks were extremely wary, due to frequent shore-shooting and commented that wild birds often obtained sanctuary on ponds in suburban parks, a habit which may have influenced their tendency to breed in these areas.

Recent breeding status

The *Atlas* demonstrates that the Mallard has a very widespread breeding range throughout the county, utilising a variety of wetland habitats (Westerberg & Bowey 2000). As Temperley states it can be found on lakes, ponds, rivers and marshy areas at all times of the year. Today it probably breeds on almost every wetland, regardless of size, from the western moorlands down to the coastal fringe, including all local reservoirs. Birds are found on almost any area of open water (Gibbons *et al.* 1993), no matter how small, although slow-moving or still water sites are preferred. Notable concentrations occur at locations where open shallow water is bordered with significant areas of marginal vegetation. The best breeding sites have islands and indented shorelines with tall vegetation and insect rich shallows (Giles 1992). Fledgling success is closely related to the abundance of hatching chironomids (Brown & Grice 2005). Many Mallards breed around the fringes of urban parks and similar small wetlands, including garden ponds. Indeed, taken together, these areas may hold more breeding pairs than any other habitat in the county (Blick 2009). The adults have an extremely varied diet, they can up-end, dabble or even dive for short periods to obtain seeds and invertebrates; they may grub around on land far from water and also take food from the hand (Brown & Grice 2005). Furthermore, some birds nest at a considerable distance from water, often among crops such as oilseed rape.

The Mallard has become notorious for nesting in bizarre locations. One of the strangest reports was of a duck successfully rearing young in an old Magpie's nest at Rainton Meadows in 2001. The DBC annual report for 2007 also documents a nest situated in a plant pot on top of a spiral staircase two metres above a pond. In the same year successful broods were raised at Hardwick Hall in a half-barrel containing bedding plants, adjacent to a noisy marquee, with another nest 3 metres above ground on top of an ivy-covered wall.

Breeding activity is widely reported from numerous sites, e.g. Acer (Dunston) Pond, Chester-le-Street Riverside, Coxhoe Pond, Durham City Riverside, Holywell Hall Lake, Hetton Lyons CP, Far Pasture, Pockerley Farm Pond, along the River Derwent, the River Team at Urpeth, Ryton Willows, Shibdon Pond, Swalwell Pond, Castle Eden Dene, Joe's Pond, Hetton Bogs, Sedgeletch, Houghton Gate, West Pastures, Hurworth Burn Reservoir and especially WWT Washington. Breeding distribution shows a bias towards locations along river valleys, from the uplands to the coast. The pattern of spread is different to that of other wildfowl which tend to have a more patchy distribution within the county. In terms of breeding density, lowland areas undoubtedly illustrate a higher level of occupation than upland ones, although this may, in part, be a reflection of observer coverage. Birds breeding in the uplands are therefore probably overlooked and under-represented in the county records.

Adaptable and fertile, the Mallard can be found breeding at any time of the year from mid-January to November, when weather conditions allow, although the majority of breeding activity is concentrated between March and July. In 1978 ducklings were noted to emerge on 27 November at WWT Washington, with a similar occurrence in the same month in 2003. In the previous year two females with ten ducklings were seen at Sedgefield on 22 October.

More than 100 broods are recorded annually throughout the county (Armstrong 1988-1993), with many more going unreported. The fact that 1,300 eggs were laid in the grounds of WWT Washington during 1981 demonstrates just how common the Mallard is as a breeding bird in County Durham. With such a potential food resource available, levels of predation can be high. The Red Fox, *Vulpes vulpes* is heavily implicated in many losses. All but one of a late brood of 11 at South Church were taken by this notorious opportunist in just one night in September 2007. Although there are many observations of second and third clutch re-lays, duckling survival rates can be low. The Grey Heron *Ardea cinerea* is another implacable predator, taking many of the 73 ducklings from 11 broods at Sedgeletch in 2007. The destruction of eggs by Magpies *Pica pica* and other corvids can also be significant. Ducklings are not the only ones at risk during the breeding season; unpaired females are often relentlessly pursued by gangs of amorous drakes until they may succumb to exhaustion, suffocation or drowning (Joynt *et al.* 2008).

The *Atlas* indicated a minimum population of 200-300 pairs for the county, though considering the estimated 300 breeding pairs around Teesmouth and North Cleveland during 1999-2006 (Joynt 2006), this figure is probably too low and should be revised upwards to around 450-500 pairs in total. At present, no reliable information is available to indicate whether the local population is undertaking the long term increase experienced nationally (Marchant *et al.* 1990).

The Mallard is the basis of all European domesticated ducks and will inter-breed happily with a range of species. In 1991, a male Mallard hybridised with a female Red-crested Pochard *Netta rufina* rearing seven young at Charlton's Pond (Bell 1992). Such inter-breeding however, especially with domestic populations, leads to genetic variations in plumage and feather structure, which affect the natural phenotype and reduce the fitness of the bearers. Parkin & Knox (2010) suggest that selective removal of such individuals *"would probably benefit the gene pool"*.

Recent non-breeding status

The Mallard is the most common species of duck in the UK with an estimated 70,000 breeding pairs. In addition, the resident population is supplemented annually by the release of hundreds of thousands of birds for shooting purposes. Harradine (1985) estimated that 400,000 birds are released by shooters every year throughout Great Britain, with around 600,000 birds being shot. Post-breeding moult gatherings usually commence in July and can result in large counts from wetland sites around the county; for instance, 830 birds were noted at WWT Washington in July 2001 and 310 at Seaton Pond in August 2001. While our breeding birds tend to be sedentary, moving only to the nearest open water or to the coast during hard weather, the wintering population is augmented by large numbers of birds arriving from the Continent. The majority of these originate from Fennoscandia, Northern Russia, the Baltic States and Germany.

Counts at Teesmouth start to build up in September and October, but the significant increase does not occur until December and January when most migrant birds arrive. The highest counts at Teesmouth were 1,750 in 1964 and 1968 and 1,150 in 1976 and 1980. At Longnewton Reservoir, 820 were seen in September 1988 where 300-500 was the usual winter maximum. At Wynyard, 350-500 is typical, but a record of 930 was present in December 1978. About 150-300 birds can also be seen during the winter at many wetland locations around the county, such as Bishop Middleham, Derwent Reservoir, Shibdon Pond, Brasside Ponds and Charlton's Pond. An interesting observation was a December count of 61 birds on South Shields Marine Park Lake, a site record. At Teesmouth, the lowest numbers occur in April, May and September with less than 100 birds recorded; the number of adults present increases significantly in June and July, as birds start to use the area as a moulting site.

During the 1980s a major shift in the distribution of wintering birds seems to have occurred with a decrease in the number of birds using the North Tees Marshes and a corresponding increase in birds using WWT Washington, with a peak of 1,960 in November 1990, the highest record for the county at a single site (Armstrong 1991). However, by the early part of the 21st century, there was a general decline in the number of Mallard wintering across the whole of the UK, a change mirrored in County Durham. On the North Tees Marshes, the wintering population had reduced to less than 500 and at WWT Washington the average had dropped to around 600 birds. Counts for recent years include 398 at WWT Washington in January 2007 and 233 in 2008. Bishop Middleham, Hetton Lyons CP, Rainton Meadows and Brasside Ponds tend to hold between one and two hundred birds. The lower sections of the Wear can also support good numbers, with 285 on 28 November 2009 at Chester-le-Street.

Passage is often noted along the coast, at sites such as Whitburn, with peaks between late August and late December, and presumably includes some birds entering the UK to spend the winter. The species is almost always outnumbered by Wigeon *Anas penelope*, Teal, Goldeneye *Bucephula clangula* and Common Scoter *Melanitta nigra* during such movements. In most years, no more than 20-50 per day are seen even on days of heavy duck passage; 2006 however, proved to be an exception, when 247 were noted flying north during gales on 1 November.

Distribution & movements

In its nominate form, the Mallard breeds across Eurasia and North America, with a range of other less numerous sub-species being represented elsewhere. It is one of the most widespread breeding birds in the world. The Mallard also occurs as a passage migrant in Durham, in reasonable numbers. Many of these birds are of Scandinavian origin, although some are northern British birds moving south and west to their wintering grounds.

Significant ringing recoveries of Mallard in Durham include one ringed in West Flanders, Belgium, on 9 June 1951, which was found dead at Teesmouth two months later (Blick 2009) and another ringed as a juvenile on 24 June 1959 at Borviksbruk (Varmland) in Sweden that was recovered at Port Clarence on 18 November 1959. Ringing recoveries from Northumberland indicate that many of the wintering birds in the north east are of Northern Eurasian origin (Kerr 2001).

Pintail (Northern Pintail)
Anas acuta

A scarce winter and passage visitor and a very rare, intermittent breeder.

Historical review

This dabbling duck has long been a winter visitor to County Durham, today in relatively modest numbers. In the main, Pintail frequent the complex of wetlands around the North Tees Marshes; at present the most favoured location is Dorman's Pool. Blick (2009) suggested that it had probably never been very common, even prior to the extensive land claim around the Tees Estuary, but that it may have been a more regular breeding species in the past. A record that hints at a previous more widespread and possibly more common status for this species, confirming its long-term presence as a Durham bird, is indicated by the excavation of Pintail remains from the medieval strata of Barnard Castle (Yalden & Albarella 2009). Nelson (1907) suggests a rather different situation to Blick, writing *"Half a century ago, the Pintail was a numerous species in the Tees where, as George Mussell tells me, it was greatly sought after by the professional gunners, who would not trouble with other fowl if they could get the pintail, and, as it was most plentiful in May, and no restrictions were at that time placed upon shooting, great numbers of this delicious duck were procured and brought into market"*, quoted by Mather (1986). Nelson thought it was so sought after because it made *"good eating"*.

Despite this, at the beginning of the 19th century, several writers considered the bird rare. Hutchinson (1840) said, *"The Pintail is one of the rarer ducks of the County"*. He mentioned one obtained by him on 26 December 1838 which was sent to the University museum for preservation and this was the first specimen that William Proctor, who had many years experience of preserving birds, had seen. In the Derwent valley, Robson considered this species a rare visitor and the only example he quoted was of one shot in March 1894 at Dunston (Robson 1896).

Temperley (1951) reported that it was a *"regular winter visitor in small but now gradually increasing numbers"*. He observed that it was *"reported each winter though in very small numbers and often single birds only, not only from the Tees mouth marshes, but from various flooded brick-fields, ponds, reservoirs and similar places. The largest flock so far recorded was one of 22 birds seen at the Tees mouth in September 1948"*.

Temperley (1951) knew of no records of it having bred in Durham in the past, despite Tristram's (1905) assertion that it was *"said to have formerly bred in the County"*. Hancock (1874) suspected it of breeding at Prestwick Carr in south Northumberland, just to the north, during the 19th century (Kerr 2001). Breeding was proven there in 1916, 1918 and again in 1945 (Kerr 2001), so it is not inconceivable that birds may have bred at Teesmouth around that time. In 1941, two pairs were present at Teesmouth throughout the summer, although breeding was not proven. In 1944, two pairs remained into May, with possible breeding taking place, but it was not until 1954 that a single pair was proved to have bred on the North Tees Marshes. Successful nesting again took place in 1961, when a brood of nine ducklings was observed; however, this latter case may not have involved truly wild individuals, since between 1959 and 1965, a pair or two of feral birds had bred at Rossmere Park, Hartlepool. Although the original birds had their wings clipped, their young were free-flying (Blick 2009). The 1961 breeding episode may well have involved the offspring of these feral birds. Summering birds have been intermittently present in the Teesmouth area ever since this time (Stead 1964). The following year, 1962, a probable nest of this species on Cowpen Marsh was trampled by cattle (Joynt *et al.* 2008). The species again bred successfully in Durham, probably at Teesmouth, when a female was seen with five young on 10 June 1963.

The wintering population at Teesmouth also grew to 40-50 by 28 March 1965, and 53 were on Dorman's Pool on 10 March 1968, the largest figure ever recorded at Teesmouth. Through 1969, birds were recorded at Teesmouth in every month and in 1970 the winter population reached a plateau at 50 in early March, though 1971 brought a record wintering total of up to 60 on Dorman's Pool in March.

Recent breeding status

In the breeding season, Pintails prefer shallow, often ephemeral, wetlands which may revert to drier, more open, semi-natural habitats (Fox & Meek 1993). In much of County Durham therefore, the species' optimal habitat is in short supply.

After the indications of breeding between the 1950s and the 1960s and the continued growth of the winter population, by 1969, birds were recorded at Teesmouth in every month of the year with further speculation of breeding having occurred.   In both 1970 and 1971 a pair was present at Teesmouth in May, but there was no confirmation of breeding in either year. Through the rest of the 1970s, occasional individuals were recorded in the summer, for example, birds at Saltholme Pool and up to five on Cowpen Marsh during the summer of 1977, but the true status of this species as a rare but possibly regular breeding bird in the county, could never be fully ascertained.

The situation changed in spring 1985, when a female with two or three ducklings was located at a moorland reservoir close to the River Greta in the Pennine uplands. This was the first proven breeding of the species in the west of the county, but regular visits to the site on subsequent occasions did not record any further nesting (Fox & Meek 1993). During the *Atlas* survey period, summering (and therefore potentially breeding) birds were recorded, on the North Tees Marshes, around WWT Washington and at one locality in the Pennine uplands (Westerberg & Bowey 2000). In the west of the county, a pair of birds was present at the western end of Derwent Reservoir, during the spring of 1995 (K. Bowey pers. obs.), and a female was seen intermittently through the summer months, then a small flock of 'female-type' birds was noted in July of that year, probably a family party of full-grown juveniles.

Another instance of possible breeding during the period of the *Atlas* survey came from the vicinity of the WWT Washington, where two feral pairs bred in 1988 (Armstrong 1989).

In 2001, there was single spring report, at Dorman's Pool on 8 April and this location brought the only summer sighting, of an eclipse drake, which was present on 10 July. According to Blick (2009), it is possible that a pair attempted to nest more recently at Teesmouth, since one or two birds are quite frequently seen in suitable habitat during May, June and July. Sometimes birds linger well into spring and two were present in June and July 2008, but it is not certain whether such records relate to genuine breeding attempts. For example, a pair was regularly seen at RSPB Saltholme in mid-June 2009, but the outcome was unknown.

Gibbons *et al.* (1993) concluded that there was a British breeding population of just 30-40 pairs, and recent work has suggested something of a decline since then, with an average of 25 pairs reported to the Rare Breeding Birds Panel between 1993 and 2006 (Parkin & Knox 2010); County Durham's contribution to this total is unlikely to exceed one pair per annum (Westerberg & Bowey 2000). In summary, as stated by the *Atlas*, given the tenuous summer presence of Pintail in Durham, and the relatively small British population, it must be concluded that at best the county *"provides only marginal breeding habitat for this attractive duck"*.

Recent non-breeding status

In Durham, the majority of sightings of this species occur between mid-September and January, with peak occurrence in November. Most wintering birds arrive from late August and September and small numbers remain in the area until March, with most of these usually gathering around the North Tees Marshes. The most important location for the species early in the year has for some time been the area encompassing Dorman's Pool and Saltholme Pools at Teesmouth. The county's annual maximum counts usually occur there, often during November to December or February and March, the numbers in most years relating to 'scores' of birds but not much more. From 1999, the annual winter maximum at Teesmouth has steadily increased, with 44 there in 1999, 50 in 2000, 63 in 2001, 70 in 2002 and 73 on 17 February 2003, but there have been no more than 62 since 2003 (Blick 2009). Prior to this period, maximum winter figures through the 1980s and the early part of the 1990s had hovered between 15 and 30 birds in most years, other than the 60 seen on 7 March 1971, as noted in the historical review (Blick 2009).

Away from the main wintering area, records are sporadic and sightings from central and inland sites are always noteworthy, although Derwent Reservoir and Crookfoot Reservoir quite regularly attract birds. Most such records concern small numbers of Pintail, which tend not to linger. Occasionally, individual birds 'adopt' a local site, such as Boldon Flats, and seven there on 7 March 2010 was a good record.

Autumn usually brings renewed activity, with a very small scale sea passage along the coast observed mostly from the Whitburn Observatory and Hartlepool Headland, but at best, numbers are in single figures. The main period for sightings tends to be from early July through to October. Even the best passage days for this species may only provide day-totals of 10 to 15 birds, as from 26th to 28 September 2007 at Hartlepool, when 24 birds were logged over a three day period. Other Hartlepool counts have included 26 on 3 November 1984 and 32 there on 15 September 2000. More rarely, birds are noted on spring passage, such as the party of 11 that flew north at Whitburn Observatory on 3 May 2010.

Hybrid drakes, showing Mallard *Anas platyrhynchos* and Pintail characteristics have been seen on Charlton's Pond from December 1980 to at least January 1987 and on Crookfoot Reservoir in January 1998 (Blick 2009).

Distribution & movements

The Pintail has a circumpolar breeding distribution, which is centred further north than most other dabbling ducks, though unlike some species of geese, it is not found on the Arctic islands. The wintering grounds are to the south of the breeding areas, largely in western and southern Europe, but extending down into western Africa.

Just one marked bird has been seen in County Durham. In March 2011, a nasal banded female was at Rainton Meadows, and had been marked at Pontevedra, Galicia, north west Spain on 12 December 2010.

Garganey
Anas querquedula

A scarce passage migrant and rare breeder.

Historical review

Temperley (1951) described the Garganey as *"a summer visitor in very small but increasing numbers, which breeds occasionally in the county"*. Increased observer coverage since Temperley's day suggests that it is a regular passage migrant in small numbers, which breeds occasionally.

The first record for the county was documented by Hogg (1845), who said that a *"fine bird was shot in the marshes near the Tees in January 1829"*, a rather strange date for this species, which is a summer migrant to Britain, though winter records are not totally unprecedented. However, as there is no evidence to indicate that the specimen was seen by an experienced ornithologist, it is more likely that the date is suspect, since the majority of the very small number of winter records for the British Isles, come from the milder west or from Ireland (Lack 1986, Brown & Grice 2005). The next record, a bird shot on the Tees in 1831, came from Edward Backhouse (1834). Hutchinson (1840) however, makes no mention of the species at all in his own work.

The earliest breeding records for Durham come from the North Tees Marshes, occurring between 1880 and 1887 (Nelson 1907), which *"gave hope of permanent colonisation of that county"* (Holloway, 1996). This was part of a relatively small group of breeding records, away from the Norfolk Broads, which occurred in Britain during the 19th century (Holloway 1996).

Breeding took place in the same area, some years later in 1914, when a breeding pair was noted at Teesmouth (Milburn 1916). Temperley (1951) confirmed breeding in the same area in 1927 and 1928 and the species bred again at Teesmouth in 1933, when Bishop saw three males together in April and felt *"sure it was nesting"*. Birds were subsequently noted at Teesmouth during spring between 1937 and 1939 and during the war years, but no nests were found.

Temperley recorded the species away from the Tees estuary in the Sunderland/South Shields area from 1945-1951. Breeding was suspected but only confirmed in 1947 when Dr. H.M.S. Blair saw a duck with downy young at Whiteleas Pond. When the young fledged, the birds flew between Whiteleas and Boldon Flats (J. Coulson pers. comm.).

A pair reared 10 young at Boldon Flats in 1952, but 14 birds were flushed there on 17 July, so it is possible that more than one pair may have bred in this locality during that year. A further isolated instance of breeding away from Teesmouth took place in 1959, when three juveniles were seen, possibly in the same area.

Stead documented breeding at Cowpen Marsh in 1963, with two pairs there in 1964, and at Saltholme on four occasions during the period 1901-1961 (Stead, 1964 & 1969). Stead (1966) also recorded that breeding had been proved on the Durham side of the Tees estuary on four occasions during the 20th century, the most recent being in 1957. There was also some evidence of breeding at Shotton Pond, when a female was seen there with four half-grown juveniles on 30 July 1963. Subsequent inland breeding records refer to the four-year period 1965-1968 when an area of flooded meadows in the south west of the county produced counts of seven young, five young, four young and three eggs in successive years. In 1967, birds again bred successfully in south west Durham, a female being noted with four young on 8th to 9 July, the pair at this site being present from the last week of March. Up to eight birds were at Teesmouth in August 1967 and two pairs were there in 1968, though breeding was not proven. In 1969, at least two pairs were at Teesmouth, from 22 March and breeding was attempted but not proven. There followed a successful breeding episode at Teesmouth in 1975, when four young were seen on Cowpen Marsh.

Since the Second World War, the species has been recorded annually in Durham, usually on both spring and autumn passage and predominantly from the North Tees Marshes, though there have been occasional years with no records. Other wetlands that have attracted the species include Barmston Ponds and WWT Washington which have hosted birds in most years, but a comparison with older records suggests that there may have been something of a decline in the number of spring Garganey in the Durham area through the 1960s, 1970s and onwards.

In 1970, birds were, as usual, at Teesmouth in the spring and many birds were seen in autumn, and in this year birds passed Hartlepool Headland on active migration on three dates in August. Furthermore, one was at Barmston from 28 May to 2 June and birds were present here over the next couple of years suggesting that breeding may have been attempted; six individuals were noted on 25 April 1973. A pair was seen many times at one site at Teesmouth in 1975, and breeding was suspected (Joynt *et al.* 2008). On 29 July 1975, 19 adults were around the North Tees Marshes, including one female with four ducklings (Bell 1976). In the north of the county, there is also a single mid-sixties record of a pair at Shibdon Pond sometime after April in 1965 (Bowey *et al.* 1993). The next report of the species in this area was not until a pair was present at Shibdon, on 5 May 1974, with another report of a pair on Derwenthaugh Meadows in the spring of 1975.

Recent breeding status

Nationally, the numbers of Garganey fluctuate markedly from year-to-year and figures indicate that there are usually less than sixty pairs breeding in the whole of Britain (Gibbons *et al.* 1993). The Garganey is shy and retiring, with a predilection for ponds whose edges have dense vegetation and this ensures that it is difficult to both locate and confirm breeding (Owen *et al.* 1986).

Typical habitat for Garganey in Durham is a small pond with shallow margins and large amounts of emergent vegetation into which birds can 'disappear' and perhaps breed. The difficulty in proving breeding, at even very well-watched locations, is considerable. An examination of the pattern of occurrences at the regularly observed Shibdon Pond during the mid-1990s serves to illustrate this point. This wetland recorded two different, long-staying males in eclipse plumage during the autumn of 1994. A few years later, in 1998, a pair was present on 12 April. The following spring, in 1999, a pair was seen again and breeding appeared to have been confirmed by the appearance of two juveniles from 7 August. This process was apparently repeated in 2000, when once again, a pair appeared on 5 May. A lone female was seen to fly out of the marshes on one occasion in June, before disappearing back in that direction, and a juvenile was once again seen in August. It is not inconceivable that birds had been breeding at the site over much of this six to seven year period, without conclusive proof.

Teesmouth produces June records in most years and the species has bred successfully in that part of the county in nine years during the 20th century (Blick 2009) and again on at least one occasion in the early part of the 21st century. In April 1998, up to three pairs were at Haverton Hole and breeding was proven on 12 June, when a duck with eight half-grown ducklings was seen, the first confirmed breeding record at Teesmouth for 23 years (Joynt *et al.* 2008). In 1999, up to four birds frequented Haverton Hole and seven drakes were noted at Saltholme Pools in May, whilst up to nine were at Dorman's Pool in July and August, though breeding was not confirmed (Joynt *et al.* 2008). During the Teesmouth breeding atlas survey, there was confirmation of breeding in 2002, when a female with a brood of ducklings was seen on the North Tees Marshes, and in 2003, when there were two adult females and two fully-grown juveniles on Dorman's Pool in early August (Joynt *et al.* 2008).

The *Atlas* (Westerberg & Bowey 2000) illustrated a low intensity of registrations and occupancy for this species, which was roughly clustered at two main sites over the six-year period of survey work. Most records came from lowland waters, with Teesmouth and the ponds in the Seaton and Hetton areas being particularly favoured. However, an examination of *Birds in Durham* (Armstrong 1988-1993) revealed that this was probably an underestimate of the species' true distribution. The major problem is drawing the distinction between records of passage birds and potential breeders. Passage birds may linger on into June, confusing the picture further and local breeding attempts may be more frequent than the *Atlas* suggests.

Birds are usually found singly or in pairs, and it is not unusual for such a pair to be noted at a wetland for only a day or two before they move on. Occasionally, birds remain for longer and breeding is attempted, though a successful outcome may only be an annual occurrence in Durham.

There are relatively few threats to the Garganey's status as a breeding bird in County Durham as most attempts occur on protected sites. However, newly arrived adults can be prone to disturbance from birdwatchers. In the longer term climate change may constrain future breeding success, as the species is clearly heavily dependent on adequate spring water levels in its favoured habitats, such as the North Tees Marshes.

Recent non-breeding status

Records of this species can come from almost any body of water in the county. Between 1980 and 2010 it was noted at over 20 locations, with a definite preponderance of sightings around the North Tees Marshes, though the Bishop Middleham area has also attracted birds in recent springs.

The Garganey can be one of the first migrants to appear, with arrivals occasionally occurring as early as late February in Britain (Brown & Grice 2005), although such occurrences are rare. The earliest date for Teesmouth is of four birds on Greatham Tank Farm on 9 March 1980 (Blick 2009). The year's first birds, typically, arrive on the

North Tees Marshes late in March or early April and passage normally continues through the county until May. Most observations refer to no more than one to three birds though rarely larger numbers have been recorded at this time of year at Teesmouth. Away from the North Tees Marshes, the Garganey remains a scarce species in Durham. A pair was at Bishop Middleham on 4 May 2003 and a male was at Hetton Lyons on 22 June. Two males were at Shibdon on 8 May 2004 and a pair was at Sunderland Academy Pools on 21 May. Several spring records refer to single drakes, as in 2006 with birds at Bishop Middleham on 22 April, Bowesfield Marsh the next day and Brasside Ponds on 9 May. Similarly in 2007 drakes were again at Bishop Middleham on 21 April and others were at Rainton Meadows on 8 May, Castle Lake on 12 May and Crookfoot Reservoir on 16 June. A pair was at Hurworth Burn Reservoir on 23 June. In 2008 a pair remained in the Seaton Pond area from 20 April to 6 May, but there was no indication of breeding. Once again there were drakes at Rainton Meadows on 25 June and Hetton Lyons on 16 July, with a female at Low Barns on 22 May. In 2009 a pair afforded good views at Wader Lake, WWT Washington remaining from 17 May to 2 June. On 21 March of the same year, a pair was on the old brick pond at Hagg Farm, Hamsterley Mill (M. J. Harbinson pers. comm.). Presumably the same pair was seen at the Far Pasture Wetland two days earlier, indicating birds prospecting for a breeding site.

Over the period 1976 to 2005, the mean reported first arrival dates of Garganey were: 1976-1985, 14 April; 1986-1995, 15 April; and, 1996-2005, 7 April - an earlier mean arrival date of seven days (Siggens 2005).

Return passage commences in July, with the final sightings of the year in late August, although a few birds are occasionally seen in September. Exceptionally birds are noted in October, such as one at Shibdon Pond in 1985. There have also been a very small number of November records, one from 1954 on upper Waskerley Reservoir was untypical in locality as well as date; another was at Washington in 1976, while the latest date for a bird at Teesmouth is of a female lingering to 11 November 2011.

In the county, there are on average, perhaps 15 to 20 records of this species each year. In most years slightly more birds are seen in the autumn than spring, but it would not be unusual for less than 10 individuals to be recorded in any year. However, it can be difficult to demonstrate such figures. For example, in 2009, one or two birds were seen on the North Tees Marshes on numerous dates from 29 March through to May, but it is difficult to gauge the turnover of birds in that area and therefore to judge the actual number of birds involved.

It is rare for more than seven birds to be seen together in Durham, though at least ten were at Teesmouth in July 1975 and on Fishburn Lake the following month of that year. The highest count for the county was of 23 birds that were scattered around the North Tees Marshes on 29 July 1975; this count included four ducklings on Cowpen Marsh (Blick 2009).

The species is very occasionally recorded off the coast during sea watches. For example, there are about 20 records of birds passing off Hartlepool Headland (Blick 2009), whilst occasional birds are also noted at Whitburn, mainly in April/May and August/September. A particularly unusual sighting was of a drake that flew north past South Shields Leas on 6 April 2001, in the company of two Tufted Ducks *Aythya fuligula*.

Distribution & movements

The Garganey breeds across Europe and northern Asia, as far east as Kamchatka. In Britain, it is a rare breeding bird and unusually for a dabbling duck, a summer visitor that winters in sub-Saharan Africa.

Blue-winged Teal
Anas discors

A very rare vagrant from North America; there are three records, including a drake seen annually from 2007 to 2011.

The first accepted record for County Durham relates to an adult female found by Chris Kehoe and Mike Gee amongst roosting Teal *Anas crecca* and Shoveler *Anas clypeata* on the Reclamation Pond, Teesmouth on 30 August 1994. Present for just five minutes before flying off towards Saltholme Pools, the initial views of the bird could not reliably rule out Cinnamon Teal *Anas cyanoptera*. It was rediscovered by the same observers on the

Reclamation Pond during the early afternoon of 3 September, allowing a more critical examination and confirming its identification as the first county record of Blue-winged Teal. It remained in the area intermittently until 29 September, visiting nearby Haverton Hole, particularly towards the end of its stay (Roger *et al.* 1995).

The second was found at Haverton Hole, Teesmouth, on 1 January 1999. Aged as a first-winter male, this bird was very elusive and spent much of its time asleep in the reeds, although it remained in the area until 11 January (*British Birds* 93: 522).

The third record was of an eclipse drake found at Haverton Hole, Teesmouth on 31 August 2007. Although mobile and often elusive, it was faithful to this area until 15 October, by which time it was starting to moult into an adult drake (*British Birds* 102: 533). Further sightings of drakes at Saltholme Pools on 14 November 2008, Haverton Hole, Saltholme Pools and Dorman's Pool from 23 September to 26 October 2009, Saltholme Pools on 8 April 2010, and Saltholme Pools from 8 August to 1 October 2011 are all thought to relate to this same bird and were accepted as such by BBRC (*British Birds* 103: 568).

An earlier record concerns a female at WWT Washington present from 14th until 21 February 1981. This bird was eventually caught in one of the collection pens on 21 February, and examined in the hand, confirming its identification. Following its capture the bird was transported to WWT Slimbridge, but what happened to it subsequently is unknown; presumably it was added to the large collection of wildfowl held there. The identification of this bird was accepted by BBRC, though it was thought to relate to an escape from captivity (*British Birds* 75: 491-492).

In addition, a drake *Anas sp.* at Haverton Hole, Teesmouth, from 11 November 2006 to 16 March 2007 was thought to be a hybrid Shoveler *Anas clypeata* x Blue-winged Teal. Although resembling a drake Australasian Shoveler *Anas rhynchotis*, this bird showed a dark eye typical of such hybrids, examples of which have been recorded in both the United Kingdom and North American previously. They are not necessarily of captive origin.

Distribution & movements

Blue-winged Teal breeds across much of North America from southern Alaska, and across much of temperate Canada to the south and central United States. It is a long distance migrant, wintering in the southern United States, through to northern South America and the Caribbean, though many birds move no further than the southern United States. It is an annual vagrant to the British Isles, with a total of 240 records by the end of 2010, and has been recorded in all months.

Shoveler (Northern Shoveler)
Anas clypeata

A regular, though local, passage migrant and winter visitor. It is uncommon in summer and a rare breeding bird that is seldom found away from its traditional breeding haunts in the county.

Historical review

National population trends indicate major increases in both the range and population of this species from the 19th century to the present day. It was during the earlier phase of this expansion that the county was first colonised by breeding birds. Hogg (1845) stated it *"visits us every summer and has bred in Cowpen Marsh"*. Tristram (1905) recorded a pair nesting in Saltholme Marsh in 1881 and perhaps their progeny were the nine young *"bagged"* on 1 August the same year by friends of Nelson (Temperley 1951). The Shoveler spread relatively rapidly as a British breeding species through the 19th and early 20th century and by the 1930s was breeding in all but eight English counties (Holloway 1996).

In the north west of the county, Robson (1896) recorded the taking of a specimen, an adult male, near Derwenthaugh on 25 May 1886, which he described as an *"accidental visitor to the valley"*, so presumably the bird was uncommon in the area at that time. However, considering the timing and location of this record it is tempting to speculate that the bird may have been one of a breeding pair, as suitable wetland was certainly present. Shoveler was certainly breeding in southern Northumberland during the 19th century (Kerr 2001), so breeding in this part of Durham was not that unlikely.

Two nests near the mouth of the Tees were found in June 1899 (Milburn 1901) and by 1907, Nelson said, *"At the Tees estuary it is of fairly frequent occurrence in spring and autumn, mainly owing to the presence of a well-protected breeding place in south-east Durham"*. In 1910 and 1912, Courtenay also documented breeding at Teesmouth with ducklings seen in both of those years (Temperley 1951). Milburn (1929) noted a significant increase in numbers with eight pairs nesting in 1916 and about a dozen pairs rearing young locally by 1927. It was suggested that it was resident and increasing in numbers at Teesmouth around this time (Richmond 1931), with four nests in one part of the marshes, and odd pairs away from Saltholme and Cowpen. Bishop reported eight pairs nesting in the Teesmouth area in 1931, but *"owing to increased molestation they have seldom succeeded in rearing young"* (Temperley 1951). Courtenay also noted *"it certainly nested at Hurworth in 1914, for in May of that year I saw one accompanied by eight youngsters"* (Courtenay 1931).

Prior to the drainage of West Rainton Marsh in 1928 Shoveler bred there annually with nests being found each year between 1924 and 1928 (Temperley 1951). The bird also bred at Whiteleas Pond, on the outskirts of South Shields, when the farmer from Newton Garth Farm found the nest in vegetation well away from the pond. By the mid-20th century, the Shoveler was considered *"a summer resident in very small numbers, a regular passage migrant, that occasionally remained over the winter"* (Temperley 1951).

After the initial increase in the early third of the 20th century, the species' breeding presence at Teesmouth declined. At least three pairs attempted to breed in the area in 1957, two being successful. Although a presence was maintained until the early 1960s, only two to four pairs were recorded annually at the North Tees Marshes, with low numbers being maintained by regular single broods (Stead 1964). Breeding was also proven at Crookfoot Reservoir in 1966 and in 1969 at Bowesfield (Blick 2009). After breeding at Teesmouth every year since 1930, during 1965 there was no proof of breeding, for the first time in over 30 years.

Nonetheless, over this period, the number of birds passing through the county was rising and in 1966, a peak count of 58 was recorded on 17 August, at Teesmouth. Parslow stated, in 1967, that the Shoveler's main phase of expansion in Britain and Ireland occurred early in the 20th century and was matched by a similar spread in Western Europe. He incorrectly stated that since the 1930s, it had ceased to breed regularly in Durham. The late 1960s were a less successful time for this species as a breeding bird in the county, though two pairs were present in summer 1967 at Teesmouth, and a pair was there in 1968, with another pair present that spring until 6 June at a different site, though breeding was not proven. Shoveler was also proven to have bred at Cowpen Marsh and Crookfoot Reservoir in 1969. The highest count of this species over the post-Temperley period occurred at Teesmouth, when 110 were present in August 1963.

Despite an increase in the British population over the latter part of the 20th century, a greater proportion of pairs now breed on protected sites, as unprotected sites have been lost (Gibbons *et al.* 1993). As increasing numbers occurred across the county through the 1970s, breeding behaviour spread to a few more sites away from the traditional south eastern stronghold.

Passage birds used to occur in two autumn peaks, in early August and then later in October, but more latterly at Teesmouth, there has been a single autumn peak in September during four out of five years in the period 2006-2010. The early 1970s also saw an increasing presence of this species at Barmston Pond, and in the first half of this decade breeding was proven there, as well as annually at Teesmouth, with up to three broods in the latter area. The peak early autumn count at Barmston rose to 35 by 8 August 1971, when 74 were at Teesmouth on the 9th of the same month. In 1973, at least six pairs were around Teesmouth in spring, but there was no confirmation of successful breeding. Throughout the decade, however, birds bred intermittently at both Barmston and Teesmouth. The species first bred at WWT Washington in 1977, possibly as a transfer of allegiance from Barmston Pond. Breeding was also confirmed at WWT Washington in 1978, the only breeding site in the county for that year, with most of the five broods documented managing to fledge. Additional breeding may have occurred at the site at this time, but the birds were very secretive.

Autumn flocks reached double figures at Barmston, Teesmouth and Shibdon Pond, with 78 at Teesmouth on 2 November 1976, but by and large the Shoveler was a scarce wintering bird in Durham during the 1970s.

Recent breeding status

The species' summer distribution in Durham is restricted to the available wetland sites in the eastern coastal plain and, to a lesser degree, along the main river valleys of the county. During the 1988-1993 survey period, there were no records of summering birds above an altitude of 130m above sea level, indicating the essential lowland character of the species in the north east of England.

Shoveler have more stringent habitat requirements than most other dabbling ducks (Owen *et al.* 1986), in that they favour relatively shallow eutrophic waters with considerable marginal vegetation, or occasionally, wet pastures and flood meadows. Consequently, more than most other ducks, the Shoveler is prone to the consequences of wetland drainage and is more likely to fall victim to agricultural improvement (Westerberg & Bowey 2000). Durham has relatively few suitable wetland sites available for this species.

In the last thirty years in County Durham, the Shoveler has been quite a rare breeding bird, even in its historical heartland, Teesmouth, where it has probably never nested in anything other than small numbers. Breeding was proven in Durham in every year since 1990 to the end of the first decade of the 21st century. Over a three decade period, the breeding strongholds of the Shoveler have been located within the complex of marshes at the mouth of the Tees, an area favoured since the 1930s; the wetlands in the Low Barmston area, including WWT Washington; and Shibdon Pond, in Gateshead (Westerberg & Bowey 2000). Survey work, during the *Atlas* period confirmed that breeding birds were recorded in all six years from 1988 to 1993, but never in more than very small numbers. Over this period, all of Durham's known breeding birds were located in wetland nature reserves or other protected sites. In June 1988, five pairs reared 44 young around the North Tees Marshes (Joynt *et al.* 2008). Breeding was confirmed for the first time at Shibdon Pond in 1991, when a female fledged nine of a brood of ten ducklings, and this success was repeated in 1992. In only two of the six years of the *Atlas* survey period were more than two successful pairs reported (Westerberg & Bowey 2000). Over this time frame, the only sites that hosted confirmed breeding pairs in more than a single year were Shibdon Pond (with pairs producing young each year from 1991 to 1995) and the complex of sites at Teesmouth, where successful breeding was noted annually (Westerberg & Bowey 2000). At least four broods were noted around the county in 1994, whilst nine broods were noted on Teesside in 1995 (all around Saltholme Pools). In 1999, nine broods were seen around the North Tees Marshes, the largest number ever documented since Milburn's record of 12 pairs in 1912, and between 2002 and 2006 the number rose to sixteen pairs (Joynt *et al.* 2008).

The key recent locations for the species at Teesmouth have been Saltholme Pools and Cowpen Marsh (Joynt *et al.* 2008). In June 2001, the species was reported from three sites at Teesmouth: the Reclamation Pond, Saltholme Pools, and Dorman's Pool, as well as at Shibdon Pond and Barmston Pond, all of these being previously utilised breeding sites. The only confirmed breeding record however, concerned the sighting of a female with five young ducklings at Rainton Meadows on 6 July (Armstrong 2005). In 2005, confirmation of breeding came from Cowpen Marsh at Teesmouth in July, whilst Bishop Middleham and Shibdon Pond also held nesting pairs, and breeding was strongly suspected at Rainton Meadows. In the following year, Bishop Middleham was the only site at which this species was confirmed as breeding successfully, a duck with a brood of eight young being noted there on 25 June. In 2007, the only proof of breeding came from the North Tees Marshes, where one pair was noted with four young on 13 July, with possibly another female with three young seen nearby two days later. Pairs were seen at several other localities during May and June, but there were no further reports of nesting attempts or ducklings being noted.

The density of breeding birds on any occupied water in the county is low as the Shoveler, unlike most ducks, is highly territorial (Gibbons *et al.* 1993). It is unusual for reports to refer to more than a single pair on any occupied site, though the exception is the Teesmouth wetland complex with up to ten pairs per annum in the Cowpen-Saltholme area (Blick 2009). Successful breeding occurred in every year of the first decade of the 2000s; sites where this took place away from Teesmouth, included: Rainton Meadows, Bishop Middleham, Bowesfield Marsh, Hurworth Burn Reservoir and Shibdon Pond.

The male Shoveler, unlike most wildfowl species, is often present throughout the incubation process and not infrequently accompanies females when they have broods of young. The presence of such a gaudily plumaged bird on breeding waters should alert the observer to search for the much less obvious female (Gibbons *et al.* 1993). However, lone females with young are shy and are much more likely to go undetected. The county's main concentration of breeding birds remains the North Tees Marshes. Nonetheless, other sites now seem to be attracting summering birds, although less frequently, with further pairs in the Bishop Middleham, Washington and Gateshead areas. The *Atlas'* estimated annual county population of around five pairs is probably still reasonably accurate (Westerberg & Bowey 2000). In good years this probably rises to double figures.

Recent non-breeding status

Today, birds are often present in the county throughout the year, with peak numbers occurring in the autumn. This probably represents a post-breeding dispersal from other breeding areas in north east England and may be supplemented later in the year by north European birds, a few of which pass along the coast in autumn (Blick 2009).

A detailed study of movements during the 1980s and early 1990s at Shibdon Pond using a comparison of daily counts revealed the emergence of a complex pattern. It appeared that presumed British breeding birds arrived there to moult during mid-August and early September, departing for southerly wintering grounds through October only to be replaced by a second autumn influx, presumably of more northerly birds. These arrived either to winter in the north east or to utilise the area as a staging post on their way further south. As the cold weather worsened most of these birds moved away, leaving only small numbers to remain throughout the winter at the site (Bowey *et al.* 1993). It is not certain that this observed pattern holds true for the general movement of the species through the county.

During the 1990s, the highest autumn counts were respectively: 225 at Teesmouth in 1995; 72 at Shibdon Pond in the same year; and 52 at Barmston in 1994; it is notable that numbers at Shibdon and in the Washington/Barmston area decreased in the latter half of this decade.

Peak numbers at Teesmouth usually occur in the months August-November (typically September) and in recent years figures have risen to between one and two hundred birds. Recent maxima at Teesmouth illustrate this process over the last three decades. There were: 139 in September 1988; 232 in October 1995; 260 in September 2000; 308 in August 2005 (Blick 2009); 309 in September 2006 and 300 in October 2009 (WeBS).

In 2007, the peak counts at Teesmouth were 176 birds in February and 157 in September. Increased winter numbers are an interesting development in the county, as the Shoveler was a scarce bird in winter as recently as the early 1970s. The five year mean of winter peaks at the Tees WeBS site has risen from 55 in 1993 to 129 in 2011. This may be due to the long run of mild winters experienced in the previous decade and a half, perhaps encouraging birds to winter further north. At Teesmouth, the five year mean of autumn peaks stood at 231 in 2010, whereas in 1994 it was 87. Whilst the North Tees Marshes still holds the bulk of the county's birds, from the perspective of the late 2000s, the long-term trend shows the number of birds in the county may no longer be rising, as it has done over the previous 17 years or so (Newsome 2008).

August usually sees the commencement of the autumn influx and numbers at regularly frequented sites rise at this time of the year. At such times birds are usually reported from Shibdon Pond, Castle Lake at Bishop Middleham, the Washington/Barmston area and various sites around Teesmouth. The species usually becomes more widespread in September, being noted at a wider range of the county's wetlands than at other times of the year.

Occasionally small numbers of birds on passage are noted off Whitburn Observatory and other coastal stations. For example, in 2007, coastal passage was noted on 12 dates through the year at Whitburn, with most occurrences taking place in the autumn, though occasional spring and even winter sightings can occur along the coast.

A bird which exhibited the characteristics of a presumed hybrid Shoveler x Blue-winged Teal *Anas discors* was present at Haverton Hole from late 2006 until 16 March 2007.

Distribution & movements

This species breeds across Iceland and eastwards across northern Eurasia, and in North America, from Newfoundland in the east across to Alaska in the west. The British breeding population winters in France and Spain, being replaced in winter by birds from Scandinavia and Russia.

Red-crested Pochard
Netta rufina

A scarce passage migrant from continental Europe, or from naturalised populations in southern England, recorded in all months, most often between August and May.

Historical Review

A single bird seen by several observers on the River Wear in Durham City from 3 February to 5 March 1963 represents the first record for County Durham. Originally published as *"doubtless an escape"*, the weather conditions at the time more likely point to a bird of wild origin. The winter of 1962/1963 was exceptionally cold, with the north east of England experiencing arctic conditions from December through to March, resulting in all of the inland waters in the county being frozen solid. This forced many waterfowl to make use of any available ice free water, including the River Wear in Durham City. The Red-crested Pochard was no doubt utilising the same resource, having presumably escaped even harsher conditions on the near Continent.

Recent status

There were 32 records of Red-crested Pochard in County Durham from 1970 to 2011, though records from WWT Washington are not included here as they may relate to birds of captive origin. There were just two records during the 1970s. The first was of a drake near Ferryhill on 16 April 1973, with presumably the same bird at Hurworth Burn Reservoir on 8 May, having visited Northumberland in the intervening period, and the second was of an eclipse drake at Low Barns NR, Witton-le-Wear on 4 September 1979.

A slight improvement during the 1980s saw a total of seven birds, including the first multiple record, a pair at Shibdon Pond, Blaydon in November 1980, the female of which remained for much of the winter. Other notable records include the first for the North Tees Marshes, a female at Saltholme Pools from 14th until 16 November 1982, and a drake flying north past Ryhope Dene on 14 November 1987, which was relocated on the River Tyne at Wylam the following day. A female first seen on Saltholme Pools, Teesmouth on 6 October 1989, later relocated to Charlton's Pond, Billingham, and remained in the area until at least December 1991, occasionally visiting Cowpen Bewley WP. Despite the arrival date suggesting a potentially wild origin, this bird hybridised with a drake Mallard *Anas platyrhynchos*, nine hybrid young being noted on 10 June 1991, and must therefore be regarded as being somewhat suspect.

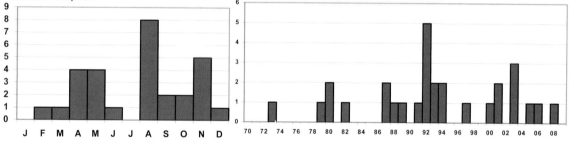

Records of Red-crested Pochard in County Durham, 1970-2009

A total of eleven birds were recorded during the 1990s, a conservative estimate, as several widely scattered records of a female in the south west of the county during 1991 and 1992 were thought to relate to a single bird. A party of four birds, aged as juveniles, at Crookfoot Reservoir from 30 August until 13 September 1992, with two remaining until the 18th, represented the largest party ever seen in the county, and had presumably been hatched somewhere in the south of the UK. Another notable record during the period is of a drake at Hetton Lyons CP, Houghton-le-Spring on 4 May 1994 which to date represents the only record for this well-watched locality. The 1990s were the best decade to date for this species, with not only the highest number of birds recorded, but also the peak of five in 1992 representing the highest total ever seen in the county in a single year.

A further nine birds were seen during the 2000s, maintaining the average of around 10 birds per decade since the 1980s. Surprisingly, five of these were around the North Tees Marshes, including a returning drake from 17

April to 13 July 2003, 12 April to 2 July 2004, and 7 May to 27 July 2005, with what may also have been the same bird at Saltholme Pools on 3 October 2005.

Away from the Tees Marshes, notable records during the period include the first record for Sunderland, with a female at Silksworth Lakes on 3 August 2005. A drake flying north past Hartlepool on 2 August 2006 was the first record for the Headland. The similarity in dates is noteworthy, although perhaps not unexpected, given that August is the peak month for the occurrence for this species in the county. Two drakes at Castle Lake, Bishop Middleham on 18th and 19 June 2010, were the first recorded at the newly created Durham Bird Club reserve here.

The true status of Red-crested Pochard within the county is clouded by the many presumed feral birds noted in the WWT Washington area, particularly during the period 1998 to 2004. These birds were apparently not part of the collection, though were frequently very tame, and must be considered of potentially dubious origin, though a female from January to March 1998 had previously been seen at nearby Boldon Flats. Present year round, though only reported intermittently, the number of birds here reached a peak in 2001 when up to five were recorded, 3 drakes and 2 females, with the final sighting in the area being of a pair in 2004, with no subsequent records. A female at Shibdon Pond from 4th to 16 April 1978 and at WWT Washington in May and June 1978, and again in May 1979, was considered to be an escape, on the basis of its behaviour. In addition to this, birds at Barmston Pond on 29 November 1986 and 17 November 1987 are presumed to relate to the free-flying birds from the collection at WWT Washington, although this may not necessarily be the case.

Distribution & movements

The Red-crested Pochard has a scattered distribution across much of continental Europe, eastwards through central Asia to north west China and western Mongolia. Largely sedentary in southern Europe, more northerly populations winter principally around the eastern Mediterranean, east to the Caspian Sea, with most of the Asian population wintering in the Indian subcontinent. First recorded in Britain as long ago as 1818, this species undoubtedly occurs in the wild state on occasions. These are most likely to be from amongst the birds breeding in France and the Netherlands, though the situation is complicated by the number of feral birds at large in the British Isles and elsewhere in Europe.

Known to have bred in a feral state in Britain sporadically since at least the 1930s, a large feral population now exists at Cotswold Water Park on the Gloucestershire/Wiltshire border, reaching a peak of 117 birds in 2001, with a smaller population in Baston and Langtoft Pits in Lincolnshire. On this basis the species was given dual categorisation by the BOU and also added to Category C of the British list in October 2005. Most recent British records, including those in County Durham, probably stem from these feral populations, though it will doubtless still occur in the wild state occasionally.

Pochard (Common Pochard)
Aythya ferina

A rare breeder, but a common passage and winter visitor.

Historical review

Hogg (1845), quoted John Grey as saying it was not uncommon *"on the Tees in the winter, but never numerous"*. Tristram (1905), referring to the later 19th century, wrote that it was *"frequently met with throughout the winter"*, stating that it was thought to have bred at Teesmouth, *"but I have no certain proof, though it breeds sometimes in North Yorkshire and Northumberland"*. Hancock (1873) gives no breeding record for Durham, referring only to eggs taken at Scarborough in 1844. In the west of the county, Robson (1896) recorded that *"an example of this common winter visitor was shot near Dunston in the year 1893"*, presumably on the Tyne.

Nelson (1907) thought it an *"extremely local breeder"*, not common in the Teesmouth area. Stead writing in 1964 concluded that breeding in the early part of the 20th century was restricted to one or two pairs at a single locality. However, in 1929 Temperley considered that the Pochard had become more common since Hancock's day with birds now breeding in both northern counties. Between 1910 and 1929, Courtenay (1931) documented flocks of up to 50 at Hurworth Burn Reservoir. The first definite breeding record of the 20th century came in 1903 from a pond on an old brick-field site in south east Durham (Temperley 1951). The nest was found by J. Bishop, who also documented a further nest with six eggs the following year. Meanwhile, Bolam (1912) reported the species as breeding at numerous localities in Northumberland, having established itself there in the previous two decades (Kerr 2001). The original Durham nesting site was near Billingham and the species bred there again in 1919 and on an annual basis for some years afterwards, with one or two pairs being present. The site was lost to development in the early 1940s.

Temperley also documents that before the draining of West Rainton Marsh in 1928, a few pairs of Pochard bred there, with young noted at this site in 1924 and nests present from 1925 to 1928. In 1946, two pairs also reared young at another brick-field site, this time in north east Durham. Temperley saw the birds himself and noted that breeding again took place in 1948. He considered the Pochard *"a bird of inland waters"*, being, *"a rare summer resident of irregular and local occurrence; fairly common as a spring and autumn visitor, with a few remaining throughout winter"* (Temperley 1951).

In spring 1959, a visit to Brasside Ponds revealed that two pairs of Pochard were present. On 23 May, the female of one of these pairs was observed sitting on a nest. At that time, these two pairs were the only birds in Durham attempting to breed (*The Vasculum* Vol. XLIV, 1959). In 1960, the Pochard bred on at least one site in County Durham, again presumably Brasside, with a brood of four ducklings noted on 21 June. In 1963, three females reared three broods at Teesmouth pools where the species was said to have bred for a number of years. However, throughout this decade the Pochard remained a rare breeding bird within the county, although summering birds were seen on several occasions (e.g. in 1964 and 1969).

This period also saw the start of a build up in wintering numbers which continued through the 1970s and into the 1980s, with a record peak count of 194 for Durham at Hurworth Burn Reservoir 15 March 1964. This site was favoured by the species for the rest of the decade, although increasing numbers were also being reported at the newly completed Derwent Reservoir.

Recent breeding status

In England, the Pochard is a rare breeding bird, but each year a small number of pairs breed within our county. Their breeding habitat needs to meet certain specific requirements, usually undisturbed, still or slow moving waters with dense marginal vegetation, and brackish fleets around saltmarshes can also be important (Brown & Grice 2005). As diving ducks, they require water with a depth of around three metres to breed, where there is a rich source of submerged and emergent vegetation. The adults are mainly vegetarian, feeding both day and night on submerged aquatic plants, but the ducklings need an abundant supply of aquatic life (Owen *et al.* 1986). Dense, bank side vegetation is required for cover, with the nest usually on a platform of reeds (Gibbons *et al.* 1993). The species is highly sensitive to recreational disturbance, including angling (Brown & Grice 2005).

The North Tees Marshes have remained the centre of the Durham breeding population for many years. Pochard bred there during the early part of the 20th century and in almost every year since 1986 (Blick 2009). Usually, only two to three pairs are present annually, in spite of the significant population just to our north, in south east Northumberland (Kerr 1995). Breeding away from the Teesmouth area is a relatively rare event, occurring regularly at only a couple of sites.

Nationally, the Pochard appears to have increased as a breeding bird in the latter quarter of the 20th century, possibly as a result of better observer coverage (RBBP), although it has declined as a wintering species since the peak of the 1970s (Gibbons *et al.* 1993). The *Atlas* determined that between three and six pairs bred annually in the county (Westerberg & Bowey 2000), though there were many more summering, non-breeding birds present (Armstrong 1988-1995, Bell 1988-1993). This represented up to 1.5% of the entire British breeding population (Gibbons *et al.* 1993), and in addition, two or three feral pairs nested at South Shields Marine Park Lake during the early to mid-1990s. At the time of the *Atlas* survey there appeared to be no strong upward trend in breeding numbers (Armstrong 1988-1993) although since the late 1990s there is some indication of an increase, especially at Teesmouth. During the 1990s it was observed that some areas of water around the North Tees Marshes had become deeper, in particular Saltholme Pools, Dorman's Pool and the Reclamation Pond, and as a consequence, more attractive to Pochard (Blick 2009). Three to five pairs nested at Teesmouth between 1996 and 2000. Saltholme Pools proved to be the centre of a concerted colonisation attempt and by 1999 the population was significant in a national context, with 20 broods being reared across the Teesmouth area during that year (Joynt *et al.* 2008).

Between 2001 and 2010, proof of breeding was obtained from Dorman's Pool, Saltholme Pools, the Reclamation Pond, Haverton Hole and the Long Drag, with regular summer sightings coming from: Rainton Meadows, Joe's Pond, Shibdon Pond, Hurworth Burn Reservoir, Bishop Middleham, Barmston Pond and Hetton Lyons CP. The species is known to have bred at five of these latter sites, but probably only at a total of 10-12 sites throughout the county. In 2003 a pair was noted with three ducklings on the nature reserve pool at Whitburn CP, a remarkable record given the size of the pond. In 2005, fourteen pairs hatched young, but by 2007, the only place where successful breeding occurred was once again in the south east of the county. In 2008 there were 13 breeding pairs on the North Tees Marshes and also broods at Billingham Beck and Charlton's Pond.

There is some potential for expansion of this species' breeding range within Durham, though sites with the required combination of habitat are relatively few in number. In addition to the disturbance caused to incubating females, ducklings are thought to be easy prey for predators such as pike *Esox lucius*, a fact which has been implicated in the poor breeding success of the species (Fox 1991). As Joynt (2008) has remarked, hopefully the development of the RSPB reserve at Saltholme will help to ensure the bird's status within the county.

Recent non-breeding status

Prior to the 1960s, flocks of more than ten Pochard anywhere in the county were unusual, but from the early part of that decade, the annual winter maximum counts at a number of sites began to rise. By the 1970s, mid-winter flocks, in January or February, of a few hundred birds could be found in several areas, especially Seal Sands. A mid-winter count of 13 birds in 1976-77 rose to 335 the following year, with 776 in 1978/1979 and a peak of 930 on 30 January 1981; the largest ever count for Durham (Bell 1982). However, such large concentrations at Teesmouth proved to be a relatively short-lived phenomenon and despite a figure of 850 in 1981-82, no more than 115 have been observed since the mid-1980s, although 214 were noted at Saltholme Pools in February 1996. The large congregations at Seal Sands are now a thing of the past and in our present century, peak counts of the species from all sites at Teesmouth have barely reached one hundred; the highest recent count was 88 in March 2010.

In the north of the county, Shibdon Pond proved to be a favoured site with flocks of more than 200 birds in most winters from the late 1970s, and a count of 500 in February 1979. In the late 1990s a large winter flock developed at WWT Washington with up to 250 birds, but at both these sites numbers have declined rapidly in parallel with the situation at Teesmouth. Many other sites across the county attract birds in varying numbers, including the reservoirs at Hurworth Burn and Crookfoot, Barmston Pond, Brasside, Bishop Middleham and the Marine Park Lake at South Shields. However, in recent years the highest counts rarely exceed 40 individuals. The best count for winter 2010 was 35 at Hurworth Burn Reservoir.

In Durham, the highest counts usually come in the first quarter of the year and freezing weather can produce good numbers at sites that have retained open water. Numbers decline rapidly throughout March, although in April birds can still be found at sites scattered throughout the county. Small gatherings of birds recur from late August onwards with numbers growing through September. As autumn progresses, birds tend to be reported from the more usual locations, but at this time of the year few sites have more than 15-20 individuals. The main body of birds arrive throughout October and November, but peak winter numbers are not usually reached until January or February. The Pochard tends to be highly mobile in winter, which leads to a variation in numbers using a given site. This was noted in 2006 when birds often moved between Hetton Lyons CP and Rainton Meadows.

Typically, there are few reports from coastal observation points. For example in 2007, coastal passage was noted on just nine dates at Whitburn, with a total of 49 birds. Sightings tend to be more frequent in autumn, especially from late August to mid-October. Hartlepool remains the key coastal watch point for sea passage with: 81 in October 1963; 64 in November 1967; 46 in November 1972; 39 in October 1977; 44 in October 1982 and 97 on 3 November 1984 (Blick 2009).

Distribution & movements

The Pochard is a Palaearctic species that breeds in Iceland and Spain and across Europe from southern Finland to the Mediterranean, and as far east as north east China and Japan (Parkin & Knox 2010). As a breeding bird, it did not become established in Britain until the early part of the 20th century, and although its numbers have increased in recent decades, it remains rare. The majority of our wintering population originates from the nations around the Baltic.

A bird caught at WWT Washington in 1994, had been ringed at the same site eleven years earlier (Armstrong 1995). A drake present at Haverton Hole from 11th to 29 May 2009, and again from 23rd to 30 April 2010, had a blue bill-tag, attached when it was a first-winter bird in Loire-Atlantique, north west France on 13 November 2006 (Blick 2009).

Ring-necked Duck
Aythya collaris

A rare vagrant from North America; there are 13 records, documented in all months, involving several returning birds.

The first record for both County Durham and Northumberland was found by Brian Little at Derwent Reservoir on 13 February 1977 amongst an exceptional influx of around 500 Pochard *Aythya farina*; remaining until 26 March (*British Birds* 71: 494). The same bird was at Hurworth Burn Reservoir from 30 April to 4 May, nearby Crookfoot Reservoir on 5 May, and was again at Derwent Reservoir on 27 August, and from 16 October until 6 November. In the intervening period this bird was also seen at Shibdon Pond and a number of sites in Northumberland, including Capheaton Reservoir 29 May to 14 June where it was apparently paired with a female Pochard *Aythya ferina*.

All records:

1977	Derwent Reservoir, male, 13 February to 26 March, 27 August, and 16 October to 6 November (*contra BB*), same Hurworth Burn Reservoir, 30 April to 4 May, and Crookfoot Reservoir, 5 May.
1988	Charlton's Pond, Billingham, male, 23 April (*British Birds* 82: 516).
1990	South Marine Park, South Shields, female, 25 February to 4 March, 11th to 12 April (*BB*: 84: 462), 8 January 1993, and 30 November 1993 (*contra BB*) (*British Birds* 87: 516).
1990	South Marine Park, South Shields, male, 6 April, 11th to 18 April, 23rd to 24 April, 7 May, 31 October, and 11th to 12 December (*British Birds* 84: 462) (*British Birds* 87: 516). Also 13th to 21 April 1991, 9th to 25 June 1991, and 5th to 28 October 1991 (*British Birds* 85: 516). Same at Shibdon Pond, Blaydon, 31 May 1990 (*British Birds* 84: 462).
1993	Hartlepool Headland, male, 8 May, same Crookfoot Reservoir, 9th to 10 May, and Hurworth Burn Reservoir, 18th to 22 May (*British Birds* 87: 515, *British Birds* 88: 502).
1999	Seaton Pond, male, 30 August.
2002	WWT Washington, female, 3rd to 13 January, same Low Barns NR, Witton-le-Wear, 16 January to 30 April.
2002	Low Barns, Witton-le-Wear, male, 16 January to 30 April, additional male, 17 January to 30 April.
2003	Reclamation Pond, Teesmouth, male, 21 July to 25 August.
2007	Low Barns NR, Witton-le-Wear, male, 28 February to 7 March, same Holywell Hall, Brancepeth, 17 March.
2010	Cowpen Bewley WP, male, intermittently, 23 January to 13 March, and intermittently from 29 December into 2011.
2011	Cowpen Bewley WP and Dorman's Pool, Teesmouth, male, from 2011 to 18 February, and again from 10 December into 2012.
2011	Bowesfield Marsh, male, 19th to 20 April.

The party of three birds at Low Barns from 17 January to 30 April 2002 is noteworthy in being one of very few multiple records in Britain, the largest of which was of a party of six at Stithians Reservoir, Cornwall in October 1979.

Distribution & movements
The Ring-necked Duck breeds largely in southern Canada and the north eastern United States, and winters mainly in the southern United States, Central America, and the Caribbean. A period of eastwards expansion in

North America during the latter part of the 20th century has also seen a large increase in the number of British records in recent years. Prior to 1977, the year of Durham's first record, there had been just 26 recorded in Britain, although almost 300 had been documented by the end of 1993, at which point the species was no longer considered by BBRC.

Ferruginous Duck
Aythya nyroca

A very rare vagrant from southern and eastern Europe; there are four records.

The first county record of this species relates to an adult drake at Hebburn Ponds from 3rd to 7 April 1948 (Temperley 1951). *"On April 3rd, 1948, Mr. J. R. Crawford, of Sunderland, observed a male of this species on one of the Hebburn ponds. It was closely associated with a pair of Tufted Duck, Aythya fuligula. He had it under observation until April 6th. On April 7th, the writer (George Temperley) visited the pond in company with M. G. Robinson, where we found the bird and were able to observe it under the most favourable conditions. It was an adult drake, bearing all the distinguishing characters of the species. It had every appearance of being a wild bird, being very wary and alert and more easily flushed on disturbance than its companions the Tufted Ducks. By the following morning, April 8th, it had gone, though the pair of Tufted Duck was still present on the same water. It was not seen again either there or elsewhere in the County."*

Commenting on the record, the editor of *British Birds* at the time noted: *"Although it is scarcely possible to be certain that Ferruginous Ducks in this country are not 'escapes', it is deserving of note that although there has been an increase in the number reliably recorded in recent years there are now excessively few in captivity in the British Isles - only two drakes are recorded in the latest summary of the Avicultural Society's Waterfowl Registery. There is thus a good prima facie case for regarding apparently wild birds as actually so"* (*British Birds* 42: 61).

All records:

1948	Hebburn Ponds, male, 3rd to 7 April
1982	Brasside Ponds, Durham, male, 22 April to 7 May.
1985	Shibdon Pond, Blaydon, female, 4th to 10 November.
1991	Shibdon Pond, Blaydon, female, 31 January to 3 February.

In addition to these records hybrid *Aythya* resembling Ferruginous Duck *Aythya nyroca* have been recorded in the county on a number of occasions. The first of these was at WWT Washington from December 1983 to early April 1984, with the same bird returning from 23 November 1984 to 29 March 1985, 9 November 1985 to 11 April 1986, and again from 12 November 1986 to 9 April 1987. This bird was often thought by some observers to be a genuine Ferruginous Duck. In November 1986, it was found roosting in one of the wildfowl collection pens allowing critical examination of its bill pattern and confirming that it was a hybrid, probably with Pochard *Aythya ferina*. Another female hybrid was at Shibdon Pond and Far Pasture, Gateshead from 26 September to 21 December 1996, with what was thought to be the same bird at Shibdon Pond again from 3 April to 21 June 1997. A bird also thought to be a Pochard x Ferruginous Duck *Aythya nyroca* hybrid was at Brasside Ponds, Durham from 23 November to 2 December 2001.

Distribution & movements
The Ferruginous Duck has a scattered and rather fragmented breeding distribution, mainly across Eastern Europe and Central Asia, though small numbers breed in Iberia and occasionally in North Africa. Sporadic breeding also occurs in the Netherlands and in Britain breeding attempts took place at Chew Valley Lake, Avon from at least 2003 until 2006, probably being successful in 2006, when a young male was seen in October of that year (Davis & Vinnicombe 2011). All populations are known to be migratory, with the majority of European birds wintering in the eastern Mediterranean, though some penetrate into sub-Saharan Africa; more easterly populations winter in the Indian subcontinent. Formerly a much more regular vagrant to Britain, particularly to the southern and eastern

counties, there was a marked decline in the number of records from the mid-1990s. The species was assessed by BBRC from 1998 to 2005, documenting an average of around 10 records per year. The lack of recent records in County Durham is no doubt related to the downward national trend.

Tufted Duck
Aythya fuligula

A common resident, passage migrant and winter visitor.

Historical review
In the 19th century, the Tufted Duck was an unusual bird to be seen on the marshes and inland waters of County Durham, and even up to the 1950s flocks rarely exceeded twenty to thirty birds. Temperley (1951) said it was a *"summer resident in small but increasing numbers"*, and a *"regular autumn and winter visitor, being chiefly a bird of inland waters"*. Hutchinson (1840) wrote that it was *"rare"* and usually single when seen on rivers, noting that one had been shot on the Wear near Durham. Hogg (1845) said it was a rare visitor near Hartlepool, giving only one record of a bird shot on the Tees in December 1823. Tristram (1905) highlighted that it was *"not a very common visitor, though breeding in Northumberland"*. Hancock first confirmed breeding there, on the Trevelyan estate at Wallington in 1858. In Yorkshire breeding was first recorded in 1849 (Mather 1986).

Most English counties had been colonised by this species around 1912 (Holloway 1996) and in Durham, during the first half of the 20th century, its numbers increased. The first absolute proof of breeding came in 1914, when J. Bishop found a nest on the same brick-field pond in south east Durham where he discovered breeding Pochard *Aythya ferina*. A few pairs bred there over the next 25 years or so, until the site was degraded by development. Breeding was also confirmed at West Rainton Marsh, from 1924 to 1927, when the pond was drained. The species showed a distinct preference for Hurworth Burn Reservoir, where it eventually bred. Courtenay, writing in 1931, commented that *"except for the Mallard no duck is to be met with at Hurworth more frequently than the Tufted"*. In the same year Richmond noted that the Tufted Duck was a resident breeding bird at Teesmouth, and Sharrock (1976) confirmed that by the early 1930s most suitable waters in Britain had been colonised. Certainly by the late 1930s birds were established as breeding in the centre of the county, with two pairs nesting at Brasside Ponds near Newton Hall in 1938 (*The Vasculum* Vol. XXIV, 1938) and in 1946, eight pairs reared young on another brick-field pond in north east Durham. A visit to Brasside Ponds during May 1959, detected *"no fewer than fourteen pairs"* (*The Vasculum* Vol. XLIV, 1959). However, in the south east of the county, following the loss of Charlton's Pond, no birds were known to breed until 1960, when a single brood of ducklings was seen on Crookfoot Reservoir (Blick 2009). By 1965, at least five broods, amounting to 29 ducklings, had hatched on the park lakes at Hartlepool and breeding birds were established in the mid-Wear valley, with four broods at Witton-le-Wear in 1969. By 1971 at least 17 broods were recorded from seven wetland sites around the county.

Recent breeding status
The Tufted Duck is the commonest breeding British diving duck. Brown and Grice (2005) estimate the English breeding population as in excess of 6,000, noting that the bird *"has benefited enormously from the creation of gravel pits and new wetlands in the English lowlands"*. In recent decades the species has expanded and consolidated its range at national and local levels. As illustrated in the *Atlas*, there are notably linear patterns of breeding distribution, along the Tyne and Wear valleys, with a particular concentration of breeding birds in the lower Tees valley (Westerberg & Bowey 2000). Many lowland ponds host a small number of pairs, but considering its oft repeated description as a lowland species (Gibbons *et al.* 1993), it is surprising how many upland breeding records are noted in Durham, particularly in the relatively exposed areas of upper Teesdale. In 2007, birds were noted at eight sites west of the A68, although their appearance on upland reservoirs and ponds was generally sporadic. The most interesting record for that year concerned two males on Burnhope Dam, near Hunstanworth, 445 metres above sea level.

In order to breed Tufted Ducks require water of around five metres deep; locations having a greater depth are less favoured. An adequate supply of aquatic animal life is essential for provisioning the ducklings. The young feed mainly on flies and aquatic larvae, usually *Chironomidae*, in the first few days of life before turning to larger crustaceans such as zebra mussel *Driessena polymorpha* and spire-snail *Potamopyrgus jenkensi* (Owen *et al.* 1986). In some areas, the slower stretches of our rivers can also support the species, although this is not a commonly available habitat in the county as all of our principal rivers have a largely upland course for much of their length. Perhaps less than 5% of the county's breeding pairs favour this habitat, compared to over 75% which can be found on open waters larger than one hectare, this being the habitat of choice (Marchant *et al.* 1990). The birds nest on islands or along the bank side, being most successful where the vegetation is tall. The highest density of occupied tetrads occurs on the North Tees Marshes, along the Wear Valley and on the Teesdale reservoirs.

According to Armstrong, by 1993, numbers for the county had reached over 30 pairs. Unlike the Pochard, the Tufted Duck is relatively tolerant of humans and breeding pairs can be found on suburban park lakes, which also serve as an important winter resource. In this context breeding has been noted at Shibdon Pond in 1993, Saltwell Park Lake in 1999, Clockburn Lake in 2000 and also Mowbray Park Lake, Sunderland. In June 2002 a pair with 6 ducklings was seen on the small pool at Whitburn Coastal Park. The Shibdon record was the beginning of a colonisation process for Gateshead and by 1997 six pairs were breeding at that site, where none had been reported in the late 1980s (Bowey *et al.* 1993). In the same area, breeding also occurred at Acer (Dunston) Pond between 1993 and 1995, Axwell Park Lake in 1994, Far Pasture in 2001 and Lamesley Pastures (Water Meadows). In 2000, a local expansion that illustrated the process occurring elsewhere. In 2000 at least 92 young were hatched from 18 broods, on seven different waters throughout the county.

Around Teesside, numbers rapidly increased from 13 ducklings produced by two broods in 1976 to 130 ducklings from 21 broods a decade later. The present estimate for the North Tees Marshes over the survey period for the Cleveland Breeding Atlas is around 50-55 breeding pairs with 16-20 pairs in any one year (Joynt *et al.* 2008).

Between 2003 and 2010 successful breeding, where more than one pair was present, was recorded at a minimum of 20 sites away from the North Tees Marshes. These sites included: Brasside Ponds, Castle Lake (Bishop Middleham), Cowpen Bewley, Crookfoot Reservoir, Herrington CP, Hurworth Burn Reservoir, Joe's Pond, Lamesley reedbeds, Low Barns, Mowbray Park Lake, Rainton Meadows, Seaton Pond, Sedgeletch, Shibdon Pond, WWT Washington and the Sedgefield village pond. Pairs were also recorded at Derwent and Grassholme Reservoirs, but the furthest west where breeding was proven was at Blackton Reservoir, in 2008. The overall breeding estimate for the county is of around 60-70 pairs per annum.

In the summer of 1995, in an unusual breeding incident, a young Tufted Duck duckling was found on a small, formal garden pond (of under 2m diameter), in Axwell Park, Gateshead. The pond was roughly 250m equidistant from two sites where the species was breeding, Shibdon Pond and Axwell Park Lake (Armstrong 1997). The duckling was uninjured and very active. There was no prospect of the species having bred at the site, or of a female Tufted Duck 'dump-laying' in a Mallard *Anas platyrhynchos* nest. The duckling was 'fostered' into a brood of ducklings of approximately similar age on Axwell Park Lake, apparently successfully based on later observations.

Recent non-breeding status

There was a small but steady increase in the number of Tufted Duck recorded as wintering in County Durham during much of the 20th century, but especially in the last 30-35 years of this period. Temperley (1951) recorded flocks upwards of 50 at Hebburn Ponds. The principle wintering areas during the 1980s and 1990s were: Brasside Ponds, Hetton Lyons, Hurworth Burn Reservoir, McNeil Bottoms, Shibdon Pond, WWT Washington and Witton-le-Wear. Shibdon produced counts of an estimated 600 birds on 3 February 1979 and 400 in January 1984. Since the peak of the 1980s, numbers have declined considerably at all of these sites, with Shibdon holding a maximum winter count of only 20-30 individuals in 2000.

During the 1990s, the Marine Park Lake at South Shields became a significant wintering site. This seems to have developed because local people were used to feeding the flock of moulting Mute Swans, *Cygnus olor*, a behavioural pattern from which the Tufted Ducks became the unintended beneficiaries. For example, 201 birds were noted in January 2001, the largest count for the whole of Durham that year. Recently many more birds have become tame enough to take bread, notably at less urbanised sites such as Shibdon Pond and Rainton Meadows.

By the middle of the first decade of the 2000s, the mid-winter maximum counts on the main waters around the county had declined dramatically from the peak of the late 1970s and early 1980s, possibly as a result of the milder winters experienced over the period in north east England. Maximum counts now tend to occur much later in the winter, or even into early spring (e.g. 235 on the North Tees Marshes in April 2007). The traditional peaks which occurred in January or February were often associated with freezing events in the Baltic Sea, where the largest winter gatherings tend to occur. During prolonged cold weather in our own region, birds often resort to rivers. Temperley (1951) noted the species using the polluted Tyne at Scotswood Bridge in hard weather and an exceptional winter gathering of 500 birds occurred again on the Tyne at Elswick in February 1942 (Kerr 2001). In 2006, freezing conditions at WWT Washington forced 92 birds on to the Wear at Cox Green.

Contrary to the observed trend of high counts occurring in late winter since the 1980s, the maximum counts for Teesmouth now largely peak in late summer. For example, 245 around the North Tees Marshes in July 2003; 276 in August 2006 and 362 on 10 September 2009 (Blick 2009), a tendency which suggests that these are post-breeding congregations of birds from our region.

Coastal passage in the region tends to be relatively light, being noted mainly in the autumn, largely between August and November, but with smaller movements from April to June. In 2007, coastal passage was noted on 44 dates from the Whitburn Observatory, between 21 April and November. A daily maximum of no more than 10-30 individuals is normal for both autumn and winter passage. The highest recorded count of birds at the coast was 175, flying north past Hartlepool on 15 December 1963 (Blick 2009).

Distribution & movements

The Tufted Duck breeds eastwards across northern Europe from Iceland to Siberia, but has only been breeding in the British Isles since the middle of the 19th century. Since that time it has spread across much of the United Kingdom, the culmination of a westwards expansion in range that started over 200 years ago. Wintering birds in Britain include British breeding stock and birds from further north and east, including Iceland, Scandinavia and Russia.

Clearly some Durham wintering birds move further to the south and west in hard weather, as demonstrated by the female ringed at Marine Park Lake, South Shields on 10 November 1996, which was shot at Lough Sheelin, in the Republic of Ireland on 14 November 1998 (Arsmtong 1999b).

Scaup (Greater Scaup)
Aythya marila

An uncommon passage and winter visitor, occasionally seen inland.

Historical review

The Scaup is a freshwater diving duck in summer and a maritime one in winter (Ogilvie 1975). It breeds on large lakes and beside rivers, where there is some cover. As a winter visitor to England it prefers shallow coastal waters and brackish estuaries, feeding at night on mussels, crustaceans, ragworms and small fish (Brown & Grice 2005). During the day the birds often fly out to sea to roost in large flocks.

Nelson writing in 1907 said that *"before the advent of steamships"* and increased industrialisation, the Scaup was *"one of the commonest of sea ducks"* at Teesmouth. Sir Cuthbert Sharp in his *List of Birds observed at Hartlepool* (1816), noted that in the winter of 1788-89, *"above a thousand were caught in a week,"* and sold for one shilling per dozen. Hutchinson (1840) confirmed the Scaup as *"a very abundant duck"*, which *"usually confines itself to the coast, especially to those parts in the vicinity of the Tees"*.

More recently in the mid-20th century, Temperley thought the Scaup a *"common winter visitor to the coast"*, which was occasionally met with on fresh water. He noted that it preferred rocky rather than sandy substrates and that one of its favourite haunts was off the mouth of Horden Dene. Temperley (1951) also remarked that small numbers occasionally entered the Tyne and could be seen at Jarrow Slake, or on the fresh water at Hebburn Ponds. He considered that the Scaup rarely ventured further inland, writing that the bird had been seen at Hurworth Burn Reservoir only once in twenty years, though W.B Alexander had reported five there on 12 October 1929. Almond *et al.* (1939) had also logged inland records of two drakes on Darlington Boating Lake in December 1936 and a duck in December 1938. They noted that the species was most numerous at Teesmouth after severe weather, when a flock of more than a hundred were observed in December 1938.

From 1946 onwards Scaup were not plentiful at Teesmouth, until a considerable influx of birds occurred along the whole of the North East coast in 1954. Two hundred and fifty were at Seal Sands in March, with three drakes remaining there throughout the summer (Blick 2009). At the same time fourteen birds were noted at Jarrow Slake. In February of the same year, three females were found on Thornley Flooded Field, near Shotton Colliery, an area to which the species continued to make occasional visits. David Simpson located two males on Shotton Pond from May to September 1963, observing the birds from his bedroom window.

The next record-breaking passage occurred off Hartlepool in 1962, when, during an eight hour watch 788 Scaup flew north in a northerly gale on 18 November. On the same date birds were noted inland on Crookfoot Reservoir (Coulson 1963). There were a further series of inland records at Hurworth Burn Reservoir, with six birds present on 29 September 1964, one in October 1967 and others in November and December 1968. During the late 1960s, inland birds were noted at Derwent Reservoir and Barmston, although by the end of the decade wintering numbers at Teesmouth had fallen significantly.

Recent breeding status

During the survey period for the *Atlas*, a number of individuals were recorded on wetland habitats during the breeding season, but almost all of these appeared to be late wintering birds. The most noteworthy were a pair on Hury Reservoir in late April 1992, three birds at Derwent Reservoir in May 1988 and a drake that remained on Shibdon Pond in early April 1988, which was seen displaying to female Tufted Ducks *Aythya fuligula* (Armstrong 1988-1993). In May 1994 a female at Shibdon Pond pair-bonded to a drake Tufted Duck and copulation was seen to be attempted. Yet in none of these cases was there any suggestion of successful breeding (Westerberg & Bowey 2000). Although there have been a number of notable summer records for the species, there is clearly no established breeding population within the county. A drake at WWT Washington in June and July 1995 was considered an unusual summer record. In 2006, a duck took up residence on the Reclamation Pond on 25 March where it remained until the 8 May. Another out of season drake on Greenabella Marsh during mid-June 2009 was considered the third breeding season record in the first decade of the 21st century. Scaup tend to wander inland most often during spring and late summer. The majority of these records occur at periods of passage, mostly March to April and August to October, but some occur in winter when birds can be found among other groups of diving ducks.

Recent non-breeding status

During the 1970s and 1980s there was a general decline in the size of the wintering population of Scaup at Teesmouth, in spite of occasional peaks, for example the 120 birds noted at Seal Sands on 18 February 1970. In fact, during 1987, inland records at Derwent Reservoir, with birds present in every month of the year, culminated with a flock of nine in September, which proved to be a greater number than the mid-winter peak at Teesmouth. The largest count for the species for many years occurred at Seal Sands with 27 birds in March 1986, but for the rest of the decade winter numbers here were less than double figures. Despite the downward trend, Whitburn observatory logged 313 birds flying north on 7 February 1991, the largest coastal movement for some years. The influx seems to have been prompted by severe weather leading to the freezing of the Baltic. The subsequent movement inland led to twelve different birds being present at Gateshead, the largest number being a flock of eight at Ryton Willows during early February (Bowey *et al.* 1993). During the rest of the decade small numbers continued to winter at Teesmouth, with a mere five birds present in 1993, a derisory total compared with the numbers of the past. However, Hartlepool Docks managed to produce thirty birds on 6 January 1997.

Today, most reports of this species come from Teesmouth, with a scattering of birds inland and, depending on weather conditions, some notable periods of passage, so that the bird's status seems to be moving ever closer to that of a pure passage migrant. During the first decade of the 2000s, winter numbers have remained consistently low and the large counts around the county's south east coast and Tees estuary are sadly now no more. On 1 January 2001, for example, there were only three birds at Greatham Creek.

There were over fifty inland records of Scaup during the 1970s, many more than in previous years, involving at least seventeen wetland sites. Some locations held birds at several different times, such as Barmston Pond, Shibdon Pond and Hurworth Burn Reservoir, with Derwent Reservoir attracting the species on at least four separate occasions. An examination of the Shibdon Pond records over a ten year period, suggest an interesting trend. There were more than fifteen records of the species from 1987 to 1996, sometimes up to four birds together. Moreover, five different birds passed through the site during November and December 1987. There were also a small number of August records, with a pair present on 11 August 1987, a female on 29 August 1991 and another female in the same month in 1992. In addition, a juvenile was at Shibdon on 10 September 1993, a male on 22 December 1993 and a drake 28 January 1996. These findings seem to indicate an autumn overland movement, perhaps via the Tyne Valley, with birds occasionally dropping off at Shibdon. The records from Derwent Reservoir during the 1970s tend to confirm this observation. Inland records also seem to have fallen, although in 2007 birds were noted on seven fresh water sites including South Shields Marine Park Lake, Hetton Lyons and Low Barns.

Passage tends to begin in autumn and is usually noted at Hartlepool Headland and Whitburn; however, it is now considerably less than during the 1970s and 1980s. In 1972, there was an exceptional winter passage off Hartlepool, with a total of 208 birds flying north on 18 January. The following year at Whitburn, on 30 November, birds were passing north at up to forty an hour. The largest sea watch total for the county was of 837 birds off Hartlepool Headland on 18 January 1962. These counts date from a period when the wintering population of Scaup in the Firth of Forth was around 25,000 birds, since which time numbers have declined dramatically. During the 1990s, the heaviest recorded passage was off Whitburn in Dec 1995, when 142 individuals flew north and 37 south, presumably a result of birds relocating to the Durham coast from other wintering sites in Britain. Throughout the 2000s, regular autumn records occurred from late August through to early November, with flocks of 30 to 40 birds on many dates. Exceptionally for recent years, 134 birds passed Hartlepool Headland on 9 November 2003.

Distribution & movements

The Scaup has an Arctic circumpolar distribution, breeding from Iceland to eastern Siberia. The species has occasionally bred in Britain, although this is well to the south of its normal range, but unfortunately not in County Durham. Birds primarily winter in north west Europe, particularly in the Baltic and Wadden Sea and around the British coast.

The origin of some of the county's wintering birds was hinted at by the bird that was ringed at Husavik, Iceland on 19 January 1930, which was recovered in Durham, presumably as a tide-line corpse, on 13 October of the same year (Witherby & Peach 1932).

Lesser Scaup
Aythya affinis

A very rare vagrant from North America; three records.

The first county record was of an adult drake found by Richard Taylor at Saltholme Pools, Teesmouth on 6 June 1999 whilst performing survey work for the *Breeding Birds of Cleveland*; there were fewer than 30 records for Britain at the time. The bird was often to be found on the nearby Reclamation Pond, particularly during the latter part of its stay. It remained in the area until 31 July, often visiting the nearby Reclamation Pond, by which time it had almost completed its moult into eclipse. Although expected to remain for the rest of the summer, it must have completed primary moult elsewhere, as it was not seen subsequently (*British Birds* 93: 53). There were fewer than 30 records for Britain at this time.

The second record was less straightforward as it involved a female at Stoneybeck Lake, Bishop Middleham from at least 22 February until 23 April 2003 (*British Birds* 97: 566). This bird was initially considered an 'odd-looking' Scaup, before being conclusively identified on 26 March. The identification of this bird was credited to Alan Wheeldon by BBRC, but the original finder remains unknown.

The third record came three years later when a female was located, amongst other diving duck, on the Reclamation Pond, Teesmouth on 7 January 2006. This bird, thought to be in its first-winter, remained until 15 January, occasionally visiting Dorman's Pool, although it was often elusive during its nine-day stay (*British Birds* 100: 698). What was thought to be the same bird, and accepted as such by BBRC, was at WWT Caerlaverock,

Dumfries and Galloway from 17 January to 10 March 2006, and again from 27 November 2006 to 13 March 2007 (Fraser *et al.* 2007, Hudson *et al.* 2008).

Distribution & movements

The Lesser Scaup is an abundant and widespread breeding species in North America. Its breeding range stretches from Canada, across central Alaska and the northern United States. In winter it can be found throughout much of the United States, southwards to northern South America. It was formally a very rare vagrant to the British Isles with the first record in 1987, though increased knowledge of key identification criteria has proven the species to be an annual and increasingly regular vagrant to Britain. Just over twenty years after the first record, there had been 157 records accepted by the BBRC by the end of 2010, although the true number of individuals is difficult to assess.

Eider (Common Eider)
Somateria mollissima

Sponsored by

NTBC

An uncommon passage migrant, winter visitor and a recently established breeding species, in very small numbers.

Historical review

The oldest record of Eider for Durham is an ancient one, with remains discovered in an archaeological dig, dating back to the medieval period, from the early settlements on Hartlepool Headland (Yalden & Albarella 2009). This is one the county's earliest recorded species, known as Cuthbert's Duck, or Cuddy's Duck, because of links with St. Cuthbert on the Farne Islands in Northumberland. The Eider is also regarded as the first bird in the world to be given conservation protection, as St. Cuthbert offered the species sanctuary on the Farnes in the 7th century.

There are several references to Eider reported by Fowler (1898), in his study of the Bursar's rolls for the Priory of Durham. The roll for 1380/1381 recorded the payment of a shilling to a painter in Newcastle for a picture of an Eider. A further reference from Gurney provides the earliest instance of the use of Eider-down for the stuffing of bedding and cushions. It relates to an inventory taken on 14 June 1397, and refers to *"two small cushioned couches of which one is made from St. Cuthbert's down"*, or as originally written in Latin, *Iternij parva pulvinaria quorum j est de Cuthbert doun*. From being specially mentioned we can conclude that this item was of particular value. Similar references occur again in the inventories for 1401 and 1418. Whether the Eiders which provided the down came from Durham, or more likely nearby Northumberland, we do not know.

In 1815, Sir Cuthbert Sharp wrote that it was *"extremely rare"* at Hartlepool, and he recorded the shooting of one there in 1789 (Sharp 1816). A specimen in Marmaduke Tunstall's collection had been shot at Hartlepool sometime before 1827, and was possibly the same bird referred to by Sharp. Lofthouse (1887) said that it has *"been met with a few times on the Tees"*. Nelson (1907) recorded just six occurrences for the Tees Bay prior to 1907 (Blick 2009). However, he did report that there was one present in the winter 1902/1903, and more significantly, a flock of *circa.* 20 birds near the Tees breakwater on 2 February 1905. Through the early part of the 20th century Bishop said he saw it at Teesmouth on odd occasions inside the estuary, but it was clearly not a common bird at that time; nor was it well documented along the rest of the Durham coast.

George Temperley (1951) regarded the Eider as a *"rare winter visitor off the coast and then they are usually in flight…. they rarely find feeding grounds along the Durham coast sufficiently attractive to retain them and at no point is the coastline suitable as a breeding haunt"*. He observed that the bird was probably *"never more than a casual visitor"* even before the Durham coast was impacted by industrialisation. Since regular birdwatching began in the 1950s, many more have been seen. This may partly be due to increased observer activity, but probably is also a result of the protection of the Farne Islands, the main breeding station to the north of the county. Certainly, a distinct increase in numbers became evident in the mid-1950s (Taverner 1959). The 1950s also saw a national increase, which was reflected in the number of birds occurring in Durham. By 1953, the species was more abundant than ever in Northumberland and reports were increasingly frequent in Durham, as far south as the Tees. The ornithological report for 1955 commented that *"birds are now wintering more frequently off the Durham coast"* (Temperley 1956). By 1957, it was more numerous than ever before at Teesmouth, with flocks of up to 30 present in late December and 34 in the Tees Bay in January 1958 (Blick 2009).

By the late 1950s, Taverner believed that the increases were due to more than simply the movement of birds from Northumberland. He wrote that *"Durham is one of the most interesting counties for it has followed the general pattern. This suggests that the Eiders wintering on the Durham coast come from the same place as those wintering further south and there is now a great deal of evidence that points to the Dutch colonies as the source of these birds"* (Taverner 1959). Furthermore, *"up to now, I have assumed that the Eiders wintering in Durham are from the Farne Islands colony, but this recent evidence suggests that such may not be the case. The breeding numbers at the Farne Islands have shown no fluctuations that could account for the pattern described above and it seems likely that some of the Durham birds are from the Netherlands"* (Taverner 1967).

This upward trend continued through the 1960s, with a new peak of 93 being reached for the Tees Bay on 12 January 1969 (Parrack *et al.* 1972). Birds lingered at sites such as the North Gare until March and a few remained throughout the summer, which was then considered unusual. By 1971, the species was recorded in the county in every month, with reports from Whitburn, Hartlepool and Seaton Carew, but numbers remained generally low, with the highest count that year being just 21 at Hartlepool on 7 November (Coulson 1972).

This period also coincided with some expansion in the species' breeding range in Northumberland, as during the 1960s Eiders began breeding increasingly along the Northumberland mainland coast (as opposed to just the Farnes), with, for example, ten mainland pairs by 1964 (Kerr 2001).

The first and only historic inland record concerned an immature drake at Hurworth Burn Reservoir on 19 November 1959 (Grey 1960).

Recent breeding status

Prior to the beginning of the 21st century, there was no suggestion of Eiders breeding in the county, unlike neighbouring Northumberland where references go back well over a thousand years to the time of Saint Cuthbert. During the early to late 1980s and through the early to mid-1990s, summering birds were increasingly noted off the Durham coast. These sightings were mainly concentrated in the South Shields to Sunderland area (Whitburn Steel in particular), with increasing regularity and in every spring and summer between 1988 and 1993 (Armstrong 1989-1994). This trend continued through the decade and occasional birds would be noted on shore at less disturbed locations. In 1995, a small number were noted off Whitburn throughout the year although they were less common during May and June. Perhaps more significantly, the following year 17 were at Sunderland Docks in July (Armstrong 1997 & 1998).

In the summer of 2001, up to 15 birds were present off Whitburn throughout June, with movement of birds past that point noted on several days. Elsewhere along the coast in summer, birds were present off South Shields, Sunderland, Blackhall Rocks and in Hartlepool Bay (Siggens 2005).

Birds were first recorded as breeding in the county in 2002, when two nests were found in Sunderland South Docks. Only a single nest was found the following year. The birds were nesting in an area which was difficult to search, and only about a third of the possible breeding area was checked. There were up to six males, five females and a single immature male present in early April 2003 (J. Coulson pers. comm.). A flock of flightless, part-grown juveniles was also seen on the sea in mid-July of that year, approximately four miles to the south of this location (K. Bowey pers. comm.), the nearest breeding site to the north being Coquet Island, Northumberland some 50 kilometres away. Both sexes were also present in 2004 and 2005, but nests were not searched for. One female was noted to be carrying a ring, probably as a breeding bird from a study on Coquet Island, but it was not possible to read the ring number to confirm this (J. Coulson pers. comm.).

There have been no known successful breeding attempts since, although nine were noted in Sunderland Harbour on 28 May 2007 (Newsome 2008). The lack of public access and hence comparatively undisturbed nature of Sunderland South Docks may have favoured breeding Eider; this also makes it more likely that breeding could take place again here without being noticed. Conversely, the increase in the number of predators, particularly Foxes *Vulpes vulpes*, as the Docks became derelict, may have contributed to the Eiders' departure, just as it did the nearby Common Tern *Sterna hirundo* colony (J. Coulson pers. comm.).

Away from the mouth of the Wear, Blick (2009) noted that a summering flock appeared in the Hartlepool area in 1991, with up to 30 birds being seen, and this flock has persisted since then, with up to 52 there in 1993; 49 in 1996; 40 in 2004 and rising to 114 in 2008. It is therefore also possible that Eider may breed in the Teesmouth area in the future (Blick 2009).

Recent non-breeding status

Today, the Eider is principally a regular winter visitor to the Durham coastline, although in the past it seems to have been a genuinely scarce species, as noted previously. From the early 1980s there has been a small summer flock present.

The species' most favoured haunt within the county includes virtually the whole of the Tees Bay from Hartlepool Headland to the mouth of the Tees (Blick 2009). Outside of the breeding season, birds might be noted anywhere along the Durham coast, particularly the stretch from South Shields to Sunderland, Whitburn Steel and just south of Hendon. The latter site produced a peak count of 346 on the sea at Ryhope on 20 March 1997 (Armstrong 1999).

Increasingly in the late 20th century, sea-ducks including Eider have built up in numbers off the formerly despoiled stretch of coastline between Dawdon, south of Seaham, and Crimdon, to the north of Hartlepool. In the 1980s hardly any birds were recorded wintering in this area, but spoil dumping on the coast ended in 1993 and during the 1990s small numbers began to appear on the sea, with 29 off Blackhall Rocks in October 1996 (Armstrong 1998). A sharp increase followed as coal waste cleared and the marine fauna recovered, by the end of the first decade of the 21st century. Blackhall Rocks became a favoured location for this wintering flock, with numbers typically peaking in late autumn. Up to 60 were noted there in late September/early October 2003, by 2007 the flock had expanded to 210 on 11 November, and a peak of 340 was reached on 26 September 2008 (Newsome 2007-2009).

There is likely to be some interchange between this gathering and the regular wintering flock on the north side of Hartlepool Headland. Although numbers there have fluctuated, there has been an upward trend since the 1980s, and a sharp increase since 2005, with the flock increasing from an annual peak of just 44 in 2004 to 358 in 2005. The highest count of birds on the sea in the county was at Hartlepool in 2008, with 413 on 5 January (Joynt 2009).

During most years, passage is noted at Whitburn Observatory in every month of the year, except perhaps July, though many records no doubt relate to birds making only local movements. Numbers peak in late autumn and the highest counts off Whitburn include 300 north on 13 November 1983, an exceptional count at that time, 325 north and 10 south on 9 December 1990, and 377 on 30 October 2008 (Baldridge 1984, Armstrong 1991, Newsome 2009). The heaviest movement in the county to date came between 11th and 13 November 1999, when 1,718 flew north past Whitburn in just 16 hours of sea watching. This equates to 150-200 birds per hour when the observatory was occupied, suggesting that the total movement may have involved up to 3,000 birds, assuming that it continued during daylight hours when the observatory was empty (Armstrong 2003).

Passage is also regularly recorded from Hartlepool Headland, where high counts have included 52 on 2 December 1978, 173 on 13 November 1983, 574 on 7 November 1995 and 1,460 on 12 November 1999, indicating a steady increase at least in the last 20 years of the 20th century (Blick 2009). The final count here was during the record movement described above at Whitburn.

Nationally, the Eider population increased markedly over the past 200 years, but stabilised in the early 1990s and remained stable up to around the year 2000 at least (JNCC 2010). The increase in numbers within the county is therefore consistent with this pattern up to the early 1990s. However, the subsequent continued increase in wintering birds on the Durham coast may be a more local phenomenon, linked to the cessation of the dumping of coal waste, as described above.

Inland records have tended to come from the mouths of the main river systems, Tyne, Wear and Tees, such as four noted feeding on crabs on the River Wear at Sunderland in January 2001. Eiders sometimes move up the Wear as far as Wearmouth Bridge, and the birds seen here in the early years of the 21st century may have been breeding in the nearby South Docks (P.T. Bell pers. obs.). Inland birds have regularly been noted at Jarrow Slake since 2001, including a maximum of 8 there on 13 January 2008. Five were also observed further up the Tyne at Hebburn on 3 March 2008. A single on the Reclamation Pond, Teesmouth on 31 October 2008, and a drake at Saltholme Pools on 30 November 2009 were also unusual (Joynt 2009 & 2010).

Genuinely inland records are extremely rare in Durham. One of the most unusual reports occurred on 14 November 1995, when a flock of ten birds, seven females and three sub-adult males, was found on the Tyne at Timber Beach, Dunston, some 14 miles inland. The following week, on 20 November, an immature male was noted at Ryton Willows and three birds were reported at various locations along the Tyne, namely Blaydon Haughs, Scotswood, Newburn and, furthest west, at Wylam. Undoubtedly, they were all part of the larger original flock. The occurrence of these inland birds would appear to be associated with the heavy passage of Eider noted on the coast at this time, 391 moving north off Whitburn in the two days prior to the discovery of the birds at Dunston (Armstrong 1997).

Prior to these sightings, there were only three inland records of the species for the county, a dead female at Longnewton Reservoir on 27 February 1983, an immature at Brasside Pond on 4 November 1993 (Baldridge 1984, Armstrong 1994), and the one historic record from 1959 noted previously.

Distribution & movements

This species exhibits a circumpolar breeding distribution, inhabiting coastal habitats in Arctic and sub-Arctic zones. The southern limit of its breeding distribution in Europe is northern England, with significant numbers breeding around the Farne Islands in Northumberland, and across to the Netherlands. There are just two known ringing recoveries within the county, but predictably, both individuals were ringed on the Farne Islands, in May 1965 and June 1980; both were found dead at Hartlepool in May 1968 and April 1991 respectively (Hickling 1970 & Bell 1992).

King Eider
Somateria spectabilis

Very rare vagrant from the Arctic: one record.

The sole record for County Durham is of a second-summer male at Hartlepool Headland on 4 July 1998 (*British Birds* 92: 567). The bird was found by Mike Gee at around 2:25pm on the rocks between the Heugh Breakwater and the Old Pier, but it flew north just after 3pm. Around three hours previous to this, at 11:20am, Martin Blick had observed the same bird at South Gare, North Yorkshire whilst counting Eider *Somateria molissima*; it was seen to fly north at around 1:30pm, presumably arriving at Hartlepool shortly afterwards. A drake, it was aged as a second-summer individual on the basis of the relatively small bulge to the orange shield and apparently small 'sails', though in all other respects it resembled an adult male. This was not only the first record for County Durham, but also only the second for Yorkshire, the first being as long ago as 1846 (Mather 1986).

Distribution & movements

King Eider has a circumpolar breeding distribution in the high Arctic, with the most northerly breeders dispersing southwards during winter to escape harsh conditions, though some remain on the breeding grounds year round. British records are thought to emanate mainly from the large Norwegian wintering population, with some birds dispersing further south into the North Sea. It is a rare but annual vagrant to Britain, though most records are from Scotland, and in particular the Shetland Isles, with very few records further south. Out of a total of 226 British records by the end of 2010, just 36 have reached England, plus a single bird in Wales.

Long-tailed Duck
Clangula hyemalis

A scarce passage and winter visitor, occurring rarely in summer.

Historical review

According to Temperley (1951), the first record of this species for Great Britain occurred in County Durham in 1678 when a pair was noted *"feeding together for several days on the River Tees below Barnard Castle"* (Willughby 1678). Tristram (1905) considered that the bird occurred *"frequently on the coast in winter"*, with *"many being shot at Teesmouth"*. The record of forty individuals seen together in the Tees Bay in October 1887 (Blick 2009) was clearly exceptional and Temperley writing in 1953 felt that as a winter visitor to the coast, the bird was *"never numerous"*. This is in strong contrast to the bird's status in Northumberland, where it has always been more common, with the area around Lindisfarne achieving nationally important status, holding the largest wintering population in England in 1981 (Day & Hodgson 2003).

As a maritime diving duck, the Long-tailed Duck tends to remain well offshore. The birds can sometimes dive to depths of more than nine metres, exceeding any other duck species. They seem to prefer waters with sandy substrates and feed on mussels, cockles, crustaceans and small fish, and exceptionally grain may be taken (Ogilvie 1975). Their plumage and 'yodelling' call has led to a host of alternative names, such as: Ice Duck, Sea Pheasant, the Calloo Duck or Old Squaw. The bird's strange cry is sometimes transcribed as *"coal and candle licht"*, Bannerman noting that the species has a good Scot's accent (Bannerman 1953). Birds tend to arrive in Durham at the end of October and usually depart in March. The species was rarely observed on inland sites in the past but its offshore status in County Durham seems to have altered little since Temperley's time.

A flock of ten, found wintering off Seaton Carew in 1962 was thought to be the largest flock at Teesmouth for over 20 years. The following year, unusually high numbers were noted along the east coast from Durham to Norfolk and a December record of nineteen on the sea at Hartlepool with a further nine flying north, was considered the highest for the county at that time (Bell 1963). For the rest of that decade the number of birds wintering around Hartlepool Bay and Teesmouth crept into double figures in most winters, with the highest count being fifteen on the Tees Estuary on 4 January 1964. Through the 1970s, wintering numbers around Teesmouth were low with just one to five birds being observed in some years. A gathering of 79 off Hartlepool on 21 October 1972 was exceptional.

Temperley (1951) thought that the duck was a very rare vagrant to inland waters. Joseph Bishop noticed a drake frequenting Charlton's Pond near Billingham in December 1931 and M. G. Robinson observed one on the Skerne at Darlington Park in January 1939. During the winter of 1943/1944, an immature bird spent nearly four months at Fulwell Waterworks Pond, Sunderland, almost a mile from the sea; it was present from 26 December to 5 April 1944. From the late 1950s onwards there were increasing reports of birds inland. The trend began in 1959 when a bird was at Fishburn Lake from 20th to 29 October. There were subsequent inland records in at least three years during the 1960s: an immature at Hurworth Burn Reservoir from 13th to 20 October 1960; one at Charlton's Pond on 25 November 1961; another at Fishburn Lake on 13th to 18 November 1962; a female/immature on Derwent Reservoir in January 1968; and, a male on a brickyard pond at Shotton Colliery on 16 April 1968. This increase in inland records continued through the 1970s with birds reported at a dozen or more sites including Crookfoot Reservoir, where a drake was noted displaying to a Goldeneye *Bucephula clangula* during its long stay there in March and April 1974. At Derwent Reservoir through the 1970s, the species became of almost annual occurrence.

Recent status

Today, the Long-tailed Duck remains a winter and passage visitor along the Durham coastline, usually between mid-October and April. A few birds winter on Seal Sands and on the sea between South Shields and Hartlepool Bay. Most records occur towards the end of the year at typical locations around Hartlepool and Teesmouth, but rarely exceed more than ten to fifteen birds. Although 21 birds flew north past Whitburn over two days in November 2000, the general impression is of a slight downward trend, contrary to the increased observer activity over recent decades. In 2001, only one bird was reported during the first quarter, a single female flying south past Whitburn Observatory, whilst three or four birds were noted during the summer months. In 2005, an unusual sighting of a pair of birds in summer plumage occurred on the sea off Seaham on 3 September.

166

During the *Atlas* survey period (Westerberg & Bowey 2000), as would be expected, there was no suggestion of successful breeding within the county, although a drake was present at Hetton Lyons Lake on 3 June 1993 (Armstrong 1988-1993). Another out-of-season record concerned a drake in summer plumage on Saltholme Pools from 22 May to 22 June 2007.

Inland records, although not common, are no longer considered that unusual, with an average of two birds every three years or so; some of these may remain for long periods. The west Gateshead area has produced a number of intriguing sightings. In early 1985, a female spent the first quarter of the year on the Tyne at Ryton Willows amongst Goldeneye, occasionally being seen as far downstream as Blaydon. The following year, a duck was seen on 9 January, at Wylam, upstream of the previous year's sightings. In late 1987, a duck was found at Shibdon Pond on 21 December, remaining in the area until 13 April 1988 and commuting regularly between Shibdon and the River Derwent at Swalwell. It is possible that these records may all refer to the same bird returning in successive winters.

Over the last thirty years (1980-2010), single birds have been seen at most of the county's main inland waters, including: Brasside Ponds, Crookfoot Reservoir, Charlton's Pond, Derwent Reservoir, Longnewton Reservoir, Low Barns, Shibdon Pond, Smiddyshaw Reservoir, WWT Washington and sites around the North Tees Marshes; 1994 was a particularly good year, when four sites held birds during the year.

During the first decade of the 21st century, inland birds have been recorded in at least six years. A bird was noted at Silksworth Lake on 2 October 2003. The following year a female toured some of the county's northern wetlands, appearing at Seaton Pond on 2 November before moving in succession to Hetton Lyons, Rainton Meadows and Brasside, before settling at Hetton Lyons and remaining until 14 December. Most unexpectedly, a male appeared in its place, staying until 14 January 2005. A female at Longnewton Reservoir on 29 November 2004 stayed until 18 March 2005. A drake present at Herrington CP on 17 November 2006 was found dead on 10 December. A juvenile female appeared at Hurworth Burn Reservoir on 7 December 2008 and stayed until 22 March 2009; before relocating to Crookfoot Reservoir from 6th to 22 April.

Sea passage is most obvious along the Durham coast in October and November, especially when onshore winds drive birds nearer land. Peak numbers are usually recorded in November and are usually of counts of birds passing Whitburn or Hartlepool. The largest day-count occurred when 79 were 'logged' off Hartlepool Headland on 21 October 1972. The only number approaching that total in more recent times was of 53 flying north, at the same location, on 9 November 2001 (Blick 2009), although observers at Whitburn logged 27 flying north on both 1 November 2006 and 31 October 2008.

Distribution & movements

Long-tailed Ducks breed throughout the Arctic, wintering along coastlines south of their breeding range. British wintering birds probably originate from Fennoscandia and Russia. The greatest number of birds is usually noted in Durham between mid-October and early April and encompasses both wintering and passage individuals. Some birds are occasionally seen in May and September and, very rarely, in the period from June to August (Blick 2009).

Common Scoter
Melanitta nigra

A common passage migrant and winter visitor to the Durham coast.

Historical review

An intriguing record, which indicates the historical presence of this species within the county, came from the discovery of 'scoter' bones during the excavation of a mediaeval settlement on Hartlepool Headland. It seems highly likely that the remains were of *Melanitta nigra* rather than any other scoter species (Yalden & Albarella 2009).

This is the most frequently encountered sea duck in Durham with birds noted throughout the year, something which has been the case for more than a century. During the 19th century this species was described as a common winter visitor off the Tees estuary, Seaton Carew and Hartlepool (Blick 2009). In the mid-20th century, Temperley

(1951) regarded it as a *"common winter visitor, often in large flocks"*, observing that these contained many immature individuals. He gave a wonderful description of the bird's herd instinct, *"They fly in dense packs, arising and alighting as a unit. When feeding the whole flock will often submerge simultaneously"*. He noted that winter flocks usually keep well out to sea and when disturbed or changing their feeding grounds, appear in flight *"like some dense black smoke screen on the horizon"*. Temperley also noted that the bird was a passage migrant, remarking that *"only very occasionally was it encountered inland"*. In this respect, Hutchinson documented seven birds on the River Wear at Durham in April 1837, most of which were shot, the confused birds trying to escape by diving rather than by flight (Hutchinson 1840). Robson (1896) writing of the Derwent valley, described the Common Scoter as *"only a winter visitor"*, noting that an immature bird had been shot near Derwenthaugh. More recent inland records include a pair of Common Scoter on a flooded brickfield near Durham on 27 June 1948, indicating the potential for a breeding site within the county. Other typical inland records for this period include four birds on Tunstall Reservoir, which remained for five weeks in October 1953 and a male on Shotton Brickworks Pond in September 1953. In 1962 four birds at Smiddyshaw Reservoir on 11 August presaged a pattern that was to emerge during the latter part of the century and the early 21st century, of birds appearing in summer and early autumn often in small parties at some of the county's westerly reservoirs.

A significant feature of the Durham coast in the 1950s and the early 1960s was the large flock, of several hundred birds that wintered off Seaton Carew (Blick 2009). Over the following decades however, this site was used only intermittently and by declining numbers of birds, so that although two hundred birds were present in April 1964, there were only fifteen in March 1970.

Recent status

Between the mid-1960s and the first decade of the 21st century the noticeable decline in number of birds wintering off Seaton Carew and Teesmouth continued, but this was mitigated by the appearance of birds further north, with a flock of 514 being noted off Hartlepool in November 2005 (Blick 2009). The relocation of birds to other Durham coastal sites seems related to the improvement in the quality of sea water, following the reduction in colliery waste despoilment, ultimately leading to the recovery of marine fauna and providing an increase in foraging opportunities for sea ducks. Large numbers of birds now winter further north from Dawdon to Crimdon Dene, with numbers building up through the 1990s and 2000s. One of the largest flocks of birds in this area comprised 550 off Ryhope in November 1991, with 400 in Marsden Bay around the same time. The area of sea between Horden and Blackhall has always been a favourite haunt of the species and 1,100 birds were present there in November 2009, with at least 800 at Blackhall Rocks in February 2010. There were 750 birds present in this area on 25 July 2010, an indication that by the 21st century, birds were able to be noted off the Durham coast in every month of the year. In the North Tees area, Hartlepool Headland has been the main location for this species, with Parton Rocks attracting 160 birds on 11 March 2000. The Common Scoter now has a year round presence, although the species does not breed.

It is not unusual for the occasional bird to be found on Seal Sands or sheltering from rough seas in Hartlepool Docks during the winter, but genuine inland records have always been considered rare in Durham. Most such records concern single birds, but groups of two or three have been seen on a few occasions, usually from August to October. During the 1960s there were five records of multiple occurrences at the western reservoirs. Records have also occurred along the county's main rivers such as the adult and a first year bird that were together on the Tyne at Ryton Willows on 30 April 1988 and a drake, again at Ryton, on 21 January 1989. There seems to have been a significant increase in such inland records during the 1990s and the early part of the 21st century.

For instance, seven birds were at McNeil Bottoms on 16 June 1990 and single drakes were on the Tyne at Gateshead in February and at Dunston in August 1996. In the first decade of the 21st century, inland birds were recorded in every year, involving at least sixty eight birds at twenty seven sites. Once again small flocks were noticed on the county's westerly reservoirs. These included seventeen birds at Pow Hill Country Park, on Derwent Reservoir on 5 July 2002 and fifteen female/immature birds at Tunstall Reservoir on 19 November 2005. The summer of 2007 brought an unprecedented series of records with a flock of nine birds, three ducks and six drakes, on Smiddyshaw Reservoir on 20 June. It would seem reasonable to conclude that these birds are involved in overland migration, analogous to the large movements that have long been observed in Yorkshire (Mather 1986). Perhaps this represents a previously unrecognised trend in Durham, with birds crossing the Pennines to wintering sites in the west. Cardigan and Liverpool Bays can hold very large numbers of this species, with an estimated

79,000 at the latter site in February 2003, *"more than the previously known total for the whole of Great Britain and Ireland"* (Parkin & Knox 2010).

In Durham, most coastal movements of the species are noted from the county's two main sea watching sites at Whitburn Observatory and Hartlepool Headland. Birds are also reported from the central coastal section, at Ryhope, Horden and Blackhall. Passage can be quite heavy at times and usually starts from late June lasting through to November. Typically, the largest numbers of birds are noted in July and August, with later peaks in November and December. The July peak is associated with birds returning to British waters from their breeding grounds from further north and post-breeding moult movements, which sometimes lead to flocks cutting across country.

Weather conditions are a key determinant of autumn passage movements, which often relate to birds relocating themselves within British coastal waters. The predominant vector of movement off the Durham coast tends to be a northerly one. The scale of passage can be high, with as many as a hundred per hour heading north during the summer peak. The heaviest documented coastal passage comes from Whitburn Observatory, with month totals including 4,664 flying north in October 1991, 3,543 birds logged during November 1998 and 3,170 birds in July 2000 (Armstrong 1992, 1999b & 2005). Large counts off Hartlepool Headland include 1,850 on 20 October 1991 and 1,600 on 3 November 1998 (Blick 2009). The highest figures for one day comprised 2,813 individuals off Whitburn on 21 October 1972, a count matched exactly on 3 November 1998. The latter count was achieved in just six hours of observation, making this the heaviest rate of passage ever noted in the county. In 2008, at Whitburn, a total of 11,886 birds were recorded on 208 dates throughout that year; including 2,166 birds in 5.3 hours on 2 July (Newsome 2009). An unusual pale coffee-coloured bird was seen off Hartlepool Headland on 14 November 1964 (Blick 2009).

Distribution & movements

In the nominate form, the Common Scoter breeds *"on fresh water on tundra and taiga bogs across the northern coasts of the Palaearctic, from Iceland, through Fennoscandia to Siberia"* (Parkin & Knox 2010). Small numbers breed in Shetland, the flow country of Scotland and Ireland, but so far as is known, never in England. In the breeding season it feeds on plants, insect larvae and freshwater crustaceans, but in winter it becomes maritime, diving to a depth of two to four metres, to feed on molluscs, hence *"the mussel eaters strong heavy bill"* (Ogilvie 1975). Birds winter in the Baltic and the Kattegatt, as well as around the coasts of Great Britain. Numbers tend to vary at different sites and aerial surveys have highlighted the limitations of land based observation, concluding that far more Common Scoters winter in British waters than previously realised (Brown & Grice 2005). The tendency of flocks to form large rafts on the sea and to congregate in flight suggests their vulnerability to oil spills and the potential for injury from offshore wind farms.

Surf Scoter
Melanitta perspicillata

A very rare vagrant from North America; nine records.

The first record for County Durham was an adult drake watched flying north past Hartlepool Headland, with a pair of Common Scoter *Melanitta nigra*, shortly after 08.00hrs on 17 September 1977 (*British Birds* 71: 495). Mike Gee initially saw the party of three birds well to the south and as these flew closer, both he and Geoff Iceton observed the bird's striking white head markings and multi-coloured bill. The birds passed within 200 yards of the shore. The same bird was seen by Bryan Armstrong and Les Rimmer, heading north past Whitburn, just over half an hour later. A drake Surf Scoter was seen off Holy Island, Northumberland the following day, presumably the same bird.

All records:
1977	Hartlepool Headland & Whitburn, male, flew north, 17 September.
1981	Whitburn Observatory, male, flew north, 12 November (*British Birds* 76: 488).

1990	Hartlepool Headland, male, flew north, 17 September (*British Birds* 84: 463).
1995	Hartlepool Headland, male, flew north, 4 November.
2000	Hartlepool Headland, male, flew north, 23 September.
2003	Whitburn, male, flew south, 31 October.
2006	Whitburn, female, flew north, 14 October.
2006	Hartlepool Headland, female or immature, flew north, 11 November.
2011	Blackhall Rocks, male, 31 May to 5 June.

Most of the county records have occurred between September and November, usually during northward passage of Common Scoter. The bird at Hartlepool in November 1995 had passed Flamborough Head, Humberside earlier that morning, and was reported passing Whitburn and Newbiggin in Northumberland later in the day, giving an indication of how far such birds can move in a day. Only two birds have lingered; the male at Hartlepool in September 2000 was present on the sea for just five minutes before heading north, whilst the drake at Blackhall Rocks in 2011 provided the only occasion for observers to study this species at leisure.

Distribution & movements

Surf Scoter breeds in Alaska and across much of western and central Canada, and winters along both the Atlantic and Pacific coasts of North America. It is a regular vagrant to the British Isles, though most frequently to Scotland where small numbers can usually be found amongst the huge Scoter flocks in the Firth of Forth each winter. There had been over 350 British records by the end of 1990, at which point the species was no longer considered by the BBRC.

Velvet Scoter
Melanitta fusca

A scarce passage migrant and winter visitor in small numbers to Durham coastal waters.

Historical review

The number of Velvet Scoter recorded off the Durham coast appears to have changed little over the years. Temperley (1951) described the bird as *"much less frequently seen than the Common Scoter and never in large flocks"*. He considered the species as a *"winter visitor in very small numbers"*, with most sightings being of *"single birds, or at the most two or three together"*. Although a sea diving duck, Temperley also observed that the bird was *"more prone to enter bays and estuaries than the Common Scoter"* and could occasionally be seen on the Tees estuary at low tide. Backhouse (1834) documented *"several specimens killed in the mouth of the Tees"* in 1829/1830; while Hutchinson (1840) considered that the bird *"only appears in winter"*. Thomas Robson (1896) recorded a single specimen of the *"Velvet Duck"* as the bird was then known, shot inland in 1888, probably at the confluence of the Derwent with the Tyne. Courtenay recorded that he occasionally saw the bird diving off the mouth of Horden Dene, between 1907 and 1929 (Courtenay 1931), while M. G. Robinson saw five birds at Teesmouth on 2 August 1931, a record that at that time was considered worthy of mention in *British Birds* (*British Birds* 25:195).

Inland records of this species are very rare. One was recorded near Cotherstone 32 miles from the sea around 1856, Robson's bird followed in 1888, while in November 1941 two were seen inland on an old brickfield pond near Boldon Colliery.

During the 1950s and 1960s, the counts for this species in Durham were considerably higher than today, with 24 noted off the Tees on 7 April 1956 and 31 birds off Seaton Carew on 1 January 1961. Further north, a flock of 25 was off Sunderland on 22 October 1960.

Recent status

Today, the Velvet Scoter can be found regularly wintering off the south east Durham coast. Most sightings take place off Hartlepool Headland and Whitburn. The species used to be frequently sighted off Seaton Carew until the flocks of Common Scoter *Melanitta nigra*, once a regular feature of that area, moved away (Blick 2009).

The Velvet Scoter is considered the least abundant sea duck to winter regularly in England, with around five hundred birds present in one year, in contrast to the much larger British wintering population of five thousand birds mostly to be found in Scotland (Brown & Grice 2005). The birds often occur as part of mixed species scoter flocks, although they tend to keep together within such groups. They feed mainly on mussels, but also take crustaceans and sand eels. Brown & Grice (2005) note *"that county avifaunas give no indication that the status of the Velvet Scoter has ever been substantially different from that of today"*. The slight increase in numbers recorded in Durham in recent decades is, therefore, probably the result of increased observer coverage. There have been no recent inland sightings.

In County Durham, birds on passage are usually in higher numbers than are birds feeding off the sea. Few birds are recorded during the early part of year, but numbers increase during autumn, with November being the peak month for sightings. In 2001 however, 24 birds flew north past Whitburn and South Shields during July. One of the highest monthly counts was achieved during October 1992 with 87 birds seen from Whitburn Observatory. The highest daily total at Whitburn for the 1990s was 40 birds on 3 November 1998, but 46 birds passed Crimdon Dene during late September 1997, indicating that such passage occurs along the whole of the Durham coastline. More recently, a monthly passage count of 51 was recorded off Whitburn in October 2004 with high day-counts of 34 off Hartlepool on 23 November 2003 (Blick 2009) and 36 at Whitburn on both 28 September 2007 and 31 October 2008. Most observations of coastal movement are however, usually low, with rarely more than single figures counted on any one day.

Sightings of birds feeding on the sea are notable and are seldom recorded, although more birds may be present than is commonly realised. In 2006, eight birds were with Common Scoter, *Melanitta nigra* off Hartlepool in early January and five birds were on the sea off Blackhall from 28 March to 17 April. In 2007, two birds were off Seaton Carew from 7th to 21 January, increasing to four on 1 February. These birds relocated to Parton Rocks, Hartlepool, remaining for six weeks and being joined by two further individuals on 15 February. Blackhall Rocks still produces regular sightings and eight birds were seen there in November 2009.

Distribution & movements

The Velvet Scoter breeds on lakes and large ponds in the Arctic and northern boreal zones, from Scandinavia to Siberia, having a circumpolar distribution. The species does not breed in Iceland or in Greenland, but there are isolated populations as far south as the Black Sea. The majority of the western Siberian and north west European population, around one million birds, winter in the Baltic around Riga and in the Kattegat (Brown & Grice 2005). Those that frequent the North Sea coast are largely birds of Scandinavian origin.

Goldeneye (Common Goldeneye)
Bucephala clangula

A common passage migrant and winter visitor.

Historical review

According to Temperley (1951) Goldeneye's historical status was somewhat *"confused or uncertain"*. Selby (1831) said it was *"common"* during the winter in the north east, whilst Hutchinson (1840) said *"in winter is very numerous"*. Commenting on its mid-20th century status Temperley (1951) described the Goldeneye as *"a common winter visitor to the coast and, in lesser number, to inland waters"* usually noted from October to March, occasionally in April. He said that it was a regular visitor to Hebburn Ponds and, on 2 October 1943, 17 of 24 ducks counted on the pond were Goldeneye. It was also regularly seen on Hurworth Burn and Crookfoot Reservoirs. The status and fortunes of this species has varied since then, though it appears to be more common than in the past.

Counts during the post-Temperley period conformed to his description, with small flocks being widely scattered, though never numerous; for example, in 1965, no flock larger than 20 was reported. A step change occurred with the construction of Derwent Reservoir in the mid- to late 1960s, and from the latter part of that decade that site routinely attracted flocks, with 40 on 11 December 1966 and 27 there on 24 November 1968. On 5 August 1960, an immature bird was near Durham, out of season. Numbers at Teesmouth rose from counts of just 41 in the

1954/1955 winter, building slightly through the 1960s, then very steeply in the late 1970s, when over 200 were present (205 in 1978/1979 and 224 in 1979/1980).

Recent breeding status

On a number of occasions since the early 1980s, birds have been recorded during the spring and summer in Durham. Such reports are usually split roughly equally between the tidal reaches of rivers and inland locations such as reservoirs. Breeding season records during the late 1980s to the early 1990s came from: upper Teesdale, Ryton Willows, Hurworth Burn and Crookfoot Reservoirs. In most years, there are a small number of such out of season records in Durham. Most intriguing in this period was a series of records at Shibdon Pond: a June sighting in 1985; two, in summer 1986; and a fledged juvenile in August 1986. In May 1990, an immature male spent over a fortnight on Shibdon Pond. Most of these seemed to refer to lingering winter visitors or late passage birds.

During the *Atlas* period, the only genuine summering records were in the area around WWT Washington in the early 1990s and most were likely to refer to escaped birds rather than genuinely wild individuals. Breeding was first confirmed in the county at Washington on 11 May 1991, when a female and four young were seen on the Wear at Chester-le-Street, upstream of Washington. A few years later, a brood of seven was hatched at WWT Washington in 1994, but none of the young survived. Both of these instances probably referred to birds of feral stock derived from the WWT (Westerberg & Bowey 2000).

In the early 1990s, small parties of sometimes displaying birds were noted staying later and later into the spring at Shibdon Pond. In 1996 and 1997, such small groups were present into late April. In both years birds were often displaying, behaviour which sometimes occurs from late winter onwards; however, in these instances birds were clearly pair-bonded and on occasion attempts at copulation were noted. In 1997, a female was seen investigating a nestbox mounted on a tern raft. After disappearing from sight for some weeks, what was presumably the same female reappeared in late May and began circling the nestbox investigated earlier in the year. She spent much time circling the box and calling up to the hole. At no stage were any young noted and on investigation during the following winter, the box did not appear to have been used.

This pattern of late 'hangers-on' is now well-established. In 2001, up to eight late birds were at Shibdon Pond well into April, with smaller numbers at Derwent Reservoir. In 2007, a late spring female was at Brasside Ponds on 11 May, whilst three birds were on Greatham Creek on 1 July 2007. Apart from such isolated incidences of breeding, there are no indications that this species is likely to breed in Durham in the near future.

Recent non-breeding status

This species winters in considerable numbers around the British Isles, *c.*25,000 being estimated in British waters (Cranswick 2005). This is only a small fraction of the 200,000 reckoned to winter in Europe as a whole (Lack 1986). Wintering birds begin to arrive in Durham during late October, remaining in most years until late March/early April. Peak numbers often coincide with the hardest weather of the winter in January and February. Small numbers of birds spend the winter on many freshwater sites around the county, birds being routinely noted from 15 to 20 sites during the winters of the 2000s, though the number doing so has declined, after peaking in the late 1980s. Most of the county's significantly sized inland waters attract small numbers, of tens of birds, in most winters. The deeper reservoirs sites are particularly attractive to this species, though favoured sites have varied through time. Some sites have proven attractive to the species over a period of years before they have moved on. Some of the rivers, particularly on their tidal lower stretches, can occasionally attract what are now considered quite large numbers of birds, often in dissipated 'flocks' over a stretch of the river.

In the early 1980s, Teesmouth held the 5th largest number of wintering Goldeneye in Britain, with the peak count of 470 during the winter of 1981/1982 being the largest number recorded in Durham. Wintering numbers there have declined considerably through subsequent decades, with no more than 153 birds observed since. The reasons for the initial increase, over a three-decade period, and subsequent decline, are unclear (Blick 2009).

In the north of the county, during the late 1970s and through the 1980s, a large flock developed on the River Tyne at Ryton Willows. Numbers here varied from year-to-year but maximum counts peaked at over 200 birds; the largest was 264 in February 1988. The site held nationally important numbers during the period 1986 to 1990 (Bowey *et al.* 1993). Numbers then declined sharply over the period 1990 to 1997. This decline occurred in two stepped phases, coincident with a reduction in the volume of sewage discharges into the Tyne. This reduction resulted from the construction of the Tyneside Interceptor Sewer system. Such effects have been documented elsewhere (Cambell 1984). Furthermore, the decline appeared to be related, in an inverse manner, to the increase in Goldeneye occurrence at Washington (Bowey *et al.* 2007). Through the mid- to late 1990s, the River Wear at Washington became the main stronghold of the species in Durham, with up to 124 birds recorded there in January 1997. Birds fed on the river during the day and roosted on the WWT reservoir at night. Numbers had declined to a mid-winter maximum of around 40 by 1999.

In the early part of the 21st century, the winter distribution of Goldeneye showed a distinct easterly bias in Durham, with birds largely located on the lower stretches of the main river systems, and small numbers scattered at a range of wetland sites, in particular on some of the upland reservoirs. By the beginning of the 21st century, the days of three-figure Goldeneye flocks in the county were gone. Early quarter maximum counts were in the order of 20-30 at key sites, maximum winter counts being in the order of 40-50 birds. Double-figure counts came from fewer locations and during the second half of the first decade of the 21st century, the highest counts of the year usually came from the North Tees Marshes, with up to 64 birds there in February 2008, but numbers more typically between 30 and 40. There were occasional larger gatherings, of up to 35 birds, on some of the reservoirs in the west.

Goldeneye arrive later than many winter wildfowl in Britain and small numbers arrive in Durham from the second half of September, though most birds appear in mid- to late October, with numbers building to peak in December, January or early February, depending on weather patterns. Sea passage is most evident in October and November, especially if strong onshore winds prevail. In 2007, observations at Whitburn Observatory revealed a total of 399 birds passing there on 25 dates between 11 September and 27 November, the highest day-counts being 17 on 18 October, 62 on 3rd, 69 on 4th, 57 on 11th and 26 on 24 November. Some of the largest day-counts occurred in the late 1960s, when 102 flew north past Whitburn on 5 December 1967 and the same number flew north off Hartlepool on 22 November 1969. A northerly gale in November 2006 produced a count of 593 birds past Whitburn Observatory in two days, including 381 on 1st. When large numbers are on the move, flocks can be often seen on the sea; up to 200 have been noted in such circumstances.

The first inland wintering birds in Durham do not usually appear at wetlands until October, most in early November. Wintering birds begin to move away from Durham in March, most having gone by early April, and just occasional stragglers staying into May, with scarce out of season birds being seen in the summer months, as noted above.

In the winters of 2005/06, 2006/07 and 2007/08, an aberrantly plumaged bird was reported in the Wear valley, at Low Barns and then at Low Butterby Ponds; the plumage was much paler than normal with a head pattern similar to a male Smew *Mergellus albellus*.

Distribution & movements

The Goldeneye has a Holarctic distribution. It is principally a winter visitor to the British Isles. Limited ringing data on the species suggests an origin in Scandinavia and Russia for the majority of British wintering birds (Lack 1986). A small breeding population has become established in Scotland since the early 1970s with range expansion and an increase in numbers since then (up to 200 egg-laying females by 2006 - Hollings *et al.* 2008); though the core British population is still centred upon Scotland (Parkin & Knox 2010).

Smew
Mergellus albellus

A scarce, irregular passage migrant and winter visitor.

Historical review

The smallest of the sawbill ducks, the Smew is recorded in small numbers in most winters in County Durham, principally between November and March. Numbers noted vary from year to year, principally depending upon the severity of the winter weather on the near Continent. In recent decades it has been recorded rather more frequently than formerly.

Temperley (1951) called it *"an irregular and uncommon winter visitor on the coast and on inland waters"*. In the 19th century, Selby (1831) described it as a *"winter visitant, but rare in the adult state in the north of England"*. One was killed in the winter of 1829/1830 near Sunderland and it seems likely that an influx of Smew had occurred that winter, as in December 1829 two males were also killed at the mouth of the Tees, with another shot shortly after that at the same place (Temperley 1951). These appear to be the first documented records of its occurrence in the county. Hutchinson (1840) said that it was occasionally seen in *"stormy winter weather"* and mentions a number of birds taken on inland stretches of the Wear. Although Hogg (1845) regarded it as a rare visitor, Nelson writing in 1907 thought it frequent at Teesmouth in the previous forty to fifty years and mentions several killed on the Tees near Yarm. In February 1919, there was a rare record of an adult drake shot on the Derwent at Chopwell, in the north of the county (Temperley 1951).

An unusual *"invasion"* of Smew occurred in February 1937, when there were at least five different birds with reports from: Charlton's Pond at Billingham; East Boldon Pond; East Rainton; Hebburn Ponds; and, Hurworth Burn Reservoir (*The Vasculum* Vol. XXIV, 1938). It seems that this species has occurred to a regular pattern, stretching back to the Victorian period with rare or low level occurrences, punctuated by intermittent influxes.

The Smew seems to have been less common in the 1950s before becoming slightly more regular in occurrence in the 1960s and 1970s. Thereafter sightings increased quite markedly from the mid-1980s onwards, probably related to increased observer activity or possibly a change in weather conditions on the Continent.

During the 1950s birds were noted in three out of ten years. In 1954, a drake was present on Hurworth Burn Reservoir on 27 December, and then in 1956 a redhead was off Seaton Carew in February, and one was on Charlton's Pond, Billingham in March. During 1957 there were only two records for Northumberland and Durham in the whole year, prompting a comment on *"the scarcity of the Smew in our two counties"* (Grey 1958). The Durham bird was a redhead flying south over the sea off Hartlepool on 13 January during a north easterly gale.

The 1960s produced 12 records, in seven out of ten years. In 1960 one was at Tanfield Ponds, Stanley on 16 October, then a drake was on Greatham Creek on 27 March 1962. The severe winter of 1962/1963 produced an influx to the north east, and although most records were in Northumberland, a redhead was on Greatham Creek on 30 December 1962 with early 1963 producing records of one in the Tees estuary from mid-January to early March, with two there on 24 January, and an adult drake again on Greatham Creek on 17 February. A very unseasonal report concerned a female on Mowbray Park Lake, Sunderland, on 1 July 1963, which stayed until 25 October. This was believed at the time to be a wild bird that became progressively tamer during its stay at this urban wetland site. What may have been the same bird appeared at Hurworth Burn Reservoir on 8 November, but it is perhaps more likely to have been a different individual occurring at the usual time of year for the arrival of new birds in the county. Each of the years from 1966 until 1969 produced just single records. These were redheads at: Hartlepool on 22 February 1966 and 24 November 1969, and a drake at Greatham Creek on 19 February 1967; all remaining for just one day. In 1968 a redhead stayed on Derwent Reservoir through January and February.

During the 1970s birds were noted on an almost annual basis, the only blank year being 1978. In 1970 a redhead flying north off Hartlepool on 23 February was the only record for that year. However, early 1971 brought a flurry of reports with a drake and two redheads on the Wear at Houghall on 9 January and one to three redheads present at Witton-le-Wear from 14 February until 7 March. There was but a single record in 1973, a redhead at Cowpen Marsh on 8 October, and then in 1974 a redhead appeared at Hurworth Burn Reservoir on 26 October and remained until 22 March 1975. The only other bird that winter was a drake at Derwent Reservoir on 1 December 1974. During 1976, two drakes were on Teesside from 1st to 7 February moving between Haverton Hole, Saltholme Pools and Charlton's Pond in Billingham. Then in 1977, a drake spent the period from 19 January until 13 February on the tidal stretch of the River Wear between South Hylton and WWT Washington. Although 1979 was a good winter nationally for the species, this was not reflected in Durham as there were just two records of three birds; a redhead was at Hart Reservoir and Hartlepool docks from 7th to 14 January with the only other sighting relating to two drakes which flew downstream at Newfield near Willington on 12 January.

Recent status

The Smew is now an annual visitor to County Durham, albeit in very small numbers, with the best years tending to coincide with severe winter weather, particularly in The Netherlands, where a large winter population exists. Both closed waters and stretches of the county's main river systems are frequented. Any location that attracts Goosander *Mergus merganser*, or Goldeneye *Bucephala clangula* may also attract Smew. Some of the most frequented sites for Smew over the last three decades have included the various ponds and stretches of the River Wear around Low Barns NR and WWT Washington, the River Tyne at Newburn/Ryton, Shibdon Pond, Crookfoot and Hurworth Burn Reservoirs, and in particular since 1998, the Dorman's Pool/Reclamation Pond/Saltholme Pools complex of wetlands on the North Tees Marshes.

For the past 30 years the usual pattern of records for the species is that one to five birds are scattered across the county in most winters; these are often highly mobile and may visit several sites. The north east, like the rest of Britain, however, experiences periodic influxes of Smew with numbers in peak years approaching or even exceeding double figures.

For the first half of the 1980s, only one to two birds were recorded annually, although the early part of 1985 brought severe winter weather and an influx of Smew. It is difficult to be certain how many individuals were involved but the total was probably around 12 birds. The first sighting concerned two redheads on Brasside Pond on 4 January, and then there were up to four birds, two adult drakes, an immature drake and a redhead, on the River Wear in the vicinity of WWT Washington from 5 January until 2 February. In the north of the county, at least five individuals were noted in Gateshead between 17 January and 9 March, whilst in the south east, a drake and a redhead were present on Seal Sands during January. Although there were no sightings in late 1985, five birds were recorded during the first quarter of 1986, and then a further notable influx totalling 10 birds took place in 1987, following another spell of severe weather. Similar locations attracted birds as in 1985 with one to two redheads around WWT Washington from 8 January until 19 February, and up to two drakes and a redhead in the Shibdon Pond area over the period from 21 January until 22 February. Teesmouth held a drake and up to three redheads between 15 January and 11 March, while a redhead also appeared at Derwent Reservoir on 15 February. There were single birds in February and March 1988, both staying for one day. Towards the end of the year, a relatively early bird was at Fishburn Lake on 31 October, with further brief visits by redheads to Seal Sands on 3rd to 4 November and WWT Washington on 5th and 11 December.

During the period 1989-1996, only one to two birds were recorded annually, but for the ten years thereafter an average of four to seven birds were recorded each winter, except for the early part of 1999 when an exceptional influx occurred. This trend for increased numbers of wintering birds in Durham correlated with the situation in the UK as a whole. National wetland birds counts indicate that the wintering numbers of Smew rose during the mid-1990s peaking at 453 during the cold winter of 1996/1997 (WeBS).

The notable 1999 influx was evident from the first few days of January and was virtually confined to the North Tees Marshes. Here eight to ten birds were regularly seen during January and February with a maximum count of 13 on 13th to 14 March; the timing of this peak perhaps indicative of a very small spring passage through the county. The Teesmouth count was the highest ever for both Cleveland and Durham, unequalled to date (Iceton 2000, Armstrong 2003). Elsewhere, a lone redhead was present at Witton-le-Wear and Escomb Lake in early January, with perhaps the same bird noted again at Escomb from 29 April until 9 May, before relocating to Beechburn Gravels Pits where it remained until the exceptionally late date of 15 May.

From 2007, the pattern of records appears to have returned to the levels experienced before the mid-1990s. This seemed to mirror the picture nationally as the peak national counts in the 2006/2007 and 2007/2008 winters were low at 100 and 109 birds respectively. The only Durham record in 2007 concerned a pair on Saltholme Pool from 8 March to 3 April. In early 2008, a drake was on Back Saltholme Pools on 13 January; a redhead was at Portrack Marsh from 27 January until 24 February, visiting Bowesfield Marsh on 20 February, and a drake was at Low Barns on 9th to 10 February and again on 24 February. A redhead was again on Portrack Marsh on 28 November, with another redhead at Hurworth Burn Reservoir on 21 December and nearby Crookfoot Reservoir on 30th, where it remained until 11 April 2009. Later in 2009, two redheads were again at Crookfoot Reservoir on 15 November and remained there or at Hurworth Burn Reservoir to the year's end. In addition a redhead was briefly at Hetton Lyons on 24 November. During the first winter period of 2010, there were perhaps four to five birds wintering, with one to three redheads regularly at Hurworth Burn Reservoir and on the North Tees Marshes, although the proximity of the two locations meant that movements between them could not be discounted. At the end of the year, one to two redheads were once again at Crookfoot/Hurworth Burn Reservoir from 13 November into 2011, and likely the same birds returned to this area in November 2011. One at Hetton Lyons CP on 7 February 2011 was also notable for the area.

An analysis of the occurrence of Smew in Durham since the mid-1980s suggests that some birds are faithful to particular wintering sites, returning each winter for several years. This may explain why there were occasional 'flurries' of records in particular areas over a number of years, as appears to have taken place from 1985 to 1988 in Gateshead, around the North Tees Marshes in the late 1990s and part of the first decade of the 21st century, and at the closely linked sites of Crookfoot and Hurworth Burn Reservoirs.

Looking more closely at occurrences in western Gateshead what might reasonably be termed the first modern record for the area was a redhead at Shibdon Pond from 23rd to 25 November 1984. The 1985 influx brought sightings from several locations, when at least five individuals were noted in the borough over that winter. The following winter brought a pair of birds to Shibdon from 4th to 8 March 1986. Then, during January to March 1987, there was a series of sightings of one to two drakes and a redhead at Shibdon, whilst in 1988 there was a single report of a redhead on the Tyne at Ryton Willows from 26th to 28 February 1988. There was then a three-year gap before the next record of a redhead in that area, presumably a new bird, on the Curling Pond at Ryton Willows on the rather early date of 20 October 1991; the next record not coming until 1994. Although the numbers involved are low this pattern suggests a degree of site fidelity, with numbers peaking as expected around the initial influx, followed by a gradual decline as returning birds either succumb to mortality or establish new wintering habits.

The number of redheads observed in Durham far outweighs the number of adult drakes, perhaps in a proportion of five to one at Teesmouth (Blick 2009), whilst in Northumberland it has been suggested that six redheads are seen for every adult drake (Kerr 2001). This is what might reasonably be expected as most adult drakes move no further south than northern Germany, with the proportion of redheads increasing the further south and west one goes in the wintering range (BWP).

Disregarding the July 1963 record, the earliest autumn bird was recorded on 7 September 2001 and the latest in spring occurred on 15 May 1999.

There are just a handful of records of birds passing along the coast in Durham, all relating to redheads; one flew south at Hartlepool on 13 January 1957; two south over Whitburn CP on 9 December 2000; singles north off Whitburn Observatory on 9 September 2001 and 12 November 2006; and one north off Blackhall on 26 January 2010.

Distribution & movements

Breeding birds can be found in the boreal zone of Eurasia, from Sweden and Finland east to Siberia, whilst Smew winter largely in Europe to the south of their breeding range. Autumn departures from breeding areas begin in September, with the main passage through Sweden and the Baltic from mid-October to November. Early arrivals reach North Sea countries in October, but the main arrivals are not until December and January following cold weather further east (BWP). UK numbers peak in January or February (WeBS), with British birds thought to be primarily of Fennoscandia origin, but also including some birds from north west Russia (BWP). Considerable numbers visit Germany and the Netherlands, with the first five years of the present century producing a small wintering population in Britain of around 250 birds, largely in southern England (Parkin & Knox 2010).

Red-breasted Merganser
Mergus serrator

An uncommon passage and winter visitor, and a very rare breeding species.

Historical review

In Durham today, this species occurs far more frequently as a winter visitor to coastal and offshore locations than inland and it appears to have become increasingly common over an extended period of time. There may be a long ago established cultural reference to its, or the Goosander's *Mergus merganser*, presence in the Pennines at the time of the Norse incursions. The name of the hamlet of Cargill, just to the south of the A66, in what was North Yorkshire prior to 1974, may be derived from the Old Norse *"strake, gill"*, meaning 'merganser valley' (Yalden & Albarella 2009), although this reference might equally have applied to its similar congener, the Goosander.

By the middle of the 20th century, Temperley (1951) observed that it was, *"a regular winter visitor to the coast, but in small numbers and usually during or after severe weather. Seldom if ever seen on inland waters"*. He noted that it was usually seen singly or occasionally two to three together and that most birds were immature with drakes only rarely occurring in spring. Hutchinson (1840), said *"it is met on the coast during winter, but not generally… and driven in severe winters to seek refuge in the bay at Hartlepool and in the estuary of the River Tees, where specimens are obtained. It is seldom found on rivers"*. Hogg (1845) said he had seen two specimens shot near Hartlepool, but does not refer to its status. Writing a little later Hancock (1874) said it was *"occasionally captured in our estuaries"* (probably mainly referring to Northumberland). By the early 20th century Tristram (1905) could refer to its annual presence at the coast and also commented that it was less numerous than the Goosander, *"nor does it go so far inland"*. Temperley thought that its status had not changed appreciably in the previous hundred years. Nelson (1907) stated that it was *"frequently found in winter and spring in the Tees estuary"*, around the middle of the 1800s, although, at the time of writing, he classed it as an uncommon winter visitor. Almond *et al.* (1939) documented that *"brown and grey birds are to be seen most winters and adult drakes are not infrequent"*, noting that it had been seen as late as 20 April 1933. On this date three parties of birds, one numbering eight individuals, were seen in the Tees estuary. A flock of 14 birds was noted here in January in the late 1930s. On 3 April 1945, an adult drake was seen on Hebburn Ponds by Fred Grey. The species was first known to have bred in England in 1950 (Holloway 1996).

Throughout the 20th century, the Red-breasted Merganser was not rare, nor particularly numerous in Durham. This continued to be the case through the 1950s and 1960s, though numbers off the Durham coast grew somewhat through the latter decade. Notable through these decades were some inland records, namely a female shot at Shotton Brickworks Pond on 10 November 1957 and a slightly bizarre record in 1961, of a drake picked up in the town centre at Stockton on 28 April 1961 and later released by the RSPCA. A long-staying inland bird was at Hurworth Burn Reservoir from 17 November to 10 December 1968, whilst the following year three were at Derwent Reservoir on 12 October 1969. Of interest, considering the growth in numbers a little further to the north through the 1960s, was the party of 25 that was seen to fly across the Tees Bay on 18 October 1958, which appears to be the largest recorded to that date. From the early 1960s onwards, numbers on the sea off Crimdon Dene grew from 17 in February 1961 to 31 on 21 February 1963. Otherwise, it remained a sparsely distributed species at coastal locations, mainly around or to the north of Teesmouth, through this period. In the early 1970s, the sea off Crimdon Dene remained the species' principle winter station in the county, though rarely more than 15 birds were present there, and a similar number were recorded around Hartlepool Docks in early 1973.

There was an intriguing run of inland records at Derwent Reservoir and Low Barns between 1970 and 1973, with three occurrences at the former site and at least three at the latter. In one instance, there were up to six birds at Low Barns, Witton-le-Wear on 28 October 1973, one of the largest gatherings on any inland location in County Durham. This may have been a family party that had managed to breed unobserved somewhere in the Wear valley. A pair had been observed at this site in late November and early December 1970.

A notable movement of birds occurred off Hartlepool Headland from 21st to 24 October 1970, when 119 birds were noted flying north. The highest known count of birds in Hartlepool Docks was of 34 in January 1979 (Blick 2009).

Recent breeding status

In most parts of the country, breeding Red-breasted Mergansers are associated with coastal localities, although in northern England they can also be found on fresh water, especially on the eastern side of the Pennines, and around the Cumbrian lakes (Gibbons *et al.* 1993). It is likely that the very few breeding records of this species in Durham emanate from these populations. The first known breeding attempt in the county comes from the early 1970s, when a pair was said to have nested on the River Tees near Piercebridge in 1975, and birds had apparently done so since 1973 (observations by G.D. Moore), (Mather 1986).

There was another unconfirmed report of breeding taking place on the River Tees in 1991 and a pair was on Cow Green Reservoir on 15 May 1990. In 1992, a pair was again noted there during May, and not long afterwards, successful breeding was confirmed in upper Teesdale. However, it is not certain whether the nest was on the Durham side of the county boundary. In 1993, pairs were noted on two inland waters during the summer (Armstrong 1994) and, subsequent to the *Atlas* survey period, the species has since been confirmed as breeding within the county boundaries in upper Teesdale (Westerberg & Bowey 2000).

There have been a number of additional breeding attempts, through the 1990s and early 2000s, but it is clear that, as yet, there is no real established breeding population within the county. The numbers are very small, probably only a single pair, but there is potential for expansion. In most recent years there have been no reports of breeding in Durham. Small populations of any species are always at risk from random factors, such as poor weather during the breeding season, and this might be of particular relevance in the known breeding areas in the extreme west of the county. Another factor to be considered in the establishment of a breeding population is the antipathy shown by anglers and fishing interests, which exercise extensive 'control' of this species elsewhere in the British Isles (Mead 2000).

Recent non-breeding status

The species can be seen, in small numbers, throughout the year in Durham, particularly around the Hartlepool and Teesmouth areas, although so far there has been no proof of breeding in those locations. Teesmouth is by far the most frequented area of the county in the winter months and the area around the mouth of the Tees usually holds the largest winter concentration of birds in both the early and latter parts of the year. Hartlepool Harbour and Docks, Hartlepool Bay and the area off North Gare are particularly favoured. The maximum counts usually occur in January or February, with small numbers of birds often in the Seaton Snook/Seal Sands area. Elsewhere along the coasts small but increasing numbers are reported in winter from the coastline between Dawdon and Crimdon, including Blackhall Rocks, and also from South Shields. The numbers around Teesmouth decline in March, but there is usually some observed passage and movement of birds north at this time.

The largest counts of Mergansers, except for those seen on sea passage, all come from the Teesmouth and Hartlepool area. Up until 1993, no more than 15 birds were recorded on the sea at the same time. This increased in November 1993 and February 1994 to 28 birds, a total surpassed in 1999 when up to 60 were recorded in the area during December.

The number of birds wintering in the county might be described as steady throughout the 2000s, with the highest concentrations occurring during the first and last quarters of the year. Teesmouth is by far the most important regularly frequented location, with up to 85 birds recorded on Seal Sands on 22 February 2005. One of the largest counts came on 13 March of that year, when 111 birds were recorded from around Teesmouth, Hartlepool Headland and Hartlepool Bay; a total which may have included birds moving north from other wintering grounds (Joynt 2006).

Mid-winter counts at Teesmouth can reach as many as 50 birds, but rarely more than this, usually in January or February. Other high counts include 66 around Teesmouth in January 2001 and 70 in November 2004 (Blick 2009). In 2007, the most regular localities in the North Tees area were Seal Sands and Seaton Snook, with single site counts of 50 at Seaton Snook on 24 February and 41 on Seal Sands two days later being the most impressive.

In most years, there are a small number of records of birds on freshwater sites in the county, largely between October and April, but not exclusively so. These occurrences are never common but a number of wetland sites have played host to birds on occasion, often those relatively close to the coast or the main river corridors of the county, perhaps indicating movement of the species along these. Typical sites over the last three decades include: Crookfoot Reservoir; Charlton's Pond; Hart Reservoir, with up to eight there in February and March 1979 (the largest number on any inland site); Longnewton Reservoir; the Tyne at Ryton Willows, and Shibdon Pond. More recently during the 2000s, inland records have come from: the River Wear at Chester le Street; Haverton Hole; Hetton Lyons; Jarrow Slake; Joe's Pond; Portrack Marsh; Selset Reservoir; Shotton Pond (four males, the first since 1957); and, Sled Lane Pond.

Autumn usually brings significant movement of this species past our coastal stations. A small number of birds are often seen in September, but it is usually October before significant numbers occur with peak sightings, perhaps of 40 to 60 birds, reached between the middle of November and early to mid-March.

Coastal movement can be quite evident between September and November, especially during strong onshore winds, with 15 to 30 per day being regularly noted in some autumns off Hartlepool and Whitburn. The highest count was 71 passing Hartlepool Headland on 21 October 1970 (Blick 2009). In the 1990s, Whitburn's highest counts were a total of 165 north and nine south, mainly during October, in 1991, and a total of 108 north in October 1993 (Armstrong 1992 & 1994).

In 2007, passage of Red-breasted Mergansers was noted on a total of 51 days during the year at Whitburn Observatory, with over 70% of these sightings occurred between June and November. A total of 51 birds were logged on seven dates between 22 September and 14 October, peaking at eight on 27 September and 14 on both 13th and 14 October.

Distribution & movements

Red-breasted Merganser has a circumpolar breeding distribution, breeding in a band from northern temperate habitats north into the low Arctic zone. The most northerly populations are migratory, wintering to the south of their breeding range. The birds that breed in more southerly locations tend to be resident.

Goosander (Common Merganser)
Mergus merganser

A common winter visitor and small-scale passage migrant, uncommon as a breeding species.

Historical review

The Goosander has increased in numbers in Durham, both as a winter visitor and as a breeding species, since the middle of the 20th century, reflecting an expansion in range and numbers nationally. The first record in the county was of a male shot near Wycliffe-on-Tees in January 1789 (Allan MS 1791-1800). In 1840, Hutchinson noted that it rarely appeared in County Durham except in the severest of weather. Since then its status and frequency of occurrence has increased greatly. There is a fascinating record cited by Nelson (1907) from March 1853, concerning a bird shot on the River Tees, near Stockton, the gizzard of which contained part of a golden ear-ring; presumably swallowed by a fish that the Goosander had then eaten (Mather 1986). Nelson (1907) said that Goosander was *"formerly not infrequent in the Tees estuary, but is now considered as being of rather rare and irregular occurrence"*. Courtenay, with more than twenty years experience of Hurworth Burn and Crookfoot Reservoirs between 1910 and 1929, had not encountered a single individual there (Courtenay 1931).

Across Britain the species went through a considerable expansion in range and, to a lesser extent, numbers, in recent decades (Owen *et al.* 1986). In broad terms, the Goosander is a relatively recent addition to the list of British breeding birds, being first proven to have nested in Perthshire, Scotland, in 1871 (Thom 1986). There was a large winter influx of the species in 1875/1876, which probably fuelled colonisation (Harrison 1988). After reaching the Borders, on the River Tweed, in the 1930s (Marchant *et al.* 1990) the expansion continued locally (Meek & Little 1977) spreading southwards across much of northern England, in the second half of the 20th century (Brown & Grice 2005). The first evidence of successful breeding in England was obtained in Upper Coquetdale, Northumberland, in 1943 (Meek & Little 1977).

Temperley (1951) considered the Goosander a winter visitor in small numbers, which preferred inland waters and which was very occasionally seen on the coast. He noted that it had been recorded from both Hurworth Burn and Waskerley Reservoirs. The species is stated to have been more numerous and widely distributed in the winters of 1947/1948 and 1948/1949 than in any previous year.

Breeding in County Durham was suspected in the early 1950s, but it was not until 1965 that confirmation was obtained, when a female with *c.*8 ducklings was watched on the Tees for about 15 minutes (Bell 1966). In 1967, towards the end of June, a farmer in the Barnard Castle area noted a large grey duck, with a brown head and long thin beak with a brood of 13 big brown and white striped ducklings on the Tees. Since 1965 nesting may well have been annual, and is almost certainly under-recorded. Throughout the late 1960s, the process of consolidation continued and in 1969 two pairs bred along the River Tees between Middleton-in-Teesdale and Barnard Castle. In the 1970s there was a steady increase in the number of broods recorded. The following year four broods were seen on the Tees above Barnard Castle, with five broods there in 1971, in which year breeding was recorded for the first time on the River Wear. Again a family party was noted on the River Tees close to Thorngate Bridge at Barnard Castle, in the summer of 1972 (K. Bowey pers. obs.). G.D. Moore reported five pairs along the Tees between High Force and Piercebridge in 1975 and in that year the population of the county was estimated to be a minimum of ten to twelve pairs. Family parties were seen on the River Tees at East Lendings, downstream of Barnard Castle in the summer of 1976. In 1977, a pair with seven young was on the Tees at Wycliffe (Mather 1986). The first confirmed breeding on the Derwent came in 1975, while broods were noted on Tunstall Reservoir in 1974 and 1984. Most of the suitable habitat on the River Tyne falls outside County Durham, but there were records of successful breeding on that watercourse in 1986 and 1987.

In the mid- to late 1960s, the British breeding population was estimated as being 500-1,000 pairs (Atkinson-Welles 1970). This was revised to 1,000-2,000 at the time of the first B.T.O Atlas survey in 1968-72 (Sharrock 1976), and more accurately to between 915 and 1,246 pairs, of which 130 to 150 were in Northumberland (Meek & Little 1977), and 2,700 adult males at the time of the *New Atlas* (Gibbons *et al*, 1993). The last figure was accompanied by caveats relating to birds nesting away from riverine habitat, and also the timing of the movement of birds between winter quarters and breeding grounds.

Through the 1950s and 1960s, the species remained scarce in Durham and its wintering status largely unchanged from Temperley's time. For example in 1966, the only party of birds reported was of five birds at Smiddyshaw on 13 November, though singles were noted at a number of locations. This changed with the establishment of Derwent Reservoir in 1966, which provided Durham with a focus for wintering birds. By the winter of 1968/1969, the site had begun to attract sizeable flocks of Goosander, which probably commuted between the reservoir and the Northumberland Loughs. Thus there were 73 at Greenlee Lough on 19 January 1969 and 63 at Derwent Reservoir on 9 March 1969. The latter count has not subsequently been exceeded at that site although there were maxima of 56 on 15 March 1970, 50 on 14 February 1971 and 50 on 25 November 1973.

Recent breeding status

As a breeding bird in Durham, the Goosander is principally an upland, riverine species, favouring fast- flowing rivers (Owen *et al.* 1986). Along the higher altitude sections of our main river systems, it is well represented (Armstrong 1988-1993). The typical nesting site for Goosander is a suitably sized nest hole, usually in mature trees, but in neighbouring Northumberland, especially on higher ground, it has been known to use rabbit holes, crevices in crags and crag ledges screened by heather (Kerr 1985). Consequently, its breeding presence is usually associated with mature riverside woodlands, usually broadleaved or mixed in nature. Its preference is for faster flowing watercourses and therefore normally the upper stretches of the county's rivers.

The number of birds increased through the 1980s, 1990s and 2000s, probably due in part to the improved water quality, and consequent improved fish stocks on the county's rivers. Successful breeding was recorded in every year throughout the 2000s and in 2008, there were eight definite breeding successes with birds noted at twelve other potential breeding sites; only a snapshot of the true position. The Goosander does not breed until it is two years old (Cramp *et al.* 1977) and first year drakes will be in 'redhead'' plumage in March/April. During the 1987 BTO survey, 27 drakes and 35 'redheads' were found on Durham rivers. The survey results indicated successful hatching at seven sites (four on the Tees and its tributaries, one on the Wear and two on the Derwent - 37 young counted), but this was believed to be a gross under-estimate of the true figure. There has been considerable consolidation of the Durham population between the late 1970s and the early 1990s, although nationally the increase appears to have begun to level off.

Birds can also be found breeding on a number of the county's upland reservoirs, where suitable nesting sites are available nearby (e.g. Derwent Reservoir); though in some instances nests are located on nearby tributaries. Moreover, it does penetrate into the lowlands where the river geography allows it to do so, for example, breeding on the lower sections of the Derwent as far down as Winlaton Mill at less than eight metres above sea level. In fact breeding occurs along almost the whole length of the well-wooded Derwent, upstream of Winlaton Mill. The species is well distributed along much of the Rivers Tees and Wear, upstream of Croft and Chester-le-Street respectively. A pair was found breeding on the Tees around Low Worsall in March 1987, the first breeding record for Cleveland, where the species is uncommon, with a total of around eight breeding pairs (Joynt *et al.* 2008). Successful breeding was recorded at Hurworth-on-Tees in 2010, the lowest such record on that river to date.

The relatively limited availability of holes in mature trees close to its riparian habitat limits the species' distribution within the county, although fish stocks, disturbance and deliberate persecution are other factors. It seems likely, as noted by the *Atlas*, that the species may be under-recorded along some of the more remote sections of the county's rivers. An early brood of ducklings was noted on the Wear near Bishop Auckland on 16 April 2006. Birds of the year usually fledge in July and disperse along the river systems away from the nesting areas (B. Little pers. obs.).

The *Atlas* (Westerberg & Bowey 2000) took Carter's figure, of one pair of Goosander per 4.6km of occupied river (Gibbons *et al* 1993) and, assuming an occupied length of 200km of river in Durham, estimated a potential population of 43 pairs for the county. This has probably risen slightly over the years, since breeding birds have penetrated further into the lowland sections of some of the county's rivers. For example in 2001, spring records came from 17 locations, along the Tyne, Wear and Derwent. Evidence of successful breeding came from the River Wear between Wolsingham and Frosterley where 14 females and four ducklings were noted on 15 June, and at Chester-le-Street, with a female with two young, and on the Derwent at Swalwell, where a female with ducklings was seen on 12th. In 2007, successful breeding was confirmed at just five localities (though these included the lower section of the river at Merrybent and Rockcliffe on the Tees). A more accurate figure for the county is now believed to stand at between 50 and 60 pairs.

The principal threat to the species, as for other large piscivorous birds, is through 'control' by anglers, although this is supposed to be undertaken under licence in England (Brown & Grice 2005). Research has shown that Goosanders can take large numbers of small game fish, but there have been no proven benefits to fish stocks after they have been controlled (Carter 1990). However, there undoubtedly remains a deep-rooted antipathy to the Goosander within the fishing community.

Recent non-breeding status

The species has a widespread distribution in winter and winter numbers have increased to well beyond the "small numbers" noted by Temperley. In 2010, birds were reported from over 100 sites. In 2007, records came from over 70 inland localities, the majority being during January to April and October to December. Many sightings come from the main river systems, the Derwent, Tyne, Greta, Tees and particularly the Wear, which can hold a good population for most of its length. Although the majority of birds leave the upper stretches during the winter months, some sightings remain as far west as the respective water heads. Most British Goosanders, like those in southern Scandinavia, are resident and only move short distances in winter, but an influx from central and northern Scandinavia, Finland, the Baltic States and Russia around the turn of the year, often related to hard weather on the Continent (Cramp et al.1977), may lead to high mid-winter counts.

Over the winter period, total numbers can be difficult to assess, as birds may be distributed along much of the length of the major rivers. However, these birds may congregate at still water roost sites at dusk, which are favoured as communal roosts. This was first noted at Derwent Reservoir in the late 1960s. Such locations are mostly at upland sites, but in spring 1978 a then unprecedented lowland roost developed at Shibdon Pond, with numbers rising from two in January to a peak of 34 in March and a dozen remaining into May (Bowey et al. 1993). In the 1980s, McNeil Bottoms was a favoured roost location with counts of 15 on 9 February 1985, 35 in January 1986, 43 in February 1987 and 46 in December 1988, though Derwent Reservoir still attracted birds regularly, with 58 there on 3 January 1982.

The main winter site for this species in the early 1990s was on the River Tyne upstream of Ryton Willows, where 78 birds were recorded in January 1992 and 74 in January the following year (Bowey et al. 1993). This figure dropped after 1993 and just over 20 birds were seen annually, with 18 recorded in February 1999. However, 40 were recorded at nearby Clara Vale on 21 December 1999. A major change seemed to occur in the middle of the decade with Brasside Ponds attracting 40 in April 1995, but the largest congregation of the decade came from Baldersdale where 95 were recorded on 14 January 1996. Occasionally during the winter, one or two birds can be found on Seal Sands or on the lower sections of the River Tees, particularly in periods of hard weather (Blick 2009).

In the 2000s the main winter roost for Goosander appeared to be Longnewton Reservoir, which held 76 birds in March 2005 and 49 in November 2007.

Pairs of birds begin to take up territory on the main river systems during January, adult and sub-adult males departing for their moulting grounds in Finmark during late May, returning to pair with the females again at the end of the year (Little & Furness 1985). Adult males moult in northern Norway, when the females are incubating, not returning until November (Wernham et al. 2002). Males normally migrate to the coast of Norway in order to moult during the summer so it was with some surprise that a moulting male was found on the lower Derwent in August 1991 (Bowey et al. 1993). Thus, during autumn reports of males are scarce, though one was at Cowpen on 27 October 2001.

During the 1970s what appeared to be a passage roost developed at Tunstall Reservoir off the Wear valley. Annual dusk maxima in late September at this site between 1973 and 1976 were: 53 on 26 September 1973; 65 on 24 September 1974; 69 on 29 September 1975; and 55 on 26 September 1976. This is somewhat surprising as the national wildfowl counts regularly produce low mid-September totals for this species, e.g. 106 in 1976 and 166 in 1977. After 1976 the increased pressure from fishing on Tunstall Reservoir made it unattractive to Goosander, but echoes of the same pattern were observed elsewhere, with 23 birds at Hisehope Reservoir on 18 September 1977, 27 on Waskerley Reservoir on 14 October 1979 and 12 at the latter site on 12 October 1980.

In most winters, a few birds are seen along the coastline, and even occasionally on the sea; the highest coastal count was of 15 passing Hartlepool Headland on 8 December 1968. In 2007, coastal passage totalled 52 birds on 20 dates at Whitburn Observatory, and records were scattered through the year, but November was the best month with 13 birds on five dates, including five on 4th.

Distribution & movements

The Goosander has a circumpolar breeding distribution covering a latitudinal band that incorporates temperate and low Arctic zones. Most populations are migratory, to a greater or lesser degree, wintering to the south of their breeding areas, but many birds do not make large-scale movements unless forced to do so by hard weather.

The few ringing recoveries indicate that the birds seen wintering in Durham are largely of local origin, such as the bird that had been ringed on 12 July 1975 on the South Tyne and which was recovered near Durham City, on 2 January 1976. Moreover, an immature in Hartlepool Docks on 29 December 1977 bore a yellow wing-tag, indicating that it had been caught as a duckling on the North Tyne in July of that year (Unwin 1978).

Ruddy Duck
Oxyura jamaicensis

A once locally common passage and winter visitor and a scarce summer breeding species; subject of a national cull and on the verge of local extinction in 2011.

Recent status

A native of North America, where it is relatively common, this species colonised the county rapidly during the late 20th century, before an extensive extermination programme saw it all but disappear from Durham and elsewhere. During the late 20th century, the Ruddy Duck arrived in Durham having spread north and east from south west England since the early 1950s. The species is an opportunist, which originally escaped from captive collections during the early to mid-1950s in Gloucestershire. It first bred in Britain in the south west of England in 1960 (Brown & Grice 2005), with at least 50 pairs being present in 10 counties by 1975. During the following decades, the colonisation of unpopulated areas continued, with expansion initially centred on the Midlands, Cheshire, Yorkshire and Anglesey (Hagemeijer & Blair 1997). In the north, it first bred in Yorkshire in 1980 (Mather 1986) and by the 2000s it had become a widespread, established and regular breeding species in many parts of the UK (Parkin & Knox 2010).

The Ruddy Duck is a highly successful species that appears to have exploited a previously vacant ecological niche in Britain, and as a result, it showed one of the fastest range expansions of any UK bird species over the final decades of the 20th century (Brown & Grice 2005). It breeds on lowland wetlands of varying size, with abundant emergent vegetation, among which it nests. The species' reproductive capacity is great, as the females lay on average, seven eggs and the ducklings grow rapidly, to become independent at the age of just three weeks. The Ruddy Duck can also 'dump-lay', depositing eggs in the nests of other species, including Tufted Duck *Aythya fuligula* and Pochard *A. farina*. Consequently, young Ruddy Ducks are often seen accompanying broods of these species after being hatched by their surrogate mothers (Gibbons *et al.* 1993).

The spread of the Ruddy Duck across the British Isles brought the first record of the species to County Durham in 1977, when three birds were seen at Derwent Reservoir on 25 April (Unwin 1978). It was another nine years before the first breeding attempt (Baldridge 1987). The species was not recorded again until 1981, when two females were noted at Brasside Pond from late March until September, with presumably one of these same birds being at this site from 14 March to 18 June 1982, and then there were three records of a single female at scattered wetland sites around the county in 1983. In overview, this species was a rarity in County Durham during the 1980s, the first breeding took place in 1986, and the 1990s saw a spectacular increase, the numbers occurring annually rising by around 60-fold over the period 1993 to 2002.

Recent breeding status

There were no further records until 1986 when the species bred successfully for the first time at Brasside Pond, three young being noted there on 15 July, with successful breeding also occurring at this location in the next two years (Armstrong 1989). The following year, two females were discovered at Joe's Pond, but no breeding attempt was documented. Birds successfully produced young here in 1990 with one pair raising two broods, an exceptional occurrence of double-brooding (Armstrong 1991). These two sites were the stronghold of Ruddy Duck during the *Atlas* period (Westerberg & Bowey 2000). Further expansion in range came in 1993, with birds reaching several new potential sites, but breeding was only proven at two.

In a number of respects, 1993 was a key year for this species in Durham, numbers increased significantly, growing to a new record of 19 individuals, and new sites continued to record birds. By 1995, breeding had been confirmed at Haverton Hole, Joe's Pond, Seaton Pond and Shibdon Pond although no more than two pairs were successful during the survey period for the national *Breeding Atlas* (Gibbons *et al.* 1993).

In 1994 the first breeding took place at Teesmouth, when three pairs were located at Haverton Hole, and a single pair was at Saltholme Pools. In 1996 at least ten broods of ducklings were seen and birds were present from 29 February to 12 December (Blick 2009). Over the next few years, success varied with between three and 14 broods produced annually across the North Tees Marshes (Joynt *et al.* 2008). Breeding birds were recorded at Saltholme Pools, Haverton Hole, the Long Drag, Cowpen Marsh and the Reclamation Pond. By the end of the 1990s the Ruddy Duck was breeding widely across the whole of County Durham, with the North Tees Marshes an established stronghold.

Breeding numbers increased rapidly from 1992 onwards, from 11 breeding season birds in 1993 to about 30 in 1994 and at least 40 in 1995, with individuals present in Durham from March to November. The breeding population was estimated to be at least forty pairs by the end of the decade. At Teesmouth, numbers remained at around 10 to 15 pairs from 1998 to 2007 at the favoured breeding sites of Haverton Hole, Saltholme Pools and the Reclamation Pond, with ducklings hatched there as late as September (Blick 2009).

The Ruddy Duck was perceived as presenting a significant conservation threat (Hughes & Grussu 1994) to its endangered European relative, the White-Headed Duck *Oxyura leucocephala*, through hybridization, and a number of control programmes were initiated at the outset of the Millennium across the country. Undoubtedly this had considerable impact at some sites, although birds managed to persist at others.

The *Atlas* (Westerberg & Bowey 2000) documented the breeding distribution and local status of the Ruddy Duck during 1988-1993 and demonstrated a moderate increase in records as its range expanded across the county. Breeding was first recorded at Shibdon Pond, in 1994, when five young were reared, with the species becoming quickly established there as an annual breeding bird. By 1997, up to six pairs were breeding at this site, amply illustrating the species' colonising potential. The maximum spring count of nine males and four females occurred on 11 May 1998, but a control programme commenced in 2000 and the number of birds declined rapidly thereafter.

In 2007, culling work, funded by Defra, began in earnest on the North Tees Marshes, much to the consternation of many local bird watchers, and by the end of June just a few birds remained. In that year, breeding was confined to just three sites in the north of the county, and five pairs produced 15 young. Only two pairs bred in 2008, although birds were recorded at 13 sites, and, by the following year, it was clear that the Ruddy Duck was becoming difficult to locate in Durham. In 2009, records came from just eight sites and the maximum count comprised eight birds in the Saltholme area during mid-August, with no breeding recorded in the county for the first time in over 20 years. Thus 2008 remained the last documented record of breeding within the county, a sad end for a bird that had bred at more than 20 sites in its 23 years as a Durham bird.

Recent non-breeding status

In 1995, a few birds started to linger at some sites through the autumn, such as WWT Washington, where birds were present on 19 November. The first wintering record occurred at Rainton Meadows the following year. By 1998, Ruddy Ducks could be observed throughout the year, with a wintering population at Teesmouth from around the late 1990s (Iceton 2001). In 1999, 26 sites held birds with Haverton Hole holding 19 birds in April. Later that year, the Teesmouth WeBS count recorded 64 birds in October, whilst away from Teesside, Brasside Ponds attracted 34 birds in August 1999. Further record counts came from the North Tees Marshes, with 78 in October 2000, and 151 in September 2002 (Blick 2009). This latter figure is the largest gathering ever recorded in the county, a count now unlikely to be exceeded, as a result of Defra's countrywide eradication programme.

Rather remarkably, this species quickly established patterns of movement in and around England, and Durham, with birds appearing at breeding sites in late March and early April. Numbers usually increased markedly during April, as birds moved back into their breeding areas. Such activity resulted in the appearance of birds at a range of sites from Shibdon Pond in the north to several locations in the south of the county. A drake on the sea off Lizard Point on the morning of the 14 April 2001 was rather unusual, but is consistent with the period of movement back to the breeding grounds.

Birds generally vacated breeding sites in the late summer and, up to the early 2000s at least, gathered at favoured lowland post-breeding locations elsewhere in the county, with the lake at Raby Castle being a favoured resort. Pre-exodus gatherings occurred each autumn, especially around Teesmouth, followed by the movement of the majority of birds out of the area. Prior to the commencement of wintering in the north east, many of the county's birds apparently moved south to winter on large reservoirs in the Midlands and south west of England (Kerr 2001).

In 2003, the species was still recorded in every month of the year, from over 40 sites and reached a pre-breeding peak in April of 74 birds. The Defra-sanctioned cull, which commenced in 1999 (Parkin & Knox 2010), had removed over 2,000 birds nationwide by September 2001 (Brown & Grice 2005), and by 2003 was apparently affecting numbers in the county, with previously well frequented sites holding fewer birds. A sharp decline in numbers was noted after 2003; however, 108 were still counted on the North Tees Marshes in August 2006 (Joynt 2007).

The last bird of the decade was one at Rainton Meadows that was present from early August 2009 until 9 January 2010. Following this record, the only other report of this species in Durham during 2010 came from the Allotment Pool at Haverton Hole, where three birds were noted from 10 April until 29 July. Only four birds were seen in 2011, including a displaying male on a moorland pool in July. It is unlikely that there will be many more records in Durham of this fascinating little American water bird.

Distribution & movements

The Ruddy Duck is a North American breeding bird that became naturalised in the UK as a consequence of escapes from wildfowl collections in the south west of England during the early to mid-1950s. It was initially brought to the UK in 1948 to WWT Slimbridge and first bred in the collection there in 1949. After escaping and breeding successfully 'in the wild', it was formally admitted to Category C of the British List in 1971 (Parkin & Knox 2010).

Red Grouse (Willow Ptarmigan)
Lagopus lagopus

A common resident in areas of upland heather moorland.

Historical review

The remains of this species have been found in the north east of England at sites along Hadrian's Wall, specifically at Corbridge a few kilometres north of the Durham county border, indicating that birds were probably common and taken by local Roman garrisons for food (Miles 1992).

The Cellarers' rolls of the Monastery of Durham indicate that the species was still common in the region around the 14th century (Fowler 1898). As might be expected, the Red Grouse figures prominently in Gurney and Ticehurst's work on the rolls (Ticehurst 1923). Records from these documents indicate that birds were usually purchased in lots of three to six or seven brace. They were generally documented under their medieval names of *"morecok"* or *"murkoke"* though here and there in the text they appear under the feminine equivalent, which the author commented on as *"unusual"*. Thus an entry for mid-July 1348 *In vj morehennes emp. November 24-30, 1348. In ix aucis et xj gallinis de mora et j capone cum pastu, iij s. xj d. ob. q.*, suggests that *"six moor hens (female Red Grouse) were bought and in November 1348 nine geese, eleven moorcock and one capon with feed were purchased for 3 shillings and 11 pence"*. *Gallus de Mora* would seem to refer to the Red Grouse, indicating that the Durham moors were being harvested for this species at the time.

Clearly, Red Grouse have held a prominent place in both the ornithological and cultural life of the county for many hundreds of years, courtesy of the high-quality habitat that has been created and preserved for the species locally. One good indicator of this is the size of shooting bags from sporting estates in the county. For example, in what now forms the south western extremity of the county, *"Lord Strathmore's keeper in 1843 was matched to shoot 40 brace on the Teesdale moors on 12th August and performed the feat with great ease, bagging 43 brace by two o-clock"* (Mather 1986). The importance of grouse estates to rural communities in northern England in the 19th century and the necessity to protect early broods of grouse and their habitat, is evidenced by this quotation from a legal charge in the Minute book of the township of Harmby in Richmondshire, North Yorkshire; a few miles south of the Durham/Yorkshire border, *"...And if any person shall burn any Ling, Heath, Furze etc. between 22nd February and the 24th June he shall be sent to the House of Correction for any time not exceeding a month, there to be whipt and kept to hard labour"* (Mather 1986).

Writing of the 19th century, Abel Chapman (1889) highlighted that the Durham Moors were '*infinitely more prolific of grouse than the more alluvial moors of Northumberland'*. The level of 'cropping' of the Durham moors was shown by the size of one season's bag from Wemmergill (this is now in Durham, but was then correctly described as being in North Yorkshire). The year 1872 was a record one here, 3,931 brace were shot by six guns and a total of 17,074 were killed in the season, 5,668 of which were claimed (i.e. shot) by Sir Frederick Millbank. This feat was documented as being even greater by George Temperley (1951), who recorded that in 1872, Sir Frederick Milbank and his friends had shot 18,231 birds in this one season. The average bag on this moor over 12 seasons was 4,133 brace with the largest single day's shoot claiming 1,035 brace from six guns, Sir Frederick claiming 96 brace of these in one drive lasting just 23 minutes. This feat is commemorated by a granite monument on Wemmergill Moors (Mather 1986). Not far away, at High Force in Teesdale, in 1886, eight guns killed 2,616 brace between 13th and 17 August. A century beforehand, in the latter part of the eighteenth century, Marmaduke Tunstall (1784), who lived at Wycliffe-on-Tees, some 30 kilometres down the Tees valley from Wemmergill, said that the species had declined considerably in the Teesdale district. He said not long before his time of writing that 20-30 brace could be shot on his moors in a day, but numbers had declined as a result of *"too much demand"* (i.e. shooting). Large variations in annual shooting bags have been apparent over the last two hundred years in County Durham, as elsewhere, with population numbers rising and falling on a roughly regular cycle.

The species undertakes movements away from heather moorland only during exceptionally cold, snowy weather and such activities were recorded in the past, but only rarely. The phenomenon was known by Marmaduke Tunstall who wrote in 1784: "*in severe weather they will come down to the vales in the neighbourhood of the moors and feed with the common fowls, and sit on the ling coverings of the poor cottages, sometimes in great numbers, the poor peasants not regarding or meddling with them*" (Nelson 1907). Both Robson (1896) and Temperley (1951) mention a number of sightings of the species in lowland areas in such circumstances. Almost invariably these were associated with severe weather conditions on the bird's moorland haunts. During 1892, a bird was shot by Colonel J. A. Cowens near Coalburns, in the west of Gateshead. The higher areas of this part of Gateshead borough attracted birds again two years later, when several were seen by Colonel Cowens' gamekeeper during January and February 1894 on and around Barlow Fell. This was part of a major altitudinal movement of Red Grouse which occurred in late January of 1894 after a period of unusual and extreme weather conditions. A severe snowstorm on 24 November 1893 was followed by a rapid thaw and then re-freezing which resulted in the glazing of the moors with ice. Birds were consequently driven off the moors by the New Year and twelve birds were found at Ferryhill, in the middle of the county and others as far away as Seaham on the coast. Two years later, in 1896, a brace of birds was shot near Winlaton, above Blaydon on Tyne (Bowey *et al.* 1993).

By the middle of the 20th century, Temperley (1951) described this species as a resident in Durham and said it was *"Plentiful on the heather moors to the west of the county, where it is encouraged...by game preservers"*. A major decline in numbers across upland areas of Britain, from around 1930, was highlighted by Hudson (1986).

There are no confirmed breeding records away from the higher moorland areas of the county in the last 60 years, though there is strong anecdotal evidence to suggest that birds were once breeding on some of the more easterly areas of mid-altitude heathland then present in the county, an example of this being on the remaining heather moorland of Silver Hills on Ravensworth Fell, in the Team valley, just prior to the Second World War. This area of habitat was lost during open cast operations, which destroyed all suitable heath in that area, but it has been reliably reported that birds were present prior to this process.

Nothing substantial seems to have been published on this species' local status during the 1950s and 1960s. Indeed, in most ornithological reports over this period there was no mention of either population figures or shooting bags for any of the years between 1952 and 1969, if indeed any mention of the species at all. It has to be presumed that the species' distribution and status remained stable in Durham over these decades, with no major fluctuations.

It has been estimated that 2.5 million Red Grouse were shot annually in the early 20th century in Great Britain and Ireland (Parkin & Knox 2010) but bag numbers have declined by a significant amount both nationally and in County Durham during the subsequent century. Despite the documented declines of the early 20th century in particular, the first half of the 1970s appeared to experience a series of productive breeding seasons and in 1973, the species was said to have had its most successful breeding season since 1883, largely founded upon a dry spell that coincided with the hatching period (Unwin 1974). The contrast in bag numbers from just under a century previous is of interest, with 58 brace shot on the first shoot of the season on Widdybank Fell in upper Teesdale on 20 August 1974, in what was described as a decent breeding season; a far cry from some of the figures posted by Sir Frederick Milbank. Aggressive behaviour by a male on Widdybank Fell was described in 1975. This bird was said to have attacked Land Rovers that were driven over the Birkdale Track that went through its territory. The excessively hot, dry summer of 1976 was the first of the decade that produced a poor number of birds, though another good season was recorded in 1977. Nationally, the numbers of Red Grouse have fluctuated enormously over time, with huge increases taking place in response to the regimes of intensive management of heather moorland that were initiated in the 19th century. This rise was followed by some considerable variation in numbers in the early 20th century, around a long downward trend, shooting bags having declined hugely since the 1940s (Brown & Grice 2005).

Recent breeding status

In County Durham, Red Grouse occur almost exclusively on upland dry heath and blanket bog, where heather *Calluna vulgaris* is the dominant vegetation (Gibbons *et al.* 1993). Any threat to the quality or extent of heather moorland in the county poses the greatest threat to this species (Battens *et al.* 1990) given its absolute dependence on this habitat (Westerberg & Bowey 2000). In the summer of 1975, a nest was found at over 770m above sea level on Mickle Fell, not in heather but in an open situation, and this is perhaps the highest elevation nest ever found in Durham. Today the species is still abundant in the uplands of the county, with most breeding birds located to the west of the A68 trunk road. Living at high altitudes can expose the species to extreme weather conditions at almost any time of the year. For example, the prolonged cold weather of the first few months of 1979 severely impacted upon the local population. By the time of the breeding season, many dead birds had been found and subsequently, clutch sizes were said to have been reduced and laying dates delayed. In slightly different circumstances but equally catastrophic terms, heavy snow in late April 1981 wiped out early nests; replacement clutches were smaller than usual as a consequence but overall production was said to be relatively good. In recent decades, the lower heather line in County Durham has receded, in some cases significantly, due to both climatic affects and excessive sheep-grazing, though stocking levels, especially within the North Pennine Special Protection Areas, are today far more sympathetically controlled.

The species thrives, in terms of numbers, where the heather is actively managed, especially where small plots are subject to rotational burning to create blocks at different stages of growth (Hudson 1992). This style of detailed, quite intensive management of such a widespread habitat results in a mosaic that provides young shoots for feeding and older, taller patches of vegetation for nesting and cover. The species is managed, in this way, on most upland shooting estates in the county, and this practice is of significant economic importance in the Durham uplands as highlighted by Westerberg & Bowey (2000). Red Grouse is so closely linked to the distribution of heather in the uplands, that its distribution is, in effect, a reflection of heather's upland distribution in the county (Clifton & Hedley 1995). Pairs may be on eggs from early April, more typically late April and early May, with young hatching from the middle of May onwards.

The essential heather moorland habitat is extensive in its distribution across the higher western ground of the county, indeed it may be the single most widespread habitat (agriculturally improved land aside), covering an area in excess of more than 20,000ha (Westerberg & Bowey 2000). It is particularly prevalent along the eastern slopes of the Pennines, at elevations over 300m above sea level. The only, albeit small, population of birds now regularly occurring to the east of the A68 is on Hedleyhope Fell, near Tow Law. Linear transects can be used to gauge breeding densities and 42 pairs were counted along the 14km of road between Stanhope and Muggleswick in spring 1984.

The population is prone to fluctuations and poor years, such as 1989 and 1993 during the *Atlas* survey period. These can lead to the curtailment or at least the restriction of shooting and such a situation occurred in both of these years on most of the county's estates. By contrast, 1992 was a notably good year, with one estate alone reporting that 3,000 brace of birds were shot during the season (Westerberg & Bowey 2000). These figures are worth contrasting with those from Wemmergill a century previous. In the south west of Northumberland, over 20,000 birds were shot on just two estates in 1997 (Kerr 2001); the number of birds present on the Durham heather moorlands will not be lower. For much of the rest of the 1990s poor summer weather was blamed for low or, at best, below average breeding success. On 2 May 1993, early season breeding success was demonstrated by the observation of a female with six young in upper Teesdale. This bird must have commenced laying a clutch of eggs by 2nd or 3 April at the latest, if not in late March, and provides a reminder that late heather burning into the first half of April may be a negative factor for some breeding grouse.

Research indicates that there is a population periodicity in this species, with high numbers occurring on a cycle of between four and ten years (Parkin & Knox 2010). The reasons for this, whether parasite-related or summer-weather-dependent are complex. Since 1911, the focus of Red Grouse research in the United Kingdom (Lovat 1911) has been to dampen these cycles This phenomenon has been extensively studied because of the species' commercial importance (Ratcliffe 1990). Significant population crashes can be caused by the nematode intestinal worm *Trichostrongylus tenuis*, which at high grouse densities reduces host fecundity (Hudson 1986) and can be lethal to individuals with high worm burdens (Wilson & Wilson 1978). As elsewhere, birds in Durham are prone to such outbreaks of infection and in 2005, birds on some moors in the county were severely affected by infestations of worms and shooting days were reduced as a result. Within the last decade the practice of providing medicated grit for birds has been widely adopted by estates to combat *Trichostrongylus*, with good results. Shooting bags in the latter part of the first decade of the 21st century, on several estates, showed a marked increase due to a combination of the grit, predator control and improved heather-burning practices following the introduction, in 2007, of a new Code of Practice for heather management.

The county's main populations of red grouse occur almost exclusively on managed driven grouse moor estates at locations such as Eggleston Common, Lune Forest, Pikestone Fell, Middleton Common, Muggleswick Common, Stainmore, Stanhope Common and Wolsingham Park Moor (Westerberg & Bowey 2000). According to the *Atlas*, the county's breeding population was numbered in tens of thousands of birds, with figures of some 25,000-30,000 pairs, in the best seasons, being postulated (Westerberg & Bowey 2000); there seems little reason to amend this estimate at the present time.

Recent non-breeding status

Red Grouse are widely distributed on the heather moorlands of the west of the county throughout the year with Knitsley Fell and Wolsingham Park Moor typically marking the regular eastern extremity of the birds' distributional limit. Occasionally birds are noted at Hedleyhope Fell. The species' non-breeding distribution in Durham is essentially the same as that of the breeding season.

Birds are often at their most obvious in cold weather and many birds were prominent by the moorland roadsides during the hard weather of February 1978. Flocks of birds collect at protected feeding niches in bad weather when there may be small-scale altitudinal movements. Birds will move to lower levels and valley bottoms, gills and other sheltered locations. They are generally driven together by hard weather, as illustrated by the flock of 400 that was at Eggleston in January 1985. A more dramatic example came in 1986, when bad weather early in the year led to concentrations of birds in various places. An estimated 500 birds were flying together at Hamsterley Common on 2 February, with noted 300 at Eggleston later that month, whilst 200 were in Langleydale and a similar number at Derwent Reservoir in the same period. At this time, birds were said to be favouring south-facing slopes when many of the moors were under a foot of frozen snow. There followed a very poor shooting season, and some family parties were still being seen in early September, indicating a late start to the breeding season.

An unusual record came from Widdybank Farm in winter 1991, where one was found in the farm flower bed. Several hundred were reported at Hisehope Reservoir in January 1995 and likewise at Bollihope in the following month. In 2009, the extensive winter snows rendered the species' camouflage useless, and on 17 February, 600 birds were noted on Bollihope Common with 270 on Lune Moor around the same time. A good example of local altitudinal movements occurred during the extreme weather of early 2010. In this instance, relatively large numbers of birds left the higher moorland to gather at lower altitude areas, with large flocks seen at Knitsley Fell on 18 January.

Amongst a party of eight birds on Bollihope Common in late January 1984 was a bird with white on its wings and under parts.

Birds intermittently appear at unexpected locations prompted by poor weather. In recent decades, there are a small number of examples, such as the report of two birds at Witton-le-Wear on 29 January 1974, which were believed to have moved downstream during strong westerly winds. This unlikely occurrence was followed by three even less likely appearances in the south east of the county, perhaps the most unusual being a bird seen at Dorman's Pool on 14 December 1997 (Blick 2009) followed, a few years later, by the amazing record of a carcass that was found on the sand dunes at Seaton Carew on 18 February 2001. The final sighting, of a single bird observed and photographed at Cowpen Bewley Woodland Park on 7th and 8 December 2010, was associated with a period of very harsh snowy weather.

Distribution & movements

The Red Grouse is a resident species with British birds belonging to the endemic sub-species *L. l. scoticus*, though Irish birds are sometimes considered separately under the trinomial *hibernicus*. The species is found, in the form of Willow Ptarmigan *Lagopus l. lagopus*, across northern Eurasia, northern Canada and Alaska. The British form was formerly considered a separate species (Parkin & Knox 2010).

Land management and conservation policy at national and European levels and the viability of sport-shooting estates in the uplands are all relevant to the long-term fortunes of Red Grouse and the heather moorlands upon which they depend.

Black Grouse
Tetrao tetrix

This is an uncommon resident species that has declined in numbers over the last century and a half. Nonetheless, the county still holds a large proportion of England's remaining breeding stock centred upon the North Pennines.

Historical review

Black Grouse were much more common and widespread in the 19th century than today and were once found throughout Britain, being native to at least 25 English counties (Brown & Grice 2005). The species has demonstrably been present in Durham from historic times, perhaps on a far more widespread geographical basis than may be inferred from its current distribution. The excavation of the remains from Roman sites at Piercebridge, in the south of the county, and Corbridge just to the north of the Durham area, bear witness to this (Miles 1992, Yalden & Albarella 2009). Black Grouse remains were also considered numerous in Roman excavations at Carlisle (Yalden & Albarella 2009), indicating that the species was not only relatively common and widespread in northern England at the time but also popular as a food item amongst the Roman garrisons along Hadrian's Wall.

The Cellarers' rolls of the Monastery of Durham make frequent mention of this species as "*Heath Cocks and Hens*" and indicate that the species was common and taken for food during the 14th century (Fowler 1898). There are least ten quoted extracts relating to Black Grouse and/or Capercaillie *Tetrao urogallus* in these. The quotations sometimes leave it open to argument as to which of the two species is being referred to. What is clear is that the Black Grouse was the pre-eminent species reflecting its then much more widespread, and presumably common, status in the county. Samples of relevant extracts that apparently relate to Black Grouse include: *1323-4. In vj gallis silvestribus nigris December, 1324. In xxvijgallis nigris*, January 1325. *In vij gall, silvestribus*, February 1325. *In x gall, nigris*, May 1325. *In xvj gall, nigris*, September 1337. *In cxxxiij gall., ij caponibus, iv duoden. pull, columb. et xxiiij gall, silvestribus empt. a diversis per parcellas, cum eorum esca, xxvs. ij d*, January1348. *In perdic, plovers, anatibus et gallin. silvestr., ij s. viij d.*, December 17, 1348. *In ij gall, de bosco cum minutis avibus empt.,vd.*, 1417. *Inj auca et jgall, silvestr. empt. vd. ob. In xxvj gallinarum silvestr. Empt., iij s. iiij d.ob*. It seems that in most of these instances the Black Cock (*gallis nigris*) was being referred to (Fowler 1898). *Gallus silvestris niger* or *gallus niger* would seem to refer to the Black Cock and *gallina silvestris* to the female or Grey Hen. Clearly good numbers were bought: twenty seven in January 1325, sixteen in May of the same year and 133 unspecified grouse, with two capons, four dozen young pigeons, 24 Grey Hens and appropriate feed for 25s and 2d. Temperley (1951) suggests that the reference to *gall.de bosco* indicated something unusual, perhaps relating to the "*Cock of the Wood*" or Capercaillie, rather than a grouse. The inference that we are dealing with a fowl of significant proportions is borne out by the phrase *cum minutis avibus,* meaning that it was purchased with birds of much lesser size.

In information taken from "*The Regulations and Establishment of the Household* of *the fifth Earl of Northumberland*, "*black game*" are described as Heath Polts and are priced at three shillings each, the prices relating to the year 1512 (Gladstone 1943). Black Grouse remains were excavated from 14th to 16th century archaeological deposits at the Castle Keep in Newcastle, less than a kilometre to the north of Durham over the Tyne (Kerr 2001). The proof that it was present and possibly abundant in upper Teesdale in prehistoric times is provided by the discovery of bones in Teesdale caves as documented by James Backhouse in the 19th century (Mather 1986).

By the late eighteenth century, some measure of decline in the north of England may already have commenced, as Tunstall (1784) wrote that it had "*grown very scarce all over the north of England*". He documented that it used to be "*on the moss*" not too far from Wycliffe-on-Tees, in Teesdale. Temperley (1951) presented evidence of the species' previously more widespread distribution. He cited Hutchinson (1840) mentioning breeding in marshy pastures in the township of Ebchester in the middle portion of the Derwent valley and also around Greencroft, near Annfield Plain. Hancock (1874), writing of the two northern counties, said that "*it was plentiful in the wild tracts of both counties*", "*especially where the birch abounds and in damp situations*".

The much wider distribution of the species in former times is supported by Thomas Robson (1896) and his observations in the Derwent valley. He recorded having seen one or two birds on Barlow Fell, between the villages of Barlow and Rowlands Gill, presumably in the latter part of the 19th century, and concluded that these were "*storm driven*" individuals. However he also told of how he was informed that the species had bred "*for several years*" at Horsegate, not far away from Barlow Fell, to the west of High Spen and also at Ash Tree Farm above Chopwell. Birds were shot at both of these sites by the Cowens (the father and son Members of Parliament for Newcastle who lived at Blaydon Burn House).

In his national review of the distribution of Black Grouse, Gladstone (1924) described the species as indigenous in the north east, but only "*local*" in County Durham, as opposed to "*numerous locally*" in neighbouring Northumberland. Temperley thought that by 1951 it was "*a resident; less common now than formerly*" and said it had a very local distribution, being found on the fringes of the western moorlands, where woodland and birch scrub provided cover. In the 'vale of Derwent', he said it was found as far east as Hisehope Burn, though it was reported as becoming scarce there at the time of his writing. In Weardale, it bred as far east as the Bollihope Burn, whilst in the Tees valley it was found above and below High Force.

To the north of Durham, populations in Northumberland were thought to be increasing somewhat from 1953, before falling away again sharply from the late 1980s (Kerr 2001). This situation seems to have been largely mirrored in Durham, despite a paucity of hard data. A summary of the situation for this species in the 1950s and 1960s is reliant on rather vague contemporary commentaries. In 1964 the numbers at leks were reported as being "*well up to normal*" and in 1965 it was said to be "*common in Hamsterley S.F. (State Forest)*". Despite the sparse published data it was considered plentiful in Weardale in 1967 and up to 20 birds were regularly seen in the Langdon Beck area. Just one year later the alarm began to ring, albeit quietly, with the statement "*Not quite so many in old haunts*". In 1960 there had been specific reports from locations such as Bollihope Common, Langdon Beck, Allen's Green and Wolsingham Park Moor, all places where birds might be seen five decades later, but intriguingly there were also reports from Warden Law, near Hetton-le-Hole, though no further details were published.

There were more systematic counts available in the 1970s and perhaps the first realisation that something more than a steady decline might be taking place. On 21 February 1971, 37 males were on Wemmergill Moor near Selset Reservoir but only 16 were counted on 22 February 1981 (Mather 1986), and this lek then soon fell into disuse. In 1970, up to 17 males had been at Langdon Beck, and for a while this lek was thought to be increasing. Numbers hovered around this maximum through the 1970s, with reports from four other smaller leks in Teesdale and Weardale.

The establishment of conifer plantations in the 20th century initially offered Black Grouse a valuable source of food and ground cover, largely because fencing kept sheep out of the plantations. Birds were found in a number of such areas in Durham, such as The Stang. As the trees grew, the ground vegetation changed and such sites became less attractive. This process of timber maturation coincided in Durham with the other factors that were impacting negatively on this species and it was through the late 1970s and early 1980s that several Durham sites, which formerly held the species, were deserted and others suffered significant reductions.

One of the largest counts of recent times came from the Langdon Beck lek in March 1980, when 51 males were noted. Smaller gatherings of up to seven birds were found at six other sites around this time, including The Stang. In the next few years the numbers of blackcock at Langdon Beck fell away. The peak count in the period 1981-1984 was of up to 27 males, declining to 24 in 1986, but rising again to 30 in April 1987 and 34 in February 1988. Twenty-five birds at Holwick in the autumn of 1982 and the 16 at Newbiggin-in-Teesdale at the same time may have involved birds dispersing from the Langdon Beck area. In 1983, 21 males had been shot during grouse drives in upper Teesdale and this may have explained the poor numbers noted at the Langdon Beck lek in the following spring, which peaked at just 14 on 16 April. Further afield, up to 30 were on Woodland Fell in January 1984 and nine males were displaying at Westernhope Moor, Weardale on 30 September 1985 the year that more systematic recording of this species in Durham commenced in earnest.

The locally iconic status and cultural significance of the Black Grouse, the male or 'black cock' in particular, is indicated by the existence of three public houses in County Durham that are named in honour of it. These are The Moorcock at Eggleston, not far from Barnard Castle, The Moorcock at Honey Hill, near Waskerley and The Bonny Moorhen at Stanhope.

Recent breeding status

Within Durham today, birds are confined to the western uplands, the greatest density of birds occurring in Weardale and Teesdale, and their tributary valleys, mainly between 300m and 450m above sea level. Most of the Durham sites are located on 'open fell' or sheep walk. It is interesting to note that in areas dominated by heather moorland or coniferous plantations, black game are, at best, present in relatively small numbers. Open birch/alder woodland on moorland edge, as favoured by the species in Scotland, is a very limited habitat in Durham although land management schemes have set out to create such areas in recent years. This species has been in general decline throughout Britain for a century or more. Nevertheless, County Durham today still holds one of the healthiest remnant populations of this species in England (Warren & Baines 2008).

The national decline of Black Grouse is well documented, falling from an estimated 25,000 males in spring at the end of the 1980s (Baines & Hudson 1995) to 6,500 males in 1995/96 (Hancock et al. 1999) and just 5,078 more recently (Sim et al. 2008).

By the mid-1990s, fluctuating and apparently declining numbers in Durham were giving considerable cause for concern based on the result of a ten-year study, between 1987 and 1997, by the Durham Upland Bird Study Group, supported by the RSPB. The numbers and distribution of this species in the county was by then reasonably well known (Westerberg & Bowey 2000). In 1987, the Durham Bird Club contributed to the first national survey and discovered 171 males across the county's western uplands, with 24 communal leks and an additional 27 scattered males; leks held between one and seventeen birds (Armstrong 1988). The data from this ongoing work suggested that the early 1990s produced the highest counts of that decade and during the ten-year study period over 60 lek sites were identified as holding displaying birds with the number of birds attending these fluctuating considerably from year to year. In 1993, a peak was reached with a total of 334 males attending 40 surveyed sites (Armstrong 1994). Subsequently, there was a steady decline in numbers. In 1994, 36 leks were recorded and these contained 199 males whilst 125 males were recorded from lek sites in 1997. In 1998, 33 sites held 187 males but by the turn of the Millennium, the number had probably stabilized. Counts at the county's key lek, at Langdon Beck in upper Teesdale, saw the highest numbers there in March 1993, when 35 males were present, falling to 17 in February 1999. In 1996, two leks in upper Weardale attracted 19 females but only two or three males were present, demonstrating the attractant power of such display leks. Black Grouse leks were recorded in all of the county's main watersheds, the Tees, Wear, and Tyne, and their tributaries, though birds were least obvious in the upper Derwent valley. At the time of the *Atlas* exercise, the most easterly record was from just east of the A68 trunk-road, in the area of Hedleyhope Fell.

Black Grouse tend to associate with habitat edges such as where heather *Calluna vulgaris* moorland adjoins marginal, poorly drained, agricultural grassland, with its damp rush-dominated patches (the areas known locally as 'white-moor'). Birds also use short, semi-improved, herb-rich, fields that are grazed by sheep (Baines 1994, Starling-Westerberg 2001). Detailed analysis of survey data and the species' seasonally collected faeces has demonstrated that the Black Grouse require a patchwork or mosaic of different habitats. Such varied habitats assume primary importance to the species at different times of the year and different stages of the breeding cycle. For instance, chicks require insect food from moorland pools and flushes whilst later in the year adults may be found in scrub and shrubs, often feeding on the berries of rowan *Sorbus acuparia* and hawthorn *Crataegus monogyna*.

Heather offers winter food and, together with rushes *Juncus spp.*, provides important cover. Rushes are especially utilized for nest sites and by young chicks, which feed primarily on sawfly *Symphyta* larvae (Starling-Westerberg 2001). Black Grouse adults are vegetarians, feeding on a wide variety of plant matter during the year. Cotton-grass *Eriophorum vaginatum*, provides protein in spring (Trinder 1973), supplemented in some areas by tree buds and catkins (notably of hawthorn and birch *Betula spp.*) and the fresh young shoots of herbs and grasses. In summer and autumn, the flowers, fruits and seeds of herbs, sedges and rushes form the main part of the diet.

Males will gather at leks throughout the year, with the most spectacular displays occurring between February and May in the early morning and again in the evening, when they indulge in 'roo-cooing', hissing and 'perform flutter-jumps', the noise of which may be heard up to a kilometre distant in calm weather (Westerberg & Bowey 2000). Away from their display grounds Black Grouse, greyhen in particular, can be easily overlooked.

Fragmentation of the existing population through habitat loss poses the main threat to the county's nationally important population. Whilst major land-use changes can have a dramatic effect on grouse numbers, smaller, more localised alterations may prove just as significant for this species. At one medium-size site in Durham in the late 1980s the birds all but disappeared when the area started to be used regularly by a local gun club. If the recovery of this species in the county is to be maintained, it is essential that the mosaic of habitats required by this bird is maintained and extended. The marginal farmland habitats favoured by this species have been increasingly 'improved' by a process of drainage and fertilization, as well as having been subjected to overgrazing by sheep. In many areas of the county these deleterious land-management practices have been reversed thanks to a process of active intervention through the national North Pennines Black Grouse Recovery Project initiative, (a partnership between the Game & Wildlife Conservation Trust, Natural England, the Ministry of Defence, RSPB and National Wind Power, later joined by Northumbrian Water and the North Pennines Area of Outstanding Natural Beauty Partnership) (Warren & Baines 2008).

During the first decade of the 21st century, numbers have been higher than for several years. In 2001 the peak casual counts from Teesdale numbered at least 45 males, including 21 males at Harwood, 16 at Langdon Beck and eight near the Cumbrian border. In early 2004, around 50 birds were counted in roadside fields near Eastgate in Weardale. Over the period 2005 to 2007, there were annual records of birds to the east of the A68 trunk road. Numbers in northern England as a whole rose over the period between the mid-1990s and the mid-point of the first decade of the 21st century, from 773 males in 1998 to an estimated 1,200 in 2007 (Parkin & Knox, 2010). Subsequent consecutive poor breeding years in 2007 and 2008, due to wet weather when the chicks hatched, followed by the severe winter of 2009/2010, have seen numbers across northern England fall once more to just 495 males (Warren 2010). This recent reversal has also been evident in Durham. Black Grouse remain severely threatened and the species' long-term future depends on improving breeding success and providing further areas of scrubby woodland to provide an emergency food source in severe winters. Despite such fluctuations in numbers, the county today still holds the largest proportion of England's remaining breeding stock.

Recent non-breeding status

The species' non-breeding distribution in the county is largely similar that in the breeding season, though its numbers are somewhat elevated post-breeding. Female Black Grouse in particular disperse away from the lek sites into the wider upland habitat, during the summer and much of the winter, though males stay quite close to the lekking areas throughout the year. As for the Red Grouse *Lagopus lagopus*, the severity of winter weather may play a key role in how successful the subsequent breeding season is, or at least, the number of males attending leks.

Local adult Black Grouse, especially males, appear to be relatively sedentary staying close to their focal leks and surrounding habitats. Juvenile females have two dispersal phases, one in October and the other in late March and early April. The average dispersal distance of hens is nine kilometres from their natal site but some have been known to have moved as far as 30km (P. Warren pers. comm.). Such movements are usually within the complex of upland Pennine habitats associated with the species. In some instances however, bad weather may have been the prompt that induced the rarely recorded movements into other areas. In the early 2000s, a hen was radio-tracked to an arable field between Darlington and Barnard Castle. In somewhat similar circumstances in March 2008, a greyhen was flushed from Barlow Fell above Rowlands Gill, in the lower Derwent valley. It flew over 1.5km, in a circuitous fashion, before landing in the top of a willow tree. The last record in this area was over 100 years previously. This female was known to be at least 15km from the nearest known Black Grouse lek, which was on the eastern edge of Edmundbyers Common at the top of the Derwent valley, demonstrating that some level of local dispersal of this species is still possible.

Distribution & movements

This species breeds extensively in the forests of the Palaearctic, from the British Isles in the west, through Scandinavia, Russia, Mongolia and eastwards to China, in a range of racial forms. The British birds belong to the subspecies *britannicus*. Most of the forms are resident, though some of the northern continental birds occasionally exhibit irruptive behaviour when conditions conspire to prompt this (Parkin & Knox 2010).

Capercaillie (Western Capercaillie)
Tetrao urogallus

Extinct, formerly a resident species.

Historical review

George Temperley (1951) reported that in County Durham, the species was "*Extinct. Formerly resident in the Forest of Teesdale*". There is strong material evidence that the species formerly inhabited Northern England as bones have been found in the region and also in Roman remains at Settle in the Yorkshire Dales. It has long been extinct as a breeding bird in the county. It is represented as a Durham bird courtesy of sub-fossil remains that were recorded during the excavation of a cave midden in the south west of the county. In 1878, numerous sub-fossil bones of Capercaillie were found in a limestone cave in the Teesdale Fissures, which are dated to the Flandrian Period, some 10,000 years ago. These were discovered on the Durham side of the Tees in Teesdale, at 488m (about 1600 feet) above sea level and were reported by James Backhouse. The bones (comprising: metacarpus, coracoid, humerus, tibia, femur and sternum) were initially identified by D. Bramwell. They have been recently compared against reference material by Louisa Gidney and their identity corroborated (Jessop unpub. 2007).

Bolam (1912) thought that this species may have been present in Northumberland in the 14th century. It was unsuccessfully introduced to the Eslington Hall area of that county in the 1870s (Kerr 2001) but no such re-introductions are known for Durham. The species was certainly extinct in England by about the middle of 17th century (Witherby 1938-1941) but Holloway (1996) indicated that it probably became extinct much earlier. It became extinct in Scotland by about 1760, only to be later re-introduced there (Thom 1986). Mather (1986) speculated that the destruction of the forests in the uplands were probably the cause of this species' extinction in England.

There is also archaeological evidence confirming the species' presence in the county until at least Mediaeval times (Yalden & Albarella 2009). This comes courtesy of bones excavated from 11th century middens in Durham City (Rackham 1979), at least six fragments of the species being excavated from the site of Anglo-Saxon tenements in Durham City across the period from the 10th to 13th centuries (Rossiter 1999).

In addition, there are some ambiguous but intriguing references in 14th century manuscripts which possibly indicate its presence in County Durham as late as that date (Jessop unpub. 2007). Ticehurst (1923) reviewed the evidence for this noting that there were 14th century grants of land in County Durham being held by tenure of paying "*One Woodhenne yearly*" to the Bishop of Durham. "*In the seventeenth year of Bishop Hatfield, 1361,*

Margaret, late wife of Robert Orleans, held of the Lord Bishop in capite one message and fifteen acres of land ... in Sokyrton by paying (inter alia) yearly . . . one hen called a wood-hen (et unam gallinam, voc. woodhen.)". It is possible that this 'Wood-hen' was the Capercaillie.

Furthermore, the Cellarers' rolls of the Monastery of Durham (Fowler 1898) contain a reference for September, 1337, to a *gallis de bosco*, a literal translation of the Capercaillie's name Cock of the Wood. Another reference in 1348 is for the purchase of two *gall. de bosco*. The species was evidently scarce in Durham, even in these earlier times, as it is mentioned much less often than is the Black Grouse *Tetrao tetrix* in these texts. If correct, it indicates that birds may still have been extant somewhere in the county at that time.

Distribution & movements
The Capercaillie breeds across a huge swathe of coniferous forest in the northern hemisphere of Europe and Asia, from Scandinavia east as far as north west Mongolia. It can be found in European woodlands from Iberia as far north as the edge of the Arctic Circle. This species occurred in Britain in historical times and there are sub-fossil remains from even earlier periods, but there is little doubt that it has long been extinct in Durham (Parkin & Knox 2010).

Red-legged Partridge
Alectoris rufa

An uncommon local resident breeding bird. Its numbers are supplemented annually by releases of hand-reared poults in locally significant numbers.

Historical review
This species was first successfully introduced to England in Suffolk around 1770 followed by many subsequent widespread releases (Mather 1986). It was not mentioned by either Hutchinson (1840) or Hancock (1874) as having been recorded in Durham, though it was known to have been the subject of a programme of releases near Wynyard sometime before 1876 according to Nelson (1907). This first documentation of Red-legged Partridge in County Durham is mentioned in J.H. Gurney's *'Rambles of a Naturalist'* (1876) where he wrote, *"Some years ago one was shot at Stockton, but this may have been one of a bevy which I learn from Mr Grey were turned out at Wynyard, and bred at Cole Hill"* (Bell 2008).

The second record was of a bird of the year that was shot at in the Kibblesworth-Ravensworth Castle area of the Team valley in 1909. Further single birds were shot at Marsden on 30 November 1911 and at an unknown locality in the county on 6 November 1918 as documented by Mr. Arthur Thew in *The Field* of November 16th, where he wrote of shooting a bird out of a covey of five which proved to be a hen Red-legged Partridge. Canon Tristram (1905) stated that some had been turned out by Prince Duleep Singh at Mulgrave Castle and that since then stragglers were occasionally shot north of the Tees, as at Elton. The exact origin of these birds is unclear as the species had still not expanded north from Yorkshire at this time (Mather 1986). In April 1933, of three displaying partridges seen at Mansfield Scar, near Darlington one was a Red-legged Partridge (*The Vasculum* Vol. XIX 1933). Temperley's most recent record, at his time of writing in 1951, was of one shot at Winston, in the Tees valley, on 13 October 1949. He called it, *"An introduced species. Never more than a rare straggler in County Durham"*, and said that it was well-established in Yorkshire, but was not so successful further north with odd specimens being recorded *"now and then"* in Durham and Northumberland.

Across the county there were no further published records until 1971, when a single bird was near Wolviston and Greatham on 8 April. Two birds were shot at Wheatley Hill on 20 October of that year. In relation to the Wheatley Hill record, D. Simpson said, *"A friend shot two at Greenhill's Farm, Wheatley Hill on 20 October 1971; these were the earliest records I have for the area"*. Local enquiries revealed that the birds had been released by shooters at Harehill Farm, Haswell Plough approximately two kilometres to the north west (Simpson 2011). It seems probable that most of the county's records at this time will have come directly from release schemes both in Durham and Northumberland where birds had been noted regularly from about 1969 (Kerr 2001). Reports may have also been aided by the slow spread northwards of the Yorkshire population (Sharrock 1976).

Two were with Grey Partridges *Perdix perdix* near Durham City on 12 December 1973 and there were singles at Shibdon Pond in 1973 and 1974 and five were shot in the Whickham area in November–December 1976. One that flew into a window at High Shincliffe on 23 June 1977 was somewhat unexpected. In the south east of the county, the species remained scarce with only three further published records up to the end of 1976 (Joynt *et al.* 2008). A small upsurge in sightings came in the late seventies with birds at Hargreaves Quarry on 30 April 1977 and Dorman's Pool on 29 April 1978. The species was being recorded in small numbers in Durham annually by the end of the 1970s (Bell 2008) and an upturn in sightings gathered pace through the 1980s. In 1979, there were six records over the period March to May, almost as many as over the previous decade, from across the county with birds being noted near Washington, at Dorman's Pool, Derwent Reservoir, near Darlington and at Langdon Beck.

Recent breeding status

Today, the Red-legged Partridge remains an uncommon bird in County Durham, though locally it is subject to regular rearing and releases and numbers fluctuate from year to year. The largest numbers are usually to be found in the upland fringe areas of the western part of the county where there are dedicated game rearing interests although numbers have tended to increase in lowland areas too in recent years. The species naturally prefers dry, arid areas with acidic soils (Bell 2008).

Red-legged Partridges are generally uncommon and remain local in their distribution across the county, as indicated by the map in the *Atlas*. The species' association with farmland habitat, which does not attract large numbers of observers results in it being somewhat under-recorded. The county lies towards the northern limit of the Red-legged Partridge's naturalised range in Britain, and many of the birds present in the area undoubtedly originate from release programmes in and around shooting estates (Westerberg & Bowey 2000). The extent to which the populations would be self-sustaining without annual releases is debatable.

The local status of Red-legged Partridge is quite complex and its distribution has been determined, to a considerable degree, by the commercial release of captive-bred stock on particular estates and by shooting syndicates (Raw 1991). For a period it had been further complicated by the rearing and release of Chukar *Alectoris chukar* x Red-legged hybrids, a practice which had all but ceased by about 1992. 'Red-legs' appear to have gained a foothold as a breeding species in the early 1980s, breeding birds at Ravensworth Grange and a female with a brood of small young at Wycliffe-on-Tees being, in 1982, the first confirmation of 'wild breeding' in Durham. In reality, it would appear that birds have been very much more widespread than documented. A synopsis of the releases of hybrid Red-legged Partridges in County Durham (Raw 1991) made reference to the release of nearly 1,500 such birds on two shooting estates alone after 1976. The Game & Wildlife Conservation Trust commissioned a study into the impact of Chukar and hybrid releases on wild stocks of 'red-legs'. This concluded that the British feral population of 'red-legs' declined rapidly in the 1980s and that this decline was attributable to the large numbers of hybrids released for shooting purposes at the time, with further hybridisation and reduced breeding success leading to a diminution of the wild, pure-bred stock (Potts 1989). Consequently, the Department of the Environment introduced a national ban on the release of non-native and naturalised wildlife species (including Chukar hybrids), which was effective from January 1992. It is only through the late 1990s and more recently that pure 'red-legs' have once again been routinely released locally.

A thumbnail sketch of the species' detailed situation in the well-studied area of Gateshead over this period is perhaps indicative of its relatively recent history across the wider county. The first 'modern' reference to birds in the Gateshead area was of one at Shibdon Pond in 1973 and 1974. During the late seventies birds began to be released in the Ravensworth Estate in the Team Valley. Breeding took place at Ravensworth Grange in 1982 and family parties were seen there two years later. During 1985, birds were noted on Kibblesworth Common and the following year, what was probably a family party was noted at Lamesley on 4 August, a site from which birds have been subsequently reported on many occasions. In July 1989, a female with three young was seen not far from Marley Hill and during 1990 a pair was noted on Ravensworth Fell. The release of birds in the Gibside Estate, most of which were probably hybrids, coincided with the appearance of birds at the Far Pasture wetland in the Derwent Walk Country Park. This site attracted birds on a number of occasions in 1991, with as many as 15 there in October (Bowey *et al.* 1993).

More generally across the county, releases of birds were documented at Pittington in 1985 and Joe's Pond in 1987 (the latter of 50 birds) and in Weardale in 1988, though it is unclear whether or not these birds were hybrids. The following year, March and April observations came from near Causey Arch, High Coniscliffe, Hury Reservoir and Middleton-in-Teesdale. It would seem that any pattern to sightings had disappeared to be replaced by widespread low-level release of small numbers of birds. Up to18 birds were at Hury Reservoir near a known release site on 17 January 1988. As the *Atlas* documented (Westerberg & Bowey 2000), in the period 1986-1990, two estates in the county released 1,200 *Alectoris* hybrid partridges, only 400 of these being 'accounted for' during shoots.

The species is reported in all months and in 2007 records came from about 30 localities across the county including Selset Reservoir, Hargreaves Quarry on Teesside and Lamesley, Ruffside and Derwent Reservoir in the north. *Alectoris* partridges can occupy a wide range of habitats, being found on lowland arable land and some cropped systems on the edge of the upland-fringe areas (Gibbons *et al.* 1993). During the survey period for the *Atlas,* groups of birds were found near Cotherstone, Baldersdale, Woodland and Muggleswick, though there was little doubt that most of these related to local releases of birds. Since the early 1980s, breeding records have been relatively few and far between, though no doubt small numbers of birds do breed as 'wild pairs', and such breeding was documented at Piercebridge on 28 July 1994 and in Baldersdale in August 1996 and again in 1997.

Around Teesside, records have been increasingly prevalent from the late 1990s with perhaps 10-15 pairs in the parts of Cleveland away from Teesmouth (Joynt *et al.* 2008). Birds were recorded in a wide range of habitats, from moorland edge to industrial wasteland. During 2002 and 2003 birds were also believed to be increasing in the South Tyneside area and there was an increase in the lower Derwent valley. These increases probably related directly to the release of birds by local shooting syndicates, although some pairs were present in areas where shooting did not occur (such as the Gibside Estate in the Derwent valley in summer 2004). Evening survey work by the Durham Bird Club in 2005 revealed this species' presence in 15 of the county's 48, whole or part, ten-kilometre squares, showing a wide distribution but with some emphasis on the north central section of the county, some of the Wear uplands and in scattered areas of the coastal strip. Two pairs held territory at Kibblesworth in 2007, an increase on 2006 and in 2008 birds were reported from 50 localities across Durham. Despite this much wider representation and documentation of breeding success, it remains doubtful as to whether the Durham population of 'red-legs' is yet self-sustaining.

In neighbouring Northumberland, huge numbers of birds were released for shooting in the 2000s. In one area alone, in summer 2007, birds were present close to a supplementary feeding station in their 'many hundreds'. This is likely to be repeated in Durham in some areas so that wild breeding birds will be swamped annually by releases. It is known that fresh releases occurred in Durham in the Lambton Estate in summer 2002. The magnitude and geographical spread of Red-legged Partridge releases continues to have a dominant effect on the species' status in Durham.

Recent non-breeding status

Outside of the breeding season the reported numbers tend to increase, especially towards the autumn and the end of the year, largely coincident with shooting estates releasing stock. Somewhat unusual records include the covey of nine that was on Hartlepool Headland on 1 May 1997, and there were two records from Sunderland gardens in May 1999. More unusual still, was the bird discovered roosting some three metres from the top of a 20m high quarry face near Sherburn Hill in 2008.

Distribution & movements

Red-legged Partridge is endemic to the Western Palaearctic and is restricted largely to France and Iberia and there is no evidence to suggest that, in the British context, it is anything but a sedentary species. It was first introduced to the UK in 1673 to Essex, Surrey and Sussex, though largely unsuccessfully and it was not thought to have become established. In the 1770s, large numbers of eggs were imported from the French population and hatched for release in Suffolk, and this large-scale release is thought to be the core origin of the feral population in the UK (Parkin & Knox 2010). There were at least another 60 recorded introductions in the period 1830 to 1958, and by the end of the 19th century the species had spread as far north as Yorkshire (Lever 1977), its penetration further north into Durham not occurring in any substantive way until the more concerted releases of the 1980s.

Grey Partridge
Perdix perdix

A widespread resident which remains quite common. Found in lowland agricultural habitats, river valley field systems and on the fringes of upland moors.

Historical review

The Partridge is listed with considerable frequency in ancient documentation relating to the county. Temperley (1951) noted that *"common partridge"* appeared to have been *"abundant"* in the 14th century, as the species was frequently recorded in the Cellarers' rolls of the Monastery of Durham. From the numbers of birds purchased at one time, the purchase of 'lots' of twelve to eighteen brace were not infrequently recorded, it was evidently quite a common species and no doubt a valued addition to the diet of the Priory. In these documents it was generally entered under its Latin name *perdix*, but several variants of its English name occurred for example, *pertrikis* and *pertrykes* (Ticehurst 1923).

In the early 19th century, Selby (1831) thought that it was *"abundant"* in the north east and that it had benefited from the then *"recent agricultural improvements"* (Kerr 2001). Writing of its presence in the Derwent valley in the late Victorian period, Robson described it as *"very common in our district"* (Robson 1896). Around the same time in the south east of the county, Nelson (1907) said this species was an abundant breeder around Teesmouth at the turn of the 20th century. Perhaps some decline took place in northern England in the next few decades because George Bolam (1932) thought that it was *"less common than in the past"* (Kerr 2001). Around the mid-20th century, Temperley (1951) documented it simply as *"a resident"*, and said it was common from the coast to the fringes of the moors adding that birds were present in the agricultural areas but absent where cultivation did not occur. He said that it was preyed upon by Carrion Crows *Corvus corone* and ground predators. Stead (1964) considered it a common breeding resident around Teesmouth through the 1950s and 1960s.

All of the national data indicate that over the last half-century there has been a quite dramatic reduction in the British breeding population of Grey Partridge, a situation which has been exacerbated in the last three decades. The causes of this are almost certainly related to the intensification of farming methods, post Second World War, though this process so far appears to have had relatively little effect in many parts of Durham although it is unlikely to be anywhere near as common as it was in historical times. Writing of his long-term study area, D. Simpson commented that the Shotton Colliery area since the 1950s has always had *"good numbers of Grey Partridge, even during the worst of years"* (Simpson 2011).

Through the early 1970s, data in *Birds in Durham* suggested that it was declining a little in areas of intensive agriculture in the county, but it was still plentiful elsewhere. In 1973, numbers were said to be recovering and this apparently continued through 1974 with birds doing well in eastern parts of the county, after some perceived local declines during the 1950s. Throughout the period, the county's westerly populations seemed as strong as ever. Farmland in the area around Barmston was good for this species through the decade, as demonstrated by the 50 at WWT Washington in the autumn of 1974, the 85 there in early 1976 and a similar number at the year's end in 1977. Winter coveys of 30-50 birds were typically reported from farmed areas in the early 1970s and through that decade small coveys were noted in upland locations such as Baldersdale, Burnhope and around the various upland reservoirs, some sites being at well over 300m above sea level. During the hard weather of February 1978, large numbers of birds were reported in snow-covered fields between West Boldon and Wardley, with around 200 birds there on 18 February and another 127 around Barmston Pond on 21st.

Despite the declines in other parts of the country, the figures of Grey Partridge in Durham through the 1970s and 1980s indicated that the species was doing reasonably well, with good numbers regularly reported. Good breeding success was the norm in the 1970s with, for example, a brood of 23 chicks in a Kelloe Law meadow in summer 1975 not being atypical. Newly hatched young were noted on Harthope Fell in late July 1977, one of the most exposed locations in the Pennines.

Recent breeding status

Nationally, the Grey Partridge is a widespread bird of open, arable farmland, field margins, grass leys and hedgerows and like many farmland birds, it is not as common as it used to be. It nests on the ground usually at the edges of fields and in hedges and banks with often quite long linear territories and relies primarily on plant food (weeds and cereals). Within lowland Durham, the Grey Partridge predominantly occupies mixed and arable farmland. It is most common on farmland throughout the eastern lowlands and extends its range to several coastal habitats including disused industrial land. County Durham has a wide variety of agricultural land practices many of which seemingly suitable for Grey Partridge. It is this variety that has perhaps buffered local populations against some of the declines experienced nationally. It is also closely associated in Durham with upland rough pasture, moorland margins and in-bye land in the west of county (Gibbons *et al.* 1993). Birds are widespread and thinly scattered but only really absent from the heather-dominated areas. Nests here are reported to be in quite similar situations to those of Black Grouse *Tetrao tetrix* with a preference for clumps of rushes *Juncus spp.* within unimproved field systems. Once again, this choice may have helped buffer the species from the worst of the population declines seen in the more intensive agricultural areas of Britain.

The Grey Partridge's critical, long-term national decline, which resulted in the species' placement on to the red list of Species of Conservation Concern, has been extensively documented (Marchant *et al.* 1990, Mead 2000). According to national monitoring data, this downward trend seems to have steepened during the late 1970s, when it was determined that over 85% of birds were lost (Parkin & Knox 2010). In 2009, Grey Partridge was found to have reached its lowest monitored level nationally on Breeding Bird Survey squares since the commencement of the scheme, declining by 50% since the start (Risely *et al.* 2010). Gibbons *et al.* (1993) remarked that the decline of the species had been much less well marked in the east of the country and this is certainly the case for County Durham.

In the 1980s, successful breeding was widely reported with high densities of breeding birds in some areas, such as the five broods produced in the Joe's Pond area in 1984 from 10-15ha, whilst birds routinely bred in the uplands, as illustrated by the family parties noted high above Stanhope in Weardale in June 1985. The largest number counted in one area in the early 1990s was 65 at Eppleton Colliery, after the breeding season, in September 1990. The long-term pattern of records has failed to indicate any serious local declines but rather points to a stable population in many lowland areas.

Certainly, Grey Partridge was relatively well recorded in the county during the survey work for the *Atlas*. This showed concentrations of birds in some eastern areas, indicating relatively high local observer activity as much as concentrations of breeding birds. Westerberg & Bowey (2000) pointed out that it was likely that surveyors may have missed some pairs on farmland and in the upland margins. The highest concentration of occupied tetrads in the *Atlas* was coincident with the outcroppings of the magnesian limestone, in the east of the county, and the predominant agricultural practices that go alongside this geological feature (Westerberg & Bowey 2000).

Birds can also be found at much lower breeding densities on reclaimed and derelict land, even in urban areas, and even the exposed hilltops of the north Pennines support pairs of birds (Westerberg & Bowey 2000). An extreme example of this was documented in the *Atlas*, when a pair with eight young was seen in July 1991 at an elevation of 620m above sea level on Chapel Fell. Birds are found throughout much of the eastern lowlands and occur extensively in coastal habitats, from the Whitburn Coastal Park in the north to those areas of the Tees estuary claimed from the sea. In 1990, there was an unusual report of two pairs that both raised young on the Jetty Peninsula at Seal Sands, in an area where birds had not been known to nest before. As winter coveys split up in March, pair-formation takes place and males begin to 'sing'. The species can breed late in the year and quite often young broods have been noted in early September.

The factors identified for the national decline might well apply to Durham at some stage, but at present numbers still seem to be holding up well. During the first decade of the 21st century, this species remained widespread across Durham, breeding successfully at many different sites in a range of locations and habitats. Up to 50 pairs were reported during the breeding season in 2006 and breeding-season birds were noted at a minimum of 180 localities in 2008. Coastal areas of Durham remain productive for this species. For example, to the north of Ryhope Dene at least six pairs bred in the coastal fields between Hendon and Ryhope, over a distance of some 4km, in the spring of 2006. In 2007, the species was widely reported from more than 140 localities, from locations as far west as Herdship Farm in upper Teesdale and Killhope in upper Weardale, to North Gare at coastal Teesside, and north to the county boundary with Northumberland. In that year, sample surveys revealed eight pairs in the Seaton Pond area, seven calling males at Brancepeth Beck, six males at East Carrside, four pairs at Trimdon and four pairs in territory at the Kibblesworth Common Bird Census site. In 2010, the species was reported as breeding from 25 sites, with ten pairs noted in the Seaton Pond area and eight pairs at Eppleton.

The range of sites holding the species is indicated by looking at where family parties were reported during the breeding season over the period 2005-2010. These included: The Stang, at over 350m above sea level in the far south of the county; Ramshaw in the upper Derwent valley, in the north of the county; Low Hardwick; Langley Moor; Brancepeth Beck; Old Quarrington, Oakenshaw and Bishop Middleham in central lowland and mid-altitude locations and, in the south east, the Castle Eden Walkway and Hurworth Burn Reservoir; though many hundreds of other sites would have also held breeding birds over this period (Newsome 2006-2012).

The relative status and distribution of this species and its closest relative Red-legged Partridge *Alectoris rufa*, was illustrated by the survey results from forty, one kilometre squares across the county in 2005. Survey data from these revealed 33 pairs of Grey Partridge, which were found in twelve of the forty squares (36%), whilst just six pairs of 'red-legs' were found, and these were restricted to just two (5%) of the squares, presumably near shooting estates.

Survey work for *The Breeding Birds of Cleveland* revealed that there were over 60 occupied tetrads in the north part of what was Cleveland, with at least 130 to 150 pairs breeding in this part of the county (Joynt *et al.* 2008). The *Atlas* came up with a 'conservative county estimate' of four pairs per occupied tetrad, based on selected surveys and the national figures in high-density occupied areas given by Gibbons *et al.* in 1993. This suggests that Durham may hold a population in the range of 1,500 to 2,000 breeding pairs and this range probably still stands as accurate today, though the current figure is now perhaps nearer the lower limit.

There were some large importations of this species into the UK from estates in Czechoslovakia and Hungary during the early part of the 20th century (Bell 2008). Small numbers continue to be released for shooting purposes, though they are thought to make up less than 0.2% of the total UK population. With several large estates in County Durham specialising in Grey Partridge shooting, such as that around Raby Castle in the Tees valley (Joynt *et al.* 2008), it is safe to assume that some birds continue to be released directly into the county; it is not known whether the level of this activity has been sufficient to significantly enhance local populations (Bell 2008).

Recent non-breeding status

In the post-breeding season and towards the end of the year, the largest counts of Grey Partridge tend to be made. The number of birds locally is at its height at this time and, assuming a reasonably successful season, there may be as many as 10,000 Grey Partridges across Durham at the outset of the autumn. The extremes of winter, shooting pressures and natural mortality rapidly impact upon this figure. Winter survival is thought to be the critical stage of the Grey Partridge's life-cycle, especially in the uplands where winter weather is harshest. On one large upland estate, a strategy of providing feeding stations is now proving effective in helping to sustain the population.

Through the 1980s this species remained a common widespread resident, and winter counts of coveys of over 20 birds routinely came from sites all the way from the coast to the edge of the moors. Its hardiness in such locations was demonstrated on 19 January 1980 when two birds were on snow-covered pasture at Widdybank Fell at more than 410m above sea level. Through the decade the largest counts came from lowland farmland such as the 50 around Joe's Pond in October 1983; 73 at Barmston on 25 January 1984; 69 at South Hylton, Sunderland in January 1985; 75 in the Pittington-Sherburn area on 14 December 1985; 57 at Greenside on 7 January 1986 and 86 at Barmston in February 1986. At WWT Washington, 63 were in one field in October 1984, the highest count there since the early 1970s. Large parties of birds were also reported from more elevated locations, such as the 52 at Derwent Reservoir in February 1986. The first slight suggestion of any local declines came in 1987 but this was somewhat countermanded by reports of large coveys the following year.

During the 1990s, despite the national declines, some still impressive numbers of birds were being recorded in the non-breeding season, such as 92 in January 1995 in the River Browney area, 86 in January 1995, around Burnhope on the slopes of the Browney valley, and at Sedgeletch, 69 in February and 84 in December 1996.

In some of the best areas for this species, for example along the easterly strip inland from the coast south of Sunderland, wintering numbers can be high. As a snapshot of such locations, on a sweep to the south and south west of the City of Sunderland, from Seaton, via Burdon west to Hetton and Houghton, there are still good numbers of breeding and wintering Grey Partridges, as evidenced by the peak monthly counts of 65 birds around Seaton Pond and 45 at Hetton Lyons in December 2008. In the Seaton area alone there were gatherings of over 70 birds in mid-November 2007 with a similar number at Farnless Farm near Bishop Middleham in September 2004, indicating that numbers in Durham were being maintained through this decade.

Smaller gatherings of between 20 and 30 birds are more routinely noted in lowland areas from around Rainton Meadows, Trimdon, and Sharpley Plantations. Reports of the largest winter gatherings often come during the first few weeks of the year, mainly from east Durham, when cold weather can concentrate numbers and, at such times, gatherings of from 20-40 in January and February are by no means unusual. In most years, smaller coveys of 10 to 20 birds coalesce to create flocks 30 to 50 strong. Numbers in the west of the county at this time of the year tend to be somewhat smaller, as food concentrations tend to be poorer, though gatherings of up to 25 birds have occurred, such as at Langdon Beck in November 2007.

One of the intuitive observations made about this species in Durham over the last decade was that it seems to have benefited from the provision of game crops and winter set-aside fields. This statement is somewhat in contradiction to the national data, which seems to suggest that such practices have only provided limited success (Baillie *et al*. 2006) and that release by shooting syndicates may now be artificially buoying up local populations (Parkin & Knox 2010).

On occasion, single birds are seen very much out of their normal context and habitat, for instance there are records of birds on Hartlepool Headland and Seaton Carew beach (Blick 2009) and a record of a pair of birds feeding at low tide on the foreshore at Trow Rocks in 2009. A pair using a garden feeder at Langley Moor in March and April 2004 was unusual for this species (Siggens 2006).

Of particular interest was the observation of a pair of birds of the '*montana*' morph, the only documentation of its occurrence in Durham, which was present on Edder Acres Farm at Shotton Colliery during one summer in the late 1970s. The farmer and his friends who first reported them believed they were Red Grouse *Lagopus scoticus* as they were very red. D. Simpson and other observers who saw them described them as very distinctive birds (Simpson 2011).

Distribution & movements
Grey Partridges occur as breeding birds from northern Iberia, north and east across Europe into the British Isles and as far east as the western provinces of China. It is a resident species that is resolutely sedentary over much of its Western European range (Parkin & Knox 2010).

Quail (Common Quail)
Coturnix coturnix

A scarce and erratic summer visitor, though perhaps under-recorded and somewhat overlooked.

Historical review
The earliest record for the county is from a manuscript by Tunstall (1784). He said that a "*few are found here, but not frequent*"; one infers from this that he meant that birds were occasionally present in the area of the Tees valley around Wycliffe and Whorlton on Tees. In the 19th century, Hogg (1827) reported that "*several*" had been taken in the "*hot summer of 1826*" and Blick (2009) reported that some birds were even being reared and released in the Wynyard Estate at some time prior to 1829. Selby (1831) recalled being told of Quail breeding near Cleadon but thought that it was "*now a bird of rather rare occurrence in the northern counties*", implying that it had at one time been otherwise. Backhouse (1834) thought that it was "*not uncommon*" whilst Hutchinson (1840) considered it "*thinly dispersed*", and recalled that he had shot one at Gilesgate Moor and a young bird in at Lanchester in October. George Temperley (1951) noted a specimen in the Hancock collection, was labelled as "*killed against telegraph wires at Cleadon Station 1850*". In the south east, it was recorded as breeding at Greatham in 1868 (Blick 2009). Hancock (1874) documented it as having bred at Fulwell and Westoe and there was also a nest in Hancock's collection that was taken at Fulwell in August 1869. In *The Bird -Life of the Borders*, Abel Chapman (1889) wrote of one that was shot on 22 September 1870 in a stubble field above Frosterley, the only one he had seen in Britain, so in that sense it was presumably scarce locally. Nonetheless, according to Holloway (1996) it was occasionally found in nearly all counties and regions of Britain during the last quarter of the 19th century, and was probably breeding annually, in small numbers, over much of the country (Holloway 1996).

Temperley highlighted its occurrence in the north east and north west parts of the county, essentially the lower Derwent valley and the Sunderland to South Shields areas (Temperley 1951) and noted that these same areas were also favoured locations for the species in the latter part of the Victorian period. There are many historical references to this species' occurrence within the westerly portion of what is now Gateshead borough, including the lower Derwent valley and the areas above Ryton and Crawcrook. Around 1876 a nest with eggs was found on Barlow Fell, above Rowlands Gill, and several were shot in the Blaydon Burn area. In 1883 one was 'taken' at High Spen, whilst the previous year a nest was found between Greenside and Ryton (Bowey *et al.* 1993). This area has produced a number of records over the best part of a century. In 1891 a clutch of eleven eggs was taken from a Greenside nest and two years later, in a summer that was obviously very good for the species, another nest was found in the Greenside area and three or four birds were frequently seen at Fellside, above the Gibside Estate (Bowey *et al.* 1993). During the 20th century there were no documented records in the favoured Gateshead area until the 1930s when the Greenside-Ryton area once again proved attractive to birds. Singing birds were noted there in the summers of 1934, 1937 and 1939. During the late 1950s two birds were flushed from damp pastures at Derwenthaugh (Bowey *et al.* 1993).

Temperley (1951) said that since Selby's time, Quail had remained "*a bird of rather rare occurrence*". At the middle of the 20th century, he described it as "*a rare and irregular summer resident; never numerous and frequently absent altogether*". The agricultural methods employed in the 19th century were possibly more suited to the bird's breeding requirements and the discovery of its nest or chicks. There have been very few confirmations of breeding records in the 20th century and beyond in the county. On 18 July 1933, Miss Anderson of West Boldon took three eggs to the Hancock Museum, which had been found on Pike's Hole Farm. A bird had been disturbed during mowing and at least 12 eggs found. Six were removed and others left in case the incubating bird returned, which it did not. They were identified as Quail eggs. At this time it was thought a rare visitor to Durham but it was noted that "*at one time it appears to have been fairly plentiful*" (*The Vasculum* Vol. XIX 1934). Some of Temperley's then more recent records included information about two nests found in Teesdale in 1949 (Temperley 1951).

So called 'Quail years' occurred in 1947 and 1964, with birds being noted in both years in Northumberland (Kerr 2001), but in neither case was there much evidence from Durham; the only report was of a bird noted in a rainstorm at Greatham Creek on 6 June 1964. As part of a national review of the species' status, Parslow (1967) commented that "*breeding has probably occurred at one time or another since 1954 in every English county except Middlesex (and perhaps Durham and Devon, where birds were present in summer in 1964 only)*".

Birds were rather rare through the 1960s, not even averaging one per year through the decade, despite the one nationally recognised Quail year in this decade. One was calling near Sadberge, Darlington on 22 June 1965, a bird was heard at Wynyard on 21 May 1966, one was singing near Fishburn on 3 June 1967 and a bird was found dead at Hartlepool on 24 May 1969.

Quail remained very scarce in Durham through the 1970s. Birds were heard at two sites near Hartlepool from 1st to 19 June and also near Hurworth Burn Reservoir on 8 June in 1970, then no birds were reported in the county for seven years, until the spring of 1977, when two were calling at Haverton Hill on 3 June and a late migrant was at Marsden on 20 September; the first birds reported in Durham since 1970. There were then no further records until spring 1981.

Recent breeding status

Due to the Quail's secretive nature, its habit of singing mainly at dusk and dawn and its preference for little observed agricultural habitats, this is a species that is rarely seen and is probably relatively poorly documented in County Durham. Nevertheless, singing birds are now recorded annually with occasional spring or autumn passage birds noted in coastal areas.

As the species is largely in the northern edge of its British range in County Durham, there are no discernible local threats to what is, in Durham, only a 'marginal' breeding bird (Westerberg & Bowey 2000). There is little doubt that Quail are rather scarce, irregular and inconspicuous breeding birds (Gibbons *et al.* 1993), though they have been recorded in Durham every year since 1992. It seems likely that absence in previous years may often have been due to a lack of observer coverage.

Quail can be found in a range of agricultural habitats, largely arable crops, especially cereals, but occasionally in grass lays or permanent pasture. Typically, most records of this secretive species come in June when calling is at its height. A combination of its scarcity and its preferred habitat being in largely under-watched farmland means that this species in Durham, as in the rest of the UK, is probably very much 'under-recorded' (Westerberg & Bowey 2000).

Birds were more prominent than for many years in 1981, being reported from three sites, one at Meadowfield on 2 June, up to six near Piercebridge through July commencing a run of records in this area that continues to the present time. Birds were noted annually across the county from 1981-1986, in very small numbers with Piercebridge again featuring in 1982 and 1984. By contrast, 1983 was a good year in the county with up to three birds calling at Piercebridge from 28 May to late July and five in August, last reported on 25th. Five birds were involved in records around Sedgefield from 16 June to 14 July with two at Houghton-le-Spring and two at Littletown. The years 1985 and 1986 reverted to type, with just a handful of records, with one to two birds in fields around Washingwell Woods at Whickham in both years, and an additional bird reported at Joe's Pond in summer 1986. In 1987, calling birds were near Longnewton Reservoir in mid-May, at Leamside in early June and at Piercebridge on 17 July. Breeding was suspected at one site in the county in 1988, when a male was calling from 31 May to 5 June near WWT Washington. Two were at Holmside through June and July, one was at Great Lumley in late July and one was near Sadberge on 7 August.

During the time of the survey work for the *Atlas*, there was a notable influx of this species to Britain in 1989; such influxes tending to occur on a roughly four to six-year cycle (Pennington *et al.* 2004). The influx of 1989 involved by far the largest numbers of birds recorded during the 20th century in Durham and this 'inflated' its local presence and the *Atlas'* mapped distribution derived mainly from this Quail influx year (Gibbons *et al.* 1993). In what might be considered a normal year, Quail is a really rather scarce species in the north east.

The *Atlas* mapping showed the species' distribution to be spread across the county, with some concentrations in the north, and especially the north west where observer effort was high and this area, as has been noted previously, is traditionally a good one for the species (Bowey *et al.* 1993). This is reflected by the pattern of records from western Gateshead, an area regularly frequented by the species in the 1930s and again in the modern period from the mid-1980s, suggesting that observer activity has a profound effect on the perception of this species' distribution and status locally. In the summer of 1989, during this special year for the species, single birds were noted calling at Clockburn Dene, Highfield, Lockhaugh Meadows, Milkwellburn Wood, Ravenside above Chopwell and at Washingwell Wood, with at least five, and probably seven, birds in the Greenside area. There is compelling research evidence that indicates a clustering of calling birds occurs because migrating birds are attracted to calling birds already present in suitable habitat, with many apparently suitable places between such aggregations of birds remaining unoccupied (Parkin & Knox 2010). At this time in Gateshead, birds were also reported to be present in the Team Valley and during August a family party was flushed by a farmer on Parson's Haugh near Ryton Willows (Bowey *et al.* 1993); an important record for a species that is so difficult to secure proof of breeding. It is likely that this pattern was matched in many other less well observed parts of the county.

The year-by-year maps published in Gibbons *et al.* (1993) indicated that there were very few records in 1988, whilst 1990 showed a small scatter of records across the county, with a slight concentration of occupation once again in the Gateshead and north west areas. There were very few reports in the early part of the 1990s, but in 1994, nine sites reported birds. Seaton Pond seemed to be developing as a regularly occupied site and birds were recorded singing from mainly oilseed rape, winter wheat and other arable crops. In 1996, at least three birds were found singing from rape crops in the Whickham to Kibblesworth area, between 8th and 15 June. In June 1997, which was a very good spring for the species, up to nine males were thought to be at Rainton Meadows alone.

This is a scarce bird in the Teesside area, with relatively few records up to the 1970s. According to Joynt *et al.* (2008), there were just 13 occurrences during 1968 to 1985, with no birds at all in ten of those years. Over the survey period for *The Breeding Birds of Cleveland*, from 2000 to 2006, there were just a handful of occupied tetrads in the north of Teesside area. In typical years just a handful of singing males are recorded, at most (Joynt *et al.* 2008).

A review of records from the first decade of the 21st century indicated that the numbers of this species occurring in Durham were as variable as ever, with a low base of just a handful of singing birds in the summers of 2001 and 2003, rising to as many as 40 singing birds in the summer of 2009. As is often the case with this species, breeding was rarely proven, such evidence being gathered in just one of these years when, in 2004, five juveniles and a female were noted at Hetton Bogs on 20 June. At the outset of the decade, 2000 was a good year with a minimum of nine birds in the county, though there were some 12 singing males in Cleveland (Blick 2009). During 2005, at least 18 singing males were noted and nocturnal survey work revealed birds in 10 ten-kilometre squares. These were distributed largely across the northern and central northern portion of the county, with some records in the central and south east areas. In what was a poor year nationally, 2006, there were reports from 12 localities in

Durham. The only record that definitely involved more than one bird came from Dalton Piercy, where five or more were heard calling from 13th to 23 June. These however, were later linked to a release of locally reared birds that had taken place here, originally in 2005, but possibly repeated later. In 2007, seven individuals were reported from six localities on four dates during June, one date in May and one date in August, while one observer checked six traditional occupied sites on 18th and 22 June, but no birds were heard or seen. In summer 2009, which was a Quail year, there were reports of 50 birds from 26 localities, including 40 singing males. Seaton Pond held up to seven birds in 2009 and this site held at least two singing birds in every year between 1994 and 2001. It is clearly an important area for this species in Durham. In 2010, singing males were reported from eleven sites.

In non-invasion years, the *Atlas* estimated the population figure to be as low as just ten pairs; not an unreasonable estimate of the breeding figures in many quiet years, though this figure will be very considerably exceeded in good years, such as 2009. Nonetheless, the number of Quail in the county probably remains very much under-recorded. This is based on the fact that the average number of reported calling Quail between 1988 and 1993, excluding the influx year of 1989, was just 2.8 per annum (Armstrong 1988-1993). In that rather exceptional year, 1989, between 39 and 49 singing males were documented. Gibbons *et al.* (1993) indicated that the national population is usually of 'up to 300 pairs', though in the exceptional year of 1989 in excess of 1,600 were believed to be present across the country.

Most records in the county come from lowland, farmed landscapes and birds are rarely heard in the uplands. However, one was calling in what was considered ideal breeding habitat at 310m above sea level near Frosterley in June 2004 (over 130 years after Abel Chapmen had heard one in that area). Another at Harwood-in-Teesdale, on the 25 June 2005, and one that flew into bracken on Bollihope Common on 4 July 2006, were at similarly elevated positions and might be thought exceptions to the general distributional pattern in the county.

Recent non-breeding status

Quail, Hartlepool Headland, May 2004 (M. Sidwell)

There are at least two records of birds in gardens and a few records of passage birds at Hartlepool Headland. There was also a very unusual record of one observed and photographed sitting on a house roof at Millfield, Sunderland, on 16 May 2010, although the origins of such urban birds must be viewed with some caution as Quail are commonly kept by aviculturalists. Most birds are reported as arriving in late May or early June though one calling at Durham City on 6th and 7 May 1987 was unusually early and almost matched the passage bird at Mere Knolls Cemetery on 7 May 1995. These early records were pre-dated by one that was singing on 3 May 2009 at Elemore. An even earlier record concerns one in a small roadside garden in Norton on 18 April 2002, giving very close views as it fed on cornflakes provided by the householder (Little 2003).

The date for this however, was noted at the time as being suspiciously early, because of the possibility that this may have been a released bird. The autumn passage bird at Marsden Quarry on 20 September 1977 remained the latest recorded for the county for many years, until this record was extended to 1 November 1997, when a bird was at Medomsley, in the Derwent valley.

Distribution & movements

Quail are highly migratory. They breed over much of Europe and north west Africa and the Middle East. European breeding birds winter in sub-Saharan Africa. In northern Europe the species occurs in varying numbers and with marked fluctuations from year to year. The scale and impact of any local release programmes, though they are known to have occurred, are thought to be minimal.

This species has experienced very considerable declines across its European range in the last three to four generations (Hagemeijer & Blair 1997). One suggested reason for these observed declines over the last century or so has been the impact of continental hunting pressures during migration (Marchant *et al.* 1990).

Sponsored by

Pheasant (Common Pheasant)
Phasianus colchicus

A common resident; populations are extensively consolidated by annual releases of birds from shooting estates.

Historical review

There is some disagreement about the earliest history of the Pheasant in Britain. Some authors have asserted that it was introduced by the Romans; others suggest that this took place around 1299 or that it was first brought to Britain by the Normans in the 11th century (Sharrock 1976). Yet others claim that this did not happen until much later, perhaps in the 14th century. What is clear is that remains of this species have been found in the Roman excavations at Corbridge, though it would seem quite possible that these remains were from birds that were brought north with the legions as domesticated stock, rather than occurring as locally wild individuals (Miles 1992). There have been various claims for its presence in Britain since Roman times, but it is known to have been documented in this country since at least 1058 (Parkin & Knox 2010).

Whatever the national situation, Pheasant has been known in County Durham since at least the 13th century, as is proven by its mention in the Bursar's Rolls of The Monastery of Durham for 1299, which mentions "*uno fesaund*". During the 13th and 14th centuries, the Pheasant may have been an uncommon bird in County Durham, for in this period there is just a single mention of it in the Bursar's Rolls, though it occupies one the earliest chronological mentions of any bird species in those texts (Ticehurst 1923).

In the Northumberland Household Book of 1512, Pheasants were entered as *Fesauntes*, or *Fessauntis* and this documentation has often been quoted as evidence of the early existence of the species in Great Britain. Mention in such a text however does not necessarily mean that these birds were obtained locally though the species was certainly present in the north east of England, in some capacity, before 1059 (Gladstone 1943). Perhaps subsequent introductions served to supplement the earlier releases as indicated by Holloway (1996)? Nonetheless, it had become well-established in Britain by the end of the 15th century and intensive Pheasant rearing was certainly underway by the mid-1800s and continues to the present (Holloway 1996).

In 1827, Hogg listed this species as being an uncommon bird in the south east of the county. By contrast, Temperley in 1951, called it "*Introduced; but now naturalised and a common resident*", though he asserted that it did not become established as a feral breeder until around the 19th century (Temperley 1951). Up to the start of the 20th century, it occurred largely as a reared and 'pampered' game bird. As game-rearing declined proportionately more wild birds survived. He noted it as being widely scattered, from the industrialised east to the edges of the western moors.

Temperley's work (1951) told of a male and female 'Brown Pheasant' *colchicus*, the originally introduced variety, which were taken at Ravensworth in the Team Valley in 1833. The birds originally brought to Britain were thought to have come from the Caucasus and consisted of the nominate race *colchicus*, which lacks the white neck ring. Later releases included birds of the race *torquatus,* the Ring-necked Pheasant which originates from China. These were largely brought to Britain from the middle of the 19th century, replacing the originally introduced *colchicus.* These races readily hybridise and recent research has indicated that at least another four races have been introduced to the UK "*over the years*" (Bell 2008). Thomas Robson documented the shooting of a pure white specimen at High Spen in 1877 (Robson 1896), at that time the species was "*strictly preserved*" in the valley.

A bird noted in a Chester-le-Street garden on 11 February 1967, was at that time quite an unusual record (*The Vasculum* Vol. LII July 1967) and this proved to be the first such documentation of this behaviour which is now commonplace in the county.

Recent breeding status

In the British context, Pheasants are largely birds of lowland farms and are rarely found very far from some level of cultivation, in particular arable fields with nearby cover. That said, in Durham, Pheasants breed in a wide range of habitats including marshland, woodlands and hedgerows, but unlike the Grey Partridge *Perdix perdix*, they are absent, or present in only relatively low numbers, from much of the easterly coastal strip and large parts of the eastern half of the Magnesian limestone plateau. Birds are most numerous in the low-lying, central and eastern portions of Durham though birds can still be found, at much lower densities, on and around the moorland fringes, to the west. Birds do not routinely breed on open moorland, but a female, which was flushed from a nest containing 12 eggs in deep heather *Calluna vulgaris*, near Eggleston on 18 May 1993, demonstrates that they can. It is widespread as a breeding bird on farmland in the south east of the county, though it is less common here than in some other parts of Durham (Joynt *et al.* 2008). Traditionally, at least in modern times, this species has not been associated with Teesmouth, so a female with a nest at Haverton Hole in 1989 (Bell 1990) was considered notable.

As in other parts of Britain, this species' numbers are heavily supplemented by the release of many tens of thousands of hand-reared birds each July and August, which are used to augment the shooting stock of local estates (Hill & Robertson 1988). Across Britain, very large numbers of birds are released annually and it has been suggested that this amounts 12 million birds annually, which would make it, after the domestic fowl, one of the most common birds in the country (Parkin & Knox 2010). Pheasants are reared and released almost everywhere that there is rough shooting in Durham. One estate in the county released 1,800 birds in 1994, which may be typical of the scale of releases elsewhere (Armstrong 1995). This means that the Pheasant is extensively distributed throughout the county and may be found in many areas where they would probably not persist 'naturally', as purely wild stock. In 1990, a bird was recorded at Racehead Plantation, in upper Weardale at an elevation of 530m above sea level.

The numbers of natural breeding birds is difficult to estimate because the rearing of birds for shooting causes very high local concentrations (Marchant *et al.* 1990). Some of these survive the shooting season to breed in the following years.

In spring, males are bold and conspicuous and fond of open places; their territories always comprise open ground next to cover, and are usually at a density of 0.6-2.2 males per ha (Gibbons *et al.* 1993). As they are polygynous, Pheasants do not lend themselves to an estimate of density by a description of breeding 'pairs'. During Breeding Bird Survey work in Durham in 2007, Pheasants were recorded in 33 of 35 sampled one-kilometre squares; an occupancy rate of 94% (Newsome 2008). At Kibblesworth, in 2007, the long-term study revealed a 'healthy' population of 16 territorial males in approximately 10 hectares. The *Atlas* estimated that even in an average spring, it is probable that there are some 10,000 breeding females in County Durham (Westerberg & Bowey 2000). In one southern part of the county, Pheasant rearing ceased in the early 2000s, to be replaced by a programme of encouraging wild-breeding Pheasants and Grey Partridge; the initial results proved encouraging.

Recent non-breeding status

The species is reported throughout the year. In 2007 for example, it was noted from over 150 localities from Herdship Farm in upper Teesdale, east to Seaton Common and north through the centre of the county, all the way to the border with Northumberland. It is probably least numerous in the extreme north east of the county, along the coastal strip from Sunderland north to South Shields, though birds are still present in this area and are noted regularly, for example, in and around Whitburn village. The majority of records refer to ring-necked birds though there is a great variety in the plumage of birds.

During winter, birds collect in scrub and woodlands for shelter and roosting, except when feeding at dawn and dusk (Gibbons *et al.* 1993). Nineteen birds roosted in hawthorns at Shibdon Pond on 12 January 1987. Twelve birds at 435m above sea level, feeding on top of Brownberry, above Grassholme Reservoir, on 3 January 1987 demonstrated the species' hardiness. Winter survey work during January and February 2007 revealed that the species was present in 80% of the 25 tetrads surveyed in one lowland ten-kilometre square, but whether this is representative of the county as a whole, it is not possible to say with any degree of certainty. In most years the largest reported counts of birds are made towards the end of the year. A gathering of 140 birds in the vicinity of Derwent Reservoir in October 1986 were almost certainly released birds from the Ruffside Estate. Most of these 'flocks' will refer to locally released birds, such as the 70 noted at Morley Lane, which were described as "*ready for the gun*", on 5 September 2007. In that autumn, similar numbers were recorded at Barningham, Croxdale, Byers Green Hall, near Derwent Reservoir and at Sherburn.

In the Derwent Walk Country Park during the 1980s, large numbers of birds occurred when shooting was prevalent in the adjacent Gibside Estate, leading to large gatherings of birds when shoots were in progress there, for example 75 birds seen together there in October 1986 (Bowey *et al.* 1993). Since the cessation of shooting there, numbers have dropped to a much lower level, indicating that most of these gatherings comprised birds released for shooting as opposed to 'wild breeding birds'.

Away from shooting estates, during the 1990s and the last decade, the highest counts have often come from WWT Washington. Birds were first noted on the bird-feeders around Hawthorn Wood there during the winter of 1980 and this proved to be the start of major congregations of this species which occurred at this site through the 1980s and into the early 1990s. Up to 40 were counted here in January 1982, 65 on 16 January 1986 and 75 on 28 October 1997. Birds happily exploited the food that was made available as a result of the feeding of the captive wildfowl over this period.

During the autumn and winter, hard weather usually forces a few birds into gardens and coastal locations. A female was in Marsden Bay in late September 1990 but more bizarrely, a bird was seen on South Shields Pier on 13 October 1999; it eventually turned around and walked back to land. An adult male was on the beach near Horden on 7 October 2007, but it had probably just strayed from nearby fields rather than being representative of any more elaborate movement. On 5 November 1983, a female was seen to fly in off the sea at North Gare, Teesmouth, presumably it had journeyed directly from across the Tees estuary.

Over recent decades, birds have increasingly taken to visiting garden feeders and the habit appears to be spreading. The earliest documented occurrence of this in the county was of 14 birds coming to a bird table at Balder Grange in Baldersdale during hard weather in December 1981; 27 were doing likewise there in March 1985.

For released birds, shooting is not the only hazard they face. Pheasants in the wild can suffer from the intensification of agriculture, and in years of high summer rainfall the species' breeding success, like many game birds in such situations, can be poor (Hill & Robertson 1988). The genetic deterioration of their breeding competence, as a result of frequent crosses with hand-reared stock, has intensified in recent years (Hill & Robertson 1988). In addition, very large numbers die on country roads but there is an even greater volume of mortality where busy main roads lie adjacent to major shooting estates, such as the A183 and the nearby A1(M), which are in close proximity to the Lambton Estate which loses hundreds of birds to road traffic collisions each autumn. In April 2007, 34 pheasant corpses were noted on one drive down this section of the A1(M) between Junctions 62 and 63 (Durham and Chester-le-Street).

Distinctly-coloured individuals such as melanistic (almost black) and tawny-coloured males are not infrequently observed in the county, and these are likely to have their origin in release programmes. Examples amongst many include an albino bird that was at Nookton Burn on 10 July 1988, an all-black bird at Low Barns in November 2006 and a melanistic female in Old Shotton in June 2008 close to where several melanistic birds had been noted in the previous year.

Being large and obvious birds, observations of interesting behaviour are more easily made and reported than for smaller species. In 1986, a female that laid eggs in a garden at Balder Grange ignored passing lawnmowers just a short distance away when incubating. In 2001, a female was seen in a confrontation with a domestic cat, and during the contretemps it sounded an alarm call not unlike that of a 'roding' Woodcock *Scolopax rusticola*. During the autumn and winter of 2004 and 2005, birds were noted in the Barlow Burn, in the north west of the county, walking around and feeding in pastures, presumably on earthworms, alongside Red Kites *Milvus milvus*, which were doing likewise, without any noticeable interaction between them. In summer 2008 a 14-day old juvenile was seen to chase an adult Jackdaw *Corvus monedula* from a garden feeder full of seeds.

Distribution & movements
The native distribution of the Pheasant is one that is based upon South East Asia, parts of China and extending westwards to the Black Sea, in a number of racial forms and types but in the British context, for all intents and purpose, this is a sedentary species. In terms of the British population, many of these races are combined to

create a melange of baffling intra-specific hybrids, which vary hugely in colour and patterning around the species' basic plumage blueprint. It is one of the most widely introduced bird species in the world and has been recorded from at least 50 countries worldwide (Parkin & Knox 2010).

Red-throated Diver (Red-throated Loon)
Gavia stellata

A very common winter visitor and passage migrant along the coastal strip.
Scarce away from marine habitats and rare on inland waters.

Historical review

It would seem that the Red-throated Diver has always been relatively common along the county's coastline, especially during the period September to April. Hutchinson (1840) said that it was common in Hartlepool Bay and he recorded one that was shot at Kepier on the River Wear, near Durham City, on 17 December 1840. Hancock (1874) observed its frequent occurrence in summer plumage along the north east coast. Nelson (1907) quoted an example of one killed at Teesmouth in 1901, which disgorged eight sand-eels, three of which were eight inches long (Mather 1986).

Temperley (1951) called it a "*common winter visitor off the coast*" at the mid-point of the 20th century, though he also noted that it was "*rarely seen on inland waters*". At that time, birds were frequently seen fishing offshore between October and March and, as Temperley said, "*Occasionally earlier and later*" than this.

Even with limited seawatching activity, there were still some interesting winter movements documented in the post-Temperley period. Counts of 40-50 birds were made in the Teesmouth area in most winters in the late 1950s and 1960s, while larger movements at Hartlepool included 80 on 11 January 1964, 74 on 16 February 1964 and 80 on 30 January 1965. Gatherings of feeding birds were frequently noted off Crimdon Dene, including over 50 on 8 February 1960 and 46 on 31 January 1965, while one observer counted 52 flying north here in 75 minutes on 25 January 1963.

Interesting inland occurrences in the 1960s included one at Hurworth Burn Reservoir on 28 January 1961, reportedly in summer plumage, although this would be a rather odd date for such a plumage state. Charlton's Pond attracted birds in April 1958 (two), April 1960 and February 1966, while others were at Hart Reservoir in March 1960 and March 1968. All the inland sightings in the 1950s and 1960s fell between 28 January and 12 April.

Confidence in identification was clearly an issue through the 1960s and into the 1970s, and this may have somewhat masked the species' true status at this time. Of 20 divers on the sea at Crimdon on 27 February 1973, it was reported that "*only three were confirmed as Red-throated*". With current knowledge of the status in the county of the three regular species of diver, we can surmise that the other 17 birds were likely to be Red-throated as well.

Recent status

This remains easily the most commonly recorded diver species along the Durham coast, judged to make up almost 90% of all of the bids identified (Blick 2009), although intensive seawatching studies at Whitburn between 2006 and 2010 showed it to be 95% (M. Newsome pers. obs.). This status is one that seems to have been maintained through time, even amongst the historical records of the species. Birds can regularly be seen offshore at many sites between South Shields and North Gare, but there are usually large winter gatherings from Hawthorn south to Hartlepool Bay. The numbers to be found off this stretch of coast have increased notably in the period between the late 1980s and the end of the first decade of the 21st century, as the coastal marine fauna has recovered after the long period of coal despoilment that was experienced off the Durham coast. The population wintering off the Durham coast is, in relative terms, not high in comparison to those found in the southern North Sea, where nearly 60% of the estimated British wintering population of 17,000 resides (O'Brien *et al.* 2008). Nonetheless, they are an important indicator species of coastal faunal and floral welfare.

Peaks in numbers on this stretch of coast generally occur in late winter and again in mid-autumn. Winter gatherings have included 65 at Crimdon Dene on 21 January 1978, 71 at Horden on 10 March 1999 and 57 off Castle Eden Dene on 14 November 2007, while 40 flew past Horden on 8 March 1981, with a further 47 still on the sea. Examples of autumn counts include 72 on the sea between Horden and Blackhall on 16 September 1989, nearly all of which were adults, and 53 at Crimdon Dene on 26 September 1999. Loose gatherings of up to 36 birds are now noted annually off localities such as Blackhall Rocks, Chourdon Point and Nose's Point at any time between early September and early April.

At Teesmouth, birds regularly use the Seal Sands area through the winter period, but numbers vary considerably from year to year. A presence of up to 12 birds might be considered normal, but 51 were counted on 1 February 1998, 24 on 12 December 2004 and five were at Seal Sands and 17 at nearby Seaton Snook on 23 November 2005. This small wintering population also utilises the mouth of the Tees and Hartlepool Bay, with occasional birds wandering into Hartlepool Docks to escape rough weather and winter seas. Gatherings at Teesmouth have included 32 in Hartlepool Bay on 21 February 1998 and 25 off North Gare on 13 January 2003, while Hartlepool North Sands attracted 65 on 21 January 1978.

In the north of the county, small numbers are frequently noted in the mouths of the Rivers Tyne and Wear, with one to three birds occasionally wandering as far inland as Jarrow Slake on the Tyne and upriver to the Wearmouth Bridge at Sunderland. Varying numbers utilise the sea off Whitburn and Marsden through the winter with local influxes occurring in response to food availability, 10 to15 birds constituting an average population. Late 2010 produced higher numbers in this area with 25 on the sea off Whitburn Coastal Park on 31 December.

Although parties of Red-throated Divers are most often encountered in autumn and winter, impressive spring gatherings have occasionally occurred, such as the 35 between Nose's Point and Fox Holes on 5 May 2007 and 17 off Crimdon Dene on 10 May 2008, while a pair was noted displaying off Hartlepool Headland on 24 June 1996. The summer presence of this species has become more prominent through the 2000s; counts such as 14 off Crimdon Dene on 28 July 2007 would have been considered most unusual 40 years earlier, but are now becoming more commonplace.

In the past four decades, there has been an average of around two birds every three years on inland fresh-water sites, over 60 birds being recorded in such situations between 1970 and 2010, making it by far the most frequent of the divers to be seen inland. All inland records in this period have fallen between 16 September and 31 May, nearly 60% occurring between the last week of December and the last week of March. Without a doubt, the most extraordinary inland report in the county's history took place on 17 May 2008, when a flock of ten birds, nine of them in summer plumage, was found on Selset Reservoir, in Lunedale, in the far south west of Durham. This is the only recent occurrence of more than one bird and only the second-ever inland occurrence in May.

Almost any inland water body can attract the occasional wanderer. Reports from the upland reservoirs in the west of the county since 1970 have included birds at Balderhead Reservoir (1), Derwent Reservoir (5), Selset Reservoir (10), and Tunstall Reservoir (2), while one at Witton-le-Wear is also notable. The reservoirs in the south east have featured well with birds at Crookfoot Reservoir. (5), Hurworth Burn Reservoir (3) and Longnewton Reservoir (2). Records at smaller wetlands in more central areas include Bishopton (1), Brasside Ponds (1) and Hetton Lyons CP (3). Slightly more expected sightings have come from the lower reaches of the main rivers and adjacent wetlands. Barmston Pond (3), Charlton's Pond (2) and Shibdon Pond (2) are all close to tidal sections of the Wear, Tees and Tyne respectively, and occasional birds have been seen on the rivers themselves, for example, as far west as Ryton on the River Tyne and Chester-le-Street on the River Wear. The freshwater habitats in the Teesmouth area have also attracted single birds, with reports coming from Dorman's Pool, Haverton Hole, Portrack Marsh, Reclamation Pond and Saltholme Pools.

Sea-watching activities from the county's main coastal watch points produce the majority of the records. The weather conditions largely determine the number of birds that are noted in coastal movements by land-based observers (Blick 2009). The number of coastal birds tends to be relatively low in the winter period, December through to early March, but periods of harsh weather on the Continent or during strong winds from between the north west and north east often induce greater movements. Significant counts under such conditions have included 119 at Hartlepool on 11 January 1974, 223 at Hartlepool over 14th to 16 January 1977 (including 103 on 15th) and 139 at Whitburn on 22 January 1979. Birds occasionally move *en-masse*, such as the 100 birds that flew north off Whitburn in three flocks on 31 January 1982. The most recent example of cold weather movements came in December 2010 as temperatures in the north east regularly fell below minus 10 degrees Centigrade as a result of severe weather in Northern Europe. Limited seawatching at Whitburn produced passages of 134 on 4th and 152 on 19th; no doubt the total movements over this period would have been very much larger had prolonged observations been undertaken.

207

Although there is a rise in the number of birds recorded in spring, passage is not usually that prominent off County Durham. Counts from seawatch points at this time of the year usually relate to single figures, with day counts barely reaching double figures in most Aprils and Mays. Regardless of this, the highest single day count was of 180 birds moving north off Hartlepool Headland on 12 April 1970, during strong onshore winds (Blick 2009). Such passage levels are clearly unusual and counts from Whitburn of 25 on 6 May 2002, 16 on 25 May 2004 and 19 on 2 May 2010 might be considered unusually high.

Mid-summer is a quiet time for this species off the Durham coast, although a small number of birds, presumably first-summer individuals, are typically noted from late May through to early August. Reports of birds at this time usually number from one to five, although increased seawatching effort at Whitburn in the second half of the 2000s highlighted occasional larger movements, such as 10 on 14 June 2008, 18 July 2010, 11 on 13 July 2008 and 31 July 2010. The summer status of this species has clearly changed between 1980 and 2010. In 1984, 98 hours of seawatching at Whitburn during July and August produced just 12 sightings, whereas in 2007, a total of 87 were logged in the equivalent period.

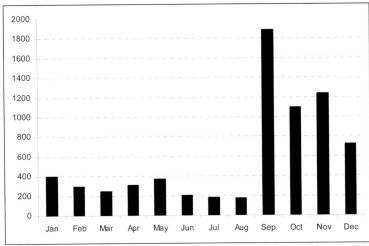

Red-throated Diver Passage off Whitburn, by month, 2006-2010

Autumn passage usually commences from late July onwards, occurring in a more concerted fashion from mid- to late August. Studies at Whitburn over 2006-2010 showed that 59% of all passage occurred in the three-month period from September to November, and the distribution of the total number of birds logged per month is illustrated. Typically, movements may involve counts of 'tens' of birds in a day and September usually brings the highest numbers, but the observed effects are very much dependant on the availability of favourable seawatching conditions for observers. Strong northerly winds can produce a flurry of passage birds, presumably displaced from locations further north around the UK coastline and *en route* to wintering areas in the southern North Sea. There is a clear peak in numbers from coastal watch points in September and again in November, although numbers vary considerably from year to year in response to conditions. Documented passages at Whitburn during periods of northerly winds include 91 on 19 November 1991, 87 on 10 October 1992 and 73 on 15 September 2002. Elsewhere, up to 80 birds were counted at Hartlepool in mid-September 2007, part of a total of 651 birds recorded over the whole month here. The scale of passage is perhaps best illustrated by looking at the cumulative totals of birds over passage periods, such as 466 noted at Whitburn over 6th to 30 September 2004 (including 162 on 25th) and 207 at Whitburn over 10th to 18 September 2009 (which included 51 on 14th).

Although the number of Red-throated Divers reported on the Durham coast has increased over the past three decades, this appears to be largely a result of increased observer effort. Seawatch data from Whitburn showed an 'average birds-per-hour' figure of 2.1 in 1991 (675 birds in 325 hours), 3.1 in 1994 (785 birds in 251 hours), 1.6 in 2007 (1,506 birds in 922 hours) and 2.6 in 2010 (2,230 birds in 851 hours). These slight fluctuations show there appears to be little in the way of any long-term upward trend.

Distribution & movements

This species has a circumpolar distribution, breeding in wetland habitats from the high Arctic to northern temperate areas. In Europe, they can be found as a breeding bird in Iceland, through much of Fennoscandia and down into northern Scotland and Ireland, the only areas where the species breeds in the UK; some of the most southerly breeding locations in the world. The British wintering population includes breeding birds from Shetland and Ireland, as well as birds from Scandinavia (Cramp *et al.* 1977).

Black-throated Diver (Black-throated Loon)
Gavia arctica

An uncommon passage migrant and winter visitor to the county's coastal waters, but rare inland.

Historical review

The first specific reference to the species in Durham is of immature birds being shot at the mouth of the Tyne and at Seaton Snook in the winter of 1829/1830 (Backhouse 1834, Selby 1831). Hutchinson (1840) said it was "*seldom found either on the coast or on the Wear, Tees or Tyne*", though one was shot on the River Wear near Durham in December 1844. Hancock (1874) intriguingly regarded it as a "*frequent winter visitant*", a statement that does not really stand up to current-day knowledge of occurrence, and he referred to a summer-plumage bird in his collection that was shot on the River Wear at Durham, "*some years ago*". The date on the specimen says 16 December 1845, which may mean that this is perhaps the record referred to as having been "*shot on the Wear in 1844*". Reliable old records include those of Robson, who noted the shooting of two birds on the Tyne, off the mouth of the River Team, during the winter of 1894 (Robson 1896). It was always considered very rare on inland waters in Durham, but Courtenay found two together on Hurworth Burn Reservoir on 18 February 1924 (Temperley 1951).

By the mid-20th century, Temperley (1951) said that this species was "*an infrequent visitor to the coast in winter and still rarer on inland waters*". The true status of this species, like some of the other diver species, is somewhat clouded by the problems of its identification in winter plumage, which may mean that many records of winter divers in the past were of unproven species identification. It would seem that there was some confusion about the critical identification features that separated this bird, in winter plumage, from Great Northern Diver *Gavia immer* and even Red-throated Diver *Gavia stellata*. Nonetheless, it would appear that it has always been very much less common than Red-throated Diver in the north east. This is well-illustrated by comparing the incidence of reports between the 1950s and the 1960s, when the number of reports in the former decade numbered in the scores to hundreds but in the latter, only in tens. There seems no reason to suggest that there was any major change in the patterns of occurrence of the species over this period, more that optics and the identification ability of observers had improved significantly, facilitating more accurate recording of the species.

Of the confirmed post-Temperley records, two inland birds are worthy of mention. One was at Hart Reservoir on 27 January 1960 and an oiled bird was seen at Brasside Ponds on 23 February 1966. The remaining observations through the 1960s were entirely typical of the current status; occasional single birds (and once, two together) were noted at Marsden, Seaham, Hartlepool and Teesmouth between late October and late March. A summer-plumage adult flying south at Hartlepool Headland on 22 June 1966 was the only out-of-season bird.

Recent status

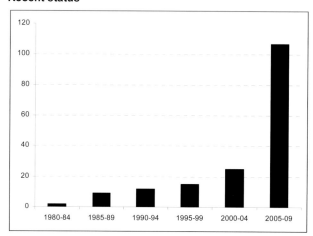

This species is an annual but scarce visitor to Durham. Typically, birds are seen in small numbers at the start of the year, with sightings increasing, occasionally into double-figures, towards the year's end. Although most of the sightings are concentrated into the winter months, it has been recorded at other times of the year with a small number of mid-summer records for Durham. It remains the scarcest of the regularly recorded divers to visit Durham, with studies at Whitburn in the period 2006-2010 indicating that 1.5% of divers passing the coast involved this species.

Black-throated Diver at Whitburn by five-year period, 1980-2010

There was an average of just seven records per year through the 1980s, increasing to ten in the 1990s, but a significant rise took place in the 2000s to an average of 29 birds per year. The increasing number of records over this 30-year period is probably indicative of observers' increased effectiveness in spotting birds, courtesy of improved optical equipment, greater observer experience and an improved understanding of the criteria used to identify the species in flight and at a distance, all contributing to this increase. Perhaps the modern understanding of its status is more truly indicative of the species' true long-term status in, and off, the county, and of the true scale of passage along the Durham coast, as opposed to the perceived historical situation which indicated that the species was a very rare visitor indeed (Newsome 2007).

In terms of geographical occurrence patterns, this species might be noted almost anywhere along the Durham coast, from September to April, but it is clear that the area around Teesmouth, Jackson's Landing at Hartlepool Docks in particular, is one of the most attractive sites for the species. The Headland has attracted from one to three birds in all but five years in the period 1980 to 2010, including a run of eight consecutive years from 1981 to 1988, and is the county's most favoured location, principally between December and February. Examples of extended stays in the shelter of the harbour and docks include 22 December 1996 to 8 January 1997 (up to three birds), 19 January to 28 March 1999, 14 December 2002 to 9 March 2003, and 27 November 2007 to 23 January 2008 (two birds).

Elsewhere on the Durham coast, records have come from several localities including Blackhall, Castle Eden Denemouth, Crimdon, Easington and Hawthorn, and occasionally the South Shields area. A total of 68 birds were reported on the central stretch of coast between 1980 and 2010; this is a relatively high proportion of all non-passage birds recorded, particularly in comparison with Great Northern Diver, which is only reported occasionally in this area. Perhaps the geography of the coast and in particular the seabed, along with the associated food sources, favours Black-throated over Great Northern.

Coastal reports have normally involved birds in the expected period of September through to March and it is rare to see a summer-plumage bird on the sea off the Durham coast. Unseasonal birds are very occasionally noted on the sea, for example off Whitburn Coastal Park on 17 July 1983, Castle Eden Denemouth on 17 August 1986, Hartlepool Headland on 20 May 2006 and Blackhall Rocks on 21 August 2010. The majority of reports of birds resting on the sea relate to singles, so three off Hartlepool Headland on 14 January 1996 and four birds at Dawdon Blast Beach on 26 March 2006 were both notable gatherings.

There have been just fifteen Black-throated Divers seen at inland sites in Durham over the period 1970 to 2010. These records largely follow the pattern of inland occurrences shown by Red-throated Diver, but they are much less common. Sightings have all fallen in the period between 2 December and 17 March, with late February and early March being a favoured time.

- Cow Green Reservoir, 15 April 1975 (two birds)
- Maiden Castle, Durham, 17th to 21 February 1979
- Billingham Pond, 10 February to 10 April 1981
- Wynyard Lake, 18 February 1981
- Crookfoot Reservoir, 1 March 1981
- South Marine Park, South Shields, 22nd to 24 March 1981
- Crookfoot Reservoir, 14th to 19 March 1983
- Longnewton Reservoir, 23 January 1985
- River Tyne, Gateshead, 16 February 1985
- Shibdon Pond, 2nd to 7 December 1985 (died)
- River Wear, Washington, 18th to 19 January 1986
- Shibdon Pond (and adjacent river Tyne), 22 January to 20 February 1986
- Hebburn Riverside, 17 March 1996
- Tunstall Reservoir, 8th to 16 December 2009

The sight of two birds at the desolate moorland locality of Cow Green Reservoir in April 1975 was particularly notable; a Great Northern Diver was at this site in October of the same year. The bird at Shibdon Pond in 1985 developed the habit of flying off to the Tyne in order to feed and on 7th, it flew into overhead cables breaking a wing, necessitating that it be destroyed (Bowey et al.1993).

At the key seawatching localities of Hartlepool and Whitburn, Black-throated Divers have been regularly logged since the commencement of greater seawatching effort, in the mid-1970s. Despite this, and other developments that aid identification and location, the species remains a scarce visitor to the Durham coast.

Autumn provides most records, particularly after northerly winds from mid-September to late November. One or two birds are the norm, but three have been noted on six dates, and four were logged at Hartlepool on 23 November 2007.

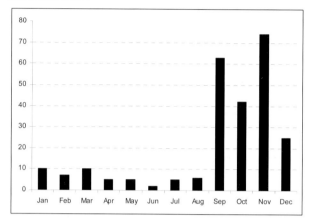

Monthly Distribution of Black-throated Diver, Hartlepool & Whitburn, 1980-2010

The monthly occurrence pattern is broadly similar to Great Northern Diver, but the November peak is not as pronounced, with the autumn passage period commencing earlier. This is perhaps to be expected, as the majority of Black-throated Divers moving off our coast probably originate from northern Eurasia and therefore have less distance to travel than Great Northerns, which may be arriving from Greenland and Baffin Island. Summer sightings are decidedly uncommon, with May to July passage records totalling just 14 birds. Several of these have been in full summer plumage, but as with Great Northern Diver, immature birds are the norm.

The best year for coastal passage was 2007. A total of 32 birds were observed off Whitburn, all but four being in the period 2 September to 16 December, while observers at Hartlepool logged a further 12 birds. Sightings included three past Whitburn on 14th and 25 November, whilst four passed Hartlepool on 23 November. In 2005 and 2009, the combined total off Whitburn and Hartlepool was 29 in each year; a figure unheard of in the 1980s and 1990s. The second half of the 2000s was by far the best period for observing this species in Durham.

Distribution & movements

This is a tundra and boreal breeding species of northern Eurasia. The majority winter coastally, just to the south of their breeding area. Most British wintering birds are believed to emanate from the Scandinavian population (Parkin & Knox 2010). There is a small British breeding population, which is largely restricted to northern and north west Scotland (Parkin & Knox 2010).

Great Northern Diver (Great Northern Loon)
Gavia immer

An uncommon passage migrant and winter visitor to the county's coastal waters, but rare inland.

Historical review

The first definite reference to the species in the county appears to be that from Hancock (1874), who documented the taking of a *"mature specimen"* on the Tyne close to the Tyne Bridge on 12 October 1824. In Willughby's *Ornithology* however, under the title of *"Gesner's Greatest Douker"*, appears the intriguing statement, *"Mr. Johnson (of Brignall, Near Greta Bridge), in his papers sent us, writes that he hath seen a bird of this kind, without any spots on its backs or wings, but yet thinks it not to differ specifically but accidentally"* (Mather 1986). This title was published in 1678 and Johnson of Brignall did much of his collecting at Teesmouth, so this specimen in all likelihood came from there.

Hogg (1827) said, that it was "*sometimes observed near Hartlepool*". Edward Backhouse (1834) noted that it was *"occasionally killed in the mouth of the Tees in the winter*", while Nelson (1907) recorded just one summer occurrence, in July 1877. Temperley (1951) believed that it was most likely to occur between October and March, but occasionally as late as May. British Birds remarked on the remarkable influx of grebes and divers into Britain at the end of January 1937, which included two Great Northern Divers near Sunderland on 2nd and 3 February (*British Birds* 30:12, 370-379). This was part of a remarkable national influx of various diving species. Although Temperley said that this was "*a rather rare winter visitor off the coast*", in reality, its true status at the time was not well known or documented.

Post-Temperley, the late 1950s and 1960s saw the start of a familiar pattern as single birds occasionally appeared at Teesmouth and Hartlepool between November and February. The winter of 1959 produced more sightings than normal, including four at Teesmouth and one at Hartlepool on 8 February. Notable spring sightings included two off Hartlepool on 30 May 1959, with further single birds there on 7 May 1968 and 10 May 1969, neatly falling into the current pattern of low-volume May passage. Much more unusual was one at the same location on 15 July 1960. Other notably productive years occurred in 1968 and 1969; 16 individuals were noted in the Hartlepool area during November and December 1968 peaking at five on 8 December, whilst eight were noted passing the Headland between 25 October and 17 November 1969.

Perhaps the most unusual historical record is of the bird that crashed into a locomotive works at Darlington on 21 February 1956 and was later successfully released, apparently unharmed, out to sea. Several other inland records came in the 1950s and 1960s, including Burnhope Reservoir on 15 March 1953, Crookfoot Reservoir on 5 November 1961, Smiddyshaw Reservoir on 17 December 1961 and Hurworth Burn Reservoir from 27 November to 15 December 1963.

Recent status

In more recent years, the majority of birds in Durham have occurred in the months between October and April, with few seen between May to September. Most of these have been in immature or winter plumage; this includes some of the birds that have very occasionally been observed in May and June (Blick 2009). There has however, been some change in status since the mid-2000s, as previous highest year totals and day counts of passage birds were exceeded by considerable margins.

Whilst the species can be noted anywhere along the Durham coastline, the only area that regularly attracts birds is around Hartlepool Bay, Hartlepool Docks and Seal Sands (Blick 2009). Seal Sands in particular can be productive in mid-winter and five birds were noted there in February 1983 (Blick 2009) and again on 13 February 2007 (Blick 2009), while extended stays have been noted here in December 2000-January 2001 and October-December 2004. For close views, Hartlepool Docks often provides opportunities and birds may remain there for prolonged periods. Examples in the late 2000s included birds from late February to 14 March 2005, 18 March to 9 April 2006, 21 November to 28 December 2008 (two) and 29 November 2009 to 12 February 2010.

Elsewhere on the coast, infrequent sightings have come from Whitburn north to Trow Rocks, and also along the county's central stretch of coast at sites such as Blackhall Rocks, Crimdon, Horden and Shippersea Bay. Observations are usually of single birds and the length of stay is normally only a day or two. The incidence of birds entering the river mouths of the Tyne and Wear is much lower than at the mouth of the Tees and the adjacent docklands of Hartlepool. The reasons for this are unclear as all provide adequate shelter from rough sea conditions, so perhaps local food availability is a factor.

The main period for coastal observations is November to February, the true passage birds witnessed in spring and autumn seemingly reluctant to land and linger. It is very unusual to see birds on the sea in the summer months and one off Seaham on 17 July 1999 was only the third-ever record for that month in the county, following birds in 1877 and 1960.

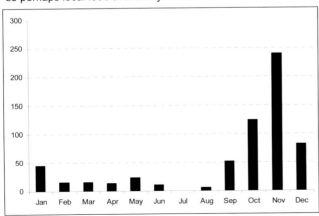

Passage of Great Northern Diver by Month at Hartlepool & Whitburn, 1980-2010

This remains by far the rarest diver species to appear inland in County Durham, with only four recent inland records to add to the historical occurrences referenced above. On 20 October 1975, one was on Cow Green Reservoir, Upper Teesdale, but was unfortunately found dead the following day. In December 1984, a bird was on Derwent Reservoir (Kerr 2001). A summer-plumaged bird was observed flying west over Barlow Fell early in the morning of 27 May 1997 - only the second record for the Gateshead area since the first in 1824. Finally, one was seen flying over Boldon Flats on 20 December 2005, only 3km from the coast. This situation contrasts with that in Yorkshire, where the appearance of this species "*on any large sheet of inland water from September to May need not occasion any surprise*" (Mather 1986). The absence of an obvious flyway between the east and west coasts of England is no doubt a major factor in County Durham.

The level of coastal passage off the north east coast is never great in comparison with the west coast of Britain and Ireland. Numbers vary greatly from year to year with the heaviest passage associated with the late autumn period. October and November often provide the peak movements, particularly when winds are from a north or north westerly direction. An extended run of strong westerly winds from fast-moving Atlantic depressions in autumn 2007 produced exceptional numbers in the county, suggesting that the birds were of North American origin, rather than from Iceland and Greenland. The last three decades have seen a rise in the frequency of passage birds in comparison to earlier years, but it is perhaps more likely that observations have increased for the same reason as other diver species, greater effort and improved techniques and technology. The totals logged at Whitburn in five-year periods are detailed below.

Spring passage tends to be light through April, peaking in early May and occasionally extending into June. It usually features singles or small numbers of birds. Up to the late nineties, there had been only three June records in the county in the previous 50 years, but the 2000s produced a further eight sightings. For example, two flew north past Whitburn on 4 June 2009, whilst three flew north in June 2010. Again, this increase is likely to be linked with changes in observer effort.

Although day counts are most often limited to just one or two birds, some impressive days of passage occurred in the late 1990s and in particular, during the second half of the 2000s. At Hartlepool, six flew north on 14 November 1999 while observers at Whitburn recorded seven moving north on 28 October and eight north on 31 October 2008. Day-counts of five birds have been made on several other dates at both Hartlepool and Whitburn through the 2000s and this level of passage is now not considered to be exceptional.

By far the best year on record was 2007, when 101 birds were logged at Whitburn alone, contributing to a county total of nearly 150 birds for the year. At Whitburn, the first autumn birds were recorded from 9 September with the majority of the passage occurring in November, during a period of sustained northerly winds. Peak days included four on 9th, 13 on 10th, nine on 11th, six on 12th and six on 14 November. Approximately 25 birds were also logged at Hartlepool during the autumn of 2007, including 9 on 10 November. By way of contrast, late autumn passage in 1992 resulted in a total of seven birds past Whitburn; at that time, the largest passage total in the county since the late 1960s (Armstrong 1993).

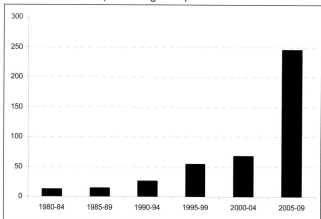

Indicative of the level of passage divers on the Durham coast, concerted observer effort at Whitburn between 2006 and 2010 amassed a total of 267 Great Northern Divers; 3.5% of all divers noted flying past. In comparison with the numbers noted during the 1950s and 1960s, these figures might seem unimaginably high.

One of the most unusual records of this species locally, was of a leucistic individual seen flying north off Whitburn on 19 December 2009. The plumage was almost wholly white except for some darker feathering on the under-wing.

Combined passage records of Great Northern Diver at Hartlepool & Whitburn, 1980-2010

#

213

Distribution & movements

This species breeds across Greenland, Iceland, and westwards over much of North America, as far south as the northernmost states of the USA. Birds wintering in Europe, and consequently Britain, appear to be derived not just from the Iceland population but also from birds breeding in Greenland and around Baffin Island (Parkin & Knox 2010).

White-billed Diver (Yellow-billed Loon)
Gavia adamsii

A rare visitor from Arctic breeding grounds, mainly in autumn: nine records.

The first record for Durham came in 1981 when an adult in winter plumage was found by Tom Francis and Rob Little in Hartlepool Docks on 14 February (*British Birds* 75: 483). It remained until 22 February and became quite a celebrity, attracting over 400 observers from around the country. The close views obtained ensured it left a lasting impression on many visiting birdwatchers.

Nineteen years passed before another was recorded in the county, but further sightings through the 2000s saw its status become 'more expected'.

All records:

1981	Hartlepool Docks, adult, 14th to 22 February
2002	Hartlepool Headland, adult, 14 September (*British Birds* 96: 545)
2002	Hartlepool Headland, adult, 18 October (*British Birds* 96: 545)
2003	Lizard Point, adult, 18 August (*British Birds* 98: 639)
2007	Whitburn, two adults, 13 October (*British Birds* 101: 526)
2007	Hartlepool Headland, adult 13 October (same as Whitburn) (*British Birds* 101: 526)
2008	Whitburn, adult, 31 October (*British Birds* 102: 536)
2008	Whitburn, adult, 1 November (*British Birds* 102: 536)
2011	Whitburn, adult, 12 March

A combination of increased seawatching effort, improved optical equipment and a greater appreciation of key flight identification features is partly responsible for the alteration in status of this species. Following the obliging 1981 bird, all records have been of 'fly-bys' on seawatches and several have been tracked passing other east coast watch points. The sight of two birds passing Whitburn within a few minutes of each other in 2007 was most unusual, though not unprecedented in England.

White-billed Diver, Hartlepool, February 1981
(Tom Francis)

Distribution & movements

The White-billed Diver breeds primarily along the coasts of the Arctic Ocean, in Russia, Canada and the United States, and winters in coastal waters of the northern Pacific and off northern Norway. Small numbers are thought to also winter in the Atlantic, to the west of Ireland, as demonstrated by a small but regular spring passage off the Outer Hebrides (Bell 2006). Northerly winds originating from the Arctic, particularly in late autumn, have been shown to move small numbers down the English and Scottish east coast, with Northumberland, North and East Yorkshire witnessing regular sightings over the past 25 years.

Fulmar (Northern Fulmar)
Fulmarus glacialis

A common, though restricted, breeding species, also occurring as a abundant passage visitor along the coast; a scarce visitor to inland areas.

Historical review

This species' well-documented spread from northern Europe commenced over two centuries ago (Harrison 1988). The first record of the species for Durham comes from Edward Backhouse (1834), who said that one was killed near Sunderland in 1814. Hogg (1827) subsequently documented a bird shot in a severe snow-storm in February 1823, at Seaton Snook. In 1845, Hogg said that this was still the only incident he knew of one being killed in the county, clearly not acknowledging, or being aware of, Backhouse's earlier record. William Backhouse told of one that was found dead on a sand bank at Teesmouth at an unknown date, presumably in the first half of the 19th century, while another was picked up on the beach at Castle Eden Dene on 16 November 1872. Hancock (1874) had one in his collection taken on 11 October 1850 from "*sands near Whitburn*", and another was found dead there in March 1850. One was found exhausted on the beach at Sunderland in September 1868. Clearly, over the first three-quarters of the 19th century, the Fulmar was a relatively scarce visitor to the county's coastline. In 1885, a bird was found dead at Hartlepool on 12 October and another was found in a field near Hesledon Dene. Nearly all records up to this date were apparently of storm-driven birds and most were found in the autumn period.

Around 1919, the pattern of occurrences changed, with occasional birds beginning to 'haunt' the cliff tops of the north east coast line of the county. That spring, birds were noted on-shore on the Farne Islands, to the north, and Bempton Cliffs, to the south of Durham (Chislet 1952). Around 1924 or 1925, birds began to frequent the cliffs at Marsden Bay and Frenchman's Bay at South Shields. Twelve birds were present in the spring of 1926 but, unfortunately, most of these were shot. In 1927, 22 were counted and a nesting attempt was noted on the mainland cliffs of Marsden Bay.

"*On July 17th, 1927, I saw a Fulmar (Fulmarus g. glacialis) brooding a nestling, still in down, on the cliffs at Marsden, near South Shields. For some time Fulmars have been frequenting Marsden Bay and occasionally Frenchman's Bay, which is a little further up the coast. Though a resident informs me that they were here three seasons ago, this is the first proof available of their actually breeding here. Last year (1926) the largest number of Fulmars I saw was about twelve in June, but as nine were reported to have been shot in mid-July and the last seen was on July 22nd, it is doubtful whether any young could have been reared. This year (1927) on February 9th a hen with a well-developed ovary was brought to me and was said to be one of four at Marsden. The largest number seen this year was on May 1st, when eight were seen at Frenchman's and fourteen at Marsden. One pair was seen mating (May 1st), though the majority seemed to be non-breeding birds. On July 17th, when the young one was seen, only four or five others still remained.*" (Noble Rollin 1928).

In 1928, 24 adults were present, eight pairs laid eggs and seven well-grown young were seen on Marsden Rock, with others also on the mainland cliffs. There then followed a period of steady growth and colony expansion. A survey in June 1945 revealed 136 adults on ledges, with 48 young being found on the same ledges in mid-August 1945. Prior to the Fulmars becoming established on the cliffs at Marsden, this area was not known to have had any nesting seabirds in recorded times. Also in the 1940s, birds were seen investigating Claxheugh near Sunderland, a rocky outcrop two miles up the River Wear, but eggs were not known to have been laid there. By the end of 1948, the species was breeding at Marsden, Ryhope, Nose's Point and Horden. In 1951, Temperley described the Fulmar as "*a resident, breeding on the coast in increasing numbers*", while surveying in 1953 showed that 72 young were produced in the Marsden area, though none by birds nesting in the newly colonised area between Ryhope and Easington. This breeding distribution was maintained through the 1960s with numbers remaining relatively stable.

Recent breeding status

At present, the breeding habitat of Fulmar in County Durham is strictly limited, being confined to the magnesian limestone cliffs of the county's coastline, and most occupied sites are in the north east of the county (Lloyd *et al.* 1991). These cliff locations are the only habitat in which the species is known to breed locally, unlike in neighbouring Northumberland, where small numbers of birds have colonised some man-made sites, such as Bamburgh Castle (Kerr 2001). The county's main colony remains at Marsden Bay and is concentrated around the upper sections of the mainland cliffs there, with the perpendicular faces being more favoured by Kittiwake *Rissa tridactyla* (J. Strowger pers. obs.). In most breeding seasons, around 75% of birds use the mainland cliffs as opposed to the offshore stack (Westerberg & Bowey 2000). Observations over many years have shown that the majority of birds tend to use the sections of Marsden Bay that lie to the north of the Grotto, with smaller numbers of birds around the Grotto itself, as well as south from there (Westerberg & Bowey 2000). From the early 1970s at least, small numbers also nested on the east-facing cliff in the adjacent working quarry.

In 1990, at the Marsden colony, 188 pairs were present, with 120 pairs reported as breeding in 1992 (Westerberg & Bowey 2000). Using the 1990 count as being the most reliable available for the main colony at that time, the *Atlas* suggested a maximum population estimate for the county of 200-210 pairs (Westerberg & Bowey 2000). More recently, detailed monitoring work undertaken in 2005 and 2006 as part of the national Seabird Monitoring Programme (SMP), revealed that there were 230 Apparently Occupied Sites in the main Fulmar colony; a decline of some 7% from the previous breeding season's figure. The local results determined that 2006 was a poor one for Fulmar productivity at Marsden, with a small improvement (0.32 fledged young per occupied site), albeit from a very low base in 2005. In terms of productivity, in both of these years, Marsden was significantly below the national average (SMP) (Newsome 2006-2007). The reasons for this low productivity are not known with any degree of certainty, but disturbance to breeding birds from coastal visitors may be a factor. In 2007, there were estimated to be between 230 and 250 breeding pairs present in the county, most of these being in the Marsden area.

The only way of arriving at an accurate assessment of the breeding population is to count the young, because adults come and go and possibly half of those present may be non-breeding birds. Indeed, a feature of Fulmar behaviour in the months prior to egg-laying is the complete desertion of the site, sometimes for a period of a few days. Students from Durham University made a study of the presence and movements of the Fulmars at Marsden in the late 1960s and early 1970s, discovering that the Fulmar has a long period at the breeding colony prior to egg-laying (Coulson & Horobin 1972). The re-occupation of the cliff starts in early November with an occasional visit by one or two birds. The main period of activity at the cliff is during the morning and, as the numbers build up, the diurnal period of occupation increases. By mid-December the first birds to arrive in the colony do so before dawn and the last to leave remain well after dark until near midnight. Almost throughout the pre-egg stage, the colony is deserted each night and re-occupied the next day; birds only stay regularly overnight just before egg-laying. A similar pattern of occupation occurs after breeding but in the reverse order. The numbers of birds at the colony in January and February exceed the breeding population and include many non-breeders. The non-breeders progressively decline in numbers until May when only the breeding birds remain with a few non-breeding birds. The daily variation in the numbers of birds at the cliff is influenced by the wind speed. In general, the birds leave the colony under freshening conditions and the number present at the colony can be interpreted in terms of the wind conditions over the last three days. It is suggested that the synchronised departures are primarily feeding trips, the birds using the strong winds to reach feeding areas, except that the departure just before egg-laying is linked to egg development and synchronised laying in the colony (Coulson & Horobin 1972).

During spring and early summer, prospecting birds occasionally make landfall at other cliff top sites away from Marsden. In this respect, a small number of birds re-colonised Dawdon Blast Beach, to the south of Seaham, from 1987 and Shot Rock, near Easington Colliery, from the early 1990s. Between 10 and 20 pairs now attempt to nest at Dawdon annually, but a severe limiting factor here is deliberate shooting of birds. In 1992, twenty pairs attempted to nest, many were shot, and none were successful (Westerberg & Bowey 2000). Persecution persists and human disturbance is still considered a limiting factor to the colony's growth. Perhaps this was the reason for its demise after the initial colonisation in the late 1940s. Prospecting birds have also been noted on the stretch of coastline between Blackhall and Crimdon since the early 1990s, though nesting attempts here appear more sporadic and prone to failure. The speed of growth of satellite colonies appears to have distinctly slowed since the days of rapid expansion, in the mid-20th century, but occasional pioneering birds will settle on suitable cliffs and ensure further extension of range within the county.

During October and November, sightings around the colonies are usually few. Birds are routinely noted returning to the Marsden Bay colony from mid-December and 100-150 have been frequently encountered on calm days between then and late February. Larger pre-breeding gatherings have included 475 in Frenchman's Bay on 30 January 1999 and 400 in Marsden Bay on 17 February 2007, while an intensive survey of the full area on 11 February 1993 produced an impressive 1,042 birds on ledges and a further 180 on the sea.

There is no record of the Fulmar breeding inland in County Durham, as it does in Northumberland, though this may happen in the future especially considering that since 1967, birds have nested sporadically on Roseberry Topping and the escarpment of the Cleveland Hills, to the south of the River Tees (Stead 1969). Inland prospectors have been noted since the late 1940s, when one or two would often be seen gliding across the cliff face and sitting on the ledges of Downhill Quarry, West Boldon. Occasional reports continued at that site up to the early 1970s. The majority of inland sightings have come between April and early August with activity most prominent in June. Unusually however, one was over Washington on 12 January 1974. Most have come within 15km of the coast, but on 22 May 1967, two circled the 'artificial cliffs' of Durham Cathedral before flying away to the north. The major river valleys in the north of the county form a more expected corridor for birds to follow and occasional sightings have featured as far west as Clara Vale, Ryton Willows, Shibdon Pond and Whickham along the Tyne, and WWT Washington on the Wear. More random sightings over inland conurbations have included singles over Billingham on 12 June 1966, Shotton Colliery on 23 and 30 June 1969, Darlington on 27 May and 31 July 1974 and over Darlington on 15 April 1978. Other reports far inland include one flying north over Burnhope Reservoir in Weardale on 3 June 1975, over Heighington, 48km inland, on 25 June 1977 and over Derwent Reservoir on 1 August 1978 and 18 May 1986.

Non-breeding status

Although large movements of Fulmar can occur at any time through the year, the 'classic' autumn passage period spans a period between mid-August and late September. By far the most impressive of these was in August 1962 when huge concentrations of sprats *Spratus spratus* at Teesmouth attracted huge numbers of Fulmars between 21st and 27th, the likes of which have not been witnessed since. Over 30,000 Fulmars were noted at Hartlepool during this period, including 7,500-8,000 on 26th and approximately 21,000 moving north between 08.50hrs and 16.50hrs on 27th. Counts from Northumberland at this time were unexceptional and the food availability in the Tees Bay appeared to be wholly responsible for such incredible numbers. Increased seawatching in the 1970s produced further evidence of large-scale passage, such as in 1973, when northerly movements were more or less sustained between July and September. The peak occurred on 10 September when a "*spectacular continuous stream*" flew north at Whitburn (Unwin 1974), associated with a movement of Sooty Shearwaters *Puffinus griseus*. At least 400 Fulmars were noted between 08.00hrs and 09.00hrs with a similar passage rate evident later that day. A further example of the scale of northerly movement came at Whitburn in September 1986. Sample counts revealed on: 10th, 1,500 per hour; 11th, 1,100 per hour; 12th, 1,800 per hour; and, on 13th, 600 per hour. Linked to this substantial presence, 1,600 birds passed Hartlepool in an hour and a half on 14th. In the same period of September 1993, after 3,600 moved north past Whitburn in three hours on 6th, a further 2,000 passed there in five hours on 13th.

These impressive passage rates were not reached in the late 1990s and 2000s, with the decline in North Sea fish populations and the associated fishing industry perhaps being contributory factors. Occasional examples of large autumn movements at Whitburn included counts of 2,000 flying north on 28 August 1998, 1,300 north on 16 September 2005 and 1,150 north on 13 August 2006.

Outside of the autumn passage period, large movements have been recorded in most months, but these are generally weather-dependent. Spring can be a productive period with several day-counts in the region of 500-1,000 birds being noted at both Hartlepool and Whitburn between early March and late May, while 1,800 flying north at Whitburn on 28 May 2007 was an impressive spring tally. The only month that consistently sees a very minimal presence off the Durham coastline is October. The rapid and far-distant dispersal of juveniles out to sea and the post-breeding primary moult of adults (BWP), means that the species is scarce during October. A rapid return to coastal waters, presumably by adults, occurs during November and heavy passage can occur, such as 541 moving north off Whitburn in 75 minutes on 25 November 1977.

The first report of a 'blue' Fulmar, the colour morph more predominant in northerly breeding populations, in Durham came on 18 March 1933, when one was seen with other Fulmars at Nose's Point, Dawdon. Further sightings came from Frenchman's Bay on 17th and 21 August 1951, and twice more in the 1960s. It wasn't until the start of the 1970s that birds were more frequently noted. More regular seawatching through the 1980s and 1990s confirmed the status as a scarce but regular visitor to Durham, the autumn bias perhaps reflecting observer effort rather than the birds' increased presence. By far the best years during this period were 1986, with 24 noted passing

217

Whitburn in four days over 10th to 13 September, and 1998, with 10 birds logged between Hartlepool and Whitburn during August. A further increase in seawatching effort from the early 2000s, particularly in winter and spring coverage at Whitburn, resulted in a swing away from the autumn bias and has produced a better appreciation of winter and early spring occurrences. Year totals of 16 in 2004, 27 in 2008 and 13 in 2010 (the majority of these being at Whitburn) indicate that occurrences are more regular than previously thought. Correlation of weather patterns and the occurrence of blue Fulmars has led to the understanding that brisk northerly winds originating from further north than Iceland, particularly during January and February and again between late May and late September, are likely to produce sightings. Most birds seen are at the pale end of the spectrum, but on 28 September 2007, two blue Fulmars out of seven logged at Whitburn approached Fisher's 'double-dark' colour description (Fisher 1984).

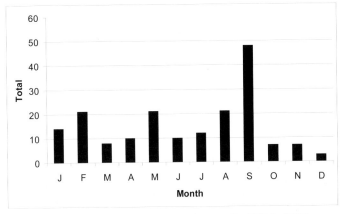

Blue Fulmars in County Durham, by month, 1980 to 2010

Very occasionally, blue Fulmars have been seen prospecting the cliffs at Marsden, most recently on 29 July 1996 and between 25th and 27 April 2010. There has never been any indication of a breeding attempt involving this colour morph.

Leucism is not uncommon in Fulmars, but wholly white birds were seen at Hartlepool on 3 September 1965 and 6 September 1982; and at Whitburn on 18 August 1986.

Distribution & movements

The Fulmar breeds around the coasts of the North Atlantic and Pacific Oceans, from the temperate through to the Arctic zone. Winter dispersal is limited to the North Sea and North Atlantic. The species has undergone a remarkable expansion in range and population size since the early part of the 19th century. This culminated in the first breeding in the British Isles - excepting the long-established colony on St. Kilda - on Foula, Shetland, during the late 1870s (Fisher 1984).

There is little information pertinent to Durham from recoveries of ringed birds. One found dead at Sunderland on 25 June 1970 had been ringed as a chick on Foula, Shetland, in 1962 and might therefore have been in our area for its first or second year of breeding. A bird ringed on Fair Isle, Shetland, as a nestling on 16 August 1971 was caught in a fishing line on 7 June 1977 at South Shields and then released. In August 1975, a colour-ringed bird seen from a boat off Sunderland had been marked in July 1960 at Eynhallow, Orkney, and in addition to this, a bird ringed in Scotland as nestling at Eday, Orkney on 26 July 1977, was found dead at Seaton Carew on 15 April 1979 (Blick 2009).

Fea's/Zino's Petrel
Pterodroma feae/madeira

An extremely rare autumn vagrant from Madeira and Cape Verde: three records.

None of the recorded birds has been identified to species level, all being accepted as 'non-specific *Pterodroma* petrels', although Fea's Petrel *Pterodroma feae* is the most likely to be involved. The first Durham record of this enigmatic seabird came on 1 September 2002, when Ian Mills watched one flying north off Whitburn CP (*British Birds* 98: 640). It is highly probable that this was the same bird as had passed Flamborough Head, East Yorkshire, almost exactly four hours earlier.

218

1. The Knotts, Teesdale (*Dave Raw*). The rugged Durham uplands play host to breeding Buzzard, Merlin and Ring Ouzel and occasionally Hen Harrier, along with a range of other moorland species.

2. Langdon Beck, Upper Teesdale (*Mark Newsome*). Rough pasture and damp grassland is the home to a healthy population of Black Grouse, as well as nationally important numbers of breeding waders such as Golden Plover, Curlew and Snipe.

3. Balderhead Reservoir (*Mark Newsome*). The upland reservoirs play host to wintering wildfowl and breeding Wigeon, while surrounding moorland attracts Hen Harrier, Short-eared Owl and the occasional Rough-legged Buzzard.

4. Pennington Plantation, Hamsterley Forest (*Mark Newsome*). Large commercial forests provide cover for Crossbill and Goshawk, and are the only regular breeding habitat for a small population of Nightjars.

5. Muggleswick Woods, Derwentside (*Mark Newsome*). The upland oak woodlands are a last refuge for Wood Warblers, while spring echoes to the song of Tree Pipits, Redstarts and Pied Flycatchers.

6. Farnless Farm, Bishop Middleham (*Mark Newsome*). Well managed farmland in central and eastern Durham provides a home for declining species such as Corn Bunting, Grey Partridge and Yellow Wagtail, whilst the purposely grown bird food crops attract numerous finches and buntings in winter.

7. Low Barns DWT, Witton-le-Wear (*John Bridges*). Natural wetland habitat is home to breeding waterfowl, and surrounding woodland holds Marsh and Willow Tits and a few pairs of Lesser Spotted Woodpecker.

8. Rainton Meadows DWT, Houghton-le-Spring (*Ray Scott*). Newly created wetlands have transformed reclaimed mining sites to bird-rich habitats, home to breeding Great Crested Grebes, Reed Warblers and a variety of wildfowl.

9. Lower Derwent Valley, Gateshead (*Mark Newsome*). As well as being the reintroduction site for Red Kites in north east England, the mature woodland and river holds healthy populations of Goosander, Dipper, Green Woodpecker and an increasing number of Buzzard.

10. Sandhaven Beach, South Shields (*Mark Newsome*). The mouth of the River Tyne supports wintering waders, gulls and divers, whilst the adjacent sandy beaches are a favoured late-summer roosting site for terns, including the rare Roseate.

11. Marsden Bay (*Mark Newsome*). The county's only seabird colony thrives here despite it being a popular tourist area, and holds thousands of breeding Kittiwakes plus small numbers of Fulmar, Cormorant and Razorbill.

12. Whitburn Coastal Park (*Mark Newsome*). Coastal scrub provides cover for grounded migrants in spring and autumn, and along with the many warblers, chats and flycatchers, occasional top class rarities have been discovered including Red-flanked Bluetail and Isabelline Shrike.

13. Castle Eden Dene mouth (*John Bridges*). Scrubby clifftops are a favoured habitat of Skylark, Whitethroat and a small population of Stonechat, whilst the coastal waters support wintering populations of seaduck, grebes and waders.

14. North Tees Marshes (*John Bridges*). Despite being surrounded by industry, the wetland habitat of sites such as RSPB Saltholme holds nationally important numbers of wintering wildfowl and waders, as well as having an outstanding history of rarer species.

15. Whooper Swans, Herrington CP (*Mark Newsome*). Large numbers pass through during spring and autumn migration.

16. Barnacle Geese, Whitburn CP (*Mark Newsome*). The Spitzbergen population occasionally passes in spectacular numbers.

17. Wigeon, Boldon Flats (*Mark Newsome*). Upland wetlands hold a significant proportion of the English breeding population.

18. Garganey, WWT Washington (*Mark Newsome*). A regular passage migrant, with a pair or two occasionally breeding.

19. Ring-necked Duck, Cowpen Bewley, Feb. 2010 (*Ian Forrest*). **20. Surf Scoter, Blackhall Rocks, June 2011** (*Mark Newsome*).
21. Eider, Hartlepool (*Ian Forrest*). Large numbers spend the winter on the coast, particularly in the south of the county.

22. Goosander, Chester-le-Street (*Dave Johnson*). A scarce breeding species on the main river systems and their tributaries.

23. Black Grouse, Upper Teesdale (*Mark Newsome*). The Durham uplands hold a healthy population of this iconic species.

24. Grey Partridge, Seaton Common (*Ian Forrest*). Still a common breeding species in some arable areas.

25. Great Northern and Black-throated Divers, Hartlepool Harbour (*Ian Forrest*). One of the most favoured sites in the county.

26. Fulmar, Marsden Bay (*Mark Newsome*). Small numbers breed along the coast, while much greater numbers pass offshore.

27. Sooty Shearwater, off South Shields (*Steve Egglestone*). A regular autumn migrant, sometimes in large numbers.

28. Storm Petrel, Whitburn CP (*Tom Tams*). Over 1000 have been tape lured and ringed in the county.

29. Double-crested Cormorant, Billingham, Jan. 1989 (*Alan Tate*) **30. Cattle Egret, Saltholme Pools, November 2008** (*G. Iceton*)

31. Grey Heron, WWT Washington (*Mark Newsome*). The largest breeding colony in the county.

32. Purple Heron, Saltholme Pools, October 2011 (*Ray Scott*). A rare spring and autumn visitor from Europe.

33. Black Stork, Gateshead, August 2009 (*John Malloy*). Of a county total of ten birds, this was the only one to linger.

34. Glossy Ibis, RSPB Saltholme, October 2011 (*Ray Scott*). The fifth for the county, this bird arrived as part of a national influx.

35. Black-necked Grebes, undisclosed locality (*Mark Newsome*). A rare and irregular breeding species.
36 & 37. Red Kites, Derwent Valley (*K.Bowey & R.Scott*). A successful reintroduction scheme returned the species to the county.

38. Hen Harrier (*Dave Johnson*). An iconic moorland raptor sadly lost as a breeding species.
39. Northern Harrier, Lunedale, February 2009 (*M.Newsome*) **40. Pallid Harrier, Dorman's Pool, October 2011** (*M.Newsome*)

41. Goshawk, Hamsterley Forest (*Greg Jack*). A small population breeds in secluded forested areas.
42. Buzzard, Teesdale (*Ian Forrest*). Once a highly localised breeder, this species has now recolonised much of the county.

43. Merlin, Seaton Common (*Ian Forrest*). A scarce winter visitor and upland breeding species.
44. Peregrine, undisclosed locality (*Ian Forrest*). A small number now breed in lowland and coastal areas.

45. Baillon's Crake, Sunderland, May 1989 (*Pete Wheeler*). This confiding individual became a celebrity during its four day stay.
46. Sandhill Crane, Marsden, September 2011 (*Ian Mills*)　　**47. Kentish Plover, Seaton Snook, May 1994** (*Jim Pattinson*)

48. Avocet, North Tees Marshes (*Ray Scott*). Following colonisation in 2006, the species now breeds in two areas.
49. Pacific Golden Plover, Whitburn, September 2002 (*Jim Pattinson*). The second county record of this rare Asiatic visitor.

50. Golden Plover, Lunedale (*Mark Newsome*). A widespread moorland breeder as well as a common winter visitor.
51. Lapwing, Upper Teesdale (*Mark Newsome*). The western uplands hold an important breeding population.

52. Great Knot, Teesmouth, October 1996 (*Jim Pattinson*). The second British record was often distant during its stay.
53. Long-toed Stint, Saltholme Pools, August 1982 (*Jeff Delve*). This confiding individual was the second British record.

54. Purple Sandpiper, Hartlepool (*John Bridges*). The rocky coastline holds a small wintering population.

55. Broad-billed Sandpiper, Teesmouth, June 2007 (*M.Sidwell*). **56. Sharp-tailed Sandpiper, Teesmouth, Sept. 2011** (*M.Sidwell*)

Surprisingly, a further sighting came just three weeks later. A bird which had initially been seen passing Flamborough Head on the morning of 23 September was seen off Whitburn by Peter Collins, Brian Unwin and other observers some five hours later (*British Birds* 97: 568). Just over four hours later, it was observed passing the Farne Islands, Northumberland. Another record came in the outstanding seabird autumn of 2007, when one was identified by Ross Ahmed as it flew north off Whitburn on 11 September (*British Birds* 101: 527). The bird was presumably associated with the large-scale movement of Great Shearwaters *Puffinus gravis* into British waters around this time.

Distribution & movements

Fea's Petrel breeds on Bugio, off Madeira, and on the Cape Verde Islands, whereas Zino's Petrel nests solely on the highest peaks of Madeira. Population levels are low, particularly of Zino's Petrel with an estimated 80 breeding pairs (Shirihai 2010). Outside of the breeding season, they disperse widely throughout the North Atlantic. The rise in the number of reports in Britain since the early 1990s is marked, although North Sea records are still rare and unpredictable in occurrence.

To the end of 2010, there have only been three accepted records of Fea's Petrel in Britain, all of which were well-photographed individuals seen from boats off the south west. However, there are almost 90 accepted records of Fea's/Zino's Petrel for Great Britain and Ireland. 'Soft-plumaged' Petrel was only recognised as a separate species by the BOU as recently as 2000, and although the chances of the highly endangered Zino's Petrel reaching British waters from its sole breeding grounds in Madeira are slight, it cannot be ruled out. There is one accepted Western Palaearctic record (from Israel) of the Southern Hemisphere piece of the jigsaw, still named Soft-plumaged Petrel. A well-photographed individual apparently of this species in northern Norway in June 2009 (still awaiting formal acceptance) clearly demonstrates that all is not clear-cut as regards sightings in British waters.

Cory's Shearwater
Calonectris diomedea

A scarce autumn passage visitor from breeding grounds on the North Atlantic islands.

Historical review

The first record for County Durham came in 1960 when one was seen by Jack Bailey 'near South Shields' on 8 August (*British Birds* 55: 565). At the time, Cory's Shearwater was still a national rarity with just a handful of sightings per year, and North Sea records were quite unusual. It was a full 12 years before the next sighting, when Tom Francis saw one flying north at Hartlepool Headland on 14 September 1972 (*British Birds* 66: 334). This turned out to be the only record in the county during the 1970s.

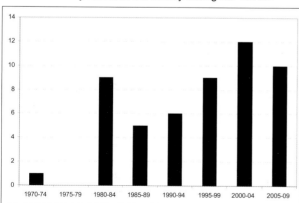

From the early 1980s, sightings became more frequent as the popularity of seawatching increased. A total of 14 birds through the 1980s was followed by 15 in the 1990s and at least 22 since 2000. This increase in numbers recorded is most likely due to an increase in observer effort over the period, although the rise in the number of sightings has been gradual, despite the large increase in the interest in seawatching. This suggests that the frequency of occurrence may have increased but not to the degree implied by a simple examination of the data without factoring in observer effort. The records since 1970 in five-year periods are illustrated.

Cory's Shearwaters in County Durham, 1970-2009

All records for County Durham fall between 29 June and 23 September. There is a fairly narrow window for this species' occurrence in the North Sea and spring records are decidedly rare, but it is surprising that none have been seen in Durham after late September. As illustrated, the distribution of sightings across late summer and autumn shows peaks in mid-July and late August, although with the limited number of individuals, it is difficult to form any firm conclusions. Nonetheless, it is fair to say that the mid-July to early September period has provided most sightings.

Nearly all sightings have come from the established seawatching locations of Hartlepool Headland and Whitburn. Apart from the first bird at South Shields in 1960, the only records away from Hartlepool and Whitburn have been at South Shields on 2 August 1984 and Seaham on 11 August 2006. Whitburn has produced 30 records, in comparison to 23 at Hartlepool Headland. There have only been two instances of two birds being seen n a day; north past Hartlepool (and later Whitburn) on 14 July 2001, and north past Hartlepool on 25 August 2002.

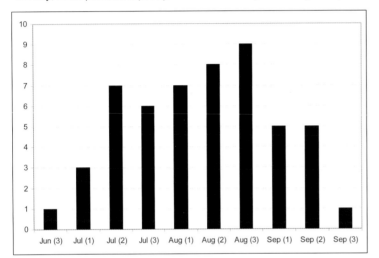

The best years for this species came in 1984 and 2001. In 1984, singles were noted at Whitburn on six dates between 16 July and 4 August and another on 10 September, with one seen at South Shields on 2 August. It is conceivable that such a series of sightings could be attributable to two or three birds feeding along the north east coast. In 2001, Hartlepool Headland produced five of the six birds seen that year. The run of four birds in the second week of July included two passing both Hartlepool and Whitburn on 14th.

Cory's Shearwaters in Durham, by ten-day period June to September, 1970-2009

Distribution & movements

Cory's Shearwaters of the race *borealis,* which is responsible for virtually all British records, nest on the North Atlantic island groups of the Canary Islands, Madeira, Cape Verde and the Azores. The winter range is predominantly off the coasts of North and South America. The nominate race *diomedea,* known as Scopoli's Shearwater, nests in the Mediterranean, with a small outpost of birds on the Biscay coast of France.

The post-breeding dispersal of Atlantic-nesting Cory's Shearwater brings regular, but varying, numbers to the inshore waters of south west England and Ireland; 1980 brought over 17,000 records, but the species is scarce in most years. North Sea sightings often follow on from large numbers in the south west, but year totals are normally in the tens of birds. To the end of 2010, there has only been one acceptable claim of Scopoli's Shearwater in British waters (off the Isles of Scilly) (Fisher & Flood 2004). It has however, been suspected on other occasions off the south west of England and in the North Sea and, as such, should be considered possible in Durham waters.

Great Shearwater
Puffinus gravis

A scarce autumn passage visitor from breeding grounds in the South Atlantic.

Historical review

This species has always been rare off the Durham coast and has probably been seen around 40 times since the first 19th century records. Temperley said that the Great Shearwater was a *"very rare casual visitor"* to Durham (Temperley 1951) and referred to two records from the 19th century. The first record for the county is of one collected at Teesmouth, probably at sometime between 1834 and 1846; this was in Edward Backhouse's collection. William Backhouse, in his "*Additions to Mr Hogg's Catalogue*" (*The Zoologist* Vol. IV 1846) mentioned this specimen from Teesmouth; it was in the collection of Edward Backhouse at the time. As the specimen is not mentioned by Edward Backhouse in his catalogue in 1834, it was probably obtained between these two dates. Interestingly, the species was not documented by Hancock (1874). Tristram (1905) recorded one as being "*captured off the Tees, January or February, 1874*" and he also stated that "*a few years ago, one was picked up dead about the same place and brought to Mr Cullingford, Durham Museum*". The reference to the January or February capture remains the only mid-winter record for the county.

The first record for the 20th century, and the only one in the first half of that century, was also the first definite record for the south east of the county (Blick 2009), and was referred to in Richmond's 1931 paper on the *Birds of Teesmouth*. On 14 July 1931, W. E. Almond observed two Great Shearwaters about four miles out from Teesmouth (Almond *et al*. 1939). Strangely, George Temperley's documented "*most recent record*" was of two birds together, three miles out from Teesmouth on 18 July 1931; presumably the same record as above. Only three further birds were seen over the following forty years, with all sightings coming from Hartlepool Headland. Single birds flew past that location on 29 August 1959, 10 October 1959 and 26 August 1969.

Recent status

A total of 54 fully documented birds were recorded in Durham between 1970 and 2010, with the species being recorded in 15 years over this period. All sightings have been between 25 July and 27 September with over 75% falling between 19 August and 11 September, noticeably peaking in the period 25th to 29 August.

Extreme annual fluctuations have resulted in some long periods without any sightings; there were no records between 1984 and 1994. The annual totals for all sites combined in five-year periods are illustrated.

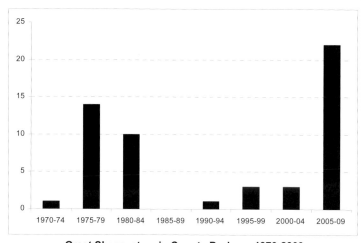

Great Shearwaters in County Durham, 1970-2009

The peak in the mid-1970s was due to a spate of sightings from fishing boats working up to 10 miles offshore from Sunderland. Fifteen late-summer birds were noted by J. Barker between 1974 and 1976, including a run of birds in August 1976 of two on 19th, four on 20th and three on 29th. Autumn 1976 was a particularly good one for the species in Britain and the North Sea, for example in Yorkshire (Mather 1986). Despite a number of pelagic trips off Durham and Northumberland since the late 1970s, there has never been a repeat pattern of such records. Other than the 'at sea' records in the 1970s, all other recent sightings have come from Hartlepool Headland and Whitburn. Through the 2000s, sightings at Whitburn took the overall total of birds recorded there to 32, while 12 have been noted off Hartlepool since 1970. A review of Great Shearwater records off the south east coast of the county resulted in three previously published sightings being removed from the county record (Joynt 2009).

The best year in County Durham came in 2007 when 17 were seen moving north between 28 August and 27 September, 16 of which were seen from Whitburn Observatory. This included four at Whitburn and one at Hartlepool on 10 September, six moving north past Whitburn on 11th and another three north there on 27th. Although these numbers were high for Durham, it is likely that many more, which were far out to sea, went unseen by observers. At the same time, observers at Flamborough Head in East Yorkshire logged a new North Sea record of 33 on 11 September and 13 passed Newbiggin, Northumberland, on the same day. The strong northerly winds in the second week of September 2007 only moved a very small proportion of these birds down to the north east coast of England.

Distribution & movements

Great Shearwaters nest in the South Atlantic, in the vast seabird colonies of Gough Island, Inaccessible Island, Nightingale Island and Tristan da Cunha. In the non-breeding season (the northern hemisphere's summer), they are long-distance oceanic wanderers undergoing a huge migration loop that takes them off the eastern coast of North America, through the south west approaches of the British Isles, then back down through the eastern Atlantic. Numbers seen in British and Irish waters vary considerably from year to year, no doubt influenced by food availability and prevailing weather systems at the critical time in late summer. The peak movements in British terms usually occur between late July and mid-September, and are most evident from the coasts of Cornwall and south west Ireland, with small numbers occasionally penetrating into the North Sea.

Sooty Shearwater
Puffinus griseus

A common passage visitor from the southern oceans mainly recorded in late summer and autumn.

Historical Review

The first record of this species for Britain was of one shot at the mouth of the Tees in mid-August 1828, by G Marwood (Mather 1986), being documented by Hancock (1874) as *Dusky Shearwater*. Its status in previous times is somewhat difficult to ascertain, due to a lack of information and the technical challenges around its identification some distance off the coast.

In the earlier part of the 20th century, it was said to be *"not uncommon in these parts in July, 2-3 miles out at sea"* (Richmond 1931). More specifically, in his article on the *Birds of Teesmouth*, Richmond documented his observations on 1 September 1927, of *"what I took to be a definite southward movement of Sooty Shearwaters, small parties of these birds passing steadily very close in-shore"* (Richmond 1931). In addition, in 1929, one or two small flocks were noted flying north in the bay off Seaton Carew on 16 November. In 1951, Temperley had summarised this species' status as a *"very scarce autumn visitor"* to the Durham coast (Temperley 1951).

Before the advent of more regular and concerted seawatching, the 1950s and 1960s produced little more than occasional sightings of 'Sooties'. A large movement was recorded along the east coast of England in 1959 (Ferguson-Lees 1959) and following sightings in East Yorkshire on 27 August, 22 were watched moving north, and 10 south, past Hartlepool on 28th, with another two there on 31st. This was the largest movement recorded locally until the early 1970s.

Recent Status

Increased interest in the documentation of seabird movements from the late 1970s resulted in a much better understanding of the appearance of Sooty Shearwaters in the North Sea. Off the Durham coast, the first passage birds are usually noted in mid-July (although late June and early July sightings are not unusual), and movements usually peak during August, continuing through September and early October. Numbers are usually at their highest from late August to mid-September when hundreds of birds can be noted in a day, especially during concerted periods of passage. Weather conditions are hugely influential on the numbers of birds seen from land-based locations, and numbers entering the North Sea vary greatly from year to year.

Hartlepool Headland and Whitburn are the prime sites for recording this species, Hartlepool attracting more birds in strong north westerly winds; birds passing Whitburn appear to be pushed further offshore in these conditions. Away from the main seawatching sites, few birds are reported but this is normally a reflection of observer effort. Almost any coastal station, such as Blackhall Rocks, Chourdon Point, Dawdon Blast Beach and Seaham, can document birds if observers are based there during a movement of the species. The geography of Durham's coastline normally means that birds pass these other watch points much further out to sea.

The highest recorded level of sea passage was of 1,954 birds flying north past Hartlepool Headland on 22 September 2002, a day which saw tremendous numbers in the North Sea, with counts of over 1,000 birds at several other sites in eastern England. At Whitburn, a total of 1,271 were counted on this date. There has been only one other comparable period of passage when, in 2005, a high pressure weather system on 16 September 2005 produced counts of 1,092 north at Hartlepool Headland and 926 at Whitburn. Other large day-counts at Whitburn have included 471 on 11 September 1976 (with 206 at Hartlepool), 253 on 27 September 1993, 565 on 17 September 2001 (with 509 at Hartlepool) and 347 on 4 September 2007.

Outside the main autumn period, sightings are rare but not unprecedented. Between 1976 and 2010, a total of seven birds were noted between December and May. Single birds were recorded at Sunderland on 9 January 1976, Hartlepool Headland on 28 December 1978, and at Whitburn on 2 March 1990, 20 January 1998, 5 December 1999, 12 December 2002 and 31 May 2006.

Distribution & movements

This species breeds in the southern oceans, on islands around New Zealand, Tasmania and the southern coasts of South America. It wanders into northern ocean areas, both Atlantic and Pacific, during its non-breeding season, performing a 'figure of eight' migration route back to the southern oceans (Cramp *et al.* 1977).

Manx Shearwater
Puffinus puffinus

A very common passage migrant and non-breeding summer visitor in fluctuating numbers.

Historical review

There are records of the remains of this species being excavated during the archaeological investigations of the Mediaeval settlements on Hartlepool Headland and at Newcastle (Yalden & Albarella 2009), hinting at a possible breeding presence of this species along some parts of the Durham and north east coast at that time.

Sharp (1816), in his catalogue for Hartlepool, said it was "*rare*". William Backhouse referred to a specimen from Teesmouth in Edward Backhouse's collection, which was taken some time "*before 1846*". Edward Backhouse in 1834, also described it as rare, and he included a reference to one being killed at Marsden Rock, rather intriguingly in the light of later information, "*a short time ago*". Tristram (1905) said that it occurred at Castle Eden Dene, Hartlepool and Seaton Carew, but he gave no dates or numbers. Nelson (1907) said that it was unusually plentiful at Teesmouth in 1876, when a flock of ten was seen there on 7 July (Mather 1986), and also in 1885, 1887 and 1904. Likewise, Temperley recorded that more were 'on the move' in 1948 than is usually the case, and in summarising the status, noted that "*the main northward migration takes place at the end of April and the beginning of May and birds are seen passing south again in July early August, though single birds have been seen much later in the year*".

Without doubt, the most intriguing record of this species in the county refers to the apparent attempted breeding at Marsden in the summer of 1939. In this instance, an egg was collected from under over-hanging vegetation on the top of the Marsden Rock. This egg was identified by Dr. H.M.S. Blair, some five years later. The full account was published in *British Birds* under the title, "*Manx Shearwater Breeding on Durham Coast*", and is repeated below.

"*In the late summer of 1944, Mr. Wilfred Robson, of South Shields informed me that he still had in his possession an egg taken on Marsden Rock, which he believed to be a Puffin's. As there is no*
record of the nesting of the Puffin on the Durham coast, I took the first opportunity of examining this egg, which was roughly end-blown, and was surprised to find it to be that of a Manx Shearwater (Puffinus p. puffinus). The details of this interesting discovery are as follows :—In June, 1939, Mr. Robson—then a school-boy— succeeded, with several companions, in scaling Marsden Rock, always a rather hazardous undertaking. While on the top of the rock, he came upon the egg in question well in under an overhanging rocky outcrop covered with grass. No bird was seen near the nest. Marsden Bay having been closed to the public from 1939 to June, 1944, it is impossible to say whether any Shearwaters returned to the locality in the interval. None were definitely identified in the summer of the latter year ; but as the beach was closed each evening, the Rock could not be kept under observation during the hours when these birds, if present, might be expected to become most active and noisy. While the North Durham coast remained closed, there was a marked increase in the numbers of sea-birds breeding around Marsden. In

1944 - besides numerous Fulmars and Kittiwakes - at least a dozen pairs of Herring-Gulls nested either on the Rock or about the adjacent cliffs. One pair of Cormorants nested on the Rock in 1939, and others may have done so since ; but as these birds resort to an inaccessible part of the sea-face of the Rock, no positive proof of this could be obtained while the restrictions continued in force" (Blair 1945).

This was the first, and is possibly still the only, documentation of this species breeding on the east coast of Britain, although there have been at least two suggestions of at least a nominal breeding season presence of birds on the Farne Islands, Northumberland, during the 20th century, in 1921 and 1998 (Kerr 2001). Temperley (1951), not unreasonably, speculated that birds might have bred on the summit of the Rock for years without its presence being known about or it being molested.

Despite limited seawatching effort through the 1950s and 1960s, the current pattern of occurrence was already in evidence. A *"remarkable passage"* was noted in the last week of August 1959, culminating in 410 to 430 birds being logged at Hartlepool Headland in a watch of 12 hours. A summer feeding presence was noted off Hartlepool at the end of June 1961 when in eight days, 464 were counted flying north and 590 flying south. A count of 133 at Seaham on 28 June 1963 was noteworthy for the location, while in 1969, totals of over 200 passed Hartlepool Headland on four days between 5 July and 2 August, peaking at 348 flying north and 141 south on 24 July. Clearly, there were some years with a considerable offshore presence. Also noteworthy in this period was the rare inland occurrence of one flying past an observer, at 20 yards distance, at Shotton Brickworks Pond on 9 June 1956, a day when coastal passage was in evidence.

Recent status

The main spring passage along the County Durham coast starts in the second half of April and peaks in late May, and there is usually a small-scale summer presence, although numbers are periodically much larger. A more pronounced autumn passage usually commences in late June and July, with numbers building through August and continuing into September. A decline occurs from late September, with only occasional birds being sighted through late October and into November. Numbers each year and in each season can vary tremendously and may be linked to food abundance off the north east coast.

Jan	Feb	Mar	Apr	May	Jun	Jul	Aug	Sep	Oct	Nov	Dec
1	2	1	48	463	1015	2900	2500	1381	597	5	3

Peak monthly day-counts of Manx Shearwater off Hartlepool or Whitburn

Although November records are not unusual, sightings in the winter period of December to March are much less frequent. There were just twelve January records of single birds between 1970 and 2010 (five at Whitburn and seven at Hartlepool), five in February (four at Whitburn, one at Hartlepool) and one in early March (Whitburn). There are single December records for Seaton Carew (found dead), Teesmouth, Sunderland and Whitburn, plus a total of six off Hartlepool in this month.

There have been 13 instances of day counts topping the one thousand mark, quite insignificant on British terms but notable in the context of the North Sea. The counts are detailed below:

- 30 June 1984, Whitburn — 1,015 north
- 16 July 1984, Whitburn — 2,900 north
- 4 August 1984, Hartlepool — 1,075 north
- 4 August 1984, Whitburn — 1,000 north
- 31 July 1987, Hartlepool — 1,150 north
- 9 September 1989, Whitburn — c.1,000 north
- 16 July 1996, Hartlepool — 1,005 north
- 10 September 1997, Lizard Point — 1,200 north
- 25 August 2003, Hartlepool — 1,500 lingering offshore
- 29 August 2003, Hartlepool — 1,644 north
- 18 August 2009, Whitburn — 1,125 north, c.1,400 in 'rafts' on the sea
- 14 September 2009, Whitburn — 1,381 north
- 5 June 2011, Whitburn — 1,667 north

The counts during the summers of 1984, 2003 and 2009 all involved a sustained presence for a week or more. The birds were not purely passage birds displaced by adverse weather conditions, but were clearly feeding offshore, perhaps stimulated by a period of local food abundance.

The main seawatching locations of Hartlepool Headland and Whitburn Observatory have provided the vast majority of data, but this species can be seen almost anywhere along the coast on days of passage. Counts of 450 off Crimdon Dene on 22 July 1996 and 315 north at Blackhall Rocks on 14 September 2009 illustrate that heavy passage can occasionally be witnessed at less prominent sites.

As with other seabirds, County Durham does not have much historical evidence of storm-driven 'Manxies' being found inland. In recent years, the first such record was of a bird picked up on a road one mile west of Billingham town centre 4 September 1974. In September 1981, an apparently uninjured individual was found in Blackfell, Washington. The bird was cared for at WWT Washington before being released, successfully, out to sea. On 16 September 1997, during a period of strong westerly winds, one was grounded at the Gateshead MetroCentre and was taken into care. After an overnight rest, it was released successfully back out to sea at Roker. Apparently healthy birds heading the 'wrong way' have also been seen flying west up Greatham Creek on 12 September 1992 and flying around the Seaton Snook and Seal Sands area on several occasions.

Distribution & movements

In Britain, Manx Shearwater breeds extensively off the western coasts, in particular around the Irish Sea, the Welsh islands, and on some of the UK's northern islands groups, e.g. the Hebrides, Orkney and Shetland. Breeding also takes place on the Azores, Canary Islands and the Faroes, but the true range is not well known. Outside the breeding season, long-range dispersal includes movement of both adults and juveniles to South America (Cramp *et al.* 1977).

Balearic Shearwater
Puffinus mauretanicus

An uncommon passage visitor from breeding grounds in the Mediterranean, mainly in late summer and autumn.

Historical review

The first documented occurrence of this species, which was then believed to be a sub-species of the Manx Shearwater *Puffinus puffinus*, came in 1959 when Jack Bailey saw a bird flying north with Manx Shearwaters at Souter Point, Whitburn, on 18 September. Despite not having been recorded previously, this first sighting was quickly followed by four others in the same autumn. Three were seen at Teesmouth by V.F. Brown on 4 October, and the late Fred Grey saw another at Souter Point during the 'great gale' of 28 October.

Nine birds were reported during the 1960s; this was quite a high total considering the technical challenges for observers at that time. Four sightings came from Hartlepool, all between 17 August and 1 September, while one at Seaham on 18 September 1962 was notable for the locality. More intriguing were the four seen at Marsden "*several times*" in mid-June 1963 (Bell 1964), especially in view of the early date and group size of what was, at that time, a very scarce bird in the North Sea.

Recent status

From a Durham perspective, the status of this species changed quite markedly over the next four decades. Through the late 1980s and 1990s, it become an annually reported passage migrant, with ever-increasing numbers documented, though it remained, at the early part of the 21st century, a scarce species.

After a single bird at Hartlepool on 8 September 1970 and two there on 14 August 1971, there were no further reports until 1983. At the time, this shearwater was still classed as a race of Manx Shearwater, and this taxonomic position may have influenced its reporting during the 1970s and 1980s. It is clear however, that few were identified along the coastline, even in times of notable shearwater passage, as in the years of 1976 and 1978.

More reports followed from the mid-1980s with single birds noted off Whitburn on 11 occasions between 1984 and 1989, along with two at Hartlepool. The re-consideration of this form as a full species, 'Mediterranean Shearwater' (comprising the forms *mauretanicus* and *yelkouan)*, by the BOU in 1991 stimulated a rise in interest amongst observers and a consequent increase in sightings. In the 1990s, 42 birds were noted, including sightings of two birds on six occasions, with three birds at Whitburn on 28 August 1995.

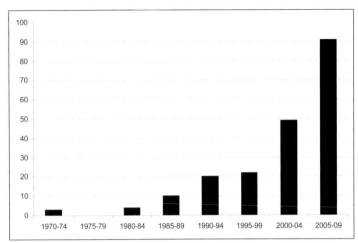

Balearic Shearwater records in County Durham, 1970-2010

Further taxonomic studies led in 2000, to the decision to treat the two forms *mauretanicus* and *yelkouan* as separate species, the former being commonly known as Balearic Shearwater. Through the 2000s, the number of birds recorded off the Durham coast continued to rise. Seawatching observations over this period in the south west of England and in the North Sea confirmed that the species was an increasingly frequent late summer and autumn visitor to UK waters. The increase in sightings in Durham over the period 1970-2010 is illustrated.

Since 1970, over 230 birds have been recorded in County Durham, with all records falling between 22 May and 29 November. Over time, there has been a subtle change in late autumn sightings; prior to 2009, the latest bird seen off the Durham coast was on 24 October (one of only three birds recorded in the second half of that month), but a total of four November birds was noted during 2009 and 2010. It is perhaps surprising that the occasional mid-winter and spring birds have not been detected, as is increasingly being demonstrated at Flamborough Head (B. Richards pers. comm.).

The peak period falls during the second half of August and first half of September (47% of records fall into this period), but the appearance of this species is very much dependent on prevailing weather conditions. Most records have been of single birds (83% of sightings), but between two and four have been recorded on 28 dates. The two best days of passage occurred in 2003, when five flew north at Hartlepool on 25 August (four of these also being noted passing Whitburn) and five were feeding offshore at Hartlepool on 22 September. Whitburn has accounted for 56% of birds recorded, with Hartlepool 43%. The only birds seen away from these locations since 1970 have been at Marsden Bay on 5 September 1996, off Dawdon Blast Beach on 15 September 2007 and Blackhall Rocks on 12 August 2010.

The best period for records was the autumn of 2007, which was notable for its movements of seabirds along the north east coast. Between 4 July and 28 September, a total of 20 birds were noted off Whitburn, 13 at Hartlepool and a single at Dawdon. Although no day-counts were greater than the three birds at Whitburn on 30 August, there was clearly a larger than usual presence of this species in the North Sea, perhaps associated with food availability, but no doubt greatly influenced by the prevailing weather systems that were feeding flows of northerly winds into the North Sea.

Distribution & movements

In world terms, this species is one of the rarest birds to be regularly recorded in Britain. Balearic Shearwater is classed as Critically Endangered with extinction (IUCN 2006) with the whole of the world population nesting on a small number of islands in the western Mediterranean. In the early 2000s, the number of mature breeding birds was estimated to be less than 10,000, with an overall population estimate of 20,000-25,000 individuals and an estimated population decline of 7.4% per annum (Arcos & Oro 2004). The main threats come from predation by cats *Felis catus*, common genets *Genetta genetta* and brown rats *Rattus norvegicus* in the breeding colonies, long-line fishing, food availability and human disturbance elsewhere (Birdlife International 2011).

The species is undertaking a change to its distributional range, which may have been set in train by a suite of conservation-related triggers since the 1990s. This may account for the species' increased occurrence in British waters, with the coincident increase in sightings along the north east England and the Durham coast. Previously a scarce visitor to Britain, with an average of 318 birds annually in UK and Irish waters during the 1980s (Wynn & Yesou 2007), subsequent studies since 2007 have suggested that approximately 10% of the world population now visit UK inshore waters in late summer and autumn (www.seawatch-sw.org). Some birds are also wintering off south west England, a new phenomenon possibly linked to elevated winter sea temperatures. As part of this overall trend, sightings in the North Sea have correspondingly risen.

Macaronesian Shearwater (Barolo Shearwater)
Puffinus baroli

An extremely rare autumn vagrant from Madeira and Cape Verde: four records involving five birds.

The first sighting of this difficult to identify pelagic species came in 1984, when Martin Blick, Tom Francis, Geoff Iceton and Andrew Robinson documented two 'Little' Shearwater (as it was known at the time) flying north at Hartlepool Headland on 24 September (*British Birds* 78: 532). These records were accepted by BBRC, contributing to a total of 25 accepted sightings in the 1980s. This was the first of four occurrences for the county.

All records:
1984	Hartlepool, two 24 September
1989	Whitburn, 15 July (*British Birds* 83: 442)
1990	Hartlepool, 7 September (*British Birds* 84: 452)
2000	Hartlepool, 15 July (*British Birds* 95: 479)

At 2012, two further Durham claims remained under consideration by BBRC, at Hartlepool Headland (and later Whitburn) on 7 August 2005 and at Whitburn, on 4 September 2007. Only four records were accepted by BBRC in the 2000s, indicating the high degree of caution being exercised in relation to the identification of this species.

Distribution & movements
The taxonomy and identification of small black-and-white shearwaters in Britain was under review during the 2000s. At 2010, the thinking was that the form *baroli,* nesting on the Azores, Canary Islands, Madeira and Salvages, was a separate monotypic species, with *boydi* (previously considered a subspecies of 'Little Shearwater' along with *baroli*), the form nesting on Cape Verde, being more closely related to Audubon's Shearwater *Puffinus lherminieri* (Onley & Schofield 2007). Past records of this species were under review by BBRC in the early part of the 21st century, with the likely outcome that many previously accepted records would be found 'no longer proven'.

Wilson's Petrel (Wilson's Storm-petrel)
Oceanites oceanicus

An extremely rare vagrant from the southern oceans: one record.

The sole record for Durham is of a bird seen passing Hartlepool Headland on 7 September 2005. Initially found by Richard Taylor and later seen by Graham Lawlor and Steve Keightley, the bird was watched for a period of five minutes as it moved north east past the Headland during a period of light seabird passage.

Distribution & movements

Wilson's Petrel is oft-quoted as one of the most abundant species in the world. It nests along the Antarctic coastline and nearby islands, dispersing widely during the non-breeding season to the North Atlantic, Pacific and Indian Oceans, though remaining wholly pelagic throughout. It has long been known to be common off the eastern coast of North America during the northern summer, but its true status in European waters was only revealed following pioneering pelagic trips in the mid-1980s.

Although now known to be regular off Cornwall and south west Ireland in late summer and early autumn, it is a truly rare bird in the North Sea. The first North Sea record came on 1 September 2002 when one was seen on a pelagic trip 14km off the Northumberland coast. The Hartlepool bird was only the second sighting on the English east coast, but was soon followed by another well documented bird at both Sheringham and Cley on the North Norfolk coast on 23 July 2010.

Storm Petrel (European Storm-petrel)
Hydrobates pelagicus

An uncommon but regular non-breeding summer visitor to coastal waters.

Historical Review

Temperley described it as "*a very rare visitor*" and "*only seen after exceptionally severe gales*". The first reference to it in the county is courtesy of Sharp (1816), who said it was "*frequently caught by the children in winter*". Temperley cast some doubt on this statement (Temperley 1951). Hogg (1845) quoted Sharp's statement, but referenced it to just one record that was known to him, a bird shot near 'Stockton on the Tees' in the winter of 1837. Hancock (1874) called it a "*casual visitant*" to the north east, on the coast, and mentioned five captures for the whole of Northumberland and Durham. Just one of these was in Durham; found dead on the beach somewhere between South Shields and Sunderland in 1835.

In the winter of 1895/1896, Tristram (1905) recorded 'many' being taken, including one found inland in the yard of an inn in Durham City. Temperley's only recent documented sighting at his time of writing was of two together flying over the mud-flats of the Tees Estuary on 29 October 1927, after a westerly gale.

Few sightings followed in the 1950s and 1960s, although several did feature birds in unusual circumstances. On 10 November 1956, five followed a ship into Hartlepool Docks, one of them later falling down the funnel of a steamer. On the same date, one was picked up alive on the beach at Seaton Carew. An unusual inland occurrence came in 1962 as one was watched at Billingham Pond for 10 minutes during a westerly gale on 29 October. At times, the bird approached as close as eight feet.

Recent Status

Until 1988, when the nocturnal tape-luring of Storm Petrels began (Milton 1995), the true status of this species in the county was unknown. Indeed, during the period 1970-1993, there were only 27 records of birds off the Durham coast (Milton 1995). Since 1989, when the first birds were mist-netted, over 1,000 have been caught along the county's coastline during the summer months at Whitburn, Sunderland Docks, Seaham, Hartlepool and North Gare. All birds ringed have been caught between 15 June and 16 September, although in most years, ringing is concentrated between early July and early August. It is believed that the birds caught are non-breeding birds and the pattern and timing of catches (M. Cubitt pers. comm.) would tie in with documented influxes of non-breeders to Shetland and other breeding areas (Fowler & Okill 1988). The numbers caught vary tremendously from year to year, with the species clearly going through periods of abundance offshore, presumably governed by food availability. The peak catches at Whitburn have included: 43, on 31 July 1990; 24, on 10 July 1991; 22, on 14 July 2006; 29, on 19 July 2008; and, 28, on 25 July 2008. The largest catches at Hartlepool include: 26, on both 12th and 26 July 2003; 20, on 26 July 2004; and, 33, on 25 July 2006.

Although the nearest known breeding colonies to Durham are in Orkney (Gibbons *et al*. 1993) with others in Shetland and down the west coast of Britain and Ireland (Lloyd *et al*. 1991), there was initial speculation as to whether there may be a breeding colony much closer than those known. The species was documented in the *Atlas* on the strength of a single female, bearing an egg almost ready to lay, caught during tape-luring sessions in July 1989 (Westerberg & Bowey 2000). To date, there has been no evidence to confirm breeding at any new site down the English or Scottish east coast.

The pool of ringing data since 1989 has led to some interesting recoveries. As would be expected, many of the recoveries have come from ringing sites with similar projects along the North Sea coast away from breeding colonies. A total of 17 recoveries have involved the Isle of May, with nearby Fife Ness featuring 15 times. However, 14 recoveries have involved birds moving to and from the breeding colony at Sanda Island, Argyll and Bute. Only one of these has been of a bird moving from Sanda to Durham, the remainder all being birds ringed in Durham and being later recovered on Sanda. Out the these 14 recoveries, only two have been during the same year; one ringed at Hartlepool on 27 July 2001 and recovered on Sanda on 9 August 2001, and one ringed at Whitburn on 2 August 2008 and recovered on Sanda six days later. These movements support the theory that the majority of birds in the southern North Sea are wandering non-breeders.

Additional discoveries have involved movements to and from north west Britain and Ireland, including the Calf of Man (5 controls), Ardglass, Down (3 controls), Harrington, Cumbria (2 controls), Tullagh Point, Donegal (2 controls) and also St Kilda. There have been multiple recoveries involving the Northern Isles, with seven from Orkney, six from Fair Isle and five from Shetland. The south west of Britain has also featured including movements from Wooltack Point, Dyfed, to Hartlepool, and from Hartlepool to Prawle Point, Devon. Nine recoveries have also featured movements to or from Norway.

Some Storm Petrels are clearly lingering off the English north east coast, as proven by multiple controls the same summer at sites between Kilnsea, East Yorkshire, and Seahouses, Northumberland. That said, some birds are clearly on the move and ringing has demonstrated some remarkably rapid journeys. The six of the most interesting recoveries as regards speed and distance are tabulated:

Ringing Site	Date	Control Location	Date
Collieston, Grampian	26 July 2003	Hartlepool same day	26 July 2003
Tarbat Ness, Highland	30 July 1990	Whitburn	31 July 1990
Whitburn	4 August 1989	Fair Isle	8 August 1989
Seaham Docks	21 July 1994	Eilean nan Ron, Highland	24 July 1994
Hartlepool	7 August 2000	Lindesnes Fyr, Vest-Agder, Norway	10 August 2000
Whitburn	2 August 2008	Craig Stirling, Grampian	4 August 2008

Selected Storm Petrels Controls to and from Durham, 1989-2008

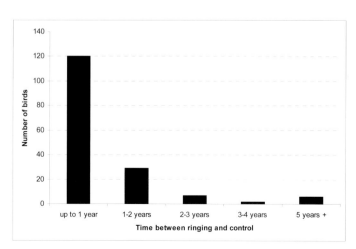

British-ringed Storm Petrels have been recovered from as far away as Namibia, South Africa and Zimbabwe, as well as several West African countries (Wernham *et al*. 2002). No Durham-ringed birds have been involved in recoveries outside the Western Palaearctic, but movement from Faro, Portugal, to Whitburn in 29 days in July 1992 gives an insight to the distribution and movements of non-breeding birds.

Storm Petrels are long-lived birds, defying their frail appearance and oceanic life style. Most Durham recoveries concern periods of less than a year between ringing and recovery, but one ringed at Tarbert Ness, Highland, on 12 August 1981 was controlled at Whitburn on 30 July 1991, approaching its tenth anniversary, while one ringed at Whitburn on 12 August 1990 was re-trapped 10 years later to the day on Sanda Island, Argyll and Bute. A breakdown of time between ringing and subsequent recovery involving Durham birds is shown below.

During 1989-91, out of 210 Storm Petrels trapped by tape-luring, three had either a whole leg or at least most of the leg below the tarsal joint missing. This constitutes over 1.4% of all those caught, a surprisingly large proportion of wild birds to be carrying such major bodily damage. Over the same period, a further four or five petrels also had toes or a part or the whole of a foot missing; if these were included in the calculation, the percentage would obviously be much higher (Bowey 1995). Other ringers working on the species have also noted this phenomenon with ringers on the Isle of Man finding that up to 15% of birds caught there had some leg or foot deformity. Studies have revealed this to be caused by attacks from predatory fish as the birds feed on the surface of the sea (Manx Ringing Group pers. comm.).

The tape luring and ringing of Storm Petrels at both Hartlepool and Whitburn revolutionised the understanding of this species' status in the county and information from ringed birds will no doubt lead to further discoveries.

All seawatch records of Storm Petrels in Durham fall between 16 May and 29 November, with the majority seen between mid-June and late September, as illustrated. As might be expected, Hartlepool Headland and Whitburn Observatory account for the vast majority of records. Single birds have been observed at just three other sites in the county: Fox Holes, South Shields and Sunderland. In contrast to Leach's Petrel *Oceanodroma leucorhoa*, Whitburn Observatory has been the most productive location for this species with 106 birds noted since 1970, with 94 off Hartlepool Headland in the same period. This is possibly due to many late summer records involving feeding birds, rather than storm-driven individuals so the geographical influence of Hartlepool Bay and Headland, as observed in Leach's Petrel occurrences, is not as manifest in the pattern of sightings.

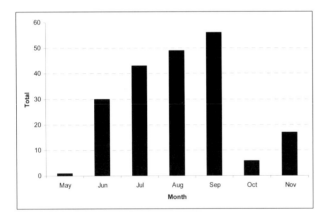

Seawatch records of Storm Petrel in County Durham, 1970-2010

Up until 2001, sightings on seawatches were sporadic and unpredictable. Indeed, there were 15 years between 1970 and 1999 when none were recorded. Apart from a remarkable (for the time) 11 birds in 1978, including six at Whitburn on 31 July, no more than five birds were seen in a year, with one or two birds being the norm. A remarkable change in this pattern was witnessed from 2000 as all previous seawatch counts were exceeded. After just 33 seawatch sightings between 1970 and 1999, a total of 165 were logged between 2000 and 2010.

Although there was a more regular frequency through the summer months as birds presumably fed offshore, all the peak day-counts were weather-related, generally being linked to strong northerly airflows.

The best days of passage at the start of the 21st century included:

- 10 September 2001   eight at Hartlepool and 11 at Whitburn
- 23 September 2004   16 at Hartlepool and three at Whitburn
- 9 August 2005   one at Hartlepool and nine at Whitburn
- 11th to 13 August 2006   totals of 13 at Hartlepool and nine at Whitburn
- 26 June 2007   three at Hartlepool and nine at Whitburn
- 21 July 2008   four at Hartlepool and five at Whitburn

Quite why there was such an upsurge in records during this period is difficult to explain. The drop in numbers to just seven sightings in 2009 and 2010 combined may hint at a return to a more 'normal' and long-established pattern of occurrence. It may be that the 2001-2008 increase in numbers was related to local, relatively short-lived phenomena, such as subtle changes in sea temperature or a temporary change in food distribution in the North Sea.

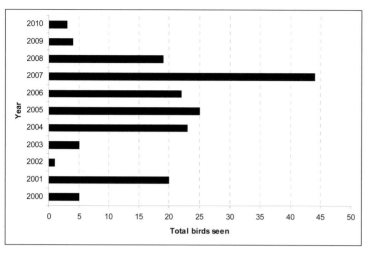

Annual totals of Storm Petrels seen during seawatches in County Durham, 2000-2010

The summer of 2011 changed the status of this species dramatically as late July brought an unprecedented influx to the north east coasts of England and Scotland. After a brisk northerly wind and heavy sea on the evening of 23 July, the morning of 24th produced a passage of 75 birds flying north at Whitburn Observatory. A full day watching on 25th resulted in an incredible total of 124 north and 15 south, whilst the passage continued unabated on 26th, with 352 north and 6 south. With a change in the weather, the final ten birds moved through on 27th and there were no subsequent summer sightings. Away from Whitburn, the influx was detected at other Durham seawatch sites including day totals of up to 82 birds at Blackhall Rocks, Crimdon Dene mouth, Hartlepool Headland, Seaham and South Shields.

Very few inland birds have been seen in recent years with the species apparently resilient enough to remain out at sea even in the roughest of conditions. The only sightings since 1970 have been single birds captured in Redheads Shipyard, South Shields, on 5 November 1973 and in Sunderland Town Centre on 29 October 1989.

Distribution & movements

This a colonial breeding species that can be found along the coasts of the north east Atlantic and the Mediterranean, with the Atlantic population (at least) moving to south and south west Africa in the non-breeding period. The largest Atlantic breeding colonies are located in Iceland, Norway, the Faeroes, northern Britain and Ireland (Cramp *et al.* 1977).

Leach's Petrel (Leach's Storm-petrel)
Oceanodroma leucorhoa

A scarce autumn passage visitor, usually after gales.

Historical review

It was called "*a very rare vagrant*" by George Temperley (1951). He highlighted four inland records along the Tyne, all of them within the modern Gateshead area, spanning a period between 1830 and 1949. The first was a bird killed on the Tyne opposite the Old Mansion House, Newcastle in 1830. Some years later, on 1 March 1886, one was shot near Blaydon, though the specimen was not found until the 10th. In the 20th century, there were two other records, both of birds shot on the Tyne. One was 'taken' below Bensham Boathouse in November 1928, and finally one was at Ryton Willows on 24 April 1949. Considering that none of these birds were documented as occurring in storm conditions, this cluster of records along the Tyne over a long time period would seem to be suggestive of some regular movement across country from Irish Sea to North Sea via the Tyne valley (Bowey *et al.* 1993).

Other than the records from along the Tyne corridor, Temperley (1951) commented that there were few other records, most of these relating to the Tees Estuary. These included one washed ashore, date uncertain but before 1905; this had been mentioned by Tristram (1905). Nelson (1907) referred to one, a female, which was shot at the mouth of the Tees on 17 September 1903. One was also shot at Cowpen Marsh in September 1914. Well inland, one was picked up in Middleton-in-Teesdale on 19 October 1935. A further inland sighting came in 1952 when one was found in the grounds of Polam Hall School, Darlington, on 29 October. This was part of an enormous wreck of petrels that occurred in Britain at this time, caused by severe weather in the south west (Boyd 1954). Several other birds were found in Northumberland at this time, presumed to have been driven across the Pennines rather than arriving from the east coast. Very few others were noted in the 1950s and 1960s; one off Whitburn on 18th and 19 June 1956 was unusual in date, while one at Hartlepool on 27 October 1960 and a total of nine there between 20th and 27 November 1965 were a little more expected.

Recent Status

This species' occurrence is unpredictable and episodic. Its annual appearance in Durham is not guaranteed and there were 15 years between 1970 and 2010 in which it was not recorded. Almost all sightings since 1970 have occurred in the period 28 August to 30 December, the one exception was a bird flying north close inshore, off Whitburn Observatory, on 12 June 2008.

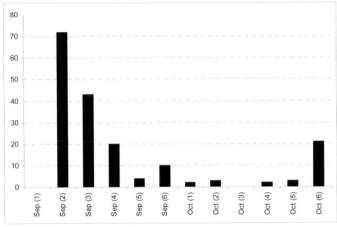

Leach's Petrel records in County Durham by five-day period in September & October, 1970-2010

The peak period for movements of this species is the second week of September; in fact 71% of all records since 1970 have fallen between 8th and 16th of that month. The breakdown of autumn records is illustrated.

The key sites of Hartlepool Headland and Whitburn Observatory have provided virtually all the records. The only bird seen away from these two locations was one that flew south past Hendon on 19 September 1990. Hartlepool has recorded most birds, with 128 logged between 1970 and 2010, compared to 56 at Whitburn in the same period. The geographical position of Hartlepool is perhaps the reason for this bias, with storm-driven birds being pushed into Hartlepool Bay and departing to the north past the Headland; there is no such channelling effect at Whitburn.

The appearance of Leach's Petrel in County Durham is very much weather-dependent and is often associated with disturbed weather patterns, such as the arrival around the British Isles of the tail-end of Atlantic storms. Comparatively large numbers have been encountered in our area at the end of October. These are usually linked to 'wrecks' of storm-driven birds on the south and west coasts, such as in 1989, 2006 and 2009. The large numbers on the south and south west coasts of England in these years were displaced by fierce gales in the south west approaches and Biscay, presumably weakening birds on their wintering grounds and pushing them back onto the English coast (Gantlett 2006).

The first large-scale movement came in 1974 when late October gales produced 20 close inshore past Hartlepool Headland, including 13 north and three south in six and half hours on 30th. It was another 15 years before any equivalent numbers were seen, despite the rise in the number of observers. A strong north easterly gale in early September 1989 resulted in unprecedented numbers of Leach's Petrels being displaced into the North Sea, with counties from Northumberland to Norfolk recording larger than usual numbers. The peak day for movement was 9th, when 29 flew north past Hartlepool Headland and 19 north past Whitburn. Many more must have been missed against the mountainous seas of this day; more than 100 were noted along the East Yorkshire at the time

(Curtis 1991). Since 1989, there have only been two further instances of more than four birds being logged during a period of seawatch observation, in both instances at Hartlepool Headland. On 13 September 1993, 25 were noted moving north and 12 likewise on 16 September 2004.

Despite the species' appearance being associated with storms, very few have been found driven inland in recent years. Of the three recent such records away from the open sea, only one has been truly inland; in a field three kilometres inland of Seaton Carew on 15 September 1972, and later successfully released out to sea. The other two sightings concern one in Hartlepool Docks on 19 December 1976, eventually flying off south west, and one over Seal Sands on 13 September 1980.

Nocturnal petrel ringing has produced occasional, but regular, records of this species in both Northumberland and Yorkshire, but surprisingly, only two birds have been detected in County Durham. After one trapped and ringed at Whitburn on the night of 9 July 2010, a second individual was trapped there the following night. The weather conditions were unremarkable at the time, as was the number of Storm Petrels being attracted, so such events are difficult to explain.

Distribution & movements

The Leach's Petrel has a very large world range, being found breeding on islands off the north Atlantic and north Pacific coasts. It winters far out to sea, generally moving south to areas of tropical convergences. British colonies are concentrated in the north and west of Scotland, but east Atlantic populations form only a tiny proportion of the world population (BWP).

Gannet (Northern Gannet)
Morus bassanus

A very common coastal passage visitor, especially in spring and autumn.

Historical review

The skeletal remains of Gannet, from an undated human midden discovered in a quarry at Whitburn Lizard, on Cleadon Hills, in the spring of 1878, implied that this species was present and hunted in former times in the county, but whether this was as a breeding bird or a passage migrant cannot be stated (Howse 1878). Hogg (1845) described it as "*frequent at Hartlepool*", but there is relatively little in the historical record about this species. Records from some of the contemporary 19th century authors seem to indicate that inland records of the species may have been somewhat more prevalent in the north east during the late 18th and 19th centuries (Kerr 2001). One such record refers to a bird shot at Stockton in January 1823 (Blick 2009).

Temperley (1951) called it a "*regular passage migrant in spring and autumn*" and said it was observed off the Durham coast in April and May, and again in August and September. The study of active seabird migration was in its infancy in Temperley's day, but he did note a number of significant movements of this species. For example, on 25 October 1947, 600 were logged flying north off Whitburn in a six-hour period, whilst on 21 September 1948, 460 flew north off Whitburn. A count of 1,400 flew north off Whitburn in nine hours on 28 October 1948 demonstrated that large numbers were occasionally displaced onto the north east coast. Post-Temperley, 640 flying north at Hartlepool on 2 September 1965 was also a noteworthy total.

In line with the species' current status, the Gannet was only rarely seen inland in the past, but a 'bird of the year' was found injured in a turnip field near Chester Moor on 5 November 1932 (Temperley 1951). A number of the birds that have been found grounded in the county have been later released back out to sea, after varying periods of recuperation. For example, one found on a cricket field in Stockton on 27 May 1955 was picked up and later released (Blick 2009). On 12 September 1966, an immature was found at Longnewton, east of Darlington, having struck overhead wires. This bird recovered and was successfully released at Seaton Carew in early October. In Gateshead, there is a single record from the Tyne corridor when, on 28 May 1968, a lone adult was observed flying west along the Tyne from Newburn Bridge (Bowey *et al.* 1993). During rough weather on 20 February 1969, a moribund individual was found on a rooftop in South Shields, and on 9 September of that year, one was noted flying over Billingham (Blick 2009). In 1970, an inland bird was on the River Tees at Croft on 15 February and another was over Saltholme Pools on 17 October, later seen heading inland (Blick 2009).

Recent status

The nearest Gannet colony to County Durham is at Bempton Cliffs, North Yorkshire, but the handful of known ringing recoveries indicate that the Bass Rock colony, in Lothian, one of the largest in Britain with over 40,000 pairs, is the likely source for many of the county's birds. Passage northwards along the Durham coast, presumably to Bass Rock, can begin as early as late February, but this only really builds up to a significant flow of birds during March, rising through April and continuing into May. Late summer and autumn remains the peak period with passage rates seemingly rising in recent years, in line with growth at the breeding colonies. Through the winter months, it remains a relatively scarce species off the North Sea coastline.

Few birds are seen off the coastline from December through to February, even in more inclement weather conditions with northerly winds, suggesting that most leave the North Sea during the winter period. A gathering of 72 following a trawler into Hartlepool Harbour on 29 December 1979 was exceptional, whilst other impressive out-of-season counts include 89 at Whitburn on 19 February 1993, 71 at Whitburn on 30 December 1996 and 120 past Hartlepool on 4 February 2001.

The return of adults to breeding colonies commences in late March, and there is usually a noticeable increase in passage activity throughout April. Spring movements peak in late April and early May, but numbers fluctuate greatly from year to year and their visibility at the coast is heavily dependent on the presence or absence of onshore winds. Evidence of passage levels at Whitburn have been provided by counts of 1,250 north in 15 hours of observation over 13th to 16 April 1994, 2,228 north in four hours on 28 April 2004, and over 3,200 moving north in the first ten days of May 2010. Unusually, an adult landed on Marsden Rock on 29 April 1995; the closest Durham has come to any breeding behaviour.

During the late spring and high summer, birds feed along the Durham coastline and parties of plunge-diving birds are noted offshore from many coastal locations between Tyne and Tees, but particularly those in the north of the county, such as South Shields and Whitburn. This species is known to forage at considerable distances from the colony at Bass Rock, up to 540km has been recorded with a mean distance of 232km (Hamer et al. 2000), which brings the Durham coast comfortably into range for many feeding birds. Notable northbound movements at Whitburn in this period have included 1,150 on 6 June 2002, 1,357 on 30 May 2006 and 2,214 on 6 June 2009.

The heaviest passage of Gannets along the county's coast usually occurs in the late summer and early autumn, mainly in late August and September, with large numbers being recorded on a daily basis from coastal watch points. As well as birds being seen in large numbers off Whitburn and Hartlepool Headland at this time, smaller numbers are routinely noted off locations such as Blackhall Rocks, Castle Eden Denemouth, Dawdon Blast Beach, Ryhope and Seaham. The highest-ever count for the county was of 4,250 flying north past Whitburn on 22 July 2010, but other impressive totals at this locality in recent years have included: 2,090, on 9 September 1989; 1,830, on 5 September 1992; 2,000, on 25 September 2004; 2,300, on 3 August 2006; and 2,450, on 4 September 2007. Passage is usually somewhat lighter at Hartlepool, but tallies during heavy movements in the 2000s have featured: 1,184 moving north in 2.5 hours, on 2 September 2000; 1,050 north, on 20 July 2008; and, 2,270 north, and 125 south, over 14th to 15 September 2009.

Although concentrated passage can still occur through to mid-October, numbers decrease significantly in late autumn. Despite a nominal presence for much of the time, northerly gales can result in a sudden reappearance of mainly juvenile birds. A count of 800 at Whitburn on 3 November 1998 was exceptional, with 108 there on 19 November 1991 and 85 on 8 November 2010 also being unusual; most year pass with November counts only just reaching double-figures.

The colony of Gannets at Bempton Cliffs, East Yorkshire, has grown from 20 pairs in 1970 to 7,859 pairs in 2009 (RSPB), whilst the Bass Rock colony has also undergone an equally impressive growth rate (JNCC). The incidence of large Gannet movements off the Durham coast has increased in line with the growth of these colonies and can be expected to continue to do so in coming years. Day-counts of 800-1,000 were unusual through the 1970s and 1980s and it was not until the early 1990s that four-figure movements became more regular.

Inland records over the last four decades are scarce but clearly clustered into the species' main passage periods, with most records occurring in either March/April or September. Two were seen in Durham City on 14 September 1990; an adult was noted flying downriver and an immature was observed being attacked by Magpies

Pica pica. Birds were found dead by the River Wear at Witton-le-Wear on 22 March 1975 and at Longnewton Reservoir on 18 January 1989, an immature circled Shotton Colliery on 6 September 1992 before heading west, and one was on the River Tyne at Jarrow Slake on 19 July 2000. Several have been noted in the vicinity of the Tees estuary in recent years, as might be expected. One picked up alive under overhead wires at Eaglescliffe was cared for by the RSPCA and released successfully, another was picked up exhausted at Cowpen Marsh on 8 April 1986 and one flew over Stockton and Norton on 29 September 1995. Unusual for the time of year was a bird flying west up Greatham Creek on 22 July 1997, but the most remarkable sight was of one found asleep at the side of the road at Greatham Creek on 18 March 1979. More recently, the autumn of 2011 brought an unusual series of inland sightings. On 13 September, an adult flew north over the cricket ground at Chester-le-Street, soon followed by two adults flying north over Sedgeletch on 27th. A juvenile was also present on Castle Lake, Bishop Middleham on 2 October.

Distribution & movements

The breeding range of the Gannet is restricted to the North Atlantic, with 67% of the world population residing in British and Irish waters and the remainder in the Faeroe Islands, France, Germany, Iceland, Norway, Russia and Canada. Twenty one colonies in Britain and Ireland hold approximately 260,000 pairs, with a steady growth rate at nearly all of these in recent decades (JNCC). The winter range covers the North Atlantic to the south of their breeding range, penetrating as far south as the coasts of West Africa.

Local ringing data for this species is limited but several birds ringed at Bass Rock have been recovered along the Durham coastline. Firstly, a bird ringed at Bass Rock on 4 July 1936 was recovered at Seaton Carew on 24 September 1936. Two other birds ringed at the same location have been recovered locally; these were ringed as pulli during the summers of 1949 and 1950. They were found dead at Hartlepool on 9 September 1953 and North Gare Sands on 10 June 1953 respectively. Most recoveries have been of birds less than five years of age. Examples of the longevity of this species also come from recoveries of nestlings ringed at Bass Rock. One ringed on 30 June 1961 that was found dead at South Shields on 17 August 1977; one ringed on 13 July 1969 was found dead at Blackhall 30 April 1980; and one ringed on 1 July 1971 was found dead at Blackhall 1 February 1986. Another interesting recovery involves a bird ringed as a pullus at Skarvklakken, Andoya, in northern Norway on 3 August 1972 and found dead at Hartlepool on 17 October 1972; this was the first Norwegian-ringed Gannet ever recovered in Britain. Most bizarrely, a nestling ringed at Bass Rock on 26 July 1960 was found injured but alive at Croxdale on 12 October 1960; it was later released at Marsden Rock, before being found in Lancaster, Lancashire, on 15 October of that year.

Cormorant (Great Cormorant)
Phalacrocorax carbo

A very common coastal resident and winter visitor, regularly recorded along river valleys and on inland waters in small numbers.

Historical review

It is likely that the broad status of the Cormorant in County Durham has remained largely unchanged for decades, if not centuries. The earliest mention of the species from the area comes courtesy of a record from 1544, by William Turner, who stated "*I have seen mergi* (meaning Cormorant or Shag *Phalacrocorax aristotelis*) *nesting on sea cliffs about the mouth of the Tyne river*" (Evans 1903, Gardner-Medwin 1985). This is not a specific reference to the Durham side of the river, and may have referred to Tynemouth, but it may also relate to the Marsden area. Whatever the birds were at the mouth of the Tyne, Shag or Cormorant, it is obvious that in those long-distant days, at least one or the other of these species was nesting '*about the mouth of the Tyne'*. There was a later assertion that Cormorants nested on Marsden Rock in 1813 and for several years afterwards. The first inland record in the county was of a bird shot near Wycliffe-on-Tees in September 1782 and Hogg (1845) said it frequented the Tees "*to a great distance inland*".

During the early 20th century, Cormorants were regularly present in Marsden Bay and off Teesmouth (Temperley 1951), large flocks resting on sandbanks there in August (Nelson 1907). The Rev. George Courtenay expressed some surprise that he only saw Cormorant twice when he was domiciled in Sunderland, on 24 January and 7 March 1907, and he stated, "*I must not be positive but I do not think that I ever saw a Cormorant in Sunderland in all the years that I spent in Sunderland except on these two occasions*" (*The Vasculum* Vol. XIX 1934). This strongly suggests that it was not that common in and around the River Wear or adjacent seas at the turn of the 20th century. In early 1932, a Cormorant that was found dead at Park Lake, Darlington, as a result of starvation - because of a damaged bill and parasite attack - attracted some public attention. At the time, observers were surprised by its inland location (*The Vasculum* Vol. XVIII 1932). At the mid-point of the 20th century, Temperley (1951) noted that it was "*a very common resident on the north-east coast*", though he stressed the point that it did not normally breed on the Durham coast; although one or two pairs had nested on Marsden Rock "*during the years of the first war*". He recorded, in the mid-20th century, that it was still found occasionally well upstream on the county's main rivers and more regularly on the lower Tyne, around Jarrow Slake and around 1950, "*a few birds*" regularly spent the winter on the Tyne near Blaydon with occasional stragglers occurring on fresh water lakes and ponds.

Over the ensuing years, inland birds became an increasingly common feature in Durham. In 1966, three were found sitting on fence posts at the newly constructed Derwent Reservoir on 2 September, while birds were noted at Hurworth Burn Reservoir and various other wetlands around the south east of the county during the late 1960s. Small numbers utilised the River Tyne to the west of Gateshead more frequently, up to seven in 1969 rising to 16 in December 1971, and counts of up to 100 in the Tees estuary also became more routine. Coastal passage was also prominent on occasion, with 230 noted flying south between Crimdon Dene and Hartlepool on 1 March 1962 being particularly noteworthy.

Recent breeding status

As a breeding bird in Durham, the Cormorant remains a strictly coastal species. For nesting purposes, it is confined to a short section of the county's north east coast, on the weathered sea-stacks at the southern end of Marsden Bay. Birds have traditionally preferred the flat plateau of the main stack of Marsden Rock, leaving the steeper cliff faces to other species such as Kittiwakes *Rissa tridactyla*. They have also nested, in much smaller numbers, on Pompey's Pillar and Jack Rock, further south along Marsden Bay (Westerberg & Bowey 2000).

Despite some confusion over the exact date that the species first bred at Marsden, the history of the colony through the latter part of the 20th century has been well documented. Breeding commenced in 1954 when three nests were built, but the outcome is not known. The colony fluctuated between six and ten pairs through the remainder of the 1950s; however, success rates were reported to be low, with just two young produced from nine nests in 1957. An increase came through the 1960s as the colony became more established, the number of active nests rising from 12 in 1962 to 25 by 1968, while the number of birds occupying the site in the non-breeding season reached a new high of 79 in November 1969. A slow and steady increase through the 1970s was evident. The colony numbered 53 nesting pairs in 1974, and the number of individual birds in the area rose accordingly. Pre-nesting gatherings in March 1974 reached approximately 250 birds, while 90 juveniles present in August of the same year gave some idea of the success rate.

The boom period for the colony's growth then came through the 1980s and into the early 1990s, in line with the national trend. A total of 153 pairs were present by the mid-1980s (Lloyd *et al*. 1991), rising to 324 pairs in 1990, but falling to 225 pairs in 1993 (Armstrong 1991-1994). British populations levelled off through the early 1990s to around 7,000 pairs (Gibbons *et al*. 1993), so the Marsden colony still formed a significant percentage of this figure. Although overall numbers increased in the north east of England coastal population, between the Humber and the Firth of Forth, some colony extinctions occurred at this time, the cause of which was not always known (Lloyd *et al*. 1991). The partial collapse of Marsden Rock in 1996 may well have affected subsequent breeding activity in the county.

Further declines of the Marsden colony took place during the 1990s and into the early part of the 21st century. In 2004, there were 145 occupied nests, compared to 248 in 1999, amounting to a 41% reduction in just five years. By 2005, there were 105 Apparently Occupied Nests (AON), but the trend appeared to have been reversed with an increase of almost 40% the following year, to 150 AONs. The following years saw a dramatic change to the colony in this location. In 2007, the traditional site on top of Marsden Rock was abandoned (for the first time since the

1960s), and the whole colony relocated to Jack Rock, just off Lizard Point close to Souter Lighthouse. With this move, the total number of nests decreased considerably compared to previous years, largely one presumes, because of the smaller available area of this stack. Nests containing juveniles numbered 91 on 24 June of that year. The reasons for the desertion of Marsden Rock were never ascertained. It is of interest that the colony moved *en masse* at the start of the breeding season and no nesting attempts were made at all on what was traditionally the species' main breeding site. Estimates of numbers around the colony at Marsden included counts of 'around 300 birds' on 24 April and 'over 400' there in late July of that year. Clearly, many birds were present as non-breeders in summer 2007. In 2008, breeding occurred only on Jack Rock, where there were 110 occupied nests and an estimated 260 juveniles produced; there were no attempts on Marsden Rock. In 2009, the Marsden breeding colony was split over several stacks, with 18 pairs back on Marsden Rock, four on Pompey's Pillar and the majority, 91 nests on Jack Rock, and a similar situation was evident in 2010.

Although successful breeding has never taken place away from Marsden, nesting behaviour has been noted at Shibdon Pond on a number of occasions (Westerberg & Bowey 2000). Observations of birds in breeding condition were made on a relatively frequent basis during the late 1980s and early 1990s, including pair-bonding behaviour and occasional short-lived nest building, such as in spring 1989 when a rudimentary nest platform was constructed. Furthermore, in 1991, birds attempted to nest at a site close to the River Tyne in the east of Gateshead, though the outcome was unknown (Bowey *et al.* 1993).

Recent non-breeding status

It would appear that during the late 1970s and through the 1980s, birds increasingly moved inland onto the county's less polluted rivers and waterways to exploit the area's recovering fish stocks. This was a trend noted as occurring elsewhere across Britain around this time (Marquiss & Carss 1997). Birds can now be noted at any time of the year along the main river corridors and larger water bodies, though they are at their most common in the non-breeding season.

In the north, inland roosting along the Tyne, for instance at Corbridge, was well-established in Northumberland by the early 1990s, and this site peaked with a count of 64 birds during 1993 (Kerr 2001). Further east, and within Durham, a nightly Cormorant roost began to establish itself at Shibdon Pond from the late 1980s, growing in size through the 1990s. Roost counts here during the winter months showed that this site regularly attracted birds throughout the rest of the 1990s and reaching 87 in December 95 and rising to 104 by 8 January 2000. Further east, 107 birds were noted on riverside cranes at Hebburn on 3 January 1995, while 111 birds were counted on the Tyne between Bill Quay and Wylam on 23 December 1995. In addition to the established roost at Shibdon Pond, 27 were roosting on the High Level Bridge in the centre of Gateshead in January 1996.

During the late summer of 1995, there was an observation of a single immature bird descending from a great height, estimated to be approximately 500m, and dropping directly in to a willow tree *Salix sp.* in the roost on the central island at Shibdon Pond. The implication of the observation was that this bird knew the roost well and the height at which it was approaching from was intriguing. The method of approach suggested an arrival from a considerable distance. Birds leaving or entering the roost at Shibdon normally fly at around eight to ten metres above ground level, 35m at a very maximum, unless having to overfly obstructions (K. Bowey pers. obs.). A similarly intriguing report came from WWT Washington, where a party of 18 birds watched on 21 August 1999 rose on a thermal to a height at which they were lost to view.

During the winter months, birds can often be found sheltering in the mouths of the rivers Tees, Wear and Tyne and also use the adjacent docks and harbour areas, often during severe weather conditions. For instance, 77 were counted at Hartlepool Docks in January 1986, 95 were there on 28 December 2005 and 101 on 16 January 2008 (Blick 2009). Along the River Tees, up to 65 have been noted roosting in trees at Preston-on-Tees, downriver from Yarm, in the winter months; this gathering has been noted regularly since 1984 (Blick 2009). The mouth of the River Tyne often holds up to 60 birds with a similar number on the south pier at the mouth of the river Wear. Counts at Whitburn, half-way between the Tyne and the Wear, are generally very low through the winter months suggesting limited interchange between the two areas, with birds utilising their respective river corridors for feeding. Harsh weather can drive large numbers further upstream from the river mouths, with 106 on the River Wear between South Hylton and Fatfield in such conditions on 17 February 1986.

Through the late autumn and winter, the number of records from inland waters increases. For example, during 2007, birds were reported from nearly 70 inland sites, from small ponds and minor river systems to the larger lakes and reservoirs. Some of the main inland sites frequented (and their peak counts) include: Brasside Ponds (65), Crookfoot Reservoir (111), the River Wear at Durham City (17), Hart Reservoir (26), Haverton Hole (76), Hurworth Burn Reservoir (59), Low Barns NR (35), Portrack Marsh (34) and WWT Washington (35). Most of the larger upland reservoirs such as Balderhead, Grassholme, Selset and Smiddyshaw also receive occasional visits from up to 14 birds at some point during the year, most frequently in the autumn when birds may be crossing the Pennines from colonies on the west coast. The use of Derwent Reservoir, where maximum counts have risen from single figures to 45, has increased dramatically over a 25-year period, as illustrated

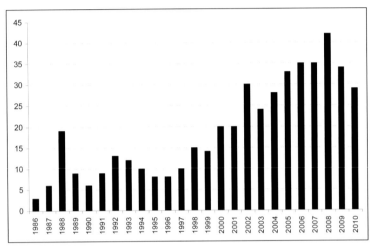

Cormorant Maximum Winter Counts (WeBS) at Derwent Reservoir, 1986-2010

The number of birds seen around the mouth of the Tees has substantially increased in recent decades and birds are present there in varying numbers throughout the year. The monthly high-tide counts carried out as part of the Wetland Bird Survey (WeBS) have shown that peak numbers usually occur in August or September. Since regular counting began, the highest counts have been 136 in September 1969, 185 in February 1977, 337 in September 1989, 480 in September 1990, 731 in August 1995 and 772 in August 2003. Most of the largest counts have involved birds roosting on the Philips Gantry at Seal Sands. Over the ten years 1996-2005, the average peak count there has been 516, which is well above the threshold (230 birds) qualifying the area for national significance. The numbers however, have fallen in recent years in line with population decreases elsewhere and, since 2006, the highest annual count has varied between 300 and 360. The autumn population at Teesmouth, at the end of 2009, ranked as being the 15th highest in the UK (Blick 2009).

Occasionally, quite large numbers of Cormorants are recorded passing along the Durham coast, usually under the influence of high winds. Such movements at Hartlepool have included 130 flying south on 9 February 1978 (Blick 2009) and 76 south on 17 October 1992, while at Whitburn, late summer movements of birds departing south from the colony have numbered 190 on 18 July 2010, 176 on 9 August 2010 and 156 on 15 August 2009. Maximum post-breeding gatherings at the Marsden colony have been 794 in mid-September 1990 and 446 on 17 September 1992. With local populations at this level, it is unsurprising that occasional large movements are encouraged under suitable weather conditions.

In the north east of England, there seem few obvious threats to the species, although falling fish stocks would affect birds very significantly indeed. On occasion, inland foraging birds strike overhead cables and small numbers of coastal birds are caught in fishing nets, but these are probably relatively minor in their effect on the local population. There is little doubt however, that Cormorants are still actively persecuted locally, presumably by fishermen. This species is very unpopular with anglers, and this disapproval sometimes manifests itself in action against the species. This was illustrated in the winter of 1971/1972 when of 11 birds feeding at Charlton's Pond, at least two were killed. This was despite the fact that the pond had been a protected nature reserve since 1968. More birds were killed here in the following winter (Blick 2009).

Continental Cormorant
Phalacrocorax carbo sinensis

Since the early 1990s, there have been regular reports of birds resembling the continental race, *P. c. sinensis*. Its true historical status is difficult to determine however, as fully valid field characteristics have only recently been established. Through the 1990s, sightings of one or two birds came from inland wetlands such as: Brasside Ponds,

Chester-le-Street, Crookfoot Reservoir, Darlington, Low Butterby and Shibdon Pond, with most falling between November and early April. An increase in observer interest in this subject from the mid-2000s led to more numerous reports and perhaps a more accurate reflection of its status in the county.

In summary of the situation between 2006 and 2010, sightings came from eight to ten localities per year with most coming in the winter and early spring period, between late November and late April. It should be noted that few observers attempt to identify all Cormorants observed down to sub-specific level and initial findings should be treated as tentative. Peak counts have included: five at Saltholme Pools, on 27 March 2006; up to four at Herrington Country Park in November and December 2006; five at Derwent Reservoir, on 5 July 2007; five on the River Wear in Durham City, between February and April 2008; seven at Brasside Ponds, on 4 June 2008; and, six at Hury Reservoir, on 22 February 2009. Other favoured sites included lowland wetlands such as: Castle Lake, Hart Reservoir, Hetton Lyons CP, Low Barns, Rainton Meadows and Shibdon Pond, while several reports have come from upland reservoirs such as Cow Green, Derwent, Hury and Waskerley. Coastal sightings were relatively few and came mainly from the well-watched areas at either end of the county's coastline, but the lack of other coastal reports is perhaps due to limited opportunities for close study.

The majority of summer birds observed have been immature and it is possible that a significant proportion of non-adult Cormorants seen well inland outside the main winter period are *sinensis*. With the increasing numbers at inland colonies in south east England (JNCC) and the increasing frequency of birds at inland lakes and reservoirs through the summer months during the first decade of the 21st century, the establishment of an inland colony in the county seems a distinct possibility.

Distribution & movements

The Cormorant has an extremely large distribution, being found on every continent except South America and Antarctica. The nominate form *P. c. carbo,* is found mainly in Atlantic waters and nearby inland areas; on Western European coasts and south to North Africa, the Faroe Islands, Iceland and Greenland and on the eastern seaboard of North America, in the Canadian maritime provinces. The subspecies *P. c. sinensis* occurs from north central Europe east to southern China, wintering in South East Asia and Indonesia. A further four subspecies occur in Africa, Japan and Australasia (Parkin & Knox 2010).

An interesting set of ringing recoveries, often involving colour-ringed birds, has given a clearer picture of the origin of some of the wintering and summering birds in the county. Over the years, at least five birds ringed as pulli on the Farne Islands, Northumberland, have been reported at Hartlepool or, just south of the Tees, in April or May of the following year (Blick 2009). At least twelve birds from the Farnes colonies, were recovered in the South Shields to Sunderland stretch of the Durham coast between 1961 and 1965; a number of these were birds drowned in fishing nets. Recoveries of birds from more northerly colonies have involved individuals ringed as nestlings in Shetland, Orkney, Highland, Dumfries & Galloway and Lothian. One early inland recovery was of a Farnes-bred bird, ringed as a pullus on 20 June 1961, and recovered at Sacriston in November 1965. Many of the recoveries have come from the south east of the county and from Shibdon Pond, but one ringed in the nest at St. Ninian's Isle, Shetland, on 22 June 1990 was found dead at Grassholme Reservoir on 12 December of the same year. Repeated patterns of movement have also been proved; for example, three birds from the breeding colonies on the Solway, one ringed in 1993 and two ringed in 1994, were noted at the Shibdon Pond roost in the winter of 1993/1994 and again in 1994/1995.

Cormorants ringed in colonies in central and south east England have been noted on several occasions. One ringed in the nest at Rutland Water in June 1998 was at Longnewton Reservoir in April 2000, and one in the Tyne Estuary in October 1998 had been ringed at an inland colony in Nottinghamshire the previous year (Kerr 2001). A bird of the race *sinensis,* ringed at Abberton Reservoir in Essex, was present at Shibdon Pond on 1 May 1995, and again during the autumn of late 1995. This bird was later found dead at the mouth of the River Derwent (K. Bowey pers. obs.). A more detailed picture of movement is available for a bird ringed as a nestling at Abberton Reservoir in Essex in May 1998. It was seen on Seal Sands in 1999, in Spain in 2001 and then back at Teesmouth in August 2002.

Rapid post-fledging movements have been demonstrated by sightings of colour-ringed birds. During the 1990s, birds ringed in the nest at Abberton Reservoir in Essex, Besthorpe in Nottinghamshire, Craigleith in Lothian, Haweswater in Cumbria and Rutland Water had reached Teesmouth by July or August of the same year, approximately one to two months after fledging.

Double-crested Cormorant
Phalacrocorax auritus

An extremely rare vagrant from North America: one record.

The county's sole record is of a first-winter at Charlton's Pond, Billingham, from at least 11 January to 26 April 1989 (*British Birds* 86: 453-454), and again on 16 June 1989, although the latter date was never published by BBRC. The bird was first noticed by Martin Blick on 11 January 1989 and initially mistaken for an unusual looking Shag *Phalacrocorax aristotelis*, although an immature 'Cormorant' had apparently been seen regularly on site since at least 8 December 1988, and these sightings may well have related to this bird. Several other local observers saw the bird during the ensuing days with opinions on its identity being divided between Cormorant *Phalacrocorax carbo* and Shag *Phalacrocorax aristotelis*. It was not until 30 January that the late Terry Williams suggested that the mystery Cormorant might be a vagrant from North America. Armed with this information, several observers saw the bird during the morning of 31 January, none of whom disagreed with the revised identification, however unlikely it seemed. By the late afternoon, the news was broadcast that the bird was a Double-crested Cormorant; a new bird for both County Durham and Britain (Blick 1989).

The bird obligingly settled into a routine for much of its protracted stay, though it became more erratic in its appearances as time wore on. On most mornings, it would fly in from the east an hour or so after daybreak and generally stay for most of the day, usually roosting on a small wooden raft at the eastern end of the pond. It usually departed to the south east during the late afternoon, presumably to roost somewhere in the Tees estuary, though it occasionally roosted overnight at the pond. During its long stay, it was also noted on nearby Haverton Hole on at least six occasions, and once on the River Tees between Middlesbrough and South Bank.

This was the first British and European record and it was estimated to have been seen by more than 1,500 observers during its stay (Blick 2009). There have been no further confirmed British records.

Distribution & movements

Double-crested Cormorant breeds across most of North America, both coastally and inland, though birds from the interior head to coastal areas during winter when it can also be found as far south as Central America and the West Indies. The species suffered a major decline between the end of the Second World War and the mid-1970s, mainly due to contamination from pesticides and persecution. Since the early 1970s, legislative protection has resulted in a strong recovery, so much so that it is now considered a pest in some areas (Coniff 1991). Despite the increase in North America, the Billingham individual remains the only British record, although a first-winter was in County Galway, Ireland, from November 1995 to January 1996, and there have been several records from the Azores.

Shag (European Shag)
Phalacrocorax aristotelis

A common coastal passage and winter visitor, and an occasional rare breeder.

Historical review

This diminutive cormorant is predominantly a winter visitor along the Durham coast and probably has been for many hundreds of years. Temperley (1951) called this species a "*resident on the north-east coast, though not breeding in County Durham*". He noted a marked increase in numbers during the winter, though he also commented that there were fewer Shags off the Durham coast than in neighbouring Northumberland. He said it may have been more common in the mid-1800s, as Sharp (1816) and Hogg (1845) both described it as "*common on our rocky coast*", although Hutchinson (1840), said it didn't breed in Durham and was "*but a wanderer on the Durham coast*" in autumn and winter. Blick (2009) noted that it was previously, in the 1880s, regarded as quite rare in the Redcar area, just to the south of Durham.

In the early part of the 20th century and in relation to Teesmouth, Richmond stated that the Shag was increasing and was occasionally noted in small numbers in winter (Richmond 1931). Blick (2009) affirmed this local status and commented that numbers had probably not changed very much through the remainder of the 20th century.

In the north east of the county, its status and numbers probably did not change very much until the last quarter of the 1900s, when there was increasing evidence of it spending more time locally during the spring and summer months. A summering pair at Marsden Rock in June 1960 was unusual, while evidence of large late autumn movements came in 1965 when northerly gales resulted in 102 flying south at Hartlepool on 21 November and 250 doing likewise the following day. By the mid-1970s, small numbers were noted joining Cormorants *Phalacrocorax carbo* roosting on Marsden Rock in late winter.

Inland records have always been exceptional, even more so when involving flocks. On 20 November 1965, following northerly gales, a party of seven birds dropped into Darlington power station. One of these had been ringed on the Farne Islands on 29 June of the same year. All seven were cared for and released at the coast the next day. In January 1969, an oiled bird was found at Charlton's Pond, Billingham, while three sightings in 1975 included birds at Grassholme Reservoir on 19 February (Dale 1975), on the Reclamation Pond on 30 April and at Tunstall Reservoir on 16th to 17 October.

Recent breeding status

Breeding was first recorded in Durham when Dr. John Coulson discovered a pair of adult Shags on a ledge on the east face of Marsden Rock on 7 June 1960. They had built a nest though this was at the time incomplete and they were not yet incubating. Unfortunately, he could not return at low tide so the outcome of the breeding attempt is unknown. The situation was repeated with the same observer in 1967, when he found two Shag nests, with birds incubating, on the south side of the pier at South Shields (alongside three Kittiwake *Rissa tridactyla* nests). They were all washed off by a heavy swell a few days later (John Coulson pers. comm.). From the early 1980s, the species was increasingly noted along the coast from May to August and in particular, around Marsden Bay in June and July. A genuine toehold of breeding activity was not established in the county until the early 1990s, but throughout its history as a breeding bird in Durham, the county's breeding population has never gone beyond a handful of pairs, probably between one and five at very most. The potential breeding distribution has always been limited by the restricted nature of suitable habitat.

The first modern breeding record occurred during the *Atlas* survey period, when a pair was seen mating and nest-building amongst Cormorants on Marsden Rock in early June 1989. Subsequent to this, birds were present in most summers, although in some years, no breeding attempts were reported. By 1993, up to six birds were noted in suitable areas at Marsden, with one of the adults being on a nest by 31 May. The first evidence of successful nesting eventually came in July 1994, when young were seen in a nest on Marsden Rock.

The species has not gone on to establish in any significant numbers. Throughout the 2000s, summer records were limited to sightings of one to three birds, mainly during seawatching activities at Whitburn. Although several have lingered around Lizard Point and Marsden Rock, they have been predominantly immature birds and there has been no indication of nesting attempts. The situation now seems to have moved back to an irregular summer presence, with no breeding activity being noted in some years.

Recent non-breeding status

This species is usually recorded in every month of the year along the Durham coast, but it tends to be rather scarce over the summer period. The vast majority of records of this species come from the expected coastal locations such as Marsden, Whitburn and Hartlepool, although it is much more widely dispersed along the Durham coastline than the reported observations might lead one to believe.

Birds are routinely present during the non-breeding season along the length of coastline north from Whitburn to the Tyne estuary, and frequently use the area between the Tyne piers in which to fish. The number of birds tends to peak in the early months of the year, tailing off through early spring as they move back to breeding colonies. Marsden Rock is the main roost site with birds arriving on the seaward side in the late afternoon, having first assembled in Marsden Bay. Wintering numbers gradually increased through the 1980s, with 80 roosting on 13 September 1990 followed by September peaks of 57 in 1991 and 50 in 1992. A modern-day high of 84 roosting on the Rock on 21 March 1993 was ironically followed by a crash in numbers through 1994 and 1995. The roost at this time attracted no more than 12 individuals and a similar decrease was noted at Flamborough Head, East Yorkshire, where the winter roost of over 1,000 birds fell by 75% in this period (Newsome & Willoughby 1995). The reasons for this decline remain unclear. Through the 2000s, numbers roosting at Marsden fluctuated and were usually at their peak during January, maximum counts varying between 18 (2009) and 78 (2005).

In the south of the county, wintering birds are usually present between Hartlepool and North Gare, but in much lower numbers than at Marsden. The largest winter gathering at Hartlepool Headland was 71 on 23 February 1977 (Blick 2009), but the 2000s produced a one-off peak of 45 in January 2004. Small congregations also occur at the mouth of the Tees, with 25 on the sea off Seaton Snook on 5 March 2007 being a notable such flock. Elsewhere along the Durham coast, Shags can be encountered at sites such as Seaham, Blackhall and Crimdon, but the numbers reported are never great. The county's overall wintering population is probably no higher than 100 birds in most years.

After a very limited summer presence in most years, the start of an extended autumn passage, probably comprising largely of birds of the year from the Farne Islands, usually commences from late August and continues until early November. Passage can be concentrated into short periods, usually coinciding with inclement weather and northerly winds to encourage birds away from breeding areas. Prior to the population crash in the mid-1990s, impressive counts included: 38 north and 178 south at Whitburn, on 20 October, with 155 on the sea there two days later; 188 north in a single flock at Marsden, on 3 September 1992; and, 122 at Whitburn on 4 November 1998. Following a general reduction in the number of sightings in the late 1990s, recent higher counts in the north of the county have included: 120 at Whitburn, on 9 November 2000; 170 birds flying south at Whitburn, in four hours on 23 November 2007; flocks of 90 and 40 moving south past South Shields Leas, on 7 September 2008; and, two flocks totalling approximately 150 birds flying south past Easington on 8 September 2008. Off Hartlepool, larger counts have included: 206, on 26 November 1977; 203, on 9 September 1989; and, 267 on 20 October 1991. The latter count was the highest day-count for the county (Blick 2009). Such movements however, particularly involving large single flocks, are sporadic, unpredictable and not annual.

Recent inland records have been few with birds normally shunning freshwater habitats in Durham. It is likely that the majority of such records relate to ill or injured birds (Blick 2009). Singles were seen at Saltholme Pools on 1 October 1989 (Blick 2009) and at Hetton Lyons CP on 29 November 2003. Also notable, although still in tidal zones, occasional birds have appeared on Seal Sands and on the River Tees as far upstream as the Tees Barrage.

Distribution & movements

The Shag has a restricted global distribution, being found only in the Western Palaearctic, where it breeds in North Atlantic coastal areas from Iceland in the north, south to Morocco, including the whole of the Norwegian coastline as far north as the Kola Peninsula. It also has a restricted and discontinuous breeding distribution along the shores of the Mediterranean and Black Seas. The British population is estimated to be around 26,600 pairs (Mitchell *et al.* 2004), of which up to 1,500 pairs nest on the Farne Islands.

It has been speculated that birds wintering off the Durham coast comprise mainly birds from the Farne Islands, or perhaps further north on the Scottish east coast. The limited ringing date for the county would seem to confirm this, at least to some degree. Birds ringed as nestlings on the Farne Islands have been recovered at various points along the Durham coast, including Easington, South Shields, Sunderland, Whitburn, as well as at Teesmouth. A high proportion of the birds discovered as tideline corpses in the south east of the county had been ringed as pulli on Bass Rock, the Farne Islands or the Isle of May. A juvenile at Hartlepool in late November 2008 had been ringed with a darvic ring 'XPI' on the Isle of May, Fife, on 5 July 2008. Regular sightings of birds with darvic rings on seawatches at Whitburn may also relate to birds making such east coast movements. Recoveries of nestlings ringed on the west coasts of Britain also demonstrate that there is interchange of immatures between the west and east coasts. One ringed on Bardsey Island, North Wales, on 13 June 1957 was found dead at Marsden on 3 September of the same year, and another ringed at Great Saltee, County Wexford, on 31 May 1973, was found dead near Hart on 5 April 1974.

Bittern (Eurasian Bittern)
Botaurus stellaris

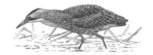

A rare migrant from Europe, with approximately 70 birds since 1970, recorded in all months, most frequently from November to February; it bred until the early part of the 19th century but not since.

Historical Review

The earliest specific reference to the Bittern in County Durham is by Tunstall in the late 18th century who wrote that *"there were many in the neighbourhood of Wycliffe-on-Tees"* (1784), though at the time of his writing this location, just over the River Tees, was part of North Yorkshire. The species was probably resident, breeding mainly in marshy areas in the south east of the county, as stated by Selby who wrote that it was *"rarely met with in the northern counties, although before the drainage of our bogs and mosses it used to be common and well known"* (Selby 1831). Records documented by Wallis (Rossiter 1999) in neighbouring Northumberland indicated that Bitterns were once much more common there and without a doubt, were breeding in the north east of England into the early 19th century, as stated by Holloway (1996). At Wallis' time, it seems likely that birds would have been breeding in similar areas of Durham, such as Bradbury, Preston and Mordon Carrs. In like fashion, it has been suggested that prior to about 1800, Bittern would probably have bred in the then still relatively untouched and extensive areas of marsh around Teesmouth (Blick 2009). The national trend towards drainage and the general destruction of wetlands during the 19th century was almost certainly, in part, responsible for the complete extinction of this bird as a local breeding species as well as for the drastic decline in the entire British population (Holloway 1996).

Writing in 1834, Edward Backhouse mentioned the species as being *"rare"*, though he added that *"several specimens have however lately been killed at Cowpen Marsh and Sunderland"*, with Hutchinson (1840) stating that *"the appearance of the Bittern is very uncertain and it has no permanent residence here. In some seasons generally in autumn, a straggler or two is taken"*. Other 19th century references to the occurrence of the Bittern in County Durham come from William Backhouse (1854) writing in his *List of Birds of Darlington*, that the species *"formerly inhabited the Four Riggs Bog"*, and from Hancock (1874) who wrote *"There are four or five entries in my journal of the capture of the Bittern in Northumberland and about the same number in Durham; and many others have occurred"*.

The available anecdotal evidence points to the Bittern having been a former resident in County Durham, though there are no proven records of the species having bred. One can only assume that breeding took place occasionally until at least the early part of the 19th century, but not since.

At the beginning of the 20th century, Tristram (1905) was able to give a reflection of the species' status at that time, writing; *"The Bittern was a resident in some marshy districts within living memory. It is now only an irregular winter visitor, but always late, generally in the month of February. An aged fowler told me some 40 years ago that in his youth a pair had always bred in Cowpen marshes, near Stockton. One was shot there in 1901 and several have been taken near the Tees."*

In the Derwent valley, there was but a single record of this species and, as is the case with so many old records, it was of a bird that had been shot. A female was killed on Derwenthaugh, close to the confluence of the River Derwent with the River Tyne, on 13 January 1947 (Bowey *et al.* 1993). On examination, its stomach was found to contain *"four frogs, various water beetles... and portions of sedge"*. Other records of the species from the middle of the 20th century include birds at Billingham Bottoms in January 1947 and Charlton's Pond in 1949. During the 1950s, when the species was at its national nadir (Brown & Grice 2005), the *Ornithological Reports for Northumberland and Durham* failed to show even one or two birds a year, other than in 1956 when a pair bred successfully at Gosforth Park Lake, Northumberland (Kerr 2001).

The status of the Bittern in County Durham changed little during the early part of the 20th century with Temperley (1951), stating that *"hardly a year passes without one or two individuals being shot in both counties"*. This confirms the species status as a scarce winter visitor to our region, though there were just a handful of published records during the 1950s. One was at Crookfoot Reservoir in September 1951, with moribund birds at Sunderland on 4 January 1957 and West Stanley on 14 January 1958. Perhaps surprisingly the species was not recorded at all during the 1960s despite still occurring regularly in Northumberland (Kerr 2001).

Recent Status

The 1970s started better than the previous decade for records but not for individual birds. One was found at Billingham Bottoms on 24th to 31 January 1970 but on 4 February, it was found shot. Unfortunately, this seemed to be an early 1970s pattern as one found *"long dead"* at Shotton Brickworks Ponds on 22 March 1970 had also been shot. These were the first Durham records since 1958.

There have been an estimated 75 Bitterns recorded in County Durham since 1970, though the exact number is difficult to assess due to the elusive nature of this species. The 1970s saw a total of eight birds, with a slight increase to ten in the 1980s, though since then records have been virtually annual. Although it has been recorded in all months, there is a distinct peak in records in January, usually associated with prolonged periods of cold weather. The best year on record was 2010 with between nine and eleven birds, five of which were between January and March, one was in June and July with the remainder from November onwards.

 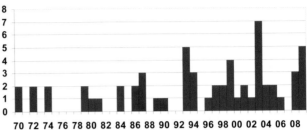

Records of Bitterns in County Durham 1970-2009

The North Tees Marshes including Haverton Hole became established as the prime location for this species in Durham from the mid-1980s onwards, particularly during the winter months, though there were seemingly no records on the marshes from 1959 until one was on Dorman's Pool on 13 April 1980 with a further four records during the 1980s. The next birds were in 1993 when probably three different birds were recorded. The species has remained an almost annual visitor in this part of the county since. The species has been recorded less than ten times between May and August with records from:

- Long Drag - from 13 August to 25 September 1986
- Dorman's Pool – 26th to 27 July 1984; on 25 July and 20 August 1999; 18 August to 4 November 2002; 18 July to 12 October 2003 and 25 May 2008
- Cowpen Marsh - on 12 May 2008 and 29 August 2009
- Haverton Hole - on 1 May 1984 and from 30 June to 6 July 2010

It is possible that the same individuals were involved in some of these sightings, such as in 2002 and 2003 (Joynt *et al.* 2008). There has been some speculation that birds may breed in the south east of the county soon (Joynt *et al.* 2008) but despite the increasingly regular occurrence of this species, breeding has yet to be proved. Most birds have been recorded from the extensive reedbeds at Haverton Hole and Dorman's Pool, though individuals on Seaton Common on 21 March 1981 and Greenabella Marsh on 28 December 1998 are noteworthy.

Bitterns have been recorded away from the North Tees Marshes less regularly, with over half of these being from the well-watched areas around Houghton-le-Spring and Washington, though there have also been five records from Low Barns, the last of which was in 2008. There have also been at least five records from the centre of the county within five miles of Durham City, with perhaps the most notable being at Brasside Pond from 13th to 17 August 1974.

Records from elsewhere are much scarcer with only five sites recording sightings on more than one occasion; Hetton Lyons CP (1994 and 2010), Low Barns (1998, 2003, 2004, 2008 and 2011), Rainton Meadows (1994, 2004 and 2010), Shotton (1970 and 1993) and WWT Washington (1987, 1990 and 2010). Other records of Bittern since 1970 include: one in the central reservation of the A1(M) close to Ferryhill on 6 October 1974; one at Shibdon Pond on 28 January 1984 intermittently until at least 14 March (Bowey *et al.* 1993); one found dead in the water by Sunderland North Dock on 26 April 1989; at Barningham in late November 1997; at Drinkfield Marsh, Darlington on 1st and 2 February 2006; flying west along the River Wear at Castletown, Sunderland on 22 March 2010 and at West Boldon on 22nd and 23 December 2010. The bird at West Boldon in December 2010 was found in an emaciated state at Mount Pleasant Marsh and taken into care before being released at East Chevington,

Northumberland on 2 January 2011. One at Castle Lake, Bishop Middleham from 25 August to 30 September 2011 was notable in being the first record for the Durham Bird Club reserve.

Distribution & movements

The Bittern breeds in suitable habitat across much of Western Europe and Asia, with an isolated population in southern Africa. Southern birds are largely sedentary, though those in the north of the species range are subject to periodical movements in response to harsh weather, seeking out areas of unfrozen water. Historically the Bittern was once a common species across much of the British Isles, though numbers declined markedly at the beginning of the 20th century and its British breeding range is now largely restricted to East Anglia, Lancashire, and Kent. The number of Bitterns wintering in Britain fluctuates each year according to the severity of the winter weather on the near Continent. This pattern fits nicely with the obvious January peak in records in County Durham.

Little Bittern
Ixobrychus minutus

A very rare vagrant from southern Europe: four records.

A bird quoted to have been shot at Blaydon on 12 May 1810, is in fact erroneous and actually refers to a bird shot at Blagdon, Northumberland on the same date (Bolam 1912). The specimen is held in the Hancock Museum, Newcastle-upon-Tyne. The first county record therefore falls to a bird reported to have been 'taken' at Stanhope on an unspecified date in 1869, with the specimen becoming part of the Rev. H.H. Slater's collection, though its whereabouts are now unknown (Tristram 1905).

All records:

1869	Stanhope, juvenile, 'taken'
1889	Gateshead, juvenile, found dead under telegraph wires, November; specimen in the Hancock Museum (Howse 1899)
1973	Hurworth Burn Reservoir, male, 1 June (*British Birds* 67: 315)
1976	Low Barns, Witton-le-Wear, male, 23rd to 24 May (*British Birds* 70: 415)

Another bird was said to have been caught near Sheriff Hill, Gateshead around November 1889, although no specimen was presented at the time (Temperley 1951). Consequently, the record remains not fully substantiated although the coincidence of timing with the other Gateshead record lends it a degree of credibility.

Distribution & movements

Little Bittern breeds patchily in wetlands across Europe, western Asia, tropical Africa, and parts of the Indian subcontinent. The nominate European race is strongly migratory and winters almost exclusively in sub-Saharan Africa. Breeding has been confirmed in Britain on two occasions, in Yorkshire in 1984 (Allport & Carroll 1989) and Somerset in 2010, though it has been suspected of occurring more regularly in East Anglia during the 19th century (Brown & Grice 2005). Little Bittern is predominantly recorded as a spring overshoot to the British Isles, although it has been recorded in all months, with a total of almost 500 records by the end of 2010. The bird at Gateshead in 1889 is unusual not only by virtue of it being an immature, but it is also one of only four British records for November, all of which occurred before 1910. Both of the 20th century records are more typical and conform to the national pattern, relating to male birds having overshot their southern breeding grounds on spring migration.

Night-heron (Black-crowned Night Heron)
Nycticorax nycticorax

A very rare vagrant from Europe: seven records.

The first record for County Durham relates to an immature found by Tim Bennett at what is now know as Low Barnes NR, Witton-le-Wear on 1 September 1977 (*British Birds* 71: 489). This site also hosted the county's seventh record almost thirty year later.

All records:

1977	Low Barns, Witton-le-Wear, juvenile, 1 September
1979	Brasside Pond, Durham, adult, 2 July (*British Birds* 73: 494)
1982	WWT Washington, juvenile, 28 September (*British Birds* 76: 479)
1988	West Boldon, adult, 24 April, found injured; died in care (*British Birds* 83: 443)
1992	Lumley Castle, Chester-le-Street, first-winter, 13 November (*British Birds* 87: 511)
2005	Cowpen Marsh and adjacent brine fields, second summer, 4th to 8 July
2007	Low Barns, Witton-le-Wear, adult, 4 February

Thomas Robson in his 'Birds of the Derwent Valley' (1896) makes reference to a bird said to have been shot close to Dunston, Tyne and Wear some forty years earlier. The specimen was said to have been held in the collection of a Mr J. H. Hedworth, having being obtained by his grandfather. However the evidence surrounding this record is at best anecdotal and it was not accepted by George Temperley in his review of the county avifauna, as he was unable to trace the original specimen 'due to the distance of time' (Temperley 1951).

An adult found roosting in trees by the River Tees at Low Coniscliffe from 23rd to 24 April 2007 bore blue and pink colour rings, and was presumed to be an escape. Different escaped birds, again bearing colour rings were at Middlesbrough and Northallerton in nearby North Yorkshire during the same year, though their origins were never traced.

With just one record in April, records in County Durham do not strictly conform to the national pattern, though July records are not unusual. It is interesting to note that the September and November records all relate to juveniles which have presumably fledged on the near Continent. The individual at Low Barns in February 2007 is of interest given the number of colour-ringed individuals sighted later that year, but this bird was not seen to have any rings. Winter records in the British Isles are not unprecedented, and include an immature in Lincolnshire in December 1979, that had been ringed as a nestling at Belyayevka on the Black Sea coast of the Soviet Union on 8 June (*British Birds* 73: 494).

Night-heron, Low Barns, March 2007 (Chris Bell)

Distribution & movements

Night Heron has an extensive range and can be found breeding on all continents except Antarctica and Australasia. It has a patchy distribution in Europe, predominantly in the south and east, though breeding regularly occurs as far north as Belgium and northern France. The majority of European birds winter in sub-Saharan Africa though small numbers remain in the vicinity of the breeding grounds throughout the winter, particularly at more southerly latitudes. It is a regular vagrant to the British Isles, most frequently as a spring overshoot from March onwards, although it has been recorded in all months. It was removed from the list of species considered by the BBRC at the end of 2001 by which time there had been over 600 records.

Cattle Egret (Western Cattle Egret)
Bubulcus ibis

A very rare vagrant from Europe: three records.

The first for County Durham was found by S. Thexton in fields adjacent to Low Barns, Witton-le-Wear on 15 April 1979. This bird spent most of its time feeding amongst the local sheep, roosting nightly in the nature reserve itself, choosing the relative security of the island in the centre of the main lake. It remained until 17 April when it was seen to fly high to the north west (*British Birds* 73: 494). A bird found feeding amongst cattle at Loch of Kinnordy, Angus, from 10th until 19 May later that year may well have been the same bird.

The second was found on the evening of 10 October 1986 at the somewhat unlikely location of Longnewton Reservoir. The bird was present until dark roosting amongst gulls on the narrow causeway between the two reservoirs, though was not present at dawn the following morning (*British Birds* 81: 542).

It was over 20 years until the next occurrence, when an adult was found at Saltholme Pools on the morning of 15 November 2008. This bird settled in fields close to Port Clarence where it fed amongst horses for much of the remainder of the day. It was seen to depart high to the north east at around 3pm and was not seen subsequently (*British Birds* 102: 543).

The County Durham records are notable for being so far north, with just a handful of occurrences from Scotland, and a single record for nearby Northumberland in September 1986 (Kerr 2001). Both the 1986 and 2008 birds were part of significant influxes into Britain.

Distribution & movements

Cattle Egret was originally found in southern Europe, Africa and Asia, though it now has an almost cosmopolitan distribution having colonised both of the Americas and Australasia since the 1930s. The European population continues to expand with breeding now occurring regularly in Italy and as far north as Brittany in northern France. It is an increasingly regular vagrant to the British Isles, particularly since the beginning of the 21st century, with almost 450 records by the end of 2008 when it was no longer considered by BBRC. More than half of these have occurred since 2007 including a record influx of around 150 birds during the winter of 2007/2008, which culminated in the UK's first breeding record, on the Somerset Levels in 2008.

Little Egret
Egretta garzetta

A former rare vagrant; it has become an increasingly frequent visitor since the beginning of the 21st century. Present year-round, its numbers peak in August following post-breeding dispersal from further south in the UK.

Recent status

The first Little Egret for County Durham was found by Andy Clements at Dorman's Pool, Teesmouth and was present on 3rd and 4 July 1979 (*British Birds* 73: 495). Still a British rarity at that time, this was one of just 14 British records in 1979 and the furthest north of that year. The second record came from the same location, and was present on 7th and 8 June 1983 (*British Birds* 77: 510); one of 21 British records in 1983, it was England's most northerly of the year, although a single bird reached Grampian in June.

It was almost ten years before another appeared in Durham, when a single bird was present by Greatham Creek, Teesmouth intermittently from 3rd until 7 May 1992. By this time, the species was no longer considered by the BBRC. There were a further six records from the North Tees Marshes during the 1990s: 12th to 14 May and 5

June 1994; 5th and 26 May 1995 (two on the latter date being the first multiple record); 8th to 12 June 1996 and on 10 May 1998. The increase in occurrences during the 1990s also brought the first records away from the North Tees Marshes, with single birds at: WWT Washington on 26 May 1995; on a small stream near Haswell from 24th until 27 April 1996 and unusually, on coastal rocks at Hartlepool Headland on 12 June 1999. This last bird was continually harassed by gulls until it flew off inland. In contrast to the previous two decades which had recorded but a single bird each, a total of 11 birds reached County Durham during the 1990s and the species has occurred annually ever since.

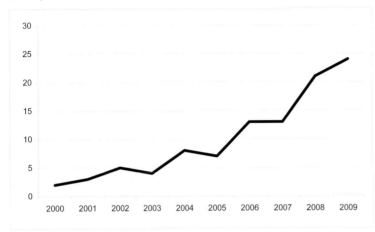

Peak Counts of Little Egrets on the North Tees Marshes, 2000–2009

The early part of the 21st century saw a dramatic change in the species' status with the annual total of Little Egrets recorded in the county reaching double-figures for the first time in 2002. The number of birds has continued to increase year on year and by the mid-point of the decade it had become very difficult to accurately estimate the true number of birds involved each year.

The vast majority of records continue to come from the North Tees Marshes, where the species can now be found year round. Although initially recorded only as a spring overshoot, the first autumn records came in 2000 with one to two birds present from August through to October, signalling the beginning of a shift in the occurrence pattern to a late summer peak. The winter of 2003/2004 brought the first wintering individual in the county, with a single bird near Dalton Piercy from late December, before it relocated to the North Tees Marshes. Although still not common, increasing numbers now spend the winter months within the county, principally around the Tees Estuary.

Coinciding with the sharp increase in numbers in the Tees Estuary, a roost was established close to Greatham Creek in August 2004. Initially, this consisted of up to seven birds but the August maximum had increased to 27 by 2010, with a record 36 on 6 September 2010. This level was maintained in 2011 with an autumn peak of 34 on 29 August. This roost is largely occupied in late summer, usually reaching a peak in mid- to late August, before numbers decrease rapidly with the onset of winter. A single bird was found roosting in Rossmere Park, Hartlepool during the winter of 2005/2006 and despite the urban location, this has become the only established winter roost site in the county, with peak counts of eight on 7 December 2009 and nine on 26 December 2011. Numbers present often exceed the number thought to be present on the North Tees Marshes during the winter, and this indicates that there are more birds wintering in the surrounding countryside. These birds no doubt utilise the many dykes and streams in the area.

Records away from the North Tees Marshes remain rather scarce, though increasingly regular, particularly so at Bishop Middleham where the species has been annual in small numbers since 2007. Further north there have been nine records for the Washington/Barmston area, and seven records for South Tyneside, with the first being at Sunderland AFC Pools in May 2004. Hetton Lyons Country Park, close to Houghton-le-Spring, has had four records, all since 2005, and Gateshead had its third record in August 2010 when up to five were at Shibdon Pond. The species is extremely rare in the west of the county with just two records of singles at Low Barns, Witton-le-Wear from 14 August until 3 September 2001 and at Thorpe Farm wetlands, close to Greta Bridge, on 31 August 2008.

The marked increase in the number of Little Egrets in County Durham since the 1990s directly correlates with the northward spread of the species in Britain; it seems likely that breeding will occur in Durham relatively soon. Little Egret nests colonially in trees, often amongst Grey Heron *Ardea cinerera*, but birds can be surprisingly difficult to detect during the breeding season. The established heronries at Wynyard and WWT Washington are likely locations for first breeding in the county.

Distribution & movements

Little Egrets are distributed widely across temperate and tropical latitudes throughout much of Europe, Africa, southern Asia and Australia. The species was heavily persecuted, particularly during the 19th century, as its decorative plumes were highly prized in the Victorian millinery trade, resulting in its European range becoming restricted to southern Europe. Conservation measures led to a reversal in fortunes to the extent that by the end of the 1960s it had spread north to Brittany, with an estimated 400 pairs breeding in north west France by 1993 (Lock & Cook 1998). The subsequent increase in the French breeding population is thought to be responsible for the dramatic increase in Little Egret sightings in Britain over the last 40 years; which brought the first breeding at Brownsea Island, Dorset, in 1996. By 2009 there was an estimated 820 pairs breeding in Britain at over 70 colonies as far north as Yorkshire. The most recent estimate of the UK winter population gave a post-breeding total of over 4,500 birds, indicating that the actual breeding population may be even higher than the above figure suggests (Holling 2011).

Several birds with colour-rings have been recorded in County Durham, all of which have been ringed as nestlings in the UK. A bird ringed in Gwent in June 2006 was present around the North Tees Marshes in August and September of the same year and was later seen at Spurn Point, Humberside in March 2007. One ringed in Norfolk in June 2008 was on the North Tees Marshes from August 2008 to at least January 2009, with another ringed in Norfolk in May 2009 being noted at Castle Lake, Bishop Middleham in July 2009, before it relocated to the North Tees Marshes in August. Another bird ringed in Norfolk in May 2010 was at Teesmouth in late August 2010, having been at Coatham Marsh, North Yorkshire in July and at Shibdon Pond earlier in the month. All of these records give a clear indication that the August peak of Little Egrets within the county corresponds directly with post-breeding dispersal from the recently established colonies in the UK. It is interesting to note that three of the birds seen in Durham had been ringed in the same colony at Terrington St. Clement, Norfolk. Birds from this colony have also been seen in France, the Netherlands and southern Spain (Clark *et al.* 2011).

Great White Egret (Great Egret)
Ardea alba

A rare vagrant from Europe: 18 records, from October to July, most often from April to June.

The first Great White Egret for Durham was found by Dave Moore, along the Hutton Beck, close to Hutton Magna on 16 May 1993, and was observed briefly in flight the following day. This bird was later relocated at Aldbrough St. John, North Yorkshire, some 12km to the east along the same beck, where it was present from 19th to 22 May. Given the number of subsequent records from the North Tees Marshes, it seems remarkable that the first bird was found inland (*British Birds* 87: 511-512).

All records:

1993	Hutton Magna, 16th to 17 May
1998	Reclamation Pond, Saltholme Pools, Haverton Hole, and Dorman's Pool, Teesmouth, 17th to 18 April (*British Birds* 92: 560)
1999	Cowpen Marsh and Greatham Creek, Teesmouth, 16th to 17 June (*British Birds* 93: 519)
2001	Saltholme Pools, Dorman's Pool, & Greatham Creek, Teesmouth, 24 June to 12 October (*BB*: 95: 483)
2001	Neasham, end of November, found dead (*British Birds* 95: 483)
2002	Portrack Marsh, Stockton-on-Tees, 11th to 12 October; same North Tees Marshes, 12 October to 9 November (*British Birds* 96: 551)
2002	Dorman's Pool, Teesmouth, 2 December (*British Birds* 96: 551)
2003	Saltholme Pools, Teesmouth, 11 July (*British Birds* 97: 570)
2005	Haverton Hole, and adjacent North Tees Marshes, Teesmouth, 24 April to 28 May (*Contra BB*), also Portrack Marsh, Stockton-on-Tees, 24 April (*British Birds* 100: 28)

2007	Saltholme Pools, Teesmouth, 6 April
2007	Greatham Creek and Cowpen Marsh, Teesmouth, 17 June
2007	Saltholme Pools and Dorman's Pool, Teesmouth, 27th to 28 June
2009	Portrack Marsh, Stockton-on-Tees, 9 May; same RSPB Saltholme, Teesmouth 9th to 12 May; same Seaton Common, Teesmouth, 18 May, and RSPB Saltholme, Teesmouth, 18th to 28 May
2009	Dorman's Pool, Teesmouth, 8 June
2009	Greenabella Marsh, Teesmouth, 4 November
2009	Reclamation Pond, Teesmouth, 26 November; same Long Drag, Teesmouth, 28 November
2010	Dorman's Pool, Teesmouth, 27th to 28 March
2011	Long Drag, Teesmouth, 17 September

The increase in records since the late 1990s corresponds with the national trend, and the species is now regarded as an annual vagrant to the North Tees Marshes, being recorded in eight of the eleven years, between 2001 and 2011.

Distribution & movements

Great White Egret, Dorman's Pool, June 2007 (Ian Forrest)

The Great White Egret has a cosmopolitan breeding distribution, throughout most temperate and tropical regions of the world. The nominate race breeds patchily across much of southern, central and Eastern Europe, and central Asia, although it is absent from the Iberian Peninsula. The taxonomy of this species is complex and evolving. It was formerly a very rare vagrant to Britain, the number of records per year has risen dramatically since the mid-1990s, correlated with the recent spread of the species into the Netherlands and northern France. It was removed from the list of species considered by the BBRC at the end of 2005, by which time there had been over 300 records.

Two different colour-ringed birds were noted on the North Tees Marshes in 2009, confirming that four different birds were involved in what was to be a record year for the species locally. The first of these was present intermittently from 9th until 28 May, had been ringed as a pullus at Lac de Grand Lieu, Loire Atlantique, France on 5 May 2007. This location is the origin of several other colour-ringed birds recorded in Britain.

Grey Heron
Ardea cinerea

A common and widespread resident and winter visitor, that breeds more sparsely than indicated by its widespread distribution.

Historical review

This is the only member of the heron family that routinely breeds throughout Britain. It can be found along many streams, rivers and wetlands across the county throughout the year. It is a resident species in Durham but additional birds are regularly noted as passage migrants with a small influx of birds in winter.

This species has clearly been present in Durham for many hundreds of years and was documented in the 14th century Cellarers' rolls of Durham Monastery, which record monetary rewards being paid in 1349/1350, (*Cuidam deferenti Heronceaus Priori, de Acley usque Dunolm., vj d, "for the bringing of a particular Heron to the Prior from*

Ackley to Durham, 6 pence") and again in 1381/1382 (Fowler 1898). Herons do not appear in the documentation as being bought for food until 1404 (Ticehurst 1923).

The British Birds' Census of heronries in 1928 (Nicholson 1929) recorded six occupied nests at a site at Gainford 28 May. Temperley (1951) noted that this particular heronry had been founded *"immemorial"* at its original site of Gainford Great Wood, as evidenced by the egg in the Hancock collection, which had been taken there *"around 1823"*. He also noted that this breeding site had moved and divided during the period 1914-1918. This census also noted that there had previously been a heronry in Ravensworth and Temperley (1951) expanded upon this, commenting that this site was largely deserted by about 1850 or 1855, although occasional pairs nested there until the mid-1890s, as observed by George Temperley himself; though the nests were frequently robbed. Hutchinson (1840) told of a small heronry at the Sands Hall, near Sedgefield, which was deserted even by his time of writing, due to the trees having been felled. He also made reference to birds nesting at Wilson's Gill, near Stockley Heugh. Tristram (1905) reported a heronry in the Park at Raby Castle, and alluded to birds nesting near Sedgefield and at Gainford.

Temperley commented at length upon the apparent disparity between the number of birds seen around the county, from lowlands to upland streams, and the limited number of heronries. He speculated that the majority of birds nested singly or in small groups and observed that Abel Chapmen (1889) had drawn attention to this habit. Hancock (1874) had indeed recorded single nesting pairs at Cleadon House in the north east of the county in 1814 and at West House with another noted near Cocken on the Wear, downstream of Durham, around 1830 (Temperley 1951). In the Derwent valley, Robson recorded birds nesting in the Gibside Estate and higher up the Derwent valley at the Hagg, near Hamsterley (Robson 1896). Into the 20th century, even higher up the Derwent, Temperley noted pairs nesting at the Sneep near Muggleswick up to 1943 and in 1946 and elsewhere, four pairs in a wood near Eastgate and a pair at Whitworth Park, near Spennymoor in 1948.

At the mid-point of the 20th century, Temperley (1951) summarised its status as *"a resident, generally distributed throughout the County and at all times of the year"* but *"only breeding in small numbers"*. He noted that it was *"Frequently met with in the Tees valley and in greater numbers around Teesmouth, equally common in the Wear, especially in its upper reaches"*. He especially focused on the scarcity of heronries found in Durham and Yorkshire during a national census, with the two counties having a breeding density of less than five breeding pairs per 100,000 acres. He was aware that autumn migrants bolstered the county's winter numbers.

The breeding of herons near Gainford, in the mid-Tees valley, is worth documenting as the data from in this area was collected over a long period. Birds were nesting there in the early decades of the 19th century although the heronry was described as 'small' in 1851. There were 19 nests before 1912 but apparently only two in 1920 and just one in 1921/1922. Subsequently, birds were found nesting at a new site nearby (Dyance Wood), with one pair there in 1922 rising to seven or eight pairs by 1926. Birds were still divided between the two sites in 1928 with five at one and six at the other (Temperley 1951). In 1934 there were was a total of 15 nests, with 16 found in 1944, declining to single figures by the late 1940s. During the 1950s and early 1960s, Stead (1964) said that the herons regularly visited Teesmouth but that the nearest heronry was over 20 miles away at Dyance, near Gainford (Joynt *et al.* 2008). The colony went through a small expansion in the 1950s, with certainly six successful nests in 1952 and a maximum for the decade of an estimated 20 nests, in 1956. In that year, on 27 April three adults and up to nine young were found shot beneath the trees; a not unusual fate for local birds. Such human interference may have been a limiting factor through the 1950s with some evidence of persecution here sadly continuing into the 1980s. Sample counts showed that there were at least 12 nests in 1961, 21 nests in 1973, 15 in 1983 and finally 13 in 1985. Scots pine *Pinus sylvestris* and especially larch trees *Larix spp.* were favoured for nesting. The plantation was lost to felling in late 1985 and pairs subsequently established two smaller colonies of three to five nests in nearby woods.

In 1967, Parslow noted that across England this species had experienced *"A fairly general decrease"* and that the two known Durham heronries were in a *"precarious state"* (Parslow 1967). Since Temperley's summary, a number of other heronries were known to have been discovered in the county. For instance, in 1952, one was found in Weardale near Harperley containing three occupied nests, with a similar small number of nests persisting there up to the 1970s. In 1973 and 1974, birds also bred or attempted to breed at three other small heronries in Weardale, sometimes failing due to human disturbance. In 1967, there was said to be six nests at another unnamed site 'further north' (Bell 1968) and through the late 1960s and early 1970s large numbers of birds were noted at Derwent Reservoir, autumn counts of over 20 birds not being atypical. This may have been related to

251

undiscovered breeding sites in that area; the overall sense being that very small, isolated and unknown heronies were probably the mainstay of the species in the county at that time.

Recent breeding status

The Grey Heron has never been a common breeding bird in Durham and this was evident from the *Atlas* fieldwork. The mapped distribution was probably more indicative of the presence of birds during the breeding season than the widespread presence of breeding colonies (Westerberg & Bowey 2000). Despite this, the actual distribution of heronries will have been under-recorded during the *Atlas* survey, as determining the exact location of single pairs, or even small numbers of breeding pairs, especially when nesting in conifers, can prove difficult (K. Bowey pers. obs.). In terms of nesting, the species tends to avoid the county's western uplands, though there are some exceptions to this. It is also less-frequently noted fishing in such areas, as fish populations tend to be lower. Some of Durham's upland reservoirs may be an exception to this rule, as many of these are actively stocked and managed as fisheries, although the absence of marginal shallows in these can leave fish out of reach of feeding Grey Herons. There was at least one small heronry close to an upland reservoir in the early 1990s and birds nested in at least nine different upland locations during the *Atlas* survey (Westerberg & Bowey 2000).

The county's largest heronry at WWT Washington has numbered double-figures of nesting pairs for over two decades. It grew from two pairs in 1988, to six in 1990 and 13 pairs by 1993. Through the 1990s, birds moved to nest in trees from their original nesting locations on the ground amongst brambles *Rubus fruticosa*. By 1999 over 20 pairs were present fledging more than 60 young, with up to 106 adults and juveniles present in August of that year. This is undoubtedly the largest colony in the area. Breeding numbers peaked in 2006, when 43 pairs nested, declining to 28 successful pairs in 2008, increasing to 37 pairs in 2009 and 2010.

During the 2000s, other than the largest at WWT Washington, at least 13 active heronries were documented in Durham, ranging in size from just one to about 30 pairs. If all heronries combined contained the maximum number of nests recorded during the decade in one year, this would give a notional county population of around 135 pairs. Active heronries were noted at: Axwell Park, Beechburn Gravel Pits, Brancepeth Beck, Croxdale & Low Butterby, the Gibside Estate, Hardwick Hall, Harperley Banks, Hurworth Burn Reservoir, Piercebridge, Sedgefield, Witton Park and the Wynyard Estate. From the mid- to late 1990s, at least one other large colony became established, at Axwell Park in the lower Derwent valley. By 2008 this had over 20 pairs nesting. Another important site was at Croxdale/Low Butterby where 30 active nests were reported in late March 2005. Colony size fluctuates from year to year and through time. For instance, birds were found breeding at Wynyard from 1978 (possibly since 1974), increasing to a maximum of 23 nests in the early part of the 2000s (Blick 2009) but numbers have declined considerably since 2006.

The Grey Heron breeds early in Durham, as it does elsewhere. This probably maximises the period between the fledging of young and their first experience of cold winter weather (North 1979), with the need to cope with the dietary impact associated with freezing temperatures. At WWT Washington, adults are present throughout the year, with display and nest building occurring from late January. Females are usually incubating by February and the first young are often hatched in late March. Occasional second clutches were noted at this site from 1995. Seasoned observers of this colony, suggest that breeding success is related to the spring weather patterns and food availability (Westerberg & Bowey 2000). Colour-ringing, of young here has shown that some young birds return to breed in their natal colony and that this usually occurs at two years of age.

From a national perspective, systematic heronry counts since 1928 have provided abundant data on this species and the national population trend has been strongly upward, though not continuous, through time (Parkin & Knox 2010). Durham has only a relatively small breeding population. During the *Atlas* survey period, the county's 'known' population was estimated at 40-50 pairs (Westerberg & Bowey 2000), but the number of birds found breeding since indicates that this was an underestimate. Today, there may be as many as 100-120 pairs scattered between 15 or so heronries with a few isolated pairs. This is a fraction of the estimated 14,000 pairs believed to be present in Britain in 2003 (Parkin & Knox 2010). This species still faces at best, disgruntlement and at worst, direct persecution from fishermen and it has been well documented that this is sometimes enacted at local nest sites (C. Jewitt pers. comm.). The scale of this is not known, but shot birds are occasionally found.

Recent non-breeding status

This wide-ranging and obvious species is well distributed across Durham; its most favoured habitats are reed-fringed ditches, rivers, ponds and marshes mostly in lowland locations along the county's main river valleys and their tributaries. Birds are also reported from the upland reservoirs and the upper stretches of the main river systems. Post-breeding congregations and dispersal leads to reports from more widespread wetland locations through the post-breeding phase of the year (birds were reported from at least 40 sites in July and August 2009). Typical late-summer and autumn sightings are of less than six birds but congregations at favoured wetlands can be larger. For example, the most frequented locality in Gateshead, Shibdon Pond has had counts of up to 30 in the late summer and other sites that regularly attract double-figure numbers include Low Barns, Hurworth Burn Reservoir, the North Tees Marshes and WWT Washington. Peak counts usually occur in July and August, and again in late winter, usually around February. The marshes around Teesmouth and Washington regularly attract more than 20 birds at individual sites and over 50 in the extended complexes of wetlands. Peak counts at Teesmouth include 71 in July 1993, 78 in July 1999, 83 in August 2000 and 75 in August 2009. Early in the year, counts of small numbers of birds can come from many wetland sites ranging from Shibdon Pond and Whitburn in the north to Brignall Banks, Whorlton and Aislaby in the south. Birds have been reported visiting gardens since the mid-1990s; garden ponds are often visited in search of fish or spawning amphibians in spring.

Locally, a wide variety of prey items have been recorded, including small mammals, Mallard ducklings *Anas platyrynchos*, Little Grebe chicks *Tachybaptus ruficollis* - it appears most small birds and animals are fair game. More unusually, a well-grown Coot *Fulica atra* was taken at Bear Park in 1994 (the Heron was noted as being 'incapable' of flight afterwards) and, in January 1987, one attempted to catch a Green Sandpiper *Tringa ochropus* near High Coniscliffe. A Starling *Sturnus vulgaris* was caught in flight at Shibdon Pond (Bowey 1997). On a number of occasions, birds have been noted hunting at night on grasslands by the lights of the A1 in western Gateshead, perhaps taking amphibians in the long grass? Land-based hunting has been noted at many sites, for example at Rainton Meadows where birds have taken small mammals on dry grassland. It tends to be only during severe freezing conditions that Grey Herons vacate inland freshwater areas and frequent inter-tidal pools along the coast or the tidal reaches of Greatham Creek and Seal Sands.

In 2007, a one-legged bird was seen fishing at South Church, Bishop Auckland. Despite its disability, it was noted as being in good health and was proficient at catching fish from a garden pond. On one occasion during the 1990s a bird that was hunting at Shibdon Pond was seen to spear its own foot (K. Bowey pers. obs.). There are a small number of records of melanistic birds, for example by Greatham Creek in 1978 and Saltholme Pools in 1981; this may have been the same bird that was first noted south of the Tees at Coatham Marsh from 23 July to 5 September 1973. Another dark bird was in this area in 1993.

Distribution & movements

In its nominate form, the Grey Heron breeds over much of Europe and the western portions of Asia, and down into Africa. In the European context, it can be found as a breeding bird from Ireland in the west, eastwards to Russia and from Spain and Italy in the south as far north as the Arctic Circle. Northerly populations are migratory (Parkin & Knox 2010)

The majority of Grey Herons present in Durham during autumn and winter are probably locally bred, but between July and mid-October there are regular sightings of birds along the coast or arrivals in off the sea. Whilst such coastal passage is reported, it is not always clear as to whether it refers to dispersing birds from inland that have moved to the coast or birds coasting from north or south or genuine immigrants from the Continent.

National ringing data indicate that many Grey Herons come to Britain from southern Scandinavia and the Netherlands (Wernham *et al.* 2002). A bird ringed at Lodbjerg Plantage, Jylland in central Denmark on 16 May 1985, found long dead on Seal Sands on 19 January 1986 indicates that some Scandinavian birds are involved in such sightings. There is normally an increase in the number of coastal records at Hartlepool through August into early September (Blick 2009). Whilst some birds have been observed arriving from far out to sea, on other occasions birds have been noted flying out to sea, only to turn around and come back to shore some considerable distance away – giving the impression of them having arrived as immigrants. Most coastal sightings refer to one to three birds, although 14 birds passed Hartlepool in July and August 1970 and a single flock of nine passed there on 28 August 1967 whilst 14 arrived at Hartlepool on 12 September 1981 (Blick 2009). During September 2003, a total of 27 birds were noted coming in off the sea at Whitburn Observatory, including one flock of 11 birds on 9

September. There is little ringing evidence to suggest that these birds are from colonies further north in Britain, although there is clear evidence from the national ringing data set that birds move between colonies (Wernham *et al.* 2002).

From a local perspective, there is little discernible pattern to the dispersal of birds from local ringing studies, or from birds controlled in the county from elsewhere. Birds ringed at WWT Washington during the early 1990s were subsequently noted at Holywell Pond, Northumberland in 1992 and within Durham in 1994 at Barmston Pond, Brasside Ponds, Shibdon Pond and Witton-le-Wear. At least four birds ringed as pulli at Washington during 1992 and 1993 were found breeding there by 1996.

The earliest ringing recovery of a bird from Durham was of one ringed as a nestling in the heronry at Gainford on 22 June 1928 that was recovered on 21 March 1929 at a small heronry on the River Glen at Ewart Newton near Wooler in north Northumberland, (Temperley 1951). Coming into the county from the west, a bird ringed as a nestling at Floriston, Cumbria on 9 May, 1925 was reported near Ferryhill on 24 June, 1925, a movement of some 100km to the east south east. A bird ringed as a pullus at WWT Washington on 16 May 1993 was found dead on 20 October 1993 at Abergavenny, Gwent in Wales 350km to the south west.

Purple Heron
Ardea purpurea

A very rare vagrant from southern Europe: seven records.

The first Purple Heron for County Durham was an adult found by Tom Francis on Cowpen Marsh, Teesmouth on 10 June 1975. It was present for only a short time on Swallow Fleet before it became nervous and flew high to the south west, being lost to view close to the Transporter Bridge, Middlesbrough (*British Birds* 70: 446).

All records:

1975	Cowpen Marsh, Teesmouth, adult, 10 June.
1983	Whitburn, juvenile, in off the sea, 21 October.
1985	Seaton Common, Teesmouth, adult, 15 May.
2006	Spion Kop Cemetery, Hartlepool, adult, in off the sea, 4 June.
2009	Haverton Hole, Teesmouth, first-summer, 29th to 31 May.
2009	Haverton Hole, RSPB Saltholme, and Dorman's Pool, juvenile, 7th to 8 September.
2011	RSPB Saltholme and Haverton Hole, juvenile, 26th to 27 October.

With four birds in spring and three in autumn the records of this species in County Durham conform to the national pattern, though the birds in 1983 and 2011 are noteworthy for the late date as there are few British records for October. Surprisingly three birds have been watched coming in off the sea including the bird in 1985 which had flown over South Gare, North Yorkshire, and was seen to drop onto Seaton Common but was not relocated.

Distribution & movements

Purple Heron breeds in Iberia, France and across much of central and southern Europe, as well as in Africa, southern and eastern Asia. The European population is migratory and winters in Africa predominantly south of the Sahara. It is a regular vagrant to Britain, typically as a spring overshoot between April and June, though small numbers also occur in late summer and autumn. The majority of records in Britain come from the southern and eastern counties of England, and a pair bred in Britain for the first time in 2010 at Dungeness, Kent.

Black Stork
Ciconia nigra

A very rare vagrant from southern and eastern Europe: ten records.

County Durham's first Black Stork was a bird shot at Greatham in August 1862 before being added to the collection of W. Christy Horsfall of Low Horsfall Hall, near Leeds, on 8 September 1862 (*The Zoologist* 8 1862). Writing in 1905, Canon H. Tristram in the *Victoria County History of Durham* recalled the event. *"One morning in August 1862, my children came running into my study at Greatham vicarage to tell me that a Black Stork was walking about in the Seaton fields. I went out and watched the bird for an hour, marching about in a swampy meadow. Next morning it was still there, but was shot in the afternoon by a man from Hartlepool. It is now in the Hartlepool Museum"* (Tristram 1905). This bird was only the eleventh record of the species in Britain at the time, all of which were captured or shot. The current whereabouts of this specimen are unknown; it is no longer in the Museum of Hartlepool.

All records:
1862	Greatham, August, shot
1880	Sunderland South Dock, October, captured alive (specimen went to the Sunderland Museum)
1989	Hamsterley Forest, adult, 13 May (*British Birds* 83: 447)
1995	Great Eggleshope Beck, upper Teesdale, adult, 28 June (*British Birds* 89: 489)
1995	Durham City, sub-adult, 26 September (*British Birds* 89: 489)
1996	Hartlepool, adult, flew west, 19 April (*British Birds* 90: 460)
2006	South Shields and Whitburn Coastal Park adult, 16 May, presumably the same Frosterley, 17 May (*British Birds* 100: 706)
2008	Greenside, juvenile, 8th to 9 August, same Clara Vale, and Crawcrook, 11th to 14 August, also in Northumberland (*British Birds* 102: 547)
2009	Castle Lake, Bishop Middleham, 16 October (*British Birds* 103: 574)
2010	Deepdale, Barnard Castle, adult, 27th to 28 May

All of the modern day records have occurred since the recovery of the European breeding population.
The long-staying juvenile in 2008 was often confiding, though frequently elusive, it was later relocated at various sites in Yorkshire before heading south east from Spurn Point on 2 September, and was last noted circling over Great Yarmouth, Norfolk the following day.

That four birds have been found in the extreme west of the county is especially noteworthy. The bird at Frosterley in 2006 had been seen flying low south over South Shields and Whitburn the preceding day, and had presumably gained height and followed the River Wear inland. This situation is mirrored in some respect by records of White Stork *Ciconia ciconia* in the county, several of which have been relocated in the dales after having earlier being seen at more easterly localities.

Distribution & movements
The breeding range of Black Stork extends across much of central and eastern Europe, with an isolated population in Iberia. Further east, its range extends across Asia as far as the Pacific. The majority of the European population winters in sub-Saharan Africa north of the equator, though some Iberian birds are present year round. The species suffered a significant decline during the early part of the 20th century, though the European population underwent significant increases, particularly since 1990 (Birdlife International 2004), with a corresponding increase in the number of vagrant birds recorded in Britain. There had been almost 200 British records by the end of 2010, most between April and September.

White Stork
Ciconia ciconia

A rare migrant from Europe and also an escape from captivity: 25 records involving 27 birds, recorded from February to June, also in September and December.

Historical Review

The first record for County Durham was listed by Hogg (1845), and related to two birds said to have been present on Cowpen Marsh, Teesmouth in the spring of 1830, one of which was shot, though what happened to the specimen is unknown.

It was over fifty years later that the county was to record its second, when a bird was shot at Morton Tinmouth, just south of Bolam, by a T. Robson on 14 February 1884; though as with the first record, it is not known what happened to the specimen (Fawcett 1890). The only record during the early part of the 20th century, relates to a bird that toured various sites in North Yorkshire, and the south east of County Durham from late October to December 1938. This bird spent the majority of its long stay at Great Ayton and North Ormesby, both in North Yorkshire, although it was also seen close to West Hartlepool on more than one occasion and at nearby Hurworth Burn Reservoir (*The Naturalist* Dec 1938: 327).

Recent Status

The next record was not until 6 May 1982, when a bird was watched feeding in an arable field for 45 minutes at Stainton Hill close to Great Stainton before flying off high west (*British Birds* 76: 483). Just one more was recorded during the 1980s, a single watched flying south east over Wingate on 16 September 1988.

There were no further records until 1997, when a single was observed circling over Crookfoot Reservoir on 17 April. After a five year gap, there were three records in 2002. A single was seen flying south over Shibdon Pond, Gateshead and the Tyne Valley on 25 March, with presumably the same bird at Brignall on 4 April. Two birds that flew south over Blackhall Rocks on 25 April were relocated inland at Eggleston on 10 May and another was flying south over Sedgefield on 25 May, although this may have been the known escaped bird previously noted at Lamesley Water Meadows on 14 May. The species has been an annual visitor to the county since, with a dramatic increase in records that saw a further ten recorded by the end of the decade. As many were noted only in flight, it is difficult to exclude known escapes.

 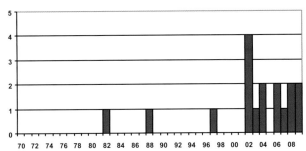

Records of White Storks in County Durham 1970-2009

On 2 May 2006, an adult was discovered at Belasis Technology Park, Billingham after dark by the on-site security guards. A number of observers' first sight of this bird was of it striding around illuminated by streetlights on the flooded grassland after 22.00hrs. It was seen to fly off to the west the following morning. This bird bore a single metal ring on its left leg and although the exact detail of the ring could not be read, it was consistent with ringing schemes used on the Continent, and was presumed to be of wild origin. Just two birds have made protracted stays in the county. The first, an un-ringed adult was at Hardwick Hall and Castle Lake, Bishop Middleham from 17th to 22 April and 9 May 2007; this bird had previously flown over Gateshead on 16 April. The second was of a bird seen in flight at various sites around Chester-le-Street and Durham City during the period from 3rd to 23 December 2009.

This is the only December record for Durham suggesting that this bird's origin may not have been a wild one, given that by this time most wild White Storks are on wintering grounds in Africa.

Two further sightings occurred in 2010, which may have related to the same bird. A single was watched 'circling' over the A66 on the eastern edge of Darlington on 17 April and was seen heading north over Coxhoe about an hour later, with it or another over Heighington on 4 May. A very similar situation was evident in 2011. After one seen flying low over East Boldon and Washington on 7 April, single birds were noted at Hardwick Hall on 16 April and Kibblesworth on 25 April. It is quite possible that all 2011 sightings related to the same individual.

Even allowing for the potential of free-flying escapes, the records within the county show a distinct peak in April and May, typical of a spring overshoot and in that respect, conform to the national pattern. This situation is clouded however by the presence of several free-flying individuals at Harewood House in North Yorkshire, which are known to wander freely during periods of sunny weather with the onset of spring, though always return home eventually. The most famous of these is known to bear a grey colour ring numbered FIU, and has been recorded in County Durham on several occasions; at Lamesley Water Meadows on 14 May 2002; and at Boldon Flats and West Boldon from 13th to 14 March 2004, relocating to Haswell near South Hetton from 16th to 19 March 2004, and finally at Springwell Farm, Darlington on 28 March. Other known escapes have been recorded as follows; south over Hartlepool on 3 April 1996 (a wide ranging escape bearing a metal ring, which was seen between Lothian and Kent); and south over Whitburn and South Shields on 30 December 2000 (later relocated in Strathclyde where it was fed by hand and found to bear a single colour ring).

In addition to these, a bird wearing a yellow colour ring bearing the letters PCC was at Low Middleton from 13th to 17 September, with presumably the same bird being seen at Hamsterley on 9 September, and flying west up the Tyne Valley at Gateshead the following day. This bird was first seen at Watton, East Yorkshire in mid-August 2006, before moving to Eaton, Nottinghamshire until early September, and then to County Durham. It was later relocated near Thirsk, North Yorkshire in late September, and was subsequently seen as far south as Essex, before re-appearing at Yearby, North Yorkshire from 1st to 2 April 2007. The exact origin of this bird was never traced, though seems likely to have been an escape from captivity, as the ring details were never traced to any European ringing scheme.

Distribution & movements

White Stork breeds in north west Africa and Europe eastwards to central Asia, with the majority of European birds wintering in Africa just south of the Sahara, though some remain to spend the winter in Iberia. In Europe the largest breeding concentrations of White Stork are to be found in the east, particularly in Poland with a population of around 52,500 pairs (Thomsen & Hötker 2006). Numbers have declined dramatically in Western Europe, though there are re-introduction schemes in France, Italy, the Netherlands and Switzerland. In the Netherlands, where the last wild pair bred in 1981, the reintroduction scheme has been particularly successful with 396 breeding pairs in 2000. This dramatic increase in numbers in the last 25 years seems likely to have influenced the pattern of records in the UK and Durham.

Glossy Ibis
Plegadis falcinellus

A very rare vagrant from Europe: five records.

The first record for County Durham was of an immature bird shot at Ryhope near Sunderland in September 1831. Writing in 1834, Edward Backhouse described the occurrence as follows: "*A beautiful specimen in the immature plumage was killed by the sea-side near Sunderland in 1831. It was brought to me immediately it was shot and is now in my collection*". The specimen went eventually to the Sunderland Museum, with the label reading, "*Shot on the shore at Ryhope, presented by E. Backhouse*" (Backhouse 1834).

All records:

1831	Ryhope, immature, September, killed
1900	Billingham Bottoms, adult, 25 November (*Zoologist* 1901: 185)
1988	Wynyard, immature, 22 October (*British Birds* 82: 513)
1992	Haverton Hole, Teesmouth, adult, 5 May (*British Birds* 86: 457)
2011	RSPB Saltholme, juvenile, 15th to 18 October (*British Birds in prep.*)

The bird at Wynyard in 1988 was seen by just one observer. The adult at Haverton Hole in 1992 was inadvertently flushed from Cowpen Marsh on the morning of 5 May, before being relocated at Haverton Hole later that morning and again at around 16.00hr; staying for much of the evening before flying off south east over Middlesbrough as dusk approached. The juvenile in 2011 was the only bird to stay for more than one day and arrived during a large national influx of the species.

Distribution & movements

The Glossy Ibis has a scattered distribution across Europe, Africa, Asia, Australasia, the southern United States, Central American and the Caribbean. In Europe the species has a southerly breeding distribution and can be found around much of the northern Mediterranean, eastwards towards Turkey and the Caspian Sea and into Russia. Most of the European birds winter in sub-Saharan Africa. It was formerly much more common in Western Europe and was once a regular vagrant to Britain, particularly during the 19th century, though it became much rarer during the latter part of the 20th century. During the early part of the 21st century, Spain's breeding population was increasing (Birdlife International 2004), correlating with increased vagrancy to Britain. There had been over 500 British records by the end of 2010.

Spoonbill (Eurasian Spoonbill)
Platalea leucorodia

A very scarce migrant from Europe, recorded from March to December, most frequently in May and June.

Historical Review

The first reference of Spoonbill occurring in County Durham is by Hogg (1845), who wrote: "*I have only heard of a single Spoonbill having been killed on the Tees marshes and this was some years ago. It was seen by Mr Hixon.*" The next was not until 1929, again from the North Tees Marshes, and it related to a first-year female shot at Seal Sands by F. W. Bulmer, though no date is attached to this record. There were a further six records relating to seven birds, by the close of the 1960s. These related to: an adult flying in off the sea at Seaton Carew Golf Course on 11 June 1941; an adult on Cowpen Marsh, Teesmouth on 19 April 1951; two adults at Cowpen Marsh and Seal Sands, Teesmouth from 15th to 17 June and 1 July 1956; an adult at Greatham Creek, Teesmouth from 14th to 24 May 1959; an adult at Greenabella Marsh, Teesmouth on 13th and 14 July 1963 and an adult watched flying up the mouth of the River Wear at Sunderland on 20 June 1967. The bird at Sunderland in 1967 is especially noteworthy in that it is one of the very few records of this species away from the North Tees Marshes.

Recent Status

Over 115 Spoonbills have been recorded in County Durham since 1970, the vast majority of which have been on the North Tees Marshes, where the species has been virtually annually reported since the mid-1990s.

Approximately 90% of all Spoonbills between 1970 and 2011 were recorded in spring, mainly in May and June, with a single March record and six in April; most records refer to adults. Formerly an irregular visitor, in many years the species went unrecorded, particularly during the 1980s and early 1990s. It become much more regular in the period 1996-2010 with only two blank years in that time, in 2005 and 2010. The increase seems to be related to the breeding success in the Netherlands over the period and it is from here that the majority of British Spoonbills are thought to originate. The best year for the species was 2002 when at least eleven birds were recorded on the North Tees Marshes, though there may have been as many as eighteen. Two separate parties of four birds were seen on

Cowpen Marsh and Dorman's Pool, Teesmouth on 25 May 2002; the largest number of birds seen in the county on a single day, though the party of five which flew south west over WWT Washington on 26 May 1976 remains the largest group. On at least one occasion in spring, long-staying adults have been seen carrying nesting material on the North Tees Marshes, but there has been no further evidence of breeding (R. M. Ward pers. comm.).

Away from the oft-visited sites around the North Tees Marshes, there have been just 15 records of 20 birds, all in spring. The Barmston Pond/WWT Washington area has produced six records in: May 1970, June 1972, May 1976, May 1992, April 1996 and May 2007. The remaining records are widely scattered across the county, with birds noted at: Cow Green Reservoir on 28 April 1977; flying north over McNeill Bottoms near Witton-le-Wear on 2 May 1998; two over Tunstall Hills, Sunderland on 21 May 1988; at Boldon Flats, East Boldon on 15 May 1992, this bird having been at Barmston Pond, Washington the previous day; at Barnard Castle on 23 May 1996; flying north at Whitburn on 27 May 2006; at Castle Lake, Bishop Middleham on 15 March 2008 and 23 May 2011; at Derwent Reservoir on 20 June 2011, and at Hurworth Burn Reservoir on 30 June 2011. The bird at Cow Green Reservoir in 1977 is the most westerly sighting in the county occurring at over 500 metres above sea level.

Extreme spring arrival dates are 15 March at Castle Lake, Bishop Middleham, and a party of four at Dorman's Pool, Teesmouth on 29th and 30 June 2000, with one remaining until 1 July.

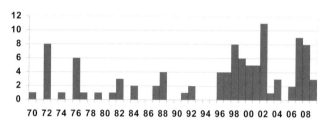

Records of Spoonbills in County Durham 1970-2009

Autumn records are rare with all sightings referring to the North Tees Marshes, where four have appeared in September. An immature bird turned up on 16 November 1984, being joined briefly by a second bird on 18th, with the original bird lingering a full month until 16 December; the latest recorded date for the species. In 1997, three were present from 19th to 25 November.

Distribution & movements

The Spoonbill has a wide but fragmented breeding range from Europe and North Africa to the Red Sea, India and China. In Europe the largest colonies are to be found in The Netherlands, Spain, Austria, Hungary and Greece. The majority of European birds winter in Africa, though small numbers remain in the milder areas of Western Europe, including southern Britain. The species is an extremely rare breeding bird in Britain, with the first colony being established in Norfolk in 2010 after an absence of nearly 300 years, although single pairs bred successfully in Lancashire in 1999, and Dumfries and Galloway in 2008.

Three different colour-ringed birds, all ringed as pulli in Holland, have been seen in south east Durham, around Teesmouth. The first had been ringed in May 1990 and subsequently visited Kent and Norfolk in June 1991, before moving south from Holland to Spain each autumn from 1991 to 2001, returning north each spring, apart from a visit to Coatham Marsh on 27 May 1999 and Dorman's Pool from 28 May until 6 June 1999. The second was ringed in July 1994 and spent the summer of 1999 touring Britain, visiting sites in East Anglia in June, before moving north to Scotland then south to Dorman's Pool from 2nd to 5 July; then shuttling back and forth between sites south and Dorman's Pool until late July. From April 2000, this bird commuted north and south in Europe, being seen in Spain in April 2001, France in September 2002 and several times in Holland. The third was ringed in the nest in June 1996 and recorded four years later on Dorman's Pool on 29th and 30 June 2000. It then visited the Ouse Washes in mid-August 2000, before being seen in Spain in February 2002 (Blick 2009).

Little Grebe

Tachybaptus ruficollis

A common but localised resident, also believed to occur as a passage migrant.

Historical review

The Little Grebe is the most numerous species of grebe that regularly breeds and winters in Durham it was noted as a breeding bird in neighbouring Yorkshire as early as 1791 (Mather 1986), but its documented history in Durham is not quite so long. In 1840, Hutchinson said it was "*a rare species*" and mentioned it as having bred in a wetland at Houghall, near Durham City from where it had been 'lost' by his time of writing. This suggests that it certainly bred in the county sometime in the first third of the 19th century. In contrast, Hogg (1845) reported that it was common in the Hartlepool and Cowpen Marsh areas. Hancock (1874) said it was present throughout the year but thought it most numerous in the winter. He recognised that it bred on the Tees near Stockton, though from his comments and in contrast to Hogg, he clearly felt that it was little more than a scarce breeding bird. Perhaps a genuine change in status had occurred during the mid-part of the 19th century? Robson (1896) knew of only one instance of it having nested in the Derwent valley; a clutch of eggs having been taken at Axwell Park, not far from Blaydon, on 2 May 1881. He supposed the species to be "*more common than generally realised*".

Temperley (1951) considered that the species was a common resident in Durham, "*more plentiful than formerly*" whilst Bolam made comment that it had increased considerably over the previous three decades in neighbouring Northumberland (Kerr, 2001). In the north of Durham, two pairs nested at Sled Lane Pond, Crawcrook in 1934, but as an example of the problems faced by the species, it was reported that between April and September, the birds built between them nearly 30 nests and that most, if not all, of these were destroyed by local people (*The Vasculum* Vol. XX 1934).

There is little quantitative information published on the status of this species during the 1950s and 1960s, though Mather (1986) noted that there was a drastic reduction in its breeding numbers in Yorkshire after the severe weather of the 1962/1963 winter. One of the largest post-breeding, gatherings was noted at Hurworth Burn Reservoir on 8th to 9 September 1967, when 29 birds were present.

The 1970s brought reports of the species as being widely but thinly scattered on suitable waters in the county. In 1974, pairs were present on at least 14 different sites during the breeding season. In addition, there were reports of twelve pairs in Cleveland in the early 1970s (Stewart 1975); the majority of these being concentrated in the North Tees Marshes wetlands and ponds. It is interesting to note that at this time it was considered a commonly occurring summer visitor to the upland reservoirs of Lunedale and Baldersdale, despite it being predominantly a bird of lowland waters. The outcome of any such breeding attempts at these more exposed locations is not documented (Bradshaw 1976). More recently a pair bred successfully at Westgate in Weardale, some 360m above sea level, the highest elevation and most westerly site at which such activity has been noted in recent decades.

Numbers of Little Grebe have varied quite considerably at some locations. The species appears to exhibit declines probably in response to winter freezes, followed by recoveries. Habitat loss or disturbance may also play a part but the real reasons seem complex and aren't fully understood. In the early 1970s, Shibdon Pond held a number of breeding pairs and winter gatherings of as many as 13 individuals. Birds at this protected site declined and it was lost as a breeding species during the 1980s only for breeding to occur again in 1992, for the first time in over a decade (Bowey *et al.* 1993); numbers then built up to a situation where five or six pairs were breeding annually by the end of the 1990s.

Recent breeding status

Little Grebes are relatively secretive and they occupy a wide range of open water and wetland sites, some of which can be surprisingly difficult to census. Birds will occupy some potential breeding sites over the winter as long as waters remain open, but by late February and early March a more concerted increase at breeding sites is apparent as birds draw attention to their presence by their distinctive 'whinnying' call.

The species requires still or slow moving water with emergent vegetation (Gibbons *et al.* 1993). Fast moving streams and rivers, as well as the larger upland reservoirs of the west of the county, which tend to be devoid of plant cover, are usually, but not always, avoided. Quite small ponds and marshes can be occupied but the main habitats in the county are the larger ponds and reservoirs in the east, especially where emergent vegetation provides nesting cover for this shy species. The mid- to lower reaches of the River Tees and especially the Wear are also favoured. Today this dainty grebe is a relatively common breeding bird at suitable sites, with the main distribution being concentrated in the county's 'central wetlands' from Gateshead south through to Sedgefield, and down to the Tees Marshes in the south east . Teesmouth is one of the main breeding areas for the species in the region and sites in this area holding breeding birds include: Charlton's Pond, Cowpen Bewley Pond, Cowpen Marsh, Dorman's Pool, Haverton Hole, the Reclamation Pond, Portrack Marsh and Saltholme Pools (Blick 2009).

The *Atlas* (Westerberg & Bowey 2000) stated that it was likely that some pairs would have been missed during the survey period, although the published map is probably a reasonably accurate reflection of the species' largely central and easterly distribution in the county; albeit that the presence around Teesmouth was underestimated at the time. The *Atlas* indicated a dearth of records from parts of the River Tees, which suggested a degree of observer bias rather than a true absence of Little Grebe from this area, especially considering their considerable presence on the North Tees Marshes, as illustrated by Joynt *et al.* (2008). There were a reported 17 pairs around the North Tees Marshes by the early 1990s, (Bell 1992) with over half of all the breeding birds reported in *The Breeding Birds of Cleveland*, of between 15-20 pairs, being in this area (Joynt *et al.* 2008).

At the onset of the *Atlas* survey period, the collation of casual records suggested a stable population of approximately 25-30 pairs on 12-18 waters. The better understanding of the situation in the Wear valley brought this estimate, in 1993, to 65 known breeding pairs at 25 localities, excluding Teesmouth. This reflected much more dedicated monitoring along the River Wear rather than any population increase. In 2007, the species was recorded at 78 sites in the spring and summer, with breeding being confirmed at 35 of these. All of these were east of grid line NZ02 and the majority of them were to the east of the A68.

This was still believed to be an under-representation of the species' true summer distribution. The wetlands around Rainton Meadows have one of the highest concentrations of breeding birds away from the North Tees Marshes, with five pairs there in 2007. Pairs now regularly occupy 50-65 sites in the county, with between one and five pairs breeding on most of them. Given the species' elusive behaviour it seems probable that the breeding population may be as high as 80 pairs in favourable years. Breeding success can be variable, depending on climatic conditions but, in good years pairs may double or even triple brood. Occasionally, dependent young may still be seen in early September.

Recent non-breeding status

Post-breeding dispersal of birds takes place from mid- to late August or earlier if water levels are low. This dispersion from breeding sites is presumed to take place nocturnally as the species is so rarely seen in flight. Dead birds have been found under overhead cables locally (Blick 2009). From early autumn until the year's end, there is a general movement away from the breeding sites to other preferred waters. The autumn build-up at Bishop Middleham ponds is now a regular feature and the numbers perhaps suggest local breeding going unnoticed. In the winter, birds will often resort to a range of wetland sites some of which are not occupied during the summer. Examples include Hartlepool Docks and Seal Sands.

Small gatherings are regularly recorded on lowland reservoirs after the breeding season, in particular on Crookfoot, Hurworth Burn and Hart Reservoirs, with between six and fifteen birds being quite usual, although 19 were on Crookfoot on 7 August 1976 (Blick 2009). In August 1993, 44 were counted on Brasside Ponds. The area's largest ever post-breeding gathering, of 152 birds was recorded on 13 September 2000, around the North Tees Marshes (Iceton 2001); this count included 98 birds on the Reclamation Pond alone (Blick 2009).

From autumn onwards, birds will often frequent the slow-moving lowland stretches of the county's rivers, where many birds spend the winter. These sites resist freezing in severe weather more so than still waters. Such well-frequented sites include the lower sections of the River Derwent, near Swalwell, which draws birds from breeding sites at Shibdon Pond and Axwell Park Lake. Similarly, 42 birds were counted along the River Wear between Chester-le-Street and Durham City in January 1993.

Birds on the sea are only rarely seen, but there is an example every few years, such as the bird seen off Whitburn on 6 October 1999 and singles at Jackson's Landing, Hartlepool in February 2005 and December 2009, during periods of cold weather.

Distribution & movements

Excepting Australasia, Little Grebes breed over much of the Old World, though they are not present in the extreme north of Eurasia. Most populations are mainly resident, the most northerly and easterly birds in Europe are migratory, moving to locations, including Britain, to the south and west of their breeding range.

In November through to January, the overall numbers in Durham shows a distinct decline and it is presumed that many of the county's Little Grebes leave the area. Nonetheless birds do winter in Durham, in relatively small numbers. It is presumed that birds that leave move south and west, though there is little concrete evidence for this statement by way of ringing data.

Great Crested Grebe
Podiceps cristatus

An uncommon breeding resident, frequently seen on passage and as a winter visitor to the coast.

Historical review

During the latter half of the 19th century, Great Crested Grebes were killed throughout Britain for their feathers, resulting in their near-extinction nationally as a breeding bird and a 'temporary absence' in our own area for over 100 years. The reason dates back to the mid-19th century's craze for 'muffs and stoles' made from the bird's silky under-plumage (Sharrock 1976). Legal protection from 'industrial scale persecution' came from 1880 onwards and this helped enlighten attitudes. The increasing availability of gravel pits and reservoirs during the 20th century assisted in the gradual re- colonisation of the species in Britain (Harrison & Hollom 1932, Prestt & Mills 1966).

Hancock (1876) described it as "*rare*" in the region and treated a summer plumaged bird in the mouth of the Tyne during Easter week 1860 as a special event. Tristram (1905) called it "*a rare straggler*", mainly to the coast and quoted just two records in the Teesmouth area. The first confirmed breeding in Northumberland was provided by Bolam in 1911 (Kerr 2001) though nest productivity varied there throughout the 20th century with frequent failures mirroring some of the experiences which would eventually emerge in Durham.

There is an intriguing historical record from the Gateshead area of an immature bird shot on Derwenthaugh in May 1907 (Bowey *et al.* 1993). The species was then still very rare as a breeding bird in Britain and the presence of an immature so early in the season could point to local breeding. The first confirmed successful breeding for County Durham would not be recorded for a further 50 years. Richmond (1931), writing of Teesmouth, stated that the Great Crested Grebe was '*chiefly a hard-weather bird*' there. The ground-breaking national survey of 1931 (Harrison & Hollom 1932) suggested that the species had never bred in Durham but that a national recovery was then evident in the Midlands although breeding records from the northern counties were still lacking. W. Eltringham studied the courtship of a pair at Crawcrook near Ryton in 1936, though he did not secure confirmation of nesting. He carefully noted many of the pair bonding behaviours so well known today and recorded that at "*some vocal signal they turn and fly directly towards each other, alighting halfway face to face on the water, where with outstretched neck and bill to bill they seem to scold one another for a brief space, and the turn round quite calmly and swim off in the opposite directions*" (The Vasculum Vol. XXII 1936).

The effects of the national recovery were slow to materialise in the northern half of England and it was not until 1944 that a pair attempted to breed in Durham at Charlton's Pond, Billingham (Stead 1964). This is the first known nesting attempt in the county although it failed because of egg-collectors; though it seems likely that some kind of attempt at breeding had taken place at Crawcrook eight years previously. The pair at Charlton's Pond was present from late May into early June 1944, with birds displaying over the period 9th to 12 June. On 27th a nest with three eggs was found, but it was robbed in mid-July and the adults deserted the site.

By the middle of the 20th century, Temperley (1951) thought that Great Crested Grebes were primarily still an "*autumn and winter visitor in small numbers to both salt and fresh waters*". There had been no more breeding attempts though it had become a more frequent autumn and winter visitor to Durham as the national breeding population continued to expand slowly. Temperley mentioned that it occasionally visited Jarrow Slake, Crookfoot and Hurworth Burn Reservoirs and in severe weather, sheltered in the Tees Estuary.

The first report of successful breeding in Durham eventually came in 1957 when four 'half-grown' young were found at a disused brick-works pond at Brasside, near Durham City. Expansion of the breeding population subsequently proved painfully slow. A nesting attempt there in 1959 failed at the egg stage, largely due to human interference but two pairs attempted to breed in 1960 with one pair rearing young. Even by 1964 the rearing of young by three pairs at the same site in the county was regarded as an outstanding achievement. During the remainder of the 1960s pairs summered at between two and five sites but in any one year only one pair was ever successful, rearing between one and three young.

Recent breeding status

The species became a frequent summer visitor, albeit in very small numbers, by the 1970s and breeding, or at least the number of territorial pairs began to increase. In 1973, two pairs reared four young at two locations and summering birds were found on five other inland waters. The first successful breeding attempt in the south east of the county was at Wynyard Park Lake in 1973 (Blick 2009) when a pair raised four young. In 1974, successful breeding again occurred at two waters, with three pairs at Brasside rearing just two young because of disturbance and one pair elsewhere rearing four young from a replacement clutch. Breeding may also have been attempted at third water. In 1975 a pair even appeared on Cow Green Reservoir, amidst the high fells of upper Teesdale, but did not attempt to breed. By 1976 there was an increase in reports of pairs holding territory with at least five pairs going on to breed. Two of these were once more at Brasside Pond, where family parties were present in mid-July. One of two young also survived on the landscaped gravel pit at Witton-le-Wear. During this period one or two birds began to appear each spring on the reservoirs of Derwent, Hurworth Burn and even exposed Smiddyshaw. The occasional attempts at breeding on such open waters tended to end in failure; as illustrated by the eight adults at Hurworth Burn in the spring of 1977, where only one pair built a nest but no young were fledged. In the following year 14 adults on the same reservoir resulted in just two pairs producing four young. By 1982, 15 young were reported on four separate waters. This did not include Hurworth Burn Reservoir where ten birds spent the spring and summer. Further range expansion, if not improved breeding success, was evident by 1987 when, excluding Teesmouth, pairs nested on eight waters and but still produced just 15 flying young from eight successful broods at five sites (Armstrong 1988).

In the south east of the county, one pair nested annually on Charlton's Pond from 1980, occasionally rearing two broods in one season (Blick 2009). Around Teesmouth, numbers had increased through the 1970s (Joynt et al. 2008), with between six and eight pairs rearing up to 14 young per year during that decade. Numbers rose further during the 1980s, from seven to ten nesting pairs, increasing to perhaps 12 pairs by the early 1990s (Iceton 2000).

Westerberg & Bowey (2000), considered that the combined results of the six years of Atlas fieldwork (1988-1993) for Durham had overstated the species' breeding distribution in the county since not all prospecting pairs went on to nest in any one year and in some years pairs may have been absent at smaller sites. The species' presence on waters during the breeding season was mapped, but not the outcome of any breeding attempts. During this period between six and 11 waters were occupied in each year by eight to 23 pairs, of which two to ten pairs were successful in rearing between just three and 17 young. Clearly there was considerable year-to-year variability. In fact 1993 was the most successful breeding year in Durham to that point, with 10 pairs raising 17 young. This record was broken in 1994 when 14 pairs, from 21 nesting attempts, raised 29 young. This number of fledged young was matched in 1999 by 12 successful pairs from 20 nests. In other years, as the century closed, only between eight and 14 young were fledged. The situation improved with newly-occupied sites such as Seaton Pond bringing welcome success for the first time in 2004. A notable improvement came in the period 2005-2010, largely as a result of habitat creation at Teesmouth especially associated with RSPB Saltholme. In 2009, the nearby Reclamation Pond held nine pairs with 12 at Saltholme, with seven and ten respectively at each site in 2010. Elsewhere in Durham the same limits on breeding success continued.

A major constraint on County Durham's breeding population is a severe lack of natural lakes and the unsuitability of many of the artificial water bodies present. Considerable difficulties are faced when trying to breed on many of the county's reservoirs, which are prone to receding water levels in early summer and a paucity of emergent vegetation upon which to anchor nests and provide cover for young. In addition, the western upland reservoirs are generally steep-sided, lack littoral vegetation and suffer severe on-shore wave action due to the 'long reach' of some of these large waters. Rivers and smaller ponds are avoided in preference to reservoirs and larger waters in the eastern half of the county. Sites with emergent vegetation for nesting and cover are ideal but the extent of this type of habitat in Durham is restricted, so it is no coincidence that the mining subsidence ponds at Brasside, which offered the species its first enduring foothold in Durham, are still used. In such sites, breeding success tends to be noticeably greater although the number of pairs here seem to have fallen inexplicably in recent years. At many more sites, pairs have no option but to build their nests close to the shoreline which can then become dry and exposed in summer and, consequently prone to attack by ground predators. Birds at Hurworth Burn Reservoir regularly experience such difficulties. In 2007 six nests occupied in early May had all failed by early July. Derwent Reservoir is a notable exception to this and birds regularly nest here. Other upland reservoirs at Smiddyshaw and in Lunedale have seen successful breeding in recent years depending on spring rainfall and water level management.

Unintentional disturbance by fishermen and recreational walkers are thought to contribute to nest failures at some sites and even deliberate egg theft has been cited. The concentration of the county's relatively few breeding pairs at a few key sites means that the population is vulnerable should these be weather affected in any one year. Habitat creation schemes have offered the clearest success story and point the way to how the species might eventually secure its place in the county.

Recent non-breeding status

The species is found in County Durham throughout the year. Adult Great Crested Grebes undergo a full flight-feather moult during July or August and become flightless for a few weeks (Newton 2010). Birds exhibit notable annual gatherings along the coast at this time where in-shore waters, with gently-shelving beaches, provide the necessary feeding and sanctuary during the wing feather moult. The timing is such that those adults still successfully rearing young on inland waters during the summer cannot be involved and these gatherings probably include birds from failed or very early breeding pairs. The immediate coastline south from Dawdon to Teesmouth is favoured, with Blackhall Rocks and Crimdon Dene seemingly key locations. Build ups occur annually and usually begin in July and extend well into September before the flocks disperse. For the remainder of the autumn and early winter the coastal flocks tend to be much smaller.

There is a degree of unpredictability, year-on-year, over exactly when these coastal moult gatherings peak and the number of birds they attract. In the 1990s annual peaks of between 45 and 55 birds occurred variously from late August to mid-September. Exceptionally 80 were reported in late August 1994 and 76 in early August 1998. In contrast, the highest count in 1997 was of a mere 23 in mid-September. The highest early autumn counts came in 2000 when 81 were found on the sea off Easington and Blackhall in July and 87 were off Castle Eden Dene a few days later. It is probable that these two counts involved much duplication. The following year 78 were seen off Blackhall in August, but more recently this gathering hasn't been so prominent and from 2005 to 2010 the autumn counts on this stretch of coast were of between 25 and 40 birds. At the same time as Durham's autumn coastal flock declined there was an increase in autumn gatherings at nearby Teesmouth, where the Reclamation Pond held 42 in August 2005 (Blick 2009) and north Teesmouth as a whole, 67 in September 2007.

There are also autumn gatherings at some inland waters; Crookfoot Reservoir was a favoured location in the 1980s with up to 14 birds present (Blick 2009) and Hurworth Burn has held 10 to 15 birds more recently. In general, numbers fall quite rapidly from late August onwards, and from October through to the early part of winter just a few locations might typically hold two to four birds as most appear to vacate our area and disperse southward (Lack 1986).

The species is quite hardy and over the last two decades has appeared regularly in mid- to late winter along the Durham coast. The area around Newburn Sewer outfall and North Gare have provided the largest late winter counts, usually peaking in February as a prelude to moving on to breeding waters. In Hartlepool Bay, 69 gathered on 15 February 1997 and the same number was seen there on 14 February 1998 (Blick 2009). February counts in the years 2005 to 2008 ranged between 35 and 45 birds but in January 2010 the Newburn Sewer outfall attracted 43, rising to an 82 by 23 February. In March of that year, 42 were counted on the North Tees WeBS, including 27 on Reclamation Pond.

Whilst the majority of this species' coastal presence is along the southern half of the county's shoreline, Great Crested Grebes can be seen in smaller numbers almost anywhere between South Shields and Hartlepool during the winter months. Inland waters are never entirely abandoned as long as they remain ice free. In the cold weather of early January 1982, four birds were present at the mouth of the Tyne and individuals are sometimes recorded at Jarrow Slake; while four birds were on the Tyne between Bill Quay and Gateshead on 17 March 1996. There are

December and January records for Crookfoot, Hurworth Burn and Longnewton Reservoirs, as well as Brasside and Joe's Pond ahead of the return of breeding birds, which is usually evident from early February.

Occasionally, passage movement is observed along the Durham coast; on 18 January 1985, for instance, a total of 26 flew north off Whitburn. Such passage may involve birds from the Continent. A few birds pass along the Durham coastline each autumn linked no doubt to the flocks that gather off our coast at that time. Very light coastal passage is noted at Whitburn Observatory, with around 25 birds being logged throughout the year, most of these being seen in late summer and early autumn.

Distribution & movements

The nominate race breeds across much of Europe eastwards into central Asia, with other races represented in Africa and Australasia. The most northerly parts of Europe are devoid of breeding birds; the majority of European birds winter to the south and west of their breeding range. Northern England is towards the north western limit of the species' range and climatic factors must impose some additional natural constraints on the local population, especially during winter.

Red-necked Grebe
Podiceps grisegena

A scarce passage migrant and winter visitor to the coast from the Baltic and Fennoscandia; often occurring as part of a more widespread influx in response to hard winter weather in the Baltic region.

Historical review

The first report for Durham was made by Fox (1827) who told of '*one killed at Jarrow Slake about three years ago*'. Edward Backhouse (1834) considered it "*common*" at Teesmouth in immature plumage and Hutchinson (1840) implied that it was the most common of the grebe species, based upon the number of specimens seen by him. These accounts are surprising in terms of the species present day status but may reflect a genuine change. Hancock (1874) recounted an apparent influx in the winter of 1830, when "*a number were killed in the neighbourhood of the River Tyne*" and Bolam (1932) obviously recognised the species' tendency to arrive as an influx in some years. He said it was 'especially numerous' in 1891 and again in the early months of 1897 and 1899. The 1891 arrival was documented by Nelson (1907) at Teesmouth, when many were in the estuary and off the mouth of the Tees on the sea. One can only speculate how many birds might have been involved with such influxes but Nelson recorded that '*there were at least 35 off Redcar (N Yorks) on 19 January 1891 and many more in the following month*' (Blick 2009). Nelson also noted an influx at Teesmouth in 1895 but this involved smaller numbers. Other notable arrivals took place in February 1922 and especially in February 1937, when birds were reported from sites across the county including: East Boldon, East Rainton, Hebburn Ponds (2 birds), Hurworth Burn Reservoir (5 birds), Jarrow Slake, Leamside, Saltwell Park Lake and Teesmouth (2 birds); a total of at least 14 birds in eight localities.

Temperley (1951) called it a "*very irregular winter visitor*", recording that it was never very common but occasionally "*a number of them*" appeared on coasts and inland waters usually remaining for a few days at a time. Temperley (1951) cited the influxes referred to above, all of them notable for occurring during the months of January and February.

In the two decades following Temperley's publication there were six years in which there were no records but overall there was a steady increase in the number of reports, perhaps as observer effort grew, particularly around Teesmouth. The most unusual record over this period was a bird in summer plumage seen on Cowpen Marsh on 6 May 1962, one of very few breeding season records for the county. In October 1962, an individual stayed for a fortnight on the harbour side of the pier at South Shields. Typically in this period, there were often no more than one or two birds seen per year, the maximum being seven during the severe winter of 1962/1963 when from the January to early March of 1963, four were reported on the Durham side of Teesmouth (two of these were observed to have oil contamination to their plumage), two were on the Wear at Chester-le-Street and one, which lingered in

Seaham Harbour, was in summer plumage by 7 March. Birds reliably showed their preference for the calmer waters of docks, harbours, creeks and pools along the immediate coastal strip and only occasionally appeared further inland.

Four birds were noted in the early spring of 1969, these included birds at Greatham Creek in mid- to late March, one in Hartlepool Dock from mid-March to 1 April and one inland at Fishburn over the period 3rd to 5 April.

Recent status

Today, this species can be considered a scarce but annual passage and winter visitor. Birds have been reported in every year but one (1975) since 1970. With the exception of obvious invasion years the numbers reported annually have steadily increased from, perhaps, typically two to six during the 1970s and 1980s to between five and 15 through the 1990s and 2000s. Increased observer effort, especially in observing sea passage past Whitburn Observatory and Hartlepool Headland has been probably most responsible for this improved standing. The majority of records come from Whitburn Observatory, Hartlepool Headland and Marina, and Seal Sands.

A period of light coastal passage is now noted most years from August to early October. Occasional birds might then appear during November and December before mid-winter influxes usually bring one or two birds to a few favoured sites from January through to March. National totals of this species in England and Wales are estimated to be in the order of 50 birds in a normal winter (Chandler 1981) and Durham usually holds no more than five or six overwintering birds. In addition to the few that do appear to be settled at coastal locations from January to early March, there are some years in which mid-winter movements are evident without necessarily being associated with any influx to local waters. For example, 13 moved north off Whitburn between 10th and 19 January 1985 with a peak of 7 on 18th and in 1987, 16 moved north there between 15th and 29 January, with a peak of nine on 16th.

Favoured winter retreats include the sheltered bays and harbours of Hartlepool, Newburn Sewer outfall and Seal Sands, where one to two birds might be present in most winters from late December onwards, with less frequent winter reports from the Tyne southwards to Seaham.

Most birds disperse by late March but occasionally they linger or are seen moving along the coast up to mid-April, exceptionally into early May. Unusually, a bird stayed at Haverton Hole, Teesmouth from 12 March to at least 6 May 2006 whilst, remarkably, another was on Dorman's Pool from 16 April until 14 May; both birds attaining the species' attractive summer plumage during their stay. The trend for more records of later staying birds continued in 2008, when a bird stayed in the area of Hartlepool Headland until 17 April, by which time it had moulted into summer plumage.

Coincident with the increase in sea-watching along the Durham coastline since the early 1980s, came the realisation that a light autumn passage of Red-necked Grebes occurs along the coast in most years. This can begin as early as July and the birds seen then are possibly associated with the development of the late summer, post-breeding flock which gathers in the Firth of Forth, and may be linked to moult. July birds are normally in breeding plumage. Early autumn passage records in Durham are by no means annual but include singles at Seaton Snook, on 27 July 1980 and Teesmouth, on 30 July 1983. In addition, there is a series of sightings from Whitburn Observatory beginning on 2 July 1984 with further singles there on: 2 July 1993, 8 July 2004, 30 July 2005, 19 July 2007 and 20 July 2008. The autumn movement is modest and not evident annually; counts normally don't exceed six or seven birds over a full season. It shows most strongly in August or in some years September, though it can continue to mid-October. Notable peaks have included 11 off Whitburn between 4 August and 19 September 1996, with 11 over the autumn of 2007 and 10 in 2009.

Red-necked Grebes are scarce on inland waters, most frequently occurring in autumn. Singles were seen at: Crookfoot Reservoir in late November 1971 and October 1991; Bishopton, in early October 1983; Longnewton Reservoir, in mid-October 1985; Brasside Ponds, in October 1993; Castle Lake, Bishop Middleham, in late November 2010; and Seaton Pond in August 2011.

Infrequently there are much larger numbers during 'invasion' years, when the total of birds in the county might exceed 20. Throughout much of the modern recording period, post 1970, the annual tally falls well short of double-figures, except for the marked influx that occurred in response to hard winter weather in the early months of 1979. This was the largest invasion to the county since 1937, and quite possibly 1891. In mid-February 1979, a high pressure system centred over Scandinavia produced gale force winds and snow on the eastern seaboard of England over 14th and 15 February and a large number of grebes, particularly Red-necked, entered the country. It was determined that a grand total of 481 birds reached Britain in the period 19th to 25 February (Chandler 1981), of which at least 19 were in Durham. These included up to nine around Teesmouth, three at both Sunderland Docks and Jarrow Slake, and possibly wandering birds at a number of other sites including inland records from Crookfoot Reservoir and on the River Wear at Maiden Castle, Durham (Unwin 1979).

The years since have not been without some notable peak counts. The years 1985, 1987, 2006 and 2007 each achieved cumulative counts, between coastal movements and overwintering birds, which probably exceeded 20 individuals. In 2009 there were reports representing a total of perhaps 28 individuals, mainly during autumn passage.

Distribution & movements

Red-necked Grebes breed in eastern Europe and across central Asia, to eastern Siberia. Their breeding range extends into Alaska and eastern Canada. European breeding birds winter to the south and west of their breeding areas, largely in marine habitats.

Slavonian Grebe (Horned Grebe)
Podiceps auritus

A scarce winter visitor and passage migrant; recorded in all months except June, most frequently between October and April.

Historical review

George Temperley (1951) called this species "*an uncommon winter visitor*", largely on the coast and, according to Temperley, rare on inland waters. It was first recorded by Hogg (1827), who said that one was shot in January 1823 in hard weather though he does not give a locality, with the same author stating in 1845 that he had seen a specimen which had been shot on the Durham coast in the winter of 1829/1830, though again no locality is given. By way of contrast Edward Backhouse writing in 1834 thought that the species was "*common in the immature state*", though it is conceivable that this reflects the challenges around its identification in winter plumage. Nelson (1907) wrote that this species was an annual visitor along the coastline and that, in relation to Teesmouth, 'several' were killed in the Tees estuary during the winter of 1874/1875, and again in the winter of 1896/1897 when it was "*more than usually abundant*". Nelson was also said to have 'procured one' that had been feeding on sprats, in a channel at 'the Teesmouth' in 1901 (Mather 1986). Other records during the early part of the 20th century are scant, though one was at Whitburn in February 1937 during a marked influx of Red-necked Grebes *Podiceps grisegena*, and another was at Tyne Dock, South Shields on 15 February 1950.

Inland records were noted by Temperley as uncommon, though one was shot on the Tyne, at the mouth of the River Team during the winter of 1894 (Bowey *et al.* 1993), and two were said to have been shot on Nicholson's Fish Pond at New Seaham, '*about 1907*'. Another was at Belasis Pond, Billingham in December 1931, with perhaps the same bird being present at the same locality in November and December 1932.

Post-Temperley, birds were reported with somewhat more frequency, though whether a genuine change in status was involved is debatable. A bird that was at Graythorp, on the Durham side of the Tees in 1956, was reported as only the fourth record in that area since 1947, indicating its comparative scarcity. The Ornithological Reports for Durham and Northumberland, during the 1950s and 1960s, indicate a small upsurge in reports during the latter decade, mentioning one on the Marine Park Lake, South Shields from 26th to 29 December 1959. The 1960s brought at least ten records in five years, with something of a bumper year in 1969. The years 1962 and 1964 had only a single record each; at North Gare on 13th to 21 October 1962 and two summer plumage birds on Hurworth Burn Reservoir on 24 July 1964, which remains one of only two July records for the county to date. Other more typical records were at Charlton's Pond, Billingham from 11th to 17 February 1966 and one found moribund at Cowpen Marsh, Teesmouth on 25 February 1968. Records in 1969, compared to the previous pattern were exceptional: two being noted off Hartlepool on 21 March, with single birds on Charlton's Pond, Billingham from 22nd to 28 March, at Brasside Ponds from 28 March to 3 April and finally at South Shields on 30 August.

Recent breeding status

Slavonian Grebe has never bred in England despite first being proven to breed in Scotland as long ago as 1908 (Thom 1986). Locally there have been occasional and increasingly frequent records of birds outside of the usual wintering or passage period. For example in 2004, two wintering birds at Teesmouth assumed summer-plumage and were still present on 20 April, one of these remaining on the Reclamation Pond until 2 May. One was also at Saltholme Pools on 15 April 2004, with what was believed to be a different bird at Dorman's Pool on 1 May 2004, though none of these birds were seen to indulge in any signs of breeding activity (Joynt *et al.* 2008).

In 2005, three wintering birds, conceivably all or some of the birds from the previous year, commuted between Seal Sands and the Reclamation Pond through February and March, and frequented the Reclamation Pond into April, by which time they were in full breeding plumage. Two of the birds were known to have displayed during May and on 10th they were seen nest-building, though unfortunately, the nest was taken over by Coots *Fulica atra* and the nesting attempt was abandoned (Joynt *et al.* 2008). This would appear to be the first time that a pure pair of this species has attempted to breed in England (Holling 2008).

Recent non-breeding status

At least 180 Slavonian Grebes have been recorded in County Durham since 1970, making it marginally the rarest of the grebes to occur within the county. The species has a markedly different occurrence pattern to that of the Black-necked Grebe *Podiceps nigricollis* in that the vast majority of birds are recorded during October to April, whereas the peak period for Black-necked Grebe is from April until September. In addition to this, Slavonian Grebe is more likely to be recorded on the open sea outside of the breeding season.

There are records along the entire stretch of coastline between South Shields and Teesmouth, though records predominate from the relatively sheltered waters of Seal Sands and the many docks and quays of the Hartlepool area. Slavonian Grebe is known to prefer relatively calm inshore waters during the winter months and in the north of the county, seawatching aside, the majority of records are from the mouth of the River Tyne at South Shields; itself sheltered by two large breakwaters.

Records of this species increased dramatically from an average of around one bird per annum during the sixties and seventies to an average of at least eight birds per year since 2000. There were four blank years during the 1970s and two during the 1980s, though Slavonian Grebe has been annual in increasing numbers since 1983, when a total of four birds were recorded. The peak year was 2005 when nineteen birds were noted, the majority of which were between August and December and included a party of four off Fox Holes Dene on 23 September; one of the largest individual counts away from the Teesmouth area.

Between 2003 and 2008, a small wintering population became established at Teesmouth with two to three birds commuting between Seal Sands and the nearby Reclamation Pond each winter, though a maximum of six birds was reached on the 18 February 2006. No birds returned for the winter of 2008/ 2009, and just one bird was recorded subsequently on the 30 January 2010. The reasons for this decline are unclear though it is possible that the same birds were responsible for the majority of sightings and have switched to a different wintering area. The count of six was the largest number of this species in County Durham to date although split between two localities with four on Seal Sands and two on the nearby Reclamation Pond.

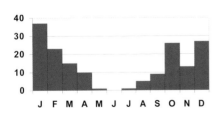
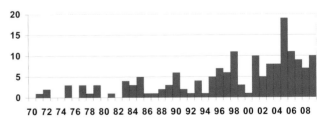

Records of Slavonian Grebes in County Durham 1970-2009

At least fourteen birds have been noted passing Whitburn Observatory between August and April, all of which have flown north; there are no records of migrating birds between January and March. Particularly early birds passed on 8th and 9 August 2005, though the peak period of passage is from September to November, when a total of nine birds have been recorded including a single party of four birds on 4 September 2010. Return passage has also been noted in April with a single on 12 April 2008 and two on 10 April 2009. There are just two records specifically referenced as flying north past Hartlepool Headland, on 10 November 2001 and 31 October 2008, with the latter bird also passing Whitburn.

Although essentially a marine species during the winter months, in excess of 25 birds have been recorded inland to some extent, excluding those recorded on the North Tees Marshes. The majority of records are from the eastern half of the county with most large water bodies having at least one record. Hetton Lyons Country Park has the distinction of being the best inland site for this species with three records of singles on: 7 December 1995; 22nd to 24 October 2005 and 7 November 2006. Four other sites have also received more than one record: Charlton's Pond, Billingham, with singles in January 2003 and September 2005; Crookfoot Reservoir, singles in April 1993 and October 2003; Derwent Reservoir, singles on 21 October 1984, 27 March 1988, and 18th to 26 July 2011; Shibdon Pond, singles in December 1983 and February 1996; and WWT Washington, singles in October 1987 and February 1996. Further west, the species is much rarer with individual birds at: Selset Reservoir, on 19 February 1975; Lambton Ponds, Witton-le-Wear, from 23rd to 25 March 1988 and at Lartington Low Pond from 26th to 27 January 1997. With the exception of the birds at Charlton's Pond in September 2005 and Derwent Reservoir in July 2011, all of those recorded inland have been between October and April, which is the peak period for this species in the county.

The only mid-summer coastal record was of a summer plumaged adult flying south past Whitburn on the 3 July 1984. This bird was presumably a failed breeder from further north (Bannerman 1959), though why it should be heading south so early is difficult to explain as autumn migration does not normally take place until late August and reaches its peak during October and November (Cramp *et al.* 1978). The individual at Derwent Reservoir in July 2011 was notable in both date and location.

The earliest bird to be recorded during autumn is of a single flying north past Whitburn on 8 August 2005, with it or another flying north past the same locality the following day. A summer plumaged adult present at Barmston Pond, Washington from 21 April to 16 May 1975 represents the latest spring date for a bird in the county, though a summer plumaged adult on Dorman's Pool, Teesmouth on the 1 May 2004 is the only bird to have actually arrived in May.

Distribution & movements

Slavonian Grebe breeds in the boreal and sub-Arctic zone across much of the Old World and in North America, wintering in predominantly coastal habitats often to the south of its breeding range, and chiefly on the open sea. In Western Europe the majority of breeding birds are to be found in Iceland and Scandinavia, though a small population has existed in northern Scotland for over a century and in 2009 was estimated to be 29 breeding pairs, having recently suffered minor declines (Holling 2010). In Britain the species can be found wintering around the entire coastline in small numbers though the majority are to be found at a number of favoured sites, in particular the Firth of Forth, the Moray Firth, the Clyde Estuary, Pagham Harbour in Sussex and the sheltered estuaries of Devon and Cornwall. The UK wintering population was estimated at around 725 birds between 1986 and 1993 (Baker *et al.* 2006).

Black-necked Grebe
Podiceps nigricollis

A scarce passage migrant and rare winter visitor, which has been recorded in all months and has bred on at least six occasions.

Historical review

Selby (1831) said that immature birds were sometimes killed upon the coasts or estuaries of the region, but this statement may have had a strong bias towards records in Northumberland. In direct relation to Durham, Hogg (1845) documented one that was killed at Teesmouth in January 1823 (The Zoologist 1845), which appears to be the first specific reference to the species in the county. Nelson (1907) recorded birds at Teesmouth on 12 April

1846 and in the winter of 1874/1875, with Hancock (1874) describing it as "a very rare winter visitant", though Tristram (1905) said he knew of only a single Durham specimen, but he gave no date or location for this.

"Normally a very rare winter visitor", and "the rarest of the grebes" to be recorded in County Durham, was how this species was described by Temperley (1951), though he also recorded a breeding record. After recent summer records in Northumberland and North Yorkshire, in the early 1940s (Kerr 2001, Mather 1986), Temperley felt that it was not a surprise when Black-necked Grebes bred in County Durham in both 1946 and 1947. The unfolding of these events began on 3 June 1946, when two adults were present on a 'flooded brick-field' in northeast Durham, and on 23 July were seen to be feeding a juvenile, with two juveniles being noted there on 28th, and a third adult present on 9 August 1946. Birds were still present at this site on 22 August, and two recently fledged juveniles seen by Dr. H.M.S. Blair several miles away on Whiteleas Pond at Newton Garth Farm, near South Shields during 13 August to 8 September 1946, were not thought to have originated from the original 'breeding site' and were presumed to have bred locally, though this was never proven (*British Birds* 40: 21-23). In 1947 a pair returned to the original site and again bred successfully rearing two young, though just a lone bird returned each summer from 1948 to 1950 and there were no further records from this site. This point is re-iterated by that fact that in 1952 it was expressly stated that the ponds where it bred "a few years ago" in County Durham were not occupied (Temperley 1953). This locality was Hebburn Ponds which was filled in during the post war period; Temperley noted that "with regret the locality is now to be reclaimed" (Temperley 1953).

In 1954, a pair was known to have summered at Thornley Flooded Field, near Shotton Colliery, being present from at least 26 June until the end of August, with an additional adult noted on 8 August. Breeding was also thought to have been attempted on this occasion but as no fledged juveniles were seen, any attempt made must have been unsuccessful. Throughout the rest of the 1950s there were just two published records of singles at Whitburn on 23 February 1959 and Jarrow Slake from 1st to 4 August 1959.

Reports of birds were scarce during the 1960s with: one at Hartlepool on 15 January 1961; two in the Tees Estuary on 2 August 1964; a bird in breeding plumage on Cowpen Marsh on 21 August 1965 and a bird on Seal Sands on Christmas Eve 1966.

It is interesting to note that the majority of Black-necked Grebes recorded in the county between 1940 and 1969, with the exception of breeding birds, tended to occur in the period from August until December, suggesting that most were dispersing from breeding sites elsewhere in Britain. This is in direct contrast to the species' more recent status, which shows a strong spring bias with most being recorded during April to June, though juvenile birds continue to be recorded in the county during August and September. This is presumably related to the increase in the number of pairs breeding in Northumberland, where breeding first occurred in 1963 and had risen to an estimated 20 pairs during the late 1990s, becoming established as the county which held the majority of breeding pairs in England (Kerr 2001

Recent breeding status

Like many other species of grebe, Black-necked Grebe require shallow, usually lowland water bodies, with abundant emergent vegetation and usually copious amounts of submerged plants such as amphibious bistort *Polygonum amphibium*. In 1971, the first successful breeding in the county since 1947 occurred when a pair reared young on a temporary pond close to Fishburn (Raw 1985). The pair eventually fledged at least two of the three young hatched, though did not return the following year as the pond had dried out and there were no subsequent reports from this site.

This species has always been a rare breeding bird in the county (Raw 1985) and this was very much evident during the *Atlas* survey period (Westerberg & Bowey 2000). Birds were noted on apparently suitable wetlands in four out of the six survey years, but only one of these produced records in more than two years consecutively. This site, McNeil Bottoms, held a pair in 1988 during late May and early June with three birds there in May and June 1989, and a single bird was present in late April 1990. Courtship display was noted in some years, but successful breeding was never confirmed (Armstrong 1988-1993).

Since then, the breeding status of the species in the county has changed somewhat. Birds bred successfully for the first time in many years at McNeil Bottoms in the mid-Wear valley in 1999, when a pair present from 3 June to 1 August successfully reared two young. Birds returned to this site from mid-June to mid-July 2000 though there was no evidence to suggest that breeding had been attempted; the birds did not return the following year.

Two breeding attempts took place in 2004, though only one of these was successful. A pair which arrived at Brasside Pond, Durham on 22 June was later seen with two young on the 28 July and subsequently hatched another young, though only two survived to fledging. One of the juveniles remained on site until the 17 September and presumably the same bird was seen at Barmston Pond, Washington around 18km away from 18th to 23 September. In addition to this, a summer plumaged adult that was on the Reclamation Pond, Teesmouth from 30 March was joined by a second bird on the 6 April with display and nest-building taking place, but a Coot *Fulica atra* ousted the birds from their nesting platform and the pair was last seen on 6 May (Joynt *et al.* 2008). This same aggressive Coot was also responsible for the failure of a pair of Slavonian Grebes *Podiceps auritus* at the same site in 2005.

Subsequent to the successful 2004 breeding attempt, Brasside Pond held birds each summer from 2005 until 2008, though breeding did not take place again until 2007. A pair which arrived on 16 June 2007 had hatched two young by 12 July, both fledging and being last observed 14 September. In the following year, what was presumably the same pair was present from 2 April 2008 and although display and mating were observed on a number of occasions, successful breeding did not take place. Despite being present each summer from 2004, successful breeding took place on just two occasions at this site, and there were no further reports from this locality after 2008, possibly related to a rise in the water levels and the consequent loss of emergent vegetation. This site was also often used for shooting and suffered from some human disturbance.

It would appear that the species has begun to establish a tentative hold as a breeding species within the county as suggested by the *Atlas* (Westerberg & Bowey, 2000), though currently there are no regular sites. Neighbouring Northumberland has long been a favoured stronghold of this species (Kerr 1995) and there seems to be no reason why birds should not continue to breed occasionally within County Durham. Part of the reason for the increased appearance of the species in the county over this period, may relate to the change in status at one of the country's' principal sites for this species, Capheaton Reservoir in Northumberland. This site, which during the 1990s held up to 17 pairs, was drained in the early 2000s, with a subsequent dispersal of what was believed to be the breeding birds from this location to other sites in the north east of England (Kerr 2001). Over this period, a number of sites in Northumberland held breeding birds after the original site's loss of suitable habitat, which could also be the reason for the species appearance as a more regular breeding bird in Durham. Threats to the species include habitat change and degradation, egg theft and human disturbance of breeding sites.

Recent non-breeding status

In excess of 200 Black-necked Grebes have been recorded in County Durham since 1970 and the species is an annual passage migrant in both spring and autumn, occasionally recorded during the winter months. There have been six years without records: 1970, 1976, 1980, 1982, 1991 and 1996, with annual records in ever increasing numbers since 1997. This is illustrated by an increase in the average annual number of records from two birds per year during 1970 to 1999 to eight or nine per annum during the period 2000 to 2009, including at least 18 during 2005, which was the peak year of occurrence to date; though 15 were recorded in 2004 and 14 in 2006. This dramatic shift is thought to be related to birds being displaced from their traditional breeding grounds in Northumberland following the drainage of the main breeding site there during the early part of the 2000s. Black-necked Grebe is most often a freshwater species during passage and is much less likely to be recorded on the open sea than its close relative the Slavonian Grebe *Podiceps auritu*s, though it does frequent inshore waters during the winter months. Records are widely scattered throughout the county with few substantial areas of freshwater yet to receive a record.

In direct contrast to the 1950s and 1960s, the period from March to June has established itself as being the peak period of passage for Black-necked Grebe within County Durham, with almost two-thirds of records since 1970 being from this period. The majority of records are from the eastern half of the county and birds have penetrated further west on just two occasions: to Derwent Reservoir on 7 May 1988 and Escomb Lake on 19 May 2006. The North Tees Marshes has recorded in excess of thirty birds, though other favoured localities during spring include Longnewton Reservoir, with six records of nine birds and Hurworth Burn Reservoir with five records totalling eight birds.

Spring passage begins in mid- to late March, with most birds' appearance being brief; all but two of the March records were of birds on single dates, highlighting an urgency to locate suitable breeding areas. The notable exception was of a summer plumaged adult at Brasside Ponds from 23 March to at least the end of June 2005, which was presumably hoping to attract a mate as the species had bred successfully there in 2004.

Passage reaches its peak during April and May when at least 85 birds have been recorded. Most do not linger, though a single bird remained on the North Tees Marshes from 10 May to 18 July 2001, and three were at Castle Lake, Bishop Middleham from 3 May until late August 2005, none of which were successful in attracting a mate. There are a number of records of two birds together, some of which pair-bonded and displayed, though most move through quite quickly, presumably to breed at sites further north in Northumberland. The highest individual count is of four birds at Saltholme Pools, Teesmouth from 22nd to 24 April 2002, though parties of three have been noted on at least four occasions: at McNeil Bottoms, in May and June 1989; at Longnewton Reservoir, on 8 May 2000; at Castle Lake, Bishop Middleham, from 3 May to late August 2005 and at Hurworth Burn Reservoir on 17 May 2008.

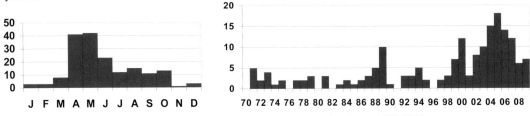

Records of Black-necked Grebes in County Durham, 1970-2009

Passage begins to tail off by June and by mid-month it is not possible to separate late spring migrants from failed breeders, which is illustrated by the fact that several of the records during June relate to obvious pairs. The most notable record during the month of June is of a juvenile present at Hetton Lyons Country Park on 26 June 2005, the earliest such bird to appear in the county, by almost a month. It had presumably been hatched in late May or early June as the incubation period for this species is said to be 20-22 days, with fledging estimated to take a further 21 days (van Ijzendoorn 1944).

Return passage begins in July and around a quarter of Black-necked Grebes recorded in County Durham have been noted between July and October. Autumn passage mirrors that of spring with the majority of birds recorded from lakes and reservoirs in the eastern half of the county with just two records from the west: at McNeill Bottoms on 4 October 1987 and Balderhead Reservoir on 22 October 1993. The North Tees Marshes has attracted a total of 12 birds during this period, followed by six at Longnewton Reservoir with four each at Crookfoot Reservoir, Hurworth Burn Reservoir, and Rainton Meadows.

Sightings in July relate to adult birds, some of which are presumably failed breeders and the occasional juvenile towards the month's end. Single adults have been recorded at: an undisclosed locality, from 12th to 20 July 1975; a 'central Durham site', on 23 July 2000; Cowpen Marsh, Teesmouth, 16th to 17 July 2000; Rainton Meadows, 21st to 22 July 2004, and from 21st to 24 July 2005 (accompanied by a juvenile); SAFC Academy Pools, 26th to 27 July 2005; and, more unusually, two adults were on the sea off Ryhope from 23rd to 27 July 1994. There have been four records of juveniles during July, the earliest of which was at Rainton Meadows from 21st to 24 July 2005. Others were at Watergate Park, Gateshead from 25th to 31 July 2004, with Longnewton Reservoir providing records from 27 July 2004 and 23rd to 29 July 2009.

A total of 31 have been recorded during August and September with all but one of those specifically aged, relating to juvenile birds. The only bird aged as an adult during this period frequented Saltholme Pools, Teesmouth from 9 August to 2 October 2010, though many adults will have lost their summer finery by mid-September and are more difficult to distinguish from juveniles. Most records are of single birds though two juveniles were at Hurworth Burn Reservoir on 11 September 1999, and two further juveniles were at Saltholme Pools, Teesmouth on 28 August 2003. A further 13 have been recorded in October, the most noteworthy of which were a long-staying bird at Haverton Hole, Teesmouth from 27 October to 14 November 1973 and two on the sea off Whitburn on 31 October 2009.

Just 18 Black-necked Grebes have been recorded in the period from November until February, most of which relate to birds seen at coastal localities, with a distinct increase in the number of records since 2000, including a total of five birds in 2010. Seal Sands, Teesmouth is the favoured locality with four records involving five birds, including a single bird present from 30 January to 16 April 2010 and two birds from 21 December 2010 increasing to three during the early part of 2011, which may indicate that the species in the process of establishing a wintering population in the area. All other coastal records from the mid-winter period have been from Hartlepool and Seaton Carew. Inland, single birds have been noted at Low Barns on 16 January 1971, and 8th to 14 February 1971; at Cox Green, Washington on 21 February 1979; at Longnewton Reservoir from 3 November to 18 December 1984; at Saltholme Pools, Teesmouth from 18 February to 20 March 2002 and at Silksworth Lakes, Sunderland from 24 January to 2 February 2010.

Birds are rarely seen on the open sea and generally occur at times of passage with just 10 records falling into this category since 1970. A bird was picked up badly oiled on the shore at Sunderland on 15 October 1973 and died three days later. Two summer plumage birds were off Ryhope from 23rd to 29 July 1994 with singles off: Ryhope, on 22 October 1995; Whitburn, on 4 October 2000; Seaton Carew, on 10 December 2000; Whitburn, on 19 September 2001; Hartlepool Headland, on 6 April 2006; and, two off Whitburn, on 31 October 2009. A bird was found dead at Hartlepool North Sands on 15 January 2010 and one was at Hartlepool Headland on 28 February 2010. In addition to these records, three singles have been noted on active coastal passage, all flying north, at Whitburn on 22 April 1981 and at Hartlepool Headland on 18 September 2002, and 20 August 2005.

Salt water sightings from the more sheltered waters of Seal Sands, Teesmouth relate to winter occurrences with single birds on: 3rd to 4 January 1973; 27 January 2002; 17 March 2002; and, from 30 January to 2 March 2010, with two birds present from 21 December 2010 into 2011. Individuals were also present in Hartlepool Marina from 4 December 1999 to 18 February 2000, and from 30 December 2000 to 3 January 2001; possibly the same bird returning.

Distribution & movements

Black-necked Grebe in its nominate form has a large though fragmented breeding range, which extends from south western Europe eastwards to central Asia, with a large though isolated population in Manchuria and north eastern China. Two other poorly defined races breed in North America and Africa. During the winter months most birds winter to the south of their normal breeding range in a range of habitats, both freshwater and marine, though African birds are thought to be chiefly resident. A small breeding population has existed in Britain since at least 1904 (Parkin & Knox 2010) with a current population estimate of around 28 to 43 pairs (Holling 2010) having recently suffered a decline from a peak of 48 confirmed pairs in 2002 (Holling 2007). Britain is at the northernmost limit of the species' wintering range and during the winter the majority of birds are to be found along the south coast of England at a number of favoured sites. The UK wintering population was estimated at around 120 birds between 1981 and 1984 (Baker *et al.* 2006).

Honey-buzzard (European Honey-buzzard)
Pernis apivorus

A generally scarce passage migrant recorded from May to November, most frequent in September, with notable influxes in 2000 and 2008; bred in the 19th century but not confirmed since.

Historical review

The first documented record for County Durham is detailed in a paper by G. T. Fox entitled *Notice of some rare Birds, recently killed in the Counties of Northumberland and Durham*, which was read to the Natural History Society in 1831. The record is quoted by Fox as: "*In the beginning of September this year (1831) a bird was killed by the gamekeeper of John Gregson, Esq., of Durham, at Burdon (south of Sunderland) in that county, where he had observed it hovering about for nearly a week. It was sent to me as an unknown species and upon examination I identified it as the young of the Honey Buzzard. I further found that another specimen of this species had been killed about a fortnight after, on the chimney of the engine belonging to the paper-mill at Shotley Bridge*" (Fox 1831). This latter specimen is also of a dark morph juvenile, and can be found in the Hancock Collection, Newcastle upon Tyne, though it is labelled as "*Chester-le-Street, about the year 1836*" (Hancock 1874).

The historical status of the Honey-buzzard in the county has been variously described by Victorian authors with Selby (1831) calling it *"one of the rarest of the Falconidae"*, and giving no Durham records, whilst in contrast Edward Backhouse (1834) stated that *"Several specimens of this elegant bird have lately been shot in various parts of this county"*. Writing in 1840, Hutchinson declared that *"In the months of September and October the Honey Buzzard is frequently met with, but its stay is short as it is seldom observed in any other period of the year"*. T. J. Bold, writing to the *Zoologist* in October 1841 stated that Honey-buzzards had been *"abundant"* in that year and in 1848 he reported that a first year male, in uniform dark brown plumage, had been shot by Sir John Eden's keeper on his estate near Beamish on 7 October of that year, adding that it was *"the first in the district since the unwonted abundance in 1841"* (*The Zoologist* 1848). This particular specimen found its way into the collection of Lord Ravensworth. Most observations at this time suggest birds on autumn passage.

In 1874, Hancock, inferred a change in status and said *"It is certainly now, according to my experience, one of the commonest large birds of prey"*. His view may have partly reflected the scarcity of other large birds of prey at the time. He documented 25 specimens between 1831 and 1868 *"all taken within the two counties"* of Northumberland and Durham and said that *"it occasionally breeds in the district"* without giving any details specific to Durham. He did go on to say that: *"Young birds very much predominate, and usually two or three are taken about the same time and near the same place, as if they belonged to the same brood."* He also mentions that: *"This species arrives on our coast in May, and takes its departure in August, September and October, the old birds leaving the district first, the immature frequently not till the middle of October"*.

Presumably based on Hancock's commentary, Holloway (1996) considered that the north east may have had a residual breeding population until around 1900 and that some pairs may have been located in Durham.

The *Zoologist* of 1911 brought news of firm evidence of breeding with a report that Honey-buzzards–were nesting in Durham:*"Mr O. V. Aplin states that Mr. Isaac Clark has given him particulars of a nest of Pernis apivorus built in some beechwoods on the banks of the Derwent in 1899 (The Zoologist 1911). The nest contained two young, early in August of that year"*. It was said shortly after this that the only other recent breeding record in Great Britain, appears to have been in Herefordshire in 1895 (*The Zoologist* 1895).

In reality, the presence of this species in the Derwent valley was more than the transient activity implied by the above extract. In 1993, Bowey *et al.* described the detailed history of this species as *"one of the most fascinating"* of any that has been documented in the Gateshead area. The species was first documented in the lower Derwent valley, in what is now Gateshead, when a "*nearly mature male*" was shot near Blaydon in 1841 during what was part of a more general influx of birds to the British Isles. The October 1848 juvenile male, near Beamish, then followed.

In the late 19th century, the Honey-buzzard was indeed discovered to be breeding in the lower Derwent valley on the Gibside Estate. Details are to be found in *The Birds of the Derwent Valley*, Thomas Robson's report to the Vale of Derwent Naturalists (Robson 1903). He described his discovery of a nest containing two half-grown young, and three pieces of a wasps comb some 60 feet up a beech tree on the Gibside Estate in the summer of 1897. This was apparently not the first time the species had bred, as R. McQueen, the then keeper of the estate told Dr H. M. S. Blair in 1932 that *"when he was a boy the birds came every year for as far back as he could remember"*, with the first nest being found in 1896. A pair of birds was once again present in 1898, though in that year they did not nest but, in 1899 the locally named "*Big Hawks*" produced two young (Bowey *et al.* 1993). This proved to be the last confirmed breeding in the area and was referred to by Thomas Thompson in an address to the Tyneside Naturalists Field Club in May 1901. This was later confirmed by George Bolam, who stated that the nest was *"on the Derwent banks and contained two young ones in the early part of August."* It is conceivable that birds may have nested a little higher up the Derwent valley in 1898; for birds were noted at Shotley Bridge and a bird of the year was shot nearby.

After this period of breeding at the end of the 19th century, Temperley recorded the Honey-buzzard less frequently in the first half of the 20th century and gave no subsequent records of breeding, describing the species only as *"a rare casual visitor in spring and autumn, formerly an occasional summer resident"* (Temperley 1951). Nonetheless, birds were occasionally shot, for example an adult female was obtained in this manner at Hylton, Sunderland on 30 September 1938.

There is little documentary evidence of this species' occurrence in the immediate post-Temperley period. It remained a resolutely rare species in Durham either on passage or as a potential summer visitor with very few documented records, and just three records of single birds during the 1960s. These being: a passage bird at

Hartlepool on 17 September 1960; a bird coming in off the sea at North Gare on 4 September 1965; and, one over Fairfield, Stockton on 26 May 1968. One over Norton on 30 July was thought to be 'probably' this species.

Recent Status

There have been at least 120 Honey-buzzards recorded in County Durham since 1970, though most of these have been since 2000, with an estimated 80 birds recorded during two notable influxes in 2000 and 2008.

There were just three sightings in the 1970s, including spring observations at Teesmouth on 15 May 1972 and 20 June 1974, and just five records in each of the last two decades of the 20th century. The species' status has apparently changed very little for a large part of the century. All confirmed records have been between May and November with extreme dates being, of one north east over Rainton Meadows, Houghton-le-Spring on 4 May 2002, and a very late bird south over Seaburn on the 13 November 1982. There is an even later and remarkable record of one in the late autumn and early winter of 1982 when, following a coastal influx of Common Buzzards *Buteo buteo*, a dark morph, juvenile Honey-buzzard was found near East Boldon, being present from at least 14th to 24 November; although according to Baldridge (1983), it was also reported into late December of that year. Roosting in thick hawthorn hedges and a small willow copse, it spent much of the day on the ground taking what few earthworms and other invertebrates were on offer. This is a significant sighting and one of the latest reports of this species in Britain (Cramp *et al.* 1977), though the lack of supporting evidence casts an element of doubt over this record.

A total of eighteen Honey-buzzards have been recorded in spring since 1970, all of which have been in May and June. The majority of records are from coastal localities, several of which involve birds seen flying in off the sea. Further inland, a number of single birds have been noted with sightings south east over Darlington on 17 May 1989, over Hamsterley Forest on 30 May 2001, north east over Rainton Meadows DWT, Houghton-le-Spring on 4 May 2002, south over WWT Washington on 27 May 2004, over Langley Moor, Durham on 28 June 2006 and at Croxdale Hall on 10 June 2009 before flying off south.

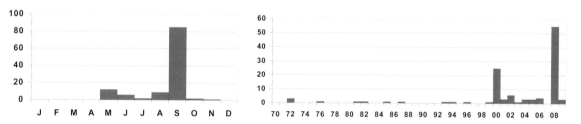

Records of Honey-buzzards in County Durham, 1970-2009

There are just two July records, the first of a moulting dark morph individual seen over Hamsterley Forest on the 22 July 2001, which was apparently a different bird to that noted there in May, and one flying west over Durham Cathedral, Durham City on 25 July 2009.

Prior to 2000, there were just six recent autumn records of Honey-buzzard for County Durham. These were of individuals: near Darlington on 3 August 1972; at South Shields Marine & Technical College on 3 September 1972 (this was a very weak and emaciated bird which later died); south over Greatham Creek, Teesmouth on 16 October 1976; south over Seaburn on 15 September 1981; one south there on 13 November 1982; flying in off the sea at Sunderland on 15 September 1985; and, west over Cowpen Marsh on 2 August 1987. The bird in 1976 is notable for being one of only two October records for the county, whilst the exceptionally late birds in 1982 remain the only November records for County Durham.

There have been more regular sightings since the turn of the Millennium which have changed the status of the species sufficiently for it now to be considered a scarce but annual passage migrant, which on rare occasions arrives in considerably larger numbers. The period 2001-2007 might be considered typical and produced sightings of 21 birds.

Exceptionally, in late September 2000 an observer at Marsden noted the first bird in what was, at that time, an unprecedented movement of Honey-buzzards through eastern England. A period of strong easterly winds and rain, associated with a weather system tracking across from the Continent, pushed a spectacular movement of over 500 birds from northern Europe across the North Sea to the east coast of Britain. This was the largest ever movement of the species recorded in Britain and in County Durham at that time, and an estimated 24 birds passed through the county between 20th and 30 September. Most of the birds were juveniles at predominantly coastal localities with several exhausted birds noted coming in off the sea before heading south. Inland sightings of single birds came from Esh Winning and Tunstall Reservoir on 20 September and WWT Washington on 27th. The final bird directly associated with this influx, a dark morph juvenile, was found 'grounded' at Marsden Hall on 30 September and taken into care but later died.

The unprecedented movement of birds in 2000 was exceeded by an even greater movement of Honey-buzzard through Durham and Britain in September 2008. Similar weather conditions prevailed with a weak low pressure system moving across the southern North Sea creating a strong south easterly airflow that funnelled many southbound Scandinavian birds to the eastern counties of England. The movement eclipsed that of September 2000 with an estimated 800 birds recorded in Britain, most of which quickly re-orientated and headed south, before leaving Britain via East Anglia and southern England. An absolute minimum of 55 birds were recorded passing through County Durham from 13th until 24 September, with an estimated 27 birds on 13 September alone, when at least nine birds passed through the North Tees Marshes in under two hours in what was quite easily the most fully documented occurrence of this species in the county. This influx was more protracted than that of 2000 with birds still being reported coming in off the sea as late as 20 September, as well as being on a broader front inland, the last of which passed south over Low Barns DWT, Witton-le-Wear on 24 September. As in the previous influx, the majority of birds specifically aged were juveniles with just four distinctive pale morph adults (Newsome 2008).

Apart from these two exceptional years, the Honey-buzzard has always been a very scarce bird in modern County Durham. There have been another fifteen autumn records since 2000, the majority of which are from coastal locations. Perhaps the most notable concerned a juvenile which spent almost four hours at Summerhill, Hartlepool on 19 August 2008, before moving south ahead of heavy rain, and a pale morph juvenile found exhausted close to the Seal Sands roundabout, Teesmouth on 28 September 2010. The latter bird was taken into care, with a broken foot. Unfortunately this injury was thought unlikely to heal by the close of the autumn migration window, and it remained in care until at least the following spring. Between 2009 and 2011, records reverted to a more typical pattern with two spring passage migrants seen in the county in the first of these years, three spring birds and one in the autumn of 2010, and just one autumn bird in 2011.

Records further inland include three in August and two in September from widely scattered localities, with a bird at Ruffside on 18th and 23 August 2003 raising hopes that it might have been locally bred. A very late bird at Dryderdale on 27 October 2000 is one of just two October records and was thought to have lingered from the influx in September.

Recent breeding status

There has been some suggestion since the turn of the 21st century or longer that birds may have summered or even bred intermittently in the county with sporadic sightings in the Hamsterley Forest area and in the upper Derwent valley, its old breeding haunt. This was illustrated when a pair were noted displaying in the north west of the county on 12 June 2006, in an area where summer birds have been reported in the recent past. It is possible that this very secretive species could be breeding undetected in the county. Honey-buzzards breed almost exclusively in well-wooded landscapes, from forests to wooded farmland, nesting in deciduous, coniferous or mixed woodland in both uplands and lowlands. They often utilise old crows' nests or squirrel drays and prefer mature trees with nests placed 10-30m above ground, preferably adjacent to clearings. Although wasp and bee pupae and larvae are their main food source, early in the season, frogs, lizards, small mammals and the nestlings of pigeons are important (Brown & Grice 2005). The availability of prey in cool springs will always be an important factor particularly for these northern birds, so the exploitation of frogs as a relatively abundant food source for newly-arrived birds as in northern Scotland could be relevant to any success in Durham (Roberts et al. 1999).

Distribution & movements

The Honey-buzzard is a widespread and common summer visitor over much of Europe and western Asia, except the extreme north. It occurs in woodlands from the Caucasus right through the Continent south into Iberia with the entire world population wintering in tropical Africa. The British breeding population is small – estimated between 33-69 pairs in 2000 (Batten 2001) - though this may be expanding (or be under-recorded). During a study in 1989-1997, three distinct breeding habitats occupied by this species were identified – lowland southern woodland, central hill country with mixed farmland/woodland and upland coniferous plantations (Roberts *et al.* 1999). A total of 52 nests were located over that period in 16 nesting areas, and 32 different nests were used; eight in the uplands, four in the central hills and 20 in the lowlands. One to three pairs have been nesting in northern Scotland since the early 2000s at least, perhaps originating from Scandinavian migrants – which are the most likely source of any birds which may settle to breed in Durham. Some of the records within County Durham may relate to birds which have bred in Britain, although most are probably of Scandinavian origin, as were those associated with the influxes in September 2000 and September 2008.

The Honey-buzzard has always been a scarce breeding bird in Britain. According to the Rare Breeding Birds Panel (RBBP), between two and 33 pairs were confirmed as breeding between 1991 and 2002 rising to a five-year mean of 40 pairs in 2009. At 2010 it featured on the amber list of Species of Conservation Concern.

Studies of radio-tagged birds from both northern Scotland and south east England have confirmed migration routes through France, Spain, Morocco/Algeria and Mauritania to sub-Saharan wintering areas around the Niger delta and in Guinea, involving distances of over 4,500km (www.roydennis.org). Although the Scottish chicks initially made their way south west into north west England before changing tack to a south easterly direction, the English nesting birds moved quickly east of due south before re-orientating. If these are typical routes, this would have a bearing on Durham sightings.

Black Kite
Milvus migrans

An extremely rare vagrant from Europe with just four records.

The first authenticated Black Kite record for County Durham was seen by Peter Collins and Dave Foster from the old railway line at Peepy Plantation, Washington on 21 May 1981 (*British Birds* 75: 496). The bird was initially flying low, south west being mobbed by corvids and Jim Pattinson, who had been alerted, managed to acquire photographs before it was lost to view.

All records:
1981	Peepy Plantation, Washington, 21 May
1985	near Stanhope, 21 April (*British Birds* 79: 539)
1988	North Gare and Hartlepool, 16 April (*British Birds* 86: 466)
1994	Belasis Technology Park, Billingham, flew north, 17 August (*British Birds* 88: 504)

The bird in 1988 was initially watched flying north west at South Gare, North Yorkshire, before crossing over to the north side of the Tees estuary. It was last seen amongst the cranes of Hartlepool Docks.

Distribution & movements

The nominate race of the Black Kite breeds across much of Europe and North Africa, eastwards to north western Pakistan, though it is absent from Scandinavia. European birds winter in sub-Saharan Africa where they mix with the resident population. There are a number of other races of Black Kite in both Africa and Asia. The Black Kite is an increasingly regular vagrant to Britain, most frequently as a spring overshoot from April to June, although it has occurred in all months. It was removed from the list of species considered by BBRC at the end of 2005 by which time there had been over 350 records, the majority of these from southern England. In 1981, when first seen in our area, there had been 53 national records.

Red Kite
Milvus milvus

Sponsored by

Go North East
www.simplygo.com

Historically, the Red Kite was a breeding bird in County Durham until, following a sustained period of persecution by man; its status was reduced to that of a rare vagrant and passage migrant. This changed following the advent of the national reintroduction programme in 1989 and in 2004, with the north east reintroduction in the lower Derwent valley.

Historical review

Historically, this species was known to be common across Great Britain as documented in many literary and cultural references (Coker & Mabey 2005). In medieval times, it fed in London, where it was given special protection for its role in cleansing the streets, and at that time was probably the most widespread large bird of prey in Britain (Carter 2001).

Three hundred years ago, kites would have been a common sight over much of northern England but their decline, commencing in earnest during the 17th century, accelerated through the subsequent hundred years. Persecution took the form of both shooting and poisoning - principally during the 18th and 19th centuries, leading to its near extinction by the late 19th century (Carter 2001). At this time, just a handful of birds remained in Wales (Evans 1991) and this was the case, in effect, until the latter part of the 20th century (Mead 2000).

The Red Kite became extinct as a breeding bird in England by the end of the 1870s, the entire UK population being restricted to just a handful of birds in the valleys of central Wales by the early 20th century. Genetic analysis established that the whole Welsh population originated from just a single female (Bowey 2007). It, therefore, remained a rare breeding bird in the UK for over 200 years, with less than twenty pairs throughout much of the 20th century. Similarly, the species has been rare in Durham for more than two centuries (Temperley 1951). Clear evidence of its former presence comes from its old-fashioned name, Glead or Glede, (from the Anglo-Saxon *glidan*, 'to glide or slip away'), being enshrined in local place names (Bowey 2007). In 1769 John Wallis wrote, "*We have the Glead, or Swallow-tailed Falcon, in the alpine and some of the vale woods*" (Rossiter 1999). Documentation of the species' local presence during the 16th century comes from the late 1970s' excavations of the Castle Ditch in Newcastle upon Tyne (Rackham & Allison 1981).

There are just a handful of documented 19th century occurrences of Red Kites in County Durham (Temperley 1951):

- 1834: Three birds were 'taken' near Bishop Auckland, Durham. It is possible that one of these birds, a juvenile, found its way to the Hancock Museum. The multiple nature of this record, and the ageing of the museum specimen, gives tentative evidence to suggest that this was the last documented breeding in the county.
- 1842: One was shot near the Whitworth Estate, Spennymoor, season unknown.
- 1883: One flew in from the sea at Teesmouth on 15 September. This was presumably the bird, which was later shot near Warrenby, to the north of Redcar and on the south side of Teesmouth. This specimen was purchased by the Dorman Museum

John Hogg (1845), described Red Kites as "*rare visitants*". In his summary of the species, George Temperley (1951) described it as, "*a rare casual visitant, of which there have been no records for over half a century. Once a resident*". Based on this evidence it seems reasonable to assume that kites, as breeding birds, had gone from County Durham by the 1840s (Lovegrove 1990).

Recent status

In the early part of the 20th century, the Red Kite was recorded in the county in only seven years between 1900 and 1989, prior to the commencement of the British reintroduction programme in the latter year. It was not recorded at all in the first seven decades of the century. Between 1975 and 1988 there were only eight occurrences involving probably just seven birds (Bowey 2007).

A Red Kite was seen flying 'in off the sea' at Ryhope on 27 April 1975, and subsequently over Grangetown, Sunderland. This was the first Durham record since 1883. There were a further six records to the end of 1988, all of which pre-dated the British re-introduction scheme. These were:

1976	On 18 September, one at Hamsterley Forest
1981	On 6 April, one in the Raby Castle area, near Staindrop for up to a week
1982	One in the Cleadon Hills area of South Shields 'for a few weeks' in February, with probably the same bird later seen flying south over the A19 towards Shotton, on 15 March Another in the Eggleston area of Teesdale from May through to July was possibly the same bird. This individual spent some considerable time in the Blackton Beck area (to the west of Eggleston village)
1985	A bird was found poisoned at Hamsterley Forest on 12 April 1985. Evidence from *post mortem* showed that the bird had ingested the pesticide Alphachloralose (the kidney tissue examined having a poison burden of 25mg of Alphachloralose per kg)
1987	One was seen moving south over Hawthorn Dene on 15 March
1988	One at West Rainton, near Durham City, on 24 March

In 1989, various national agencies began a programme of reintroductions designed to return the species as a breeding bird to the whole of the UK. The initial releases were on the Black Isle in Scotland and the Chilterns, in England. Despite these operations, Red Kites remained very scarce birds in Durham, with just 15 reported in the 13 years between 1989 and 2001. The observed pattern of sightings, with nine reports in spring (between March and June), indicates a spring dispersal of first-summer or non-breeding birds away from the re-introduction areas. There is another cluster of reports (six sightings) over the months October to December, which may have involved some continental migrants visiting Durham, or wandering kites from the British release programme, perhaps prospecting for potential breeding habitat or mates. It was only after successful breeding was established in both northern Scotland and the Chilterns that Red Kites began to be recorded with significantly increased frequency in the county, from around 1997. Records have been almost annual in Durham since 1993 (apart from 1998 and 2002), though it was not until 1997 that a wing-tagged bird was seen, confirming that birds from the reintroduction scheme were reaching County Durham. Subsequently, Durham records increased considerably.

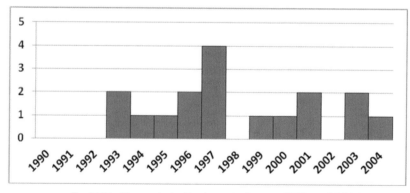

Red Kite Records in Durham between 1989 and 2004

The series of four sightings in 1997, all of them in the south west of the county, prior to the commencement of the Yorkshire releases (in 1999) is intriguing. It represented the most concerted pattern of Red Kite 'activity' in Durham for over 150 years. Considering the paucity of records in the past, it seems unlikely that four birds were involved in these sightings – especially considering the pattern of records both before and immediately after this cluster of activity. Partial wing-tag combinations were noted on two birds, confirming that at least two individuals were involved; it seems more likely that two birds only accounted for these records – despite their spread over a nine-month period.

In 2004, a bird was noted at Boldon Flats on 10 June and it or another was seen at Ruffside in the Derwent valley on 13th; the same bird being in the Townfield/Ramshaw area of the upper Derwent valley on 15th of the month. This was the last observation of a Red Kite in Durham prior to the commencement of the Northern Kites' release programme, just one month later, in the summer of 2004.

The Red Kite's local status as a passage migrant in Durham had changed quite markedly over the last two decades of the 20th century with an increase in sightings occurring long before the national, reintroduction work began. There were probably two main factors responsible for this change. Firstly, a genuine increase in the number of continental immigrants to the area during the 1980s in a process that was supplemented later, by the appearance of birds from reintroduction schemes elsewhere in the UK (Bowey 2007). This pattern of records is not isolated and was noted elsewhere in England, most evidently in Norfolk (Brown & Grice 2005). The driving force behind this shift in status can probably be traced to the increase in the migratory Swedish population, which took place from the 1970s onwards (Hagemeijer & Blair 1997).

Release of the first Red Kite, 'Speedy' (courtesy Ken Sanderson)

Recent breeding status

In the early 2000s, the north east of England was chosen for a Red Kite release project as part of the national programme. The resultant Northern Kites Project, released 94 birds between 2004 and 2006 at two 'secret locations' in the lower Derwent valley; these were the National Trust's Gibside Estate and Northumbrian Water's Lockhaugh Sewage Treatment Works, at Rowlands Gill, less than six miles from Newcastle City Centre. This was a world first in large bird of prey conservation with a reintroduction exercise based in a semi-urban location. The chicks originated from the re-established UK population in the Chilterns.

In 2005, birds attempted to breed, unsuccessfully, near Sherburn Tower Farm, Rowlands Gill, about one kilometre from the main release sites. In 2006 two pairs raised young, one pair in the Derwent valley and one pair in Langleydale, near Kinninvie, marking the first breeding in the county for around 170 years. By 2007, the first pair had crossed the county's northern boundary to breed successfully near Wylam, in Northumberland.

	Breeding Pairs	Pairs Fledging Young	Young Fledged
2005	1	0	0
2006	4	2	3
2007	10	7	11
2008	22	15	23
2009	25	13	20
2010	27	13	24

By 2008, the first pair to breed successfully had bred for the third year in succession and they had also became grandparents, one of their 2007 female offspring becoming the first Durham-bred generation of kites, to produce young (in a nest to the west of Derwent Reservoir, near Blanchland). This year also brought the area's first successful nest of three young, whilst one first-time nesting female managed to lay four eggs, though only one youngster was successfully reared. Analysis showed that her other three eggs were infertile. This female was a Yorkshire-bred bird.

In 2008, birds bred in four well-differentiated areas of the north east, despite poor summer weather. Over the years 2005-2010, there was a steady growth in the number of breeding pairs.

The 2009 breeding season continued the upward trend of previous years, despite the indifferent weather during the crucial early summer breeding period. More significantly perhaps, in 2009, a pair bred successfully in a large Rowlands Gill garden, with at least one youngster being fledged; the nest was in a Scots Pine *Pinus sylvestris* less than 45m from the house.

The rise in breeding numbers was less pronounced between 2009 and 2010 than in previous years. In part, this may have been an artefact resulting from the fact that, as birds become more widespread, they were more difficult to monitor accurately. There were however, some checks on the population's growth partly through the impact of poor weather in the breeding season and the hard winter of 2009/2010. As at 2010, the stronghold for breeding Red Kites in Durham remained the Derwent valley although there were some signs that it was spreading slowly into other areas.

Away from the Gateshead concentration, several other key areas are beginning to develop. Shortly after their release, birds quickly established a breeding location in Langleydale in the south west of the county, but these birds were lost around 2010. In late April 2010, the male of this pair was found dead (along with a Buzzard *Buteo buteo*) next to a rabbit carcass close to the nest. The female was incubating at the time and presumably deserted. The Toxicology Report showed that the kite had died from ingesting carbofuron, sadly the first known instance of poisoning in Durham since the species' return.

The Northern Kites monitoring programme, which was continued by Friends of Red Kites (FoRK) from 2010, revealed a number of fascinating details. For instance, post-breeding inspection of nests showed a range of unusual items used as decoration including: a white teddy bear's head; a discoloured white sports sock; plastic supermarket bags; a range of soft toys; once white gloves; a sponge ball; an England football flag; and, a pair of underpants - confirming Autolycus' admonition from Shakespeare's *The Winter's Tale* (1610), "*when the kite builds, look to lesser linen.*"

The Northern Kites Project strove to show Red Kites to the public and from 2006 to 2008 a public viewing point close to nesting birds by the Derwent Walk attracted over 24,000 visitors, with 10,805 in 2006, the year of the first nest. The involvement of schools, Red Kite 'branded' buses, the promulgation of Red Kite health walks and the bird's conspicuous presence ensured that people in the Derwent valley took the birds 'to their hearts'. This once reviled 'poultry thief' rose on a tide of public goodwill to become the iconic symbol of the valley.

Recent non-breeding status

By the latter part of the first decade of the 21st century, the Durham Bird Club was receiving well over 200 Red Kite records per annum, mainly from the core areas that had been so central to the species since its return. These were the lower Derwent valley and Barlow Burn, parts of the Causey Arch/Beamish Burn area of the Team valley; Clara Vale, Wylam and around Eggleston and Kinninvie (in Langleydale and Teesdale). By 2006, in Rowlands Gill and the surrounding area, Red Kites had become common garden birds, with numerous instances of birds visiting back-garden feeding stations. It is a true opportunist, exploiting a variety of food sources. Live vertebrate prey has not been recorded as being routinely taken by local birds, with a notable exception, the remains of young corvids have often been be found at nest sites. Since the first releases, Red Kites have been seen routinely over the high-density housing estates of Winlaton, Swalwell and, after the large 'car-boot sales' held at Blaydon Rugby Club, birds have often been seen foraging on the waste left by visitors – having clearly reverted to their mediaeval role as urban scavengers.

In winter 2005/2006, large-scale communal roosting by Red Kites commenced in the Derwent valley, firstly at the well-used Sherburn Towers site, at Rowlands Gill, and later, in the Gibside Estate. These roosts grew over subsequent winters, with up to 40 individual birds noted during a Northern Kites coordinated roost count on 12 December 2007. The ongoing monitoring work revealed that over 55 different kites were using the lower Derwent valley and Gateshead over the 2008/2009 winter period; this was probably a considerable underestimate of the real numbers.

In 2009 and 2010, the highest counts came from the communal roosts in the lower Derwent valley, at Sherburn Tower, Rowlands Gill, and in the Snipe's Dene area of the Gibside Estate, slightly lower down the valley. Birds routinely commuted between these two main locations, depending on the prevailing weather conditions and group dynamics.

Distribution & movements

From a global perspective, the Red Kite is confined almost entirely to Europe. It is sparsely distributed over much of the Continent, being somewhat more widespread in the east. The largest populations are found in Germany, France and Spain, with smaller numbers in Portugal, and a tiny remnant population in Morocco (Hagemeijer & Blair 1997). Approximately 75% of the European population, an estimated 19,000-25,000 pairs in 2004 was found in Germany, France and Spain, with significant numbers also present in Sweden. The British population was estimated to be a minimum of 1,156 pairs by 2009 (Hollings 2011). As a bird of passage, the species is scarce but quite widely recorded along the eastern counties of England. There are a small number of records of German-ringed birds being recovered in eastern England in spring (Wernham *et al.* 2002). Occasional continental migrants reach the British east coast in autumn and winter, but the numbers involved in any such movements are relatively small (Brown & Grice 2005).

British kites are apparently, largely non-migratory but young birds wander during their first two winters (Carter 2001). In Durham, late spring and early summer is a time of dispersal for first summer and non-breeding kites and this period usually produces sightings of single birds away from the core areas. Nonetheless, by 2010, birds were not infrequently reported from locations across the county. Observation of the wing tags of birds has revealed considerable movement between the various re-introduction areas in Britain. This was evident from 2004 onwards when one of the first birds to be released in the Derwent valley, moved to Wales, then to the Chilterns before returning to Gateshead and then heading south to Yorkshire, where it eventually bred. Over the period 2005-2010, birds from Dumfries & Galloway, northern Scotland and Yorkshire were all recorded in Durham.

White-tailed Eagle
Haliaeetus albicilla

An extremely rare vagrant limited to historical records only with none seen for almost a century

Historical review

This species probably once bred widely throughout lowland Britain and Ireland (Brown & Grice 2005) but by the 18th century there were just a handful of birds left in England; most of these probably being in the Lake District until the 1790s (Holloway 1996); though breeding in the Keswick area was said to persist until the close of the 19th century (Macpherson 1901).

Historically, this species may well have been present in County Durham based on the presence of the name 'erne' which was highlighted by Yalden (2007) as being embedded in the place name *Yarnspath Law*, which was listed by him as being in Durham. The name can be translated as eagle's path (Yalden 2007). In fact, this location is close to Kidland Forest in the Cheviots, in Northumberland some 40km to the north of Durham. Nonetheless, this does indicate a presence of the species in the north east of England at the time of the development of this place name. There are less ambiguous records of the species' presence in Durham from Roman times, with the excavation of its bones from the Roman site of Binchester, in the central south of County Durham (Yalden 2007). These birds may or may not have been associated with what would have been the once extensive wetlands associated with the River Wear or the River Skerne Carrs.

Drawing on other sources, Rossiter (1999) speculated that this species was very much more common in the past, probably being widespread across England, and perhaps breeding in the north east, into Anglo-Saxon times, at least until the 8th century. Perhaps they were also present in Durham? There is a suggestion of breeding in the Hexham area around 700 AD, according to the '*history of Erneshou*' (Kerr 2001), so such activity is not completely improbable.

Temperley (1951) called it "*a very rare casual visitor*". Selby (1831) stated that several had been killed in Northumberland and neighbouring counties. Hogg (1845) recalled that one was shot at Teesmouth on 5 November 1823 by Mr. L. Rudd. This bird was actually shot on the Yorkshire side (Mather 1986) but Edward Backhouse stated that it had been previously recorded at Norton, near Hartlepool which qualifies it as Durham's first documented occurrence. Originally, this bird was reported as a Golden Eagle *Aquila chrysaetos* but it was re-identified from the corpse. This record may have referred to a bird reared in the British Isles, as the species was probably still relatively, widespread across the country and around this time, some 200 pairs were thought to be still breeding (Holloway 1996). A possible 'sea eagle' was reported in September 1837 near Cocken on the Wear...."*but it could never be approached within gunshot*" (Temperley 1951), in all probability it was a White-tailed Eagle. Hancock (1874) gave a Durham record from Ravensworth Park. M. Richardson's *Local Historian's Table Book* said this occurrence took place in 1835. This fascinating reference to the species relates to a bird that wintered on the Ravensworth Estate, in the Team valley, possibly in 1835. The bird was first noted by Thomas Hancock whilst he was out walking near the Lambton Estate in the Wear valley, not far from Chester-le-Street. It later took up residence in the Ravensworth Estate where "*it made a safe retreat*" (Hancock 1874), and Lord Ravensworth, writing to Hancock, said of the bird, "*we treated him with hospitality....I have seen him a score of times*" (Temperley 1951).

In a letter to George Bolam, in January 1916, C.E. Milburn talked of two White-tailed Eagles at Teesmouth in October 1915, which *"for a few nights"* roosted on a headland-like slag heap near Grangetown, on the Yorkshire side of the Tees. Milburn said he watched one of these cross over to Greatham without it once flapping its wings (Temperley 1951). One of these birds was killed at Guisborough (Yorkshire) on 17 November, the other bird *"probably went inland to be seen on the Yorkshire moors"*, near Great Ayton on 1 December 1915. A continental origin for these birds is perhaps most likely at this time, especially as the species was by then almost extinct in the British Isles (Brown & Grice 2005). Temperley was cautious of sight-only records of 'white-tails', some of which when shot proved to be Rough-legged Buzzards *Buteo lagopus*.

Distribution & movements

Today, 'sea eagles' are birds of the north. They breed in Iceland, Greenland, northern, central and eastern Europe, across Asia to the Pacific. The most northerly nesting birds are migratory and juveniles and first-winter individuals may wander, though many populations are sedentary. The White-tailed Eagle became extinct in the British Isles in 1918 and subsequent endeavours to re-introduce the species in Argyll in 1959 proved unsuccessful. Further attempts took place on Fair Isle, Shetland, in 1968 and then in a more concerted way on Rhum from 1975 in the Inner Hebrides; the first chick being hatched in the wild in 1985. There are now over thirty pairs breeding in Scotland (Parkin & Knox 2010). Following the commencement in 2008 of a re-introduction scheme in Fife, with over 60 young birds released there, it seems probable that in the future some of these will wander up and down the coast and may give rise to sightings of these magnificent eagles in Durham.

This huge bird of prey was recorded more often in the 19th century than the 20th century and since it became extinct in Britain, the last British bird being shot in Shetland in 1918 (Parkin & Knox 2010), it has not visited Durham. However, it continues to occur in England as a scarce vagrant from the Continent. In the period 1990-2011, it was recorded in both North Yorkshire (Blick 2009) and Northumberland (Kerr 2001), but not in Durham. Nonetheless, with the steady increase in the Scottish population, derived from the re-introduction programmes commencing in the 1970s on Rhum (Bainbridge *et al.* 2003), it might one day soon be seen again within the county.

Marsh Harrier (Western Marsh Harrier)
Circus aeruginosus

A once rare visitor and very rare breeder, the Marsh Harrier has become far more frequent in recent years and is now regarded a scarce but increasingly regular visitor and extremely rare breeder.

Historical review

In common with most birds of prey, the Marsh Harrier was likely to have been present in the county in greater numbers prior to the advent of extensive gamekeeping activity and in all probability bred in the more extensive marshy areas around the Tees Estuary, the Skerne Carrs and even on some of the county's upland, moorland bogs until at least the 18th century (Hogg 1845).

In 1831, Selby described it as not uncommon through the northern counties *"in low and marshy districts"*. At that time, it bred annually in neighbouring Northumberland (Kerr 2001). One was shot on Cowpen Marsh in 1829 (Backhouse 1834), this is probably the specimen labelled from that site that was in the Sunderland Museum, although with no date assigned to it. The species declined over the next half century and by 1874, Hancock said that it had *"almost disappeared"*, as a result of game preservation, becoming, as he put it, nothing more than a *"casual visitant"*.

According to Temperley (1951) this was a *"rare vagrant"* to Durham. He described the fact that it was once a regular summer visitor, and a summer resident in the county. Temperley cited six examples of birds being 'collected' over the period 1829 to 1930, the last possibly being of a pair of birds prospecting for a breeding site.

Evidence that the Marsh Harrier once bred in Durham is given by John Hancock (1874) who found a nest with four eggs on Wemmergill near Middleton-in-Teesdale in 1823 in the same year and possibly on the same trip as the then young Hancock found the nest of a Hen Harrier *Circus cyaneus*. In 1905, Tristram, summarising his long experience in the county wrote, "*in my youth I have several times taken the nest*", "*the last bird of which I have heard was in 1840*" (Tristram 1905). This was probably the female shot at Hartington, near Durham, in August 1840, a record also mentioned by Hancock. Another bird was shot at Whitworth on 8 April 1848. A severe national decline in the species occurred during the late 19th century and it eventually became extinct as a breeding bird in Britain (Sharrock 1976). Consistent with this, there were very few records in Durham in the second half of the 19th century though a bird was shot between Etherley and Bishop Auckland in 1880 (Temperley 1951).

Records remained very rare in the county into the 20th century. One was seen at Seaton Bank near Seaham on 29 to 30 October 1930, a rather late date for the north of England. Interestingly a female shot at Prior's Close Bog, Leamside, on 29 May 1913, proved on examination to have an incubation patch and therefore must have been preparing to breed. Another bird, described as a 'blue' hawk, was in the area at the same time, and presumably was the male of the pair.

Holloway (1996) noted that following prolonged persecution and wetland destruction during the late 18th and 19th centuries, the Marsh Harrier became extinct as a British breeding bird between 1899 and 1911, after which re-colonisation occurred. Increases in the 1970s throughout most of Europe including Britain were largely attributed to the banning of pesticides including DDT, which allowed populations to recuperate. A reduction in persecution levels in this country and reduced hunting pressures in southern Europe also assisted the recovery. This was further helped by the creation of huge reedbeds in the Dutch polders. In England the flooding of large eastern coastal areas for defensive purposes in World War 2 had also created suitable wetlands for the species.

Conservation efforts in the 1920s in the Norfolk Broads had led to up to four pairs nesting annually from 1927 onwards (Sharrock 1976). By the 1950s, up to 12 pairs were nesting in East Anglia, with occasional pairs elsewhere in the south of England; however, by 1971 numbers had again declined with only one pair clinging on.

A bird on the North Tees Marshes from 17th to 27 May 1959 was the first recorded at Teesmouth since 1829 and the first Durham record since the one shot near Leamside in 1913. Subsequently, there was a steady increase in reports of Marsh Harriers on passage in Durham during the following decades and a minor upsurge in records took place during the 1960s. In particular, there was a series of sightings from the still quite extensive area of Cowpen Marsh, Teesmouth (Blick 2009). In that period, one to three birds were noted in every spring except 1969. The most unusual record over this decade was the presence of a wintering bird at Teesmouth in January 1964. Up to the winter of 2010/2011 this remains the only such record for Durham though overwintering in East Anglia and elsewhere is now common-place.

In summary, since 1962, Marsh Harriers have been noted in County Durham in every year excepting 1969. This slow and steady increase in the number of birds continued through the 1970s, when the national population was still on a very slow upwards trajectory. This annual figure rose to around five per year through the 1980s, reaching double-figures by the early 1990s. The annual pattern of occurrence had a strong emphasis on the Teesmouth area, though passage birds increasingly appeared elsewhere across the county, such as at Barmston Pond on 14 May 1971 and the 'moor buzzard' which was watched on 2nd and 3 June 1973, in Weardale.

The strong seasonal emphasis over the 1960s and 1970s was on spring passage birds, so worthy of note were the two autumn records of 1973, one at Barmston on 9 September and one near Hawthorn Dene on 6 October. This trend continued through the 1980s, as breeding success further south produced numbers of young birds that returned in the following years and wandered across Britain, supplemented by continental overshoots. One of the best years on record for this species was 1986 was, when birds were at Teesmouth from 19 April, and others were recorded elsewhere around the county, at Boldon Flats and Shibdon Pond, and most notably on 12 May when a group of three birds hunted over a field of oil-seed rape between Whitburn and Cleadon.

Recent breeding status

That this species once nested in upland areas of the county is well documented. Although the bird will nest in arable crops, such as winter wheat and oilseed rape, its preferred habitat of marshland with reedbeds and vegetation is not well-represented in County Durham, although in recent decades there have been some major wetland and reedbed creation schemes.

In 1979, a pair was seen in suitable breeding habitat at Teesmouth from mid-May to mid-June (Blick 2009) and then during the main survey period for the *Atlas*, the North Tees Marshes attracted passage Marsh Harriers in every year from 1988 to 1993. The numbers of birds and the period over which birds stayed tended to increase and by 1992 and 1993 birds were regularly present from late April through to early July (Bell 1988-1993). It was anticipated from this level of activity that the species would soon breed again in the county and successful nesting occurred in 1996 when one pair reared three young, in the reedbed at Haverton Hole (Joynt *et al.* 2008).

Whilst the species has become an ever more common passage migrant in the north east as the national population has increased, it is hoped that it will breed regularly once again with the habitat improvements that have taken place around the county, particularly at Teesmouth in recent years. At present, successful breeding has not been repeated. Improved security and land management provided by the RSPB at Saltholme may offer the best prospect for this. Single birds, mostly immature males, have occasionally been seen over moorland in late spring without any suggestion of breeding.

Recent non-breeding status

At 2011, the Marsh Harrier is once again an annual passage migrant in small numbers, largely in the months April to September. Although more usually associated with reedbeds, migrants can appear anywhere in Durham, both near the coast and well inland, with birds sometimes lingering over undisturbed pasture and arable fields. The species has for many years been regularly attracted to the Teesmouth area and many of the spring records over the last three decades, perhaps 60% or more, occurred in that area, with many of the birds hunting briefly around the complex of Teesmouth wetland sites in late April or May. The majority of the autumn records involve slightly longer-staying individuals.

By the mid-1990s, it was increasingly difficult to ascertain just how many birds were occurring in the county, particularly in the Teesmouth area. This was due to longer-staying birds, becoming mixed with passage migrants. It is clear that in 1994, five or six individuals were recorded around the North Tees Marshes on 24 April and 6 May alone (Blick 2009), in marked contrast to previous years. In the first decade of the 21st century, 15-25 birds have been recorded annually.

Most recently, between 2006 and 2011, Marsh Harriers were largely an ever present species during May-July around the North Tees Marshes. Across the county a well-distributed spread of spring passage migrants occurs mainly in May, but there is also an upsurge of late summer migrants in August, with birds being seen at coastal locations and also well inland, including the western valleys and moors. Through the first decade of the 21st century, the autumn period produced a steady decline in sightings and few birds were noted into October. However, the winter of 2010/11 saw an individual remain on the North Tees Marshes for the full winter period; this situation is likely to be repeated as the species consolidates its now healthy population in England.

An unusually plumaged juvenile, showing broad white tips to nearly all of its feathers, was noted around the North Tees Marshes from 4 August to 15 September 2001 (Blick 2009).

Distribution & movements

This species breeds in western and central Eurasia and north west Africa. It is largely migratory, wintering in Africa, though small numbers do winter in Britain and Europe. As these are mainly females, improved survival rates, arising from avoiding the risks inherent in migration, together with the fact that many males are polygamous, should assist increased breeding success in the long term (Gibbons *et al.* 1993).

Following an average annual increase in the British breeding population of 19.6 % from 1971, when just one pair bred, by 1990 there were at least 87 nests, fledging 213 young and by 1995 this had increased to 157 to 160 females being present in suitable habitat. The population has continued to expand since then (Stone *et al.* 1997). In 2001 it was downgraded from Red to Amber as a species of conservation concern. By 2005, it was estimated that 360 females were present in Britain (Eaton *et al.* 2006) and the Rare Breeding Birds Panel cited a five-year average of 404 pairs as at 2009 (Holling 2011). The earliest and latest dates for birds in Durham are 5 March (2007) and 26 October (2009).

A wing-tagged bird that was present at Dorman's Pool in August 2002 had been ringed in the River Tay area of Scotland in 2000 where nesting began in 1990. A juvenile that was seen around the North Tees Marshes on 11th to 12 August 2005, had been wing-tagged in the same area in June 2005 (Blick 2009). The problems of interpreting data from individuals migrating through the county are well illustrated by the following example. In July 2004 a young Marsh Harrier was radio-tagged on the Tay estuary. Beginning its migration on 8 August it first flew south west to the Isle of Bute; on 22 August it had reached Wigton, Cumbria. By 08.00hrs on 23 August it had flown 89km east to Blaydon but only paused here briefly as by noon it was at Throckley. Later that day it moved into south east Northumberland where it remained until 14 September. On that day it was in the Stannington area in the morning but by 15.45hrs it was near Danby, North Yorkshire, presumably passing through Durham on the way, and at 17.30hrs it was by the Humber Bridge – a distance of 170km in one afternoon at a mean speed of 42.5km per hour. It continued steadily south from there reaching its African wintering area near the River Senegal by 14 October.

Hen Harrier (Northern Harrier)
Circus cyaneus

Sponsored by

Leengate /

Industrial & Welding Supplies

An uncommon passage migrant and winter visitor that breeds rarely and sporadically.

Historical Review

The Hen Harrier was deemed to be quite widespread as a breeding bird in suitable areas of Britain up to the early part of the 19th century (Holloway 1986). Hancock (1874) listed a much older report of it even having bred on the Town Moor, Newcastle (Northumberland). Selby (1831), writing of both Northumberland and Durham, described the Hen Harrier as being "*not uncommon*" and he was familiar enough with the species to mention its habit of "*communal roosting in rank heather in parties of 5 to 6*". Rossiter speculated that this species was probably common on the moorlands of Northumberland in the late 18th and early 19th centuries (Rossiter 1999). Robson (1896) had been told of a pair on the Ravensworth Estate, in what is now Gateshead, some time before 1836; both birds were shot and their eggs taken. Hutchinson (1840) said they "*inhabited the moors, occasionally descending to lower ground but their breeding is kept down by gamekeepers*". Hancock had been informed of a pair breeding at Hedley Edge, near Tow Law, in about 1845 and he had himself taken eggs from a nest on the Wemmergill Moors, Lunedale (then part of North Yorkshire) when just 15 years of age in 1823. This appears to be the first documented account of breeding for the present Durham recording area. By 1874, Hancock had reported a decline in the species' fortunes and stated that it had become "*merely a casual visitant*" after "*almost succumbing to the zeal of the gamekeeper*". Tristram (1905) thought the last breeding in Durham had occurred in about 1876 and even by the mid-20th century Temperley (1951) still considered the Hen Harrier to be a rare winter visitor and "*unlikely to re-establish itself until the rearing of game was abandoned*".

These descriptions agree closely with the broader national picture. In the face of persecution from gamebird interests and marked changes in land use, the Hen Harrier had been completely lost as a breeding bird to mainland Britain by the end of the 19th century. Tentative signs of a recovery came only after World War II when sightings increased. Northumberland recorded its first breeding success for more than 60 years in 1958 and by the mid-1960s the population there had reached at least eight pairs (Galloway & Meek 1978). A re-colonisation of northern England appeared to be underway as successive records of "*first breeding in the 20th century*" came from Bowland, Lancashire in 1969 (White *et al.* 2008), from Durham in 1970 (McAndrew 1972) and from Yorkshire in 1971 (Mather 1986). The hope was that this would prove to be merely the start of a sustained and full recovery across northern England.

Recent breeding status

The discovery of a nest with four almost fledged young, on a central Durham heather moor 18 July 1970 was a very welcome event after an absence of almost a century. A pair returned to the same site in 1971 but sadly the adult male was said to have been shot and the nest rumoured to have been destroyed.

Optimism in Durham waned and it was to be 1983 before breeding was once again confirmed. A pair raised two young not far from the 1970/1971 site and in 1984 four young were raised at another nearby location (F. G. Grey pers. comm.). Breeding probably occurred at yet another moorland site in 1986. If an increase in breeding numbers was underway it was proving to be a slow and intermittent process. The monitoring under licence of breeding Merlin *Falco columbarius*, Hen Harrier and other species became more coordinated and expansive in 1992 with the formation of the Durham Upland Bird Study Group (DUBSG). The graph shows the number of breeding pairs of Hen Harrier found in Durham and the nest outcomes from 1990 to the 2010 season, inclusive.

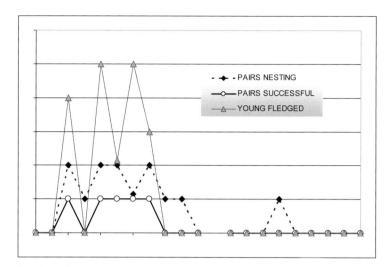

Number of Hen Harrier nests and their outcomes in Durham .1990-2010

The data excludes 2001 when Foot and Mouth Disease restricted access. A significant majority of Durham's prime heather moorland is surveyed each year by DUBSG fieldworkers and several large areas of moor have been the subject of constant annual monitoring for the last 20 years. Initially, the additional fieldwork brought some reward and from 1992 to 1999, one or two pairs were found nesting each year although never more than one pair was successful. In all, 19 young were fledged in the years 1992-1997 from five successful nests but crucially, 62% of known nests failed over the full survey period. There has been only one known nest in the last decade but the hen bird deserted at the egg stage after the adult male disappeared. It should be borne in mind that the data does not include sites where pairs appeared early in the season but then failed to proceed to nest.

Most summers still manage to produce a few records of lone birds and in 2005 a pair displayed at a second site, although they failed to nest. DUBSG members participated in the National Hen Harrier Survey of 2010 and collectively spent over 380 hours in the field surveying suitable habitat between early March and late July. No pairs and no display were seen and just five individual birds were recorded. The expectation fostered in the 1990s that the Hen Harrier might at least have a tenuous foothold as a breeding bird in Durham had once again all but evaporated. The *Atlas* (Westerberg & Bowey 2000) stated that "*there seems little likelihood of this most graceful and charismatic raptor becoming a more regular sight floating across Durham's moorlands*". A decade later, nothing seems to have changed and there seems little hope of a breeding population becoming established without the cooperation of those who own the grouse moors. The present position resonates with Temperley's closing comments on the species made in 1951, yet the perceived difference now is that far from game rearing needing to be "*abandoned*", the future of the Hen Harrier plainly depends on the active and sympathetic management of driven grouse moors which provide the birds' optimal habitat.

All nesting attempts in Durham have been in rank heather on managed grouse moorland. They have also all been within the North Pennine Special Protection Area (SPA) which was designated in the late 1990s, under European Union legislation, as a conservation area for which the Government has responsibilities to develop and protect the biodiversity through good land management practices. The quality of the habitat is formally classified as 'favourable' and there has been no significant deterioration over the period to explain the marked decline in the breeding status of the Hen Harrier in Durham.

The Hen Harrier's current standing in Durham can only be understood within the wider national context. The species has an unfavourable conservation status in Europe and is on the UK's priority red list of Birds of Conservation Concern (Eaton *et al.* 2009) because of its historic population decline to critically low breeding numbers. A full national survey in 2004 (Sim *et al.* 2007) gave an estimated 806 pairs for the UK and the Isle of Man, the vast majority of which were in Scotland. A repeat national survey in 2010 recorded an estimated 646 pairs, a 20% decline (Anon 2011). The perilously small English population has been centred over the last 30 years or more on the Forest of Bowland, Lancashire where the heather moorland is mostly managed under a "*special protection*" scheme through sympathetic land owners. A few pairs have occasionally nested each year elsewhere in the North Pennines and Cumbria. The position for England has been recently summarised by Natural England (2008) with the results of their Hen Harrier Recovery Project for 2002 to 2008. The comparatively small area of Bowland accounted for over two-thirds of all the 127 Hen Harrier breeding attempts in England. Only 19 attempts were recorded on other driven grouse moors; this despite the far greater extent of equivalent habitat available across the rest of the uplands of northern England. Overall, there was an average of just 18 breeding attempts per year (range 10 -23) from 2002 –2008 (most recently, the number of breeding birds in England has fallen to an unsustainable and alarming four pairs). Nesting attempts on 'other' grouse moors were found in the study to be twice as likely to fail compared to Bowland (74% compared to 35% failure rate, respectively). Earlier studies in Scotland (Etheridge *et al.* 1997) and northern England (Stott 1998) had also both demonstrated a far lower success rate for nesting females on commercially managed grouse moors than on other types of moorland. Interestingly the Natural England study showed that, if successful, nests on grouse moors were the most productive, indicating that the habitat was very favourable. The study concluded that there was unequivocal evidence (direct and indirect) that illegal persecution was limiting the natural distribution and breeding density of Hen Harriers in England. Persecution was thought to account for nearly all of the failed breeding attempts outside of Bowland. Potts (1998) acknowledged that without such constraints the moors of northern England could naturally support over 230 pairs.

It is accepted that the presence of raptors on moorland does have some impact on game bags and there has been considerable dialogue over the years between the conservation bodies and those representing commercial grouse moor interests to try to resolve the issues. Initiatives have included the Government's Raptor Working Group in the mid-1990s, the Langholm Study (Redpath & Thirgood 1997) and its various follow ups, the Hen Harrier Recovery Project already mentioned and a dialogue currently being facilitated by the Environment Council. Meanwhile, the majority of Durham's moors carry their SPA designation yet singularly fail to meet the stated objectives in species biodiversity. It is hoped that sooner, rather than later, a route will be found to allow this iconic species to be regularly seen displaying and breeding successfully across Durham's uplands. Without Hen Harriers Durham's magnificent moorlands will continue to lack an important dimension and be deprived of a key aspect of their natural heritage.

Recent non-breeding status

Wintering Hen Harriers favour heather moorland, adjoining rough pastures, patches of early first and second rotation planting within coniferous forests, marshes, estuaries and open farmland. Recent agricultural schemes to provide set-aside land for winter seed bearing crops and game cover show signs of proving particularly attractive.

In the 20 years between Temperley (1951) and the 1970 Durham breeding record, there were a mere eight reports of overwintering birds in Durham. Notably, a ringtail spent three weeks at Crookfoot Reservoir from late November 1953 and Derwent Reservoir had ringtails in November 1966 and February 1969, after an adult male had been seen there in January of that year. Winter records from both upland and lowland sites then became a little more regular, especially from the late 1970s. The uplands have since provided the majority of winter records although lowland sites do support birds in most years, particularly in hard weather. Hen Harriers are prone to wander quite widely and no one location will hold birds for any substantial length of time. Given this, it is difficult to estimate wintering numbers on a consistent basis.

Birds are typically seen in the uplands from mid-October through to early March with a slight peak evident in December and January. Favoured locations here include the reservoir areas of the Derwent valley, Muggleswick Common, Lunedale and Baldersdale and the moorland fringes of the Stang, Bowes, Hamsterley and Stanhope Commons. Records from 1970 to the present day show that only about 5% of upland sightings have involved adult males. In the two decades beginning in 1980, the winters have typically brought reports of one to two birds from six to ten upland locations but from about 1999 there have been noticeably fewer locations and smaller numbers recorded. Satellite tagged and tracked birds reared in Bowland, Lancashire are known to occasionally pass through the Durham uplands in late autumn or winter; indeed, tagged birds were sighted in five years during the first decade of the 21st century, nearly all in the south west of the county.

Lowland records show a similar pattern of occupancy over the winter months with the favoured locations being the North Tees Marshes, the WWT Washington, Whitburn and latterly, Hetton Bogs and Rainton Meadows where schemes for set-aside land have attracted birds. WWT Washington drew in birds in the winters of 1978/1979 to 1984/1985 inclusive, with a maximum of five. A bird initially tagged at a nest in Cumbria spent almost three months at Teesmouth in the final quarter of 2005. Typically, each winter, there might be reports of one to two birds over a period of weeks from three to four lowland locations. In the past 40 years the DBC archive shows that approximately 25% of lowland winter birds seen in Durham have been males; comparison with the much lower equivalent figure for the uplands reflects the gender size dimorphism and different food selection found within Hen Harriers (Newton 1979).

The habit of forming communal winter roosts can provide an opportunity to study this elegant raptor in greater detail. Roosts were found and monitored in the Durham uplands from the mid-1980s though the birds at these sites are vulnerable to persecution and specific roost locations must be withheld for this reason. Typical counts have been of one to three birds, usually ringtails, but exceptionally six birds have been seen. Roosting occurs in rank heather or sedge and rushes (*Juncus spp.*). There are perhaps three or four traditional 'roost centres' spread throughout the county's uplands though by no means all may be occupied in every winter and those that are may only hold birds intermittently. A combination of weather enforced movement and a natural tendency to range widely, appear to maintain a state of flux and change at roosts. Even within a given area a number of alternative roost sites will often be used within the same season. In one well watched area seven alternative roost sites are known ranging from 500 to 2500 metres from a notional centre. Alarmingly, the frequency and number of birds appearing at upland winter roost sites has fallen significantly in the last decade.

Given that birds from Scotland and northern England will wander during the winter months to coastal and inland locations in our region it can be difficult to discern movements. The DBC Annual Reports' archive for the last 30 years (1980-2009) contains just six reports to suggest any early autumn movement along the coastal strip. These few sightings probably involve Scottish birds, 'coasting' south from the breeding grounds, rather than birds of continental origin. First dates for these early autumn sightings of ringtails have been at Teesmouth on 14 August 1996 and at Whitburn on the same August date in 2003; the latest date was at South Shields on 9 September 1983. Autumn records then show a distinct lull until they become more general in mid-October when arrival of continental birds may be involved. Direct evidence of autumn or winter influxes from the Continent is scant; individual ringtails have been seen coming in off the sea and making landfall at North Gare on 26 October 1979, at Marsden on 16 October 1992 and in midwinter there on 3 January 1982.

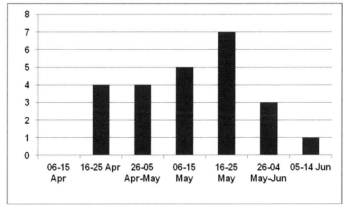

Number of spring coastal records of Hen Harrier by 10-day periods over 30 Years (1980 -2009)

Spring passage along Durham's coastal strip is more clearly marked and occurs at a time when any possible breeders in the North Pennines would be already settled on territory. The inference is that at least some of these birds are departing continental migrants although proof of their origin, through ringing or tagging, has yet to be established. Over the 30 year period (1980-2009), the DBC archive reveals 24 records of individuals along the coastal strip between 16 April and the end May with one outlying example in early June. The distribution, divided by 10-day periods, is shown in the bar-chart.

Thirteen of these years showed a complete absence of late spring records. The earliest dates were of ringtails at Teesmouth on 16 April 1987 and Sunderland on 16 April 2003 and the very latest was a second calendar-year male over Cleadon Hill on 7 June 2008. The peak years were in 1994 when 4 different birds were seen around WWT Washington from late April through to mid-May and in 2005 when singles were seen in May at Shotton Pond, East Boldon and Whitburn. Teesmouth and the northern coastal area around Cleadon, Marsden and South Shields have produced the main body of reports. Most involve ringtail Hen Harriers but proportionally more immature males have been positively identified with the reports in late May.

Distribution & movements

This is the most northerly of the European harrier species. It breeds extensively in France and northern Spain but discontinuously across much of the rest of northern Europe until substantial populations are met with in Sweden, Finland and Russia. The breeding distribution in the British Isles is largely confined to upland areas in the north and west, including the Isle of Man and Wales, but principally in Scotland. British breeders are partial migrants in that many birds tend to spend the winter wandering upland areas although some, especially first-year males may travel far greater distances. Northerly populations on the Continent are wholly migratory, moving to the south and west each autumn with some, probably from Scandinavia and the Netherlands, arriving in Britain. They account, in part, for birds wintering in south and east Britain where, joined by some of the UK breeders, they occupy a variety of habitats including coastal marshes, fens and open agricultural land.

A bird that was wing-tagged as part of an RSPB study in Grampian during summer 1993 was present at Derwent Reservoir on 16 November 1993. Tagged individuals from Geltsdale, Cumbria were seen in the county's uplands on several occasions between 2000 and 2009.

Northern Harrier
Circus cyaneus hudsonius

An extremely rare vagrant from North America.

In North America, the Hen Harrier is represented by the racial form, *Circus cyaneus hudsonius*, the Northern Harrier (previously known as the Marsh Hawk). On 22 February 2009, a series of photographs of an unusual subadult male harrier was taken by Mark Newsome on moorland at Cocklake Side, adjacent to Selset Reservoir. The bird showed several distinct features at odds with Hen Harrier and after further expert research, its identification was confirmed as a Northern Harrier, only the second for Britain and preceding the third seen in Norfolk in October 2010 (Hudson 2011).

The identification and taxonomy of the Northern Harrier in relation to the nominate race, *Circus c. cyaneus* of Europe and Asia, has been the subject of much debate since Britain's first record, a juvenile, was found on the Isles of Scilly in October 1982 (Martin 2008). Once the taxonomic situation is further evaluated by the BOU, it is possible that Northern Harrier may be elevated to the status of a separate species, as has already happened with several other authorities.

Pallid Harrier
Circus macrourus

Very rare vagrant from south eastern Europe or Asia: One record.

The sole record of Pallid Harrier for County Durham is of a juvenile found by Ian Forrest at Dorman's Pool, Teesmouth during the afternoon of 20 October 2011. It roosted in the extensive reed beds at the north end of the pool that evening and again on 21st, and although typically faithful to Dorman's Pool, it roamed more widely during the latter part of its stay, visiting both RSPB Saltholme and Greenabella Marsh. It was last seen at Cowpen Bewley Woodland Park on the evening of 23rd (*British Birds* in prep.). By this time, many hundreds of admirers had seen this stunning bird, which often afforded excellent views as it hunted at close range around the edge of Dorman's Pool, sometimes just a few metres distant from the gathered crowd.

Distribution & movements
Pallid Harrier breeds in south eastern Europe from the Ukraine, through Central Asia to north western China, and winters predominantly in Sub-Saharan Africa and the Indian Subcontinent, with small numbers also regularly wintering in south western Asia. It was formerly an extremely rare vagrant to Britain with just 3 three records prior to 1993, but there had been a total of 29 records by the end of 2010, the majority of which have come from the south eastern counties of England and the Northern Isles, particularly Shetland. There are three accepted records for the neighbouring county of Yorkshire.

There are a number of spring records between late March and June, but more recently, the majority of records have involved juvenile birds between August and October, several of which have lingered and exceptionally over wintered on two occasions (Norfolk 2002/03, and Cornwall 2009/10). The species has been undergoing a marked westward range expansion in recent years with several pairs confirmed breeding in Finland and European Russian during the summer of 2011, and were doubtless responsible for the unprecedented number of juvenile birds recorded in Western Europe during the autumn of 2011. A record 43 were logged at the migration watchpoint at

Falsterbo, Sweden (Naturhistoriska riksmuseet online), in comparison with the mean annual average of just two birds per autumn there between 1973 and 2010. This marked influx was also reflected in Britain with an incredible 30-40 juvenile birds recorded between August and November 2011, with many counties like County Durham receiving their first record of this striking vagrant. Assuming that the species will continue to increase in the western part of its range and consolidate its expansion, then it is entirely reasonable to speculate that the county will receive further records in the not too distant future.

Montagu's Harrier
Circus pygargus

An occasional and very rare breeder and scarce passage visitor.

Historical review
Selby (1831) knew of only one Durham occurrence, a bird shot some years earlier near the Northumberland border at Allenheads. Hancock (1874) considered the species to be a casual visitor to the north east but did have in his collection two birds in juvenile, "*nest*" plumage which were shot on Wolsingham Park Moor in 1835 where they had "*undoubtedly bred*". Calvert (1884) included a pair of Montagu's Harriers in his collection, said to have been shot at Shull, near Bedburn, Hamsterley in 1846 by Henry Gornall a well known provider of specimens to local 'gentlemen-collectors' such as Calvert and Joseph Duff. According to Robson (1896) the species was "*not very common*" in the Derwent Valley though eggs had been obtained from birds nesting on Hedley Fell, above Chopwell. He gave no date and the location may well have fallen on the Northumberland side of the bisecting boundary. In 1868, a second-calendar year male was shot at Axwell Park in the lower Derwent Valley (Hancock 1874) though there was no suggestion of breeding.

The first significant 20th century record came from Wolsingham Park Moor in 1929, when a lone adult male was caught in a trap set by a keeper near a 'cock nest' it had been seen to form in the heather (Temperley 1951).

Apart from an immature bird shot at Port Clarence, on the North Tees Marshes on 2 September 1932 (*The Vasculum* Vol. XIX 1932), there were no further reports until a quite remarkable series of events unfolded in the mid-1940s and over the subsequent decade. At the time of Temperley's review (1951), the story of local breeding had only partly developed and it is now worth documenting in some detail. It relates to one of the most exciting events in the region's ornithological history and reflects the considerable skill and determination of the observers of the day, revealing along the way some important behaviour. Sadly, it also shows the effect of persecution on a vulnerable population.

The national population of Montagu's Harrier had seen something of an upsurge by the late 1940s and throughout the 1950s, the number of breeding birds continued to grow to a peak numbering at least 40-50 pairs (Nicholson 1957). Several reasons for this have been postulated, including a reduction in keepering activity during and after the Second World War, favourable climatic factors and an increase in coniferous afforestation with ground cover perhaps reaching an optimum for the species at that time. In neighbouring Northumberland young plantations proved particularly attractive and breeding was recorded there from 1952 and then almost annually from 1959 to 1966 (Galloway & Meek 1978). On the North Yorkshire Moors breeding was evident in young conifer plantations and occasionally on adjacent moorland from as early as 1937 and continued through to the early 1950s.

The prelude to renewed breeding in Durham came in May 1944 when an adult male was reported on Muggleswick Common (F. G. Grey pers. comm.). In 1947, at the same location, when travelling home on 6 August, M. G. Robinson saw a large bird fly across the Edmundbyers to Stanhope Road, he recognised this as a male Montagu's Harrier. Thus began the difficult and protracted saga of the breeding of these birds on the North Durham moors. The gamekeeper thought two pairs were probably present early in the season and on the 13 August, G. W. Temperley and M. G. Robinson saw two adult cocks and three recently fledged young on the wing – the first confirmed breeding in Durham since 1835. A pair returned to the site in the following spring and gradually through the summer of 1948 it was realised that a loose, social colony of three pairs was breeding in an area of about two square miles. At one nest, the adult female and three of the five young were shot soon after leaving the nest. At another, one of four young was shot just after fledging and at the other, three young fledged. Fred Grey described the incongruous site of several of these rare and magnificent birds in flight over the moor with the then active Consett blast furnace and steel works set as a backdrop. Even further surprises came that year with news from a forester, G. H. Longstaff, at Hamsterley Forest that a pair had raised four young in a ride within conifers planted in the early 1930s. This then set in train breeding attempts at each of these two locations in the following few years. The outcomes are summarised below.

Location		1947	1948	1949	1950	1951	1952	1953	1954	1955	1956	1957
Muggleswick Common	No. of pairs	1	3	1	1	1	1	1	0	0	0	0
	Successful nests	1	3	1	1	0	0	1	0	0	0	0
	Fledged young	3	8	3	4	0	0	2	0	0	0	0
Hamsterley Forest	No. of pairs	0	1	1	1	0	2	2	1	2	1	0
	Successful nests	0	1	1	1	0	2	2	1	1	1	0
	Fledged young	0	4	4	2	0	6	6	4	2	3	0
Combined	No. of pairs	1	4	2	2	1	3	3	1	2	1	0
	Successful nests	1	4	2	2	0	2	3	1	1	1	0
	Fledged young	3	12	7	6	0	6	8	4	2	3	0

Summary of breeding Montagu's Harrier at two County Durham locations, 1947-1957

Though the peak total of three pairs in 1948 was never surpassed at Muggleswick Common the site did support single successful nests in three of the next five years. In 1949, two adult males were present and in 1951 no less than three cock birds were seen together on 13 May when a 'cock nest' was found. The four young from the one successful nest that year were seen on the wing on 13 August and one or two lingered until mid-September. In 1951, a pair appeared in late April but the female was not seen again. In 1952 a male arrived on the early date of 21 April but a female was not seen until a full month later and no nest was built. The final breeding took place here in 1953, when a clutch of five eggs was seen on 1 June from which two young later fledged. There was no reported breeding in later years although single males were seen in the general area in the late spring of 1957 and 1958. Odd birds were very occasionally noted elsewhere in the county during the 1950s.

Meanwhile at Hamsterley Forest, nests in the plantation area produced young in every year, except 1951, up to and including 1956. Between 1952 and 1955 a second site on the adjacent heather moor was established and this was successful in all but one year. Observations were made mainly by Miss C. Greenwell and her colleague Miss D. N. Bell. In the early years the duo needed to rely on public transport and undertook a long walk on the many occasions they visited the forest area. In 1954, they noted that nesting duties were being shared by two female birds before one was driven off by the male soon after 4 July. Four young were seen on the wing by 30 July. In 1955 a nest with four eggs once again had two females in attendance but this was deserted and at a neighbouring site a late clutch produced two flying young by the end of August. The female of the 1956 pair went missing, probably shot, before the young had fledged but the cock continued to feed the young unaided and later three were seen with him on the wing. The following two years only produced sightings of single males before records ceased entirely. All records fell away quite dramatically by the end of the 1950s compared to a greater persistence of the species in Northumberland where it bred until 1966, and again in 1992 (Kerr 2001).

Recent breeding status

There is a single recent record of breeding in the county. During local breeding atlas survey work, a pair and a nest containing three eggs was found on heather moorland on 2 June 1992 but by 10 June, the nest was empty, presumed predated. The birds did not return in 1993.

On current evidence there is no suggestion that the species will re-establish a breeding population, although successful nesting on the North Yorkshire Moors in 2009 provides some cause for optimism. Away from its favoured habitat of heather moorland and young conifer plantations, Montagu's Harriers elsewhere in Britain have made use of cereal crops, gorse heath-land and reed beds. The current national population is estimated to be about 15 pairs (Holling 2011).

Recent non-breeding status

The collapse of the breeding population is highlighted by the fact that there were a mere three records of passage birds in the county between 1959 and 1983. An immature was present on Cowpen Marsh, North Tees in August 1959 and single ringtails were seen there in May 1964 and May 1968.

The period 1984 to 2011 saw rather more records of passage birds, although in these 28 years, there were 13 years without any records. A total of 25 birds have been reported; the earliest being an adult male at Hart on 25 April 1998 and the latest, the sole juvenile amongst reports, was seen at Crookfoot Reservoir on 25 August 2008. Overall, there is: one April record; 15 in May; three in early June; one in very late July; and, five in August. There have been seven full adult males and four second–calendar year males in spring time. Most birds have been seen at Teesmouth or along the northern coastal strip but a ringtail flew over Aislaby in May 1997 and an adult male passed near Hutton Magna in August 1998. Peak years have been: 1995, 1997, 2001 and 2002, each with three records.

Distribution and movements

Montagu's Harriers are found breeding across Europe and into Asia as migratory summer visitors from sub-Saharan Africa. Britain lies on the north western edge of their European breeding range. Their occasional presence in Northern England only during the later spring and summer is consistent with this.

There are a small number of documented movements of this species, courtesy of the birds ringed at nest sites in Durham during the 1950s. A bird ringed as a juvenile in the nest at Hamsterley on 10 July 1954 was recovered in Aigre, Charente, France on 24 August 1959. In addition, a bird ringed as a juvenile in a nest at Hamsterley two years later, on 15 July, was picked up dead or dying at Bouzy, Marne, France on or about 10 September 1956.

Goshawk (Northern Goshawk)
Accipiter gentilis

A scarce resident breeder and possibly a rare passage visitor.

Historical Review

The Goshawk is considered to have been widespread historically in the county, as it was in much of the rest of the UK, with a decline taking place long before the earliest avifaunas were being compiled (Brown & Grice 2005), although there is little documentary evidence to bolster this statement. The presence of this species in the north east, around 1,000 years ago, is demonstrated by the finding of its remains during the excavation of a medieval site just to the south of the Durham border, in North Yorkshire (Yalden & Albarella 2009). The Goshawk became extinct as a breeding species in Britain in the 19th century. The last known breeding birds in northern England were shot in Westerdale North Yorkshire in 1893 (Batten 1990).

The oldest known direct reference to the species in Durham relates to its apparent use in falconry and occurs in the Bursar's rolls for 1378 to 1379, which states *Will'o Cotom portanti j Goshauk de d'no Joh'e de Lilborn d'no Priori, xij d.* (Fowler 1898). This alludes to the payment of twelve pence to William of Coatham for bringing one Goshawk from John de Lilburn to the Prior. Whether this bird was sourced locally is impossible to say.

Goshawk were known to be breeding in Scotland into the late 19th century (Baxter & Rintoul 1953), and Rossiter speculated that it might have done so in Northumberland, and therefore possibly Durham, for example in the heavily-wooded Derwent valley, at least until the 17th century (Rossiter 1999). It was already believed to be "*very scarce*" across Britain by the beginning of the 19th century, with many of the major declines probably taking place prior to the sixteenth century (Holloway 1996).

Hancock (1874) reported that there was one well-authenticated record, a mature female which had been shot near Castle Eden Dene "*a few years ago*". Tristram gave the date as 1872, and nominally this is the first and only well-documented pre-20th century record of Goshawk in County Durham, though the species must have been present long before this time. This shot individual pre-dates the extinction of the Goshawk as a breeding species in the UK, though the coastal locality may suggest a continental immigrant. A specimen of an immature female in the collection of Birmingham Museum has a labelled origin of "*County Durham, January 1884*", but no other details are available. Throughout the first part of the 20th century, the Goshawk was still considered a scarce straggler, with Temperley (1951) describing it as a "*rare vagrant*". Indeed the immature that flew off the sea and up the Tees estuary on 2 January 1934 was considered by its observers to be one of the rarest birds they had ever seen (Almond *et al.* 1939).

Recent breeding status

Goshawks evolved in boreal and temperate forest in the Northern Hemisphere. However, the development of high density, 20th century coniferous plantations led to the re-colonization of much of Britain, from those woodlands; especially in places where there was nearby moorland or open agricultural land, upon which birds were able to hunt the medium-sized mammals and birds it favours as prey (Petty 1989).

Nationally, since 1968 Goshawks have nested annually, with exponential population increases following peaks of the importation of birds in 1973 and 1975 (Marquiss & Newton 1981). It is considered that the British population is probably derived entirely from imported birds which have escaped, or which have been deliberately released, as this is a difficult species to breed in captivity, with an estimated 20 birds per annum 'entering the wild' between 1970 and 1989 (Kenward *et al.* 1991).

In the north east, modern records, especially in Northumberland, increased after the release of birds in the Border forests by hawk keepers in the late 1960s and early 1970s (Day *et al.* 1995). In Durham, records became centred on the Derwent valley and Hamsterley Forest. A record prior to this, of a bird at Fellside in the Derwent valley from 15th to 18 February 1953 was probably an escape, but could also have been a continental migrant (Bowey *et al.* 1993).

The Border releases were thought to involve 15-20 birds. Despite persecution, the first modern record of breeding in Northumberland occurred in 1973, and although young were hatched both parents were shot. Breeding success continued through the 1970s and during the Northumberland Breeding Atlas work in the early 1990s, up to 24 occupied home ranges were recorded (Day *et al.* 1995). The latest monitoring work in that county has identified *c.*35 territories of which, half are in Kielder Forest (M. Davison pers. com.). This well-established population is the most likely source of Goshawk recruits to Durham, although hawk keeper's releases do probably still occur.

The first records of Goshawk in Hamsterley Forest date from around 1979/80, and breeding is known to have taken place at that time. Thereafter a well known nest in a large Larch *Larix decidua* was robbed two years running and it was decided to remove the tree to encourage the birds to relocate elsewhere in the forest (G. Simpson pers. comm.). Although subsequent nests were monitored by a local ringing group, records from this area were not usually reported.

The first modern sighting of the species in the Gateshead district was of a bird watched near Washingwell Woods on 9 February 1972. The next was of a female at the same site being mobbed by a female Sparrowhawk *Accipiter nisus* and Carrion Crows *Corvus corone* on 14 October 1985. The two spring records in 1988 no doubt refer to the same bird, a male at Shibdon Pond on 3 April and 1 May. During the autumn and winter of 1990 there were a series of sightings of a male again at Shibdon Pond and also in the Derwent Walk Country Park, followed by a male at Ryton Willows on 9 December 1991.

In County Durham, the Goshawk has remained a rare breeding bird principally restricted to the larger blocks of upland conifer forest, although through time, birds have begun to be seen in a wider range of wooded habitats, particularly where there is extensive contiguous woodland in the context of a wider landscape, such as in the Derwent valley. By the early 1980s there were credible and persistent reports of birds in the lower Derwent valley,

294

with Gibside being reported as holding birds not infrequently between 1980 and 1985 and birds were noted displaying there on at least one occasion in the early 1990s.

Breeding of this secretive species is difficult to prove and its presence is most often confirmed by display flights over nesting territories in early spring. By 1991, such display was noted at eight sites in the county, and although breeding was not confirmed at any of these, it seems that by this time the Goshawk had become firmly re-established as a breeding species in Durham. During the first decade of the 21st century, the species seemed to make some headway in penetrating the lowland parts of Durham and it was not unusual to see some display in the smaller woodland areas of the east of the county and two lowland central localities produced several sightings in spring 2007.

Hamsterley Forest for many years, provided most sightings at the peak time for display in March and April, and this large area held a number of pairs in the early 21st century; with at least two pairs present there in 2007, and displaying birds at another three sites in that season. In 2007 birds were regularly noted at a site in the south west of the county, two being there on 21 January, with two seen in the same area again on 17 February. In spring one was over the Burnt Houses area on 19 February, a displaying male was at a central Durham site and other birds were noted at Eggleston Common, and the Derwent Gorge.

Although Petty (1989) recorded an average national productivity of 2.3 young per breeding pair, because of the resident and largely sedentary nature of Goshawks, persecution appears to ensure that the population in established breeding woods, including that in our larger forests, rarely reaches levels at which natural expansion can occur. However, it is possible that new territories in Durham have been founded by established breeding pairs relocating in response to deliberate harassment (A.L. Armstrong pers. obs.).

In 2010, most current estimates put the UK population at around 100 to 200 pairs although Baker *et al.* (2006) suggested that it was as high as 400 pairs and the RBBP place the 5-year average at 430 breeding pairs. It is now considered that the national breeding population is large enough, and sufficiently located in such extensive woodlands, that it is buffered against the impact of the theft of eggs and young, although these activities can still have a dramatic effect on local populations. In Durham, game keepers are aware of the increased numbers of Goshawks and do not hide their dislike of the species. It is likely that illegal disturbance will continue to limit the species' local breeding success and range for the foreseeable future. Across its range the Goshawk can breed in small woods, as small as 3ha in extent, but in such places their nests are more easily found and robbed, or destroyed. Productivity is usually greater in larger woodlands (Petty 1989) and this appears to be the case in Durham.

Although sightings of Goshawk have increased in recent years, with widely scattered records from both coastal and urban localities, the local breeding population seems to have remained stable at between six and eight pairs and this is mainly restricted to three core sites. Although successful breeding is considered to be infrequent (DUBSG data), it is more likely to occur in areas of large coniferous forest; hence the availability of such habitat within the county could be a factor in limiting the expansion of the species.

Some local observers have commented on the apparent size differences between same gender individuals. The majority of Durham's birds are larger than those from continental populations which might be expected to provide natural immigration. This relates to the source of the imported birds in the 1960s and 1979s. The former were from central Europe and are smaller than the later releases from north east Europe particularly Finland. Regressive genes from the original releases may be at play or new releases by hawk-keepers could still be involved.

Recent non-breeding status

There is some movement of young birds away from breeding areas during autumn and winter, occasionally into more open habitats. Many sightings during the latter part of the year in Durham fit with the established breeding distribution, in particular around the west and north west of the county. Dispersing juveniles and immature birds can appear more widely.

Distribution & movements

This species has a huge global range that wraps around the northern hemisphere, penetrating south into central Europe. It is found across northern Europe, eastwards through Fennoscandia, Siberia and throughout North

America. By and large, most populations are resident, though small numbers of northern birds move south in winter (BWP).

Immigration from the Continent does occur, as illustrated by a Norwegian bird trapped at Theddlethorpe, Lincolnshire on 17 October 1994 (Wernham *et al.* 2002) and in the decades 1980 to 2000, there was some suggestion of a limited, natural immigration into the county (Armstrong 1995).

Sparrowhawk (Eurasian Sparrowhawk)
Accipiter nisus

A widespread relatively common breeding species, as well as a passage migrant and winter visitor, in some numbers.

Historical review

In the early Victorian period, Hutchinson (1840) described it as "*well dispersed, though not very abundant*", its increase "*prevented as much as possible by the shooter and gamekeeper*". Hancock (1874) said of it, "*not nearly so plentiful as formerly*", which implies a great number and good distribution in the region, perhaps in the late 18th and early 19th centuries. The Sparrowhawk was described as the commonest bird of prey in the Derwent valley in the late 19th century (Robson 1896); unsurprisingly in an area that has retained much of its original woodland cover throughout recent centuries. He noted unusual local references to the species, including a bird without any breast bands taken at Blaydon in February 1854 and an unusually large clutch of eleven eggs that was discovered in the lower Derwent valley in the late 19th century. The species was described as a common and generally distributed resident at Teesmouth around the turn of the 20th century (Nelson 1907).

By 1905, Tristram was saying that it was "*very rarely to be seen*" but that a few pairs had, "*escaped destruction … in upper Weardale and the woods of the Tees*". George Temperley (1951) wrote that, as a young man, the only place to see a Sparrowhawk was on a gamekeeper's "*rack*" and "*it was seldom to be seen even there*". However, he recorded a steady increase during the first part of the 20th century, to the point where it was no longer considered an unusual sight within the county. Temperley also speculated on the regularity of the arrival of immigrants into the county, citing Nelson (1907), who had on "*several occasions observed individuals freshly arrived in the neighbourhood of the Tees mouth*", occasionally in good numbers.

During the two World Wars, and in the respective periods immediately following the conflicts, populations of Sparrowhawks were seen to increase, as significant numbers of gamekeepers were on active service. Locally, this led Temperley to describe it by that time as a "*well distributed resident*". The statement was made however, prior to the period when Sparrowhawk numbers fell dramatically in Durham, during the 1950s and 1960s, a decline directly attributable to the introduction of agricultural organochlorines. These chemicals are absorbed readily in the body fat of birds and as a bird-eating raptor, Sparrowhawk accumulates such large quantities as to be directly poisoned. The situation was worsened by shell-thinning attributable to D.D.T. residues in the birds' eggs from the late 1940s onwards. Numbers fell so dramatically that in 1961 the Sparrowhawk was given the same legal protection enjoyed by other birds of prey (Newton 2008).

Locally, by the late 1950s and early 1960s, the species was almost unknown in the lower Tees valley, there being just single records in the 'Tees plain' in both 1962 and 1963 (Stead 1964). In 1964 there were about 30 reports for the year across the region but no breeding was proven in either Durham or Northumberland. By 1969 at least one pair bred successfully in Durham – the first since the D.D.T. - induced crash. The year 1971 saw an increase in the breeding population with three to four pairs in one area of south west Durham and several other records in spring and summer, including a pair in the north east of the county. In 1973 five pairs were known to have bred in west Durham, with four young reared at one site and apparent success at one or two other sites. By 1974 it seems that the local breeding population was starting to recover slowly - indicated by increases in the number of sightings/reports in the autumn. Breeding included two pairs in upper Weardale, two young fledged from a nest in upper Teesdale and a pair in Hamsterley Forest. In the east of the county, one pair's nest was robbed at Easington Lane.

The recovery accelerated through the 1980s and the Sparrowhawk is now considered to have fully regained its former position. Nationally, the estimated population was 39,000 pairs in 2000 (BTO-BBS) and this was also reflected in its local status. After this period of growth and stability however, the Breeding Bird Survey 2009 recorded a decline of 7% from 1995-2008 with most of this occurring since 2005. It is not yet clear whether the local population has followed this national trend.

Recent breeding status

Compared with other birds of prey Sparrowhawks are not particularly long lived, although first breeding can occur at one to three years of age. Newton (2008) found in one national study that 72% of all females died before they could breed, while 15% of the most productive individuals in one generation produced over 50% of the young in the next generation. Age and habitat quality interact to have a direct bearing on breeding density and success.

Following the reduction in chemical poisoning, the maturing of amenity woodlands led to a surge in the availability of suitable nesting habitat through the 1980s and early 1990s, as plantations sown on old spoil heaps and opencast restoration sites began to enter the growth phase making them most attractive to Sparrowhawks. The county's population by the late 1990s probably stood at its highest level for well over a century (Westerberg & Bowey 2000). Today, breeding densities can be quite high where there is sufficient suitable woodland, even close to quite large settlements, such as in the lower Team valley. Newton's study (2008) confirmed that they favour and breed most successfully in conifer stands aged 25-30 years, with tree spacing at ground level between two and four metres apart. The birds tend to build a new nest in a different tree every year, perhaps as an anti- predator device. Stability in local populations was best achieved where woodland is managed on a long-term rotation (of 40-60 years), providing a mosaic of tree stands of various ages. Where such rotation results in marked alteration in suitable habitat and/or small bird densities, through these and other developments in land use, then the numbers of Sparrowhawks will also change accordingly.

In the first decade of the 21st century, the Sparrowhawk could be found relatively abundantly across the whole of County Durham, from the east coast denes to the woodlands that fringe the western uplands and, in all likelihood, the population is at an all-time high. In the west, breeding birds are largely concentrated in the river valleys, probably because of the lack of suitable nesting woodlands at higher altitudes. Amongst general observers, it is the most reported raptor species and is now a regular sight over all parts of the county, though it is probably not our most common diurnal bird of prey. It can now be seen, quite routinely, in suburban gardens and the county's urban centres, hunting the small birds that are attracted to the food that is now made widely available by so many householders. The BTO Garden Birdwatch scheme confirmed that in 2010 in the north east an average of 16.5% of gardens were visited by Sparrowhawks, the highest rate of any region in England. Although the Sparrowhawk seems to prefer coniferous woods for nesting locally, it does happily and commonly nest in broad-leaved woods, in tall scrub, shelterbelts and even in overgrown hedgerows. Unlike some species however, such as the Kestrel *Falco tinnunculus*, it very rarely nests in single or small groups of isolated trees.

The numbers breeding in any area depends primarily on the availability of suitable woodland. The constraints on this species include the continuing threat to individual birds from game preservation interests and the availability of prey associated with the reduction in numbers of formerly common, small birds, linked with changes in agricultural practices. This may be somewhat compensated, on a local basis, in the suburban fringes, as Sparrowhawks seek to take advantage of the significant growth of garden feeding stations and the bird populations they support. This has led some to consider, erroneously, that the Sparrowhawk is responsible for the decline of some songbird species. Although clearly the impact of agricultural change and, in some urban situations, domestic cats are much more heavily involved in their decline. According to *The Breeding Birds of Cleveland*, there are now between 40 and 45 pairs north of the Tees (Joynt *et al.* 2008). By extrapolating the densities from the well-studied, densely populated, river valleys such as the extensively wooded lower Derwent valley and the mid-Wear valley, and factoring in the species known densities in mixed habitat areas, for instance the Hetton/Houghton area, the *Atlas* postulated a county population range of 200-400 pairs for Durham (Westerberg & Bowey 2000). This estimate is probably still a reasonable assessment of the number of breeding pairs in the county, with the upper figure probably representing the area's capacity.

In spring 2007, birds were noted displaying at 12 locations, some of which had multiple pairs present, and no doubt many other woodland locations went unrecorded. Single birds were observed at a further 34 widespread sites where breeding was considered likely. Breeding success appeared high in 2009 with successful broods of five at both Cowpen Bewley and South Tyneside College, and four in Backhouse Park, Sunderland. During the early summer period reports came from sixty one-kilometre squares, indicating a healthy and at least stable population (Newsome 2010).

Recent non-breeding status

Outside the breeding season some dispersal away from breeding areas is evident and in January and February birds are commonly seen around many parks and gardens and also over the large estates, nature reserves and country parks with Croxdale Estate, Hardwick Hall, Hetton Bogs and Lyons, Rainton Meadows, Shibdon Pond, Thornley Woods and WWT Washington producing some of the most regular sightings.

In 2007, over 600 records were received, increasing to over 800 in 2008 for this widespread species, many of which came from suburban and urban areas, particularly during the winter months (Newsome 2009).

Birds can be seen across the county, though they are much more common in the well-wooded lowlands or in the upland fringe, where woodlands are extensive. Winter sees an influx of birds into the area and ringing information suggests that some of these are of local origin, although some continental migrants may also linger. In September 2008 the species was part of a movement of raptors along the coastal strip. At least 42 birds were recorded moving south from mid-month with three to eight birds seen at various sites including: Boldon Flats, Dorman's Pool, Fellgate, Penshaw Monument and Ravensworth Grange. The movement continued into early October when eight were noted over Houghton on 4th.

On 21 May 2008, a female was rescued from a warehouse in Birtley where it had resided for two days. Another incident of a bird being trapped in a building, though not for such a protracted period, includes a male that had chased a House Sparrow *Passer domesticus* into a greengrocer's shop at Rowlands Gill in the winter of 1990.

Distribution & movements

In its nominate form, the Sparrowhawk can be found breeding across Europe as far east as western Siberia. Other races can be found in Siberia, central Asia and in north west Africa. Although more southerly populations are sedentary, northern birds migrate south and considerable numbers of Scandinavian birds visit Britain during the winter (Cramp *et al.* 1977).

Most ringing returns indicate that locally bred birds undertake only small scale movements, with birds ringed as nestlings at Jarrow being recovered at Tynemouth, Northumberland and Low Fell, Gateshead. One of the longest distance recoveries of a locally ringed bird is of one caught at Clara Vale on 20 February 1994 and recovered at Fulwell Quarry in the June of that year.

An incident, which occurred at Clara Vale on 3 March 1999, when two birds were caught in a mist-net, may have been indicative of an example of co-operative hunting, where one bird flushes prey and the second follows close behind. The first bird flew into the mist-net in front of the hide and was immediately extracted, but the ringers were amazed to see that within seconds a second bird had also flown into the net. Sparrowhawks are known for the ease for which they can escape from mist-nests, and one morning in 2008, an individual was seen to escape on five occasions. Only one bird of 64 ringed by the Durham Ringing Group between the late 1990s and early 2000s has been recaptured. It had been ringed on 4 May 2001 at Clara Vale and was re-trapped there on 31 December 2005, illustrating both the sedentary nature of local birds and their low likelihood of being re-captured once previously caught.

Buzzard (Common Buzzard)
Buteo buteo

An uncommon resident, passage and winter visitor, that is now more common than at any time in the past 150 years.

Historical review

The early recorded history of the Buzzard in Britain is linked most closely with its persecution. In 1457 an Act of James the Second of Scotland included the Buzzard in a list of vermin to be destroyed. Thereafter it figures in the records of churchwardens who were concerned with any species that might compete for man's food. Under a 1566 Act they were obliged to pay two pence *"for everie head of Busarde [and Ryngtale]"* (Tubbs 1974). Yet by the late 18th century it was still described as *"the commonest of the hawk kind we have in England"* (Pennant 1776).

Land enclosure and agricultural improvements from the 18th century onwards, with the associated control of vermin; followed by the development of breach loading guns and percussion caps, which allowed game to be driven to shooters, leading to the emergence of sporting estates – were all factors which combined to produce a retraction in the bird's national range. The Buzzard sought refuge in the higher, unimproved land to the west and this was reflected in the local situation. The reduction in keepering during, and following, the two World Wars led to a growth in numbers and an eastward expansion, which continued up to the catastrophic effects of *Myxamatosis* upon the rabbit *Oryctolagus cuniculus* population in 1954. This was despite the fact that rabbits did not become a pest species until the 19th century; hence the association of Buzzards with them is a relatively modern phenomenon (Tubbs 1974).

The Buzzard's national recovery, starting in the 1960s, was slow. Although re-colonisation of the south and east of England and eastern Scotland, together with more enlightened attitudes among lowland gamekeepers and better practices designed to minimise the impact of raptors around game release pens, all led to a steadily increasing national population.

This species was present and apparently widespread in England in 1800 (Moore 1957). Hutchinson (1840) however, highlighted the contrast in its status in Durham, saying *"This bird, although common in many parts of England, is by no means so here, specimens of it being less frequently obtained as of either the Rough-legged or Honey Buzzard. They are generally brought from the vicinity of the moors. In the eastern parts of the County they are hardly ever seen. They rarely breed here, probably in consequence of the persecution they experience from gamekeepers"*. Hancock's writings (1874) draw further attention to the persecution of this species. He highlighted several specimens in the collection of Mr Smurthwaite of Staindrop, which had been taken in that neighbourhood. Hancock had one, a female, taken at Ravensworth in February 1837. On 26 March 1856, Hancock himself found one at Whitburn Sands and Thomas Robson listed three records in the lower Derwent valley over the later part of the 19th century, with birds being shot at Lockhaugh around 1836, Hagg Hill in October 1852 and at nearby Dunston Haughs on 1 September 1883 (Robson 1896). According to Holloway (1996), this species bred in almost all British counties at the outset of the 19th century. Even at that time, however, there is little documentation to indicate that the Buzzard was breeding widely, if at all, in County Durham.

Nelson recounts the tale of W. Walton of Middleton-in-Teesdale, whose grandfather in the late 19th century, kept a pet *buzzard-hawk* which *"cock-fighting being then in vogue, he pitted it against the game-cocks, when it came off victorious"* (Nelson1907). The origin of this individual is not documented. The only suggestion that the Buzzard was ever common in the area comes from Canon Tristram (1905) who said *"Within living memory it regularly bred in many parts of the County but has been exterminated by game preservers aided by egg-collectors"*. He recalled, as a boy, taking three nests of four eggs each, around 1834, all of the nests being within a mile of each other. Backhouse (1885), largely writing of upper Teesdale, said the Buzzard had declined from 20-30 years previously. Abel Chapman however, writing in 1889, considered the status much more critical and recalled Hutchinson's earlier comments *"not a single pair now nest in Durham or Northumberland. The few Buzzards that do occur are generally met with during ….October but these are usually …. Rough-legged Buzzards"* (Chapman 1889).

According to Mather (1986), Nelson said that there was authenticated evidence of Buzzard breeding in north west Yorkshire in the late 19th and even early 20th century. Presumably this was towards the Durham border, even within what is now considered part of Durham, with breeding in this undisclosed area continuing until around 1906 (Mather 1986).

Temperley (1951) thought the Buzzard a *"rare occasional visitor, except in upper Teesdale, where of late years, it has occurred fairly regularly and would probably breed if not molested"*. He quotes a gamekeeper in Teesdale regularly shooting Buzzards that were coming in from the Lake District. He also reported however, that it was not an uncommon sight in the dale for a single or even a pair of birds to be seen in the breeding season. In this context, a bird was noted in upper Teesdale in 1936 when George Temperley found moulted feathers below Falcon Clints (*The Vasculum* Vol. XXIV 1938). Indeed, nest building had taken place but this did not continue to the egg-laying or incubation stage. Records in lowland and east Durham, during the mid-20th century appeared to be rare with a bird remaining for four weeks on the Lambton Estate, in October 1949, whilst a mid-summer record of a bird over the Team Valley on 20 June 1937, was, at that time, most unusual (*The Vasculum* Vol. XXIV 1938, Bowey *et al.* 1993). In autumn 1944, one was noted for two weeks at High Reservoir between Waskerley and Stanhope. The species' decline at the time was lugubriously remarked upon, "*this hawk, although once common, is a vanishing species and it is impossible for anyone who loves wildlife to mention the buzzard without feeling profound melancholy*" (*The Vasculum* Vol. XXIX 1944).

Post-1951 records remained generally sporadic. In 1952, however a pair did nest on the then Yorkshire side of the Tees in Teesdale, probably within vice-county 66, when three eggs were found on 30 April, which were subsequently taken.

Despite the assertion in the *Breeding Atlas* (Sharrock 1976) about its absence from the county, breeding took place in 1968, when two young were raised in the west and the pair returned the following year, although no young were subsequently recorded. During the 1970s sightings began to slowly increase, for example in 1974 a total of 10 was recorded, although these included five coastal birds.

Recent breeding status

At the time of the 1968-1972 Breeding Atlas (Sharrock 1976), this species was absent from most of south eastern England and large parts of central and eastern England. In northern England, the Lake District was its stronghold, but County Durham was considered to be "*devoid of the species*"-(Clements 2000). However, the 1952 and 1968 records prove that this was not quite correct.

Some years after the event, it was revealed that in the late 1970s Buzzards had bred in Lambton Estate woods "*for several years*" (T. Dunn pers. comm.). This appears to be the first modern breeding record outside the western dales and represents a rare example of the enlightened attitude of gamekeepers.

Few species have seen such a dramatic turn-around in local breeding status within recent times as the Buzzard. A review of *Birds in Durham 1981-1990* (Baldridge 1982-1986, Armstrong 1987-1991) confirmed an average of 8.2 sightings per annum with the records equally divided between the eastern and western parts of the county. Although a few of these were noted in suitable locations for breeding, this was never confirmed. Following the formation of the Durham Upland Bird Study Group in 1992 a more systematic effort was made to determine the species' true breeding status, and this confirmed that it remained a rare breeding bird within the county. The pair that bred in the south of the county in 1991 was the first fully documented successful breeding since 1968 (DUBSG). National and local evidence gathered apace through the 1990s and confirmed that an eastward range expansion was occurring (Gibbons *et al.* 1993). During the 1988 to 1993 period, and especially from 1993 onwards, there was a steady growth in the number of Buzzards seen in Durham. By 1995, the species' range had expanded dramatically, with 14 territories occupied and breeding proven at two sites in the south of the county and in Weardale, and breeding probable at a further site in the south and in Teesdale. Two other pairs were suspected of having bred. It is believed that this expansionist phase was driven, largely, through recruitment from the well-established neighbouring population in Cumbria, which in 1995 stood at 400-500 pairs (Cumbria Raptor Group pers. comm.) and later, from Northumberland. In the spring of 1995, birds were noted displaying in the Crookfoot Reservoir area (Bell 1996) long before the concerted expansion into lowland areas took place (Joynt *et al.* 2008).

On publication of the *Atlas* (Westerberg & Bowey 2000), the species' distribution was mapped on a ten-kilometre square basis, for 'security' reasons. Within seven years it had spread across the west of the county and into many lowland areas. By 1997, it was estimated that between 15 and 20 territorial pairs were present (DUBSG). The once westerly bias to the species' local distribution pattern disappeared in the early part of the 21st century. From the perspective of 2006, it was legitimate to say that just 15 years ago, finding a Buzzard in Durham was a noteworthy event, but since then, the species has seen a dramatic turn-around in its fortunes.

Apart from the urban areas, most of County Durham contains suitable habitat for the nesting and hunting requirements of Buzzards and the opportunistic nature of the species has meant that it has rapidly begun to colonise all such suitable habitat in recent years.

Buzzards have the capacity to absorb energy from very poor food due to a proportionally long gut compared to other raptors. This allows birds not only to live on worms, which many other raptors could not, but also to survive more readily in periods of frost without eating much. This undoubtedly has assisted in the expansion.

Nationally, the Buzzard is now considered to be probably the most common diurnal bird of prey. As yet, this is not quite true for Durham. Its expansion into the lowland parts of the county was even more evident by 2000. Data in The New Atlas (Gibbons *et al.* 1993) gave some 31,000-44,000 territories nationally and these figures were reviewed in 2000 using CBC/BBS trends (Robinson 2005), meanwhile Clements (2002) put the figure at 44,000-61,000 territorial pairs in Britain.

Durham Bird Club data for 2007 confirmed its expansion eastwards into the county's lowlands was ever more evident, and recorded pair-bonding and displaying males in suitable habitat at a minimum of 37 sites as far east as the coastal denes. In 2009, there were 34 occupied territories in the western part of the county, largely west of the NZ00 grid line, with 17 pairs fledging a total of 33 young. The area can clearly sustain many more pairs than this if allowed. The total county population was then estimated to be in the range 77–100 pairs (NERF 2010). Common Buzzard was the most reported raptor to the Durham Bird Club in spring 2010, with 210 sightings.

Recent non-breeding status

Birds are now widely distributed across the county during autumn and winter. Well wooded areas tend to be best favoured, but birds can be observed in almost any location, barring heavily developed areas and the conurbations, though occasional birds are even noted over-flying these. In autumn 2010, Buzzards were recorded in 70 locations in the county. Typical woodland sites for the species are found in the whole of the Derwent valley, the Wear valley from Chester-le-Street upstream to the county boundary, the Tees valley from Langdon Beck down to open farmland in the Gainford area and across Hamsterley Forest. Often quite small woods are used where there is a suitable mosaic of habitat. The northern section of the mid-county now also plays host to birds, with regular sightings from localities such as: Cornsay, Lanchester, Sawmill Wood and Waldridge Fell. Further east, birds are also becoming more regularly observed, right down to the mid- to southern section of the coastal strip, in areas such as Hurworth Burn Reservoir and the coastal denes.

Evidence of migration was apparent in autumn 2008, associated with a large passage of Honey-buzzards *Pernis apivorus* in eastern England. Groups of up to four were noted at many coastal sites between 13th and 29 September, and passage peaked on 20 September, when ten flew south over Penshaw between 09.50 and 13.30 (Newsome 2009). Early September 2010 also produced evidence of coastal passage with six birds over Dorman's Pool on the 2nd, four moving south over West Boldon on 12th and a single bird flying in off the sea at Whitburn on the 14th (Newsome 2011).

The sudden disappearance of birds from a hitherto established territory is strongly suggestive of human interference and still happens all too frequently in western parts of the county. It is clear however, that Buzzards are now well-established throughout Durham and a continued slight increase in their population can be expected.

Distribution & movements

Buzzards breed throughout most of Europe and northern Asia, most populations are largely sedentary, though some northern birds, including many from Scandinavian and northern Asia, do move south in hard weather. It is the former that probably account for winter migrants to Britain.

There is a single recovery of a ringed bird in the Gateshead area, on 9 December 1975, when a bird ringed as a nestling near Penrith was found dead in Chopwell Woods (Bowey *et al.* 1993). A bird which had been wing-tagged in Cheshire was present at Crimdon Dene in autumn 2006.

Rough-legged Buzzard
Buteo lagopus

A scarce and erratic winter visitor from northern Europe with more than 58 records involving at least 61 individuals between 1970 and 2010, recorded from October to April, most frequently from the uplands in the west of the county.

Historical Review

The first record of Rough-legged Buzzard in County Durham is difficult to trace, with the earliest reference being made by Selby (1831) who listed the species as "*an occasional and rare visitant*" stating that "*In the autumn and winter of 1815 several visited the northern counties, and I was fortunate in procuring several specimens.*" However, none of these specimens are now known to exist, with the earliest adequately documented specimen being that of a bird shot at Marsden Rocks, near Sunderland in 1823, which Selby (1831) stated was held in "*the Museum of the Society*", though it is not listed in Hancock's catalogue (Hancock 1874), having presumably been destroyed at some point.

Writing in 1834, Edward Backhouse made reference to "*a very large one killed at Whitburn in the spring of 1830, and another shortly afterwards taken in the marshes near Greatham*". Hutchinson (1840) said that "*Some are captured almost every year on the moors in winter. In December 1839 and January 1840 several were shot in Weardale*". Over thirty years later Hancock (1874) listed the Rough-legged Buzzard as a "*rare and casual visitant*", though he provided details of only one further specimen in addition to that from Marsden, which involved an adult female shot at Bishop Auckland in 1840. He did, however, also state that "*about that time (1840) several specimens occurred on both sides of the Tees*". Other 19th century references to the species are of a bird shot near Sunderland in 1844, a female shot at Boldon Flats in 1848, the specimen going into the Hancock Collection in Newcastle-upon-Tyne, and a female shot at Castle Eden Dene on 10 January 1876 having been in the area for around six weeks, with another seen in the Dene in April of the same year (Temperley 1951). Writing in 1885, J. Backhouse could give only one record for upper Teesdale, a single bird shot at Millbeck, near Newbiggin, "*several years since*".

There was a period when the relative fortunes of Rough-legs and Common Buzzards in the north east were very different from that of today. Abel Chapman (1889) wrote that "*the few Buzzards that do occur, are generally met with during….October, but these are usually of the northern type, or Rough-legged Buzzard of which I have examined 3 or 4 shot at this season*".

In summarising the situation for Durham at the mid-point of the 20th century, George Temperley (1951) described these birds as "*rare and irregular winter visitors*". There were just six accepted records for the early part of the 20th century, the first of which was at Teesmouth in October 1915. This bird is recorded as having attempted to roost on a huge slag heap close to Grangetown, before being driven off by two White-tailed Eagles *Haliaeetus albicilla*, which were temporarily using a crevice in the same artificial crag (*The Vasculum* Vol. II 1916). Nelson (1907) recorded a large influx to Yorkshire in 1903 and Mather (1986) commented that birds were recorded at that time along the east coast from Holy Island to Spurn (Mather 1986). Subsequent records are of: one shot at Port Clarence on 30 October 1926, the specimen residing in the Dorman Museum, Middlesbrough; at Teesmouth on 23 October 1936; and, two at Teesmouth in October 1938.

Temperley (1951) made reference, without apparent question, to several reports of this species from R. Martinson of Wolsingham, who stated that Rough-legged Buzzards "*had been coming to the moors in that district on and off for years*". All sightings related to Weardale and the records included: two on or around 11 August 1942; two from 19 August until 30 September 1943; a single from 8 August until 30 September 1944; a single from 12 August until 30 September 1945; and, a single from 7th to 26 September 1946. In terms of geography, these records tally with Hutchinson's remarks of over a century before, but they seem remarkably early for this species in the modern context. Such August records of Rough-legged Buzzard are extremely rare in Britain, with Thom (1986) listing just a single such record for Scotland, in 1982. However, without further evidence to the contrary, it is difficult to challenge Temperley's decision to include these records given the then current knowledge of Rough-legged Buzzard movements, in spite of the early date of their occurrence.

During the 1950s and 1960s records of this species were few and far between. One was noted over Eggleston, in Teesdale on 13 April 1955 and another flew in off the sea at South Shields on 23 December 1958, and then headed inland. There were no further authenticated reports in the county until 1975. Major national influxes of this species were once believed to follow good lemming *Lemmus lemmus* years in the northern breeding areas. This has however, not proven to be an entirely consistent correlation. National influxes occurred during the winters of 1966, 1973 and 1974, but Durham had no documented records in any of these periods.

Recent Status

The Rough-legged Buzzard is a late autumn and winter visitor to Durham, with October and November being the principal months for its first arrival in the county. Over the last 20 years, sightings have tended to persist through mid-winter and into the early months of the following calendar year. Most records come from the county's western uplands where it is almost certainly an under-recorded annual visitor, in small numbers.

A total of 52 Rough-legged Buzzards have been recorded in the county since 1970, all of which have been between October and April. Extreme dates have been one flying south offshore at Hartlepool on 3 October 1981 and north west over Seaton Common, Teesmouth on 18 April 1999.

The first record during the recent review period is of a juvenile seen over Wolsingham Park Moor on 18 February 1975, with presumably the same bird at nearby Tunstall Reservoir later in the month. This was the first for seventeen years with the next coming some four years later when a bird was watched flying over Peepy Plantation, Washington on 25 March 1979.

During the 1980s, the years 1982 and 1988 with four records each were the best of the decade; 1982 had three published records with another recently coming to light of a bird in the upper Derwent valley in November of that year. In 1988, there was a single coastal passage record and the rest from typical upland areas in Weardale and Teesdale. The species has since become a more frequent visitor, but was still wholly absent in the eleven years between 1980 and 2009.

Most records occur in the western uplands, in particular Teesdale and Lunedale, and to a lesser extent Weardale. There are also records from The Stang and the Hamsterley area. Birds have been noted on the uplands in twelve winters since 1980 and average about eight records per decade in this area. A number of birds were recorded in the county during the major influx of 1994 which involved around 140 birds across Britain. This was the best winter on record locally when seven birds were reported including at least three at the Stang Forest. Many have been recorded on just one date, although there were wintering birds in Teesdale from November 1982 until 3 March 1983, around Lunedale and Eggleston from 9 November 1985 to 18 January 1986, two different juveniles in Teesdale from 9 November 1988 to 25 March 1989 and 9 December 1988 until 7 January 1989 respectively, an adult and a juvenile in Teesdale on 6 November 1991 with one remaining until the year end, in Teesdale from late October until 6 December 1994, west of Hamsterley Forest from January until 19 February 1995 and at Langdon Beck in January and February 1998.

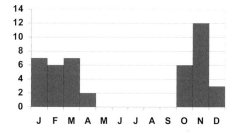 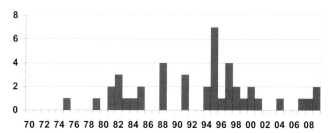

Records of Rough-legged Buzzards in County Durham, 1970-2009

Most birds are noted between November and March although they occasionally linger until early April. An individual at Harthope, St. John's Chapel on 7 October 1998 was noteworthy in being particularly early, so far inland. More recently, birds have occasionally appeared in Teesdale and Lunedale in February and early March after an apparent absence during mid-winter; this may point to evidence of return passage as they move north again having perhaps overwintered in more southerly or westerly locations in Britain.

Records from the coast are much scarcer, with just ten reports since 1980, all of which have been in October-November and March-April suggesting that these birds had just crossed or were about to cross the North Sea. The well watched area of the North Tees Marshes has recorded just five birds in the review period. These were at Teesmouth on 24 October 1982, west over Seaton Carew on 17 November 1984, west over Seal Sands and Cowpen Marsh on 6 October 1991, north west over Seaton Common on 18 April 1999 and in-off the sea over Seaton Common on 4 November 2007. Further north along our coastline there are four further records of birds, south over Whitburn on 10 November 1982, south over Marsden on 26 October 1988, flushed from trees at Ryhope Dene on 15 November 1994 and over Hawthorn Dene on 18 March 1995.

Further inland in the lowlands, the species is very scarce and there are just ten records, all relating to single birds, between 1970 and 2011. A juvenile was over Hetton Lyons Country Park on 2 January 1996, over Rainton Meadows on 1 December 1997, an adult north over Hetton Lyons Country Park on 18 November 2000 and north over Longnewton Reservoir on 21 February 2004. The widely scattered dates suggest that most of these birds were not recent immigrants but passing through en route to a more suitable wintering area. More recently, single juveniles over Hurworth Burn Reservoir on 22 October 2010, and Preston Farm, Stockton-on-Tees on 24 October 2010 were part of a large influx of the species into Britain that winter, and a precursor to at least three birds being found in Teesdale and Lunedale during the early part of 2011. Another small autumn influx was witnessed nationally in 2011, and at least four birds passed through eastern areas of County Durham between 20 October and 28 November, including one which lingered at between Bishop Middleham and Hardwick Hall from 24th to 26 October.

The vast majority of birds recorded in the county have been aged as juveniles and there have been just five modern day records of adult (or near adult) birds. These were in Teesdale on at least 6 November 1991, over Hetton Lyons Country Park on 18 November 2000, at Cocklake Side, Lunedale on 25 November 2000, in Lundale during February/March 2011 and near Woodland on 29 October 2011. Up to three, or possibly four birds, present with a number of Common Buzzards *Buteo buteo* around the Stang Forest at the end of February 1995, is the largest group recorded in the county to date.

In the 1997 influx, 43 were seen across Britain, of which 33 were in England, nine in Scotland and one in Wales. At that time, the north east shared a large proportion of the English total with five birds noted in both Durham and Northumberland (Fraser *et al.* 1999). These represented something of a glut over previous decades. After at least 13 birds in the 1980s and a minimum of 20 in the 1990s however, records undertook something of a downturn in the first decade of the 21st century, a minimum of eight and perhaps eleven birds were seen to the end of 2009, but there were four years in this period in which none were recorded. In 2009, at least two and possibly three different first-winter birds were observed in Lunedale and upper Teesdale. In the winter of 2010/2011, birds were once again wintering in this area with at least three individuals again involved.

Distribution & movements

The Rough-legged Buzzard has a circumpolar breeding distribution across much of the low Arctic and northern temperate zones, though in Europe the breeding range is restricted to northern Scandinavia. It winters to the south of its breeding range, though the largest numbers are well to the east of Britain, the majority in eastern Europe, particularly in Poland and the Baltic States, though small numbers, of mainly first-winter birds, reach the British Isles annually; these being largely from Scandinavia (Lack 1986). The species is however, subject to periodic influxes further west, usually in response to food shortages hence the numbers wintering in Britain fluctuate from year-to-year.

Golden Eagle
Aquila chrysaetos

An extremely rare passage visitor with just six fully substantiated occurrences since 1970.

Historical review

Temperley (1951) believed that in the 18th century, when Golden Eagle bred on Cheviot (as per Wallis 1769) and in the Lake District, wandering birds would have occasionally visited Durham. He mentioned several examples of this but was uncertain as to whether 'the eagle' referred to in such records was the Golden Eagle or White-tailed Eagle *Haliaeetus albicilla*. Some early examples were proven to be White-tailed Eagles, for example that of Hogg at Teesmouth on 25 November 1823. There is one documented record, confirmed by the dimensions given in the Gentleman's Magazine[1] for 1756, which reads, "*1765, October Thursday 17th. A Golden Eagle of an enormous size was shot at Ryhope, near Sunderland. It measured from the extremities of its wings 7 feet 6 inches, from the bill to the tail 3 feet*". Canon Tristram, writing in 1905, recounted the tale of some 30 years previous when, in November, he was crossing from Teesdale to Nenthead on foot, over Killhope Fell, when fog came down and he sat down to rest. On rising as the fog lifted, he realised that "*a Golden Eagle in young plumage with its white tail*" was perched on a post within feet of him (Temperley 1951).

It is interesting to speculate on the possibility of Golden Eagle breeding in County Durham in times past. It has been suggested that the species was breeding in the Derbyshire, Peak District in 1668 and around North Yorkshire before 1790 (Holloway 1996). So, it is not inconceivable that birds might have been present in habitats that are now located in Durham's remote south west, such as Mickle Fell, on what used to be the Yorkshire side of the River Tees, prior to the 1974 boundary changes.

Recent status

In 1967, an immature bird was seen flying north at Wellhope Burn in upper Weardale on 29 April. At the time Golden Eagle had been breeding in the Lake District for some years and a number of young were produced that adopted, in some instances, a nomadic existence. This was the first confirmed county record of the 20th century; though it seems highly likely that some would have occasionally wandered over the Cumbrian/Durham border, prior to this.

A number of birds were seen in the north west of Yorkshire through the 1970s and into the early 1980s, and these records were associated with the birds then breeding in the Lake District (Mather 1986). In November 1972, an immature bird was noted in both Weardale and Teesdale. Less specifically, Golden Eagle was one of the species that was referred to as having been "*seen on the moorlands*" of upper Teesdale "*in recent years*" by Bradshaw (1976), though no specific records were listed. It is conceivable that birds noted in the early 1970s would have been derived from the breeding birds then present in the Lake District on the western side of the Pennines. In 1978, a wandering adult Golden Eagle was reported at Lune Head Moss in the south west of the county on 26 May, although the record was not published in the county ornithological report of that year. This was probably linked to the two established pairs breeding in the Lake District over the period 1976 to 1982 (Parkin & Knox 2010). The following year, an adult bird was seen at Long Grains Beck, Lunedale on 20 May 1979 (Mather 1986) in a similar area to the previous year's report, raising the possibility of prospecting for territory within the county by this species, in an area that would not be unsuitable for a breeding attempt. Indeed, a bird was seen on Mickle Fell on 8 August 1984 in the very same area of the upper Lune valley.

Later in the 1990s, an immature bird was at Swinhope on 29 June 1995 and then presumably the same individual was seen at Edmundbyers on 3 July. The last confirmed record of this species in the county was of a bird that was seen at The Stang and over Gilmonby Moor on 13th and 14 April 2002.

In addition, in March 1991, experienced raptor worker B. Walker, reported a juvenile Golden Eagle flying down the Euden/Sharnberry valley whilst he was sat at Brown Law, Hamsterley Forest. However, due to no formal submission of a record for what is an extremely rare bird in the county, the record is omitted from the totals.

Distribution & movements

This species breeds over a wide area around the Holarctic though it is usually sparsely distributed; nonetheless, it can be found across much of Europe, northwest Africa, the Middle East, Asia and North America. It is essentially sedentary though young birds from Scotland may wander, though most usually within the established breeding range (Brown & Grice 2005).

Osprey (Western Osprey)
Pandion haliaetus

Sponsored by

NORTHUMBRIAN WATER

A once rare but now a scarce and increasingly recorded passage migrant, recorded from March to November.

Historical Review

This species is believed once to have been a locally common breeding bird in England (Brown & Grice 2005), with a suggested 500-1,000 pairs across Britain (Dennis 1987). It apparently bred on the eastern most side of Ullswater in the Lake District to the end of the 18th century (Holloway 1996) and based on this broad evidence it is tempting to speculate as to whether the species ever bred in the county. It seems possible that, many hundreds of years ago, prior to the extensive drainage of some of the larger wetland areas, in the central southern part of the county, that it may well have done so.

The first documented record of Osprey for County Durham is by Edward Backhouse who wrote; "*When at Seaton Carew in the autumn of 1828 I frequently saw one perched on the mast of a wreck on the sands at Teesmouth.*" (*MS*: 1834). It was reputed to be fairly frequent at Teesmouth in the early 19th century (Stead 1964) but there is little documentary evidence of records at this time.

In addition to the above record, Backhouse (1834) also reported a bird which had been shot near Durham in the spring of 1883, which was then in his possession. Hutchinson, writing in 1840, stated that "*The Osprey has not infrequently been killed both on the coast and in the inland parts of the County. This has usually happened in autumn and winter*", though he also makes reference to a bird shot near Darlington in June 1840, presumably because of the summer date. J. Duff of Bishop Auckland wrote of a bird trapped near Windlestone Hall, Rushyford in the spring of 1849 (*The Zoologist* Vol. V 1849) and another said to have been taken on board a ship off Hartlepool, during a late autumn storm in the same year (*The Zoologist* Vol. II 1850). Over twenty years later, Hancock (1874) said that it "*occurs in our district not infrequently*" and he listed eleven specimens as having been taken in Northumberland and County Durham between 1830 and 1860 with just one definitely attributable to County Durham, a female shot at Heworth on 23 September 1841. This was one of two historical records for the Gateshead area in the north of the county, the second being documented by Robson (1896) who told of another killed at Nabs End near Blaydon Burn around 1865 (Bowey *et al.* 1993). J. W. Fawcett (1890) gave an additional two records relating to an undated specimen held in Sunderland Museum, which was shot at Marsden, and one shot near Durham on 2 October 1883. This is doubtless the same bird recorded by Tristram (1905) as having been shot at Aldin Grange near Durham on 22 October 1883, though it is listed by Cullingford as having been shot on the 23rd (*The Zoologist* Vol. VII 1884); this was described as a male bird of the year.

The early 20th century saw the nadir of this species in Britain and then, around the middle of the century, its resurgence, as it first became extinct as a breeding bird in 1916, and then re-commenced breeding in Scotland in 1955.

The species remained a rare migrant throughout the first half of the 20th century. One was seen over Stockton High Street on 5 July 1933 (Almond *et al.*1939). Temperley (1951) described it as, "*An occasional visitor of very irregular occurrence*" but he listed just three further records by 1951, of a male shot near the Swing Bridge at Stanhope as it fed on a trout on 6 May 1939, another shot at Seaham Harbour on 17 October 1949 and a very emaciated female found dead but still warm, in woods near the vicarage at Muggleswick on 10 May 1951. Through the rest of the 1950s, despite events in Scotland, there was little sign of this species in Durham, with just two more documented records. Stead (1964) gives no records at all for Teesside between 1850 and 1954, making reference to just one record at Seal Sands, Teesmouth on 31 August 1955. The only other published record during the 1950's is of a bird at Cauldron Snout in upper Teesdale on 12 May 1957.

The 1960s brought some local records and the establishment of a pattern of low but annual occurrence that was to be consolidated over successive decades. There was a dramatic increase in the number of records, resulting in a further 20 birds at widely scattered localities by the close of the decade, although none at all were seen in 1961, 1963, 1964, and 1966. A remarkable series of records in September 1960 saw a total of three flying in off the sea at Hartlepool and Crimdon on 17 September with a further four watched flying south offshore at Hartlepool on 29 September. The latter record remains to this day the maximum count for the species in the county. A bird at Hamsterley Sewage Farm from 11th to 13 June 1967 is notable for being the first 'modern day' bird to linger in the county. With the exception of eight in 1960, annual counts through the 1960s were typically of one to four birds, and this pattern continued through the 1970s.

On 26 September 1976 there was the most unusual sight of an obviously exhausted Osprey sitting on a domestic TV aerial in the Fulwell area of Sunderland.

Recent Status

The Osprey is today a scarce but regular migrant through Durham, with one or two birds occasionally lingering for variable amounts of time during the summer and autumn. It is regularly noted on passage, especially during the autumn when birds tend to frequent inland water bodies and pass down the coast. This species is now established as a breeding bird at several sites in England and Wales and still increasing within its Scottish stronghold.

The typical pattern of records in Durham over the last two decades is that the first early spring migrant is noted in late March or early April , as in 2009, when the year's first bird was seen over Eggleston, Teesdale on 25 March. Early birds are usually followed by a few sightings through April, with more birds, perhaps mainly non-breeding or immature birds, arriving in May. Through the last four decades, most records are of single birds seen on one day, or flying through, although since the mid-1990s, there has been an increasing tendency for birds to summer in our area with ever greater regularity, a very encouraging sign for the future.

Overall, there have been in excess of 320 Ospreys recorded in County Durham since 1970, and the species has failed to be recorded in just the two years at the beginning of the period, 1971 and 1972. There has been a progressive increase in numbers during the review period with each new decade seeing the rates of occurrence almost doubling. In the 1970s there were a total of 28 birds, yet three decades later, the period 2000 to 2009 saw an estimated 142 birds recorded, including a peak of 28 in 2008 alone. A record 33 birds were seen in 2010, though as ever the true figure is likely to be much higher with a number of unsubstantiated reports coming from local fishermen. Records have been widely scattered throughout the county and the species can occur just about anywhere though they seem to favour passage routes along our river valley systems and occasionally birds linger at some of our larger reservoirs, particularly in late summer and autumn.

The majority of birds are recorded in spring, usually from late March to early June, with a distinct peak of records in May. There have been just ten March records, the earliest of which flew north over Garmondsway Moor, Coxhoe on 13 March 2007. These 'early' birds are thought to represent birds en route to their breeding grounds in the Scottish Highlands, whilst those recorded in late April and May are presumed to be either immature non-breeding birds or of continental origin, perhaps heading further north to Scandinavia.

There can be little doubt that the upsurge in sightings in Durham is directly related to the ever-growing number of Ospreys now breeding in Scotland, though it seems likely that a proportion of the birds passing through our region will be migrants of a Scandinavian origin.

Given the national range extension and growth in numbers, with seemingly ever more passage birds in Durham and successful breeding in neighbouring Northumberland, it is likely to be only a matter of time before the species does likewise in County Durham. Favoured localities over recent years have included the Derwent, Crookfoot and Hurworth Reservoirs, which between them have probably attracted more birds, for longer periods of time, than any other sites in the county, and when breeding does occur, it seems likely that one of these sites will feature. These sites would certainly benefit from targeted conservation measures aimed specifically at attracting Ospreys and would no doubt prove an excellent beacon for public conservation interests within our region

There have now been a number of records of Osprey throughout the summer period with several July sightings. One notable July record concerned a bird seen from a fishing-boat half a mile offshore from Ryhope on 15 July 1975 as it flew south west carrying a fish. During the survey period for the *Atlas,* a pair of birds spent the early part of June 1992 around Crookfoot Reservoir. At the time, some courtship behaviour was noted but beyond this, there was no concerted attempt at breeding (Westerberg & Bowey 2000). Six birds have made protracted stays during the summer months at potential breeding locations. At Derwent Reservoir, birds have lingered from 10 May until at least 4 August 1977, from 30 June until 24 July 2007, from 21 July until 3 August 2010 and during July and August 2011. At Crookfoot and Hurworth Burn Reservoirs, single birds have stayed from 23 June until 2 September 2009 and again through August 2011. None of these birds showed any evidence of breeding activity, suggesting they may have been immature, but as the species bred in neighbouring Northumberland for the first time in 2009, the first breeding pair in County Durham is unlikely to be too far in the future.

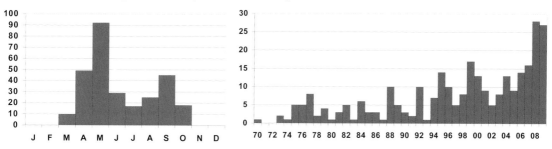

Records of Ospreys in County Durham, 1970-2009

Return migration probably begins as early as July, and continues right through until late October, although September is the peak month. The 2008 September influx of Honey-buzzards *Pernis apivorus* also brought a number of Ospreys to the county, at least fourteen were recorded between 4th and 27th of that month; easily the largest number ever seen during one September. At the height of this passage an exhausted juvenile spent several hours sat on an electricity pylon next to the Seal Sands roundabout, Teesmouth on 13 September, with at least one more passing through the North Tees Marshes and another down the coast south of Sunderland on the same day. The latest bird ever recorded in County Durham is a juvenile which lingered in the Derwent Valley around Gibside and Far Pasture from 21 October until at least 17 November 1994, which remains the only November record of Osprey in the county and one of the longest staying autumn birds (Bowey *et al.* 1993). The maximum count during the review period is of three birds together, which has occurred twice, both times in autumn, over Greatham Creek, Teesmouth on 2 October 1999 and at Burnhope Reservoir, Weardale on 4 September 2008.

Nationally, the species has been in a colonising phase since the 1980s, expanding both its range and numbers (Dennis 1987). Scotland has for many years held the UK breeding population but a re-introduction project in Rutland during the early 2000s and the natural colonisation of the Borders, Cumbria and then Wales confirms that the species is slowly gaining a foothold in England and beyond. In Kielder Forest a pair built a foundation nest in 2008 and returned in 2009 to fledge three young.

Distribution & movements

This species breeds, at variable densities across Eurasia, North and Central America and down into Australasia. In Europe the species is highly migratory, with birds wintering south into Africa, though some spend the winter months around the Mediterranean. The Osprey has a patchy breeding distribution across most of Eurasia, North and Central America and Australia. The Osprey became extinct as a breeding bird in Britain in 1916 and did not breed again until 1955, being largely restricted to the Scottish Highlands, especially the Spey Valley. Recently, there has been a large increase in the Scottish population and in addition to a re-introduction project at Rutland Water, Leicestershire in 2001, the species re-colonised Cumbria in 1999/2000 (Brown & Grice 2005). This was the first time that Osprey had bred in England for 160 years; Northumberland has subsequently been re-colonised. Despite the increases in the UK, the Osprey remains Amber listed as a species of conservation concern. The national population is estimated at about 180 breeding pairs (Holling 2011).

During the late summer of 2008 there was an interesting series of records involving juveniles from the Loch Garten satellite-tagging project. Two young birds passed through Durham, over the River Wear on 1 August, with a third tagged juvenile passing through on the 14th. A highly publicised bird from the Loch Garten nest, 'Nethy', settled for a short period at the Oakenshaw Local Nature Reserve on 22nd before moving on. In 2009, the process was repeated, with on 30 August a satellite-tagged juvenile, 'Beatrice', once again from Loch Garten, passing over West Pelton, near Chester-le-Street, at 09.00hrs and then over Bishop Auckland at 11.00hrs, later the same day. These discoveries confirm that many birds must pass through Durham un-noticed.

Kestrel (Common Kestrel)
Falco tinnunculus

Sponsored by
Durham County Council
(Countryside Service)

A common resident breeder and an uncommon passage visitor.

Historical review

The bones of this species, or possibly those of the Merlin *Falco columbarius*, were found in a quarry at Whitburn Lizard, on Cleadon Hills, in the spring of 1878, implying that the species was present and perhaps hunted in former times in the county (Jackson 1953).

In the late 19th century, Robson (1896) thought that the Kestrel was scarcer than the Sparrowhawk *Accipiter nisus* in the Derwent valley, a situation that, with the rise in Sparrowhawk numbers in recent years, is probably still true today. In 1875, John Sclater noted that nine pairs had reared broods in Castle Eden Dene, indicating the number that could be present at that time when birds went unmolested (Temperley 1951). Nelson (1907) described it as "*common and generally distributed*" around Teesmouth around the end of the 19th century.

This was a "*common and well distributed resident*" midway through the 20th century according to Temperley (1951). It was evidently common in the west of the county during the first part of the 20th century as illustrated when one day at Swinhope Head, George Temperley put up some young Kestrels eating a thrush with four adults hovering. On hearing the clamour three more adults arrived and he had the *"unusual sight of 7 adult kestrels flying within close range and glimpses of the younger hawks flitting amongst the trees"* (*The Vasculum* Vol. XXV 1939).

Temperley recorded that it was met with in every type of country in Durham and even observed it hovering over gardens, allotments and other open spaces well within towns and villages. He also noted that for several years birds occupied the sea cliffs at South Shields, without giving any further details on this, and that a pair had bred on a slagheap at Teesmouth around the turn of the 20th century.

In most years since 1950, two to three pairs per annum have bred in the Shotton Colliery area, primarily at Edder Acres Farm, Shotton Colliery and Castle Eden Dene. In the latter site, a pair nested every year from 1954 to 1965 on a ledge on the cliff face at Gunner's Pool. Many hours were spent watching the adults feeding the young by one observer, who noted that initially this was a quiet spot but had become increasingly popular as a place for recreational visits and after 1965, seemed to suffer from too much disturbance for the birds (Simpson 2011).

In common with many birds of prey, Kestrel numbers declined both locally and nationally during the 1950s and 1960s following the introduction and widespread use of organochlorine pesticides. These were used most extensively in the predominantly agricultural eastern part of the county where the greatest density of Kestrels occurred. Egg thinning was a particular symptom of DDT poisoning, which affected raptors at the top of the food-chain.

Local reports in the early 1970s reflected the national trend and the changing impact of pesticides, although this was largely implicit, as most records came from urban locations or the west of the county. Successful breeding was noted on Durham Cathedral between 1971 and 1974 and concentrations of five to eight pairs were recorded respectively in the areas of upper Teesdale and Upper Weardale, the latter decreasing to five pairs in 1974 perhaps arising from the increasing 'improvement' of traditional hay meadows. In 1975 three young were fledged from a nest on the ledge of a sea cliff at Hendon, mirroring Temperley's earlier information.

The downward trend was reversed following a phased voluntary withdrawal of the pernicious chemicals from 1962 with practically a complete ban in 1982. However, following the resultant recovery of the population through the 1980s, significant declines have since been recorded especially in lowland eastern England possibly linked to the diminution in rodent prey following agricultural intensification (Baillie *et al.* 2007). High levels of residues from anticoagulant rodenticides in the liver of Kestrels is a concern which is currently being monitored (Predatory Bird Monitoring Scheme).

Recent breeding status

Recent estimates of the UK population, range from around 37,000 pairs in 2000 (Baker *et al.* 2006) up to 58,000 pairs in 2007 (BTO-Breeding Bird Survey). However, the latter work also estimated a significant 20% decline during the period 1995-2008 with most of this occurring post-2005. Clements (2008) considered that earlier estimates of the national population were too low, being based on an underestimate of the maximum density in good habitat, which should be around 30 pairs per hectare (100 per square kilometre). He suggested that higher densities are common in areas of mixed farmland in eastern and southern England, and projected a British population of 53,000 to 57,500 pairs, which is close to the BTO/BBS estimate. The survey also indicated an alarming 54% decline in Scotland over the same 1995-2008 period. The Kestrel, at 2010, remained an amber-listed bird of conservation concern (Eaton *et al.* 2009), largely on the basis of its European status.

Today, hovering Kestrels are a familiar sight across Durham and it remains the most common of County Durham's diurnal birds of prey. It is a widespread species, in terms of broad habitat selection, being present from town centres to the coast and on to the remote uplands of the west. That this species can be considered common locally is well illustrated by the fact that the Durham Bird Club received nearly 1,000 reports of Kestrels in 2007 and 2009, and almost 1,500 in 2008.

In Durham, as elsewhere, Kestrels need open country for hunting. They are, largely, small mammal specialists but are capable of taking a variety of other prey, including small birds and many invertebrates, especially earthworms (Village 1990). The species' favoured habitat is permanent pasture land, which supports high levels of small mammals, in particular field voles *Microtus agrestis*. Locally it is also often found in open woodland and along the woodland edge, especially where these areas lie adjacent to grassland or farmland, although it tends to be less common in such habitats (Westerberg & Bowey 2000).

Like all falcons, Kestrels do not construct nests but need a horizontal ledge or platform upon which to lay their eggs (Village 1990). They are highly adaptable and will use a wide variety of nest sites, often close to human habitation. Within the county nesting has been recorded in old Carrion Crow *Corvus corone* nests, in both woodlands and on pylons, and often located close to some of the county's busiest roads (Westerberg & Bowey 2000). Kestrels also use factory air-vents, building ledges and recesses, natural crags and quarries and holes in mature trees, with Ash *Fraxinus excelsior* being most favoured. Most of the county's town centres, assuming that there is open habitat for hunting nearby, have a pair or two of nesting Kestrels. Along the length of the Durham coast birds can be found wherever hard or soft cliffs provide suitable nest cavities and ledges. Nestboxes are used increasingly, including tea chests at WWT Washington, specialist Kestrel boxes at Lockhaugh and Rowlands Gill and Tawny Owl *Strix aluco* chimney-style boxes at Gibside. Indeed, in 2008 a pair was evicted from a box near Haswell by a Barn Owl *Tyto alba*. Normally only one brood is attempted each year, laid in April or May, and the typical clutch size is four to five eggs, but occasionally successful broods of six are noted.

Birds can regularly be found in the western uplands of the county and it is probably our most common raptor in such situations, although it would appear that suitable nest sites are limited over 350m above sea level.

Local populations can fluctuate considerably from year-to-year and it is likely that this is driven by the fluctuation in the number of small mammals upon which the species preys. This local variation was well illustrated by detailed monitoring work in Gateshead during the late 1980s and early 1990s. In 1988, eight pairs nested in approximately 20 square kilometres of the lower Derwent valley and adjoining Tyne valley (Bowey *et al.* 1993). This density was surpassed in 1992, when six pairs nested in an area less than half this size (Bowey *et al.* 1993). In the west Gateshead area in 1991, 23 nestlings were ringed from 14 nests (Armstrong 1992). Some of these densities are considerably higher than those suggested by Village (1990), who quoted an upper limit of 36 pairs per 100 square kilometres in good vole years on grassland, down to 10 pairs per 100 square kilometres on intensive arable farmland.

In the *Atlas*, Westerberg & Bowey (2000) considered that using Village's criteria, allowing for quality and availability of habitat and nest sites, would have indicated a county population of around 200 pairs. They considered this to be "*on the low side*", based on monitoring work undertaken in the north west of the county, which in good breeding years was estimated to hold over 40 pairs (Bowey *et al.* 1993). Using the Gateshead survey data, as a basis for extrapolating an estimate of the county population, the *Atlas* stated that over 400 pairs might be present in the county in the best breeding seasons. According to Joynt (2008), there were probably 45-55 pairs in the old Cleveland area, north of the Tees, with the lowest densities observed in the arable areas to the west and north of Stockton.

Illustrative of the recent situation in Durham, in 2007, birds were noted at over 60 localities during the breeding season and populations in the east of the county were considered 'particularly healthy'. Three pairs reared young at Rainton Meadows, with two pairs raising young at nearby Hetton Lyons Country Park. Further family parties were noted at twelve additional sites from Blackhall at the coast, to Eggleston Common and Ruffside in the Pennine uplands and upland fringes (Newsome 2008).

At present, local trends are uncertain. There appears to be relatively little evidence that collisions with cars, especially involving young inexperienced birds, the ingestion of rodenticides or the loss of permanent grassland, all of which have been implicated in declines of the species nationally, have had any significant effect in County Durham. In 2008, it was felt that the local population was, at least to some degree, bucking the trend of an observed national decline of the species, with potential nesting pairs being reported from at least 150 sites (Newsome 2009). Declines noted in some parts of western and northern Britain linked to agricultural intensification, overstocking in upland areas and increased competition for food from the growing Buzzard *Buteo buteo* population could yet become significant in the western part of the county. Surprisingly, given its typical prey items, this species also suffers from persecution, although at a lower rate than most raptors. In 2008 one confirmed and two probable shooting incidents involving Kestrels were recorded in Durham (RSPB Birdcrime 2008).

Recent non-breeding status

There is a general dispersal of birds after the breeding season. In autumn loose post-breeding groups of up to 10 are frequently noted over farmland in the east, taking advantage of recently cut fields or on moorland hillsides in the west, hunting voles. This phenomenon is well illustrated by the distinct increase in numbers around the North Tees Marshes and other good hunting areas. In August of some years, up to 15 birds have been seen at one time over Greenabella Marsh, whilst up to 20 have been recorded on the stretch of land between Port Clarence and Seaton Carew at this time of the year (Blick 2009). Similarly there was a report of 15 around WWT Washington during July 1990 (Armstrong 1991). Where feeding conditions are good, then some sites can retain such groupings through the winter as evidenced by counts of 10 on the rough grasslands of Murton and Rainton Meadows in February 2008. Although most often reported in spring and summer, Kestrels are resident and present in every month. On the higher ground of the west in mid-winter birds exploit road-kill carrion as a more readily available food source than their usual prey. It is however, most numerous in the eastern lowland areas with one to six birds typically noted at many locations. In January 2010, a Kestrel visited a bird table for several days at Norman's Riding Farm, Winlaton during severe weather. The bird regularly took small pieces of duck, on one occasion staying to feed for several minutes (D. McCutcheon pers. comm.). Records of Kestrels visiting bird tables are rare.

The mortality rate of this species in the first year is high at 50-60% (Village 1990), though an adult female ringed at Lockhaugh on 11 May 2004 was recaptured on 4 January 2009 at the same site, demonstrating the relative site-faithfulness and longevity of some mature birds; this bird was in at least its 7th calendar year at recapture.

There have been a number of reports of birds roosting on the window ledges of houses, for example at Balder Grange in Baldersdale in January 1997 and a juvenile that undertook such behaviour in Dipton, near Consett, in July 2008. Unusual behaviour was noted when a bird was seen caching prey presumably for future retrieval on South Shields Leas in 2008 (Newsome 2009). An almost completely white male was observed in the Seaton Common and Greenabella Marsh area from August 1997 to March 2000 (Blick 2009), whilst a leucistic bird was at Marsden Lime Kilns in 2000 (Armstrong 2005).

Distribution & movements

The Kestrel is a widespread breeding bird across Europe, Asia and Africa. It is a largely resident species over much of its range, though northern and eastern populations do winter further south in Europe and in N. Africa (BWP). The UK wintering population is supplemented by birds from Fennoscandia, which often use oil installations in movements across the North Sea (Wernham et al. 2022).

There is clear recent evidence in *Birds in Durham* of the regular autumn passage of birds along the Durham coast, first noted by Nelson (1907) and commented upon by Joynt (2008). In the period between late July and early November, Kestrels are regularly seen arriving from over the sea or moving south along the coast. Specific examples of such movements include the 20 birds that passed Hartlepool Headland between August and November 1970 and the five that arrived at Hartlepool in one hour on 27 September 2000 (Blick 2009). The origin of these birds is debatable. Some of the ringing recoveries confirm that Scandinavian birds are involved. For example, two birds ringed in the Swedish province of Norrbotten, the first in July 1961 and the second in July 1976, were recovered three months later in the county, the first in Durham City on 20 October 1961 and the second at Hartlepool (Blick 2009). Furthermore, a bird ringed as a nestling at Kolari, Lappi, Finland on 7 July 1998 was controlled 97 days later at North Sands, Hartlepool, on 12 October 1998, just under 2,000km to the south east of its original ringing location.

It is likely that birds breeding further north in Britain are also involved in these coastal movements. There are a number of records of birds being ringed in Northumberland and recovered later in the year in Durham. A first-year bird ringed in Kielder Forest on 3 September 1987 was controlled at Lockhaugh 64km to the east south east on 27 November 1988. If such southward dispersing birds reach a coastline the inclination will be to move along it. Most of the nestlings ringed in Gateshead over the period from the mid-1980s to the mid-1990s, fitted into a regional pattern of the southerly dispersal of juveniles with recoveries of birds coming from Norfolk and Oxfordshire. A similar movement was shown by the bird ringed as a pullus on 2 July 1965 at Kimblesworth near Sacriston, which was shot at Moreton Morrell, Warwickshire, on 29 November 1965. Clearly there is also some random dispersal of juveniles from natal sites and the bird ringed at Blaydon Station, which was later found in Galloway, south west Scotland fitted into this pattern (Bowey et al. 1993).

An analysis of ringing controls and recoveries in the Teesmouth area by Blick (2009) indicated that Scandinavian Kestrels appeared to fledge about a month later than British bred birds, suggesting that birds moving along the coast in July and August are most likely to be British birds, and that some of those seen later in the autumn, from September to November, for example, may have arrived from Scandinavia (Blick 2009).

Over 300 Kestrels have been ringed in County Durham, the majority of these being young in the nest. For example, from 25 January 1989 to 1 February 2011, 166 pulli and 17 adults were ringed in the north of the county alone, with just one of the latter being recaptured. Some birds ringed further south have been noted in the Durham area. For example, the bird ringed at Newby, North Yorkshire, in June 1980 moved 12km north to Billingham, in January 1981, and one that had moved north west from New Holland, on Humberside, in June 1957 was again found at Billingham in August 1958. Most spectacularly, a bird ringed on 6 October 1966 at Sint Kruis-Winkel, Belgium, was found dead at Barmpton, to the north east of Darlington, on 27 September 1969.

Red-footed Falcon
Falco vespertinus

A rare vagrant from eastern Europe or western Asia: 15 records, most often May and June.

Historical review

In the 60 years since the publication of *A History of the Birds of Durham* (Temperley 1951), the Red-footed Falcon has been recorded with slightly greater frequency than prior to Temperley's review. He assessed its status then as a very rare vagrant, on the basis of the two records up to that time. The first county record dates from October 1836, when an adult male was shot at Trow Rocks, South Shields, and was obtained from a Mr Clarke (Hancock 1874). This bird was just the seventh record for Britain and came just six years after the first British record (Brown & Grice 2005).

Over one hundred years later, on 30 October 1949, a female or immature bird was seen near the North Gare, at Seaton Carew by Miss C. Greenwell and Miss D. Bell; a fortnight earlier, two Red-footed Falcons had been present on Holy Island in Northumberland (Kerr 2001). Not until 25 October 1969, did the third occurrence of this distinctively handsome raptor from eastern Europe take place in Durham; this was a male that was found at Saltholme Pools, Teesmouth.

All records:

1836	Trow Rocks, South Shields, male, October, shot
1949	Seaton Carew, female, 30 October (Temperley 1951)
1969	Saltholme Pools, Teesmouth, adult male, 25 October (*British Birds* 63: 274)
1971	Backhouse Park, Sunderland, female, 24th to 29 April (*British Birds* 65: 330)
1974	Barmston Ponds, Washington, first summer male, 29th to 30 May (*British Birds* 68: 315)
1983	Cowshill, Weardale, female, 2 August (*British Birds* 78: 541)
1987	Longnewton, adult male, 1 May (*British Birds* 81: 551)
1990	Seal Sands and Greenabella Marsh, Teesmouth, adult male, 10th to 21 May *(contra BB)* (*British Birds* 84: 466, *British Birds* 85: 520)
1992	Killhope, Weardale, first summer male, 8 June (*British Birds* 86: 466)
1992	Haverton Hole and area, North Tees Marshes, female, 10th to 12 June (*British Birds* 87: 519)
1992	Haverton Hole, Teesmouth, male, 12 June (*British Birds* 86: 466)
1994	Hartlepool Headland, female, in off the sea, 24 May (*British Birds* 88: 505)
1995	Hartlepool Headland, first summer female, in off the sea, 24 May (*British Birds* 89: 496)
2000	Hargreaves Quarry, Teesmouth, female, 9th to 10 June (*British Birds* 94: 467)
2003	Baydale Farm, Darlington, female, 8 June (*British Birds* 100: 31, published as North Yorkshire)

That the first three county records should all be in October is unusual, as there are less than 30 October records for Britain as a whole. With the exception of the bird in Weardale in August 1983, all of the subsequent records have been in the period from late April to mid-June. This is the classic time for the species and conforms to the national pattern. The influx in 1992 saw three birds in County Durham, all of which were in a five-day period in early June and included two together at Haverton Hole, Teesmouth on 12 June. The bird at Killhope on the Cumbrian border around the same time had penetrated far inland. The two females flying in off the sea in 1994 and

1995 are remarkable not only for being on the same date, but that they were also seen by the same observer. The bird in 1994 even perched on top of the lighthouse for a short time before heading off inland.

Distribution & movements

Red-footed Falcon, Greenabella, 1990 (M Sidwell)

Red-footed Falcons breed with a highly fragmented distribution, across wooded steppe from east Hungary to temperate Russia, via central Eurasia and eastward to Mongolia. They are highly migratory and most birds winter in Africa, south of the Sahara. The number breeding in Europe is relatively small and said to be declining, though the number of records in Britain over the last three decades of the 20th century increased quite considerably (Parkin & Knox 2010). That said, numbers vary annually and are largely weather dependant, as illustrated by the prolonged period of warm temperatures and light easterly winds in late May and early June 1992 that resulted in the influx of 120 birds into Britain (Nightingale & Allsop 1994). The Red-footed Falcon is now a regular vagrant to Britain, most notably as a spring overshoot in May and June, although it has occurred as late as November. The species was removed from the list of species considered by the BBRC at the end of 2005, by which time there had been over 750 records.

Merlin
Falco columbarius

Sponsored by

NORTH PENNINES

Area of Outstanding Natural Beauty

GLOBAL GEOPARKS NETWORK

An uncommon resident, regular but scarce in winter and on passage.

Historical review

The oldest known reference to the Merlin in County Durham suggests the species' use in falconry. An extract from the Bursar's rolls of the Monastery of Durham for the year 1333 states *"Henry Hunter and Walter Maughan rode forth from the priory with Merlins to the Earl Warren"* (*Heruico Hunter et Waltero Mauyhan euntibus cum emerlionibus ad Com. Warenn. ex parte Prioris*) (Ticehurst 1923). Presumably this alludes to a presentation of Merlins made to Earl Warren from members of the priory manor. It might be assumed that the birds were bred on one of the Monastery hunting estates in the west of the county.

Records from the 19th and early 20th century consistently mention the widespread persecution of this iconic moorland falcon in its breeding range. Gamekeepers of that era seemed determined to dispose of any predator that might impact on their stocks of Red Grouse *Lagopus lagopus* and appear to have shown an irrational prejudice against Merlins much as they did other 'vermin'. Hancock (1874) was particularly concerned and claimed that *"This beautiful little Falcon is rapidly disappearing by the hand of the gamekeeper"*. James Backhouse (1885) talked of it still being shot or trapped in Teesdale despite presenting his own evidence that its main summer prey was the Meadow Pipit *Anthus pratensis* and not young grouse. Mather (1986) repeated details of 16 nests and 44 adult birds that were destroyed on the moors around Bowes between 1881 and 1890. Tristram (1905), not surprisingly, described it as *"now being rarely seen owing to the gamekeeper"*. Attitudes were very slow to change; Almond *et al.* (1939) said that in Teesdale *"it received no mercy from gamekeepers"* and Temperley (1951) related a conversation he had had with a gamekeeper who despite knowing the main prey not to be Red Grouse had still shot birds anyway and destroyed their nests on his moor. On the other hand, Temperley mentioned estates in Weardale and the Derwent valley that had by then a far more enlightened attitude to Merlins and gave them sanctuary.

313

The habit of birds wintering near the coast was known from Backhouse's time and Temperley said that it was regularly reported from Teesmouth and around South Shields.

Recent breeding status

The breeding population of Merlins in Durham continues to be closely associated with expanses of open, heather moorland invariably managed for the benefit of Red Grouse. Indeed the interests of Merlins are directly linked to the grouse shooting economy centred on the moorland estates and the management of heather *Calluna vulgaris*. In a reversal of previous mindsets, the shooting estates and gamekeepers of today actively promote the wellbeing of Merlins on their moors and are keenly interested and protective towards them. Grouse moorland management provides a mosaic of heather blocks of differing age structures which is favoured by Merlins. The control of predators such as fox *Vulpes vulpes*, stoat *Mustela erminea* and corvids additionally helps boost nest site productivity as it does for nesting waders (Fletcher *et al.* 2010). Lower levels of sheep grazing and the practise of rotational heather burning both help improve habitat quality and are to the long term benefit of Merlins. Over 70% of Durham's heather moorland now falls within the designated North Pennine Special Protection Area (SPA) which was scheduled in the late 1990s. This will hopefully serve to safeguard the birds' future through a range of appropriate land management practices although the longer term threat of receding heather lines will remain important.

Returning pairs can appear on or near the breeding sites by late February but full residency is not normally taken up until late March. Pairs often show an affinity to a particular breeding territory from year to year although the exact nest site is normally varied within this home range. Ground nesting in blocks of mature heather is the preferred choice and nests have been found at elevations between 320m and 640m above sea level. There have been isolated reports of pairs nesting in cotton grass *Eriophorum angustifolium* and bent grass *Agrostis* palustris (L. Waddell & G. Wall pers. comm.) although these sites retained vestiges of heather. Tree nesting is almost unknown in Durham with no sites reported for the last 40 years. This contrasts with the species' habit of tree nesting at the edge of conifer blocks adjacent to moorland in Northumberland (Newton *et al.* 1986). There are no records of crag nest sites.

In an earlier breeding study from one area of Weardale, the number of breeding pairs varied between 11 and 14 from 1969 to 1972. In 1973 a large number of infertile eggs were reported. This may have been the result of organochloride pesticides then present in the environment.

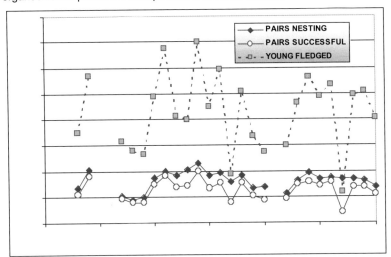

Summary of Merlin Breeding Success in County Durham

In 1983 and 1984 the RSPB covered the county as part of the first national Merlin breeding survey (Bibby & Nattrass 1986). The fieldwork in Durham was undertaken by M. Nattrass who later went on to co-ordinate extensive annual surveys in Durham, recommencing in 1987 and continuing to this day as part of the monitoring work of the Durham Upland Bird Study Group. The majority of suitable habitat has been surveyed annually from about 1990 with never less than 75% of traditional sites visited each year and more typical coverage of about 85%. Pairs are followed from the first signs of occupancy through to ringing the young birds just before fledging. The study has provided one of the most complete set of breeding data for any species in the Durham recording area. The graph summarises the results over this study period.

National surveys were undertaken in the years 1994 (Rebecca & Bainbridge 1998) and 2008 (Ewing *et al.* 2011). These revealed a slow decline in the UK's breeding population to an estimated 1,130 pairs by 2008. The Durham population has seen a slight decline, though breeding success is quite variable and the 2008 national survey did follow a year of particularly poor local productivity. The Durham data shows that one of the greatest threats to productivity is prolonged spells of heavy rain in mid-June which limits the adults' ability to hunt and chills the young birds in the nest at a vulnerable time in their development. Heavy continuous rain occurred for four to five days in both June 1997 and 2007 and led to the widespread death of nestlings.

The Durham population is presently stable and in the range 32-36 pairs. Breeding productivity has been measured at 2.7 young per pair and 3.6 young per successful nest.

Amongst a total of 40 avian prey species found near Durham nest sites, the commonest are: Meadow Pipit *Anthus pratensis*, Starling *Sturnus vulgaris*, Skylark *Alauda arvensis*, Wheatear *Oenanthe oenanthe* and Chaffinch *Fringilla coelebs*; whilst Snipe *Gallinago gallinago,* Barn Swallow *Hirundo rustica* and Great Spotted Woodpecker *Dendrocopos major* have also featured at some sites.

Recent non-breeding status

Durham's breeding Merlins are partial migrants. Some, especially juveniles, may leave the area in winter when local numbers are supplemented by visitors from other parts of the UK. Overall it remains uncommon during the winter period. The autumn and winter distribution is usually considered to be on the low-lying ground to the east where single birds frequent wetlands, open farmland and coastal habitats. Harsher weather and the absence of small birds on which it preys are bound to cause movement away from its upland breeding haunts to lower altitudes. The evidence however, suggests that this is an overly simplistic picture. Coastal areas have a strong observer bias in the winter months so whilst they are probably genuinely favoured by Merlins, the reported effect is likely to be exaggerated. Birds may winter on any open farmland which can attract small passerine flocks but such records will tend to go under-reported. Reports over the last two decades also show a small winter presence in upland valleys and on moorland edge in western areas. Clearly some adults do not wander too far from their breeding territory. Reports of single birds in the uplands in December or January have come from The Stang, Bowes, Lunedale, upper Teesdale, Wolsingham and Rookhope. Some of these sightings have been when there was complete snow cover and freezing temperatures. Notable was an apparent territorial dispute between three birds witnessed over high ground on 2 January 2001.

Merlins are thinly scattered in the east of Durham from August through to March. Typically there will be reports from 10-15 locations each winter involving one to two birds at each site, though exceptionally, four male Merlins hunted on Boldon Flats in November 1986. Favoured locations have included Boldon Flats, Low Barns, Mordon Carr, McNeil Bottoms, Rainton Meadows, Sedgeletch, South Hetton, South Shields, Teesmouth, WWT Washington and West Auckland Flats. There is a marked association with water bodies and presumably the congregation of the potential prey species that these support.

Where hunting has been witnessed, autumn prey items have included House Martin *Delichon urbicum* and Swallow *Hirundo rustica* and in winter targeted species have included finch flocks, Pied Wagtail *Motacilla alba*, Starling *Sturnus vulgaris*, Dunlin *Calidris alpina* and Turnstone *Arenaria interpres*.

The winter population in Britain has been put at between 2,000 and 3,000. It is difficult to arrive at a figure for Durham's winter numbers, especially since they probably vary quite considerably even within a single winter, but an estimate of between 30-50 individuals can be postulated.

Distribution & movements

Merlins have an almost complete circumpolar distribution across northern Europe, Asia and North America. Northern populations are migratory, wintering to the south of their breeding grounds. The majority of the British breeding population winters within the country (Lack 1986) with a general movement towards lower altitudes, which support the core of the overwintering population. Young birds of the year show greater dispersion from the natal area with a strong southerly bias to their movements (Heavisides 1987).

Since the present long-term Durham study began in 1983, a total of 1,586 young birds have been ringed in the nest, inclusive to the year 2010. There have been 101 (6.4%) recoveries. The vast majority of these are the result of fatal collisions with windows, other man-made objects or road vehicles. Almost half involve recently fledged birds in their first three months of life with August being the peak month for recoveries. Selected examples of birds in their first year ringed in the nest in Durham and later recovered in the same calendar year are shown below:

	Ringed	Date	Recovered	Date	Days	Km
A	Central Durham	20/6/89	Billingham, Teesside	12/07/89	22	53
B	Central Durham	24/6/90	North Shields, Tynemouth	31/07/90	37	34
C	North Durham	06/7/92	Seahouses, Northumberland	06/08/92	31	94
D	North Durham	06/7/92	Chollerford, Northumberland	08/08/92	33	34
E	North Durham	01/7/93	Weston, Warwickshire	17/08/93	47	264
F	Central Durham	20/7/93	Scunthorpe, Humberside	13/08/93	23	159
G	North Durham	20/6/94	Boulmer, Northumberland	27/07/94	37	64
H	Central Durham	03/7/95	Auldtoun, Strathclyde	01/08/95	29	148
I	North Durham	06/7/95	Eskdalemuir, Dumfries	02/09/95	58	87
J	Central Durham	18/7/95	Hull, Humberside	21/08/95	34	165
K	North Durham	07/7/96	Bridlington, Humberside	04/08/96	28	143
L	South Durham	01/7/99	Gulf de Morbihan, Brittany, France	05/12/99	157	762
M	South Durham	04/7/03	Seahouses, Northumberland	13/08/03	40	120
N	Central Durham	23/6/05	Les Sables d'Olonne, Vendee, France	14/10/05	113	903
O	North Durham	17/6/06	Ely, Cambridgeshire	18/08/06	62	295
P	South Durham	30/6/08	Lulworth, Dorset	19/08/08	50	464
Q	Central Durham	25/6/09	Stapleford, Nottinghamshire	08/08/09	44	196
R	North Durham	25/6/09	Shrivenham, Oxfordshire	22/10/09	119	364

After fledging in early July most young birds appear to stay in their natal area for a mere two or three weeks before dispersing independently. Examples A, C, D, F, G and H show just how rapid and far this initial movement can be, especially considering that the young birds are still a few days away from being able to fly at the time they are ringed. Coastal sites appear to be favoured, at least with these very early dispersals. Most recoveries show a distinct movement on a bearing near to due south (for example: E, L, N, O, P, Q & R) but there are also cases of dispersal to the north or north west (C, H, I & M). Both of the French recoveries (L & N) lie relatively close to one another in similar coastal lagoon habitats on the Atlantic seaboard.

Selected examples of Durham-ringed nestlings recovered when in their second calendar year or older.

	Ringed	Date	Recovered	Date	Comments
A2	Central Durham	23/6/87	Swindon, Wiltshire	12/1/88	Wintering; 2nd CY
B2	South Durham	25/6/88	Yatton Keynall, Wiltshire	20/02/00	Wintering; over 11years
C2	North Durham	25/6/89	Stockton-on-Tees	07/09/00	Aged over 11years
D2	North Durham	27/6/90	Lindisfarne, Northumberland	11/02/96	Wintering; over 5 years
E2	Central Durham	01/7/93	Seaton Ross Humberside	07/02/96	Wintering; over 2 years
F2	North Durham	29/6/94	North Berwick, Lothian	11/01/95	Wintering; 2nd CY
G2	North Durham	01/7/96	Yorks Dales moorland	15/06/99	Breeding; over 3 years
H2	South Durham	30/6/96	South Durham moorland	18/04/04	Breeding ; over 8 years
I2	North Durham	26/6/98	North Durham moorland	18/06/99	Breeding; 2nd CY
J2	Central Durham	29/6/09	Powys, Wales moorland	18/08/10	Breeding; 2nd CY

Two birds (B2 & C2) were over 11 years old when found, though the longevity record for the county is 13 years and relates to a much earlier recovery. The data indicates where some Durham birds may overwinter as adults (A2, B2, D2, E2 & F2). It also provides not unexpected evidence that young birds later return to local sites to breed as adults; H2 & I2 were both found within 10km of their natal area. Alternatively, birds sometimes move to entirely new breeding populations in other parts of the country as demonstrated by G2 & J2.

There are two recoveries at lowland locations in Durham of birds ringed outside of the county, but within the north east of England. Firstly, one that was ringed in a nest on Danby Moors, Yorkshire on 29 June 1970 and which was found dead at Billingham, 45km away, on 6 March 1978; and, a nestling ringed in the Cheviots,

Northumberland on 22 June 1977 found dead, at a similar distance from its natal site, at Seaham on 29 August 1978.

Against this background, the occasional reports of birds in Durham coastal locations in July are quite readily explained as being the early dispersal of young birds of the year. These are by no means regular but recent records include sightings of singles in 2006, at Seaham on 22nd, Whitburn on 23rd and 31st and South Shields on 23rd; and, in July 2009, at Whitburn on 18th, Langley Moor on 27th and Greenabella Marsh on the 30th. August produces more regular reports of one to two birds in widespread coastal and lowland locations. An interesting report concerned an adult male Merlin that was seen catching dragonflies at Seaham Pond in August 2004.

Spring passage at the coast is discernible in some years. The period from mid-April into the first week of May has occasionally produced a number of records, which probably relate to returning Scandinavian or Icelandic birds although there has been no direct evidence to confirm such an origin. Single birds were seen in early May 1986 at Cleadon Hill, South Shields and Darlington and in late April or early May in 2001 at Hartlepool, Rainton Meadows, Whitburn and Teesmouth. The period of 20th to 29 April produced records of singles at Seaton Pond, Whitburn and Seaton Carew in 2008 and at Bishop Middleham, Rainton Meadows and Cowpen Marsh in 2009.

There are a few June records of birds away from breeding grounds. These presumably relate to wandering immature birds. Most Merlins do not breed until they are two-years old (Lieske *et al.* 1997). Singles were seen on Boldon Flats and at Great Lumley in June 1986, at Teesmouth in June 1991 and at Whitburn in June 1994.

Hobby (Eurasian Hobby)
Falco subbuteo

A scarce passage migrant and summer visitor that is being recorded with increasing regularity. Birds have summered and it was first proven to have bred in 2009.

Historical review

Reports of this handsome long-winged falcon have steadily increased in recent years, but at the outset of the 20th century it was a genuinely rare bird in County Durham. Temperley (1951) documented the Hobby as a "*rare vagrant, of which there are no recent records*", indeed up to his time of writing, there were just five documented records. One, the first record for the county, was of a bird that was shot near Stockton at some time "*prior to 1827*" (Hogg 1827). The second record was described by Selby (1831) who thought the Hobby "*a rare species in the north*". He gave a single record for Northumberland and Durham, which was one shot at Streatlam Park, near Barnard Castle, but he did not give a date for this occurrence. This specimen eventually went to Hancock's collection.

According to Hutchinson (1840), a young male was killed at Thornley in October in 1832, a late date for birds in the north of England. However Hancock (1874) appeared to correct this and said that one was shot there in September 1832, which would be more typical for a passage bird. Hancock also said that he had obtained an adult female Hobby "*several years ago*" (before 1874) which he thought was a breeding bird that had had its nest in the locality. Finally for the century, Tristram (1905) mentioned a bird that was obtained at Grangetown in 1868, but gave no further details.

Holloway (1996) reported that in the 19th century the species was breeding in 19 English counties, unsurprisingly none of these were in the north east of England.

The first county record of the 20th century occurred at Hartlepool on 18 September 1960. There was then an eight-year gap until the next, a spring bird that was over Dorman's Pool on 4 May 1968. One was found dead below wires at Middleton St. George on 9 June 1969 and this was followed by records of single juveniles at Teesmouth on 16 August and Hurworth Burn Reservoir on 1 September; the best year for records of the species in the county at that time.

Through the 1970s birds were observed with greater regularity and they were recorded on average in two years out of every three throughout the decade, with birds in 1970, 1971, 1973, 1974, 1975, 1976, 1977 and 1979. One or two birds were usually noted in these years, with most records coming from the south east of the county, often from the Teesmouth area. On 2 October 1971, two adults together over Teesmouth were mobbed by a Merlin *Falco columbarius*, the first multiple occurrence of the species in Durham. Most intriguingly over this period was the occurrence of a pair in the Piercebridge area from 23 May to 2 June, with just the male noted on 5th and 6 June 1976 (Sharrock *et al.* 1978). In 1976, an unprecedented number of passage birds were reported with four birds being noted in addition to the Piercebridge pair. The next best year for sightings was 1979, when singles were noted on five dates from 2 June to 29 August, including a juvenile in the area of Barmston Pond on 9th to 10 August, which may well have indicated local breeding in northern England.

Recent breeding status

The Hobby remains a rather scarce species in County Durham; sightings are generally few and irregular and they largely relate to passage or, perhaps increasingly, to wandering non-breeders from the expanding population in the south of England (Brown & Grice 2005). However, the period 2007 to 2011 saw clear year on year increases and it is likely that this trend will continue. Breeding was first successfully proven in 2009, but it may well have been occurring intermittently, at a low density for a decade or more prior to this. Until recently, this species was thought to be a rare bird of southern heathland, but over the last thirty years various studies have demonstrated that in Britain, it is more abundant than previously imagined and that most birds breed in lowland farmland habitats (Brown & Grice 2005).

During work on the *Atlas* (Westerberg & Bowey 2000), in 1994 and 1995, around the lower Derwent valley, sightings of individual Hobbies were frequent enough during the late spring and summer to consider that a territorial pair might have been present. In 1994 for example, sightings occurred from late May until September with a possible juvenile seen in early September. These records occurred in an area of mixed woodland and farmland (Westerberg & Bowey 2000).

Hobbies were seen on an annual basis up to 1998 at some of the lower Derwent valley sites, including Axwell Park, Hollinside, Highfield, Shibdon Pond and Strother Hills. It seems reasonable to assume that breeding was taking place or at least being attempted somewhere in this area at that time, as it did to the north in neighbouring Northumberland (Kerr 2001). Breeding was first documented there as early as 1966 and up to four pairs bred in the south west of Northumberland from 1997 to 1999 (Kerr 2001).

Nationally, the population has undergone a considerable increase over recent decades, rising by 23% between 1995 and 2008 (Parkin & Knox 2010). Mirroring this locally, there would appear to have been a genuine increase in sightings over the last two decades of the 20th century and especially at the outset of the 21st century.

Low breeding density, unobtrusive behaviour and large hunting ranges make this one of the most difficult raptors to census (Parkin & Knox 2010). In the period 2005 to 2010, there appeared to be a genuine increase in the number of records of this mercurial falcon, to the extent that proof of successful breeding was highly anticipated, especially given the national increase in the UK breeding population both in terms of numbers and distribution. Though, by 2008, it had still not been proven to have bred in Durham. The long-expected first proof of successful breeding came in 2009 when careful, long-distance observation of a site in the south of Durham confirmed that at least two young had been reared. Breeding was also suspected at one or two other localities, but firm evidence was lacking. The same pair was again successful in 2010 when three juveniles fledged and there were also frequent sightings of birds in the Bishop Middleham area. In 2010, further summer season sightings were received from 13 locations around the county, including: the Hetton/Houghton & Rainton Meadows area, Hamsterley Forest, Cocken, Hurworth Burn, Darlington and Piercebridge, and the upper Derwent valley in the north, with sightings of juvenile birds in the mid-Wear valley in late summer 2010. Signs of breeding activity occurred at three of these sites and the presence of additional breeding pairs in the county now seems likely. The summer of 2011 saw the original breeding pair once again successfully fledge three young, and sightings of further pairs came from at least two other southern Durham farmland sites. It is quite likely that the county breeding population in 2011 stood at 2-4 pairs.

Most sightings in Durham are in the central to eastern lowlands of the county, though birds can be found at almost any elevation, as was shown by the fact that in 2008 one woodland that contained territorial birds, was located in one of the county's western dales, but the woodland itself was located just thirty metres outside of the county.

Since the early part of the 21st century birds have bred as far north as Strathspey in Scotland (Parkin & Knox 2010), so confirmation of breeding in Durham was something of an inevitability. The species was recorded as breeding in Cumbria in the 1930s (Brown & Grice 2005).

Nationally, Hobby populations decreased in each of the periods 1800-1849 and 1850-1899, were relatively stable from 1900 to 1969, but have increased markedly since then (Gibbons *et al.*1996). With an estimated breeding population of 500-900 pairs in 1988-1991, compared to only 60-90 pairs in the 1950s, the New Atlas records a remarkable growth in population and an extension of range (Gibbons et al. 1993). National summaries suggest breeding populations of 20 and then 27 pairs respectively in Yorkshire in 2005 and 2006, but it is accepted that the overall figure of up to 970 pairs nationally (Holling 2010) was too low, based upon the UK population being estimated to be 2,200 pairs (Clements 2001). The Rare Breeding Birds Panel analysis for 2009 suggested that almost 1,200 breeding pairs were now present in the UK (Holling 2011).

Recent non-breeding status

Birds have been recorded annually on passage from 1980 onwards, with up to five birds noted in 1980, between 29 May and 4 September. However, for much of the rest of the decade sightings remained scarce with just single records in: 1981, 1982, 1983 and 1988. Up to four were noted in spring and summer 1984, all of them in the Whitburn area, though at least three different birds were involved. In 1985, three records occurred between 31 July and 17 September, suggesting the birds involved were autumn migrants. The period 1986 to 1987, produced the best cluster of records of the decade, with five in 1986, these falling between 7 May and 3 September, and six in 1987; with a cluster of sightings in the west Gateshead area between late June and early August. These sightings were spread over the longest period in which the species had then been recorded, 24 April to 21 September.

During the 1980s, there were between one and six reports per annum, with an average of around three each year. This number rose through the 1990s, the mean for that decade being about six per annum, but the numbers were relentlessly rising over the period, and into the 21st century. The most concentrated period for records ever recorded in the county, probably relating to at least 13 birds, was during September 2008. A high proportion of these passage migrants were juveniles and were associated with a large passage of Honey-buzzards *Pernis apivorus*, Common Buzzards *Buteo buteo* and Ospreys *Pandion haliaetus* down the east coast of England. It is highly likely that their origin was continental rather than from British breeding populations.

Up to the late 1990s, the majority of records referred to birds seen on only one date, but over the last decade and a half there has been a tendency for birds to be seen in certain areas over longer periods of time, though individual sightings were typically fleeting. The location of the species' favoured food items, which are usually clustered at wetland sites, or in river valleys, has the effect of concentrating Hobby sightings. The species specialises in hunting large insect prey, often damsel and dragonflies *Odonata* but various butterflies and moths are also taken, whilst small birds, most notably the aerial insectivores such as hirundines and Swift *Apus apus*, are often fed to the nestlings (Brown & Grice 2005). As observers also tend to congregate at wetland sites, this probably serves to excessively skew the number of Hobbies recorded in such areas. In truth, birds in the county's wider countryside are probably considerably and disproportionally under-recorded.

The first birds of the spring were most often seen in late May or early June through the 1980s and 1990s, but as birds are now more regularly observed, they are more usually first seen during the first week or ten days of May; as in 2009 and 2010. June is often the busiest month for sightings in the county, and such records might include amongst them locally prospecting birds, non-breeding birds and late spring passage migrants, bound for further north or east. For some time the earliest record in the county was the bird noted over Dorman's Pool on 4 May 1968, but this was superseded by an adult at WWT Washington on 24 April 1987, which remained the earliest ever for Durham until the bird seen at Boldon Flats on 23 April 1997. April records for the county are uncommon, there being just a handful of these during the 1990s and 2000s. Most late records occur in September or even earlier when the species was rarer in the county and October records are scarce. There was a late record of a bird over Teesmouth on 2 October 1971, which was the latest for the county, until surpassed by a bird on 13 October 1990 at High Spen, which now constitutes the latest documented record for Durham. Other October sightings include: one at Whitburn, on the 4th in 1992; a bird flying through Seaton Carew Churchyard, on the 2nd in 1994; a bird over Hartlepool Headland on the 7th in 1998; in 2000, one on the 2nd at Hargreaves Quarry and one was catching insects over Castle Lake, Bishop Middleham, on 6 October 2009.

One of the suggested reasons for the growth in the Hobby population through the last three decades is the greater availability of large dragonfly species for young birds in the post-fledging phase, leading to better provisioning of the young and improved juvenile survival (Brown & Grice 2005). As there now seems to be a greater availability and wider range of dragonfly prey species in Durham over the first decade of the 21st century, a trend possibly associated with climate change, then the prospects for this, the most elegant of British falcons, look good within the county.

Distribution & movements

The Hobby breeds across much of the Palaearctic, from Portugal and Spain, east and north to the British Isles, across Europe and eastwards. It can be found as a breeding species over most of Europe, through Asia and east to Kamchatka, the Japanese islands and down into China. It is a long-distant summer migrant, reaching the United Kingdom by May and leaving by mid-October, or in the case of the north east of England, usually sooner. Birds spend the winter in southern Africa, being one of the most long-distance migrants amongst British birds of prey (BWP).

Gyr Falcon (Gyrfalcon)
Falco rusticolus

An extremely rare vagrant from the Arctic: five records of six birds.

Historical review

Considering both the rarity and elusive nature of this species in the British Isles and the tendency for sightings to predominantly occur in the north west of Britain, on remote Scottish islands or in south west England, Durham has, over the latter years of the 20th century, attracted a good number of records of this, the largest and perhaps most dramatic of all the falcons.

The first record for County Durham was of two grey morph birds, present on Wemmergill Moors, Lunedale in the spring of 1846, both of which were shot (Duff 1851). A specimen of one of these was obtained by Duff and is now held as a mounted exhibit in the collection at the Dorman Museum, Middlesbrough, having been sold in 1901 by Duff's son to Thomas Nelson. Duff's initially entry in *The Zoologist* confusingly places Wemer Gill (sic) in Northumberland but the exhibit itself is correctly labelled '*Wemmergill, North Yorks 1848*'. It became the first for County Durham's present recording area as a result of the boundary changes imposed in 1974.

Temperley (1951) noted an undated 19th century report of a "*Greenland Falcon*" chasing terns at Teesmouth but J.H. Gurney (1876), who originally drew attention to the claim, clearly had some doubts over its authenticity and Temperley thought the evidence "*too slender to permit it to be added to the Durham list*".

All records:

1846	Wemmergill Moors, two, grey phase, spring, both shot
1987	Pikestone Fell, grey phase, 7 February (*British Birds* 81: 552)
1991	Bollihope Common, first-winter, white morph, 8 December to 19 January 1992 (*British Birds* 86: 474)
1994	Hunstanworth Moor, white morph, 21 November (*contra BB*) (*British Birds* 89: 497)
2006	Startforth, Barnard Castle, adult white morph, 12 January (*British Birds* 101: 534)

Despite the complexities of large falcon identification, especially with regards to falconers' hybrids, the bird at Pikeston Fell was well described (Raw 1988), being just reward for the many hours that the observer had spent surveying Durham's moorland fells. The relatively long stay of the individual in the winter of 1991/1992, on Bollihope Common, around Eggleston, on the edges of Hamsterley Forest, and over into Teesdale, is noteworthy. The white phase bird seen on a fence-post near Hunstanworth in November 1994 then became the third example of a sighting from Durham's heather rich moorland expanses in just eight years. More unusually, a white morph bird, was photographed as it perched on a bird bath in a garden at Startforth, Barnard Castle on 12 January 2006 after it had struck a window. This was most probably the same bird that had been reported just three days earlier a little further down the Tees valley at Hell's Kettles ponds, Hurworth Place near Darlington.

Distribution & movements

The Gyr Falcon breeds in the Arctic in northern Europe, North America, and Asia, with most birds wintering just to the south of their breeding range, though others disperse more widely. There are three recognised colour morphs, though only two of these have been recorded in Britain, grey phase birds have their origins in northern Europe, whereas the prized white phase birds originate in the high Arctic, predominantly Greenland. It is an annual vagrant to the British Isles, though the majority of records are from northern Scotland, in particular the Western Isles and Shetland, though there are also a number of records from Devon and Cornwall. Most frequently recorded during the period from October to April, there had been over 380 British records by the close of 2010.

Gyr Falcon, Startforth, January 2006 (courtesy J. Wood)

Sponsored by

chromazone

Peregrine (Peregrine Falcon)
Falco peregrinus

A scarce breeding resident, a scarce passage and winter visitor

Historical review

There is archaeological evidence of the Peregrine's historical presence in Durham from the excavation of a mediaeval site on Hartlepool Headland, though whether this was a wild bird or one used for falconry cannot be determined (Yalden & Albarella 2009). There are three extracts relating to 'hawking' (all from the Bursar's rolls of the Monastery of Durham). One in 1335 seems to refer to the purchase of meat for the falcons, which were considered the property of the king – *In carnibus recent, empt. pro falconibus d'ni Regis, iiij*. Only the nobility were supposed to use Peregrines and perhaps one of the Priors also practised the art of falconry (Ticehurst 1923).

Temperley (1951) described the Peregrine as "*once a regular resident, now a rare persecuted vagrant*". In earlier times, there seemed to be some uncertainty about its status in parts of Durham. Hutchinson (1840) did not refer to it, but Edward Backhouse (1834) recorded that "*several specimens have been killed at Marsden and Dalton near Sunderland*". Peregrines were recorded as breeding at 'Shawnberry' in Weardale until 1843 and birds were seen there again in 1884 when they were also breeding at Wearhead. There was an even earlier record of breeding on the coast in Castle Eden Dene, up to 1810. Hancock (1874) stated that it used to breed annually in one or other of the two counties, but at his time of writing "*can scarcely be said to do so*". A previous stronghold was upper Teesdale and writing in 1885, J. Backhouse said, "*it one time inhabited Falcon Clints and various other suitable places*". Falcon Clints, apparently taking its name from the species, was an important historic site and is the grandest section of cliff topography in the Durham Pennines. In 1900 alone, seven birds were reported as being shot there and in 1907 a further two pairs were shot. Pairs were present in June 1934; and again in May 1938 both of the latter birds were shot. In 1938 Temperley saw a pair there on 8 May but later reported that both birds were shot off the nest (*The Vasculum* Vol. XXIV 1938). The authors of *The Birds of the Tees Valley* could therefore justifiably conclude: "*It tries yearly to nest in Upper Teesdale, but the birds are usually shot*" (Almond *et al.*1939). In 1944, a pair was present from February to mid-June, but successful breeding was 'not accomplished'. In 1947, two clutches of eggs were 'collected'. Temperley also noted the Peregrine, as being seen on an annual basis at Teesmouth in winter.

Peregrines were known to be breeding in north west Yorkshire, in small numbers but with some success, up to the Second World War, within the area now encompassed by County Durham (Mather 1986).

In June 1952 a pair had a very late nest on a crag in southwest Durham which eventually failed at the egg stage. This pair was thought to have already made two unsuccessful nesting attempts on the Yorkshire side of the Tees earlier in the same season.

The widespread use of DDT in the environment during the 1950s and 1960s caused the well documented decline of the Peregrine as a breeding bird both nationally and locally (Ratcliffe 1963). A very gradual recovery coincided with the phasing out of chlorinated hydrocarbons and other harmful pesticides and from a low of 385 pairs in 1961, the National Peregrine Survey of 2002 found 1402 breeding pairs. Similar increases in Europe have resulted in a downgrading of conservation listing from 'rare' to 'secure' (Birdlife International 2004) and consequently the species has been moved from the amber to the green list in the UK's list of Birds of Conservation Concern. This encouraging position is not reflected locally although the population did show tentative signs of recovery by the 1980s when one or two pairs began to breed successfully in most years.

Recent breeding status

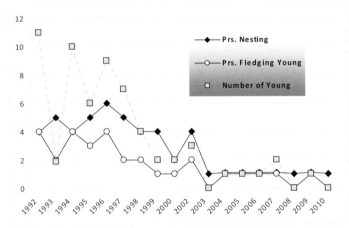

Breeding success of Peregrine in the Durham uplands, 1992-2010

Sadly, it is striking that the advances seen nationally over the last few decades have not been realised within Durham and despite its status as a Schedule 1 protected species the local Peregrine population remains fragile and vulnerable with few signs of a sustained recovery. The quality and extent of suitable habitat are both more than adequate and these factors do not pose any impediment to population growth. Regrettably, Temperley's comments of 1951 still find an echo today as the Peregrine continues to face illegal persecution and undue disturbance from factions within the game shooting, pigeon-fancier, falconry and egg-collecting interest groups (Holmes *et al.* 2000, Holmes *et al.* 2003, Court *et al.* 2004, RSPB 2006). The problem manifests itself in frequent breeding failures and in unexpectedly low occupancy at eyries early in the season. In the north of England a strong correlation has been found between failed eyries and their proximity to managed grouse moorland (Amar 2011).

The county's uplands do not have a large number of high crag sites that might be considered as prime breeding sites for Peregrines, but disused quarries can provide an attractive alternative. Breeding pairs have also resorted to what might be considered 'sub-optimal' nesting situations almost equivalent to ground-nesting. The sites are all within a mosaic of upland habitats, incorporating extensive sheep walk, heather moorland and river valley systems. Annual monitoring, under licence, of all traditional nest sites has been undertaken by the Durham Upland Bird Study Group for more than 18 years and the results reveal the very considerable challenges still faced by this species.

The 'upland data' shows the number of nesting pairs and the overall outcome from the seven traditional Durham upland eyries which have been comprehensively surveyed each year from 1992 to 2010. No data was collected in 2001 when Foot & Mouth Disease prevented observer coverage. Four of these sites fall within the North Pennine, Special Protection Area (SPA) which has EU designation because of its importance for upland bird communities and where the UK Government has an obligation to foster and broaden the biodiversity. The graph shows a starkly deteriorating situation. In the 1990's, between four and six sites regularly attracted nesting pairs although four or fewer of these ever managed to fledge young. This gave some cause for optimism for a recovery but the number of successful pairs then fell to no more than two each year from 1997 and from 2003 this number fell to no more than one nest per year. Even then, the productivity was unsustainably low with typically and most unusually, just one youngster fledgling per nest. Over the last eight years, only six young are known to have fledged and there have been three years when none at all were raised. The situation for the four sites within the Durham SPA has become critical. In what is meant to be the region's flagship upland conservation area, no Peregrines have fledged since 1999 and none have nested since 2002. In all upland sites, 41% of nests have failed to produce any fledged young over the 18 year study period. This figure takes no account of the many cases where adults were present at traditional eyries early in the season but for unexplained reasons did not go on to nest. Neither does it account for nests which were successful only after first or sometimes even second clutches were lost. At the same time that the population in the Durham uplands has seen this marked deterioration some pairs have begun to tentatively establish themselves in the lowlands and these do at least bring one source of hope for the future.

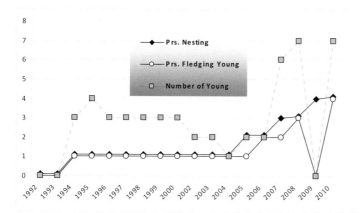

Breeding success of Peregrine in the Durham lowlands, 1992-2010

The 'lowland data' illustrates the fortunes of lowland sites since the first breeding in recent times was recorded in 1994 when a single pair became established in a quarry in central Durham. This one site has gone on to produce a total of 37 young with success from first clutches in every year bar one for the last 16 years (ignores 2001's 'FMD' restrictions). In all lowland sites (excluding Teesmouth) 21% of nests have failed to produce any fledged young during the study period (compared with 41% for upland sites). The number of occupied sites began to increase in 2005 and by 2006 a pair had raised young at a coastal location probably for the first time since the 19th century. In 2009, after a very encouraging spring produced four lowland pairs with nests, reported human disturbance and persecution once again took their toll. Shockingly, the adult female and chick at one nest were found poisoned with contaminated pigeon bait.

The above figures exclude breeding birds on the north side of Teesmouth in Cleveland where one site has enjoyed regular success since 1999 (Joynt *et al.* 2008) and another well publicised pair raised young on a chemical site tower in 2009 but failed in 2010.

Whilst lowland sites perhaps offer the best route to some future success, as 2009 showed, they too are not immune to persecution. The rightful heritage of the Durham uplands will never be realised unless the emblematic Peregrine is allowed to prosper in its natural environment. The UK Government's legal obligations towards the promotion of species conservation and protection within the EU designated North Pennine SPA are clearly not being met as we witness the increasingly perilous state of breeding Peregrines here and elsewhere.

Recent non-breeding status

During the winter period birds can be found over the lowland and coastal areas of the county whilst still retaining a presence in the uplands. The North Tees Marshes produce regular sightings, favoured areas include Seal Sands, RSPB Saltholme, Seaton Snook and around the Nuclear Power Station. Further records often come from sites spread across the whole county including Castle Lake, Crookfoot Reservoir, Derwent Reservoir, Eggleston, and the Marsden/Whitburn area. Birds do occur in open farmland habitats, away from the normal observer coverage and go largely unrecorded. It is difficult to gauge precisely how many birds are present in the recording area in winter. Breeding birds begin returning to upland nesting territories during February and there is usually a local comparable decrease in coastal and lowland sightings, especially over the period between March and July.

Distribution & movements

This is a cosmopolitan breeding species, occurring on all continents except Antarctica. Some populations are resident and many are highly migratory. The nominate form is widespread in Europe and there are at least another 15 races represented around the world. Some birds in the north east may very well be immigrants from further north or dispersing first-winter birds.

There is a single ringing recovery of the species in the county. A bird, ringed as a nestling during June 1983 in the Lake District, was found, with an injured wing on 10 April 1987 in Central Gateshead and was taken into care (Bowey *et al.* 1993).

Impediments to further increases in the local population centre largely on continued persecution since both upland and lowland terrains have shown a capacity to support greater numbers. The ability of Peregrines to adapt to urban breeding appears to offer the best hope for a step-change but a solution must be found in the uplands.

Sponsored by

Glead Ecological & Environmental Service

Water Rail
Rallus aquaticus

An uncommon winter visitor, passage migrant and a scarce breeding resident.

Historical review

Records of this species in the Teesmouth area date back to 1827, when Hogg described it as "*not uncommon*" by the side of becks and similar habitat (Hogg 1927). Backhouse (1834) also said that it was, "*not infrequent in the marshes near the mouth of the Tees*". In 1840 Hutchinson noted that it "*frequents the carrs in summer and brooks in winter*", inherent in this observation was a hint that there was a difference in habitat selection and behaviour between resident birds and winter immigrants (Hutchinson 1840). It was over this period that the species ceased to breed in many parts of the British Isles (Holloway 1996) although Proctor (1846) still said that it was "*not uncommon*" and across the region, Hancock (1874), thought it "*not uncommon though rarely seen*". In 1896, Robson wrote of the Derwent valley "*a pair of these birds were shot on Shibdon Flats a few years ago*", he also mentioned single birds shot at Dunston in 1889 and Dunston Lodge in 1894. He made no mention of the season when these were taken but considering the large tracts of marsh present in that part of the county at the time, it seems likely that birds were breeding locally (Bowey *et al.* 1993).

Three-quarters of a century after Hogg, Nelson (1907) described Water Rail at Teesmouth during the breeding season. A pair or two were known to have bred in this area around the beginning of the 20th century (Blick 2009) and there is little doubt that birds must have bred in some numbers prior to the major losses in wetland habitat from around the estuary. Stead (1964) did not document any subsequent breeding records for Teesmouth.

Temperley (1951) had thought it "*a resident: not uncommon in summer, more frequent in autumn and winter*" and said that it was more likely to be heard than seen. There is little detailed information about the status of this species post-Temperley. It was 'recorded at Teesmouth' in 1966 (Bell 1967) and a few years later, in 1968, there were scattered records of the species, mainly in winter at Teesmouth, Hurworth Burn Reservoir, Witton-le-Wear and most notably Shotton Colliery Pond, where there were up to five birds in December of that year. That the species occurs as a winter visitor to the north east coast was confirmed in 1967, when a bird was found with a broken leg in South Shields on 2 November. Up to the early 1970s the species was known primarily as a winter visitor, but it was almost certainly being overlooked in the breeding season. In the early 1970s, there were clearer signs that the species had a year-round presence, with regular spring and summer sightings and territorial calling at some sites, such as Brasside Ponds, Low Barns, Low Butterby and Shibdon Pond. Confirmation of breeding in Durham did not come until 1975, though it clearly must have been occurring well before this and probably at least as far back to the 19th century

Recent breeding status

In Durham, as elsewhere in Britain, the Water Rail is largely a lowland species restricted, for breeding purposes, to wetlands with large amounts of vegetative cover. Habitat suitable for Water Rail is limited in its extent in County Durham though quite widely distributed. Sites with suitable habitat tend to be concentrated along the main river corridors of the county. The species favours fen or tall fringing vegetation around open water, habitats dominated by plants such as reed sweet-grass *Glyceria maxima*, common reed *Phragmites australis*, great reedmace *Typha latifolia* and bulrush *Scirpus lacustris*. Where appropriate habitat is found birds tend to aggregate, the result being clusters in the species' distribution. This was well illustrated by the mapped distribution in the *Atlas*, which showed groups of occupied tetrads around Teesmouth.

Being a skulking, largely crepuscular species, Water Rail is renowned for being difficult to study (Jenkins 1995) breeding birds are best located by their calls in early spring (Bayliss 1985). As a consequence, it is probable that it remains under-recorded to this day, particularly from small, relatively isolated wetlands. Proof of successful breeding, may only be confirmed at between one and three sites in any year, if indeed, at all. Such sites are largely in the lowlands but occasionally birds are recorded at higher elevations; for example, the bird at Tunstall Reservoir in the summer of 2004.

It was only with the advent of more detailed scrutiny at wetland sites that confirmation of breeding in Durham occurred. The first came at Low Barns in 1975, when two young were noted on 13 June. Breeding was repeated at the same site in 1976, when a nest discovered in May eventually produced three young. Success came again at Low Barns in 1981, when a pair was seen with four young. It was not until 1982 that Teesmouth had confirmation of breeding and this was repeated in 1986. More detailed studies in the 1980s, most notably at Shibdon Pond (where Robson in 1896 had noted the presence of birds) and Teesmouth eventually proved that birds were breeding in the county on a regular basis. Downy young were noted at Shibdon Pond in June 1988. The following year well-grown juveniles were seen there again, with birds present in two territories and breeding was also confirmed in 1990.

During the *Atlas* survey, 1988-1993, breeding was proven most regularly at Shibdon Pond. The species was closely studied there and birds were confirmed as breeding in five of the six survey years, and were strongly suspected of doing so in the sixth (Westerberg & Bowey 2000). In each successful year, a minimum of two territories were established, with up to five singing, territorial males mapped in a marsh area that amounted to around five hectares. Subsequent work at this site indicated that in the best years, such as 1997, up to seven territories could be occupied (Westerberg & Bowey 2000). On 23 September 1996, 16 birds, largely juveniles, were scattered around the edge of Shibdon Pond, an indication of the success of that season's breeding. Up to five other localities held birds in the summer, some of these probably hosting breeding pairs. In the middle of the 1990s, in 1996, breeding was thought to have occurred at nine sites across Durham, with at least 13 pairs involved at these locations in that year (Armstrong 1998).

The largest area of available Water Rail habitat in the county is around the North Tees Marshes and they host a large proportion of the county's breeding population (Joynt *et al.* 2008). Breeding was confirmed at Teesmouth in 1982, 1986, 1989 and 1991, and in every year from 1994 at sites such as Bowesfield, Cowpen Bewley Pond, Cowpen Marsh, Dorman's Pool, Hargreaves Quarry area, Greenabella Marsh, Haverton Hole, the Graythorp area, the Long Drag, Portrack Marsh and Saltholme Pools. Teesmouth had at least five breeding pairs in 1994 (Joynt *et al.* 2008). Elsewhere, there is a relatively small amount of suitable habitat scattered across the county, the pattern of wetland occupation undoubtedly being related to observer coverage. Significant areas of habitat that harbour the species include Brasside Ponds and the string of wetlands along the mid-Wear valley, from Brancepeth Beck to Low Barns.

The fact that this species remains somewhat overlooked in Durham was illustrated by a nocturnal survey conducted in 2005 that located birds at a number of sites where it was previously unrecorded. These were spread across seven ten-kilometre squares and in two of these squares it had not previously registered a breeding season presence. In 2006, birds were recorded at 28 sites, with breeding confirmed at three of these, including Burnhope Pond, at 250m above sea level. Territory mapping around the North Tees Marshes in 2006 indicated the presence of ten territories. In 2007, records of Water Rail were reported from 26 sites during the year, all of them east of the A68 and the majority in the central third of the county.

Through the 1990s and 2000s, the species was confirmed as having bred successfully at a minimum of 11 sites across Durham. These included: six sites around Teesmouth, Rainton Meadows, Hetton Bogs, Low Barns, Birtley Sewage Treatment Works and Shibdon Pond.

A minimum county population estimate of 15 to 20 pairs was suggested by Westerberg & Bowey (2000), with "*as many as 30 pairs*" being stated as a better working estimate. Recent work at Teesmouth has suggested a population there of up to 30 pairs (Joynt *et al.* 2008, Blick 2009). In 2010, an estimated 35 pairs were believed to be present across the county at all locations. With this additional knowledge, a more realistic guide figure for the county's current breeding population is 40-50 pairs. Nationally, the species declined through the late 20th century due to drainage and perhaps later in the century, the severe winter weather of the late 1970s and 1980s (Marchant *et al.* 1990, Mead 2000) whilst there was also a contraction in distribution of around 35% between the two national Atlas surveys (Parkin & Knox 2010).

Recent non-breeding status

Water Rail is probably more common than realised for apart from wintering birds seen at now well recognised wetland sites and a few passage birds, many must go unseen at smaller wetland habitats in Durham. There is little doubt that it is most common as a winter visitor in the county. The bulk of records usually relate to the winter period, when birds are more noticeable. The majority of sightings of Water Rail are of single birds in late autumn or winter but occasionally two to three birds are seen together. Locations that regularly attract wintering birds include Brasside Ponds, Cowpen Bewley, Far Pasture, Hetton Lyons, Low Barns, Low Butterby, Rainton Meadows, Sedgeletch, Seaton Pond, Shibdon Pond and WWT Washington, as well as the many sites around the North Tees Marshes. In winter 2009/2010, up to six were at RSPB Saltholme during freezing conditions. Many winter sightings are of birds emboldened by hard weather, hence the number of reports is somewhat weather dependent. Large counts are relatively unusual but not unknown. As many as ten were present at Shibdon in February 1985 (Bowey *et al.* 1993) and eleven were in the Dorman's Pool and Reclamation Pond area during freezing conditions in early December 1998 (Blick 2009). The highest winter count of the 1990s came from Brasside Ponds when 16 were seen in December 1994. The largest known count came in icy conditions on 1 December 2008, when 19 were distributed around the adjoining wetland complex of Rainton Meadows, Joe's Pond and Chilton Moor.

As well as birds spending the winter at typical marsh sites across the county, it seems likely that unknown numbers of birds will spend the winter beside secluded stretches of ditches, overgrown streams and small ponds in the lowland, eastern half of the county. The numbers actually wintering in the county in such circumstances can only be speculated upon.

Wintering birds are occasionally found in more unusual situations, such as the birds that were present in Roker Park, Sunderland from early February through to late March 2003 (Newsome 2007). A bird that visited a suburban garden in Low Fell between December 1995 and 24 March 1996 was fed on fish scraps, and most extraordinarily, it returned to spend the following winter in the same location. This record was not unprecedented, as one had spent over a month in a suburban South Shields garden in March to April 1983 and also one was in a Whitburn garden in December 1988. Some of the unusual food stuffs taken by Water rail locally include bread, at Gilesgate, Durham in November 1990, a rabbit carcase at WWT Washington in the same winter and, in December 1997, two birds at Hetton Bogs that enjoyed the remains of a Christmas turkey.

Small numbers of passage migrants are seen in most autumns along the Durham coast, usually between late September and December at sites ranging from South Shields in the north to Hartlepool Headland in the south. Small numbers of migrant Water Rails are most often noted in coastal locations with the bulk of records being in October and early November. Evidence for these often comes from those that have experienced the greatest peril or are in the weakest state. For example, the bird that hit a television aerial in Dormanstown on 11 October 1955, one rescued from a Sunderland City centre window ledge on 6 December 1978, one dead in a car park at Chester-le-Street in autumn 1981, one killed against wires at Tunstall Hills, Sunderland on 1 November 1988, one found in similar circumstances at Whitburn, on 30 October 1995 and, one rescued from a cat in a Whitburn garden on 17 October 1999. Movement often continues into November as shown by the coastal migrant that was seen at Whitburn on 22 November 1998. Even later arrivals are known, for instance one was on Hartlepool Golf Curse on 25 November 2000 and in 1970 one that was seen to arrive from over the sea at Hartlepool Headland, on 26 December but was then killed by a Great Black-backed Gull *Larus marinus*. One of the most unusual records was of a bird found on a girder under a fridge in the Glaxo Smith Kline factory at Barnard Castle on 20 October 1987. Initially reported as a young Curlew *Numenius arquata*, it was caught and released into a marshy area near the factory, apparently fit and well.

Spring birds are much less often noted in coastal locations; a bird flushed at Trow Quarry on 1 April 2001 was just the second site record there since 1994. One on the edge of the beach at Ryhope on 11 March 2006 was another example of this scarce spring migrant.

Distribution & movements

Water Rails breed across much of Western Europe and north west Africa and into central Asia as far east as Japan. Birds are resident across much of Britain and Europe, though northern European populations are migratory. These, alongside birds from the Low Countries and northern Germany, as demonstrated by the national ringing database, are the probable source of most of the additional birds that visit Britain, and Durham, during the winter months. There is one recovery from south Durham, of a bird ringed in northern Germany (Wernham *et al.* 2002).

Spotted Crake
Porzana porzana

Formerly a very rare breeding bird and a scarce passage migrant. Now a scarce but almost annual passage migrant and possible rare breeder.

Historical review

Today this species is considered predominantly a scarce passage migrant in Durham but in the 19th century it was recorded as an irregular summer visitor and was known to have bred in the county. Alpin (1890), in an early review of the species' status, concluded that Spotted Crake probably bred in every county of England and Wales where habitat was available. Tunstall (1784) said that it was *"not infrequently shot here"*, though he commented upon its rarity elsewhere in England.

Nationally, there was probably a significant decline in numbers following the widespread land drainage and agricultural improvement schemes of the 18th and 19th centuries (Holloway 1996). Numbers reported have always been low with a distribution reflecting the availability of extensive herb rich wetlands. One interesting report comes from Lord Lilford, writing of the *Birds of Northamptonshire*, who tells how *"John Hancock of Northumberland had visited in 1843 and in collecting eggs at Whittlesey Mere and Yaxley, took almost as many eggs of this bird as the water-rail"* (Alpin 1890).

Temperley (1951) documented a number of historical sources that stated that the species appeared to be more common in the county during the 19th century. Hutchinson (1840) had reports of it from the "*Carrs of Morden and Mainsforth*". He thought it was usually present until the end of September or October whilst Proctor (1846) reported that it had "*formerly resorted to Framwellgate Carr, where several have been shot in the month of March, the time of its arrival here*". According to Hogg, (1845), it had indeed been shot in Mordon Carrs, to the south west of Sedgefield and more specifically, in the old bed of the Tees, near Mandale Mill on 4 October 1832. Hancock (1874) reported Proctor's assertion that it bred some years ago at Framwellgate Carrs; but in his document he considered that it was by then *"rather rare"*.

Nelson (1907), wrote, "*Formerly, the Tees marshes were favourite resorts of this bird and certain spots in its haunts appear to have special attractions, for if one is killed, its place is soon occupied by another*". He shot two at Saltholme Marsh, on 16 September 1882. It seems certain that, before 1900, the species was more common in Durham than at present. Temperley listed at least six birds shot on the North Tees Marshes in the final two decades of the 19th century. The fact that it was breeding there seems in little doubt, as indicated by the number of mid-summer records, one of these was an adult female, "*acquired*" on 25 May 1899, which had taken from her a fully-formed egg. In relation to this record, Milburn stated that he "*saw a bird of this species at Tees-mouth on August 23rd*" going on to say "*Mr. C. Braithwaite caught a female, which was injured on May 25th, and, upon our skinning it, imagine our surprise in taking from the body a full-formed and well-marked egg. This was also at the mouth of the river, where this bird breeds*" (Milburn 1901). In 1900, a pair with young was seen at Teesmouth (Nelson 1907).

Other records around this time include one "*taken*" at Haughton-le-Skerne and exhibited on 20 September 1892, one shot on Shibdon Flats (the site of Shibdon Pond) in 1891 and another shot, by Colonel Cowens, in Blaydon Burn "*some years ago*" (Robson 1896). The past status of this species in the north of the county around west Gateshead is uncertain; no doubt it was rare but Robson (1896) suggested that it may have been more common than at present. According to Thomas Robson (1896), Col. Cowens had three Spotted Crakes in his collection and, since the gentleman was a well known local sportsman, it seems possible, even likely, that they were taken close to the Blaydon Burn area in which he lived (Bowey *et al.* 1993).

Between 1900 and 1951 only two birds were recorded, a male killed against telegraph wires on the north side of Teesmouth on 21 May 1933. Joseph Bishop, the official 'watcher' for the Tees Marshes at that time wrote, "*a male Spotted Crake struck the wires on May 21st near a point that I had guessed as the possible nesting site. In fact last year I spent a night here in hopes of hearing the bird*". This suggests that the species may have still been breeding around Teesmouth into the 1930s or was, at least, a much more frequent spring visitor than at present. The second report was of one was at Hurworth Burn in October 1948.

George Temperley (1951) listed this species as, *"a rare summer visitor, which has been suspected, of breeding occasionally"*. At the end of his summary on the species he commented that it was *"fairly certain"* that the species had bred in the county although there had been no recent, direct proof of it having done so.

Post-Temperley there were relatively few records of the species:

1952	A bird was at Boldon Flats on 27 April, intriguingly about a month earlier a bird had been heard calling there
1952	One seen at Primrose, Jarrow on 30 July
1953	One at Tanfield Ponds, near Stanley, on 18 October, was flushed on a number of occasions and seen on the ground
1962	One found dead at Greatham Creek, Teesmouth, on 6 October
1968	One at Hartlepool on 27 August
1968	One at Barmston, Washington on 17 November, at that time, the latest recorded date for the species in Durham

None were recorded between 1969 and 1976. Spotted Crake remained a resolutely rare bird in Durham during the 1970s with just two records, both in 1976. The first was a bird found dead in North Sands Shipyard, Sunderland, on 3 September. The second was a very obliging bird that commenced a three-day stay at WWT Washington from 26 September; it was often seen running around at observers' feet.

Recent breeding status

Through the 1990s and into the 2000s, there was a small but discernible increase in the number of records in Durham, which included singing males in spring at a number of sites around Teesmouth and at two other sites elsewhere in the county.

During the *Atlas* survey period, birds were noted on three occasions (in 1988, 1989 and 1991) in suitable breeding habitat (Bowey 1999). In all cases, these birds were more probably late summer/autumn passage individuals, the sightings falling within the period 12 August to 21 September. Two males found singing at a site between late May and July 1994 were of greater interest, providing the first concerted evidence that breeding might have been attempted since the 1930s; although the relationship between singing birds and breeding is not clear. Subsequent to this, spring birds were seen or heard calling at the same site in 1996, 1997 (singing on 10 June and 26 July), 1998 (an adult being chased by a Water Rail *Rallus aquaticus* on 11 April) and 1999 (singing in late April and early May). In at least one of these years, a juvenile was observed later in the autumn. In addition, a spring bird was at a different site, in the south of the county, in May 1996 and a singing bird was at a site in the centre of the county from 18th to 21 May 1998.

Over the following decade, singing birds were noted in June 2000 and May 2006 and one was seen in early June 2002, at sites around the North Tees Marshes (Joynt *et al.* 2008). This is a pattern of occurrence that is strongly indicative of breeding behaviour. Elsewhere, during the decade, adults were present in spring/summer at least twice at the site that attracted birds on a number of occasions in the 1990s. This body of evidence suggests that there may be a small, perhaps intermittent, breeding presence of Spotted Crake in at least two favoured parts of the county. The population is unlikely to number more than two or three singing males in even the best years. The latest assessment of the national population suggests 30-80 singing males at up to 30 sites nationwide (Stroud *et al.* 2012).

Recent Non-breeding status

In Durham the species is more common in the autumn than in spring and it cannot be considered as anything other than a scarce species. It occurs principally as an autumn passage migrant, normally after periods of easterly winds, at wetland sites or rarely, elsewhere along the coast.

The increase in records since 1990 is consistent with evidence of an increase in the number of passage and non-breeding birds nationally; most of these being reported between September and early November (Stroud *et al.* 2012). Shy and difficult to observe in heavily vegetated marshland vegetation, it is possible that Spotted Crake is an annual passage migrant in small numbers but with cryptic plumage and largely nocturnal habits it seems destined to remain poorly known. Many of the pre-1900 records relate to specimens taken by shooters often in pursuit of Snipe *Gallinago gallinago*. The species appeared more than four times as often (37 documented records) in the period 1971-2010, than in the previous forty years 1931-1970 (eight records). No doubt this relates to greater observer effort as well as a genuine change in status. It was recorded in only sixteen years between 1900 and 1996, with a total of 23 birds documented in the literature (Bowey 1999). In reality, the species will of course occur more frequently than these figures suggest, birds are undoubtedly overlooked but the species in real terms remains scarce.

During the 1980s a change in recorded status commenced. There were two records in 1981. Unusually, the first was a bird in spring at Dorman's Pool, Teesmouth, from 27 April until 3 May. This was the first reported in spring since 1952. The year's second record was of a bird that flew into wire mesh fence at WWT Washington, on 15 October; it recovered, was ringed and then released. A gap of five years ensued until 1986, when one was noted on the Long Drag at Teesmouth, over the period 9th to 18 August. In 1988 there were two records, a single at Dorman's Pool from 3rd to 21 September and more unusually, an obvious passage bird flushed from dry coastal, grassland in Trow Quarry, on South Shields Leas on 12 October. The following autumn a juvenile was at Shibdon Pond, from 11th to 19 September, the first time the species had been recorded in consecutive years in Durham since 1952/1953.

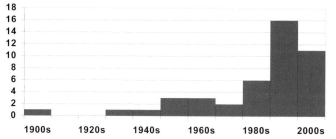

Records of Spotted Crake in County Durham 1900-2009, by decade

Records increased quite markedly during the 1990s and included a series of spring reports from a number of sites in both the north and south of the county. The first of the decade came in 1991, with an adult was at Shibdon Pond from 12 August until 7 September; at that time, the longest stay of a bird in the county during the 20th century. Birds were recorded on sixteen occasions during the decade and in every year from 1994 onwards. Since 1994, Spotted Crake has been recorded in Durham in most years, principally at sites around Teesmouth, there being just four years without records: 2001, 2004, 2007 and 2008. The best year of the 20th century for records of this species in Durham was 1995, with three, possibly four, records. These included one singing on 13th and 14 May at Dorman's Pool, Teesmouth and a bird at Whitburn on 8 September, presumably a coastal passage bird. The third was once again at Dorman's Pool, on 27 September, which may have been the same as one seen there on 11 November. These autumn records coincided with what was considered one of the best autumn influxes of the species to Britain, with up to 80 birds reported across the country (Gantlet 1995).

The trend of more regular sightings from the 1990s continued during the 2000s, though with some reduction in frequency from around the middle of the latter decade. The over-riding trend was in an increase in the number of singing birds in spring and summer and the preponderance of sightings around Teesmouth. Barring records at Hetton Bogs on 5 August 2002 and 24 August 2003 and a slightly odd record of one found behind refuse bins at Souter Lighthouse on 22 September 2000, eight of the county's eleven documented records to the end of 2009 were from sites around the estuary. In 2010, there were two records, one at Dorman's Pool on 9 August and a juvenile at Shibdon Pond from 15 August to 12 September 2010.

Birds have been recorded in every month of the year in Durham, except February. The majority of the county's records since 1933 (almost 60%) refer to autumn birds found between the beginning of August and the end of October (August and September being the months with most records). Autumn birds have been found between 5 August 2002 and 29 October 2000. Around 21% of records have occurred in April and May, with a spread of spring records falling between 11 April 1998 and 9 June 2002, though this pattern is somewhat confused by the increased presence of singing birds. There are a handful of records that do not fit into these broad spring or autumn passage periods, these include single birds at: Dorman's Pool, on 11 November 1995; Haverton Hole, on 2 January 1999; and, Low Barns on 28 December 1999.

Distribution and movements

Spotted Crake breeds rather sporadically across much of Europe and east as far as south west Siberia and the north western provenances of China. It is a highly migratory species, with very small numbers wintering in Europe, the vast majority of the European population winter in sub-Saharan Africa.

Three Spotted Crakes have been ringed in the county. These were one at WWT Washington, in 1981, one on the Long Drag, on 29 August 1997 and, one at Haverton Hole, on 4 May 2006, which was heard calling after dusk on the same day. These have generated no recoveries or controls.

Baillon's Crake
Porzana pusilla

An extremely rare vagrant from Europe: three records.

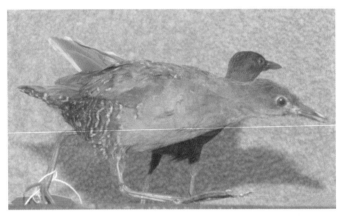

Baillon's Crake, Swalwell, July 1874 (K. Bowey, courtesy NHSN)

Baillon's Crake is an extremely rare visitor to County Durham, for which there are two historical records and one modern occurrence. The first record for the county was one shot, by Thomas Thompson of Winlaton, on the banks of the Derwent, near Swalwell, on or about 12 July 1874. George Temperley, in "*A History of the Birds of Durham*" (1951), commented, "*from the state of its plumage and the time of year when it was shot, in all probability it was breeding in the neighbourhood*". There were at least three, and perhaps five, records of breeding in England in the latter part of the 19th century; nests were found in East Anglia in 1858, 1866 and 1889, when a nest was discovered at Sutton Broad, Norfolk (Holloway 1996). Such evidence adds weight to the suggestion that the bird shot 'on the banks of the Derwent', might have been breeding or at least holding territory. There is another specimen in the Dorman Museum, Middlesbrough, which was shot by Thomas H. Nelson, along with two Spotted Crakes *Porzana porzana*, at Saltholme Marsh, on 16 September 1882.

The scene in Mowbray Park, May 1989 (David Tipling)

The only modern record concerns one found on 17 May 1989 in Sunderland (*British Birds* 83: 459). This was a quite extraordinarily extrovert Baillon's Crake, which was initially discovered by a member of the public as it fed around the paved edges of the small Mowbray Park lake, situated directly behind the Sunderland Museum and Winter Gardens. This bird was unusually bold in its behaviour and, lacking any apparent fear of humans, it allowed a very close approach, routinely feeding within feet, and sometimes even inches, of observers, occasionally pecking at shoelaces in the apparent mistaken identity of them as worms. It delighted the hundreds of bird watchers and even larger numbers of the general public who lined up to watch it behind the balustrade of the ornamental lake. For the first three nights, it roosted in low vegetation on a small island, but at dusk on 20 May was seen by the observer who first identified it, to clamber to the topmost branches of a small tree in what was an obvious preparation for departure, as it was not present the following morning. There have been no subsequent records of the species in the county.

Distribution and movements

Baillon's Crake is patchily distributed from south west Europe and north west Africa across Asia, east to Japan and south to southern Asia. It is believed that European birds migrate south to Africa, but as the species is so secretive, the full extent of the breeding and wintering ranges is not well known. The species is an extremely rare vagrant to Britain with a total of 81 records to the end of 2010, the majority of these during the 19th century. There have been just 16 post-war British records. Records are widely scattered throughout Britain, though many of the more recent records have been in areas with scant vegetation indicating that more probably pass through Britain undetected.

Corncrake (Corn Crake)
Crex crex

A scarce spring and autumn migrant from Europe, with 45 records since 1960. A former widespread summer visitor, though it may not have bred since 1958.

Historical review

The Corncrake was once a common summer visitor to much of County Durham, being said to have bred in almost every meadow, clover-field, and rough pasture, and occasionally cornfields, right up to the outskirts of towns and villages. In 1861, Hogg noted a scarcity of the species around Norton, Teesside, over a three to four-year period. In the late 19th century and early in the 20th century, Temperley (1951) said that it was to be found breeding in "*almost every meadow*" of Durham, even on the outskirts of towns, though numbers fluctuated from year to year. The lower Derwent valley and the marshlands around the confluence of the River Derwent and the Tyne were a traditional stronghold for this species, even after it had been lost from many other lowland areas. Late in the 19th century, Robson (1896) found a nest with 12 eggs near Rowlands Gill on 28 May 1895 and at the time birds were regularly reported around Blaydon, Rowlands Gill and Winlaton Mill (Bowey *et al.* 1993).

The change in agricultural practices and farming methods over the period from about 1885 quite dramatically impacted upon the fortunes of this species across England (Brown & Grice 2005). Nationally, the major decline commenced earlier further south and, around 1907, this became more marked and widespread. According to Tristram, in the *Victoria County History of Durham,* the numbers of Corncrakes "*have much diminished lately*" (Tristram 1905). Around this time, Nelson (1907) recorded migrant Corncrakes as late as the first week of November at Teesmouth.

In line with the rest of the English Corncrake population, a serious decline took place in Durham from around 1915, and continued up to 1929. From then, it became an increasingly scarce bird over much of Durham. Nonetheless, on 22 May 1926, M. G. Robinson observed birds mating at Neasham Lane, not far from Darlington. Around this time, Temperley said that it was quite usual to "*live in the County and not to hear the call of the corn-crake once in a season*" (Temperley 1951). This calibrates his then 'current' experience, with his memories and knowledge of the species' past status. Temperley said that the only area of the county in which the reduction in birds had been entirely absent was Upper Teesdale; he noted that this may have been due to the later hay cuts in that area.

In 1929, it was said that the Corncrake was "*common throughout the district*" (*The Vasculum* Vol. XV 1929), as it had been according to Hancock, but now it could not now be so described. George Temperley questioned whether the Corncrake had become scarce because of the conditions here or in its winter quarters. Occasionally still noted at the coast, one was on top of the river wall at Teesmouth on 10 August 1933 (*The Vasculum* Vol. XIX 1933). The confiding nature of local birds was illustrated by the report of J. Greenwell (*The Vasculum* Vol. XXIV 1938) some years previously, who told about his encounter with a Corncrake near Bishop Auckland. The bird was so close he could almost touch it through a hole in a hedge. He described it displaying and "*cuttering*" and its voice, as it faced him, as being almost deafening.

It was reported to be increasing in numbers near Wolsingham in Weardale, where four birds were heard 'craking' at one time (*The Vasculum* Vol. XXIV 1938). In the same issue of *The Vasculum*, the 'Records section' stated that only three reports had been received up to 28 June 1938: one at Blackwell, one on the road from Scotch Corner to Greta Bridge and one near Winston Old Colliery on the Barnard Castle road. Clearly, the declines across the county were patchy. The species managed to survive as a breeding bird in the west of the county, particularly in Upper Teesdale, until at least the late 1940s, being aided no doubt by the fact that hay harvests there were cut later than those in the lowlands, allowing the chicks time to fledge.

In the early 1940s, a national review of its status stated that the "*Corn-Crake is still met with in most suitable areas in Durham and in Northumberland, where it is generally distributed and breeds commonly*". This review highlighted the fact that of 14 reports from County Durham in 1938, seven of these reported the 'presence' of the bird, with as many as four being heard in the Stanley area. In 1939, there were 15 reports of the birds' presence including five in the Darlington area (Norris 1945). In 1943, Corncrakes were reported as being more in evidence than usual. By contrast, Corncrakes appeared to have been less in evidence in 1947 than in 1946 and "*evidence of breeding*" was scanty. In 1948, "*rather more . . . than in previous years*" were reported (Temperley 1951). It appears to have been quite a regular passage migrant through the Teesmouth area up to the late 1950s (Blick 2009).

Temperley (1951), writing in the middle of the 20th century, when the decline was already well underway, described it as "*a summer resident, once abundant, now scarce*" and stated that "*most had been lost in recent years*". Temperley reported on the ongoing decline in the number of reports of the species, especially through the early 1930s and into the 1940s, until the species became a scarce summer visitor. There was then a slight increase, from 1946, into the early 1950s. For example, in 1951, one was singing in fields at the Wynyard Estate; this was a very good year for the species with many records in the north east, confirmed breeding records included a nest that was found in a Jarrow school playground and at Hunwick. This good year was followed by the 'corncrake year' of 1952. In that year, there was a large increase in the number of reports including many records of birds from around County Durham. Birds were reported at Blaydon (where young were noted), Derwenthaugh, Hamsterley, Witton-le-Wear and near Barnard Castle. A pair reared young at Darlington Sewage Farm and several pairs were calling in Teesdale, around Middleton.

The lower Derwent valley held birds long after many Durham locations had lost the species. In 1938, the species could still be found around Blaydon and Rowlands Gill, birds persisting at the former site into the early 1940s. By the early 1950s, Temperley noted that Durham had lost nearly all of its breeding Corncrakes but that birds persisted in the north west of the county. In 1952, birds were present at Blaydon (i.e. Shibdon Pond), where young were seen, and also at Derwenthaugh. Birds were noted there in the following year and in 1954, one was in territory in the area now encompassed by Shibdon Pond Nature Reserve. That same year, birds were in the lower Derwent valley but not on the by then spoil-tipped Goodshields Haugh, which had been one of its favourite local haunts in the 1930s (Bowey *et al.* 1993). As the fifties wore on, sightings in the Gateshead area became increasingly scarce, though 1958 brought a report from Old Ravensworth on 3 August.

In 1957, there were said to be no calling birds in Durham, which was indicative of the general decline of the species over the decades. The last confirmed record of breeding in the county took place at Satley in summer 1958, though unfortunately, the eggs were destroyed by a mowing machine. The intensification of grassland management is cited as being the single major factor responsible for the rapid decline of the Corncrake in England, and across much of Europe (Green 1995). Nonetheless, the mid-1960s still saw a few birds persisting in the dales, including a bird near Stanhope throughout June 1966.

Just a handful of singing birds were recorded during the 1960s, including singles at Warden Law and Ovington in 1963, between Barnard Castle and Cotherstone in 1965, Stanhope in June 1966 and at Portrack near Stockton-on-Tees on the 5 June 1967. This serves as a sad testament to the demise of this species in the county and across the British Isles in general. In 1967, Parslow said that "*very few pairs still breed fairly regularly (but are not known to do so annually) in west Yorkshire, Durham, and perhaps some counties in central Scotland*" (Parslow 1967). There were also five records from the coastal plain in the 1960s: one was at Hartlepool Observatory on 14 May 1967, singles were at Hartlepool and Greenabella Marsh on 3 May 1969, one was found dead near Seaton Carew in mid-

June 1969 and one was at Saltholme on 7 October 1969. The bird at Saltholme Pools is notable in being the latest of just two October records for the county in recent times.

Recent Status

A total of just 38 Corncrakes have been recorded between April and June since 1970, with most relating to singing birds. During the 1970s, there were three records of singing birds from the western uplands of the county. It was described as still attempting to nest occasionally in the early 1970s in the meadows of upper Teesdale by Bradshaw (1976). For example, one was present from 30 June 1973 in meadows at Broadley's Gate Farm, Newbiggin-in-Teesdale, at 393m above sea level. It was still present on 7 July. Further singles were at Grassholme Reservoir from 14th to 28 May 1975 and near Dent Bank, Middleton-in-Teesdale, from 17 May to 12 June. Traditionally, Upper Teesdale was one of the last breeding strongholds of the species in the county and given the protracted stay of these individuals, it is tempting to speculate that breeding may still have taken place occasionally as late as the early 1970s, although it wasn't proven in any of these cases.

Elsewhere around this time, one was singing at Middleton St. George on 21 May 1971 and in 1973, one was reported at Kelloe Law during July and August. The only confirmed records of the species from its previously favoured area near Shibdon Pond were of birds in both 1973 and 1974. In the former year, a male held territory for at least a week during May in the Derwenthaugh Meadows, where it was repeatedly heard by an observer undertaking a botanical survey, in a site that was once a traditional location for the species. This is believed to be the last record of a territorial Corncrake in the lower Tyne Valley (Bowey *et al.* 1993), whilst in the following year, one was heard calling close by, in the 'twelve-score' fields at Swalwell, on 23 May 1974.

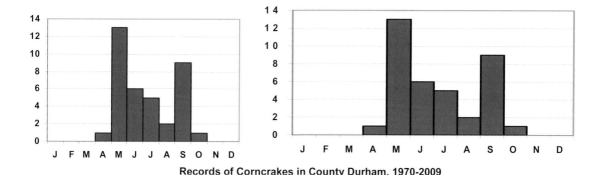

Records of Corncrakes in County Durham. 1970-2009

Records are widely scattered across the county, with most being recorded on just a single date, and not heard or seen subsequently. The exceptions to this are: a bird calling from a field of growing corn in the Tees valley for several weeks from late May 1984; one singing near Carlton, Stockton-on-Tees, from 27 May to 1 June 1988; a bird singing at Cleadon Hill, from 14th to 15 June 1992, one singing at Portrack Marsh, Stockton-on-Tees, from 31 May to 10 June 1998; one singing near West Boldon, from 15th to 17 May 1999; and, one singing at Sadberge from 22nd to 29 June 2009, with remarkably two singing males there on the 29 June, though not subsequently.

There are just three spring records of coastal migrants since 1970, the first of which was a bird that was found stunned beneath a window at Hartlepool Fire Station on the 21 May 1975. This bird was revived with oxygen by local fire fighters, before being kept overnight by the RSPCA and released at Seaton Carew the next day. Another, on the 15 May 1985, spent much of the day in the bowling green at Hartlepool Headland during a spectacular fall of Scandinavian migrants. One at Cowpen Marsh on 16 April 2004 was the earliest spring record for the county.

The species is much scarcer from July onwards, as it is much more difficult to detect when not singing. There have been just eighteen records since 1970, the majority of which have been in September. Five singing males were recorded in mid-summer during this recent period, all of which were in July. These were: at Kelloe Law during July and August 1973; by the Teesmouth Field Centre on 25 July 1981; at Sacriston, on 12 July 1994; at RSPB Saltholme, from 3rd to 6 July 2009; and, near Copley on 30 July 2009.

In stark contrast to the spring records, most autumn birds are recorded from the coast, having been inadvertently flushed from grassland, and are usually not seen again. There are just three August records, two of which are from Whitburn CP, on 27 August 2000 and 29 August 2001. The first of these was flushed from the cliff top and gave observers good views of it on the short sward of the Whitburn Rifle Range, before moving into a nearby garden. The most recent record is of a bird seen briefly at Bowesfield Marsh, Stockton-on-Tees, on the 29 August 2010.

Of the nine September records since 1970, four are from coastal localities in Tyne and Wear, with the exceptions being single birds: at Kelloe in late September 1972; WWT Washington on 6 September 1976; Shibdon Pond on 7 September 1990; Dunston on 16 September 1990, which was found in a garage and later released at WWT Washington; and, at Bowesfield Marsh, Stockton-on-Tees on 16 September 2006. All dated records in September have fallen in the period from the 6th to the 17th, highlighting the peak passage for the species in Durham, from late August to mid-September.

The only October record during the recent review period is of a bird found injured at Teesmouth Field Centre on 2 October 1981, within yards of where a bird had been singing in July of the same year, though there were no reports in the intervening period and it was thought to be a different individual.

Distribution & movements

The Corncrake breeds across much of central and north west Europe, eastwards into western Asia, with the entire population wintering in southern and eastern Africa. In Britain, since the decline, the breeding range is largely restricted to north western Scotland and the Western Isles. An English re-introduction scheme began in 2003 with 52 chicks being released at the RSPB Nene Washes, in Cambridgeshire. This has met with some initial limited success, with 23 singing males in the summer of 2009. Elsewhere, targeted conservation measures have seen the Scottish population recover to over 1,100 singing males by 2009 (Holling 2011).

Moorhen (Common Moorhen)
Gallinula chloropus

A very common and widespread resident in all suitable habitats.

Historical review

The Moorhen has a long pedigree as a Durham species, its remains being described from amongst the excavated bones taken from the mediaeval strata of Barnard Castle (Yalden & Albarella 2009). These were presumably birds that had been captured along the River Tees, below the escarpment upon which the castle is built, and along which section of river birds still nest today.

The Moorhen was described by Temperley (1951) as our commonest resident water bird. The same may be implied from the various 19th century chroniclers, although ornithologists of that era perhaps found it so widespread that very few specific records are mentioned and indeed some, undoubtedly due to over-familiarity, failed to list the species at all. Temperley considered it abundant and widespread and noted that it was to be found on nearly all water bodies, from the largest reservoirs to the smallest field ditch, so long as there was enough nearby cover for nesting. An early 20th century account recorded an instance of exceptionally late nesting when the brother of Mr R. H. Brown saw a brood of young *'still in down'* on a pond near Durham City, on 29 October 1924 (Brown 1925). Temperley was in a position to observe first-hand the drop in numbers as a result of severe winter weather in the infamously cold winter of 1946/1947, but he also judged that a full recovery of the resident population had occurred by at least the spring of 1951.

None of these early recorders commented on any evidence of passage or winter visitors.

Recent breeding status

The Moorhen is common, less reliant on large water bodies than the Coot *Fulica atra*, and it has shown a remarkable consistency over the last two centuries. It is still today regarded as our commonest resident water bird with a widespread distribution wherever there is cover for refuge and breeding. The North Tees Marshes and WWT

Washington hold good numbers throughout year but the species uses a wide array of wetland habitats, including ponds, streams, marshes, wet meadows, damp or muddy farmland margins, ditches, drainage culverts, large gardens, parks, artificial water features and sewage works. Larger lowland water bodies are extensively occupied, but numbers at upland waters are never high and these can be abandoned entirely in winter. It also thrives along the banks of all the major river systems from the Derwent to the Team, Tyne, Wear, Browney, Gaunless and Tees especially in their middle and lower reaches. Its wide range is shown by a presence at Whitburn CP, close to the north east coastal corner, and south and west to the River Tees below the raging torrent of Cauldron Snout in Upper Teesdale (Newsome 2009).

There are few areas of Durham where the species is not represented in wet or damp habitats, except perhaps on the exposed upland expanses. Nevertheless, birds can be routinely found at altitude in the upper Tees and Wear valleys and on some water bodies lying on the fringes of moorlands and commons, such as at Fish Lake, Lunedale, and adjacent to Muggleswick Park in the upper Derwent valley. Breeding was confirmed in 1960 at a reservoir on Middleton Common at an elevation of 485m above sea level and during the *Atlas* survey (Westerberg & Bowey 2000), successful breeding was recorded at sites over 420m above sea level, in the Bowes Moor and Harwood-in-Teesdale areas. The species' absence from large areas of open heather and rush moorland was amply demonstrated in the *Atlas'* mapping exercise (Westerberg & Bowey 2000), which showed this largely sedentary species is restricted at elevation, presumably by a lack of adequate food supply and potential for winter freezing. Small numbers can nevertheless be found in the west, as in 2009, when up to eight were noted at Derwent Reservoir and up to three at Blackton Reservoir, Harelaw Quarry, Hudeshope Beck and Gilmonby.

It was apparent from the *Atlas* work, that the River Tees basin had not been adequately surveyed, especially along the river's middle and lower sections. To an extent, this situation was corrected by the work on *The Breeding Birds of Cleveland* (Joynt et al. 2008), which showed that Moorhens were widely distributed along the north side of the lower Tees valley, being particularly abundant around parts of the North Tees Marshes, but less well distributed in the Hartlepool area. Indeed, apart from this concentration of birds around North Tees Marshes, the species is present, but not necessarily common, along the whole Durham coastal strip, largely because of an absence of suitable habitat (Westerberg & Bowey 2000).

As well as the North Tees Marshes, an obvious major centre of population for Moorhens is at WWT Washington, where a complex of artificial ponds and 'pens' for its captive wildfowl collections offers excellent habitat and feeding opportunities (Westerberg & Bowey 2000). From its creation in the mid-1970s, WWT Washington rapidly became a breeding stronghold for Moorhens. Bolstered by the supplementary feeding of the captive wildfowl, Moorhen benefited considerably, resulting in a particularly high density of breeding pairs there. This particular population has been closely monitored and during the *Atlas* survey period, the site produced peak spring counts of between 130 and 180 birds, with a maximum number of 38 broods totalling 118 young in 1991 (Armstrong 1992). A near identical breeding success was recorded in 1996 and the situation remains similar, if not slightly increased, to the present day.

Other lowland sites with notable breeding concentrations include: Bishop Middleham, Brasside Pond, Hetton Lyons CP and Seaton Pond, Low Barns NR, Joe's Pond, Rainton Meadows NR and Shibdon Pond. At Rainton Meadows, habitat management led to a steady increase in numbers through the late 1990s and early 2000s with 24 pairs raising 4.1 young per successful nest there by 2006. The middle to lower reaches of the Wear and Tees also hold significant numbers of breeding birds along their banks and smaller tributaries.

As important as these recognised wetland sites are for the species, the majority of the county's breeding population is probably far more dispersed and occurs at much lower densities (Westerberg & Bowey 2000) and birds rapidly exploit new opportunities. For instance, the species quickly colonised the Sunderland Football Academy pools, just months after their creation in 2002.

The first chicks are usually noted at WWT Washington from the first week of April, exceptionally by 13 March in 1993. The extended breeding season, with double or even triple broods, means relatively recently hatched young may be observed into the second half of October; older young often assisting with the rearing of siblings from different broods. The 1924 record for the latest brood still stands, though pairs with small young have also been seen at East Boldon on 11 October 1978, and on the River Don, Jarrow on 5 October 1986. The strategy of multiple broods allows for the loss of nests or young from predation or flooding and provides the species with the ability to recover quickly from such losses, or those due to hard winters. A pair on the River Gaunless at South

Church, Bishop Auckland, in 2007 had three broods in all, with the full-grown young from the first brood seen to help with feeding the small chicks from their parents' second (Newsome 2007).

Predation, especially by Mink *Mustela vison*, was once thought to be significant in some parts of the county, for example along the lower River Derwent (Westerberg & Bowey 2000). Courtesy of the recovery of the county's Otter *Lutra lutra* population however, the mink is an increasingly scarce species along the county's rivers.

The *Atlas* estimated that the county's breeding population during the early 1990s was in excess of 400 pairs and probably less than 700. There is evidence that habitat creation at some sites has further benefited the species and there are also suggestions that the general population continues to do well with no perceived threats; it is now probably in the range 600-900 pairs.

Recent non-breeding status

Counts of 90 at Boldon Flats in November 1954 and of 100 at Crookfoot Reservoir in October 1959 were both considered unusually large at the time and also happen to be the first specific quantitative counts documented for Moorhen in the county.

The largest counts predictably come at the end of the breeding season in late autumn and early winter, with increases noted at many sites over the period September to November. These usually reflect breeding success for the season since Moorhens seldom move more than 20km from their natal area (Lack 1986). Cold weather can stimulate further peaks at sites that retain open water during the months of December and January. The winter counts at WWT Washington are amongst the best documented; a site record of 145 was set in January 1987, but by December of the same year, 200 were present and, in the winter of 1988/1989, the site held 180-190 birds. Moving a decade on and numbers at WWT Washington had continued to increase with 255-265 in November 1997 and 1998 and an all-time site record of 315 in November 1999. Late autumn numbers have remained in the range 240-280 since.

Elsewhere, Shibdon Pond, Joe's Pond and Rainton Meadows all typically hold 70-90 birds in late autumn. The extensive North Tees Marshes population also regularly produces winter counts of up to 100. The importance of our river systems in winter is shown by a count of 150 birds noted on the banks of the Wear between Durham City and Chester-le Street in January 1993. Smaller counts of 25-45 have come from Cobey Carr, Derwent Walk CP, Hardwick Hall, Herrington CP, Hetton Bogs, Hetton Lyons CP, Lamesley, Low Barns NR and Ramside Hall.

Moorhen are renowned for their catholic diet and illustrative of this were the birds seen feeding on Hawthorn *Crataegus monogyna* berries near Bishop Auckland on 7 February 1996 during bad weather. Stranger though were the detailed observations between 1983 and 1995 of a small number of Moorhens, which developed the habit of feeding on the pollen of reedmace *Typha latifolia*. This plant grows to almost 2m in height and has a spiky male flower. The Moorhens would climb up the stem and, on reaching the flower, strip the anthers of the yellow pollen, becoming dusted over the head and neck in the process. On a few occasions, they were observed dropping beakfuls of material from the flowers to the water below, where it was consumed by their offspring. It was presumed that the pollen formed an easily available source of protein for the growing young (Bowey 1995). A partially leucistic bird was at Herrington CP in 2008; it produced young in 2009, some of these showing some leucistic traits of their parent but to a diminished degree. Some birds that inhabit parks and similar locations, such as the South Marine Park Lake in South Shields, can become quite tame, especially when food is regularly available.

Moorhens are resident along the coastal strip throughout the year, so direct evidence of passage or winter influxes remains scant. Some occurrences along the coast probably involve no more than very local movement though occasionally, what are believed to be migrant birds are seen along the coastline, with most records being in the months April and August-November (Blick 2009). From the few such records spread over several decades, it is possible to construct a case for occasional and light autumn and spring passage. In 1961, a bird settled on the sea off Hartlepool Headland on 22 October and, in 1965, single birds again settled on the sea off North Gare and Hartlepool on 8th and 22 November respectively. On 2 October 1976, one was seen perched on top of a poplar tree in Seaton Carew church yard. Records suggesting spring movement include one found on Hartlepool Headland on 24 April 1971 during a fall of migrants. In 1987, single birds, taken to be migrants, appeared on the Whitburn coast on three dates during April and, in 1990, one was found in Mere Knolls Cemetery on 6 May. There is some suggestion of the occupied upland areas being vacated during the winter and the species' local upland populations may undertake a degree of altitudinal migration into lower river valley sites.

Distribution & movements

The Moorhen has a huge global range and is found breeding on all continents, Australasia and Antarctica excepted, though on the former it is replaced by a closely related congener. In Europe, they are largely sedentary, though some birds from north west Europe regularly move into Britain in autumn and winter.

The increase in some winter gatherings, especially during cold spells near the coast probably includes a proportion of visitors from outside Durham, though there are no significant ringing recoveries to demonstrate this categorically, or to suggest their possible origins.

Coot (Eurasian Coot)
Fulica atra

A very common resident and winter visitor on open waters in the east of the county.

Historical review

In 1829, Hogg described the coot as *"very rare"* when writing of it in the Stockton area, noting in 1845, that the only location where it could be considered common in the south east of the county was the Wynyard Estate. A half century later and Nelson (1907) said that it was a *"resident, generally distributed and common"* and Tristram (1905) considered it to be *"not uncommon"* on larger ponds. At a more local level, in the north of the county, the species was summed up by Robson as "*not so much distributed as the Water Hen*" (Robson 1896), and this, in many respects, accurately describes its status today. The earliest record of breeding at Teesmouth was of at least three pairs on Cowpen Marsh in 1912 (Blick 2009).

In the mid-20th century, Temperley (1951) regarded it as common on open waters but less abundant than the Moorhen *Gallinula chloropus*; he said that it was seldom found on rivers or streams and that it rarely wandered. He believed it had expanded its range somewhat through the formation of subsidence ponds associated with historic mining activity. Large flocks of Coot gathered on wetlands in winter, indicating that there may have been some winter influxes, either from further north or the Continent. Being so common, it appears to have attracted few if any specific records and the county's first quantitative reports were of 80 on Hurworth Burn Reservoir in January 1946, 69 at Fishburn Lake in October 1962 and a large count of 153 on Hurworth Burn Reservoir in September 1964.

By 1964 Stead, writing of the Teesmouth area, said that Coot was locally common as a breeding resident and by 1969 *"increasingly common"*, highlighting that it then bred on Cowpen Marsh. The 1960s and 1970s saw an increase in the numbers around Teesside, which Blick (2009) asserts was not related to observer coverage. By 1973, there were 15-20 pairs at Cowpen and Haverton Hole, six to eight pairs Saltholme and two to four pairs at Dorman's Pool and Wynyard, with one or two pairs at Charlton's Pond (Joynt *et al.* 2008).

Recent breeding status

The Coot requires shallow fresh water, preferably with ample marginal vegetation for cover and nesting. The species prospers at sites where there is a particular combination of cover and open eutrophic water, shallow enough to allow bottom feeding on the macrophytic plants that form a large part of its diet. The absence of marginal cover is not a complete deterrent to breeding and it is quite common to see nest platforms built out on open water.

Nesting occurs on ponds, reservoirs, gravel pits, marshes with expanses of water, ornamental lakes and occasionally on the slow moving sections of our main rivers. The *Atlas* (Westerberg & Bowey 2000) demonstrated an expected strong easterly bias to breeding in the county, since conditions at our lowland waters best match the species' habitat requirements. Given its preferences, the Coot's distribution would seem to be restricted by the availability of waters offering the right conditions. In recent decades however, the species' fortunes have benefited from conservation and recreation-driven improvements at several wetland sites. The birds are generally tolerant of modest human disturbance and have been able to exploit these new opportunities, some of which are on the urban fringes. The species' range has not expanded greatly, but it has been able to progressively consolidate its position in the east of Durham. All of this a far cry from the position described between 1972 and 1974 (Coulson 1973-1974, Unwin 1975) when egg collecting and deliberate human disturbance at the then Washington Ponds resulted in little breeding success for the 20 or so pairs established there each spring. Thankfully, attitudes and habitat management have improved greatly since.

Over the period 1990-2005, there were several excellent examples where initiatives to provide new ponds or to improve the quality of existing ones yielded significant benefits in the numbers of breeding pairs and wintering birds. Example sites include: Seaton and Hetton Lyons Ponds, the Joe's Pond and Rainton Meadows complex, Bishop Middleham lakes, Hardwick Hall and McNeil Bottoms.

Brasside Ponds, whose origins lie in mining subsidence, possess the classic combination of shallow open water with well-vegetated margins and they have traditionally enjoyed the status of the county's premier breeding site. Shibdon Pond in the north of the county was an important site in recent decades, although breeding numbers at both Brasside and Shibdon have fallen over the last few years for reasons that are not clear. Hurworth Burn and Crookfoot Reservoirs and sites at Teesmouth, such as Charlton's Pond, Hart Reservoir and Saltholme Pools have also featured strongly. The Reclamation Pond at Teesmouth has been particularly interesting. Here, breeding birds are now out-numbered in spring by non-breeders. Counts have exceeded 250 in the month of May in recent years. This flock then acts as a nucleus attracting an even larger summer gathering of moulting birds. However, the recent in-filling of this waterbody has resulted in a drastic change in numbers here.

The *Atlas* recorded pairs in 87 tetrads so beyond the major sites, given there are a good number of minor ones, the species is dispersed chiefly throughout the east of the county. Virtually all lowland bodies of water, of any significant size, attract birds at some time in the year, even some small relatively coastal ponds, such as that at Whitburn CP. Coot appear to prefer larger, relatively open waters, than Moorhen *Gallinula chloropus* and the species seems much less partial to running water than that species, only rarely being noted on rivers.

The Coot is essentially a lowland species, with some exceptions. The upland reservoirs in Durham's western parts have never supported breeding Coot to any significant degree. The main controlling factor appears to be that the waters are mostly man-made reservoirs which are generally too steep sided and too deep for them to support the growth of aquatic weeds upon which the birds would depend. Derwent Reservoir has held a few pairs regularly, but Grassholme Reservoir in the south west only recorded its first breeding pairs in 1995, and neighbouring Hury Reservoir in the following year. Populations here remain small. Altitude alone is not the limiting factor in such situations, as Fish Lake in Lunedale, at 450m above sea level, holds a few pairs in most years and in 1975, breeding was recorded on a small water at Cowshill in Weardale, at an elevation of 415m.

Young broods usually first appear from mid-April, very occasionally at the end of March, as demonstrated by the birds that were feeding chicks on 29 March 2005, whilst frequently produced second broods extend the breeding season through until July at many sites.

The number of breeding pairs can vary quite considerably from year to year, possibly affected by spring water levels. At 2010, some sites, such as Castle Lake and Rainton Meadows, appeared to be in the ascendency at the apparent expense of others, such as Brasside and Shibdon Ponds. Not all sites follow the same trend each year and success at one might coincide with a poor season at another. The maximum breeding count at any single site has been from Brasside, where in 1993, 46 pairs raised 118 young. However, this had declined to fewer than ten pairs by 2006. At Teesmouth, the number of breeding pairs has declined with 44 pairs on Cowpen Marsh in 1976 to only nine in 1986 and about 20 pairs in 2004 (Blick 2009). Other notable breeding counts have been at Joe's Pond and Rainton Meadows, with about 30 pairs in 2007 and 2008, although this number fell to about half this level in 2009. The wetlands at Bishop Middleham held 17 pairs in 2009, whilst Seaton and Hetton Lyons Ponds each typically hold six to eight breeding pairs. In a good year, both Hurworth Burn and Crookfoot Reservoirs have supported up to 15 pairs. McNeil Bottoms might hold 12-15 pairs with the sites at Teesmouth showing similar numbers. The present whole Teesmouth population is about 85-95 breeding pairs (Blick 2009). In addition, there are many smaller waters holding one to three pairs. In 1994, casual observation revealed 150 pairs breeding at 35 locations (Armstrong 1995).

The Coot has undoubtedly benefited from wetland habitat improvements and continues to do so. There are some threats to the species, including loss of habitat at some sites that have been important for the species in the past, such as the Reclamation Pond at Teesmouth. At some sites, dry springs can lead to nests becoming stranded on the dry banks, making them prone to predation by Foxes *Vulpes vulpes*. Fishing interests at some waters such as Lambton Ponds and Witton-le-Wear, can sometimes sit awkwardly with the species, due to disturbance.

Despite fluctuations, the evidence points to a general increase in breeding numbers due to habitat improvement from new conservation and recreation projects. The current breeding population is estimated to be about 200-230 pairs in all but the poorest years and perhaps in excess of 280 in the very best.

Recent non-breeding status

Adult Coots undergo a full moult between July and early September during which time they become flightless for a few weeks. Breeding pairs will tend to remain on territory, but many moulting birds are known to assemble in large flocks at selected sites across the country. The Reclamation Pond at Teesmouth became established through the first decade of the 2000s in attracting a large moult gathering during the summer months. Little is known about the birds' origins, but they presumably include a high proportion of one-year-old non-breeding birds (Cramp *et al.* 1977).

The county's core, resident population appears to be largely sedentary, apart from those in the western uplands. Breeding waters continue to be occupied throughout the year unless hard weather prompts local movement to open waters that may include estuaries, coastal marshes and rivers. Continental migrants arrive in autumn to supplement flocks and most sites will see peaks sometime between November and January.

Lambton Ponds, Shibdon Pond and Brasside regularly attracted winter flocks of about 250 during the 1970s and continued to do so for many years, with other sites typically hosting flocks of between 40 and 80. Crookfoot Reservoir can hold as many as 150-160 at times. Charlton's Pond has recently held 100-130 in the winter months and waters such as Cowpen Bewley WP and Wynyard have held up to 50-60 birds (Blick 2009). Far fewer are now found at Low Barns NR, with nearby McNeil Bottoms now favoured, and the balance has fallen away from Brasside and Shibdon Ponds. The most remarkable change has been seen at Teesmouth largely as a result of increased water levels in the Reclamation Pond from the late 1990s. The WeBS counts for the North Tees area demonstrate the onset of large gatherings from that time.

Year	Jan	Feb	Mar	Apr	May	Jun	Jul	Aug	Sep	Oct	Nov	Dec
2000	546	373	423	313	260	324	505	654	630	518	548	728
2003	346	318	320	287	229	359	541	853	1156	1070	1234	1331
2006	1588	807	523	512	356	366	1037	1115	1357	1499	1940	1307
2009	837	516	403	242	265	258	526	981	1361	976	1095	471

WeBS counts for Coot on the North Tees Marshes at 3 year Intervals, 2000-2009

Elsewhere during the 2000s, Rainton Meadows winter counts were up to 250 and Brasside Pond's numbers still exceeded 100, with several other sites holding 40-80 birds. The overall peak winter population in Durham is now probably between 1,500 and 1,800 birds.

At Teesmouth, during very cold weather, it is not unusual for a bird or two to move to Hartlepool Docks and at such times, birds from the North Tees Marshes move to Greatham Creek and the adjacent south west corner of Seal Sands (Blick 2009). Nonetheless, it is unusual to see a Coot on the sea off the Durham coastline, but single birds were noted at North Gare on 8 January 1995 and on the sea at Whitburn, accompanied by an Eider *Somateria mollissima*, on 6 July of the same year. Birds with aberrant plumage are most unusual, and a pale grey individual was at Haverton Hole in late 2009, at Saltholme Pool in March and April 2010, and then on Dorman's Pool from May to August (Blick 2009).

Distribution & movements

The Coot is widespread over much of Europe and Russia, though absent from cooler and higher latitudes. In the UK, it is scarce or absent from the higher Pennine chain and in upland areas of Wales, the West Country and Scotland. Those in southern and western Europe are essentially sedentary, but continental populations in the north and east abandon their breeding areas in autumn and move generally in a south westerly direction. These account for the migrants arriving in Britain to bolster resident flocks during October or November. Some 20,000 Coots are thought to visit this country from the Continent each winter (Newton 2010). Spells of severe winter weather on the

Continent might induce further influxes in mid-winter, but normally, winter flocks disperse by late February or early March.

There are few direct records of passage or ringing recoveries to indicate the origin of winter movements into Durham, but this is perhaps not too surprising given that the species is mainly a nocturnal migrant (Newton 2010). An early significant record was of a bird ringed at Abberton, Essex, on 19 December 1957 that was recovered at Barnard Castle, 220 miles to the north west on 2 September 1960, suggesting that this may have been a northerly breeding bird wintering in the south of England. Providing some corroboration for this was the bird ringed at Shibdon Pond on 18 March 1989, which was found dead at Druridge Bay, Northumberland, just 33 days later; perhaps it was moving to the coast to make a subsequent North Sea crossing? The only continental recovery concerns a first-year bird ringed at Woumen, Belgium, on 10 August 1974 and found long dead at Durham City on 21 August 1976. The winter influxes of continental migrants must be inferred by the size of the winter flocks that gather at some sites, especially Teesmouth, where numbers far exceed those that might be accounted for by local breeders and their surviving young. There is also recent evidence of cross-Pennine movements, provided by a colour-ringing scheme in Lancashire and Merseyside. A Coot ringed at Marine Lake, Southport, on 11 November 2008 was seen at Hardwick Hall on 15 April 2010, and another ringed at Farnworth, Greater Manchester, on 26 November 2009 was at Hurworth Burn Reservoir on 3 April 2010.

Crane (Common Crane)
Grus grus

A scarce passage migrant from Europe, mainly in spring: 40 records involving 66 individuals since 1959. Probably bred in the county prior to 1600.

Historical Review
Records of the Crane in Durham can be traced back to the 14th century, courtesy of the Cellarers' rolls of the Monastery of Durham (Fowler 1898), but its local presence actually goes much further back, to the Roman period of occupation of northern England, as shown by the discovery of the remains of this species in archaeological excavations at Corbridge and Housteads (Miles 1992). Further archaeological evidence of its presence in the county comes from the Roman site at Piercebridge (Yalden & Albarella, 2009) and, more recently, from both medieval and post-medieval (the latter at Durham Cathedral) sites within Durham (Yalden & Albarella, 2009). There is also evidence from a medieval excavation just over the northern boundary of the county at Newcastle Quayside.

Cultural reference points are an often useful tool in mapping the former range and occurrence of such an obvious and spectacular species as this. There are at least two County Durham place names that make 'specific' allusion to Cranes and which, therefore, hint at its former presence in the area, perhaps in the late medieval period. The first of these is Cornsay, which derives from the Old English *corn, hoh* meaning "*crane height*". The second location is Cornforth, which again, comes from the Old English *corn, ford* unsurprisingly meaning "*crane ford*" (Yalden & Albarella 2009). Such obvious reference points indicate that this was probably a well-known bird in the county and, despite the subsequent transference of the term 'crane' to the Grey Heron *Ardea cinerea*, this did not happen until after the loss of the Crane as a breeding bird and the establishment of both of these place names. A little to the north of Durham, in southern Northumberland, lies the hamlet of Tranwell, which derives its name from another Crane reference from the Old Norse word '*trani*', meaning Crane, and the Old English '*wella*' meaning "spring". Clearly in past times, this species had a significant enough cultural profile in Durham, and the wider north east, for settlements to be named after it.

As stated above, it was quoted in the Bursar's and Cellarers' rolls of the Monastery of Durham in the 14th century, the name *grue* appearing in the accounts relating to the purchase of wild game, suggesting that birds had been taken locally. Gurney (1923) identified a number of references to the Crane occurring in these accounts in the years: 1312/1313, 1358, *c.*1375 and again in 1390. Unfortunately, from the material quoted, the exact time of year of these occurrences cannot be determined and the full context of only one entry is given, it states: *In iii signis, uno grue, xvj perdicibus, ij curlows, xiiij pluvers, xiij s. iiij d.* ("*three swans, one crane, sixteen partridges, two curlew,* *fourteen plover,13 shillings and 4 pence*"), and clearly refers to the purchase of one *grue* or Crane alongside more expected food items such as partridge and plover (Fowler 1898).

The first well-documented record of Crane for County Durham is of a bird shot at Dyke House Farm, then part of West Hartlepool, in May 1865. This was obtained by a Jonathan Smith, the then gamekeeper of the Hart Estate. The specimen was passed to the Hartlepool Museum, before being transferred to the Hancock Museum (Hancock 1874), where it can still be found in the NHSN collection to date.

There were just two further records by the end of the 1960s, both of which occurred in 1959. An adult was present on Cowpen Marsh, Teesmouth, from 4 August until 1 November, occasionally roosting on the Reclamation Pond, and two were present between Sedgefield and Coxhoe from 28 November to 1 December (Grey 1960).

Recent Status

The next record came almost twenty-five years later, when a party of four was seen soaring over Cleadon Hills on 4 April 1984 during a period of light south easterly winds. The next, was an adult seen flying north over Marsden Quarry on 5 May 1985, with two further records in 1987, both of adults, near Ludworth on 12 June and at Stillington on 16 August. The final record of the 1980s is of two birds present close to Durham Tees Valley Airport from 25 May until at least 29 July 1988. Initially found by an airline pilot, these birds were seen by many observers during their unusually protracted stay.

The 1990s produced a further six records although the first was not until 1995 when a single bird was watched flying north over Whitburn village on 29 May. A first-summer initially seen over Cleadon Hill on the morning of the 10 May 1996 was later seen at Boldon Flats and then Barmston Pond, Washington until 12 May, when it departed to the south, later being seen over Finchale Priory. The following year, a single was watched circling over Haverton Hole, Teesmouth, on 9 May 1997, before departing to the north. This record was noteworthy in that it was the first for the North Tees Marshes since 1959. There were three records in 1999: two adults on Cowpen Marsh, Teesmouth, on 14th and 15 May, an adult at Castle Lake, Bishop Middleham on 13 September and another over Greatham Creek, Teesmouth, on 2 October.

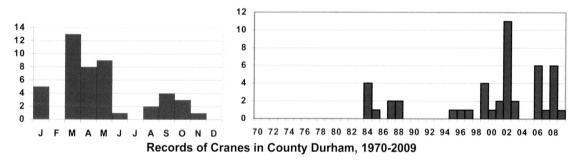

Records of Cranes in County Durham, 1970-2009

Records of Crane in the county have increased dramatically since 2000, with 23 records involving 37 birds during the decade. The species was recorded virtually annually, with just two blank years, 2004 and 2005. The best year was 2002, with five records involving eleven birds, including two parties on 17 March, the first of five that flew east over the North Tees Marshes, after being initially located over Belasis Technology Park, Billingham, and two flying south over Shotton Colliery later that afternoon. The decade has also produced the only two January records, a single coming in off the sea at Hart Village on 5 January 2003, which was later over Hartlepool Town Centre before heading south, and a party of four tracked flying north through the North Tees Marshes and later Hartlepool on 7 January 2006. These birds had been at Scarborough, North Yorkshire, just over two and a half hours earlier, and were later seen at various points throughout England before finally settling in south Devon in February.

A further seven birds were recorded in 2010 in what was a good year for the species. Sightings included: two at Crookfoot Reservoir on 18th and 19 April; a single flew north past Nose's Point, Seaham, and 15 minutes later at Whitburn, on 24 April; one was over Eppleton on 22 May; one circled over Hartlepool Town Centre on 28 May; and finally, two flew east over Coxhoe on 16 August. A similar situation was evident in 2011 when a total of nine birds was recorded between 25 March and 21 May. Following two at Saltholme Pools on 25 March (including a ringed individual), a new bird was present there the following day. One 10 April, two were intially seen on Cowpen Marsh

341

in the morning and were tracked flying over Raintonpark Wood, Low Barns and Witton-le-Wear later the same day. Another two were seen flying over Great Lumley, and then Durham City, on 1 May, whilst the final sighting for the spring was of two flyng over RSPB Saltholme on 21 May.

There is a distinct spring bias to records of Crane within County Durham with most recorded between March and May, although autumn records are not unusual. It is interesting to note that during the 1980s and 1990s, the peak month of occurrence was May yet, since 2000 there have been just five May records, with the peak passage period now being from late March to early April. Records are widely scattered, though typically most are not far from the coast, though the bird soaring over Abbey Bridge, Barnard Castle, on 3 April 2001 had penetrated a long way west. The separate groups of five over Billingham and the Tees Marshes in 2002, and in fields at Seaton Moor, inland of Seaham, on 31 March and 1 April 2008 are the largest parties to be recorded in the county in recent times.

Distribution & movements

The Common Crane breeds from Scandinavia and Eastern Europe eastwards to eastern Siberia, with smaller populations in Turkey and Transcaucasia. A noted long-distance migrant, birds from northern Europe winter in southern Europe, North Africa and the Middle East, often in huge numbers. This was once a widespread species across the British Isles, probably becoming extinct as a breeding bird around 1600. A very small breeding population has been present in the fenlands of East Anglia since the early 1980s and a crane re-introduction programme commenced on the Somerset Levels in the autumn of 2010. An increasingly regular migrant in the British Isles, records in our area are very likely to involve birds *en route* to and from their Scandinavian breeding grounds.

Sandhill Crane
Grus canadensis

Very rare vagrant from North America: one record.

The sole record of Sandhill Crane for County Durham is of an adult watched flying south along the coast on 29 September 2011 (*British Birds* in prep.). It was initially seen at 11:00 by Paul Hindess, Ian Mills and Mark Newsome at Marsden Quarry as it flew low over the centre of South Shields and was watched circling over Cleadon Hill before it headed off south over Sunderland. After an anxious wait, the same bird was picked up from the railway bridge on the Hart-Haswell Walkway, just to the north of Hartlepool at 12:15 where the assembled crowd of 10 observers watched it arrive from the sea before gaining height and heading off south over Hartlepool. After a brief sighting from Newburn bridge, the bird was seen again from the dunes at North Gare by six fortunate observers and was watched flying low over the sea towards South Gare when it left County Durham airspace at around 12:35. Just over 1.5 hours from start to finish, this 'one in a lifetime' event was witnessed by just 20 observers and will be remembered by those fortunate enough to be present for many years to come. The Sandhill Crane was later logged heading south east at several sites along the North Yorkshire coastline between South Gare and Kettleness between 12:35 and 14:00, before being last reported circling over Whitby at around 14:10, by which time it had also become the first record for Yorkshire.

This same bird had been present at Loch of Strathbeg, Aberdeenshire and the surrounding area from 22nd to 26 September 2011 before being seen to fly high south around midday on 26th. After having being missing for two days, it was seen again in Northumberland flying low south over Newbiggin Ash Lagoon, at 09:10 on the morning of 29 September. From here, it was seen at several sites in south east Northumberland, even alighting briefly close to Holywell Pond, before last being seen heading south over Whitley Bay at 10:40. After leaving Durham and then Yorkshire airspace, there were no further sightings until 1 October when it was reported flying south over Rimac, Lincolnshire, before being photographed on the salt marsh at Snettisham, Norfolk later that day. The following day, it was watched flying south over Kessingland, Suffolk, and after a brief stop at RSPB North Warren, finally settled at RSPB Boyton Marshes where it remained until 7 October. What was presumably the same bird was seen with Common Cranes *Grus grus* in southern Spain in November.

Distribution & movements

Sandhill Crane breeds from north eastern Siberia, through northern Alaska to Arctic Canada, as well as the western and central United States, Florida and Cuba. The nominate form which breeds in the Arctic is thought to be responsible for all of the European records and is a noted long-distance migrant with the entire breeding population heading south to winter in the southern United States and northern Mexico often in flocks in excess of 10,000 birds. There have been just three previous records in Britain, all of which were in the northern isles of Scotland; a first summer on Fair Isle, Shetland from 26th to 27 April 1981; a first summer at Exnaboe, Shetland from 17th to 27 September 1991; and an adult on South Ronaldsay, Orkney from 22nd to 29 September 2009. Elsewhere in Europe, there are single records from Ireland in September 1905 (shot), the Faeroe Island in October 1980 (shot), the Azores in June and July 2000, and in Finland and Estonia in September 2011. The species is also a regular vagrant to Japan in the east.

Little Bustard
Tetrax tetrax

An extremely rare vagrant from Europe: one record.

The sole record of Little Bustard for County Durham is of a bird said to have been shot at Harton, South Shields, in December 1876. This specimen was sent to John Hancock for examination and to *"improve the stuffing"*, on 9 March 1877 (Temperley 1951). Eventually, the specimen was presented to the Natural History Society of Northumberland, Durham and Newcastle-upon-Tyne, in 1935, though Hancock's details attached to this did not record by whom it had been shot. Other accounts however, almost certainly relating to this specimen, indicated that it had been *"shot at Marsden, by William Sisterson about 1860"* and that it was a female.

There have been no subsequent records of this species in County Durham. The species has suffered a major decline across much of its former breeding range, and with just ten British records during the last 30 years it seems unlikely to occur again.

Distribution & movements

Little Bustard breeds from the Russian steppes westwards to Iberia, although it is now absent from large tracts of south east Europe where it formerly bred. It is largely sedentary in the western parts of its range, and it is believed that the more migratory eastern birds, stimulated by adverse weather in their favoured habitat of steppe grassland, were responsible for the majority of British records (Parkin & Knox 2010). There have been 111 records of the species in the Britain, although the vast majority of these were during the Victorian era. There have been only 19 records since 1958, the last being on the Isles of Scilly in March 2002. Little Bustard suffered a major decline across much of its former breeding range during the latter part of the 19th century (Cramp & Simmons 1979), and the subsequent decrease in the number of British sightings is doubtless a direct result of this decline.

Oystercatcher (Eurasian Oystercatcher)
Haematopus ostralegus

A very common passage and winter visitor, and a locally common breeding species.

Historical review

In modern terms, this species' distribution is primarily a coastal one through much of the year except during the breeding season, when birds are present from locations as far west as the highest reaches of upper Teesdale and in scattered fashion down to the coast. The first documented record of the species in Durham dates back over 400

years and comes from Benett's records emanating from the Bursar's rolls of the Durham Monastery, which stated that *a 'seepye' was bought in 1533 for 1d* (Gardiner-Medwin 1985).

Three centuries later and Hutchinson (1840) described the species as breeding at Teesmouth. Other local historical chroniclers of the time mentioned little or nothing of its nesting, though at the end of the Victorian period, Tristram (1905) documented its occasional breeding there. In the 19th century this species was almost exclusively a bird of the coast, and the first inland nest in Britain, was found along the River Nith, Dumfries and Galloway, in 1896 (Holloway 1996). By the 1920s, birds were to be found along the river corridors of northern Northumberland (Kerr 2001).

Into the 20th century and its status had clearly changed, for Temperley (1951) reported its first breeding in Durham in more modern times. This occurred in spring 1933, when a pair nested at Teesmouth, as reported by Milburn but the eggs were stolen. *"As there seems to be no previous record of the Oyster-Catcher having nested in Co. Durham, a nest containing three eggs which I saw on May 27th, 1933, in rather an unusual kind of site, may be worthy of mention.... The nest was on one of many small mounds of coal shale refuse from the adjacent Clarence Steelworks and the eggs showed up most glaringly against the dark surroundings. Furthermore, the place is only about three miles from the Town Hall of Middlesbrough"* (Milburn 1933). Bishop claimed that two pairs nested in 1933. Breeding attempts took place in the next six years but with little success, until 1939 when at least two pairs reared young (Temperley 1951). Birds nested again in 1951 at Greatham Creek and nearby in 1959 (Stead 1964).

In terms of its general status Temperley (1951), observed that the Oystercatcher was *"a summer resident, passage migrant and winter visitor"*, recording that non-breeding birds were at Teesmouth throughout the summer. He considered it less common in winter than previously although small flocks were present. He thought the largest numbers were present during March and April and again over the period August to September; clearly passage birds moving through Teesmouth.

In the period between 1960 and 2010, the Oystercatcher's breeding range in the county spread inland from its original coastal haunts. Inland breeding was occurring in Northumberland by the early 1950s (Kerr 2001), but at that stage, not yet in Durham. From the 1950s onwards, birds began nesting on riverine shingle beds, reservoir shores and in fields along river valleys. The inland colonization was part of a national range extension, which commenced in the early 19th century (Marchant *et al.* 1990). In 1960, inland breeding was reported in Teesdale, where one pair was noted, and over the next few years the range and inland population increased. Birds attempted to breed at two sites in the south east of the county in 1964, and the following year, a pair bred in a field not far from Hart Reservoir (Stead 1969).

In the early 1970s, two pairs nested at Teesmouth in 1971, but breeding was considered irregular at the time, although small numbers of inland breeding birds were then present in the Tees, Wear and Derwent valleys. In the following year, a pair nested at Witton-le-Wear in 1972 and birds bred as far west as the Eastgate-Westgate area of Weardale. In1974 a pair bred in an arable field near Darlington for the fourth year in succession. At this time two pairs nested at Derwent Reservoir and there were a few other inland reports, mainly in the Wear and Tees valleys. Three pairs bred around Cow Green Reservoir in 1975, marking the limit of the westward expansion. Two adults were seen with young near Greatham Creek in June 1976 (Bell 1977) and since then a small number of pairs have bred annually around the North Tees Marshes. Birds bred for the first time at WWT Washington in 1987. During the summer of 1988 Oystercatchers were reported to be more common than ever in the western dales and breeding was confirmed at Hurworth Burn Reservoir for the first time in 1989.

Recent breeding status

Nationally, there has been a large increase in numbers in recent decades; from between 19,000-33,000 breeding pairs in the early 1960s to 33,000-43,000 pairs by the mid-1980s (Piersma 1986). O'Brien (2005) estimated the British summer population to be 113,000 pairs in 1985-1998. The results of a joint BTO/RSPB/Defra survey found that in Durham, between 1982 and 2002, there was a decrease of 50% across 375ha of nine monitored sites; this compared to a decline of 5% in Northumberland and a 52% increase in England and Wales as a whole; most of the latter in north west and eastern England. Oystercatcher features on the amber list of Species of Conservation Concern and Britain holds a significant percentage of the European breeding population (Birdlife International 2011).

In County Durham breeding was confirmed at 18 sites in 2000, but no doubt occurred at many more locations. Through the 2000s it became a common breeding species in the west of the county, and by 2007 it was reported from over 110 inland locations. The majority of the county's breeding birds are to be found in the Pennine Uplands and spread across the three watersheds of the Derwent (Tyne), Wear and Tees. Upland reservoirs, moorland pools, quarries and river systems all hold numbers of breeding birds, from Derwent Reservoir in the north to The Stang in the south.

The upland distribution of Oystercatchers was well illustrated by a detailed study of the breeding waders of Baldersdale and Lunedale, in the Tees watershed, by the RSPB (Shepherd 1993). At the time, 97 breeding pairs were located and it was subsequently confirmed that this level of occupation was representative of much of the county's upland habitat (Westerberg & Bowey 2000). A notable breeding concentration also exists on arable land in the south of the Wear lowlands. Elsewhere, lowland birds are largely restricted to reservoirs, nature reserves and along the River Tyne (Westerberg & Bowey 2000).

Coastal Oystercatchers nest on shingle beaches, salt marsh, dunes and rocky shores. Where such suitable habitat exists in the county, largely between Sunderland and South Shields, and Crimdon and Teesmouth, it is subject to comparatively high levels of human disturbance. As a consequence, these locations are largely avoided by breeding birds. Four pairs attempted breeding on the coast in 1994 at three sites and at least one pair of Oystercatchers nested each year within the Sunderland Docks in the early 1990s (A. L. Armstrong pers. obs.).

Despite the decrease noted above, inland habitats now clearly support the majority of the county's breeding population, though breeding birds can be found in relatively small numbers around Teesmouth, with just ten to twelve pairs being present on an annual basis during much of the first decade of the 21st century (Joynt et al. 2008). Birds bred near Longnewton Reservoir in 2004 and have been recorded breeding at Hart Reservoir and in Wynyard and breeding may also have occurred at Crookfoot Reservoir in the recent past (Joynt et al. 2008). Most nesting pairs in this area however, are concentrated around the mouth of the Tees, principally on areas of tipped slag near Greatham Creek (Blick 2009). Occasional breeding pairs are scattered along the ribbon of the Durham coast at sites such as Crimdon and Whitburn, though in such instances coastal fields are more likely to be used than the disturbed beaches. Between 2007 and 2009, c.30 pairs were documented nesting at around 20 different lowland locations, though not all sites were visited, or occupied, in any one year. In 2010, breeding was confirmed at only 12 lowland sites. Allowing for under-recording in the coastal lowland areas, between 25 and 50 pairs are probably present across this part of the county.

Of the relatively small number of coastal pairs that do attempt breeding, many do so within the protection of industrial compounds, such as those around the Tees Estuary, where human disturbance is minimized, or on isolated offshore stacks e.g. at Marsden Rock (Westerberg & Bowey 2000). One of the most interesting recent breeding reports came from Carrville, Durham, in the summer of 2007, where birds were seen carrying worms onto a factory roof where they appeared to be feeding young. In the same season, roof-nesting, a phenomenon noted elsewhere in Britain (Duncan et al. 2001), was also believed to have taken place on the Rainton Bridge Industrial Estate, Durham (Newsome 2008).

Local breeding birds begin to move inland during February, depending on weather conditions, and such movements increase through March. At this time of the year, birds may be found concentrated in flocks prior to settling down into breeding territories, often in areas close to the breeding grounds such as around some of the upland reservoirs. This phenomenon was well illustrated by figures from 2010 when, in mid-March, around 120 birds were present at Hury and 110 at Derwent Reservoir.

Westerberg and Bowey (2000) suggested a minimum breeding population for Durham in excess of 330 pairs, which, given current knowledge, represented 0.3% of the British population. Noting the significant national increase in the usage of wet meadows in the first decade of this century, at the same time as the survey work found a decrease of similar scale in Durham, it is difficult to accurately suggest what the current county population might be.

There are no obvious or immediate threats to this species and the long-term trend of what is probably still an expanding population of inland breeding Oystercatchers in Durham, seems likely to continue for at least the immediate future, as predicted in Westerberg & Bowey (2000). Any expansion in the county's coastal breeding population would appear to be reliant upon an unlikely reduction in recreational pressure at suitable coastal locations.

Recent non-breeding status

Teesmouth provides the main concentration for non-breeding birds in Durham and, although numbers fluctuate throughout the year, there are peaks in autumn and winter with the latter sometimes occurring late in the season in February as birds are moving back to their breeding grounds, presumably to the north.

The seasonal mean peak low water counts (the average highest count at low tide) at Seal Sands in the period 1990 to 2001 were: winter, 238; spring, 118; and, autumn, 310. The peak counts (the highest counts) for the same period are: 425, 195 and 460 respectively. The picture is much the same when projected through to March 2010 (R. M. Ward pers. comm.). Individual counts however, can considerably exceed these mean peak counts. Although the trend for local wintering numbers showed a significant increase between 1969 and 2000, the numbers remained below the qualifying threshold for national importance (Ward *et al* 2003).

During the summer months, small numbers of non-breeding immature birds remain at the coast; sometimes these number many hundreds of birds. This gathering is usually in the order of 200-400 birds, but up to 600 were present in July 1985 and again in July 2006. It is thought that the majority of these birds are immature, non-breeders. Many Oystercatchers breeding to the north of the British Isles join local birds on the British coast in late summer to moult and subsequently remain through the winter. Those along the British North Sea coast are predominantly from Norway and the Low Countries (Wernham *et al.* 2002). Norway is the predominant breeding season origin for recoveries of autumn passage and wintering birds at Teesmouth (Ward *et al.* 2003).

A moult flock is present at Teesmouth and there have been some breeding season recoveries associated with this group, including some ringed as adults and others as pulli, from Shetland and North Wales. Where the latter go to as adults is not known, although some have subsequently wintered at Teesmouth and singles have been recovered in the Netherlands and in mid-September in Suffolk (Ward *et al.* 2003).

In most years, autumn numbers at Teesmouth exceed the winter total, and this was the trend for WeBS counts between 1969 and 1973, and again in the period 1988-2001. Typical WeBS counts on the North Tees wetlands in autumn would be around 1,200 birds, for example 1,164 in August 2005 and 1,264 in October 2007. Between 700 and 900 birds would be typical of winter passage and wintering totals although 1,082 were noted in February 2007 and 1,053 in December of that year.

In the winter the species' distribution is almost exclusively coastal and the largest wintering numbers are concentrated around the Tees estuary, with peak counts coming towards the beginning and end of the year; often in February and December, when over 1,000 birds may be present, though as previously mentioned 700-900 birds would be nearer the usual winter maximum. The largest Tees count of the 1990s came in December 1999, when the total for this area was 1,741. In the north of Durham, Sunderland Harbour and Whitburn Steel usually hold smaller numbers, and figures here are in the order of a few hundred during early January and February or December. There is a frequent interchange between birds at these sites, facilitated by their relative proximity. Further north, the South Shields area holds a good wintering population, numbering over 100 birds and these often use The Leas, as well as the foreshore, as feeding grounds. Along the coastal strip many grassland sites hold small flocks of up to 20 birds, when rough weather or high tides drives them off the shoreline, for example at the Whitburn Rifle Range and Seaton Carew. A high autumn count for South Shields occurred in October 1992, when 710 birds were in the Trow area. Along the central section of the Durham coast, flocks of 50-100 are occasionally noted at sites such as: Blackhall Rocks, Crimdon Dene, Dawdon Blast Beach, Featherbed Rocks, Hawthorn Hive, Hendon, Seaham, and Salterfen Rocks. At Teesmouth the species' favourite high-tide roost sites are on the rocks at Seaton Snook and Seal Sands peninsular, and to a lesser extent at Hartlepool, though spring high tide sites are limited here. The roosts within the estuary include birds from Hartlepool and south of the estuary.

Three birds at Hury Reservoir on 11 December 2008 and a single at Derwent Reservoir in December 1993 suggest that inland wintering, in mild winters, may occasionally occur in the Durham uplands. On 19 May 1989, a melanistic bird was at McNeil Bottoms, and in 2000 a partial albino returned to the coast for the sixth year running.

Birds appear in the county as passage migrants in both the spring and autumn. The spring passage period spans mid-March to mid-May, with most birds being noted in the first half of this latter month. Counts from well-watched localities illustrate the trend, such as the gathering of 156 at Whitburn Steel on 9 March 2010 and 534 birds on the North Tees WeBS count in March 2010.

Return in autumn is evident throughout August and September as most breeding birds vacate their territories from late July, and a steady movement of birds past Whitburn Observatory and through Teesmouth is recorded. Up to 200 birds are seen passing Hartlepool Headland on some days. Visible migration past Whitburn amassed 469 flying south in 3 hours on 7 August 1981 and 230 on 8th (Baldridge 1982). In August 2007, a typical year on the Durham coast, southerly passage was evident at the Whitburn Observatory, where almost 500 were counted passing this point over the period from 8th to 12 August, with a considerable movement of 836 birds on 21st of the month. In 2008, early August produced a concentrated period of southward bound passage, with 897 logged on four dates between 2nd and 9th, including 196 on 3rd and 344 on 8th of the month (Newsome 2008 & 2009). In August 1999, five flocks in two afternoons, totalling 270 birds, were seen departing from Seal Sands in a south westerly direction. The departure of diurnal migrant flocks from Seal Sands moving southwest along the trans-Pennine migratory pathway between Teesmouth and Merseyside (Evans 1968) has been regularly observed in autumn (Ward *et al.* 2003). Similar movement along a south westerly vector was also recorded over Shibdon Pond in August 1986 when 43 were noted (Armstrong 1987).

Distribution & movements

This species has a discontinuous distribution across Europe, eastwards to Siberia and central Asia. Many European birds are short distance migrants along their near coastlines, though some populations winter further to the south of their breeding areas, mainly around the North and Irish Seas.

At least 1,100 birds have been ringed on Teesside and there are at least 30 recoveries or controls of birds moving to or from Teesmouth. Ringed individuals from Teesmouth's non-breeding, summer population have been found in Shetland in the following summers. One ringed on Seal Sands on 2 May 1988 was found dead on the Isle of Sheppey, Kent, in December 1996, whilst another ringed on the same day was killed by a Peregrine on Seal Sands on 8 February 1998 and a third bird ringed on that date, was found dead at Saltburn on 2 March 2000 (Blick 2009).

Birds ringed at Teesmouth in the autumn, mostly by Durham University teams, have been recovered on the Mersey and in France in the following winter, as well as in Norfolk and France during the following autumn and in the Shetlands in summer. One ringed in July 1983 was seen in the Netherlands in November-December 1989, August-September 1990, December 1992, October 1996, 1997 and 2000, December 2001 and September 2003. Another bird ringed in July 1983 was found dead at Wells, Norfolk, in January 2002, having reached the age of 20, a similar age to that of the previous bird. Another long-lived individual, ringed in August 1986, was recorded at Wrangle, Lincolnshire, in August 1995. High site fidelity of wintering birds at Teesmouth is indicated by the single November recovery, in the Netherlands, of a bird ringed at Teesmouth in the December of a previous year (Ward *et al* 2003).

Black-winged Stilt
Himantopus himantopus

A very rare vagrant from southern Europe: one record.

The county's sole record relates to a first-summer male initially found on Cowpen Marsh, Teesmouth by T. Farooqi on 23 April 1993, remaining in the area intermittently until 4 May, often feeding and roosting on Seal Sands, it visited Hurworth Burn Reservoir on 30 April and 1 May. This bird was part of a notable influx of 11 birds into the UK at the time (*British Birds* 87: 521).

Distribution & movements

Black-winged Stilt breeds in suitable wetlands along the Atlantic coast of France and locally throughout the Mediterranean basin through to the Black Sea; most European birds winter in sub-Saharan Africa and increasingly in south west Iberia. By the end of 2009, there had been a total of 389 British records, mainly from the southern counties of England. It is a rare bird further north with only four records for Northumberland and eight birds recorded in Scotland.

Black-winged Stilt, Cowpen Marsh, April 1993 (photographer unknown)

Avocet (Pied Avocet)
Recurvirostra avosetta

A scarce but increasingly common summer visitor and passage migrant, first breeding in 2006.

Historical review

The Avocet became extinct as a British breeding bird around 1842 (Holloway 1996). Having been absent for over 100 years, they successfully re-established a breeding population at Havergate Island on the Alde-Ore estuary in Suffolk from 1947 (Brown & Grice 2005). This came about after the deliberate flooding of coastal marsh areas for defence purposes and reduced human disturbance in these during World War II. Prior to this re-establishment phase, the species was described by Temperley, in the mid-20th century, as, "*a very rare casual visitor*" (Temperley 1951).

The species' documented history in the county goes back to Backhouse (1834), who said it was "*very rare*", he did not know of any then recent captures in the county, which implied that there were older records. His allusions to this species, "*this bird, a few years ago was not uncommon at the mouth of the Tees*" infers that it was previously more common in the county. Hogg (1845) said one was shot at Teesmouth in the winter of 1827/1828. J. H. Gurney stated (*The Zoologist* 1876) that it had been recorded in Durham on two or three occasions, with one in Duff's Bishop Auckland collection, having been shot at Teesmouth in the spring of 1849 (Temperley 1951).

Nelson (1907) recorded three single birds shot at Teesmouth in the 19th century. The last of these was shot by a Stockton gunner (then roasted for supper) on the Tees, about 1870 (Mather 1986). The historical documents however, do not specify from which side of the River Tees these birds were taken. Nonetheless, during the late 19th century and first half of the 20th century, Avocets were clearly rare visitors to Durham and there were no definite records at all after the dinner table bird of 1870, until May 1931. The latter was the only then 'modern' record quoted by Temperley (1951) and it concerned three birds together at near Greatham Creek, Teesmouth, on 22nd and 23 May 1931, all were present to 23rd, but just one remained on 24th to 25th (Richmond 1931).

Post-war records include one that was on a pool near Dormanstown on 9 September 1947 (Blick 2009). It was almost 13 years until the next visit, a single being recorded from Seal Sands and the Reclamation Pond over 21st to 30 May 1960; it was joined by a second on 24th (Blick 2009).

Birds were recorded again in April 1966, over June and July 1968 and with a party of seven birds that arrived at Greatham Creek on 21 March 1969, two remaining until 26th, birds had been recorded in Durham in successive years for the first time in over a century (Parrack & Bell 1970, Parrack *et al.* 1972).

Recent Status

The majority of Durham's records of this species have come from Teesmouth, largely between March and May. With the establishment of the slowly increasing population in East Anglia, and more particularly the expansion of the North Sea population (Hagemeijer & Blair 1997) visits to County Durham, and Teesmouth in particular, became more frequent as the 1970s and 1980s progressed. After the 1969 records however, there were no further sightings until a single bird was present at Seal Sands and Dorman's Pool from 8 May 1973 until 13th, the seventh year since 1930 that birds had appeared at Teesmouth; perhaps it was this bird that remained on Seal Sands from early July until at least 16 September of that year. There followed an unprecedented influx on 4th to 5 May 1974 with 17 birds on the Reclamation Pond on the former date and a different group of 10 birds on Seal Sands the next day. Both groups of birds were described as restless and they quickly moved on, nonetheless the number of birds involved in this influx totalled more than had been recorded in the county for the whole of the previous century. Singles appeared again at Teesmouth in March 1976 and March 1979.

During the 1980s, the number of Avocet records increased considerably, the only blank years in this decade being 1982, 1985 and 1986. Since then, years in which Avocets were not seen became the exception rather than the rule and the only blank year since 1986 was 1990. During the 1980s, there were just four records away from the North Tees Marshes; these came on 15th to 17 June 1980, when two were at Castle Eden Denemouth, and on 30 July 1980, when a single flew south off Whitburn. There was also a single bird inland on the Newburn-Ryton stretch of the River Tyne in early April 1984 and at Hurworth Burn Reservoir on 22 January 1989. The following day this latter bird was relocated at Teesmouth, where it remained until 14 May, being joined by a second for the last four weeks of its stay.

Over the 1990s and 2000s a pattern of occurrence was established with very occasional winter sightings but most records continued to fall into the period March-May with the latter month recording most occurrences. Numbers remained low, with one or two birds involved in the majority of occurrences, although a group of 14 flew north off Seaton Snook on 26 April 1998 and 20 appeared on Back Saltholme on 30 September 2000. In 2004 however, birds were present on the North Tees Marshes intermittently between 9 March and 12 April with a maximum of nine together on 14 March, at this stage however there was no sign of breeding activity.

Recent breeding status

There was no evidence of breeding by this species during the *Atlas* survey period, although spring sightings at Teesmouth occurred in all of the survey years excepting 1990. On some occasions these involved apparently paired birds. There was no further evidence of breeding but, as the trend of spring presence continued, the possibility that this might take place increased. Prior to 2006, this striking wader remained merely a passage migrant to County Durham, despite its change in status at Teesmouth through the late 1980s, 1990s and early 21st century (Joynt *et al.* 2008). Eventually, during April and May 2005, nest-building activity was recorded by birds around the Tees estuary (Joynt 2006). The predominantly spring passage pattern changed in 2006 when birds bred for the first time at WWT Washington, then from 2008 breeding also took place on the North Tees Marshes (Blick 2009); a major change in the species' local status since the 19th century (Anderson 2007).

On 7 June 2006, a pair of Avocets arrived on Wader Lake at WWT Washington, adjacent to the River Wear. All but five of the previous records in Durham had come from the North Tees Marshes and there had only ever been three previous occurrences at inland sites. It quickly became apparent that this was a pair and they settled down to breeding activity, copulation being noted over the next few days. By 12 June, the pair had established a nest scrape, on the edge of a shingle island housing a breeding colony of Common Terns *Sterna hirundo*, and over the next days it was determined that there were three eggs in this. On 7 July two young had hatched. Over the next few weeks adults and young were seen daily, but in early August one youngster had disappeared. By 9 August, one of the adults had left and it may have visited the North Tees Marshes around this time. The other adult remained with the juvenile, which on 18 August, 42 days after hatching, made its maiden flight and as it became stronger it began to feed on the River Wear at low tide. The last sighting of these birds came on 28 August. It is thought that a failed breeding attempt may have taken place at Teesmouth in May 2006 (J. Grieveson pers. comm.) and it is possible that it was the pair involved in this, which subsequently relocated to Washington. Successful breeding at WWT Washington occurred annually from 2007 to 2011, with two young raised in 2007, two in 2008, two in 2009, when a second breeding pair briefly appeared before leaving, and two again in 2010, after the adults lost an early clutch of eggs. The summer of 2011 was significant in there being two pairs nesting sccessfully, raising three and two young respectively.

At Teesmouth, in spring 2008, birds nested at Greenabella Marsh, but failed early in the season. They moved to the nearby saline lagoon, south of Greatham Creek, and were joined by a second pair. Three pairs bred in the Greatham Creek area, two pairs hatching eight chicks, but one of these disappeared around 13 June (Joynt *et al.* 2008). In 2009, ten pairs were in the same area, and after the breeding season at least 35 birds were around Teesmouth. In 2010, the first North Tees birds returned on 2 March, with the WWT Washington birds returning by 11th. Numbers at Teesmouth had increased to 16 by 19 March with 30 there by 28 March. Fifteen pairs settled to breed, mostly on the Saline Lagoon. The first chicks were noted on 8 May and a minimum of 27 young were raised (Joynt 2011). Predation and natural losses were high, but breeding productivity was of just over one chick per pair (INCA 2010). Up to 39 birds were present on the North Tees Marshes in June 2010, with at least 14 juveniles present in July of that year; the colonisation of the county going from strength-to-strength.

Recent non-breeding status

Apart from the breeding birds at Washington Wildfowl and Wetlands Trust, the only inland records concern a lone bird found on the Newburn to Ryton stretch of the River Tyne in early April 1984, one at Hurworth Burn Reservoir in January 1989 and one at Longnewton Reservoir from 2nd to 4 April 2001 (Blick 2009). The first of these was noted on a few occasions during the first week of the month and it seems likely that this was the bird that had been frequenting Teesmouth in late March of that year. It had been present for three days before disappearing in early April, re-appearing on the Tees on 11th to 13th of the month (Bowey *et al.* 1993). A party of three seen flying north along the coast at Whitburn Coastal Park on 9 April 2011 was also significant in being the first record for the South Tyneside area.

Distribution & movements

The Avocet has a discontinuous and fragmented range globally. It breeds across temperate Eurasia eastward as far as Mongolia, as well as in east and southern Africa. In Europe, breeding occurs around the coasts of the southern North Sea, the most northerly breeding populations in the world, the west of France and locally along the north coast of the Mediterranean Sea from Portugal to Turkey. Inland breeding occurs in Spain, Austria and the former Yugoslavia, and Avocets also breed along the north coast of the Black Sea. The north European populations are partly migratory with more birds moving south in colder winters, to southern Europe or North Africa. The majority of the English breeding population is believed to winter on the estuaries of southern, and in particular, south west England.

At WWT Washington, two young were colour-ringed on 19 June 2010. Both were together at Cley in Norfolk on 27 July that year, then one of them moved to Spain, being sighted at Salinas la Tapa in Cadiz province on 7 October 2010 (Newsome 2011).

Stone-curlew (Eurasian Stone-curlew)
Burhinus oedicnemus

A very rare vagrant from Europe: five records.

The first record is of a bird in the collection of the eminent ornithologist and collector Marmaduke Tunstall, which was housed at Wycliffe-on-Tees. The specimen was said to have been taken in the neighbourhood in August 1782. At the time the species bred in considerable numbers on the Yorkshire Wolds (Nelson 1907) and might have been expected to stray further north on occasion. This specimen remains in the collection of the Hancock at the Great North Museum.

All records:

1782	Wycliffe-on-Tees, August, obtained (Tunstall 1784)
1843	Saltholme, Teesmouth, killed (Hogg 1845)
1864	Frenchman's Bay, South Shields, 4 February, shot (Hancock 1874)
1997	Piercebridge, 19 May
2011	Calor Gas Pools, Teesmouth, 7th to 8 May

Stone-curlew, Wycliffe-on-Tees, August 1782 (K. Bowey, courtesy NHSN)

The only 20th century record was of a bird watched bathing in the River Tees at Piercebridge on 19 May 1997, before being disturbed by gunshots. By remarkable coincidence, this was only six miles east along the River Tees from the location of the first in 1782. The sporadic appearance of this species was continued into the 21st century, with one on rough ground at Teesmouth providing observers with a rare opportunity to see this species in Durham.

Distribution & movements

Stone-curlews breed from Western Europe and north west Africa across to the Indian subcontinent. The majority of birds are migratory, wintering south into North Africa, though small numbers stay in southern Europe. Formerly more widespread, the British breeding population is now largely confined to East Anglia and the southern counties, having declined considerably during the latter part of the 20th century, although this had been somewhat reversed, by concerted conservation efforts through the 1990s and 2000s, resulting in 370 pairs breeding by 2010.

Collared Pratincole
Glareola pratincola

A very rare vagrant from southern Europe: one record.

The only record for County Durham was seen briefly by Jim Perfitt, Keith Robson and others at Barmston Ponds, Washington, on 17 June 1973 (*British Birds* 67: 325). Initially seen flying around what was then the east pond; it was harassed by the Black-headed Gulls *Larus ridibundus* and despite settling briefly, it quickly flew towards the west pond where it was not relocated. Just seven observers were present to witness the event.

Distribution & movements
Collared Pratincole breeds locally, in a highly fragmented fashion, from North Africa and southern Iberia eastwards through the Caspian Sea area to Kazakhstan and Pakistan. It also breeds in scattered colonies across sub-Saharan Africa where the majority of the European population winters. There had been a total of 101 British records by the end of 2009, only nine have been recorded further north than Durham.

Little Ringed Plover
Charadrius dubius

An uncommon but regular passage migrant in small numbers and a scarce local breeder.

Historical review
This wader is now a regular summer visitor to Durham, with birds being seen mainly between early April and mid-August. Echoing it's national status, this is a relatively new species to the county, being unknown as a Durham bird prior to the early 1950s. It is now an established though scarce breeding species, and represents something of a rare success as a recent colonist.

Nationally, Little Ringed Plovers were first proven to have bred in Britain at Tring in 1938 (Ledley & Pedlar 1938) and in Yorkshire in 1948 (Mather 1986). The first Durham record and the first for the north east, came in 1954, when on 1 May, David Simpson found one at Shotton Brickworks Pond. It was under observation for over three hours but was not present the next day (Temperley 1955). This alongside Lancashire's first record at Leigh Flash in May 1953, are amongst the earliest for the north of England (Brown & Grice 2005). In Durham, it remained very rare through the 1950s, with no further records to the end of that decade.

This status changed in the early 1960s, when in 1962, the first breeding record for the north east occurred (Bell 1967). In the spring of that year a nest was found with four eggs at the working gravel extraction site at what is today Low Barns at Witton-le-Wear but strangely no adults were ever seen. Little Ringed Plover bred at this site for at least three years after the first nesting but as the gravel became quickly colonised by plants, the birds appeared to move further up-river to the west of Witton-le-Wear onto active gravel working in the area for at least five further years (J. Coulson pers. comm.). Success was proven in 1963, once again at Low Barns where a pair was present with an additional single female also noted. A nest scrape was made, five eggs laid and one youngster hatched, and was presumed to have been reared successfully (Bell 1967). A clutch of five eggs is most unusual (Cramp *et al.* 1983) and this may have been the result of both females laying in the same nest; the fact that just one egg hatched lends some weight to this theory - the normal-sized clutch, of four eggs, perhaps being infertile. In addition, two adults noted at Coxhoe on 24 July 1963 perhaps indicated that a second breeding attempt was made in Durham that year. A juvenile noted at Teesmouth over the period 27th to 31 July, 1963, the first record of the species for Teesmouth, may have been derived from one of these.

In 1964, breeding birds were again present at Witton-le-Wear, with four eggs laid and hatched by mid-June, but less than a month later, no birds were present. It was commented erroneously in a national review of the species' status in 1964, that the only counties with an appreciable gravel production industry where nesting had not occurred by 1962 were Lancashire, Northumberland and Durham (Parrinder 1964).

Subsequent developments during the early 1960s, at Low Barns, the Witton-le-Wear site, and elsewhere, included: in 1965, two pairs, of which one hatched two young and the other three young; in 1966, of two pairs, one reared three young, the other pair laid four eggs which were accidentally destroyed. A newly-hatched chick was found two months later, probably from a replacement clutch. In 1967, three birds were at the usual site on 30 April, and three nests were later found there, with a fourth a few hundred years away. In 1968 a pair bred near Darlington.

Away from the breeding location in the 1960s passage birds were noted at a handful of sites in the months of June, July and August, including sightings at Teesmouth (first record on Cowpen Marsh in late July 1963) and Hurworth Burn Reservoir. In 1968 single passage birds were noted in May at Barmston and Teesmouth. More passage records from these sites in 1969 totalled 10 over five dates at Teesmouth between July and September, and singles at Barmston on 4th and 10 August.

There were no confirmed records of breeding in 1969 although two adults and two juveniles at Teesmouth over 13 June to 4 August indicated that this had taken place somewhere in the area. Three to four birds at Witton-le-Wear in early May 1970 suggested likewise, but there was no proof of breeding, although breeding by one pair was suspected at Washington. In 1971, single pairs were noted at four sites, with breeding proven at two, successfully at one of these.

The first confirmed breeding of Little Ringed Plovers at Teesside occurred in 1972, and subsequently in that area birds have nested around the North Tees Marshes at: Billingham Beck Country Park, Portrack Marsh, and also at a development site at Stockton (Joynt *at al.* 2008) and breeding has been attempted every year since then (Blick 2009).

By the early 1970s the species was firmly established as a breeding bird in the county. Eleven widely spread pairs were located in 1973 and it was present at ten locations in spring 1974, with young noted at six of these. By 1973, the former breeding area at Witton-le-Wear (from 1963-1968) was too overgrown for nesting. In the mid-1970s birds were nesting in the extreme north of the county, but this site now lies under the MetroCentre; the last occasion upon which birds were known to have hatched young there was 1976 (Bowey *et al.* 1993). It was appropriate that, in 1979, a pair nested, rearing three young, at Shotton Brickworks Pond, the site of the county's first record twenty years previously. The finder of both was able to watch this breeding attempt from his bedroom window (Simpson 2011).

Recent breeding status

In Durham, as in the rest of Britain, Little Ringed Plovers are largely dependent upon man-made habitats, that include shallow-edged water bodies with 'open' foreshore that are in the early stages of the succession process, i.e. relatively free from vegetation. Birds breed and forage across such level, barely vegetated ground where it fringes still, shallow water. Occupied sites usually comprise quarries, gravel pits, mine workings, land cleared for development or reservoirs; the latter with their varying levels of exposed substrate can be ideal, in some seasons. A national survey in 1984 revealed that, across the country, only around 3% of Little Ringed Plover breeding sites could be considered 'natural' (Parrinder 1989).

Since its first nesting attempts in the early 1960s, its occupation of such breeding sites in the county has been both variable and unpredictable. Through time, it tends to move, 'tracking' these, as they are developed or reclaimed either by vegetation or restoration. In any year during the *Atlas* period up to nine sites were occupied in the county with young being noted at anything between one and four of these (Westerberg & Bowey 2000). Over the period, breeding occurred on at least eleven sites, not all of these being regularly occupied and, in the majority of instances, just one pair was present.

Most occupied breeding sites in Durham, have been concentrated in the east and north of the county, below the 150m contour; although, occasionally, birds appear and breed at sites as far west as the upland reservoirs.

Today, the majority of occupied sites remain in the central and coastal parts of Durham, though regularly occupied westerly sites include Derwent Reservoir and Eastgate Quarry, in Weardale. Over almost 20 years, birds have regularly produced young at Derwent Reservoir, at an elevation of around 225m above sea level, so a lowland location is not a pre-requisite for success. At Teesmouth, the best breeding year for this species came in 2005, when there were a total of five pairs breeding, spread between two sites, Saltholme and the Greatham Creek to Brinefields area (Joynt *et al.* 2008).

Although human activity on working sites can deter many predators it can also lead to disturbance and act as a constraint to breeding success. Adverse weather in June, particularly heavy rain and flooding can also have an impact. In such situations the higher elevation sites in the west tend to suffer most (Westerberg & Bowey 2000).

Little Ringed Plover (Mark Newsome)

The national population trend has been one of continued increase since the species first nested in the UK. The national population was estimated at 825-1,070 pairs by Gibbons *et al.* (1993). The latest BTO estimate gives a population range of 1,046–1,181 pairs in 2007 (Conway *at al.* 2008). The trend has been reflected locally. The continued establishment of the species' breeding population in the county was evident in 2007, with reports from 24 potential breeding localities during spring and summer. Breeding was only known to have been attempted at six of these but one of them held five pairs; at the other locations, just single pairs were present. In 2010, birds were noted at 13 potential breeding sites with two to three pairs at a number of these, including: Castle Lake at Bishop Middleham, Lamesley Pastures, around the North Tees Marshes and at Sedgeletch. Successful breeding was reported from at least five sites.

The potential for population growth in Durham is limited by the availability of suitable habitats and it is unlikely that the county will see any further major increases in the number of breeding pairs in the future without active management that specifically sets out to maintain and increase the breeding population.

There is some overlap in the Little Ringed Plover's habitat preference, with those of its larger congener, Ringed Plover *Charadrius hiaticula* (Marchant *et al.* 1990). Nevertheless, there is no evidence of this competition limiting either species' range or numbers in County Durham, although aggressive interactions between birds of both species have been noted at sites where both are in territory (K. Bowey pers. comm.) although Mead (2000) noted some displacement of this species by the larger plover as the latter has spread inland.

Recent non-breeding status

Most Little Ringed Plovers return to Durham from their sub-Saharan wintering grounds in Africa, during early April, although since the late 1990s the first birds have increasingly been noted from mid- to late March. For example, March arrival dates over the period 2005-2009 were respectively: 21st, 26th, 17th, 25th, and 16th; the last being the earliest for the county at the time. After the next arrival in 2009, at Lamesley Water Meadows on 17 March, birds were seen at a further twelve sites during the spring – a typical showing (Newsome 2010). Spring 2011 pushed the first arrival date even earlier, as one was seen at Sedgeletch on 14 March.

Autumn dispersal and passage commences early in Durham, with successful birds leaving their breeding sites by late June. Post-breeding gatherings can occur from mid-June, for example the 12 birds at Castle Lake, Bishop Middleham, on 18 June 2007 and the 11 around the North Tees Marshes on 15 July of that year. In both instances, some of these may have been passage birds, as well as local breeders. Small groups of adults and juveniles are often seen at Teesmouth in July and August. The largest known gathering in the county was of 15 on the Brinefields' tidal pool on 14th to 15 July 2009 (Newsome 2010). Most birds have usually left the Durham area by the third week of July, although occasional birds will linger until late August and, exceptionally, into mid- to late September. Such sightings include the latest record, a single bird on Dorman's Pool on 27 September 2006 (Newsome 2007), two that were at Bishop Middleham on 25th in 1999 (Armstrong 2003) and two at the same site on 23rd in 2007 (Newsome 2008).

Distribution & movements

This species reaches the northern limit of its breeding range in Britain, breeding southwards as far as North Africa, and eastwards across Europe and central Asia into south west Asia. It is a migratory species, the European populations wintering in north and West Africa, with some of them penetrating to areas south of the Sahara.

Ringed Plover (Common Ringed Plover)
Charadrius hiaticula

An uncommon breeding species, more common as a winter visitor and passage migrant.

Historical review

The first traceable mention of this species in the county comes from the Allen manuscript, Tunstall Museum, in 1791, which stated that the "*Sea-lark - frequents our shores in summer but are not numerous...*" as quoted by Mather (1986). This was an old name for the Ringed Plover, not as might be thought, the Shore Lark *Eremophila alpestris*. In the early part of the Victorian period, Hutchinson (1840) and Hogg (1845) had noted Ringing Plovers as breeding on sandbanks near Seaton, Teesmouth, and they were also reported in this area by Nelson in 1907; though there was little suggestion of how common it was as a breeding bird during this almost 70-year span. Around the late Victorian period, some birds were noted as nesting on iron-slag, perhaps one of the first examples of the species breeding on a man-made habitat (Joynt *et al.* 2008). Holloway (1996) documented the fact that some decline in numbers was noted at Teesmouth towards the end of the 19th century. Though at this time, Nelson (1907) wrote that passage birds were numerous at Teesmouth during mid-July to October, and common through the winter.

Into the 20th century and Milburn considered that breeding numbers had declined to just a couple of pairs at Teesmouth by around 1916 (Milburn 1917). Temperley (1951) noted that Milburn, writing to George Bolam of the 1920 season, said that this had risen to three successful pairs by 1920 and, by 1933, Bishop reported increased breeding numbers. Temperley also quoted Milburn as reporting in 1920 that a few pairs were nesting on the shingle bank side of a reservoir 'a few miles inland' from Teesmouth. This was the first documented instance of inland breeding in the county (Temperley 1951). Furthermore, up to 1919 two pairs had nested there regularly "*for the last ten years at least*". Perhaps the shingle bank reference suggests that this may have been Hurworth Burn Reservoir. In 1947, Fred Grey found a Ringed Plover's nest with four eggs in a turnip crop near a flooded field in north east Durham. Such inland breeding was still uncommon in national terms as the main inland expansion did not occur until after the *Breeding Atlas* (Sharrock 1976).

Temperley (1951) described Ringed Plover as "*a resident, breeding in diminishing numbers*" and he also documented it as a passage migrant and a "*common winter visitor*". Breeding declines, he associated with the increased industrialisation of Teesmouth and parts of the Durham coast. He recorded its winter presence at Jarrow Slake, on the River Don and also on the Wear, between Pallion and Hylton. Spring migration here, he noted, took place from March to May, though with some build-up prior to this.

In both 1951 and 1952 a pair attempted to nest on the Durham coast near Sunderland. Their initial attempts were unsuccessful as the four eggs were washed away by a high tide, a not infrequent hazard. However, a recently hatched youngster was there in August 1952, so a re-lay must have taken place. Two pairs attempted to breed there in 1953. Around this time at least four pairs attempted to breed at Teesmouth. There is little other data through the 1950s. Stead (1964) classed this species as a common breeding bird around Teesmouth and up to 15 pairs nested at Teesmouth in the early 1960s, with four pairs at Hartlepool, and three at Whitburn around this time.

A typical period of spring passage was noted at Teesmouth in 1964, with up to 450 birds there on 21 May. No doubt these numbered mainly Scandinavian birds, which were re-locating to northerly breeding grounds. In 1967, the peak passage occurred at what was later understood to be the usual time at Teesmouth, with 460 birds there on 24 May.

In 1970, over 20 pairs nested on the Durham side of the estuary, including three or four pairs in Hartlepool Docks (Stewart 1970) and by the late 1970s, it was estimated that area held between 25 and 40 breeding pairs. In 1974, pairs were also noted at Whitburn and Jarrow Slake.

Recent breeding status

In many counties, Ringed Plover are found principally along open coasts and shorelines, but in Durham today they inhabit a wider range of locations, both inland and at the coast. Most coastal breeding birds are still centred upon the Teesmouth area with occasional outliers scattered northwards as far as Sunderland, and with odd pairs appearing in places such as unused or derelict areas within Sunderland Docks where up to four pairs nested annually during the early 1990s (A. L. Armstrong pers. obs.). During the late 1980s and in to the early 1990s, a small population of at least eight pairs developed on derelict land along the River Tyne in Gateshead (Bowey *et al.* 1993). These birds were largely lost as most of this land was redeveloped. Nonetheless, it indicates this species' ability to quickly realise a habitat opportunity, where a lack of disturbance facilitates nesting and the safe rearing of young on suitable ground (Westerberg & Bowey 2000).

The *Atlas* distribution map was considered to accurately represent this distribution, giving a good overall picture of the Ringed Plover's range in Durham at that time, which largely holds true today. The species is scattered from east to west in a manner that reflects its principal nesting habitat requirement of sparsely vegetated sand, shingle or gravel, rather than the geography of rivers or altitude. It is found along the open shorelines of still water bodies and undisturbed beaches, though there are relatively few of these in Durham. Riverine shingle, a habitat frequented elsewhere (Prater 1989) is also largely absent in Durham. Birds breeding in the Pennine uplands, in the west of the county, are found congregated around the considerable number of large and small upland reservoirs.

The species' lowland, inland distribution is rather limited to the availability of man-made water bodies, which provide it with the required stretches of sparse vegetation and open foreshore for nesting and foraging. Its use of sites with the latter habitat assemblage does sometimes overlap with its close relative the Little Ringed Plover *Charadrius dubius*, which is both a rarer and less extensively distributed breeding species in Durham.

At the species' principal lowland stronghold, around the Tees estuary, breeding numbers peaked in the mid-1980s with for example 32 territories and 22 nests on the Brinefields in 1986. Since then numbers have declined to around 12 to 15 pairs on the north side of the estuary (Joynt *et al.* 2008). The reasons for this are unclear. Habitat management by raking gravel and removing vegetation on the relatively undisturbed Brinefields appears to have brought little improvement, and perhaps predation by corvids or the increasing population of foxes *Vulpes vulpes* is implicated.

Away from protected sites human disturbance at lowland sites, particularly increased recreational pressure, is one of the principal threats to the county's breeding population. In some areas, beach users have innocently trampled eggs, young and/or inadvertently kept adults away from nests and their contents (Westerberg & Bowey 2000). Clearly, many parts of the county's open coast, with apparently suitable breeding habitat, are not occupied by the species. Over the years, birds have attempted to nest on the shore immediately to the south of the South Shields Pier but have never been known to succeed; likewise the birds which have attempted to nest at Jackie's Beach, Whitburn. In most instances, successful beach-breeding of this species is restricted to the larger more open sand beaches and dunes surrounding Teesmouth, such as in the area of Seaton Common, and within the seasonally fenced compound at Crimdon, which protects the colony of Little Terns *Sternula albifrons*. However, the asynchronous nesting of the two species means that earlier plover nests are still subject to disturbance, and more widely this is likely to continue to limit the county's population, to undisturbed or protected sites; is a reflection of the national situation (Prater 1989).

In 2000, inland breeding was confirmed at Monkton Fell and then in 2005 at the Lunedale, Baldersdale and Waskerley Reservoirs. At the latter site in 2007, a pair reared at least one juvenile; and at Selset Reservoir three pairs were present in early June, whilst a single bird in the extreme west, around Cow Green Reservoir, in May was possibly a breeder. A pair was in territory on derelict land near the Federation Brewery at Gateshead in May 2007. In 2008, coastal breeding was confirmed at three sites including at Crimdon Dene, with at least seven successful breeding pairs at various upland reservoirs. In 2009, the county's breeding population was split between upland reservoirs (9 pairs), coastal beaches (9 pairs), and docklands (4 pairs) (Newsome 2008-2010).

The *Atlas* work indicated a population for the county of around 25 to 50 pairs (Westerberg & Bowey 2000). This figure probably remains in the right order of magnitude today. However, if just two pairs were present each year around all of the upland reservoirs at which it has been recorded in territory in recent years, then this figure alone would be around 24 pairs. Adding this to the numbers from Teesmouth and elsewhere would bring the current county population figure close to the upper estimate limit of 50 pairs.

Recent non-breeding status

During autumn and spring passage, Teesmouth, and in particular Seal Sands, supports large numbers of Ringed Plovers, with the latter season's population reaching levels of national importance of over 300 birds (Ward *et al.* 2003).

Migration is evident from late July onwards. Most birds seen locally are likely to be from Greenland and Iceland *en route* to winter in West Africa (Taylor 1980). Some local breeders are known to remain while others disperse. It is August/September when the greatest autumn numbers occur with counts peaking at Seal Sands where over 400 can be present and at Whitburn where 232 were noted in 2009 (Newsome 2010).

During the winter this species is a regular feature of the sandy beaches between Sunderland and South Shields. In winter 2009/2010, Sandhaven Beach, attracted up to 70 birds during January 2010, with up to 91 birds at Jackie's Beach, Whitburn, a little way to the south. During the last quarter of the year numbers tend to build up at sites such as Whitburn Steel, with as many as 120 being recorded in roosts there. In this situation, birds can often be found feeding on the Steel at low tide and in the fields to the south of Whitburn Observatory when the tide is in. At Teesmouth, between 50 and 100 birds are usually present in winter, although considerably more have been counted on occasion, perhaps in response to cold weather in Europe. Birds ringed at Teesmouth in early winter have been noted in Kent in early December and Dyfed in early February (Ward *et al.* 2003).

The early year maximum counts usually occur in January at both Teesmouth and Whitburn, after which there is slow decline through subsequent months into late April as British breeding birds move away from wintering sites. Corresponding with this coastal decline, there is an increase in sightings at inland locations.

The month of May is when most Ringed Plovers are seen in the county, again mainly around Teesmouth. Around 400 to 600 are the usual highest counts in May, though 1,137 were counted there in May 1997 (Armstrong 1998) and an exceptional 1,250 birds, presumed to be of the race *tundrae* were noted near Hartlepool Power Station in May 2006 (Newsome 2007). It is thought that these are mainly migrants heading for Iceland and Greenland.

Distribution & movements

Three weakly defined sub-species of Ringed Plover occur in Britain; the native race *C. h. haticula* also extends to Scandinavia, the Baltic, Ireland and France and is largely resident although some migrate to south west Europe for winter; *C. h. tundrae* from Northern Scandinavia and Finland and high Arctic areas of Russia which migrate through in spring and autumn *en route* to and from East and Southern Africa; and *C. h. psammodroma* from north east Canada, Greenland, Iceland and the Faroes which also pass through Britain in both seasons and winter in Western and Southern Africa.

Approximately 3,000 birds, almost all full grown birds, have been ringed at Teesmouth, largely by Durham University. Birds ringed in autumn at Teesmouth have been recovered in Cornwall (September), France (August and early October), Western Sahara (April), and Iceland (May). Recoveries from Cornwall and Cumbria suggest that birds may migrate down the British east coast in one year and the west coast in other years (Ward *et al.* 2003) and recoveries show that this can occur for northerly spring migration too. Other birds are thought to use the trans-Pennine migration pathway between Teesmouth and Merseyside (Evans 1968).

Several birds ringed as nestlings at other east coast sites have been seen around Teesmouth in the following autumns, indicating that not all birds passing through this area in spring and autumn are necessarily bound for, or have come from, nesting grounds outside Britain. Birds in this category have been ringed at Holy Island and Beadnell in Northumberland, Spurn in Humberside and Gibraltar Point in Lincolnshire. At least three birds ringed in Norfolk moved north to Seal Sands by the autumn of the same year.

Birds ringed at Teesmouth during spring or autumn passage have been recorded during the following years at: Walney Island, Portsmouth, Canvey Island, near Bristol, Cornwall, northern France, southern Norway, Sweden, Portugal and Iceland. However, relatively little has yet been determined with regard to the wintering grounds or the breeding areas of birds passing through the estuary.

Birds controlled at Teesmouth during spring or autumn passage have been ringed in previous springs or autumns in: Finland, Norway, Poland, Denmark, France, Spain, Highland, Dumfries, Essex, Cornwall and Cumbria. The only bird known to move any great distance during the same period of migration was a juvenile ringed at Teesmouth on 8 September 1980 and recovered near Bangor, Gwynedd, North Wales, just 20 days later.

The considerable age which this species can attain was attested to by the bird ringed at Barmston Ponds, Tyne & Wear, on 5 October 1974 which was found dead 'near Middlesbrough' over 17 years later on 5 June 1991 (Blick 2009).

Killdeer
Charadrius vociferus

A very rare vagrant from North America: two records.

Ian Mills found the first for County Durham in fields close to Tilesheds Farm, East Boldon on 31 March 1976. Only the 16th British record, it remained in the area until 9 April (*British Birds* 70: 420). A bird had been at Bainton Gravel Pits, Cambridgeshire from 6th until 27 March, so the Boldon bird may have been the same individual on its way north.

The second record was some fourteen years later at Cronkley Farm, upper Teesdale on 25 March 1990 (*British Birds* 85: 522). Found by a visiting birdwatcher, David Sharrod, it showed very well down to 20 metres amongst breeding Lapwing *Vanellus vanellus* and Redshank *Tringa totanus*. Like the preceding bird, it was presumably migrating northwards having wintered to the south. The similarity in the dates of the two records is noteworthy.

Distribution & movements
Killdeer is a highly migratory species found across the Americas, from southern Alaska, Canada and throughout the USA south to Mexico. Northern populations spend the winter around the Caribbean and into the northern parts of South America. It is a rare but regular vagrant to Britain, with 51 records by the end of 2009.

Kentish Plover
Charadrius alexandrinus

Rare vagrant from Europe: 12 records involving 13 individuals, between April and June.

The first record for County Durham relates a bird observed at Teesmouth on 8 June 1902 by Joseph Bishop and Charles Milburn. No mention is made of which side of the River Tees the bird was seen, though the record is published as the first county record in George Temperley's *A History of the Birds of Durham* (1951), so was presumably seen north of the river. One of the observers, Joseph Bishop, also found the second record, an adult female at Seaton Carew on 20 May 1904, which was found dead close the breakwater at North Gare the same day. Remarkably, Bishop and Milburn were also responsible for the third record in 1924.

All records:

1902	Teesmouth, 8 June (Temperley 1951)
1904	North Gare, adult female, 20 May, found dead, specimen in the Dorman Museum (*The Naturalist* 1904: 283)
1924	Teesmouth, two, 11 May
1954	Cowpen Marsh, Teesmouth, 28th to 29 May (Stead 1964)
1961	Saltholme Pools, Teesmouth, female, 30 April to 1 May (Stead 1964)
1964	Reclamation Pond, Teesmouth, 24 August (Bell 1965)
1968	Barmston Pond, Washington, 27 May
1979	Seal Sands, Teesmouth, female, 20th to 21 May
1980	Long Drag, Teesmouth, male, 13th to 15 May

1981	Long Drag, Teesmouth, male, 20th to 21 May
1987	Tidal Pool, Greatham Creek, Teesmouth, male, 26 May
1994	Seaton Snook, Teesmouth, male, 16th to 23 May

Distribution & movements

Kentish Plover breeds across most of temperate Europe and Asia, and North Africa. Birds in their northern European range are migratory and winter south into Africa. The birds of the Americas are considered by some authorities to be a separate species, the Snowy Plover *Charadrius nivosus* (IOC 2010). Formerly a rare breeding species in Britain, primarily in Kent and Sussex, it had become all but extinct as a breeding species by the end of the 1920s. The species also suffered a major decline across much of Europe during the 20th century. As a direct result, the number of passage birds seen in Britain annually has declined and now averages just 20 birds per year, mainly in spring. Most records are from the south east of England, in particular Norfolk and Kent.

Dotterel (Eurasian Dotterel)
Charadrius morinellus

A scarce but almost annual passage migrant, largely in spring less commonly in autumn; it has been proven to have bred once.

Historical review

The species was possibly quite widespread as a breeding bird in the uplands of northern England prior to the 19th century. It certainly bred on many hills in the Lake District at this time (Holloway 1996). Eggs were taken on Skiddaw, north of Keswick, in 1784 and many nests were found in the Lake District in the second half of the 19th century (Holloway 1996). Hutchinson (1840) reported its passage in autumn as rapid, but more lingering in spring, with the greatest numbers in May. He said it had not been recorded breeding in Durham, but he speculated on locations where successful breeding might occur, these included Burnhope, Killhope and Wellhope Moors. The Dotterel was certainly known to be breeding in the extreme east part of the Cumbrian north Pennines during the 19th century (Macpherson 1892), although it was uncommon there. It seems that at a time of elevated Dotterel populations, locations such as Cross Fell, Great Dun Fell in Cumbria and a number of the highest fells on the Durham side of the Pennines might have been used as breeding locations. Hancock (1874) considered it an annual passage bird in spring and autumn and in particular noted its occasional visits to areas near Sunderland, on spring passage. He also refers to eggs which had been obtained in the north Pennines. In 1951 however, Temperley observed that since 1840, the Durham Moors had been extensively searched "*in vain*" by ornithologists, including himself, but that no Dotterel were found breeding, and moorland passage birds were considered rare by his time of writing. Nonetheless, it has long been accepted that breeding used to occur on Mickle Fell in upper Teesdale (Mather 1986), and in a letter to John Hancock in April 1875, J.E. Anderson stated that '*they build*' on another fell top a few miles south of Cross Fell (which may have been Dufton Fell or Mickle Fell)? Mather (1986) tells of a Riley Fortune finding a pair on a hilltop in the north west of Yorkshire and young birds were seen there in the summers of 1895, 1902, and 1904 but the exact location was unstated, it is conceivable that it may have been in this area of upper Teesdale? An old Durham breeding record, dating from between 1910 and 1926, was also mentioned by Harrison (1988) but the details of this cannot be traced.

At Teesmouth, Nelson (1907) knew its haunts, and he recorded up to 30 birds in May 1903, whilst in the same area, in 1937 and 1938, flocks of 12-18 birds were reported. Three birds noted at the Darlington Sewage Farm on 3 September 1929 were unusual as it was then, and still is, a rare bird in the county during autumn passage.

At the mid-point of the 20th century, Temperley called it "*a passage migrant in very small numbers*". He said that it was mainly a spring migrant, annually recorded in the late 1940s and early 1950s at Teesmouth, though records elsewhere in County Durham, were he said, 'exceptional'. The Ornithological Reports for Northumberland and Durham since 1932 however provide no details, except for a 'trip' of 14 at Teesmouth in early May 1937. In mid-May 1946 a pair was seen by a local gamekeeper at "*the head of Allendale*", which was presumably close to Rookhope Head, in the upper reaches of the Rookhope valley.

No further birds were reported until 1961, when a passage bird was in a potato field between Cleadon and Whitburn on 14 June, an exceptionally late date for spring migrants. There is a record of this species in the Gateshead area in the late 1960s. A small group of five birds was found on Ravensworth Fell on 20 May 1967. Birds were not reported again until 1973 when two males and a female were on coastal grassland at the Leas, South Shields on 21 May, remaining until 24th. In the autumn of that year, two birds were at Whitburn Golf Course on the morning of 12 September.

Recent breeding status

There is only one modern breeding record for County Durham, a clutch of eggs, with attendant male, being discovered at just over 600m above sea level, on a fell top above upper Teesdale in spring 1993 (Strowger 1998). Although more associated with Scottish peaks, breeding does occasionally take place on high ground in England, and indeed, there were 41 documented breeding attempts in England between 1972 and 1995. All the known nests in this period occurred at 576 and 890m above sea level.

Recent non-breeding status

Since the 1967 sighting on Ravensworth Fell, in Gateshead, there have been at least two further sightings in that area. On two occasions between 1982 and 1985 parties of birds were located in May, once again on Ravensworth Fell (Bowey *et al.* 1993). The first sighting concerned a group of about a dozen birds with the second referring to a single bird. It would seem likely that this locality may once have been a traditional stopping-off site for migrating Dotterel and the occasional presence of 'northern' Golden Plover *Pluvialis apricarius altifrons* with which they sometimes migrate, might lend some support to this theory. Confirmation of this behaviour occurred on 10 May 2008, on a fell above Weardale, when two Dotterel were seen in the company of four northern Golden Plover (J. Strowger pers. obs.).

During 1970-2011, Dotterel were seen with some increasing regularity in Durham, but there were at least 17 years over this period in which birds were not recorded. There were seven blanks years in the 1970s, two in the 1980s, four in the 1990s (but with many fewer birds recorded in that decade than in the previous one) and years without sightings in the 2000s. After being a genuinely rare bird at the outset of the 1970s, the species' status very much increased in the 1980s, before it became rarer again in the 1990s and 2000s (Siggens 2005). The spate of occurrences in the 1980s, when it was recorded in eight years out of ten, is probably the best period for this species in the county, at least in recent times. Over the years, Cleadon Hills attracted birds on a number of occasions, particularly in spring, most frequently between 1975 and 1986. Birds were recorded there in: 1975, 1980, 1981, 1983 (twice), 1986 and 2008 (in flight).

The majority of records, 80%, were in the spring with early May, particularly the first two weeks of the month, producing most sightings. This is traditionally the period of greatest spring passage in England (Strowger 1998). Most records over this period came from coastal areas but birds were also seen near: Bowes, over The Stang, Stanhope, Cow Green and Smiddyshaw Reservoirs. 'Trips' consisted of two to six birds generally, but there were larger parties of 14 in 1980, 16 in 1988 and nine in 1996. In 2005, after a rather early bird on Dawlaw Fell, south east of Cow Green Reservoir, on 29 April, eight birds was seen on Widdybank Fell and later by Cow Green Reservoir on 17 May. On 21 May, a party of five were near the edge of Cow Green Reservoir, and were presumably part of the earlier flock in that area. The species is probably an annual passage visitor to the upper Teesdale area, but the days of passage are unpredictable, so they can easily be missed. Two at Hetton Lyons Country Park on 13 May 2011 were a rare sighting in lowland Durham.

Autumn passage, confined to the coast, has been recorded, between 14 August and 18 November, on eight occasions. A juvenile bird that was briefly at Barmston on 18 August 1999 was the 35th species of shorebird to be recorded there over the years. The bird involved in the latest of these autumn records was unusually late for Durham, it was a first-winter found in 2001, amongst Golden Plover, at Back Saltholme. Three of these autumn passage birds were juveniles, but the last, at Teesmouth in August 2010, was a brightly plumaged female.

Distribution & movements

The Dotterel is a species that inhabits high montane habitats, from Scotland through Scandinavia and eastwards across northern Eurasia. Also in mountain habitats, south into southern Siberia, Mongolia and parts of Russia. Birds are long-distance migrants wintering in Africa (Cramp *et al.* 1983).

American Golden Plover
Pluvialis dominica

A very rare vagrant from North America: nine records.

The first record for County Durham refers to a summer plumaged adult found by Chris Clark on the Long Drag, Teesmouth on 2 July 1979. It remained in the area until 9 July mainly frequenting Seal Sands and the adjacent Greenabella Marsh. It was judged to be an adult male on the basis of the solidly black underparts, particularly the flanks and undertail coverts. At the time of this record, birds were still considered conspecific with Pacific Golden Plover *Pluvialis fulva* and known as Lesser Golden Plover (*British Birds* 73: 506).

American Golden Plover, Greenabella Marsh, July 1979 (P. Wheeler)

All records:

1979	Long Drag, Seal Sands & Greenabella Marsh, Teesmouth, adult summer, 2nd to 9 July
1984	Greatham Creek, Dorman's Pool, & Reclamation Pond, Teesmouth, juvenile, 23 August to 4 October (*British Birds* 78: 545)
1988	Langley Moor, Durham, juvenile, 29 September (*British Birds* 83: 461), presumed same Greatham Creek and Saltholme Pools, Teesmouth, 30 September to 1 October (*British Birds* 83: 461)
1994	Whitburn, adult, 18th to 21 September (*British Birds* 88: 511)
1994	Whitburn and Boldon Flats, juvenile, 8 October to 1 December (*contra BB*) (*British Birds* 88: 512)
1995	Greatham Creek, Seal Sands, and Seaton Snook, Teesmouth, first-summer, 11th to 13 June (*British Birds* 89: 498)
1996	Cowpen Marsh, Greatham Creek and Saltholme Pools, Teesmouth, first-summer, 31 August to 2 September (*British Birds* 90: 469)
2011	Whitburn, adult, 4th to 13 September
2011	Whitburn Steel, juvenile, 13 October

In addition, a bird seen and heard calling several times as it flew over Greenabella Marsh, Teesmouth on 22 August 1999 was accepted by BBRC as 'either American or Pacific Golden Plover', although the observers thought the call more resembled that of American Golden Plover (*British Birds* 94: 470).

Distribution & movements

American Golden Plover breeds on coastal tundra across Canada and North America as far west as Alaska, and into extreme north east Siberia. A long-distance migrant, its migration route takes it over the western Atlantic to spend the winter in South America. It is one of the most frequently recorded Nearctic waders in Britain, principally in autumn, with many being displaced by Atlantic storm systems. There had been over 270 British records by the end of 2005 at which point the species was no longer considered by the BBRC.

Pacific Golden Plover
Pluvialis fulva

A very rare vagrant from Siberia: two records.

The first for Durham was an adult in summer plumage that was found by Tom Francis and Graeme Joynt by Greatham Creek, Teesmouth, at around 10.00hrs on 5 August 1995. Initially located by virtue of its solid black undertail coverts amongst a large post-breeding flock of Golden Plover *Pluvialis apricaria*, it was over an hour before the bird's key identification features could be fully observed. Shortly afterwards, it flew towards Seal Sands though it returned on that evening's high tide to roost to the Reclamation Pond; a pattern which was to be repeated until it was last seen on 15 August (*British Birds* 89: 498).

The second record was also of an adult in summer plumage found by Dave Foster in a large flock of Golden Plover *Pluvialis apricaria* roosting on Whitburn Steel in the late afternoon of 18 September 2002. It remained in the area until 29 September and was often seen at high tide on nearby arable fields or the rifle range adjacent to Whitburn Coastal Park (*British Birds* 97: 578)

Distribution & movements
Pacific Golden Plover breeds across the tundra of northern Siberia and western Alaska, wintering from north east Africa, through southern and South East Asia to Australasia and Polynesia. The majority of British records (over 85%) relate to summer plumage adults, with July and August being the peak months of occurrence. To some extent the County Durham records mirror this pattern. There had been 75 British records by the end of 2010, predominantly from the southern and eastern counties.

Golden Plover (European Golden Plover)
Pluvialis apricaria

An abundant wintering species though with a restricted winter range.
Fairly common as a breeding bird in the county's western uplands.

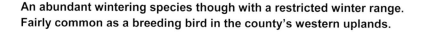

Historical review
The first documentation of the species in Durham probably dates back over 450 years, *"plover"* being mentioned several times in the accounts of the Monastery of Durham from the first third of the 16th century (Benett 1844). These were shown as *'costing about 1 and half old pennies each'* (Gardner-Medwin 1985), though it is not absolutely certain as to whether this refers to Golden Plover or Grey Plover *Pluvialis squatarola*. Benett distinguished between Lapwing *Vanellus vanellus* and 'plover', by using the old name *wype* for the former species, for example in the purchase during 1531 of *"3 plovers and 1 wype, 5d"*.

There is little specific information about the species' status from the 19th century texts, though Robson (1896) said of it in the Derwent valley *"in some winters it is common at Gibside especially around Fellside"*. In times past there was evidence that birds bred at much lower elevations in the county, with a 19th century record of a nest with eggs being found on the fell above Fellside Road, near Whickham, at around 210m above sea level.

Into the mid-20th century and George Temperley (1951) called it *"a summer resident, breeding on the moorlands in the west of the County"* but possibly also a passage migrant and winter visitor. He observed that the northern race *altifrons*, with its stronger black front appeared as both a passage migrant and winter visitor, although how he discerned this in their drabber winter plumage is unclear.

In terms of its breeding behaviour, Temperley noted that Golden Plover preferred burnt areas of exposed moorland and that few remained on the moors after the end of July, a situation which still pertains today. According to Temperley, the coastal wintering flocks comprised both races, in the thousands of birds.

In the early 1970s the Jarrow Slake mudflats were the pre-eminent site in Durham with maximum counts of 2,000 noted in autumn 1972 and early winter 1973 and counts of at least 400 in every other year of the decade. By

1975 reclamation of the 'slake' was underway and it was feared that this would result in the loss of the site. Although there was a general reduction, peak counts of 900 in October 1977 and 1,000 in January 1979 were achieved. During the decade, counts at Teesmouth were usually below those of Jarrow, typically of 300-600 in autumn with a peak of 1,150 in October 1976, with winter maxima of 800 in 1972 and 1977. Elsewhere, large autumn passage flocks of 3,000 birds were noted sporadically near Wolviston and Witton-le-Wear, the latter perhaps involving birds moving down from higher ground.

This picture continued into the 1980s with numbers at Jarrow generally in the mid-hundreds, with occasional peaks of 1,000. Flocks at Teesmouth were normally of comparable size although 1,570 were there in November 1980 (Baldridge 1981). Longnewton Reservoir and the fields nearby was an important site for this species during the decade with up to 1,000 birds there in some years, but this site appears to have decreased in significance more recently.

Recent breeding status

In County Durham, as a breeding bird, it is almost entirely restricted to unenclosed moorland above 300m, west of a line drawn between Smiddyshaw Reservoir in the north and Barningham Moor in the south, though a few pairs are occasionally found breeding at lower elevations (Westerberg & Bowey 2000). Wolsingham North Moor is typically the easternmost breeding location in most years. The species is widespread in distribution across such areas. Breeding density is generally greatest on the tops of hills that feature gently sloping ground and nests are most usually located in tussocks of heather Calluna vulgaris or cotton-grass Eriophorum vaginatum. Chicks preferentially forage in areas of mixed grassland or wet flushes, with nearby patches of heather or tussocks of grass providing cover from predators. Golden Plover nesting on heather moorland and bog lay, on average, 11 days earlier than those on grass moor and, in these situations, mean clutch size is also slightly larger (Crick 1992), possibly due to grass moor providing reduced food supplies for this species (Ratcliffe 1976). As the species has one of the largest eggs of its genus (over 16% of female weight), a typical clutch of four eggs can amount to 73% the female's weight (Nethersole-Thompson & Nethersole-Thompson 1986), access to good feeding grounds prior to laying is most important and females spend up to 90% of their time feeding prior to laying (Byrkjedal 1985).

In County Durham, much of the farmland close to where Golden Plover breed, in Weardale and Teesdale, is designated under the Environmentally Sensitive Area scheme, and any intensification of farmland within these areas would almost certainly be detrimental to the species. Potential threats to breeding birds in Durham come from a number of sources. The moors are now managed more intensively and access tracks proliferate. Although they favour short-rotation burned moors, excessive heather burning can result in less protective vegetation, leaving nests more prone to predation by corvids and mustelids. Higher stocking levels can increase the chances of trampling by sheep. There is no evidence that increased recreational activity under 'right to roam' has had any impact locally. Changes in husbandry regimes, which led to a reduction in the management of land for Red Grouse Lagopus lagopus in the Pennine uplands, might have negative implications for this species in Durham. Across Britain these factors, together with increased winter mortality, have been cited as causes in the national decline of the breeding population, although locally there does not appear to have been any such effect. The British population fell from 30,000 pairs in the early 1970s (Ratcliffe 1976) to 23,000 pairs in the late 1980s (Marchant et al. 1990) although BBS work since 1994 has indicated some later increase.

Although nesting almost exclusively outside enclosed land, the pastures and the hay meadows located within the upper slopes of the county's upper river valleys, are used for feeding by off-duty breeding birds, and are important during the species' pre-breeding period (Whittingham 1996). During the breeding season, birds can be found on most of the higher slopes of the county's moorlands such as: Bollihope, Grains o' the Beck, Killhope, Tan Hill and Widdybank Fell. Population levels of over five pairs per square kilometre have been noted in parts of Teesdale and Weardale (Whittingham 1996) and these figures are typical of much of the county's uplands. A further indication of the numbers present in some areas comes from the Rookhope Burn, where in spring 2007, an observer estimated 40 pairs along a seven kilometre stretch of moorland (Newsome 2008). These breeding densities compare favourably with the low densities encountered over large areas of the West Highlands of Scotland and in Ireland (Gibbons et al. 1993). Assuming a lower figure of two pairs per occupied square kilometre, to give a minimum population figure, then a population of over 800 pairs was estimated for Durham in the Atlas (Westerberg & Bowey 2000). The 2007 Breeding Birds Survey results gave a total of 329 pairs in five out of forty kilometre squares surveyed, a rather high density, which if extrapolated across the county would suggest that the population was several times higher than estimated by Westerberg & Bowey (2000).

Recent non-breeding status

This species is a common winter visitor to County Durham, occurring in large numbers though it is rarely seen away from its relatively limited number of favoured sites. A county-wide Golden Plover survey undertaken by Durham Bird Club over the period 08.00-12.00hrs on 27 November 1994 was designed to capture a snapshot of its local winter distribution. The area covered was 66% of the Vice-County below the 300m contour and c.75% of the non-urban land. Only one known site was not surveyed, Jarrow Slake which had by that time lost much of its former importance.

The total county population was estimated at between 9,497 and 11,756 birds. A large component of this was the flock of 2,000 on the Whitburn foreshore, which may have constituted part of the 3,500 seen on Boldon Flats earlier in the morning. Taking the total estimate as a minimum, it represented 3.7% of the whole British wintering population and over 0.5% of the North West European breeding population. Boldon Flats alone supported over 1 % of the estimated British wintering population, whilst six other sites held 600 or more birds. All bar one of the flocks were located to the east of '1° 51' west' and below 200m altitude, with the exception of one that was at no more than 260m. Large gatherings, in addition to that in the Whitburn area, were: 2,000 along the Tyne in Gateshead; almost 1,300 at Teesmouth; and, 1,220 around Derwent Reservoir, the most westerly and at the highest elevation. The principal habitats used by the surveyed flocks were grazed pasture (42.3% of birds), rocky coast (17%), flooded waste ground (17%) and winter crop land (6.4%).

Today, post-breeding gatherings of 150-450 birds may occur in moorland edge locations such as at Derwent Reservoir from early August or even be noted at easterly and truly lowland sites from late June onwards. Such flocks comprise presumably failed breeders, supplemented increasingly by juveniles. Birds returning to coastal wintering locations, such as Teesmouth and Whitburn, increase through June into July and may have reached as many as 1,000 by the end of July, as in 2008 (Newsome 2009), or the 1,500 present at Whitburn by late August 2010. Numbers continue to rise into September and usually peak from October until the end of the year.

Towards the end of the first decade of the 21st century, the fields around the Whitburn Rifle Range consistently provided single most important site in the county, the threshold for national importance, 2,500, being exceeded in every winter since 2000, excepting 2004; numbers peaked at 5,000 in late autumn 2007 (Newsome 2008). Birds from this wintering population regularly move north of the Tyne and to Boldon Flats and other nearby sites, particularly if there is disturbance at Whitburn.

The growing importance of Whitburn for this species can be given context by counts from the national BTO WeBS and Winter Farmland Bird Survey which confirmed that the winter distribution of this species is more biased towards eastern Britain than it was in the 1980s and 1990s. They also show that arable farmland is increasingly important for feeding birds. As demonstrated by a study in south Norfolk that found considerable numbers feeding on cereal and bare till fields, day and night, with birds dispersed across many more fields and habitat types in darkness and feeding actively regardless of the phase of the moon. Prey items on arable fields consisted mainly of small beetles, millipedes and earthworms (Gilling 2003).

The North Tees Marshes is the other main wintering area in the county. The species' favoured site in this area is around Saltholme Pools. Numbers here are usually lower than at Whitburn, counts over the period 2005-2010 usually peaking at about 1,500-2,000. The highest recent counts at Saltholme Pools have been: 2,350, in October 1994; 2,650, in November 1996; 3,000, in December 2001; 4,000, on 23 November 2002; and, 3,050 in December 2009. Seal Sands tends to be used mainly in autumn, to a lesser extent in spring passage, or by wintering birds in response to disturbance inland or severe weather (Ward et al. 2003).

Away from Whitburn and Teesmouth, locally significant winter flocks can be found at a number of widely scattered eastern sites; most of these are long established. Counts at sites such as: Bishop Middleham, Bowesfield, Chester Moor, Greenside, Langley Moor/Meadowfield, Longnewton, Ravensworth Fell and Sherburn are usually below 1,000 birds but can occasionally reach up to 2,500. Changes to the drainage of damp or poor grassland, such as at Meadowfield and Chapman's Well can reduce numbers present or result in the complete loss as at the MetroCentre pools. Most of the latter flock appear to roost on the banks and flats of the Tyne at Derwenthaugh/Dunston. The latter comprised over 3,000 birds in 1998/1999 (Armstrong 2003) but only c.1,000 by

February 2008 (Newsome 2009). Smaller flocks of up to 300 birds can be found at many more lowland locations with much site fidelity evident over the decades.

If weather conditions are favourable, birds can be found back on the fringes of breeding areas or even within territories from January, for example on Bowes Moor in 1989, Hamsterley in 1992, Wolsingham North Moor in 2000, Baldersdale and Gilmonby in 2005. Exceptionally in 1975, up to 21 birds remained on Widdybank Fell through the whole winter period into 1976 (Unwin 1977). If conditions remain good then numbers in such areas can increase rapidly in February, evidenced by 300 at Muggleswick in 1993, and in 2000, the 120 birds at Sleightholme and 200 on Barningham Moor (Armstrong 1994 & 2001). Wintering flocks in the east, decline quickly from late February/early March. For example, in 2009 numbers at Whitburn, after a winter peak of 3,000 in February, fell to just 180 in March. At this time pre-breeding assemblages continued to build, typically on pasture fields adjoining moorland, such as: 300 at Waskerley, on 15 March; 600 at East Stoney Keld, on 27 March; 300 on Lartington Moor, on 11 April; and, 150-290 at Barningham Moor, Grains o' th' Beck and Selset Reservoir (Newsome 2010).

Distribution & movements

Golden Plover breeds across much of northern Europe, from eastern Iceland and eastern Greenland westwards through Britain and Scandinavia across much of northern Arctic Russia to central and northern Siberia. All populations are migratory and some of the northerly birds replace more southerly birds in winter, as is the case in Durham, where different cohorts of birds are present through the year.

The British race *P. a. apricaria* mostly winters in Britain but a few move to the Iberian Peninsula and North Africa (Cramp & Simmons 1983). The 'northern' race *P. a. altifrons* makes up the majority of the migrants in Britain and often occur in the same localities which have held small winter gatherings. Inland during March, there is usually a build-up of birds, prior to breeding locally or further north in Britain. At this time northern birds make up an increasing proportion of birds moving through the county. Typically, such flocks number a few hundred. Spring passage has two distinct but over-lapping phases as mainly British birds move 'upwards and outwards' from their county wintering sites in February to early April. There is then a smaller but distinct passage of *altifrons* in late April or early May, largely comprising birds returning through Britain from wintering in Iberia and North Africa (Byrkjedal & Thompson 1998). These are usually in their black-fronted breeding finery and thus attract reliable identification. They also appear to favour traditional stopover sites such as near Tow Law Fell, Ravensworth Fell, Waskerley and Woodland Fell. In some springs at sites in the upland fringe *altifrons* may be noted on pasture when locally nesting birds are already holding territory nearby.

Local ringing recoveries have demonstrate that interchange between the county and the Netherlands is probably frequent, as shown by a bird ringed at Burum (Friesland) in the Netherlands on 12 March 1959 that was recovered at Jarrow Slake on 17 January the following year, and an adult ringed at Drente, Holland in late December 1990 that was on Seal Sands on 14 July 1993. Locally wintering birds may move to the coast or even away from the county with the onset of harsh winter weather in late December or January, sometimes in the company of local Lapwings. For example, 3,000 were at Whitburn in November 2010, this fell to 1,500 after hard weather arrived in December, and around the North Tees Marshes the number fell from 3,700 to 47 over the same period. South westerly movement of birds that have been in Durham was illustrated by an adult ringed on Seal Sands on 14 July 1993 that was killed in Biera Litoral, Portugal, in February 1996 (Blick 2009).

Grey Plover
Pluvialis squatarola

A common passage migrant and winter visitor from northern Europe, mainly to Teesmouth.

Historical review

The first reference to the species in the county is courtesy of Sharp's list for Hartlepool (1816), who thought it 'rare', as did Hogg (1827), though Edward Backhouse (1834) described it as 'common' at Teesmouth in autumn and winter.

Temperley (1951) recorded that it was *"a regular passage migrant and winter visitor in small numbers"*, which usually arrived in August and September, occasionally in July; with spring passage being noted through the county in April and May. In the 1920s and 1930s Bishop noted flocks as being present at Teesmouth from mid-March and as late as mid-June; clearly, these were passage and perhaps non-breeding birds. Around 1950, Temperley thought that records had been fewer 'of late', citing one war-time observer, with two years duty at Seaton Carew, saying it is *"a bird I rarely see at Teesmouth"*.

The main location for the species in the county during the first half of the 20th century was Teesmouth, but birds also used to frequent Jarrow Slake with some degree of regularity before most of the mudflats in that area were lost following land-claim, which began in the early 1970s. Although it was hardly ever seen inland according to Temperley (1951), he did document a handful of records from October at Darlington Sewage Farm.

In the post-Temperley period there are few reports. However in 1960 one was present at Teesmouth on the late date of 21 May, with two there on the 24th and one of these lingered until the end of the month. At this time, winter peak counts at Teesmouth were of 340 on 28 September 1957, 368 on 1 November 1964, and 367 in mid-February 1977. Since then, changes to the feeding areas resulting from increased silting may have reduced the carrying capacity of the site (Blick 2009).

At this time there were a number of inland reports from wetlands, though such occurrences were and still are never common in the county. One near Coxhoe on 17 May 1960 was an isolated report. They more often comprise juvenile birds in late August and through September, though may be somewhat overlooked. In the early 1970s such inland records included: three birds with Golden Plover *Pluvialis apricaria* at Low Barns on 17 February 1971; one at Hurworth Burn Reservoir on 7 October 1971; up to six at Barmston Ponds through September 1973; one at Hurworth Burn Reservoir on 31 August and 18 September 1974; and, two at Barmston Ponds on 25 August and 11 September 1974.

Recent status

The 1980s saw generally higher counts at Teesmouth than earlier decades but the picture continued to be variable from year to year. Notable totals were, in 1980: 234 in February and 243 in December; 288 in February 1982; 272 in December 1988; and, 257 in March 1999. However, some very low winter counts also occurred, for example: 65 in January 1983; 62 in December 1986; and, 24 in December 1987. Typical autumn influxes were noted from late August in most years. Throughout this period sporadic single figure counts were recorded at Boldon Flats in most years but exceptional records were of two at Smiddyshaw Reservoir on 17 October 1982 and one at Derwent Reservoir on 21 May 1989 (Armstrong 1990).

This pattern continued through the 1990s with Teesmouth's winter maxima of around 200 recorded in most but not all years. The peak count of the decade was of 342 in February 1996. Inland one to three birds were noted at scattered sites in autumn. In September 1990 passage brought six to Hurworth Burn Reservoir, four to Derwent Reservoir and then six to Smiddyshaw Reservoir. In October 1991 one to three were recorded at Hetton Lyons and at Shibdon Pond/Dunston (Armstrong 1991 & 1992).

Teesmouth remains the species' most-favoured location in Durham. There, the first returning birds, nearly all of which are adults, generally appear in late July or early August. Some of these early-returning birds may stay to moult, but others move further south, as far as western France. Here they are joined by some of the birds which have moved on after moulting, as suggested by one ringed bird being recovered in north west Spain in January. Juvenile birds begin to arrive at Teesmouth in September, some staying, but others move south again as far as western France. Observations of colour-ringed birds on Seal Sands have shown some juveniles remain to compete for feeding territories, whilst the unsuccessful birds either leave the estuary or stay to feed non-territorially (Ward *et al.* 2003).

By November most of the adults that spend the winter at Teesmouth have returned after moulting on the Wadden See shores. These birds may displace some of the juveniles that have established territories on Seal Sands which then migrate further south. The juveniles which do remain become part of the wintering adult population in subsequent years (Ward *et al.* 2003).

It has been suggested that Grey Plover have become more vulnerable to mortality in severe winter weather following population increases of some temperate zone wader species leading to greater competition in feeding areas (Clark 2002). Within Britain this may have been exacerbated by the degradation of some sites, such as Teesmouth. When freezing conditions occur on the Wadden See as early as December, more adults move west to

Teesmouth. In some years a marked influx is noted in late January and February although the wintering area origin of these birds is unknown (Townsend 1982).

Most birds leave for the Siberian breeding grounds in March. Some juveniles linger until late April or early May when a few adults in breeding plumage pass through quickly; their sun-bleached flight feathers a confirmation of having wintered further south (Ward *et al.* 2003). Very few Grey Plover are seen in the summer in Durham; recent WeBS maximum counts were of 15 on 11 June 2006 and 8 on 13 July 2008.

Inland records of Grey Plover remain by no means common in Durham but they do occur on an almost annual basis, primarily in the autumn and sightings largely involve juvenile birds. Unusual in winter, a single was at Hurworth Burn Reservoir in January 2010; whilst a summer plumage bird, inland at Hetton Lyons on 29 May 2010 was exceptional, only the second such occurrence in almost 50 years.

The seasonal mean peak counts of Grey Plover on Seal Sands over the period 1990–2001 show: 160 in winter; 13 in spring; and, 72 in autumn (Ward *et al.* 2003). Within these figures, analysis of WeBS data shows that while a total of 342 was achieved in winter 1995/1996 numbers have declined somewhat since then, with the five-year mean of winter peaks standing at 157 (as at 2011). The overwintering flock appears to have fluctuated throughout the last forty years, but has not exceeded 200 birds since winter 1998/1999. The current importance of Seal Sands for this species is exemplified by records of the species in 2010, when this site almost had a monopoly on winter records in the early part of the year with a minimum of 127 being recorded there on 28 February.

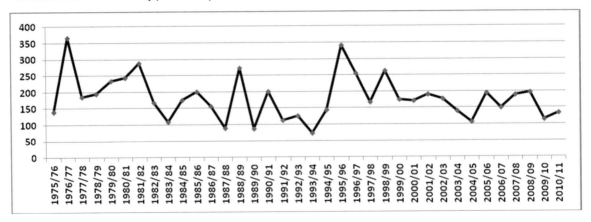

Tees WeBS Winter Maxima, Grey Plover 1975/76-2010/11

The only other site to record birds regularly is at Whitburn Steel where it is found annually but in single figures. It is probably birds from this site which occur infrequently at Boldon Flats, brought there in mixed wader flocks from the coast.

Birds are recorded regularly in small numbers on autumn sea watches at Whitburn in August and birds are noted passing Hartlepool Headland in most autumns, as they return to the British Isles from their breeding grounds. Passage counts at Whitburn associated with strong northerly winds have included 23 on 12 August 1990, 45 on 13 August 2006 and 15 on 14 August 2010. All days of notable autumn passage at Whitburn between 1990 and 2011 occurred between 11th and 22 August.

Distribution & movements

Grey Plovers breed in high arctic habitats, and have a circumpolar distribution, which encompasses three continents. Wintering birds, which can be found around the coasts of Europe, come from the northerly summer range, with many birds penetrating south into Africa and even South America. Grey Plovers migrating through, or wintering in Britain, originate from western Siberia and migrate via the Gulf of Finland and southern Baltic (Branson & Minton 1976, Evans *et al.* 1984). Moult largely takes place on the Wash and German Wadden See before birds disperse to wintering or breeding grounds. Most migrants winter in southern Europe and West Africa (Byrkjedal & Thompson 1998). It has been suggested that the distribution of wader populations within Britain may be shifting north and east (Clark 2002). As global climate change leads to reduced severe winter weather events, this trend may become more evident, with implications for the wintering Grey Plovers in the county.

About 500 birds have been ringed at Teesmouth since the mid-1970s. A juvenile ringed on Seal Sands on 16 December 1975 was seen again on Seal Sands in October 1997, aged 22, and one ringed on Seal Sands on 19 September 1978 was in Luneberg, Germany, in April 1988. One ringed at North Gare on 28 January 1979 was found in western Siberia, in July 1985 (Blick 2009). An example of extreme site fidelity was demonstrated by the colour-ringed bird that was at Greatham Creek and the tidal pools in 1996 and 1997, which had been ringed there as a juvenile 22 years previously, in the autumn of 1975.

White-tailed Plover (White-tailed Lapwing)
Vanellus leucurus

A very rare vagrant from central Asia: one record.

The only County Durham record is of a bird found by Brian Bates on 21 May 1984, on newly-sown farmland between Cleadon and Sunderland, adjacent to what is now Sunderland AFC's Academy of Light (*British Birds* 78: 545). At 16.50hrs, he was cycling around the area when he noticed a crouching plover in a distant field. After dismissing his first thought of Dotterel *Charadrius morinellus* a closer approach revealed that it was a species with which he was unfamiliar. He described it as "*a rather plain plover, with an almost white head and extremely long, lemon-yellow legs*". Telephone discussions with other birdwatchers led to the conclusion that it was a White-tailed Plover, a species with just two previous occurrences in Britain at the time (Dymond *et al.* 1989). A handful of local birdwatchers, Ian Mills, David Constantine and Peter Hogg, arrived to view the bird for the next hour or so. At about 18.10hrs it took flight, circled high over the three observers and flew off to the south west. During its stay it had been constantly harassed by the local territorial Lapwings. What was presumably the same bird was relocated at a private site in Shropshire on 24th and 25 May, although this bird was accepted as a different individual for the national totals.

Prior to the bird in County Durham there had been just two British records at Packington Gravel Pits, Warwickshire from 12th until 18 July 1975, and at Chesil Beach, Dorset briefly on 3 July 1979. Since then, there have two further occurrences, in Dumfries & Galloway and then Lancashire in June 2007, and a widely travelled bird that visited several sites in south east England in 2010 after having been initially found in late May on Merseyside.

Distribution & movements
White-tailed Plover breeds discontinuously from the Middle East, through Iran to Kazakhstan. Although birds breeding in the marshes of Iraq are mainly resident, those from the former USSR vacate their breeding areas to winter in the Middle East, north east Africa, Pakistan and north west India.

Lapwing (Northern Lapwing)
Vanellus vanellus

An abundant passage and winter visitor and a very common breeder.

Historical review
Plovers, of what species it is impossible to determine, were evidently a favourite dish and to be had in considerable quantity in Durham during the 14th century. Entries of such in the Bursar's Rolls of the Durham Monastery are numerous and birds appear to have been purchased in lots of several dozen. The earliest record is of a lot of fifty that was purchased in 1312/1313. Their price in 1390 was two pence each, which seems rather high when compared with other goods at the time and would seem to indicate the high esteem in which they were held. Records were always entered as "*pluvers*" or in a Latinized form of the same word. These records probably often referred to Lapwing, but Golden Plover *Pluvialis apricaria* is not out of the question, or it may have been a generic

term that applied to both species (Ticehurst 1923). A century and a half later, and Lapwing or *wype* were being purchased for consumption at the Durham Monastery, for example in 1531, a purchase of *"3 plovers and 1 wype, 5d"* (Benett 1844). Lapwings must have been a very well known bird to societies through the ages as references to it in both Chaucer and Shakespeare are couched in terms which expect familiarity from their audiences (Cocker & Mabey 2005).

In the Victorian period and earlier times, this was a common resident species locally, and inhabited rough and uncultivated lands, as said of it by Thomas Robson (1896), late in the 19th century, in relation to the Derwent valley. This remains the case today, though increasingly birds are forced to nest in spring cereals as other more optimal habitats are lost. That birds were then present in the region in numbers is confirmed by the observation that, on an unspecified date in October 1899, approximately 1,000 Lapwings arrived at Teesmouth in fifteen minutes (Nelson 1907).

As early as 1929 Temperley (1929) noted the demise of the *"peewit"* in the north east; this after Hancock had once stated that *"its eggs were so plentiful that it did not pay to gather them for market."* Despite such observed declines, it was, in 1931, still cited as breeding at Teesmouth by Richmond (1931). By the mid-20th century, Temperley (1951) described it in general terms as *"a resident, passage migrant and winter visitor"*; whilst stating that at the time of writing it was best known as a winter visitor; this after major declines in the breeding population over the first three to four decades of the 20th century. He identified that that it was no longer breeding in many areas where it was previously abundant, with the main decline in numbers occurring in the lowlands.

Temperley (1951) also observed that the birds that were winter visitors used ploughed fields and damp places, and that large flocks sometimes occurred on passage in October and November. He also noted that birds ringed as pulli in Durham had been recovered in Louth and Galway in the Republic of Ireland, indicating a south west vector of migration for locally-reared birds, to their milder wintering quarters across the Irish Sea.

Post-Temperley these declines appear to have continued and in the early 1960s, it was clear that this species had experienced something of a decline in its breeding strongholds of western Durham. In 1964, it was reported as being *"still scarce or absent"* from *"many areas where usually present"* (Bell 1965). However, this species was considered common and widespread in the north Teesside area throughout the 20th century (Joynt *et al.* 2008) with particularly high breeding densities occurring around Cowpen/Greatham Creek (Stead 1964).

Recent breeding status

With its plaintive penetrating song and calls, and looping song flight Lapwing is the characteristic bird of Durham's flat or gently sloping heather, grass moorland and bog. Nationally the species has suffered a marked decline. A BTO/RSPB survey carried out in 1998 suggested a population of about 63,000 breeding pairs in England and Wales (Wilson *et al.* 2001). This represents a decrease of 49% in the eleven years since the surveys for the *New Atlas* (Gibbons *et al.* 1993). In the 1998 survey, the second lowest decline was found within the north east of England and Cumbria (42%), only Yorkshire's was lower, at 28%. One study indicated that in order to replace annual adult losses, Lapwings need to produce 0.83–0.97 fledglings per pair each year. A literature review indicated that Lapwing productivity was only sufficient to do this, in eight of 24 studies (Peach *et al.* 1994). This poor state of affairs led Mead to conclude *"Outlook seems to be fairly grim but should be ok on reserves"* (Mead 2000).

Despite these large declines, Lapwings are still to be found breeding across much of Durham in good numbers. At the national level the decline have been associated with changes to agricultural management regimes, from spring tillage to the planting of autumn cereals and intensively managed grassland. Partly as a result of this, nesting Lapwings are now more dependent upon less productive grassland habitats with an increased dependence on upland grasslands across the UK. The ideal breeding habitats are produced by a mixed farming system providing some open ground (Shrubb 1990, Shrubb & Lack 1991). Silage systems, with dense sward and early mowing, are least favoured and this is a regime whose acreage has increased in Durham over recent decades. Lapwing's increased reliance on grassland has coincided with increased livestock levels in the uplands with concomitant risk of trampling and nest desertion. Nest record data between 1962 and 1999 indicated that marginal upland had relatively lower rates of reproduction than ungrazed grassland and suggested that population changes may have related to changes in clutch failure influenced by increases in grazing regimes and predation associated with habitat change (Chamberlain & Crick 2003). A study in 2007 indicated that 88% of nest predation took place under darkness, suggesting mammalian causes (Bolton *et al.* 2007).

368

Notwithstanding these concerns, Lapwing probably breed in a greater variety of habitats than any other wader species that nest in Durham. They can be found breeding alongside Redshank *Tringa totanus* and Snipe *Gallinago gallinago* in lowland wet grassland areas, such as the Teesmouth marshes; they nest on tipped slag and 'waste ground'; in old quarries and on sand and gravel extractions with Ringed Plovers *Charadrius hiaticula* and Oystercatchers *Haemantopus ostralegus*; and very extensively with Curlew *Numenius arquata* on the upland pastures in the west, and alongside Golden Plovers *Pluvialis apricaria* on heather and high moors, as well as suitable farmland across the whole county.

Locally, the return to breeding territories takes place relatively early in spring, but this is weather dependent. For example four birds on Bowes Moor on 8 January 1989 were clearly early pioneers. More concerted returns occur from mid-February in most years, by which time display flights are often evident on lowland sites. Today, many pairs are still found along the main river systems and it remains quite widespread in the eastern arable areas though nowhere in the same numbers as in the uplands. Examples of lowland concentrations have included: in 2002, 50 pairs at Bishop Middleham; 35 pairs at Houghton Gate; 18 pairs at both South Hetton and between Seaham and Murton; and, five pairs at Brancepeth Beck. In 2003, 18 pairs were at Boldon Flats and 15 pairs at Rainton Meadows. In 2005, several nests were lost due to heavy rain in early March and numbers on the Kibblesworth CBC fell to a five-year low of nine pairs (Newsome 2006, Siggens 2006, Newsome 2007). During 2007 work on the BTO BBS revealed a total of 272 pairs in 21 of 40 one-kilometre squares surveyed.

With very few exceptions, the species is absent as a breeding bird from the heavily built-up areas of the county. Gaps in its local breeding distribution are also found on the highest moorland tops, and within solid forestry plantations. In the west, most pasture and moorland fringe areas hold several pairs. The concentration of breeding Lapwing in the less disturbed Pennine uplands is evident from the *Atlas* map (Westerberg & Bowey 2000). Example densities in upland areas were: in 2006, 48 pairs in Eastgate Quarry; in 2008, 42 pairs at Thimbleby Hill Quarry; 20 pairs at Middle End; 15 pairs at Rimey Law; 13 pairs on Lintzgarth Common; and, 10 pairs in Pennington Plantation, Hamsterley.

Using some of Shrubb & Lack's (1991) north of England breeding densities extrapolated across Durham, Westerberg & Bowey (2000) estimated the breeding population to be 11,240 breeding pairs, which at the time was approximately 5% of the estimated British & Irish breeding population (Gibbons *et al.* 1993). Shrubb & Lack's (1991) survey found, on the Lapwing's preferred habitat of spring tillage in northern England, an average of seven pairs per kilometre square. Dedicated survey work by the RSPB for upland waders in south west Durham during the early 1990s, found an estimated 1,165 pairs in Lunedale, Baldersdale and upper Teesdale and in some parts, densities were between 34.3 and 46.4 pairs per square kilometre (Shepherd 1993). At the time, this was amongst the highest breeding densities in Britain and confirmed the importance of the county for the species, in the national context.

The highest density of Lapwings breeding around Teesmouth, used to be found in the fields and marshes between Seaton Carew and Port Clarence. It has been estimated that there were over 100 pairs in this area in the 1980s (Blick 2009). A survey in 1982 found 53 pairs around the Tees Marshes (Bell 1983) and in this area, numbers remain similar today (Joynt *et al.* 2008). In 2009 and 2010 it was estimated that 57 and 49 pairs respectively bred on the managed grassland at RSPB Saltholme.

The major threat to Lapwing populations throughout Britain is the continuing intensification of agriculture, together with a move away from a mixed farming system, as described above. Though, in northern England, Lapwing numbers have been comparatively stable over the past decade, changing agricultural practices are resulting in a marked downward trend in the south (Marchant *et al.* 1993).

Recent non-breeding status

The first juveniles are usually noted in early May but can be as early as late April. Survival rates vary from year to year and most juveniles have fledged by mid- to late June and post-breeding flocks of birds have started gathering by the second half of that month, with parties of several hundred strong noted in some lowland areas by late July. At coastal and wintering sites, such as around Saltholme Pools and Cowpen Marsh at Teesmouth birds are present from July, from which times flocks of up to 1,000 and more might be present (Blick 2009). Subsequently, through August and September, numbers may decline before rising again through the autumn. By this time, flocks probably include birds form further north in Britain and from the near Continent. The main wintering flocks are largely coastally based and centred on two centres, one in the south east of the county, Teesmouth, and

369

one in the north east, Whitburn. The peak counts at some inland sites occur in September/October as birds move from upland areas down to the coast, for example at Bishop Middleham with: 1,800 in 2008 and 1,200 in 2009, and at Derwent Reservoir with 200 in 1982 and 1,260 in 2009.

Around the two main wintering areas, at Whitburn, on the coastal fields and intertidal rocks, and the freshwater complex around Teesmouth, 3,000-5,000 birds are quite routinely counted at the height of winter, but much higher numbers have occasionally been noted. For example: WeBS counts at Teesmouth were of 9,800 in December 1990; 10,500 in December 1995 of which 9,361 were on the north side of the estuary; and 9,000-10,000 in November 2002 (Blick 2009). Today, the North Tees area holds by far the biggest winter concentration in the county, with the RSPB Saltholme reserve being of particular importance, holding winter peak counts of 3,163 in December 2008 and over 5,300 birds in December 2009, but less than 400 during the severe weather of December 2010.

On 27 November 1994, the Durham Bird Club surveyed the vice-county, between 08.00hrs and 12.00 noon, for Golden Plover, but in the process all Lapwing seen were also counted. This gave a snapshot of the total winter population in the county, as all of its principle winter resorts were covered. The number of birds counted was 10,143, the count probably achieving 90-95% efficiency of detection for this species. The results indicated that in the mid-nineties around 11,000 Lapwings were wintering in the county, about 0.5% of the national total at the time (Armstrong 1995). This total might be severely reduced in the onset of hard weather. According to the analysis of this count, over one-fifth of all birds seen were located on grazed pasture with other important habitats noted during the count being arable winter crops and flooded 'waste ground'.

Away from the main areas, flocks of some three to six hundred can be encountered at as many as twenty or so sites around the county, including: Bishop Middleham, Boldon Flats, Bowesfield Marsh, Derwent Reservoir, near Greta Bridge, Houghton Gate, Longnewton Reservoir, Marwood, Portrack Marsh, Preston-le-Skerne, Ravensworth Fell, Ryhope, the SAFC Academy Pools, Sedgeletch, Stanley, Stonechester and the Tyne at Dunston. Flocks of c.200 birds have been noted over several years roosting on factory roofs at Birtley, Darlington, and Belmont.

National winter WeBS counts, mainly at coastal sites, after increasing in Britain through the 1980s and into the 1990s are now decreasing steeply (Calbrade *et al.* 2010). These comprise primarily birds of continental origin and across Europe this is one of the most rapidly declining species; consequently it is now a 'red-listed' Species of Conservation Concern.

Distribution & movements

From Ireland and Britain, Lapwings breed across a range extending east through north and central Europe through Asia to China. It is a largely migratory species, though this trait is less pronounced in western populations and the southern part of its breeding range. Birds from the north winter as a far south as southern Europe, through the Middle East and into North Africa. Scandinavian birds appear to winter in Britain, as do many British breeding birds. This species is often associated with hard weather movements, triggered by long periods of frosts or snow cover.

In October and November there is an influx of birds to Durham, sometimes appearing as flocks, which arrive from over the sea at any point along the Durham coast. Some of these birds are known from ringing work to have come from northern Europe. Cold weather triggers local movements with birds seeking out better feeding and milder conditions. In such conditions, local birds often move south, some flying as far as France and Spain. Periods of prolonged hard weather leads to a marked exodus of local birds, as in the winters of 1983, 1984 and 2010, when the species was largely absent for a number of months.

Well over 1,200 Lapwings have been ringed in Durham, many as pulli and some in cannon-net catches at Teesside. There are over 30 recoveries of birds to the county that were ringed elsewhere, mainly as pulli. Birds ringed in Durham in summer have been recovered in the following winters in France, Spain and Portugal. For example, the bird ringed at Hamsterley Forest on 4 June 1953 that was recovered at Etan de Lacanau, Gironde, France, on 4 March 1956; the juvenile ringed at Stockton-on-Tees on 4 July 1953, which was recovered at Gijon, Asturias, Spain on 10 February 1954; and, the bird ringed at West Boldon on 2 June 1955 that was found dead near Celorico da Beira (Beira-Baixa), Portugal on 20 February 1956, over 1,000 miles to the south west.

A number of birds ringed as pulli on the Durham moors have been recovered in Ireland in subsequent winters. For example, a bird ringed on 11 June 1927 at Hamsterley was recovered at Dundalk Bagenalstown, Ireland on 15 February 1929.

Some of the birds recovered in Durham, and ringed elsewhere, have originated to the east and north of the county, with birds from Shetland, the Netherlands and Norway. For instance, a bird found injured at West Hartlepool on 20 December 1954 had been ringed as a juvenile at Herikstad, near Stavanger, Norway on 30 May of that year and a bird recovered on 20 October 1958 at Hetton-le-Hole had been ringed on 18 June 1958 at Sola (Rogaland) Norway. A bird found stunned by the car park at WWT Washington on 16 December 2010 had been ringed as a pullus at Kverneland, Rogaland, Norway on 16 May 2004.

Some locally-bred birds don't appear to move that far, at most up and down an altitudinal gradient from the Durham uplands to the Durham coast, as exemplified by a colour-ringed bird at Saltholme Pools in August 1993 that had bred in upper Teesdale in 1992 and 1993.

Whilst many local Lapwing return to breed in their natal areas, as indicated by the bird ringed as a pullus at Wolsingham on 21 May 1949 that was found dead near Bedburn Beck only a few miles away on 2 May 1952 and one ringed at Langdon Beck on 6 June 1958, which was recovered at Newbiggin-in–Teesdale on 15 May 1965. That said, some clearly do not and birds ringed in the county as pulli have later been found in summer, between late April and mid-June, in places as distant as Norway, Sweden, Finland and even Russia. For example, one ringed near Consett on 1 June 1957 was recovered at Sokol (Vologda) Russia on 6 May 1960, indicating that it was breeding in this location.

Great Knot
Calidris tenuirostris

A very rare vagrant from Siberia: one record.

The sole County Durham record relates to an adult found by Mike Gee on 13 October 1996, whilst undertaking a routine WeBS count on Greenabella Marsh, Teesmouth (*British Birds* 90: 471). Unable to confirm its identification that evening due to failing light, this was achieved when the bird was seen on Seal Sands the following day. It was not until 15th that the bird was well observed by a number of observers. Located at dawn in the high tide roost on Greenabella Marsh, it soon flew off but was relocated on Seal Sands shortly afterwards. This pattern of behaviour was repeated until it was last noted on 5 November, though during the latter part of its stay it often fed on Bran Sands on the Yorkshire side of the estuary. As views were often distant, it became known as the "*Great Dot*" by many observers, though it did eventually show down to a few feet on Bran Sands, allowing good photographs to be taken. This was only the second record for Britain following a bird for one day at Pool of Virkie, Shetland on 15 September 1989.

Distribution & movements
Great Knot breeds in north east Siberia and is a long distance migrant wintering from southern Asia to Australia. It is a very rare vagrant to Britain, with just three records to date, all of them adults, the most recent of which was at Skippool Creek, Lancashire on 31 July, and again on 16th and 17 August 2004 and previously in Ireland.

Knot (Red Knot)
Calidris canutus

A fairly common passage migrant and common winter visitor to coastal areas, in particular the Tees Estuary.

Historical review
The earliest documented record for Durham is of a bird shot at Sunderland in January 1814, during a severe winter, which was later used by Thomas Bewick as subject for an engraving of the species (Jessop 1999). The next

reference to the occurrence of Knot in County Durham is by Edward Backhouse (1834), who observed that it was *"very common at the Tees mouth where immense flocks usually arrive early in the autumn and continue through the winter"*, whilst some years later Hogg (1845) said that it was *"not unfrequent' upon sandy coasts"*, and *"numerous on the shores of the Tees"*.Strangely, Sharp (1816) did not include this species in his list of the birds of Hartlepool, though it will doubtless have been present.

During the early part of the 20th century, Nelson (1907) documented that there were large numbers on passage at Teesmouth, with *"enormous flights"* in October and November, with some summer plumage birds persisting into June, indicating that there had been little change in status during at least the previous 60 years. Courtenay (1933) however, observed a reduction in numbers from the *"immense flocks"* noted during 1907 to 1914 to what he described as a *"fairly large company"* during eleven visits between 1921 and 1929.

In 1931, W.K. Richmond, writing of the birdlife of Teesmouth, stated, *"Though it is most difficult to be certain of visible migration, large hosts of waders, in August especially, can be seen apparently coming in direct from the sea. Godwit, Knot, etc., arrive from about July 23rd onward, but the most considerable numbers of Dunlin, Ringed Plover and Turnstone are to be seen in mid-September"*. He went on to state that the "*enormous flights*" of Knots mentioned by Nelson (1907) were apparently a thing of the past, for although large flocks may be seen flying up and down the coast in winter, the species was poorly represented in the estuary at this time. This is further evidence of the decline first noted by the Rev. Courtenay some ten years earlier, though the true extent of which will never be known as there are no accurate counts at this time.

There is little information regarding this species from the 1940s, but Temperley (1951) still regarded it as *"a common winter visitor and passage migrant, often in large flocks"*, indicating that it remained a frequent visitor throughout, though he gave no specific counts. Writing little over a decade later Stead (1964) still considered the species an often abundant passage migrant and winter visitor, giving an average winter maximum of around 4,000-5,000 birds with numbers peaking in January and February; most having departed by March. These figures however, represent a significant decrease in numbers from the early 1950s, when flocks of 10,000 birds were considered not unusual. Stead (1964) opined that the pollution of Bran Sands on the Yorkshire side of the estuary in 1953 may have been a major factor in the decline. In this respect, a count of 6,000 at Teesmouth on 15 February 1958 was published as being *"the largest flock noted in the estuary in recent years"*. It was evidence that the winter population had recovered somewhat towards the end of the 1950s, this being double the peak count of 3,000 recorded in 1956.

The wintering population at Teesmouth remained high throughout the 1960s with an estimated 6,000 wintering in an average year, though in some years the totals were much higher. A count of 10,000-12,000 birds on Seal Sands on 14 January 1962 was the highest since the early 1950s, and a consolidation of the 6,000-10,000 present in January 1961. These high totals were maintained through the decade, but there just two further six-figure counts, an exceptional 15,000 on Seal Sands in February 1964 (Stead 1964) and 10,000 at the same locality in February 1965, with numbers declining to around 6,000 by the winter of 1968/1969.

Inland occurrences are scarce, though according to Temperley (1951) flocks occasionally visited Jarrow Slake during the winter months, with around 100 noted there during January to March 1955, and a few were also said to visit nearby Hebburn Ponds at high tide before these were drained in the late 1950s. In addition, Blick (1978) writing of the period from 1968-1973 made reference to Knot occasionally visiting freshwater localities within the Tees estuary such as Dorman's Pool; a situation that doubtless occurred historically as it does today. Further inland, the only published records prior to 1970 were of: a party of 12 at Darlington Sewage Farm, on 10 September 1928; one at Crookfoot Reservoir, from 4th to 8 September 1960; and, singles at Washington Ponds in July and August 1969.

Another noteworthy record involves a wholly white bird at North Gare on 19 December 1964, though it was not a true albino as its legs and bill were said to be of normal colouration. Another leucistic bird was at Hartlepool Headland in the winter of 2011/12.

Recent status

The main wintering areas for Knot in County Durham are still overwhelmingly the Tees estuary (including Hartlepool Bay), and to a lesser extent the coastline immediately to the north from Hartlepool Headland to Crimdon Denemouth. In addition small numbers occur annually along the entire coastline to the north with both Blackhall Rocks and Whitburn Steel being favoured localities although counts at these sites are rarely high. Peak numbers usually occur between November and March, though in most years the species is recorded year-round, especially at Teesmouth where a handful of stragglers are usually present in June.

The first returning birds usually appear at Teesmouth from mid-July, when small flocks of adult birds are to be found either feeding on Seal Sands or roosting upon Seaton Snook at high tide. Numbers are generally low at this time and rarely reaching three-figures, though 500 were at 'Teesmouth' in late July 1971 (McAndrew 1972), and more recently 109 were on Seal Sands on 25 July 2003 (Joynt 2004). Numbers continue to build into late August when the first juvenile birds appear and this continues into September with most birds at this point in the autumn using Teesmouth as a staging post *en route* to their moulting grounds on the Wash (Ward *et al.* 2003). The first wintering birds begin to arrive in late October and November and it is usually during January that the winter population reaches its height. Return migration takes place from early March onwards with just a few birds remaining in April, May and into June.

Teesmouth was once considered of international importance for its wintering Knot which during at least the 1950s was estimated at around 10,000-12,000 though the peak count, of approximately 19,000 birds came on 13 December 1970 (Blick 1978). The in-filling of large parts of Seal Sands for industry and the resultant loss of inter-tidal mudflat during the early 1970s saw the species suffer a serious decline, from which it has never fully recovered. This is most evident by the sharp drop in mid-winter peak numbers from 19,000 in 1970 to 2,000 during 1972/1973; as Seal Sands was reduced to a third the size it had been in 1969 (Evans 1979). Despite signs of some recovery during the latter part of the seventies, with numbers gradually building from 2,500 in the winter of 1974/1975 to a peak of 9,500 on 16 December 1979 (Bell 1980), this increase was not maintained and the species has been in gradual decline ever since.

Around 4,000-6,000 birds were still a regular winter feature through much of the 1980s, though mid-winter numbers had fallen to less than a quarter of this by the close of the next decade with a peak count of just 1,250 during the winter of 1999/2000 (Armstrong 2005). Some cause for optimism was provided by counts of 6,830 in January 1993 (Bell 1994) and 6,022 in December 1995 (Bell 1996), these, to date, were the last counts in excess of 5,000 birds. Despite a minor recovery having taken place again since 2000, with 1,000 to 4,000 peaks, the species remained in gradual decline reaching an all time low mid-winter count of 700 in January 2010 (Joynt 2011). The reasons for this decline are unknown but may be linked to the increasingly mild winters experienced in the UK through the 1990s and 2000s, with more birds choosing to remain on the Continent (Maclean *et al.* 2008).

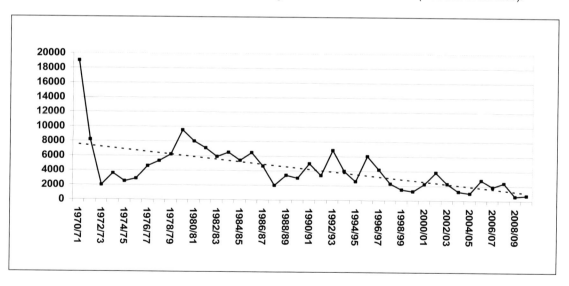

Peak winter counts (November to March) of Knot at Teesmouth 1970-2009 (with trend line)

Since at least 1985 when 200 were noted at Crimdon Denemouth on 17 February, the stretch of coastline from Hartlepool Headland to Crimdon Dene has become increasingly attractive to this species. A high count of 5,000 birds roosting on Hartlepool on 17 January 1991 (Armstrong 1992) doubtless included some of this population, as did presumably the 4,000 roosting in Hartlepool's West Harbour on 18 February of the same year (Bell 1992). Other particularly noteworthy counts include 2,000 on 1st and 2 December 1994 and 1,700 on 18 November 1998, though more recently numbers have rarely exceeded a couple of hundred birds.

In most years a handful of birds are recorded from freshwater localities around the North Tees Marshes, most frequently during early autumn, though numbers rarely enter double figures and the majority of records involve just one or two birds. Particularly good counts include: an exceptional 152 at Saltholme Pools, on 1 August 2008 (Joynt 2009); 12 at Saltholme Pools, on 1 September 2008 (Joynt 2009); and, 37 on the Calor Gas Pool, on 1 August 2009 (Joynt 2010).

At Whitburn the species is largely noted during southbound migration between July and September with peak numbers passing through between late July and the first two weeks of August. Most are noted passing offshore and typically counts rarely exceed 50 birds per day though an exceptional 325 flew south on 12 August 2004 (Siggens 2006) and the following year 119 flew south on the 29 July 2005. A record of 150 flying north on 11 November 1983 indicates that passage can continue into November though in this instance it probably referred to birds heading for wintering grounds to the north, having presumably completed their post-breeding moult further south.

Just a handful of birds regularly winter at Whitburn Steel each year, numbers rarely reaching double figures; exceptionally 330 were there in March 1997, 280 in March 1999 and 220 on 22 January 2000. In addition, an unprecedented 720 were counted during March 1996 following a period of harsh weather on the Continent which brought increased numbers of this species, Grey Plover *Pluvialis squatarola* and Bar-tailed Godwit *Limosa lapponica* to the county (Armstrong 1998).

There are very few significant wintering populations away from the Tees estuary though three-figure counts are occasionally reported from the central Durham stretch of coastline between Sunderland and Crimdon Dene with Blackhall and Ryhope being the favoured localities. Around 100 were known to regularly winter at Blackhall during the 1980s, though exceptionally 500 were noted in January 1995 (Armstrong 1997), with other notable counts being 400 in February 1998 and 440 in February 1999; numbers have rarely exceeded 100 since. At Ryhope, counts during the early 1990s suggested a typical winter population of 100-150 birds, but a cold weather movement in February 1996 saw a count of 604, in-line with greatly increased numbers elsewhere in the county (Armstrong 1998), the only notable count since is of 280 in January 1999. In addition to this small numbers used to regularly winter along the Tyne at Jarrow Slake until at least the 1970s though there have been few records since. Around 100 were present during the early part of 1973 (Coulson 1974), though by the close of the decade, 20 on 20 February 1979 was considered noteworthy and there have been no subsequent published reports. The decline of this species and other wading birds at this site can be directly attributed to land reclamation for industry, commencing in the 1950s and continuing into the mid-1970s.

Further inland, reports of Knot are scarce though they have been just about annual in occurrence since the 1970s, with an average of around three occurrences per annum since 2000. The majority of these have occurred from July to October and to a lesser extent in April and May. Records are widely scattered across the county though the well-watched wetlands surrounding Washington and Boldon Flats have accounted for over half of the sightings. Elsewhere there are eight from the reservoirs in the western uplands, including five from Derwent and two from Smiddyshaw, five from Longnewton Reservoir and four from the Houghton-le-Spring area. Perhaps surprisingly there are just three instances from the Gateshead area, two of which are from Shibdon Pond. Most observations relate to just one or two birds though exceptionally eight were at WWT Washington on 2 September 1990, and a party of seven were at Longnewton Reservoir on 28 August 2000, whilst four flying east over Shibdon Pond on 25 September 1994 and up to four at Boldon Flats during early 1996 were noteworthy.

Outside of the main migration periods there are few inland records between November and March, though exceptionally a cold weather movement during the early part of 1996 brought a total of 10 birds to Boldon Flats, Silksworth and WWT Washington, whilst just inland, 26 were near Cleadon during the same period. This same movement also brought unprecedented numbers to the coast with a combined WeBS total of 5,582 birds during March (Armstrong 1998). There are just four other mid-winter inland records: at Witton-le-Wear, on 23 January 1972; Hart Reservoir, on 16 January 1988; Boldon Flats, from 15 to 19 January 2003; and, at Seaton Pond in February 2006. Perhaps more noteworthy is the record of an adult in summer plumage at Washington from 30 June to 1 July 1973, which remains the only bird to have occurred inland during mid-summer to date.

Distribution & movements

The Knot has a circumpolar breeding distribution in the high Arctic and is a noted long-distance migrant wintering in Western Europe, West Africa, Australasia, and South America. Six races are recognised worldwide all of which have distinct breeding and wintering ranges. The vast majority of birds recorded in Britain are thought to be of the race *"islandica"* which breeds in Arctic Canada and Greenland and winters in Western Europe, though small numbers of nominate birds from north eastern Siberia have also been proven to pass through eastern England *en route* to the their wintering grounds in West Africa (Piersma *et al* 1992).

All of the birds which utilise the Tees estuary are thought to be of the race *'islandica'* and this has been quantified by using bill measurements (Tomkovich 1992, Koopman 2002) and by a number of ringing recoveries. An adult female colour-ringed at Alert, Ellesmere Island, Canada on 12 June 2003 was seen at Saltholme Pools on 1 August 2008 (Joynt 2009). There are a number of recoveries and controls from Iceland and Norway, both countries are important staging posts during migration; an example was the seven Teesmouth ringed birds controlled at Balsfjord, Norway during 12 to 26 May 1985 as part of a Durham University expedition (Bell 1986, Davidson *et al.* 1986).

Perhaps the most remarkable recovery from Teesmouth concerns a juvenile ringed at North Gare on 23 September 1983, which was controlled at Langebaan Lagoon, Cape Province, South Africa, on 14 April 1985 (Bell 1986); a distance of some 9,921km and the longest distance recovery of a British ringed Knot to date. Birds from the Siberian breeding population winter in southern and western Africa, but do not normally pass through Britain on migration, and as this bird was trapped alongside birds trapped in France and Norway in subsequent years and assumed to be of Nearctic origin, it would appear unlikely that this bird was of the nominate race *'canutus'*. However assuming it to be from the Greenland population, then South Africa was far outside its normal winter range (Ward *et al.* 2003).

Large-scale ringing of waders at Teesmouth since at least 1976 by Durham University has provided considerable insight into the sometimes complex moments of Knot, and its use of the Tees estuary in particular. The first adults arrive in late July and early August and quickly move on, heading towards the Wash to undergo a post-breeding moult. Juveniles begin to arrive during late August, and like the early returning adults do not stay long with recoveries from northern and western France suggesting that these birds quickly head south. From November and December a series of fresh influxes bring many adults to Teesmouth that have completed moult elsewhere, with some passing through before moving further north and west as winter progresses. Others remain at Teesmouth for the winter, peak numbers usually occurring in January (Ward *et al.* 2003).

There is little evidence to support any degree of winter site fidelity in this species, with many birds switching sites during an individual winter and others noted elsewhere in subsequent winters. Examples of this include: an adult ringed in Norfolk on 10 February 1990 that was controlled at Hartlepool on 7 January 1997 (Bell 1998); an adult ringed in Tayside on 11 December 1988, which was controlled at Hartlepool on 8 December 2001 (Iceton 2002); and, an adult colour-ringed at Hartlepool on 7 January 1997, which was seen in Holland on 12 January 2003 (Joynt 2004). An adult ringed on the Cromarty Firth on 17 January 1987 and re-trapped at Hartlepool on 1 December 1994 and 7 January 1997 (Bell 1998) however, provides evidence that some will winter at the same site in subsequent years. There has been some suggestion that the species performs a 'migratory circuit' beginning at the Wash with birds travelling earlier and further along the route in some winters more than others. Birds trapped at Teesmouth though have not provided evidence to support this theory (Ward *et al.* 2003).

Almost all of the Knot wintering at Teesmouth leave at the beginning of March and are known via ringing recoveries to head to the Wash, Morecambe Bay and the Wadden Sea (particularly the northern German part) where they undergo a pre-nuptial moult before migrating back to Greenland and Arctic Canada via Iceland and northern Norway during May. Birds which have wintered at Teesmouth have been controlled from both of these staging posts (Ward *et al.* 2003).

Two notable longevity records include an adult ringed on Bran Sands on the Yorkshire side of the estuary on 22 November 1985 which was killed by a cat at Hartlepool on 20 December 2001 (Iceton 2002) and an adult colour-ringed at Hartlepool on 7 January 1987, which was later seen at Den Helder in the Netherlands on 12 January 2003 (Joynt 2004). Both of these birds had reached at least 17 years of age, against a British longevity record of 27 years and four months (BTO online).

Sanderling
Calidris alba

A common passage migrant and winter visitor to coastal areas.

Historical review

The earliest documented record of Sanderling in County Durham comes somewhat inadvertently from George Temperley's text on the Wood Sandpiper *Tringa glareola* (Temperley 1951). This tells of how Durham County's first Wood Sandpiper was shot at White Mare Pool; a locality close to Easington on the Durham coast. The account tells how *"it was running around on the edge of the pool in company with a Sanderling when both were killed with one discharge."* These birds were said to have been shot by a Jon. Richardson of Newcastle-upon-Tyne in 1826 and were retained in the private collection of Edward Backhouse.

Further historical records of Sanderling are scant though Tristram (1905) wrote that *"it had been shot several times in June in full summer plumage at Seaton and the Tees mouth"*, whilst Nelson (1907) noted that, *"during its migration down the coast in autumn, it was at times quite as numerous as the Dunlin and Ringed Plover in the Tees district."*, indicating that it was by no means a scarce visitor to the region at the time. The few early counts that do exist include a party of 120 birds at Teesmouth by J. Bishop on 8 June 1933 and a flock of up to 150 on the shore at Seaton Carew by R. D. Sistern during the winter of 1945/1946. Although neither of these records is exceptional by modern standards, both represent significant individual parties of birds.

Temperley (1951) described the Sanderling as *"a common passage migrant and winter visitor in small numbers"* noting that the species exhibited strong southward passage through the county in August and September with a less marked return movement in May and June. In addition to this he commented that the species was most numerous at Teesmouth, being less common along the rest of the Durham coast and that there were no records from Jarrow Slake, just upriver of the mouth of the Tyne.

Stead (1964) gave the normal winter population at Teesmouth as around 200-300 birds though did stipulate that the majority were to be found on Coatham Sands on the Yorkshire side of the estuary. Like Temperley (1951) before him, he also made reference to the strongest passage occurring in September when flocks of up to 500 birds could be found including many juveniles. Impressive counts of up to 500 at Teesmouth on 26 May 1957, and 22 May 1960 were especially noteworthy at the time for being in spring, though modern day records have shown that peak numbers of Sanderling today pass through Teesmouth during late May, not September as documented by Temperley and others.

In 1969, Stead considered the Sanderling to be *"rather less numerous than in the years prior to 1962"*, which is borne out by a peak count of just 85 at Seaton Snook on 21 July 1963. Just two years later numbers had doubled to a maximum count of 159 at Seaton Carew on 5 December 1965, a significant increase though still less than half of what would normally be expected to winter around the Tees Estuary. These reduced numbers at Teesmouth were to continue until the end of the decade with annual maxima of 155 at Seaton Carew on 25 December 1967 and 150 between Seaton Carew and North Gare on 16 November 1968. The reasons for this decline are not known though may have been linked to increased human disturbance at this popular seaside resort.

Historically, the only published records of Sanderling to have occurred inland within County Durham were of an individual shot near Newbiggin-in-Teesdale sometime prior to 1885 (Backhouse 1885), and a party of three with Ringed Plover *Charadrius hiaticula* at Boldon Flats on 16 April 1953. Temperley (1951) however, wrote that it had *"rarely been observed inland"* indicating that these were unlikely to have been the only records. Writing of the period from 1968 to 1973, Blick (1978) also commented that *"in May each year, a few birds are seen on the marshes principally Dorman's Pool"*; a situation which persists to date and there is no reason to doubt that this would have occurred in the past.

Recent status

Particularly during the winter months this charismatic wader can still be found on almost any section of sandy beach between the mouth of the Tyne and the mouth of the Tees, although the beaches at Hartlepool North Sands, Seaton Carew, and between South Shields and Whitburn support the bulk of the wintering population in County Durham. Sanderling is now recorded through much of the year in the county. Although it is typically absent for five to six weeks during mid-summer, June records are not exceptional; particularly during the early part of the month.

The first adults normally appear during July, with the population supplemented by many juvenile birds from September onwards, a large proportion of these spend the winter in the area. Numbers generally reach their peak in February before the birds begin to depart in March. In addition a secondary peak frequently occurs during May when the remainder of the local population is supplemented by birds passing northwards through the county *en route* to their breeding grounds in the Arctic, and it is often this month which sees the highest counts of the year.

The beaches to the north and south of Hartlepool have long been the most favoured localities for this species and this is a situation remains unchanged, though numbers here, formerly considered of national importance, have declined markedly in recent years; particularly since 2000. A record low maximum of just 60 occurred during the winter of 2009/2010 (WeBS). The reasons for this are not known, although the world population of the species is estimated to have declined by up to 30% during the last ten years (Birdlife 2009). Climate effects and an increase in human disturbance at favoured sites may all be contributory factors behind the decline.

The exact numbers of Sanderling present during the winter months is difficult to assess in some years as many of the published counts simply refer to 'Teesmouth', particularly during the 1970s when a peak count of 705 was made on 13 December 1970 (Blick 1978). From a County Durham perspective these counts do not represent an accurate picture as in most years more than half of this total will have comprised birds counted on the Yorkshire side of the estuary with Coatham Sands being an important wintering area for this species. A more representative estimate would be of a winter population of 200-300 birds throughout much of the 1970s and 1980s with 350 at Seaton Carew on 26 February 1972 (Coulson 1973). Similarly, many of the counts made since are unrepresentative with most published records listing BOEE and WeBS totals for the given month; the majority of these counts being undertaken on a Sunday morning when human disturbance on the county's beaches is at its height and the number of birds recorded are correspondingly low. A real decline is however evident with a sharp drop in the numbers wintering in Hartlepool Bay since 1991 from a peak of 275 at North Gare Sands on 17 February (Bell 1992) to around 80 in the winter of 1992/1993; a position which has been maintained to date with a typical winter population of around 100 birds.

There is however some evidence to suggest a minor change in wintering areas in the North Tees with increasing numbers to be found wintering along the beach between Crimdon Dene and Hartlepool Headland since the late 1980s. Earlier reports suggest a population of around 100 birds here in a typical winter during the 1970s (Coulson 1974) though during the 1980s counts of 230 on 23 March 1985 (Bell 1986) and 245 on 27 February 1988 suggest an increase. This was maintained through much of the 1990s with a typical wintering population of around 250-300 birds though 610 were present on 29 November 1990 (Bell 1991). This is the largest single winter count to date. Since 2000 numbers have been fewer, in line with the general decline of the species locally though typically around 80-100 still winter in the area.

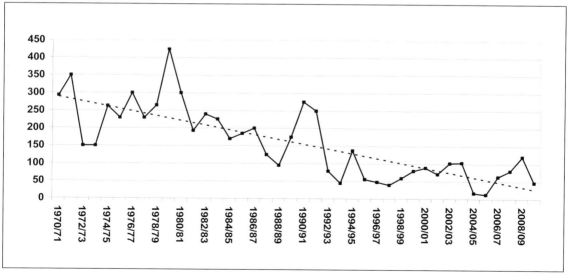

Maximum winter counts (November to March) of Sanderling in Hartlepool Bay, 1970-2009 (with trend line)

The increased human disturbance along the beach at Seaton Carew following the renovation of the seafront and the removal of several sewerage outflow pipes may have influenced this marked shift in wintering area in recent years, with relatively small numbers to be found at Seaton Carew, the majority of which congregate around the outflow pipe at Newburn.

In most years an occasional bird is reported from the numerous freshwater pools around the North Tees Marshes, most frequently during May when passage is at its height. The majority of these records involve single birds, however noteworthy exceptions include: 54 on Dorman's Pool, on the late date of 7 June 1975 (Bell 1976); 50 on the Long Drag, on 17 May 1992 (Bell 1993); and, 12 on the North Tees Marshes, on 12 May 1996 - comprising eight on the Reclamation Pond and four on Saltholme Pools (Bell 1997).

The county's coastline between South Shields and Sunderland also supports a healthy population of Sanderling particularly during the winter months. Throughout much of the 1970s and 1980s Whitburn supported a wintering population of around 100 birds though in some years such as the winters of 1973/1974 and 1976/1977 numbers were much lower with no counts in excess of 60 being made. Particularly high winter counts from Whitburn during this period included: 180 on 19 November 1972; 150 between December 1978 and February 1979; and, 190 on 29 November 1980. These were eclipsed by a total of 325 on 2 January 1987 (Armstrong 1988), the highest mid-winter count from the site to date.

Numbers have declined in recent years, particularly since 1990 though there is limited evidence to suggest that some of the wintering population may have switched to nearby South Shields, where numbers have generally increased since the late 1980s. Evidence of such local movement is provided by one or two colour-dyed birds noted at both sites during November and December 1984 (Baldridge 1985). Prior to 1980, a wintering population of around 60-80 was considered average though counts in excess of 100 birds have been made on five occasions since: 150 on 2 January 1987; 112 on 26 January 1991; 126 on 9 January 1994; 125 in January 2000; and, 150 in March 2009. The latter count could conceivably have involved some passage birds. The count of 150 in January 1987 is especially noteworthy in that it was made on the same date that a record 325 was counted at nearby Whitburn though intriguingly there were no greatly increased counts from the Hartlepool area around the same time.

To the south, there are few noteworthy records from the central Durham coast between Sunderland and Blackhall, despite an apparent abundance of suitable habitat. The paucity of records may in part be related to poor observer coverage along this under-watched stretch of coastline especially during the winter months. Notable counts from this stretch include: 140 at Seaham, during the early part of 1986; 51 at Horden, in November and December 1994; 70 at Hendon, in February 1999; and, 101 at Horden, on 26 February 1999.

Sanderling is the rarest predominantly coastal wader to occur inland in County Durham with just 21 records since 1970, the majority of which involve single birds in May and July. Records are widely scattered throughout the county with five sightings coming from the reservoirs of the western uplands including two from Derwent Reservoir, four from the Washington area, three from both Hetton Lyons and Longnewton Reservoir and two from the well-watched Hurworth Burn Reservoir. Just two records have involved more than one individual, these being two at Blackton Reservoir in mid-May 1993 and two at Selset Reservoir on 24 May 2009. Just one bird at Barmston Pond, Washington on 3 August 1995 was specifically aged as a juvenile, which is an exceptional date given that most juveniles do not arrive until September, though it is likely that the individual at Hetton Lyons on 20 September 1994 also related to this age class. Just two birds are known to have penetrated significantly upriver along the Tyne, being on the banks of the River Don at Jarrow on the unusual date of 17 February 1985 and at Shibdon Pond on 6 August 2011, whilst a single at Crookfoot Reservoir on 30 November 1993 remains the only other record out with the main migration period.

Distribution & movements

Sanderling has a circumpolar breeding distribution in the high Arctic which includes northern Canada, Greenland and north eastern Siberia. An extremely long-distance migrant it spends the winter around the coastlines of Western Europe, The Americas, Africa, southern Asia and Australasia, with those wintering in Britain thought to originate from both the Greenland and Siberian populations.

Over 2,000 Sanderling have been trapped and ringed at Teesmouth since the mid-1970s predominantly as part of ongoing research by both Durham University and the WWT Advisory Trust (Blick 2009). This research has suggested that the vast majority of the birds which utilise the Tees Estuary are probably of Nearctic origin and there is as yet no data to indicate that any of those trapped, to date, have their origins in Siberia, either by way of ringing recoveries within the breeding area or from along the proposed migration route (Ward *et al.* 2003).

Sanderling which breed in the Nearctic are known to pass through Iceland and southern Norway on migration each spring and autumn, and it is this migration route which provides much of the evidence regarding the origin of the birds seen in County Durham. One of the earliest recoveries involves a bird ringed at Jaeroen, 30km south east of Stavanger, Norway on 24 September 1939 which was found dead at Hartlepool 7 October 1939 (*The Vasculum* Vol. 26: 1940), also giving indication of the pace of the species' southbound migration. There are also at least five records of birds having been ringed or controlled at Sandgerdi in south western Iceland and being subsequently ringed or controlled on the coastline of County Durham. These involve records in both spring and autumn from the beaches of South Shields (Newsome 2010), Crimdon Dene (Bell 1997, Iceton 1999), and Hartlepool North Sands (Joynt *et al.* 2008) all of which support a healthy population of the species during winter and on migration.

The first adult birds arrive at Teesmouth from mid-July, some of which undergo a post-breeding moult, whilst others move quickly through heading for more distant wintering areas, which extend as far south as Ghana.

Juvenile birds begin arriving from August, many of these have been proven to be wintering at Teesmouth in subsequent years, and a further influx of birds usually takes place from late October. These birds arrive having undertaken a post-breeding moult on the Wash before heading north to winter along the coasts of north east England (Ward *et al.* 2003).

Colour-ringing studies indicate high site fidelity for both moulting and wintering Sanderling at Teesmouth with a high degree of consistency between the distribution of individual birds (Gilbert 1991), though at least two birds sighted on the Wash during their first year have been noted at Teesmouth in subsequent winters (Cooper 1987). Further indication of this is given by an adult ringed at Teesmouth on 7 November 1980 which was controlled at the same locality on 5 December 1993 by which time it was at least 14 years old (Bell 1994).

The majority of wintering birds begin to depart from March onwards having undergone a pre-nuptial moult before heading northwards to their breeding grounds in the Nearctic, although there is some evidence to suggest that some of these birds also utilise the Wash during their spring migration having wintered further north at Teesmouth (Ward *et al.* 2003). Admixed with the departing winter population, a strong spring passage occurs from April usually reaching its peak in late May, consisting of birds which have wintered well to the south. A bird which was trapped at Teesmouth on 21 May 1985 was seen just three days later on Handa Island off north west Scotland (Bell 1986) giving clear indication of both the speed and direction of the species spring migration. Interestingly this bird had been originally trapped and colour-ringed as a juvenile at Seaton Carew on 10 September 1984.

Spring passage reaches its height during May and it is this month which usually provides the highest count of the year. There are at least five records in excess of 500 birds, all from Hartlepool Bay and Seal Sands. These were: 640 on North Gare Sands, on 15 May 1976 (Bell 1977); 650 at Seaton Snook, on 13 May 1979 (Bell 1980); 795 at Seaton Snook, on 17 May 1992 (Bell 1993); 566 on Seal Sands, on 18 May 1993 (Bell 1994); and, a record flock of 925 at North Gare Sands, on 12 May 1995 (Bell 1996). The latter is the highest individual count of Sanderling recorded in Durham and was counted by the Durham University Wader Study group during darkness using an image intensifier. That these peak counts should come within a seven-day period neatly illustrates the relatively narrow time frame in which peak northbound migration occurs.

Outside the winter period there is clear evidence to suggest that substantial numbers of Sanderling pass through the Whitburn and South Shields area on their northbound migration particularly during late March and early April. A striking example is provided by the count of 360 made at Whitburn on 30 March 1982 when just 10-20 had been present in the preceding days (Baldridge 1983). Further examples include, an increased count of 144 at Whitburn on 7 April 1989 (Armstrong 1990) and an influx of 140 to Whitburn on 2 April 1992, following three days of wind and rain (Armstrong 1993).

Autumn passage is less intense and usually begins in July when flocks of birds can be seen passing southwards along the coast at Hartlepool often in some numbers with 135 south on 22 July 1972 culminating in a particularly large count of 481 in the estuary the following day (Coulson 1973) and in 1978 a count involved 209 south at Hartlepool on 31 July (Unwin 1979). More recently, 205 at Seaton Carew on 29 July 1984 (Bell 1985) is the only July count to have exceeded 200 birds. Numbers continue to increase as the autumn progresses with an estimated 600 at 'Teesmouth' in August 1977 (Bell 1978) providing a good example of this. A reduction typically occurs in October when the majority of migrating birds have departed to the south, until in November the population increases again as new birds arrive to spend the winter. Autumn passage in the north of the county is less well documented, presumably as it is difficult to differentiate between birds merely passing through and those arriving to winter in the area, though a count of 181 was made at South Shields on 31 October 1981 with just 51 present at the year's end (Baldridge 1982).

Semipalmated Sandpiper
Calidris pusilla

A very rare vagrant from North America: three records.

The first Durham record was identified by Tom Francis on 8 May 1989 at Saltholme Pools, Teesmouth; it had been seen the previous day, but had been believed to be a Little Stint *Calidris minuta*. Based on his experience of the species in the USA he realised that it was a summer plumage adult, not only the first for County Durham, but the first spring record in Britain. It remained in the area until 10 May and occasionally visited Dorman's Pool. An adult was also seen briefly on the Long Drag, Teesmouth on 23 July 1989, before being relocated on nearby Greenabella Marsh, where it was present until 25th. This bird was accepted by BBRC as being the same bird as in spring on return passage, an event which has never been replicated by any other vagrant Nearctic wader in Britain (*British Birds* 85: 526).

All records:
1989	Saltholme Pools and Dorman's Pool, Teesmouth, adult, 7th to 10 May, same Long Drag and Greenabella Marsh, Teesmouth, 23rd to 25 July
2002	Seal Sands and Dorman's Pool, Teesmouth, adult, 14 July (*British Birds* 96:567)
2006	Saltholme Pools, Teesmouth, adult, 5th to 11 July (*British Birds* 100: 711)
2011	Saltholme Pools, Teesmouth, adult, 31st July to 5 August (*British Birds* in prep.)
2011	Saltholme Pools and Greatham Creek, Teesmouth, juvenile, 22 October to at least 26 November (*British Birds* in prep.)

Distribution & movements

Semipalmated Sandpiper, Saltholme Pools, October 2011 (Ian Forrest)

Semipalmated Sandpiper breeds on the tundra of north Alaska and Canada, migrating southwards through the Great Plains and down the eastern seaboard of America to winter in the Caribbean and South America. It is a regular vagrant to the British Isles amassing a total of 99 records by the end of 2010. Most of these relate to juveniles in September and these have a strong westerly bias. Many of the records from the east coast of Britain are of adults which have presumably crossed the Atlantic in a previous autumn. This pattern is replicated in County Durham with four records being of adults found at a time of year when a direct Atlantic crossing was unlikely. The only juvenile was part of an unprecedented influx of over 20 birds to the Britain in the autumn of 2011 following a number of Atlantic storms, with many more recorded in Ireland.

Little Stint
Calidris minuta

A scarce passage migrant, typically recorded from May to October, though most numerous in August and September.

Historical review

The species has a long association with County Durham based on the evidence of historical information. A large number of 'stints' were purchased for consuming at the Monastery of Durham, with a number of entries referring to the period 1531-1532 amounting to a total of 330 birds (Benett 1844). It is distinctly possible, that this name covered a range of small wading birds, but not Dunlin *Calidris alpina*, as these were documented as costing twice as much as a 'stint'. Unfortunately, no precise season for these purchases was documented, so this cannot be used to infer an identification of the species referred to (Gardner-Medwin 1985).

The earliest specific reference to the occurrence of Little Stint in County Durham is by Sharp (1816) who included it in his list of the birds of Hartlepool though he did not comment upon its status. Several years later Selby (1831) said that it was *"seldom met with"*, whilst Backhouse (1834) gave two records, singles shot at Hartlepool in the autumn of 1830, and at Sunderland *"a few years ago"*. Hutchinson (1840) however, was the first to make reference to the species occurring at Teesmouth, stating that it was, *"occasionally shot on the sands at the entrance to the river Tees"*.

Nelson (1907) was the first to describe the species' status as a passage migrant through Teesmouth noting that *"it occasionally occurred in the Tees area in May and June, generally singly and never in flocks as it autumn"*; a pattern which exists to date. He noted particularly good passage in 1881, 1887, 1889, 1892, 1894, and again in 1903, when flocks of 40-50 birds were recorded around the Tees Estuary. Conversely the Rev. G. F. Courtenay recorded it just twice at Teesmouth between 1907 and 1929, a single on 20 September 1922 and three on 27 August 1928, though J. Bishop referred to it as an *"irregular passage migrant"*, recording a flock of eleven in August 1931.

At the mid-point of the 20th century Temperley (1951) said that it was *"an irregular autumn passage migrant in small numbers"*; *"very occasionally met with in spring"*, indicating that little change in status had occurred during the fifty years before his commentary. He summarised that September was the peak month for passage and that the majority of birds at this time of year were juveniles, usually in small numbers amongst Dunlin *Calidris alpina* and Ringed Plover *Charadrius hiaticula*, though occasionally in much larger flocks.

Over a decade later Stead (1964) described the species as *"a regular passage-migrant to the estuary and marshes"* usually in small numbers from August to October and occasionally in April and May. Writing of the period from 1954 to 1961 he made reference to the species having occurred in spring *"in 4 out of the last 8 years"* (Stead 1964) and some years later that it was annual in spring at Teesmouth, from 1962 to 1967 (Stead 1969). The maximum count was a flock of 12 on Cowpen Marsh on 6 June 1960; still the largest spring count to date, it coincided with a record autumn for the species.

Autumn 1960 brought an unprecedented influx of almost 300 birds into the county, an event yet to be repeated. From 18 August to 14 September, an estimated 18 to 30 birds were present, predominantly at Teesmouth. Easterly winds, with fog and low cloud, on 17 September saw numbers increase dramatically with an estimated 200 birds at Teesmouth with another 88 reported from inland sites around the county. This same weather system also produced a huge fall of migrant passerines along the coastline. Numbers of Little Stints remained high until late September with the largest individual flocks being 113 at Saltholme Pools on 24th and 75 at Coxhoe Sewage Treatment Works on 17th. The flock at Saltholme Pools is easily the largest recorded in the county to date, whilst the party at Coxhoe is the greatest number recorded at an inland locality.

Two further influxes occurred during the late 1960s though neither was on such a grand scale as that of 1960. In 1967 a peak of 50 were present on the North Tees Marshes on 19 September, including a single flock of 43 on Dorman's Pool, with up to 12 birds noted at Washington Ponds around the same time. This event was repeated in 1969 with 47 at Saltholme Pools, Teesmouth on 14 September, and five at Washington Ponds from 19th to 21 September. To put these influxes into perspective, the peak count at Teesmouth during September 1966 was of just six birds.

Outside of the period from May to October the species is extremely rare in the county with just a handful of mid-winter records. The first was of a party of three at Hartlepool on 19 February 1950 (Stead 1964), followed by singles on Seal Sands, Teesmouth on 2 January 1966, and Greenabella Marsh on 26th to 27 February 1966. The latter two presumably relate to the same bird, which was also thought to be responsible for an early record at Dorman's Pool, Teesmouth on 3 April 1966.

Temperley (1951) considered the species to be rare inland stating that there were no records for the well-watched Jarrow Slake, although it had been seen at nearby Hebburn Ponds on occasion. In addition to this he noted that it had occurred at Darlington Sewage Farm occasionally, quoting a record of a single bird on 30 September 1927. Post-Temperley single birds were at Primrose Ponds, Jarrow on 1st to 7 September and 16 September 1953, and at both Crookfoot and Hurworth Burn Reservoirs during the autumn of 1959. The influx of September 1960 saw many more recorded inland as well as the record flock at Coxhoe. Multiple records at Washington Ponds during the Septembers of 1967 and 1969 coincided with significant influxes to the North Tees Marshes.

Recent status

There appears to have been little change in the status of Little Stint within County Durham during the last 100 years, it remains a scarce passage migrant to the region, chiefly in autumn though a handful of birds are usually noted each spring in an average year. It is subject to periodic influxes during autumn which occasionally deliver three-figure counts though no more than thirty are recorded in a typical year. The North Tees Marshes provides the bulk of sightings in the county, a situation which has existed since at least the Victorian era, though the species can be encountered on almost any suitable wetland particularly during times of influx.

The vast majority of spring sightings come from the North Tees Marshes where it is recorded annually in May and June, being absent in just one spring, 1983, since 1970. In most years just one or two are noted, though flocks of up to four have been recorded on seven occasions since 1970, with the peak spring counts being of five at 'Teesmouth' on 20 May 1972 (Coulson 1973) and five on the Long Drag on 16 June 1986 (Bell 1987). A single bird on the Long Drag on 3 May 1987 (Bell 1988) was the earliest spring migrant during the period 1970-2009 but most are recorded between late May and the first half of June, although records persist until the month end. Some of these 'late' records involve small flocks, with the largest being a party of four on the Long Drag from 23rd to 26 June 1992 (Bell 1993), and it is difficult to ascertain whether these are late stragglers or the vanguard of the autumn migration, which typically begins in July.

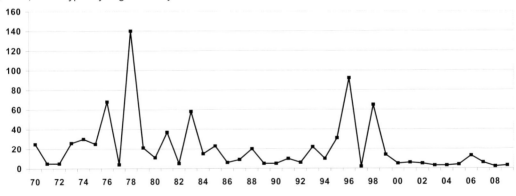

Peak autumn counts of Little Stint at Teesmouth 1970-2009

Away from the coast, Little Stint is very scarce during spring with the most records occurring in the well-watched Washington area in the 1970s. The species was almost annual there at this time, being absent in the springs of 1976 and 1978, the majority of records relating to single birds in May and June; although four were present from 18th to 21 May 1973 (Coulson 1974) and 27 May 1977 (Unwin 1978). Since the loss of Barmston Pond there has been just one spring record in this area, from WWT Washington on 13 June 1989. Elsewhere there have been just five such records, all since 2000. These were singles at: Hetton Lyons, on 22 May 2001, 5 June 2002 and 25 May 2003; Seaton Pond, on 28 June 2006, with two there on 26 May 2001, one remaining until the 28th.

382

During autumn migration the Little Stint is much more numerous than it is in spring with records spanning from July through to December, though peak numbers usually occur in August and September when numbers are bolstered by many juvenile birds. Peak numbers frequently occur during periods of inclement weather and light easterly winds, and are often associated with influxes of other passage waders and to some extent migrant passerines on the coast. The North Tees Marshes is the most important area for this species, with a handful of birds typically found elsewhere in most autumns, not infrequently well inland of the coast. July usually sees the first adult birds arrive at Teesmouth, most frequently from mid-month, although in many years the first does not appear until early August.

The majority of records relate to single birds present for just a day or two before they move on. Exceptions include 12 on 8 August 1975, the largest number of presumably adult birds to be recorded in the county. Juvenile birds typically appear at Teesmouth during mid- to late August and continue to build in numbers into September when the annual peak is usually reached. In an average year around 15-20 birds are recorded, often loosely associating with both Dunlin *Calidris alpina* and Curlew Sandpiper *Calidris ferruginea*. In common with the latter species, Little Stint passage has been low since 2000 with just one count in excess of ten birds, 13 at Saltholme and Dorman's Pools on 27 September 2006.

September 1978 saw a considerable influx to Teesmouth, the largest since autumn 1960. One or two were present from 24 July until 7 September when 12 arrived at Saltholme Pools rising to 140 on the Reclamation Pond on 9 September (Bell 1979). More than 100 remained in the area until the 13th, after which numbers declined quickly, few remaining into October; the last on Saltholme Pools was on the 28th. This influx was associated with a with a low pressure system moving through the North Sea, that brought a small fall of passerines to the coast and large numbers of other waders at Teesmouth, including Curlew Sandpipers *Calidris ferruginea*, Ruff *Philomachus pugnax* and Spotted Redshank *Tringa erythropus* (Unwin 1979). There have been just four further counts in excess of 50 since 1970, these being: 68 at Seal Sands, on 4 October 1976; 58 on the Reclamation Pond, on 13 September 1983; 92 there on 27 September 2006; and, a combined count of 65 on the North Tees Marshes on 6 September 1998.

Relatively few birds linger beyond October and there have been just ten November records since 1970, the majority of which have been in the first few days of the month. Notable exceptions are single birds at Saltholme Pools on 12th and 16th November 1996. The most outstanding record was a first-winter bird found at Monsanto on 18 November 1984 that later moved to Dorman's Pool, remaining until 31 December, though was not seen the following year. This was the first December record since 1975 and one of a few winter records for the county as a whole.

Coastal records away from Teesmouth are scarce though typically one or two are seen each autumn amongst other waders at Whitburn, most frequently on the beach at Whitburn Steel where a peak count of five was made in September 1984 (Baldridge 1985). Records elsewhere are few and far between, though singles were at Castle Eden Denemouth on 25 September 1975 and Blackhall Rocks on 29 August 2008. Unusually a party of seven frequented the beach adjacent to Roker Pier for much of September 2004 and a few kilometres inland six different birds were on floodwater along Cleadon Lane, Whitburn between 6th and 30 September 2008.

Records inland are more frequent during autumn though there are rarely more than half a dozen individual records in any given year. In a direct parallel to the North Tees Marshes the first adult birds typically appear in mid-July with the earliest being at Washington on 12 July 1977 (Unwin 1978) with peak passage occurring from late August until mid-October and mainly consisting of juvenile birds. Just one bird has lingered into November, at Washington in 1975 with undoubtedly the same bird at nearby Barmston Pond from 2nd to 4 December (Unwin 1976). An exceptional mid-winter occurrence involved two birds at Boldon Flats on 27 December 2005, the first December record in the county since 1984.

During the 1970s the vast majority of inland records came from the wetlands immediately surrounding Washington where the species occurred regularly. Elsewhere, records are fairly evenly spread throughout much of the eastern portion of the county with Bowesfield Marsh, Bishop Middleham, Hurworth Burn Reservoir, Longnewton Reservoir and the Houghton-le-Spring area all attracting birds. Most records involve one or two birds though occasionally parties of up to five have been recorded. Much higher counts include: 16 at WWT Washington, from 23rd to 29 September 1973; 11 at Hurworth Burn Reservoir, on 7 October 1973; 12 at Barmston Pond, in late August 1975; eight at Longnewton Reservoir, from 25th to 26 September 199; and, 13 at Barmston Pond, on 13 September 1995. All of these were eclipsed by a count of 37 at Barmston Pond during an influx in September 1996, which brought an estimated 160 birds to a minimum of twelve sites throughout the county (Armstrong 1998) including at least eight at nearby WWT Washington.

Further inland in the western half of the county records are much scarcer with only four sightings concerning single birds at Derwent Reservoir in September 1977, 1984 and 1990, and an adult and a juvenile at Smiddyshaw Reservoir on 17 September 1995. Inland along the River Tyne, the species is also surprisingly scarce with just two records from Gateshead Borough, both of which are from Shibdon Pond with a single on 11 September 1989 and up to five during the influx of September 1996, whilst closer to the coast singles were at Jarrow Slake in September 1989 and 2001.

Distribution & movements

Little Stint is a fairly common breeding species of the arctic tundra from northern Scandinavia eastwards across northern Russia to Siberia as far east as the Taimyr Peninsula. It is a long-distance migrant with the majority of the population wintering in Africa, chiefly along the coasts of the Indian Ocean and to a lesser extent the Indian subcontinent itself, though small numbers also regularly winter in southern Europe.

A juvenile trapped and ringed on the extreme southern tip of Norway at Tjorveneset, Vest Agder on 22 September 1984 and re-trapped just seven days later at Dorman's Pool, Teesmouth on 29 September 1984 (Bell 1986), gives a clear indication of the rapid southward dispersal of this species from its breeding grounds in northern Scandinavia *en route* to winter in Africa. This is the only Little Stint to have been trapped within the county to date.

Temminck's Stint
Calidris temminckii

A scarce but annual passage migrant from north east Europe, recorded from April to October, most frequently in May.

Historical review

The first documented record comes from the first half of the 19th century from Edward Backhouse (1834), who said it was rare, though one had been shot at Seaton Snook in the autumn of 1833, having been *"in the company of a small flock of grey plover"*. There is also an old record, from what is now Gateshead; a single bird was shot on the King's Meadows, an island in the middle of the Tyne below Blaydon, on 25 May 1843, a typical date for a spring passage bird. There is then a long gap in documented records until a single bird was present at Saltholme Pools, Teesmouth from 30 August to 3 September 1954, though Tristram (1907) had described the species as *"a very rare autumn visitor"*. Through the 1960s, further sightings came from Teesmouth: at Cowpen Marsh, on 17th to 20 May 1963; Cowpen Marsh, on 19 June 1963; Seaton Snook, on 15 September 1963; Dorman's Pool, on 9 July 1966; Dorman's Pool, on 22 July 1967; and, at Saltholme Pools, from 5th until 24 October 1969. This last record remains the latest ever for the county. Barmston Pond near Washington was also established as a regularly visited site in the late 1960s with single birds on 22nd to 23 May 1967, 17 August 1968 and from 3rd until 10 August 1969, with two birds present on 5th.

Recent status

Between 1970 and 2011, a total of approximately 180 individuals were recorded in County Durham. Of these, 135 were in spring with the earliest records being singles at Saltholme Pools from 22 April to 6 May 2003, and at the same location on 22 April 2004. There is only one additional April record, again from Saltholme Pools, on 27 April 2002. It is tempting to speculate that all of these records might relate to the same annually returning bird.

The vast majority of records come from the North Tees Marshes where it is virtually annual in spring, though records are influenced by suitable weather conditions, in May and June. The species has failed to appear in spring on the North Tees Marshes in just seven years since 1970, these were: 1970, 1984, 1994, 1995, 1997, 1998 and 2001. The peak passage occurs in mid- to late May and birds often occur in small parties, the peak count being four on the Long Drag on 14 May 1987. Records continue until late June with the latest being at Monsanto Option on 27th to 28 June 1982.

Temminck's Stint was recorded regularly in the Barmston Pond/WWT Washington area, particularly in the 1970s and 1980s when there were just eight blank years out of 20. Since then, however, it has only been recorded twice, both times at WWT Washington; on 13th to 14 May 1992 and 6th to 8 June 2010. The reason for the decline in records can be attributed to the scarcity of suitable habitat in the area, particularly at Barmston Pond, where an increase in water levels has resulted in a lack of suitable feeding areas.

The species is much less frequent on autumn passage with a total of 46 individuals recorded since 1970, although with 'spring' birds having occurred as late as 30 June, the cut-off point is somewhat arbitrary. The earliest 'autumn' record relates to a single bird on the Long Drag, Teesmouth, on 2 July 1982. Peak passage occurs in July and August, and unlike spring records, usually relate to single birds, although two were at Barmston Pond on 12 August 1974. There have only been five September records, the latest of which was on the Long Drag, Teesmouth on 17 September 1993, though a winter plumaged adult lingered at Saltholme Pools, Teesmouth from 4 September until 4 October 1999.

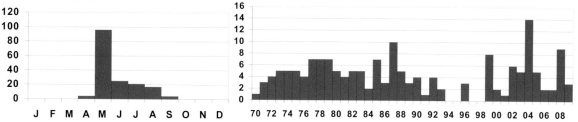

Records of Temminck's Stints in County Durham 1970-2009

Temminck's Stint is a rare bird away from the now established sites of the North Tees Marshes and to a lesser extent the Barmston/Washington area. During the review period (1970-2011), there have been just 12 records away from these favoured localities. These were at: Hurworth Burn Reservoir, on 12th to 14 May 1973 and 19 June 1988; Crookfoot Reservoir, on 10th to 12 August 1984; flying north at South Shields Pier, on 9 May 2001; Seaton Pond, on 16 May 2002; two at Bishop Middleham, on 19th to 20 May 2004; Longnewton Reservoir, on 30 June 2004; Hetton Lyons, on 24 May 2005; Bowesfield Marsh near Stockton-on-Tees, on 13 May 2006; Houghton Gate, on 9th to 10 May 2009; Castle Lake, Bishop Middleham on 7 May 2011; and three at Rainton Meadows from 16th to 20 May 2011, one of which was seen at nearby Houghton Gate on 19th to 20th.

There has been an average of 48 birds per decade during the review period, though there were just 22 during the 1990s, there being four years since 1970 without records. The peak year was 2004 with 14 birds, the majority of which were recorded in spring. Other good years were 1987 with ten and 2008 when nine were recorded.

Distribution & movements

Temminck's Stint breeds from Scandinavia eastwards to eastern Siberia, with occasional pairs breeding in the Scottish Highlands. Its wintering range includes Africa, the Indian subcontinent and South East Asia. Passage birds in County Durham are presumably *en-route* to and from their Scandinavian breeding grounds. Just one Temminck's Stint has been trapped and ringed in County Durham, during the night-time ringing of waders on Dorman's Pool in August 1999 (Iceton 2000).

Long-toed Stint
Calidris subminuta

Very rare vagrant from Siberia: one record.

The sole record for County Durham is of a juvenile, discovered independently by John Dunnett and Tom Francis, at Saltholme Pools, Teesmouth during the afternoon of 28 August 1982. Hampered by distance and strong heat haze, it was initially thought to be an unusual White-rumped Sandpiper *Calidris fuscicollis*, especially as one observer thought they had noticed a white rump in a distant flight view. The bird was relocated at much closer range in the evening and its identity became more mysterious. It appeared 'stint-like' in proportion and clearly did not have a white rump. Despite extensive searching, the bird was not present the following day, though a tentative identification as Long-toed Stint had been suggested from the limited literature available at the time. The bird was re-found on Saltholme Pools on the morning of 30 August when the identity was confirmed. It remained faithful to Saltholme Pools until 1 September by which time it had been seen by many hundreds of observers. Often very confiding, it provided exceptional views enabling many photographs to be taken (*British Birds* 78: 546, *British Birds* 85: 431-436).

At the time of the occurrence this was considered the first record of the species in Britain and only the second record for Europe, following a bird at Ottenby, Sweden in October to November 1977. Subsequently, the first record was found to be of a bird at Marazion Marsh, Cornwall on 7th to 8 June 1970, following a review in 1995, when it was accepted as the first for Britain and the Western Palaearctic (*British Birds* 88:512, Round 1996). This bird had been previously accepted as a Least Sandpiper *Calidris minutilla*, despite doubts about this conclusion. This meant the Saltholme bird was the second for Britain, it has not occurred since, and it remains one of the signature rarities of the Tees Marshes.

Distribution & movements

Long-toed Stint has a disjunctive breeding range across much of southern and eastern Siberia in a wide variety of boreal and arctic habitats. It is a long-distance migrant with the majority of the population wintering in South East Asia and The Philippines, although small numbers also reach the Indian subcontinent and Australia. It is an extremely rare vagrant to Britain with just two records, and one in Ireland in June 1996. There are just a handful of records from elsewhere in Europe, the most recent of which was a bird in Holland in October 2009.

White-rumped Sandpiper
Calidris fuscicollis

A rare vagrant from North America: 18 records, involving 21 individuals between July and October.

The first record for County Durham was an adult at West Hartlepool Corporation rubbish tip, Seaton Carew, from 13th until 17 August 1963. Found by the late Phil Stead, amongst several Wood Sandpipers *Tringa glareola* and Dunlin *Calidris alpina*, the pool it was frequenting has long since been filled in, though it was a magnet for migrant waders at the time (*British Birds* 57: 269).

All records:

1963	Seaton Carew tip, Teesmouth, 13th to 17 August
1973	Dorman's Pool, Teesmouth, 4th to 14 August (*British Birds* 67:323)
1975	Reclamation Pond, Teesmouth, 16 August (*British Birds* 69: 337)
1977	Reclamation Pond, Teesmouth, 10 September (*British Birds* 72: 521)
1980	Long Drag, Teesmouth, adult, 26th to 27 July (*British Birds* 75: 501)
1980	Long Drag, Dorman's Pool, & Saltholme Pools, Teesmouth, up to three, 4th to 21 October, adult and two juveniles from 5th until at least 14 October (*contra BB*) (*British Birds* 75: 501)
1983	Dorman's Pool & Reclamation Pond, adult, 30 July to 9 August (*British Birds* 77: 522)
1986	Seaton Snook, adult, intermittently, 7th to 17 August (*contra BB*) (*British Birds* 80: 536)
1989	Long Drag & Reclamation Pond, Teesmouth, adult, 5th to 14 August (*British Birds* 83: 461-462)
1990	Saltholme Pools and Greatham Creek, Teesmouth, adult, 23 July to 3 August (*British Birds* 85: 526), presumed same, Whitburn Steel, 15th to 29 August (*contra BB*) (*British Birds* 84: 469)
1995	Saltholme Pools, Dorman's Pool and Reclamation Pond, Teesmouth, adult, 13th to 19 October (*British Birds* 89: 500)
1998	Dorman's Pool, Teesmouth, adult, intermittently 13th to 27 August (*British Birds* 92: 574)

2001	Saltholme Pools, Teesmouth, adult, intermittently 29 September to 11 October (*British Birds* 96: 568)
2002	Cowpen Marsh and Seal Sands, Teesmouth, adult, 13 July (*British Birds* 96:568)
2002	Seaton Snook and Seal Sands, Teesmouth, adult, 8th to 19 August (*British Birds* 96: 568)
2006	Seaton Snook, Teesmouth, adult, 26 July, joined by a second adult on 30 July with both birds remaining to 3 August, with one until 5th
2010	Saltholme Pools, Greenabella Marsh and Greatham Creek, Teesmouth, adult, 28 July.
2011	Saltholme Pools, adult, 20th to 24 July.

White-rumped Sandpiper, Saltholme Pools, October 1980 (P. Wheeler)
One of only two juveniles recorded in the county.

Distribution & movements

White-rumped Sandpiper breeds on the tundra of the Canadian Arctic, and is a long distance migrant wintering in southern South America. It is one of the commonest American waders noted in Britain with over 400 records by the end of 2005, at which point it was no longer considered by the BBRC. Chiefly an autumn vagrant, it is unusual amongst vagrant Nearctic waders in that more than half the British records relate to adults. It occasionally occurs in small parties, chiefly of juveniles in October, so the party of three in October 1980 is not entirely unprecedented, though not surprisingly remains the largest party recorded in the county. This party also included two juvenile birds, which are to date the only juveniles recorded in County Durham despite there being a county total of 20 birds. This probably reflects the easterly location of County Durham with most records relating to adults amongst migrating parties of Dunlin *Calidris alpina*, in July and August. It is assumed that these birds had made transatlantic crossings in previous years.

Baird's Sandpiper
Calidris bairdii

A very rare vagrant from North America: three records.

The first record for County Durham was of a bird found by Martin Blick, John Dunnett and Chris Sharp at Saltholme Pools, Teesmouth, on 5 May 1979. Present for just one day amongst migrating Dunlin *Calidris alpina*, it was the first spring record for Britain (*British Birds* 73: 507).

The second bird was a juvenile, initially seen briefly on Greenabella Marsh, Teesmouth, on 29 September 1986; it was relocated on the nearby Long Drag the following day, and was seen by large numbers of local observers. A few days later it was on the Yorkshire side of the River Tees, on the lagoons at South Gare, from 3rd to 15 October, where it was often accompanied by a Little Stint *Calidris minuta* (*British Birds* 80: 537).

Nearly 20 years passed before the third record, a worn summer plumage adult at Back Saltholme Pool, Teesmouth, from 2nd until 8 September 2005 (*British Birds* 100: 40).

Distribution & movements

Baird's Sandpiper is a long-distance migrant breeding in northern Canada, north west Greenland, and north eastern Siberia, and wintering in South America as far south as Tierra del Fuego. It is a regular vagrant to the British Isles with almost 230 records by the end of 2010, the majority of which relate to juveniles in September. Spring records are rare and presumably relate to birds that have crossed the Atlantic in a previous autumn. With 80% of British records relating to autumn juveniles, it is notable that two of the Teesmouth records relate to adults, although the two records in September conform to the national trend.

Pectoral Sandpiper
Calidris melanotos

A rare migrant from North America or Siberia: at least 125 individuals from April to December, though most frequent in autumn from July to September.

Historical review

The first record of Pectoral Sandpiper for County Durham relates to a bird said to have been shot by Dr Edward Clarke *'very near Hartlepool'* in October 1841. This record is documented by William Yarrell in *A History of British Birds* (1843), although the original specimen cannot now be traced. At the time of occurrence this was believed to have been the first British record, although this is now considered to have been a bird shot at Breydon Water, Norfolk, on 17 October 1830 (Parkin & Knox 2010).

There are just two subsequent 19th century records. A bird shot *"in or near the Tees mouth"* on 30 August 1853 (*The Naturalist* Vol. III 1853) and another said to have been shot by a Mr Henry Gornall close to Bishop Auckland, *"a few years before 1873"* (Hancock 1874). Temperley expressed some concern regarding this specimen, in particular with regard as to where it was obtained, but he concluded, albeit with a modicum of reservation, that it was an acceptable record (Temperley 1951). The specimen is held in the Hancock Museum, Newcastle-upon-Tyne.

The first 20th century record was of a bird found by Eric Shearer frequenting a sewage bed near Coxhoe from 12th to 14 September 1961, before what was presumably the same bird was relocated in a nearby marshy field from 28th to 30 September 1961. It was seen by several eminent local observers of the time including the late Fred Grey and the late Phil Stead (*British Birds* 55: 573). There were a further five records during the 1960s, all of which came from the North Tees Marshes. These were at: Cowpen Marsh, from 12th to 17 September 1962 (*British Birds* 56: 400); and, from 20th to 30 July 1963; at Seaton Carew Tip, on 15 October 1969; and, at Saltholme Pools, from 4th to 7 October; and, 18th to 24 October 1969.

Recent status

Between 1970 and 2011, a total of 116 individuals were recorded in County Durham. Of these, 24 (20%) were in spring, with the earliest record being of a single on pools at Seaton Common, Teesmouth on 19th and 20 April 2004. With the exception of a bird at Barmston Pond, Washington on 29 May 1976, all of the spring records for the county come from the North Tees Marshes, where the species has been recorded in 18 of the last 40 years. Extreme spring dates for Durham are 19 April and 22 June, and there have been eleven records in May and eight in June. The best spring was 1995 with two birds, singles on the Long Drag on 3 May, and at Cowpen Marsh on 28th and 29 May. Most birds stay for just short periods but single birds lingered on the Long Drag, from 20 June to 2 July 1979, and around the Tees Marshes intermittently from 6 June to 30 July 1983. A bird at Dorman's Pool on 10-12 June 1973 was seen to be the subject of display by several male Ruff *Philomachus pugnax*.

The species is much more frequent on autumn passage with a total of 92 individuals recorded since 1970, the vast majority of which come from the North Tees Marshes. The earliest 'autumn' record relates to a single bird at Saltholme Pools, Teesmouth from 3rd to 7 July 1976, though as some 'spring' birds have lingered into July, the cut-off point between the seasons is far from clear.

Return passage of adult birds continues throughout July and August, with the first juvenile birds beginning to appear in late August. Passage reaches an obvious peak in September and consists mainly of juvenile birds. There have been several small parties recorded on the North Tees Marshes during September, with the largest being a group of six birds on Dorman's Pool on 19 September 1970 which included a long-staying adult present from late August. Other notable groups have included: four at Dorman's and Saltholme Pools on 4 September 1971; up to three birds on Dorman's Pool, in September 2003; and, three on Dorman's and Saltholme Pools, in September 2006.

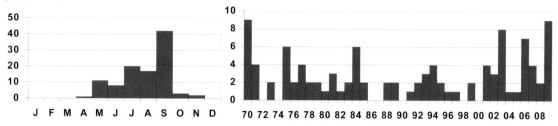

Records of Pectoral Sandpipers in County Durham, 1970-2009

There have been just six October records, with two in November, the latest of these being with a party of Redshank *Tringa totanus*, on the A1 flashes at Bishop Middleham on 11 November 2001. The sole December record, from Cowpen Marsh, Teesmouth on 25 December 1975, is thought to relate to a bird which had been present at the same locality from 1st to 8 November that year.

The Pectoral Sandpiper is a very scarce bird away from the North Tees Marshes, with just 19 records elsewhere, nine of which have been in the Barmston/Washington area. The Bishop Middleham area has attracted birds in recent years with six records since 2000, including two in both 2001 and 2009. The remaining records are of singles at: Hurworth Burn Reservoir on 15th to 23 July 1975; Boldon Flats on 1st to 5 September 1985; Bowesfield Marsh, Stockton-on-Tees, on 3 August 2001; and, Whitburn on 26 September 2010. This last bird, a juvenile was present very briefly on the small pool immediately behind the observatory and was at Whitburn on 26 September 2010, only the second record for South Tyneside, and came some 25 years after the first.

The best decade for Pectoral Sandpipers in the review period was the 2000s with a total of 39 birds, though there were 30 in the 1970s. In contrast to this, the 1980s and 1990s were relatively poor, with just 35 birds through that twenty-year period; including just 16 during the 1990s. There have been seven blank years since 1970, with the peak years of 1970 and 2009 both recording nine birds, the majority in autumn. Other good years include 2003 and 2006, when annual totals of eight and seven respectively were recorded.

Distribution & movements

Pectoral Sandpiper breeds on the tundra from Alaska and northern Canada, across the Pacific to the Taymyr Peninsula of eastern Siberia. A long-distance migrant, both populations winter in southern South America. The species is a regular migrant to the British Isles in both spring and autumn, being the commonest 'North American' wader to reach Britain, particularly in September and October when westerly storms displace many juveniles across the Atlantic. Many of these birds must migrate southwards to spend the winter in Africa, returning northwards the following year and accounting for the numerous spring sightings on this side of the Atlantic. There are also several recent records of displaying birds in suitable habitat in northern Europe including a single pair that bred in Scotland in 2004.

Two Pectoral Sandpipers have been ringed in County Durham. The first of these was present on Saltholme Pools from 30 September to 22 October 1989 and its identity was uncertain, at one stage being suggested as a potential Long-toed Stint *Calidris subminuta*, until trapped on the evening of 11 October. It was identified in the hand as being a juvenile female at the small end of the range for this species. The second was trapped and ringed during a six-day stay at Portrack Marsh, Stockton-on-Tees from 4th until 9 September 1994, and was later seen at nearby Saltholme Pools, Teesmouth from 11th until 19 September.

Sharp-tailed Sandpiper
Calidris acuminata

A very rare vagrant from north eastern Siberia: five records.

This rare Siberian wader was first found in Durham by the late Phil Stead during the early evening of the 21 August 1963 on a small pool at the rear of Cowpen Marsh, Teesmouth (*British Birds* 57: 269). An adult, still largely in breeding plumage, it was watched for some two hours as it fed the company of around 40 Ruff *Philomachus pugnax*. It remained until 24 August though was often absent for long periods. Described as quite tame at first it become more wary, mobile and elusive towards the end of its stay. At the time of its discovery, it was just the seventh record for Britain.

All records:

1963	Cowpen Marsh, Teesmouth, adult, 21st to 24 August
1977	Long Drag, Teesmouth, adult, 3 September (*British Birds* 71: 501)
1997	Long Drag, Teesmouth, adult, 26 August (*British Birds* 91: 475)
2010	Greatham Creek, Teesmouth, adult, 20th to 21 September (*British Birds* 104: 581)
2011	Greatham Creek, Teesmouth, adult, 9 September (*British Birds* in prep.)

Distribution & movements

Sharp-tailed Sandpiper breeds on the tundra of eastern Siberia from the Yana River eastwards to the Kolyma River Delta, and occasionally further east. The entire population migrates south to spend the winter in the islands of the western Pacific, through New Guinea to Australia, and New Zealand. The species is a very rare vagrant to the British Isles with just 30 records by the end of 2010. Most records refer to adults in late summer, with those in County Durham conforming to this pattern.

Sharp-tailed Sandpiper, Long Drag, September 1997 (the late Jeff Youngs)

Curlew Sandpiper
Calidris ferruginea

A scarce passage migrant, typically recorded from May to October, though most numerous in August and September, very rare in winter.

Historical review

The first documented record of Curlew Sandpiper for County Durham comes courtesy of Thomas Bewick's *'History of British Birds'* (sixth edition, 1826), the figured specimen being said to have been shot near Sunderland at the end of January 1814. Several years later Backhouse (1834) said that he had shot several birds at Seaton Carew during the autumn of 1830, including one from a flock of 20-30 birds; some of which are doubtless the birds listed by Selby (1831) as having been *"killed near Hartlepool"* and forming part of Backhouse's collection. Backhouse (1834) also makes reference to others having been killed at both Hartlepool and Sunderland around the same, with Fox (1831) mentioning two specimens that had been *"lately"* shot from a flock of five at Hartlepool, presumably referring to the autumn of 1830. Hutchinson (1840) wrote that although it was rare the Curlew Sandpiper had been shot several times on the Durham coast, with a number of specimens having been sent to William Proctor of Durham for preservation. The majority of these specimens were said to have been taken from the sands between Hartlepool and Seaton Carew, and were generally obtained in January. Such mid-winter occurrences would be considered extremely unusual in a modern day context, but Hutchinson himself had examined one such specimen in January 1841 and its identity was not in doubt.

Nelson (1907) considered the species an irregular passage migrant, being more plentiful in some years than others, with extensive migrations in 1873, 1881 and 1887, when it was *"abundant at the Tees mouth"*. In 1890, he writes of an *"immense flight of Pigmy Curlews"*, stating that in excess of 100 birds had been shot at Teesmouth, during August of that year, indicating that a very considerable influx had taken place, perhaps the largest in the county's history. It is interesting to note that Nelson (1907) did not record any spring birds at Teesmouth with his earliest record being on 27 July 1894.

Almost half a century later Temperley (1951) referred to the species as being *"a passage migrant in small and varying numbers on the coast"*, stating that it was usually met with in September and October, and that small numbers were also recorded during spring. Temperley is the first author to specifically mention the occurrence of this species on spring migration in the county though he does not give any specific records. He is himself sceptical of many of the historical records stating that *"in early times it seems to have been confused with the Dunlin"*, perhaps referring to the numerous mid-winter records prior to 1850, though several of these involve specimens.

Stead (1964) painted a similar picture to that of Temperley stating that a few adults passed through Teesmouth in early August, though that passage reached its height at the end of the month when the majority of those recorded were juveniles. Writing of the period from 1951-1964 he also noted that the species had been recorded at Teesmouth in May in just three out of the last eight years, suggesting that spring passage was not as regular as Temperley (1951) had implied. Like many previous authors Stead also makes reference to the numbers fluctuating from year to year; he considered flocks of more than 10 unusual, quoting a flock of 55 at Saltholme Pools, Teesmouth on 2 September 1954 as being *"the largest concentration noted in recent years"*.

During the 1960s the maximum counts at Teesmouth typically reached around 15 per autumn, though 1964 (three birds) and 1968 (seven between July and October) occurred. By way of contrast, in 1963 birds were present at Teesmouth from 8 August, peaking at 44 on the Reclamation Pond on 28th. In 1967, 26 were present on 19 September.

Autumn of 1969 saw a huge influx during late August, with a record count of 230 at Teesmouth on 28 August (Blick 1978); 143 were still present on the 13 September, virtually all of them juveniles. During this influx others were reported inland, including a flock of up to 29 birds at Washington Ponds between 28h August and 1 September, with birds present there from 27 August to 7 September. These are the highest numbers recorded in Durham.

Historically, inland occurrences in the county have been scarce, with Temperley (1951) stating that it had been occasionally recorded at Jarrow Slake and on the River Don, quoting a record of three summer plumaged birds there on 27 July 1938. He noted the species as being *"unusually plentiful"* during the autumn of 1946 with at least 36 on the River Don and five at nearby Jarrow Slake. Small numbers were also recorded at Jarrow Slake during the autumns of 1952 and 1953 though did not reach double-figures on either occasion. Further inland it was noted occasionally at Darlington Sewage Farm, a peak count of 33 there on 11 September 1928 and a flock of about a dozen birds on 9 December 1928, following a period of very cold weather (Almond 1930), a remarkable mid-winter occurrence. The only other mid-winter record in the period was also inland in the Team Valley, Gateshead on 23 January 1943, again noteworthy for both location and date. Many more were recorded inland during the 1960s including a series of records from near Stanley with singles on 31 August to 1 September 1960, and 18 August 1965, two on 28 August 1965 and an exceptional spring bird on 8 June 1969. Birds were said to be 'regular' at Hurworth Burn Reservoir during the autumn of 1961 and three were at Washington Ponds on 15 September 1967.

Recent status

The status of Curlew Sandpiper in County Durham has changed little during the last forty or fifty years, it remains a scarce passage migrant during autumn that is subject to periodic influxes and remains rare in spring.

The majority of records come from the North Tees Marshes, though it might be encountered on any suitable area of wetland particularly during autumn, usually in the company of Dunlin *Calidris alpina*.

The North Tees Marshes have provided the bulk of spring sightings, the majority in May and June, though there have been three records in late April: on 28 April 1971; at Saltholme Pools, on 26th and 30 April 1981; and, two on Saltholme Pools, on 27 April 2008. In most years between one and four are found but few make extended stays and in some years it is absent in spring. Few spring counts exceed five, exceptions being: six at Teesmouth, on 6 June 1971; six on Dorman's Pool, from 1st to 2 June 1980; nine on the Long Drag, on 26 May 1986 (Bell 1987); and, six on Saltholme Pools from 19th to 24 May 2001. After mid-June it is difficult to determine whether any birds seen are heading north or have failed as breeders and are on southbound migration. An unusual record concerns a 'winter-plumaged adult', more likely a bird in its first summer, which lingered around Saltholme Pools from 20 June to 21 July 1993. Inland spring records are decidedly uncommon. All but one of the seven records since 1970 have been in May, the exception being a remarkably early bird over SAFC Academy Pools, Sunderland on 15 April 2003, the earliest date for the county (Newsome 2007). The well-watched Washington area has attracted four records with singles in: May 1974, May 1977 and May 2004, with an exceptional flock of four from 24th to 28 May 1972 (Coulson 1973). Other inland birds were near Bishopton on 13 May 1980 and at Low Barns, Witton-le-Wear on 8 May 2004 (Siggens 2006).

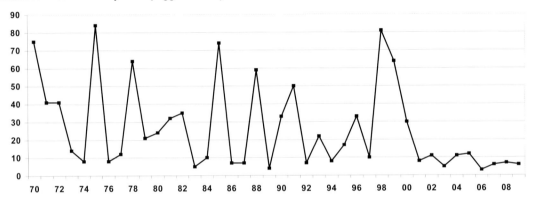

Peak autumn counts of Curlew Sandpiper at Teesmouth 1970-2009

Curlew Sandpiper is much more numerous during autumn than it is in spring, with records spanning from July until November, though peak numbers usually occur from late August until mid-September, these typically involve mainly juvenile birds. The numbers fluctuate, being dependent upon prevailing weather conditions and to some extent, breeding success. The North Tees Marshes generally provides the majority of records and the highest counts, although small numbers are recorded annually elsewhere; more so, during 'influx' years.

The first adult birds typically appear at Teesmouth from early July, most records relating to single birds, with an exceptional flock of 29 by Greatham Creek on 30 July 1990. Adult birds are genuinely scarce in autumn and this is the highest count in the county to date.

Juveniles begin to arrive from late August with numbers usually reaching a peak during early September. In an average year on the North Tees Marshes perhaps 15-25 birds might be recorded but during influxes this can increase three to four fold; no individual count has exceeded 90 birds since 1970. The highest count was 84 around the North Tees Marshes on 31 August 1975 (Bell 1976). Other notable counts during the period 1970-2009 were: 75 at 'Teesmouth', autumn 1970; 74 on the North Tees Marshes, on 12 September 1985; and, 81 on the North Tees Marshes, on 11 September 1998, this included a flock of 77 by Greatham Creek. In contrast, particularly poor years were: 1983, 1989 2003 and 2006, when no count exceeded five birds, with just three recorded in 1989 (Bell 1990). There have been no notable influxes since 2000 when a peak of 33 was reached on 30 August. A party of 17 were counted on Seal Sands at low tide on 12 September 2010 suggested that some birds might be passing through undetected, using the more poorly observed inter-tidal habitats of the estuary, as the species did in times past.

Relatively few birds remain after mid-September, though in most years odd birds stray into October, occasionally until quite late in the month. There are just five November records, all bar one relating to single birds: at Saltholme Pools, throughout much of November'1975; at Saltholme Pools, on 12 November 1978; on the

Reclamation Pond, on 5 November 1988; and, Saltholme Pools on 1 November 1999. The exception was two by Greatham Creek on 2 November 1987 (Bell 1988). That none should have occurred between December and March during the last fifty years is at odds with the species' historical status when prior to 1900 it was said to be *'generally obtained in January'*, (Hutchinson 1840). This perhaps confirms Temperley's (1951) initial suspicions that some of these non-specimen winter records may have been erroneous.

Coastal records away from Teesmouth are rare though not exceptional and mainly refer to the Whitburn area, where the species is often found in small numbers feeding with Dunlin at Whitburn Steel. The majority of these records relate to juveniles in August and September with counts not usually exceeding two to three birds, although an exceptional 18 were present on 21 August 1990 with a flock of 20 from 25th to 31 August 1999 (Armstrong 2003). Additionally, nine flew north past Whitburn Observatory on 10 September 2000 (Armstrong 2005). Slightly further inland up to 10 birds visited Boldon Flats during September and October 1985 and up to three were on floodwater along Cleadon Lane, Whitburn from 11th to 22 September 2008, occasionally visiting nearby Boldon Flats. One at Whitburn on 5th and 10 November 2005 is noteworthy for the late date.

Inland, there are three July records, the earliest at Hurworth Burn Reservoir on 9 July 1992, the latest, two at Washington on 7 November 1974. The majority of inland records come from the Washington area and to a lesser extent Hurworth Burn Reservoir with a few from the extreme west, these coming mainly from Derwent Reservoir, where there were: three in September 1990; nine, on 26 August 1991; and, singles in September 1996 and 1997. There are just a handful of occurrences from the Gateshead area including four records from Shibdon Pond and a party of six on the River Tyne at Dunston from 26th to 30 August 2002. The Houghton-le-Spring area has attracted birds more often with at least six records from Seaton Pond and four each from both Hetton Lyons and Rainton Meadows. Elsewhere there have been at least six records at Bishop Middleham, all since 1999, at least five from Longnewton Reservoir and several from Bowesfield Marsh near Stockton-on-Tees, including nine from 2nd to 3 September 2005 (Joynt 2006). The majority of inland records involve small parties of up to five birds. Exceptionally, 12 were at Washington Ponds, from 1st to 4 September 1973, 17 were at Barmston Pond on 30 August 1975, and 17 were at WWT Washington on 8 September 1988.

Distribution & movements

The Curlew Sandpiper breeds in Arctic Siberia, and is a long-distance migrant wintering in Africa, southern Asia, Indonesia and Australia. It is a regular migrant to Western Europe including Britain, though autumn numbers fluctuate.

An indication of the wintering range of birds that pause at Teesmouth is neatly illustrated by the sole recovery of a bird ringed there. An adult ringed at Seal Sands on 12 August 1989 was later re-trapped at Merja Zerga, Morocco on 7 December 1989; a distance of 2,232km to the south (Bell 1991). A colour-ringed juvenile seen on Seal Sands, Teesmouth in September 1991 had been ringed at the Elbe Estuary, Germany in August of the same year (Armstrong 1992). Limited ringing recoveries elsewhere in the UK suggest that at least some juvenile birds move west out of the Baltic and cross the North Sea to Britain for a brief stop-over before continuing their southbound migration.

Purple Sandpiper
Calidris maritima

A fairly common passage migrant and winter visitor, recorded in all months, most frequently from September to May

Historical review

One of the earliest references to Purple Sandpiper in County Durham is by John Sclater of Castle Eden, who writing during the 1870s described the species as being *"rarely met with on this part of the coast"* (*The Zoologist* 1875). Almost a century later Temperley (1951) described the Purple Sandpiper as a *"common passage migrant and winter visitor in small numbers"*; stating that a few birds in winter plumage sometimes remained throughout the summer. He recorded passage as taking place along the coast in August and particularly September, with small numbers remaining throughout the winter, with a northward spring passage occurring again in April.

Temperley (1951) gave no records of the species having occurred inland, making specific reference to the fact that it had not occurred at either Jarrow Slake or Hebburn Ponds where other small waders occurred. As its scientific binomial suggests this species has a strong association with the coast, in particular weed-strewn rocky foreshores. Temperley (1951) also made reference to this strong habitat preference, stating that the species was *"rarely seen in tidal estuaries"*; a situation which persists to date. Temperley drew attention to the fact that Joseph Bishop had seen birds on the slag breakwaters at Teesmouth during the 1930s, an early reference to the species' use of man-made structures.

Stead (1964) considered the Purple Sandpiper to be a *"regular passage migrant and winter visitor"* in his review of the birds of Teesside, noting the species was locally common from September to May. He gave the typical winter count at Hartlepool as being of around 50 birds and mentioned small flocks of up to five birds being present on the slag outcrops at North Gare during the winter months. Similar sized counts at Teesmouth can be traced back as far as 1893, indicating that there seems to have been little change in its status during at least the last 70 years (Blick 2009).

Hartlepool has long been one of the most favoured sites for this species within the county, with a high-tide roost located in the town's West Harbour since sometime in the mid-1960s. A count of 70 there on 16 April 1960 was particularly high and doubtless influenced by passage birds on their way north. Winter counts of: 57, on 10 March 1963; 65, on 10 January 1965; and, 56, on 23 November 1965, were typical for this site at the time. An increase in numbers as the decade came to a close saw counts of 100 on 31 March 1969 and 106 on 23 November 1969; new county maxima at the time.

Away from Hartlepool, small numbers are found elsewhere along the county's winter coastline, particularly at Whitburn, where the rocky foreshore of Whitburn Steel provides winter feeding habitat. There are few historical counts from this area, though 22 were present on 20 December 1969, in line with the increase in numbers wintering at Hartlepool that year.

The species is extremely rare inland and a single bird on the Reclamation Pond, Teesmouth on 14 August 1961 (Blick 2009) remains the only record from the North Tees Marshes, despite being only a few kilometres from the coast.

Recent status

Purple Sandpiper is still a fairly common winter visitor and passage migrant to the Durham coast, though as a winter visitor this rather nondescript species has declined in recent years. It is almost entirely restricted to its chosen habitat of tidal rocky coastline and is rarely recorded away from this specific habitat. The two most favoured localities are Hartlepool Headland and Whitburn Steel. Away from these core areas, locations that have produced significant counts include: Ryhope, Seaham Harbour, South Shields and Sunderland Dock. The species can be encountered anywhere there is suitable habitat along the Durham coast. In this respect, it is probably under-recorded particularly along the central stretch of coastline where there are relatively few observers.

The first birds to return are usually noted from mid-July onwards, with numbers building to a winter peak in January and February. Return passage to the breeding grounds begins in March, though in most years a slight increase in numbers occurs locally in April and May, as more southerly wintering birds pass north through the county. June records are extremely rare, having occurred on just six occasions since 1970, these were: two at Hartlepool, on 1 June 1970; one at Hartlepool, on 11 June 1977; one Seaton Snook, on 4 June 1989; one Whitburn, on 29 June 1991; one Sunderland Docks, on 27 June 1996; and, two flying north at Whitburn Observatory, on 24 June 2004.

The majority of the county's wintering population of Purple Sandpiper even today remains centred around Hartlepool and in particular the rocky coastline surrounding Hartlepool Headland, though many are also known to feed on the exposed reefs offshore from Seaton Carew at low tide. The majority if not all of these birds congregate at high tide to roost in the West Harbour, where typically the highest count of the winter is made during January or February. Occasionally birds resort to roosting at the nearby Newburn Sewer, where a peak count of 109 was made on 3 February 2007 (Joynt 2008). The roost in the West Harbour has existed since at least the 1960s, though it rose to prominence during the 1980s when erosion by the sea rendered the walls unsafe and inaccessible to people at high tide, making it an ideal roosting locality for this and other wader species. Restoration of the harbour and its surrounds was completed by the local council in 1992, though a piece of bold mitigation included the creation of an island in the centre of the dock designed specifically for roosting waders, which continues to be utilised (Blick 2009).

During much of the 1980s and early 1990s, the Teesmouth/Hartlepool area was considered a nationally important site for this species, as it regularly held over 1% of the British wintering population of Purple Sandpiper (Blick 2009). Following a decline, that status was lost in 1995. This decline has been mirrored nationally with an estimated 32% drop in the English wintering population. The reasons for this are not yet fully understood but may be linked to climate change and the redistribution of wintering birds across north west Europe, along with reduced organic input from sewage outflows. This is part-way illustrated by an almost three-fold increase in numbers wintering on the Outer Hebrides since the winter of 1984/1985 and increased wintering numbers on Shetland (NEWS 1997/1998).

Building upon the increases noted during the late sixties the Hartlepool population continued to grow during the 1970s. Mid-winter peak counts grew in the 1980s: 178, 1 January 1976 (Bell 1977); 210, on 19 March 1981 (Bell 1982); 246, on 16 January 1982 (Bell 1983); and, 278 on 2 April 1983 (Bell 1984), the last count swollen by passage birds. Relatively high numbers, approaching 150, were maintained throughout much of the remainder of the decade, although 245 were present in Hartlepool Bay in the winter of 1989/1990. Another very high count of 277 was made in West Harbour in March 1991 (Bell 1992), just one short of the 1983 record, although this signalled the start of the decline, with no count of in excess of 200 birds since December of 1991. The decline continued into the 2000s with just four three-figure counts since 2000, the highest being 136 in Hartlepool Bay in February 2000 (Iceton 2001), with 118 in November 2000 and February 2004. Towards the end of the first century of the 21st decade, the wintering population at Hartlepool seemed to have stabilised at around 80-100 birds.

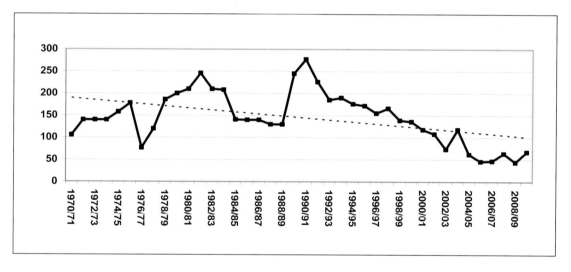

Peak winter counts (November-March) of Purple Sandpipers at Hartlepool 1970-2009 (with trend line)

In addition to the data surrounding the wintering population, a particularly noteworthy record involves two birds seen to indulge in prolonged song-flight around Hartlepool Bay on 5 May 1984 (Bell 1985).

Whitburn has traditionally supported a healthy population of this species during the winter months particularly during the 1980s though numbers have been in steady decline since; the reasons for this are not clear. Throughout much of the 1970's around 30-40 birds regularly wintered at Whitburn Steel though numbers increased dramatically in the winter of 1978/1979 rising from a record of 83 birds on 8 December 1978 to a peak of 112 during the first week of February 1979. This increase was consolidated with around 120 during the early part of 1980, though in the next three winters no counts in excess of 100 were made. Around 130 were counted during three successive winters from 1983/1984, followed by a count of 146 on 7 January 1987 (Armstrong 1988), the highest count from this site. This count neatly corresponds with the 1980s dip in the Hartlepool population though there is no evidence to support any switch in winter site usage. This peak has not been sustained since with the only other Whitburn count in excess of one hundred being of 131 on 4 March 1988; numbers have declined since. An average of just 40 per winter was maintained throughout much of the 1990s and since 2000 numbers have struggled to reach double-figures, 24 on 23 January 2009 was the peak count of that decade.

Close to Whitburn, small numbers are regularly present during the winter months at both South Shields and Sunderland, where typically they roost on the piers at high tide. Numbers at these sites fluctuate and there is doubtless some interchange with birds at Whitburn. Few counts at South Shields have exceeded 40, though at least 60 were present on 30 March 1983 and a site maximum of 62 was recorded in February 2001 (Siggens 2005). In the Sunderland area numbers rose to as many as 80 during November 1986 and peaked at 114 during January 1987. To the south, foraging parties of: 53 at Ryhope, in November 1987; 78 at Hendon, on 17 February 1988; and, 60 at Ryhope, on 10 February 1991, were noteworthy. Seaham Harbour was established as a roosting site for this species during the 1990s with 160 there on 3 March 1991 and 206 during March 1992 (Armstrong 1993). Later counts of over one hundred included 135 in February 1993 and 131 in March 1994; although these numbers have not been maintained, with an average of around 80 birds per winter between 1995 and 2003 and just 20 on 14 January 2006, in-line with declining numbers elsewhere within the county.

Inland records are extremely rare with only three truly inland occurrences within the county of single birds at: Barmston Pond, on 14 August 1971 (McAndrew 1972); Boldon Flats, on 19 February 1989 (Armstrong 1990); and, at Grassholme Reservoir on 18 October 1991 (Armstrong 1992). In addition, there are four additional records from the banks of the Tyne within Gateshead Borough. The first of these was a single on exposed riverside rocks at Bill Quay sometime during the winter of 1989/1990 (Bowey *et al.* 1993), followed by a party of four birds at the MetroCentre Pools briefly on 22nd November 1994 before they flew east along the Tyne (Armstrong 1996). The most recent ones are both from Timber Beach, Dunston, a single bird on 20 November 2005 (Newsome 2006), and two on 8 April 2007 (Newsome 2008).

Distribution & movements

Purple Sandpipers breed in the Arctic and sub-Arctic from north eastern Canada and Greenland eastwards across northern Russia to Siberia as far east as the Taimyr Peninsula. In Europe the largest breeding populations are to be found in Iceland and Arctic Norway, though it penetrates as far south as the Faeroe Islands and the mountains of southern Scandinavia. In Britain, a tiny population of one to three pairs has existed in the Cairngorms National Park, Scotland since at least the 1970s, though the last confirmed breeding was in 2003 (Holling 2007). A migratory species, most winter to the south of their breeding range along the eastern seaboard of North America and around the coastlines of Western Europe as far south as Iberia, although some Icelandic birds move no further than the coast.

An interesting record involved two adults ringed in North Wales at Rhos-on-Sea on 10 November 1979 and re-trapped at Hartlepool on 29 December 1988 (Bell 1990). Purple Sandpipers are known to display marked site fidelity during the winter months (Burton & Evans 1997, Craggs 1990). Another adult which had been colour-ringed in Finland on 13 December 1972 was seen at Hartlepool on 14 April 1992, 11 May 1992, and 15th and 28 April 1993 (Bell 1995). By the time this bird was last sighted at Hartlepool in 1993, it was known to be in excess of 20 years old; just short of the European longevity record of 20 years and 9 months (Euring online). At the time, this was the only foreign ringed Purple Sandpiper to be recovered in the county and suggested a Scandinavian origin for those wintering along the Durham coast. However, sightings of three colour ringed individuals at Whitburn in March and November 2011, and at Hartlepool Headland in November 2011, confirmed that at least some of our wintering birds originated in Spitzbergen.

The only other recovery of this species in the county is of an adult colour-ringed at Hartlepool on 4 March 1993, which was seen on the Farne Islands, Northumberland on 20 July 1995 (Bell 1996) giving a clear indication of the rapid southward dispersal from their breeding grounds to the north.

Dunlin
Calidris alpina

**A very common passage migrant, a common wintering species and a scarce,
but probably under-recorded, breeding species.**

Historical review

This species has a long-documented association with County Durham that can be traced back to the 16th century. The oldest documented reference to the species named as Dunlin relates to Durham according to a letter published in British Birds in 1912, by J.E. Harting. In this correspondence it is stated that, "*The meaning of the name dunlin is the 'little dun thing', a diminutive akin to grayling, titling, Sanderling, duckling, and gosling, and this is the spelling to be found in the oldest mention of the name, which occurs in the Durham, Household Book, which contains the accounts of the Bursar of the Monastery of Durham, A.D. 1530-4. The price then paid for these little birds, known elsewhere as stint, purre, sand-lark, and ox-bird, was at the rate of 4d. a dozen*". The first written description of the species, was made by Ralph Johnson of Brignall near Greta Bridge, to John Ray for Willughby's Ornithology, published in 1678, this was presumably taken from a bird obtained at Teesmouth by Johnson, who routinely hunted for specimens at the mouth of that river (Mather 1986).

Temperley (1951) said that this species was "*a summer resident in very small numbers*", the southern race *schinzii* breeding in a few locations in Durham. He said that there were very few breeding records, possibly because of the difficulty of finding breeding birds on the large moorland areas of the west. A century before this, Hutchinson (1840), thought it did not nest in Durham. Backhouse (1885) said that he had found birds in the summer on the watershed between Weardale and Teesdale in 1884, but he did not confirm breeding. Tristram (1905) said that it "*used to breed on the moors*" and speculated that it might still. Birds were found in territory on the moors between Teesdale and Weardale between 1944 and 1949. In 1952, ten birds were displaying near the Weel, upper Teesdale, on 26 May (K. Baldridge pers. comm.). Elsewhere the same year, three-day old young were noted in upper Teesdale and at another site a deserted nest with eggs was found. In 1964 a successful breeding season was revealed by the fact that some 30 young were noted in the western uplands. This was probably an indication of the size of the Durham breeding population during the mid-1960s and there is no reason to think that the situation now is much different. In 1965 Dunlin were again considered to be "*plentiful*" and young were seen. In neighbouring Northumberland during the 1970s, at least 90 breeding pairs of Dunlin were discovered, nearly all on the south west moors and the population there is believed to routinely number between 75 and 90 pairs (Kerr 2001).

By way of contrast, Nelson (1907) reported that "*it had nested...for many years past in the marshland near the Teesmouth*", this breeding population, numbering just a few pairs. A nest with four eggs was found on the North Tees Marshes (in the Cowpen Marsh area) on 23 May 1899 (Milburn 1901) with another nest found in 1902. From this latter nest two eggs were taken, which are now in the Dorman Museum (Stead 1964). No breeding was confirmed at Teesmouth after 1907, though breeding was suspected of taking place in the first decade of the 20th century.

Temperley (1951) said that the northern race of Dunlin was "*a very common passage migrant and winter visitor*". Large numbers were seen on the coast in autumn and at Teesmouth in winter, and he added that it was noted occasionally along the shore of the Rivers Tyne and Wear, including at Jarrow Slake. Teesmouth was very much the winter stronghold of the species up to the early 1980s when winter flocks of up to 6,000 were not uncommon. A count of 10,000 was recorded in November 1973. During this period relatively high numbers were also recorded at other sites with 1,000 at Jarrow Slake in October 1971, 700 at Boldon Flats in January 1979 and 450 at Whitburn in 1984.

Recent breeding status

As a breeding species, this bird is both more widespread and abundant in the Durham uplands than many casual observers realise, though it is by no means common (Westerberg & Bowey 2000). It was referred to as occurring on the hilltops of Mickle Fell, presumably as a breeding bird, in the early 1970s but also being present in areas on the high fells or where heather had been recently burnt, alongside Golden Plover *Pluvialis apricaria* (Bradshaw 1976). The Dunlin is mainly a bird of the high moorlands (Gibbons *et al.* 1993) and the species' relatively shy nature and unassuming habits after the end of the song period lead to it being easily missed. To quote Ken Baldridge who studied the species in Durham for many years, *"I've always loved Dunlin, such a charming little character, confident in its camouflage and quite willing to let you walk through its territory without a call or fluttering wing to help you see it"*. It is found especially where wet moorland abounds, usually above the 500m contour in Durham. The greatest concentration of breeding birds is located on the high, western tops that give rise to the county's principal river systems. This is where its favoured habitat is found; it prefers wet ground, usually encompassing open pools and *Sphagnum* bogs, these often being interspersed with sheep walk and short heather (Nethersole-Thompson & Nethersole-Thompson 1986).

In the early to mid-1990s territorial birds were found in: the upper Wear valley, Baldersdale, upper Teesdale, at the head of the Rookhope Burn, above Rookhope and in small numbers in the Derwent valley. Birds were recorded at up to nine sites with up to 10 pairs being seen. Reports continued intermittently during the period up to the early 21st century with a maximum of six pairs in any one year, though this is probably a considerable under-estimate of the true numbers. During the breeding season, birds are often noted near Cow Green Reservoir in late May and on Bowes Moor throughout the breeding season. In 2009 it was felt that at least 10-12 pairs had nested in the county.

There is little recent evidence of lowland breeding in the county, though small numbers of birds are noted around Teesmouth throughout the breeding season and it is certainly possible that breeding occasionally occurs in some of the peripheral wetlands around the estuary. There have been no known breeding records since the early part of the 20th century at Teesmouth, despite a number of summering birds being present at various locations round Teesmouth (Joynt *et al.* 2008). However, in 1992, three juveniles were on the Long Drag, Teesmouth on 1 July and this may have indicated local breeding.

If Durham birds nest at even half the densities of the lower figure found by Stroud *et al.* (1987) in Caithness and Sutherland, i.e. one to three pairs per occupied square kilometre, then a county population of 39 pairs could be estimated from the 20 occupied tetrads that were mapped in the *Atlas* (Westerberg & Bowey 2000). Without doubt it is an uncommon breeding species in northern England and as a result, Durham can be considered to hold an important proportion of England's total of breeding birds.

It is known that dry summers can impact negatively upon the species' breeding success locally (J. Strowger pers. obs.) especially when moorland pools and the surrounding wet areas dry out. It remains an unknown as to how global climate changes will impact upon this species.

Recent non-breeding status

The number of wintering Dunlin at Teesmouth declined significantly over the period 1969 to 2000 (Ward *et al.* 2003). The number of spring and autumn passage birds did likewise over the same period, in broad terms, though less obviously so for autumn birds during the period 1988 and 2000 (Ward *et al.* 2003).

It would appear that the long-term decline of winter numbers at Teesmouth, which commenced in the 1970s, continued apace through the mid-1990s and into the 2000s. The Teesmouth mid-winter peak count was about 50% less than in the previous three winters. Adult Dunlin tend to be site-faithful to their wintering sites, but unusually mild or harsh weather can influence some birds' choice of wintering sites, making them stray further north, east, south or west depending on conditions (Blick 2009). An influx of long-billed *alpina* birds to the Durham coastal area in early 1997, in response to severe weather on the near Continent, led to the largest count of Dunlin at Teesmouth since the early 1980s, with 3,371 there in January 1997; very much against the trend. In the period 1995-2010, the winter counts at Teesmouth were in the range of 2,500 to 6,000 birds, and although 8,000 were recorded in January 1974 and just over 10,000 in November 1973, such numbers have not been present for over two decades (Blick 2009). The late 1990s saw further decreases in the Teesmouth wintering population and by 2001, the mid-winter peak was at approximately one quarter of what it was only 15 years previously. Seasonal mean peak low water counts of Dunlin on Seal Sands between 1991 and 2001 were 1,126 in winter, and 1,443 in spring (Ward *et al.* 2003). Nationally, the species has declined over the last forty years or so (Eaton *et al.* 2009). The long-term trend for wintering Dunlin is a decrease of 25% and the ten-year trend shows a decline of 41%. By the end of the first decade of the 21st century, winter flocks of up to 6,000 birds at Teesmouth are now only a memory. The largest flocks in recent years, probably of the race *schinzii*, appeared in August with 2,235 in 2003, 1,270 in 2002 and 1,496 in 1998. In 1996 a flock of 993 birds in January had increased to 1,859 in February due, it is thought, to extremely cold weather affecting other sites on the east coast.

It is believed that a consolidation of sediment and the impact of a rapid expansion of algae mats on to open mud may provide part of the explanation for the decline in Dunlin winter numbers at Teesmouth (Ward *et al.* 2003).

Small numbers of birds spend the winter away from the major aggregations of Teesmouth, on estuaries such as the Tyne and the Wear, with flocks of as many as fifty being recorded at Jarrow Slake and further inland around the Gateshead area, on the Tyne, where as many as 300, numbers swollen by early passage birds, have been noted in late winter and early spring. In recent years, such as in 2005 and 2009, South Shields held as many as 279 in the winter, and by 2009, the numbers in the north of the county were sometimes exceeding those around Teesmouth, though few of these were feeding in that area. In 2010, up to 350 were at Teesmouth in January, peaking at 474 in February.

Distribution & movements

The Dunlin has a circumpolar distribution, breeding in Arctic and sub-Arctic zones in Europe, Russia and Siberia, Greenland and Canada. Some birds also breed in moorland habitats at lower latitudes, such as those in Britain and Ireland. This species winters far south of most of its breeding range in Western Europe, around the Mediterranean, and down to the coasts of West Africa; as well as eastern China and North America. Birds of a variety of races winter around the coasts of Britain.

Small numbers of Dunlin are noted in both spring and autumn at suitable passage sites around the county with peak numbers occurring on autumn passage and smaller numbers in spring. Away from the main site of Seal Sands at Teesmouth, passage birds occur on a wide front across the county and birds appear in small numbers at many established wetland sites such as: Bishop Middleham, Derwent Reservoir, Rainton Meadows, Seaton Pond, Shibdon Pond, Smiddyshaw Reservoir and WWT Washington. Inland, 105 at Derwent Reservoir on 22 October 1995 was an unusually high count away from the coastal strip.

Extensive study and bird ringing by Durham University research groups since autumn 1976, involving the ringing of over 10,000 Dunlin at Teesmouth, has revealed the complexities of Dunlin movements through the Tees estuary area. This work has shown that three races of Dunlin visit Teesmouth: *C. a. alpina*, breeding in northern Scandinavia and Russia, occurs in winter; *C. a. schinzii*, from Iceland, is seen on migration in spring and autumn; and, *C. a. arctica*, from Greenland, is uncommonly recorded on passage.

Autumn passage usually starts in July at Teesmouth (a flock of 2,500 was at Greatham Creek on 17 July 2008) and Whitburn, this building through August and into September. At this time, active migration is sometimes noted along the coast, for instance on 12 October 1995, at least 1,645 birds flew south past Ryhope in less than three hours and 1,205 flew past Horden in an hour and 40 minutes on the same date.

Adult *alpina* and *schinzii* arrive at Teesmouth in July and early August, passing through *en route* for wintering areas in north west Africa. Ringing returns have shown that these birds follow a route along the French, Spanish and Portuguese coastlines, as illustrated by two recoveries: one ringed as an adult at Saltholme Pool on 26 June 1964 was recovered at Foz do Arelho, Estramadura, Portugal on 23 August 1964, and a juvenile bird ringed at Seal Sands on 24 August 1996 that was caught and released at Puerto Real, Cadiz, Spain some 2,045km south on 12 September 1999. Juvenile *schinzii* pass through in August and early September, followed by longer-billed *alpina* juveniles in September and October, some of these remain through the winter, others move south and west; some to the Irish Sea coasts (Ward *et al.* 2003). A bird ringed at Revtangen, Norway in October 1957 and recovered at Teesmouth in January 1958 along with three birds, ringed at Sappi, Turku and Pori, in Finland, that were controlled at North Gare between mid-January and mid-February 1978, indicated the origin of some of Teesmouth's wintering birds. Additionally a bird that had been ringed in Denmark, at Aflandshaga, Sjaelland, Denmark on 13 July 1976, was found dead at North Gare in February 1978. During 1978, an additional five birds ringed at Teesmouth in 1977 were controlled around other British estuaries, as far dispersed as Belfast Lough and the Isle of Sheppey, Kent.

Juvenile *alpina* follow a route from Russia around north and western Norway or down through the southern Baltic regions of Poland and Germany. Adult *alpina* usually only occur at Teesmouth from late October, after they have moulted, which they do in the Wadden Sea area or around The Wash. Large numbers of these birds move to Teesmouth from The Wash sometimes creating a late autumn peak count in October or November, so that numbers at this time of the year may be higher than in the spring or mid-winter. Additional arrivals of birds from the Continent, or other sites from elsewhere in Britain can occur through November and December, so the peak numbers across the Tees Estuary are often achieved in January.

In spring, most wintering *alpina* leave Teesmouth during early March, then passage birds swell numbers on the North Tees Marshes, particularly during April, a few staying until late May. A flock of 120 birds, presumably migrant *alpina*, in a field to the west of Rainton Meadows on 17 March 1999 being in non-wetland habitat was most unusual. Other birds, including some of those known to have wintered in Ireland, pass through in May. Passage of the shorter-billed *schinzii* and more northerly-bound *arctica* occurs in mid-May (Blick 2009). Some birds controlled at Teesmouth had been previously ringed in Morocco and western France in April but the main passage of *alpina* is along the Irish Sea coasts in spring, as indicated by birds ringed at Teesmouth in July and August and controlled in subsequent springs on the Severn estuary, in Morecambe Bay, Walney Island or the Solway Firth.

An adult ringed on South Uist in the Western Isles, on 4 May 1986 was found dead on Cowpen Marsh on 24 July 1986, whilst the ring from a bird ringed as a juvenile at South Bents, Whitburn on 27 August 1990 was found in an owl pellet on 10 March 1994 on the Isle of Lewis, Western Isles. Another bird ringed at North Gare in August 1979 was found dead at sea off Durafjordur, Iceland, on 5 June 1986. A few *schinzii* migrate in a southerly direction in autumn through western Britain, but move north to Iceland via the Tees, as shown by two juveniles that were ringed in August, one in County Cork, Ireland and the other on the Dee estuary and found at Teesmouth the following May (Blick 2009). A bird ringed at Teesmouth in February 1976, as a one-year old bird, was killed at Estremadura, Portugal, on 3 May 1993, making it over 17 years old.

Broad-billed Sandpiper
Limicola falcinellus

**A rare vagrant from Scandinavia: 17 records, May to September,
principally in May and June.**

The first Broad-billed Sandpiper for County Durham was found by Jack Bailey and the late Phil Stead amongst over 2,000 Dunlin *Calidris alpina* on Seal Sands, Teesmouth during the late afternoon of 13 August 1961, it was flushed shortly afterwards by youths. Fortunately the bird was again present on Seal Sands on the evening of 14th before relocating to Dorman's Pool at high tide in a pattern that would be repeated daily until last seen on the afternoon of 19 August (*British Birds* 55: 573). This bird remains the only August record for Durham and one of only four 'autumn' records, two of which have been juveniles in September.

All records:
1961	Seal Sands & Dorman's Pool, Teesmouth, adult, 13th to 19 August	
1973	Barmston Ponds, adult, Washington, 25 May (*British Birds* 67: 324)	
1974	Greenabella Marsh, Teesmouth, adult, 22nd to 29 June *(contra BB)* (*British Birds* 68: 319)	
1981	Long Drag, Teesmouth, adult, 31 May to 1 June (*British Birds* 76: 494)	
1986	Long Drag, Teesmouth, adult, 13 June (*British Birds* 80: 537)	
1987	Greatham Creek & Seaton Snook, Teesmouth, adult, 3rd to 4 May (*British Birds* 82: 525)	
1990	Tidal Pool, Greatham Creek, Teesmouth, adult, 23 June (*British Birds* 85: 526)	
1992	Seal Sands & Greatham Creek, adult, 3rd to 4 June *(contra BB)* (*British Birds* 86: 480)	
1992	Greenabella Marsh, Teesmouth, adult, 26 June (*British Birds* 88: 514)	
1994	Saltholme Pools, Teesmouth, adult, 5th to 7 May (*British Birds* 88: 514)	
1994	Tidal Pool, Greatham Creek, Teesmouth, adult, 22nd to 26 July (*British Birds* 88: 514)	
1994	Seal Sands & Greatham Creek, juvenile, 27 September to 3 October (*British Birds* 88: 514)	
2002	Dorman's Pool, Teesmouth, adult, 15 June (*British Birds* 96: 569)	
2003	Saltholme Pools, Teesmouth, adult, 6 June (*British Birds* 97: 580)	
2007	Saltholme Pools, Teesmouth, adult, 27 May to 1 June (*British Birds* 101: 539)	
2010	Saltholme Pools, Teesmouth, adult, 6th to 15 June (*British Birds* 104: 581)	
2010	Seaton Snook, Teesmouth, juvenile, 13 September (*British Birds* 104: 581)	

There is just the one record away from the North Tees Marshes, an adult at Barmston Ponds, Washington, on 25 May 1977, which is especially noteworthy for being at an inland locality.

Distribution & movements

Broad-billed Sandpiper breeds in the forest bogs of northern Norway, Sweden and Finland, eastwards into adjacent north western parts of Russia, with other scattered populations further east in Siberia. European birds migrate south through the eastern Mediterranean to winter from the coasts of East Africa to the Indian subcontinent. It is a regular vagrant to the British Isles, most frequently in May and June. There had been over 230 records in Britain by the end of 2010. County Durham is second only to Norfolk in the number of this species recorded.

Buff-breasted Sandpiper
Tryngites subruficollis

A rare vagrant from North America: 15 records, May to October, most often from August to October.

County Durham's first was found by Andy Goodwin and Jonathon Guest at Saltholme Pools, Teesmouth on 10 September 1977 (*British Birds* 71: 501). This bird formed part of a record influx of the species into Britain during the autumn of 1977, which involved an estimated 52 birds, most of which were in September.

All records:

1977	Saltholme Pools, Teesmouth, 10 September
1980	Saltholme Pools, Teesmouth, 18th to 21 September (*British Birds* 75: 502, *British Birds* 76: 495)
1985	Long Drag & Saltholme Pool, Teesmouth, adult, 17th to 18 August, and 26 August to 4 September
1989	Haverton Hole, Teesmouth, adult, 1st to 3 June
1990	Whitburn Steel, adult, 4th to 8 August
1994	Saltholme Pools, Teesmouth, juvenile, 3rd to 4 September
1995	Saltholme Pools, Teesmouth, juvenile, 25 September, same Belasis Technology Park, Billingham, 26 September
1996	Whitburn Steel, juvenile, 16 September
1997	Rainton Meadows DWT, adult, 7 May
2007	Greatham Creek, Teesmouth, adult, 13 August
2007	Seaton Carew Golf Course and Hartlepool Power Station, juvenile, 23rd to 25 September
2007	Saltholme Pools, Teesmouth, juvenile, 2nd and 6 October
2009	Saltholme Pools, Teesmouth, juvenile, 24 September to 12 October
2009	Whitburn Rifle Range, juvenile, 4 October
2011	Whitburn, juvenile, 14 September

Durham records are typical of east coast sightings in England. They include five birds aged as adults, two in spring, and three in autumn. All of the latter were in early to mid-August and fall outside of the typical occurrence pattern in Britain. Hence they are presumed to probably relate to birds that have made a Transatlantic crossing in a previous autumn.

Distribution & movements

Buff-breasted Sandpiper breeds in the high Arctic of North America from northern Canada to Alaska, with a localised population in extreme north eastern Siberia. A noted long distance migrant, most birds migrate southwards through the North American prairies to winter on the grasslands of central South America as far south as Argentina. Some birds are also known to take a more easterly migration route, crossing the Great Lakes to New England, and then heading out over the sea to north eastern South America, and it is these birds that are assumed to be the source of the vast majority if not all of the vagrants in Western Europe. It is one of the most regular North American waders to be recorded in Britain. The majority of British records are of juveniles in late September and early October, predominantly from western localities.

Ruff
Philomachus pugnax

An uncommon but regular passage migrant, scarce in winter and historically a rare breeder.

Historical review

A visit to Teesmouth today, at almost any time of the year, might reveal individuals of this species. It occurs as both a spring and autumn passage migrant, and in small numbers both as a wintering bird and during the summer months. It has a very long association with Durham, and the hinterland of the Tees estuary has probably attracted it for many centuries.

Benett recorded the payment of "*1d*" to Willelm o' Sherston, for the purchase of a "*ree*" for consumption at the Monastery at Durham, in the 14th century (Gardner-Medwin 1985). "*At one time the Ruff probably nested in many suitable places from Northumberland to the south-west, though ornithological history records its disappearance only from its later breeding areas in such counties as Durham, Yorkshire, Lincolnshire, Huntingdonshire, Cambridgeshire, Norfolk and Suffolk*" (Cottier & Lea 1969). According to Holloway (1996), this was a widespread breeding species, though uncommon, along the east coast counties of England up to the end of the 18th century, and breeding took place in Northumberland, Durham, and other counties along the English eastern seaboard into the early 19th century. By 1825, it was said to have ceased to breed on the marshes about the mouth of the Tees (Holloway 1996), though it bred at Prestwick Carr, Northumberland until at least 1853 (Kerr 2001) and, on an occasional basis, until 1865 (Holloway 1996).

Sharp (1816) described it as a "*very rare*" bird in the Hartlepool area, whilst Hogg (1827) said it was "*extremely rare*", though he noted that it occurred as a passage migrant at Teesmouth. By contrast, Edward Backhouse (1834) described it as "*not uncommon*" at Teesmouth, though he said that it was rarely met with in breeding plumage. He did note however that one was shot at Whitburn Moors, "*a few years ago*". Hutchinson (1840) noted it as being occasionally found in small parties in September and October in Durham, also that it was met sparingly at the carrs of the interior. For example, he wrote of one being got at Framwellgate Carr one June, which may even hint at breeding, and of large numbers brought to market in the autumn of September/October 1840.

Temperley (1951) described it as being a "*fairly regular autumn passage migrant, in small numbers*", and commented that it "*has been known to breed*" however, he gave no evidence that breeding was ever regular in the past.

Breeding occurred at Teesmouth at the turn of the 20th century, 'probably in 1901', definitely in 1902, and possibly in 1903 (Nelson 1907, Temperley 1951) but not apparently in 1904, as was suggested by Bannerman (1961). These breeding attempts took place in an area of marsh where Hartlepool Power Station is now situated (Joynt *et al.* 2008); in 1901 a Ruff and a Reeve were seen behaving as if they had young there, though none were seen. In 1902 two pairs arrived and two nests with eggs were found, but it was said both were inadvertently destroyed. Millburn (1915) however stated that in 1902 after one of the nests was flooded, this female laid a replacement clutch within ten yards of the original site and that both females reared young. A clutch of eggs in the Dorman Museum, Middlesbrough labelled North Tees Marshes 1902 is probably from the flooded nest. Birds were present in 1903, but no nest was found. In addition to birds breeding at Teesmouth, Tristram (1905) had it on "*good authority*" that Ruff formerly bred at Boldon Flats (see Backhouse's reference to the bird shot at Whitburn Moors sometime before 1834).

Temperley (1951) observed that autumn passage birds were irregular in timing, but occurred mainly from mid-August, though they could be present right through to the end of September. He noted a wintering record, a bird being at Hebburn Ponds on 15 December 1929 and likewise in December 1928 at the Darlington Sewage Farm. On this subject W.E. Almond wrote, "*The majority of the passage waders, which were to be seen on the Darlington Sewage Farm in the autumn of 1928, departed by the beginning of October. On December 9th, however, one Ruff (Philomachus pugnax) and about a dozen Curlew-Sandpipers (Calidris testacea) were seen. This was during a spell of very severe cold. The Curlew-Sandpipers were not seen again, but the Ruff evidently stayed the winter, for it was seen in January, February, April, May and July of 1929. By the end of April it was assuming breeding plumage. I believe this bird remained until it was joined by others of the same species on passage in August*" (Almond 1930).

Little was documented about this species through the 1950s. However, as well as passage birds, an occasional wintering individual was found, for example one at Jarrow Slake on 20th and 21 February 1954. The species continued a small breeding season presence through the 1960s at Teesmouth. In this respect at least four, including two males in breeding plumage, were at Teesmouth in April 1960, with a pair there on 25 June, the male being present to at least 29th of the month. Birds wintered at Teesmouth in 1961 and birds were present in the county throughout the year, excepting June, with some breeding activity recorded, again around Teesmouth. Numbers of spring passage birds were noted on Cowpen Marsh from 12th to 18 April 1964, when eight birds were present. This pattern of presence continued through the decade. In 1967 birds were at Teesmouth throughout much of the year, with five there on 31 January. Up to eight were at Cowpen Marsh in May with at least one remaining into June. Barmston Pond also attracted birds in the autumn, with up to 35 there on 8 September 1967. A similar pattern occurred to the end of the decade, with spring males displaying in April, and odd birds being essentially 'ever present'.

Through the early 1970s, breeding plumage males were recorded at Teesmouth annually and in 1973 and 1974 at Barmston Pond, with up to 11 at the latter site on 18 May 1974. In 1973 two or three of the males remained through June and July at Teesmouth, and four or five birds were there through the summer. Over this period, wintering birds were routinely recorded in small numbers and the species occurred in the county during every month of the year.

Recent breeding status

Since Temperley's report of breeding (1951), at a time when there was more extensive suitable habitat, there has been no definite proof of success, despite the strong suggestion of birds showing an interest in breeding around Cowpen Marsh in the 1960s and early 1970s, and possibly even at Barmston Pond, in the 1973-1975 period.

In recent years, the only significant breeding season records or breeding behaviour have come from the North Tees Marshes. In this extended complex of wetland sites, passage birds are present in most springs and in many years (for example, in every year between 1988 and 1992), summer plumage males were noted as present, often in May and quite regularly into June, with birds displaying in some years. In 1992, at least two males and one female were noted together over a considerable period of time in these critical months and breeding cannot be discounted. It is worth noting that birds have summered on a regular basis around the Druridge Bay area, in neighbouring Northumberland, since the early 1990s, and up to twelve males displayed in that area, with breeding suspected on a number of occasions (Kerr 2001).

Britain lies at the western edge of the breeding range of Ruff so breeding season numbers will always be small. Nationally, probably fewer than five females breed in any one year (Gibbons et al. 1993). However, it is feasible, that this species could nest again in the county, since lekking males are quite regularly reported in June, though proving this is likely to be difficult; females when nesting or with young are very secretive. Their preferred breeding habitat is low lying, wet grassy meadows which are usually grazed and flooded for part of the year.

Recent non-breeding status

The bulk of spring passage at the sites around Teesmouth occurs between mid-March and late April. Though this is typically in single figures and has remained so over recent decades, up to 13 were present at this time in 2009. Small numbers are scattered elsewhere and some often linger through to July.

During the 1990s peak passage counts at Teesmouth demonstrated a bi-modal peak during the decade, with numbers rising to maximum daily counts of around 90 in 1993 and 72 in 1997; suggesting some level of cyclical breeding success on the main breeding grounds, or weather related displacement which then affected passage numbers at Teesmouth (Armstrong 2003).

Most of the Ruff passing through Britain in autumn are heading to wintering grounds in central and southern Africa. Locally, autumn passage is much more pronounced than it is in spring. The movement of adults through the county is underway by late July. The first juveniles arrive by early August and their numbers peak later that month or in early September. Counts of 60-90 birds are noted in most years, for example in mid-August 2003 around 70 autumn passage birds were recorded across the county; though 176 were recorded on 8 August 1980 (Baldridge 1981). Analysis of figures from Teesmouth in the five years leading up to 2009 indicated an average autumn peak of 39, with a maximum count of 58 birds.

An excellent passage occurred in 2008, particularly in the north east of Durham, where extensive floodwaters due to heavy rain brought up to 44 were at Boldon Flats on 11 September. A total of 73 were in this general area on 17 September 2008. Numbers usually decline rapidly in late September with only a few birds present in October and a rump of single figures into the last two months of the year. Whilst the North Tees Marshes are easily the most important suite of sites for this species, autumn passage Ruff are often recorded at wetland sites across the whole county including in recent years: Barmston Pond, Bishop Middleham, Boldon Flats, Bowesfield Marsh (up to 28), Derwent Reservoir, Hetton Lyons, Hurworth Burn Reservoir, Longnewton Reservoir (up to 55 in 1985), Rainton Meadows, Seaton Pond, Shibdon Pond, Smiddyshaw Reservoir, Whitburn Steel, and, WWT Washington. Almost any water or wet field in the county can attract some passage individuals or small flocks.

A few birds linger into the winter in the south east, largely around Teesmouth. These birds are part of a larger sub-population, which winters in the north east portion of North Yorkshire and has done so since at least the early 1980s (Baldridge 1980, Mather 1986). In the last three decades there has rarely been more than three or four Ruff wintering at Teesmouth, though around 10 were noted in 1985 (Baldridge 1986), and nine in 1989 (Armstrong 1990). More recently 11 were present in January 1999 and seven in January 2000 (Armstrong 2003 & 2005). Around 20 birds were wintering at four localities on the North Tees Marshes in 2001 (Siggens 2005) and Bowesfield Marsh was a key winter site in January 2009, with nine birds present there (Newsome 2010). Elsewhere in Durham, wintering birds have always been quite rare and usually only singles. Examples include: a female at Shibdon Pond, on 31 December 1983 (Bowey *et al.* 1993); singles at Longnewton Reservoir in January 1986; on the River Tyne at Blaydon, on 1 January 1999 (Armstrong 1987 & 2003); and, in 2010, two birds were at Castle Lake, Bishop Middleham in February (Newsome 2012).

Distribution & movements

Ruff breed across northern and north east Europe, eastwards into Russia and further east still, to Siberia. Their distribution is a patchy one. Wintering grounds are spread between Western Europe, Africa, east to India, though small numbers of birds, usually males, remain much further north, within the southern portion of their breeding range. These wintering areas are highlighted by the scant ringing data for this species. For example, a first-year bird ringed at Saltholme Pools on 19 August 1974 was killed at Matera in Italy on 25 March 1975; whilst much further to the south, a first-year bird that had been ringed at Saltholme Pools on 25 August 1980 was reported in the West Flanders part of Belgium on 31 July 1981 and then at Dialloube in Mali, on 15 April 1982 (Robinson & Clark 2011).

Jack Snipe
Lymnocryptes minimus

An uncommon passage migrant and a regular winter visitor in small numbers.

Historical review

The 'judcock' or 'half-snipe' was described by Robson (1896) as a regular winter visitor to the Derwent valley (Bowey *et al.* 1993). Temperley (1951) called it a "*passage migrant and winter visitor of irregular occurrence*", "*usually in small numbers*". Temperley noted that it usually arrived in October staying until March and he told of an impressive flock and one of the largest such gatherings recorded in Durham, of 13 birds that was flushed from a marsh at Birtley on 16 January 1932. Nelson (1907) recorded one as early as 1 October and the highest number recorded by him at Teesmouth was six, on 23 October 1900. One at Darlington Sewage Farm on 13 August 1931 was a very early autumn bird. The latest historical spring for Durham, as documented by Temperley, was of one at Boldon Flats on 4 May 1934.

Through the 1950s and 1960s, as observed by David Simpson, this species was a regular winter visitor to the Shotton Brickworks Pond, with birds arriving in September and leaving in April. He commented that at one spot where he "*still sees them today*" he used to see them in the 1950s, illustrating this species' use of traditionally occupied wintering locations. He commented upon the species' behaviour. "*During the day they would rest on stubble fields, potato fields and a part of Edder Acres Plantation and then they would 'flight' into the pond area to feed at dusk".* On 6 March 1962, ten birds were present at the Shotton Brickyard Ponds and on 8 March of that year, 15 were counted there (Simpson 2011).

Birds were widely reported in small number through the 1960s, but the detail was relatively poorly documented. In 1969, there were reports of singles from many localities around Durham in the autumn, with nine at Shotton Brickyard Ponds on 5 November and a similar number there on 4 January 1970.

The Shibdon Pond/Derwenthaugh Meadows complex was for many years, during the 1970s and 1980s, one of the principal sites for this bird in the county, with for example 10 there in October 1980, though numbers declined during the 1990s and the early 21st century following habitat changes. The largest number counted there during the last three decades of the 20th century was 15 on 3 November 1973. Chilton Moor also attracted this species on a regular basis throughout the winters of the 1980s.

In 1985, it was estimated that *c*.30 birds were scattered across Durham (Baldridge 1986). However, the Derwenthaugh counts and the 12 on Portrack Marsh in December 1987 (Armstrong 1988) are more indicative of their true status in Durham.

Recent status

There is little doubt, that this remarkably subtle and beautifully marked snipe, remains one of the most under-recorded, least known and least understood of wintering birds in County Durham. It is very difficult to see when on the ground and will only fly up when approached to within a metre or two. A bird injured by a mechanical grass trimmer during management work in October 1996, at Shibdon Pond, demonstrated how confiding this species can be.

In recent winters, small numbers have been noted in rushy flushes on open moorland, provided they don't freeze over, such as the six with 13 Snipe *Gallinago gallinago* on 8 November 2010 on Wolsingham North Moor (M. Passant pers. comm.). It is likely that a significant total population could be present across the county in such under-visited locations. This would appear to have been confirmed by an experience in the early 1990s, when tens of birds were caught or seen in the lamps whilst wader research workers were attempting to net Golden Plover *Pluvialis apricaria* and Woodcock *Scolopax rusticola* on the Durham moors (R. M. Ward pers. com.). Consequently, although the majority of records are of birds in the eastern half of the county, this may as much reflect the pattern of observational activity as it does truly match the actual winter distribution of the species.

This bias was apparent in 2008 when no birds were reported further west than Far Pasture and Watergate Park in Gateshead and were recorded at 21 locations in central and east Durham, and in 2009 when reports came from nearly 30 such widely scattered sites. Favoured areas, in which birds have been regularly recorded over recent decades include: Cassop Vale, around Chilton Moor, Daisy Hill, Hetton Bogs, Hetton Lyons, Rainton Meadows, Seaton Pond, Sedgeletch, Shibdon Pond, Shotton Ponds, Whitburn Coastal Park and WWT Washington.

A clutch of sites around Teesmouth are frequented by Jack Snipe, in particular the Dorman's Pool-Saltholme-Cowpen Marsh area, which attracted 19 birds on 10 November 1973. There were 20 at Portrack Marsh on 27 November 1994 and again in late January 2003, whilst 20 were at Haverton Hole on 26 October 2005 and 18 October 2010 (Blick 2009).

The highest counts usually occur in January and February with a small peak sometimes noted in March/April, which probably comprises passage birds. The last birds are normally noted in mid-April, rarely in early May. In central Durham, sightings tend to decline from the second week of March. There is sometimes a small influx in April, indicative of passage through the county. The last birds of spring are usually seen around early to mid-April, late records making it into the period 13th to 19th of that month. Exceptionally late recent records in May include birds in 1973 at Teesmouth on 2nd, in 1983 at Whitburn on 14th, in 1995 at Seaton Pond on 13th and at Haverton Hole on 7 May 2008.

The first returning autumn birds are usually reported in the last week or few days of September or more regularly in early to mid-October through to early November. At this period birds are often noted from sites in the coastal strip, as birds pitch down after an arduous sea crossing. It is not unusual to see a bird or two arriving from the sea along the coastline in October and November. Exceptional counts from Hartlepool Headland include: six, 13 and 20 birds recorded over three consecutive dates in October 2005 (Newsome 2006).

Allowing for the biased distribution of observers in the county, several have commented that there seems to have been a trend towards lower numbers of wintering Jack Snipe locally over the last 15 years or so. This trend may be related to the run of mild winters from the mid-1980s, though there was an increase in the number of sites from which the species was recorded in the hard winter of 2010, a total of 30 sites, when this figure had been no more than 20 for rest of that decade.

Distribution & movements

This species can be found throughout most of Scandinavia as a breeding bird, and to the east across the northern and mid-latitudes of Russia and Siberia. European Jack Snipe winter to the south west of their breeding areas, largely in southern and Western Europe. Most Durham birds are presumed to be from the Scandinavian population. The latest estimate of the British winter population in the period 2004-2009 was 100,000 (Musgrove *et al.* 2011). This seems high compared to the known numbers in Durham and is perhaps further confirmation that it is significantly under-recorded locally.

Over a period of many years, from the late 1970s to the early 1990s, three birds were caught and ringed at Shibdon Pond and Derwenthaugh Meadows; they generated no controls or recoveries.

Snipe (Common Snipe)
Gallinago gallinago

A common breeding bird in the county's western uplands but a rare and local breeder in lowland locations; also a common winter visitor to wetland areas.

Historical review

This species has been present in Durham for many centuries. It appears in the Bursar's account rolls for Durham in 1430 (Ticehurst 1923). Similarly, Benett (1844) in the *Durham Household Book*, recorded the purchase of '*snype*' on two occasions, one bought along with a '*plover*' for 2½d in 1533, and four more bought for *1d.* (Gardner-Medwin 1985). The fact that Snipe were recorded so rarely, would suggest that they were either uncommon, or difficult to catch.

"A resident" visiting marshy and wet grounds in winter, were Robson's comments on this species' local status in the Derwent valley in 1896, probably a fitting description of its status over much of County Durham during the 19th century. Nelson (1907) described Snipe as a local resident and reported that it was *"very numerous"* in some seasons at Teesmouth, noting *"I have flushed a large 'whisp', which might almost have been called a flock, of fully one hundred, evidently newly arrived"*. This was presumably in the autumn or winter. Snipe was mentioned by Richmond (1931) as a resident breeding bird at Teesmouth a quarter of the way into the 20th century.

In 1951, Temperley called it a *"common resident, passage migrant and winter visitor"*. He said the species bred throughout the county in moist places, bogs and marshes, noting that it was found in all suitable places in the western moorlands valleys, while in the lowlands, *"every bog or marsh of any size has a few pairs"*. In winter it was found in considerable flocks around favoured feeding sites, Temperley recording 40 to 60 birds at Hebburn Ponds for example, during September, October and November.

In the south east of the county, the local breeding population seemed to have fallen considerably by the 1960s, largely due to industrialisation and drainage of wetland habitats around Teesmouth (Stead 1964). As a result, just ten pairs were found, restricted to the area around Saltholme and Cowpen Marsh, in the period 1962-1967. Following further declines towards the end of the 1960s, there were probably not more than five pairs present towards the end of that decade (Stead 1969).

In contrast, the wintering populations around this time were more impressive. On 25 October 1970, for example, a total of 196 were counted at Teesmouth, at Brenda Road Pool and Saltholme Pool. Large numbers were also present in the county in the autumn of 1973, with 171 at Witton-le-Wear on 1 October, up to 200 in the Barmston Ponds area in early November, and a combined total of 440 around the North Tees Marshes on 1 November (Stewart 1971 & 1974).

Recent breeding status

Snipe are well-represented as breeding birds in Durham and, as noted in Westerberg & Bowey (2000), they do not seem to have declined locally to the same degree as has been recorded nationally, especially in the more southerly parts of England (Parslow 1973, Gibbons *et al.* 1993). The distribution mapped in the *Atlas* is still believed to be a largely accurate representation of the species' breeding range in the county (Westerberg & Bowey 2000).

The majority of the county's breeding population can be found west of a line running north to south west of Hedleyhope Fell, with one regular easterly breeding outpost being at Chapman's Well. The Snipe remains a common and widespread breeding bird in the Durham uplands, especially in the wetter areas, where moorland flushes and wet pastures provide muddy feeding areas and tussocky vegetation for nesting (Westerberg & Bowey 2000). The sound of the species' characteristic drumming is a familiar backdrop to a spring walk in the county's uplands, and can be heard at any time from early to mid-March through into early to mid-June.

An RSPB survey in 1993, covering 3,391 hectares of Baldersdale and Lunedale, revealed 165 drumming males and this was believed to be an underestimate of the true numbers (Shepherd 1993). Today, numerous breeding season reports come from the western uplands, with display noted typically from the second week of March. This was illustrated by at least eight birds being recorded in just 100 square metres of pasture at Langdon Beck in April 2007. The following year, concentrations of displaying birds during April and May included over 30 drumming birds at Balderhead Reservoir, 20 at Gilmondby Moor, 50 near Grassholme Reservoir and more than 20 at Langdon Beck, but birds were present in almost all upland areas from Derwent Reservoir in the north to The Stang in the south (Newsome 2008 & 2009). Rough pasture and 'white moor' are often the most productive habitats and hold the highest breeding densities. Birds are only absent from the highest fells, such as Chapel and Cronkley Fell.

It is possible that any programme of extensive improvement of the upland meadows and pastures that adjoin the county's western moorlands might have significant negative impacts on some local populations of Snipe (Baines 1988). A shorter, more even sward and drier surface would make fields less suitable for the species, and higher stocking rates may result in more nests being trampled (Westerberg & Bowey 2000). Nevertheless, Snipe can still be found extensively across our uplands.

'Drumming' is probably more difficult to detect when only one or two pairs of birds are present, especially where arable farming predominates, as in the eastern parts of the county. In such areas breeding Snipe are only to be found on a few wetland fringes near some rivers and ponds, or adjacent to more extensive wetland complexes (Westerberg & Bowey 2000), such as around the mouth of the Tees.

During the 1970s, drumming Snipe were present on the North Tees Marshes each spring and this continued, at a relatively low level, through the 1980s and into the 1990s, though breeding was rarely confirmed during this period (Joynt *et al.* 2008). The survey work relating to *The Breeding Birds of Cleveland* located a cluster of territorial birds around the North Tees Marshes in the period 1999-2006, with one or two pairs in other wetland locations in that area. This suggested some level of increase since the 1970s and, based on the number of occupied territories and estimated numbers within these, there are likely to be around 20 territories around the North Tees Marshes and other local areas today (Joynt *et al.* 2008). This apparent increase is in contrast to the national trend over the same period.

Elsewhere in the east of the county, in the first decade of the 21st century, the numbers of occupied lowland sites have ranged from eleven in 2000 to just four in 2001, but the species is likely to be under-recorded in lowland areas. In 2008, although breeding was only recorded at four lowland sites, display was noted at a further two sites, and mid-summer records came from another six sites. The following year, breeding was confirmed at RSPB Saltholme, with two pairs, and also in the east at Moorsley, with display being noted at Elemore and Shotton Pond during April and May. Wetland creation and restoration at sites such as Burdon Moor and Lamesley Pastures in Gateshead during this period has created lowland breeding habitat for waders, including Snipe overall however, to the east of the A1 trunk road, this species is not a common breeding bird in the county.

'Drumming' is often noted from early March and birds may have chicks by mid-April, especially in years with mild late winter and early spring weather. A bird was noted back in breeding territory at St. John's Chapel in upper Weardale on 15 March 1996, despite much of the ground in the area still being frozen (Armstrong 1998). The last 'drumming' birds are usually heard around early to mid-June.

In southern England, the length of the Snipe's breeding period is determined by the time for which the ground remains soft enough for the birds to probe for invertebrates (Green 1988). The drying out of foraging areas is not as critical in western County Durham, as the Pennines receive considerable rainfall throughout the spring and summer, as illustrated by the finding of a bird on eggs in Northumberland as late as mid-July (A. N. Hoodless pers. comm.). Gibbons *et al.* (1993) recorded that the Pennines held some of the highest densities of breeding Snipe in Britain, and this remains so today.

The very best areas may hold in the region of 15 pairs per square kilometre. The *Atlas* determined that a population based on this maximum density, but assuming a low level of occupancy of one tenanted km^2 per occupied tetrad, would equate to an estimate of 2,500 breeding pairs in the county (Westerberg & Bowey 2000). This figure is probably still reasonably realistic today. The latest national breeding population estimate was 52,000 (Baker *et al.* 2006).

Nationally, a rapid decline in breeding Snipe numbers has been recorded in lowland England since the 1970s as a result of agricultural drainage (Baillie *et. al.* 2010). However, the overall position is unclear, with a suggested population rise in the UK as a whole in the period from 1995 to 2008, but not a statistically significant one, and virtually no change in the English population in the same period (Risley *et al.* 2010). Snipe is on the amber list of Birds of Conservation Concern, because of an overall decline in its European population (Robinson 2005).

Recent non-breeding status

There is a noticeable decrease in lowland sightings in spring and summer, but by early autumn flocks begin to appear at suitable wetland sites, and these may well be locally or regionally-based breeding birds, descending to more lowland sites for the winter, or prior to moving further south. Birds begin to gather at favoured wetlands from mid-August, and counts of scores of birds from such sites are not unusual by the end of the month. For example, in 2010, 46 birds had already returned to Castle Lake, Bishop Middleham, by 6 August, with 65 counted there just eleven days later on 17th (Newsome 2012). Other high early autumn counts include c.100 in the Mordon Carrs area at the end of August 2004 and 96 at Barmston Pond on 18 September 2005 (Siggens 2006, Newsome 2006).

Most birds have left the county's exposed uplands by the end of August, with very few remaining in such locations by the end of September. Post-breeding gatherings also sometimes occur in western areas, however, with records to date including a high count of 48 at Gilmondby Moor on 3 September 2009 (Newsome 2010).

This lowland build-up takes place prior to any major coastal influxes, which tend to be in the period from late September to mid-November, but often concentrated in October. Numbers then continue to increase through the autumn as immigrants arrive, peaking from November to January. This inward movement in autumn probably contains a fairly large percentage of Scandinavian and northern European birds, as discussed in the distribution and movements section.

Birds can be found scattered across the county during the winter, when small numbers might be flushed from any damp fields or wet patches of ground, though the largest concentrations tend to occur at traditionally occupied wetlands, such as the suite of wetland sites around the mouth of the Tees. Larger flocks are less common, and confined to a small number of sites. The breadth of wintering Snipe distribution across the lowlands of the county was illustrated in 2007, when records came from over 90 locations during the year, the majority of which were of wintering birds (Newsome 2008).

Peak winter counts across the county in the early 2000s were lower than in the past at some sites, and this may relate to the milder winters in both the UK and on the Continent, which were experienced at the time. This may have led to smaller numbers of birds coming into the county from overseas during the autumn. It is also possible, however, that the decline in the European Snipe population, as noted previously, may be having some impact on numbers here. There have been particularly noticeable declines in the reports of some formerly regular large winter gatherings since the late 1970s and 1980s.

For example, Shibdon Pond and its immediate surroundings once held very large numbers of Snipe from October to late February. The largest number estimated there was a gathering of 300 during the early winter of 1977. During the early 1980s, this winter peak began to decline to between 150 and 200 birds, whilst by the late 1980s, mid-winter gatherings of 100-120 birds were more usual (Bowey *et al.* 1993). The flight patterns and feeding areas of Snipe at Shibdon Pond were closely studied from 1983 to 1987 and nets were set in feeding areas, resulting in catches of over 80 birds at this site and a number of notable ringing recoveries, which are discussed in the distribution and movements section. Further declines in peak counts occurred here during the 1990s and today, counts of tens of birds rather than hundreds would be a more usual mid-winter peak at this site.

Many lowland wetlands around the county hold small numbers of birds during the winter, with frequent reports of from 10 to 30 birds from a number of sites. Higher recent winter counts, though not matching those recorded historically at Shibdon include 150 at Bowesfield Marsh, Preston-on-Tees, on 14 February 1998 and 124 at Billingham Beck Valley on 27 November 2002. In the north east of the county, Rainton Meadows and Sedgeletch, near Houghton-le-Spring, are favoured wintering sites. Rainton Meadows had a peak count of 125 birds on 28 December 2002 (Armstrong 1999b, Newsome 2007).

In most winters, about 100-150 birds around Teesmouth would be considered a typical maximum count, although in view of its skulking habits, the species is probably under-recorded. By far the highest count of this species in many years came from Portrack Marsh, with c.400 on 9 March 2003, then a new record count for Cleveland. This was matched by 400 at Saltholme Pools during December 2008 (Joynt 2004 & 2009).

Although most wintering birds are usually in low-lying areas of the county, small numbers occasionally remain in the uplands during winter, especially when mild weather has dominated, such as in the early part of the 21st century. For example, a single bird was at Harthorpe Head Quarry, at 620m above sea level, on 20 January 2001, twelve were found on Wolsingham North Moor in January 2007 and three were near Blackton Reservoir, at 350m above sea level, on 11 December 2008 (Siggens 2005, Newsome 2008 & 2009).

The uplands tend to be less frequented during cold periods of the winter months, or abandoned completely when freezing conditions prevail. When the ground is frozen in hard weather, birds are forced to move, even in the lowlands, and can be found feeding around springs and the edges of streams. In severe winter weather, Snipe abandon wetland habitats and may occasionally resort to coastal fields along the county's eastern seaboard.

A severe freeze in December 1981, for example, drove many birds 'out of habitat', and several Snipe corpses were found on the beach at Seaburn. Similar conditions in January 1984 resulted in an exceptional 138 in the Whitburn area on 31st, as birds were forced towards the coast. A cold weather influx on 23 November 1993 produced a site record count of 151 at Boldon Flats. More recently, 11 were in clifftop stubble at Seaham and three were on South Shields Leas on 29 December 2005, coinciding with other hard weather movements (Newsome 2006). Cold weather at both the beginning and end of 2009 produced two notable hard weather records at Cowpen Bewley, with 36 seen following a tractor in a ploughed field during March, and 15 feeding behind sheep in a frozen field during December (Joynt 2010).

The species is undoubtedly more numerous in some autumns than others and it is not unusual to see Snipe arriving over the sea in October and November, indicating the overseas origin of many of our coastal and wintering birds. Records of such visible migration include nine flying south at Horden on 13 November 1994 (Armstrong 1996), seven in off the sea at Hartlepool Headland on 29 December 2000 (Iceton 2001), ten in off the sea at Dawdon on 18 August 2003 (Newsome 2007), and ten south at Hartlepool Headland on 8 September 2009 (Joynt 2010). Another notable passage record concerns a single migrant seen feeding along the rocky shore at Whitburn with Ringed Plover *Charadrius hiaticula* and Dunlin *Calidris alpina* in October 1991 (Armstrong 1992). Snipe are less often seen returning north on spring migration, so approximately 20 flushed from an area of playing fields on the edge of South Shields in March 1980, probably on passage, was a notable record (Baldridge 1981).

Other notable non-breeding Snipe records include a partial albino, with white primaries, at Saltholme on 31 July 2004, and a melanistic bird in the same area from 22 July to 29 August 2008 (Siggens 2006, Joynt 2009).

Distribution & movements

Snipe breed over much of northern Europe, Asia and North America, as well as in South America, and south and east Asia, though a number of races are involved in this wide breeding range. To a large extent, breeding birds winter to the south of their breeding range, though in Europe and particularly in Britain, the breeding and wintering ranges overlap.

Movements of birds between Durham and Scandinavia are confirmed by the following ringing records. A bird ringed at Pity Me on 10 September 1959 was found dead on 5 September 1964 at Lausviken, Gotland, Sweden. Another ringed near Pori in Finland in August 1975 was re-trapped at Teesmouth in February 1976 (Blick 2009).

Between the mid-1970s and the early 1990s over 200 Snipe were ringed at Shibdon Pond and this generated a number of interesting recoveries, the most spectacular of which were of two birds which had travelled far further than the expected Scandinavian breeding grounds, being found over 2,000 kilometres to the east, in what was the USSR. The first was ringed at Shibdon on 3 September 1975 and reported, presumably shot, at Goretskiy, USSR in July 1977. The second was ringed on 25 February 1983 and was shot at Smolensk, Belarus on 17 September 1983 (L. J. Milton pers. comm.).

Some probably locally-bred birds move south west as the weather worsens, as shown by the bird ringed at Billingham in August 1971, which was in Co. Kerry in Ireland in January 1973 (Blick 2009). Likewise a wintering adult ringed at Wareham in Dorset on 21 January 1982 was re-trapped at Saltholme on 14 August 1983, illustrating movements within the UK (Bell 1984).

Great Snipe
Gallinago media

A rare vagrant from Scandinavia and eastern Europe: at least 9 records, August to October.

The first authenticated record for County Durham is of "*several*" said to have been shot near Sedgefield in the autumn of 1826 (Selby 1831). Hutchinson (1840) wrote 'a few have been killed in Mordon Carrs', which could have related to these birds, as the site is close to Sedgefield, and was a popular shooting locality at the time. In similar vein there was a reference to "*five or six shot in marshes in Durham*" in a review of the species printed in the Northern Echo in 1891 (Northern Echo 1891). These records were referenced by Tristram (1905), who wrote "*Rarely an autumn passes without one or more specimens being recorded, I possess a specimen shot at Sedgefield in 1826*". This record is accepted by BBRC though the current whereabouts of the specimen(s) are unknown, a peculiarity of all of the county's Victorian records.

All records:

1826	near Sedgefield, autumn, several shot (Selby 1831)
1830	Thornley, September, shot (Hancock 1874)
1830	Witton-le-Wear, October, shot (Hancock 1874)
1833	near Bishop Auckland, shot (Hancock 1874)
1885	Burnledge-in-Weardale, early October, shot (The Field 1855)
1896	Romaldkirk, 30 September, shot (The Field 1896)
1901	Teesmouth, 1 September, shot, (Milburn 1903)
1976	Cowpen Marsh, Teesmouth, 21 August (*British Birds* 70: 421)
1976	Hartlepool Headland, 23 September (*British Birds* 70: 421)

In addition to these records, a bird was said to have been flushed from vegetation between Saltholme Pools and Dorman's Pool on 3 September 1994 (Armstrong 1995), but as this record wasn't submitted to the BBRC, it remains 'unproven'.

The Great Snipe was seemingly recorded more regularly in Durham during the 19th century though many records were poorly documented and offer largely anecdotal evidence. An example of this is the first known reference to the species in the county, which was made by Hogg (1827) who wrote, "*the only specimen know to have been seen here was killed by a Mr John Grey as it was flying over the Tees about four years ago*". This record is not included in the national totals though there is little reason to doubt its authenticity, although by the standards of modern scrutiny it must be considered outside of the formal county record. Selby (1831) regarded the species as "*a rare and occasional visitant*", though he offered no records other than those at Sedgefield in 1826. Hutchinson (1840) also made reference to "*solitary individuals have been shot in other parts of the county in autumn and winter*", but again offering no detail regarding these. In addition Tristram (1905) stated that "*single birds have been very occasionally seen in September on the Darlington Sewage Farm*", with no supporting documentary evidence of this.

The weight of anecdote indicates that the species was a rare but regular vagrant to Durham during that era, most frequently shot during Snipe *Gallinago gallinago* shooting and it is interesting to note both the references to 'autumn', and in particular 'September' with regards to these records. This timing is typical of the modern vagrancy pattern, which has an obvious peak in records in September.

That both of the late 20th century records were in the same year is remarkable. The bird at Hartlepool in September 1976 occurred during a prolonged spell of south easterly winds and rain. An obviously tired migrant it was found sat between the goalposts of the football pitch on the Town Moor, by an understandably incredulous Geoff Iceton, though it later moved into the bowling green before eventually being lost flying up Gladstone Street.

Distribution & movements

The breeding range of Great Snipe extends from Scandinavia eastwards through Poland and into Russia as far east as western Siberia. It winters largely in sub-Saharan Africa, the winter range extending as far south as South Africa. The species has experienced large declines through the 19th and 20th centuries and it may have been considerably more common in the past in the north east of England, as it assuredly was around the rest of the British Isles. Many past records were generated by the intense shooting of Snipe along the east coast of England, a practice which has declined very significantly in modern times (Parkin & Knox 2010). Formerly a regular vagrant to the British Isles, there are over 700 records, though only 155 since the formation of the BBRC in 1958, mirroring the species' marked decline in Europe. Most recent records have come from the Northern Isles and it is virtually an annual visitor on the east coast of England, though many must evade detection.

Short-billed Dowitcher
Limnodromus griseus

A very rare vagrant from North America: one record.

The sole record for County Durham is of a juvenile found by Bernie Beck and Richard Taylor amongst roosting Redshank *Tringa totanus* on Greenabella Marsh, Teesmouth during the late evening of 29 September 1999. The bird remained in the area until at least 30 October, often feeding in nearby Greatham Creek (Beck 2000). The same bird, identified by a missing tertial feather in the left wing, had previously been at Rosehearty, Grampian from 11th to 24 September 1999, and was the first record of this Nearctic species in Britain (*British Birds* 94: 472, *British Birds* 95: 354-360).

Distribution & movements

Short-billed Dowitchers breed largely in three discreet populations in northern North America, two of replace these with which are in Canada, in Labrador/Quebec and the Canadian interior, and the other in Alaska. A long distance migrant, it migrates southwards through Atlantic Canada and along the eastern seaboard of the USA (Parkin & Knox 2010) wintering along the eastern and western coasts of South America, as far south as Brazil and Peru. It is an extremely rare vagrant to Britain, with the bird at Rosehearty and the North Tees Marshes in 1999 being the only record, although a further three have been recorded in Ireland.

Long-billed Dowitcher
Limnodromus scolopaceus

A very rare vagrant from North America: one record.

The first Long-billed Dowitcher for County Durham concerned an elusive first-winter bird found on 13 November 2007 at Seal Sands and on the adjacent Seaton Snook; it was also seen on 14th (*British Birds* 102: 556). The bird was found by Robin Ward and Phil Shepherd during the former's final low water count of waders at Teesmouth, one of 341 he had conducted over an 18-year period; a long awaited and much anticipated addition to the county's list of birds (Ward & Joynt 2009).

Distribution & movements

Long-billed Dowitcher breeds primarily in Arctic Siberia, having a more northerly breeding distribution than its close relative the Short-billed Dowitcher *Limnodromus griseus*, and in this area its range has been expanding in recent decades. Its North American range is restricted to the coastal tundra of western and northern Alaska, eastwards to the Mackenzie River. In autumn, it migrates southwards through the interior and along the eastern seaboard of the USA, wintering from the coastal areas of the southern states of the USA south into northern Central America.

It is unclear whether it is Siberian or North American birds, or both, reaching Britain. This species is one of the commonest 'Nearctic' waders in this country, often being located amongst Redshank *Tringa totanus* at coastal localities, although it has been recorded inland. Records have come from all months of the year but mostly between September and November; there had been almost 210 by the end of 2010.

Woodcock (Eurasian Woodcock)
Scolopax rusticola

A common migrant and winter visitor, and locally resident breeding species.

Historical review

There are relatively few entries of *"Wodekokes"* in the accounts of the Cellarers' Rolls of the Monastery of Durham and it is perhaps surprising, for an edible species, that it is not mentioned until 1347, and that the numbers purchased were relatively small (Ticehurst 1923). Perhaps they were rather uncommon, or the inhabitants of County Durham in the 14th century were not very adept at snaring them. The comparatively high price (1½d. to 3½d. each) paid for them indicates that they may have been somewhat unusual luxuries. Two centuries later, according to Benett (1844) in the *Durham Household Book*, two were bought for 3d on 26 October 1533 (Gardner-Medwin 1985). The first reference to this species breeding in County Durham comes from Tunstall's manuscript, which recorded the shooting of a young bird, two-thirds grown, in September 1782 near Durham.

Breeding numbers of Woodcock appeared to be on the increase in the 19th century, from around 1820 onwards, probably associated with land management changes. These included an increase in the area of lowland plantations, and the planting of game cover and shelter belts following the Enclosure Acts of the mid- to late 18th century, and the consequent changes in the intensity of hunting patterns. The species was breeding throughout the country by the 1860s, but prior to this was largely a winter visitor to Britain (Holloway 1996).

In Durham, Selby (1833) thought it mainly, or only, a winter visitor to Durham, though it was regular at other times of the year, indicating perhaps a sparse breeding presence in the region. By 1840, Hutchinson was documenting it as having bred in Durham, with three young out of a brood being shot near Allensford on the Derwent, in May 1830. In the 1860s and 1870s, Temperley (1951) recorded that several nests had been found in the Derwent valley and elsewhere. By 1874, Hancock said it was breeding in the area of the Tyne, around the confluence with the Derwent and that it was not as 'uncommon' as previously supposed. This may have been a real change in status, or it may simply reflect a change in knowledge.

The heavily-wooded Derwent valley has apparently always been a stronghold for the species, confirmed by Temperley's reference to early clutches of eggs having been taken at Medomsley in April 1872 and Lintz Green on 8 April 1893. In Robson's time, the species was widespread as a breeding bird in the lower Derwent Valley, Chopwell Wood being noted as a *"favourite resort"*, and there were records of clutches of eggs being taken at a number of sites (Robson 1896).

By 1951, Temperley described Woodcock as a *"resident and winter visitor"* nesting in all suitable woodlands in the county, often close to centres of population, for example at Ravensworth Castle in the Team valley, and at Gibside. A BTO enquiry into its status as a breeding bird in Britain, carried out in 1934/1935, reported that twelve pairs had been recorded in 7,000 acres near Stockton-on-Tees and 10 to12 pairs annually in a 2,000 acre estate near Chester-le-Street. The Stockton site may have been the Wynyard estate, while the Chester-le-Street location was presumably the Lambton Estate.

In relation to its winter status, Temperley wrote that wintering birds arrived in October, many making landfall on the Durham coast before moving inland. He noted that they were seen in daylight flying in from the sea, usually singly, but occasionally in small parties, and that *"the shore shooters then reap a harvest"*. Hancock (1874), reported that one shooter at Marsden Rock, in October 1865, shot 31 Woodcock in a day. Temperley added that although the return passage in spring was less obvious, Woodcock could again be seen on the coast at the end of March or the beginning of April.

A study of long-term activity in one area is provided by David Simpson, who has been birding and recording his findings in the Shotton area for over fifty years. In May 1957, he noted at least three pairs of Woodcock nesting in Edder Acres Plantation, Edder Acres Farm and in Calfpasture Dene near Shotton Colliery, giving an indication of the local breeding density at that time. Over the period 1952 to 2000, he found four nests containing just hatched eggs and several times he saw chicks with their mother. Over the years he commented that he watched Woodcock fly out of Edder Acres Plantation at dusk on winter evenings, on their way to feed on the nearby marshy meadows. Sometimes up to ten birds were counted in just a few minutes flighting out of a ride in the woods (Simpson 2011).

Recent breeding status

Inevitably, due to the largely crepuscular nature of the species, any pattern of records collected by daytime observers is likely to give only an incomplete picture of the species' true distribution and breeding status. 'Roding' males, during April-June, are thought to be reliable indicators of the presence of breeding birds (Hoodless *et al.* 2008), but nesting Woodcock are notoriously difficult to find. Furthermore, Woodcock are prone to desert their nests if disturbed and it is believed that desertion can account for as many failed nests as predation does (Hoodless & Coulson 1998). As a result, systematic ground searches are not an appropriate method for assessing numbers of this species.

The Woodcock's display flight, 'roding', is usually first noted in the latter part of February in Durham, and increases in frequency during March and into April. It is widely reported from many sites from mid-April to late July, particularly around the well-wooded dales and upland fringes of the Derwent valley, Weardale and Teesdale. The vast majority of presumed breeding birds are in the west of the county, but the apparent distribution probably partly reflects observer habits.

In reality, breeding Woodcock can be found in many of the county's woodlands, from the coastal denes in the east all the way up to the western valleys. As an example, they breed in all of the lower Derwent Valley woodlands, and are present in relatively large numbers in some of the larger areas of tree cover, such as Chopwell Woods, the Derwent Walk Country Park and the Gibside and Ravensworth Estates. Survey work in this area during the late 1980s and early 1990s indicated that there were over 50 displaying males in Gateshead borough, and it was suggested that, in such ideal habitat, there may well be more (Bowey *et al.* 1993). The possible density of breeding birds was illustrated by the presence of two females nesting within 30m of each other at Edmundbyers in spring 1984 (Baldridge 1985).

According to Fuller (1982), Woodcock are found mainly in large deciduous woodlands that are greater than 80 ha in area, and they apparently prefer areas with an understorey and dense ground vegetation that provide shelter from avian predators (Hirons & Johnson 1987).

The species is known to nest in conifer plantations up to the thicket stage, as well as more open woodlands, but avoids dense, mature plantations, especially where the canopy is closed. In Durham, and the north east generally, it would appear that Woodcock routinely also breeds out on open moors and along moorland fringes, especially where there is bracken. This behaviour has been noted at Widdybank Fell (A.N. Hoodless pers. comm.) and in the upper Derwent valley. It is highly likely that this is an under-recorded phenomenon, and confirmed breeding records include an adult found with five or six flying young on high pasture near Middleton in Teesdale in 1991 (Armstrong 1992). This is an unusual number of young, as the typical clutch is four or exceptionally five eggs. It is possible that two broods were therefore involved here (A.N. Hoodless pers. comm.). Breeding season records in 2009 came from as far west as Langdon Beck and Herdship Farm in Teesdale and from Lunehead (Newsome 2010). These are exposed localities with limited tree cover, demonstrating that this is not necessarily a bird that breeds solely in local woodlands.

Nonetheless, in most years, the majority of reports came from relatively well-wooded sites around the centre of the county. Hamsterley Forest, because of its sheer scale, is a highly favoured locality with many breeding birds present. In spring and summer 2009 for example, there were a total of 21 roding males reported in this area, comprising 10 at Low Redford Plantation, 6 at Shipley Moss and 5 at Windy Bank (Newsome 2010).

It is rather difficult to establish the size of Woodcock populations because roding males establish a hierarchy, with a few dominant birds securing most of the mating opportunities with receptive females (Hirons 1980). Hence, counts of roding males may represent only about a quarter of the total breeding population, but they are useful for estimating the size of the male population (Westerberg & Bowey 2000, Hoodless *et al.* 2008). The *Atlas* stated that the density of breeding Woodcock was likely to be highest in the county's deciduous woodlands, with up to 16 birds per square km, but lower in the west of the county, where habitat is of poorer quality. A population estimate of 1,160 pairs was calculated, using a figure of four birds per square kilometre in the documented 62 lowland occupied tetrads and an estimated two birds per square kilometre in 21 occupied upland tetrads (Westerberg & Bowey 2000). In the south east of the county, breeding numbers are believed to be low, with very small numbers being present around the Wynyard estate and the Castle Eden walkway (Joynt *et al.* 2008).

The survey work for the *Atlas*, whilst believed to have provided a reasonably accurate overview of the species' distribution in the county probably missed some of the detail. This is simply because of the variation in observer coverage between well-recorded areas such as around Durham City and Sacriston, and less visited sites elsewhere. It is highly likely that Woodcock is more consistently widespread in the county's woodlands than the detail of the *Atlas* map indicated. Subsequent records have shown that, as suggested in the *Atlas*, the species is indeed present in such large woodland blocks as those around Raby Castle and Weardale Forest, although there were no breeding records available at that time (Westerberg & Bowey 2000).

Confirmation of breeding is relatively rare as nest sites are so well hidden, but notable records include one in the east of the county at Elemore Hall, where an adult was seen with a chick in May 1986, and a well-developed chick found on the early date of 17 April 1988 at Ravensworth in the Team valley. A party of five were noted flying over Waldridge Fell on 26 May 1994, with those following behind the leader giving begging calls, suggesting that this was an adult with young (Baldridge 1987, Armstrong 1989 &1996).

A small number of confirmed breeding records have been reported since 2000. An adult with two immatures was at Grassholme Reservoir on 10 June 2000, and in 2007 two recently fledged juveniles were at Chopwell Woods on 12 May, and two begging juveniles were seen pursuing a 'ragged' adult in flight at Shipley Moss on 25 July. In 2009, a pair was noted with three young at Dryderdale on 8 May and the following year a pair with four young was at Waskerley Reservoir on 16 May, and family parties were reported in July at Cow Green, Hamsterley and Low Redford.

Nationally, Woodcock declined rapidly and significantly on Common Bird Census (CBC) plots for the three decades up to 2000. However, because the CBC did not include many coniferous forests and its plots were concentrated in lowland Britain, it is unclear how well this trend represented the overall UK population. The BBS is also inefficient at recording this scarce, elusive species and no longer provides an index for Woodcock. However, range contractions were also recorded elsewhere, concurrent with the CBC decline (Gibbons *et al.* 1993). Possible causes of a national decline could include recreational disturbance, the drying out of natural woodlands, overgrazing by deer, declining woodland management, and the fact that new plantations are maturing. However, there is no strong hypothesis to date (Fuller *et al.* 2005).

In any event, there are no clear trends from data in Durham, and the issues raised as possible causes of a national decline may not be significant factors in the county. It is notable that a TBC survey in 1997 concluded that the species' status in Cleveland had changed little in the past 40 years (Bell 1998).

The first survey aimed at monitoring the UK's breeding Woodcock took place in 2003 and has provided a new, much higher baseline population for future monitoring (Hoodless *et al.* 2009). However, the upward revision is due to new methodology and carries no information about population trends (Baillie *et al.* 2010).

This species still has an amber listing as a Bird of Conservation Concern, but this is based on declines in the European population. Note also that numbers shot in the UK have increased threefold since 1945 and are currently running at a historically high level (Baillie *et al.* 2010).

Recent non-breeding status

Outside of the breeding season, Woodcock records in the county come from a wide range of locations. However, there is an extensive migration of presumably mainly Scandinavian birds to the Durham coastline in the autumn, and this accounts for many of Durham's non-breeding records. This will also include some individuals from Finland, western Russia and the Baltic states (A.N. Hoodless pers. comm.).

The first of these late autumn migrants are usually noted around the end of September, and the major influx of birds takes place from mid-October to early November. At this time of year, high numbers can occur in the right weather conditions. For example, a major arrival took place in early November 1994, with classic 'fall' conditions of south easterlies and rain grounding large numbers of Woodcock on the coast on 4th and 5th, including at least 20 at Hartlepool Headland, 27 between Whitburn and South Shields, and birds strung out along the whole of the Durham coastline. A similar arrival took place in early November 2008 when at least 17 were in the Whitburn area on 1st of the month, 21 in Marsden Quarry alone on 6th and 20 on Hartlepool Headland on the same date (Bell 1995, Armstrong 1996, Newsome 2009).

Birds typically become more widespread inland after such influxes. In 1984, for example, after a similar coastal arrival, a party of 13 birds were seen flying over Sedgefield on 9 November. It is unusual to see a flock of Woodcock of this size, and these birds had presumably been flushed from nearby (Baldridge 1985).

Woodcock is very much a visible migrant and arrivals are observed far more frequently than in most land bird species. There are no recent records of the "*occasional small parties*" noted coming in off the sea by Temperley, however, despite the numbers grounded on the coast in fall conditions. All recent records of Woodcock flying in off the sea are of single birds. In the period 2000-2009, for example, singles have been noted coming in off the sea at Hartlepool, Teesmouth, Whitburn and Marsden in the months of January, February, September, October, November and December, reflecting the months of continental immigration. Notable amongst these records was one which flew in from the east and landed unexpectedly on Seal Sands on 18 November 2006. Another single in off the sea at Whitburn on the early autumn date of 4 September 2007 was well ahead of the main autumn arrival. One at Hartlepool Headland on 31 December 2000 failed to make land fall, and was seen trying unsuccessfully to take off from the surface of the water.

Further continental arrivals sometimes boost local populations during late November and December, and sometimes there is evidence of continued immigration during the mid-winter period, with occasional birds being noted arriving 'in-off the sea' at coastal stations such as Marsden and Hartlepool, even into January. Exhausted individuals are often flushed from coastal grasslands, such as South Shields Leas, or areas of scrubby cover, at this time of the year.

An exceptional hard weather movement took place in very severe conditions in December 2010, on a hitherto unprecedented scale. Counts included 31 in Whitburn Coastal Park on 5th, 30 at South Shields South Marine Park on 7th, 40 completely out of habitat at Whitburn Steel on 8th, 30 at Seaton Common on 9th, 20 at Horden on 11th, and a remarkable count of 71 at Easington Colliery on Boxing Day. Not surprisingly, there were also many reports of fatalities during this period (Charlton 2011).

Some of these Woodcock may have been newly arrived migrants, retreating from the even colder Continent, but others were probably locally wandering birds, searching for unfrozen ground in what was a period of low temperatures and severe winter weather. At such times, despite the exceptional counts of December 2010, many records are of single birds, often inadvertently flushed by walkers.

Cold weather, or fatigue in the case of newly arrived migrants, sometimes forces birds into more urban areas and occasionally into gardens, largely in autumn or less often in mid-winter. Woodcock can often then be found in unexpected places, particularly in the east of the county.

Severe weather in December 1981, for example, drove a number of birds to the coast to seek feeding areas and as a result there were three victims of collisions with windows at Sunderland Museum alone during this period. More recently, one was noted in a Durham City garden on 30 January 2009 (Baldridge 1982, Newsome 2010). A single, small Whitburn garden with little cover has produced five Woodcock records in 13 years between October and January, including three autumn migrants hitting windows, and one storm-driven after heavy snow, on 9 January 2010 (P.T. Bell pers. obs.).

Other hard weather records include one at Balder Grange, Teesdale, on 28 December 1995, which came into a house to retreat from the bitter cold outside when a door was left ajar. Another Woodcock was noted 'out of habitat' feeding in a ploughed field behind a tractor at Cowpen Bewley in March 2009, with Snipe *Gallinago gallinago* (Armstrong 1997, Joynt 2010).

In more typical winter conditions there are usually widespread reports of birds across the county between November and February, not just from woodland sites, but a range of scrub and more open habitats. The majority of these come from lowland woodlands and damp scrub right down to the coastal strip; however, some are noted even from the moorland fringes. This pattern will to some extent reflect observer coverage. Gatherings of birds are occasionally reported from prime wet woodland locations at this time of the year.

In the early months of 1986, for example, 46 birds were reported from over 30 locations around the county (Baldridge 1987). In early 1995, a better estimate of the true numbers wintering in parts of the county was provided by the record of 20-25 birds noted on a local Pheasant *Phasianus colchicus* shoot at Cotherstone (Armstrong 1997). Another high winter site count came from WWT Washington, with 23 on 20 January 2002 (Newsome 2007). Most recently, in the winter of 2009/2010, the species was again reported from over 30 sites (Newsome 2012). However, it seems likely that any published pattern of records, and our consequent understanding of this, represents a very considerable under-estimate of the species' true winter presence in the county.

Much smaller numbers are also occasionally noted at coastal sites in March and April, as presumed continental birds begin to drift north and east on their return migration (Blick 2009). In 1988, for example, one was noted in a South Shields garden in mid-March. In 2001, an influx between 3rd and 5 March brought six to the South Shields area and in 2008, single birds were noted on five dates in Marsden and Whitburn between 17th and 30 March (Armstrong 1989, Siggens 2005, Newsome 2009). One seen flying out to sea from Whitburn CP on 26 March 2005 was a rare example of return migration actually being witnessed (Newsome 2006).

Distribution & movements

This species' breeding range encompasses northern and central Europe, the whole of Scandinavia and, moving eastwards, a band across the middle and upper latitudes of Russia, as far east as Japan.

Few Woodcock are ringed in the county, and the only Scandinavian recovery from a Durham-ringed bird to date is of a full-grown bird (age unknown) ringed at Graythorp, Hartlepool on 8 November 1975, which was shot at Soderhamn, Sweden on 2 July 1977, presumably whilst breeding (Bell 1980).

Two other birds ringed at Hartlepool Headland in autumn, presumably Scandinavian or Russian arrivals, were later recovered having continued heading south west to winter. The first was a full-grown bird ringed there on 28 October 2004 and killed at Mount Sion, Gwynedd, Wales, just 23 days later. The second bird was an adult ringed on 6 November 2008 and shot at Cumragh River, Dromid, County Kerry, Ireland, on 24 January 2009 (Joynt 2005 & 2010).

Chicks ringed in the north east of England have also been recorded in Cumbria, the Isle of Man and Ireland during following winters. They include a bird ringed as a nestling at Hamsterley on 15 June 1958 which was recovered at Lepertown, Waterford, Ireland, 430 kilometres to the south west on 31 January 1960, indicating that locally breeding birds may also head further south or west during the winter (Coulson 1961). Another ringed as an adult at Balder Grange, Teesdale on 28 December 1995 was shot near Shap in Cumbria two years later on 27 December 1997 (Armstrong 1999).

Black-tailed Godwit
Limosa limosa

A fairly common passage migrant and scarce winter visitor, most frequent in April and May, and from July to September.

Historical review

The earliest reference of the Black-tailed Godwit in County Durham is by Selby (1931) who said it was *"a rare species in the northern counties"* but made no specific mention of Durham. Edward Backhouse (1834) said *"occasionally killed in marshes near Greatham"*. Neither Hogg (1845) nor Hutchinson (1840) mentioned any records of the species but William Backhouse (1846) said that it was *"occasionally met with about the Teesmouth in autumn"*; whilst Hancock (1874) said it was a *"rare, casual visitant"*.

Nelson (1907) quoting old sources, indicated that it was more common in the early part of the 1800s than it was *"at present"*, though he personally had limited experience of it. There are very few well documented records from the early part of the 20th century though a specimen held in the Dorman Museum, Middlesbrough is said to have been shot on Cowpen Marsh, Teesmouth on 14 September 1914, and the late W. B. Alexander recorded single birds at Holme Fleet, Teesmouth on 28 August 1929, and in the Tees Estuary on 4 September 1929. Joseph Bishop was the first to document the annual spring passage through Teesmouth, albeit in small numbers, stating in correspondence to W. K. Richmond in 1931 *"that Black-tailed Godwits (Limosa limosa) are fairly regular at Greatham in May"*, with subsequent observations by himself of a single bird in 1931, four in May 1933, and up to six birds during the spring of 1939. In 1944, up to four birds in summer plumage were present from 22nd to 30 April, with 13 on 3 May, and in 1945, small parties were recorded from 16th to 25 April, the largest flock being of ten birds.

At the middle of the 20th century, Temperley (1951) described this elegant wader as *"an uncommon spring and autumn migrant, of more regular occurrence now than formerly, but only in small numbers"*. He quoted no specific records, though did indicate that the general increase in records since the 1940s had also been replicated in Northumberland. Stead (1964) attributed this to an increase in the European breeding population and considered it *"a more frequent visitor to the estuary than it was 50 years ago"*. He quotes an average of *"3-5 records annually, mostly of solitary individuals or parties of up to five birds with occurrences in both spring and autumn"*, detailing an especially noteworthy flock of 12 birds on the Reclamation Pond, Teesmouth on 25 July 1954.

The 1960s saw the species further consolidate its status as a scarce migrant, in particular to the North Tees Marshes, being recorded annually in both spring and autumn. Just two birds were noted in 1960, though 1961 saw one or two birds on Cowpen Marsh from 12 March for six weeks. Many were present during the period 23 July to 1 October, including a peak count of 41 on 7 August. This was almost three times the previous maximum count, in 1954, and included a single flock of 35 birds on Seal Sands. This dramatic influx was not repeated in subsequent years despite a general increase in the number of spring records; peak counts being, seven in May 1962 and 1963, 11 on 10 May 1964 and 13 over Seal Sands on 28 April 1969. As the decade came to a close, one of the most notable events of 1969 was the successful breeding of a pair of Black-tailed Godwits on Cowpen Marsh, Teesmouth with a single youngster being raised (Blick 1978).

Records away from the North Tees Marshes appear to have been genuinely rare, with the first well documented record coming from Darlington Sewage Farm in 1929. This event was noteworthy enough for the observation to be printed in *British Birds*, "on August 28th, 1929, eight Black-tailed Godwits {Limosa limosa) appeared on the Darlington Sewage Farm. Some were in partial summer plumage and in all the white wing-bar was conspicuous. At first, they were very tame and could be approached to within twenty yards, but later became exceedingly wary. Three were still present on September 12th and the last was seen on October 5th" (Richmond 1931).

Other early records away from Teesmouth were of: a summer plumaged bird at Cleadon on 30 April 1934; six on a flooded field near Cleadon Village, on 21 April 1945; four at Jarrow Slake, on 7 May 1950; three at Boldon Flats, on 8 September 1950; and, two at Jarrow Slake on 21 September 1950. An increase in records during the 1960s, in line with increased numbers passing through the Tees Marshes, saw single birds at: Darlington, on 12 April 1961; Crookfoot Reservoir, on 7 May 1961; Washington, on 4 May 1969; Hartlepool, on 20 August 1969; Washington, on 14 September 1969; and, Hurworth Burn Reservoir, on 20 September 1969. The final four records highlighting that the species was becoming an increasingly regular migrant through the county during autumn as well as in spring.

Recent breeding status

Despite a pair having successfully reared one young on Cowpen Marsh in 1969 (Blick 1978) there have been no further records of confirmed breeding in the county although birds have been recorded annually at Teesmouth during the breeding season since the 1970s, with display noted on a number of occasions, though relatively few linger into the summer months. During the *Atlas* survey period, there were summer records of birds during June and July 1988 and again during spring/summer 1991 and 1992. In May 2003 a small colony appeared to have established a territory on Cowpen Marsh, but despite at least one bird being present throughout the summer, there were no confirmation of breeding having taken place (Joynt 2004). At present it seems unlikely that a population will become established in the near future.

The UK breeding population was estimated at 59-66 pairs in 2007, the majority of which are on the Ouse Washes in Cambridgeshire (Holling 2010), these belong to the nominate race *limosa*, which breeds in continental Europe. A handful of pairs have also nested in Orkney and Shetland since at least 1973 (Sharrock 1976); these are of the race *islandica* which breeds predominantly in Iceland and the Faeroe Islands. This is the commonest form to occur on passage through the North Tees Marshes.

Recent non-breeding Status

Despite continuing to be a scarce but annual passage migrant to the county throughout much of the 1970s and 1980s in both spring and autumn, the numbers visiting each year have increased since late 1990s and the species can now be found throughout the year with small numbers regularly wintering on the North Tees Marshes. This area provides the majority of the county records in both spring and autumn, though the number of passage birds elsewhere has increased in line with the Teesmouth population.

Spring passage typically begins in mid-March and continues into June, though the peak numbers are usually recorded in late April and early May. Early returning birds have been more difficult to detect in recent years due to the presence of the overwintering population, though single birds on Cowpen Marsh on 14 March 1978 and on Seal Sands on 10 March 1993 are presumed to represent early migrants, as was the party of eight on Dorman's Pool on 14 March 1999. Typically, the peak spring count is achieved in late April, though prior to 2000 this had exceeded thirty birds on just two occasions; 32 at Teesmouth on 29 April and 2 May 1975 (Bell 1976), and 31 on Dorman's Pool on 21 April 1983 (Bell 1984). The peak spring count in 2000 was 37 on Saltholme Pools, 109 were at the same site on 29 April 2001 and then 216 were around the Tees Marshes on 23 April 2006 (Joynt 2007).

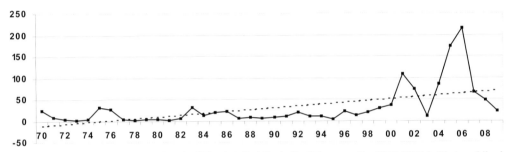

Maximum spring counts (April to June) of Black-tailed Godwit at Teesmouth, 1970-2009 (with trend line)

Away from the North Tees Marshes this species can appear at almost any wetland during spring though it is typically scarce with just a few records each year between mid-March and early June. The wetlands in the Washington area have attracted birds almost annually since 1970, though records have decreased in recent years. Typically most records involve parties of one to three birds though five were present on 17 April 1970 and six were at WWT Washington in the first week of May 1990; the highest spring counts for that area. Elsewhere, Hurworth Burn Reservoir has recorded birds on five occasions, including a party of 12 on 10 April 1993 and there is a scatter of sightings from the eastern half of the county, typically involving single birds. Further west, there are just two spring records, a party of five at Derwent Reservoir on 19 April 1994 and an exceptional flock of 32 at Grassholme Reservoir on 27 April 1996 (Armstrong 1998). Light coastal passage takes place during this period with occasional records from the Whitburn area. Intensive seawatching here in recent years has produced spring counts of 26 north on 22 April 2006 (Newsome 2007) and 41 north on 26 April 2008 (Newsome 2009).

Autumn migration begins in late June and continues through to early September though in most years the peak numbers are recorded in July. Typically the number of birds involved is higher than in spring though the true number is difficult to assess given the more protracted nature of the autumn migration and the peak autumn count each year is probably not representative of the numbers involved. The first birds to arrive are adults, though the population is supplemented by juveniles from mid-August onwards, these reach a peak in numbers in September.

The North Tees Marshes is the prime location for this species in the autumn. Few counts in excess of 30 birds were made prior to 1990, excepting 40 on Dorman's Pool on 2 July 1983 (Bell 1984). A new high count of 61 along the Long Drag came on 24 August 1991 but this was exceeded by a count of 70 around the North Tees Marshes on 3 July 1993. There was just one count in excess of 50 in the next five years, this being 60 on Dorman's Pool on 2 August 1997. The first three-figure count came in 1999, when 153 were at Dorman's Pool on 21 July 1999 (Iceton 2000). Subsequently, numbers continued to rise: 163, on 30 July 2001; 174, on 28 June 2005; and, 281, on 13 July 2006 (Joynt 2007).

418

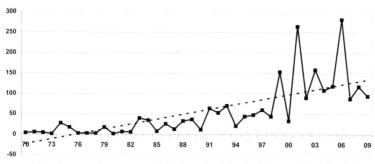

Maximum autumn counts (July to October) of Black-tailed Godwit at Teesmouth, 1970-2009 (with trend line)

Black-tailed Godwits are more numerous inland during autumn, particularly during late August and early September. The Washington area has traditionally held the best sites away from the Tees Marshes with records being almost annual there from 1972 to 1997, though with few since. Most records involve parties of up to five birds though occasionally more, 16 at WWT Washington on 22 July 1975, 21 in September 1996 and an exceptional 31 on 23 August 1997 (Armstrong 1999). More recently, Bowesfield Marsh near Stockton-on-Tees has established itself as a site that regularly produces double-figure counts; numbers here rising from three in August and September 2000 to a high count of 59 in September 2009 (Joynt 2010). Other favoured localities have included Hurworth Burn Reservoir during the late seventies and early eighties and in the far west Derwent Reservoir. Elsewhere, other notable records include: 45 over Pennington, Hamsterley Forest, on 24 August 1997 (Armstrong 1999); 18 over Shotton Colliery, on 25 August 1997; 17 at Longnewton Reservoir, on 10 August 2000; 21 at SAFC Academy Pools, Sunderland, on 30 August 2008; and, 26 at Castle Lake, Bishop Middleham on 8 August 2009. Coastal passage can be evident during July and August with small numbers noted each year passing offshore at Whitburn Observatory and other points along the coast. Noteworthy counts of southbound birds include: 13 at Whitburn, on 23 July 1984; 21 at Whitburn, on 29 July 1991; 46 at South Shields on 14 August 1998; 32 there, on 2 July 2006; and, 21 at Whitburn, on 3 July 2009. The latter two records indicating how early this species begins its southbound migration.

Since the early 1980s, Black-tailed Godwits have become an increasingly frequent winter visitor to the Tees estuary where they feed on Seal Sands at low tide and disperse around the marshes in small groups at high water. The first December record was as recently as 1979 when one was on Seaton Common on 2nd (Bell 1980), the first overwintering bird being present on Seal Sands from 5 October 1980 until March 1981 (Bell 1982). Between one and five wintered through much of the 1980s though none were recorded in the winters of 1987/1988 or 1988/1989. Records were at best sporadic during the 1990s being absent in five winters and typically just one or two birds involved. Exceptions were a count of 22 in January 1996 (WeBS) and 10 during the winter of 1996/1997 (WeBS). Despite there being no records in the winter of 1998/1999, the species has overwintered every year since, with a highest count of 43 in January 2004 (WeBS), the typical winter population being around 30 birds. This relatively recent phenomenon has been noted in neighbouring Northumberland (Kerr 2001) and has increased in line with trends around the rest of UK (Musgrove *et al.* 2011). It is thought that the majority of these wintering birds are of the Icelandic race '*islandica*', which traditionally wintered in Ireland and south west England but have now spread to other parts of the British Isles. The UK wintering population, including September and October, counts, has recently been estimated at around 43,000 birds (Musgrove *et al.* 2011).

Winter records away from Teesmouth are decidedly scarce, with the first as recently as 22 December 1983 (Baldridge 1984) when one was at Jarrow Slake. Over a decade later single birds were at Boldon Flats in February 1991 and WWT Washington on 5 December 1996. The following winter saw two small groups overwinter, with two at Boldon Flats from 30 November 1997 until February 1998 and three at WWT Washington from 17 December until February 1998. There have been a handful of other winter records, mainly during the 2000s and a number of them from the Tees Valley area, the exception being eight north over Rainton Meadows on 22 January 2005 and a party of eight at Castle Lake, Bishop Middleham from 12 November to 5 December 2005.

There is a long series of records, mainly from nocturnal westerly flights of birds along the Tyne Valley during the early spring. This phenomenon was first noted in 1963 when birds were heard on 26th and 28 April as they flew up river at Blaydon. In the following year, a small party was heard at 23.30hr on 3 May. Birds were noted overhead at Stella, Blaydon in 1981, with birds heard again on 1st and 8 May, and again at the same site in 1984 when they were heard on 10th and 27 March. The fact that these movements were still being noted after 20 years, strongly suggests that this is a traditional cross-country movement and that it could be occurring on an annual basis (Bowey *et al*. 1993). Kerr (2001) highlighted the occurrence of birds at Grindon Lough, Northumberland, in April 1939 and April 1947, which would seem to indicate a long-term spring movement of birds east to west, using the Tyne valley as a flyway. That such over-land movements in the county may be more regular than previously supposed is suggested by the previously mentioned 32 birds at Grassholme Reservoir on 27 April 1996, the two there 12 April 2005 and 11 at Selset Reservoir on 24 April 2007.

Distribution & movements

Black-tailed Godwit breeds in a localised fashion from Iceland to eastern Siberia. The nominate race breeds from western and central Europe eastwards into central Asia, with western populations wintering predominantly in western and central Africa from Senegal to Chad. The race *'islandica'* breeds abundantly in Iceland and to a lesser extent, the Faeroe Islands and the Lofoten Islands of Norway. It winters in the United Kingdom and Ireland, France and the Netherlands though some penetrate as far south as Iberia.

Detailed studies by Durham University and the WWT Advisory Service have shown that the majority of Black-tailed Godwits which utilise the Tees Estuary are of the race *'islandica'*, and this is borne out by a number of ringing recoveries. During July and early August the first birds to arrive are adults which are thought to be *en route* to the Wash to moult, but by mid-August the first juveniles arrive, peaking in September, before heading south with any remaining adults to winter to the south of the Teesmouth by October. Relatively few birds remain during the winter months, though close observation reveals that those that do are invariably first-winter birds (Ward *et al.* 2003).

Two colour-ringed birds were present on Seal Sands in August 1994. The first, on 1 August, had been ringed on the Wash in the autumn of 1993; the second on 9 August had been ringed in Fife in the spring of 1994 (Bell 1995). This pattern of occurrence is further illustrated by an adult on Greenabella Marsh on 31 July 2010 that had been ringed at Iken Marsh, Suffolk on 29 August 2008 (Joynt 2011).

Evidence that some of these birds continue south to winter in continental Europe is provided by a bird ringed at Golf du Morbihan, France on 26 October 2001, which remained in France during November and December of that year before being seen again on 26 July 2002 at Dorman's Pool, Teesmouth during its southbound migration (Blick 2009). Not all birds head so far south, one ringed in its first summer on the Ribble Estuary, Lancashire on 3 April 1988, was later seen on Seal Sands on 5 February 1992, the only mid-winter recovery for Teesmouth (Bell 1993).

A colour-ringed bird seen at Dorman's Pool, Teesmouth on 28 July 1999 had been marked as a chick in north east Iceland earlier that yea, the only Icelandic recovery to date. This bird was seen at the Eden Estuary, Fife on 5 November 1999, near Belfast, Northern Island on 4 May 2000, and was back at Teesmouth on 27th to 28 May 2000 (Blick 2009); proving that not all birds conform to expected movement patterns.

Bar-tailed Godwit

Limosa lapponica

A fairly common passage migrant and winter visitor to coastal areas, in particular the Tees Estuary.

Historical review

The earliest reference to the occurrence of Bar-tailed Godwit in County Durham is by Hogg (1845) who said that birds were present in small flocks in the latter part of the year at Teesmouth, suggesting that the Tees estuary has attracted this species for many hundreds of years. Over 60 years later Nelson (1907) recorded *"large flocks in some seasons"*, making specific reference to the years of 1876, 1881, 1887, 1890, 1892, 1895 and 1899. Bishop however did not consider it to be an abundant bird of Teesmouth, indicating that a decline in numbers may have taken place. Nonetheless during the winter of 1943/1944 when the coast was closed to human access during World War II, larger numbers than usual were reported, with R. D. Sistern recording a flock of up to 200 birds on the sands at Seaton Carew (Temperley 1951), hinting that human disturbance may have been partly responsible for this decline.

Temperley (1951) described the Bar-tailed Godwit as *"a winter visitor, sometimes in large flocks, and also a regular spring and autumn passage migrant"*. He said that its only regular haunt in the county, was at Teesmouth, with a few occasionally being noted at Jarrow Slake, a situation that pertains into the early 21st century. Temperley believed that it was not as abundant in Durham as it was in neighbouring Northumberland and this statement is still true. As a passage migrant he considered it much less common in spring than in autumn when the flocks contained many juvenile birds giving September as the peak month of passage. Temperley had however personally noted birds at Teesmouth in May in *"full red plumage"*, indicating that light spring passage was not unusual.

Stead (1964) considered the species to be *"a regular passage migrant and winter visitor to the estuary and coast"* being 'abundant' from August to April with a few non-breeding birds remaining for the summer months. The largest concentrations at Teesmouth being found at Seaton Snook at high tide, spending low tide feeding on the mudflats of Seal Sands; a situation which remains much the same today. Flocks of around 300 birds during the winter months were not considered unusual, with 600 at Seaton Snook on 12 September 1953, and 420 at the same locality on 26 January 1955 being given as the highest counts in recent years.

Records indicate that the winter average of around 300 birds at Teesmouth was maintained until the mid-1960s, with an estimated 300-350 being present in January 1958, 300 in both March and October 1963, and 280 in September 1965. However, in 1967 the peak count had almost halved to just 160 on 29 October, with subsequent annual peaks being 142 on 15 December 1968 and 120 on 20 September 1969. The reasons for this decline are unclear but could be related to the disturbance created by the ongoing land-claim on Seal Sands.

Essentially a coastal species, there are very few historical records of Bar-tailed Godwit straying inland, though Temperley (1951) made reference to the occasional bird visiting Hebburn Ponds. The first dated record is of an adult in summer plumage at Darlington Sewage Farm on 20 August 1936, though the next was not until 1962, when two were at Hurworth Burn Reservoir on 21 September, with one remaining until 1 October. The only other published record was of three at Washington Ponds on 3 August 1969. In *Birds of Teesside* Stead (1964) made reference to the fact *"that a few flight inland with the Curlew"*.

Recent status

Teesmouth and in particular the mudflats of Seal Sands remain the stronghold of this species in County Durham though the species has been in steady decline since the 1970s, the reasons for which are not fully understood. Essentially a winter visitor the first birds typically arrive in July, with most having departed by late March, though numbers are at their peak between November and February. In addition it is not unusual for occasional birds to be present around the Tees estuary in May and June, essentially giving the species a year-round tenure within the county (Blick 2009).

The highest number recorded is of 927 at Teesmouth on 19 January 1970 (Blick 1978), though how many of these were within County Durham is not clear as there is considerable interchange with birds on Bran Sands on the Yorkshire side of the estuary. This total has not been even closely approached since as it was around three times greater than the typical 1960s winter population at Teesmouth. Numbers remained high during the early part of the 1970's with counts of 485 on Seal Sands on 13 February 1971 and 450 on 27 February 1972, though just 74 were present during the winter of 1974/1975, which was attributable to the disturbance of large parts of Seal Sands, the mudflats being reduced from around 400ha in 1969 to just 140ha by February 1974 (Evans *et al.* 1979). A marginal recovery was evident in 1976, when 380 were counted during February, numbers staying stable at around 250-300 through much of the 1970s and 1980s, save for a peak of 456 on 13 January 1985, the highest count since 1971. Numbers have declined steadily since, reaching a peak of just 79 in February 1998 (WeBS), a situation that has been maintained since with few counts in excess of 100 birds. The exception was during the winter of 1995/1996 when cold weather on the Continent was deemed responsible for an exceptional count of 526 on 18 February 1996 (Bell 1997), at a time when many other wintering waders were also displaced including increased numbers of both Knot *Calidris canutus* and Grey Plover *Pluvialis squatarola* locally.

This marked decline has been mirrored at a national level with Musgrove *et al* (2011) stating that the Bar-tailed to Black-tailed ratio during the winter months nationally has fallen from 4:1 to less than 1:1 over the first decade of the 21st century. This situation is doubtless clouded by the increased numbers of wintering Black-tailed Godwits *Limosa limosa* nationally. This recent decline is thought to be linked to increasingly milder winters in the UK, with many wading birds wintering on the Continent, chiefly around the Wadden Sea (Maclean *et al.* 2008). The peak of

526 birds at Teesmouth in February 1996, in response to cold weather on the Continent, would seem to lend support to this theory.

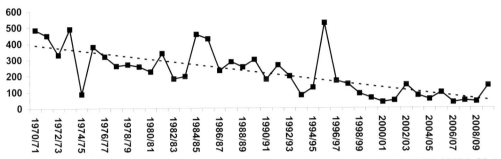

Peak winter counts (November to March) of Bar-tailed Godwit at Teesmouth, 1970-2009 (with trend line)

Each year, a few are at freshwater sites on Teesmouth though typically only one two birds are involved, usually during spring and autumn migration; exceptional counts were of 72 roosting at high tide on the Long Drag in late May and early June 1982 (Bell 1983) and, to a lesser extent, a party of nine at Saltholme Pools on 18 September 2010. In addition, Blick (2009) considered it not unusual to see birds heading away from Teesmouth on autumn evenings with Curlew *Numenius arquata* and Whimbrel *Numenius phaeopus,* a movement highlighted by single birds over Charlton's Pond, Billingham on 27 October 1985 and Norton on 15 July 1982 (Bell 1983); the latter accompanied by Whimbrel. An interesting record was of a largely white bird by Greatham Creek on 18 November 1990 (Bell 1991) with presumably the same around the Tees estuary from 12th to 30 January 1991 (Bell 1992). Away from Teesmouth the only area to produce regular winter sightings is Whitburn where a handful of birds regularly spend the winter at Whitburn Steel. Numbers rarely reach double figures, though exceptionally 30 flew north on 7 December 1980 and 33 flew north on 6 November 2000, the latter being the largest single day movement at that site.

Marginally inland along the River Tyne, Jarrow Slake was also formerly a regular winter haunt for this species, though the mudflats here were in filled from the 1950s; five were there on 22 February 1974, with one two per winter during the 1980s and up to three in August and September 1989. A single bird present from 19 December 2002 to February 2003 was said to be first to winter at the site for almost fifteen years, the only other recent record relating to passage birds on 4 April 2005 and 10 September 2006.

Inland the species is a scarce bird in County Durham, averaging one or two records per year, though occasionally more are recorded for example, five in 1996 (Armstrong 1998). Most occur between July and October and, to a lesser extent, during May, with occasional mid-winter records, the majority of which involve birds associating with wintering flocks of Curlew. Over half of all records come from the well-watched Barmston/Washington area, where it was regularly noted during the 1970s and 1980s, though records have declined in that area in recently decades. Elsewhere there are few sites that have produced multiple occurrences though Gateshead Borough has eight records all since 1989. These include three from Shibdon Pond, with the others recorded from the banks of the Tyne as far upriver as Scotswood. Further inland, there are four records from Hurworth Burn Reservoir, three from Derwent Reservoir and three recent records from the Bishop Middleham area. There is just one record from the well-watched Houghton-le-Spring area, at Hetton Lyons on 31 October 1999 (Armstrong 2003). Most records involve single birds or occasionally small parties of up to four, though exceptionally 43 flew east along the Tyne at Scotswood on 14 December 1989 (Bowey *et al.* 1993) and 45 were at Birtley on 17 September 1995 (Armstrong 1997). Other notable counts include nine west over Hurworth Burn Reservoir on 8 September 1979, six at Derwent Reservoir on 1 December 1985, seven heading down the Tyne valley at Shibdon Pond on 18 October 1997 and 12 south over WWT Washington on 31 August 2000. Some suggestion has been made that this species may utilise the Tyne corridor on migration between the west and east coasts, based upon occurrences in Northumberland from 1968 to 1980 (Kerr 2001), and the records from Scotswood in December 1989 and Shibdon Pond in October 1997 might offer some support to this theory.

The species is often seen at Whitburn during autumn migration from late June until September, when it is regularly reported passing offshore. Notable counts were; 25 north, on 22 July 1982; 22 north, on 20 June 2000;

and, 22 north, on 21 July 2007. Why these birds are heading north at this time of year is unclear. Occasionally migratory parties pause briefly at Whitburn Steel with the highest count to date being of 24 on 6 September 2006.

Distribution & movements

Bar-tailed Godwit breeds in the Arctic from Scandinavia to western Alaska, with birds of the nominate race breeding from Scandinavia to central Siberia and wintering around the coasts of southern and western Europe, Africa, the Arabian Peninsula, and parts of the Indian subcontinent. This is this race that occurs annually in the British Isles, including County Durham. An extremely long-distance migrant; birds from these eastern populations have been proven via satellite tracking to undertake the longest non-stop migration of any bird on earth, taking just nine days to cover the 11,000km between New Zealand and the Yellow Sea in China (Gill *et al* 2005).

Studies by Durham University and the WWT Advisory Service show that small numbers of post-breeding adults and a few over-summering juveniles undergo a post-nuptial moult at Teesmouth from July, though most have moved on by September. Limited ringing recoveries suggest that these birds, especially immatures, head further south to winter in southern Britain. A further influx largely consisting of juveniles takes place in September and October, with a few staying to winter locally, though the majority move on to winter in southern Britain and locations as far south as West Africa (Ward *et al.* 2003). Relatively few have been recovered/controlled in the county, but those that have suggest that at least some pass through southern Norway *en route* to Teesmouth (Bell 1987 & 1996), and from there can reach the western edge of the Sahara in as little as five days; as demonstrated by a young male ringed at South Gare on the Yorkshire side of the estuary on 13 October 1982 which was dead at Largoub, Western Sahara on 18th, some 3,660km to the south south west (Bell 1984).

Numbers at Teesmouth begin to build up from late October, boosted by the arrival of adult birds that have completed their post-breeding moult on the Continent at both the Wadden Sea and Westerschelde, rising to a late winter peak. Most depart for their breeding grounds in Siberia in March and April via a pre-nuptial moult on the Wadden Sea (Ward *et al.* 2003). There are a number of recoveries of birds ringed at Teesmouth from the Frisian Islands surrounding the Wadden Sea (Bell 1982, 1987 & 1988) in both spring and autumn, including a bird ringed as first-summer on Seal Sands on 2 July 1992 that was re-trapped at Sylt, Schleswig-Holstein, Germany on 12 March 1994 and 6 April 1997, highlighting the close links between this important staging post and the Tees Estuary.

Other birds ringed at Teesmouth have been found wintering in France (Bell 1992), Morocco (Bell 1987), the Channel Islands (Joynt 2004), on the Humber (Goodyer & Evans 1980) and in Devon (Bell 1998). The latter bird demonstrated remarkable site-faithfulness during winter, being trapped at Dawlish Warren on 14 November 1992 and 28 January 1997, having been originally ringed as a first-summer on Seal Sands on 2 July 1992 (Bell 1998). Another notable record involves a juvenile ringed at North Gare, Teesmouth on 22 October 1980 which was seen in Schleswig-Holstein, Germany, in August 1992 and subsequently wintered on Jersey every year to 2003, by which time it had reached the considerable age of 22 (Joynt 2004).

Whimbrel
Numenius phaeopus

A relatively common coastal passage visitor, but scarce inland.

Historical review

As '*Whympernell*', Whimbrel was mentioned twice by Benett (1844) in relation to provisioning the Durham Monastery during the 16th century. The first of these was on 8 August 1530 when "*3 curlews et 1 whympernell*" were purchased for *13d*, whilst in 1531-1532, two were bought for *4d*.

Willughby (1678) documented a report by Ralph Johnson of Brignall that the Whimbrel 'is found upon the Sands in the Teez mouth'. He made the first description of the species, saying, "*it is less by half than a Curlew, hath a crooked bill but shorter by an inch or more; the Crown deep brown with speckles. The Back under the Wings white, which the Curlew hath not...it is found upon the sands in the Teesmouth*" (Willughby 1678).

Some centuries later, Temperley (1951) described this 'small curlew' as a regular passage migrant, "*observed in spring and autumn, usually on the coast*". Richmond (1931) referred to its occurrence at Teesmouth on passage migration as one of the birds that 'keep to the foreshore'. He went on to say, 'Though most of the passage migrants do not apparently stay long, the movement is protracted, and birds such as the Whimbrel, arriving as early as 17 July, may be seen well into October (Richmond 1931).

Temperley (1951) noted that spring passage occurred in April and through early May, with a few birds even being seen in high summer. He noted that two were at Teesmouth on 7 June 1933 and an unusual winter or exceptionally early spring record occurred in 1945, when one was at Teesmouth on 3 February. He observed that autumn passage extended from July to September and into early October. Temperley thought that inland records were few, but he did document the occurrence of a small flock near Wolsingham, Weardale, from 12th to 15 September 1944.

A flock of 100 birds seen in a turnip field near Wylam, a village which straddles the Tyne, over 15 miles from the sea, on 21 September 1947, remains unprecedented as an inland record in the north east (Kerr 2001). However, it is unclear whether this occurrence was on the north or south side of the River Tyne, so it may not have been in County Durham but in Northumberland.

Large flocks of Whimbrel were not a feature in Durham in the 1950s and 1960s, though Blick documented a notably early spring migrant at Teesmouth on 10 March 1957 (Blick 2009). A flock of more than 50 birds flew in over the Tees estuary on 11 August 1961 and 40 were on the Reclamation Pond, Teesmouth on 10 September of that year. Over 75 birds were in the area around the Reclamation Pond on 20 July 1963. A couple of years later passage was thought to involve larger than usual numbers at Teesmouth over July to September, with a flock of 70 birds flying north over the North Gare on 4 September 1965 being, at the time, the largest single flock ever recorded at Teesmouth. In 1967, record numbers were noted at Seal Sands in late summer, including 76 counted on 30 July (Bell 1968).

Through the 1970s, this species' usual pattern of passage and movements was maintained, and the highest count in Durham to date came in this period, when 212 birds flew south over Seal Sands on 1 August 1970 (Blick 1978). A number of interesting but scarce inland records occurred, such as the single at Derwent Reservoir and two at Witton-le-Wear in autumn 1970 (McAndrew 1972).

Recent status

Today, the Whimbrel is a regular passage migrant through Durham, over-flying the county during spring and late summer. More specifically, it regularly feeds around the mouth of the Tees and its associated marshes, and at some shoreline locations between Sunderland and Lizard Point, Whitburn, in the north of the county, mainly during the late summer and autumn.

Spring passage through Durham usually occurs from mid-April until early June, with the first of the year occasionally appearing even earlier, in late March. Such spring migrants are mainly on the coast, often including birds feeding on the mudflats at Teesmouth, as well as birds seen on active northward migration past coastal watch points such as Hartlepool Headland and Whitburn Observatory. The earliest spring records, the 1957 occurrence aside, are of singles on Seal Sands on 16 March 1995, Greatham Creek on 15 March 1996, flying north over Fence Houses on 13 March 2002 and north over Sunderland Harbour on 17 March 2005 (Bell 1996 & 1997, Newsome 2007a & 2006).

More usual first dates are in early to mid-April, however, and the Whimbrel often lives up to its old name of May Bird, with spring passage peaking in late April or early May. Peak spring counts on the North Tees Marshes are typically of 20-40 birds. The ten-year mean spring peak count, 2000-2009, was 25 birds. There was significant variation in this period, however, from an exceptionally low six in 2003, to 54 on Cowpen Marsh on 26 April.

Inland spring records have included one feeding in pastures with Curlew *Numenius arquata* on Ravensworth Fell in early May 1992 (Armstrong 1993), and five at Hurworth Burn Reservoir on 8 May 2005, followed by five there on 23rd and 30 April 2006 (Newsome 2006 & 2007b). An exceptional inland count of nine was recorded at Castle Lake, Bishop Middleham on 9 May 2010 (Newsome 2012).

Late spring birds, or perhaps early returning migrants, which may be failed breeders, are quite regularly noted in June at sites around Teesmouth, such as Greatham Creek, or flying past Whitburn Observatory.

One to three birds summered in 1990 at Teesmouth and there have been occasional mid-summer records since, such as two on Seal Sands in mid-June 1995. However, the birds were not in breeding habitat, and there is no indication of a breeding population of this species being present in the county (Bell 1988-1993, Armstrong 1997). Given that this species is at the south western extremity of its European range in the Northern Isles and the far north of mainland Scotland, where it favours heathland and bog (JNCC 2009), this is very far removed from Teesmouth, and breeding is unlikely ever to occur there.

In the autumn, return passage often starts in late June, lasting through to September. A more noticeable arrival of migrants usually takes place from the first week of July and reaches a peak in late July and early August, with much of the autumn passage taking place during the first three weeks of August along the county's coastal strip. Passage typically declines through September, the last reports of the year often coming around mid-month.

The North Tees Marshes is again the area of the county which holds the largest numbers of birds, with an autumn peak of 30-50 being typical, occasionally building up to as many as 70 to 100 birds. The ten-year autumn average peak for the period 2000-2009 was 42 birds. As in spring, there was significant variation in this period, however, from a low of 21 in autumn 2008, when the spring peak was unusually higher than the autumn one, to a high of 95 on the North Tees Marshes on 10 August 2007.

Other high autumn Teesmouth counts have included 108 flying over the Brinefields on 3 August 1986 and 125 roosting on Seal Sands on 24 July 1990 (Bell 1987 & 1991). As the numbers at sites such as Seal Sands build through August, flocks numbering tens of birds are also often noted off Hartlepool Headland, such as 21 noted on the shore there in August 2000 (Armstrong 2005). Radar studies identified a narrow direct migratory route for waders from Teesmouth south west to the Ribble estuary during autumn (Evans 1968). Not all Whimbrel would appear to stop at Teesmouth, some continuing straight overhead. This regularly includes flocks of 10 or more birds on this trans-Pennine route, heading over Teesmouth by day (R.M. Ward pers. comm.).

Even in autumn, there are relatively few inland records, so singles noted over upper Teesdale on 5th and 12 September 1975, five far to the west at Balderhead Reservoir on 23 July 1994, 10 at Derwent Reservoir in July/August 1998, and 15 at Longnewton Reservoir on 19 August 2007, were notable (Unwin 1976, Armstrong 1996 & 1999b, Joynt et al. 2008).

Visible migration of Whimbrel is more obvious at the coast than in other wader species, helped by the birds' distinctive, often seven-note flight call. In spring and autumn birds are often noted passing Hartlepool Headland and Whitburn Observatory, when numbers mirror the build up of birds at Teesmouth. There is an apparent east-west split in Whimbrel migration in the UK, with the highest counts in western locations in the spring, and in eastern ones in autumn (Parkin & Knox 2010). The generally higher autumn peaks in the county are consistent with this national pattern.

Spring coastal passage is typically far lighter than in autumn and peaks in early May. Higher counts off Whitburn have included 37 north on 4 May 2005 and 21 on 7 May 2009. However, single figure day counts are more typical in spring. The predominance of autumn passage was illustrated in 2009, when only 67 out of 243 birds (28%) recorded passing Whitburn during the year were in spring (Newsome 2010).

Inland spring movements have included at least 11 heard flying north near Langley Park during the night of 23 April 1994, two in the Pennines over Barningham on 4 May 1998, a flock of four flying up the Tyne at Dunston on 16 May 1998, and one over Langdon Beck on 29 April 2006 (Armstrong 1996 & 1999b, Newsome 2007b). Inland passage through Durham may be under-recorded and this sample of records gives indication of the movement through the county in spring. Autumn movements have also been observed inland from the well-recorded Gateshead area, where most reports of this species have been of birds flying over. Records have included a flock of over 70 flying west over Sunniside on 26 July 1985 (Bowey et al. 1993), 13 north over Lamesley in August 1994, and 21 east over Shibdon Pond on 12 August 1995 (Armstrong 1996 & 1997).

The scale of passage along the coast during the autumn migration is illustrated by the 110 which flew south past Whitburn in two hours on 29 July 1979. Two years later, observers along the coast between Holy Island in north Northumberland and Hartlepool Headland noted heavy passage on 8 August 1981, including 100 flying south in just over an hour at Whitburn. Other heavy movements have included: 111 south past Hartlepool, on 9 August 1989; 120 there, on 2 August 1995; and, 105 off Whitburn, on 5 August 2007. Peak passage is usually recorded between the last few days of July and mid-August, as reflected in these high day counts.

Summarising movements through July and August 1991, a combined total of 255 birds flew past Whitburn Observatory on 27 dates. In 2007, higher counts resulted in a total of 115 birds being logged on 16 dates off Whitburn during July, with another 308 birds recorded on 16 dates in August (Armstrong 1992, Newsome 2008).

After feeding within the estuary and around the Tees Marshes in autumn, Whimbrel leave Teesmouth and often head south west over the centre of Teesside. In this part of the county, it is therefore not unusual to hear Whimbrel calling overhead as they fly over urban areas such as Stockton and Billingham, often at night (Blick 2009).

Whimbrels sometimes linger late into the autumn and there is a record of a single overwintering bird; in the severe winter of 2010/2011. A bird noted at Teesmouth until 26 November 2010 was believed to be the same individual as was detected in early 2011. Another very late individual with a slightly damaged wing, which may have delayed its departure, was at Cowpen Marsh on 9 December 1992. Other late records for the county include a single on Dorman's Pool, Teesmouth on 11 November 1984, one again with a slightly damaged wing in the Greenabella/Cowpen Marsh area on 11th to 14 November 1989, and singles on the North Tees Marshes on 12 November 1994 and north over Whitburn on 7 November 1995.

Distribution & movements

Whimbrel has a circumpolar, though patchy, breeding distribution. Small numbers of birds breed in Britain, provisionally estimated at a minimum of c.300 pairs in 2009, largely on islands in northern Scotland (Holling 2011). The species is made up of a number of races. Those occurring as breeding birds in northern Europe, i.e. Iceland, Scandinavia, and north Britain, are of the nominate form, which breeds as far east as western Siberia. Birds from across this range winter far to the south of the breeding areas. In the European context this means sub-Saharan Africa, though a few remain in the southernmost parts of Europe. This species is only recorded in tiny numbers wintering in Britain, with generally less than 20 per year, and similar numbers in Ireland (JNCC 2009).

The origin of some of Durham's passage birds is indicated by the small amount of ringing information relating to this species. There are only two ringing recoveries for the county and both were ringed as breeding adult birds on Fetlar, Shetland, then recorded on autumn passage on Seal Sands. A female ringed on Fetlar on 16 June 1988 was on Seal Sands on 31 July 1993, whilst another colour-ringed there in summer 1986 was seen on Seal Sands on 6 July 1994 (Armstrong 1994 & 1996).

These ringing records indicate predictable southward movement in autumn, but there is not enough evidence here to be confident of the origin of the majority of the county's passage Whimbrel. However, it is likely to comprise birds from Iceland, the Faeroes, Fennoscandia and north west Russia, as well as Scotland (Wernham *et al.* 2002).

Overall, numbers of passage Whimbrel in the county seem to be stable, with no clear trends, and this reflects an apparently stable European population (Robinson 2005), although there appears to have been a recent decline in the small Scottish population (Holling 2011).

Curlew (Eurasian Curlew)
Numenius arquata

Sponsored by

NORTH PENNINES
Area of Outstanding Natural Beauty

A common resident, passage migrant and winter visitor in large numbers.

Historical review

The Curlew was plentiful in the county as long ago as the 14th century, and commonly found and used as food. It was frequently mentioned in the Cellarers' rolls of the Monastery of Durham, as noted by Fowler (1898). The earliest mention in the county is as "*Curleus*" in 1311-1312 and it would appear that Curlews, spelt in a variety of ways, were numerous, though less easy to find than 'Plover', as they were more expensive. They were bought in small lots of up to about half a dozen and their price varied from three halfpence in 1338 to eight pence each in 1390. Several centuries later, Benett (1844) in the *Durham Household Book*, recorded Curlew as being bought for 3d, two for 6d and three for 18*d* during the 16th century (Gardner-Medwin 1985).

In early Victorian times it was a breeding bird of the uplands. Hogg (1845) noted that it fed on grasslands in August and September, then at the coast in the winter. Robson (1896) recorded that the species bred sparingly with a scattered distribution in the upper Derwent valley. The Curlew appears to have been exclusively an upland breeding bird in the county until the end of the 19th century, when it began to breed increasingly in grass fields in the lowlands, even down to the coast. Colonisation of the lowlands continued through the 20th century (Alexander

& Lack 1944, Parslow 1973). During the mid-20th century, breeding Curlew spread into lowland agricultural areas, but this expansion in range had stopped by the latter part of the century (Gibbons *et al.* 1993).

Temperley first noted Curlew breeding in the lower Team valley in 1932 and at the same time Bishop, based at Teesmouth, said that it was *"breeding on grazing tracts where it had never previously been known"*. Heslop-Harrison noted that in his early days this species was rarely, if ever, seen in the Team Valley, but in 1944 and 1945 a pair raised young at Urpeth Bottoms, just above Bewick Main (*The Vasculum* Vol. XXX, 1945). Prior to this time, breeding birds had not been known to be present in that area, although they would probably have been breeding on the nearby higher ground of Ravensworth Fell (Bowey *et al.* 1993).

In the middle of the 20th century, Temperley (1951) described Curlew as a summer resident, passage migrant and winter visitor. He observed that its status had changed considerably *"in recent years"*, and by the late 1940s and early 1950s, Curlews were even nesting on the outskirts of towns in industrialised areas. He also reported that some were remaining in tidal areas throughout the year, recognising that these summer flocks probably consisted of non-breeding birds.

Temperley stated that Teesmouth was the main wintering site for Curlew on the Durham coast. This was confirmed by Blick (2009), who noted that prior to the 1950s over 1,000 had frequently been recorded in that area.

Stead (1964) noted that the species bred in several areas within ten miles of the coastal plain in the early 1960s, but not around the Tees Marshes, where breeding was not confirmed until the late 1960s, despite the fact that small numbers of birds were already summering around the estuary (Stead 1969, Blick 1978).

A spring party of at least 110 birds at Hurworth Burn Reservoir on 16 April 1964 was considered exceptional at the time (Bell 1964). High Teesmouth counts of non-breeding birds in the 1960s included a mid-winter total of 450 near Greatham Creek on 9 January 1966, rising slightly to 501 on 12 March (Bell 1966).

The partial reclamation of Seal Sands from 1970 to 1974, reducing the intertidal area by about two thirds, had a dramatic adverse effect on the numbers of wintering and passage waders, including Curlew (Evans 1978/1979, Evans *et al.* 1979 & Evans 1997). Hence, numbers of wintering and passage Curlew at Teesmouth fell sharply. The TBC report recorded a highest winter count of just 70 birds on the whole estuary in January and February 1973. However, numbers recovered quickly to a winter maximum of 350 in February 1974 (Stewart 1973, Unwin 1975), and have since increased significantly, as discussed in the following section.

Recent breeding status

Curlew is a common large wader of Durham's western uplands, and there is a smaller but significant population that breeds, rather unsuccessfully, on lowland farmland and pastures. Where breeding, their characteristic display flight and haunting call makes them a prominent and noticeable part of the county's fauna. The seasonal distribution of Curlew in Durham is strikingly different. The majority of the breeding season records come from the west of the county, though not exclusively so, and the wintering population is largely in the eastern and central lowlands, along the coastal strip and in the south east of the county.

In Durham, the species breeds mostly on upland heather moorlands and in rough grazing areas, with smaller numbers of birds on lowland wet pastures, some on permanent pasture on areas of extensively managed farmland and, occasionally, on 'set-aside' (Westerberg & Bowey 2000). It is a widespread breeding species, for example, in 2007 it was reported from more than 160 locations across the county (Newsome 2008). The highest densities of breeding birds occur in the county's Pennine uplands, along the Gaunless uplands and on the Durham plateau (Westerberg & Bowey 2000).

The species' favoured upland habitats are generally on acidic or leached soils (Marchant *et al.* 1993), which are particularly common in western and central regions of the county as a result of the altitude and climate (Westerberg & Bowey 2000). During the breeding season, much of the open moors and grasslands of the western portion of the county are populated by this species. Curlew can be found breeding at significant altitudes in Durham, commonly up to and over 550m above sea level, with extreme examples at significantly higher elevations. Surveys conducted in Weardale and Teesdale in the early 1990s found breeding birds at up to 630m above sea level in various habitats, including hay meadows, rough grazing pastures and heather moorlands which were managed for Red Grouse *Lagopus lagopus* shooting (Westerberg & Bowey 2000).

Birds move from their coastal wintering areas into upland breeding locations as early as late February, though the timing of this move is very much weather-dependant. In 1975 for example, mild weather in late February saw Curlew return to Wolsingham Park Moor by 18 February and Cow Green Reservoir by 25th. A mild January in 1989 resulted in an exceptionally early return, with birds noted in upper Weardale by 26th (Unwin 1976, Armstrong 1990). In 2008, birds had returned to a breeding area near Langley Moor by 25 February, but display wasn't noted in upper Teesdale until late March, demonstrating the disparity between upland and more lowland breeding sites. On arrival, pre-breeding flocks may be found gathering on the fringes of the main breeding areas. In 2009, for example, 100 birds were at Grains o' th' Beck in the extreme south west of the county, on 5 March, whilst 275 were scattered around the Wellhope Burn, in upper Weardale, on 8 March (Newsome 2009 & 2010).

Birds can still be found widely in lowland areas of the county, although breeding densities in such areas are considerably lower than in the uplands. The lowland breeding population appears to be very much in decline, but this is difficult to quantify, as the species is under-recorded in breeding areas. Since 1990, lowland breeding estimates have ranged from 22 pairs in 15 locations in 1996, to probable breeding in just five locations in 2007, but both of these figures are likely to be under-estimates.

Elsewhere in the county's lowlands there are scattered population of birds, for example, in areas around Crookfoot Reservoir alone, there were perhaps 20 pairs during the first decade of the 21st century (Joynt et al. 2008). The contradiction between this and the above records for Durham as a whole underlines the inconsistency of lowland breeding data across the county.

The breeding density of Curlew varies greatly from location to location, because of the relatively wide range of habitats used for nesting. Even between areas of apparently similar habitat, large variations in breeding densities are sometimes evident, such as in Robson's study of the early 1990s, when he found that on two relatively close Red Grouse moors in Teesdale, the breeding densities of Curlew were 4.3 pairs per square kilometre and 9.2 pairs per kilometre respectively (Westerberg & Bowey 2000). Such large variations in breeding density create problems when trying to accurately estimate the county's breeding population. For the purposes of the *Atlas*, this was resolved by averaging the breeding densities according to the areas of habitat, leading to an estimated population range of between 2,600 and 6,100 breeding pairs (Westerberg & Bowey 2000).

According to Gibbons et al. (1993), County Durham was an area of high abundance for this species, and therefore it is of considerable importance in the national context. The mid-eighties estimates of the national population indicated that there were between 33,000 and 38,000 pairs of Curlew in Britain (Reed 1985). In Durham, the *Atlas* concluded that the county's breeding population formed a significant proportion of the country's total, between 7% and 18.5 % of the national figure (Westerberg & Bowey 2000).

However, a more recent estimate of the UK breeding population placed this far higher, in the range of 99,500 to 125,000 pairs, for the period 1985-1999 (Thorup 2006). This would suggest that the breeding population of Curlew in the Britain represented approximately 50% of the European population.

In any event, whilst it remains a common passage, winter visitor and breeding bird in Durham and across the north east region, its breeding numbers have declined in the UK by about 42% between 1995 and 2008. This decline is mirrored in the north east region, with an estimated 38% decline in the same period (Risely et al. 2010).

Curlew is on the amber list of Birds of Conservation Concern, and near-threatened on a global scale. The national and international causes of this decline are habitat fragmentation, loss of moorlands and agricultural intensification in breeding areas, as well as disturbance and land reclamation on wintering sites and migration staging posts (Robinson 2005).

In Durham, upland afforestation has not had a major impact, but agricultural intensification, including drainage of wetlands and over-grazing of breeding habitats by sheep, is likely to have had an adverse effect (Westerberg & Bowey 2000). The negative impacts will probably have been experienced most acutely in the county's lowlands. Impacts on non-breeding sites in the county are discussed in the following section.

An unusual record concerned a white individual, which held territory in May 1985 near Tow Law, and was seen again there in 1986 (Baldridge 1986).

Recent non-breeding status

At the end of the breeding season, once young have fledged, the breeding grounds are rapidly deserted and from late June to August the county's wintering areas are once again increasingly populated by some local birds, which are joined by others from more northern breeding grounds. Small numbers are widely reported on passage during this period, before flocks are settled into established areas for the winter. As autumn progresses, many local breeding birds leave our area for the winter, heading south west, to be replaced by others which have bred in Scandinavia.

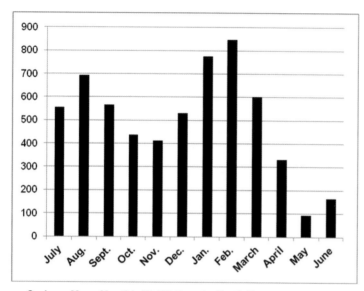

This is reflected at Teesmouth, where from late June adults begin to arrive for post-breeding moult, numbers peaking during August and September. The main immigration of juveniles follows later, supplementing numbers in the second half of September (Ward *et al.* 2003).

The seasonal pattern of Curlew numbers at Teesmouth, as illustrated, covers the ten-year period from 2000 to 2009, but Ward *et al.* (2003) also found the same seasonal pattern over the longer term, from 1975 to 2000. The autumn peak typically numbers 500-700 birds, but far higher counts have occurred, such as 1,211 on the WeBS count on 21 August 1994 (Armstrong 1996).

Curlew - Mean Monthly WeBS Counts, North Teesmouth, 2000-2009, (no counts March to August 2001)

In Durham, Curlew winter on lowland farmland and at suitable coastal sites. Preferred winter habitats are mud flats and areas of sand at low tide, but birds also use rocky pools, salt marshes, and damp pastures both on the coast, and increasingly inland.

Teesmouth is the main wintering area for this species in the county and it still attracts large numbers of birds. After a slight fall in numbers following autumn passage, numbers typically rise again to their highest level of the year in mid-winter. On 12 February 1984, 945 were counted on the north side of the estuary. This was at the time the highest count for at least 35 years (Baldridge 1985). Such a total would now be considered unremarkable however, as from 500 to 1,000 birds typically winter on the North Tees Marshes, and counts exceed this in some years. On the WeBS count of January 2003, 1,604 birds were present, probably the highest recorded total in Durham to date (Joynt 2004). Numbers here have recovered and increased significantly since the setback of the major land reclamation at Teesmouth in the early 1970s.

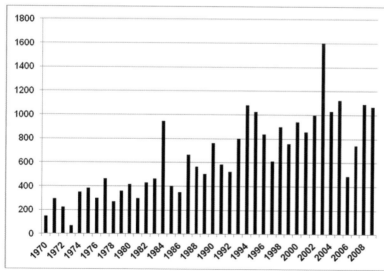

Curlew - winter maxima, North Teesmouth (January & February), 1970-2009

The recovery of the non-breeding Teesmouth Curlew population from this large-scale habitat loss reflects the fact that this species is a 'generalist' feeder, both in terms of habitats used and prey items taken. This has enabled Curlew to adapt to the significant changes at Teesmouth over the years. It has not only moved in response to habitat losses to industrial development, but also to losses of feeding areas due to the spread of mats of the green algae *Enteromorpha* over parts of the mud flats in the 1990s. The other Teesmouth waders which are 'generalised' feeders, Oystercatcher *Haematopus ostralegus,* Black-tailed Godwit *Limosa limosa* and Redshank *Tringa totanus,* have also shown significant population increases at Teesmouth in this period (Ward *et al.* 2003).

There is considerable variation in mid-winter counts at Teesmouth, however, such as in 1985, when the January peak at North Teesmouth was reduced to 184 birds. This may have been associated with a movement of birds out of the area because of severe winter weather. Winter numbers that year recovered to 402 in February (Baldridge 1986).

During late winter, an influx of birds occurs at Teesmouth prior to departure to breeding areas. This shows as a significant average rise between December and February in Figure 1. Many individuals which have been absent from the estuary since post-breeding moult re-appear around this time (Ward *et al.* 2003).

Birds can also be routinely found in autumn and winter at many other sites in the county, including the tidal mudflats along the River Tyne. In the past, these mudflats included Jarrow Slake, which suffered substantial habitat loss through gradual infilling during the 1950s and 1960s, before eventually being completely lost to development during the 1990s. Despite being reduced to a fraction of its former size, this site still attracted 140 Curlew on 2 January 1979 and 130 on 26 October 1980 (Unwin & Sowerbutts 1980, Baldridge 1981).

There has been an increasing trend in the county towards birds feeding on inland fields in winter in recent decades. Such birds undertake daily flights between their feeding areas and roosting sites, feeding on worm-rich pastures during the day and roosting on tidal or other wetland habitats at night. This was first noted in the late 1980s and early 1990s. By the late 1990s, most of the county's wintering birds were behaving in this way.

The number of Curlew feeding on estuarine mud during the day at Teesmouth declined hugely during the 1990s, reflecting this change in behaviour. This was noted from the beginning of the decade, as increasing numbers of birds began regularly feeding on fields near Saltholme. On the January 1994 WeBS count, for example, 810 Curlew were on the Saltholme fields, while only 201 were foraging across the intertidal mudflats on 13 January, underlining the extent of this change (Armstrong 1996).

Similar behaviour has been noted around Hurworth Burn Reservoir, where, during the last quarter of 1992 for example, a south easterly movement of birds at dusk suggested that up to 100 were regularly feeding in nearby fields and roosting overnight at Teesmouth (Armstrong 1993). Radio telemetry studies undertaken by Durham University during winter 1999 have confirmed this behaviour, with daily commuting of the Hurworth Burn/Fishburn wintering population to roost and feed overnight on Seal Sands (Tsai unpubl. 2003, R. M. Ward pers. comm.). Some of Teesside's wintering Curlew now feed well inland, with as many as 100 having been noted at Upsall Carr, Wynyard Park and Preston Park, flying on to the North Tees Marshes almost at dark, during the first decade of the 21st century (Blick 2009).

Away from Teesmouth, the most important of these inland feeding areas is just outside Sunderland in the Houghton Gate area, most of these birds commuting to roost, nightly, in large numbers at WWT Washington. This roost grew during the 1990s, increasing from around 50 birds in 1990 to over 600 by January 2007 and 2008, making it second only to, and in some years comparable with, the Teesmouth mid-winter roost. Smaller winter flocks of up to 100 are also noted at Fellgate, Follingsby Lane, Lambton and Sedgeletch. The majority of these satellite flocks probably follow a daily pattern of movements to and from the Washington roost. This roost is also used by birds feeding on the nearby River Wear inter-tidal mudflats.

Examples of such gatherings also occur in western Gateshead, with up to 100 birds feeding on pastures around Greenside and Barlow/Blaydon Burn, then flying to the River Tyne mudflats to roost at night (Armstrong 1993 & 1996). This behaviour was also noted in the Cleadon area in 1996, with Curlew gathering to roost at Boldon Flats (Armstrong 1998). These birds may have been those that formerly fed around Jarrow Slake during the 1980s before its reclamation, and subsequently moved inland. The Boldon Flats roost reached a peak of 120 birds in February 2009 (Newsome 2010).

The county's mid-winter population grew through the 1990s, and this may have been linked to this behavioural change from feeding on inter-tidal mudflats to inland fields. Overall wintering numbers in the county increased by approximately 30% in the 15 years 1987 to 2001 (Siggens 2005).

Wintering in the uplands is unusual, so up to 12 remaining at Derwent Reservoir throughout the winter in 1977 (Unwin 1978) and one on Cotherstone Moor on 2 December 1990 (Armstrong 1991), are notable county records.

Numbers decline at non-breeding sites after wintering birds disperse and spring passage is over, as birds return to breeding areas. However, small numbers of immature/non-breeding birds remain on the Tees estuary through the summer, beginning to moult by late May (Ward *et al.* 2003). Over-summering by non-breeding birds in estuarine habitats was first recorded in the county by Temperley, as noted in the historical review. The current trend at Teesmouth is typically for 50-100 birds to remain through the summer.

A few birds pass along the Durham coastline in most months of the year, because many locally wintering Curlews make minor movements up and down the coast between feeding grounds, as is evidenced by regular records of birds moving north and south. However, more significant movements of birds on active migration occur during passage periods. Such observations occasionally come from coastal watch points such as Whitburn Observatory and Hartlepool Headland, and are clustered in spring and autumn. Counts usually number tens or at most scores of birds, so the totals of 111 seen flying south off Hartlepool on 24 September 1981 (Bell 1982) and 122 north off Whitburn on 25 June 2000 were somewhat exceptional (Armstrong 2005).

Visible migration can also occur at Teesmouth, such as in March and April 1995 and again in April 1996, when small flocks were noted departing in a north easterly direction, presumably heading back to their Scandinavian breeding grounds. An exceptionally heavy inland movement of Curlew was also noted on 18 March 1999, when 1,500 flew west into the Pennines over Dryerdale, near Wolsingham, in just two hours (Armstrong 2003).

The overall position for non-breeding Curlew in the county is favourable, as illustrated by the trends identified above. This is despite the decline both regionally, nationally and internationally in the species' breeding population. This reflects the Curlew's ability to adapt in winter, colonising new feeding habitats in the form of inland fields, which are far more plentiful in the county than estuarine mud flats, and coping with significant changes at its wintering stronghold of Teesmouth.

Distribution & movements

The nominate race of Curlew breeds over much of northern and eastern Europe and east through western and central Siberia, though it is absent from some parts of northern Europe, such as Iceland. The wintering range encompasses the milder parts of western and southern Europe, and south into the African tropics and sub-tropics.

Autumn and winter populations at Teesmouth, as elsewhere in northern Britain, originate primarily from breeding grounds in Scandinavia, although some local breeding birds are also involved (Ward *et al.* 2003). This is substantiated by a significant number of ringing recoveries of birds that were ringed and colour-ringed at Teesmouth in autumn and winter, largely as part of a programme undertaken by Durham University. They include, for example, one ringed as an adult on Seal Sands on 23 July 1994 then seen on the nest at Balso, Bjurholm, Sweden on 29 May 1997 (Bell 1998). A number of Finnish recoveries include one ringed as an adult at Seal Sands on 21 August 1991 and found dead after flying into a window at Keskusta, Oulu, Finland on 26 May 1999 (Iceton 2000). Another ringed as a first year bird on Seal Sands on 26 November 1994 was re-trapped at Kiusala, Turku-Pori, Finland on 26 June 1996 then re-trapped again back on Seal Sands on 22 August 1998 (Armstrong 2003).

Many of the August and September arrivals at Teesmouth are juveniles from Scandinavia, as indicated by ringing recoveries. For example, four birds ringed as nestlings in Norway and Finland have been recovered at Teesmouth in the following August, September and October (Blick 2009).

Few migration stop-overs occur between the northern breeding grounds and Teesmouth. There is only a single ringing recovery to date to indicate that such behaviour may occur, this being of an adult in Denmark at the beginning of September (Ward *et al.* 2003).

The majority of northern British breeding birds are known to move to Ireland in winter (Bainbridge & Minton 1978). This movement of locally bred birds to the south west has been substantiated by a small number of birds being recovered or controlled in Ireland. They include one ringed as a nestling on the moors near Edmundbyers on 6 August 1988 then shot at Milltown, Tuam, Ireland on 26 December 1988 and another ringed as a nestling at Langdon Beck, upper Teesdale on 24 June 1990 and also shot at Tuam, on 5 September 1994 (Armstrong 1989 & 1996).

The only Teesmouth spring or autumn passage birds recovered in winter have been individuals in Lancashire, Humberside and Northumberland, as well as more local recoveries (Ward *et al.* 2003). The Humberside record, for example, concerned one ringed as an adult on Seal Sands on 17 March 1977 then re-trapped at Cherry Cob Sands, Keyingham, Humberside on 26 October 1987. More local winter observations include another adult ringed on Seal Sands on 9 August 2000 then noted just 36km to the north at Washington, on 11 December of the same year (Bell 1989 & Iceton 2001).

Other more local movements of Curlew include one ringed as a nestling at Newton Mulgrave Moor, North Yorkshire, on 31 May 1991 and re-trapped on Seal Sands on 2 July 1992. This was only the second recovery of a North Yorkshire Moors-bred Curlew on the Tees (Bell 1993).

Curlews can be long-lived, as illustrated by a bird colour-ringed as an adult on Seal Sands in September 1981 then seen again there on 17 July 2007. It was therefore a minimum of 27 years old, just four years short of the BTO longevity record for this species (Joynt *et al.* 2008).

Terek Sandpiper
Xenus cinereus

A very rare vagrant from Siberia and eastern Europe: four records.

The first record was of a bird seen by Peter Evans on Cowpen Marsh, Teesmouth on 27 September 1952. Initially located amongst a large flock of Redshank *Tringa totanus* and Grey Plover *Pluvialis squatarola* on a small pool behind the sea wall; it remained until the 28th (*British Birds* 46: 188).

All records:

1952	Cowpen Marsh, Teesmouth, 27th to 28 September
1979	Long Drag, Teesmouth, 20th to 22 June *(contra BB)* (*British Birds* 73: 510)
2008	Saltholme Pools, and adjacent River Tees, Teesmouth, 5th to 13th, and 17 July (*British Birds* 102: 556)
2009	Saltholme Pools, Teesmouth, 16th to 17 June (*British Birds* 103: 589)

Distribution & movements

Terek Sandpiper, Long Drag, June 1979 (Tom Francis)

The European range of Terek Sandpiper is restricted to a small population in Finland centred around the northern Gulf of Bothnia, and into Belarus. To the east the species breeds more widely, but locally throughout northern Russia to eastern Siberia. Birds winter far to the south, along the coasts of southern and eastern Africa, the Middle East and most of southern Asia including the Indian subcontinent through Indonesia to Australia. The species is an annual vagrant to Britain with a total of 73 records by the end of 2010. Most records relate to adults in May and June, although July records are not unusual. In this respect, the pattern of occurrence in County Durham is typical, although the first record is unusual; being one of only five September records for Britain.

Common Sandpiper
Actitis hypoleucos

A fairly common breeding summer visitor and passage migrant, typically recorded from April to September, very rare in winter.

Historical review

There is little historical documentation surrounding the occurrence of Common Sandpiper in County Durham, though it appears to have been common throughout much of its existing modern day distribution. One of the earliest references to this species comes from the 19th century, with Thomas Robson describing it as *"one of the commonest waders that visit the Derwent"* and going on to describe finding a clutch of eggs near Winlaton Mill in May 1895 (Robson 1896).

George Temperley (1951) described the species as *"a common spring and summer resident"*, which was most frequently found breeding on moorland streams in the western uplands, and less so in the industrialised east of the county. Summarising its status on each of the major rivers systems within the county, Temperley said that it bred as far down stream on the Tyne as Ryton, Gibside on the Derwent and below Durham on the Wear, and that it could be found along the entire length of the Tees, with a few pairs breeding annually at Teesmouth. In addition, Temperley noted that the species occurred at other wetlands on passage.

Regarding the species at Teesmouth Stead (1964) stated that *"a few birds frequent the slag revetment walls and marshy pools in the spring and summer"*, with most records relating to solitary birds not lingering long in the estuary. He gave just one instance of breeding at Teesmouth; a pair that bred on Cowpen Marsh from 1927 to 1931 (Milburn 1932). Unusually for such an urbanised location a pair bred at Charlton's Pond, Billingham in 1966 (Stead 1969).

Both Temperley (1951) and Stead (1964) gave little indication of the numbers involved on migration, though double-figure counts were reached on at least four occasions during the 1950s and 1960s. These were 20 at Tanfield Ponds on 28 July 1953, 18 at Witton-le-Wear on 19 August 1969, 19 at Greatham Creek, Teesmouth on 12 August 1969 and 21 at Saltholme Pools on 15 August 1969. That three of these counts should be made during an eight-day period in August of the same year indicates significant passage through the region at that time, though some overlap will no doubt have occurred between the two counts from the Tees estuary.

Temperley (1951) stated that the first birds typically arrived between 11th to 13 April, with the majority departing in August and early September. This status has changed little since. The earliest recorded bird between 1951 and 1970 was one at Langdon Beck, upper Teesdale on 4 April 1961.

As Temperley (1951) suggested most have departed by early September, and post-Temperley there was just a handful of October and November records. What was probably the same bird was at Greatham Creek on 24th and 31 October, and again on 14 November 1954, with others: at Hurworth Burn Reservoir, on 26 October 1959; on the River Tyne at Wylam, on 7 November 1959; at Swalwell on 16 October 1962; and, at the mouth of Crimdon Dene on 10 October 1964. The bird at Wylam in November 1959 was a particularly late migrant, or perhaps it was wintering in the area. A single bird on the River Wear at on 12 January 1950 is the only mid-winter record prior to 1970, and one of very few winter records for the county to date.

Recent breeding status

Common Sandpiper remains a common summer visitor to County Durham. As a breeding species it is very scarce in the south east lowlands and is almost entirely restricted to the west of the county where it can be found along most watercourses throughout the summer, particularly those in the western uplands. It is particularly prevalent along the stony shores of rivers and upland becks, though reservoir foreshores are also an important breeding habitat for the species particularly in the west where this habitat is common. There are few if any upland reservoirs and their feeder streams that do not have at least one breeding pair of Common Sandpipers in residence between May and early July.

The habitat and geographical features favoured by Common Sandpipers are most abundant in the Pennine uplands, where the species is extensively distributed across three watersheds, the Derwent (Tyne), the Wear and the Tees. The highest located territories in the county are found at over 500m above sea level in locations above Cow Green Reservoir in upper Teesdale; some of the highest altitude territories of this species in England (Westerberg & Bowey 2000). Below an altitude of around 130m above sea level, breeding territories become largely restricted to the three main rivers, with breeding birds extending down to Darlington on the Tees, at 30m above sea level, to Durham City on the Wear and to Swalwell on the River Derwent at less than 5m above sea level (Westerberg & Bowey 2000). Most of the upland reservoirs have breeding birds and examples of such concentrations include six pairs between the car park and the dam at Cow Green Reservoir in spring 2008 and six pairs along a 1.3km stretch of the Balder Beck, near Balderhead Reservoir in the same year. During June and July 2009, breeding birds were reported from over 30 one-kilometre squares, the majority of these being west of grid line NZ06.

In the east few, if any, breeding birds are present in the lowlands or along the coastal strip and it is a rare breeder around Teesmouth, not having documented as breeding successfully there for many years (Joynt *et al.* 2008), although in the past it has bred at Charlton's Pond, Cowpen Marsh, Portrack Marsh and Crookfoot Reservoir, as well as along the lower reaches of the River Tees (Blick 2009). There is only one published record of breeding taking place in this part of the county since 1970, a pair bred in the old shipyard at Haverton Hill in 1985 (Bell 1986).

At the time of the *Atlas* survey work, lowland breeding records were scarce but notable exceptions came from the wetlands at Far Pasture, near Rowlands Gill, which supported up to two breeding pairs, and the pools at Gateshead MetroCentre, with one or two pairs. Both sites have now been lost, the former due to vegetational succession rendering it unsuitable for breeding and the latter to development (Westerberg & Bowey 2000). In such instances, the Common Sandpiper mirrors the 'pioneering' behaviour of Little Ringed Plover *Charadrius dubius*, moving into newly created wetland sites and breeding until vegetational changes militate against ongoing success. Another site to have lost the species as a regular breeder during recent decades is WWT Washington, where breeding first commenced in 1974 and was last noted in 1984 (Baldridge 1985). Additional breeding records of interest from the lowland east of the county include a pair that reared young at Chester-le-Street in 1999, and a pair seen copulating on the Wear at Sunderland Bridge near Croxdale during the same year; a locality where it had bred previously (Armstrong 2003).

The density of breeding birds varies enormously across the county, being largely dependent upon the geography of the watercourse, which they inhabit. One of the earliest references to population density was the 14 pairs counted at Witton-le-Wear in 1972 (Coulson 1973). In the spring of 1980, census work revealed 48 birds along seven miles of upland water way in early May, including 8km of the Sleightholme Beck and 3.5km of the River Greta, the observer commenting that birds seemed less densely distributed above 400m above sea level (Baldridge 1981). In early May 1983, up to 28 birds were at Derwent Reservoir indicating the size of the breeding population there at that time (Baldridge 1984). Three territories were found along less than a kilometre of the Tees at Wycliffe in spring 1989 (Armstrong 1990); six pairs were in a 5km stretch of the River Greta over Bowes Moor in spring 1991 (Armstrong 1992); at least seven pairs along this stretch in 1992 and 1993 (Armstrong 1993 & 1994); and, two to three pairs were along each of four, one-kilometre stretches of the Wear between Witton Park and Stanhope in 1993 (Armstrong 1994). Regular observation in 1996 suggested a population of at least 20 birds on territory along the 25km stretch of the River Tees between Eggleston and Cow Green Reservoir (Armstrong 1998), a relatively low count although this area is popular with tourists and suffers from disturbance. By contrast 16 pairs were distributed along just 8km of the Sleightholme Beck in the spring of 1998 (Armstrong 1999). More recently during the early spring of 2007, eight pairs were counted at Waskerley Reservoir, with between four and six territories located at Bollihope, Bowless and Derwent Reservoir. In June of the same year at least 11 at Selset Reservoir, seven at Cow Green Reservoir and six at Balderhead Reservoir doubtless involved breeding birds. In June of the same year, at least 10 were along the River Greta near Bowes, indicating a continuing strong presence in this area (Newsome 2008). Breeding densities in the Peak District are reported to range from 0.7 to 4.7 pairs per kilometre of watercourse (Gibbons *et al.* 1993) and working on the assumption that each occupied tetrad in Durham held an average of two pairs per kilometre of occupied watercourse, the *Atlas* calculated that the county population was around 400 pairs (Westerberg & Bowey 2000). This assertion was verified using data gathered from Waterway Bird Surveys conducted along the Rivers Wear and Greta.

Concerns have been expressed about the increasing amount of recreational disturbance to the uplands, and in particular the river systems and reservoirs favoured by this species; both are increasingly popular with hikers and anglers (Yalden 1992) especially in the light of changes in access policy to the uplands in the early 21st century. A number of local studies have attributed the nationally observed population declines to the species' vulnerability to such recreational disturbance (Gibbons *et al* 1993), though the effects on the species in County Durham from such factors are not known.

Recent non-breeding status

Common Sandpiper occurs commonly as a passage migrant throughout the region in both spring and autumn though is usually more numerous during autumn when it can occur in some numbers. It can be encountered on almost any body of water from rivers and streams to freshwater lakes and reservoirs right down to tidal rocks at the seashore and is often numerous at Teesmouth during autumn where it tends to favour the tidal waters of Greatham Creek.

The spring's first returning birds typically appear around the middle of April, though there has been a tendency for this to have become earlier in recent years with March records in four years since 2000. Some of these birds may be relocating wintering birds rather than genuine arrivals from the south. Despite this, March records are still rare with just ten since 1970, two in the 1970s, five in the 1990s and three in the 2000s. The earliest being one at Lamesley on 11 March 2006 (Newsome 2007).

In most years there is typically an early bird or two during the first half of April, followed by a more concerted influx of birds that usually occurs from around the 23rd onwards with records becoming much more widespread and numerous as the month progresses. By the end of April, birds have usually penetrated to breeding sites as far west as Cow Green Reservoir, in upper Teesdale, though occasionally they arrive sooner, such as the pair displaying at Waskerley Reservoir on 18 April 2009 (Newsome 2010). Passage continues into May, though is short-lived and is usually over by mid-month, with the majority of the county's breeding population, including those in the extreme west back on territory by early to mid-May. Throughout the duration of spring migration passage is typically light and often involves just one or two birds at any given locality, many of which have moved on by the following day, though the larger upland reservoirs often produce higher counts, such as the 28 birds at Derwent Reservoir in May 1983 (Baldridge 1984. Away from these numbers rarely reach double-figures.

Post-breeding dispersal takes place rapidly after young birds have fledged and may be noted from as early as the third week of June, when small numbers of birds begin to appear away from their breeding haunts. There are usually few observations from upland breeding areas after mid-July, although birds often linger at some of the larger reservoirs such as Derwent until early August. Notable post-breeding counts from there include 34 on 16 July 1989 (Armstrong 1990) and an exceptional 50 on 10 July 1994 (Armstrong 1996).

Late July and early August is when the species is at its most numerous in the lowland east of the county when records are widespread and individual counts frequently reach double-figures. The highest count ever made at a single site was 78 birds on the River Wear at Washington on 7 August 1981 (Baldridge 1982) - 40 were at Gosforth Park in Northumberland on the same date (Kerr 2001) - these were obviously part of large influx to the region, though they moved quickly through with just six present at Washington the next day. The River Wear at Washington remains one of the best sites for autumn gatherings of this species though numbers have never approached this total since. A peak count of 10-15 birds is more typical, with 19 on 31 July 1977 (Unwin 1978) being the only other count to exceed 15 birds since 1970.

The highest count of the year typically comes from the North Tees Marshes where a peak count of 15 to 25 birds is typical with occasional larger gatherings, such as 41 by the Long Drag on 8 August 1980 (Bell 1981) and 65 to 70 at the mouth of Greatham Creek on 11 August 1997 (Bell 1998). On this latter date a number of high counts were made across the county, with a total of 82 at Teesmouth (Bell 1998); this being the highest total recorded there. On the same date, 23 were at Whitburn and 16 at Rainton Meadows (Armstrong 1999), with 34 at Hart Reservoir the following day. Later in the month, two flocks totalling 33 birds at Ryhope on 25 August (Armstrong 1999) were noteworthy for their coastal location. A similar influx during August 2004 brought a minimum of 62 birds to the North Tees Marshes on 13th. Other notable counts away from both the North Tees Marshes and WWT include: 18 at Wynyard, on 10 August 1975 (Bell 1976); 16 at Charlton's Pond, 27 July 1980 Bell 1981); 15 at Norton Bottoms, on 11 August 1986 (Bell 1987); 16 at Ryton, on 23 August 1987 (Armstrong 1988); and, 21 at Crookfoot Reservoir on 6 August 1996 (Bell 1997).

Additionally, passage is frequently detected along the Durham coast with for example a party of 12 flying south off Whitburn Observatory on 30 August 2008 (Newsome 2009). In late summer it is not unusual to hear Common Sandpipers calling as they fly overhead at night mainly along the coast but sometimes along river valley corridors.

Passage slows through September and in most years few birds are noted after mid-month with the highest September count being 11 at Greatham Creek, Teesmouth on 4 September 2008 (Joynt 2009). October records are uncommon, less than annual and the majority refer to single birds, exceptions being: three at WWT Washington, from 6 October 1974, one remaining until 3 November (Unwin 1975); two at Hartlepool Docks, on 2 October 1994 (Armstrong 1996); two at Saltholme Pools, on 15 October 2005 (Joynt 2006); two at Shibdon Pond, on 15 October 2006 (Newsome 2007); and, two at Cobey Carr Ponds, Willington on 22 October 2006 (Newsome 2007). Other than obvious wintering birds, November records are exceptional with just three occurrences since 1970 at: Witton-le-Wear, on 1 November 1973 (Coulson 1974); WWT Washington, three dates to 10 November 1977 (Unwin 1978); and, Blackhall Rocks on 5 November 1997 (Armstrong 1999). The latter record is particularly noteworthy for its coastal location.

Wintering birds are rarely recorded in the county and there have been few such occurrences since the first in January 1950. An isolated example involved a single bird on the River Wear at WWT Washington during January to March 1976 (Unwin 1976), being arguably the first to successfully overwinter in the county. More recently, at least one bird wintered at the mouth of the River Don and on Jarrow Slake in the winter of 1984/1985 remaining until at least 21 February. Remarkably, both Green *Tringa ochropus* and Wood Sandpiper *Tringa glareola* were present in the same area during January and early February presenting a unique mid-winter treble (Baldridge 1985). Three years later, a bird was noted on the River Wear at WWT Washington on 17th and 22 December 1988 (Armstrong 1989). Clearly, overwintering is not a common occurrence in the north east of England and the clusters of such records in certain areas seem to be associated with the presence of returning birds, year-on-year. This phenomenon appeared to be confirmed in the winters of 1988/1989, 1989/1990 and probably also in 1990/1991 by occurrences in western Gateshead. In the first of these winters, at least two, and probably three, different birds spent the winter on the Tyne between Blaydon, Ryton Willows and Wylam, with two being seen together on the Tyne at Blaydon Station in the last week of January 1989. The following year what may have been one of the same birds was noted between November and February at Ryton Willows, Shibdon Pond and on the River Derwent at Eelshaugh where there was also a single winter report in February 1991 (Bowey *et al* 1993). There were just three further mid-winter records in Durham by the close of the decade, these were single birds: at Croxdale, on 25 December 1992 (Armstrong 1993); on the River Wear at Prebend's Bridge, Durham, on 9 January 1993 (Armstrong 1994); and, on Sandhaven Beach, South Shields, on 5 February 1996 (Armstrong 1998). The latter bird is especially unusual for its coastal location during winter, though it was presumably wintering undetected on the adjacent River Tyne. At the beginning of the 21st century, Jarrow Slake again attracted a regular wintering bird during the winters of 2000/2001 and 2001/2002, it was present from November to March and gave further evidence that occasional birds may regularly winter along the Tyne. Two further records involved singles at Brancepeth Beck on 2 March 2002 (Newsome 2007) and Dalton Piercy Burn on 1 January 2004 (Newsome 2006). The former could conceivably represent a very early migrant although it is unlikely.

Distribution & movements

The Common Sandpiper breeds across much of northern and eastern Europe, eastwards through central Asia to Siberia, Kamchatka and Japan. A few spend the winter in the milder parts of Europe including the south coast of England, though the majority of the European population winters in Africa, both north and south of the Sahara.

Interesting coastal observations during spring include four birds flushed from the face of Marsden Quarry during thick mist and drizzle on 21 May 1973 (Coulson 1974), and three seen to fly in off the sea at Whitburn on 14 May 1985 (Baldridge 1986). Both of these records giving clear indication that passage is still ongoing well into May.

Just three Common Sandpipers are known to have been trapped and ringed in County Durham making it all the more remarkable that there is a single foreign recovery. A first-year bird ringed at Seal Sands on 20 August 1988 was controlled at Chenquiette, Morocco on 2 May 1989; a distance of 2,563km to the south south west (Bell 1990).

Spotted Sandpiper
Actitis macularius

Very rare vagrant from North America: three records.

The first reference to this species in the county dates back to the early 19th century, and occurs in Sir Cuthbert Sharp's *History of Hartlepool* published in 1816, which noted the occurrence of Spotted Sandpiper near Hartlepool. No doubt this record was of either suspect origin or perhaps a confusion of nomenclature, Temperley (1951) did not list it. The first fully authenticated record was of a juvenile, found by Tom Francis, at the northern end of the Long Drag, Teesmouth on 16 September 1997. It remained until 29 September (*British Birds* 91: 478). This was the only record

Spotted Sandpiper, Long Drag, September 1997 (Tom Francis)

436

for Britain that year.

The second was a summer plumage adult at Pow Hill Country Park on the southern shore of Derwent Reservoir from 19 June to 6 July 2002. This bird was holding territory and displayed on a number of occasions to Common Sandpipers *Actitis hypoleucos*. It ranged widely around the reservoir, also visiting the Northumberland shore during its extended stay (*British Birds* 96: 572).

The third record was also an adult in summer plumage found at Saltholme Pools, Teesmouth on 11 August 2004. Heavy rain during the late evening brought an arrival of 14 Black Tern *Chlidonias niger* and a flock of at least 30 Common Sandpiper to Back Saltholme Pool, which included this bird. Anxious and mobile these birds could be heard leaving as darkness approached; unsurprisingly there was no sign of the bird the following day (*British Birds* 98: 659).

Distribution & movements

Spotted Sandpiper is the Nearctic counterpart of the Common Sandpiper, with a breeding range that extends over much of North America. During the winter some birds migrate southwards to coastal areas of the southern United States, though the vast majority spend the winter in Central America, the Caribbean, and northern South America. Formerly a very rare vagrant, an increased knowledge of the identification criteria of non-breeding adults and juveniles, particularly since the 1970s, has led to this species being recognised as a regular vagrant to the British Isles with a total of almost 170 records by the end of 2010. Recorded in both spring and autumn, the species also has a penchant for overwintering in Britain and now does so just about annually. Several spring birds have been noted displaying to Common Sandpiper, including the bird at Derwent Reservoir in June 2002; a pair attempted to breed in Scotland in 1975.

Green Sandpiper
Tringa ochropus

An uncommon passage migrant and a scarce but regular visitor in winter.

Historical review

The earliest reference to this species in Durham comes from Selby (1831) who recorded that one was shot in October 1830 near Hylton Castle, Sunderland, and he described it, at that time, as a '*rare species*'. One in summer plumage was also shot at Streatlam Park in the summer of 1838, and the specimen is now in the Hancock Museum, but there is no other detail about this record. If this was a genuine mid-summer record, it is intriguing, raising the possibility that it could have been a bird in territory. This is conceivable, as birds were said to have bred in Cumbria in 1917 (Brown & Grice 2005). The species was documented by Hancock (1874), who thought it a 'rare autumn visitant'. The first record in the Gateshead area was in the lower Derwent valley, where Robson told of one that was shot near Whickham in the 19th century, but he gave no precise date or locality (Robson 1896).

It seems that there has been a real increase in the number of records of Green Sandpiper since Hancock's time and this process continued into the 20th century, as stated by George Temperley (1951). He said that it was recorded annually at South Shields, in the Team valley and at Darlington Sewage Farm, where a party of twelve birds was present on 23 August 1927, some of these staying to mid-October 1927. A further bird was seen from December 1928 to January 1929, at this same site.

During the 1930s, Temperley noted that birds were often seen in the Team valley, (this would have been prior to the development of the Team Valley Trading Estate). Birds were overwintering there, such as in 1935-36, with up to three being present in the autumn of 1936 (Bowey *et al.* 1993). In 1944, Dr. H. M. S. Blair saw two birds near South Shields on 15 January 1944, so clearly the phenomenon of wintering in the north east of England is a long established one.

In overview, Temperley described Green Sandpiper as a "*regular passage migrant in small numbers*" and he reported that it occurred with greater frequency at his time of writing (1951) than previously. He noted that a few remained into the winter and this quite accurately describes its modern status in the county. He also commented that spring passage through the county was less marked than in autumn, a statement which also stands the test of time.

Through the 1950s and 1960s, birds were reported regularly at many wetland sites, but rarely in large numbers. Exceptional counts included 12 at Tanfield Ponds on 25 August 1953, up to eight at Hurworth Burn Reservoir in August 1961, ten at Coxhoe Sewage Farm on 5 August 1962 and again in July/August 1963, and 11 spread around Teesmouth on 14 August 1964. Tanfield Ponds proved one of the most attractive sites for this species through the 1960s and in 1964, with regular watching, based on the timing and throughput of birds at the site, an estimated minimum of 18 individuals visited between 26 July and 4 September (Bell 1965). Some more unusual sightings included the bird flushed from the River Tees at Langdon Beck on 24 April 1955 and one at Smiddyshaw Reservoir on 26 October 1958, which was a long way west, and late for such a high altitude record.

A small number of wintering records occurred over this period, for example the winter of 1958/1959 brought three birds, two near Stockton in the south of the county, and one at South Shields in the north, this latter bird staying into 1959. Witton-le-Wear attracted wintering birds in four consecutive years from 1963/1964 through to 1966/1967 (Bell 1964-1968, Parrack & Bell 1970).

Through the 1970s small numbers of wintering birds were noted on an annual basis, from at least thirteen sites overall. During October to December 1973, single birds were noted at five sites, including West Boldon, where birds had wintered in the previous three years, and during the first quarter of 1974 birds were noted at Witton-le-Wear, Durham City and Rowlands Gill. By the middle of the decade, local overwintering by Green Sandpipers was described as *"now a regular occurrence"* by Unwin (1975). Passage birds were also by then routinely and widely reported, including up to seven at Sedgeletch in July 1975 and widespread reports in August that year, when it was thought to be more numerous than usual (Unwin 1976).

Recent status

In the last quarter of the 20th century and the early 21st century, Green Sandpiper has been a regularly recorded spring and autumn passage migrant and a less common winter visitor. The species is as often found feeding by inland streams or an isolated pond or small wetland, as it is on one of the county's larger wetland sites. It is the passage wader most likely to be seen away from established wader habitats. Though it is recorded regularly in spring, and occasionally in winter, Green Sandpipers are most common on autumn migration.

Spring passage is evident in most years from late March, with an increase in sightings in April. Spring migrants are typically noted at a handful of sites, between early April and early May, the last of the spring usually being noted by the middle of that month. The spring peak for Green Sandpiper passage comes in April, earlier than the peak for many other spring passage waders, which occurs in May. This reflects the shorter distance between wintering and breeding grounds for this than other later returning wader species (R.M. Ward pers. comm.). Numbers are low, however, with typically just five-ten in spring in the first decade of the 21st century, although it is sometimes difficult to distinguish late winterers from passage birds at this time. Although generally far commoner than Wood Sandpipers *Tringa glareola*, spring numbers are comparable in this period.

This is one of the earliest migrant waders to return from its breeding grounds, re-appearing in the UK in mid-June. This is reflected in the county, with the first autumn birds appearing around this time. The first young birds are then recorded in late July or early August in the UK (Wernham *et al.* 2002). This is again reflected locally, resulting in a peak in numbers in late August. The autumn passage then continues through to late October. The number of birds noted in autumn between 2000 and 2009, was typically 30 to 50 individuals; far higher than in spring. Numbers significantly exceed this in some years. In 1994, for example, as many as 63 individuals were noted at 21 sites in July to September (Armstrong 1996), and in autumn 2001, over 70 birds were recorded (Siggens 2005, Iceton 2002). In autumn 2008 around 50 birds were noted, and the following year the county saw a comparable influx in July and August, totalling a minimum of 45 birds (Newsome 2009 & 2010).

A moderately large gathering of Green Sandpipers in Durham is of four to five birds at any one site, as noted at Boldon Flats, Bishop Middleham, Hurworth Burn Reservoir and Seaton Pond in July and August 2009 (Newsome 2010). Higher autumn counts can occur, however, such as the 12 birds at Shibdon Pond on 16 August 1981 (Bowey *et al.* 1993) and 25 were at Boldon Flats on 19 August 1987 (Armstrong 1988). Seven were at Bradley Hall Pond in the north west corner of the county on 21 August 2000. On 19 August 2001, coinciding with a good passage of this species, 25 were at Barmston Pond and 17 were at Teesmouth (Siggens 2005, Iceton 2002), the birds being grounded by heavy thundery showers suggesting that in a typical year many more birds pass over the county than are recorded. The Barmston count is a particularly high one for a site of this size, and followed the deliberate lowering of water levels at the pond to create ideal conditions for passage waders. It is notable that all of these high counts are within a six-day period between 16 and 21 August.

The pattern of occurrence for this species is very much more widespread than that of many other waders, with one or two birds being noted at many sites around the county in most years. There is also much less of a focus on the Tees wetlands than there is for many wader species, although birds are routinely noted in that area in all seasons. For example, in 2000, between early July and mid-October, 17 widely scattered sites held from one to seven birds, comprising a total of at least 58 individuals (Armstrong 2005). From June to August 2010, the species was reported from 25 locations (Newsome 2012). Typical sites away from Teesmouth which regularly attract birds include: Castle Lake at Bishop Middleham, Hetton Lyons CP, Hurworth Burn Reservoir, Lamesley Pastures (Water Meadows) and Reed Beds, Seaton Pond, Sedgeletch, Shibdon Pond and WWT Washington. There is a strong lowland bias to autumn passage distribution with few upland records. Two at Smiddyshaw Reservoir on 13 August 2009 was therefore a notable westerly record (Newsome 2010). The only site in this part of Durham which attracts birds with any degree of regularity is Derwent Reservoir, though birds are never common here. Birds are usually noted until the second week of September, with passage being over by mid-October.

As was noted by Kerr (2001), small numbers winter in the north east of England even in the harshest of weather, and through the 1970s and 1980s up to four were present in winter at Greatham Creek (Blick 2009). Wintering birds have been present every winter in Durham from 1975 onwards, and numbers wintering were thought to be on the increase in the county through the mid- to late 1980s. Favoured wintering sites in the 1970s and 1980s included the River Tees at High Coniscliffe, which attracted birds every year between 1977 and 1988, with usually one or two, but including up to four on 25 February 1988, and Sedgeletch/Houghton, which had birds every year from 1983-1998, including up to three in December 1993. Wintering birds were also regularly at Langley Moor in the 1980s and early 1990s.

High winter site counts have also included four at Witton Park late in 1994 (Armstrong 1996) and five at Haverton Hole from January to March 1999 (Iceton 2000). A more unusual sighting was of a coastal winter bird at Hendon on 29 January 1992 (Armstrong 1993).

During the winter, if the pattern of records from the first decade of the 21st century is now typical, ten or more individuals may be expected at a handful of sites, scattered around the county; most of these sightings relate to single birds. In 2001 for example, small numbers were reported to be wintering at favoured locations, with records from Jarrow Slake, Joe's Pond, Herrington County Park and Sedgeletch (Siggens 2005). In 2002 and 2003, single wintering birds were recorded regularly from 15 sites in both years, and Sedgeletch held an exceptional five birds on 5 December 2002 (Newsome 2007a). From 2007 to 2009, the number of Green Sandpiper wintering sites ranged from nine to 15. One at Tunstall Reservoir intermittently from 28 October 2009 to the year's end was in a most unusual upland location for a wintering bird (Newsome 2008-2010).

In late autumn, UK research has shown that those remaining to winter become more solitary in their habits, as demonstrated above, and may even become territorial. Green Sandpipers are also known to be highly site faithful to passage and wintering sites, with an 80% return rate between years (Smith *et al.* 1992). This is also reflected locally, with what would appear to be the same birds returning to winter at sites year after year. For example, wintering birds were noted in the Lockhaugh area in 1987/1988 and again in 1988/1989 (Bowey *et al.* 1993). These preferred wintering locations are often small marshes or lengths of streams, rather than the larger wetland sites. In just such a circumstance a bird was found on a tiny farm pond at Reeley Mires, Barlow Burn, in November 2004, kilometres from the nearest large wetland (K. Bowey pers. comm.).

Green Sandpipers are uncommon on the coast, and rarely seen during coastal movements, so eight flying in off the sea at Hartlepool on 18 July 1987 was notable seawatching records (Bell 1988, Joynt 2010). Normally records at the seawatch sites of Whitburn and Hartlepool Headland are limited to sporadic sightings of single birds.

Distribution & movements

Green Sandpiper breeds in north central Europe and Scandinavia, as well as to the east, into the mid-latitudes of Russia and Siberia. Birds winter in west and southern Europe, with small numbers in Britain, and down into Africa and on the Arabian Peninsula. There have been no ringing recoveries of this species in the county up to the end of 2010 (R.M. Ward pers. comm.).

Spotted Redshank
Tringa erythropus

A scarce to uncommon passage visitor, rare in winter.

Historical review
During Victorian times this species was considered to be a very rare straggler to the county (Temperley 1951). Hancock (1874) gave just two occurrences for Durham, an immature at Jarrow Slake on an unknown date, but before 1831, as it was mentioned by Selby (1831), and one from Blanchland on the Durham/Northumberland border, on 12 August 1840. Nelson (1907) said that it was rare at Teesmouth, having been seen on six or seven occasions in that area, but he suggested that it might have been overlooked. He listed the following records: one in William Backhouse's collection, date unstated but presumably in the early part of the 19th century; one in 1876; one on 15 September 1881; one on 21 September 1899; and one in September 1902.

Temperley observed that records of this species had increased during the first half of the 20th century, and this may have been a result of greater observer knowledge and better optical equipment. Over this period he said most sightings occurred in July and August. This species was considered 'fairly regular' at Teesmouth by the 1930s, as noted by Richmond, writing of the breeding birds of Teesmouth, who said that "*Mr. J. Bishop writes to me....that the Spotted Redshank is regular – usually in July and August.*" (Richmond 1931).

A series of early records of single birds came from Darlington Sewage Farm, beginning with one on the late date of 4 November 1928, other records following there on 14 August 1930, 24th and 25 August 1931, 15 September 1934, from 7th to 9 September 1935 and on 22nd and 24 August 1939. This indicated that passage birds occurred more widely, and that regularly watched sites would produce records. Spring birds, rare at the time, were noted at Teesmouth, with one on 24 April 1933, and two there on 4 May 1939.

By his time of writing, Temperley (1951) called this a "*fairly regular autumn passage migrant in very small numbers*", but said it that was "*less often noted in spring*". He reported that the first record of an adult in full breeding plumage came from Teesmouth on 5 July 1947.

A very unusual occurrence took place in 1959, when a flock of up to eleven birds was present at Hurworth Burn Reservoir on 22nd and 23 September, part of a large influx of birds in the region that autumn. This was repeated in 1968, when up to 12 were at Hurworth Burn Reservoir over the period 7th to 11 September. These remain the largest numbers of birds recorded in Durham away from Teesmouth (Bell 1959, Stewart 1968).

The first winter record came in 1960, with probably the same bird at Billingham Ponds on 12 December and by Greatham Creek on 27 December and last seen on 25 April 1961 (Bell 1960 & 1961). Wintering birds became a regular feature at Teesmouth during the 1960s, and in 1967/1968 one lingered very late into the spring, being last seen on 3 June 1968, by which time it was in full summer plumage (Bell 1967, Stewart 1968).

The 13 at Teesmouth from 20-28 August 1964 were the largest single flock recorded in the county to that point, being matched in 1973, when 13 to 14 birds were noted there in August and September (Bell 1964, Stewart 1973).

Recent status
In its immaculate, almost black, breeding plumage there are few more splendid birds to be seen than the Spotted Redshank, though such a sight in Durham is uncommon. Only a few such birds appear in May or June, for this species is a scarce, mainly autumn passage migrant, as it would appear to have been for many years. Today, Spotted Redshank is largely a scarce passage visitor to County Durham. Occasional birds stay into the winter, though this is a rare winter visitor, which is recorded almost annually. Regardless of the time of year, Teesmouth is the most favoured area for this species in Durham, though a few sightings come from other scattered wetland sites around the county in most years.

Spring passage through Durham has always tended to be rather sparse, with one to five individuals per year being typical, and birds are only ever seen regularly in spring at Teesmouth. Larger numbers do occasionally occur, however, such as in 2002, when a higher spring passage produced around ten birds, but all records were confined to Teesmouth, between 8 March and 4 May (Little 2003, Newsome 2007). In 2007, spring migrants were on the North Tees Marshes from 4 April to 9 May, but no more than five birds were seen in total. Another on the River Tyne near Newburn on 18 April 2007 was a notable local record (Joynt *et al.* 2008, Newsome 2008). This is one of the earliest waders to appear on spring passage, peaking in April, earlier than the peak for many other spring passage waders, which occurs in May. This is likely to reflect the shorter distance between some parts of the wintering and breeding grounds for Spotted Redshank than other later returning wader species.

Birds in full summer plumage, either as spring migrants or returning adults in early autumn, are scarce in the county, being recorded barely annually. Such records have included one at Seal Sands on 29 April 1988 and, away from Teesmouth, one at Houghton Gate from 16 to 20 April 2004 (Armstrong 1989, Siggens 2006). In 2009, a wintering bird lingered on the North Tees Marshes until 19 April, by which time it was in virtually full breeding plumage (Joynt 2010). Returning birds retaining summer plumage have included several on the North Tees Marshes in 2007, when adults were regularly recorded from mid-June onwards, with a maximum of three at Greatham Creek on 7 July (Joynt 2008).

Most Spotted Redshanks are seen in the autumn, when a trickle of birds passes through the county from June to October. The autumn's first returning birds are adults, often appearing around mid-June in low numbers, but these build through July and into August, as adults are joined and then outnumbered by juveniles. There is usually a presence of this species, albeit in small numbers, around the county's wetlands from mid-August through to mid-October, with some of the later birds occasionally remaining into early November.

Although Teesmouth predominates, birds can appear at a wide range of wetlands from Shibdon Pond in the north to Crookfoot Reservoir in the south. Other sites that regularly attract the species include Castle Lake (Bishop Middleham), Derwent Reservoir, WWT Washington and nearby Barmston Pond. Although Barmston Pond's heyday was in the historical review period in the late 1960s and 1970s, prior to the draining of the eastern of two ponds, the remaining pond enjoyed something of a revival in the 1990s and early 2000s, when water levels were deliberately lowered in autumn for passage waders. Spotted Redshanks were often prominent amongst them, including five juveniles together there on 22nd and 23 August 2001, during a particularly heavy and diverse wader passage (Siggens 2005).

Peak site counts in Durham rarely exceed single figures, and the higher counts almost invariably come from Teesmouth. In some years the maximum counts are of only three or four birds. Larger numbers can occur, however, as noted under the historical review. They have included ten at WWT Washington on 7 September 1978 and 13 at Teesmouth on 24 August 1986, which was the largest autumn passage there since 1973 (Unwin 1979, Bell 1987).

It is difficult to estimate the total number of birds involved in a year's autumn passage, because of the constant throughput of birds, and conversely, the tendency of some Spotted Redshank to linger and even overwinter. If the first decade of the 21st century is representative, this revealed numbers typically in the range of 10-20 per autumn.

There is considerable variation, however, with higher numbers for example in 1992, when late August brought up to seven to Barmston Pond, five to Teesmouth, four to WWT Washington, two to Whitburn Steel and a single to Shibdon Pond, out of a total autumn passage of at least 25 birds (Armstrong 1993). Conversely, 2004 fell far below this level, with only an estimated 11 birds in the entire year, of which perhaps as few as six were involved in autumn passage (Joynt 2005, Siggens 2006).

In many years, quite a rapid decline in sightings takes place from around mid-September, though in 1994, three lingered into early November and one remained to winter into 1995 in the Greatham Creek area (Bell 1995 & 1996). In the early years of the 21st century, up to four were at WWT Washington between 4th and 12 October 2007, with a late bird there on 13 November, and an exceptional eight were still together on Seal Sands as late as 7 November 2008 (Newsome 2008 & 2009).

Most wintering birds have been present around the North Tees Marshes. Blick (2009) noted that there was a wintering bird at Teesmouth in almost every winter from the early 1960s to January 1995. The most favoured location for birds was in the Seal Sands to Greatham Creek and Greenabella Marsh area. Usually there were just single birds, and presumably the same individual(s) were often involved year after year, but in the winter of 1975 up to three were noted, this figure rising to five by 31 March, by which time wintering birds may have been supplemented by early migrants (Unwin 1975b). Likewise in 1980, more than one bird was recorded. The winter of 1986 was believed to be the first since 1966 when no birds wintered at Teesmouth (Bell 1981 & 1987). Birds quickly reappeared, however, subsequently occurring in most winters. Wintering Spotted Redshanks in Durham remained a scarce but regularly recorded feature through the 1990s and into the 2000s. Wintering birds at Teesmouth have occasionally lingered late enough into the spring to develop full summer plumage, as noted previously.

Wintering birds are rare away from Teesmouth, so singles at Boldon Flats until mid-December 1992, on 10 December 1993, and at WWT Washington 18th to 19 December 1996 are notable records (Armstrong 1994 & 1998). Wintering birds have also been noted increasingly regularly at Jarrow Slake, in early 1987 for example, and in each winter between 2002 and 2005, then again in 2008. In 2005, what was thought to be the same bird, an adult female returning from the breeding grounds, re-appeared in July (Armstrong 1988, Siggens 2006, Newsome 2006, 2007a & 2009).

Other winter records include single birds on the coast, on the beach at Hartlepool Headland on 26 December 2000 and at Whitburn Steel on 13 February 2007. What may have been the Whitburn individual then appeared at WWT Washington on 19th and 20 February 2007 (Armstrong 2005, Newsome 2008).

Spotted Redshank pass along the coastline in small numbers, sometimes at night when their distinctive call betrays their onward movement. Coastal records include one heard calling over Souter Point in July 1992, one flying south at Whitburn on 24 August 1993, and occasional singles passing there in August 1998, August 1999 and July and August 2010 (Armstrong 1993-2003, Newsome 2011). A single also flew south over Fulwell on 17 July 2003 (Newsome 2007). Visible spring migration is less often noted, but on 9 May 2010, a summer plumage bird flew north at Whitburn Observatory (Newsome 2012).

Distribution & movements

Spotted Redshank has a breeding distribution that stretches discontinuously from Scandinavia, eastwards, though Russia and on to Siberia. The wintering grounds stretch from south western Europe, where a small number winter (including a very small number in the British Isles) across to West Africa and as far east as Vietnam and south east China. There have been no ringing recoveries of this species in the county to date.

There are no clear trends in the county's records for passage or wintering Spotted Redshank, and this reflects an apparently stable international breeding population (Butchart *et al.* 2011).

Greenshank (Common Greenshank)
Tringa nebularia

An uncommon passage migrant, rare as a wintering species.

Historical review

The earliest reference to Greenshank in the county is from a passage in Ray's posthumously published *Synopsis* of 1713. Ray wrote that this species (the '*Greater plover of Aldrovandus*') occurred sometime before 1689, when the reporter of the record died. Translated from Latin, the passage indicates that '*D. Ralph Johnson (of Brignall, in the south of County Durham) saw it on the coast of the Bishopric of Durham*'. This was perhaps the first record for Britain, as noted by Gardner-Medwin (1985).

Temperley (1951), reviewing the 19th century records, noted that Hogg (1845) said that it was '*not uncommon*' but never numerous, though Hancock (1874) described it as '*a rare autumn visitant*'.

At his time of writing, Temperley (1951) described the Greenshank as a "*passage migrant in small numbers*" occurring more regularly in autumn than in spring, with few spring records. He also noted that it occurred mainly on the coast (i.e. at Teesmouth), and occasionally at inland sites. Temperley mentioned no wintering records for Durham, despite this having been noted in Northumberland (Kerr 2001). He thought that its status in the county had not altered over the previous 100 years or so.

Temperley listed a number of sites where Greenshank had been recorded, including Jarrow Slake, the River Don at Jarrow, Hebburn Ponds, Darlington Sewage Farm, Boldon Flats, and Hurworth Burn Reservoir, but noted that the main 'resort' was Teesmouth.

The 1950s and 1960s saw relatively little information published on the status of Greenshank, except when larger numbers were recorded, as in 1963, when a peak of 23 birds were on the North Tees Marshes on 17 August. The county's first apparent wintering bird was recorded at Teesmouth the following winter, from 10 December 1963 until 9 February 1964, when it was found dead under power lines at the Reclamation Pond (Bell 1963 & 1964).

Recent status

This elegant wader is a regular spring and autumn passage migrant through Durham. It is quite frequently seen in autumn at a wide range of sites on inland waters, and very occasionally it is present right through the winter, mainly at Teesmouth. Fly-over records come from many areas, particularly during the peak autumn passage period, courtesy of its distinctive call.

The North Tees Marshes have for many years been consistently the best area in the county for this species. Birds can and do appear at all of the county's main wetland sites however, especially when there has been a significant summer drawdown of water levels. In most years, apart from a few weeks in June, the species is almost continuously present from April through to October, as passage periods overlap, but birds do not breed. The lowland and easterly predominance of Greenshank records in the county was illustrated in 2007, when out of a total of 97 records in the year, just seven came from sites west of the A1M (Newsome 2008). Greenshank remains rare in winter.

The first for the year usually appear from early to mid-April, most often, but not exclusively, on the North Tees Marshes. Small numbers can occur even earlier than this, in March, such as the four on the Tees at Low Coniscliffe on 26 March 2007. Such isolated early records may perhaps involve birds which have wintered further south in Britain (Newsome 2008). An unusual upland spring record came from Langdon Beck, upper Teesdale, where two birds were present on 29 April 2006 (Newsome 2007).

The spring passage of this species through Durham tends to be small and relatively concentrated, with typically under 20 birds recorded in total across the county, as they head north. The ten-year spring average for the period 2000-2009 was just 14 individuals. Higher numbers can occur, however, such as in April and May 1976, when spring migrants amounted to around 20 birds across the county (Unwin 1976). The heaviest spring passage in the period 2000-2009 was in 2001, when approximately 27 birds were seen in the county, including 15 on the North Tees Marshes, with a single site maximum of eight at Haverton Hole on 8 May (Iceton 2002, Siggens 2005).

An intriguing mid-summer record came in the mid-1970s, when a bird was present in upland habitat above Cow Green Reservoir over the period 7th to 12 June 1975, in an area that is not dissimilar to the species' favoured breeding habitats in locations such as north west Scotland. This would appear to be the only such record in the county, but considering the vast area of the Durham uplands, it would be surprising if parts of this area were not, at least occasionally, investigated by passage birds (Unwin 1975b).

Other mid-summer records are more predictably from lowland passage sites, and probably refer to wandering individuals which have never reached the breeding grounds, or early failed breeders. They have included birds noted on three dates at Joe's Pond in June 1989, and another single which spent much of June 1990 on Wader Lake at WWT Washington. Another wanderer appeared at Rainton Meadows and nearby Redburn Marsh on 7th and 13 June 1997 respectively (Armstrong 1989b, 1990 & 1999a).

The first of any early returning birds usually appear around the last week of June, with numbers building up through July and August, typically peaking at 25-50 birds across the county in the latter part of August or early September. A far higher autumn peak is possible, as in 2004, when combined records suggest that around 100 birds were in the county at the same time in mid- to late August (Joynt 2005, Siggens 2006). The increase in numbers from mid-August is the result of the later arriving juveniles joining the passage of adult birds. This can produce two peaks in Greenshank passage at Teesmouth, as noted in 2006, when a late July peak was followed by a lull in early August after most of the adults had moved through, then a second wave from mid-August as the juveniles arrived (Joynt 2007).

The autumn passage is far larger than that experienced in Durham during spring. Regardless of the fact that even at the height of this movement the number of birds in the county may not exceed 25-50, the number moving through the area during the whole autumn passage period, in from the north and out to the south, is many more than this. Overall autumn numbers are far more difficult to estimate than in spring, but considering figures for the 1990s and 2000s, totals in the range of 50-100 birds per autumn are typical. The autumn average for the period 2000-2009 was 67 individuals per annum. The migration period is protracted, extending from late June to early November.

It is likely that in the best years, because of the throughput of birds, the true figures are even higher than this. Blick (2009) estimated that the number of birds passing through the Tees estuary during the autumn may reach the low hundreds in years of heavier passage.

In autumn, birds are usually widespread at wetland sites from west to east, though they are far more common in the lowlands. The North Tees Marshes remain dominant, with around 20-50 birds passing through in a typical autumn, but possibly far more, as suggested by Blick (2009). However, there are also records scattered across many inland sites in most years. For example in 2000, in August alone, birds were recorded at 22 sites. Away from Teesmouth, the majority of sightings refer to single birds, but larger numbers are often noted rising to three or four birds at favoured sites, with some higher counts, such as seven at Shibdon Pond on 8 August 1990, and more unusually seven birds together on the River Tyne at Clara Vale on 21 August 2000 (Armstrong 1990 & 2005).

Occasionally there are far larger influxes, as in August and September 1979, when unprecedented numbers occurred. This was most obvious and spectacular at Teesmouth, when 16 birds on 1 August rose to 26 on 19th, and a peak of 61 on 9 September, when most of the flock flew off south west. This is still the highest ever count in Durham to date. Around the same time 12 were at Hurworth Burn Reservoir on 12 August and 10 were at WWT Washington on 9 September. In 1981, 23 birds were recorded at WWT Washington on 23 August, a large number for an inland site, eight remaining there into early October (Unwin & Sowerbutts 1980, Baldridge 1982).

Another large influx came in 1987, when 20 birds were at Boldon Flats on 18th and 19 July, during a passage of northern waders that included numbers of Wood Sandpiper *Tringa glareola* and Green Sandpiper *Tringa ochropus* at the same site. In addition, 16 were around Teesmouth on 12 August of that year, and on 12th and 23 August very high counts of 25 birds came unusually from the coast at Whitburn (Armstrong 1988).

A particularly heavy autumn passage was recorded in 1999, when at least 108 birds were noted from 16 sites across the county, with a typical August peak which included up to 30 on the North Tees Marshes, 18 at WWT Washington and 16 at nearby Barmston Pond, probably including some of the birds from WWT Washington (Armstrong 2003). An even heavier passage followed in autumn 2000, amounting to c.120 birds (Iceton 2001, Armstrong 2005).

Of the county's inland wetlands, the only one which has routinely produced high counts in recent years is Castle Lake at Bishop Middleham. The highest count there to date was of up to 17 in the final days of August 2009. Other high inland counts have included 15 at Barmston Pond in August 2000, and 16 at each of Mordon Carr and Longnewton Reservoir in August 2004 (Newsome 2010, Armstrong 2005, Joynt 2005 & Siggens 2006).

Few are recorded from the uplands, so a peak of six at Derwent Reservoir in autumn 1998 was notable, and this was the only site at over 200 metres above sea level to record passage birds that year (Armstrong 1999b).

Birds become less widespread and decline in numbers as September progresses, and are much less common in October, with a few late stragglers lingering into November and even very occasionally December. Particularly late records include singles at Seal Sands on 15 November and 24 December 2003, and two on Greatham Creek on 12 December 2005, although no birds stayed on to winter at Teesmouth in either of these years (Joynt 2004 & 2006). A late single at Barmston Pond on 17 November 2007 was the only record from this reliable site that year (Newsome 2008). However, passage wader numbers have declined dramatically there since the deliberate autumn drawdown of water levels ceased at Barmston Pond in the mid-2000s.

Overall, numbers of passage Greenshank appear to be stable in the county, with no clear trends, and this reflects an apparently stable European breeding population (Robinson 2005).

One unique autumn record concerned a juvenile showing physical and plumage characteristics suggestive of hybridisation with Marsh Sandpiper *Tringa stagnatilis* at Back Saltholme in September 2000. This individual unfortunately flew off after being disturbed by Little Egrets *Egretta garzetta* and was never relocated (Armstrong 2005).

Greenshank remains a comparatively rare wintering bird in the county, with most winter records formerly coming from Teesmouth, such as in the period 1982/1983 to 1985/1986 when a bird first noted on 26 February 1983 presumably returned in consecutive winters. This was the first winter record for Teesmouth for eleven years (Baldridge 1984-1986). In 1989, two autumn birds were in the Gateshead area until late November, one of them remaining throughout the winter until April, and often being present at Shibdon Pond and along the River Tyne west to Newburn (Bowey *et al.* 1993).

In recent years, wintering birds on the Tyne at Jarrow Slake have become more frequent, with single birds noted there in successive winters from 2002/2003 to 2004/2005, and again in January 2008. More unusual was a wintering bird at Daisy Hill in 2007, an intriguing inland record. In contrast at Teesmouth, apart from the isolated December birds noted earlier, which did not linger, there were no wintering records at all in the period 1987-2009. It is therefore apparent that overwintering Greenshank remain rare in the north east, just as they were 30 years ago, involving only a few isolated individuals.

It is not unusual for a few birds to be seen moving south over the sea during the autumn, most regularly off Whitburn and Hartlepool Headland. Such visible migration has included at least 15 birds flying south off Hartlepool on 19 August 2001 (Iceton 2002), whilst recent Whitburn counts have included five on both 13 August 2004 and 18 August 2006.

Distribution & movements

There is a small but stable Scottish breeding population, and beyond there, breeding birds can be found across Scandinavia and northern Russia, and east and north across central and southern Siberia. Small numbers winter in Western Europe, including Britain, but most are long-distance migrants that winter in sub-tropical and tropical regions of Africa and Asia.

There is only one ringing recovery from the county to date, as few Greenshank have been ringed. An adult ringed at Saltholme Pools on 26 July 1999 was killed in 2003 in Calvados, France, on the later autumn date of 21 September. This is consistent with a predictable annual southward autumn migration (Joynt 2004).

Lesser Yellowlegs
Tringa flavipes

A very rare vagrant from North America: six records.

The first Lesser Yellowlegs for County Durham was an adult in summer plumage seen by Eric Shearer at Barmston Ponds, Washington on 10 May 1967 (*British Birds* 61: 341). The Washington area also hosted the county's second and third birds including the only spring record, a summer plumage adult that spent three days at WWT Washington in May 2002.

All records:

1967	Barmston Ponds, Washington, adult, 10 May
1976	Barmston Pond, Washington, adult, 17th to 25 August (*British Birds* 70: 422)
2002	WWT Washington, adult, 18th to 20 May (*British Birds* 97: 582)
2003	Dorman's Pool & Saltholme Pools, Teesmouth, adult, 13th to 14 August (*British Birds* 97: 581)
2003	Greatham Creek, Cowpen Marsh, Greatham Tank Farm & Saltholme Pools, Teesmouth, juvenile, 22 September to 19 October (*British Birds* 97: 581)
2008	Saltholme Pools, Teesmouth, juvenile, 13th to 14 October (*British Birds* 102: 557)

Distribution & movements

Lesser Yellowlegs breeds across much of Canada and eastern Alaska, and winters in the southern United States, Central America, the Caribbean, and much of South America. It is one of the most frequently recorded transatlantic waders to the British Isles with over 310 records by the end of 2010, most of which have been in autumn. The high proportion of adults typifies the easterly location of County Durham suggesting that most of these birds have crossed the Atlantic in a previous autumn. Given the vagrant waders recorded on the North Tees Marshes, it is surprising that the first record for such a regular vagrant did not come until 2003, with second just over a month later.

Marsh Sandpiper
Tringa stagnatilis

A very rare vagrant from eastern Europe: two records.

The county's first record of Marsh Sandpiper was of a summer plumage adult found by the late Edgar Gatenby on Cowpen Marsh, Teesmouth during the afternoon of 25 May 1963. It remained faithful to this area until 29 May and was only the ninth record for Britain at that time (*British Birds* 57: 268). The second was also an adult in summer plumage on the North Tees Marshes. This bird, found by John Dunnett and Graeme Lawlor, was present at Saltholme Pools on 5th and 6 May 2003 (*British Birds* 97: 581), close to forty years after the first.

Distribution & movements

Marsh Sandpiper breeds across middle latitudes from Eastern Europe, through Central Asia to Mongolia and north east China. The species extensive wintering range encompasses much of sub-Saharan Africa, the Indian subcontinent, South East Asia and Australia. It is a regular vagrant to Britain with 136 records by the end of 2010, although the majority of these come from the south east of England, the species being much rarer further north with just nine records for Scotland.

Wood Sandpiper
Tringa glareola

An uncommon passage migrant.

Historical review

The first record of this species for County Durham came from a coastal wetland, the White Mare Pool near Easington, and was highlighted by Selby (1831) who documented it being shot there in 1826. Further details came from Edward Backhouse who said that "*a beautiful specimen of this very rare bird*" was shot by Mr Jon. Richardson of Newcastle upon Tyne. He went on to say that, about the same time, another was shot at Seaton Snook, Teesmouth by Chas. Janson, also from Newcastle. The first breeding record of the species in Britain followed, outside the county at Prestwick Carr in Northumberland in 1853, with birds displaying again at Gosforth Park in 1857 (Kerr 2001).

George Temperley (1951) said that this was a "*very rare and irregular passage migrant*", much more frequently noted in Northumberland than Durham, reflecting the above historic breeding record. It was clearly a much rarer bird in the county as a passage migrant in the 19th century than in the following century, and a bird that was recorded with greater frequency as the 20th century wore on. Records from the first half of the 20th century comprised one at Urpeth Bottoms, in the Team valley, on 1 September 1923, then two singles at the Darlington Sewage Farm, on 8th to 9 September 1935 and 22nd to 24 August 1939 respectively. A marked increase in records followed in the 1940s, with a total of six records between 1945 and 1948, three in spring, from mid-May to early June, and three in autumn, between mid-August and mid-September, all of single birds. These records came from flooded fields in the north east and south east of the county, the River Don, and Darlington Sewage Farm.

Temperley felt that this apparent sudden increase in records, from just two in the 19th century, then nine up to the mid-point of the 20th century, might be due to the increased number and competence of field observers rather than any significant change in the status of the species as a passage migrant (Temperley 1951). Through the 1950s and 1960s the species remained a relatively rare passage migrant that was recorded only intermittently in Durham. However, there were two significant influxes, in 1952 and 1963.

What was, at the time, an unusually large passage in early autumn 1952 (Temperley 1953) was reflected in Durham, with at least eight autumn birds, between three and five in July and August at Boldon Flats, and two to five on the North Tees Marshes over the period 13th to 24 August. There were also four spring birds, at Primrose Ponds, Jarrow, with up to three from 12th to 14 May, and Boldon Flats. This was to this point the largest passage of the species in Durham and in one year, the total exceeded the sum total for all records in the first half of the 20th century, though there seems little doubt that it was previously overlooked.

The 1952 influx marked a turning point, as more frequent records followed, with single spring birds then recorded in 1953, 1954, 1957 and 1959, and slightly larger numbers, from two to seven birds, being noted in most years in autumn in Durham. Favoured sites were Boldon Flats, Cowpen Marsh, Saltholme Marsh and Teesmouth more generally, as well as Darlington Sewage Farm and Hurworth Burn Reservoir.

In the early 1960s, numbers increased slightly further, with perhaps seven in the autumn of 1960, birds being noted from at least five sites, including one at Coxhoe Sewage Farm on 11 October, still a very late date for this species. This was followed by small numbers in spring 1961 and 1962, with a larger but still relatively small autumn presence. The autumn of 1963 then saw an exceptional passage, with a total of 38 birds, including one flock of 25, on the North Tees Marshes on 8 August (Bell 1963). This movement was also well represented in Northumberland (Kerr 2001). Through the rest of the decade birds were much less regularly noted, with just a handful each year. The exception was 1965, when 13 were reported near Stanley on 25 August, presumably at Tanfield Ponds (Bell 1966).

Recent status

Today, Wood Sandpiper is a regular though scarce spring and autumn visitor to the county. Passage is typically more pronounced in autumn than in spring. It is both locally and nationally much less common than Green Sandpiper *Tringa ochropus* as a passage migrant (Parkin & Knox 2010).

Records can come from almost any wetland in the county, though most sightings tend to be congregated in the eastern half, with a noticeable concentration of occurrences in the south east of the county. As with many wader species, there is a particular concentration of records from Teesmouth, with its complex of wetlands. WWT Washington has also been favoured over the years.

In Durham, this is a rather rare spring migrant with just five to ten birds in most years, these usually appearing between early May and mid-June and moving on quickly, staying little more than two to four days. Spring passage can be more pronounced, however, as in 1977, when a total of 14 birds were noted (Unwin 1978). The earliest spring record to date came far ahead of typical spring arrivals, on 10 April 1997, when two were at Saltholme Pools (Bell 1998).

Spring 2008 was a particularly good year for this species locally, matching the spring passage of 1977, with around 14 birds recorded across a suite of wetlands from south to north (Newsome 2009). The following spring's passage involved around ten birds, including an influx of about six in a short period around the second week of May 2009 (Newsome 2010).

More unusually, there are at least four records of birds singing and displaying over Saltholme Pools, Dorman's Pool and Cowpen Marsh in late May and June (Blick 2009). The first such records came in 1978 and 1979 (Bell 1979 & 1980). More recently, one was noted singing continuously at Saltholme Pools on the evening of 13 June 2003, having lingered in the area for 17 days from the beginning of the month. Two were then observed singing and displaying on Seaton Common in late May and early June 2004 (Joynt 2004 & 2005). However, there is very little likelihood of this wader breeding in Durham. Its breeding range is far to the north where it nests in secluded marshy moorland. A few pairs breed in northern Scotland but the bird's main breeding area is northern Europe (Parkin & Knox 2010). This species is already on the extreme south western edge of its range in northern Scotland, so climate change may make its future occurrence as a breeding species in the county even less likely.

Returning autumn passage birds can appear from mid- to late June, but the main autumn passage occurs from early July, when the number of reports tends to increase, usually peaking in August. This peak coincides with the arrival of juveniles joining the adults. This is reflected in the fact that August has more Wood Sandpiper sightings than any other month, with around 225 birds recorded in August over a 25-year period between 1985 and 2009. Autumn numbers are almost invariably higher than those noted in spring, though this is never a common passage migrant in Durham. Birds can be seen almost continuously from late July to mid-September, although there is probably a steady trickle of birds moving in and out of the area.

In most years there are 15 to 25 records over approximately two months of autumn passage, but larger gatherings can occur, such as the 24 around the North Tees Marshes on 2 August 1980, when there were also approximately seven elsewhere around the county at four other sites (Baldridge 1981, Bell 1981). In 1987 there was an unprecedented July influx, during which 20 birds were at Boldon Flats on 19th, probably the largest single site count in the county, whilst the previous day 13 birds had been on Cleadon Hill Pond just a few kilometres away (Armstrong 1988).

Autumn 2000 saw another good passage of this species, with 22 birds recorded, and this was the first of a number of good years in the first decade of the 21st century (Armstrong 2005). Exceptional numbers arrived in autumn 2002, when around 60 birds were seen in the county (Newsome 2007a), with up to 25 on the North Tees Marshes, including 15 on Seaton Common, on 5 August (Little 2003). This was certainly the best year since 1963, and probably the highest total in the county to date.

Birds can be quite widely dispersed and rarely occur in large numbers at any one location. In 2008, for example, birds were reported from 11 locations across the county, with most records of just one or two birds present at each site (Newsome 2009). Relatively few are recorded far inland, but Low Barns (Witton-le-Wear) is a favoured locality, and the count of seven there as long ago as 13 August 1975 remains a notable record (Unwin 1975b).

Numbers seem to have risen over the last 20 years or so, certainly the average number recorded on autumn passage has increased. In the ten years from 2000 to 2009, annual totals in the county have never fallen below 20 birds, the lowest being around 23 in 2001. In contrast, between 1988 (the year following the 1987 influx) and 1999, ten out of the twelve years failed to reach an annual figure of 20 birds (Siggens 2005). This is also reflected in the peak month of August, with a total of 135 birds in the ten years from 2000-2009, and just 90 in the 15 years from 1985 to 1999. The August average has thus more than doubled from around six in the period 1985-1999 to 13.5 in 2000-2009. Although the exceptional August of 2002 (when approximately 52 were seen) inflates the average for the recent period, this does not however, account for what is a genuine upward trend over the period 1985 to 2009.

This trend may simply be the result of increased observer coverage. However, Wood Sandpipers have also probably benefited from increased habitat management for waders, and for autumn passage waders in particular, at sites such as RSPB Saltholme, Castle Lake (Bishop Middleham), Barmston Pond, Lamesley Pastures and Shibdon Pond. At Barmston, deliberately lowering water levels for autumn passage produced an impressive suite of waders on 19 August 2001, including four Wood Sandpipers, as well as an exceptional 25 Green Sandpipers (Siggens 2005).

The number of birds in Durham declines in early September, and records have usually ceased by the middle to end of that month. September records include two taking advantage of flood water which attracted a wide range of waders, on Cleadon Lane, Whitburn from 12th to 15 September 2008 (Newsome 2009). There are relatively few October records, by far the latest being reported from Witton-le-Wear on 27 October 1974 (Unwin 1975). Other late records include one or two on the North Tees Marshes until 14 October 1978 and a single at Dorman's Pool on 9th and 10 October 1999 (Bell 1979, Iceton 2000).

Movement of this species at the coast is relatively rarely observed, so at least one heard calling flying north over Whitburn CP on 30 May 1991 is a notable record (Armstrong 1991). Other coastal records include singles flying south in autumn, at Whitburn Observatory on 7 August 2004, off Hartlepool on 11 August 2004, and over Hartlepool Headland on 20 August 2006 (Siggens 2006, Joynt 2005 & 2007).

Wood Sandpipers are described as 'very rare indeed in winter' in the UK (Parkin & Knox 2010) but one was on the River Don and Jarrow Slake in January and early February 1985, where Green Sandpiper and Common Sandpiper *Actitis hypoleucos* were also present in this period, a unique trio of unlikely wintering waders in Durham (Baldridge 1986).

Distribution & movements

Wood Sandpiper breeds in swamps and bogs, across north-central Europe and Scandinavia, and eastwards across temperate and Arctic Eurasia to Kamchatka. It also breeds in tiny numbers in the British Isles, in Scotland (Parkin & Knox 2010). The species is a long distance migrant, wintering in tropical Africa as far south as the Cape and the Arabian Peninsula. However, there have been no ringing recoveries of Wood Sandpiper, and only four ringing recoveries involving birds in the whole of the UK to date (R.M. Ward pers. comm.).

Redshank (Common Redshank)
Tringa totanus

Sponsored by
Council for the Preservation of
Rural England

A very common winter visitor and passage migrant; also a widespread breeding species in the west of the county.

Historical review

There is evidence of the Redshank's presence in the county extending back many centuries. The remains of this species have been identified in mediaeval excavations at Barnard Castle (Yalden & Albarella 2009). These were perhaps birds that had been hunted from local meadows and moorlands. The purchase of *"reydshanks"* in the 16th century is mentioned three times in Benett's *Durham Household Book* of 1844, at *"four for 2d. two for 1d. and eight for 4d"* (Gardner-Medwin 1985).

Reviewing the historical records, Temperley (1951) noted that in 1840, Hutchinson said that it was largely a bird of passage in Durham, stating that 'at these periods a few specimens are occasionally taken on the coast'. Temperley thought this record suggested that a large change in the breeding status of Redshank had taken place over the subsequent century.

Hancock (1874) noted an increase in its breeding numbers during his lifetime. In 1885, J. Backhouse made no mention of Redshank breeding in upper Teesdale, but in the second edition of his manuscript in 1898, he highlighted a pair of Redshank which had frequented a large marsh in upper Teesdale in June 1893, saying that they were 'probably breeding there'. Just fourteen years later in 1907, Temperley himself found 'many pairs' breeding in the damp meadows between Langdon Beck and Widdybank Farm, and commented that they had become 'very numerous' since then. Robson (1896) mentioned the taking of a number of clutches of eggs from Shibdon Flats, close to the Derwent's confluence with the Tyne, late in the 19th century.

Traditionally, the area around the North Tees Marshes was a breeding stronghold for this species (Joynt *et al.* 2008), although numbers there have fluctuated. At the turn of the century they were present in increasing numbers (Nelson 1907). However, with the loss of the intertidal habitats and the surrounding salt marshes during the early 20th century, Redshank suffered major declines in the area, Temperley commenting that increasing industrialisation seemed to be 'driving the birds away'. In 1916, Milburn estimated that only about a dozen pairs were breeding at Teesmouth. By 1931, numbers had recovered again to an estimated 50 pairs at Greatham Marshes alone (Richmond 1931). As Milburn later commented, *"the slag reclamations which have absorbed large stretches of the estuarine mud flats on both sides of the Tees estuary in the late 1920s and early 1930s now provide breeding places for such birds as Redshank"*, suggesting some benefit from industrialisation (Milburn 1933). However, by the late 1930s, Bishop reported that the number breeding successfully was again decreasing steadily and Temperley (1951) wrote that by 1940-1942 none were breeding there.

Around this time, in contrast, Redshank was noted to be spreading as a breeding species in Weardale, and also in lowland damp grasslands at Dunstonhaugh around the mouth of the Derwent and also near South Shields. Outside the breeding season, Temperley noted that the largest flocks occurred on the coast in August and September, with the greatest concentrations at Teesmouth in winter.

By 1951, Temperley said that Redshank was *"a common resident, passage migrant and winter visitor"*. He reported that there had been an increase in its breeding numbers from the mid-19th century up to the middle of the 20th century, writing that it was 'probably more abundant and widely distributed now than at any previous time on record'.

Post-Temperley, breeding Redshank returned to Teesmouth and by the 1960s were once again widespread over the North Tees Marshes, with around 20-30 pairs through to the early 1970s (Stead 1964 & Blick 1978).

The partial reclamation of Seal Sands from 1970 to 1974, reducing the area of mud flats by about two thirds, had a dramatic adverse effect on the numbers of wintering and passage waders, including Redshank (Evans 1978/1979, Evans *et al.* 1979 & Evans 1997). Hence, wintering and passage Redshank at Teesmouth crashed from annual peaks of 1,265 on Seal Sands on 13 September 1970 to just 250 at Billingham Bottoms in January 1974 (Stewart 1970 & 1975). The following year's autumn maximum was just 333 in October (Unwin 1976). However, numbers have since recovered, as discussed in the following recent non-breeding status section.

Jarrow Slake on the River Tyne is another area of mud flats which suffered substantial habitat loss through gradual infilling during the 1950s and 1960s, (before eventually being completely lost to development during the 1990s). When the area of mud flats was still extensive this site also regularly attracted high numbers of wintering and passage birds, including 900 Redshank in the early months of 1974, in contrast to the low numbers at that time at Teesmouth (Unwin 1975).

Recent breeding status

In Durham, the 'sentinel of the marshes' is a widespread resident, breeding in large numbers, particularly in the western uplands. It is an abundant passage migrant in spring and autumn and a very common winter visitor. In the breeding season, the three-note alarm call of the Redshank is a frequently heard sound of the county's western moors. Over two-thirds of the Redshank's breeding range in Durham extends over the Pennine upland marginal land in the west, and it occurs on many parts of the less well-drained moorlands. They can also be found in a variety of lowland habitats, from salt marshes to freshwater marshes and damp pastures. In County Durham, as elsewhere in Britain (Gibbons *et al.* 1993), damp pastures and poorly drained, rough grazing land are the Redshank's principal breeding habitats. In the county, these are largely utilised at altitudes of below 400 metres above sea level (Westerberg & Bowey 2000), along the lower sections of watercourses and, in some considerable numbers, around the county's upland reservoirs. The upland dales of Lunedale and Baldersdale, off the Tees valley, and upper Teesdale itself are considered to be of national importance for their breeding wader populations, including Redshank (Shepherd 1993), but large numbers of breeding birds are also found in Weardale and the upper Derwent valley.

The return to breeding areas in the uplands typically occurs in early March, but birds can return earlier, as in 1990, when two were back in upland territory by 24 February on Bowes Moor, probably due to very mild weather (Armstrong 1991). Redshanks begin to disperse from breeding sites in early July.

Survey work in the early 1990s in Baldersdale and Lunedale, and in upper Teesdale, came up with an estimate of 5.5 and 4.7 pairs per square kilometre respectively for these two areas, and at least 66 pairs were known to have raised young (Shepherd 1993). The *Atlas* calculated that if such figures were extrapolated to all occupied tetrads in the Pennine uplands, this would equate to a population of between 1,316 and 1,540 pairs for the uplands of the county. However, the *Atlas* considered that a more realistic figure for the Pennine uplands was probably in the range 500 to 1,000 pairs, as occupied tetrads outside the above mentioned areas would vary in terms of habitat quality and land management practices. Away from the Pennine uplands, data suggested that the lowland breeding population was approximately 50 pairs. Adding this to the mean of the above range (750 pairs) gave a total county breeding population of approximately 800 pairs (Westerberg & Bowey 2000), 2.41% of the estimated UK population (Thorup 2006).

As for many other breeding waders of the Pennine uplands, predator control is believed to benefit Redshank's breeding productivity (Shepherd 1993). Any long-term decline in shooting interests, and hence predator control, over the county's uplands might therefore be to the detriment of this species. However, from the 1970s to the early 1990s, the primary identifiable threat to breeding Redshank at both a national (Marchant *et al.* 1993) and county level was habitat degradation. The main causes of such degradation were land drainage and agricultural improvements, resulting in changes to grassland management practices. A reduction in grazing density would, in some Pennine areas, be considered beneficial to the species (Westerberg & Bowey 2000).

To the east of the Pennines, many areas that would have once held Redshank no longer do so as a consequence of more intensive management of grassland and in particular land drainage, as well as an increased presence of arable land since the Second World War (Lack 1992). This has resulted in the species being lost from many previously occupied lowland locations. It is now confined to pockets of still appropriate habitats and 'unkempt' margins, whether these be on managed wetland nature reserves, or around the western reservoirs, sewage treatment works, gravel pits and other 'waste' ground habitats.

The now severely limited area of wet meadows in the central and eastern parts of the county usually restricts the lowland breeding populations to fewer than five pairs per site. An exception to this 'rule of thumb' is the large population associated with the complex of extensive coastal grazing marshes, wetlands and brine fields at the mouth of the Tees (Westerberg & Bowey 2000).

A breeding survey at Teesmouth revealed 29 pairs on the North Tees Marshes in April and May 1982 (Bell 1983), whilst estimates at the same location in the mid-1990s placed the breeding Redshank population at over 20 pairs (Bell 1995, Armstrong 1993). The most recent survey work indicates a slight but not significant reduction from the 1982 estimate, to around 25 pairs at Teesmouth during the early part of the first decade of the 21st century (Joynt *et al.* 2008).

The breeding stronghold at Teesmouth is the various marshland sites between Seaton Common and Hargreaves Quarry. RSPB Saltholme is now a key site, with 23 pairs confirmed as breeding in 2009 and 2010, an increase of 10 pairs from 2008, as a result of careful habitat management. This is likely to have boosted the overall Teesmouth breeding population. Further pairs bred successfully in 2009 at Greatham Creek, on the Saline Lagoon and on Seaton Common (Newsome 2010 & 2012, Joynt 2010).

Although the Redshank is a scarce lowland breeder relative to its status in the uplands, it remains quite widespread. Away from Teesmouth, a small number of pairs breed in the Sunderland area, where damp grasslands provide suitable undisturbed habitat. In spring 2006, for example, four pairs were in territory around Rainton Bridge, and there were further breeding pairs at Sedgeletch and Houghton Gate. From 2007 to 2009, up to three pairs were noted at Rainton Meadows, and one to two breeding pairs were recorded at Herrington Country Park, Houghton Gate, Lamesley Pastures, Sedgeletch and WWT Washington. Display was also noted at Castle Lake, Bishop Middleham, where Redshank bred in 2006. Thus, apart from the North Tees Marshes, breeding in this three-year period was only recorded from seven lowland sites (Newsome 2007b-2010).

An unusual breeding record came in 1990, when at least two and probably three pairs bred on the old Dunston Power Station site next to the Tyne (Armstrong 1991). Habitat creation and restoration at sites such as Burdon Moor and Lamesley Pastures in Gateshead from the early years of the 21st century, has created breeding habitat for waders including Redshank. This may have offset the gradual loss of 'brownfield' sites in that area, such as the former Dunston Power Station, to re-development (P.T. Bell pers. obs.).

Monitoring programs at the time of the *Atlas* show no clear national population trend, though a decline was strongly suspected, and was certainly evident in the south (Marchant *et al.* 1993). Since then, the BTO Breeding Bird Survey has indicated a 33% national decline between 1995 and 2008 (Risely *et al.* 2010). The European population has also been described as being in 'moderate decline' from 1980 to 2005 (Robinson 2005). However, this is much less obvious in the Durham context.

Recent non-breeding status

In the winter, the resonant alarm call of the Redshank moves down from the county's western moors to the marshes in the south east of the county, and the north eastern coastal strip.

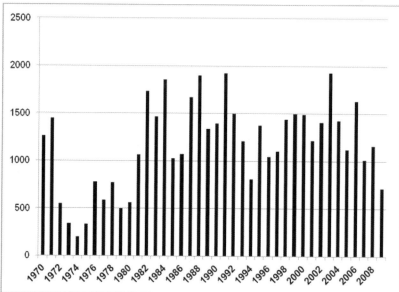

Redshank - autumn maxima, North Teesmouth (July-October)
(N.B. only the 'whole estuary count 'was available for 1971, no July-October counts for 1974; the figure shown is for November-December)

Durham has a large autumn and wintering population of Redshank centred upon Teesmouth, which is of national importance for this species (Holt *et al.* 2011), and has been close to qualifying as internationally important during some autumns. Large numbers arrive in the north east through July, August and September. Some of these birds move on as passage migrants, presumably to estuaries to the south and west of Teesmouth, others remain through the winter.

The largest numbers at Teesmouth are usually noted through August and September. Examples of large

451

counts from the north side of the estuary at this time of year include: 1,857 in September 1984; 1,901 in October 1988; and 1,932 in September 2003 (Baldridge 1985, Armstrong 1989 & Joynt 2004). These counts reflect a significant recovery of this population following the major land reclamation at Teesmouth in the early 1970s, as illustrated by the autumn (July-October) maxima for 1970-2009). The recovery of the Redshank population from this large-scale habitat loss reflects the resilience of this species. As a 'generalist' feeder, it has been able to adapt to significant changes at Teesmouth over the years. It has not only moved in response to habitat losses to industrial development, but also to losses of feeding areas due to the spread of mats of the green algae *Enteromorpha* over parts of the mud flats in the 1990s. The other Teesmouth waders which are generalised feeders, Oystercatcher *Haematopus ostralegus*, Black-tailed Godwit *Limosa limosa* and Curlew *Numenius arquata* have also shown significant population increases at Teesmouth in this period (Ward *et al.* 2003).

Numbers fluctuate widely from year to year, and the high 2003 count has not been matched to date in subsequent years. Although the overall trend in these North Teesmouth counts is comparable to that across the whole Tees estuary, some of the annual fluctuations are artificial, resulting from varying numbers of birds roosting outside of the county on the south side, and not being included in the North Teesmouth totals. This local variation may be caused by, for example, disturbance, or different tide heights (R.M. Ward pers. comm.).

In most autumns, numbers at Teesmouth decline from October to December, as birds move south to locations such as the Wash, as indicated by ringing recoveries. The seasonal pattern of Redshank numbers at Teesmouth is illustrated in the graph below, for the ten-year period 2000-2009. A similar pattern has been found over the longer term, from 1975-2000 (Ward *et al.* 2003). However, the spring peak has been less pronounced in this most recent ten-year period than in the previous decade (R. M. Ward pers. comm.).

High autumn numbers can also be found in the north east of the county around Whitburn Steel, an important feeding and gathering location for this species, with birds feeding on the rocky shoreline. Large numbers also occur at WWT Washington, Timber Beach on the River Wear, and Jarrow Slake on the Tyne. Further upstream on the Tyne, 130-140 are regularly seen feeding on the mud flats in Gateshead, as in October 2009, birds moving from the river onto Shibdon Pond at high tide (Newsome 2008-2010). Small numbers of inland, passage birds are also recorded at a wide variety of ponds and wet areas, particularly in autumn, and some sites such as Castle Lake at Bishop Middleham attract birds throughout the year.

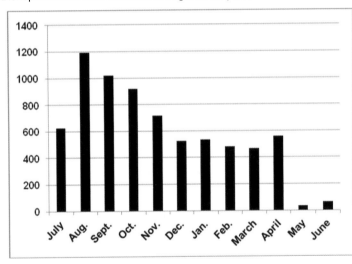

Redshank - Mean Monthly WeBS Counts, North Teesmouth, 2000-2009

By December, numbers at Teesmouth will typically have fallen from their autumn peak. Mid-winter numbers have, like autumn counts, recovered at Teesmouth following the land reclamation of the early 1970s. January WeBS counts on the North Tees Marshes have ranged from 163 in 1979 to 1,008 in January 2001, the highest January total to date (Unwin & Sowerbutts 1980, Iceton 2002). However, the recovery in winter numbers took longer than in autumn and spring, probably because wintering birds have increasingly switched to foraging inland, and thus not appeared in these estuary counts (Ward *et al.* 2003).

Winter counts also vary widely and birds are susceptible to freezing conditions, which can dramatically reduce numbers. Redshank are particularly vulnerable to increased mortality in severe weather, as confirmed by high numbers of ringing recoveries from north east England in severe winters (Clark 2004).

Freezing conditions in January in both 1985 and 1987, for example, resulted in hard weather movement of birds out of the area, with the Teesmouth recovery checked, and North Teesmouth WeBS counts of just 312 and 185 respectively (Baldridge 1986, Armstrong 1988). In winter 1991, although 632 were at Teesmouth in January, this number fell to 154 in February, and the icing over of some of the mudflats may have led to the deaths of some birds and the exodus of many others. In early February, 40 dead birds were found, including 10 which had been ringed in previous winters at Teesmouth (Armstrong 1992). This 1991 mortality rate was unusually high, and most dead birds were found on Seal Sands (R.M. Ward pers. comm.).

Between 1975 and 2000, national and Teesmouth numbers both showed a significant increase, highly significant in the case of Teesmouth. However, statistical analysis showed no significant relationship between the national and site trends (Ward *et al.* 2003).

Nationally, while Redshank and many other wader species have historically tended to winter in larger numbers on milder western coasts, there is an apparent shift towards the east, at least on non-estuary sites. This is as a result of rising temperatures due to climate change improving the survival prospects of east coast wintering birds (Parkin & Knox 2010). However, it should be noted that, in the case of Teesmouth, there have still been some severe winters with high mortality in the past 25 years, such as 1990/1991, as noted above.

Away from Teesmouth, Whitburn Steel has also seen recent high mid-winter counts, including 620 in December 2005 and 700 during January and February 2007, perhaps reflecting the trend noted by Parkin & Knox. The whole coastal stretch between Sunderland and South Shields is important for wintering birds, and 301 Redshank roosted at South Shields Pier on 14 February 2006 (Newsome 2006-2008). It is noticeable that whilst the typical annual peak at Teesmouth comes in the autumn, the highest counts at Whitburn are usually in mid-winter. This may reflect the fact that the open coast is less susceptible to freezing than the Teesmouth mudflats. The majority of birds leaving Teesmouth in the autumn are thought to disperse locally inland, or join the open coast flock centred upon Hartlepool Headland. This is supported by observations of colour ringed birds, and the fact that numbers at Hartlepool typically increase during the winter period (Ward *et al.* 2003). Recent high WeBS counts there include 565 at North Hartlepool in January 2006, and 350 in Hartlepool Bay in February 2007 (Joynt 2007 & 2008).

It is also notable that a number of wader species, including Redshank, made markedly increased use of the coast between Blackhall Rocks and Hartlepool Headland during the 1990s. This coincided with the reduction in colliery spoil on the Durham beaches (R.M. Ward pers. comm.).

The mudflats along the Wear and the Tyne, and some inland wetlands, also attract high winter numbers, especially at high tide roosts. Regular reports come from WWT Washington, where Wader Lake held high-tide counts in the range 100-150 from January to April 2007 and 100-200 during the last quarter of that year. Counts here have included 170 roosting in January 2008 (Newsome 2008 & 2009). Occasional three-figure counts have also come from Boldon Flats, Marsden and Sunderland, but it is likely that many of these birds were the same as those more regularly seen at Whitburn, as there appears to be much movement of birds around the various wetland sites in the north east of the county.

After the turn of the year, there is a build up in numbers again at Teesmouth through March and April, as southerly wintering birds move back north. This spring increase is far lower, however, than the autumn peak (see graph). The Teesmouth flock at this time typically peaks at 700-800 birds in April, and again numbers have recovered following the land reclamation of the early 1970s (Ward *et al.* 2003). The fact that passage birds swell the numbers at Teesmouth in spring as well as autumn has long been known, as was illustrated in 1977, when up to 775 were present on 17 April. That same spring local breeding birds were already in upper Teesdale by 12 March (Unwin 1978). These later spring passage birds are moving on to other breeding areas further north.

From mid-April, coastal numbers decline dramatically as most of the remaining birds leave for breeding areas, and the numbers around Teesmouth fall to their lowest level of the year in May and June, with scores, as opposed to hundreds, of birds present.

Active passage is routinely noted along the whole of the Durham coastline, in particular at Whitburn and Hartlepool during July and August. One of the highest counts of such movement to date was of 268 flying south off Hartlepool Headland on 19 July 1980 (Bell 1981), whilst day counts at Whitburn have included 131 on 13 July 2005 and 134 on 17 July 2011.

An interesting record concerns a partial albino bird, which was present on Seal Sands on 8 October 2006 (Newsome 2007b).

Distribution & movements

Redshank breeds from Western Europe, throughout much of Fennoscandia and discontinuously east to central Asia. Wintering birds are found along the coasts of Western Europe, around the Mediterranean and into north west Africa, and eastwards to the Middle East and South East Asia. It is the nominate race *T. t. totanus* which breeds across much of Europe.

A number of authorities, including WeBS and the International Wader Study Group, accept that three races of Redshank visit Britain, with the British, Irish, and southern Scandinavian population now treated as a race *T. t. britannica*, separate from the nominate race. The third form is the larger Icelandic race *T. t. robusta* which winters around British coasts.

There are a number of records of birds of the Icelandic race ringed at Teesmouth and then recovered in Iceland in subsequent springs and summers. For example, an adult ringed at Seal Sands on 2 September 2000 was seen at Nordur-Thingeyjar, Iceland on 24 May 2005 (Joynt *et al.* 2008). Conversely, a bird ringed as a chick in at Nordur-Thingeyjar, Iceland on 3 July 2008 was on Seal Sands on 15 October 2008 (Joynt 2009). By late winter, up to 75% of the wintering Redshank at Teesmouth have come from Icelandic breeding grounds (Mitchell *et al.* 2000).

Teesmouth also provides wintering grounds for breeding birds from the eastern half of northern Britain (Ward *et al.* 2003). Ringing recoveries also suggest that some birds wintering at Teesmouth make stop-overs during migration/moult at other east coast sites in Scotland and northern England. For example, a bird ringed at Teesmouth in September 1982 was found on the Ythan Estuary, Aberdeen, on 22 July 1983, but was back on Seal Sands on 30 December 1988 (Blick 2009).

The British breeding population is known to be resident, although some migrate as far as Iberia in winter (Parkin & Knox 2010). The proportion of this race on British wintering sites declines with latitude, with British breeding birds being replaced by Icelandic breeders at more northerly wintering sites (Furness & Baillie 1981).

Some ringing recoveries of local breeding birds show southerly movement in winter. For example, a bird ringed as a nestling on 3 June 1962 at Langdon Beck was shot on 15 November 1964 in St. Mary's Bay, Kent, 400 kilometres to the south east. Conversely, a bird ringed on 26 November 1967 at Walton, Suffolk was found dead on 16 June 1968, presumably as a breeding bird, at Mordon on the Skerne Carrs, near Sedgefield, 340 kilometres to the north west.

However, not all locally-ringed birds make such long-distance movements, some local breeding birds remaining to winter in the county. For example, one ringed as a chick near Edmundbyers on 12 June 1965 was recovered at Wardley, Gateshead on 24 November 1965; a modest movement for a wading bird (Bell 1966). Likewise, one ringed on Seal Sands on 1 April 1990 was found dead at Lealholm, near Whitby, in North Yorkshire, on 26 June 1990, presumably whilst it was in breeding territory (Bell 1991).

Many Redshank return to winter at Teesmouth year after year, as is shown by re-trapping of ringed birds. However, ringing has also shown that some adults switch wintering grounds from year to year, moving between different British and Irish estuaries (Ward *et al.* 2003). Mid-winter movement between estuaries is also possible, as demonstrated by a bird ringed at Teesmouth in December 1975 which appeared a month later at Eyemouth on the Scottish border.

Most autumn juveniles behave differently from adults by not remaining to winter at Teesmouth. This has been suggested by observations of colour-ringed birds (I. Mitchell pers. comm. in Ward *et al.* 2003). These young birds disperse, exhibiting 'exploratory' behaviour, and there are recoveries from France and elsewhere in Britain (Ward *et al.* 2003). For example, a first year bird ringed at Seal Sands on 24 August 1988 was found dead at Leverton Marsh near Boston, Lincs., on 19 March 1989 (Bell 1990). Another first year ringed at Seal Sands on 3 September 1989 was found dead exactly a year later at Raie de Somme, France (Bell 1991).

Such exploratory behaviour may also result in juveniles which do remain to winter at Teesmouth moving to different wintering sites when adult. For example, a first year bird ringed at Seal Sands on 16 January 1981 was shot on the Malahide Estuary, Dublin on 15 November 1987 (Bell 1989).

Birds of the nominate race from northern Scandinavia are also known to pass along the east coast of Britain (Prater 1981), although most winter outside the UK. However, there is just a single local recovery to date to support this assertion. This involved a juvenile ringed on Seal Sands on 12 October 1978 and found dead at sea near Statfjord, off south western Norway, on 25 August 1986 (Bell 1988).

Redshank can reach a considerable age, as is illustrated by a bird with white primaries that wintered in the Brinefields area of Teesmouth every year between December 1970 and April 1978 (Bell 1979). The oldest local recovery to date was of an adult ringed at Seal Sands on 9 August 1988 and found dead at Saltburn-by-the-Sea on 11 November 2001 (Robinson & Clark 2011).

Turnstone (Ruddy Turnstone)
Arenaria interpres

Sponsored by

South Tyneside Council

A common winter visitor and passage migrant along the coastal strip, small numbers of non-breeding birds occasionally summer. Scarce at inland locations.

Historical review

Turnstone remains have been described from the excavated strata of mediaeval Barnard Castle (Yalden & Albarella 2009), an unexpected location for this largely coastal species. The mediaeval remains of the species have also been recorded at Castletown, Sunderland, which is just seven kilometres from the modern winter stronghold of this species, Whitburn Steel. Perhaps birds were traded from coastal locations in the region as food items?

Turnstone was described by Temperley (1951) it as "*a regular winter visitor and passage migrant*", coastal in its occurrence and he considered it to be more numerous at his time of writing than previously. Sharp, in the area of Hartlepool, writing in 1816 considered it "*rare*". Hogg (1845) described it as a winter visitor and "*certainly rare*". He gave records of birds being shot near Seaton in the autumn of 1829 and of one at Seaton Snook in February 1837. By contrast, Edward Backhouse (1834) described it as 'common at Seaton Carew in the immature state'; he also observed that it sometimes remained at the coast in 'summer plumage', documenting a pair killed at Cowpen Marsh in June 1829, in summer plumage. Some observers considered it rare on some sections of the coast, e.g. at Castle Eden Denemouth (Sclater 1874). Temperley observed its passage as most evident in August and September and also in April, but rarely was it totally absent from the coastal strip at any time of the year, even in high summer. In 1933, Bishop told of "*many turnstones at Teesmouth from 8 May to June 14th*".

Despite the perception that Turnstones are rare birds inland this is not strictly correct. Inland, it was recorded at least 21 times, at a minimum of eight different sites, in thirteen of the twenty years between 1960 and 1979; followed by seven records in the next ten years. 'Regular' sightings came from Hurworth Burn Reservoir in the 1960s and 1970s, on at least four occasions, and at Barmston Pond through the 1970s, with birds there on at least five occasions during that decade. Most of these occurrences were of one or two birds, most often late in the summer or early autumn period, between 3rd June and 15 September, sometimes the birds involved were summer plumage adults. Three were at Barmston 21 August 1970 and seven were there on 9 September 1971, but one there on 8 February 1979 was an unusual inland winter record.

During the 1970s there was a developing awareness of the presence of a small number of birds along the Durham coast, especially north of Teesmouth, during the summer, as occurred in 1977 and again in 1978, when a party of 24 were in the Tees Estuary in mid-June.

Numbers wintering at Whitburn rose through the 1980s. A count of 130 on the rocky shore at Whitburn on 11 January 1979 was considered a large flock for that site. However, by 1985, up to 224 were there in the first quarter of the year, and by 1988 there was a maximum count of 250 in mid-March, though this total probably included some passage birds. September 1989 brought a site record, 346. The highest comparable count from Teesmouth around this time was 446 in the winter of 1985.

Recent status

In the county, this species has a widespread distribution along the coastline, with concentrations of birds at key sites, and birds might be counted anywhere between the mouth of the Tyne and the mouth of the Tees, in small numbers.

The 1990s saw a decline in counts at Hartlepool Bay from a maximum of 695 on 13 October 1991. This appears to be despite the creation of an artificial island within Hartlepool Harbour which can attract high tide roosts of several wader species. For example 560 Turnstones were recorded on the island on 28 September 1996, but this had dropped to 305 by September 1999 (Armstrong 1999).

Today, Wetland Bird Survey records for Teesmouth show a regular wintering population of some 150-250 birds from both sides of the estuary, and it has been suggested that this population has remained more or less stable since the 1950s (Blick 2009). Small numbers of birds are found at several locations along the Durham coast between the months of October and March.

The mud flats at Seal Sands are predominantly the domain of passage migrants, with counts typically in the 20-50 range. Sandhaven Beach in South Shields also attracts passage birds with for example up to 100 roosting there in spring 1999 (Armstrong 1999).

Over the first decade of the 21st century, Whitburn Steel was the single most important wintering site for the species, although large fluctuations in count numbers here over the winter period indicates that there is a good deal of movement between sites to the north and south of this key area. The seaweed covered rocky shoreline at Hartlepool also holds good numbers of wintering birds. Up to 240 were at Whitburn in January 2007, whilst 274 were in Harbour Hartlepool in January 2000. The amount of seaweed present at Whitburn appears to be a direct factor in the level of presence, with feeding frenzies created after local sea conditions have dumped additional seaweed on the tide line. In addition to the population in the Whitburn and Hartlepool/Teesmouth areas, small numbers of birds winter at the mouths of the Rivers Tyne and Wear. Whilst along the rest of the Durham coastline birds can be noted at Hendon, Salterfen Rocks, Seaham, Blackhall Rocks and Crimdon Denemouth. Birds from Teesmouth regularly forage as far north as Castle Eden Denemouth. The species is no doubt more regular along rocky stretches than records suggest and many such lengths of coastline probably hold small numbers of birds. The total wintering population of Durham was assessed as being of around 450 birds in 2005. Passage birds can exceed such numbers. For example, the highest number of Turnstone to have been seen in the Hartlepool area during autumn passage periods include 686 counted in September 1991 and 1993, with a combined total of 843 birds at Hartlepool and Teesmouth on 17 October (BOEE/WeBS counts).

While the main habitat is the coastal rocks and beaches around the tide line, a few birds quite frequently visit the Tees estuary's marshes in spring including Dorman's Pool; likewise at Whitburn, where up to four were on flood-water near Cleadon Lane between 8th and 22 September. At nearby Boldon Flats, seven were reported on 11 March 2008. There was a good presence of this species at RSPB Saltholme during early April 2009, peaking at 18 on 6th. High tides and occasional storms will drive birds onto the cliff tops, grassy areas and fields close to Whitburn Rifle Range, South Shields' Leas and Seaton Carew where up to 150 have been recorded.

Inland records have continued at scattered sites with for example up to 5 birds at WWT Washington in August 1997, and singles noted well inland at Selset Reservoir on 5 May 1990 and Derwent Reservoir on 11 September 1999. As might be expected several locations along the Tyne Valley flyway have recorded one to four birds including Birtley, Blaydon Haughs, Dunston mud flats, Shibdon Pond, Bill Quay and Jarrow Slake from the late 1980s through to the present day.

Genuinely inland birds are much scarcer but do occur on a reasonably regular basis in small numbers, largely at passage periods. A slight increase in recent decades may reflect greater observer presence at inland waters. Examples in 2009 included, in May, two at Castle Lake on 15th and singles at Longnewton Res. on 25th and at Hetton Lyons on 5 December.

A few birds summer at Teesmouth but numbers start increasing from July onwards beginning with non-breeding immatures. These birds usually moult then remain locally for the winter period (Ward et al. 2003). A large proportion of the returning birds noted in late July and August are adult, and a small number of non-breeding birds remain in the county throughout the summer. In this respect, small numbers are regularly logged at Whitburn Observatory and on the rocks near Hartlepool Headland through the year. Birds are often seen moving past coastal watch points, in small numbers, but 170 past Whitburn Observatory on 21 September 1982 was exceptional.

Distribution & movements

This species has a circumpolar breeding range, which encompasses widespread locations around the coasts of Scandinavia, the Baltic and White Sea and along the Arctic coasts of Siberia and around North America and Greenland. In winter, it is very widespread indeed, being found along the coasts of Eastern Europe, Africa, Australia and North and South America.

Birds originating from Greenland and north east Canada moult, winter and occur on passage in Britain. This origin has been confirmed locally by ringing studies at Teesmouth carried out by Durham University which have led over the years to the capture and ringing of over 1000 Turnstones.

Numbers decrease in April as the majority of birds leave the Durham coast to return to their northerly breeding areas, in part to be replaced in late April and May by passage birds from further south. Evidence for this comes from the observation of a bird that had been ringed at Morecambe Bay in August 1970 and was seen at Hartlepool in May 1972, as well as one that wintered on the Wirral, for several years after it had been ringed at Teesmouth in May 1977. Other recoveries include a bird ringed in Mauritania (West Africa) in April 1985 and which was at Teesmouth in May 1985 and another bird that was ringed on Jersey on 11 January 1989, which was found dead at Teesmouth on 14 May 1996. An adult that was caught and colour-ringed at North Gare on 5 May 1977 was subsequently noted at Seaforth, Lancashire in March 1978, New Brighton, Cheshire, later that month and was then controlled at Boston, Cheshire on 9 April.

Colour-ringing of wintering Teesmouth birds has shown the species to be subsequently faithful to the moulting and wintering site used in their first-year (Ward *et al.* 2003). For example, a bird ringed at South Gare on 23 November 1976 returned to that area in many winters until it was found dead on North Gare on 12 January 1986 (Blick 2009).

One colour-ringed at Redcar on 24 November 1976 was present at Whitburn on 16 May 1977. The oldest ringed bird recorded in Durham was an individual that was ringed at North Gare on 5 May 1977 and controlled in Finistere, France, on 1 May 1997 making it at least 21-years old. Another bird ringed at Teesmouth, in August 1976, was recorded in the Netherlands on 6 August 1977; perhaps it was part of the moulting flock of Greenland birds that utilise the Wadden Sea and then migrate to winter in Britain during late autumn.

A juvenile bird that was ringed in western Norway on 18 August 1970 was present at Teesmouth five years later and was then seen regularly in the Tees estuary area in a number of subsequent winters. It would appear that this individual was part of the Greenland population, which reaches the North Sea coastlines via Norway. Some of the population that winters at Teesmouth moult on the Dutch coast as confirmed by several recoveries from Vlieland (Ward *et al.* 2003).

Wilson's Phalarope
Phalaropus tricolor

A rare vagrant from North America: 13 records involving 14 birds, June to October, also January.

The first Wilson's Phalarope for County Durham was found by the late Phil Stead on the Reclamation Pond Teesmouth, at 16.00hrs on 12 October 1963. Described as being in winter plumage it was presumably an adult and was watched feeding for around 20 minutes before flying off towards Seal Sands as the tide receded. It had returned to the Reclamation Pond by the early morning of 13th and was watched intermittently by several observers, including the late George Temperley, until it was last seen flying towards the estuary at around 14.30hr (*British Birds* 57: 270). At that time, this was just the twelfth record for Britain.

All records:
1963	Reclamation Pond, Teesmouth, 12th to 13 October	
1971	Dorman's Pool, Teesmouth, adult female, 5 June (*British Birds* 65: 335)	
1971	Dorman's Pool, Teesmouth, 4th to 8 September (*British Birds* 65: 335)	
1973	Barmston Ponds, Washington, adult female, 20th to 22 July (*British Birds* 68: 335)	

1974	Barmston Ponds, Washington, adult female, 17th to 22 July (*British Birds* 68: 319)
1977	Saltholme Pools, Teesmouth, 7th to 10 September, and 17th to 28 September (*British Birds* 71: 504)
1979	Long Drag & Dorman's Pool, Teesmouth, adult & juvenile, 31 August to 3 September (*contra BB*) (*British Birds* 73: 510)
1980	Long Drag & Dorman's Pool, Teesmouth, adult, 13th to 20 September (*British Birds* 74: 473)
1983	Reclamation Pond, Teesmouth, juvenile, 15th to 27 September (*British Birds* 77: 527)
1985	Reclamation Pond, Long Drag & Dorman's Pool, Teesmouth, juvenile, 26 August to 11 September (*contra BB*) (*British Birds* 79: 551)
2005	Seaton Carew Golf Course, first winter, 9 January (*British Birds* 100: 45)
2007	Castle Lake, Bishop Middleham, adult female, 15th to 18 August (*British Birds* 101: 545)
2011	Greatham Creek, adult, 13th to 14 August (*British Birds* in prep.)

The duo in 1979 remains the only multiple occurrence in Britain, with presumably the same birds at Pagham Harbour, Sussex on 26th and 27 August, and at Dungeness, Kent on 27 August. The bird at Seaton Carew in 2005 represents the only mid-winter record in Britain. This bird was presumably heading south but got caught up in a storm system which brought hurricane force winds to northern England.

Wilson's Phalarope, Seaton Carew, January 2003
(Tom Francis)

Distribution & movements

Wilson's Phalarope breeds in the interior of western Canada south to California, and throughout much of the western United States. Most birds migrate through the interior of the USA to winter in South America, from Peru south to Argentina and Chile. The species is an annual vagrant to the British with almost 240 records by the end of 2010, chiefly in autumn with spring records being more unusual.

Red-necked Phalarope
Phalaropus lobatus

A rare vagrant from Scandinavia and the Arctic: 20 records, most often from August to October, also in May and June.

The first fully documented record of this species relates to an undated specimen held in the Hancock Museum (Howse 1899). The bird, listed as being in first plumage, is labelled as having being taken at South Shields, though no date is given. This record was not listed by Hancock in his *'Catalogue of the Birds of Northumberland and Durham'* (Hancock 1874) so presumably it must have been shot sometime after 1874. It is however included by George Temperley in his *'A History of the Birds of Durham'* (Temperley 1951), who quotes the first dated record as being the bird picked up dead at Teesmouth on 23 October 1891, which was in Nelson's collection (Nelson 1907).

In the last quarter of the 17th century, Mr. Johnson of Brignall, was reported to have shown famous botanist and ornithologist John Ray *'a bird of the Coot kind scollop-toed not much bigger than a Blackbird'*. Ray went on to describe the specimen, the description of which there is no doubt refers to a Red-necked Phalarope; Willughby's account also making reference to *'Mr Johnson's small cloven-footed gull'* (Willughby 1678). The place of the bird's capture was not specifically stated, but as the owner Ralph Johnson routinely hunted on the Tees Marshes, it was speculated that this bird was caught there around that time (Gardner-Medwin 1985). This record, of considerable historical interest, cannot be stated as the first for the county, as there is a doubt as to whether it was taken in County Durham or not, and the specimen seemingly no longer exists.

All records:

- -	South Shields, juvenile, undated specimen in Hancock Museum
1891	Teesmouth, 23 October, picked up dead (Nelson 1907)
1901	Seaton Carew Golf Course, 6 September, shot (Milburn 1903)
1943	Graythorp Shipyard, Teesmouth, 2 October (Temperley 1951)
1968	Reclamation Pond, Teesmouth, 21st to 23 September
1969	Hartlepool Headland, 14 September
1969	Dorman's Pool, Teesmouth, 15 September
1970	Dorman's Pool, Teesmouth, 11 August
1973	Dorman's Pool, Teesmouth, adult female, 14 August
1981	Dorman's Pool, Teesmouth, adult, 7 September
1986	Long Drag, Teesmouth, juvenile, 26 August
1988	Long Drag, Teesmouth, adult female, 29 June
1988	Hartlepool Headland, 23rd to 24 September
1988	Hartlepool Headland, 13 October
1997	Crookfoot Reservoir, 14 September
2001	South Shields, juvenile, 9th to 14 November
2002	Hartlepool Headland, juvenile, 13th and 15 September
2007	Saltholme Pools, Teesmouth, adult female, 14 June
2008	Hartlepool Headland, juvenile, 17 August
2011	Saltholme Pools, adult male, 29 May

In addition to these records, single unidentified phalaropes, at Seaham Harbour on 14 December 1962 and flying south at Hartlepool on 3 December 2006, were considered to possibly relate to this species, though given the dates, Grey Phalarope *Phalaropus fulicarius* may have been more likely.

The three spring records relate to adults in late May and June, typical of a species which does not normally reach its breeding grounds until that month. The November record in 2001 is noteworthy, especially as most birds leave the breeding areas in August, and serves as a reminder that all phalaropes at sea in late autumn are not necessarily Grey Phalaropes.

Distribution & movements

Red-necked Phalarope has a circumpolar breeding distribution which includes a relict breeding population in Scotland, though the majority of passage birds in Britain presumably originate from Iceland and Scandinavia. In winter the species is wholly pelagic, spending the non-breeding season at sea in three well defined areas, clustered off the Arabian Peninsula, the coasts of The Philippines and the coast of Peru.

Grey Phalarope (Red Phalarope)
Phalaropus fulicarius

A rare migrant from the Arctic: at least 71 individuals between August and June, most frequently from September to November.

Historical review

The first authenticated record of Grey Phalarope for County Durham is of two birds said to have been shot by the River Tees at Haverton Hill in the autumn of 1824 (Hogg 1827). This is the only specific record included in Hogg's *'Catalogue'* though he did list the species as *"Occasionally seen in our marshes"* (Hogg 1827). Prior to this record, the 'Grey Phalarope' was listed by Cuthbert Sharp in his *'List of Birds Observed at Hartlepool'* (Sharp 1816), as being *"rare"*, though he gave the scientific name of the species as *'Tringa lobata'*, which must cast doubt as to which species of phalarope was involved, though based on current knowledge Grey Phalarope is most likely.

There are a number of further Victorian records but many of these are poorly documented. Edward Backhouse (1834) listed two records. One specimen held in his collection had been shot near Sunderland *"a few years ago"*, and the second was *"seen running along the highroad near Cleadon during a tremendous storm on the 14 October 1829".* Hutchinson (1840) said *"it had been killed a few times in the County"*, although he also listed just two records, one shot near Thorp, near Easington on 27 October 1832 and another on a pond near Easington in the latter end of October 1836. One of the most notable records is of a bird said to have been *'on the Tees'* 'in the first week of June 1850', and reported by T J Bold in September 1855 (*The Zoologist* Vol. XIII 1855). More typical subsequent records were of a single shot *"in the channel leading to Hartlepool"* on 15 November 1850 (Hancock 1874) and one obtained at Teesmouth on the 19 November 1899 (Milburn 1901).

There were a further 17 records in the period between 1900 and the close of 1969, but the first of these was not until 8 October 1932, when a male was shot, from a party of three birds off Whitburn by J R Crawford. Crawford was also responsible for sightings off Whitburn on 28 October 1948, and 7 November 1948, both of which were flying north into a strong north easterly wind (Temperley 1951). A record of four birds swimming in South Shields Harbour on 30 October 1949 were believed to be phalaropes and probably of this species although they were not positively identified as such at the time (Temperley 1949). Described as *"grey, thrush-sized gulls"*, they were seen to take flight and exit the harbour *"swiftly with rapid wing-beats"*.

A bird at Teesmouth on 13 August 1950 (Temperley 1950) is noteworthy for the early date and being one of only two August records for the county. There were a further seven birds recorded by the end of that decade, all of which were in the Seaton Carew/North Gare area in January or October, with the exceptions being single birds at Cowpen Marsh on 11 September 1954 and at South Shields from 14th until 19 January 1959. Only five birds in total were recorded during the 1960s with singles: inside the Tees Estuary, on 15 October 1960; at Seaham Harbour, on 2 November 1960; and, at Hartlepool on 17 November 1966, 24 November 1969 and 6 December 1969.

Recent status

The scarcity of this species in the county is indicated by the fact that during the review period from 1970 to 2011, a total of just 41 Grey Phalaropes had been recorded in the county, maintaining an average of around ten birds per decade. However, the 1990s were particularly poor, with just three records, all of which were in the period December to February.

Noteworthy coastal records include: two together off Hartlepool, on 19th and 20 November 1972; a bird photographed from a boat some 14 miles off Sunderland on 25 September 1974; and, two together off Hartlepool on 9 November 2001. Single birds at Greatham Creek, Teesmouth on 6th and 7 March 1978 and 12 September 1983 are noteworthy as they were the first on the North Tees Marshes since 1954. Perhaps the most interesting record was bird photographed on a small pool at North Gare on 14 May 1977, only the second spring record for the county at the time. A further spring record came in 2011 when a summer plumaged bird flew south off Whitburn on 2 May.

Autumn gales during late September 2007 brought many to the English east cost and provided the largest ever influx of this species to the county. A run of records commenced with one at Whitburn on 14 September, four were at Whitburn on 27 September, two passed Hartlepool on the same date and then another two past Hartlepool and a single past Whitburn on 29 September. All of these were flying north, though several paused briefly on the sea, particularly off Whitburn Observatory (Newsome 2008). Assuming all of these to have been different birds, the total of ten birds is the highest annual total in the county, the previous best years of 1972 and 1977 producing just three birds each. A further six unidentified phalaropes were seen during the autumn of 2007, most of which were probably Grey Phalaropes. The four birds off Whitburn on 27 September 2007 is the highest confirmed day-total for the county, although it is worth noting the record from 30 October 1949. Few Grey Phalaropes linger on our coastline, so one or two birds present at Hartlepool Headland for nearly three weeks in November/December 2011 is notable.

The species is rare inland in Britain and five such birds have been recorded in Durham. These probably refer to birds that have travelled upriver after being blown inshore rather than having come overland from the west. These records were at: Piercebridge, on 21 November 1971; Barmston Ponds, Washington, on 12 August 1974; on the River Wear between WWT Washington and Hylton Bridge, from 24th until 26 September 1989; on the River Tyne between Ryton and Newburn on 22 October 1987; and, on the River Tyne at Wylam, on 12 October 1988.

Flight identification of phalaropes at sea is difficult, especially during poor weather conditions when birds are most likely to occur. Unidentified phalaropes have been noting heading north during sea watches on a number of occasions. Birds have been seen at Hartlepool on: 11 November 2006, 27 September 2007, 28 September 2007, 29 September 2007 (two birds), 9 November 2007 (two birds) and 30 October 2008. In like fashion at Whitburn: on 31 October 2008, 17 October 2009 and 7 November 2010. All of these were thought most likely to refer to this species.

Distribution & movements

The Grey Phalarope has a circumpolar breeding distribution in the high Arctic, and is one of the most widespread breeding waders of the northern hemisphere. Outside of the breeding season, the species is essentially oceanic, only being driven inshore after gales, most frequently off the western coasts of Britain and Ireland. It is scarce in the North Sea. The main wintering grounds and migration routes are imperfectly known although large numbers are known to frequent the plankton-rich waters off western Africa and western South America.

Pomarine Skua
Stercorarius pomarinus

An uncommon passage migrant, mainly in late summer and autumn, but scarce in spring.

Historical review

This skua is a passage migrant along the Durham coastline, being seen in quite large numbers in some years whilst being very scarce in others. The majority of birds are recorded between August and November, with the highest numbers usually in late autumn. Today, it is recorded with much greater regularity and frequency than it was in historical times.

John Hancock had in his possession, specimens of the *'Pomatorhine Skua'* that were shot on the Tyne in 1830 and on 14 September 1846, both of them adults (Temperley, 1951). Hancock (1874) detailed two records in his collection that had been shot off the Durham coast, one of them probably shot off Seaton Carew on 24 October 1837. Up to Hancock's time, it was considered a rather rare visitor though large numbers were noted at times, often alongside other skua species, such as those observed in autumn 1879 (Temperley 1951). Following 50 Pomarines moving north west off Teesmouth on 8 October, a further 100 were counted travelling in the same direction the next day. On 14th, about an hour after the onset of a north easterly gale, large numbers of exhausted birds were driven into the Tees Bay and onto the surrounding sands. An estimated five to six thousand moved through the bay that day and over 150 were shot (Nelson 1907, Wallace & Bourne 1981). A year later, during a gale on 28 October 1880, another large movement of skuas, including Pomarines, took place off the north east coast; it was noted as prominent off both Teesmouth and Yorkshire. Further significant skua movements involving Pomarines occurred on 14 October 1881 and again in mid-October 1886. After a 25-year gap, the next major movement documented by Nelson was at Teesmouth on 30 September 1911 when, during a gale, about 200 Arctic Skuas S. *parasiticus* and many Pomarines, in parties of threes and fours, moved south along the coast (Nelson 1911). All of the Pomarines identified at the time were adults (Temperley 1951), the inference from this being that there were actually many more that were immature birds.

Temperley (1951) gave the status in the mid-20th century as "*an uncommon autumn passage migrant*". He said that it was mainly noted in August, September and October, was of irregular occurrence, being very much dependent upon weather conditions and, when it was recorded, the majority noted were immature birds. In the earlier years of the 20th century, prior to 1950 and the 1911 influx aside, this species was a very irregular passage migrant in the autumn. For example, in the ten years prior to 1953, it was only recorded in two years, these being 1945 and 1948. In both of these years however, large passage movements, which are now known to be intermittently characteristic of this species, were noted off Whitburn when 65 flew north in six hours on 9 November 1945, 49 flew south in nine hours on 28 October 1948 and 18 north in eight hours on 7 November 1948. The 1945 movement was associated with a much larger movement of Kittiwakes *Rissa tridactyla* that was judged to have involved about 2,000 birds (Temperley 1951).

Post-Temperley, there were few reports in the 1950s, but better passages, involving at least 40 birds each year, were reported in 1961, 1963 and 1966. The totals were, in each case, largely due to brief periods of passage activity, when good numbers might have passed a coastal watch point on a single day. For example, 42 were seen off Whitburn on 18 October 1962 and 195 flying south at Hartlepool during a northerly gale on 26 October 1962; the latter featured flocks of up to 22 individuals, including three dark morph birds. In a north westerly gale on 28 October 1962, five flew up the Tees Estuary past the North Gare and disappeared over Middlesbrough (Blick 2009). In autumn 1963, 18 were noted off Whitburn on 14 October and 'dozens' flew north at Teesmouth on 27 October 1966.

Recent status

The species has been much more regularly recorded off County Durham since 1969, except for a rather lean spell during the period 1979-1982. Sporadic movements noted in the 1970s included at least 18 attracted by a wreck of sprats *Sprattus sprattus* at Teesmouth on 22 July 1971, 22 at Whitburn on 21 October 1972, 76 north past Hartlepool Headland on 10 October 1973 and 43 off Whitburn on 15 October 1976.

Two years stand out as prominent in the recent history of the species, 1985 and 1992. In autumn 1985, exceptional numbers of Pomarine Skuas were noted along North Sea coasts (Fox & Aspinall 1987). This movement was precipitated by gale force north westerly winds which were believed to have pushed many Pomarine Skuas south into the North Sea, from their intended journey vector into the Atlantic. On 29 October, 103 flew north in five hours at Whitburn, and this proved to be one of the heaviest movements of this species since 1962. Deteriorating conditions on 31st produced further passage when at least 51 flew north in 90 minutes off Lizard Point, Whitburn. This passage continued during the first week of November and then peaked during a north westerly gale on 10th when a further 110 passed Whitburn. On the same date, eleven were present in the Tees Estuary and most unusually, two were in the Tyne Estuary. On 19 November, more gale force winds produced a passage of 80 birds north at Whitburn. At Hendon, on 17 November, two patrolled the promenade searching for anything edible, including the discarded bait of local sea-anglers.

The 1985 influx was exceeded in 1992, when on 9 October; the largest passage of the 20th century was noted at Seaton Carew. On this day, at least 1,060 birds were observed moving predominantly north, whilst counts of 147 and 156 were made at nearby Hartlepool on the following two days. This movement was also logged at Whitburn, where observers noted 839 flying north and 60 south during the period 9th to 12 October. Birds quickly navigated back out of the North Sea and there were no further sightings after 20 October. The count at Seaton Carew was the largest documented on the east coast of England during this period of passage (*Birding World* 5: 369).

Movements from the mid-1980s and through the decade of the 1990s generally fell into three time periods. Late summer influxes of mainly adult birds were rarely noted, with stand-out years being: 1987, with 19 at Hartlepool on 27 July; 1995, with 120 at Hartlepool on 6 August; 92 at Whitburn on 7 August and 32 at Ryhope the same day. More frequent were mid-autumn passages, generally at the end of August and through the first ten days of September, and involving a mix of adults and the first influx of juveniles. Counts of up to 51 birds were achieved in several years, with larger movements including 60 at Hartlepool and 43 at Whitburn on 9 September 1989 and 51 at Hartlepool on 4 September 1993 with 64 at Whitburn the following day. Movements associated with late autumn storms were also infrequently noted, usually involving a high proportion of juveniles between late October and mid-November. Day counts of 99 at Hartlepool on 1 November 1986 and 61 at Whitburn on 20 October 1997 were notable, but a period in November 1999 produced 115 at Whitburn and 71 at Hartlepool on 14th, with 41 at Whitburn five days later.

The trend over the first decade of the 21st century was of decreased passage and double-figure counts became much less common, despite increased observer effort and favourable seawatching conditions; no doubt contributors to the better documentation of this species over the latter quarter of the 20th century. Episodic movements produced no day-counts of three figures, although 61 at Hartlepool on 6 October 2002 (with 29 at Whitburn on the same day), 30 at Hartlepool on 31 August 2003 and 25 at Whitburn on 3 September 2007 (with 19 at Hartlepool on the same day) were noteworthy. Year totals at Whitburn were: 12 (2006), 137 (2007), 30 (2008), 53 (2009), 91 (2010) reflecting the overall scarcity of the species, considering the total of more than 3,900 hours observation time at Whitburn over these five years.

Records between mid-December and March are unusual in the north east, but in 1974, there were several winter sightings, all of singles, and perhaps relating to the same bird. One was off Whitburn on 5 January, off Hartlepool on 9th and 10 February and again on 5th and 8 March. Reports were almost non-existent through the

1980s and early 1990s, but the occasional winter bird became almost expected from 1996 onwards, perhaps as a result of increased observer effort during the winter months. Month totals of eight for December, five for January, two for February and one for March came from Hartlepool and Whitburn combined between 1996 and 2010, sightings being split fairly evenly between the two localities.

Although a regular spring passage migrant off the north west of Scotland and Ireland during late April to early June, sometimes in large numbers (Davenport 1992), few have been seen in County Durham at this time of year. Sightings have come in ten springs since 1970, all between 6 May and 3 June, with a concentration in the last ten days of May, the majority of sightings relating to full adults. The most productive spring was 1991 when seven passed Whitburn on 16 May followed by three the next day; three at Whitburn on 3 June 2001 was also notable. An adult at Hartlepool on 6 May 2002 was the first May record for the site, whilst adults at South Shields in 1998 and 2004 were of interest.

Inland records are very rare indeed and there are only a handful of recent sightings of live birds inland, all from the Teesmouth area or nearby sites. An immature flew south west over Wynyard on 21 November 1985, no doubt associated with the large coastal movement of the time. Two adults flew west over Dorman's Pool on 12 September 1987 and a juvenile was noted attacking waders in the Saltholme-Dorman's Pool area over 16th to 17 September 2000. In addition, an immature was found dead in Hargreave's Quarry on 19 January 1986 (Blick 2009).

Distribution & movements

This is a high Arctic tundra breeding species, which is found from northern Europe east to Siberia and across North America. It winters largely out to sea, mainly in coastal tropical waters in the nutrient-rich waters off the west coast of Africa, just north of the equator (Parkin & Knox 2010). Movement from the northern seas and breeding areas occurs from early August, but this probably involves failed or non-breeders. The main movement through the North Atlantic and North Sea occurs during September and October, with adult birds preceding the juveniles. Fluctuations in skua numbers along the British coast have long been believed to be related to cyclic variations in the population of Norway and Arctic lemmings *Lemmus lemmus* and *Dicrostonyx torquatus* (Kokorev & Kuksov 2002), and it may be that this is one of the factors, along with prevailing weather patterns, contributing to the huge degree of variability in the presence of this species off the Durham coast.

Arctic Skua (Parasitic Jaeger)
Stercorarius parasiticus

A common passage visitor, the largest numbers being recorded in autumn.

Historical review

This species is the commonest skua on passage in British waters particularly in the North Sea, and it is the most frequently observed skua species off the north east coast, as it accompanies, and harries, the large flocks of post-breeding and migrating terns that feed off the Durham coast in late summer and early autumn. It occurs in its maximum numbers from late June to late September, the passage period usually extending, in a somewhat lesser volume, through October and occasionally into early November. This modern pattern appears to mirror the historically observed situation, although numbers are now in decline and the exciting passages witnessed in the 1960s, 1970s and 1980s now occur much less frequently.

In modern times, it is ironic that the species has only rarely been reported inland, but one of the earliest documented records for the county was a bird in just such a situation, recorded by Abel Chapman (Chapman 1889) as having been shot at Ireshopeburn in Upper Weardale on 12 September 1874. Nelson (1907) noted large numbers of this species in the Teesmouth area, amongst the mixed skua movements in the 'skua years' of 1881, 1885, 1886, 1887 and 1891, whilst in the mid-20th century this species was viewed by Temperley (1951) as a "*common passage migrant in autumn, seldom recorded in spring*".

Subsequent to the publication of Temperley, records through the 1950s illustrated that this species was not uncommon, but noted in variable numbers on an annual basis. It was documented in much larger numbers in some years. For example, in 1959, the observed passage was heavy throughout August, and the total for Teesmouth over this period was estimated at about 500 birds. On 28 August 1959, at the time of a great northward passage of shearwaters, about 180 Arctic Skua were recorded flying south off Hartlepool. Passage continued throughout September on a reduced scale, although on 26th, 127 flew south and 23 north off Hartlepool. Skuas were being driven before the tremendous gale from the north on 27th to 28 October 1959, and 60 to 70 Arctics were seen off Hartlepool on both days. Even in November of that year, of the 40 skuas which flew south on 15th, most were thought to be Arctics.

At Hartlepool in 1962, totals in the hundreds were recorded on at least seven days in August. This was prompted by a huge concentration of sprats *Sprattus sprattus* at Teesmouth, and the numbers of Arctic Skuas associated with this gradually built up in the estuary until 80 birds were there on 25th and 150 on 26th and 28th (Blick 2009). Although most of these skuas were intent on harrying the terns, which were feeding on the sprats, several skuas were noted foraging on the mud-flats and feeding on the stranded fish. The county's largest documented passage of this species occurred at this time when 750 Arctic Skuas moved south past Hartlepool Headland on 23rd of the month. There was a further significant passage of birds on 26 October, when 195 flew south past Hartlepool, and two days later, when during a north westerly gale, passage included five birds flying inland up the River Tees.

Outside the established autumn passage period, reports were infrequent and it is difficult to be sure of the species' true status through the 1950s and 1960s. Winter encounters were unusual but there were several coastal reports between December 1964 and February 1965, possibly involving a handful of wide-ranging birds. Up to 12 off Hartlepool in December 1961 was also a significant early winter presence. Records away from the coast were very infrequent through the mid-20th century; a status which has changed little. One was found near Boldon during a period of severe fog on the very unusual date of 7 March 1943. On 26 June 1960, an adult was at Hurworth Burn Reservoir, again unusual in date and ageing, as most inland birds are immature. During late August 1962, 35 were on the Reclamation Pond, obviously associated with the huge gathering of birds in the estuary at that time and, in similar fashion, 16 were on Dorman's Pool and Cowpen Marsh on the North Tees Marshes at the same time.

Recent status

The general pattern of occurrence has not changed much over the period 1970 to 2010, although there has been an obvious reduction in the level of passage through the second half of this period. In the 1970s and 1980s, passage occurred in differing levels in all weather conditions, whereas movements through the 1990s and 2000s were more weather-dependent. Even in seemingly ideal weather systems, peak migration days were harder to predict and more observer effort was required to produce equivalent results.

Records of this species off the Durham coast outside the main passage periods are unusual, and rather rare in winter. Since 1970, there have only been 15 years featuring sightings between mid-December and late March, with the frequency reducing through the period. Regular reports during the 1970s included five at Hartlepool on 21 January 1973, eight at Hartlepool on 8 December 1973 and again on 12 January 1974 and eight at Whitburn on 2 February 1974, while one to two birds were noted on a further nine occasions. The sightings between December 1973 and February 1974 coincided with south westerly gales and the occurrence of large shoals of sprats and Atlantic Herrings *Clupea harengus* in the North Sea.

Through the 1980s and 1990s, one or two appeared infrequently at Hartlepool, Ryhope, Seaton Carew and Whitburn during the winter, with three at Hartlepool on 6 February 1983 the best day count; these were involved with the large influx of pelagic species onto north east coasts at the time. Since 1996 however, there have only been two further mid-winter reports of Arctic Skuas, two at Hartlepool on 14 January 2000 and one at Whitburn on 16 January 2004.

Spring passage generally takes place between the end of April and the first few days of June, although the intensity varies greatly from year to year. Most reports in the period concern just one to three birds, but movements of 18 at Hartlepool on 27 April 1985 and seven at Whitburn on 25 May 2004 were more notable. A more complete picture of spring occurrence was revealed with increasing observer effort at Whitburn in the second half of the 2000s. Between 2005 and 2010, the earliest spring passage bird was on 23 April and totals varied between four in 2010 to 39 in 2005, with an average of 18 birds per spring. Significant day counts included seven on 10 May 2005, 18 on 16 May 2005 and six on 31 May 2006. Away from Whitburn, noteworthy passage counts at Hartlepool in this period included five on 20 May 2006 and 16 on 28 May 2007.

Although there have been occasional weather-triggered appearances through late June and early July, it is normally late July before there is any significant increase in numbers. The mouth of the Tees estuary used to be the prime site for large numbers of Arctic Skuas, drawn in by the concentrations of gulls and terns. Gatherings of 15-30 were commonplace through the 1970s and 1980s; for example, approximately 80 were in the river mouth on 15 August 1985 and impressive counts of 230-264 were made at Teesmouth on several dates in mid-August 1988 (Blick 2009). Following congregations of up to 70 in early August 1998 and 47 on 20 July 2003, there have been no equivalent assemblies and gatherings into double-figures became a rarity in the latter half of the 2000s. The reason for this decline may be two-fold. Firstly, tern numbers utilising the estuary have declined through the 2000s and secondly, breeding success of Arctic Skuas in this period has led to far fewer birds being off the north east coast in general.

True passage of Arctic Skua peaks in the second half of August and first half of September. Day counts of 225 to 440 were documented on 12 occasions between 1970 and 1986, all such movements falling between 14 August and 22 September, and with five of the largest movements coming in the last five days of August. Of these, 440 at Hartlepool on 30 August 1980 and 400 there on 26 August 1986 are the greatest. Since 1986 however, there have been no day-counts greater than the 178 at Whitburn, on 13 September 1993. Between 2000 and 2010, there were only four years with counts in the range 119-137, all other years falling in the range of 36-84. As a measure of the volume of passage at Whitburn, yearly totals between 2005 and 2010 varied between 385 birds (2009) and 1,175 birds (2007), with an average of 731 per annum; a far-cry from thirty-years previous.

Late summer movements generally involve adult birds, possibly failed breeders, with only a small percentage of immature birds. The first juveniles usually appear from the third week of August, and by mid-September, this age class is dominant. Immature birds remain in the majority through the remainder of the autumn until passage ends in early November.

Although the majority of significant data relates to the established seawatching localities of Hartlepool and Whitburn, Arctic Skuas can be encountered at any coastal watch-point. Vantage points include South Shields, Seaham Harbour, Chourdon Point and Blackhall Rocks, although the numbers observed are smaller due to birds generally being further out to sea.

Inland skuas are a true rarity in County Durham and such instances were only recorded in ten years between 1970 and 2011. The year 1976 was remarkable for a series of four inland records in September, which in all probability referred to the same bird. A single bird was noted at Witton-le-Wear on 5th, at Stanhope on 11th, in Upper Teesdale, at Langdon Beck, on 18th and then at Grassholme Reservoir on 21st. The bird was picked up at the latter site in a weakened state on this last date and died a week later. The 1970s also produced single birds at Hurworth Burn Reservoir on 12 September 1974 (found dead six days later), Bowlees on the unusually early date of 23 July 1978 and Derwent Reservoir on 15 October 1978. Despite being a productive period for coastal passage, there were only three inland records through the 1980s. One at Windlestone on 25 August 1980 was followed by two sightings in 1984; a single at Morton Tinmouth on 30 September and two flying west up the Tyne Valley, seen from the west end of Newcastle, on 4 November (Kerr 2001). The 1990s followed in a similar vein with inland sightings in three years. Two were at Longnewton Reservoir on 6 September 1991, with a juvenile found dead there on 22 November of the same year. One was at Hury Reservoir (and later Grassholme Reservoir) on 14 August 1995 and one was at Crookfoot Reservoir on 29 November 1999. The 2000s however, saw a distinct downturn in events, no doubt linked to the ever-decreasing Scottish population. The only inland records were of a dark morph bird in a mixed gull flock at Hetton Lyons CP on 21 August 2007 and another which flew north low over a housing estate at Hedworth on 28 September 2011.

In addition to the inland sightings detailed above, the appearance of single birds, or exceptionally small flocks, on the North Tees Marshes is not unusual. Dorman's Pool, the Reclamation Pond and Saltholme Pools have all attracted occasional birds between late May and late October, though their appearance is not annual. Of more interest, 44 gathered on Cowpen Bewley landfill site on 26 August 1986 and at the same time, approximately 25 Arctic Skuas flew inland over Cowpen Marsh. In 1998, up to 18 were observed gaining height over the Tees estuary and heading off strongly south west, whilst on 7 August 1999, 16 birds in two flocks headed off to the south west over Greenabella Marsh. Whether these birds continued their journey cross-country to the west coast, or re-orientated back to the east coast is not known.

Interesting aspects of behaviour noted include an adult seen to chase a Woodpigeon *Columba palumbus* down a hedgerow at Cleadon Hill on 21 June 1981, and one seen pursuing a large bat *Chiroptera sp.* out at sea off Hartlepool Headland on 1 September 2000.

Distribution & movements

This species breeds throughout most of the Arctic and sub-Arctic, and has an almost circumpolar distribution. The British populations in northern Scotland, confined as they are to the Hebrides, Orkney, Shetland and the northern Scottish mainland, are towards the southern limit of the species' breeding range. Arctic Skuas are long-distance migrants and the European breeding birds are to be found wintering off the southern coasts of West Africa, the earliest returning adult birds reaching their wintering quarters as soon as early October. Northern populations, from elsewhere in the species' breeding range, winter on a much broader front, though at latitudes similar to north European birds.

Britain's Arctic Skua population was in quite steep decline over the 25 years or so to 2010 (Parkin & Knox, 2010). Presuming that many of the passage birds seen off Durham originate from this population, it would require a considerable recovery in the British breeding numbers before Durham again experienced the levels of passage which were so regularly witnessed during the 1960s and 1970s.

Long-tailed Skua (Long-tailed Jaeger)
Stercorarius longicaudus

An uncommon passage migrant, mainly in late summer and autumn, but scarce in spring.

Historical review

During the vast majority of the 19th and the 20th centuries, this species was considered a rare, casual visitor to County Durham, but since the late 1980s, numbers have increased dramatically, with several notable influxes being documented.

The first unequivocal record of the species in Durham was of one shot at Whitburn on 24 October 1837 (Hancock 1874). Earlier recorders however, such as Hutchinson (1840) and Hogg (1845), failed to document its presence in County Durham at all. Hancock also stated, with reference to the whole of the north east coast, it was

Long-tailed Skua, Whitburn, October 1837 (K. Bowey, courtesy NHSN)

"*the rarest of the skuas and a mere casual visitant*", and referred to an adult male that was 'taken' at Seaham Harbour on 25 October 1879. The autumn of 1879 saw a huge influx of skuas along the English east coast including, according to Nelson (1907) "*a considerable number*" of Long-taileds. These were said to have accompanied the many thousands of Pomarine *Stercorarius pomarinus* and other skuas recorded in the mouth of the Tees over 14th and 15 October, in particular.

Temperley (1951), at the mid-point of the 20th century, described Long-tailed, or *'Buffon's Skua'* as it was often then known, as a "*very rare and irregular autumn visitor*", only rarely met with off the Durham coast. He was able to quote the handful of 19th century records, including those during the skua influx of October 1879, but said it had only rarely been recorded off the Durham coast since that exceptional event. He also documented one at Teesmouth on 5 November 1881 and referred to a single inland record; an immature bird found long dead at Shotley Bridge on 31 October 1891. This was actually the last record noted up to Temperley's time of publication. The fact that there were no known 20th century records up to 1951 emphasises the then rare nature of the species.

The first documented records for the 20th century came in 1955 when one was off North Gare on 9 October, followed by an immature off Teesmouth on 22 October. Then, rather remarkably, considering the previous scarcity of the species, a total of 13 Long-tailed Skuas were reported in five consecutive years, from 1958 to 1962 inclusive. More specifically, these included: three past Marsden on 18 October 1958; an adult north at Hartlepool, followed by possibly two immature birds on 30 August 1959; an adult at Whitburn on 6 September 1959; a 'tired-looking' adult which flew over the Hartlepool Promenade on 28 October 1959; an adult at Hartlepool on 8th to 9 October 1960; immature birds at Hartlepool on 1st and 21 October 1960, two at Hartlepool on 17 October 1961; an adult at Hartlepool on 26 August 1962; a juvenile at North Gare on 2 September 1962; a juvenile off Seaham on 12 September 1962 and an adult there on 20 September 1962.

Later in the 1960s, sightings came in both 1964 and 1969. In the first of these years, an adult, accompanying eight Arctic Skuas *Stercorarius parasiticus* flew up Greatham Creek on 7 August and away over Cowpen Marsh, later returning the same way. A good showing in 1969 included one at Souter Point on 1 June, another at Marsden on 25 June, two off at Hartlepool on 26 July and then four there on 21 August.

Recent status

The rise in the popularity of seawatching through the 1970s inevitably led to more records of Long-tailed Skuas, despite remaining a BBRC rarity up to 1979. All of the 27 records in this decade fell between 20 July and 1 October and Hartlepool and Teesmouth accounted for the majority. Three different adults frequented this area over 20th to 22 July 1971, while seven birds in autumn 1976 and eight birds in autumn 1978 comprised the best tallies. Both of these years were notable for the increased numbers of Long-tailed Skuas seen from North Sea coastlines (Wallace & Bourne 1981). A sickly juvenile on Seaton Common on 18 September 1976 unfortunately died three days later.

An increased frequency of occurrence was evident through the early and mid-1980s, with year totals varying from two in 1981 to 19 in 1984. These generally occurred in autumn seawatches between mid-July and mid-October, with significant passages of note (for the time) including four at Hartlepool on 13 August 1983 and four at Whitburn on 12 September 1984. The first status-changing year came in 1988. After a quiet early autumn period, a total of 1,224 were documented around the coast of Britain between 28 August and 2 November, the bulk of which were in the North Sea (Dunn & Herschfeld 1991). An impressive total of approximately 100 birds were recorded in County Durham, including September counts of 21 adults at Hartlepool on 24th, 13-17 at Hartlepool on 29th and 34 at Whitburn over 29th to 30th. Passage continued up to 12 October and included three at Whitburn on 12th. Putting Durham's counts into context, east coast peaks at Flamborough Head included 41 on 23 September and 43 on 29 September, contributing to an autumn total there of 367 birds. The influx may have been linked to the low availability of sand eels *Ammodytes sp.* in 1988, which in turn led to a total breeding failure in Shetland's Arctic Terns *Sterna paradisaea*. The summer of 1988 was also reported to be a 'lemming year', although Long-tailed Skua breeding success in Scandinavia was low, so perhaps the birds originated from further east (Dunn & Herschfeld 1991).

While the passage of 1988 had been the greatest witnessed since 1879, the numbers in 1991 comfortably exceeded this. It is difficult to assess the level of duplication in the county, but a total of 750-800 birds were noted during the autumn. After an uneventful August, a huge arrival took place in early September. This commenced with 30 at Hartlepool and 15 at Whitburn late on 5th. An increase came on 6th as more birds travelled south, with passage totals of 54 flying north and 139 flying south at Hartlepool, with 10 north and 82 south at Whitburn. The peak of passage occurred as birds returned back north on 7th; observers at Whitburn logged a total of 215 north and 40 south in 12.5 hours' watching, whilst 46 birds were noted at Hartlepool. Lesser numbers followed over subsequent days, but included impressive day tallies of 45 at Hartlepool and 20 at Whitburn on 11th. County Durham's numbers contributed to a total of some 3,000 birds on the English north east coast over this period. Small numbers continued to be seen through the rest of the autumn, with 32 at Whitburn on 18 October being the only count of importance that month. Sightings continued at Whitburn until 4 November.

Reports of Long-tailed Skuas declined sharply in the next two years with county totals of just 14 birds in 1992 and 68 in 1993, although 17 adults and 11 juveniles at Whitburn between 24 August and 5 September 1993 were notable, as were the 15 past Whitburn on 13 September of the same year. Fitting in with the three-yearly cycle of lemming abundance were a county total of 195 birds in 1994, which included an early influx of adults between 8th and 18 August (including 32 at Whitburn over 13th to 15th). A later passage of mainly juveniles featured Whitburn totals of 42 on 14 September, 13 on 3 October (when 11 were noted at Hartlepool) and 12 on 16 October. The year 1995 also proved to be a good one, with a county total of over 250 birds. A rush of August adults provided the bulk of the year's passage with approximately 100 at Hartlepool on 7th, 60 at Whitburn over 7th to 8th and daily counts of up to 11 at both sites to the end of August.

Year totals in the remainder of the 1990s and through the full decade of the 2000s varied between eight and 57 birds, with no repeat of a dramatic influx. There were few periods of concentrated passage, with 17 at Whitburn and 14 at Hartlepool on 18 September 2001, 12 at Hartlepool on 22 September 2003 and 29 at Hartlepool on 9 September 2005 being isolated movements. This period did however produce several late autumn records, including a total of 19 November sightings. Eight were at Whitburn on 14 November 1999 and four there five days later were significant, as was the latest-ever county record of a juvenile at Dorman's Pool over 26th to 27 November 2005.

Spring sightings in County Durham have been very few. The earliest on record is of two past Whitburn on 21 May 2004, and a total of 19 have been logged between 21 May and 8 June. Birds have been seen in only six springs since 1980, the best year being 2004 with eight birds, including five at Hartlepool on 22 May. All spring sightings have involved adults (or near adults). There are no winter records for the county, which is to be expected for a pelagic species wintering in the South Atlantic.

This species is very rare inland and there are only two recent sightings. One was seen over Muggleswick Common on 11 September 1970 (surprisingly, the only record of the species for that year), and a juvenile was photographed as it fed on a dead rabbit on Barningham Moor on 14 August 2011. Several birds have been seen marginally inland on the North Tees Marshes, such as at Cowpen Marsh, Dorman's Pool, Greenabella Marsh and Seaton Common, with most in September being immature birds (Blick 2009). A juvenile found dead at the Reclamation Pond on 11 November 1990 had damage consistent with an attack from a falcon *Falco sp.*

Although it is now recorded every year in the county, numbers remain unpredictable compared to the other skua species. It is the most pelagic of the skuas and probably many more individuals pass through the North Sea than are recorded by coastal watchers. Undoubtedly, improved optical equipment used by dedicated and skilful seawatchers is in part responsible for some of the increase in the records of this once very rare species, but to what extent this increased incidence is real or apparent cannot be stated with confidence. What is abundantly clear is that from the late 1980s and into the 21st century, this species was recorded in numbers that were far greater than in any previous decades.

Distribution & movements

This is an Arctic breeding species nesting on tundra up to 1,300m elevation. It is found across northern Scandinavia east to Siberia and at similar latitudes across North America. The Long-tailed Skua is a less well-understood species, its full distribution and movements being imperfectly known. It is apparently a more pelagic species than other skuas, being observed much less frequently by coastal watchers when on passage. It winters, largely out to sea, south of the equator on both sides of the South Atlantic.

Great Skua
Stercorarius skua

A common passage migrant, mainly in summer and autumn.

Historical review

The Great Skua is often known as the 'Bonxie'; a bastardisation of the original Norn (old Shetland language) word for *bunksi* or *bonksi* meaning a stout or thick-set person (Jakobsen 1928). The sight of these impressive pirates of the sea has become much more frequent in County Durham since the 1970s.

Backhouse (1834) gave its status as "*rare*", it was not documented at all by Hutchinson (1840), whilst Hogg (1845) said that it was "*sometimes killed on our coast*". Hancock (1874) summed up the status as a "*Rare autumn and winter visitant*". Writing of Teesmouth in 1931, Richmond (1931) said that "*the Great Skua (Stercorarius skua), which should not normally occur inshore, has been observed in the estuary in clear weather*", giving examples of such occurrences on 8 August 1928 and 2 August 1931. Clearly, some birds were present just offshore and occasionally inshore in Durham, during the late summers of the early 20th century, although the species was rather poorly documented in the county over this period. Temperley (1951) said that at his time of writing, its true status had not changed much from previous times. He called it an "*uncommon autumn passage migrant, less often seen in winter*", and noted that birds were off the coast from time to time, but usually well off-shore, and only in small numbers.

The increase in seabird studies from the late 1950s onwards proved the species to be a much more regular passage migrant than was previously envisaged, particularly through the early 1960s, with over 100 birds being seen in most years. An example of more pronounced passage in this period included 95 north and 45 south at Hartlepool on 22 August 1962, but such numbers were then exceptional.

Coastal observers also noted interesting behaviour in this period. On 14 September 1959, at Souter Point, an observer saw one strike a Great Black-backed Gull *Larus marinus,* which fell into the sea, whilst at Hartlepool in 1962, a pair of Great Skuas killed and partly ate a Kittiwake *Rissa tridactyla.*

Recent status

Improved optics and a greater awareness of seabird movements, and identification criteria, has led to a much improved understanding of the status of this species in Durham. Temperley's summary remains valid but the increased observer effort and the rise in population levels of the species means that Great Skua is now routinely encountered through much of the late summer and the autumn, mainly at the seawatching points of Hartlepool and Whitburn, but also quite frequently off other coastal vantage points.

Winter sightings in the North Sea are usually few and unpredictable, but it is clear from the data accumulated in recent decades that there is a small wintering presence. Occasional records of one or two birds occur at Hartlepool and Whitburn in most years between mid-December and March, with winter storms having produced better movements including three at Hartlepool on 22 January 1978, eight at Whitburn on 21 December 1986, four at Whitburn on 24 March 1992 and three at Whitburn on 25 January 1995. A more pronounced winter presence was noted at Teesmouth in 1976 when four were noted on 24 January, five the following day and another on 31st. This was part of an unusual winter movement of seabirds that involved Arctic Skua *Stercorarius parasiticus,* Pomarine Skua *Stercorarius pomarinus,* Sooty Shearwater *Puffinus griseus* and Manx Shearwater *Puffinus puffinus.* The majority of winter birds have been seen at Hartlepool and Whitburn, but sporadic sightings have also been made close inshore at Crimdon, Marsden, North Gare and South Shields.

A small spring passage is detected in most years, but this is usually limited to a handful of sightings. After infrequent singles in mid-April, a more discernable passage takes place in May, often peaking in the second half of the month. Compared to autumn movements, the numbers concerned are very small suggesting that the northbound return to breeding grounds is funnelled to the west of Britain with few utilising the North Sea. Volumes of passage can be judged from the second half of the 2000s, when spring seawatching effort was high. In 2005, spring passage was concentrated into just two days, 16th to 17 May, when 10 flew north off Whitburn. In 2006, there was a low-level passage featuring one to three past Whitburn on five dates plus eight moving north on 31st, and a total of six at Hartlepool; including three on 22nd. Five flying north at Whitburn on 28 May was the peak for 2007, but subsequent years produced only the occasional single. As with all movements of pelagic species, sightings close inshore are heavily weather-dependent and subject to observer effort in the peak conditions.

Return migration of presumably failed breeders and immature birds commences from late June and small numbers are regularly seen through July and the first half of August. Larger movements are exceptional in this period; better totals at Whitburn in the 2000s included 47 on 12 August 2006, 21 on 22 July 2005 and 26 on 20 June 2010, with a high total of 156 at Hartlepool on 1 August 1986. From late August and through the rest of the autumn, winds from the north westerly to north easterly sector can produce large-scale passages and, with the decline in breeding numbers of Arctic Skuas through the 1990s and 2000s (Furness & Ratcliffe 2004), the 'Bonxie' has increasingly been the most frequently encountered skua during seawatches. Large movements were unusual in the 1970s with 235 at Hartlepool on 11 September 1976 being exceptional; no other count in that decade topped 48. Growth in the Shetland population through the 1980s and 1990s resulted in an increased presence off the north east coast and day-counts in the region of 50 to 95 became more regular. Exceptionally, 121 flew past Hartlepool on 6 September 1988 with *c.*200 there on 22 September 1988. On 13 September 1993, 261 flew past Whitburn (the highest day-count in County Durham) and 111 passed that site on 30 September 1998. The relatively constant seawatching effort at Whitburn through the latter half of the 2000s gave some indication of the current volume in passage numbers. Between 2006 and 2010, yearly totals varied from a low of 224 in 2006 to a high of 510 in 2007, with an average total of 385.

Although passage peaks in September, late autumn gales can infrequently result in a large displacement of this species into the North Sea. In 1986, 95 at Hartlepool on 1 November was the highest-ever count for that month, but this was exceeded by 111 at Whitburn on 3 November 1998 and 254 at Hartlepool on 6 November 2000. In most years however, November day-counts struggle to reach double figures, whilst nine at Hartlepool on 8 December 1973 and eight at Whitburn on 30 December 2009 were impressive early winter counts.

There appears to be only a single truly inland record of the species in the second half of the 20th century, when one was at Derwent Reservoir on 9 September 1978. One at Boldon Flats on 28 August 1986 is notable however for being 3km from the coast. According to Blick (2009), birds very rarely venture into the mouth of the Tees in the same manner as Arctic Skuas, although, over recent decades, one or two birds have been seen on Seal Sands, as well as flying over Cowpen Marsh, Dorman's Pool and the Long Drag during severe weather conditions (Blick 2009).

Distribution & movements

The 'great skua' group is notable for its bi-polar distribution, but the breeding range of the Great Skua encompasses Scotland, the Faeroes and Iceland, with smaller number in Ireland, and across northern Scandinavia and into Russia. In Britain, nesting is restricted to sub-arctic habitats in Shetland, Orkney, Caithness, Sutherland and the Outer Hebrides. Wintering birds disperse widely across the North Atlantic, with mainly juveniles moving as far south as Brazil, Guyana and the Cape Verde Islands.

There have been at least six ringing recoveries involving birds ringed as pulli on Shetland and recovered in County Durham. Of interest, one ringed at Foula on 21 July 1960 was recovered off Sunderland on 12 August 1964, whilst birds found dead at Hartlepool on 25 June 1981 and in June 1986 had been ringed on Foula in July 1974 and July 1984 respectively. The world population of Great Skuas is only around 13,600 pairs, of which 9,600 nest in Britain and Ireland (Mitchell *et al.* 2004, Reeves & Furness 2002). Foula holds the largest colony in Britain with approximately 3,000 pairs, so it is to be expected that the locality features well as a source for ringing recoveries.

Ivory Gull
Pagophila eburnea

An extremely rare vagrant from the Arctic: two records.

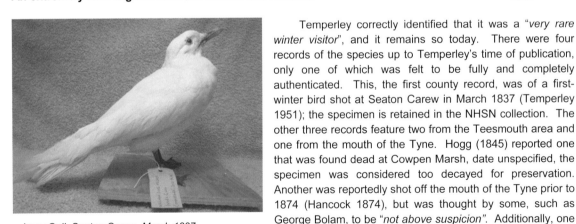

Ivory Gull, Seaton Carew, March 1837
(K. Bowey, courtesy NHSN)

Temperley correctly identified that it was a "*very rare winter visitor*", and it remains so today. There were four records of the species up to Temperley's time of publication, only one of which was felt to be fully and completely authenticated. This, the first county record, was of a first-winter bird shot at Seaton Carew in March 1837 (Temperley 1951); the specimen is retained in the NHSN collection. The other three records feature two from the Teesmouth area and one from the mouth of the Tyne. Hogg (1845) reported one that was found dead at Cowpen Marsh, date unspecified, the specimen was considered too decayed for preservation. Another was reportedly shot off the mouth of the Tyne prior to 1874 (Hancock 1874), but was thought by some, such as George Bolam, to be "*not above suspicion*". Additionally, one was reported by Lofthouse (1887) as being shot at Teesmouth on 14 February 1880, but there were doubts in each of these cases regarding date, location or authenticity of capture. Consequently, these three reports are excluded from the county's official totals.

The only modern record of Ivory Gull was in the winter of 1970/1971. A first-winter bird frequented the River Tyne, moving between North Shields and South Shields, from 18 December 1970 to 23 February 1971 (*British Birds* 64: 353, *British Birds* 65: 336). At the time, this was only the eighth British record since 1958. Initially found by Fred Grey and John Strowger, the bird attracted many visiting birdwatchers, even more so when a Ross's Gull *Rhodostethia rosea* was found in the same area.

Distribution & movements

In Europe, Ivory Gulls breed only in Svalbard, inside the Arctic Circle. Elsewhere, the range is restricted to islands in the high Arctic between Franz Josef Land and Arctic Canada, with small numbers in north and south east Greenland. The wintering range is thought to be restricted to within or close to the edge of the pack ice. The population in Canada and Svalbard has been in steep decline since the 1980s and this is probably the only gull to visit Britain

Ivory Gull, River Tyne, winter 1970/71 (Bryan Galloway)

as a vagrant that has suffered a decline in records since the beginning of the 20th century. Ivory Gull is an iconic rarity in Britain, appearing only sporadically and often at remote Scottish locations. A total of 137 had occurred to the end of 2010, but only 53 since 1950. The majority have been found between November and February, although there are records for every month of the year. The last sightings in adjacent counties were in North Yorkshire in 1986 and Northumberland in 1979.

Sabine's Gull
Xema sabini

A scarce but almost annual visitor in late summer and autumn from Arctic breeding grounds.

Historical review

Temperley (1951) called it a *"very rare straggler"*. The first record for County Durham came on 11 October 1879, when the specimen of a juvenile bird came into the possession of Mr Fred Raine of Durham, having been shot by a fisherman at Seaham Harbour the same day. The specimen was retained in the Hancock Museum (Howse 1899) and is in the NHSN collection to date. The second Durham record concerns an immature shot in the Tees Bay on 6 October 1889, the skin of which is in the Dorman Museum, Middlesbrough (Blick 2009). Very unusually, the next two occurrences were both in spring. On 9 May 1932, N.K. Duncan saw two at Greatham Creek, described as "*one adult in summer dress and the other not quite mature*". The same observer saw two at the same

Sabine's Gull, Seaham, October 1879 (NHSN)

location on 17 May 1944, this time both adults in full summer plumage. The birds were apparently fearless and allowed a close approach, with the encounter being described in *British Birds* as follows.

"On May 17th, 1944, as I was nearing the bridge over Greatham Creek on the Stockton to Hartlepool road, in company with a friend, we observed two Sabine's Gulls. They were in adult plumage. The entire head to below the nape and down in front to the upper breast was slate blue, circumscribed by a distinct narrow border of deep black. The lower neck down to the upper back was white, washed with pale lavender, while the back and wing coverts were darker lavender. The forked tail was very conspicuous in flight, as were the black primaries and primary coverts. The black feet and legs were also noted. They were busily feeding, hovering gracefully over the water and taking their prey from the surface with scarcely a ripple, or, alighting for an instant in the water to make a capture and rising again with ease. One bird only uttered an occasional harsh tern-like cry, grating but not loud. We watched them for over two hours, in flight and resting on the shore, with good glasses; during which time they often came within a dozen yards of us, as they showed no fear. On a previous occasion, May 9th, 1932, I had seen a couple of these birds at exactly the same spot; but only one of them was in adult plumage, the other being immature. They also allowed me to approach to within a few yards, so that I was able to observe every detail of their plumage", Norman K. Duncan (Temperley 1945). These remain the only spring records for County Durham.

Other records mentioned by Temperley (1951) include birds that were probably on the Yorkshire side of the Tees Estuary. No further records followed until 1960, when adults were seen at Seaton Snook on 24 July and Hartlepool Headland on 2 August. A further four birds were noted during the 1960s, all of these between 29 September and 24 November, Hartlepool having three sightings and Seaham a single. In summary, this was a rare visitor that during 1932-1968 was only reported in Durham on seven occasions, although probably 10 birds were involved.

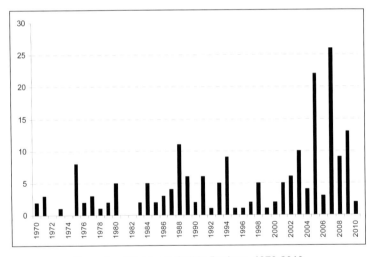

Records of Sabine's Gull in Durham, 1970-2010

Recent status

Since 1970 however, only four years are without records and nearly 200 Sabine's Gulls have been seen in County Durham. The total number of birds seen each year between 1970 and 2010 is detailed.

Two years stand out as being exceptional for Sabine's Gulls in Durham, 2005 and 2007. In both years, favourable weather conditions in September brought a flurry of sightings at both Hartlepool Headland and Whitburn. In 2005, a strong north easterly wind on 9 September pushed four birds past Hartlepool and nine past Whitburn, all but one of which were juveniles. A further six flew north at Hartlepool on 16 September, contributing to a county year total of 22 birds. In 2007, good seawatching conditions through much of the autumn contributed to birds being recorded on nine dates (Hartlepool and Whitburn combined). The year's total of 25 birds included three at Hartlepool on both 23 August and 29 September, and five juveniles flying north past Whitburn on 28 September.

Almost all records fall between 20 June and 25 November, with the exception of the county's only winter record; an immature reported flying down river at South Shields on 1 February 1978 (Unwin 1979). Summer sightings are decidedly rare, with only two birds seen between 20 June and 20 July, but there is a gradual increase evident from the last week of July. In line with records in the south west of England, the peak period is mid-September, but the species' appearance off the Durham coast is very much weather-dependent. All summer and autumn records in the period 1970 to 2010 are summarised by ten-day periods in the graph.

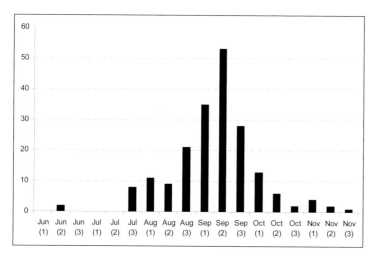

Most visits by this species are brief; indeed, approximately 90% of all birds since 1970 have been seen during seawatches. On rare occasions however, Sabine's Gulls can linger in a favoured area, often providing close views. Recent examples have included an adult at Seaton Snook over 10 July to 7 August 1971, a second-summer bird at the same locality over 7th to 10 August 1998, four different individuals around Hartlepool Headland between 13th and 22 September 2002, and a juvenile around South Shields and the River Tyne from 25 October to 9 November 2009.

Summer/autumn records of Sabine's Gull in County Durham, 1970-2010, by ten-day periods

As to be expected, the key seawatching localities of Hartlepool Headland and Whitburn Observatory have accounted for the bulk of the records. Hartlepool records total 107 birds and Whitburn's 74. The only other birds noted during sea passage have been single birds at Seaham in August and October 1988. Occasionally, individuals have turned up in more sheltered areas of beach and coastline, such as at Newburn, Seaton Carew and Sunderland Docks. Inland birds are truly exceptional in the north east, so particularly noteworthy are the records at Jarrow Slake on 28 August 1978, Felling on 9 August 1979 (adult moulting into winter plumage) and Dorman's Pool on 14 October 2001 (juvenile).

Distribution & movements

Sabine's Gull has a circumpolar distribution, nesting on the Arctic tundra throughout northernmost North America and Siberia. Most winter in the Pacific off western South America, but the Greenland and east Canadian population move through the Atlantic to winter off south west Africa. In Britain, the species is mainly encountered as an autumn migrant from early August to late October. The volume of passage varies greatly each year, with autumn gales occasionally bringing large numbers to south west England and Ireland. Far fewer birds penetrate the North Sea, occurrences usually coinciding with periods of northerly winds.

Sponsored by

INTERNATIONAL PAINT

Kittiwake (Black-legged Kittiwake)
Rissa tridactyla

An abundant coastal passage visitor and restricted breeding species, with smaller numbers in winter.

Historical review

In 1840, Hutchinson said that they did not breed in Durham and there is no evidence of Kittiwakes breeding in the county until the 1930s. The relatively recent arrival of breeding Kittiwakes here may seem surprising as large numbers have been recorded nesting in neighbouring Yorkshire as far back as 1770, as noted by Nelson (1907). Of the early 19th century, Nelson wrote that *"Charles Waterton, in 1834, found the nests so numerous as to totally to defy any attempt to count them"*. A national decline followed as demand then rose for Kittiwake feathers for use in fashion, as happened with some tern species. This was curtailed by the Seabird Preservation Act of 1869, which Nelson noted *"put an end to the butchery"*.

Kittiwakes were first seen on Marsden Rock by David Rollin in 1930, but he did not record breeding. The colony at Marsden probably became established in 1931, and ciné footage was taken in 1932 showing birds on the south side of Marsden Rock (J. Coulson pers. comm.). The colony spread to other ledges and faces and on to the mainland cliffs as it grew in subsequent years. By 1937, Fred Grey counted 308 nests.

Temperley (1951) suggested that the closure of the coastal cliffs and beaches during the Second World War may have assisted the growth of the colony. Reliable data however, would be difficult to obtain in this period, and the figures do not bear this out, showing consistent growth with a slowing, not an increase in the rate of growth during the war (J. Coulson pers. comm.). In 1945, when access to the cliffs was restored, Fred Grey counted 750 nests (Temperley 1951).

By the middle of the 20th century, Temperley wrote that the Kittiwake was *"a common summer resident and passage migrant"*, and a *"winter visitor in small numbers"*. He said that locally breeding birds usually returned in March, occasionally as early as mid-February, leaving the colony in August. He considered Durham's coastal passage of Kittiwakes to be significant, including counts of up to 6,000 north in a gale on 28 October 1948 over a nine-hour period, and 4,200 north in eight hours on 7 November 1948. Temperley also documented some, then rare, inland records, for example one at Darlington Sewage Farm on 18 August 1937.

In 1952, the Marsden Rock colony had expanded to 1,344 nests, including nests on the mainland cliffs. By 1960, it had grown to contain 2,878 nests, increasing further to 3,887 nests by 1970 (J. Coulson pers. comm.).

The earliest record of the species in the Gateshead area concerned three birds feeding under the Tyne Bridge on 10 May 1948 (Bowey *et al.* 1993). This was the prelude to the remarkable establishment of inland Kittiwake colonies on the River Tyne. In 1949 a colony became established just outside the county on buildings on the Tyne at North Shields, some three kilometres from the sea (Temperley 1950).

In 1960, a change in behaviour led to birds being noted up river as far as Newcastle during the spring and summer. The following breeding season, Kittiwakes could be seen almost daily between February and June, within sight of the Tyne Bridge (Coulson & Macdonald 1962). In 1962, this led to the establishment of a breeding colony on the Sheet Metal Works just west of the Baltic Flour Mill on Gateshead Quayside. One bird breeding here had been reared as a chick at North Shields (J. Coulson pers. comm.). From 1962 to 1964, up to 10 pairs nested, before the building was demolished in 1965. In 1964, the first three nests appeared on the Baltic Flour Mill, over 14 kilometres inland, increasing to 35 pairs by 1970.

Site of Kittiwake colony, Gateshead, 1970 (John Coulson)

The River Tyne was severely polluted at this time, and Kittiwakes were occasionally noted feeding on the river, taking material from the surface, and following the Shields Ferry, probably feeding on fish or crustaceans associated with sewage. They were also seen taking fish from just below the surface from fresh water coming in from the River Derwent, a few hundred metres from the confluence with the Tyne (Coulson & Macdonald 1962). This suggested the possibility of a change in food supply being responsible for birds moving on to the river to breed; this however, was not borne out by subsequent events. By 1973, Fitzgerald and Coulson noted that *"Kittiwakes have not fully adapted to the river environment and appear to be strongly influenced by their natural pelagic and sea coast environments"*. As the river became cleaner, wintering gulls on the Tyne, particularly Common Gull *Larus canus* and Great Black-backed Gull *Larus marinus*, which depended on the food supply provided by sewage, declined dramatically in numbers. In contrast, Kittiwakes breeding on the river did not decline, but nor did they show any tendency to fish on the, by now, far cleaner Tyne (Raven & Coulson 2001).

Kittiwakes rarely feed on the river, and if they do so, only by picking small objects from the surface, as noted above. Raven and Coulson recorded that *"it is clear the Kittiwake does not use riverside nesting sites because food is available within the river limits"*. Coulson studied food regurgitated by birds breeding at the Baltic, and found that this was entirely marine, mainly Sandeels *Ammodytes sp.* Birds breeding there flew down the river and out to sea to collect food. The main advantage to birds breeding on the river is likely to come from shelter from strong winds and possibly higher spring temperatures (J. Coulson pers. comm.), and this may have triggered their move inland.

Turner (2010) also suggested that birds nesting on riverside structures may benefit from reduced risk of loss of eggs or chicks to some avian predators. He noted however, conversely that the 'commute' to feed at sea was a potential disadvantage, that some riverside structures, such as high ledges on the Baltic, could be more exposed to bad weather than cliffs, and that unintentional human disturbance may be greater at some riverside sites.

Elsewhere in the county, at West Hartlepool, a colony was formed on a dock warehouse, with five pairs nesting in 1958. One adult had been ringed as a chick at Marsden five years earlier (J. Coulson pers. comm.). Again however, the building was demolished in 1960 (Coulson 1963). In 1967, seven pairs also nested at North Sands, Hartlepool, with 10 pairs on the pier there in 1970 (McAndrew 1971).

Significant historical non-breeding records include 10,000 flying north off Hartlepool on 9 November 1957 (Blick 2009). In August 1962, huge concentrations of sprats *Sprattus sprattus* appeared in the bay and estuary of the River Tees, attracting large numbers of gulls and terns. By mid-August, some 6,000 Kittiwakes were present and on 21st, they had reached huge numbers off Hartlepool, the count certainly running into five figures. A similar incident occurred in 1971, when a wreck of sprats at Teesmouth in the third week of July attracted thousands of Kittiwakes. Estimates on 22nd ranged from 25,000 to at least 100,000 (McAndrew 1972).

Away from the Tyne breeding sites, the species remained scarce inland, so records of 24 at Smiddyshaw Reservoir on 27 March 1965 (Bell 1966), and four or five at Witton-le-Wear on 7 September 1971 (McAndrew 1972) were notable.

Recent breeding status

In Durham, Kittiwakes nest on both natural coastal cliffs and, in the context of inland nesting locations, quite routinely on man-made structures. Breeding colonies of this species are largely restricted to the coast, mainly at Marsden, with the exception of the famous colonies along the River Tyne from the mouth, inland as far as Gateshead.

The Marsden colony remains by far the largest in the county, and numbers continued to increase steadily there for some years, with 4,465 nests in 1977 (J. Coulson pers. comm.), and peaking at 5,763 nests in 1992 (Strowger 1993). This gradual expansion of the colony was in marked contrast to the situation in Scotland where numbers fell dramatically during the 1980s. The trend at the Marsden colony however, was in line with the overall population change in Britain and Ireland during 1969-1987, which showed an increase of *c.*22% (Carter 1995). The number of nests at Marsden increased at approximately 100 per year for the first 60 years of the colony's existence, with no marked period of more rapid growth (Coulson 2011).

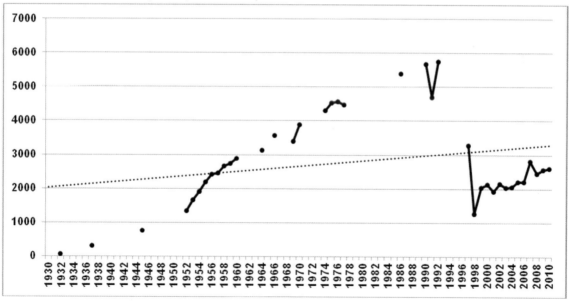

Kittiwakes at Marsden: no. of apparently occupied nests by year (data supplied by J. Coulson) with trendline

475

A dramatic decline followed in the late 1990s, however. A survey in 1997 discovered that numbers had fallen by 24% in a single year and the colony collapsed to a low point of just 1,259 nests in 1998 (J. Coulson pers. comm.). This decline was not part of a national trend, but was believed to be caused by a toxin produced by algal blooms. This resulted in the death of approximately 13,000 birds. This occurred at sea approximately seven kilometres offshore, in or close to an area used to dump human sewage. Nutrients from the sewage may have been responsible for the unusually large algal blooms. Many dead Kittiwakes floated ashore in this period during onshore winds. Female birds were killed in larger numbers, possibly because of differential feeding areas or techniques. As a result, in 1998, over a quarter of the nests at Marsden were occupied by only a male bird. The extensive dumping at sea of raw sewage ceased after 1998 (Coulson & Strowger 1999) and a gradual recovery followed, with 2,158 nests in 2002, 2,821 nests in 2007, and 2,610 nests in 2010 (J. Coulson pers. comm.). The variation in numbers at the Marsden colony is illustrated above.

Kittiwakes at Marsden are noted for out-competing Fulmars *Fulmarus glacialis* on nesting ledges. Fulmars return earlier, but then leave the ledges for extended periods, during which time Kittiwakes move in. Returning Fulmars are then unable to reclaim their ledges (Coulson & Horobin 1972).

Although in general there has been little persecution of the Marsden colony over the years (J. Coulson pers. comm.), a serious incident occurred in 2001, when more than 100 birds were shot (Siggens 2005), and a smaller number were killed in the summer of 2009.

The Marsden colony may be the most numerically significant in the county, but the inland Kittiwake colonies on the River Tyne in Gateshead and Newcastle are unique. They form one of the few urban seabird colonies in Britain and the furthest inland Kittiwake colony in the world (Coulson 2011). On the River Tyne at Gateshead, Felling and Tyne Dock by 2010, there were five separate colonies on the ledges of large riverside warehouses/bridges and along the river frontage, as well as a very successful colony on a specifically designed 'Kittiwake Tower'.

By the early 1970s, the initial colony at the Baltic Flour Mill was well-established and another colony had formed even further upstream at Dunston. Up to 32 pairs nested at the Dunston site for about 10 years from 1971, prior to the demolition of the building upon which the birds nested. At over 18 kilometres from the river's mouth, this is the furthest up river that breeding has ever been noted. On 19 July 1972, counts revealed the presence of a total of 47 nests; 30 at Dunston and 17 on the Baltic Flour Mill ledges at Gateshead (Bowey *et al.* 1993). In the early 1970s, a few pairs of Kittiwakes also nested downstream on the end of South Shields Pier, but these nests were washed off (J. Coulson pers. comm.).

During the 1980s, further small colonies became established along both banks of the River Tyne, although some of these were on old warehouses which have subsequently been redeveloped. By the early 1980s, there were some 103 pairs nesting, split between Gateshead and a developing colony just outside the county, on the Newcastle side of the river (Bowey *et al.* 1993). Despite the loss of the Dunston site, numbers gradually increased through the 1980s, reaching 148 pairs at the Baltic Flour Mill in 1988 (Armstrong 1989). A handful of pairs also established a foothold in yet another new colony, on the International Paints building, downstream at Felling from 1985 (J. Coulson pers. comm.).

At the beginning of the next decade, in summer 1990, 188 pairs were nesting on the Baltic Flour Mill (Armstrong 1990) and this colony continued to increase to a peak of 310 nests in 1996 (J. Coulson pers. comm.). By 1991, the new Felling site had expanded to some 30 pairs, rising to 100 pairs by 1994 (Armstrong 1996) and peaking at 215 apparently occupied nests in 2002 (Turner 2010). Numbers increased rapidly after wooden ledges were installed there in 1996, but predation by Carrion Crows *Corvus corone* in 2007, the placing of metal spikes on one ledge, and the removal of the wooden ledges after the 2008 breeding season all probably contributed to a subsequent decline to 116 apparently occupied nests in 2009 (Turner 2010). Counting is difficult at this site as it is only easily visible by boat or from the north side of the Tyne.

In the late 1990s, the Baltic Flour Mill was re-developed into the Baltic Centre for Contemporary Arts, funded through an Arts Lottery Award. During the development phase, Durham Bird Club worked with Gateshead Council to design and create an alternative nesting structure. This comprised a 16m high tower topped with three 'cliff-faces' with artificial ledges, constructed in heavy-section timber. The design was made to accommodate up to 200 nesting pairs and the tower was installed close to the Baltic in time for the 1998 breeding season. Both 'decoy' Kittiwakes, and old nests collected from below the cliffs at Marsden, were used to attract birds. A small number of Kittiwakes moved on to the structure in the first season, with 40 pairs occupying it that year, rising to 128 pairs by 2000 (Armstrong 1999b & 2005).

Kittiwake Tower, Gateshead (M. Newsome)

The structure subsequently had to be moved 1.5km downstream to Saltmeadows in 2001. A prominent location was chosen to be visible to birds heading up-river, and Kittiwakes returned to the tower immediately, numbers falling only slightly to 105 pairs that same year (Siggens 2005). Numbers gradually recovered following the tower's relocation, peaking at 143 apparently occupied nests in 2007 (Turner 2010), when the colony produced at least 140 young. Numbers dipped slightly to 110 apparently occupied nests in 2009, following the collapse of a ledge (Newsome 2010, Turner 2010).

Kittiwakes originally ringed on the Baltic Flour Mill have been re-trapped on a number of occasions at the tower, confirming the relocation of some birds from the Baltic (Northumbria Ringing Group pers. comm.). However, the structure has not been able to accommodate all the birds originally displaced from the building, in part because the south face pointing away from the river has proved far less popular, and never been even close to capacity. Based on observations of colour-ringed birds, the tower has probably never attracted more than about 25% of birds from the Baltic, with the remainder either crossing the river, or moving to International Paints, and in one case Tynemouth (J. Coulson pers. comm.). A more gradual phased closing down of existing ledges, prevented by the pace of development at the Baltic, might have increased the number of birds relocating. The tower has nevertheless been a success and this bold, innovative design should provide a model for displaced birds in other development sites. The Kittiwake Tower was designated as a Local Nature Reserve by Gateshead Council in 2003 and by the end of the first decade of the 21st century it had been used successfully for well over ten years.

Kittiwakes have shown the same order of preference for the three faces of the Tower in every season since its construction, until the colony was disrupted by the ledge collapse noted above. The face pointing north west has been consistently the most favoured, with most apparently occupied nests. This may be because the other river face, pointing towards the coast, is very exposed to cold east and north easterly winds. The south face, pointing away from the river, and directly into the mid-day sun, has not surprisingly held far fewer birds than either of the other two faces, as noted above (P. T. Bell pers. obs.).

Birds have also returned to nest on the Baltic, despite an attempt to 'design them out' in the original redevelopment by grading off flat ledges. Kittiwakes may never have been eliminated completely from the site, with at least two pairs remaining throughout (J. Coulson pers. comm.). More birds quickly returned, as a small number of pairs settled on the scaffolding used for the building work during re-construction in 1999, underlining the resilience and adaptability of Kittiwakes (Armstrong 2003). Numbers have slowly built up, with 18 pairs noted in 2007 and 30 pairs present in 2009 (Newsome 2008 & 2010).

Although nesting birds had been present close by for some time, breeding did not actually occur on the Tyne Bridge at Gateshead-Newcastle until 1996. Numbers have remained relatively small on the south side of the bridge, but 43 nests were occupied in July 2009. Another small colony on the Tyne was established by 1997, downstream at McNulty's (formerly Redheads) slipway, Tyne Dock. The highest count to date at this colony was 42 apparently occupied nests in 2008 (Turner 2010).

The presence of Kittiwakes on structures in central Newcastle and Gateshead has divided public opinion. The droppings, smell, and even the sound of the colonies has caused some concern, and was the reason for the original relocation of birds from the Baltic. A number of buildings on the Newcastle side of the river have also been fitted with spikes and netting as deterrents.

Elsewhere along the Durham coast, smaller colonies also use artificial structures, mainly dockside buildings at Seaham and Hartlepool. A colony at the Steetley Magnesite Pier, Hartlepool, was in existence from about 1969, with 16 pairs noted nesting in 1973. Numbers reached 72 pairs in 1995 but birds suffered from regular disturbance and persecution, resulting in the colony being abandoned in 2003 (Joynt 2008).

At Seaham Harbour, nesting was first noted in 1985, with three pairs, building up to 24 in 1990 and a high of 85 in 2001 (Baldridge 1986, Armstrong 1991, Siggens 2005). Numbers have dropped slightly since then, with 52 pairs in 2007, but the colony appears to be well-established (Newsome 2008). Birds here nest on a natural cliff in the fish dock and on remains of wooden piles. Several wooden ledges were placed between the piles, and ledges were cut into the cliff on the north side of the dock, to attract more birds. Kittiwakes have also nested on the roofs of sheds on the other side of the docks, making two quite distinct colonies (J. Coulson pers. comm.).

In 1992, 51 nests were noted at Hart Warren, but this was an isolated report, as there have been no subsequent breeding records from this site (Armstrong 1993). A colony centred on buildings at Hartlepool Fish Quay in the Victoria Harbour has also existed since 1992, although there are records of Kittiwakes nesting in the area back to 1959 (Stead 1964). This colony numbered 35 pairs in 1994, rising to 64 pairs in 1998 and a gradual increase has continued since with up to 160 pairs by 2006 (Joynt et al. 2008). Birds have expanded slightly from the harbour with nine pairs noted on the adjacent New Fleece public house on Northgate in 2006.

Another Teesside colony is located on the jetties of the Conoco Phillips oil terminal at Seal Sands. This is a relatively new, but rapidly expanding, colony situated at the mouth of the Tees. Five or six pairs nested for the first time in 2001, soon expanding to 162 pairs in 2006 (Joynt et al. 2008). Numbers have declined since then, with 99 pairs in 2007 and 113 in 2008, but this colony also appears to be well-established (Newsome 2008 & 2009).

Combining figures from the main colony at Marsden, the colonies on the south side of the Tyne, the other coastal sites and those at Teesside, the Atlas suggested a breeding population for the county of some 6,200 pairs in 1992 (Westerberg & Bowey 2000). Taking into account the subsequent decline of the main colony at Marsden and the increases in some of the smaller colonies in the following decade, this figure should be revised downwards to the range of 3,200 to 3,300 pairs by 2010.

Birds arrive at the nesting ledges at any time from late February, but mainly through April and into May. By early June the majority of birds are incubating and most young have fledged by mid-August. There is a trend towards later return in recent years. Coulson & White (1956) recorded that the annual return at Marsden started in mid-January, earlier than recorded by Temperley. Coulson (2011) noted that the time of return has become progressively later, so that by the first decade of the 21st century, the first birds often did not arrive until early March.

Although the Kittiwake population has declined nationally since the 1980s, the major loss has occurred in Scotland, particularly in Shetland. This is attributed to declines in abundance of sand eels, linked in some areas to rising surface sea temperatures due to climate change. Species such as Kittiwakes are particularly vulnerable to such changes as they can only take prey from near the surface of the sea, unlike diving species such as auks, which have access to a greater variety of prey in the water column (JNCC 2010). There was however, little evidence of a widespread decline in Kittiwakes in England up to 2005, and Kittiwakes have been successful in the county over the past eighty years, including during the more recent period of rising sea temperature (J. Coulson pers. comm.). Their progress in Durham has, as yet, only been checked by the local phenomenon of a sudden decline at Marsden in the late 1990s, as previously described.

Other factors linked to food availability are also likely to impact upon Kittiwakes, including industrial pollution in the coastal waters where breeding birds forage. Changes to breeding sites, including local impacts through the erosion of the limestone cliffs at Marsden and the changing uses of artificial structures, are also likely to influence the number of available nest sites annually (Westerberg & Bowey 2000).

Recent non-breeding status

Small numbers of birds, typically single-figure counts, are usually seen in coastal locations around January-February and at the year end, with favoured sites being South Shields, Seaham, Sunderland and Hartlepool, often around harbours and fish quays. Larger numbers have occurred at this time, sometimes in response to storms. An exceptional 10,000 were noted offshore at Hartlepool on 12 January 1978 (Bell 1979). The following year, up to 400 were storm-driven into the mouth of the River Wear in mid-February (Unwin & Sowerbutts 1980). Gatherings in the mouth of the Tyne were a regular feature in the early 1980s, including c.300 in February 1982 (Baldridge 1983). It is noticeable that all of these winter records are from the early 1980s or before. To date, there are no more records of comparable winter numbers. This may reflect the trend towards a later return of birds to breeding ledges, as noted previously.

Coastal movements tend to peak as birds return to breeding colonies in spring, and again with post-breeding dispersal in autumn. At Whitburn, it is sometimes difficult to determine how much of this is local movement involving

Marsden birds, but the highest counts clearly transcend this. High spring totals include 11,500 flying north on 28 May 2007 and 7,000 north in 7.6 hours on 12 May 2009. Autumn totals include 9,500 moving north in 6 hours on 19 September 2005 and at least 11,700 north in a full-day watch on 15 September 2009 (Newsome 2006, 2008 & 2010).

Large post-breeding flocks have also congregated on the coast, particularly at Teesmouth, during the late summer. This included 11,400 on the Tees Estuary WeBS count on 14 August 1988 (BOEE 1988). Elsewhere on the coast, during the 1970s and 1980s, flocks of over 2,000 were regular at South Shields in August and September, and included first-year birds, which are rarely seen in colonies, and probably many birds which had not yet bred (J. Coulson pers. comm.). Other high post-breeding counts have included 2,200 at Sunderland on 9 August 1981, 2,000 at Seaham in August 1986 and 2,000 at South Shields on 27 August 2000 (Baldridge 1982 & 1987, Armstrong 2005). Again, it is noticeable that these high counts are mainly from the 1980s or earlier, and there are no comparable post-breeding figures to date after the year 2000. Even the high South Shields count noted is exceptional, as the maximum North Tees WeBS count that year was just 93 in August (Iceton 2001). Lower post-breeding numbers may reflect the decline in the Scottish breeding population experienced by this species since the 1980s, as noted previously.

Kittiwakes are scarce but annual inland, with a few records each year of mostly single birds. They are reported in most months, but the period April-August, and particularly post-breeding dispersal in late summer, produces slightly more frequent records. No doubt the success of Kittiwakes nesting inland on the River Tyne is responsible for increased numbers of sightings further upstream, as far west as Ryton Willows. Inland records adjacent to the River Tyne are therefore more frequent, and in slightly higher numbers. At Shibdon Pond, it has become a scarce but annual summer visitor, with five in July 1985 (Bowey *et al.* 1993) and 12 present on 9 June 2006 (Newsome 2007b), plus occasional mid-winter records, mainly of first-year birds.

Away from the Tyne, regular inland records have come from Longnewton Reservoir, including three on 28 March 2006 (Newsome 2007b) and WWT Washington, with two or three between 22 July and 8 August 1993 (Armstrong 1994). Further inland, Derwent Reservoir has also amassed a few Kittiwake records, including two on 29 October 1978 (Unwin 1979). Storm-driven birds can also appear inland, as on 11 November 2007, when gales brought an adult and a juvenile to Hetton Lyons CP and another juvenile to nearby Sedgeletch. An unusual inland winter record concerned an adult found freshly dead at Highfield, Rowland's Gill, well away from rivers or wetlands, on 17 January 1995 (Armstrong 1997).

There is evidence of east-west movements of Kittiwakes in Durham, which may also explain some of the inland records. Examples include: a flock of 127 flying in from the west on the Tyne at Dunston, on 17 March 1996, before continuing east; four flying off high to the west from Shibdon Pond on 29 June 1998; and, a flock of 56 flying inland over the Long Drag, Teesmouth, on 22 February 1999 (Armstrong 1998 & 1999b, Blick 2009). All of these records, however, are from a short period from the mid- to late 1990s, there having been no further similar observations to date.

A wholly white leucistic Kittiwake was noted flying north past Whitburn on 13 August 2006, having been seen at Filey Brigg, North Yorkshire, the previous day (Newsome 2007b).

Distribution & movements

Nominate race Kittiwake breed around the coasts of the North Atlantic, and in the adjacent Arctic seas. The race *pollicaris* is found in the North Pacific. On the European side of the Atlantic, birds breed from Portugal in the south, to Svalbard in the far north. Wintering birds have a pelagic habit and a wide distribution across the Atlantic, young British birds routinely visiting Greenland and Newfoundland.

The earliest ringing recovery is of a bird found inland at Waskerley Reservoir in April 1941, which had been ringed near Murmansk, Russia on 28 July 1940 (*British Birds* 38: 13, 246). More recent records include the following recoveries of dead birds at Hartlepool. One ringed at Vindfarholmen in central Norway on 28 June 1978 was found on 8 October 1978, while another ringed at Cruden Bay, Grampian, in July 1979 was recovered in January 1980 (Blick 2009). One ringed as a nestling on the west coast of Iceland on 16 July 1987 was found on 8 June 1989 (Armstrong 1993). Another ringed as a nestling on Inchkeith, Firth of Forth, on 14 July 2000 was recovered on 26 May 2008 (Blick 2009).

There are also the following notable recoveries from birds ringed at Marsden. One ringed as a nestling on 25 June 1991 was found freshly dead at Houtribdijk, Isselmeerpolder, the Netherlands, on 29 January 1993. More surprisingly, an adult ringed on 17 June 1992 and seen again there at a nest the following summer, was shot at Narsak, Nuuk, Greenland on 18 September 1993. A large number of ringing recoveries of Kittiwakes have been made from Greenland as a result of birds that feed in the North Atlantic being shot there for food. Virtually all of these have been young birds; this was thought at the time to be only the second transatlantic recovery of an adult (Armstrong 1994).

At the Kittiwake Tower, a nestling colour-ringed on 10 July 2006 was found and photographed breeding at Boulogne, France on 8 April and 19 June 2011. Another ringed as a nestling on 3 July 2008 was found dead on the shore at North Berwick, Lothian, on 27 September the same year (Northumbria Ringing Group pers. comm.).

There are also a number of recoveries in the county of birds ringed on the Farne Islands. Most have been recovered on the coast between June and September, but a notable record far inland concerned one ringed as a nestling on 6 July 1960 and recovered at Wolsingham on 11 April 1966. Another ringed on 30 June 1958 was recovered at Jarrow on 8 May 1961. Southward dispersal from the Farnes in autumn is indicated by, for example, one ringed as a nestling on 14 July 1966 and recovered at Horden on 29 September the same year, and another ringed as a nestling on 6 July 1968, and recovered at Seaton Carew three weeks later on 27th.

These records suggest significant movements between breeding colonies, and also a predictable wide dispersal, as well as some southward movement, in autumn and winter.

Bonaparte's Gull
Chroicocephalus philadelphia

An extremely rare vagrant from North America: four records.

Bonaparte's Gull, Saltholme Pools, autumn 1977 (Tom Francis)

The first record of this North American vagrant for Durham occurred in 1977, when a moulting first-summer bird was discovered by Dave Britton at Saltholme Pools on 12 August (*British Birds* 71: 506). At a time when there had been fewer than 30 previous records, this individual was probably seen by more observers than any other Bonaparte's Gull in Britain (Britton 1978). It remained until 2 October, during which time it completed its moult into second-winter plumage. The next came ten years later, when Dave Foster found a winter-plumage adult feeding at the sewage outfall at Whitburn on 17 January 1987 (*British Birds* 81: 562). After being watched by several observers, the bird moved off and was not relocated.

The most recent example was a moulting adult photographed by Ray Scott at Whitburn Steel on 28 August 2010, but the bird's identity was not discovered until the photographs had been uploaded onto the internet. The bird was still in the area, being seen again at Whitburn Steel on 31 August, Seaham and Ryhope on 3rd and 4 September and again at Whitburn Steel on 11 September (*British Birds* 104: 586).

What was possibly the same adult was seen again at Whitburn in 2011. Initially found by Mark Newsome in ploughed fields by Lizard Lane on 5 August, it remained in the area until 11 September, occasionally feeding in fields between Cleadon and Whitburn and also resting on the beach as far south as Roker pier (*British Birds* in prep.). The bird was in almost full summer plumage when found, but had moulted into full winter pluamge by the time it left the area.

At least one other bird must have moved through the county unnoticed, as the long-staying adult at Chevington and Newbiggin, Northumberland, in September/October 2006 was seen at Saltburn, North Yorkshire two weeks after it had left Northumberland.

Distribution & movements

Bonaparte's Gull breeds widely across western Canada and Alaska. The wintering range covers rivers and lakes in the northern states of USA, and south to Mexico and the Caribbean. An infrequent vagrant to Britain up until the late 1970s, sightings of this species increased considerably over the period 1990 to 2010. A total of 178 had occurred to the end of 2010, with the late winter and early spring period being the most productive time of the year.

Black-headed Gull
Chroicocephalus ridibundus

An abundant and widespread winter visitor and passage migrant; also a locally common breeder.

Historical review

For generations, this species has been, for most people, the commonest 'seagull' and this would appear to have been its sustained status for many years in the north east of England. It is in Durham, as elsewhere, the main species that is observed following the plough at inland sites in the autumn and it is also inclined to feed in grass fields, regularly coming to the bread thrown by members of the public in parks and at ponds across the region.

Temperley (1951) called it an "*abundant winter visitor and passage migrant*", and a "*summer resident in varying numbers*". He referred to its natural variations in numbers over the previous 200 years, with the species being widespread in the 17th and 18th centuries, the severe declines that occurred in the 19th century being followed by re-colonisation of many parts of the county by the end of the 19th century and the beginning of the 20th. Apparently the species was not common in Durham during the early part of the 19th century, as it was not mentioned by a number of the early chroniclers of the county's birds, such as Hogg (1827); though the same author later called it "*not uncommon*" in his 1845 manuscript (Hogg 1845) dealing with the south east of the county, suggesting a marked change in status in a relatively short period of time. Hutchinson (1840) indicated that there were no known colonies but it occurred, though not abundantly, in other seasons. By Hancock's time of writing (1874), he said it was a resident, breeding in meres and loughs in "*all wild moorland districts*". The status of this species changed significantly in the Derwent valley between the late Victorian period, when Robson described it as "*an irregular visitant to the valley*" (Robson 1896), and the modern period. Gurney (1918) wrote of Durham, in his national survey summary, "*No regular gulleries are known*". Since that time, George Temperley (1951) recorded that many colonies had been documented in Durham, but that none were permanent and many moved site from year to year, according to conditions.

Notwithstanding Gurney's observation, it is known that in the first half of the 20th century, birds were recorded breeding at a number of locations primarily in the uplands, illustrating re-colonisation of the county. In 1918 or earlier, a colony was established at Sunniside Moss near Tow Law, with about 100 pairs there by the time of the British Trust for Ornithology's Survey in 1938. However, this was one of only two active colonies in the county discovered by surveyors that year, with the other at the Mines Reservoirs above Howden Burn, near Frosterley, holding just one occupied nest. The year 1938 was reportedly a poor year due to a very dry spring making several moorland locations unsuitable. This site had held 25 nests in 1937, illustrating the temporary nature of some moorland colonies. Previous upland sites also noted as holding breeding birds during the period between the two world wars but not in 1938 were on the moors of the Wear-Derwent watershed between Blanchland and Wolsingham, which was colonised about 1925; the old Mines Reservoirs near White Hill, off the Stanhope to Eggleston road, colonised in 1930 or earlier; and near Tunstall Reservoir, which was colonised by 70 to 80 pairs in 1937, but was vacant the following year (Temperley 1951). In the lowlands, a large colony was established at Priors' Close Bog, near West Rainton, but this site was drained in 1928. In addition, there were 40 nests in 1945 at a long-established site on moorland west of Edmundbyers, whilst a new site was also colonised, probably during the

481

late 1930s or early 1940s. Numbers quickly built up here to about 200 birds in 1944, following which the ponds were reclaimed. Those birds that did nest in 1945 lost their eggs to collectors and in 1946, none at all bred. In the south east of the county, a few pairs were noted breeding at the Darlington Sewage Farm. A few pairs were also recorded as nesting sporadically around the North Tees Marshes between 1930 and 1938 (Stead 1964, Blick 2009), but they were subject to much disturbance and did not attempt breeding here again until 1971 and 1979, but none were successful in either year. A survey by the Tyneside Bird Club was carried out in 1973 and found four colonies in County Durham, the smallest consisting of ten to twelve pairs and the largest up to 400 pairs. All of these were in moorland areas (Unwin 1974).

Temperley (1951) reported wide-ranging numbers of this species in the county through spring and autumn, feeding on agricultural land. He reported that it had special roosting locations in winter where large numbers could be seen, and these included Jarrow Slake and Teesmouth. In August 1962, a huge concentration of sprats *Sprattus sprattus* appeared in the Tees Bay and the estuary of the River Tees. This concentration of food attracted approximately 50,000 gulls, the majority of which, perhaps over half, were Black-headed Gulls.

Recent breeding status

Black-headed Gulls, as a breeding species, are surprisingly widely distributed in Durham. The main breeding sites remain on moorland bogs and by reservoirs in the uplands. Some of these traditional sites are used annually whilst at others, breeding is more sporadic. It has been noted that the numbers in breeding colonies can vary quite considerably from year to year (Kerr 2001). Some of the factors influencing such variations include the levels of predation experienced at the colony in the previous year, human disturbance and water levels (Kerr 2001). Occasionally, breeding sites are deserted for no apparent reason, as happened at Smiddyshaw Reservoir in 2002, where the 145 nests present in May had all been abandoned by June.

As most Black-headed Gulls nest in a relatively few discrete colonies, it should be possible to carry out an accurate census of the county's breeding population, but this has not been attempted in County Durham for many years. Despite the fact that this is a noisy and conspicuous bird when breeding, it is not known exactly how many extant breeding colonies there are in the county, although clearly there are many present in the western uplands and attached to both small and larger bodies of water. In 1983, 800 adult birds were present and presumably breeding on Widdybank Fell. The largest and best monitored colony was in Lunedale and generally held between 600 and 1,300 pairs during the decade from the mid-1980s to the mid-1990s, with possibly as many as 2,000 pairs in 1995. Of 1,200 nests here in 1993, as many as 500 were washed out by heavy rains in May, illustrating the precarious state of some of the breeding sites; too little water and birds may not settle to breed, too much and flooding follows. There is little detailed data for other sites, although up to 500 pairs were known to be breeding on Bowes Moor in 1993. During the 2000s, Grassholme Reservoir became abandoned as the county's largest breeding site and birds moved the relatively small distance to nearby Selset Reservoir. The *Atlas* (Westerberg & Bowey 2000) determined that the county population might be in the region of 2,000 pairs, or more, yet this may well be a gross under-estimate of the true number. Following the publication of the *Atlas*, several breeding colonies of up to 70 pairs were found in the west of the county during the early 2000s, although other moorland ones no doubt went unreported. In 2002 and 2003, seven upland locations held breeding colonies numbering over 400 pairs in total, with another 50 pairs at Hisehope Reservoir and 42 pairs at Fish Lake, to the south west of Barnard Castle. In total though, this was an unknown proportion of the actual numbers present in the county. In 2005, 150 pairs were in one mid-Wear valley colony at McNeil Bottoms, although birds had temporarily and completely abandoned this site for an unknown reason during 2000.

Whilst the bulk of the breeding population is upland in nature, smaller numbers nest in the lowlands. There is a well-established colony on two pools at Smallways, on either side of the A66 - only half of the colony lies within County Durham - although no recent estimates of the numbers here have been made. A rapidly increasing colony on the North Tees Marshes was established from 2004, although breeding was suspected in 2003 because of the presence of recently fledged juveniles. The tern islands at Saltholme attracted birds and, by 2006, the number of breeding birds had reached 20-30 pairs, producing 34 young (Joynt *et al.* 2008). A further increase followed with 66 pairs in 2008 (Blick 2009), whilst by 2010, a total of 191 nests were occupied, although breeding success did not appear to be high with 56 young fledged (Joynt 2010).

Recent non-breeding status

The European population has increased markedly since the mid-19th century, with an expansion of range into Iceland and Scandinavia in the north and a southward spread through much of France and into northern parts of Spain, Italy and Greece. In Britain, there was a dramatic increase in the 20th century following on from a low point in the 1880s when there was a fear of extinction (Cramp *et al.* 1983).

Today, non-breeding birds can be seen across the county throughout the year, though few adults are noted far away from their breeding colonies during the breeding season. Following the end of the breeding season, the first fledged juveniles make their way to lowland and coastal sites usually by the last few days of June, with a rapid build-up in numbers following. Hence, by late July, flocks of several hundred can be noted at lowland wetland sites and some coastal locations. Some of these birds may not be local breeders, but already have migrated from the western Baltic. On 11 August, a juvenile was captured in Durham that had been ringed in southern Norway (J. Coulson pers. comm.). During August, there is often a late summer peak and sometimes very large gatherings noted in the Tees estuary as some birds from Scandinavia and Eastern Europe arrive in the UK. Most years see counts of several thousand, although there are very occasionally much larger flocks; for example over 12,000 birds were present at Teesmouth on 14 August 1988 (Bell 1988)

By October, many wintering birds have arrived in the county and this familiar, common species is thereafter very widely distributed in all parts of the county to the early spring period. Gatherings of 200 or 300 birds are a routine sight at many minor wetlands, along river systems, at sewage treatment works and on short grass swards, such as urban playing fields. Favoured landfill sites and larger ponds and lakes may produce much larger counts often numbering many hundreds or even thousands of birds. Many birds roost on some of the inland reservoirs and wetlands, for example Longnewton Reservoir (Blick 2009) where 4,400 were present in 2010, but an exceptional count of some 15,000-17,000 birds was made on 7 October 1996. Further inland, the largest known wintering roost site is at Derwent Reservoir, which regularly attracts thousands of birds; 3,000-4,000 is the regular winter peak count here, usually in January, but it is not uncommon for numbers to reach 6,000-8,000 birds. Likewise, some of the other upland reservoirs such as Balderhead, Hury and Selset attract several thousand roosting birds at times. A particularly large count of 10,000 was on Hury Reservoir on 15 March 1980 which, given the date, no doubt included some birds moving through *en route* to breeding grounds.

There is also a large offshore roost in Trow Bay, South Shields, which regularly holds many thousands of birds. Peak counts here can occur from October through to March and it is tempting to speculate that those birds feeding nearer the coast tend to roost on the sea, whilst those inland head to the upland reservoirs. Some of the larger counts here include 15,000 in October 1983, 10,000 in January 1985 and 7,000 in February 1999. Favoured bathing localities that are close to landfill sites, such as Rainton Meadows and Shibdon Pond, regularly attract winter roost and pre-roost gatherings in excess of 1,000 birds, although up to 8,000 have been noted at the latter site and 6,200 at the former. These would probably include birds feeding at the Birtley, Greenside and Houghton landfill sites, where counts of up to 4,000 birds have been made. Whilst it is located in the Northumberland recording area, the large roost in the Scotswood area adjacent to the river Tyne draws many of its birds from the Durham side of the river, with birds drinking and bathing at Shibdon Pond after feeding at landfill sites in west Gateshead. The peak count from this location, courtesy of a long-term study, was of 21,000 birds in January 1997 (Kerr 2001).

Feeding birds also occur along the whole of the coastline although there are rarely any real big concentrations, but 6,000 fed on the shore in rough seas at Seaburn in January 1984 and there were 2,600 off Castle Eden Denemouth early in 1993. At Teesmouth, the Wetland Birds Survey maximum counts have shown that up to 4,000 birds gather in the autumn and winter periods (Blick 2009). Occasionally, much larger gatherings occur as in both January 1977 and 1978, when 11,000 to 12,000 birds were noted (Blick 2009). In an average winter, it is probable that well over 25,000 birds use Durham's coastline, inland waters and landfill sites between the months of August and March.

In some respects, this species seems more prone to extreme cold and perhaps other winter-related mortality factors than a number of other gull species. This may simply be a physiological result of its smaller size than its congeners. It is not unusual to find dead birds in winter and over 200 corpses were located between Hartlepool and Crimdon Dene in early January 1971, during an exceptionally cold spell (Blick 2009), although the cause of death was not determined. Occasionally, there are disturbing reports, such as that from Seaham Docks in winter 2001, where over 50 ill or dying birds were noted on the 19 November, and at least 20 were dead the following day. A similar report occurred on 13 December and on both occasions poisoning was suspected. In some instances, such poisoning is not directed at the birds but may result from the birds foraging on bacterially-infected food items. In any event, these factors have probably influenced many of the recoveries of ringed birds, which have revealed so much information about the origin of the county's wintering birds.

Among feeding birds, several techniques are employed. Walking and picking up items like earthworms is common on short grassy habitats, and this species is often seen following the plough. Over some areas, a slow searching flight one to two metres above the surface is employed with the feeding bird dipping down to pick up food items, whilst not infrequently birds search over hedges occasionally picking off berries or insects. Such feeding techniques have been noted at WWT Washington in the 1970s and 1980s and on 17 August 2006, when 40 were seen picking insects from tree tops in Bishop Auckland. Flocks are also noted flying high and catching swarming insects such as flying ants. Such feeding behaviour was noted on 21 July 2006, when approximately 500 birds were hawking insects with Swifts *Apus apus* over the South Shields ferry. The species also feeds on the surface of lakes or ponds in the manner of a phalarope *Phalaropus* sp., or up-ends like a Mallard *Anas platyrhynchos*, or occasionally surface-plunges to take food. In this manner, one bird was seen to plunge dive into the River Wear at Durham City and emerge with a four-inch long trout *Salmo* sp. on 11 May 1995 (Armstrong 1997).

This species often exhibits plumage abnormalities and individuals with a faint pink flush on their breast are quite regularly noted amongst winter gull flocks. Sometimes, this can be quite extreme and a bright pink bird was at Shibdon Pond on 14 January 1987 and another bird was at the same locality in the early 1990s (Bowey *et al.* 1993), whilst an adult seen at Charlton's Pond on 27 December 1992 likewise had the entire underparts and head suffused with a bright pink colouration. Records of leucistic or partly leucistic parts are also fairly frequently recorded, including birds with completely white wing tips such as that which wintered at Shibdon Pond in 1989/1990. All-white birds were noted at Longnewton Reservoir on 16 September 2002 and perhaps the same individual flew past Whitburn Observatory on 21 September 2002. More unusual was the melanistic individual, a dark sooty grey bird with paler carpal bars, which was on Seaton Common on 31 October 2006 (Blick 2009). An adult bird at Dorman's Pool on 29 May 2002 had a curved beak that was adjudged to be three or four times its normal length, another such bird was at Shibdon Pond during one winter in the late 1980s.

Distribution & movements

This species breeds across much of northern Europe and eastwards in a wide band through central Asia. Wintering birds are widely distributed across Europe and into North Africa.

During the five years 1909-1913, over 11,000 nestling Black-headed Gulls were ringed under the auspices of the British Birds Ringing Scheme, and many of these were marked at Ravenglass on the Cumberland coast. Of the total of 414 recoveries, thirteen came from Durham. Eleven of these were recovered within six months with the others recovered after 15 and 20 months respectively (*British Birds* 8:9, 209-213).

Over 7,000 birds ringed in Britain have been recovered abroad, with over 7,500 foreign-ringed birds recovered in the UK. Recoveries have shown that a large proportion of the birds present in Britain in the autumn and winter have bred or been hatched in northern and central Europe, particularly Norway, Sweden, Finland and Denmark, but also in Belgium, the Netherlands, northern Germany, and further east in Poland, Lithuania and Russia (Wernham *et al.* 2002).

In County Durham, birds have been recovered that support the national picture with recoveries of birds originally ringed in Denmark (3), Finland, Latvia, Norway (5), Russia and Sweden. Of those birds ringed within the county, recoveries have come from Denmark, Finland and Germany (3), whilst a bird ringed at the Saltholme breeding colony in July 2007 was sighted in Lugo, Spain, in November 2008 and found dead at Donnington, Shropshire, in November 2009 (Blick 2009). Three ringed at Teesmouth in September 1975, October 1981 and September 1989, were later found dead in Germany in May 1976, June 1982 and July 1990 respectively (Blick 2009).

A summary of some of the most intriguing ringing recoveries from, and to, Durham over more than a century (including one of the earliest recoveries of a ringed bird in the county) are listed below:

Ringed (Durham)	Date	Recovered	Date	Notes
Denton Fell, Cumberland	04/06/1910	Ryhope	12/09/1910	
Durisdeer, Dumfriesshire	26/06/1910	Croft Spa, near Darlington	07/02/1911	
Ravenglass, Cumberland	06/06/1912	Blaydon Race Course	14/09/1912	
Roskilde Fjord, near Frederikssund, Denmark	14/06/1922	Blaydon	21/11/1922	
Russia	not known	Teesmouth	02/09/1953	
Jokijärvi, Hauho, Finland	24/06/1962	West Hartlepool	03/02/1963	
Foteviken, Malmöhus, Sweden	21/06/1964	Greatham Creek, Teesmouth	18/01/1965	
Hognestad, near Stavanger, Norway	10/6/1965	Gateshead	-/01/1966	
Babite Lake, Latvia	03/06/1967	Dunston, Gateshead	14/01/1968	
Coxhoe Tip	02/12/1989	Vassa, Finland	15/07/1991	
Hett Hills	09/12/1982	Galveborg, Sweden	16/07/1991	1,255km the north east
Grassholme Reservoir	23/06/1991	Rodovre, Denmark	01/04/1993	
Stainmoor	01/07/1983	Lothian, Scotland	20/06/2000	17 years old

Little Gull
Hydrocoloeus minutus

A common passage visitor, mainly in late summer and autumn, but rare in winter.

Historical review

Temperley (1951) described the Little Gull as "*an autumn and winter visitor usually in very small numbers and of rather irregular occurrence*". He noted that its occurrence in the first half of the 20th century was more frequent than in the past and this trend may have presaged the events that followed in the 1950s and 1960s.

The species was not mentioned by many of the early chroniclers of bird life in the county, which may mean that it was either scarce or not readily recognised. John Hancock (1874) said though it was formerly considered a rare bird in the northern counties, it frequented the coast "*with considerable regularity*". He documented a number of birds that were shot in the 19th century, including two juveniles at Whitburn on 6th and 7 October 1847, one at Hebburn, an adult male, on 7 October 1849 and two winter-plumage birds at Whitburn on 26th and 27 December 1876. About 100 years ago, it was considered a passage migrant along the coast "*in small numbers only*" (Nelson 1907). The majority of the pre-1900s records occurred between September and March (Brown & Grice 2005). Temperley (1951) said that seldom a year passed without a few being reported, usually in the period October to December, and that these were mostly immature birds of one description or another, though an adult in winter plumage was at South Shields on 19 June 1929. This was the 'single bird' observed by C. N. Rollin over the River Tyne which had a dark and nearly complete hood (*The Vasculum* Vol. XV 1929). Clearly the species was very much less common in Durham in the past, when all sightings of the species were notable and annual totals numbered perhaps just a bird or two each year.

The status of this species along the north east coast, the Durham section of this in particular, began to change in the early 1950s, and this process of change accelerated through that decade and into the 1960s. Over this period, as the number of sightings increased almost year-on-year, record counts and observations exceeded each other with increasing regularity.

The 1950s started quietly for this species with a couple of autumn records at South Shields and Whitburn in 1953. This was followed by a prescient comment from George Temperley in 1955, following the occurrence of a second-year bird at Teesmouth from 10th to 18 July and a summer plumage adult at Teesmouth on 28th, that "*in no year had so many been seen in the north east*" (Temperley 1956). It was the following year when, in retrospect, the enormity of the change that was taking place can be seen to have commenced. In 1956, a party of 13, eight adults and five juveniles, was at Crimdon on 25 August, whilst on 29th, four immatures were at Teesmouth and these flew west up the estuary. Subsequently, flocks of up to 35 began to appear annually in autumn in Durham, both on the coast and at inland reservoirs (Hutchinson & Neath 1978). Over the next three years, numbers were out of all

485

proportion to previous years. In 1957, more were at Teesmouth than ever before with probably 22 there from 10 July to 26 September, including ten adults that were to the north at Crimdon on 26 August. In 1958 and 1959, autumn peak counts were respectively 19 about the Hartlepool to Teesmouth coast in autumn and probably five at Teesmouth during July. In 1958, 12 were at Hart Reservoir on 4 September and the following year, there were *"more than ever before"* (Grey 1960), with the then largest-ever passage along the Durham coast, in October, when 87 passed north off Hartlepool between 17th and 31st.

Numbers were lower in 1960, just one at Teesmouth between 17th and 31 July, but up to four were at Hart Reservoir in August, with birds at Teesmouth and Hartlepool in early October. It was in 1961 when the main change in status occurred. That year, birds were noted at Hurworth Burn Reservoir from mid-June, the main arrival there taking place from mid-July on four dates and, in August, more than 20 were counted. Birds were noted coming and going 'all day' on 3 August, and it was speculated that a roost was nearby, but it seems in the light of observations from later years, that birds were moving in from the north, the Durham coast, and leaving to the south west. On 30 July, there were as many as eleven at Hurworth Burn Reservoir, this number rising as high as 22 by 15 August, and up to 28 were recorded on 14 September. The maximum total of different birds passing through this site this year was 45 individuals, but there were almost certainly many more than this (Coulson 1962).

In 1962, Hurworth Burn Reservoir was drained through the summer so fewer Little Gulls were recorded, though up to 35 were present on 10 September. Bell (1963) speculated on the whereabouts of these Little Gulls prior to their appearance at Hurworth Burn; presumably they had been somewhere along the Durham coast. Through the rest of the decade, this pattern of occurrence persisted with birds appearing at Hurworth Burn Reservoir in late July and being noted intermittently through August and into early September, a regular passing through of birds being evident in most years, with birds arriving from the north and east and then departing to the west and south west. Peak counts in this period included 27 on 8 September 1963, at least 200, mainly juveniles, on 24 September 1964 and 18 on 18 September 1966. Elsewhere during the mid-1960s, Teesmouth attracted up to 16 birds during July and August, and 31 were off Hartlepool on 14 September 1968. During the summer of 1969, an unprecedented flock of first-year birds was present around the North Tees Marshes, rising from two birds on 11 May to a peak of 20 on 3 July. This was the commencement of what was to become a regular pattern of summering birds around the Tees Estuary, which persists to the present.

Through the early 1970s, these summering first-summer birds became a regular feature at Teesmouth and as the decade wore on, it was clear that this species was visiting the county with ever greater regularity. In 1973, birds were regularly around Teesmouth in summer and autumn, and a passage of up to 211 birds flying north was noted off Hartlepool on 19 September. The following year, as was becoming usual, birds were present in small numbers throughout the summer at Teesmouth, from May to September. The summer of 1975 saw a further movement of birds through Hurworth Burn Reservoir, 17 being noted there on 19 July, but more significantly this year brought the realisation that there was a considerable concentration of Little Gulls on a regular basis off the Durham coast, largely centred on Castle Eden Denemouth. In 1975, this build-up took place from the end of July and reached a peak of 103 birds on 30 August, the number declining during early September (Unwin 1979). A similar pattern of occurrence was reported in 1976, with up to 65 in July of that year, and these were mainly adult birds. In 1978, at least 150 birds were found, strung along the coast between Seaham Harbour, Horden and Castle Eden Denemouth. In 1979, the regular roost between Blackhall Rocks and Horden Colliery reached a new peak. A total of 158 there on 20 July was three times more than the previous best July count, and was followed by 226 on 1 August. At least 80% of these were adults. It was speculated at the time that one of the attractant features in this area was the combination of a sewage outfall and a coastal pool, formed by the damming of the Dene's stream by colliery waste, although such features were not scarce along the coast at that time.

Cleary, there had been a very significant growth in the numbers of this species in England, and the British Isles, post-Second World War, as outlined by Brown & Grice (2005). Hutchinson & Neath (1978) indicated that such an agglomeration of birds may have existed since the early 1960s, citing the passage of birds inland at Hurworth Burn Reservoir as a possible consequence of the coastal build-up. In reality, this build-up had probably been occurring, albeit at a somewhat smaller scale, since the mid-1950s, when flocks of up to 35 birds began to appear in autumn in Durham, both on the coast and at inland reservoirs (Temperley 1957). The increase in the number of passage birds recorded moving through Britain & Ireland since the 1970s far exceeds that which might be accounted for by the increase in seawatching over the same period (Wernham *et al.* 2002), but it does correspond in timing with the westwards expansion of the Baltic breeding population into Holland and Denmark (Hagemeijer & Blair 1997).

Recent Status

Through the early 1980s, the numbers of the coastal Little Gull flock remained high. As the decade wore on, the numbers rose but the flock became more mobile and dissipated. In 1980, up to 230 birds were at Castle Eden Denemouth on 16 August, whilst a count of 305 along the coast on 6 September, represented the largest flock of Little Gulls recorded in Durham at the time (Baldridge 1980), and only 23 of these were at the denemouth. Peak counts in subsequent years along this stretch of the Durham coast were: 158 at Castle Eden Denemouth on 20 August 1981; with 220 in the first half of August, the vast majority of these being adults; up to 150 on 28 August 1982; 137, mainly adults, on 11 September 1983; and, 70 at Castle Eden Denemouth on 16 July 1984.

In 1986, a large coastal gathering of 221 birds between Ryhope and Seaham on 26 July was presumably the Durham coast congregation that had drifted north from its usual area off Castle Eden Dene and, by the late 1980s, the focus for this species was swinging more towards Seaham. Indeed, the flock there reached a peak of 248 on 28 August 1988, then the second largest gathering in Durham. Subsequently, most of the large numbers noted tended to be in the Seaham area, with a likely factor being the sewage outfall, just south of the harbour which was a regular gathering point for birds.

From summer 1989, flocks of birds, often comprising exclusively adults, began to be noted at inland wetland sites, in a pattern that seemed analogous to that which occurred in the early 1960s at Hurworth Burn Reservoir. This commenced when a flock of 143 adults was found at Hetton Lyons CP on 16 July. Birds involved in this series of sightings often stayed for just brief periods after bathing and drinking, before heading back to the coast. Birds were noted at Hetton Lyons CP on a relatively regular basis over the next decade, including 75 there on 17 July 1994 and 26 on 1 August 1999. Other sites that attracted such flocks through the period included Murton Moor pool with 41 on 30 July 1996, and WWT Washington with peaks of 45 adults on 19 July 1996 and 61 on 10 July 1998.

Most reports from within the peak late summer period during the 1990s concerned birds around the North Tees Marshes and the Castle Eden Dene - Blackhall area. The numbers off the Durham coast grew to a peak around 1992, before entering into a steady decline through the rest of the decade. In 1992, the late August maximum occurred on 25th, when a group of 484 birds was in the Dawdon Blast Beach area, comprising four separate smaller flocks of 100, 109, 193 and 82; the largest gathering recorded in the county to that year. High numbers remained in 1993, when on 16 August, a total of 394 were between Seaham Harbour and Seaham Hall. A similar peak was reached in 1994 when 377 were between Sunderland and Seaham on 18 August. However, numbers declined from this point of the decade. Subsequent annual peak counts were: 100 at Castle Eden Denemouth, on 17 September 1995; 198 off Hartlepool, on 29 August 1996; 104 at Seaham, on 12 August 1997; and, 117 at Dawdon, on 16 August 1998. In 1999, when in relative terms there was said to be to be a "paucity of sightings", the peak count was just 32 at Castle Eden Denemouth in late July.

It was at one time, postulated that the Durham coast appeared to be a major migration staging post for this species after leaving their breeding grounds in northern Europe (Unwin 1979) and it may be that adult birds were using the sheltered mid-section of the Durham coast to undertake a post-breeding moult, prior to migrating to their wintering grounds, as suggested by Hutchinson & Neath (1978). During the 1970s, 1980s and 1990s, flocks generally formed, either on a beach, around a sewage outfall or on the sea, on a rising tide. Then on the approach of high tide, they would lift off and fly inland, i.e. westwards. With no coinciding reports of inland gatherings, it is not possible to say for certain whether they were setting off on the next stage of their migration, or simply going to an inland roost and that they might be back on the sea as the tide ebbed a few hours later (B. Unwin pers. comm.). This observation would be supported by the behaviour of birds visiting Hetton Lyons CP and other sites during the 1990s.

The numbers of birds noted in the county went into steep decline after the highs of the 1990s, from which they have never recovered. The appearance of birds through the first decade of the 21st century was more sporadic and, in some instances, in lower numbers than during the 1970s and 1980s. Birds did not stay as long but arrived in greater numbers, then moved on – largely first-year birds rather than adults, as in the 1970s and 1980s. Counts on the once-favoured central stretch of coastline included 41 off Seaham in 2001 and c.200 off there on 7 September 2004, but just five through the summer of 2006. The emphasis of the species' activity in the county seemed to have changed again from the over-summering, non-breeding or moulting flocks of the 1980s to a much larger but more ephemeral presence, based on passage along the county's coast. Sewage treatment improvements now mean the outfalls are no longer as attractive to gulls as in the past and the cleaning up of the Durham coast, from pit waste tipping, means that most cliff-base pools have now disappeared. These changes may be part of the reason why significant Little Gull gatherings between Seaham and Blackhall no longer occur. The only area to have a maintained, or possibly slightly increased, presence is the North Tees Marshes, where up to 32 birds continue to summer.

Apart from the aforementioned gatherings at favoured wetlands, inland sightings have always been relatively infrequent, with some years producing few or even no such records. As in the past, the closely allied Crookfoot and Hurworth Burn Reservoirs, still occasionally attract birds, but nowhere near the previous numbers. At Bowesfield Marsh, the highest recent count was of up to 16 in June and July 2003 (Blick 2009); perhaps these were some of the summering Teesmouth birds being slightly displaced. Other inland records have included up to eight birds at Barmston Pond, Bishop Middleham, Boldon Flats, Brasside Pond, Herrington CP, Houghton Gate, Longnewton Reservoir, Rainton Meadows, Seaton Pond, Sedgeletch and Shibdon Pond. Much more unusual were single birds at Derwent Reservoir in March 1988 (Kerr 2001) and at Low Barns NR from 5th to 8 December 2002.

During the 1980s, it was realised that birds were increasingly regularly noted moving off the coast between July and October, and this may have been the case since the 1960s. In some years, very few coastal passage birds are documented whilst in other years, large numbers are noted. Along the Durham coast, passage is usually recorded between late July and late October, although up to six birds have also been recorded passing land-based stations such as Whitburn Observatory and Hartlepool Headland during the winter period, particularly during periods of strong on-shore winds. Summer and autumn passage normally commences in July, and is dominated by adult birds. For example, passage in July 2007 consisted of mainly summer-plumaged adults from 11 July, with 73 birds being logged at the Whitburn Observatory during that month.

August and September are usually the peak months for passage along the Durham coast and 'birds of the year' often figure prominently in movements; the first juveniles are usually seen in the latter half of August or early September. Significant counts in the 1990s included 160 north past Horden on 24 September 1995 and 533 north in two hours past Hartlepool on 27 September 1995, whilst 300 were feeding offshore at Hartlepool on 27 September of the same year. A more recent measure of the scale of passage is indicated by counts at Whitburn Observatory in autumn 2007. During August of that year, a total of 530 birds were logged on 18 dates including peaks of 157 on 11th, 75 on 12th, and 110 on 29th. A total of 954 birds was logged over 24 dates in September, with as many as 144 being seen in one day.

Clearly, there is a strong dependency upon the prevailing weather conditions for high counts, and the largest movements tend to occur during strong northerly or north easterly winds. During the autumn of 2003, record numbers of Little Gulls were noted off the Yorkshire and north east coasts of England. Hartlepool Headland featured prominently in movements and day counts in early October included; 1,106 on 1st, 900 on 2nd, 3,891 on 3rd and 2,500 on 4th. Surprisingly, no more than 350 were noted at Whitburn in this period. Other high seawatch counts in recent years include a movement of 1,175 flying north off Whitburn on 15 October 1983 and 455 there on 8 October 2004. Other large day-counts at Hartlepool include: 711, on 23 September 1988; 655, on 27 September 1995; 840, on 16 September 2000; 750, on 23 September 2004; and, 585, on 8 November 2005 (Blick 2009).

A completely white bird passed Hartlepool on 3 September 2000 and one with a white mantle, rump and under-wing was on Saltholme Pools on 12 June 2003 (Blick 2009).

Recent breeding status

The summering of mainly one-year-old birds at Teesmouth was first noted in the county around 1970 (Blick 2009). In most springs, the first birds are seen in early to mid-April and numbers build up to a peak in late June or early July, these perhaps numbering 15-20 birds in most years. The largest number noted in Durham was the 42 birds that were present on 3 July 1986. During work on the *Atlas*, birds were recorded summering at Teesmouth in most years of the survey period. In particular, up to 18 were noted in July 1990, 12 in early July 1992 and 17 in June and July 1993. Subsequent to the *Atlas*, up to 32 were noted on the Tees Marshes on 25 July 1999 and 23 in the summer of 2008. Almost all these birds were first or second-summer individuals and, whilst they frequented areas in which breeding could have taken place, there was no suggestion of it having done so (Bell 1988-1993).

The change in status in Durham coincided with the extension westwards of the Baltic breeding population and the tentative colonisation of the British Isles, which culminated in four breeding attempts around England between 1975 and 1987 (Brown & Grice 2005). The higher numbers that have been recorded in locations such as Liverpool Bay in recent decades are a relatively recent development (Brown & Grice 2005).

The numbers of summering birds at Teesmouth declines as birds depart in late July and all have usually vacated the area by late August. Amongst this summer gathering, it is not unusual for a small number of summer-plumage adults to be present. It is not readily apparent why one-year old Little Gulls, and a few adults, should summer so consistently in the same area of Teesmouth for a period of over 40 years, without there being any suggestion of concerted attempts at breeding (Blick 2009). A lingering adult at Thorpe Farm, near Greta Bridge, in the Tees valley, from 23 May to 12 June 2008 at a small wetland site, was an exceptional and intriguing 'breeding season' record.

Distribution & movements

Little Gulls breed from the Baltic to western Siberia, with another population in eastern Siberia. Small numbers of birds also breed, intermittently, around the North American Great Lakes. Wintering birds are found offshore, off the coasts of southern and western Europe and around the Mediterranean. Small numbers are also regularly noted off north eastern North America.

The origin of the summering birds along the Durham coast is uncertain, but the supposition is that they come from the Scandinavian and Baltic breeding populations, the nearest breeding grounds to Britain being in Holland and Denmark (Hagemeijer & Blair, 1997). The only known ringing recovery for the county tends to confirm this assertion; a one-year bird was found dead at Easington on 10 August 1965 and had been ringed as a nestling at Oulu, Finland, on 4 July 1964.

Ross's Gull
Rhodostethia rosea

A rare vagrant from the Arctic: nine records.

The first record for County Durham was an adult that frequented the River Tyne, moving between North Shields and South Shields, between 24th and 31 December 1970 (*British Birds* 64: 354). This was, at the time, only the seventh for Britain; an occurrence noteworthy enough in itself, but throughout its stay, the bird shared the river with an Ivory Gull *Pagophila eburnea*. For such a rare bird historically, the 1970s brought three more to County Durham, a relatively large proportion of the national total of 16 birds in Britain during the decade. At the time, the North Sea fishing industry was centred on east coast ports such as North Shields and Hartlepool, which no doubt supported the appearance of birds in the county. The recent decline of this industry may have played a part in the subsequent downturn in occurrence after the mid-1990s.

Of the nine records in County Durham, three were mid-winter occurrences, four were in early spring and singles have been seen in mid-summer and late autumn. This broadly reflects the national pattern of occurrence, although the cluster of three records between 6th and 11 April is unexpected. The summer-plumaged adult which frequented Greatham Creek in June 1995 was presumably the bird that had been at Filey Brigg, North Yorkshire the day before and which was subsequently seen in Northumberland. This beautiful example was the last to be seen in the county.

Ross's Gull, Sunderland, February 1994 (Paul Cook)

All records:

1970	South Shields, adult, 24th to 31 December
1975	Seaton Carew, adult, 6th and 9 April (*British Birds* 69: 341)
1976	South Shields, adult, 9th to 11 April (*British Birds* 70: 425)
1976	Hartlepool Headland, adult, 7 May (*British Birds* 70: 425)
1983	Seaton Carew, adult, 9 February (*British Birds* 77: 534)
1992	Hartlepool Docks, adult, 11 October (*British Birds* 86: 490)
1994	Sunderland, adult, 26 February to 7 March (*British Birds* 88: 519)
1994	Hartlepool, first-summer, 11 April (*British Birds* 88: 519)
1995	Greatham Creek, adult, 12th to 27 June (*British Birds* 89: 505)

Distribution & movements

This enigmatic gull breeds locally on the tundra of north east Siberia, from the Lena River east to at least the Kolyma River. In Canada, it is rare and local in the western Hudson Bay region. Siberian birds migrate east in September to unknown wintering areas, which are assumed to lie near the edge of the pack ice, and perhaps in the Bering Sea or North Pacific. A total of 92 Ross's Gulls had occurred in Britain to the end of 2010. The north east of England, from East Yorkshire to Northumberland, has accounted for a comparatively high proportion of these, but occurrence rates have declined somewhat since the 1980s and early 1990s.

Laughing Gull
Larus atricilla

A rare vagrant from North America: six records, one involving a long-staying individual.

The first record of Laughing Gull for County Durham came in 1981, when Martin Blick located an adult, still mostly in summer plumage, on Seal Sands on the evening of 21 June. The bird was still present the next day and was seen by many observers. At the time, Laughing Gull was still a very rare vagrant to Britain: this record constituted the 21st occurrence.

Laughing Gull, Silksworth, February 1996 (Paul Cook)

The most famous Laughing Gull to visit the north east of England was the individual that spent over three years (albeit discontinuously) in the Tyne valley, between Blaydon and South Shields. Initially discovered as a first-winter bird at Shibdon Pond in January 1984, it frequented Newcastle General Hospital on the north side of the river Tyne for long periods, but also paid regular visits to sites on the County Durham side of the river. It often roosted with other gulls off South Shields, being seen in the pre-roost gathering on Sandhaven Beach on many occasions. It was last seen in County Durham on 19 February 1987, and was last observed in neighbouring Northumberland on 31 March 1987 (Kerr 2001).

The county's two most recent sightings occurred in December 2005 following Britain's best period for this species. Weather systems associated with the remains of Hurricane Wilma displaced large numbers of Laughing Gulls across the Atlantic with over 50 birds recorded in Britain. The two recorded in County Durham were notable for being part of only a handful of sightings in the east and north of England.

All records:

1981	Seal Sands, adult, 21st to 22 June 1981 (*British Birds* 76: 498)
1984	Blaydon, first-winter, 22nd to 25 January 1984 (*British Birds* 79: 552)
1984	South Shields, same as Blaydon, 29 February discontinuously to 9 March, 13 July and 21 December 1984 (*British Birds* 80: 541)
1985	Shibdon Pond, first seen in 1984, 25 January and 8 March 1985 (*British Birds* 79: 552)
1985	South Shields, first seen in 1984, 5th, 6th & 17 March, & 21st to 31 December 1985 (*British Birds* 80: 541)
1986	South Shields, first seen in 1984, 1 January discontinuously to 30 April, and 4 August to 31 December 1986 (*British Birds* 80: 541)
1986	Shibdon Pond, first seen in 1984, 29 November and 13 December 1986 (*British Birds* 80: 541)
1987	South Shields, first seen in 1984, 1 January to 24 March 1987 (*British Birds* 80: 541)
1987	Shibdon Pond, first seen in 1984, 24 January 1987 (*British Birds* 80: 541)
1996	Silksworth Lake, first-winter, 19 February to 12 March 1996, 5th to 11 April 1996 (*British Birds* 90: 483)
1996	Mowbray Park, first-winter, same as Silksworth, 21st to 22 February 1996 (*British Birds* 90: 483)
2003	North Tees Marshes, second-summer, 16th to 17 July 2003 (*British Birds* 97: 582)
2005	Shibdon Pond, second-winter, 4 December 2005 (*British Birds* 100: 46)
2005	Wellhouse Farm, Whitburn, adult, 14 December 2005 (*British Birds* 100: 717)

Distribution & movements

Laughing Gulls are a locally common breeding species in North America, from Nova Scotia south along the eastern seaboard of the USA, to Florida and the Gulf coast and in the Caribbean, Central America to northern Venezuela. The northern breeding populations migrate south to winter within the southern breeding population. Only 99 birds had been accepted in Britain to the end of 2004 with an even spread of occurrence across the year, but the 2005 occurrences substantially changed this pattern (Ahmad 2005); there had been 191 accepted records to the end of 2010.

Franklin's Gull

Larus pipixcan

An extremely rare vagrant from North America: two records.

The first record of this rare North American visitor, an adult in summer plumage, was found on 24 July 1977 by Martin Blick and Don Griss at North Gare (*British Birds* 71: 506). The bird lingered a short while before flying off to the west, and was not seen again. This was only the third to be seen in Britain following two birds in 1970.

A spate of sightings around Britain in 1991 brought the county's second record when John Dunnett discovered an adult at the Reclamation Pond on 19 June. It remained in the area for a further four days being seen at various locations around the North Tees Marshes and roosting each evening on the Reclamation Pond. After disappearing for two months, what was almost certainly the same bird was again at the Reclamation Pond between 28 August and 1 September 1991. What was presumed to be the same bird was seen in the gull roost at Longnewton Reservoir on 4 September.

Franklin's Gull, Reclamation Pond, June 1991 (Tom Francis)

Distribution & movements

Franklin's Gull breeds in the central provinces of Canada and adjacent states of the northern USA, migrating south to winter in the Caribbean, Peru, Chile, and Argentina. It is a rare visitor to north west Europe. The majority of European sightings have been in Britain and Ireland. To the end of 2010, there have been a total of 66 records in Britain since the first in 1970, with a distinct south westerly bias to occurrences.

Mediterranean Gull
Larus melanocephalus

**An uncommon but increasing winter and passage visitor, mainly in winter.
Occasional in the summer months and a likely future breeding species.**

Historical review

This species was unknown as a Durham bird at Temperley's time of writing, yet today it is an annual visitor in small numbers, since its first occurrences in the winter of 1955/1956. The first county record was of an adult which visited the South Shields ferry landing stage, during its wintering stay based at the North Shields Fish Quay. This bird was present in the area from around 12 November 1955 to at least 17 March 1956. Later that same year, the second county record came with an adult found at Hartlepool on 29 October that was then present to the year's end. This individual spent fifteen subsequent winters in the Hartlepool area, usually arriving in late July and remaining until late March. Assuming that this was the same bird throughout the period of observation, which seems self-evident based on its behaviour patterns and its favoured locations, it had reached at least 17 years of age by the time of its last observation in March 1971 (Blick 2009).

Very few additional birds were seen during the 1960s. In 1964, a sub-adult was at Seaton Snook from 8th to 19 August and later at Hartlepool and Graythorp Pond, while an immature was present at Hartlepool in March 1969.

Recent status

A distinct change in status took place through the early part of the 1970s. Although there had only been four individuals seen between 1955 and 1969, the year 1970 produced three 'new' birds in the Hartlepool area, with a further two seen in 1971, including a first-summer at Teesmouth intermittently between early May and early July. Reports in 1973 brought the first sightings at Crimdon, Seaham and Whitburn, contributing to perhaps six individuals during the year. After two or three new birds seen in 1974 and 1975, a new high was reached in 1976 as a total of approximately 15 different birds were seen; an unimaginable figure ten years earlier. These included an adult and a second-winter at South Shields and up to four together at Teesmouth in late June.

Through the rest of the 1970s and into the early 1980s, yearly totals varied between three and eight birds and several regularly wintering individuals became a feature, including one at Seaham Harbour between 1977 and 1982. A distinct change in occurrence pattern came in the mid-1980s as Mediterranean Gulls started to become noted year-round in Durham. Some 15 birds in 1983 included four sightings through the summer months, but this total was eclipsed by approximately 32 birds in 1984. South Shields became the centre of the species' activity, as at least six individuals were located in the first winter period, four through the high-summer months and five in the second winter period. A fresh juvenile at Dorman's Pool on 4 August was also the first of its kind to be seen in County Durham, whilst an adult at Whitburn in May of the following year constituted the first-ever sighting for that month.

A similar level of occurrence was evident through the remainder of the 1980s and during the 1990s. Year totals were difficult to be sure of due to interchange at favoured sites and likely relocation of birds along the coastline, but between 20 and 35 individuals were noted annually. Notable landmarks included up to ten different birds on the northern stretch of coastline in January 1987 and five birds together at Newburn on 28 February 1993. Inland sightings started to become more frequent as observers realised that Mediterranean Gulls might visit almost any water body in the east of the county. Occasional sightings came from Crookfoot and Longnewton Reservoirs, but these were followed in 1993 by birds at Brasside Ponds, Hetton Lyons CP and Rainton Meadows, while an adult appeared on the River Wear in the centre of Durham City in January 1994. A pair displaying to each other at Hartlepool Headland in both 1994 and 1995 intimated that birds were quite settled on this side of the North Sea.

An analysis of records in the west Gateshead area through this period sheds some light on the changing status of this species at inland localities in the county. Prior to the mid-1980s, this species had not been recorded in the area. From 1986 onwards, there were a number of sightings, primarily at Shibdon Pond but also elsewhere in the borough, and from that time the species was recorded annually. The first record for Gateshead was a first-winter bird on 8th and 15 March 1986, with second-winter birds then noted on the River Tyne at Felling and at Shibdon in late 1986. During January 1987, four individuals were at Shibdon Pond, three of them second-winter birds and one in first winter plumage. One of the second-winter birds, having only one leg, was christened 'Monopod' and this individual returned to winter at Shibdon every year up to the winter of 1993/1994 (Bowey et al. 1993). Further birds appeared at Shibdon and adjacent sites such as the MetroCentre Pools through subsequent winters, with up to three individuals noted in some years. The landfill site at Birtley also produced winter sightings between 1995 and 2001, hinting at where local birds may have been feeding. Passage birds were noted in early spring on a number of occasions, these often being first- summer birds, such as on the River Tyne at Felling on 11 April 1987 and at Shibdon Pond in April 1988, April 1989, March 2000 and March 2001. The only summer record was of an adult on the Tyne at Scotswood on 3 August 1989. In summary, at the very minimum, 21 different birds were involved over a 16-year period, with none having been recorded prior to 1986.

In line with national increases, the situation through the 2000s was one of further increases and new trends became established. The wintering situation through this decade remained much as in the 1990s, with the majority of records from the coastal strip and most falling into the two main geographical areas of the county; between the rivers Tyne and Wear in the north, and around Teesmouth and the North Tees Marshes in the south east. Sightings are however, dependent as much upon the distribution of observers as they are the on presence of the birds themselves. The highest counts for the county have reached no more than five birds together (at South Shields on 22 November 2006 and between Newburn and Seaton Common on 20 January 2007), but each winter, it is likely that 10-15 birds are present at the coast. Although this is an increase on the previous decade, County Durham still lags some way behind neighbouring Northumberland, where there are now regular autumn gatherings of over 20 birds at Newbiggin-by-the-Sea (Holliday 2011).

Today, returning birds re-appear in Durham en masse around mid-July, although numbers vary from year to year. Studies in the north of the county in 2006 showed that at least 15 different individuals frequented the coast and inland fields between South Shields and Seaburn during July, with a maximum of four birds together, including several juveniles. A similar situation was evident in this area in 2009, with at least eight birds seen between 18 July and 5 August, comprising at least six adults and two first-summer birds. These frequented the beaches at South Shields and Whitburn Steel and were also found feeding in arable fields at Boldon Flats/Cleadon and on South Shields Leas. There is annual evidence of a light autumn passage through the county, with a small turnover of birds at coastal watch points, such as Whitburn and Hartlepool, at this time of the year. In 2009, up to four were seen together at Whitburn Steel and on South Shields Leas in September and October, whilst a similar situation was evident in the south east of the county, with more regular sightings on the North Tees Marshes and in the Hartlepool/Seaton Carew area. The incidence of inland sightings also increases through this period as gatherings of the commoner gull species grow in size.

Whilst the coastline between South Shields and North Gare encompasses most of this species' favoured locations, birds have been seen at a growing number of inland localities over recent years, including many wetland locations that attract bathing gulls, particularly if these are located within easy flying distance of local landfill sites. Inland birds tend to be much less frequently noted than coastal individuals, suggesting that few remain site-faithful for long periods and utilise a number of localities in an area for feeding and bathing. More favoured sites through the 2000s include Boldon Flats, Hetton Lyons CP, Rainton Meadows, Shibdon Pond and WWT Washington, with an increasing number of sightings coming from the less established wetlands such as Castle Lake (Bishop Middleham) and Herrington. Most reports concern single birds, although up to three have been seen together at Hurworth Burn Reservoir and WWT Washington. Longnewton Reservoir is the most important inland water, with the gull roost there producing a regular turn-over of birds, particularly in winter. Between five and ten individuals has been an average yearly total through the 2000s, but 12 different birds were seen during January-March 2003 and a 29 different individuals between August and December 2003. The appearance of birds here is likely to be a function of observer effort and many inland birds must be overlooked amongst gull flocks on inland tips and arable areas.

Very few birds have been detected further west than the wetlands around Gateshead, Houghton-le-Spring and in central Durham. One at Willington on 29 January 2000 and two adults at Eastgate in spring 2007 were unusual, but occasional birds are likely to join in with the huge numbers of Common Gulls *Larus canus* roosting and feeding in Teesdale and Lunedale during the winter.

Recent breeding status

In the spring of 2007, a pair of summer-plumage adults was recorded in a Black-headed Gull *Chroicocephalus ridibundus* colony near Eastgate in Weardale. It is unknown whether breeding was attempted or, if it was, whether any such attempt was successful. One or two adults and second- summer birds have also been seen amongst the Black-headed Gulls in June-July at RSPB Saltholme, since the summer of 2008 (Blick 2009), whilst a fresh juvenile on Saltholme Pools on the early date of 29 June 2006 indicated that nesting may have taken place not too far away. Also of interest, an adult at Hetton Lyons CP from 30 March until 10 April 2000 was often seen displaying to Common Gulls during its stay.

Breeding first took place on Coquet Island, Northumberland, in 2009 and had occurred sporadically in Yorkshire prior to this, so it seems clear that if this species has not already bred in one of the county's Black-headed Gull colonies by 2010, then it probably will in the near future.

Distribution & movements

The Mediterranean Gull breeds almost entirely in Europe, mainly on the Black Sea coast of Ukraine, with a recent spread to the northern Caucasian Plains and Azerbaijan. It also breeds at scattered localities throughout Europe, including the Netherlands, southern France, Italy, Greece, Turkey, southern England, Belgium, Germany and Spain, with further range expansion being noted. It winters in the Mediterranean, the Black Sea, north west Europe and north west Africa. In Britain, at the north western limit of the species' world range, breeding first occurred as recently as 1968 on the south coast of England (Lloyd *et al.* 1991); the population currently numbers 600-700 pairs (JNCC).

An extensive colour-ringing scheme in various colonies in Europe has led to some interesting insights into the movement of birds. The most comprehensive in our area concerns a second-summer male colour-ringed (PAU2) at Paczkowski Reservoir in southern Poland on 7 May 2006. It was seen at: Crimdon Dene Beach in July 2006; South Shields in December 2006; Seaton Carew in January 2007; South Shields in October 2007; Seaton Carew in December 2007; Scalby Mills, North Yorkshire, in March 2008; Whitley Bay, Northumberland in October 2008; Seaton Carew in November and December 2008; North Shields in October 2009; Mietkow Reservoir, Poland, in May 2010; and Whitley Bay in October 2010. Studies of colour-ringed birds seen in Northumberland has shown that many seen in the north east have originated in north west European countries such as The Netherlands, Belgium and Germany, but also in more eastern countries such as Serbia.

Common Gull (Mew Gull)
Larus canus

An abundant passage and winter visitor, and occasional scarce breeder.

Historical review

Parslow suggested that the numbers of Common Gulls in Britain had increased in the 19th century (Parslow 1973), though gulls were poorly known at this time and it may be that the species was not that easily differentiated from others by many observers, with consequent errors in the documentary record of the species (Brown & Grice 2005).

There is almost no documented information about this species' status in the county prior to the outset of the 20th century. It would seem that it was a common and regular part of the area's avifauna at least in winter, as it was across much of the British Isles (Holloway 1996), but there is little evidence to assert this, especially during the 19th century or earlier.

In the middle of the 20th century, Temperley (1951) said this was "*a common winter visitor and passage migrant*", but he stated that the Common Gull did not nest in the north east of England; however that situation has changed since his time. He noted that non-breeding birds were often "*present in the summer*". Considerable flocks were reported to gather at the coast in harbours and throughout the winter. The species was also noted by Temperley to feed inland on fields, roosting on the sea at night. He noted the post-breeding passage of young birds along the coast in the autumn and that, in the spring, there was a northbound passage, which was "*more noticeable inland*", as it is today.

Old breeding records of Common Gull in the north east of England were documented by Bolam (1932), this occurring on the Farne Islands during the period 1910-1914. These were the first instances of documented breeding in England, but after this, breeding did not occur again in Northumberland until the late 1960s, on the south west moors of Northumberland (Kerr 2001). This neatly ties in with the development of breeding activity in nearby Durham, especially as it would appear that a number of these records may have been located over the county boundary into Durham (K. Baldridge pers. comm.).

Recent breeding status

Despite its name, this species is anything but a common breeding bird in England, with less than 100 pairs being noted as breeding in most years (Brown & Grice 2005). The species has a well-documented preference for upland areas in British mainland locations (Gibbons *et al.* 1993), consequently most breeding birds, though small in number, are to be found in the west of the county.

A small number of Common Gulls have nested in County Durham on the moorland adjacent to Northumberland since 1967, according to Brown & Grice (2005). Birds were confirmed as breeding in Durham in 1967, on moorland where breeding had been suspected in 1965 and 1966. In 1967, two pairs bred, one reared three young and the other two, in the Smiddyshaw Reservoir area in association with Black–headed Gull *Chroicocephalus ridibundus*. Birds had also been noted in association with Black–headed Gull colonies around Hisehope Reservoir at this time (K. Baldridge pers. comm.). In 1968, there were five nests, and four pairs bred in the area where birds bred in 1967, at least three, possibly five, young being fledged. Four pairs with three nests were found on the usual site in 1969. Birds continued breeding at this site to at least 1971, but none were noted here in 1973. One pair nested in 1974, but the young disappeared before fledging. Four pairs reared 10 young in 1976, whilst four pairs nested but only two hatched young in 1978. Birds have probably attempted to breed in Durham on an annual basis since then, with no more than six pairs at three localities in any one year.

Breeding has occurred on the Pennines from the north of Weardale, south to Bowes Moor. In this respect, one or two pairs of Common Gulls associated with the Black–headed Gull colony on Bowes Moor probably moved with the colony as it relocated northwards and in to Lunedale, at Grassholme Reservoir and, at present, to the shores of Selset Reservoir (K. Baldridge pers. comm.). Common Gulls tend to be associated with the edges of Black-headed Gull colonies in Durham and they are mercilessly harried by the resident 'Black-heads'. As a consequence, they may nest but without much success (K. Baldridge pers. comm.). Birds are very site-faithful and the dry site on open moorland, to the west of Consett, provided the bulk of the county's nesting reports from 1970 to 2010. The species has not been immune from human interference. In 1980 in the far south west of the county, three clutches of eggs were laid, and one of these was destroyed as was one of the adults, and one pair bred at Stainmore at this time (Mather 1986). Two pairs attempted to breed in 1982, but both nests were deserted, presumably due to human disturbance. Other, single pairs have attempted to breed amongst Black-headed Gulls on a number of occasions, such as in 1998, often beside small tarns or old mining reservoirs (Westerberg & Bowey 2000). By the mid-1980s, pairs began to be present with greater regularity in the south west of the county and one pair bred there in 1982, to the north of the A66 and west of Bowes. During 1986, breeding took place in at least three sites, two pairs were in the south west, one pair was in the north of the county - rearing three young - and one pair was in the north west, producing one fledged youngster. Through the 1990s, breeding was confirmed in some years but many of the possible breeding sites were not checked on an annual basis, so the actual situation is not fully documented. Nonetheless, in 1993, limited survey information suggested that five pairs bred in the county at two sites in the west. In 1997, the long-established site referred to above was no longer occupied. At that stage, this species' status as a breeding bird in the county was deemed tenuous, but published information suggested that no pairs were successful (Westerberg & Bowey 2000). However, it may be that in the latter part of that decade, birds were present in colonies in the south west of the county.

Through the 2000s, reports of breeding were intermittent, largely because of a lack of observer coverage, though one or two pairs attempted to breed in most years. After Black-headed Gulls had moved onto the Lunedale Reservoirs, Common Gulls followed them and one pair attempted to nest there, unsuccessfully, on the top of a dry-stone wall at Grassholme Reservoir for five consecutive years. Three successful pairs were present in the county in 2004, and in 2006, there were five pairs at one site in Weardale, which raised two young. Birds at this now regularly occupied quarry site sometimes nest on the cliff-type faces of the old quarry or in amongst boulders at the base of the cliffs on the westerly pool side of the site (K. Baldridge pers. comm.). By 2005, the species was felt to be 'just about holding on' as a breeding bird in Durham, though that year was typical in that, at the only known breeding site, four pairs settled but it was believed that only two pairs progressed to the egg laying stage, and no young were known to have been reared. In 2007, breeding took place at two sites in the west of the county. At one site, two pairs fledged six young and at the other, two pairs fledged one. The peak number breeding occurred in 2008, when up to six pairs bred, though in the previous year, one often-occupied site held two pairs, producing six young and another site had two pairs with young. In 2010, nesting birds were noted at three locations, amounting to between six and eight pairs, but as ever, success was variable; this species is something of sporadic breeder in Durham, at least in terms of producing young. Today, the county's breeding population amounts to barely a handful of pairs and all of those that attempt to breed do so in the Pennine uplands. Lack of observer coverage, and the scale of the landscape and available habitat, however, means that some breeding attempts probably go unrecorded. At the moment, there is little evidence of expansion to other sites, as occurred in neighbouring Northumberland, where lowland open-cast and gravel extraction sites have both been utilised, with up to 23 pairs there in 1993 (Kerr 1994), although this habit has subsequently declined in that county.

Nationally, there has been a decline in the number of inland breeding sites for Common Gull in Scotland (Gibbons *et al.* 1993). Nesting Common Gulls in exposed upland places in the county are prone to periods of cold, wet weather particularly early in the breeding season. Over a number of years, the principal breeding sites in Durham have suffered from a combination of human disturbance and predation from larger gull species, in particular Lesser Black-backed Gulls *Larus fuscus* (Westerberg & Bowey 2000).

Recent non-breeding status

During the summer, it is a relatively uncommon species, though a small number of immature birds might be present at sites along the rivers or local wetlands around the county. These largely comprise one-year-old birds, which spend the summer drifting around the county, commuting between inland wetland areas and the coast.

Common Gull is a very common winter visitor to the UK, with an estimated 700,000 wintering birds in Britain (Parkin & Knox 2010) and numbers appear to have been increasing in England over the second half of the 20th century, with a marked increase in the number of birds wintering inland (Brown & Grice 2005). Although frequently outnumbered by some other gulls, this can be the commonest gull around river mouths in Durham. Over the years, the Wetland Birds Survey (WeBS) counts have provided the most reliable monitor of numbers of this species around some of the county's main wetland sites and estuaries. The highest counts of this species usually occur in January and February and number in the tens of thousands.

Birds can be seen across much of the county during the winter, with roving and scattered winter flocks usually numbering 300-500 birds, occasionally in the low thousands, especially when visiting landfill sites. Common Gulls often gather at the same roost sites as Black-heads Gulls, but not normally in such large numbers. They feed alongside other gull species, but on average seem to spend a greater proportion of their time searching pastures for invertebrates. They return regularly to traditional feeding pasture, often in the western half of the county, and many such sites attract gatherings of up to 200 or more birds. Winter birds more often roost at the coast, but not exclusively so. During the winter, many wetlands are visited by large numbers of Common Gulls on a daily basis, particularly as birds drink and bathe before joining into flocks to travel to night-time roosts. Birds often arrive *en masse* and spend short periods of time at the respective wetland before heading off to the main winter roosts on inland reservoirs. Such flights of birds create spectacular ribbons of birds as dusk approaches, and some of the larger roost sites, such as Derwent Reservoir, hold many thousands of roosting birds. Especially when the spring passage coincides with the tail-end of the wintering population, truly huge numbers of birds can build up in westerly locations, for example the roost count of 20,300 birds at Balderhead Reservoir on 13 March 2007.

In 1978, there were some huge gatherings that included 13,000 birds at Derwent Reservoir in January and 25,950 at Teesmouth at the same time, suggesting a wintering population in the county of well over 40,000 birds at the time. A count of 11,000 at South Shields the following month may have been in addition to these, or part of the movement of birds through from further south (Unwin 1979). From the 1970s through to the 2000s, the county's principle winter roost site for the species was at Derwent Reservoir, with many thousands of birds present there on a nightly basis between November and late February each winter. Particularly large counts occurred here: in December 1982, when 12,000 were present; in November and December 1985 when 15,000 were there; and, in January and February 1989, when up to 14,000 were present. The largest ever gathering in the county occurred in January 1998, when 48,000 birds were counted at this site (Kerr 2001). The largest numbers counted at Teesmouth were the 12,000 present in February 1977, the congregation of 15,000 that occurred in August 1962 and an estimated 34,000 on 22 January 1978 (Blick 2009).

Large winter roosts have also occurred, for many decades, on the sea off South Shields, with many thousands of birds sometimes in the Tyne Estuary area during the first quarter of the year. For example, a gathering of 10,000 birds was in Whitburn Bay over 19th to 21 January 1976. These figures are typical of the number of wintering birds that were noted in this area through the 1970s and 1980s. It may be that a slight southerly shift of some of these birds was responsible for the huge roost of Common and Black-headed Gulls that developed in Sunderland Docks in the early 1990s, though up to 3,000 were still roosting at South Shields in winter period 1999.

The totals of wintering birds usually decline during late February and early March, only to increase again with the influx of passage birds heading north through the county. Spring movements through the north east are substantial and these often commence in late February. The diminution of the county's large winter population is followed by a rapid passage of birds through the area. A huge overland passage movement, numbering scores - if not hundreds - of thousands of birds, occurs through Durham and Northumberland in late March and early April. At this time, birds occur inland on open countryside, often feeding on fields during the day and roosting on reservoirs and inland waters at night. The roosts that develop in association with this movement are considerable. For example, the spring passage as observed from flocks in fields near Wolsingham in March 2001, numbered 9,000 birds as they paused on their way to more northerly breeding grounds. The spring passage of birds through the county in 1998 brought counts of 1,100 at Longnewton Reservoir, 2,000 at Smiddyshaw Reservoir and a mixed flock of 16,600 gulls, which largely comprised this species, at Balderhead Reservoir. On 24 March 2002, 25,500 smaller gulls were roosting on Balderhead Reservoir, and around 90% of these were thought to be Common Gulls (Newsome 2007). At this time of the year, coastal passage is usually noted at Whitburn Observatory but this is relatively light, compared to the inland movements, although a count of 3,440 along the coast between Blackhall and Castle Eden Denemouth on 21 March 1993 almost certainly involved northbound passage birds.

Returning birds can re-appear from late June or early July, these probably consisting of post-breeding birds from northern Britain or southern Scandinavia, at least initially. For a number of years, adult Common Gulls have been observed as returning to western Durham in late summer. Flocks of 30-40 birds can be seen feeding in newly cut hay fields in upper Teesdale and other westerly sites from the last week of July, with 300-500 birds present in such locations by the beginning of August (K. Baldridge pers. comm.). It would appear that these have arrived to moult in Britain after breeding in northern Europe. More concerted returns across the county usually bring the arrival of the first juveniles along the county's coastline from early August. One bird caught in east Durham on 11 August had been ringed in southern Norway as a chick, presumably in June or July; an example of how quickly such birds can move south. The majority of early returning coastal birds comprise adults, these being joined by ever more 'birds of the year', as early autumn arrives. Larger build-ups take place in September and on through the autumn, with many birds in roosts by mid-October, and a further late autumn influx of birds in November (Kerr 2001) consisting, presumably of bird from further afield. The autumn counts from Teesmouth in 1984 well illustrate this progressive build-up in numbers. In mid-September of that year, 700 birds were gathered, and counts rose to 5,000 in mid- October, 10,000 in November and 12,000 in December.

Leucistic birds were at South Shields on 31 January 1984 and Teesmouth on 24 April 1985. There was a most unusual observation of a bird in full juvenile plumage at Shibdon Pond over 20th to 23 February 1994; this species has normally undertaken its full post-juvenile moult by late September (Bowey et al. 1995).

Distribution & movements

This species breeds throughout the boreal and temperate zones of Eurasia, from Ireland in the west to the Pacific and, in different racial (possibly specific) forms, in Alaska and north west Canada. It is principally a migratory species, wintering in Western Europe, the Mediterranean, the Middle East and eastern Asia, as well as western North America. In some areas of Europe, Britain being one of these, birds are present throughout the year, though the winter and summer populations probably contain different groups of birds (Parkin & Knox 2010).

It is presumed that most of the wintering birds in the county originate from more northerly parts of Europe and this is certainly borne out by the national ringing data (Wernham *et al.* 2002). The degree of severity of winter weather in Norway and Sweden would appear to be important in determining the numbers of birds wintering along the Durham and north east coast more generally.

Despite capturing several thousand gulls at landfill sites for marking, Durham University caught only five Common Gulls. It was clear that they were rarely attracted to landfill sites where the cannon-netting took place, contrasting with the behaviour of other gulls (J. Coulson pers. comm.). There have been many birds ringed in northern Europe and recovered along the east coast of Britain, including Durham, over a number of decades. These included seven birds ringed in breeding colonies as nestlings, such as: one from Denmark in June 1928; birds from Norway in July 1963, July 1988 and June 1991; one from near Pori, Finland, in June 1971; a bird from the Murmansk region of Russia in June 1982; and one from Yell, Shetland, in July 1993. Most of these were recovered in the year after ringing in the Hartlepool-Teesmouth area (Blick 2009). Amongst other interesting recoveries illustrating this north to south movement were:

- A bird ringed in Finland in June 1931 found in Durham in February 1933
- One ringed at Krokane, Kinn, Sogn og Fjordane, Norway, on 7 July 1963, which was recovered at West Hartlepool on 3 August 1964
- A bird ringed as a nestling at Vest Agder, Norway, on 18 July 1978, controlled at Whickham on 17 February 1979
- One ringed in a colony at Serikstad, Rogaland, Norway on 1 June 1979, controlled at Greenside Tip on 2 February 1980
- One ringed in Norway in July 1985, found dead at Teesmouth in December 1990.

Two Common Gulls ringed by Durham University at landfill sites in the winter were later recovered in Finland during the breeding season (J. Coulson pers. comm.). In addition, there are three recoveries of birds that were already ringed when they were found wintering in County Durham, which were later recovered in Norway and Finland, demonstrating that birds return north to breed.

Ring-billed Gull
Larus delawarensis

A rare vagrant from North America: 17 records.

The first record of this species in County Durham came from Teesmouth, when John Dunnett, Tom Francis and Geoff Iceton found an adult on the Long Drag pools on 20 April 1982 (*British Birds* 77: 531). Although sightings had been rapidly increasing of this species first recorded in Britain in 1973, especially in south west England through the late 1970s, it was still a rare bird in the north east. Yorkshire had its first in 1978 (Mather 1986), with Northumberland following in 1985 (Kerr 2001).

Only two other birds were seen in Durham during the 1980s, but a distinct upturn came in the following two decades with totals of seven in the 1990s and eight in the 2000s. Some of these may have been attributable to returning birds, such as that at Billingham during the period 2002 to 2005. Records have come in all months of the year except November. A fairly even spread of sightings up to 2001 showed no real pattern of occurrence, but the regular appearance of late winter birds in the Billingham and Stockton area in the first half of the 2000s formed a distinct February/March peak. Despite sightings seemingly gathering pace, the lack of records since 2005 has seen a reversion in its status and at the end of the first decade of the 21st century, it could still be considered as a rare visitor to the county.

All records:

1982	Long Drag, adult, 20 April
1986	Whitburn, first-winter, 12 December (*British Birds* 82: 528)
1987	Boldon Flats, second-winter, 3 February (*British Birds* 82: 529)
1990	North Tees Marshes, first-summer, 24 June to 15 July
1991	North Tees Marshes, second-summer, 5th to 13 May
1992	North Tees Marshes, adult, 9th to 26 September
1994	South Shields, adult, 9 March
1994	Shibdon Pond and then in Gateshead area, second-winter, 24 August intermittently to 27 February 1995
1995	Seaton Snook, second-summer, 19th to 20 July
1995	North Tees Marshes, first-winter, 28 October
2001	Stockton-on-Tees, second-winter, 16th to 19 January
2001	Portrack Marsh, second-winter, same at Stockton, 13th to 19 February
2002	Billingham Tech. Park, adult, 19 February to 26 March
2002	Billingham Tech. Park, second adult, 12th to 14 March
2003	Billingham Tech. Park, adult, 18 January
2005	Billingham Tech. Park, adult, 21st to 28 February
2005	South Shields, adult, 22 February
2005	Longnewton Reservoir, adult, same as Billingham, 2 March
2005	Billingham Tech. Park, adult (different from February adult), 13 March
2005	Billingham Tech. Park, adult (one of the previous two), 23 March

Distribution & movements

Ring-billed Gull is a widespread breeding species in North America, wintering south to northern Central America. An increase in both occurrence and understanding of the species' identification features led to a huge rise in sightings through the 1970s and 1980s. The species was no longer considered by BBRC after 1987, and attempted to breed (in a mixed pairing with a Common Gull *Larus canus*) in Scotland in 2009 (Holling 2011). Winter is the peak for sightings in the UK, with the south and west of the country being favoured.

Lesser Black-backed Gull
Larus fuscus

A common summer visitor, increasing as a breeding species, and a scarce but increasing wintering species.

Historical review

Prior to Temperley's summary of this species (Temperley 1951), there is little documented information referring to this species' presence or status in the county, which does not mean that it was absent. It would appear to have always occurred, largely at coastal locations in Durham. Of birds of the race *graellsii* which breed in Britain, Temperley (1951) said that this was a "*summer visitor, though not normally breeding in the County*" and he also observed that it was a regular passage migrant. Such birds arrived in March and April at the coast, staying until September or October; as they do today. He reported at that time that it was not usually present in the winter, though late birds were occasionally noted in November. In terms of range in the county, he said that it was widely distributed along the Durham coast, along rivers and even on to the moorlands of the west of the county (Temperley 1951).

Towards the middle of the 20th century, birds were noted in the Herring Gull *Larus argentatus* colony on Marsden Rock during the 1940s, though breeding was not confirmed here until 1949, when a nest with three young was on the summit of Marsden Rock. Temperley (1951) speculated that it may have bred along the Durham coast in "*times before industrialisation*". Temperley could find no records of moorland colonies, but postulated that this might have been occurring as it was taking place in Northumberland at the time (Kerr 2001). In fact, there was evidence of this having occurred on a moorland site on Edmundbyers Common in 1948, the year before the nest was found on Marsden Rock, when a nest with two egg shells was found there (Temperley 1949). The birds were present the following year at this site, but no nest was found.

Through the 1960s, this was considered a rather scarce and irregular bird in County Durham. Regular visits to the Newcastle/Gateshead Quays area in 1964 revealed the presence of this species between 19 February and 4 November. During winter 1964/1965, there were occasional winter reports from South Shields and Marsden, but four at Derwent Reservoir on 25 January 1969 was most unusual, as at this time it was then a decidedly scarce winter visitor. Towards the extreme southern edge of the county, there was a breeding colony at Tan Hill, at the head of Arkengarthdale (then in Yorkshire) of 20 pairs in 1960, but these later moved over the border to Westmorland. In 1963, a few pairs were found on 'either side of the border'; 40 nests were on the Yorkshire (later the Durham) side in 1968 and 1969. Towards the end of the decade, in 1968, a pair nested on a stack at Marsden Bay, South Shields, where the last Durham nest had been in 1962. This established a pattern that would persist in the north east of the county for the next two decades and what was presumed to be the same pair was present through to the late 1970s. Through much of the first part of the decade, they were thought to be the only breeding pair in the county, though a pair was suspected of breeding on a building in South Shields in 1971. Adults, seen well inland during the summer of 1974, were believed to have come from the Stainmore colony that was technically at that time in Yorkshire, but now in Durham County. At least 300 birds were at the gullery on 30 May 1976. This was spread over 0.8km over a flat-topped moor at Bog Moss, to the south of the A66 (Mather 1986), with at least 60 in this colony in late June 1978; less than one fifth of the number noted by the same observer in 1969 (Unwin 1979). Roof-top nesting in Durham was first confirmed in 1976, with two pairs in South Shields and three pairs in Sunderland (Unwin 1977), and this had spread to Gateshead by 1978, and Hartlepool by 1979. In this latter year, 146 had gathered at Teesmouth on 9 July which was then considered to be a very large number for Durham.

Recent breeding status

As a result of a process that started in the early part of the mid-20th century, increasing numbers of pairs are now nesting each year in the county. Today, the species can be found across the county during the summer, its breeding range split between the urban centres and a small number of pairs in both upland and coastal areas. The majority of breeding birds are found nesting alongside Herring Gulls *Larus argentatus* on buildings in urban Tyneside, Wearside and Teesside, and in selected inland industrial areas. A few pairs do occasionally nest in more traditional sites at Marsden Bay and on the moorland of the west of the county, not infrequently in this latter context, in association with Black-headed Gulls *Chroicocephalus ridibundus*. Records from other sites during the breeding season are usually of passage birds or non-breeders. The majority of mid-summer records come from the eastern half of the county, although birds are regularly noted in Derwent Reservoir area and near Barnard Castle.

Through the summer months, birds are usually widespread and relatively common in the east of the county and industrial estate rooftops are now often used as nesting sites. Small numbers arrive back on their breeding sites in the main inland colonies from early March. The fact that most of the county's nesting Lesser Black-backed Gulls are now found on buildings in urban areas means that they are exposed to the control programmes operated by some local authorities in an attempt to reduce the numbers of gulls (mainly Herring Gulls) that nest in towns. The numbers of moorland nesting birds are also limited, as they are perceived as a potential threat to nesting grouse. The start of the roof-top nesting habit cannot be pinpointed, but the colonisation of buildings in the Teesmouth area commenced in 1979, when a pair reared young on a factory roof in the north of Hartlepool (Bell 1980). From this location, birds spread to a range of other buildings in the area, numbers rising through the 1980s and early 1990s. Birds certainly bred in Sunderland City centre in the mid- to late 1980s and through the 1990s, spreading along the River Wear industrial corridor as far upstream as Southwick (C. Gibbins pers. comm.). A pair laid eggs on the Brinefields, Teesmouth, in 1987, but failed to raise any young and at least two pairs nested at Hartlepool Docks in 1990 and 1991 (Blick 2009). In 1994, birds were found breeding at Jarrow alongside the Tyne and up to five pairs bred on the ground in Sunderland Docks during the early 1990s. The national survey of gulls nesting on buildings undertaken in 1994 found just 39 pairs of roof-nesting Lesser Black-backed Gulls in the county (Raven & Coulson 1997). Today, many urban sites within ten miles of the coast, or the estuarine section of the main rivers, hold breeding birds, including Gateshead, Peterlee, South Shields and Sunderland in the north of the county. At Teesside, by 1997, the largest colonies were to the south of the docks. None of the mobile Hartlepool colonies has ever exceeded 25 pairs. One or two pairs have attempted to breed at Longnewton Reservoir since 1994 (Blick

2009), with two pairs nesting there in 1996 and one pair bred at the Tees barrage in 1998. The number of pairs in Hartlepool had increased to at least 22 by 1997 and there were between 20 and 30 pairs at Portrack in 1998, with 47 pairs there in 1999 (Blick 2009). Seven pairs were found on a factory at Billingham in 1998, numbers rising to 54 pairs by 2004. Overall, there are probably some 150 nesting pairs in and around Teesmouth (Joynt *et al.* 2008), though it is possible that small numbers could be nesting unseen on any of the many warehouses and industrial complexes in the area (Blick 2009).

Along the Tyne, by the summer of 2009, very considerable numbers, perhaps 35 pairs, were breeding on roof-tops in the Chain Bridge Industrial Estate at Blaydon, in west Gateshead. This phenomenon was first suspected in the early to mid-1990s, probably first occurring on, and spreading from, the roof-tops of the nearby armaments factory on the Newcastle side of the river, which has been the site of a very considerable mixed species winter gull roost for many years (Kerr 2001). A similar inland colony was established at the Belmont Industrial Estate in Durham City during the early part of the 21st century, with tens of birds present there by 2002, and at least five pairs were at the Dragonville Industrial Estate, Durham City, by the middle of the first decade of the 21st century. In 2006, there were over 200 nests at the Darlington Industrial Estate, a reduction from the more than 300 nests recorded there in 2005 due, it was thought, to painting, which discouraged breeding. Other roof-top colonies are known at Low Willington in the Wear valley (with 34 adults there in summer 2007) and one or two pairs on the Caterpillar Factory in Peterlee. Small amounts of breeding activity may still also occur in the uplands for, in 2004, single pairs were known to have bred in moorland Black-headed Gull colonies, and four were in the gull colony at Selset Reservoir in summer 2009.

Westerberg & Bowey (2000) suggested a population estimate not exceeding 75 pairs at the mid-1990s. Clearly, there has been a very substantial increase in the county's breeding population since then, and the number today is probably in excess of 450 pairs, and possibly exceeds 600 pairs. Despite this, there have been some local declines, for example just three pairs were found nesting in Sunderland City in 2004 as part of a gull survey undertaken in that area (J. Coulson pers. comm.), where tens of pairs had nested 15 years previously.

Recent non-breeding status

This species is still a scarce wintering bird in the county, with only a handful of birds being noted in most winters during December, January and February. During much of the early part of the 21st century, the number of wintering birds in the north of the county was probably in single figures. Across the county, records of one or two birds tend to come from 10 to 12 widespread localities over the winter months. Up to five were seen in the Trow Rocks gull roost in some winter periods, and inland birds visit a range of wetland sites, including Barmston Pond, Shibdon Pond and Watergate Park, amongst others. Even this number is a considerable increase over two to three decades ago, when one or two birds at most would be seen through the winter, and the phenomenon of local overwintering did not become regular until the late 1970s or early 1980s. Favourite winter haunts tend to be amongst other large gulls, often along the river corridors a little way inland. They visit landfill sites and similar locations on or around the county's main rivers and estuaries. In 2008, records of one or two birds came from 18 widespread localities across the county, but none from the coast; this was probably the highest number of wintering birds on record. During most winters of the 2000s, between four and twelve individuals were probably present. Most sightings came from wetlands in central and inland locations and the number of birds increased as the decade wore on.

This is one of the earliest migrants to arrive in the county, with the first birds of the spring being noted from early February. A trickle of migrants joins the small number of locally wintering birds, with greater numbers moving in through March. For example, summer arrival at South Shields in 2001 started with 15 birds in February, increasing to 63 by the end of March. A more noticeable arrival takes place anytime between mid-February and early March with counts of ten to fifteen birds coming from some sites as more immigrants move in; birds then become noticeably more widespread from early April onwards. The numbers in the county rise through the latter part of April, though passage movements are rarely large in Durham. Coastal passage in the county is usually light but is occasionally noted off Whitburn Observatory, perhaps involving 50-100 individuals in total in spring. Nonetheless, at some well-watched wetland sites, such as Castle Lake and Shibdon Pond, small numbers of passage birds might be present, though detecting these is complicated by the presence of inland breeding colonies. In some years, during spring and early summer, there is a build-up of non-breeding birds, mostly immatures, at Teesmouth. At such times, 20 to 30 can be seen loafing alongside terns around the river mouth and estuary (Blick

2009).

In the north of the county, a more concentrated local influx occurred during the late spring and summer of 1989. The small flock of birds, usually present at Shibdon Pond at this time of the year, built up steadily through May and by 11 June, 104 were counted. The gathering went on to peak at 194 on 31 July (Bowey *et al.* 1993). On this latter date, a number of dark-mantled birds were present in the flock. Four of these were adjudged to be of the race *intermedius*, which occurs as a scarce but annual passage migrant in Durham, whilst two others were darker still and considered to be possibly the much rarer north eastern race *fuscus* (Bowey *et al.* 1993). This sub-species is only on the British list courtesy of a ringing recovery of a bird from the breeding grounds, and is currently judged to be only reliably identifiable by in-hand examination or ringing evidence (C. Gibbins pers. comm.). It may be that *fuscus* once occurred regularly in Durham, before it declined in northern Europe.

Post-breeding, fledged juveniles from within and outside the county, usually produce an increase in numbers in the coastal strip by mid-July. Numbers increase further as birds move south but almost any wetland in the county may attract birds from around early August. In August 1988, a large influx of gulls and terns brought an exceptional number of Lesser Black-backed Gulls to the Teesmouth area. A count of 106 on 3 August was nearly twice the previous maximum count and numbers steadily increased, peaking at about 920 on 19 August, the birds being spread over a large area involving most of Teesmouth. Birds were seen moving off inland between these two dates, so well over 1,000 birds were estimated to have been involved in this influx (Blick 2009).

Perhaps the most visible evidence of movement through the county in recent years is the series of large autumn counts made at the Longnewton Reservoir roost. This site is considered of national importance for this species (Banks *et al.* 2007). An autumn, night-time roost developed here through the 1990s and 2000s. Example counts include: 490 in late June and July 1993; 780 in September 1997; 2,600 in September 1998; 2,765 in September 2000; 3,000-3,500 in late August and early September 2002; 2,930 in September 2004 and 3,310 in September 2005 (Blick 2009). A few hundred were also seen at this site during March in the early 2000s, but spring passage is much less pronounced. At 2009, the BTO threshold for national importance was 500 birds and in the latter part of the first decade of the 21st century, Longnewton Reservoir ranked as the eighth most important site for this species in the UK (Blick 2009).

Autumn influxes can also occur at other favoured inland wetlands, such as Bishop Middleham, Hardwick Hall, Hetton Lyons CP and Shibdon Pond, but their appearance is much less predictable. Most birds have left the county by early September and, after early October, there is usually a rapid decrease in any remaining numbers. The last double-figure counts of the year are usually reported in the latter part of this month, as birds move south.

Scandinavian Lesser Black-backed Gull
Larus fuscus intermedius

Dark-backed Scandinavian birds of the race *intermedius* are recorded in most years, with many more probably going undetected. Of these, Temperley (1951) said "*this is a winter visitor in small numbers*". Temperley told of five dark-backed birds with 48 'British' race birds at Jarrow Slake on 26 October 1947. Such dark birds are now noted annually, often off Whitburn Observatory and in the North Tees area. At the former site during the period 2003 to 2011, birds were recorded between 15 January and 1 June, with a distinct peak from late March to late April. Autumn passage birds were also occasionally recorded. Several sightings have also come from other coastal areas where gulls gather, such as Boldon Flats. Undoubtedly, this race is under-recorded in the county.

Distribution & movements
This species breeds in north and west Europe, from Iceland to the Taymyr Peninsula and as far south as Portugal. There are three clearly defined races, the British one, *graellsii*, is the one most regularly noted in the north east, though passage migrants of the darker backed, Scaninavian race *intermedius* are noted on a regular basis. The species is highly migratory, the British race spending its winter off Iberia, south to the coasts off north west Africa, though an increasing proportion of bids has been wintering in Britain, particularly in southern counties, over recent decades (Parkin & Knox 2010).

Temperley (1951) documented a number of ringing recoveries in Durham of birds from the Cumbrian (then Westmorland) breeding colonies. These included some of the earliest-ever ringing recoveries in the county:
- One ringed as a nestling at Foulshaw, Westmorland on 1 July 1911, recovered at Deptford, Sunderland,

29 August 1911

- One ringed same site as a young bird on 12 July 1922 and reported at Marsden Rock on 29 September 1922
- One ringed same site in July 1920 that was found at Seaton Carew five weeks later

Another early recovery was of a bird ringed as a nestling on the Fame Islands, Northumberland, on 2 August 1913 and reported at Sunderland on 12 September 1914. Subsequent to this early recovery, there have been a number of other Farnes-ringed birds found in Durham (*British Birds* 13: 12, 307-312), including:

- One ringed on 4 August 1959, recovered at Ryton-on-Tyne on 10 September 1961
- One ringed on 13 August 1962, recovered at Middlestone, Spennymoor, in July 1966
- One ringed on 13 August 1962, recovered at Chester-le-Street on 14 June 1964
- One ringed on 4 August 1965, recovered near Stockton-on-Tees, on 28 June 1966.

According to Blick (2009), at least ten ringed birds have been recovered in Cleveland, five of which were ringed as pulli on the Farne Islands in August and recovered around Teesmouth up to seven years later, mirroring the earlier referred to recoveries.

The only known foreign-ringed recoveries of Lesser Black-backed Gulls in Durham are one ringed as a nestling on 30 June 1963 at Lyngoya, in Hordaland, Norway and found dead at Sunderland on 15 February 1964, and a ringed (and wing-tagged) bird from Badajoz, Spain, in January 2007, which was at Portrack on 31 August 2007 (Blick 2009).

Sponsored by

Herring Gull (European Herring Gull)
Larus argentatus

An abundant passage and winter visitor, and a common breeding species at an increasing number of locations, both at the coast and inland.

Historical review

Hutchinson (1840) mentioned the wide-ranging nature of this species in the county, occurring even at that time, in both coastal locations and inland. It was not known to be breeding in pre-Victorian times according to Hutchinson (1840) nor for another hundred years. Temperley (1951) reiterated Hutchinson's point about its being seen at *"all times of the year"*, emphasising this point, at least for immature birds.

George Temperley (1951) called this species a *"common visitor ... but particularly in winter"*. He also noted that a *"few pairs"* had then, recently begun to breed in Durham. He noted that it was common on the coast but could also be seen across the county *"even as far inland as the western moorlands"*. He speculated that birds may have bred at the Durham coast prior to industrialisation.

The closing of the coast to public access during the 1939 to 1945 period provided security for cliff-breeding species and gave a breeding opportunity to the Herring Gull. The documented breeding took place in 1943 on Marsden Rock, in the north east of the county. In 1943, there was at least one nest, but possibly more. Subsequent to this, birds bred annually on Marsden Rock, at least from 1944 until 1951. Fifty young were reared on the Rock in 1948 and 105 in 1949. At Temperley's time of writing, there was no documentation of breeding on Durham's western moorlands, although birds were breeding in such situations in neighbouring Northumberland at that time (Kerr 2001).

As the 20th century progressed, this species increased significantly in Durham, almost certainly benefiting from the increased fishing activity along the coast, following fishing vessels into the main harbours, and in particular, from the practice of tipping large amounts of domestic refuse into open landfill sites.

As the number of birds increased at Marsden, overspill of young birds during the mid-1960s led to the species beginning to resort to roof-top nesting in South Shields and Sunderland. This behavioural trait had been noted in London from around 1960, when a pair attempted to nest in London Zoo, being successful from 1962 (*British Birds* 57: 80-81). The first nesting in the Hartlepool area appears to have occurred in 1959, when two pairs attempted to nest on mooring buoys at Hartlepool Docks, but birds were not known to be successful in this area until 1961 (Stead 1964). Later in the decade, this behaviour spread south along the Durham coast and birds were nesting on buildings in 'old Hartlepool' by 1967, with between eight and eleven pairs on hotels, shops and warehouses in and around the docks by 1970 (McAndrew 1972). There has been a continuing expansion of this breeding colony since then (Joynt *et al.* 2008).

Meanwhile, in the far south westof what would become part of the county's recording area, a gull colony near Tan Hill, at Bog Moss, contained ten pairs of Herring Gulls, as well as many Lesser Black-backed Gulls *Larus fuscus*, in May 1969. By 1970, young were reared from at least eight 'urban' nests in Hartlepool, the species was also nesting on buildings in Sunderland, and there were scores of pairs in central South Shields. In this area by 1971 and 1972, a number of pairs were nesting on house roof-tops, close to the old South Shields Girls' Grammar School, in Westoe. By the former year, at least 50 pairs were nesting on buildings in Sunderland, where the habit was said to have been known for at least ten years. At this time, there was probably twice this number nesting on buildings in and around the South Shields town centre. By 1974, the balance had shifted from the coast and the majority of breeding birds in the county were nesting on roof-tops in South Shields, Sunderland and Hartlepool. A survey in 1974 revealed at least 700 roof-top nesting pairs; 250 pairs at Marsden, 236 in South Shields, 166 in Sunderland, and probably around 50 pairs in Hartlepool (Unwin 1975). By the middle of the decade, these birds in South Shields and Sunderland were perceived to be such a problem for local people and businesses that the first efforts to control their numbers were instituted by local authorities and other concerns. These persisted into the 1980s and in some instances were amplified through that decade. Control methods employed included shooting, the provision of poisoned baits at nests and the removal of eggs from nests. Even after 'culling' efforts in 1976, 209 and 189 pairs nested in South Shields and Sunderland respectively, though the figure for Shields had fallen to 150 pairs by 1977, no doubt as a result of the cull. Studies of the town-nesting habit in South Shields and Sunderland by Durham University revealed that many of the birds breeding in these locations around the late 1960s and early 1970s had been ringed as chicks on the Isle of May, in the Firth of Forth, Scotland (Fitzgerald & Coulson 1973).

Recent breeding status

A decade into the 21st century, the majority of breeding Herring Gulls in Durham are still concentrated towards the coast, or are located around the urban conurbations clustered at the mouths of the county's principal river systems, namely the Tees, the Wear and the Tyne. Many Herring Gulls now breed well inland, often along these river corridors and, in small numbers, on a more scattered basis, in the western uplands.

In recent decades, the number of Herring Gulls nesting on buildings in Durham has increased, despite the national decline in the total number of breeding birds (Lloyd *et al.* 1991). During a national survey of inland-breeding gulls, a pair that nested on a chimney stack in Durham City in 1994 was described as the furthest inland breeding record on buildings at that time. Small numbers, associating with Black-headed Gull *Chroicocephalus ridibundus* colonies around the western moors, may also have been breeding at this time. The species is seen in small numbers around most of the larger uplands reservoirs during the breeding season, and it seems probable that a small amount of breeding activity takes place in the county's uplands. In May 2007, single pairs were noted at two moorland sites in the far west, but it is uncertain whether breeding attempts were made. Many breeding season observations from sites further inland are of non-breeding birds, many of which penetrate along the county's rivers or move inland to feed at rubbish tips.

In 1994, 1,348 pairs of Herring Gulls were found nesting on buildings in County Durham (Raven & Coulson 1997). Factoring in the numbers breeding at Marsden Bay, approximately 150 pairs at the time, the *Atlas* estimated that there was a breeding population of approximately 1,500 pairs (Westerberg & Bowey 2000). There has certainly been a decline in the coastal numbers since the mid-1990s, though the numbers on buildings have probably proven stable, or grown somewhat. Breeding surveys in South Shields and Jarrow in 2001 and Sunderland in 2004 showed that in South Shields, 160 properties had nesting gulls, with 14 having multiple nests, giving a total of over 500 pairs. In Jarrow, 16 properties totalled some 50 pairs. In Sunderland in 2004, a total of 482 nests were found, with 321 on industrial areas and 161 on residences, offices and shops, with another 50 nests outside the Roker study area, giving a total of about 550 pairs. At this time, there were also four nests in Whitburn village (J. Coulson pers. comm.). At the mid-point of the first decade of the 21st century, the county probably held over 1,750 breeding pairs.

Today, only at Marsden are birds found nesting in what might be considered traditional, natural sites on the sea cliffs and rocky stacks. Breeding Herring Gulls colour-ringed by Durham University at Lizard Point during the 1980s and 1990s were found to return repeatedly to the same nesting sites. Birds had a low average annual mortality rate of fewer than 10% per annum, giving them an average expectation of adult life of between 10 and 12 years (J. Coulson pers. comm.). In 2005, the number of coastal pairs at Marsden had fallen to 49, from 97 pairs in 1997, and there was a further drop in the 2006 breeding season to 47 apparently occupied sites. In 2007, the long-established colony on Jack Rock, off Lizard Point, was almost fully displaced by the relocation to that site of the Cormorant *Phalacrocorax carbo* colony from nearby Marsden Rock. Up to eight pairs of Herring Gulls also nested on the ground in Sunderland Docks in 1990-1995 and about 40 pairs in a similar situation in South Shields during the late 1980s. Over the years, some of these dockside sites have been lost to demolition, but the birds tend to quickly relocate to alternative sites.

At the turn of the 21st century, 532 nests were counted around Hartlepool (Mitchell *et al.* 2004), one of the ten largest roof-nesting populations in Britain (Joynt *et al.* 2008). Nesting occurred at Longnewton Reservoir from 1991 (one pair in 1991, three pairs in 1992, 11 pairs in 1994 and 28 pairs in 1997) (Blick 2009), and there were 60 nests at the Terra factory in Billingham by 1999 (Iceton 2000). The main population in the south east of the county is centred on north Teesside, in Hartlepool and on industrial buildings around the Tees estuary (Joynt *et al.* 2008). It is likely that unrecorded pairs nest on other high buildings in this area (Blick 2009).

Non-breeding Status

The number of Herring Gulls in the county increases from the end of July, birds gathering in post-breeding congregations on ploughed fields and at wetlands around the county. In the immediate post-breeding period, there is a general dispersal of birds from breeding sites, with quite large numbers sometimes building up around Teesmouth and other coastal stations in August. Maximum numbers are reached during the period from October to early February, at the end of which time most of the adults have returned to their breeding areas.

Non-breeding birds, of a variety of age groups, can be seen across the county throughout the year, but numbers are highest in winter, when large numbers congregate at favoured inland refuse tips, particularly in the east of the county. A Durham University colour-ringing enquiry during the late 1980s and through the 1990s discovered that outside of the breeding season, many hundreds of Herring Gulls feed at landfill sites in County Durham. Birds were not entirely dependent on the food found at landfill sites though. On successive days only 25-30% of birds feeding at these sites were the same, most fed on only one or two days at landfills during a week. At other times, the birds, which largely roost at night on the coast between Sunderland and Hartlepool, fed in fields on grain, worms and insects, or along the shore line and at sea, following fishing boats. Some of the most frequented landfill sites include Greenside, Houghton-le-Spring, Wardley and when open, Kibblesworth, with c.8,500 birds utilising the latter site in early January 2001. At Cowpen Bewley landfill site, c.8,000 were estimated to be present in December 2008. In 2009, an estimated 6,000 were present there on 3 January, while Houghton held up to 1,500 in November of that year. Many of the Houghton birds bathe at Rainton Meadows where counts included up to 1,200 in December 2009. These counts are not unprecedented for wetland sites that are situated close to major landfills. The beaches and rocks around Whitburn and to the north of the Tees Estuary frequently hold large numbers of Herring Gulls, particularly at low tide in autumn and winter.

Other locations once well-frequented by this species included the various sewage outfalls along the major rivers such as the Tyne, although this habit has fallen into abeyance with the development of Interceptor Sewer Systems that reduce point source sewage (Raven & Coulson 2001). Due to the development of the SITA Incinerator, in the south east of the county, many of the local landfills had closed by the early 1990s, but some sites, such as those at Hargreaves Quarry and Cowpen Marsh, attracted several hundred Herring Gulls each winter from 2000 to 2008 (Blick 2009). Lesser numbers are present at many inland wetlands throughout the year, with small numbers also feeding and bathing along the county's main river corridors throughout the year. During the winter, as well as at other times of the year, birds follow fishing boats, from a few kilometres out to sea, into harbour at the Tyne, Sunderland, Seaham and Hartlepool, and birds can be found loafing around the fish quay areas through much of the day. Large numbers can gather around the Tees Estuary in mid-winter, particularly during severe weather conditions, and some 17,000-18,000 were present there in both January 1977 and January 1978 (Blick 2009).

Most of the county's wintering Herring Gulls roost on the coast and there are sizeable roosts on the sea and in the docks at Sunderland, off South Shields and Teesmouth. For example, 4,500 were roosting off South Shields in January 2000, but there is evidence that this roost declined towards the end of the first decade of the 21st century. Large inland winter roosts also develop at a small number of sites in the interior of the county, a good example being the one at Derwent Reservoir. This usually persists from October to March and may hold many thousands of birds, such as the 6,300 counted in January 1997 (Kerr 2001). The average wintering population of the county is probably in the region of 20,000 to 30,000 birds.

As well as racial variability, the Herring Gull seems particularly prone to plumage aberrations. Partly white or coffee-coloured birds are seen on an almost annual basis in the county. On 20 December 1982, a pure white bird with pink legs and a black, yellow-tipped bill was present at Shibdon Pond, and another white bird with normal beak and leg colouration was there from 3rd to 15 February 1986. A strikingly pale leucistic first-winter bird was in the Marsden and Whitburn area on several dates up to 22 March 2007, having first been noted there in early November 2006. Individuals with white wings or white primaries have been seen on more than one occasion in the Teesmouth area, and at least one completely white bird has been recorded there (Blick 2009). A leucistic adult with pure white wing-tips was noted at Whitburn Steel on 5 July 2008.

On 6 February 1986, a bird exhibiting the characteristics of a hybrid Glaucous *L. hyperboreus* x Herring Gull was noted at Shibdon Pond (Bowey *et al.* 1993). Another possible hybrid juvenile Herring Gull x Glaucous Gull was seen at Stella on 16 December 2007, and further possible hybrids with Glaucous Gull were noted at Cowpen Bewley landfill in late August and at Saltholme Pools in late December 2009. Such hybrids are not uncommon within the breeding range of Glaucous Gull and it is likely that other such birds go undetected in their wintering areas.

Scandinavian Herring Gull
Larus argentatus argentatus

Coastal passage can be noted off Whitburn Observatory on many dates throughout the year, but the highest numbers usually come in both winter periods. A significant proportion of the birds present in the county during the winter are probably of the larger Scandinavian race *argentatus* and these mix freely with local birds. By way of example, in 2007, birds of the Scandinavian race were regularly noted at Whitburn Observatory in both winter periods. In the autumn, there were regular sightings from late September onwards and up to ten birds were seen on many dates; the peak movement logged was of 62 on 10 December. On the main days of gull passage, Scandinavian birds were estimated to have made up between three and five percent of the Herring Gulls seen, although this is likely to be an underestimate of true numbers, as many immature birds are more difficult to assign racially. Movements of the Scandinavian race at Whitburn in January 2009 included 46 on 11th and up to 12 on many other dates, whilst in 2010, the winter peak was 21 on 12 December. This race is without doubt under-recorded locally. At nearby Sunderland harbour, 20 *argentatus* present on 31 December 2010 were estimated to be 10% of the total number of Herring Gulls present. A study by Durham University discovered that large Herring Gulls with characteristically different wing-tip patterns and darker plumage arrived in Durham from late September and through October, after having completed their wing moult. Subsequent sightings and ringing recoveries of these birds showed that they came from northern Norway (as far as the North Cape) and from the north coast of Russia, as far east as Murmansk, but none came from southern Norway or Denmark. In November and December, about 24% of the gulls caught at the county's landfill sites comprised these continental immigrants. Such birds synchronously leave the area and return to their northern breeding grounds over just a few days in late January or early February (Coulson *et al.* 1984).

On 2 October 1990, a very large dark-backed bird with yellow legs was noted at Shibdon Pond. The size and colouration of this bird were consistent with it being of the type *omissus*, which is considered an extreme variant of the Scandinavian race (Bowey *et al.* 1993). In 2007, adults with yellow legs, possibly from the north east European populations of *argentatus*, were noted at Greenside landfill site on 16 February, Hartlepool on 7 May and at Stella on 16 December.

Distribution & movements

The nominate race of this species, *argentatus* breeds mainly in Scandinavia, south to Denmark. Its close relative, *argenteus*, essentially the 'British' race, nests around Britain and Ireland, on the Faeroes, Iceland, and from western parts of France along the North Sea coasts to Germany (Parkin & Knox 2010). Britain is home to a large number of wintering Herring Gulls from the north and east, including many birds of differing, or indefinable racial origin. British birds appear to winter locally, though this depends on age and gender, with young birds tending to move greater distances than do adults.

During the 1980s and early 1990s, Dr. John Coulson's research group at Durham University undertook a large-scale study on the movements of Herring Gulls and the species' use of landfill sites within the county, during which over 5,000 Herring Gulls were ringed and colour-ringed. Most of these were cannon-netted at the following landfill sites: Coxhoe (the main site), Consett, Greenside, Seaton Carew and Wingate. In addition, a number of locally breeding Herring Gulls at Lizard Point, were similarly marked. The local breeding birds were found to rarely use landfill sites, most remaining within the county throughout the year, feeding along the coast and at sea, behind fishing boats. The exception was one adult female which moved annually to winter near Scarborough, returning each year to breed on Jack Rock (Coulson & Butterfield 1986).

Adult Herring Gulls visiting Durham County during this study were subsequently found breeding along the whole east coast of Scotland, from Eyemouth to Orkney, but none were shown to breed on the west coast of Scotland. One of the surprising features of the movements of these birds was that most did not remain in the county for the whole non-breeding season, but were present only for a few days or weeks, moving on to winter further south. In one year, over 2,000 Herring Gulls were caught at Coxhoe between September and December, but only 2% of these were later detected amongst about a thousand gulls present there in the following January. Individual birds repeated the same pattern of movements in successive years and birds captured and marked in November were most likely to be present in Durham in the following November, but not in any of the intervening months. Males from Scotland arrived in Durham from July to December, while the females were later, mainly arriving in October and even as late as January. In fact, some males were already returning to Scotland in mid-winter, as some females were still moving south into the county (Coulson & Butterfield 1986).

During this study, swabs were taken to measure the extent to which gulls carried food-poisoning organisms. Some 20% of individuals carried *Salmonella* and 40% carried *Campylobacter* bacteria. Adults visiting Coxhoe were implicated in the spread of a strain of *Salmonella* that was present at Coxhoe and which later caused abortions in cattle along the east coast of Scotland.

A number of earlier ringing recoveries and controls support the Durham University's findings, for instance:

- A bird ringed in west Norway on 8 July 1938 was recovered at Stockton-on Tees on 25 August 1939
- An immature, found dead at Jarrow Slake on 15 February 1957 had been ringed as a juvenile at Baresta, near Floro, Norway on 6 July 1956

The origin of many of the wintering birds elsewhere in the county is illustrated by a number of recoveries of ringed birds, about 20 of which were ringed as pulli in northern Britain, at locations such as Fair Isle, the Bass Rock, the Isle of May and the Firth of Forth. Another four were ringed as pulli in western Norway, Denmark, and northern Russia. More intriguing was the adult ringed at Wingate on 20 December 1979 and later controlled on Kharlov Island, on the Kola Peninsula, Russia, 2,535km to the north east on 28 June 1984. Another bird found in Durham had been ringed in County Down, Ireland, demonstrating that Durham birds come from the west as well as the east. Two birds ringed in County Durham in winter were located in Iceland in April, two and five years later, indicating a more north-south migration (Blick 2009). There is clearly some interchange with birds breeding elsewhere in the north east of England too, as a number of birds ringed in the Farne Islands, Northumberland, have been later controlled or recovered in the county. For example:

- One ringed 22 August 1961 was recovered at Woodland on 23 December 1962
- One ringed 18 July 1963 was recovered at Jarrow Slake on 7 March 1964
- One ringed 9 August 1963 was recovered at Blackhall Rocks on 17 April 1964

A bird ringed at Lumley on 18 February 1986, which at that time was at least four-years old, demonstrated the longevity of the species. It was found dead at Washington just 9km away on 7 April 1996, making it over 14 when it died.

Yellow-legged Gull
Larus michahellis

A scarce but increasingly recorded annual passage migrant and summer visitor from continental Europe.

Recent status

This distinctive large gull, which prior to 2004 was treated as a race of the Herring Gull *Larus argentatus,* hails from the western Mediterranean. Since the late 1980s, Yellow-legged Gull has been observed regularly in Durham. The numbers increased through the 1990s, in line with the considerable upturn in its pattern of occurrence across the whole of the British Isles (Collinson *et al.* 2008). Today, it is an annual though uncommon passage migrant in Durham and a scarce visitor at other times of the year, mainly in late winter.

The first fully documented Durham record occurred in 1985 when what was then described as a 'Yellow-legged Herring Gull' was at Seaton Snook, Teesmouth from 18th to 20 July and on 3 August (Bell 1986, Baldridge 1986). The following year a bird was at WWT Washington on 10 December. There then commenced an interesting series of sightings at Shibdon Pond, from 1987 through to the early 1990s, starting with an adult in the company of two Lesser Black-backed Gulls *Larus fuscus* on 27 July 1987 (Bowey *et al.* 1993). In 1988, two different adults were at this site from 18 March to 1 April and another from 18th and 19 August.

Over the period 1988-1991, a minimum of eighteen birds were recorded in the county; 11 of these records were at Shibdon Pond, with others at Marsden, Sunderland Docks, Brasside Pond and Whitburn. The summer of 1989 saw an influx of this species to the area around Shibdon, associated with the large numbers of second-summer Lesser Black-backed Gulls present at the time. Four different birds, three adults and a second-summer were recorded at Shibdon between 22 May and 31 July of that year (Bowey *et al.* 1993). This pattern of records in the north of the county was unlikely to have been indicative of the species' true occurrence across the county as a whole and was more to do with targeted observer effort, and raised awareness of this species' possible presence in the area.

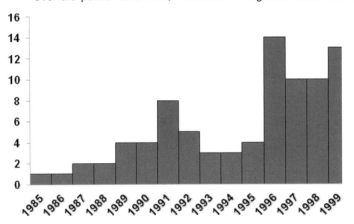

Yellow-legged Gull Records in Durham, 1985-1999 (minimum numbers)

It was from around 1991 that a significant increase in the number of sightings of Yellow-legged Gull commenced in Durham, with the realisation that close scrutiny of parties of Lesser Black-backed Gulls might reveal this species. Through the early 1990s, records came from an ever-growing suite of sites where gulls congregated. As more birds were recorded the complexity of the numbers meant interpretation became more challenging, though the general trend is indicated by the pattern of records per annum between 1985 and 1999. The number of sites at which the species had been reported from across Durham grew from eight at the end of 1991 to 16 by the end of 1996.

The rise in the number of records over this period was associated with a greater observer awareness of the bird, the possible review of its taxonomic status and, no doubt, a real increase in its numbers in the county. That this species was genuinely scarce in Durham prior to the early 1990s is illustrated by the fact that despite being present when over 7,000 large gulls were caught and ringed by Durham University research groups at all times of the year, over the period 1970 to 1992, Prof. John Coulson did not observe a single record of Yellow-legged Gull at any landfill site within the county. This was despite frequent visits to such locations and the inspection of tens of thousands of gulls for the presence of leg rings over this period (J. Coulson pers. comm.).

During the 1990s as well as Shibdon Pond, the shallow flood-water pools close to the Gateshead MetroCentre proved attractive, as many thousands of gulls drank and bathed there prior to roosting on the roofs of nearby industrial buildings. As time passed, the number of records from these sites, as a proportion of the total recorded each year, fell as birds were discovered more widely across the county. Other favoured locations for this species in the late 1990s included South Shields and Hartlepool. By 1999 up to three different birds were noted in spring at South Shields from late February through to May.

Favoured sites in the 2000s included: Longnewton Reservoir, the Tees Barrage and the refuse tips and pools around the North Tees Marshes (Blick 2009), as well as many sites elsewhere in the county. After the turn of the Millennium it became increasingly difficult to accurately interpret the total number of birds seen each year and whether birds, adults at least, were newly arrived or returning birds from previous years. Some birds are known to have returned year-on-year, such as the adult that frequented the area around the River Tees by the Tees Barrage/Portrack Marsh and the Belasis Technology Park and was present from spring 2006 until at least the summer of 2010. This bird was often recorded from high summer through to late March and may have been responsible for sightings of other adults in the North Tees area over a number of years. In 2007, at least four different birds were seen at Longnewton Reservoir during January and February, whilst later in the year an adult was there on 10 November and a first-winter on 22nd and 24 December.

Through the period 2000-2010, the coastline between South Shields and Sunderland Harbour produced occasional reports throughout the year and anything between four and ten birds were reported in the county per annum over this period, but there were probably more than this. Over the decade, there was an emphasis on winter and spring passage records, but birds were recorded in all months of the year. The total of over 13 different birds in 2009 and at least 13 in just the south east of the county in 2010 perhaps reflects the increasing numbers now being seen, and breeding, in south east England (Holling 2009).

Distribution & movements

This species is essentially the west Mediterranean equivalent of the Herring Gull, which it replaces in that area. Yellow-legged Gull was once considered a sub-species of Herring Gull (Grant 1982), but was elevated to species status based on phenotypic and genetic evidence (Collinson *et al.* 2008). It breeds around the western Mediterranean coasts and up the Atlantic seaboard islands of Spain and France. It was recorded as breeding in southern England in 1997 (Parkin & Knox 2010). It largely winters within its breeding range. Its status in the UK is changing, as a result of population increase and range expansion; it is nowadays a more frequent visitor, primarily in the late summer as a result of post-breeding dispersal.

There are two peaks of occurrence in the county; one in spring , as birds - mostly adults - move north in the company of Lesser-black Backed Gulls, and a second late in the summer or early autumn, when immature birds tend to be more numerous. Despite the large increase in the number of sightings in recent decades, this species can be difficult to identify at times as some races of Herring Gulls exhibit yellowish legs and Herring x Lesser Black-backed Gull hybrids have been noted in the county on several occasions.

Caspian Gull
Larus cachinnans

A rare visitor from Eastern Europe: at least 30 records.

Caspian Gull was recognised as a full species by the British Ornithological Union in November 2007 and the early 2000s heralded a much better understanding of its identification and occurrence patterns. The first record for County Durham occurred in 2001 when Chris Bell, Jamie Duffie and Richard Taylor found a first-winter bird in the gull roost at Longnewton Reservoir on 21 February. It re-appeared in the roost three days later.

Since this first record, birds have been identified in a further seven years up to 2011, although their appearance has remained unpredictable. Ten individuals were seen in the gull roost at Longnewton Reservoir between 2001 and 2009 and this has remained the most productive inland site. Similar levels of scrutiny at other reservoir roosts in the county may reveal occasional birds at those sites. The North Tees area, in particular the day time roosting areas for gulls adjacent to the landfills at Cowpen Bewley and Seaton Meadows, have produced at least 17 birds, and the results achieved draws particular attention to these hot spots.

There is a clear peak of occurrences in late winter, matching a pattern of arrival in south east England early in the winter with a slow spread northwards as the season progresses. The small cluster of late summer sightings indicates that the species can also be expected outside the traditional winter 'gulling' season.

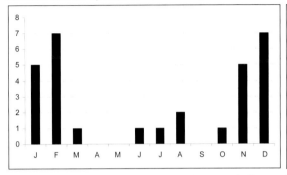 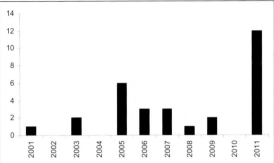

Records of Caspian Gull record in County Durham, 2001-2011

The best year came in 2011 when at least 12 birds were recorded in the county, 11 of which were in the North Tees area. Following sightings of two different first winters and a second winter between 22 January and 15 March, the last two months of 2011 produced a minimum of four first winters, three second winters and one third winter. The turnover of gulls around Cowpen Bewley and Seaton Common is vast, so the true numbers could have easily been more.

Away from the south east of the county, Caspian Gull has remained a rarely detected species. A second winter bird at South Shields and Marsden on 23 December 2006 was the first record for the north of the county, soon followed by a first winter at Stella on 17 December 2007. An adult at Rainton Meadows on 27 January 2011

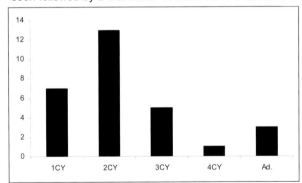 was the first for this well watched area. It is likely that landfill sites and adjacent wetlands in the Gateshead and Houghton-le-Spring areas will produced further sightings with continuing study.

The chart shows the age range of Caspian Gulls in County Durham. The majority of birds identified have been in their first or second calendar years and very few birds older than their fourth calendar year have been detected. This perhaps to be expected and mirrors the age breakdown of other scarce wintering gulls in the north east of England, such as Glaucous *Larus hyperboreus* and Iceland *Larus glaucoides*.

Ages breakdown of Caspian Gulls in County Durham, 2001-2011

Distribution & movements

Like the Yellow-legged Gull *Larus michahellis*, this taxon has had a problematical status but in essence, it is the eastern equivalent of that species. Its breeding range encompasses the Black and Caspian Seas, east to Kazakhstan and increasingly north west into Poland. Birds normally disperse south and south east during the winter, but small numbers also move north west to winter around the southern North Sea; in the UK, Netherlands and northern France. The first British record is credited as being in Essex in 1995, but its true past status is confused. The first decade of the 2000s saw advances in gull identification studies which led to the determination that the Caspian Gull is now a regular winter visitor in small numbers to Britain, mainly south east England and the Midlands.

Iceland Gull
Larus glaucoides

An uncommon winter visitor and passage migrant.

Historical review

Iceland Gull is the least common of the 'white-winged gulls' to occur in Durham, although its status has changed somewhat in more recent years. Through the 1970s and 1980s, it occurred less regularly than Glaucous Gull *Larus hyperboreus,* but in recent years, annual totals have equalled and in some years, exceeded, this species.

Hancock (1874) said that it was a rare and casual visitant to Northumberland and Durham, but he documented no specific records for County Durham. Up to 1951, Temperley (1951) knew of only four records, the first being at Teesmouth on 1 February 1902. Nelson (1907) documented this Teesmouth record, which consequently appears to be the first record of the species for the county. Temperley listed more recent records, at his time of writing, as being one at the River Don at Jarrow on 11 May 1939, a third-winter at Cleadon on 25 December 1939 and one at Whiteleas Pond, near South Shields, from 15 March to 29 April 1948. He also highlighted the difficulties of distinguishing between Iceland and Glaucous Gulls, something which no doubt affected an accurate assessment of its status in these earlier years.

Subsequent to Temperley's summary of status, an increase in observer activity resulted in more regular sightings through the 1950s and 1960s, although the species remained a rare bird locally. Approximately 21 birds were recorded in these two decades, although it was unrecorded in seven of those years. Of the more interesting sightings, one at Whitburn on 13 April 1951 was possibly the same as was found dead at Seaton Carew in July 1951.

Recent status

Iceland Gulls have been recorded annually in County Durham from 1973, with possibly between 550 and 600 birds being involved, although an accurate figure is impossible to calculate. Its status has changed from an average of six birds per year in the 1970s, 15 per year in the 1980s and 1990s, to an average of approximately 25 birds per year through the 2000s.

Nonetheless, it remained a scarce bird in the county in the 1970s. For instance, in 1973, there were just two 'acceptable' records, at Marsden on 2 March and at South Shields on 1 December, followed by just one sighting the following year, at Seaton Carew in November. This was clearly an indication of the challenge of gull identification at the time, prior to advances in field identification techniques (Hume 1980). A much improved showing in 1976 totalled some 12 birds, of which four together at South Shields on 10 April were accompanied by two Mediterranean Gulls *Larus melanocephalus* and a Ross's Gull *Rhodostethia rosea*; an extraordinary gathering of gull species for the time.

Through the remainder of the 1970s and into the 1980s, small numbers were detected each winter with most sightings coming from the much-frequented gull-watching areas around the Tyne estuary and Hartlepool Headland. Notable clusters of records included three at Seaton Carew in March 1981 and four around the mouth of the Tyne in the first quarter of 1983.

As a consequence of the activities of dedicated gull-watchers and a better appreciation of both field characters and occurrence patterns, a shift in status of this species was evident in the mid-1980s. Up to five were present around South Shields in the first three months of 1984, with five present there in early May likely to be transient birds, whilst further early spring passage was detected the following year, with five individuals at South Shields in late April and early May. Further small influxes occurred in early 1987, when six birds were detected utilising Sunderland Docks, and February 1989, when four were at South Shields and two at Sunderland. Overall, the general picture remained relatively constant; the coastal strip held a small number of wintering birds from November to March and there was evidence of a limited, early spring passage, whilst inland records remained few.

Through the 2000s, numbers recorded were higher than ever before. Whether this reflected a genuine increase in birds wintering in the county or merely an increase in observer activity and reporting is uncertain. One of the best years on record was 2001 when the county total of some 31 birds was boosted by 18 individuals using the roost at Longnewton Reservoir between January and April. This included six first-winter birds, three second-winters and nine adults. Longnewton Reservoir also attracted Iceland Gulls in subsequent years including seven different individuals in the late winter/early spring of 2002, and three or four individuals in most other winters.

The North Tees area featured more strongly through the 2000s, particularly during the early spring period as birds moved through the county from wintering sites to the south. Eight different immature birds were attracted to Dorman's Pool between 20 February and 3 May 2002, while approximately 28 different birds were seen in the first four months of 2008, favouring sites such as Cowpen Bewley, Hartlepool and Seaton Common. A further 15 new birds appeared in March of that year. In 2009, there was a conservative estimate of 20-25 birds detected during the first four months of the year, most sightings coming from the landfill sites at Cowpen Bewley and Seaton Common.

Although the number of sightings along the northern stretch of coastline through the 2000s declined, up to five at South Shields and three at Hendon in the first quarter of 2000 was still impressive in relation to previous occurrence patterns. Further use of Sunderland Harbour and in particular, the South Pier for roosting, was demonstrated by five birds together there on 14 March 2004, with at least eight individuals moving through this area during March 2005. One adult bird was noted in this area each winter from December 2007 to at least 2010. Much interchange between coastal and inland sites, the rivers Wear and Tyne was evident, with a number of birds noted at Sunderland also being seen at Boldon Flats, in the Houghton-le-Spring area and at South Shields.

Unlike Glaucous Gull, summer records of this species are exceptional. Spring passage tends to persist until the first week of May and sightings up to the 10th are not infrequent. There have however, only been a handful of records between late May and mid-September in Durham, including birds at Hartlepool on 17 August 1980 and South Shields eleven days later, WWT Washington on 17 May 1983, Marsden on 20 May 1995, MetroCentre Pools on 14 July 1996, Marsden on 17th and South Shields on 23 August 1996 (these last three sightings presumably involved the same individual), Marsden on 12 August and Whitburn on 30 August 2000, Sunderland harbour on 25 May 2006 and Dorman's Pool on 6th to 7 June 2008. The majority of these summer individuals have been of non-adult age classes.

The first sightings for the autumn tend to come much later than Glaucous Gull, as would be expected for a species breeding no closer than Greenland. The earliest autumn record was at Whitburn on 29 September 2001, whilst others at Seaham from 19 October to 4 November 2006 and at South Shields Leas on 27 October 2007 are notable. Through early November, sightings become more expected, although these are not annual.

Truly inland records, away from the river valleys, are rare. Most inland sightings are confined to lowland reservoirs or freshwater wetlands in the east of the county. The extensive gull roost at Derwent Reservoir has attracted birds on several occasions, including in February 1984 (two birds), March 1984, December 1984, January 1985, May 1989, January 1996, January 1998 and November 2007 (Kerr 2001). Other notable inland records consist of birds at Stillington (January 1975), Coxhoe (February 1981 and December 2007), Willington (two in January 2005), and the River Wear north east of Durham City (March 2006 and April 2008). More frequent sightings have come from Barmston Pond, Boldon Flats, Brasside Ponds, Crookfoot Reservoir, Hetton Lyons CP, Hurworth Burn Reservoir, Lambton, Mowbray Park, Rainton Meadows, Ryton Willows, Shibdon Pond, Silksworth Lake, Stella, Watergate Park and WWT Washington. The incidence of inland birds tends to be fluid and dependent on local feeding and roosting conditions. For example, at least three different juvenile birds utilised waste land at the site of the old power station at Stella through December 2007, but this land was later developed and the gull roost ceased.

The current distribution of records centres more on the North Tees than in the past, and there have been several events that may have led to these changes in the focal point of sightings. The changing location of landfill sites through the 2000s appears to been a key part of this, as such sites have traditionally been productive feeding sites for gulls. The expansion of sites at Cowpen Bewley and Seaton Common have drawn in huge numbers of gulls, while the cessation of the landfills at Birtley, Greenside, Rainton Meadows and Willington reduced numbers in those areas. Linked to the cleaning of the coastline, sewage outflows ceased during the late 1990s and 2000s, the cessation of the outflow at Hendon in 2001, no doubt lessening the attraction of this area to gulls species both here and at nearby Sunderland Docks. The decline of the local fishing industry probably also played a part, particularly at South Shields (McDiarmid 1990). The Tyne used to be the most prominent locality for 'white-winged gulls' in north east England, as large numbers of gulls were drawn into the river mouth and harbour by the numerous returning trawlers. This now happens only to a limited extent, and a similar situation is evident at Hartlepool and Sunderland.

Distribution & movements

The Iceland Gull breeds in the Arctic regions of Canada and Greenland, and outside the breeding season can be found wintering in the northernmost states of the eastern USA as far inland as the great lakes, in Iceland, Ireland, the United Kingdom, the north coast of Norway, the southern tip of Scandinavia and the northern tip of Germany

The only evidence of the origin of Iceland Gulls on the Durham coast is provided by a single ringing recovery of a bird found dead at Seaton Carew in July 1951. It had been ringed as a juvenile in West Greenland on 14 August 1949.

Kumlien's Gull
Larus glaucoides kumlieni

There are just two accepted records of this form in the county. The first came on 5 February 1996 when a third-winter was identified at the MetroCentre Pools in Gateshead. Nine years later, an adult was studied in the gull roost at Longnewton Reservoir on 12 February 2005. Although there have been several other claims of Kumlien's Gull, the individuals concerned were considered to have not shown the full suite of characters needed to be fully confident of the identification. Such birds are nevertheless worth documenting and include: a third-winter at South Shields Leas on 23 March 1998; an adult at Longnewton Reservoir on 6th to 7 April 2001; a first-winter at Longnewton Reservoir on 9 March 2002 and a first-winter at Seaton Common on 9 April 2008.

Distribution & movements

Kumlien's Gull has variously been considered a subspecies of Iceland Gull, a subspecies of Thayer's Gull *Larus thayeri,* or more probably part of a 'hybrid swarm'; a highly-variable population of hybrids that survives beyond the initial hybrid generation and back-crosses with its parent types, accounting for the high degree of variability in field characters, such as wingtip pattern and iris colouration (Baxter & Gibbins 2007). It breeds on Baffin Island and Southampton Island in the Canadian Territory of Nunavut and in north west Quebec. Outside the breeding season, it is found on the North American east coast from the Gulf of Saint Lawrence and Newfoundland south to Virginia, and inland to the Great Lakes region. Vagrants have wandered west to Britain and in winter, they regularly occur in Iceland and the Faeroe Islands. It is a rare visitor to the remainder of Western Europe as far south as the Azores and Madeira.

Glaucous-winged Gull
Larus glaucescens

An extremely rare vagrant from the North Pacific: one record.

On 31 December 2008, Toby Collett located an adult 'Herring Gull-type' bird with grey primary tips, rather than the usual contrasting black, on the main pool at RSPB Saltholme. Photographs of the bird obtained on this first day played a crucial part in its identification over the next 36 hours. Discussion around these raised the possibility of the bird being a Glaucous-winged Gull; a species only seen three times previously in the Western Palaearctic. The bird was re-found on New Year's Day, close to the Cowpen Bewley landfill site and by 2 January the identification had been confirmed (Collett 2009). The bird lingered in the Cowpen Bewley tip area, often frequenting the ice-covered Saltholme and Dorman's Pools, until 10 January 2009 (*British Birds* 102: 560) and several thousand observers were believed to have seen the bird by the time it left the area, making it one of the most watched rarities seen in the county.

Distribution & movements

Glaucous-winged Gull is a Pacific species, with a range extending along the north westcoast of North America, and adjacent north east coast of Asia. The winter range extends south to the coasts of California and northern Japan. By the end of 2010, there had been just five records in the Western Palaearctic, these being in the Canary

Islands, Morocco, Denmark and Britain (two). The first British record was of a third-winter in Gloucestershire (and subsequently Carmarthen and Berkshire) between December 2006 and April 2007 (Allen 2007). This was one of the most unexpected species to have been recorded in County Durham.

Glaucous Gull
Larus hyperboreus

An uncommon but regular winter visitor and passage migrant, occasionally seen in the summer months.

Historical review

In the 19th century, Backhouse (1834) said that Glaucous Gull was "*not uncommon*" on the coast near Sunderland while Hutchinson (1840) reported that "*an individual is now and then obtained*", also noting that one was shot near Croxdale in 1837, well inland. Hogg (1845) documented other reports, saying that it "*is a rare bird*" in the Teesmouth area, though specimens were obtained every winter. Hancock (1874) summarised the 19th century status as being "*not uncommon*" in autumn and winter, but adults were rare. In 1951, George Temperley collated the records of one Whitburn shore-shooter for the beginning of the 20th century. These notes indicated that birds were seen or shot on 7 October 1910 (an immature female), 2 December 1910 (an immature male) and 3 January 1911 (several seen, one adult female shot).

Temperley (1951) brought the status of Glaucous Gull up to date by calling it a "*winter visitor in small numbers, rather irregular in its occurrence*". He said that its status had been pretty much constant during 'recorded times' and noted that adult birds were still rarely seen. He also noted that birds were seen in most winters in his time, however, up to 20 together at a gull roost at Hendon, Sunderland on 16 January 1945, was exceptional, and it still remains the largest-ever gathering recorded in the county. Significant passage was noted at Hendon over 26th to 27 November 1947, when 23 flew north. He recorded occasional birds late into April, for example an adult at Jarrow Slake on 1 April 1939 which stayed until 15 April and was later seen at South Shields on 20 April. Possibly this same bird was then noted at Seaham on 18 June and reported locally on a number of other dates, causing Temperley to speculate that the bird may have been summering locally.

Glaucous Gulls were recorded on a regular basis through the 1950s and 1960s including, in 1969, at least four birds in the Hartlepool area during January to April. In no winter was it particularly numerous, although on 9 December 1967, a total of 19 birds were seen at Hartlepool Headland, 16 of which flew north (Blick 2009). One might speculate that at this time, many birds may not have been identified by observers and that it was perhaps more frequent than the historical record suggests. There are also two more interesting inland records in this period, both of them in the mid-sixties. Firstly, an immature bird was seen in the Team Valley, Gateshead, on 19 August 1964, a rather unusual date for the species, and one was seen well inland, on the River Tyne, below the Tyne Bridge, on 21 December 1966.

Recent status

Although the number of Glaucous Gulls recorded in the county has fluctuated over the past four decades, and perhaps more are recorded in current times than in the 1970s, there has been quite a change in the general pattern of occurrence. An average of some 20-25 birds per year through the 1970s and early 1980s included a regular wintering presence in both the Tees and Tyne estuaries. At the mouth of the River Tyne, for example, up to three were present in January 1980, five through the first quarter of 1982, eight during the same period in 1983, and at least ten birds through January and February 1984. In the south of the county, five were at Hartlepool on 31 January 1976 and up to six in the Teesmouth area during January 1980, as well as small numbers at both sites through most other winters. Hartlepool Headland, with its associated fish quay and boats, has long been a favoured area, and an adult was noted here every winter from 1985 to 2003, returning in late August or September and leaving in March (Blick 2009). The Sunderland Docks area was also productive in this period, with at least four birds through the winter of 1986/1987. It is quite clear from the records of this period that birds were routinely encountered at these preferred sites.

From the late 1980s, this wintering presence became less constant and these favoured sites less reliable; many sightings of Glaucous Gull were now of short-staying or transient birds. This downward trend in wintering numbers seems likely to be linked to the decline in the North Sea fishing industry. There are now far fewer trawlers arriving back in port at North Shields and Hartlepool, and consequently, far fewer gulls are drawn back to the estuaries in the trawlers' wakes. Very occasional local influxes do still occur however, such as the eight different birds at Hartlepool in January 1998.

During the 2000s, as few as ten birds were seen in the county in some years. The years 2007 and 2008 however, were more productive with year totals of approximately 40 and 57 birds respectively. Eight passed Whitburn Observatory between 18 November and 28 December 2007 and a further seven were logged there between January and April 2008, while an impressive total of some 25 birds were detected moving through the North Tees area between January and May 2008. Weather conditions elsewhere in the species' wintering range may be a factor behind such influxes and it is also notable that in these more productive years, the number of birds detected in the once-favoured Tyne estuary was very low. Linked with such influxes, there has been a change in the feeding habits of Glaucous Gulls, which in turn reflects the locations where they are now found. As the fish quays and river mouths are no longer a source of abundant fish waste, landfill sites close to the coastline provide more productive feeding grounds and these, as a consequence, receive more attention from observers.

Although generally thought of as a winter visitor, summer sightings were regular through the whole decade of the 1970s and into the early 1980s, with reports of this species every year between June and August. Teesmouth and the Tyne estuary were the most favoured localities and frequently attracted one or two birds, which often spent the whole summer with the local gull gatherings. Occasional summer sightings also came from Hartlepool, Sunderland and Whitburn, but more unusual were birds at East Boldon on 13th and 16 June 1977 and at WWT Washington on 31 May 1978, 21 August 1980 and in July 1981. The best year was 1976 when one or two were present at Teesmouth through June and July, increasing to at least three in August; one also summering on the Tyne estuary. Regular summer records ceased after 1985 and the only out-of-season sightings since have been at Haverton Hole in June 1988, Hartlepool in July 1995, Whitburn on 27 August 2000 and Whitburn on 28 May 2007. The reason for this cessation in summer sightings is hard to determine, but correlates with the end of protracted winter stays by small numbers of birds at the county's river mouths.

Within the current trend of occurrence, Glaucous Gulls have been reported in every month of the year, although adults are almost entirely confined to the period between late November and late April. Numbers have a tendency to gradually increase during the last two months of the year and peak in January/February. Typically, there is a slight fall in numbers in late February followed by an upsurge in numbers around late March and through April. This is presumably indicative of a return spring passage, northwards to Arctic breeding grounds, of birds wintering further south than Durham. Such passage can lead to occasional late spring birds, such as at Marsden on 24 May 1994, Whitburn on 28 May 2007 and Saltholme Pools on 8 May 2008. It has been calculated that in Durham, about seventy per cent of birds reported are first or second-winter birds, just under twenty per cent are third or fourth calendar-year birds, and a little over ten per cent adults. In neighbouring Northumberland, post-1980, it was determined that immature birds out-numbered adults by a ratio of roughly eight to one (Kerr 2001).

The furthest inland records for the county have come from Derwent Reservoir, which has frequently attracted birds to the large winter gull roost over the last three to four decades. One was there in February 1976, with two roosting in March of the same year (Kerr 2001), followed by another in March 1981. Subsequently, birds were seen in March and April 1985, February 1987, February 1988, January 1994 and February 1996, with two there again in May 1997 (Kerr 2001). The last records suggest that there may be some movement west through the county in early spring. Longnewton Reservoir in the south of the county has also provided many records and interchange between here and the North Tees Marshes is a likely reason for its popularity. At least five different birds roosted there in February 1986, while nine birds were detected in March and April 2001. One to three birds have been recorded there in most other winters.

Most other inland records have come from wetland sites in the east of the county, such as Charlton's Pond, Crookfoot Reservoir, Hart Reservoir, Hetton Lyons CP, Hurworth Burn Reservoir and Rainton Meadows. More unusual were birds at Shotton Colliery on 17 November 1972, Durham-Tees Valley Airport in January 1987, High Urpeth on 19 March 1988, Langley Moor on 3 March 2001 and Houghton Gate on 17 March 2007. Inland landfill sites have also attracted occasional birds, such as at Coxhoe in February 1994, May 1995 and February 1998, whilst one was at Willington on 25 February 2007.

In the Gateshead study area, it is surprising that this species occurs less regularly than the Iceland Gull *Larus glaucoides,* even though it is generally the more frequent winter visitor to the coast. It is more restricted to the mid-winter months there than is Iceland Gull and like that species, the majority of records refer to immature birds. Shibdon Pond was the most favoured locality through the 1980s and 1990s, but this is probably as a result of the concentration of observers at that site. In more recent years, the River Tyne and adjacent roost sites have been more productive, such as the derelict land by the old power station at Stella which attracted at least two in December 2007 and four different birds in January and February 2008. As with other localities, a small spring passage has been detected in some years, such as second-summer and juvenile birds on 3rd and 10 April 1988 respectively and a juvenile on 14 March 1992 (Bowey *et al.* 1993). Nearby landfill sites of Birtley and Greenside attracted feeding birds on several occasions in the 2000s, but with the closure of these, a decline in gull activity in the area become apparent.

Over the years 1970-2010, there have been a number of credible reports of Glaucous x Herring Gull hybrids. This pairing is not infrequent in breeding areas and it is quite likely that such birds occur in the county on a fairly frequent basis. Birds have been observed most regularly at the expected gull hot-spots of the Tees and Tyne estuaries. With the variations in immature gull plumages, it is not always possible to be sure of a bird's true parentage and hence, few conclusions can be drawn as regards occurrence patterns.

Distribution & movements

This species breeds extensively around the Arctic Circle, from northern Europe (though not in Scandinavia), through Northern Norway (Svalbard) and Siberia, and across to North America. In Europe, the population winters just to the south of the breeding range down the Atlantic coastline as far south as Brittany, with most of the British wintering population thought to emanate from Iceland.

Great Black-backed Gull
Larus marinus

Mainly a very common passage and winter visitor, but a few non-breeders are present during the summer months.

Historical review

There is little mention of this species in the chronicles of the early Durham ornithologists and it can only be surmised that this species has long been present in Durham as it was elsewhere in Britain (Holloway 1996), though its true local status in the past remains undocumented.

Temperley (1951) said that this was "*a winter visitor and passage migrant*", noting that it was much less common than Herring *Larus argentatus* and Lesser Black-backed Gulls *Larus fuscus*, and that it moved well inland in the winter. It has never been known to breed in the county although there was a small colony of birds breeding to the south west of County Durham, on moorland around the northern Yorkshire/Lancashire border, during the late 1940s and early 1950s (Mather 1986). During the early part of the 20th century, Jarrow Slake was noted as a main winter roost, with up to 110 estimated to be present there in early December 1929. The fact that birds moved along the coast at certain times of the year was known during the middle of the 20th century, and an example of such movement came on 24 August 1950, when 600 passed Teesmouth (Mather 1986).

Post-Temperley, there was little attention paid to this species by the Ornithological Reports for Northumberland and Durham. What little information was forthcoming largely related to winter flocks, such as the 2,000 birds in the Tees Estuary on 14 September 1969. Clearly, large numbers were along the coastal strip of the county at the time, as numbers had risen to 2,400 by 14 December and 2,500 on 18 January 1970.

Recent status

During the summer months, aside from small numbers of immature birds that can be seen along the county's river systems and along the coast, the species is a relatively scarce one in Durham. It can be found across the county during the winter and, like the Herring Gull, it is well distributed from South Shields to Hartlepool. It also gathers, often in large numbers, at inland refuse tips, sewage outfalls and other areas that provide easily available food. In broad terms, it most common along the coastal strip with smaller numbers reported from many inland sites throughout the non-breeding season. The majority of large counts come from several favoured areas, such as Hartlepool Bay, the North Tees Marshes, Seal Sands and Sunderland. Birds are also frequently seen at the mouth of the River Tyne and down the central stretch of the county's coastline, but few counts of any significance are made away from such locations.

Like the Herring Gull, it can be found bathing in wetlands and at reservoirs, and can frequently be seen following fishing boats both along the Durham coastline and further offshore. During the winter, many birds feed at sea off the coast of Durham, many only coming inland when sea conditions are stormy (J. Coulson pers. comm.). Inland gatherings are scarcer than on the coast, and numbers in these tend to be lower. Nonetheless some landfill sites, particularly those in the east of the county, occasionally attract larger numbers, especially towards the year's end. For instance, up to 800 were noted at Cowpen Bewley in early January 2009. Less regularly, quite large winter roosts can develop at some inland sites, such as Derwent Reservoir. Almost invariably, this species roosts in concert with Herring Gulls in such situations; for example, 900 were roosting at Derwent Reservoir in January 1991, with 720 birds there in January 1997 (Kerr 2001).

Peak numbers of this large gull are usually recorded between August and January, but it is generally outnumbered by the other common gulls and in some years the peak count in the county's main area, Teesmouth, is under 1,000 birds. In the west of the county, there are usually scattered records through the year with most reports being of one to eight birds, though it is not unusual to see a Great Black-backed Gull, drifting low over the western uplands in the depths of winter, or quartering purposefully across the moors, clearly hunting.

The occasional autumn wreck of sprat in the mouths of local rivers attracts very large numbers of gulls including Great Black-backed and strong onshore winds in the months between August and December can bring birds inshore to seek shelter around locations such as Hartlepool and Teesmouth. At such times, 2,000-3,000 Great Black-backs have been noted. The highest counts in the county are of 4,200 on 14 October 1973, 4,860 on 16 November 1975 and 4,700 on 22 January 1978. Such high counts make Teesmouth one of the three most important waters for this species in the UK (Blick 2009).

Further inland, birds are less frequently reported in large gatherings, but exceptions occur for example, 800 were on flood-water at Preston-le-Skerne on 20 July 2009. At Longnewton Reservoir, 270 were roosting on 31 December 2009. Peak late autumn and winter counts in the north of the county during the first decade of the 21st century included c.1000 at South Shields in October 2001 and 1,130 on 24 November 2006. It is probable that in the 'average' winter, there are between 1,200 and 2,000 Great Black-backed Gulls in Durham.

Large numbers of birds arrive in the north east of England from early autumn and over this period, considerable coastal passage movements may be noted, though this species tends to be less well recorded than other more glamorous species. Coastal passage is usually evident off Whitburn Observatory and Hartlepool Headland during several periods of the year, most evidently from late summer and peaking through late autumn. For example, up to 157 passed Whitburn on several August dates in 2007; 550 flew north in two hours on 23 November 2007; 250 flew south on 8 December 2007 and 680 flew north in nine hours on 3 October 2008. Also at Whitburn, in response to strong southerly winds, 400 flew south in 2.5 hours on 10 January 2009 with 500 doing likewise in 2.5 hours the following day. On 14 November 2009, 226 flew south in just 15 minutes. There was evidence of a large overland passage on 29 November 2007, when 240 flew west over Ruffside Moor, presumably heading cross-country for The Solway.

The number of adults to be seen locally declines dramatically during February, when birds migrate to their northerly breeding locations, although some immature birds and non-breeding sub-adults remain in the county during the spring and summer. For example, 95 were feeding in the Tyne Estuary in July 1994 (Kerr 2001). Coastal passage in late winter is occasionally noted, and counts of such activity at Whitburn Observatory in 2007 included 350 flying south in just over two hours on 13 January and 200 south on 11 February.

The scavenging nature of this gull is well documented but occasionally individuals exhibit their truly predatory nature. It has been observed to attack Wigeon *Anas penelope* and Pochard *Aythya ferina* on Seal Sands and been seen chasing migrant thrushes as they arrive over the sea, as well as attacking auks close inshore (Blick 2009). Birds observed at Shibdon Pond have been recorded taking Moorhen *Gallinula chloropus,* attacking Teal *Anas crecca*, Coot *Fulica atra*, and, on at least one occasion, a fully-grown Mallard *Anas platyrhynchos* (Bowey *et al.* 1993).

517

At present, there is no sign of potential breeding in Durham although the Great Black-backed Gull spread, as a breeding species, to the east coast of Scotland as recently as the 1960s. Currently, it does not regularly breed closer to County Durham than the Farne Islands, Northumberland, and Dumfries & Galloway, with other breeding colonies in the Firth of Forth and Abbeystead, Lancashire (Brown & Grice 2005).

As with many other gull species, aberrant plumaged individuals are frequently noted. A leucistic immature was seen around Hartlepool and Seaton Carew in October 1979 and March 1980, an unusual pale brown-backed adult on Seaton Tip in February 1979 and December 1980 and a pale, coffee-coloured juvenile was at Hartlepool on 14 November 2002 (Blick 2009).

Distribution & movements

As a breeding bird, this species is restricted to the North Atlantic, nesting from eastern Canada and the USA west to Svalbard and north Russia, and from there, as far south as the British Isles and northern France. Northerly birds tend to winter to the south of their breeding range, though some of the southern populations, whilst wandering as immature birds, once they reach adulthood have a tendency to winter close to their breeding areas (Parkin & Knox 2010).

Most of the Great Black-backed Gulls coming to the Durham area in autumn and winter, as shown by ringing controls and recoveries, originate from Norway, and these, unlike Herring Gulls, breed along the whole coastline of Norway. One early ringing recovery lending credence to this was the bird ringed as a pullus on 29 June 1956 at Sola in Norway, which was shot at Bishop Auckland on 5 December 1960.

Many Great Black-backed Gulls were ringed by Durham University over the period 1988-2005 and these were subsequently reported in the breeding season along the whole western coast of Norway, but only one bird ringed in Durham was subsequently found breeding in Scotland (J. Coulson pers. comm.). Blick (2009) highlighted the fact that there were some 30 known recoveries of 'Great Black-backs' in Cleveland, although some may have been recovered to the south of the River Tees. All of these had been ringed as pulli, and these included five from Hoy, in Orkney, two from the Moray Firth, 14 from the west coast of Norway. The Norwegian recoveries included a bird ringed as a pullus on 11 July 1963 at Klepp, Norway, found dead at Teesmouth in March 1964, and one ringed as pullus in–Rogaland, Norway, on 25 June 1992 and found dead at Port Clarence on 18 March 1993. In addition, there are three local recoveries of birds from the Great Ainov Islands, Murmansk, in northern Russia. One ringed there on 28 June 1960 was recovered at Teesmouth on 3 December 1961; another, ringed on 26 June 1962 was recovered at Chester-le-Street on 26 November 1962, and a third, ringed in July 1966, was recovered at Beamish on 24 May 1967.

Two birds ringed as full-grown birds near Thorpe Thewles were recovered less than a year later in Norway. An adult ringed on Seaton Tip in December 1978 was found dead on the west coast of Iceland on 30 March 1990 (Blick 2009). Birds ringed in Scotland have been subsequently recorded at Teesmouth as early as mid-August and those from Norway, as early as mid-September, although one individual ringed as a pullus in Norway during summer was found dead at Teesmouth in the following May (Blick 2009). Based on the re-capture of birds in years after they were ringed, it was shown by Durham University that many birds demonstrate winter site-faithfulness, returning to the same wintering area in successive years (J. Coulson pers. comm.).

Bridled Tern
Onychoprion anaethetus

An extremely rare vagrant from the tropics: one record.

County Durham's sole record of Bridled Tern came in 1988. On the evening of 9 August, Keith Cowton and the late Doug Cowton were seawatching at Hartlepool Headland and found an adult feeding offshore. It lingered for approximately 45 minutes, coming to within 50m of the shore, before drifting south into Hartlepool Bay; it was not seen subsequently. During July 1988, a Bridled Tern had made intermittent appearances at Coquet Island and Hauxley, Northumberland, with presumably the same bird at Sands of Forvie, north east Scotland, on 2 August, and at Coquet Island again on 13th and 28 August. All sightings were accepted as referring to the same bird; it was the twelfth record for Britain at the time (*British Birds* 82: 531).

Distribution & movements
Bridled Tern is a bird of the tropical oceans, nesting off the Pacific and Atlantic coasts of Central America including the Caribbean, some areas of West Africa, around Arabia and East Africa down to South Africa, and into parts of the Indian Ocean and south east to northern Australasia (Birdlife International 2011). It is dispersive and pelagic through the non-breeding season, and is a rare visitor to north westEurope. A total of 23 have been recorded in Britain since the first in 1931, the majority of which have occurred in early summer.

Little Tern
Sternula albifrons

A common passage migrant and restricted summer visitor.

Historical review
Temperley (1951) described the Little Tern as a *"summer resident in small numbers"*. He said it had maintained a *"precarious"* breeding station at Teesmouth *"for some years"* and he speculated on it being more common in the past, though he said earlier records and listings made little reference to it.

In the historical context, Little Terns have attempted to nest around Teesmouth's sandy beaches since at least the early 19th century, as documented by Sharp (1816), who described it as being common around Seaton Snook (Joynt *et al.* 2008). Hutchinson (1840) said that it bred at Teesmouth, and Hogg (1845) said that it inhabited Teesmouth and Seaton Point, indicating that it was breeding. More specifically, this was highlighted by Stanley Duncan in his note in *British Birds* in 1911, in which he stated that *"J. Hogg in the Zoologist (1845, p. 1187) refers to the Lesser Tern inhabiting in summer the sandy beach near the Tees Mouth, but no note of nesting is mentioned"* and he went on to say that it was now doing so (Duncan 1911). Nelson (1907) however, did not mention its breeding, though he recorded it as occurring in both spring and autumn. According to Blick (2009), the first documented nesting of this species at Teesmouth was in 1908, but a pair had attempted to breed on the Durham side of the Tees in 1905 but failed to produce young (Temperley 1951). By contrast, Stead (1964) stated that the species *"first bred at Teesmouth in 1910, when three pairs nested on Coatham Sands"*.

The first proven successful breeding was not until 1922, when birds moved from the south side of the Tees Estuary to nest in the Durham area. This colony was found by Joseph Bishop in June, after birds were flooded out from a colony on the south side of the estuary. It was established in a derelict ship yard at Graythorp adjacent to Seal Sands and because it was enclosed and safe from disturbance, it was successful and persisted until 1932, with a maximum of 30 nests in 1931. The site was eventually deserted because of vegetation growth on the area where the birds nested (Temperley 1951). In 1932, a few pairs, no doubt the remnants of the re-located colony from the shipyard, nested on an unprotected site and this colony thrived during the war years, 1939-1945. Temperley (1951) referred to seeing nesting birds here, and their young, but he was doubtful about the species' continued future as breeding bird in the county.

From the early 1950s to the 1970s, the species' breeding hold in the Teesmouth area was rather tenuous, with up to ten pairs nesting in various locations around the estuary (Joynt *et al.* 2008). In the early 1950s, breeding success was intermittent, though attempts by small numbers of pairs were regularly noted. In 1952, there were several scattered colonies around Teesmouth, the largest of these contained 18 nests, but many were destroyed, though at least two young were hatched at one of the sites. Small colonies attempted to breed over the next few years, but no successful breeding was proven until 1955, when half a dozen young were reared at Teesmouth. In 1956, the main colony was washed out by a high tide and the birds relayed only to be washed out again, though a few birds may have reared young. In the following year, there was the best success of the decade with at least 15 young reared at Teesmouth in 1957. This same roller-coaster pattern continued more or less unchecked through the 1960s and early 1970s. Between 1960 and 1969, only in 1962 were any birds successful in rearing young, but at least two pairs bred at Teesmouth in 1969. One publicised breeding success in a 1967 colony, which was listed under Durham in a British Birds paper was, in fact, on the south side of the estuary (*contra.* Norman & Saunders 1969).

Up to four pairs nested successfully at Teesmouth in the early 1970s, but then from 1973, the species' fortunes in Durham took a decided downturn and breeding did not occur again for five years, when two pairs fledged three chicks in 1976, once again at Teesmouth. In 1978, an adult with two juveniles was seen on 16 July at Teesmouth, but it was not known whether these had been reared locally or on the south side of the estuary. The 1970s was not a good time for the species in Britain with noted declines being widespread and a very marked decline in both passage birds and summering individuals in the Durham area (Sharrock 1976). The next documented breeding evidence came in 1981, when at least one pair was thought to have bred at Teesmouth, young being fed by an adult in early August. Numbers were poor through the mid-1980s, with a few birds noted annually around the Teesmouth area but no further evidence of successful breeding, though more and more birds were being noted on the south side of the estuary by 1986. During the mid-1980s, birds bred at North Gare, as well as at Coatham Sands, on the south side of the Tees estuary (Joynt *et al.* 2008). From this time and on through the 1980s and 1990s, the history of this colony was rather chequered. For instance, in 1986 about 30 pairs nested, rising to a peak of 45 pairs in 1987, and numbers then fell away to about 25 pairs in the period 1989-1992. Prior to 1996, most pairs nested in the Coatham Sands-South Gare area and a very few pairs sometimes laid eggs in the Seaton Sands and North Gare area (Blick 2009).

Recent breeding status

This species has a highly restricted breeding distribution in County Durham, and it has had for generations. Through modern times, as a breeding species, it has been largely confined to the Teesmouth area, though birds frequently forage further up the coast and they are more widely distributed during passage periods. The Little Tern, as a breeding bird, is prone to erratic breeding outcomes and the recent success of colonies along the Durham coast in the last two decades or so has been secured only through the implementation of successful wardening schemes.

Nationally, there was a considerable decline in the number of Little Tern colonies between the late 1960s and 2000, down from 154 in 1969/1970 to just 130 in 2000, after rising to as many as 168 in the mid-1980s (Parkin & Knox 2010). The decline in the number of colonies was matched by a reduction in the number of 'apparently occupied nest' sites over the same period, from around 2,800 pairs in the mid-1970s to 2,153 pairs in 2000 (Parkin & Knox 2010), with some probable slight recovery by the middle of the first decade of the 21st century.

For many years, from the early to mid-1980s through to 1996, Teesmouth's main colony occupied a site on the south side of the estuary (i.e. outside the recording area of County Durham), with small numbers of pairs attempting to breed at smaller 'satellite' sites within the county in most years over that period. Both breeding sites in terms of habitat consisted of relatively broad sand/shingle beaches fronting extensive dune systems. Occasional pairs have displayed, and even possibly attempted to nest, at one or two other locations on the Durham side of the estuary, for example on a slag-topped sea wall.

During the *Atlas* survey period (1988-1993), the 'Teesmouth' breeding population varied between 24 and 45 pairs (at a mean of 36 pairs), with up to seven pairs of these breeding on the Durham side of the estuary in at least three of these years (Westerberg & Bowey 2000). In 1995, after the end of the *Atlas* survey period, some 15 pairs of Little Terns colonised an entirely new site well to the north of Hartlepool, which was centred on the beach at the mouth of Crimdon Dene, just to the north of Hart Warren. In 1996, the majority of the birds, which had previously nested to the south of the Tees estuary, moved north to the beach at Crimdon. In this first year, 33 pairs reared 52 young (Joynt *et al.* 2008). From 1997 onwards, almost all of the 'Teesmouth' population of this species, nested at Crimdon Dene some 10km to the north of the mouth of the Tees, although over this period, a small number of pairs continued to attempt to breed around the North Gare Sands, and also, on occasion, to the south of the estuary (Blick 2009). This colony increased to 65 pairs in 1997 and continued to grow. In 1999, however, the colony was ravaged by human egg-collectors who stole 27 of the 65 clutches of eggs that had been laid (Joynt *et al.* 2008). Through the 1990s, County Durham's breeding population increased from seven pairs in 1994 to 30 in 1995, 35 in 1996 and 63 pairs in 1997 (Wilson 1992-1997), all of these being based at Crimdon. The Crimdon Beach location comprises a mixed sand/shingle beach, adjacent to a narrow dune system next to a stream outflow of the Crimdon Burn. The birds' utilisation of this area was all the more unexpected in view of the intense recreational pressures prevalent in this area.

In 2000, birds moved away from this site, the first time that birds had failed to breed there since 1995, although two pairs nested successfully at another site in the south of the county. Birds returned to Crimdon, albeit late in the season, in 2001, and 32 chicks were fledged, whilst eight chicks were produced from 15 nests at a second colony based on North Gare. At Crimdon, 61 pairs nested in 2004, but a total colony failure occurred in 2005, largely as a result of fox *Vulpes vulpes* predation, then 25 pairs reared 37 young in 2006. In 2007, nesting took place at two sites and record numbers of young were fledged. The success rate of these was much higher than in other recent years. The main colony at Crimdon Denemouth held 51 successful nests and a total of 105 chicks were fledged (95 of these were ringed), despite some predation of nests early in the season. At the small colony established at Seaton Snook, four pairs attempted to nest, but success rates were not as high as at Crimdon. The following year saw further success with around 70 pairs producing 67 young, despite early season predation of eggs by hedgehogs *Erinaceus europaeus* and foxes. In 2009, 120 young were raised by 60 pairs making Crimdon the second most productive colony in the whole of the UK in that year. Other pairs elsewhere in the North Tees area appeared not to have been successful. Such high levels of breeding productivity led to the astonishing sight of almost 240 Little Terns being seen off Hartlepool and in the Tees Bay area for a short period of time in the late summer of that year (Blick 2009); probably the largest ever congregation of this species in the county.

After 2009's success, the species' propensity for boom and bust returned in 2010. The year started well when 30 had returned to Crimdon by 25 April, with over 70 pairs present by late May. Shortly afterwards 91 pairs were incubating, almost 50% up on the previous highest number of pairs. This turned into a disastrous year for the colony when predation by a male Kestrel *Falco tinnunculus* reduced productivity to almost zero; only a single chick was fledged. The Kestrel's outstandingly successful tactic was to hover over the nearby golf course, some distance away from the colony, then swoop in and pick up a chick, before the colony of terns was aware of the attack. Almost every chick, Little Tern or Ringed Plover *Charadrius hiaticula*, on the beach was predated in this way (INCA 2010). The summer of 2011 however saw a return to successful breeding, as 84 pairs managed to fledge a record 147 young.

Fox, human predation (at least in 1999), high tides and avian predation (in 2010) have been responsible for part or complete failure of the colony in recent years. In some instances, a few losses at the Crimdon colony have been attributable to other human activities (including accidental trampling and crushing by vehicles). In most years, predation by corvids, gulls or raptors, whilst it may have occurred, has not proven significant. The threats to the county's breeding population of Little Terns have been well documented (Wilson 1992-1997) and this was exemplified by the disastrous breeding season of 1999. Tidal inundation of colonies has also led to numerous failures and desertions. Efficient wardening of the colony has minimised some of these threats (especially from people), but the fox and kestrel problem remains virtually intractable. There is some suggestion that availability of the optimum nesting substrate – large stones, shingle and beach debris on sand, with vegetation absent – plays a positive role in concealing incubating and brooding adults along with their eggs and chicks from detection by predators. The dynamic nature of the coastal habitats in some areas makes for uncertain provision of such a substrate, and considerable between-year variation in habitat condition at the same site (G. Barber pers. comm.).

Recent non-breeding status

After young have fledged from the main colony at Crimdon, in mid- to late July, both adults and juveniles relatively quickly move away from the area, drifting south and sometimes north, with numbers occasionally congregating in the area of Hartlepool Bay and around Teesmouth, though never in the large numbers of the other tern species that use that area in late summer. At this time of the year, and through August, small numbers of birds can quite often be seen passing the county's coastal watch points at Whitburn and Hartlepool. In the later years of the first decade of the 21st century, small number of birds were increasingly noted in late summer loafing on some of the less disturbed beach areas in the north of the county; for instance, up to three birds were regularly present at South Shields and Whitburn during the second half of July in both 2008 and 2009.

Small numbers of passage birds are routinely noted on seawatches during the spring and summer months at locations away from the colonies, such as Whitburn. Between 2006 and 2010, a total of 122 birds were logged at Whitburn Observatory on 63 dates, the extreme dates being 30 April and 18 September. No more than five birds were noted on any one day, and the peak period was between the second week of July and the first week of August, with 46% of all sightings.

The first returning Little Terns in Durham are usually seen in the last week of April or in early May, with birds then present through to mid-August and, more occasionally, into early September. In 2000, the species undertook a typical early autumn departure from the county, by and large through late July, the final records coming in August, with single adults being in the roosts at Seaton Snook on 5th and at Dawdon on 17th. The earliest bird recorded in the county was on 18 April 2008 at Crimdon and the latest was the exceptionally late bird that was noted at South Shields on 9th to 11 October 1964.

A thirty-year analysis between 1976 and 2005 of the first arrival dates for Little Tern revealed that the mean arrival date of 8 May, over the period 1976 to 1985 was two days later than that for 1986 to 1995, and that the mean arrival date for the period from 1996 to 2005 was six days earlier than this, i.e. 20 April. This is an advance of nine days over the whole study period, with most of the change occurring in the second ten-year period (Siggens 2005).

This species is largely a 'sea-bird' in Durham, with relatively few inland records, though it has been recorded over Cowpen Marsh, at Haverton Hole and Longnewton Reservoir, and eight were at Crookfoot Reservoir on 7 July 2000 (Blick 2009). Another rare inland record came at Hurworth Burn Reservoir on 24 June 2007 when two birds were seen, and this record presaged a cluster of such sightings over the next few years. On 27 July 2008, a flock of 30 flew west over Hurworth Burn Reservoir, with 15 being seen later the same day over Castle Lake, Bishop Middleham, then flying off west over the A1; a remarkable inland occurrence. In 2009, several parties were seen at Hurworth Burn Reservoir in early July. On 5th, a party of 10 stayed for about an hour before flying off north east, while the following day groups of nine, four and two lingered briefly, before continuing south. In 2010, sightings at Hurworth Burn Reservoir included five on 27 June, five on 11 July and seven on 25 July 2010; the third consecutive summer that birds had been noted at this site. Further away from the coast and the Teesmouth area, two were at Castle Lake, Bishop Middleham, on 10 July 2010. The furthest inland record for Durham however, was the bird at Derwent Reservoir in July 1991 (Kerr 2001).

Distribution & movements

This species has a discontinuous breeding distribution that encompasses much of Western Europe, as far north as southern Sweden and Scotland, through Eastern Europe and into central Asia. Other populations are to be found in eastern and southern Asia, and down into the western Indian Ocean and Australasia. It is highly migratory, wintering off the coasts of Africa and Arabia (Parkin & Knox 2010).

Many Little Terns were ringed at the colonies based around Teesmouth and Crimdon over the period 1980-2010 and these generated a number of interesting recoveries and controls. One of the oldest local ringing recoveries of this species is of a bird that was ringed at Tentsmuir, Fifeshire, in July 1949 as a pullus and was present at Teesmouth in August of that year. An adult ringed at Inverkeilor, Tayside, in July 1979 was found breeding at Teesmouth in June 1990. Three adults that were breeding on the south side of the Tees estuary in 1991 had being ringed respectively as pulli: at Teesmouth in July 1976, Carnoustie, Tayside, in June 1980 and near Spurn, Humberside, in June 1987. The Carnoustie-ringed bird was still at Teesmouth on 13 May 1995, aged 15. This age was also reached by a bird ringed in the nest at South Gare in July 1976, which was still at Teesmouth in June 1991. A bird ringed as a pullus at Brora, Highland, in June 1991 was at Teesmouth on 13 May 1995.

That there is some interchange of birds between colonies along the east coast of England was demonstrated by the bird ringed as a pullus at Crimdon on 19 June 1996 and found dead as a breeding adult in the Little Tern colony at Great Yarmouth, Norfolk, on 5 June 1999. Most impressive in terms of the distance travelled, were the two pulli ringed in the Durham colony on 11 July 1997 that were recorded in Senegal, West Africa, the first on 8 October 1997 and the other on 16 May 1998 (Blick 2009).

Gull-billed Tern
Gelochelidon nilotica

An extremely rare vagrant from southern and eastern Europe: four records involving five individuals.

The first Durham sighting of this rare visitor came in 1978. At 10.10hrs on 6 July, Peter Bell saw a summer plumaged adult flying north with Sandwich Terns *Sterna sandvicensis* close inshore at Whitburn. It was one of seven records in Britain during that year, and on a very typical date of occurrence for east coast sightings (*British Birds* 72: 527).

A further four birds have been seen in Durham, but all have been rather brief visits. The two in 2006 were notable in that they were tracked flying up the north east coast during the evening, first being seen at Hartlepool Headland at 18.50hrs, Lizard Point at 19.40hrs and then in Northumberland, at St Mary's Island, at 20.05hrs.

All records:

1978	Whitburn, adult, 6 July
1991	Reclamation Pond, adult, 5 June (*British Birds* 85: 530)
2006	Hartlepool Headland, two adults, 9 May (*British Birds* 100: 719)
2006	Lizard Point, two adults, same as Hartlepool, 9 May (*British Birds* 101: 549)
2007	Hartlepool Headland, juvenile, 29 September (*British Birds* 101: 549)

Distribution & movements

The North European population of Gull-billed Tern is restricted to small numbers in northern Germany and Denmark. It is more widespread though local in Southern Europe, breeding discontinuously through Turkey and south west Russia to Kazakhstan, Mongolia and China. The European population winters in coastal West Africa south to the Gulf of Guinea, while the Asian populations winter from the Persian Gulf to the Indian subcontinent and South East Asia. Formerly a more regular visitor to Britain with 335 accepted records to the end of 2010, there has been a steady decline in appearances over the past 30 years. It is now considered a very rare vagrant to the UK, mainly in spring and early summer.

Caspian Tern
Hydroprogne caspia

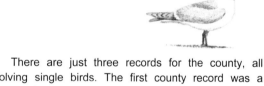

An extremely rare vagrant: three records

Caspian Tern, Seaton Snook, August 2000 (Iain Leach)

There are just three records for the county, all involving single birds. The first county record was a short-staying adult, initially found by Geoff Iceton on the Long Drag pools on 12 July (*British Birds* 76: 501). A total of eight Caspian Terns were seen in Britain during 1981, all between 6 June and 6 August, and the Long Drag bird was very typical in date, location and brevity of stay. The second record for the county came just short of a decade later and concerned a much more obliging adult found on Saltholme Pools by Graeme Lawlor on 2 August 2000 (*British Birds* 94: 479). It stayed in the Teesmouth area until 5 August, commuting between Saltholme and Seaton Snook, but also visited Bran Sands on the south side of the River Tees. Surprisingly, this was the only Caspian Tern seen in Britain in 2000. The most recent record was in 2006. On 2 July, an adult was seen flying south at Whitburn at 06.40 by Paul Hindess and Mark Newsome and was seen passing Hartlepool Headland at 07.45 (*British Birds* 100: 719).

Distribution & movements

The European population of Caspian Terns breeds on the Baltic coasts of Estonia, Sweden and Finland to head of the Gulf of Bothnia. Further east, there are fragmented populations from the Black Sea coast of the Ukraine across the steppe-lake region of central Asia to north westMongolia and east China. European birds are migratory and winter off West Africa south to the Gulf of Guinea; the Asian population winters on the coast to the south of their breeding range. Other populations of this species occur in Australia, South Africa and North America. In Britain,

Caspian Tern is a rare but annual vagrant, of which there were over 290 records of this species in Britain to the end of 2010 (Hudson *et al.* 2011). The European population is currently in decline and has consequently contributed to a decrease in sightings through the 2000s.

Whiskered Tern
Chlidonias hybrida

An extremely rare vagrant from southern and eastern Europe: two records involving three individuals.

A run of warm southerly winds in April 2009 brought a record-breaking influx of an estimated 24 Whiskered Terns to Britain, including a flock of 11 in Derbyshire on 24th. On 26 April, John Dunnett found two adults on Back Saltholme Pool, on the North Tees Marshes; the first for County Durham (*British Birds* 103: 598). The birds moved around various wetlands in the area before one of the birds was spooked by a Peregrine *Falco peregrinus* and disappeared shortly before dusk, the other remaining until dark.

On 25 July 2010, the county's second, a juvenile, was found on Saltholme Pools by Chris Sharp, and it remained loyal to this area until 25 August (*British Birds* 104: 588). It often fed close to the main road that bisects Saltholme Pools east and west. This was the earliest-ever juvenile to be recorded in Britain, perhaps indicating a breeding site close to the English Channel coast.

Distribution & movements
Whiskered Terns nest in scattered colonies through southern and eastern Europe, with the largest numbers further to the east from the Volga delta into Asia. European birds winter in tropical Africa. To the end of 2010, a total of 200 Whiskered Terns had occurred in Britain, the majority in May and June.

Black Tern
Chlidonias niger

An uncommon spring and autumn passage migrant.

Historical review
George Temperley (1951) described the Black Tern as "*a fairly regular passage migrant in very small numbers*". He said that it was most usually observed in May and August, but that there were records from June and July and as late as October. Temperley observed that it seemed to be occurring more frequently in his time than it had formerly.

Looking further back, Backhouse (1834) called it "*very rare*", noting that one was killed a few years ago "*at the pond on Sunderland Moor*" and this bird would appear to constitute the first record for the county. This was an abundant breeding bird in England during the 17th and 18th centuries, and it was still common as a breeding bird in some parts of East Anglia, for instance, up to the middle of the 19th century (Parkin & Knox 2010). It would therefore seem likely that this species would have been more regular in appearance in times past prior to the drainage of the county's larger wetlands, such as the Skerne Carrs, yet Hutchinson (1840) didn't mention it in his list of the birds of the county. There is a record of one from Stockton in 1837 and another that was shot near the Batts at Bishop Auckland in June 1850, the latter an interesting time of the year for one to occur inland. Hancock (1874) listed no records at all for Durham and just four for Northumberland, suggesting that it was indeed rare across the region in the Victorian period. One, described as a male in second-year plumage, was shot near Castle Eden in late October 1872, whilst other 19th century records for Durham include young birds that were obtained at Teesmouth in 1867 and three that were shot about four miles from Darlington in August 1868. It was said to occasionally occur in the adult state during the latter half of the 19th century (Temperley 1951). Tristram (1905) noted that birds were occasionally taken in the Tees estuary whilst Nelson (1907) said that individuals occasionally lingered at Teesmouth into late May. Chapman (1889) reported that one was at Waskerley Reservoir on 30 May 1887.

After this time, particularly in the early decades of the 20th century, Temperley recorded its becoming more frequent, though not apparently common; for example, two were in the Tees estuary on 21 September 1901 and a specimen was taken at Saltholme in September 1908. A few years later, one was present at Teesmouth on 2 May 1914 and two were at the same locality for over a fortnight in May 1917. During the 1930s, the species was observed a number of times at Darlington Sewage Farm; it was noted as visiting this location almost annually over a short period. On 1 May 1930, one was at Holme Fleet, Teesmouth and, by 1938, Bishop was reporting that the species had become an annual spring visitor to the Tees Marshes. On 16 September 1945, two birds were at Seaton Sands, alongside other tern species. In 1950, one was at Teesmouth on 18 August and the species was also noted on 21st and 25 August, and 3 September, with two there on the last date when two were also at Hurworth Burn Reservoir.

In the north of the county, birds were much less well reported than in the south east, so an immature bird on the River Don near Hebburn on 20th to 21 July 1929 was notable, whilst one on a pond at Brockley Whins, South Shields, on 10 August 1937 was of similar interest. One was seen by Dr. H.M.S. Blair on a flooded field, perhaps Whiteleas Pond, in north east Durham on 22 May 1945.

This low level, but regular presence was maintained through the 1950s and, although the spring passage of 1954 was larger than usual, numbers of birds in Durham were very much less than further south in the UK. Between four and eight birds in total were recorded on Teesmouth that spring, with two or three birds regularly at Cowpen Marsh between 5th and 12 May and, in the north, one was at Boldon Flats on 10th. In the autumn of the same year, passage was again more evident than previously. Birds were present in the Tees estuary from 8 August when three adults in breeding plumage were noted, then from 15th, birds were present throughout the remainder of the month and into September, numbers peaking at 12 birds on 1 September, before declining to seven by 5th. In the north of the county, one was at Boldon Flats on 12 September with three there on 15th, then two from 18th to 20th. The last of the autumn was one at Hurworth Burn Reservoir on 26th. These numbers paled in comparison with the gathering of 50-55 on Teesmouth following a torrential thunderstorm on 28 August 1958; at the time this was the largest flock recorded in the county.

In the 1960s, passage birds became more regularly recorded and it was realised that birds moved along the Durham coast, sometimes in reasonable numbers. Hence, 39 birds were counted flying past Hartlepool on 28 August 1960, a previously unheard of number of birds on the move. In 1962, at Teesmouth, a record number of 65 Black Terns was present on 20 August, no doubt detained on their migration by the concentrations of fish that attracted so many seabirds to the area at the time (Blick 2009). Autumn counts during the rest of the decade were typically much lower, but did include peaks of 19 on 11 September 1963 and 27 over the period 23rd to 28 August 1964, whilst 13 flew south off Hartlepool on 28 August 1966.

During the 1970s, a small number of spring birds was noted, almost on an annual basis around Teesmouth during May, with occasional sightings elsewhere in the county. Whilst counts typically featured just a handful of birds, there was stronger passage in 1973 and particularly in 1974. In the former year, a total of perhaps 14 birds was noted in May, with one to three at Teesmouth, one at Washington and three at Shibdon (Blaydon) Pond. In 1974, one of the most notable spring passages on record at the time took place from 10th until 17 May. On the former date, five were at Hurworth Burn Reservoir, with seven there and on the North Tees Marshes next day, then on 16th, five to seven birds were present at Shibdon Pond, on Teesside and at Washington, with six at the North Tees Marshes on 17th. The autumn of 1974 was also an outstanding one for this species, commencing in the last week of August with the usual handful of records. On 2 September however, a total of 87 birds flew south off Hartlepool in three hours. Thereafter birds were at Teesmouth throughout that month, with at least 21 present on 8th, building to a peak of 30 on 14th. By 22nd, 22 were still present, but only one remained on the last day of the month. Inland sightings included one to three at Washington, with two well inland at Witton-le-Wear on 7th. The remainder of the decade saw a return to normal with the general pattern being of a scattering of records through May and somewhat more regular sightings of very small numbers of autumn birds being typical. Teesmouth remained the prime site for this species, but others were noted at inland sites such as Blaydon, Brasside and

Washington. Singles were also noted well offshore on two dates; about 12 miles out to sea off Sunderland on 7th and 8 June 1975 and seven miles out on 31 May 1976.

Recent status

In Durham over the period 1980-2010, this delicate species had an annual presence as a scarce passage migrant. The number of birds seen varied markedly from year-to-year, and appeared to be closely linked to weather patterns, influxes often being associated with sultry or thundery continental weather, with east or south easterly winds. In general terms, it is scarcer during spring and although more regular in autumn, numbers are never particularly high. Almost any wetland of reasonable size in the county can attract the species, although adult birds in spring rarely stay for long. Autumn juveniles are much more likely to linger, particularly on the marshes and estuary around the mouth of the Tees, which is the species' principle site in the county. Other regularly frequented locations are Hartlepool and Whitburn, where birds are noted passing along the coast. Inland ponds and reservoirs including the western sites such as Derwent and Smiddyshaw Reservoirs can also attract passage birds in the right conditions.

During the spring migration period, May is undoubtedly the peak month for sightings but in most years fewer than 10 birds in total are recorded. There are occasional years when this species does not appear at all in Durham during spring passage, for example 1995 and 1998; likewise, there are other years when influxes occur. By way of example, in 1980, 12 birds were reported as passing through the county over 11th to 13 May, whilst between 1st and 3 May 1990, the best spring passage on record took place, with perhaps 70 birds in total, including 53 at Longnewton Reservoir. Even so, these numbers were low compared to those noted further south on the eastern side of the UK. The only other recent years to produce spring totals into double figures were 2008 and 2009. In the former year, up to 24 birds were reported between 4th and 9 May including nine together at Castle Lake on the latter date, with a further three birds at Teesside later in the month, whilst in 2009, strong south easterly winds produced 12 birds between 14th and 17 May.

Although May is the time of greatest passage, there have been a number of earlier records over the past three decades. April birds have occurred in 1987, 1991, 1992, 1994, 1996, 2007 and 2009, the majority occurring during the last week. Particularly early birds were noted at Saltholme Pools on 20 April 1987 and Crookfoot Reservoir on 21 April 1996, with the earliest on record being one at Derwent Reservoir on 13 April 1991 (Kerr 2001). Late spring migrants are not uncommonly recorded during the first week of June, whilst in late June and early July, birds occasionally appear; for example at Blackhall on 26 June 1995, Shibdon Pond on 24 June 1997, the North Tees Marshes on 1 July 2000 and Barmston Pond on 2 July 2006. These are presumably wandering non-breeding birds or failed breeders from the near Continent. In the past, occasional birds have lingered at Teesmouth, but not in any concerted fashion, indicative of breeding behaviour. There is no suggestion of a breeding population of this species having been present in Durham although a summer plumage bird was observed sitting amongst the breeding gulls in a Black-headed Gull *Chroicocephalus ridibundus* colony at a wetland in the mid-Wear valley on a single date in early July 1991 (Westerberg & Bowey 2000).

Autumn passage birds appear from the last week of July, with most birds being recorded during August or September but, as in spring, numbers are variable and influenced by prevailing weather patterns. In most years, autumn totals are in the range of 10-12 birds. Some years, they are very much poorer; for example in 1985 and 1997, when only four to five birds were reported, and in the exceptionally lean year of 1989, no birds were noted after 23 May. By way of contrast, there have been a number of good years, beginning in 1980 when almost 30 were recorded at Hartlepool and Teesmouth between 15th and 17 August. It was 1987 before the next comparable count, after up to four during August and up to six during early September, 18 were present on the WeBS count at Teesmouth on 13 September. In 1992, peak passage occurred over 11th to 13 September with approximately 30 birds noted, including seven at Shibdon Pond and 11 passing south off Whitburn on the former date. In 1999 and 2001, birds gathered at Seaton Snook at the mouth of the Tees in late August, with maxima of 23 and 25 respectively, whilst in 2004, there were 35 autumn records including up to 27 at Saltholme Pools in August. Active passage was again noted in 2009 and 2010 with the former year producing a count of 18 birds passing Hartlepool on 8 September, whilst 18 were logged at Whitburn Observatory over the period 19th to 29 August 2010.

October birds have always been scarce in Durham, but such occurrences in the two decades 1990-2010 became even rarer. There were sightings at Teesmouth on 19th and 22 October 1980, 24 October 1982 and 9 October 1983, with another late bird at Whitburn on 15 October 1983. However since then, the only October reports have concerned a single at Shibdon Pond on 26 October 1996 and three north past Hartlepool on 13 October 2005. There is one November record, of a bird on the North Tees Marshes on 10 November 1984.

Distribution & movements

Black Terns of the nominate form breed over much of Europe, from Iberia to southern Sweden, and east through Europe and eastern Asia to central Siberia. The North American race *surinamensis* has also been recorded in Britain as a vagrant. European and Asian birds winter principally along the tropical coasts of West Africa, although a few Russian birds winter on the Black and Caspian Seas (BWP).

White-winged Black Tern (White-winged Tern)
Chlidonias leucopterus

A rare vagrant from southern and eastern Europe: 25 records.

Historical review

George Temperley unsurprisingly described this species as "*a very rare vagrant*" with just one record to his time of publication. The first County Durham record came in 1869, when an adult was "*killed by G. Mussell in the marsh at Port Clarence on the north side of the Tees on 15 May*" (Nelson 1907). The specimen is referenced by Hancock (1874), who said he obtained the example "*from the collection of Mr Oxley of Redcar, April 1871*". It was almost a century before the county had its second, in August 1967, when an adult frequented the North Tees Marshes from 13th to 18 August. A major change in status then followed, illustrated by the fact that the species was encountered in a further 15 years to 2010.

All records:

1869	Port Clarence, adult, 15 May 1869
1967	Reclamation Pond, adult, 13th to 18 August (*British Birds* 61: 344)
1968	Washington Ponds, adult, 30 May (*British Birds* 62: 472)
1969	Washington Ponds, juvenile, 12 August (*British Birds* 63: 279)
1972	Saltholme Pools and Greenabella, adult, 13th to 16 May (*British Birds* 66: 343)
1977	Saltholme Pools, juvenile, 10th to 11 August (*British Birds* 71: 509)
1977	Seal Sands, juvenile, 22 August (*British Birds* 71: 509)
1979	Dorman's Pool, adult, 11 August (*British Birds* 73: 514)
1979	Jarrow Slake, adult, 2 September (*British Birds* 73: 514)
1979	Dorman's Pool, juvenile, 3 September (*British Birds* 73: 514)
1980	Saltholme Pools and Haverton Hole, juvenile, 17 August (*British Birds* 75: 511)
1982	Seal Sands, juvenile, 30 September - 5 October (*British Birds* 77: 537)
1984	Long Drag Pools, adult, 31 July - 5 August (*British Birds* 80: 546)
1985	Long Drag Pools, adult, 29 June (*British Birds* 79: 557)
1985	Dorman's Pool, juvenile, 1 September (*British Birds* 79: 557)
1995	Hartlepool Headland, adult, 10 July
1996	North Tees Marshes, second-summer, 29 June - 7 July
1999	Hartlepool Headland, juvenile, 7 August
2000	North Tees Marshes, adult, 25 June
2001	Dorman's and Saltholme Pools, adult, 4th to 10 June
2002	Dorman's and Saltholme Pools, second-summer, 10 June
2002	Whitburn Steel, adult, 10 July
2002	Greatham Creek and Seaton Common, adult, 3 August
2011	Back Saltholme Pool, adult, 1 July
2011	Saltholme Pools and Haverton, adult, 19th to 22 August

All records have fallen between 13 May and 5 October, with distinct clusters in late May/early June, late June/early July and the first two weeks of August. The earliest juvenile arrival date is 7 August; the first week of August is quite a productive period nationally for early dispersing juveniles.

Overall, adults and second-summer birds have outnumbered juveniles by almost two to one, although there is a much stronger bias towards non-adult birds in autumn. The North Tees Marshes is clearly the most favoured area with only six birds being seen away from there; two each at Hartlepool and Washington, plus singles at Jarrow Slake and Whitburn.

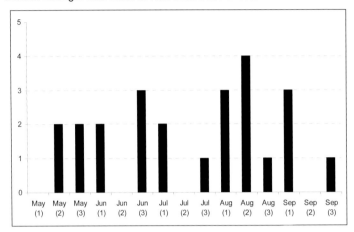

**White-winged Black Tern (all records)
by ten-day period
(based on arrival date)**

Distribution & movements

The European range of White-winged Black Tern includes Poland and Hungary and sporadically further west, but the main population is scattered through Asia as far east as north east China and the Russian Far East. The wintering range includes sub-Saharan Africa, the Indian subcontinent, South East Asia and Australia. Formerly a national rarity, it was no longer considered so after 2006 as national records had then exceeded 800. Although it is now an expected spring and autumn visitor in small numbers to the UK, drainage of east European marshes might see it revert to its previous status of rarity in the future (Fraser & Rogers, 2007).

Sandwich Tern
Sterna sandvicensis

A very common passage migrant and summer visitor.

Historical review

The first reference to Sandwich Terns in the county was made by Hutchinson in 1840. He stated that it frequented the mouth of the Tees and noted that it nested amongst the "*maritime plants which grow near the beach*". He implied that they bred at Teesmouth, rather than stating this categorically. Nelson (1907) said it was "*not infrequent*" in summer.

As with many tern species, populations in north western Europe were brought to the brink of extinction at the end of the 19th century by egg collecting for food and the hunting of adults for the millinery trade, but the species recovered in response to protective legislation in the early 20th century (Mitchell *et al.* 2004). There were nesting attempts in the Tees estuary during the 1930s, of which at least one was successful. Bishop informed Temperley that two pairs attempted to breed in the estuary in 1929, but that the eggs were lost to gulls. Four pairs attempted to breed in 1930 and one of these pairs hatched young; in 1931, six pairs laid eggs, but none reared young. What was presumably the same series of records was also documented by Stead (1964), who noted that up to six pairs attempted to nest in the North Gare area between 1929 and 1931, but with very little success. Sandwich Terns were scarcely established at the time, and there were apparently only about twenty pairs of Common Terns *Sterna hirundo* and Little Terns *Sterna albifrons* respectively, in this period. In 1937, breeding was suspected but spoiled by flooding, then no further breeding activity was recorded until it was suspected again in 1943, when the coast was closed off to the public for the duration of the war.

Temperley (1951) described the Sandwich Tern as "*a common passage migrant*", which had been known to breed. He noted that as passage migrants, the species passed north in May, with return movement occurring in August and September. Temperley highlighted a bird ringed as a nestling at Blakeney Point in Norfolk on 30 June 1928 and recovered on 2 September of the same year at North Sands, Hartlepool (*The Vasculum* Vol. XV 1929, Witherby 1929). He cited this as evidence that young terns may move northwards after fledging, before beginning their southward migration. This is supported by later ringing evidence.

In 1964, up to 490 birds were at Seaton Snook on 8 August, including a bird that had been ringed in Aberdeenshire in June of that year, indicating a more typical southward post-breeding dispersal (Bell 1965). In 1973, a notable inland record came from Jarrow Slake, where three were present on 25 August (Unwin 1974).

Recent breeding status

Along the Durham coast, small numbers of birds are regularly seen off Whitburn, Salterfen Rocks, Seaham, Hartlepool and other coastal sites throughout the spring and summer. The only modern breeding attempt in the county occurred at Saltholme Pools in 2006. On 18 June, a remarkable 82 were present and a few settled to breed, for the first time since 1931. A total of five pairs laid six eggs and hatched at least two young, but only one survived to fledge. This bird, which was ringed on 30 July, was later recovered after being shot in northern Spain, on 3 February 2007. Up to eight returned to Saltholme Pools in June 2007 and up to 33 were recorded the following summer. In spring 2009 up to nine were again present, including a mating pair, but no further breeding attempts have occurred to date (Joynt 2006-2009).

Recent non-breeding status

The Sandwich Tern is a common passage migrant along the Durham coastline from mid-April through to early September, with small numbers of birds apparently spending the summer months along the coast, in particular around the mouth of the River Tees (Blick 2009).

The first spring records of what is an early migrant in Durham typically come towards the end of March, usually in the final 10 days of the month. Earlier spring arrivals can occur, however, the earliest to date being two flying south off Whitburn on 7 March 1992 (Armstrong 1993). A more exceptional early record concerns a presumed wintering bird flying south off Whitburn Observatory on 17 January 2009 (Newsome 2010). This is only the second winter record for the county. Birds become more widespread from early April, with light northerly passage usually evident off the coast during the month. A thirty-year analysis of first arrival dates for Sandwich Tern between 1976 and 2005 revealed a slightly earlier arrival for the species in Durham over the period, on average, three to four days earlier than in the mid-1970s (Siggens 2005).

Sometimes, gatherings of up to a hundred birds occur at locations such as the mouth of the Tyne, especially in the last few days of April or the first week of May, before they are fully established in breeding colonies to the north. The key period for spring migration tends to be from the last week of April through until the first week or two of May. For example in spring 2007, there was a distinct passage off Whitburn in early May, with counts of 118, 286 and 240 documented between 5th and 7th of the month (Newsome 2007). Spring passage is generally much less conspicuous than it is in the autumn, being more rapid, involving fewer birds (Wernham *et al.* 2002) and when less time is spent watching by land-based observers. Spring trans-Pennine movement is also likely to occur overnight, in the same way as in autumn, as noted below (R. M. Ward pers. comm.).

The return passage begins as early as mid-June, virtually overlapping the spring passage, but it is early to mid-July before the first juveniles appear. The early summer dispersal from breeding grounds further north sees large coastal post-breeding gatherings at a number of sites along the coast. Regularly frequented locations include Sandhaven Beach at South Shields, Whitburn Steel, Nose's Point, Seaham, Dawdon Blast Beach, Chourdon Point, Blackhall Rocks and Hartlepool Headland. This autumn build-up usually begins in June, and is often evident, initially, at Whitburn Steel where 150-200 birds might have gathered by mid-month. The highest counts to date here were of 350 on both 23 July 1991 and 15 July 2003 (Armstrong 1992, Newsome 2006). Other high counts have included 420 at Sunderland South Docks, where large post-breeding tern flocks were regular during the 1980s, on 19 August 1985 (Baldridge 1986).

As the summer progresses, the number of terns off the Durham coast increases and by August, hundreds of birds have usually gathered around the mouth of the Tees, with favoured roost sites in this area being North Gare Sands and Seal Sands. In most years, between 1,000 and 1,500 birds can be present in these roosts between late July and mid-August, but occasionally larger numbers occur (Blick 2009). Numbers in the estuary including those birds on the south side outside the county, typically peak during early August when in excess of 3,500 birds were recorded during the 1990s (Ward 2000). Late summer counts for the whole estuary between 1990 and 1997 averaged 1,835, representing over 1% of the international population (Ward 2000). The highest count to date from exclusively the north side of the estuary was of c.4000 at Seaton Snook on 16 July 2002 (Little 2002).

As in other tern species, numbers at Teesmouth fluctuate widely, increasing significantly when there are sprat *Sprattus sprattus* wrecks in the estuary. Durham counts are also influenced by the fact that in some years, more birds roost just outside the county on the south side of the Tees. Increased observer coverage in later decades may also have boosted more recent counts. There was a very marked increase however, in the eighties, and a sharp decline from around 2004/2005, as illustrated, which shows the five-year mean peak counts from 1975 to 2009. The data is taken from DBC and TBC Annual Reports and WeBS counts for the period. This may reflect national trends, as described below.

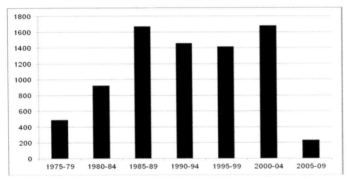

**Sandwich Tern - post-breeding records, North Tees Marshes
5-Year mean peak counts, 1975-2009**

Nationally, Sandwich Terns exhibit the most erratic population trends and distribution of any breeding seabird in the UK. The population fluctuates dramatically between years owing to large variations in the proportion of mature birds attempting to breed, and distribution varies owing to mass movements between colonies. The species increased in the UK from the 1920s to the mid-1980s however, as a result of protection from increasing recreational disturbance on beaches as well as from persecution. This included a 40% increase in the UK population between the national censuses of 1969-70 and 1985-1988. The UK population peaked in 1987 and has fluctuated since, declining until the mid-1990s then increasing until 2002, before declining again. A more marked decline in breeding productivity is apparent since 2000. Predation on eggs and chicks by foxes *Vulpes vulpes* is probably the most important factor determining breeding productivity, and fox populations are thought to be increasing due to reduced human control. Sandwich Terns nest on low ground close to the tideline, and are therefore also vulnerable to tidal inundation, which may increase in future as a result of increased storminess and sea level rise due to climate change (JNCC 2010).

Although the post-breeding counts at North Teesmouth are variable as previously noted, it is at least possible that the marked increase in numbers there in the 1980s is a real one, reflecting this national trend. Equally, the sharp recent decline in that area may reflect reduced productivity. In 2004 for example, both the North Teesmouth and Whitburn Steel counts fell dramatically from high figures in 2003 of 1,615 and 350 respectively, to peaks of just 306 and 30 in July, and birds dispersed early. At the time, this was attributed to diminishing local food supplies and a poor breeding season nationally (Joynt 2003, Newsome 2006 & Siggens 2005). It is entirely consistent with the national trend.

Peak sea passage in the county predictably tends to coincide with this post-breeding build-up. A particularly heavy movement was noted off Hartlepool Headland on 14 September 2006, when 1550 flew south in just one hour. Many of these birds presumably headed into the river, as 700 were present on Seaton Snook on 22nd (Joynt 2006). The following year, autumn passage peaked more typically in July, with a heavy southbound movement at Whitburn early in the month, perhaps in response to adverse weather. After 538 were noted on 5th, 1,000 flew south in just over four and a half hours on 7th, but the following week produced no more than 253 birds. At Whitburn Observatory, 5,477 birds were logged in just under 87.5 hours of recording in July, an average of 62.7 birds per hour compared to an August average of just 13.8 birds per hour (Newsome 2008).

By September, numbers at roosts have started to decline, and drop rapidly after mid-month. By the end of that month there are few birds locally, with fewer still in October. The occasional Sandwich Tern lingers into November, such as the bird off Hartlepool Headland on 25 November 2000, the latest autumn record for the county to date. Presumably the same individual was also seen on Crimdon Beach that same day (Armstrong 2005, Iceton 2001). There is an even later record of an adult, unseasonably in summer plumage, inland at Charlton's Pond, Billingham on 10 December 1996 (Armstrong 1997). This was the first winter record for the county.

The scale of movements north and south along the Durham coast over a whole season is indicated by, for example, the cumulative total of birds logged off Whitburn between April and October 2008. A total of 9,375 birds were recorded during this period (Newsome 2009). Many of these however, will have been local movements, with the same birds probably logged on several occasions, presumably including 'commuters' from the large colonies in Northumberland. This 'commuting' behaviour was illustrated by the 180 birds flying north off South Shields on the evening of 3 July 2000, many of which were carrying fish, presumably returning to feed young in the Northumberland colonies (Armstrong 2005).

Apart from presumed migrating birds flying overhead into the Tees estuary, this tern is relatively rarely encountered inland, making the December record described above even more notable. At Teesmouth, there are a few other inland records from Dorman's Pool, Coatham Marsh and Hartlepool Docks, but not many locations further inland. Along the Tees, four were at Portrack Marsh on 24 July 2000 (Iceton 2000). The furthest inland record in the county to date concerns one flying west over Blackton Reservoir, Teesdale on 14 June 1996 (Armstrong 1997).

A few probably wander into the mouths of, and up, all of the county's main river systems in the late summer of most years. A series of observations from the Gateshead stretch of the River Tyne suggests this, for birds are seen regularly as far west as Timber Beach, Dunston, at this time of year. The first record in Gateshead dates from 12 August 1979 when two adults were at Shibdon Pond, and two days after this, a party of ten were on the River Tyne near Bill Quay. On 22 August 1984, two were noted heading west and upstream at the Tyne Bridge, with up to six being seen there during the month. Birds occurred at their most westerly documented site on the river in 1989, with a party of five being seen at Ryton Willows on 5 September (Bowey et al. 1993). Other higher counts on the Tyne include 18 flying west over Friar's Goose on 31 July 1994, eight at Dunston on 29 August 1995, and 25 at Jarrow Slake, where smaller numbers regularly occur, on 21 July 2001 (Armstrong 1995 & 1995, Siggens 2005).

On the River Wear, the species regularly occurs as far inland as WWT Washington, where records include an exceptional 41 on 23 August 1989, and most recently two there on 6 August 2007 (Armstrong 1990, Newsome 2008). Other inland sightings, away from the main river valleys, include two flying south over Waldridge Fell on 20 July 1997 and up to 20 roosting at Boldon Flats in mid-July 1999 (Armstrong 1998 & 2003). Inland records have also come from Seaton Pond and Rainton Meadows (Siggens 2005). There have been few inland records since 2004, coinciding with a fall in numbers at coastal post-breeding roosts, as previously noted. However, a party of 29 watched flying inland at Longnewton Reservoir at dusk on 19 July 210 was exceptional.

Other notable records include a leucistic bird north past Whitburn on 24 August 1991, a partially leucistic individual at Seaton Snook on 4 August 2000 and a first-summer, an unusual plumage in the UK, at South Shields on 16 May 2001 (Armstrong 1992 & 2005, Siggens 2005). A leucistic juvenile was also seen at Sandhaven Beach, South Shields, on 11 August 2010. The same individual had moved north to Bellhaven Bay, East Lothian, on 12 August, but was at Seaton Carew (and then South Gare) on 18 August. This illustrates the pattern of post-breeding dispersal of juveniles, presumably from the colony on the Farne Islands. At Teesmouth, some Sandwich Terns were noted to be stained pink, along with many gulls, on 14 August 1998, perhaps as a result of iron oxide picked up by bathing off the steel works (Armstrong 1999).

Distribution & movements

Sandwich Terns breed along various coasts in the northern hemisphere, in a scattered fashion. In Europe, this breeding range stretches as far north as the Baltic and east to the Caspian Sea. Other populations are found in North America, the West Indies and parts of eastern South America. European birds winter in coastal habitats along the Mediterranean coasts, around the Middle East and from north west Africa, south to southern Africa.

It is apparent from ringing data that several thousand birds use Teesmouth as a staging post each summer and early autumn. While the flock size may stay almost static over a period of time, the turnover of birds can be considerable, with many just staying a few days or at most a small number of weeks. It has been estimated that as many as 10,000 birds may move through the estuary in the autumn (Blick 2009). Flocks of both Sandwich and

Common Terns are regularly seen leaving Teesmouth during August evenings and nights, and heading high south west inland, on a presumed trans-Pennine migratory path (Ward 2000, R.M. Ward pers. comm.). A similar movement was suggested by birds heard calling while passing south west over Sunderland late at night on 17 August 1978 (Unwin 1979). Conversely, smaller numbers have also been observed arriving at Teesmouth from the south west in spring (Ward 2000).

Colour-ringing of birds at their nesting sites and re-trapping on passage has shown that large numbers of birds using Teesmouth in autumn originate from the Farne Islands and Coquet Island in Northumberland. In addition to Teesmouth, ringing recoveries of birds from the Farnes have come from a number of locations along the Durham coastline including Blackhall Rocks, Crimdon and Whitburn. In 2000, Ward analysed records of birds ringed as nestlings in colonies elsewhere and re-trapped in late summer flocks at Teesmouth, and found that 68 out of a total of 85 birds were from Northumberland. Birds from Aberdeen, Fife, Orkney, Norfolk, Essex, Holland, Belgium, Ireland, Northern Ireland and Denmark have also been seen or re-trapped here in autumn. Birds ringed or noted at Teesmouth have been known to move north in autumn to Coquet Island and Fife (Blick 2009). This northward movement, also noted in birds ringed as nestlings from Essex and Norfolk, probably forms part of a general dispersal of birds after fledging (Ward 2000). This corroborates Temperley's earlier assertion (Temperley 1951) that such northward movement could occur before birds begin their southward migration.

There are a number of ringing recoveries from Teesmouth of birds ringed as nestlings on the Continent. They include one ringed at Fjando, Jyland, Denmark on 23 June 1979 and re-trapped at North Gare on 8 August 1979, and one ringed in the Netherlands on 31 May 1998, re-trapped at Seal Sands on 20 July 1998. More recently, two ringed at Zeebrugge, Belgium on 6 June 1996 and 26 May 2003 were re-trapped at Seal Sands on 8 and 30 August 2009 respectively, and two more ringed at Heist, Belgium on 11 and 21 June 2004 were also both re-trapped at Seal Sands on 8 August 2009. From Ireland, two ringed as nestlings on Lady's Island Lake, Wexford, on 22 June 2001 and 14 June 2006 were both re-trapped at Seal Sands on 8 August 2009 (Unwin & Sowerbutts 1980, Armstrong 1999, Joynt 2010).

Conversely, continental recoveries of birds ringed at Seal Sands include the following: an adult ringed on 1 August 1993 seen at Griend, the Netherlands on 26 April 1998; an adult ringed on 20 July 1998 seen at Zeebrugge on 23 April 2007; and a first-year ringed on 19 July 2000 and re-trapped at Westkapelle, the Netherlands, on 29 August 2007 (Iceton 1999, Joynt 2008).

The migration routes and wintering grounds used by Sandwich Terns are illustrated by the following data, from Ward's study in 2000. By October, juvenile Sandwich Terns ringed at Teesmouth post-breeding roosts had reached the Gulf of Guinea (one ringing recovery) and Senegal (two recoveries). Other recoveries of juveniles comprised four in March in Senegal and the Ivory Coast, and one in Spain in February. All remaining recoveries in Ward's study were of birds ringed as adults at Teesmouth roosts. Wintering areas for these birds extended along the West African coast (Senegal to Ghana), with five recoveries, plus single recoveries from the Congo, Angola and South Africa, and a more surprising January recovery from the Netherlands. Recoveries of birds en route consisted of single records from France in April and Portugal in September (Ward 2000). The Spanish recovery in February is at the northern limit of the wintering grounds of UK breeding birds, which extends around the Iberian Peninsula (Wernham et al. 2002).

Unseasonal ringing recoveries have also come from Senegal on 27 June and 15 August, suggesting that birds sometimes stay late or return very early to their wintering areas (Blick 2009). The June recovery was of a bird ringed as an adult at Seal Sands on 28 July 1996 and found freshly dead or dying in Senegal on 27 June 1998. It may have remained in its wintering quarters as a result of being in poor condition (Armstrong 1999).

Sandwich Terns, like most seabirds, are relatively long-lived, as illustrated by the following recoveries: two ringed in the nest on Coquet Island in 1967 and two others from Farne Island nests were all re-trapped at Teesmouth in 1986, having reached the age of 19 years; another ringed on Coquet Island in 1970 was re-trapped at Teesmouth in 1993, aged 23 years; and another ringed at the Sands of Forvie, Newburgh, Grampian, in 1978 was re-trapped in 1999, aged 21 years (Blick 2009).

Lesser Crested Tern
Sterna bengalensis

An extremely rare vagrant from the Mediterranean: one individual responsible for records spanning 11 years.

The identification of orange-billed terns was still a developing art in the early 1980s, and a summering bird on Blakeney Point, Norfolk, in 1983, eventually identified as Lesser Crested, had helped clarify many of the problems. Presumably the same bird relocated to north east England in 1984, first being seen at Seaton Snook by Russell McAndrew on the morning of 17 June. The bird had relocated to the Long Drag pools that evening, but was back at Seaton Snook early morning on 18th. It then remained in the area until 20th, providing many observers with the chance to see this exceptionally rare visitor to the UK. It was subsequently seen at South Gare and Redcar in North Yorkshire in late June and visited the Fame Islands for the first time in August. 'Elsie', as she was named, then returned to the Fame Islands every summer until 1997 and was responsible for numerous sightings on the English and Scottish east coast, including an excellent run in County Durham up to 1995.

A juvenile tern with orange bill and legs, but otherwise similar to Sandwich Tern, was seen at Seaton Snook on 30 July 1993, and what was considered to be another similar individual was seen on 9 August (Bell 1993). There was speculation at the time that these were the hybrid offspring of 'Elsie', for she was known to have paired and successfully reared young in that year with a Sandwich Tern on Inner Farne, Northumberland (Kerr 2001).

All records:

1984	Seaton Snook, 17th to 20 June (*British Birds* 79: 556)
1987	Dorman's Pool, 13 June (*British Birds* 81: 567)
1987	Hartlepool Docks, 23 August (*British Birds* 82: 531)
1988	Whitburn, 21 August (*British Birds* 83: 468)
1990	Hartlepool Headland, 10 June (*British Birds* 84: 476)
1990	Whitburn, 6 July (*British Birds* 85: 531)
1990	Reclamation Pond and Seaton Snook, 9th to 13 July (*British Birds* 84: 476)
1991	Reclamation Pond, 16 June (*British Birds* 89: 507)
1994	Hartlepool Headland, 12 May (*British Birds* 88: 520)
1994	Seaton Snook, 18th to 28 May (*British Birds* 88: 520)
1995	Hartlepool, 11 June and 15 July (*British Birds* 89: 507)
1995	Seal Sands and Seaton Snook, 12th to 16 June and 15 July (*British Birds* 89: 507, BB91: 482)

Distribution & movements

Vagrant Lesser Crested Terns in Europe originate from the Mediterranean population, which breed on islands off Libya and winter in coastal West Africa. Other populations exist in the Red Sea, Arabian Gulf and coastal Northern Australia, dispersing throughout the tropical Indian Ocean. It is a vagrant to several European countries and has bred in either pure or mixed pairs (with Sandwich Tern) in Italy and Spain, as well as England. Apart from the summering bird on the Fame Islands, there have been very few vagrants in Britain. Since the first UK record on Anglesey in 1982, a total of nine birds had occurred to the end of 2010.

Common Tern
Sterna hirundo

A very common passage and summer visitor, breeding in small numbers.

Historical review

The only historical breeding records of this species are from Teesmouth. Temperley referred to large numbers of terns breeding on sandbanks at the mouth of the river prior to industrialisation. He highlighted the Cott manuscript (1670), which recorded "*a shelf of sand, raised above the highe water marke that entertaines an infinite number of sea-fowle, which lay their Egges here and there*". The nesting terns here were likely to be Common Tern, though little distinction was made at that time between Arctic Tern *Sterna paradisaea* and Common Tern (Fletcher 2010).

The first half of the 19th century saw sporadic references. Sharp (1816) described the species as "*common*", though Edward Backhouse (1834) referred to just one specimen being taken at Seaton Carew, "*a few years ago*", and Hogg (1845) mentioned it, but did not refer to its breeding. Hutchinson (1840) was the first to refer unequivocally to breeding Common Terns, stating that it "*breeds among the gravel at Teesmouth*". Another breeding reference comes from Morris (1855) in which, Thomas Bedlington wrote, "*Sea swallows lay their eggs on the shore of the Tees*". He reported seeing more than 100 nests there in 1852, though some years later "*not a nest*" was to be found. Nelson (1907), referring to the south side of the Tees, noted that there were "*old inhabitants now living* (1906) *who can remember terns breeding near the estuary*".

All tern populations in north western Europe were brought to the brink of extinction at the end of the 19th century by the hunting of adults for the millinery trade, but recovered in response to protective legislation in the early 20th century (Mitchell *et al.* 2004).

There was no further breeding information in the county until 1922. The species' subsequent breeding history around the Tees estuary is well documented. A colony of two pairs was established, and found by Bishop, in 1922. Eleven pairs were there in 1923 and this rose to 50 pairs over the period 1928-1930 (Stead 1964 & 1969), dropping to 10 pairs in 1931, after flooding disrupted the colony. Richmond (1931) stated that the scattered ternery that "*sprang up in recent years*" was apparently declining. He said that Sandwich Terns *Sterna sandvicensis* were scarcely established and that there were apparently only about twenty pairs of Common and Little Terns *Sterna albifrons* respectively. Neither of these species was breeding there in Nelson's time. Breeding Common Tern numbers rose in 1941, because of the war restrictions to human access, and over this period more than one site in the area was occupied. Pairs nested regularly over the period 1944-1950, but no young were known to have been reared because of the reinstatement of human disturbance. The fortunes of this colony declined over the next few years and pairs were scattered, egg-collecting was rife and numbers fluctuated subsequently, between 10 and 25 pairs, right up to the early 1960s.

Temperley (1951) said the species was rarely seen inland, though two were noted at Hurworth Burn Reservoir on 25 May 1928 and one was at Darlington Sewage Farm on 17 September 1930. At his time of writing, Temperley described Common Tern as a "*common passage migrant*" and "*summer resident, now in very small and decreasing numbers*". He observed that it attempted to breed at Teesmouth and reported that it was then the most common species of tern along the Durham coast, with the main passage period northwards being in May, and the main return being in August.

By 1952, 24 pairs were nesting in three colonies at Teesmouth, but the largest of these, with 11 pairs, was washed out that year (Temperley 1952). There are spasmodic records thereafter from Teesmouth in the 1950s and 1960s, indicating a small breeding population. In 1955, *c.*20 pairs attempted to breed in a ploughed field there, with five or six young being found in July (Temperley 1955). In 1957, at least 20 young were fledged at Teesmouth (Grey 1958) and, in 1964, at least ten pairs were noted at what was described as the 'regular' Teesmouth site (Bell 1965). In 1967, at least six pairs nested there, where "*a few do nest most years*" (Bell & Parrack 1967). However, by 1971, there was very little breeding success at Teesmouth (McAndrew 1972).

A notable record came in August 1962, when huge concentrations of sprats *Sprattus sprattus* appeared in the bay and estuary of the River Tees, attracting some 2,000 Common and Arctic Terns, all fishing in the bay (Bell 1963).

Inland records came from Barmston Pond, Billingham and Brasside Pond in 1968. The first inland breeding in the county followed in 1969, when a pair took to a raft at Charlton's Pond, Billingham, increasing to 12 pairs by 1974 (Joynt 2008). A pair successfully reared one young at Darlington in 1971. Much overland movement was also recorded in late July and early August 1971 and, in 1973, a post-breeding gathering held up to 110 birds at Barmston Pond on 14 August (McAndrew 1972, Unwin 1974). This presaged a large increase in the number of inland sightings from the mid-1970s onwards, which probably resulted from the gradual cleaning up of the major river systems, such as the Tees, Tyne and Wear, with a consequent recovery of freshwater fish stocks (Westerberg & Bowey 2000).

Recent breeding status

The distribution of this species in Durham reflects its ability to feed on both marine and freshwater fish. Consequently, colonies are to be found at the coast in the south east of the area, or at wetlands along the main river systems. Most breeding Common Terns in Durham nest at protected or undisturbed sites. Traditionally, colonies have persisted at the mouth of the River Tees, and in the 1990s, the mouth of the River Wear.

On the North Tees Marshes, the breeding population has recovered and Teesmouth has again become the most important breeding area for Common Terns in the county. From 20 or so pairs present in the early 1990s, numbers had risen to over 200 pairs by the early 2000s. In 1986, nine pairs bred on the North Tees Marshes, with 27 pairs in 1987 and 15 pairs in 1989 (Joynt et al. 2008). Rafts in Cowpen Marsh held four pairs in 1994, rising to 24 pairs by 1998, and single pairs bred at Haverton Hole during the period 1996-1998 (Joynt et al. 2008). A large colony (of up to 300 pairs) which had developed on the south side of the estuary, just outside the county, was lost to development in 2000. A saline lagoon with tern-friendly features was built on the north side of the river, not far from Greatham Creek, to try to relocate this colony in time for the 2001 breeding season. This was used by 88 pairs in that first year (Joynt et al. 2008). By 2004, this co-operative effort between industry and nature conservation had resulted in the successful relocation of almost 480 pairs to protected sites north of the Tees, which became part of RSPB Saltholme from 2008 (Blick 2009). Birds were present at a number of sites, but most were at the saline lagoon. A count on the North Tees Marshes in June 2007 revealed a total of 490 pairs (Newsome 2007). This fell in 2008 to 300 pairs, probably due to predation by breeding Black-headed Gulls Larus ridibundus at the Saltholme islands. By 2010 however, numbers had recovered to 372 pairs (A. Snape pers. comm.). Locally breeding birds have been seen perching on wires at Teesmouth, and unusually one was feeding by plunge-diving from wires at Dorman's Pool on 16 July 2002 (Little 2002).

In 1989, Common Terns also began breeding on the Wear, at a disused site in Sunderland's South Dock. This colony became established after most working activities in the Docks had ceased. Numbers increased rapidly and during the nineties, this became the county's only large colony, with 120 pairs fledging 170 young in 1995 (Armstrong 1996). A peak of 200 pairs was reached in 1997 (Armstrong 1998). Wet weather in early summer 1997 however, resulted in low breeding success, with only c.50 young surviving to fledge. The original site of the colony became increasingly vegetated and less suitable, particularly so in wet weather. The following year saw more heavy rain early in the breeding season and as a result, birds abandoned the traditional site and moved to a less favourable, more exposed area nearby, where their breeding output was virtually zero. Ironically, this happened soon after Sunderland Council had designated the original area as a Site of Nature Conservation Importance. An attempt to manage the vegetation on the original site failed to entice the terns back, and after the 1997 peak, numbers declined steadily, until only 25 pairs attempted to breed in 2005 (Newsome 2006), and there have been no subsequent breeding records from this area. Breeding success had remained low after 1997 and there were regularly years of total breeding failure prior to the colony being abandoned. The birds moved to several derelict areas in the port complex, but did not return to the original protected site. In the final years of the colony, birds moved, unsuccessfully, to the southern end of the docks (J. Coulson pers. comm.). The decline and subsequent abandonment of the colony was believed to be as a result of predation of eggs and young, particularly by foxes Vulpes vulpes but also by rats Rattus norvegicus, possibly stoats Mus ermine and latterly Kestrels Falco tinnunculus (J. Coulson pers. comm.). Occasional human disturbance may have contributed, particularly after the relaxation of security within the port, and the sites favoured by terns became deserted. UK breeding terns are known to show a low degree of site faithfulness from year to year, in response to predation or habitat change (JNCC 2009), so the abandonment of the Sunderland colony was predictable.

Inland, Common Terns have also made significant gains since the early breeding efforts in the 1970s. There is, at present, a relatively small but increasing number of inland breeding sites, mainly in the north of the county. There is a considerable potential for increase in the number of these, particularly where artificial provision is made for the species by the creation of nesting areas such as islands and rafts. The majority of these inland birds occur in the east, where the rivers are wider and slow-moving providing the ideal fishing conditions for this species.

The Charlton's Pond colony eventually rose to 20 pairs, though it fluctuated over the period 1975-1998 (Joynt et al. 2008). Successful breeding of Common Terns at WWT Washington followed, beginning with a single pair in 1990. The first breeding attempt in Gateshead came in 1986 at Shibdon Pond (Bowey et al.1993), but birds did not succeed here until the 1990s, after the installation of artificial nesting rafts. After further unsuccessful breeding attempts in 1993 and 1994, birds bred successfully from 1995 (Armstrong 1995). At the time of the Atlas survey, inland sites such as WWT Washington and Shibdon Pond held in total less than 15 pairs per annum. The importance of these inland sites continued to grow in the early 21st century, the Shibdon Pond colony by then increasing to over 25 pairs. In 2007, inland breeding was confirmed at three sites in the county, with Shibdon Pond and WWT Washington holding around 30 pairs each, and Joe's Pond a single pair (Newsome 2008). By 2010, the colony at WWT Washington had grown to 67 pairs, which went on to produce 65 young, and a new peak was reached in 2011 when 101 pairs bred.

Today, the county population is likely to number over 550 breeding pairs, an almost 100% increase from the estimated number since publication of the *Atlas* (Westerberg & Bowey 2000). In part, this growth resulted from the re-location of breeding birds from the south side of the Tees estuary, in the Bran Sands area, to the North Tees Marshes. It also reflects the small but steady growth of the inland breeding population.

Many inland sites, such as Barmston Pond, Bishop Middleham, Brasside Ponds, Herrington CP, Hetton Lyons CP, Hurworth Burn Reservoir, Longnewton Reservoir, Rainton Meadows and Watergate Park, now receive regular visits from birds, especially during June and July (Newsome 2008). Several of these sites had attracted occasional breeding birds by the early 21st century, and more may follow. This is likely to depend upon appropriate, undisturbed nesting facilities being provided and maintained.

Common Terns in Durham nest successfully in areas that suffer little disturbance, low rates of predation, and offer the adults a nearby, good food supply. The ultimate failure of the Sunderland South Docks colony, as previously described, underlines the importance of this. Durham birds breeding inland nest on poorly vegetated islands or rafts in small water bodies near rivers. Freshwater fish caught, for example three-spined sticklebacks *Gasterosteus aculateus*, may be of lower nutritive value than marine fish (Massias & Becker 1990), but are easier to catch on days when poor weather conditions at the coast might reduce fishing success (Frank 1992). However, it should be noted that inland nesters also continue to collect food at the coast, and may be seen commuting between the coast and inland sites carrying fish.

Nationally, Common Tern numbers have increased slightly since 1986, and in the longer term, between the first census in 1969/1970 and 1985-1988, a 9% increase occurred. Local trends however, vary considerably, reflecting differing pressures facing Common Terns in different habitats across their wide geographical range (JNCC 2009). The positive trends in Durham, noting the local distortion in the county's population caused by birds moving from south to north across the Tees, are consistent with this favourable national picture.

This species has benefited nationally from habitat creation in the form of gravel pits, tern rafts, islets in industrial lagoons, and positive conservation management to control vegetation succession, gull competition, and predation. Maintaining the population is likely to depend on the continuation of such management in perpetuity (JNCC 2009). Again, the Durham population reflects the national position, particularly in response to positive human intervention at Teesmouth, and at inland sites such as Shibdon Pond and WWT Washington.

Recent non-breeding status

The first spring arrivals in Durham are typically in late April, when single birds often appear around estuaries or the established breeding colonies. Birds can arrive earlier than this, however: for example, one flying north past Whitburn Observatory on 15 April 2007 was a week earlier than the recent '10-year mean arrival date' (Newsome 2007), whilst the earliest spring sighting in 2011 was at the same locality on 9 April. An analysis of first arrival dates, between 1976 and 2005, in ten-year periods, revealed mean arrival dates of: 1976-1985, 20 April; 1986-1995, 27 April; and 1996-2005, 24 April (Siggens 2005). The first birds usually appear at inland breeding sites in the last few days of April or the first few days of May. There is also one exceptional far earlier spring record, concerning a bird not yet in breeding plumage at Seaham Harbour on 3 March 1978 (Unwin 1978).

Occasionally, one or two-year-old birds, with darker, heavily-worn flight feathers are seen at Teesmouth in summer (Blick 2009). This is referred to as *portlandica* plumage. Such birds are scarce in the UK as they generally spend their first summer and second winter in the northern tropics. One such recent record concerns two first-summer birds at Saltholme Pools on 5 July 2009 (Joynt 2009). Away from Teesmouth, there is a single record of such a bird on the River Tyne at Ryton Willows on 11 June 1992 (Bowey *et al.* 1993), whilst seawatching activities at Whitburn in the early part of the 21st century provide several further examples.

Post-breeding dispersal brings an influx of birds from outside of the area, supplementing the local breeding population. This results in a build up in numbers at coastal roost and loafing sites from late July onwards. This build up continues into August, when peak numbers are typically recorded. The highest counts ever recorded in the

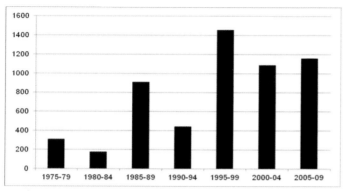

**Common Tern - post-breeding records, North Tees Marshes
5-Year mean peak counts, 1975-2009**

county are *c.*5,000 on the North Tees Marshes on 30 August 1999 and 22 August 2006 (Iceton 1999, Joynt 2006). The numbers at Teesmouth fluctuate widely, increasing significantly when there are influxes of Sprats into the estuary, and also depending upon prevailing weather conditions. Durham counts may also be influenced to a lesser extent by the fact that in some years, more birds roost just outside the county on the south side of the Tees. There is however, a slight upward trend in the 1990s and 2000s, as illustrated. This shows the five-year mean peak counts from 1975 to 2009, taken from *Birds in Durham*

and the Teesmouth Bird Club Annual Reports for the period. The ten-year mean peak counts show this more markedly, with 596 in the eighties, increasing to 951 in the nineties, and 1,126 in the 2000s; these perhaps reflect the favourable national picture for breeding Common Terns at the time.

Away from Teesmouth, the largest post-breeding roost is usually on Whitburn Steel, which became established as a regular roost site from 1982 onwards. The highest count here was 600 on 24 August 2007 (Newsome 2008), but counts in the low hundreds are more typical. Other regular roosts included Sunderland Harbour and South Docks from the early eighties to the early nineties, where a combined count of 652 Common and Arctic Terns was recorded on 10 August 1985 (Baldridge 1986). In the early years of the 21st century, Sandhaven Beach, South Shields, became a favoured location, with a maximum of 420 on 23 July 2006 (Newsome 2007).

Inland, from late July through August, large numbers of birds can sometimes be seen along the Tyne, the Wear and around the mouth of the Tees, with sometimes as many as 200 birds being noted on locations relatively high up these rivers, for example on the Tyne, at Bill Quay. As water quality has improved, more of these birds are being seen further up river and in late July 1987, a flock of 105 birds was present as far west as Scotswood Bridge (Armstrong 1988). Post-breeding counts at WWT Washington are also significant in some years, with the highest recorded total being 240 on 19 August 1990 (Armstrong 1990).

The heaviest sea passage typically coincides with the post-breeding peak at coastal roosts in late August, extending into the beginning of September. The highest passages ever recorded in the county have come since 2007, probably reflecting an increasing level of observer coverage at Whitburn Observatory. These movements included over 1,300 flying north past Whitburn in 13.75 hours in northerly winds on 4 September 2007 (Newsome 2008). On 29 August 2010, a large high- pressure system and northerly winds gusting at up to 50 miles per hour, displaced large numbers of terns from further north. The following day produced a very large movement of terns returning north in the afternoon, including 2,040 Common Terns. The fact that terns attempt to move north in the autumn is difficult to explain, but in such strong northerly winds they have little alternative, as they are only be able to fly into the wind; attempting to fly with strong tail-winds can result in them being blown into the sea.

The post-breeding peak is followed by a rapid decline in September as the species exits the county, the final birds usually being recorded in October. The latest coastal record was of a single off Hartlepool Headland on 5 November 1997 (Bell 1997). There are two late inland autumn records, with a juvenile on the Tyne at Newburn on 10 November 2008 (Newsome 2009), and one found freshly dead at a Willington gravel pit on the exceptionally late date of 24 December 1994 (Armstrong 1995).

Distribution & movements

The Common Tern is a temperate nesting species, which is found throughout Europe, Asia and also in Central North America. European birds winter off the coasts of Africa, with other populations off southern Asia and Australasia (BWP).

Count and ringing data suggest that over 1% of the UK breeding population, with their young, migrate through Teesmouth in late summer (Ward 2000). This study also found that 55% of adults using Teesmouth have yet to initiate primary moult. This is consistent with the involvement of a substantial Baltic contingent of non-moulting adults amongst Teesmouth's late summer population as most adults from the Baltic undergo primary moult later, after arrival on their wintering grounds. This is confirmed by ringing recoveries. For example, birds mist-netted at night at Seal Sands on 7th to 8 August 1999 included two juveniles from Finland and a single juvenile from Estonia (Armstrong 2003).

There are therefore two different populations involved in post-breeding roosts at Teesmouth; birds which breed locally, and those which breed in the Baltic. Local breeding birds winter mainly in West Africa, from Sierra Leone to Ghana, as evidenced by a number of ringing recoveries from Teesmouth. They include, for example, one ringed as a nestling at Seal Sands on 2 July 2003, then re-trapped at Malan, on the Ivory Coast on 23 February 2004 (A. Snape pers. comm.).

The Baltic birds leapfrog over UK birds to winter in Namibia and South Africa. The furthest travelling locally-ringed Common Tern was one ringed as an adult at Seal Sands on 11 August 2008 and re-trapped at Port Alfred in Cape Province, on the east coast of South Africa, on 19 September 2009; a distance of 10,171 kilometres. Conversely, an adult ringed at Port Alfred on 29 November 2002 was re-trapped at Seal Sands on 3 September 2006 (Newsome 2009). There have also been a number of Teesmouth ringing recoveries of birds from Mile 4 Saltworks in Namibia, as the South African Ringing scheme (SAFRING) ring birds there (A. Snape pers. comm.).

A selection of notable Common Terns, from or to Durham, includes:
- An adult ringed at North Gare on 19 July 1977, caught on a fishing boat in Portuguese Guinea on or around 15 December 1978
- A bird ringed as nestling in Taurage, Lithuania, on 27 June 1988, controlled at Seal Sands had travelled 1,501km west in 53 days, on 19 August 1988
- A bird ringed at Seal Sands on 3 September 1994 was controlled 10,124km to the south east at Cape Province, South Africa, on 29 April 1995
- A juvenile ringed at Seal Sands on 3 September 1994 was killed 2,662km to the south west off Essaouira, Morocco, fifteen days later on 18 September 1995
- A bird ringed as pullus on 21 June 1987 at Hame, Finland was controlled over 12 years later at Seal Sands on 4 September 1999

The local breeding birds appear to leave earlier in autumn, the last local recapture of young ringed at RSPB Saltholme in 2010 coming on 18 August. This was also when Baltic birds began to appear in numbers in 2010, with five from Finland and singles from Stavanger and Stockholm on the ringing nights of 18th to 19 August and 17 September. On 3 September 2010, the first ever Russian-ringed Common Tern for the UK was caught at Teesmouth (A. Snape pers. comm.).

There is a rapid turnover of birds at Teesmouth during this period. This results in fluctuating numbers, for example a peak of 1,470 in mid-August 1995 declining to less than 100, before rising again to a second peak of over 750 in early September. This turnover is also reflected in the fact that in 1995, none of the 552 Common Terns mist-netted over five nights were recaptured (Ward 2000).

Ringing data from late summer at Teesmouth, combined with records from rings read in the field at Seaforth Docks, Merseyside has identified a trans-Pennine autumn migration flyway between these two sites. This was confirmed when an adult Common Tern was observed at Seaforth in August 1995, seven days after it was dye-marked at Teesmouth. This is further backed up by regular observations of Common and Sandwich Terns departing high to the south west from Teesmouth at dusk in autumn. Smaller numbers have also been observed arriving from the south west in spring (Ward 2000).

A bird ringed as a nestling at WWT Washington on 29 June 1994 was observed at Seaforth Docks on 28 May 1998 and again on 4 June 2000, indicating that an interchange of birds can occur between these two breeding colonies (Armstrong 1999 & 2005). A bird ringed as a chick in Belgium was re-trapped in the Sunderland South Dock colony in the 1990s (J. Coulson pers. comm.), illustrating that birds fledged on the near Continent can also transfer to UK breeding colonies.

Roseate Tern
Sterna dougallii

An uncommon passage visitor and one time, very rare breeder.

Historical review

In his catalogue of the birds of the county, Edward Backhouse (1834) listed it but recorded no status for the species. Hutchinson (1840) called it the most numerous of the terns at Teesmouth in the summer, which, in retrospect and in the light of its national status in the 19th century (Yarrell 1885), seems highly unlikely. Temperley clearly disbelieved Hutchinson's assertion, which was presumably based upon a misidentification. By contrast, William Backhouse (1846) said it occurred rarely in the Tees Bay and Proctor (1846) said that it was noted occasionally on the coast. Yarrell (1885) noted that nationally "*there is no doubt that numerically this species has undergone considerable diminution*", citing egg collecting and the collection of plumes for hats as the reasons for this. All tern species endured big declines in the 1800s due to trapping, their tail feathers being highly prized for hat decoration (JNCC 2009).

The species was not listed by Tristram (1905), while Nelson (1907), said that there were no authenticated records of its capture at Teesmouth "*in recent years*", but he presumed that it passed up and down the coast. Bishop, in a letter quoted by Temperley from 1928, stated that it "*makes its appearance at odd times*", but gave no specific examples or occurrences.

At the mid-point of the 20th century, Temperley said it was "*a rare passage migrant, of which there are few definite records*". The most recent records at that time were of one at the Velvet Beds, Marsden, on 23 July 1949 and an adult at the mouth of the Tees on 8 September 1950.

Numbers remained low through the 1950s, with the species being described as rare along the Durham coast in 1955, when a small passage of birds was noted in July/August off Teesmouth. The decade's highest recorded count was just three, at Seaton Snook on 8 July 1956. The first spring records at Teesmouth followed in 1958, with singles on 31 May and 1 June (Grey 1959).

Slightly higher counts and more frequent records followed in the 1960s, including eight flying south at Seaham on 25 July 1962. The majority of the records in this period again came from Teesmouth, including an adult noted feeding a juvenile at North Gare on 26 September 1964, these two birds lingering until the late date of 4 October. The highest counts in the remainder of the decade were six at Seaton Snook on 25 May 1967, and eight at Teesmouth on 4 September 1969.

High numbers of Roseate Terns arrived on Coquet Island in Northumberland in the 1960s, with over 150 pairs nesting in 1968, some moving from the Forth Islands (J. Coulson pers. comm., Kerr 2001). These breeding numbers in the north east are unsurpassed to date. It is therefore likely that post-breeding Roseate Terns were also more common in the county at this time, and were overlooked.

There was an exceptional gathering of birds at South Shields in late August 1971, with up to 25 birds being recorded between 28 August and 24 September, the maximum count coming on 30 August (Unwin 1973). This was the prelude to more frequent records, and higher late summer counts, as observer coverage increased in subsequent decades.

Recent breeding status

This 'red-listed' species of conservation concern experienced the most dramatic decline in the UK of any seabird species between Operation Seafarer (1969-1970) and the Seabird Colony Register Census (1985-1988) (JNCC 2009). This decline was attributed to the trapping of adult birds on the wintering grounds off West Africa, as the birds that visit our coastline mainly winter in Ghana. Other factors, such as predation and nesting habitat loss may also have played a part. The species has undergone a gradual recovery since, in response to education programmes in the wintering areas in West Africa, and conservation management at breeding sites (JNCC 2009).

In the late 1980s and early 1990s, occasional late spring and early summer sightings at coastal localities in the north east of the county began to occur. The usual pattern of Durham summer records, which had been established through the 1970s, involved birds appearing locally between late July and early August. From the late 1980s, this 'window' was extended, as both the number of birds and the frequency of late spring and early summer sightings increased. This trend was typified in 1993, with a single record in May, and two in June (Armstrong 1993). This presaged the development of a small breeding colony in the county in the mid-1990s.

Most unusually, this colony was at a mainland site, Roseate Terns breeding within an established Common Tern *Sterna hirundo* colony at a disused area in Sunderland's South Dock between 1995 and 1999. This was, at that time, the county's only large Common Tern colony. The first successful Roseate Tern breeding attempt for the county came here in 1995, when two pairs successfully reared two young. The following year, a remarkable 11 pairs were present in July. Birds arrived late that year however, perhaps including failed breeders from Coquet Island, and it is believed that no young fledged in 1996 (Armstrong 1997). In 1997, four pairs nested and it is believed two young were raised, but this proved to be the final successful breeding effort by Roseate Terns at Sunderland South Dock. The large Common Tern colony here, which peaked at 200 pairs in 1997, declined and dwindled to zero pairs over a period of eight years from this 1997 peak. As the Common Terns declined, the breeding of Roseate Terns ceased. The last record of Roseate Tern attempting (unsuccessfully) to breed anywhere in County Durham came from Sunderland South Dock in 1999 (Armstrong 2003). There were never more than two successful breeding pairs at Sunderland and never more than two young fledged in a season. This is the only location at which Roseate Tern is known to have successfully nested in Durham. Elsewhere in the county, a pair was noted displaying and mating at Hartlepool on 10 July 1994, but nesting was not proven (Joynt *et al*. 2008). In 2006, two visited Saltholme Pools intermittently between 9 and 13 July and were seen on the artificial tern island, perhaps prospecting for a nest site (Joynt *et al*. 2008).

Recent non-breeding status

The first spring arrivals tend to come later than for other tern species, from late May onwards. The earliest record in the county was of a one fishing off South Shields on 3 May 2005 (Newsome 2006).
Numbers in Durham typically remain low until late summer. Then, as with other tern species, post-breeding dispersal brings an influx of birds from breeding colonies outside the area, resulting in peak numbers locally from late July to early September.

From the mid-1970s to the late 1980s, even the late summer peaks were relatively low, with the highest counts being 14 moving south off Hartlepool on 2 August 1975, up to 14 in a tern roost at Sunderland Harbour in August 1982, and more than ten at South Shields on 18th to 19 August 1983. This period coincided with the major international decline of the species, which probably explains low numbers and relatively few records despite increasing observer coverage.

During the mid-1990s, Whitburn Steel and South Shields became the most regular post-breeding gathering sites in Durham, with the highest count being 16 at Sandhaven Beach, South Shields and in the mouth of the Tyne on 20 August 1998. The following year, 30 were at Seaton Snook, Teesmouth on 30 August. This late summer period, as well as producing peak counts at coastal roost sites, inevitably produces the heaviest sea passage. This included 52 flying north off Whitburn Observatory in four hours on 4 September 2001 (Siggens 2005).

The pattern of sightings along the Durham coast varied over the period from the late 1990s through to the 2000s, with the species being somewhat less frequently reported from 2004 to 2008, but this period did produce another high count of 30 at Seaton Snook on 20 August 2006 (Newsome 2007). During the first decade of the 21st century, post-breeding birds tended to use Sandhaven Beach, South Shields most regularly, from late July to early September. This pattern was particularly evident in 2009, with birds present here from the last week of July and the first juvenile being noted on 4 August. The highest count was of 22 birds on nearby South Shields Pier on 12 September (Newsome 2010).

The pattern of records in the county is much influenced by the size of the nearest Roseate Tern colony, on Coquet Island, Northumberland. This colony grew strongly in the first few years of the 21st century, almost tripling from 34 pairs in 2000 to 94 pairs in 2006 (Morrison & Gurney 2007), after the introduction of nestboxes to shelter eggs and chicks from nest predators. As a result, by 2006, it had become the largest colony in the British Isles outside of south west Ireland. In 2009, high local figures for non-breeding birds reflected a good breeding season on Coquet Island, where 90 pairs raised 101 young. The Coquet Island colony enjoyed another good summer in 2010, with 76 young being ringed (M. Kitching pers. comm.). This provided a significant 'pool' of birds, which resulted in an unprecedented passage off the Durham coast in 2010 (Newsome 2010).

The late summer 2010 was the best-ever period for Roseate Terns in County Durham, particularly along the coast between South Shields and Whitburn Steel. Observers at Whitburn Observatory recorded high counts of passing birds, and there was a continued presence of the species around South Shields Pier, Sandhaven Beach and Whitburn Steel from mid-July to early September. Sightings peaked in July with 11 at Whitburn on 18th, and 17 on Sandhaven Beach on 31st. In August, numbers gradually increased with daily sightings of up to 15 at Whitburn Observatory and up to 20 resting on South Shields Pier. On 12 August, 44 flew past the Observatory, with 54 the next day.

At the end of August 2010, a large high-pressure system and northerly winds gusting at up to 50 mph displaced large numbers of terns from further north. On the afternoon of 29th, tern passage included 69 Roseates past Whitburn. The following day produced a considerable passage of terns northward, including 115 Roseate Terns, the largest number yet recorded off the Durham coast (Newsome 2010). This unprecedented event was likely to have been at least partly a reflection of the success of this species on Coquet Island. Plentiful food off the South Tyneside coast during August 2010 must also have contributed, as large numbers of terns lingered all month.

As with other tern species, Roseate Terns exit the county rapidly in September, and there are few October records, the latest being an adult feeding a juvenile at South Shields on 14 October 1983 (Baldridge 1984).

An interesting local observation took place on 26 June 1983, concerning a bird seen by Barry Stewart and Chris Gibbins at the mouth of the River Tees. Whilst watching a party of three Roseate Terns, two of which were adults in summer plumage, they realised the third bird was in an unfamiliar plumage, to which the observers could find no reference. It transpired that this was a Roseate Tern in first-summer plumage, equivalent to the so-called *portlandia* plumages of Common and Arctic *Sterna paradisaea* Terns (Sharrock 1980). At the time, this was one of the few records of a bird in this stage of plumage in Britain and Ireland (Stewart 1984).

Distribution and movements

Roseate Terns breed in scattered colonies that are located over a wide area of Europe, Africa and the Indian Ocean. Only small numbers breed in north westEurope, most of these being around Britain and Ireland, with some in the Azores and other populations in the north eastern USA and the West Indies. North westEuropean birds winter off the west coast of Africa.

The predictable link between Coquet Island and southward post-breeding dispersal into the county is confirmed by ringing recoveries at Seal Sands. Out of nine Roseate Terns re-trapped there between 2001 and 2010, six were ringed as young on Coquet Island. Four of these birds were re-trapped on their first autumn journey south, within a few weeks of fledging. For example, one ringed on Coquet Island on 2 July 2009 was re-trapped at Seal Sands on 30 August 2009 (A. Snape pers. comm.).

The remaining three re-trapped Roseate Terns at Seal Sands were all ringed as young at Rockabill, Ireland. For example, one ringed at Rockabill on 4 July 2001 was re-trapped as an adult at Teesmouth on the same day as the above juvenile (A. Snape pers. comm.). The interchange of Roseate Terns between colonies on opposite sides of the British Isles has long been understood, but this is direct evidence of such movement (Newsome 2009).

Arctic Tern
Sterna paradisaea

A common passage and summer visitor, though very rare as a breeding bird.

Historical review

Edward Backhouse (1834) called the Arctic Tern *"very common"*, but said Common Tern *Sterna hirundo* was only occasionally met with. Hutchinson (1840) said that it bred at Teesmouth, but Hogg (1845) said he had never seen it or heard of it being observed in Durham. More recently, Nelson (1907) said that small numbers were occasionally at Teesmouth in early May and also in autumn.

Arctic Terns, in common with other tern species, were probably reduced to low levels internationally by hunting for the millinery trade and egging, but are likely to have increased since the 1930s owing to legal protection (Mitchell *et al.* 2004). Bishop, the official watcher at Teesmouth in the 1930s, reported that small numbers of Arctic Terns were present from August to September in the Tees estuary between 1937 and 1939, and Sistern reported them there in 1943. There was an unusual inland record at Hebburn Ponds in 1947, a bird being reported there on a daily basis over the period 24 April to 7 May.

Temperley (1951) described the Arctic Tern as a "*passage migrant off the coast*". He felt the reports of the various early recorders were somewhat contradictory, reflecting the difficulty of separating Arctic and Common Tern. By 1951, he said that there were relatively few definite records of the species from the Durham coast excepting at Teesmouth, though he believed they were present in huge mixed flocks of terns recorded off the coast in May and June, and from August to September each year.

There is little information post-Temperley until the 1970s, although the maximum count of birds in the county came in August 1962, when 1,200 were reported to have gathered at Teesmouth (Blick 2009). When large numbers of terns are gathered together however, it is difficult to discern the real proportions of Common and Arctic Terns in such flocks. Up to 810 were at Seaton Snook on 3 August 1964 demonstrating the scale of autumn passage at the time (Bell 1965). In 1970, there were two unusual inland sightings. Two birds were at Witton-le-Wear on 17 May and one was at Shotton on 20 July (McAndrew 1971). The following year brought another record at Witton-le-Wear, one being seen on 24 May, suggesting some overland migration, as had already been noted in neighbouring Yorkshire (Mather 1986). By contrast, a flock of seven birds at Barmston Pond in late July 1971 was perhaps more indicative of post-breeding dispersal to valley-located wetland sites. The year also brought a late bird, present off Whitburn on 31 October (McAndrew 1972). In 1970, a pair spent most of May on Saltholme Pools, a prelude to subsequent breeding attempts in the county many years later (Blick 1978).

Recent breeding status

Occasional summer sightings in coastal localities occur in Durham, but these are usually well away from potential breeding habitat. The first known breeding record for the county came during the A*tlas* survey period (Westerberg & Bowey 2000). This concerned a single Arctic Tern apparently paired with a Common Tern at Teesmouth in the summer of 1992. It was noted feeding a Common Tern's chick, either as part of a mixed pair or acting as a helper (Armstrong 1993). The following year, one was paired with a Common Tern on the Long Drag in June, but the birds failed to lay eggs (Bell 1993). In 1994, an adult paired with a Common Tern fed four fledged young at Greatham Creek on 13 August (Bell 1994), but as the young had already fledged when observed, these birds could have bred outside the county.

Subsequently, up to five pairs bred in 1997 in the large colony of Common Terns at Sunderland South Docks, but only one chick was found during ringing of the Common Terns, in what was a year of low breeding success in the tern colony. Arctic Terns never bred there again (J. Coulson pers. comm.). One was in the Common Tern and gull colony at Saltholme Pools throughout the summer of 2006, with one or two birds there again in late June 2007 (Joynt 2006 & 2007).

Recent non-breeding status

In relative terms, only small numbers of Arctic Tern compared to Common Tern are usually reported in Durham, despite the fact that almost five times as many Arctic Terns breed in Britain (Mitchell *et al.* 2004). A much larger percentage of this species' colonies, and a greater number of birds, are also located close to the county's borders. There are large breeding colonies on the Farne Islands (Kerr 2001) and along the east coast of Scotland, as well as in Orkney and Shetland (JNCC 2010). Despite this, relatively small, although not insignificant numbers of Arctic Terns pass along the Durham coastline. It is difficult to tell whether this has always been the case, or is a more recent pattern, because of the problem of separating the two species, particularly when considering older records. Earlier records frequently lump the species together as 'commic' terns.

The first birds in spring are usually observed around the third week of April, or slightly earlier, and by the last week of April, birds are regularly passing along the Durham coastline. One north at Whitburn Observatory on 10 April 2009 was the earliest ever recorded in Durham, the previous earliest record being two at Whitburn on 11 April 1988 (Newsome 2010, Armstrong 1989). An analysis of first arrival dates between 1976 and 2005 revealed that there had been little change in this in Durham (Siggens 2005).

Spring passage is usually light, rising to double figures on many days, but rarely peaking above 30 to 40 birds per day. If higher rates of passage do occur at this time of the year, these tend to be concentrated into isolated days and associated with onshore winds, for example c.550 meandering slowly north off Hartlepool Headland on 8 May 2006, and a movement of c.130 birds per hour there on 28 May 2007 (Joynt 2007 & 2008). The main seawatching sites of Hartlepool and Whitburn tend to be the only places where birds are regularly noted for much of the summer period, though the species must be present along the length of the coast between these locations. During high summer, a light daily passage of birds is regularly noted north and southbound, often fishing: presumably these are foraging adults from the Farne Islands' colonies.

Arctic Tern is principally a bird of autumn passage in Durham. At both Whitburn Steel and Teesmouth, there are significant post-breeding gatherings of this species, along with other terns, from late June onwards and, in the latter area, birds are routinely recorded between mid-July and mid-October. Away from Hartlepool, Teesmouth and Whitburn, just a handful of birds tend to be noted on several autumn dates at locations such as Blackhall Rocks, Castle Eden Denemouth, Chourdon Point, Dawdon and Seaham, but all coastal locations must occasionally attract birds at this time of the year.

Large-scale movement south tends to take place from July onwards. Towards the end of this month, juveniles begin to be noted amongst the flocks of adults. More sightings come in late July, as birds disperse from breeding colonies outside the area, particularly Northumberland. At such times, passage off coastal watch points can be quite high, such as the 801 which flew north in 10 hours off Whitburn Observatory on 21 July 2007, a record single-day passage for the county to date. This heavy movement was the result of birds being displaced following northerly winds. A similar passage was noted on 24 July 2005, when 670 flew north past Whitburn in seven hours in strong northerly winds (Newsome 2008 & 2006). Birds routinely roost on the rocks of Whitburn Steel at this time of the year. The highest count to date here was an exceptional 209 on 30 July 1992, when 100 were also feeding offshore (Armstrong 1993). However, the roost of just 20 here on 26 August 2001 was more typical (Siggens 2005).

In terms of active passage, in some years only a few hundred birds might be noted close inshore, and it may be that birds pass along the coast further out to sea than other terns, or alternatively that their main movements north and south are concentrated along the west coast of Britain (Blick 2009).

Over the five-year period 2005-2009, which includes the two high counts noted above in 2005 and 2007, the average annual maximum single-day passage off Whitburn was just 394 birds (Newsome 2006-2010). All of these maxima predictably occurred between mid-July and mid-August. Similarly, when a spectacular tern passage took place on 30 August 2010, Arctic Terns were typically outnumbered, when just 250 flew north, compared with 2,050 Common and 115 Roseate Terns *Sterna dougallii* (Newsome 2010). Given that the Farnes population alone was 2,199 pairs in 2010 (National Trust 2011), this underlines the point that the majority of these birds are not seen passing along our coast.

Some of the most significant gatherings of Arctic Terns in the county come in late summer at Teesmouth, where the Seaton Snook tern roost can attract 100 or more in late July and through August, such as the 100 that were there on 24 July 2009 (Joynt 2009). The highest count of roosting birds in the county to date was 710 at Seaton Snook on 14 August 1988, almost matched by *c*.650 there on 31 July 2006 (Bell 1989, Joynt 2007). However, these numbers are far lower than those of Common and Sandwich Terns *Sterna sandvicensis*, and in some years, less than 50 are recorded at the North Teesmouth roosts, reflecting the comparatively light sea passage noted above.

Nationally, the Arctic Tern population increased through the 1970s and early 1980s, peaking around 1988, perhaps as a result of increasing sand-eel *Ammodytes sp.* stocks around Shetland. A collapse in the sand-eel stock between 1984 and 1990 however, resulted in a reversal in fortunes. The population fell to a low point in 2004, but has recovered slightly since. Arctic Tern suffers from amongst the lowest breeding productivity of any UK seabird, due to food shortages, gull predation, and poor weather hampering foraging and chilling eggs and chicks. Declining sand-eel stocks around the Northern Isles may be linked in some areas to rising surface sea temperatures due to climate change (JNCC 2010). Sand-eels have generally remained plentiful around the Farnes in recent years however, where the Arctic Tern population was stable in 2010 (National Trust 2011).

As the numbers at the North Teesmouth post-breeding roosts are relatively low and erratic, there are no clear long-term trends. However, it is notable that the highest ever recorded roost count, in 1988, coincided with the peak in the UK breeding population. Conversely, a very low peak of just 14 birds at Seaton Snook in 2000 (Armstrong 2005) coincided with a period of decline in the UK population. Low breeding productivity was also reflected in one of the Seaton Snook roost counts when, out of a high total of 320 birds, just two were juveniles, on 25 July 1998 (Iceton 1999).

Smaller numbers remain in local inshore waters through most Septembers and sometimes linger into the first week or so of October, with a few birds usually present intermittently through that month, but numbers decline rapidly as October progresses. Arctic Terns have lingered into November more frequently than other tern species in recent years, with November records every year between 2004 and 2009. They include a series of records at Hartlepool Headland in 2008, where one flew south on 7 November, one flew north on 12th and finally a first-winter was at the fish quay on 21st, the latest autumn record to date (Joynt 2009).

Inland records may be more regular than records indicate even though such observations of small numbers of birds have been made on a regular basis since the mig-1960s. The largest inland count to date is of 21 birds at Longnewton Reservoir on 27 April 2004 (Joynt 2005). Amongst the large numbers of Common and Sandwich Terns witnessed setting off from Seal Sands after dusk on late summer nights on a trans-Pennine migration route, there will undoubtedly be Arctic Terns (R. M. Ward pers. comm.). A bird found dead on in the Moorhouse and Teesdale NNR was presumably indicative of such overland migration, as may have been parties of three, 15 and 16 birds flying high to the west over Whitburn village on the evening of 3 August 2006 (Newsome 2007).

There are a number of inland sightings over the modern period from wetlands in Gateshead, such as along the River Tyne and Shibdon Pond. These commenced with a report of one at Shibdon Pond on 10 August 1977. In the period 1980-2010, birds were noted rarely at Ryton Willows, whilst Shibdon attracted occasional birds. In the eastern half of Gateshead, on the Tyne by Felling and Bill Quay, large flocks of 'commic terns' gather in late summer and no doubt these regularly contain small numbers of Arctics. Small numbers drift up river with Common Terns in late summer, into areas around the Tyne at Dunston (Bowey et al. 1993), and similarly on the Wear at WWT Washington. A flock of 41 terns flying west up the Derwent valley on 22 August 1997 included 'some' Arctics, and it may have consisted solely of this species, as only Arctic Terns were positively identified (Armstrong 1998). This record again suggests overland migration.

Other records further inland include two at Derwent Reservoir on 25 May 1993, and two flying east over Brandon on 30 April 1998 (Armstrong 1994 & 1999). The species is also occasionally recorded at sites such as Hetton Lyons CP and at Hurworth Burn Reservoir, where two pairs were displaying on 20 May 1999, with one pair again present on 15 June 1999 (Armstrong 2003). In 2007, a typical year, the only inland bird was one a long way west at Waskerley Reservoir on 6 May, but a juvenile flying upriver at Jarrow Slake on 15 September was more predictable, as such birds have not infrequently been noted at this site in late summer and autumn (Newsome 2008).

There are also a number of notable records of first-summer birds, unusual in the UK, the first being one inland at Crookfoot Reservoir on 7 July 1993. Many of the other records of first-summers have been at Teesmouth including one or two lingering in the mouth of the Tees in late June 2000, at Saltholme Pools on 14 July 2006, at North Gare on 13 June 2007 and at Saltholme on four dates in July 2009 (Bell 1994, Iceton 2001, Joynt 2007-2010). Occasional birds have also been detected at Whitburn during seawatch activities in the first decade of the 21st century.

Distribution & movements

The Arctic Tern breeds throughout the Arctic, sub-Arctic and northern temperate zone of Eurasia and North America. It winters in the southern oceans, as far south as the Antarctic pack ice.

Ringing records include one ringed at Maklappen Island, Scania, Sweden as a nestling on 29 May 1933 and recovered at the mouth of the Tyne on 24 July 1949, aged 16. Others ringed as nestlings include one ringed on 26 June 1954 near Heiligenhafen, Schleswig-Holstein, Germany, found dead at Marsden, South Shields on 16 July 1968, and one ringed at Fjortoft, Norway, on 24 July 1970, found dead at Darlington on 13 September 1970 (McAndrew 1972).

Another bird from Schleswig-Holstein in Germany was recovered at Teesmouth 13 years after being ringed. Other European-ringed birds include an adult ringed in Jylland, Denmark, on 1 July 1974 and re-trapped at North Gare on 22 July 1986. Two birds ringed at the same colony at Turku-Pori, in Finland, have been recaptured at Seal Sands. These include a bird ringed as pullus on 24 July 1989 that was controlled just 64 days later on 26 September 1989, having travelled 1,520km to the west south west. A decade later, another pullus ringed there on 28 June 1999 was controlled just 53 days later on 20 August 1999. One ringed in Norway in July 2009 was controlled at Teesmouth in the autumn, just two months after being ringed (Blick 2009). An Estonian bird ringed as a nestling on 19 June 2007 was re-trapped at Seal Sands on 9 August 2007 (Joynt 2008).

One ringed on the Farne Islands in the summer of 1957 was found dead at Crookfoot Reservoir on 6 May 1960. Birds ringed on the Farne Islands between 1963 and 1965 were subsequently recovered: near Easington, at

Seaburn, Seaton Carew and South Shields, illustrating one of the major points of origin for birds occurring along the Durham coast. One ringed as a nestling on the Farnes was reported at Teesmouth 14 years later, while a nestling ringed on Coquet Island was reported at Teesmouth later the same year (Blick 2009). Three adults ringed on Seal Sands, two on 2 September 1995 and one on 26 July 2003, were later recorded at Groningen, The Netherlands, with two present there on 2 June 2008 and one on 21 May 2009 (Blick 2009).

These ringing records illustrate a wide range of origin of birds in the Teesmouth post-breeding roosts, and show a predictable southward dispersal from both continental and northern UK breeding colonies in autumn. There are no recoveries of birds ringed in the county from the species' remote wintering grounds in the southern oceans.

Guillemot (Common Murre)
Uria aalge

A very common passage migrant and non-breeding summer visitor, but scarcer in winter.

Historical review
This species was recorded in large numbers during the 19th century, with Sharp (1816) describing it as "*very common*" at Hartlepool, and Hogg (1845) also giving its status as "*very common*". Hutchinson (1840) said it was "*abundant along the whole length of the coast*" about late July.

By 1931, Richmond wrote that "Guillemots, Razorbills Alca torda, Manx Puffinus puffinus and Sooty Shearwaters Puffinus griseus, seem not uncommon in these parts", i.e. around Teesmouth in July. In 1951 Temperley observed that Guillemots did not occur in such great numbers, prompting him to speculate that the species was probably much more common "a century ago". However, he did add that it had been noted "taking part in the occasional northerly passages of sea birds along the Durham coast in times of north easterly gales".

More generally, Temperley observed that in the mid-20th century, the Guillemot was "a regular visitor off the Durham coast at all times of the year", but chiefly in late summer, when "young birds are passing". He said that there were no records of it ever having bred on Durham's sea cliffs, but did comment that he thought that both races i.e. Northern Guillemot U. aalge aalge, and Southern Guillemot U. aalge albionis were probably recorded off the Durham coast.

Recent breeding status
Guillemots have never been proven to have nested successfully in the county though non-breeding birds are often present, loafing offshore during the summer or even present on the cliffs at Marsden. For example, the species was seen on a regular basis in the Marsden Bay area in the summer months of most years during the 1988-1993 *Atlas* survey (Westerberg & Bowey 2000).

Observations of occasional birds amongst the other breeding species on the cliffs at the Marsden seabird colony increased during the early 1990s. It was thought that this might be a precursor to breeding (Westerberg & Bowey 2000). In 2008, a single bird lingered on cliffs around Lizard Point in mid-June, then the following year at least two birds visited the cliffs there in June (Newsome 2009 & 2010). They may have been inexperienced, prospecting birds. Another more isolated breeding season record concerned a single, apparently healthy individual perched on the cliff top at Chourdon Point on 19 April 1996 (Armstrong 1997). To date, however, there has been no confirmation of breeding for the county. The closest breeding colonies to Marsden are on the Farne Islands in Northumberland (Fisher & Holliday 2008), some 75 kilometres to the north.

Recent non-breeding status
The first months of the year tend to be relatively quiet off the coast, with low numbers of birds noted off, for example, Hartlepool Headland and Whitburn. Only small numbers winter inshore in most years, and birds are scattered in small numbers down the whole of the Durham coast. Modest counts of 38 off Whitburn-Marsden and 24 off Hendon–Seaham on 17 January 1993 were still higher than in some years (Armstrong 1994).

Occasionally, however, far higher numbers of Guillemots are found lingering inshore during the winter, usually in response to adverse weather conditions. In 2000, for example, 1,200 were on the sea off Whitburn on 15

January, with a combined total of 1,400 including some Razorbills off South Shields the following day (Armstrong 2005). This may have resulted from hard weather further afield or possibly an abundance of small fish close inshore. Around 1,000 Guillemots remained on the sea between Whitburn Observatory and Whitburn Steel up to 3 February 2000, but numbers declined rapidly thereafter. A very heavy Guillemot passage was also noted in this period, with 1,000 flying south in just 10 minutes off Whitburn Observatory on the early morning of 31 January 2000, followed later that day by 859 flying north and 19 south in an hour. On the same date, birds were estimated to be passing north at Hendon, Sunderland, at the rate of 600 per hour.

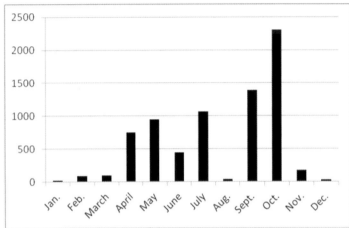

Guillemot movements off Whitburn Observatory
5-year monthly mean peak day-counts, 2005-2009

The heaviest sea passage of auk species (a combined Guillemot and Razorbill count) coincided with adverse winter weather, with 13,400 south off Hartlepool Headland in six and a half hours on 31 January 1976 in easterly winds (Bell 1976). All auk species, including Guillemots, suffer in such extreme conditions, and this movement was followed by a 'wreck' of oiled birds, with *c.*750 Guillemots and Razorbills on the Tyne-Tees coast over 6th and 7 February (Unwin 1977). January storms, with strong northerly or easterly winds, also forced large numbers inshore and caused many oil pollution victims in both 1978 and 1979, but numbers did not reach the spectacular 1976 totals (Unwin 1978, Unwin & Sowerbutts 1979). In February 1983, all auk species, including Guillemots, suffered again in a prolonged spell of strong north easterly winds in what *British Birds* described at the time as *"the largest seabird kill ever recorded in Britain"* (*British Birds* 76: 240). In this case, few were oiled, and starvation was the likely cause of the losses (Baldridge 1984). Another large wreck of auks occurred along the north east coast in February 1994, with 47 dead Guillemots in the Whitburn area alone, including a small number of oiled birds (Armstrong 1996).

The typical seasonal pattern of Guillemot movements in the county is illustrated which shows the five-year monthly mean peak day counts off Whitburn Bird Observatory for the period 2005 to 2009 (figures derived from Newsome 2006-2010). There are significant variations from year to year, largely resulting from changes in weather conditions, as peak movements invariably occur during northerly or north easterly winds, hence the value of considering averaged monthly maximum day counts over a five-year period. Prior to 2005, it is more difficult to assess the size of Guillemot movements, as observers tended to group counts of Guillemot and the usually less numerous Razorbill together as 'auk species'. Guillemots typically outnumber Razorbills, and did so over this five-year period by on average approximately two to one. A higher average ratio of Guillemots to Razorbills of approximately four or five to one however, has been noted in the Teesmouth area and may be more typical (Blick 2009).

An obvious increase in sightings tends to occur in early March, when birds begin returning to breeding colonies further north. Spring passage does not usually begin in earnest until late April off the Durham coast, however, and peaks in May. Spring movements have included 1,500 north past Whitburn in 5.2 hours on 13 May 2006. Larger numbers begin to move again in early July, and have included a high count of 1,800 moving north past Whitburn in four hours on 21 July 2008 (Newsome 2007 & 2009). This post-breeding peak may involve non-breeding birds or failed breeders, or adults still able to fly with chicks at sea, moving to safe areas to moult.

The year's first juveniles, presumably from the Northumberland colonies, are usually noted off the Durham coast around the first week of July. A lull invariably follows in August, as adult birds are rendered temporarily flightless by moult and accompany their flightless young, drifting south on the sea. This lull is clearly evident in the dramatic August drop shown in the monthly mean peak day counts.

Although passage stops in August, and therefore this appears as a gap in the monthly mean peak day count figure, summer 'rafts' of birds are to be found loafing on the sea off the Durham coast during this flightless period.

Such rafts are made up of family parties from further north. They have included 306 off Whitburn on 23 August 1990 and 350 at Blackhall Rocks on 29 August 2008 (Armstrong 1991, Newsome 2009). Rafts of over 100 are frequent and occur most years.

Far larger gatherings occasionally occur in late summer, often in response to the presence of fish shoals. An exceptional c.2,000 Guillemots were noted feeding on large shoals of Herring *Clupea harengus* and sprats *Sprattus sprattus* in an area of about one square mile, from a boat six miles north east of Sunderland on 27 July 1975 (Unwin 1976). Vast shoals of sprats were probably also the attraction for a gathering of c.2,000 auks, of which an estimated 85% were thought to be Guillemots, off Sunderland in August 1981 (Baldridge 1982). High numbers also occurred at Teesmouth in 2006, when c.1,400 were in the mouth of the Tees in late July, many lingering throughout August, with c.820 still present in early September (Joynt 2007).

The August lull in movement is typically followed by the heaviest sea passage of the year in the autumn, as illustrated. The record sea passages of Guillemot off Whitburn Observatory have come within this period, with 4,000 north on 11 October 2004 and c.4,000 flying north on 16 September 2009, the latter surpassed a few weeks later by 5,700 north on 17 October (Siggens 2006, Newsome 2010).

Most 'wrecks' of auks have tended to come in extreme weather in the early months of the year. Guillemots however, were noted starving in Scottish and Irish waters in September 2006. Fortunately, there was little evidence of this from our coastline, although 15 close inshore in Sunderland Harbour on 19th were obviously weak and at least one dead bird was found nearby. Similar numbers were close inshore in the Newburn and Hartlepool areas in the following week and some birds also entered the Tyne during this period (Newsome 2007).

The final two months of most years tend to produce relatively small numbers of passage birds. Small numbers of birds tend to linger close inshore in the autumn, and there is a small residual population off the Durham coast which probably stays through the winter.

Away from the North Sea, Sunderland Harbour and Jarrow Slake on the Tyne are also favoured localities. Along the River Wear, birds are regularly recorded as far upstream as Timber Beach. On the Tees, birds are routinely present around Hartlepool Bay, in the nearby Docks, and on the river itself as far west as the Tees Barrage.

In Durham, the species is rare inland, although it may have been under-recorded along rivers in the past. In the well-studied Gateshead area, the first documented inland record was of one on the Tyne above the Scotswood Bridge on 1 February 1985. In early 1986, there was a series of sightings of birds, with at least two, and perhaps four individuals involved. A bird further inland at Ryton Willows on 17 February was picked up uninjured after crashing into a tree and two days later, two birds were on the River Tyne at Felling, whilst on 2 March, one was at the same locality. This cluster of sightings was associated with a period of hard weather in the North Sea, which had driven many auks inshore during February 1986 (Bowey *et al.* 1993). A bird was found long-dead, under power lines at Shibdon Pond on 12 June 1994. This bird had presumably found its way inland during the large wreck of auks that had occurred along the north east coast in February 1994 (Armstrong 1996). There were a number of inland records in 2001, including up to 20 at Jarrow Slake in February (Siggens 2005).

The long-term pattern of Guillemot records in the county is difficult to interpret, given a large increase in observer coverage, and a willingness to regularly separate Guillemot and Razorbill counts when seawatching only from around 2005 onwards. There has been a significant long-term national increase in the breeding population of 30% between the late 1980s and 2009. Declines in productivity however, have been found at breeding colonies in the north and east of the UK since 2000, as a result of food shortages which may be caused by warming sea temperatures due to climate change (JNCC 2009). This makes future population trends uncertain. Given that these northern colonies are likely to account for most of our records, future non-breeding numbers in the county are difficult to predict.

Distribution and movements

Guillemots breed around the coasts of northern Europe, North America and the north Pacific, largely on the boreal and lower Arctic zones. In the north Atlantic, which includes the British breeding population, their range extends from Iberia as far north as Svalbard, and eastward into the Baltic. Wintering birds are pelagic and European birds are distributed over a large portion of the North Atlantic in winter, although some remain in the inshore waters of the North Sea (BWP).

Ringing recoveries include a breeding adult of the race *U. aalge hyperborea* marked on Kharlov Island, Murmansk, Russia, on 27 July 1940, and found at Hartlepool in May 1950. It was thus at least 12 years old. Another ringed as a nestling on Helgoland, Germany, on 18 July 1967, was recovered at Seaham Harbour on 3 August 1968.

There are a number of ringing recoveries of birds from breeding colonies on the Northern Isles, and one from the Western Isles. A bird ringed as a nestling on Fair Isle 25 July 1962 was recovered at Whitburn 504 kilometres to the south on 9 February 1963 (Bell 1964). One of the bridled form was ringed on Fair Isle on 24 June 1972 and recovered at Easington on 28 February 1986, 528 kilometres to the south. It was at least two years old when ringed, and was therefore at least 15 years old when recovered (Armstrong 1987). One ringed on Canna in the Inner Hebrides as a nestling on 2 June 1976 was found dead at Hartlepool on 10 February 1983, and another ringed as a nestling on Fair Isle on 29 June 1985 was picked up at Seaham on 26 February the following year (Armstrong 1993, Armstrong 1987). Finally, one ringed at Sumburgh, Shetland, on 25 June 1988 was picked up oiled but alive, for treatment, at Marsden on 13 September 1989 (Armstrong 1990).

During the 1960s and early 1970s, a series of recoveries of Guillemots ringed on the Farne Islands in Northumberland, were discovered along the Durham coastline from South Shields to Teesmouth. Many of these were found as tide-line corpses, in an oiled or partly-oiled state, a sad reflection of the times in relation to maritime and shipping activities off the county's coastline. Examples of these recoveries include:

- One ringed as a pullus on the Farnes on 11 July 1959 and drowned in fishing net 10 miles off Hartlepool on 12 August 1966
- A bird ringed on 23 June 1962 was found oiled at Seaham Harbour on 10 March 1963
- A bird ringed on 24 June 1961, found oiled at Seaham on 3 September 1964
- One ringed on 29 June 1964, found oiled at Sunderland on 6 September 1964
- A bird ringed as an adult on 20 June 1960, found Hartlepool on 1 February 1965
- A bird ringed as a pullus 2 July 1966 recovered oiled at Teesmouth 6 November 1968
- One ringed as a pullus on 29 June 1961 recovered oiled at South Shields 31 December 1970

These records confirm a predictable southerly dispersal down the east coast from the Northern Isles in the autumn and winter. They also suggest that birds from the Northern Isles may form a significant component of the movements seen off the Durham coast in spring and autumn, though clearly many birds from the large colonies on the relatively nearby Farne Islands are also involved in sightings off the Durham coast.

Razorbill
Alca torda

**A very common passage migrant, though not as common in winter.
Also now established as an uncommon breeding species.**

Historical review
The bones of this species were found in a quarry in an area described as Whitburn Lizards, on Cleadon Hills, in the spring of 1878, implying that the species was present and hunted in former times in the county (Howse 1878). Although it is not possible to tell whether the remains referred to breeding birds or passage migrants, it is more likely that birds would have been caught while perched on cliffs, suggesting that it may have been a breeding bird in earlier times.

Backhouse (1834) said that the Razorbill was common in the county, and Hogg (1845) said it was "*very common*" in spring and summer. He further stated that it could be seen fishing off South Shields in summer and that it was "*not uncommon*" in the mouth of the Tees. Bewick (1804) stated that his woodcut, which featured in *A History of British Birds (Volume 2, first edition 1804)*, was of a bird that was shot at Jarrow Slake in May, but no year was documented.

In the mid-20th century, Temperley (1951) described it as *"an autumn visitor off the coast; less often seen in winter"*. He said that it bred on the Yorkshire coast and in Northumberland at the Farnes, but he reported that it had never been known to have bred in Durham.

Recent breeding status

The Razorbill is present off the Durham coast throughout the year in varying numbers. The range of the county's breeding birds is severely limited by a lack of suitable nesting habitat. For this, Razorbills require wide ledges or boulder beaches overlooking the sea, usually in relatively secluded inaccessible locations. There are a number of such sites around Marsden Rock in Marsden Bay, South Shields, but nowhere else in the county. Consequently, the species is completely confined to the cliffs at this site (Westerberg & Bowey 2000). Nesting birds prefer the seaward cliff faces of Marsden Rock, but small numbers of birds also nest on mainland cliffs to the south of the Rock, in the area around Lizard Point. Elsewhere, other non-breeding birds may be encountered in coastal waters during the breeding season, but it is rare to see such birds on land and it would appear that the rest of the Durham coast is not suitable for further colonisation by this species.

During the 1980s, a small number of birds were regularly seen close inshore during the summer months, largely in the Marsden area. For a number of years, however, there was no evidence of breeding. Two Razorbills attempted to land on Marsden Rock on 23 June 1981 and two were described as apparently prospecting a sea stack at Marsden in the summer of 1984 (Baldridge 1982 & 1985). This was the prelude to the colonisation which followed.

Razorbills were first definitely established as a breeding species in the county in 1987, when two pairs nested at Marsden. In 1988, the number had increased to seven pairs, although only one chick was believed to have been reared. Consolidation occurred during 1989 and 1990, when nine pairs were present, increasing to 11 pairs during 1991-1993 and, in this latter year, eight chicks were reared (Westerberg & Bowey 2000). By 2000, the breeding Razorbill population had continued to increase steadily at Marsden, to reach a new peak of 28 apparently occupied nests (Armstrong 2005).

Coinciding with the arrival of breeding Razorbills at Marsden, a new colony was also established in north Northumberland in 1988 (Bradshaw *et al.* 1989). Between 1969 and 1970, and again during 1985-1987, there was a substantial increase in the British population (Gibbons *et al.* 1993). These new local colonies therefore fit into this later identified period of population growth.

Accurate counting of the local Razorbill population can be difficult, due to the nature of the cliffs at Marsden, but in 2005 and 2006, the colony here was monitored as part of the national Seabird Monitoring Programme. This revealed that there were identical numbers of 56 pairs breeding in both seasons. Given that there are established large colonies further north, such as around the Farne Islands in Northumberland (Fisher & Holliday 2008), the Marsden colony is comparatively small and might be considered to be still in the relatively early stages of establishment. Unfortunately in 2007-2009, no accurate counts were made, but numbers were considered to be 'fairly stable' at between 50 and 60 pairs (Newsome 2007). The species' habit of leaving the breeding ledges when chicks are only three weeks old, usually at dusk, and immediately taking the young out to sea, makes it difficult to establish the outcome of breeding (BWP). Whilst it is heartening to see additional species establishing themselves as breeding birds in the county, with an estimated UK population of 187,000 birds in the Seabird 2000 census (JNCC 2010), Durham's breeding population of Razorbills is of little significance in the national context. It is notable however, as a colony close to a large urban population and at a site frequented by large numbers of people.

Nationally, the breeding population continued to increase significantly from over 132,000 during Operation Seafarer in 1969-70, to the above Seabird 2000 census figure. The continuing, albeit small-scale, success of Razorbills in Durham is consistent with this pattern. It should be noted however, that Razorbill productivity declined nationally in some years in the first decade of the 21st century, as with Guillemot *Uria aalge*. This coincided with food shortages which may be linked to declining sandeel *Ammodytes spp.* productivity due to rising sea temperatures (JNCC 2010). This makes future population trends difficult to predict.

Recent non-breeding status

The first winter quarter tends to be relatively quiet for this species off the Durham coast, the majority of counts being in single figures. Exceptions do occur, however, usually in response to extreme weather conditions. In February 1983, for example, prolonged north easterly winds caused widespread starvation in what was at the time described as 'the largest seabird kill ever recorded in Britain' (British Birds 76: 240). In the county, many corpses were found along the shoreline, with twice as many Razorbills as Guillemots; 28 Razorbills being found dead on one stretch of less than a hundred metres at Marsden on 19 February 1983 (Baldridge 1984).

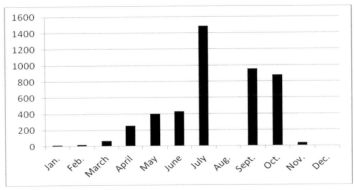

Razorbill movements off Whitburn Observatory
5-year monthly mean peak day-counts, 2005-2009

The typical seasonal pattern of movements in the county is illustrated (figures derived from Newsome 2005-2009). There are significant variations from year to year, largely resulting from changes in weather conditions, as peak movements invariably occur during periods of northerly or north easterly winds, hence the value of considering mean monthly maximum day counts over a five-year period. Prior to 2005 it is more difficult to assess the size of Razorbill movements, as observers tended to group counts of Razorbill and the usually more numerous Guillemot together as 'auk species'.

A noticeable increase in reports usually occurs from around the last week of March, when counts rise as birds begin returning to breeding colonies, both locally at Marsden and further north. This was typified by records in 2009 when, after mainly single-figure counts in the first two months of the year, a passage of 144 north past Whitburn Observatory was noted on 28 March (Newsome 2010). Movement tends to increase through April, as more birds return to breeding colonies.

Passage then peaks with post-breeding dispersal in the second half of July. Such movements have included 2,600 north off Whitburn in 8 hours on 20 July 2008 (Newsome 2009). As with Guillemot, this peak may involve non-breeding birds or failed breeders, or adults still able to fly with chicks at sea, moving to safe areas to moult.

A lull invariably follows in August as adult birds are rendered temporarily flightless by moult and accompany their flightless young, before passage resumes in September. This is clearly evident in the illustrated monthly mean peak day counts. During this late summer lull in passage, auk flocks including Razorbills are often seen loafing on the sea in 'rafts', close inshore. These are family parties with newly-fledged young, dispersing from colonies further north. A notably large raft of 350 was off South Shields on 5 August 2006. In September of that year, when Guillemots were reported starving in Scottish and Irish waters and small numbers were affected locally, there was little evidence of any local impact on Razorbills (Newsome 2007).

The autumn then produces a second peak passage period at both Hartlepool and Whitburn during September, sometimes extending into October, mirroring the pattern of Guillemot movements. This period includes the record passage of Razorbill for the county at the time of writing, with 17,000 passing south off Hartlepool Headland on the morning of 1 October 2006 (Joynt 2006). In 2007, a marked increase in the number of Razorbills relative to Guillemots was noted in the autumn passage period, with an exceptional 92% of auks passing on 15 September being Razorbills (Newsome 2008). However, this does not apply every year, as the annual Whitburn passage peak for Razorbill was earlier on average than that of Guillemot during this five-year period, as comparison of the two monthly mean peak day count figures shows.

After autumn movements have ended, numbers typically decline during the final quarter. This period sees a return to a low residual winter population, with few records and single-figure counts. As in the first quarter, this is influenced by weather conditions. Small numbers were noted off North Gare during the autumn gales in 2007, several tired birds appearing close inshore (Newsome 2007).

Increased observer coverage, and a willingness to separately count Razorbills when seawatching, has undoubtedly played a large part in the observed increase in the scale of Razorbill movements recorded in the first decade of the 21st century. Both an increase in the national breeding population and the establishment of the local colony at Marsden, may also have contributed to this upward trend.

Distribution & movements

Razorbills breed along the coasts of Scandinavia, Iceland and the Faeroe Isles, Britain, the Baltic and across the North Atlantic on the west coast of Greenland and north east Canada. The wintering grounds are to the south of the breeding areas, with some going as far south as the Mediterranean and even north west Africa.

The county's first ringing recovery is an early record, of a bird ringed as a 'full grown' at Skokholm Bird Observatory on 1 July 1938, then recovered at Sunderland on 2 April 1939 (*British Birds* 33: 161). One ringed as a first-year bird on Handa on 10 July 1974 was found dead at Blackhall Rocks on 8 September 1980 (Baldridge 1981). Another ringed as an adult at Grimsey in northern Iceland on 27 July 1994 was found dead at Hartlepool on 6 February 1996 (Blick 2009). The two later ringing recoveries indicate southward dispersal from breeding colonies further north in the autumn and winter, similar to the pattern noted in Guillemot.

Great Auk
Pinguinus impennis

Extinct

Historical review

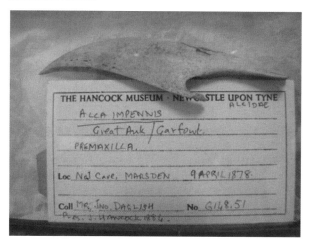

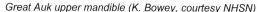

Great Auk upper mandible (K. Bowey, courtesy NHSN)

Temperley (1951) noted that it was "*extinct*", not surprising as the last known birds had been recorded off Iceland over 70 years previously. An upper mandible of a specimen of a Great Auk was discovered alongside other bird and mammal remains "*in some old sea caves in the limestone cliffs at Whitburn Lizard*". In fact, in the spring of 1878, workmen employed in a quarry on Cleadon Hills discovered a collection of both human and animal bones in a cave at Whitburn Lizards, some 46m above sea level (Howse 1878). The cave contained a midden of human food remains. Amongst the bird bones identified by John Hancock was the upper mandible of a Great Auk, the specimen of which was held in the Hancock Museum (Jackson 1953) and is still retained in the NHSN collection. Alongside the Great Auk mandible were the remains of red deer *Cervus elephus*, roe deer *Capreolus capreolus*, badger *Meles meles*, marten *Martes martes* amongst others. It was presumed that the local inhabitants had been in the practice of catching Great Auk along the local coastline. As this species would have been only available when it was on shore during the breeding season, it is interesting to speculate on when and from where these birds would have been obtained. Other bones found in the same place included those of Kestrel *Falco tinunnculus* or Merlin *Falco columbarius*, Gannet *Morus bassanus* and Razorbill *Alca torda*. Similar results were determined by the British Museum for remains found in excavation of 9th and 10th century settlements on Lindisfarne, which also revealed Great Auk remains (Kerr 2001).

Rossiter (1999) speculated on whether this species was once a breeding bird on the Farne Islands, as the profile of the islands matches areas in which it once bred further north, around Orkney and Shetland until the 17th and 18th centuries. There is a well documented record of a bird being taken at the Farnes during the period 1763-1767, by John William Bacon, of Etherston in Northumberland, who domesticated it and kept it as a 'pet' (Rossiter 1999). Remains of the species have been excavated from sites on Lindisfarne, and it is tempting to suggest that the species might have been much more common off the north east coast around the 9th and 10th centuries (Rossiter 1999). Perhaps it is from around this period that the Durham specimen dates.

Distribution & movements

The Great Auk became extinct in 1844, when the last known pair was killed on the island of Elday off the Iceland coast on 3 June 1844 (Fuller 1999); it is the only Palaearctic species to have become extinct in the last four hundred years. Prior to this it bred in British waters, on St. Kilda off north west Scotland, until at least 1698, probably to 1821 (Fuller 1999), with possible breeding off Papa Westray, Orkney in 1816 (Yalden & Albarella 2009).

At one time, this large flightless auk bred over large areas of coastal northern Europe including some of the more remote British Isles, including Orkney and other sites. Wintering birds would have presumably moved south into North Sea waters during the non-breeding season (Fuller 1999) and may have been, at one time, a regular winter visitor off the Durham coast.

Black Guillemot
Cepphus grylle

A scarce passage visitor from more northerly breeding grounds, most frequently encountered in late summer and autumn.

Historical review

Sharp (1816) described it as "*very rare*" with Hogg (1845) repeating this status without additional comment. Edward Backhouse (1845) also described it as rare, mentioning "*an individual in winter plumage was killed not long ago at Whitburn near Sunderland*". Hutchinson (1840), however, gave no records for County Durham, while Hancock (1874) described it as "*not infrequent in severe weather*", but did not refer to any specific records. By contrast, Tristram (1905) described it as *"not uncommon"* but occurring only in winter, while Nelson, writing in 1907, recorded that immature birds were noted sparingly in autumn and winter (Blick 2009). Temperley (1951) summarised the past status as *"an autumn visitor of very rare occurrence"*.

The only confirmed sighting during the early part of the 20th century concerned two flying into a north easterly gale at Whitburn on 28 October 1948 (Temperley 1951). With increased observation during the 1950s and 1960s, a further six birds were seen; at Hartlepool in November 1953, April 1957, October 1959 and August 1969, with one at Seaton Snook in January 1954 and another close to South Shields pier during the third week of November 1959. The Hartlepool bird in 1969 was notable for being the only bird in living memory at the time to stay for more than a day, being present from 27th to 31 August. There was also a remarkable report of a party of ten flying south off Whitburn on 2 January 1955, but with the benefit of current knowledge and improved optics, the validity of such a record must be questionable.

Recent status

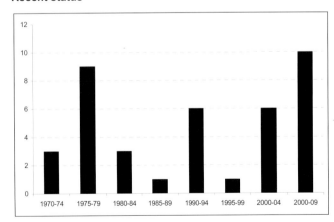

Black Guillemot records in Durham by five-year period, 1970-2009

A total of 44 birds was seen between 1970 and 2011, although the frequency has been very erratic. An adult in breeding plumage was found dead at Hartlepool on 25 May 1974; a rare instance of a bird in this plumage being noted locally. Up to four birds have been recorded in some years, but there have been 16 years without sightings in this period. The spread of records in five-year periods during the last four decades is shown in the chart.

The late 1970s and late 2000s were productive periods for sightings. Both periods included good passages of other sea birds, so perhaps the increased observer effort associated with these were responsible for securing more Black Guillemot sightings.

The species has been recorded in all months of the year but mid-winter is the time with fewest records; only five birds have been noted in the four-month period from December to March since 1970. Clusters of occurrence come in the last two weeks of August/first week of September (eight birds) and again in the last two weeks of November (five birds). These correspond with periods of increased movement in other auk species (particularly Little Auks *Alle alle* in late November), suggesting that Black Guillemots are pushed further south in the North Sea alongside other auk species.

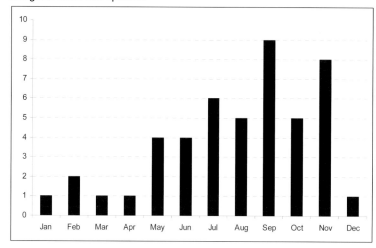

Black Guillemot records in County Durham by month, 1970-2009

As expected, the seawatching hubs of Hartlepool Headland and Whitburn have accounted for the majority of sightings, with 19 at the former and 28 at the latter. Two have been seen along the central stretch of coast (Blackhall in September 1992 and Fox Holes in August 2005), while one in Marsden Bay on 4 July 1978 is a significant early summer sighting close to an established seabird colony.

Only one bird in recent history has made a prolonged stay in the area. A juvenile was present of Hartlepool Headland from 10 September to 17 October 2011 and provided many observers with their first chance to see the species in the county.

Distribution and movements

The Black Guillemot has a nearly circumpolar breeding distribution, from the Canadian Arctic via Greenland and Iceland to the British Isles and the Scandinavian coast, through to the Russian Arctic. Birds also nest on the Atlantic coast of North America as far south as Maine. Approximately 20% of the European population breeds on the British coastline, mainly around Shetland, Orkney and the Western Isles, but extending south to Anglesey and south west Ireland. The species is rather sedentary and remains close to breeding grounds through much of the year. It is only an infrequent visitor to the English east and south coasts.

Little Auk
Alle alle

A regular autumn passage migrant in varying numbers, also infrequently seen during winter and early spring.

Historical review

The specimen illustrated by Thomas Bewick in his *British Birds* (1804) appears to be the first documented record of this species for the county, occurring at sometime around the turn of the 19th century. The account states that it was a bird *"taken alive"* on the Durham coast and fed for a while on grain (Bewick 1804). Other than this, the first mention of Little Auk comes courtesy of Sharp (1816) who considered it *"rare at Hartlepool"*, while Hutchinson (1840) noted *"the Little Auk is only met with occasionally in winter"*. George Temperley, in his overview (1951), described this species as *"a winter visitor of rather irregular occurrence"*. He documented large numbers being seen in the winters of 1841, 1876, 1878/1879, 1891, 1895, 1910/1911, 1912, 1920/1921, 1923/1924, 1936/1937 and 1948. Clearly late autumn and winter influxes took place on a fairly regular basis. Some of these movements were clearly vast in terms of the scale of the numbers of birds passing along the coast. As Blick stated, one can only guess at the numbers of Little Auk that must have passed across the Tees Bay in January 1895, when birds were said to have been seen continuously for a fortnight (Blick 2009). Invariably at such times, birds were recorded inland in the county, and most inland records are associated with coastal 'wrecks' of the species. Robson (1896) told of one that was shot near Blaydon in the winter of 1894/1895, when there had been "*many on the coast*"; no doubt referring to the above movement of birds. The influx of late January and early February 1912, the largest influx since 1895, was notable in that large numbers of dead or starving birds were reported as summarised in *British Birds* magazine (Nelson 1912). Likewise, in January-February 1912, there was a considerable wreck, when hundreds of birds were washed ashore and birds were found inland over most of northern and eastern England (Chislett 1952).

There is little information from the first half of the 20th century, but in November 1948, a large-scale passage northwards was seen along the Northumberland and Durham shore. J. R. Crawford, who routinely watched seabirds between South Shields and Sunderland (J. Coulson pers. comm.), counted 3,500 birds flying 'up the coast' in less than eight hours on 7 November (The Naturalist 1949). Temperley made clear the significance of this movement at the time saying, "*We frequently witness such northerly passages, no doubt caused by the N.E. gale sweeping the birds, engaged in fishing in the North Sea, up against the coast when they can only escape being driven inland by following the shore in a N. N. Westerly direction, only a few degrees off the wind*". He reported that some birds were recorded in-shore at the time. No doubt associated with this movement, one was picked up dead at Marley Hill on 17 November, over 14 miles from the sea (Bowey *et al.* 1993). Several inland records were noted in Temperley (1951), all in the typical mid-November period. Canon Tristram picked up one in Durham City on 26 November 1852, while one was taken in an unfinished house at Chester-le-Street on 25 November 1875, and another was found in a pond near Cleadon on 16 November 1878. In the 20th century, a single Little Auk was on the River Greta near Barnard Castle during the influx of January 1912. Post-Temperley, further inland occurrences included the following single birds: one picked up in a garden in Washington after a blizzard, on 29 February 1958; one killed against power lines at Elton in November 1961; birds at Stockton-on-Tees and Thornaby 1961 and a moribund bird picked up in a car park in Peterlee on 6 November 1969.

Recent status

This high-Arctic breeding species is the smallest and most maritime of the auks that are regularly recorded along the Durham coastline. It is intermittently noted in large numbers, usually during strong north westerly, northerly or north easterly winds between mid-October and late February. The appearance of large numbers off Durham's coast is erratic, with some autumns passing with no more than an occasional wanderer being seen. Suitable weather patterns however, coinciding with the migration of large numbers of the species through the northern North Sea, can displace thousands of birds towards the county's coastline. Large movements of the species along the Durham coast have been well documented for over a century and a half, but the more regular appearance of large numbers in autumn is a fairly recent phenomenon. Indeed, "*exceptional numbers*" in late October 1974 (Unwin 1975) featured a peak of just 98 flying north at Hartlepool on 30th; a figure which, twenty years later, would be considered little more than typical. Within the last thirty years, the first major autumn passage recorded was over 21st to 25 November 1987 when counts included 758 at Whitburn on 22nd and 1,692 Hartlepool on 23rd. Three years later, a new high count was made when 3,750 flew north at Whitburn in five hours on 12 December 1990. Another extended presence came in mid-November 1995 as 1,850 were logged at Hartlepool Headland on 12th and 1,147 at Whitburn on 18th. Drawn-out periods of passage are quite unusual, as birds are normally keen to exit the North Sea and seldom linger for more than a few days.

The decade of the 2000s saw several very large autumn movements. On 9 November 2001, a total of 2,008 passed Hartlepool, with 1,300 also at Whitburn on the same date, while counts at Whitburn in 2004 included 1,500 on 14 October, 2,933 on 13 November and 4,500 moving north in eight hours the following day. Only a year later, heavy passage on 2 November 2006 amassed 3,154 flying north in six and a half hours at Whitburn and more than 1,400 at Hartlepool Headland. The best year on record was 2007 when 18,669 were counted off Whitburn between 9th and 14 November, including a peak day-count of 11,261 (in nine and a half hours) on 11th. Over 3,000 also passed Hartlepool Headland on this date with many seeking shelter in the harbour here. Many more birds were present just to the north of County Durham, culminating in a new highest British count of 28,803 moving past the Farne Islands on 11th (Fisher & Holliday 2008).

Whilst autumn birds generally seem to be fit and healthy, mid-winter gales can result in a 'wreck' of Little Auks, with many birds dying and small numbers even pushed far inland. A period of severe weather in February 1983 resulted in such a wreck, and several thousand Little Auks were pushed onto the coastline in the period 7th to 11th. Peak counts at Whitburn included 1,105 on 7th and 3,500 on 11th, with at least 1,000 seen at Hartlepool on 11th along with 350 resting in Tees Bay. At least 166 Little Auks were found dead along the south east coast of the county at this time (Blick 2009). February 1983 was also exceptional for the number of birds blown well inland. In the south of the county, birds were found stranded at Charlton's Pond, Darlington, Dorman's Pool, Newton Aycliffe, ICI Billingham and Wynyard Lake (Blick 2009). It was a similar position in the north of the county as birds were found at Blaydon and Wylam, and as many as ten were scattered around the grasslands on the Team Valley industrial estate (Bowey *et al.* 1993). Elsewhere, exhausted or dying birds were noted at Durham City, Houghton and Thornley Hall, while one was also found in the boiler house of the pithead baths at Eppleton. Further years featuring strong mid-winter passages have included 1995, when 7,510 passed Whitburn on 12 January, and 2003, when counts included 1,350 at Whitburn on 31 January. Mid-winter wrecks are of relatively rare occurrence, there being just these three major examples between 1980 and 2010.

Little Auks only start to move away from their breeding areas in Svalbard from mid-August, arriving on the western Greenland coast from mid-September, so early autumn birds in the North Sea are exceptional. The earliest dates for County Durham are 13 September 1993 at Lizard Point, 16 September 2007 at Seaham and 16 September 2009 at Whitburn. Birds have been seen on a further eleven occasions in the final eight days of September, including 29 north at Whitburn on 24 September 2001 and four past Hartlepool Headland on 28 September 2007. Many of these birds are still in summer plumage. At the end of winter, birds start their return migration to breeding grounds during late February and March, arriving back at Svalbard in April. Very few birds have been seen in County Durham after late March, with only five sightings of single birds during April and an exceptionally late bird at Hartlepool on 22 May 2004 (Blick 2009).

Inland birds always coincide with periods of heavy coastal passage and birds have often joined flocks of Starlings *Sternus vulgaris*, ending up making landfall in fields or on house roofs. Several have been seen on the North Tees Marshes in areas such as Dorman's Pool and Seal Sands, while the suburban areas of Billingham, South Shields and Sunderland have also seen their share of back garden birds.

Occasionally, lone birds head upstream along the Rivers Tyne, Wear and Tees, with birds noted at WWT Washington on 29 January 1995 and 24 February 1996. Other more exceptional inland sightings have included birds on the River Tyne at Ryton Willows on 28 January 1996, at Derwent Reservoir on 1 December 1996, on Middlestone Moor on 4 December 1996 and on moorland in Lunedale in early December 2003.

Distribution and movements

The Little Auk is a high-arctic species that breeds on eastern Baffin Island (Canada), Greenland, Jan Mayen, Svalbard, Franz Josef Land, Novaya Zemlya and Severnaya Zemlya. Most of the Svalbard population migrates to winter off south westGreenland with some remaining off the Norwegian coast. The world population is possibly as high as 30 million pairs, one and a half to three million of which nest in the Barents Sea, to the north of Norway and Russia. The number of Little Auks seen in British waters varies greatly from year to year, usually being dependent on weather conditions. Late autumn is the peak period when movements can run into the thousands, with severe mid-winter storms also responsible for large numbers

Little Auk, Hartlepool Headland, November 2007 (Ian Forrest)

Puffin (Atlantic Puffin)
Fratercula arctica

A locally common non-breeding summer visitor and passage migrant.

Historical review

There is relatively little historical documentation of this species in the Durham, though it has clearly been known as a bird of the north east of England for many centuries (Kerr 2001). The earliest reference to the species in relation to Durham appears to be in the context of a consignment of birds that was apparently sent south from the Farne Isles, for use in a banquet at Durham Priory in the summer of 1532 (Gardner-Medwin 1985).

Sharp (1816) noted it as *"rare"* at Hartlepool, whilst Hogg (1845) said it was not infrequently noted off Hartlepool in the summer. Proctor (1846) reported that it was not uncommon off Durham, whilst Hancock (1874) described it as *"frequent"* off the Northumberland and Durham coastal strip, though his overt Northumberland perspective, and the species' high profile on the Farne Islands, may have 'skewed' this view somewhat. Nelson (1907) said that he had never seen it in the Tees estuary, but subsequent to his period of documentation, there were a couple of winter records from that area; one on 12 January 1929 and another on 2 January 1934. From this, it is safe to say that it was a rare winter visitor in the estuary.

Temperley (1951) summed up the known status by describing it as "*a summer visitor off the coast in small numbers*", the largest numbers were reported to occur in April and August when birds were moving along the coast. He noted that it did not breed off the Durham coast and recounted large movements off Durham in 1948, when on 21 September during north easterly gales at least 300 birds were noted flying north. Little was reported of its status through the 1950s and 1960s, largely due to the difficulty in separating the auk species on seawatches at the time. Movements of 70 at Hartlepool on 28 June 1963 and 45 there on 3 October 1964 were thought notable, as *"such numbers are exceptional away from the Farnes"*.

Recent status

Although Razorbills *Alca torda* and Guillemots *Uria aalge* regularly feed close inshore, it is much less usual to see Puffins in numbers on the sea away from its breeding stations in the counties north and south of Durham, i.e. on Coquet Island and the Farne Islands in Northumberland and at Flamborough Head in Yorkshire. In Durham, whilst Puffins might be been seen offshore in almost every month of the year, most sightings occur between April and September with birds being particularly evident during June and July. The majority of birds are logged from Whitburn Observatory, although they occur in in-shore waters along the full length of the Durham coast.

Sightings in the winter months are rare. Although a small proportion of Puffins nesting on North Sea coasts spend the winter in the mid-North Sea, many move out into the Atlantic for at least part of the non-breeding season. Winter records are by no means annual in Durham, but have included nine at Hartlepool on 31 January 1976, six at Whitburn on 19 February 1983 and five at Whitburn on 22 January 2007. The Beached Birds Surveys, which have been operating since 1978, have discovered few tideline corpses in the winter months, except for two notable years. In 1983, following an exceptional period of adverse weather which resulted in large numbers of Little Auks *Alle alle* in the southern North Sea, a total of 95 Puffins were found on the south east Durham coast, only seven of which were oiled (Blick 2009). In February 1993, following the grounding of the tanker *Freja Svea* off Teesmouth, a total of ten Puffins were found dead on adjacent beaches.

Birds return to breeding colonies in the latter part of February and through early March (Kerr 2001), but few are detected in Durham in this period. Numbers only increase noticeably through April with movements of up to 50 birds becoming more frequent by the month's end. There has never been any suspicion of Puffins attempting to breed in Durham, although up to four were present around the cliffs at Marsden in May 1995, possibly prospecting by non-breeding birds.

The peak period for Puffin activity is from late May to mid-July, with mid-June being particularly productive. The highest day count in the county was 1075 flying north past Whitburn on 5 June 2011, whilst other notable movements at Whitburn have included; 665 on 22 June 1987, 880 on 10 June 1988, 486 on 31 May 1991, 579 on 12 June 1999, 996 on 11 June 2011, and 996 on 26 July 2011. Numbers passing Hartlepool in early summer are always much lighter, possibly due to the increased distance from the Northumberland colonies, or perhaps due to differing local feeding conditions in the north of the county. The largest counts here in recent years have been 129 on 10 June 1972 and 137 on 6 June 1986, with fewer than 20 being more the norm.

Post-breeding dispersal from colonies takes place from late July, with young birds heading straight out to sea rather than remaining close inshore, as happens with Guillemots and Razorbills (Kerr 2001). The consequence of this is a rapid decline in passage from mid-July and few juvenile birds recorded on seawatches, typical August totals being under 20 birds per day. Occasionally, autumn gales produce an increase in passage with peaks at Whitburn including 70 on 30 September 1998 and 45 on 20 September 2002, while 115 were logged at Hartlepool on 4 October 2003 and 76 were there on 13 September 2009. Late autumn movements are even more sporadic, suggesting a gradual shift in distribution out of the North Sea and into the Atlantic. Early November 2007 was exceptional at Whitburn with 52 birds being seen on eight dates including 11 on 3rd and 13 on 10th and 11th.

There are only two inland records of Puffins in Durham. In 1976, one was found dead under wires beside the Reclamation Pond (Blick 2009) and on 21 February 2007, one flew west up the River Tyne over Newburn Bridge (Fisher & Holliday 2008).

Distribution & movements

This species breeds around the North Atlantic, from the north east states of the USA, north west France and as far north as the high Arctic. Its wintering grounds are poorly understood, with evidence suggesting use of the North Sea, as well as the more expected areas of the North Atlantic as far south as the Canary Islands.

Ringing data from the Farne Isles has shown that Puffins disperse widely in the North Sea, with birds from Northumberland being found in Scandinavia and south west France (Kerr 2001). There have been over 20 ringing recoveries of Puffin in County Durham, mainly involving birds found as tideline corpses in the winter months. By way of example, five of the birds picked up dead along the south east coastline of the county during the February 1983 wreck of auks had been ringed. Three of these had been ringed as pulli on the Farne Islands, Northumberland, in 1970, 1981 and 1982; another was ringed as a pullus at Runde, Romsdal, Norway in June 1982; and the fifth bird had been ringed as an adult on the Isle of May, Fife, on 5 April 1974 (Blick 2009).

As might be expected, the majority of Durham recovered birds have been ringed as pulli on the Fame Isles. Other recoveries have involved birds from Fair Isle, Craigleith (Lothian) and the Isle of May (Fife). A bird ringed as a pullus on the Farne Islands on 9 July 1959 was caught and released at Easington Colliery on 13 December 1966; an unusual winter record of a live bird. Interesting recoveries in terms of longevity have included: an adult ringed on the Isle of May on 19 July 1974 and found dead at Horden on 21 February 1983; an adult ringed on the Isle of May on 18 July 1977 and found dead at Blackhall on 6 February 1986; and, a nestling ringed at Craigleith, Fife, on 22 June 1974 and found dead at Roker on 2 March 1986. A bird ringed as an adult in 1975 on the Isle of May was killed by gulls in Marsden Bay on 19 April 1998; a sad end for such a long-lived bird.

Pallas's Sandgrouse
Syrrhaptes paradoxus

A very rare vagrant and possible rare historic breeder; represented only by 19th century records.

This irruptive species invaded Britain from its breeding range on the Asian Steppes on a number of occasions between 1863 and 1909; the largest invasions being in the years 1863 and 1888 (Brown & Grice 2005). Temperley (1951) called it "*a very rare irregular summer visitor*" to County Durham and documented records from the two main invasions along with a small number of more minor ones. In both of the major influxes birds were present in Durham, as well as in Northumberland to the north (Kerr 2001) and Yorkshire to the south (Mather 1986). In the first invasion of 1863, a flock originally numbering 17 birds arrived at Port Clarence on 13 May and remained for several weeks to constitute the first record for Durham. At Ryton, close to the River Tyne, a female was shot out of a flock of sixteen in a pea field on 2 June and a male was shot out of a flock of around 12 at Cowpen, Teesmouth on 30 June of that year. On 24 November, the corpse of a female was taken to William Proctor in Durham, long after it had flown into wires; it was said to have been from of one of two flocks that were near Whitburn in the middle of June. One might speculate that between these northerly and southerly locations, many other birds would have been scattered around the interior of the county during the truly amazing events of the summer of 1863. Tristram

(1905) described the 1863 invasion, "*from the months of May to July many more were seen and taken on the coast, and on the sand hills of Seaton and Cowpen Marshes. I saw a flock of very nearly twenty for several days, but I regret to say that most of them were shot*", these birds were referred to by Blick (2009).

The next occurrence in the county came just over a decade later. Nelson (1907) recorded that three birds were 'taken' "*on the sand near the Teesmouth*" in late August 1876, when several flocks were also seen on the Continent.

Birds re-appeared in force in May 1888 and this proved to be an even larger influx than that of 1863. Detailed records included six at Teesmouth between mid-May and mid-June, two that were shot at Whickham in May, two seen about half a mile out of Durham City on 25 May, two shot at Warden Law, date undetermined, and six in a field between Bishop Auckland and Byers Green on 3 June. Tristram (1905) said of this invasion that "*numbers were shot all over the county, one taxidermist, Mr Cullerford, of Durham, had over 60 specimens brought to him*". During the 1888 invasion breeding occurred in North Yorkshire (at least two pairs) and also in Northumberland and it is conceivable that birds may have bred in Durham. A note in the Zoologist for 1888, told of "*sand-grouse breeding in Durham*", "*a nest with three young is near here*". Temperley (1951) had no confirmation of this breeding record but it seems credible given the reports further to the north and south. The 1888 influx was the largest appearance of Pallas's Sandgrouse in Britain. It started in May of that year and involved perhaps 800-900 birds in Yorkshire alone (Blick 2009) with perhaps 3,000 across Britain as a whole (Brown & Grice 2005).

In 1890, a further small invasion to Durham occurred. This had a distinctly northern focus with all records coming from a small area of what is now Gateshead District. On 25 May a male was shot at Whickham and around the same time a badly injured female was picked up near Blaydon and another was shot at Swalwell on 28 May. There have been no records in the county since.

Distribution & movements

This species breeds on the steppes of central Asia. It is a largely resident species though occasionally high breeding success followed by poor weather conditions elicits irruptive behaviour. The most recent record in the north east of England was on 5 September 1969 when a male was shot at Seahouses on the Northumberland coast, reminiscent of records from a century earlier; another was seen at Elwick the following day (Kerr 2001). Only three birds have been seen in Britain since these Northumberland records; two on the Isle of May in May 1975 and one on Shetland in May 1990.

Feral Pigeon/Rock Dove (Common Pigeon)
Columba livia

A very common urban and farmland resident.

Historical review

The Romans had domesticated the Rock Dove and were probably rearing it on an almost industrial scale for fresh meat (Yalden & Albarella 2009), with up to 5,000 in one dove-house. The species' remains are found in the majority of Roman site excavations in Britain, including those across the north east of England (Yalden & Albarella 2009).

The cliff-nesting Rock Dove is the original ancestor of the Feral Pigeon, the colloquial name given to the pigeons that frequent our streets today. The species was first brought into captivity 5,000-10,000 years ago and was the probably first bird to be domesticated by humans (Johnson & Janiga 1995). Pigeons have been domesticated in the UK since at least the 14th century (Holloway 1996). As escapes from such stock, Feral Pigeons have steadily replaced wild Rock Doves from the UK largely through a process of interbreeding (Cramp *et al.* 1985), though the 'two' birds are genetically one and the same. The exact point at which the 'change-over' occurred locally is unclear, though by the end of the 19th century it would appear that pure Rock Doves were close to extinction in England, lingering only on the Yorkshire coast. They rapidly died out as a race in England and Wales early in the 20th century (Holloway 1996).

References to pure Rock Doves are hard to find in the literature. In 1834, Edward Backhouse wrote, "*The Rock Dove, I am informed inhabits the rocks near Hawthorn Hive near Easington*", though by 1874, Hancock considered it impossible to determine whether there were any pure Rock Doves left along the County Durham coastline. Even at that time, he doubted the truly wild origins of the birds breeding along the north east coast. Pure Rock Doves were said to breed occasionally at Marsden, though they are little mentioned by other authors.

The colonisation of urban centres by Feral Pigeons probably took place during the 19th century, but this was poorly documented and consequently there is little background information on the process, as noted by Holloway (1996). It is rather speculative, but it seems likely that 'pure Rock Dove' bred on Durham's coastal cliffs until at least the early part of the 19th century, if not later than this (Bell 2008). Temperley (1951) said that it was, *"a doubtful resident, probably now quite extinct in its original form, though doubtless at one time breeding in the caves along the shore"*.

Recent breeding status

Today, in its purest 'wild' form, the Rock Dove is restricted to the coastal areas of north west Scotland and the Scottish Islands, having been replaced everywhere else within its UK range by the Feral Pigeon (Parkin & Knox 2010). In Durham the Feral Pigeon, i.e. the 'feral' Rock Dove, is a numerous breeding species though it is more sparsely distributed in the rural west. It is principally a bird of varied, lowland habitats in Durham. It is most common in urban areas where there is contact with people although many are also present in rural areas on farmed landscapes with outbuildings that offer nesting sites. Substantial scattered colonies of birds are present in all of the county's larger towns and cities and around industrial complexes where derelict buildings predominate. Good numbers also occur in many villages, with some Feral Pigeons penetrating along the county's river valleys, as far west as the Cumbrian border. Birds have spread furthest west in the south of the county, along the line of the A66, though the species tends to be much less numerous in such situations and at such elevations.

Birds along the county's north eastern coastal strip still inhabit cliff top habitats, often in plumages that very much resemble those of the wild Rock Dove (Armstrong 1988) and some such birds can be found using the species' typical sea cliff habitats at all locations from South Shields to Hartlepool.

The Feral Pigeon has the ability to produce young in any month of the year, which is one trait that may have initiated human interest in this as a domesticated species for fresh meat. Early human selection will also in itself have increased reproductive vigour in the species (Johnson & Janiga 1995). This continuous breeding behaviour is illustrated by the presence of a nest with young at Dunston, Gateshead during the first week of January 1993, and at the other end of the calendar, a recently fledged juvenile at Shotton Colliery on 30 December 2001 (Armstrong 1994, Siggens 2005).

In the county's urban environment it nests mainly in derelict buildings or artificial cavities around rooftops, whilst at the coast, caves in limestone cliffs or rock crevices are still used (Westerberg & Bowey 2000). It seems likely that the species' nesting density is, at least in part, limited by the availability of suitable nest sites (Gibbons *et al.* 1993) and the density it achieves in 'natural' locations in the county appears to be very much lower than that in town centres.

The species does not show a wide altitudinal tolerance in Durham and it is principally a lowland bird. In 2007, no records were received from west of grid line NZ12 (this grid line lies east of Consett and passes through Tow Law), although this may be a result of under-recording of an 'overly familiar' species (Newsome 2008).

There are few of the county's larger built-up areas which do not host a large population of Feral Pigeons, though the breeding densities appear to be greatest in the largest conurbations, such as southern Tyne & Wear (Westerberg & Bowey 2000) and around Teesside (Joynt *et al.* 2008), where waste food from shopping centres and other outlets provides ample 'forage' for what is a highly opportunistic town-dwelling species. Large 'hotspots' occur in derelict buildings in urban areas, and many scores of pairs can nest in such areas, for example around Sunderland Docks. In its more 'natural' habitats, i.e. along the coastal cliffs from Ryhope north to South Shields, the species feeds on seeds and plant materials like cabbage leaves, in cliff top crops and grasslands (Westerberg & Bowey 2000).

Between the early 1970s and the early 1990s, the national population expanded by an estimated 39% (Gibbons *et al.* 1993). It was believed that a major factor in this upward trend was the increase in urban food supply from fast food outlets (Westerberg & Bowey 2000). Based on a *pro rata* extrapolation from the national figures, the *Atlas* made a population estimate of 1,100 pairs. However, population trends within the county are unknown, as no monitoring work has been attempted. At the time of the *Atlas'* publication, it was felt that this might be something of an underestimate. It now seems likely that the true figure is closer to 2,000-5,000 pairs. Figures derived by subsequent work on *The Breeding Birds of Cleveland* however, suggest a significantly higher population. In that area, Iceton (2000) called it 'abundant' and there may be over 2,500 pairs in the area to the north of the Tees alone. This work confirmed that the majority of birds locally are in urban areas, especially in industrial Teesside (Joynt *et al.* 2008). Justification for this re-assessment is based partly on Gibbons *et al.'s* (1993), depiction of the north east as an area of high population density for this species.

As stated in the *Atlas*, few birdwatchers deem this species worthy of study and consequently, it is probably under-recorded and not really well understood. The *Atlas* almost certainly underestimated its true lowland distribution in the county, in particular in the urban heartlands of the north and east (Westerberg & Bowey 2000). This widespread lack of interest is illustrated by the fact that the Feral Pigeon was not fully included in *Birds in Durham*, the county's annual ornithological reports, until 1988.

Recent non-breeding status

t is estimated that 2.25 million adult and 2.5 million young 'racing pigeons' are raced in Britain each year during the main racing season, from April through to September (Yalden & Albarella 2009) and the north east has a long tradition of involvement in the sport. Escapes from pigeon fanciers' lofts help supplement feral numbers.

Throughout the year, birds can be seen in town centres and other urban areas. Large flocks gather in many suburban fringes in late summer and early autumn. Numbers become concentrated in agricultural areas on the edges of towns, particularly during crop harvesting and on newly sown fields. Locally high concentrations also occur around cereal storage and processing facilities sited in rural areas, such as at Piercebridge. In winter, like the Woodpigeon *Columba palumbus,* it forms large flocks at suitable feeding sites on farmland where cereals or oilseed rape is grown. Such flocks frequently number hundreds of individuals. An increase in available winter forage, in the shape of oilseed rape and other autumn sown cereal crops has occurred over the last 30 years. Feral Pigeons also favour *Brassica* crops (Westerberg & Bowey 2000).

The highest recorded count of Feral Pigeons to date was 1,500 near Seaton Pond on 11 January 2008 (Newsome 2009). Other large gatherings have occurred in autumn and winter on the agricultural urban fringe, with 1,000 at Kibblesworth in January 1991; a similar number in cliff top stubble at Grangetown, Sunderland in December the same year; 1,000 feeding in fields at Offerton in November 1996; and, at least 1,000 in the Ryhope/Hendon area in October 2004 (Armstrong 1992 & 1996, Siggens 2006). Recently-sown crop fields attracted large flocks in October 2007, with over 600 near Cleadon on 10th and 500 at Fellgate on 24th (Newsome 2008). A winter gathering on roofs in Hartlepool Docks reached 800 on 13 December 1998 (Armstrong 1999); presumably a communal roost site.

Distribution & movements

Both the original wild Rock Dove and the derived Feral Pigeon have a cosmopolitan distribution. It is principally a resident species, which spends the winter in its breeding areas, though birds often move to locally abundant food sources such as agricultural habitats.

The species is still widely bred domestically as the racing pigeon, with large numbers of competition birds travelling through, and covering considerable distances both within and far beyond, the county. Hundreds of birds can sometimes be seen flying in off the sea or over other areas, and tired racing pigeons do sometimes occur on the coast (Blick 2009). Loft-ringed birds are also often noted in town populations of Feral Pigeon, indicating the regular addition of lost competition birds to the wild stock. It has been estimated that 3.6% of all town-based Feral Pigeons carry racing rings, indicating an ongoing source of recruitment to the urban populations of this species from lost captive-bred stock (Yalden & Albarella 2009).

Stock Dove
Columba oenas

A common resident.

Historical review

The national range of this species increased dramatically during the 19th century, partly as a result of an increase in arable farming and partly an increase in the number of trees, following the switch from wood to coal during the industrial revolution (Harrison 1988). The extent of the recorded increase however, was also exaggerated to some extent by an improved understanding of its identification, as it was not even identified as a separate species from Rock Dove *Columba livia* until the late 18th century (Holloway 1996).

The earliest references to Stock Doves in the county relate to the 1860s. It was first documented by Tristram (1905) as having been recorded in 1862 or 1863 at Elton, and definitely breeding there by 1867. Gurney (1876) said that he discovered it at Castle Eden Dene in May 1866; describing the taking of specimens at Darlington and High Coniscliffe, he speculated on it becoming abundant in the county "*before long*" which suggests an expansion was well underway by then. Hancock (1874) found a nest at Castle Eden Dene in 1871, and it was noted breeding at Brancepeth and the Ravensworth Estate in 1874. By 1877, it was considered quite common in Castle Eden Dene. In Robson's time, Stock Dove was known to nest regularly in Axwell Park and Gibside, though it was believed by him to be "*less common than the Wood Pigeon*" in the Derwent valley (Robson 1896).

Tristram said that it nested regularly in the Banks at Durham City. He also documented Stock Doves nesting around Durham Cathedral, when two nests were found in drains that were entered via stone gargoyles, one on the Prebend's Bridge and another in the Cathedral (Tristram 1905).

Teesside had been colonised by this species by the second half of the 19th century, as illustrated by Nelson (1907) who noted that "*near Redcar, in February 1888, 14 were killed at a single shot*". By 1907, he recognised it as a resident in Yorkshire "*now breeding more or less abundantly in most districts*". He also recorded it as an immigrant, noting "*a flight coming from seaward*" at Teesmouth on 5 October 1901.

In 1929, Temperley (Temperley 1929) noted that the Stock Dove was a relative newcomer and "*very local fifty years ago. It is now generally distributed in both counties*". By 1951, he called it a "*common resident*", though he felt that it was still "*rather local*" as a breeding species. He also considered Stock Doves to be more numerous in winter, stating that flocks arrived from overseas in autumn, often mixed in with Woodpigeon *Columba palumbus*. This suggestion of large-scale winter immigration boosting winter numbers is not corroborated by more recent understandings of its status. Through the late 1950s and 1960s there was some considerable decline in numbers nationally, largely it is believed to the impacts of the ingestion of organochlorine seed dressings (Harrison 1988).

Temperley said that the species nested in sea-cliffs at South Shields and Marsden, and that during the Second World War, in rabbit burrows in the dunes at Teesmouth, when the coast was closed to public access. In 1964, Stead described it as a breeding resident in small numbers on the marshes and in the deciduous woods of Teesside (Stead 1964).

Recent breeding status

Although it is considered widespread in the county today, the Stock Dove is by no means a numerous bird and it is infrequently seen in large numbers, though it is thought that this may, in part, reflect the "*low profile*" nature of the species rather than its real distribution (Bowey *et al.* 1993). Generally, this species has been rather poorly studied and documented in the county in modern decades, and in some respects we lack both a detailed understanding of its local distribution and an insight into its broader ecological requirements.

This is a bird of varied habitats in Durham. It can be found in locations ranging from the coast, where it nests in caves in the limestone cliffs, westwards to the moorland edges. Along the coastal strip and across the centre of the county the species seems to prefer farmland habitats, often favouring areas with arable crops. However, it can also be found in parkland and woodland edge, where mature trees provide nesting holes. Regardless of local specifics, it is usually associated with trees (Gibbons *et al.* 1993).

Stock Doves nest in holes in trees, although when suitable trees are not available they will use old buildings and crevices in cliffs (Cramp & Simmons 1985), as well as artificial sites such as owl boxes (Westerberg & Bowey 2000). Many of the birds in the uplands utilise old or abandoned buildings, with disused quarries and cliff faces also providing nesting sites. Birds still nest in the area around Durham Cathedral and Castle for example, as noted in the early historic records (Siggens 2005 & 2006). The density of breeding birds depends, in part, upon the availability of suitable nesting sites (Gibbons *et al.* 1993). Pairs tend to breed more or less in isolation from one another, though there are some apparent 'colonies', where nesting sites are clustered and locally abundant, as in the old castle at Ravensworth (Westerberg & Bowey 2000).

A 1984 study revealed just ten pairs in the coastal strip north of the Tees (Bell 1985). Survey work during 1988-1994 showed significant concentrations in the mixed agricultural areas of central and eastern County Durham, along the well-wooded Derwent valley in the north west and the Tees valley to the south, as well as around the Tees Estuary, which is largely grazing land (Westerberg & Bowey 2000). The species as a breeding bird has a wide altitudinal tolerance. Notable examples of the breadth of this come from Bowes Moor at 350m above sea level in the west, down to Greatham Creek at around 5m above sea level in the east (Westerberg & Bowey 2000).

The *Atlas* probably provided a reasonably accurate pattern of the species' breeding distribution across the county, but it may have under-represented its presence in some areas, particularly in the west (Westerberg & Bowey 2000). The *Atlas* map indicated an apparent absence of the species from the relatively poorly wooded areas to the east of Darlington. However, such gaps may simply reflect lower observer coverage and the field work for *The Breeding Birds of Cleveland* showed Stock Doves to be widespread, in generally small numbers, over much of this area (Joynt *et al.* 2008). The best breeding areas around the Tees are to the west of Stockton, Elton, Longnewton and Eaglescliffe, with other lesser concentrations around the Wynyard Estate.

Assuming the county to be typical of Britain as a whole, the *Atlas* arrived at an estimated *pro rata* population figure of around 2,700 pairs for Durham. Tempering this in the light of the fact that the abundance map in the '*New Atlas*' (Gibbons *et al.* 1993) indicated that Durham harbours lower breeding densities of Stock Dove than areas further south, the population was re-estimated to be 2,000 pairs (Westerberg & Bowey 2000). There is, however, evidence of a slight decline in the south of the county in recent years (Blick 2009), which may pertain, more broadly, to the rest of the county, suggesting a small downwards revision of this estimate may be appropriate.

In 2007, most reports came from the eastern half of the county but birds are by no means uncommon in the upland areas of the west, with small numbers noted even in the upper reaches of Weardale, Teesdale and Lunedale. In reality, this is a widespread breeding species across the county, though it is somewhat local in the west, being found near upland woodlands and fringes. It is present from Grains 'o the' Beck in the far south west and Harwood in the west, to the eastern coastal cliffs of Dawdon Blast Beach and Marsden. In 2008, Stock Dove records came from approximately 140 sites across the county (Newsome 2009).

Nationally, from the 1960s, the population went through a slow increase (Marchant *et al.* 1990) until the '*New Atlas*' highlighted the fact that this trend had by then stabilised (Gibbons *et al.* 1993). The current situation, as reflected in the national Breeding Bird Survey, indicates no significant change in the UK Stock Dove population between 1995 and 2009 (BBS 2009), although the species is on the Amber List of Birds of Conservation Concern, because of the international importance of the UK population (Robinson 2005).

As with other pigeons, the breeding season tends to be protracted. For example, in 2008 song or display was observed at Hardwick Hall from as early as 15 January to as late as 9 December (Newsome 2009).

Recent non-breeding status

Birds begin flocking from late summer and early autumn and by October communal roosts and feeding gatherings are well-established and widespread. Over the last three months of the year, flocks, including young of the year, can reach as many as two hundred birds in favoured areas. The largest gatherings in recent decades have been recorded either in this final quarter, or soon after the turn of the year. For example, at least 130 were at Pittington on 28 December 2000 (Armstrong 2005) and 335 were at Haverton Hole on 23 November 2001 (Siggens 2005). The largest feeding flock recorded to date was in the Cowpen and Greenabella Marsh areas of Teesmouth, peaking at 680 between November 2003 and February 2004 (Blick 2009). Counting the regular and often large Greenabella flock is made easier by the tendency of birds to perch up on wires in this area. A communal roost site has also been noted nearby, with at least 80 roosting on the bridge by the Seal Sands hide on 15 October 2000 (Armstrong 2005). In the Bishop Middleham area in 2009, significant numbers were attracted to crops deliberately designed and planted to conserve Corn Buntings *Miliaria calandra*, including 90 birds at Farnless Farm on 23 November (Newsome 2010).

During winter, flocks of 20-50 birds gather at sites across the county, particularly in the agricultural lowlands of the east, and in the Tees lowlands. Parties of up to 10 birds are more widespread and can be seen at many sites, particularly in the first half of the year. In recent decades more birds have been recorded using artificial feeding stations such as those at Thornley Wood in the Derwent valley, and WWT Washington, where numbers have sometimes reached 'scores of birds' in hard weather.

During the 1980s and early 1990s the largest reported wintering flocks, at least early in the year, tended to come from WWT Washington and nearby Barmston Pond, where records included 272 at Barmston Pond on 6 February 1987 (Armstrong 1988). The species could be noted at many other widely scattered localities during the first winter quarters of most years over this period, ranging from Seal Sands in the east to Consett in the west, but with some concentration of birds across the eastern, arable-dominated parts of the county. The highest counts in the first quarter are more recent, and have come from two areas which have regularly held large wintering flocks, with 400 at Salter's Gate, near Tow Law on 15 January 2003 (Newsome 2006) and repeated high counts at Greenabella Marsh, as noted above.

Stock Dove is generally not thought of as a continental immigrant, but references indicate that it was very much seen as this in the past, as noted previously. Recently, a pattern of both spring and autumn records has emerged, which may support the theory that such movements occur. The most surprising record is of a massive movement of 945 flying south over Dorman's Pool, Teesmouth on 5 November 2005, by far the highest count of Stock Doves in the county to date (Newsome 2006). However, this may simply have been a local movement, although the number of birds involved far exceeds even the highest of the local Greenabella Marsh counts. Possible coastal migrants include birds seen at Hartlepool Headland where this species is unusual, with two on 15 February 1997 (Bell 1997) and a single on 9 September 2001 (Siggens 2005).

Such records have become regular at South Shields and Whitburn since 2006, beginning at South Shields Leas, with four flying north on 23 March and two north on 1 April 2006. This was repeated at the Leas the following spring and autumn, with single birds moving south on 25 March, 15th and 20 September and 24 October 2007. Possible migrants were then noted apparently arriving 'in-off' the sea at Whitburn Observatory on single dates in February, March, April and August 2008. A further bird was recorded arriving in like fashion at Whitburn on 21 February 2009 (Newsome 2007-2010). As the species routinely nests along the coast in the Marsden area however, the possibility of local birds swinging out to sea and then returning to land as an explanation for such occurrences cannot be discounted. So, despite this series of interesting records, there is nothing to date to conclusively corroborate Temperley's assertion of large-scale winter immigration.

Distribution & movements

Stock Doves breed over most of Europe, North Africa and some parts of the Middle East. Another race is found in central Asia. Birds are largely sedentary in the more southerly parts of their range, though some of the northern and eastern populations are at least partly migratory.

The only ringing records also concern local movements, with one ringed at Blaydon in April 1945 being found dead at Hartlepool in December of the same year (Blick 2009) and another ringed as a nestling at Stainmore, near Bowes on 9 June 1993, being found dead at Whixley, North Yorkshire on 31 January 1994 (Armstrong 1995).

Woodpigeon (Common Wood Pigeon)
Columba palumbus

An abundant resident and winter visitor.

Historical review

There is evidence of this species' presence in Britain since the last Ice Age (Robinson 2005). Writing of the Woodpigeon in 1784, Marmaduke Tunstall of Wycliffe-on-Tees, said, "*have many here, and what is singular, more in the winter than in summer, even in the severest weather. Are very mischievous in gardens, destroying all sorts of grains, cabbages etc...They usually begin cooing in March, though I have heard them in January, in mild warm weather*" (Mather 1986); which amounts to a rather accurate summation of the species in County Durham today.

Locally, it was considered abundant in Durham and Yorkshire by the end of the 19th century (Holloway 1996). In 1840, Hutchinson wrote, "*wherever the county is interspersed with woods and plantations the Cushat is found*". Woodpigeon was considered a common resident in the heavily wooded Derwent valley in the 19th century (Robson 1896) and it remains so today.

By 1907, Nelson said that the species was increasing wherever suitable woods were found. He documented a series of large winter movements, stating that migration of this species was 'very pronounced' in 1881, 1884, 1889, 1894, 1898, 1899 and 1901; an exceptional influx of birds at Teesmouth, in late October and November 1884, being particularly notable. Referring to 1884, he noted that the influx "*was observed at its full strength at Redcar, where one gunner on the sands shot 50 in three days, all birds coming in from the sea*". Nelson himself observed "*great flights come in at Redcar*" in November 1901. He attributed these variable movements to hard weather or food availability (Nelson 1907).

There is little doubt that the large-scale planting of conifers in some parts of the county, after the First World War, provided much emergent habitat from which this species benefited particularly through the latter half of the 20th century, at locations such as The Stang and Hamsterley Forest.

Temperley (1951) called it "*a very common resident and most abundant winter visitor*", breeding in almost every wood and plantation. Large flocks in autumn and winter were "*often very numerous*". With the increased ploughing associated with the drive for home-produced food during the Second World War, he wrote that numbers of winter flocks "*increased alarmingly and became a veritable plague*".

Post-Temperley there was little substantive documentation of this species' status in the county, though it was noted as still common in the south east of the county through the 1960s and 1970s. Blick (1978) observed that there had been some decline over the late 1960s and early 1970s. In the early 1970s, it was increasingly noted as feeding in gardens in the Durham City area. It was also observed to be very common in the Lambton Estate area, near Chester-le-Street, and c.3,000 flew over Witton-le-Wear on 29 November 1973, an indication of its winter abundance in the county at that time (Unwin 1974).

Recent breeding status

The Woodpigeon is one of the commonest and most obvious species in County Durham with a very widespread distribution from urban parks and gardens to moorland plantations. It is conspicuous throughout the year and there are few areas of woodland and scrub in the county where it does not breed. The species has undergone a very considerable increase in numbers in the north east over the last thirty years or so and there were some quite marked changes in the species' habitat selection over the latter decades of the 20th century. During this period it expanded away from its traditional woodland strongholds into urban gardens and habitats composed as much of mature scrub as of woodland (Westerberg & Bowey 2000). At this time it became well represented in suburban and even urban locations, although the highest densities remained in arable farming areas (Gibbons *et al.* 1993). Some local studies appear to confirm this trend. For example, at Shibdon Pond birds did not breed at all in the early to late 1980s, but 'invaded' the scrub habitats at this site during the early 1990s, rapidly expanding to become a common nesting species (Westerberg & Bowey 2000). The apparent local increase was in line with figures from the national Common Bird Census data through the 1980s, which also indicated an upwards trend in numbers (Marchant *et al.* 1990).

By the time of the *Atlas*, Woodpigeon had become one of the most widespread species in the county (Westerberg & Bowey 2000). Allowing for gaps likely to be due to under-recording, it was only absent as a breeding bird from the higher moorland areas in the west of the county, where scrub and woodland are also absent. Taking into account the 'abundance mapping exercise' illustrated in the '*New Atlas*' (Gibbons *et al.* 1993), which indicated that the Durham area held relatively large populations, and working on the assumption that the county was typical of Britain as a whole, the *Atlas* calculated the county's breeding population to be in excess of 57,000 pairs (Westerberg & Bowey 2000). In some local studies, in Gateshead for example, this widespread distribution was confirmed by its presence in almost 70% of all surveyed squares (Bowey *et al.* 1993).

The *Atlas* illustrated a marked concentration across the more rural, central and eastern parts of the county (Clifton & Hedley 1995). The arable farmland in the eastern third of central Durham still held the highest numbers, both in winter and during the breeding season. While records also came from many western areas, the upland pastures were clearly not as productive for this species as lowland oilseed rape and cereal crops. Subsequent records do not suggest any change in distribution, although the range of habitats used by this species was illustrated by the presence of a male proclaiming territory 500 metres above sea level at Killhope on 25 April 2005 (Newsome 2005).

The gaps in recording noted by the *Atlas* in the east and south of the county were addressed by the work of *The Breeding Birds of Cleveland*. This determined that there were probably 1,500-2,000 pairs in the northern part of what had been Cleveland in the 2000-2006 survey period, when the species was described as widespread and common. In this area, particularly high breeding densities are to be found in some of the woods adjacent to suburban housing areas (Joynt *et al.* 2008).

Despite the fact that shooting takes a toll on Woodpigeons in the county, it would appear that the move into suburban and urban areas over recent decades has more than offset any negative impacts upon the species in its rural strongholds. This is reflected in the healthy position of the national Woodpigeon population, which showed a 35% increase in the period 1995-2008. In this dataset, Woodpigeon was the most widespread British species, being recorded from 94% of the 3,243 surveyed squares across the UK in 2009 (Risely *et al.* 2010); 79,383 individual birds were recorded in these.

Birds are dispersed and widespread during spring and summer, consequently few flocks exceed 30 to 50 birds at this time and these usually comprise locally breeding birds, or the recently independent young of early broods. Breeding data submitted from year to year tends to be limited, but available information shows that, like other pigeon species, the Woodpigeon has an extended, virtually continuous breeding season. Song and display is noted from January onwards, being widespread by March, and nest building continues into early September. Examples of particularly early and late breeding include: nest building at Byland Lodge, Durham City on 4 January 2005 and a newly-fledged juvenile at Houghall on 22 November 1996 (Newsome 2006, Armstrong 1997). No doubt other early and late breeding activity goes unrecorded. Most local birds nest in trees and tall shrubs, but single pairs nested on top of a girder inside the Glaxo Smith Kline factory, Barnard Castle in May 1991, and 15 metres up on a steel building support at Sunderland Enterprise Park in July 2008 (Armstrong 1992, Newsome 2009). Most unusually, a pair also bred in a Tawny Owl box at Winlaton Mill in April 2001 (Siggens 2005).

Leucistic and partial albino birds are occasionally recorded, including a leucistic individual in Westoe Cemetery, South Shields from 24 March until at least 5 May 2001 (Siggens 2005).

Recent non-breeding status

During the species' relatively short non-breeding season, birds form very large winter flocks, to feed on suitable aggregations of food, particularly in agricultural habitats. These large flocks gather on suitable farmland in the winter and roost in nearby woodlands. In hard weather they undertake very obvious local movements to avoid the worst weather and to ensure reliable foraging. The majority of such reports relate to large feeding flocks that develop during the first and last quarters of the year. Gatherings of between 300 and 500 birds are commonplace across the county at this time, and flocks of up to 4,000 have been recorded. Arable farmland in the eastern third of the county probably holds the largest numbers in winter. Habitats attracting numbers of such birds include autumn sown oilseed rape, cereal crops and, where available, stubble. Prior to the trend for increased planting of oilseed rape, there was evidence that an increase in the use of herbicides and a switch from spring tillage was leading to a reduction of winter feeding habitat for this species (Gibbons *et al.* 1993). The introduction of rape as a widespread crop in Durham over the last thirty years appears to have reversed this trend, at least locally, as illustrated by the following high counts. Some winter concentrations can be very large indeed, particularly in the north east of the county. For example, 4,000 were noted over a field of bean stubble at Great Lumley on 28 November 1998, 4,000 were recorded feeding near Seaton Pond in the first quarter of 2008 and 3,800 were at Sharpley in late September 2009 (Armstrong 1999, Newsome 2009 & 2010).

The largest winter counts often relate to flocks moving to or from roost sites. These include the highest counts in the county to date, with 5,000 going to roost in woodland at Lambton Park on 3 January 1974 and 5,000 at Crookfoot Reservoir on 2 January 1997 (Unwin 1975, Armstrong 1998). More than 4,000 birds flew over the Nissan Car Plant between Washington and Sunderland in February 2008, coming from a regular roost in Peepy Plantation. Similarly, more than 3,000 birds flying over Houghton in December 2008 were believed to be roosting in the Lambton Estate (Newsome 2009).

More recent decades have shown a trend for a slightly increased presence of birds in suburban settings, as illustrated by the BTO Garden Birdwatch results for the north east, which show that the number of gardens visited by this species increased by 3% per annum over the period 1995 to 2007.

Singing and display of local birds routinely continues into September, and so breeding birds' activities sometimes overlap with the arrival of the county's incoming continental migrants. Continental migrants no doubt arrive in the county each winter, though there is little recent documentary evidence for this phenomenon, unlike in times past (Nelson 1907) when large flocks were said to be involved. Recent records suggest immigrants arrive mainly in November. These include a flock of 420 flying west over Seal Sands on 4 November 1975 (Blick 2009) and more than 2,000 birds heading south west over Shotton Pond, in a series of flocks in one and a half hours, on the early morning of 11 November 2001 (Siggens 2005). Similarly, 104 were noted flying high, in a southerly direction, over Whitburn village in 25 minutes in the early morning of 1 November 2005, and a passage of 1,200 went over Dorman's Pool just a few days later on 5th. In none of these cases however, were birds actually seen to cross the coastline. Presumed departing continental migrants were noted over the sea at South Shields on several dates in March 2006 (Newsome 2006 & 2007). Perhaps the best piece of recent evidence to suggest that this species migrates across the North Sea comes from the observation of a bird seen from a Newcastle upon Tyne bound ferry, in mid-North sea, that was flying strongly east, towards Denmark (Kerr 2001). Hard weather movement may have explained the passage of 174 south along the coast at Horden in 20 minutes in January 1996, and smaller numbers were also noted moving south there the following month (Armstrong 1997).

Distribution & movements

Nominate race Woodpigeons breed over much of Europe and western Siberia. Other races are present in central and southwest Asia and on the various Atlantic Islands, such as the Canaries. In the south and west, they are largely resident, though birds from the north and east are migratory.

None of the ringing recoveries in the county indicate significant movements. One, ringed as a nestling at Norton-on-Tees on 10 May 1927, was recovered at Stainton-in-Cleveland on 10 May 1929. Another, ringed as a nestling in the Langdon Estate (Teesdale) during the summer of 1948, was shot at Usworth near Sunderland in February 1952. One ringed as a nestling at Nunthorpe in June 1969 was in Darlington in June 1973, and another ringed as a first-year bird at Marske, North Yorkshire in August 1980 was also near Darlington in December 1987 (Blick 2009). A bird ringed as a pullus at Shibdon Pond on 24 July 1993 was recovered (presumably shot) at Fell House Farm, Walbottle on 16 June 1996. The final ringing recoveries are of birds ringed as nestlings at Teesmouth. One ringed at Saltholme in September 1989 was shot at Easby, North Yorkshire in February the following year (Armstrong 1990), whilst another ringed on the North Tees Marshes in June 1996 was shot on nearby Cowpen Marsh in September 2003 (Blick 2009).

Collared Dove (Eurasian Collared Dove)
Streptopelia decaocto

A common resident found mainly in suburban areas; gathering on agricultural land particularly in autumn where these lie close to suburban fringes.

Historical review

A relatively recent addition to the British list, Collared Dove was first recorded in Norfolk in 1955, and spread rapidly throughout the country (Hudson 1965 & 1972). The first breeding in the north east occurred at Ponteland in Northumberland (Kerr 2001) just three years after its first appearance in Britain and a year before the first breeding in Yorkshire (Mather 1986). The species reached Teesside as a breeding bird in March 1960, when two pairs bred at Linthorpe, Middlesbrough, just south of the Durham border. It reached Durham as a breeding bird in 1961, when two pairs were found in areas just outside of West Hartlepool (Stead 1964) and birds were noted at Hurworth Burn and Thorpe Thewles (Bell 1961).

By 1962, up to 27 birds were resident in the West Hartlepool area in July, and two were noted in Sunderland (Bell 1962). In the winter of 1962/1963 Collared Doves were present at Houghton-le-Side, near Darlington, with one pair apparently breeding there in 1963 (Hudson 1965). The West Hartlepool population expanded rapidly, to between twenty and twenty-five breeding pairs by 1963 (Bell 1963). In the following year, the species was described as 'common' in Sunderland, with flocks of up to 24 birds being noted, and six pairs in the Cleadon area.

At this time birds probably also bred at Norton-on-Tees, as they were known to have summered in that area and 25 were noted in the autumn (Hudson 1965).

After the 1964 breeding season, a flock of 201 Collared Doves was recorded in Burn Valley Gardens, Hartlepool in October (Blick 2009) and 32 were feeding in stubble at Cleadon Hills on 28 December (Bell 1964). By 1965, the Hartlepool breeding population had reached approximately 120 pairs, but there appeared to be some withdrawal from town gardens, possibly due to a lack of nest sites, with birds moving out to the edges of town, to sites such as Ward Jackson Park (Bell 1965). By the following year, birds were breeding across the eastern strip of the county, though the main population was still in the south east. Young were seen at South Shields, Durham and Darlington, and twenty birds were noted as far as west as Wolsingham (Bell 1966). The species was documented for the first time in the Chester-le-Street area in 1967, a single bird being noted at the Isolation Hospital there on April 30th (*The Vasculum* Vol. LII, 1967). Collared Dove was seen for the first time at Hartlepool Observatory in May 1967, then several times in November that year, and again on 1 June 1968 (Bell 1967 & 1968). Such coastal records may have involved newly arriving migrants.

The south east and east coast of the county remained the species' stronghold for some years, though it may have been under-recorded elsewhere, as observers were already becoming blasé about the species. In 1969, McAndrew noted that the initial surge in the Hartlepool population appeared to be over, reporting no increase in numbers there since 1965 (Bell 1970). This is not consistent with Hudson, who in his second analysis on this species in 1972 gave a maximum in 1970 of over 150 pairs in West Hartlepool alone. At this time, birds had spread west to Stockton-on-Tees and Bishop Auckland. It had become an established resident in Sunderland, South Shields and Gateshead, but had still not significantly penetrated the inland areas of much of the county (Hudson 1972).

Whatever the precise timing and size of the population peak in Hartlepool, by 1971 this population had declined to about 90 pairs, presumably as birds dispersed more widely. Elsewhere the spread continued, but Collared Doves remained scarce in the centre of the county with just two pairs reported in Sedgefield, two in Durham City and one in Chester-le-Street in 1971 (McAndrew 1972).

In 1973, it was regarded as plentiful in the urban areas along Durham's coastal belt, such as Hawthorn, though Hartlepool was still the stronghold, with 135 there on 25 December. It was by now increasing in Durham City, and also noted at Finchale for the first time. Collared Doves remained scarce in the west and north west at this time however, so 10 roosting at Hamsterley Mill Park during the winter of 1973 was notable (Unwin 1974). There are no published breeding records in the north west of the county prior to 1973, when displaying birds were reported at Rowlands Gill (Bowey *et al.* 1993). In the following year evidence of breeding came from Bishop Auckland, and 12 were reported in a garden there in December. Up to 150 were also in the north east of the county at West Boldon during 1974 (Unwin 1975).

Collared Doves have since spread to occupy suitable habitat in virtually all areas of Durham, with particularly high densities from the north east to the centre of the county. The mapping work for the *Atlas* did not indicate that the spread into the county was only from the south east. Given that Collared Dove colonised Northumberland before reaching Durham, some may also have arrived from that direction (Westerberg & Bowey 2000).

Recent breeding status

This is a common and widespread breeding species with records coming from most of the county, except the most exposed, treeless areas in the west. In Durham, as nationally, the Collared Dove's primary habitat preferences appear to be village and suburban gardens (Marchant *et al.* 1990), though it also breeds in the centre of towns where green space and scrub is available. It forages for grain and weed seeds, readily taking to garden feeding stations, and it uses garden shrubs and trees, bushes and hedges for nesting. Telegraph poles, overhead wires and television aerials are routinely used for resting, roosting (Cramp *et al.* 1985) and also as song posts. Collared Doves have also been recorded utilising a wide range of nest sites, including buildings on occasions. One pair nested successfully inside a satellite dish at Sherburn in 2006 (Newsome 2007), and a similar breeding attempt was noted behind a satellite dish in Whitburn (P.T. Bell pers. obs.).

The westward spread observed through the 1960s continued in the 1970s and 1980s, with a first record coming from High Force, upper Teesdale on 6 June 1976 (Unwin 1977). Unwin then reported in 1978 that "*records suggesting continuing penetration into new areas came from Eggleston, Whorlton, Cotherstone, Willington and Kirk*

Merrington" (Unwin 1979). The first records for Balder Grange, at almost 200m feet above sea level, followed in 1980 (Baldridge 1981).

By 1982, a study in the Teesside area concluded that the breeding habitat in that area was by then 'almost saturated' (Bell 1993). Observers in Seaham considered that the local population had more than doubled in a single breeding season between 1982 and 1983, up to 30 pairs (Baldridge 1984). In Durham City, the population was still increasing in the early 1980s and this trend was probably still progressing at the time of the *Atlas* survey (Westerberg & Bowey 2000). In the early 1990s, a slight decline in numbers was noted nationally from a 1982 peak, based on studies on Common Bird Census plots, but this was documented with the caveat that this survey method did not accurately measure changes in villages and suburban areas (Marchant *et al.* 1990), where the species is most prevalent in Durham.

Current trends are unclear, with populations in many areas now apparently stable. In the south east it is widespread and very common north of the Tees, though it is less common in the most densely populated urban areas. There are probably 1,000 pairs breeding in the north Cleveland area (Joynt *et al.* 2008). However, there are still some parts of the county where the species occurs at very lower densities or is even absent.

Even in the early 21st century, the Collared Dove remained absent from many parts of the Durham uplands. The range of this species has not altered significantly since the mid-1990s, though it started penetrating some of the county's more westerly dales during this period. In this part of the county, severe winter weather is likely to more significantly constrain its ability to expand its range. The species' upland distribution therefore tends to follow patterns of human settlement along the river valleys, and away from villages it remains thinly distributed in the uplands. By 2008, birds were breeding as far west as the villages of Bowlees, Gilmonby and Hunstanworth and the furthest west it has been noted was in 2007, when remarkably single birds were close to Cow Green Reservoir in February and May (Newsome 2008 & 2009). Currently, in general terms, it can be found from the eastern areas of the county, north and south along the coast, and westwards through the central lowlands, and thence up the valleys and dales, as broadly illustrated by the *Atlas* (Westerberg & Bowey 2000).

The *Atlas* offered a breeding population estimate of 'around 2,500 pairs' (Westerberg & Bowey 2000), though this may now be considered a little low. Subsequent growth and expansion into low density or previously unoccupied areas suggests that this should be increased by some 10-20% to approximately 2,800 pairs. This is consistent with the continuing growth of the national Collared Dove population, which increased by 26% in the period 1995-2008 (Risely *et al.* 2010).

This species can breed in most months of the year given suitable weather, an ability which has helped the local Collared Dove population to maintain and increase its numbers. Mating behaviour between pairs has been noted in a Middle Herrington garden on Christmas Eve (Armstrong 2005), with nest building at Houghton in mid-January (Newsome 2009). Hatched egg shells have been found at Cotherstone on 21 February, fledged young there in mid-March 2007, and males singing 'again' at Bowburn in mid-November 2008 (Newsome 2008 & 2009). Large, fledged young have been observed in gardens in Rowlands Gill in early October, as the parents were settling down to their next, and presumably final, breeding attempt of the year (K. Bowey pers. obs.). Such impressive production of young is useful, as Collared Doves are frequently predated by Sparrowhawk *Accipiter nisus* in Durham (Westerberg & Bowey 2000), though this is unlikely to reach levels that would adversely affect the population of this productive species.

Other observations include a bird noted calling under streetlights in the early morning darkness of 17 February 2000 at Great Lumley. Leucistic individuals occur occasionally and an aberrant 'yellow' female was in a garden at Shotton Colliery in July 2000 (Armstrong 2005).

Recent non-breeding status

Post-breeding flocks typically begin to appear from early July onwards. Most involve fewer than 20 birds, although several large post-breeding and winter gatherings are noted in most years, with some exceeding 100 birds. Flocks occur mainly in the agricultural fringes, close to villages and smaller towns. Larger flocks are sometimes noted during the autumn and winter, as birds gather to feed around farms and on harvested crops. These have included 200 at Witton Gilbert in December 1994 feeding on spilt grain (Armstrong 1995) and 220 in the Eppleton area through October 2009 (Newsome 2010). More notably, 200 were at the surprisingly western location of Woodland Fell, outside of the typical peak period for Collared Dove flocks, in April 1987 (Armstrong 1988).

Birds often become concentrated around locally abundant food supplies, such as farmyard grain stores, and up to 100 birds have been recorded at Whickham and near Clara Vale in such situations. The species' preference for feeding on spilt grain in the autumn could bring it into conflict with people for reasons of hygiene, though locally Collared Doves are not usually sufficiently abundant to be regarded as a pest. Nonetheless, reports of flocks of over 150 in these circumstances 'outside the breeding season' are not uncommon (Armstrong 1989-95). The highest counts of the species in the county to date have resulted from just such a concentration of food, at the Hartlepool Docks grain terminal, where spilt grain was a regular food source until 1981. The record count was 520 on 8 February 1976 (Blick 2009).

Communal winter roosting is usually noted at a number of sites around the county. This has included a maximum of 145 at a factory near Hetton Lyons Country Park in November 2001 (Siggens 2005). Smaller roosts are more typical, such as 25 birds together at Hardwick Hall on 15 January 2008 (Newsome 2009).

Distribution & movements

The Collared Dove originated from southern and eastern Asia, but the species went through an extraordinary range expansion during the 20th century, though this process may have started as early as the late 19th century (Parkin & Knox 2010). Despite this historic westward movement, birds are otherwise sedentary.

Coastal records have been noted at Whitburn and on the coast on Teesside in both spring and autumn, although it is unclear whether these sightings involve continuing immigration or are merely local movements. Occasionally sightings occur during sea watches, including one noted flying in off the sea at Whitburn on 3 May 2004 that was first seen approximately three kilometres out to sea, suggesting that it was a genuine newly-arriving migrant (Siggens 2006). Another bird was noted apparently arriving low over the sea at Whitburn Observatory on 3 May 2008, this has been noted in previous springs but it is unclear whether some such sightings involve actual immigration or whether these are merely 'spring forays' by local birds (Newsome 2009). That genuine migrants, both Britain-derived ones and those from further afield, have reached the county is beyond dispute. An adult ringed in Herford in German Westphalia on 26 December 1963 was found dead at Sunderland on 20 November 1964. This bird was presumably part of the original influx. Another adult ringed at Hartlepool on 6 November 1965 was found dead in Southend, Essex in 1969 and a full-grown bird ringed at Twyford, Hampshire on 23 July 1983 was killed at Egglescliffe on 9 May 1985 (Blick 2009).

Turtle Dove (European Turtle Dove)
Streptopelia turtur

A scarce summer visitor and passage migrant. This species occurs in the area principally as a passage migrant, with sporadic breeding records.

Historical review

According to Gibbons *et al.* (1993), the Turtle Dove reaches the northern edge of its regular British breeding limit in Yorkshire so Durham has probably always been on the northern periphery of its range. Consequently, it seems not unlikely that many of the records, even of apparently territorial birds may refer to over-shooting passage migrants, rather than truly breeding birds; it has always, at best, been a scarce summer visitor and rare breeding bird in Durham.

Edward Backhouse (1834) described it as "*rare*". He reported that it was occasionally killed near Darlington and he documented that one was "*lately caught*" alive at West Hendon, Sunderland, so this presumably occurred in the early 1830s. Hogg (1845) said that it was "*very rare*" in south east Durham. He had in his possession a female that had been shot on 14 September 1829 at Norton and he recorded one that had been shot 26 years before this date in the same area, which would appear to make this the first documented record for the county. Hancock (1874) categorically stated that it had never been known to breed in Northumberland or Durham.

Tristram (1905), wrote of a few found every spring and he believed that it may have bred at Castle Eden, near Sedgefield, and at Wolsingham but, according to Temperley (1951), Tristram gave little evidence to support this assertion,. Nelson (1907) described it as "*extremely local but breeding in several districts*", though a number of these may have been south of the Tees. There was an expansion of this species' range into Yorkshire through the 1880s and 1890s (Mather 1986) and in 1912 the first breeding took place in Northumberland (Holloway 1996). As a result a small breeding population was established to the north of County Durham, in Northumberland, by 1932 (Kerr 2001). Temperley (1951) reported a slight increase in occurrence in Durham at this time, with not just individuals but pairs being more frequently reported and also greater numbers as passage migrants in July, August and September.

By the early 1950s, Temperley (1951) considered it "*a rare and irregular passage migrant*" and said it was suspected of breeding occasionally in the south of the county, though at that time it was not yet proven. Somewhat by the way of contrast, Chislet (1952) said it was "*far from rare*" in lower Teesdale, though presumably on what was then the Yorkshire side of the Tees.

Regardless, the species remained rare in Durham through the 1950s and 1960s. It was recorded as present in both Northumberland and Durham in 1950, but there was almost no published detail about its exact status and no nests were found. In 1952, birds were recorded during the summer at Piercebridge and Hurworth Burn Reservoir without proof of breeding. One near Middleton in Teesdale on 11 July 1952 was the first noted there by an experienced observer. Birds occurred in north Durham in 1955 but breeding was not proved. Breeding probably became established in the county sometime during the 1950s, if not sooner. By 1960 up to eight were seen together in the area around Hurworth Burn Reservoir in early June, and this was apparently developing into a local stronghold for the species in the county as it was to remain for many years subsequently along the old railway line.

Birds were also calling at Neasham on 2 July 1960 and through the 1960s, small numbers of birds were regularly present at Darlington, Hamsterley Forest, Shotton Colliery, Teesmouth and Wynyard Park. Breeding was suspected or confirmed in all of these areas through that decade. In 1962, three pairs were at Hurworth Burn, one female was shot there and two others reared three young. Breeding was suspected or proven at this site until at least 1968. Breeding was proven in Wynyard Park in 1966. In 1968, up to six were near Darlington in June and July, with a nest found there on 18 June, but this was later destroyed. In his national review, Murton (1968) said that a few pairs had by that time colonised Durham. In 1969, breeding was confirmed in south east Durham. Two late birds were noted at Darlington on 23 September 1967 and single migrants were noted at Hartlepool and South Shields, in spring 1968.

Elsewhere, the species was first recorded in the Shotton Colliery area in Edder Acres Wood on 19 May 1957. Subsequently, birds were seen regularly in spring and summer and on 14 August 1962, a nest containing two young was found. After this the species "*increased greatly*" in that part of the county and over the next few years birds could be heard singing in Calf Pasture Dene, Edder Acres Dene and Edder Acres Wood. At its zenith in this area, in the mid- to late 1960s, up to eight singing birds could be heard in a walk of just a few kilometres, and singing birds (up to three) were also present in Castle Eden Dene, with another three in woodlands near Wheatley Hill. In 1965, four pairs nested in the area but with little success. Nonetheless, a party of eight was seen in August of that year (Simpson 2011).

In 1976, a few birds were in the Tees valley above Yarm in July and August, indicating that breeding may have taken place there and again in the same area in 1978. In 1979 a pair at Neasham may have been nesting. Other breeding season reports over this period came from Barmston, the Castle Eden Walkway, Durham City, near Ferryhill, on the North Tees Marshes and in the Washington area.

Reports of largely single migrants occurred in most springs and autumns, largely from the south east of the county or the coastal strip between Hartlepool and South Shields, though occasional passage birds appeared elsewhere in the county; six flying south over Cowpen Marsh on 6 June 1972 was the largest count (Blick 2009).

Recent breeding status

Most sightings of this species in Durham occur during the period May to July, with a few reports of autumn birds usually in August, occasionally in September. In essence, this is a bird of the lowlands in Britain (Parkin & Knox 2010). Typical habitats for this species are agricultural, with strong elements of arable, and a heavy reliance on hedgerows. For breeding purposes, Turtle Doves require open ground for feeding and mature shrubs or hedgerows for nesting (Sharrock 1976).

As summarised in the *Atlas*, in recent decades there are probably too few meaningful breeding season records across the county to establish any significant distributional pattern (Westerberg & Bowey 2000). For a number of decades, from the mid- to late 1960s to the early 1990s, there was an intermittent breeding season presence of the species in a few favoured areas of the county, with some concentration in the lowland south and east of the county. Local population hot spots persisted, sometimes, for quite long periods of time. The histories of these clusters, such as the one around Shotton Colliery, illustrate the pattern through time and in terms of geographical distribution.

During the 1980s, the Castle Eden Walkway, in the south east of the county, was one of the most reliable locations for this species in the county, with up to six or seven territories being documented in 1986 (Joynt *et al.* 2008). Numbers there have crashed since and it is now only a scarce, and increasingly rare, passage migrant in the south east of the county. In 1995, two birds were calling along the Walkway from mid-May, but breeding wasn't proven (Joynt *et al.* 2008); there are no reports of breeding or even territorial birds there since around 1996 (Blick, 2009).

Through the 1980s, which probably saw the peak of this species' breeding activity in Durham, birds were regularly noted at Hurworth Burn, though less regularly as the decade came to a close. This area, in relative terms, had been particularly well-favoured by the species from the mid- to late 1950s. Other areas that regularly attracted birds over this period largely conformed to the south easterly distribution pattern and were at: Neasham, Piercebridge and Castle Eden Dene. One feature of the 1980s distribution however, was the increasingly widespread pattern of breeding activity across the county. It appears that small populations were occasionally established away from the south east. For instance, in the north of the county, on 10 July 1981, four birds were present together in the Hollinside area, close to Whickham, and it seems likely that birds may have bred in that part of Gateshead that year. Indeed, the species was recorded in small numbers in this area in 1983, 1986 and again in 1987, when it seems likely that birds bred in the Clockburn area, adjacent to Hollinside, with birds present there from 8 May through to at least 14 July (Bowey *et al.* 1993).

The species will have been somewhat under-recorded in some rural parts of the county's southern strip, with intermittent records of singing birds in this area, from Wycliffe-on-Tees, downstream to Piercebridge and Darlington, from the early 1980s to the early 1990s and into the *Atlas* survey period. For example, in May and June 1986, there were two pairs near Teesside Airport. By the end of the 1980s, Turtle Doves were becoming increasingly scarce as both a breeding bird and passage migrant as was evident from its former stronghold near Shotton Colliery. By 1990 numbers in this area had dropped rapidly, the last one that was heard singing in Edder Acres Wood was around 1994. In summary, the first local bird was noted in this part of the county in 1957 then birds bred regularly from the early 1960s, persisting at a low level through the 1970s and 1980s, then the decline came and they were 'all gone' by 1995. It was believed that these birds had come north and north west from the Hurworth Burn Reservoir area, possibly following the railway lines as a conduit of dispersal (Simpson 2011).

The overall downward trend continued in the 1990s, and by 1994, only the Hurworth Burn Reservoir area seemed to be regularly holding birds in the summer. Thereafter, reports even from there became spasmodic, ending with a final long-staying single bird in May and June 1997 (Armstrong 1995-98). In the old Cleveland county, there were no breeding records at all by 1997 in what was described as a 'disastrous year' for this declining species (Bell 1998). A particularly low point was reached by 1999, with a combined total of just five records in the county in the DBC and TBC Annual Reports, all relating to passage birds (Armstrong 2003 & Iceton 2000).

Into the early 21st century the only suggestion of a breeding population over the decade was of a handful of pairs establishing in the extreme south of the county, at The Stang. Occupancy of this surprisingly elevated and westerly location began in 2000, when a single was noted calling there on 15 June (Armstrong 2005). In 2005, it was felt that the species showed 'signs of consolidation' at this site (Newsome 2006), when up to five birds were present in June of that year, including birds singing and displaying. Breeding success however, was unknown, as no juveniles were reported. In 2006, a bird was at this site on 18 May and single birds were noted there on several dates until 2 July, but there was no proof of breeding. By contrast, 2007 was apparently a poor year with just one there on 7 May and two there on 26 June. It is likely that one or two birds were present there in the intervening period. A bird was at this site on 24 May 2008, it sang and remained until at least 16 June but there was no suggestion of breeding. Two were seen here on 28 May 2009 but unfortunately were not seen again and a singing male was present on 16 June 2010. A slightly more encouraging situation was evident in 2011 with up to four birds seen including two singing males, the final sightings being on 20 August.

Perhaps this pattern of scattered, short-burst activity indicates the establishment of small, sub-populations, by a pair or even a handful of birds, in particularly favourable areas for the species. These seem to persist over short spans of time, before a downturn occurs perhaps associated with a run of poor summers, a change of local foraging opportunities or decline in habitat quality, leading to a local extinction of whatever small population had become established.

It is still unlikely that in any one year, more than a handful of breeding pairs are successful in the county, perhaps five pairs or even less (Westerberg & Bowey 2000). In general terms, records in the early 1990s, became even scarcer. In the south east of the county, birds were largely restricted to a handful of passage birds on an annual basis. Accordingly, there were no known territorial birds in the north Tees area from 1995 to 2008 (Joynt *et al.* 2008). Nationally, there was a marked contraction in range (Parkin & Knox 2010) between the two national breeding atlases (Sharrock 1976, Gibbons *et al.* 1993). With some degree of irony, this period probably marked the zenith of this species' breeding presence in Durham. Considering that the species is declining nationally and internationally (Parkin & Knox 2010), it seems most unlikely that this it will consolidate its status as a breeding bird in the county, and at the end of 2010, it is more likely to revert to its status prior to the mid-1950s, as merely a scarce passage migrant.

Recent non-breeding status

Over many years this species has occurred largely as a spring and autumn passage migrant, with most sightings in spring. The majority of birds over the years have appeared in the east and along the coastal strip of the county, with occasional inland records, some of which may have been prospecting breeding birds.

It is usually early to mid-May before any birds are seen, with occasional late April records. The earliest known record for the county is of a bird at Darlington on 12 April 1984. The majority of records have occurred in late May and early June, but there appears to have been a small drift of birds occurring into July. Autumn birds are largely noted in late August, with a small peak in September and a few records over the years into October.

There was an increase in the number of passage records through the late 1950s and from the 1960s through to the 1980s, the period when the species was establishing itself as a regular but scarce breeding bird in the county. Coincident with the declines noted from the late 1980s the number of passage migrants has also decreased. A notable record came in 1994, however, when six were seen together on Cleadon Hills on 9 May (Armstrong 1994). In contrast to much of the first decade of the 21st century, 2006 was a good year with five spring and five autumn sightings, but by the second half of this decade, the number of sightings in most years was once again down to single figures per annum. In 2007, there were just three records, in 2008, there were two records and, in 2009, a single bird at Westfield Pastures, Ryton, on 29 May, was the year's only report. More birds were recorded in 2010, but were still rather uncommon, whilst just two migrants were noted in spring 2011.

Occasionally single birds have appeared at the coast amongst migrants during fall conditions, suggesting a continental origin for such individuals. Just such a late individual spent 27 October 2005 commuting between Well House Farm and the Charlie Hurley Centre, Whitburn. Other late migrants include one at Barmston Pond on 21 October 1975, one at Marsden Quarry on 24 October 1976; and a bird at Cleadon Hills on 29 October 1983. One of the most unusual records of this species for the county was of the bird that was found freshly dead in Hartlepool Town Centre on 29 January 1972, a most unusual occurrence. It was commented that the observer could not rule out the possibility of it being an escaped cage-bird (Blick 2009), but this seems rather unlikely.

Over the thirty years between 1976 and 2005, the reported first arrival dates of Turtle Dove were analysed, in ten-year groupings. The mean first arrival dates were: 1976-1985, 18 May; 1986-1995, 7 May; and, 1996-2005, 14 May. This is an advance in first arrival date of four days from the first ten-year study period, to the last, and this is perhaps indicative of the rare and sporadic occurrence pattern of the species in the county over the period, which has seen it declining considerably from what was a relatively low base even at the outset (Siggens 2005).

Distribution & movements

This species is a long-distance, trans-Saharan migrant. It breeds over much of Europe, reaching its northern breeding limit in the north east of England, though it does extend into southern Sweden. Other races breed in North Africa and across into Asia and the Middle East. A male bird was caught and ringed at Shibdon Pond on 28 May 1994 but other than this, there is little ringing data relating to this species in the county.

Ring-necked Parakeet (Rose-ringed Parakeet)
Psittacula krameri

An introduced exotic species with probably a small localised breeding population.

An exotic feral species in southern English counties, most if not all of the birds recorded in County Durham at least until the onset of the 21st century, are likely to have been escapes from captivity, rather than vagrants from the established population in the south east of England (Bell 2008). A perhaps controversial addition to the county avifauna, a self-sustaining population has probably existed at Hartlepool since about 2006.

The first documented records of Ring-necked Parakeet in County Durham occurred during the 1970s when a single bird flew over Cowpen Marsh, Teesmouth on 14 October 1972 and another was at Crookfoot Reservoir on 13 February 1977 (Blick 2009). Neither of these early records was published in local bird reports as they were both considered to relate to escapes from captivity, and the species did not form part of the official British list at the time of their occurrence. Ring-necked Parakeet was added to Category C of the British List in 1983 on the basis of a large self-supporting population in London and the south east of England estimated at that time as being around 1,000 birds (Lack 1986).

The first well documented record of this species after its admission to the British List is of a female in the High Sharpley area from 30 August 1989 to at least the 13 January 1990 (Armstrong 1991), though single parakeets over Shotton Colliery on 20 August 1986 and 13 July 1989 were also considered to relate to this species but were unfortunately never published as such. All three of these birds were undoubtedly local escapes and were the only records during the 1980s.

Recent breeding status

Since the early 1990s the species has been of almost annual in occurrence in County Durham though many records are typically brief in nature and do not refer to any suggestion of breeding. The majority of sightings are from urban areas in the eastern half of the county though birds are occasionally noted further west with singles at McNeil Bottoms on 14 October 1999; Willington on 6 April 2005 and Bishop Auckland on 11 September 2008. There have been just four years which have failed to produce a record since 1989, these being: 1992, 1995, 2000, and 2002.

At least 17 birds were recorded during the 1990s including a long-staying male at Charlton's Pond, Billingham from 16 October to 18 December 1990. Records came from widely scattered localities throughout the county, though most were recorded to the east, with a cluster of records surrounding Durham City between 1994 and 1999 perhaps relating to the same wandering individual. Two to three birds were noted in most years through the rest of the decade. Records were submitted in every month of the year without any discernible pattern. With the exception of the bird at Charlton's Pond in 1990 and singles at Marsden between 24 April to 7 May 1994, and at Norton intermittently from 29 January to 7 April 1997, all records referred to just single dates though undoubtedly some will have stayed longer, undetected in urban locations. Two which flew north over Peterlee on 21 June 1996 and two at Chester Moor on 3 August 1997 were the first multiple records for the county, and could conceivably have been the same pair.

Although there were no records in 2000 a further nine birds were recorded between 2001 and 2004, including five in 2003. Again, like most of those recorded previously all were recorded on single dates, with records from Whitburn, Shotton Colliery and Washington, where singles at two different localities in December 2003 and January 2004 probably related to the same bird. Whitburn Coastal Park attracted three birds during the period with singles there in March 2003, September 2004, and May 2005, whilst Hetton Lyons was the only other site with more than one record with singles in March 2003 and December 2004.

The recent emergence of a breeding population in Hartlepool possibly has its origins in birds present in Acklam Park in Middlesbrough (Blick 2009), on the south side of the Tees, where two pairs bred in 2006, raising at least three young (Joynt *et al.* 2008). There had been little real pattern to records in County Durham prior to this but two birds were found in Ward Jackson Park, Hartlepool, in late December 2005, increasing to three by 11 February

2006. It then became apparent that a pair had settled and were regularly visiting a hole in a mature tree, though there was no confirmation of breeding in that year from Ward Jackson Park. In 2007 up to three birds were again present for much of the year and were seen to be prospecting nest-holes, though as in 2006 breeding wasn't proven (Bell 2008). However a party of six birds over Torquay Avenue, Hartlepool on 22 October might well have been a family party indicating that breeding had occurred for the first time in the county that year. Throughout 2008 sightings continued at Hartlepool, with up to seven in Ward Jackson Park in January. Numbers fluctuated as birds presumably moved to and from feeding areas, but they returned to the park to roost each evening. They were more secretive through the following spring and summer and although, once again, there was no direct proof of breeding, they almost certainly did breed in the area. One suggested location was the nearby Tunstall Beck, where sightings were frequent. During the second half of the year, fourteen birds were seen flying over Burn Valley Gardens on 24 August, thirteen were near Tunstall Hall, just before dusk on 13 December and up to eleven in Ward Jackson Park from December onwards; all presumably relating to the same roving flock. The increase in numbers offers circumstantial evidence that the species had bred again in the area. Up to fifteen remained at Hartlepool during 2009, though there was again no confirmed report of breeding having taken place.

The harsh winter weather of 2009/2010 had no obvious affect on this exotic species. Courtship activity was noted from early February 2010 and breeding again seemed likely as the population made a dramatic jump with 32 being noted over Burn Valley Gardens on 16 July 2010. To date, breeding has yet to be confirmed directly, though it seems increasingly likely that a small population of birds is now established in the Hartlepool area. Whilst this location continues to be the chief focal point for this species, birds have been increasingly recorded from other locations across County Durham. There was a distinct upsurge in sightings away from the Hartlepool area from 2005 onwards, prompting speculation that other breeding populations might soon be established in the county.

Recent non-breeding status

There were records from five different parts of the county in 2005 ranging from Hartlepool to Whitburn in the north and Willington in the Wear valley. Birds were seen in Saltwell Park, Gateshead in September 2007 and at Seaham in October with unidentified parakeets, probably of this species, seen at another four sites during the year. In 2008, two were near Newton Aycliffe in January, with singles at Old Quarrington and Eppleton in June, in the Hetton area from August to November and in Sunderland in October. Two were at Bishop Auckland in September. In 2009 single birds were at Hetton Park in the first half of the year, over Dorman's Pool, Teesmouth on 6 May and in Gateshead on 7 July.

At the present time the Ring-necked Parakeet remains a scarce bird in the north east of England. Nationally, the future of this exotic species is uncertain with studies being undertaken to determine the impact of this and other non-native species on native wildlife and agriculture, with enquiries also being made into the feasibility of controlling numbers should this be deemed necessary. This will no doubt affect any future spread of this species within our region and the British Isles as a whole, though it is at present unclear if any culling will actually take place. Current estimates indicate that there 20,000 to 30,000 birds now resident in the UK, mainly in the south east of England but with signs that the population is still expanding. Despite large areas of suitable habitat being available, there is little evidence to suggest that the localised, 'possible' breeding population in the environs of Hartlepool is showing any signs of expansion, though the establishment of further colonies is possible with time.

Distribution & movements

Ring-necked or Rose-ringed Parakeet as it is often known, is a native of Asia and Africa, covering a vast sub-tropical range, from the west coast of Africa, east to Burma, in a variety of racial forms. It is a very common cage bird in the UK with many thousands having been imported each year since the 1960s and many more being bred in captivity (Bell 2008). Birds have been seen regularly in the wild in England since the late 1960s, and possibly much longer, with breeding confirmed in Britain in 1971. The current feral population stems from a combination of escaped birds and intentional releases. Comparative biometric analysis of 'wild' British birds and museums specimens by Pithon & Dytham (2001) was unable to definitely identify the British breeding birds to race, though it would appear that they are nearer to the Indian subspecies than any other (Parkin & Knox 2010).

Great Spotted Cuckoo
Clamator glandarius

A very rare vagrant from Southern Europe: one record.

On 2 July 1995, Ian Foster found a juvenile Great Spotted Cuckoo close to Dorman's Pool. It moved around the area, being seen at Greenabella Marsh and Hargreave's Quarry, but spent much of its time in the Brinefields area close to the Long Drag (*British Birds* 89: 509). This was only the fourth mid-summer record for Britain at the time, and is still the earliest ever juvenile in Britain by five days. The bird remained in the area until 9th and was seen by several hundred observers during its stay. It spent most of its time in small scattered trees and bushes adjacent to the Long Drag, often feeding on the ground where it would remain out of sight for long periods and only be seen when accidentally flushed.

Great Spotted Cuckoo, Long Drag, July 1995 (M. Sidwell)

Distribution & movements

Great Spotted Cuckoos are a widespread summer migrant to south east and south west Europe, arriving on breeding grounds from late February. A smaller western Asian population breeds discontinuously from Central Turkey to northern Iraq and south west Iran. The wintering grounds are in sub-Saharan Africa. There had been a total of 46 British records to the end of 2010, the majority of which in the coastal counties of southern and south east England during March and April.

Cuckoo (Common Cuckoo)
Cuculus canorus

A formerly common and widespread summer visitor, also a passage migrant, apparently declining and now considered uncommon.

Historical review

The earliest reference to the Cuckoo in the area appears in Willughby's Ornithology in 1683, from a document by Ralph Johnson of Brignall, near Greta Bridge. He was presumably referring to birds he observed in his area, which sits astride the modern southern boundary between Durham and North Yorkshire (Mather 1986). Other than this, there is little from the early writers on Durham's birds, which may reflect the almost mundane way that they viewed its summer presence. Hutchinson (1840) recorded young birds on the moors in August, which would today be considered quite a late date for this species.

Its widespread and common occurrence during the late 19th century is illustrated by the fact that Robson (1896) reported taking sixteen clutches of eggs in the Derwent Valley, all of which contained one Cuckoo egg. He documented eight different host species including Whinchat *Saxicola rubetra*, Greenfinch *Chloris chloris* and Tree Pipit *Anthus trivialis* (Bowey *et al.* 1993). Around the turn of the century, Nelson (1907) called it 'common and generally distributed' in the Teesmouth area. Temperley (1951) described it as a "common summer resident" which was "well distributed throughout the county". He noted that it was most common on the moors of the western areas, where it parasitized Meadow Pipit *Anthus pratensis*; he observed no changes in its status during his long period of interest.

Post-Temperley, the species remained a common summer visitor, but little was written on its relative status over the following decades. One of the references we do have comes from Stead (1964) who, reviewing the status of Cuckoo in the Teesmouth area from the period of Temperley's publication to the early 1960s, thought that it was less numerous than 10 years earlier.

Perhaps the most insightful observations on this species in Durham over the latter half of the 20th century come from David Simpson, a long-term observer around Shotton Colliery in the eastern half of the county. In 2010, on reviewing his 50 years of records, he commented that from being a common summer visitor in the 1950s and 1960s, the Cuckoo had declined to such an extent that for the first time in his memory, none were seen locally in spring 2007, and likewise each subsequent year to spring 2010. His notes tell of him finding, on 21 May 1953, five Meadow Pipit nests in the area, two of which contained Cuckoo eggs (Simpson 2011).

One of the most interesting records from the 1950s was of an albino cuckoo nestling that was found in a Meadow Pipit's nest on moors near Wolsingham in the summer of 1952. This distinctive bird was later seen as a fledged juvenile (Temperley 1953). In 1954, a rufous or 'hepatic' female was shot at Edder Acres Farm, Shotton Colliery (Simpson 2011).

The early 1970s produced reports from across Durham, including what is still one of the latest autumn records, of one at Billingham on 28 September 1972.

Recent breeding status

The species has been found breeding in a wide range of open country habitats across the county (Armstrong 1988-1993) but is mostly concentrated in open mixed farmland in the lowlands, wetlands and the upland fringes. Birds are regularly recorded on the high, exposed areas of the county such as on upland commons and fells. They traditionally could be heard around the county wherever suitable host species were present, though they tended to avoid built up areas and dense woodland.

The *Atlas* map showed two main areas where Cuckoos were recorded. One of these was in the central lowlands of the county and the other on the eastern edge of the Pennine uplands. These two groupings suggested that the best breeding areas for Cuckoos were, at least in part, dependent upon the presence of their 'fixation hosts' (Gibbons *et al.* 1993).

In Durham, the two principle host species for the Cuckoo are probably Dunnock *Prunella modularis* and Meadow Pipit *Anthus pratensis*. Birds breeding in the central lowlands are likely to parasitize hedge and woodland edge species such as Dunnock and Robin *Erithacus rubecula*. At Shibdon Pond, in July 1986, a juvenile was being fed by both Dunnock and Meadow Pipit foster parents (Bowey *et al.* 1993). Birds in the westerly cluster, on the edge of the Pennines, are more reliant on ground-nesting species as hosts, principally Meadow Pipits (Westerberg & Bowey 2000). In wetland areas Sedge Warbler *Acrocephalus schoenobaenus* and perhaps the generally much less common, Reed Warbler *Acrocephalus scirpaceus*, may be parasitized as elsewhere in Britain (Glue & Morgan 1972). Certainly in 2007, interactions between adult Cuckoos and Sedge Warblers and Barn Swallows *Hirundo rustica* were noted, but the only confirmed foster parents were Meadow Pipits. In the following year on the North Tees Marshes, evidence suggested that Cuckoos were not choosing the Reed Warblers, which are now quite numerous there, preferring Meadow Pipits and Dunnocks instead (Newsome 2008 & 2009). Across Britain it is has been estimated that almost half of all female Cuckoos lay in Dunnock nests (Davies 1987). One juvenile Cuckoo in the county was still being fed by Dunnocks on the very late date of 6 August 2006.

Previous work has acknowledged the difficulty of estimating the Cuckoo's breeding population. Most 'sightings' of Cuckoo refer to calling males. The male's unmistakable call, which is audible over considerable distances, provides a guide to male distribution. Conversely, many mated females will go unnoticed, resulting in a possible under-estimate of breeding numbers (Westerberg & Bowey 2000). However, as the species appears to be largely polygamous (Marchant *et al.* 1990), with several males attached to each female, a simple assessment of the number of breeding 'pairs' is always likely to be difficult. In the lowlands, where there is plenty of cover for birds, females can be difficult to observe, but in upland areas, as many as three or four males have been observed calling to a single female (L.J. Milton pers. comm.). Allowing for the fact that such males can travel quite large distances as they accompany females, calling frequently, it is possible to over-estimate the number of males (Marchant *et al.* 1990).

In view of the above difficulties, the *Atlas* placed strong caveats on its representation of the species' breeding distribution in the county, considering that the maps may have been an under-estimate of the position at that time (Westerberg & Bowey 2000). Working from suggested national population densities of between five and ten pairs per occupied 10km square (Sharrock 1976) would have given a county estimate of 145-290 pairs. However, based on the number of occupied tetrads in the county, allowing for two breeding females for each one of these, the county breeding population was estimated as likely to be higher than this, at around 340 'pairs' at the time of the *Atlas* (Westerberg & Bowey 2000).

Nationally, in 1993 the *'New Atlas'* suggested that there was some evidence of a decline across the British Isles over the previous two decades (Gibbons *et al.* 1993). Since then, trends have been unremittingly downwards, with an estimated 65% decline in the UK population since the early 1980s (Risely *et al.* 2010). However, studies in Cleveland in 1985 and 1999 suggested there had been little change in local populations of *c.*40 pairs in the north of Tees area in that period (Joynt *et al.* 2009).

During the 1990s the number of locations where birds were recorded in June was logged separately in some DBC Annual Reports, as a sample indication of the strength of the breeding population. This period produced records from 14-16 locations in June 1992, 1995, 1998, and 1999, with only 1993 and 1997 falling short at just eleven and eight June locations respectively. This therefore failed to demonstrate any clear trend, probably merely illustrating the variation in records submitted from year to year. High site counts in the period included six at Muggleswick Common in May/June 1991, at least five calling males at Waldridge Fell in May 1994 and five at Horden in May 1995 (Armstrong 1990-1999).

Trends in the county since the *Atlas* are difficult to assess. In earlier decades, the species was considered ubiquitous and was therefore under-recorded, except for extreme dates. However, because of a combination of its known scarcity, and far greater observer coverage, the likely decline is to some extent masked by more extensive recording in recent years. Nevertheless, anecdotal evidence suggests that birds have been getting ever scarcer through the early years of the 21st century. This species was still considered to be 'a common summer visitor and passage migrant' in Durham in 1990, but that status had been changed to that of 'an uncommon and apparently declining summer visitor' by 2009 (Armstrong 1991, Newsome 2010).

This decline may have been particularly severe near the coast and in the south of the county, where records are now mainly of migrants. The species continues to occur regularly in the North Tees Marshes where a high total of eight were noted on the Long Drag on 24 May 2000 (Iceton 2001). In 2001, the worst year to date in Durham for Cuckoo records, the species was reported to DBC from just 12 locations in the whole county through an entire season (Siggens 2005). However, the fact that there was just a single record from the uplands that year, from Muggleswick, a regular location, is a clear sign of severe under-recording. In 2005, Joynt considered that the Cuckoo's status as a breeding bird in Cleveland was under threat, and the following year no juveniles at all were seen there (Joynt 2006). In the whole of 2007, there were just three birds reported in South Tyneside, one each in May, July and August, a decline that has been noted as evident in that area for some years (Newsome 2008).

Even in this period, however, the Cuckoo remained widespread in the county. The Hetton/Houghton area has remained something of a stronghold in the east, as illustrated by the presence of five between South Hetton and Elemore Hall on 19 June 2002. The breadth of 2005 reports underlines the range of the species in the county, with records from Hamsterley Forest, Tunstall Reservoir and Knitsley Fell in the west, the Hetton/Houghton area, Shotton Colliery and Whitburn in the east, Bishop Middleham in the centre, and Seaton Common and Teesmouth in the south east. Not all would have been breeding birds with the Whitburn records, at least, certainly being only migrants. Note also the Shotton Colliery record in this list, which was one of the last from this location, as previously recorded by David Simpson in the historical background section (Newsome 2006).

There was the hint of an improvement in Durham from 2007 to 2009, but this may simply reflect the current level of interest in, and hence reporting of, this iconic and declining species. It is not supported by any change in the national downward trend at the time of writing. In 2007, Cuckoos were recorded from 26 areas to the west of the A68, 25 between the A68 and the A19, and just 6 to the east of the A19. The following year, Cuckoos were noted in over 100 1km grid squares during spring and summer. The majority of these records came from the western half of the county on moorland fringes, with Meadow Pipit being predominantly the favoured host. This appeared to be a much better year for the species than recent preceding years, but may simply reflect better recording, as noted above. The pattern was similar in 2009 with records from at least 90 1km squares, approximately two thirds of which were in upland areas west of grid line NZ20, which lies to the west of Durham, passing through Stanley and Sunniside (Newsome 2008-2010).

These three years of more concentrated recording effort probably provide a more realistic assessment of the current status of the species in the county. They suggest that, though declining, particularly as a breeding bird in the east and south, to date the Cuckoo remains sparsely but widely distributed across Durham, with the west still favoured. Only further monitoring will show whether the Durham upland population is following the national decline, though the intuitive answer to the query is that it is.

As a long distance migrant, the Cuckoo is susceptible to any degradation of habitats in its wintering areas in Africa and potential problems along its migration routes to Northern Europe (Westerberg & Bowey 2000). The principal causes of the decline in the UK population are considered likely to be a combination of deteriorating conditions along these migration routes, or in the overwintering grounds in sub-Saharan Africa, and a shortage of larger insect prey (mainly large caterpillars) during the breeding season in the UK (Risely et al. 2010). This being the case, it may mean that at least part of the cause of its population decline in the UK is beyond our control. In 2009, a number of declining species were found to have reached their lowest levels since the start of the national Breeding Bird Survey and amongst this group was the Cuckoo, which had experienced a 44% decline over a fourteen-year period (Risley et al. 2010); the present trend for this culturally resonant species seems to be relentlessly downwards.

Cuckoos may be noted in Durham from late April, usually from around the 21st of the month, with most birds arriving in the latter half of May. There is a sprinkling of earlier records in the county however, the earliest modern record being of one on 2 April 1991 at Teesmouth (Armstrong 1991). Further early arrivals were noted on 5 April 1985 at Hartburn and 5 April 2002 at Greenabella Marsh, Teesmouth (Bell 1985, Little 2002). All of these are pre-dated by an older record of a bird that was heard singing at Durham by the River Wear on 1 April 1946 (*The Vasculum* Vol. XXXI 1946).

Reports of the species decline through July, as adults vacate the breeding grounds from mid-June onwards. The first juveniles appear around the first few days of July, and most records from then on are of juveniles. By July in 2009, for example, the majority of records in the eastern third of the county (Newsome 2009) were of juveniles. Numbers decline through August with the final records of the year usually coming in the second half of the month. However, there are a small number of September records, including a very late bird at Marsden on 29 September 1976. What may have been the same bird was found dead at Harton Cemetery, South Shields on 6 October (Unwin 1977). The latest ever in the county is the only record of a live bird in October, one at Cowpen Bewley Pond on 15 October 1993 (Bell 1994).

The only modern county record of the rare rufous morph concerns a female at Dorman's Pool on 6 July 2002 (Little 2003). Another notable record was of an individual with an unusual three note call at the Grove, Hamsterley throughout June 1990. This species does sometimes call under cover of darkness, as was the case with one heard in the middle of the night at Brasside in May 1998 (Armstrong 1991 & 1999).

Distribution & movements

Cuckoos breed across much of Eurasia, the extreme north excepted. They are long distance migrants that winter in Africa, to the south of the Sahara, largely in equatorial regions. Asian breeding birds winter in southern Asia.

At the time of writing, five adult Cuckoos had been satellite tagged by the BTO, to try to improve our limited understanding of this species' migration, which may assist international conservation efforts (BTO 2011). It is notable that there is just a single Cuckoo ringing recovery to date in the county; one ringed at Bedburn, Hamsterley on 17 July 1965 was recovered at Pool, Otley, Yorkshire on 22 August 1965.

Barn Owl (Western Barn Owl)
Tyto alba

A widespread though relatively uncommon resident. Numbers have generally increased in recent years but remain prone to the impact of hard winters.

Historical review

The species' long time presence in the north east of England was demonstrated by the excavation of skeletal remains from the Roman site at Piercebridge, in the south of the county (Yalden & Albarella 2009). No doubt, it persisted through the millennia and into the last two centuries of ornithological recording. It was believed to be the most common owl in Britain during the 19th century (Holloway 1996).

By late in the 19th century Robson (1896) said of this species "*now rarely met with in our district*", though in the late 1940s Temperley (1951) documented that it was breeding in the Derwent valley and surrounding district in the late 1940s. Temperley said of it, "*a resident, locally distributed and nowhere common, formerly more numerous*".

At the turn of the 20th century it was said by Nelson to be "*the most generally distributed of the owls*" around Teesmouth (Joynt *et al.* 2008). There was a serious national decline of this species over the first 80 years of the 20th century, with some recovery in later decades. This species was probably in long term decline in Durham for over a century, perhaps reaching its low point in the mid-1980s. In the first two-thirds of that period its national population declined by over fifty percent. In Victorian times, the Barn Owl was considered the commonest owl in Durham and in the 1930s there was an estimated 150 pairs (Blaker 1934). By 1985, this figure had declined to an estimated 18 pairs (Bowey 1993) but there is some evidence of a stabilisation in numbers followed by a more recent recovery.

During the late 1950s and 1960s the Barn Owl was breeding in a number of areas in the south east of the county, including Wolviston, Cowpen Bewley, Hartlepool, Hartburn, Graythorp, Port Clarence and Portrack (Stead 1964). In 1967 a pair bred, apparently successfully, in Durham Cathedral. Nesting locations in the late 1960s and early 1970s included near Port Clarence in 1968 with breeding being reported in eight locations around the county in 1971.

By the mid-1970s the breeding population at Teesmouth was thought to be eight pairs in that area but by the mid-1980s it had fallen to no more than three or four pairs. Industrialisation and urbanisation of the lower Tees valley, was believed to have contributed to its local decline (Joynt *et al.* 2008).

Recent breeding status

The Barn Owl is a well-known crepuscular farmland bird, which feeds largely on small mammals. It is found in rough grassland areas and in extensively farmed countryside, roosting and nesting in undisturbed cavities in trees, barns and other similar locations (Taylor 1994).

The *Atlas* documented it as "*sparse, but widespread*" (Westerberg & Bowey 2000) and nowhere was the species present at high densities. For many years, its stronghold was the central lowlands and the north east of the county, areas that are less exposed to the extremes of winter weather experienced in the west of the county. The distribution, as revealed by the county survey of 1992 (Bowey 1993), indicated that pairs were found largely from around Durham City, eastward to the coastal strip, with pockets of birds in the north west and scattered pairs to the south and south east of the county. Traditionally, the species avoided the county's western uplands (Westerberg & Bowey 2000). The *Atlas* highlighted the disparity between the information revealed about the species by general survey techniques and more dedicated work (Bowey 1993). It is clear that it is difficult to map Barn Owl distribution without targeted effort.

Prior to the 1992 survey, the numbers of pairs of Barn Owls in Durham was believed to be in double-figures, but probably not more than 20-30 pairs. The Hawk & Owl Trust survey in the early 1980s (Shawyer 1987) indicated the presence of 18 territories. The 1992 survey suggested a county population of 35 to 40 pairs (Bowey 1993). Through the late 1990s and beginning of the 21st century birds were increasingly found in areas previously believed unoccupied. It was clear that an upturn in numbers and a westward extension of range was taking place over this period. Through this latter period, reports of birds came from 111 one-kilometre squares across the Durham recording area. This process of expansion was supported by the species' increasing use of nestboxes, the large-scale provision of which was initiated by the Durham Biodiversity Partnership around the turn of the Millennium. In 2007, birds were noted in almost 64% of the county's ten-kilometre squares, despite an absence of reports from some traditionally occupied squares, which suggests an element of under-recording rather than an absence of birds. Two of these squares held six breeding pairs each, indicating the density of territories that can occur in some parts of the county.

By the middle of the 2000s, birds were present in territories from the coast to as far west as Langdon Beck. In lowland Durham, birds were spread from Gateshead in the north to Piercebridge in the south, and along the full stretch of the coastline from Marsden to the North Tees Marshes. Particular concentrations were noted in the area around Houghton and Hetton and to the north west of Bishop Auckland. Over this period, there were increased sightings on the North Tees Marshes and across the north of Teesside area and it was described as "*A rare but welcome sight*" with four to five pairs (Joynt *et al.* 2008). During the Durham Bird Club 2005 owl survey, breeding was confirmed in two 10km squares in the south west of the county (Evans 2006). The following year, the importance of quarries (both working and disused) on the Magnesian Limestone escarpment became apparent, with at least three such sites being occupied by birds (Evans 2006). Most Barn Owl territories in Durham are still located in the lowlands, a high proportion of them to the east of the A1 (M) motorway. For example, of 16 territories examined in 2004, all but one was to the east of the motorway; the exception was in Weardale.

Nonetheless, between 1990 and 2010, there was a large increase in the species' use of locations in the west of the county, presumably linked to the run of relatively mild winters over this period. In 2008 there were records of birds from 11 one-kilometre squares west of the grid line that passes north-south through Barnard Castle. Some birds were present almost as far west as the Cumbrian border, such as at Langdon Beck (at 420m above sea level). A similar westerly limit was noted in Weardale, with mid-summer reports of birds at Cowshill. In Lunedale, one was hunting in daylight by Selset Reservoir on 2 August 2009. The severe weather conditions of December 2009 and early 2010, and a repeat of this situation in winter 2010/2011, appears to have had an impact on birds in such areas and winter weather of this nature is likely to check any increase in range and growth in numbers.

Nationally, by the mid- to late 1990s, there were believed to be around 4,000 pairs in Britain (Toms *et al.* 2003) and by 2009, there were estimated to be over 60 pairs across Durham. Many pairs however, would have gone unregistered so it is conceivable that at 2010, there may have been close to 100 pairs of Barn Owl in the county; a rise of some 150% since the 1992 survey work (Bowey 1993).

The threats to this species are well-documented and considerable (Shawyer 1987). The principal risks to the Barn Owl in Durham, in descending order of probable local impact are:

- Harsh winter weather, particularly the length of any periods of snow cover
- Loss of suitable nest sites, due to the re-development of farm buildings and the decline in the number of hedgerow trees
- Loss of suitable habitat where barn owls can feed, because of the intensification of farming
- Increased deaths of Barn Owls from road accidents
- Pesticides and other agricultural poisons. These are known to have suppressed breeding performance in some cases and in some instances, birds have inadvertently succumbed to the use of various poisons

All of these, but particularly harsh winter weather, probably contributed to the species' long-term decline in the county (Bowey 1993), recent recoveries have probably been attributable to the milder winters experienced from the late 1980s until near the end of the first decade of the 21st century. The provision of nestboxes has in part offset the loss of some nest sites, but hard winter weather is always likely to be the determining factor for this species' distribution in the north east of England. What has been evident over the last decade is that as numbers rose across Durham there were more reports of road casualties. Three deaths were reported in 2002 and 2003; six in 2005 and the same number in 2007; the A1 (M) and A19 claimed several victims over this period.

At Haswell in May 2000, a captive bird was visited by a wild bird on several occasions. Over the course of the 2000s there were a number of reports of escapees from captivity. One such bird was found dead at Boldon in 2006, after its leg-straps were snagged on a branch. Another was found dead at East Boldon in the first week of May 2008. This is a trend that has been prevalent for many years, with similar fatalities coming to light during the 1992 survey; what is clear is that escaped, captive-bred birds do not fare well when in the wild.

The species is clearly adaptable and at times can be quite resilient, for example in 2005, a pair in Teesdale raised young despite a gas-powered bird scarcer being positioned near their nest tree. An indication of its sometimes prolonged breeding season was illustrated by the 2007 report of a pair that fledged three young near Satley in November.

Recent non-breeding status

Being effectively a sedentary species, disregarding juvenile dispersal, the non-breeding situation of this species in Durham is essentially equivalent to its breeding range. Hard weather can alter birds' nocturnal hunting patterns and many records from casual observations come in the period November to February, when days are short and prey scarce. At such times, birds often hunt in the late afternoon. In some instances, additional winter competitors in birds' territories may be challenging and birds can be aggressive to similar species. For example, on 13 November 2007, a Wear valley pair vociferously drove off an intruding Long-eared Owl *Asio otus* that had strayed into 'their' farmyard. Elsewhere, birds have been observed defending their nests against Sparrowhawk *Accipiter nisus* and Fox *Vulpes vulpes*.

Dark-breasted Brn Owl
Tyto alba guttata

Temperley called this race of Barn Owl "*a rare vagrant from abroad*". On 25 February 1927, one was shot at Prior's Close Bog, Leamside and sent to the British Museum where in 1945, it was confirmed as being of the continental, dark-breasted race. Two further birds have been seen in the North Tees area (found dead at Hartlepool Docks in autumn 1957, and at Hargreave's Quarry on 18 December 1983). However, all records of this race must be treated with some degree of caution as the species' racial forms are genetically complex, for occasionally even pale-breasted forms can produce dark-breasted young (Parkin & Knox 2010). A seemingly dark-breasted bird at Whitburn on 4 August 2006 was unlikely to have been of continental origin.

Distribution & movements

Barn Owl is one of the most widely distributed and cosmopolitan of bird species, it is found on all continents excepting Antarctica and the Arctic. Consequent to this wide world range, there are many racial forms, with over 30 documented subspecies. British breeding birds, in the north of the islands, are amongst the most northerly breeding Barn Owls in the world, as the species is largely absent from northern Europe (Parkin & Knox 2010).

Barn Owls are largely resident, though there is usually some dispersal of young birds. For a largely sedentary bird there are a surprising number of interesting ringing recoveries from Durham. In November 1937 a bird ringed as a nestling at Stocksfield the previous July was recovered at Winlaton, Gateshead. An adult ringed at South Shields on 9 November 1949 was recovered at Cocksburnspath, Berwick upon Tweed, Northumberland on 26 April 1950. A bird ringed as a pullus on 1 July 1962 at Sacriston was recovered near Driffield, Yorkshire on 14 February 1963. One ringed in Kielder, Northumberland on 27 June 1965 was recovered at Witton-le-Wear on 4 December 1965. In addition, there are two recoveries of birds ringed as pulli at Stocksfield in Northumberland and the other at Barford near Coventry, recovered in their first winter at Stockton and Wynyard respectively (Blick 2009). In November 2007, a dead bird was found Cowpen Bewley mid-month, it had been ringed as a pullus at Burstwick, Humberside on 17 June 2007.

Snowy Owl
Bubo scandiacus

A very rare vagrant from the Arctic: two records.

The Snowy Owl is an extremely rare winter visitor and there are just two records in Durham. A nomadic species, breeding in the Arctic tundra regions, it occasionally wanders south in response to food availability and climatic effects, but on mainland Britain this species remains a major rarity. Hancock (1874) documented "*a fine specimen*" that had been shot at Helmington Row, near Bishop Auckland, on 7 November 1848. The corpse went into the possession of Henry Gornall, a professional "*collector and bird preserver*" (1851 Population Census) who lived at Bishop Auckland and who had himself quite possibly shot the bird. He was a close working associate of the 'gentleman-collector', Joseph Duff, also of Bishop Auckland and between them they had supplied Hancock with several specimens and were in regular correspondence with him. Temperley (1951) considered the evidence of this record to be reliable given that a contemporary account (*The Zoologist* Vol. 71, 1848) was published, and that the area around Helmington was largely rough pasture and sheep-walk, suitable if not entirely characteristic habitat for the species. Snowy Owls were clearly much more frequently encountered in Britain in the 1800s, and the Durham record is one of 140 records for the century, with frequent encounters in eastern and southern England, as well as Scotland. The fate of the specimen is not know with certainty but an unlabelled and clearly old mounted Snowy Owl was at 2010 held in the Bishop's Palace, Auckland Castle.

There is also a more recent record of this dramatic owl. On 26 November 1981, Mr D. Bell of Wolsingham was acting as a beater to a grouse shoot on moorland to the north of the town when he flushed a large white and speckled owl which had been perched upright in the heather. The owl flew back low over the beater and away. Despite the absence of optical aids, the views were close enough to note the big yellow eyes and an odd facial expression produced by a moustachial arrangement of feathers above the beak. Mr Bell conferred with another beater who had also seen the strange bird and they soon found the remains of a Red Grouse *Lagopus lagopus* where the owl had first been seen. Having some ornithological experience, including a familiarity with Barn Owl *Tyto alba*, the observer wrote a description which he forwarded direct to the British Birds Rarities Committee which accepted the record as a probable female Snowy Owl. Mr Bell was apparently fortunate to see the bird again a few days later (*British Birds* 77: 538).

Distribution & movements

Snowy Owls breed across the tundra and islands of Arctic Russia, Siberia, Canada and northern Greenland, and also occasionally in Iceland and northern Scandinavia, depending on food availability. Most birds disperse south in the winter, but some are resident or nomadic if food is available. A very small population was resident in Shetland, Scotland, from the mid-1960s to the early 1980s, breeding there between 1967 and 1975. Away from the Northern and Western Isles, where occasional birds are still seen with some frequency, it is a very rare vagrant with few recent English records.

Little Owl
Athene noctua

Sponsored by
The Original Bakehouse

A widespread and fairly common resident, primarily in agricultural habitats.

Historical review

The Little Owl is an introduced species in Britain. Prior to the introductions, the species was considered a rare visitor to Britain from the Continent (Holloway 1996). Releases occurred in a number of areas and places, many of them unsuccessful (Lever 2005); including in Yorkshire in the 1840s. Concerted release programmes, using birds from the Netherlands, took place in Kent in the 1870s and Northamptonshire in the 1880s (Holloway 1996) and these were eventually responsible for the establishment of the species in Britain. The first nest was found in Kent in 1879. Most of England, to the south of Harrogate, Yorkshire, had been colonised by the species, by the 1950s (Holloway 1996). Mather stated that prior to its introduction this species was an accidental visitor from the Continent, of extremely rare occurrence. Nelson listed four such records for Yorkshire, but none are known for Durham (Mather 1986).

The species reached County Durham in the early 20th century, the first specimen being shot at Chilton Moor near Fence Houses in April 1918 by T. Harrison, a gamekeeper of New Seaham; it was identified by Dr Hugh Blair. It had presumably reached the southern part of Durham, undetected, prior to this and the first records in that part of the county came when was one was shot in Teesdale in 1923 (Mather 1986) and another was recorded near

Darlington in the same year. In 1931, F. J. Burlenson and Joseph Bishop (the RSPB Watcher at Teesmouth during the 1930s) reported that a pair had attempted to nest near Rushyford, "*two to three years previously*", though the actual date is now thought to be 1926. Further colonisation and consolidation followed through the early 1930s. A male was shot at Stonechester Farm near Toft Hill on 26 October 1932. Joseph Bishop reported that young birds, still bearing down, had been knocked down by a car near Stockton on 8 August 1933. Considerable interest was paid to a pair nesting near Shildon in 1935, with an analysis of their pellets and prey remains being undertaken; concern was expressed at the time about the possible impact that the species would have on native fauna (*The Vasculum* Vol. XXII 1936). In 1934, a pair bred in a disused quarry at Fulwell, Sunderland and the first breeding of this species took place in Northumberland, near Corbridge, in 1935 (Kerr 2001).

Colonisation and spread through the county continued apace and by 1949 the species was well-established in southeast Durham, had penetrated up the Wear valley as far west as Wolsingham and was present along the coast as far north as South Shields.

By 1951, George Temperley thought that the Little Owl was "*a resident in increasing numbers, as yet chiefly confined to the east of the County*" and that it was "*spreading rapidly*". There was little information documented on this species through the 1950s, though frequent reports in the Stanley area in 1958 indicated an extension into that part of Durham. In his national review of the species in 1967, Parslow stated that by the 1930s, the species had reached north west Wales, Durham and Northumberland and that, through the 1950s, it continued its spread northwards (Parslow 1967). By the late 1950s, Little Owls were nesting in various locations in the north easterly coastal strip of the county; Trow Quarry at South Shields was by then a favoured locality for nesting birds (J. Edwardson pers. comm.). Up to the 1960s, the Little Owl nested at several places in the south east of the county including Wynyard (Blick 2009).

Records from the 1970s and 1980s indicate that this species was still relatively sparsely distributed in County Durham being most numerous in the eastern half of the county. Although Little Owl's expanded their range in subsequent decades they were still largely absent from some large areas of apparently suitable habitat, even into the early to mid-1990s, particularly in the west of the county. An illustration of the relatively slow spread of the species away from the eastern portion of the county is that the first Gateshead records, in the north west of our recording area, did not occur until 1975. In the western valleys, the first confirmed breeding from near Wolsingham wasn't until 1973, despite the species having been recorded in that area more than 20 years prior to this.

With expansion there were also local losses. For instance, during the late 1970s birds were known to have bred in Clockburn Dene, not far from Whickham, and in the early 1980s they were present at Clara Vale, in western Gateshead, but by the mid-1990s both of these nest-sites had been abandoned.

Recent breeding status

Little Owls generally favour open farmland with hedges and mature hedgerow trees. Open rural areas, particularly where more traditional land management practices are adopted, are often occupied. For nesting purposes they will use disused farm buildings and out-houses, working or disused quarries, and rocky outcrops but tree holes, particularly in mature ash *Fraxinus excelsior*, the most common mature hedgerow tree in the county, are favoured. There is also a definite concentration of occupied tetrads, as noted by Westerberg & Bowey (2000), on the magnesia limestone escarpment, in eastern and north eastern parts of the county where several limestone quarries hold birds, some of them with multiple pairs, often in close proximity. In 2006, successful nest sites were discovered inside a house roof and on a coastal cliff. Established pairs can be noisy and prominent in their territories during March.

As the *Atlas* showed, birds are well represented at coastal sites and around the North Tees Marshes (Westerberg & Bowey 2000). Most of the occupied tetrads in this latter area are in the farmland, to the west and south west of Stockton, north towards Wynyard and north east to the coast and in the rural areas to the north east of Hartlepool (Joynt *et al.* 2008). They are well represented in farm land hedgerows in the mid–Tees valley around Sadberge, Piercebridge and Staindrop.

The species' strongest populations are to be found in the easternmost parts of the county with sightings coming from coastal cliffs, for example at Dawdon and Trow Quarry, and from inland farmland. Concentrations are particularly evident in the Houghton area with sightings in most of the local quarries and even in suburban locations towards the edges of the town during the mid-2000s.

In more westerly areas birds can be routinely found along the main river valley bottoms as far as Mickleton, in the Tees valley, Frosterley on the Wear, and Crawcrook, on the Tyne. The county's larger woodlands and conifer plantations do not provide the correct conditions for the species. Nationally, birds have tended to avoid higher ground (Gibbons *et al.* 1993), and upland habitats in Durham proved to be no exception until quite recently. Over the last ten or more years birds have been noted near Muggleswick and above Edmundbyers, in the upper part of the Derwent valley; the latter records being of birds at over 280m above sea level. The trend for penetrating into the county's upland fringes has continued, aided perhaps by mild winters. By 2008, small numbers of birds were being seen in quite exposed moorland edge habitats from such widespread sites as Balderhead, Grassholme and Hury Reservoirs, Gilmonby, Hill End, Cowshill and Salter's Gate. In the most extreme situation, one hardy pair was present at around 400m above sea level in The Stang in 2008; birds having also been seen there in 2003. It remains a scarce species in the upper Tees valley, to the west of Eggleston, so records at Newbiggin-in-Teesdale in December 2001 were noteworthy.

In 2005, it was believed that that there were over 180 territories in the county and successful breeding was proven from at least 25 pairs, which had raised 38 young. In 2006, 270 known territories were reported with the highest density being in the ten-kilometre grid square in the centre of the county, around Brancepeth and Spennymoor, which held at least 50 pairs. In 2009 an adult at Cowshill in the Upper Wear valley was feeding a fledged juvenile in July, and in the very exposed location of Weatherhill Engine above Stanhope, a pair was utilising rabbit burrows for shelter at 430m above sea level.

Little Owls can be extremely site faithful (Glue & Scott 1980) and, because of this trait, many observers' reports refer to birds being 'in the usual place' or at 'a regular site'. Nonetheless, in what is a relatively under-watched habitat, i.e. farmland, they can be discrete and their preference for such locations probably means that many pairs will be overlooked. The species can be relatively common in the most favourable habitat. For example, around one farmstead near Trimdon in February 2007, ten calling birds were noted, with five pairs going on to breed. In 2006, breeding was noted in South Tyneside at Marsden within sight of an active Sparrowhawk *Accipter nisus* nest. Over the last decade, between Kibblesworth and Lamesley, there has been a cluster of territories that was thought to number a minimum of 10 pairs, with up to four pairs on one farmstead. At nearby High Thornley there were two pairs of Little Owls before one pair was usurped from their nestbox by Starlings *Sturnus vulgaris*. A few kilometres away, in the Garesfield area, there were seven pairs and here, one natural site in a tree cavity produced broods of three and four in consecutive years before a Tawny Owl *Strix aluco* nestbox was erected nearby. This almost immediately attracted that species and Little Owls began falling as occasional prey to the larger species before they moved away. This behaviour was emphasised by the licensed attempt to tape lure Little Owls at Kibblesworth in the early 2000s, when a Tawny Owl showed an unhealthy interest in proceedings, perching on top of the mist net poles and diving at the tape recorder.

At the time of the *Atlas* survey work (Westerberg & Bowey 2000), County Durham was still relatively close to the northern limit of the species' distribution in the UK (Gibbons *et al.* 1993), although birds have been present in Northumberland, to the north, since the 1930s (Kerr 2001). As a result, it was felt that some fluctuation in population numbers might be expected, particularly after severe winters. The *Atlas* stated that, '*Little Owls had probably slightly strengthened their position over the last decade*' (Westerberg & Bowey 2000) and this trend is probably continuing. During the 2005 Durham Bird Club owl survey, Little Owl was recorded in 19 ten kilometre squares, breeding took place in 16 of these. It was distributed chiefly across the central and eastern part of the county, though birds did penetrate up to the moorland edges in some areas, and it was present in two squares where it was not logged as being present during the *Atlas* survey period a little over a decade previously. The population estimate from the *Atlas* fieldwork indicated that there were over 100, perhaps 150, pairs in the county (Westerberg & Bowey 2000). Targeted field work in the early part of the 21st century suggested that this species is not as scarce as previously envisaged. Because of this, and what has probably been a genuine consolidation in numbers (which may be related to climate amelioration over recent decades), the *Atlas* population estimate can be revised upwards. It is known that there are between 35 and 40 pairs in the North Tees area alone, according to the number of occupied tetrads, in *The Breeding Birds of Cleveland* (Joynt *et al.* 2008). Previous estimates in that area (Blick 1978) had been lower, being less than two fifths of this total. In 2005, around 180 pairs were located across the county, with 275 being reported in the following two years (Evans 2006). Continued extensive work suggests that there are now at least 330 known territories in the county. This postulated increase seems unlikely to be attributable solely to a population expansion and in large part is probably due to greater knowledge resulting from the intensive survey work carried out during the early 21st century.

Recent non-breeding status

Outside of the breeding season, birds are chiefly found in their breeding territories, with the exception of dispersing juveniles, which may not move that far from their place of birth (Wernham *et al.* 2002). In 2010, the Durham Bird Club received 232 records of the species, from 104 locations the geographical spread of which had a strong easterly bias as does the species' breeding distribution.

The species is a generalist feeder, often taking many large insects and small mammals, but one at Lumley, in 2006, routinely plundered Swallows' *Hirundo rustica* nests in the local farm and on one occasion was caught 'red-handed' by the local farmer, trying to make off with an almost fully grown nestling.

Distribution & movements

Little Owls occur across the Palaearctic from North Africa and Arabia, across Asia to the Himalayas and westwards up into Europe, to the fringes of the North Sea. It is largely resident with some dispersal though apparently no set migration patterns. It is not a native British species and there appears to be no conclusive evidence to say that it has ever occurred, naturally in a wild state in the British Isles (Parkin & Knox 2010), despite some claims to the contrary.

Most ringing information confirms the sedentary nature of Little Owls, with very little movement recorded between sites and islands around the British Isles (Wernham *et al.* 2002). Local ringing efforts in the Gateshead area secured 28 birds but the only movement of note was a bird ringed as a breeding female in May 1994 that was found dead just two kilometres away in mid-September 1999. There are just two known ringing recoveries for the county. A bird that was ringed at Brasside near Durham on 7 June 1964 was recovered at Newcastle on 22 July 1964 and, around the same time, but moving in the opposite direction, one ringed at Kirkley, near Ponteland, Northumberland on 8 June 1964, was recovered on 26 April 1965 at Wardley, Gateshead.

Tawny Owl
Strix aluco

Sponsored by
The Original Bakehouse

A fairly common and widespread resident.

Historical review

This species is Britain's most numerous owl (Marchant *et al.* 1990) and probably the most common local bird of prey. Tawny Owls can be found across the county in both summer and winter. They are resident and sedentary, being found in most types of woodland, including blanket forestry, small copses and even along sparsely timbered hedgerows, where there are mature hedgerow trees that possess appropriately-sized cavities. Their range in the county extends into the suburban environment in large urban parks, mature gardens, allotments and occasionally disused industrial settings.

Around the middle of the 20th century, Temperley (1951) said that it was a "*common resident*". In the 19th century however, Tawny Owls nationally were regarded as uncommon despite the greater availability of deciduous woodland at that time. In the last quarter of the 19th century, the Long-eared Owl *Asio flammeus*, was said to be the more common of these two species in northern England (Holloway 1996), but the true situation for Durham at the time remains unclear.

Locally, there are some indications from the historical texts, that it may have been somewhat less widely distributed during the mid-19th century. For instance, Hogg (1827) described it as "*rare*" in south east of Durham. A few years later, Hutchinson (1845), who had a more central overview of the county, described it as "*not uncommon*". Late in the 19th century, Robson (1896), writing of its status in the well-wooded Derwent valley, described the species as "*regularly nesting in Chopwell Woods*" and he documented a gathering of 14 birds in Blaydon Burn during October 1882, as well as an unusual nest site, down a rabbit burrow at Rowlands Gill (Bowey *et al.* 1993).

Through the middle and latter parts of the 20th century, the species benefited from the maturation of the county's coniferous plantations, such as those at Hamsterley Forest, Stobgreen and Chopwell, and in such areas any evening visit in early spring is usually accompanied by a multiple chorus of many territorial birds.

Recent breeding status

Tawny Owls are essentially a woodland species in Durham, as elsewhere, although they can also be found inhabiting gardens, parks (Gibbons *et al.* 1993) and areas of dense scrub. Nesting in the latter habitat is problematic for the species unless boxes are provided as nest sites are limited, although it will, not infrequently, use old Carrion Crow *Corvus corone* nests, a regular occurrence in conifer plantations (Avery & Leslie 1990). In some instances there is competition, not just from avian competitors for nesting cavities, for instance where nesting cavities have been in-filled with grey squirrel *Sciurus carolinensis* dreys. Birds nest in greater numbers in mature woodlands although they are also known to breed in parks, cemeteries and in large gardens (Westerberg & Bowey 2000). The well-wooded valleys and afforested areas hold the largest populations and highest densities of breeding birds. The most widely utilised local habitat is broad-leaved woodland but birds are also found in a wide range of other woodland types especially.

Due to its nocturnal nature, this species may be overlooked in some areas and, in all probability the extent of its true range in County Durham may, in some instances, be under-estimated. Artificial nestboxes have been used with some success in a number of areas over the years, starting at Low Barns Nature Reserve in the 1970s, but the species more commonly uses tree cavities, particularly in old ash trees. Mather (1986) reported a nest found on the crags of Cronkley Scar cliffs, in upper Teesdale, during spring 1962.

Tawny Owls breed relatively early in the year; in most years they are still vocal in March though many pairs will be incubating eggs by the end of the month. Evidence of very early nesting, was provided by the discovery of two nestlings at East Herrington on the remarkably early date of 11 March in 2005. Other examples include two pulli ringed in a nestbox at Washington Sewage Treatment Works, in early February 2001, the earliest in the UK that year. On a similar timetable a pair was discovered incubating eggs at Joe's Pond on 18 January, with the same pair unusually going on to lay a third clutch after the earlier two had failed. The first young of the year may be seen out of their nests by early April.

This is by far the most common of the owl species locally, with a widespread distribution wherever its basic requirements are met. Given the largely nocturnal habits, it is perhaps not too surprising that many reports of this species relate to calling birds. Most locations from where it is reported are in reasonably well-wooded areas, though reports do come from suburban locations.

In the early 1980s, breeding densities were studied at Hamsterley Forest. At that time, 31 pairs occupied the 250 hectare study site. In the mid- to late 1980s, mixed woodland at Thornley Woods held seven to eight pairs most years in an area of 80 hectares, whilst studies from a smaller mixed woodland area at Elemore Hall, reflected the cyclic changes in the number of young produced, which typically results from annual fluctuations in the abundance of the species' rodent prey. Such variations mean that isolated census work based on one season's work, is of limited indicative value; as numbers can vary so much between years; in some years of poor prey availability, some pairs opt not to breed.

It is a large and hardy species and current knowledge suggests that it is to be routinely found in the higher parts of the dales' woodland corridors (Westerberg & Bowey 2000). Some pairs inhabit conifer stands above the 300m contour e.g. in woodland at Nookton Burn in the north west of the county (Westerberg & Bowey 2000). In 2009, the species was reported from locations as far west as Selset Reservoir, in Lunedale. At Grassholme, a bird was utilising an exposed barn for roosting, the location being at 300m above sea level in an area largely devoid of trees, but presumably holding a good food supply. In 2009, in The Stang, 11 young were noted in four nestboxes and a pair of ground-nesting birds produced young, though the nest failed due to predation, as is often the case with such activity (Mikkola 1983). Over the last decade birds appeared to become yet more widespread across the western dales with pairs found at Brownberry Plantation near Grassholme Reservoir, High Force, Selset Reservoir and Westgate in Weardale, all of these locations being at around 350m above sea level. This trend may be related to the run of mild winters experienced between the early 1990s and the late 2000s.

The *Atlas* (Westerberg & Bowey 2000) indicated a concentration of occupied tetrads to the north of Durham City and in the north west of the county where there was keen nocturnal observer survey effort, and it is likely that this pattern more accurately reflects the true abundance of the species over the rest of Durham (Westerberg & Bowey 2000). In 2005, it was reported as being well represented throughout the year from 24 of the county's 48 ten-kilometre grid squares. Gaps in the lowland distribution may be related to a limited availability of woodland in some areas (Percival 1990), and this may be a factor in some parts of Durham, particularly in the south east. There are probably around 50-55 pairs in the north of the Tees area, and it is considered a sparse but widespread species there. The main distribution in this area is to the west of the urban areas of Teesmouth, concentrated in the wooded areas to the west of Stockton and around Wynyard, as well as having some representation in the mature, suburban areas (Joynt *et al.* 2008).

Some areas that have been the subject of more intensive study, such as the Derwent valley woodlands, during the late 1980s and 1990s, indicate that they support high population densities. For example, in heavily-wooded western Gateshead, the species is almost ubiquitous. In that area it was believed that there were between 10 and 15 pairs in each of the Derwent Walk Country Park, Chopwell Woods and the Gibside Estate, with up to 90 pairs in the whole of Gateshead District (Bowey *et al.* 1993). Working on the assumption that this north west corner was 'typical' of the rest of County Durham, extrapolation of the number of pairs per unit area i.e. occupied tetrad, suggested a population in excess of 400 pairs (Westerberg & Bowey 2000). Although this might be considered somewhat high, Percival (1990) suggested an average density of 10 pairs per occupied 10km square which, extrapolated across Durham, gave a largely similar figure of *c.*420 pairs for the county.

Tawny Owls, by and large, continue to thrive in Durham although they face threats in particular from collisions with road traffic and in some areas, breeding pairs are still persecuted by gamekeepers. At the end of the first decade of the 21st century, it is considered to be still relatively common and its population stable.

In 2004, pairs at nine monitored sites, including Cathedral Banks in the centre of Durham City, averaged 2.2 young per nest. In 2006, the Hetton, Elmore and Houghton areas held 16 pairs and these produced 31 young; an indicative snapshot of local breeding productivity. On the evening of 22 June 2007, several family parties were found at each of the following sites: Brancepeth Station, Morley Lane, Pit House, South and West Brandon and Stanley Crook, amounting to over 12 pairs. At least 25 birds were calling in the Beamish area in spring 2009, giving some indication of just how common this species can be where ideal habitat requirements are met. A further indication of the carrying capacity of local woodlands came in 2007, when ten territorial birds were heard calling in less than 40ha at Elemore Woods on 28 March, with 12 birds in the same area in 2009. Local breeding success is heavily dependent on the availability of small mammal prey and the variations in the populations of these. Food items identified from pellets collected at nests and roosts include: short-tailed (field) vole *Microtus agrestis*, Greenfinch *Chloris chloris*, rabbit *Oryctolagus cuniculus* and beetles *Coleoptera spp.* The species' dietary range is catholic, and sometimes even quite bizarre. In the mid-1990s returning home from a night shift, close to Barmoor, Ryton, an observer saw a Tawny Owl on the ground in the middle of the road; he slowed down to obtain a better view and discovered that the bird was feeding on the remains of a pizza.

Recent non-breeding status

It should be emphasised that this is a very widespread species in the county with sightings being reported from over 100 locations in spring 2007 and 2009, but this probably grossly underestimates the number of birds actually present. As this species is essentially sedentary once it reaches maturity (Parkin & Knox 2010), the species' non-breeding distribution is largely similar to its breeding situation. Any differences between these are probably largely a consequence of the dispersal of the young of each breeding season's efforts.

Autumn records can be almost as numerous as in spring, as birds commence vocal periods as nights lengthen towards the winter (Mikkola 1983). Birds can sometimes be encountered during daylight hours in quiet, well-forested areas, with such daytime activity regularly noted at Elemore and Hamsterley. In early 2006, a bird was discovered roosting in a cramped cavity in a limestone quarry, made all the more surprising as it was at least 800m from the nearest semi-mature trees.

Distribution & movements

This species is a widely distributed woodland owl, which can be found across much of Eurasia in a small range of racial forms. It breeds in north and Eastern Europe and it is a generally a resident and rather sedentary species that usually only undertakes short-distance movements (Parkin & Knox 2010). The availability of woodland is a key determining factor in the number of Tawny Owls to be found in any particular area (Percival 1990). Across Britain, there was some evidence of a contraction of range between the two national breeding atlases (Sharrock 1976, Gibbons *et al.* 1993) but this has not been evident locally.

There are few records to suggest migration or even local movement but very occasionally, wandering individuals can be seen at the coast and also around the coastal marshes of Teesmouth, though birds are not common in this area (Blick 2009). Most dispersing young move only a very short distance from their natal territories with the usual movement between the territory in which they were fledged and the location of their first breeding attempt being less than five kilometres (Parkin & Knox 2010) and there is local ringing evidence that bears out this fact.

A study in the Gibside Estate, in the lower Derwent valley, which commenced in 1991, to the end of 2010, had ringed 45 adults and 138 nestlings. The most young ringed, from six nestboxes, in any one year was 15 in 1996, with 14 in 2011, the only blank year in which none were ringed was 1992, demonstrating the wide variability in breeding success between seasons. In most years one to three adults per annum were caught on the nest, rising to six when capture techniques were refined. In most instances such captures demonstrated that birds were both site and nestbox faithful between years. Despite the numbers ringed no significant movements out of the area, or even away from the study site, have been proven, further indicating the species' sedentary nature. For instance, two Tawny Owls re-trapped in Gibside in spring 2007 had both been ringed there as nestlings, with the eldest subsequently becoming the most successful female breeding on-site. She adopted a nestbox only 200 yards from where she was hatched. A bird re-caught in a nestbox in Gibside in 2010, had previously been caught in the same nestbox in 2003, 2004, 2006 and 2008, after it was initially ringed there in 2002.

Long-eared Owl

Asio otus

Sponsored by
The Original Bakehouse

An uncommon but widespread resident and a regular winter visitor.

Historical review

Despite much study during the first decade of the 21st century, this remains a seriously understudied and rather enigmatic species in Durham; that said, there is now a better understanding of its true status than ever before. Its long-time occurrence in the area is attested to by the fact that the undated remains of Long-eared Owl, from many centuries ago, were found in the Teesdale Caves at the top of the Tees valley (Yalden & Albarella 2009).

Temperley (1951) considered that Long-eared Owls had once bred more commonly across Durham but he claimed that, at his time of writing, it was more common than was widely supposed. It was thought to be most numerous in coniferous woods and in the western parts of the county but more widely distributed in the winter. Hancock (1874) characterised the species as "*common in wooded districts*" in Northumberland and Durham. It was persecuted in the 1800s but it recovered in numbers, as game preserving reduced during and after the First World War (Temperley 1951). Towards the end of the 19th century, it was breeding in the Derwent valley from where Robson reported three clutches of eggs taken at Ravensworth, in the Team valley, and another from Chopwell Woods, the latter on 4 April 1884 (Robson 1896).

Temperley recorded that the species was a regular winter visitor to the Ravensworth Estate during the 1930s and there is anecdotal evidence to suggest that it was also regularly breeding there at that time (Temperley 1951).

In broad terms, Temperley (1951) felt that it was "*a resident in well-wooded localities but nowhere common*"; also recorded as an "*immigrant in small numbers*". It is of interest that Temperley did not seem to fully appreciate the Long-eared Owls' winter status in Durham, due no doubt to inadequate observer coverage. In the south east of the county, Nelson (1907) wrote that it was observed annually at Teesmouth, usually in November or December. In the context of breeding he described it as local and confined to wooded districts. Despite the fact that Nelson (1907) had recorded the species as being quite frequently seen in autumn on the coast, there were, very surprisingly, no such reports, between 1911 and 1962. In the south east of the county, Stead (1964) commented that there had been a general decline in the species' numbers since about 1935, though in 1965 a juvenile was found near Crookfoot Reservoir on 8 July, indicating local breeding.

Records of summering birds came from Hamsterley in 1972, 1977 and 1978, with an isolated occurrence at Hurworth Burn in 1971, which is of interest in the light of later events. Other records around this time came from Derwentside and Washington in 1976 and Wycliffe-on-Tees in 1983. Over the period from 1970 to the mid-1980s, successful nesting was only recorded from near Stanhope, in 1975 and 1976, and near Murton, in the east of the county, in 1987 and 1988, and possibly in 1986.

From the time of Temperley's review, from the perspective of the late 1980s, it seemed as if the breeding population of this species had suffered some decline. At that time, it was felt that this probably paralleled the national position, for which a general reduction of Long-eared Owl numbers had been noted and a comparison made with the increase in Tawny Owls in mutually preferred habitat over that time period (Mead 2000).

Recent breeding status

Proof of breeding for Long-eared Owl has always been difficult to establish and the handful of confirmed breeding records prior to the 1990s tempted speculation that the species persisted in Durham as a rare but annual breeder; in reality this assessment was to prove far from accurate.

Fieldwork associated with the second national breeding atlas (Gibbons *et al.* 1993), supplemented with fieldwork for the Durham *Atlas* provided a total of just three confirmed breeding records with a further three probable breeders (Westerberg & Bowey 2000). This highlighted either the genuine rarity of the species as a breeding bird in the county, or the difficulties of determining its true status. It is now known that the latter held true. Adopting a 'minimalist approach' to estimating the county breeding population at the time, the *Atlas* indicated a 'baseline estimate' of at least 15 breeding pairs, but it also highlighted the fact that this was quite possibly a gross under-estimate, placing the estimated number within very wide caveats.

The distribution and nesting habitats of the species in the county have, at best, always been 'imperfectly known', as confirmed by Westerberg & Bowey (2000). Because of the species' shy and extremely nocturnal nature (Mikkola 1983) it is often overlooked. Its status and distribution in our area are now much better understood thanks to a targeted and ground breaking survey, organised by the Durham Bird Club, which took place early in the 21st century.

Long-eared Owls will use conifer blocks but in recent years it had become increasingly appreciated that mixed scrub and broad-leaved woodland strips are occupied (Evans 2005, Evans 2006). Birds take advantage of hunting opportunities provided by tree planting on recently reclaimed opencast sites and this habitat is widely used in the species' stronghold in Nottinghamshire and surrounding counties (Scott 1997). The key determining influence on the species' presence seems to be the nearby availability of rough grassland, rich in small mammals, the field vole *Microtus agrestis* being the species' principal prey item (Glue & Hammond 1974). The larger areas of coniferous forest and smaller mixed woodlands abutting areas of rough grassland, which support prey species, remain as suitable breeding habitats.

Scott (1997) estimated that there were between 2,000 and 4,000 pairs of Long-eared Owls in Britain, though probably nearer the lower figure and it was suggested that there had been a 50% decline in numbers since around 1950. This figure was set at a lower level by Brown & Grice (2005), who estimated the English breeding population to be between 600 and 2,000 pairs, saying that there is good information that the species' range has indeed contracted in England over recent decades.

As highlighted in the *Atlas* (Westerberg & Bowey 2000), the distribution map suggested that most breeding season registrations were in the central to eastern part of the county; it is now known that birds are distributed across large areas of the county (Evans 2005, Evans 2006). The observed concentration of occupied tetrads to the north of Durham City, shown in the *Atlas* was probably more indicative of the species' actual breeding density across much of the lowland parts of County Durham.

Since the publication of the *Atlas* the species has been confirmed as breeding in many additional tetrads across the county. For instance, in 2005 the Durham Bird Club survey checked at least 59 sites for the species, twenty-one 10km squares were covered and a total area of around 2,600 square kilometres. This confirmed successful breeding for 15 widespread pairs by mid-June. By mid-July, 45 young were known to have left 21 nests, giving an average production of 2.25 young per successful pair and an estimated population for the number of breeding birds of 27 to 39 pairs (Evans 2005). Breeding pairs were found from the edges of the uplands, on heathland and moorland fringes, down to the coast and lowland sites. There were two instances of successful pairs being in close proximity to each other, with Tawny Owl *Strix aluco* nearby. At one lowland site, Rainton Meadows, two pairs of Long-eared were just 500m apart with Short-eared Owls *Asio flammeus* also in the area.

Early March sees pairs calling and displaying at breeding sites from dusk. By mid-month, females are showing an interest in likely nest sites. Most nest sites used were old corvid nests, 12 of the 13 identified nests were either of Magpie *Pica pica* or Carrion Crow *Corvus corone*, though seven nests could not be identified. The principal habitats utilised were farmland, with nine nests, woodland with six nests and heath/moorlands with five nests. Breeding success was estimated to be below the British average, though this was estimated from the number of fledged young (Mead 1987). The average altitude of nests was 160m above sea level, though the species is known to favour low-lying areas, below 150m above sea level for breeding (Scott 1997). Notwithstanding this, several nests were located at over 200m above sea level in woodlands, and 240m for the moorland sites.

In 2006, ten active nests had been found by the end of April and young were out of the nest by June at one regularly occupied central site, though some regularly breeding pairs did not attempt to breed. July brought confirmation of breeding at The Stang, which had produced a nil return during the 2005 survey. Five pairs bred successfully at Rainton Meadows in 2006. One Long-eared Owl was found on a reclaimed opencast site, hunting from tree shelters at dusk (Evans 2006). This habitat is widely used in Nottinghamshire (Scott 1997). During 2005 and 2006 it was found that Tawny Owls were present at almost 60% of Long Eared Owl breeding sites (Evans 2006).

In 2007, further work revealed 42 successful pairs, one of these being found on moorland at an elevation of 360m above sea level; 14 pairs were in new locations, where there had been no previous indication of breeding. This highlights the fact that birds are potentially present and breeding at many undetected sites. In the summer of 2007, the Hurworth Burn area held six pairs. Several summer reports of hunting adults came from widespread locations where breeding was not confirmed, including Waskerley Station. At this site, the observer initially took the bird to be a Short-eared Owl *Asio flammeus* as it hunted over moorland at 360m above sea level until it came closer, demonstrating how the species may be overlooked in such habitats.

In the south east of the county, the species is considered to be a scarce breeder (Joynt *et al.* 2008), with probably just three to four pairs located during the survey work for *The Breeding Birds of Cleveland* (Joynt et al. 2008). Recent breeding has occurred at or near Cowpen Bewley, Dorman's Pool, Urlay Nook and North Gare (Blick 2009) and almost certainly in one or two other coniferous woods in the south east of the county. At the end of 2008, the county population was estimated to be in the region of 90 breeding pairs (Newsome 2009).

Recent non-breeding status

Winter visitors usually arrive in small but varying numbers from early autumn onward, sometimes exhibiting sporadic, notably larger influxes. Autumn arrivals at the coast have been regularly noted from mid-September but tend to peak between late October and early November and there are several accounts of daytime flying birds seen coming in off the sea, which indicates a continental origin for many of our winter visitors.

During the winter months, birds form small or occasionally larger communal daytime roosts. A variety of habitats are used for such roosts. Hawthorn *Crataeugus monogyna* thickets or overgrown hedges are particularly favoured, whether isolated or within more wooded areas. Mature trees, usually on a woodland edge are also used with larch *Larix spp.*, Scots pine *Pinus sylvatcia* and spruce *Picea spp.* and even gorse *Ulex europaeus* all having been employed. Many of these roost sites are traditional so that by careful and sympathetic observation a better understanding of population changes from year to year has unfolded.

In such locations, Long-eared Owls are prone to disturbance but localities at: Tunstall Hills (Sunderland), Cleadon, Jarrow, Washington, Joe's Pond, Kelloe, Murton, Haswell, Shibdon Pond, Teesmouth and Neasham have all regularly held roosts over the years; though not necessarily in every year, or for the full duration of any season. Typically, these roosts hold between one to three birds, sometimes four to eight, and occasionally even greater numbers. The north easterly bias in the geographical distribution of roosts may be based more on observer coverage rather than the true distribution of roosts, although particularly in the early to late autumn, there may be a genuine coastal bias to the location of these.

Low numbers of birds were reported in the late autumns of 1976, 1978 and 1979 and yet later in these winters sizeable gatherings had developed at several sites. A large influx was known to have occurred along the north east coast of Britain in January 1979 and in the following March, 13 birds were found together at Washington with at least 10 others present in the county. This same cold weather spell resulted in reports of four birds being picked up dead and emaciated. In this instance, the scale of the national influx was not fully reflected by numbers of Long-eared Owls in Durham. In contrast the influx of 1975/1976 was felt equally on a national and a county scale, it remains one of the largest documented. By November 1975 in excess of 60 birds had been found locally, with no doubt many more undiscovered. On Tunstall Hills there were 15, 14 and four were at two Washington sites, nine at Kelloe, seven at Cleadon and five at Seaburn. There were still at least 30 in the county during March 1976 with the county's largest single roost at Peepy Plantation, Washington having grown to 25 birds.

The winter of 1986/1987 produced another large influx with approximately 45 birds present by February. A group of 14 were discovered in a clump of hawthorns on Boldon Flats in January but following disturbance the majority moved to a large tree in a nearby garden. These events demonstrate that Long-eared Owls will readily move roost in such circumstances. Another very notable influx occurred in the winter of 1994/1995 when 71 birds were counted at 11 sites in January with a maximum gathering of 18. Apart from those winters, there have been influxes of lesser proportions in 1974/1975, 1985/1986 and 1988/1989 during the last 40 years.

In the south east of the county the highest number at winter roosts in recent years were of up to 13 birds at one Teesmouth site in December 1994 and 1999, with a combined total of at least 26 birds at five locations north of the Tees in January 1995 (Blick 2009).

Over the last decade one winter roost at a northern site held ten birds in January 2000, with another ten at Newton Aycliffe. Eight birds were at a winter roost in November 2003, but human disturbance saw the break-up of this roost by December. In 2005 five roosts held 23 birds, the largest holding 10 birds and in 2007, 17 birds were spread between nine locations during the first quarter whilst separately up to 18 birds were present at Rainton Meadows in January. In the final quarter of 2007 there were four roosts totalling 20 birds in the east of the county on 20 December. Through 2008, winter roosts were very much at a premium and few were discovered; by far the largest concentration being at Rainton Meadows where up to 18 were found in three separate roosts in January 2008. Today the winter population of Long-eared Owls is typically probably 20-30 birds above the base-line breeding population in the poorer years, with as many as 60 additional birds in some of the better winters.

Roosts may not become fully established until as late in the winter as January or February, suggesting that there can often be some mid-winter influxes from the Continent in response to hard weather. The winter roosts usually break up by late March or early April but there have been reports of a few birds lingering until late April.

Up to six coastal migrants were recorded in both 2005 and 2008, largely in late September and through October. An exceptionally early bird was seen flying in over Tees Bay on 12 August 1988, mobbed by 30-40 skuas *Stercorarius* spp. Single figures are more or less typical, in any one autumn with birds being mainly recorded from the north east of the county, at sites such as Whitburn Coastal Park and Marsden, though they occur right along Durham's coastal strip. Most recently, between three to five coastal birds have been seen annually in October and November at places such as Hartlepool and one or two occasionally appear in April and early May. These are presumably birds heading for breeding areas in northern Europe (Blick 2009).

Distribution & movements

This species breeds over much of Eurasia, though it is not present in northernmost areas or in southern Asia. In different racial form, it also breeds in North America and some parts of the African continent. Over much of its range, the species is resident, though northern birds are migratory and dispersive, depending on weather conditions, population densities and the state of local small mammal populations. This owl is a regular spring and autumn passage migrant along the coast, in small numbers (Blick 2009).

Ringing data gives clear evidence for the immigration of foreign-ringed birds into the county during the autumn and winter. For example, one ringed as chick on 9 June 1963 at Zwonitz, Sachsen in Germany, was found dead at South Hetton on 27 December 1964. From a similar direction was one ringed on Helgoland, Germany, on 7 April 2000 and found at Thorpe Thewles on 4 January 2002. Other locally recovered birds of foreign origin include one that was ringed as a nestling on 1 June 1962 near Pirkkala (Hame) Finland and which was shot on 12 January 1964 near Sunderland. Most spectacular was the bird ringed at Rybachiy, Kaliningrad, Russia, on 9 October 1979 and found dead 1,405km to the west at Port Clarence on 10 January 1983 (Blick 2009). More local movements are

591

indicated by the bird that was found exhausted at Sunderland in autumn 1970, which was rested, ringed and then released at Hurworth Burn Reservoir. This bird was later found dead at Market Weighton (East Yorkshire) in March 1971. Perhaps it was a continental bird that had moved further south through the winter. Of known local origin was the bird ringed as a juvenile male at a breeding site near Newton Hall, Durham, on 25 May 1993, which was found dead 86km to the south, in North Yorkshire, on 15 March 1995.

Short-eared Owl
Asio flammeus

Sponsored by
The Original Bakehouse

Present all year. A regular passage and winter visitor; also a scarce breeder, mainly in upland areas.

Historical review

The Short-eared Owl's prominence on autumn passage and during the winter months in Durham was well known to 19th century chroniclers but its status as a breeding species was perhaps not so well understood. Hutchinson (1840) thought that "*it had not bred in Durham*" and Hogg (1845) referred only to birds on passage. Hancock (1874) however, did recognise it as an occasional breeder and Tristram (1905) made reference to some remaining on the moors throughout the year where "*it had been known to occasionally breed*". Interestingly for such an early account, Tristram also drew attention to the predator-prey link between cycles in field vole abundance and the appearance of Short-eared Owls. By the final decades of the 19th century, breeding was being regularly recorded in the north west of Yorkshire, close to Durham's southern boundary, in Cumberland, in Northumberland and also in Durham itself (Holloway 1996). Temperley (1951) considered it "*a rather irregular resident but fairly consistent in winter in small numbers*". He referred to a record from Thomas Thompson of Winlaton who had a clutch of five eggs in his collection that was taken from a nest found under a bramble bush in Chopwell Woods in 1868. The account is important because it not only gives the first specific account of breeding in the county but also relates to a site which today would not be considered strictly as the species' favoured upland terrain.

Mather (1986) has suggested that in the adjacent county of Yorkshire the Short-eared Owl was genuinely no more than an occasional breeder up to the early part of the 20th century and this may well have also held true for County Durham over that period. Records did subsequently slowly increase. It was believed that breeding occurred on rough grassland in the Lamesley area of the then undeveloped Team Valley in 1942 (Bowey *et al.* 1993); this again being notably in lowland terrain. The first fully documented occurrence of moorland breeding came as late as 1948 when a pair raised three young in the upper Derwent valley. The joint annual ornithological reports for Northumberland and Durham included comments in 1951 that the species was by then "*much more common*" on Durham's north west moors and in 1953 that it appeared to be "*expanding its range*".

The habit of immigrant birds arriving at the coast in autumn, especially during October coincident with Woodcock *Scolopax rusticola,* was well documented from the earliest records as was the tendency to overwinter along the coastal strip in fluctuating numbers at sites such as those around the North Tees Marshes. It was considered "*very numerous*" there in the winter of 1957/1958 when at least 14 birds were present over the North Tees Marshes on 29 September (Grey 1958). Return passage in springtime is not recorded within the historic accounts.

Recent breeding status

Today, this is undoubtedly the county's scarcest breeding owl species (Evans 2006). The breeding population is chiefly confined to the western uplands on open, undisturbed moorland with blocks of mature heather *Calluna vulgaris,* unimproved acid grassland and rough pasture, (the so called 'white moor') or in young conifer plantations. New and in some circumstances, restocked conifer plantations in their early growth stage are keenly exploited and afforestation in the first part of the 20th century probably contributed greatly to the increase in breeding numbers seen in Durham and more generally across northern Britain. Birds are known to favour conifer blocks in the first 10 years of growth until the thicket stage is reached when the trees reach about 1.5 metres tall. After this, suitable open nest sites and the quality of rough grassland supporting prey species tend to diminish. Today there is less acreage planted as new growth conifer blocks than was once the case.

This is a challenging species to monitor and several features of its breeding ecology present difficulties in developing even a modestly accurate picture of its status. Firstly, Short-eared Owls are migrant and nomadic. Their presence and abundance is determined by the availability of small mammal prey items, especially field voles *Microtus agrestis* whose numbers are prone to cycles ranging in the extreme between plagues and crashes. An area supporting breeding owls in one year may be entirely deserted in the next such that an annual census of an area can produce quite erratic results. Working in southern Scotland, Village (1987) tagged 21 breeding birds in one season but only one returned to the same survey area the following year after a crash in the vole population. Two of the tagged birds were controlled in the following year, each having moved over 420km to other breeding territories. Birds show more site-faithfulness whilst vole numbers are stable or increasing. The species' overall tendency to wander was referred to by Lack (1986) who reported that over 50% of recoveries of UK-ringed Short-eared Owls of all ages had travelled more than 100km. Further monitoring difficulties arise because although this owl species can often be seen quartering the ground in daylight hours it is actually mainly crepuscular in its hunting and display habits and can remain unobtrusive for long periods. It also displays and nests as early as March, often in remote locations such as on white moor landscapes, which are not well covered by observers, particularly early in the season. Finally, the hen bird will sit very tight on the nest and the young disperse and hide in surrounding vegetation at only 12-14 days old.

Against these challenges any estimate of the breeding population needs to be made with caution. The variability in local populations was demonstrated by G. Wall who in the early 1970s surveyed the same area of moorland in upper Weardale over a number of seasons. He found six breeding pairs in 1972, an impressive 12 pairs in 1973 but just three pairs the following year. To add to this varied picture, in neighbouring Northumberland the local peak in the vole cycle was considered to have occurred in 1972 with a crash evident in 1973 (Galloway & Meek 1978). The *Birds in Durham* annual reports suggest breeding numbers reaching a peak in 1980 and 1981, 1988, 1992, 1999, 2004 and 2009; though these must be taken as very simple and qualitative assessments. The *Atlas* (Westerberg & Bowey 1999) cautioned against the risk of giving an exaggerated distribution by showing the combined results from several years of survey work. At its peak the population in the Durham uplands was then thought to reach 50-60 pairs but this now seems overly optimistic, even for its time. Today it is thought that in a very good year the upland population may no more than 20-25 pairs and in poorer years 10-12 pairs or fewer. During the 2005 owl survey the species was recorded in 17 ten-kilometre squares during the breeding season in Durham and likely breeding took place in just eight of these, but there was little confirmation of such breeding attempts. At this time, it was distributed largely between the uplands and the coastal strip, with all of the county's coastal 10km squares registering birds during the survey, but many of these will have been related to passage or lingering winter birds, rather than territorial pairs.

In the past 25 years, Short-eared Owls have bred at one time or another on all of the county's major tracts of moorland and sometimes on adjoining rough pasture and forestry plantations including: Baldersdale, Barningham, Edmundbyers, Hamsterley Common, Langleydale, Lunedale, Muggleswick Common, Rookhope, Stainmore, The Stang, Stanhope Common, Teesdale up to Cow Green Reservoir, Weardale and in the upper Derwent Valley. Pairs have nested at elevations greater than 600m above sea level on Mickle Fell as well as on other exposed fell tops. Birds feeding fledged young on Cotherstone Moor by late May in 1986 and 1987 confirm just how early nesting can take place. Equally, in 2009 a nest with three young aged six to ten days was found on 9 July near Hamsterley, suggesting that double broods can be raised in favourable conditions. Nests are not infrequently found close to Merlin *Falco columbarius* nests indicating that some mutual advantage may be gained from such close association, even though juvenile Merlins are on rare occasions predated by Short-eared Owls (M. Nattrass pers. comm.). The quality of breeding data year-on-year does not allow trends to be determined precisely but anecdotal comment from several experienced field-workers through the 2000s all point to a recent decline. In 2010 this owl was added to the list of species considered by the UK's Rare Breeding Birds Panel on the basis that the national population is poorly understood and may have fallen below the monitoring threshold of 1,500 pairs.

Active management of heather moorland for Red Grouse *Lagopus lagopus* provides an excellent mosaic of habitats for breeding Short-eared Owls but with this process come the attendant risk to any species that might be perceived as a threat to grouse stocks, as Short-eared Owls sometimes are. An old report concerned five Short-eared Owls killed during a grouse shoot in Teesdale in 1947 but as recently as 1999 one was caught on an illegal pole trap on a Weardale moor.

The phenomenon of breeding at lowland sites in Britain is considered a relatively recent development, associated mainly with coastal marshes, dunes, rough pasture and once again new plantations. Not discounting the earlier historic records for lowland Durham, a pattern of limited breeding in the Durham lowlands has emerged over the last decade or so. The situation can be confused by the appearance of late spring migrants, especially towards the coast. Over-summering, without evidence of breeding, was noted at Teesmouth and WWT Washington in 1979 and again at Shibdon Pond in 1989 (Bowey *et al.* 1993). In 1992, 1993 and 1997 various agricultural set-aside and new woodland planting schemes attracted wintering birds and some lingered into the early summer each year, again without clear proof of breeding. In 2004 pairs were present in late spring at four widespread sites in the east of the county, where breeding was subsequently confirmed or thought probable. Three sites were occupied in the early summers of 2005 and 2007 but the outcomes were uncertain. In 2006, birds were confirmed as breeding at two lowland community forest sites away from the coast (Evans 2006). A pair went on to summer each year from 2008 to 2010 on rough grassland near Shotton and almost certainly bred there – a recently dead, three-quarters grown youngster was found in long grass on Shotton Air-field one summer (D. Simpson pers. comm.). Whilst the exact status of lowland breeding is still be clarified it would seem that one to three pairs could now be breeding each year in such situations in the county.

Recent non-breeding status

There is a strong coastal bias to winter sightings of Short-eared Owl, but the uplands are not necessarily vacated even in hard weather. Immigration is evident from August, occasionally even in late July, when modest numbers begin to be seen along the coastal strip. An obvious peak occurs with some regularity each year during October although, as ever with this species, numbers can vary. Winter occurrence is then usually quite widespread through until early March. In a typical winter reports of one to four birds may come from 15 or more localities. Sites which have featured in the last decade include: Blackhall, Daisy Hill Nature Reserve, Elemore Hall, Hedley Hall, Hedleyhope Fell, Hetton Lyons, Longnewton, Ravensworth, Sedgeletch and Usworth. The North Tees Marshes and other favoured locations might on average hold five to seven wintering birds.

Hard winter weather often produces higher counts and the development of sizable communal roosts can occur especially in the first quarter of the year. In the notably cold weather spell of February 1979, which also resulted in the development of Long-eared Owl *Asio otus* roosts, a roost of up to 20 gathered near Washington whilst eight were found at Boldon and 10 at Teesmouth. In December 1980, Greenabella Marsh at Teesmouth held 21 birds with nine at Jarrow Slake the next month and in January 1984 a group of 11 was seen at Follingsby in the north of the county. High numbers were again apparent in the early months of 1989 with 13 on the North Tees Marshes, seven at South Shields Leas and four at Shibdon Pond. In the final months of 1993, the North Tees Marshes held a roost of sixteen. In the late winter of 1996, the then newly reclaimed and landscaped old pit heaps at Rainton attracted a very modest count of two Short-eared Owls; after further years of sympathetic land management the site, by then a Durham Wildlife Trust Nature Reserve, and the nearby Seaton Pond area could boast no fewer than 25 birds after an influx in October 2007. The spectacle was maintained throughout the winter and afforded observers the opportunity to witness the birds skirmishing with Long-eared Owl, Kestrel *Falco tinnunculus* and Stoat *Mustela erminea*. Wing clapping was seen, apparently as a method of flushing prey, and by February some pair bonding was evident.

The species is a clear beneficiary of set-aside and game cover planting schemes both as a base for winter feeding and an incentive to linger in spring for possible breeding. Upland sightings during the winter months have included nine on Barningham Moor in November 1987 and eight on Cockfield Fell in the early months of 2002 but one to two birds might be expected in any upland location in all but the most severe of weather.

There are ample examples showing the arrival of continental birds to Durham's shores in the autumn months with several reports of birds arriving "in off the sea". Most notably two were seen flying west towards the coast on 9 October 1975 still some nine miles offshore and in 1979 a total of five birds were seen making landfall at various points between Marsden and Ryhope over the period 3rd to 17 October. Whitburn Observatory recorded single birds arriving on three dates during October 1991 and three individuals arrived on the Marsden shore on the last day of October 1999. An early season example concerned one, flying west, that was seen from a ferry two miles off Sunderland on 1 September 2001 and a late example was one that arrived at Whitburn Observatory on 11

November 2007. The immediate coastal strip seems to be favoured by these new arrivals and birds are typically seen in these circumstances on The Leas at South Shields, Whitburn, Blackhall and Teesmouth. Evidence of spring migration is usually less obvious but records of one or two birds are normal from late March until early May at coastal sites with a general northward movement evident.

Distribution & Movement

This owl breeds intermittently across north west Europe, Scandinavia and Russia and winters in north west and southern Europe. It is prone to long distance movements.

Tengmalm's Owl (Boreal Owl)
Aegolius funereus

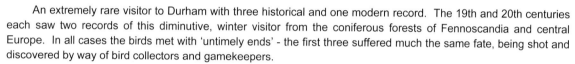

A very rare vagrant from Northern Europe and Russia: four records.

An extremely rare visitor to Durham with three historical and one modern record. The 19th and 20th centuries each saw two records of this diminutive, winter visitor from the coniferous forests of Fennoscandia and central Europe. In all cases the birds met with 'untimely ends' - the first three suffered much the same fate, being shot and discovered by way of bird collectors and gamekeepers.

Tengmalm's Owl, Whitburn,
October 1848 (K. Bowey, courtesy NHSN)

Temperley (1951) detailed the first three occurrences. The first county record dates from October 1848, when John Hancock reported that he "*bought a fresh specimen of it (Tengmalm's Owl) in a poulterer's shop in Newcastle. It was shot near Whitburn on the 11th or 12th October, 1848*"; this was the first record of the species for the British Isles. This record was reported by T J Bold as being killed on the sea coast near Marsden in October 1848 (*The Zoologist* Vol. 8 1850). The mounted specimen remains in the Natural History Society's collection in Newcastle upon Tyne. Hogg (1845) recorded another that was shot in November 1861 near Cowpen to the north east of Norton, near Hartlepool. A third was shot by a gamekeeper named Young, at North Hylton, Sunderland, on 4 October 1929.

The fourth and only modern record for the county concerns a bird found dead, and identified in fascinating circumstances. It constitutes one of the county's most unusual records for any species. The 'minimal remains' of a bird, just the leg, were found under a line of overhead electricity cables near Fishburn on 10 January 1981 by Mr. W. R. Lawton, who was out walking his dog. The leg, which was small but sturdy, featured strongly hooked talons and bore a ring. The ring was forwarded to the Ringing Office at the BTO. Subsequent enquiries with the Oslo Museum, Norway revealed that the leg belonged to a Tengmalm's Owl. The bird had been ringed as a nestling, at Greflen, near Vang, in Hedmark County, Norway, on 10 June 1980 (*British Birds* 78: 562, *British Birds* 79: 559). It was judged to have moved over 800km in the autumn of its first year, before expiring in late 1980. This, together with the two adults that were found in Orkney, Scotland in mid-October 1980, proved to be the first British records of this species for almost 20 years.

Distribution & movements

Tengmalm's Owl breeds discontinuously through southern and central Europe eastwards to the Urals, with the European stronghold being from Fennoscandia east through Russia. Birds are largely resident and mostly sedentary but the northerly populations are sometimes 'irruptive' in nature, depending on fluctuations in the numbers of their prey, forest rodents. Since the discovery of the Fishburn remains, there have only been two further

Tengmalm's Owls in Britain; in East Yorkshire in 1983 and Orkney in 1986. The overall British total stands at 49, only seven of which have been since 1959; three of these being in the winter of 1980/1981, when the remains of the Fishburn bird were discovered (Parkin & Knox 2010).

Nightjar (European Nightjar)
Caprimulgus europaeus

An uncommon summer visitor and rare passage migrant.

Historical review

This summer visitor to Britain is a long-time occasional breeder in Durham and a rare passage migrant, although as the Nightjar is a strictly nocturnal species, it is likely that many birds have been in the past, and still are, overlooked by observers.

Temperley (1951) said that it was a *"summer resident; not numerous and very local in its distribution"*. He noted that it *"Breeds about the fringes of the moorlands in the western part of the county"*, *"less frequently on scrubby open ground, amongst heather, gorse or young trees"*, in the east. All of which suggests the species was more widely distributed in the county that it is at present.

The Nightjar was probably much more common in Durham in Hutchinson's time (1840) and he gave a detailed account of its inhabiting abandoned agricultural habitats in the west of the county, where heather and gorse were taking over and *'straggling trees of larch and fir had been wafted in* from neighbouring plantations. George Temperley (1951) thought it unlikely to have been all that much more common in the east of the county in the previous 150 years than at his time of writing, i.e. as far back as 1800. He noted that the species was not included in some of the notable bird lists of the Teesmouth area, though Hogg (1845) did mention it in his text. In 1862, it was recorded as breeding in a corner of Greatham Churchyard (Tristram 1905). Thomas Robson told of a male and a female in the collection of Mr Thomas Thompson of Winlaton, which had been shot at Lockhaugh, near Rowlands Gill, but no date was given (Robson 1896). Nelson, writing in 1907, recorded it almost annually on passage at Redcar in May and September, just to the south of the Durham recording area, but whether birds were such regular visitors to Teesmouth cannot be ascertained. At the end of the 19th century, this species was believed to breed in every British county and region, aside from some of the smaller off-shore islands and the Outer Hebrides (Holloway 1996).

At the time of Temperley's review, one or two pairs were still breeding annually in south east Durham during the late 1940s and at the outset of the 1950s. Temperley said, "*On the Derwent there are still favoured spots, not so many miles above its confluence with the Tyne, where the Nightjar might still be found*". It is presumed that this refers to the complex of woods around the lower Derwent valley, which would have included such sites as Chopwell Woods (Bowey *et al.* 1993).

Nightjars were breeding on the outskirts of South Shields in 1933 and in 1943 it was seen every summer night on a farm within three miles of the town (Blair 1944). What is clear is that there was a substantial reduction in the numbers of Nightjar throughout Britain from around 1930. In the post-Temperley period, it became scarcer even as a passage migrant within the county. During the 1950s there were just a handful of records, one was a male found dead near Hart Station on 7 June 1954 and one was flushed from sand dunes at North Gare on 8 September 1956. Perhaps more interesting was the fact that three pairs were recorded as breeding in one square mile of Weardale in the summer of 1958. This record most likely to refers to the Dryderdale-Knitsley Fell areas, on the eastern edge of the current Hamsterley Forest.

This species was noted in territory at Wynyard from about 1960 to at least 1966, and almost certainly it bred there over this period, and maybe again in 1968; however, there were probably no more than one or two breeding pairs over this period (Blick 2009). At around the same time, in 1962, three pairs and a few single males were noted as being present in Hamsterley Forest. The species probably maintained a low-level breeding presence at this site over the next few years and up to 1967 the site was said to hold the 'usual numbers' (Bell 1968), although there was no documentary information that quantified what these numbers were.

There was a significant decline in this species between Temperley's time and the modern period, even though, as a breeding bird, it appears to have persisted in small numbers during most of the decades between 1950 and the early 1980s. Anecdotal reports suggest that birds may have been present in some areas where it was previously present, such as in the lower Derwent valley, into the mid-1970s. Some greater effort was expended by local observers to better understand the species' true status in Durham during the 1970s and this resulted in greater knowledge of birds, in particular, in Hamsterley Forest.

Recent breeding status

At no time in the last forty years has the species been anything other than a scarce breeding bird and a rare passage migrant in Durham. The *Atlas* (Westerberg & Bowey 2000) stated that, as most suitable habitat had been examined over the survey period, 1988-1993, that the mapped minimal distribution for this species was likely to be a 'a fairly accurate representation of its distribution' in the county at that time, though in reality many potential, though still unlikely areas that could have held the species remained unchecked from year-to-year.

In Durham, over recent decades, this species has been found almost exclusively on clear-felled areas within the large forestry area of Hamsterley Forest at altitudes between 250-350m above sea level (Westerberg & Bowey 2000); though across most of its range in Britain, it is traditionally a species of lowland heathland (Gibbons *et al.* 1993). Such forestry areas are now amongst the most important habitats nationally for this species (Gibbons *et al.* 1993). In Durham, it tends to occupy recently felled, clear-fells, for a period of between five and ten years after re-stocking, after which period the maturing tree cover renders the habitat unsuitable for Nightjars (Westerberg & Bowey 2000). Elsewhere in the county, records are rare and odd pairs may perhaps be overlooked?

Lowland heathland was once quite widespread over parts of mid-Durham, but opencast coal extraction and agricultural improvements have virtually removed all extensive examples of this habitat. It is estimated that only 115 hectares of lowland heath remain in Durham (North East Biodiversity Forum, 2001). At only a few sites, such as at Waldridge Fell, near Chester-le-Street, do significant areas of such lowland heath remain, which might attract the species. This area has not been used recently by Nightjar and this may be because of the fragmented nature of the habitat.

Detailed work in Hamsterley Forest over the period 1985-1995, showed that there had been a gradual increase in the numbers of birds, from the mid-1980s, after a low point around 1989-1991, to an established population of between 15 to 20 'churring' males by the late 1990s (Westerberg & Bowey 2000). According to the published record from the mid-1980s through until the early 1990s (Armstrong 1988-1994), this species was on its way to, or had already become extinct as a breeding bird in Durham. It was believed that territorial birds had been absent from Hamsterley Forest over the period between 1985 and 1991, in fact this was not the case. Birds probably reached their lowest ebb around 1983-1985, when very few were present, but even through this period birds bred successfully. For instance, in 1985, when the Durham Bird Club record said that no birds had been reported, a nest with young was found (G. Simpson pers. comm.). Birds were breeding, sometimes in small numbers throughout this period. Numbers rose steadily from two singing males in 1986, to six in 1988, eight in 1990, 10 in 1991 and then 19 in 1992, reported as part of the national survey. The re-colonisation of Hamsterley Forest, or the re-discovery of birds there, and the later increase in the population, coincided with a greater availability of suitable habitat resulting from the first cropping of mature conifers, which provided much open ground (Westerberg & Bowey 2000).

Ongoing re-structuring of the forest estate at Hamsterley continues to provide more suitable habitat but the locations where Nightjars first re-colonised, in the 1980s, were lost by the late 1990s, as a result of re-stocking (Westerberg & Bowey 2000). In 2002, Forestry Commission sources claimed up to 40 pairs present in Hamsterley Forest by 28 June.

Intermittently, birds have been noted in habitat and territory at other locations in the county, away from Hamsterley. For instance in the first decade of the 21st century, when the species nationally was on an upward trajectory, birds were heard calling at Urlay Nook, to the south of Longnewton, in June 2004 and June 2005 (Joynt *et al.* 2008); indicating that at least a small breeding presence was present in that part of the county. There were suggestions of a small presence of birds at a site in east Durham around 2002-2003, but no details were ever forthcoming about these birds. A singing male near Horden Dene for at least two weeks in late May and early June 2011 was a notable coastal territory, but there was no indication of a second bird or a breeding attempt.

There have been three relatively recent national surveys of Nightjar. The first, in 1981, indicated that it was absent from the county (Gribble 1983), although two were noted in Hamsterley Forest that summer and birds were found displaying at Wynyard in both 1982 and 1985 (Joynt *et al.* 2008). The survey of 1992, identified 19 'churring' males, almost all of them in Hamsterley Forest (Morris *et al.* 1994) and in 2004 during the survey, 16 males were located, again largely in the vicinity of Hamsterley Forest (Conway *et al.* 2007). This small reduction in numbers from 1992 compares with a 5% increase in Northern England and 36% increase nationally over the same period.

Through the first decade of the 21st century birds had a regular and more obvious summer presence in Hamsterley Forest, with up to sixteen birds 'churring' there in 2009, being the peak confirmed count for this period. This compares with just six in 2004, the year of the repeat national survey. In recent years, the first singing birds have been usually noted during the last ten days of May, for instance over a five-year period these were: 2006, 28 May; 2007, 25 May; 2008, 29 May; 2009, 21 May; and, 2010, 22 May. In 2006, three separate areas of Hamsterley Forest produced 18 records between 28 May and 5 August; in recent years birds have usually been noted in territorial areas into very late July or early August. Around the mid-point of this decade, birds began to be noted away from this core population with a group of breeding birds being discovered in similar habitat at The Stang, on the county's southern border. Up to three birds were there in 2008, but greater observer effort would likely reveal more birds there. Early June 2005 saw two sites away from the Hamsterley stronghold produce reports; one of these was in Weardale, with a minimum of three 'churring' birds on 7 June, whilst a male was flushed from heather at a site to the east of the A68 on the same date. The latter site also held a hunting bird on 27 July.

Detailed surveying in Hamsterley Forest in 2008 revealed frequent reports from late May until the last was heard on 30 July. In the western sector, birds were recorded in three areas with perhaps five singing males present while the central section of the forest produced four males. The Black Bank Plantation/Shipley Moss area also held a maximum of five singing males in June. A conservative estimate of at least 14 singing males could be made for Hamsterley, but there may well be more in other less accessible areas of the forest. A further three males were noted in The Stang, in the south of the county. In 2009, of the 16 pairs located at Hamsterley Forest, the western sectors around Neighbour Moor and Pennington proved particularly attractive. A more comprehensive survey in 2011 revealed a larger population than estimated in previous years. Following the first male back on territory on 21 May, at least 39 territorial males were found during the summer months. There were notable concentrations in the western side of the forest and also in the north east, but as with the nature of commercial forest, favoured areas will change from year to year. Productive survey nights combined bright moonlit conditions and little wind, and during one such night on 13 July, at least 20 birds were encountered between Black Midding Plantation and Neighbour Moor. In summary, the maximum counts of singing birds in Hamsterley Forest have included: 16 in 2004, 11 in 2005, 10 in 2006, 7 in 2007, 14 in 2008, 16 in 2009, 15 in 2010 and 39 in 2011.

The main threat to Nightjar in the county appears to relate to the cycle of forestry management but this is also one of the major beneficial factors for Nightjar in the county, as the cycle of felling and replanting creates suitable habitat on a rotational basis. One informed commentator stated that birds in Hamsterley appear to be 'hefted' towards the type of site in which they were reared (Brian Walker pers. comm.). The continuous working of the forest for timber will no doubt result in the Nightjar population being mobile in the coming years as birds respond to the changes in habitat. This is one of the species that has been identified which might benefit from the effects of climate change and nationally it has seen a marked increase in recent years, and it is conceivable that it will colonise new areas in the county in the near future.

Non-breeding Status
Away from its breeding strongholds, this species has always been a very scarce passage migrant in Durham. Its occurrence doesn't really exhibit a strong pattern, either to coastal locations or to simple time periods. In a forty-year period since 1970, there were probably no more than 17 records of passage Nightjars in the county. These came from a wide variety of locations, largely between July and early October, most in September (4) and October (5). Exceptionally there have been a few records in spring, but with just two May birds away from breeding sites since 1970, this is a real rarity as a spring passage migrant in Durham. The latest autumn passage record is the bird at South Shields in 1973.

Passage birds (all known records):

1973	South Shields, one, possibly, two were in the Harton area in October 1973. One was perched on a garden cold frame on 7th, then it or another was noted a few hundred yards away during mid-October, then it was flushed from nearby trees on 24th, still present on 25th
1986	Teesmouth, one on 10 September
1987	Saltholme Pools, Teesmouth one on 9 July 1987
1992	Hargreaves Quarry, singing male on 6th and 7 June 1992, trapped and ringed on the latter date
1993	Miller's Wood, near Chopwell Woods, one flushed in daylight, on 12 October
1995	Eppleton, one on 13 September
1996	Hetton Bogs, one on 4 October
1997	Whitburn Coastal Park, one on 19 May
1997	Witton-le-Wear, one on 11 October
2000	Hetton Lyons, a male on 15 September
2001	Tunstall Village on 12 August 2001.
2003	Biddick, Washington, one flew in front of a car on 23 September 2003
2004	Whitburn Coastal Park, one on 23 June 2004
2005	Deerness Valley Walk near Stanley Crook, one hunting after dark on 5 October
2010	Hartlepool Headland, a roosting bird on 23 May
2010	WWT Washington, female on 12 August

Distributions & movements

Nightjars are found breeding over much of Europe, as far north as southern Scandinavia, in the Middle East and east into central Asia. In Britain, populations are fewer in number and size towards the north of the British range. The species is a long-distance migrant, the British populations believed to winter in sub-Saharan Africa. A small number of birds were caught and ringed in Hamsterley Forest in the late 1980s and early 1990s, but there were no recoveries of these birds.

Nightjar, Hartlepool Headland, May 2010 (Tom Francis)

Common Nighthawk
Chordeiles minor

A very rare vagrant from North America: one record.

On 11 October 2010, construction manager Mark Penrose was working on a site by waste ground adjoining Warren House Gill, Horden. He noticed an unfamiliar bird and thought it could be some sort of owl, due to its colouration. It was sat in a lone tree on the site and allowed very close approach, so he took several digital photos. It also flew around the area, at one point even attempting to land on his car roof. At this stage, he thought little more of the bird, other than it was 'unusual'. The following day, Ian Smith was visiting the site. During the course of conversation, Mark said he had seen a 'funny looking bird' and showed Ian the photos. Ian immediately recognised the bird as a Common Nighthawk and asked Mark to show him the location where the bird had been, but there was no sign of the bird. Ian waited until dusk but forlornly, it was not seen again.

Distribution & movements

Common Nighthawks are a widespread breeding species throughout temperate North America south to Panama and the Caribbean. The wintering range covers South America south to central Argentina, with regular migration occurring over the west Atlantic, where it is frequently encountered on Bermuda and the Lesser Antilles.

All British records have fallen in the period 9 September to 4 November and are often linked to fast moving depressions crossing the Atlantic. Of the 21 previously accepted records for Britain, all but seven have been in Cornwall and the Isles of Scilly.

Swift (Common Swift)
Apus apus

A common summer visitor and passage migrant.

Historical review

The Swift is a common summer visitor to Durham and probably has been for many hundreds of years. It was considered a widespread and common bird in the 19th century, though some have claimed that it declined in industrial areas, because of increasing pollution, although this was never fully substantiated (Holloway 1996) and was unlikely to have been a factor in the Durham area. Hutchinson (1840) observed that it favoured older buildings for nesting such as churches, church towers and in villages it used the areas under the eaves of cottages. On their spring arrival, he recorded seeing birds on 3rd or 4 May (Hutchinson 1840).

In fact there is little documented reference to this species or its status from the county's early lists of birdlife. One of the few references to its actual status in the county comes from Thomas Robson, who considered this species to be scarce in the Derwent valley and was he was not aware of any nesting sites in the valley late in the 19th century (Robson 1896). This was perhaps as much to do with the valley at the time being still heavily wooded and relatively little developed, without suitable building stock for Swift's nests.

There were some suggestions of increase at the outset of the 20th century (Holloway 1996) and Nelson (1907) said that it was generally distributed in the south east of the county, at least in the Teesmouth area, and fairly common. Joseph. Bishop, the official RSPB Watcher at Teesmouth in the 1930s, reported that in 1929 he observed large southerly movements of the species in that area during June and July (Temperley 1951).

In 1951, Temperley said that it was a "*common summer resident*" and recorded its arrival time as being from the last few days of April, with most birds arriving in the first and second weeks of May. Dr Hugh Blair recorded birds nesting along the cliffs of Marsden in June 1933 (Temperley 1951) and in 1946 Temperley himself saw birds entering cracks in Marsden Rock. Stead (1964) noted that it was more common in the older residential parts of West Hartlepool and Stockton than in the newly built areas, suggesting the species has a penchant for older housing stock.

Little of significance relating to the species was published through the 1950s and 1960s, though in 1958 there was late record of three birds at Hartlepool on 8 October, perhaps the latest for Durham to that point, though a bird at Teesside on 8 November 1961 extended this record. At the other end of the season, early dates for the occurrence of Swift in the north east during the 1960s included a bird at Haverton Hole, on 16 April 1966, which was then the earliest Teesside record (Bell 1970) and one at Barmston on 19 April 1968. A flock of up to 500 feeding at Shibdon Pond on July evenings in 1973, during fine weather, probably comprised locally breeding birds and family parties, suggesting that numbers were buoyant around the mid-point of that decade, and this was reflected in comments about there being '*good numbers of breeding birds in the area*' in 1977.

Recent breeding status

Between May and July, breeding birds are present across much of the county, from upper Teesdale in the west to the coast, though breeding densities vary according to factors such as local micro-climate and the availability of nest sites, which in itself is a function of housing age and style. Today Swifts nest entirely in man-made structures, particularly older buildings (Gibbons *et al.* 1993), and the *Atlas* showed them to be present in all of the main built-up areas of the county, from the cityscape of the Tyne & Wear and Teesside conurbations, to the more isolated market towns, such as Barnard Castle in the south west (Westerberg & Bowey 2000). In general terms the greatest number of birds tends to be in the eastern third of the county with birds present, often at lower densities, in the west of the county; although even here there are local hotspots. Obvious concentrations of birds occur on Wearside and on Tyneside, through the populated central corridor of the county, including Chester-le-

Street and Durham City, as well as in the smaller towns and mining villages, and then also around the Hartlepool area (Westerberg & Bowey 2000, Joynt *et al.* 2008).

Most of the smaller rural villages in the west of the county have a few breeding pairs. The species' favourite breeding haunts are old buildings, often in locations such as churches and large houses. Settlements such as Durham, Hartlepool, and Barnard Castle all have good-sized breeding colonies, as do some areas with more modern-built houses that have tiles and soffit boards that are attractive to the species; a good example is Whickham on Tyneside. Up to 30 pairs nested in the village hall roof at Clara Vale in the extreme north of the county in the summer of 1986, indicating that large numbers can be concentrated in relatively compact suitable locations; consequently, such birds are susceptible if these sites are re-developed, as this location was during the late 1990s.

Towards the end of the first decade of the 21st century, the breeding densities appeared to be greatest in and around some of the towns and villages of the western dales and uplands rather than in lowland Durham. This may be related to the availability of suitable nesting cavities in houses in these areas. In 2009, counts of between ten and forty nesting pairs came from a number of locations in Weardale, including: Cowshill, Daddry Shield, Frosterley, Ireshopeburn, St John's Chapel, Stanhope, Wearhead and Westgate. This local population probably numbered over 150 pairs, which were distributed ribbon-like up the dale. By contrast in late July 2009, just twenty pairs were estimated to be present in Durham City centre, once something of stronghold for the species.

Nationally, the population levels were believed to be stable up to the early 1990s (Marchant *et al.* 1990), though numbers appear to have declined since then (Mead 2000). Reliable local estimates of the population across the county have been difficult to determine because whilst birds are present in many places, being a supremely mobile species, they are not always breeding in the areas in which they are routinely seen. Furthermore, relatively little in the way of breeding information tends to be submitted from casual observers. Some of the county's urban areas where there are suitable nesting sites clearly hold many pairs but there is little reliable local data on breeding density in such areas. By way of contrast to this statement, in 2007, some indication of breeding density was provided by counts of *c.*20 pairs nesting in Walter Street, Houghton, and *c.*15 pairs utilising a 1950s housing estate close to Hetton Downs. In addition, significant numbers of birds were breeding around the similarly aged housing stock located at Swalwell and Whickham, in Gateshead, where some 40-50 pairs were obvious through much of the summer; these birds nesting in houses in the estates below Whickham Shopping Centre. In 2008, the suburban housing estate at Hetton which had previously supported a productive nesting colony was demolished. The loss of such colonies is a real threat to the species in many areas, though there are few other local threats to Swifts, excepting competition for nesting sites (Kaiser 1997), which can be at premium where there is a lack of availability of suitable buildings with nesting crevices, and this is likely to limit numbers in some areas (Gibbons *et al.* 1993).

Swifts are present in Durham for only a few months of the year, nest preparation commencing just a few days after arrival in the first week of May. Birds occupying a nest hole under house eaves in the village of Heighington, for least eight successive years managed to oust nesting Starlings *Sturnus vulgaris* from the site within a few days of their return each year. Most breeding activity is complete by late July or very early August, at the latest, when sites are quite suddenly vacated. One very unusual breeding record concerns an adult that was seen leaving a nest cavity in Sunderland, which still contained young, on 14 September 1977 (Unwin 1978).

During the breeding season, flocks of several hundred can be regularly observed over prime wetland sites such as Shibdon Pond, in Gateshead, and around Teesmouth. In 2000, around 1,500 birds were over Saltholme on the 5 June, brought down by prolonged rain. These birds presumably comprise local breeding birds on feeding forays, although the records of 1,800 over Hargreaves Quarry on 16 June 1982 (Blick 2009) and up to 5,000 at Teesmouth in June 1982, beg the question as to where they originated. Gibbons *et al.* (1993) estimated the national population to be in the region of 80,000 pairs and the *Atlas*, working on the assumption that Durham was 'typical' of the country as a whole, suggested a county population estimate of '*in the region of 900 pairs*' (Westerberg & Bowey 2000). In the south east of the county, key colonies are located in the outskirts of Stockton and in south west Hartlepool, where there are up to 60 pairs. There are probably around 250 pairs in the north Tees area in total (Joynt *et al.* 2008). Considering that large flocks numbering over 1,000 birds are sometimes reported during poor weather at favoured sites (Armstrong 1988-1993, Newsome, 2006-2009) however, this estimate may seem on the low side and perhaps the real figure is nearer 1,500 to 2,000 pairs. Nationally, there may have been some declines in this species over the last 25 years or so, especially since around 1994 (Parkin & Knox 2010) but numbers fluctuate considerably from year to year (Brown & Grice 2005).

Non-Breeding Status

Large mid-summer concentrations of Swifts can be seen, especially during poor weather, over many wetland areas, sewage treatment works and other habitats where there may be concentrations of insect food (Westerberg & Bowey 2000). Overcast or thundery conditions can result in spectacular feeding concentrations over such sites. For example, an estimated 1,000 birds were over Shibdon Pond in the summer of 1967, though more usual numbers there in such conditions might be 100 to 200 birds. Other high counts include 500 at Shibdon in July 1973 and again on 29 June 1980 when a further 500 were feeding over Birtley Sewage Treatment Works. A gathering of 1,000 birds at Derwent Reservoir in mid-June 1978 was unusual for a summer upland gathering. There were concentrations in 1993 of 1,500 over Durham on 28 July and 2,000 at Bearpark on 3 August. At Birtley Sewage Treatment Works, 2,500 were reported during June 1998. What was described as a 'feeding frenzy' of around 2,000 birds was over the Reclamation Pond on 11 June 2005. In 2007, large counts from about mid-June to mid-July included c.1,000 over the Reclamation Pond on 9 June, c.2,000 there on 14 June and 700 at Whitburn on 2 July. The summer months of 2009 brought c.2,000 in the Saltholme area on 20 June, 200 over Whitburn Coastal Park on 4 July and c.900 feeding over Whitburn Village on 2 August, with c.1,100 at Greatham on 3 August.

Most local birds arrive in early May and can soon be found nesting across Durham. The earliest record for Swift in the county is of a single bird over Hartlepool on 12 March 1995; the next earliest, one over Durham City on 30 March 2001. Traditionally, it is the last week of April or the first few days of May before the first Swifts are seen and within a week many birds are usually screaming over their nesting colonies and feeding over marshes and ponds. By the end of the first decade of the 21st century, birds were returning much earlier to County Durham than they were 30 to 40 years earlier. In the late 2000s, the earliest arriving birds were more typically noted in mid-April, as in 2008 when the first bird was seen on 19 April, rather than the first few days of May as 30 years previously. The pattern of the first birds arriving well in advance of the main arrival of birds however, persisted; the main influx sometimes being up to a week behind the first birds. For example, the first spring arrivals in 2009 were two birds at Saltholme Pools on 18 April, followed by one at Lamesley Reedbeds on 19th. A more widespread arrival came between 25th and 30th of the month. Between 1976 and 2005, the first arrival dates of Swift were analysed in ten-year groupings, these were: 1976-1985, 1 May; 1986-1995, 22 April; and, 1996-2005, 23 April, an advance of over a week (Siggens 2005).

In 2007, after the first bird at Saltholme Pools on 20 April, an obvious arrival occurred on 21 April, with one to two birds noted at four other sites and subsequent days saw birds at a number of other locations. By 29th, up to 50 had gathered over Saltholme Pools and birds had reached western upland localities, such as Muggleswick (Newsome 2008). The 2007 arrival date was considerably earlier than the mean arrival date at Teesside between 1985 and 1995, which was 27 April (Joynt et al. 2008). These pieces of evidence suggest that this earlier arrival is a trend rather than fluctuation. The 300 birds over Saltholme Pools on 27 April 2009 was a remarkably high number for so early in the spring. In 2010, a 'swift species' was reported in the lower Derwent Valley on 3 April, this was early for Common Swift and the bird was seen during an influx of Pallid Swifts Apus pallidus into southern England. In that year, the first definite birds of the spring were at RSPB Saltholme and Haverton Hole on 19 April, with four at Saltholme the following day. Birds were widespread by the last week of April, including 150 at Hetton Lyons and 200 at Castle Lake. Active spring migration is not normally observed so an estimated 230 moving west in an hour at Ryton Willows on 17 May 1987, which was preceded by 146 west up the Tyne valley on 10 May, were of interest as they are probably indicative of the scale of movement that is typical at this time of the year.

This is one of the earliest of the summer visitors to leave Durham and in most years, birds start moving south in late July. Heavy passage was noted at Sunderland on 15 July 1975, when c.300 went over in 30 minutes, flocks of up to 100 were moving over South Shields on 21st and 22 July of that year; whilst 420 flew south in two hours over South Shields on 31 July 1979. Despite evidence of such an early exodus from the county many other adults will still be attending nests in July when some of their neighbours had probably already commenced their journey south. Adults start to leave the area entirely during late July or early August and by the latter half of the month numbers are normally very much reduced. At this time of the year large southerly movements of birds are sometimes noted, for example the 160 moving south over Shibdon Pond in ten minutes on 11 August 1991; this was just a small proportion of the many birds that were moving over on that day. A steady southerly movement of 600 birds per hour passed along the lower Derwent valley on 31 July 1997. Significant movements in August are illustrated by the 550 moving south in an hour at Hawthorn Dene on 5 August 2000 followed by 'about 1,000' south at Dorman's Pool on 12th.

Birds can often be seen passing over the sea and along the coast in early autumn from locations such as Whitburn Observatory and Hartlepool Headland, though such movements probably take place on a much broader front. On occasion, these movements can be quite marked with hundreds of birds flying south on some days at this season. On 18 August 1985, 850 passed south in just 30 minutes at South Shields. That figure however, was dwarfed by the 4,200 that flew over Seaton Snook on 24 July 1961 (Blick 2009), easily the largest documented movement in the county. A flock of 1,500 to 2,000 birds over Boldon Flats on the evening of 28 July 1987 were presumably birds gathering prior to them leaving.

The distinct decline in numbers from the first week of August onwards is very evident from the upland fringes with few counts of more than ten birds in the latter part of the month. In most years just a few birds remain with us into September so a gathering of 200 at Washington on 2 September 1977 was a large number for so late in the season. Flocks of about 50 over Teesmouth on 8 September 1977 and South Shields on 9 September 1974 are the latest 'large' gatherings on record. In 2008, most September reports were noted within ten miles of the coast and no counts exceeded six birds, a typical pattern. Regular sightings up to 19th were followed by six singles over the period 20th to 21st with the final bird at Hartlepool Headland on the late date of 3 October. On occasion, a few birds are noted into early October. Such occurrences seem to have been more common during the late 1970s and early 1980s and less frequently recorded recently, though the last records of the year have tended to occur much later.

In 2000, the final reports involved two at Saltholme Pool and one at Cleadon Hills on 1 October, one at Hebburn the next day, and one flying south at Fulwell on 11 October 2000. Such dates are not unprecedented, one was over Durham on 28 October 1976; and one was at South Shields on 6 November 1975. Late 'swifts' at South Shields on 6 November 2005 were noted but these occurred at the time of a small national influx of Pallid Swifts, one of which was at Newbiggin, Northumberland on the same day, so identification was uncertain. Other late stragglers were at Marsden over 5th to 6 November 1982 and at Whickham on 9 November 1982. For many years the latest record for the county was the exceptionally late wanderer that was at Seaburn on 2 December 1984. However, all of these late dates are exceeded by the sighting in late 1993, when a Swift was noted heading south east over Winlaton on 5 December.

In the south east of the county, there have been several records of individuals showing some white in their plumage (Blick 2009). Such birds can be problematic, occasionally giving the impression of rarer species. A partial albino bird was present at Shibdon Pond on 28 June 1975 and more initially challenging to the observers was a Swift with a partially white-rump at Shibdon Pond that was seen in the early summer of 1985.

Distribution & movements

Swifts breed over most of Europe and North Africa, as well as in the Middle East and Asia, to the north of the Himalayas. This is a highly migratory and supremely aerial, species, with all northern populations wintering in sub-Saharan Africa.

Swift is a difficult species to trap and consequently the numbers ringed in Durham have never been high. There have been a number of exchanges between ringing sites in the north east. For example, one ringed at Seaton Burn on 6 August 1962 was recovered at Blaydon-on-Tyne on 6 July 1965 whilst one caught as an adult bird at Birtley Sewage Treatment Works on 29 June 1980 was controlled at Seaton Burn, Northumberland on 23 June 1986. This bird was at least seven years old and must have travelled a prodigious distance in its lifetime. Other local recoveries include one caught as an adult bird ringed at Birtley Sewage Treatment Works on 27 June 1987 that was found dead at West Boldon on 17 May 1988. One caught at Portrack Marsh in June 1987 had been ringed near Sutton Coldfield, Warwickshire, in May 1984 (Blick 2009). A bird ringed as an adult female on 8 July 1988 at Lamesley was found freshly dead at Howick in the Scottish Borders on 29 June 1991, where it may have been breeding or perhaps just on an extended feeding trip. Perhaps the most interesting recovery of a locally ringed Swift concerns a bird that was ringed at Shibdon Pond on 27 May 1979 and which was recovered on 20 May 1984, five years later in Finistere, France.

Pallid Swift
Apus pallidus

A very rare vagrant from Southern and Eastern Europe and North Africa: three records.

Early on the morning 25 October 1999, Martin Blick found a Pallid Swift flying around the lighthouse at Hartlepool Headland. The bird lingered in the area until approximately 14:30hrs, which allowed over two hundred observers to see an extremely rare and mobile visitor (*British Birds* 93: 541). This bird formed part of what was then an unprecedented influx of 12 individuals to the east coast of Britain over the period, 24 October to 5 November. Prior to this, there had only been 12 records of Pallid Swift in Britain since 1978.

It was just two years before the species was recorded in the county again. On 23 October 2001, Paul Cook found a Pallid Swift over Church Lane, Whitburn, but it soon flew towards the coast. At 10:30hrs the following morning, the same observer found another bird over Marsden Quarry. This moved north over South Shields Leas and then lingered around Gypsy's Green, South Shields (*British Birds* 96: 583). Unfortunately, it was seen to be taken by a Sparrowhawk *Accipiter nisus*. During the afternoon of 24th, what was presumed to be the original Pallid Swift from Church Lane, was relocated to the south of the Whitburn Rifle Range. This bird moved slowly north up the coast, eventually crossing the River Tyne to Tynemouth (*British Birds* 96: 583). The late autumn of 2001 had produced a second influx of this species to Britain, with ten birds found during 2nd to 31 October, eight of them along the English east coast between Northumberland and Kent.

Distribution & movements

Pallid Swifts are locally common through the Mediterranean, and also from Mauretania and the Canary Islands across north west Africa and the Middle East to the Arabian Peninsula and south Iran. Some are resident in southern Europe but most winter in the North African tropics. Since the first, a total of 74 have been recorded in Britain to the end of 2010, a high proportion of which having been seen from the late 1990s onwards. Whilst several birds have occurred in spring and summer, late autumn has become firmly established as the peak time with most sightings coming from the east and south east coasts of England.

Alpine Swift
Apus melba

A very rare vagrant from Southern and Eastern Europe: five records involving six individuals.

On 24 July 1871, Abel Chapman saw a single Alpine Swift on the coast near Souter Point. He was accompanied by his uncle George Crawhall, who was familiar with the species abroad and therefore had no difficulty in identifying it. The time of year is typical for this species and there had apparently been others seen in Norfolk, Essex and Kent in the preceding month (Temperley 1951).

It was nearly 130 years before the next Alpine Swift in the county, but the species has been recorded in four years of the last decade. Three of the records came in the spring period, but two birds flying around Roker Park and the adjacent terraced streets in late October 2006 was unexpected. The 2010 bird at High Spen was the first record away from the county's coastal strip.

All records:

1871	Souter Point, Whitburn, 24 July (Temperley 1951)
2000	Greenabella Marsh, Teesmouth, 21 June (*British Birds* 94: 482)
2002	Hartlepool and Seaton Common, 9 June (*British Birds* 96: 582)
2006	Roker, Sunderland, two, 22 October
2010	High Spen, Gateshead, 9 May

Distribution & movements

Alpine Swifts breed discontinuously throughout southern Europe and north west Africa, and also locally through Turkey east to Iran, Afghanistan and Pakistan. The wintering range is poorly understood but thought to be in the Afro-tropics. A total of 560 records had been accepted up to the species' removal as a BBRC rarity in 2006.

Kingfisher (Common Kingfisher)
Alcedo atthis

Sponsored by

www.simplygo.com

An uncommon though quite widespread resident.

Historical review

Historical records tell of Kingfisher being resident along the Derwent, with a nest found there in 1870 (Robson 1896). In addition, one was shot at Lockhaugh on 9 April 1884 (Bowey *et al.* 1993). Temperley (1951) thought that Kingfishers may have increased in number during the late 19th and early 20th centuries and based this view upon a statement by Hutchinson in 1840, who said it "*is scarce; a few pairs only breeding on the banks of the Wear and the brooks of the county*".

In 1905, Nelson stated that the species appeared annually in the tidal pool and marsh streams around the North Tees Marshes. Illustrating its scarce nature in the county in the early part of the 20th century, Courtenay speaking of his time in the Sunderland area, said that "*in 1906 (October 26th) one flashed past him on the Wear just above Hylton. It was a long time ago and I doubt if one is to be found thereabouts today*", (*The Vasculum* Vol. XIX 1934). A record of one at Chester-le-Street, on Good Friday 1936, however suggest that occasional birds did venture downstream on the Wear, though J. Heslop-Harrison elicited the following comment in response to his observation, "*this is the first I have ever seen in this district and I can hear of no one else who has noticed it locally*" (*The Vasculum* Vol. XXII, 1936).

Further west in Durham, the Rev J. Greenwell noted that the Kingfisher was attracted to Weardale, particularly around Bishop Auckland where it was 'fairly common'. He mentioned once being at Pixley Hill, some two miles from the River Wear when, in a horseshoe in the river, he was surprised to see a Kingfisher fly past. "*I have never before or since seen a Kingfisher so far from water, and this short cut was fully two miles long*" (*The Vasculum* Vol. XXI 1935).

Temperley (1951) thought it, "*A resident, more common than is usually supposed*" and he highlighted that the species was more abundant on the Wear, saying it nested within a few miles of Bishop Auckland and on that river's tributaries, such as the Bedburn Beck, Rookhope Burn and Bollihope Burn. He confirmed that it bred on the Tees and its tributaries, on the Tyne almost as far down as Gateshead, and also on the Derwent almost to that river's confluence with the Tyne. Breeding was recorded on the Don at East Boldon. He also documented Kingfishers moving to the coast in the autumn and winter which may be compared to today when coastal records are rather scarce. In this context, Temperley mentioned records, presumably of dispersing juveniles, in autumn, in unusual locations such as Hebburn Ponds, Saltwell Park Lake (Gateshead) and Darlington South Park Lake.

There was an interesting record of a bird being taken into custody on 16 August 1953, when one was picked up alive by a policeman in Stockton High Street, it was 'quite unhurt' and later 'discharged' on to the River Tame (Temperley 1954).

In the early 1950s, this was considered a rather scarce species in the county, the low numbers at the time presumably being as a result of declines experienced during the severe winter of 1947 and probable industrial pollution of some local streams. In 1952, it was reported more than usually in the first three months, including one along the Wear at Durham Banks, but by and large its status as almost a rarity remained through the decade. This was further exacerbated by the major national and local decline inflicted upon it in the severe winter of 1962/1963 (Brown & Grice 2005, Bell 1964). A good example of the impact of this winter comes from Teesmouth where, subsequently, no birds were recorded in the north part of Cleveland, until December 1967, when one was at Billingham Beck. Breeding was not confirmed again in the area until 1984 (Joynt *et al.* 2008). In 1967, there were just two reports for the whole region and the following year only two pairs were known to have nested, one on the

River Wear and one on the Tees. An increase in reports was noted in 1969, when pairs were at two localities on the Wear and three on the Tees, with one on the River Derwent, at Swalwell, the first to be seen there since 1956. A bird occupied a stretch of the River Skerne in central Darlington from October 1968 to February 1969. The fact that individual records of this nature were documented indicates just how scarce it was still at this time.

By the early 1970s, this species was described as making a welcome come-back to the lower stretches of the Tees after being absent throughout the late 1960s (Bradshaw 1976). Two pairs were known to have nested on the Wear in 1970. The following year saw a marked increase in the breeding population, with nine pairs on the Tees, at least four pairs on the Wear and single pairs on two streams in the south east of the county. It was also presumed to be present on the Derwent at that time. The recovery continued steadily through 1973 when six pairs bred along the Wear and probably at four other locations in the county. Illustrative of the species' poor winter survival were the observations from 1976 that indicated that there were, at that time, at least 11 pairs reported to be breeding successfully in the county compared to just a single pair after the very cold winter of 1978/1979.

The relative importance of water quality and climate is shown by the River Browney, an intensively watched tributary of the Wear. A regularly observed pair of Kingfishers disappeared on a stretch of this river as a result of the cold winter of 1978/1979 but a new pair took up territory in 1980. The phenomenally cold weather of early 1982 froze the Browney and the kingfishers disappeared again. Later that year the river was badly polluted by a phenol spill and it was three years before Kingfishers were seen again, though some five years later they still had not bred.

During the late 1970s and 1980s, the estimated maximum breeding population for Durham was only of about 17 pairs which is much lower than is now known to be the case after the long run of mild winters through the 1980s and 1990s.

Recent breeding status

Across Durham, Kingfishers regularly breed on the wooded middle sections of the main river valleys, the Derwent, the Wear and the Tees and their tributaries. Some sites, such as Ebchester on the Derwent, Witton-le-Wear on the Wear and Piercebridge on the Tees, must be especially suitable for Kingfishers as they have consistently provided reports even when populations have been at low ebb elsewhere.

The last two decades of the 20th century and the first part of the 21st century saw something of a change in status of Kingfisher in Durham. From a low in the early 1960s, it seems that by 2008, most of the county's streams and rivers, at least those below an altitude of 250m above sea level, had been colonised, or re-colonised (Newsome 2009). This is in part, a reflection of the improved water quality, and the consequent increased prey availability, and the very much milder winters that the north east had experienced (Brown & Grice 2005).

Kingfishers are found only near to water and along watercourses. This was very evident from the *Atlas* map (Westerberg & Bowey 2000), which illustrated concentrations of birds along all of Durham's main river systems as well as their smaller tributaries. Given the species' well defined habitat requirements, the *Atlas'* mapped distribution was felt to be broadly representative (Westerberg & Bowey 2000) though some pairs within the main areas will have been missed. The species requires slow-moving, clean water with sufficient fish stocks to sustain breeding. The glacial deposits along the banks of the Derwent provide many breeding sites (Westerberg & Bowey 2000) and here it can be found nesting from just above its confluence with the Tyne to upstream above Derwent Reservoir, almost as far west as Blanchland.

A key habitat requirement for Kingfisher is a bank side that has just the right type of ground sediment of a soil particle size and structure that makes for an easily excavated but stable nest burrow (BWP). Such situations are limited in the county, and tend to be concentrated along lowland stretches of river, or where the river geography provides a more meandering profile, with attendant slack waters and water-cut vertical bank sides.

Only detailed examination of river banks efficiently locates nest sites revealed by the dark slime of droppings and disgorged fish offal that stains the nesting bank, outside the entrance hole (Sharrock 1976). An occupied nest can be located by the smell, on close examination, but only if the observer has a Schedule 1 Species Disturbance Licence.

Kingfishers can be surprisingly difficult to spot despite their distinctively bright plumage. Their small size, mercurial nature and periods of relative inactivity as they perch looking for fish can make them difficult to locate. Crucial to revealing its occupation of a stretch of river, is a familiarity on behalf of the observer with the species' loud and distinctive call, which readily draws attention to it (Westerberg & Bowey 2000). Overall, it is probably an overlooked species in Durham. For example, intensive observer coverage over many years in Gateshead, revealed a very healthy population with birds nesting on all three river systems in this relatively small area; the Derwent, the Tyne and the River Team, with somewhere between eight and ten pairs breeding in most years (Bowey *et al.* 1993).

The Kingfisher is also regularly reported in the areas of Durham City, Langley Moor, Billingham Park, Shibdon Pond, Billingham Beck and Hurworth Burn, the lower Tees around Eryholme and during winter periods occasionally on coastal marshes (Armstrong 1988-1993). It is not restricted to lowland waters, with sightings around St. John's Chapel at over 300m above sea level. In the south east of the county, it is largely a winter visitor to the Teesside area. There were just a handful of breeding attempts in this part of the county during the 1980s and 1990s, and today there are probably only three to four pairs breeding in what used to be the north Cleveland area (Joynt *et al.* 2008). It is known to breed, in some years, around Billingham, Wynyard, and the River Tees, from Portrack to Yarm and occasionally in several other localities (Blick 2009).

Confirmed breeding locations in County Durham, during the period 1988 to 1993, fluctuated between two and fifteen sites per annum (Armstrong 1988-1993) but clearly such results were dependent upon observer effort, and the real number of breeding birds would have been very much higher than this. Birds were noted as nesting at 20 breeding sites in the county in spring 1994 (Armstrong 1995). Over the most recent decade the population appears to be still expanding into previously unoccupied areas and the species is said to be enjoying breeding success at a number of sites.

The *Atlas* showed Kingfisher present in 40 tetrads, but not all nesting sites would have been recorded. Indeed, Westerberg & Bowey (2000) speculated that as many as one in four might have been missed. Working on an occupancy rate of one to two pairs, per occupied tetrad, the *Atlas* suggested that the county held a population of between 50 and 100 pairs. Perhaps the true figure present is nearer the upper limit of this range rather than the lower one. During 2007 the Durham Bird Club received 246 reports from 73 sites and by 2008 there were 650 reports from 93 sites. Nevertheless, the Kingfisher population remains relatively vulnerable from the possible effects of harsh winters (Gibbons *et al.* 1993). Being a largely riparian species, Kingfishers may suffer from deteriorations in water quality, particularly around breeding and feeding sites. Such loss of quality might come about from industrial or agricultural spillages, river engineering works, the depletion of fish stocks or in the most agriculturally intensive areas, 'diffuse source' pollution, from agriculture run-off (Marchant *et al.* 1993). In fact, general improvements to water quality and a series of mild winters have contributed to continued expansion and the majority of Durham's river systems are now populated.

The breeding data from 2007, suggested that local pairs start nest-hole excavation during the first half of March and can be feeding their first brood of youngsters by the end of April; with these being on the wing by early to mid-May. The species often double-broods and brood sizes of up to five have been noted locally in good breeding seasons. By 2008, even the river systems with historically poorer water quality, such as the River Team in Gateshead and the Don in Jarrow, supported one or two pairs.

Collisions with windows have been a significant cause of Kingfisher mortality in some areas of the county, where houses lie close to water courses, such as by the River Gaunless at South Church near Bishop Auckland. Three birds were killed here in 2007. On the Tees in 2009, a pair was noted feeding young near Eggleston Abbey in June, where the observer had the chance to show his son his first Kingfisher's nest, some 35 years after finding his own first nest on the same stream.

In 2010, Kingfishers were noted at 22 breeding locations across the county, although there was little information about successful breeding. The previous harsh winter may have affected populations. At Cowpen Bewley, a pair reared a second brood with the young fledging on 6 August.

Non-breeding status

Soon after breeding the young birds disperse from their natal areas during July and August and can account for some irregular sightings such as one in the dunes at Sandhaven Beach, South Shields, on 19 July 2007 and a bird seen fishing in rock pools on the coast at Whitburn for a day in August 1973. Family parties, or the young from several dispersed pairings, may gather and up to four birds having been noted together at Shibdon Pond and the Far Pasture wetland (Bowey *et al.* 1993), feeding on fish made more easily available by low summer water levels.

Birds are most frequently seen on smaller streams, away from the larger rivers, during the autumn and winter periods especially if the main rivers are in flood. The general winter distribution in the county is more easterly and perhaps more local than during the summer months.

The DBC total of 246 reports for the whole of 2007 included locations during the winter, even as far west as Langdon Beck in late January at almost 400m above sea level, with others at Derwent and Hury Reservoirs. In most winters there are usually a few wintering birds reported from sites in the south east of the county, such as Billingham Beck, parts of the River Tees and occasionally, the North Tees Marshes. Few sites tend to produce regular sightings in the early part of the year but exceptions include the River Derwent at Swalwell, Far Pasture wetland and the Thornley Woods Feeding Station hides, all in the Derwent valley. The latter location, especially when water levels in local streams and rivers are high, is hugely attractive to the species. It was in such circumstances at this site that allowed prolonged observations to be made of feeding birds, leading to the publication of observations on the winter feeding rate of the species, which were observed in these instances to be consuming approximately 63% and 90% of their respective body weights (Rewcastle *et al.* 1997).

At places such as Hurworth Burn, Sudburn Beck near Staindrop and the Durham Light Infantry Museum in Durham, the movement of the water keeps it ice-free and can encourage birds. At Washington Waterfowl Park some water is kept open artificially for the collection and Shibdon Pond is fed by constant-temperature water from a drift mine.

If the weather worsens in late autumn and winter then birds will move downstream to warmer, ice-free localities and even appear in small numbers at the coast (Brown & Grice 2005). A bird at Castle Eden on 20 August 1993, was probably illustrative of dispersal from a breeding site further west in the county rather than being a locally bred bird. A bird seen flying over an oil-seed rape crop at Bishop Auckland on 21 June 1993, mirrored a similar sighting in the same part of the county almost 60 years previously. A most unusual observation came from the A1300 near South Shields in August 1994, where a bird was seen flying south along the road, a long way from the nearest water course. Similarly, a bird was seen flying along the A66 near Longnewton Reservoir on 15 October 1992. Dispersal of young was illustrated in the south east of the county, where young birds were at Greatham Tidal Pools on 12 July 1993 and Hurworth Burn Reservoir on 31st. Other dispersing birds were at WWT Washington in June and late July 1995. A bird was at Mere Knolls Cemetery, Seaburn on 18 October 1998 and another at Saltwell Park Lake in Gateshead on 7 October 2007, the first known record there since August 1932.

In 2007, a number of coastal birds were seen during the second half of the year. One flew over the sea 100m off Hartlepool Headland on 16 October and another was at Hartlepool Fish Quay on 2 December. One notable historic record was of a bird on a boat off the mouth of the Tees on 14 November 1887 (Nelson, 1907). Blick (2009) also documented a number of other coastal records from the south east of the county, including: singles at Hartlepool Headland on 22 April 1975, 5 July 1975, 8 August 1985, 23rd and 26 October 2002, 21 December 2003 and 7 September 2008 and noted that there are at least five records of birds seen around Hartlepool Docks (Blick 2009).

Distribution & movements

Kingfishers can be found breeding over much of Europe, north to southern Scandinavia, with birds also present in north west Africa, southern Asia and Australasia. Northern populations are migratory, but they winter within the species' breeding range. British birds are essentially sedentary, although there is some juvenile dispersal.

The period from early July through to the winter months of the following calendar year is the principal time when young birds disperse. Most local lowland pairs tend to remain in their territories throughout the year. The only significant recovery of a ringed Kingfisher dates from the first half of the 20th century, when on 6 September 1943, a bird that had been ringed on 11 July at Kirkley Mill near Ponteland, Northumberland, was found on Back Redheugh Road, Gateshead (Bowey *et al.* 1993).

Bee-eater (European Bee-eater)

Merops apiaster

Very rare vagrant from southern Europe: nine records, involving at least 51 individuals to 2009. A pair bred in 2002.

Recent breeding status

One of the true ornithological highlights in the county's history unfolded in 2002; prior to this few if any would have envisaged that this species might ever breed in the north east of England. The surprise happened when a pair nested at the Durham Wildlife Trust nature reserve at Bishop Middleham Quarry. The birds were originally discovered by Mike Hunter on 2 June 2002 when studying butterflies. The pair soon settled, started to undertake courtship feeding and copulation was noted. A nest hole was excavated in one of the sandy bank faces of the quarry and it was clear that the birds were attempting to nest. The RSPB and DWT then set up a warden's post to cope with the flood of attention from both birdwatchers and potentially, egg thieves. At this stage, round-the-clock security was provided by a team of volunteer wardens from a range of local organisations. Five eggs were laid and on July 24th, news broke that the parents were carrying food to their nest hole. The birds became instant media celebrities. All of the eggs hatched but one nestling died in the nest, one died just prior to fledging and of the three fledglings seen on the wing, one disappeared almost immediately. It is estimated that some 15,000 people came to see the birds during their tenure at Bishop Middleham. The family party of four, the adults and two juveniles, was seen in the company of Swallows *Hirundo rustica* on August 28th, and were last reported in the area on 2 September. The pair at Bishop Middleham were the first nesting Bee-eaters in Britain for almost 50 years; the species does normally nest further north than Paris. The only previous occasion that Bee-eaters bred successfully in the UK, was in 1955, in a sand quarry at Streat, East Sussex, when two pairs reared seven young (*British Birds* 95: 468).

Recent non-breeding status

In Durham, the Bee-eater is a rare vagrant that had been recorded on nine occasions, involving at least 51 birds, to the end of 2010. The first sighting of this colourful Mediterranean migrant in the county occurred as recently as 15 July 1979, when Michael Duncton and Steve Howat saw one flying over the Wear valley at South Hylton, close to Sunderland (*British Birds* 73: 516). The national count of 15 birds during that year was the second highest year-total on record. It is perhaps surprising that this species had not been recorded in the county prior to this. The second record came from the North Tees Marshes just two years later, but was typically brief, when a single passed over Cowpen Marsh on 21 May 1981 (Blick 2009).

A series of records in 1997 was remarkable for the north east. During an unprecedented nationwide influx involving at least 123 Bee-eaters across England, observers in Durham were able to claim several sightings. This commenced with a singleton at Cotherstone in late May that presaged a party of six birds over Cornthwaite Park, Whitburn on 30th. This group paused for about 20 minutes before heading south east over the sea. Incredibly, this flock was exceeded in number by a group found in the lower Derwent valley during August. Fourteen wet and bedraggled birds were seen going to roost in heavy rain in the Derwent Walk Country Park on 21st and these were followed by a flock of ten birds circling over Shibdon Pond on the evening of 29th. It seems likely that some of these birds were mobile in this part of the county for a period of nine or ten days, for around this time several others were reported, including one claimed sighting by an experienced observer, of a flock of 27 birds high over Shibdon Pond. Unfortunately this record was never formally submitted.

Extended opportunities to see this delightful species came in 1999 when ten paused for three days at Cleadon.

All records:
- 1979   South Hylton, Sunderland, 15 July
- 1981   Cowpen Marsh, 21 May (*British Birds* 75: 513)
- 1997   Cotherstone, 21st to 22nd May
- 1997   Whitburn, six, 30 May
- 1997   Derwent Valley, fourteen, 21 August

- 1997   Shibdon Pond, Gateshead, ten, 29 August
- 1999   Cleadon Hills, ten, 20th to 22 May
- 2002   Bishop Middleham, 2 June to 2 September, pair bred, raising five young, three of which fledged
- 2003   SAFC Academy Pools, Sunderland, 15 July

Distribution & movements

Bee-eaters breed discontinuously throughout the Mediterranean, north west Africa and the Middle East, and across into Central Asia. Sporadic breeding takes place outside the usual distribution, sometimes leading to permanent colonisation. The population winters in two separate areas of tropical Africa; from West Africa and Senegal to Ghana, and in eastern and southern Africa. The species was formerly much rarer in Britain but was dropped as a BBRC species at the end of 1990 when British record totals approached 500. Five breeding attempts have taken place in Britain since the first in 1920, although the Bishop Middleham nest was the last to be successful.

Roller (European Roller)
Coracias garrulus

A very rare vagrant from Southern and Eastern Europe: four records.

Unfortunately, this brilliantly colourful species seems to occur less often today in Britain, than it did a century ago. There are three records for the 19th century. The county's first, according to Selby (1831), was shot near South Shields, sometime before 1831. He recalled that "*a fine specimen now in the possession of Edward Backhouse Jun. Esq. of Sunderland was shot lately near South Shields*". The second record in Durham was of a bird detailed by Bolam (1912) as having been acquired at Ravensworth Castle in 1868. Again, the exact date is unknown. Curiously, this record was not mentioned by Temperley. The third record has more detail and some tangible evidence. Temperley (1951) noted that Hancock had in his possession, a Roller shot near Willington in the Wear valley "*on the Hunwick Estate, Durham, on May 25th or 26th, 1872, by Mr. H. Gornall of Bishop Auckland*". The specimen was presented, probably sold, to Hancock by Gornall and was housed in the Hancock Museum; the specimen still exists today in the same collection.

The only modern record is of a bird that was found in the area of Mundles Farm, near East Boldon from 13th to 30 July 2000 (*British Birds* 94: 482). It was initially found by non-birdwatchers and later identified as a Roller by the local farmer; this identification not being 'confirmed' until 25th. Over the next five days, it was watched for long periods by many observers before, on 30th, it grew restless, drifted further east and disappeared. This bird represented the fourth occurrence of the species for Durham.

There is an additional published record from the mid-20th century. In the autumn of 1958, a very late bird was reported just to the south of Sunderland, on 11 October, remaining in the area until 23rd (Grey 1960). It is perhaps significant that a Roller had been reported in the Bamburgh area of Northumberland on 24 September (Kerr 2001), and it is conceivable that this was the same bird on its way south. However, the Northumberland record was not accepted by BBRC and the Sunderland bird never formally submitted and as such, it cannot be included in the county record.

Distribution & movements

The breeding range of Roller covers north west Africa, Spain, and Eastern Europe east through Turkey, southern Russia, south west Siberia, Kazakhstan and western China. Most winter in East Africa, from Kenya to Zimbabwe, with smaller numbers in equatorial West Africa. The peak time for occurrence in Britain is late spring, with a secondary peak in September. A total of 309 had been accepted as occurring in Britain to the end of 2010, but the rate of occurrence has been in decline corresponding with a decrease in its European breeding population (Hagemeijer & Blair 1997).

Hoopoe (Eurasian Hoopoe)
Upupa epops

A scarce spring and autumn passage migrant: 33 records in the last one hundred years.

Historical review

The ostentatious plumage of this south European bird makes it conspicuous and it is likely that a high proportion of the Hoopoes occurring in Durham have been recorded. Their ornamental appearance has also once made them attractive to collectors and taxidermists and a high proportion of early records were of shot birds.

In the Tunstall manuscript of 1784, there is a reference to *'many Hoopoes'* being seen in Yorkshire, last year (presumably 1783), and of one that was shot within a few miles of this place, Wycliffe-on-Tees in September of that year (Mather 1986). As Wycliffe-on-Tees, where Tunstall lived at Wycliffe Hall, is now in County Durham then this probably constitutes the first record for Hoopoe in the county.

Aside from this above report, the first fully documented local record of Hoopoe came in 1827, when a female was shot near Westoe, South Shields, which was presented to the Newcastle Museum by G.T. Fox. Further references in Temperley (1951) include one "*in the possession of a Mr Bennett, killed at Gibside, in the county of Durham*" as quoted by Selby (1831), but there is no date for this record. Edward Backhouse (1834) gave the status as "*rare*" but added that "*several specimens have at different times been killed in the neighbourhood of Sunderland*". Hogg (1845) quoted others as saying that several had been shot in the Tees Marshes, including one at Coatham Marsh. Temperley (1951) had some doubts about this, believing that Hogg's source, a wildfowler, had meant "*Whooper*" i.e. Whooper Swan *Cygnus cygnus*, and not 'Hoopoe'. William Proctor, in 1846, recorded one as being shot at Seaham about 1840 and another killed at Aycliffe in the same year, while Hancock (1874) wrote of seven entries of the capture of this species in his journal, but did not say how many were taken in Durham. In 1848, one was caught in the rigging of a ship coming into the Tyne with the specimen later ending up in the Hancock collection. On 11 June of the same year, Robert Calvert recounted that he saw a Hoopoe flying back and forth over the sand hills between Sunderland and Ryhope, being pursued by a gang of boys. Canon Tristram (1905) recorded two further specimens, one taken near Bishop Auckland around 1850 and deposited in the Dorman Museum at Middlesbrough, and "*another in the possession of Mr Cullingford which was killed near Durham twenty years ago*" i.e. around 1885.

In the first half of the 20th century George Temperley (1951) said that it was *"Now a rare straggler, previously an uncommon spring and autumn passage migrant"*. A review of the records by him indicated that it was a rare visitor in Victorian times, perhaps more common in the first half of the 1800s. In line with the national trend at the time, it became much rarer in the first few decades of the 20th century. Subsequent to Tristram's record of '1885', there was a long absence of the species from the county, with the next record not occurring until 1933, when one was found near Croxdale on 14 September. It had been seen to emerge from a rabbit hole and was duly shot. It came into the hands of S.E. Cook for preservation (*The Vasculum* Vol. XIX 1934). This was the only record in the county during the first half of the 20th century. Temperley surmised that the absence of sightings was associated with the decline of the species in north west Europe, noting that the disappearance from northern France, Scandinavia and the Low Countries may have been linked with the rapid spread of the Starling *Sturnus vulgaris* and the associated competition for nest holes (Temperley 1951).

Post-Temperley, just three birds were recorded between 1951 and 1970. Single birds were seen at: Neasham Hall on 3 May 1958, Seaton Carew on 17 May 1963 and at North Gare over 19th to 20 April 1964.

Recent status

Over the last 40 years there has been a considerable upturn in the number of sightings of this species in Durham with the spring and autumn periods being when most records have occurred, although birds have been recorded in many different months. There were 29 records of Hoopoe between 1970 and 2010. All recent spring birds arrived between 15 March and 30 June, while autumn reports span the period 15 August to 29 November.

By and large, the rate of occurrence of this species has altered little over the last 30 years or so, with an average of around six per decade since the 1980s. Although Hoopoes have been recorded in 23 years since 1970, most visits have been rather brief. Of the 29 recent records, 20 have been seen on just one day, with two staying for two days and seven lingering for three days or longer. A high proportion of the single-day birds have moved quickly through the area in which they were found, spending only an hour or two feeding in suitable habitat. The North Tees area, north to Hartlepool, has been the most favoured part of the county, with seven birds seen here. Elsewhere, sightings have been widely scattered with 17 inland birds as far west as Hamsterley Forest, Shotley Bridge and Wolsingham, three birds on the coastline between the Rivers Tyne and Wear, and two on the central stretch of coast.

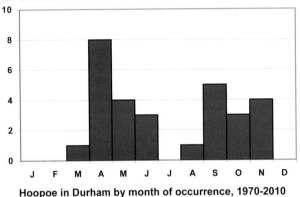

Hoopoe in Durham by month of occurrence, 1970-2010

Although over half of the March and April records have been within a mile or two of the coast, only two out of the seven in late spring (May and June) have been at the coast. Unlike the situation in Northumberland (Kerr 2001), many of Durham's records have been inland. June birds in particular have appeared well inland, with birds at Low Barns on 27th to 28 June 1971 and Muggleswick on 30 June 1991 being especially notable. The late autumn arrivals are of interest, with four birds appearing between 28 October and 29 November, possibly indicating a more easterly origin. Two of the November birds have appeared in the environs of Durham City, including one which found itself in the grounds of County Hall over 29th to 30 November 2010 during a period of heavy snowfall. This bird was weak and emaciated and despite being taken into care on 30th, died the same day. The specimen is lodged in the NHSN Hancock collection in Newcastle upon Tyne.

Distribution and movements

Hoopoes have a very widespread breeding range through much of southern Europe, Africa and Asia. In the Western Palaearctic, they breed from Iberia and most of France, across to the southern Urals, Caspian and Caucasus, and north to Poland and the Baltic States. Many Mediterranean birds are resident or partially migratory, but they are strong migrants elsewhere, wintering in sub-Saharan Africa. It is a scarce spring and autumn migrant in Britain with less than 130 individuals seen in most years, mainly in southern England, but it is frequently recorded north to Shetland.

Wryneck (Eurasian Wryneck)
Jynx torquilla

A former breeding species, now a scarce but regular spring and autumn migrant.

Historical review

This beautifully marked 'woodpecker' is today a scarce but regular passage migrant in autumn seen less often in spring. It is clear from historical records, that the Wryneck was formerly a quite widespread summer visitor in the county and commonly bred in many areas. The earliest reference is by Selby (1831), who noted that "*About Durham it used to be pretty common, and I have frequently heard the peculiar and loud notes of three or four individuals at the same time, upon their first arrival in spring*". Evidence of persecution and a decrease in numbers was noted by Hutchinson in 1840; "*The banks and woods near Durham are favourite resorts of the Wryneck, as they find in the fine old trees abundance of food and holes for their nests. Unfortunately for the bird, its distinct notes proclaim its arrival when some idle boy sallies out with his gun bent on its destruction. From this cause the numbers that formerly resorted to the fine old trees in the banks have lamentably diminished*". The writings of Backhouse, Hogg and Proctor confirmed the status in the middle of the 19th century as "*not uncommon*", while Hancock (1874) noted that "*a pair bred in 1813, and for four or five years after, in the garden at Cleadon House, though this is probably about its northern limit*".

From the middle of the 19th century the species was much less common and a rapid decline appears to have then taken place. Incidental records included a bird which was observed by Rev. G. Sowden at Houghton-le-Spring for three days in October 1856, who then shot it on 3 October. In relation to the Derwent valley, Robson (1896) commented that it was "*common in the lower part of the vale up to the year 1842*". In 1834 it was found nesting near Swalwell and on 16 May 1836 a male was shot in Axwell Park. Almost 40 years later, in the diary of Thomas Thompson of Winlaton, it is recorded that an early migrant was shot in Gibside on 4 April 1872 and this proved to be the last local record in this area for almost a century (Bowey *et al.* 1993). The Alfred Chapman collection of wild bird eggs, now housed in the Hancock Museum, held two eggs of the Wryneck marked "Durham 1872". These eggs are the last evidence of breeding in the county although according to Holloway, this species bred in Durham, in an outlying breeding cluster into the 1890s (Holloway 1996). Wrynecks clearly bred in the county regularly up to the mid-19th century and not infrequently over the latter half of the century, this is in line with the known decline of the species over England as a whole, the reasons for which are not readily apparent.

Only two records are quoted by Temperley for the first half of the 20th century. One was picked up dead by F.T. Tristram in Close Gill at Castle Eden Dene on 3 September 1905, the specimen ending up in the Hancock collection. Another was reported in the BOC *Bulletin* for 1910, being "in Durham" on 13 May 1909. This bird was apparently the only one recorded in any county north of Lincoln in that year, a clear reflection of the scarcity in Britain at the time. Temperley (1951) refers to its loss as a breeding bird in Cumberland, Westmorland and Lancashire and of its being much less common even further south. The lowest ebb for this species in Durham seems to have been in the period from 1910 to around 1950 and this pattern was mirrored in Northumberland, where there were no reported sightings between 1923 and 1951 (Galloway & Meek 1980).

Subsequently, the number of records increased somewhat through the late 1950s and 1960s, though the species remained a genuine rarity locally at this time; early autumn arrivals becoming the established pattern. The first record for the 20th century in the south east of the county was of a coastal passage bird near Graythorp on 28th and 29 August 1954 (Blick 2009). Better years for coastal migrants, particularly in the Hartlepool and North Tees areas, included 1958, 1966 and 1969, while one inland at Hurworth Burn Res. on 12 September 1965 was notable. Unusually, four were also seen between 5th and 10 May 1969; three at Graythorp/Seaton Carew and one at Marsden. Spring arrivals were much more unusual in this period. Records further increased until it was once again an annual, though scarce, passage migrant in Durham.

Recent status

At least 166 Wrynecks have been seen in the county since 1970 and its status has not changed much over the 40 years. It remains a scarce passage migrant, more frequent in autumn, and prone to larger arrivals with suitable weather conditions. Nearly 30% of recent records (1970-2010) have been in spring, the earliest being at Hawthorn Dene on 1 April 1988 and the latest at High Sharpley on 31 May 1991. Of the more than 119 autumn birds, the earliest was at Marsden Quarry on 11 August 2004, with the latest arrival being at Whitburn on 16 October 1994. The appearance of migrant Wrynecks is far from guaranteed. There were seven blank years between 1970 and 2011, while just a single bird was noted in a further ten years.

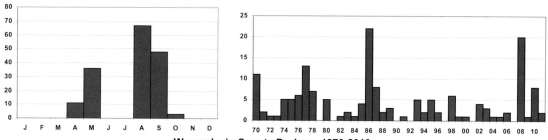

Wrynecks in County Durham, 1970-2010

Spring sightings have shown something of a decline over the period 1970 to 2010. After decade totals of 21 for the 1970s (including seven in 1978) and 12 for the 1980s, only nine were seen in the 1990s and 5 in the 2000s. The reason for this decline is uncertain and is in some ways mirrored by that of Bluethroat *Luscinia svecica*. Although Wrynecks have been suffering from a long term reduction in numbers and range in western and central Europe (del Hoyo *et al.* 2002), the reason for fewer spring migrants may be down to changing weather patterns and migration routes.

In contrast, autumn totals have not changed significantly since 1970. The best recent autumns were 1986 and 2008. In 1986, a total of 22 arrived between 19th and 31 August, 15 of which were found between Sunderland and South Shields. The peak day was 28 August when 14 arrived during rain and onshore winds, with 7 in the Whitburn area alone. In 2008, 20 birds were found between 26 August and 23 September, the main 'fall' period being 6th to 9 September with 15 arrivals, including four in the South Shields area on 6th.

The well-watched northern section of coastline has been the most productive area for Wrynecks. A total of at least 76 have been found in migrant coastal habitat between the Rivers Tyne and Wear, with over 70% of these in the Marsden/Whitburn area. The southern section of coast, from Hartlepool to the mouth of the River Tees, has accounted for more than 56 birds with Hartlepool Headland claiming at least 37. Surprisingly, only 11 have been found on the central stretch of coastline between Sunderland and Hartlepool, although the abundance of habitat and relative lack of observers does not bode well for locating a quite secretive species.

A total of 21 Wrynecks have been found inland. Of these, six have been only marginally inland (less than 5km from the coast); while a further six have occurred between five and 12km from the coast. Of the more interesting sightings, three have been found in the Houghton-le-Spring area (Fence Houses, Hetton Bogs and Hetton Lyons CP), with singles at Boldon Colliery, Hurworth Burn Res. and Norton. A further nine birds have been found more than 12km from the coast. In the Gateshead area, a male was in song at Stella, Blaydon from 7th to 11 May 1970 then singles were at Low Fell on 30 April 1989 and Shibdon Pond on 11 May 1989, while in the Durham City area, birds were at Great High Wood on 7 May 1974 and Langley Park on 21 April 1987. More significantly inland, the upper reaches of the River Wear has produced singles at Wolsingham on 1 May 1995, Upper Weardale on 16-17 May 2005, Low Barns on 24 August 2006 (caught and ringed, but unfortunately killed by a cat the following day) and Hamsterley Forest on 13-14 September 2010. There has been just one sighting in upper Teesdale, at Harwood-in-Teesdale on 27 August 1976. It is interesting to note that three of the spring birds seen well inland have been in suitable breeding habitat. Indeed, the bird at Wolsingham was observed emerging from an already excavated nest hole, but the hole was found to be occupied by a Great Spotted Woodpecker *Dendrocopus major* and the Wryneck was not seen subsequently. Perhaps one day, birds may return to the areas where they were once common.

Distribution and movements

Wrynecks breed widely across Europe, from Iberia east to the Urals, Caspian and north Caucasus, and across temperate Asia to China and northern Japan. The wintering grounds include tropical Africa, India, South East Asia, southern China and southern Japan, but some birds in the south of the breeding range are resident or only partially migratory. It was formerly a regular breeding bird in Britain, but is now a scarce spring and autumn passage migrant, mainly to the east and south coasts, with just occasional breeding records.

Green Woodpecker (European Green Woodpecker)
Picus viridis

A common resident that despite its somewhat localised distribution, is now much more widespread than in times past.

Historical review

The status of the Green Woodpecker in the north east of England has changed considerably during recorded times. Rossiter (1999), indicated that Green Woodpeckers were reasonably widespread but not common in the north east during the eighteenth and early 19th centuries, a decline seemed to take place from around 1840, leading to it becoming much rarer thereafter. In the early part of the 19th century it seems to have been fairly common and well distributed in both Durham and Northumberland. In 1824 Hogg described it as 'rather frequent' around Wynyard and Elton, to the north of Stockton. Selby (1831) wrote: "*The Green Woodpecker is seldom seen in the northern parts of Northumberland. I have, however, met with it in the woods about Hulne Abbey, near Alnwick, and upon the banks of the Wansbeck. It is common about Durham and indeed in all localities where timber abounds and has attained an advanced age*". E. Backhouse said it was more common than the Great Spotted Woodpecker *Dendrocopos major* but by 1840 the first observer to record any falling off in numbers was Hutchinson (1840), who wrote that it was "*formerly much more common than it is at present; the cause being apparently the scarcity of decaying trees and the persecution which every bird considered somewhat rare encounters at the present day. They are occasionally met with near Durham City, but I have never found a nest*". Proctor (1846), considered it 'rare' and Sclater (1884) recorded one in Castle Eden on 16 October 1874, only the third he had noted in that area in 15 years. Subsequent writers consistently described it as: "*a rare bird*" or "*an occasional visitor*". Hancock (1874) wrote: "*not by any means common; occasionally breeds in the district*"; but he quoted only one instance of breeding at Minsteracres in the valley of the Tyne, date not recorded. He gave no records of it breeding in County Durham. Sclater (1884) said, "*it used to breed regularly at Woodlands (near Lanchester) where I have several times found its nest*". J. Backhouse (1885) said of it in upper Teesdale, "*occasionally met with in Park End Wood and has once been seen in High Force Wood*". It was only thought to be an occasional species, even in the well-wooded Derwent valley. Indeed Robson thought it rather rare in the lower valley in the late Victorian period, he documented only two records, specimens obtained from Lockhaugh and Axwell Park (Robson 1896).

Temperley (1951) described it as a "*not uncommon resident in wooded districts*" but did note a change in status during his 'recorded times'. Until the early 1800s, Temperley thought that it must have been a common and widespread species, this was followed by a severe decline, after which it was considered 'rare' for quite a long period. Temperley thought a significant increase in numbers occurred during the first two or three decades of the 20th century and from then the species spread quickly after its first burst of population expansion.

Tristram (1905) had given its status as "*formerly common, now rare.*" It was documented as being present at Wynyard by Nelson (1907). The recovery in numbers seemed to start a few years after this, with year-on-year more records from different areas of woodland. A few years afterwards observers began to note an increase in numbers in the few localities where it had previously been located and a slow but gradual extension of its range followed. It moved back up the Tees valley, George Bolam recorded it near Eggleston in September 1913, and it spread from larger woods in Weardale up and down that valley. In the late 1920s it had crossed into the Derwent and established itself at Hamsterley Mill, Muggleswick, Chopwell, where this species was first heard in 1927, Gibside and the Ravensworth Estate at roughly the same time. The species had spread between several of the county's river valleys by 1927 (Temperley 1951) and by the mid-1930s was generally distributed. According to Brown & Grice (2005), the main increase occurred after 1940, and by 1945 birds were breeding in most of the deciduous woodlands in Durham from the coast to the fringes of the western moorlands.

This same process of spread and establishment was said to have occurred in neighbouring Northumberland, alongside a similar northward movement of Great Spotted Woodpeckers *Dendrocopos major* over the same period (Kerr 2001).

Temperley (1951) had speculated that the widespread planting of conifers may have led to an increase in the number and availability of wood ants *Formica rufa*, a major source of food for Green Woodpecker. Ants can be abundant in some conifer plantations such as Hamsterley Forest and Chopwell Woods, where today Green Woodpecker is common. It does seem likely that the British afforestation programme, post the First World War, facilitated the expansion of the Green Woodpecker in the north east of England in the first third of the 20th century (Temperley & Blezzard 1951), although the species does tend to prefer mature, broad-leaved trees for nesting (Glue & Boswell 1994).

In 1967, Parslow noted that the Green Woodpecker had increased and spread quite considerably in Durham and Northumberland in the east, and in Lancashire, Westmorland and Cumberland in the west. The species further expanded its British breeding range during the 20th century, colonising parts of southern Scotland from which it was previously absent (Parkin & Knox 2010). It is estimated that there was an overall doubling of the national population in the forty-year period between 1964 and 2004 (Parkin & Knox 2010). Nonetheless it remained a relatively uncommon species in Durham through the late 1960s and early 1970s, although there was an increase in the breeding population at the outset of this decade, and a range extension was illustrated by, in 1973, two reports from the county's north eastern coastal areas, the first there for several years. More concrete evidence of this trend came with a bird in Castle Eden Dene in 1974, the first there since about 1960. Through the second half of that decade birds were mainly confined to western valley woodlands, with sightings regularly noted in the Bedburn and Hamsterley area and just a few isolated easterly records in Barnes Park, Sunderland in February 1975, at Washington and Leam Lane, Gateshead.

Recent breeding status

The Green Woodpecker prefers mature or even ancient woodlands with plenty of open ground on which it can forage (Gibbons *et al.* 1993). It is now widespread across the county, although it remains relatively localised. During the late 1980s and early 1990s, *'Birds in Durham'*, (Armstrong 1988-1993) routinely listed reports of this species from around 70 'breeding' sites in the county on an annual basis. In 2007, records came from over 80 such sites and, with this being a relatively sedentary species, many of these will be in the vicinity of breeding territories, regardless of the time of the year.

The Atlas (Westerberg & Bowey 2000), mapped the main concentrations of breeding Green Woodpeckers along the Derwent valley and in the woodlands around Durham City, although it was felt that this pattern partly resulted from better observer coverage in those well-studied areas. It seems probable that whilst the distribution, as shown in the *Atlas*, was illustrative of the overall distributional picture, it may not have been comprehensive. The bulk of the county's population is concentrated in the west, particularly in the northern section covering the well-wooded valleys of the Derwent and mid- to upper Weardale. The lower Derwent Valley was a stronghold for the species for much of the 20th century and it remains so today.

In 2008, nearly 300 records came from more than 80 localities, many of these from the central lowland or eastern portion of the county. Further west, birds penetrate up the valleys and into the well-wooded upland fringes of all of the county's main valley systems, including at locations such as Blackton Reservoir, around Derwent Reservoir and Snaisgill right up to moorland edges at 300m above sea level. Some of the most the most productive areas of the county for this species are the well-wooded upland fringes, for example, the Hamsterley Forest area, and in the more lowland and central area, mature parkland and open woodland, such as around Bishop Auckland and Durham City. Breeding birds can be found right down to the coast in the denes at Blackhall Rocks, Crimdon Dene, Castle Eden Dene, Hawthorn Hive and Warren Gill.

The *Atlas* showed only 64 occupied tetrads during the survey period (Westerberg & Bowey 2000), and considering the sometimes elusive nature of the bird, it is possible that a significant percentage of genuinely occupied sites at that time were not located. In 2009, the species was recorded from fifty-two, one-kilometre grid squares around the county during April, May and June. Breeding densities are highest in areas of woodlands where populations of ants are high, or on more open habitats, for example nine birds were noted around Waldridge Fell in April 2001.

Drumming Green Woodpeckers, unlike Great Spotted Woodpecker *Dendrocopos major*, are seldom noted; however its loud springtime 'yaffling' makes it easily located and April, the peak month for calling, produces more records than in any other month of the year. Yaffling by territorial birds is usually noted from late February onwards, but it has been occasionally noted as early as January, especially in fine weather, despite the species' reputation as the 'rain bird'. This folk-lore name, based on a belief that it is noisiest before rain, has been long established, the old Roman name was *Pluviae aves*. In 2007, nearly half of all observations came between March and May, the period of nesting activity, when birds are calling; birds clearly become much quieter and elusive at other times of the year. Despite being a large and obvious bird, it can be secretive when breeding, and the main proof of breeding usually comes from the observation of juveniles (Glue & Boswell 1994) in summer and early autumn.

One of the more unusual local breeding reports of the species came from Paddock Hill Woods in spring 1987, where a pair bred in the same branch of a tree as a pair of Great Spotted Woodpeckers (Bowey *et al.* 1993).

The main areas without birds are those where large stands of woodland are lacking, such as in the north east of the county, South Tyneside in particular, and in the south east around Teesmouth. There are perhaps eight to ten pairs in the north of Cleveland area (Joynt *et al.* 2008). Allowing for possible under-recording, the *Atlas* estimated the Durham population to be between 100 and 200 pairs, and further consideration would suggest that the higher limit is almost certainly closer to the real figure, especially as there appears to be a continued spread and consolidation in a number of areas of Durham into the first decade of the 21st century. During 2002 and 2003, there was confirmation of local range expansion, with first observations of the species coming from Middle Barmston Farm and Cowpen Bewley by one observer who has visited both sites since the 1950s.

The species' taste for ants occasionally brings it into some interesting feeding locations. In the well populated Derwent valley it has been seen taking ants from fence posts, where the insects were climbing prior to swarming and, most intriguingly, on a regular basis between Winlaton Mill and Thornley Woods, it has been observed feeding on the ants from nests that are situated behind the kerbstones on the south-facing side of A694 road. The birds stand on the tarmac of the road, or occasionally on the kerbstones themselves, just 30-45cm from the wheels of the traffic as they feed.

Recent non-breeding status
The species' winter distribution in the county effectively mirrors its breeding distribution, though there may be some local movements out of core woodlands sites, to forage more widely at such times. The species is a lot less common than Great Spotted Woodpecker and it is not usually found in the suburban, or even urban, areas into which that much more common species sometimes penetrates, particularly in winter. Over two-thirds of all records tend to fall in the period between January and late May. The sometimes reported easterly bias in the distribution of this species in Durham, from the coastal denes, is perhaps more of an artefact of the distribution of observers rather than a true reflection of the species' actual distribution in the county.

Sightings away from the species' frequented locations usually occur in August, when presumably dispersing juveniles, wander and are responsible for such observations, such as those at Burnside, Houghton-le-Spring, on 9 August 2008. A juvenile found dead at Hartlepool Civic Centre on 31 July 2009 was clearly unusual but indicative of this dispersal trend. There is a most unusual mid-winter record of the species from Hargreaves Quarry on the North Tees Marshes on 11 December 2005. The closest regularly occupied sites to this observation, in the south east of Durham, are probably the coastal denes to the north of Crimdon or Urlay Nook and Stillington Forest Park to the west. One on Cleadon Hills on 19 March 1984, even with the passage of time, remains a most unusual record.

Distribution & movements
This species breeds over much of Europe, from central Scandinavia to the Mediterranean coastal areas and the northern parts of Africa, east to the Caucasus and Caspian Sea. Although in some areas it is more locally distributed. Generally it is non-migratory across its range (Parkin & Knox 2010).

Great Spotted Woodpecker
Dendrocopos major

A common and widely distributed resident. A passage migrant and winter visitor in uncertain numbers.

Historical review
This is England's and the Palaearctic's most widely distributed and numerous woodpecker species (Brown & Grice 2005) and, today, it is quite easily the most common of the three regularly recorded in County Durham. It is present in almost all suitable woodlands of the county and is increasingly recorded in well-wooded suburban areas and even gardens.

Amazingly, given its status today, the *'great spotted'* became extinct as a breeding bird in northern England, during the early years of the 19th century. Bolam (1912) recorded that the species suffered a dramatic decline in Northumberland during the first half of the 19th century (Kerr, 2001). Its fortunes turned around quite dramatically in the second half of the century. The spread into and re-colonisation of the north east appeared to have begun round 1870 and it had reached the Scottish borders around Duns Castle by 1887 (Holloway 1996). These increases in northern England continued until the 1950s (Holloway, 1996).

Temperley (1951) said that it was rare and local in the early part of the 19th century and noted that the species had not been included in some of the earlier catalogues of species for the county, for example that of Hogg (1824); who had never seen one in north Cleveland, indicating that it was genuinely scarce in that area. Edward Backhouse (1834) thought that it was less plentiful than Green Woodpecker *Picus viridis*, which was then not uncommon. Hogg (1840) stated that this was *"an exceedingly rare bird in the county"*. In the national context, the

Great Spotted Woodpecker seems to have almost disappeared from most areas of the northern England counties by the early part of the 19th century (Brown & Grice 2005). Sclater, at Castle Eden, saw one there on 10 December 1876, which was only the third time that he had noted it in 16 years of observation and considering the time of year it is tempting to think that it could have been a continental immigrant.

By Hancock's time of writing (1874), it appeared to be somewhat more common and he thought it more common than the Green Woodpecker. From specimens held in the Hancock Museum, Temperley (1951) concluded that it had bred at Maiden Castle, Durham City in 1863 and at Cleadon by 1880, for a juvenile was obtained there. By 1905, Tristram reported that it was *"occasionally met with at all times of the year in the wooded parts of the County and breeds regularly in some localities"*.   At the end of the 19th century, Robson still considered this species to be a rare resident in the Derwent valley, although it was breeding regularly at Gibside in the lower valley by this time (Robson, 1896).   Temperley (1951) considered that this was the beginning of the spread of the British race into County Durham.   Like the Green Woodpecker, they spread significantly in the early 20th century and by 1921 had reached both Chopwell Wood and the Ravensworth Estate, in the Team valley, from its 'beach-head' in Gibside (Bowey *et al.* 1993).   George Temperley first saw it at Ravensworth, one of his main bird watching locations, in 1921.   By the early 1950s, it was to be found in all of the county's principal woodlands such that in the autumn of that year, Temperley wrote: "*The same period has also witnessed a most decided increase and an extension of range of the Great Spotted Woodpecker. It never became as scarce as the Green Woodpecker and its recovery began rather earlier and was more gradual. In numbers it has kept ahead of the Green Woodpecker and is now a very common woodland species.*"

In 1950, a female was reported visiting a 'tit-bell' in a garden in South Shields, during the early winter. This bird was joined by a male after April (Temperley 1951), and these would appear to be the first records of birds using garden feeders in the county, a habitat that was to become so prevalent in subsequent decades. Little substantive information was published on this species' status in the county during the 1950s and 1960s, though research in the south east indicates that birds were scarce there in the past but also probably much under-recorded (Joynt *et al.* 2008).

During the 1970s, this species did not routinely penetrate the western valley woodlands, so one at Dent Bank near Middleton-in-Teesdale in September 1973 was considered a notable record at the time.   Considering that the predation of nestboxes would become a regularly reported practice of this species through the 1980s and 1990s, the 1979 report of a bird predating a nest of Tree Sparrows *Passer montanus* was somewhat prescient. A measure of this species' change in status in three to four decades is indicated by the fact that by the late 1980s it was being described as increasingly urban, and was noted at many garden and other feeding stations around the county during the winter months.

Recent breeding status

The *Atlas* indicated that the county's main areas of occupancy, during the survey period 1988-1993, were around densely wooded zones, such as: the coastal denes; Hamsterley Forest; the woodlands around Durham City; and, the contiguous woodland ribbon of the Derwent valley (Westerberg & Bowey 2000). Today, Great Spotted Woodpeckers are widespread across the county, being found in almost all woodland blocks and types, the availability of suitable woodland habitat being the only limiting factor. In lowland Durham, few areas do not hold nesting birds. It is sparsely distributed in the south east of the county, with around 50 to 60 pairs, largely away from the urban centres, in the woodland areas west of Stockton and north to Wynyard (Joynt *et al.* 2008). The only parts of Durham where it is absent are the treeless portions of the western uplands, though tendrils of 'great spotted' breeding populations have now penetrated up the wooded upper dales into some of these areas. Traditionally, the Great Spotted Woodpecker was less often reported from high ground than the Green Woodpecker (Westerberg & Bowey 2000) though this has changed somewhat in recent years. In 2008, the species' westerly boundaries were the Beldon Burn in the north, Langdon Beck in upper Teesdale, Balderhead Reservoir and Gilmonby in the south. These are close to the limits of mature woodland areas along the county's main river systems.

The species prefers mixed or deciduous woodland, where dying trees supply easily excavated nest sites. It lives at high densities in broadleaved woodland where, on average, there is a pair per 20 acres, whilst in pure conifer stands it has been estimated that birds require over 80 acres of woodland to maintain a territory (Bowey *et al.* 1993). There is also a trend for it to use nestboxes in suitable urban fringe areas (Gibbons *et al.* 1993) although this is not a habit noted locally. Both locally and nationally however, it is becoming more evident in urban gardens

(Parslow 1973). Suburban breeding has been occurring in Durham, even in some of the larger conurbations, for around 20 years or even longer. In 2009, such breeding birds were recorded in the Sunderland area, with three pairs in the centrally located Backhouse Park and another pair in Barnes Park, and also in Saltwell Park, Gateshead. Birds have for many generations been found in the woodlands along the riverbanks in the well-wooded heart of Durham City.

Territorial drumming males are usually noted from early February, though it can commence from the second week of January. This sound becomes more familiar in woodland visits through March and into early April, and often continues until late April. In addition to traditional woodland drumming posts, of dead or hollow branches, a variety of other sites have been noted locally as being used when birds proclaim their territories. These have included various metal posts, telegraph poles, the plastic cover of a streetlight, an electricity pylon, a church spire's weather vane and the main electricity supply pole for Hardwick Hall. The first fledged juveniles are usually noted from mid-June, but this can vary from late May to late June, and a rise in reports through late June and July no doubt will include many dispersing young.

Various bodies of data from national surveys gathered by the British Trust for Ornithology indicate that there were periods of population increase through the 1970s that could be ascribed to the results of the Dutch Elm disease epidemic of the time, which created increased availability of standing dead timber (Parkin & Knox 2010) and again through the 1990s. The causal reasons for this second phase of increase are less clear, but there is a suggestion that better winter survival as result of the provision of supplementary food in gardens and milder weather may be factors. Westerberg & Bowey (2000) stated that there had been a 50% increase of reported nesting locations in Durham over the period 1988 to 1993, with the caveat that these figures were heavily dependent upon observer effort, though probably indicative of a genuine trend at the time. It is clear that at the present time the Great Spotted Woodpecker was very much a thriving woodland species in Durham. The *Atlas* estimated the population at delete of between 250 and 750 pairs for the county (Westerberg & Bowey 2000), but it would seem that the true number now lies nearer, or even beyond, the upper limit of that range.

There appear to be few threats to the species locally, other than large-scale deforestation, which has not occurred in the county in recent times and seems unlikely to do so in the near future. Some observers consider the damage that these woodpeckers can do to nestboxes, when predating the eggs and nestlings of Blue and Great Tit *Cyanistes caeruleus* and *Parus major*, intolerable and may even take action to exclude or eradicate the species from their gardens (Westerberg & Bowey 2000).

Recent non-breeding status

During the non-breeding season sightings come from almost all areas and in winter the species ranges widely, often accompanying flocks of titmice and other woodlands birds, and visiting garden feeding stations. In 2007, the Durham Bird Club received 763 records of this species, the spread of records covering the majority of the county; even small stands of trees in the western uplands received the occasional bird. The westerly extremes were upper Teesdale and St John's Chapel, at 350m above sea level.

During the winter periods, birds readily come into urban areas and take advantage of garden feeding stations as well visiting stations at nature reserves such as those at Clara Vale, Cowpen Bewley Woodland Park, Low Barns, the Thornley Woodland Centre and WWT Washington. It has been postulated that the spread of this winter behaviour has resulted in a better overall survival rate and supported the growth in local and national breeding numbers (Marchant *et al.* 1990). A bird was seen feeding on fat on a Bishop Auckland garden feeder every day from 20 December 1973 to 13 May 1974, and then again in December 1974. Meanwhile, the first documentation of bird table feeding in the south east of the county came in November 1974 (Blick 2009). In 1975 birds regularly visited a garden feeder in Hamsterley Mill and from then on this habit became increasingly reported.

An interesting observation from the middle of the 20th century commented on the species' foraging behaviour. *"During the past winter, the oak marble galls produced by gall wasps have been plentiful in Lambton Park. These however have now been thinned out to a very remarkable extent by the attacks of woodpeckers. The birds peck a deep hole in one side of the gall, and through it extract the larva of the wasp. Galls so treated were strewn on the ground under the oaks in all directions"* (*The Vasculum* Vol. XXXIV, 1949).

A melanistic bird, all dark except for a small white shoulder patch, was seen at Shotton Colliery on 27 August 2007 such melanism is rarely reported in woodpeckers (Newsome 2008).

Northern Great Spotted Woodpecker
Dendrocopos major major

The Scandinavian and east European race of Great Spotted Woodpecker *Dendrocopos major major* is a migrant which reaches the east coast of Britain in some years, occasionally in quite good numbers, largely in September, sometimes in October, and remains through the winter. Consequently, locally breeding birds are supplemented by an unknown number of northern breeding birds, and it seems likely that small numbers of these large, subtly darker birds, probably reach our area each winter.

The earliest record of this race in Durham is mentioned by Abel Chapman who observed these "*migrants from Northern Europe*" at irregular intervals in Northumberland and Durham (Chapman, 1889). There was a possible influx of northern birds to the area in the winter of 1849/1850, with more numbers than is usual on Tyneside (Temperley, 1951), though he thought the numbers of his race were almost impossible to determine. He believed, probably quite justifiably, that coastal birds in the autumn were more likely to be of this race, but not exclusively so, given how uncommon local breeding was at that time. An adult female shot at Winlaton Mill on 31 January 1931 (Bowey *et al.* 1993), was confirmed as being of this form. Other confirmed records include a male that hit telegraph cables at Cleadon on 4 February 1931. Three months previously Dr. H.M.S. Blair had observed one in West Park, South Shields, with others reported to him as having been in the neighbourhood. Presumably a year when there was something of an influx of birds that winter.

Occasionally large influxes occur and just such one took place in 1957, when birds were recorded in many coastal areas between Northumberland and Dorset (Brown & Grice 2005), though none were noted in Durham at this time. During the 20th century, notable irruptions of this race took place in 1901, 1929, 1949 and 1962 and also September in 1968 (Parkin & Knox 2010). Birds were noted in Durham in at least some of these years.

One in an allotment garden in the centre of Sunderland on 1 December 1952, might have been a continental immigrant, for at this time the species was much less common than at present and rarely seen away from mature woodlands.

In many instances apparent records of this race have to be qualified within caveats of 'probably referring to the race *major* ', excepting where birds are caught by ringers. An irruption of continental birds in late 1972, meant that there was a number of birds in coastal areas during early 1973, for example four or five birds in the Hartlepool area, three at Sunderland and two at Cleadon, were probably largely of this race. There was a suggestion of a small influx of continental birds in autumn 1974; one at Seaton Carew churchyard on 10 October and one at Hartlepool on 11th and 12th were probably of this form. Likewise in 1977, birds at Marsden on 26 September and Crimdon on 9 October may have been continental birds, as was one at Hartlepool on 8th to 10 October 1982, and possible continental immigrants at Mere Knoll's Cemetery in early September and 3 October 1987. The following year, possible migrants were reported at Grangetown cemetery and Mere Knoll's cemetery and at Whitburn in October 1988 and at Whitburn, the first in that area for ten years. Another at Boldon Flats on 7 November may have been a continental bird.

In the late 1980s and 1990s, there were two definite records of this race, when two birds were caught and ringed in western Gateshead. A bird spent much of the winter of 1989/1990 in the area around the Thornley Woodlands Centre, its identity being confirmed by repeat capture and ringing over this period, on more than one occasion alongside local breeding birds. Another bird was at the Clara Vale feeding station in November 1991, again this bird was caught and ringed and its racial identity confirmed by biometric measurement (Bowey *et al.* 1993).

Around the south east of the county, there have been several records of such birds at North Gare and Hartlepool Headland. These include three on the coast in September 1968 during an influx to England (when 17 were at Spurn) (Brown & Grice 2005); six on the coast between 18 September and 3 October 1972 (one of which was found dead at Teesmouth Field Centre and described as *Dendrocopos m. major*); three between 30 August and 12 October 1974; seven or eight in the latter half of October 1990; and eight between 5th and 27 September 2003 (Blick 2009).

It should be borne in mind that not all Great Spotted Woodpeckers observed near the coast are necessarily from northern Europe. As birds breed in a number of coastal situations and young birds may disperse along the coast from their natal sites at Castle Eden, Crimdon and Hawthorn Denes, then genuine autumn immigration may be almost impossible to prove. One seen flying in-off the sea at Whitburn on 6 April 2007 however was a good contender for such, whilst another was at Hartlepool Headland on 24 October 2007. In 2008, possible coastal migrants were reported at Hartlepool on 13th and Greenabella Marsh on 28 September. Then in 2009 singles were at Marsden Lime Kilns, Roker Park, Marsden Quarry, and Whitburn between mid-September and mid-October, and one flew over the Long Drag late in November.

Distribution & movements

This is the most widespread species of woodpecker in the British Isles (Parkin & Knox 2010). It is polytypic and various taxa are found across the Palaearctic from the Canary Islands, through North Africa, around the Mediterranean basin, and north through Europe to Scandinavia east to Siberia. Locally restricted populations occur in North Africa and the Middle East. Though northern populations are highly dispersive or irruptive, wintering to the south within the overall breeding range, many populations of birds, including the British stock, are essentially sedentary (Parkin & Knox 2010).

Autumn coastal birds may be continental migrants although genuine immigration is difficult to prove as local birds now occur right down to the mouths' of the county's coastal denes. The sedentary nature of local birds is confirmed by the fact that the few ringed locally are rarely recovered excepting at or near their ringing stations. The mean distance at which ringed British birds are found from their ringing site is just two kilometres (Wernham *et al.* 2002).

Lesser Spotted Woodpecker
Dendrocopos minor

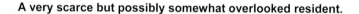

A very scarce but possibly somewhat overlooked resident.

Historical review

The northern limit of regular breeding for this unobtrusive species in the 19th century was considered to be Yorkshire, almost as far north as the Durham border (Holloway 1996). By the early 21st century this had advanced northwards to Northumberland and south Cumbria (Brown & Grice 2005). This is the smallest, and least abundant, of the British woodpeckers, and a small population would appear to have been present in County Durham since the 19th century, though due to its secretive nature and preferred habitat, they are rarely recorded. Temperley (1951) recorded it as a "*very rare visitor*" to Durham and he documented one record of breeding in the county.

The first known record of the species in Durham relates to one that was said to have been "*shot 'some years ago' at Stockton Bridge*", according to Hogg (1845). Robson recorded one that had been obtained in Gibside in the Derwent valley but he gave no date for this occurrence though from his text and publication date, it must have been around 1888 (Robson 1896). From the 19th century there were anecdotal reports of birds in the Gibside Estate and at Bradley Park, near Crawcrook, in the Tyne valley (Bowey *et al.* 1993). Reports of them being seen 'frequently' in the Gibside Estate were considered by Temperley (1951) to be less than reliable but there is no doubting the specimen obtained in Gibside around 1888.

A bird was noted near Sockburn, to the south of Darlington, by M. G. Robinson in the late 1920s and Temperley's (1951) only documented breeding record was of a pair nesting in that part of the county, in West Cemetery, Darlington, in 1934. In this instance, the corpse of a young bird was found. Single birds were seen in Darlington on 12 March and 16 November 1948. A pair was in Whitworth Park near Spennymoor on 25 March 1950. In March 1952, one was at the same location on the 27th, with two there on 30th and the male was seen again on April 9th, but breeding, though suspected, was not proven.

In 1953, there was a most unusual record of a bird in a garden at Cleadon on 4 April, this remains one of the few records from the north east of the county. On 10 September 1954, a bird was watched down to a distance of some five yards in a suburban garden in Broom Lane, Whickham (Bowey *et al.* 1993). During the 1950s, birds were noted regularly in the Wylam area in spring and drumming was heard there in April 1960. Many years later, in 1997, a bird was drumming in woodland near Crawcrook on 7 April.

In 1959, a pair excavated a nest hole at a site in south Durham on April 25th but there was no confirmation of breeding success. During the 1960s sightings were few, with one at a site in Durham in 1961 and one at Winston, on the banks of the Tees, in March 1965.

Birds have a habit of being recorded in the same area sometimes many years apart, indicating that small numbers may persist without being regularly located. For instance, birds were noted near Hurworth Burn Reservoir on 27 November 1960, again on 24 December 1966 and once more in January 1979. A similar pattern occurred in the area around Houghall on the outskirts of Durham City, with records there in April 1976, April 1981 and at nearby Great High Wood in January and March of 1987.

This situation was mirrored in the south of the county where, in 1973, the Darlington area was reported as becoming a regular haunt of the species. One was noted there in the grounds of a hotel on 21 January 1973 then one was noted in a town garden in mid-July. This started a long run of almost annual records in this area, which lasted until 1983. Most of these occurred in winter and autumn but not exclusively so. In 1979, one was calling at Darlington in a garden in early May and it or another was there again in late August. Birds were present in Darlington South Park in January and April 1981, not that far from the area where birds bred in 1934. Also, just to the south of Durham, around the border of Durham and North Yorkshire, from 1975 to the mid-1980s, birds were seen with some regularity in the Yarm to Kirklevington area of the Tees valley. In early May 1976, a male excavated a nest-hole there, and breeding was confirmed in 1983 and 1985 (Blick 2009).

In the north of the county, birds were present in woodland in the lower Derwent valley through the 1970s, most being regularly noted at Hamsterley Mill, initially from February 1975, in an observer's garden, and then annually in that area until 1979; with a pair present in 1977, and on a number of occasions, drumming was heard. Around western Gateshead, on 19 November 1978, a bird was seen in a garden at Stella, whilst other autumn records around that area included, a male at Ryton Willows on 4 September 1986 and in the Derwent valley, about one kilometre from Hamsterley Mill, a female was at Chopwell Woods on 28 November 1987.

Through the rest of the 1970s, a series of apparently haphazard, largely spring, records came from scattered sites in the county's main valley woodland, such as Rosa Shaftoe reserve near Spennymoor, Shincliffe village and Sadberge.

Recent status

It has been shown that Lesser Spotted Woodpeckers, forage in a relatively small area around the nest site throughout the year, perhaps utilising as little as 50ha of woodland. The presence of abundant decayed wood is more important than tree species (Cramp et al. 1985) but some preference for river valleys with alders *Alnus glutinosa* has been recorded (Gibbons et al. 1993). The preferred habitat of this species, which includes open forest, parks, orchards and gardens with large mature trees (Winkler et al. 1990), is found in the county but not on a scale that is likely to make this species a common one in the area. Lesser Spotted Woodpeckers are exclusively associated with broadleaves (Gibbons et al. 1993) and avoid conifers and dense stands of timber (Cramp et al. 1985). It seems likely that birds do, and have bred in small numbers for many years.

In the early and mid-1980s, there was something of an upsurge of observed breeding activity. For example, a pair was on the outskirts of Durham City in early spring 1981. In 1983, for the third year running a locality near Barnard Castle held birds in spring, two drumming males had been seen at this site in 1981 and in 1983 a female was present but no males were noted. A pair was at Beamish Hall in the second week of July 1984. Spring birds were noted at Castle Eden Dene in both 1985 and 1986, and in 1986 a pair excavated a nest hole near Lartington, but this was later deserted, it was thought, as a result of pressure from Great Spotted Woodpeckers, *D. major*. Other sightings later in the decade came from: Finchale, Hardwick Hall, High Coniscliffe and Witton-le-Wear. In 1988 there were April and May sightings at Clockburn Dene in the Derwent Walk Country Park, and again in December; and at McNeil Bottoms and Darlington in September. Around this time successful breeding was reported in both 1983 and 1985 in the Yarm to Kirklevington area of the Tees valley (Blick 2009), though this was on the south side of the Tees and therefore just outside of the county area.

The *Atlas* map (Westerberg & Bowey 2000) showed that the species was found largely along the central sections of the county's main river valleys, where mature broad-leaved woodland is most common. For the *Atlas*, a full review of all Gateshead's spring-time records, over the long-term, revealed that it seems likely to have bred, at least occasionally, in the extensive woodlands located in the western half of Gateshead (Bowey et al. 1993). Birds were probably present in suitable breeding habitat, during the spring, at three different sites in Gateshead during the late 1970s, the 1980s and early 1990s. There were records of a male at Ryton Willows on 7 March 1990 and single birds in the Derwent Walk Country Park during March 1992. The distribution illustrated in the *Atlas*, was derived from all 'breeding season' records collected during the survey period, 1988-1993. These amounted to just ten registrations in total, illustrating its rarity and its elusive nature (Winkler *et al.* l995). In one instance, in summer 1989, birds were watched at Pont Burn Woods in the lower Derwent valley carrying food for young.

Subsequent to the publication of the *Atlas*, the species has been proven to have bred in a number of the county's woodlands, two of these being in the Tees valley as well as in the lower Derwent valley. Winter visitors, which have delighted observers at winter feeding stations, have also indicated that breeding is likely to have taken place in locations close to these facilities. The feeding stations at Clara Vale, near Wylam, and the Thornley Woodlands Centre are close to areas that have produced regular spring records of the species over a number of decades.

The small size, secretive nature and the bird's habit of feeding on small, vertical twigs high in the canopy makes this a difficult species to find and observe. It is on the edge of its European range in north east England, a range within which it is found at low densities across Europe (Cramp 1985), and clearly, it is not a numerous species in the county; because of these factors, the real population of Durham is difficult to determine. The *Atlas* worked on the assumption that all areas where the species had been found during the survey period probably held a breeding territory (Westerberg & Bowey 2000). Using these criteria, it was stated that at least five pairs were probably nesting in the county. Taking into account winter records over the survey period for the *Atlas*, as the species is sedentary (Winkler *et al.* 1995), then the county population might 'rise' to "*at least seven pairs*" (Westerberg & Bowey 2000); it seems highly likely that other pairs are missed and that the county's population is higher than previously postulated.

Gibbons *et al.* (1993) stated that there had been a small expansion of this species in the north east of England since the first breeding Atlas (Sharrock 1976) but more recently there has been a reduction of numbers in Britain. The initial increase in the north east may have been due to better knowledge from an increased number of observers, whilst the national decline is thought to result from the loss of standing dead timber, which had been provided by the trees killed by Dutch elm disease and had facilitated the earlier rise in numbers after 1970 (Brown & Grice 2005). The Lesser Spotted Woodpecker population grew as more dead timber became available with the arrival of this disease. However, as more of the dead elms were felled for infection control reasons or collapsed naturally, the feeding and nest sites they had provided were lost, with a consequent fall in population numbers. During the period of the *Atlas* survey in Durham, there were still good numbers of standing dead elms, which may have accounted, in part, for the perceived local, small increase. Since this time they appear to have disappeared from some areas in the north of the county and nationally, the gains in range and numbers of the 1970s have now been reversed (Brown & Grice 2005). It is now apparent that the national population has declined dramatically with a drop of 76% between 1970 and 2008 with the number of breeding pairs within the range 1,400-2,900.

In the last decade of the 20th century birds were noted at four sites in spring 1992, including one in Durham City and one drumming in a Darlington. In 1993 an autumn bird was roving with the local tit flock in Bishop's Park, Bishop Auckland. The Durham City area produced another two in 1994 and another autumn single was in Darlington. Hamsterley Forest held one on 9 March 1996 and another male was in Durham City in May. Croxdale Hall recorded one with a winter tit flock in January 1997 as was another in Weardale in March.

About the same period, there was a notable increase in the breeding season presence of birds at a number of sites in Teesdale, principal amongst these were Deepdale Woods near Barnard Castle and at Cotherstone, with birds regularly present at both sites from 1990 to 2005. Birds were also seen occasionally over the last two decades at Wynyard Park (Blick 2009) in the south east of the county.

The first decade of the 21st century has seen a marked increase in the number of records for this species and in 2002 and 2003, there were 42 records of the species from eight sites over the two-year period, probably amounting to the greatest registered activity of this species in the county's history. In 2004, reports came from Durham City, the Derwent Gorge, Hetton Bogs, Hetton House Wood, Merrybent and Morton Wood near Fence Houses. In 2005, there were five records through the year with sightings at Cotherstone, Clara Vale and Thornley Woods. Singles at Winston and Tees Bank Woods near Barnard Castle in 2007 hinted at what would appear to be a regular small population on the banks of the River Tees. The species was recorded at 21 different sites between 2004 and 2009, with a cluster of records concentrated in the central section of the county around the mid-Wear Valley, with sites such as Low Barns attracting birds. Birds were regularly seen in the upper Tees valley, as far downstream as

Winston, where records date back as far as 1965, and other birds were present in both the lower and upper Derwent valley, such as at Rabbit Bank Woods near Consett. One in the River Team valley, near Urpeth, on 4 February 2008, was not that far from the location where the pair was noted over two decades earlier at Beamish Hall. This pattern in effect, describes what is a low-level, but widespread breeding presence of the species, that has never been adequately or properly documented but which must be centred on the best woodland habitats in the county. There is little doubt that this secretive species is not as scarce or as restricted in Durham as documented records or perception suggests. Association with winter tit flocks noted at Bishop's Park, Croxdale Hall, Low Barns and Pont Burn Woods suggests that scrutiny of such roving bands would reveal more of this species.

Distribution & movements

This species breeds from Iberia in the south west of Europe, north to the British Isles and Scandinavia and east across to China and north Japan. Some of the northern populations are winter dispersive, with some suggestion of irruptive behaviour, that is not well documented, but the British breeding stock appears to be largely sedentary (Parkin & Knox 2010) and its winter range effectively mirrors its breeding range (Brown & Grice 2005).

Despite the fact that it occurs in the far north of Europe, in Durham this species has always been at the northern edge of its distribution in the British Isles and it therefore remains very scarce (Gibbons *et al.* 1993). There is just a single known ringing record, one that was caught in the winter of 2001 at Low Barns by the Durham Dales Ringing Group.

Sympathetic management of woodland is likely to be a key factor in ensuring a breeding population of birds in the county in the long term. Maintaining mature woodland with dead limbs and standing dead timber is of prime importance. As this small woodpecker is on the northern edge of its range, without a concerted national increase in numbers, which is currently improbable, its tenuous foothold in the county is likely to remain just that and reported declines in the national population perhaps bode ill for the future in Durham.

Red-eyed Vireo
Vireo olivaceus

A very rare vagrant from North America: four records.

The first for County Durham was found by Brian Bates in Mere Knolls Cemetery, Seaburn on 27 October 1990 (Bates 1991). Always high in the canopy, the bird was often elusive. It remained until 29 October by which time it had been seen by hundreds of observers (*British Birds* 85: 548).

All records:

1990	Mere Knolls Cemetery, Seaburn, 27th to 29 October
1991	North Gare, Seaton Carew, 12th to 13 October (*British Birds* 86: 525, *British Birds* 88: 547)
2004	Marsden Quarry, 27 October (*British Birds* 98: 686)
2010	North Gare, Seaton Carew, 11 October, trapped (*British Birds* 104: 592)

With a total of four records each, County Durham and Suffolk–both hold almost one-third of reports of this species from the east coast of the British Isles, with a total of just five recorded elsewhere. Remarkably two birds have been found in the coastal buckthorn at North Gare, within 100 yards of each other, although nine years apart. The second of these in 2010 was trapped and ringed, one of only 21 ringed in Britain (Clark *et al.* 2011). More remarkable perhaps is the fact that this bird was found on the same date as the county's first Common Nighthawk *Chordeiles minor*, some 15km to the north.

The paucity of records further north in Britain and Scandinavia is somewhat unexpected and it may be that some of these east coast birds are ship assisted. The fact that all four county records have been found during falls of migrants supports a natural occurrence and is more probably related to increased observer coverage at such times.

Distribution & movements

Red-eyed Vireo breeds across much of North and South America. The North American population is strongly migratory and winters mainly in Cuba and central South America. It is the commonest American passerine recorded in Britain with 119 records accepted by BBRC to the end of 2010, chiefly in late September and October; following westerly gales. Over two thirds of the records have come from south-western England, with just twelve from the east coast of England, all since 1988.

Golden Oriole (Eurasian Golden Oriole)
Oriolus oriolus

A scarce migrant from Europe: 27 records involving 29 birds, recorded from April to June, though most in May and June.

Historical Review

Golden Oriole, Hebburn, prior to 1831
(K. Bowey, courtesy NHSN)

The first record of Golden Oriole in County Durham concerned a female shot near Hebburn some time prior to 1831. Labelled as "*Shot at Hebburn near Newcastle about the year 1831*", this specimen was originally part of Hancock's personal collection (Howse 1899) it is now housed in the NHSN collection.

Almost a century later, Mrs A. M. Golson recorded the second, a male by the roadside between Greatham and Newton Bewley on 17 May 1929 (*The Vasculum* Vol. XVI 1930), though the date was incorrectly given as the 4 May. This species remained rare throughout the next quarter of the 20th century; there were just two further records by the close of the 1960s, with none at all noted until a 'yellow' male was discovered singing in 'Bluebell Wood' near Shotton Colliery on 23 June 1956, followed by one at Hamsterley Forest between 1st and 5 June 1959.

Recent Status

This scarce species has bred in quite small, but now in ever decreasing numbers in the south east of England (Brown & Grice 2005). However, in the latter part of the 20th century it was recorded with slightly greater regularity as a passage migrant from sites in the northern half of England.

This species is largely a scarce passage migrant in Durham, with records normally being confined to the coastal strip. In parallel with the national picture (Frazer & Ryan 1994), sightings in Durham, over the last 30 years or so, have become a little more regular with most records occurring in late spring. At least 26 Golden Orioles have occurred in County Durham since 1970, all of which have been in spring and confined to the period between 16 April and 24 June, with the majority of records in late May and mostly along the coastal strip.

Just two birds were recorded during the 1970s; singing males at WWT Washington on 30th and 31 May 1978 and Castle Eden Dene on 24 June 1978. An increase in sightings came during the 1980s, as seven birds were recorded in five different years, including three birds in 1989. The first coastal migrants appeared in this period, setting the trend for the years ahead with a female at Whitburn on 17 May 1983, a tide line corpse of a male at North Gare on 19 May 1984, a female on the Long Drag, Teesmouth on 27 May 1985 and another female at Cherry Knowle Dene, Ryhope on 25 May 1989. Away from the coast, a male which died flying into a window at Langley Moor, Durham on 16 April 1987 is the earliest record of this species in the county.

After no records for four years, the 1990s proved to be the best ever decade for this species. The female at WWT Washington on 26 May 1993, the only inland occurrence of the decade, was obviously a migrant and quickly moved on. Eleven coastal migrants were recorded between 1992 and 1999, with peaks of three birds in both 1994 and 1995. Two males were reported singing on the coast briefly during the last weekend of June 1992; a full adult male was on the Long Drag, Teesmouth on 30 April 1994, and a singing first-summer male was in Hargreaves

Quarry from 14th until 16 May 1999. Elsewhere, two different males were in Mere Knolls Cemetery, Seaburn on 28th and 30 May 1994 and both adult and immature males were seen on the Long Drag, Teesmouth during the evening of 29 May 1995. Around three weeks later, a female type bird was watched flying north over Greatham Creek on 19 June, highlighting the potential of this area to attract the species. This relatively small area, just inland of the coast has recorded five Golden Orioles, more than anywhere else in the county. In 1997, an anecdotal report was received of one calling in woodland in upper Teesdale on 30 May.

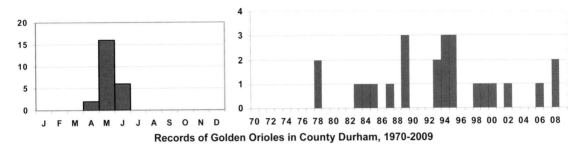

Records of Golden Orioles in County Durham, 1970-2009

Just six birds have been recorded since 2000, though in contrast to the previous decade just two of these were coastal migrants, with single first-summer males at Moor Farm, Whitburn from 30th to 31 May 2000, and on nearby Cleadon Hill on 25 May 2006. Elsewhere, there was a singing male at Shibdon Pond on 30 May 2001, the first fully authenticated record for the Gateshead, which was followed by a second in 2002, when another singing male was at Winlaton Mill on 3 June before flying off south east. An adult male was found in a distressed state at Rockwell Nature Reserve, Darlington on 3 May 2008, this was taken to an animal welfare centre, where it made a full recovery allowing it to be released the following day at Low Barns, Witton-le-Wear. Gateshead's third singing male of the decade was at Birtley Sewage Treatment Works on 28 May 2008. Extreme arrival dates during the period are at Langley Moor on 16 April 1987 and Castle Eden Dene on 24 June 1978.

There are several reference to this species in the annual county bird reports that are not recognised here; an unconfirmed report of a pair near Barnard Castle from May to July 1978 (*Birds in Durham* 1978); two males reported singing on the coast during the last weekend of June 1992 (*Birds in Durham* 1992); and a male reported 'calling' in Upper Teesdale on 30 May 1997 (*Birds in Durham* 1997). The evidence presented in these cases is anecdotal and as such unacceptable, though particularly the latter two reports may have been genuine. In addition to these records, a male was heard singing at Green Lane Plantation on the south eastern outskirts of Darlington on at least 24 May 1988 (Bell pers comm.) and another was reported from a garden in Whickham on 8 July 1989 (Bowey 1993). Unfortunately, neither of these birds was ever submitted for consideration.

Recent breeding status

On 3 June 1989, during survey work for the *Atlas*, a singing male was detected in an area of mixed woodland at Pont Burn Woods in the Derwent valley, where a few days later it was seen to be accompanied by a female. The pair remained throughout the rest of June but was not seen thereafter, despite several subsequent visits to the site (Westerberg & Bowey 2000). Unfortunately successful breeding was never proven, the consensus of opinion being that it may have been attempted but failed at an early stage. In Britain, breeding of this species has largely been confined to the East Anglian fen basin (Gibbons *et al.* 1993) where it has nested almost exclusively in hybrid poplar *Populus* spp. plantations, though numbers here have declined dramatically in recent decades (Parkin & Knox 2010).

Distribution & movements

Golden Oriole breeds across much of Europe and western and central Asia, though is absent as a breeding species from most of Scandinavia except southern Finland with a handful of pairs in Sweden and in Britain each year, the latter chiefly in East Anglia. It winters in sub-Saharan Africa southwards from Cameroon and Kenya to South Africa. A typical spring overshoot, most frequently in May and June, around 100 migrant birds per annum are recorded in Britain (Fraser & Rogers 2006), the majority of which are from the south coast of England, in particular the Isles of Scilly which typically provides the most records each spring.

Isabelline Shrike
Lanius isabellinus

A very rare vagrant from Asia: two records; two subspecies have occurred.

Daurian Shrike
Lanius isabellinus isabellinus

The sole record for County Durham of this form is of the first-winter bird found by Paul Cook and Andrew Hetherington in mature hawthorns at the rear of Marsden Quarry on 21 October 1999. It quickly disappeared in strong wind and rain and was seen just briefly on 22nd before seemingly flying north. Fortunately the same bird was re-located in a small hawthorn bush close to the main A183 coast road at Marsden Grotto on 26 October where remained until 9 November, often showing at very close range. It was seen to catch a number of wasps and bees (*British Birds* 93: 560).

This is thought to be the commonest form of the Isabelline Shrike complex recorded in Britain with most records relating to first winter birds in October and November. The bird at Marsden was a typically pale first-winter individual and was presumed to be of this form. It occurred on a typical date and was one of three recorded in Britain that autumn.

Turkestan Shrike
Lanius isabellinus phoenicuroides

The only record of this form for County Durham is of a male found by Paul Hindess at Whitburn during the early morning of 14 May 2006. This bird was initially seen from the sea-watching observatory sat atop one of the warning signs for the rifle range, having presumably just arrived in off the sea. It quickly relocated to the scrub behind the hide before moving northwards into Whitburn Coastal Park, where it showed well around the vegetated mounds for around an hour before being last seen close to the lighthouse at Souter Point. From here it was watched to fly inland towards the Lime Kilns at Marsden, but was not relocated subsequently (*British Birds* 100: 744). The six 'Isabelline Shrikes' so far recorded in Britain in May and June were all identified as belonging to this form.

Distribution & movements

The taxonomy of the Isabelline Shrike complex and is currently thought to consist of four races, three of which are sometimes considered to represent distinct species. Two of these have been recorded as vagrants to Britain. Isabelline Shrike is an annual vagrant to Britain, chiefly in late autumn, and there had been 80 recorded in Britain by the end of 2010, most of which are thought to relate to 'Daurian Shrike'. As would be expected, the majority of records are from eastern and southern coasts of England, although there are 21 records for Scotland, over two-thirds of which are from the Northern Isles.

Red-backed Shrike
Lanius collurio

A scarce migrant from Europe and a very rare breeder prior to 1885.

Historical Review

Through the second half of the 19th century and the first half of the 20th, there was a steady decline in the breeding status of this species throughout England, due in part to habitat destruction, eventually leading to its extinction as breeding species in England the late 1980s. This species probably bred sparsely in a number of locations in Durham prior to and during the 19th century. It is evident that it was declining rapidly during the late 19th century, but was probably not uncommon in Northumberland and Durham to around 1850 (Holloway 1996).

The earliest specific reference to this species in County Durham is by Selby (1831) who reported that a pair was "*killed by Lord Ravensworth's gamekeeper at Ravensworth Castle, in the County of Durham, where they had their nest*", though unfortunately he gave no date for this occurrence. Writing in 1840 Hutchinson gives the first date record for County Durham stating "*The bird which Bewick figured and described as the Woodchat Shrike was a female Red-backed Shrike shot near Ushaw by Mr Wm. Proctor sub-curator of the University Museum (Durham) on 10th September, 1824*". He added that this was to his knowledge the only capture of this species in County Durham, so this record may well pre-date the reference by Selby (1831). The current whereabouts of this specimen is unknown having presumably been lost or destroyed.

Subsequent 19th century records are a little vague, though the species was undoubtedly rare, with Hogg (1845) writing that he had "*never seen this species in the S.E. corner of the County*", whilst just one year later Proctor (1846) described its status as "*an occasional summer visitant*". The only other documented records around this time are both by Joseph Duff of Bishop Auckland who recorded in the *Zoologist* in 1849 that a female had been shot near Bishop Auckland in November 1844, and some two years later (*The Zoologist* 1851) wrote on 16 December 1851, that a female had been taken three weeks earlier near Barnard Castle, i.e. in late November. Both of these records are exceptionally late compared to the modern pattern of occurrence, with no subsequent November reports in the county excepting a juvenile at South Shields on 1 November 1981; one might ponder whether all of the details relating to these records were correct at the time.

The only well document record of Red-backed Shrike breeding in County Durham comes from James Backhouse Jnr. who writing in the *Naturalist* in 1885 described the event as follows, "*An instance of the breeding of this south country species in Teesdale came to my knowledge four or five years ago. The nest was found by my friend, Mr Wearmouth, of Newbiggin by the Teesside near to that village, and was built in a low thorn bush. It contained six eggs, one of them now being in my collection. This season (1885) the cock Red-back has again been observed near the old breeding place*". Presumably the same male was seen again in 1887, but no nest was found (*The Naturalist* 1888: 37).

Temperley (1951) gives no further records of this species, stating that: "*there are no recent records of it being observed even as a passing migrant*", though it seems hard to imagine that Red-backed Shrike did not occur in the county occasionally during the late Victorian era at least. At the mid-20th century it was called a "*very rare summer visitor, of which breeding records are few*" by George Temperley.

A male at Cleadon on 5 June 1954 was seemingly the first 20th century record, and one of just seven birds recorded during the 1950s, all of which were obviously coastal migrants. A further three birds in 1954 included a juvenile at Seaton Carew on 22 August and up to two birds at Hartlepool Headland from 25th to 27 August, with subsequent records being a male at Jarrow on 17 May 1957 and two at Crimdon Dene on 6 September 1958.

By the early 1960s, coastal migrants, particularly in the autumn, were more regularly recorded in Durham. Whilst the number of passage birds increased at this time, it could never be described as other than a scarce passage migrant. This change in status may have been connected to the development of an expanding Scandinavian population through the 1960s into the 1970s, though numbers have declined there since (Parkin & Knox 2010). In this sense, the 1960s saw the Red-backed Shrike consolidate its status as a scarce passage migrant, being absent in just four of the ten years. Twelve individuals were recorded in the decade, all but one in autumn, with two in August and nine in September, the sole spring record was an individual at Hartlepool Headland on 13 May 1960. Indeed, Hartlepool Headland produced the most with six records, whilst nearby North Gare/Seaton Common recorded four, with two juveniles at North Gare from 17th to 21 September 1960 the only multiple occurrence. A single at Marsden Hall on 21 September 1963 was surprisingly the only bird recorded away from the Teesside area.

Recent Status

Today in Durham, the Red-backed Shrike is a scarce but annual passage migrant in both spring and autumn. Records of the species are divided roughly three-fifths to two fifths between the two migration periods, with the weighting towards spring. Most spring birds appear between late May and early June and the autumn sightings fall into a period from mid-August to around the third week of September, occasional late birds are noted in early or even mid-October.

A total of 161 Red-backed Shrikes were formally recorded in County Durham between 1970 and 2011, and the species is now best regarded as a scarce migrant in both spring and autumn, though spring records have predominated in recent years. Records have been almost annual since 1970 with just five years without records: 1975, 1990, 1999, 2003 and 2007. The best individual years were 1977 and 1998 with 17 birds each. These two peak years were distinctly different in that during 1977 the records were spread evenly between spring and autumn, whereas in 1998 the vast majority were in spring, with just one bird recorded during the autumn.

At least 95 Red-backed Shrikes have occurred in spring since 1970, all of which have been in May or June. The majority of records are concentrated along the coastline from South Shields to Teesmouth, in particular the well-watched areas around Hartlepool and Whitburn though a number have also penetrated further inland. The species has been recorded in 24 springs since 1970 though numbers vary; each year's total being dependent on weather conditions and other factors. There have been two significant 'falls' of Red-backed Shrike during the review period; both were largely confined to May, coinciding with the arrival of many, mainly Scandinavian migrants.

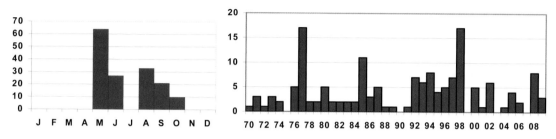

Records of Red-backed Shrikes in County Durham, 1970-2009

Suitable conditions from 7th to 25 May 1985 brought eleven birds to Durham. They were associated with a spectacular fall of migrants along the east coast of England following a period of north easterly winds and rain mid-month. At least five were between Hartlepool and Seaton Carew over 10th to 17 May and three more were in the Whitburn/Cleadon area around the same time, with at least one lingering until 25th. At the same time the county's first Thrush Nightingale *Luscinia luscinia* and at least 30 Bluethroats *Luscinia svecica* were recorded.

The 15 recorded between 11 May and 4 June 1998 represents the best spring for Red-backed Shrike in County Durham, with two phases of arrival. The first, from 10th to 16 May, was associated with a period of easterly winds and coastal drizzle and involved 10 birds, including four on Cleadon Hills on 12th. Later in the month much stronger north easterly winds and rain saw six more arrive between 29 May and 4 June, five on the former date.

Several birds have appeared well inland during May and June with records scattered as far west as Wolsingham. Particularly notable records are from Salter's Gate on 26 May 1974; Shibdon Pond on 20 May 1977 (see below); Charlton's Pond, Billingham on 7 May 1985; Drinkfield Marsh, Darlington on 16 May 1993; Ravensworth Fell, Gateshead on 10 June 1994; a Wolsingham garden on 15 June 1996; Lamesley on 16 June 1997; Newfield, Bishop Auckland on 5 June 2000; Crow Trees, Quarrington Hill on 7 June 2005; and, near Urlay Nook on 29 May 2008. It is particularly significant that all of these records related to adult males.

A male at Charlton's Pond, Billingham on 7 May 1985 is the earliest spring record in the county, with the latest being a female in scrub at the end of the Zinc Works Road, Teesmouth on 30 June 2009.

A total of 66 Red-backed Shrikes have been recorded in autumn and although the species is statistically less frequent than it is in spring, birds at this time of the year are less prone to episodic influxes. It has been unrecorded in 13 autumns since 1970; eight of these were between 1990 and 2009. Autumn migration is more protracted than in spring and records span the entire period from August to October with the majority occurring from late August until mid-September. Most records relate to juvenile or first-winter birds and only five birds have been aged as adults during the period. These were: a male at Hartlepool Headland on 30 August 1980; a female at East Boldon on 7th to 8 September 1983; a male at Whitburn Coastal Park from 24 August until 4 September 2001; a male at Hartlepool Headland on 3 August 2002; and, a female in scrub by Seaton Snook, Teesmouth from 12th to 15 August 2002. The best autumn for records was 1977 when nine birds were recorded between 14th and 29 August, including two together at Marsden on 20th, one of which remained until the following day.

Inland records are scarcer during autumn and just two birds have been recorded significantly inland, a juvenile at Shibdon Pond from 6th to 8 September 1986 and one at Crookfoot Reservoir on 10 October 1981. Elsewhere, single birds were a short distance inland at: Boldon from 1st to 6 October 1976, Boldon Flats on 27 August 1977 and, East Boldon from 7th to 8 September 1983. Also near the coast, single juveniles have been recorded on the

North Tees Marshes on four occasions. These were by the Reclamation Pond from 26th to 27 August 1977, on the Long Drag on 10 September 1995 and from 13th to 18 September 2008, and at Dorman's Pool from 20 September until 1 October 2009.

The earliest arrival date during autumn is of an adult male at Hartlepool Headland on 3 August 2002, whilst the latest record by a good margin is of a juvenile at South Shields from 26 October until 1 November 1981. This bird arrived almost two weeks later than all of the other October records for the county. A pale, sandy plumaged juvenile shrike in Cornthwaite Park, Whitburn on 8 November 2000 was thought to be a Red-backed though Isabelline Shrike *L. isabellinus* could not be ruled out before it disappeared into nearby gardens; this would be the latest modern record for the county.

Recent breeding status

Not fully reported in the literature at the time, one of the most fascinating modern occurrences of this species was recorded in Bowey *et al.* (1993). On 18 May 1977, a male was found at Shibdon Pond. The bird was heard singing and the following day a female was found to be present. The pair was noted prospecting for nest sites and they were present until the 22nd, when it was believed they were disturbed by birdwatchers. Anecdotal evidence suggested that a pair bred successfully elsewhere in the area that year and this may have referred to the re-located Shibdon birds but this could never be proven. Spring 1977 was a good one for the species and birds were suspected that summer of breeding at Backworth north of the Tyne (Kerr 2001). More recently a pair was seen in suitable breeding habitat near Beamish on 31 May 1997 though they did not linger and there were no subsequent signs of breeding having taken place.

Distribution & movements

Red-backed Shrike breeds across much of Europe and western Asia though is absent from much of Iberia and northern Scandinavia, and winters in sub-Saharan Africa, chiefly from Zambia and Malawi southwards, and also in south-eastern and coastal Kenya. This species was once a widespread summer visitor to Britain, regularly breeding as far north as Yorkshire until around 1850. By the middle of the 20th century a major decline saw the species restricted to southern counties of England and by the late 1980s it was essentially extinct as a regular breeding species. The full reasons for this decline are not completely understood. The Red-backed Shrike is now best regarded as a scarce passage migrant to Britain usually appearing after anticyclonic conditions bringing winds from the east. Appearing in both spring and autumn in equal numbers an average of around 200 birds per annum are recorded in Britain, though numbers are on the decline (Fraser & Rogers 2006).

Despite the number of records, very few Red-backed Shrikes have been trapped and ringed in County Durham. The most recent of these were females ringed at Castle Eden Denemouth on 5 June 1982 and at Hartlepool Headland on 8 June 2002. There have been no relevant controls or recoveries of this species.

Lesser Grey Shrike
Lanius minor

A very rare vagrant from Europe: four records.

The first record of Lesser Grey Shrike for County Durham was a first-winter bird found by E. Shearer at Tursdale Sewage Farm near Coxhoe on 24 October 1960 (*British Birds* 54: 194). There are several inaccuracies surrounding the publication of this bird, with *British Birds* listing the locality incorrectly as 'Tinsdale' Sewage Farm, whilst the Ornithological Report for 1960 (Coulson 1961), gave the date as 25 October, presumably in error.

All records:

1960	Tursdale Sewage Farm, Coxhoe, first winter, 24 October
1974	Port Clarence, adult, 29 June to 3 July (*British Birds* 68: 329)
1981	Barmston Pond, Washington, adult, 20 May (*British Birds* 75: 525)
1984	Jarrow, adult, 17th to 25 November, trapped on 22nd, found long dead on 3 March 1985 (*contra British Birds* 78: 579)

In broad terms, the records within County Durham conform to the national pattern, excepting that most of those recorded have been significantly inland. The bird near Port Clarence in 1974 is noteworthy in being one of only a handful of mid-summer records in Britain, whilst the individual at Barmston Pond in May 1981 was one of two recorded in the north east that spring, the other being in Northumberland in June.

Present along the main A185 approach to the Tyne Tunnel, the bird at Jarrow in late November 1984, attracted significant debate as to its identity, before being trapped and ringed on 22 November. Lacking the black forehead typically associated with this species it was aged in the hand as being an adult female and following its release was seen daily until 25th. This remains the latest date this species has been recorded in Britain although there are four other November records. Unfortunately, it did not travel far and was found long dead at the same location on 3 March 1985.

Distribution & movements

Lesser Grey Shrike breeds predominantly in south east Europe eastwards into western Asia, although small numbers breed further west along the northern Mediterranean to southern France and north eastern Spain. A noted long-distance migrant, the entire population winters in southern Africa from Namibia to southern Mozambique and northern South Africa. It is an annual vagrant to Britain between May and November, with the majority of records being of adults in May and June, and mainly first-year birds in September. There had been over 180 records by the end of 2010, though the species has become much rarer over the period 1990-2010, in line with a long-term decline in the European breeding population, which is estimated to have fallen by at least 10% during 1990-2000 (Birdlife International 2004).

Great Grey Shrike
Lanius excubitor

A very scarce migrant from northern Europe, recorded from September to May, but with most in April, and from October to November.

Historical Review

The Great Grey Shrike has always been a scarce winter visitor to Britain. The wintering birds that reach the UK are presumed to be Scandinavian breeding birds that have moved south west to escape the harsh northern winters (Bell 2010).

The earliest documented records of Great Grey Shrike are referenced by Hogg (1827) who recorded three specimens, though the accounts are rather vague. Hogg stated "*J. H. shot one of these rare birds December 11th, 1824, and about the same time two others were observed near Stockton*". Some four years later, Selby (1831) described the status of the species in the county, "*The visits of this species appear to be very irregular, some winters passing without the appearance of a single individual, whereas in others, it is by no means uncommon*". The species' status as a scarce winter visitor during the 19th century is further quantified by Hutchinson (1840), who described it as "*only seen in winter and then but only occasionally*" and Hancock (1874) who said it was a "*rare winter migrant*". At the time Hancock had four specimens in his collection, although only two had been obtained in County Durham; a male shot near Hebburn in 1836 and an immature male caught on lime twigs near South Shields in the winter of 1837.

The Teesmouth area attracted a number of birds through the middle of the 19th century. In October 1841, one was killed near Cowpen and in 1865 John Hogg wrote, "*On September 26th, 1865, I shot a Great Cinerous Shrike (Lanius excubitor) ...this species is rarely observed in this district*", he was presumably referring to the south east of the county but the specific location was not stated (Temperley 1951).

In 1876, Hancock noted, that a male *Grey Shrike* was shot on 21 November 1876 near the Parson's Field, Durham. Temperley (1951) thought that it had become more common over the latter part of the 19th century. Robson, writing of the Derwent valley, told of a number of birds being "*obtained*", including: a female at Blaydon

Haughs in 1836; a pair late in the 19th century in Whickham; and, in 1880, a bird that was shot by the river at Dunston Haughs (Robson 1896).

A similar pattern of occurrence continued into the early part of the 20th century with Tristram (1905) writing that "*A winter seldom passes without one or more captures being reported*", though conversely Temperley (1951) makes reference to the species being a "*rare and irregular autumn and winter visitor, less frequently recorded now than formerly*". Although there would doubtless have been others, he listed just three records; at Seaham Hall on 16 January 1908, at Haverton Hill, Teesmouth on 31 October 1936 and two together at Folly Plantation, near Birtley on 4 November 1944, with one still being present on 8th.

Eight birds were recorded during the 1950s, with the species becoming regarded as more of a scarce passage migrant in spring and autumn rather than a rare winter visitor. There were five October records during the period, all of which were at coastal localities, including singles at Marsden Quarry and South Shields on the 19 October 1952. Elsewhere the two 'spring' records comprised singles at Houghton-le-Spring on 31 March 1954 and at Greatham on 15 April 1956, with the sole mid-winter record being of a single at Warden Law on 31 December 1953.

In addition to these, records of single birds were submitted to the Yorkshire Naturalists' Society of one '*near Barnard Castle*' on 4 June 1954; and '*near Middleton-in-Teesdale*' on 5 June 1954 (Mather 1986). That both these records should be on consecutive days in the extreme west of the county and on unusual dates, suggests that only one bird was likely have been involved, despite the records being at least 15km apart. These records are exceptionally late according to present day status but there are a handful of mid-summer records of Great Grey Shrike in Britain, the most recent of which was one in Cumbria in July 2009 (Hartley 2009).

A further 28 were recorded during the 1960s, with the species being annual in occurrence throughout the decade, averaging about two to three birds per annum. In line with the previous decade more than three-quarters of these were in October and November at predominantly coastal localities, further consolidating the species' status as a scarce passage migrant. A noted influx of six birds over 21st to 23 October 1960 included three at Stanley on 22nd and two at Hartlepool Headland on 23rd. The three at Stanley remains the highest number recorded at a single site in the county. Other noteworthy 'autumn' records include a single at Low Barns, Witton-le-Wear on 11th and 12 November 1967 and two at Hamsterley Sewage Farm in November 1969, though none appeared to have lingered into December. Just three birds were recorded during the mid-winter period, at Hurworth Burn Reservoir on 27 December 1960, Thorpe Thewles on 16 February 1963 and Charlton's Pond, Billingham from 7th to 19 December 1965 though none were seen subsequently. Birds on spring passage were noted on just three occasions during the decade at: Hurworth Burn Reservoir from 31 March to 7 April 1960, Graythorp from 9th to 10 April 1967 and at Frosterley on 15 April 1967.

Recent Status

The Great Grey Shrike remains an annual visitor to Durham, in very small numbers, mainly between late October and mid-April (two to three in most years); though numbers have varied widely.

A total of 117 Great Grey Shrikes have been recorded in County Durham since 1970 and the species is chiefly regarded as a scarce migrant in both spring and autumn, though it is more numerous in autumn. Single birds have overwintered in the county on a number of occasions particularly during the 1970s though in recent years winter records have been scarcer. Reports are widely scattered though the majority of migrant birds, particularly in autumn, are from predominantly coastal localities. The species was annual in occurrence during the 1970s but there have been eight blank years since 1985 and the current average is just one bird every other year. The best individual year was 1976 when 11 were recorded, but a remarkable influx in 2010 saw a total of 10 arrive between late September and early November.

At least 64 Great Grey Shrikes were recorded during the 1970s in what was to be a golden era for this species in County Durham. To put this into perspective, this total is more than would be recorded in the county in the next thirty years. Almost half of these were autumn migrants found during September to November, although 18 during March and April represents a healthy total for spring passage and is indicative of this species being a much more regular winter visitor to Britain in this era. Amongst the many mid-winter records during this period single overwintering birds were at Low Barns, Witton-le-Wear from 17 December 1970 until 10 January 1971, Croft-on-Tees from 20 February until 3 April 1972 and Shibdon Pond each winter from February 1972 until March 1975. Most notable amongst these wintering birds was what was almost certainly the same bird, first discovered on 25

February 1972 at Shibdon Pond, and seen annually thereafter to 26 March 1975; it frequented the scrub and hedgerows of Shibdon Pond and Derwenthaugh, with occasional forays into Axwell Park and to Swalwell.

Several high annual totals included ten in 1974 and nine in 1975, though the total of 11 in 1976 remains the best annual total to date. In 1976, five coastal migrants arrived between Marsden and the Tees Estuary during the period from 26 September to 3 October with two further birds at Marsden and Hartlepool Headland during late October. That five should arrive so early in the autumn is unusual though the weather during this period was exceptional producing what were described as "*mammoth falls of migrants*" that autumn (Unwin 1977).

The 1980s saw numbers decline with just 18 birds recorded in the entire decade, the majority of which were prior to 1985. The peak year was 1982 with seven, of which five were coastal migrants between 9th and 19 October. These birds were found during what a classic autumn for Siberian vagrants in the county. Just three birds were seen during the latter part of the decade, these were at Lartington on 16 March 1986, at Hargreaves Quarry, Teesmouth from 23rd to 25 April 1986 and at Whitburn on 25 October 1988. There were no records of overwintering though a single bird was at Cleadon Hill from 15th to 20 January 1982.

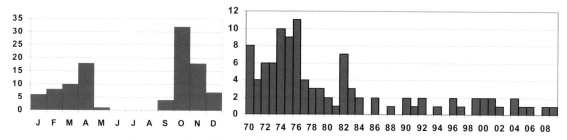

Records of Great Grey Shrikes in County Durham, 1970-2009

Just eleven Great Grey Shrikes were recorded during the 1990s which was a particularly poor decade for the species in the county. There were three blank years and four years which each recorded two birds. Seven coastal migrants were reported, the most notable of which were two together at Whitburn on 23 October 1990. Elsewhere single birds were at Clara Vale, near Crawcrook on 14 January 1996, Lamesley from 29th to 31 December 1997, High Urpeth on 3 May 1999 and Killhope on 21 November 1999. The individual at High Urpeth is very late and is one of only two May birds recorded in County Durham to date.

Like the previous decade, just eleven were recorded between 2000 and 2009, though records were more evenly spread with just two blank years. Of these eleven, five were coastal migrants found between September and November, with Hartlepool Headland having most records with four birds, including two in 2001 on 25 September and 14 October respectively. The remaining bird was one at Dawdon Blast Beach from 3rd to 7 October 2002. There were no records from the well-watched Whitburn/Marsden area throughout the entire period. Elsewhere, singles were at the Long Drag, Teesmouth from 15th to 16 April 2000, Hurworth Burn Reservoir from 26 November 2004 until 10 April 2005, with perhaps the same bird at nearby Quarrington Hill from 10th to 12 April 2005, at Bowesfield Marsh on 29 October 2006, Knitsley Fell on 12 November 2008, and at the west end of Hamsterley Forest from 10 March to 15 April 2009. One at Neighbour Moor, Hamsterley from 21 Jnuary to 4 April 2011 could well have been the bird from 2009 returning to the area, as this species can show a high degree of site fidelity. There have been just two records of wintering birds in the last 20 years. The individual at Hurworth Burn Reservoir was the first time a Great Grey Shrike had spent the entire winter period in County Durham since the 1970s. A bird that lingered around the Northumberland/Durham border in the Wylam area from late 1995 through to early 1996 however, intermittently visited the Durham side of the river, with occasional reports through the winter at Eels Wood, Wylam and from Ryton Willows and Clara Vale on 14 January 1996.

In contrast to these recent trends, a minimum of 11 Great Grey Shrikes were recorded in County Durham between 29 September and 11 December 2010 in the best ever influx of this species in the county. The first was at Hendon on 29th and 30 September, although the main arrival took place eight days later when a total of six birds were recorded between 8th and 10 October at coastal localities from Whitburn to Seaton Snook at Teesmouth, including a minimum of four birds on 9th. Birds were associated with a large fall of continental migrants and quickly moved on, just a single bird at Hartlepool Headland, from 9th to 11 October, remaining for more than one day. Outside of the main arrival period, singles were also at Seaton Pond, Seaham from 13th to 15 October, the Long

Drag and RSPB Saltholme on 15 October, at Hetton Lyons Country Park on 17 November; and near Tunstall Reservoir on 11 December.

The earliest Great Grey Shrike to be recorded in County Durham was at Hartlepool Headland on 25 September 2001, whilst the latest was a bird which lingered at Barmston Pond, Washington until 5 May 1973.

Distribution & movements

Great Grey Shrike breeds in central and northern Europe from France and Scandinavia eastwards across northern Asia, and in similar latitudes across North America. As a relatively short distance migrant, northerly populations normally winter as far south as the southerly limit of the species' breeding range. A number of races are recognised though British records are thought to relate to the nominate *excubitor*, with most birds probably originating in Scandinavia. The species was formerly a scarce winter visitor to Britain though it is perhaps now best regarded as a scarce autumn migrant; records have declined markedly in recent years. A small return passage in usually evident in late March and early April, presumably indicating that some might spend the winter undetected in the British Isles. Around 100-130 birds are estimated to occur each year in Britain (Fraser & Rogers 2006). Just two Great Grey Shrikes have been trapped and ringed in County Durham, at Whitburn Country Park on 8 October 2010 and at Hartlepool Headland on 10 October 2010, with no controls or recoveries.

Woodchat Shrike
Lanius senator

A very rare vagrant from Europe: ten records.

County Durham's first record of Woodchat Shrike was of an adult at Billingham Bottoms on 24th and 25 April 1971 (*British Birds* 65: 346). This bird was photographed by Adrian and David Allen though unfortunately, they did not realise the significance of their find and the bird's presence only became known when the photographs were exhibited at a meeting of the Teesmouth Bird Club; the bird had long since departed (Smith *et al.* 1972).

All records:

1971	Billingham Bottoms, adult, 24th to 25 April
1983	Marsden Hall, first winter, 8th to 12 September, trapped 11th (*British Birds* 77: 554-555)
1994	Boldon Flats, adult, 24 September to 14 October
1997	Mere Knolls Cemetery, Seaburn, adult male, 22 May
1998	Westoe Colliery, South Shields, adult, 28 May to 7 June
2000	Whitburn Coastal Park, adult male, 19 June
2006	Whitburn Coastal Park, adult male, 21st to 23 May
2008	Greatham Creek, Teesmouth, first winter, 18 September
2010	Hartlepool Headland, first winter, 26 September to 17 October, trapped on 27 September
2011	Trow Quarry, South Shields, adult, 21 August

Of the ten records for the county, five have appeared in spring during the period from 24 April until 19 June, with the majority in late May. This might be considered an expected spread of dates given County Durham's northerly location, although the species typically occurs earlier in southern England; even as early as late March. The first, at Billingham in 1971, is especially noteworthy for being the only April record, and 10km inland. The bird at Whitburn Coastal Park in 2000 is very late for a coastal spring migrant and remains the only June record to date.

The five autumn records have arrived between 21 August and 26 September, with several making protracted stays. The bird at Boldon Flats is unusual not only for being significantly inland, but also for being one of just two adults recorded during the autumn period. Adult birds are extremely rare in Britain in this season. The bird at Hartlepool during autumn 2010 spent the longest period in the county, 22 days. This bird fed predominantly on insects but was seen to kill a Blue Tit *Cyanistes caeruleus* on one occasion.

Two birds have been trapped and ringed in the county, both of which were first-winter individuals; at Marsden Hall on 11 September 1983 and Hartlepool Headland on 27 September 2010. All of the adults recorded in County Durham have been identified as being of the nominate race *senator* and there is no evidence to suggest that any of the first-winter birds are any different.

Distribution & movements

Woodchat Shrike breeds throughout southern Europe, North Africa and Asia Minor, and winters predominantly in the Sahel belt of sub-Saharan Africa. Four races are recognised though differences are slight. Woodchat Shrike is an increasingly regular vagrant to Britain in both spring and autumn with almost 600 records by 1990 at which point the species was no longer considered by BBRC. The ten year average for 1990-1999 was of 21 birds per annum, with a peak of 36 in 1997 (Fraser & Rogers 2006).

Chough (Red-billed Chough)
Pyrrhocorax pyrrhocorax

A very rare vagrant: four records involving five birds.

Historical Review

Today, this species is restricted in Britain largely to coastal localities in the west of Scotland, Wales and Cornwall, but in historic times it was very much more widespread in England (Brown & Grice 2005). In the past it was a very rare vagrant to the county and up that to the end of 2010 it had not been recorded in Durham for over 40 years.

Certainly, in the early 1800s, *'red-legged daws'* were said to be present to both the south and north of Durham. There are records on cliffs at Boulby (Blick 2009) and they certainly bred more generally on coastal cliffs in Yorkshire up to the beginning of the 19th century (Mather 1986). The Chough also used to breed near St. Abb's Head and Fast Castle, just north of the Scottish border, but this population had died out by 1851 (Holloway 1996). The first documented record of Chough for County Durham came from the banks of the River Derwent on 20 May 1905. This bird was seen well by Messrs. W. Johnson, H.S. Wallace and T.C. Fortune, all of whom were members of the Vale of Derwent Naturalists' Field Club, and the sighting was written up by Johnson (Johnson 1905). They later consulted E. L. Gill, curator of the Hancock Museum, and with his help examined the specimens of this species held in the museum, confirming their identification.

Prior to this record Hogg (1845) had written in his catalogue that *"stragglers are occasionally killed along the Durham coast",* though he gave no documented records. He did allude to the fact that these birds were probably stragglers from the small population at St. Abb's Head. Hancock (1874) also made reference to this population, but was not satisfied that the species had ever occurred in County Durham, although he did concede that it may well have occurred in Northumberland.

All records:

1905	River Derwent, 20 May (Johnson 1905)
1927	Whorlton, 23 April (*The Vasculum* Vol. XIV 1928)
1967	Seaburn on 24 October 1967 (Bell & Parrack 1968)
1969	Hartlepool, two, late March (Blick 1978)
1969	Seal Sands, Teesmouth, 18 September (Blick 1978)

The second in 1927 was seen by a Mr F. Williams near Whorlton, to the west of Darlington, on the 23 April and was reported to the Darlington Naturalist Club at the time. The only confirmed record along the northern section of Durham's coastline was of a bird at Seaburn on 24 October 1967. There is little background to the two most recent records, of two birds at Hartlepool, in late March 1969, and later that year, one near Seal Sands on 18 September (Blick 2009). Perhaps they and the 1967 report all refer to the same birds? All of these records have been considered as possible escapes from captivity, even at the time of occurrence. Temperley (1951) regarded both of the early 20th century occurrences as being probably escapes but what he based his view on other than the species rarity or how regularly this species was kept as a cage bird in the region at that time are unknown. Overall, there seems, little direct evidence to support the various assertions of the records relating to escapees. The nearest breeding population to County Durham lies 180km away on the Isle of Man, and there are several records of Welsh birds dispersing up to 130km from breeding sites. In the light of this, records in Durham are not that outlandish and may represent genuine vagrants.

Distribution & movements

The Chough has a disjunct range, breeding on mountains and sea cliffs from southern Europe and North Africa eastwards into central Asia, northern India and China. In north west Europe the species is entirely restricted to the French region of Brittany and the British Isles where it is essentially a rare and localised resident of south western Ireland, north western Scotland, Wales and the Isle of Man. In the early 21st century, within the species' very limited, westerly range in the British Isles, numbers have been increasing in recent decades after two centuries of decline. It was absent as a breeding species in England for over 50 years, a pair bred in Cornwall in 2002, increasing to at least six pairs in 2010, with 49 young being raised during the period from 2002 to 2009 (Parkin & Knox, 2010). The Chough is chiefly resident throughout its range, though juveniles are known to disperse more widely. Ringed birds from North Wales have been recovered in the Isle of Man in 2005 and Lancashire in 2007, representing movements of over 100km from their natal area, whilst one found dead in Buckinghamshire, December 1991, had been ringed on Islay, Strathclyde in May 1986, over 604km to the north west (BTO).

Magpie (Eurasian Magpie)
Pica pica

A very common and widespread breeding resident.

Historical review

There are three quotations from the Bursar's rolls of the Durham Monastery that mention Crows *Corvus corone*, Rooks *Corvus frugilegus* and Magpies; these were documented in 1348, 1368/1369 and 1378/1379. The first of these indicated that the three species, assuming '*corvos*' included the Rook and the Crow, were sufficiently numerous on the manors of the Priory to require 'thinning down' (Fowler 1898). In consequence of such a long history of human persecution in Durham, this species has unsurprisingly experienced some ebb and flow in its numbers in the northeast of England over the last two hundred years. Hancock, in 1874, wrote that it had become rare, when it had once been "*so abundant*". Less than forty years earlier, Hutchinson (1840) had reported on its common status in Durham, '*no bird better known than the magpie*'. In the 19th century, Robson, commenting from the perspective of the Derwent valley, thought that it was decreasing due to persecution, although he did document a number of nests in the lower Derwent valley in both 1895 and 1896 (Robson 1896). At the beginning of the 20th century, the Magpie, like other corvids, was heavily persecuted by man (Birkhead 1991). Decline appeared to continue until the First World War, and then a change came with the cessation of game preservation that was associated with the conflict.

After the Great War, the number of keepers employed declined and with that social change came a reduction in game rearing practices, allowing the species to recover and re-colonise areas from which it had been excluded by persecution; the numbers of Magpies began to increase (Temperley 1951). In an article about his time living in Sunderland, the Rev. George Courtenay said that a Magpie was recorded on October 28th 1920 "*on outlying fields within my three-mile limit, the first and only time I have seen one so near the town*" (*The Vasculum* Vol. XIX 1934). This indicated that the species was moving in towards built-up areas, even as early as the first quarter of the 20th century.

The species' recovery continued through the 20th century, and by Temperley's time of writing, its status had changed dramatically since the outset of the century; he described it then as "*a resident, increasing in numbers and rapidly becoming common*". He did note that birds had still not fully penetrated into the coastal fringe of the northeast of the county, though the population was increasing around Hurworth Burn and Crookfoot Reservoirs. By the 1960s, the population was relatively stable, increasing again during the 1970s; a trend that, in some areas, continued into the 1990s (Marchant *et al.* 1990). In 1973 it was described as 'widespread, if thinly distributed', still somewhat different from the modern situation. In relation to Teesside, Stead (1964) and Blick (1978), noted that it was becoming more common, though both said that it was 'less common' in the built-up areas.

Recent breeding status

Originally, the Magpie was a bird of open farmland. It began to spread into more suburban areas during the 1940s (Marchant *et al.* 1990). In the mid-20th century, Temperley (1951) stated that, "*The Magpie rarely appears in the coastal district between the mouths of the rivers Wear and Tyne, due to the lack of woodland and tall hedgerows.*" This statement is no longer true and birds are now common in the Cleadon and Whitburn areas (Westerberg & Bowey 2000) and more generally in the 40 years or so since 1970, there has been a very significant increase in the numbers of Magpies in Durham.

Survey work in the Gateshead area during the mid- to late 1980s found that birds were present in over two-thirds of the surveyed one kilometre squares examined at that time. Numbers of this species increased rapidly through the 1980s both locally and nationally (Marchant *et. al.* 1990). Today, the Magpie is a common and ever more widespread resident across the county. Its range stretches from the coast including the well-wooded coastal denes, such as Castle Eden and Hawthorn; along the ribbons of disused railway lines and the linear scrubland that has sprung up since their closure in the 1960s; westwards as far as the moorland fringes and hinterlands. It now breeds in a wide variety of 'scrub and woodland' habitats, including: woodlands, hedgerows with mature trees, parks and large gardens (Westerberg & Bowey 2000).

The species' spread probably accelerated over the last fifteen years of the 20th century, expanding west up Teesdale over this period with birds being recorded for the first time at Balderhead Reservoir in 1988 and at Selset Reservoir in 1989 (Armstrong 1988-1993). At present, it can be found in the far reaches of the west of the county, with birds to be noted at elevations as high as St. John's Chapel in upper Weardale, and Selset Reservoir, in Lunedale where it is linked with conifer plantations.

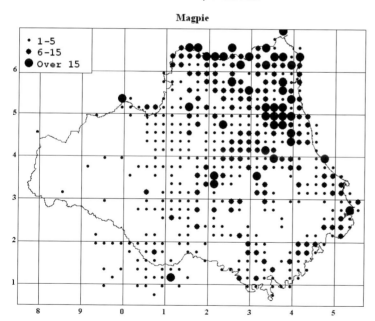

Magpie

- • 1–5
- • 6–15
- ● Over 15

Distribution of Magpie in County Durham by Tetrad, 2008-2010

Although the *Atlas* mapping exercise indicated that it was absent from the Hartlepool area in the period 1988-1994, this was incorrect. In fact, it is a common and widespread bird in that locality as documented in *The Breeding Birds of Cleveland* (Joynt et al. 2008), which showed it to be present in almost every surveyed tetrad of the area, with perhaps as many as 400 pairs in the area north of the River Tees (Joynt *et al.* 2008). The impression that Magpies are somewhat less common in the open farmland of southern Durham than it is in the former coalfield areas of the east of the county, appears to be borne out by the mapped distribution and timed counts of 'birds seen per hour' across Durham during the period 2008-2010, as indicated by the map. The lower densities and absences in some of the westerly tetrads may indicate an element of predator control in those parts of the county.

Nationally, over recent decades, the Magpie has become increasingly common in our major towns and cities (Marchant *et al.* 1990). By way of local example, forty birds were present at Boldon Business Park on 2 October

2005, while 40 came in to a roost site in urban Hebburn in winter 2008 (Newsome 2006 & 2009) and a considerable colonisation of the urban edges and then the centres of the urban areas around Teesside occurred over the years 1975 to 2005 (Joynt *et al*. 2008). Once a rural bird, suburban areas now probably hold greater densities of Magpies than farmland in Durham, with up to 16 pairs per square kilometre (Gibbons *et al*. 1993). During the *Atlas* survey, the Magpie was found to be present in 363 tetrads in the county (Westerberg & Bowey 2000). Using the results of species-specific census work undertaken in Gateshead in the early 1990s, which suggested a breeding density of 8.47 nests per occupied tetrad, a population estimate of 3,075 pairs was calculated for Durham. In 2007, in a relatively smaller survey, ten pairs were found nesting around Rainton Meadows and Joe's Pond and seven pairs were at nearby Hetton Lyons, indicting the number of birds that can persist in close proximity. Local census work indicated that over the period 1987 to 2007 the population was still increasing in parts of the county. For instance, at the Kibblesworth CBC site a figure of 14 pairs in 2007 represented an all-time high, but despite this high density, there was little evidence from the census that the species was having any detrimental effect on local populations of other bird species in the area (Newsome 2008). Elsewhere, numbers may have stabilised or even dropped in some parts of the county. The most recent national survey put the British population at 590,000 territories (Gregory & Marchant 1996). Public feeling against the species is sometimes rather negative and it is often targeted as a scapegoat for declines in song bird numbers, despite the fact that studies have shown numbers of breeding passerines are little different in areas where there are many Magpies in comparison to those areas where there are few (Mead 2000).

In 2007, a pair was observed building a nest on the edge of a rookery at Carrville in March, unusual behaviour for this territorially nesting corvid. The principal threats to Magpies have always come from man and it is still persecuted today.

Recent non-breeding status

In winter, it is commonly encountered across Durham and at this time of the year birds roost communally, often at traditional spots located in clumps of hawthorn and similar bushes. Roosts normally contain a maximum of 30-50 birds and may be reported from all parts of the county; though the numbers are lower in the western uplands. Information gathered in the Hetton to Houghton area in 2007 well illustrates the species' winter roosting behaviour with several large roosts were the area. At Rainton Meadows, the winter maximum was 92 on 7 February, whilst November roosts included 75 at Hetton Bogs and 45 at Hetton Lyons. The following winter saw the Rainton Meadows roost peak at 135 during December 2008. In lowland areas, counts of over forty birds at roosts routinely come from heavily wooded areas, such as those in Gateshead at Blaydon Burn, Lockhaugh, and Ryton Willows. The species is often reported as being scarcer in areas of the dales, such as upper Teesdale, where 'vermin control' is still practised; although 21 were at Derwent Reservoir in August 2009. The Magpie was once a scarce bird at Hartlepool Headland but it is now seen quite regularly there (Blick 2009). The species was probably more numerous in County Durham in the autumns at the outset of the 21st century than at any time in the previous 200 years.

Distribution & movements

Magpies are widespread in Eurasia, from Japan to Ireland, from North Africa to the edges of the boreal forests in northern Europe, and penetrating in the south, into the Middle East and as far as south East Asia. It is largely sedentary. As a resident it seems capable of withstanding the very harshest of winter weather. Even in the north of its European range in Sweden, most dispersive movements are within 50km (Cramp *et al*. 1994).

Local birds are rather sedentary, and even those ringed as nestlings, when controlled or recovered tend not to have moved large distances. This is illustrated by the fact that the mean distance moved by eleven birds ringed in the west of Gateshead area between 1988 and 1997 and later controlled or recovered, was just over 2.5km; with over one-third of these being recovered at or close to their site of ringing. An exception to this trend was the nestling ringed at Blaydon Burn on 24 May 1992 that was shot on 17 March 1993 at Netherwitton, Northumberland, some 27km to the north west.

Jay (Eurasian Jay)
Garrulus glandarius

A widely distributed and common breeding species.

Historical review

The history of this woodland species in the county is poorly documented and there is little that can be said about its past status other than it has been the victim of persecution in Durham for well over 150 years. Holloway mapped the species' status in the county as common, unlike some parts of the north of England, where numbers had already been suppressed by game-rearing interests (Holloway 1996) and for fly-tying and the millinery trade (Mead 2000).

This species was considered a common resident in the county in the 19th century. References to its past status by Hogg (1827) and Hutchinson (1840), suggested that the Jay was "*frequent in the woods of south-east Durham*" and a "*common inhabitant of woods, plantations and thickets*". Just 34 years later Hancock (1874) wrote "*the Jay has shared the fate of the Magpie and is now nearly annihilated in the two counties* (Durham and Northumberland) *where a few years ago it was by no means uncommon*". He believed that it would have been lost to the region if it were not for its roving habit, which meant that some birds always escaped persecution. A few years later, Tristram (1905) concluded that "*the mis-directed energies of the gamekeeper have all but exterminated the jay in the eastern and central parts of the county, where in the memory of man it was not uncommon*".

In the 20th century, Temperley (1951) recalled its once common status, in the early 19th century, and its decline as a consequence of game rearing and the associated persecution. He reported that this rather sedentary species tended to be "*heard and not seen*" although it was common in most local woods. A decline in the number of gamekeepers after the Great War allowed the species to slowly increase in Durham (Temperley 1951). In the Birtley area, it was reported in *The Vasculum* that the Jay had nearly disappeared locally through the early part of the century. By the early 1950s, it was becoming more common "*chiefly because, during the war, the keeper's gun has not been quite so active*". It was however, still persecuted "*... one relates with horror that dead jays have been found recently, lying just where they had been shot*" (*The Vasculum* Vol. XXXIX 1954).

In the south east of the area, Blick (2009) felt that there had probably been quite a major increase in this species' numbers during the 20th century. One flying north over Hartlepool Headland on 25 April 1962 was, at that time, an unusual sight and somewhat out of context in an area largely devoid of woodland habitats.

Recent breeding status

This colourful, though well camouflaged species, at least in the context of its woodland habitat, is relatively common as a breeding bird along all of the county's wooded river valleys and in the major conifer plantations. In particular the Derwent valley in the north west, the Browney and the Wear between Durham City and Chester le Street extending to Sacriston and the mid-Wear, all support relatively high numbers as illustrated by Westerberg & Bowey (2000). Further west, Jays track the River Wear westwards at least as high as Stanhope, at 219m above sea level. The species is especially common in Hamsterley Forest and the surrounding woodlands.

Unlike many other areas of Britain, Durham's uplands have not been extensively afforested so, because of a lack of suitable woodland habitats at such elevations, Jays are not usually found on land much over 300m in the county. The most westerly reports usually come from the wooded area to the west of Derwent Reservoir in the north or in the extensive conifer plantations of The Stang in the south. In summary, this is a fairly common and widespread resident in County Durham, in both coniferous and mixed woodland as well as parkland. The highest densities are found in broad-leaved or mixed woodland, though it can quite routinely be found in pure conifer stands; though, usually in lesser numbers (Marchant *et al.* 1990).

In most breeding seasons, small numbers are reported from all the traditional wooded localities, with sites that attract larger numbers of observers, such as Croxdale, Brasside Pond, Thornley Woods and the lower Derwent valley more generally, providing the bulk of observations, and typical counts of four to eight birds. The only areas of the county that are routinely devoid of breeding birds are the north east quadrant, around South Tyneside, and the open habitats of the North Tees Marshes. For instance in 2007, no birds at all were reported in South Shields or surrounding area. Coastal reports of the species tend to be scarce, excepting where breeding populations inhabit the major wooded denes that stretch down to the coast, such as Hawthorn and Castle Eden Denes.

In Durham's south east, this species is a sparsely dispersed breeder with perhaps 30-35 pairs in the north Tees area. Concentrations of birds can be found around Wynyard and north towards Nesbit and Crimdon Denes. Effectively, the Jay is absent as a breeding bird over much of the Teesmouth area, including the urban centres of Teesside and north to Hartlepool (Joynt *et al.* 2008).

Following a peak around the mid-1980s, the national population declined to a low in the late 1990s. Thereafter, numbers quickly recovered to an estimated 160,000 territories in 2000 (BTO), this being broadly equivalent to 1960s population. The CBS/BBS index records show some further increases up to 2010 which is again approaching the peak noted in the mid-1980s.

Locally, the *Atlas* (Westerberg & Bowey 2000), working on the assumption that the county population was typical of the national one as illustrated by Gibbons *et al.* (1993), calculated that Durham held some 1,700 territories. Although Westerberg & Bowey (2000) stated that this, considering the recorded occupation of tetrads, might have been 'on the high side'. Counterbalancing this, the *Atlas* also highlighted that its distribution maps, may have underestimated Jays' true status, because of the relatively secretive nature of birds during the breeding season and this probably still holds true today. Although this species is a very noisy and obvious resident whilst pair-bonding, they can be remarkably secretive once nesting is underway and locating breeding birds is surprisingly difficult. This coupled with the fact that it is a species of sometimes quite dense woodlands that benefit from little observer coverage, means that the population could well be under-recorded.

Recent non-breeding status

The species' profile is higher during the post-breeding season as birds of the year disperse out of their natal territories and at such times it is most often seen flitting between autumn stands of oaks as it collects acorns for winter provisioning. This practice of gathering seeds is usually witnessed in woodland areas but in some instances birds have been noted doing less usual things. For instance, at Castle Eden Dene, birds have been observed bringing acorns from the local woodland onto the golf course to cache in the short turf. In Durham City, Jays regularly cache both acorns and conkers in the plant pots and tubs of gardens adjoining mixed deciduous woodland (A.L. Armstrong pers. obs.). In recent years, birds have been increasingly noted feeding on bird tables and wire peanut holders in both woodland feeding stations and gardens.

In autumn and winter, most gatherings of birds number fewer than six birds, it is not a social species in the way a number of other corvid species are. Nonetheless, autumn can bring concentrations of birds to some areas. Counts of from eight to eleven are at the upper end of typical numbers. The feeding station at Thornley Woods often attracts numbers of birds, with 14 recorded there on 29 August 2000. Jays are more obvious during the winter months with nearly two thirds of all reported sightings usually coming in the periods January to March and October to December.

Whilst this species, in spring, occasionally takes the eggs and young of small birds, it almost certainly has little or no effect on game species. Nonetheless, it still regularly features in the gibbets of game keepers, though such a sight is becoming less common in the County Durham countryside of the early 21st century.

Jays are a mainly sedentary though intermittently, in response to food shortages, northerly breeding birds exhibit eruptive behaviour, and large numbers of birds can move south west across the Continent . One such influx occurred in autumn 1983, when large numbers of Jays occurred in southern Britain and birds were recorded elsewhere in areas where the species was not normally present. In fact, this influx took place after the failure of the continental acorn crop, and birds wandered extensively. In the north east of England, birds appeared in many locations where they were previously only rarely noted or even unrecorded. Such influxes have been recorded in the region previously, but only very rarely, for example in 1897, 1901 and 1917 (Kerr 2001).

During the 1983 influx, small numbers of birds were seen coming in over the sea at several points along the east coast of England during October and November, including 28 birds flying in off the sea at Whitby, Yorkshire on October 9th (Brown & Grice 2005). Birds were not seen in such numbers along the Durham coast although there were several sightings of coastal birds from areas where the species is not usually present. For example, four were on Cleadon Hills on 9 October, two flew south over South Shields on 16th and another five were at Whitburn on 22 October whilst between 26 October and 12 November, five single birds flew south west over Hartlepool. At Castle Eden Denemouth on October 23rd, 13 birds was an exceptionally high count and a very large count of 35 came from the Wynyard Estate on 19 October (Blick 2009).

A further, smaller influx took place in 1985 and in this instance birds were noted in a number of unusual sites were they were previously rare, such as Shibdon Pond and in Sunniside Village, where a party of six were observed flying west on 25 October 1983 (*Bowey et al.* 1993). Perhaps more indicative of local movement from a breeding location slightly further north, such as Nesbit or Crimdon Dene, was the bird that was seen to fly over the Long Drag and Dorman's Pool on 22 September 2005 and one that was on the Long Drag on 31 October 2009 (Blick 2009).

Distribution & movements

The Jay breeds throughout Eurasia, from Ireland in the west to Japan in the east. Its range extends to the edges of the northern forests and as far south as northwest Africa. Most local Jays are rather sedentary, probably moving very little during the year or their lifetime and there is no known, relevant ringing data for the species in Durham.

Jackdaw (Western Jackdaw)
Corvus monedula

An abundant and widespread resident.

Historical review

Of Durham's three common black crows, the 'little crow', is probably the most widespread and may have been so, for many centuries. This species was mentioned, along with other corvids, in the 1410 cellarer's rolls of the Monastery of Durham (Ticehurst 1923), confirming that it has been present for as long as any aspect of the county's fauna has been recorded and it was presumably used as a food item in times in the past. Firm evidence on the size of the county's Victorian population is difficult to come by, although the adjective '*abundant*' was used by both Hutchinson (1840) and Tristram (1905). Late in the Victorian period, Jackdaws were recorded as nesting in the slopes of Scar Bank, on the banks of the Derwent, in the lower Derwent valley (Bowey *et al.* 1993). Nelson (1907) also reported it as an immigrant to the Durham coast, being '*often in great numbers*' at Teesmouth, along with Rooks *Corvus frugilegus*.

During the 1920s it was noted as nesting on the cliffs at Marsden and Frenchman's Bay, South Shields (Noble-Rollin 1928). It was described by George Temperley (1951) as widely distributed, resident and very common. He recorded nesting locations on crags, in quarries and on sea-cliffs, as well as in buildings of various forms. Temperley felt that numbers were steadily increasing, partly as consequence of agricultural changes after the Second World War.

This observed trend continued through the 1950s as highlighted in this report from the north of the county, "*In the Birtley area, one of the most remarkable features to attract the attention of the bird lover is the enormous increase in numbers of nesting pairs of jackdaw. Indeed, this bird has become a perfect pest*" (*The Vasculum* Vol. XXXIX 1954). There have been few attempts to systematically census the county's nesting population, but counts around Durham City revealed that there were about 200 pairs nesting on Durham Cathedral in 1953, most of them on the West Tower (Temperley 1954).

There is little substantive or qualitative information about this species' status locally during the 1960s and through much of the 1970s. In the winter of 1970 however, a mixed roost of Rooks and Jackdaws, which contained about 10,000 birds, was reported from Heighington (Coulson 1972) but during the remainder of the 1970s the documented largest flocks of Jackdaws alone tended not to exceed the range of 250 to 500 birds.

Recent breeding status

Across the county, Jackdaws can be found breeding in many habitats, including the centres of towns and villages, along coastal cliffs, in the industrial areas, in deciduous woodland, even in single dwellings such as farms and old houses. This common and widespread species is almost as common in the upland areas of Teesdale, Weardale and the Derwent valley as it is on the coastal cliffs at Marsden, as Westerberg & Bowey (2000) illustrated. It is one of the most widespread of breeding birds in the area, being particularly common in the central lowlands, the central valleys of the Deerness and Browney and across the eastern fringe of the county. In the south east, the population is more sparsely distributed in the more agricultural areas and over the coastal plain, directly north of the Tees Estuary (Joynt *et al.* 2008).

The nest sites are frequented throughout the year and consequently, the year-round distribution of the species is, in effect, the breeding distribution. It nests in small, sometimes large, colonies often on the outskirts of towns and villages and in quite large numbers on coastal cliffs. By contrast, in woodland, open parkland and agricultural habitats pairs are more solitary, being spread across the habitat, according to the availability of nest sites, which are usually cavities in trees. In rural areas birds routinely use farm buildings. It has been recorded using a wide variety of man-made and natural nest sites including owl nestboxes, chimneys and, in undisturbed areas, even down rabbit burrows (Westerberg & Bowey 2000).

Larger colonies are found in towns, especially where older housing stock provide chimney nesting sites, and because of a lack of this feature in some of the more modern housing, gaps in colony distribution have been artificially created in some areas (Westerberg & Bowey 2000). Today, a high proportion of Durham's birds probably still nest in the disused chimneys of houses in local villages and settlements. An illustration of the county-wide situation can be found in the lower Derwent valley where birds nest in abundance in older villages such as Chopwell, Rowlands Gill and Winlaton Mill but are largely absent from newer housing areas. The habit of chimney nesting is widespread wherever houses have been converted to central heating from their original coal fires. A reduction in nest site availability may very well be the most significant threat to this species in the county (Westerberg & Bowey 2000). In the largest urban areas, birds are found mainly on the periphery of the developed area, where nesting locations tend to be in closer proximity to farmland, and the feeding opportunities offered by that habitat.

Jackdaws, like Rooks *C. frugilegus*, usually feed within a few kilometres of their breeding sites (Cramp & Perrins 1994). Further west in Durham, colonies can be found at relatively high altitudes, in both the Wear and Tees valleys, where birds often use disused quarries, and feed on the surrounding hill pastures (Westerberg & Bowey 2000). In the south east of the county, this species has long been a common resident, but it has become even more numerous over the years (Joynt *et al.* 2008).

The BTO Garden Birdwatch shows an extraordinarily regular seasonal cycle year on year in the population from 1995 to 2010. The 2010 national results confirmed that the North East region had an average of 35.15 % of its gardens visited each week, the highest for any region. Numbers from this study show a distinct peak in late May as young fledge from their urban natal sites (BTO Bird Trends).

The *Atlas*, considered the county population to be in the region of 40,000 pairs (Westerberg & Bowey 2000). This number was based on detailed survey work undertaken at Durham University around winter roost sites, and thereby may have somewhat inflated the estimate of breeding pairs. Until the early 1990s, the national population level was believed to be stable, after increases during the 1970s (Marchant *et al.* 1990). At a local level, this seems to be the case in Durham, though national census data indicate a fairly steady increase in numbers since the 1970s (Parkin & Knox 2010). This sustained increase was probably associated with improved breeding performance linked to its generalist feeding habits. Other than direct persecution from land managers, the species faces few direct threats in Durham although, in the wider context, large-scale agricultural change might affect the species' breeding density and success in some areas. The sheer number of birds that inhabit Durham makes it difficult to determine the population and fluctuations in it, though it was determined that there were perhaps as many as 700-800 pairs of this species in the north of Tees area over the period 2000-2006 (Joynt *et al.* 2008). The BTO estimated the number nationally as 555,000 territories in 2000, based on the *New Breeding Atlas* (Gibbons *et al.* 1993) data, updated using BBS trend data (Baillie *et al.* 2010).

Recent non-breeding status

During the non-breeding season, birds forage widely on rural grassland for invertebrates. In urban-fringe areas, they often investigate waste paper bins for scraps and birds have frequently been seen searching dung for edible items (Bowey *et al.* 1993). Large post-breeding flocks of birds can be seen in the most productive areas after the young have fledged, a flock of this nature 300 was counted at Greatham on 18 September 1983 (Joynt *et al.* 2008).

The 1980s saw reports of some large single species flocks, with the highest figure noted being of 2,000 birds at Ryton Willows on 23 November 1986. The species is often found in close association with Rooks, it will feed in the same fields and roosts communally with that species (Cramp & Perrins 1994) through the winter, sometimes forming huge pre-roost aggregations. These mixed winter flocks can be very large and tend to occur towards dusk, though it is often impossible to determine the relative proportions of the two species in the make-up of such flocks.

Aggregations numbering many thousands are routinely noted in the Deerness and Browney valleys to the west of Durham City. The flock of c.3,500 that was at Langley Park in December 1985 might be deemed typical, as it is an area where this phenomenon is frequently recorded, but there are many such examples from elsewhere in the county. For example, of more than 2,000 corvids counted at Hurworth Burn Reservoir in late January and early February 2007 a high proportion were Jackdaws. The mixed Rook & Jackdaw roost mentioned by Coulson (1972) still occurs every winter but has rarely exceeded 3,000 birds in recent decades.

In 1990, birds from the Rowlands Gill area were investigated after they were suspected of contaminating doorstep milk with a bacterium causing food poisoning (Bowey et al. 1993). This phenomenon was investigated by researchers at Durham University, studying birds at the Lockhaugh STW, as part of a study into environmental health and outbreaks of Campylobacter. The infective agent was proven to be being carried by local Jackdaws, feeding as they were around the filter beds of the treatment works and also on the cream of local milk bottles. Almost every Jackdaw captured during this study and from which beak and cloacal swabs were taken, proved to be infected with Campylobacter (J. Coulson pers. comm.). The practice and problem was reported elsewhere in the county but it seems to have died out as householders took precautionary measures.

In May 1993, there was an interesting observation of one of the few documented incidents of direct avian predation by this species. This took place in Bowburn, near Durham City where House Martins Delichon urbica were collecting mud for their nests when a Jackdaw suddenly swooped down picked up a martin and took it to a roof top, where it began to eat it (Evans 1997).

Nordic Jackdaw
Corvus monedula monedula

Although Parkin and Knox (2010) refer to birds regularly migrating along the east coast, there seems little reliable evidence for this phenomenon from local modern observations and there are few reliable recent reports of coastal visible migration, though Nelson's old assertions are acknowledged (Nelson 1907). Nonetheless, it seems likely that the winter numbers may include some continental migrants and birds showing the characteristics of the Scandinavian race, *C. m. monedula*, were noted on at least eight occasions in the period 2000 to 2011. Most sightings were in early spring, between 29 March and 4 May, and concerned one or two birds at localities close to the coast, such as Boldon Flats, Cleadon and Whitburn. A tight party of 36 Jackdaws arriving from the sea on 29 March 2008 may well have been from Northern Europe. Inland reports in winter, such as at Sedgeletch on 26 January 2008 and Herrington County Park on 1 December 2009, also hint at autumn arrivals from the Continent spending the winter in our area, with direct evidence of autumn arrivals provided by a Nordic bird at Trow Quarry on 29 October 2005. It is quite conceivable that coastal parties in spring and autumn probably involve more continental birds than is sometimes appreciated.

Distribution & movements

This species breeds across much of Europe, Iceland and Scandinavia, northern Fennoscandia excepted. It is also found as a breeding bird in eastern Asia and into northwest Africa. Some populations, such as those in Western Europe, are partially migratory wintering within the breeding range of the more southerly located birds, but the majority of birds are more or less sedentary. The northern races, found in Scandinavia are much more migratory, wintering well to the south of their breeding range.

The largely sedentary nature of local birds is evidenced by the sparse ringing data relating to it in the county. One that was found dead at Cowpen Marsh in August 1994 and another that was shot on agricultural land in Wynyard on 27 March 2000 had both been ringed as nestlings at Crookfoot Reservoir, the first in May 1985 and the second in May 1995 (Blick 2009). An aberrant, coffee-coloured bird (exhibiting non-eumelanic schizochromaticity), hatched in Chopwell in 1988, was observed nesting building in 1991 only 400 metres from its birth place, and was still present in 1992 (Bowey et al. 1993). This species is normally rather wary and can consequently be difficult to catch for the purposes of ringing but during May and June birds become bolder as they seek food for growing broods. In the months of May and June in 2008, 2009 and 2010, 38 birds were ringed at Barmoor, near Ryton. At this site, on 18 December 2010, a bird was re-caught that had been ringed at the same location on 25 January 2002. The greatest known distance moved by a bird ringed at this site was of one caught on 16 June 2002 and that was found dead three kilometres to the north at Heddon-on-the-Wall on 9 April 2004.

Rook
Corvus frugilegus

An abundant and widespread resident.

Historical review

This is probably the most numerous of the crow family in many parts of County Durham, although it may be locally out-numbered by Jackdaws *Corvus monedula* in some areas. Cultural evidence gives a strong hint to this species' long-term presence in the west of County Durham. Rookhope, just off the upper Wear valley, is derived from the Old English, "*hroca, hop*" meaning 'rook valley', presumably indicating the species' presence in that area before the Norman invasion (Yalden & Albarella 2009). Rooks would appear to have always been common and widespread. There are three references within the 14th century Bursar's Rolls referring to crows, including Rooks and Magpies *Pica pica*. In 1348 and 1368-1369 the '*corvos*' were believed sufficiently numerous to require thinning down, while in 1378-1379 young Rooks were regarded as an article of food (Ticehurst 1923).

In the late 19th century, Hancock (1874) observed that there was "*scarcely any well wooded area without a colony*". He suggested that the persecution of raptors in the north east had led to a consequential increase in this species' numbers. Temperley (1951) noted that Rooks were abundant, widespread and increasing in number, a trend which he considered had commenced locally during the first part of the 19th century. The attraction of certain areas to birds over considerable periods of time is illustrated by the fact that rookeries at Axwell Park and Gibside in the lower Derwent valley, which were still active in the 1960s, were mentioned as having Rookeries by Robson in his text in the late 19th century (Robson 1896).

A number of local surveys have been conducted in Durham since the beginning of the 20th century. Temperley (1951) reported that in 1931, a census of rookeries in a sixty square mile area around Darlington had revealed a total of 36 rookeries, which contained a total of 2,376 breeding pairs, equating to almost 40 pairs per square mile. Surveys in Teesdale by the pupils and staff of Barnard Castle School, during the 1930s revealed an increase in number there of more than a third between 1933-1935 and 1939-1941, from an average of 877 nests per year to 1,214 nests (Temperley 1951). An April census of rookeries within a two-mile radius of Durham City in 1953, revealed 14 rookeries containing 1,022 nests, a density of about 85 breeding pairs per square mile. Over 200 Rooks were known to have been shot in this area during that month (Temperley 1954).

Recent breeding status

Across Durham, Rooks favour the agricultural mosaic of arable, pastoral and woodland found along the county's main river valley systems. It is less common in the county's upland areas, although there are colonies in the west and northwest of the county and birds, probably based in Cumbria can be regularly seen high on the county boundary with Durham on the A66 trans-Pennine route. It extends as a breeding species up to elevations of 360m above sea level for example, at Cowshill, in upper Weardale. The combination of harsher environmental conditions at higher altitudes, the presence of dense heather, bracken and tall grass–restrict ground feeding opportunities, and the lack of suitable nest trees–contribute to the species' relative scarcity, but not its absence, in such areas. Birds will scavenge across open upland pasture with rushes and sedges.

From central Durham eastward towards the coast, colonies tend to become smaller with gaps in the distribution in some lowland areas perhaps due to the more extensive tracts of urbanised, tilled or densely forested land. On the magnesian limestone, a paucity of suitable trees may limit numbers. Within and around towns, such as Durham City, Herrington and Barnard Castle, scattered complexes of small and large colonies have formed. These tend to be within visual and behavioural contact of each other and might actually be considered to constitute a single colonial unit. In 1986, counts at such clustered colonies revealed at least 1,608 nests around some 15 rookeries, largely in four clusters in the north east of the county.

Rookeries are relatively easy to census for a number of reasons. Sites tend to be traditionally used so they can be easily re-surveyed, although small new colonies do spring up between breeding seasons. Birds also often attend colonies throughout the year, being especially obvious in late winter and early spring, when nest refurbishment commences. Finally, intensive breeding activity occurs early, with maximum nest numbers being attained by mid- to late April (Marshall & Coombs 1957, Griffin 1999, Westerberg & Bowey 2000). With the exception of colonies in pines *Pinus* spp., nests can easily be easily counted by surveyors, as long as this is done before leaves erupt in the spring.

The main baseline data for the species locally come from the BTO's National Survey of Rookeries, carried out in 1975/1976. The results of this survey showed that Durham's colonies had the highest mean size of those in any English county, forming part of a pattern of larger rookeries in more northerly areas (Sage & Vernon 1978). In Durham, counts in excess of 300 nests came from rookeries at: Whitehill Hall Wood, Chester-le-Street (543), Coalpark Gill (356), Greatham Beck (337) and Coppy, near Beamish (319).

Survey coverage in 1980 and 1996 was not as complete as the full national survey of 1975/1976. Counts at sites assessed as directly comparable with those surveyed in 1975/1976 showed an overall increase of 4.5% in 1980. Additional sites with over 300 nests were found at Cole Hill, near Crookfoot Reservoir (334) and High Coniscliffe (305, which showed an increase from 250 nests in 1975). The more limited survey in 1996 indicated an overall decrease of 5.2% at rookeries that were comparable with those surveyed in 1975/1976. The decrease in the mean colony size was explained, in part, by examination of information from the four rookeries with over 300 nests in 1975/1976. Coalpark Gill and Greatham Beck were not covered in 1996; Whitehill Hall Wood had declined to 134 nests, probably largely due to tree-felling for housing development, and Coppy to 123. At Coppy, there was evidence of shooting in 1975 (Sowerbutts 1998).

Rookery Survey Results in Durham, 1975-1996

Date	No. of Rookeries Counted	Number of Nests Found	Mean No. of Nests per Colony in Survey	% Change from Previous Survey
1975/76	284	15,345	54.0	
1980	236	13,234	56.1	4.5% increase
1996	137	6,112	44.6	decrease of 5.2%

The sample surveys suggest that the population in Durham remained fairly stable between 1975 and 1996. Nationally, an increase until the 1950s was followed by a decline in the 1960s (Sage & Vernon 1978) and then a recovery, assessed at around 40%, although only 9% in Northern England, by 1996 (Marchant & Gregory 1999).

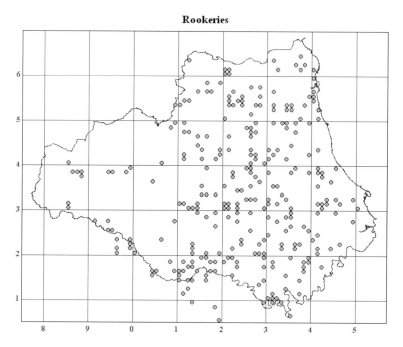

Rookeries

Rookeries in County Durham by one-kilometre Square

In the north Teesside area, declines amounting to 38% in the numbers of nests in monitored rookeries were noted over the period 1975-2005; from 556 to 334 nests (McAndrew & Clayton 2005). During work on *The Breeding Birds of Cleveland*, rookeries were recorded in some 25 tetrads, north of the Tees, probably amounting to around 600-650 pairs of birds in the period 1999-2006. There have been some considerable declines in this part of the county the last 20-30 years and it can no longer be considered an abundant species in this area. As an illustration, at Greatham—numbers fell from 404 nests in 1975 to 127 by 2005 (Joynt *et al.* 2008).

Between 1970 and 2010, over 500 colonies have been occupied across the county, some of these were small and transitory others large and have persisted for decades. Many of Durham's rookeries can be found in urban fringe locations. In these situations, they occur in often occur in the grounds of hospitals, churches and old houses where stands of mature trees remain in relatively open and somewhat protected surroundings. It has been suggested that the use of such locations may represent a historical response to the persecution of colonies in farmland copses (Bannerman 1953). It is believed that shooting now routinely affects very few rookeries in Durham, though in 1996, a local study revealed that around 5% of those investigated had been 'shot out' (Westerberg & Bowey 2000). Like many farmland birds, Rooks have been adversely affected over recent decades by land-use changes and the increased use of pesticides (Marchant *et al.* 1990).

The *Atlas* distribution map suggested that for most resident species, atlas sightings legitimately related to breeding distribution but for a wide-ranging species like Rook this may not hold good. The consequence being an unintentional overstating of the species' breeding status; the bird's presence in a tetrad does not mean that it breed there. Nonetheless, many of the mapped tetrads in the *Atlas* would have held active rookeries (Westerberg & Bowey 2000).

By contrast, the accompanying map is of Durham rookeries and is based on data from a number of sources, allowing the mapping of all known recent breeding colonies at a one-kilometre square resolution (D. Sowerbutts pers. comm.).

A 2005 review of the monitoring work undertaken around the rookeries in the Houghton-le-Spring and Hetton area over a 20-year period revealed a dynamic situation. This area held 754 Rook nests in 1986, 610 nests in 1998 and 545 nests in 2002. At that latter point, there had been a between-year reduction of 113 nests from the previous season. The 2005 review showed that two rookeries had been lost in the area since 2003, one of these probably due to disturbance, the other as a result of tree-felling. The remaining five colonies counted regularly since the mid-1980s, showed a combined total decline of 384 nests or 51% over a 19-year period, though this was not necessarily typical of the county position as a whole. This local trend continued into 2007 when there were just 250 nests, a further reduction of 116 nests. Other declines were noted at Eppleton Hall, where a maximum number of 711 nests in 1996, had declined to 139 by 1999; the former count being the largest count at any rookery for many years in Durham.

This remains one of the most widespread, numerous and visible species in Britain and in 2009, on all Breeding Bird Survey plots, 37,822 Rooks were recorded; this number was only exceeded by the numbers of Woodpigeon and Blackbirds recorded (Risely *et al.* 2010). The local breeding population varies with evidence pointing to some overall decline. The *Atlas* estimated the population to be around 18,000 pairs (Westerberg & Bowey 2000) but this may now be too high an estimate considering observed local declines.

Recent non-breeding status

Feeding flocks in autumn and winter are widespread on arable farmland and grassland, even into the west of the county. They are often in the company of Jackdaws *Corvus monedula*. Numbers can be high, particularly in late afternoon pre-roost gatherings. Records of up to 3,500 birds were received annually in the 1970s and early 1980s with a mixed flock of 10,000 reported at Heighington at the beginning of 1970 though today that particular mixed, mid-winter flock is typically less than 3,000 (D. Raw pers. comm.).

Over this two decade period, and into the 1990s, large flocks were particularly obvious in the lower Browney valley, around Witton Gilbert, Lanchester and Burnhope. They were also prominent in the Crookfoot Reservoir area, in the south east of the county. Wintering numbers in the area of the Derwent Reservoir rose from 463 in February 1986, to 700 at the beginning of March and 1,500 a fortnight later. In 1987, 3,000 birds at Crookfoot reservoir on 10 February had declined to 100 by the end of the month, presumably as wintering birds moved into breeding colonies. A mixed flock of Rooks and Jackdaws at Heighington contained 2,000 Rooks in December 1987, despite the fact that the nearest rookery held only about 200 nests; indicating some level of winter influx to that area.

Through the first decade of the 21st century, reports of large winter gatherings notably declined, though the species remains common and flocks of 300-500 birds are a routine sight. The regularly observed flocks of thousands of birds, so noticeable through the 1960s, 1970s and into the 1980s were much less often reported. Whether this is related to a reduction in numbers, a change of foraging strategy or, less likely, a change in observation pattern is not known. Larger recent counts included: up to 1,000 at Brancepetth Beck, in November

2001; 1,500 on organic farmland at Houghhall, in January 2003; and, a mixed flock of 3,000 Rooks and Jackdaws at Annapoorna Farm, near Crook, on 13 September 2003.

The fact that Rooks travel prodigious distances to their winter roosts was demonstrated in early survey work by J.M. Dewar in during 1904-1906. He noted that the Dilston roost, located in the Tyne valley towards Corbridge, Northumberland, drew birds from as far away as Ryton in the north west of County Durham. He also noted the behaviour of birds at "*a small rookery at Stella-on-Tyne. It was situated about midway between the large Blaydon and Ryton rookeries. The Stella rookery belonged to about two dozen birds, and was situated in quite small trees. Each morning, in winter, the nests were visited by a dozen birds that came from Ryton rookery (Dilston roost), and a similar dozen, which came from a rookery at Axwell Park (Blaydon) (Durham roost). The two parties met apparently on good terms with each other. Both the Ryton and Blaydon rookeries were large and probably of ancient origin. In the evenings the Stella rookery was revisited by the same number of birds, half of which went on to the Ryton rookery, the other half to the Blaydon rookery. These observations, which were often repeated, suggest that the colonization of the Stella rookery took place from two different stocks, since the Ryton Rooks roosted at Dilston and belonged to a feeding territory distinct from that of the Blaydon Rooks which roosted near Durham. The Stella Rooks must, in part at least, have been hatched at Stella, and it was interesting to find that they were still bound by a Blaydon and a Ryton tradition*" (Dewar 1934).

Detailed study of this species by Durham University during the 1990s indicated that tens of thousands of Rooks were roosting in the winter at Coalpark Gill, near Langley Park at that time, and that these were coming to the roost from considerable distances around the roost site, despite the fact that a number of major rookeries were located within a 12km radius of the roost site, these were bypassed by the birds on their way to the roost (Griffin 1999 & 2000).

An all-brown bird accompanied normal coloured birds at Hamsterley Hall in the Derwent valley on 14 August 1964 and birds with white wing feathers or other white plumage markings have been noted on a number of occasions over the years.

Distribution & movements

This temperate and boreal species breeds across much of the Palaearctic; it is largely absent from northern treeless areas. The most northerly populations are partial migrants, particularly in hard winters, though most birds to the south of the range are sedentary (Parkin & Knox 2010).

Local Rooks appear to be largely sedentary, other than their sometimes extensive daily movements to and from roosts or to feeding sites. Temperley (1951) believed that continental visitors augmented the county's resident birds during the winter. Although there is little evidence of this in the form of ringing returns or visible migration, the size of winter flocks is suggestive of an influx from elsewhere.

A few birds are occasionally seen passing along the coast or even apparently arriving from over the sea during migration periods, these birds may originate from northern Britain or the Continent, although there is little independent evidence for this (Wernham *et al.* 2002). There are some foreign-ringed recoveries in Britain, but a more local origin for such Durham sightings seems to be indicated by the sparse ringing recoveries. A bird recovered in Guisborough on 21 March 1981, had been ringed at Consett on 5 February 1981 and a bird ringed at Durham on 9 February 1963 was recovered on 23 October 1964 at Longframlington in Northumberland.

Carrion Crow
Corvus corone

A common and widespread resident.

Historical review

There are old references to the species in the county, three of these coming courtesy of quotations from the cellarer's rolls, which refer to corvids, more specifically Crows, Rooks *Corvus frugilegus* and Magpies *Pica pica*, in documentation for 1348, 1368-1369 and 1378-1379. The first of these indicates that all three species were numerous on the manors of the Priory (Ticehurst 1923). In the 1368/1369 reference, the phrase *In uno capucio coruino* is somewhat obscure but it has been interpreted as meaning a 'hood' made of crow-skins. If this is correct it is difficult to imagine the utility of such a garment.

As early as 1533, the Carrion Crow was believed to be so numerous and considered such a destructive and 'evil' bird, that Parliament passed an act for its elimination (Temperley 1951). It was not until the more widespread rearing of game that Hancock (1874) was able to write, "*it is rapidly disappearing under the persecution of the game preservers*". This seemed to be confirmed by Tristram (1905), who said that it was rare except on the moors, '*where it may be occasionally seen*'.

During the first half of the 20th century, following the decline in the number of gamekeepers after the two World Wars, the Carrion Crow became a widespread resident, breeding in nearly all woodland and in isolated hedgerow trees (Temperley 1951). It thus reclaimed much of its former range as recorded for example by Hutchinson (1840) who confirmed that it "*frequents all the cultivated districts and the borders of the moors*".

After the First World War, the reduction in persecution of the species, coupled with changes in agriculture between the wars, allowed the population to increase. By the late 1930s, populations had recovered in most of the eastern parts of County Durham, and H.M.S. Blair recorded the species increasing in Harton, South Shields. Post World War 2, several large communal roosts were recorded by Temperley, such as the flock of seventy at Ravensworth in the Team valley on the rather late spring date of 5 May 1946 (Temperley 1951). Attitudes towards the species however, particularly amongst the agricultural community, remained very negative and have changed little since the Ministry of Agriculture Fisheries and Food (1948) concluded of the Carrion Crow that, "*on balance, it is a harmful bird, and is especially an enemy of the poultry farmer. The good work done by it in the destruction of insect and other pests by no means compensates for the bad, and farmers generally are advised to do what they can to destroy (it) when found on (their) land.*" Nonetheless, the species' expansion has continued to the present both nationally and locally, as indicated by CBC data for the species since 1962 (Marchant *et al.* 1990).

Recent breeding status

Today, this species is a widespread and common breeding resident in the county and probably significantly more numerous in Durham than it was in the 19th century. This is most likely to be due to a reduction in the total number of gamekeepers particularly in the eastern half of Durham, although it is still shot on some farms. Baillie *et al.* (2010) reported that shooting bag returns show little change in the number of crows killed by keepers since 1960, suggesting that control on shooting estates has stabilised numbers there and that increases have occurred largely on un-keepered farmland and other habitats.

From east to west, it can be found in a wide range of different habitats, from seashore to woodland and farmland (Westerberg & Bowey 2000) and up to the moorland fringes and upland heath. Since the beginning of the 1900s birds have adapted to live in more urban situations alongside man. Recent examples of such behaviour include pairs that nested on top of car-park lighting at the Gateshead MetroCentre and on electricity pylons at Shibdon Pond (Westerberg & Bowey 2000). A most unusual nest site was found in the Tees Estuary in 1994, when a pair nested on a navigation buoy in the middle of the river mouth (Joynt *et al.* 2008).

In the Pennine uplands, birds can be found on land well over 300m above sea level as far west as Forest-in-Weardale, albeit in lower numbers here than in other areas of the county, due no doubt, to local persecution.

It is a widespread but scattered species in the north Tees area, with around 350-400 pairs (Joynt *et al.* 2008). During the *Atlas* survey period, it was found to be present in over 420 tetrads across the county and from this presence, using the nationally derived breeding density of 9 territories per occupied km^2 (Gibbons *et al.* 1993), a county population estimate of around 8,800 pairs was made.

Nationally, the species has increased steadily since the 1960s. The New Breeding Atlas (Gibbons *et al.* 1993) recorded 790,000 territories in 1990, increasing to 987,500 pairs in 2000 (BTO - CBC/BBS trend data). It has been suggested (Crick *et al.* 1997) that the increase is linked to improving nesting success and earlier laying, perhaps arising from climate change, and the successful exploitation of changing food resources. Nesting begins in March and can continue through to June, though clutches are laid later in the west of the county. Typical clutch size is from four to five eggs, with incubation lasting 17-22 days and young fledging at 29-30 days (Coombs 1978). Picozzi (1975) reported 2.94 young per successful pair, but across all attempts only 1.1 to 1.7 young.

Home ranges are vital to this species. In the breeding season these vary in size from 14 to 49 ha in extent and they are defended most vigorously, particularly near the nest. Flocks of non-breeders may also be present, these are mostly immature birds although some adults, of up to five-years of age, and paired, may be involved. They may even be in a suitable physiological condition to breed but without being able to establish a territory, it is impossible for them to do so (Coombs 1978).

One of the most striking characteristics of the Crow is its adaptability to different food sources. Prey includes young and adult birds, small mammals, carrion generally, grains and other vegetable matter, fish, amphibians and invertebrates of grassland and seashore (Coombs1978). Coastal Crows are regularly seen dropping mussel shells in flight in order to smash them open. In urban areas they will scavenge any discarded food. Land-fill rubbish disposal sites and sewage treatment works are particular haunts of the species both locally and nationally. Given this species' exploitation of a wide variety of food resources, and its adaptability to nesting in diverse locations, the continued presence of Carrion Crows in the county, probably close to its present population levels, seems assured.

Recent non-breeding status

Unlike Rooks *Corvus frugilegus* and Jackdaws *Corvus monedula*, this corvid does not usually form large flocks, so reports of large gatherings tend to be few. Nonetheless most years produce a number of such gatherings and these are often associated with abundant food sources, such as landfill sites or occasionally, recently ploughed fields, cut hay meadows or other particular feeding opportunities. For example, in 2007, gatherings included around 200 birds noted near Crook in August and September, and a similar number noted at South Church, Bishop Auckland, in the same season. Alternatively, such associations may relate to pre-roost gatherings as illustrated in the winter of 1989, when over 100 birds would regularly gather at dusk on Goodshields Haugh by the River Derwent (Bowey *et al.* 1993). In the south east, smaller groups of 30-60 have occasionally been recorded at sites such as Cowpen Marsh and Preston Park (Blick 2009).

Local, and indeed most, Carrion Crows are considered resident and sedentary (Parkin & Knox 2010) though in 2007 small numbers were noted flying over the sea from Whitburn Observatory on several dates late in March. It was once considered not that unusual to see individuals or even small groups moving along the Durham coast or arriving from the sea at coastal watch points. Such activity predominantly occurs in early spring, for example when 22 were observed at Hartlepool Headland on 8 April 1972 (Blick 2009). The actual occurrence of coastal passage is difficult to judge, because of the presence of local birds, but in autumn 2007, the single flock of 36 birds that flew south down South Shields' Leas, prompted the observer to consider whether these might have been migrants; perhaps from further north in the UK (Newsome 2008).

Distribution & movements

Carrion Crows breed in Western Europe, from Scotland south to Iberia, east to Germany and also in the Alps. Another closely related race breeding in Asia may be a separate species. Until recently, the Carrion Crow was long considered, by most authors, as conspecific with the closely related Hooded Crow *Corvus cornix* but these are now thought to be best treated as two species (Parkin & Knox 2010).

There is little ringing data referring to this species in Durham , but an adult ringed at Lockhaugh STW on 23 May 1995 was found dead over five years later, at Leazes, Northumberland on 20 September 2000. This remains one of the few recoveries for this species which is only rarely ringed in the county.

Hooded Crow
Corvus cornix

A scarce, irregular and declining winter visitor and passage migrant, which is very much less common today than it was 30-40 years ago.

Historical review

Though clearly very closely related, Hooded and Carrion Crows *Corvus corone* are now thought to be best treated as two distinct species (Parkin & Knox 2010).

The Hooded Crow was first documented in the county by Willughby, who said it was frequently recorded around the mouth of the Tees, presumably in winter (Willughby 1678). It was known as a winter visitor to Hutchinson (1840), mainly singly but occasionally in small parties. By Hancock's time, it was described (Hancock

1874) as a common winter visitor, most frequently noted along the sea shores; hence its old name of 'sea crow'. Backhouse (1885) stated that it was common in upper Teesdale in the winter but further north in the county, Robson (1896) believed that it occurred only "*sparingly*" in the Derwent valley. Holloway (1996) noted that it had bred intermittently in the Borders and down into north east England as far south as Flamborough Head in Yorkshire. In the light of this it is possible to speculate that at some time in the past it may have bred occasionally in County Durham.

Nelson ascribed an average arrival date for winter migrants to Teesmouth to be around 6 October, with the earliest report being on 10 September 1880 (Nelson 1907). He noted that overwintering birds occasionally remained into spring, for example the bird that was noted on 17 May 1902. In the early part of the 20th century, the Rev. George Courtenay commenting upon its status around Teesside noted, "*The vanishing status of the Hooded Crow in Durham (so far as my observation goes) is borne out by my experience at Teesmouth. Between 1909 and 1913 I came across it five times. The first time there were four or five together, about a pool on the links; on the other three occasions there was just a single bird. And since 1913 I have not seen another*" (*The Vasculum* Vol. XIX 1933).

Temperley's own observations indicate that it was a regular winter visitor to Durham, at least until the early years of the 20th century; becoming less common and more irregular in its appearance as time wore on. He highlighted the paucity of records from 1925-1950, with the exceptions of the winters of 1941/1942 and 1943/1944, when small flocks were noted along the coast (Temperley 1951). Around this time, Dr. H.M.S. Blair, from his perspective at South Shields and his exceptional knowledge of Scandinavian birds, commented that he thought that its declining status may have been in part determined by a change in its behaviour across the North Sea. It had become, as he had noted when on expedition, more of an urban-based scavenger during the winter in Scandinavia, which may have affected its need to migrate and consequently its status along the northeast coast as a winter visitor.

David Simpson reported that in the 1950s he regularly saw small parties of 'hoodies' flying southward over Shotton Colliery during October, but since then they had become very scarce. On the 27 October 1971, a Hooded Crow was seen at Edder Acres Farm, Shotton Colliery with a party of Carrion Crows and this bird stayed all winter, until March 1972; his most recent observation of the species in that part of the county (Simpson 2011).

In the early part of 1959, three wintering birds were at Seaton Carew (Grey 1960) and the following year, four were near Rowlands Gill in March (Bowey *et al.* 1993), when one or two birds were also at Teesmouth on dates in February, March and in April. The gathering up to 34 birds at Graythorp, at Teesmouth, in February 1961, seemed to be a return to the sort of numbers experienced in the region in earlier times and was part of an increasingly rare phenomenon for the county, an influx of Hooded Crows. Certainly the species appears to have occurred in much larger numbers in the north east in the past, as evidenced by the report of up to 28 seen near Ponteland, Northumberland, in early 1964 (Kerr 2001).

Influxes occurred to the south east of the county, during the winters of 1947/1948, 1960/1961, 1969/1970 and 1976/1977. In these, about 40 to 60 birds comprised the maximum totals, although it is likely that over 100 birds passed through the whole of Cleveland in some of these winters (Blick 2009). The numbers recorded from neighbouring Northumberland, indicate the nature of the species' decline in the north east of England over the latter half of the 20th century, particular in the period 1960-1980 (Kerr 2001). Over this period, the species changed from being a regularly recorded wintering species, sometimes in large numbers, to an increasingly scarce passage migrant.

During the 1970s, even fewer birds were recorded in the county, but at least one bird was noted on a regular basis at Shibdon Pond over the middle part of this decade, possibly the same returning individual being seen in the winters of 1973, 1974, 1976 and 1978 (Bowey *et al.* 1993).

Recent status

The Hooded Crow used to be recorded each winter in Durham, certainly up to the mid-1980s, with anything from three to ten individuals per annum, but occasionally in considerably much larger numbers than this. Hooded Crow then declined quite dramatically in the Durham area over the last quarter of the 20th century, and even more so in the first decade of the twenty-first century. There were few sightings in the decades of the 1980s and 1990s, with a marked decline being apparent over this period. From a maximum of 12 in 1980 (a figure which may have included multiple sightings of wandering individuals) to six in 1981 and seven in 1984, the average subsequently fell to 2.2 birds per annum.

During the period covered by the *Atlas*, a Hooded Crow was found, paired with a Carrion Crow *C. corone* at Warden Law in 1991 and was present until at least July of that year (Westerberg & Bowey 2000). In 1994, a bird was found at Low Westwood on the unusual date of 24 May, but breeding evidence was not forthcoming (Armstrong 1994). That a bird may have made a local mixed pairing attempt was indicated by the appearance of a hybrid Carrion Crow x Hooded Crow, which first appeared at Low Barns, Witton-le-Wear, in 2000 and itself bred, subsequently being observed feeding young. Later sightings that year probably referred to the original bird, but could have involved one or more descendants (Armstrong 2005). During 2006, there were several reports of birds from the Low Barns area but all of these probably referred to the long present hybrid birds. The future picture in this area looks likely to be further complicated by the fact that six birds were seen here in September, suggesting that the long-staying hybrid bird raised a family earlier in the year (Newsome 2007). Birds and the descendants of the originals were noted in this area at least until 2008 (Newsome 2009). Another hybrid was present in Deepdale, in Teesdale, on 26 June 2005.

This species' modern low level of occurrence continued through to 2011, the pattern showing a slight preponderance for sightings in the year's first quarter. In recent years, its local status has become established as a scarce winter visitor or passage migrant. In 2005, there were probably two birds present in the east of the county in February and on the coast in March as well as, rather unseasonably, in July at Hawthorn. In the autumn a migrant was at Marsden on 11 September 2005. In 2006, reports came from the Houghton-le-Spring landfill site at Sedgeletch and the old power station site at Stella from 13 December to the year's end. In late December 2007, a single bird found at the Path Head landfill site, near Blaydon, lingered until mid-January 2008. Then further singles were on at Ragpathside Farm, near Cornsay on 1 March and at Barningham on the rather unusual date of 1 June. In 2009, a bird was at Seaton Common in early June, the first at Teesmouth for a number of years, whilst one flew north over Cleadon Hills on 15 March. Further birds were seen at Seaton Common in late April 2010 and in February 2011, whilst migrants were at Hetton Lyons County Park on 25 March 2011 and Rainton Meadows on 23 November 2011.

Distribution & movements

Hooded Crows breed throughout northern, central and eastern Europe, and in Britain, mainly in western Scotland. They are largely sedentary although continental birds are partial migrants or fully migratory, moving to the south and west wintering within the breeding range of the species, or that of the Carrion Crow. In Scotland, although a few pairs are scattered along the east coast, the main population is concentrated to the north and west of the Great Glen. Whilst this might suggest that Hooded Crows are birds of highland areas, this is not the case in Ireland or throughout Europe and the Middle East (Parkin & Knox 2010).

Where Hooded Crows and Carrion Crows meet in Europe there is an extensive zone of hybridisation varying from 24 to 170km in width (Coombs 1978) and extending over a distance of about 1300km. Since the late 1940s there has been a steady extension north and west of the range of Carrion Crows across Scotland and a roughly equivalent contraction in the Hooded Crow's range, which receded in a roughly complementary manner. The reasons behind this change in status are far from clear although it has been speculated that the shift may relate to climate change. The decline in numbers recorded in Durham may relate to the northwards shift of this hybridisation zone and the 'squeezing out of Scottish hoodies. It is possible that the long run of milder winters from the mid-1980s to the early 2000s, meant that Scandinavian birds did not have to move south and west to meet their winter energetic requirements, as they foraged in new urban locations, as intimated by Hugh Blair's observations in the 1950s.

Raven (Northern Raven)
Corvus corax

An irregular but increasing, mainly winter visitor, and an extremely rare breeder.

Historical review

The earliest records of this species in the north east of England comes from remains excavated from Roman sites such as those at Corbridge, Northumberland (Miles 1992), just a few miles north of the current Durham border. There are a number of records of Raven remains having been found in Durham archaeological sites, from the mediaeval and post-Mediaeval periods (between 1000 and 1400 years before the present) as well as there being a number of Raven-themed place names (Yalden & Albarella 2009). The cultural resonance of this species locally is confirmed by several place names in the county's uplands such as Raven Hills and Raven Seat on Middleton Common, and in the east at Ravensworth Castle and Estate in the Team valley and another Ravensworth just over the border in North Yorkshire. Its etymological origin indicates the species' local prominence and its presence in the period before the Norman invasion. The name itself derives from the Old Norse, *"hrafn"* (Yalden & Albarella 2009).

Marmaduke Tunstall recorded that the species nested in the area of Wycliffe-on-Tees '*some years ago*' in a manuscript written in the late 18th century (Tunstall 1784). By the end of that century Raven was recorded as breeding in urban locations, such as the lantern tower of St. Nicholas' Cathedral, Newcastle upon Tyne; just a few hundred metres north of County Durham (Kerr 2001). Around the year 1800, according to Yalden & Albarella (2009), the Raven bred in every English county but it had gone from much of the lowlands and was ever more restricted to the northern and western counties by the end of the century (Holloway 1996).

Temperley (1951) quotes Backhouse (MS. 1834) recording that Ravens were then breeding *"in the rocks at Datton* (Dalton?) *near Sunderland and also at Marsden",* and there is some suggestion that they may have lingered on until the latter part of the Victorian period in this area. Hutchinson (1840) recorded that inland breeding sites at Wolsingham Park and along the Derwent Banks were lost by 1840. In the far south west of the county, illustrating many land managers' long-term antipathy towards the species in the area, eleven birds were killed on Bowes Moor in 1880 and the corpses of these were seen by Thomas Nelson on the wall of the local keeper's cottage (Mather 1986). By 1885, it was reported by Backhouse as to becoming, *"rapidly extinct"* in the Tees valley, though it was still present in the Falcon Clints and Cronkley Scar area.

The early 20th century was an apparently bleak period for the species in Durham and relatively little is documented of its status at this time despite Temperley's summary work. It was somewhat surprising then that in the summer of 1947, rather intriguingly considering its then local status and range, one was noted by Charles Hutchinson, initially being mobbed by Swallows *Hirundo rustica*, over Lockhaugh in the lower Derwent valley, on 28 June. What was presumably the same bird was noted on at least two other occasions over the subsequent week flying over from the nearby Gibside Estate (Bowey *et al.* 1993). A pair was present on the Lambton Estate, during 1948, and this was perhaps the last lowland breeding presence in the county for almost half a century. At this time, these two instances would have been have been very unusual lowland sightings in the county.

By the middle of the 20th century Temperley (1951) was describing it as a resident of the west of the county, where one or two pairs then still attempted to breed. He noted that it had suffered years of persecution and it would, no doubt, have been widespread before the advent of game preserving. In 1952, a pair attempted to breed in upper Teesdale, a nest was built and five eggs were laid, but the young were not reared presumably as a result of human interference. This same pair had previously attempted to nest on the then 'Yorkshire side' of the Tees, but the nest had been destroyed (Temperley 1953).

A few years later, in 1955, a gamekeeper in the Teesdale area boasted of trapping several Ravens during the previous winter and 'several' nests were destroyed during that spring. One or two nests were built in trees, an unusual practice in the county at that time. This pattern of occasional attempts to breed, only to be thwarted by persecution, continued through the next decade, with a nesting pair in Teesdale in 1961 having their young destroyed in the nest (Bell 1962). This was a lean time for the species in the county and a bird over Hartburn, Stockton-on-Tees on 14 April 1973 (Blick 2009) was the first Teesside record since 1936. Two birds over Widdybank Fell on 2 July 1973 may have been wanderers from Cumbria or failed local breeders (Unwin 1974).

Although, the species has undergone considerable decline and retraction of its national distribution in recent centuries there was, however, a marked re-occupation of its former range during the latter part of the 20th century

(Gibbons *et al.* 1993) and the early part of the 21st century. However, this has not occurred in County Durham over this period, and is unlikely to do so unless it is granted a stable presence in the west of the county from which it could expand.

Recent breeding status

The Raven is a bird of wide distribution in the northern hemisphere and it is adapted to a remarkable range of climates (Ratcliffe 1997). In northeast England it has long been considered a bird of the uplands but this represents a contraction of range to a refuge from persecution, and a not very successful one at that. The Durham uplands are essentially either grouse-moor or sheep-stocking areas, or both, and the species attracts the attention of both sets of land 'managers'.

This attention is well illustrated by an analysis of the species' known breeding performance in the county between 1990 and 2009. This period began well when a pair in Teesdale finally succeeded in rearing two young in 1991, only for the juveniles to disappear soon after fledging. Over the whole of this period there were a total of 14 known nesting pairs which reared just 15 young across five successful breeding seasons. No more than two pairs were known to attempt nesting in any one year and never more than a single pair was known to be successful in these twenty breeding seasons.

In 1996, a pair laid one egg at a site in Teesdale but the attempt came to nothing. However, in the same season, a pair in Lunedale successfully fledged four young from a very prominent nest. The success was not repeated in 1997. In 2005, the nine reported records of the species, between March and September, mainly referred to single birds and were all confined to the south west of the county. The notable exception was of two birds soaring over Bishop Middleham Quarry on 15 May. This was remarkably the same date as two birds at this location in 2004 and suggests that the species may be prospecting for a foothold in the east as has occurred in many southern counties of England in recent years (Brown & Grice 2005). In 2006, pairs were noted at up to six sites early in the year with display evident, and at one site in upper Teesdale four young were successfully fledged. Breeding was also attempted in this area in 2007 but the outcome was unknown.

Birds are usually now present in small numbers in the Dales and the upper Derwent valley during the breeding season without nesting necessarily taking place, for example five birds were in Lunedale in March 2002 and 13 were on Newbiggin Common in March 2009, when birds were noted at 10 other sites. A family party of five birds south of Blanchland in August 2003 could have originated on either side of the county boundary (A. L. Armstrong pers. obs.) and an adult in heavy wing-moult in June 2007, which was observed flying from Northumberland into Durham over Blanchland village, suggest that birds may be breeding in this part of the Derwent valley, perhaps as an over-flow from the north. Certainly, by the end of the first decade of the 21st century, the Northumberland population was burgeoning with 15 pairs in 2009 fledging at least 29 young (NERF Annual Report 2010) and in future this might provide another recruitment corridor to Durham's population similar to the western dales.

For their rare local nesting attempts, Durham Ravens mainly utilise crag locations, with an occasional pair resorting to tree top sites (Westerberg & Bowey 2000). They usually build substantial nests, which gives them some latitude in site selection, but can, where necessary, build surprisingly small structures in trees (Ratcliffe 1997).

The persecution which has led to such a perilous population situation for this species locally, has been exacted upon Ravens for a number of 'purported reasons', amongst them: damage to livestock and through the activities of egg-collectors. Although persecution is still a problem in the uplands, Mearns (1983) identified other issues, including improved sheep husbandry and the consequent reduction in available carrion to the species, as being crucial in some local declines up to that point. Nonetheless, the national recovery of the species, referred to above, in parts of lowland England (Parkin & Knox 2010) during the late 1990s and early part of the 21st century, suggests that, if persecution can be reduced, it should only be a matter of time before a sustained breeding presence is established in Durham. The brightest hope for the future of a long term viable breeding population in the county is probably for birds to become established in lowland quarries in its central and eastern areas.

Recent non-breeding status

Most reports of this species come from the county's western uplands and are largely concentrated in the first and final quarters of the year, and comprise birds that have drifted over the county boundaries from the larger population centres of Cumbria and Northumberland. Birds tend to be present in very small numbers at a number of 'traditional' upland localities over these periods. In the first half of 2007, records of Ravens came from 14 sites in the west of the county, including birds in all of the main river valleys. The largest gathering was of 14 birds in the Rookhope area on 21 February but around this time, parties of three were also seen on Stanhope Common in Weardale and at Ruffside, in the Derwent valley (Newsome 2008).

As one of our earliest nester, with egg laying possible as early as the first week in February (Ratcliffe 1997) the presence of birds, particularly pairs, in the late winter season can denote both lingering winterers and early breeders. So the position in the upper dales in that season is a little unclear.

In the years 2005 to 2010, encouraging numbers of records were received with western birds present during the first quarter. In 2006, pairs were noted at up to half a dozen locations in the late winter and at one site a flock of 11 birds was seen. It is considered that many of the birds that spend the winter around the western periphery of the county are immature individuals searching for suitable territories, probably wandering birds that have been reared in the preceding year, in neighbouring Cumbria. It is probable that such birds comprised most of the 13 that were noted around Newbiggin on 12 March 2009.

Interaction of Ravens with local birds of prey is often noted and in 2006 such sightings involved two birds mobbing a Common Buzzard *Buteo buteo* on 2 February and one mobbing a wing-tagged Red Kite, *Milvus milvus* in Teesdale on 21 June.

Distribution & movements

This species has a huge global breeding range, which encompasses a circumpolar distribution around all northern latitudes and penetrating south well into north Africa, North America and the whole of Europe and central Asia. Most populations are resident, with young birds dispersing in the autumn. It is believed that this later phenomenon is responsible for the influx to Durham each Britain in recent years, from the thriving Cumbrian population.

Goldcrest
Regulus regulus

A common and widespread resident. It also occurs in large numbers as an autumn passage migrant and can be common in winter, particularly in coniferous woodlands.

Historical review

According to Holloway (1996) the breeding range of the Goldcrest in Britain during the 19th century was difficult to determine with any degree of accuracy, though it appears to have been generally distributed in southern England and Wales; he observed that as conifer plantations increased in their extent then the Goldcrest increased in number. In the early part of the 19th century, writing of Northumberland and Durham, Selby said that it *"abounded in conifer areas and that vast numbers arrived on the coast in autumn"* (Selby 1831). The species is one of those that would have benefited from the formation of the Forestry Commission, and its subsequent work, after the First World War (Brown & Grice 2005).

Goldcrests were probably less common as breeding birds in Durham in the past, before large areas of conifers were planted during the mid-19th and early 20th centuries. Blick commented that it may have been more of a garden bird in the past than it is today, as the species is known to have nested in a Norton garden for several years around 1812 (Blick 2009).

At his time of writing Temperley (1951) described the Goldcrest as a *"common resident, winter visitor and passage migrant"*. Dr. H.M.S. Blair noted that it was occasionally seen at the coast until December, though October was the main time of its appearance. Despite its preference for conifers the species will feed in an array of other habitats, particularly when on migration. At such times it can regularly be encountered in mixed woodland and scrub along the coastal strip as well at inland locations across Durham. At the midpoint of the 20th century, it was known to breed in conifer woods and on the fringes of the thickest plantations in Durham (Temperley 1951). In the south east of the county, local populations were determined to have declined after the severe weather of the 1962/1963 winter (Stead 1964).

Recent breeding status

This species is a common resident in suitable woodland throughout the county, breeding in many types of woodland, though it is present at much higher densities in coniferous plantations such as those at Hamsterley Forest and The Stang. It is also widespread in mixed woodland, penetrating the upper reaches of the Tees, the Wear and the Derwent valleys during spring and summer. There is little doubt that it is one of the most common birds in such woodlands, where the birds glean the tree foliage for tiny insects and arthropods (Gibbons *et al.* 1993). From the coast to the Pennine woodlands, principally where conifers have been planted, these tiny birds can be found at almost any altitude in the county. It is probably able to persist in the higher altitude locations, because of the dense sheltering nature of these plantations. Such sites develop their own microclimate, which is not, necessarily, as exposed or extreme as other more open habitats at similar elevations.

The *Atlas* indicated that Goldcrests occurred throughout Durham during the breeding season (Westerberg & Bowey 2000), but coverage for this species during that survey process was believed to have been somewhat uneven and the full extent of its breeding distribution was probably underestimated.

Unsurprisingly, Goldcrests are absent as breeding birds from areas of open moorland, densely populated urban areas and most of the county's lowland farmland except for shelter-belts. They occur in the greatest numbers in the plantations and scattered shelter-belts of the west of the county, though they are also widely dispersed in the now mature plantations that were used to landscape old colliery spoil heaps over much of the county's eastern to central ground during the 1960s and 1970s.

Some Goldcrests do breed in mature, purely broad-leaved woodland, but where this habit occurs, the species tends to be present at much lower densities, as pointed out by Gibbons *et al.* (1993). This activity is probably not that common over much of the county and tends to be most prominent when populations are high (Brown & Grice 2005). In the well-studied, Paddock Hill Woods in the Derwent valley, during the late 1980s and early 1990s, one pair per fifteen acres was found to be present in the deciduous parts of that woodland, whereas one pair per two and a half acres occurred in the coniferous plantations of the same sites (Westerberg & Bowey 2000). Scattered pairs can also be found in suburban parks and even large gardens, particularly where species such as Lawson's cypress or *Leylandii* and yew *Taxus baccata* are found (A. L. Armstrong pers. obs.). A pair in South Shields Marine Park in May 2001 was notable.

The felling of mature woodlands or the overall reduction of plantation size, for example reducing woodland block size to less than ten hectares, can bring about a local decrease in numbers (Fuller 1982). It is rare for significant counts of Goldcrest to be made during the breeding season however in April 2007 at least 16 singing males were found at Elemore Hall Woods, with the breeding population here slightly increasing to 20 singing males in 2011. This gives an indication of the population density in relatively small lowland mixed woodland of less than 20 hectares in extent. Larger forested areas such as Hamsterley Forest, The Stang and around Derwent Reservoir clearly hold much greater numbers than recorded here. Eighteen singing males in a 2km stretch of Hamsterley Forest in May 2011 show how abundant the species can be in prime habitat. Breeding density in the best habitat can be very high, with as many as 400 to 600 territories per square kilometre having been recorded (Parkin & Knox 2010).

The *Breeding Birds of Cleveland* survey work determined that there were probably between 50 and 100 pairs in the North Teesside area, with most of these being concentrated in the Wynyard Estate's coniferous plantations (Joynt *et al.* 2008). Really detailed information on numbers and breeding densities, not to mention distribution, across the whole county is still somewhat lacking. Breeding birds survey work in 2006 found birds in only five out of 31 surveyed kilometre squares, totalling just 10 pairs. This figure is perhaps more reflective of the difficulty of finding birds in their nesting woods, where they often spend large amounts of time high in the upper reaches of tall conifers, rather than the actual numbers present.

The *Atlas*, estimated the county population, based on a 'by area' extrapolation of the national figures published in Gibbons *et al.* (1993), to be between 5,000 to 10, 000 pairs, but it was suggested that the real number was probably towards the bottom end of this range (Westerberg & Bowey 2000). Nationally, numbers seem have remained stable since the 1980s (Mead 2000) and one might expect this, in terms of population trends, to be the case in Durham, although exceptionally two successive cold winters over the period 2009-2010, appear to have impacted upon these numbers. Goldcrest, which had in 2008 reached its highest level since the start of the Breeding Bird Survey, declined by 56% between 2008 and 2009 (Risely *et al.* 2010). The data on the species underlying national trends is confusing and contradictory, with some survey methods indicating declines and others an increase (Parkin & Knox 2010), though most agree that for quite some time lowland populations do not seem to have been be prospering (Mead 2000). Accounting for such information, the lower estimated figure from the *Atlas* is probably best used as an indicative guide for the number of Goldcrests that breed in Durham today.

Recent non-breeding status

One of the environmental factors that most seriously impacts upon Goldcrest populations is winter survival (Marchant *et al.* 1990), as the species, not surprisingly considering its diminutive size, is susceptible to prolonged cold. It has been estimated that a single severe winter can halve the number of Goldcrests in the subsequent breeding season (Marchant *et al.* 1990).

On 27 January 2000, during a cold spell, 14 were observed seeking food in the filter beds at Sunderland Bridge STW. The importance of this industrial habitat to this species during the winter should not be underestimated. The combination of large numbers of invertebrates courtesy of the treatment work's filter beds, in locations that are often surrounded by screen-planting comprising coniferous tree species, such as can be seen at the Fishburn STW, provide sheltered, relatively food–rich, if localised, overwintering conditions for this species. The species' use of such sites is illustrated by the ringing capture totals from two sites in the north of the county. At Lockhaugh STW in the lower Derwent valley, invertebrates rising from the filter beds are blown into the surrounding hedges. At this site, 19 Goldcrests were caught and ringed in four visits over two months of the severe winter weather in the winter of 2010/2011; in addition 21 were caught in two visits in November 2008 and 10 on the morning of 30 December 2009 at Birtley STW.

Small numbers are usually noted in tit flocks at many other localities, often visiting suburban gardens and parkland, with impressive gatherings sometimes noted at more heavily wooded sites during the winter. At Hamsterley, in the winter of 2007, over fifty birds were counted between The Grove and the visitor centre on 13 October and 25 were at Croxdale on 6 December. On 17 November 2000, large counts were noted at many inland sites with 80 at Elemore Woods and 50 around Lumley Castle. Singing is occasionally noted during the winter months; this is often witnessed at Elemore Woods, Hamsterley Forest and other sites that hold high density populations, although more often in the early part of the year and usually associated with milder conditions.

This tiny species is a regular though surprisingly scarce passage migrant along the Durham coastline in spring, largely between late March and late April. Spring counts in 2007, by way of example, were limited to just single figures at the main watch points during late March and April, but 20 were in the Whitburn area and eight were on Hartlepool Headland on 28 March. In some springs, it is not recorded and even when it is, rarely are more than ten birds per season involved in reports, these usually being seen at sites such as Marsden Quarry, Hartlepool Headland and North Gare.

In the autumn, however, from early September to mid-November, large numbers of birds usually arrive along the Durham coast, in some Octobers in spectacularly large figures. Some of the largest falls to have been recorded include 500 birds between Hartlepool and South Gare (North Yorkshire) on 1 October 1965, at least 220 between Hartlepool and Seaton Carew over the 2nd and 3 October 1983, about 200 between Hartlepool Headland and West View Cemetery on 23 October 1990 and on 5th and 6 October 1998, and about 200 at Hartlepool on 7th to 9 November 2000 and again on 21st to 22 October 2001 (Blick 2009). In autumn 2000, hundreds of birds were present in the South Shields Marine Park, in Mere Knolls Cemetery and Marsden Quarry on 8 November. Similarly, in 2005 there was a large arrival of birds over the period from 15th to 19 October. At this time, along the whole of the county's coastline, double-figure counts came from twelve different sites and there were estimates of 350 birds at Hartlepool Headland and 400 in the Marsden to Whitburn area. Brown & Grice (2005) highlighted the influxes of Goldcrest that occurred along the English North Sea coast in the autumns of 1975, 1990 and 1992 – with *'massive numbers'* along the English coast in 1990, at that time there were *'several thousands'* at Marsden Quarry on 20th to 21 October. At such times the whole length of the Durham coast can be 'deluged' by many hundreds of this diminutive species and on the morning of 21 October 1990, 148 were ringed in Marsden Quarry.

By way of contrast, in years with an absence of what are considered classic fall conditions in either late September or early October when onshore winds and drizzle combine, then the highest site counts might barely reach double-figures. Nonetheless, at such times the total number of birds collectively along the coast can still be significant. For instance, 493 were noted on the 1992 Co-ordinated Migrant Count on 4 October, and the Hartlepool Headland area had over 90 of these on the day. In 1994, 1995 and 1996 however, it did not figure at all in the lists of the commonest species recorded on the respective Co-ordinated Counts (Bowey 1999c).

During the late autumn numbers also increase in the county's woodlands and large numbers may appear after easterly winds, as migrants filter inland. In such conditions Goldcrest might be found frequenting gardens, ornamental shrubberies and patches of scrub across the county's urban locations.

Distribution & movements

This is a species of boreal and temperate forests that breeds throughout Europe and, in a patchy fashion, east across Asia as far as Japan. Wintering birds from northerly populations are migratory and the winter range extends south of the main breeding areas to the Mediterranean zone of Europe where it is scarce as breeding bird. At this time of the year, the sedentary birds of temperate areas are joined by migratory visitors from the northern boreal populations (Parkin & Knox 2010).

The origin of some of Durham's autumn Goldcrests is well demonstrated by an analysis of the ringing data for this species. Cleary autumn-arriving birds are originating from countries to the north and east of the British Isles as shown by the first-year male bird ringed at Christanso, Bornholm, in southern Denmark, on 8 October 1990 that was at Graythorp on Teesside just 14 days later, on 22 October. A similarly rapid movement was shown by a first-year bird ringed at Falsterbo, Sweden, on 17 October 2001 which was controlled at Hartlepool five days later (Blick 2009). These distances are by no means small but even more impressive movements have been documented, for example the bird found dead at Whitburn on 31 October 2010, which had been ringed in Finland some 1600km to the east.

What may be local birds or autumn migrants have been documented making onward southern journeys from Durham as shown by the juvenile male ringed at Hargreaves Quarry on 12 September 1998 that was in Kelton Quarries, Leicestershire on 15 October 1998. Internal movements in England are not isolated, as confirmed by the national dataset (Wernham *et al.* 2002). That such tiny waifs are capable of making the return journey across the North Sea is illustrated by the adult female ringed at Crimdon Dene on 18 September 1993 that was found dead on 29 January 1994 at Jylland, Denmark, 805km to the east; presumably it was returning along its vector of travel to its point of origin.

Firecrest
Regulus ignicapilla

A scarce migrant from Europe, recorded from September to May, but with most in April, and from October to November.

Historical Review

This species, now an annual passage migrant along the Durham coast, with increasing numbers noted at inland locations during the winter, was previously, up to the early to mid-1990s, a decidedly scarce passage migrant, being a genuine rarity in Durham until the early 1970s.

The first record of Firecrest for County Durham was listed by Tristram (1905) who wrote: "*I possess a specimen shot at Brancepeth by Mr Dale, keeper to Lord Boyne in April, 1852.*" The current whereabouts of this specimen is unknown, and this bird was to remain the only fully documented record of this species in the county for over a century. Prior to this record, as related by Temperley (1951), Hutchinson (1840) made reference to a bird having been shot by George Proctor at Bow Bank near Middleton-in-Teesdale in October 1835, though Proctor himself did not include the record in his list of the *Birds of Durham* published in 1846. Proctor was said to have been something of a careless recorder, but even so one would presume that had he obtained a specimen of such merit he would have been unlikely to have not included it in his documentation.

Recent Status

At least 133. Firecrests have been recorded in County Durham since 1972 and the species has become an increasingly regular migrant since the beginning of the 1990's with only one year, 1998, when it has not been recorded at all. The recent increase in the number of Firecrest records in the north east of England corresponds well with the northwards expansion of the species' breeding range through Europe; it is known to have bred in Denmark since 1961 and in Britain from 1962 (Hagemeijer & Blair 1997).

The majority of records have come from coastal localities such as Dawdon Blast Beach, Hartlepool Headland, and the Marsden/Whitburn area, though occasional birds have penetrated inland and there is an increasing trend for single birds to winter in the county. Hartlepool Headland is the single site with most records, a total of 32 during 1970-2009, while 46 were recorded in the Marsden and Whitburn areas during the same period. The majority of passage birds are noted during autumn, from late September until early November, with birds also occasionally seen in spring from March until early May.

After an absence of Firecrests in Durham for over 120 years, a large high pressure system centred over Europe brought an influx of at least five birds to the county during early October 1972. The first was at Marsden on 1 October, with two there on 2nd to 3rd, whilst up to three were at Hartlepool Headland from 2nd to 6 October. The first spring record during the modern era was at North Gare from 19th until 21 April 1975, with a bird inland, near Washington, on 20 April of the same year. The one further spring record before the decade came to a close was at Hartlepool Headland on 4th to 5 April 1978. During autumn, singles at North Gare from 25 September until 2 October 1977 and Marsden from 26 September until 1 October were noteworthy for being the first September records, with a further five post-1972 records roughly split between Hartlepool Headland and the Marsden area. The one exception to this developing pattern was an individual well inland in Darlington's South Park on the 9 October 1973. The decade closed with a respectable total of 15 birds, the species failing to have been recorded only in 1974 and 1976.

The 1980s saw a further 16 birds recorded when, largely as a result of an influx in April 1983, the species was more numerous in spring than it was in autumn. Like the previous decade there were only two blank years, 1984 and 1986. A total of eight were recorded during spring including the first March record at Hartlepool Headland on 28 March 1981, and the first May record at Cleadon Hill on 6 May 1983. The six April records included an influx of four birds during 5th to 18 April 1983. Three of these were coastal migrants at typical localities, but the fourth was well inland at Shibdon Pond, Blaydon on 14th; this represented the first record for Gateshead (Bowey *et al.* 1993). Seven autumn birds occurred in the 1980s, with dates ranging between 29 September and 1 November; the majority being, typically, in the Hartlepool and Marsden/Whitburn areas. Dawdon Blast Beach received its first record on 20th to 21 October 1988. A single bird in a garden at Brancepeth on 16 February 1981 was noteworthy not only for being significantly inland but also for being the first winter record for the county, and in the location of the first record in 1852.

 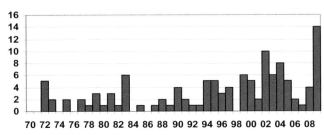

Records of Firecrest in County Durham, 1970-2009

Firecrest became established as a scarce but regular passage migrant in Durham during the 1990s with an increase to 31 records, a number equal to that of the previous two decades combined. It was unrecorded in just one year, 1998, and has been noted annually since then, in ever increasing numbers. Eight birds were recorded in spring, six of which were in April except for single birds at Hartlepool Headland on 4 May 1994 and at Marsden Hall on 1 May 1999. The most noteworthy spring record was a single, 2km inland at Tunstall Hills, Sunderland on 4 April 1992. A further nineteen coastal migrants were recorded during autumn with extreme dates of 20 September 1996 and 9 November 1990. An average of two to three birds were recorded each autumn throughout the period, with the best year being 1999 when five birds were recorded between Whitburn and Hartlepool from 22 September until 24 October. There were three 'late autumn' records during the period involving four birds with one bird wintering. Up to two birds were present in Hargreaves Quarry, Teesmouth from 25 November until 2 December 1995 with one remaining until at least 24 February 1996. The others were singles at Cowpen Bewley Woodland Park on 23rd and 24 November 1997 and at Mount Pleasant Marsh near West Boldon from 4th to 6 December 1997.

At least 57 Firecrests were recorded in County Durham during 2000-2009, representing an almost two-fold increase on the previous decade. Unusually just three were recorded in spring, at Hartlepool Headland on 20th to 21 April 2001 and 17th to 18 April 2004, and at Whitburn on 29 March 2007. A total of 46 autumn birds were reported from coastal localities, including influxes of nine birds in October 2002 and 13 in October 2009. The influx in 2002 included a peak count of three at Whitburn Coastal Park on 13 October, the largest single site count. On 22 October 2009, five birds were present between Whitburn and Hartlepool, the highest single day count of this species in the county. A total of eight birds were seen inland between November and March through the 2000s, although only one bird, at Mount Pleasant Marsh, West Boldon, present from 12 January until 28 March 2008, could be regarded as having overwintered. Other inla nd birds were at: Hetton Lyons Country Park, Houghton-le-Spring on 3 November 2003; Fence Houses from 26th until 29 November 2004, Sedgeletch on 28 November 2004, on the Long Drag, Teesmouth from 26 November until 23 December 2006, Sedgeletch on 2 December 2006, Portrack Marsh, Stockton-on-Tees on 30 November 2008, and at Lamesley on 1 March 2009.

The best year on record was 2009, when 14 were seen, most of which were in October, whilst the maximum count for a single site is of three birds in the Whitburn Coastal Park on 13 October 2002. The total of eight birds recorded in 2010 represented an average year by recent standards, with a single spring record at Whitburn from 31 March until 3 April, and an influx of five birds between 7th and 11 September, when two were at both Hartlepool and South Shields, with a single at North Gare, Teesmouth. These are the earliest autumn Firecrests to be recorded in County Durham. Inland birds were at Far Pasture NR from 20 February until 24 March and at Tunstall Reservoir, on 14 November. A similar situation was evident in 2011 as a total of six birds was recorded. Notable amongst these were another individual found wintering at an inland sewage treatment works (Fishburn during late January and early February) and one at Hetton Bogs on 29 September. It seems increasingly likely that a few birds now regularly winter and are overlooked as they move around in the company of Goldcrests *Regulus regulus*, Long-tailed Tits *Aegithalos caudatus* and mixed tit flocks.

Extreme arrival dates for migrant birds during spring are 28 March 1981 at Hartlepool Headland, and 6 May 1983 on Cleadon Hill, and during autumn, 7 September 2010 at Hartlepool Headland and South Shields and 24 November at Hartlepool Headland.

Recent breeding status

One of the most unusual records to date is of a singing male between Tunstall Reservoir and Wolsingham in Upper Weardale from 9 April until 6 June 1995. This bird had established a territory and was frequently seen in close association with a female Goldcrest though it is not known if breeding was attempted. Hybrids between Firecrest and Goldcrest are extremely rare though not unheard of (Holling 2009).

Distribution & movements

Firecrest breeds across much of temperate Europe and north western Africa and is partially migratory, birds from central Europe wintering to the south and west of their breeding range. A small breeding population, of around 600 pairs (Holling 2009), exists largely in southern England; although these birds are thought to be largely resident. Elsewhere the species is a scarce migrant chiefly in autumn from September to November, and to a lesser extent in March and April when numbers are lower. This species is much commoner as a migrant to the south coast of England and further north records are much scarcer, though generally on the increase in line with the westward expansion of the species in continental Europe.

A number of Firecrests have been trapped and ringed in County Durham since the first at Marsden on 1 November 1980, including at least five at Hartlepool Headland, but there have been no subsequent recoveries.

Penduline Tit (Eurasian Penduline Tit)
Remiz pendulinus

A very rare vagrant from Europe: three records, involving four birds.

The first record of Penduline Tit for County Durham was of an adult female trapped and ringed at Haverton Hole, Teesmouth by Derek Clayton on 17 July 1992; initially discovered in the net by Russell McAndrew at around 9:00pm. As the light was fading, the decision to roost it overnight was taken. The bird was an adult female undergoing post-breeding moult, an extensive brood patch was present but this had already begun to re-feather. The bird was released at dawn the following morning, promptly flew into the extensive reedbeds and was not seen subsequently (*British Birds* 87: 560). The second record came less than a year later when Chris Brown trapped and ringed an adult male in Hargreaves Quarry, Teesmouth on 8 May 1993. This bird was seen to construct a nest, though it gradually became more elusive and was not seen after 9 May (*British Birds* 88: 543).

Almost fifteen years later, two males were discovered at Portrack Marsh, Stockton-on-Tees on 23 March 2006 and remained until 3 April. They were often elusive and ranged over a wide area. What were presumably the same birds were seen 30km to the west at Drinkfield Marsh, Darlington on the evening of 4 April, before being seen again at Portrack Marsh on 5th and 6 April. They were last seen below the upper car park at Dorman's Pool, Teesmouth on 9 April (*British Birds* 100: 744).

The three records from County Durham remain the most northerly records of this species in Britain, the nearest record to the south being a male at Knaresborough, North Yorkshire on 19th and 20 September 1992. The bird at Haverton Hole in July 1992 is especially noteworthy in being one of only two July records in Britain. Even though the Haverton bird bore the remnants of a brood patch, it presumably represented an example of rapid post-breeding dispersal from the near Continent.

Distribution & movements
Penduline Tit has a scattered distribution across much of Europe, although it is much less common in the north western part of its range. To the east it breeds from southern Russia into central Asia in several distinct forms. The species underwent a significant range expansion during 1975 to 1985, although there was a slight decline in the early 21st century, when the Dutch population was estimated at between 140-210 pairs, with a further 75-150 pairs in southern Scandinavia. Since the first at Spurn Point, Yorkshire in 1966, the species has become an increasingly regular vagrant to Britain with almost 250 records by the end of 2010. Recorded in all months, but with a peak in October, the majority of records come from south east England, in particular from East Anglia and Kent.

Blue Tit (Eurasian Blue Tit)
Cyanistes caeruleus

An abundant and widespread breeding resident.

Historical review
During the 19th century this species was believed to be the most common of its family across the British Isles (Holloway 1996), and this was almost certainly the case in Durham, though relatively little comment was made on its range and status within the county by early recorders of the area's bird life. In the absence of evidence to the contrary, the assumption is made that this species was sufficiently widespread and common in the county as not to warrant special comment.

Temperley (1951) stated that the Blue Tit was "*a resident*" and "*very common*" and believed it more common than the Great Tit *Parus major*. Dr. H.M.S. Blair, who was resident at South Shields, believed that some of the birds seen in the area in October during the 1940s represented continental migrants. This was based on his observations of plumage difference in the birds seen there in late autumn compared to those noted early in the summer. Likewise, Temperley (1951) thought that there were influxes of immigrants to Durham in September and October.

Recent breeding status

This is one of the county's most familiar and widespread species, in 2006 Breeding Bird Survey work found it to be present in 23 of 31 (74%) sample one-kilometre squares surveyed. Birds are present from the county's western boundaries in upper Teesdale and near Derwent Reservoir all the way down to the North Tees Marshes (Newsome 2009). It is the most common of Durham's breeding tits and one of the most widespread and common of any species; it will nest almost anywhere there are trees with natural holes or where nestboxes are provided, regardless of whether these are situated in woodlands or in gardens. The *Atlas* (Westerberg & Bowey 2000) showed it to be present in almost every available 10km square of County Durham where appropriate nesting habitat is available during the breeding season. An indication of the numbers in a prime woodland site was given by the count of 42 singing males made in Castle Eden Dene in late May 1979.

The density of breeding birds is highest in broadleaved woodland and lowest in the upland valleys, where it tends to be somewhat replaced by the Coal Tit *Periparus ater*. There are a number of nestbox schemes in the county, for example at Hamsterley Forest, woodlands in Teesdale and in the Derwent valley. In these schemes, Blue Tits have traditionally occupied a high proportion of the nestboxes. In Hamsterley Forest, they have been found to breed at altitudes as high as 365m above sea level (Westerberg & Bowey 2000). The absence of trees over much of the higher ground in the west of Durham effectively precludes the species from nesting in those areas. Across the county, Blue Tit breeds abundantly in gardens, although generally with less success, because of a lack of available suitable food for provisioning nestlings at the correct time of the year (Parkin & Knox 2010). Populations of Blue Tits over the last 30-40 years have benefited enormously through the increased provision of nestboxes, particularly in urban areas. These have allowed the species to expand its range into the heart of towns and villages across the county. A study in Durham City in 1974, showed the variety of clutch sizes laid in such situations, from four to ten eggs per clutch.

Data from monitored breeding sites over the years has demonstrates graphically how this species is very much dependent on the weather for breeding success. For example, in 1991, despite a good start to the breeding season in the Derwent Walk Country Park, when a maximum clutch size of 14 eggs was noted, subsequent poor weather during incubation led to the desertion of about 60% of nest sites. The poor breeding season was confirmed by the proportion of adults to juveniles that were caught at ringing sites later that autumn. The usual ratio for this was 3:1 in favour of first-winter birds to adults but in 1991 this was 43% young birds to 57% adults during the months of November and December. There followed a series of poor breeding seasons in the mid- to late 1990s, with evidence for these once again coming from the lower Derwent valley. Particularly poor seasons were experienced in 1996 and the cold wet spring of 1998, when just 10% of monitored nestboxes in the area managed to fledge young. Most of the young birds in boxes starved to death just a day or two short of fledging. Nestboxes elsewhere in the county fared better in the 1998 spring, though were less productive than usual. Another poor season followed in 1999 and such a run of bad breeding seasons can conflate to negatively affect local population numbers. A poor breeding season was experienced in 2001, was exemplified by data from the Kibblesworth CBC site, where only 12 out of 64 boxes were occupied.

In woodlands, this is a species that coordinates its breeding season so that the eggs hatch to coincide with the availability of the maximum number of small caterpillars upon which the young are to be fed (Perrins 1979). To a large extent this includes the suite of caterpillar species that predominate upon on oak species *Quercus spp*. Accordingly, most pairs lay eggs at roughly the same time. As in April 2007, which saw some warm, fine weather, most pairs bred early (Newsome 2007). To some extent, this observation is borne out by a pair that was feeding young at Ramside Hall on 26 April, but two days later, another pair was still nest building at Hawthorn Dene. Early season singing was noted on 12 February 2004, but this more usually commences in March, building in intensity through April. In similar fashion, a pair was investigating a nestbox on 8 February 1987 during mild weather, but this activity increases in frequency prior to territory establishment around a suitable nest site, in April.

In coniferous woods where nestboxes are provided, population densities can be high. For example, 1.9 pairs per ha. amongst birds utilising boxes in parts of Hamsterley Forest (Westerberg & Bowey 2000). Comparable, or even greater, densities may be achieved in high quality, mature oak woodland even without boxes. When additional nest sites, in the shape of nestboxes, are provided, the population density in such areas grows considerably; indicating that in the best quality woodlands, it is the availability of nest sites that limits the population. A combination of natural and nestbox-utilising pairs, made up the more than 50 pairs found nesting in the 23ha of Thornley Woods, during the early 1990s; a density of 2.1 pairs per hectare. A very high figure of more than 16 pairs in one acre (almost 20 pairs per hectare) was present in the nearby Paddock Hill Woods at the same time, a figure that was artificially boosted by the provision of nestboxes (Bowey *et al.* 1993). High density does not always lead to success and in 1988, in the Derwent Walk Country Park, just 3% of nestboxes were occupied, the average brood size being just seven, possibly as a result of the wet spring weather that year. Even in mixed habitat sites, large numbers can breed where boxes are provided, for example up to 29 pairs have been noted on the Kibblesworth CBC site, but birds here appear to prefer natural nest holes over the numerous artificial nestboxes available.

According to the *Atlas*, a county extrapolation from the national population estimate of 2,000 breeding pairs per occupied 10km square from national data (Gibbons *et al.* 1993) suggested an approximate population for Durham of 52,000 breeding birds (Westerberg & Bowey 2000). There are probably around 2,200 pairs in the north Tees area, although there are some gaps in that area's breeding distribution, mainly around the Teesmouth NNR area, where there are few trees or garden nestboxes to be found (Joynt *et al.* 2008).

Hard winters can be a threat to small birds such as the Blue Tit, but the ever-increasing popularity of garden feeding stations, is likely to have offset any serious threats to this abundant and adaptable species. Over time, the population nationally is holding steady (Brown & Grice 2005) though according to national Breeding Bird Survey statistics, the Blue Tit experienced a decline, of 4%, between 2008 and 2009, due to the prolonged freezing temperatures in January and February 2009 (Risely *et al.* 2010). Perhaps the greatest threat to the species is the risk from cold, wet weather in May and June, during the incubation and nestling phases of the breeding season. This can severely impact upon breeding success, as in the wet summers of 2007 and 2008, and subsequently, the breeding population of the following spring.

The species has been recorded using a number of unusual nesting localities across the county, including: ornate lamp standards in Whickham and Durham City, a traffic sign pole, the roof of the visitor centre at Hamsterley Forest, a functioning drainpipe at the Thornley Woodlands Centre, a nestbox on the ground and a security alarm box at Ramside Hall Golf Club. An example of nesting displacement took place in 1982, a pair laid three eggs in a nestbox at Edmundbyers, the box was taken over by a pair of Pied Flycatchers *Ficedula hypoleuca* which covered the eggs with its own nest and successfully reared young. On a number of occasions mixed clutches of eggs have been recorded in titmouse nestboxes.

Ringing studies have also revealed a numbers of instances where Blue and Great Tit *Parus major* eggs have been laid in the same boxes, for example at Thornley Woods and at Burnhope near Edmundbyers in the late 1980s and at Shibdon Pond in 1994. In addition, a female ringed as an adult at Thornley Wood on 15 December 1988 was found long dead in a nestbox occupied by Great Tits at Thornley Wood on 26 May 1990.

Recent non-breeding status

Outside of the breeding season, Blue Tits gather in loose feeding flocks that rove along hedgerows, through woodlands and visit garden feeding stations in a sequential fashion throughout the autumn and winter. During the mid-winter at the best feeding sties, the numbers in these flocks can grow to scores of birds and in most winters flocks of 40 birds and more are reported from dozens of sites around the county. Most significant counts of this species in Durham relate to observations at such times.

Higher point counts lead to the conclusion that considerable numbers of birds visit garden feeding stations over short periods of time. Ostensibly, these roving flocks usually number 20 to 40 bids but the total number of birds actually visiting suburban feeders can be much greater than is sometimes realised. This was well evidenced by the example from an Ebchester garden on 14 December 1988, when 134 different Blue Tits were ringed in just one morning. Observations at the feeders the next day, however, confirmed that only about 25% of birds visiting the feeding station had rings on their legs (Armstrong 1989), indicating that, at a conservative estimate, many hundreds of birds were using that one garden each day. Further evidence was provided by the 90 birds caught and ringed in one morning at the Thornley Woods feeding area and the 236 caught there during the winter of 1995/1996. In the autumn and winter of 1994, 188 birds were ringed at the Thornley Woods feeding station, 68% of which were first-winter birds, the oldest bird being six-years old.

These winter flocks begin to develop soon after the breeding season when family parties gather in ever larger numbers, coalescing into quite large roving bands (e.g. the 130 birds at Thornley Woods in July 1988). These can mix with a range of other tit and warbler species such as Chiffchaff *Phylloscopus collybita* and Willow Warbler *Phylloscopus trochilus*, feeding on the move.

This species is capable of apparent 'explosions' in numbers in good breeding seasons as in 1985, when 120 birds at Thornley Woods on 30 July had risen to 250 the next day, courtesy of the coordinated fledging of young and the joining up of family parties into feeding flocks. These flocks, minus the warblers, form bands of wandering tits intermixed with a few other woodland species, such as Nuthatch *Sitta europaea* and Treecreeper *Certhia familiaris*, which persist through autumn and winter. They can be found in almost every area of woodland and travel *en masse* along adjoining hedgerows and gardens. In such flocks Blue Tits generally outnumber all of the other tit species by a figure of two to three to one.

Occasionally, birds wander into areas that are in habitat terms, largely alien to them, though in terms of distances travelled, these might only be a few kilometres away from regularly inhabited woodlands. For example, in 1986 one was reported from upper Teesdale above the 450m contour on 20 July, and a bird was by the Harwood Beck in upper Teesdale on 28 August 1990, at 420m above sea level, both very high elevations for the species. More extreme was the flock of 12 birds were seen sheltering from a snow storm in the lee of a dry-stone wall at over 370m above sea level near Selset Reservoir in the early part of 1994.

An odd report in the winter of 1986 referred to a bird in Sunderland AFC's then stadium at Roker Park, which flew the entire length of the pitch during a match and landed on the crossbar. At the start of the year in 1977, a bird with fawn wing coverts was at Balder Grange in the south west of the county. Another aberrantly coloured individual, entirely grey and white, visited a Durham bird table in February 1983.

Distribution & movements

This species is widespread and common over most of Europe; its breeding range extends east into Asia Minor. Over most of its western range, Blue Tits are largely sedentary, though some of the northern and continental birds undertake winter movements, sometimes in an eruptive fashion, in large numbers (Parkin & Knox 2010). British birds are largely sedentary and there is little evidence of birds from the Continent reaching County Durham, though it seems likely that some occasionally do.

Evidence for the presence of winter visitors in the area comes from ringing studies in the Derwent valley. Several birds ringed in nestboxes on the edge of the Durham moors have been re-trapped at winter feeding stations at lower elevations. The reverse of this has also been proven, with a bird ringed during the winter in a garden in the lower Derwent valley, being found breeding in a nestbox near Edmundbyers the following spring. This suggests that some Blue Tits nesting on the edges of the uplands behave as short range, altitudinal migrants.

Further evidence for this activity is provided by a series of ringing results, such as: the adult ringed in a Barnard Castle garden on 4 April 1976 and controlled at Wallsend on 20 December 1980; a nestling ringed at Edmundbyers on 9 June 1988 controlled at Rowlands Gill on 7 December 1988, had only moved some 7km, but it had descended by 150m in terms of elevation. Complementary cases of young birds exhibiting altitudinal migration come from a pullus ringed in a nestbox at Hamsterley Forest on 12 June 1988 that was controlled at Hart Station on 7 January 1989, a distance of 42km and a juvenile female ringed on 30 September 1987 at Saltholme that was controlled at Swarland, Northumberland on 21 April 1988. Meanwhile, an adult female ringed in a Whickham garden on 21 February 1988 was controlled as a breeding female on a nestbox in Backstone Bank Wood, Wolsingham on 14 June of that year, and an adult female nesting at Burnhope near Edmundbyers in 1989 had been ringed the previous winter lower down the Derwent valley at Blackhall Mill.

The phenomenon of coastal passage seemed to be more regularly reported in the 1970s, although these birds were usually thought to be local breeding birds moving along the coast rather than immigrants. Considerable numbers of birds are present along the Durham coast in the autumn, as shown by the 306 counted on the 1992 Coordinated Migrant Count on 4 October. This high figure included over 100 in the Mere Knolls Cemetery area alone, demonstrating how many titmice find their way to such coastal migration points in the autumn. Such numbers throw in to some doubt claims for the presence of coastal or continental migrants and national data sets indicated that most movements made by British birds are undertaken by juveniles, which later return to their natal area (Wernham *et al.* 2002). This figure was exceeded on the 1994 Durham Co-ordinated Migrant Count, when 606 were noted between the Tyne's mouth and Teesmouth on 2 October (Armstrong 1995). A bird ringed at Dipton Woods in June 1971 near Hexham, Northumberland, was re-trapped at Hartlepool on 23 December of that year, demonstrating that some coastal birds are from the county, or the region's, interior rather than drifting down the coast.

Clearly some birds do move up and down the north east coast. For instance, the bird that was ringed in Gosforth Park, Northumberland on 13 June 1965 and was recovered at Hart in the south east of Durham on 29 January 1966. In the opposite direction, a nestling ringed on 1 June 1980 at Helmsley, North Yorkshire was controlled at Marsden Hall on 19 October 1980. A pullus ringed at Elwick at Teesmouth on 20 June 1978 was controlled at Castle Eden Denemouth 34km away on 19 February 1979. A first-year bird from Shibdon Pond, ringed on 8 October 1978 had found its way to the coast at Castle Eden Dene by 23 March 1979. A juvenile bird ringed at Hauxley in Northumberland on 7 October 1988 was controlled at WWT Washington on 8 December 1988: an illustration of a bird coasting south to find a wintering site, perhaps had been bred further north rather than being a continental migrant.

Over 15,000 Blue Tits have been ringed in Durham - more than any other species. Local ringing studies indicate that some birds can occasionally survive to over seven years old. For example, a bird ringed at Ravensworth on 8 June 1978 as a nestling was killed by a cat in Gateshead on 30 April 1986. Even older was the bird ringed as a juvenile at Shibdon Pond on 13 June 1987, which was found dead at Blaydon at the exceptional age of over nine years, on 28 June 1996.

An analysis of the controls and recoveries of locally ringed Blue Tits indicate that movements can be quite complex, though these are rarely long. Of the birds ringed by the Durham Ringing Group, during the late 1980s and the 1990s, birds on the coast had a tendency to move further than those ringed inland and when birds do move these movements are not simply south in the autumn and north in spring. A good example of such complexity was illustrated by the first-year bird ringed on 2 January 1976 at Barnard Castle in Teesdale that travelled north to be controlled in Weardale, at High Pond, on 8 July 1979. In an unlikely reciprocation, a nestling ringed at High Pond on 19 June 1978 was controlled in a Barnard Castle garden on 11 March 1979; indicating the movements that can take place amongst the breeding birds of adjacent valleys.

There is a raft of altitudinal and other local effects at play in local movements and some east-west and west-east movements have been documented. Such as the bird ringed at Grune Point, Cumbria on 4 October 1964 and recovered at Hawthorn, near Seaham some 110km to the east just eight weeks later on 5 December. In the opposite direction, a bird controlled at Satterthwaite in Cumbria on 26 March 2003 had originally been ringed four months earlier in December 2002, 117km to the east, at Shotton Colliery. In summary, most ringed Blue Tits that have been found or re-trapped in the county do not move large distances from their breeding or natal territories and many birds nest close to their natal area.

Great Tit
Parus major

An abundant and widespread breeding resident.

Historical review
Holloway (1996) noted that this would have been a common and familiar species during the 19th century, though much less common than Blue Tit *Cyanistes caeruleus* away from broadleaved woodlands. This seems likely to have been the situation for this species in County Durham. In terms of data or pronouncements upon its status locally, almost nothing substantive was documented by the earliest chroniclers of Durham's ornithology. It is assumed that it was present and common in Durham.

At the mid-point of the 20th century, Temperley described it as, "*a resident; generally distributed and common wherever there are woods and copses, gardens, orchards and hedgerows*". He also seemed confident in his assertion that many of these flocks must consist of immigrants (Temperley 1951).

Recent breeding status
This is a widespread breeding species in most woodland, many parks and occasionally in gardens. It is very common across the county nesting primarily in woodland but also breeding in parks, shelterbelts and farmland and is well distributed from coastal fringe to the most westerly treed locations. In Durham, it is associated with woodland, mature hedgerows and, to a lesser extent, coniferous plantations. Although it appears to have more of a

preference for broadleaved trees than the closely related Blue Tit *Cyanistes caeruleus*, it still occurs in a wide range of woodland types and situations (Cramp & Perrins 1993) and in 2010, it was reported from more than 200 locations across the county (Newsome 2009). Like Blue Tit it will routinely use nestboxes.

During the late 1980s, local survey work in the well-wooded lower Derwent valley recorded it in 57% of all surveyed one-kilometre squares (Bowey *et al.* 1993). As a breeding bird, it was present in almost all 10km squares in the county during the *Atlas* work between 1988 and 1994 (Westerberg & Bowey 2000) and little has changed since then as judged by survey work from the latter part of the first decade of the 21st century. In May 2007 records of birds came from as far west as Cow Green Reservoir, though it seems unlikely that birds present were actually breeding at this extremely exposed location, given the paucity of suitable food for nestlings.

Detailed studies in Hamsterley Forest, during the early 1990s, showed that breeding was largely confined to streamside broadleaves (Westerberg & Bowey 2000). In this extended location, some broadleaved plantations can be found at relatively high altitudes and in such situations the species has been recorded breeding in both nestboxes and natural sites at 365m above sea level. The number of breeding pairs in Hamsterley Forest can be as high as 1.5 breeding pairs per hectare but this is unusual and even in the best areas 0.4 pairs might be considered a good density (Westerberg & Bowey 2000). Local breeding numbers was illustrated the total of 21 pairs on the Kibblesworth CBC area in 2007.

Observations over many years in the Derwent valley suggest that it is somewhat more inclined to choose natural nest sites in deciduous and mixed woodlands than its close relative. Yet published work indicates that Great Tits are particularly keen on nestboxes and, in some instances, almost entire populations can 'be moved' into nestboxes if sufficient of these are provided (Gosler 1993); all nine nestboxes erected at Heighington in March 1977 were occupied during the first breeding season and four boxes at Edmundbyers in 1980 produced 40 young. Successful breeding was recorded for the first time at Shibdon Pond in 1994, an incident indicating that birds were expanding out of their more traditional woodland habitats. More interesting in relation to this breeding attempt was the fact that the Great Tits took over a Blue Tit occupied nestbox and they proceeded to successfully rear three of the young Blue Tits, the eggs of which had been laid before the larger species took up residency.

As this species tends to breed slightly earlier than the Blue Tit, it sometimes misses bad weather such as the series of cool, wet springs leading to the poor breeding seasons experienced in the late 1990s. Whilst this species did suffer to some extent between 1996 and 1999, it did not take the full impact of the worst of the poor summer seasons, unlike Blue Tits.

In Durham, birds have utilised many non-natural sites, such as rotten fence posts, drainage pipes, footpath signposts and holes in walls. In May 2000 a pair nested 30cm away from a wasp nest at Houghall. The species' stoic and determined nature was illustrated in 1977 when a pair at Castle Eden Dene continued to feed a brood after the nesting tree was felled; the nesting limb was placed upright and the young fledged successfully. A poor season at the Kibblesworth CBC site in 2001 was demonstrated by the fact that just eight out of 64 nestboxes available were occupied by this species. From BBS work nationally, between 2008 and 2009, the Great Tit was determined to have declined by up to 5%, mainly as a consequence of the severe weather of January and February 2009 (Risely *et al.* 2010) though across England the species has experienced a long term increase in numbers, sustained over a number of decades (Brown & Grice 2005).

Nationally, 1,000 breeding pairs per occupied 10km square has been used to estimate populations (Gibbons *et al.* 1993) and the *Atlas* concluded that this would give an estimated County Durham population of approximately 25,000 breeding pairs (Westerberg & Bowey 2000). There are probably around 1,000 pairs in the north Tees area, where it is almost as widespread as the Blue Tit, though not as numerous. In this area, it is very much less numerous around Teesmouth and up the coastal strip towards Crimdon (Joynt *et al.* 2008).

Recent non-breeding status

In winter behaviour and habitat choice, this species is very similar to the Blue Tit, it is however less common than that species, in a ratio of perhaps one to two. The size of the congregations of birds that can develop in the post-breeding period was illustrated by the 39 birds caught and ringed in just an hour at the Ravensworth Estate on 3 November 1977. In Auckland Castle Park during a period of deep snow in early 1981, 40 to 50 birds were noted in a single flock. In 1984, flocks of 45 and 34 were observed, respectively, in Thornley Woods and Paddock Hill Woods and these proved the largest counts of the species in that year, but such numbers were more likely indicative of the scale of winter gathering that might be found in most of the county's high quality valley woodlands.

Using 2007 as a typical year, most of the reports of any significance were of winter feeding flocks, although at least 40 birds were present in Elemore Woods during the breeding season. Typical winter gatherings of up to 20 birds were reported from many widespread locations during the first and last quarters of the year with somewhat larger totals, of more than 30 birds, counted at Hawthorn Dene in December.

This species is easily attracted to garden feeding stations and this behaviour has been prevalent in the county for over 50 years. Mixed winter tit flocks usually include Great Tits and feeding flocks of birds outside the breeding season occasionally rise to as many as 40 birds. Sample high counts include, in 1985, the 57 at Thornley Woods (Baldridge 1986), and in 1989, 60 at WWT Washington (Armstrong 1990). The numbers of this species do not tend to build up to the very large numbers sometimes seen with Blue Tits, so a flock of 85 in Wynyard on 1 January 1986 was exceptional (Blick 2009).

Hard winter weather, with prolonged snow cover, is a major natural threat to Great Tits, though snow cover has much decreased in the latter part of the 20th century and early 21st century. In poor weather, Great Tits have greater difficulty in finding food than the smaller Blue Tit, which is better adapted to feeding on tree seeds whilst Great Tits prefer to spend more time feeding on the ground (Gosler 1993). The numbers of Great Tit breeding from year to year fluctuate more than for Blue Tit, presumably for this reason. The provision of food in gardens no doubt helps many more birds to overwinter successfully than would be otherwise the case. During the winter months considerable numbers of birds can visit winter feeding stations over relatively short periods of time. For example 30 birds were caught and ringed in three weeks at the Thornley Woodlands Centre (Bowey et al. 1993) and in the winter of 2002/2003, 157 new birds were ringed at this site plus an additional 24 were recaptured from previous winters, including one eight-year old bird.

A partially melanistic bird with almost completely black under parts was noted at Whitburn in 2009.

Distribution & movements

Across European, northwest Africa, west, and central Asia, Great Tits are common breeding birds, the range extending east to China and into eastern Siberia. In Europe, they are resident over much of the range, with northern bids exhibiting irregular and irruptive movements that seem related to food supply, most notably in 1957 (Brown & Grice 2005). Occasional birds on the east coast of County Durham might relate to birds involved in such movements, though there is to date no documentary evidence for this assertion.

In the post-breeding phase of the year, it is usual for young birds to wander and this brings birds that have bred at sites close to the coast to coastal locations in variable numbers and when they do this they have a tendency to move north and south, 'coasting' as it is termed, and this sometimes concentrates numbers in locations more associated with truly migratory species.

Slight coastal passage sometimes seems to be evident in September and October but assessing the provenance of birds at the coast in autumn is difficult. Typical increases in numbers were noted in 2006 at Whitburn, with 13 birds on 24 September, and Tunstall Hill, with 23 on 15 October, indicative of most years at the coast. However, the total of 66 counted along the Durham coast on the 1992 Co-ordinated Migrant Count, on 4 October (Armstrong 1994), indicates that there may be many more present than observers sometimes realise. In such situations it is almost always less numerous than Blue Tit. Occasionally birds turn up in more out of the way locations, such as the two that were noted on the Seal Sands Peninsula in March 1986 (Blick 2009). Most local birds appear to be sedentary and most of the movements noted are of juvenile and first-winter birds, as per the national situation (Wernham et al. 2002).

Well over 10,000 Great Tits have been ringed in Durham over the years and the most substantive information gleaned from this is that most local birds do not move more than a few kilometres. As illustrated by an early ringing recovery. An adult ringed at South Shields in November 1952, was re-trapped at the same spot on 20 February 1956, then found dead there on 26 July or that year. If it was one-year old on first ringing, this made it to at least four and a half when it died (Temperley 1957) in the same area that it was ringed.

There is some ringing evidence of birds moving up and down the county's altitudinal gradient between the breeding and wintering seasons. Birds moving to the coast include a nestling from a nestbox in the lower Derwent valley in June 1988 that was caught 20km away in Marsden Quarry on 23 October; an adult female that was ringed at Thornley Woods on 20 March 1988 and controlled to the north, at Hauxley, Northumberland on 8 October 1988, a nestling ringed in Barnard Castle on 19 June 1979 that was controlled at Whickham on 27 January 1980 and, nestlings from Hamsterley Forest that have been trapped at Thornley Woods.

57. Buff-breasted Sandpiper, Seaton Carew GC, September 2007 (*Paul Macklam*). A rare transatlantic visitor in autumn.

58. Snipe, Upper Teesdale (*Mark Newsome*). A widespread breeding species in upland moorland areas.

59. Short-billed Dowitcher, Greatham Creek, Sept. 1999 (*Wayne Richardson*). The first British record.
60. Curlew, Teesdale (*John Bridges*). Upland pastures and moorland hold a healthy breeding population.

61. Terek Sandpiper, Saltholme Pools, June 2009 (*Chris Bell*)   **62. Spotted Sandpiper, Derwent Res., June 2002** (*Jim Pattinson*)

63. Lesser Yellowlegs, WWT Washington, May 2002 (*M.Hunter*)   **64. Wilson's Phalarope, Castle Lake, August 2007** (*John Malloy*)

65. Grey Phalarope, Hartlepool, December 2011 (*Ian Forrest*). An uncommon autumn and winter visitor, but rarely lingering.

66. Pomarine Skua, Roker, October 2011 (*Dave Johnson*). An uncommon passage migrant, occasionally seen in large numbers.

67. Long-tailed Skua, Seaton Snook, July 2003 (*Wayne Richardson*). Several impressive sea passages took place in the 1990s.
68. Sabine's Gull, River Tyne, October 2009 (*John Malloy*). This dainty Arctic gull is a rare but annual visitor to the coast.

69 & 70. Kittiwakes (*Mark Newsome*). Both the 'Kittiwake Tower' and Marsden's natural sea cliffs provide nesting sites.

71. Ross's Gull, Greatham Creek, June 1995 (*Tom Francis*)   **72. Laughing Gull, South Shields, 1984-1987** (*Ian Mills*)

73. Glaucous-winged Gull, North Tees, January 2010 (*Richard Dunn*). The second British record of this Pacific species.

74. Glaucous Gull, Hartlepool (*Martyn Sidwell*). The county's rivermouths and harbours are productive areas for white-winged gulls.

75. Little Tern, Crimdon (*Ray Scott*). The Crimdon colony is one of the most productive in the country.

76. Whiskered Tern, Saltholme Pools, April 2009 (*Martyn Sidwell*). The first county record came during a record national influx.

77. White-winged Black Tern, North Tees, August 2011 (*I.Forrest*) **78. Lesser Crested Tern, Teesmouth, July 1990** (*Jeff Youngs*)
79. Roseate Tern, South Shields (*Glyn Sellors*). Post-breeding dispersal from Coquet Island can bring good numbers in late summer.

80. Razorbill, Lizard Point (*Ray Scott*). The county's only breeding colony is at Marsden.

81. Little Auk, Whitburn, November 2007 (*Mark Newsome*). Autumn can bring spectacular numbers of this Arctic auk.

82. Great Spotted Cuckoo, North Tees, July 1995 (*Jim Pattinson*) **83.Alpine Swift, Roker, October 2006** (*Mark Newsome*)
84. Common Nighthawk, Horden, October 2010 (*Mark Penrose*). One of the most unexpected vagrants to the county.

85. Barn Owl (*Ian Forrest*). Conservation efforts have helped boost the breeding population.

86. Long-eared Owl (*Dave Johnson*). Survey work has revealed a healthy and widespread distribution.

87. Bee-eater, Bishop Middleham, summer 2002 (*Andy Hay, rspb-images.com*). One of the famour breeding pair.

88. Roller, East Boldon, July 2000 (*Jim Pattinson*)   **89. Lesser Spotted Woodpecker, Merrybent** (*Chris Bell*)

90. Isabelline Shrike (Turkestan), Whitburn CP, May 2007 (*Mark Newsome*). This rare visitor from Asia stayed for just a few hours.

91. Isabelline Shrike (Daurian), Marsden, Oct. 1999 (*Paul Cook*)　　　　**92. Lesser Grey Shrike, Jarrow, November 1984** (*Ian Mills*)

93. Hooded Crow, Seaton Common, May 2009 (*Derek Clayton*). Once a common sight in winter, it is now a scarce migrant.

94. Red-eyed Vireo, Seaton Carew, October 2010 (*Chris Bell*)   **95. Penduline Tit, Portrack, March 2006** (*Wayne Richardson*)

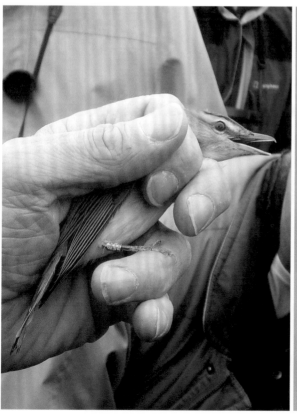

96. Willow Tit, Tilery Wood (*Ian Forrest*). Still a common bird is some areas, despite a widespread national decline.
97. Skylark (*John Bridges*). Quite common on both upland pasture and in lowland arable areas.

98. Cetti's Warbler, Dorman's Pool (*Ian Forrest*). The species was first recorded in 2005, and bred successfully in 2010.

99. Eastern Crowned Warbler, Trow Quarry, October 2009 (*Ian Fisher*). The first British record of this vagrant from the far east.

100. Greenish Warbler, South Shields Leas, August 2009 (*Steve Egglestone*). An expected early autumn scarce migrant.

101. Arctic Warbler, Hartlepool, September 1996 (*Iain Leach*). With only seven records, this species has remained a rare visitor.

102. Wood Warbler, Muggleswick (*Mark Newsome*). A characteristic but declining sight in the Oak woodlands.

103. Subalpine Warbler, Trow Quarry, September 2008 (*John Malloy*). Of six county records, this is the only autumn bird.

104. Grasshopper Warbler, Whitburn (*Mark Newsome*). A common sight and sound throughout eastern Durham.
105. Booted Warbler, Hartlepool, September 1999 (*Jim Pattinson*) **106. Melodious Warbler, Whitburn, June 2003** (*John Malloy*)

107. Aquatic Warbler, Long Drag, August 1986 (*Chris Brown*). The second of just two county records.

108. Paddyfield Warbler, Hartlepool, Sept. 1969 (*Tom Francis*)   **109. Rose-coloured Starling, Hetton, Aug. 1995** (*J.Pattinson*)

110. Waxwing, Castle Eden Walkway, December 2008 (*Ian Forrest*). An irruptive and at times common winter visitor.

111. Dipper, River Wear Valley (*Dave Johnson*). A characteristic resident of the county's rivers and streams.

112. Ring Ouzel, Bollihope (Dave Johnson). A summer visitor to upland crags and gullies.

113. Black-throated Thrush, Hartlepool, April 2010 (*Ian Forrest*). The sole county record of this Siberian vagrant.

114. Fieldfare, Upper Teesdale (*Mark Newsome*). The upland pastures attract large numbers in early spring.

115. White-throated Robin, Hartlepool, June 2011 (*Gary Crowder*). The third British record and the first for the mainland.

116. Thrush Nightingale, Hartlepool, May 1997 (*Jim Pattinson*)   **117. Red-flanked Bluetail, Whitburn, October 2011** (*S. Howat*)
118. Siberian Rubythroat, Roker, October 2006 (*Ian Fisher*). This rare visitor stayed for three days in a suburban garden (I.Fisher).

119. Pied Flycatcher, Muggleswick (*John Bridges*). The Oak woodlands hold a small breeding population.

120. Redstart, Weardale (*David Kay*). A common summer visitor to the west of the county.

121. White-spotted Bluethroat, Dorman's, Apr. 1983 (*T.Francis*) **122. Siberian Stonechat, Teesmouth, Dec. 2011** (*R Stephenson*)
123. Wheatear, Castle Lake (*Ian Forrest*). A characteristic breeding species of the western uplands.

124. Yellow Wagtail, North Tees Marshes (*John Bridges*). Small numbers breed on well managed farmland.

125. Tree Sparrow, Eppleton (*John Bridges*). Conservation efforts have helped to keep Durham populations quite healthy.

126. Amur Wagtail, Seaham, April 2004 (*Chris Bell*). One of the longest-travelled vagrants seen in Durham.

127. Richard's Pipit, Hartlepool, October 2005 (*W.Richardson*)   **128. Water Pipit, Haverton, January 2007** (*Steven Clifton*)

129. Twite, Greenabella Marsh (*Ian Forrest*). The North Tees Marshes holds the only regular wintering flock in the county.

130. Common Crossbill, The Stang (*John Bridges*). A common, though irruptive, breeding species in the commercial forests.

131. Arctic Redpoll, Hargreave's Quarry, Nov. 1995 (*Jeff Youngs*)   **132. Pine Bunting, Langley Moor, Mar. 1998** (*Norman Urwin*)
133. Ortolan Buntings, Marsden Quarry, Sept. 1995 (Jim Pattinson) **134. Little Bunting, Hartlepool, Sept. 2005** (Chris Bell)

135. Corn Bunting, Sherburn (*David Kay*). Once a common farmland bird, few strongholds remain.

136. Brown-headed Cowbird, Seaburn, May 2010 (*John Pillans*) The fifth British record of this North American vagrant.

A bird ringed at Catterick Garrison on 6 June 2004 was controlled at Low Barns on 22 January 2005, a movement of 34km to the north, north west but a significant one for this species, which is not renowned for long-distance forays. This movement was matched by the bird ringed at the same location on 17 September 2006 and controlled as a breeding female at Low Barns on 9 May 2009. From a different direction but to the same location, a bird ringed as a pullus at Apperly Dene, Northumberland on 5 June 1988 was controlled and identified as a female at Low Barns on 27 November 1991, and controlled again on 14 April 1993.

Crested Tit (European Crested Tit)
Lophophanes cristatus

A very rare vagrant: one record of one, possibly two birds.

The first and only fully authenticated record of this species for County Durham was of a bird, or birds, shot at Sunderland Moor in the middle of the 19th century (Temperley 1951). Joseph Duff published a note in *The Zoologist* for 1850 (*The Zoologist* Vol. VIII) saying "*In the second week of January (1850) a male specimen of the Crested Tit was shot on Sunderland Moor and is now in the possession of Mr. Calvert of that place*". Duff and Robert Calvert were friends, the latter being the author of *The Geology and Natural History of Durham* (1884), in which he stated "*In 1850 I saw a specimen of this little bird, which had just been shot on Sunderland Moor*".

The species was included in William Proctor's list of Birds of Durham for Ornsby's *Sketches of Durham (1846)*; in which he described Crested Tit as "*very rare*". This work however, was written four years before the bird on Sunderland Moor and it was based upon reports of a man called Farrow who claimed to have once seen three or four birds near Witton Gilbert, "*some twenty years ago*" i.e. around 1825.

Hancock (1874) thought this was not enough to establish the authenticity of Crested Tit as a bird of Durham. Later Hancock must have had some contact with Joseph Duff as, in a hand-written note in one of his ornithological volumes, he said, "*Mr. Duff says he got one of the Crested Tits shot on Sunderland Moor, but cannot recollect how it came into his hands*". This was the first information that there had been more than one bird at Sunderland. This record was referred to by Gurney in his article on the status of Crested Tit in England (Gurney 1890). Thomas Nelson knew Calvert and contacted him at sometime around 1890/1891. Calvert informed him that two birds had been shot and he presented one of the specimens to Nelson (Nelson 1891). It is believed that this went, via Nelson, to the Hancock Museum, and as at 2010, this specimen still existed in the NHSN collection in Newcastle upon Tyne.

Distribution & movements

Essentially endemic to Europe, this species has a wide distribution, largely in coniferous forests, across the Continent from Iberia in the south west, up through north west Europe, as far as northern Fennoscandia and east through the Balkans and into Western Russia. It is a largely sedentary species (Parkin & Knox 2010). Birds nesting in Britain are of the Scottish race *scoticus*, which has never been found south of the Scottish central belt. This is a rare species in England, especially in the modern context, with around ten records in the 20th century (Brown & Grice 2005). Those vagrants that have been racially identified, have proven to be of the race *mitratus*, indicating a continental origin, but perhaps the nominate race from Scandinavia/Eastern Europe might be more likely for the Durham bird(s). The most recent record of this species in the north east of England was of one at Hauxley, Northumberland from late August and through September 1984 (Kerr 2001).

Coal Tit
Periparus ater

An abundant and widespread breeding resident.

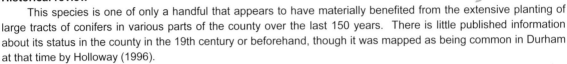

Historical review

This species is one of only a handful that appears to have materially benefited from the extensive planting of large tracts of conifers in various parts of the county over the last 150 years. There is little published information about its status in the county in the 19th century or beforehand, though it was mapped as being common in Durham at that time by Holloway (1996).

Hutchinson (1840), identified very early on that the planting of Durham's "*poor quality*" land with conifers, after the enclosure of the Commons, around 1750 onwards, offered up new opportunities for this species as this habitat matured, so it is likely that it underwent an increase in numbers in the early to mid-1800s. At the mid-point of the 20th century, Temperley (1951) described it as a "*resident, but much less numerous and less widely distributed*" than Blue *Cyanistes caeruleus* and Great Tits *Parus major*.

In an article on the irruption of titmice in autumn 1957, the authors expressed their view that the appearance in mid-September of small numbers of Coal Tits, almost simultaneously at many places on the south and east coasts of England, indicated the start of an influx of birds from the Continent. One of the first birds reported as part of this was at South Shields on the 16 September, amongst others noted at Spurn (Yorkshire), Dungeness (Kent), Gibraltar Point, Portland (Dorset) and Farnington Marshes (Hampshire), over the previous two days (Cramp *et al.*1957).

The 1950s and 1960s saw little information published about this rather over-looked species, despite the fact that it was in all likelihood increasing in numbers at the time, benefitting from the landscaping, with conifers, of the many areas of mining reclamation in the county. In contrast to this likelihood, reports suggested some local declines during the early to mid-1970s. However, an unusual sighting of this species came from the centre of Hartlepool on 13 September 1971. It was increasingly noted at bird tables during the mid-decade and winter reports suggested an ever wider distribution across the county. Singles at bird tables in Sunderland on 25 January at Croft 23 February 1980, still remained notable. Meanwhile, one at Balder Grange near Cotherstone on 23 July was significant because it was the first there since the winter storms of 1979, which had resulted in a marked decrease in numbers, particularly in the west, in the early 1980s. The true number of Coal Tits in suitable habitat was indicated by the 80 to 100 birds feeding in just one area of Hamsterley Forest in early 1981 and again in November 1982.

Recent breeding status

This species is found widely across the county, in all types of woodland habitat, although it reaches its highest densities in mixed or pure conifer woodlands. During the *Atlas* survey period, Coal Tits were relatively poorly recorded in the southeast of Durham-probably as a function of the relative scarcity of suitable woodland in that part of the county. This was later confirmed by the work of the Cleveland Breeding Atlas, which recorded that it was relatively poorly distributed species around the northern part of Teesside, though there are some concentrations of birds around Wynyard and also to the west of Stockton, with a total population of perhaps 150-200 pairs (Joynt *et al.* 2008).

Coal Tits are most often associated with coniferous woodlands and the species, with its thinner beaks and relatively longer claws, is adapted for feeding in such habitats (Perrins 1979). Its habit of nesting in holes in the ground, where there is a lack of suitable tree-nesting holes (Cramp & Perrins 1993), is well documented and is recorded for Durham (Armstrong 1988-1993). The species is also widespread in other woodlands types and will also regularly nest in gardens in both rural and suburban areas, holes in walls being oft-used nesting locations in such situations (Westerberg & Bowey 2000). However, in the breeding season it is very rarely noted in the built-up zones. That said, by the late 1990s, this species was reported to be breeding in greater numbers than in the past in the south east of the county (Iceton 2001), probably facilitated by the increased provision of garden nestboxes. Breeding season data are usually relatively poorly reported but at a small scale level, between 2006 and 2007, there was an increase, from three to five pairs, in the Hetton area. As well as holes in walls, in 2007, birds were noted breeding in a hole in an old elder *Sambuca nigra* stump and also inside a metal gatepost (Newsome 2008).

The species preference for conifers was clear in the distribution maps in the *Atlas*, which showed a more evident westerly distribution than most other woodland species (Westerberg & Bowey 2000). This pattern closely matches the distribution of the largest coniferous woods, particularly in the upper Wear and Derwent valleys. In pure conifer stands, it is easily the most common tit. CBC survey work in the Derwent valley indicated that there it occurred at densities of around 0.1 pairs per ha. in deciduous woodland and one pair per ha. in coniferous woodland (Westerberg & Bowey 2000). In the largely coniferous Elemore Woods, in the east of the county, a linear survey produced a count of 68 birds on 6 May 2008 (Newsome 2009).

Gibbons *et al.* (1993) stated that the national population of Coal Tit was believed to be around 16.5% of that of the Blue Tit *Cyanistes caeruleus*. This statement led to the *Atlas*, through extrapolation, to state the local population was around 8,500 pairs (Westerberg & Bowey 2000). It is not thought that there has been any significant increase or decline since this figure was derived. As a consequence of the availability of increasing areas of maturing forestry, reclamation planting schemes and farmland shelter belts, all of which had, by the end of the 20th century, matured to provide extensive habitat for this species in the county, Coal Tit was probably at its the most numerous that it had ever been in Durham.

Recent non-breeding status

Coal Tits begin to disperse from their main breeding areas from late August onwards, frequently being seen in urban gardens away from the breeding areas around this time. During the winter many areas record small numbers of this species, with slightly larger groups of ten to twelve birds being reported from widespread woodland sites from Brownberry Plantation, Lunedale, through Hamsterley up to the Derwent Walk Country Park and at the coast in the shape of Castle Eden Dene. Counts of 64 came from Castle Eden Dene and Sacriston in January/February 1994. At the end of winter 1994, 44 birds had been ringed at the Thornley Woods feeding station, 61% of which were first-winter birds. The largest flocks recorded in the county in the early 1990s were of 200 birds in Bishop's Park, Bishop Auckland on 13 August 1993, demonstrating the size to which post-breeding flocks can build as family parties join together. Over 100 birds were counted in Hamsterley Forest in mid-August 1997, but this extensive site must contain many times this number.

Although Coal Tits regularly visit garden feeding stations along with other tits, they often cache whole food items away from the site, rather than compete with the larger more aggressive species present (Perrins 1979). There were reports of such behaviour from suburban gardens in Sunderland and South Shields in early 1982, when it was still considered unusual.

It occurs at feeding stations in greater numbers than is sometimes realised; at least 24 birds were ringed at the Thornley Woods feeding stations in November and December 1991. One of those caught there was already five years of age (Armstrong 1992). In one morning on 28 July 2008, 47 birds were caught at the Gibside Estate feeding station. The species' habit of storing large quantities of seed during the autumn (Cramp & Perrins 1993) makes them less susceptible to starvation than a number of other small woodland species. The winter survival of coal tit, is heavily influenced by the success or otherwise of the autumn seed crops, especially that of beech *Fagus sylvatica*. As with many 'garden species', the increased provision of food in the winter over recent decades has probably helped many birds survive poor weather than would otherwise (Brown & Grice 2005).

During the early part of 2007, there were no gatherings larger than the 20s reported from Breckon Hill, Croxdale and Elemore Woods. More were noted during the autumn months when its presence in westerly plantations was indicated by the 34 birds noted at Stanhope on 23 September 2007 and at least 40 birds along on a two-mile walk through Hamsterley Forest in mid-October 2007. Such figures are clearly 'just the tip of the iceberg' in terms of the county's total autumn and winter population.

Reports of coastal Coal Tits in Durham probably increased somewhat through the 1980s, such as the small coastal passage reported at Seaburn and Whitburn during September and October 1981 and then subsequently through many of the autumns of the 1980s and 1990s. Eight were noted on the 1992 Coordinated Migrant Count on 4 October, and both the scatter and numbers of birds, often in association with other tit species, perhaps suggests that these were more likely to be relocated local birds, birds from further north in Britain or even from higher elevations in Durham and Northumberland, rather than birds from over the North Sea.

Continental Coal Tit
Periparus ater ater

Of more interest perhaps in relation to the origin of birds was the observation during an influx in 2003 when Coal Tits became unusually numerous at coastal watchpoints. These included a party of six observed to arrive 'in-off' the sea at Hartlepool on 4 October, and numbers swelled to 30 at Whitburn Coastal Park by 11th. Although none were confirmed at the time as such, it seems possible that some were of the continental race *P. a. ater*, as had occurred in the winters of 1957 and 1996 (Brown & Grice 2005); the latter influx bringing birds ringed in both Germany and the Netherlands to East Anglia (Wernham *et al.* 2020). Subsequent to this, continental birds were identified at Marsden and Whitburn on 12 September and 17 October 2005 respectively, and at Hawthorn Quarry on 30 April 2008. There was also an apparent influx of passage birds to coastal locations in Durham in autumn 2008, particularly during the second half of September. At this time, a single bird was seen to come in off the sea at Whitburn on 20th and later in the month, seven were at Hartlepool Headland, on 26th to 27 September (Blick 2009). In addition to these, small groups were present in several other coastal areas and nine were trapped at Hartlepool Headland in the period 5th to 11 October. Most were presumed to have wandered from local breeding sites but, associated with the above pattern of records, one showing the characteristics of the continental race stayed in a Whitburn village garden from 10 December until at least 6 January 2009 (Newsome 2010).

Distribution & movements

This species breeds throughout most of Europe, in North Africa, and eastwards through the Siberia and central Asia to Manchuria and China. The birds in south and west Europe are sedentary but the northerly and easterly populations tend to be short distance migrants. However, some of these are irrruptive, occasionally in large numbers. British birds are very sedentary, though some local birds may make short distance altitudinal movements within the region.

In most years, a few Coal Tits are noted along the Durham coast between mid-September and early November, such as one noted at Blackhall Rocks on the coast on 4 November 1999, but the origin of these birds has never been adequately proven. The significant ringing recoveries of this species in Durham however, do not bear out a continental origin for birds observed at the coast in the autumn. The data rather indicates that a small volume of British-derived birds appear to coast north and south through the county at this time. Examples of ringed birds moving south include one ringed at Fairfield, Stockton, on 25 October 1985 that was found dead 107km to the south at Almondsbury near Huddersfield, West Yorkshire, on 30 April 1986. Another ringed in Stockton in October 1986, had moved 41km east south east to Grosmont, North Yorkshire, by April 1987. Birds moving north include one ringed at Rowlands Gill on 14 February 1988, which was found dead at Wideopen, Newcastle on 15 September 1988 and a bird ringed as a juvenile at Catterick Garrison in North Yorkshire on 20 September 1998, which was controlled four and a half years later at Low Barns on 11 January 2003. Though this is a movement of just 34km to the north, north west, the bird was of a considerable age for a Coal Tit. There is also ringing evidence to illustrate east to west and west to east movements, which may be related to altitudinal migration. For example a bird ringed as a nestling at Hamsterley Forest on 13 June 1993 was controlled at Hexham on 24 October 1993 and one ringed at Skelton Castle in February 1998 had moved 58km west to Redford, Hamsterley by January 1999 (Blick 2009). A bird caught at Thornley Woods in November 1993, had been ringed there seven years previously.

Willow Tit
Poecile montana

A once common but increasingly uncommon and local resident.

Historical review

This species was not officially recognised in Britain until a specimen was collected in 1897 (Holloway, 1996). In summarising its status, Temperley (1951) described it as a *"resident in small numbers"*. At Temperley's time, the distribution of this species was very imperfectly known. W.B. Alexander, who saw two birds at Crimdon Dene on 17

November 1929, first identified it in the field in the county and it was also recorded near Neasham on 1 August 1930 by M.G. Robinson (Temperley 1951). The first documented breeding of this species in Durham was in April 1934 at Ravensworth, in the Team valley, when birds were observed by George Temperley as they excavated a nest hole. *"On April 18th, 1934, I was rambling through what remains of the woodlands of Ravensworth Park, near Gateshead, on the look out for early spring migrants. Approaching a glade, usually frequented by Wood-Warblers (Phylloscopus sibilatrix), whose arrival was almost due, I was listening intently for their accustomed song. Suddenly the opening notes of the expected song were heard; but the full phrase was not completed ...* (Temperley 1934). His attention had been attracted by the calls of a nesting pair and his extensive note in *British Birds* went on to describe the circumstances of the successful breeding attempt.

There is however, earlier evidence which proves that birds were nesting locally long before this discovery, as specimens of Willow Tit, labelled as Marsh Tit *P. palustris*, have been found in a number of museum collections. For example, there is one in Liverpool Museum, which was labelled as having been collected at Sherburn, County Durham, in October 1885. Temperley (1951) felt that it had bred, *"undetected in the County for an unknown period"*; eggs having been collected under the name of Marsh Tit. For example, a clutch of eight eggs taken at Wild's Hill, Rowlands Gill on 20 June 1888, which came from a 'bored hole', was undoubtedly Willow Tit and not of Marsh Tit, as originally catalogued. George Temperley noted that *"it is quite possible that the Willow Tit will prove to be a regular resident in our counties, but that it is being overlooked on account of the difficulty of differentiating it and the Marsh Tit"* (*The Vasculum* Vol. XX, 1934). Confirming Temperley's assertion was the observation that this species had been noted in the Stockton District for the last eight years (*The Vasculum* Vol. XXIV 1938).

In April 1944, a pair was found nesting near Crookfoot Reservoir and birds returned to the same tree and excavated a new nest hole the following year. It had not been recorded in Teesdale at Temperley's time of publication (1951).

By 1952, the Durham distribution was still a matter of speculation. At this time birds were recorded at Boldon on 12 October, the first record for the South Shields area, and also at Durham Banks on 15 October, whilst in the south east two were at Billingham Bottoms on 26 July 1952. It was described as 'well distributed' in the wetter woodlands of the south east of the county by Stead (1964) but little substantive information on its status was added through the 1960s.

During the 1970s it was considered a widespread resident, especially in the eastern half of the county, where it was as likely to be seen in open scrub as it was in mature woods. At this time, resident birds could be noted in various locations including Brasside Ponds, Hurworth Burn Reservoir, Low Barns near Witton- le-Wear, Washington Ponds and, in good numbers, around Bedburn.

Recent breeding status

Through the 1980s, the majority of records came from the east of Durham. An apparent stronghold of the species in the Washington area, especially around Peepy Plantation, was probably more a result of observers' concentration in that area. Through the decade breeding reports came from Joe's Pond and Darlington in 1982, Hetton in 1983 and Wycliffe-on-Tees, the most westerly location, to this time. This boundary was superseded by birds at Balder Grange, in Baldersdale, in April 1985. In the late 1980s, it became clear that there was a considerable number of breeding birds in the lower Derwent valley, with 20 birds noted around Paddock Hill Wood in November 1986. Breeding was again confirmed at six sites in the Hetton area in July 1989.

Today, despite the well publicised national declines, the species remains reasonably well distributed throughout the county, from the coastal denes as far west as Hamsterley Forest. Willow Tits rely upon the presence of standing, rotten timber in which to excavate nest holes, so any activity that impacts upon such habitat will be a threat to the species. However the observed declines in the national population up to the early 1990s were thought to be associated with a loss of damp habitats through improved drainage (Marchant *et al* 1990). Pinpointing the causes of the severe declines experienced over the last twenty years or so, are more problematical (Lewis *et al.* 2009).

Durham lies within 100km. and 150km. respectively of the northern limit of both Willow and Marsh Tit distribution in Britain and this is reflected in the largely lowland distribution of both species, as mapped in the *Atlas* (Westerberg & Bowey 2000). The distribution maps for these two species indicated that Willow Tit had, overall, a more westerly distribution than its close congener. Over the last forty years or so, it has also been more widely distributed than the Marsh Tit in Durham. This has been attributed to Willow's more catholic habitat preferences, which include damp woodlands, scrubby areas around wetlands and its ability, in some degree, to use conifer woods (Cramp & Perrins 1993). The Willow Tit's almost exclusive use of early stage succession woodland habitats (Lewis *et al.* 2009) led to the species being becoming relatively common around colliery reclamation sites in the county. In recent years, it is in some of the county's more central areas, in such habitats, where the bulk of the population has been reported, and there is clearly a strong centre of population around Houghton, though this may be emphasised by the concentrated observer effort in this area.

Areas where Willow Tits still prosper include Joe's Pond/Rainton Meadows, the Hetton area and WWT Washington, where the species, during the 1990s, was relatively common. In Thornley Woods during the 1980s and early 1990s, up to four pairs bred in most years. Collectively, these mixed scrub/woody habitats are more widespread across the county, than the Marsh Tit's preferred pure broadleaved woodland (Fuller 1982). The *Atlas* map (Westerberg & Bowey 2000) illustrated a clear distributional association with the county's main river valleys, unsurprisingly, as this is where most of Durham's habitat for this species is situated.

Conversely, in the period since the publication of the *Atlas*, there have been relatively few records of the species west of Ebchester in the Derwent Valley and Witton-le-Wear in the Wear Valley, indicating that the Willow Tit has a more easterly distribution than the Marsh Tit locally. In the extreme east of the county, birds can now be found in scrub, virtually on the beach in some places for example at Blackhall Rocks, Hawthorn Dene and Ryhope; the furthest west that regular sightings emanate from today are around Stanhope and Derwent Reservoir.

The *Atlas* records, by and large, were a full 10km further west than most other recent local records. This core range contraction eastwards may be an indication of the continuing decline of this species both locally and in Britain (Parkin & Knox 2010). Conversely, in Cumbria there has been a huge retraction west from the Pennines. In this area there was a small population near to the Durham border before 2001 (Stott *et al.* 2002) but no longer. By the start of the survey work for the new BTO Atlas, in 2008, the nearest Cumbrian Willow Tits to the Durham border were over 50km away, this distance, no doubt, preventing any movement between populations (S. Westerberg pers. comm.).

This can be a very unobtrusive species when it is incubating eggs. However, in 2007, breeding data came from many of the species' 'regular' locations, because observers 'knew' that birds would be present. For example, in 2007, nine pairs raised a total of 33 young in the Houghton/Hetton area; up to six males held territory around Tunstall Hill in Sunderland and successful breeding was also noted at: Brasside Ponds, Elemore Woods, Escomb, Hurworth Burn Reservoir, Sedgeletch, Shotton Colliery and WWT Washington. More central locations that hold birds include Low Barns and the Old Quarrington area. Despite this north easterly emphasis, there is still evidence of birds rather further west than has sometimes been documented in the early part of the 21st century. For instance, in 2008, as well as regular sightings from locations such as Blackhall Mill, Clara Vale and Hedleyhope, birds were noted at Rabbit Bank Wood, near Consett, from several locations around Hamsterley Forest and as far west as Rookhope. In the absence of concerted survey work to establish the population in the western half of the county, these records may suggest that observer bias heavily influences the perceived distribution of this species in Durham.

During the 1980s and 1990s, CBC work in the lower Derwent Valley produced estimates of up to one pair of Willow Tit per 15ha. of local woodland (Westerberg & Bowey 2000). As the species is more widely distributed and has been recorded at higher densities than Marsh Tit was at that time, the breeding population was almost certainly greater. At an average breeding density of three to four pairs per occupied tetrad, the *Atlas* calculated the Durham population at around 300 pairs (Westerberg & Bowey 2000), though, in retrospect, this now seems slightly low for that period. However, given the subsequent very much evident national declines (Brown & Grice 2005) this is probably now an over-estimate of the county's true population. Today, in the south east of the county, it is considered slightly less common than Marsh Tit *Poecile palustris*, with around 15 pairs, certainly less than 20 pairs, in the northern part of what was Cleveland. These were mainly found around the Wynyard Estate, and along the Castle Eden Walkway; where there were perhaps four to six territories (Joynt *et al.* 2008). In this general area, singing birds, or successful breeding has been recorded recently at Urlay Nook, Hardwick Hall Country Park and Hurworth Burn Res, though no doubt, it goes unrecorded at many other locations where birds are present.

In 2009, a number of declining species were found to have reached their lowest levels since the start of the BBS, and included in this list was the Willow Tit, which was found to have experienced a 73% decline since the outset of the work in 1996 (Risely *et al.* 2010).

In 2010, breeding was confirmed at Bishop Middleham, Blackhall Rocks, Duncombe Moor, Elemore GC, Elemore Woods, Hetton Bogs, Hetton Lyons CP, Rainton Meadows, Thornley Woods, Shincliffe and along the Deerness Valley Walk. Reports in that year, indicated some local declines, but in relative terms, this species seems to be maintaining itself relatively well in the county, somewhat against national trends.

Recent non-breeding status

Despite the national decline, this species remains a widespread species in Durham, for example there were almost 400 reports of it from over 100 localities in 2008 and it was very much more widespread than the Marsh Tit. By way of contrast, on the other side of the Pennines, in 2008 there were just 20 records from three sites in Cumbria (Hartley 2009).

The majority of reports to the DBC are non-breeding season sightings. These show a north easterly distribution in the county during the winter months, mirroring its breeding range, but heavily skewed by observer bias. In 2008, large early winter counts came from Rainton Meadows, with up to 18 birds, and Hetton Bogs with 15 birds.

Reports of birds visiting winter feeding stations for sunflower seed or peanuts have increased in recent years, though this habit has been noted in the lower Derwent valley, at Thornley Woods for decades, as it has at Hurworth Burn Reservoir, Low Barns and WWT Washington, with several others coming to suburban gardens, for example at Langley Moor and Shotton Colliery. Concerted bird ringing activities at some of the county's feeding stations has shown that Willow Tits (and also Marsh Tits) can be extremely faithful to one site. By way of illustrating this point, when ringing resumed at WWT Washington after a four year gap in 1990, the only ringed bird that was still present from previous ringing activities there was a Willow Tit.

Unlike other tit species, there is not usually an autumn increase in numbers at coastal sites, though two were noted on the 1992 Coordinated Migrant Count on 4 October and again on the equivalent watch in early October 1995. A bird at the Marine Park at South Shields on 9 November 1974 was the first known sighting there, but the only indications of possible coastal passage in recent years involve single birds at Whitburn village on 11 October 2008 and in Whitburn Coastal Park on 25 August 2011, although the presence of resident, breeding birds in coastal areas to the south of the River Wear, and possible dispersal of young birds from such areas, makes the reliable detection of passage migrants almost impossible.

Distribution & movements

Willow Tits breed throughout much of Europe, in North Africa, and eastwards across the Palaearctic, east to Siberia and south into to Manchuria and China. The birds in the west and south, including most of Western Europe, are sedentary but the northerly and easterly populations show occasional irruptive movements probably to avoid severe winter conditions.

The British population is very sedentary (Parkin & Knox 2010). At just one per cent, this species has the lowest percentage of ringing recoveries more than 1km from the ringing site of any British bird (Clark *et al.* 2010). Though some local birds may make short distance movements within the county, for example a bird ringed at Clara Vale in 1998 moved 2km to Ryton three months later, the essentially sedentary nature of local birds is hinted at by the small number of recoveries and controls of locally ringed birds. For example, a bird ringed as a juvenile at Lockhaugh STW on 28 August 1988 was found dead two years later at the same place. The exception is a bird ringed at Low Barns on 26 June 2005 as a juvenile, and presumably fledged in that area. It was found dead, after hitting a window at Barnard Castle 17km to the south west, a considerable for this species, and in the relatively short period of just four days; easily the longest known movement of a Durham-ringed bird.

Marsh Tit
Poecile palustris

An uncommon and local breeding resident.

Historical review

There is little doubt that this species is under-recorded in Durham and it would appear to have been for as long as bird recording has featured. There is little information about the species' historical status in the county, other than it was known to be present in Victorian times, and there is little documentary evidence of its presence prior to the early 1900s, though it was undoubtedly present, as museum specimens indicate. Hutchinson (1840) did refer to it in passing, commenting that 'the Marsh-Titmouse', like Coal Tit *Periparus ater*, often had its nest '*at the bottom, among the roots'*.

It would appear to be considerably less common than its close relative the Willow Tit *P. montanus* preferring purer stands of mature, deciduous woodland. Temperley (1951) described it as "*a resident, much less numerous than the great, blue or coal tits*". He thought it more numerous than Willow Tit, a few being noted in winter mixed tit flocks. It was described by Stead (1964) as common in the south east of the county, but locally distributed in woodland, including around Wolviston and in Crimdon Dene, but through the 1950s and 1960s this species, other than being recognised as present was little studied and its status imperfectly known.

Through the 1970s, Castle Eden Dene was perennially noted as the species' main locality in the county and at least ten pairs were present there in late May 1979. It was also plentiful in the Bedburn area and around Durham City. Its status might be summarised as locally distributed in well-established woodland from the coast to the Pennine dales, though somewhat localised. Up to three birds were noted at a bird table at Balder Grange, in Baldersdale in February and October 1977.

Recent breeding status

In broad terms, despite its name, for breeding purposes the Marsh Tit prefers drier habitats than does the Willow Tit. However, in winter, roving tit flocks containing this species might be seen almost anywhere around the county's woodlands, and Marsh Tits are more often seen in such flocks than their close relatives.

This Marsh Tit has long been associated with the woodlands found in Durham's coastal denes, such as at Castle Eden and Hawthorn but it is a species with a sparse distribution across much of the county. It is a bird of mature deciduous woodland, with a particular penchant for oak *Quercus petrea/Q. robur* or beech *Fagus sylvatica* (Harrap & Quinn 1996). Hence, it is found in many of the ancient semi-natural woodlands, such as the western oak woods in much of Teesdale, and the Muggleswick NNR. Accounts in *Birds in Durham* for the 1980s very much mirror this disjunct distribution in Durham, though a pair nested in a nestbox in the south east of the county near Hartburn in 1980 (Bell 1981). Towards the end of that decade it was reported from around 20 sites per annum across the county. The species was noted as nesting near Rowlands Gill in May 1978 and breeding occurred in the Derwent Walk Country Park for the first time in ten years in 1997.

The species' visual similarities with Willow Tit have always caused some identification problems, however its strident '*pitchu'* call helps resolve this problem. In any mapping exercise for this species a modicum of reserve about the results needs to be adopted, as inevitably some errors will have crept in to the mapping process. A survey around Teesmouth in 1989 indicated a presence of the species only along the Castle Eden Walkway (Joynt *et al.* 2008).

There is little doubt that this species was under-recorded locally for many years. Whilst during the 1970s and 1980s, it was considered less common than the Willow Tit this relative position is now less clear, largely because of the major decline in numbers of the latter.

A comparison of the *Atlas* mapping of this species, with its known distribution from the local submitted records, indicated that the map in Westerberg & Bowey (2000) was largely an accurate one. In the east of the county, the largest populations were typically recorded in the denes at Castle Eden and Hawthorn, where maximum counts were of 22 and 12 birds respectively in 2008, with birds also present in Nesbitt Dene, further south. In the more western portion of its local range the population in recent years has been centred in and around the mid-Wear valley, in particular Low Barns, with birds recorded east to Bishop Middleham and west into Hamsterley Forest. Other strongholds are found around Barnard Castle, the Derwent Gorge and Muggleswick.

It has been determined that the Marsh Tit requires up to six hectares of woodland per successful territory, so the breeding density per occupied tetrad is nowhere very high in Durham. Only in Castle Eden Dene does the population in a single wood routinely exceed six pairs (Westerberg & Bowey 2000). At a mean density of just two pairs per occupied tetrad, the thinly spread county population, probably still only amounts to around 100 pairs, as stated in the *Atlas* (Westerberg & Bowey 2000). It is generally significantly less common in the north of England than it is further south (Brown & Grice 2005).

Breeding occurs in a number of scattered small areas of woodland in the south east of the county, including Wynyard, Elton, Newton Woods, Dalton Piercy, Hart Station and at Wolviston (Blick 2009). Today, there are estimated to be around 20 pairs in the northern part of what was Cleveland (Joynt *et al.* 2008).

In the well-studied, lower Derwent valley woodlands, Marsh Tits disappeared from some woodland for several years during the 1980s and 1990s, only to re-appear in years of good beech nut crops (Armstrong 1988-1993). A bird at the Thornley Woods feeding station hide in November 1995 was the first recorded at that location since its construction in 1991 and the first record in the area since the winter of 1987/1988. The locality producing most regular sightings over the years in this area is Paddock Hill Woods, where birds bred successfully in 1988 and again intermittently during the late 1990s and early 2000s. The ultimate determining factor for such local fluctuations in breeding activity is not fully understood. In 2007, there were no confirmed breeding records of the species anywhere in Durham, though clearly birds would have fledged young in many of the 'usually occupied sites'. In 2008, 129 records of the species were reported from 27 locations. Two birds near Thwaite Hall in July were a little further west than usual but a single at Park End Wood near Holwick in upper Teesdale on the 16 June was the most westerly record for many years. In 2010, adults and juveniles noted at Bearpark, Crimdon and Newsham Wood, were all noteworthy sightings as these are locations in which the species is not regularly recorded.

Recent non-breeding status

In the early 1980s, it was thought that the local population of Marsh Tits had been reduced as a consequence of the hard winters of 1978 and 1979, so a gathering of 10 birds in Castle Eden Dene, in November 1982, was exceptional even for this well frequented site. In autumn 1997, birds were noted at Baybridge in the upper Derwent valley and at Low Force in Teesdale, which was the first time that the species had been recorded west of the NY/NZ dividing grid line in Durham.

Numbers recorded across the county are generally small, nonetheless, a count of 21 in the whole of Castle Eden Dene in December 2009, illustrated that in the right habitat, this can be a reasonably common species.. When beech trees were felled as part of habitat improvements in the Muggleswick NNR in the late 1990s, Marsh Tits became far less abundant at this site, presumably due to the loss of an important winter food resource.

The winter numbers, or more precisely the number of birds reported, in the county fluctuate, apparently in relation to the availability of beech mast (Westerberg & Bowey 2000). In winter some feeding stations are regularly frequented, such as those at Hawthorn Dene and Low Barns.

Although Marsh Tit is one of the most sedentary of woodland species, for example one seen at the Kibblesworth CBC site on 11 March 2000 was only the second record there in over 30 years, they do occasionally wander away from their more usual sites; Ryton Willows in the Tyne valley has held birds on a number of occasions but they have never been known to breed there. Singles noted at Shibdon Pond on 30 June 1985, at the Metro Centre in autumn 2002, unsuccessfully attempting several times to fly across the Tyne, and at Axwell Park on 17 March 2004, were all presumably wanderers originating from the small populations in the lower Tyne and Derwent valleys. The two that were noted on Hartlepool Headland on 11 October 1985 (Blick 2009) however, are slightly more difficult to reconcile with nearby breeding birds. A single bird reported from South Shields on 18 September 1988, was well out of its normal range and suggested a movement along the coast.

Distribution & movements

The Marsh Tit is a largely sedentary species in Western Europe. Some Durham birds may become locally nomadic, at a small scale, during the winter, as they get caught up in mixed tit flocks. It breeds throughout much of Europe apart from N Scandinavia, Ireland and most of Scotland. The northerly and eastern European populations may wander in winter (Parkin & Knox 2010).

The species' sedentary nature is illustrated by local ringing work. Between 1995 and 2010, the Durham Dales Ringing Group ringed a total of 81 Marsh Tits at Low Barns Nature Reserve, but there were no significant controls or movements from these. One of the few recoveries was of a bird found freshly dead at Witton-le-Wear on 11 November 2007, it had been ringed five days previously at Low Barns on 16 November, and had moved two kilometres.

Bearded Tit (Bearded Reedling)
Panurus biarmicus

A rare migrant from England and continental Europe, 30 records involving at least 77 birds, recorded in all months, though most frequently from October to March.

Historical Review

The first Bearded Tits for County Durham were two males seen by the late Edgar Gatenby at Dorman's Pool, Teesmouth on 3 December 1966. Just over two weeks later a party of four birds at the same locality on 18 December 1966 comprised one male and three females and were presumably different birds. These are the only pre-1970 records of the species, although three at Ingleby Barwick on 6th and 7 December 1948 were less than a mile outside of the county boundary.

Recent Status
A total of 71 Bearded Tits have been recorded in County Durham since 1970, with the vast majority of records coming from the North Tees Marshes, in particular from the large reedbeds at both Haverton Hole and the Long Drag. The peak months of occurrence are October and November although the species has been recorded in all months, and it has frequently wintered in the region particularly since the mid-1990s. The best year on record was 2004, when an estimated 17 birds were recorded, boosted by an influx of 14 birds during October and November. Involved with this influx was a group of 11 that were present on the Long Drag on 18 November 2004; still the largest party in County Durham.

Since 1970, the North Tees Marshes have accounted for an estimated 62 birds with most of these occurring in the ten-year period between 1997 and 2006. The first record since 1966 involved a female at Dorman's Pool from 1 April until 7 July 1983, later joined by a male from 25 April until 2 October, which raised hopes of a breeding attempt though there was no proof of this. A long gap ensued until the next at Teesmouth in 1995 when a mobile male toured the North Tees Marshes between 2nd and 15 May, it was trapped at Haverton Hole on 6 May. Both this and the previous male in 1983 had previously been ringed as nestlings at different sites within Britain, indicating that spring records of this species are more likely to relate to dispersing British birds rather than continental immigrants.

The period between 1997 and 2006 saw the Bearded Tit establish itself as a regular passage migrant and winter visitor to the North Tees Marshes with a total of 59 birds recorded including several small flocks. Most sightings came from the extensive reedbeds at Haverton Hole and the Long Drag with occasional records from Dorman's Pool. Overwintering was first recorded in the south east of the county in 1997, when a party of three birds was present on the Long Drag from 25 October 1997 until 15 March 1998, and this occurred again in the winters of 1998/1999, 1999/2000, 2002/2003, 2003/2004 and 2004/2005. The largest party was the 11 birds on the Long Drag during the winter of 2004/2005, but at least eight were at the same location during the winter of 2002/2003 and another group of eight was at Dorman's Pool on 19 September 2006. A party of three on Greenabella Marsh on 18 October 1998 and another seen nearby, on Cowpen Marsh on 3 November 1998, are the only records away from the established 'core areas' around Teesmouth. The most recent record was on 19 September 2006.

The species is extremely rare away from the North Tees Marshes and there have been just nine records. All related to singles and occurred at Charlton's Pond, Billingham on 13 January 1988, Shibdon Pond (male) from 24 November 1992 until 12 March 1993 (the first overwintering record for Durham), Portrack Marsh, Stockton-on-Tees on 15 October 1998, SAFC Academy Pools, Sunderland on 30th and 31 October 2004, Dawdon Blast Beach on 31 October 2004, Cowpen Bewley Woodland Park on 3 November 2004, Billingham Bottoms from 16th to 20 February and 27 March 2005 (male), Hetton Lyons Country Park on 24 October 2007 (female), and a female at Hartlepool Headland on 9 October 2010.

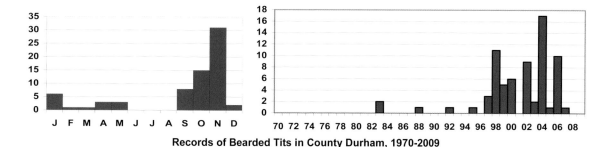

Records of Bearded Tits in County Durham, 1970-2009

It is interesting to note that the majority of the records away from Teesmouth related to birds seen on just a single date, with just two making protracted stays. Most occurred in late October and early November. The records from Dawdon Blast Beach and Hartlepool Headland suggested possible immigration from the near Continent as both were found during large falls of continental migrants. The individual watched flying west and calling over Charlton's Pond in Ja nuary 1988 suggests that movements of this species can occur well into the mid-winter, not always in response to harsh weather conditions.

Recent breeding status

The species has often been cited as a potential colonist. Thus far, a pair present at Haverton Hole until late April 2000 are the only birds to have shown any conclusive signs of breeding activity within the county. The male of this pair was seen carrying nesting material in March and was also seen to chase off a rival pair during early May, but heavy rainfall later that month flooded the reedbed and unfortunately there were no subsequent sightings (Joynt *et al.* 2008). Successful breeding took place for the first time in Northumberland in 1999 (Kerr 2001).

Distribution & movements

Bearded Tit has a scattered breeding distribution across much of Europe, including parts of the British Isles eastwards through central Asia to western China. The European populations are largely sedentary though many populations do undertake dispersive movements in times of food stress probably brought about by failure of the *Phragmites* seed crop. Ringing recoveries have proven that these movements occur both within the British population and from the near Continent with several recoveries from both Belgium and the Netherlands and from as far afield as Norway and Switzerland.

Several Bearded Tits have been trapped and ringed on the North Tees Marshes including a party of three on the Long Drag in 1997, with further birds trapped in 1999, 2000, 2002 (3), and 2004 (2). In addition to these birds, the long-staying male at Shibdon Pond during the winter of 1982/1983, was trapped and ringed on 9 February. This is the only Bearded Tit to be trapped and ringed away from the North Tees Marshes and when examined in the hand was exhibiting a cloacal protuberance indicating that it was coming into breeding condition.

Two birds have been controlled on Teesside; a male trapped at Saltholme Pools on 3 August 1983 had been ringed as a nestling at Ousefleet, East Yorkshire on 25 July 1982, some 104km to the south, whilst a male caught at Haverton Hole on 6 May 1995 had been ringed as a nestling at Leighton Moss, Lancashire on 9 May 1994 where it had remained until at least 19 September 1994. The most remarkable record involved a male trapped and ringed on the Long Drag on 7 November 2004, which was later controlled at Borrevannet, Vesfold, Norway on 29 October 2005; a distance of 882km to the north east. This was the first British ringed Bearded Tit to be recovered in Scandinavia, although a bird ringed in Norway was controlled in Hampshire in 1998.

The British population is currently estimated at around 589 pairs (Holling 2010), the majority of which are to be found in East Anglia and southern Britain, though isolated populations exist at Leighton Moss, Lancashire and at a single locality in Perthshire, Scotland. This represents a dramatic increase from between two and four pairs in 1948 and is widely attributed to the expansion of this species from the Netherlands in the mid-1960s (Brown & Grice 2005).

677

Short-toed Lark (Greater Short-toed Lark)
Calandrella brachydactyla

A very rare vagrant from southern Europe: three records.

*Short-toed Lark, Reclamation Pond, May 2008
(Martyn Sidwell)*

All of Durham's records have been found in the North Tees area, though only one has lingered. The first was found by Peter Bell and S. P. Warwick in rough grass at the western end of Saltholme Pools on 17 April 1983. It lingered here until 20 April, being seen by large numbers of visiting birdwatchers (*British Birds* 77: 545). This was only the sixth April record for Britain at the time. The second record came eleven years later, when one was found on the Town Moor on Hartlepool Headland on 20 May 1997. The final sighting was in 2008, when one was located adjacent to the Reclamation Pond on the afternoon of 30 May, showing well until the evening. This sighting was part of a late spring fall which involved a Red-throated Pipit *Anthus cervinus* within a few hundred metres of the lark. Durham's total of three birds is surprisingly low compared to that of the neighbouring counties of Northumberland and North Yorkshire, which have attracted regular sightings between 1990 and 2010.

Distribution & movements

Short-toed Lark breeds in southern Europe and north west Africa and from Turkey across Central Asia to Mongolia. The western population winters to the south of the Sahara across to the Middle East, whilst eastern birds move to south west Asia and the Indian subcontinent. They are a regular, scarce migrant to Britain and north west Europe in both spring and autumn, most frequently encountered in the southeast of England, the Northern Isles and the Isles of Scilly. A BBRC species until 1993, Short-toed Lark was removed as a rarity once records topped the 430 mark.

Woodlark
Lullula arborea

A rare vagrant from continental Europe: 11 records, involving 13 birds.

Historical review

George Temperley (1951) summarised its status as a *"rare vagrant…few records"*. Several Woodlarks were seen in historical times and certainly the earlier records suggest that the species could have been more regular in the county, but presumably was overlooked. The first reference is by Backhouse (1834) who noted that the Woodlark was "*occasionally killed near Durham*", while Hancock had in his collection, two specimens out of three killed by Thomas Robson at Swalwell in March 1844. It seems likely that these were the birds which had been described as wintering locally by Robson (1896). As no earlier records, including Backhouse's assertion, are known, this record constitutes the first for the county. A more intriguing sighting came in 1884 when George Bolam noted, "*On 3rd August 1884, when in company with my late friend, Mr Henry Wallace, I saw a Wood-Lark in the grounds of Ravensworth Castle, Co. Durham, where it might, not unnaturally, have been supposed to have passed the summer...*" Certainly early August is an unusual time for a migrant and this remains the only summer record for the county.

The next record took over forty years to appear, when two spring migrants were seen in 1929. The first report came on 3 May 1929, when George Bolam and H.M.S. Blair saw a Woodlark along the cliffs at South Shields, noting that there were many Skylarks *Alauda arvensis* in the seed and corn fields but the "*Little Lark*" was always alone and kept to itself. The same observers saw it several times over the next two days, "*sketched it and made quite sure of it*". What was presumably a second bird was found by the same observers the following day, 4 May 1929, some ten miles away at Hebburn Ponds.

Post-Temperley, there were no records until the latter part of the 1950s and into the 1960s. The first in County Durham since spring 1929 was found at Fence Houses on 22 April 1958. Singles, fitting neatly with current occurrence patterns, were at Cleadon Hills on 1 January 1964, then only the second winter record for the county, and Hartlepool on 16 November 1968. Subsequent to this sighting, there were then no records until more than half-way through the 1990s, emphasising just how rare this species was in Durham for much of the 20th century.

Recent status

There have only been five records of Woodlark since 1970, only one of which involved a bird seen for more than one day. After a gap of 28 years, the first sighting was of a wintering bird found on waste land at Norton on 26 November 1996; it remained in the area until 21 February 1997. All other sightings have involved short staying or birds flying over. November is clearly a month when the species is on the move, with four of the five recent birds being found between 10th and 26th, whilst the mid-winter Whitburn bird was involved in a large cold-weather movement of Skylarks.

All records since 1970:

1997	Norton, 26 November to 21 February 1997
1999	Shotton Colliery, 23 November
2000	Mere Knolls Cemetery, 10 November
2006	Whitburn Village, 2 January
2008	Lambton, 23 November

Distribution and movements

The Woodlark is a widespread breeding species through much of Europe, southern Scandinavia and north west Africa, spreading east into Russia, the Caucasus and the Middle East. Many birds in the eastern part of its range move south during the winter, but milder winters in north west Europe through the 1990s and 2000s appear to have encouraged more to remain further north. In 2010, the British population was estimated to be approximately 3,000 pairs, with expansion from strongholds in southern England leading to regular breeding as far north as Yorkshire. However, in England, it is a rare migrant outside its normal breeding range.

Sponsored by

NORTHUMBRIAN WATER

Skylark (Eurasian Skylark)
Alauda arvensis

An abundant resident and common passage migrant that has shown some local declines in recent years.

Historical review

The Skylark is a well-known farmland bird, associated with open fields and landscapes. It has an attractive aerial display and song-flight that has given it a certain cultural iconic status. It is one of the most widespread of British birds. This species has a long history in the county and its presence in mediaeval Durham was confirmed by the excavation of remains from the mediaeval strata of Barnard Castle (Yalden & Albarella 2009).

Skylark has a single mention in the accounts of the Bursars' rolls of the Monastery of Durham in the 14th century. At such a time it might be imagined that Skylarks would have been a fairly frequent article of diet, so it is a little surprising to find that there is just one mention of them in these accounts. It was documented in the cellarer's roll for Christmas week 1344, when "*six dozen larks*" (*vi duoden alaud*) were purchased along with ducks (Ticehurst 1923).

There is relatively little documentation of its status in the county up to the end of the 19th century. Robson (1896) mentions the taking of a white individual locally and described the species as "*commonly distributed*". In 1905, Tristram thought it "*rapidly diminishing in numbers*" but Temperley (1951) stated that he thought this only temporary, and probably within the bound of local fluctuation. Nelson (1907) recorded that they were the most numerous migrant in autumn at Teesmouth.

In the Scottish Naturalist for March-April, 1931, Noble Rollin wrote on the length of lark song. He timed over a thousand songs, half or more in the district of South Shields, and demonstrated that the average length of song was greater in March than in April; lengthening again in May and June, and becoming more truncated in July, presumably affected by moult. It was found to be of longest duration in September; greater in October than in July, and much lower again in November.

Temperley (1951) wrote that the Skylark was a "*very common resident*" and a "*winter visitor in considerable numbers*". He described it as generally distributed over the county, that there had been no real change in its status during recorded times, and that it was "*seldom seen on heathy moors*", which seems a decidedly odd statement considering the species' modern distribution in the uplands of the county. J.J. McKinney noted that as fires were lit on Victory Night June 6th 1946, a Skylark was singing at 10:15 p.m. as if enjoying it all (*The Vasculum* Vol. XXXI 1946).

Through the 1950s, 1960s and early 1970s the species appeared to remain widespread and common, though little substantive information about its status in Durham was documented over the period. It was described as probably the 'commonest breeding bird of the area' by Stead (1964). Blick (1978) observed that it was a 'regular breeder in quite good numbers', and about 1,000 were recorded beside Seal Sands on 15 December 1976 (Blick 2009). One of the farmers in the farmland survey area of Kibblesworth noticed a Skylark nest with young whilst working in May 1981. After marking the nest and contacting a local ringer the result was the only Skylark pulli to be ringed in that area during a twelve-year period of study.

Recent breeding status

This is still a common resident and passage migrant in Durham, though it has shown some local declines in numbers in the late 1990s and in the first decade of the 21st century. In the north east region as a whole, numbers declined by about 38% between 1994 and 2005 (Risely *et al.* 2010). Overall, in Durham the Skylark population has not fared as badly as elsewhere in the UK, and it is still abundant on the western uplands. It has declined mainly in the lowland, agricultural areas, though it is still common. There are still good numbers of Skylark in the coastal strip, where rough grassland prevails.

Research by a raft of agencies, including the RSPB, BTO, the Game Conservancy Trust and researchers at Oxford University, has shown that the intensification of cereal growing and other agricultural changes in farming practices, such as cropping patterns, in the second half of the 20th century, were the principal reasons for a major national decline of the species (Marchant *et al.* 1990). It has been estimated that this decline amounts to 58% between the years 1969 and 1994 (Fuller *et al.* 1995). The national population was relatively stable until around 1980 (Marchant *et al.* 1990), when the loss of birds began in northern and eastern England.

Although the British population has declined, its range has changed relatively little and it remains widespread in Durham. Where local losses have occurred then these have been primarily in the eastern arable areas of the county, at least over the last 30 years, as indicated by reports in *Birds in Durham* over that period.

The *Atlas* well illustrated that the species was widespread in its spring and summer range, reaching as far west as the upper reaches of the Tees, Wear and Derwent valley watersheds and a presence over most of the county, from the coastal strips to open arable farmland and through the varied central portion of the county. Any gaps in its distribution tended to coincide with extensive areas of unsuitable habitat, in the main large-scale forestry, such as Hamsterley Forest and urban areas (Westerberg & Bowey 2000).

Skylarks can be seen and very clearly heard in virtually all open habitats in the county, such as farmland, marshland, rough grassland and extensive open moorland. A singing bird, however, does not necessarily represent a breeding pair early in the season it may be a male that is still 'advertising' for a mate. Song is usually heard from the start of February in Durham and is always a welcome early sign of spring. Territorial Skylarks sing for much of the breeding season. Consequently, frequently visited sites can produce large numbers of documented birds. For example in spring 2000, there was a high concentration of birds breeding in the open grasslands around Rainton Meadows with over 40 pairs thought to be in territory there.

In May 2003 a survey of South Shields Leas' coastal grasslands revealed 39 singing birds. In 2007, which might be seen as typical, breeding densities in some of the prime lowland breeding locations included 82 singing birds at Hedleyhope Fell in early March, about 50 at Farnless Farm, Bishop Middleham, over 40 territories at Brancepeth Beck in March, 40 at Park House Farm, Oakenshaw, and 20 singing over set-aside fields at Pittington Hills in June. Also in June, over 150 territories were found on another four separate farms in the Bishop Middleham area. Such survey work indicated the importance of managed farmland areas in central Durham. In the north of the county, the long-running, Kibblesworth CBC site revealed 16 pairs, an increase of five pairs from the 2006 total. Although more limited by the availability of habitat, some coastal areas also hold good breeding populations. Fledged juveniles are usually first noted in the last week or so of May, though singing continues in many localities until the second week of July.

The survey in the north of Teesside for *The Breeding Birds of Cleveland* showed that Skylarks were still widespread and common. Although absent from the urban areas, it was found extensively around the fringes of Hartlepool. There are very high numbers of birds breeding around the Teesmouth NNR, with up to 55 pairs in that one area. Tetrad occupancy across north Teesside was around 70%, and there were probably around 850 breeding pairs in this whole area (Joynt *et al.* 2008).

Although it is usually poorly documented, largely because of its sheer scale, the breeding population of the county's uplands considerably dwarfs that of every other part of the county. The *Atlas* postulated that the county held as many as 30,000 pairs of Skylark (Westerberg & Bowey 2000). It is difficult to challenge this figure today given its continuing strongholds in the west.

Non-breeding Status

Post-breeding gatherings occur during late July and build through August, and these may be noted at favoured inland farmland sites such as in September 2007 when 110 were at Farnless Farm, near Bishop Middleham on 17th and 100 at Brancepeth Beck on 22nd.

Autumn brings an extensive emigration from the county, with most significant movements usually occurring between the last week of September and through October. Throughout October there are usually many reports of 'visible migration', often from coastal locations such as Beacon Hill and Hawthorn Dene. Over this period, flocks numbering from small numbers of tens, up to constant streams of overhead birds, are often noted. Grounded, feeding birds are routinely found in grassland and farmed areas particularly along the coastal strip. In the winter months, numbers in the county are usually relatively low as birds move out of our area. The exception is of some decent-sized flocks that tend to linger mainly on coastal stubble fields or *Brassica* crops. However, prolonged periods of snow cover can lead to a movements away even from these sites.

In some years, if there is little undisturbed stubble across the county to attract the species, small gatherings, usually counted in scores of birds at most, are found scattered rather haphazardly across the eastern lowland portion of the county, with occasional groupings in places such as Seaton Common and the Teesmouth NNR.

When stubble is left in such areas, then the winter congregations can be considerable. For example in December 2000, up to 250 were around the Hart village area and 400 at Urlay Nook, and then 500 at Chourdon Point in January 2001. The severest weather can drive local birds away but these can be replaced by birds from elsewhere and mid-winter, hard frosts often precipitate complex movements of birds through or over the county. The 800 birds counted flying south over the Long Drag on 23 December 2001, was a good example of such a re-location, as was the easterly movement, during snow showers, of 500 birds over Shibdon Pond in less than three hours on 9 February 1991 and the 550 birds noted flying across South Shields Leas in one hour in December 2000. In 2005, heavy snow in December led to a mass exodus from inland areas resulting in 855 flying south over South Shields Leas on 29th and 30th and a huge flock of 1000 in a stubble field at Dawdon on 29th, plus an even larger one, estimated to amount to more than 2,000 birds, at Cold Hesledon on the last day of the year. Larger than all of these was the movement observed on 1 January 2010, during very severe weather, when several flocks up to 200 strong, totalling 2,610, were noted heading south over Sunderland South Dock between 10.30 and 11.40. Very large numbers were witnessed moving south at several other coastal sites on this date, and this was probably the largest ever documented movement of this species through Durham.

In the first quarter of most years, birds are relatively sparsely reported, although some breeding sites such as Rainton Meadows retain a few birds throughout the winter, and they are sometimes noted singing from as early as mid-January. Birds are largely absent from the more exposed areas during the winter months, including their summer strongholds in the west but will re-occupy even some elevated sites by late February

Distribution & movements

The species is found as a breeding bird across Europe, North Africa and Asia Minor, and from there, across a broad band through central Asia to the Pacific via Japan. In Europe, northern populations are migratory, often wintering within the species range but to the south. Patterns of movement are often determined by weather conditions, particularly the extent and duration of winter snow cover.

Spring passage occurs from around the middle of February, with flocks moving north and north west from wintering areas further south. Singing birds are usually first noted back in their breeding areas by the last week of February, with more concerted arrival taking place through March. In the core of its westerly breeding range in the county, the extensive uplands, which are usually vacated by the end of August, emigration usually commences around the middle of July.

There is little relevant ringing information for Durham relating to this species, but a bird present at Haverton Hole in December 1959 (Blick 2009) had been ringed as a pullus at South Shields in June 1953.

Shore Lark (Horned Lark)
Eremophila alpestris

A scarce winter visitor and spring and autumn passage migrant to the coast; formerly more common.

Historical review

The Shore Lark was only recognised as a visitor to Britain in 1830, so understandably there is no reference to its occurrence in Durham in the writings of Selby (1831), Backhouse (1934) or Hutchinson (1840). The first mention is by Hancock (1874), who noted that it was "*a casual visitant and is seldom taken in our district*". He recorded two occurrences in Durham, the first mentioning "*two specimens came into my possession in 1855 which were killed near Rock Lodge, Roker, a few years before*". The second related to notes published by J. H. Gurney stating that "*four specimens were shot in the county of Durham by Mr. G.C. Pecket on the 26th of July, 1867*"; on such an unusual date for this species, it seems likely that the date was a printing error in the communication of the date. Further evidence in the late 19th century was provided by Tristram (1905), who stated that four specimens were taken at Seaton Snook and several others seen in the winter of 1870/1871. The winter of 1879/1880 featured the first documented large-scale arrival on the east coast of England, and Hancock referred to one caught by Mr. Wm. Yellowly using a clap-net at South Shields in the week before Christmas 1879.

In 1907, Nelson wrote of the increasing frequency of Shore Larks visiting Teesmouth. "*In 1900, a party of twelve appeared on the 12th of October and others continued to arrive until the 22nd of December, while a flock of two hundred individuals frequented a piece of reclaimed land from November until February of the following year. At the Tees mouth it usually haunts the foot of the sand hills near high-water mark, where it feeds along the debris cast up by the tide or on the short herbage at the edges of the tidal pools*" (Nelson 1907). Following Nelson's observations, no other observer recorded the species with anything like such regularity or in such numbers and this remains the largest gathering in the county. Temperley (1951) considered it to be very rare on the Durham coast, away from Teesmouth.

In the early part of the 20th century, C.E. Milburn recorded a dozen at Teesmouth in late November 1916 *(The Vasculum* Vol. III 1917), and Joseph Bishop noted "*a fair showing*" at Teesmouth in the winter of 1937 *(The Vasculum* Vol. XXIV 1938). Few were reported away from Teesmouth in this period, one of the exceptions being a juvenile shot at Whitburn Bents on 29 October 1924. Temperley noted that there were no further records from 1939 until the publication of *A History* in 1951.

Post-Temperley, and with the rise in popularity of bird recording, the 1950s and 1960s produced more sightings, on an almost annual basis after 1958, mainly from the coastline between Hartlepool and North Gare.

Most sightings concerned one or two birds, but seven were at North Gare in February 1961 and five were present at Teesmouth between 24 December 1961 and 4 March 1962. Away from Teesmouth, occasional sightings of one or two birds came from Marsden, South Shields and Whitburn.

Temperley noted two rare inland occurrences. He observed two were in a field between Gateshead and Low Fell during severe snow storms in 1890, and four were on Tunstall Hill on 10 November 1908. A further inland sighting came in 1958 when one was at Jarrow Slake on 26 December during a period of wintry weather.

Recent status

This species is almost exclusively recorded along the coastline, most regularly in winter but also, occasionally, as a spring or, more often, autumn passage migrant; the majority of records falling between mid-October and late February. It would seem that this species was somewhat more common in the early part of the 20th century and the latter half of the 19th century than it was in the last few decades of the 20th century. The most favoured localities in the county are Hartlepool Headland, South Shields Leas, North Gare and Seaton Carew seafront. Prior to the late 1960s and early 1970s, it was a regular but still scarce visitor however, over the last forty years its occurrence in the county has become both much less frequent and predictable.

The early 1970s was the best period for this species since the influx of November 1900. Just prior to this period a flock at Whitburn over the winter of 1969/1970, peaked at 29 on 30 January (Coulson 1972). Between October 1971 and April 1973, sizable flocks were seen in the winter months (Blick 2009). An early spring showing at Teesmouth in 1970 featured 25 at Seal Sands on 12 April, with 15 still present on 2 May, while singles were seen at Hartlepool in April and December. Following nine near Horden on 16 January 1971, a large arrival took place in autumn 1971 with Teesmouth receiving the lion's share. One at North Gare on 18 September was the precursor and remains the earliest ever autumn record, one of only a handful of September sightings in recent times. The peak on the south side of the estuary was 105 in October, with a few wandering parties occurring on the Durham side, but 17 were at North Gare in late October. Up to nine were at Whitburn around this time, with occasional sightings thereafter. Through the winter of 1971/1972 part of the wintering flock at South Gare moved north, leading to North Tees peaks of 46 in January and 63 on 9 April 1972. Two remained in the area until 7 May, with two to three also at Whitburn in the first three months. Following seven at North Gare on 8 October, another arrival took place from mid-December 1972, this included 45 at Seal Sands on 23rd. The flock remained at Teesmouth during the winter, increasing to 50 by 24 February 1973, and up to 40 lingered into April. Elsewhere, sightings in 1973 included seven at Horden on 10 March and three early returning migrants at Whitburn on 25 September. The loss of favoured wintering habitat on the south side of the Tees estuary in 1974/1975 had a profound effect on the species' subsequent wintering presence in the county.

Since 1973, there has been no count greater than nine birds together and there have been ten years where Shore Lark was not recorded. Through the late 1970s and early 1980s, records followed a typical pattern of occasional short-staying October and November arrivals, mainly in the North Tees and Hartlepool area, and infrequent longer staying birds, with flocks of up to eight birds around Seal Sands. More unusual sightings include four at Seaburn from 29 March into early April 1980, up to four at South Shields between 6 January and 10 April 1982, three at Horden on 21 February 1982, while one at Teesmouth on 10 May 1981 was a very late bird.

Following a gap of six years with no records in the county, sightings resumed in a similar vein from 1991 onwards. A total of 84 birds were recorded in the twenty one years from 1991 to 2011, over 80% of which arrived between 27 September and 30 November. Most of these were once again along the coastline between Hartlepool and North Gare, with occasional sightings of up to five birds at: Blackhall, Crimdon Dene, Dawdon, Hawthorn, Hendon, Marsden, South Shields Leas and Whitburn. Nine at South Shields on 8 November 1998 was the highest count for the north of the county in recent decades. Prolonged winter stays are now unusual, exceptions being one at Whitburn from 13 November 1993 to 7 January 1994 and up to three at North Gare from 6 January to 19 March 1995.

Occasionally, birds are detected in March and April on spring passage, with Hartlepool Headland being particularly favoured; three at Chourdon Point on 21 April 1998 were of note. May sightings are much more infrequent, but not unprecedented in terms of the English east coast. Three at Hartlepool Headland from 9th to 17 May 1999 form the latest spring record, whilst singles at Hartlepool on 7 May 2000, by the Reclamation Pond on 6 May 2006 and at North Gare from 12th to 15 May 2011 are also significant.

Only one inland bird has been encountered since 1970, and this was as far inland as is possible in County Durham; one frequented a bare hillside by Cow Green Reservoir in upper Teesdale from 9th to 22 January 1976.

Distribution and movements

There are three major breeding distributions for this species. These are in Northern America, as far south as Mexico, northwest Africa, south east Europe and Asia Minor and central Asia east to Mongolia; and most relevant to Britain, from Scandinavia east through the Siberian Arctic. It is found in 46 sub-specific forms across this distribution. The race *E. a. flava,* responsible for the British wintering population, breeds in Arctic Eurasia and migrates south to the southern coastlines of the North Sea and Baltic Sea. Numbers seen in Britain vary from year to year, but are most regular on the east and southeast coasts of England.

Sand Martin
Riparia riparia

Sponsored by
BANKSGroup
development with care

A common summer visitor and also a passage migrant

Historical review

There seems little doubt that this species was common and widespread during the 19th century though there is little documentary evidence to prove this. Robson described "*a good breeding station*" at Winlaton Mill, in the late 19th century, as well as others that were located higher up the Derwent valley (Robson 1896). Nelson (1907), speaking of Teesmouth, said that it was abundant in suitable locations.

In the mid-20th century it was described by George Temperley (1951) as, "*a summer resident*", and said to be more local in its distribution than Swallow *Hirundo rustica* and House Martin *Delichon urbicum* because of its association with water. In the early part of the 20th century there was a coastal colony in sandy cliffs "*a few miles north of Hartlepool*" presumably in the Hart Warren area, and there were also a few pairs nesting in sandy areas of cliffs at Marsden in the 1930s and 1940s; the use of this habitat persists locally today. Temperley said it arrived in early April and left Durham during August and September when coastal migrants were seen. An exception to this pattern came courtesy of Nelson, who recorded a party of 30 birds at Teesmouth on 22nd October 1907, a very late date for this species in the north east (Mather 1986).

In the early 1950s birds were recorded as breeding in the spoil heaps left by the crusher, from the limestone quarries, in upper Teesdale 150m upstream of the High Force Hotel; a good example of the opportunistic nature of the species and its use of industrial by-products even in the mid- 20th century (Bradshaw 1976).

Recent breeding status

The map in the *Atlas* illustrated a patchy distribution of breeding colonies across Durham that were effectively concentrated along the county's three main river systems with a dearth of sites in the more industrialised east (Westerberg & Bowey 2000). Sand Martins choose sites for breeding where the substrate is easily excavated, and they will return annually to traditional breeding colonies, although they can quickly establish new ones nearby should the former become unavailable as a result of erosive forces or human intervention. For example, the small colony that used to be present close to Strother Hills Wood in the Derwent valley became unsuitable for nesting due to the erosion of the sandy cliff which once held birds.

Nationally, the species has an association with artificial cliff nest sites, such as sand and gravel workings and this is part of a national trend which started post-Second World War (Marchant *et al.* 1990). Locally such sites have been occupied since at least the mid-20th century. Some Durham colonies became deserted during the population crash of the early 1980s and remain unoccupied to the present. The glacial sandstone cliffs that are found along the Rivers Derwent, Wear and Tees and the Tyne, at the county's northern boundary, are the species' preferred local habitat. As documented in the *Atlas* (Westerberg & Bowey 2000), the species' main area of occupation is located in the extreme north west of the county where extensive colonies persist often in working sand and gravel quarries, though similar colonies are scattered across the county.

Sand Martins do not need water at their breeding sites (Gibbons *et al.* 1993), though they no doubt favour such locations. This allows them to colonise both current and old quarry workings often at some distance from water. Where local sand quarries exist the nests tend to be associated with the harder substrate layers. Exceptions to this are the Sand Martin colonies now to be found in sand deposits associated with some of the boulder clay cliffs along the county's coastline. A small colony has been present in the soft cliff-top at the north end of Marsden Bay since at least the late 1990s, but numbers fluctuate greatly here as a result of the dynamics of cliff erosion, and the consequent availability of accessible substrates. In 2010, additional cliff colonies were noted along the coast at Crimdon and Horden and at least 50 pairs nested in three small colonies between Hendon, Sunderland and Seaham Hall at that time.

Occasionally, pairs use artificial sites, such as a pair that nested in a pipe at Hartlepool Headland in 2003, 2004 and 2006 (Joynt *et al.* 2008). In Durham City centre there is a long established colony using drainage pipe holes in the concrete retaining wall along the Wear, and, in the late 1990s, 3-4 pairs from there opportunistically nested in gaps in the eroded sandstone wall of a building exposed in a nearby major development site; the works being re-programmed to allow fledging to take place at the end of their second season using the site (A. L. Armstrong pers. comm.). Specially designed artificial nesting banks, complete with holes partially filled with sand to satisfy the birds' urge to excavate, have been constructed at Low Barns in the Wear valley and at RSPB Saltholme; both attracted the interest of birds shortly after construction.

Sand Martin population trends have fluctuated hugely in the last fifty years and in particular the WeBS data available for the last 30 years or so shows fluctuations in numbers from year-to-year but with an overall stable population since 1978. One major cause for this would appear to be varying adverse weather conditions in the species' wintering area; the Sahel region of Africa (Winstanley *et al.* 1974). Szep (1995) found that rainfall in the trans-Saharan wintering grounds prior to the birds' arrival influences annual survival rates and thus their abundance in the following breeding season. Further research conducted by Robinson *et al.* (2008) confirmed a positive correlation between minimum monthly rain-fall during the wet season in West Africa and subsequent breeding success. In addition, it was recently discovered that increased summer rainfall on the breeding grounds also has a negative influence on survival rates through the following winter (Cowley & Siriwardena 2005).

Before the major declines of the late 1960s and early 1970s (Winstanley *et al.* 1974), this communally nesting species was to be found quite widely across Durham but there was a general fall in the number of breeding Sand Martins in Durham in the 20-year period prior to 1995. Since then the species appears to have staged something of recovery and numbers are probably now higher than at any time in the last forty years. The *Atlas* stated that with the colonisation of working sand and gravel quarries (Westerberg & Bowey 2000) Durham's population was increasing. At that time the best sites for the species, such as those around Blaydon Burn and Crawcrook, in western Gateshead, probably each held around 600 pairs on a regular basis with a maximum count at Crawcrook Quarry of some 739 active nests on 27 May 1996. Whilst these colonies still exist, numbers are probably now somewhat lower. It was thought then that the county held around another 700 pairs at other sites, giving a total population estimate of around 1,300 pairs. The accuracy of counting holes at colonies to determine breeding numbers has been raised (Mead 1979, Cowley 1979), as it can be difficult to distinguish between active and inactive holes; hence, there is a tendency for some non-occupied sites to be included in the estimated numbers and colony-size estimates to be somewhat inflated as a consequence.

Since the publication of the *Atlas* a number of other large colonies have been identified elsewhere in the county, which in total probably hold over 600 pairs; this includes a colony at Wynyard which held over 200 pairs around the turn of the Millennium and in 2006 (Iceton 2000, Joynt *et al.* 2008). In the same year there were perhaps 250 nest-holes at the long-inhabited colony at Hayberries near Romaldkirk, a site which held 220 pairs back in 1990. In 2007, there were 250 nest holes in High Haining Quarry, up to 60 in the cliffs at Whitburn, and more typical-sized colonies of 80-90 in riverside locations along the River Wear at Brancepeth Beck and Frosterley, whilst in 2010 inland quarry colonies of over 100 pairs were present at Bishop Middleham and Sherburn. During the first decade of the 21st century a colony at Easington Lane illustrated the fluctuations in fortunes that occur at Sand Martin colonies as numbers ebbed and flowed, and eventually flowed away altogether. It started with 132 nests in 2001, fell to 90 in 2002, there was no nesting in 2003, 100 were present in 2004 and then the colony was deserted by 2006. This is the repeated story of many colonies in the county over the years. Over the most recent decade there were additional colonies holding over 100 pairs at Croxdale Viaduct, Eggleston Burn, Page Bank and Seaham Beach, whilst one at Aycliffe had 200 pairs in 2007, with many other small colonies, each having 10 to 15 pairs, scattered around the county. In the south some birds nest along the lower River Tees, especially around Portrack, Bowesfield and Preston Park, where about 80-100 pairs have been counted in recent years (Blick 2009). In total there are probably 220-250 pairs breeding around the north of Teesside (Joynt *et al.* 2008). The current population estimate for the whole of the county is around 2,200 pairs, with possibly slightly more than this present.

Apart from adverse weather conditions to which Sand Martins are susceptible and the drainage of wetland areas that affects feeding and post-breeding roost sites, there are few local threats to the species. Over the years, many small colonies that became established in working quarries have been lost as those locations were worked out or the nesting faces lost to new excavations. Somewhat ironic however, is the fact that a greater threat to this species now is the restoration and landscaping of such disused quarry sites. The species' readiness to use artificial sites means that there should consideration for incorporating such features into all construction projects involving bridges, embankments and river bank management.

Recent non- breeding status

The species congregates in large numbers, alongside other hirundines and Swift *Apus apus,* in favoured post-breeding feeding localities. These are often along river valleys or over wetland complexes or, especially in poor early spring weather, at working sewage treatment works, where birds flock in numbers to feed on the swarms of insects that provide a relative glut of food. Some of the largest gatherings over the past 20 years include 500 at Saltholme Pools on 13 July 1996, 600 at Dorman's Pool on 13 August 1999, 800 at Haverton Hole on 2 August 2001, 500 at Bishop Middleham on 4 July 2004 and 900 at Saltholme Pools on 25 May 2005.

Pre-emigration, birds sometimes roost at wetland sites, such as those at Haverton Hole and others on Teesside, and occasionally large numbers of Sand Martin joined the large Swallow roost that was at Shibdon Pond, with as many as 500-1,000 birds being recorded there during the late 1970s (Bowey *et al.* 1993).

Most of the county's breeding birds move out of the area during late August with only a few staying into early September. Numbers decline rapidly during the first half of September, with most birds having left by mid-month and just late stragglers hanging on until the end of the month. However, it is not unusual for some late breeding, perhaps triple-brooded, birds to have young in the nest even into the first week of September; the latest recorded instance of this came from the banks of the Wear close to WWT Washington on 17 September 1978.

An analysis of departure dates over the past 40 years shows an average last date of 27 September but with a tendency for slightly later departures during the periods 1971-1980 and 2001-2010 when it was 2nd to 3 October before the last bird left. This of course masks the extremes but October birds have only been recorded in 11 out of 40 years with particularly early departures in 1986 and 1988 (last date 7 September), whilst very late birds have been noted on dates between 26th and 29 October in 1980, 1983 and 2006. The latest ever record was a bird at North Marine Park in South Shields on 8 November 2005: a year that also produced a number of November records of Swallows and House Martins.

The species is one of the earliest of the summer migrants to arrive although data shows that during the 1970s and the 1980s this was not always the case. Whilst the first arrival date of this species since 1988 has always been in March there was not a single March record during the five year period from 1983-87 with the first bird in 1986 not until the remarkably late date of 19 April. There were March arrivals in just three out of 10 years from 1971-1980 although the average arrival date was 30 March, whilst the late arrivals in the mid-1980s pushed the first date back to 5 April in the period 1981-1990. Since then however, arrival has become progressively earlier. From 1991 to 2000 first arrivals were on average on 20 March and during the first decade of the 21st century this advanced a further three days to an average arrival date of 17 March. Whilst mid-March birds are now expected, the earliest on record was a single at Teesmouth on 10 March 1977.

Distribution & movements

Sand Martins breed across much of Eurasia and are long distance migrants wintering to the south of the Sahara. North American breeding populations winter in South America. An abundance of data means that migration routes for this species are particularly well known with birds concentrating at short sea crossings. UK breeding birds move into France then follow the coast SSW to Biscay skirting the north west fringes of the Pyrenees before some pass overland to the Coto Doñana where they join others following the Ebro valley to the Mediterranean coast of Spain. From here they pass into Morocco and move to the primary wintering zone of The Sahel from Senegal eastwards. Many birds move east within this zone and pass back north in spring via the Niger inundation in Mali. Some other birds, presumably non-UK breeders, winter in East Africa south to Mozambique (Cramp *et al.* 1988).

The migration routes outlined above come from the wealth of ringing data on UK birds. Of over 1,600 UK ringed Sand Martins recovered abroad more than 1000 recoveries have come from France and Spain with 372 (almost 23%) from Senegal and 21 from Algeria. A similar situation applies with those birds that were ringed abroad and recovered in the UK i.e. of 573 recoveries in the UK some 430 or 74% of birds were ringed in France and Spain and 88 (15%) in Senegal (Wernham *et al.* 2002).

Among Durham ringed birds there are two recoveries from overseas which follow the general pattern; a bird ringed at Shincliffe on 24 August 1962 was recovered at Mer in France on 18 September 1962, and a juvenile ringed at Crawcrook on 30 July 1966 was controlled at Ploegsteert, West Flanders, Belgium on 6 September 1966. One bird caught in Gateshead during May 1992 had been previously ringed in the Channel Isles.

A trio of juveniles ringed at Crawcrook in June and July 1966, demonstrated the rapidity which local youngsters head south. One was in Lincolnshire on 7 July 1966, after being ringed on 26 June; another was controlled at Brough, Cumbria on 6 August 1966, after ringing on 26 July 1966; and finally one was controlled at Boston, Lincolnshire on 12 August 1966 after being ringed on the same date as the previous bird.

There are a number of other interesting ringing recoveries within the UK from the ringing of local birds. A total of 12 birds ringed in the county have been controlled in Hampshire, Lincolnshire, Kent, Norfolk and Sussex, whilst birds ringed in Essex, Kent, Norfolk, Northamptonshire and Sussex have been controlled or recovered in Durham. These birds are illustrative of movement from this region to and from the south east of England. For example, six birds ringed in Gateshead in the late 1980s and early 1990s, were later caught in Hampshire, Kent and Sussex suggesting that they were gathering there to make the shortest possible crossing of the Channel.

Long-term ringing data has also demonstrated that the county's breeding birds return to their colonies year-on-year, once they have settled on a breeding site and whilst that site remains viable. A number of birds ringed as migrating juveniles in East Sussex and Kent in the autumns of 1989/1990 were subsequently controlled in the breeding colony at Hayberries Quarry, Romaldkirk in 1991. More spectacular in terms of distance travelled, was the bird ringed by the BTO expedition to Djoudj national park in Senegal that was controlled at Hayberries Quarry Romaldkirk, 4,500km to the north on 19 August 1992.

Swallow (Barn Swallow)
Hirundo rustica

A very common summer visitor and passage migrant.

Historical review

This species' long-term presence in the county is illustrated by its cultural referencing in the place name of Swalwell, located at the north end of the Derwent valley just to the east of Blaydon upon Tyne. This name is derived from the Old English "*swealwe, well*" meaning 'swallow spring' (Yalden & Albarella 2009).

J. Backhouse (1834) found it nesting in upper Teesdale at 2,000 feet (over 600m) above sea level during the early part of the 19th century and in the late 19th century it was described as "*arriving in the middle of April*" in the Derwent valley (Robson 1896). What were considered to be relatively early arrival dates at Teesmouth, of 23 April 1899 and 19 April 1901, were recorded by Milburn (1932). Nelson (1907) commented that it was an abundant summer visitor.

George Temperley (1951) described it as "*a summer resident*", but knew it also a passage migrant that it was "*widely distributed throughout the county.*" In Temperley's time the first arrivals of Swallows usually occurred during the second or third week of April, and birds remained to August or September, with a few lingering individuals being noted in October in mild autumns (Temperley 1951). Variation in the arrival and departure dates of this species has always fascinated people and during an unusually mild spell of weather, a Swallow was seen at Whickham on March 27th, 1933 (*The Vasculum* Vol. XIX 1933); a very early date and at that time probably the earliest ever documented in the county. In the summer of 1942, an albino bird was found at Vince Moor Farm, Croft Spa near Darlington. It was a juvenile and others from the nest were normal in appearance (*The Vasculum* Vol. XXVII 1942).

Stead (1964), described the species as a common breeding bird outside of the industrial belt of the south east of the county. There is little documentation of the status of this species post-Temperley, suggesting it remained a common and widespread bird, and as such was poorly reported. One exception was the observation of a huge, presumably pre-roost gathering of birds, numbering 1400-1600 birds, which were seen resting on wires at Stanley on 19 August 1964 (Bell 1965).

Recent breeding status

Gibbons *et al.* (1993) estimated the national population to be 570,000 pairs, although this figure was revised upwards to 726,000 territories in 2000 (Birdlife International 2004). It remains one of Britain's most common summer migrants, as it is in Durham where, aside from small scale annual fluctuations, this species' status does not appear to have materially altered through the course of the 20th century, or into the first decade of the 21st century. It is usually considered a common breeding species in the Teesmouth area, but some declines were noted over the period 1995-1998 (Joynt *et al.* 2008).

As the *Atlas* documented, breeding Swallows are well represented in both urban-fringe and rural areas across Durham. The highest densities are found in the central lowlands (Westerberg & Bowey 2000) and whilst nesting in towns occurs, it is very much less common here, mirroring the national situation; urban nesting occurring only at low densities (Gibbons *et al.* 1993). Breeding, as mapped by the *Atlas,* occurs right across Durham from the coastal strip west to the high moorland edges.

It is absent from large areas of the Pennine uplands, because suitable nesting sites are not available, although birds will willingly take advantage of the shelter provided by isolated uninhabited old or ruined farm buildings even on open moorland. In 2006, one pair nested in the wild and very exposed location of Cauldron Snout proving how adaptable this species can be. Across the county it is very much dependent on man-made sites for nesting with many of these locations found on farms or other undisturbed rural sites (Moller 1983).

There is one record of a tree-nesting pair in Durham. This occurred during the early 1970s on the outskirts of Sunderland, a nest with five eggs was found resting against a large, main horizontal branch of an enormous spreading beech tree *Fagus sylvaticus,* and supported from below by a similar branch. The use of natural sites, including caves, is now very rare in Britain (Gibbons *et al.* 1993) but in Durham there are recent examples of such behaviour as it has been found breeding along the rocky northern cliffs around Marsden.

In urban-fringe locations, derelict buildings are often used and birds will sometimes use garages and other structures such as church porches for nesting. A frequently utilised site is on supporting beams or girders under bridges, and during the 1970s a pair nested in the end of a box girder used in the construction of the A19 Hylton Bridge. This occurred whilst work on the bridge was suspended for several months. The local Swallows, of which several pairs bred in a pig farm on the river bank below the bridge, were quick to take advantage of this extremely safe nest site suspended high above the River Wear. Around Teesmouth occasional pairs nest in concrete blockhouses, old sheds and even in some bird-watching hides (Blick 2009). Wherever they nest a slight support is all that is required to start the building process; hence at Balder Grange in 1975 birds built on top of a hanging basket and also on a hula hoop hanging from a nail in the wall. Three broods are frequently reared with some birds still breeding late into the summer; a late active nest with young was reported from the Tees Marshes on 14 September 1974 (Joynt *et al.* 2008), and an exceptionally late record was of a pair with three young, presumably just about to leave the nest, at Bollihope on 9 October 1983.

The Breeding Birds of Cleveland showed Swallows to be widespread but as expected to be more sparsely distributed in the urban areas. Overall, there was probably a 60% occupation of tetrads in North Teesside with a population of over 450 pairs (Joynt *et al.* 2008). In County Durham, there are well over 1,200 farmhouses and similar establishments (K. Bowey pers. obs.) from the coast to the uplands and almost every one of these will have a handful of breeding Swallows in and around the out-buildings, and feeding over the farm yard. The *Atlas* map showed that it was recorded breeding in 549 tetrads and based upon this, and a very conservative population density of 12 pairs per occupied tetrad, the *Atlas* estimated a breeding population of around 6,600 pairs in Durham (Westerberg & Bowey 2000). This estimate is probably still a reasonably accurate figure.

In the UK as a whole there have been fluctuations in numbers but no evidence of an overall decline over the last 40 years (Robinson *et al.* 2003). The BTO's national census and survey datasets suggests a rise of over 20% from 1967 until 2008, most of these increases have occurred since 1994 (Risely *et al.* 2010). There has been however something of a shift in population with a general decline in the lowland east, brought about by increasingly intensive agricultural methods, being offset by increases in the west and north, as more land is turned over to pasture (Evans & Robson 2004). Prior to this, population declines had been recorded in Britain in the 1980s (Marchant *et al.* 1990). This was thought to be primarily a result of declining levels of rainfall across parts of Africa. During the same period, however, reduced egg production by adults was also implicated in declines in Denmark; perhaps as a result of poor body condition of returning migrants (Moller 1989). Further studies (Robinson *et al.* 2008) have shown that rainfall levels in Africa do have a significant impact on annual fluctuations in the numbers of birds in Britain. By the end of the first decade of the 21st century, the impact of climatic warming might also have been beginning to have an adverse effect on numbers. One of the possible effects of climatic warming is that birds may have earlier starts and later finishes to the breeding season, leading to increased chick mortality in hot dry summers and possibly, reduced fledging survival success because of the poor condition of birds migrating through North Africa (Turner 2009). Whilst Swallows may face considerable threats both on their wintering grounds and on migration, in Durham the loss of nest sites as a result of the renovation of old buildings in rural districts and the reduction in food supply because of agricultural intensification and increased rural hygiene (Voous 1960) may also be negative factors.

Recent non-breeding status

Large gatherings occasionally gravitate towards wetland locations in adverse weather, for example 1,000 were at Derwent Reservoir on 20 May 1979, but it is usually late summer when most flocks occur as birds began to gather in increasingly large numbers prior to migration. Post-breeding, family parties gather together to feed over wetlands and to utilise the large communal evening roosts.

Large roosts accumulate during August and September mainly at wetland sites with reed beds and other similar tall vegetation. Locations that have regularly been frequented in recent include the Long Drag area of Teesmouth, Shibdon Pond, Birtley STW, Lockhaugh STW, Bishop Middleham and Wingate, although the numbers that have occurred in these sites and the consistency of their occupation can vary from year to year. These roosts are usually first formed in July and persist through August and on a diminishing scale through September. Numbers utilising roosts vary from just a few hundred birds to many thousands of individuals and they fluctuate, probably with a constant flow-through of birds. Significant roost counts over the years have included those at Shibdon Pond which regularly held 3,000 to 4,000 birds through the 1970s and early 1980s, and Wingate where 2,000 were present on 1 September 1976 and 23 September 1979 and 3,000 in early September 1988. In the south of the county, Haverton Hole held up to 500 in the 1970s but no more than about 200 in the 1980s, whilst Dorman's Pool hosted up to 3,000 in 1999 (Blick 2009). Several thousands were also gathered at Great Lumley in September 1990 and August 1991 where they went to roost in reedmace *Typha latifolia,* but sadly this site was drained in 1992. In 2007, an impressive roost built up at Lamesley reed beds, with an estimated 3,000 birds in late August, building up to a peak of 5,000 on 14 September. This declined to 3,500 by 21st and all birds had had left by the end of the month. A total of 2,916 birds were ringed at the roost during this period. There is no doubt that some of the birds involved were local breeders but some also come from just further north, for example in Northumberland, and others from much further afield.

For many people this is the species that best symbolises spring and summer; its arrival announces the commencement of fair weather and summer being on its way. However, the first Swallows usually arrive in periods of unsettled weather which often, amongst the April showers might include at least some sleet or even snow. The main arrival of Swallows in Durham takes place from mid-April to mid-May.

Analysis of first arrival dates of Swallows over the last 40 years clearly shows that the mean first arrival date of the species is becoming progressively earlier. Taking the 10-year means running from 1971-1980, 1981-1990, 1991-2000 and 2001-2010, the respective arrival dates were 13 April, 4 April, 30 March and 27 March. From 1971 to 1987 there was just one record of a March Swallow, at Cleadon on 29 March 1981. In the subsequent 23 years, there have only been four years when the first Swallow did not appear until April, the last of these in 2000. The earliest bird on record was a single at Far Pasture on 16 March 1995.

Birds begin to move south both from, and through, Durham from mid-August onwards. Most birds depart the area in September though a few stay on into early October, with stragglers often being reported later than this. Occasional odd late birds move through the county in early November. A similar analysis of departure dates over the last 40 years shows a trend for later departures although the differences are not nearly as marked as for spring arrival dates. From 1971-80 the mean departure date was 30 October yet in the last 30 years this has been stable at 4th to 5 November, with November birds noted in two out of every three years during the latter period. The latest bird on record was a juvenile present at Hartlepool Headland on 5 December 2002.

Scores or sometimes hundreds of birds per day regularly pass along the Durham coastline. By way of example, on 1 May 2005 birds were recorded at North Gare passing north over the River Tees into Durham at a rate of 200 per hour; the same location recorded 555 and 1,100 north on passage on 9th to 10th and 15th to 16 May 2009 respectively. Movements can be more prevalent in the autumn with southerly passage on 6 August 2006 producing 275 birds per hour over Dorman's Pool, Teesmouth. Movement can also be seen inland, for example, birds moved south over Far Pasture on 22 September 2000, when 1,750 passed in under three hours (Armstrong 2005) and 900 moved through Hurworth Burn Reservoir on 14 September 2008.

An interesting and very striking individual with rich rufous under parts, reminiscent of the eastern Mediterranean race, *H. r. transitiva*, was noted at Saltholme Pool between 22 April and at least 13 May 2007. A completely white nestling was ringed near Greatham Creek in late June 1990 and it was then seen feeding around the North Tees Marshes throughout July and August 1990 (Blick 2009). A partially white bird was at Hurworth Burn Reservoir on 3 June 1997 and another completely white individual was seen at Hetton Lyons on 14 August 2001.

On 22 September 1979, whilst trapping birds going to roost at Shibdon Pond, a bird was caught that superficially resembled a Red-rumped Swallow *Cecropis daurica*; after careful examination it was determined that it was a hybrid Swallow-House Martin. Another bird thought to be of this mixed parentage was trapped at Lockhaugh STW in May 1990 (Bowey *et al.* 1993), with other similar birds recorded at Hartlepool on 11 September 2002 (Blick 2009), Barmston Pond on 19 May 2005 and at Roker Park, Sunderland on 1 October 2010.

Distribution & movements

Swallows breed over most of Eurasia, North Africa and North America. All populations are long-distance migrants and the European birds winter in sub-Saharan Africa.

At least 1,324 UK-ringed Swallows have been controlled or recovered abroad. Of these 383 (28.9%) have been on passage in France and Spain, and 176 (13.3%) in Algeria and Morocco in North Africa. A further 343 (26%) have been recovered on the wintering grounds in South Africa with smaller numbers in Nigeria, Zaire (Congo) and Namibia. A similar pattern is shown from the 135 Swallows recovered in the UK that were ringed abroad; of these 59 (43.7%) were ringed in France and Spain and 27 (20%) in South Africa (Wernham *et al.* 2002).

There have been four foreign recoveries of Durham-ringed Swallows. A juvenile ringed at Haverton Hole on 22 August 1974 was found dead at Halfway House, Transvaal, South Africa, on 16 January 1976. A bird ringed at Seal Sands, as a nestling, on 17 August 1987 was caught and identified as a female at Aglou, Morocco, on 11 January 1988, a distance of over 2,800km. Remarkably two juveniles ringed at Wingate on 31 August 1988 were recovered in Africa; one was found dead on 26 March 1990 at River Kraai, Cape Province, South Africa, a distance of 9,920km and the second was also found dead, nearly three years later, at Smara, Sahara Morocan, a distance of 2,569km. In addition a first- year female which was ringed in Zambia on 23 November 2009 was controlled at Birtley on 30 May 2010, a distance of 8,274km to the north north west. Other notable recoveries include:

- A bird ringed as pullus at Bowes on 21 June 1992 was found dead 6,667km away at Dollisie Congo Brazzaville on 22 April 1993
- A bird ringed as pullus at Hurworth Burn on 23 August 1994 was found dead on 29 December 1994 at Alger Algeria, a distance of 2,024km
- A pullus ringed at Harehill Farm, Haswell Plough on 14 August 1998 was controlled at Bloemfontein, Orange Free State, South Africa on 10 March 1999, almost 9,700km to the south

Birds ringed at Shibdon Pond in the late 1980s, Barnard Castle in 1987, and at Saltholme and Wingate in 1988, were all controlled later in their onward autumn movement at Icklesham in East Sussex, a distance of 446km to the south, in some instances as little as seven days elapsed between ringing and recapture on the English south coast. Typical north-south movements of Swallows to and from Durham include two birds ringed at Shibdon caught in roosts further south within a fortnight of the birds being ringed, and:

- A first-year, ringed at Shibdon Pond 28 July 1975, found dead at Ryhope on 19 July 1979
- One ringed at Shibdon on 28 August as a juvenile, controlled at Pitsea Marshes, Essex on 20 September 1988

Much less typical records include the bird ringed in Bolle di Magadino, Swtizerland on 29 April 2004 on its way north, which was controlled at Lockhaugh on 21 May 2005, 1,226km north of its ringing station and the juvenile ringed at Wingate on 18 August 1990 that was apparently controlled in Northampton on 13 January 1991 (Armstrong 1992).

This species is well known for its year-to-year site fidelity and evidence of this in Durham comes from the Derwent valley. Nestlings which were ringed here have on a number of occasions returned from African winter quarters to breed within a kilometre of their hatching site (Bowey *et al.* 1993).

House Martin (Common House Martin)
Delichon urbicum

A very common summer visitor and passage migrant.

Historical review
This regular summer visitor to Durham is usually present from mid- to late April until the middle of October. It breeds in small, widely scattered colonies throughout most of the towns and villages of the county, though in earlier times no doubt it would have utilised caves and coastal cliff locations along Durham's northern coastline, as it was recorded doing so regularly in Northumberland, at least until the 1990s (Kerr 2001).

Bolam (1912) considered that it was more numerous but not as widespread as the Swallow *Hirundo rustica* in the north east. Temperley (1951) described it as a "*summer resident*" and "*passage migrant*" and considered it to be not as numerous or widespread as the Swallow. J. Backhouse in 1885 recorded it nesting in Teesdale at elevations as high as 1150 feet (*c.*350m) above sea level. In the modern context, birds were noted as nesting at the Langdon Beck Hotel in upper Teesdale in the summer of 2002 at around 390m above sea level. Declines in the east of the county during the 19th century were thought to be as a result of industrialisation (Temperley 1951). It was reported that, up to the end of the 1800s, the species nested about the Durham Cathedral windows and extensively in the City of Durham, however, by 1905 Canon Tristram noted that "*not one was left*" there (Tristram 1905). It arrived in the last fortnight of April, a week to ten days later than the Swallow, and departed in August and September, according to Temperley (1951). By the early 1950s Temperley described it as "*increasing in County Durham*". This species was probably helped considerably, in terms of breeding success, by the introduction of the Clean Air Acts of the late 1950s and early 1960s (Mather 1986), which would have resulted in a greater availability of its aerial insect food.

Recent breeding status
This is a highly visible species of the county's urban areas and rural settlements. The *Atlas'* distribution map (Westerberg & Bowey 2000) clearly showed that it was found widely in the urban areas of Durham, though it was not as widely distributed as the Swallow in rural locations and habitats. The impression of some observers recently has been that the bulk of the county's population is located in Durham's suburban lowlands, though this may be in part be attributable to an observer bias towards those areas.

The *Atlas* concluded that House Martins were more common than Swallows as a breeding species and much more likely to be encountered in the centre of built-up areas. In such situations, they tend to prefer older housing stock, with its wider eaves to provide nesting localities. In places where the preferred house styles predominate, quite large, extended colonies of birds can develop. In County Durham, it is almost completely reliant upon man-made structures for nesting, and nests are usually located on the outside of buildings (Clark & McNeil 1980), although in 1974 in Weardale 12 pairs nested inside a cattle-shed. Also in 1974, a colony was found utilising a natural site, where birds were nesting on a cliff-face in a flooded quarry at Cowshill in Weardale. This small colony never exceeded 12 pairs but birds were present at least until 1979 and probably 1982, but no birds were found the following year.

Its distribution in the county is a widespread one, but with a strong easterly and central focus, although it is commonly found in the dales, even in some of the more exposed of the county's upland fringes. For instance, in 2009 breeding took place on buildings at Tunstall Reservoir, and at the very exposed location of Smiddyshaw Reservoir in the upper Derwent valley.

The House Martin's breeding season can be a protracted one. They arrive in April or early May and parents are often seen feeding late broods in late September, much later than most other species, even resident ones. In 2007, a late brood, presumably the third of the year, was still in a nest at Rowlands Gill on 1 October, just 25 days before the last record of the species in the county for that year. Even later than this, young were still in the nest at Barnard Castle on 12 October 1987 and at Woodland on 20 October 1975. At the other end of the bird's breeding season a pair was observed nest building on 14 April 2003, at Darlington.

At the outset of the 1990s, Marchant et al. (1990) reported that there had been no recent, nationally detectable changes in population level, but since then there has been some evidence of a decline. Although at the last national atlas (Gibbons et al. 1993), these trends were believed to be more perceived than proven yet recently published data from the BTO's CBC/BBS suggests that a very significant national decline of up to 60% took place between the late 1960s and 2008, with the majority of losses noted between 1983 and 1992. Since 1994 the BBS data shows an overall slight increase of 4% (Risely et al. 2010).

Unfortunately, the House Martin is not a bird that has been intensively studied in Durham. Anecdotal reports around the time of the survey work for the Atlas suggested that the distribution was more restricted than it once was, but to date, these assertions are still not substantiated; though clearly birds have been lost from some areas where they used to be present. It is unclear whether this is part of a long-term periodicity in numbers or a downward trend. Local fluctuations do occur in populations but these appear to be predominantly in a downward direction. The demise of one colony however, may be offset by the establishment of small colonies or single nests elsewhere. Hence there is some suggestion of a resurgence of numbers in some areas that is at least in part associated with the crop of new-built housing of the 1950s and 1960s, particularly where this abutted favoured wetland feeding areas, such as around Teesmouth and in sites in the lower Derwent valley, such as Rowlands Gill.

Notwithstanding such fluctuations, local evidence confirms that there have been significant losses in some parts of the county over the last few decades. Some thriving local colonies have been lost to re-developments (such as the once large colony located at the old Blaydon School, demolished around 1995) whilst some have declined with no apparent local causes, as for example those in some residential parts of Barnard Castle in the early 1990s. Some even disappeared, for example in Highfield, Rowlands Gill, between the start and the end of the 1990s. More recently there were 67 occupied nests in the Homelands estate, Houghton, on 1 June 2005 but just 19 the following year; the decline caused by housing demolition. The overwhelming impression by many observers and commentators was that House Martins were 'thin on the ground' in 2008, and this was felt to be the case more generally towards the end of that decade. This perception, however, as outlined above does not fully reflect the national picture.

This species is less widely distributed around Teesmouth than the Swallow and it is also less numerous. More closely associated with urban areas for breeding, it is much less common in the coastal strip. There are probably around 300 pairs in the North Teesside area (Joynt et al. 2008). The Atlas applied Tatner's (1978) population figures to the county's occupied tetrads (i.e. eight pairs per occupied tetrad), to provide a population estimate of 2,784 pairs for Durham. This was believed to be too low and a figure of "nearer 5,000" was recommended as the guide population figure (Westerberg & Bowey 2000). The Atlas' estimate probably remains a broadly reliable figure to represent the current population. However, a lower limit should now be applied to this to accommodate any downturns that may have occurred in the last decade or so; hence the suggested present estimate of breeding numbers would be between 3,500 and 5,000 pairs.

Non-breeding Status

For feeding purposes, the species chooses sites where there are abundant flying invertebrates, but they can also be widely encountered in urban, mainly suburban, and rural areas where they catch prey over hedges and gardens, as well as higher in the air column, on warmer days. Birds also gather in very large flocks at wetland sites, particularly during poor weather. The species' normal foraging behaviour is generally at high altitude, and is more akin to Swift Apus apus in this regard than other hirundines, but rain and wind often force birds to lower levels and in such situations it seems particularly fond of feeding over sewage treatment works where huge populations of emerging insects can be found. Other favoured sites include ponds, wetlands, sheltered river valleys and in the lea of hedges and tree lines (Armstrong 1988-1993). Larger gatherings recorded over the last two decades include: 800 birds at Ebchester STW on 31 May 1991; 800 at Far Pasture on 11 September 1992; 900 at Brasside on 23

August 1993, 800 at Birtley STW on 11 May 1997 and almost 900 birds at Hetton Lyons in June 2001. Adverse weather conditions led to a gathering of 850 over the Reclamation Pond on 12 June 2005.

The main departure of the species usually takes place from late August and throughout September, with flocks of several hundred sometimes gathering over wetland sites during these months; for instance in 2000, pre-migration flocks were noted at Hetton Lyons, where there were up to 700 birds, whilst 13 other sites attracted flocks of between 100 and 500 birds at the time. Birds are regularly seen until the latter part of September, with sometimes considerable numbers remaining into October. A particularly early exodus from the county was apparent in 2003 when the last bird was noted on 2 October, but the last few stragglers usually hang on into the second or third week of October. The average date for the last bird during the period 1991-2010 was 19 October, with November birds recorded in just 4 out of 20 years. This is a full 15 days earlier than that recorded during 1971-1990 when November birds were seen in 12 out of 20 autumns. Why such a pattern of occurrence has occurred is unclear. The latest recorded bird in the north of county was one that was at Whitburn on 29-30 November 1986 but the latest was at Teesside on 2 December 1984 (Blick 2009).

Conversely the first House Martins of the year used to arrive in the county around mid-April but during the last 20 years there has been a progressively earlier arrival. During 1971-1990 the average first arrival date was 14 April, but between 1991 and 2000 the average first date was 9 April and in the period from 2001-2010 the first bird was noted on average on 4 April; indicating a tendency for this species to both arrive and depart earlier than had previously been the case. March records have also increased significantly, particularly during the last decade. There was just one bird during the 1970s, an exceptionally early bird, which remains the earliest ever recorded in the county, being noted at Branksome, Darlington on 13 March 1977. The 1990s produced a single record at Whitburn on 22 March 1997, yet in the period from 2001-2010 March birds were seen in five out of 10 years. It is still the case however that in some years there are very few birds present before the beginning of May.

Distribution & movements

This species breeds over much of Eurasia and in northwest Africa. The European population winters far south of the breeding areas, in Africa to the south of the Sahara. In its winter quarters and on migration House Martins are subject to similar pressures to other hirundines with rainfall in Africa prior to the birds arrival a major factor influencing winter survival rates and hence the relative abundance of the species the following breeding season. Szep (1995) determined that from British population monitoring data there were no perceived impacts upon the species from droughts in Africa during the 1960s and 1980s.

In comparison to other hirundines, House Martins are rarely ringed so the amount of data from ringing schemes is very limited. Of the recoveries of UK ringed birds almost 50% are from France and 12% from Algeria, however, given that these come from a very small sized sample, this may not be as significant as first appears (Wernham *et al.* 2002). Nonetheless, ringing studies at colonies in the region have shown that the same birds regularly return, year after year, to their breeding sites (Kerr 2001). There have been no records of Durham ringed birds having been recovered abroad and just a couple of Durham ringed birds recovered elsewhere in the UK. One ringed in the nest at Hartlepool on 11 July 1990, was at Icklesham, East Sussex on 9 September 1990 and an adult ringed at Birtley STW on 2 June 1998, was caught at Tinsley STW in South Yorkshire, just eleven days later.

Red-rumped Swallow
Cecropis daurica

A rare vagrant from southern Europe: nine records involving ten individuals.

The first record for County Durham, and the only autumn sighting, came in 1995 when Rob Little saw one flying south down the Long Drag on the morning of 10 September, accompanied by other hirundines (*British Birds* 89: 511). A further nine birds have been seen in the county since, including five during a notable spring influx in 2009.

All records:

1995	Long Drag, 10 September
1998	Whitburn village, 31 May (*British Birds* 92: 586)
2002	WWT Washington, 18th to 19 April (*British Birds* 97: 591)
2003	Seaton Common, Teesmouth, 29 April (*British Birds* 97: 590)
2004	Low Barns, Witton-le-Wear, 17 May (*British Birds* 98: 666)
2009	Castle Lake, Bishop Middleham, 9 April
2009	Far Pasture, Gateshead, 27th to 30 April
2009	Long Drag, Teesmouth, two, 2 May
2009	Cowpen Bewley, 20 May

Just two have remained for more than a few minutes. The individual at WWT Washington in 2002 was seen briefly by the same observer on two occasions, and the Far Pasture bird in 2009 which, although intermittent in its appearance, was seen by many observers.

Distribution & movements

Red-rumped Swallows are a widespread and locally common breeding species in North West Africa and around the Mediterranean, with a discontinuous range to the east through Turkey and the Middle East. This western population is assumed to winter in northern, equatorial Africa. Other populations exist through eastern Asia and equatorial Africa. In Britain, it is a rare but regular visitor, mainly in spring, chiefly to southern England but it is prone to occasional influxes in both spring and autumn. It was removed as a species considered by BBRC in 2006 when records had passed the 500 mark.

Cetti's Warbler
Cettia cetti

A rare vagrant from southern England: ten records, involving an estimated sixteen birds all since 2005, including a pair which bred in 2010 and 2011.

The first Cetti's Warbler for County Durham was initially heard calling several times by Martin Blick from dense bramble scrub and adjacent reeds at Dorman's Pool, Teesmouth on 2 November 2005. Although present until 23 April 2006, it was extremely elusive throughout its stay and was heard calling much more frequently than it was seen, most often around dawn and dusk, and often in response to the calls of nearby Wrens *Troglodytes troglodytes*. It was trapped and ringed by Derek Clayton on 9 November and was thought to be a female based on wing length and weight (Blick 2006).

All records:

2005	Dorman's Pool, Teesmouth, female, 2 November to 23 April 2006, trapped on 9 November
2009	Roker, Sunderland, 15 October
2009	Bowesfield Marsh, Stockton-on-Tees, male, 7 November to 27 March 2010
2009	Hurworth Burn Reservoir, male, 9 November
2009	Long Drag, Teesmouth, female, 9 November to 12 February 2010
2009	Haverton Hole, Teesmouth, male, 16 November
2009	Dorman's Pool, Teesmouth, male, 21 November until 2011 (ringed in Hertfordshire)
2010	Lamesley Reedbeds, male, 7 February to 7 March
2010	Dorman's Pool, Teesmouth, pair bred involving original male, raising five young
2010	Haverton Hole, Teesmouth, male, 15 October to 2 December (singing on 18 October)
2011	Dorman's Pool, Teesmouth, pair bred, raising at least one young

A remarkable influx during the winter of 2009/10 brought at least six birds to County Durham. The first of these apparently struck a kitchen window in Roker, Sunderland on 15 October, and was photographed as it recovered before disappearing into the undergrowth never to be seen again. This same garden in Edgeworth Crescent had previously hosted County Durham's first Siberian Rubythroat *Luscinia calliope* in October 2006. Three of the six Cetti's Warbler records were from the North Tees Marshes and included a long staying male, which was still present in 2011. Another male at Hurworth Burn Reservoir on 9 November was noteworthy for both the isolated inland location and for being the most northerly record in England at the time. This status did not last as a singing male was present further north in reedbeds at Lamesley in February and March 2010, the first record for Gateshead. Despite expectations there was no further influx in the autumn of 2010, although the bird at Haverton Hole in October was probably a new arrival.

Surprisingly, a number of birds survived the harsh winter of 2010-11 and further colonisation seemed a possibility. During February and early March 2010, birds were singing at Bowesfield Marsh, Dorman's Pool and in the Lamesley reedbeds, though only the bird at Dorman's Pool was known to have attracted a mate. The original male from 2009 was very active throughout the early part of 2010 and frequently sang at both Dorman's Pool and nearby Hargreaves Quarry. The presence of two birds was suspected but not proven until 31 May, when one was seen to be carrying food and on 6 June three fledglings were seen; representing the first successful breeding of this species in County Durham. The following day a total of five fledglings were counted, four of which were trapped and ringed in Hargreaves Quarry during the period from 23 June until 11 July. These quickly dispersed although a single bird was heard calling on the Long Drag in early December, which was assumed to be from this brood, as the original male remained in residence on Dorman's Pool at this time. One of these juveniles was retrapped in October 2011 in Northumberland. Having gained a toehold as a breeding species in County Durham, and with large areas of suitable habitat to be found throughout the south east of the county, there seems every reason to assume that the species will consolidate this early success and eventually colonise.

Distribution & movements

Cetti's Warbler inhabits dense cover and scrub, usually close to water and it is found in North Africa, throughout much of southern Europe northwards to France and southern England, and eastwards through south western Asia to Afghanistan and north western Pakistan. The European population is largely resident though dispersive. The first British record was in Hampshire in March 1961, and then following an influx into southern England in 1971, breeding was confirmed for the first time in 1972, and by 2006 the British population was estimated at around 2,257 pairs (Holling 2010). Cetti's Warbler has been gradually spreading northwards throughout England and Wales, especially since the beginning of the 21st century, and the first record in Northumberland in 2010 meant that all English counties had provided records by the end of that year. At this time however, it remained an extreme rarity in Scotland, with just one record from Lothian in October 1993.

The male at Dorman's Pool in November 2009 was controlled on 18 May 2010 and was found to have been originally ringed as a juvenile, at Rye Meads, Hertfordshire, on 20 June 2009. Interestingly, another juvenile ringed at Rye Meads in July 2009 was later controlled at Marston Sewage Works, Lincolnshire in late September 2009 representing another notable movement. These recoveries give some indication of the origin of these pioneering birds. In addition to this, and following the successful breeding on the North Tees Marshes, a total of five juveniles were trapped and ringed in Hargreaves Quarry, Teesmouth during late summer and autumn 2010. A juvenile from this brood was controlled at Newton Pool in Northumberland on 2 October 2011.

Long-tailed Tit
Aegithalos caudatus

A common breeding resident.

Sponsored by
Wm. C. Harvey Limited

Historical review

According to Holloway the Long-tailed Tit was distributed 'more or less commonly' across much of the British Isles in the 19th century (Holloway 1996). But there is little or no substantive information about the species locally

695

prior to the work of Temperley (1951) and even that provides only sparse detail about its status in County Durham. Mead (2000) noted that the species was well distributed across Britain and Ireland "*100 years ago*", presumably it was also in Durham.

Bolam, with reference to Northumberland, (1932) described this species as "*seldom numerous, but nesting in many places and forming roving bands in winter*', his comments may also have been indicative of its status in Durham. Halfway through the 20th century, Temperley (1951) said that this was a "*resident, well distributed in wooded areas*". He commented that it was noticeably numerous in 1944 and 1945 and that it was most "*frequently noticed in autumn and winter when it wanders around in flocks*".

A correspondent writing of observations of birds in the garden of the Isolation Hospital at Chester-le-Street in the autumn and winter of 1954 said, "*As I generally do, this autumn I continued my observations on birds and insects in the Hospital garden. This year seems to have been exceptional in respect to the species seen. In the first place the long-tailed tit put in an appearance*" (*The Vasculum* Vol. XXXIX 1954). This would appear to be the first local report of this species utilising gardens, though the habit may have gone previously undocumented.

One of the few notable reports of this species prior to the modern recording period was of a flock of 40 birds noted near Rowlands Gill in February 1960 (Bowey *et al.* 1993). A review of reports in the north west of the county, suggested that local populations were high during the early seventies when a number of large flocks were reported, for example 120 birds together at Ryton in winter 1974, (Bowey *et al.* 1993) and 100 at Witton-le Wear on 13 February 1972; although these numbers had possibly developed in part as a consequence of a series of mild-winters at that time. Declines were noted during the cold winters of the late 1970s with reports of groups of 30 in December 1975 being reduced to seven or so birds by February-March 1976. Such catastrophic declines in numbers can be widespread and are often associated with the severe winter weather, such as that of winter 1963-1964 (Parkin & Knox 2010). For instance, in 1980 the species was locally reported to have "*not recovered from the severe winter of 1979*" (Unwin & Sowerbutts 1980) with peak counts of nine birds being reported, though the species was recovering through the early 1980s.

Recent breeding status

This is one of the most commonly recorded species in the county. It breeds widely in woodland and scrub, in small copses and along overgrown, disused railway lines. It is frequently found where broom dominates, the dense vegetation offering relative safety in which to build the dome-shaped nests. Though it tends to be only numerous in a number of key areas, such as the Derwent valley, nonetheless, this species is widespread and can be found in many areas throughout the year.

Long-tailed Tits frequent a variety of lowland habitats with tree or scrub cover. In the best of such areas, the population can reach a density of 15 pairs per tetrad, but this is exceptional. The species' preference for woodland was seen in its mapped distribution in the *Atlas*. This distribution, in the main, followed the county's river valleys upstream, as this is where much of the woodland occurs. The species has a sparse distribution on higher ground, in the west of the county, and it has not been recorded breeding successfully above 300m, though there is woodland habitat above this altitude. Consequently, Long-tailed Tits tend to more widely distributed in the east of the area than are some of the other woodland species, partly because of their avoidance of exposed upland situations and also down to the species' willingness to breed in scrubland and hedgerows.

Probably because of the series of hard winters in the late seventies and early eighties, local population levels declined but they rose again through the later part of that period, and this trend continued through the 1990s (the period when the survey work on the *Atlas* was conducted) and on through the first decade of the 21st century. Other than severe winter weather, there are few threats to this species locally, other than possible habitat loss; especially the removal of woodland, scrub or hedgerows. This species is so widely distributed across the county, and its nesting habitat is so common, it is difficult to envisage a process that would occur at such a scale as to affect significantly the Long-tailed Tit population (Westerberg & Bowey 2000).

The species' nest sites are most commonly found in low shrubs such as, gorse *Ulex europaeus*, broom *Cytisus scoparius* or hawthorn *Crataegus monogyna*. Usually at a height of one to two metres although in mature woodland, nests have been found at heights of more than 10m, for example in the fork of a tree (Westerberg & Bowey 2000). Breeding pairs are usually in territory by March but early nest building has been noted on 4 March 1989 and this is not atypical, nest-building routinely commences from midway through to late March, earlier in mild winters (Brown & Grice 2005). In 1987, one monitored nest at Clara Vale was started on 9 March and was complete by the 18th. During the 2000s some birds have been noted adapting to more suburban locations, with breeding birds building in garden cypresses in Rowlands Gill at the end of the first decade of the 21st century. At Finchale on 2 April 1975, two pairs were found nesting within 70m of each other. Once settled, breeding birds can be surprisingly difficult to locate, but as soon as the young have fledged, noisy family parties are quite easily located.

For the *Atlas*, it was assumed that an average of four to six pairs of Long-tailed Tits were found in each occupied tetrad, which gave an estimated county population of around 600 pairs. There is no reason to suggest that this estimate will have declined since that work was complete, indeed courtesy of the exceptional run of mild winters in the first few years of the new century, up to winter 2009, it is likely that numbers would have risen. Around 120 pairs were identified by the survey work of 2000 to 2006 for *The Breeding Birds of Cleveland*, to the north of the Tees (Joynt *et al.* 2008). A current estimate of 'more than' 1000 pairs for the county is likely to realistic although this may still be somewhat on the low side, however the impacts of consecutive severe winters in 2009 and 2010, was not apparent at the time of writing.

Recent non-breeding status

Flocks of family parties start to appear around woodlands and along hedgerows from early June and through July, and these grow in size during the late summer, as they join up with roving, mixed titmouse flocks during autumn and through the winter.

During the winter of 2007, illustrative of the county-wide position, over 600 records of the species were received from local observers from almost 200 locations across the county, an indication of how common this species had become by the early 21st century. Birds were reported from locations as far west as Upper Teesdale, although the vast majority of observations come from lowland sites. Severe winter weather, which is exacerbated at higher altitudes, is an important limiting factor for this small species and national population declines of up to 80% have been recorded in extreme winters (Gibbons *et al.* 1993), particularly when trees became glazed with ice. Being a small insectivorous species, Long-tailed Tits suffers badly in such hard circumstances but population levels can recover quickly given good breeding seasons and subsequent milder winter weather (Brown & Grice 2005); this propensity for population booms and busts around hard winters in the county had been commented upon by Temperley (1951).

During the final quarter of most years, flocks of 15 to 30 birds, usually merged family parties, are noted from many widespread locations across the county. Reports of larger numbers (of between 40 and 80 birds) are less common but are not infrequent. Most of the observations submitted of this species are of flocking birds during the winter periods. In the south east quadrant of the county, this species is rather scarce in the Teesmouth and coastal strip area, so a flock of around 50 birds that was at Hartlepool Headland on 8 October 2000 was noteworthy (Blick 2009).

The run of mild winters through the 1990s and early 2000s led to high population levels, and large winter flocks of the species were one of the expected features of a walk through any type of woodland across the county by the winter of 2006/2007. Indicative of the locally high numbers was the observation of one flock containing over 100 birds at Rainton Meadows during the final quarter of 2008. By 2008, at a national level, this species had reached its highest monitored numbers since the commencement of the Breeding Bird Survey but, like a number of other diminutive resident species, it was found to have declined due to the prolonged freezing weather of January and February 2009; in this instance by some 12% between 2009 and the previous year (Risely *et al.* 2010).

Despite a number of woodland birds routinely using garden food and accessing garden feeding stations, for many decades Long-tailed Tits were once rare visitors to such locations. However, in the period from the mid-1980s to the middle of the first decade of the 21st century, this habit grew hugely in its observed frequency and it is now widely and routinely noted from many places in the county. Whether this is as a result of an increase in garden feeding-stations or the species learning to follow titmice to available food-sources remains unknown. It was certainly a common sight by the late-nineties to see Long-tailed Tits descend on garden feeders. The Garden Bird Feeding survey indicated a slight increase in the number of gardens in which the species has been recorded since the mid-1990s, this is corroborated by Breeding Bird Survey data, and elsewhere in Britain, like Durham, it has taken to visiting garden feeders (Parkin & Knox 2010).

The apparent attractiveness of Sewage Treatment Works, as demonstrated by local ringing totals, to this species during the winter is of interest. For example over 150 were ringed adjacent to the Lockhaugh STW in the lower Derwent valley over two months of the severe winter of 2010-2011. The attraction was probably the large numbers of invertebrates emanating from the treatment work's filter beds. These become available to the birds after the invertebrates are blown on to the surrounding scrub and hedgerows, from which they can be gleaned by foraging flocks of birds.

Communal winter roosting in this species is well documented (Brown & Grice 2005), but the report of 20-30 birds utilising a squirrel dray as a roost in the winter of 2003 is noteworthy (Newsome 2007).

This species is not usually regarded as a migrant but in September and October, wandering birds can be observed anywhere in the county, including along the coastline. Over the years, small parties have been seen on Hartlepool Headland, at Seaton Carew, in the North Gare bushes and along the Long Drag (Blick 2009) as well as many other coastal locations between Hartlepool and South Shields in the north of the county. The frequency of sightings along the coast has risen markedly in the 21st century. Such occurrences were previously not common, as illustrated by the fact that the four seen at Hartlepool Headland on 30 October 1971, were the first records there since 1957; and the three to ten birds at Marsden in October 1988 were the first there for ten years. Contrast these numbers with a rather different picture provided by ringing operations at Whitburn Coastal Park during autumn 2011, when 90 birds were trapped and ringed. Coastal movements have been particularly prominent in most years over the last decade with small parties of up to a dozen birds a common sight in the Whitburn/Marsden area during autumn (T. I. Mills pers. obs.). A large flock of 40 passed through a Hartlepool garden, on 21 September 2007 (Blick 2009). Perhaps this is further evidence of a current high population. More intriguing perhaps was the party of birds flying high over the sea at Marsden on 14 October 1975 then heading south and likewise the bird seen to fly in from over the sea at Hartlepool on 25 October 1989 (Blick 2009). It is almost certain however, that most coastal records involve local wanderers rather than continental birds, though birds have been recorded, rarely, on the Farne Islands in Northumberland (Kerr 2001). Most locally observed birds in coastal areas are probably locally bred or at most short-distance internal migrants either from inland locations in the county or perhaps further north in Britain. This assertion is at least partly borne out by the record of two birds ringed at Saltholme on 4 December 1997, which were later re-trapped inland west of Stockton, at Urlay Nook on 12 November 1999 and trapped there again on 2 October 2000.

Distribution & movements

The species' breeding range encompasses most of Europe, Asia Minor, Siberia and Central Asia, as far east as Kamchatka and parts of China. Most birds across the range are sedentary, though some of the northerly populations (the 'white-headed' races) exhibit a migratory nature in winter, occasionally undertaking long-distance irruptive movements, sometimes in large numbers though few reach the British Isles (Parkin & Knox 2010).

Durham Long-tailed Tits, like most British ones, are largely sedentary, roving in groups around a home range (Brown & Grice 2005). They move on average only short distances and the mean movement between ringing site and recovery site for birds ringed in Britain being just two kilometres (Wernham et al. 2002). This would appear to be confirmed by the details of birds in Durham. Most locally ringed birds are controlled or recovered locally, indicating that whilst they may rove in mobile flocks in the local context, these flocks move only on a restricted local beat, which serves the purpose of maximising the foraging efficiency of this small insect-eating species, particularly in winter. These group-based movements probably account for the five birds that were ringed in 2010 at either Rowlands Gill or Lockhaugh Sewage Treatment Works and were re-trapped to show movement between the sites, in 2010 and early 2011; all movements of less than three kilometres. Likewise a bird ringed at Rowlands Gill in July 2010 was controlled at Chopwell Woods in February 2011.

'Long-distance' movements of this species include one ringed at Ebchester on 27 September 1989, which was at Cattersty Pond at Carlin How, North Yorkshire on 29 October 1990, a distance of 69km to the south east. More unexpected was the bird that was ringed as a fully-grown bird at Grune Point near Silloth, Cumbria on 16 October 1988, which was controlled 99km to the east, at Clara Vale on 13 May 1989. This is a very significant movement for this species. It was perhaps predictable that this individual, when re-trapped, proved to be a breeding female, for it is females of this species that are most likely to disperse away from their natal areas (Brown & Grice 2005). Another female bird that was ringed at Lockhaugh STW on 21 August 1993, as a juvenile, was later re-trapped at Bolgam, Northumberland in October 1994, 29km to the north west, where it had presumably spent the summer and reared a family.

The above details relate to the British race *Aegithalos caudatus rosaceus*, but there have been a number of claims over the years of 'white-headed' birds, which may have been individuals of the north European race *Aegithalos caudatus caudatus*, but there are no fully authenticated records of this race for the county. It has been suggested that some of the birds that appear in eastern and southern coastal counties may be genuine migrants of the race *A. c. europaeus*, which has been recorded from the English south coast (Brown & Grice 2005). It is conceivable that such birds, if they are reaching Britain, might penetrate as far north as Durham but at present there is no evidence to support such an assertion.

Eastern Crowned Warbler
Phylloscopus coronatus

A very rare vagrant from the Far East: one record.

The appearance of Britain and Ireland's first Eastern Crowned Warbler in a small clump of sycamores at Trow Quarry, South Shields on 22 October 2009 will go down as one of the most sensational discoveries in County Durham's ornithological history (*British Birds* 103: 609). This bird was found by Dougie Holden and Derek Bilton and was initially identified as a Yellow-browed Warbler *Phylloscopus inornatus* before being re-identified by Mark Newsome from photographs posted on the internet on the evening of 22 October. The bird was estimated to have been seen by 2,500 observers from across the British Isles before its departure on the night of 24 October. Restricted to the south eastern corner of the quarry it frequently showed well and was accompanied by a single Yellow-browed Warbler throughout its three day stay.

Distribution & movements

Eastern Crowned Warbler breeds in the hill forests of far eastern Asia from south eastern Russia to north eastern and central China, Korea and Japan. A long distance migrant it spends the winter in South East Asia from Assam southwards to Sumatra and Java in Indonesia. A true far-eastern vagrant the individual at Trow Quarry represents the first record of this species in Britain, though there were four previous records for Europe, the most recent of which was at Katwijk ann Zee, Netherlands on 5 October 2007. All of the European records have been in the period from 30 September until 23 October, with only the Durham bird remaining for more than a day.

Eastern Crowned Warbler, Trow Quarry, October 2009 (D. Holden)

Greenish Warbler
Phylloscopus trochiloides

A rare migrant from north eastern Europe: 28 records, involving 29 birds, most frequently from August to October; one spring record, in May.

Recent Status

The first Greenish Warbler for County Durham was found by Cliff Waller and Alan Wheeldon at Hartlepool Headland on 16 May 1970. It spent much of its time in the gardens surrounding St Mary's Church and was heard singing on occasions. This remains the only spring occurrence of the species in County Durham to date and it was the first May record of Greenish Warbler for Britain, although there have been several others since (*British Birds* 64: 360).

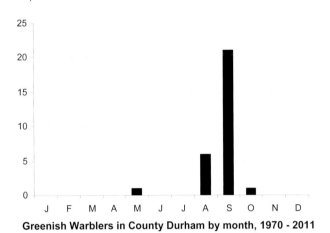

Greenish Warblers in County Durham by month, 1970 - 2011

Since the first in 1970, a further 28 Greenish Warblers have been recorded in the vice-county all of these have been in autumn between August and October, with the vast majority coming in the first half of September. Formerly a very rare vagrant, more than two-thirds of the county total has occurred since 2000, including a remarkable influx of eight birds in 2002. This dramatic change in status has been mirrored throughout the rest of Britain and is thought to be linked to the westward expansion of this species in continental Europe.

Greenish Warbler is a typical 'drift' migrant; most if not all of the records of this species have been associated with periods of easterly winds and all have been from coastal localities. Hartlepool Headland has recorded a total of eight birds and Whitburn Coastal Park seven individuals. The only multiple occurrences were of two birds at Hartlepool Headland from 13th to 15 September 2002 and two at Whitburn from 4th to 6 September 2005.

A total of five birds were recorded during the 1980s, although after the first record 14 years passed before the county's second at Marsden Quarry from 22nd to 25 August 1984 (*British Birds* 78: 576). This bird was found during a small arrival of continental migrants, including Wood Warbler *Phylloscopus sibilatrix* and Red-backed Shrike *Lanius collurio*. It was trapped and ringed during the morning of the 23 August. The four subsequent records were all at Hartlepool Headland on: 25 August 1987 (*British Birds* 82: 549); 7 September 1988 (*British Birds* 82: 549); 31 August 1989 (*British Birds* 83: 483) and 13th to 14 September 1989 (*British Birds* 83: 483).

Three further birds occurred during the 1990s with the first seen briefly at Spion Kop Cemetery, Hartlepool on 18 September 1995 (*British Birds* 89: 518). This bird seemed to have just arrived and was watched flitting around the gravestones and weeds before flying off inland shortly afterwards. Subsequent to this, a particularly vocal but elusive bird was at Bents Park, South Shields from 7th to 8 September 1996 (*British Birds* 91: 507), and another was at Marsden Quarry on 25 August 1999 (*British Birds* 94:491).

A remarkable total of 18 Greenish Warblers was found in County Durham during 2000-2009, double the total number of birds seen during the preceding 30 years. This dramatic change in status saw the species establish itself as a rare but almost annual passage migrant to the county, being recorded in five out of those ten years.

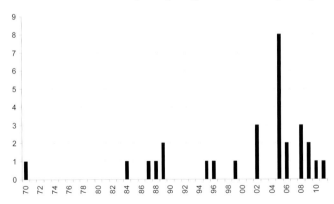

Greenish Warblers in County Durham by year, 1970-2011

During early September 2002, advantageous weather conditions resulted in a spectacular fall of migrants at the Durham coast, and as the rain associated with the weather system eased, three Greenish Warblers were found at Hartlepool in the ensuing days. These were at: Spion Kop Cemetery on 10th and 11 September; Hartlepool Headland from 10th until 17 September, with two birds present from 12th to 15th, both of which were trapped and ringed during their stay (*British Birds* 96: 596). A total of 21 Greenish Warblers were recorded in Britain during 2002.

A record eight birds were seen in 2005, the majority of which arrived in just two days in the first half of September. The first were two at Whitburn Coastal Park from 4th to 6 September, after a period of light winds and fine weather, which had produced few other migrants. Amongst other migrants grounded by a weather system moving northwards through the North Sea from 8 September, were five Greenish Warblers discovered on 10th and 11 September. These were at: Seaham Hall Dene on 10 September; Seaton Snook, Teesmouth from 10th to 11 September; Trow Quarry, South Shields from 10th to 12 September; Bruntoft Avenue, Hartlepool 11 September and Whitburn Coastal Park on 11 September (*British Birds* 74:86, *British Birds* 75: 740). The bird at Seaham Hall Dene was the first to be recorded away from the favoured areas of Hartlepool and South Tyneside. The final bird of 2005, at Marsden Quarry on 15th and 16 October, arrived during a record influx of Yellow-browed Warblers *Phylloscopus inornatus*. This is the only October record of the species to date.

Subsequently, there were two birds in 2006; at Hartlepool Headland from 18th to 20 August and Whitburn Coastal Park on 24th to 25 September. In 2008, three birds were discovered between Whitburn and Hartlepool on 7th and 8 September, the most noteworthy of these being at Blackhall Rocks on 8th; one of just two records from the stretch of coastline between Sunderland and Hartlepool. Further birds were noted at Whitburn Coastal Park on 5-6 September 2009, at Marsden Quarry on 6-7 September 2009, on South Shields Leas on 2-3 September 2010, and at Whitburn Coastal Park on 24-25 August 2011. The earliest Greenish Warbler to be recorded in County Durham was at Hartlepool Headland from 18th to 20 August 2006, while the latest was at Marsden Quarry from 15th to 16 October 2005.

One published record of a Greenish Warbler, at Hartlepool Headland on 16th and 17 September 1967 (Bell 1968), previously accepted by BBRC, was later deemed unacceptable following a national review of records between 1958 and 1982 (Dean 1985, *British Birds* 78: 577).

Distribution & movements

The taxonomy of the Greenish Warbler is complex. The western race *viridanus* breeds from north Eastern Europe east to western Siberian and Kashmir; it winters almost exclusively in the Indian subcontinent. This race, which is responsible for the majority of British records, has been expanding westwards over the late 20th century, breeding as close to the British Isles as Poland, the Baltic countries and southern Finland by 2010. To the east it is replaced by the race *plumbeitarsus*, which is considered by some authorities to represent a distinct species, the 'Two-barred Greenish Warbler' (Collinson *et al.* 2003). The Greenish Warbler is an increasingly regular vagrant to Britain with over 450 records by the end of 2005 at which point the western race *'viridanus'* ceased to be considered by BBRC. Records span the period from May to November although spring records are the exception, and more than three-quarters recorded in Britain have been in August and September. Three Greenish Warblers have been ringed in County Durham, all of them first-year birds. These were at Marsden Quarry on 23 August 1984, and two at Hartlepool Headland on 13 September 2002. No controls or recoveries have been generated from these.

Arctic Warbler
Phylloscopus borealis

A very rare vagrant from north eastern Europe: seven records.

County Durham's first Arctic Warbler was found by Martin Blick and John Crussell at West View Cemetery, Hartlepool on 4 October 1979; it was still present the following day (*British Birds* 73: 524). A large high pressure system was centred over Scandinavia at this time creating a prolonged period of south easterly winds that brought several Yellow-browed Warblers *Phylloscopus inornatus* and Firecrests *Regulus ignicapillus* to the area.

All records:
       1979   West View Cemetery, Hartlepool, 4th and 5 October
       1984   Hartlepool Headland, 6 September (*British Birds* 78: 577)
       1984   Whitburn, 12th and 13 November *(contra BB)* (*British Birds* 78: 577)
       1984   Seaburn, 17 November (*British Birds* 78: 577)

1991   Hartlepool Headland, 31 August to 2 September (*British Birds* 85: 540)

1991   Grangetown Cemetery, Sunderland, 12 October (*British Birds* 86: 517)

1996   Hartlepool Headland, 19th to 23 September (*British Birds* 90: 502)

The records in County Durham broadly conform to the national pattern although the two November records from 1984 are especially noteworthy as there is just one other November record for Britain - from Unst, Shetland on 10 November 2007 - making the bird at Seaburn on 17 November the latest to be recorded in Britain. It is conceivable that this was the same bird which frequented high trees in Church Lane, Whitburn, and disappeared three days before a bird was discovered in Mere Knolls Cemetery, Seaburn less than two kilometres distant across open fields. Hartlepool Headland has attracted most with three records. The bird in 1984 favoured the 'Doctor's Garden'; the next in 1991 was mobile in gardens along the seafront. In contrast to this, the bird in September 1996 remained faithful to the large sycamore in front of the Borough Hall throughout its five-day stay.

Distribution & movements

Arctic Warbler breeds locally in northern Scandinavia and is widespread across much of northern Russia eastwards through Siberia to Japan and western Alaska, being the only species of *Phylloscopus* Warbler to breed in the New World. It is an extremely long distance migrant with the entire population wintering in South East Asia and Indonesia as far south as Sulawesi and the Lesser Sundas. Arctic Warbler is an annual and increasingly regular vagrant to Britain and has been recorded from late June until mid-November, though the majority of records are from September, typically occurring slightly later than Greenish Warbler *Phylloscopus trochiloides*. There had been over 300 British records by the close of 2010, more than half of these have been on Shetland. Elsewhere, the majority of records come from the coastline of eastern Britain, though several have been recorded as far west as the Isles of Scilly.

Pallas's Warbler (Pallas's Leaf Warbler)
Phylloscopus proregulus

A rare migrant from Siberia: 69 records, involving 74 birds, all in October and November.

Historical Review

The first Pallas's Warbler for County Durham was at Hartlepool Headland on 12th and 13 October 1962. This diminutive *Phylloscopus* was initially found by Jack Bailey in flowerbeds surrounding the outer bowling green during the early afternoon, and was seen shortly afterwards by several other observers including Graham Bell and the late Phil Stead, as it fed actively by gleaning insects from leaves in typical manner. It was trapped at 5:00pm, although due to the failing light and the bird's apparent hunger, it was quickly photographed to verify the record and released without being ringed; it represented only the seventh record for Britain at the time. Still present the next day, it was last seen flying down the street around midday (*British Birds* 56: 112-113, *British Birds* 56: 405). At the time of this occurrence there had been just two records of Yellow-browed Warbler *Phylloscopus inornatus* in County Durham, both of which were in South Tyneside in the 1950s.

Recent Status

A total of 73 Pallas's Warblers have been recorded in County Durham since 1970 with more than three quarters of these since 1990 and the species is becoming an increasingly regular visitor in late autumn. There have only been four years without records since 1990. A vagrant from Siberia, all but two of the records are from the coast between South Shields and Hartlepool Power Station, with birds typically appearing during periods of onshore wind and rain from mid-October until mid-November. November records are not unusual with a total of 21 records, by which time its close relative the Yellow-browed Warbler *Phylloscopus inornatus* is very scarce indeed with just six November records by comparison. The site with most records is Hartlepool Headland with a total of 19 birds, but 41 birds have been found on the stretch of coastline between South Shields and Sunderland; the majority of which

702

have been in the Marsden/Whitburn area. Just three birds have been found along the stretch of coastline between Sunderland and Hartlepool, all of these were at Dawdon Blast Beach; no doubt this pattern relates to the amount of observer activity.

During the 1970s there was only a single bird, at Hartlepool Headland on 9 October 1974. This was one of thirteen in Britain that year and part of a notable influx into Europe; just eight were recorded in Britain during 1970-1973 (*British Birds* 68: 327).

A total of eleven were recorded in the county during the 1980s, the first of which was at Hartlepool Headland on 19th and 20 October 1981 (*British Birds* 76: 517). During early October 1982, favourable weather features resulted in classic fall conditions to the east coast of Britain, especially during the period from 10th to 12 October. At least seven Pallas's Warblers were found in County Durham over this period, part of a large influx into Britain at the time involving over 120 birds (Howe & Bell 1985). The first at Marsden Quarry on 7 October was followed two days later by another at Hartlepool Headland which stayed until 10th when two more were found, a second bird at Hartlepool Headland and one at Mere Knolls Cemetery, Seaburn. The influx reached its peak the following day, when a further three birds were found amongst large numbers of Goldcrests *Regulus regulus* within 200 yards of Souter Point Lighthouse, Whitburn, and there were possibly two different birds seen nearby on 12 October (*British Birds* 76: 517). This peak count of three birds at Whitburn on 11 October 1982, remains the highest individual site total on a single day in the county.

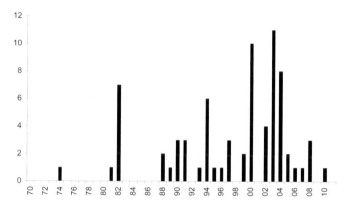

Pallas's Warblers in County Durham by year, 1970-2011

It was five years before the species was recorded again, with singles at Marsden Hall on 25 October 1988 and inland at Low Fell, Gateshead, on 27 October 1988 (*British Birds* 82: 550). A single bird at Dawdon Blast Beach on 29 October 1989, brought the decade to a close (*British Birds* 83: 484).

A total of 20 birds between 1990 and 1999 signalled the start of Pallas's Warbler being regarded as an almost annual species in the county; it was absent in just two autumns in the period. The decade started with a total of three birds in the Marsden/Whitburn area in 1990. These were at Marsden Quarry on 18 October (*British Birds* 86: 520), with further individuals at Marsden Hall and St Mary's Church, Whitburn on 21 October (*BB*: 85: 540). A further three birds were recorded on 12th and 13 October 1991, including a bird at Dawdon Blast Beach on the former date. This bird shared a single hawthorn and one tiny patch of bramble on the beach with a Yellow-browed Warbler *Phylloscopus inornatus* and a Red-breasted Flycatcher *Ficedula parva*. Just one bird was recorded in 1993, at Hartlepool Headland from 14th to 17 October, though six were recorded between 13 October and 7 November 1994, the peak year of occurrence during the 1990s. These included a bird inland at Shibdon Pond on 22 October and the first November record of this species in the county, a bird at Hart Warren, Hartlepool for two days from 7 November. A further nine were recorded before the decade closed with singles at Hartlepool Headland on 10th and 11 November 1995, and at Whitburn from 15th to 18 November 1996. Three birds were between Whitburn and Hartlepool Headland from 27 October until 21 November 1997 with two further singles at Hartlepool Headland and Marsden during the period from 19th to 23 October 1999.

An exceptional 40 individuals were recorded between 2000 and 2009, including record influxes of 10 birds in 2000 and 11 birds in 2003. The influx in 2000 was entirely restricted to November with all 10 birds appearing during the five days from 7th to 11 November. The first were at Seaton Snook and Whitburn on 7th, with five birds arriving on 8th when three were in the South Shields area and singles were at West View Cemetery, Hartlepool and at Hartlepool Headland. Subsequent to this, singles were at Marsden Quarry on 9 November, North Sands, Sunderland on 10th and finally at Mere Knolls Cemetery, Seaburn on 11 November. This was, at the time, the largest influx of Pallas's Warblers recorded in County Durham and the first to be restricted to November.

Surprisingly there were no records in 2001, although four were recorded in 2002, all of which were on the coast between South Shields and Sunderland on 12th and 13 October. A prolonged spell of easterly winds during the latter part of October 2003 saw a record total of eleven birds reach the Durham coastline between 16 October and 1 November, although most occurred in the eleven day period between 16th and 26 October, with the final bird appearing at Seaton Snook from 1st to 3 November. The following year saw the last of the bumper influxes during the decade with a total of eight birds recorded between 20th and 28 October, five of which were in the Marsden/Whitburn area, two at Hartlepool Headland, and a single at Dawdon Blast Beach. Just two were recorded in 2005, both of which were in the Sunderland area in October, with 2006 and 2007 producing single records, both in the Marsden/Whitburn area, including a remarkably late bird at Marsden on 21 November 2007. Three birds between 1st and 7 November 2008 included two birds at Hartlepool Headland on 7th; there were no records in 2009. A single bird at Marsden Hall on 11 November 2010 was the only bird recorded that year. With only seven birds recorded between 2005 and 2009, perhaps one or two birds per autumn represent a return to normal after the glut of records between 2000 and 2004.

Just two Pallas's Warblers have strayed significantly inland in County Durham, both of which reached the Borough of Gateshead. The first was in a garden at Low Fell on 27 October 1988, and the other briefly at Shibdon Pond, Blaydon on 22 October 1994. The only other record away from the immediate coastal strip was of a single on Cleadon Hills, some two kilometres inland, on 9 November 1997.

The earliest Pallas's Warbler to be recorded in County Durham was at Marsden Quarry on 7 October 1982, while the latest were at Whitburn on 21 November 1997 and along Quarry Lane, Marsden on 21 November 2007.

Distribution & movements

Pallas's Warbler breeds in central and eastern Siberia southwards to northern Mongolia and north eastern China, wintering chiefly in southern China and north eastern Indochina. It is an extremely long distance migrant with birds at the western end of the species' range having to undertake a migration of around to 5,000km to reach their wintering ground in southern China, a distance that is broadly equivalent to that travelled to reach Western Europe. This is an increasing regularly vagrant to Britain with around 100 Pallas's Warblers reaching Britain in a typical autumn (Fraser & Rogers 2006), although numbers fluctuate from year to year and are subject to periodic influxes dependant on weather conditions. Most arrivals are associated with anti-cyclonic weather in late autumn, ideally with an easterly airflow extending well to the east of the high pressure system creating ideal conditions for westward migration. The best year on record is 2003 when over 300 were recorded in Britain (Fraser & Rogers 2006). This species was removed from the list of species considered by BBRC at the end of 1990. Six Pallas's Warblers have been ringed in County Durham, at: Hartlepool Headland on 12 October 1962; Marsden Hall on 18 October 1990; North Gare, Teesmouth on 26 October 2003; Seaton Snook, Teesmouth on 1 November 2003 and two at Hartlepool Headland on 7 November 2008. They have generated no controls or recoveries.

Yellow-browed Warbler
Phylloscopus inornatus

A scarce migrant from Siberia, recorded from September to November, and once from January to March.

Historical Review

County Durham's first Yellow-browed Warbler was found by John Coulson at Westoe, South Shields on 4 October 1951. At the time John was a sixth-form pupil, who had permission to watch birds in one of the large gardens in the village. What was presumed to be the same bird was also seen by Dr Hugh Blair and the late Fred Grey between 8th and 28 October, though given the large range in dates it seems highly likely with hindsight that more than one individual was involved. The following year produced the county's second, just yards away from where the first had been, in Westoe, South Shields and found by the same observer. A particularly early individual, it was present from 14th until 17 September.

A decade would pass until the next sighting which involved a long-staying bird along West View Road, Hartlepool from 14 October until 6 November 1962, although like the first in 1951, the protracted stay suggests that more than one individual was involved during this period. This bird was found two days after the county's first Pallas's Warbler *Phylloscopus proregulus* at Hartlepool Headland a little over 3km away (*British Birds* 56: 405).

The species was to remain annual in occurrence for the rest of the 1960s with at least 17 birds recorded between 1963 and 1969, all of which arrived between 16 September and 7 November. The Hartlepool area accounted for the majority of the records with 15 birds, though away from there, singles were at Marsden Hall on 12th and 13 October 1964, at Graythorp, Teesmouth on 9th and 10 October 1965 and near North Gare, Teesmouth on 11 October 1969. The best individual year was 1967 when at least five birds were thought to have visited Hartlepool Headland during the period from 16th to 24 September, including peaks of three birds on 16th to 17th and 23 September.

The two November records in the period at Hartlepool Headland on 7th and 8 November 1963, and 2 November 1966 are unusual and one might speculate that these birds may have been the rarer Hume's Warbler *Phylloscopus humei* given the dates of occurrence; at that time however, Hume's was believed to be a race of Yellow-browed Warbler and would probably not have been considered. There have subsequently been four modern records of Yellow-browed Warbler in November.

Two birds were trapped and ringed during this period - one at Marsden Hall on 12 October 1964 and another at Graythorp on 9 October 1965.

Recent Status

Over 380 Yellow-browed Warblers have been recorded in County Durham since 1970 with more than half of these since 2000. Formerly a rare passage migrant with a total of 17 birds recorded during the entire 1970s, the considerable rise in numbers has seen that total exceeded in each autumn since 2005 and the species is now best regarded as an annual passage migrant, with just three years without records over the period, the last of which was in 1983. With the exception of a single mid-winter bird, all of the records have been in autumn from September until November, though the vast majority of birds have occurred between the last week of September and mid-October, with just a handful of records in November. The species typically appears during periods of anti-cyclonic weather and is usually associated with onshore winds, though it does occasionally occur in relatively calm conditions. Almost all of the records of this species are from the coastline between South Shields and Teesmouth, although not surprisingly most are recorded from the well-watched locations of Hartlepool Headland and the Marsden/Whitburn area, with just a few penetrating further inland. The site with the largest number is Hartlepool Headland, with in excess of 90 birds since 1970, but the Marsden/Whitburn area has also hosted an impressive total of around 130. The best years on record were 2005, when an estimated 39 birds were recorded, and 2011, when approximately 40 were seen, although in both years, there were probably many more. A count of nine birds at Hartlepool Headland on 15 October represents the highest single site day count.

The 1970s saw little indication of a change in status for this species with a total of seventeen birds recorded, which was actually one less than in the 1960s. The peak year was 1975 when four birds were seen between 9th and 25 October, though three which arrived between 21 September and 7 October 1977 indicated the wide range in dates upon which this species can occur. The latter year also saw the first bird to be recorded significantly inland in the county, when a single was in a Norton garden on 25 September 1977.

The 1980s brought a dramatic change in status with a total of 96 birds recorded, including three annual peaks in excess of 10 birds. The best year was 1986 when an estimated 22 birds were noted between 27 September and 2 November, including peaks of four birds at Whitburn on 27 September and five at Hartlepool Headland from 5th to 8 October. A single at Ryhope Dene on 2 November 1986 was the first November record since 1970. A further 20 were recorded in 1987, all of which were between 28 September and 11 October, with 10 birds in the autumn of 1988 and just three in 1989, giving an indication of how numbers can fluctuate between years.

The 1990s represented a downturn in real terms with just 66 birds recorded, two thirds the total of the previous decade, though numbers were still well above average in comparison to those noted during the 1960s and 1970s. The best year was 1994 when a total of 12 birds were recorded between 17 September and 21 October, but 10 were also found in 1991. The highest individual total recorded during this period was of three birds at Hartlepool Headland on 13 October 1990, and there were three further counts of more than one bird throughout the entire decade. Two birds were at Mere Knolls Cemetery, Seaburn on 30 September 1991, two at Holy Trinity Churchyard, Seaton Carew on 18 September 1994 and two at West View Cemetery, Hartlepool on 7 October 1998.

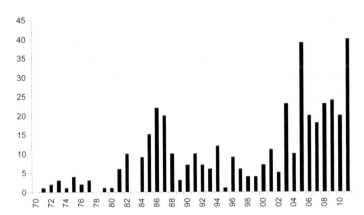

Yellow-browed Warblers in County Durham by year, 1970-2011

In excess of 240 Yellow-browed Warblers were recorded between 2000 and 2011, though more than three-quarters of these were subsequent to 2003, with annual totals prior to this being seven (2000), 11 (2001) and five (2002). A new record total of 23 birds were recorded in the autumn of 2003, although this was quickly surpassed by a huge arrival of at least 39 birds between 5th and 25 October 2005. The first to arrive was at Whitburn Coastal Park on 5 October, followed by occasional daily reports of singles until 9th, although the main arrival took place from 13th and reached a peak on 15 October, when an estimated 16 birds were thought to be present between South Shields and Sunderland, with at least nine at Hartlepool Headland amongst others reported elsewhere. Records gradually tailed off towards the end of October with the final bird being at Sunderland North Dock on 25 October. Subsequent to this influx, the species maintained high numbers each autumn until the end of the decade, with annual totals being; 20 (2006), 18 (2007), 23 (2008), 24 (2009) and, 20 (2010). A new high was reached in 2011 when a prolonged October arrival resulted in over 40 birds being found. The average number of birds per annum since 2000 was 20, a huge increase on just one to two during the 1970s.

The earliest Yellow-browed Warbler to be recorded in County Durham was at Marsden Quarry on 8 September 1993, with the next earliest being the second record for the county on 14 September 1952. There have been just seven birds recorded in November since 1970, with five of these occurring in 2008, when two were at Hartlepool Headland from 6th to 8 November, at Marsden Quarry on 8th and again at Hartlepool Headland on 15th, the latest to be recorded in the county.

Despite increasingly regular sightings of wintering birds in Britain, just one has been found in Durham; a mobile and often elusive individual in a mixed tit flock at Jarrow Cemetery from 29 January until 9 March 2008.

A total of 22 Yellow-browed Warblers have been recorded inland to some extent during September and October since 1970, but considerably more must go undetected. The most significant of these were at: Norton, Stockton-on-Tees on 25 September 1977; Charlton's Pond, Billingham on 13th to 14 October 1981 and from 30 September until 2 October 1990; Lambton Lane, Fence Houses on 19th to 21 October 1994; Hetton Lyons Country Park, Houghton-le-Spring on 11 October 2004 and Barmston Pond, Washington on 12 October 2008. Elsewhere most have been within five kilometres of the coast including seven records from the North Tees Marshes and four from the Cleadon area.

Distribution & movements

Yellow-browed Warbler breeds across much of Siberia from the northern Urals eastwards to Kamchatka and as far south as Afghanistan and northern India. A long distance migrant, it winters from central Nepal and Bangladesh to the Malay Peninsula and is one of the commonest *Phylloscopus* warblers to be found in South East Asia during the winter months. It is recorded in ever increasing numbers in Western Europe each autumn and is now perhaps regarded as a scarce migrant despite its origins in Siberia. The vast majority of birds occur from early September until mid-October though records frequently continue into November and an increasing number of birds have been found wintering in Britain in recent years. The reasons for this increase are not yet fully understood though one theory is that these birds are undertaking normal migration and are the vanguard of this species establishing a new winter range within Western Europe taking advantage of the increasingly mild oceanic climate (Gilroy & Lees 2003). The exact number recorded in Britain each autumn is difficult to estimate as numbers fluctuate, though these have increased considerably during the latter part of the 20th century. The best year on record was 2005 when a record influx in October brought an estimated 1,250 birds to the British Isles. To put this into perspective the average number of birds per annum during the 1970s had been just 79 (Fraser & Rogers 2006). In excess of 20 Yellow-browed Warblers have been trapped and ringed in County Durham since 1970, the majority of which have been at Hartlepool Headland and the Marsden/Whitburn area, with six ringed on the North Tees Marshes; they have generated no controls or recoveries.

Hume's Warbler (Hume's Leaf Warbler)
Phylloscopus humei

A very rare vagrant from central Asia: five records.

The first Hume's Warbler for County Durham was found by Bernie Beck and Chris Kehoe at Hartlepool Headland on 11 November 1994. Initially discovered in sycamores *Acer pseudoplatanus* by Borough Hall, it quickly flew off in the direction of the Fish Quay, though it was relocated in nearby gardens where it gave excellent but brief views. This was one of six Hume's Warblers recorded in Britain during November 1994 (*British Birds* 91: 508).

All records:
1994	Hartlepool Headland, 11 November
1999	Whitburn, 17th to 18 October (*British Birds* 94: 493)
2000	North Gare, Seaton Carew, 10th to 12 November (*British Birds* 94: 493)
2008	St Mary's Cemetery, Norton, 10 February until 24 March (*British Birds* 102: 584)
2008	North Gare, Seaton Carew, 1 November (*British Birds* 102: 584)

The second record was of an elusive bird, which appeared in gardens on a Whitburn housing estate and is notably early for this species, although it has occurred in Britain as early as 13 October. The bird at North Gare in 2000 was very mobile in the extensive sea buckthorn *Rhamnus catharticus* there. Like all of those subsequently this bird was very vocal, constantly giving its characteristic disyllabic call, making it easy to locate. The wintering bird at Norton in 2008 was one of four birds present in Britain that winter. Remarkably this bird coincided with the first wintering Yellow-browed Warbler *Phylloscopus inornatus* for County Durham at Jarrow some 50km to the north. Later the same year the bird at North Gare was on a typical date and although vocal, it quickly became elusive.

Distribution & movements
The nominate race of Hume's Warbler breeds from the Altai mountains to western Mongolia and southwards to Afghanistan, the north western Himalayas and the mountains of north east China. A relatively short distance migrant in winter it is found from southern Afghanistan to northern India as far east as West Bengal. The Hume's Warbler was long considered a distinct form of the more familiar Yellow-browed Warbler *Phylloscopus inornatus*, though was split by the BOU in 1996, with the first acceptable report from Britain published the following year as having occurred in Sussex in November 1966 (BOU: 24th report). The species has subsequently been proven to be an annual vagrant to Britain, chiefly in late autumn with November being the favoured month although coastal migrants have been found as late as December and several birds have been found wintering in Britain. There had been over 100 records by the close of 2010, though almost all had occurred since 1989 with the majority from the eastern and southern coasts of Britain.

Radde's Warbler
Phylloscopus schwarzi

A very rare vagrant from Siberia: eight records.

County Durham's first Radde's Warbler was found by Colin Freeman at Marsden Quarry on 19 October 1976 and its identity confirmed by the late Fred Grey. This bird was associated with a fall of late migrants in the Marsden area. The 21st British record, it was one of four recorded in Britain that year (*British Birds* 70: 436).

All records:

1976	Marsden Hall, 19 October
1982	Whitburn, 12 October (*British Birds* 76: 518)
1988	Hartlepool Headland, 12 October (*British Birds* 82: 581)
1989	Marsden Quarry, 28th to 30 October (*British Birds* 83: 485)
1990	Mere Knolls Cemetery, Seaburn, 18 October (*British Birds* 85: 545)
1999	Marsden Quarry, 23 October (*British Birds* 100: 742)
2000	The Leas, South Shields, 14 October (*British Birds* 95: 515)
2005	Hawthorn Dene, 16 October (*British Birds* 100: 742)

The second record came during an influx of migrants in the second week of October 1982, when a bird was discovered at dawn on 12th in a weed-choked gully on the cliff tops at Whitburn, but it disappeared inland after just ten minutes and was seen by only two observers. A bird which was found in St. Mary's Churchyard at Hartlepool Headland on 12 October 1988, remains the only record for this well-watched site. The five subsequent records were all discovered during significant falls of migrants. The last record at Hawthorn Dene in October 2005, seen by just a few observers, highlighted the potential of this under-watched area of the county's coastline. That five out of the county's seven records have been along a five kilometres stretch of coastline between Seaburn and South Shields is remarkable given the paucity of records further south. However the elusive nature of this species no doubt militates against its discovery.

A published record of Radde's Warbler, present at Marsden Hall on 19 September 1996 (Armstrong 1998) that was known to have been seen by a number of experienced observers was never formally submitted to BBRC and therefore does not form part of the official county record. Had it been, it would be one of only a handful of September British records and the earliest ever.

Distribution & movements

Radde's Warbler breeds in southern Siberia eastwards to Ussuriland and north eastern China, and is a noted long distance migrant wintering in south East Asia from northern Myanmar to central Thailand. The species is an increasingly regular vagrant to Britain with over 270 records by the end of 2005 at which point it was no longer considered by BBRC. More than three-quarters of these records have been since the mid-1980s, with influxes of 25 in 1991, and 28 in 2000. All of the British records fall in the period from late September until early November, though the majority are recorded in October, typically appearing slightly earlier than the similar Dusky Warbler *Phylloscopus fuscatus*, which frequently occurs into November. As would be expected the majority of records are from the coastline of eastern and southern Britain, though records do have a distinct southerly bias with less than twenty being recorded on the Scottish mainland but over 50 from the Isles of Scilly in the far south west.

Dusky Warbler
Phylloscopus fuscatus

A very rare vagrant from Siberia: five records.

The first record of Dusky Warbler for County Durham was a bird found by Martin Blick and Rob Innes at Holy Trinity Churchyard, Seaton Carew on 24 October 1981; it was still present the next day. Unusually for this species, this individual spent much of its time high in the sycamore *Acer pseudoplatanus* canopy, but being typically vocal it was generally easy to locate. The 30th record for Britain, this was only the third record for north east England, following two birds in Northumberland in 1980 (*British Birds* 76: 518).

All records:

1981	Holy Trinity Churchyard, Seaton Carew, 24th to 25 October
1982	Hartlepool Headland, 11 October, trapped (*British Birds* 76: 518)
1982	Mere Knolls Cemetery, Seaburn, 11 October (*British Birds* 76: 518)
1984	Marsden Quarry, 8th to 14 November, trapped on 9th (*British Birds* 78: 578)
2010	Spion Kop Cemetery, Hartlepool, 10 October

During a spectacular fall of eastern migrants on 11 October 1982, two individuals were found. One was in the shrubs surrounding the outer bowling green at Hartlepool Headland and was trapped and ringed, while the other was seen briefly in Mere Knolls Cemetery, Seaburn. The next at Marsden Quarry on the 8 November 1984, posed some identification problems and was initially thought to be a Radde's Warbler *Phylloscopus fuscatus*, but it was trapped and ringed the following day and its identity resolved. Some 26 years elapsed before the next, a vocal individual present in weeds alongside Hartlepool Rovers RFC ground, opposite Spion Kop Cemetery on 10 October 2010. Given Durham's east coast location, the twenty-six year absence in records is difficult to explain, as there were many records of this species along other parts of the east coast of England over that period.

Distribution & movements

Dusky Warbler, Hartlepool, October 2010 (Steven Clifton)

Dusky Warbler breeds in Siberia from the Ural Mountains eastwards to the Sea of Okhotsk and southwards through the Russian Altai to China and the eastern Himalayas. A long distance migrant, it winters from Nepal and northern India to South East Asia as far south as Singapore. Like the Radde's Warbler this is an increasingly regular vagrant to Britain from late September onwards, though it typically occurs later than Radde's with records well into November, and it has wintered on occasion. Over a third of British records have been in November, in direct contrast just to nine for Radde's Warblers. Like Radde's, this species has seen a dramatic increase in the number of records since the late 1980s having formerly been a very rare vagrant. There had been over 300 recorded in Britain by the close of 2005, after which it was no longer considered by BBRC. The majority of records come from the southern and eastern coasts of Britain, although several have penetrated as far west as Wales and the Isle of Man.

Wood Warbler
Phylloscopus sibilatrix

A localised and increasingly scarce breeding summer visitor and a scarce passage migrant.

Historical review

Temperley (1951) gave little information on this summer visitor, merely saying it was, *"much less plentiful and more locally distributed than the willow warbler, but not uncommon in suitable localities"*. Its arrival date was stated as the *"last days of April or early May"*; very similar to the position in modern times. It is apparant from the lack of historical detail that the species must have been a relatively widespread and familiar species.

Recent breeding status

Wood Warblers are largely restricted to the county's broad-leaved woodland with a limited under- storey, tending to favour those dominated by oak *Quercus* spp. or, occasionally, birch *Betula* spp. (Simms 1985). In Durham, as in many situations in Britain, Wood Warblers prefer relatively elevated locations (Gibbons *et al.* 1993) and the combination of these two characteristics effectively restricts them to the westerly parts of the county. Traditional key sites for this species have been Muggleswick Woods, in the Derwent Valley, where up to 18 singing males were noted during the 1980s (Baldridge 1981-1987); Tees Wood near Barnard Castle; Tunstall Reservoir and Backstone Bank Woods in the Wear valley; Pont Burn Woods in the lower Derwent valley (Armstrong 1988-1993). These inland woodlands are largely still intact but Wood Warblers are very thinly spread now. Away from these traditional inland sites breeding season occasional records in suitable woodland are received - at Brancepeth, two males singing in May 1999 and single birds singing in suitable habitat on the coast in late May at Castle Eden Dene in 1992 and 1993 (Armstrong 1993-1994).

Data from the BTO's Breeding Bird Survey showed a dramatic decline in this species from mid-1994 and in the next 15 years, there was a 60% reduction in the UK population. Contrastingly, there was only a moderate decrease across Europe (Pan European Common Bird Monitoring Scheme 2010). At 2010, the species was classified as 'red' on the list of species of conservation concern - the highest level. This triggered BTO-supported survey work in the species' wintering area south of the Sahara in the Sahel region. BTO studies on fledglings produced per breeding attempt showed little variation over the period 1970-2010 (Eaton *et al.* 2009). Sadly the situation in County Durham mirrors the national picture and its once-healthy population of Wood Warblers is now in rapid and possibly terminal decline.

Wood Warblers have always had a propensity for wide, between-year population fluctuations and variation in breeding success (Simms 1985). This was evidenced by work in the Pont Burn Woods' Common Bird Census during the late 1980s and early 1990s, where breeding pairs fluctuated between zero and five pairs over a five-year period between (Westerberg & Bowey 2000).

There is also a tendency in males towards polygamy (Simms 1985) and, in some instances, to rapid territory establishment, resulting in a shortened period of song that makes detection difficult. Nevertheless, information from *Atlas* survey work and a species-specific survey by the Durham Bird Club, in 1990-1992, indicated a county population in the range of 70-100 pairs. This range was estimated from the number of recorded territorial males, the greatest concentration of which was located in the Derwent valley, in the north west of the county. It has always been a rare breeding bird in the south east of Durham, with possibly just one territory revealed by the survey work between 2000 and 2006 for *The Breeding Birds of Cleveland* (Joynt *et al.* 2008).

An indication of the sudden change in fortunes of this species is evident from a comparison of the healthy population levels in 1992 with the suddenly transformed situation in 1995. By 7 May 1992, five birds were in song at Muggleswick and in the first two weeks of May reports came from a further 15 sites, with an impressive total of 16 singing males in the Derwent Gorge area over a two-day period from 10th. In 1995, a sharp downturn in birds occupying traditional breeding woodlands was evident with one to two birds appearing at Thornley Woods on 5 May, two singing at Muggleswick from 7th when more than six were in song at Tunstall Reservoir and five were singing at Hamsterley Forest on 7 June. By 2008, the species was 'clinging on' as a breeding species in Durham: a single bird singing at Muggleswick, possibly two males holding territory in the Bedburn area of Hamsterley with up to three birds at Tunstall Reservoir. Singles were seen at Thornley Woods and Glbside on 19 May.

In the early to mid-1990s, birds bred at many sites in the lower Derwent Valley, including: Thornley Woods, Clockburn Woods, Low Horse Close Woods, Strother Hills Woods, Chopwell Woods and Victoria Garesfield, amongst others. In this area, during an average late 1980s breeding season, it was considered that there would be between ten and twelve pairs of Wood Warbler breeding in Gateshead, largely in the lower Derwent valley. In good years, this would rise to as many as 20 pairs. The species' stronghold in Gateshead at the time was undoubtedly the lower Derwent Valley, with up to eleven pairs often being present in the Derwent Walk Country Park alone, with smaller numbers of birds being heard singing at Washingwell Woods, along the Stanley Burn, around Chopwell Wood and occasionally at Ryton Willows. In 2009, there were no Wood Warblers breeding at any of these sites. Over the same period, Tree Pipit *Anthus trivialis* disappeared from many, if not all of the same woodlands. By the end of the first decade of the 21st century, there were routinely no records of Wood Warbler in Gateshead at all.

Despite the fact that, in County Durham, many of its most favoured sites have been managed by conservation-minded organisations, for many years, this has not stopped its loss from many sites, where it was formerly a regular if not common breeding species. Today, other than a meagre handful of birds in the few remaining 'strongholds', the species can be a difficult one to see in Durham. Based upon breeding season records collected over the period 2008-2010, it seems that the county's breeding population may now be approaching as few as 10-15 pairs; the future looks most uncertain for this oak woodland songster.

Wood Warblers arrive in our area very consistently in the third week of April, with the first appearing usually two or three days either side of the 24th. A bird considered exceptionally early appeared in the Bedburn area of Hamsterley Forest on 18 April 2007, bettered only by one at the same locality on 17 April 2011. Over a thirty-year period, between 1976 and 2005, the reported first arrival date of Wood Warblers was analysed in ten-year groupings. These were found to be: 1976-1985, 30 April; 1986-1995, 24 April and 1996-2005, 23 April. It was found that at the end of the period the mean first arrival date in Durham was a week earlier than at the outset (Siggens 2005).

Recent non-breeding status

Sightings of Wood Warblers away from their breeding grounds have always been sporadic even when breeding numbers were more plentiful. There are still, very occasional records of birds in lowland areas in the county, though these tend to be few and far between. An example came in 2007, when three birds were noted at Old Quarrington on 9 July and 5 August, with several singles there on dates in between. In this year, further singles were near Trimdon on 10 July, at Hedleyhope Fell on 6 August and at Haverton Hole on 15 August. Birds of the year might be seen until the latter part of July and, very rarely, into early August, although adult birds are increasingly difficult to find in the county after mid-June.

Spring passage at the coast is infrequently recorded with many blank years and even when recorded, only small numbers of birds are involved. Examples of typical reports include singles often in song during May - at Cleadon in 1995, at Sunderland in 1997, two birds at Marsden and South Shields in 2005, on 3 May at Whitburn, and at Mere Knolls Cemetery in 2007. In 2001, a spring bird on a CBC plot at Kibblesworth was the first in 15 years.

Autumn passage is similarly uncommon, though somewhat more prevalent with sightings occurring on average less than every two years since the early 1990s, though three birds were at Hartlepool Headland on 14 August 1969 (Blick 2009). Birds appear to move south early – a very early bird was at Whitburn on 24 Jul 1992, with more typical reports relating to singles along the coast from early August or into early September and these are largely noted at the main coastal migration points.

Examples of sightings include:

- 1995: two at Marsden Hall on 8 Sep and singles at Whitburn on 9th and 14 September
- 2002: two to three at Hartlepool and Mere Knolls Cemetery in September
- 2004: singles were seen at Whitburn and Hartlepool in August
- 2008: an exceptional number of coastal sightings involving seven birds. Typical early movement occurred in the first week of Aug with singles from three coastal sites followed by further individuals later in the month with the final report from Hartlepool on 7 September.

Rarely do birds appear after late September so migrants at Whitburn Coastal Park on 16 October 2004 and 13 October 2009 were particularly noteworthy.

Since the number of sightings of passage birds has continued since the mid-1990s at a similar level, despite declines in the numbers of local breeding birds, it would seem that continental migrants are responsible for most of the reports along the Durham coast.

Distribution & movements
This species breeds from the Pyrenees to central Siberia, as far north as the sub-Arctic zone and south to southern Italy and the Balkans. They winter in Africa south of the Sahara, largely on the eastern side of the continent (BWP). In the late 1980s, some effort was expended in trying to ring Wood Warblers in the Derwent valley, but relatively few were caught. These included: five adults and four nestlings in the lower valley in spring 1987, single adults near Edmundbyers and in Thornley Woods in May 1989, and two broods of six young at Pont Burn Woods in June 1989. Unfortunately this small number of ringed birds generated no data on the origins or movements of the county's birds.

Chiffchaff (Common Chiffchaff)
Phylloscopus collybita

A very common breeding summer visitor and passage migrant, uncommon ir winter, though very much more common than in times past.

Historical review
At the end of the nineteenth century, in the Derwent valley, Robson considered it "*by no means common*", although he did find it nesting near Winlaton Mill (Robson 1896). A "*summer resident*", "*much less numerous and more local in its distribution than the willow warbler*" is how it was described by George Temperley in 1951. He said, it "*arrives early in April, sometimes even in March*", "*departs in September*". Temperley thought it had "*never, at any time, been a common species in Durham*" and he felt that the then replacement of deciduous woodlands with conifers had been detrimental to this species. He did say that it was most frequently found in the centre and south east of the county at locations such as: Raby Park, Barnard Castle, Bishop Auckland, Brancepeth Park and Harperley (Temperley 1951). He mentions the species as "*reported from Chopwell Woods.*" and as formerly occurring, occasionally, in Ravensworth Park, in the Team valley. In the south east of Durham, it was scarce during the 1950s, with increases occurring in the 1960s and 1970s (Stead 1964). Some early wintering records include:

- 1967: One at Hartlepool on 8 January
- 1968: One at the Starling roost in Hartlepool on 2 January with another in a Hartlepool garden on 14 December
- 1970: One thought to be this species was at Teesmouth on 20 December
- 1971: Single birds at Hartlepool on 4 November and 28 December were believed to be Chiffchaff

Recent breeding status
Today it can be found in most local woodlands, both coniferous and mixed, with the highest densities in mature broadleaved areas. It inhabits a range of woodlands, principally broad-leaved or mixed (Avery & Leslie 1990) although it does also occur, at lower densities, in pure conifer stands, where it may be the only warbler species present (Westerberg & Bowey 2000). Large populations of Chiffchaff are established in the species' county strongholds of the mid-valley woodlands of the Wear, the Derwent and the Tees. Although the species is found effectively across the county in appropriate habitats, from the wooded denes that run down to the coastal strip, through mature scrub and woodland habitats from the northern boundary to the abutting boundary with North Yorkshire. In 2007, the spring months saw birds spread across much of the county, with birds reported at over 80 sites during April and more than 65 sites during May. Territorial birds were seen as far west as Beldon Burn in the north and The Stang in the south, but the highest concentrations were in lowland woodlands and along the central stretches of the main river valleys. Two summers later, birds were noted as holding territory in moorland fringe

areas as far west as the western edges of Hamsterley Forest in 2009.

Most of the county's significant areas of woodlands or mature scrub, whether these are coastal or inland, support the species. Some three to four decades ago, it was considered very much a woodland bird but in recent years it has become increasingly widespread and common in sites that better fit the description of scrub rather than woodlands. At the time of the publication of the *Atlas* it was stated that this species was less widespread and less catholic in its habitat choice than Blackcap *Sylvia atricapilla*. There has probably been some extension in the range of this species locally and it is now found in many more scrubby habitats. Nonetheless, the highest densities still tend to occur in mature, broad-leaved in the river valleys of the county.

In spring 2010, birds were noted for the first time in suburban garden habitats in the lower Derwent valley, another extension of habitat for this increasing species, in the county. A strong sign of a healthy population is birds occupying isolated/fringe habitats such as in the Whitburn Coastal Park where fledged young have recently been noted, or birds singing in a small area of trees at Roker Marina; more surprisingly, a singing bird in a large suburban garden in South Shields with another apparently holding territory in a Cleadon garden (T. I. Mills *pers. obs.*).

National trends are well documented by the BTO, the national CBC results showed a large increase following the cold winter of 1962/63, although most winter in the Mediterranean. The Chiffchaff's population crashed in the late 1960s/early1970s, a fate shared by other trans-Saharan warblers (Marchant *et al.* 1990). Increases occurred through the 1980s (Marchant *et al.* 1990) and numbers have continued to increase. Long term CBC, BBS and CES data all illustrate this strong recovery (Mead 2000). BTO data also emphasises the importance of overwinter survival as a critical factor responsible for changes in abundance. Numbers of Chiffchaff have shown a widespread, moderate increase across Europe since 1980 (PECBMS) and in England there was a 32% increase in the five-year BBS index up to 2008 (Risely *et al.* 2010).

The national picture is mirrored by local survey work which indicates that Durham's populations were stable or rising slightly through the 1980s and early 1990s (Westerberg & Bowey 2000). There was estimated to be a 25% increase in the Cleveland area over the period 1981-1991 (Iceton 1999). By 2008, a measure of the high population levels of this species locally was indicated by the receipt of 1200 records for the county for that year. Breeding population increases were reported from several localities with Hetton Bogs holding 15 pairs (up from 11 in 2007) and at Rainton Meadows territories doubled to 10 compared to the previous year. This species' range extends towards the western county boundary with six singing at Hunstanworth and Baybridge.

Working on the assumption that County Durham was typical of Britain as a whole, and using national population figures as a baseline (Gibbons *et al.* 1993), the *Atlas* calculated the county population to be around 7,100 pairs (Westerberg & Bowey 2000). It would appear that levels have risen somewhat since then (Newsome 2005-2009). The survey work for the *Breeding Birds of Cleveland* indicated that there was a population of 300-350 pairs in the north Teesside area (Joynt *et al.* 2008).

Recent non-breeding status

Overwintering birds are now recorded annually, although these can be easily overlooked when associating with tit flocks. As an overwintering species it appears to be less numerous than Blackcaps, though this is sometimes difficult to judge, as the latter species is routinely noted at garden feeding stations, whereas the Chiffchaff appears to prefer more natural situations. Wintering birds are quite frequently noted in scrub habitats in and around sewage treatment works, where winter insect populations are relatively high. In some years, Far Pasture, adjacent to the Lockhaugh STW in the lower Derwent valley, has held a number of birds; up to three individuals being there together and seven winter birds being ringed at the site between 1989 and 1991; five of them in November and December 1990 with one still being present in the thickest snow of February 1991. The 2010 winter had what were considered at that time to be unprecedented numbers, with at least nine birds present in the Gateshead area alone.

The picture of overwintering in the county over the latter half of the first decade of the 21st century is indicated by the following statistics:

- ten sites held possible wintering birds in November and December 2005
- there were reports of at least 11 wintering birds in January 2006
- in 2007, January and February records came from Birtley STW, Cowpen Bewley Woodland Park, Drinkfield Marsh, Hetton Bogs, Long Drag and Sedgeletch, and from Brasside Ponds and Shibdon Pond in the last five weeks of the year – a rather small number of wintering sites compared with

recent years
- in the winter of 2008, wintering birds were noted at ten sites
- five were at Sled Lane, near Crawcrook, on 6 December 2009

The Chiffchaff is one of the county's earliest arriving summer migrants appearing from mid-March although the principal local arrival occurs during early April and passage birds can appear almost anywhere in the county during this month. First appearances are usually in their most favoured woodlands but a novel site for the first report of the year in early April 1974 was on a fishing boat eight miles off Sunderland, where eight birds alighted (Unwin 1975).

An example of an early arrival and build up was in 2007 when individuals first appeared from 7 March (the earliest date for the county), with birds seen or heard at a total of 13 sites between 7th and 15 March. One at Low Barns on 8 March was the earliest-ever arrival there by 14 days, while one at Langley Moor on 17th was the observer's second earliest-ever at this site. A rapid build up through March resulted in many localities having singing birds by the month's end. The last week in March 2008 saw an exceptional arrival with birds reported from 45 widespread sites across the county. As a spring passage migrant, relatively few are seen along the coast in most years with birds arriving directly onto breeding grounds, although the arrival period is quite protracted with birds at times, still appearing along the coast into late May.

By contrast, autumn passage normally involves considerably more birds and inland sites, such as Lockhaugh and Shibdon Pond, amongst many others, regularly attract scores of birds. For example, between July and September 2010, over 290 were ringed at Lockhaugh, indicating the numbers that move through such sites when breeding conditions have been good; more than 40 birds have been ringed in one morning at this site, as on 20 July 2004. During late August and early September, single birds not infrequently appear in suburban gardens, often in association with tit flocks. By mid-September however, many inland birds, principally the adults, have already moved south; the offspring of local breeding birds however, may still be present in late September and early October. The rate of exodus of birds of the year is indicated by counts from specific sites, for example in September 2007 the count of birds at Hetton Bogs decreased from 15 in the first half of September to just one or two birds by the last week.

During September and October passage along the coast is usually light with daily counts of one to three birds at coastal watch points, though parties of as many as 30-40 have been recorded from around Hartlepool Headland and Marsden Quarry in the right conditions. The Co-ordinated Coastal Migrant count on 8 October 1995 revealed 20 birds along the coastal strip, but as with many autumn passage migrants, onshore winds have a big impact in grounding birds at the coast. For example, in 1998 a substantial fall of up to 75 birds appeared in easterly winds on 1 October at Hartlepool and birds continued to arrive along the coast for several days with 17 at South Shields and 25 at Marsden Hall on 3rd and 20 at Hartlepool on 6th. In contrast, 2005 saw low numbers reported at the coast with only 18 birds reported in the peak arrival from 9th to 19 September. Similar numbers were recorded during the third week of October with passage petering out in early November.

Scandinavian Chiffchaff
Phylloscopus collybita abietinus

As well as the locally breeding race *collybita*, two other distinctive eastern races of Chiffchaff are reported, usually during late autumn and on occasions in winter. A few birds showing features of *P. c. abietinus*, from north and north east Scandinavia are recorded in most years, for example:
- 1948: probably the first report of a bird with the relevant characteristics of this race in Durham was identified by Dr. H. M. S. Blair in his garden at South Shields on the rather early date of 9 August 1948. The observer's wide experience of Scandinavian birds was well known and he described the species' characteristic shrill call-note *"like the cheeping of a domestic chicken in distress"*
- 1979: the bird noted at Marsden on 19th to 20 October was the first modern documentation of this race in the county
- 1990/1991: in November up to six birds showing characteristics of 'eastern' races were seen on 21st at Marsden and another 'eastern' Chiffchaff was singing an unusual song in the following spring at Shibdon Pond from 1st to 15 April 1991
- 1992: one to three birds were present during October at Cornthwaite Park, Whitburn

- 1995: two in November were at Far Pasture and Shibdon Pond, where two were seen in December
- 1996: one to two birds were at Whitburn on 12 November and one at South Shields on 27th, with another at Whitburn on 6 December.

Between 2000 and 2011, small numbers of birds ascribed to this race were seen at coastal locations in nine of the 12 years. It is clear that a significant proportion of late autumn migrant Chiffchaffs are of north eastern origin and assignable to this subspecies.

Siberian Chiffchaff
Phylloscopus collybita tristis

Birds showing characteristics of the rarer 'Siberian' race, which originates in and around the Urals, have been recorded on the following occasions:

1987	a striking individual was at Seaham Hall on 3 October
2004	one was seen at Whickham on 31 October and a bird at Sedgeletch, first reported in November, spent the winter there and was last seen on 16 March 2005
2005	one in Whitburn CP on 6 October
2008	one noted at Billingham Green between 29 November and 2 December
2009	a single bird was trapped and ringed at Whitburn Coastal Park on 21st to 22 October with another on 23rd in scrub opposite the Old Cemetery in Hartlepool.

The criteria for identification of this race has been in a state of flux for some time and following a review of previous claims in Durham, only birds fitting the currently applied national standard have been retained in the record.

Distribution & movements

Birds, in various racial forms, breed from southwest Europe through Europe and most of northern Asia to eastern Siberia. They inhabit almost the whole of the boreal and temperate forest area as far south as the Mediterranean and Black Seas. The nominate race occurs in the west of this range with other, not always easily distinguishable, races further to the east (BWP). Birds winter largely in Africa, though increasing numbers winter further north, some around the Mediterranean and increasingly, but still in relatively small numbers, in the British Isles. The origin of these latter birds is unclear, but they may come from more easterly locations in Europe as is the case for wintering Blackcaps (Parkin & Knox 2010).

Many Chiffchaffs have been ringed in County Durham and these have generated some information about the origins and destinations of the birds breeding and passing through the county. A number of juvenile birds ringed at Lockhaugh in the lower Derwent valley have been controlled on the south east or south coast of England (at Beachy Head and Essex). Another juvenile, ringed there on 19 August 2000, was recovered at Katwijk, De pan van Persijn, Zuid-Holland, Netherlands on 1 May 2001, when its remains were found in the pellet of a Long-eared Owl *Asio otus*. In a reverse of some of the above movements, an adult ringed at Sandwich Bay, Kent, on 24 July 1978 was controlled at Beamish Hall on 20 April 1980.

A number of 'foreign-ringed' Chiffchaffs have been recovered or controlled in the county, these include: a bird ringed on Sark, Channel Isles, on 1 May 1976 and controlled near Port Clarence, Teesside seven days later; one ringed in southern Sweden on 28 September 1996, controlled in Hargreaves Quarry on 13 April 1997 and a bird ringed in Pori, Finland, on 27 June 2004 and controlled at Hartlepool on 20 October 2004.

Unsurprisingly, it would appear that some birds passing along the Durham coast in autumn have been bred, and will return to breed in subsequent years, to countries to the north and east, as illustrated by the bird ringed at Hartlepool on 26 September 2000 and controlled at Vest-Agder, Norway on 4 April 2001 (Blick 2009).

Willow Warbler
Phylloscopus trochilus

A still abundant summer visitor, and passage migrant, which has undergone considerable declines in recent years.

Historical review

In the nineteenth century, this species was considered by Robson to be "*very common*" in the Derwent valley (Robson 1896) and no doubt this would have adequately described its status over much of Durham. Temperley (1951) commented that it was very occasionally met with in winter. This possibly refers to misidentified records of Chiffchaff *Phylloscopus collybita* but there is at least one genuine mid-winter record of a Willow Warbler in the county. This is one of the most unusual records of this species in Durham, referring to a bird found overwintering near Shibdon Pond that was shot on 16 January 1890, somewhere between Blaydon and the Scotswood Bridge; the specimen was sent to the NHSN collection at the Hancock Museum. At his time of writing, 1951, George Temperley described this species as a "*common summer resident*", "*the commonest of our warblers*", and "*the most widely distributed*". He said that it arrived in the second or third week of April remaining until August or September. In times past, certainly during the 1940s and 1950s, it was considered the most widespread of the warblers in the south east of Durham (Stead 1964).

Recent breeding status

This species, with its distinctive descending song, is perhaps the most obvious early spring migrant, and despite a steady decline in recent decades, there is little doubt that it is still our most common and widespread summer visitor (Westerberg & Bowey 2000). Breeding season records are usually widespread and come from virtually all parts of the county, which have scrub or suitable woodland habitats. Breeding birds can be found from locations as far west as Herdship Farm, high in Upper Teesdale, down to the coastal denes. Willow Warbler has probably the least demanding habitat requirements of any of our breeding warblers (Simms 1986). It utilises shrubberies and amenity planting, in the centre of towns, mature woodlands with dense under-storey (Westerberg & Bowey 2000) and the edges of conifer plantations high in the western dales. Even in areas of heavily maintained agricultural land, a few Willow Warblers maintain territories, alongside Whitethroats *Sylvia communis* in sparse hedgerows (Westerberg & Bowey 2000). After arrival, in many places, its song is the principal spring signature sound of the county's countryside. Individuals appear to be site faithful between years as shown in May 1984, when no fewer than 12 birds ringed in 1983 across the west Gateshead area were re-trapped in the same breeding territories; amazing that birds weighing between nine and 12 grams can travel thousands of miles to their wintering grounds and return to the same spot where they bred previously.

Typically, birds arrive in Durham from early April (Armstrong 1988-1993, Newsome 2007-2009) and quite rapidly fill up most available habitats. A typical pattern of arrival occurred in 2007, when the first at Hetton Bogs was noted on 3 April. This was the forerunner of a swift return to breeding areas in central Durham and by 10th of the month birds had been recorded at a further 10 localities and typical counts by the third week of April included 42 singing birds at Hetton Lyons Country Park, 37 around Hurworth Burn Reservoir, 31 at Hetton Bogs and 24 at Shotton Pond.

Breeding season survey work in Gateshead, over the period 1986/87, found the species in over 70% of all visited kilometre squares and, in suitable habitat, it was found to occur at very high densities. For example, 25 territories were found in around 18 hectares at Shibdon Pond in 1984 (Bowey *et al.* 1993). Since that time it has undergone some quite severe declines but it is still the most common warbler in the Teesside area, having an occupancy rate of over 70% in tetrads in the Cleveland atlas (Joynt *et al.* 2008), with notable populations along the coastal strip. It was estimated that there are probably around 700-900 pairs in the north Teesside area (Joynt *et al.* 2008). Birds tend to occur at lower densities in the oak woodlands of the upper dales, where woodland with sparse under-storey was traditionally more attractive to its congener, the Wood Warbler *Phylloscopus sibilatrix*. However, birds can still be found at such sites and more than any other of our breeding warblers, they are present on the 'high ground' of the west of the county (Westerberg & Bowey 2000).

The species is now classified by the BTO as category amber - the second highest level of concern in species conservation – on the strength of a 31% decline on CBC plots between 1974 and 1999. Willow Warbler abundance has shown regionally different trends within the UK (Morrison *et al.* 2010). In fact the latest BBS data for 1995 – 2008 highlights the north east of England as one of the worst hit areas with an alarming 24% fall in breeding numbers over the 14 year period. Interestingly figures for the same period show big variations in other parts of the UK – in Scotland there has been 16% increase over the period while southern England's fall is only 8%. A ray of hope comes in the form of an increase in 15% in breeding birds from 2008 to 2009. Detailed research indicates that these falls in numbers were due to poor adult survival and thus changes on migration or in Africa might have been to blame (Mead 2000). A further pressure on the species is a reduction in habitat quality on the breeding grounds (Fuller *et al.* 2005). The recent population decline is associated with a shallow decline in productivity as measured by CES and with a substantial increase in failure rates at the egg stage, which raises NRS concern (Leech & Barimore 2008). Illustrative of this trend, 82 birds, largely juveniles, were ringed at the Lockhaugh study site on 13 August 1988 and 191 during July of that year but two decades later, just 12 were caught there in August 2008; though catch numbers did increase to 107 in August 2009. There is a small but significant decrease in the number of fledglings per breeding attempt. Average laying dates have become a week earlier, perhaps in response to recent climatic warming (Crick & Sparks 1999). Numbers have shown widespread moderate decrease across Europe since 1980 (PECBMS 2010). From local Breeding Bird Survey data in 2006, the Willow Warbler was, despite large increases in the number of Chiffchaff in recent decades, still more widespread as a breeding bird in County Durham; 25 occupied BBS kilometre-squares held at least 163 singing birds.

A county population estimate of almost 28,000 pairs (Westerberg & Bowey 2000) was derived using nationally available figures (Gibbons *et al.* 1993), and working on the assumption that County Durham was 'typical' of Britain as a whole. Recent data however shows that the population levels within our area are not typical of UK levels. This level of population estimate indicated at that time that it was one of our most common breeding birds and our most common summer visitor. Despite well-documented declines, it probably remains so. An indication of the population range and occupancy of sites was provided by the results from the county's BBS sites for 2007, when a total of 231 singing birds were found in 24 of the forty one-kilometre squares. More specifically, in 2007, breeding data submitted from the Houghton and Hetton areas suggested that populations of this species remained more or less stable in comparison with the previous year. At Hetton Lyons CP, 32 pairs was down 'three pairs' on 2006 while at Hetton Bogs, 29 pairs was three up on 2006.

Interesting data was received on breeding during 1998 when nest-building was observed at Durham on 5 May and four nests discovered at Hetton each had three eggs or three young. An unusual nest site was in a potted shrub growing in a tunnel of garden netting, with a late nest on 14 July at Cotherstone in ivy *Hedera helix* at the side of a garage. A bird exhibiting mixed song patterns of both this species and Chiffchaff was at Barmston in spring 2009, and such birds have been noted on occasions in the past, notably at Church Dene, Ryton in the spring of 1987.

Recent non-breeding status

Spring usually sees very little coastal passage of this species, with most birds heading straight into breeding territories. Over a thirty-year period between 1976 and 2005, the reported first arrival dates of Willow Warblers were analysed in ten-year groupings and the mean dates for these were: 1976-1985, 7 April; 1986-1995, 5 April and 1996-2005, 3 April; an advance of four days over the period. Arrival dates in the latter part of the first decade of the 21st century were, on average, even earlier still, with more first arrival dates occurring in the last week of March. The earliest ever spring migrant in the county was at Hamsterley on 4 March 1972.

Widespread reports of roaming juveniles usually come after mid-June and many birds are loosely associated with late summer tit flocks. Following the breeding season in 1992, good numbers were trapped for ringing with 33 at Ebchester on 28 June and 34 at Far Pasture on 26 July. Further evidence of a good breeding season came with reports in July of up to 44 at Hetton, at least 40 at Bishop Auckland, 36 at Joe's Pond and *c.*35 near Darlington. While in 1995, further census work in the Hetton area gave an indication of breeding season populations with July numbers of *c.*60 at Hetton Lyons, *c.*75 at Hetton Bogs and *c.*45 at Joe's Pond. Regular ringing at Far Pasture resulted in a total of 109 birds caught from May to Aug, with 88% juveniles. Comparative post breeding season counts from recent years show numbers of 50 at Sedgeletch, 55 at Hetton Bogs and up to 60 at Rainton Meadows in 2008.

Late summer usually brings a general dispersal from breeding areas and birds start trickling through coastal sites in the first half of August. During late August and the first two weeks of September greater numbers are noted along the coast with the extent of records very much weather dependent. Good numbers of passage birds were observed along the coast on 17 September 1960, when many birds were counted between Crimdon Dene and Teesmouth (Blick 2009). Some other notable arrivals over recent decades include:

- 1979: on 17th to 19 August 81 birds were counted along the coastline
- 1980: heavy rain and strong northerly winds at midday on 30 August grounded c.30 birds at Marsden Hall and c.50 at Hartlepool Headland
- 1995: this species formed the bulk of a large fall of drift migrants on 8 September with spectacular numbers noted. Estimated totals of birds included 100 at Marsden and 80 in the Whitburn Coastal Park
- 2008: there was marked coastal movement on 6 September with 40 between Hawthorn and Nose's Point, and over 90 birds between Trow Quarry and Seaburn, including 30 in Marsden Quarry.

Most birds have moved south by mid-September but birds occasionally linger into October, for example the three that were at Mere Knolls Cemetery on 24 October 1981, while an individual was at South Shields on 17 October 1999 and two were there on 26 October 1985. November sightings are rare with just three known records - one was at Mere Knolls Cemetery, Seaburn on 6 November 2004; one or two extremely late birds were at Whitburn in 1990 with the final report on 11 November, and the latest ever was at South Shields on 12 November 2000.

Northern Willow Warbler
Phylloscopus trochilus acredula

A bird showing characteristics of the paler north eastern race, which breeds in northern Scandinavia, was found dead at South Shields on 12 October 1974. Further recent examples were noted at Marsden Quarry on 26 October 2011 and Trow Quarry on 9 November 2011. This subtle subspecies is surely overlooked and a proportion of migrant Willow Warblers in both late spring and late autumn are likely to involve this race. A bird considered showing some characteristics of the eastern race *P. t. yakutensis* was reported at Mere Knolls Cemetery, Seaburn on the very late date of 6 November 2004. However, this subspecies has never been proven to occur to Britain.

Distribution & movements
From the Bering Straits in the Far East, through Siberia and west as far as Western Europe, the Willow Warbler is found as a breeding bird across an extensive range that reaches as far north as southern Sweden and south to southern France, though it is not common around the Mediterranean. Birds winter in Africa, south of the Sahara, eastern populations travelling huge distances of over 10,000km to reach their wintering grounds (BWP).

Ringing has shown that birds from breeding sites further north, use sites in Durham as staging posts on their southerly migration. For example, one ringed at Fenwick, Northumberland on 10 August 1957 was recovered at Ryton a month later, where it died having flown into a house on 10 September. This is further illustrated by the fact that all birds ringed in Gateshead and subsequently caught on autumn migration were trapped due south of their original capture point. The furthest of these was a bird ringed at Shibdon Pond on 27 July 1989 and later controlled at Le Chaume, Vendee, in France on 24 August 1989, on the coast of the Bay of Biscay. This direction of movement was emphasised by the information of a bird ringed as a nestling at Barnard Castle on 15 June 1986 that was controlled near Bruges, Belgium on 26 April 1987.

Reciprocal movements have also been recorded with birds ringed in southern England; one ringed at Wansford, Northamptonshire, was controlled at Lockhaugh, whilst a first-year bird ringed at Steyning, Sussex on 30 August 1978 was controlled at Marsden Hall, 457 km to the north, on 30 August 1979, exactly one year late, but with a journey to Africa in between captures. Also, a juvenile ringed at Marston Sewage Farm, Grantham, in Lincolnshire on 20 August 1998 was controlled on 11 September 1999 at Wolsingham, 213km to the north, it had presumably been breeding somewhere in the north east.

Birds ringed overseas and caught locally, include a juvenile that had been ringed at Eigersund, in Rogaland, Norway, on 4 August 2001, controlled at Hartlepool 15 days later plus an adult ringed on the island of Jersey on 28 April 1980 that was controlled at Neasham on 18 May 1980 and again the following year, on 23 May; indicating that it was breeding in the area.

Cross-country movements, in opposite directions, undertaken by young birds presumably before the full migratory urge took over, were shown by the juvenile ringed at Glencaple, Dumfries & Galloway, on 4 September 2002, which was at Billingham Beck four days later and a similarly aged bird, ringed at Shibdon Pond on 24 July

1976 that was killed by a cat at Rossendale, Lancashire just over a month later. The speed at which spring migrating birds move was indicated by the adult ringed at Hartlepool on 16 May 2006 that was 148km to the north at St. Abb's Head in the Borders the next day.

Blackcap (Eurasian Blackcap)
Sylvia atricapilla

Sponsored by
Wm. C. Harvey Limited

A common breeding summer visitor and passage migrant.
Also a scarce, but increasingly common winter visitor.

Historical review

This is the most widespread *Sylvia* warbler in the county. It breeds mainly in woodland or dense scrub, largely of a deciduous nature, with a well-developed scrub layer but may also be found in mature hedgerows and today, in areas of scrub. Whilst this warbler is mainly a summer visitor, it is regularly recorded in small numbers during the months November-March, most usually in suburban gardens.

In the mid-20th century, this species was thought "*less plentiful*" than the Garden Warbler *Sylvia borin* (Temperley 1951) but this is certainly not the case today. It is more widespread than that species, being present in over a third of surveyed squares 1988-1993 and it is also present in greater numbers than that species. It can be found nesting in a wide range of woodlands, parks and large gardens with dense shrub layers.

George Temperley (1951) described it as a "*summer resident*", "*well distributed and not uncommon*". He said that it arrived in the last week of April or early in May and that a few were occasionally recorded as wintering (Temperley 1951), an example of this being the male that was trapped and ringed on 9 March 1950 in a Westoe Garden. One was at the same place in November 1950. The year previously, wintering Blackcaps were seen in Northumberland and Durham in December 1949 and on 9 March, 12 November and 4 December 1950. Wintering birds had however, been recorded in Northumberland as early as 1909, when a bird was noted at Powburn, Northumberland (Bolam 1912). However, it was 1948 before regular wintering was recorded to the north of Durham (Kerr 2001). In fact this habit in Durham had been documented in Durham in the early Victorian period. As stated by John Hancock, who wrote: "'*Mr. Dale, of Brancepeth, Durham shot a male on the 5th December,1848, in his garden, when it was feeding on the berries of the privet; and about two years afterwards the same gentleman killed a female, likewise in December, and near the same place*" (Hancock 1874). This suggests that Blackcaps have been wintering in the county for some considerable time and poses the question as to what extent the increase in wintering numbers is related to the increase in observer awareness and the practice of feeding birds in gardens?

The developing habit of wintering in Durham was ever more obvious through the 1960s. Records of late or wintering birds comprised: a male in a Stockton garden on 16 January 1966, one at Rowlands Gill from 24th to 27 December 1966, singles in 1967 at Hartlepool on 9 November and Castle Eden Dene on 10 November with another at Hartlepool in a Starling roost on 31 December. One was seen in garden in Chester-le-Street on 22 December 1968 by Judith Dunn. She had heard of them overwintering in the south and west of England but considered "*the sighting of an individual in this locality at this time of year must be exceptional*" (*The Vasculum* Vol. LIII, 1968). This species was said to have a rather local distribution in the coastal plain in the south east of the county (Stead 1964), with scattered pairs breeding in the more extensively wooded areas, in locations such as Wynyard.

Recent breeding status

In County Durham, as in the British Isles more generally, the species has increased in its range and number in the last 50 years or so (Mead 2000, Parkin & Knox 2010); this is now one of the most common and obvious of our summer visitors. The species increased by 51% in the period 1995-2008 across England (BBS), though the north east of England saw a much smaller increase of 32% in that time (Risely *et al.* 2010). In 2010, 202 individuals were ringed in Gateshead, the highest total in this area and this probably reflects both a bumper year for the species in the UK and the relatively high population levels of this species locally, compared to three decades ago.

The *Atlas* map gave a broadly accurate and still pertinent, picture of the species' distribution across the county, although that publication may have underestimated the total number of tetrads occupied by the species. There has been a large increase in numbers between the publication of the two national breeding atlases, with active colonisation of parts of northern England and Scotland over this period (Parkin & Knox 2010).

The early spring birds appear from late March, with the main arrival of migrants occurring through April, and these are consolidated by the arrival of birds, in more marginal habitats, in May (Westerberg & Bowey 2000). By the middle of that month Blackcaps are to be seen and heard across much of the county in suitable breeding habitat , from small wooded valleys in the west, to the upper reaches of the Rivers Derwent, Wear, Tees and Lune, down through woodland and scrub areas in central Durham to the coastal denes between Seaham and Crimdon. The favoured breeding habitat of the species in the county appears to be mature woodland sites, the species optimal local habitat, which tend to attract the first arriving males with later birds tending to 'fill in' a wide range of mature and less mature areas of scrub (Simms 1985), though these latter habitats have been increasingly utilised in recent decades.

The species is now a frequent component of the county's urban bird life, routinely holding territories in parkland, where this has dense under-planting, in amenity landscape planting and even large mature gardens. An interesting breeding report concerned three pairs in suburban Darlington in the summer of 1974, which fed young on blackcurrants from large gardens. All of the county's woodlands, from the coastal denes to the dales' oak woodlands are tenanted by Blackcap, though densities of birds to be found in upland-fringe areas are much lower than in the east and valley woodlands. Nonetheless, in 2008 birds were recorded at Derwent Reservoir and Eggleston in the west, over to the western limit of High Force, and up to six birds were also noted at The Stang in the far south of the county. In 1983, twelve pairs were counted in two Derwent Valley woodlands with six pairs in the Hetton area, where counts of up to 38 birds were made following the 1986 breeding season. There are relatively few breeding birds around Teesmouth with most birds in this part of the county concentrated to the north and west of Stockton. There may be as many as 200-250 pairs in the northern part of what was the northern portion of Cleveland (Joynt *et al.* 2008). Populations in the well-studied Houghton/Hetton area in 2007 were very similar to 2006, with 12 pairs at Hetton Bogs being an increase of one pair over the former year, while eight pairs at Hetton Lyons CP was a decrease of one territory; suggesting a largely stable situation in at least this part of the county during the first decade of the 21st century.

The species is less disposed to coniferous woodlands and may avoid pure stands of conifers if under storey or thicket stages of woodland are not well represented. As the county's woodlands are concentrated along the river valleys, these consequently, form linear strongholds for the species. Assuming Durham conforms to Britain as a whole, using national figures (Gibbons *et al.* 1993) a county population estimate of almost 6,500 pairs can be calculated (Westerberg & Bowey 2000). Blackcap was recorded from 61% of the 3,243 of surveyed BBS squares across the UK in 2009 (Risely *et al.* 2010), making it one of the most widespread of British warbler species.

Recent non-breeding status

Very late March or the first week of April usually brings the first genuine spring arrivals to Durham with birds most often being noted in lowland breeding locations before any are noted along the coastal strip. One on board a fishing boat off Sunderland on 19 April 1975 was an interesting record of migration in action. Comparative studies have shown that the arrival dates of spring migrants have become earlier since the mid-1970s (Brown & Grice 2005). The local picture is somewhat clouded by the presence of late wintering birds, which throw into doubt the earliest arrival dates. An obvious arrival of spring migrants usually occurs in many areas during the first ten days of April, with spring passage peaking in May. The coastal denes usually have singing birds by early May, the lowland woodlands slowly fill with birds during the first fortnight, but some sites in upper Teesdale may not be occupied until the second half of May. By the end of the month sightings come from many smaller woodlands even in the far west of the county. Small numbers of coastal migrants in May are probably bound for locations further north, and many migrants probably pass through the area but it is probably only along the county's coastal strip that these are readily distinguishable from locally breeding birds.

In the post-breeding phase of the summer, many birds, especially juveniles, get caught up in roving tit flocks, which seem to disperse randomly from their natal territories, as indicated by local ringing studies. Up to 37 juveniles were noted feeding on raspberries at Hetton Bogs at the end of July 2004 in just such circumstances. Regular ringing at Far Pasture from May-Aug 1995 showed that 81% of the 47 birds ringed were juveniles.

In general terms, the number of reports from inland locations becomes increasingly less through August and into early September; this trend accelerates during September. For example, in 2008 between 15 and 20 birds were noted as being present at Rainton Meadows at the start of the month, but this had declined to just one or two by the final week of the month. This exodus often coincides with an increase in sightings at coastal sites in September, which, presumably is bolstered by the arrival of some continental birds, but perhaps also birds from further north in the British Isles. A few local birds might still be present in favoured areas into early October but most have left by late September. Much larger numbers of birds are recorded in coastal areas during September and October.

As a coastal passage migrant, the Blackcap is more numerous in autumn than spring, and it is quite often the most common warbler recorded in the county during the autumn. October tends to be the peak month for migration through the county, with the number of birds often rising to double-figures in the well-watched areas of Hartlepool and Whitburn/Marsden. A good autumn passage in October 1976 was illustrated by the monthly total of 40 birds caught and ringed in Marsden Quarry. Thirty-two were ringed at Marsden Hall in October 1979, many staying around for over a week to feed on elderberries, a favourite autumn food. A total of 43 were ringed at Marsden Quarry through October 1987, which is indicative of the scale of passage through that site.

This was the most common warbler noted on the Durham Bird Club's Coordinated Migrant Counts between 1992 and 1996. A maximum of 59 birds were noted in the coastal strip, between the mouth of the River Tyne and Hartlepool Headland, on the 1992 Coordinated Migrant Count on 4 October; with 47 noted the following year, of which 45 were between the mouths of the Tyne and the Wear.

The weather and wind direction in particular, has a strong bearing on the number of Blackcaps grounded on the Durham coast in autumn. For example, persistent westerly winds through that season in 1983 led to just three records for the whole month of October, whereas 50 were counted between Sunderland and Marsden on 11 October 1982 during a period of strong onshore winds. Other notable one-day counts recorded include more than 18 at Marsden Quarry on 5 September 1981 and 22 at Whitburn on 13 November 1984, in a year that was notable for late migrants, both common and rare, in the north east of England.

Small numbers of birds are quite often still being reported at coastal sites in late October or early November; such late birds are believed to originate from northern Europe as they often arrive alongside migrating northern thrushes. It is uncertain whether these are newly-arrived birds from the east, set to stay in Britain for the winter, or the last of northern European and British breeding stock, feeding up before heading south to Europe for the winter. These birds are often seen in areas rich in elder berries with some appearing in gardens, where they often visit feeding stations. Brown and Grice (2005) highlighted the fact that this species was now occurring more often as a late autumn migrant compared to in the past and that nationally, the number of wintering birds rose hugely from just an average of 18 per annum in the period 1945-1954 and as high as 3,000 per annum by the early 1980s. This growth was certainly reflected in the numbers wintering in County Durham. Most wintering birds in the county are noted at bird tables, or other feeding devices.

Prior to the 1960s, the Blackcap was rare as a wintering bird in England (Parkin & Knox 2010), though the habit had been recorded some time before this in Durham (Temperley 1951). Through the 1970s the species' winter status in the county changed from an intermittent, rare winter visitor to something of a scarce but increasingly regular presence between November and March; starting with a male eating bread on a South Shields lawn on 5 December 1971 and winter records in almost every year of that decade. In 1974, there were at least three wintering birds, from Billingham in the south to Ryton in the north. The winter of 1974/1975 was particularly good one, with at least 10 wintering birds in the county, the first time numbers reached double-figures. Most of these records came, and still do, from garden feeding stations. It seems likely that the increasing popularity of this activity has been a significant factor in helping birds winter in the north east of England and the rest of the country.

It is believed that these wintering birds largely consist of birds from Central Europe, developing a new migratory route, west as opposed to south, bringing them to the milder oceanic climates of the British Isles. Ringing studies have indicated that most of the Blackcaps wintering in the British Isles originate in southern Germany and northern Austria (Berthold 1995).

During the late 1970s and 1980s, an average of about five to eight birds per annum were reported each winter but in hard weather, as experienced in the winter of 1976, none were recorded. By contrast, during the mild winter of 1985/1986, about 15 different individuals were noted at sites scattered around the county. An indication of the number of Blackcaps, which can be present, was given by the seven individuals counted between Ryton and

Blaydon in early 1975, with another two at Hamsterley Mill and one at Witton-le-Wear in March. In 1987, up to a dozen birds were found wintering around the county over the first few months of the year, with up to six birds during March in the Lockhaugh area of Gateshead alone. The winter of 1989/90 saw at least 11 different birds present in the west of Gateshead (Bowey *et al.* 1993) in just one small part of the north west of the county. In similar fashion, at least five different birds were known to be wintering in an area of about three square kilometres in the Blaydon area, during the winter of 2000/2001. More widely, in 2007, wintering birds were noted at 15 locations between the start of year and the second week of March. In both 2008 and 2009, reports of wintering birds came from ten localities, all in suburban localities in lowland Durham. Most overwintering records are from lowland or coastal locations, but one was discovered relatively high up Teesdale, at Barnard Castle in January 2008, though this was trumped by a bird even further west and at higher elevation that visited a garden feeder in Baldersdale from 11th to 28 December 1975.

Between 1989 and 1998, 13 wintering birds were ringed in west Gateshead, in the months of November and December, at just four sites, Thornley Woods, Lockhaugh, Whickham and Shibdon Pond; four birds were ringed at Lockhaugh on 12 November 1989. The spread of birds at these sites over this period indicates that wintering birds are probably under-recorded and quite widespread at inland locations in the county.

Distribution & movements

The Blackcap breeds in Europe, Western Siberia, North Africa and south west Asia. Most populations are highly migratory. The north westerly populations winter around the Mediterranean within the breeding range of the more southerly breeding birds, but many populations winter in sub-Saharan Africa. There is growing evidence that this species' wintering areas, and their migration routes to these, are changing rapidly, possibly in response to changes in climatic conditions (Parkin & Knox 2010).

Many Blackcaps have been ringed in Durham over the decades and the recoveries and control of these indicate that the county acts as staging post for birds from further north on their journey south. For example one ringed at Vest-Agder, Norway on 12 September 2001 was at Hartlepool on 27 September 2001. Coastal ringed birds from Durham have been later found to the south, in France and Spain, such as one ringed at Seaton Carew, 12 October 1982 and recovered in Belgium on the 16 October 1982. The fact that birds reach such continental locations by staging at points further south in England is illustrated by some of the birds caught in Durham, for example: a juvenile ringed at Lockhaugh STW, on 26 June 1999 and controlled at Icklesham Sussex on 2 October 2000; one ringed on the Long Drag 21 September 2002 and recovered at Worplesdon, Surrey 3 October 2003; one caught at Saltholme on 3 September 1989 and controlled at St Albans Head, Dorset on 25 September 1989; another ringed at Seaton Carew on 19 September 2009 and then controlled at Portland Bill, Dorset on 1 October 2000, presumably preparing to cross the English Channel.

Sometimes locally breeding birds have been ringed at locations further south before being caught in the north east, such as: a breeding bird at Shibdon Pond in 1992 that had been ringed the previous autumn on passage at Sandwich Bay Bird Observatory, in Kent; one that was ringed at Saltholme on 28 September 1999 and found at Gosforth Park, Tyne & Wear, on 16 July 2000; likewise the male that was ringed at Lemsford, Hertfordshire on 17 September 2005 and found breeding at Low Barns on 2 April 2007, with another breeding at the same site in spring 2006, which had been ringed near Ripon, North Yorkshire on 22 July 2005. Alternatively, birds ringed locally in their breeding or natal territories have been recovered far to the south, such as the juvenile male ringed at Ravensworth on 26 August 1996 that was found dead, after flying into a window, at St. Peter, Guernsey, Channel Islands on 14 April 1998. Even more spectacular, in terms of the distance travelled, was the adult ringed at Ravensworth on 28 August 1995, which was found dead three-and-a-half years later, 2067km to the south at Alger, in Algeria, North Africa, on 18 February 1999.

Of real interest was the young male bird ringed at Monkwearmouth, Sunderland on 27 March 1986, which was controlled at Linthorpe, Middlesbrough on 23 December 1987, presumably a bird moving its wintering station to a location slightly further south in a subsequent winter; where had it been during the breeding season?

Garden Warbler
Sylvia borin

A common breeding summer visitor and passage migrant.

Historical review

The Garden Warbler is found throughout the county in mature scrubby areas. In a number of ways it broadly matches the Blackcap *Sylvia atricapilla* in its choice of habitat and the distribution of the two species is very similar. Its evenly-flowing warbling song emanating from dense cover is often the first indication of its presence and it is more often heard than seen, and as such it can be a relatively unobtrusive species though the song is both loud and obvious, if not always readily separable from that of the Blackcap.

Temperley (1951), provided little detailed commentary on the status of this species and almost no information from previous documentors of the county's bird life, though he did say that it was, "*A summer resident; well distributed and common in woodlands, dense thickets and gardens in the lowlands*", and that it arrives in the last week of April or early May. In 1892, the species was found nesting at both Gibside and Spen Banks and according to Robson they were "*somewhat common*" in the breeding season (Robson, 1896). Today they are rather more restricted in their distribution in the county than both the Blackcap and Whitethroat *Sylvia communis*, seeming to occupy a habitat niche somewhere between that of those species.

Recent breeding status

Garden Warblers can be found in tall scrub, scrubby woodland and along woodland edge, tending to avoid more mature patches of woodland, which it tends to leave to the Blackcap. They usually prefer habitats that have a dense under storey (Simms 1985) that facilitates their unobtrusive nature. They are one of the last of our breeding warblers to arrive in the spring, often not appearing in numbers, until well into May (Westerberg & Bowey 2000). It occupies a more transitional habitat, somewhere between woodland and scrub, than those preferred by the Lesser Whitethroat *Sylvia curruca* and the Blackcap, although there can be significant overlap between these species, and in some instances all three can be found breeding alongside each other, for example at Shibdon Pond (Westerberg & Bowey 2000). In farmed areas, Garden Warblers often choose overgrown copses of dense hawthorn *Crataegus monogyna* or blackthorn *Prunus spinosa* scrub for their territories (Gibbons *et al.* 1993), and while it does not wholly avoid commercial forestry and shelter-belts, it is much less often found in pure coniferous areas than is the Blackcap.

The *Atlas* showed their principal distribution in Durham to be concentrated around the lower to mid-altitudes of the county, with major gaps in distribution in the west, on the highest ground. The species is well represented in wooded habitats around the centre and north west of the county, and along the river valleys, where much of the county's woodland can still be found. At the coast, it is relatively poorly represented even where suitable habitat is available, exceptions being in the coastal denes, which attract good numbers of the species. In the west of the county, Garden Warblers are largely confined to the river valleys, although they penetrate further up these than some other warblers, for example Whitethroat. Breeding Garden Warblers are routinely reported from sites as far west as Stanhope Dene and Wolsingham in the Wear valley, and Muggleswick in the Derwent valley. In some studies in the county, population densities have been found to be high, for example ten pairs in 150 acres of scrub and woodland habitats in the Derwent Walk Country Park during the early 1990s (Bowey *et al.* 1993). More recent survey work has revealed that birds were in territory in nearly 81 kilometre squares across the county in 2009, and emphasised the species' breeding season distribution around the wooded areas of central Durham. Small numbers of birds are present in the western upland fringes, with some territorial birds found in the Beldon Burn and around Derwent Reservoir in the north, Snaisgill in Teesdale and at The Stang in the south. Woodlands around Hamsterley Forest, and Tunstall Reservoir also held a small number of birds. In the south east of the county, most breeding birds are around the Wynyard and Crookfoot areas (Joynt *et al.* 2008). In this part of Durham, this species is thought to be much less common than Blackcap; there may be as few as 30-40 pairs in what was the northern part of Cleveland and almost none around the coastal plain or in the Teesmouth area (Joynt *et al.* 2008).

The breeding grounds are not usually fully occupied until after mid-May, which is the peak month for activity for this species. In 2008, reports came from around 60 potential breeding sites, with up to four singing males being recorded at over 30 localities in 2007; these were mainly located in the central and eastern portions of Durham but birds were noted as far west as Muggleswick. Despite being relatively widespread, counts of singing males from individual sites are usually small and the majority of records from observers tend to be of single singing birds. A decline in the species' singing activity through the summer months usually leads to few reports of the species in the county after mid-June; in 2007 there were only three reports of single birds during July.

Westerberg & Bowey (2000) felt that the pattern of distribution of this species, as mapped in the *Atlas*, did not give a fully complete picture of the species' true range across the county. This was demonstrated by the fact the species was much better represented in the areas that were known to have been intensively surveyed, i.e. those to the north of Durham City and in the north west of the county. These appeared to be more heavily populated than other, less intensively surveyed areas with similar habitat. The mapped pattern in the *Atlas*, it was felt, indicated that at least some of the gaps in distribution were due to a lack of recording, as opposed to a genuine absence of the species (Westerberg & Bowey 2000).

Nationally, in the late 1980s and early 1990s, there was some recovery in numbers after a recorded decline during the 1970s (Marchant *et al.* 1990). Moreover, this seems to have been evident locally, with the numbers of breeding birds increasing through the 1990s and the 2000s. However, the national data from the Breeding Bird Survey showed a species decline of 13% over the period 1995-2008 in the UK but a greater decline, of 19%, in England; the only place that this species has increased over the last two decades being in south west England (Risely *et al.* 2010).

Using an extrapolation to all occupied tetrads of the county, of the highest breeding densities, suggested that there may be up to 3,100 pairs present locally. Westerberg & Bowey, (2000), offered up a slightly more conservative population estimate of somewhere between 2,000 and 3,000. This probably still adequately encompasses the breeding population of this species in the county.

Recent non-breeding status

Arriving distinctly later than the Blackcap and most of the other summer visitors, this species tends to filter away relatively early in the autumn, with few being seen locally after late August or the first week of September. The first arriving migrants might be noted in the last few days of April but the main influx is usually in early May. In most years birds are noted on the breeding grounds before any passage birds are seen along the coastal strip. For example, in 2007 the first recorded was at Hamsterley Forest on 18 April, typical in terms of location, but perhaps a little earlier than usual. Subsequently, there were no further reports until the final week of April, when an obvious arrival resulted in birds at 15 localities between 24th and 30th of the month. In 2008 and 2009, the first singing birds were noted on 13th and 14 April respectively, in both instances at inland sites and preceding a more general arrival by six days or more.

The earliest ever recorded arrival dates for Garden Warbler in Durham are 10 April 1968 at Hartlepool (Blick 2009) and 11 April 1972 at Low Barns in the Wear valley. Over the years there have been only a relatively small number of records of birds occurring before 20 April, including singles at Houghton on 18 April 1984 and at Finchale on 16 April 1993. The reported first arrival dates of Garden Warblers were analysed, in ten-year groupings between 1976 and 2005. The mean arrival dates for these were: 1976-1985, 4 May; 1986-1995, 26 April; and, 1996-2005, 25 April; a nine day change over the period (Siggens 2005). By 2006-2009 this had advanced further to around 15th to 16 April.

Birds were at one time routinely noted along the coast in late May, but in recent decades coastal arrivals have been more and more limited in spring. Consequently, six birds at Hartlepool Headland, five at North Gare and three at Whitburn Country Park on 16 May 2009, was considered one of the best spring passages on the Durham coast in many years. By late July, inland reports have started decreasing, largely because birds are not singing and many observations at this time of the year are of birds in mixed tit flocks with a range of other warblers and woodland species. It leaves the county earlier than the Blackcap, birds appearing on the coast from mid-August. Coastal passage peaks in late August, most birds have left the county by mid- to late September but movements continue into October; arrivals of migrants in the county in early October would usually include small numbers of Garden Warblers.

During autumn migration this species never approaches the number of Blackcaps occurring as passage migrants; there are relatively few times when double-figure counts have been recorded and in many years the total autumn coastal sightings would not exceed 15-20 birds. Nonetheless, traditionally, the greatest numbers of Garden Warblers have been observed on passage during autumn, and these have mainly occurred amongst falls of passerines along the coast between mid-August and early September. One of the most notable of these occurred in wet easterly weather on 26 August 1986, when 30 Garden Warblers were found along a mile stretch of coastline between Marsden and Whitburn; presumably just a fraction of the number which must have been present along the whole of the coastline from the River Tyne to Tees. A fall of migrants on 8 September 1995 resulted in 20 birds in the Marsden-Whitburn area including ten in Whitburn Coastal Park. Another good autumn showing occurred on 20 Sep 2000 with 14 birds at South Shields, including eight in Trow Quarry. In recent years however, coastal autumn migrants have become very much scarcer than in the mid- to later part of the 20th century. Hence the influx of migrants which brought 40 birds to Hartlepool Headland, ten at Seaton Common and one to three birds at another six coastal sites further north in the county between 9th and 13 September 2002 was of note. By contrast, and more typical of recent times, was the half-a-dozen birds noted between 8 September and 14 October 2007 at sites between Marsden Quarry and Hartlepool Headland. That the right conditions are vital to the appearance of coastal passage birds was shown in 2008. In this year, on-shore winds and poor visibility on 6 September brought a flurry of coastal records including more than 20 birds scattered around Hartlepool Headland and eight at Whitburn Country Park as well as another 13 birds, which were distributed around north east coastal locations.

The last birds of the year are usually noted in early to mid-October; a late bird was at Marsden Quarry on 27 October 1980 and in 1970 one was at Beamish Wood, near Stanley up to 26 October, a very late record indeed for an inland bird. An exceptionally late record was at Hartlepool Headland on 11 November 2000. But this date was exceeded in 1984, a year that was exceptional for late migrants, when the latest record for the county occurred at Whitburn on 13 November.

Distribution & movements

Garden Warblers breed in western, northern and central Europe as far east as western Siberia. Birds winter in sub-Saharan Africa (Parkin & Knox 2010).

Many Garden Warblers have been ringed in the county, though these have generated relatively few interesting recoveries or controls. A juvenile ringed at Shibdon Pond on 19 July 1989, was killed against a window at Eston, Middlesbrough just over a fortnight later on 4 August 1989. It was presumably making its gradual way south through England, but it had only travelled 56km before it met its untimely end. A juvenile, ringed at Gosforth Park Lake, Newcastle upon Tyne, on 20 July 2003 was controlled as a breeding female at Low Barns, Witton-le-Wear, on 28 May 2006, 40km to the south west. Of more significance was the record of one that was ringed at Telemark, Norway on 17 August 2008 and which was caught at Hartlepool Headland four weeks later.

The most extraordinary ringing recovery of this species in Durham, concerns the bird ringed at Spurn, Humberside on 9 October 1974, which was discovered dead at Castle Eden Dene on 2 January 1975. This was only the second British January record of a species normally wintering in tropical Africa; it had travelled over 55 kilometres north west after ringing, presumably in the wrong direction. During the field work for the *Winter Atlas* (Lack 1986), there were just three records of wintering Garden Warblers in England.

Barred Warbler
Sylvia nisoria

A very scarce migrant from eastern Europe, recorded from August to November.

Historical Review

The first record of Barred Warbler for County Durham came surprisingly late, the bird being found by the late Phil Stead at Hartlepool Headland on 21 September 1960. A huge fall over 16th to 17 September deposited large numbers of migrants along the east coast of England and it seems likely that this bird had arrived earlier and remained undetected. Why the first record for County Durham came so late is puzzling as it was first recorded in Northumberland in 1913, and was virtually annual there during the late 1950s (Kerr 2001).

The next in Durham was at Teesmouth on 3 September 1964, and the species became a virtually annual visitor during the late 1960s, with a further six recorded, though it did not occur in either 1966 or 1969. The best year was 1968 with three records, all singles at Hartlepool Headland on 25th to 26 August, 21st to 22 September and a very late bird on 2 November.

Just one bird was trapped and ringed during the period, an adult at Graythorp on 17 September 1967. The vast majority of Barred Warblers occurring in Britain are first-year birds; adults are extremely rare with only a handful of records and this remains the only one recorded in the county despite over 100 records to date.

Recent Status

A total of 101 Barred Warblers have been recorded in County Durham since 1970, all of which have been in autumn between August and November but with more than half of these in September. There have been seven blank years during the period but only one in the last 20 years. The 1980s were a particularly poor decade for this species locally with just 13 birds recorded, though across the entire period the numbers recorded each autumn have remained static at around two to three birds per year since 1970. One of the classic 'drift' migrants, this species typically occurs at coastal watch points during periods of easterly winds associated with a high pressure system on the continent.

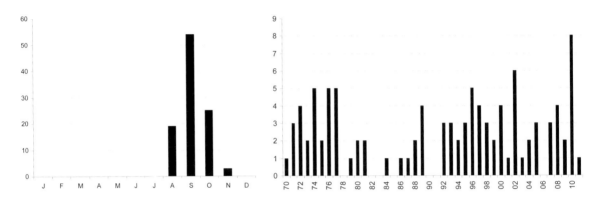

Barred Warblers in County Durham, 1970-2011

Records are evenly distributed between the northern and southern sections of the coastline with the Hartlepool/Seaton Carew area having a total of 47 birds, whilst the area between South Shields and Sunderland recorded a total of 39. Just eight have been recorded from the under-watched central section of Durham's coastline, between these two areas. These were at: Hendon from 14th to 16 September 1996 and 30 September 2010; Crimdon Denemouth on 21st to 22 September 2000; Dawdon Blast Beach on 3 September 2002, 17 September 2005 and 7 September 2008; Castle Eden Denemouth on 10 September 2002 and Beacon Hill, Easington on 7 September 2008.

Hartlepool Headland has recorded most with a total of 41 birds since 1970, including a peak count of three birds on 2 October 1974, which remains the highest individual day-count for a single site. The total of eight birds recorded in 2010 represents the best year for this species in the county, and is more than double the total that might be expected to occur during an average year. Records spanned the period from 7 September to 17 October, though more than half were in the ten day period from 27 September to 7 October. Two were trapped and ringed on Hartlepool Headland during September, though the remaining six birds were all found along the 10km stretch of coast between Marsden and Hendon, including a bird in Sunderland North Dock from 30 September until 16 October. This site previously hosted birds in the autumns of 2004 and 2005.

The earliest Barred Warbler ever recorded in County Durham was at Hartlepool Headland on 16 August 1987 whilst the latest is of an individual at Whitburn on 13 November 1984. There have been just two other November records since 1970, at Graythorp, Teesmouth on 9 November 2000 and at Hartlepool Headland on 2 November 2008.

Only six Barred Warblers have strayed inland during the review period though most of these were within five kilometres of the coast. They were noted at: East Boldon on 2 October 1976; well inland at Shibdon Pond, Blaydon on 26 August 1979; Hargreaves Quarry, Teesmouth on 22 September 1992 and 11 September 1993; Cleadon Hills on 17 September 1993 and at Graythorp, Teesmouth on 9 November 2000. The most significant of these was the bird trapped and ringed at Shibdon Pond in August 1979, some 23km inland.

Distribution & movements

Barred Warbler breeds throughout much of central and eastern Europe from southern Scandinavia to northern Italy and as far west as Denmark and Germany. To the east the range extends into central Asia. The entire population winters in southern Arabia and north eastern Africa as far south as Kenya. It is a scarce migrant to

Britain and is typically associated with periods of anti-cyclonic weather and light easterly winds, which create ideal conditions for 'drift' migration particularly during August and September although records occasionally continue as late as November. Spring records of Barred Warbler are extremely rare in Britain with just a handful of occurrences in May and June, the majority of which have been birds in their first summer. Similarly, during autumn adults are extremely rare, with almost all of the records relating to birds in their first-winter. As would be expected, the majority of records come from the eastern and southern coasts of Britain with the Northern Isles usually faring particularly well. Around 100 birds per year are recorded in Britain, though almost 300 were logged during 2002 which remains the best year to date for this species (Fraser & Rogers 2006).

Barred Warbler, Hartlepool Headland, September 2010 (Chris Bell)

A minimum of 10 Barred Warblers have been trapped and ringed in County Durham. These were at: Graythorp on 17 September 1967; Shibdon Pond, Blaydon on 26 August 1979; Marsden Quarry on 27 September 1986; Hargreaves Quarry, Teesmouth on 22 September 1992 and 11 September 1993; Hartlepool Headland on 10 October 2002, 2 November 2008, plus 7th and 27 September 2010; and at Whitburn Coastal Park on 2 October 2010. These have generated no controls or recoveries.

Lesser Whitethroat
Sylvia curruca

A scarce to widespread summer visitor and passage migrant.

Historical review

The skulking nature of this species, its relatively short song period (Simms 1986) and unfamiliarity with its song, has probably led to a significant long term under-recording of it in the county. Selby (1831) said that this species was "*rare in the north of England*", but he recorded two nests found near Durham by William Proctor. Hutchinson, 1840, recorded it as far west as Lanchester and north to Chester Dene. More generally, this has always been a scarce species in the north east of England and it was considered rare in Durham in Victorian times (Holloway 1996). A. C. Chapman in 1893 called it "*one of the rarest warblers in the northeast corner of England*". During the 19th century, this species was considered to be much scarcer than Common Whitethroat *Sylvia communis,* though it was documented as breeding in the "*district between the Tyne and Derwent*". In the late 19th century a pair was shot in Axwell Park and during 1911 birds were reported in Gibside and near Sunniside, but the species was not noted at Chopwell until 1930 (Bowey *et al.* 1993). In May 1928, M. G. Robinson, found a pair of birds nesting in gorse at Neasham Lane, near Darlington.

In the very early 20th century, it was described by Bolam (1912) as being scarce as a breeding bird in "*northern districts*". In 1937 birds were found nesting near Ryton and also in that year a nest with four eggs was found near Rowlands Gill (Bowey *et al.* 1993). In 1939 it was sufficiently scarce for George Temperley to note "*The* distribution of Lesser Whitethroat in Northumberland and Durham is very local and irregular. Readers are asked to report on any birds which they may have observed this summer and to make a point of identifying the species and recording its occurrence next year" (*The Vasculum* Vol. XXV 1939). Around the middle of the 20th century, Temperley described it as "*an uncommon summer resident*", "*variable in its numbers and very local in its distribution*". He said that in areas where it bred, often two or three pairs were found within a short distance of each other. He reported its having increased in the years prior to the publication of the '*History*' (Temperley 1951). At his time of writing, he said that that it usually arrived in early May and bred fairly regularly but in small numbers, with the best areas for it between the Tyne and the Derwent, in a restricted area a few miles west of Hartlepool and in the Darlington, Stockton and Gainford areas.

Prior to 1962, the only breeding records for the northern side of the Teesside coastal plain were from Stockton, at sometime around 1900 and subsequently. There is anecdotal evidence to suggest that there has been an increase in the numbers of Lesser Whitethroat in the south east of the county since the 1960s (Blick 2009).

Recent breeding status

This species, which has the most restricted range of any of the regularly breeding British warblers, excepting Dartford *Sylvia undata* (Marchant *et al.* 1990) and, more latterly, Cetti's Warbler *Cettia cetti*, breeds relatively sparsely in County Durham. It has never been a common bird in the northeast of England, although it has been noted as 'spreading' in Northumberland since the early 1960s (Marchant *et al.* 1990). Lesser Whitethroats are elusive, the males are most easily located by their rattling song; however, once territories and pairs are established, they become largely silent (Simms 1986). Even in well-watched areas, the species has a propensity to 'melt away' at the end of their song period (K. Bowey *pers. comm.*), often remaining undetected until they are feeding young, or announcing a possible second brood with another burst of late-season song.

During the *Atlas* survey period, there were no records of this species above the 120m contour and many of the occupied tetrads were along the river valleys (Westerberg & Bowey 2000). It is obviously a species that prefers lowland locations. In most years, the majority of records of the species come from sites located in the eastern half of the county (Newsome 2007-2009). Its preferred local habitats include stands of tall, overgrown scrub with a dense understorey of rank vegetation. Such suites of vegetation may often be found along disused railway lines and, increasingly, in areas of untended, mature amenity planting, in urban or suburban areas.

Localities that are well favoured by Lesser Whitethroat include the scrubby areas surrounding Brasside and Shibdon Ponds. Indeed, the lower Derwent valley has always been something of a stronghold for this species, though it is probably much overlooked in many localities elsewhere in the county. At Shibdon Pond, the species uses dense hawthorn *Crataegus monogyna* and willow *Salix* scrub, adjacent to marshland (Bowey *et al.* 1993) though extensive scrub clearance at this locality has reduced the viable habitats for this species. Breeding also occurs in isolated, relatively small patches of scrub on a regular basis. Hawthorn and blackthorn *Prunus spinosa* scrub on Cleadon Hills supports one or two pairs in most years as does Temple Park in South Shields (T. I. Mills *pers. obs.*). Fledged juveniles usually appeared from early June but little in the way of meaningful breeding data is usually submitted for this species (Newsome 2007).

The species is relatively well distributed in the northwest of the county, though only patchily spread in many other areas. Gateshead borough, which includes the favoured sites in the lower Derwent valley, is something of a stronghold for it, with birds noted annually at a number of sites; although even here, it is probably much overlooked (Bowey *et al.* 1993). The species' distribution as observed in this area, well covered by observers as it is, may be a better indication of the species' true presence across much of the rest of the county. In the area north of the River Tees, it remains uncommon (Joynt *et al.* 2008). The survey work for *The Breeding Birds of Cleveland* however, indicated that there were much larger numbers present than had been previously realised, with possibly 50-60 territories in the north Teesside area (Joynt *et al.* 2008); although these levels may have been reduced by the distinct national dip in numbers since 1998.

The *New Atlas* (Gibbons *et al.* 1993) indicated a low occupancy figure of 0.65 territories per occupied square kilometre. Assuming a similar occupancy rate for Durham and taking into account that Durham is towards the northern edge of the species' range in Britain, the *Atlas* suggested that the county supported a population of around 150 pairs (Westerberg & Bowey 2000). Combining the new, updated estimates for the area to the north of Teesside, would perhaps suggest that a review upwards of this figure, to 200 pairs, would be appropriate. Considering the amount of available, suitable habitat in the county, this species remains sparsely distributed, and it may be that even this revised figure might be too low figure. Results from Breeding Birds Survey sites around the county, confirm the scarcity of this species as a breeding bird, with only three territories being detected in each of three 1km squares surveyed from a countywide total of forty squares. In spring 2010, singing males were reported from just 16 potential breeding sites during May and the only locations to hold two singing males or more were Bolton Garths, Cowpen Bewley, Elemore, Hurworth Burn Reservoir and Willington.

Nationally, Lesser Whitethroat abundance was roughly stable, with some short-term fluctuations, from the 1960s until the late 1980s, but the CBC/BBS and CES trends provide evidence for a moderate decline that lasted into the late 1990s. These changes were significant and large enough over the relevant periods to trigger BTO concerns. BBS has subsequently shown a significant sharp upturn, but this contrasts strongly with the continued decrease recorded by CES ringers. Wide fluctuations in survival and productivity have been recorded by CES ringers, and may be influencing population change, but pressures during migration and in winter are the most likely causes of decline (Fuller *et al.* 2005).

A measure of the very restricted distribution of this species in Durham and neighbouring counties is given by BBS data spanning 1995–2008. It was only recorded in a maximum of five one-kilometre squares in eight separate years throughout Northumberland, Durham and Cleveland during that period. The other seven years produced breeding season records from one to four squares. Whilst fluctuations have occurred the population has remained relatively stable, with a 3% drop in figures in England but 2% rise for the UK overall.

A graphic example of the fluctuations in this species' fortunes during the early to mid-1990s can be seen from the following breeding season summaries (Armstrong 1992, 1995):

- 1992: While a good spring arrival was apparent with reports from 18 sites during the month, several negative comments were received concerning breeding. Near Piercebridge numbers were considered down with two regular sites unoccupied. A pair bred on Cleadon Hills but apparently abandoned their nestlings; the first non-productive year there since 1984. There was no confirmation of breeding at Shibdon Pond, where three to four pairs held territory in 1991. Birds with young were noted at Hetton, Pittington and Shotton (two pairs) in July.
- 1994: This was an exceptional year for reports of this, the first being on 23 April followed by multiple arrivals at Marsden, Witton Gilbert and New Lambton. A further 10 sites provided reports before the month's end. A good spread of records from inland sites prompted several observers to comment that the species was noticeably more common than the previous year. In May, up to eight birds were singing in Sacriston Woods by 6th, with eight localities occupied in Gateshead borough. including up to five at Shibdon throughout the month. A total of 21 singing males were recorded from 11 central sites including four to five at Sherburn Hospital; with breeding proven at 12 sites.

The dip in fortunes in 1992 was reflected in national data (CBC/BBS), which also pinpointed 1997 and 1998 as the last big dip years - during May 1998, singles were seen at five sites in the Hetton–Murton area and two were regular at Washington WWT. In June, five were in the Moorsley–Pittington area and a brood of four were raised at Sedgeletch. Around Lamesley, three former breeding sites appeared to be unoccupied, but since then there has been a gradual increase in numbers locally (Armstrong 1998, Newsome 2003-2009). Breeding Birds Survey returns for 2007 indicated the scarcity of this species as a breeding bird with only three pairs being detected in 40 one-kilometre grid squares surveyed. Singing males were reported from 30 locations during May 2008. In 2009, distribution during spring and summer showed a lowland bias and breeding was confirmed at several sites, with two pairs at Hetton Bogs raising 13 young.

Recent non-breeding status

The first birds of spring usually arrive from mid-April, with the major influx in the last week and into the first week of May. Early arrival dates include those in April 1991, when birds arrived in Flass Vale, Durham on 12th and Blackhall Mill on 14th with two at Hartlepool Headland on 13 April 2007 being notable; and the earliest for the county was at Hurworth Burn Reservoir on 9 April 2009.

Coastal migrants, at least in spring, tend to be noticeable by their absence, the vast majority of birds in the region heading straight onto breeding territories, with a typical arrival in 1996 - in the last week of April, four sites provided records with the first at Hetton Lyons Park on 26th, followed by singles in the Derwent Walk Country Park on 25th, Langley Moor on 27th and four in song at Shotton Pond on 29th. Birds trickled in during early May with one to two at Marsden, Tunstall Hills, Washington WWT and Gateshead. Of these reports, only the Marsden birds appeared in non-breeding habitat. In May 1998 however, two distinct periods of passage were noted on the coast in May: at Whitburn-Marsden three appeared on 13th to 14th with four birds from 29th to 31st, when 15 were at Hartlepool (Blick 2009). Of note in 2001, the first was recorded at Hartlepool on 22 April, when it appeared on the rocks in front of the observatory.

Post-breeding dispersal means that birds can be seen away from breeding location by late July, for example one at Dawdon Blast Beach on 29 July 2007 (Newsome, 2007); at such time some birds may still be breeding, possibly rearing second broods. The latest inland birds tend to be seen up to early or mid-September, while autumn coastal migrants tend to begin to appear from late August into September.

Autumn passage is usually light. A typical set of autumn records was received in 1998 with a trickle of eight coastal records from the well-watched sites of Hartlepool and Seaburn to South Shields. Final reports were on 3 October, at South Shields (2), Hartlepool (2) and at Hawthorn Dene. More notable examples of passage unfolded in 2005, when an early movement was witnessed with 12 birds logged at coastal sites as southerly migration got under way. The species featured prominently in the September 2008 fall of migrants, with a minimum of 36 birds being recorded at coastal sites, including: four at Hartlepool; six at Marsden Quarry; while four at Trow Quarry, South Shields on 6th was exceptional. The following day five were seen at Cleadon Hills and nine were counted along the coastal strip between Dawdon and Easington.

Birds usually stay into early September, with some individuals turning up well into October, for example in 1988 easterly winds saw one to two at Seaburn, Whitburn and Marsden until 22nd with one on 25 October at Whitburn in 1981. One unusual sighting concerned a bird feeding on a fat-ball in a garden on Hartlepool Headland on 30 October 2004. The county's latest reports concern individuals lingering into November:

- 1984: in an exceptional autumn for late migrants, singles were at Whitburn on 17th and South Shields on 20 November
- 1994: one was reported on 6 November at Sunderland
- 1995: a very late bird was at Roker Park, Sunderland from 13th to 18 November, which constitutes the latest record
- 2008: a single was at Marsden Quarry on 1 November.

Distribution & movements

Birds are found breeding across the mid-latitudes of Europe and Asia, from Britain in the west to eastern Siberia in the east. There are a number of racial forms, the nominate one being that which is encountered in Western Europe and the British Isles. Wintering birds from western populations are found wintering in northern eastern Africa, the easterly races in the Indian sub-continent.

In May 1989, a bird that had been ringed as an adult male on 21 April 1989 at Kampen in the Netherlands was controlled at Ryton Willows on 10 June 1989, where it was breeding. This is a distance of 567km to the west north west and it illustrates the well-documented south easterly migration route taken by this species (Wernham *et al.* 2002). Meanwhile, a bird ringed as a juvenile at Seaton Burn, Northumberland on 20 August 1985 was controlled nearly five years later at Clara Vale on 27 July 1990.

Whitethroat (Common Whitethroat)
Sylvia communis

A common summer visitor and passage migrant. It remains widespread and relatively common in areas that are not built up.

Historical review

In a letter dated March 1672, Ralph Johnson of Brignall, wrote to John Ray, "*it is like enough our whitethroat Curruca cenera is of the Ficedulae, for its her manner with us to fall upon a ripe cherry, whose skin when she hath broken with a chirp she invites her young brood, who devour it in a moment*". This was the first documentation of

Whitethroat in the region, as pointed out by Mather (1986) writing of Yorkshire, though this area is now part of Durham.

This species remains a common visitor to the county, as it was in the 19th and 20th centuries. It is one of our most common warblers, exceeded in number only by the Willow Warbler, *Phylloscopus trochilus*. In the 19th century it may have been the most common and abundant summer visitor in the region, exploiting the hedgerows that grew out of the Enclosure Acts of the 17th and 18th centuries (Holloway 1996). Bolam (1912) considered it one of the most common lowland visitors to the region, a situation which was maintained in the north east until the 1969 crash which saw numbers decline across the UK. An estimated 77% of breeding pairs did not return to Britain in the spring of 1969; this extremely high level of mortality being attributed to the continuing drought in the Sahel region to the south of the Sahara Desert, the species' principal wintering area (Mead 2000). In the mid-twentieth century, Temperley called it *"one of our commoner summer residents"*, arriving in late April and departing during August and September. Temperley considered that they were more widely distributed than Garden Warbler *Sylvia borin* and Blackcap *Sylvia atricapilla* and preferred more open habitats. In the 1950s and early 1960s, it was said to be the most common warbler in the south east of Durham by Stead (1964).

Recent breeding status

Following the late 1960s population crash (Winstanley *et al.* 1974), by the mid-1970s, there had been a slight improvement in our county. However, in 1974 numbers of breeding birds remained low and although there were signs of some improvement, it was clear that this species was still nowhere near returning to its pre-1969 status. A year later in 1975 a further improvement was noted with breeding season reports of birds at eight sites at Washington Riverside Park, five pairs in Castle Eden Denemouth, four at both Wynyard and a site in the Browney valley, three at Witton-le-Wear and one or two in at least 10 other localities (Unwin 1976).

Numbers fluctuated nationally around their lower level until the mid-1980s, since when they have sustained a consistent shallow recovery. It seems likely that habitat loss since the 1960s, particularly on farmland, will eventually limit the degree of recovery. A shallow upturn has been detected widely across Europe since 1980 (PECBMS 2010). The national recovery in numbers, which occurred after that 1980s decline (Merchant *et al.* 1990), has been confirmed locally. For example, evidence of such a recovery came from observations at Sunniside, where annual counts in the same relatively small area, showed that where just one pair was present in 1982, there were nine pairs by 1989 (Westerberg & Bowey 2000). Another illustration of the recovery was the results of an extensive CBC plot, conducted on Cleadon Hills, in the north east of the county, during the late 1970s. This site returned around four pairs per annum in the 1970s, whereas by 1990 population levels were so high that several birds held territory along the edge of a Whitburn housing estate, in an apparent overflow from more suitable adjacent habitat (T. I. Mills *pers. obs.*).

Compare the situation in the 1970s with a much healthier picture 20 years later in 1995, when the first sighting was at Hetton Lyons Park on 25 April, with an influx of eight birds on 28th to 29th. By 30 April, a bird was already nest-building at Blackhall. Passage birds in May were obvious, with 16 birds in the first two weeks at Marsden-Cleadon Hills, where an estimated 20 pairs held territory by the month's end. A further nine pairs were recorded between Cleadon and Seaburn. A two kilometre walk along hedgerows at Great Lumley revealed seven singing males, with 12 at Tunstall Hills and at Dawdon, seven to nine pairs were in territory by mid-month. Extensive census work in the Gateshead area produced over 100 singing birds. Regular counts came from the well-watched Hetton area gave a snapshot of breeding season populations:

	May	Jun	Jul	Aug
Hetton Lyons	19	18	42	25
Joe's Pond	17	11	28	20
Seaton	25	15	35	

At Far Pasture, Lockhaugh, a constant effort ringing site, 37 birds were ringed from May to Aug, with 81% juveniles and trends also suggested second broods were more successful than first.

BBS data for 1995 to 2008 indicate that a slow recovery was still tasking place locally, although national levels were nowhere near those of pre-1969. Over the 14-year period of the BBS, to 2008, the population of this species had grown by 34% in the north east of England. Numbers at the Kibblesworth CBC site reflected this trend as 15 pairs were counted in 2006 – a continued increase from three pairs in 1999. Interestingly, figures for the 14-year period show large variations across the UK with Scotland's increase being 86% and England's overall at only 18%.

Across Durham, this species can be found widely in low scrub and hedgerow habitats. Whitethroats are, essentially, lowland birds (Gibbons *et al.* 1993), being found mainly to the east of the county's 130m contour line, although the species does penetrate further west in Durham, but in such situations it tends to be confined to the river valleys of the Derwent, the Browney, the Wear, the Gaunless and the Tees (Westerberg & Bowey 2000). In 2009, the only locality to hold breeding birds west of grid line NZ03 was the western end of Derwent Reservoir, but small numbers of birds were present in several moorland fringes areas such as Hedleyhope Fell, Muggleswick and Wolsingham.

The greatest number of birds, though not necessarily the highest breeding densities, is probably to be found in the mixed agricultural landscape, which predominates over much of the county's central lowlands and south eastern parts (Westerberg & Bowey 2000). This is especially so, where hedgerows have been retained and there is a well-developed, associated ground flora along hedge bottom and ditch, with low woody herbaceous plants such as bramble *Rubus fruticosa* (Lack 1987). Across Durham, Whitethroats are found in many marginal areas, where there is tall, rough herbage, with such herb species as hogweed *Heraculum sphondylium* and rosebay willowherb *Chameniron anguistifolium*. The rapidity, with which territorial birds can settle into their breeding cycle, was evidenced by the fact that on 13 May 2007, adults were noted carrying food to nestlings at Hawthorn Quarry, after their arrival in the latter half of April. A bird caught at Lockhaugh on 7 May 2011 had originally been trapped there on 20 May 2007, demonstrating that birds can and do return to their original breeding sites. This bird was at least four and a half years old, having been hatched sometime before the autumn of 2006.

In spring 1995 in the Gateshead Borough, over 70 territories were recorded around 50 surveyed farm fields. If this density, of 90 territories per surveyed tetrad, were extrapolated across all occupied tetrads in the county, the resultant countywide population would amount to more than 24,000 territories. The *Atlas*, calculating the population using national figures (Gibbons *et al.* 1993), however, felt the 24,000 figure was not representative and the final determined figures was "*in the region of 7,400 pairs*" (Westerberg & Bowey 2000). *The Breeding Birds of Cleveland* work, between 1999 and 2006, indicated the presence of a relatively large population of perhaps 600-700 pairs in this area (Joynt *et al.* 2008).

Counts during the 2007 breeding season emphasised once again the common nature of this species, demonstrating its abundance in high quality habitat: 60 along the River Wear at Croxdale on 30 June; up to 27 at Hetton Bogs in early August; 25 singing males between Nose's Point and Foxholes Dene; 19 pairs at Rainton Meadows, 17 pairs at nearby Hetton Lyons CP and 15 pairs at Hetton Bogs with 16 singing males at Tunstall Hills. A decrease was noted on the Kibblesworth CBC, where a count of 10 pairs was a decrease of 5 from 2006. By the end of the first decade of the 21st century, the species was once again doing well in the county, especially compared to the late 1960s or mid-1980s and in 2008, potential breeding pairs were reported from over 200 localities.

Recent non-breeding status

Small numbers of birds are regularly seen along the Durham coastline in both spring and autumn. In the former season birds usually arrive in a very narrow corridor of dates in late April so that early arrival dates are few and far between. Very few birds turn up before 20 April, with the earliest date of arrival being 9 April 2010 at Hetton Lyons; spring passage probably peaks in early to mid-May. Between 1976 and 2005, the first reported arrival dates of Whitethroats were analysed in ten-year groupings. These were found to be: 1976-1985, 25 April; 1986-1995, 24 April; 1996-2005, 19 April. Over the period, the mean first arrival date of the species in Durham had advanced by six days, most of the change occurring in the latter period (Siggens 2005).

Breeding sites are increasingly deserted from early August, birds joining up with roving tit flocks from around early July and in the post-breeding phase the species is commonly noted at many widespread localities. The frequency of reports declines throughout August as adults move south and dispersal takes juveniles further afield.

Unlike some species, falls of migrants along the Durham coast seldom contain large numbers of Whitethroats, rarely more than 15-25 birds are recorded at any one time. A handful of birds tend to linger on into the first ten days of September but autumn migration is usually over for this species by mid- to late September. October records are exceptional: singles at Durham on 23 October 1976, at Cleadon on 25 October 1994 and the latest recorded at Hartlepool on 29 October 1982.

Distribution & movements

Whitethroats breed over much of Europe, into north west Africa and east as far as western Siberia and central south west Asia. All populations winter in Africa to the south of the Sahara, largely in the Sahel zone (BWP).

There is little substantive ringing information in relation to this species in Durham, though a juvenile ringed at Lockhaugh on 27 June 1998, which was controlled at Radipole Lake Dorset on 19 August 1998, illustrated what is no doubt the general southward vector of movement for autumn birds from the county.

Subalpine Warbler
Sylvia cantillans

A very rare vagrant from southern Europe: six records.

The first Subalpine Warbler for County Durham was a male found by Martin Blick, William Fletcher, and Tom Francis at Hartlepool Headland on 8 May 1975. This bird was found in St Mary's Churchyard following a period of heavy rain and light north easterly winds, although it later moved to the Croft gardens where it was frequently heard singing and occasionally indulging in 'display' flight (*British Birds* 69: 349).

All records:

1975	Hartlepool Headland, male, 8 May (*British Birds* 69: 349)
1985	Whitburn, female, 14 May (*British Birds* 80: 560)
1986	Finchale Priory, Durham, male in song, 11 May (*British Birds* 80: 560)
1999	Hartlepool Headland, male *'cantillans'*, 7 May (*British Birds* 93: 557)
2008	Hartlepool Headland, female, 28 May
2008	Trow Quarry, South Shields, female or first winter, 6-14 September

All five spring records have fallen into the period between 7th and 28 May which is entirely typical for this species. The singing male at Finchale Priory, downstream of Durham City, on 11 May 1986 is out of context with this usually coastal vagrant, though another singing male was in Bedfordshire at Whipsnade just four days earlier. All of the other Durham records have been associated with typical fall conditions involving periods of light easterly winds and rain.

The sole autumn record was found close to South Shields on 6 September 2008 during a good period for Scandinavian migrants. This bird remained faithful to an area of scrubby vegetation and a few small hawthorns for nine days in the south eastern corner of Trow Quarry, which was fenced off due to nearby excavations, allowing the bird an undisturbed feeding area, no doubt contributing to the length of its stay. Opinions differed as to the age of this bird, though advice sought from experienced ringers suggested it may have been an adult female based on the rather fresh plumage and iris colour. Just one of the Durham birds has been identified to race, a male of the western form *cantillans* that was at Hartlepool Headland on 7 May 1999.

Distribution & movements

Subalpine Warbler breeds in southern Europe and north western Africa and winters in sub-Saharan Africa in the Sahel belt from Mauritania eastwards to southern Egypt and Sudan. Four races are recognised, three of which have been identified in Britain; nominate *cantillans* which breeds from Iberia to Italy; *albistriata* which breeds in south eastern Europe and western Turkey; and *moltoni* which breeds on Corsica and Sardinia as well as adjacent parts of the Italian mainland. The Subalpine Warbler is a regular vagrant to Britain, chiefly in spring from April to June, and to a lesser extent in September and October. Records are widely scattered along the coastline of the British Isles, although the majority of records in early spring come from southern England, in particular Cornwall, and Shetland where there are over 150 records. The species had amassed a total of over 600 British records by the end of 2005 by which time it was no longer considered by BBRC.

Pallas's Grasshopper Warbler
Locustella certhiola

A very rare vagrant from Siberia: one record.

The only record of Pallas's Grasshopper Warbler for County Durham is of a bird seen briefly but well by Paul Cook in thick cover in the 'Doctor's Garden' at Church Lane, Whitburn on 9 October 2010; despite much searching the bird was not seen again (*British Birds* 104: 660). This individual was one of four recorded in Britain in autumn 2010 with the other three being on Shetland between 8th and 9 October. The similarity in arrival dates to the Whitburn bird is noteworthy.

Distribution & movements
Pallas's Grasshopper Warbler breeds in central and eastern Siberia eastwards across central Asia to north eastern China, and winters from north east India and southern China southwards throughout much of South East Asia. Four sub-species are recognised, with all of the birds recorded in Western Europe thought to be of the race *rubescens* which breeds in northern Siberia and has the longest distance to travel to reach its wintering grounds. It is a rare but annual vagrant to Britain chiefly during late September and early October, though has been recorded as early as the 13 September, with a total of 46 records by the end of 2010. More than three-quarters of all records have been found on the Shetland Isles, where the species is somewhat of a speciality. Further south, it remains a prized find with just seven records for England including the bird at Whitburn, though its skulking nature no doubt mitigates against its discovery and many must pass through undetected.

Grasshopper Warbler (Common Grasshopper Warbler)
Locustella naevia

A scarce to widespread summer visitor and passage migrant.

Historical review
Grasshopper Warblers are skulking and secretive and the vast majority of observations of this enigmatic species in Durham are actually of singing males, which are heard rather than seen. Its distinctive song however, has meant that it has been a relatively well documented species over the decades and even the centuries.

Abel Chapman, in *Birdlife of the Borders* (1889) said that he first encountered Grasshopper Warbler at Silksworth, Sunderland, on 22 April 1882, finding a nest with four eggs there on 15 May. One of the species' most favoured districts has long been the lower Derwent valley, this was acknowledged in Seebohm's '*History of British Birds*' (1883) in which he quoted... "*is especially numerous in the County of Durham, perhaps in no locality more so than in the warm and sheltered valley of the Derwent*". On this same subject, "*Somewhat common along the banks of the Derwent*", was how Robson described the species in the late 19th century and he mentioned finding a nest as early as 19 May (Robson 1896).

At the mid-point of the 20th century, Temperley (1951) reported that it was a "*summer resident in small numbers*", that "*arrives...towards the end of April or early May*"; he noted that it was distributed over most of the county though "*local and variable*" and he too related how the lower Derwent valley was locally renowned for this species (Temperley 1951).

Between the publication of Temperley (1951) and the commencement of recording by the Durham Bird Club in 1974, relatively little substantive information was documented on this species in the Durham ornithological record. One of the few notable records over this period came in 1953, when, in a marshy woodland area close to Greencroft Ponds near Stanley, six birds were heard reeling during May of that year. Otherwise most years, saw little more than the merest presence of the species being put on record until 1970s.

In this decade, Low Barns, Witton-le-Wear, appeared to have been something of a stronghold for this species, with eight birds there in the summer of 1972. The numbers of records of Grasshopper Warblers generated however is strongly linked to the amount of observer effort. In 1971 birds were recorded as being present at around a dozen locations across the county, a typical representation in the county at that time. In 1973 there was a suggestion of more birds than usual being noted in Durham, though this may have been a result of growing observer coverage. In that year, up to seven sang at Low Barns whilst one singing at Tunstall Hills in Sunderland was published as the first known territory to be held within the city (Unwin 1974), at least since Abel Chapman's time.

Recent breeding status
This is a species that in Durham is largely restricted to lowland sites with dense, rank vegetation, but these can be very varied and may include young plantations, scrub, marshland and overgrown waste ground. Numbers present can vary considerably from year to year and periodically this species appears in larger numbers on an unpredictable, but roughly three to four-year cycle. For example, six were singing at Castle Eden Denemouth in spring 1979 in what was considered a good year for the species in a number of areas across the county. By contrast in 1984 the species was said to be more sparsely distributed than usual. They can be found in a variety of rank habitats across Durham and a common factor in these habitats is low, dense vegetation. Many apparently suitable areas are not occupied every year. They undoubtedly prefer damp situations although they have been noted singing from rape crops locally (Westerberg & Bowey 2000).

The map that was shown in the *Atlas* illustrated the breeding range for this species as extending sparsely across much of lowland Durham, but no doubt that publication underestimated the number of tetrads in which the species actually occurred. The map in the *Atlas* recorded its presence in 48 tetrads. In most breeding seasons, birds are most strongly represented from the north, central and coastal areas of the county and usually absent from the upland and upland fringing areas. Nonetheless, despite its largely lowland status, birds are sometimes noted in the west of the county, for example on 11 May 1974, six birds were singing around Derwent Reservoir. More recently, a singing bird was present at Stanhope Gorge in May 2007 and two males sang at Hedleyhope Fell in 2008 but both of these occurrences were 'trumped' in 2009 by a singing bird at Selset Reservoir; this is the most westerly report of the species in Durham, a very unusual and exposed location for this species.

Grasshopper Warblers are notoriously difficult to detect during daytime as they can be very secretive, singing principally around dawn and dusk (Cramp 1986). This habit can lead to significant observer biases in mapping and the understanding of its distribution.   Highly motivated observers in some areas have documented what approximates more to the species 'real' distribution of the species through targeted fieldwork. In 2005, evening and nocturnal survey work by the Durham Bird Club, when searching for breeding owls, revealed this species in 12 of the county's 48 ten-kilometre grid squares, five of which were not noted as being occupied during the survey work for *The Summer Atlas of Breeding Birds in Durham*.

The high-pitched, soft 'reeling' song is almost inaudible to some observers and after pairing; the male tends to sing only in short erratic bursts or stops singing altogether. Late-season singing, associated with double-brooding, is sometimes more prolonged and obvious (Westerberg & Bowey 2000). Unusually, five were heard 'reeling' along the old railway line at Hurworth Burn on 13 August 1978 and in 2007 singing was noted as continuing into the second week of August.

Illustrative of the species' status in Durham during the 2000s, singing birds were reported in May and June 2004 from 19 widespread locations including: Barmston Pond, Brancepeth Beck, Burdon Moor, Cox Green, Far Pasture, Greenabella Marsh, Hetton Bogs, Hetton Lyons, Houghton Dene, Portrack Marsh, Rainton Meadows, Reclamation Pond, Seaton Pond, Sedgeletch/Philadelphia, Shibdon Pond, Shotton Colliery, South Shields Leas, Waldridge Fell and Whitburn. Proof of breeding however, came only from Shibdon Pond and Hetton Bogs. Typically, there is little information about the success of breeding of this elusive species in Durham and proof of breeding is difficult to obtain due to its extremely skulking habits. Significant breeding data for the county came from Low Barns in 1981, when three pairs were proven to have reared 15 young (Baldridge 1982).

Even in the best areas of the county for the species, Grasshopper Warblers tend to occur at low densities and it is unlikely that many tetrads hold more than a few pairs. Although the species is more common than some observers realise, in the right habitats, it does tend to occur in aggregated clusters of suitable habitats, at a macro-level. This was illustrated in the well-studied Gateshead area, from the mid-1980s to the mid-1990s, when birds were regularly recorded from: Blaydon Burn, Lockhaugh STW/Far Pasture, Ryton Willows and Shibdon Pond, with birds often also present at Blackhall Mill, Chopwell and Clockburn Dene (Bowey *et al.* 1993); some of these sites holding a small number of territories. In 2007, which might be considered typical for the purpose of illustration, singing birds were noted at 42 localities during the spring and early summer, with a few sites having up to four territorial birds. An insight into the densities that occur in the best habitats for the species however, came from the Rainton, Leamside and Chilton Moor areas, in the north east of the county, where at least 20 singing males were found during June 2007. In the nearby Murton and Sharpley areas, another 13 singing birds were present. In 2008, the species was reported from 40 localities across the county, which was felt to represent around 65 territories. Amongst these, five territories were located in the Chapman's Well area, whilst up to six males were noted between Rainton Meadows and Joe's Pond. In 2009, four at Dawdon Blast Beach and at least eight singing on South Shields Leas were particularly noteworthy concentrations in the late April period; in the same year four were at Birtley Brickworks on 1 June (Newsome 2005-2009). Temple Park, South Shields held four territories in May-June 2009.

The population estimate made in the *Atlas* was in the region of 50 to 200 pairs (Westerberg & Bowey 2000). There were around 30 territories of this sparsely distributed species around the northern part of what was Cleveland (Joynt *et al.* 2008), most of these being clustered in the Teesmouth area. It is now believed that the *Atlas* estimate was something of an underestimate of the species' true status, and perhaps the upper limit for this range might be extended upwards to 300 pairs.

It would appear that there has been some decline in numbers in some parts of the county, for example in the south east it was felt that numbers have declined since the 1970s, this decline was particularly evident over the period 1979 to 1985 (Joynt *et al.* 2008); as borne out by the national situation. That this species is neither widespread nor that common nationally, is illustrated by the fact that Grasshopper Warblers were recorded from just 3% of the 3,243 surveyed Breeding Birds Survey squares across the UK in 2009 (Risely *et al.* 2010). The species, which is a red-listed Species of Conservation Concern and a UK BAP species, increased by 24% over the period 1995-2008 across the whole of the UK, but showed a decline of 23% for England (Risely *et al.* 2010).

Recent non-breeding status

Male Grasshopper Warblers usually arrive a week or two earlier than the females, with the first singing birds being reported from mid-April. The main spring arrival of this species in the north east of England is usually concentrated in late April and the first week of May. The arrival and presence of females, unless they are trapped for ringing, goes almost undocumented in most years. Unusually for such an inconspicuous species, cause of mortality was noted in June 1984, when one was killed by a car at High Urpeth.

The earliest migrant arrival of the 1970s was on 17 April 1979 at Witton-le-Wear and this arrival date was not equalled until a bird was at Shibdon Pond on the same date in 1987; the first arrival dates in 1980s largely fell in the period 22nd to 28 April. Over thirty years between 1976 and 2005, the reported first arrival dates of Grasshopper Warblers were analysed, in ten-year groupings. The mean first arrival date moved from 25 April in the first ten-year period, to 19 April in the last (Siggens 2005). Through the last part of the first decade of the 21st century this date was, on average earlier still, 14 April. In the four-year period 2006-2010, there were three years in which some of the earliest ever arrival dates for the county were recorded. In 2009 the first 'reeling' male was at Houghton Dene on 13 April, the same date as the earliest bird in 2007, when one was at Shotton Pond. The first arrival of 2010 was at Whitburn on 11 April, the second earliest record at the time following one at Clara Vale on 10 April 1998, but the earliest date was bettered in 2011 with one reeling at Rainton Meadows on 7 April. The general trend clearly indicates that this species' occurrence in Durham is becoming progressively earlier.

Reports of Grasshopper Warblers usually tail-off quite dramatically through August, with relatively few autumn migrants being recorded in the county in most years. September birds are relatively uncommon in Durham. By way of example, a late bird was at Marsden on 27 to 28 September 1992, and in 2008, such records included single birds at Teesmouth on 7 September and Trow Quarry on 22 September. October birds are decidedly scarce in the county, with just four or five recorded in this month between 1990 and 2009. In 2007, two late autumn migrants were seen, one was on Cleadon Hills on 22 September and one was at Marsden Quarry on 1 October. In 2000, the last of the year was a bird ringed on 1 October on Hartlepool Headland. Other late birds were at Hartlepool Headland on 2 October 1976, on the same date in 1998 with another there on 8 October 1995. A *Locustella* warbler seen in Marsden Quarry on 11 October 1993 was presumably this species, whilst a very late bird, the latest ever, was caught and ringed at Castle Eden Denemouth on 22 October 1982. The fact that a number of these late occurring records were of birds trapped for ringing, suggests that some late passage migrants may go through the county unrecorded at this time of the year, perhaps even on an annual basis.

Distribution & movements
This species can be found breeding over much of Europe, from southern Scandinavia down to northern Spain and Italy. Its range extends eastwards through Russia, to central Asia and south west Siberia. Wintering birds are to be found in the northern tropical areas of sub-Saharan Africa and, for more easterly breeding birds, the northern parts of the Indian subcontinent (Parkin & Knox 2010).

Savi's Warbler
Locustella luscinioides

A very rare vagrant from southern Europe: four records, involving at least five birds. Probably bred in 1994.

The first record of Savi's Warbler in County Durham is of a singing male initially found by Ken Smith at Haverton Hole, Teesmouth on 21 May 1982 and then independently found by Derek Clayton and Russell McAndrew on 23 May. The bird sang from the large reedbed on the northern shore and was seen by numerous local observers before last being noted on 29 May. A second bird was said to have been present on 28th, and a brief mating attempt noted before being interrupted by a rather aggressive Reed Warbler *Acrocephalus scirpaceus*, although only one bird is accepted by BBRC. This is presumably due to the lack of a description of the second bird as it is clearly mentioned in the original record submission (*British Birds* 78: 574).

All records:

1982	Haverton Hole, Teesmouth, male in song, 21st to 29 May, second bird seen on 28th May, brief mating attempt noted
1994	Haverton Hole, Teesmouth, male in song, 23rd to 26 April
1994	Long Drag, Teesmouth, pair, mid-May until 30 August, probably bred, male trapped 28 July and 30 August, female trapped 29 July
1995	Long Drag, Teesmouth, male in song, 1st to 4 May

The remarkable series of records in 1994 almost certainly culminated in breeding taking place in the extensive reedbeds on the Long Drag, Teesmouth, though unfortunately this could not be proven conclusively. Having been noted singing from at least mid-May a male was trapped during the routine ringing of hirundines on 28 July, with it or another adult seen carrying food on the same date. The following day another bird was trapped and was seen to be a female with an extensive brood patch, reinforcing opinion that these birds were a breeding pair, though no young were ever seen or trapped at this regular ringing site. The last sign of the birds came on 30 August when the male was re-trapped in the same area by which time it was in heavy wing moult. The following spring, a singing male was again present along the Long Drag from 1st to 4 May but not subsequently and presumably it did not attract a mate (Joynt *et al.* 2008). Due to the recent decline of this species in Britain and adjacent parts of continental Europe it seems unlikely that these events will be repeated again in the near future. However, this species is notoriously secretive in nature and with a wealth of good breeding habitat on the North Tees Marshes this possibility cannot be discounted entirely.

Distribution & movements
Savi's Warbler has a discontinuous breeding distribution across much of Europe, although it is absent from Scandinavia and is most numerous in the southern part of the range. Further east it can be found breeding throughout much of temperate Russia, and the Ukraine, and in central Asia from the Caspian Sea eastwards through Kazakhstan to north western China. European birds are migratory and winter in West Africa from Senegal to northern Nigeria. A small breeding population formerly existed in East Anglia and south east England, reaching a peak of 30 pairs in 1979, though this had reduced to an estimated 10 pairs by 1993, with just one record of proven breeding between 1994 and 2004 (RBBP). The decline of the British breeding population also saw a marked

decline in records of migrant birds within the British Isles and on this basis the species was re-admitted to the list of species considered by BBRC in 1999.

Booted Warbler
Iduna caligata

A very rare vagrant from north east Europe and Asia: five records.

The first Booted Warbler for County Durham was identified by Tom Francis at Hartlepool Headland on 7 June 1992. Located during the early evening in scrub surrounding the bowling green, it was also present during the early morning of 8 June, but was not seen subsequently. This bird arrived during a prolonged period of south easterly winds and represented the first spring record of the species for Britain, although it was quickly followed by singing males at Spurn Point in Humberside, from 10-22 June 1992 and at South Walney, Cumbria on 17 June 1992. Despite the recent range extension into southern Finland, this species is extremely rare in Britain in spring and there have been just two other spring records since 1992, both of which related to singing males in early June.

All records:
1992	Hartlepool Headland, 7th to 8 June (*British Birds* 86: 513)	
1995	Seaham Hall Dene, 21 September (*British Birds* 89: 517)	
1999	Hartlepool Headland, 20th to 21 September (*British Birds* 93: 555)	
2003	Whitburn, 2 October (*British Birds* 100: 85)	
2004	Spion Kop Cemetery, Hartlepool, 10 August (*British Birds* 100: 738)	

Booted Warbler *Iduna caligata* or Sykes's Warbler *Iduna rama*
2009   Marsden Quarry, 29 September (*British Birds* 103: 615)

This bird was considered to be a Booted Warbler by the observer, though the BBRC felt that the description did not adequately eliminate the very similar Sykes's Warbler and accepted it as one or the other. Sykes's Warbler was, until 2002, considered a race of Booted Warbler (*Ibis* 144: 708-709).

Distribution & movements
Booted Warbler breeds in central Russia and western Siberia, southwards to Kazakhstan, western Mongolia and the Xingjian province of China, though the range is expanding westwards and it is now breeding regularly in southern Finland. It winters entirely on the Indian subcontinent, as far south as Karnataka but occasionally it reaches Sri Lanka. The species is a fairly regular vagrant to Britain with 123 records by the end of 2010, chiefly from late August to mid-October, though with a distinct peak in September. In this respect all records of this species in County Durham fit the established vagrancy pattern for this species, though the bird at Hartlepool in 2004 was exceptionally early and is the earliest autumn record of this species in Britain by five days.

Icterine Warbler
Hippolais icterina

A rare migrant from northern and eastern Europe: 82 records involving 95 birds, recorded in May and June, and from August to September, with one October record.

Historical Review
It is somewhat remarkable that the first confirmed Icterine Warbler in County Durham was as recently as 1966, although what was almost certainly an Icterine Warbler was seen at Hartlepool Headland in September 1960, but the identification criteria for this species were still being developed at that time (Stead 1969). The first well

documented record for County Durham was an individual trapped and ringed at Marsden on 29 August 1966 by M. L. Chambers and the late Fred Grey; it was associated with a large fall of Scandinavian migrants and was still present the following day. The same weather pattern saw another two birds discovered on 30 August, with one trapped and ringed at Graythorp and another at Hartlepool Headland. These were all found during a period of light easterly winds providing ideal conditions for 'drift' migration; four birds were found in Northumberland during the same period (Kerr 2001).

Four more birds were found by the end of the 1960s, all of them at Hartlepool Headland. These were on 4 September 1967, two on 17 August 1968, and on 24 September 1969. All of these birds were associated with an arrival of continental migrants during easterly winds.

Recent Status

At least 88 Icterine Warblers were recorded in County Durham between 1970 and 2011, and the species is perhaps best regarded as a scarce migrant in both spring and autumn, although autumn records predominate. Despite being just about annual in occurrence since 1970 there have still been 14 years without records. The best year was 2008 with a total of 11 birds, although there were ten in 1994 and nine in 1995. The 1994 peak is particularly noteworthy in that eight of these birds were recorded in spring.

To date there have been 24 Icterine Warblers recorded in spring, most of which have been in the relatively narrow period from late May to early June. As might be expected the majority have been concentrated along the coastline from South Shields to Teesmouth, in particular at the migrant hotspots around Hartlepool and Whitburn. The first spring record was of a singing male at Seaburn from 23rd until 29 May 1984, though the species has subsequently appeared in nine springs since 1990, with most records unsurprisingly relating to singing males.

There have been two notable falls of this species in spring both of which occurred in the early 1990s. Suitable weather conditions saw a total of six birds recorded during the period from 29 May to 8 June 1992, with a peak of three in the Marsden area over 29th to 31 May. Two years later strong north easterly winds and rain during 22nd to 25 May 1994 brought a record arrival of at least eight birds. The Marsden/Whitburn area held an estimated three to five birds during this period including a singing male along Quarry Lane, Marsden which lingered until 27 May, and a further five were discovered in the Hartlepool/Teesmouth area. As in 1992, the arrival of these birds was associated with a large fall of Scandinavian migrants to the east coast of England.

In contrast to autumn, a number of spring birds have penetrated well inland; the most noteworthy being singing males at Eels Wood near Clara Vale on 7 June 1992 and at Shibdon Pond, Blaydon on 11 May 1993. These are the only records for the Gateshead area. The influx of 1994 also produced three records of inland birds albeit all the locations were within a few kilometres of the coast. Singles were found in the centre of Hartlepool at North Cemetery on 23rd to 24 May, on the Long Drag, Teesmouth on 24 May, and trapped and ringed in Hargreaves Quarry on 25 May. These latter two birds are the only records for the Borough of Stockton-on-Tees.

The bird at Shibdon Pond on 11 May 1993 is the earliest spring record in the county with the latest being at Whitburn Coastal Park from 8th to 13 June 2002.

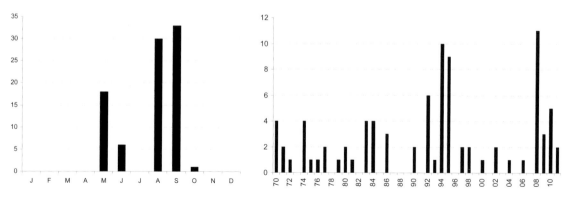

Icterine Warblers in County Durham, 1970-2011

A total of 66 Icterine Warbler have been recorded during autumn since 1970, making the species more than twice as numerous as it is in spring. All of the records have been at coastal locations with the majority coming from late August to mid-September, though the species has been recorded as late as 1 November. As in spring most occurrences are associated with typical 'fall conditions' during periods of easterly winds and rain and they often arrive in association with other scarce migrants.

Despite several blank autumns, with suitable weather conditions birds can arrive in numbers and nine were recorded in two years, between 7th and 9 September 1995 and from 6th until 8 September 2008. Multiple arrivals are not uncommon with the best single count being of four birds at Hartlepool Headland on 30 August 1970, one of which remained the following day.

The earliest autumn arrival is of a bird trapped and ringed at Hartlepool Headland on 3 August 2002. Only four other Icterine Warblers have been recorded before mid-August, these being at Hartlepool Headland on 9 August 1975, 9th to 10 August 1983, 12 August 2011, and at Seaton Snook, Teesmouth on 10 August 2004. Similarly, only three birds have been recorded later than mid-September, these being at Holy Trinity Churchyard, Seaton Carew from 17th to 18 September 1994, at Hawthorn Dene on 30 September 2000 and at Cleadon from 31 October until 1 November 1984. The 1984 bird at Cleadon was exceptionally late and remains the only October or November record to date.

Distribution & movements

Icterine Warbler breeds in central and eastern Europe from southern Norway and eastern France eastwards to western Siberia, and winters in sub-Saharan Africa, chiefly south of the Equator. It is a scarce migrant to Britain, one of the classic 'drift' migrants, usually occurring after light easterly winds in both spring and autumn, chiefly in late May and early June, and again from late August to mid-September. As would be expected the majority of records come from the east coast of Britain from the Northern Isles southwards to Kent, though small numbers occur annually along the south coast of England as far west as the Isles of Scilly. Numbers vary annually and are highly dependent on the occurrence of suitable weather conditions, with the ten year average from 1990 to 1999 being around 140 birds per annum. Eleven Icterine Warblers have been trapped and ringed in County Durham. These were at: Marsden on 29 August 1966; Graythorp on 30 August 1966; Whitburn on 28 August 1986; Hargreaves Quarry, Teesmouth on 25 May 1994; Hartlepool Headland on 3 August 2002, with single birds ringed there on 24th and 28 May 2008 respectively and two on 7 September 2008, and one on 12 August 2011. Finally, one was ringed at North Gare, Teesmouth on 8 September 2009. There have been no subsequent recoveries.

Melodious Warbler
Hippolais polyglotta

A very rare vagrant from southern Europe: one record.

The sole record of Melodious Warbler for County Durham is of a singing male at Whitburn Coastal Park from 27 June until 7 July 2003. This bird was initially heard singing by Peter Collins though it would be a week before it was seen well enough to be identified, due to poor weather causing it to remain hidden amongst deep scrub. The late arrival date of this bird initially appears noteworthy as most birds have arrived on their breeding grounds by late May, though spring migration is known to be especially protracted in this species and can continue into June (BWP). In this respect, this bird was more than likely a late spring overshoot and not the failed breeder that it might first appear. This bird came less than a month after a Melodious Warbler had been present on 31 May 2003 at South Gare, just across the River Tees in North Yorkshire and less than a mile outside of County Durham.

Distribution & movements

Melodious Warbler is very much the south western counterpart of Icterine Warbler *Hippolais icterina* and breeds throughout much of south west Europe and parts of north western Africa. In Europe the majority are to be found in France and the Iberian Peninsula, though to the east its range extends along the shores of the Mediterranean to both Italy and Croatia. A relatively short distance migrant the entire population winters in West Africa from Gambia and Sierra Leone eastwards to Nigeria and Cameroon. It is a scarce passage migrant to Britain in both spring and autumn though the vast majority of birds are recorded in autumn from July through to October. As would be expected there is a distinct south westerly bias to records in the British Isles with most recorded along the south coast of England as far west as the Isles of Scilly. Records further north are extremely rare and the species has yet to be recorded in Northumberland, though there are over 40 records for Scotland, most from the Northern Isles. Much rarer than its close relative the Icterine Warbler, around 30 birds are recorded on passage in Britain each year, with a peak of 60 in 1981 (Fraser & Rogers 2006).

Aquatic Warbler
Acrocephalus paludicola

A very rare vagrant from eastern Europe: two records.

The first Aquatic Warbler for County Durham was found by John Coulson in an oat crop at Frenchman's Bay, South Shields on 28 August 1951, although at the time he was unsure as to its identity. He reported the sighting immediately to the late Fred Grey who was able to return with him shortly afterwards and who concluded that the bird in question was an Aquatic Warbler. It was the first record for the English east coast north of Norfolk. Said to have been seen at close quarters and kept under observation for several hours, it was also seen later by the late Hugh Blair, who concurred with the initial identification (Temperley 1951). Unfortunately the oat crop this bird frequented no longer exists as the farmland was converted to recreational land in the 1960s and forms part of what is now known as The Leas.

The second record was almost 35 years later when a juvenile was caught and ringed by the Tees Ringing Group in reedbeds along the Long Drag, Teesmouth on 12 August 1986 during a routine ringing session. It was photographed and released shortly afterwards but was not seen subsequently.

Distribution & movements
Aquatic Warbler is listed as globally threatened with a world population of just 15,000 pairs. It is a rare, localised breeder in eastern Europe from Germany eastwards to western Siberia. The main wintering grounds were only discovered in 2007 and are now known to be in West Africa south of the Sahara, principally at Djoudj National Park in Senegal. Birds from eastern Europe are known to head west south west to staging posts in Western Europe, before heading south into West Africa for the winter, with the south coast of England falling on the edge of this migration route. Essentially a scarce passage migrant to the south coast of England during August, the Aquatic Warbler is a genuinely rare vagrant elsewhere, with very few records further north. Whilst there have been over 1,200 Aquatic Warblers recorded in Britain since 1958, there has been a marked downturn in the number of records since 1997, with a current average of just 10-15 birds per annum, well down from the peak of 102 in 1976 (Fraser & Rogers 2006). The recent sharp decline in numbers appears to be related to the drainage of wetland areas in the species' core breeding range.

Sponsored by

Sedge Warbler
Acrocephalus schoenobaenus

Glead Ecological & Environmental Service

A common breeding summer visitor and passage migrant.

Historical review
There is little substantive information about this species' status in the county from the north east's historical records. One is left to presume that it was, as stated by Holloway (1996), a *'familiar bird'* in Durham *'breeding along the banks of all streams and rivers'*, as it was said to be across the rest of Britain and Ireland. In the north of the county, it was described as "*somewhat common in the vale*" by Robson

(1896), presumably this was the case before many wetland areas in the Derwent valley were drained though the 19th and early 20th centuries. Holloway stated that nationally, its population appeared stable from the beginning of the 19th century to the late 1960s (Holloway 1996).

A "*common summer resident in the lowlands*", was how it was described by Temperley in 1951, which, according to him, arrived at the end of April and in early May. He mentioned that it was sometimes found in drier habitats and Dr. Hugh Blair, of South Shields, observed it nesting in hedgerows and similarly bushy places in the north east of the county around this time.

Relatively little detailed information was published on this species in Durham over the period from Temperley (1951) until the commencement of recording by the Durham Bird Club in the early 1970s. An exception came in 1952 with the documentation of a late date for the occurrence of this species, on 13 October at Tanfield Ponds, near Stanley, when two were heard, most unusually, singing with one remaining to 20th; one of the latest records for the county (Temperley 1953). Scant reference to its breeding status came in May 1953 when at a wetland near Greencroft Ponds, near Stanley, 21 were heard singing, a high density of singing birds in the county (Temperley 1954) and a few years later, when 16 singing males were revealed by a survey at Tanfield Ponds, in May 1960.

In the south east of the county, there was a good population of Sedge Warblers throughout the 1950s and early 1960s, in '*suitable habitat*' around Teesmouth (Stead 1964). Blick (1978) summarised the species' situation a decade and a half later, noting that there were small numbers around several marshland sites in the area.

During the 1970s, detailed breeding information was scant but up to 12 territories were at Low Barns, Witton-le-Wear, in 1975 and 12 birds were singing along the River Wear between Washington and Chester-le-Street on 19 May 1976. The first breeding records at WWT Washington took place in 1977 in wetland habitats that were created from around the middle of the decade. In 1974, birds bred in Darlington on an old allotment site; at that time a rare documentation of birds breeding in a drier habitat.

Recent breeding status

The Sedge Warbler is mainly a lowland bird in County Durham, with many of the occupied sites being clustered, ribbon-like, along the river valleys. Away from the species' traditional reed bed and marshes, it also uses bramble/scrub habitat and is often found along disused railway lines. Concentrations of occupied sites are found to the north, west and south of Durham City, these being associated with the River Wear and its tributaries the Deerness, Browney and Gaunless along with their associated wetlands. Relatively large concentrations of birds are present at Teesmouth and from there up the Tees valley to Croft, as well as up the River Skerne towards Darlington. The species has been recorded from a wide range of other habitats, including dry scrub and farmland hedgerow, at altitudes from five to 350m above sea level. These latter upland records most likely refer to unmated males, the species being effectively absent as a breeding bird in the county's uplands. A singing male on Bowes Moor in 2002 was an extreme example of such behaviour as were birds at Langdon Beck on 23 May 2007 and one that was singing high in Weardale at almost 430m above sea level at Lanehead, on 4 June 1975.

From the late eighties, the species began to be recorded on a number of occasions in oilseed rape *Brassica napus oleifera* crops. The first documentation of this activity locally was of a bird singing in a crop at Whitburn on 26 May 1987 and three such singing birds were present in one such crop on Fellside in 1989, and this habit has been noted regularly around the county ever since.

In 1980, there were up to 25 territories at Shibdon Pond (with 21 singing there in 1987, even after the crash in numbers) and four of these birds holding territory had been ringed there the previous year as juveniles, emphasising the species' propensity to return to its natal area to breed; males have a well documented site-fidelity between years (Wernham *et al.* 2002). This was illustrated by an incident in the spring of 1994, when two breeding males were caught, about half a metre apart, in a mist-net at Shibdon Pond. These two birds had been ringed the previous year, in the same mist-net location, presumably during a territorial dispute, over their adjacent territories. Subsequently, these birds had travelled to Africa for the winter, before returning to the same territories and repeating their capture in an almost identical dispute a year later. In similar fashion, some Teesmouth-ringed birds have returned to their natal sites; for example, one ringed in May 1972 was in that area again in July 1977 (Blick 2009).

Detailed local survey work in Gateshead Borough during the 1990s indicated that this species was substantially more widely distributed in arable habitats in the county than previously realised, particularly in oilseed rape crops (Bowey 1999b). The first habitats in which birds usually appear however, are the traditionally occupied wetland areas. During the initial song period males are easily located but once females arrive and pairs become established, the males usually fall silent (Catchpole 1976), unless practising bigamy or indulging in a second brood (K. Bowey *pers. obs.*). By the second week of May, singing males are often present in rape fields. All of the county's principal wetlands hold birds but it is conceivable that birds using rape crops may outnumber these populations (Bowey 1999b).

Gibbons *et al.* (1993) suggested a national population of 250,000 pairs, a decline of around 15% from the first national *Breeding Atlas* (Sharrock 1976); this fall in numbers was due to serious declines in the 1970s and 1980s, followed by considerable recovery from around 1985 (Marchant *et al.* 1990). In recent years, the North Tees Marshes have held the county's largest number of breeding birds in wetland areas, but, as always, it is difficult to gauge the true population size in such a complex of closely associated sites. The fact that this species is a relatively common breeding bird around this area was evidenced by the fact that in spring and summer 2006 at RSPB Saltholme, 21 adults and 100 juveniles were ringed. Sample densities of breeding birds at other sites in the county during 2007, included: 28 singing males at Brancepeth Beck; nine pairs at Rainton Meadows and Joe's Pond and eight pairs at Hetton Lyons. In spring 2009, the species was reported from nearly 40 lowland sites away from the North Tees Marshes; the largest concentrations of singing males were at Rainton Meadows, Mordon Carrs and Low Barns.

Sedge Warblers were recorded from 11% of surveyed Breeding Bird Survey squares across the UK in 2009 (Risely *et al.* 2010). The species increased by 9% over the period 1995-2008 in the whole of the UK but showed a decline of 7% for BBS squares in England (Risely *et al.* 2010). The loss of wetland sites to drainage adversely affects this species, as might changes to the Common Agricultural Policy that served to reduce the amount of oilseed rape grown (Westerberg & Bowey 2000). Much of the decline noted in the early eighties was linked to poor adult survival in the drought-ridden Sahel wintering grounds in Africa (Peach *et al.* 1991), indicating that this might be a species prone to the future effects of climate change resulting from global warming.

Based upon the intensive work in Gateshead, extrapolated across the area of appropriate habitats for the species in Durham, Westerberg & Bowey (2000) estimated the county population to be in the range of 520 to 660 pairs. Assuming the 1990s Gateshead data to be representative of the whole of the county, as many as 60% of birds were thought to be utilising rape habitats as oppose to those thought more typical of the species. After major declines associated with problems in this species' wintering grounds, the population around Teesmouth was estimated to have declined to around 50 pairs over the period 1977-1980 but *The Breeding Birds of Cleveland* indicated a much increased population two decades later, with an estimated 200 pairs in the of the North Tees Marshes and along the River Tees (Joynt *et al.* 2008). Balancing all of these factors and current knowledge about distribution and habitat usage, the upper limit of the figures postulated in the *Atlas*, of around 650 pairs, is probably an appropriate working figure for the number of pairs inhabiting the county at the end of the first decade of the 21st century.

Recent non-breeding status

Small numbers off birds are seen in most springs and autumns at the county's main coastal watch points, though birds can appear at any point along the coastal strip. From ringing data generated at the county's main wetland sites, it would appear that some of the birds that pass through Durham are birds that may have bred, or reared, further north in Britain and move through sites such as Shibdon Pond plus the various North Tees Marshes sites on their onward migration south.

The first birds of the spring usually appear in the county during late April; the principal arrival of males occurring in early May, with females arriving eight to ten days later (Westerberg & Bowey 2000). Early records of migrants occurred on 10 April 2010 at Haverton Hole and 11th at Hartlepool Headland in 1981, but the earliest date is of one at Dorman's Pool on 29 March 2003. Between 1976 and 2005, reported first arrival dates were analysed in ten-year groupings. These were: 1976-1985, 29 April; 1986-1995, 22 April and 1996-2005, 22 April; and advance of one week (Siggens 2005). The first observed arrival dates of birds through the second half of the first decade of the 21st century was, on average, even earlier still; 19 April in the period 2006-2009.

Many adults have already left breeding sites in the county by the end of July and these will be exiting the region during August. Birds that are still present at breeding locations, wetlands in particular, in the early part of September are almost invariably birds of the year, either locally bred or from further north in the British Isles. Coastal migrants are occasionally recorded later, through late September and rarely into early October. The latest bird recorded at the much-watched Shibdon Pond is 28 September 1992, and this site regularly retains birds into mid-September, which is probably indicative of the situation at many other less frequently observed wetlands. An unusually late bird, particularly for an upland location, was at Derwent Reservoir on 29 September 1985. Later birds have been noted at Teesmouth, with one there on 5 October, and at Tanfield Ponds, where two birds were noted as singing on 13 October 1952, one still being present on 20th.

There is one winter record of this species for Durham, which is both an exceptional record for the county and one of only two known records for England. This bird was noted in the reed bed area at WWT Washington on 7th and 8 December 1984; up to 1973 there were no later records of this species in Britain than 24 November (Baldridge 1984). The only winter record documented by Brown & Grice (2005) in England was of a single bird referred to as occurring during the survey period of the *Winter Atlas* (Lack 1986).

Distribution & movements

This is a common breeding species across much of the Western Palaearctic, occurring from the edges of the high Arctic in parts of Fennoscandia to the more southerly climes of Greece and Turkey. East to west it is found in a broad zone, from Central Siberia, through the Urals and west as far as Ireland. Its wintering grounds are in Africa, south of the Sahara and it is in this, the Sahel zone that droughts seem to have been the cause of major declines in the British population between 1965 and 1985 (Parkin & Knox 2010).

There has been relatively extensive ringing of Sedge Warblers in the county, particularly around Teesmouth, from the late 1970s to the 2000s, and this has revealed that birds migrating, stage at wetland sites further south in England (Blick 2009). By way of example, a bird was at Sandwich Bay, Kent, five days after being ringed at Teesmouth. Another was at Goole, Humberside, seven days after ringing; one visited Radipole, Dorset, nine days after being on Teesside; one was at Chichester; West Sussex, 12 days after ringing while four others were caught at Icklesham, East Sussex, some two weeks after ringing (Blick 2009). Most spectacular in terms of onward movements, was the juvenile ringed at Billingham on 28 August 2003 and controlled at Kings Lynn, Norfolk the next day.

Ringing returns from sites in the north of the county have shown similar, if not quite as dramatic results. For example, one ringed at Shibdon Pond on 14 May 1978 was controlled on 18 August 1979 at Wraysbury, Buckinghamshire with another ringed there on 25 May 1981 controlled at Ilkley in Yorkshire on 14 July 1983. In contrast to this expected southerly movement, an adult ringed on the North Tees Marshes in May 1994 was later found breeding at Gosforth Park, Northumberland, in June 1995; whilst young birds ringed at Gosforth have been controlled as breeding birds at Shibdon on a number of occasions and on two occasions birds caught at Iklesham on the south coast in the autumn, were controlled at Shibdon in subsequent springs.

Demonstrating this species' onwards passage to the continent, eleven other birds ringed at Teesmouth, between August 1977 and July 2007, were later re-trapped on the west coast of France, anything between seven and 31 days after being at Teesmouth (Blick 2009). A bird caught at Lockhaugh on 6 August 2002 was later controlled at La Masereau, Frossay, in the Loire-Atlantique, France, almost exactly a year later. A nestling ringed at Teesside in June 1994 was later trapped in France in August 2000, when it was six years old. Meanwhile, in reciprocal movements, at least six birds that were French-ringed have been re-trapped in Cleveland in the spring or summer following the autumn in which they had been ringed (Blick 2009). Similarly, a French ringed bird, caught at Oudalle, Seine-Maritime in the previous August was caught at Shibdon Pond in spring 1994. The most interesting ringing report was of a bird caught and ringed as a first winter bird on Sark, Channel Isles, on 6 December 1984 that was controlled at Shibdon Pond on 14 July 1985. A further five birds have been caught on passage at a wide range of sites in southern England, between Somerset and Sussex. Two breeding females from Shibdon Pond were caught in subsequent years breeding in Yorkshire while two birds controlled at Shibdon Pond during the summer had originally been ringed at locations on the English south coast. A juvenile ringed at Haverton Hill on 19 August 1992 was controlled at West-Vlaanderein, Belgium on 26 August 1992 and going in the opposite direction, a juvenile ringed at Brabant, Belgium on 11 August 1992 was controlled at Saltholme on 13 May 1993. One of the most spectacular controls was of a bird ringed in Senegal, West Africa in January 1993, which was caught on the North Tees Marshes in May of that year as breeding bird.

On a small number of occasions, birds ringed as juveniles further north have been controlled in the county, for example one ringed at Inchture on Tayside, Scotland 20 August 1992, which was at Haverton Hole on 5 September 1992; another juvenile ringed at St. Abb's Head, in the Borders, on 1 August 1999 that was re-trapped at Billingham on 28 August 1999 and the juvenile caught at Loch Spynie, Grampian on 27 July 2003, which was controlled at Haverton Hole on 6 August 2003.

The longevity of the species is demonstrated by a bird ringed at Shibdon Pond in 1979 and re-trapped there in 1984 while another ringed at Saltholme in May 1972 was re-trapped there in both 1973 and again in July 1977; both birds being more than six years of age.

Paddyfield Warbler
Acrocephalus agricola

A very rare vagrant from eastern Europe and Asia: three records.

The first Paddyfield Warbler for County Durham was found during the morning of the 18 September 1969, when John Burlison, Geoff Iceton and Russell McAndrew had brief views of an unusual *Acrocephalus* warbler in bushes bordering the outer bowling green on Hartlepool Headland. It was trapped later that evening and roosted overnight, before being ringed and released in its original location early the next morning. It was to remain until the 22 September, though it was frequently elusive. A first-year bird, Russell McAndrew was the first to suggest that the mystery bird was a Paddyfield Warbler, although one aspect of the wing formula did not appear to fit and seemed to favour Blyth's Reed Warbler *Acrocephalus dumetorum*, leading to it being submitted to BBRC as such. After much correspondence, the Hartlepool bird was eventually accepted as the third Paddyfield Warbler for Britain. Both of the previous records had been on Fair Isle, Shetland (*British Birds* 65: 349, *British Birds* 72: 348-351).

Remarkably, the second record was found in exactly the same location some 15 years later, when Chris Bell, Stephen Ryan and Mark Wallace discovered a mystery warbler in the outer bowling green on Hartlepool Headland on the rather late date of 27 October 1984. Observer opinions were split between Booted Warbler *Hippolais caligata* and Paddyfield Warbler; the bird was trapped and proved to be County Durham's second Paddyfield Warbler. Ringed and released back into the bowling-green, it remained until 28 October (*British Birds* 78: 574, *British Birds* 79: 571).

The third bird was just inland of Newburn Bridge, Hartlepool during the late afternoon of 17 September 1994. Initially skulking and difficult to view, it was still present the following morning when it was trapped and ringed. Once in the hand, it was confirmed a Paddyfield Warbler. Although often elusive, this bird showed well on occasion in a tiny hawthorn and was seen by many observers on 18th, though was not seen subsequently (*British Birds* 88: 538).

All three birds in the hand were aged as being in their first calendar year; this is not unexpected given the timing of their occurrence.

Distribution & movements
Paddyfield Warbler breeds from the northern coast of the Black Sea eastwards through central Asia to southern Siberia, Mongolia and western China, and winters entirely in the Indian subcontinent. A recent population increase has seen the species spread north west with breeding first recorded in Latvia in 1987 and in Finland in 1991 (Nankinov 2000) and possibly the Netherlands in 2007 (*Dutch Birding* 29: 343-344). It is a rare vagrant to Britain and has been recorded from May to December, although the majority of records have been in September and October. There had been 82 British records by the end of 2010, with more than half occurring on offshore islands.

Blyth's Reed Warbler
Acrocephalus dumetorum

A very rare vagrant from eastern Europe and western Asia: one record.

County Durham's only Blyth's Reed Warbler was discovered by Paul Cook in gardens at the north end of Cornthwaite Park, Whitburn on 2 October 2007. Initially located by its incessant harsh "*tek*" call, it took several minutes before the bird emerged and could be seen to be an un-streaked *Acrocephalus* warbler. It would be two hours before the observer managed long enough views to confirm the identification. Frustratingly elusive in the dense vegetation, it proved difficult to see during that first afternoon, though showed better the next morning and remained skulking in the same area for another two days, calling almost constantly, enabling it to be easily tracked through the thick cover. This bird formed part of a record influx of twelve birds to Britain during the autumn of 2007 (*British Birds* 101: 564).

Distribution & movements
Blyth's Reed Warbler breeds widely across the Palaearctic from Finland eastwards through European Russia and central Siberia to Lake Baikal, and southwards through Kazakhstan and Tajikistan to northern Pakistan. It winters almost entirely in the Indian subcontinent as far south as Sri Lanka, though there are a number of winter records from north western Burma where it is presumably regular in occurrence. Formerly a great rarity in Britain, this species has been gradually extending its range westwards with an estimated 5,000-8,000 pairs now breeding in Finland, as well as a further 5,000 to 10,000 pairs in the Baltic States. This, coupled with a greater awareness and understanding of its key identification criteria, has resulted in Blyth's Reed Warbler become an increasingly recorded vagrant to Britain. There had been a total of 115 British records by the end of 2010 with just 16 prior to 1990, most of which were on Scottish islands. The majority of British records are concentrated into the period from late September to mid-October.

Marsh Warbler
Acrocephalus palustris

A rare migrant from Europe: 31 records of 35 birds, recorded from May to September, though most frequently in May and June.

Recent Status
The first fully documented record involved a bird that was first found at the edge of the old Whitburn Colliery buildings, and then skulked around a manure heap in the arable field immediately south of what is now Whitburn Coastal Park over 23rd to 24 May 1984. Found by Dave Foster, it was trapped, closely examined in the hand by Ian Mills and ringed before being released. Its arrival coincided with the county's first spring Icterine Warbler *Hippolais icterina* at nearby Seaburn later the same day.

A bird heard singing at Low Barns, Witton-le-Wear on 7 June 1973 and seen on 12th was originally published as the first county record of Marsh Warbler in *Birds of Durham* (Coulson 1974). However, this was before the launch of the Durham Bird Club and, significantly, during a year when there was no Records Committee. Consequently, the report did not receive the due consideration that has become standard for a potential county 'first', especially bearing in mind the difficulties involved with Marsh Warbler identification. As there is uncertainty over whether an adequate description was supplied in support of the report, the 1984 occurrence must be regarded as the first well documented record (B.Unwin pers. comm.).

A total of 31 Marsh Warblers have been recorded in May and June, with most records being of singing males at coastal localities between South Shields and Teesmouth following periods of easterly winds and rain. The species is a late migrant with most records being concentrated into the period from late May until mid-June, often appearing with other typical late migrants en route to Scandinavia. It is becoming increasingly frequent in occurrence, reflecting the spread of this species into Scandinavia and has been absent during spring in just eight of the last twenty years.

In addition to the Whitburn bird there were two further records during the 1980s. A singing male was on Cleadon Hill on 3 June 1985 with another singing male at Haverton Hole, Teesmouth from 27th to 30 June 1987, both of which were trapped and ringed.

The next two decades saw a marked increase in the number of records with a number of multiple arrivals. These included: a single at Whitburn and four different birds at Hartlepool Headland during the period 29 May until 10 June 1992; a single at Marsden and three in the Hartlepool/Seaton Carew area during 29th to 31 May 1998; three birds at both Hartlepool and Whitburn in the period 6th to 11 June 2002 and two at North Gare, Teesmouth, along with singles at both Hartlepool Headland and Sunderland AFC Academy Pools, from 28 May until 7 June 2008.

Away from the immediate coastal strip, Haverton Hole, Teesmouth has recorded three birds, on 27th to 30 June 1987, from 13 June to 18 July 1993 and over 11th to 15 July 2006, with the only other records from the North Tees Marshes being of a singing males on the Long Drag from 8th to 10 July 2009, and 20 June 2011. All of these records are consistent with unpaired males attempting to establish a territory and attract a mate. Further inland, singing males at Shibdon Pond, Blaydon on 11 June 1996 and at Far Pasture on 4 June 1997 represent the only records for the Borough of Gateshead. Most notably, three birds, including two singing males, were at Herrington CP, Houghton-le-Spring from 15th until 25 June 2009.

The peak years for this species were 1992 and 2009, five birds occurring in each of these years during May and June, with the earliest arrival date on record being of a singing male at Marsden Quarry from 22nd to 23 May 1993.

There have been just four records of this species outside of May and June, though only the latter two of these could be regarded as autumn migrants. Two singing males have been found in July both of which were located on the North Tees Marshes and the two autumn passage birds were both in South Tyneside. The first was trapped and ringed at Whitburn on 26 August 1990 and the other was present on The Leas, South Shields on 10 and 11 September 2005. This species is notoriously difficult to separate from Reed Warbler *Acrocephalus scirpaceus* in autumn and identification difficulties no doubt mask the true status of this species in the county.

Marsh Warblers in County Durham, 1970-2011

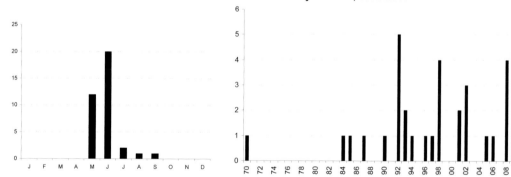

Recent Breeding status

During the *Atlas* survey period (Westerberg & Bowey 2000), there was a single record of apparent breeding behaviour in suitable habitat in the county. In 1993, a singing male was found in wetland edge habitat at Haverton Hole and held territory for around four weeks from 13 June until at least 18 July. There was a strong suggestion of a pair being present, with the singing male being seen to display to a bird believed to be another Marsh Warbler (Joynt *et al.* 2008), though there was no conclusive evidence of breeding having taken place.

In 2009, a male was found singing from low vegetation by the fishing lake at Herrington Country Park, on 15 June and remained on territory until at least 24th. Incredibly, it was joined by a second singing male over the period 20th to 21 June, with the two birds often noted singing in competition only 20m apart. Subsequent reports confirmed the presence of a third non-vocal bird in the same area, which was presumed to be a female. Reports

747

ceased after 25 June and unfortunately, there was no substantive evidence of breeding having taken place or being attempted.

At present, Marsh Warbler is a very rare breeding species in Britain with a tiny core population in southern England and just occasional sporadically breeding pairs elsewhere as far north as Shetland. Considering the increasingly restricted British range (Gibbons *et al.* 1993) this species appears highly unlikely to establish a breeding population in Durham in the future, but isolated breeding attempts could take place.

Distribution & movements

Marsh Warbler breeds in temperate Europe from France eastwards into western Siberia, though the species has been extending its range recently further north west into Scandinavia and north western Russia. It winters predominantly in south east Africa, from Cape Province north to Zambia and Malawi. Although a handful of pairs breed in Britain each year, it is largely a scarce migrant to Britain, usually as a spring overshoot in late May and early June, though birds may occur as late as July. The majority of records come from the Northern Isles and the east coast of Britain south to Yorkshire and the Lincolnshire coast. The species is much rarer further south despite the close proximity of breeding birds in both France and the Netherlands. This pattern of occurrence seems to indicate that the majority of migrant birds recorded in Britain are of Scandinavian origin. Numbers vary annually and are dependent on the occurrence of suitable weather conditions, with the ten year national average from 1990 to 1999 around 50 per annum. A total of six Marsh Warblers have been trapped and ringed in County Durham. These were at: Whitburn on 23 May 1984; Cleadon Hill on 3 June 1985; at Haverton Hole between 27th and 30 June 1987; Whitburn on 26 August 1990 and two birds at Hartlepool Headland on 6 June 2002. There have been no subsequent recoveries.

Reed Warbler (Eurasian Reed Warbler)
Acrocephalus scirpaceus

A localised summer visitor and uncommon passage migrant.

Historical review

The breeding profile of the Reed Warbler in Durham has always mirrored the patchy distribution of its favoured habitat, common reed *Phragmites australis*. The interest shown in a bird that, even in the 21st century, was still in the process of fully colonising the county, has focused observers upon it and as a result, the data available on this species compared to many others in the county is considerable.

While it was nesting quite commonly in suitable marshy areas in the south of North Yorkshire, through much of the 20th century, this species was considered to have reached the northerly limit of its British breeding range in Durham for much of that period (Mather 1986, Brown & Grice 2005). Yet, in the late nineteenth century, this species bred in North Yorkshire, as far north as a site near Redcar, which was thought, at that time, to be the northern boundary of the species' breeding activity in the British Isles (Holloway 1996).

In 1951, Temperley summarised the status of Reed Warbler locally as a "*rare casual summer visitor*", and he listed only three records for the county; all of which referred to breeding occurrences. Firstly, around 1830, a nest of this species, with eggs, was discovered built into a bed of willows in a railway cutting between Blaydon and Derwenthaugh (Temperley 1951), from an area that was close to the present day breeding site of Shibdon Pond. This was the first known nesting of the species for the vice-county of Durham. The second recorded breeding also comes from what is now part of Gateshead Borough; a nest with four eggs, being found near Ravensworth in the Team Valley on 8 June 1885 (Bowey *et al.* 1993). The third instance was from near Hell Kettles at Darlington in 1894 and the nest was exhibited at a meeting of the Darlington and Teesdale Naturalists' Field Club, on 12 June 1894 (Temperley 1951). At his time of writing, Temperley (1951) stated that Reed Warblers did not usually breed further north than Yorkshire.

In truth, we can probably glean relatively little of the species' true status in Durham up to the 1950s from Temperley's work, other than the certainty of these three breeding records. Species that were difficult to identify in the field at that time may, as a consequence, have not been fully or authoritatively documented. In Temperley's account (1951) he mentioned the "*systematic draining of lowland swamps and marshes*", which may indicate that the potential breeding habitat for this species, and therefore perhaps its distribution, had been more widespread in years gone by.

After a 64-year gap, during which no birds were recorded, a bird was at Jarrow on 3 September 1958 and around this time, regular breeding was established in the north east of England at Gosforth Park, in southern Northumberland (Kerr 2001). The species was still considered rather rare in Durham, when three birds appeared at Teesmouth in the autumn of 1965; firstly one was found dead at Hartlepool on 3 September; another was there two days later on 5 September and finally one was trapped and ringed at Graythorp on 2 October. From that time on, birds began to be recorded with much greater regularity. In 1966, single birds, probably different individuals, were at Hartlepool on 16th, 17th and 19 September and one was trapped at Hartlepool on 15 October. On reviewing this species' national status, Parslow said that the Reed Warbler bred in every English county 'except Durham', but only occasionally in Cumberland and at single localities in Westmorland and Northumberland (Parslow 1967). The next indication of breeding in the county came from Billingham Bottoms, where birds were seen in May 1969 and on 11 August 1970, whilst another autumn passage bird appeared at Hartlepool in September 1969; but it was through the 1970s that the colonisation, or perhaps re-colonisation, of the county took place in earnest.

Recent breeding status

Today, this species is a regular nesting bird around the Tees Marshes and a suite of other wetlands around the county, though it remains virtually confined to lowland stands of common reed *Phragmites australis* for breeding. It also occurs as an occasional coastal migrant. As a breeding species, the Reed Warbler breeds semi-colonially, where conditions permit, but single pairs occur especially at some smaller sites. The species' stronghold remains the reed beds and other wetlands around the north side of the Tees Estuary and while dense colonies of Reed Warblers are difficult to census accurately, this problem really only applies as a serious challenge at a handful of sites around Teesmouth (Westerberg & Bowey 2000).

The relatively recent increase of this species in the county started in earnest when birds began to establish themselves at Teesmouth during the early part of the 1970s, a male singing at Haverton Hole on 25 June 1972. Two spring birds appeared there in 1973, when a rare spring passage bird was at Hartlepool on 21 May. Meanwhile, in the centre of the county in 1973, a pair was reported to be present at a site in Weardale. Whilst this location was not published at the time, it was Low Butterby Ponds, the ox-bow pools left after a man-made change to the river's course during the early 19th century, located between Croxdale and Shincliffe. It was said that birds had been present and breeding for '*the past three years*' (Unwin 1974). These were the first known breeding records since 1894, though there were no reports of breeding at this location in 1974. In that summer a bird was heard singing at Low Barns on 19 June. Furthermore, a bird believed to be this species was at Seaton Carew on 25th to 26 September. A juvenile was mist-netted at Haverton Hole on 19 August 1974 and a bird was at Hartlepool on 31st. Breeding was suspected at Haverton Hole in the summer of 1975, when one or two birds were in song from 8 June and a pair held territory into July. Reports continued into September but no confirmation of breeding was received. In 1976, four to five birds held territory at this site and breeding was proven for the first time; ten adults were ringed there between late June and early August (Blick 2009). It would seem that from this point the species' status in Durham had fundamentally changed from that of rare visitor. From 1976, the Haverton Hole breeding population slowly increased to an estimated ten pairs by 1982, and around this time new sites began to be colonised. This process continued through the 1980s and 1990s, both up and down river of Haverton Hole (Joynt *et al.* 2008) reaching a total of some 45-50 pairs by the end of the first decade of the 21st century (Blick 2009).

Nonetheless, during the 1980s and early 1990s, the species was still a relatively scarce summer visitor to the county. At that time, it bred at only at a handful of sites away from the main Teesmouth colonies and at some of these, only occasionally. Breeding was recorded with increasing regularity during the latter part of the 1980s and some of the main sites for the species over this period were WWT Washington and Shibdon Pond. At WWT Washington, attempts to establish a reed bed took place through the late 1970s. This succeeded in attracting a bird late in that decade, when one sang there from 15 June to 17 July 1979. Another individual summered there in 1981, with three present the following year and at least three pairs breeding annually thereafter. During the early to mid-1980s, several spring records came from Shibdon Pond and in 1983 two to three birds were present from 5 June to 30 August. Breeding was confirmed there in 1986, 1987, 1991 and 1992. In the latter year, four males appeared to be in territory (Bowey *et al.* 1993). It is likely that breeding took place at this site prior to 1986, at least on occasion, as single birds were ringed there in August 1979 and May 1980, with another singing there in 1982. By 1993, up to six territories were recorded at this site.

In 1985, two singing birds were discovered at Wingate, near Durham, on 10 May and one was present on 1 June 1986. By 1988, away from Teesmouth, there were probably four small breeding colonies in existence with a total of not more than 15 to 20 pairs summering each year. *Birds in Durham 1996* (Armstrong 1997) recorded between one and four singing birds at three sites away from the Teesmouth stronghold; at the WWT Washington, Shibdon Pond and at Low Butterby, which attracted fewer birds due to a reduction in size of the reedbed. Only three to four birds were singing on 13 Jun, in contrast to more than 10 the previous year. Single summer sightings came from a further four sites. At the end of the 1990s, the Durham population was estimated to be in the region of 120 pairs, of which, around 80% were located around the Tees Marshes (Westerberg & Bowey 2000). The spread of the species into new sites continued after the turn of the Millennium with, for example, birds breeding in a small reed bed at Swalwell, in the Derwent Walk Country Park for the first time in 2000.

By 2001, further consolidation had taken place, with breeding reports coming from the four established sites, two new ones and with singing birds at an additional eight sites. At Teesmouth, towards the end of the first decade of the 21st century, singing birds were to be found in the Dorman's Pool area, in common reeds on the land-claimed parts of Seal Sands, alongside the Long Drag, around Greenabella Marsh, Cowpen Marsh, Portrack, Bowesfield and at Teesside Park (Blick 2009). In 2006, 14 breeding sites were occupied around the county and the numbers breeding at Teesmouth, were hinted at by the 48 adults and 118 juveniles ringed at Dorman's Pool that year, with another 94 adults and 147 juveniles ringed at Saltholme. In 2007, breeding was strongly suspected at six Durham sites away from Teesmouth as well as the established breeding birds at Shibdon Pond and WWT Washington. By 2010, singing males were noted at a minimum of 18 sites in the county; a very considerable increase from the six breeding sites reported just a decade before. These sites included the specially created reed beds at Birtley Sewage Treatment Works. The numbers breeding in the core area around Teesmouth were large compared to the relatively recent past, with around 50 singing birds noted in the Dorman's Pool area alone on 17 June 2010.

Today, the majority of the county's breeding population is to be found around Teesmouth, though it is now very much more widespread at other wetland sites, large and small, across the county. Surveys at Teesmouth revealed: in 1987, 27 singing birds at eight localities; in 1997, about 40 pairs at 10 localities; in 2004, at least 60 pairs and by 2007 about 200 pairs (Blick 2009). There was thought to be approximately 400 pairs across Teesmouth during the work for *The Breeding Birds of Cleveland* Atlas, 2000-2006 (Joynt *et al.* 2008).

At the mid-point of the first decade of the 21st century, it was legitimate to say that there had been a large increase in numbers, a major extension in range and a consolidation of established populations of Reed Warblers, over the previous ten years; this being the continuation of a trend that had been underway over the previous thirty to forty years. The ongoing increase in numbers and range extension that has taken place in Durham, is part of a general expansion of range in Britain and Ireland (Gibbons *et al.* 1993), which continued through the 1990s and 2000s (Parkin & Knox 2010). Overall, Reed Warblers increased by 28% over the period 1995-2008 in the whole of the UK and by 22% in English Breeding Bird Survey squares alone (Risely *et al.* 2010).

Drainage, infilling and vegetational succession are all threats to the reed beds on which this species largely depends but as many occupied sites are designated as nature reserves, these threats are unlikely to be significant in Durham. Indeed, there is potential for further expansion as *Phragmites* occurs in over 70 tetrads in vice county 66 (Graham 1988) but it remains to be seen how many of these are capable of supporting breeding Reed Warblers.

Recent non-breeding status

The earliest spring arrival of a Reed Warbler during the establishment phase of the species was on 20 April 1987, when a bird at Washington beat that site's previous earliest date by six days. In 2007, singing birds were noted on the North Tees Marshes from 18 April, a little earlier than the average arrival date for the species. In recent years, more birds have been recorded during the last week of April, though most singing males still arrive in the first half of May. The earliest date for a bird in the county is one at Haveton Hole on 9 April 2011. Over the period 1976 to 2005, the reported first arrival dates were analysed, in ten-year groupings, these were: 1976-1985, 7 May; 1986-1995, 3 May and 1996-2005, 23 April. The mean first arrival date of the species advanced by 14 days between 1996 and 2005 (Siggens 2005), in line with the population expansion in Durham over that period.

In Durham, most birds have left breeding sites by August and this is an uncommon spring and autumn passage migrant in the county, occurring largely on the coast, usually in the well-watched areas of Hartlepool and

Whitburn/Marsden. Spring passage birds are not common, with rarely more than a bird or two being noted. Autumn records are more numerous with birds occurring at any time between mid-August and mid-October, though mainly in September; although in 2000, coastal migrants were noted in the Trow Quarry to Sunderland area between September and November but even in autumn, coastal birds are generally scarce. A good autumn passage might number between five and ten records along the county's coastal strip, although in some years more birds might be recorded; for example, well over 20 were present during September and October 1998. The documentation of coastal migrants became very much more regular from the early 1970s, with autumn passage birds becoming effectively an annual feature from around 1973. One bird seen bathing in a garden pond in Whitburn on 25 October 2004 was most unusual, whilst a bird inland at WWT Washington on 27 October of that year, was very late for an inland bird. There are two sightings of the species relating to November, one was at Marsden Hall on 5th in 1984 and later still, one was ringed at Graythorp on 15 November 1975 (Blick 2009).

Distribution & movements

Reed Warblers, of the nominate race, can be found breeding in wetland habitats across much of Europe, from Spain and Portugal in the southwest, north to Fennoscandia and east to the Volga, with some birds also present in northwest Africa. To the east, the eastern race is present in a wide area covering Asia Minor, through Iran and east to Kazakhstan. Birds winter in Africa to the south of the Sahara (Parkin & Knox 2010).

Since the mid-1970s regular ringing at Haverton Hole and along the Long Drag has shown that breeding Reed Warblers at Teesmouth are relatively site-faithful (Blick 2009). There was probably a small-scale developing inland passage through the county as breeding populations moved north and south, as indicated by birds being ringed at Shibdon Pond in August 1979 and May 1980. The migratory ability of this species was amply demonstrated by the bird ringed as a juvenile at Teesmouth in July 1980 that was last re-trapped in August 1991, presumably having crossed the Sahara 22 times (Joynt et al. 2008). This longevity was almost matched by a bird originally ringed on the North Tees Marshes on 5 July 1993 that was controlled at Gosforth Park, Northumberland in summer 2003, making it at least 10 years old. There are many examples of birds ringed in the north east and controlled further south in England, or ringed further south and found breeding in the north east. Occasionally more local movements have been documented, such as one of two males ringed at Shibdon in the summer of 1987 that was re-caught at a breeding site on the Tees Marshes in the summer of 1991 (Bowey et al. 1993), and one ringed at Dorman's Pool in 2006 that was controlled at Rainton Meadows in 2008.

Sample Ringing Controls and Recoveries Relating to Durham

Ringed	Date	Recovered	Date
Rostherne Mere, Cheshire (as nestling)	Summer 1975	Teesmouth	26/06/1976
(same bird as above)		Rostherne Mere	August 1977
Corsham Lake, Chippenham, Wiltshire	09/08/1976	Saltholme	03/06/1977
Staffordshire	Summer 1981	Shibdon Pond	Summer 1982
Shotton, Clywdd, Wales	September 1990	Shibdon Pond (breeding male)	Spring 1993
Haverton Hole (as juvenile)	08/08/1988	Icklesham, Sussex	05/09/1988
Long Drag (as juvenile)	02/08/1998	Slimbridge, Gloucestershire	27/08/1998
Haverton Hole (as juvenile)	26/08/2004	Portugal	22/09/2004
East Chevington, Northumberland	16/08/2006	Haverton Hole	31/05/2007
Haverton Hole (as juvenile)	24/08/2007	Sandwich Bay Kent	04/09/ 2007
Morocco	11/04/1989	Haverton Hole	14/07/1990
(same bird as above)		Haverton Hole (again)	Aug. 1991

Some of the most significant recoveries for the county include: the juvenile ringed at Haverton Hole on 28 August 1981 and found dead in an army barracks in the Sahara at Tan-Tan, Morocco on 19 October 1982; an adult female from Saltholme May 2008 controlled at Hondaribbia Guipuzcoa, Spain August 2008 that was presumably heading to its wintering grounds in West Africa and which was later re-trapped at Rainton Meadows on 14 May 2009.

Great Reed Warbler
Acrocephalus arundinaceus

Sponsored by

Gateshead Council

A very rare vagrant from southern Europe: two records.

The first record of Great Reed Warbler for County Durham is of a bird shot by Thomas Robson close to Swalwell on 28 May 1847. Said to have been skulking in the low herbage close to a mill dam, Robson was initially attracted to the bird by *"its note"* which he did not recognise. This no doubt refers to the species characteristic loud song, by which many subsequent vagrants to Britain have been located. The specimen is retained in the Great North Museum collection, Newcastle-upon-Tyne. Notably, this also represented the first record of the species in Britain & Ireland (Palmer 2000).

The second record was over a century later when a singing male was present at Haverton Hole, Teesmouth from 22 June to 4 July 1995. This bird was initially found by Russell McAndrew in the south eastern corner of the main pool, and although occasionally elusive, particularly on windy days, it frequently showed well with its loud song audible from several hundred metres away (*British Birds* 89: 517). That both records in the county should refer to singing males in May and June is entirely typical of this species' occurrence in the British Isles.

Distribution & movements

The nominate race of Great Reed Warbler breeds in North Africa and across much of Europe eastwards into western Asia, though is absent from Great Britain and much of Scandinavia. Strongly migratory, it winters in tropical Africa from Sierra Leone to southern Ethiopia, southwards to Namibia and South Africa. It is a regular vagrant to British Isles, predominantly as a spring overshoot in May and June, though it has been recorded from late March until mid-November. There had been over 240 British records by the close of 2010, the majority of which were singing males at widely scattered localities. Several have established territories and remained for long periods, although the species has yet to be proven to have bred in Britain.

Great Reed Warbler, Swalwell, May 1847 (K. Bowey, courtesy NHSN)

Waxwing (Bohemian Waxwing)
Bombycilla garrulus

An irregular winter visitor and passage migrant, that is sometimes recorded in large numbers during 'irruption years'.

Historical review

In the modern context, the Waxwing is renowned as a northern breeding species that exhibits an 'irruptive pattern' of occurrence. Large numbers of birds occasionally move south and west from Scandinavia and Russia, primarily as a result of a failure in the berry crop in these areas. It would appear, from the historical record, that this pattern of behaviour is long-established.

The first fully documented record of Waxwing for Durham comes from the 'Allan' Manuscript, which stated that *"a number of these birds appeared about Darlington in 1787"*. This is pre-dated however, by information from Ray and Willughby's correspondent Mr D. Ralph Johnson, who lived in the extreme south of modern Durham, at Brignall.

He reported an influx of Waxwings in the previous March when writing to Ray and Willughby on 7 May 1686. He said "*a new English bird...They came near us in great flocks, like fieldfares, and fed upon Haws, as they do*" (Gardner-Medwin 1985). It is presumed that this influx was noted at Brignall, which is in the modern recording area of County Durham. This is one of the earliest British records, with just one Yorkshire record preceding it in the national context (Mather 1986).

In the nineteenth century, Hogg (1827) said that it was an occasional visitor in winter, one being shot at Norton in the winter of 1790-1791. Thomas Robson, author of *Birds of the Derwent Valley*, said that the 'Bohemian Chatterer' was of "*rare occurrence in the valley*", though when it did occur it was sometimes seen in large numbers. Records from the 19th century in this area included an unusually late bird that was recorded at High Spen in May 1889 (Robson 1896).

One of the few early 20th century records, prior to the early 1930s, came on 19 November 1927, when one was seen about two miles from the coast near Sunderland, after strong easterly winds had been blowing for some days. It was said to have flown off inland, with "*a kind of whistling trill*" (Byers 1927).

Temperley (1951) described it as a "*winter visitor in small numbers every year, but periodically considerable flocks arrive and may remain for some weeks in the County*". He commented upon their favoured haunts, their liking for the berries of the guelder rose *Viburnum opulus* and that they frequently appeared where that shrub was common such as at High Horse Close Wood in the Lockhaugh area of the lower Derwent valley. In the first half of the 20th century this area around Rowlands Gill became renowned for its wintering Waxwing. During the 1930s Waxwing were noted on an almost annual basis here. In 1931, George Temperley corresponded to *BB*, "*On November 1st, 1931, a small flock of seven Waxwings .... was observed on a patch of scrub consisting largely of Guelder-rose bushes near Rowlands Gill in the Vale of Derwent. Within a week the flock had increased to 30, and on November 10th, when I visited the spot, I counted 75 birds. The maximum was reached on November 24th, when there were well over 100 birds. From that time onwards the flock dwindled in numbers, and on the last two days on which they were observed, December 5th and 6th, only three were seen*" (Temperley 1932). In 1951, Temperley documented the series of sightings from that location during the 1930s and 1940s where they fed on guelder rose berries, which formed a significant component of the local woodland understorey. Temperley himself visited the garden of Charles Hutchinson at Lockhaugh to view the birds and he commented that flocks were seen there in 1931/1932, 1932/1933, 1935/1936, 1936/1937, 1937/1938, 1941/1942, 1946/1947 and 1949/1950 (Temperley 1951).

Mirroring the reports mentioned above, in the early 1930s, in the Teesmouth area, Joseph Bishop reported considerable autumn numbers in the Stockton-on-Tees district. In 1943, a year in which birds were not recorded at Lockhaugh, one was noted in King George Road, South Shields, on November 25th. Also in 1943 there was a party of about six birds, near West Stanley railway station on December 17th, with four there again on December 20th.

Large influxes were noted in Yorkshire in the winter of 1946/1947 (Mather 1986). With reference to this winter the single bird at Cleadon on March 3rd was said in *British Birds* to "*carry the record a little later than any of the notes published*" (Temperley 1948), the latest spring record at the time. Large numbers were also recorded in Durham and Northumberland during the winter of 1949/1950 (Cornwallis 1961).

During the 1950s, small numbers of birds were again noted in the High Horse Close area in the winters of 1951 and 1953, with a dozen birds also reported in the Ravensworth Estate in the former year (Bowey *et al.* 1993). In the national context there was a series of invasions of Waxwings in the four successive winters from 1956/1957 to 1959/1960 (Cornwallis 1961). During the influxes in February and March 1957 and from November 1957 to March 1958, birds reached Durham (Cornwallis 1961), for example, there were 12 at Stanley in early 1958. Later that year nine to ten were at South Shields on 9th to 10 December, but the influx into Durham was completely overshadowed by that in neighbouring Northumberland where over 1,000 birds were noted (Kerr 2001). In the next winter period, a further influx was recorded but only relatively small numbers, with a maximum of 25 at Chester-le-Street on 29 October 1959. Many parties of less than 10 birds were reported during November but it was a poor berry crop locally and by December most birds had moved on (Grey 1960).

There were very few reports of occasional birds during 1960-1962 although these did include an interesting observation of birds eating snow at Sunderland in the harsh conditions of November 1962 (Bell 1963). There was a widespread invasion again in the autumn of 1963 with birds at many localities in Durham (and Northumberland) from early November but only in ones and twos and small parties; the largest gathering concerned 20 at Witton Park. Widespread small parties remained into 1964 with a maximum of 15 at Darlington in January of that year.

The autumn of 1965 saw one of the largest documented influxes to that time, with the first birds appearing on the early date of 13 October when seven were at South Shields. From 16 October records were widespread and numbers built up with the largest flock recorded in the county being of "*several hundred*" in Castle Eden Dene around the last week of November. Birds dispersed over a wide area and there were no large flocks reported during December. Surprisingly the only Durham report early in 1966 was of a single bird at Darlington on 27 February, but later in the year small numbers were widespread with a maximum of 20 birds at Rowlands Gill over 12th to 14 December. A similar pattern prevailed in 1967 with 38 at Darlington by far the largest party reported. The next two years produced just a few reports.

The autumn of 1970 saw the largest numbers since 1965. The first birds arrived on 27 October and were quickly followed by flocks, and during November there were gatherings of 300 in Spennymoor; 150 in Sunderland, 70 in Hartlepool and up to 40 at many other localities. A rapid decline in numbers followed in early December. A smaller-scale invasion occurred during 1971 with widespread records from 11 November but no flock greater than the 83 present at Cleadon on 5 December was noted. The following year saw fewer, with perhaps less than 100 birds in total later in the year. Amongst the late visitors was a group of 17 birds which fed on Swedish whitebeam *Sorbus intermedia* berries in the Market Square, Sunderland on 3 December, unconcerned by the Christmas shoppers passing by metres away.

The smallest numbers for years came in 1973 and 1974, with none at all in the first part of the latter year. The situation at the end of the year however, could not have been more different as, after the first report at South Shields on 25 October, Waxwings poured into Durham. By the end of the month 20-25 were present in the coastal locations of Hartlepool, South Shields and Sunderland, then in November there were gatherings of 100 at Felling and Jarrow, 30-50 at Durham, Gateshead and Hartlepool, and 10-20 at several other locations. December produced flocks of 100 at South Shields and Wrekenton and 80 in Sunderland, the increase in numbers at the coast suggesting continued immigration. In the following year small numbers were noted up until early April with a few birds until the end of that month, then autumn produced a lesser influx but still sufficient for numbers to build to 100 at Hamsterley Mill on 4 December although this was more than the rest of the county combined.

A review of the position up to 1975 appeared to show something of a developing pattern with a very good year followed by a less good one, then numbers becoming progressively smaller until the next big influx some four to five years after the last. Hence there were major influxes noted in 1965, 1970 and 1974 with lesser numbers in 1966, 1971 and 1975 and smaller numbers during the in-between years.

However, the remainder of the 1970s failed to produce any substantial flocks at all with by far the largest counts being of up to 21 at South Shields and Jarrow respectively in 1976 and 1977. In the decade that followed Waxwing became a very scarce, almost rare, winter visitor to Durham with the period 1977-1985 producing only one flock of more than four birds: 20 at Peterlee on 5th to 6 December 1981. There were only two to three records per annum in 1978, 1980, 1982, 1984 and 1985 and no records at all in 1983; this would appear to be the only blank year on record in the last 50 years.

Recent status

Following on from the relatively barren years of 1975-1985 this species has been recorded with a greater degree of regularity and in larger numbers over the last 25 years compared to the 30-year period from after the Second World War. It seems likely that part of the reason for this trend relates to the greater number of active observers now. From late October to late February, small numbers of Waxwing can appear almost anywhere in the county, usually where berries are in good supply. Some birds appear in town and village centres but they also have a penchant for disused railway lines and the 'amenity planting blocks' of suburban estates, depending on the supply of berries available. There is sometimes a secondary peak in numbers in late winter or even early spring. It is thought this involves birds that have penetrated further south, and are moving northwards again, stopping over in Durham prior to onward migration through the east coast of Scotland and on to Scandinavia, as was demonstrated by colour-ringing exercises based in Aberdeen in the early 1990s (Kerr 2001).

In 1986 and 1987 there were relatively small influxes of birds early in January rather than in autumn, presumably as birds only had to move from Scandinavia as food became scarce. In 1986 the first birds appeared on 5 January with relatively modest peak counts of 13 at Washington in February, and 18 there in March when 15 were also noted at Whickham. There were just singles at three sites in December 1986 but on 5 January 1987 a flock of 12 appeared at Washington with parties of one to ten at seven other locations between there and Sunderland during January, the proximity to the coast again suggestive of recent immigration. The Washington flock increased to 28 by 7 January and 51 by the end of the month, the first count in excess of 50 birds anywhere in Durham since December 1975. Up to 45 were present in February but numbers declined as the berry crop became depleted and just six remained in March.

In 1988 there was a return to the former pattern of late autumn arrivals with a massive influx of birds into County Durham as part of a national irruption of over 3,000 birds. The first in Durham came early, on 12 October, when 25 were at Washington, but it was November when most arrived. During this month there were many widespread sightings with substantial flocks of 200 close to the A19 at South Hylton, and 80-100 at Houghton, Stockton and Washington. By December more reports came from inland locations with a December maximum of 60-70 birds at Hamsterley Mill. The first part of 1989 saw birds still in the area with 50-60 around the berry-rich suburbs of Washington in January and 30-35 in several areas during February and March. April saw an obvious increase as birds headed back towards their breeding grounds, sightings being concentrated in the north east of the county. At this time, Washington held up to 69 birds until 20th, Blaydon a maximum of 67 from 9th until 24th and Tunstall a flock of 40 on 13th. There were very few birds during the winter of 1989-90 but Washington again proved attractive to those that did occur with a maximum of 12 birds there on 24 February, the last sighting coming on 17 April.

Late 1990 saw probably the biggest influx since 1970 with the earliest ever county record concerning a bird at Washington on 2 October, the start of an extensive invasion. A few more were noted from 18th into early November before the major influx occurred from 10 November. The Borough of Gateshead was particularly favoured during this irruption with November flocks of up to 92 birds at a number of locations, 50-60 also being noted at Houghton and in the Washington area. Even more were seen during December with estimates of well over 500 birds in Gateshead which included flocks of up to 300 at Swalwell, 160 at Wylam and 90 at Dunston (Bowey *et al.* 1993).

Not surprisingly, 1991 produced some good-sized flocks during the first quarter with a couple of flocks of 70 in January, but a decline during February and then a notable increase during March as birds returned north. On 3 March, the Washington flock peaked at 100 with 64 at East Boldon and 27 at Newton Hall staying through until mid-April. Late 1991 saw a further influx but although on a smaller scale than the previous year it did produce the largest ever flock in the county at Southwick in Sunderland. On 6 December a flock of 224 was present here, increasing to 300 the next day and then 400 on 14th, but such numbers quickly consumed the local berry crop and only 150 remained on 16th before birds moved on. Flocks of 50-65 were also noted at Peterlee and Swalwell in December. The early part of 1992 produced a few small flocks of 20-30 but the prime site of Washington held 40 in January increasing to 60 in February and March with *c.*50 still present there as late as 22 April.

A lean spell then ensued with just two records in late 1992, and a handful of sightings in both 1993 and 1994, and then just a single record of two birds at Whickham on 25 February 1995. During 1996 however, there were 257 records when a mid-winter influx led to estimates of 10,000 birds being present in the UK (Armstrong 1998). The influx began in the county with two to three at Blaydon and Gateshead on 1 January. A few more appeared in the next 10 days or so and then 84 were at Washington on 12th rising to 109 by 18th but declining to 80 by the end of the month. Flocks of 30-60 were noted in the far north east at Boldon and Cleadon, suggesting they may have been newly-arrived, but there were many reports of two to three birds as far west as Ireshopeburn. A flock of 120 was present in Darlington on 23 February with more reports during March when 110 were at Wardley and 40-60 at Darlington and Newton Hall. Good numbers were in the Gateshead area in mid-April with a maximum of 66 in Gateshead and 50 at Rowlands Gill. There was also a strong presence late in 1996 but a relatively late arrival, the first sighting not coming until 13 November. By 27th, 127 were present at Houghall with 56 at Castle Eden Dene and 40 at Washington. Blaydon held 123 in December when the Washington flock increased to 300 by 10th and a record-equalling flock of 400 was there on 29 December but the flock dispersed before the New Year.

The period from 1997 onwards has seen birds present in the county in reasonable if variable numbers on an annual basis. In the north of the county the most favoured locations have tended to be around Birtley, Gateshead, South Tyneside and Washington, with Billingham and Stockton perhaps the most reliable areas in the south although, in the better years, birds were found in any location with a good supply of berries. From 1997-2000 there were only four flocks in excess of 50 birds with 60 at Consett on 2 February 1997, 200 at Blaydon on 2 December 1999 and 70 at Waldridge and Washington in February 2000.

The beginning of 2001 saw an obvious influx with many reports of up to 20 birds in January but 25-50 at six widespread locations and a couple of big flocks; 240 were present at South Shields and 164 at Stockton, but no more than 50 were seen in February. In autumn 2001 and during 2002 the highest counts were of 30. Another winter irruption occurred in January 2003 but numbers peaked in February, when 180-190 were present in Stockton and Washington, 109 in Jarrow and 85 in Darlington. A further influx occurred at the end of the year with 200-220 at both Stockton and Washington and 300 in a Jarrow housing estate during the last week of November. December 2003 to January 2004 saw the Jarrow flock drop to around 90, which was the largest gathering until yet another impressive showing in November-December 2004. At this time, probably over 1,500 birds were present in the county, numbers peaking in the first week of December. Several locations held flocks of 50-100, 150-200 were at Framwellgate Moor and Spennymoor, with 300 at Houghall and Durham's third 400-strong flock was in the same Jarrow location as that in 2003.

The first week of January 2005 produced flocks of 100-200 at Brandon, Sherburn and Spennymoor but numbers declined rapidly and there were few significant flocks for the remainder of the winter. Despite a fairly moderate influx at the end of 2005 a gathering at Port Clarence exceeded three figures with a maximum count of 115 on 15 December, then the first quarter of 2006 produced 80 at Jarrow in January, and a northbound gathering of 78 at Washington in mid-March with further travelling flocks of 50-70 in the north east of the county, apparently held up on their northbound journey by northerly winds and some snow during the second week of April. The winters of 2006/2007 and 2007/2008 had few records but produced annual peaks of 47 and 33 respectively.

The 2008/2009 winter produced another substantial invasion. The first birds appeared on 28 October and during the first half of November there were parties of up to 30 at 18 locations, with Jarrow holding 90 birds by 10 November. The housing estate centred on Saxon Way in Jarrow was one of the county's main sites for Waxwing over the first decade of the 21st century; this location again held 400 birds on 21 November 2008. Elsewhere, 185 were present at Stockton, 100 at Wynyard and 82 at Billingham. Washington's numbers peaked at 103 on 20 December, but increased further to 116 during January and 200 in the first week of February, when 150 at Birtley may well have included many of the same birds. On 26 February a flock of 100 were at Chester-le-Street. Smaller flocks of up to 55 birds were reported from Bishop Auckland, Durham, Gateshead, South Shields, Stockton and Sunderland during the first quarter of 2009. There was just a single record in the autumn of 2009 and a quiet start to 2010 with a maximum count of 26 birds at Jarrow on 14 February.

The autumn of 2010 produced an enormous and very early influx to Britain, particularly into Scotland. During November, flocks reached 230 in Jarrow, 140-150 at Great Lumley and South Shields and 60-70 at Elemore and Sedgeletch. Most were dispersed by December when only three flocks of 25-50 were noted, all in the south of the county.

Distribution & movements

The Waxwing has a circumpolar distribution and breeds at high latitudes in both Eurasia and North America. It is an irruptive species, occurring in the British Isles in small numbers on an annual basis in winter, but occasionally appearing in spectacular influxes. Birds found in Britain are of the nominate race, which breeds in Europe and western Siberia. The reasons behind the irruptions of this attractive and often confiding species into the UK have been established for many years and largely relate to seasonal food shortages in Scandinavia and points east (Svardson 1957). The earliest and latest dates for Waxwing in Durham are 2 October 1966 and 1990 and 17 May 1975.

The national ringing data-set shows clear evidence for the origins of birds reaching the UK (Wernham *et al.* 2002). Of 32 birds ringed abroad and recovered in Britain most were ringed in Norway (18) and Finland (9), the others coming from Sweden (2), Denmark, Belgium and The Netherlands. There is one Durham recovery of a foreign-ringed Waxwing, a bird found dead at New Brancepeth in January 1975 which had been ringed in Norway on 13 November 1974, a movement of 630km.

Of those birds ringed in Britain and recovered abroad (21 in total) five were found in Sweden, Norway (4), Denmark and Finland (3), Russia (2) and one each in Belgium, France, Germany and The Netherlands.

A number of birds ringed in Durham have been controlled or recovered elsewhere within Britain. Two birds out of six ringed at Hartlepool on 3 November 1974 were recovered, one near Helmsley, North Yorkshire on 15 March 1975 and one at Tonbridge, Co. Antrim, Northern Ireland on 16 April 1975. A bird ringed at Acklam on 19 November 1988 was present at Monkwearmouth, Sunderland on 1 May 1989 (Blick 2009). A bird, which had been colour-ringed in 1990 at Aberdeen, was found at Swalwell in December 1990 and through careful observation it was later tracked around the country also being seen near York and at Manchester (Bowey *et al.* 1993). A female ringed at Swalwell, Gateshead on 30 December 1990 was killed by a cat at Thirsk, North Yorkshire forty days later on 8 February 1991. Another female ringed at Swalwell on 24 December 1990 was found dead having flown into a glass window in Stockport just 24 days later on 17 January 1991. A number of birds present at Newton Hall, Durham City in March and April 1991 had been colour-ringed in Aberdeen in the autumn of 1990 and were seen as far south as Manchester during the winter months.

Nuthatch (Eurasian Nuthatch)
Sitta europaea

A widespread and locally common breeding resident in suitable woodland habitats.

Historical review

Mather (1986) makes reference to what is the first apparent record of the species from the modern Durham area. It was made in a letter from Ralph Johnson of Brignall, to John Ray in 1678, "*The Nuthatch or Nut Jobber... hath not a long tongue as the other [the Woodpecker kind] because she feeds not on Cossi as they do, but on other insects and especially on nut-kernels*". He then described the action of a Nuthatch breaking open a wedged nut or acorn, something that he was presumably seeing in woodlands such as Brignall Banks, along the River Greta, a tributary of the Tees in what is today the southernmost portion of County Durham.

The next reference came courtesy of Selby (1831), who said "*I have frequently seen the Nuthatch*", "*in the neighbourhood of Durham*". Edward Backhouse (1834) said it was "*not very common*". Hutchinson (1840) considered it "*very scarce*" but he recorded its occurrence around Durham woodlands, including Marsden Castle, Houghall, Great High and Little High Woods. He also said that it was still found in Auckland Park and where trees of a "*great age*" occur. Selby (1831) recorded that it had been noted on the "*banks of the Tyne*", indicating a relatively northerly location for the species at that time, despite the fact that later in the Victorian period it had retrenched to positions much further south (Hancock 1874).

Hogg (1845) wrote "*for many years I have observed it near the City of Durham*". He thought the species was at its northern limit in Britain in County Durham. Hancock (1847) documented its regular breeding in Auckland Park "*30-40 years ago*" (i.e. around 1805), but that it was now "*a rare casual visitant*". From then until the early part of the 20th century, it remained a "*casual visitant*", with no reports of breeding, most sightings being of single birds in the south of the county. In 1901 at least one, possibly two pairs bred near Wynyard (Temperley 1951). From 1925, records became more frequent. Joseph Bishop noted Nuthatch nesting in south Durham in the 1930s and in commenting upon its "*Jack in the box*" behaviour, illustrated the then status of the species. He concluded "*it gives me great pleasure to forward the above notes and thus to throw a little further light upon the status of one of County Durham's rarest nesting species*" (*The Vasculum*, Vol. XXIII, 1937).

It would appear that sporadic breeding was taking place in Durham up to about 1845; the species then retrenched and possibly even became extinct as a breeding bird. The main breeding range of this species over the second half of the 19th century was further south in Yorkshire. Subsequently, there was re-colonisation of Durham, this taking place from around the turn of the 20th century. The first 20th century nest was found near Wynyard, but regular breeding was not established or noted as occurring there until around 1927 with birds spreading to the rest of southern Durham, along the Tees valley by the 1940s (Holloway 1996). Maurice Robinson recorded Nuthatch with young on his way to school in Darlington in 1926.

In 1951, Temperley regarded it as "*a resident in very small but increasing numbers*". The two or three decades preceding Temperley's time of writing saw some very considerable changes in the status of this species in Durham and this change accelerated over subsequent decades.

In 1932, there was a report in *British Birds* of birds being noted on the Durham side of the Tees, at Eggleston, in August 1927, and again in 1928, 1930 and 1931 (Clegg 1932). Birds were nesting near Barnard Castle in April 1930 and from 1927 to 1931 a pair bred near or in Darlington South Park. In 1936, birds were nesting in the wooded outskirts of a village in south Durham. In 1937, Joseph Bishop said that there were six pairs breeding in the lower Teesdale area. In 1940, a new pair was found nesting on the outskirts of Darlington and a pair was at Middleton-one-Row. In 1935 and 1942, Nuthatches were observed in Middleton-in-Teesdale, and they were found breeding there in 1943 and 1947. This indicates a healthy population in the upper Tees valley by the late 1930s and early 1940s, possibly expanding up the valley from Darlington and the lower areas after becoming established in that area from the 1920s or possibly reaching Darlington after coming down the Tees valley. By the late 1940s, birds were back and breeding around the Durham City area, with a nest discovered at Durham Banks in 1949.

There was little information available about the species' status in the county during the 1950s, though in 1952 it possibly bred at Durham Banks, when two birds were seen with a juvenile. Two pairs bred near Middleton-in-Teesdale and a pair was building at Whitworth Park in April but breeding was not proven. At the other end of the decade, birds bred successfully at a site in the south of Durham.

In the early 1960s, Stead said that this species was not present as a breeding bird in north Cleveland, and only sparsely distributed in south east Durham (Stead 1964). In 1960, five pairs were estimated to be present along the Tees in the Eggleston area upstream of Barnard Castle and it was also reported around Durham Banks, where birds were present the following year, and Darlington, near Chester-le-Street, Spennymoor, Wynyard and Shotley Bridge were all noted as breeding locations. Through the rest of the decade breeding reports were sparse coming from Barnard Castle (1962), Durham City (1963 & 1967), Plawsworth, near Chester-le-Street (1964), Croxdale Hall and Hamsterley (1967). In 1969, five birds were singing in Castle Eden Dene and birds were also reported from Durham City, Hamsterley Forest and the Tees valley (Parrack *et al.* 1972).

In the north of the county, the first record of this species in the lower Derwent valley dates back to 1964 when birds where noted at Hamsterley Mill. Birds were regularly present in that area over the next 15 years but, apart from occasional reports from Lintzford and an isolated report from Thornley Woods in 1976, birds did not seem to be colonising Gateshead over this period (Bowey *et al.* 1993). The species was beginning to establish as a breeding bird in Northumberland, in the Tyne valley from around 1970, breeding having been first recorded at Bywell in 1964 (Kerr 2001).

Through the seventies the number of sightings rose considerably and the range of sites from which the species was reported did likewise. Over the first few years of the decade it was noted regularly from: Castle Eden Dene; the Tees valley, up to Low Force and downstream of Barnard Castle, in the East Lendings area; sites along the Wear, upstream to Low Butterby and down to Finchale, including Durham City, where breeding was recorded; and in the Bedburn area, where numbers were said to be increasing. Darlington was a good area for the species at this time and 13 were counted in South Park in February 1973. From around 1973 birds were noted at Wynyard and Wolviston in the lower Tees valley and in the late 1970s, there were regularly one to three pairs in the Wynyard Estate (Blick 1978).

Recent breeding status

Thirty years ago, despite being rather common in southern England, the Nuthatch was still a relatively uncommon and rather localised species in County Durham. Finding one away from the handful of mature deciduous woodlands where it was established was quite unusual. The local population of Nuthatch has expanded quite dramatically since the 1970s, with new woodlands being colonised in every year during the *Atlas* period (Armstrong 1988-1993) and in every year since. Nuthatch numbers have increased substantially in England since the mid-1970s, expanding its range across northern England and into the Scottish Borders (Brown & Grice 2005); the BBS index over the period 1994-2006 showing an increase of 62% (Raven *et al.* 2007).

Today it is widespread and increasingly common in woodlands in many parts of the county. At the time of the *Atlas*, birds were routinely found in Castle Eden Dene, Hawthorn Dene and Wynyard Park (Westerberg & Bowey 2000), but there has been a major local range expansion since then and the population of Nuthatch is still expanding locally in Durham, as it is across much of north west Europe (Hagemeijer & Blair 1997). Nationally, there has been a steady and consistent increase in numbers and range since the early 1970s, with an apparent doubling in numbers since 1967 (Parkin & Knox 2010).

The expansion through the county in its initial growth phase, during the sixties and seventies, was in part linked to the spread of Dutch elm disease, and the consequent increase in the number of standing dead trees, which provided an abundance of new nesting sites. The co-incidental increase in garden bird feeding at this time probably led to an increase in winter survival rates. During the 1980s, the species' expansion, in both range and numbers, gathered pace due to this two-fold advantage. For example, from the autumn of 1981, birds were increasingly seen in the lower Derwent valley woodlands. Nuthatches first started breeding in the Derwent Walk Country Park in 1983, a year after the first elm died as a result of the disease. The first confirmed successful breeding there took place in 1984, with no fewer than eight pairs present by 1986, and at least ten pairs by 1992 (Bowey et al.1993). That decade saw birds increasingly reported in areas of deciduous woodland where hitherto they had been only occasional visitors or completely unknown.

Birds in Durham (Armstrong 1990) described the Nuthatch as being *"an uncommon resident of deciduous woodland"*, with records for that year from just 47 sites, including several new ones in the county, such as Flass Vale, in the centre of Durham City. Clearly, a process of 'in-filling' was occurring. Since then, the range expansion has continued unabated. This has included a movement into suitable habitats in urban/suburban locations such as Saltwell Park, Gateshead and, most recently, birds have started penetrating into the north east of the county. Birds are now increasingly reported from suburban localities, such as Jarrow Cemetery and Morton Wood, though the isolated incident of a bird singing near the coast at Whitburn village in July 1984 has, to date, not been repeated.

As stated in the *Atlas*, the distribution of Nuthatch, more so than for most other local woodland species, tracks the county's river valleys, along which the best examples of mature woodland occur. These include the lower Wear and Derwent valleys, and Teesdale in the Barnard Castle area, which are all still major strongholds of this species. In the west, the species' presence is limited by altitude, with the altitudinal limit of the species in the county occurring at around 260m (Westerberg & Bowey 2000).

Even in the late 1990s this species was still extending its range in the county, with four new breeding tetrads logged in 1999 alone (Armstrong 2003). A bird singing at Baybridge in the Derwent valley in late April 1999 was the most westerly location for the species at that time; the individual moved west, out of Durham into Northumberland.

The woodlands in the west of Durham which most regularly record birds include those at Balder Grange in Baldersdale, the valley woodlands around Stanhope in Weardale (Armstrong 1988-1993) and Baybridge in the upper Derwent valley (Westerberg & Bowey 2000). The species frequents mature deciduous woodlands, especially those with abundant oak *Quercus* spp. or beech *Fagus sylvatica* (Harrap & Quinn 1996); locally, elm *Ulmus* spp., and ash *Fraxinus excelsior* are also frequently used as nest sites (Westerberg & Bowey 2000).

In 2006, the Durham Bird Club undertook a Nuthatch survey to increase knowledge of the species' distribution and abundance and the changes in these over recent decades. Further expansion in range was revealed after a review of the electronic records for 2005, which provided breeding season records for five tetrads not mapped in the *Atlas*, whilst survey results from 2006 identified an additional twelve occupied tetrads. Colonisation of these may have taken place previously, but clearly the trend is one of continued expansion and infilling of previously unoccupied areas. Some of these 'newly' occupied tetrads appear to have been colonised from adjacent areas, for example at Croxdale Oxbow and Arlaw Banks in the Wear valley. By way of contrast, a number of the newly documented sites, such as Bishop Middleham, Elemore Hall School, Hardwick Hall, Satley, Stanley and Wingate are relatively isolated from the large blocks of mature deciduous woodland which are the traditional strongholds of the species.

The lack of records during 2006 from much of the south of the county, especially the Tees Valley, appeared to reflect a lack of observer coverage rather than a genuine absence of birds from areas which had been first re-colonised some 60-70 years previously.

In 2007, there were 429 records of the species, mainly from well-established locations, but the frequency of occurrence in some particularly well watched areas, such as Hetton, suggests that this species is continuing to expand its population most notably towards the north and east. Today there are probably around eight pairs in the area north of the Tees (Joynt et al. 2008) and it has nested in Wynyard, from about 1970 and possibly since 1961, Stockton in 1976, Hartburn in 1986 and Ward Jackson Park in Hartlepool from 2007 (Blick 2009).

During fieldwork covering 1988–1993, Nuthatches were recorded from 98 of the possible 825 tetrads in the county (Westerberg & Bowey 2000). The density per occupied tetrad varied from one to eight pairs, with an average of 2-3, leading to a population estimate of around 200–300 pairs. Surveys in April 2006 indicated the difference in territorial densities of this species in woodlands at coastal and inland locations. The first in Hawthorn Dene found just one calling bird in a 4.3km transect, whilst in the Durham City area, a 1km transect at Cocken Woods produced 7 calling birds. A further survey found five calling birds in an area of just 0.5km through Durham Banks and five calling along a 1.5km route in Great High Wood, on the outskirts of Durham City. Clearly there are greater numbers in the inland areas of the large coastal denes, where 22 birds were noted in the western end of Castle Eden Dene during March 2003. Today the population has probably grown by at least 20% over the figures estimated in the *Atlas*, which would suggest a current population of around 350 pairs. By 2009, almost all suitable woodland in the county held birds, as did a number of town parks, though not those in the extreme north east of the county. The westerly limit of this species in Durham was delineated by a line traceable from Brignall in the south through Barnard Castle to the western edge of Hamsterley Forest and north up to the Derwent Gorge, and the species continued to expand north and east through the county.

The first documented use of nestboxes in the county occurred in 1989, close to the Thornley Woodlands Centre, and ever since birds have used boxes in the lower Derwent valley on an annual basis; three pairs used boxes near the Thornley Woodlands Centre in spring 1995. The Nuthatch survey of 2006 produced two examples of nestbox use close to human habitation. At Shincliffe a pair occupied a tit nestbox and fledged at least three young on 28 May, whilst at Satley a pair nested in a nestbox in an orchard in 2006. The temporal difference between lowland and higher altitude nesting birds was evidenced by the fact that a box with young in Brignall Banks in spring 1995 contained five young on 31 May, by which time three boxes at the lower elevation Paddock Hill Woods, in the Derwent valley, had already fledged a total of 19 young birds (Armstrong 1997).

The species' dependence on 'dead wood' for nesting, and in some instances its adoption of nest- boxes, as well as the supplementing of its winter food supply because of the widespread development of garden feeding stations, has doubtless speeded up its spread and helped it consolidate its numbers across the county.

Recent non-breeding status

The distribution of Nuthatch in the non-breeding season is largely similar to that of the breeding season, but with added mobility of their joining local tit flocks to rove through woodlands and along hedgerows. A post-breeding flock of titmice at Bishop's Park, near Bishop Auckland in late summer 1992 contained 19, mainly juvenile, birds indicating the strength of the local breeding population in that area at that time, and their affinity for joining such roving bands at this time of the year (Armstrong 1993).

The species' readiness to exploit winter feeding stations probably commenced in the 1980s and was noted at the Thornley Woodlands Centre from around 1984. Birds were noted on garden feeders in several areas during 1990, with four to five birds on the feeders at the Thornley Woodlands Centre. Their occurrence at such feeding stations increased dramatically through the 1990s, as birds found these locations whilst accompanying mixed tit flocks. There was a count of 22 during March 2003 in the species' traditional eastern stronghold of Castle Eden Dene whilst later that year, in September, WWT Washington had its first in 10 years. A juvenile at Westgate, at 275m above sea level on 1 July 2000, is one of the most westerly records in the county as well as being at a relatively high elevation. One of the most interesting records of dispersing juveniles occurred on 2 July 1998, when two birds were noted in a small plantation on the moorland edge at Sleightholme in the extreme south west of the county. This upland area was at least 5km west of the species' nearest breeding site in the Greta valley at the time and about 3.5km from the nearest breeding woodland in Deepdale, situated to the south across the moors. This is the highest altitude occurrence of the species in the county, at 320m above sea level (Armstrong 1999b).

British Nuthatches are largely regarded as sedentary (Wernham *et al.* 2002), though they sometimes join mixed flocks of titmice and this may provide a mechanism for dispersal and expansion of the range. Wandering birds are occasionally seen at locations away from breeding sites, primarily dispersing juveniles from late June to July. A bird was in Ward Jackson Park, Hartlepool on 10 July 1997, a typical date for a juvenile dispersing from its breeding area, and only the second time the species had been recorded in the Hartlepool area to that time, the previous record being August 1974.

A record of a juvenile on Ravensworth Fell in gorse in the late summer of 1995, over half a kilometre from the nearest tree, illustrates the fact that young birds regularly wander outside of their usual woodland habitat. Likewise, a number of sightings occurred at Shibdon Pond through the latter half of the 1990s, usually in the company of roving titmice. An exceptional winter record came from near Sunderland City Centre on 23 February 1996, well away from the species' normal range in the county. The origins of a bird at Dawdon Blast Beach on 19 August 2006 is uncertain, though it could easily have moved along the coast from Hawthorn Dene. Two birds in Jarrow Cemetery in August 2006 remained until the end of the year, but this species remains rare in South Tyneside. Single birds at West Hall Wood, Whitburn in July 1984, Marsden Quarry on 31 August 2005, Marsden Quarry and Whitburn Coastal Park on 1 October 2011 and Whitburn village on 3 October 2011 demonstate the local scarcity in this well-watched coastal area. Apparently migrating birds have been noted at Hartlepool Headland on 9 August 2004, 5 May 2007 and 20 August 2008 (Blick 2009).

Distribution & movements

This species breeds across the Palaearctic in a predominantly deciduous belt of woodland, in a wide range of geographically, subtly different forms, from Iberia north to southern Scandinavia, east to Kamchatka and into China. It is largely a non-migratory species (Parkin & Knox 2010).

The first local Nuthatch was not ringed until 1986, at Thornley Woods, but many have been ringed at various sites since. These have not however generated a huge amount of data though a notable record was of a bird of over five years in age that had been ringed at Thornley Woods on 29 November 1991, and was controlled there on 28 December 1996.

Treecreeper (Eurasian Treecreeper)
Certhia familiaris

A common and widespread breeding resident.

Historical review

This is a widespread, though in relative terms unseen resident across County Durham and it has probably always been thus. Once observers have learnt the call, they quickly realise that it is both very vocal and a common woodland species. It is present in most areas of woodland, especially in all of the county's main valley woodlands, such as the Tees, Wear, Derwent, Tyne and upper Team valleys. Past writers on the county's birdlife documented little detail about this species but, in relation to its distribution across England in the 19th century, it was described as *"ubiquitous"* by Holloway (1996), and it is assumed that this statement applied to its situation in the county.

In the mid-twentieth century, Temperley (1951) said that it was a, *"well-distributed and not uncommon resident in suitably wooded districts"*. Traditionally, the species had been considered as a scarce one, as is evidenced by the observer who, when a Treecreeper killed itself against his window in 1931 near Birtley, commented that he had never before seen one in that area (Bowey *et al.* 1993).

In 1952, renowned observer Fred Grey described his observations of a *"white looking bird"* at South Shields on 19 October, which occurred during a large influx of thrushes and other migrants. It was assumed that this was of the northern race *familiaris* and an immigrant, as no Treecreepers were then known to breed in the South Shields area (Temperley 1953). In like fashion, 'clean looking birds' were at Hartlepool on 9th and 24 September 1961, and these were also believed to have been of the northern race (Stead 1964).

Recent breeding status

In County Durham, as in the rest of Britain (Gibbons *et al.* 1993), mature woodland, almost to the exclusion of all others, forms this species' main breeding habitat. The species is found in a wide variety of woodland locations, its range extending along the main river valleys into the upland west of the county as far as Newbiggin-in-Teesdale, Eastgate in the Wear valley and up to Nookton Burn in the Derwent valley in the north of the county. It shows a preference for the mature woodland areas, with a strong presence in the well-wooded Derwent valley and in the valley woodlands of the Wear and in Teesdale. It is absent from the central parts the main urban areas, though, in modern times, it can be increasingly seen in the thickly-wooded mature parks and cemeteries of suburban, and even urban, areas such as Saltwell Park in Gateshead and South Park in Darlington, as well as from 'urban woodlands', such as Flass Vale in the centre of Durham City. Most of the county's lowland woodland habitats hold small numbers with the only area where it is seldom encountered being the far north east of the county, where the vicinity of St. Paul's and the cemetery in Jarrow are one of the few woodland habitats in the area.

It is more widespread than the Nuthatch *Sitta europaea*, utilising conifer woods to a greater degree though, as it is less noisy and more secretive than that species, it tends to be less well recorded or noted, despite the fact that it is perhaps two to three times more numerous. In 2009, an early nesting pair was feeding a fledged juvenile by 2 May at Adder Wood, in the Wear valley, with most other reports of juveniles on the wing not coming until the last week of the month (Newsome 2010).

Breeding Treecreepers are dependent on mature trees and woodlands. As the *Atlas* illustrated, the species' distribution in the county is similar to that of other woodland species, in that it is fundamentally linked to the presence of woodlands. Distribution is probably most similar to that of the Long-tailed Tit *Aegithalos caudatus*, another small, insectivorous, woodland species that shows susceptibility to cold weather (Westerberg & Bowey 2000). Work by the BTO has shown low temperatures have a greater effect on Treecreeper populations rather than long periods of snow, frozen water in bark crevices being the critical factor (Marchant *et al.* 1993). Between 2008 and 2009, largely due to the prolonged freezing temperatures in January and February 2009, Treecreeper experienced a 27% decline in the numbers monitored through the Breeding Bird Survey (Risely *et al.* 2010). Although birds can be heard calling frequently on woodland visits, Treecreepers can be quite inconspicuous and their camouflage probably leads to them being under-recorded both casually and even during more detailed survey work, by inexperienced observers. It was felt at the time of the *Atlas* that the mapped distribution shown was a largely accurate one (Westerberg & Bowey 2000).

In the UK, the species is more numerous in deciduous woodlands, although locally it does occur in extensive conifer woodlands, such as those at Hamsterley Forest (Westerberg & Bowey 2000). By contrast, on the Continent, the species is found mainly in coniferous woodland (Harrap & Quinn 1996) with the Short-toed Treecreeper *Certhia brachydactyla*, occupying deciduous woodland.

As the *Atlas* documented, in mature deciduous woodland, such as that of the lower Derwent valley, the breeding population can be as high as 25 pairs per occupied tetrad. However, a more typical population density was believed to be of between three and four pairs per occupied tetrad (Westerberg & Bowey 2000). More recent Atlas survey work in the south east of the county, between 2000 and 2006, suggested that the species was *"generally distributed in woodland"* but *"nowhere very numerous"* in *"north Cleveland"* (Joynt *et al.* 2008), as was indicated by statements a century before (Nelson 1907). More specifically, it is estimated that there are around 55 pairs in the north of Tees area (Joynt *et al.* 2008). Extrapolation of the above breeding densities to all of the occupied tetrads in the *Atlas,* gave a population estimate of at least 400 pairs in the county, however this seems to be on the low side considering the area of woodland in the county and the fact that most woodlands hold populations of Treecreepers. Perhaps a more accurate estimate for County Durham would be around 1000 breeding pairs.

Recent non-breeding status

As it is sedentary (Flegg 1973) and vulnerable to harsh winters (Harrap & Quinn 1996). This species is largely limited to lower altitudes in the county and cold winter weather is one of the main limiting factors for this species (Brown & Grice 2005). Low Force, in Teesdale, is the furthest west that the species is usually reported with any degree of regularity, though in winter 2009 a bird was present at Cronkley Pasture, in upper Teesdale; such birds are likely to suffer disproportionately in hard winters (Newsome 2010).

Being rather sedentary, its local winter range is largely equivalent to its breeding distribution. Outside of the breeding season it is regularly reported in small numbers from many wooded locations, mainly from the eastern lowland areas, but penetrating west along the valley woodlands. Winter reports usually concern small groups or counts in the range of five to ten birds, most often in mixed tit flocks in locations such as: Causey Arch Woods, Croxdale, Deepdale, Elemore Woods, Hetton Bogs, Low Barns and Rabbit Bank Wood. Winter counts are occasionally higher than those noted in spring and summer, such as the 16 birds in Thornley Wood in March 1991, and such gatherings are usually associated with nomadic titmouse flocks during the winter period. In 1994, it was observed as becoming increasingly common at winter feeding stations, though this habit has been noticed in the county since the late 1980s, spreading through the early part of the 1990s. This is a specialised species in terms of its foraging strategy.

What was at the time considered unusual behaviour from this species was observed in the mid-1980s at a bird feeding station at the Thornley Woodlands Centre. This was the habit of birds, initially, picking at fragments of peanuts from the ground below bird feeders. Such behaviour became established and spread locally through the late 1980s (Bowey *et al.* 1993), becoming ever more commonplace through the nineties, and was recorded at many such locations by the beginning of the 21st century. Today the species is now a not infrequently noted visitor, especially amongst winter titmouse flocks, to suburban garden feeding stations and individuals are routinely seen picking up small food fragments from the ground below feeding stations at Thornley Woodlands Centre and WWT

Washington. Opportunistic birds were noted taking sunflower seeds and peanuts from the Cowpen Bewley Country Park feeding station during November 2008 (Newsome 2009).

Evidence of post-breeding dispersal of this species often comes in the way of late summer records from atypical locations, where it is not usually reported, for example in late summer 1992, when birds were noted at both Shibdon Pond and Whitburn, both well-watched sites where this species is rare. A total of 22 birds were counted on the Coordinated Migrant Count of 10 October 1993, demonstrating that much larger numbers of birds than are sometimes realised are present along the Durham coastal strip. On 17 September 1997, five birds were in Burn Valley Gardens, Hartlepool, the first records there for several years, but this absence may really be due more to a lack of observers in this area than to a true absence of the species here. A bird was noted in Barnes Park, Sunderland during the breeding season in 1999, an area where it is not usually noted. Perhaps there was a small previously unrecognised population of birds in this area at the time (Armstrong 1993, 1994, 1999 & 2003).

An unusual report came on 31 July 1989, when an exhausted bird was on the patio at Marsden Grotto. Another at Bill Quay on the Tyne on 21 December 1990 was also well away from its usual haunts, as was one in Whitburn Churchyard which appeared in October 1989 and remained until 11 March 1990. Several birds were in coastal locations around the Whitburn area in the autumn of 1990, including two in Whitburn Churchyard in November 1990. In January 1991, a bird was found wintering in Mere Knolls Cemetery at Seaburn, for the first time. The most unusual recent sightings of the species concerned three birds seen at Jackson's Landing, Hartlepool in December 2007. A single bird was noted in the Church Lane area of Whitburn on 1 November 2008, which may have been a passage migrant. Another notable coastal bird was one at Hartlepool Headland on 20 September 2008 (Armstrong 1990-1992, Newsome 2008 & 2009).

Distribution & movements

This species breeds over much of Europe, northern Asia and North America. Further south it is replaced by the very similar Short-toed Treecreeper. Over much of its range, the species is sedentary but some of the northern, paler races are partial migrants, sometimes exhibiting irruptive behaviour. Occasional migrants on the Durham coast might relate to birds from these northerly and easterly populations but there is little empirical data to confirm this. The little available ringing data for local birds indicates that birds are essentially sedentary. Any birds that have been re-trapped are near their site of ringing, or have made small local movements, such as the bird found dead in Rowlands Gill in September 1997, which had moved about one kilometre from Lockhaugh, where it had been ringed in the spring of that year.

Wren (Eurasian Wren)
Troglodytes troglodytes

An abundant resident and apparently rare passage migrant.

Historical review

This little bird is very common across Britain and throughout Durham. It is a familiar sight in gardens and can be recorded breeding virtually in every habitat in the county, even in some parts of the open western moorlands and along the bases of cliffs and beaches of the northern section of the county's coastline.

In the 19th century, Holloway described this species as being commonly and widely distributed over the whole of Britain and Ireland (Holloway 1996). In Durham it conformed to this profile, as it was a very common species across the county during Victorian times though its population, then as now, varied greatly from year to year, according to the severity of the previous winter's weather. Temperley (1951) called it *"a common resident"* and said that *"in autumn they are occasionally seen along the shore under circumstances that suggest passage migration"*. Nelson (1907) thought that the species *"passed through"* Teesmouth as a migrant.

Recent breeding status

Although a small species, of unassuming habit, the Wren is widespread and vociferous and this diminutive bird is one of the most common of all in the county. Records have come from all types of habitat, from the open moorlands, next to Cauldron Snout in Teesdale in 2008, down to the largely concrete environment of Seaham Harbour in the same year (Newsome 2009). The Wren was adjudged the most widespread in Britain from Breeding Bird Survey results in 2009, occurring in 91% of surveyed one kilometre squares (Risely *et al.* 2010).

As this indicates, this species is a widespread and adaptable bird, which is effectively ubiquitous in lowland habitats across the county. The mapped distribution, as shown in the *Atlas*, is largely an accurate one, genuinely reflecting the species' widespread presence across most of the county (Westerberg & Bowey 2000). Any gaps in that mapping exercise, e.g. in the area to the east and north east of Darlington, were almost certainly due to a lack of observer coverage in those areas. Subsequent work has proven that this is also a widely distributed breeding bird across the south east of the county (Joynt *et al.* 2008).

The Wren's choice of habitat is broad and in Durham this encompasses town gardens, hedgerows and riparian corridors as well as the lower to mid-altitudes of the uplands, where there is scrub cover for nesting purposes (Westerberg & Bowey 2000). It favours woodland thickets and dense scrub, though it can be found from the moorland fringes and farmland in upper Teesdale, down to bramble scrub in the cliff-top gullies in the central stretch of the Durham coastline. It is found in suburban gardens and in green-space plantings of all of the major settlements, though breeding populations are less dense in the urban areas (Joynt *et al.* 2008). In Durham it breeds from sea cliffs and shoreline to the higher, sheltered nooks and crevices of the western uplands, an altitudinal breeding range, at its most extreme of over 400m, from east to west, as shown by the bird in song at Falcon Clints at an elevation of over 420m above sea level, in March 1980. It often chooses unusual sites to nest in and by way of example in spring 1997, birds were recorded nesting in a nestbox just one metre from the ground at Swalwell, a cave entrance at St. John's Chapel in Weardale and the roots of a rowan *Sorbus acuparia* in Hudeshope (Armstrong 1999b).

In early April 2007, there were 33 singing males at Tunstall Hills and 70 singing birds in Elemore Woods in the first week of May, which gives some indication of the sheer numbers present in lowland areas. Wrens are most common in dense, lowland woodland and in such habitat can reach very high densities in some years (particularly after a run of mild winters); for example, the 28 territories in around 10ha noted in Paddock Hill Woods during spring 1989 (Bowey *et al.* 1993). More recent studies in the Houghton/Hetton area revealed 21 pairs at Hetton Lyons CP in 2007 (up three pairs over the 2006 figures), 31 pairs at Hetton Bogs (up one on 2006) and 28 pairs at Rainton Meadows/Joe's Pond. On the Kibblesworth Common Bird Census, 34 pairs were found in 2007, an increase of five pairs from 2006. Surveys of prime habitat in spring 2009 produced some high counts of territorial males with 35 singing males being found in a survey of Castle Eden Dene on 6 May, 25 males in Elemore Woods on 20 March, 19 at Rosa Shafto on 2 June and 8 on the western fringe of Hamsterley Forest at Sharnberry Flats on 18 May.

The species can undergo considerable between-year fluctuations in numbers, as evidenced by the 400% rise in the number of territories at the Paddock Hill Woods study site, in the lower Derwent valley, between the years 1985 and 1989 (Bowey *et al.* 1993). Some large gaps appeared on the species' distribution map in parts of the Pennine uplands (Westerberg & Bowey 2000) and these were apparently due to a lack of suitable breeding habitat on the open moors in those areas.

There were probably over 2,000 breeding pairs in the south east of the county during 2000-2006, occupying over 80% of the tetrads in this area (Joynt *et al.* 2008). The *Atlas* made an estimate of the county's breeding population of Wrens, based upon this being assumed to be 'typical of Britain as a whole', and this translated to an estimated 79,200 pairs (Westerberg & Bowey 2000). There is little that has happened since that estimate was made to suggest that this will have altered dramatically. Nationally, the population trend for Wren is one of a very slow decline, since about 1976. The reasons for this are not fully understood (Marchant *et al.* 1993) but there appears to be no obvious manifestation of this in the local context.

Recent non-breeding status

The species suffers badly during periods of prolonged, severe winter weather (Marchant 1983). The Wren's winter distribution is essentially the same as the breeding distribution, but the gaps in upland distribution mapped by the *Atlas* in parts of the uplands, in broad terms, persist through much of the year, though birds can sometimes be noted in some of the most exposed locations in the county, even during mid-winter. The species has been observed in tall heather on Muggleswick Common in severe weather in January during the mid-1980s (Westerberg & Bowey 2000).

Upland Wrens survive in many moorland fringe and exposed areas in milder winters in Durham, readily taking advantage of cover provided by any human habitation or artificial planting (Newsome 2009). A bird at Newhouse Moor on 31 August 1989 was at 580m above sea level, possibly the highest elevation that this species has been recorded in County Durham. Birds can remain in the extreme uplands of the county's western moors, with birds lost in the deep cover and milder micro-climates of the world under the heather and even under snow cover. One hardy bird was on a snow-covered Hamsterley Common at 430m above sea level on 1 January 1986 (Baldridge 1987). The species' more typical distributional limits in west Durham range from Beldon Burn and Cowshill in the Derwent valley and Weardale to Bowlees and Harwood in Teesdale and Spital Grange in the far south. In the winter, at lowland sites, birds can be found in relatively large numbers in the dense fen and reedbed areas of local wetlands, such as around the North Tees Marshes, WWT Washington and Shibdon Pond. An illustration of the numbers present in such habitats was given by the 17 birds that were ringed at Lamesley in one morning on 17 October 2010 (Richard Barnes pers. comm.).

In Northumberland, Galloway & Meek (1983) noted that mortality could be high in severe winters but that milder years could produce large population increases (Kerr 2001). In some local studies, such as that in Gateshead in the late 1970s, in the spring of 1978 it was estimated that in the woodland of Clockburn Dene there had been a 50% decline after the severe snow of the previous winter (Bowey *et al.* 1993). It remains to be determined at the time of writing, what affect the severe weather of 2010 had on local wren populations.

In March 1984, at Balder Grange, a bird roosted in an old Swallow's nest, despite the curious deaths of three other birds that had used the same roost site in January. A similar site was used by roosting birds in a porch in Ebchester through the early 1990s and another pair in Cotherstone in winter 1999. More of a tight squeeze was the 12 birds that roosted in a nestbox at Ingleton in January 1987 (Armstrong 1988).

The numbers of birds caught at the Lockhaugh and Ravensworth ringing study sites in Gateshead were down in 1996 from the previous year, 39 birds being compared to 63 in 1995, largely due to the reduced adult survival over the winter. A bird was in thick snow at Cronkley on 23 December 1996 (Armstrong 1998).

Distribution & movements

Wren has a wide distribution that encompasses the whole of Europe, much of Asia and North America. Some populations, such as those in Scandinavia are at least partial migrants, winter to the south, within the breeding range of the species. The majority of the European population, including that of the British Isles, is essentially sedentary. Local birds would appear to be resident, as illustrated by the limited dispersal of one that was ringed at Elwick on 11 October 1986, which was near Saltholme Pools just 15 days later (Blick 2009). The available data for birds ringed locally suggests that most of them are essentially sedentary. Most ringed birds have been trapped and re-trapped near their site of ringing at long-term study sites, or have, at most, made small local movements. There is little local evidence to indicate that the seasonal altitudinal movements noted in some small, largely insectivorous birds, from areas of great altitudinal elevation to lower ground, e.g. Blue Tits *Cyanistes caeruleus* and Great Tits *Parus major*, occurs in Wrens in Durham.

Fifty-nine Wrens were recorded along the Durham coast as part of the Coordinated Migrant Count in early October 1995 (Armstrong 1996) and a count of 23 birds at Hartlepool Headland on 6 October 1998 was higher than usual at this site (Blick 2009). Small numbers are often seen at the coastal watch points in autumn, but in most instances they could have flown from the nearby urban/agricultural areas, rather than crossing the North Sea, although one bird ringed at South Gare, just a few kilometres south of Durham, on 9 October 1974 was controlled at Spurn, Humberside, five days later (Blick 2009).

The Wren is a small-scale but regular spring and autumn migrant in Britain, with ringing recoveries involving the Scandinavian countries confirming that the species does cross the North Sea (Wernham *et al.* 2002). Nonetheless, as it is such a widespread and common resident species in the county's coastal denes and shrubberies, passage migrants can be difficult to detect in the autumn on the Durham coast. One individual bird was seen to fly 'in off' the sea at Whitburn on 27 October 1989. An amazing record concerns a bird ringed on 19 October 1999 at Kroon's Polder, Vlieland, Netherlands, that was controlled only four days later at Marsden Quarry. At the time this was only the fourth foreign-ringed Wren to be found in Britain, and it is possible that it may have been ship assisted (Armstrong 2003). It is conceivable that October 2007 counts of 41 birds between Easington and Beacon Hill on 1st and 26 at Crimdon Denemouth on 27th may have involved a number of overseas arrivals, but it is almost impossible to confirm or refute this. No doubt many more arrive in County Durham than are detected through the autumn and the fact that 19 were ringed at Dawdon Blast Beach on 27 September 1990 during a fall of migrants might be significant (Armstrong 1991).

Starling (Common Starling)
Sturnus vulgaris

An abundant resident and winter visitor.

Historical review

The first Durham reference to the Starling comes courtesy of correspondence between Ralph Johnson of Brignall and John Ray in the late seventeenth century (Mather 1986). A measure of the species' relative scarcity as a breeding bird even a century later, is illustrated by the fact that the engraver Thomas Bewick, was known to have expressed how delighted he would be should Starlings ever nest in his roof; which, at that time, was located in what is now central Gateshead. Brown & Grice (2005) documented the fact that it was absent as a breeding bird from many northern English counties in the early nineteenth century. This status is confirmed by the review in *The Breeding Birds of Cleveland,* which stated that during the first half of the 19th century, it was described as rare as a breeding bird in northern England (Joynt *et al.* 2008); Nelson (1907) reported its increasing "*enormously within the past half-century"* in Yorkshire.

Over the subsequent 60-70 years there were was clearly some change in local status and writing in 1874, Hancock, believed it to be common everywhere, though he reported that it had not been so just a view years previous. He ascribed the increase in Starlings to the destruction of birds of prey. The phenomena of winter influx of Starlings appears to have commenced only during the late 19th century as it was said to be scarcer in winter around 1840 (Hutchinson 1840). Nelson (1907) recorded an immigration of a huge number of Starlings off the coast of Redcar, south of the Tees, in early November 1881, so clearly autumn migrants were crossing the North Sea to enter north east England at that time. In the late 19th century, Robson documented a number of unusually coloured birds being taken around the Derwent valley, including a cinnamon-toned one (Robson 1896) and he further commented that the species had "*increased immensely of late years*".

Exceptionally, Starlings have been found breeding in County Durham from mid-January onwards and the earliest documentation of this took place at a nest site at Vigo, near Birtley, on 4 January 1931, where a pair was seen feeding young at a nest hole 'in hard frosty conditions', ice being on the ground but '*judged from the noise the young were making they were quite lively'* (*The Vasculum* Vol. XVII 1 1931).

By the mid-20th century, Starling was described as being an abundant and generally distributed resident, whose numbers were supplemented by a large winter immigration from the Continent (Temperley 1951).

Recent breeding status

Today, the Starling is a widespread breeder across the county in a wide variety of habitats (Westerberg & Bowey 2000). Its distribution corresponds strongly with the presence of tree cover, although they will nest in walls, roof cavities, drainpipes, old woodpecker holes and even Sand Martin *Riparia riparia* burrows (Westerberg & Bowey 2000). It is more widely distributed and more abundant in the lowlands than it is towards the western uplands, though breeding birds will occupy suitable sites right up to the edge of these. It is also largely absent as a nesting species from the coastal cliffs. Pairs breed in many habitats and the main limiting factor is often the availability of nest sites. The greatest number of birds can be found in the eastern half of the county, giving a distributional skew to the pattern of occurrence with the highest breeding densities occurring in the mosaic of urban gardens and open spaces around housing estates and the town centres (Parkin & Knox 2010), where parks and gardens provide food in close proximity to nesting sites.

The species has always been adept at exploiting nesting cavities, but nestboxes were not routinely used in Durham until the 1980s. Successful fledging was noted from one attached to the front housing of a static caravan at East Lendings near Barnard Castle in the summer of 1977 just 1.5 metres above the ground (K. Bowey pers. obs.).

Using an extrapolation of the national population figures and habitat specific densities, to the appropriate habitat areas of Durham, Westerberg & Bowey (2000) estimated a population range of between 20,000 and 35,000 breeding pairs for the Durham recording area as a whole but considering the well documented national decline in numbers, particularly in woodland populations (Parkin & Knox 2010), perhaps the real figure today lies nearer the lower limit of this range. The survey work for *The Breeding Birds of Cleveland* indicated approximately 4,000 pairs nesting in the northern part of what was Cleveland County, during the period 1999-2006 (Joynt *et al.* 2008).

By the end of the 20th century there was significant evidence of a decline of this species across northern Europe (Tiainen *et al.* 1989) and Britain (Marchant *et al.* 1990, Feare 1995); numbers on English Common Bird Census plots fell by 78% between 1967 and 1999 (Brown & Grice 2005). The main threats being the loss of foraging grounds, especially the ploughing up of permanent pasture and a change in the sowing time of arable land, much of which occurred through the 1950s to 1970s (O'Connor & Shrubb 1986). In some areas, the conversion of land to coniferous plantation has impacted the species (Gibbons *et al.* 1993). Over the period 1996-2009, Breeding Bird Survey data showed a 38% decline (Risely *et al.* 2010) and in 2009 the Common Starling was added to the red list of the UK's Birds of Conservation Concern because of its severe decline over 25 years and more (Eaton *et al.* 2009). That said local evidence of a decline in breeding numbers in Durham is difficult to assess other than anecdotally; a growing number of reports of fewer birds being seen in suburban gardens. One seasoned observer noted in mid-2011 that he had not seen a Starling in his garden at Shincliffe for over three years; a site less than four kilometres from the location where half a million birds roosted in the autumn of 1951.

Recent non-breeding status

The first significant post-breeding gatherings typically come in early June when roving flocks of perhaps 100-300 birds, including many juveniles, can often be found feeding on insects at sites such as sewage treatment works and farmland hedgerows. Emerging insect populations on the county's moorland areas also attract quite large post-breeding flocks in high summer to moors or fringing pastures. These flocks largely consisting of juveniles have often moved up the dales, to exploit the cropped pastures and meadows in late summer, only to disperse again in autumn. In 1999, there was a roost of around 10,000 birds at Baldersdale in late August possibly comprising a large portion of the Tees valley's breeding stock.

Numbers generally increase in lowland areas during the late summer months with peak counts at this time of the year often reaching into the low thousands at some sites. During the mid- to late 1980s and the early 1990s, mid-summer roosts were regularly present at Shibdon Pond with, for example, up to 4,000 birds in the reeds there in June and July 1987. Late post-breeding and early autumn gatherings of several thousand birds were also recorded going to roost in the dense willows of the main island at Shibdon Pond. Such roosting behaviour must have been repeated in many locations across the county. In June 1988, hundreds of juvenile birds were noted feeding, presumably on kelp fly larvae *Coelopa sp.*, in piles of rotting seaweed on the shore at Whitburn.

Large winter roosts are an obvious and often spectacular aspect of Starling biology. Temperley (1951) noted that numbers were supplemented by a large winter immigration from the Continent and referred to gatherings at West Dunston Staithes on the Tyne and on the Transporter Bridge over the Tees.

The roost at West Dunston Staithes was estimated to contain a million birds nightly during the winter of 1941/1942 and may well be the origin of the central Newcastle roost that attracted large numbers of birds in subsequent decades.

Through the winter months of the 1950s and 1960s, the county played host to some enormous night-time roosts of Starlings and these appeared to be a particular feature in Durham City. From the early 1950s through until the 1960s, there was a huge roost of birds on the outskirts of Sherburn Village to the east of the city. In 1952, birds were recorded as flying from 12 to 15 miles away to this roost, from the direction of both Sunderland and Hartlepool. It was estimated to contain at least 100,000 birds on 18 January 1953 and during the winter of 1959 the birds, which numbered 'not less than half a million', were concentrated in an area of hawthorn scrub of about 15 acres in extent. In this instance, birds were reported coming to the roost from locations far as 13 miles distant. In 1960, this roost was estimated to contain between a half a million and a million birds. One dead bird found here had been ringed in Moscow. About half a million were again estimated to be present the following winter when it was said that the roost had been in this location for a few years (*The Vasculum* Vol. XLVI 1961).

The roost on the Transporter Bridge in 1946 (Temperley 1951) is one example of the use of an industrial structure. In the winter 1955 a huge roost formed at the ICI works at Billingham, debatably estimated by the local press to contain 'a million and a half' birds. In a similar industrial situation, the Consett Steel Works attracted a roost of around 50,000 birds on 31 October 1959, but just a few hundred were roosting there by the end of the year (*The*

Vasculum Vol. XLV 1960). Perhaps these birds were largely continental immigrants, which subsequently moved on to other localities in the British Isles.

In West Hartlepool, a modest roost of just over 3,000 birds in the autumn of 1966, shared with smaller numbers of Song Thrush *Turdus philomelos*, Blackbird *Turdus merula* and House Sparrow *Passer domestica*, had grown to almost a quarter of a million Starlings by December. Blick (1978) noted that during the late 1960s and early 1970s, large numbers of birds also roosted in Rossmere Park, in Hartlepool.

A large roost developed at Lambton Castle towards the end of 1971 and this grew to around half a million birds at its peak in December of that year. This roost was seen each winter until at least 1974. Subsequently, a roost of 100,000 birds was at Chester-le-Street on 5 January 1977 which was probably a relocation of the birds that had previously roosted at nearby Lambton. In 1978, a large roost developed in Sunderland Town Centre, and this persisted for some years on the iconic Wearmouth Bridge and in later years on the Crowtree Leisure Centre in central Sunderland, peaking at around 20,000 birds in December 1985.

In Newton Aycliffe in 1984, a roost of around 150,000 birds was present in a conifer block in late February, though this had been abandoned in 1985, in favour of a new roost about a mile and a half away. These large roosts tend to break up and disperse through March, as wintering birds move away (Brown & Grice 2005).

Through the 1980s and 1990s, birds from Gateshead and South Tyneside travelled large distances late on winter afternoons to the huge roost of birds that was located around St. Nicholas' Cathedral and the Central Station in Newcastle upon Tyne until concerted efforts were made in the late 1990s to protect the city's building heritage and they moved them away. At that time there were important pre-roost gatherings of birds that numbered tens of thousands in the East Cemetery, Gateshead and in Saltwell Park, where birds congregated in late afternoon prior to flying to the central City roost.

More recently, there appear to have been fewer and far smaller winter roosts in the county and certainly nothing that compares to the huge roosts that were present at so many sites over the period 1950 to 1980. It has been suggested that the number of birds wintering in England has declined in recent decades (Feare 1995, Brown & Grice 2005). Reports of larger roosts have been the exception rather than the rule, though a gathering at Haverton Hole numbered 41,000 in November 2007, 84,000 the following year and 48,000 in 2009. A more modest roost of 12,500 was at Low Barns in November 2008.

Most non-breeding season sightings relate to large feeding groups during the winter months or to post-breeding period roosts. During the winter, roving flocks of birds can be seen feeding on arable farmland and many urban gardens in Durham are visited by smaller groups during the period November to March. By January, the county will host many flocks that number in the range of 100 to 300 birds, with occasional reports of much greater numbers. Around 5,000 birds were noted feeding on snow-covered fields near the then Sunderland Airport on 30 November 1973. Over the years, as garden feeding stations became more popular birds became ever more adept at their exploitation and on 4 December 1987, at Roker, a bird was documented feeding on a peanut hanger for the first time in the county. Many households now put out waste food scraps, as well as bird food, during the winter months and this obviously helps Starlings survive cold periods of weather. On a number of occasions, birds have been noted collecting nesting materials during the winter months, such as on 29 December 1987, at Langley Moor.

More than perhaps any other passerine species, Starlings seems to produce individuals with plumage aberrations. As early as 1896, Thomas Robson documented a number of unusually coloured birds taken in and around the Derwent valley, including a cinnamon-toned bird. Albino birds have been recorded in the Sherburn roost (1951), and at Stockton (1968), on the North Tees Marshes (1974, 1977, 1978, 1995), at Marsden (1975), Boldon Colliery (1977), Seaton Common (1977), Chester-le-Street (1981), Elwick (2000) and Hartlepool (2000, 2003). There is also a record of a leucistic bird from Teesmouth. An individual with a white body but dark wings, tail and head was at Billingham in 1980, whilst birds with pale grey or brown feathers were on the North Tees Marshes in 2003 (Blick 2009).

Distribution & movements

Starlings breed over most of Europe, though are less common or absent from many southern parts of the Continent. Their range extends east as far as Pakistan and central Asia. In Europe, in the north and east, the species is a mainly migratory. British birds are principally resident but are joined by large numbers of continental individuals during the winter.

The final quarter of the year usually produces evidence of significant autumn immigration. There was a most definite and concerted autumn immigration of Starlings to County Durham throughout the 1950s to the 1970s, though in more recent times this would appear to have reduced considerably or even abated completely in some years. During the autumn of 1955 many large flocks, some numbering thousands of birds, were seen flying in from over the sea over the period 21st to 24 October at various points along the coast. Smaller numbers of migrants were observed arriving at the Durham coast in the October and November of 1972. Such reports declined through the 1970s, and became even less common in the 1980s, though some birds were noted 'on migration' in autumn 1981. The modern numbers pale into insignificance in comparison with the numbers noted during the 1950s and 1960s. Nonetheless, this species was the most commonly noted on four of the five Coordinated Migrant Counts along the Durham coast between 1992 and 1996, in the remaining year 1993, it was the third most numerous (Bowey 1999c).

Coordinated Migrant Counts, Durham Coast 1992-96

Date/Year	Number Seen	Position in list of most Numerous Species Recorded
4 October 1992	1,718	1st
10 October 1993	1,483	3rd
2 October 1994	1,782	1st
8 October 1995	3,429	1st
13 October 1996	2,572	1st

Durham's resident Starlings are joined by continental birds, believed to be predominantly from Scandinavia and Northern Russia. This immigration is affected by wind direction but, in general, peak arrival is in November. There is a prodigious amount of ringing information generated by the many thousands of Starlings that have been ringed in the county, which demonstrates this. Ringing at winter roosts, in the north east of England particularly during the 1950s showed that large numbers of the county's wintering Starlings come from Scandinavia and northern Russia. For example the adult ringed at South Shields on 2 February 1950 which was recovered at Nyborg, Island of Fyen, off the coast of Denmark on July 7th, 1951 and another adult ringed at the same place and time which was recovered at Larsnes, Norway on May 21st 1952. This situation pertained through time as illustrated by the pattern of subsequent recoveries from such areas.

Birds Ringed in Winter in Durham and Recovered In Countries to the North and North East

Ringing Site	Date	Recovery/Control Location	Date
Pity Me	26 Dec 1961	near Pskov, Russia	14 Aug 1964
Hartlepool	29 Jan 1967	Vadde Island, Stockholm, Sweden	15 Jun 1967
Whickham	25 Jan 1969	Smolensk, Russia	01 Apr 1974
Stockton in	Nov 1992	Norway	25 Feb 1995

The origin of many of the county's wintering Starlings has been known for a long time as some of the county's earliest ringing recoveries indicated this; examples include one ringed on Helgoland on 30 March 1925 and reported at East Boldon on 31 December 1925, one ringed near Viborg, Denmark on 30 May 1924 and recovered in East Boldon on 20 February 1929, and another ringed at North Jylland, Denmark on 30 May 1924 and recovered in County Durham on 20 February 1929.

Birds Ringed in Winter in Durham and Recovered in Countries to the East

Ringing Site	Date	Recovery/Control Location	Date
Stockton	7 Dec 1952	Altenbruich, Nieder Elbe, Germany	21 Jun 1953
Norton on Tees	Feb 1954	Denmark	June 1954
Wolviston	1 Jan 1962	Koed, Kolind, Jutland, Denmark	02 Apr 1962
South Shields	5 Aug 1967	Gjol, Jutland, Denmark	01 Apr 1968

Some birds ringed in northern and eastern countries have been recovered in County Durham.

Birds Ringed in Countries to the North and North East and Recovered in Durham

Ringing Site	Date	Recovery/Control Location	Date
Tallinn, Estonia	6 Jun 1958	Sherburn near Durham City on	02 Jan 1960 (roost)
Baltic Sea coast	spring 1932	Darlington	March 1935
South Holland	14 Jul 1937	Chilton, near Fence Houses	20 Feb 1938.

Not all recovered birds have travelled large distances, for example one ringed in Newcastle on 1 February 1956 was found dead in an empty house just across the river at Blaydon on 6 November 1956.

Birds Ringed in North East England and Recovered in Durham

Ringing Site	Date	Recovery/Control Location	Date
Newcastle	01 Feb 1956	Blaydon	06 Nov 1956
Newcastle	18 Mar 1956	Sunniside	05 Dec 1956
Newcastle	31 Jan 1957	Witton Gilbert	10 Feb 1957
Newcastle	27 Dec 1956	Darlington	15 Feb 1959
Wallsend	03 Sep 1961	Stockton on Tees	02 May 1966
Beal, Northumberland	29 Nov 1958	Norton on Tees	09 Jun 1963.

Some birds travel through the county to winter further south, moving along the eastern seaboard of Britain, such as recorded movements from South Shields to Boston, Lincolnshire, and Aberdeen to Billingham. Others move further south as winter progresses and perhaps adopt these as their breeding location such as the individual ringed at Pity Me as an adult on 10 November 1960 and recovered at Somercotes, Alfreton, Derbyshire on 1 July 1963.

Birds Ringed in the North and Recovered Further South in England

Ringing Site	Date	Recovery/Control Location	Date
South Shields	09 Jan 1958	Boston, Lincolnshire	05 Aug1959
Whickham	15 Jan 1983.	Aylesbury	20 Mar 1984
Aberdeen	January 1984	Billingham	July 1985
Whickham	12 Jan 1995	Scunthorpe, Humberside	11 Apr 1997

In spring, only a very few are seen at the Durham coast returning to the north and east from their wintering areas, as found nationally (Brown & Grice 2005).

No doubt many birds come to untimely ends. An adult ringed at Whickham on 25 April 1956 was found drowned in same area on 29 June 1956. Some Starlings however, can reach quite considerable ages. For example, one that was ringed as a fully grown bird, at Whickham on 1 June 1969 was found dead in Gateshead on 24 August 1978 and another also ringed at Whickham on 1 June 1978 was found there on 1 March 1988, when it was at least nine years old. A bird ringed as an adult female at Low Thornley, Rowlands Gill on 12 May 1990, was found dead at Winlaton on 17 March 1998 and a bird controlled at Low Barns on 16 October 2005, had been originally ringed as an adult female over ten years earlier and 124km away to the west, at Bispham, Blackpool, Lancashire on 6 July 1995. The bird ringed at Whitburn in December 1979 and that found its way down a chimney at Roker three months later is not an isolated incident. It is not known when the bird had died, for which just the leg was discovered in a loft at Whickham on 2 April 1998, almost 18 years after being originally ringed as an adult male nearby, on 24 May 1980.

Rose-coloured Starling (Rosy Starling)
Pastor roseus

A rare vagrant from south east Europe and adjacent south western Asia: 13 records predominantly from June to October.

The first Rose-coloured Starling for County Durham was a bird shot at Newton Hall near Durham on 14 October 1829. It is recorded by Hutchinson in his manuscripts as follows: *"On 14th October, 1829, the ground having a slight covering of snow, a Pastor was shot at Newton Hall near the City of Durham, which I purchased and preserved and it is now in the University Museum. When first seen it was searching for food amongst the litter cast from the stables"* (Hutchinson 1840). Unfortunately, he did not describe the age of this bird, though it was presumably an adult.

All records:

1829	Newton Hall, Durham, 14 October (Hutchinson 1840)
1856	Whitburn, adult, July (Hancock 1874)
1856	Black Fell, Gateshead, adult, 15 September (Hancock 1874)
1860	Shotley Hall, Shotley Bridge, no date given (Hancock 1874)
1949	Durham City, adult, 20 February (*The Vasculum* Vol. XXXXV 1949)
1973	Billingham, adult, 27th to 30 June (*British Birds* 67: 336)
1975	East Boldon, adult, 12 July (*British Birds* 69: 354)
1987	Sunderland, adult, 3rd to 5 July (*British Birds* 81: 588)
1995	Heighington, adult, 21st and 22 June (*British Birds* 89: 521)
1995	Hetton Downs, Hetton-le-Hole, adult, 28 August to 10 September (*British Birds* 89: 521)
2002	Whickham, Gateshead, adult, 10th to 12 June
2002	Saltholme Pools and Dorman's Pool, Teesmouth, adult, 17 June
2005	Whitburn Coastal Park, adult, flying inland, 2 June

All of the birds specifically aged in County Durham have been adults, with most falling into the classic overshoot pattern for this species, being recorded in June and July. The two birds in 1995 coincided with large numbers breeding in Bulgaria and Hungary and were amongst 32 recorded in Britain that year (*BB*: 89: 264). The two in 2002 were part of an unprecedented influx into Britain with an estimated 128 birds being recorded in June and a further 45, or more, in July. This influx coincided with the largest breeding numbers in Bulgaria and Hungary since 1995, precipitating some exceptional movements in eastern parts of Europe (Gantlett 2002). The December 1949 record is unusual in that it falls outside the established vagrancy pattern for this species, though winter sightings are not unheard of, with several instances of birds amongst flocks of Starlings *Sturnus vulgaris* in Britain in recent years. The bird in Durham City was said to be feeding amongst Starlings and described as being "*in winter plumage*"; it may have been a first-winter having made landfall elsewhere in the previous autumn.

Distribution & movements
Rose-coloured Starling breeds from south east Europe, eastwards to Kazakhstan and southern Iran; it winters predominantly in the Indian subcontinent, as far south as Sri Lanka. This species is known for its erratic and highly variable westward movements, during which it may breed in large numbers in south east Europe, although in other years it may be absent from the region as a breeding species. These periodic irruptions often produce influxes into the British Isles, primarily in June and July, though a small number of adults usually reach Britain on an annual basis. A second peak in records during September and October relates to juveniles, usually amongst flocks of Starlings, with the majority of these being recorded from the south coast and south west of England. An increasingly regular but rare visitor to the British Isles, it was removed from the list of species considered by BBRC at the end of 2001, by which time there had been over 600 records.

Dipper (White-throated Dipper)
Cinclus cinclus

A common but habitat-restricted resident.

Historical review
The major rivers and most of the minor streams and becks in the uplands of County Durham support good populations of this species. Commenting from the perspective of the whole of the north east, Hancock (1874) noted that it was a *"constant resident of rocky burns"*. The current range of Dippers in Durham seems remarkably similar to that quoted by earlier authorities. Robert Calvert (1884) in *Notes on the Geology and Natural History of the County of Durham* said that they were *"common on the retired rocky channels of the quiet burns of the west"*. Canon Tristram (1905) agreed in *The Victorian History of County Durham* though he added that they were *"occasionally by streams near the coast"* and *"much persecuted through the ignorance of anglers"*.

An article by Temperley in *The Vasculum* in 1933 (Vol. XIX), noted that the status of the Dipper in County Durham was incorrectly recorded as scarce in *The Practical Handbook of British Birds*. He stated that the Dipper was plentiful on many Durham streams and it was only in the lowest reaches of the rivers and their tributaries that had been excessively polluted by colliery water that the species was absent.

In his county review, Temperley (1951) said that Dippers were *"common on all our streams and rivers and occasionally on shallow ponds and on the margins of reservoirs"*. He said that *"On the Derwent the Dipper is to be found all along the banks of the mainstream, and on practically every tributary, to a point within three miles of its junction with the Tyne"*. This is a point that lies less than three miles from the City of Newcastle. At the end of the 19th century Robson recorded it as nesting a little way upstream, near Rowlands Gill (Robson 1896). Temperley went on to say that it was *"common"* on the Wear, to within a mile or two of Bishop Auckland, with smaller numbers found below and above the City of Durham. Nelson (1907) stated that birds had been seen on the rocks on the shoreline of Teesside during severe weather, but there are no recent records of this habit.

On the Tees, Dippers were described as *"abundant"* and as present on all of the main streams and tributaries to within a few miles of Darlington. Temperley (1951) reported that the species clung on to its old haunts long after increased industrialisation had driven off Grey Wagtail *Motacilla cinerea* and Common Sandpiper *Actitis hypoleucos* though in the modern context it is the Grey Wagtail that appears more tolerant of pollution and disturbance. Temperley observed that there was a propensity for the species, presumably juvenile birds, to be seen away from their usual haunts in late summer. Clearly some birds do move around the region, presumably using watercourses as conduits to dispersal. For example, a bird ringed as a first summer male on 18 May 1964, at Rowley Burn near Hexham (Northumberland) was found dead on 18 January 1965 at Cockfield Fell, not far from Barnard Castle (Kerr 2001), some 39km to the south east in a straight line, but probably much further as the Dipper travels, i.e. along watercourses. Presumably by this time of the year, the bird would have been holding territory.

Recent breeding status
Breeding records of this species come from all of the county's main river systems, as well as most of their tributaries, with an emphasis on the fast-flowing sections and areas where the river geography is varied. There is a concentration of birds, because of the topography, in the west of the county. The species' riverine range lies within three watersheds, the Derwent (Tyne), the Wear and the Tees. Its regular breeding range extends down to Swalwell on the Derwent, where the River Tees intersects with the A1(M) trunk road and Croxdale on the Wear. In 2005 however, successful breeding was noted at Chester-le-Street, perhaps the furthest downstream breeding location recorded on the Wear watershed (Newsome 2006).

The lower limits for Dippers probably also indicate the absence of white water stretches on lower reaches, reflecting the distributions of the aquatic invertebrates on which they feed though, in each case, suitable prey seems to extend a mile or two downstream of the lowest breeding Dippers. It is essentially a bird of upland streams and rivers, or at least of those rivers that demonstrate, regardless of altitude, the geographical profile of such water courses. An exception is found on the lowest stretch of the River Derwent at Swalwell, where birds nest on the tidal section of the river (Wiffen 2003). Mapping of colour-ring sightings of this pair illustrated how their feeding territory moved upstream with the tidal cycle.

Dippers breed remarkably early. They regularly start singing in October and are usually holding territory by the end of December. In 2000, nest building was noted at Langley Moor on 6 February, though at higher altitudes it is usually not noted until much later, as much as six weeks later at extreme elevations. Birds have generally laid their first eggs by mid-March with the earliest recorded clutch at Eastgate at the end of February in 1975. Fledged juveniles are usually seen from late April, the 30th of this month in 2007 being a not untypical date (Newsome 2008). A small percentage of pairs attempt second broods in May and June (Wiffen 2003).

Dipper, Weardale, March 2011 (Ray Scott)

The combination of early nesting and the Dipper's choice of nest sites mean that many initial attempts are washed away, though most pairs are quick to rebuild. At least one nest with eggs on the Derwent survived a period of flooding; the nest in a hole in the river bank was below the flood water level for several hours, but was in an air-pocket and survived without the eggs chilling (Wiffen 2003). Nests are usually associated with waterfalls, culverts and bridges though at least three have been found in disused, partially-flooded lead mines in Weardale and also at Eller Beck, near The Stang in the south of the county. Artificial nestboxes located under bridges along watercourses have been successfully used in two locations in the county. In times when the population is at a high density, some first-year birds will resort to building in less usual locations. In 2003, two nests were found that had been constructed in the thin branches of small willows *Salix sp.* growing in the River Derwent. These nests had been built by colour-ringed first-year birds and both nests were approximately 100m from well-established nests with adult pairs. Both nests in the willows were washed away before the eggs hatched (Wiffen 2003). In 1992 an unusual nest site was discovered by bat workers 100m up a tunnel in upper Weardale. Bigamous males were at Winlaton Mill in spring 1990 and Lintzford in 2003, their identity and attendance at two nest sites proven by colour-ringing. At least two birds in the Rowlands Gill area, both females, were still in their territories seven years after ringing (S. S. Westerberg pers. comm.).

Since Dippers are found along linear habitats it is a relatively simple task to locate breeding birds by walking riverbanks. Only in June and July, when birds are in moult, do they become difficult to locate, often hiding in bank-side holes when disturbed (Westerberg & Bowey 2000). The density of breeding Dippers varies considerably, and is largely dependent upon the nature of the watercourse. Clean, fast-flowing rivers and streams in the uplands, with high levels of oxygen and high densities of oxygen-dependent invertebrates, hold up to one pair per kilometre of river course. However, on the lower stretches of rivers, nesting birds are more typically located at around 2.5km apart. In good quality water courses the greater volume of water in lower stretches, and consequently a greater quantity of prey, has some influence on the density of birds. Another factor which may affect breeding density is acidification of the watercourse due to afforestation, as has been widely documented, including by Ormerod *et al.* (1985). However, the *Atlas* recognised that as there has not been large-scale forestation of Durham's uplands, acidification would not seem to be a likely problem for Durham's birds.

Studies in the Derwent catchment during the 1980s and 1990s showed that one of that river's tributaries, the Pont Burn, had pairs approximately every 400m along its length. Using an extrapolation from these breeding densities, manipulated to reflect the known densities along the county's different river systems, the *Atlas* determined that the county population was around 210 pairs (Westerberg & Bowey 2000), and there is little reason to review this at present. In 2010, 40 breeding pairs were located at typical locations. This is just a snapshot of the true population during the spring, but these range from Harwood in Teesdale, in the far west, down to Sedgeletch on the Wear watershed, in the east. Nationally, the Dipper population would appear to have remained fairly stable over a 35-year period (Brown & Grice 2005).

Detailed studies on the Derwent revealed that nesting birds could be found from 360m down to 10m above sea level, with birds regularly found nesting on the tidal stretch of the Derwent. During the early 1990s, it would appear that this species was more common along the Derwent than in Robson's time, with at least eleven nesting pairs in a

7km stretch, and as many as thirteen in the best of years (Bowey *et al.* 1993). A colour-ringing study between 2002 and 2004 found 16 pairs regularly breeding on the 11.7km stretch of the Derwent between Swalwell and Chopwell Woods. This is a density of 13.5 pairs per 10km of river, which is higher than the 9.7 pairs per 10km recorded in Perthshire, Scotland, the highest density recorded in Britain (Ormerod *et al.* 1985). The main factor responsible for this density appears to be the total hardness of the water (mg/l CaCO3), the density of Dippers on the Derwent fitting on to the linear relationship between Dipper density and water hardness (Wiffen 2003). Birds' sensitivity to environmental conditions was demonstrated by the desertion of the River Balder by this species between September 1990 and April 1991, due to the excessive turbidity of the water in the river, as a consequence of work in the upstream Balderhead Reservoir (Armstrong 1992).

On the River Wear, territories have been located in the far west in the Weardale Forest at 430m above sea level, while traditionally birds were also found as far down the river as Shincliffe, to the east of Durham City (Westerberg & Bowey 2000). A Dipper seen at WWT Washington in July 1974 was thought to indicate breeding on the lower reaches of the Wear though this has never been proven. During the first decade of the 21st century, birds have been found on sections of the R. Wear as far down as Chester-le-Street. Birds apparently bred on the Red Burn, near Rainton Meadows, Chilton Moor, in 2004, at just less than 50m above sea level, one of the species' lower altitude breeding records for the watershed (Siggens 2006). The species' pattern of presence on other small streams in the east of County Durham is not yet fully clear.

In 2007, reports from the most eastern localities such as Rainton Bridge and Sedgeletch were relatively infrequent, and the only site in this part of the county from which breeding was confirmed was at New Lambton, on the Wear (Newsome 2008). Higher up the Wear, for many years the Rookhope Burn has held far fewer birds than its geography would suggest it should. This is believed to be due to residual heavy metal pollution and leachates associated with mining wastes higher up that valley. During the *Atlas* survey, birds were recorded along the Rookhope Burn's tributaries, indicating that these were not as badly affected. Of the Wear's lower tributaries, typically the total breeding population of the River Gaunless is between seven and nine pairs, with lower numbers on the Deerness and Browney rivers, though three territories were present in three miles along the Browney, up and down stream of Malton, during spring 1987 (Armstrong 1988).

In the north of the county, breeding birds were on the Beamish Burn, a tributary of the River Team, during the mid-1990s. These were the most isolated breeding Dippers in the county, at an estimated 16km from the nearest other breeding birds, on the Pont Burn or the Browney valley. Two pairs were present in this area in 1997 and these birds persisted into the 2000s (Armstrong 1999).

In the far west, the Tees watershed holds the highest altitude territories in the county, with birds sometimes breeding at over 500m above sea level, as in 1994 when a pair nested to the west of Cow Green Reservoir, at 520m, which may have been the highest elevation at which the species has nested in the county. In nearby Lunedale, a bird was on Arngill Beck on 4 July 1987, again at 520m above sea level. Breeding densities in such areas are illustrated by the four pairs in a 4.5km stretch of the Sleightholme Beck at Stainmore on 10 May 1980 and three pairs along 4km of the River Greta, where it crosses Stainmore, on the same date; both of these stretches of watercourse are more than 300m above sea level (Baldridge 1981). From there, birds occur all the way downstream on the Tees watershed, though in lower densities with diminishing altitude, to Darlington, where they nest at 30m above sea level.

Dippers can be relatively tolerant of the close presence of humans, as long as any disturbance is not directly targeted at them. Examples of where birds have bred in disturbed locations include the Malton picnic site on the River Browney and at Carrick's Picnic Site above the head of Derwent Reservoir. In both locations, Dippers have bred successfully oblivious to the large numbers of visitors that are sometimes present. Fortunately the gross persecution that this species used to suffer from, apparently at the hands of anglers, has largely subsided though there are instances of such activity in the last 20 years in the county.

A young Dipper, hatched near Wolsingham on the Wear, indulged in prolonged food-begging from a Common Sandpiper *Actitis hypoleucos* in spring 1994, following that bird for 60m before giving up.

Recent non-breeding status

The species' winter and non-breeding season distribution has a familiar feel to it and throughout the winter birds are present across the county in a pattern that is reminiscent of their breeding distribution. A bird flying across

Grassholme Reservoir in Lunedale on 7 May 1990 was an unusual sight, but perhaps it was a local breeding bird relocating around the reservoir.

In December 1990, wintering birds were on the Tyne downstream of Wylam Station, well below the tidal limit of the river. Demonstrating the species' hardiness, a Dipper was watched at Arngill in Lunedale (at 440m above sea level) in 1979 feeding through holes in the drifted snow that covered the beck. Another was watched in 1985 feeding among ice along the frozen Tees at Widdybank (Baldridge 1986).

Dippers often retire to small becks when their main nesting rivers are in flood, for example six birds were found on the Ulnaby Beck, High Coniscliffe when the Tees was in spate on 4 January 1980, where none were found during the breeding season. These movements in flood conditions often result in the birds moving across country well away from the river course. In some urban situations, such as in the Team Valley in Gateshead, this may be necessary as long sections of streams are culverted.

There is some evidence of local movements and occasional movements from further north in England, as indicated by records of birds in the east of the county, mostly during the winter months or in the post-breeding dispersal phase of the year. By and large though, Dippers are relatively sedentary (Brown & Grice 2005). The species was noted at Billingham Beck in January 1974, at Finchale in early 1979 and a bird at the River Don at Jarrow, in August 2001 was the first known record there, whilst one at Sedgeletch that October was the first there in a decade. A bird at Herrington Country Park on 17 September 2008 was unusual, as was another at the same place in 2009 on exactly the same date as the previous year (Newsome 2010).

Black-bellied Dipper
Cinclus cinclus cinclus

Birds of the continental form, *C. c. cinclus*, are partial migrants and very small numbers winter on an annual basis in eastern England (Brown & Grice 2005). There is a single record of this race in Durham, a bird fitting the description was on the River Wear at Witton-le-Wear on 24 December 1970 (McAndrew 1972). That Scandinavian birds can reach north east England was demonstrated by the Norwegian bird ringed as a pullus on 31 May 1993, which was found dead in North Yorkshire in October 1993 (Blick 2009). There have also been several other sight records of this race in Yorkshire, plus birds in Northumberland in 1987 and 1989.

Distribution & movements
This species breeds across mainland Europe, with isolated populations occurring in north west Africa and central Asia. In general, the species is rather sedentary, though some of the northern and eastern populations undertake altitudinal movements, down the watercourse upon which they breed. The nominate race occurs as a rare migrant to Britain, the source of these dark-bellied birds possibly being twofold, the first being continental Europe, the second Scandinavia, the population of these being partially migratory (Cramp *et al.* 1988).

The Derwent watershed holds a high density of breeding birds and colour-ringing work on local birds has identified two records of male bigamy and also highlighted a juvenile dispersal of first-year females. Observations of colour-ringed pulli indicate that only males stayed within the study area and their natal areas. The closest record of a female setting up a territory was 8km from its hatching place. A bird ringed at Edmundbyers as a juvenile on 10 June 1985 was controlled 13km away, as a breeding bird down river at Hamsterley Mill on 18 February 1989. A bird ringed as a nestling at Carrick Hill, above Derwent Reservoir on 21 May 1991, was found 9km away at Combfield House on 12 March 1992 (Armstrong 1993).

White's Thrush (Scaly Thrush)
Zoothera dauma

A very rare vagrant from Siberia: two records.

There is one historical record of this impressive Asian thrush. On 31 January 1872, one was shot and wounded by Rowland Burdon at Castle Eden Dene. It was caught alive approximately two weeks later, with the

specimen remaining in the possession of Mr Burdon. Some of the feathers were sent to Alfred Newton, who confirmed the identification, and John Sclater later gave a very detailed description of the bird in the *Zoologist*, 1872 (Temperley 1951). The specimen is now in the NHSN collection.

The best portion of a century had passed before Durham's second bird appeared. The only 20th century record of this species came on 7 November 1959 when, on a morning of rain and low cloud, Stephen Hayes found one in the Harton area, a well-populated part of South Shields (*British Birds* 53: 423), later tracking it to the West Park, where Fred Grey was taken to see it, but it was not seen after this date. The Ornithological Report for Northumberland and Durham for 1959 listed it by the beautiful name of 'Golden Mountain-Thrush' (Grey 1960).

White's Thrush, Castle Eden Dene, January 1872 (K. Bowey, courtesy NHSN)

Distribution & movements

White's Thrush is a widespread breeding species throughout central and southern Siberia, from the Yenisey River to Ussuriland. An isolated population also exists in the foothills of the European Urals. The wintering range extends throughout southern China and Taiwan to Indochina and central Thailand. It is a very rare autumn and winter vagrant in Britain with a total of 72 accepted records to the end of 2009, a high proportion of which have occurred in the Northern Isles.

Grey-cheeked Thrush
Catharus minimus

A very rare vagrant from North America: one record.

The only record for the county, and one of only a handful of records for mainland Britain, came in quite amazing circumstances. On 17 October 1968, Eric Meek was teaching at Horden Junior School. Eric had given a talk about birds in an attempt to enliven his class, and had clearly made an impression on one of the pupils as he was presented with a small bird that the schoolboy had found in the gutter of the A1068 outside the school. Eric was quite astounded by what had been presented to him; he identified it as a Grey-cheeked Thrush, then only the sixth British record (*British Birds* 62: 476).

The tale took a further bizarre twist. The specimen was forwarded to the Hancock Museum at Newcastle for preservation, where it was prepared as a freeze-fried specimen, rather than being stuffed and mounted in the traditional fashion. In unknown circumstances, the specimen was subsequently eaten by a rat and is no longer in existence (*British Birds* 89: 6). The autumn of 1968 was notable for the number of Siberian vagrants in Britain, but also featured a higher than normal number of trans-Atlantic waifs, including another Grey-cheeked Thrush on Bardsey Island, Gwynedd, on 31 October (Dymond *et al.* 1989).

Distribution & movements

Grey-cheeked Thrushes breed in the extreme north east of Siberia, throughout Alaska and northern Canada, across to Labrador and Newfoundland. The population migrates through the eastern United States to winter in northern South America. The vast majority of the 51 British records have occurred in October following fast-moving depressions crossing the Atlantic, with extreme dates of 22 September and 26 November. The Isles of Scilly have attracted most of the sightings, with the Durham bird being the only record on the English east coast.

Ring Ouzel
Turdus torquatus

A common summer visitor to our western uplands, with small numbers of passage migrants usually noted, mainly along the coastal strip.

Historical review

During the 19th century this species was said to have bred in its greatest numbers in the Pennines, amongst other upland areas in England, its population being stable and well-established (Holloway 1996). Backhouse (1885) in his *Avifauna of Upper Teesdale,* reported a pair nesting at an elevation of over 2,000 feet "*under Crossfell*" in 1844, so clearly this species has been a breeding bird in County Durham for many years. This nest, though, was almost certainly south of the county boundary. Hogg (1845) only documented one sighting in his work on the south east of Durham, no doubt an October passage migrant seen in a hedge near Newton Bewley.

Temperley (1951) reported that this was a "*summer resident in some numbers in the moorland districts*" and also a scarce lowland migrant. He said it "*arrives in March, departs in October*". He believed that it was not uncommon on the upper reaches of the Derwent as far down as Muggleswick and Weardale nesting "*as far east as Wolsingham*". Fifty or so years earlier, Thomas Robson (1896) had considered it an uncommon local species in the Derwent valley, though he did document the shooting of a bird at Broad Oak, a little way up the valley from Blackhall Mill. It was, Temperley stated, more scattered in upper Teesdale, but it did extend "*far up the valley*" and that "*later in the season flocks of young birds are to be found feeding on bilberry, cowberry and crowberry on the moors*".

Temperley (1951) said that it was "*very rare on spring migration*" highlighting a record at Teesmouth in April-May 1899 of a bird which "*frequented the slag wall*". Other early 20th century reports of passage birds included birds at Cleadon in 1933 and Hebburn Ponds 1944. Late autumn records included one at Teesmouth on 7 November 1937 and one that was seen at Herrington, near Houghton-le-Spring on 13 November 1853 (Temperley 1951), which was considered a very late date by George Temperley.

Over the past fifty years the shortage of organised data collection on the species has resulted in little progress in our knowledge of the bird loved by so many ornithologists but studied by so few. The current situation, in 2011, is slightly better. The ornithological records for Northumberland and Durham rarely reported the species prior to Temperley (1951). Subsequently, records appeared more frequently but only in nine of the following 20 years. In 1965 there was a large coastal influx in early October with hundreds of birds being reported. This had been preceded by a considerable southerly passage inland at Smiddyshaw Reservoir. On 19 August flocks of up to 10 were passing through, with 29 being flushed from one patch of heather. In 1968 and 1969, clutches were reported in which there were two normal sized eggs and a third that was the size of a Meadow Pipit's *Anthus pratensis* egg. Coastal passage was significant in 1969 when, between 3rd and 8 May, 30 were reported on the coast, with 10 of these at Marsden. The date suggests that these may well have been Scandinavian birds. Local breeding birds would, by this date, have been incubating their first clutches.

More detail relating to the species emerged through the 1970s, with the commencement of more systematic recording from around that time. Some of the more interesting records that arose from that decade include the landing of a tired migrant bird on a fishing boat eight miles out into the North Sea, off Sunderland, in April 1974. Clearly the Tees valley around Widdybank and up to Cauldron Snout has long been a favoured haunt of this species and those watching it. The 1975 breeding season was considered a good one for this species in that area, and six pairs were found around Falcon Clints and Cronkley Scar, as well as another three family parties between Arngill Force and Fish Lake in Lunedale, in early July.

Passage periods tended to produce some of the most interesting records, with a notable fall of what were presumably birds of Scandinavian origin along the Durham coast, on 26 September 1976. The true numbers involved can only be guessed at but around 20 birds were in Marsden Quarry at the north end of the coast and around 30 birds were at Graythorp, Teesmouth, in the south east of the county. Many hundreds must have been present between these points. Probably the earliest arriving bird of the decade was one that made it, rather unusually to a coastal site, Marsden Quarry, on 19 March 78, though this was matched by a bird at Falcon Clints on the same date in the previous year. Late records during the decade included one at Marsden over 9th to 10 November 1977 and not quite so late, but a long way to the west, one at Falcon Clints in upper Teesdale on 8 November 1978 (Unwin 1979).

Recent breeding status

Breeding Ring Ouzels occupy a very specific habitat niche in Durham's uplands. As a breeding species, they are rarely found below an altitude of 300m above sea level but this is by no means their only limitation to territory selection (Westerberg & Bowey 2000). Birds favour moorland streams and gullies, where the bank-sides provide sheltered nest sites. The species tends to be concentrated along the minor tributaries of the main river valleys, where these cut into the upland block as steep-sided valleys, gills and gullies. The main river valley basins of the county are mostly too low-lying for Ring Ouzel, even along their courses in the far west. Worked-out quarries are also attractive, as are old farm buildings and, in some instances, dry stone walls (Gibbons et al. 1993). Access to areas of sheep-grazed pasture or wet flushes, within 300-500m of their nest, has been shown by many studies to be important for feeding (Rebecca 2001). The species' distribution, as mapped in the *Atlas* (Westerberg & Bowey 2000) clearly reflected this reliance on upland areas for breeding, hence its distinct distribution in the west of Durham. Access, for surveying purposes, to the species' favoured upland domain, is always likely to have been incomplete in the past and, as a consequence, many pairs will have been missed.

Between the *Breeding Atlas* (Sharrock 1976) and the *New Atlas* (Gibbons et al. 1993) there was a decrease of 27% in the Ring Ouzel's breeding range, including the northern Pennines. A national survey carried out in 1999 (Wotton et al. 1999), which compared the situation then and in 1991, implied a further 56-61% decline, which resulted in the species being added to the red-listed of Birds of Conservation Concern (Eaton et al. 2009). It is believed that this species' breeding range contracted quite considerably in the last two decades of the 20th century (Kerr 2001), with birds disappearing from some of their formerly inhabited lower-lying territories, there being a suggestion of an up-the-hill movement by Blackbirds *Turdus merula*, with a resultant possibility of some inter-specific competition. The reasons for this are complex, but are thought to be related to changes in the habitat, both in the breeding areas on migration and in the wintering areas in North Africa.

The area between Bollihope and Eggleston has apparently been much favoured by birds over the years in Durham, but this is probably, in part, a reflection of observer bias and a tendency for birdwatchers to focus on sites where birds are regularly and fairly easily seen in late March and April. Other sites where the species is routinely present and breeds include: the Cauldron Snout/Widdybank Fell area of upper Teesdale (in spring 2003, nine singing males held territory between Widdybank Farm and Cauldron Snout), upper Weardale, the Rookhope valley, Muggleswick Common and around Waskerley and Hisehope Reservoirs. Evidence from upper Teesdale, where regular counts have been made over many years, suggests a stable population, albeit with annual fluctuations (I. Findlay pers. comm.). A further indication of Durham's breeding population is that in 1999, 32 upland sites were identified as having breeding Ring Ouzel and in 2009, 37 one-kilometre squares were similarly identified (Newsome 2010). In 2008, records came from 20 moorland locations between April and July and referred to a minimum of 30 pairs in these areas. Clearly these figures are not directly comparable, the 2009 ones being the result of largely casual observations. The *Atlas* estimated the population size in the county to be in the order of 250-300 pairs (Westerberg & Bowey 2000).

In 2000, three area studies were undertaken in upper Teesdale, in the Eggleshope Burn area and in Upper Weardale (Cowshill/Killhope) respectively. The upper Teesdale area has been studied for many years and has consistently produced counts of between nine and twelve singing males. In the Eggleshope Burn area, including Sharnberry Beck, ten territories were located in the altitude range 260-460m. Confirmation of breeding was, however, confirmed in only two cases. In eight of the territories, rough grassland comprised 55-90% of the area used by the birds, with grazed pasture, spoil heaps and heather moorland making up the rest. In two cases, 75-90% of the territory area was heather moorland with 10-25% rough pasture. All were in stream valleys. In Upper Weardale, ten territories were located between 410-510m. In one case a nest with three eggs was located and the fledged young were later seen accompanied by the adults. The main habitat type here was grazed pasture with spoil heaps, quarries and standing water also featuring. These, and other records, appear to confirm that areas of short grass are an important feature of the landscape in the vicinity of nests. The adults used these areas for feeding, often in excess of 250 metres from the nest site (J. Strowger pers. obs.). The newly fledged young are also often taken to these areas of short grass and rough pasture where they can feed, but also remain hidden in areas of longer grasses and bracken (J. Strowger pers. obs.).

Prior to this work, the county's birds had never been adequately or systematically surveyed, nor had the threats they face been properly assessed. A subjective view would be that this is a species which may be suffering a shallow decline in Durham, in keeping with the national situation (Mead 2000), but whatever factors are at play, they may not be applying as stringently in Durham as elsewhere.

Recent non-breeding status

Ring Ouzels are regularly recorded into November in Durham, but are rarely recorded as a wintering species in the county, or elsewhere in England (Brown & Grice 2005), though there were a number of records in the mid-1990s. In 1995, a bird was at Hetton Lyons on 26 December and was believed to be the first December record for Durham. In 1996, a female and three males wintered in widespread locations around County Durham. A female was at Sedling Rake in Weardale on 15 January, and another female was at Horden from 13 January, remaining until 12 February. In this latter month, a male was at Ryhope on 8 February and during February and March another male, probably present from late January, was found at Wolsingham, defending a crab apple *Malus sylvestris* tree (Armstrong 1997 & 1998).

As well as being a breeding species, this is a regular spring and autumn migrant, with light coastal passage occurring from March. Birds typically arrive in upland areas before being seen on the coast, in mid- to late March. The earliest record, of what was considered to be a passage bird, was on 3 March 1994 in Teesdale. The ten-year average first arrival date in 2009 was 22 March (Newsome 2010). There is some evidence that birds are arriving earlier with the ten-year mean for 1982-1991 being 26 March and that for 1992-2002 being 24 March (Siggens 2005).

Spring coastal migrants tend to be relatively scarce in most years, and generally follow the arrival of birds on the breeding grounds in mid-March to early April. Coastal migrants are more usually noted from mid-April to early May and rarely involved more than four or five birds together. Mid-April 2005 brought an unusually large fall of spring migrants to the Durham coast, with 22 birds together on Seaton Common over 16th and 17 April (Newsome 2006). Further large spring arrivals have included 22 on Seaton Common on 1 May 1978, 22 at Whitburn on 21 April 1983 and 20 on Cleadon Hill on 20 April 1986. Many coastal sites can potentially hold birds, but sightings reflect observer bias. However, inland lowland sites have also featured in records, including Hetton Lyons, Sedgeletch, Follingsbury Lane, Shotton and Longnewton Reservoir.

Further, more detailed evidence from the Gateshead area (Bowey *et al.* 1993) over a number of years supports the hypothesis that small numbers of birds pass through the county's inland sites on a broad front, on an annual basis. Spring passage birds have been noted at Shibdon Pond on 16 April 1989 and 19 March 1990, on Kibblesworth Common and Ravensworth Fell areas in the late 1980s and early 1990s, on one occasion briefly holding territory. Birds were at Dunston and the Derwent Walk Country Park on 10 April 1996 and a singing male was at Clara Vale from 14th to 15 April 1996 (Armstrong 1997).

In late summer and early autumn small flocks of Ring Ouzels, possibly family parties, can be seen in the upland, their location being largely unpredictable. However, Ring Ouzels characteristically feed on berries, particularly those of the Rowan *Sorbus aucuparia*, so these are a great attraction. On 7 September 1976, 27 were at Falcon Clints with 16 still being present on 16th. A large post-breeding flock of 30 birds was seen in the Derwent Valley in late July 1992. In 1999 seven family parties were seen in Lunedale, including a party of eight birds which were mobbing a Kestrel *Falco tinnunculus*. During this post-breeding period, flocks of Mistle Thrushes *Turdus viscivorus* frequently occupy the same habitat. On 20 September 1998, eight birds were in the company of Mistle Thrushes in Teesdale; in the autumn of 2000, parties of nine, eight and 15 were with Mistle Thrushes in the Eggleshope Burn area (Armstrong 1999b, Armstrong 2005). Areas of bracken are also used by Ring Ouzel and other thrushes *Turdus sp.* in the autumn, presumably for concealment and as feeding areas (J. Strowger pers. obs.).

Autumn passage is annually documented on the Durham coast, usually from mid-September to late October and on occasions to mid-November, but the scale of this is very much weather-dependent. Marsden and Trow Quarries, Cleadon Hills, Whitburn, the coastal denes and Hartlepool Headland are all likely places to attract birds and the process probably involves a large proportion of Scandinavian birds blown on easterly winds on to the north east coastline. Notable autumn gatherings have included 12 at Marsden Quarry on 21 October 1990 and 15 on Cleadon Hill on 9 October 1993, with a further 12 at Marsden Quarry on the latter date. Parties of between five and eight birds were also noted at coastal watchpoints in the autumns of 1994, 1997, 1998 and 2005. The latest sighting of a fresh coastal migrant was at the Long Drag on 21 November 2010. Lingering birds are also seen inland, for example at Waskerley Reservoir on 12 November 2000 and Hetton Lyons on 23 October 2009. In 2007, birds passed through Rainton Meadows on 7 October and Hetton Bogs on 12 October, and then a bird spent two weeks on Elemore Golf Course from 19 October. Another late individual at Rainton Meadows on 7 November was the last bird of 2007 (Newsome 2008).

Distribution & movements

Ring Ouzel breed across the mountains of northwest and central Europe, and southwest Asia. The nominate race breeds in Britain and other parts of northern Europe. The species winters around the Mediterranean, especially to the south, along the coasts of northwest Africa (Parkin & Knox 2010).

Information about birds ringed in the county is sparse, though two birds ringed on the same date at Graythorp on 2 October 1965 were recovered abroad; both were shot, one on 24 October 1965 in France and the other in Spain on 15 October 1968. Another bird ringed in Marsden Quarry on 21 October 1988 was found dead on 29 November 1988 in Reignac near Barbezieuk, Charente, in France, confirming the implied migration route of the previous two recoveries. A bird ringed in Hargreaves Quarry on 17 October 1987 was believed to be showing the characteristics of the Alpine race *Turdus torquatus alpestris* (Blick 2009), implying that this bird had headed north from its breeding grounds in a dispersal pattern that was perhaps analogous to that of Water Pipit *Anthus spinoletta*.

Blackbird (Common Blackbird)
Turdus merula

Sponsored by
Wm. C. Harvey Limited

An abundant resident, passage migrant and winter visitor.

Historical review

There is little historical documentation of this common species, though Nelson (1907) credited the earliest reference of it in Yorkshire to Marmaduke Tunstall, from birds found at Wycliffe on Tees, which was once considered part of that county, but is now in County Durham.

From about the middle of the 19th century, the Blackbird began to spread out of its original, typical woodland habitats into the suburban gardens of Britain's expanding towns and cities (Holloway 1996, Brown & Grice 2005). In Durham it was widespread and common, but there was little detail documented about its status at this time. Late in the 19th century this was said to be a very common species in the Derwent valley (Robson 1896) and it still is there today.

In reviewing its status in the middle of the 20th century, George Temperley (1951) said that it was a *"very common and well distributed resident and, in small numbers, a winter visitor."* He documented the species as nesting at altitudes as high as 1300 feet (about 420m) above sea level, which was recorded by Backhouse (1885) in upper Teesdale, and Temperley (1951) noted its westward penetration into the uplands, along the river valleys.

Nelson (1907) recorded immigrant blackbirds arriving at the coast around Teesmouth from late September to the end of November, but he also noted that they were not as obvious on return passage, in spring. As early as the mid-20th century, it was realised, from ringing data, that some of the Blackbirds ringed in Durham progressed south west to spend the winter in Ireland (Temperley 1951).

The spread of the species into the centres of cities, into the smaller parks, city centres and gardens was apparent by the 1930s, outstripping the Song Thrush *Turdus philomelos* locally during the 1939-1945 war (Holloway 1996). During the first half of the 20th century, the Blackbird was out-numbered by the Song Thrush (Westerberg & Bowey 2000), but by the 1940s the situation had changed, partly due to a milder climate (Marchant *et al.* 1990) and also because the Blackbird was quicker at adapting to the then rapidly spreading, man-made environments. Nationally, the population continued its increase until the 1950s and then remained relatively stable (Marchant *et al.* 1990). The Blackbird seems better able to survive severe winter weather than the Song Thrush (Gibbons *et al.* 1993) as was observed of Durham's birds, by Temperley (1951).

In the south east of the county during the 1950s and 1960s, it was described as *"abundant"* by Stead, (1964) and *"common"* by Blick (1978). Nonetheless, during much of the latter part of the 20th century the species was thought to be in decline, though numbers recovered somewhat in the last few years of the century and the years directly after the Millennium (Brown & Grice 2005).

Recent breeding status

Today the Blackbird is one of the best-known and most-recognised garden birds as, in spring and summer, the males deliver their rich, fluty song from high vantage points, such as garden trees, shrubs and television aerials. It is a very common and widespread breeding resident. The Blackbird is one of the most widespread species in Britain, only the Woodpigeon *Columba palumbus* being more widespread (Risely *et al.* 2010). Across the UK, it was recorded in 93% of the 3,243 surveyed Breeding Bird Survey squares in 2009 (Risely *et al.* 2010) with a reported 38,121 individual birds documented in these.

Across Britain, it can be found from the centre of the built-up areas to the middle of upland valley woodlands. Whilst records of the species come from almost all parts of the county, it tends to be more thinly distributed in the western uplands, as observed by Brown & Grice (2005). Birds are present from the coastal denes in the east through to Grains o' th' Beck, The Stang and Langdon Beck in the far south and west. It can be found breeding commonly in nearly every scrub and woodland habitat, from the county's coastal denes through mixed woodland, plantation, gardens, parks and hedgerows around farmed land and it is routinely seen feeding on the lawns of parks and gardens in urban areas. In Durham, it is more widespread than the House Sparrow *Passer domesticus,* though less numerous in many areas. In 2009, the Durham Bird Club received over 1625 records of this species, which were estimated to refer to over 12,500 birds (Newsome 2010).

The species is known as one that occasionally exhibits advanced breeding behaviour early in the year, usually in situations where the local climate has been inadvertently influenced by man's activities (Westerberg & Bowey 2000). The length of the species' breeding season is indicated by the fact that recently fledged juveniles might be seen at almost any time between February and July. In the early 1980s, the species' local breeding fortunes varied from year to year. For instance in 1980 the breeding population appeared high (Baldridge 1981) but in 1981, it was a poor breeding season with many nests reported as being abandoned or overturned due to the weight of snow that fell during late April (Baldridge 1982). During the survey period for the *Atlas*, on 19 January 1988, a female was filmed by a television crew nesting on the wording of a shop fascia in Stockton (Westerberg & Bowey 2000). Another example of early breeding was documented in 1999 when, on 15 February, a female was feeding young in the nest in the Washington Galleries shopping centre. These chicks had fledged by 26th of the month. These data can be placed into context by the following detail from the same observer. Of 500 Blackbird nests documented between 1974 and 1982, egg-laying began around 7 April and the earliest ever recorded was 19 March. A further record of even earlier breeding came with a report of a fledged juvenile near the Arnison Centre, Pity Me on 9 February 1997.

More usually, the first young of the year might be present in gardens around mid-April, with 'early birds', that are noted in March, or even late February, not being that unusual. Twenty-two juvenile birds together in a single Hamsterley Mill garden in the summer of 1978 indicated a good breeding season in that part of the Derwent valley. At the other end of the season, in 2007 for example, the last reported new brood of recently fledged young was seen on 12 August (Newsome 2008). Interesting behaviour was noted at Billingham in June 2006, where a pair was feeding their juveniles with scavenged meat from a hedgehog *Erinaceus europaeus* corpse (Newsome 2007). This is possibly a result of dry weather forcing the species to seek any available food source due to the lack of its preferred worms and molluscs. The species has a degree of adaptability in being able to sometimes breed in unusual local situations. This was illustrated by a local population in the north west of the county, which took to nesting on the top of light housings, in the front porches of houses, in a high-density housing estate in Winlaton, during the late 1980s (Westerberg & Bowey 2000).

In May 2000, survey work revealed 37 territories around Hetton Bogs, 28 at Hetton Lyons and 8 at Shibdon Pond. In June 2000, a pair in a Chester-le-Street garden was feeding young at a nest in Clematis, after two previous breeding attempts had failed due to predation by Sparrowhawk *Accipiter nisus* and domestic cat. Spring 2000 brought two records of birds establishing territory at higher than usual elevations; one was singing at Grouse Rake, in the upper Derwent valley, in April at over 400m above sea level and another in July, at Crawleyside Cottage, north of Stanhope, was at an elevation of around 330m above sea level. Such high-level nesting is by no means unprecedented, as in 1975 a pair nested at Falcon Clints, in upper Teesdale, at close to 500m above sea level and in 1979 birds bred at old mine workings at Cleve Beck in Lunedale, at around 450m elevation. Where birds do penetrate the western uplands areas, they tend to gravitate towards the vegetated valleys and gills, for example small numbers of birds tend to be concentrated around buildings or in sheltered valleys where some shrubbery or scrub offers a modicum of cover, for example a pair of birds at Bollihope in 2007, was present in a fenced-off juniper plantation whilst the local Ring Ouzels *Turdus torquatus*, preferred the adjacent open moorland habitat. In 2008, three singing males were located at Rimey Law above Rookhope at almost 500m above sea level, showing the adaptability of the species (Newsome 2008).

Over the early part of the twenty-first century, the number of territories at Elemore Hall Woods varied from 49 in 2000 rising to 52 in 2003 then falling to 40 pairs by 2007. In 2006, survey work revealed that the species was present in 26 out of 31 surveyed, one-kilometre squares and there were an estimated 279 pairs in these. Assuming a county-wide percentage occupation of one-kilometre squares based on this figure, and a mean occupation rate of 10.7 pairs per occupied square, would give a county population of around 28,400 pairs. This is lower than the estimated figure derived from the *Atlas* work and modified from subsequent trends but, based as it is on a much smaller survey it roughly corroborates the scale of the county's population of this species. Breeding information gathered during 2007 indicated just how common this species is as a breeding bird locally, with 55 pairs located in the area around Hetton, at least 40 pairs around Elemore and 120 singing birds at Spring Gardens. Other reports included 34 pairs at Hetton Bogs, 30 pairs at Hetton Lyons and 34 pairs at Rainton Meadows, all of which were increases on 2007 survey counts. Reports from the 2008 breeding season included 48 pairs at Castle Eden Dene in May, with the observer noting it as being the most abundant species there. Interestingly, some 30 years previously, in mid-June 1979, the same area had held 41 territorial males.

In the last couple of decades of the 20th century, the species experienced some considerable declines, which were attributed to the relatively cold winters of the late 1970s and the 1980s (Marchant *et al.* 1990). Nationally, the species increased by 26% over the period 1995-2008 across the whole of the UK, but showed a lower increase of 23% for just England, whilst in the north east of England the increase was calculated as 38% (Risely *et al.* 2010). Work for the *Atlas* revealed that birds were recorded in 63% of the tetrads in the county (Westerberg & Bowey 2000). More recently, probably over 3500 pairs were estimated to be present in the northern part of what was Cleveland, these being located by the survey work of 2000-2006 for the *Breeding Birds of Cleveland* (Joynt *et al.* 2008). Particularly high concentrations of breeding birds were evident in the suburban fringes of Hartlepool and Stockton, and it remains a common bird around the Teesmouth area today (Joynt *et al.* 2008).

Using national figures (Gibbons *et al.* 1993) and assuming Durham to be typical of Britain as a whole, the *Atlas* calculated that there was estimated to be in the region of 49,000 pairs of Blackbird in the county, but considering the recent national declines and recoveries (Brown & Grice 2005), then an estimate of between 40,000 and 45,000 pairs for Durham is probably an appropriate figure at present.

Recent non-breeding status

As in the rest of England, there are three groupings of Blackbirds that winter in Durham; rather sedentary local birds, internal migrants from elsewhere in the British Isles and those from the Continent (Brown & Grice 2005). Winter records generally refer to feeding concentrations of birds with just a few reports in the early part of the year, though birds are widespread in the winter. In the winter of 2007, the largest counts were of 50 at Boldon Flats and 38 at Middleton St. George, although later in the year, survey work in the mid-Wear valley revealed counts of between 30 and 90 birds at 14 locations (Newsome 2008). Most typically today, winter gatherings are reported from locations such as the Clara Vale and Thornley Woods Feeding Stations, in the north of the county, whilst almost as many can sometimes visit gardens where specially provided apples, or other favoured food items, are made available.

In the middle of winter, birds tend to be largely absent from the upland areas; locally breeding pairs in such areas perform some degree of altitudinal migration, at the very least dropping down to lower valley elevation. The reverse process can be startling, when birds suddenly re-appear around garden territories after a winter's absence, for example the mid-March 'influx' of birds to a Baldersdale garden in 2000 (Armstrong 2005).

Birds roost communally during the winter months and such gatherings frequently number scores of birds, occasionally hundreds, but the largest such gathering was the 3,000 birds counted roosting in Darlington West Cemetery in winter 1978 (Unwin & Sowerbutts 1979).

Autumn movements, at least local ones, usually begin to be reported from late July, when birds higher up the dales start becoming mobile, and are often noted moving in an easterly direction, the contrary direction of continental migrants arriving a month or two later. Large numbers of continental birds reach the county each autumn, from mid-September to late November, often arriving just before the first Redwing *Turdus iliacus* of the winter. Despite the fact that, in most years, no more than 100 are seen in the vicinity of the main coastal watch points in any one day, it is certain that many more birds pass through these sites, effectively unseen. At suitable concentrations of food, such as ripe Hawthorn *Crataegus monogyna* berries, flocks of up to fifty birds might be routinely noted in locations such as Marsden Quarry. Autumn influxes and passage often results in large concentrations of birds from the latter part of September through October and into early November.

In some years the scale of passage can be very large, especially if weather conditions concentrate the arrival of birds. For example, during August and September 1999 there were no reports of the species along the coast, although clearly local resident birds would have been present, then by contrast in October, coastal passage was very evident. On 16th of this month, 345 were counted at South Shields and around 600 were at Hawthorn Dene. The following day at least 100 were in Marsden Quarry, 55 remaining to 22nd (Armstrong 2003). Clearly, large numbers of birds were involved in a significant influx along the length of the Durham coastline, and the total must have numbered many thousands of birds.

The Coordinated Migrant Count on 4 October 1992 illustrated a restricted arrival of Blackbirds along a relatively short length of coast. During the three-hour count, 611 Blackbirds were recorded along the Durham coast but 55% of these were found in just three out of 29 zones of the count, in the north of the county, between Seaburn and Sunderland Docks. An analysis of this occurrence pattern indicated that most of these birds made landfall prior to the count over a stretch of the county's coastline no more than 3km in length. Whether this concentration of birds indicated a limited departure point for these birds or migration conditions which concentrated them into a short landfall arc, was unknown. The restricted arrival theory was supported by a similar but less pronounced concentration of Song Thrushes into the same areas (Bowey 1999c).

The largest such counts recorded in the south east of the county include 1,055 birds at Hartlepool Headland on 15 November 1969 and around 1,000 at Graythorp on 24 October 1965 (Blick 2009). On 19 November 1972, the main influx off the autumn saw 1,000 at Hartlepool and 500 Whitburn. A similarly-sized, large autumn influx brought 1,000 birds to the Whitburn area on 4 November 1984. In 2000, a large arrival of birds occurred on 22 November with *c.*600 birds at South Shields and *c.*500 coming 'in off' the sea in one hour at Whitburn. In autumn 2004, a period of heavy passage during October resulted in birds coming in off the sea and building up to around 1,000 in Marsden Quarry on the 28th. These birds can rapidly penetrate inland and good feeding areas, where berry-bearing shrubs predominate, can quickly attract large numbers, such as the 215 birds that were noted at Rainton Meadows on 13 November 2007. Immigration is sometimes evident on the coast in early November, as in 2008, when grounded migrants were also present in the same week, including 2,000 at Hartlepool Headland on 6th, and on 9 November 1992, when 300 flew over Whitburn in 90 minutes.

Coastal spring passage through Durham is usually much less evident than is the autumn movement into the county. Most typically this consists of small loose feeding flocks of birds, 15 to 20-strong, in areas of scrub in coastal areas, from late March through to mid-April. It is quite unusual for larger numbers than this to be involved at the main coastal sites on any one day, although up to 70 were present at Graythorp in mid-April 1974 (Blick 2009) and 20 were at Marsden Hall on 11 April 1979. The largest ever documented spring exodus occurred in 1984, with a return migration of Scandinavian birds that was noted at Whitburn, during which 1,113 birds were counted leaving the county on 3 April, with a further 150 noted on other dates through that month (Baldridge 1985).

Over the years, there have been many instances of birds with abnormal plumage being observed, from piebald birds to fully white leucistic individuals. These birds can sometimes live a number of years and are readily identifiable where they inhabit observer's gardens or local areas. Examples include the two males with white heads that were reported in the Hetton/Houghton area during November 1985, one of these remaining to October 1986. Birds with partial albinism were noted at WWT Washington and Charlton's Pond in 1987 and in 2001 a male that was all white excepting for four black tail feathers was noted near Darlington, with a similar bird near Clara Vale in 2008.

Distribution & movements

This species breeds across Europe, excepting the extreme north, and down into north west Africa, the Middle East and the extreme portion of western Siberia, and in central and southern Asia. Although it is present in large portions of its breeding range through the year, many populations are migratory or partly migratory, with more northerly breeders wintering within the breeding range of the more sedentary western and southern populations, such as those in Britain.

One of the earliest recoveries of a ringed Blackbird in the county was of one ringed as a nestling at Stocksfield, Northumberland, on 31 May 1922 and later reported at Dunston, Gateshead on 21 February 1923. Many Blackbirds have been ringed in the region and the mass of recoveries demonstrate that birds move south and west in the autumn to locations in Ireland, south west England and beyond this, into France and Spain, and back in spring northwards to Scandinavia, the Baltic countries, and even further east in some instances. Demonstrating this are birds ringed at Whickham during the winter, which have later been recovered from as far afield as Estonia in April and on an oil-rig in the Ekofisk Oil Field in the North Sea on 9 October 1982 (L. J. Milton pers. comm.).

Ringing has also shown, over a 40-year period, that many local Blackbirds remain in the vicinity of their birthplace through much of their lives. The oldest known ringed bird from such work in the Teesmouth area was seven years of age (Blick 2009) but in the north of the county a bird ringed at Whickham as a first-year on 29 October 1972 was found dead there in on 1 June 1983, when it was eleven years old. Ringing studies in the urban fringe areas of the county suggest that over a quarter of all Blackbirds killed fall prey to cats and road traffic (Durham Ringing Group database). One ringed in a Whickham garden on 28 August 1974, was killed by a cat in the same location on 13 March 1980, but it had presumably experienced a productive six years in between times. Such mortality factors know no boundaries, as indicated by the adult male ringed at Whickham on 13 January 1979 that was killed by a cat on Humberside on 11 January 1980. Further afield, the same fate befell a male ringed at Marsden Quarry on 25 October 1990, which was taken by a cat 129 days later on 3 March 1991 in Veltheim Germany, some 855km to the south east, whilst a first-year male ringed at Marsden Hall on 26 October 1979 was a road casualty at Weserens, Helgoland, Germany on 4 March 1980 (Baldridge 1981).

Many birds arriving in Durham in the autumn would appear to have their origins in Scandinavia. Others pass through Germany and the Netherlands on their journeys to and from the north east of England, as illustrated by the bird ringed at Helgoland, on 7 November 1934 that was recovered near Sedgefield, in December 1937. An adult female ringed at South Shields on 9 January 1951 was trapped and released at Torslanda, Göteborg, in southern Sweden on 31 March 1952. One that was ringed at South Shields on 30 December 1962 was recovered at Åbybro, Jutland, Denmark on 13 January 1964 and another, ringed as a first-winter bird at South Shields on 24 December 1966 was recovered in a similar area, at Hirtshals, Jutland three years later, on 15 March 1967. Some of our wintering birds come from the east rather than the north or the north east, as illustrated by the record of a first-year male ringed at Graythorp on 21 November 1976, which was found dead at Spols, Poghausen, Aurich, Germany on 11 June 1978 and the bird ringed at Low Barns, Witton-le-Wear, on 22 January 2005 that was controlled at Schiermonnikoog in the Netherlands on 21 November 2005, 534km to the east (J. Hawes pers. comm.).

Some birds that arrive in the autumn do not stay through the winter in Durham. There are a number of records of locally-ringed birds that have been recovered or controlled down the east coast of England, clearly moving south from their first autumn landfall, as shown by the first-winter bird ringed at South Shields on 19 October 1964 that was controlled at Cleethorpes, Lincolnshire on 3 December 1967, the adult ringed at South Shields on 26 December 1966 and recovered at Great Yarmouth, Norfolk, on 17 November 1968 and the first-winter female ringed in West Cemetery, Darlington on 2 February 1985, which was controlled at Landguard Point, Suffolk on 9 March 1988. Some of the presumed autumn migrants ringed in the county have a tendency to head south and west, as shown by the adult male ringed at Marsden Quarry on 20 October 1988, which was controlled at Murlock, Dyfed two years later on 5 November 1990, a distance of 439km to the south south west. Other birds have been recovered wintering in Ireland and the south west of Britain later in the same winter or in subsequent years, such as the bird ringed at West Hartlepool on 10 March 1967 and recovered at Athboy, County Donegal, Ireland on 3 March 1968. Confirming the direction and sometimes prodigious distance travelled, was one ringed at Graythorp in October 1965, which was in the Gironde area of France just one month later (Blcik 2009).

It is hardly surprising that migrant birds head north again in spring, but confirmation of this comes from the first-year male ringed at Morston, Norfolk on 26 October 1972, found dead at Hartlepool on 4 March 1977. Clearly, many autumn migrants manage to make it 'home' to their northerly breeding areas. There are a number of birds ringed in Durham that have been later recovered in Norway, such as one ringed at West Hartlepool on 28 November 1965, recovered at Nordkvingevåg, Masfjorden, Norway on 6 October 1966 and another ringed there on 22 January 1967, which was shot at Overdal, Norway on 30 August 1969. From the north of the county, one ringed as a first-winter bird at South Shields on 24 December 1966 was recovered at Fedje, Hordaland, Norway on 11 April

1967. Further east and north were a first-year female ringed at Graythorp, Hartlepool, on 8 November 1975, found dead at Pikonkorpi, Tammela, Finland on 23 June 1977, and a male ringed at Marsden Quarry on 22 October 1990 that was controlled at Nidingen, Halland, Sweden on 15 March 1991. An adult female ringed at the same site on the previous day was found dead on 17 March 1992 at Alvsborg Sweden. This part of northern Europe is obviously a rich breeding ground for birds wintering in the north east, for a bird ringed at Ryton in October 2003 was recaptured 1,007km away from its ringing location in Kristiansand, Sweden in April 2004, and other Swedish controls relate to birds ringed at Whickham and Urlay Nook, to the west of Stockton. Other birds ringed locally have been recovered even further east, such as the adult bird caught at Sheraton on 24 December 1966 that was recovered at Kustavi, Pori, Finland, on 27 August 1968.

Dusky Thrush
Turdus eunomus

A very rare vagrant from Siberia: one record.

The sole Durham record relates to a first-winter male discovered by Russell McAndrew, along with G. Proctor and B.J. Coates, at 10.45hrs on 12 December 1959 at Hartlepool Headland. The bird was initially in view for some thirty minutes and when first found, it was accompanied by a Mistle Thrush *Turdus viscivorus*, two Fieldfares *Turdus pilaris*, a Redwing *Turdus iliacus* and a male Blackbird *Turdus merula*, but it was, essentially, solitary. For much of its stay the bird frequented the cliff top bowling-green and an adjoining school rugby pitch. On 10 January 1960, it was caught and ringed and it was last seen on 24 February, after a stay of over ten weeks. It was only the second record in Britain following one shot in Nottinghamshire in 1905 (*British Birds* 53: 423, *British Birds* 54: 190). Unsurprisingly it generated considerable interest amongst the active birdwatchers of the time and during its stay it was seen by many observers.

Dusky Thrush, Hartlepool, winter 1959/60 (E.S. Skinner)

Distribution & movements

The breeding distribution of Dusky Thrush covers northern and central Siberia east to Kamchatka, with the wintering range spreading from Japan south to Taiwan, southern China and Myanmar. There were only nine British records to the end of 2010. All sightings have come between 24 September and 23 March.

Black-throated Thrush
Turdus atrogularis

A very rare vagrant from Siberia: one record.

On 3 April 2010, Chris Sharp found a first-winter male in gardens on Hartlepool Headland, as it fed on the lawns and flower beds near the putting green, though it roamed widely. Initially thought to be a female, it was heard singing during the early evening. Following a clear and cloudless night giving good migration conditions, it was not seen again the following day (*British Birds* 104: 609).

Distribution & movements

Black-throated Thrush breeds in the central and northern Urals, east across Siberia and Kazakhstan, to north west China. The wintering range includes Iraq and northern India, east through the Himalayan foothills to Bhutan. It is a rare but regular visitor to Britain, being encountered as both an autumn vagrant (particularly in the Northern Isles) and alongside other winter thrushes during the winter months. A total of 70 had been recorded to the end of 2010. There had been only five previous April records (based on the date of finding) and spring birds such as this one are likely to have wintered somewhere in Britain and be leaving, via the east coast, for Scandinavia.

Fieldfare
Turdus pilaris

A common winter visitor and passage migrant.

Historical review

This handsome visitor from the north may derive its name from the Middle English word *feldefare*, which has been interpreted by some to mean 'a traveller through the fields', though this is disputed by others (Lockwood 1984). No doubt it was a welcome winter addition to the diet of the county's monks in the fourteenth century, documented as it was in the Cellarers' rolls of the Monastery of Durham. There are two quotations referring to its purchase for the table from these old texts while there are only two earlier mentions of this species known. The first of these is from April 1343, during the species' northward passage through the county, with reference to "*one dozen Fieldfare being obtained*" ("*J duoden. de feldfar*"). Secondly, in 1347 "*various ducks*", "*two curlews and three dozen Fieldfare were bought for 3 shillings and 11 pence*" ("*In iij copul. anat., vij anat. domesticis, ij corleus, et iij duoden. de feldefares emp., iij s. xj d. ob*"). The species is also referenced in a later entry for mid-December 1430 (Ticehurst 1923).

During the 19th century, John Hancock (1874) called it an "*abundant winter visitor, mainly in October*". In the 20th century it was described, in a manner that quite neatly summarises the modern situation, by George Temperley (1951) as a "*common winter visitor*". He went on to say, "*Flocks usually arrive about the end of October or occasionally even earlier*", and "*they move further south when snow covers the ground*". "*Spring emigration takes place in March and April*". These summary statements remain largely accurate for the species' current status in County Durham, though Temperley could not have anticipated the later occasional breeding season records.

Through the 1959s and 1960s, there was little information documented on this species in the county, thought an early-arriving autumn record, of a bird on 19 August 1964 at South Shields, was of interest in that it presaged the usual arrival period of even the earliest autumn birds, by the best part of three weeks (Bell 1965). The most notable 1960s occurrence undoubtedly relates to one of the first-documented breeding records of Fieldfare in Britain in 1967, which occurred in Scotland (Parslow 1967) and was mirrored in County Durham (Sharrock 1976). At this time a pair was documented as nesting in Orkney and a pair also probably bred in County Durham, where on 17 July a pair was noted in the company of three fledged juveniles in the Wynyard Estate, in the south east of the county. These birds, based upon their age and dependency on the adults, must have been fledged locally. Since that time the species has remained a very rare breeding bird in Britain, England in particular, with no more than a handful of pairs and summering birds recorded in most years (Brown & Grice 2005).

Flocks in late spring are by no means unprecedented in Durham but these seem to have become more common during the 1990s and 2000s, so a flock of 200 at Middleton-in-Teesdale on 3 May 1976 was notable for the time of the year (Unwin 1977).

Recent breeding status

Since the species first bred in Britain in the 1960s, and the observation of birds at Wynyard in 1967, the only established British breeding populations of Fieldfare have been in northern Scotland and then, in the 1980s, in central England (Batten *et al.* 1990, Brown & Grice 2005). Over the last four decades the species has bred in Staffordshire, Derby, Northumberland and various sites in Scotland. The national population was thought to amount to fewer than 25 pairs in the early 1990s (Gibbons *et al.* 1993) but this has fallen in recent years.

In Scandinavia, the species prefers open, lightly wooded country, often along streams and rivers (Simms 1978) but it is difficult to ascribe this landscape to many of the summer occurrences in County Durham, excepting perhaps the most recent in 2004. Subsequent to the 1967 apparent breeding record there have been a number of instances where birds have been present during the breeding season in Durham and it seems highly likely that at least some of these refer to attempted breeding. This started in the 1970s, when a bird was noted near Castleside, in the Derwent valley, on 16 June 1975, though in this instance there was no further suggestion of breeding other than the one-off sighting (Unwin 1976). This was followed by two birds that were noted on Cleadon Hills on 29 May 1981, which were perhaps more likely to be late migrants, though it is possible that such birds might occasionally linger and attempt to breed without being found by observers. Likewise, a bird in 1983, that was in Marsden Quarry on 2 June. Of a different character though was a bird that was at Salter's Gate on 20 July 1984, which was conceivably a very early migrant, though it seems more likely that this was a summering bird present in an upland fringe area of the county where breeding might easily be considered possible (Baldridge 1985).

It was considered, at the time that the minimally mapped information from the *Atlas* gave an accurate representation of this species' very sparse, i.e. barely represented, breeding distribution in the county during the years of the breeding survey work between 1988 and 1994 (Westerberg & Bowey 2000). Somewhat earlier in the year than most summering occurrences of this species in the area, but undoubtedly displaying breeding behaviour, on 19 April 1990, a male was observed at Charlton's Pond attempting to mate with a Song Thrush (Blick 2009). The outcome of this behaviour was unknown. An incident of perhaps greater interest took place in summer 1991, when a bird was noted singing and apparently holding territory during both June and July, in an area of open, mixed farmland at 180m above sea level at Kibblesworth, in the north west of the county (Bowey *et al.* 1993). The following spring a bird, perhaps the same individual, was present in the same area into July, although no further evidence of breeding was obtained (Westerberg & Bowey 2000). In an upland-fringe area, near Bowes in the south west of the county, in June 2004, a farmer and his wife watched a Fieldfare, over a three-week period, 'carrying worms' to a nearby area of scrub. After securing a Schedule 1 licence and within days of becoming aware of it, the site was visited but no nest or birds were seen. After conversations with the farmer, the observer had no doubt that the landowner's identification was correct and that the situation referred to a breeding attempt by the species (R. Barnes pers. comm.).

With only a very rare breeding season presence of birds in the county, as stated in Westerberg & Bowey 2000, it cannot be stated that Durham has a breeding population. Considering recent successful breeding episodes in Northumberland (Bradshaw *et al.* 1990, Kerr & Johnston 1996) however, it is conceivable that the species might be overlooked in some of the county's upland and moorland-edge habitats. Summering birds have been recorded much more regularly over the last 30 years in Northumberland than in Durham (Kerr 2001), with three confirmed breeding records in that county between 1988 and 1995 (Kerr 2001, Brown & Grice 2005).

Recent non-breeding status

While it is considered largely a winter visitor to Britain, in good years the species can be one of the most recorded species in the county during the winter period. More birds however, are usually seen in Durham during the autumn and spring passage periods than at any other time. In the autumn, this passage migrant usually arrives in the county from the last few days of September, as in 2009, when the first returning autumn migrants came in late September, with 25 at Forest-in-Teesdale on 29th, or during early October, though the main influx largely occurs in the second half of the latter month or in early November. Many hundreds of birds are frequently recorded in favourable conditions, as they arrive from over the sea and many thousands must pass over along the Durham coastline during these two key arrival months.

Major arrivals can take place in Durham anytime from the first week of October, but the principal passage period, which depends very much on prevailing weather conditions, usually takes place from the middle to the latter half of the month. The number of birds seen at the coast is also very much dependent upon the conditions there, with clear weather allowing birds to penetrate straight inland, not tarrying along the coastal strip. In just such a situation about 2,000 flew over Marsden and Whitburn on 27 October 2009. At times, the numbers recorded can be impressive, for example an estimated 20,000 arrived at Hartlepool Headland during 18 October 2001 and 2,500 were grounded there on 15 November 1969 (Blick 2009). In the north of the county, autumn migration brought 1,274 to Whitburn on the 1 November 1984, with 1,000 at Cleadon on the 3rd and 600 to Mere Knolls Cemetery on the 4th.

Of this species, with considerable foresight, Temperley said that it wintered in open country, or on wild fruit and berries in the hedgerows and trees. He further noted that in autumn arriving flocks could be seen on the coast but did not remain there, passing inland to field and hedgerows (Temperley 1951). Any area of scrub or woodland with abundant berry-bearing species, such as hawthorn *Crataegus monogyna*, can attract hundreds of newly-arrived autumn birds, but these rapidly disperse to feed mainly on agricultural land after the local berry stocks are depleted.

Birds then follow a more nomadic lifestyle roving around the county in large flocks, before moving onto open fields and up to the fringes of the county's western fells, where they feed on invertebrates gleaned from ploughed soil or permanent pastures. In many years, a few linger near the coast but often many birds quickly head west to inland feeding locations and ultimately to higher pastures and even moorland, and this has been the pattern for decades, as evidenced by the 800 that were at Derwent Reservoir on 21 November 1972 and the 300 at Barnard Castle on 14th and 280 in upper Teesdale on the 17 October 2009. The oft-seen pattern of distribution on arrival was illustrated in the autumn of 1979, when the largest party at the coast was just 15 on 28 October, whilst around 100 birds flew in off the sea at Hartlepool on 3 November. Meanwhile mid-November flocks further inland included an estimated 500 at Baldersdale on 17th and over 600 birds at Gaunless Flats, West Auckland on 18th, with other gatherings of over 150 birds inland at Bollihope, Bishopton, Hutton Magna and Spennymoor through the month. In similar fashion, few birds were seen along the county's coast during the first ten days of October 1987, making the sight of 2,000 passing over Middleton in Teesdale on the 11th even more contrasting (Armstrong 1988).

The fact that birds tend to arrive in the north east on a broad front was illustrated in 2008, when the main autumn influx occurred in early November and the larger concentrations in the first fortnight included 3,500 grounded at Hartlepool Headland on 6th in the south east of the county, 500 at Hill End, Bollihope, on 7th, 360 at Bishop Middleham, in the central west and east, and 300 passing over Derwent Reservoir in the north. At the county's eastern seaboard, in autumn 1971, there was a large immigration of birds on 23 October that was noted all along the Durham coast and on this date 660 birds passed over North Gare in just two hours, illustrating the scale of movement into the county at such times (Newsome 2009).

Depending upon the weather, this species may be well or poorly represented in the county during the first period of the year. In some years, flocks of 100-200 birds are frequent, with dozens of such roving flocks being recorded, for example the 700 estimated to be near Willington on 1 February 2008, but in relative terms, compared to the number of birds that arrive during autumn influxes, few remain locally throughout the winter. Birds frequently gather in such large numbers, but flocks numbering 80 to 150 birds are more typically seen both in lowland, central and the upland fringes of Durham. Fieldfares can be much less common in the county during late January and February, especially if the weather has been hard, which tends to drive birds further south and west. Hence, a flock of 50 birds foraging in 20cm of snow at 250m above sea level in December 2005 was a sight not often observed in Durham. In broad terms, Durham's mid-winter population is usually relatively small, for hard weather tends to drive birds further afield into areas that experience milder winter weather patterns. One significant exception was December 1995 when there was a huge influx during hard weather. Many large flocks were noted during this time including 5,000 in the Lamesley to Ravensworth area on 24th and 1,000 at Craghead on 26th. There was a continuation into January 1996 with 3,500 at Joe's Pond on 1st and 5,000 estimated to be in the Houghton area on 3rd (Armstrong 1997 & 1998).

In most years, the majority of large flocks have dispersed although small numbers of birds do remain in woodlands, where they forage amongst leaf litter, and occasionally birds visit garden feeding stations. In such situations, small numbers of very aggressive birds occasionally move into towns and suburban gardens, especially in freezing conditions and where apples, a favoured food, are provided they will readily defend these against all comers. Overall though, garden visitors are quite scarce in Durham, except in extreme weather and when they do occur, these records normally relate to just one or two birds at a time. Consequently, the 22 that visited a garden in Whickham on 28 February 1979 was exceptional. A few years later in the same area, during hard weather in January 1984, a small flock frequented another back garden at Whickham to feed on provided food and during one week 15 birds were caught and ringed; nine of these in just one day (L.J. Milton pers. comm.).

It has been suggested that in milder winters many of the autumn-arriving birds stay in the north east (Kerr 2001), but in harder weather they mainly pass through on a south westerly vector. Unusually, mild weather in the early part of 1980 allowed Fieldfare flocks to frequent the dales and moors in the west of the county and 1,300 were noted from elevated positions around Selset Reservoir in Lunedale on January 6th (Baldridge 1981).

The principal spring emigration from, and passage through, Durham occurs through March and early April. The large flocks which are so often noted in the second half of March are probably genuine birds of passage, in as much as they are returning to and then passing through the county, on their northwards migration from locations further south. It is often at this time of the year that the species most aptly lives up to its name, as it feeds in open situations on ground-dwelling invertebrates as the soil warms up. The largest numbers in the county often occur at this time and flocks of hundreds, and even thousands, are routinely noted. On the northwards passage such large flocks are indeed typical and there is a tendency for these to be recorded from westerly or inland locations. Early to mid-April sees the traditional build-up of flocks in the western dales, prior to their quitting the county for more northerly climes. At the coast, the return in spring is less noticeable than it is in the autumn, although inland flocks of several hundred are usually very evident in late March and April. For instance, in 2010, flocks of over 200 birds were reported from Barnard Castle, Burdon Moor and Foxton, as well as from two or three sites in the Hamsterley area, whilst the largest number noted at this time was 1,500 at Lunton Hill on 25 March. A major exodus took place during April 2001, with one flock estimated to contain 3,000 birds being on the edge of the Browney valley, between Tow Law and Satley; similarly a flock of 2,000 birds was noted near Blanchland on 26 April 1981, while 1,600 were noted at Bowlees on 10 April 1994.

As spring approaches, birds start heading back across the North Sea and there is a rapid decline in sightings from mid-April. Outgoing sightings of birds tend to dwindle through the later part of April and the last 'birds of the winter' are usually noted in late April, with an occasional straggler into the first few days of May. Coastal representation of the species at this time tends to be of small numbers of birds or late stragglers. In 2009 the last birds reported were, not atypically, a flock of 25 at Elemore Woods in the east of the county on 21 April whilst singles were well inland near Hury Reservoir on 22 April and in the west, at Swinhopehead on 6 May. Occasionally, this smooth exodus is interrupted and in mid-April 1975, bad weather grounded departing flocks before they managed to fly out over the North Sea. At this time around 1,000 birds were in a field at Cleadon Hills with up to 400 elsewhere between the Tyne and the Tees, whilst out to sea a flock of 60 was seen heading back to land due to the bad weather. Very exceptionally indeed, larger numbers are noted in May. For example, in early May 1974, adverse weather conditions grounded much larger numbers of birds than is normal during this month, leading to flocks of 200 at Marsden on 3rd, 635 at Graythorp on 4th, 200 at East Boldon on 7th, 150 at Washington on 8th and up to 500 at Hart Village over this period. One at Teesmouth 22 May 1972 was a very late spring bird.

As the majority of birds have left the county by mid-April, with just small numbers usually reported in early May, later records are always of interest, such as the bird at Norton on 26 May 1970. It is perhaps difficult to categorise some of these 'out of season' reports. For example, considering the location, one at Hartlepool Headland on 29 July 2005 was most likely to have been an extremely early autumn migrant rather than a potential breeding bird; but as to the bird at Hartlepool Power Station on 11 June 2008, perhaps this was an exceptionally late spring migrant. Early arrivals in the autumn include birds at Hartlepool on 20 August 1970 and Witton-le-Wear a long way inland on 25 August in the same year.

Unlike the Blackbird *Turdus merula*, which frequently illustrates plumage abnormalities, aberrant plumaged Fieldfares are by no means common. So an almost completely white bird, except for some black along the front edge of each wing, which frequented the Shield Row area of Stanley from 21 December 1969 to 4 January 1970, which was not seen to "*keep the company of other fieldfares*" (*The Vasculum* Vol. LV 1970) was of interest as was the partial albino seen at Bishopton on 5 May 1980 (Baldridge 1981).

Distribution & movements

Fieldfares breed across central and northern Europe, east to central Siberia. The northern populations are highly migratory and these usually winter to the south west of their breeding range, in southern and western Europe (Parkin & Knox 2010).

There is relatively little ringing data on Fieldfares ringed in or found in Durham. The most significant recovery to date was one that was ringed as a first-year bird at Kivach Reserve, Karelia, in north east Russia in June 1980 and found dead at Norton in January 1984, indicating the origin of at least some of the region's winter visitors. Meanwhile, the bird found dead on Bollihope Common on 21 December 1986 had been ringed in Friesland, Netherlands on 24 October 1981, indicating a more southerly, though still easterly, port of departure for some of Durham's birds.

Song Thrush
Turdus philomelos

A very common, but declining, resident, passage migrant and winter visitor.

Historical review

The presence of this species in the north east of England, some 2,000 years ago, is indicated by the remains of thrushes, including Song Thrush, Redwing *Turdus iliacus* and Mistle Thrush *Turdus viscivorus*, which have been excavated from the Roman sites, in the south west of Northumberland, at Housesteads, and Birdoswald in south east Cumbria (Yalden & Albarella 2009). No doubt birds were also present in Durham at this time. The species' first documented occurrence in Yorkshire, at Wycliffe-on-Tees (now Country Durham), came about courtesy of Marmaduke Tunstall, according to Nelson (1907).

In the 19th century Tristram (1905) wrote of it, "*Abundant, except in winter when most migrate. A few remain even in the severest seasons*". At this time, flocks were noted as arriving on the coast in October and November. Almost half a century later, Temperley (1951), at the mid-point of the 20th century, described this species as, "*a common and well distributed resident, and a winter visitor in small numbers*". He observed, like Tristram, that the species was an altitudinal migrant, being absent from the higher ground of the west of the county in the winter, though some remained in the river valleys and were also present in lowlands areas.

In terms of its habitat choice and distribution, Temperley (1951) described the bird as "*A common and well-distributed resident, found in woods, thickets, plantations and hedgerows, but most abundant in parks and gardens and near human habitation*". This statement still holds true to the present day, birds being present in most of those habitats in the county, principally on land below 300m. Locally, during the first half of the 20th century, the Song Thrush far outnumbered the Blackbird *Turdus merula* (Temperley 1951). This situation changed however, in the 1930s and 1940s, as Blackbird numbers increased, due in part, it is believed, to climatic change (Marchant *et al.* 1990). The Song Thrush is particularly susceptible to severe winter weather and a slow decline in numbers in the early part of the 20th century was exacerbated by the severe winters of 1939 and 1946/1947, from which it had difficulty recovering. This was also shown by the large declines that were noted in the south east of the county during the poor winters of 1940 and 1947 and again in the winter of 1962/1963 (Joynt *et al.* 2008).

An exceptionally large influx of 2,000 migrants was noted, along with other thrushes, at Hartlepool Headland on 6 October 1966 when large numbers were also noted in North Yorkshire in the following week (Mather 1986).

In Northumberland, it has been felt that the establishment of commercial forestry may have helped this species extend its range into some of the upland areas from which it was previously excluded by a lack of nest sites (Kerr 2001), and it is conceivable that a similar process occurred in Durham over the first 50-70 years of the 20th century. For a short time, through much of the 1950s and 1960s, the population appeared relatively stable, though by the 1970s the numbers of this species were once again in decline. This trend continued through the 1980s (Marchant *et al.* 1990), for reasons not yet fully understood, and in less dramatic form into the first decade of the 21st century.

Recent breeding status

The Song Thrush is typically a bird of scrub and woodland edge, although it is now widely found in suburban and urban habitats, such as gardens and parks. Song Thrushes are widespread in the county and will readily nest in urban gardens as long as disturbance is not too great. Indeed, anywhere with sufficient damp ground for feeding and scrub for nesting is capable of holding breeding birds, although they are often distributed rather thinly, with reports usually being of one to two singing males.

Over the latter period of the 20th century, there was a considerable reduction in the numbers of Song Thrush across the UK, amounting to a decline of around 50% in a 25-year period up to the early 1990s (Mead 2000). The causal factors behind this, over a three to four-decade period, are not fully understood though the severe winters of the early 1980s were implicated as were changes in agricultural practices including hedgerow destruction and the increased use of molluscicides on both farmland and in suburban gardens ((Marchant *et al.* 1990). In the north east of England, Song Thrushes remained fairly common birds even through the worst of this decline, though there were some local reductions. In 2009, Song Thrush was recorded from 76% of the 3,243 Breeding Bird Survey squares surveyed across the UK (Risely *et al.* 2010) illustrating its still widespread nature. Over the period 1995-2008 the species increased by 27% across the whole of the UK, from a low base and the major declines of the previous two decades, but it showed a slightly lower increase of 25% across England alone. In the north east of England the increase was much lower, at just 4%, but this is partly based on the fact that the original declines in the north east were never as dramatic as elsewhere in the country (Risely *et al.* 2010). In much of Durham, the species is still widespread and common, particularly in suburban areas, where garden mosaics, and fringing open habitats, such as large parks, provide a range of less intensively managed habitats that favour the species.

This species is well known as the 'choir leader' of the dawn chorus and the Song Thrush has perhaps the longest breeding season of any thrush (Marchant *et al.* 1990). Its beautiful, repetitive song is usually delivered from a high vantage point, such as a tree or tall building and birds are, therefore, easy to locate. Given mild weather, especially in the frost-free, east of the county, singing may commence in January. Indeed, birds can be in full song in many lowland areas by New Year's Day during mild weather, as in 1999, when the first report of this behaviour was of one singing at dusk on 1 January (Armstrong 2003). Singing may continue, almost unabated, throughout the late winter period into spring and much of the summer.

Extreme examples of singing birds include a bird that was noted at 480m above sea level in the west of the county in June 1984, a bird in full song at Sherburn on 19 December, despite freezing overnight temperatures, and one singing at Gateshead on Christmas Day 1984 (Baldridge 1985). The first fledged juveniles are usually noted around the latter half of May, within the last ten days, indicating that, for a clutch of four eggs, the first egg dates would be around the 20th to 25 April, depending upon fledgling date (Westerberg & Bowey 2000). That said, by the end of March, local lowland birds can be well advanced with breeding. For example, a nest with eggs was discovered in Blaydon Dene on 24 March 1999 (Armstrong 2003). In the west of the county, where the species is largely a summer visitor (Lack 1986), singing does not usually occur until March or early April, for example in upper Teesdale (Westerberg & Bowey 2000), though there is some evidence of this occurring earlier after the mild winters of the 1990s and early 2000s.

Successful breeding usually occurs widely across the county, but is not usually well reported by observers. In 2007, survey work in the Hetton area, revealed that 15 pairs raised a total of 50 young, an average of over 3.3 fledged young per successful nest (Newsome 2008). There is some evidence, from Common Bird Census studies in the county, to suggest that there was a decline in local numbers during the late 1980s and early 1990s, complying with the national trend (Westerberg & Bowey, 2000). At the Kibblesworth Common Bird Census site, the fact that there was just one pair of birds noted in 2004, when ten pairs was the norm here during the 1970s, was illustrative of the population trend for this species and the scale of losses suffered in some areas of the county since the 1960s. There was some subsequent recovery at this long-term study site however, five territories there in 2007 representing the highest total since 2000. It was thought that the species was benefiting from the maturation of newly-planted farm woodlands there (Newsome 2008).

At Elemore Woods, the number of singing males remained steady at 16 to 18 between 2006 and 2010, with birds favouring the dampest areas within which to breed. Around 75% of singing males were congregated around two hectares of woodland dominated by yew *Taxus baccata* and Rhododendron *Rhododendron ponticum*. These results indicate that locally, birds were breeding reasonably successfully during the second-half of the first decade of the 21st century. This would appear to comply with recent national data, for example of 20 red-listed species of conservation concern monitored by the Breeding Bird Survey since 1995, the numbers of just two, including the Song Thrush, increased significantly (Risely *et al.* 2010). There was some suggestion that locally this species may have benefited from the wet summer of 2008 (Newsome 2009). It has been shown that numbers in upland-fringe areas in the north east increase after a series of mild winters (Kerr 2001), and this may have been the case for this species in Durham during the 1990s and early part of the first decade of the 21st century.

Whilst numbers of this species have always been greatest in the lowlands of the county, it does breed in the far west, as was shown in 1982, when successful breeding was noted at Grains o' th' Beck at the very limit of available suitable habitat for the species in the county. During the survey period for the *Atlas*, breeding in upland areas was proven at 340m in Baldersdale and 385m at Sleightholme Beck in 1988 and at Bollihope Common in March 1990. Ten singing birds at Langdon Beck on 24 March 2008 demonstrated the widespread nature of the species in the county, but Song Thrush is not present at these elevations throughout the year. In the west, the species' distribution follows the well-wooded river valleys up into Weardale and Teesdale, though they avoid open moorland habitats. Unusually, a pair was discovered nesting on one of Sunderland University's buildings in mid-January 1992, the young having fledged by 27th of that month (Westerberg & Bowey 2000). In 2006, what might be best described as winter breeding was suspected in the county, with one bird at a newly built nest at Brasside on November 22nd (Newsome 2007b).

Work for *The Breeding Birds of Cleveland* indicated that there were probably in the region of 700-800 pairs in the northern part of what was Cleveland. It is scarce as breeding bird around the North Tees Marshes, largely as a result of a relative dearth of nesting habitat there and it is less common in the urban heartlands of Teesside (Joynt *et al.* 2008). During the *Atlas* work, Song Thrushes were found to be present in 353 tetrads (46.2%) in the county, compared with 483 (63.3%) for the Blackbird (Westerberg & Bowey 2000). The *Atlas*, assuming the Durham population to be typical of the national situation (Gibbons *et al.* 1993), calculated that at a density of 1.2 pairs per occupied kilometre square, the county held some 11,000 pairs (Westerberg & Bowey 2000) and this number, balancing declines and recent recoveries, probably remains a reasonable working estimate today. More recent survey work in 2006 revealed that Song Thrushes were present in 20 of 31 surveyed one-kilometre squares, with an estimated 51 pairs in these (Newsome 2007). Assuming this was representative of the county as a whole, then a mean occupation rate of 2.55 pairs per occupied square would give a county population of around 5,304 pairs; perhaps somewhat on the low side, given the above *Atlas* estimate. Blick (2009) thought that as a breeding bird in the south east of the county, this species was out-numbered by the Blackbird *Turdus merula* by a factor of about three to one, but in its turn, that it was out-numbered by the Mistle Thrush *Turdus viscivorus* by about six to one (Blick 2009). This is not the case for the rest of the county, where the Song Thrush remains very much more common than the Mistle Thrush, though not the Blackbird, by which it appears to be outnumbered by about five to one.

Recent non-breeding status

Resident numbers are swollen in the autumn by an influx of continental birds which often arrive in September and early October. The autumn influx of continental or northerly British birds is usually very much more evident, and tends to presage influxes of Redwing *Turdus iliacus* and Fieldfare *Turdus pilaris*, though it is often a 'fellow-traveller' alongside both of those species. Numbers involved in autumn thrush movements are in relative terms low, and coastal counts of note tend to be in the tens rather than the hundreds, unlike other migrant thrush species. For example, in 1979, coastal movement occurred from 30 September until late October and generally no more than 20 birds were noted at any locality, though around 200 were feeding on yew berries at Castle Eden Dene on 2 October and *c.*50 flew in off the sea in the Marsden area in 30 minutes on 13th. Cumulatively, however, birds reaching the Durham coastline in autumn and over the period from mid-September to early November must number thousands, though rarely more than one to two hundred in a day. Sixty at Hartlepool on 2 October 1971 were part of a large influx of hundreds of birds on the Yorkshire side of the Tees over the period from 2nd to 7th. Similar large flocks were noted along the Northumberland coast at this time. In the scrub and trees of Mere Knolls Cemetery there were up to 250 during late October 1987. Continental arrivals from September 2001, with an influx on the 18th, led to a large concentration of 834 birds at Sedgeletch, one of the largest ever inland gatherings of this species in the county (Newsome 2007).

Autumn migration throughout the 1990s was similar to previous decades; mostly small numbers recorded with occasional three-figure counts the exceptions. There was a notable group of 150 present at Marsden Quarry on 30 September 1990, while the late-1990s saw several large flocks during October. A total of 152 were on the South Shields Leas on 23 October 1996, 150 were at Hartlepool on 1 October 1998, while 100 were in Hawthorn Dene on 16 October 1999. A total of 163 birds were recorded along the Durham coastal strip during the 1992 Coordinated Migrant Count on 4 October (Bowey 1999c).

Exceptionally large numbers included more than 1,000 birds recorded at Hartlepool on 18 October 2001 and again in October 2002, when 500 birds were also at Marsden. There were also good numbers on migration along the coast during October 2005, with 250 birds in Marsden Quarry on the 17th. Most of these autumn migrants are considered to be of Scandinavian origin, but six birds seen at the mouth of the Tyne in late November 1983 were considered to be of the very dark and heavily spotted race *hebridensis* (Baldridge 1984).

The majority of lowland-nesting Song Thrushes in Durham probably remain in or around their territories through much of the year, though this does not apply to birds in the west of the county. By the late autumn of most years, birds have vacated the upper dales in the west and in many years none are to be found in such areas during the winter months. For instance, in February 1981, one at Balder Grange was the first seen there since the previous autumn and two birds at Westgate on the 26 February 1984 were the only ones seen in that part of

Weardale during three winters' survey work for the national winter atlas (Lack 1986). Even lowland sites sometimes register an absence of birds in the winter months with a group of 12 together at Elemore Woods in November 2008 being a rare occurrence at that site (Newsome 2009).

By mid-March, the majority of territorial birds have usually arrived back in their high dales breeding locations, as was evidenced by the two singing birds at the Stang on 13 March 1999. An unusual gathering of 27 birds at Derwent Reservoir on 19 March 1987 were probably altitudinal migrants returning to the upper valley.

In some years, Song Thrushes can be markedly scarce in mid-winter, indicating that many of our local birds move further south at the behest of cold weather. During the early part of the year this might often be considered the scarcest of the county's resident thrush species and, as it is generally less social than the other thrushes, there are rarely any reports of substantial gatherings.

Spring passage through the county is usually rather slight and can occur, if at all, relatively early in the year, from late February onwards. One on a fishing boat off Sunderland on 3 March 1975 was presumably an example of such movement (Unwin 1976). The origin for the county's autumn arriving birds has always been believed to have been largely Scandinavia, as indicated by the national data (Wernham *et al.* 2002), though in fact there is little independent evidence to demonstrate this in relation to ringing returns.

Distribution & movements

Song Thrushes are common breeding birds across much of Europe, the extreme south excepted. Their breeding range also extends eastward into central Siberia. Most populations are highly migratory, though the species is present year-round in the milder climatic conditions of Western Europe. Birds from the northern part of the range winter around the Mediterranean and Western Europe. Many of the British breeding populations are resident throughout the year, though birds in more elevated locations exhibit altitudinal migration.

Very few of the young Song Thrushes that have been ringed as pulli in the county have subsequently been proven to have travelled any great distance. There are records of Song Thrushes ringed in Durham moving south west to winter in Ireland and station further south. For example, a nestling ringed at South Shields 30 May 1953, was caught and released at Wexford, 275 miles to the south west, on 6 February 1954. A first–year bird ringed at High Urpeth near Chester-le-Street, on 21 November 1976 was found dead in Llanon, Dyfed, Wales, on 11 June 1978. An adult female ringed at Shibdon Pond on 27 April 1986 was found dead, killed by a cat, almost three years later 352km to the west south west at Ratfarnham, Dublin on 26 January 1989. Blick (2009) noted that there were several recoveries of birds ringed in autumn in the Cleveland area, which had moved south or west. These included individuals going to the Calf of Man, the Algarve in Portugal in January, County Down in Ireland, Bath in Somerset, Seville in Spain in February and Amares, Portugal, in March. One of a brood of four ringed at South Shields on 14 May 1951 was recovered at Cire d'Aunis near La Rochelle, France on December 24th of the same year. In a not-dissimilar vein was the bird ringed in autumn at Ravensworth on 18 October 1981 which was found dead on 21 March at Santander, in northern Spain, three years later (L. J. Milton pers. comm.).

Birds from further north have been recovered in Durham. For instance, the oldest known recovery of this species in the county is of a bird that was ringed as a nestling at Gosforth, Northumberland, on July 14th, 1910, and which was recovered near Easington, at Castle Eden on 5 November 1910. From further afield, a bird ringed on the Isle of May on 18 October 1951 was recovered at Sunderland on 3 February 1952, 100 miles to the south south east. The following year an adult ringed on the Isle of May, Scotland, on 19 March 1952 was recovered at Penshaw on 4 December 1952, again 100 miles to the south south east.

There are however, other recoveries that do not seem to follow either of these patterns, for instance a bird that was ringed at Haverton Hole in August 1977 that was later recorded near Perth, Tayside, Scotland, on 7 March 1979 (Blick 2009). A bird that was ringed at Graythorp in October 1968, and might have been expected to be a Scandinavian, considering the time of the year, was later killed by a model aircraft, a few miles to the south at Nunthorpe in June 1969, so this was presumably a local movement by a locally-bred bird (Blick 2009).

Perhaps the most amazing record of a ringed bird in the county was of the skeleton of a bird ringed as a juvenile at Toothill West Yorkshire on 20 May 1931, found beneath floorboards of a house in South Shields fifty-nine years later exactly, on 20 May 1990 (Armstrong 1992).

Redwing
Turdus iliacus

A very common winter visitor and passage migrant.

Historical review

The remains of a variety of thrushes, including Redwing, have been excavated from a number of the Roman sites in the south west of Northumberland, at Housesteads, and Birdoswald in south east Cumbria (Yalden & Albarella 2009), indicating that this species was widespread in the North of England, and there is little doubt that it would have been present in Durham at the time.

There is little documentary representation of this species in the county's historical ornithological records, but Redwing is known to have been a winter visitor to the area, as evidenced by the presence of specimens that were taken at various Durham sites during the 19th century in the region's museums.

Robson thought of it as a "*common winter visitor*" to the Derwent valley woodlands (Robson 1896) and it still is today. In 1951, Temperley said it was a "*common visitor in winter*", and he noted that flocks arrived in October, usually before the Fieldfare *Turdus pilaris*, and that in some years, further flocks arrived in December and January.

There are few out of season records amongst the older documentation for this species however, at 20.45hrs on 29 May 1949, a Redwing was found singing in a tree near the River Wear at Durham City (Temperley 1950).

Recent breeding status

Very small numbers breed in the Scottish highlands, but this species has never been proven to have bred in County Durham. Birds have been noted on several occasions in June though, with an intriguing record of a bird singing and in territory, in the Sunderland/Whitburn area in 1994. Singles were seen on several dates into June when a bird was reportedly collecting food (Armstrong 1996). Also, in 2008, a male in full song was found on the edge of Cleadon Hill on 17th and 18 July, but soon moved on and there was no indication of a female present.,

Recent status

On arrival on the east coast in the autumn, Redwing usually gravitate to areas of scrub or berry-rich woodland, as stated by Temperley over 60 years ago, who said, "*On its arrival it resorts to woodlands and hedgerows to feed on the berries of the rowan, guelder-rose and hawthorn, but as supplies run short it visits the open fields*" (Temperley 1951). Initially, its feeding behaviour seems to be related more to a need for immediate refuelling requirements before flocks move slightly further inland, where berry-bearing areas of scrub such as those around Rainton Meadows are favoured by the species. The observation of birds feeding in open fields is relevant at all times, but much more pertinently applied to birds that are passing north and east, through the county, on spring migration, usually in February and March, by which time almost all berries have long since been consumed.

Birds arrive in the county anytime from late September, usually the latter half, through to early October depending on the prevailing weather conditions. Probably the majority of the Redwings that arrive in Durham eventually pass on through the county, heading for wintering areas far to the south and west. Nonetheless, flocks of up to 100 to 200 birds are quite routinely reported in well-wooded urban areas, such as parks, though the numbers usually depend on the severity of the winter weather. Like Fieldfare, they frequent areas with many berry trees, or a range of tree species, but they tend to be less nomadic, remaining more in one area. Also, unlike the Fieldfare, they are more often found in small dissipated flocks, or singly, rather than the large cohesive flocks so often associated with the larger species. Once all of the local berry stocks have been exhausted then birds will move into local woodlands and, occasionally, gardens, where they forage amongst leaf litter. They rarely gather in such large numbers as Fieldfare, the biggest gatherings invariably being noted when they first arrive and when birds are concentrated on berry crops. In such circumstances up to 300 birds might sometimes be noted together.

The winter months of November and December tend to bring fewer more scattered reports of the species as birds disperse more widely, after the main bulk of arrival in October. A typical year was 2004, when this period saw many reports of flocks numbering 20-80 birds from more widely dispersed sites with only one record of a flock exceeding 100 birds. Exceptional however, was a count of 470 at Rainton Meadows in mid-December (Siggens 2006). The largest concentrations over these months are often reported from areas where observers are active in local studies or in watching their immediate local areas, though in reality, birds are probably much more evenly distributed than such patchy recording implies.

In years where there have been large influxes in the previous autumn, the year often begins with large feeding flocks being reported from sites across the county, for example, in January 2009, the 250 noted on fields at Houghton Gate on 2nd and 110 at Page Bank on 4th. In 2008, there were reports of flocks of over 100 birds from 22 locations during the first three months of the year (Newsome 2009). Numbers, however, vary considerably from year to year and it is equally not unusual, as the New Year dawns, to find that this species is relatively rather scarce in the county, the maximum numbers reported perhaps never rising to 100 individuals, though no doubt many birds are scattered around the county. Sometimes birds can be difficult to find during this period, as many birds that have arrived in the previous autumn have exhausted the easily available local food supplies and simply moved south and west.

In 1985, a count of 1,000 birds at WWT Washington on 8 November was probably composed of newly arrived immigrants. Very large mid-winter flocks are rare with the most notable occurrences of large winter gatherings recorded in the following years:

- 1987: an exceptionally large flock of 1,200 birds at Joe's Pond on 7 January
- 1998: a large mid-winter concentration of 550 at Fence Houses on 6 January

Coastal reports during mid-winter indicate that generally birds are well spread in small flocks focusing on garden shrubs and trees with remaining berries but the onset of hard weather can see a distinct influx. During a period of heavy snow in December 2009, dozens of birds were driven to feeding along the edges of a busy road in South Shields, attempting to feed on small patches of uncovered grass (T .I. Mills pers. obs.).

The return of birds in spring is much less conspicuous in the county than is their autumn arrival. Night-time emigration from the county can occasionally commence as early as late February, especially in mild years, though it is March that usually brings the marked decline in the wintering population. Occasionally small flocks of birds can appear along the coast in April, and rarely a few are known to stay into May. By and large, the last birds are generally seen during the first week of April, when small flocks of up to 20 birds might be reported from scattered sites across the county. Occasionally birds are seen later than this, such as the three at Shibdon Pond on 3 May 2007 (Newsome 2008). The principal period of spring passage tends to occur in March, and through April birds become increasingly scarce, there usually being just single-figure records and decreasing numbers of sightings as the month wears on. Redwings have a tendency to slip away out of the county and wintering birds departing for the Continent in late March are often heard at night overflying coastal locations.

After the spring exodus, in most years, none are reported in the county from early May, although birds have lingered into June on rare occasions:

- 1978: most unusually a bird was at South Shields all through May until 5 June, two days after another straggler was at Washington
- 1986: a single on the late date of 15 June was in Mere Knolls Cemetery, Sunderland
- 2008: two birds were well inland at Adder Wood near Hamsterley on 11 May and these were, somewhat surprisingly not the last of that spring, as a male was found in full song at Well House Farm, Whitburn, over 4th to 6 June; the same site that hosted a singing male in June 1985

The first autumn birds usually arrive in late of September, typically in scattered small flocks and most often in the last few days of the month, but occasionally earlier than this. The bulk of the immigrants appear during October and sometimes into early November. There are few reports of birds returning before the middle of September with the earliest reported arrival date of a single at Shotton Ponds on 1 September 1999 (Armstrong 2003).

In 2007, the first autumn immigrant was noted at Old Quarrington on 22 September, the first rush of incoming birds occurring on 27th when more than 1,000 were counted moving west well inland at Hill End. In 2008, the first autumn birds reported were 30 at both Cleadon Park and Hetton Lyons Country Park on 23 September, with two at Seaton Pond the same day. Numbers continued to trickle through during the beginning of October, with 250 at Crook, well inland, on 10th and 150 at Elemore Woods the same day. In 2009, the first returning bird was noted at Rainton Meadows on the comparatively early date of 14 September, a week in advance of the next ones at Hurworth Burn Reservoir and Whitburn on 21st of the month.

On the east coast of Britain, one of the joys of the autumn is the sound of migrant Redwings calling from dark skies overhead. However, the scale of autumn passage, especially in October, varies from year to year, and a summary of movements for this key month serves to illustrate the punctuated nature of this species' arrival patterns in the county. For example, in 2007 passage was generally considered 'light', despite 1,100 heading west over Hartlepool in two hours on 9 October 2007, and a westerly movement of over 1,000 birds at Cowpen Bewley on 20 October 2007. By contrast in October 2008, birds were numerous. In misty conditions, over 250 arrived at Whitburn on 15th, whilst at least 500 descended into hawthorns *Crataegus monogyna* at Elemore Golf Course on the same afternoon. A large-scale arrival took place on 17 October and this produced 450 birds at Elemore Golf Course, with 700 at Rainton Meadows and over 1,000 over Houghton-le-Spring, while 250 were at Cowpen Bewley WP on 18th. In 2009, a long way inland, 920 were in upper Teesdale on 17 October but this was surpassed on 27th when over 5,000 birds moved through the Marsden/Whitburn area during one morning.

One of the most spectacular ornithological events in the county occurred in the period 7th to 8 October 1977 when an influx described as phenomenal took place, with vast numbers pouring in all along the coast. A deep depression centred on southern England created strong easterly winds with rain and mist. Flocks were verging on the countless, with a conservative estimate of *c.*25,000 along the Sunderland-South Shields stretch of coast and 11,000 between Hartlepool and Teesmouth on 7th; the largest autumn influx witnessed for many years (Unwin 1978). During this, in excess of 1,000 birds were grounded in Marsden Quarry, with every bush laden with this species (T. I. Mills pers. obs.).

Distribution & movements

Redwing of the nominate race breed throughout much of northern Eurasia, with the darker, slightly larger Icelandic race *coburni* being present in Iceland and the Faeroes. Redwing winter in southern and western Europe, parts of northwest Africa and the Middle East (Parkin & Knox 2010).

Redwing ringing data relevant to Durham demonstrates the fact that, after arriving in Britain during the autumn, birds may continue to move in a southerly direction to southern or south west Britain or the Continent. Two locally ringed birds made it to the Gironde region of France, over 1,000km to the south, before succumbing; a bird ringed at Shibdon Pond on 6 December 1984 was shot on 15 February 1987, and a first-winter ringed at Marsden Quarry on 18 October 1990 was found freshly dead on 11 January 1991. Other French recoveries include a first-year ringed on 30 January 1983 at Beamish Hall near Stanley that was found dead on 20 December 1985 at Durban Corbieres in France, a movement of 1,360km to the south south east, and another first-year bird ringed on 18 October 1990 at Marsden Quarry, that was shot on 11 January 1991 in France at Etauliers, over 1,000km to the south. A number of Durham-ringed birds have been recovered in Portugal, including a bird ringed at Graythorp near Hartlepool on 24 October 1965 that was recovered in Pòvoa de Lanhoso, Minho on 3 March 1966, an adult ringed on 15 January 1983 at Darlington, shot on 12 December 1983 at Viseu, a movement of 1,615km and a first-year ringed on 4 October 1987 at Crimdon Park, near Blackhall that was shot on 27 December 1987 at Varzea - a movement of 1,806km to the south south west.

The origin of Durham's wintering Redwing is indicated by the first-year bird ringed at Epse in central Netherlands, October 1962 and which was found dead at West Hartlepool in December 1962; one ringed at West Hartlepool, on 17 December 1960 and recovered at Iggesund, Gavleborg, Sweden on 26 June 1961 and a first-year bird ringed on 5 November 1984 at Tveitsund, Nissedal in Norway, which was killed by a bird of prey on 6 February 1985 at Jarrow Slake, 753km to the south west.

Birds ringed elsewhere in Britain that have been found locally include one ringed at Whitley Bay on 2 March 1965 that was found at Jarrow on 18 February 1966 and one ringed at Kippax, near Leeds, Yorkshire, in November 1969, which was found dead near Wolviston in January 1973 (Blick 2009).

Mistle Thrush
Turdus viscivorus

A fairly common and widespread resident and small-scale passage migrant.

Historical review

Hogg (1827) documented that Thomas Bewick had reported that this species was rare in the north around 1800, and it is known to have been extending its range through the 1800s (Harrison 1988) - numbers had very much increased by Hogg's time of writing. During the 19th and 20th centuries this species spread northwards through Britain into mainland Scotland, as far as the Inner Hebrides, and through Ireland (Simms 1978).

Temperley (1951) said that it was then, "*a common resident*", and "*In autumn and winter, flocks visit more open country to feed on the fields*". This is not a habit that is particularly prevalent today, most gatherings occurring in late summer to feed on berries, in particular on early ripening crops such as rowan *Sorbus acuparia*. As a mainly ground-feeding bird, it is badly affected by cold winters but recovery, even after a 75% loss in 1962/1963, was fairly quick (Mead 2000). This species also suffered a significant decline during the winter of 1947, being almost wiped out in some districts of the south east of the county, although this was followed by a rapid recovery in this area (Stead 1964).

Today, the British distribution is similar to that of 100 years ago and little different from the modern maps (Mead 2000). Both here and in Europe, the species has expanded its range from woodland and continental upland forest into park-like lowland habitats and then into towns (Simms 1978, Marchant *et al*. 1990).

Recent breeding status

This is our largest and most dominant thrush, but it is a much scarcer species than Blackbird *Turdus merula* and not as common as Song Thrush *Turdus philomelos* although it is relatively common but thinly spread across the county. Despite its well-known song, the Mistle Thrush can easily be overlooked (Gibbons *et al*. 1993); nonetheless, it is a widely distributed species in Durham, with especially good numbers present in the western dales. It is a well reported species, as illustrated by the more than 1,000 records submitted to the Durham Bird Club in 2008 (Newsome 2009).

Tall trees are important to the species, as both song-posts and nest-sites, but open ground is needed for foraging; hence, the Mistle Thrush avoids dense forest (Cramp *et al*. 1988). They tend to nest in larger trees than the other thrushes, which favour scrub and shrub species, and the species is not so reliant on such dense shrub cover. Cemeteries, municipal parks and large gardens around the county are much favoured by Mistle Thrushes, the short grassland providing good foraging opportunities whilst nearby ornamental trees offer nest sites. As a breeding bird it is widespread across much of Durham with something of a westerly and upland bias compared to the Song Thrush, as illustrated by Westerberg & Bowey (2000), which appears consistent with a preference for the species' original habitat. The first singing birds of the year are usually heard in January, though birds routinely begin to sing at the tail-end of the previous calendar year. It will often sing at night and singing males are usually reported from across the county from March onwards. This species breeds, on average, earlier than its congeners and can nest surprisingly early in the year – not unusual were two records in 1985 of nest building in February at Chester-le-Street and Hetton, but there are isolated records of exceptionally early nesting - a pair feeding young at Blaydon on 31 January 1975 was an indication of that year's mild winter, whilst an egg found predated by Magpie in Durham City on 28 February 2010 was still on the early side for this species. More typical of early breeding reports were the three birds on nests at Hetton Bogs by 20 March 1998.

The first fledged young of the breeding season may be noted as early as mid-April but more typically the first half of May or into late May, depending upon altitude and the location of the nest site. This species has a very protracted breeding season, for example fledged young were noted at two upper Teesdale sites in late July 2009. While unusual nesting locations are selected less often than by the Blackbird, some examples of such nest sites include a pair that nested on a window ledge at Readhead Park, South Shields reared two broods in 1979; a pair that successfully reared young from a nest situated on a pipe bridge in the centre of industrial Billingham in 1972 and again in 1983 (Bell 1984); a bird incubating on a nest in a set of traffic lights in Sunderland in March 2004 (Siggens 2006) and, in 2010, a pair that nested in the rollercoaster at the fairground in the Marine Park at South Shields (T. I. Mills pers. comm.).

The eastern lowlands hold a considerable population of birds and successful breeding is confirmed at numerous sites in such areas in most years. As stated in the *Atlas* (Westerberg & Bowey 2000) in Durham, while it may be more numerous, rather than merely more conspicuous, in the relatively open ground of its favoured western haunts, the extent of this difference is difficult to quantify. The species' ability to penetrate further up the dales is demonstrated by the pair nesting below White Force, Cronkley Fell at 424m above sea level in 1975; the bird singing in January 1976 at 480m elevation; the three noted at Killhope in the extreme west of Weardale in June 1978 and the successful breeding of birds in the far south west of the county, at Grains o' th' Beck in 1983 (Baldridge 1984). The *Atlas* stated, "*A subjective estimate at a County population of 1,000-2,000 pairs would be in line with the data in the 'New Atlas' (Gibbons et al. 1993). The national estimate of 230,000 territories gives, proportionate to land area, a figure of 2,566 for Durham. This may require some downward adjustment as the species is regarded as "most frequent in South-east England"*" (Gibbons *et al.* 1993). Today, it is widely distributed around the northern part of 'old Cleveland', as shown by Joynt *et al.* (2008), but relatively sparsely so, with perhaps 100 pairs holding territory, mainly away from Teesmouth.

Overall, the national picture is of a decline in the species' numbers. In the 25 years from 1972 the CBC indicated a considerable drop of population levels by more than a third, and by over 50% on farmland plots. This overview is complicated by local variations associated with regional differences in the changes in farming practices. The species is also susceptible to the effects of severe winters (Marchant *et al.* 1990). Loss of habitat due to changing agricultural practices may also pose a threat in some areas of southern England, but is not at present a significant problem in Durham. More recently, there has been an estimated 28% decline across England, between 1994 and 2006 (Raven *et al.* 2007). In 2009, the species was moved from the green to the amber list of Birds of Conservation Concern because of population decline, and recent BBS data suggest that this decline is continuing. Studies of the number of fledglings produced per breeding attempt show an increase, with a minor increase in clutch size; population decline is thus likely to have been driven by reduced annual survival (Siriwardena *et al.* 1999). Numbers have fallen widely in Europe since 1980 (PECBMS 2010). The BTO's Breeding Birds Survey has shown an 11% decline in numbers in Northumberland and Durham between 1995 and 2009, with a 13% national decline over the same period, numbers reaching their lowest level since the start of the survey (Risely *et al.* 2010).

Recent non-breeding status

In 2007, early in the year, the species was reported from more than 200 locations in the county, from the coast to the western uplands. Many of these reports were of small numbers, but in January, at least 30 were present in a Jarrow housing estate which held a good supply of berry-bearing trees and bushes, the same area attracting 20 in December of that year. Late winter and early spring usually bring many reports of singing birds, this hardy species living up to its old country name of *storm-cock*, for example at Sherburn on 7 February 2007, a male was in full song despite a temperature of -2°C with snow lying (Newsome 2008).

Throughout much of the year, birds are normally noted in pairs or small parties. By late June family parties and small, post-breeding flocks have usually started to build. Gatherings of up to 30 birds are usually noted during the post-breeding period, when family parties join together to form loose feeding flocks, often aggregated around fruiting berry-bearing shrubs and trees, in particular early-fruiting rowans *Sorbus acuparia*. In the uplands, family parties may coalesce by mid-June to form flocks of 15-30 birds, which feed on emerging cranefly *Tipula paludosa* on heather moorland. Such gatherings commence in late July or early August. In August 2009, double-figure flocks were recorded at seven such sites, with a maximum report of 74 birds at Houghall on 18th. Other notable autumn gatherings include 70 at Hunderthwaite above Baldersdale on 9 October 1985 and 80-90 there on 22 September 1991.

Groups usually remain intact through September with many flocks scattered around the county. As October wears on, reported flock size begins to diminish, as berries are consumed, though the break-up of these is not always as simple as this suggests, exceptions persisting as long as concentrations of berries remain. As berries become less available and more clumped however, such as on female holly trees *Ilex aquifolia*, the Mistle Thrush changes its tactics and many birds move to a strategy of protecting their own 'private' berry resources. In December, records of the species are usually fewer but the now well documented habit of berry-defending is often widely observed; in Durham this practice has been extended to include crab apple *Malus sylvestris* at Fence Houses. Defence of favoured trees involving colourful skirmishes with Waxwings *Bombycilla garrulus* have often brightened up a winter's day. At Houghall on 29 November 1988, a bird unsuccessfully defended its berry horde from 11 Waxwings, while a single bird at Jarrow in late 2008 was unable to keep at bay a flock of 40 Waxwings intent on plundering the highly desirable yellow berries of *Sorbus aucuparia* '*Joseph Rock*'. Despite its pugnacious approach it was out-manoeuvred by sheer weight of numbers (T. I. Mills pers. comm.).

The coastal migrant count on 13 October 1996 suggested a distinct coastal movement with a total of 81 birds and evidence of immigration to the county comes courtesy of a party of four birds that was observed flying in off the sea at Whitburn on 23 October 1980 (Baldridge 1981).

Even in winter, Mistle Thrushes can sometimes be found close to the limit of tree cover in the west of the county, long after most of these areas and habitats have been deserted by the Song Thrush after its seasonal, altitudinal migrations have been completed. Flocks have usually dispersed by the second half of the winter period but in late 2010, the year closed with a count of 120 birds at Saxon Way, Jarrow, in late December; this is believed to be the largest-ever flock recorded in Durham (Newsome 2012).

Mistle Thrush, Billingham, December 2007 (Ian Forrest)

Distribution & movements

This species breeds throughout Europe as well as in north west Africa and central Asia. It is a largely resident species over much of its southern range, though some of the more northerly populations are partly migratory.

There is relatively little ringing information relating to this species in Durham though there are a handful of intriguing ringing controls and recoveries. A bird ringed as a nestling on 21 April 1960 at Pity Me was found dead on 29 January 1963 at Poullaouen in France, a movement of 734km to the south south west. This movement may have been triggered by very severe weather, with heavy snowfall at the time. Intriguingly, the remains of a bird ringed as an adult at Whickham on 23 April were discovered in an owl pellet at the same place on 20 August 1982 of the same. At the other end of the scale of longevity for this species, was a bird ringed as a first-winter on 31 December 1983 in Whickham and re-trapped there on 16 February 1991 making it eight years old.

Spotted Flycatcher
Muscicapa striata

Sponsored by

A locally common but declining summer breeder and passage migrant.

Historical review

According to Thomas Robson (1896), the Spotted Flycatcher was a "*common spring and autumn migrant*" in the heavily wooded Derwent valley in the 19th century. At the mid-point of the 20th century, George Temperley (1951) called this species a "*summer resident, widely distributed over the eastern part of the County, wherever there are open woodlands, parks and gardens*". He also noted that its status did not appear to have changed during the last century, as confirmed by Hutchinson (1840) who had described it, over one hundred years before Temperley, as "*a common bird with us, found in woods orchards and gardens*". Historically this species has never been common in the south east of the county, i.e. north of the Tees, due to a lack of woodland in this area (Joynt *et al.* 2008). By way of contrast, in the areas of the Derwent valley that Robson thought it so common, it had undergone a very significant decline in numbers by the early 1990s (Bowey *et al.* 1993).

One at Hartlepool from 19th to 25 October 1968 was, at that time, the latest record for County Durham. This species was documented by David Simpson in his local patch records around Shotton Colliery. Through five decades of observation, Spotted Flycatcher went from being a common and widespread summer visitor in the Shotton Colliery area, in the eastern part of Durham, to a situation in the latter part of the first decade of the 21st century, where no birds were seen in the area at all. It used to nest in the local denes, mainly along the stream edges, and it also bred in the local churchyards. In 2010, for the first time none were noted in that area and it last nested in the local churchyard, at Shotton Colliery, in 2008 (Simpson 2011).

The species was described as plentiful in the county's woodland during the 1970s (Unwin 1974-1978). More specifically, in 1975, breeding birds were widely reported across the county, from the coastal denes to the dales woodlands, with birds being especially well represented along the River Wear from Wolsingham to Lanehead. Further evidence of this species' retreat from its previous easterly breeding range in Durham is provided by the fact that nine pairs were found nesting at Jarrow in 1973, in an area which today has long since been deserted by Spotted Flycatchers as a breeding bird. Successful breeding regularly occurred in Sunderland's parks during the seventies with eight fledged young noted in Barnes Park in 1979, and breeding continuing here into the early 1980s.

Recent breeding status

Today, there is little doubt that this species' breeding range has contracted considerably since the days of George Temperley, and even more noticeably so since the late 1980s and early 1990s. There was a rapid decline in the UK breeding population of Spotted Flycatchers, commencing in the 1960s (Parkin & Knox 2010), of up to 75% during the 25 years or so up to 2007, down to around a national figure of an estimated 130,000 territories. Numbers fell dramatically in the county over the *Atlas* survey period (Westerberg & Bowey 2000).

As recently as the early 1980s, this species was to be found breeding much further east in Durham than it was by the end of the first decade of the 21st century. Previously it was very much capable of exploiting novel nesting opportunities, a trait which seems to have become less apparent at monitored nests. For example, in 1978, birds bred in a plant pot in a house porch at Hamsterley Mill whilst in 1981, during the ringing of birds in the Ravensworth Estate in the Team valley, a bird was noted flying out from about ten feet up in a yew tree. On climbing the tree a nest that appeared to be that of a Song Thrush *Turdus philomelos* or Blackbird *T. merula* was discovered. This had slipped sideways to form a shelf upon which a pair of Spotted Flycatchers had nested successfully; two pulli were ringed and these fledged. In 2000, a pair in the Wear valley nested in a two-year old nest located in an outside light fitting, in the porch of a dwelling (Armstrong 2005).

During the early 1990s, with reference to Gateshead Borough, in the north west of the county, Bowey *et al.* (1993) stated, "*The species breeds widely across the borough in open woodlands, parks, large suburban gardens and even along railway cuttings in the centre of Gateshead.*" Since that time, the species has all but disappeared as a breeding bird from this area. Up to the late 1980s and early 1990s, most woodland in that area held one or two pairs. Birds were found in some numbers in such places as Saltwell Park and the well-wooded lower Derwent valley was the local stronghold for the species. Spotted Flycatcher was recorded from one-sixth of all surveyed one-kilometre squares in Gateshead during breeding bird survey work over the period 1986/1987. Twenty years later, by the middle of the first decade of the 21st century, it was recorded on a regular basis in none. A similar decline was noted in the Houghton-le Spring area at this time with the final breeding at such eastern locations as Hetton Bogs and Elemore Woods occurring in the mid- to late 1990s and at Eppleton Hall in 2004 (Siggens 2006). These declines may be due to climate change in their winter quarters, but also cooler, wetter summers in Britain (Gibbons *et al.* 1993, Mead 2000). By 2005, the national population figures were indicating a decrease of 85% in the number of breeding pairs between 1967 and 2003 (Brown & Grice 2005).

As could be seen from the mapped distribution portrayed in the *Atlas* (Westerberg & Bowey 2000), the main strongholds for this species were in the county's middle and western dales, especially where these become more upland in character. Moreover, this remains the case today. Declines notwithstanding, up to 2007, Spotted Flycatchers were still relatively common in the west of the county and they are frequently encountered in gardens in settlements such as Barnard Castle, Cotherstone and Romaldkirk. In such areas they could once quite easily be induced to nest in less suitable areas by the provision of open-fronted nestboxes, as was done at a series of garden sites in the south east of the county at Lartington, Thorp and two locations at Wycliffe-on-Tees between 1989 and 1994 (K. Bowey pers. obs.). Today, the majority of reports come from the main western valleys, including the upper Derwent valley, with up to six in the valley of the Beldon Burn in 2007, two in the plantation adjacent to Waskerley Reservoir and numbers of birds in the woodlands of the Derwent Gorge and Muggleswick (Newsome 2008). Early survey visits to breeding areas may miss a considerable part of the Spotted Flycatcher's population, as birds arrive late in the north east.

As much as anywhere does today, the county's western valleys provide ideal habitat for Spotted Flycatchers - open woodland with glades and open areas featuring streams and abundant flying insects. On 28 July 2002, 14 breeding pairs were located between Low Force and Middleton in Teesdale and in 2006, on 19 May an observer located a minimum of 22 birds along the River Wear and its tributaries of the Rookhope Burn and the Horsley Burn, indicating that despite the declines, the species retains a strong presence in the county's western riverine woodlands. From May to August 2007, sightings came from over 35 localities across the county, the majority of these being in the upper sections of the main river systems; the furthest east was at Hardwick Hall Country Park. Densities were never high, but the River Tees valley in the Bowlees area held several pairs, as did the River Wear valley around Stanhope. In 2010, birds were in breeding habitat at Bowes, Crawleyside, Gilmonby, Harthope Beck, Rookhope, St John's Hall, Waskerley Reservoir and Wolsingham, with birds from a good spread of at least 25 other sites.

During the work on the *Atlas*, this species was well recorded, except in the south east of the county (Westerberg & Bowey 2000). It was felt at the time that the lack of registrations on the distribution map in that area may have been in part a consequence of reduced observer cover, but more recent survey work for *The Breeding Birds of Cleveland* (Joynt *et al.* 2008) has shown that the species is today genuinely scarce as a breeding bird in that area. In the south east of the county, numbers of this summer visitor have declined quite markedly since the early 1990s; there was an estimated breeding population of about 150 pairs in the 1970s and 1980s (Blick 2009) but today there are probably less than ten territories in the north Tees area in most years and possibly less than this figure (Joynt *et al.* 2008).

During the late 1980s, local survey work in the Barnard Castle area recorded up to 20 pairs per occupied tetrad (Westerberg & Bowey 2000). This however, was carried out during June in close to what was optimal habitat, where some nestboxes had been provided. It would seem that an average density of around 5 pairs per occupied tetrad would be more likely to occur over much of the rest of the county, giving a breeding population estimate of some 1,000 pairs at that time (Westerberg & Bowey 2000). Since this estimate was made, it seems likely that numbers have declined quite markedly, perhaps by as much 50%. It is not so many years ago that they were present in the suburban, and even urban, areas of the county: in such situations they could be found breeding in parks, cemeteries and gardens. During spring and summer 2008, potential breeding birds were reported from 105 one-kilometre grid squares around the county and breeding was proven in a high proportion of these. Nonetheless, at the end of the first decade of the 21st century, the species was barely hanging on as a breeding bird in the east of the county, indeed in the lowlands parts of the county breeding had almost ceased except for sporadic instances,. In 2010, pairs and juveniles were noted at only a handful of such locations such as Breckon Hill, Eppleton, Hardwick Hall, Sunderland Bridge and Tilery Wood, Wynyard; the most easterly extant breeding location in the county.

Recent non-breeding status

Spotted Flycatcher is one of the last summer migrants to arrive, usually not returning to the county until late May or even early June. The main arrival occurs during the middle of the former month, breeding birds establishing themselves in territory during the last week of May. In 2007, the first spring arrival was near St John's Chapel on 29 April but no others were detected for nearly two weeks after this date, likewise an early migrant was found at Breckon Hill on 30 April 2010.

The reported first arrival dates of Spotted Flycatcher in Durham were analysed, in ten-year groupings, over the thirty-year period between 1976 and 2005 and it was found that the mean first arrival had advanced by eight days, with the greatest degree of change occurring in the period between 1996 and 2005 (Siggens 2005). The earliest arrival date for the county is on 4 April 1999, a bird at Fence Houses. This exceptionally early report is 13 days earlier than the second earliest sighting, which was at Winlaton Mill on 17 April 1995 (Armstrong 1997).

The first fledged juveniles are usually noted in early to mid-July, though first-brood young have been noted out of the nest as early as late June, and family parties and dispersing juveniles are no doubt responsible for an increase in sightings in late July and early August. Second broods can sometimes still be present in the county into early September, but they rapidly vacate the area as soon as fledging occurs.

Autumn passage birds are normally noted at inland sites through late August or early September but rarely much later than this, though the scale of such movement varies from season to season. The last records of the year are usually seen in mid- to late September, with birds sometimes hanging on to the last days of the month or even into the first week of October. This trait seemed to occur more often in the past. October records became less frequent through the 1990s and fewer still in the 2000s, especially in comparison to the 1970s and 1980s, when a small number of sightings was routine in the first week of October; for example, the bird at Hartlepool on 13 October 1972. Some of the latest records came in the early 1980s, with the latest ever being during the exceptional arrival of late migrants in November 1984, when a bird was present in Whitburn Village on 12 November, six days later than the second latest on record, which was at Seaburn two years previously (Baldridge 1985).

As a passage migrant, Spotted Flycatcher is usually seen at coastal migration points in both spring and autumn but in ever smaller numbers compared to some of the larger influxes that were once noted, such as the 12 at Hartlepool and 10 at Teesmouth in early September 1974 (Unwin 1975). Most coastal birds occur between early May and early June and at the other end of the season, between mid-August and early October. Often small coastal influxes occurred during mid- to late August through the 1970s. More than normal were on passage in 1973, with higher numbers again on autumn passage in 1974, when during late August/early September up to 12 were at Hartlepool, 5 at Seaton Carew and Marsden, 10 scattered about Teesmouth and 6 in Hartlepool away from the Headland.

In 2007, inland passage birds were typically scarce, with single birds at Hetton Bogs on 2nd to 9 September, at Ryton on 3rd and Langley Moor on 4th, Thornley Woods on 11th and Hetton Bogs again on the late date of 2 October, this being an unusually good crop of sightings for this species in the modern context. The small numbers involved in the fall of migrants around 6 September 2008 was the best autumn showing for many years (Newsome 2009).

Distribution & movements

Spotted Flycatchers can be found breeding over most of Europe, north to Fennoscandia, in the north west parts of Africa, the Middle East and through Siberia into central Asia as far east as Mongolia. It is a long-distance migrant that winters far to the south of its breeding areas, in sub-Saharan Africa (Parkin & Knox 2010).

There is relatively little ringing information relating to Durham's Spotted Flycatchers. The most spectacular recovery was of a bird ringed as a nestling at South Shields on 28 June 1952, when birds still nested in that part of the county, which was recovered at Coimbra, Arganil, Portugal on 13 September 1953. Following probably a not-dissimilar route was the individual ringed at Hartlepool on 31 July 1969, recovered in Bordeaux, France, in October 1969. Finally, heading in the other direction was one that was ringed as a pullus on 5 August 1967 at Knaresborough in Yorkshire, but found breeding in Wynyard Park, near Billingham on 30 June 1968 (Blick 2009).

Robin (European Robin)
Erithacus rubecula

An abundant resident, passage migrant and winter visitor.

Historical review

In the Derwent valley of the 19th century, Robins were considered common by Robson (Robson 1896) and they remain so today. George Temperley (1951) said that it was a *"very common and well distributed resident"*, and *"also a passage migrant"*. Nelson (1907) commented that, *"for many years, know the robin as a regular autumn migrant between September and November to the Teesmouth district…"* Stead (1964) said it was common, breeding in gardens, hedgerows and in mixed woodland in the middle of the 20th century.

Its status as an autumn migrant has long been known and a big influx was noted along the east coast of England in October 1951. Although there was little documentation of this over much of the coast of Durham, at Teesmouth hundreds were seen during the first week of the month (Mather 1986). Further evidence of its migratory status in the region came in 1965, when there was a large influx in the autumn, with over 200 in the Hartlepool area alone on 1st and 2 October (Stead 1969). Most migratory activity of this species in Durham is associated with the autumn, but an influx in late April 1973 brought 71 birds to three sites in the Hartlepool area on 24th. In late October of that year, a bird which was washed off its perch on the anchor chain of a ship that was anchored off the River Tyne by a wave, rested briefly on the sea before flying off, apparently unharmed (Coulson 1974).

Through the 1960s and 1970s this species' status or population trends were rarely commented upon, or even fully documented, though it was clearly widespread as a breeding bird, including in the western dales. This was demonstrated by the birds that bred at Cronkley Farm high in Teesdale in 1975 and those that were present at Widdybank Farm the following year. The only aspect of this species' annual cycle that regularly attracted attention was migration. A large number of immigrants were noted along the coast between mid-September and the end of October 1976, with 50 at Marsden Quarry on 2 October, 43 being ringed there through this period (Unwin 1977).

Recent breeding status

The Robin's preferred habitat of woodland, scrub or hedgerow can be found widely across much of the central area of the county. As a result, it is a widespread species and the distribution effectively covers the whole of the county, the more open upland areas excepted. Densities generally fall away in more open, arable farmland, where birds are restricted to hedgerows and wooded shelterbelts. This limitation was in some measure reflected in the patchiness of the distribution map for the species shown in the *Atlas* (Westerberg & Bowey 2000). In the west of the county, Robins are largely absent from the open uplands, though they may be found in some conifer plantations at quite high elevations, for example a bird was singing at The Stang in June 1984, at over 480m above sea level, and in 2009 a pair raised young right at the limit of available habitat in the Bollihope Burn (Newsome 2010).

Illustrative of the situation in the wider county were the figures for Robin in 2007 in a part of the north east of the county, close to Houghton-le-Spring. In this area there were 43 singing males in the Elemore Hall area on 8 April and studies in the Houghton/Hetton area suggested that there had been a small rise in the local population, with an increase at Hetton Lyons of five pairs over 2006, and a decline of one pair to 21 pairs at Hetton Bogs, from the previous year. A total of 26 pairs were detected at nearby Rainton Meadows/Joe's Pond. The first fledged juvenile of 2007 was noted on 5 May, at Hurworth-on-Tees, but young were still being fed in the nest in mid-July at several localities; these were presumably second broods (Newsome 2008).

Recent survey work in the south east of the county has shown it to be common there, occupying over 75% of all tetrads surveyed (Joynt *et al.* 2008), with over 1,700 pairs in the northern area of what was Cleveland county. The documented breeding densities of Robin territories from several local Common Bird Census plots during the late 1980s and early 1990s, which covered a variety of habitats in the county, ranged from 0.09 to 3.3 pairs per hectare. This data, used in conjunction with the distributional data provided by Gibbons *et al.* (1993), led to a minimal estimate for the county's breeding population of 14,500 pairs in the *Atlas* (Westerberg & Bowey 2000). Considering that the estimated British population during 1989-91 was 4,200,000 territories then this local estimate, today, may seem somewhat low and may be in need of revision upwards. Robin is amongst the most widespread breeding species in Britain, as demonstrated by the Breeding Bird Survey results of 2009, when the species was recorded in 90% of all kilometre squares surveyed (Risely *et al.* 2010).

Birds can show quite rapid increases in breeding numbers between years, when conditions allow, for instance the increase in the number of territories from five to 13 at Shibdon Pond between 1993 and 1994. But this is usually only evident where detailed studies are being undertaken. For example, the ringing catches at two Gateshead study sites in 1996 indicated a poor breeding season for this species, with numbers caught in 1996 being down to 54 from 85 in 1995, and whilst adult numbers were reduced by about 30% the juvenile catch was down by over 45%, hinting at an unproductive breeding season as a result of the cool spring (Armstrong 1997). Just a few years later, and 23 juveniles were ringed at one of these, the Lockhaugh study site in 2000, indicating a good breeding season in that area compared to the previous few years. By contrast, in 2005, the population that was monitored at the Kibblesworth Common Bird Census plot fell to 30 pairs, the fourth consecutive year of decline at this site (Newsome 2006).

The major concentrations of breeding birds occur in parkland and woodland areas and a local study in 2008 revealed 51 singing birds at Hetton Bogs and a maximum count of 122 in Elemore Woods, making it the second commonest breeding bird there, after Chaffinch *Fringilla coelebs*. Other indications of 2008 population came from Hunwick and Willington where 13 to 16 birds per linear kilometre were noted, and in the Houghton area, where three sites held a total of 71 breeding pairs (Newsome 2009). Another indication of local abundance was given by the count of 80 birds along a ten kilometre stretch of the Bishop Auckland to Brandon walkway in spring 2003 (Newsome 2007).

It has been commented upon that the species faces a long-term threat in the county from the continued, piecemeal destruction or detrimental management of its habitat. However, habitat lost to the continual encroachment of new building developments in the countryside can, after a period of time, be somewhat offset by the development of new, suburban habitats as gardens around housing developments mature and provide new locations for the species to nest. During the early 1980s, birds were successfully encouraged to nest around Joe's Pond by the placing of old cleaned-out paint tins in the hedgerows. In 1994 a pair successfully fledged young when nesting in one of the hides at WWT Washington. An unseasonal breeding record was of a nest with five eggs at Barnard Castle in mid-January 2003 (Newsome 2007).

One of the most unusual breeding observations in Durham was of a brood of young Robins being fed by adult Blackbirds *Turdus merula* at Spennymoor in the spring of 1994. The female Blackbird was also seen removing the youngsters' faecal sacs (Armstrong 1996).

There are numerous reports of birds singing at night in urban locations from the early 1980s onwards, particularly close to artificial lighting, it becoming an increasingly widely reported phenomenon. For example, one was at Sherburn on 1 January 2008, illustrating the theory that such urban birds are singing more at night to avoid competition with increasing traffic noise during the later singing period (Newsome 2009).

Recent non-breeding status

Singing takes place almost all year round. At least 25 singing birds were noted between The Grove and the information centre at Hamsterley on 13 October 2007, a reflection of the mild settled weather experienced at this time of year and perhaps of birds establishing their winter territories (Newsome 2008).

The pattern of post-breeding and winter reports of this species does not reveal any new information that adds to our knowledge of what is one of our most familiar birds. It has a winter range that extends from the coast inland as far as suitable habitat allows. Counts along transect walks of known length, such as in 2007, when 46 were noted in a couple of kilometres at Hetton Bogs on 26 October, give some clue as to its lowland autumn and winter abundance in the county. In the winter, birds are abundant in the garden habitats of the county and persist in these areas throughout the coldest months of the year in many areas, though it is known from ringing data that many of Durham's Robins move south and west, to winter in less harsh local climates. Such movement is not uniform across local populations but is biased between genders, females being more likely to move, and move further, than males (Wernham *et al.* 2002).

The species very routinely takes advantage of the food put out by householders in gardens, yards and even balconies and window boxes of urban high-density houses and flats. In January 1991, a bird at WWT Washington was regularly observed to be diving for barley, provided for the wildfowl, in one of the ponds. Such a close relationship between bird and people leads to observations of fascinating behaviour and individual variations, and often birds with plumage aberrations are reported from garden feeding stations. For example, a leucistic bird with a white head and chin was present in a Shotton Colliery garden from 28 November 2008 through to 2011 (Newsome 2009). An individual with a white head was at Norton in January-February 1997 (Blick 2009).

Continental Robin
Erithacus rubecula rubecula

The autumn period can often bring a large arrival of continental birds, *E. r. rubecula*, to the coast. Influxes traditionally would take place from late September, though they might occur with waves of new arrivals, presumably from further afield, through October. However, as is increasingly the case, for example in 2009, autumn migrants were largely notable by their absence. The largest arrival of birds in this year concerned up to twenty around Dawdon Blast Beach on 22 October, while smaller numbers of migrants were also noted at Marsden Quarry, Sunderland South Dock, Trow Quarry and Whitburn Coastal Park in this period (Newsome 2010). These figures are now much more typical of the scale of arrival of continental Robins on the Durham coast, when at one time, during the 1950s, 1960s and 1970s in particular, large autumn influxes were very much more the norm.

As a passage migrant, Robin is regularly observed at the county's coastal watch points in both spring and autumn. The principal periods of passage for the species are March to April and September to October. On 26 March 1984, 24 northbound migrants were grounded by rain at Whitburn and in 1986 spring passage birds were noted at South Shields Pier, Westoe Cemetery and Whitburn Allotments during April. The numbers involved in spring tend to be relatively small, but occasionally much greater numbers occur, particularly in the autumn. A total of 285 was noted on the 1992 Coordinated Migrant Count on 4 October, rising to 363 during the 1994 count on 2 October, but the following year this was down to 228 birds with 219 on the 1996 count. It was the most widespread, but not most numerous, species along the Durham coast in four out of the five years of the count between 1992 and 1996 (Bowey 1999c).

In the past, coastal falls of Robin that numbered more than one hundred birds have occurred in late March 1958; mid-April 1966, 1973 and 1986; in mid-September 1969 and in 2001; and in early to mid-October 1951, 1965, 1979, 1982, 1988 and 2004. The autumn influx during October 1988 led to the ringing of 73 birds through the month at Marsden Quarry (Armstrong 1989). At least 400-600 birds were involved in spectacular falls in October 1965, April 1986 and September 1998 (Blick 2009). In both October and early November, these arrivals and movements of birds often penetrate quickly inland, especially at favoured sites. For example, daily counts at Shibdon Pond during the autumn periods during the late 1980s and early 1990s, suggested that small numbers of birds rapidly moved into that area, some 15 miles inland, during October influxes; the individuals involved may have been coming from the Continent or from further north in Britain; 20 were there on 27 September 1990 (Armstrong 1991).

Continental birds often remain for the winter and at times, choose habitats quite different to our resident garden birds. One was observed sheltering in caves and foraging in seaweed at the foot of Marsden Cliffs in company with a Black Redstart *Phoenicurus ochruros* during January 2004 (Siggens 2006).

Distribution & movements

Robins breed over much of Europe, Iceland excepted, eastwards as far as western Siberia, into south west Asia and also in north west Africa. The western and southern populations are non-migratory, though in some areas, such as the British Isles, there may be differential gender-related migratory strategies. The northern and eastern birds are highly migratory. These move south and west to winter mainly within the areas inhabited by the more southerly breeding populations.

Autumn arrivals of Robins also usually involve a range of other species that are caught up by the same weather systems and conditions. Some of these 'fellow-travellers' include the common thrushes, a number of warbler species and some scarce migrants such as Black Redstart and Bluethroat *Luscinia svecica*. This pattern of species indicates, in all probability, a Scandinavian origin for many birds involved in such falls, and this is certainly borne out by some of the ringing data for the county. For example:

- One ringed at Hartlepool on 25 April 1973 was controlled in Jutland, Denmark, three days later
- One ringed at Graythorp, West Hartlepool on 24 April 1973 was found dead in the Netherlands in October 1973
- A first-winter bird ringed on 6 November 1976 at Graythorp was controlled on 10 April 1977 at Hanstholm, Jutland, Denmark
- An adult bird ringed at WWT Washington on 24 January 1991, one of 50 in the park at the time, was found freshly dead on a ship, somewhere between Mongstad and Tranmeer, 548km north of its ringing site, in the middle of the North Sea on 1 May 1991. The bird was presumably returning to its Scandinavian breeding grounds after wintering in the United Kingdom
- One ringed at Sappi, Finland, on 20 September 1998 was present at Hartlepool 14 days later.

Galloway and Meek asserted that ringing in Northumberland had demonstrated that during hard weather birds moved from higher ground to lower elevations, an example of altitudinal migration (Kerr 2001), and this is probably also true for birds in Durham, but there is no direct proof to support the assertion.

Locally generated ringing data indicates a slow southern or westerly movement of British-bred birds in the autumn (Blick 2009), such as one ringed at South Shields on 19 September 1951 and recovered at Darlington on 10 July 1952; one ringed at Holywell Dene on 2 June 1962 which was recovered at Newton Aycliffe on 23 January

1963; and a bird ringed as an adult at Holywell, Northumberland on 2 September 1964 and recovered at Hetton-le-Hole on 25 October 1964 (Bell 1965).

One ringed at Marsden Quarry in September 1992 was found dead at Houghton Hall, near Kings Lynn, Norfolk in March 1994, 251km to the south. A bird ringed as juvenile at Marsden Quarry on 17 September 1993 was killed by a cat in South Woodford London on 14 October 1993. Reaching much further south was the first-winter bird ringed on 2 October 1976 at Graythorp, West Hartlepool which was controlled on 29 October 1977 at Ighil Imoula, Grande Kabylie, Algeria.

Birds in the north east clearly head south, but birds ringed in the south have also been recovered in the north, in spring or in later years. For example, the bird ringed at Selsey, Sussex, on 25 February 1963, was found at Gainford, near Darlington 270 miles to the north on 15 March 1964; and a first-year ringed on 21 July 1979 at Wilton, Redcar was found dead at Sunderland, 41km to the north west on 4 October 1979 (Blick 2009).

There is also a westerly or south westerly element to some birds' movements, as shown by one ringed at Ravensworth on 30 May 1983 that was found dead at Poyton, Cheshire on 25 February 1985 and a first-year bird that was ringed at Saltholme in September 1999, which was at Penrith, Cumbria, in September 2000. The speed at which birds can move was shown by the bird ringed on an October morning in Marske Fox Covert, which had reached Seaton Carew by the same afternoon (Blick 2009). Such a pattern of movements should not hide the fact that many locally-bred birds, as has been shown from wider ringing data locally, do not move far; a matter of a few kilometres at most and, in many instances, barely away from their natal sites.

The age to which Robins can live was indicated by the six-year-old bird trapped in a Shotton Colliery garden in the winter of 2003 ringed there in 1997.

Siberian Rubythroat
Larvivora calliope

A very rare vagrant from Siberia: one record.

One of the most remarkable occurrences of a rare bird in Durham was the first-winter female Siberian Rubythroat seen at Roker, Sunderland over 26th to 28 October 2006 (*British Birds* 100: 730). The bird was present in an enclosed suburban back garden in Edgeworth Crescent, but its identity was not confirmed until the final day of its stay. Michael Williams had been watching the bird with his two sons for two days but was uncertain of its identification. A chain of events on 28th led to Ian Fisher, the ex-Northumberland County Ornithological Recorder, identifying the bird but, with the identity established, the house owner decided that news of its presence should not be released due to the nature of the viewing conditions. The bird was last seen at dusk on 28th.

Distribution & movements

Siberian Rubythroat breeds throughout Siberia from the Ob River east to Kamchatka, with small numbers in the European foothills of the Urals. The wintering range covers Nepal east through the Himalayan foothills to north east India and Burma and south to central Thailand, southern China and Taiwan. To the end of 2010, there had been seven records for Britain, all being in October and all but two being on Shetland or Fair Isle.

White-throated Robin
Irania gutturalis

Very rare vagrant from south western Asia: one record.

The first and only record of White-throated Robin for County Durham is of a first summer female trapped and ringed by Chris Brown at Hartlepool Headland during the morning of 6 June 2011. He had initially seen a mystery passerine with orange flanks briefly in Olive Street which he had assumed to be a Red-flanked Bluetail *Tarsiger cyanurus*, before routinely trapping the bird at his ringing site in the allotment adjacent to the Doctors Garden. Despite some initial confusion regarding its identity, Tom Francis arrived shortly afterwards and was able to confirm the identification as a White-throated Robin, only the third record for Britain. It was released shortly afterwards into the nearby Bowling Green where it continued to show well on and off until 10 June, though it often spent long periods in the Doctors Garden especially during the middle of the day when its favoured feeding areas became too disturbed by pedestrians. During the early morning, it was often to be found feeding along the kerbsides of Olive Street where it had originally been found and could be remarkably confiding, though soon sought retreat in the Doctors Garden when accidentally disturbed.

With news released of this exciting find by 08:30 on the morning of 6 June, many observers were able to reach Hartlepool that same day and most enjoyed good views of the bird feeding around the bowling greens and in nearby Olive Street, although by mid-afternoon, it had become more elusive and disappeared around 3:00pm. With more and more observers arriving by the hour, tensions were beginning to mount, but fortunately for those present, it was relocated around 17:30 feeding in the herbaceous borders of the 'Doctor's Garden'. What happened during the next 30 minutes will long be remembered by local residents and birdwatchers alike and was referred to by the press as "the siege of Hartlepool". Surrounded by a ten foot wall and with the bird currently on show from the roof of a small

van, hundreds of people clambered for a view; ladders arrived from nowhere, local builders offered a vantage point from the roofs of their vans, lampposts were scaled, and local residents allowed birders into properties overlooking the large well vegetated garden. A real community spirit helped everyone get a view, with some residents even providing free tea and cakes for the crowd. By the end of the day, many hundreds of observers had seen the bird and went home happy, in no large part thanks to the help of the people of Hartlepool.

The scene at Hartlepool Headland, 6 June 2011 (M. Sidwell)

This was the first White-throated Robin for mainland Britain and inevitably attracted large numbers of observers to Hartlepool Headland, with an estimated 2,500 people having enjoyed its five stay, though many more were disappointed it did not remain until Saturday the 11th.

Distribution & movements

White-throated Robin breeds across much of much of south western Asia, from southern Turkey eastwards through northern Iraq and Iran, to western Afghanistan and south eastern Kazakhstan. A long distance migrant, the entire population winters in East Africa, chiefly in Kenya and Tanzania. It is an extremely rare vagrant to Britain with just two previous records; a male on the Calf of Man, Isle of Man on 22 June 1983, and a female on Skokholm, Dyfed from 27th to 30 May 1990. Elsewhere in Europe, it has occurred as a vagrant as far north west as Belgium, the Netherlands, Norway and Sweden, where the majority of records have been in May and June, though at least six have occurred between August and November.

Red-flanked Bluetail
Tarsiger cyanurus

A very rare vagrant from north east Europe and Siberia: five records.

The first county record came on 7 October 2002 when Paul Cook found a first-winter or female in Marsden Quarry (*British Birds* 96: 587). The bird was elusive but was eventually seen by over 150 observers.

All records:

2002	Marsden Quarry, first-winter or female, 7 October
2009	Whitburn CP, first-winter, 15 October, trapped (*British Birds* 103: 619)
2011	Whitburn CP, first-winter male, 13th to 19 October, trapped 13th (*British Birds* in prep.)
2011	Church Lane, Whitburn, first-winter or female, 13 October (*British Birds* in prep.)
2011	Whitburn CP, first-winter or female, 10 November (*British Birds* in prep.)

Formerly an extremely rare vagrant, the continued westward expansion of this species range into eastern Scandinavia has seen British records increase dramatically during the last decade. That three should be found at Whitburn during the autumn of 2011 is remarkable, especially as the well-watched Hartlepool area awaits its first. November records are unusual with the bird at Whitburn in 2011 being the second-latest to be recorded in Britain, outdone only by a bird a Gibraltar Point, Lincolnshire from 15th to 16 November 2002.

Distribution & movements
The main breeding range of Red-flanked Bluetail extends from eastern Russia and Siberian to Kamchatka, northern Japan and north east China. A small but expanding population also exists in north east Finland. The species winters in southern China, Taiwan and southern Japan, and through South East Asia to peninsular Thailand. After only 11 British records up to the end of the 1980s, the last twenty years have seen a steady increase in the number of Red-flanked Bluetails being encountered, and the number of accepted British records climbed to 68 by the end of 2009, but 2010 produced over 30 autumn records. All autumn records have been in the period 16 September to 16 November, with only a handful of spring sightings.

Thrush Nightingale
Luscinia luscinia

A very rare vagrant from eastern Europe: four records.

All four Durham records of this skulking species have been at Hartlepool Headland. The first came in 1985 when a singing male was found near the tennis courts by Chris Sharp and Tom Francis; it later moved to the bowling green and the 'Doctor's Garden' (*British Birds* 79: 569). The bird arrived during a classic spring fall of north east European passerines. Despite being rather secretive, it remained around the garden for three days. The Hartlepool bird was one of six British records in the spring of 1985, and occurred on what might be considered a classic date for the species.

All records:

1985	Hartlepool Headland, 13-15 May
1989	Hartlepool Headland, 23 May (*British Birds* 84: 482)
1996	Hartlepool Headland, 13 May (*British Birds* 90: 495)
1997	Hartlepool Headland, 18-19 May (*British Birds* 91: 501)

All of Durham's birds have arrived between 13th and 23 May; very much in keeping with spring records elsewhere in Britain. The mean spring arrival date on Fair Isle, where it is a regular vagrant, was 21 May during the 1980s and 25 May during the 1990s (Slack 2009).

Distribution & movements
The breeding range of Thrush Nightingale covers southern Scandinavia, from Denmark south east to Romania and the Ukraine, and eastward through temperate European Russia and southern Siberia. The wintering range is in eastern Africa, from southern Kenya to Zimbabwe. There are a total of 187 records for Britain to the end of 2010, a high proportion of which have been on the Northern Isles. May is by the far the peak month for occurrences, with a lesser peak in September.

Nightingale (Common Nightingale)
Luscinia megarhynchos

A very rare vagrant from Europe: six records and an old unconfirmed report of a breeding attempt.

Historical review
In the British Isles, the species has long been noted as a breeding species to the south of the Humber and for its scarcity further north.   The historical status of Nightingale in Durham is fraught with some confusion and uncertainty, and in the north east of England it has always been a rare visitor. Temperley (1951) noted that "*there are no sufficiently well authenticated records to justify its inclusion in the Durham list*", but he went on to give an example from Edward Backhouse of two birds at West Lodge, Darlington, in the spring of 1803.   Temperley's hesitance in accepting the record was presumably based on a lack of tangible evidence of the birds, which were reported to have been 'taken', but were not readily traceable to a known collection. The species was certainly rare so far north at that time.   Reviewing its status in the early part of the 20th century, Ticehurst & Jourdain (1912) noted that the species was regularly noted as far north as Normanby in North Yorkshire, a few miles to the south of Teesmouth, but they made no mention of regular occurrences further north in Durham.

Nonetheless, the evidence of Nightingale occurring in Durham and indeed its apparent attempted breeding in 1931 is well documented, but the background to this record makes its inclusion on the official Durham list of breeding species inappropriate. In the summer of 1931, a Nightingale was found singing at Stampley Moss, close to Thornley Woods, Rowlands Gill in the Derwent valley.   In summary 'it' was heard singing from 24 April to 23 June, but such cursory detail misses what is a fascinating story. Russell Goddard reported in *British Birds* in 1931 that, "*Having heard that a bird reputed to be a Nightingale was singing nightly at Stampley Moss, a tract of woodland country between Winlaton and Rowlands Gill, County Durham, I determined to investigate. Accordingly, on May 21st, 1931, I visited Stampley Moss accompanied by G. W. Temperley and B. P. Hill, two members of the Natural History Society of Northumberland, Durham and Newcastle-upon-Tyne. We arrived at the spot at 10.30 p.m. and waited patiently until 11.20 p.m., when the bird commenced to sing. Much to our delight, we found that it was really a Nightingale.   On previous occasions when I have been called out to hear reputed Nightingales in Durham, they have turned out to be Sedge-Warblers (Acrocephalus schoenobaenus). The song of the Stampley Moss bird was characteristic of the Nightingale. The oft repeated note, "piou", in crescendo and the short phrases of bubbling notes which followed were unmistakable*" (Goddard 1931).

The singing bird was seen by many observers and on a few occasions, a pair was noted. Indeed, one rumour suggested that the birds had nested but that the eggs were taken. Over 100 people were said to have come to hear the birds, which led to rowdy scenes and stones being thrown at the birds. The above incident probably constitutes the north east's first bird watching 'twitch' (Bowey *et al.* 1993).   Subsequent investigations revealed that the Reverend Father F. Kyute of St. Agnes of Bradley Hall, in Crawcrook – less than 6km from Stampley Moss – had kept Nightingales as cage birds for many years, and on more than one occasion, he had lost or deliberately released captive-bred Nightingales in the north east. In a letter written in 1932 to George Bolam, he stated that he had released two pairs at Alnwick, Northumberland, in March 1931. Later in 1931, a bird was reported to have been heard singing at Bradley Hall and Father Kyute said that he had released birds there in both 1928 and 1929. Bolam knew the rarity of this species in the north east and he documented that two, thought to be males, had escaped from Bradley Hall much earlier, in the 1890s (Bolam 1932), and that Father Kyute had deliberately released 'six pairs' of birds at Bradley Hall and Alnwick between 1928 and 1931 (Kerr 2001).

This is still the only apparent breeding attempt by this species in Durham, but whether the birds were wild or something to do with Father Kyute, remains unknown; the circumstantial evidence for what was then a very rare bird suggests that the Stampley Moss birds were probably not of wild origin.

Recent status

Despite now breeding as close as South Yorkshire, the Nightingale has remained an extremely rare visitor to County Durham. The first confirmed modern record was in 1981, when George Tuthill found a singing male in Peepy Plantation, Washington, on 20 May. It stayed until the following day and made up part of an impressive array of rare birds in the Washington area at the time, including a Black Kite *Milvus migrans* and a Lesser Grey Shrike *Lanius minor*. The remaining five records concern four short-staying coastal migrants, plus a long-staying territorial male. The October and November records are particularly interesting as the weather conditions at the time suggest they may have had an easterly origin, although plumage detals noted did not point to one of the eastern races *africana* or h*afizi*.

All records:

1981	Washington, 20th to 21 May
1984	Hartlepool Headland, 7 October
1985	Whitburn, 17 May
2010	Whitburn Coastal Park, 12 October
2011	Cowpen Bewley WP, singing male, 30 April to 18 May
2011	Whitburn, 12 November

Distribution & movements

The nominate race of Nightingale breeds across most of Europe and North Africa, and east to central Turkey, with an estimated European population of 3.2 to 7 million pairs. The eastern race breeds in central Asia through Kazakhstan to western China. It is a long-distance migrant, wintering in tropical Africa, to the south of the Sahara, mainly in central and West Africa. The species is a scarce and declining breeding species through much of south east England (Holt *et al.* 2012). Occasional birds are seen in migration periods to the north of its range, but it is genuinely rare in northern England and Scotland.

Bluethroat
Luscinia svecica

The red-spotted form is a scarce spring and autumn migrant from north east Europe, prone to larger arrivals in some springs.
The white-spotted form is a very rare spring vagrant.

Historical review

Sightings of this species increased quite dramatically around the 1960s and 1970s, only to fall away again by the late 1990s and early 2000s. Prior to 1958, it was something of a rarity in County Durham.

The first British record was of one shot a few miles north of the Durham boundary, on Newcastle Town Moor in May 1826 (Kerr 2001), but it was more than fifty years before Durham's first bird was recorded. The first Durham record of Bluethroat was referred to in Hancock's "Catalogue", reading as follows; "*Another specimen was found dead, killed by the telegraph wires, in the County of Durham on the 25th Sept., 1883, it is now in the possession of Mr. H. Walton, of Birtley, and from the description given me by Mr Walton I take the specimen to be in the very same state of plumage as that mentioned above. The specimen was left at St Marys Terrace by Mr W. for me to examine and, just as I thought, it is in the same plumage as that killed at St Mary's Island (Northumberland) on the 15th Sept.*" The specimen was deposited at the Hancock Museum.

Further examples referred to by Temperley (1951) included one obtained in 1893 by H. G. Stobart at Wolsingham on 26 September; one at Chester-le-Street about the same date and another at Chester-le-Street in the autumn of 1902 or 1903. These unusual inland occurrences were followed by more typical observations of autumn coastal migrants. An immature was shot and another seen by J. Bishop near Seaton Carew on 19 September 1903, and a young bird was observed by W.G. Alexander "*skulking amongst tussocks*" at Saltholme marshes on 16 August 1929. As a reflection of the clear changes in occurrence patterns compared with modern day understanding, Temperley also noted that all the records had been seen on autumn migration; "*none has yet occurred in spring*".

Post-Temperley, there was a distinct up-turn in records and it became an almost-expected autumn visitor, particularly for the migrant hunters at Hartlepool Headland. Birds were seen in all but one autumn between 1958 and 1969 and often in numbers. Early September 1958 brought daily sightings of up to four birds at Hartlepool between 2nd and 7th, while an impressive total of nine were found at Hartlepool Headland on 17 September 1960. Sightings at Hartlepool, Graythorp and North Gare through the 1960s were mainly of one to two birds, but another exciting migration period in mid-September 1969 resulted in one to two daily between 16th and 21st, plus four on 18th. All birds in this period arrived between 29 August and 12 October, and again, there was no sign of a spring adult.

Recent status

A minimum of 107 birds have been recorded since 1970, although nearly 70% of these occurred in just four of the years. Away from these influx years, it is much rarer. There have been 15 blank years since 1970 and a further 17 years with just one sighting in the county, indicating a distinct downturn in frequency in a short space of time.

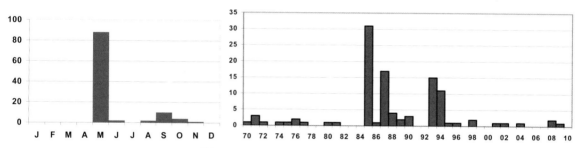

Bluethroats in County Durham, 1970-2010

Nearly 85% of recent sightings have been in spring, with arrivals coming between 9 May and 8 June; a contrast to the previous twenty years. All spring birds which have been sub-specifically identified have been of the Red-spotted subspecies *L. s. svecica*, this being the race most expected to occur during May migration (Millington 2010). The key years for spring Bluethroats in Durham were 1985, 1987, 1993 and 1994 when north easterly winds and rain in the critical migration periods led to good numbers being deposited at coastal watch points. The year 1985 provided the greatest numbers with at least 31 birds recorded between 10th and 20 May. Locations were limited to Hartlepool Headland and the Marsden/Whitburn area, but featured peaks of six at Hartlepool on 14th and 18 different birds around Whitburn over 14th to 15th (Baldridge 1986). The actual numbers arriving along the Durham coastline must have been considerably more than these figures indicate.

Another arrival took place two years later, when 17 birds were found between 22nd and 26 May 1987. Again, sightings were limited to Hartlepool Headland and Whitburn, with eight at the former site on 22nd and seven in the Whitburn area over 22nd to 24th. Lesser arrivals came between 9th and 27 May 1993 (14 birds) and 15th to 26 May 1994 (nine birds) with migrants being found in very similar localities.

The remainder of the 1990s produced just four further spring birds, and the whole of the 2000s decade just two (in 2002 and 2004). This situation was reflected elsewhere on the English east coast. The reason behind the decline is difficult to determine. A change in migration path is a possible explanation (Millington 2010).

Few spring birds have been found away from Hartlepool and Whitburn/Marsden, although this is probably influenced by observer coverage during peak arrival times. Singles have been encountered at Castle Eden

Denemouth, Roker, Ryhope and Sunderland, while the coastal areas to the south of Hartlepool, such as Seaton Carew and North Gare, have produced a handful.

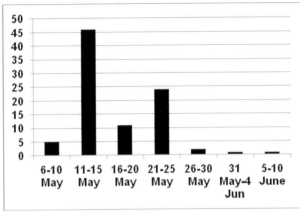

Bluethroats in County Durham, by five-day periods in spring, 1970-2010

All autumn records have fallen between 23 August and 16 November. Sub-specific identity is difficult in autumn birds, but Red-spotted Bluethroat is the expected visitor and there have been no birds giving reason to suspect White-spotted. The majority of the 17 recent autumn sightings have come in the middle two weeks of September and have been associated with falls of north European migrants, with just two birds seen in late August and four in early October. The latest sighting for the county was of a first year male found dead at Dawdon Business Park, Seaham on 16 November 2011, during a period of onshore winds and rain which also produced late records of Nightingale and Red-flanked Bluetail. The well-watched gardens and parks of Hartlepool Headland have attracted most autumn birds, with six being found there, with four at Marsden/Cleadon Hills, two at both North Gare and Hawthorn Quarry and one at Hart Warren. Although the rate of spring migrants has plummeted in recent years, autumn sightings have remained steady with no more than two in any one year, and no more than four in any of the five-year periods since 1970. However, there were 28 autumns without sightings between 1970 and 2011.

Inland sightings are unusual, although during spring arrivals at the coast, some must surely escape detection. The only records more than 2km away from the coast in recent years have been at Shibdon Pond over 12th to 15 September 1989, Kibblesworth on 9 May 1993, Thornhill School, Sunderland, on 8 June 1994 and Portrack Marsh on 23rd to 24 May 1995.

White-spotted Bluethroat
Luscinia svecica cyanecula

There has been just one confirmed record of the white-spotted form of Bluethroat *L. s. cyanecula*. A male was found by Graham Bell in reed and scrub on the edge of Dorman's Pool on 7 April 1983, where it stayed until the following day. However, female Bluethroats at Marsden on 25 March 1957 and Whitburn on 4 April 1958 were not attributed to *cyanecula* at the time, but from the very early dates and what is now known of Bluethroat occurrence patterns, it is likely that both were of this subspecies.

Distribution & movements

Red-spotted Bluethroats have a widespread breeding range extending from Norway, east across Asia to easternmost Siberia and north western Alaska. The western population spend the winter in the Mediterranean basin south to the northern Afro-tropics. It is a regular and at times numerous migrant in late spring, particularly to the Northern Isles, with lesser numbers seen during the autumn.

The breeding range of White-spotted Bluethroat is fragmented between northern Spain and Denmark and across central Europe. The wintering grounds are poorly known but include the Mediterranean basin and North Africa. Although an infrequent but regular early spring visitor to the south east of England, sightings of this form of Bluethroat are distinctly rare in the north of England.

Bluethroat, Hartlepool, October 2009, (J. Duffie)

Red-breasted Flycatcher
Ficedula parva

A very scarce migrant from eastern Europe, recorded in May, and from August to November, most frequently in September and October.

Historical Review

The first Red-breasted Flycatcher for County Durham was found by J. R. Crawford at Whitburn on 24 October 1947, during *"a day of continuous drizzle, a cold east wind and rough seas"*. Perched in the observation slits of a war time pill box situated just above the tide line and frequently engaging in fly-catching sallies, it was described as being a *"bird of the year"* (Temperley 1951).

The second record came almost ten years later when an adult male was at Cleadon on 11 November 1956. This is the latest record of this species for County Durham, and one of only two November records in the county to date. Just two more were recorded during the 1950's, single first-winter birds at Hartlepool Headland on 2nd and 6 October 1959, which were considered to be different birds (Grey 1960).

A further nine birds were recorded during the 1960s with sightings in six out of the ten years but none in 1964, 1966, 1968 and 1969. The site with most records was Hartlepool Headland with six, including two birds in both 1960 and 1961. Elsewhere, singles were at South Shields on 12th and 13 October 1962, at Marsden Hall on 21 September 1963 and at Graythorp on 23 October 1965.

All of the records during the 1960s were from September and October, with extreme arrival dates being 17 September 1967 and 23 October 1965 respectively.

Recent status

A total of 101 Red-breasted Flycatchers were recorded in County Durham between 1970 and 2010, and the species is best regarded as a scarce migrant in both spring and autumn, though autumn records predominate by far. As can be seen from the graph, it has been of almost annual in occurrence since 1970, with just six blank years since then. The vast majority of records come from the coast between South Shields and Teesmouth; only two birds have strayed inland. Most, if not all, of the records are associated with periods of easterly winds and rain, ideal conditions for 'drift' migration. The best year was 2001 when a total of 11 birds were recorded, though six were noted in both 1976 and 1996.

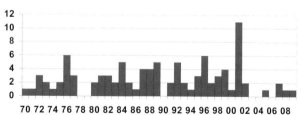

Records of Red-breasted Flycatchers in County Durham, 1970-2009

Just seven spring Red-breasted Flycatchers have been recorded in County Durham, all of these in the period from 9th to 30 May. Almost all of the records relate to first-summer or female birds with two being identified as males; only one was considered to be an adult.

The first spring record was of a first-summer/female in a Whitburn garden on 20 May 1986 which unusually was associated with a predominantly south westerly air stream and few other migrants. The six remaining records came from Whitburn on 9 May 1988 and 30 May 1992, Marsden Hall and nearby Marsden Quarry on 30 May 1992 (adult male), 11th and 12 May 1998 and 12 May 1999 and Hartlepool Headland on 11 May 2001. The two birds in close proximity in 1992 were found during a significant arrival of typical Scandinavian spring migrants whilst, in direct contrast, the only spring record of the species for Hartlepool Headland was one of few migrants recorded that day.

At least 94 Red-breasted Flycatchers have been recorded during autumn, the vast majority of which have been in September and October. All but two have been on the coast between South Shields and Seaton Carew, though records are centred on two distinct areas. The coastline between South Shields and Sunderland has recorded a total of 47 birds, whilst further south the well-watched Hartlepool/Seaton Carew area has hosted a further 42, with Hartlepool Headland attracting a total of 31. The coastline between Sunderland and Hartlepool has in contrast attracted just three birds at Dawdon Blast Beach on 12 October 1991, Crimdon Dene on 25 September 2001 and Hawthorn Dene on 9 October 2010. This reflects the far lower levels of observer coverage in this area.

The number of individuals recorded has remained fairly static throughout the review period with the forty-year average being of around two to three birds each autumn.

A single bird at Hartlepool Headland on 23 August 1970 is the earliest to be recorded in County Durham and there has been just one other August record, again from Hartlepool Headland on 30 August 1972 (McAndrew 1972, Coulson 1973).

With a total of 52 birds, September is the peak month of arrival for this species, with almost two-thirds of the records confined to the last ten days of the month. The largest arrival to date involved an influx of nine birds over 25th to 26 September 2001, as part of a large fall of common migrants following a period of strong onshore winds and rain. During this period the total of at least five at Hartlepool Headland on 25 September is the largest daily single site count of this species in the county to date. Other multiple records during September are of two on Hartlepool Headland on 27 September 1972, two which alighted on a fishing boat off Sunderland on 17 September 1976 and two in Holy Trinity Churchyard, Seaton Carew on 12 September 1989.

A further 39 have been recorded in October though records are scarce after mid-month with just four birds during the last week of the month, the latest of which was at South Shields from 29th to 30 October 1989. Other noteworthy records during the period include another bird which alighted on a fishing boat off Sunderland on 20 October 1976 and two at Hartlepool Headland from 2nd until 4 October 1987.

An individual at West Hall, Whitburn on 9th and 10 November 1984 is the latest to be recorded in the county and is the only November record to date (Baldridge 1985). It occurred during an exceptionally late run of other passerine migrants. Most autumn reports relate to first-winter birds with just three records of adult males since 1970: at Hartlepool Headland from 12th to 15 September 1973 and 25 September 2001, and at Holy Trinity Churchyard, Seaton Carew on 12 September 1989.

Only two birds were recorded inland during the period. These were at Jarrow Cemetery on 6 September 1975 and in a Norton garden on 4 October 1985.

Distribution & movements

Red-breasted Flycatcher breeds from central Europe to western Siberia, north to southern Scandinavia and south eastwards to Turkey and Iran. The main wintering grounds are well to the south of the breeding range in southern Asia, chiefly in Pakistan and the Indian subcontinent. The Red-breasted Flycatcher is a regular but scarce migrant to Britain in both spring and autumn, though autumn records predominate. Spring records are from mid-April until June with records having a distinctly easterly bias. Autumn records are more widespread, most being in the period from late August until mid-October. Around 100 birds are recorded in Britain each year, though almost double that number occurred in 1984 with an annual total of 196 birds (Fraser & Rogers 2006).

Just four Red-breasted Flycatchers have been trapped and ringed in County Durham, three of which were at Hartlepool Headland. A first-winter was ringed there on 20 September 1961, a first summer male on 11 May 2001 and an adult male on 25 September 2001. The other bird was a first-winter ringed at Whitburn Coastal Park on 9 October 2010 (Newsome 2012). They have generated no controls or recoveries.

Pied Flycatcher (European Pied Flycatcher)
Ficedula hypoleuca

A locally distributed, scarce summer visitor as a breeding species and an uncommon passage migrant.

Historical review

Temperley (1951) reported that Pied Flycatcher was a "*summer resident and passage migrant*", and that it was more local in its distribution than the Spotted Flycatcher *Muscicapa striata*, and a "*rare bird in the County*",

814

presumably he meant in terms of the number of breeding birds. Strange then to realise that the first breeding record for the county was in Axwell Park, not far from Blaydon, as long ago as 1801 (Bowey *et al.* 1993), discovered by no lesser an observer than Thomas Bewick. In the 19th century, Selby (1831) considered that it was rare in Northumberland and Durham whilst Hogg (1845) thought it scarce in his area of documentation, the south east of the county. Some years later, Hancock (1874) believed that it was, "*not by any means common in Northumberland and Durham*", though he did refer to a nest taken at Stella Park, to the west of Blaydon upon Tyne, which was found "*about 1874*". The species was, about that time, also recorded as nesting at Wolsingham, in June 1876, and at St. John's Chapel higher up the Wear valley, at the same time. In 1885, J. Backhouse stated that it was only noted as nesting occasionally in the lower "*more cultivated*" parts of upper Teesdale but it was by no means common. It was not noted by Abel Chapman in either Northumberland or Durham until 1885, when a male was found at Silksworth, Sunderland on 7 May, presumably a spring passage bird. In 1885 it was also recorded breeding at both Winlaton and Ravensworth Park, in the lower Derwent and Team valleys, as well as higher up the Derwent valley at Shotley Bridge (The Naturalist 1887).

In contrast to some of the above commentators, Tristram (1905) thought it not "*so rare as generally supposed*"; and it was felt by him to be breeding in several parts of the county. Clearly a process of colonisation and consolidation of this species in Durham was occurring over the later part of the 19th century. By 1905, Tristram could state that the species was well-established in the Derwent valley although rare in the Team valley (Bowey *et al.* 1993). By the last quarter of the 19th century, the species was breeding regularly in both Durham and Northumberland (Holloway 1996).

By 1951, Temperley recorded it as "*well-established*" in the Derwent valley, in Weardale, the Browney valley and in mid-Teesdale, though it was rare in the neighbourhood of Darlington and in the Team valley. He also documented it as a passage migrant in spring, during April and May, and from August into September, in autumn, but only in small numbers. He highlighted the occurrence of one in Westoe Village, South Shields, at the end of August 1949 (Temperley 1951) with up to five being present in 1952 (Temperley 1953).

In the early 1940s, a pioneering nestbox scheme for Pied Flycatchers was established by James Alder, George Temperley and Fred Grey in Hamsterley Forest, with the Forestry Commission's blessing, in an attempt to control the numbers of harmful insect pests. At this site, there were seven pairs in 1944, 17 pairs in 1945, at least 30 pairs in 1946 and 38 in 1947. Of the 300 boxes present in 1949, 87 of these were occupied by Pied Flycatchers; all of these '*away from the thickest plantations*' (Temperley 1951).

There is little documentation of this species' status in the county through the 1950s and 1960s, though it is known that ringing work on this species continued through these decades in Hamsterley Forest during the early 1960s; 63 young were ringed from 14 nestboxes in 1963 and 73 young from 16 boxes in 1964.

Some more detail emerged in relation to this species through the 1970s, with the more systematic recording that was associated with the early work of the Durham Bird Club. Nonetheless, breeding data was still relatively sparse at this time. Autumn influxes of varying magnitude were regularly recorded throughout the decade, from late August to late September. Some of the more intriguing records that arose during the period included the 20 birds that were in Marsden Quarry on 24 August 1971, a typical early autumn fall on a typical date, setting the trend for the decade. In relation to breeding populations, it was said to be "*quite common*" in the Bedburn and Hamsterley Forest areas in 1972 and, in 1977, 12 males were found along a one-mile stretch of the River Greta. One autumn bird at Washington on 24 October 1972, was the latest record of the species during the decade. The first bird of the year in 1976 was not even seen in Durham, but on board a fishing boat off Sunderland on 6 May (Unwin 1977).

Recent breeding status

The majority of the British breeding population of Pied Flycatchers can be found in the western half of the country, from Devon in the south west peninsula, north through Wales, via Cumbria, Dumfries and Galloway, northwards into Scotland, with the areas around Loch Lomond and the Trossachs also holding good numbers (Gibbons *et al.* 1993). Smaller, often scattered populations can be found from the west-Midlands, along the foothills of the full length of the Pennines and the North York Moors and those in Durham fall into this category.

It occurs mainly, and rather sparsely, in the middle to upper reaches of the main river systems, and along their tributaries, largely on land falling between 100m and 300m above sea level, in the western portion of the county, and largely to the west of the A68 trunk road. They are effectively absent as breeding birds in the eastern, lowland parts of County Durham. Consequently, the female found dead in a nestbox containing a clutch of Great Tit *Parus major* eggs at Crookfoot Reservoir in May 1985, was of interest in this respect (Blick 2009). The reasons for the species' absence from apparently suitable habitat at lower elevations are not clear, but may relate to competition for nest sites with other resident, hole-nesting, woodland species that have already taken up occupancy of the best nest sites, prior to the migrant flycatchers' arrival. Birds usually arrive back in breeding territory around the middle to latter part of April, with further arrivals occurring on the breeding grounds through early May. It is believed that the westerly distribution nationally largely results from the species' requirement for open, mature woodlands with a sparse shrub layer and a good supply of naturally occurring tree holes for nesting (Lundberg & Alatalo 1992). Such characteristics are found extensively in the county's western sessile oak woodlands, and these habitats are concentrated along watercourses, thereby determining the species' local distribution.

The distribution map published in the *Atlas* gave a good picture of the species' broad breeding distribution in the county, though inevitably with such mapping exercises, some of the local detail was lost (Westerberg & Bowey 2000). A snapshot of the species' breeding distribution in 2007 confirmed the *Atlas'* overall picture with breeding birds noted in western woodlands at Muggleswick and Burnhope Wood in the Derwent watershed; Brignall Banks, Tees Bank Woods and Deepdale in Teesdale; and in Dryderdale, Stanhope Dene and around Tunstall Reservoir in the Wear valley (Newsome 2008).

The readiness with which Pied Flycatchers take to nestboxes is well documented and this can result in substantial increases in their local numbers (Lundberg & Alatalo1992). It is believed that the widespread provision of such boxes in some of the county's woodlands has been a significant driving force for an increase in both absolute numbers, breeding density and, from a national perspective, range-expansion (Westerberg & Bowey 2000). In Durham, birds have been attracted into locations that might not be otherwise attractive to them by the provision of boxes, for example into parts of the largely coniferous Hamsterley Forest, mainly where more open, broadleaved trees occur along the streamside (Westerberg & Bowey 2000). Remarkably, high breeding densities of birds have been documented in some areas where nestboxes have been provided and adopted by the species. Densities of up to 200 pairs per occupied ten-kilometre square are documented (Gibbons *et al.* 1993) and in such situations, the species is effectively 'semi-colonial'. It is in such situations that there has been an increased occurrence of the species' well-documented polygynous breeding behaviour (Lundberg & Alatalo 1992).

Despite the male Pied Flycatcher's black and white plumage and its distinctive song, the species is not always easily detected, especially where it relies principally on natural nesting sites in relatively isolated woodlands. This effect is compounded by the fact that after birds have 'paired', the males cease to sing and spend much of their time foraging in the canopy of the woodlands (Lundberg & Alatalo 1992) and at local sites, breeding densities are typically rather low. Also, once the young have fledged, mostly in June, the species becomes virtually invisible. Pied Flycatchers may, therefore, be under-recorded, for example in the Wear valley above Wolsingham and the Tees above Eggleston (Westerberg & Bowey 2000). The breeding schedule of this species locally is indicated by observations at one nesting site in the south of the county, Gill Beck at Barningham, where in 2000, the first eggs were noted on 4 May and the fledged juveniles at this site were observed on 18 June (J. Strowger pers. obs.).

The areas of the county in which the species utilises nestbox schemes, give a strong indication of the numbers in the prime areas, and these provide a clue as to how to determine the nesting densities and population numbers across the whole of the species' breeding range in Durham (Westerberg & Bowey 2000). It was by using such a method that the *Atlas* estimated the breeding population of Durham to lie between 100 and 150 pairs. National census data across the UK since around 1994, however, indicate that there was a steep decline in numbers, of around 34%, leading to the species' Amber listing on the schedule of Birds of Conservation Concern in the UK (Parkin & Knox 2010). Local observations, however, do not always concur with this observed national trend. The longest running nestbox scheme in the county is operated in upper Deepdale and at Brignall Banks near Barnard Castle in the south west of the county. The scheme commenced in 1978 and was still running at 2010. The number of occupied boxes in upper Deepdale peaked in 1989 at 17, and at Brignall Banks in 1990 at 33. At both sites there has been a decline and in 2010 each had only five occupied boxes. Another nestbox scheme operating in Gill Beck, near Barningham, was started in 1997 and at the time of writing was still running. During 1997-2007, there were seven pairs in three years and eight pairs in one year. In the three years 2008-2010, eight pairs bred in 2008, nine pairs in each of 2009 and 2010. These sites are not far apart, in particular Brignall Banks and Gill Beck, but they are clearly experiencing different fortunes. Historically, from the 1950s, there seems to have been a frequent falling off in numbers at nestbox sites with time (J. Wood pers. comm.), possibly this effect is now being observed at

Brignall Banks in the River Greta. A newly refurbished nestbox scheme, near Edmundbyers, attracted seven pairs in 2010 (J. Strowger pers. obs.).

During spring 2008, the species was documented as being present in 32 different one-kilometre squares, all occupied sites being in mature woodland along the main river valleys and their tributaries. Today, core areas for breeding birds include: the Derwent valley, sites such as Muggleswick, Edmundbyers, and higher up the dale Nookton Burn; the Wear valley and its tributaries, with birds in suitable woodlands around Wolsingham, Tunstall Reservoir and across the extensive woodland complex of Hamsterley Forest; and in the Tees valley woodlands around Barnard Castle, down along the River Greta and south to Brignall Banks and the Gill Beck, Barningham as well as up the Tees valley to High Force and beyond. Birds are occasionally noted away from these principal areas and in 2009 breeding birds were present at Pontburn Woods, a former stronghold during the 1980s and, in 2004, a pair was discovered at Greencroft Park near Annfield Plain, the first recorded there in 20 years (Siggens 2006).

It has been suggested that the pattern of casual reports may seriously under-estimate the strength of the county's breeding population, but this needs to be balanced against the fact that there is some indication that the species has declined since the publication of the *Atlas*, and that it is known to have been lost locally from some previously occupied sites. So the above population estimate probably remains a reasonable working figure, until more detailed information supplants it.

Recent non-breeding status

Migrant Pied Flycatchers can appear at widely scattered localities in both spring and autumn, although they are most common along the coastal strip. Birds are sometimes seen along the coast in spring, albeit in relatively small numbers as spring passage tends to be light in Durham and birds are not recorded in every year in such situations. Inland passage birds, probably representing local movements of British breeding birds, usually occur between the last week of April and the third week of May. Autumn movements are usually much more evident along the Durham coastline, mainly through August and into the first half of September. Even at this time of the year, their occurrence pattern, and the number reported, is very much dependent on the prevailing weather conditions. Those on the coast frequently arrive in 'drift conditions' possibly indicating a continental origin. When the weather conditions produce 'falls' of migrants, usually about 25 to 30 birds per annum might be recorded between early August and early October, though numbers can be much higher in the best of falls, as on 17th and 18 September 1969, when at least 100 were recorded around Hartlepool and Teesmouth (Blick 2009), with about 100 along the county's coastline on 8 September 1995 (Armstrong 1997).

There has been some suggestion from analyses of autumn migrant reports over the last four decades that the species is now less regularly noted at the coast than formerly. This appears to be borne out with some evidence in relation to small numbers, i.e. one to two birds, which occurred more frequently during the autumn in the past. More significant passage however, admittedly weather-influenced, shows little change over time. In the last decade there have been some falls significantly larger than those between 1980 and 1999. In 2002, during 9th to 10 September, 33 were at Hartlepool and 14 were in the Marsden/Whitburn area. In 2004, over 20 were at Hartlepool on 9 August and on the following day 32 were at Whitburn. Up to 11 were at South Shields/Whitburn and Hartlepool during 18-21 August 2006, with c.20 at Hartlepool over 20 days. In the same year, a bird was seen in a Durham City garden on 11 September and in autumn 2007, unusually two inland migrants were noted in September, one at Hetton Bogs on 4th and another at Hetton Lyons on 12th. During 2008, 40 were between South Shields and Seaburn and 30 were at Hartlepool. Other locations to attract autumn passage birds through the years include Bearpark, Castleside, Cornsay Colliery, Friars Goose, Hetton Bogs, Hetton Lyons, Malton and Shibdon Pond. Rarely, a few late passage birds are noted at coastal locations into early October (Newsome 2009). The latest autumn sighting in the county was of a bird on South Shields Leas on 27 October 2011.

Over thirty years between 1976 and 2005, the reported first-arrival dates of Pied Flycatchers were analysed, in ten-year grouping, giving mean first arrival dates of: 1976-1985, 28 April, 1986-1995, 22 April; and 1996-2005, 24 April. It was therefore found that over the period, the mean first-arrival date had advanced by an average of four days, but there was much variation in the actual date from year to year, and between ten-year periods (Siggens 2005). Typical earliest recorded dates in Durham during the first decade of the 21st century fell between 15th and 23 April, but males at Low Barns on 10 April 2008 and 8 April 2011 proved to be the earliest ever for the county (Newsome 2009, Newsome in prep.).

Distribution & movements

Pied Flycatchers are birds of open woodland habitats and they breed in western, central and northern Europe, with some populations in northern West Africa and to the east, Siberia. The majority of the British breeding population of Pied Flycatchers can be found in the western half of the country. Wintering birds are to be found south of the Sahara in Africa, with large numbers in West Africa, north of the Gulf of Guinea.

The origin or destination of the birds observed along the Durham coast is not known, though it seems likely that many autumn birds may have crossed the North Sea from Scandinavia when appropriate weather conditions prevail, although some birds may be moving to or from their British breeding areas to the north (Blick 2009). This theory is borne out by the record of a bird that was ringed as a pullus in Aust-Agder, Norway in June 2007 and was at Hartlepool on 25 May 2008. Onward movement and the route taken by locally bred birds was indicated by one ringed as a nestling on 19 June 1964 at Hamsterley Forest, which was found dead on 30 August 1964 at Cabo Espichel, Estremadura, Portugal (Bell 1965).

That locally bred birds return to breed in the area was highlighted by one ringed on 18 June 1927 at Hamsterley and recovered at Witton-le-Eggleston, near Darlington on 25 May 1928, and one ringed in June 1985 at Barnard Castle, Durham, which was controlled near Loftus on 5 May 1986.

Due to the large number of Pied Flycatcher pulli ringed, locally and nationally, there have been many controls reported, where birds are recaptured by other ringers. There has been considerable interchange between local populations based at: Barningham, Brignall, Hamsterley, Lartington and Burnhope near Edmundbyers. One ringed as pullus in a nestbox at Tunstall Reservoir on 16 June 2002, was controlled as a breeding female in a nestbox at Low Crag near Lartington on 6 June 2004, illustrating this pattern of exchange of birds between woodlands in the county's western dales.

There has also been movement of birds between County Durham sites and those in Northumberland and North Yorkshire. There is a series of records showing movements of birds from: south west Scotland to Durham (a bird ringed in 1991 at Sanquer, Dumfries and Galloway was controlled at Healyfield in 1993); from Durham to Wales (one ringed on 12 June 1990 at Hamsterley, controlled in Clwyd on 31 May 1992, and another ringed on 16 June 1993 near Consett, and controlled in Clwyd in May 1994); and, to the Welsh borders (one ringed on 1 June 1999 at Brignall and controlled in Shropshire on 29 May 2000). In these, and the majority of other instances, the date of control was in the year immediately following ringing.

An interesting post-breeding record, suggesting a westward route south in autumn, was of one ringed as a pullus at Barningham in June 1999 and controlled at Little Crosthwaite, Cumbria on 1 August 1999. There have been several international records including: one ringed at Hamsterley Forest on 16 June 1957 and recovered at Safi in Morocco on 24 September 1959, which was the first British-ringed Pied Flycatcher to be recovered in Africa; one ringed on 13 June 1992 as a pullus at Hamsterley and recovered at Al Hoceima in Morocco on 16 April 1993; one ringed Hamsterley Forest 16 June 1962 which was recovered at La Guardia, Pontevedra, Spain on 8 September 1962; a bird ringed in Hamsterley Forest on 16 June 1965 and recovered between Astudillo and Fromista, Palencia, Spain on 23 August 1966; and, unusually, one ringed in north west Spain on 24 September 2000 was controlled at Barningham 22 May 2002. A particularly interesting sequence of records occurred in the 1980s and the early 1990s, when six Welsh-ringed pulli were controlled as breeding birds in Teesdale (J. Wood pers. comm.). A bird ringed as a nestling on 19 June 1983 at Lartington was subsequently re-trapped and identified as a breeding female at the same place between 1986 and 1990, making this bird at least seven years old by the time of its last handling.

Black Redstart
Phoenicurus ochruros

An uncommon passage migrant, occasional wintering bird and a very rare breeding species.

Historical review

Prior to Selby (1831), there were no known records of Black Redstart in Northumberland or Durham. Hancock (1874) thought it extremely rare as a winter visitor, and he listed no records of passage or wintering birds for Durham, though he documented an astonishing and historically significant breeding record.

In 1923, A. Coward found Black Redstart breeding on the Sussex coast, in what was then thought to be the first British record. In 1916, seven years before this, F. C. R. Jourdain had written an article in *The Zoologist* on the Black Redstart's status as a breeding species in England. In that article, he reviewed all claims of breeding to that point and showed that in every case an error of identification had been made, "*with the possible exception of Hancock's ... which however requires confirmation before it can possibly be accepted*" (Temperley 1946). What he was making reference to was John Hancock's documentation of the first breeding record of this species for the British Isles.

In his *Catalogue of the Birds of Northumberland and Durham* (1874) John Hancock described it as follows: "*In 1845, a pair* [of Black Redstarts] *nested in the garden of the late Rev. James Raine, the historian of Durham, in that City; and I am indebted to Mr. Wm. Proctor for their nest, which is now in my collection. An egg belonging to it was kindly presented to me by the Rev. James Raine, son of the above named gentleman.*"

Temperley (1951) told the story. William Proctor, who gave the nest to Hancock, was the curator and taxidermist of the Durham University Museum. He himself had already published the record for, in a "List of Birds found in the County," that was published in 1846 as an Appendix to the Rev. G. Ornsby's Sketches of Durham, Proctor had written of the Black Redstart, "*Very rare; a nest with five eggs was taken near Crook Hall in the summer of 1845.*" Temperley drew attention to the fact that this account was written just one year after the event and twenty-eight years before the publication of Hancock's Catalogue. Proctor's account gave more detail. He went on to say, "*In the year 1845 a pair* [of Black Redstarts] *built their nest on a cherry tree trained on a wall in the garden of the Rev. Dr. Raine, at Crook Hall, in the suburbs of Durham City. I regret to say that the birds were shot. The male is in the Durham Museum; the nest and an egg were given to the late John Hancock.*" Crook Hall lies on the north bank of the Wear, downstream of Framwellgate Bridge. As Temperley (1951) recounted, the nest and extant egg had then, recently (in February 1945) been sent to the British Museum (Natural History) to be critically examined. The Hon. Guy Charteris, who saw them, considered that they were undoubtedly those of the Black Redstart. His comment on the nest was that it could not be that of any other bird.

Other records up to Temperley's time of publication and summarised by him, included one that was shot on the north bank of the Tees on 28 October 1903, one shot at South Shields on 7 November 1926, a male seen by Temperley at Seaburn from 17 February to 18 March 1935 and a female on 23 February 1942 seen in a garden in Seaburn, not far from the previous record. A female was around the cliffs at Marsden from 20 January to 1 February 1945, with a bird at the same place in March 1946, from 16th to 20th, and again from 10th to 14 February 1947, and once more at Marsden from 26th to 30 January 1948. This was presumably the same bird coming back to winter in the same area year-on-year.

In 1951, Temperley summarised the species' status in Durham by saying that it was "*Once a rare vagrant, now a regular winter visitor*", which contrasts, through an interesting shift in status, with the situation in modern times, when it would be more correctly considered largely a passage migrant.

Through the 1950s it remained an intermittent, scarce visitor to Durham. One was on the coast at South Shields on 20 March 1954, with the most intriguing record being of a pair at Marsden Bay in spring 1957. The male was heard singing on May 5th and the birds were present in the area for at least a fortnight. It is therefore possible that a breeding attempt took place (Grey 1958).

The occasional scale of spring passage was illustrated by the fact that in spring 1958, at Teesmouth and Marsden, birds were seen on almost every day between 29 March and 9 April, with four together at Marsden Quarry on 1 April. The following year, three spring migrants were in the Hartlepool area in the last two days of March and early April. Rather intriguingly, an adult was inland at Durham City on 8 July, a most unusual time and location, raising the tentative thought that local breeding might have been attempted just over one hundred years after the nation's first example of this, and in the same area (Grey 1960).

Through the 1960s the species remained a reasonably regular but scarce migrant, year totals rarely being out of single figures, with the balance of spring migrants to autumn weighted towards spring. Occasionally birds appeared in out-of-the-way areas, for example one at Fairfield, Stockton on 7th to 11 April 1963. A small influx took place along the Durham coast on 26 March 1964 and this included a bird that lingered at Hartlepool until 17 April. It became scarcer through the middle of the decade, particularly in 1965 to 1967.

Recent status

The 1970s saw an increase in the number of annual passage birds from three in 1970/1971 to 11 in 1975 and 10 in both 1976 and 1978, with a continuing bias towards spring dates. This increase continued into the 1980s, apart from 1981 when only one was seen. The annual average was 9.4 birds with a peak of 20 in 1987. Once again, the great majority occurred in spring although, in 1988 and 1989, autumn predominated with 12 and 8 birds respectively.

In the 1990s, the annual average rose slightly to 9.8 birds per annum with a maximum of 22 in 1990, of which 19 were in the autumn, although over the whole decade spring continued to dominate. Migrants are largely concentrated in favoured locations along the coast. Most occur between the months of late March and mid-May in spring and in the autumn, between mid-October and early November.

Passage in the 2000s, in the main, continued the pattern established in previous decades, although overall numbers were slightly down, with an annual maximum count of up to 13 birds in 2008 (Newsome 2009).

Whilst most of the county's records concern birds along the coastal strip and at coastal watch points during passage periods, winter records have become more common. The majority of the birds observed are in immature or female plumage, although occasional adult males are noted.

Overwintering is not a new phenomenon and does not occur every year but has been noted more regularly in recent decades. The Marsden/Whitburn area is the most favoured locality with birds recorded there in 1935, 1942, 1945/1948, 1985, 1986/1987, 1995/1996, 1999, 2004, 2005 and 2007. Other frequented winter locations have included WWT Washington in 1982, Sunderland Docks in 1993, Hartlepool Old Cemetery in 1993, the Team Valley in 1995 and Newton Aycliffe Industrial Estate in 2000.

Inland records are by no means unheard of and these have become somewhat more prevalent over the last three decades. Away from the coast, a bird was at Billingham in May 1984 and 1986, and in April 1989. There were two inland Gateshead records in spring, in Clockburn Dene in April, and at Shibdon Pond in May (Bowey *et al.* 1993). Other inland records include singles at Crookfoot Reservoir in April 1993, Hetton Lyons in April 2005, Longnewton Reservoir in August 2006 and Eppleton in October 2009.

There has been some suggestion of an inland breeding presence in recent years. This has been stimulated by a series of records that would have previously been considered unusual in terms of timing and locations, and these have occurred over a relatively short space of time. Breeding has also been confirmed as having taken place in Northumberland as early as the 1960s and again in the 1970s (Kerr 2001). The possibility of local breeding was hinted at in the *Atlas* (Westerberg & Bowey 2000) which indicated that there was tantalisingly sparse, teasing evidence that there was some kind of 'breeding presence' of this nationally rare species in the county. In Durham during the *Atlas* period, the only suggestion of breeding activity that did not relate to what could be ascribed to the usual pattern of this scarce passage migrant, was of a male, in potential nesting habitat at the coast, in Trow Quarry, on 17 June 1991 (Armstrong 1988-1993). Then in 2007, a surprise came mid-summer when a first-summer male was found feeding in a suburban garden close to West Park, South Shields, on 1 August. This recalled records of singles at Souter Lighthouse from 22 July into August 2004, and then at Trow Quarry on 26 July 2006 (Siggens 2006, Newsome 2007).

The records of birds noted in Teesdale in the autumns of 2001 and 2006, at Bollihope in July 2008, Hedleyhope Fell in April 2009 and near Hamsterley Forest in the summer of 2009, all in or close to suitable breeding habitat, are also strongly suggestive of breeding. In 2010, an outstanding run of inland records from Waskerley Reservoir and Chapman's Well in April, Eppleton in May and near Hamsterley Forest in mid-June strengthened this possibility (Newsome 2012). This situation was further consolidated in 2011 when summer sightings came from Whitwell in July and on moorland edge above Stanhope in August. Whilst a coastal breeding pair would probably be noted, given the large number of coast-based observers, the county has huge amounts of active and disused industrial land along the Tyne and inland, which, together with the many quarries, provide ideal nesting sites. With the problems of surveying such sites, it is not inconceivable that the events of 1845 have been repeated and breeding may have recently taken place in Durham.

Distribution & movements

Black Redstarts breed across much of central and southern Europe, particularly in alpine and mountainous areas, with the eastern races extending east into the Caucasus, Asia Minor, the northern Middle East, northern

Afghanistan and eastwards. Small numbers of breeding birds penetrate as far north as England. The Rare Breeding Bird Panel (Holling 2007) reported a ten-year mean of 67 pairs in the UK, a decrease from the more than 100 pairs reported in the 1980s. The species is migratory with the largest numbers wintering in Africa, but some European birds do remain much further north within the breeding range during winter, with small numbers present in England in most years.

Over the thirty years between 1976 and 2005, the reported first arrival dates of Black Redstart in Durham were analysed, in ten-year groupings, and the mean first arrival date was found not to have changed significantly, being in 1976-1985, 31 March; 1986-1995, 7 April; and 1996-2005, 1 April. These dates might be somewhat confused by the possibility that some 'early migrants' might actually be undiscovered wintering birds that have made local relocations (Siggens 2005). The highest count at one site in the county is of seven birds, which were at Hartlepool Headland on 23 October 2004 (Blick 2009) although up to 16 birds were between Marsden and Sunderland on 15th and 19 April 1986 (Baldridge 1987).

Redstart (Common Redstart)
Phoenicurus phoenicurus

Sponsored by
Wm. C. Harvey Limited

A fairly common summer visitor, but a declining passage migrant.

Historical review

This species has clearly been present in the county for many centuries, as demonstrated by its excavation from the mediaeval strata of Barnard Castle (Yalden & Albarella 2009); one can visualise its breeding in the woodlands of the upper Tees valley at that time, in the shadow of the great Norman castle, just as it does in the 21st century. Nelson (1907) credited its first reference in Yorkshire to Marmaduke Tunstall, from birds at Wycliffe on Tees, now part of County Durham. At the end of the 19th century it bred in every mainland county, including Durham (Holloway 1996), though it was suggested that the species had experienced some fluctuations in numbers through that century.

Temperley (1951) commented upon its presence higher up in the Durham dales during the first half of the 20th century. By way of example he noted that it could be found breeding between Middleton-in Teesdale and High Force. A century beforehand, in the early part of the 19th century, Backhouse (1834) noted its breeding at elevations of up to 1050 feet (318m) above sea level in Teesdale. In the north west of the county, Robson considered this species common in the Derwent valley in the late 19th century (Robson 1896). So clearly the species has been widespread, at least in the west of the county, as a breeding species in Durham for some time.

Around the middle of the 20th century, Temperley (1951) said that it was "*a summer resident in suitable locations*". He thought it not uncommon in decaying deciduous trees, parkland and the sheltered woodland valleys of the western uplands. In 1944, it was noted as numerous but in 1946 he said that it was missing from many of its usual haunts, an early observation of the fluctuations in numbers that this species sometimes experiences. Indeed the Redstart has a history of such periodic fluctuations, which goes back into the early 19th century (Mead 2000). It was noted that pairs were frequently found breeding in unusual places, such as the pair which used a hole in a wall of a garden in Darlington Training College in 1943. Birds were recorded using nestboxes relatively soon after their first provision in any number, alongside Pied Flycatcher *Ficedula hypoleuca*, in Hamsterley Forest in the late 1930s and early 1940s.

At the mid-20th century Redstart could be observed fairly regularly on migration along the Durham coast, from August through to September, particularly at Teesmouth but also at South Shields. On 17 September 1960, easterly winds, with fog and low cloud, produced a huge fall of passerine migrants to the coastline, tha included more than 230 Redstarts between Crimdon Dene and the North Gare (Stead 1964) Some years later, the autumn of 1969 provided another good example of such phenomena when, during a large occurrence of migrants on the north east coast on 17 September, over 100 Redstarts were counted around Hartlepool Headland, with another 60 birds noted at nearby Seaton Carew (Blick 2009).

Through the 1970s a few pairs of Redstart could still be found nesting in coastal areas, most notably in Castle Eden Dene, which attracted up to five pairs to the end of the decade. During the mid-1970s it could be said that the species bred from High Force in the west to Castle Eden Dene in the east, tracing their way down the valleys and burns, as illustrated by the four pairs in one mile along the River Deerness near East Hedleyhope in May 1979 (Unwin& Sowerbutts 1980).

Recent breeding status

This species experienced major declines in numbers nationally and across Europe, because of droughts in their Sahel wintering grounds during the 1960s and 1980s, although it was not quite as severely affected as some other species (Gibbons *et al.* 1993). There is little doubt that Durham's population was likewise impacted upon. There appear to be significantly fewer birds in the county's valley woodlands today compared to the past, indicating some considerable change in status.

The *Atlas* map (Westerberg & Bowey 2000) gave a good indication of Redstart's breeding distribution at the level of the 10km square, in relation to the county 'footprint', but there were some areas, mainly in the western valley areas, such as the River Wear around and above Wolsingham, where the detailed representation of the species' presence was inadequately represented on the tetrad map.

As across Britain (Gibbons *et al.* 1993), in County Durham the Redstart as a breeding bird, very much has a westerly distribution. In Durham, all but a small number of the county's breeding pairs occur in the Pennine foothills, largely within the valleys of the three main river systems and along their smaller tributary streams. The greatest number of territories lie between the altitudinal limits of 100m and 300m above sea level although in some instances, small numbers of birds can be found nesting down to altitudes as low as 60m above sea level, for example in the complex of woods in the Derwent valley around the Pont Burn, though birds have declined here in the first decade of the 21st century. In some areas that might be expected to hold birds, such as the lower Derwent valley, there are few regular breeding sites, although the species has been known to nest in the Derwent Walk Country Park and in Chopwell Woods. It is only as the altitude rises that this species persists.

Occasionally pairs breed in genuinely lowland situations, such as the birds that bred at Elemore Hall in 2004 (Siggens 2006). Its sparsely distributed nature in the east of the county is well illustrated by an examination of its status as a breeding in bird in the south east of the county, around Teesside. In this portion of the county, the species has long been a scarce breeding bird. It was recorded breeding in the Wynyard Estate in 1981 but there is some suggestion of declines even since the early 1990s. Work on *The Breeding Birds of Cleveland* revealed just three pairs, which were widely scattered across the north Teesside area (Joynt *et al.* 2008).

Most birds usually begin appearing in their breeding areas during the third and fourth weeks of April, with sightings coming typically mainly from westerly locations, such as Blackton, Brignall Banks, Dryderdale, Hamsterley Forest, Stanhope Dene, Thwaite and Tunstall Reservoir. Almost all of the county's breeding birds are found to the west of the A68, although Low Barns and some sites at lower elevations occasionally hold a small number of breeding pairs. There are also occasional records, at the other end of the species' range, with breeding having been noted at up to 400m above sea level in the extreme west, for example at Cowshill in Weardale.

The mapped distribution in the *Atlas* reflected the availability of the species' preferred local habitat. This consists of mature, open broadleaved woodlands, often dominated by oak *Quercus petraea,* usually with a sparse shrub layer (Westerberg & Bowey 2000). Such conditions fulfil its requirements for nesting cavities and an abundant invertebrate food supply. Sometimes the species can be found along narrow strips or even single lines of trees, often along waterways, and hedgerows with significantly mature trees.

Redstarts will utilize suitable nesting cavities in a range of situations if in close proximity to their favoured habitats, including buildings and dry-stone walls in the absence of trees. This enables a few pairs to be supported around the meadows and pastures of the upper dales and even heather moorlands away from the valley woodlands. The species does not take to nestboxes quite as readily as a number of the titmice *Parus* spp. and Pied Flycatchers, *Ficedula hypoleuca.* Nonetheless, Redstarts will occupy both hole and open-fronted boxes mounted in suitable locations, such as at Burnhope Woods, Edmundbyers, in the Derwent valley, and Brignall Banks, along the River Greta, where the species has been breeding in such situations for decades. During the *Atlas* survey period, one pair was even noted successfully nesting in the corner of a box designed for Kestrels *Falco tinninculus,* which was situated overlooking Hury Reservoir in Baldersdale (Westerberg & Bowey 2000).

As the distribution of this species in the county is patchy, it is rather difficult to estimate the likely population with any degree of accuracy. During 2008, the species was reported in breeding situations in the west of the county from a total of 77 one-kilometre squares, with the greatest densities at Muggleswick in the Derwent valley, Dryderdale and around Tunstall Reservoir off the Wear valley. The furthest east that territories were recorded during that spring was at Lintzford and Spen Banks Wood in the Derwent valley. In 2010, territorial males were reported singing from 30 locations with seven in Deepdale Wood, six at Tunstall Reservoir, Dryderdale and Muggleswick, and five males at Spurlswood Beck and Wool-ho-wood near Barnard Castle (Newsome 2012).

Using the figures developed by Gibbons *et al.* (1993) for the national population as a basis for estimation, the *Atlas* formulated a maximum population estimate of between 250 and 300 pairs for County Durham (Westerberg & Bowey 2000). During in the last 30 years of 20th century the species underwent a general retreat from many lowland breeding areas in Britain (Mead 2000). If there has been any perceptible population trend over the first decade of the 21st century, then this has probably been marginally downwards and a current estimate closer to the lower figure of the *Atlas'* stated range might be appropriate. The BTO's Breeding Birds Survey index indicated a 59% decrease in numbers between 1994 and 2006, albeit from a low base (Raven *et al.* 2007).

Recent non-breeding status

Spring passage in Durham usually begins from the middle of April, and birds are sometimes in territory and singing by the middle of that month. Spring arriving birds now appear somewhat earlier than in the past, especially since the 1960s, at a national level, by an average of some seven days (Baillie *et al.* 2007). The dates for such observed activity have become somewhat earlier during the first decade of the 21st century, typical arrival dates over this period being between 14th and 21st. Spring birds were established in territory in the upper Derwent valley by 10 April in 2010, whilst others were noted at Dryburnside and Tunstall Reservoir around the same time, although the majority of birds were not back on territory until the final week of April (Newsome 2012). Over thirty years, between 1976 and 2005, the reported first arrival dates of Redstart were analysed, in ten-year grouping, giving mean arrival dates of: 1976-1985, 16 April; 1986-1995, 19 April; and 1996-2005, 16 April. Over this whole period, the mean first arrival date had not changed, but varied around a point in mid-April (Siggens 2005). A bird in Sunderland Docks on 9 April 1995 was at that time, the earliest in Durham since 1981. The earliest ever records for the county to date were on 3 April 1976 at Marsden, a coastal passage migrant, and a bird on territory at Hamsterley Forest, on the same date in 1981 (Unwin 1977, Baldridge 1982).

Small spring falls used to occasionally occur, usually in May, such as in the years 1985 and 1993, but this is a phenomenon that has been observed somewhat less frequently in recent decades. In recent year's coastal and inland lowland, spring migrants have become increasingly scarce and in most recent years, there have been just a handful of coastal migrants seen during spring.

Inland, post-breeding birds occasionally join tit flocks and juveniles may be reported from unusual locations, for example gardens in the Winlaton area of Gateshead (Bowey *et al.* 1993). Autumn passage has traditionally been much more evident than spring passage along the Durham coast and usually occurred from the last week of August through into early or mid-September, with occasional birds noted into early October, such as the small influx that occurred on 3 October 1998, when 15 were noted at Hartlepool. More dramatic examples include the three to four hundred birds noted in the Teesside area on 17 September 1960 with similar numbers noted again on 17 September 1969 (Blick 2009).

During the 1990s and the first decade of the 21st century, autumn migration brought fewer Redstarts, almost year upon year. It would appear that the falls of Redstarts which were once so much a September phenomenon, involving many scores of birds, sometimes hundreds, around coastal migration points such as Marsden Quarry, Whitburn and Hartlepool Headland, had become a thing of the past. Exceptions to this trend occurred in 1993 and 2008. At the peak of autumn passage in 1993, around 14th to 16 September, well over 100 birds were distributed amongst some eight or nine sites in the north east of the county, with up to 35 in the Marsden Quarry area alone (Armstrong 1994). Up to 20 were in Mere Knolls Cemetery, Seaburn on 17 September 1995, suggesting a relatively good year for this species on passage; up to 11 had been in the Marsden and Whitburn area on 19 May of that year (Armstrong 1997).

The best autumn of the 2000s came in 2008 when, on 6 September, approximately 110 birds were reported between the rivers Tyne and Wear, with concentrations of 40 birds at Marsden Quarry, 25 at Trow Quarry, 15 at Whitburn and an additional 60 in the Hartlepool area (Newsome 2009). A more typical example of the scale of passage over these decades was 2005, when autumn passage for the Durham coastline for the whole autumn amounted to eight recorded birds; the figures for 2007 were lower still and were considered perhaps the worst year on record, with just three birds reported for the whole season. A passage bird on a Fulwell bird table in autumn 1981 was something of a novelty. There was also a most unusual passage record in 1998, when a juvenile bird, judged to be about three-weeks old, was trapped and ringed at the Long Drag, Teesmouth on the very early date of 25 July (Armstrong 1999b).

Birds beyond the middle of October have always been considered late in the Durham context, so one at Hartlepool on 20 October 1973 was notable. This record was exceeded by one at Seaton Carew on 27 October 1979, one at Whitburn Rifle Range until 28 October 1988 and one at Marsden on 31 October 1995. The latest, however, was the bird at Whitburn Churchyard on 5 November 1984 the only November record for the county, in what was an extraordinary year for late-occurring migrant species (Baldridge 1985).

Distribution & movements

The species breeds over much of Europe, eastwards to west and central Siberia, down to the Middle East and across to north west Africa. The nominate race, which is represented over most of Europe, is replaced in Turkey, the Crimea and Iran by the eastern race *P. p. samamisicus*. Redstarts are highly migratory with European birds wintering in Africa, north of the Equator but south of the Sahara.

A suite of birds ringed in the Wear watershed over the years clearly illustrates the south westerly route through Iberia that some Durham-bred birds take on their way to African wintering grounds. For instance, one ringed in a nestbox at Hamsterley Forest on 10 June 1966 was recovered at St. Andrew, Guernsey, Channel Islands on or about 18 September 1966. Further south west, were birds ringed as pulli, on 16 June 1962 at Hamsterley Forest and found at Tras os Montes in Portugal on 28 September 1962 and another, from the same population, on 4 July 1965 found dead on 14 October 1965 in Paderne, on the Algarve coast of Portugal. Others that made it to this area include one ringed as a nestling at Tunstall Reservoir on 2 June 2002, found dead at Beira Litoral, Portugal on 30 September 2003, whilst another ringed at Hamsterley Forest on 21 June 1963 was recovered at Sanlucar de Barrameda, Cadiz, Spain on 14 October, 1963 (Bell 1964).

All of this indicates the generally south westerly vector, towards Iberia, which birds from the north east take on their journey to their wintering quarters. Notwithstanding this, some birds from further east would appear to follow a more northerly path during the autumn. This was illustrated by the full-grown bird, ringed at Neustrelitz in northern Germany, on 13 October 1969, which was controlled in Hartlepool 12 days later, a late occurring bird in the Durham context (25 October) (Blick 2009).

Whinchat
Saxicola rubetra

A widespread but uncommon and declining summer visitor and passage migrant.

Historical review

One hundred years ago, over much of mainland Britain the Whinchat was a very common bird breeding in rough pasture, on waste ground roadside verges and embankments (Mead 2000). This attractive summer migrant, which still breeds in small but declining amounts, once did so in much larger numbers in many more locations across the county. Traditionally, it was a regular passage migrant on the coast, particularly in the early autumn and to a lesser extent in spring, but even in this capacity it is now a much diminished presence in Durham.

Around the turn of the 20th century, Nelson (1907) described it as a "*summer visitant*", that was "*common and generally distributed*". Whinchat was described by Temperley (1951) as a "*widely-distributed and, in places, fairly common summer resident; also a passage migrant*". He said that it bred on forest clear-fells, thickets and along hedgerows in the low country and scrubby land on the fringes of the moors. At his time of writing, Temperley described it is as being particularly numerous in the middle and upper portions of Teesdale, where it bred at altitudes of up to 1,000 feet (300m) above sea level. At that time spring migrants were said to arrive at the end of April or in early May, and it stayed into September. In this respect, little has changed excepting the number of birds.

Dr. H. M. S. Blair thought it rare in the area about South Shields, with only one report of it having nested near the town in the early to mid-20th century, though it was regularly noted on passage in that area (Temperley 1951). A decline started around 1930-1940 and this continued through the 20th (Mead 2000) and into the beginning of the 21st century (Risely *et al.* 2010).

In the south east of the county, Stead (1964) reported a very local distribution on the coastal plain. In 1967, a late passage bird was at Shotton Colliery on 31 October, a very late occurrence from an inland location. Such late records are not without precedent, other late passage records include one at Crimdon Dene on the 4 November 1960 and two at Seaton Carew on 2 November 1973.

In the area of South Shields, the species was regularly noted on passage in the late 1970s and early 1980s around the scrubby grasslands of Temple Park, and the species nested there in the early to mid-1980s (K. Bowey pers. obs.). They are now gone from this area.

Recent breeding status

Nationally, Whinchat experienced a very considerable contraction in range between the early 1970s and the early 1990s, as indicated by its documented losses between the two national Atlas surveys; these loses were particularly severe in the lowland areas of England (Parkin & Knox 2010). This species would appear to have declined locally over the last two decades, as traditional lowland sites have been lost to development or become scrubbed over, and the nationally impacting factors have played out in the local context. In Durham, the Whinchat occurs largely on marginal land as has been highlighted elsewhere (Marchant *et al.* 1990).

Today, it favours moorland margins, lowland heath, derelict land and occasionally, but ever less commonly, coastal areas and marshes. All of these habitats may be selected in Durham, though many of the coastal records in Durham, in recent years, may refer to passage birds rather than breeding pairs. In the local context, it now mainly occupies habitats in the upland fringes, though it was once extensively distributed in the lowlands on rough and scrubbing over grasslands. In the western parts of the county it is often associated with areas of bracken *Pteridium aquilinum* and tends not to favour open, heather moorland, where there is an absence of posts or bushes for use as song and hunting perches (Westerberg & Bowey 2000). In its earlier years, the national forestation programme, which followed the First and Second World Wars, probably benefited the Whinchat (Fuller 1982), but new forest planting rapidly grows beyond the stage at which this species can use it (Gibbons *et al.* 1993) and this is not a habitat that is now often utilised by the species in Durham.

At the time of the fieldwork for the first national breeding Atlas (Sharrock 1976), the species was recorded in all 48 ten-kilometre squares in Durham, but its status has changed dramatically between 1970 and the first part of the 21st century. Nationally, it has undergone a marked decline since the early 1950s (Marchant *et al.* 1990) and by the 2000s the species had been amber-listed on the Birds of Conservation Concern listings (Eaton *et al.* 2009). The local distribution of Whinchat, as indicated by the *Atlas* map (Westerberg & Bowey, 2000), showed a species that occurred widely across the county, but only in relatively small numbers, and which exercised rather selective options in its choice of habitats. This species still breeds in suitable habitats around the North Tees Marshes and also in the fringes of the county's western uplands, but it has disappeared from many of its traditional hinterland habitats over the last four decades.

Three pairs reared young around the North Tees Marshes in 1997 and this area remained attractive for the species into the early 2000s, though by the end of that decade breeding birds there were increasingly scarce (Joynt *et al.* 2008). At the other end of its range in the county, successful breeding took place in 1999 at Barningham Moor and on Hamsterley Common. Then through the 2000s, birds were noted at several potential breeding sites, but the very fact that individual breeding sites were now being identified by observers indicates the level of decline that this species has undergone in the county. These sites included, in the lowlands: Chapman's Well and Rainton Meadows; at Teesmouth, Saltholme Pools and Greenabella Marsh, where they bred regularly; and, in the upland fringes and uplands: Edmundbyers, Great Eggleston Beck, Hedleyhope Fell, Langleydale Common, Leadgate, The Stang and Widdybank Fell. This species has become particularly scarce in the lowlands, as illustrated by the absence of any breeding records at all from the east of the county in 1999 and 2000, except at Teesmouth (Armstrong 2003 & 2005, Iceton 2000 & 2001).

By 2009, Whinchat was found to have reached its lowest represented level on the national Breeding Birds Survey since the start of that scheme in 1995, having experienced a calculated 57% decline since the outset of monitoring (Risely *et al.* 2010). Much of the reason for this decline, up to the 1990s at least, was attributed to a loss of habitat, particularly the improvement of agricultural land (Marchant *et al.* 1990), though this is a factor which should have had less effect in Durham than in more intensive cereal-growing areas in the south and east of England, and such effects may also be as a result of changes in its wintering quarters.

Locally, there is anecdotal evidence that the species has become increasingly difficult to find in favoured locations, even since the fieldwork for the *Atlas* (Westerberg & Bowey 2000). Whilst the *Atlas* map showed breeding season records from 98 tetrads, in 30 ten-kilometre squares, this does not reflect the situation at 2010, with birds currently being recorded in considerably less than half this number of one-kilometre squares in the breeding season.

Incidents of proven breeding became ever fewer as the 21st century's first decade wore on. In 2007, a single pair bred on Greenabella Marsh and a pair was noted at a mid-county site early in July, with two juveniles there in August, indicating that breeding had taken place. In 2009, considerable effort was expended in trying to locate birds in parts of the upper Wear watershed and the results indicated that the species was present, though in severe decline. Nonetheless, five pairs were found to the south west of Hamsterley with pairs noted at three other locations in that general area.

The population density of the species in the county, as the *Atlas* recorded, was nowhere high and what was then believed to be a 'reasonable estimation' of the Durham breeding population at 200-300 pairs, would probably now be thought of as very much on the high side. A case could be made for reducing this by as much as 50%, or perhaps even more, at the end of the first decade of the 21st century. Such a profound reduction is probably borne out by the fact that in 2010, there was no confirmation of breeding anywhere in the county. It is likely however, that undiscovered pairs were present and breeding in some of the western dales, for example four birds that were seen to the west of Linburn Head near Hamsterley on 21 August were probably 'locally produced'.

Recent non-breeding status

In most years, the first birds of the spring are usually around the last week of April or the first few days of May, birds usually being seen at inland sites before they are noted along the coastal strip. A female at Marsden Quarry on 6 April 1992 (Armstrong 1993) was the earliest for Durham, the second earliest being one at Burdon Moor, on 11 April 2009 (Newsome 2010). The average reported spring arrival date during the three ten-year periods 1976-1985, 1986-1995 and 1996-2005 remained unchanged as 25 April (Siggens 2005). Passage birds can be met with almost anywhere in the county, in any location where there is open ground with rough herbage.

Post-breeding dispersal means that small numbers of birds are occasionally seen away from breeding locations by late July. Many of the county's adult birds seem to slip away after the breeding season, leaving just a few birds of the year and whatever passage birds filter through the county from further north or across the North Sea.

Autumn passage may start as early as late July and is most evident through August and early September, but stragglers sometimes occur into mid-September. Birds from the Continent are probably responsible for later records, particularly those occasionally noted in October, even to the middle of that month, such as the bird at Haverton Hole on 17 October 2007.

Numbers vary from year to year and coastal passage has declined in terms of the overall numbers of birds recorded over recent decades. Concentrations of 15-20 birds in the past were not unusual, particularly in the second half of August if the weather conditions produced falls of migrants along the east coast of England. Such passage is now rarely evident. For many years there was a post-breeding gathering of this species around the North Tees Marshes during August and early September. This probably combined locally-produced birds and some coastal passage migrants and often totalled between 15 and 20 birds, although 49 were present on 8 September 1974 (Blick 2009), at a time when this species was very much more numerous in the county. Such coastal gatherings occasionally still occur; at least 33 birds were in the South Shields to Whitburn area in mid-August 2006.

The largest ever concentration of Whinchat probably occurred on 17th to 18 September 1960 in fall conditions that brought birds to the whole of the north east coastline. At this time, as well as several in the Sunderland area, at least 57 were present between Crimdon Dene and Teesmouth (Coulson 1961). In similar fashion, between 8th and 10 September 1995, about 115 were seen along the Cleveland coastline (Bell 1996). On the former date, 30 were in Marsden Quarry and 13 were at Whitburn (Armstrong 1997).

November birds in Durham are by no means common, but equally they are not unprecedented. One was at Crimdon Dene on 4 November 1960 (Coulson 1961) and over 40 years later, a late migrant remained at South

Shields from 5th to 9 November 2000 (Armstrong 2005). One was at Seaton Common from 12th to 15 November 2005 (Newsome 2006). The latest autumn record was of one at Dawdon Blast Beach on 27 November 1994 (Armstrong 1996).

Much more unusual, are wintering records in England, which are very scarce indeed; there were just eight such records during the three winters of fieldwork for the Winter Atlas during the early 1980s (Grice & Brown 2005). There is only one known instance of this in the county, with a male reported at Blackhall Rocks on 7 January 1985 (Baldridge 1986).

Distribution & movements

Whinchat breed over much of Europe, though they are, at best, only locally distributed in the southern parts of the Continent. Birds are also to be found breeding in Western and south-central Siberia and the Caucasus. Birds winter in sub-Saharan Africa, to the north of the Equator, with small numbers occasionally being found to the north of this zone. As a trans-Saharan migrant, Whinchat may have been affected by the droughts of the 1980s, in areas such as the Sahel, which had such sustained and drastic effects upon the sympatrically wintering Sand Martin *Riparia riparia* and Whitethroat *Sylvia communis*, although Marchant *et al.* (1990) believed this factor had a lesser effect on Whinchat than on those species.

Siberian Stonechat
Saxicola maurus

A rare autumn vagrant from Siberia: seven records.

The first record for County Durham was on 26 October 1960 when Alan Vittery saw a female or first-winter bird at Hartlepool Headland (*British Birds* 69: 360). The eastern 'races' of Stonechat were little understood at the time and the documentation of the various subspecies as vagrants was limited. The record was only published in 1977 as part of a review of Siberian Stonechat occurrence (Robertson 1977). This bird constituted the second British record of this then subspecies, the first being shot on the Isle of May in 1913.

A further six Siberian Stonechats have been seen in County Durham, four of which have fallen into the typical late September/October period, which accounts for the majority of British records. The spring bird at Hartlepool in 1985 is one of less than 20 for Britain. It occurred during a good period for north east European migrants along the Durham coast, with several Bluethroats *Luscinia svecica* and a Thrush Nightingale *Luscinia luscinia* present at the time. Unfortunately, this beautiful bird met an untimely end, being killed by a cat.

All records:

1960	Hartlepool Headland, 26 October, female/first-winter
1985	Hartlepool Headland, 13th to 14 May, male, killed by cat, remains retained by Russell McAndrew (*British Birds* 79: 569)
1990	Quarry Lane, Marsden, 7th to 10 October, first-winter (*British Birds* 85: 537)
1991	Seaton Carew, 9th to 12 October, female/first-winter (*British Birds* 88: 531)
1994	Whitburn CP, 16th to 22 September, first-winter male (*British Birds* 90: 496)
2005	Whitburn CP, 19th to 21 October, first-winter (*British Birds* 100: 79)
2011	Seaton Common, 3rd to 4 December, first winter male (*British Birds* in prep.)

In addition, two records of birds in October 1991 were published in *Birds in Durham* but were not submitted to BBRC at the time, so they do not appear in the formal record. Both birds were seen by many observers and identification was not in doubt. These were at Castle Eden Denemouth on 12th to 13th, and at Dawdon Blast Beach from 10th to 13th.

Distribution & movements

The subspecies *maurus,* to which the majority of British records are assignable, breeds widely across northern Asia, from the Urals east to the Kolyma basin, Okhotsk coast and northern Japan and south to the Caspian Sea, Mongolia and northern China. The wintering range covers the northern Indian subcontinent to southern China and South East Asia. Over 340 Siberian Stonechats have occurred in Britain, the vast majority in late autumn.

Stonechat (European Stonechat)
Saxicola torquatus

A regular winter visitor and passage migrant, also breeding in increasing numbers both on the coast and inland, a rapid change from its recently rare breeding status.

Historical review

The first mention of this species for Durham comes from 1791 and Allan's manuscript relating to Tunstall's Museum, which called the species "*common in the summer in the heather*", and "*in winter in the marshes*". Hogg, writing of the south east of the county in 1827, wrote that it was "*frequently seen in mossy fields near Elswick and Brearton*", and that it "*breeds in whin and gorse bushes*". Older accounts (Hogg 1827, Nelson 1907), described the species as breeding regularly in the south east of the county. Temperley (1951) interpreted this as indicating that the species was more plentiful a century prior to his time of writing. Hutchinson (1840) however, said that it was "*rather a scarce bird and partially distributed*". He identified the "*dual stations*" of its distribution as being along the coast and in the uplands. By 1874, Sclater stated it "*appears to have entirely abandoned this part of the coast*". He would have been referring to the central section of the county's coastline, around Castle Eden Denemouth. Of Teesdale, Backhouse (1885) said that he could find no evidence of it breeding though he knew it regularly bred in Weardale. Late in the 19th century the species was "*not common locally*" in the Derwent valley, though as an example of when it did occur, Robson reported a nesting pair near Burnopfield and he documented the taking of specimens from Axwell Park and Blaydon Burn (Robson 1896).

Temperley said that it occurred irregularly as a passage migrant and Nelson (1907) said that it was occasionally noted at Teesmouth, though it was plentiful there in the September of 1903. Early in the 20th century the species seems to have fared somewhat better in the north of the county, with breeding reported more widely. For example, birds bred intermittently near Chopwell in the Derwent valley in the early part of the 20th century, a brood was found at Muggleswick in 1925, and then it was discovered nesting near Greenside in 1937 (Bowey *et al.*1993). By contrast, C.E. Milburn (*The Vasculum* Vol. III 1917) said that it was scarce in the Middlesbrough area (just to the south of Durham), and was primarily an autumn migrant, that was never noted to breed in south east Durham. In 1925 however, it was found breeding at Blackhall Rocks and in 1922 a pair was found at the mouth of Hordon Dene, after an absence of nine or ten years from that area. It bred in Teesdale on Falcon Clints in 1925, with birds there again in 1926 and 1931, while another pair was below High Force at that time (Temperley 1951).

In the middle of the 20th century, Temperley (1951) described it as a "*resident in very small and dwindling numbers*", which was also noted as a passage migrant. He said that it was less numerous and more local than the Whinchat *Saxicola rubetra* and in the early 1950s, he described it as disappearing from many of its former haunts. He postulated that this was in part as a result of the hard winters of the 1940s, 1947 in particular. By 1951, it was said to no longer nest in the South Shields and Sunderland areas, though it was still to be found in the coastal denes further south, such as Horden and Castle Eden Denes. Temperley called it rare inland, but said that small numbers were present in the upper valleys of the Derwent, Wear and the Tees and on the fringes of the moors, where it was known to nest amongst gorse, juniper and heather.

In 1952, a pair reared two broods of three and six young at South Shields, in the basin of Trow Quarry, just inland from Trow Rocks at the end of the main beach at South Shields (Temperley 1953, J. Coulson pers. comm.). On 21 April 1955 a male was near Middleton in Teesdale (Temperley 1956) and on 8 July in one summer during the late 1950s, a nest with five eggs was found on the moors close to the Durham/Westmorland (now the Cumbria) border. A pair was on Cowpen Marsh on 29 May 1963 (Bell 1964) but by the mid-sixties, the Stonechat was virtually extinct as a breeding bird in the south east of the county, courtesy of the exceptionally hard winter of

1962/1963 (Stead 1964). In May 1967 however, a male defended a territory in the north of the county, the location of which was not fully documented.

A pair that wintered at Teesmouth and reared young at Dorman's Pool in 1969 were said to be the first pair ever to be documented breeding at Teesmouth, and subsequently birds bred regularly in that area until 1977 (Joynt *et al.* 2008). In 1973, one pair reared two broods near Jarrow, in the north of the county. In 1974, two pairs reared seven young at Teesmouth but after this run of breeding activity the species declined in this area until in 1997, when an adult was seen throughout the year (Bell 1998).

Blick (2009) thought that it had never been common as a breeding bird in the area to the north of the Tees, though there were known to have been breeding, or summering birds, in the Elwick and Hartlepool district in the early 1800s, at Cowpen Bewley up to 1949 and between Dorman's Pool and Port Clarence from 1969 to 1977 (Blick 2009). During the 1970s, the species was a regular passage migrant at the coast. In 1971 there were good numbers in the autumn, and inland it was a winter visitor to Derwenthaugh, Ryton Willows and to Shibdon Pond in the north of the county, anecdotal evidence suggesting that birds were breeding at the first two of these sites over this period (Bowey *et al.* 1993).

In summary, there was probably a considerable contraction of range and number of Stonechats in Durham during the first part of the 20th century, despite the fact that it was still at that time quite widely distributed in the county's 'rough' habitats. This decline probably continued into the middle and latter part of the 20th century as indicated by the species' well-documented decline between the two national Atlas survey periods (Parkin & Knox 2010), to the point where by the 1980s in Durham it had become an increasingly scarce breeding bird and was best known as a passage and winter visitor.

Recent breeding status

Across much of Britain this species utilises lowland heath and marginal land with scrub for breeding. Up to the mid-1990s, this was still largely a scarce passage migrant in County Durham. Even as recently as the survey work for the *Atlas* (Westerberg & Bowey 2000) during the late 1980s and the early 1990s there was little evidence to indicate that this species was a regularly breeding bird in the county. The text in that volume highlighted individual examples of breeding activity, for example, the holding of a territory by a lone male during May 1989 in what was deemed suitable habitat, between Stella and Stargate in western Gateshead. In August of that year a female or juvenile was seen in the same location, suggesting that breeding had taken place (Bowey *et al.* 1993).

As the 1990s progressed, there were indications of a change of status for this species in the area. For example, in 1990, at Ryton Willows, a pair was found in extensive gorse *Ulex europaeus* scrub during March, remaining for ten days, the male singing, but breeding was not thought to have taken place. The only confirmed breeding record during the *Atlas* period came in 1992, when two recently fledged juveniles were noted in the upper Derwent valley. At that time, Westerberg & Bowey (2000) highlighted that this was the first successful breeding of Stonechat in the county for over a decade (Armstrong 1988-1993). As milder winters began to predominate through the 1990s and the first decade of the 20th century, birds recovered locally and increasingly were noted both in winter and on passage (Olley 2006). The subsequent decade and a half saw a considerable change in this species' local breeding status from the point where, in the early 1990s, it was believed that small numbers of birds may have been breeding undetected in the county (Westerberg & Bowey 2000). In 1999 a pair bred successfully at Quaking Houses Fell, south of Stanley and Olley (2006) commented that this appeared to be the catalyst for a re-population of the county by the species. The following year a pair was at Seaton Pond on 1 May with spring singles at a couple of upland sites. In late summer, two juveniles were at Quaking Houses but it was unclear as to whether these were passage birds or the result of another breeding attempt, as earlier searches of the Fell had failed to produce any evidence of birds. On 19 July 2000, in the south of the county, a pair with three fledged young was found at Barningham Moor (Armstrong 2005).

The increase in breeding numbers continued apace through the decade with there being seven occupied sites in 2002 and 2003, in both coastal marshes sites and the western moorland, where breeding was 'highly likely'. In 2004 there were at least 15 pairs, producing at least 27 young and through the next few years the increase continued. In 2005, three pairs bred at Hedleyhope Fell, one or two pairs at Edmundbyers Common, Hill End, Muggleswick Common and Ruffside Moor with other successful pairs reported from Langleydale, as well as along the coast at Hawthorn and Easington. Recently fledged juveniles were present at Greenabella Marsh in late July, suggesting that breeding had occurred in that area. It seems likely that nesting went unreported at a number of other sites, as recently fledged juveniles were seen in several coastal areas where pairs had not been noted previously. In 2007, presumed breeding pairs were recorded at a minimum of nine inland and seven coastal locations with up to 14, of what were probably locally-bred birds, in the Easington area during August and September (Siggens 2006, Newsome 2006-2008).

Atlas survey results for *The Breeding Birds of Cleveland* indicated that there were territorial pairs in at least four locations north of the Tees in the early 2000s, three of these around Teesmouth, in particular the North Tees Marshes (Joynt *et al.* 2008). This is perhaps the reason that 13 birds, including juveniles, were noted at Greenabella Marsh in September 2006, strongly suggestive of local breeding (Joynt et al. 2008). Since 2002, Stonechats are known to have bred at a number of sites around the north Teesside area, including at Dorman's Pool, Greenabella Marsh, Hart Warren, the Hartlepool Power Station site, Haverton Hole and Seaton Common (Blick 2009).

In 2008, a minimum of 36 pairs were reported around the county during spring and summer (Newsome 2009). Breeding birds were found in upland locations, stretching from Edmundbyers in the north to Sleightholme Moor in the south, and as far west as Cauldron Snout, in upper Teesdale. This probably translated to a minimum of 20 to 25 breeding pairs but, given the abundance of suitable habitat in the upland fringes in particular, an upland population of 40 to 50 pairs in Durham might easily have existed at this time. The winters of 2009/2010 and 2010/2011 were undoubtedly hard for this species and it is thought that the county breeding population may have suffered a severe reverse over this period. Nonetheless, since the early 1990s, the species' status changed out of all recognition from its previous rarity as a breeding bird in Durham. In the six-year period 1999 to 2004 confirmed breeding came from 34 sites around the county, which had increased to 50 sites by the end of 2005 (Olley 2006). Around 1993, it was thought that the county population would not number more than one or two pairs (Westerberg & Bowey 2000); in 2010 the number probably stood close to, or exceeded, 50 pairs.

Recent non-breeding status

In the recent past this species was probably best categorised as a scarce passage migrant and an uncommon winter visitor. Inland wintering birds in particular, became increasingly scarce through the hard winters of the late 1970s and early 1980s; by way of example, the males at Ryton Willows and Shibdon in January 1984 were the first in those areas for a number of years (Bowey *et al.* 1993). This species is extremely susceptible to cold winters (Brown & Grice 2005), hence its change in status in Durham, an east coast county that has traditionally experienced relatively large numbers of winter frosts (Westerberg & Bowey 2000). Such hard winters were probably the principal limiting factor for this species in Durham around this time, as there is an abundance of apparently suitable habitat.

Records through the 1990s and 2000s indicated that the species was gradually becoming more frequent in Durham during the winter and it became increasingly less uncommon through a series of mild winters, as intimated by Olley (2006) in his review of the species in County Durham. One seen in Lunedale on 30 November 2000 was the first moorland bird seen by that observer for over 10 years, a portent of things to come. Early 21st century records suggested that this species was gradually, and persistently, becoming more frequent in the county. In 2006, the majority of wintering birds were found within 10 miles of the coast, most birds moving out of the moorland breeding areas as they did in 2008, although one or two birds could be seen at Bollihope, Knitsley Fell, Skaylock Hill and Tow Law (all of these at elevations greater than 300m above sea level) during November and December, demonstrating that this species is sometimes hardy and can cope with a little frost and cold weather, despite its reputation for winter vulnerability. In 2007, there were around 250 reports of birds from over 80 locations across the county, a spread of distribution that would have been inconceivable just two decades previously (Newsome 2007b-2009). This is also a dramatic contrast with just ten years earlier in 1997, a particularly poor year for the species in the county, when there were only three wintering records, and records from just eight locations in total, involving nine individuals (Armstrong 1997, Bell 1997).

During the first quarter of 2007, around 40 birds were reported as wintering along the coast from South Shields to the mouth of the River Tees (Newsome 2008), a huge increase in comparison with the sparse figures from just ten years previously. In recent decades, spring and summer records have been roughly split between upland areas in western Durham and the coast, with a few scattered pairs found in other lowland areas. At the end of the first decade of the 21st century, wintering and locally territorial birds can be routinely present all along the county's coastal strip.

In autumn 2007, passage and wintering birds accounted for sightings of up to eight birds coming from at least 35 coastal localities during the last quarter of the year (Newsome 2008). Today the status of birds seen along the coastal strip is difficult to determine. The position, particularly at the coast, is clouded somewhat during March, and again in October and November, when passage birds move through; these are relatively easily detected at non-breeding sites but difficult to assess where the species is now resident throughout the year.

In 2009, winter counts in the uplands included seven at Pennington Plantation (Hamsterley) on 19 October and seven at Rookhope on 7 November. At the coast, large concentrations included 22 at Hendon on 18 October and up to eight at Blackhall (Newsome 2010). What the hard winters at the end of the first decade of the 21st century did to these recently re-established local populations is not fully understood though it is known that nationally, a number of small-bodied resident birds, amongst them Stonechat, declined significantly between 2008 and 2009, presumably due to the prolonged freezing temperatures in January and February 2009, a reduction of 38% in returns from the Breeding Bird Survey being experienced by the species between those years (Risely *et al.* 2010).

Distribution & movements

Stonechats have a broad distribution in the 'Old World' and this includes Europe, northern Asia, and sub-Saharan Africa. There are a large number of named races across this breeding range and the situation is complicated by the fact that in parts of its range the species is migratory, whilst in others it is sedentary, and in others a partial migrant. Across Europe, birds are mainly found in the southern and central portions, extending as far north as the UK and into southern parts of Scandinavia as a breeding bird.

Wheatear (Northern Wheatear)
Oenanthe oenanthe

A common summer visitor and passage migrant

Historical review

The species does not feature prominently in the lists of the early documenters of the birds of Durham, presumably because it was both familiar and common and consequently was felt to warrant relatively little commentary. Holloway (1996) said that the species bred commonly across England in the 19th century, and he mapped it as being common in Durham. He also noted that at that time it occupied a much wider range of habitat than in the modern day, being found in many '*uncultivated areas of many different habitat types*' (Holloway 1996).

Historical breeding records of Wheatear in Durham include birds that nested in rabbit burrows at Seaton Snook prior to 1829 (Blick 2009) though it is likely to have been widespread in both upland and lowland locations. In the lower Derwent valley, the species was rarely recorded by Robson during the 19th century, though birds did nest in a small quarry on Blaydon Bank and, in 1892, in one of the spoil heaps of the Blaydon Main Colliery (Robson 1896), close to the modern site of Shibdon Pond. Such sites, according to Holloway, were not atypical at the time (Holloway 1996).

Temperley (1951) thought that this was a "*common summer resident and passage migrant*". He mentioned its breeding in sand dunes along the coast, presumably at Hart Warren and Teesmouth. It was said to be plentiful at Teesmouth where breeding took place in holes in the ground on the slag breakwaters and in this context Temperley said that "*it has increased of late years*". Further north though, he observed that it was "*being ousted from many of its old haunts along the coast as a result of industrialisation*". Richmond (1931) observed that it was a resident breeding bird at Teesmouth before and around 1931. Breeding was first noted in this part of the county in the first half of the 20th century, birds using the slag heaps and piles associated with the estuary's industry.

At Temperley's time, it "*until recently*", nested in the old dry stone walls about Marsden, these were presumably those that meander over Cleadon Hills and along Lizard Lane towards Whitburn. He observed that it was very local about the central strip of the county, but occasionally it could be found on scrubland, where there were old walls or rabbit holes. He said it was much more common in areas of the west, where "*mortar-less walls*" predominated (Temperley 1951). In relation to migration, he documented that arrivals occurred from the last fortnight in March and through April, as a spring passage migrant at the coast, with return passage in autumn, from

August through to October. On 27 November 1930, Guy Robinson and Bill Almond saw a Wheatear at Sadberge (Robinson & Almond 1931), which was, and still is, an exceptionally late date for the species and as such was, as stated at the time, "*worthy of special record*".

Dr. H. M. S. Blair noted that on an annual basis there was a very obvious spring passage in the South Shields area in most springs and noting in one year, 1943, that there were "*hundreds about one farm area on 23rd April*". The spring passage period noted at that time was described as being between 23 April and 2 May. Autumn gatherings have been a not infrequent occurrence over the decades and around a hundred birds were scattered around Teesmouth on 1 September 1963 (Blick 2009).

Unseasonably late, or early, birds were noted at Hartlepool Headland and North Gare in respectively, December 1959 and January 1960 (Blick 2009); this was presumably the same individual working its way along the coast. At the other end of the migratory season, one at Hartlepool on 12 March 1967 was an early record for this time, though the earliest arrival dates for the county are one day earlier than this, with birds being noted on 11 March in both 1957 and 1977 (Blick 2009).

Urbanisation around Teesmouth probably drove the breeding numbers down in that area, until just eight pairs were present by the mid-1960s (Stead 1964). This decline continued, with sporadic coastal breeding to the late 1960s, in the area of Hart Warren (Joynt *et al.* 2008).

Recent breeding status

In modern times, the majority of the county's breeding Wheatears have been, and still are, found in the west of the county, where suitable habitat is abundant on the upland grazing lands and the high moors of the Pennines (Westerberg & Bowey 2000). For nesting purposes, birds use the abundant cavities in dry stone walls, as well as disused farm outbuildings, lead mines and rabbit burrows in short-grazed pasture (Conder 1989). The Wheatear also has a history of lowland breeding in the region (Temperley 1951), albeit in much smaller numbers than are found in the county's uplands. During the survey period for the *Atlas,* there was evidence of confirmed or possible lowland breeding at Dunston, Great Lumley and Whitburn, although generally the number of lowland breeding territories appears to be in decline (Westerberg & Bowey 2000). More recent examples of such activity have become ever scarcer. In recent years, there have probably been just two pairs on an annual basis, at best, on the north side of the River Tees (Joynt *et al.* 2008), these being located in some years, in small numbers, in the slag walls of Seal Sands (Blick 2009). This has always been something of a marginal breeding species around Teesmouth (Joynt *et al.* 2008).

Wheatears favour a mosaic of heather moorland with short-grazed turf (Conder 1989) and, on the whole, most Wheatear territories in the county are found above 330m above sea level, with large numbers of occupied territories in upper Teesdale, Weardale and the upper Derwent valley. In 2007, local breeding birds were feeding young by 22 May at Killhope and fledged juveniles were seen at Bollihope in early June. In 2008, breeding birds were reported from 60 one-kilometre grid squares in Durham's western uplands and from 92 in 2009. In 2010, the largest reported congregation of birds in an upland breeding area was of 22 at Linburn Head, but birds are present across most of the county's Pennine area (Newsome 2008-2012).

Nationally, the species has exhibited a long-term decline over much of the southern half of England (Marchant *et al.* 1990) and this has continued in recent decades retreating from lowland areas between the two national breeding atlas surveys (Parkin & Knox 2010), but this is just the end of a long period of decline in England during the 20th century (Brown & Grice 2005). There is some suggestion that numbers in Durham may have dwindled in some moorland sites (Westerberg & Bowey 2000). Although the species is widespread across the county's extensive moorlands, it is thought that few sites hold more than a handful of birds. The estimate of the breeding population made in the *Atlas* was of between 200 and 650 pairs, based on a number of estimation methods (Westerberg & Bowey 2000), but it seems likely that the true figure lies much closer to the upper limit of this range than the lower and even this may be something of an underestimate based on the number of occupied upland squares in recent years.

Recent non-breeding status

The Wheatear is one of the few summer visitors to be routinely seen during March in Durham and it is quite frequently the first summer migrant to arrive in the county. Birds arrive on the moors as early as mid-March and they can be back in territory, as they were at Wolsingham, Harthope Quarry, Cow Green and Falcon Clints in 2000, by the end of March (Armstrong 2005). Today, spring passage migrants might be seen any time from mid-March through to the middle of May, but it is usually the first week of April before Wheatears appear in any great numbers. A fairly typical date for the first bird of the year to be reported would be 16 March, but this may rarely occur as late as the third week of March. The majority of British breeding birds are on territory by late April (Wareham *et al.* 2002).

Over thirty years between 1976 and 2005, the reported first arrival dates of Wheatear were analysed, in ten-year groupings. From 1976-1985 the mean was 20 March; during 1986-1995, 19 March; 1996-2005, 20 March. It was therefore found that over this whole period, the mean first arrival date of the species in Durham had not changed to any great degree (Siggens 2005). However, it varies considerably over some periods. For instance, in the first decade of the 21st century, first arrival dates ranged between 10 March in 2001, equalling the earliest ever arrival for the county, and 30 March in 2009, the latest first arrival date in many years, which was attributed to poor weather conditions for migration (Siggens 2005, Newsome 2010).

On 18 April 2004, there was an arrival of around 200 birds on the coast between North Gare and Seal Sands, an unprecedented spring presence in this area, which was not reflected by counts further north in the county. Occasionally, visible migration is noted along the county's coast, as with the 14 birds seen passing along the cliff edge at Whitburn Observatory in a period of just under seven hours on 22 April 2007, indicating that much larger numbers were moving through northern coastal areas at this time. This date saw the main day of arrival of that spring and at least 100 birds were on Seaton Common with smaller numbers around other parts of the North Tees Marshes, perhaps the best spring influx for some years (Siggens 2006, Newsome 2008).

On passage, birds frequent open country often in the more elevated parts of the county where they pitch down on locally prominent hills. A good example of this phenomenon comes from studies in Gateshead borough, where the species can sometimes be seen in quite large numbers around the open landscapes of Ravensworth Fell. For example, there were at least 25 birds in just one field at Kibblesworth, on the Fell, on one date in May 1991 whilst up to 65 were scattered around the area in early May 1992 (Bowey *et al.* 1993). Many of the later passage birds in Durham are of the large, long-winged form and undoubtedly belong to the northern race *leucorhoa* and these are bound for Iceland and Greenland.

Birds often start their return to the coast from July onwards (Armstrong 1988-2002). The scale of autumn movements in Durham is usually less marked than in spring, though on occasion larger numbers do occur, for example the 92 birds at South Shields on 20 September 1999 (Armstrong 2003). Some of the early sightings of juveniles in the eastern lowlands at the outset of the autumn passage period may be explained by the early dispersal of birds away from local moorland habitats. Such early movements may have the effect of erroneously inflating the impression of a lowland breeding season presence in the east of the county. That locally reared young very quickly achieve independence and move south is witnessed by the fact that juveniles were noted at Hartlepool North Sands, Seaton Snook and Whitburn in the first ten days of July in 2007 (Newsome 2008). More typically however, the first coastal birds of the autumn are noted at locations such as Trow Quarry, South Shields, or along the county's beaches, through the first week or so of August. Such records include 14 at South Shields on 21 and 31 August 1999 (Armstrong 2003). Passage birds are mainly noted between early August and early September with, typically, just small numbers being noted at mainly coastal locations over this period, and a scatter of inland reports filtering through across the county. Most of this southbound passage is over by late September or the first half of October with odd stragglers being recorded as late as the first week or so of November, such as the individual at Hartlepool on 12 November 2007. There are also occasional late records in the uplands, including two at Cowshill on 17 October 1996 (Newsome 2008, Armstrong 1998).

One late migrant was observed exhibiting unusual feeding behaviour, as it attempted to cling on to a peanut feeder in Whitburn in October 1991, after apparently watching Greenfinches *Chloris chloris* (Armstrong 1992).

Greenland Wheatear
Oenanthe oenanthe leucorhoa

Birds breeding in Durham are of the nominate race, but northerly birds of the 'Greenland' form *O. o. leucorhoa* routinely pass through the area, on the way to their breeding grounds in Iceland and Greenland.

Temperley (1951) reported that this race of Wheatear occurred annually in spring at Teesmouth, the information coming from Joseph Bishop the official RSPB watcher for that area, during the 1930s. But there is little information on its Durham status prior to this. Inland during the early part of the 20th century, it was also recorded at Darlington Sewage Farm on a regular basis, with a bird there as Temperley put it *"as late as 10th May"* in 1950. During the spring of 1939, it was reported to be *"more plentiful than usual"* on the Durham coast (Temperley 1951), but in reality this probably more reflects observers' then lack of experience of the sub-species' main identification features.

In Durham today, birds of the larger 'Greenland' race of Northern Wheatear are regularly reported in spring, generally in late April and through to mid-May, though it is undoubtedly much under-recorded. Many, most or even all, of the later spring passage birds that are noted in the county are probably of this larger, long-winged, race particularly those seen from mid-May into the second half of that month. Coastal passage migrants moving through at this time almost certainly refer to such populations and ringing data from other English east coast sites prove that *leucorhoa* is the dominate form present in such areas during May (Wernham *et al.* 2002). This is underlined by the numbers of Greenland Wheatears ringed just outside of the Durham area, at South Gare, which have included 71 between 20 April and 10 May 1998, a significant proportion of the total number of Wheatears ringed in this period. An adult ringed at Kvisker, Iceland, on 9 May 1996 and recovered at South Gare on 20 April 1997, is also notable as this was the first Iceland-ringed Wheatear to be found in Britain (Iceton 1999). The true scale of passage of this race through Durham is not known, but it is probably appreciable. The lack of data is largely ascribable to the fact that, as in Temperley's time, many observers are not fully aware of the relevant identification features, which reduces the number of reports of this racial form.

Nonetheless it does occur on an annual basis as evidenced by the pattern of records through the first decade of the 21st century. For example, in 2004 on 2 May, two birds of the 'Greenland race' were on Jackie's Beach Whitburn, in May 2005 up to twelve birds were in Whitburn Coastal Park, and in 2007, a little way inland, several were noted at Longnewton Reservoir in May. This latter report suggests that the northbound movement of this race takes place on a broader front through the county than is sometimes appreciated (Siggens 2006, Newsome 2006 & 2008).

Distribution & movements

This species breeds throughout Europe, in north west Africa, south west and central Asia, northern Siberia, Alaska, northeast Canada and Greenland. It is a long-distance migrant, and all populations winter in tropical Africa. Birds breeding in the county are of the nominate races, but northerly birds of the 'Greenland' race *O. o. leucorhoa* routinely pass through our area, on the way to breeding grounds in Iceland and Greenland.

A male ringed on Fair Isle, Shetland on 1 May 1977 had flown 535km south to Hartlepool by 21 May 1977 (Blick 2009), intriguingly, at that time of the year, in what appears to be the wrong direction. A colour ringed juvenile, ringed as a fledgling on Fair Isle, Shetland on 21 June 2010, was seen again on South Shields Leas on 16 August 2010.

Pied Wheatear
Oenanthe pleschanka

A very rare autumn vagrant from south east Europe and south west Asia: one record.

The sole record relates to a first-winter male found by Bernie Beck and Graeme Joynt at Seaton Snook just before mid-day, on 6 November 1994 (*British Birds* 88: 534). It showed well to visiting birdwatchers as it fed above the tide line for the remainder of the afternoon. It is perhaps surprising that this constitutes the only record for County Durham, as the adjacent counties of North Yorkshire and Northumberland attracted a total of six birds between them from 1979 to 2010.

Distribution & movements

The European range of Pied Wheatear is centred on the Black Sea, reaching eastern Romania and Bulgaria. It occurs more widely to the east, extending across southern Russia, Kazakhstan and Mongolia to northern China. The wintering range covers north east and east Africa along with the south west Arabian Peninsula. To the end of

2010, there have been 61 accepted records in Britain, the majority of which have been in late autumn on the east coast of England and Scotland.

Desert Wheatear
Oenanthe deserti

A very rare vagrant from south west Asia or North Africa: one record.

A male was found by David Watson on the edge of a timber yard at Jarrow Slake on 4 December 1955 and remained in the area until 18 December. The bird was trapped for ringing on 17 December, with photographs and a full account being published by Fred Grey in *British Birds* (*British Birds* 50: 77-79). During the next twelve days the identification was confirmed by a number of observers, amongst them some of the region's best known ornithologists including: George W. Temperley, Dr. H. M. S. Blair, H. Alder, James Alder, A. Baldridge, Brian Little and Philip Stead. The in-the-hand inspection indicated that it was a first-winter male, and the warm buff tones on the back and scapulars pointed to the western race, *O. d. homochroa*, (though subsequent work has suggested that

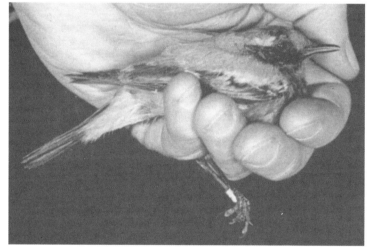

most vagrants to Britain are probably of the eastern races *atrogularis*). Although still feeding well on 18 December, the severe weather conditions with snow on the ground and temperatures of 26°F (-3°C) were thought to have caused the bird's overnight disappearance.

This was only the tenth British record, three of which were pre-1900, and a truly rare bird for the time. There have been no further records for Durham, despite the species' change in national status. It is now an almost expected late autumn vagrant to Britain.

Desert Wheatear, Jarrow Slake, December 1955 (James Alder)

Distribution & movements
Desert Wheatears breed discontinuously across the arid regions of North Africa, from Morocco to the Middle East, and across Central Asia from Iran and northern Pakistan to Mongolia and China. Some of the African birds are resident, but many northerly and easterly birds winter in the Sahara and Sahel regions of Africa. The Asian birds winter from the Arabian Peninsula across to north west India. There are a total of 114 accepted records up to the end of 2010, the majority of which have arrived in late autumn and on the east and south coasts of Britain.

Dunnock
Prunella modularis

An abundant resident and a passage migrant.

Historical review
This 'small brown bird', the perfect description and the meaning of its Anglo-Saxon derived common name (Lockwood 1984), appears always to have been common and widespread, though it is poorly documented. There is a record of the remains of an accentor species, most probably a Dunnock, excavated from the remains of Roman Birdoswald, in south east Cumbria (Yalden & Albarella 2009). Dunnocks were widespread and common during the

Victorian period (Brown & Grice 2005) though they are poorly documented in Durham. Robson (1896) recorded it as a bird which "*may be seen in any garden*", while Temperley (1951) regarded it as "*a very common and well-distributed resident, only absent from the moorland areas*". In 1964 Stead described it as abundant in the south east of the county.

For much of the 20th century the species was poorly documented, through the 1950s, 1960s and 1970s it was rarely identified as more than "*present*" in reports, with few details about its status or distribution being recorded. BTO data indicates a substantial national decline between the mid-1970s and mid-1980s based on Common Birds Census and Breeding Bird Survey data (Parkin & Knox 2010), but the reasons for this are unclear and there is insufficient data for County Durham to know whether this pattern is reflected locally. The degree of scrutiny afforded this species however, is unlikely to reveal subtle changes, only drastic ones. Nonetheless, the species' resilience was attested to by one that was singing at White Force, Cronkley Fell on 17 April 1975, at an elevation of over 400m above sea level, and also by the fact that it survived well in 1979 after the severe weather of 1978, even in high altitude exposed situations such as upper Teesdale. Likewise, the birds at Widdybank Farm and Grains o' th' Beck, which effectively defined the limit of suitable habitat for this species in the county, in 1980, suggested that there had been a recovery following the hard weather of the previous two winters.

Recent breeding status

This is a common, widespread and familiar breeding species that can be found throughout the year in a variety of dense scrubby habitats across the county. The species is most common in areas of low scrubby cover, including shrubberies in parks and gardens, and it will use this type of habitat where it is available from the coastal denes across to the moorland edge of the west of the county. Birds have been noted at Killhope, more than 480m above sea level, on 8 August 1987. There are also records from: The Stang (where four males were singing on 28 February 2007), Blackton Reservoir, Gilmonby, High Force, Newbiggin-in-Teesdale and Rookhope.

Whilst they nest in a variety of habitats, they are not usually present in the centre of mature woodlands, preferring hedgerows and woodland edge. Suitable habitat, such as that in the Ravensworth Estate in Gateshead, can hold large numbers of birds, as evidenced by the fact that 26 individual birds were ringed in a small trapping area, amounting to little more than two hectares, during a three-month period of the winter in 1980/1981.

The heather and bracken of the western uplands do not provide an attractive habitat for this species, an effect which can be seen in the distribution maps in the *Atlas* (Westerberg & Bowey 2000). Away from scrub in the valleys and gills of this area, its absence is a genuine one. Lowland gaps in the distribution maps were more likely to be a result of limited observer coverage. Preliminary maps, based on the first three years of the new national atlas, show a very similar pattern and also little difference between the winter and summer distributions of the species (D. Sowerbutts pers. comm.). The Dunnock was found in over 80% of tetrads in the northern part of Teesside at the outset of the 21st century, which probably amounted to 2,000 pairs.

Song is usually first noted from early February, though it can be noted occasionally earlier, for example birds in full song 12 January 2006. This species' song forms a significant component of the county's suburban dawn chorus until late June. Many local birds have first brood young by late May (Newsome 2009).

Local Dunnock populations can fluctuate considerably from year to year and this was illustrated from ringing returns at study sites. For instance, in 1997, a combined total of 61 were ringed at Lockhaugh and Ravensworth compared to 101 the previous year, largely because juvenile numbers had fallen by around 50% (Armstrong 1998). Looking at the Derwent valley, Lockhaugh study site in more detail, in the best breeding years early in the 1990s up to 50 juveniles were caught. This figure was down to 24 and 16 in the poor summers of 1996 and 1998 respectively, but 31 young birds were caught in 1999, the second highest seasonal total here, indicating a good breeding season. Variation in numbers between years was also revealed by data gathered in 2007. Local studies in the Houghton/Hetton area revealed 14 pairs at Hetton Lyons, four more than in the same study area in 2006, whilst 18 pairs at Hetton Bogs was two up on 2006. Even more extreme was the 100% rise in the number of territories at Shibdon Pond between 1991 and 1992, from six to twelve (Armstrong 1993).

Careful scrutiny reveals the number of this species at some sites, for example 27 pairs in the Brasside Ponds area in 1992 (Armstrong 1993), which is probably the equivalent of some 108 pairs per occupied tetrad (Westerberg & Bowey 2000). Further evidence of population density was shown by the more than 20 singing males in an area amounting to around 40 hectares on Tunstall Hills on 20 April 2007.

Concern was expressed nationally about decline from the mid-1970s through to the 1980s, which was followed by some recovery (Parkin & Knox, 2010). There has been some variable success in different aspects of the species' breeding cycle, and this was reflected by further concerns about its numbers in the late 1990s and early 2000s (Brown & Grice 2005), resulting in it being placed upon the amber list of Birds of Conservation Concern in 2009.

The Dunnock has a complex social organisation, employing a number of interwoven mating strategies, including polyandry, polygyny and polygynandry (Gibbons *et al.* 1993) and this can make it difficult to census accurately. The *Atlas* (Westerberg & Bowey 2000) provided a population estimate for the county of 22,300, based on the assumption that Durham's population of Dunnocks was typical of Britain as a whole, and there is little reason to suggest that the overall figure has changed significantly since then. At the end of the first decade of the 21st century, population levels in the county's lowland areas, both urban and rural, remain high and it is one of the most familiar songs heard in many parks and woodlands (Newsome 2010).

In Durham, this species is probably one of the two principal host species for Cuckoo *Cuculus canorus*. Across Britain it has been estimated that almost half of all female Cuckoos lay their eggs in Dunnock nests (Davies 1987). In July 1986, at Shibdon Pond, a juvenile Cuckoo was present, which was being fed by both Dunnock and Meadow Pipit *Anthus pratensis* foster parents (Bowey *et al.* 1993). In 2006, one juvenile Cuckoo was still being fed by Dunnocks on 6 August, a very late date.

British breeding Dunnocks of the race *occidentalis* are considered to be very sedentary (Wernham *et al.* 2002), most birds remaining within a 0.1 to 1km radius of their territory throughout their lives (Brown & Grice 2005). Locally, in a long-term study over the period 1986-1996 on the Ravensworth Estate, 359 Dunnocks were ringed and not a single bird was trapped that had been ringed outside of Ravensworth and there were no reports of Dunnocks ringed in Ravensworth being found elsewhere. Furthermore, over 28% of all Dunnocks caught over the study period were of birds that had been ringed in a previous year.

Longevity and site faithfulness are illustrated by a bird ringed as an adult at Shotton Colliery in 1996 and controlled there in 2003 at an age of at least eight years, and an adult ringed at Ravensworth in 1986 and re-trapped there in 1992, aged at least seven years.

A bird was noted singing from the middle of an oilseed rape crop in Gateshead during the summer of 1995.

Recent non-breeding status

Being largely sedentary and inconspicuous, the post-breeding and non-breeding distribution of Dunnocks would appear to be the same as that in the breeding season, with the possible exception of some birds at higher altitudes. During the post-breeding stage of the season, birds of the year move away from their natal territories, but sometimes not that far, as evidenced by the Ravensworth study.

The small upland gullies with scant vegetation and the areas near human habitation along the county's western fringes, where upland Dunnocks breed in the summer, appear to be vacated by this species during the harsher winter months. This may indicate some level of local altitudinal migration, with birds moving lower down the dales, or perhaps it is a consequence of there being fewer observers to record them in such places, at that time.

Through the winter, the species is regularly reported at garden feeding stations, a practice observed in Durham for well over thirty years. One of the early observations of such behaviour came in the hard weather of December 1981 when four birds frequented a bird table at Balder Grange, Baldersdale, during heavy snow. In similar fashion, the hard weather of January and February 1986 forced birds to feed on seeds and peanut fragments in the gardens of Durham City and Newton Aycliffe. In the early 1980s, 26 birds were caught and ringed in a hundred square metre trapping area, within which Pheasants *Phasianus colchicus* were being fed, between November 1980 and February 1981, in the Ravensworth Estate. Today, up to six birds are frequently noted in many gardens but counts of more than double-figures are less usual. Ten were at feeders at Elemore Golf Course on 20 December 2009 and 31 were at Hetton Bogs on 26 October 2007.

The national data set indicates that there is some immigration of continental birds (Wernham *et al.* 2002). Based upon this, it seems likely that some continental birds occasionally occur in Durham but there is little empirical evidence for this, despite assertions to the contrary. The best evidence is provided by small flocks flying west past a fishing boat ten miles off Sunderland, over the period 26th to 29 October 1976, when there was said to be a 'definite' influx to the Marsden area, and by reports of birds landing on offshore oil rigs.

Birds included in smaller autumn counts by observers looking for migrants of other species can be difficult to distinguish from resident birds which are not normally counted. The Coordinated Migrant Count on 4 October 1994 amassed a total of 287 Dunnock along the county's coastal strip, making it one of the most widespread species along the coast on this date. This demonstrates the number of birds that can be present in these coastal areas but their origin, local or immigrant is almost impossible to determine. Four birds flying south west, high over Hartlepool on 3 October 1981, seem likely to have been genuine migrants. Large numbers at migrant watch points are unusual, although 100 were counted at Hartlepool Headland on 1 October 1998, perhaps genuine continental migrants. There is no known ringing data that confirms immigration locally. Passage birds are difficult to separate from local breeding birds, or any dispersing birds from further north that might be moving along the coast. It seems most likely that many of these birds have come from not very far away, but one bird ringed at Hartlepool on 5 November 2000 was recovered at Stannington, Northumberland, on 30 March 2001.

A fully albino bird was present at Drinkfield Marsh in November 2006, a most unusual occurrence.

Distribution & movements

The Dunnock breeds throughout northern and central Europe, as far east as the Ural Mountains. Most populations are sedentary, though some of the northerly birds, such as those from Scandinavia, of the race *modularis*, do move south into wintering grounds that are largely located within the breeding range of more southerly birds (Parkin & Knox 2010). No more than a few tens of these individuals usually occur at any site on any given day, though exceptional influxes can occur (Brown & Grice 2005).

House Sparrow
Passer domesticus

An abundant but, in recent years, quite rapidly declining resident.

Historical review

The House Sparrow is the avian species most closely associated with people and to the majority of modern town dwellers it is probably the most familiar of birds, partly because it is substantially confined to built-up and suburban areas; though it may move out from these to the surrounding countryside for a few weeks at the end of the breeding season in order to feed on ripening grain and weed seeds (Summers-Smith 1963).

In the Victorian period, J. Backhouse (1898) commented on its scarcity in the uncultivated west of the county, saying that numbers could be seen at Middleton in Teesdale, but just a few miles away birds were rarely seen even as high up the dale as Newbiggin (approximately 4km to the west). Robson, writing from the perspective of the lower Derwent valley, thought it *"one of the commonest, if not the commonest resident of the vale"* (Robson 1896). Nelson (1907), commenting upon its status at Teesside, said that in spring and autumn, considerable numbers crossed the North Sea, with large flocks of clean-looking birds, which he believed were *"undoubtedly migratory"* but these may have been freshly moulted, locally-bred birds. These, he said, fed on the *"reclaimed lands"* at Teesmouth. Nelson regularly recorded migrants on the coast but there are no recent observations of such movements (Blick 2009), despite Stead's assertion that it was *"a regular passage-migrant on the coast: in small numbers"* (Stead 1964).

Echoing Nelson's earlier comments, Temperley (1951) called it an *"abundant resident and winter visitor"*; perhaps these were birds from further north in Britain? He identified some declines in numbers, which he associated with a change from horse-drawn to motor vehicles in the first part of the 20th century. He said that *"vast flocks visit corn fields"* in autumn and winter, indeed flocks of over a thousand birds used to be seen quite frequently in many parts of Durham at this time of the year (Blick 2009), the only time that birds moved away from their breeding areas.

One of the most fascinating series of records of this species in Durham relates to information gathered by Denis Summers-Smith in the early 1960s. On 21 March 1962, a report on the BBC Northern News told of a pair of House Sparrows that had built a nest at the foot of the shaft of the Waldridge D Pit, at Chester Moor Colliery. The location of the nest was at a depth of 182m (596 feet) below ground level. Subsequent investigation revealed that a pair of birds had been seen underground on 12 March, and that since then they had been feeding on crushed oats from the pit stables, as well as bread crumbs provided by the miners; the birds were also seen drinking water that had collected at the bottom of the shaft. The female of the pair was caught on the night of 29th to 30 March, on the instructions of the mine manager, who did not consider that the birds could survive, and it was released above ground; the male was not seen again after that date. These birds remained within 150m of the shaft bottom, which was lit continuously by electric light, except for the periods from 06.00hrs on Saturdays to 06.00hrs on Sundays and again from 12.00 noon to 22.00hrs on Sundays. The manager informed Denis Summers-Smith that, contrary to the news report, although there was a certain amount of hay and straw lying around, there was no evidence of the birds having attempted to build a nest. In addition to this incident, the manager was aware of two previous occasions when House Sparrows had been found down the pit, but he could give no further details. Another instance of such behaviour occurred in March 1963, when a single House Sparrow was apparently living quite happily at a depth of 300m in Horden Colliery. This bird had been adopted by the miners, who provided it with food, and it was reported that it had become quite tame (Summers-Smith 1980).

Post-Temperley there was little published information about the status of this species, the few records tending to refer to roosts or large post-breeding flocks. An example of the former was of about four hundred birds roosting amongst Starlings *Sturnus vulgaris* in West Hartlepool in autumn 1966, typical of the number of birds that were routinely present in all parts of the county at the time. Through the 1970s, local post-breeding flocks could still be of a considerable size, though by no means as large as those of three to four decades previously. For instance, even into the late 1970s, large numbers were noted in agricultural areas in the late summer. The flock of 1,000 at Bamston in July to August 1976 was just one of numerous such examples.

The trend towards decline was first indicated in the late 1980s, despite the fact that at the outset of that decade this species was a 'pest' which could be legally killed under the stipulations of the Wildlife & Countryside Act (1981). Through this decade some of its more remote breeding locations began to be deserted (Joynt *et al.* 2008), though often such information only became obvious in retrospect, the actual process of decline going almost unnoticed by many observers. Through much of the 1980s it was little studied but routinely described as abundant and widely distributed. In 1980 during the late summer flocks of 300-400 birds were noted at Bishop Auckland and *c.*200 fed in a wheat field on Cleadon Hills, with similar numbers there in late summer in 1983. Flocks reported through much of the mid-1980s were slightly smaller in scale until a flock of around 150 on a cereal crop at Washington 3 August 1987. The following year the largest gatherings of the year came from Ryton Willows (300) and Whitburn (500) in late summer. The largest count of the decade, and for some time to come, was of an estimated 1,000 birds at Whitburn in September 1989. This was the last recorded four-figure count of the species in Durham.

An unusual late breeding record concerned a pair at Washington that still had five eggs in a nest on 8 October 1976. In 1988, a nest under roof tiles of a house in Rowlands Gill was inadvertently disturbed during building work. The nest was removed and placed in a plastic bucket and relocated a few metres away. Within 20 minutes both parents were once again feeding their young, which successfully fledged.

Recent breeding status

In the past, this species was considered 'almost too familiar to be worth study'; consequently, the House Sparrow has been all too rarely censused with any degree of accuracy, until recent years, when its conservation status has focused attention upon it. Today the species is by no means ubiquitous in Durham, but it remains very common and can be seen in a wide range of both urban and urban-fringe situations. Despite major declines across the British Isles (Eaton *et al.* 2009, the House Sparrow is still a common and familiar bird in the north east of England, but there is compelling evidence that local numbers have decreased quite significantly over the past 30-40 years (Summers-Smith 1999). Some of the decreases are readily apparent in town centres, where birds were once ubiquitous. The decline appears to have been less evident, or severe, in suburban areas, which still support considerable populations of House Sparrows in many parts of the county. Birds usually nest in buildings, but at times of higher population density, the species used to be found nesting in more unusual locations and habitats, such as the colony that was present in a dense patch of holly *Ilex aquifolia* at Kibblesworth in the early 1980s. Some authors have noted that a small number of pairs also nest in woodland and parkland, generally in holes in trees (Blick 2009), but this is not a common habit locally.

In terms of distribution, the species is usually reported from locations in the coastal strip, and westwards across the county, into upper Teesdale, as far west as Forest-in-Teesdale. It has always been more local in distribution in the west of the county, so two pairs breeding at Balder Grange in Baldersdale in 1982, in the south west of County Durham, at an elevation of around 195m above sea level, was at that time exceptional. In broad terms, the House Sparrow remains a widespread species of the densely populated lowlands of Durham, but, as might be expected, it is more sparsely distributed but present in the less-populated Pennine uplands. The fine detail of its modern distribution was somewhat masked by the larger scale mapping exercise of the *Atlas*. Closer scrutiny of the species' local distribution indicates that it is not a uniformly distributed bird even in all of the county's built-up areas, but it tends to form loose breeding colonies, of some 10-20 pairs, with the intervening areas between 'colonies' being almost sparrow-free. Although frequently associated with farms, numbers there are also decreasing and some of the more isolated farms and hamlets have, in the last decade and a half, been abandoned by sparrows.

Very little substantive breeding information was generated through the 1990s; a female nest-building in the snow at Great Lumley in February 1991 was notable. As the species appeared to be largely absent from much of the south west dales as a breeding bird in the first part of the decade, a small colony of nine to ten pairs at Valley Farm, at over 350m above sea level on Bowes Moor in 1992, was of considerable interest; such breeding colonies in the west of the county are often associated with the presence of stock in and around farm buildings.

In line with its declining status, the species became increasingly well reported through the first decade of the 21st century but, whilst there were many reports, for example over 1,100 records in 2008 submitted to the Durham Bird Club, in most years casual observations by birdwatchers produce virtually no usable breeding data.

Whilst information concerning a few juveniles at widespread locations is often provided, this is usually too scant, or haphazard, for meaningful interpretation. The development of post-breeding flocks in late summer is more useful in indicating where congregations of colonies, or numbers of breeding birds, are located. In this sense, 2007 saw significant numbers of birds in the following areas: 200 at Clara Vale, 100 at Seaham and 40 at Fellgate. Flocks noted at the end of 2007 may also have been largely comprised of locally-bred birds, including 100 at Escomb in December, with gatherings of 30-45 birds at: Bear Park, Coxhoe, Houghall, Langley Moor, Meadowfield, New Brancepeth, Old Quarrington, Tow Law, Trimdon Colliery, Waterhouses and Witton.

There is little doubt that the radical changes in farming after the Second World War, and the disappearance of practices such as the widespread, small-scale keeping of free-range hens and the consequent reduction in the availability of food for rural sparrow populations, has contributed to the species' decline. The driving force for the reduction in urban populations, where the provision of garden feeding is now very common, is less clear. It seems most likely that cleaner gardening and an increased use of insecticides has led to a reduced number of invertebrates, which are essential for the successful rearing of young (Mead 2000).

Survey work undertaken in 2008 indicated that some of the largest and most dense populations were now around some of the county's smaller towns and settlements. At 2009, in Durham, unlike its cousin the Tree Sparrow *Passer montanus*, the House Sparrow seemed to be showing little sign of recovery. At that time reports suggested that in relative terms, there were few birds in the county's major conurbations (Newsome 2010). Good numbers however, were still to be found in most lowland settlements, the suburbs of the larger towns and even in the ribbons of rural villages and farmstead groupings in the west, such as: Bowes, Daddry Shield, Eastgate, Harwood-in-Teesdale, Ireshopeburn, Kinninvie, St. John's Chapel, Wearhead and Westgate.

Despite such declines, in the first decade of the 21st century, the House Sparrow was still abundant in Durham, with most breeding birds concentrated in the larger settlements and conurbations. In the south east of the county, birds were concentrated in the built-up areas, and the fringes of these, around Hartlepool, Teesside, west to Billingham and Stockton (Joynt *et al.* 2008). In some areas it was very thinly spread, as around the North Tees Marshes, which have so few breeding sites for the species; nonetheless there were probably as many as 6,000 pairs breeding (Joynt *et al.* 2008). Based upon the population estimates made during the *Atlas* work (Westerberg & Bowey 2000), the population of Durham was believed to lie in the range 25,000 to 50,000 pairs and this broad range probably still encompasses the population figure at 2010; although it is likely that the actual population is now moving quite firmly in the direction of the lower figure.

Recent non-breeding status

Nest building is usually first observed in March, but it can be earlier in mild winters, first brood young are noted in May and by June fledged juveniles are usually widespread. In rural and urban-fringe areas, July and early August sees the build up of post-breeding flocks – these usually provide the largest counts of the species each year. The diminution in the size of such post-breeding flocks, which commenced in earnest during the 1980s, continued through the 1990s. A better understanding of the species' local decline has been hampered by a lack of data (for example, the 'zero counts' on the graphic of the largest flocks below are actually 'no data' points) and a reliance upon anecdotal reports of how common the species once was. Three hundred birds at Ouston in August 1990 proved the largest gathering noted that year and for much of the rest of the decade the largest flocks numbered no more than this. For example, 300 were close to the Whitburn Observatory in August 1993 and 200 were at Witton Gilbert that month. The largest flock recorded later in the decade was 150 at Birtley in July 1998, but the species was poorly reported through much of the 1990s. That it was declining was evident from the fact that two birds at WWT Washington on 5 March 1998 were said to be the first recorded at that site since the previous April; in earlier decades it was usually present there. Reports of large post-breeding flocks through the 2000s were few but in 2008, 250 were noted at Whitburn in August.

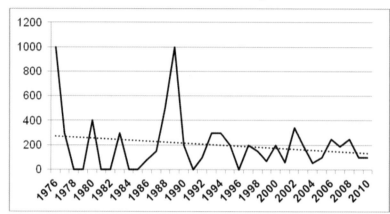

Largest post-breeding/winter flock of House Sparrow in County Durham, 1976-2010 (with trend line)

Most reports of House Sparrows concern feeding flocks during the winter periods, with early year counts numbering up to 50 being reported from many locations. Larger winter gatherings of up to 300 were noted over the period 1990-2010, especially where food was abundant, such as on stubble fields and grain stores for example, but in the second half of the first decade of the 21st century, winter flocks generally amounted to no more than 100-200 birds, and there were fewer flocks to be seen. The species still seemed to be thriving in villages further west, however, with good numbers noted in 2010 at: Crook, near Derwent Reservoir, Hury Reservoir, Kinninvie and Romaldkirk. A study of a wintering flock of 120 birds that were coming to a Whitburn feeding station in early 2008 revealed that this comprised of 75% male birds (Newsome 2009). This single sex winter flocking is well known in for example Chaffinch *Fringilla coelebs* (Newton 1972) but may also occur in the House Sparrow more frequently than previously realised.

House Sparrow seems to be a species that often produces birds with aberrant plumage, or at least its close proximity to man allows these to be documented. For instance, a coal-black example with two or three normal individuals was seen at the Coach and Horses Inn Birtley "*towards the end of the great frost*" in winter 1929. The late Chas Robson had in his possession a beautiful cinnamon specimen caught in the same area (*The Vasculum* Vol. XVII 1929). Other interesting plumage aberrations over the years have included an apparently melanistic male at WWT Washington on 22 January in 1975 and in the same year, an albino bird present in a flock of around 100 at Blackhall Colliery in August. An aberrantly plumaged male, lacking any black bib and with a broad chestnut-coloured breast band, was at Sunderland on 13 July 1994. An almost totally white bird was at Crookfoot Reservoir in July 1981 and a similar individual was at East Rainton on 17 September 1995.

Distribution & movements

A very common and widespread species breeding across Europe and Asia, which has also been widely introduced to locations from North America to Australasia. Birds are extremely sedentary.

A detailed study conducted in the 1950s by Dr J.D. Summers-Smith, partly in Stockton, Durham and partly in Hampshire, concluded that the vast majority of House Sparrows didn't move more than a kilometre or two from their birthplace (Summers-Smith 1963). It is absent from mature woodlands and most open ground although small isolated colonies occur around farm buildings in such places.

The fact that this species is very sedentary is illustrated by its behaviour around the Thornley Woodlands Centre. Surprisingly, it was not recorded visiting the Centre's feeding station, during thousands of observer hours between 1983 and 1992. This despite the fact that adults nesting little more than 550 metres away, at Lockhaugh, occasionally visited the car park of the Centre, less than 200m away from the feeding station, during the breeding season to collect caterpillars from the trees for their young. Similar local foraging behaviour was recorded in May and June at Shibdon Pond annually between 1983 and 1999 (K. Bowey pers. obs.). Here birds came to collect insects to feed their young in nests located in the housing estate approximately 400-500m to the south; for the rest of the year the species was usually absent.

The ringing data from this species in Durham confirms its relatively sedentary status. For example, a fully-grown bird ringed at Hamsterley Village in Weardale on 11 January 2003 was found injured there on 12 August 2004. Between ringing station and recovery point, there was no greater distance than a few hundred metres, and one can speculate on whether it moved more than this in the 579 days since it had been ringed. Other apparently typical ringing returns were of:

- A bird ringed at Stockton on Tees on 10 May 1958 which was recovered there on 3 July 1964
- One ringed at Billingham on 15 May 1960, recovered there on 31 May 1964
- A bird ringed at Billingham as an adult on 3 December 1960 which was still there on 8 March 1966
- One ringed as an adult at Billingham on 21 January 1961 that was still in the same place on 4 February 1967.

Of seven House Sparrows ringed and recovered in the north of the county or around Durham City in the period 1990-1999, the average distance to recovery point from ringing sites was just 1.1km; the longest movement was an exceptional four kilometres. The second furthest travelled bird was discovered by Professor John Coulson, in his garden at Shincliffe village, outside Durham City, in June 1999. It is with some irony that such is the decline in this species in that area, that he commented during summer 2011 that he had not seen a House Sparrow in his garden there for over three years (J. Coulson pers. comm.).

Tree Sparrow (Eurasian Tree Sparrow)
Passer montanus

A locally common but sparsely distributed resident, this species used to be very much more widespread but is now much reduced in number.

Historical review

Today, the Tree Sparrow remains a quite widespread species in Durham but nowhere is it as numerous as it once was. It breeds largely in agricultural areas and probably always has done where suitable nest sites are available, but local numbers are known to fluctuate quite dramatically over a period of years.

The species was first identified on Tyneside by John Hancock in 1831 when he killed one at Rabbit Banks, by the side of the Tyne, near central Gateshead (Temperley 1951). In 1840, Hutchinson wrote, "*not common*", and he observed that small numbers of pairs nested in the Banks at Durham. In the south east of Durham, Hogg (1845) recorded birds breeding around Norton, as early as 1829. Hancock (1874) documented it as breeding at Whitburn and near Durham; a clutch of nine eggs being taken at Durham in 1864. In the north of the county, Hancock found six to eight pairs at Heworth Hall, in what is now east Gateshead, on 22 June 1885. In the Derwent valley, Robson thought it much less common than the House Sparrow *Passer domesticus*, though he did know of its nesting in Axwell Park and a few other localities (Robson 1896).

Into the 20th century and Temperley (1951) observed that the coast between Marsden and South Shields, had *"long been one of its strongholds"*. Bolam (1915) reported that a few pairs nested around Eggleston Hall and Newbiggin (in Teesdale). In the 1920s and 1930s, Darlington Sewage Farm was noted as a favourite haunt in autumn and winter. In his mid-century review, Temperley described its status across the county as a *"resident in small numbers"*, with a very local distribution.

During the 1950s and 1960s, there were considerable increases in the British population, and these higher levels were maintained for a decade or so, before decline commenced (Joynt *et al.* 2008). During the 1960s the population appears to have reached a peak countrywide. At this time it was considered quite numerous as both a spring and autumn passage migrant along parts of the Durham coast, for example, 46 were counted passing Hartlepool on 8 October 1970 (Blick 2009).

During much of the 1970s it was felt that this species was not as common as House Sparrow and that its distribution was very much more localised. In 1971 there was a report of a small colony of birds nesting in a most unusual situation, a sea-cliff to the north of Seaham. During the 1970s and early 1980s, a significantly-sized colony was established around the Water Tower on Cleadon Hills, South Shields, but by the end of that decade these birds had been lost.

Over this period maximum reported winter flock size hovered around the 150 to 200 mark and these tended to be concentrated in the east of the county and in coastal areas, where arable practices predominated. For example: 150 were at Cowpen Bewley, on 30 January 1972; in 1973, Hawthorn Dene was considered the species' main stronghold. Also in 1973, there was a small scale coastal passage after mid-May that was noted at Hartlepool Headland and Marsden. At the start of 1975, *c.*100 birds were at Durham City and near Darlington but the largest flock of that year was of around 200 birds adjacent to Peepy Plantation, Washington in October. Through 1976 most of the reports were in the east of the county, 200 were at Darlington on 9 February 1977, and flocks of this size persisted through the decade at scattered locations across the county.

From the 1950s to the 1970s/1980s, the British population increased fourfold, before a major decline from the 1990s led to the species reverting to earlier levels (Summers-Smith 1989). Some local, long-term Common Bird Census studies have indicated that the number of Tree Sparrows reached a peak in the late 1960s and a trough in the late 1970s and early 1980s (Blick 2009). Post-Second World War agricultural practices, including the increased usage of agrochemicals and a move to increased ploughing regimes, combined to significantly decrease the availability of weed seeds for adults and invertebrates for nestlings. In addition, the change from the spring to autumn sowing of arable crops has led to a reduction in the availability of winter food amongst arable stubbles. Adult Tree Sparrows feed mainly on the seed of small herbs, but the nestlings require animal food in the first few days of life (Summers-Smith 1988).

Recent breeding status

As in the rest of the British Isles, the Tree Sparrow in Durham is predominantly a bird of lowland agricultural landscapes. It nests mainly in cavities in hedgerow trees and, where it occurs close to farms, occasionally in outbuildings (Summers-Smith 1988). The species is largely absent from the county's major conurbations, settlements and associated suburban areas. Hence, birds are absent from much of the 'built-up' areas from Gateshead to the mouth of the Tyne, from Chester-le-Street through to Sunderland and from Stockton east to Teesmouth, although there are small isolated pockets of birds in some of the small agricultural islands, in between these areas (Westerberg & Bowey 2000). In 2007 and 2008, observers submitted over 800 records of the species and the vast majority of these concerned birds in the eastern and central parts of the county. It is thought that competition from the Tree Sparrow's larger, dominant congener, the House Sparrow, is partly responsible for determining some of this distribution (Marchant *et al.* 1990). The species is not uniformly distributed in what appears to be suitable farmed habitat, and is absent from many areas which seem to be eminently suitable. It is mainly found on the central, lower-lying land, with fewer birds and colonies along, or more accurately, just inland from the coastal strip.

Despite its scientific name *montanus*, Tree Sparrows are not found to any great degree in the county's Pennine uplands. The most westerly location where the species can be regularly noted in the south of Durham is around Barnard Castle, where a significantly-sized colony exists along Green Lane, to the east of the town, where boxes have been provided for the birds. In the centre of the county, birds can be found as far west as Satley in the headwaters of the Browney valley, and in the north at Waskerley and Blackhall Mill.

By the early 1990s, concerns were being expressed by local observers about an apparent contraction in the species' local range. In 1993, breeding was reported from eight sites, but this was just a fraction of its true status, and there was undoubtedly an observer bias that skewed its perceived distribution. A 1994 targeted survey by Durham Bird Club, in response to concerns over its decline, led to the species being reported from 77 locations across the county and produced a population estimate of between 300 and 500 breeding pairs. In this survey, breeding birds were concentrated in an area to the north west of Durham City and across the agricultural belt running from Sherburn towards the coast.

In many places in Durham, Tree Sparrows nest as isolated pairs, though by and large, they form loose colonies of two to three pairs, where tree cavities are abundant enough to facilitate this (Summers-Smith 1988). Use of nestboxes by this species developed through the 1990s and this appears to have been first documented in the county at Kibblesworth in 1992, when two pairs used boxes. Six pairs used boxes there in 1993, rising to seven in 1994 and 14 in 1995, reaching 17 occupied boxes by 1996; at that time, these produced 47 young from 14 successful nests.

The provision of nestboxes probably helped sustain local increases through the first decade of the 21st century and a number of new colonies became established where these were provided, nesting even taking place in some people's gardens at the edge of agricultural landscapes, such as at Easington Village in the east of the county. Many hundreds of boxes have been located since the late 1990s, birds readily taking to them, and they can nest very successfully in such artificial situations. In the period 2007-2010, birds were breeding in at least 15 sites across Durham where nestboxes had been provided, such as Wallworth Moor Farm, to the north west of Darlington, where 30 pairs used boxes in 2007, with smaller numbers of birds at sites such as Byers Green and Hurworth Burn Reservoir. In this period successful breeding was confirmed at: Barlow, Burdon, Bishop Middleham, near Crook, Crookfoot Reservoir, East Rainton, High Grange, Hurworth Burn Reservoir, Hurworth-on-Tees, Langley Moor, Low Thornley, Peartree, Preston-le-Skerne, Old Durham, Ramside Hall, Sherburn and Stotfold Farm. Undoubtedly it also occurred elsewhere, as witnessed by a large post-breeding concentration of 100 birds at Clara Vale in 2007, in the county's north west.

Since the late 1990s, there is some evidence of an increase in numbers in parts of the county, for example in some areas birds have taken over farm buildings abandoned by House Sparrows, as has been observed at Clara Vale. Reports over recent years have been encouraging, showing signs of recovery after a long period of decline, with some local observers commenting that they were seeing the largest numbers of birds in their areas for a long time.

Extrapolating a breeding density derived from a sample survey, the *Atlas* suggested a county population of between 400 and 700 pairs (Westerberg & Bowey 2000). This may have been something of an underestimate, for between 100 and 120 pairs are now known to be scattered across the agricultural landscape north and west of the Tees estuary (an area poorly coved in the *Atlas*), with birds largely found well away from the urban and developed areas in that part of the county (Joynt *et al.* 2008). It seems likely that with the recovery in numbers since the *Atlas* population estimate was made, and with the greater detail now available, the population figure at 2010 should be considered to be in the range of 800 to 1,000 pairs across the county.

Recent non-breeding status

Post-breeding flocks are usually recorded from late July and through August; later in the year, these flocks coalesce into larger winter gatherings and during the 1980s and early 1990s, large winter flocks were not infrequently found on farm stubble, where the species often gathered with finches and buntings. Through the 1980s, non-breeding flocks had a tendency to be declining in size compared to those recorded through the previous decade. Small post-breeding and wintering flocks of 40–50 birds, reported from South Hylton, Cleadon Hills, Ryhope Denemouth and Marsden, were typical. Highest counts of the year were usually in the order of 90-150 birds between 1981 and 1988, although in a small number of years, no reported flocks reached three-figure counts. In such a situation, up to 100 birds were present at Marshall Lands Farm, Whickham on 27 December 1983. Some of the larger counts often came from the Houghton/Hetton area to the south of Sunderland and the farmland from around Whickham, but there would inevitably have been some element of observer bias in this pattern of reporting. Towards the end of the decade there were increasing anecdotal comments such as 'numbers well down' on previous years. The largest numbers of the decade came in 1989 with large flocks at Tantobie (near Burnopfield) at both ends of the year, peak counts being 85 in January and 200 in December.

Through the 1990s, this trend continued, as indicated by the diminution of winter flock sizes around Durham. In 1990, 95 birds were at Houghton Dene in January and February, with 60 at Langley Moor in November and December of that year and during the winter of 1991/1992 a flock of 60-70 birds visited the Clara Vale feeding station; possibly indicating that winter forage was less available in the wider countryside. In 1992, seventeen wintering flocks of between 10 and 50 birds were reported during the early part of the year, with flocks of up to 70 birds strong being noted at an additional seven sites at the end of the year. This pattern and number of birds seemed to indicate that Tree sparrow remained relatively widespread across the main agricultural areas of the county, though it was nowhere numerous. An analysis of the 1994 Tree Sparrow survey results showed that at least 705 individuals were present across the county, but no records were received from the southern part of Durham, south of a line between Bishop Auckland and Seaton Carew. Birds were known to be in this gap at the time, such as around Newton Aycliffe and Owen Grange, Hartlepool. At this time, the October and November totals for the whole county, inflated by the young of the year, were as high as 723 birds, with one flock of up to 120 at Hurworth Burn Reservoir, a traditional breeding site for the species, being one of the largest of the decade (Sowerbutts 1996). The largest other recorded winter flocks of the 1990s were 250 at Sedgeletch on 19 March 1993 and, later in the decade, 120 at Newton Aycliffe on 1 November and 82 at White Leas Farm near Easington on 26 November 1999.

During the early 2000s, numbers saw some gradual recovery and areas such as around Byers Green, in the Wear valley, began to attract slightly larger numbers of birds in the winter period. More heartening perhaps was the increasing number of winter flocks that were being noted towards the end of the decade. For instance, in the first quarter of 2009, 25 sites held flocks containing between 10 and 90 birds. Assuming an average number of 30 per flock would give a theoretical county total of 750 birds, comparing favourably with the numbers from the 1994 survey. The largest flock of 2009 came on 23 November when 150 were at Farnless Farm, benefiting from crop systems introduced there specifically for farmland birds. Most winter flocks break up through March, or early April at the latest, although breeding season activity can often be noted from February onwards, as males start proclaiming ownership of nest sites, whether these are nestboxes or natural sites.

This species was once considered a regular coastal migrant. Small numbers of birds were regularly seen moving past Durham coastal watch points and along the coastline during April and May in the 1950s and 1960s, but by the late 1970s and 1980s very few birds were recorded undertaking such activity. Something of an exception was the autumn of 1989, when up to 10 at Whitburn and six at Mere Knolls Cemetery, Seaburn were considered to be passage migrants. In 1994, possible immigrants were noted at Marsden on 2 October and at Dawdon Blast Beach on 24th. In 2008, presumed migrants were noted at Whitburn on just three dates; most interestingly, on 17 October, high-flying groups of seven and eight birds flew south. Since the mid-1990s, the number of years in which no birds were seen at the county's main coastal migration watch points has increased. This may indicate a change in status or behaviour of this species over the last four decades of the 20th century, for such visible migration activity is now very rarely noted.

Distribution & movements

The species is common and widespread through much of Eurasia and is almost entirely sedentary in the core of its range, though some northern European birds move south and west, occasionally exhibiting eruptive behaviour.

Sponsored by

NORTHUMBRIAN WATER

Yellow Wagtail (Western Yellow Wagtail)
Motacilla flava

A common, though very much declining, summer visitor and passage migrant.

Historical review

There is relatively little documentation about this species in Durham other than that it is known, in the 19th century, to have bred in some numbers in upper Teesdale, especially in the meadows between Middleton and Langdon Beck (Temperley 1951). The race

flavissima has been known from Northumberland and Durham since it was first recognised in the 1830s (Holloway 1996), but some of the most significant early records of this species relate to races other than the British one (see Blue-headed Wagtail *M f. flava*).

Hancock (1874), considering the whole of Northumberland and Durham, thought it was once rather a common species in the north east but *"of late it has become somewhat less plentiful"*. It did not appear at Teesmouth until the third week of April according to Nelson (1907), but he told of large numbers of migrating birds that were noted along the coast in the latter half of August and the first fortnight of September. In the late 19th century, Teesmouth was *"something of a northern stronghold for the species"* (Joynt *et al.* 2008). In the lower Tees valley, it was once well distributed in the area around Billingham, but by 1950 much of the habitat that it once used in this area had been built upon.

"A summer resident and passage migrant, locally common" was how this species was described by Temperley (1951), he talked about its breeding *"with great regularity"* in the coastal strip and up the county's river valleys. At his time, Temperley thought that it was still extending its breeding range in County Durham. In the 19th century the species could be found in the marshes around Dunston Haugh and in the early 20th century it was reported to have nested under cabbages in an allotment at the mouth of the Derwent (Bowey *et al.* 1993). The fact that one of its then strongholds in the county, was Dunston Haugh, an area of flat, flood-influenced grasslands adjacent to the Tyne near the mouth of the Derwent, is somewhat ironic, as in the rest of the Derwent valley, where there was so much woodland, it was not common. Elsewhere in the south of the county, flocks of up to 30 were noted inland at Darlington Sewage Farm in early May. During the early to mid-20th century, Hugh Blair stated that it was common in the area of South Shields (Temperley 1951). On 2 September 1947, Fred Grey found a pre-roost gathering near a pond in north east Durham. Along the lower Tyne valley, it was present near Jarrow in 1943 and also in the lower Team valley during 1948.

Little was recorded on this species during the 1950s and 1960s, other than that it was widespread as a lowland breeding species in Durham, and indicative of this was the statement that in 1952 it was considered in south west Northumberland, *"nearly as common as pied"* as a breeding bird (Temperley 1953). In that year, three pairs were recorded breeding in an oat-field near Boldon Flats, the first local record of its using an arable crop, cabbages aside, for breeding. In the same summer, on July 21st at least 100 birds had gathered with Pieds *Motacilla alba* at Tanfield Ponds in a mixed species, pre-roost gathering.

In the south east of the county, Stead (1964) said that many pairs were still nesting on 'slag tips' adjacent to Grangetown and Port Clarence, as well as along the sea wall on the North Tees Marshes, and birds were still using these habitats in the late 1960s (Stead 1969). A very early migrant was at Greatham Creek on 23 March 1966 and this remains one of the earliest ever records for the county. At the other end of the season, a relatively late bird was present at Dorman's Pool at Teesmouth on 23 September 1968. Of considerable interest from the point of view of its racial identity was the 'yellow-headed' male that was at Teesmouth on 12 May 1969. Considering the range of Yellow Wagtail subspecies that this part of the county has attracted through the decades (see 'Other Yellow Wagtails'), one might speculate whether this might have been a bird of the race *lutea*.

In the late 1960s and through the 1970s local numbers were appreciably higher than at present. In Gateshead for example, many pairs were breeding, with four to five pairs on Derwenthaugh Meadows alone, and up to 200 passage birds were recorded as roosting in the reed beds at Shibdon Pond in the autumn. This late summer Yellow Wagtail roost persisted into the 1980s, and its size is illustrated by the fact that 46 birds were ringed there in 1977 (Bowey *et al.* 1993). By contrast, today it is unusual to record a Yellow Wagtail in this area at any time during the spring to autumn period. This turnaround in fortunes for this species, in an area that was formally a stronghold for it, is symptomatic of its position in many lowland parts of the county. Today, it is very much a declining species both in the local context, across County Durham, and nationally (Mead 2000), as many of its traditional wet pasture breeding sites have been reclaimed for industry or drained.

Recent breeding status

The vivid, canary-yellow males with their shrill call today have a relatively sparse and scattered presence across much of County Durham. The Yellow Wagtail's range contracted in the British Isles over the two decades between the production of the national breeding atlases (Gibbons *et al.* 1993) and this has also been the story in County Durham, particularly in the last 15 years or so. The species has always had, and still has, a wide altitudinal range in the county. It can still be found breeding from sea level in the south east to around 450m above sea level, high up on the most exposed pastures of the western uplands, but it is now much more sparsely distributed in-between. Targeted species survey work undertaken by the Durham Bird Club in 1991 (Armstrong 1992) showed that the Yellow Wagtail was mainly distributed along the river valleys of the county, excepting the Derwent. It was strongly associated with its preferred habitat of damp pasture and meadows in these situations.

Almost invariably, the species nests close to fresh water (Mason & Lyczynski 1980) and traditionally, Teesmouth has been a stronghold for this species. Around the North Tees Marshes, about 20 pairs nested in the 1980s and 1990s (Blick 2009). The *Atlas* map (Westerberg & Bowey 2000) showed that there were distinct concentrations of tetrads occupied by Yellow Wagtails along the upper stretches of the Rivers Wear and Tees, as well as the tributaries of the latter river, the Balder, the Lune and the River Greta. Birds prosper in the extensively managed pastures and damp meadows of these areas, where traditional grassland management techniques have persisted to a greater degree than in many other parts of England (English Nature 2001).

The mapped distribution of the *Atlas* (Westerberg & Bowey 2000) showed them to be relatively widely distributed in the Durham lowlands, to the east of the A1, having a presence along the coastal strip and also being present in significant numbers along the central reaches of the Wear. This latter distribution included birds found along the Wear tributaries, the Deerness and Browney. The *Atlas* showed that the species was well represented in the south east of the county, across the Tees grazing marshes, and its considerable preference for this area has been confirmed through more recent survey work (Joynt *et al.* 2008). Compared to the Wear and the Tees catchments, the county's other major river system, the Derwent-Tyne complex, is very much less suitable for this species, as it is characterised by deeper and more heavily wooded valleys, with fewer open grasslands that fringe the rivers, or are situated in the floodplain.

As was observed by Westerberg & Bowey (2000), the change map in Gibbons *et al.* (1993) showed net losses in three of Durham's 10-kilometre squares since the publication of the first national *Breeding Atlas* (Sharrock 1976). This decline has continued in the subsequent 15 years or so. The *Atlas* recorded the species' presence in 99 tetrads, but today it is probably only present in 60-70 across the county (Newsome 2008).

The *Breeding Birds of Cleveland* survey work indicated that there were around 30 pairs breeding around the Tees Marshes (Joynt *et al.* 2008), with a few of these (25-30%) scattered across agricultural habitats to the west of the area. Since about the year 2000 however, the numbers breeding around the North Tees Marshes may have declined to between five to ten pairs, when just a few years ago there were over 20 pairs nesting (Blick 2009).

Early 1990s surveys in upper Teesdale and Weardale found 59 pairs split between those dales, whilst surveys in Baldersdale and Lunedale recorded between 52 and 100 additional pairs (Shepherd 1993). A crude extrapolation from these figures, using occupied tetrad data, gave an estimated County Durham population in the range 500-750 pairs (Westerberg & Bowey 2000).

An interesting insight into the state of a part of the county's breeding population came in 2002, with a repeat of part of the early 1990s 'dales surveys' of Yellow Wagtails, targeting the Weardale Environmentally Sensitive Area (RSPB 2003). This, in part, repeated the survey work undertaken in similar fashion between 1991 and 1995. This work found birds back near Ireshopeburn in Weardale from 4 May, with an influx to the area sometime between this date and the 9th. The survey found 16 nests, 87% of which were in hay meadows, but pairs were present in at least six other areas, indicating a breeding population in this part of the dale of upwards of 20 pairs. The conclusion of the survey at this point was that Yellow Wagtail numbers in Weardale were stable between the two surveys, though across the ESA there had been a contraction of range with birds being more concentrated into a smaller area. Losses of the species in some areas were noted between the two survey periods. It seems likely that optimum hay meadow management regimes were required to maintain Yellow Wagtails numbers in this area of Durham. In summary, during 1991, 25 nests were located in upper Weardale; in 1992, 11 nests with 13 additional locations holding juveniles (i.e. probably 24 pairs); in 1993, 20 nests were located, and in 1995 there were 17 nests plus eight locations with juveniles (probably 25 pairs). In 2002, the contrasting situation was thus: 16 nests and six additional locations with juveniles were found (i.e. probably 22 pairs) (RSPB 2003).

Moller (1980) identified the conversion of permanent pasture to arable, and the drainage of damp grasslands and wetlands generally, as factors that could be detrimental to the species. In the Durham context, the earlier cutting of grass for silage rather than the traditionally later hay cut is of perhaps more relevance, despite the efforts to maintain those traditional practices in the county's westerly dales. Conversely, Westerberg & Bowey (2000) noted that in recent decades some birds have been observed using new habitats, in particular arable crops, including oilseed rape.

Compared to the numbers present in the early part, and even the middle of the 20th century, only small numbers of pairs breed around the county today and these are largely in the west and around Teesmouth, with occasional pairs at other scattered localities, often in crop fields associated with nearby wetlands, for example at Seaton Pond. In 2007, there were a total of 125 submitted records of this species and it was recorded from almost 30 locations. The bulk of the reports (71%) however, came from just three main sites: Bishop Middleham (52 records), the North Tees Marshes (24) and Hurworth Burn Reservoir (13). In late July 2007, there were many juveniles amongst a total of 38 birds around Bishop Middleham, indicating successful local breeding (Newsome 2008). The year 2009 appeared to have been a relatively good breeding season for this species with birds on territory at many central Durham sites and 39 breeding pairs confirmed. Notable amongst these was the first successful breeding at Sedgeletch for 13 years and more than 15 pairs scattered across the Mordon Carrs area (Newsome 2010).

At present, the total number of breeding pairs probably falls between 350 and 400 pairs, but is now perhaps nearer the lower figure in most years. The national declines in this species mirror what has occurred in Durham, with observed declines of 65% between 1967 and 2003 according to various BTO data-sets and by up to 90% according to the Waterbirds Survey information since 1975 (Parkin & Knox 2010).

Recent non-breeding status

The arrival pattern of this migrant is fairly predictable, the first bird of the year usually appearing between the 10th and 16 April each year, but most birds arrive after mid-April. The early birds are most often noted when feeding around wetland sites, where early season congregations of insects are available. In recent years, spring passage has become lighter in the county, with just a scattering of birds along the coastal strip, often between Whitburn and Seaham, during late April and early May. The species is usually more frequent on the North Tees Marshes. After the first few birds arrive the majority come along over the next week or two. It was, at one time, not unusual for gatherings of from 30 to 50 birds to be present in the Saltholme Pools, Dorman's Pool and Cowpen Marsh area in the first week or so of May (Blick 2009), but such large numbers are now rarely seen. In 2006, spring arrivals started as early as 7 April at the Sunderland Football Club Academy Pools, followed by the largest recent count of 77 birds on Seaton Common on the 25th (Joynt 2007).

By mid-July, reports of juveniles are usually widespread and gatherings of post-breeding birds have begun to build up at favoured sites. In the summer of 1990, Longnewton Reservoir was drained and up to 30 Yellow Wagtails were seen there in August (Blick 2009), a very high count for an inland location. Birds are usually visiting roost sites by the end of July. During the 1970s and 1980s, some very large autumn roosts developed around the North Tees Marshes. These averaged around 150 in August of each year but at least 247 were roosting at Haverton Hole on 3 August 1975 with a maximum count of more than 300 at Dorman's Pool in 1981 (Blick 2009). These roosts declined through the late 1980s and early 1990s, with 70 in 1992 and 150 in 1994 being the only significant counts from the last two decades (Joynt et al. 2008).

As elsewhere in Durham, there has been a considerable reduction in the number of birds noted around Teesmouth in both spring and autumn over the last two decades (Blick 2009). Nonetheless there is still usually a build-up of birds here in the early autumn and in 2007, by way of example, the county's highest counts came from here, with 31 birds on 1st and 26 on 2 September. The autumn departure takes place through August, with a few birds still to be seen in mid-September, mainly in the south east of the county. The earliest and latest dates of this species have both come from this area, Teesmouth. The earliest was at Saltholme Pools on 16 March 2002 (Newsome 2005) and the latest recorded occurred on 3 November 1975 (Blick 2009). Unlike some migratory species that visit Durham, an analysis of the species' first arrival dates over a thirty-year period, from 1976 to 2005, revealed no substantive change in the mean first registration date of this species in the county, which is 13 April (Siggens 2005).

Distribution & movements

Yellow Wagtails belong to the complex of subspecies that make up the 'Yellow Wagtail group', which, as breeding birds, spread across much of Eurasia, south of the Arctic and to a limited extent in North Africa. At least 13 sub-species are currently recognised worldwide (Alstrom et al. 2003), of which at least eight of which have occurred in Europe. The Yellow Wagtail *M. f. flavissima* breeds in the British Isles and adjacent areas of the North Sea Basin. The nominate or Blue-headed race *M. f. flava* breeds across much of Europe, and occurs in Durham as a scarce though annual passage migrant (though it has bred), whilst other races are more akin to rare vagrants in Durham. The Grey-headed Wagtail *M. f. thunbergi* breeds across much of Scandinavia, northern Russia and western Siberia, whilst Black-headed Wagtail *M. f. feldegg* is restricted to the Balkans, Turkey and the Middle-East. Most races of Yellow Wagtail winter in Africa, to the south of the Sahara, though the most easterly races winter in South East Asia. The taxonomy of this species is complex and has not been much clarified by the advent of the analysis of mitochondrial DNA techniques and, at 2010 further research is required to clarify the relationships between races of this species (Parkin & Knox 2010).

The regular use of the North Tees Marshes during spring passage and the large late-summer/autumn roosts in that area once made Teesmouth important for Yellow Wagtails, but as the species has declined so has the area's significance for them. Ringing data attests to the fact that the large Teesmouth roosts, and those at Shibdon Pond, in the late 1970s and 1980s, drew birds from much further afield than the immediate area or even the wider county, as shown by a selection of controls and recoveries.

Firstly, those ringed in Durham and recovered further south in England:

- A bird ringed at Shibdon Pond on 20 August 1972, controlled at Christchurch, Hampshire on 6 September 1972
- An adult caught at Shibdon Pond in summer 1976 and controlled at Brough, Cumbria on 6 September 1976
- One caught by the Long Drag on 2 August 1987 that was in South Wales on 1 September 1987

Birds ringed in the north east and recovered abroad:

- A bird ringed as a nestling on Boldon Flats on 9 June 1954 was recovered at Vieux Boucau, Landes, France, on 2 September 1954, about 1,200km to the south
- One caught at Teesmouth in September 1983 was in Morocco as late as 8 May 1984
- An adult bird ringed at Saltholme on 7 September 1987 was found dead just over three years later on 17 October 1990 near Casablanca, Morocco, a distance of 2,475km to the south south west
- A juvenile caught at Haverton Hole on 5 August 1989 was present in Guadalajara in central Spain on 14 April 1991
- A bird ringed as a juvenile at Haverton Hill on 11 August 1990 was controlled at Banjul, Gambia 4,786km south west on 23 October 1992.

These snippets neatly illustrate a simplified migratory route for north east Yellow wagtails, of heading down through England on a south or south westerly vector, gathering on the south coast and then heading south through France and Iberia into North Africa, at the western end of the Mediterranean and thence south and on into West Africa. This hypothesised route is supported by the record of one ringed at St. Ouen, Jersey, in the Channel Islands on 14 September 1958 and recovered the following autumn in Durham, on 4 August 1959 and by the national ringing data (Wernham *et al.* 2002). Of course, some birds that pass through the county are breeding birds from further north in the north east of England, as indicated by the juvenile that was ringed at Haverton Hole on 24 August 1985 and found dead at Allenheads, Northumberland, on 10 May 1991.

Identifiable sub-species of Yellow Wagtail

The above account relates to the British race, *M. f. flavissima* but at least another seven races of this species have been reported in Durham. All have occurred principally as passage migrants in the lowland east of the county, many of them from Teesmouth, and are typically recorded on spring passage, often amongst arrivals of the 'British' breeding form. The complex taxonomy of this suite of races is subject to frequent review and revision and a number of the forms breed with each other producing offspring of intergraded forms. Each subspecies is given its own account below.

Blue-headed Wagtail
Motacilla flava flava

Historical review

This is the most frequently recorded of the 'Continental' races in Durham and it has bred historically. The first record for Durham of this, the nominate race of Yellow Wagtail, is of a male shot by Thomas Robson at Dunston Haughs on 1 May 1836, which was accompanied by another thought to be a female. This was one of the earliest records of the race in Britain. Blue-headed Wagtail was suspected of breeding at the same locality in 1868 and the following year breeding was confirmed, in May 1869, with two nests being found, the first confirmed breeding of this race in the British Isles.

In 1870 another nest was found on 13 June but it had been crushed by a horse and the eggs destroyed. On 5 July however, John Hancock visited the site, observed the crushed nest and saw a brood of young that were being fed by the parents. Two of these were subsequently shot and the specimens went to the Hancock Collection. There were no further records of confirmed breeding from this site, though a female was shot on 21 May 1872. Some 20 years later a pair was noted breeding at nearby Shibdon Flats in 1892, the female of which was shot off the nest and the clutch of five eggs taken. This is the last confirmed breeding record for the county, though it is difficult to be certain if all of these historical occurrences involved pure pairs and some may have related to hybrid pairings.

Over 50 years later Temperley (1951) described it as a "*Very rare summer resident*", presumably based upon the above sightings, though he gave only one further record of three males noted by C. E. Milburn amongst Yellow Wagtails *Motacilla f. flavissima* at Teesmouth on 6 May 1900. There are no other published records until 14 May 1952 when a male was at Boldon Flats, with another nine records, chiefly from Teesmouth by the close of the 1960s. The majority of these records were in April and May, though single males were: in Gateshead, on 2 August 1959; at Dorman's Pool, Teesmouth, on 23 September 1968; and, Seaton Carew, on 1 July 1969. The latter bird was said to have been displaying and may have been breeding in the area.

Recent breeding status

Birds have been noted feeding young on at least four occasions. These were a male feeding four young on Whitburn Rifle Range on 17 July 1994 (thought to be the same bird that had apparently bred on nearby Cleadon Hill); a male, seen feeding young on the Fire Station Pool, Teesmouth in June 1995 and still present in August; a female at Saltholme Pools from 27 April 2003, which raised two broods; and, a male seen carrying food at SAFC Academy Pools from 11th to 24 June 2006.

Recent non-breeding status

In excess of 100 Blue-headed Wagtails have been noted on passage since 1970, the majority of which have been at Teesmouth, with relatively few birds penetrating inland and no records from the extreme west of the county. Away from the North Tees Marshes, there are records from: Longnewton Reservoir (5); Whitburn (4); Bishop Middleham (3); Boldon Flats (3); Hetton Lyons (3); Bishop Middleham (2); Lamesley (2); MetroCentre Pools (2); WWT Washington (2); and a number of other localities including Castle Eden Denemouth, Ryhope and Rockliffe-on-Tees.

Records span the period from April to September, though most occur in late April and early May. Almost all of the records come from April and May, with the earliest being of two males in the Saltholme Pools/Dorman's Pool area, Teesmouth from 11th to 18 April 1995. Peak numbers are typically recorded in the last week of April. Significant counts include: seven (perhaps as many as 10) males on Seaton Common, Teesmouth from 24th to 26 April 2006 and six on Cowpen Marsh, Teesmouth from 22nd to 27 April 1996, with three there at the same site on 22 April 2000. Passage continues into May though there are few records of multiple arrivals, the best being a peak of four males at Cowpen Marsh, Teesmouth on 2 May 1979; this was a good spring for this form with at least eight recorded at Teesmouth between 2nd and 18 May, though none were recorded elsewhere. All records away from the North Tees Marshes relate to single birds. Few passage birds have been identified after the end of June, with single males at Dorman's Pool, Teesmouth on 4 August 1982; Teesmouth from 3rd to 11 August 1983; Dorman's Pool, Teesmouth from 8th to 11 September 1982; and, at Teesmouth on 15 September 1983. A male at Saltholme Pools from at least 16th to 20 July 1994 was perhaps breeding in the area. The paucity of autumn records is doubtless attributable to the fact that juveniles and most adult females are indistinguishable from *flavissima*. Just one bird has been ringed in the county, a male at Lamesley Sewage Farm on 29 April 1989.

In addition to the above there have been a number of published records of birds showing characteristics of the Asian race *beema*, dating back to 1970, all of which are now considered to relate to intergrades between *flava* and *flavissima* colloquially referred to as 'Channel' Wagtails. Many of the records, particularly the more recent ones refer to birds establishing territory though there are at least a dozen records of presumed passage birds, again predominantly from the North Tees Marshes. Notable exceptions include single males at: Washington, on 31 May 1978; Derwent Reservoir, on 28 May 1979; Whitburn, on 11 May 1991; and, perhaps the same male at Rockliffe-on-Tees from 29 May-3 June 2008 and 12th to 20 May 2009. The bird at Derwent Reservoir is especially noteworthy in being the most westerly record of any non-British Yellow Wagtail in the county.

Additionally, there are a few published records of *beema*-like males holding territory in the county, with at least three such birds breeding at Teesmouth between 1970 and 1974, 1990, and at Castle Eden Denemouth in 1971. All of these records are now presumed to refer to 'Channel' Wagtails. Records of this distinctive intergrade have increased in recent years given increased knowledge of its identification. Further such territorial males have been identified at: Diamond Hall Farm, Sedgefield in May and June 2000; Castle Lake, Bishop Middleham in the summers of 2010 and 2011; Hurworth Burn Reservoir, from 25 April to 12 June 2010; and, at Butterwick from April to June 2011. All of the birds since 2000 have been at inland localities.

Grey-headed Wagtail
Motacilla flava thunbergi

All records:

1969	Barmston Pond, Washington, 9 May
1976	Piercebridge, 8th to 9 May
1977	Washington, 5th to 6 June
1979	Seaton Common, Teesmouth, 15 May
1982	Monsanto Option, Teesmouth, 29 June
1985	Saltholme Pools, Teesmouth, 4 May
1985	Holmes Farm, Whitburn, 11 May
1985	Whitburn, 14th to15 May
1986	Cleadon, 18th to 19 May, (2)
1994	WWT Washington, 12th to 18 May
1995	Fire Station Pool, Teesmouth, 16th to17 July
1997	Saltholme Pools, Teesmouth, 4 May
2003	Sunderland AFC Pools, 13 July
2006	Seaton Common, Teesmouth, 23 May
2010	Cowpen Marsh, Teesmouth, 16th to 17 May

In addition, what would have been the first county record of this form was a male at a pond at Usworth, Washington on 11 May 1958 (Williamson & Ferguson-Lees 1958). This record was not locally verified or published and as such it does not form part of the formal record, though it is believed to have been genuine. Grey-headed Wagtail occurs slightly later in Britain than most other forms; typically it is associated with high pressure systems over Scandinavia in May and early June. The two July records are noteworthy, with the bird at Teesmouth seen to be carrying food though breeding was never proven. All records refer to males.

Black-headed Wagtail
Motacilla flava feldegg

A male showing the characteristics of this striking form was found by Ian Forrest at RSPB Saltholme on 28 April 2010, sometimes elusive it showed well in the early evening (Newsome 2012). At the time of press, this record was still under consideration by BBRC.

The following races have also been published as having occurred in County Durham, though all are now found to be unacceptable by modern standards (Joynt 2007, Bell 2009):

Spanish Wagtail *Motacilla flava iberiae* (Washington, 16 July 1977; Dorman's Pool, Teesmouth, 21 June 1979; Saltholme Pools, Teesmouth, 19 and 23 April 1989; Cowpen Marsh, Teesmouth, 26 April 1996);
Ashy-headed Wagtail *Motacilla flava cinereocapilla* (Greenabella Marsh, Teesmouth, 17th to 18 June 1977; Washington, 19th to 20 May 1978; Cowpen Marsh, Teesmouth, 17 May 1979; Saltholme Pools, Teesmouth, 15th to 23 May 1981; Boldon Flats, 10th to 15 May 1983; Hartlepool Headland, 12 May 2001);
Sykes's Wagtail *Motacilla flava beema* (numerous published records all of which are now thought to relate to 'Channel' Wagtail, an intergrade between nominate *flava* and *flavissima*);
White-headed Wagtail *Motacilla flava leucocephala* (Longnewton Reservoir, 4 May 1985).

The racial identity of individual birds is fraught with difficulty and there is a significant degree of integration between forms.

Citrine Wagtail
Motacilla citreola

A very rare vagrant from eastern Europe and Siberia: three records.

The first County Durham record came in 1984, when Martin Blick found a first-winter bird at Haverton Hole on 14 August. Initial views were brief but the bird was present again in the same area over 18th to 19 August (*British Birds* 88: 530). It was another 15 years before the next Citrine Wagtail in County Durham, and again it was on the North Tees Marshes. An elusive first-winter bird was found by Richard Taylor at Dorman's Pool on 4 September 1999 and remained in the Dorman's and Saltholme area until 7th, often roosting with other wagtails in the reeds below the Dorman's Pool car park (*British Birds* 93: 548). Ten years later and the North Tees Marshes again attracted the species. A first-winter bird was on the pool of the then new visitor centre at RSPB Saltholme on 23 August 2009. It remained until the afternoon of the following day (*British Birds* 103: 625), even being visible from the cafe veranda.

Distribution & movements

Citrine Wagtails breed in northern and central Siberia, with small numbers recently expanding westward to Belarus, the Baltic countries and southern Finland. The wintering range covers the Indian subcontinent, southern China and South East Asia. Since the first British record in 1954, a total of 241 records have been accepted to the end of 2010, with a high proportion of these being in the Northern Isles and Scilly. Late August to early October is the key period, but recent years have seen an increasing number of spring sightings.

Citrine Wagtail, RSPB Saltholme, August 2009
(Steven Clifton)

Grey Wagtail
Motacilla cinerea

A widespread and common resident, though restricted due to habitat preference. Also a widespread winter visitor and passage migrant.

Historical review

Little is said of this species in the historical lists and chronicles of Durham's birds. "*Less common than Pied*", was how Robson described the status of this species in the 19th century in the Derwent valley (Robson 1896), and this remains an accurate description of its presence there today. Nelson (1907) recorded it as an autumn migrant in small numbers at Teesmouth and he suspected that more over-flew Teesmouth than ever landed.

At the mid-point of the 20th century, George Temperley (1951) called it "*a resident. Common on stream sides, ascending to the upper waters to breed in summer.*" Temperley believed that it left the uplands in autumn, but was present all the year round in the lowlands, and in this he may have been correct as many more birds are present in the lowland east of the county in winter than during the spring and summer.

There is little available detailed information about the status and distribution of this species in the county during the 1950s and 1960s, but the variable water quality that was experienced in some of the county's river systems over that period may very well have impacted upon the species' distribution and breeding success over these decades. Certainly winter weather conditions can make a difference to survival patterns and the number of breeding birds in the Cleveland area were reported to be much depleted by the hard winter of 1962/1963 (Stead 1964). By 1965 it was still considered to be scarce in the lower Tees valley after the deprivations caused by that winter (Blick 2009). That this is a hardy species used to exposed locations both for feeding and nesting was well illustrated by the interesting example of a high-altitude nesting record in 1974, when a pair was found nesting at an elevation of 427m above sea level on Cronkley Fell in upper Teesdale (Mather 1986).

Recent breeding status

This attractive species is usually well reported by modern observers and in both 2008 and 2009, over 500 records of the species were received by the Durham Bird Club, but such observations, in the main, merely serve to confirm what is already known about the species rather than shedding light on new facts. What such evidence does show, however, is that the Grey Wagtail might be considered characteristic of many of the county's waterways and can be found breeding along all of Durham's main river valleys. In the main, the species prefers stream sides with broad-leaved tree cover, and watercourses with a series of in-stream rocks, riffles and areas of shingle (Gibbons *et al.* 1993); in other words, a complexity of river geography.

The *Atlas* map (Westerberg & Bowey 2000) accurately showed how the species' distribution followed the county's main waterways. Today, birds can be found along most stretches of river in the county, even the relatively minor, slow-moving ones such as the lower stretches of the Don and the Team. Nonetheless, this species is most numerous in the western uplands, and the fringing dales areas, with the rivers and streams in the west of the county being ideal. This upland bias was evident from the *Atlas* map, which clearly delineated the watercourses in the west of the county by the species' mapped distribution (Westerberg & Bowey 2000).

At the time of the *Atlas* survey work, just a few pairs nested in the east of the county, the coastal denes at Hawthorn, Crimdon and Castle Eden being favoured at that time. There are now believed to be some eight to ten pairs of Grey Wagtails breeding around the North Teesside area (Joynt *et al.* 2008). Traditionally they have been scarce as breeding birds in this south eastern part of the county but now they are scattered along some of the main water courses, particularly the Tees itself (Joynt *et al.* 2008). Numbers in many of the eastern locations have increased in the last decade and a half and this may have been brought about by an increase in overall breeding numbers. Such an increase is likely to be associated with cleaner and therefore more productive rivers and, perhaps, a change in behaviour in the species over time and not least the mild winters experienced through much of the 1990s and 2000s.

Previous research has shown that breeding performance is little affected by water quality of the watercourses where they live (Ormerod & Tyler 1991), as birds can find a great deal of their food in the foliage of trees overhanging the rivers and streams where they nest. In some instances however, perhaps birds do not even need a watercourse. Genuinely urban breeding was documented in the county, for the first time, in 2007. This took place in the very busy Sunderland City Centre, close to Sunderland Empire Theatre and the Magistrate's Court buildings, one pair raising three broods of young. Perhaps this indicates that there may be other undiscovered breeding Grey Wagtails in the county's towns and cities. In 2008, birds bred again in Sunderland City Centre and a pair of birds on waste ground in central Gateshead during spring hinted that the Sunderland birds may not be the county's only town-dwelling breeders. Most nests sites utilised locally are natural riverside sites or are located under culverts and bridges, but an enterprising pair utilised a wall-mounted planter at Hardwick Hall in 2008 (Newsome 2009).

During the 2007 breeding season, birds, as in most years, were found at many upland sites and a number of lowland riverside sites, including the River Wear at Durham City. In 2009, nesting occurred along all of the county's main river systems as well as their tributaries, down to within sight of the coast. Birds were widely distributed during the breeding season with numerous sightings on the upper reaches of the rivers Derwent, Wear, Lune and Tees. The county's upland reservoirs also attract breeding birds, particularly around Balderhead, Blackton and Hury in Baldersdale (Newsome 2010).

The highest densities of breeding birds still tend to be found along the upland stretches of rivers, or on rivers that essentially have an 'upland river' character, whether they are actually in the uplands or not, such as the River Greta in the county's extreme south west, or even largely lowland rivers, such as the Derwent, which flows into the Tyne, in the north. For both nesting and feeding, birds can often be found in association with weirs, sluices and reservoir culverts (Westerberg & Bowey 2000). First-brood juveniles can be noted from mid- to late May and through June at many sites along the main water courses on Derwent, Wear and Tees, and most of their tributaries.

During the late 1980s and 1990s, breeding densities of two pairs per kilometre of waterway were revealed by local Waterways Bird Surveys in some parts of the county (Westerberg & Bowey 2000). An indication of population levels in prime habitat in the county came from the River Wear at Wolsingham, where ten territorial birds were counted along a 1.5km stretch of the river in 2007. Further west, breeding data in 2008 included four pairs along a 1.5km stretch of the Balder Beck and, in the lowland part of the county, three pairs were on the Wear at Chester-le-Street (Newsome 2009). Some of the above documented figures approach the highest densities encountered for this species elsewhere in the country, where two to four pairs per occupied tetrad have been noted (Gibbons *et al.* 1993). Working on a 'conservative' breeding density of at least two pairs per occupied tetrad, the *Atlas* estimated that the county population was in the order of 400 pairs (Westerberg & Bowey 2000). It is thought that this might now be too conservative an estimate and that a more realistic, up-to-date figure would be in the region of 25% greater than this, i.e. a total of 500 pairs.

Recent non-breeding status

Illustrative of its modern, non-breeding distribution, in 2008, reports of this species were received from all across the county from Cowgreen and Killhope in the extreme western dales, as far east as the streams of the county's coastal denes, which run down to the North Sea (Newsome 2009).

Post-breeding dispersal of juveniles tends to occur from late May onwards, and this initially takes place immediately down and upstream of the birds' natal territory, with young birds moving further afield as the season progresses and often appearing around the fringes of wetlands as late summer draw-down reveals new feeding areas for them. This was illustrated by the bird ringed at Ebchester Sewage Treatment Works in the Derwent valley on 28 June 1992 and controlled at Shibdon Pond, 13km away, on 9 August of that year. In some of the most favoured areas for this species over the period from late June through to August, such as Sedgeletch, the Wear at Chester-le-Street and Birtley Sewage Treatment Works, numbers can rise into double figures, especially in the period August to September.

During the winter, birds tend to move downstream somewhat and the vast majority of winter reports refer to sightings from along the species' favoured breeding water courses. Birds are however, also attracted to permanent ponds and waterways, temporary floods, garden ponds and even rain-water puddles on city centre rooftops in the middle of some of the county's largest settlements, even into the extensive conurbations of Tyne and Wear and Teesside. In the urban context, birds have been noted regularly around sites such as the Saltwell Park Lake, Gateshead, and they are often seen feeding around the rooftops of office buildings in sites such as central Gateshead during cold spells. In recent years, many winter reports have come from urban areas, with regular wintering reports coming from Chester-le-Street, Durham City, Hartlepool Fish Quay, Jarrow, Peterlee, Seaham, South Shields, Sunderland and Washington. These sightings mainly refer to one or two birds together but in the winter of 2007, four were noted in Durham City and ten birds were counted around rooftop pools in Sunderland City

Centre (Newsome 2008). Extrapolating from such observations, one might speculate that such locations now hold a very significant number of the county's Grey Wagtails during the winter.

In recent decades, small numbers of birds have been increasingly reported at unusual sites such as garden feeding stations, for example in Chester-le-Street, an allotment in Houghton-le-Spring, also visiting small ponds in suburban gardens and reports of this latter behaviour have become ever more widespread in the last decade. In 2009 alone, such sightings came from locations as wide ranging as Whitburn Village in January, Old Quarrington in November and Easington Lane in December. Since the early to mid-1990s Grey Wagtails have also been, on occasion, noted using garden feeding stations, but whilst this habit has increased through the years, it has not yet become widespread in County Durham.

After moving downstream, or to more urban locations, for the winter period, most local breeding birds have dispersed back to their breeding areas by late February, though many adults will have remained in such locations throughout the winter.

Spring passage through Durham, from late February though to May, is usually rather light, if it is noted at all. August sees the build-up of post-breeding congregations and there is usually at least a suggestion of light autumn passage, mainly along the coast. Occasionally numbers of inland passage birds are noted as in August 1988, when a gathering of around 30 birds was present at Lockhaugh Sewage Treatment Works, at Rowlands Gill, despite the fact that most of the local breeding pairs were still distributed along the nearby River Derwent (Bowey *et al.* 1993). Likewise, the nine birds together at Hurworth Burn Reservoir on 17 August 2008 (Newsome 2009), which might have been locally dispersing youngsters, may also have included some short-distance migrants from higher elevations or from further north in Britain.

In most autumns, small-scale passage is noted mainly, but not exclusively, at the coast during September and the first half of October, with one or two birds reported from a number of coastal stations. In 2008, for example, four flew south at Blackhall on 9 September, six flew south at Seaham on 26 September and four flew south at Ryhope the following day, indicating that a concerted movement was underway over this period. The numbers noted were probably just a fraction of the birds actually on the move through the county (Newsome 2009). That such passage may be occurring over a period of time was shown by the apparent coastal passage noted at Whitburn Observatory on many dates between 30 August and 28 October 2009, with between one and four birds being regularly logged, up to five on 17 September and seven on 20 September. Further examples of concentrated passage include eight south over Cornthwaite Park, Whitburn on 23 October 2006 and nine flying south over Hendon in one hour on 11 September 2010. Whether these birds are locally-bred, from further north in Britain or continental immigrants is not known, and as yet there is no locally generated ringing data which can shed light on this subject.

Distribution & movements

This species breeds in most of Europe east to the Caucasus and Iran, as well as in parts of North Africa and Asia, to the east of the Urals. Over much of its range, it is resident, though more northerly populations are migratory. British birds are largely resident, though they are to a small degree supplemented in winter by an influx of birds, presumably from Scandinavia (Parkin & Knox 2010).

Pied Wagtail (White Wagtail)
Motacilla alba

A common resident passage migrant and winter visitor.

Historical review

Selby (1831) confirmed that in the early 19th century, *"a few remain with us the whole year"* and Temperley commented that this was also true in his day, more than a century later, with most resident birds leaving by the end of October. The species used to nest widely in the slag heaps located around Teesmouth around the turn of the 20th century (Nelson 1907). Nelson also mentioned the migration of Pied Wagtails along the Durham coast in large numbers in February or early March, through until late April. Migration movements were also noted at Darlington

Sewage Farm during the early 20th century, with spring passage birds noted there from the end of February through to April.

Temperley (1951) described the bird's distribution in the county as, *"Not so restricted to water as the Grey Wagtail"*. He called it a *"fairly common summer resident"* with a few remaining over for the winter, but only in *"the east of the County"*. He believed that it was a passage migrant in some numbers in both spring and autumn. Certainly the statement in *The Vasculum* in 1936 bore out this assertion, *"Further reports on the winter distribution of this species show that it may be met with over most of the low lying portions of the counties and frequently in unexpected places. This suggests that these birds are strangers from further north and not our own normal population. There is some evidence that these birds collect together regularly for roosting purposes each evening, distributing themselves over a wide area in search of food during the day"* (*The Vasculum* Vol. XXII 1936); a neat summary observation, and in many respects still pertinent to the species' modern status in Durham.

This is, at least in part however, a somewhat different pattern of occurrence from what we see today, when many more birds are present throughout the year, although there is clearly some emigration of birds from the north east, as indicated by ringing recoveries.

Temperley (1951) reported autumn roost gatherings in withy beds and similar locations. More recently, it was described as common across the Cleveland area in the early 1960s (Stead 1964). It is now a year-round resident, with very significant winter roosts at a number of locations.

Recent breeding status

Pied Wagtail occurs in Durham throughout the year, but in terms of its status, it is a complex species being a spring and autumn passage migrant, a resident breeding species and in part, both a summer and winter visitor.

Today, it is a common and widespread breeding species, with such records coming from across the county from the coastal cliffs in the east to grasslands and dry stone walls of the far west and the uplands. Consequently, it can be seen across the county in a wide variety of habitats from the built-up areas of towns and conurbations to the muck-strewn farmyards of the upland fringes. The spring and summer distribution covers the largest proportion of the county, though the numbers reported in the exposed areas of the western uplands are much smaller than in lowland areas. Nonetheless, birds do occur here, particularly around human habitation, in old quarries and along the river courses. In this respect, the species' distribution in the county follows the principal river valleys, the Wear, Tees and the Derwent, as far west as sites such as Nag's Head and Harwood. In the extreme west, only the open moors and land above 300m are routinely avoided (Westerberg & Bowey 2000). From the Pennines, it can be found eastwards, across the county, all the way to, and along, the coastline. It tends to avoid woodland areas but it is often noted in short grassland areas amongst cattle and around industrial estates in urban areas. The *Breeding Birds of Cleveland* work showed that Pied Wagtails were common around Teesside, though there appears to have been some decline in the number of breeding birds in recent years, especially around the estuary itself (Joynt *et al.* 2008).

During the breeding season, quarries, farm buildings, stone walls and derelict mine workings all provide suitable but clustered nesting cover and nest sites in westerly situations. Many pairs in the county nest around farms in and around rural villages and it is frequently found around farms and associated buildings during the breeding season, where it nests in a variety of crevices and holes, amongst rocks, in old walls and in buildings (Gibbons *et al.* 1993). There is little work that indicates current breeding densities, but a 2008 study in the Hetton-le-Hole and Houghton-le-Spring area noted nine pairs in that small part of the county; a small increase in number over 2007 (Newsome 2009).

Elsewhere in Durham, over the Magnesian limestone escarpment for instance, quarries and cliffs along the coast all routinely provide nesting niches for Pied Wagtails and, in industrial sites, stacked wooden pallets and similar situations are not infrequently used (Westerberg & Bowey 2000). In 1999, a pair reared young in a disused public toilet at Leeholme in June (Armstrong 2003). It is less common as a breeding bird in the urban centres, though it is frequently noted in such locations. In 2008, the first fledged juveniles of the year were noted in the first week of June, a typical date for the species locally (Newsome 2009).

At the time of the first national breeding Atlas work, Sharrock (1976) pointed out that the species had become scarcer in eastern England, compared to Scotland and Ireland. At the time he attributed this to an increase in cereal production and a decline in the number of cattle, stocking levels and an associated loss of farm ponds (Marchant *et al.* 1990). After a period of national increase during the 1970s the population was believed to be relatively stable by the early 1990s (Marchant *et al.* 1990). Birds, however, are susceptible in winter to severe ice and frost, which reduces their feeding efficiency (Brown & Grice 2005). Over the period 1994 to 2006 however, the population was thought have declined by 12% nationally (Raven *et al.* 2007). The *Atlas* work in Durham (Westerberg & Bowey 2000) registered the species in 341 of the county's 825 tetrads and, extrapolating from the national population data of 3.4 pairs per square kilometre (Gibbons *et al.* 1993), Durham's breeding population was estimated to be around 3,330 pairs (Westerberg & Bowey 2000). Taking on board the known national decline, the county population may now be best considered, in a rounded-down fashion, to amount to around 3,000 pairs.

The future for the species looks reasonably good in the county, though major changes to agricultural practices, such as further switches from pasture to cereal production, somewhat unlikely in the Durham landscape, or any other significant land use changes, might act to limit feeding areas currently available to birds. Today, birds are now so widespread and common even in the region's townscapes that such changes are unlikely to have a major impact.

Recent non-breeding status

In the post-breeding phase of the year, fledged young and other family parties start to join together into post-breeding flocks from around mid- to late June. Indicative of such gatherings were the 25 birds at the Birtley Sewage Treatment Works on 14 June 2008, whilst 100 were feeding on insects over recently sown fields at Hurworth Burn Reservoir on 27 July the same year. In 2009, family parties started to gather from early July, and such congregations of between 15 and 35 birds were noted at Bishop Middleham's Castle Lake, Hardwick Hall County Park, Hetton-le-Hill, Hurworth Burn Reservoir, Rainton Meadows, Roker and RSPB Saltholme (Newsome 2010).

At this time of the year, many Pied Wagtails can be found feeding in open habitats and especially on short grasslands in urban-fringe, industrial and agricultural areas, which means that scores if not hundreds of birds can gather at favoured sites during the June to September period. An example of such post-breeding flocks on short-sward grassland locations is the 114 birds that were feeding on bowling greens at Bowes Museum, Barnard Castle on 19 September 1999 (Armstrong 2003).

In autumn and winter, Pied Wagtails roost communally, often in large numbers and during the autumn of 2007, the reed beds of Low Barns, in the Wear valley, held three-figure counts throughout the autumn period, with a maximum of 275 in late September. During the 1970s and 1980s, the reed beds at Shibdon Pond were a traditional site for such roosts and in October 1975 it was estimated that between 400 and 500 birds were present there. At various times, roosts of Pied Wagtails have occurred in many contrasting locations in the county. Those numbering up to a few hundred birds have been noted at places such as Billingham Bottoms and Haverton Hole on Teesside, a culvert of the River Team at the Team Valley Trading Estate in the early 1990s, Blaydon Shopping Precinct throughout the 1980s and 1990s, and Carrville, Durham in the mid-1990s. The most spectacular, in terms of the numbers involved, was the estimated 1,500 birds that were seen roosting on the pipe-work of the Tar Distillation Plant at Port Clarence in December 1980 (Blick 2009).

Today, winter roosts seem to be either more dissipated or less well reported and in recent years, that at the Chester-le-Street Sewage Treatment Works has provided some of the county's largest winter counts. For example, in 2009, winter counts there numbered up to 200 in January. It is also clear that sewage treatment works are productive and important feeding grounds for this species, as was demonstrated by the 170 that were counted at such a facility at Birtley on 24 January 2009, the highest number ever recorded at that site, and the 120 noted at Beechburn on the same date (Newsome 2010). Winter feeding concentrations, away from sewage treatment works, tend to be more limited with 20 to 40 birds being typical of larger counts.

This is, at least in large part, a resident species in Durham, but one in which there is a pronounced southerly movement through the region in the autumn (Joynt *et al.* 2008). A large proportion of Durham's birds are probably resident in the county through the year, but there is also a passage of birds that move south in the autumn and north in spring, and at least a proportion of locally breeding birds head west and south to warmer climes in the winter. Such passage was demonstrated in 2009, when local increases in numbers of birds were noted during October. Other recent examples include up to 40 birds at Hetton Lyons and over 20 foraging on seaweed at Whitburn Steel on 29 September 2008, many of these were almost certainly migrants from further north in Britain (Newsome 2009).

This species was a regular feature of all of the annual Durham Coordinated Migrant Counts between 1992 and 1996, though never in large numbers, with 48 birds being recorded between the Rivers Tyne and Tees on 4 October 1992 (Bowey 1999).

Spring passage occurs from around the later part of February, through March and into early April and an apparent example of such spring influxes was noted in March in the north east of Durham County, when groups of six, ten and 11 were at Whitburn, South Shields and The Leas respectively between 13th and 31 March 1999, and when 87 were at the Zinc Road at Teesmouth on 26 March 2006. Occasionally, large pre-breeding season roosts develop, such as the 'more than 300 birds', which were in reeds at Low Barmston on 12 April 2008. It is conceivable that many of the birds in such situations are actually northbound spring migrants; as may have been some of the 120 birds noted around the Gateshead IKEA car park on 1 March 2006 (Newsome 2007b).

White Wagtail
Motacilla alba alba

Historical review
This, the nominate race of the species, has probably always occurred in the county, though it was little documented by the county's early ornithologists. In the 19th century, Hancock (1874) referred to the White Wagtail in his catalogue of birds of Northumberland and Durham but he ascribed it no particular status and gave no indication of numbers or documented any records, with one exception. This was of a record of one handed to him on 11 July 1878, when a specimen was brought to him of one that had been shot on the Sands at Durham City on 3 May 1875; this therefore is the first definite record of the sub-species in County Durham. Additional Victorian records include two that were shot near Durham on 29 April 1880 and another from 30 April 1880. Tristram (1905) reported it as having been seen near Durham in 1904. Slightly before this, a pair was reported as having bred in slagheaps at Teesmouth in 1899, but on the Yorkshire side of the estuary (Mather 1986).

Temperley (1951) said that this bird was usually met with singly or in small parties, and he described it as a *"passage migrant in small numbers, chiefly in spring"* a still accurate summary of its county status in the 21st century. He thought it was overlooked, as only the males, at that time, were believed to be readily identifiable, and this explains in part why autumn records are so few. He said that most reports of it were from around Teesmouth, or along the coastal strip, and that these occurred in April and May. It was also said to occasionally be recorded in September. At Darlington Sewage Farm, Almond noted birds occasionally in the first quarter of the 20th century (Almond 1929). A particularly high count was of 23 birds at Teesmouth on 29 April 1946 (Blick 2009).

The occurrence patterns of this bird were poorly documented in the post-Temperley period, though in 1952 a party of five birds was noted at Teesmouth on 16 April, with eight there on 25th of the month. That spring, birds were also observed on several occasions at Boldon Flats from 9th to the end of April, with eight there on 24th. Over the period from the 1970s to the 1990s, better documentation of its annual occurrence and improved understanding of its status in Durham resulted in birds being noted every year between 1970 and 2010.

Recent status
Today, birds showing the characteristics of the nominate, continental race, *M. a. alba* are regular spring passage migrants in relatively small numbers in both spring and autumn in Durham. It is a relatively widespread and sometimes not uncommon passage migrant, which tends to favour coastal and wetland locations when on passage. It is probably more common in autumn than realised, though less reported then as its plumage is less obvious at that time of the year. It has almost certainly bred on a number of occasions and it is perhaps somewhat overlooked in this capacity. The first of the year are typically noted in mid-March with most records of this race occurring between early April and mid-May, when one or two birds may be noted at between five and fifteen localities each year. The majority of sightings are in the coastal strip or a little way inland, though birds might be seen at any suitable site across the county. Occasionally, small flocks drop in for a few hours, and typical passage locations that attract this bird are wetland fringes and damp fields where open grassland habitats abut damp or wet areas. Concerted observation of such sites sometimes reveals a movement through of birds of this race. For instance, in spring 1991 there was a series of sightings at the Far Pasture Ponds adjacent to Lockhaugh Sewage Works, which referred to at least 10 different birds between mid-March and May (Armstrong 1992). Around the

North Tees Marshes, birds are regularly reported in the Saltholme Pools/Dorman's Pools area, and sometimes in good numbers, such as the 15 together there in spring 2005 (Blick 2009).

Typical numbers in a spring passage period vary from 'a handful of birds' up to about 15, with larger passages involving as many as 35-40 birds through the season. The best spring passages occurred in 1975, including a party of ten on a fishing boat off Sunderland on 19 April; 1977, with perhaps 30 birds in an extended spring movement; 1994, with over 30 birds and 1996, with over 40 birds. By way of example, 2001 was a reasonably good passage year, and records of from one to three birds came from 11 sites during April but only two during May. In that year, there no reported autumn records, and in total the spring passage amounted to between 20 and 30 birds. In 2007, records were fewer and again, largely restricted to the spring, with reports over the March to May period coming from Bishop Middleham, Bowesfield Marsh, Darlington, Derwent Reservoir, Saltholme Pools, Seaham, Sedgeletch, Tunstall Hill and Whitburn. In that year, autumn reports were restricted to one to three birds at Saltholme Pools, Whitburn and WWT Washington, all of these during August.

Clearly, the levels of passage vary from season to season and year to year, with some springs bringing many more reports than others. In some poor years the spring total does not exceed single figures. The largest numbers recorded at individual sites, away from Teesmouth, are 15 at Boldon Flats on 18 April 1986 and the following year, when 16 were at the same site on 19 April.

Many fewer are reported in Durham in the autumn than in the spring but it is likely that in most autumns many more pass through the county than are realised or are actually recorded. Between the years 1970 and 2010, there was evidence of autumn passage in Durham in 17 years. The largest ever autumn passage was in 2008, when at least 14 birds were noted; up to 10 were reported at SAFC Academy Pools between 14th and 17 September, and one to four birds were noted at Cleadon, Greatham Creek and Whitburn up to 29th of the month (Newsome 2009). By way of contrast, autumn 2009 was the first year that there had been no autumn records since 2001 – indicating that the previous situation in autumn, with birds less commonly noted, may have been due to birds being missed amongst migrating Pied Wagtails. Unusually late records were noted on 8 October 1989 at Saltholme and 21 October 2006 at Langley Moor. To date, the only winter record of the race in Durham was at Whitburn Steel on 17 December 2005 (Newsome 2006).

The earliest date for White Wagtail in Durham is 10 March (in 1978 and 2007), and in most years birds are recorded from late March to early May, with a smaller number of records in August and early September. Over the years there have been a small number of June reports, such as the two in 1993 (Armstrong 1994).

Recent breeding status

This race may have nested in both 1952 and 1953 at Shotton Brickworks Pond (Simpson 2011), when a male 'White' was seen to be paired with a female Pied Wagtail and it was present in each of these summers. Whilst the pair was watched on a daily basis, a successful breeding outcome was never proven. In 1988 there was a mid-summer record of a bird at Shibdon Pond and the following year a bird was found to have bred with a Pied Wagtail at an industrial site in Birtley (Bowey *et al.* 1993) in the north of the county. During the work on the *Atlas*, a pair of the nominate race was discovered feeding three young in a garden at Seaham in July 1988 (Westerberg & Bowey 2000), and a bird was recorded displaying to female Pied Wagtail in Durham City in May 1992 (Armstrong 1993).

Amur Wagtail
Motacilla alba leucopsis

On the afternoon of 5 April 2005, Steve Addinall found an 'Amur' Wagtail, the White Wagtail *M. alba* of the Far Eastern subspecies *leucopsis*, at Dawdon Blast Beach, Seaham on the Durham coast. He found the bird in an area of waste and scrub land on the site of the old Vane Tempest Pit, together with a small number of Pied Wagtails *M. a. yarrellii*. This bird was subsequently identified as Amur, 'White-faced' or 'Chinese Black-backed Wagtail', as it is variously named (Addinall 2005). This race of White Wagtail had not been previously recorded in the Western Palaearctic. The bird spent most of the following day alongside the small pool where it was discovered, being seen by a large number of visiting observers, but it was gone the next morning (*British Birds* 102: 571-572, *British Birds* 103: 260-267).

Distribution & movements

Pied Wagtails are restricted in their breeding range to Britain and Ireland, and in some local areas along the northern coasts of Western Europe. Many birds winter in the British Isles, but some move south and south west into continental Europe. White Wagtails breed across most of continental Europe north to the Faeroe Islands, Iceland and Asia Minor, wintering in southern Europe and North Africa (Parkin & Knox 2010). Amur Wagtail breeds in the eastern Palaearctic in Japan and eastern China (Parkin & Knox 2010). Like many of the White Wagtail 'group' they are long-distant migrants, wintering to the south of their breeding range, in South East Asia. Other races of the *Motacilla alba* wagtail group breed in Africa and eastwards across Asia, as far east as Mongolia.

From the evidence provided by ringing work in the north of the county, from birds caught in the autumn at Shibdon Pond during the 1970s and 1980s, it would seem that a proportion of our local birds head south and westwards for the winter. The ringing data also indicate that some of the county's wintering birds originate from breeding grounds well to the north of the north east of England, as shown by the bird ringed near Hamsterley Forest on 5 June 1954 that was recovered at Sunde, Hordaland in Norway on 26 April 1956.

Furthermore, southward movement of local birds has been demonstrated by an analysis of birds ringed in the north east which have been recovered wintering further south in England, as shown by two examples of birds ringed at Shibdon Pond in 1977. One of these, a first-year, was found dead in Manchester on 17 January 1977, after it had been ringed in September 1976; the second, an adult, was controlled at Barnsley in South Yorkshire on 8 October 1977, a year after it was ringed in September 1976. Other recoveries have come from Snow Hornby, Lancashire in December 1978, Kingston upon Thames in December 1979, Wolverhampton in June 1980 (in a building), Lewes, Sussex in October 1982, whilst one ringed at Wingate in August 1989 was recovered in January 1990 at Northampton. Some birds have been found further afield, including France and Iberia. The most interesting winter recoveries of such birds include one recovered on the island of Jersey and two adults which had made their way as far as Portugal, to Aletjo and Beira, being recovered in, respectively, April and November 1974, one of them only five weeks after ringing (L. J. Milton pers. comm.).

Richard's Pipit
Anthus richardi

A rare but regular autumn visitor from Siberia: 55 records involving 58 individuals.

The first Richard's Pipit for County Durham came as recently as 1970, though it was not confirmed until some years later. On 11 October 1970, the late Edgar Gatenby located a large pipit on waste ground by the Reclamation Pond. The intricacies of large pipit identification were still in their infancy and the bird was identified as Tawny Pipit, a fairly well-established scarce visitor to southern England. It stayed in the area until 24 October, was seen by a comparatively large number of observers and subsequently accepted as a Tawny Pipit by the British Birds Rarities Committee (*British Birds* 64: 363). Following a review of records of both species by the BBRC in 1996 however, the sighting was reassessed and accepted as a Richard's Pipit (Bell 1996). Remarkably, for what is now considered a regular scarce migrant on the east coast, the next sighting was not until 1985. Since then it has been recorded in a further 21 years. The best years for the species in Durham were 2002 and 2003, with a combined total of 14 birds, including two in the Easington Colliery area from 9 November to 7 December 2002 (Newsome 2007).

Richard's Pipits in County Durham, 1970-2010

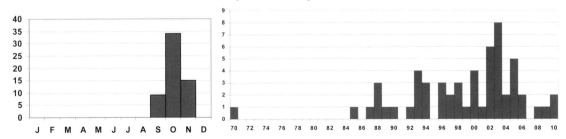

All records fall in the period 16 September to 7 December, with peaks in the third week of October and again in the second week of November. This conforms to the general national picture, with few records in the first half of September and birds sometimes remaining into the early winter period.

As might be expected for an eastern vagrant, the majority of birds have been at coastal migration 'hotspots'. The coastline between Sunderland and South Shields has attracted 34 birds, more than anywhere else, with 13 in the Whitburn area alone, the Coastal Park and Rifle Range being particularly favoured. Eight birds have been noted around Marsden Quarry, while the adjacent grassland of South Shields Leas has attracted seven. Notable for being over a mile inland, three birds have been seen at the SAFC Academy Pools, two at Fulwell and another at Sunderland South Dock.

In the North Tees area, Richard's Pipits have been seen at Haverton Hole and Greenabella Marsh twice each, and also at the Long Drag, North Gare and Reclamation Pond. More regular sightings have occurred in the Hartlepool area, with six birds at various locations around the Headland.

The coastal fields in the mid-section of the county's coastline have attracted seven birds, the most favoured area being around Easington Colliery. Singles birds have also been seen at Chourdon Point and Dawdon Blast Beach to the north and Blackhall Rocks to the south.

There have only been three inland occurrences. The second-ever record for the county was of a bird at Teesside Airport on 2 November 1985, finding the short turf to its liking for the day. Another was at Belasis Technology Park, Billingham, on 7 November 2003, while in 2002, two birds were found in rough pasture at Rainton Meadows on 21 September, with one remaining until 25th.

There seems little doubt that this species had been missed in County Durham prior to the mid-1980s, when the upsurge in autumn records commenced. However, from the early 1990s there was also a very considerable increase in the number seen in north east England and Britain as a whole.

Distribution & movements

Richard's Pipits breed in southern Siberia, Mongolia, China and central Asia, wintering in the Indian subcontinent and South East Asia. A small part of the population moves west each autumn, and it is a regular but scarce autumn and winter visitor to north west Europe and the western Mediterranean. It was formerly a rare visitor to Britain and was a BBRC rarity until 1982; it is now found in small numbers each autumn in Britain. Occasional birds have been found wintering in Britain, with small numbers also seen on spring passage.

Tawny Pipit
Anthus campestris

A very rare spring and autumn vagrant from Europe: six records involving eight individuals.

The first record for County Durham came in 1968, when C.S. Allen, I. Lawson and D. Marshall found two together at Cowpen Marsh on 12 April, with one bird lingering until the following day. The second record, and the first in autumn, came soon afterwards when one was found on Hartlepool Headland on 12 October 1969. Since then, birds have been seen in four further years, but the fact that there were no records in Durham between autumn 1992 and the end of 2010, is testament to how scarce this species is in the north east of England.

All records:
1968	Cowpen Marsh, two, 12 April, one to 13th (*British Birds* 62: 483)
1969	Hartlepool Headland, 12 October (*British Birds* 63: 288)
1974	Saltholme Pools, two adults, 10 July (*British Birds* 69: 364)
1976	Whitburn GC, 17 September (*British Birds* 70: 438)
1986	Mere Knolls Cemetery, adult, 17 May
1992	Cleadon Hills, adult, 14 September

The records are evenly split between the North Tees and the Whitburn area, with two in spring, one in mid-summer and three in autumn.

Distribution & movements

Tawny Pipits have a widespread range covering much of Europe, Asia and North Africa. The western race *A. c. campestris,* responsible for British records, breeds from Iberia and north west Africa east to the Middle East and central Asia, north to the Baltic States, and winters in Africa and India. It is a scarce spring and autumn migrant to Britain, mainly along the south coast from Scilly to Kent, also regularly in East Anglia, but it is distinctly rarer further north. The species was removed as a BBRC rarity in 1982 after records exceeded 620, but the following three decades have seen a decline in the number of birds reported.

Olive-backed Pipit
Anthus hodgsoni

A very rare autumn vagrant from Siberia: one record.

The sole record for County Durham is of a bird found by Richard Taylor on wasteland adjacent to Hartlepool Docks on 26 November 2000 (*British Birds* 94: 483). It was initially found in rough grass and gave good views before flying into a nearby housing estate, not to be seen again. This was the final British record in a very good autumn for the species, 17 birds being recorded nationally.

Distribution & movements

Olive-backed Pipits are widespread across central and eastern Siberia to northern China, Kamchatka and Japan, but the European range is restricted to the northern Urals. The winter range covers China, Taiwan and the northern part of South East Asia. Formerly much rarer, by the end of the first decade of the 21st century it had become an expected autumn vagrant, particularly to the Northern Isles, a total of 349 having been recorded in Britain to the end of 2010.

Tree Pipit
Anthus trivialis

An increasingly scarce summer visitor, more common in the west of the county, and a declining passage migrant.

Historical review

In neighbouring Northumberland, Selby and Bolam thought of this species as one of well-wooded parkland and upland valleys (Kerr 2001), a description that not inaccurately describes its habitat choices through much of the modern period, although its numbers and lowland presence have both declined in recent years.

"*Summer resident*", was how it featured in Temperley's (1951) consciousness, the author going on to say that it was "*well distributed and numerous in areas where there are deciduous trees*" and that it arrived in the latter half of April.

Post-Temperley, the species appears to have been little recorded and poorly documented, leading one to the conclusion that it was routinely observed, not often studied or written about, although probably common. Stead (1964) reported that birds bred in Crimdon Dene in the early 1960s and it was also a regular breeder in Castle Eden Dene, in the east of the county, around this time.

Recent breeding status

Tree Pipit was for a very long time considered a traditional bird of the western half of the county, though it was also found in many lowlands areas during the 1960s, 1970s and 1980s. It was usually thought of as being common, particularly in some of the more open-structured, ancient woodlands, being especially prevalent in the upland oak woods of the upper reaches of the county's river valley systems. It was also to be found in some more open areas with trees, such as those situated on the lower slopes of the uplands and down to the valley floors. For example a one-mile section of the Ayhope/North Grain Beck area of Hamsterley Forest held nine pairs in 1979 (Unwin & Sowerbutts 1980).

In the national context, Marchant *et al.* (1990) and Gibbons *et al.* (1993) noted a gradual decline in numbers of this species, most noticeably in central and southern England, and there is little doubt that this decline has now reached the north east, accelerating through the 1990s and into the early part of the 21st century. Figures from the two national breeding atlases (Sharrock 1976, Gibbons *et al.* 1993) indicated that this species has experienced some considerable decline in numbers and range over a long period, with a population decline of 15% noted between the two publications. Over a longer period, BTO data indicates an 86% drop in numbers between 1967 and 2008, most of this apparently taking place during the period from 1985-1995, but numbers are still currently falling. In 2009, BBS data showed a 25% reduction in the number of breeding Tree Pipits in England from 2003 to 2008, although across the UK this decline is less severe, but still of concern at 9% (Risely *et al.* 2010). Tree Pipit was listed as an amber-list Species of Conservation Concern, being upgraded from green, as recently as 2002, and then further upgraded to 'red' in 2008.

As recently as the mid-1980s, Tree Pipits occurred widely across the western half of the county as breeding birds. As time has passed, this westerly distribution has become ever more pronounced. For instance, in 1999 all summer reports of the species came from territories that were located to the west of the A1 trunk road; the majority were situated even further west, beyond the A68. By 2010, Tree Pipits were largely absent from the eastern half of the county, except as passage migrants, furthermore they avoid some of the more disturbed areas of apparently suitable habitat. Territorial birds were present in Castle Eden Dene from 1977 to 1979, and two pairs bred in 1983, but the species has not done so since. In the late 1980s there were still up to five pairs located along the Castle Eden Walkway but there were no reports of territorial birds in this area or around north Teesside after 1994, which is indicative of the general decline in the Tree Pipit's representation in the county's eastern half. Traditionally, this species' presence or absence was most reliant on the structure of the woodlands, as opposed to the species of tree present. The scale and rapidity with which it has retreated from some of its previous local strongholds is exemplified by a comparative analysis of the situation in one of the county's best woodland areas, the lower Derwent valley.

During the late 1980s and early 1990s, birds could be found along the rides of pure conifer stands or in clearings in the ancient oak woods of much of the Derwent Walk Country Park, between Rowlands Gill and Swalwell. At that time, birds bred at many other sites in the lower valley, including Chopwell Woods, Clockburn Woods, Low Horse Close Woods, Thornley Woods, Strother Hills Woods and Victoria Garesfield, amongst others (Bowey *et al.* 1993). In this relatively compact area of the lower Derwent valley during an average late 1980s breeding season, it was considered that there would be well over 25 pairs of Tree Pipit breeding, with smaller numbers of additional birds singing at outlying but nearby sites such as the Gibside Estate, Washingwell Woods, along the Stanley Burn, Guard's Wood and Ryton Willows. In 2009, there were no Tree Pipits known to be in territory in any of these sites (K. Bowey pers. obs.), though small numbers were present a little higher up the valley around Pont Burn Woods; a quite dramatic *volte face*.

For some time now, the species has also been known to occur in areas of new forestry and on clear fells and, more recently in some areas, this has formed some of its most favoured habitat and more densely populated locations (Gibbons *et al.* 1993). For example, 20 territories were in such habitat in the Sharnberry area of Hamsterley Forest in 2006 (Newsome 2007). Whilst habitat change has been implicated in some of the species' decline (Mead 2000), there does not seem to be significant evidence indicating that permanent expansion into areas made suitable during the early years of a forestation has occurred, so the principal reasons for the fall in numbers may lie elsewhere (Gibbons *et al.* 1993). The stronghold in the county is now probably around the large blocks of coniferous forest in the western uplands, although distribution here can be patchy, with birds present in some areas for a number of years before moving to more suitable habitat. Burton (2007) highlighted this behaviour in a study of this species in Thetford Forest in Norfolk, and a number of his conclusions would appear to hold true for Durham. Tree Pipits require open habitat for feeding and nesting but Burton commented that a major limiting factor in population was the availability of song posts. Hence, he discovered that 'prime habitat' was in areas of restocked forest between two and five years after planting, or along the edges of forest blocks where suitable song posts were available. Newly-planted areas are initially avoided as they lack song posts but also because the ground cover is poorly developed, leading to a limited food source and increased predation, reducing breeding success.

In Durham, many formerly occupied sites have become vacant in recent years, in some cases since the fieldwork for the *Atlas* (Westerberg & Bowey 2000). Although it did once breed with some regularity in the east of the county, such occurrences are now, barring odd outliers or sporadic territory holders, effectively non-existent. An isolated pair succeeded in raising two young at Hetton Lyons in 2004, and in the same year a single bird held territory at the Kibblesworth CBC site for the first time since 1975 (Siggens 2006). The vast majority of modern day records from the eastern lowland parts of the county relate to passage migrants, some of which may be delayed and singing for a few days before moving on.

The territorial stronghold through the first decade of the 2000s described an arc around the county's upland fringe, from sites such as Muggleswick and Derwent Reservoir in the north, through Bollihope, Hedleyhope Fell, Knitsley Fell, Tunstall Reservoir, Hamsterley Forest and Low Force, to Baldersdale, Bowes, Gilmonby and The Stang towards the southern boundary of the county.

A distillation of breeding reports from the second half of the first decade of the 21st century, illustrates the recent distribution for this formally widespread species. The vast bulk of records now come from known breeding areas in valleys in the west of the county. In 2005, the species was thought to be *"poorly reported in breeding areas"* (Newsome 2006). The first spring migrant of that year was typically a male in song at Burnopside on 12 April and during that season, birds were noted in territory at just nine sites, though there is little doubt that birds were also present in many other unchecked upland-fringe areas. In 2007, birds were back on territory in some of their traditional areas from 21 April, and were subsequently found at just 13 potential breeding localities. All but three of these were west of the A68, a sad indication of how much Tree Pipits had declined in Durham by that point. The majority of sightings involved just one or two birds, but four were noted on Knitsley Fell on 24 April. Particularly notable, because of their relatively eastern locations, were a pair which were seen with a juvenile on 11 July at Old Quarrington, and a single bird at Observatory Hill, Durham City, between May and July (Newsome 2008).

In 2008, between early May and mid-July, up to six birds were reported from 29 one-kilometre grid squares in the county; the majority of these were in the west, largely on the upland fringes. The largest concentration of birds came from the Hamsterley Forest area, with relatively good numbers also in The Stang in the south, and Tunstall Reservoir, in a westerly central position. Of note in the north of the county were the three singing birds in Pont Burn Woods and nearby Lintzford, as only a handful of sites remain for this species in the lower Derwent valley. In central Durham, six singing males were found at Pit House, Brandon and nearby Standalone, while further east, single birds were noted at Daisy Hill and again at Old Quarrington, the species' last easterly breeding toe-hold in Durham. In 2009, territorial birds were reported from 35 one-kilometre grid squares during May and June. The majority of these were in open woodland and woodland edge habitats west of the NZ10 grid-line. Breeding territories further east in Durham at this time, included five at Taylor's Hill, Lanchester, four at Hedleyhope Fell, two at Hedley Hill and one at Pontburn Wood. Small 'hotspots' still exist in the westerly hinterlands but this may be a reflection of the prime suitability of the habitat as described by Burton. For example, in spring 2010, up to eight were found around the Pennington area with a further five males not too far away at Dryderdale in late June; both locations are on the fringes of Hamsterley Forest (Newsome 2012).

By contrast to the above statistics, the map shown in the *Atlas* (Westerberg & Bowey 2000) illustrated a breeding season occupation by Tree Pipit of 123 tetrads across the county, with occupied tetrads being extensively distributed into the county's central heartlands and a scattering in the east. At that time, population densities were thought to be *"nowhere high"* within this mapped range and an estimate of the county's breeding population at the time was put in the region of 250-500 pairs (Westerberg & Bowey 2000). Clearly, the figure today is very much less than the lower figure in this range and it is unlikely that many more than 100-150 pairs now regularly breed in the county.

The causes for both the short-term fluctuations which the species experiences between years, and the longer-term declines, are not well understood and the threats to the species' preferred habitat in Durham would appear few. The species is a trans-Saharan migrant, but it spends the winter far to the south of those parts of Africa which were seriously affected by drought during the 1980s (Marchant *et al.* 1990). With a 58% decline across Europe during 1990-2000 however, this is not a problem restricted to the UK. Sanderson *et al.* (2006) suggested a number of causal factors although it is accepted that further study is required. Habitat degradation on the breeding grounds may not be a major influence in Durham, but other factors, such as hunting in southern Europe and North Africa, loss of staging areas and reduced overwinter survival in dry open habitats, are thought to have had a major impact on this and some other seriously declining migratory species breeding in northern Europe. Furthermore, Tilman *et al.* (2002) identified that the threat is likely to continue, as in the Sahel zone, where many species winter or stop off on migration, there was an increase of only 4% in the land used for agriculture, but the usage of fertiliser increased threefold and pesticides five-fold in the thirty years between 1970 and 2000.

Recent non-breeding status

Small numbers of birds are recorded at non-breeding sites on passage in both spring and autumn, though never in high numbers. Indeed it has almost always been an unobtrusive coastal migrant, with spring birds perhaps likely to be more obvious than autumn migrants as they occasionally sing, but they are fewer in number, as they are less inclined to linger in coastal locations. The 'great fall' of September 1960, which featured many species of continental origin was probably the biggest on record for Tree Pipit as, although numbers were not quantified, they were described as "*common*" between Blyth in Northumberland and Sunderland, although only two birds were noted between the Wear and Tees (Coulson 1961).

During the 1970s and 1980s, spring coastal migrants averaged under 10 birds per annum with peak counts of five at Mere Knolls Cemetery over 3rd to 7 May 1978, a loose flock of 10 at Newton Aycliffe on 23 April 1983 and 11 in the Whitburn area on 14th to 15 May 1985. The 1990s produced 15 birds passing through Shibdon Pond over 16th to 21 April 1994 and an exceptional flock of 21 birds at Chourdon Point on 6 May 1995. Since that date, the best year on record for spring passage was 2001, when a meagre total of five birds was recorded between 1st and 26 April (Siggens 2005).

Autumn passage birds are somewhat more regularly recorded, but there were no autumn birds at all in 1975, 1977, 1981, 1988, 1991, 1994 and 2007. During most autumns fewer than 10 birds are recorded; for example, totals of four in 1973 and nine in 2010, but a few years have seen greater numbers. Hence, during fall conditions over 8th to 9 September 1995, a total of up to 38 birds were noted in the Marsden area with an even split between grounded birds and those flying overhead. Other 'good' autumns included 1998 (16 birds), 2002 (10 birds) and, in particular, 2008 when there were at least 31 birds recorded in the middle two weeks of September, with 17 between South Shields and Seaburn over the period 6th to 9th (Newsome 2009).

Birds are usually present in the county from the second week of April, but March birds have been recorded in eight of the past 40 years, with a concentration from 1992-97 when March birds were recorded in five out of six years. Almost all of these sightings have come in the last week of the month, but an exceptionally early bird was noted at Far Pasture on 11 March 1997 (Armstrong 1999). Over thirty years between 1976 and 2005, the reported first arrival dates of Tree Pipits were analysed in ten-year grouping and the mean first arrival date of the species in Durham had advanced by over a week: 1976-1985, 15 April; 1986-1995, 7 April; 1996-2005, 9 April (Siggens 2005).

Many of Durham's breeding birds are on the move by late July, but in most years there are some records into September. October records are regular with 10 years in the last 40 producing such reports, usually in the first half of the month. However, later birds occurred between 16th and 21 October in 1982, 1996, 1997 and 2004, with an exceptionally late bird during the arrival of late migrants in November 1984, at Mere Knolls Cemetery on 7th (Baldridge 1985).

Distribution & movements

Tree Pipit, in its nominate form, breeds over much of Europe, with the exception of Ireland and the southernmost parts of France and Spain. Birds also breed in Asia Minor and in a band from Russia to Siberia, with two other races represented in the mountains of central Asia. The European birds winter in Africa, south of the Sahara, though largely north of the Equator.

One of the few significant controls of a locally-ringed Tree Pipit came from Chopwell Woods, where a bird ringed in June 1979 was re-trapped there four years later, in May 1983. There are no recoveries of Durham-ringed Tree Pipits that demonstrate where or how local birds migrate, and indeed very few from the UK. The route followed by British birds however, is clearly indicated by the fact that out of 19 UK-ringed birds, 10 were recovered in Portugal and two in Morocco, with one in Mauritania. Birds heading north have been recovered in Iceland and Norway, and a Norwegian- ringed bird has also been recovered in the UK (Wernham *et al.* 2002).

Meadow Pipit
Anthus pratensis

Sponsored by

NORTH PENNINES
Area of Outstanding Natural Beauty
GLOBAL GEOPARKS NETWORK

An abundant resident breeding across the county, also occurring as a passage migrant. It is much less abundant in winter, after an autumn emigration from the region.

Historical review

In past times this inconspicuously plumaged species was clearly a very common bird within the county. Nelson (1907) wrote of spectacular passages of birds during spring and autumn as follows *"The migration of the Meadow Pipit at the coast is a very noticeable feature of bird life; the first-comers in the latter part of February, but the main body in the middle of March, and at this period, at the Tees estuary, the sand-hills, marshes and fields are absolutely swarming with the light-coloured immigrants….In the autumn immense flocks pass south from August to mid-October. At the Tees there is frequently a great arrival of these Pipits coming in direct from seaward, the majority of which move on after a short stay"*.

Temperley (1951) called this a *"common summer resident and passage migrant, a few remaining over the winter in the lowlands"*. The numbers remaining over the winter depend on the severity of the season. In spring and summer he said that it was *"well distributed everywhere in open country"*, but was *"most abundant on the moorlands"*. In autumn, *"large flocks appear in the lower country"*, with birds leaving the moors at this time of the year. More specifically, flocks of passage migrants were noted at Darlington Sewage Farm at the end of September and in spring, at the end of March during the 1920s (Temperley 1951).

This remains a fairly accurate description of the species' modern situation in County Durham as there seems little evidence of any major changes to its status or distribution during the 20th century. It remains the most plentiful species on the county's extensive western uplands and it is well distributed across much of the lowlands and the eastern agricultural areas, though there have clearly been some local losses in these latter situations.

Recent breeding status

The Meadow Pipit is predominantly, though by no means exclusively, a bird of moorland in the north east of England; it is by far the most widespread and commonest breeding species in upland habitats. Despite this description, it is found nesting at all altitudes across the county. The species is common across most of the county wherever there is open ground and extensive grassland. It does occur close to human habitation, though it is almost absent from heavily built-up areas. The greatest number of breeding birds is to be found on the rough grasslands of the county's western fells, with all rough pasture and heather moorland holding many breeding territories (Westerberg & Bowey 2000). Examples of the numbers present can be seen by the count of *c.*100 birds around the Bowlees/Ettersgill Moor/Western Beck area in July 2000 in Teesdale, and of 300 at Bollihope on 4 April 1993. A report of 'hundreds' on the moors above Edmundbyers came on 1 April 1995, whilst birds were present at a concentration of eight pairs per square kilometre on Barningham Moor in 1997.

As a breeding species, the Durham population is, in broad terms, divided between the western uplands and those in the lowlands, where it occurs in all types of uncultivated grassland, including grazing marsh, coastal banks and the limestone and rough grasslands of the coastal strip. Close to and right along the coast, rough grasslands, pastures and cliff-tops from South Shields to North Gare hold many hundreds of pairs. For example, in 2000, 45 pairs were noted in the extensive grasslands around Rainton Meadows and in 2007, at least twelve pairs were detected in cliff top fields between Ryhope and Pincushion Rocks, while the species was very common in many other coastal grassland areas such as on South Shields Leas (Newsome 2008).

Breeding localities in central Durham are widespread but birds are both fewer and more dispersed in this area of the county. In the mid-Wear valley, Daisy Hill and Waldridge Fell hold some reasonably large populations as does the Brandon/Brancepeth area west of Durham City where a combined total of 65 pairs were noted in 2001, with 60 pairs at Brandon alone in 2007. Also in 2007, an example of a mid-altitude site holding a relatively large number of birds comes from Hedleyhope Fell, where on 7 March there were 54 singing birds, whilst the same year saw lowland population densities of 14 pairs at Hetton Lyons CP and eight pairs at Hetton Bogs (Newsome 2008).

Isolated pairs can even occasionally be found breeding on the unsown verges of trunk roads and vacant plots on industrial estates (Westerberg & Bowey 2000).

Though it is not absent from areas of intensive agriculture, it tends to occur in such areas clustered in pockets of more suitable habitat and at very much lower densities than in optimal habitats. The *Atlas* well illustrated its relative scarcity in the south east of the county. This is the area of most intensive agriculture though, as well shown by the Cleveland Breeding Bird Atlas, there are considerable numbers of birds around the North Tees Marshes. In this area, most breeding Meadow Pipits are concentrated around the grassland and dune areas to the north of the mouth of the Tees, in particular the National Nature Reserve (Joynt *et al.* 2008).

There are probably over 400 pairs breeding in the northern part of what was Cleveland; these are sparsely scattered on the agricultural landscapes to the west of the Tees estuary and the Teesside conurbation (Joynt *et al.* 2008). Other than the Cleveland Atlas work, there is little new accurate population data for much of the species' breeding range in Durham. The *Atlas'* extrapolated population density figures, using figures published in Gibbons *et al.* (1993) as a foundation, probably stand up as a working guide. There is little evidence to suggest that there have been any radical changes to the Durham population estimate of around 20,000 pairs, which was made at that time (Westerberg & Bowey 2000).

The species' ability to rapidly occupy appropriate rough grassland was illustrated in the early part of the 21st century, when birds bred in rough grassland around early-stage woodland planting on the long-established Kibblesworth CBC sites. However, the reverse is also the case, with an observed decline at this site, between 2006 and 2009, due to the maturing of the recently planted woodlands (Newsome 2010). Whilst there are no apparent large-scale threats to the overall population of the county, local extinctions could occur when breeding sites are taken into cultivation, or become afforested as exampled above. In 2009, according to national Breeding Bird Survey data, the Meadow Pipit was found to have undergone a 20% decline on study sites since the start of the BBS in the mid-1990s (Risely *et al.* 2010).

This corresponds well with the long-term UK trend of a slow decline in numbers. In the period between the two breeding Atlases a range contraction of 3.3% was evident, the majority of this reduction in range and numbers coming from the predominantly lowland areas in the southern half of England (Sharrock 1976, Gibbons *et al.* 1993).

Recent non-breeding status

Whilst this is a common breeding bird, particularly so on the moorlands of the western uplands, many birds leave Durham prior to the cold weather months and in many respects the Meadow Pipit is largely a summer visitor to the Durham area, though some small numbers of birds spend the winter along the coastal strip and on the south east coastal plain. In general terms, there are usually few reports of any large numbers during the winter period, with occasional observations of large congregations at the start or the end of the year. In some years, there are very few birds present during the mid-winter months particularly when the weather is severe. Hence in December 1981 the only report was of two birds in a Sunderland market garden, whilst the only flock, numbering just 30 birds, during the very deep snow of December 2010 was present at Haverton Hole. Conversely when the weather is particularly mild birds may be found on the moors as early as January, for example in 1980 and again in 1981, when Meadow Pipits were present at over 300m above sea level near Hamsterley (Baldridge 1981 & 1982).

On relatively rare occasions, small numbers of wintering birds can be fairly widespread through some of the Durham lowlands in the final two months of the year. For example, 70 were in the Castle Lake, Bishop Middleham area in December 2009 and in the last two weeks of December 2009, a cold snap brought what were, presumably, continental birds to the county, with up to 200 right on the coast, feeding at Whitburn Steel, on 22 December. This was by far the highest winter count for many years, but the highest number recorded during the winter period in modern times concerned a flock of 245 on Parson's Haugh, Ryton Willows on 9 January 1991 (Armstrong 1992).

Birds are more or less absent from the uplands in the first two months of most years, though early birds occasionally put in an appearance in late February, for example the 130 birds at South Shields Leas on the 21 February 2005 (Newsome 2006). Typically, the first returning birds arriving in the upper dales are on the fells around the first to second week of March. They then arrive *en masse* from around the third week of March, which usually brings an obvious increase in many areas of Teesdale and Weardale and the rest of the Durham uplands, in a spring movement through the county that persists into early or mid-April. From mid-March to early May, birds can be seen and heard moving north and west almost on a daily basis.

There are often several hundred birds per day and at this time they are frequently observed at coastal localities and, on northward or north westerly vectors, passage occurs through the county on a broad-front; such spring movements routinely number many thousands of birds. Peak counts at a rate of c.400 birds per hour have been noted on several occasions, for example 384 flying north in one hour at Hartlepool on 6 April 1961 and 480 north on 24 March 1968 at Whitburn. Groups pausing on passage have included several hundred at Whitburn on 8 April 1987 and 450 on Seaton Common on 26 March 2006 with many smaller groups elsewhere. Over 300 which came 'in-off' the sea at Whitburn on 27 March 2008 were carried on an easterly wind and may well have been birds bound for Scandinavian breeding grounds (Newsome 2008). Mather (1986) pointed out that locally breeding birds in northern England tend to be of an olive-coloured tone, whereas many of the grounded passage migrants that are seen in spring are greyer in hue and paler; these are presumably birds which are heading up to northern Europe to breed.

Autumn passage can be quite protracted from August to November but peak counts are generally recorded in September or early October. By this time many of the county's breeding birds have already left, but concentrations of birds occasionally occur at some upland sites before the last birds leave the moors. For example, counts on 6 September 2007 included 200 on Barningham Moor, 120 on Eggleston Common and 100 on Gilmonby Moor, whereas these concentrations had halved by the following week. Most have left moorland areas by early or the middle of October.

Autumn passage is usually particularly well documented in coastal areas when counts of 400 birds per hour moving south are not unusual, as illustrated in autumn 2007, when active passage was underway during the second week of September. The peak site count was of 438 flying south over Harton Downhill, South Shields. Some of the more spectacular passages recorded include 'hundreds' moving south on a broad front over Hartlepool on 14 September 1974 and grounded at Whitburn on 3 October 1976, with 1,133 south at Whitburn in 3.5 hours on 3 October 1982. The Co-ordinated Coastal Migrant Count of 2 October 1994 produced a total of 2,033 birds between the Tyne and the Tees, but numbers must far exceed this on days of good passage. For example, over 900 birds flew south at Ryhope in 1 hour on 9 September 1993, 1,123 birds flew south over South Shields Leas in 1.5 hours on 9 September 2000, whilst on 12 September 2008 over 1,000 moved south prior to 8am at Whitburn with another 800 south there in 90 minutes two days later.

Detailed analysis of the movement of Meadow Pipits through Durham during the Coordinated Migrant Count on 2 October 1994 revealed that this was probably a movement of British or more northerly arriving Meadow Pipits heading south. Birds apparently entered Durham at an inland northern location, progressed to the east coast, meeting it around the central portion of the county's coastline and then coasted south from there. This study demonstrated that migratory processes do not always occur across a broad front. Perhaps these birds were moving off the western uplands of Northumberland, or points further north, initially moving in a south easterly direction, before turning south along the Durham coast (Bowey 1999c).

Aberrant individuals have been noted occasionally in the county, and an all-white bird at Chester-le-Street in April 1970, "which looked like an albino Meadow Pipit, was observed feeding on the river bank" (The Vasculum Vol. LV 1970). An individual that had pale creamy-brown plumage similar to that of a first-year Glaucous Gull Larus hyperboreus was present and presumably breeding at Whitburn for five successive summers during 1984-1988 (Armstrong 1989). There was also a pale brown bird with pale buff wings and tail on Cowpen Marsh in May-June 1988 and April 1989 (Blick 2009). The most remarkable record of all, however, comes from the Ornithological Report for Northumberland and Durham in 1963 "Inspector R Roberts of the Darlington RSPCA was called when a Meadow Pipit was seen in difficulty near Middleton-in-Teesdale. On catching the bird he found it to have four perfectly developed legs and accordingly put it to sleep. It was also seen by VFB (V.F.Brown), who identified it as a juvenile" (Bell 1964).

Distribution & movements

This species breeds over much of northern Europe, as far west as south east Greenland, as well as down into Italy and the Balkans. Most populations are migratory, wintering in south west Europe, North Africa and the Middle East, though some of the breeding areas are populated all year round, northerly birds wintering to the south of their breeding range, but within the breeding areas of birds that winter to the south (Parkin & Knox 2010).

There is a considerable amount of national ringing data which confirms that many of the Meadow Pipits breeding in the UK winter in south western parts of Europe (Wernham et al. 2002), as indicated for Northumberland by Kerr (2001). Of 520 recoveries of British-ringed birds abroad almost 60% have come from Spain (179), France (129) and Portugal (98), with a small percentage in Belgium (42) and others moving further south into North Africa with recoveries from Morocco (37) and Algeria (3).

It is clear that some of the north east's birds do likewise, as borne out by the record of one that was ringed as a nestling at Graythorp Shipyard on 7 June 1930 and was later recovered at Alange, Extremadura, Spain, on 11 January 1931. In addition, birds ringed at Seahouses in Northumberland have been recovered in France and Portugal (Kerr 2001), and a bird ringed at Saltholme on 25 September 1999 was found dead in the Charente-Maritime region of France on 16 March 2000 (Blick 2009). Other significant ringing recoveries include a pullus that was ringed near Muggleswick on 4 June 1967 and recovered in October of that year in Empoli, Firenze, Italy, and one ringed at Saltholme on 17 September 2009 which was controlled at Lebbeke, Belgium six days later (Blick 2009). Foreign-ringed birds that have been recovered in the UK have come from The Netherlands, Belgium, Norway and Iceland (Wernham *et al.* 2002).

Red-throated Pipit
Anthus cervinus

A very rare vagrant from Northern Europe and Siberia: six records.

On the morning of 26 May 1963, Graham Bell flushed a Red-throated Pipit from the edge of one of the 'fleets' on Cowpen Marsh. It was seen later in the day by several observers including Russell McAndrew, the late Phil Stead and the late Edgar Gatenby, but it was not seen subsequently. A further five spring birds have been seen in the county, all in the period 19th to 30 May. It is quite surprising that there has only been one acceptable autumn record, despite several claims over the years.

All records:

1963	Cowpen Marsh, 26 May (*British Birds* 57: 276)	
1985	Barmston Pond, 19th to 22 May (*British Birds* 79: 568)	
1991	Ryhope, 20 May (*British Birds* 86: 505)	
1992	Cowpen Marsh, 24 May (*British Birds* 86: 504)	
2008	Dorman's Pool, 30 May	
2010	Seaton Common, 12 September	

Distribution & movements
Red-throated Pipits breed in Arctic Eurasia, from northern Norway, Sweden and Finland, across to the far east of Russia on to the Chukotskiy Peninsular. The wintering range covers equatorial Africa, south China and South East Asia. It is a regular but very scarce migrant to Britain and was removed as a BBRC species in 2006, when the total number of records exceeded 400; the Northern Isles and the Isles of Scilly are the prime spots for the species in both spring and autumn.

Rock Pipit (Eurasian Rock Pipit)
Anthus petrosus

A scarce resident, winter visitor and passage migrant, usually restricted to the coast.

Historical review
This species has a long-documented presence in the north east of England, courtesy of a description of birds on the Farne Islands by Pennant in 1769 (Kerr 2001). Its coastal occupation of Durham is likely to be similar in pedigree. It was viewed as common along the Durham coast by Hutchinson (1840), particularly on those sections that were "*girt with rocks*". Tristram (1905), although recognising the Rock Pipit as a bird of the coast, was not

aware of "it breeding in County Durham". However, Hogg (1845) had said that it was "very frequent on the Magnesian limestone rocks of the Durham coast" and Hancock (1874) had described it as "a resident, breeding plentifully on our rocky sea shores, and remaining with us the whole year".

This was confirmed by Nelson (1907) who knew of it breeding in the county, although not at Teesmouth. The lack of rocky coastline in the Tees estuary appears to have deterred it from breeding there until the early years of the 20th century. Then the building of sea-walls composed of slag provided appropriate nest sites, and so in 1909 a pair was watched building a nest "in a hole in a slag wall" on the Durham side of the estuary. Industrialisation of the County Durham coastline at the same time impacted considerably on the contiguity of its distribution, however it remained as a breeding bird, at least at Teesmouth (Richmond 1931). Pairs continued to breed at Teesmouth in similar situations in 1933, with several pairs there over the period 1940 to 1944.

According to Temperley (1951) it was "not uncommon along the sea-coast, but never penetrating inland. Breeds close to the shore in the most exposed places amongst the rocks, often just above the high water mark". This description remains accurate today except for the occasional record away from the coast. In both 1951 and 1960, two pairs were known to have bred in the South Shields area. The first record of the species away from the coast was a single bird at a pond near Stanley in September 1960 (Coulson 1961).

Nelson (1907) reported the migration of the species at Teesmouth in September and in spring, in March and April; he documented seeing flocks of 12 to 14 birds at these times. There were no records of the Scandinavian race, littoralis at Temperley's time of writing, though presumably the passage of birds, as noted by Nelson, would have contained many birds of this race, as well as birds from further north in their British breeding range. In 1957, five were in the Tees Estuary on 26 September (Grey 1958). However, the species was rarely reported in the ornithological reports, it not being until 1971, and subsequently, that records occurred annually.

Recent breeding status

Throughout the year, the occurrence of this species in Durham is almost exclusively coastal; inland records are rare and intermittent, but such an unassuming bird may possibly be overlooked. The majority of reports, though not necessarily the majority of breeding birds, come from the northern stretch of the county's coastline, between South Shields and Sunderland. As a breeding bird in Durham, the Rock Pipit is entirely restricted to the coast (Strowger et al. 1995), mainly to the splash and littoral zones of the local rocky shores. It is rarely seen more than a few metres from the seashore. Its stronghold in the county is undoubtedly the stretch of coastline between the mouth of the River Tyne and the mouth of the River Wear, particularly the low limestone cliffs north of Marsden Bay (Westerberg & Bowey 2000). In 1985 eight pairs were reported to be breeding along the coast line between South Shields and Whitburn. There were however, no confirmed breeding records for the Atlas survey period 1988-1993, although several pairs were present along the coastline north of Marsden during the summer months (Westerberg & Bowey 2000).

Typical breeding locations that have been identified in the county since then include Byers Hole at Whitburn, Camel Island at Marsden and the South Shields Pier. The dedicated 1995 Rock Pipit survey (Strowger et al. 1995) indicated that as many as 30 pairs might be found breeding along the Durham coastline, between the Tyne and Teesmouth, though around Teesmouth this species is considered mainly a wintering bird (Joynt et al. 2008). The survey also provided information on mean egg laying dates, 3 May for first clutches and 7 June for second; mean clutch size, 4.00 for first clutches and 4.25 for second; and hatching success of 95% for first clutches and 71% for second. Post-fledging observations of ringed birds in the Marsden area suggest that the young of the year remain in the immediate area for at least a few months after fledging, often being seen feeding nearby, into their first autumn and winter. The ringing study also produced some evidence which suggests that at least some of the juveniles remain in their natal area to breed in subsequent years (P. Collins pers. comm.).

On 20 February 2005, a bird was already singing at Camel Island, Marsden Bay, with three to four pairs in that area during that month. A total of 21 were counted between Lizard Point and South Shields Pier on 17 February 2007 and 11 were between Seaburn and Whitburn Steel on 24 February, though the true status of these birds, winter visitors, passage migrants or local breeding birds is unknown (Newsome 2008).

Gibbons et al. (1993) suggested that nationally there had been some decline in the breeding population of Rock Pipit since 1968-1972, but there are no local data to suggest that this is the case in Durham. Indeed in 2007, it was observed that Rock Pipit is 'more regular' in the northern section of coast than anywhere else in Durham and that perhaps ten to fifteen pairs were nesting between Whitburn and South Shields, and the resident population along this section of the northern coast is probably stable (Newsome 2008). In the summer period, the species is also regularly seen in the Hartlepool area, with occasional sightings at Roker, Hendon and in Sunderland Docks, and further south at Seaham. Man-made structures such as docks, piers and jetties can prove attractive in addition to the natural rocky coastline. The extensive cleaning of the Durham beaches, which has occurred since the loss of the deep mining industry, may well produce a less polluted environment, which could be conducive to an increase in numbers of this species in some areas.

Recent non-breeding status

An analysis of the Durham Bird Club's database for this species indicated that its principal winter resorts along the Durham coastline are, in the north, South Shields Pier and Whitburn Steel and, in the south, around Hartlepool Headland, Parton Rocks in particular. Birds are also regularly reported from Roker, the Sunderland piers, Sunderland Docks and Hartlepool Docks, Seaham and Ryhope. The numbers of birds present in the county invariably reaches its highest in the winter period, November to March, and probably involves both local birds and birds from elsewhere in Britain and the Continent (Strowger *et al.* 1995). In most years, the highest counts of the year come in mid-March, when up to 20 birds congregate at passage locations, such as Whitburn Steel, or are noted between Camel Island and South Shields Pier, although some of these may be passage birds.

The ample piles of rotting seaweed that collect on the foreshore from the off-shore beds of *Laminaria* and *Fucus* wracks, to the south of Whitburn, usually attract the largest gatherings of birds in the first quarter of the year and especially in late winter and early spring, when passage birds are thought to swell the numbers of birds present. Up to 40 birds were seen in 1975, 30 in 1976, 25 in 1979, and up to 20 in 1983. The latter part of the year also produces significant counts, such as in 1992 when 35 were at Whitburn on 30 September, increasing to 40 birds in October. More unusual were reports of passage, as far inland as Cleadon village, involving up to 40 birds in October 1996. In late 2009, a large influx took place at Whitburn Steel, and counts rose from 12 on 14 November to 20 on 1 December and then to more than 50 on 25 December. Forty to 50 of these remained in the Whitburn Steel area into early January 2010 and this is one of the largest ever gatherings of the species in the county.

Whitburn, however, does not hold a monopoly on gatherings of birds, for example at least 15 were present on the beach at Roker on 29 December 2005. It may be that this species is overlooked in other locations, as it is rarely sought out by birdwatchers. In the south of the county, there are usually a number of reports of one to three birds around Hartlepool Headland, Newburn and North Gare, to the north of the Tees estuary, from January to late March and again from early October to the year's end. In the River Tyne, similar numbers have been reported from Jarrow Slake in 1987 and 1991 and at Dunstan Staithes in 2008. In January 2010, up to eight birds frequented the Newburn area of the River Tyne. Unusually, in November and December 2008, one or two birds were present on the western edge of Whitburn village, possibly feeding and roosting in fields just inland of the coast (Newsome 2009).

In some instances active migration is noted through the county, for instance the ten birds that flew south over Whitburn Coastal Park on 23 October 2005 (Newsome 2006). The peak periods for reports of the species inland are in March/April and September/October, though there are also a number of winter sightings. The inland sighting 'peaks' coincide with the time when Scandinavian birds, or those from further north in Britain, are likely to be passing through the county on spring passage or arriving to spend the winter around the British coast. There is some suggestion from work along the Northumberland coast, that local breeding birds are largely sedentary (Kerr 2001) and this has, at least in part, been borne out by observations of colour-ringed birds in the Marsden area, which both bred and wintered in the immediate area around their natal sites.

Since the record in 1960, Rock Pipits have appeared well inland on at least 13 occasions, the furthest inland of these being one at Derwent Reservoir in October 1975. In 1987 and 1991 singles were at Jarrow Slake, a single bird was at Shibdon Pond in 1989 and in 2008, a bird was seen on three dates in January-February and once in October at Dunston. Away from the Tyne, records have come from Boldon Flats in 1987 and 1997, Hetton Lyons in 1994 and 1997, one at Longnewton Reservoir on 26th to 27 September 2005, Little Stanton, near Bishopton in 2006 and one was observed in Sunderland City Centre on 10 January 2007, this bird perhaps feeding on a flooded building roof. In 2008, the only inland record, of a presumed British bird, was at the Tees Barrage on 4 February. In 2009, one at Greatham Creek on 12 February and one at Bowesfield Marsh on 10 October were the only birds reported inland.

Scandinavian Rock Pipit
Anthus petrosus littoralis

An analysis of records has revealed that the months of March and April are the best time for the recording of the distinctively plumaged Scandinavian race of the species, *littoralis*, which becomes more easily distinguishable from local birds, *petrosus*, in spring. The extreme dates in this late winter/early spring period are 6 February to 24 April, although others have been noted at Cowpen Marsh on 15 January 2000 and Whitburn Range on 31 May 2004. Whitburn Bents/Steel is the most favoured locality with the North Tees Marshes and Seaton Common also contributing records. On average, between one and four birds are seen in most years, but no doubt many others are missed, or cannot be differentiated from local birds. The best counts for Durham have been six at Seaton Common on 26 March 2006 and four or five at Whitburn Steel on 23 March 2008. Inland sightings are very few, the first documented being two at Boldon Flats on 30th to 31 March 1987. Further birds were at Sunderland Academy Pools in March and April 2003, Boldon Flats in March 2005 and Longnewton Reservoir on 2nd to 3 April 2008.

Distribution & movements
The Rock Pipit breeds around the coasts of northern France, Britain, Ireland, Fennoscandia, the Faeroes and northern Russia. The northernmost birds, of the race *littoralis*, are migratory and winter to the south of their breeding areas, as far south as north west Africa, but many of them within the wintering area of the British race *petrosus*. Most of the British breeding birds are resident (Parkin & Knox 2010).

Water Pipit
Anthus spinoletta

A scarce and localised winter visitor and a rare passage migrant.

Historical review
The first occurrence of this species in Durham is rather clouded in confusion by the fact that for many years it was considered as conspecific with Rock Pipit *Anthus petrosus*. It is not inconceivable that it had occurred before the first fully documented record in the mid-fifties. The first county record was of one on 29 March 1956 at Greatham Creek, but as stated by Blick (2009), it seems likely that this species had been an at least scarce, sporadic visitor to the county prior to its first recognition. It was not documented by Temperley (1951) or any of the earlier writers on Durham's bird life. Interestingly, Northumberland also recorded its first in March 1956 near St. Mary's Island (Galloway & Meek 1983).

The location of the first record set a blueprint for many subsequent observations of the species in the county, as over fifty years later the open marshes and wet grasslands around Teesmouth remain the stronghold of this species. Through the 1960s birds remained scarce and the second record was five years in coming, with a bright spring-plumage bird at Crimdon Dene on 13 April 1961 (Bell 1962). Over a decade came and went before a small cluster of records occurred at either end of the county. In the south east one was on the North Tees Marshes on 1 April 1973 (Coulson 1974), but in the north the first record of this species for Gateshead, and the first wintering record in the county, was of two birds found together at Shibdon Pond on Boxing Day 1973. One of these remained until 24 March 1974 and then, what was presumed to be one of the same birds returned to winter in this area, occasionally being noted as far away as Swalwell. This pattern repeated until the winter of 1975/1976, when it was last recorded on 14 April 1976 on the nearby Derwenthaugh Meadows. It is possible that the same bird was then seen at Shibdon Pond on 27 March to 2 April 1977, not having been seen in the previous autumn/winter period. It seems most likely that, after the initial two, these records all refer to a single bird returning to winter in the same area in successive years (Bowey *et al.* 1993). The only other records of the 1970s were of one in the Tees Estuary on 21 October 1976 and one at Hart Reservoir on 12 April 1979.

Recent status

During the first decade of the 21st century the North Tees Marshes held a regular winter population of between three and 18 birds each winter. Today the Water Pipit is an annual winter visitor and a small-scale passage migrant to the county. Most observations fall into the broad period from late October to mid-April and within this there are subsumed two patterns of records, which split quite neatly into wintering birds and spring passage birds, though this may over-simplify a slightly more complex situation. Passage birds in spring are occasionally seen in their attractive summer plumage, in late March and April.

The North Tees Marshes are the most important location for this species and favoured locations in this area include Cowpen Marsh, Haverton Hole and the Dorman's Pool to Saltholme Pools area, though birds have also been noted at Seaton Common, Greenabella Marsh and further afield at Hart Reservoir. The regular wintering of birds in the county, between November and March, is a quite recently developed phenomenon. According to Blick (2009), birds have been observed at Teesmouth in most years since around 1974, but this will include some observations south of the river. The wintering trend seemed to increase significantly following two birds at Teesmouth in November-December 1984, and the number of wintering birds there rose steadily thereafter through the early to mid-1990s. The pattern of occurrence here is that the first few birds arrive in late October and through early November, but peak counts are not normally reached until the New Year, or even late winter, when it is likely that early passage migrants have swelled the numbers of wintering birds.

In December 1995, up to five birds were present at Teesmouth and just a few years later this figure had risen to double figures, with 17 being there in December 2000. The maximum count recorded is 22, which were seen in February 2001. Over the remainder of the 2000s, the maximum counts declined slightly, with 18 being noted in January to February 2003 and 17 in March 2008 (Blick 2009).

Away from Teesmouth, in both winter and passage periods, the species is a genuinely scarce migrant across the rest of Durham, though birds have turned up at a wide variety of sites in the last thirty years or so.

Wintering birds away from Teesmouth, from 1980 to 2010 can be summarised as:

1983	Birds were on the River Wear at South Hylton, Sunderland in late November to mid-December, with a maximum of four reported on 12 December
1985	Two birds were at Boldon Flats from 27th to 28 November, then one was at Whitburn Steel from 15 December until 2 April 1986
1986	A bird was once again at Whitburn Steel from December to 8 April 1987
1987	For the third winter running, a wintering bird was at Whitburn Steel, from December 1987 to 29 February 1988, and was presumably the same bird as in previous years
1992	One was at Bowburn STW on 21 November
2001	One was at Sedgefield on 7 January
2003	One was at Whitburn Steel on 4th and 5 January
2009	Two were at Whitburn Steel from 10 November, one of which remained into early 2010

In some instances, it would appear that the same birds were responsible for sequential wintering records.

Birds seen on spring passage, away from Teesmouth, between 1980 and 2010 include:

1986	Blackhall Rocks on 19 March and a passage bird was seen in mid-March alongside the Whitburn wintering bird, which remained until April
1987	On 6 April, two additional passage birds were noted alongside the wintering bird at Whitburn Steel and by the 8th all had departed
1994	A summer-plumage bird was at Shibdon Pond on 20 April
1996	Single birds were at Trow Rocks on 16 April and at MetroCentre Pools on 12 May.
1997	One at Hetton Bogs on 11 March
1998	One at Hetton Lyons CP on 20 March
2003	One at Sunderland AFC Academy Pools on 11 March
2004	One at Hetton Lyons CP on 25 March

Distribution & movements

This is a widely distributed species of the mountainous areas in the Palaearctic, from southern and central Europe, northwards and eastwards. The western populations, from locations such as the European Alps, winter on the lowlands to both north and south of their upland breeding areas. British birds are largely found in the south and east of England, becoming progressively scarcer further north. It is mainly a winter visitor to the British Isles and wintering birds favour wetland habitats, seeming to avoid, by and large, saline habitats, though brackish marshes are sometimes used (Parkin & Knox 2010). The British wintering population is estimated at less than 100 birds, although this may be a slight underestimate arising from misidentification. The total European population is within the range 540,000 to 1.8 million pairs (Hagemeijer & Blair 1997).

Chaffinch (Common Chaffinch)
Fringilla coelebs

An abundant resident, very common winter visitor and passage migrant.

Historical review

The presence of finches, presumably Chaffinch, in the woodlands of the Wear valley downstream of the modern Durham City must have led to the designation, during the Middle Ages, of the area known as Finchale, which is a name derived from Old English "*finc , halh*" which means 'finch nook'(Yalden & Albarella 2009).

Hancock (1874) described it as "*resident and probably the most abundant bird in the district*". Cannon Tristram (1905) considered the Chaffinch to be a "*common resident and frequent winter visitor*" believing that females from the breeding population preferentially left the area in winter with a predominance of males remaining. This belief perpetuated a concept first mentioned by *Linneus* when proposing the species' scientific binomial and continued by Selby and Gilbert White. Joseph Duff of Bishop Auckland writing in 1847 (*The Zoologist* 5, 1847) and 1848 (*The Zoologist* 6, 1848) sought to challenge this rather rigid view but was taken to task by the accepted authorities of the day. Today we understand that there is some evidence for differential winter movement between the sexes from continental Europe (Lack 1986) but not to the extremes imagined by Tristram (1905) and others. Almond *et al.* (1939) gave the Chaffinch's status as "*a common resident and winter visitor*" in the south east of the county. Temperley (1951) said, of this widespread common species, it is, "*an abundant resident*", "*well-distributed*" around the county. He further commented that at South Shields it is occasionally met with on the coastal cliffs, indicating a migratory origin for wintering birds.

Recent breeding status

There is little evidence of change in status since the historical assessments of a century or more ago. Breeding Chaffinches can be found in almost all wooded or mature scrub habitats across the county and in high quality woodland it is one of the most abundant, widespread and frequently reported species in Durham. This is reflective of the national situation; the 2009 Breeding Bird Survey data indicated that the Chaffinch was the third most widespread species in Britain, being present in 92% of surveyed sites (Risely *et al.* 2010). Birds penetrate as far west into the dales as available habitat allows, with only the exposed moorland devoid of trees or bushes not having breeding birds. In 1982 a pair successfully fledged young at Grains o' th' Beck, Lunedale at 365m above sea level. It is less common or sometimes entirely absent in the most built-up urban areas and industrial environments (Westerberg & Bowey 2000). The greatest density of breeding birds are found along broadleaved woodland edge, though appreciable numbers are also present in conifer plantations, parks, gardens and in mature hedgerows (Gibbons *et al.* 1993). In the *Atlas*, only Blackbird *Turdus merula*, House Sparrow *Passer domesticus*, Wren *Troglodytes troglodytes* and Willow Warbler *Phylloscopus trochilus* were illustrated as being more widely distributed (Westerberg & Bowey 2000). In the early part of the 21st century, some of the largest breeding numbers were reported from Elemore Woods, although many sites in the county, such as Hamsterley Forest, the extensive Derwent valley woodlands and The Stang, will have held similarly high densities, of breeding birds.

The less populated areas, illustrated in the *Atlas* map, especially in the south east of the county, were almost certainly due to a lack of data as opposed to a genuine absence of this species. A more precise representation of this species' status in this part of the county was illustrated in Joynt (2008). This showed that Chaffinch was both widespread and common as a breeding species across the 'old Cleveland area' with, perhaps, as many as 1,500 to 2,000 pairs in the area to the north of Teesmouth (Joynt *et al.* 2008). Nonetheless, it is sparsely distributed in the urban areas of the south east, except in the town parks and mature suburbs of some of the 'leafier' areas'.

Nationally, Chaffinches make up a substantial proportion of the total woodland bird population (Gibbons *et al.* 1993). Locally, this is probably correct for the coniferous woods in the west of the county but less so for lowland woodlands further east. Detailed studies in the Derwent valley found that Chaffinches in this habitat occurred at relatively low densities, of around one pair per five to eight hectares (Westerberg & Bowey 2000). On farmland, the species is found at lower densities relative to those in woodland; their breeding densities in such locations being limited by the quantity and quality of trees and large shrubs. Using national figures, of 30 pairs per square kilometre as a 'typical breeding density' in a variety of local habitats, as indicated by the *New Atlas* (Gibbons *et al.* 1993), the *Atlas* determined the Durham population to be in the region of 60,000-70,000 breeding pairs.

The first indication of birds re-establishing territory usually occurs at a time when wintering continental birds will still be in flocks in Durham. In 2007, the first singing male was noted at Hardwick Hall on 1 February. By late March local birds are normally paired off and setting up territory. At the very local level, area-based studies have indicated the number of birds that can be present, particularly in the county's lowland woodlands. For example, a study area at Elemore Woods, in the eastern part of the county, produced counts of singing males each spring over the period 1998 to 2009 varying between 66 and 82, average 76, with an apparently stable long term trend. The Chaffinch here was noted as '*the commonest species in the area*'. Other indications of the scale of the county's breeding population size came from the Joe's Pond/Rainton Meadows, Hetton Lyons and Hetton Bogs 'complex'. These three sites, which are located within close proximity of each other, held over 110 breeding pairs in 2007 with an increasing trend year on year (Newsome 2008). A study area at Kibblesworth recorded 47 pairs in 2005, the second highest count there since the early 1970s. Overall, the county's population in recent decades has been at least stable and perhaps slightly increasing as indicated by these year-to-year comparisons of local breeding populations. While long-term Breeding Bird Surveys have shown populations in north east England to have increased by around 20% over the last 15 years, farmland CBC indices have suggested a slight decline in that habitat (Stroud & Glue 1991), but one not nearly as serious as for many other farmland species.

Recent non-breeding status
The many birds that can be found nesting at parks and picnic sites around the county frequently become very tame and will closely approach humans that they perceive as potential sources of food, this trait being often most obvious during the winter.

In most years, it is the latter part of October before any sizeable wintering gatherings are evident with the arrival of continental birds generally coinciding with the appearance of Redwing *Turdus iliacus* and Fieldfares *Turdus pilaris*. Post-breeding and wintering flocks not unusually exceed 100 and sometimes rise to a few hundred strong through November and December. By far the largest flock reported in recent times was on the southern edge of Hamsterley Forest when 600 were present in December 1978, growing to an estimated 1,000 by March 1979. In January 1981, 400 were again at Hamsterley whilst 750 roosted in Darlington's West Cemetery. Other notable gatherings include: 450 at Ruffside in December 1988; 480 at Elemore Hall in November 1992; 430 at Swainston in December 2001 and 400 at Seaton Pond in December 2009. The largest single flock recorded in the south east of the county was 380 in Wynyard on 6 March 1982 (Blick 2009). Land set aside for game cover such as at Seaton Pond can be locally important for this species in winter.

Scattered flocks of 50-80 birds are not that unusual at any time between October and early March. During hard winter weather, larger flocks of 150-200 birds, usually associating with other finches and buntings, might be found at many sites where food is plentiful, with birds gathering under beech trees *Fagus sylvatica* in 'mast years' and at grain stores in farmyards.

This species is considered a regular passage migrant along the coast in both spring and autumn (Joynt *et al.* 2008). Given the considerable numbers involved in the regular winter influx of continental birds, the Chaffinch does not have an especially strong presence as a passage bird on the coast; though 110 birds were noted along the length of the county's coast during the Coordinated Migrant Count on 4 October 1992 (Bowey 1999c). Clearly the majority of winter visitors move directly inland. Coastal autumn passage is evident from late September to early November; while lighter return spring passage is evident from late March to very early May. In some years, more birds may be present along the coastal strip during the final months of the year and this may be accounted for by larger than usual influxes of continental migrants.

Most recent observations at the coast in October typically involve single day counts of small groups of 10-40 birds at locations such as Crimdon, Hartlepool, Marsden, Seaham and South Shields. During a period of evident passage on land in late September 1976 *"large numbers"* were noted flying west over a fishing boat off Sunderland on the 26th and 27th of the month. Similarly, at Marsden over 7th to 8 November 1993, with only light evidence of birds grounded on the coast, over 200 immigrants were seen passing west directly overhead.

By March, most lowland flocks have dispersed, as local birds have moved into breeding territories. Any flocks still persisting at this time of year, and into early April, probably consist largely of continental birds wintering here. These are almost invariably comprised of migrants from further north. Ringing data suggests the origin of these in Durham is primarily Scandinavian, as detailed below.

Distribution & movements

Chaffinches breed over much of Europe, western Siberia and, in various racial forms, down into Africa and the Middle East. Most British birds are largely resident, though large numbers of northern and eastern European birds are migratory, wintering to the south and west, including the British Isles.

Ringing data linking Durham birds with northern Europe and Scandinavian origin includes:

- One ringed at Helgoland, Germany in September 1960 was recovered at Hartlepool in October 1961
- One ringed at Helgoland on 23 September 1969 was found dead at Hartlepool about two weeks later
- An adult ringed at Barnard Castle on 15 January 1978 was controlled in Bleckinge, Sweden on 1 April 1980.
- A bird ringed as an adult female at Shibdon Pond on 13 October 1985 was killed by a cat almost four years later at Fjerritslev, Jylland in Denmark on 28 June 1989, a distance of 722km to the north east
- A bird ringed as a juvenile female at Vest-Agder, in southern Norway on 28 July 1990 was at the Thornley Woods feeding station on 1 December 1991

Evidence of movement to and from sites within northern England, west in the autumn and east in the spring, comes from other local ringing data. For example, a bird ringed as a juvenile female on 31 October 1986 at the Calf of Man, in the Irish Sea, was controlled at Hamsterley Forest on 7 May 1988. Moving in the opposite direction, an adult male ringed at Wolsingham on 26 March 1989 was controlled at Calf of Man on 18 October 1990. Another ringed near Blackpool in November 1985 was recovered at Barnard Castle in November 1989. An adult male ringed at Hamsterley Forest on 14 April 1991 was controlled to the south west at Pontypridd, Glamorgan on 3 March 1994. These records indicate the movement west or south west of some locally breeding birds in autumn and their subsequent return east to breeding grounds in spring. Probably undertaking such a movement was the first-year bird, ringed at Barnard Castle on 13 December 1979, which was killed by a cat at Alston, in the Pennines on 11 February 1980.

Some birds in the north east undergo a more southerly movement, such as the bird ringed at Kielder Forest in June 1993 that was recovered at Hamsterley Forest in March 1995, and the bird ringed at Hartlepool on 10 September 2002 that was controlled at Froggat, Derbyshire on 14 February 2003; although this latter bird may have been a continental immigrant.

What is also clear from local ringing studies is that some local birds are rather sedentary, or at least site-faithful between seasons; as illustrated by the bird that was ringed in the Ravensworth Estate in the Team Valley in December 1973 and that was still there on 18 February 1978. Likewise, the ring from a bird ringed as an adult female on 31 July 1988 at Thornley Woods, which was found in a Tawny Owl pellet one kilometre to the east of that sites on 18 October 1990.

Brambling
Fringilla montifringilla

A common winter visitor and passage migrant in variable numbers and an extremely rare and sporadic breeding species.

Historical review

In relation to the Derwent valley, this species was thought to be a "*common winter visitant*" by Robson (1896), with numbers that varied quite considerably from year to year. Robson recognised the species' reliance on beech mast, mentioning that their favourite local spot was under tall beech trees in Gibside (Bowey *et al.* 1993). Its status as a winter visitor and passage migrant was recognised by Nelson (1907), who documented the species' regular arrival "*at the Tees estuary*", in October. Whilst Canon Tristram (1905) and Almond *et al.* (1939) also agreed that it was a fairly common winter visitor in varying numbers. Temperley (1951) mentioned Dr. Blair's observation of a male singing, presumably at South Shields, on 17 March 1929. In 1929, a Brambling that had been ringed on Helgoland, Germany in 1928 was recovered in County Durham.

It was noted as being particularly numerous in the winter of 1934/1935, when a flock of at least 500 was near Blanchland in December 1934 (Temperley 1951). By contrast, in poor mast years, for example 1945, very few birds were seen. According to Temperley (1951), it was a regular autumn and winter visitor, in varying numbers, often to be found where mast was most available. It would also consort with Chaffinches *Fringilla coelebs* on stubble and other seed sources if the mast was exhausted.

Specific early records of flocks greater than 50 remained few until the years 1965 to 1967. In January 1965, 200 were present on Cleadon Hills and in December of that year an impressive flock thought to number 1,000 was at Graythorp, Teesmouth for a few days. In late December 1966 the largest flock ever recorded and exceeding 1,400 birds, was seen at Hurworth Burn Reservoir with 800 still present the following January and 250 remaining until 3 April 1967. Meanwhile 'around 300' were recorded at Neville's Cross, Durham City, on 29 January, with another flock, of 200, a few miles away at Brancepeth Park on 22 February.

Recent breeding status

Occasionally wintering birds stay on in Durham into late April and May and on at least four occasions there has been evidence to suggest attempted breeding in the county. In one instance this was known to be successful, though at the time the information was not published.

In 1978 a pair of birds lingered together in Thornley Woods, in the lower Derwent valley, until at least 23 April and in May the male was found singing. The following month a female was discovered feeding young in a nest in a different part of the same wood and it is believed that the young were successfully fledged (Bowey *et al.* 1993).

A female was observed in oak woodland at Edmundbyers on 20 June 1982. It was seen to indulge in wing-shivering on several occasions but no other birds were detected. (T. I. Mills pers. obs.). In May 1986, a lone male was discovered singing in Buck's Hill Plantation, Washingwell Woods, near Whickham, in the north west of the county. On at least one occasion it attempted to copulate with a female Chaffinch, to which it seemed to be paired. This latter bird was known to have nested and produced a family of young, which appeared to be pure-bred Chaffinch (Bowey *et al.* 1993).

The Gateshead area again attracted singing birds in the spring of 1998, after the large winter influx of 1997/1998. Over this period, flocks of around 200 birds had frequented the Thornley Feeding Station in the lower Derwent valley in early April. Rather bizarrely, at least eleven singing birds were scattered around the birch woods there and several of these stayed well into the month. That same spring, up to ten singing birds, were briefly present at Shibdon Pond. At least one, full summer plumage male, stayed in territory in Thornley Woods into the middle of May, with a strong suggestion that it was actually holding territory. However, no further proof of breeding was secured.

In 2006, the fourth suggestion of possible local breeding came with a pair seen on one day only at Satley on 9 July. Birds are obviously capable of making occasional summer appearances as witnessed by a male in full summer plumage seen at Crimdon Dene on 4 July 1963.

Recent non-breeding status

Birds usually arrive on the Durham coast and inland in the autumn, the first trickle of birds being noted in the last few days of September, but the bulk of the movement occurs usually through mid-October with light coastal passage usually apparent. Only small numbers are typically noted as grounded coastal migrants, with more passing overhead and most going unseen straight inland probably as night time migrants (Lack 1986). Coastal records usually involve a handful of birds at a few sites but exceptionally in 1987, Mere Knolls Cemetery hosted a fall of 300 Brambling on 20 October and on the same day in 2004, some 200 were grounded at Whitburn.

Larger numbers are not usually noted inland until early November or even into December, when birds start exploiting beech mast, in the years when that resource is available. This pattern has been repeated consistently since the era of more detailed observation and reporting began in the 1960s.

During the mid-winter period, flocks typically numbering 20-80 are the norm and these can be quite widespread, sometimes occurring in rural agricultural habitats but more often in small woodlands, where a few beech trees have produced seeds the previous autumn. January often yields the largest flocks of the winter season presumably derived from smaller, more scattered groups tending to gather where there is the last of the beech mast or other food source. In hard weather small numbers often accompany Chaffinches to garden feeding stations with bigger numbers visiting the larger, formal feeding stations in County Durham, such as those at Clara Vale, Rainton Meadows and Thornley Woods. Birds can be quite widely scattered across the county in winter but the Derwent valley woodlands, the areas around: Durham City, Darlington West Cemetery, Hunstanworth, Shincliffe, Sunniside, Wemmergill and Whickham, as well as coastal sites at Teesmouth have all held flocks in excess of 150. An area of set-aside land near Salter's Gate has produced the largest flocks of recent times, with 350 there in February 2002 and 300 in March 2004.

Records indicate a quite definite trend over the years 1995-2010, for the species to have become generally scarcer with fewer and smaller winter flocks, though still quite widespread. There has also been less evidence of strong autumn passage at the coast. The change appears to have occurred in the mid-1990s despite ever increasing observer interest and effort. Scrutiny of records shows that after the 1965 to 1967 period, there were above average numbers in the winter periods of 1981 to 1983, 1987 to 1989 and in 2002 to 2004. At other times, especially recently, the species has more often been described as scarce. Wintering birds are usually present well into late February and, often in diminishing numbers, through March, with the final exodus being through late March and April. Passage birds can occasionally still occur on the coast in early May. Overall, spring return passage is much lighter than in autumn.

Distribution & movements

Bramblings are the boreal breeding equivalent of Chaffinch breeding in Eurasia from Scandinavia eastwards to Siberia. European breeding birds winter to the south of their breeding areas, in southern, western and central Europe, birds wintering in a peripatetic fashion, changing their wintering area from year-to-year in search of their favoured survival food source. If the weather is severe, birds sometimes move toward the coast, for example the 250 that were at Hartlepool Docks in January 1984 (Blick 2009).

There was a sorry end for an adult female ringed on 12 March 1988 in a Whickham garden and killed by a cat three years later on 30 April 1991, rather unusually, in the same area, indicating that it had perhaps developed a habit for a specific year-on-year wintering area. In a bizarre mirror of this instance, a bird ringed in the same Whickham garden in March 1998, was killed by a cat at Jylland on its return journey to the east coast of Denmark just 34 days later on 17 April 1998. The wandering nature of the species was shown by the adult male wintering at Thornley Wood that was ringed on 23 February 1999 and was controlled wintering in Antwerp, Belgium the same year on 29 October 1999. More akin to the returning Whickham bird was the first-winter female ringed at Thornley Woods on 31 January 1998, which was re-trapped there on 23 February 1999; 388 days later, where had it been in the interim? Blick highlighted the fact that birds ringed in a winter roost in Cleveland, generated data from France, Holland, Belgium and Germany as well as in locations further south in England (Blick 2009).

Serin (European Serin)
Serinus serinus

A very rare vagrant from Europe: six records.

The first record of Serin for County Durham was of an adult male first seen by John Coulson in a garden at Westoe, South Shields on the 12 November 1950, with further sightings on 19th and 26th by several other observers including the late Fred Grey. Being the first record of this species for the north east of England, there was some speculation at the time that it might have been an escape from captivity, though a party of eight birds at North Otterington, North Yorkshire from 8th to 9 December 1950 helped verify it as a genuine vagrant (Temperley 1951).

All records:

1950	Westoe, South Shields, male, 12th to 26 November (Temperley 1951)
1955	Westoe, South Shields, 27 August to 15 September, found dead 15 September (*Transactions* Vol. XI No. 8)
1968	Hartlepool Headland, 14 September (*British Birds* 62: 487)
1969	Hartlepool Headland, female, 5 May (*British Birds* 63: 290)
2002	Seaton Common, Teesmouth, flew inland, 29 June
2011	Cowpen Marsh, 11 April

Remarkably the second record was also found by John Coulson in the same area of South Shields, some five years after the first. Present in gardens at the junction of St. George's Road, and Readhead Avenue, it was found dead on 15 September with the specimen being retained in the Hancock Museum, Newcastle-upon-Tyne. The records for Durham cover widely scattered dates and do not fit neatly with occurrence patterns of the species in Britain. After an absence of over 30 years the bird watched flying inland over Seaton Common on the 29 June 2002, was particularly unusual for its mid-summer appearance.

Distribution & movements
Serin breeds in North Africa and throughout much of Europe, excluding Iceland, Scandinavia and European Russia. Northern populations are migratory and winter south to the Mediterranean basin, though southern populations are largely sedentary. It is most frequently recorded as a scarce migrant to southern England, chiefly in April and May, and again from October to November, averaging around 70 records per annum. Records are extremely rare north of the River Humber, with just five records for Northumberland and a handful of occurrences in Scotland, most of which are from the Northern Isles.

Greenfinch (European Greenfinch)
Chloris chloris

A very common and well distributed resident, passage migrant and winter visitor.

Historical review
The oldest known reference to this species in County Durham dates from Marmaduke Tunstall's 1784 manuscript. He was presumably familiar with the species from his portion of the Tees valley (Mather 1986). Rather little is known about the species' former status in the county but the Greenfinch appears to have been a familiar enough species to 19th century chroniclers that it received little specific mention.

Nelson (1907) recorded this finch as a regular passage migrant on the coast, "*many*" being seen arriving in October 1883, 1887 and 1901. *The Victoria History of County Durham* (Tristram 1905) described its status as a common resident, often to be seen in flocks in winter, while in *Birds of the Tees Valley* (Almond *et al.* 1939) it was recognised to be a passage migrant with flocks appearing in spring at Darlington Sewage Farm towards the end of April. Temperley (1951) confirmed the species as a common resident, passage migrant and winter visitor. With keen insight he went on to describe large migrant flocks arriving in October and November, and how he believed some of these birds remained to winter in Durham whilst others passed further south; something that is not often witnessed today. Flocks arriving on the coast and at Darlington Sewage Farm in spring were judged to be moving northward after the local population had settled down to breed.

Post-Second World War, it was felt that there was some considerable decline in the region, partly in response to the introduction of more aggressive, persistent pesticides; for example, many were found dead near Haltwhistle in the Tyne valley, Northumberland in 1961 (Kerr 2001). The colonisation of urban gardens seems to have begun in Durham during the late 1960s, as elsewhere in the north east (Kerr 2001), and this continued apace through the 1970s and into the 1980s.

Recent breeding status

This attractive finch is strongly associated with densely-leafed trees for nesting and shows a dietary dependence on seeds found underneath trees and bushes, amongst crops and from favoured weeds (Newton 1985). The Greenfinch is a familiar bird to many people as it is often a resident of urban and suburban areas. It occupies gardens, parkland, arable and pastoral field systems, woodland edge and hedgerows provided groups of trees or mature bushes are present. Breeding often occurs in loose, dispersed colonies each involving a few pairs. As a breeding species, a large proportion of Durham's birds seem to be concentrated in the urban-fringe and parkland habitats, with relatively fewer in deciduous woodland and farmland areas (Westerberg & Bowey 2000). An increase in suburban garden settings was noted in the *Birds in Durham* records during the 1970s, with a fondness for the increasingly popular *Leylandii* trees, cited by several observers as one reason for the increase.

The *Atlas* (Bowey & Westerberg 2000) found the species to be a common and familiar resident across the eastern half of the county with a presence that traced the lower ground of wooded valleys higher up into the western dales. It was generally absent from high ground though much more recently, despite the core breeding population being under stress, there appears to be evidence of a slight westward expansion with a few pairs recorded at Bowes, Grassholme and even at an altitude of 310m above sea level, near Selset Reservoir, Newbiggin-in-Teesdale and Cronkley during 2008 and 2009. Many of these western birds were noted associating with, and no doubt nesting in, garden conifers and the species' penetration into more exposed westerly locations may have been facilitated by the milder winters of the 1990s and 2000s.

The male's characteristic display flight can be noted as early as February in the north east, for example in Houghton on 12 February 2008, but the main song period usually commences in mid- to late March, with breeding pairs established by late March. The effect of a long-run of mild winters from 1990 to 2008 appears to have resulted in the onset of earlier egg-laying in the county. For example, a pair at Middle Herrington, Sunderland had fledged young by the exceptionally early date of 12 April in 2001 and that same year a pair at Winlaton Mill had fledged young by 20 April, approximately one month earlier than is considered usual for the area. Second broods are normal and extend the breeding season until early July. In 1997, a survey of a 1.6km strip of woodland edge and suburban gardens in the Derwent valley produced a count of 17 territories. At Hetton Lyons Park 18 pairs in 2007 fell to 10 pairs the following year.

Up to the end of the 20th century the BTO Common Bird Census data showed a steady but modest increase in the Greenfinch's breeding population in north east England; a position supported by casual reports from Durham. The most recent BTO Breeding Bird Survey data for the north east of England showed a distinct reversal with a 20% decline over the first decade of the 21st century. This has been attributed (Robinson *et al.* 2010) to the emergence of *Trichomonosis*, an infectious and potentially fatal disease, affecting Greenfinches especially in eastern England. This protozoan parasite is carried by pigeons *Columbiforms,* but in an unusual example of inter-species transfer between avian hosts, it appears to have jumped to Greenfinches and to a lesser extent Chaffinches, *Fringilla coelebs*. Evidence first emerged in 2005, since when the problem has been described as having reached epidemic proportions, its effects continuing at the time of writing. Several observers in Barnard Castle and Durham City noted reduced numbers or the complete absence of Greenfinches in the late 2000s, at previously regularly occupied garden feeding stations. In one local study in the Hetton area, 25 pairs in 2004 had declined to ten pairs by 2008, with a loss of eight pairs between 2007 and 2008 alone (Newsome 2009). Further apparent impact, came from a comparison of the capture totals (for ringing) of this species at Clara Vale over the period 2005-2008. In 2005, 95 were ringed; falling to 86 in 2006; 76 in 2007 and then *Trichomonosis* struck and by 2008, the total had fallen to just five. Despite this, the Greenfinch remains not uncommon across its Durham range but the true impact of the disease locally is yet to be fully assessed.

Breeding densities never approach those of the Chaffinch. The *Atlas* (Westerberg & Bowey 2000) suggested a maximum number of 6,000 pairs for the area. Balancing the better knowledge of birds in the south east (Joynt *et al.* 2008) against the undoubted local declines over the last decade, when the *Trichomonosis* epidemic will have exerted a significant but unquantifiable impact on this, the current breeding population may have been reduced by as much as one third, perhaps even more at a local level, and the number of breeding pairs in Durham may have fallen to below 4,000.

Recent Non-Breeding Status

The Greenfinch quickly becomes gregarious after the breeding season and from early autumn until spring it forms flocks, usually on weedy waste ground and in some areas it begins to appear on pre-harvest oil-seed rape, but birds can be found in most suitable habitat. As with other *Carduelis* finches, flocks show a tendency to consolidate further in late winter when seed is scarce and there are advantages to being in a group when searching for food.

Winter gatherings can be large and they are often seen in the company of other finches and buntings at good feeding sites. The *Winter Atlas* (Lack 1986) showed a distinct easterly bias across the whole of the UK and the annual *Birds in Durham* record reflects the same bias in the county. The winter distribution in Durham is concentrated in the lowlands and in coastal areas; birds are associated with arable farming and artificial feeding opportunities, such as game cover planting or garden feeders. Large winter flocks in the west are unknown but smaller numbers can linger.

As the breeding season ends, flocks presumably consisting of mostly local birds, appear from late July to early September with an exceptional early gathering of 200 at Washingwell Wood, Derwent Valley on 16 July 1987. Flocks of 30-50 are by no means unusual but more notable reports have included 200 in a barley field on Cleadon Hill in August 1980, 500 at Sedgefield in early September 1989, 400 at Great Lumley in September 1992, 350 feeding on oil seed rape at Sedgefield in August 2001 and 220 at East Boldon in August 2005.

As more wintry conditions prevail larger flocks are often seen by late November or through December and these may be sustained until early March. It is likely that passage migrants will have joined with local birds to form these larger groupings. On Coatham Marsh on 10 January 1971 about 1,000 were recorded (Blick 2009) and a flock of 800 was noted on the Tees Marshes in December 1972; these remain some of the largest groups recorded in the county, with 450 still present there January 1973. Other notable records include 500 on the Tees Marshes in December 1975, 550 roosting at Chester-le-Street in January 1986, 500 at Sedgeletch during January 1993 falling to 300 by March and 250 Hetton Lyons in March 1995. The record winter count of this species in the south east of the county is of 1,400 birds in the Seal Sands area in November 1976 (Bell 1977). There is evidence from the last decade that larger flocks have not been seen with anything like their previous regularity and that overall the species has become a little scarcer during the winter months. Favoured winter food sources include yew *Taxus baccata* berries with several hundred birds feeding on these in Castle Eden Dene in December 1973 and in Darlington West Cemetery during January to March 1977. A garden feeding station in Gilesgate, Durham City attracted impressive counts of up to 300 birds in the first few months of each year between 1989 and 1992, with a remarkable peak of 500 in February 1991. The birds were supplied a diet of sunflower seeds and peanuts. Game-cover planting schemes proved attractive to the species towards the end of the first decade of the 21st century and are likely to prove highly beneficial if they can be sustained in the longer term; such initiatives attracted 350 at Seaton Pond on 30 November 2009 and 340 at nearby Sharpley in December 2010.

As in the breeding season, birds often associate with coniferous gardens trees and shrubs for roosting purposes. Roost gatherings can number hundreds of birds, the largest yet recorded in the north west of the county being of 256 at Sunniside on 24 October 1985, although up to 200 were also reported as roosting along the Banks of the Tyne in February 1988 (Bowey *et al.* 1993). Around 250 were found roosting in Darlington West Cemetery in January to March 1977 and again in 1978. A roost of 560, in Chester-le-Street during January 1986, was the largest such gathering of the 1980s. By mid-March, most of these winter flocks have dispersed with the first reports of nesting usually following during April.

Distribution and movements

The species breeds throughout Europe to the south of the Arctic Circle and is a partial migrant over much of its range with movement most evident during October and November. The population in the UK exhibits local movements after breeding. Very few British Greenfinches winter abroad and only a very small proportion of continental birds winter in Britain (Wernham *et al.* 2002). As a consequence, the autumn passage and winter visitor numbers seen in Durham are now better understood to be largely British birds moving or 'coasting' south west in autumn and retracing their steps in spring.

Ringing studies and recoveries in Durham provide data that demonstrates this situation, although the movements are perhaps a little more complex and dynamic with interchange apparent between several areas in northern England and southern Scotland. In his summary on this species Blick (2009) pointed out that ringing revealed that birds largely move either north or south in times of migration but are not faithful to any particular wintering area, and that hard weather may influence these movements.

Ringed	Date	Recovered	Date
Cartmel, Lancashire	Oct 1964	Sunderland	Mar 1965
Crimdon , Durham	Jan 1988	Thorne, S. Yorkshire	Apr 1988
Dumfries, Dumfries Galloway	Nov 1991	Wolsingham, Durham	Feb 1992
Glencaple, Dumfries Galloway	Feb 1995	Barnard Castle, Durham	Mar 1995
Heysham , Lancashire	Nov 1998	Barnard Castle, Durham	Apr 1999
Barnard Castle , Durham	Mar 1999	Sheffield	Sep 1999

Winter Movements of Greenfinch to & from Durham

Most movements are recorded within the same winter period. The bird ringed at Glencaple in February 1995 was controlled just 35 days later in Barnard Castle after an easterly movement of 117km.

Ringed	Date	Recovered	Date
Gosforth, Northumberland	Feb 1966	Shincliffe, Durham	Jul 1966
Hamsterley Mill, Durham	Jan 1979	Blackpool, Lancashire	May 1979
Crimdon , Durham	Dec 1986	Lobley Hill, Gateshead	May 1988
Saltburn, N Yorkshire	Feb 1990	Witton le Wear, Durham	May 1991
Hart Station, Durham	Feb 1993	Stamfordham, Northumberland	Jul 1993
Rowlands Gill, Durham	Feb 1993	Willington, Durham	Jul 1994

Winter to Breeding Season Movements of Greenfinch to & from Durham

In recent years, in terms of coastal passage, this species has declined significantly, few birds being observed at the coastal watch points such as Hartlepool Headland, although some Greenfinches are sometimes noted in company with other finches (Blick 2009). There seems to have been little recent evidence of autumn immigration though six birds were at Hartlepool Headland on 7 October 2009, in that year these were the only incoming 'migrants' reported.

Direct evidence of active passage has been recorded. In 1975 a flock of 150 were noted on the coast at Whitburn on 8 October and in December 1985, a flock of 100 was seen in Seaham Docks. In 1993 about 35 were present with other finches at Marsden on 7 November and on 18 October 2000, small groups totalling 100 were noted there flying south in a two-hour period, with 103 moving south at Marsden Quarry on 18 October 2006. In 2005 an impressive 650 moved south in two hours over Seaton Common, Teesmouth; the largest documented passage of this species in the county for many years (Blick 2009). The true extent of otherwise unrecorded passage of this species was perhaps revealed by the total of almost 340 Greenfinches, which were ringed on the Tees Marshes during the autumn of that year.

Five Greenfinches ringed in the Witton-le-Wear to Hamsterley area of the Wear valley between 2001 and 2006, were recovered dead either at their ringing site or very close to it within one to three years after ringing, suggesting a largely sedentary or site faithful characteristic of local birds. The greatest movement of these five birds was of one that had been ringed at Low Barns and was found about 3km away on the other side of Witton-le-Wear village.

Evidence of birds moving relatively modest distances, up and down the county's coast, has been generated over many decades from ringing data (other than that tabulated), with at least five Northumberland-ringed birds subsequently controlled or recovered in County Durham during the autumn or winter, and at least three Durham-ringed birds ringed in the winter or late spring, controlled or recovered in Northumberland. In similar fashion, at least five birds ringed at Teesmouth were later controlled or recovered further south in Yorkshire or Humberside, and three moved in the opposite direction; while one bird ringed at Billingham on 10 February 2008 was recovered at Thorndon, Suffolk on 18 August 2008. A bird that moved eastward from the west coast was ringed as a juvenile at Maryport, Cumbria on 10 November 1996 and controlled at Newton Hall on 13 April 1997. There is only one known foreign control of a Durham bird; one that was ringed at Teesmouth on 6 December 1975 recovered at Overijssel, in the Netherlands in January 1977 (Blick 2009).

Goldfinch (European Goldfinch)
Carduelis carduelis

A very common and widely distributed resident and passage migrant.

Historical review
Temperley (1951) summarised the fluctuating fortunes of the Goldfinch during the 19th and first half of the 20th century. The earliest catalogues, Hogg (1827), Selby (1831) and Hutchinson (1840) mostly described the species as being resident but by no means common. Writing in 1874 Hancock, surprisingly but no doubt accurately, described the Goldfinch as then merely being a casual visitor in autumn and winter, having been seen by him on only two or three occasions. Backhouse (1885) reported a not infrequent presence in the upper Tees valley in the winter months but when Robson (1896) saw three birds at Low Thornley, Derwent Valley in January 1890, he said these were the first he'd ever seen and Tristram (1905) considered it only an occasional autumn migrant and had no proof of breeding. Fawcett (1890) however, reporting on the coastal strip, said that it bred in Castle Eden and Seaham Denes; the first reference to its confirmed breeding in the county.

This colourful species was a favourite cage bird in the 19th century and in many parts of the country it was quite scarce as a consequence. Bolam (1912) reported that it had been almost exterminated as a breeding species in Northumberland, due to trapping for captivity. The decline in the latter half of the 19th century has been attributed to the practice of catching finches by 'liming' but climatic or food supply factors may also have played a part. Legislation was passed in 1881 and 1933 making liming illegal and attitudes and practices appear to have gradually improved as a result. Resident numbers in the Tees valley were said to have already increased greatly when Almond *et al.* (1939) presented their species list for that area. Temperley (1951) also noted a gradual, general increase "*under protection*" and mentioned breeding in parks, gardens and orchards in small colonies especially in south east Durham.

The annual ornithological reports for Northumberland and Durham during the 1930s, 1940s and on through to the 1960s, indicated a continued increase in the species locally (Kerr 2001). Numbers and distribution across the region continued to increase after Temperley's comments, although there were probable setbacks in the 1960s and 1970s as a result of more intensive farming methods and the application of herbicides, both of which were designed to eradicate some of the species' favoured food plant sources. Since at least the 1960s, flocks of up to 100 individuals have been increasingly found in places where weeds prevail, such as the edges of industrial areas, fields and housing estates. Through the 1970s the increases continued, when the colonisation of urban parks and larger suburban gardens commenced in earnest; a process already underway in the Greenfinch *Chloris chloris* by that time. It continued to increase, with extensions of its range in the county, breeding in urban locations such as Barnes Park in Sunderland and also in Teesdale in the west. This increase was most evident outside the breeding season, when three-figure flocks were becoming increasingly widespread in lowland areas of the county. The largest autumn flock in the early part of the decade was over 400 birds in the lower Derwent valley in September 1976.

A general expansion of the Goldfinch's breeding range continued during the 1980s and 1990s; in 1985 a pair successfully raising young in a tree near a multi-storey car park in Stockton town centre and such nesting locations became increasingly common as the trend continued. The spread into more northerly and westerly parts of Britain continued, as it does to the present (Parkin & Knox 2010).

Recent breeding status

If modern bird watchers are surprised by the apparent scarcity of breeding Goldfinch in the north east region at the turn of the 20th century, then ornithologists of that era would have been equally surprised by its current widespread abundance just over a century later. The Goldfinch is probably exceeded only by the Collared Dove *Streptopelia decaocto,* as the species exhibiting the greatest expansion in breeding numbers and distribution across the county over this period.

Breeding Goldfinches are now found widely across Durham, as was illustrated by the distribution map in the *Atlas*. Breeding reports come from urban sites in the east and centre of the county, for example around the Federation Brewery in Dunston, Gateshead in 2009, to the edges of some of our western moorlands in the far west of the county, for example at Harwood-in-Teesdale (Newsome 2009). Where the species occurs in the west, towards the moorland fringes, birds tend to be confined to the valleys along the main rivers; in such situations they opt for lightly wooded areas as nesting locations (Westerberg & Bowey 2000), and many parts of the western uplands, particularly above the 300m contour, do not hold breeding birds.

At lower elevations, Goldfinches prefers gardens, allotments, scrub, woodland edge, conifer plantations, arable fields, road and rail-side verges, stream courses and even industrial land but in all cases only where patches of tall weeds are allowed to grow within rather open areas. It is found at coastal sites but it is not present over much of the North Tees Marshes area, largely due to a lack of suitable scrub and nesting habitats; though post-breeding flocks are routinely seen there (Joynt *et al.* 2008).

The species is a relatively late breeder, with the first nests typically not being noted until mid-May and second broods running into late July. The nest is usually quite high in a tree or tall bushes, birds often making use of ornamental cherries *Prunus spp.* (Joynt et al. 2008). Breeding success varies considerably from year-to-year, depending on weather and seed abundance. An interesting spring congregation of birds occurred at Balder Grange with 181 there on 25 May 1986 but most flocks gather at the other end of the breeding season; it is interesting to speculate on the fact that these might have been recently arrived altitudinal migrants.

Breeding reports are not well-documented in *Birds in Durham*, but the BTO's Breeding Birds Survey index for north east England shows a remarkable two-fold increase over the period 1994 to 2010 (Risely *et al.* 2010), after a dramatic decline in Britain during the early 1980s (Marchant *et al.* 1990). The increase from 1995 to 2009 was 73% for the UK and 63% for England; while the trend continued from 2009 to 2010 with a further 11% rise in the UK. Such an impressive recent increase seems to be supported by the trend in post-breeding flock sizes recorded in Durham and described below. While few specific breeding records of interest are collated annually by observers, the fact remains that taking 2007 as a typical year, 750 records were submitted by Durham Bird Club members from 185 locations, many of which will have involved small breeding parties.

In 2008, the Hetton Lyons area had 14 breeding pairs and Rainton Meadows 10 pairs. The *Atlas* showed an easterly bias to its distribution, a little more patchy than the Greenfinch and suggested the breeding population in Durham to be in the range 1,800-2,500 pairs. With more recent evidence, including from *The Breeding Birds of Cleveland* (Joynt *et al.* 2008) that the species continues to thrive and has few real threats, it seems likely that the current breeding population in Durham now exceeds 3,000 pairs.

Recent non-breeding status

The pattern of records in most years follows a predictable pattern. In spring, birds are in breeding territories, gathering in family groups after fledging first and second broods; these 'families' gather to form post-breeding flocks in late summer, merging with others to form much larger autumn flocks. This all takes place prior to the movement south of many of our locally breeding birds, with just smaller numbers remaining to winter in the county.

Ringing studies have confirmed that a very large portion of the UK breeding population emigrates to western parts of Belgium, France and Spain in the winter months (Newton 1972). The post-breeding flocks which gather locally in September and October before the main departure in late October or early November provide a convenient indicator of overall status. That is not to say that the species is entirely absent during the winter months but the number remaining in the UK is thought to be less than 10% of the immediate post-breeding population (Lack 1986) and the main return is not apparent until late March.

Goldfinches are gregarious outside of the breeding season when thistle *Cirsium* and *Carduus* spp., ragwort *Senecio* spp., burdock *Arctium minus*, teasel *Dipsacus fullonum* and other seed-bearing weeds form important food sources. Counts of post-breeding flocks offer one means of judging population trends and variation between years. Concentrating solely on examples of flock size counts made in the peak month of September is instructive as shown by the following series of records. In 1976 a flock of 200 was noted at East Boldon, with 170 on Seaton Common, Teesmouth and in 1979 near Bishopton, 200 were present. These were amongst the largest flocks recorded for the decade at any time of year and in *Birds in Durham* 1978 (Unwin 1979), comment was made expressing concern at the then falling flock sizes. The 400 birds in the Houghton area in September 1984, was reported as the '*highest county concentration for 15 years*' (Baldridge 1985). There was no apparent uplift in numbers during the following decade and a flock of 250 at Marsden Quarry on 1 October 1988 was by far the largest count; given its location it may have involved passage rather than local birds. In September 1990, a flock of 210 was found at Chilton Moor when 280 were also at Eppleton. By 1993, when 390 were at Hetton Lyons, there had begun to emerge real evidence of a general increase in post-breeding flock sizes. Two years later Hetton Lyons held 300 whilst 350 appeared at Sedgeletch. In 1999 Hetton Lyons again held 300 and a flock at Rainton Meadows totalled 450. Hetton Lyons attracted 480 in 2002 and coastal fields at Easington supported 450 in 2003. Late September counts in 2005 recorded flocks of 480 at Dalton, 520 at Hawthorn Hive and 430 at Saltholme and in 2008 Dalton held 680 birds. Most of these flocks decline rapidly by early November.

Given the species' large-scale emigration from Britain, winter records are less frequent and flocks involve far fewer numbers, typically less than 20, exceptionally 60. A group of 360 at Billingham in late December 1972 was a particularly striking but isolated mid-winter record. Several records in the latter part of the first decade of the 21st century have run contrary to this picture, perhaps indicating an increasing tolerance to overwintering. In 2004 a flock of 400 was found at Salter's Gate near Wolsingham in early January and in 2007, after 150 were seen at Butterby on 1 February, 500 were present at Seaton Pond in December, when 300 were wintering at Rainton Meadows. The flock at Seaton Pond had risen to 530 by January 2008. Land deliberately planted with 'wild bird' seed plants or artificial feeding stations seem key to supporting these significant winter flocks. The species' popularity with observers, and its widespread distribution in Durham, is indicated by the fact that over the winter period of 2009/2010 this was one of the most reported species to the Durham Bird Club; in the first quarter of 2009, over 1200 records were received.

In late winter and early spring, when their main food has become scarce, birds often resort to conifers, in sites such as Washingwell Woods, where they feed on larch *Larix sp.* seeds sometimes in the company of Siskins *Carduelis spinus* and Lesser Redpolls *Carduelis cabaret*. At such times they have also adapted, like many species, to feed in gardens with many reports referring to garden feeding stations, where Nyger seeds are a particular attraction for this species. An impressive 60 visited a garden feeding station in Shotton Colliery during December 2008. Today this species is a regular visitor to garden feeders but it has not always been the case; such behaviour was unknown in the county only 30 years ago and only became a frequent occurrence during the 1990s.

Like other *Cardueline* finches, Goldfinches roost communally outside of the breeding season, often in scrub and in some winters, quite sizeable roosts develop. Just such a roost was present in Billingham from the late sixties to the present, with the maximum count there being of 360 on 31 December 1972. Blick (2009) speculated that this "*probably contained most of Cleveland's then wintering Goldfinches*". One of the more unusual winter records of this species was of four birds on snow-covered moorland at over 300m elevation at Grassholme Reservoir on 31 December 1979.

In County Durham, spring passage can be quite evident along the coastline from April through to early June; the later birds are presumably those from breeding locations much further north. The return of local breeders in spring is not clearly marked but light daily passage was noted on Cleadon Hill during April 1997.

Autumn passage is usually less obvious than spring movement in Durham, although a few birds have occasionally been seen apparently arriving from over the sea (Blick 2009) but there is little hard evidence of continental immigration in the autumn to winter months. Goldfinch was the third most recorded species on the series of Coordinated Migrant Counts along the Durham coast each October from 1992 to 1996 - 968 being noted along the Durham coast on 2 October 1994 (Bowey 1999). No doubt some part of Durham's autumn or even winter flocks involve breeding birds from further north in Britain.

Distribution and movements

Goldfinches breed over most of Europe into southern Fennoscandia, and as far south as North Africa, the Middle East and western Asia. In many parts of the range, particularly in the south, the species is a resident one, but in the northerly parts of its range, including some of the British population, it is at least a partial migrant. A considerable proportion of the British population winters in France, Belgium and Spain (Newton 1972).

The little substantive ringing data for this species in Durham confirms the fact that some local birds head to points further south to spend the winter, as per the national data set (Wernham *et al.* 2002). For example: a bird ringed as a pullus in Stockton in August 1984 was found dead near Doncaster, South Yorkshire, in March 1985; one ringed at Ravensworth in June 1995 was recovered at Winchester, Hampshire in March 1997 and one ringed at Hartlepool on 20 April 2007 was in Suffolk on 15 March 2009 (Blick 2009). The bird ringed at Orfordness, Suffolk on 13 October 2005 and which was at Hartlepool on 13 April 2008, had presumably been a northern bird that had moved south prior to its initial capture and ringing.

Siskin (Eurasian Siskin)
Carduelis spinus

A common passage migrant and winter visitor; a scarce local breeder.

Historical Review

In Durham, this species has radically changed both its status and habits in the last 30 to 40 years. The oldest reference to Siskins in Durham is in the manuscript of the naturalist and antiquarian George Allan (1791) of Darlington, who knew it chiefly as a winter visitor and said that it did not breed here. Hutchinson (1840) noted its irregular winter appearance and gave a knowledgeable account of its preference for larch, alder and birch seeds, as well as mentioning his own observations of a flock along the banks of the River Wear feeding on the seed heads of groundsel *Senecio vulgaris* and tansy *Tanacetum vulgare*. Temperley (1951) summarised three old reports of breeding: a nest with four eggs found in a spruce tree at Brancepeth on 7 May 1848; a fledgling male taken from a nest in a fir tree at Tudhoe in 1874 and a nest with four eggs taken in Weardale in 1874, which was in Canon Tristram's own collection (Tristram 1905). It was to be almost another century before the next breeding was reliably reported, once more in Weardale, in 1971.

It would seem from Robson's account of this species in the north of the county last century (1896), that it is much more common in the Derwent valley today than it was then, and no doubt this is a relevant statement for the whole of the county. Robson documented a number of occurrences, including flocks at Swalwell in July 1848 and a pair in Blaydon Burn in November 1888 (Bowey *et al.* 1993).

Nelson (1907) noted the irregular autumn influxes on the coast at Teesmouth, describing it as "*abundant*" there in October 1881 and again in September and late October 1901. Winter records were poorly documented but Howard Watson saw 50 feeding on a bare ground scrape in otherwise snow covered Hudeshope, Middleton-in-Teesdale on 22 March 1947.

By the first half of the twentieth century Temperley (1951) said that it was present "*In winter in varying numbers, it has bred*"; "*a pair or two, occasionally remains to breed.*" He noted that in winter it often associated with Lesser Redpoll *Carduelis cabaret*.

Recent breeding status

Up to the mid- to late 1970s, this species was very much considered a spring and autumn passage migrant and a winter visitor, often feeding in larch *Larix sp.* and alder *Alnus sp.* woodlands, in variable numbers. Over the period 1980 to 2010 across much of Durham it became a widespread breeding resident, while retaining many of the other aspects of its previously described status.

Breeding is strongly linked with conifer plantations and in particular the availability of spruce and pine seeds, which are the birds' main food in spring and early summer. Siskins are therefore to be found in commercial conifer forest blocks where they have also adapted to exploit the non-native Sitka spruce *Picea sitchensis* seed crop. Afforestation has generally helped the species expand across northern Britain during the latter half of the 20th century and a gradual colonisation of northern England, including Durham, has occurred over that period.

A few birds were reported from Hamsterley Forest in the springs of 1958 and 1959 ahead of the Weardale breeding record of 1971. Hamsterley Forest and Tunstall Reservoir, Wolsingham each produced a series of early summer sightings of pairs in most years between 1972 and 1978, although breeding could not be confirmed. By 1978 there were also reports of possible breeding in The Stang Forest and at Derwent Reservoir, and in the late spring of 1983, two pairs were present at Newbiggin-in-Teesdale. More extensive breeding was finally proven in 1985, when 30 adults and juveniles were seen in Hamsterley Forest on 16 June, growing to a flock of 100 by early August. This had followed a notable influx in the winter of 1984/1985. Breeding was also confirmed in 1988 at Strother Hills near Rowlands Gill.

The slow but definite development of a small breeding population fits well with similar expansions noted in adjoining counties over the same period. Since the mid- to late 1980s it has been increasingly recorded as a breeding species. More widespread breeding has been recorded in the last two decades. In 1992, as well as the almost regular sites of Hamsterley Forest and Derwent Reservoir, there were reports of probable or confirmed breeding at Ravensworth, Thornley Woodlands Centre, Paddock Hill, High Force and Ireshopeburn. Between 2004 and 2008 breeding was confirmed at Blanchland, Killhope, Waskerley, Frosterley, Eggleston, Copley, Selset Reservoir and Bowes. Easterly breeding sites have included Sacriston in 1993, Kibblesworth in 1998 and Wynard Park in 2009.

The breeding status is certainly now more favourable and stable than at the time of the *Atlas* (Westerberg & Bowey 2000) but the scattered distribution, mainly in the west and north west of the county, remains similar. Despite obvious advances, caution is needed since the species is still prone to annual variations linked to the health of the seed crop. Breeding may be near-regular at some sites but is probably only occasional at others. Peaks in Durham's breeding population tend to follow years of winter invasions which in themselves are determined by natural events far beyond our boundaries. The *Atlas* gave an estimate of 50-100 pairs and despite the undoubted later gains, in poor years the figure of 50 pairs may still not be reached. In good years, which are probably achieved now on a more consistent basis than previously, the figure may reach 100-120 pairs.

Recent non-breeding status

The Siskin's winter distribution is determined by the availability of larch, alder and birch *Betula sp.* seeds. It occupies open conifer plantations and wooded stream sides where its favoured trees are present, often within a mixed broad-leafed mosaic, with wintering flocks usually found at a wide range of sites often located in the county's river valleys. It is sometimes found in the loose company of Redpolls. Scottish breeding birds supported by long distance migrants from the Continent are thought to be mainly responsible for the winter population in England (Lack 1986) and this is confirmed for Durham by the listed ringing recoveries. The species' local distribution and abundance each winter can vary quite considerably depending, as ever, on the availability of key seed crops here and elsewhere.

Immigrants normally arrive at our coast from mid-September to November with winter flocks reaching a peak during December and January before dispersing in early March. The Siskin is far more widespread and numerous in Durham during the autumn and winter months than it is during spring and summer.

There were few specific reports of wintering during the 20 years after Temperley (1951) but flocks of 15-30 birds were seen in alders next to the River Tees at Darlington in the winters of 1951, 1956, 1962 and 1969. The species was said to be generally more plentiful than for many years in the winter of 1956/1957. Flocks of 30 at Norton in 1962 and 20 at Beamish in 1963 were thought notable and give some indication that the species was still by no means common. Records subsequently began to increase. Low Barns hosted flocks of 50-120 in most winters during the 1970s and from 1973 there were frequent reports from Hamsterley Forest of up to 100 birds. Other records at this time mostly involved flocks of 30-60 at a handful of sites each winter but 220 were found at Ryton Willows on 9 December 1974.

A then record flock of 300 occupied Hamsterley Forest in February 1980; while an early influx in 1984 produced counts of 60 at Balder Grange and 100 at Hamsterley in September, with 100 at Washingwell Wood and 195 at Chopwell in the Derwent Valley in December. There were still 150 in Hamsterley Forest in January 1985, which will have been the precursor to breeding there that year. The first months of 1988 produced several quite large flocks with 150 at Shibdon Pond and 100 in Deepdale both eclipsed by 350 in Chopwell Woods.

In a significant shift in status, the species began to appear in even greater numbers and consistency for a few years from 1990 onwards. In December 1990 there were 200 at Garesfield, Derwent Valley and in the first quarter of 1991 flocks of 100-200 appeared at Strother Hills, Chopwell, Derwent Reservoir, Waterhouses, Barnard Castle and Hamsterley with further groups of 50-60 noted at several other locations including Low Barnes, McNeil Bottoms and Jarrow.

In January 1992, WWT Washington held 300 and Ryton Willows 200, with notable October flocks of that year including 340 at Strother Hills, 200 at both Hamsterley and Chopwell. The main sites once again reported similar numbers in the early months of 1993 and 1994 but something of decline then occurred. This was temporarily reversed in 1997, when a very early invasion brought 400 to Paddock Hill Wood by 22 August and 200-300 to Chopwell and Thornley. WWT Washington held 360 in January 1998 when 200 were at Low Butterby. Flock counts in most recent winters have been limited to 100-200 at sites such as Joe's Pond, Low Barns, Hamsterley and WWT Washington but in January 2006 up to 500 birds were present at the Rainton Meadows and Joe's Pond complex, attracted no doubt by supplementary feeding. The flock size fell to 70 by the end of March. A flock of 430 had re-built at Rainton Meadows by January 2008 when several flocks of 100-200 were seen elsewhere in the county. The species is now clearly more abundant in winter than ever before.

The species' tendency to make use of garden bird feeders is a recent phenomenon; it was first noticed in the 1960s (Lack 1986) and, perhaps belatedly, in Durham in 1987 (Armstrong 1988), but by 2000 the habit was clearly widespread and common amongst birds in the county.

On the Continent, autumn movements are initially triggered by the condition of the birch seed crop. In years of crop failure the birds will irrupt and migrate earlier, more rapidly and in greater numbers to produce invasion conditions (Newton 2010). Coastal records occur most years from mid-September and in invasion years these can be pronounced. Winter movements are erratic and depend on the availability of its favoured seeds, with some sites holding flocks for much of the winter and others being only briefly occupied as seed stocks diminish.

Autumn influxes on the coast have been well documented for Durham. In most years coastal passage is light with just small numbers noted between September and early November, with a peak in October. Invasion years are more notable; in 1960 over 100 were seen at Whitburn on 18 September when 40 were also at Hartlepool. In 1979, a heavy influx in early October led to 70 at Marsden and Cleadon with 30 at Castle Eden. An unusually early report concerned separate flocks of 20 and 30 birds seen coming in off the sea at Whitburn on 10 July 1991, in a year when the species had generally a strong presence throughout the country. In 1992, 26 migrants were counted at Whitburn on 7 dates in September and 32 there on 8 dates in October. A Durham Bird Club Coordinated Migrant Watch along the Durham coast on 10 October 1993 produced a total count of 248 Siskin (Bowey 1999c) and over two days in early November some 450 were counted moving south with other finches at Whitburn. In September 1997, birds appeared on 19 dates on Cleadon Hill with 120 south there on the 30th. In 2007, as part of a nationwide invasion, 36 arrived at Marsden over 2nd to 3 October with 64 at South Shields on 24 October, and in that year's Coastal Coordinated Migrant Watch some 300 birds were logged on 27 October. Even on the late date of 2 December that year, 28 were seen coming in off the sea at Whitburn Observatory.

Spring passage is usually noted from the end of March through to the end of April with most birds moving through the area on a broad front, as indicated by ringing data; though it is less obvious at the coast at such times, with small groups of normally fewer than 10 birds appearing there during April.

Distribution and movements

This species breeds principally in the boreal zone of Eurasia. It has become increasingly common as far south as central Europe with a considerable increase in numbers in recent decades. There are populations in the mountainous areas of central Europe. Most northerly populations move south and west during the winter; in some years large sections of the northerly population exhibit eruptive behaviour and may behave in a nomadic fashion. In this sense the Siskin is a classic, seed-eating, migrant finch which shares a similar strategy with other northern finches such Mealy Redpoll *Carduelis flammea* and Crossbill *Loxia curvirostra*. The UK breeding population is mainly in Scotland.

Significant 'irruption' years occurred in 1983 and 1986 with large-scale ringing efforts in 1986, resulting in many birds being caught in the spring near feeding stations, for example, in a suburban garden in Whickham. These captures generated some interesting ringing data. Most Siskins ringed or controlled in Durham appear to be of Scottish origin either heading to the south coast for the winter or returning north in spring. The tables show, respectively, selected recoveries and the movements of Durham ringed individuals and the origin of birds controlled in Durham. The data confirms the source of many of our wintering birds to be north east Scotland. The bird ringed in Poland (near the Lithuanian border) in the autumn of 1993 was part of a mass westward exodus from the Baltic area and its control at Whickham was only the fourth of a Polish bird in the UK. Interestingly, the bird ringed at Whickham and controlled at Sheringham, Norfolk just 13 days later in April 1994, was probably part of the return mass movement. It was the only south bound spring recovery of a Durham ringed bird. There are two controls of birds to or from the north east and Belgium, with the first ringed at Antwerp on 16 October 1982 and controlled at Whickham on 11 April 1983, when it was presumably northbound; the second was a bird ringed at Whickham on 17 April 1986 and controlled in Liege, Belgium on 19 January 1988. Two adult female birds ringed near Wolsingham on 2 April 1999 were controlled on consecutive days, almost exactly two years after ringing, at Oye, Kvinesdal, Vest-Agder, Norway 667km to the north east of their ringing location; a Hartlepool ringed bird was controlled in the same area in 2001.

Ringed (Durham)	Date	Recovered	Date	Days
Whickham	05/04/1986	Stronsay, Orkney	25/04/1986	20
Whickham	13/04/1986	Oykel Bridge, Highlands	15/05/1987	397
Whickham	13/04/1986	Inverness, Highland	10/04/1987	362
Whickham	19/04/1986	Ottery St. Mary, Devon &	11/03/1987	326 &
		Drumnadrochit, Highland	10/08/1988	844
Whickham	19/04/1986	Forres, Morayshire	11/04/1987	357
Whickham	19/04/1986	Barnard Castle, Co Durham &	04/05/1986	15 &
		Torphins, Grampian	15/04/1989	1097
Whickham	23/04/1986	Guisborough	25/04/1986	2
Clara Vale	15/01/1992	Maryburgh, Highland	09/05/1992	115
Whickham	19/03/1994	Sheringham, Norfolk	01/04/1994	13
Whickham	13/02/1994	Logie Hill, Highland	23/03/1994	38
Thornley Woodlands	06/04/2000	Muirshearlich, Highland	21/04/2000	15
Barnard Castle	14/04/2000	Selkirk, Borders	18/05/2000	34
Hartlepool	27/09 2000	Vest-Agder, Norway	28/04/2001	213
Rowlands Gill	29/02/2004	Kilmichael Glassary near Lochoilhead, Strathclyde	28/03/2005	393
Low Barns	19/11/2005	Stocksfield, Northumberland	03/04/2006	134
Rowlands Gill	05/02/2006	Ae village, Dumfries & Galloway	15/04/2006	68

Recoveries of Durham Ringed Siskins

The adult male ringed at Clara Vale on 15 January 1992 was controlled as a breeding bird in Highland in May 1992. The bird ringed at Rowlands Gill on 5 February 2006 was at Ae village in Dumfries and Galloway on 15 April 2006, more or less due west of the ringing site; perhaps this was an example of random dispersal, or a south west Scottish bird returning to its breeding area. One ringed at Low Barns on 5 November 2005 was controlled on 19 February 2006 at Sharow Grange, Ripon in North Yorkshire, demonstrating the direction of movement of birds passing south through the county during the autumn; although in this case just 62km to the south.

Ringed	Date	Recovered (Durham)	Date	Days
Fleet, Hampshire	14/02/1986	Whickham	19/04/1986	64
Harlow, Essex	06/03/1986	Whickham	19/04/1986	44
Guisborough, North Yorkshire	04/04/1992	Whickham	07/04/1992	3
Ashtead, Surrey	01/02/1992	Hamsterley Forest	20/04/1992	79
Marley Common, Sussex	03/04/1992	Hamsterley Forest	03/05/1992	30
Elblag, Poland	24/09/1993	Whickham	05/01/1994	103
Portsmouth, Hampshire	19/01/1994	Hamsterley Forest	03/02/1996	745
Amberley, Gloucestershire	14/03/2002	Rowlands Gill	17/01/2004	671

Durham Recoveries of Siskins Ringed Outside of the County

In contrast to the abundant evidence of the movement of birds to, from and through the county, a bird ringed near Wolsingham on 30 January 2000 as a first-winter female, was found dead just 6km away at White Kirkley, Frosterley, in Weardale on 8 July 2005, presumably after breeding in that part of the valley. At its demise, it must have been close to being six years since hatching, a considerable age for a small finch species in the wild.

Linnet (Common Linnet)
Carduelis cannabina

Sponsored by
Tunstall Hills Protection Group

A very common and well distributed resident and passage migrant.

Historical review

Due mostly to its beautiful twittering song, the Linnet was widely trapped and kept as a cage bird in the past. This practice continued, illegally, in Durham well into 1980s and even the 1990s in some areas; though apparently this species did not suffer from trapping as heavily as did the Goldfinch *Carduelis carduelis*. It is considered to be less common now than it was in the 19th century, when its favoured rough agricultural habitats were more widespread and when there was much more suitable habitat for it (Holloway 1996).

This species is, as it was in the past, a very common farmland bird across the county, though there is little specific reference to it in the historical record. Temperley (1951) said of it, "*a common resident and winter visitor, breeding in uncultivated areas, such as gorse-clad slopes*". The flocks traditionally noted on the coast in September and October, were thought by Temperley to be passage migrants. The scale of the local breeding populations of Linnet, at least in the south east of the area, through the 1950s and 1960s, was indicated by the fact that birds noted feeding on undeveloped ground around Teesmouth, where weed plants grew in abundance, attracted large flocks including one flock of over 1,000 at Graythorp in September-October 1965 (Blick 2000).

An interesting record in May 1979, concerns a pair of Linnets nesting in a suburban garden on Humbledon Hill, Sunderland. The birds hatched young in a nest in a *Fuchsia* bush but did not succeed in fledging these; this was previously undocumented local breeding behaviour.

Recent breeding status

Linnets are found across the county, from farmland habitats, to the more urbanised areas. The *Atlas* map showed a well-distributed occupancy of tetrads across the county, from the upland fringes to greater concentrations in the eastern, farmed areas (Westerberg & Bowey 2000). The large apparent absence of birds in the *Atlas*' mapped distribution of this species, in the county's south east quadrant, has been shown by the subsequent work on *The Breeding Birds of Cleveland* to have been due to a lack of observer coverage in that area during the *Atlas* survey phase. In reality, this is a widespread species in the 'north of the Tees area', particularly in the coastal strip and in the agricultural areas to the west of Stockton. There were believed to be between 500 and 700 pairs of this species locally, according to the fieldwork undertaken for the above work (Joynt *et al.* 2008), which found that it was only absent from the most truly urban of areas in this south eastern part of the county.

Linnet are found mainly in the more rural parts of the county although good numbers are also present around the suburban rural interface, in all parts of the county. Small numbers often breed in roadside scrub in the central and eastern parts of the county and colonies have been known to develop along riverside scrub, such as on the Gateshead sections of the Banks of the Tyne, while in some areas a few pairs penetrate into truly urban areas, for example in amenity planted scrub in parts of Gateshead town centre (Bowey *et al.* 1993). The species' main breeding strongholds however, are on farmland areas. Large breeding colonies amongst gorse are present along the county's coastal strip such as at Beacon Hill (50 pairs in 2005), Easington and Blackhall Rocks; while further inland locations around Elemore, Seaton and Hetton Lyons Country Park hold scores of pairs in May and June, for example, 35 pairs at Hetton Lyons in 1994. There is strong distributional emphasis on the eastern part of the county, many birds being present in habitats located on the magnesian limestone areas of Durham. In such areas post-breeding flocks begin to assemble on crops during July, and these frequently number up to 250 birds in places such as Eppleton and South Hetton.

There are many fewer linnets in the upland areas in the west of the county, though birds can be found around the upland fringes in all of the upper dales, with successful breeding noted as far west as Ruffside in 2010. The widespread nature of the species is indicated by the fact that in 2007, Linnets were recorded from 116 scattered locations across the county.

The Linnet is a semi-colonial species (Newton 1985) and the sub-population size varies considerably, from area to area, from just a handful of pairs in urban settings, to quite considerable colonies of 30 pairs or more in agricultural land, where the species often chooses mature stands of gorse *Ulex europaeus* within which to nest. A long-running farmland CBC plot, covering an area of 48ha, at Kibblesworth in the north of the county, has seen much variation in numbers over the lifetime of its study period, with considerable increases during the 1990s. In the early to mid-1990s, this site held, on average, 25 pairs in two main colonies (Westerberg & Bowey 2000), the mean population level at this site being 0.52 pairs per hectare. Working on the assumption that the figures from Kibblesworth were typical of the county as a whole, the *Atlas* estimated that each occupied tetrad would contain an average of 20.8 pairs. Extrapolating this density of birds to all of the occupied tetrads across the county led to the creation of a population estimation of approximately 5,000 pairs in the county (Westerberg & Bowey 2000).

Nationally, the species has been in decline for a number of decades (Parkin & Knox 2010) although this was not as apparent in Durham. There was a countrywide decline of 76% between 1967 and 2008 (Baillie *et al.* 2010), most noticeably from the mid-1970s through to the mid-1980s, with low productivity due to a high rate of nest failure thought to be one of the primary factors influencing the decline (Siriwardena *et al.* 2000). There was however, no noticeable decline in numbers in well-studied parts of the county up to the early 1990s (Bowey *et al.* 1993). This may be in part, due to the fact that in the north east generally, and over much of Durham in particular, agriculture tends to be somewhat less intensive than in many parts of the country, partly because the productivity of the soils tends to be that much lower (Clifton & Hedley 1995). Although there was considerable evidence of such declines in the national context, particularly during the 1980s and 1990s (Marchant *et al.* 1990), they were less obvious in County Durham. It would seem though that there have been some apparent falls in numbers (Westerberg & Bowey 2000), though the *Atlas* asserted that this species was not faring as badly locally, as it has been across the rest of the British Isles.

In 2007, the Kibblesworth CBC found 11 pairs nesting, the lowest total since 1998. Thereafter, only relatively small numbers were reported until late summer with no other significant breeding records submitted. As Linnets are semi-colonial, localised habitat destruction, particularly the removal of gorse for agricultural purposes, can have a catastrophic effect on local colonies (F.S. Milton pers. obs.).

The adoption of set-aside and other environmentally sensitive schemes by some arable farmers during the late 1990s and first part of the 21st century seems to have considerably benefited this species. Perhaps the most important factor is the greater availability of undisturbed ground and arable weeds in their breeding areas (Milton 1992). Unfortunately, much set-aside land was taken back into production towards the latter part of the first decade of the 21st century, and overall the policy may not be that beneficial for the species in the long-term. The removal of gorse on the magnesian limestone plateau to encourage botanical diversity, such as at High Moorsley and Quarrington Hill, has had a detrimental effect on local breeding numbers, but such local impacts are minimal compared to the over-riding influence of agricultural policy for this species.

Recent non-breeding status

Post-breeding, small to medium-sized flocks of Linnet gather in field margins and around waste ground ,with larger flocks gathering on farmland from mid-summer to feed on grain and arable weed seeds; typical flocks across Durham might involve up to 150 birds. The flock of *c.*1,500 at Fishburn on 29 August 1981 was exceptional and at that time, was probably the largest flock documented in the county. The maximum counts of post-breeding flocks began to decline through the late 1980s and 1990s. Although flocks of 150-300 were still not infrequent in the late 1980s, widespread flocks of over one hundred were becoming less well reported.

The spread of oil seed rape has benefited the species and this crop provides excellent feeding at the crucial breeding and fledging stages and large flocks are recorded in rape fields from late June (Westerberg & Bowey 2000). In 2007, post-breeding flocks were evident from around 30 June when 60 were at Whitburn, with 90 at Great Eppleton the following day. More typically, it was the second half of August and into September before the largest numbers were noted. The biggest gathering by far was a flock of 750 birds at Great Eppleton, with groups of 200-350 at Bishop Middleham, Dalton Moor, Elemore Golf Course, Hetton Lyons and Whitburn; further counts of 100 plus came from Easington, Langley Moor and Trimdon Grange. Flocks of up to 750 were noted at a number of sites through the 2000s; the largest in 2009 being 1,500 at Elmore on 21 December. Other significant gatherings were 1,200 at Longfield House, Gateshead in January 2001 and 646 in coastal fields between Noses Point and Shot Rock in autumn 2003. Set aside fields were important for this species through the first half of the decade, illustrated by such habitat near Hetton Wind Farm holding 950 birds in September 2006.

The final quarter of the year tends to see a reduction in numbers locally, as some birds move further south and west for the winter. Nonetheless some flocks numbering up to a few hundred tend to linger into October, and through the winter on farmland habitats that provide sufficient winter forage on stubble, or where spilt grain is available. While not exclusively located in the east of the county at this time of the year, there is tendency for an easterly focus of birds during the winter period. The count of 2,500 at Blackhall on 31 December 2000 was huge and the largest documented count of this species in Durham; perhaps such a gathering indicated a greater clumping of birds on fewer suitable areas, rather than a growth in numbers. In January 2001, 1,200 birds were at Marley Hill. More recently some of the largest congregations included that at Elemore Woods where 800 fed on rape in December 2009, and this flock had increased to around 2,000 birds by early January 2010.

Occasionally birds with unusual plumaged features are observed, for example a pale buff individual was at Hartlepool in August 1991 and two other localities. An almost completely white bird was at Bowesfield from October 2006 to April 2007 (Blick 2009).

Distribution & movements

The Linnet breeds over much of Europe, as far north as Northern Scandinavia, and then westwards into western Siberia. To the south, the species extends, in differing racial forms, into southern Europe and across the Mediterranean into northwest Africa and even the Middle East. Many of the populations are migratory, the northerly groups of birds tending to move south to winter within the range of more sedentary populations of the species.

While the Linnet can be found in Durham throughout the year, its true status, as for many small passerines is more complex than can be summarised in a simple statement in relation to its movements in and out of the county. It is a resident but also a passage migrant in both spring and autumn, a summer and a winter visitor. Furthermore, it is conceivable, indeed probable that some of the same birds may behave somewhat differently from year to year, depending upon local conditions.

Spring passage is fairly conspicuous along the coastline between mid-April and late May, when thousands of birds pass through, with occasionally several hundred on single days; a light influx of passage birds was noted at Mere Knolls cemetery in early April 1985. In 1976, return passage, presumably from northern Europe, was noted from mid-March through to mid-April but one on a fishing boat 11 miles off Sunderland on 6 May suggests such passage could continue much later. Autumn passage is less obvious or predictable. However, good numbers are sometimes seen during 'falls' in August or September, but equally there are occasions when very few are noted (Blick 2009). On 4 October 1992, 760 birds were noted on the Coordinated Migrant Count along the Durham coast (Bowey 1999), but it is almost impossible to discern the difference between resident and passage birds in such situations in coastal habitats.

Many of Durham's Linnets remain in their local breeding areas, year round, though this is probably dependent upon a number of factors including weather conditions and the local abundance of food. Ringing data has shown that some north east breeding birds move south and west for the winter (Kerr 2000); for example the bird ringed at Marsden, on 26 January 1963 that was caught at Spurn Bird Observatory, Yorkshire on 9 February 1963, and again there on 17th of that month was perhaps stimulated to move by the extreme wintery conditions of that winter. Ringing data shows that both nestlings and some full-grown, autumn caught birds from the county spend the winter as far south as southern France. There are a number of long-distance ringing recoveries that indicate that at least some of Durham's 'local birds' sometimes travel further afield. For instance, a female ringed at Saltholme on 19 August 1990 was found long-dead on 3 March 1991 at Noord-Holland in the Netherlands. At the time this was only the tenth record of a British-ringed Linnet to be recovered in the Netherlands; most continental recoveries of this species come from points south or south west, largely in France and Belgium. As illustrated by the birds ringed: at South Shields on 18 June 1957 that was recovered at Mezos, Landes, France on 31 October 1959; in the nest at Crawcrook on 2 June 1967, which was found in Sore, Landes, France on 14 December 1967; the nestling ringed at

Graythorp, Hartlepool on 18 September 1965 that was recovered at St. Mèdard-en-Jalles, Gironde, France on 15 November 1965; the juvenile ringed near Hurworth Burn Reservoir on 12 August 1970 found dead in south west France on 30 December of that year and the adult male that was ringed at Saltholme on 15 August 1976 and found dead at Biarritz, in the Pyrenees-Atlantique, France on 9 November 1977.

Some birds ringed in the area do not fit into this pattern, travelling in the opposite direction to the crop of birds noted above, such as the first-year male ringed at Graythorp on 26 October 1978 which was heading north, reaching Marsden Hall, 39km away, where it was controlled on 25 August 1979. Another ringed at Teesmouth on 26 August 1983 was recovered at Dunvagel, Strathclyde, on 15 September 1984; perhaps this was a Scottish bird that had been passing through the county when originally caught (Blick 2009).

Twite
Carduelis flavirostris

An uncommon winter visitor mainly to the coast. Also a very local and decreasing breeding species.

Historical review

Given the upland nature of the county there has probably been some presence of Twite in Durham for many years, but there is relatively little detail of its status or patterns of occurrence in the county's early ornithological record. In the nineteenth century, Hancock (1874) said that Twite was a not uncommon breeder on heather moorlands in both Durham and Northumberland. Backhouse, writing in 1885, documented the finding of a nest with five eggs near the summit of one of the hills on the Durham side of the Tees, but he did not say when or on exactly which hill. Robson (1896) gave no documentation of its breeding in the Derwent valley. Nelson (1907) said it only really occurred as a winter visitor to Teesmouth and Temperley (1951) noted flocks there in 1931 and 1936. One was at Teesmouth on 21 October 1947; then a flock of around 40 birds was seen at Whitburn on 28 October 1948. Temperley concluded that its status had changed since Hancock's time, becoming less common and regular in winter and rarer as a breeding bird.

In the 20 years following the publication of Temperley, only 11 birds were reported in five of those years in the annual ornithological reports for Northumberland and Durham. Amongst the records however, was one of a pair which bred at High Force in 1952 (Temperley 1953). This is believed to be the first confirmed breeding record for County Durham, certainly since Backhouse's assertion of such in the previous century. Other records in this period were at Graythorp, Frosterley and the Tees Marshes, but no more than three birds were seen at any one site.

Since 1972, records have been annual in Durham. During 1971 to 1980 records referred to wintering birds on the Tees Marshes, but never more than 12 at any one time. Pairs, that may have been breeding, were recorded in different years at Frosterley (1972), Widdybank Fell (1978), Balderhead Reservoir (1979), and Grassholme Reservoir and Grains o' th' Beck (both in 1980). The first record of a large, late summer flock occurred on 14 September 1980 when six were at Grassholme Reservoir. Interestingly, just over the border into Cumbria, a flock of 70 was reported at Nenthead on 25 August 1980. This poses the question as to the origin of these birds - they may have been of Cumbrian or Northumbrian derivation, as there have been several breeding records just over the Durham border in both of these counties (Day et al. 1995, Stott et al. 2002).

Recent breeding status

During the breeding season Twite are difficult to locate, as they inhabit large areas of open habitat, typically, between 250-400m above sea level. They can commute large distances between nesting areas, on the moorland edge, and feeding areas, which tend to be in marginal and agricultural pastures on the upland fringes (Gibbons et al. 1993, Westerberg & Bowey 2000). The Twite also requires rich areas of seed-bearing plants, typically in the traditionally managed grasslands of the upper western dales and locally they seem to favour sites in close proximity to water, such as the upland reservoirs. As a result of its predilection for such challenging habitats for the observer, our current state of knowledge is unlikely to reflect the true status of the species in the county.

The period, 1981 to 1990, was most productive for summer records of birds in Durham, both of likely breeding birds and post-breeding flocks. Over this period summer birds were reported from Lunedale (at Grains o' th' Beck and Grassholme Reservoir), upper Teesdale (at Harwood, Widdybank Fell and Cronkley Fell), Bowes Moor, Cotherstone Moor, Eggleston and Bollihope. In the north of the county, in the late 1980s at least one singing bird was noted in the Harehope Burn off the Burnhope Burn part of the Derwent valley watershed, and one was seen flying up on to Muggleswick Park, after feeding in fields on the edge of Burnhope Woods, in June; possibly feeding young. Eight late summer records of flocks, of up to 60 birds, were reported in different years during this period from upper Teesdale. These were surpassed, however, by 150 at Harwood on 24 August 1983 and 100 in late summer 1989 in upper Teesdale. Spring records, which may have referred to local breeders, consisted of several flocks of up to seven birds in Teesdale and 20 at Sleightholme on 29 May 1983.

The decade, 1991-2000, was another highly productive period for records due to the efforts of the Durham Upland Bird Study Group. In 1996, based on the number of birds observed both before and after the breeding season, it was estimated that as many as 34 pairs may have made up the breeding population in Teesdale and Lunedale. The uncertainty over the source of birds seen in late summer and autumn, however, suggests that this figure now seems a considerable overestimate. A more likely breeding population, using only spring records as an indicator, is in the region of five to seven pairs (I. Findlay pers. comm.). During the pre-breeding period birds were mainly recorded in hay meadow, pasture and alongside roads. In addition, post-breeding flocks were well reported, for example, up to 100 on Widdybank Fell in late September/early October 1996 with 60-90 during August and September 1997 and 1998 in other areas of upper Teesdale. In 1992 and 1993 a series of records came from the Teesdale reservoirs, Hury, Blackton and Grassholme, where flocks of up to 40 birds were observed feeding during August, often in the company of Linnets. The post-breeding flocks were mainly found in the same habitats as the pre-breeding flocks with the addition of reservoir banks and spillways. Flocks have been noted as feeding on the seeds of thistle *Cirsium sp,* meadow grass *Poa sp,* docks *Rumex sp.* and meadowsweet *Filipendula ulmaria.* There is, though, no evidence of the origin of these birds.

The breeding population of Twite in Northumbria and Cumbria may have contributed to these late summer flocks. However the situation regarding post-breeding dispersal is not clear. In Raine *et al.* (2006) in a study in the South Pennines, it was noted that birds in feeding flocks may have originated from as far away as 28.2km. It was also found that 65.2% of nestlings moved less than five kilometres in the period between fledging and their first winter migration. There was also a tendency during this period for birds to be reported in October in upper Teesdale, a date which was considered by an experienced local observer, to be late for local birds. Small groups of Twite seen just outside the county boundary, on the watershed between Teesdale and Arkengarthdale, in October 1997 (J. Strowger pers. obs.), may suggest autumn passage through our area.

During 2001-2010, there was only one summer record, that of a single bird at Harwood on 30 May 2005. In addition, three were at the roadside near Little Eggleshope Beck on 15 March 2009 and three were at Widdybank Farm on 29 August 2009. In total, these records still suggest a spring to late summer presence of the species in Teesdale. This is probably not a true reflection of the species status, but rather indicates a lack of observations. There is, however, clear evidence for a significant fall in Twite numbers in the UK (Eaton *et al.* 2010).

Recent non-breeding status

Although a regular wintering bird to the north and south of our area, records suggest that Twite is a relatively localised wintering species in County Durham. The species favoured wintering habitat, salt-marsh, is restricted to the mouth of the Tees Estuary. Undisturbed coastal stubble is also used.

For many decades, it has been a regular wintering bird around Teesmouth but in the 1970s and early 1980s, counts were usually in single figures. From 1976, wintering numbers were slightly higher, with counts of: eight near Seal Sands in November 1976; nine in February 1977, 12 in February 1978; up to 25 in December 1980 and January 1981; 22 in January and February 1988 and in the 1988/1989 winter between 30 and 70 on Greenabella Marsh. This signalled the start of the period when flocks of Twite were almost annually in excess of 30-40 birds, often in the Long Drag to Seal Sands area (Blick 2009). In 1984, 50 were at Hartlepool and 60 were on the Long drag, Teesmouth in February 1990. Similar numbers were on Greenabella Marsh in November 1997 and up to 80 were there in November and December 2000; 84 were at Greatham Creek on 17 November 2001 and 90 in the same area on 18 January 2003 (Blick 2009). In 2005 the peak count was 85 at Seaton Snook in November, where they were feeding on glasswort seeds *Salicornia europaea*. In January 2006 up to 80 were present on the North

Tees Marshes and in the winter of 2006/2007 up to 30 birds were present in the North Gare/Seaton Snook area. In the period 2005-2010, this was the most favoured wintering area in the Tyne to Tees coastal strip. The year 2007 might be considered a typical year for wintering Twite with the only regular flock of wintering birds present commuting between Greenabella Marsh and Seaton Snook, on the North Tees Marshes; here the maximum count was 57 in late October. These counts were exceeded in 2009, when 100 were present in January to March in the Greenabella Marsh/Long Drag area and 141 were on the North Tees Marshes in December. In 2010, 123 were on Greenabella Marsh in February, with smaller groups moving between RSPB Saltholme, Long Drag and Seaton Snook.

The presence of winter birds over the more northerly sections of County Durham's coastline has always been somewhat erratic but in several winters since the late 1990s, the cliff-top stubble in the stretch of coast between Seaham and Crimdon attracted flocks of up to 115 birds, as in April 2001. Previously 100 were at Dawdon in December 1998, 100 at Blackhall in December 1999 and 38 were at Dawdon in December 2000. In 2010, 20 in Murton village was an unusual sight, while three at Dawdon and two at Blackhall Rocks were of interest. A lack of suitable habitat in this part of the county appears to be the limiting factor in these areas.

Inland winter records have occurred from many locations, with little obvious pattern. The Salter's Gate area appeared to be a magnet for the species during 2002-2006 with up to 30 there in February 2002 and in November 2005, when they were in a mixed finch flock on a set-aside field. Other notable counts were 40 at Edmundbyers on 4 February 1992 and 20 at Derwent Reservoir on 20 December 1998. Ten near the Cumbrian border on 2 January 2001 were unusual. The first record for Hetton Lyons Country Park was three on 2 February 2004, following a single at nearby Hetton Bog on 25 January. Other locations which have occasionally attracted small numbers have been Bishop Middleham, Brancepeth Beck, Brasside, Frankland Farm, Waldridge Fell and West Pasture.

A series of inland reports from the Gateshead area during the 1980s and early 1990s, show that this species may very well be overlooked as a wintering bird in some areas. Twite may occasionally appear at non-traditional sites, often associated with poor weather. Examples of this include the records of single birds at Shibdon Pond on 24 January 1984 and one on the edge of Washingwell Woods on 23 November 1984. Another at Shibdon Pond during hard, snowy weather in 1991 was easily approachable to within one metre, when feeding on curled dock *Rumex crispus* seeds, and was present from 11th to 13 February.

More intriguingly, in the winter of 1987, during a period of hard weather a flock of some 30 to 40 birds was reported as being present on Barlow Fell in the lower Derwent valley. Two years later birds were once again found in this area, with six birds also being present on farmland at Reely Mires, near Greenside, on 15 October 1989, though it is believed that as many as 30 had been there the previous day. Shortly after this, on 25 October, a lone male bird was in the company of Linnet, at nearby Ryton Willows. The following month a flock of approximately 30 birds was noted near Sherburn Towers Farm, Rowlands Gill (Bowey *et al.* 1993).

It seems likely these examples of inland records were produced by small flocks of birds wandering around during the winter period. Perhaps such incidents of what is essentially a nomadic winter species are more common away from traditional coastal wintering areas than is generally realised.

Distribution & movements

Twite has a well-documented, disjointed and widely separated bi-nodal distribution. Its major populations are located, effectively, at the eastern- and western-most parts of the European/Asian landmass (Cramp *et al.* 1994). The species' principle populations are in central and south west Asia, but two races occur along the maritime fringe of Western Europe, largely in Britain, especially Scotland and the Northern Isles, and also in Norway. The northern European birds are presumed to be migratory, wintering around the Baltic, but they do not appear to be responsible for the winter influxes of birds that occur along the coastal parts of eastern Britain in autumn and winter. It is believed that many of the northern British birds winter further south within the British Isles, and it is these that are thought responsible for coastal migrants in Britain. Raine *et al.* (2006) showed that birds breeding in the south Pennines mainly wintered on the east and south east coasts of England; some migrated to the west coast and others overwintered in the south Pennines.

Lesser Redpoll
Carduelis cabaret

**A relatively common though declining resident in suitable habitat.
Also a winter visitor and passage migrant in varying numbers**

Historical review

In the latter part of the Victorian period, Thomas Robson (1896) considered 'Redpoll' a resident species in the Derwent valley with a *'small breeding population'* that exhibited a noticeable build up in numbers during the autumn and winter. This presumably summarises its status across much of the county during that century, but there is relatively little detailed documentation in the earlier ornithological records. Almond *et al.* (1939) identified an increase in breeding numbers in the Tees valley with the species said to be benefitting from new plantations in the dale up to 303m above sea level. George Temperley (1951) believed it to be *"well distributed in suitable localities; a resident, and also a regular and numerous winter visitor"*. He stated it bred in thickets, comprising birch *Betula spp.* and alder *Alnus glutinosa* scrub, gardens and along streams *"often far up towards the fringes of the moors"*. He also drew attention to it being often noted on the coast in October, but said that these birds might not necessarily be from overseas.

This species was described as a common woodland resident in south east Durham by Stead (1964) during the 1950s and early 1960s. In the same two decades a study area around Shotton Colliery (Simpson 2011) held two small but stable colonies; one of these had about five pairs in a hawthorn hedge of an allotment, while the other contained about ten pairs in Edder Acres Wood. By 2010, this was a species that was rarely seen in that area.

Recent breeding status

Nationally, Lesser Redpoll breeding populations were considered stable and perhaps even increasing in the early 1970s (Marchant *et al.* 1990). By 2002, Lesser Redpoll had been placed on the Amber List of Birds of Conservation Concern because of signs of a steep decline in breeding numbers over the late 20th century (Parkin & Knox 2010). In 2009 the species was added to the Red List (Eaton 2009) because of a very considerable 88% reduction in the national breeding population over the preceding 25 years. The national decline probably commenced locally in the 1980s and it continued with some considerable losses of the species locally through the 1990s and this trend continues into the 21st century.

Birds nest in broadleaved woodlands either dominated by, or with an abundance of birch *Betula* sp. and alder *Alnus glutinosa,* as well as in hawthorn *Crataegus monogyna* scrub, but they will also breed in young conifer plantations and mixed landscape trees, usually of about 20 to 30 years in age (Marchant *et al.* 1990). One of the principal threats to the species is from the process of natural, vegetational succession that occurs in scrub and woodlands. Ultimately, this leads to a loss of the species' favoured trees and the seeds upon which they depend for much of the year, as the longer-lived, canopy dominant species, such as oak *Quercus robur* and ash *Fraxinus excelsior* (Marchant *et al.* 1990) take over. The gradual maturing of conifer plantations with lower levels of replanting has been blamed for part of the long term national decline, although where cyclical clearance and replanting takes place then numbers can be sustained. Also cited in the decline is possible increased nest predation pressure from grey squirrel *Sciurus carolinensis* and the availability of winter food supplies (Fuller *et al.* 2005).

Just two decades ago Lesser Redpoll could be considered quite widespread as a breeding bird in woodland habitats in Durham but it has since become increasingly elusive. It remains relatively well-represented along the upper sections of the county's main river systems and is still often found in riverine woodlands to the north and west of the county. It can also be a bird of upland edge and higher altitude woodlands, wherever favoured tree species are present. This pattern mirrors the distribution of the species' favoured woodland type.

The decline was highlighted in the *Atlas,* with an example from the well-studied Shibdon Pond (Westerberg & Bowey, 2000), where it was noted that breeding ceased in 1996 after, at least 15 successive years of successful breeding, when the 'colony' there numbered well into double figures at its peak during the mid-1980s. This decline was illustrated by the returns from ringing studies at the site, where between the late 1970s and mid-1980s, 443 adults and 15 nestlings were ringed during the breeding season. For example in 1980, 11 adults and 17 juveniles were ringed during the breeding season in a typical year during the peak phase; it was said that nests were easy to find in the hawthorn bushes. By the early 1990s none were being ringed at the site. Survey work in Gateshead during the late 1980s (Bowey *et al.* 1993), showed that Lesser Redpoll were then present in 19% of all surveyed one-kilometre squares in the borough. This probably mirrored the species' representation in similar, suitable woodlands and scrub across the whole of the county at the time, but such a level of occupation would be decidedly atypical today. The *Atlas* quite accurately mapped the distribution of Lesser Redpoll across the county. This showed its main distribution to be a northerly and westerly one, but that the species was still also well-represented in some lowland areas, particularly to the north of Durham City and, somewhat less abundantly in the east and south west of the county.

Like many other small, *Cardueline* finches, Lesser Redpoll are semi-colonial in breeding habit (Newton 1985), which leads to a 'clumped distribution' (Westerberg & Bowey 2000). Consequently, birds occupying small or restricted areas of habitat might easily be missed by casual observers, or even detailed survey processes, a fact that may still contribute to some under-recording of breeding colonies.

Very little substantive breeding data has been reported in the last two decades but its spring and summer presence, though now much more sparse than previously, remains primarily in western locations, where breeding takes place in locations such as Hamsterley Forest, around Derwent Reservoir and The Stang. Typically in 2007, a count of at least 20 singing males in a one kilometre square of The Stang during early April, highlighted the species' semi-colonial nesting habit, as well as its penchant for upland fringe locations and willingness to use both conifers and broadleaves. Further east, scattered small, lowland colonies still exist. These are often associated with secondary plantation woodlands and scrub along railway paths; for example, at least three pairs nested at Brandon in spring 2008. In spring 2009, song and display was noted in early May at Derwent Reservoir, near Hedleyhope Fell, Knitsley Fell, Muggleswick and West Butsfield, with 'healthy' breeding numbers noted at Pennington Plantation and Neighbour Moor in the Hamsterley Forest area.

Today, in the south east of the county, as over much of the rest of it, there has been a dramatic decline in the lowland breeding population of the species; these losses being concentrated over the last 30 years (Joynt *et al.* 2008). The only area that registered more than a 'couple of breeding pairs' during the work on *The Breeding Birds of Cleveland* was an area of scrub and new plantation at Urlay Nook, to the west of Eaglescliffe (Joynt *et al.* 2008). The species now appears to be absent as a breeding bird from most of the northern parts of what was previously Cleveland. This may be considered an extension of the earlier contraction in range from the Vale of York as noted in the *New Atlas* (Gibbons *et al.* 1993).

Extrapolating from the national population figures (Gibbons *et al.* 1993) and Thom (1986), the *Atlas* made a then 'conservative estimate' for the county population across all of the then occupied tetrads of some 2,300 pairs (Westerberg & Bowey 2000). Considering recent and current population trends, this might now be considered an unduly high figure and the current population is probably little more than half, indeed possibly lower, perhaps as few as 1,000 to 1,200 pairs.

Recent non-breeding status

Winter brings large gatherings of the species to suitable feeding areas; birch seed being one the species' principal foods at this time of the year, though they also frequently feed on the seeds of taller ruderal weed species, such as those of docks *Rumex spp.* and mugwort *Artemis*. Hence, woodlands with large amounts of birch attract many Lesser Redpolls. Flocks of up to 100 are often present along the Derwent Walk and similar locations. In good seed producing years, larch *Larix spp.* also attracts many birds, as shown by the flock of some 300 present in a larch plantation in Chopwell Woods during the winter of 1987.

By late autumn, recently arrived or nomadic small flocks have combined at inland sites, leading to more impressive numbers such as the 200 at West Moor Plantation on 22 October 2009. Some autumns bring evidence of more notable influxes into the county and in recent times winter flocks exceeding 200 were reported at Lanchester and several Derwent Valley sites. For example, in November 1984, 400 were present in the Derwent Valley and 150 at Hamsterley in November 1995.

In the first winter quarter of 2008, almost 300 reports of the species were received by Durham Bird Club from more than 100 locations, so this species remains widespread in winter. The largest numbers are likely to be reported from locations in the east of the county; often from sites with post-industrial woodland habitats, where the favoured birch and alder predominate. Many sites with mixed scrub and early-stage succession woodland routinely attract small flocks of 20-40 birds. Rainton Meadows, has become one of the best locations for attracting the species in winter, and flocks of up to, and over 100 are regularly present here throughout mid- to late winter, encouraged by the provision of feeders with Nyjer seed. More mature river valley woodlands and conifer plantations can attract three-figure flocks but by the very nature of the geography of the county, these sites tend to be further into the county's interior and consequently probably less well recorded by observers. In winter 2010, a number of sizeable flocks were reported, the largest being 150 birds near the Hamsterley Forest Visitor Centre in December, along with 150 close to the Deerness Valley Walk; other three-figure counts were at Rainton Meadows and Sharnberry in Hamsterley Forest.

Dispersal of winter flocks is usually complete by mid-April, when breeding birds will be displaying and song-flighting. Historical perspective is provided by the long-term Shotton Colliery study area, where this species used to be a very common as a winter visitor in the alders and silver birch trees; today it is very rarely seen in that area.

There is a small amount of passage through Durham in both spring and autumn, the former during April and early May, when rarely more than 20 birds per day are observed. Autumn passage tends to occur in the period mid-September through to late October, for example in 2008 this movement was concentrated in the east of the county during the second half of September, but often most coastal strip birds are noted in October. Inevitably, many of the birds noted during such movements are recorded as 'Redpoll species.' but it is likely that Lesser predominate in such activity. In 2008, at Whitburn, a total of 80 moved south between 18th and 21st, with 22 flying south at Seaham on 26th and 102 south in an hour at Ryhope on 27th. On 5 November 2005 a larger volume of coastal passage was noted at Dorman's Pool, when 275 flew south.

Distribution & movements

The Lesser Redpoll breeds mainly in Britain and Ireland, up into southern Scandinavia with other European populations in the alpine areas of Austria, France, Germany, Switzerland and Italy. It was formally much more restricted in range, but experienced major population growth and expansion of range in the mid-part of the 20th century (Parkin & Knox 2010). Populations are largely sedentary, though some make local and altitudinal movements, in a somewhat nomadic fashion.

The origin of overwintering birds in Durham is likely to be complex with immigration from northern Britain and perhaps the near Continent forming part of a generally southerly movement, the extent of which will be influenced by weather conditions and seed crop availability, as per the national situation (Wernham et al. 2002). That some wintering Lesser Redpoll come from the northern part of the Continent is shown by the adult bird ringed at Whickham on 15 December 1984 that was found dead at Sodermanland, Sweden on 12 January 1986. The pattern of reports in Northumberland during the 1960s showed that first-year birds ringed on the coast in September moved south to be controlled in Belgium later the following October or November (Kerr 2001). Birds ringed in the south east of the county during May-September confirm this southward movement, as they have been re-trapped between November and March at Fleet in Hampshire, Beachy Head in East Sussex, Chertsey and Virginia Water in Surrey (Blick 2009). This movement is further emphasised by recoveries and controls in Nottingham and Surrey of birds ringed at Hurworth Burn Reservoir and Sheraton in 1971. Some locally-bred birds travel beyond the Channel, such as the young bird ringed by the late Fred Grey, at South Shields on 7 July 1949 that was recovered at Maransart (Brabant), in Belgium on 25 October 1949. Confirming this, one ringed at Sheraton on 30 August 1971 was found dead a little further south in France, at Dewscartes, Abille on 17 January 1973, with a juvenile ringed near Hartlepool in August 1971 also found in France in January 1973. More recently two birds ringed at the Long Drag on 20 September 2008 were both controlled in Belgium in October 2008, just 30km apart. This suite of recoveries and controls indicates that this direction of movement and wintering area is still utilised today by birds from the north east of England. A juvenile male ringed at Saltholme on 17 September 1998 was controlled at Icklesham, Sussex, 32 days later on 19 October 1998; it was presumably heading to the Continent for the winter. This pattern suggests that as locally breeding birds move south for the winter, the birds seen in the county at that time may have travelled into Durham from further north in Britain.

Mealy Redpoll (Common Redpoll)
Carduelis flammea

A scarce passage migrant and winter visitor, prone to large influxes in some years.

Historical review

Early ornithologists of the 19th century, setting out to catalogue the region's birds, were certainly very well aware of the distinctive *flammea* form of Redpoll from Fennoscandia but its status as a winter visitor to our area was probably not then fully understood. William Backhouse provided the first record, writing in the Zoologist (1846), as having *"met with specimens near Darlington"*. In contrast it was not mentioned in either of the respective lists from Hutchinson (1840) or Edward Backhouse (1834). Hancock (1874) said that it was a *"common winter visitor, sometimes appearing in large flocks"* but gave no further evidence to support this assessment. His judgements on identification and status have consistently stood the test of time but on this occasion it appears that his view of it being *"common"*, with hindsight, may be regarded as a little too optimistic. Thomas Robson (1896) documented just a single record of one in the Derwent valley, an individual shot in Axwell Park, on 13 March 1892, and indicated that it was then considered a rare bird in the interior of the county. Canon Tristram (1905) however, considered the Mealy Redpoll as a *"frequent winter visitor"* but may have been reliant on Hancock's view. For neighbouring Northumberland, Bolam (1932) described the species as *"a winter visitant and passage migrant, not rare, but very irregular, sometimes coming in considerable flocks"*. Almond *et al.* in *The Birds of the Tees Valley* (1939) mention just two specific records of birds being present with Siskins *Carduelis spinus* and Lesser Redpoll *Carduelis cabaret* on the banks of the River Tees at Blackwell, Darlington in each of the winters 1926/1927 and 1927/1928. Almond did nevertheless describe it as *"an irregular winter visitor in small number"*.

According to Temperley (1951) at the mid-point of the 20th century, it was a *"rare and very irregular winter visitor"*, which he himself had never seen in the county and he thought it less common then than previously. He could only add two documented occurrences to those already mentioned; one at East Boldon on 19 November 1945 and four at Jarrow Slake on 15 March 1947. Commenting in 1956 in the Ornithological Report for Northumberland & Durham, with no further records forthcoming, Temperley seemingly had good cause to challenge Hancock's assessment (Temperley 1957). In 1959 however, a small flock of Siskins and Redpolls at Bedburn, Hamsterley on 14 March contained at least four Mealy Redpolls and in 1962 a small influx brought 12 to Shotton Colliery on 18 February and 10 to Port Clarence six days later. These records were the first signals of a subtle change in status or perhaps more accurately, began to show how the species may have been previously overlooked. According to David Simpson (Simpson 2011), who studied birds in the area around Shotton Colliery from the late 1950s to the first decade of the 21st century, this species was a regular winter visitor to his study patch. It was seen by him every year from 1956 to 1968, usually along the railway embankment at Edder Acres Wood, where up to 13 birds were recorded usually with Lesser Redpolls and Siskins.

Recent non-breeding status

Today, the species can be considered a scarce, but regular, passage and winter visitor to Durham chiefly to the eastern half of the county. Numbers vary annually and no doubt reflect the availability of food supplies and cold-weather conditions on the Continent. Most sightings tend to relate to a few individuals found amongst flocks of Lesser Redpolls but larger flocks occur occasionally at times of irruptive dispersion. Open birch, alder and willow woodland and scrub are preferred. Three significant invasions are documented in recent decades. During the winter of 1984/1985, there was a large influx of northern Mealy Redpolls into the British Isles, with considerable numbers being present in the county. On 17 February 1985 a flock of approximately 200 birds, consisting almost entirely of 'Mealies' was present in Washingwell Woods near Whickham, whilst another 50 were in Milkwellburn Woods through much of this period, with many other small parties scattered across the rest of Gateshead (Bowey *et al.* 1993). This year was clearly important in re-evaluating the species' local status.

Another notable influx came in the winter of 1995/1996 as part of one of the largest documented irruptions to the British Isles (Riddington *et al.* 2000). Up to 100 were seen in the lower Derwent valley in December 1995 mixed with Lesser Redpolls and by January 1996 four sites here held an estimated total of 500 birds, most of which were considered to be 'Mealies'. Elsewhere, mixed Redpoll species flocks of between 50 and 80 birds, were seen as far west as Barnard Castle and Derwent Reservoir, at Brandon, Croxdale, Ravensworth and Waldridge Fell. In each case 'Mealies' were judged to be present as the majority proportion of observed flocks.

A third major influx occurred with the onset of harsh winter weather at the end of 2010. From December 2010 to mid-March 2011, a feeding station at Rainton Meadows attracted up to 90 Mealy Redpoll, smaller numbers of Lesser Redpoll and, at least, four Arctic Redpoll *Carduelis hornemanni*. At the same time, 80 Mealy Redpoll were noted at Elemore golfcourse, 40 were at Boldon Colliery, 30 in the lower Derwent Valley and smaller numbers around numerous other localities.

Apart from these years of definite irruptive movement from the northern Continent, the Mealy Redpoll has occurred annually over the last decade or more, with typically groups of two to five being seen at a handful of predominantly eastern woodlands. Most are expected to be found amongst wintering flocks of Lesser Redpoll. Exceptionally, 25 were at Cowpen Bewley in January 2002, 10 at Drinkfield Marsh, Darlington in April 2006 and 15 at Coatham Stob in March 2009. All of which brings the analysis full circle to Hancock's 1874 judgement, which may have had some validity and foresight after all.

The Mealy Redpoll is highly migratory with northern birds wintering as far south as central Europe. More recently, at least since the early 1970s, small numbers of 'Common Redpoll' types have been recorded breeding in parts of northern Britain; in particular, the Highland region and some of the northern and western Scottish islands (Parkin & Knox 2010). After the influx of 1984/1985, small numbers remained into spring amongst breeding Lesser Redpoll in Washingwell Woods and likewise in the Victoria Garesfield area in the Derwent valley in spring 1996. A similar situation was evident in spring 2011 as small numbers remained in possible breeding habitat at Derwent Reservoir until early April. However, there has never been any indication of a breeding attempt.

Distribution & movements

Mealy Redpoll is classified as a sub-species of the wide ranging Common Redpoll *Carduelis flammea* that breeds across much of northern Europe and the boreal zones of Eurasia and North America. It is replaced as a breeding species in Great Britain and the near Continent by the Lesser Redpoll *Carduelis cabaret*. In 2001 the BOU designated the Lesser Redpoll as a separate species from the Common Redpoll group. Mealy Redpoll *C. flammea flammea* describes birds of the northern Scandinavian population, with Iceland Redpolls *C. flammea islandica* also sometimes claimed in Britain; despite that race being believed to be largely sedentary (Parkin & Knox 2010). There seems little doubt that the separation of Mealy Redpoll as a species distinct from Lesser Redpoll has provided an added emphasis to the field identification of 'Mealies' and contributed to a greater understanding of the status of this attractive winter visitor.

Arctic Redpoll
Carduelis hornemanni

A rare vagrant from the Arctic: seven records, involving at least ten individuals of two distinct subspecies.

Hornemann's Arctic Redpoll
Carduelis hornemanni hornemanni

The only record of this rare visitor from arctic Greenland is described by Hancock (1874): "*I have only seen a single example of this species; it was knocked down on the 24th April 1855, with a clod of earth, on the sea-banks near Whitburn, where it had been observed flying about for a few days*". The specimen is still retained in the NHSN collection at the Great North Museum, Newcastle-upon-Tyne; it represents the first record of Arctic Redpoll for Britain and Ireland (Palmer 2000).

Coues's Arctic Redpoll
Carduelis hornemanni exilipes

Following the bird at Whitburn in 1855, all of the subsequent records of Arctic Redpoll in the county are believed to have belonged to this more widespread form, which breeds as close as northern Scandinavia. The first of these was a bird seen by Dave Simpson at Shotton Colliery on 3 December 1962 (*British Birds* 57:279).

All records:

1962	Shotton Colliery, 3 December (*British Birds* 57: 279)
1985	Souter Point, Whitburn, 23 November (*British Birds* 80: 565)
1991	Dunston, 13 February (*British Birds* 86: 526)
1995	Hargreaves Quarry, Teesmouth, male, 2 November, trapped (*British Birds* 89: 522)
2006	Drinkfield Marsh, Darlington, 15th to 17 April
2010	Rainton Meadows DWT, at least four, 19 December 2010 to 3 March 2011

The county's second at Whitburn in 1985 was found by visitors from Ayrshire and was watched from the cliff top close to Souter Lighthouse, having presumably just arrived in off the sea. The third was equally brief and seen by just one observer near Timber Beach, Dunston and was accompanied by a single Lesser Redpoll *Carduelis cabaret* and another 'pale' bird which may well have been another Arctic Redpoll. The winter of 1990/1991 saw an influx of at least 50 Arctic Redpolls to eastern Britain, and although the majority were to be found in Lincolnshire and Norfolk, the Durham bird was doubtless associated with this influx. The male trapped and ringed at Hargreaves Quarry on 2 November 1995 was the sole representative found in County Durham during the winter of 1995/1996, which saw the largest ever influx of this species into the British Isles; at least 400 birds. Just across the River Tees in North Yorkshire at least seven were amongst a large flock of Redpolls near Guisborough on 31 March 1996.

One at Drinkfield Marsh on the outskirts of Darlington in 2006, loosely associated with a mixed flock of around 20 Lesser *Carduelis cabaret* and Mealy Redpolls *Carduelis flammea*, was the first to be seen in the county by many observers.

In the December 2010, a large mixed flock of Redpolls was found at Rainton Meadows in December; this was estimated to contain at least 90 birds, including many Mealy Redpolls, several of which were 'pale'. Careful scrutiny, led to the discovery of an Arctic Redpoll on 19 December, with two elusive birds present on 21st. At least four different individuals were eventually identified before the end of the winter. This large influx of Redpolls was replicated elsewhere in Britain and a number of other Arctic Redpolls were discovered mostly in South Yorkshire and Norfolk.

Distribution & movements

Arctic Redpoll has a circumpolar breeding distribution in the Arctic and sub-Arctic, with the race *exilipes* occurring over much of the species' range except for Greenland and some Canadian islands where it is replaced by the nominate *hornemanni*. During the winter months it is usually found close to its breeding areas, though it is essentially nomadic, occasionally moving much further south. It is an annual vagrant to Britain, with over 800 records by the close of 2005 when the species stopped being considered by the BBRC. Numbers fluctuate markedly from year to year and in most there are just a handful of records. Both subspecies have occurred in Britain though each has a different vagrancy pattern. Nominate *hornemanni* is essentially a rare passage migrant to northern Scotland, particularly the Shetland Islands in late September and October, very rarely occurring further south - there are just a handful of English records. The more widespread form *exilipes* typically occurs during the mid-winter period and is usually detected among flocks of Redpolls; particularly those that contain numbers of nominate Mealy Redpolls from Scandinavia.

Hornemann's Arctic Redpoll, Whitburn, April 1855 (Phil Palmer, courtesy NHSN)

Two-barred Crossbill
Loxia leucoptera

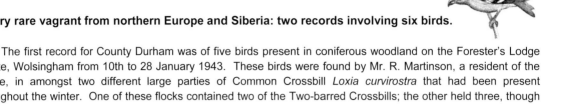

A very rare vagrant from northern Europe and Siberia: two records involving six birds.

The first record for County Durham was of five birds present in coniferous woodland on the Forester's Lodge Estate, Wolsingham from 10th to 28 January 1943. These birds were found by Mr. R. Martinson, a resident of the estate, in amongst two different large parties of Common Crossbill *Loxia curvirostra* that had been present throughout the winter. One of these flocks contained two of the Two-barred Crossbills; the other held three, though all five birds were seemingly never seen together. Writing at the time Mr. Martinson described the birds in *British Birds*, an extract of which is reproduced below: *"Two of the birds were adult males and showed a lot of crimson; this crimson in a good light appeared to be more of a pink colour and was much lighter than the crimson on the Common species. The other three showed no crimson at all and were a darkish green colour, the breast feathers being a little lighter, and showed pale yellow on their rumps. I did not notice any of the birds being streaked. The bills of all five were crossed and all five had two distinct white wing bars. The birds appeared to be of the same size as the Common species; the only difference I noticed, as far as size goes, being the bill, which did not appear to be so strong"* (*British Birds* 36: 225).

All records:
>   1943   Wolsingham, five, 10th to 28 January (*British Birds* 36: 225)
>   2008   Old Cassop, juvenile, 17 September (*British Birds* 102: 590)

The second record came some 65 years later when a juvenile was photographed by Steve Percival as it fed on peanuts from a feeder in his garden at Old Cassop near Durham City on 17 September 2008. Present for little over an hour during the late afternoon, it was not seen subsequently (*British Birds* 102: 590). This bird came after an unprecedented influx of this species into the Northern Isles during the summer of 2008, which involved over 50 individuals between July and August (Harrop & Fray 2008), but just three were seen further south, including the bird in County Durham.

Distribution & movements
Two-barred Crossbill breeds in the taiga zones of both Asia and North America. Breeding within Europe is sporadic, even in Russia, though in common with other species of Crossbill it is prone to periodic irruptions west of its normal range and it occasionally breeds in Scandinavia following such events. It is an irregular vagrant to Britain with almost 250 records by the end of 2010, although a large proportion of these are linked to notable invasions in: 1987, 1990, 2002 and 2008, all of which were largely restricted to the Northern Isles. These typically occur in July and August, and are not usually associated with Common Crossbills *Loxia curvirostra*. In contrast to this, most records further south into England are detected during October to March and are usually of birds found amongst flocks of Common Crossbill and, as such, they are perhaps not directly linked to the invasions. All of the British records are thought to relate to birds of the Eurasian race *bifasciata*.

Common Crossbill (Red Crossbill)
Loxia curvirostra

A scarce and local breeder. Also an irruptive species, which occurs in variable numbers and is occasionally widespread across the county.

Historical review
The first reference to the species in Durham comes from Hogg (1827), who recorded that a *'hen bird'* was taken sometime *'before 1810'*, in a garden at Norton on Teesside. Some years later, on 14 July 1810, a male and a female were shot in a garden at the same place. Hutchinson (1840) said, *"flocks…not infrequently met with from*

July to September". In July of 1828 the species was said to be abundant in woodlands near Lanchester and birds were '*met with*' there until the following March. A female was killed there on 23 April and three pairs were said to be there on 25 April. Hutchinson himself stated that he saw five birds, including three young, fly over him in this area on 8 June 1828, perhaps indicating that breeding had taken place locally (Hutchinson 1840) but there was no conclusive proof of breeding in the county until some 27 years later.

In 1911, Witherby listed around some 20 English counties in which the species had nested to that time (Witherby 1911). The earliest breeding record came from Sussex in 1791, followed by Suffolk in 1815, and both Norfolk and Yorkshire in 1829. This paper stated that breeding had occurred in Durham in 1838, which presumably related to the birds in the Lanchester area. This would appear to be an incorrect date and should have read 1828. The evidence would seem to be inconclusive, but if Hutchinson's account was accepted as the first breeding confirmation for Durham, it would mean that Durham was, chronologically, the third English county to document breeding by this species. The first mention of the species being conclusively proven to have bred in the north east came from Hancock's catalogue of the birds of Northumberland and Durham (1874). This account tells of a nest found by Thomas Grundy on 24 February 1856 at Coalburns, near Greenside, in what is now west Gateshead; by 1 March the nest contained three eggs. Later that year, on 14 July, the same Thomas Grundy found another pair with four eggs in a nest in a plantation at High Marley Hill (Temperley 1951). While this is a late date for a nest with eggs, Crossbill is very much influenced by local conditions in respect of cone crops and can breed almost continuously for much of the year if conditions remain favourable. In June 1881, an adult and four juveniles were watched drinking at Spen Banks above Rowlands Gill. Nelson said of records of this species at Teesmouth, it is a "*rare visitant only, in winter*". In 1890 Robson, with reference to the Derwent valley, commented that "*many of these birds were in the lower valley*" and he took eight specimens in the neighbourhood (Robson 1896).

In 1909/1910 there was a major irruption of Crossbills into Britain. The first record during this influx came from Fair Isle on 23 June 1909, quickly followed by others from Shetland, Orkney and the Outer Hebrides; before the end of the month further records came from Durham and Merionethshire (Witherby & Alexander 1911). The first English evidence of this influx was noted in Durham, where birds were noticed in the east of the county during the last week of June 1909 and by the middle of July birds were fairly frequently encountered in the Wear valley near Durham City and Wolsingham. In the last week of July birds were also seen in the Upper Browney valley and the Derwent valley (Fawcett 1909) and from January to March 1910, about forty birds were present in the Darlington area (Forder 1910).

There was another large influx of birds in 1929/1930, following which breeding was again suspected. At this time Dr. H.M.S. Blair reported several in his garden at South Shields around 26 July 1930. A note published in *British Birds* (Temperley 1931) highlighted George Temperley's summary of records in Northumberland and Durham. His 'careful enquiries' revealed that there were definitely fresh arrivals of Crossbills in small numbers in 1930 at various sites in Northumberland and also at South Shields on 16 July, mid-August and mid-September, with sightings also in the Tees Valley on August 1st and 24th.

The next proof of breeding came in April 1943 when a nest was found at Hawkside, near Middleton-in-Teesdale. Howard Watson saw the first paired birds late in 1942, on 6 December–and continued watching Crossbills in Teesdale throughout the winter period. A tall spruce, in which he had often seen a pair, was blown down in a gale on 6 April. In a branch about three-quarter's up the tree and well out from the trunk he found a nest with one broken egg, which appeared fresh, stuck to the nest lining. There evidently had been other eggs and from Mr. Watson's detailed description of the nest it was no doubt that of Crossbills. On April 10th he saw a pair mating near this spot, but there was no further evidence of breeding (Watson 1944), although there were a few parties of possibly locally-bred birds in the area later that year (Temperley 1951).

At the mid-point of the 20th century, Temperley said that Crossbill was a regular winter visitor, with a few occasionally remaining to breed. During the next three decades, the status of the species could perhaps more accurately be described as an erratically occurring scarce visitor, as most years brought reports of very small numbers primarily to coastal locations during the late summer and autumn periods. There were some years, for example 1961, 1967 to 1969, 1971 and 1974 when none were reported but in keeping with the unpredictable nature of the species there were also 'invasion years'. In 1956 flocks of as many as 'almost 100' were recorded in County Durham up until late August. In 1962 there were numerous reports from inland and coastal locations from September until the end of the year. Evidence of immigration was apparent in 1963, with flocks of birds noted over the North Sea, observations coming from a ship several hours out of Newcastle on 7th and 22 August respectively.

Another influx took place in 1972/1973 with flocks in excess of 100 in Hamsterley Forest in both years during the January to April periods. Although breeding was not confirmed, with so many birds in suitable habitat it seems likely that it occurred. Another possible incidence of breeding came in 1960 at Middleton-in Teesdale, where birds were seen and heard throughout July. On 21st of that month, three birds were seen, one of which fed the other two, undoubtedly a family party. Furthermore, single family parties were noted in Weardale and in a plantation in the coastal area; while an immature, found dead at Thornley, in July may have been bred locally.

Recent breeding status

Crossbills, as a breeding species in County Durham, are sparsely scattered in both time and geography. They are effectively confined to the larger coniferous woodlands; although they will also breed in mixed woodlands, where significantly-sized, coniferous stands abut broadleaved areas. For example, during the early part of 1998, recently fledged juveniles were discovered in Thornley Woods, indicating local breeding. The species breeds early in the year (Gibbons *et al.* 1993), very often before birdwatchers and surveyors are active and consequently it is likely that some breeding records are missed. The main strongholds for the species in the county are Hamsterley Forest, The Stang in the south and Chopwell Woods in the north (Westerberg & Bowey 2000). The presence of the species in The Stang was not mapped during the *Atlas* work, and this was down to a lack of observer coverage, but it is clearly evidenced from other work in that area.

All of the chief strongholds have extensive stands of coniferous trees, in the main, these are spruce *Picea* spp. and larch *Larix* spp., but pines *Pinus spp.* are also widely present. The seeds of the first two of these groups, particularly those of the Norway spruce *Picea abeis*, form the favoured food of Common Crossbill (Clements *et al.* 1993). As the cone crops of these trees vary dramatically from one year to another and in some species, seeds are only produced every few years, the Crossbill population tends towards mobility and in times of poor food availability they leave home ranges in search of food (Gibbons *et al.* 1993). Birds which arrive in the UK as a result of these irruptions may stay on and breed for a year or two if conditions allow before they and their offspring return to more northern and eastern parts of Europe (Newton 2006).

Scattered pairs can be found at various locations around the county according to the influx status of the species. The breeding status is difficult to untangle from the migration strategy of the species, which occurs as an irruptive migrant to the British Isles (Newton 1972). It is likely that there is a small, residual, permanent presence of Crossbills in the county around some of the larger coniferous woodlands blocks, though without specific dedicated survey work at an appropriate time of the year, it is difficult to determine the scale of this. As indicated by the *Atlas*, a breeding population of 50 to 100 pairs for Durham might be a reasonable assumption in an average year, but the species may be virtually absent in years of poor cone production (Armstrong 1988-1993). *The Breeding Birds of Cleveland* survey work recorded it as being occasionally present at Wynyard as a possible breeding species, but probably as few as 'one or two pairs' were thought to be regularly present (Joynt *et al.* 2008).

In the 1988 influx, breeding was confirmed at Ravensworth and Chopwell, but nowhere else. In 'good' years, the species can be found breeding in many stands of conifers, from shelters belts to small conifer woodlands. In irruption years, hundreds of pairs can occupy large stands of conifers. For example, in the winter of 1990/1991, there was probably the largest irruption witnessed in the county and several thousand birds were considered present in Hamsterley Forest, with some individual flocks numbering many hundreds of birds (K Bowey pers. obs.). At this time, hundreds of birds were present across Gateshead with records of many flocks of between 20 and 60 from widespread sites. Many pairs bred in Chopwell Woods that year with possible breeding noted in a number of other localities including Gibside and the Derwent Walk Country Park (Bowey *et al.* 1993). These birds presumably originated from northern and Eastern Europe (Armstrong 1991) but several hundreds of pairs stayed in the county in the following season to breed. The population at this time was clearly very much higher than the figures outlined in the *Atlas* as a baseline breeding population. At the time, based on Northumberland data for the same period (Day *et al.* 1995), it is believed that more than 1,000 pairs may have been present in Durham (Westerberg & Bowey 2000). In Kielder Forest, Northumberland, unprecedented numbers were involved during this same influx and over 40,000 birds were estimated to have been present there (Jardine 1991), with thousands of pairs staying on to breed in 1991 (Kerr 2001).

The next good breeding year would appear to have been 1998, when relatively large numbers of birds were again present, particularly in Chopwell Woods and Hamsterley Forest. At the start of the year there were up to 200 birds at both locations, with several other areas holding 30 to 60 birds. Although flock size reduced after a January

904

peak, which is to be expected as birds paired up and spread around the forest; the position in March being of "*many pairs and small flocks*" (Armstrong 1999). During the early part of 1998, recently fledged juveniles were discovered in Thornley Woods, indicating very local breeding; there seems little doubt that many more pairs bred across the county at the time.

A similar situation prevailed a decade later with the greatest concentrations of up to 150 birds noted at The Stang in February 2008. Smaller numbers were elsewhere in suitable habitat and good numbers of birds probably bred. Nest-building was recorded from Brownberry Plantation near Grassholme Reservoir on 12 January; the early date and relatively remote location, particularly in the early months of the year, giving some indication of why breeding is so rarely fully documented (Newsome 2009).

The only perceived threats that might affect this species' population are the failure of spruce cone crops and habitat loss due to the clear-felling of conifers. The latter is an unlikely event given the continuing need for commercial timber production in Britain.

Recent non-breeding status

The incidence of breeding in the UK is inextricably linked to the presence or otherwise of continental immigrants and occurs wherever birds choose to settle having found a suitable cone crop. Migrants presumably from the Continent with others from parts of northern Britain, not infrequently occur in Durham in varying numbers, although the pattern of these occurrences is unpredictable. Such is the nomadic nature of this species, birds concentrating in parts of occupied woodlands wherever suitable cones are available, that migrants are difficult to distinguish from any established resident birds. Only records from obvious, non-breeding habitats can be readily identified as such.

Over the period 1980 to 2010, the established pattern of occurrence was for 'quiet periods' of anywhere between one and six years followed by an 'active' year or two, followed by another quiet spell with few records. The numbers of birds involved in influxes and the length of time between these is very variable and not easy to predict. In chronological review, the 'best years' were probably: 1982, 1990, 1997 and 2002. Other years, 1985, 1988, 2005 and 2009, saw some immigration on a more local scale, whilst in some years there were very few birds at all. For example, in 1989 there was but one record for the whole year, on 12 May, and just five reports in 1995.

Immigration can occur over quite a protracted period of time from May right through until December as birds work their way through the cone crops of northern Europe, then down into northern Britain, the rate and extent of southerly and westerly progress is determined by food availability. June and July are often the best months for coastal migrants, as in 1999, with post-breeding gatherings sometimes evident around the same time, such as the flocks of 30-60 at Black Bank Plantation, Dryderdale and Ruffside in 2005.

The 1997 influx is the one of the best documented in the county with the first evidence of this coming on 23 May when 10 birds were noted at Cleadon, but it was more typically late June and July before the influx gathered momentum. In July, parties of one to six birds were noted in coastal areas, while inland there were birds noted at 16 locations with a maximum of 28 at Medomsley on 9th. By August, numerous small flocks were reported throughout the county and the influx appeared to reach a peak in terms of numbers during September when 80-100 were gathered at Wolsingham and Hudeshope Beck in Teesdale. Visible migration reached a peak during October and November, when small parties of one to ten birds were noted flying south and west over a number of widespread locations both inland and at the coast. The total numbers involved must have been considerable as during this period, one observer logged 36 parties passing over Cleadon with a maximum of 22 on 21 November.

In 1982, the influx of birds did not occur until October, whereas in 1990 movement was evident from June onwards although visible migration accounted for but a small proportion of birds, which settled inland later in the autumn. During early September 2002, a low pressure system which moved north out of the Bay of Biscay on 8 September and across the north east of England the following day brought with it torrential rain and easterly winds particularly during the late afternoon. The consequence was many hundreds of grounded passerines, and on 9 September 2002, 315 were in the Hartlepool area alone. Many of these were in poor condition and were noted feeding on the ground, presumably on grass and weed seeds. In similar circumstances in 2009, birds fell prey to Kestrel *Falco tinnunculus* and domestic cat. In the 1990 influx a male was noted feeding on peanuts in a Darlington garden, whilst in 1997 migrants fed on thistle *Cirsium* sp. and sycamore *Acer pseudoplatanus* in the absence of their normal spruce cones.

Distribution & movements

This species breeds extensively in coniferous forest across North America, Europe and Northern Asia, with highest densities occurring in the taiga zone forests, but birds also occur much further south in the European alpine woodlands and even into Mediterranean pine woods (BWP). Populations are famously nomadic and irruptive; with large numbers of birds sometimes being displaced by the failure of cone crops, they emigrate and settle to breed in new areas. This phenomenon is a regular but sporadically observed one in the British Isles.

Parrot Crossbill
Loxia pytyopsittacus

A very rare vagrant from northern Europe and Siberia: seven records involving at least 56 individuals, most of which occurred during a notable influx into Britain in 1990.

The first records of Parrot Crossbills for County Durham were both from Hamsterley Forest, with a male from 23 December 1982 to 2 January 1983, and a female from 28 December 1982 to 2 January 1983. These birds, found independently by Dr. Colin Bradshaw and Martine Eccles, Dave and Richard Thomas, were loosely associating with Common Crossbills *Loxia curvirostra* in the area surrounding Grove House. These were part of an influx of around 100 birds into the British Isles at the time.

All records:

1982	Hamsterley Forest, immature male, 23 December to 2 January 1983 (*British Birds* 77: 556-557, *British Birds* 79: 579)
1982	Hamsterley Forest, female, 28 December to 2 January 1983 (*British Birds* 77: 556-557)
1983	Peepy Plantation, Washington, female, 24 March (*British Birds* 79: 579)
1990	Hamsterley Forest, ten, 7 November to 10 March 1991 (*British Birds* 86: 528)
1990	Chopwell Woods, 27 from 14 November to 26 December; male and female trapped 20 November; three males and female trapped 21 November; two males trapped 5 December (*contra BB*) (*British Birds* 86: 528)
1990	Castle Eden Dene, seven, 16 December to 10 March 1991 (*British Birds* 85: 550)
1990	Castle Eden Dene, additional party of nine, 19 December to 1 March 1991 (*British Birds* 85: 550)

There were several invasions of Parrot Crossbills into Britain during the latter half of the 20th century, most notably during the winters of 1982/1983 and 1990/1991; in both of these, birds were recorded in County Durham. The female photographed at Peepy Plantation, Barmston on 24 March 1983 was presumably a returning migrant heading back north.

Another much larger influx into Britain began in October 1990 and by April of the following year an estimated 264 birds had been recorded, most having departed by mid-March. The majority of birds made landfall on the east coast of England, with records spread from Northumberland to Kent, although the largest parties were to be found inland later in the year. During this time several large flocks were discovered wintering in Britain, though the exact number present at any one site was extremely difficult to estimate with any degree of accuracy.

Parrot Crossbill, Chopwell Woods, December 1990 (S. Westerberg)

Unsurprisingly, the largest numbers of this species recorded in the county were part of this invasion when birds were present in large numbers at three large wooded areas: Chopwell Woods, Castle Eden Dene and Hamsterley Forest. The numbers present made simple interpretation of the pattern of occurrence difficult; totals accepted by BBRC were clearly statements of the absolute minimum number of birds present at these sites. This is especially true of those present in Chopwell Woods with BBRC accepting a total of 27 individuals from this site, though some observers estimated that there may have been as many as 89 present on 20 November 1990 (Armstrong 1992). These birds were initially discovered drinking on a farmyard track at Heavygate Farm on 14 November 1990, although they fed through the extensive conifer plantations in the surrounding area. The eight birds trapped and ringed by Durham Ringing Group at Chopwell Woods between 20 November and 5 December 1990 are the only examples of this species to be trapped and ringed in County Durham (Bowey & Westerberg 1994).

Recent Breeding status

After the large influx of birds to Britain in autumn 1990 a number of birds stayed in the region into 1991, all of which were in suitable areas for breeding. In Hamsterley Forest, on 26 January 1991, at least three pairs were noted in extended breeding displays in the tall, more open canopy trees near The Grove, including locking feet in flights and courtship feeding. Interestingly they were seen to chase Common Crossbills well away from 'their' trees before commencing display (A. L. Armstrong pers. comm.). At least one male was still in the area until 10 March and it seems a reasonable assumption that birds may have attempted to breed locally (Armstrong 1992).

Distribution & movements

Parrot Crossbill breeds in Scandinavia and northern Russia, though it was recently discovered regularly breeding in Scotland (Summers 2002). Like many species of Crossbill it is subject to periodic irruptions in response to food shortages, although with much less frequency than the Common Crossbill and typically much later in the year, with October being the favoured month. Most influxes of Common Crossbill occur between June and August. Over 500 migrants have been recorded in Britain, the vast majority of which were in the invasion years of 1962, 1982 and 1990; there have been very few since. Following these invasions, sporadic breeding occasionally takes place, for example in Norfolk in 1984 and again in 1985, and probably in Yorkshire in 1983 (Lunn & Dale 1993).

Common Rosefinch
Carpodacus erythrinus

A rare migrant from Scandinavia and Eastern Europe: 22 records, involving at least 26 birds, recorded in May and June, and again from August to November, with one record of successful overwintering.

Recent Status

The first Common Rosefinch for County Durham came as recently as 1977, when one was found at Hartlepool Headland by Tom Francis on 19 August. A juvenile, it frequented back gardens in the residential parts of the Headland and was seen by many observers. A number of other scarce migrants were also present at Hartlepool Headland, following a period of light easterly winds.

From the first in 1977, a minimum of 26 Common Rosefinches have been recorded in County Durham. It is an irregular passage migrant in both spring and autumn with records equally spread between both seasons, though there is a distinct peak in records in late May and early June, with autumn migration much more protracted from August until November. The paucity of records from County Durham is difficult to explain as the species is annual in small numbers in both Northumberland and Yorkshire, and it has bred in Yorkshire on at least one occasion, though the drab plumage of non-adult males may mean some go unnoticed.

At least 14 Common Rosefinches have been recorded in spring in County Durham since 1977, all of which have been in May and June, predominantly at coastal localities. The majority of records relate to birds in their first summer, often located by their simple but distinctive song. Notable late May influxes occurred in both 1998 and 2008, with these years accounting for more than half of all spring records. The peak count is of at least three singing males at North Gare, Teesmouth on 29 May 2008. Extreme dates are 19 May 1996 at Hartlepool Headland and 18 June 2009 at Elemore Golf Course. A first-summer bird at Hartlepool Headland on 8th and 9 June 1992 was the first to be recorded during spring, only the second county record. Another first summer at Hartlepool Headland on 19 May 1996 is the only other spring occurrence from this locality. A period of light easterly winds and coastal mist during late May 1998 brought a record arrival of at least four birds, with at least three first-summer males singing in the Marsden area over 30th to 31 May, and another at nearby Mere Knolls Cemetery, Seaburn on the former date. This same weather system brought a numbers of scarce migrants to Durham, including the county's fourth Woodchat Shrike *Lanius senator*. Despite this relative glut of records, a long wait ensued until the next, when a first-summer bird visited a garden on the Shearwater Estate, Whitburn on 27 May 2005. Three years later weather conditions similar to those of 1998 brought an arrival of four birds on 30 May 2008, but this time to the south of the county. The influx included three first-summer males at North Gare, Teesmouth, all of which were in song.

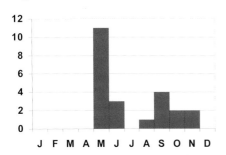 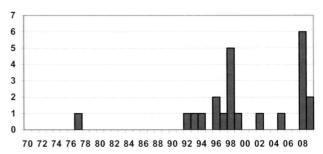

Common Rosefinches in County Durham, 1970-2009

Four birds have been recorded inland during spring, all singing males. The first was at Chopwell on 28 May 1993 (the third for Durham and first inland); followed by a first-summer male at Far Pasture, in the lower Derwent valley on 7 June 1998; an adult male on the Long Drag, Teesmouth on 30 May 2008 and an adult male at Elemore Golf Course on 18 June 2009. The latter two birds are the only red males to have occurred in the county to date.

A total of 12 Common Rosefinches have been recorded during autumn in County Durham since 1977, all of which have been juveniles between August and November, with September producing the majority of sightings. Most records are from coastal localities, in particular Hartlepool Headland and the Marsden/Whitburn area. Typically they have been found amongst 'falls' of continental migrants during periods of onshore wind and rain. Extreme arrival dates are 19 August 1977 at Hartlepool Headland and 17 November 2009 in Whitburn Village, although this last bird may have arrived much earlier and gone undetected for some time, until it was attracted to a garden feeding station.

The locality with the highest number of autumn records is Hartlepool Headland with four single birds on: 19 August 1977, 12th to 13 September 1996, 24 September 1999 and 22 October 2009. Elsewhere, single coastal migrants have been recorded at: Ryhope on 16 September 1994; Hartlepool Golf Course on 5 October 2002 and two at Marsden Quarry on 7 September 2010. This latter occurrence is the only time when more than one Common Rosefinch has been present at the same location during autumn.

Further inland, single juveniles were at Charlton's Pond, Billingham on 25 September 1997, Cleadon Village on 1 November 2008 and one that was trapped and ringed at Far Pasture, in the lower Derwent valley on 5 September 2010. This latter bird is the only example of this species to be trapped and ringed inland in County Durham; it was the third inland record for Gateshead and the second at this site.

A juvenile in gardens in Whitburn village from 17 November 2008 until 15 January 2009 was the first record of this species attempting to overwinter within County Durham, and one of only a handful of such occurrences in Britain as a whole. It had presumably arrived earlier in the autumn, and was judged to be a different individual to that present in nearby Cleadon village on 1 November 2008.

Distribution & movements

Common Rosefinch breeds from central Europe through Scandinavia across northern Asia as far south as the Himalayas and south western Asia. Strongly migratory, the species spends the winter throughout much of southern Asia, from the Indian subcontinent through southern eastern China to Burma and Indochina. In Europe, the species' stronghold has always been in Russia and the Baltic States, though has it undergone a considerable range expansion since the 1930s and now breeds as far west as south western Norway, Germany and the Netherlands. Breeding was first confirmed in Scotland in 1982 when a nest and eggs were found in Highland (Mullins 1984), though it was not documented again until 1990 when a pair fledged four young in Sutherland. A sizeable influx during 1992 saw the first confirmed breeding in England (Wallace 1999). Since then further instances of breeding in the British Isles have been sporadic. A scarce migrant in both spring and autumn, around 150 Common Rosefinches are recorded each year in Britain, which represents a significant increase on the 37 birds per annum during the 1970s. The record influx of 1992 saw a total of 248 birds across Britain (Fraser & Rogers 2006).

Pine Grosbeak
Pinicola enucleator

Sponsored by

A very rare vagrant from Scandinavia or Siberia: one record.

The sole record of this species in County Durham is of a 'female-type' bird shot at Bill Quay, Pelaw in east Gateshead, *"sometime before 1831"* (Selby 1831). Initially listed in Selby's catalogue, as having been in the collection of a Mr. Clapham, the specimen was later passed to the late William Backhouse, and eventually the specimen found its way to the Hancock Museum.

Writing in 1877 (*Zoologist* 1877: 242), J. H. Gurney reviewed all the British records of this species, and remarked that he had been loaned the specimen by Backhouse, describing it as being of a *"greenish yellow colour and in moult"*, and he expressed no doubt as to its identification. There was some initial discussion regarding the locality of Bill Quay, and as to whether the bird might have been an escaped cage bird bought in a local market, which were often referred to as 'quays' at the time, though the locality was later found to refer to a small boat mooring close to a chemical works on the banks of the Tyne. During a more extensive review of British records in 1890 (*Zoologist* 1890: 125) Gurney found only four to be acceptable *"as probably the most authentic specimens"*, including the bird at Bill Quay in 1831. This record is regarded as the first fully acceptable British record, pre-dating a first-winter female shot at Harrow-on-the-Hill, Middlesex some time prior to 1843 (Palmer 2000). Examination of the specimen, which still exists in the NHSN collection in 2011, indicated that the bird was a first-winter male in moult.

Pine Grosbeak, Bill Quay, prior to 1831 (K.Bowey, courtesy NHSN)

Distribution & movements

Pine Grosbeak breeds in the taiga zone of both Eurasia and North America. The nominate race is a scarce breeding species in northern Scandinavia, though is much commoner further east into Siberia. It is largely resident, though like many boreal species is subject to periodic irruptions further west, often in large numbers. These movements are thought to have their origins in western Siberia and regularly reach southern Scandinavia in response to food shortage, though the species seems very reluctant to cross the North Sea and remains an extremely rare vagrant to Britain. Currently, there have been just 11 British records; the most recent in East Yorkshire in November 2004. Despite the passage of more than 180 years, there is still only a single county record.

Bullfinch (Eurasian Bullfinch)
Pyrrhula pyrrhula

A widespread, common resident, with small numbers of passage birds occurring on a regular basis.

Historical Review

In the mid-20th century, this species was a common resident; Temperley referred to a steady increase in numbers "*during recent times*" (Temperley 1951). A century earlier, Hutchinson (1840) had reported that the species was thinly scattered and that seldom more than three or four were seen together. Hancock (1874) said it was a 'constant resident' but not abundant anywhere, except in the Derwent valley, where it is 'somewhat common'. As confirmed by Thomas Robson, who described it as "*to be found nesting on both sides of the Derwent*" (Robson 1896). Temperley noted a marked decline through the earlier part of the 20th century, which he attributed to the actions of finch-trappers - numbers of Bullfinch were believed to be very seriously impacted upon by trapping and bird-limers, until those practices were made illegal. Whilst there is little in the way of "facts and figures" regarding the population in the 20 years immediately post-Temperley the general impression was that the population was increasing in County Durham with one report of 10 pairs in one square kilometre at Rowlands Gill in 1960 and comments of it "*increasing*" in 1963, 1967 and 1969 (Bell 1964, Bell & Parrack 1968, Parrack *et al.* 1972). This is somewhat at odds with the national picture over this period, which showed declines commencing across Britain, despite some previous increases as predation was reduced during the fall in bird of prey numbers (Sharrock 1976, Mead 2000).

In early 1970s, the species was described as being widespread in small numbers in woodlands, rarely more than ten together, but that the population was increasing generally, particularly around Hamsterley Forest. After a short cold spell late November 1973, none were noted around Castle Eden Dene, where the species had been plentiful a month before; indicating local dispersal or major mortality. Coastal passage was occasionally suggested by birds on the coast, such as the three at Seaham on 21 October and five or six birds at Marsden Quarry on 28th in 1979.

Recent breeding status

The *Atlas* map (Westerberg & Bowey 2000) indicated that Bullfinch was widely distributed across the county, if a little under-represented in the south and east. It is sometimes quoted as being, "*Present everywhere, but rarely seen*" (Gibbons *et al.* 1993). The species' retiring habits and, in particular, the subdued song of the male means that they can often go unrecorded at some sites, and this may lead to something of an under recording during casual bird watching visits and even focused survey work. Recent reports however, suggest it is very widespread indeed with records in 2007 coming from almost 200 locations, ranging from the coastal denes to as far west as High Force in Teesdale and Rookhope in Weardale.

The species is found in most wooded areas of the county, regardless of the size and species composition of the woodland. The exception to this are the densest of coniferous blocks (Heinzel *et al.* 1995), which are relatively abundant in some parts of the county, especially where landscaping as part of the reclamation of former coal extraction areas reached maturity during the late 1990s and the 2000s. Nonetheless, in such situations, the species does utilise the edges of such habitats, particularly where they abut or merge into more natural linear features such as hedgerows, wooded riverside, streams and burns.

Bullfinches are quite frequently found in suburban habitats, and it has become a regular garden visitor to some areas; birds also occur in farmland, albeit at much lower densities than in woodland or in less intensively managed habitats (Westerberg & Bowey 2000). The first breeding record at Shibdon Pond was reported in 1988, suggesting that the species was expanding its range from local woodlands into scrub habitats around that time. The *Atlas* indicated that there were significant concentrations of the species around the Derwent valley woodlands, and also around the Durham City area, both up- and down -stream along the River Wear. Both of these areas contain high-quality, lowland woodland. The *Atlas* map also indicated that in Durham, Bullfinches are not birds of high altitude and, taking as an example the population along the well-wooded Derwent valley, as this rises in altitude the concentration of occupied territories declines along its length (Westerberg & Bowey 2000). While at these higher sites there is likely to be reduced observer coverage, there is probably a genuine effect on population that is related to elevation.

Due to its retiring nature, accurate population figures for this species can be difficult to determine. Detailed survey work in the mixed scrub and secondary woodlands around the area of Brasside Ponds in the early 1990s, indicated a local population of nine pairs (Armstrong 1988-1993). Using this figure, and extrapolating from it across the number of occupied tetrads in County Durham, led to a population estimate of around 1,000 pairs for the county (Westerberg & Bowey 2000), though this may be something of an underestimate. In the south east of the county, it is a sparsely distributed breeding bird over much of what was north Cleveland (Joynt *et al.* 2008). In this area, it occurs mainly in the more heavily wooded areas to the west and north of Stockton, such as Urlay Nook and along the Castle Eden Walkway, as well as around the Wynyard Estate and Crimdon Dene. There are believed to be around 70-80 pairs in this area (Joynt *et al.* 2008).

Nationally, Bullfinch is an amber-listed Species of Conservation Concern, as a result of concerns about recent declines in the breeding population; estimated at 158,000 territories in 2000 (Baker *et al.* 2006). The Breeding Birds Survey index between 1994 and 2006 (Raven *et al.* 2007), indicated a national decline of around 28%. This is not a phenomenon however, that is particularly evident in the context of Durham. Local studies have shown that the species is still common and widespread, for example 14 pairs were noted in the relatively small area around Hetton and Houghton in 2000. A most unusual breeding record occurred in the lower Derwent valley in 1995, when a pair of birds successfully fledged young, on 25 June, from a nest located in a honeysuckle behind a wooden shutter just 1.5m from the main entrance of the Thornley Woodlands Centre (Armstrong 1997).

Recent non-breeding status

In winter, this handsome species enjoys a widespread distribution across the county and during the winter of 2007, it was recorded at almost 200 locations, from easterly locations such as the coastal dene woodlands at Hawthorn and Castle Eden, to westerly outposts such as Deepdale Wood and Derwent Reservoir. Birds relatively rarely penetrate the extreme west of the county, though three birds fed on heather seed at Grains o' th' Beck virtually on the county boundary with Cumbria in early 1991; a lone bird was at High Force during the winter of 2007 and eight were noted at The Stang in December 2009, at almost 300m above sea level.

In essence, Bullfinches are found across the county in woodland and parks, along hedgerows, sometimes joining tit parties and increasingly in recent years, feeding at garden feeding stations. In the mid-1980s the species was described as widespread but nowhere abundant. They are at their most numerous in heavily wooded area such as the valley woodlands of the Derwent and the mid-Wear, though even in such areas winter flocks of more than 10 to 12 birds are only occasionally encountered. The bird feeding station at WWT Washington is the main exception to this observation. For some years, this site has attracted the county's largest gatherings of this species. Indeed, in the local context, this is where the species' use of bird feeders appeared to start during the late 1980s. By 1998, a site record count of 24 birds was noted though by 2008 this had risen to as many as 51 individuals utilising the feeding station during the early months of the year. Given that the actual number of birds visiting a feeding station at any one time is likely to be an underestimate of the total numbers using that facility, then clearly there is a healthy population in this part of the Wear valley. In like fashion, during the winter of 2000/2001, 62 bullfinches were caught and ringed at the feeding station at Thornley Woods, with a further six being noted without rings after this work, being indicative of the number of birds that can be present where high quality habitat abuts clustered food sources.

The BTO organised Garden Watch scheme over the period 1996-2011 showed a distinct peak in numbers in June, presumably arising from post-breeding dispersal, rather than the perhaps more expected influx in winter in response to harsher conditions.

In 2006, birds were noted feeding on dandelion *Taraxicum officinale* seed heads and ash *Fraxinus excelsior* keys in May, whilst two were seen feeding on rowan *Sorbus acuparia* berries on the central reservation of the A690 at Gilesgate, Durham. Other interesting observations of birds taking advantage of available food sources came from Seaton Pond in December 2007, where up to 12 birds were seen feeding on the unharvested seed heads of a rape crop. Likewise, up to 45 birds were feeding on heather seed at Sharnberry in 2009, close to Hamsterley Forest, where there is a large breeding population.

Even today, this species is sometimes illegally trapped by finch keepers, with known incidents of this occurring in the early part of the first decade of the 21st century in the Derwent valley. Other than hard winter weather and the effects of cold springs, to which it is prone (Gibbons *et al.* 1993), there are few perceivable threats to Bullfinches in County Durham.

Northern Bullfinch
Pyrrhula pyrrhula pyrrhula

Historical review
The 'Northern Bullfinch' is a race that occasionally appears as a passage migrant and rarely during winter, it was mentioned by Temperley (1951) as having been suspected of occurring but there were no known specimens at that time.

Recent status
While a single bird at Marsden Quarry on 19th to 20 October 1975 was described as a 'possible', the first definitive sighting of this race was an individual at Mere Knolls Cemetery on 28 October 1988. There were one or two birds at Marsden Quarry on 21st to 22 October 1990 and a single bird that showed the characteristics of the northern race at Shibdon Pond on 16 November 1992. A notable influx took place during the autumn of 1994, when one to three birds were recorded at Dawdon Blast Beach, Hartlepool and Marsden Quarry between 16th and 19 October, but judging from the number of migrants recorded elsewhere in Britain, it seems highly likely that others went unrecorded.

The first spring record related to a female that was at Whitburn on 15 March 2001. A few years later a number of Northern Bullfinches were recorded as part of the major influx to Britain during autumn 2004. One to three birds were noted at a number of coastal locations between 16 October and 8 November. The first records of this national influx occurred in Northumberland and County Durham on 16 October, with a wider arrival on 17th, with further influxes through November and birds present through the subsequent winter. The largest numbers on the British mainland were recorded from north east England, especially Northumberland, where most records came from the Farne Islands and Holy Island. Many of the birds involved in this influx were identified by their distinctive 'toy trumpet' calls and at least some were believed to have originated from populations in the far north east of Asia. Returning birds were recorded in the spring of 2005; with three at Marsden Quarry on 28 March, a single at the same site on 2 April and two birds at Sunderland North Dock on 4 April. A passage migrant in Whitburn in February 2005 was only the second record there in 15 years.

In 2006, four birds that showed the features of this race were reported from Belasis Technology Park, Billingham in mid-February. Another significant influx was apparent in 2009 with one noted at Whitburn on 15 November, two at Dorman's Pool on 21 November and one at Brasside Pond on 27 November. In addition, at least one male and a probable female were present at Middleton One Row from 24 November 2009 into 2010. In that year, a female was at Elemore Woods on 3rd and 19 December 2010, a male at Whitburn on 5 December and another at Brasside Pond on 27 December. One aberrant bird showed a large amount of white on the mantle and wings, but gave the characteristic call of this race (Newsome 2009).

Distribution & movements
Bullfinches breed over the much of northern Eurasia; they are partial migrants with much of the population wintering within the breeding range. The more northerly birds of the nominate race occasionally behave irruptively and in such situations may occur in some considerable numbers in Britain (Parkin & Knox 2010).

There is little ringing data of interest pertaining to this species in Durham. A bird ringed in the Ravensworth Estate on 16 March 1997 was recovered a year and a half later, in August 1998 at Stanley, a distance of around 7km. One ringed at Low Barns on 25 August 2007 was found dead after flying into a window at Witton-le-Wear three kilometres away on 4 March 2008; the fact that this was notable, perhaps illustrates the species' tendency to sedentary behaviour. Other recoveries of locally ringed birds tend to be within two to three kilometres of their ringing sites.

Hawfinch
Coccothraustes coccothraustes

An uncommon local resident and a scarce winter visitor and passage migrant.

Historical Review

Despite its striking appearance and large size, the Hawfinch is poorly monitored and rarely seen, owing to its rather wary nature, weak song and low population levels (Snow *et al.* 1998). Some sources suggest that this species did not colonise Britain until the nineteenth century, and although early modern British writers knew the species, it is possible that it went unrecorded for long periods of time (Marchant *et al.* 1990, Holloway 1996). Snow *et al.* (1998) stated that the species was believed to have spread north and westwards through Britain during the nineteenth and early twentieth century; breeding being first observed in Britain in the early nineteenth century (Langston *et al.* 2002). Indeed, as one contributor to the Darlington-printed *Northern Echo* in 1890 suggested, *"these birds of late years have become much more numerous"*, although the writer did not provide any records for County Durham and suggested that *'if they are rare anywhere in England, it is in the north, near the borde*r'. That said, the fact that this popular newspaper was carrying an article devoted to Hawfinches suggests a certain public familiarity with the species.

A possible factor for this sudden range expansion was that until the mid-nineteenth century Hawfinches had been widely persecuted for their supposed depredations upon fruit and garden peas (Cocker & Mabey 2005). Given that fruit growing has never been widespread in County Durham, it is unlikely that Hawfinches would have fallen foul of northern horticulturalists. However, the species and its eggs was a prized specimen for the growing band of nineteenth-century naturalists, who could count among their number both 'gentlemen' and young boys. Individuals wishing to add Hawfinch eggs to their collection could purchase them from numerous egg-dealers, although the high price of 10*d* each surely reflected the breeding scarcity of the species and tempted only the wealthiest of collectors (*Routledge Magazine for Boys* 1868). Given the vogue for collecting, it is not surprising then that the first record of the species in the county, in 1831, was of two specimens 'some time ago shot near Stockton on Tees' recorded by Selby. Hogg, in 1845, contradicted Selby, by placing the Stockton birds as being shot at Ormsby - just outside of the Durham area. Both Backhouse in 1834 and then Hutchinson in 1840 however, described the species as being present in the county, but 'rare' (Temperley 1951). Writing in 1874, Hancock confirmed this scarce status, noting that he had three specimens in his collection that had been shot at Streatham Park during the winter of 1837 (Hancock 1874). As for breeding, Hancock (1874) was of the belief that it had not been confirmed in the county. He may have been correct, although the record of a juvenile that killed itself against conservatory glass in Gibside in August 1879 tentatively suggests possible breeding in the lower Derwent valley; it was not until 1884 that this was proven, when a clutch of two eggs was taken at Winlaton. Coincidently, in the same year, breeding was also confirmed for the first time in Northumberland (Frankis 1997, Temperley 1951). The following year, and perhaps indicative of a real increase in numbers, three more clutches of eggs were 'taken' in the Winlaton area, and a pair nesting in Axwell Park was predated by Magpies (Bowey *et al.* 1993). By 1905, Temperley noted that the species had spread to Ravensworth (Temperley 1951). In the south west of the county, Backhouse suggested that Middleton-in-Teesdale was 'another centre of distribution' in the 1880s, and the newly-fledged bird that dashed itself against windows in Bishop Auckland in 1898 prompted the finder to suggest it had *"evidently bred in the neighbourhood"* (Temperley 1951). In south east Durham, from 1897 there was believed to be *'a large colony in Cleveland, where some twenty or thirty pairs breed'*, although the exact location remains unknown (Mather 1986). More precisely, by the early part of the 20th century Milburn described breeding at Yarm, Castle Eden, Wolviston, Wynyard and Dalton Piercy (Temperley 1951). There is good evidence to indicate that the species underwent a large expansion of range and numbers, which began somewhere before the middle of the 19th century. It bred on the northern boundary of Yorkshire with Durham in 1897 (Holloway 1996) and Fawcett (1890) wrote *"In various parts of Durham, where it used to be only a rare visitor, it is now resident and breeding"*.

During the inter-war years, there is a complete hiatus of records. This may reflect the demise of the collector-naturalist and a downturn in public interest in natural history rather than any real change in the status of Hawfinch.

According to Temperley (1951) the species remained "*a resident of somewhat local distribution, breeding in woods, parks and gardens*". He added that a pair had 'recently' bred within the South Shields area and suggested that due to its "*wary disposition*" Hawfinch was "*frequently overlooked*". Post-Temperley, this remark remained valid, as there were just six records from three sites. In 1959 and 1960, the species was recorded at Ravensworth in the Team valley (Grey 1960, Coulson 1961); a pair and a juvenile were reported from Chester-le-Street during the summer of 1960 and a pair was noted from the same site in 1964 (Bell 1965).

Recent breeding status

It is something of an understatement to say that the precise breeding status of this species in the county is imperfectly known and it is almost certainly, quite considerably under-reported. The singing male is a very quiet and stealthy beast (Gibbons *et al.* 1993) and it is not well known to bird watchers. The map as illustrated in the *Atlas* (Westerberg & Bowey 2000) bore testament to this. It showed a very sparse and scattered county distribution with little obvious pattern, other than a linkage to areas of high woodland density, which in the county tend to be concentrated along the main river valleys.

Birds are found in a variety of woodland types (Gibbons *et al.* 1993), but even in 1890, commentators were sure that the species '*are frequenters, in summer time, of park-like meadow land*' and that '*yew trees anywhere are a source of much attraction for them*' (*Northern Echo* 1890). Recent research by Frankis (1995) confirms the validity of this observation, suggesting that the species tends to favour woodland with an ornamental aspect. Such habitat in the county tends to be associated with old estates, churchyards and even public parkland. It is here that they can find one of their favoured spring foods, the growing buds of yew *Taxus baccata* (Milton 2005). Considering the secretive habits of the species, Hawfinch has been surprisingly well recorded in Gateshead, providing the first breeding record in the 'modern' era, when an adult with three young were seen at Hamsterley Mill in 1978. Records here suggest a small resident population, although it is at Ravensworth where the population has been at its highest during the 'modern' era of recording; breeding evidence including individuals observed singing and indulging in courtship behaviour, with a newly fledged bird ringed in 1985 and a female with a brood patch caught in 1991. Birds breeding at this private site are subject to little disturbance and may be more successful than at other sites. Over the period 1983 to the mid-1990s, breeding was proven at a number of sites in Gateshead including: Chopwell Woods; the Ravensworth Estate; Ryton Willows (two sites); Thornley Woods; Strother Hills; Victoria Garesfield and Washingwell Woods. There is debate as to whether breeding Hawfinches are more susceptible to disturbance than other carduline finches (Gibbons *et al.* 1993). In addition, '*Hawfinch colonies*' are 'worked' by Jays *Garrulus glandarius*, whilst grey squirrels *Sciurus carolinensis* are known predators of nests. Both of these species are heavily controlled on the keepered Ravensworth Estate (Langston *et al.* 2002, Gibbons *et al.* 1993, Milton 2005).

The *Atlas* suggested a distinct northerly bias to the distribution, with clumps of occupied tetrads around Washington, Chester-le-Street and Gateshead, although breeding was only confirmed at two sites during the *Atlas* survey period (Westerberg & Bowey 2000). Work in the Gateshead area suggested that up to 15 pairs breed there annually (Bowey *et al.* 1993), although this estimate was made when there was still a sizeable Ravensworth flock. As the species is so elusive it is difficult to give a precise number for a county population estimate, but based on the number of sightings registered annually, the numbers of breeding pairs is not large. Extrapolating the Gateshead estimate to a county footprint is problematic. The *Atlas* indicated that 20-25 pairs might be a reasonable 'lower limit' for a county population estimate, with perhaps as many 40 pairs in good years. Although this is comparable with the Northumberland estimate of 31 pairs (Frankis 1995), in reality, the Durham figure is probably still on the conservative side. Breeding must take place unnoticed and increases in numbers at selected sites suggest a greater population. For instance, in recent years the species has had a regular presence in the breeding season at Croxdale Hall, indicating that breeding might be taking place undetected. In 2008 the first confirmation of breeding for 22 years was recorded at Middleton-in-Teesdale where a pair with three recently fledged young was seen in July. Gibbons *et al.* (1993) thought Hawfinch had undergone a '*marked but uneven reduction in numbers*' especially of populations in southern England. Recent calculations by Langston *et al.* (2002) support this. This suggests that UK numbers of Hawfinches have declined by 37-45% during the period of 1985 to 1989, and 1995 to 1999. Although the same analysis estimates the population trend for Durham to be within the value of a plus 8% and minus 22% over the same period, the wider 20-year estimates (1975/1979 to 1995/1999) for the county shows an increase of between 16 and 23%. Given its elusive nature, potential influences on Hawfinch numbers are not fully understood, but loss of habitat and predation have been suggested as factors (Langston *et al.* 2002).

Recent non-breeding status

In the 'modern era' of recording, from 1970 onwards, the first records for the species are of five records in 1973 concerning birds at Wynyard, the outskirts of Durham City and a party of four to six at Blaydon, with breeding reported at Thorpe Thewles. The compiler of *Birds of Durham 1973* (Unwin 1974) wondered whether due to the lack of reports *"it has been absent all this time"* in the county. That said, the following year it was thought that the species was *"gaining a stronger hold"* in the county, despite fewer reports. A single noted at Darlington was the first seen by that observer since 1955, and one to two were present throughout November and December into 1975, leading to a maximum count of 14 in February in the Cox Green/Washington area. A slow increase is then perceptible with one to two birds present in suitable breeding habitat in the Derwent valley, Wynyard, Ravensworth, Cox Green and Darlington in 1976 and 1977. It was mapped as being present during the breeding season in upper Teesdale in the early 1970s (Mather 1986). This species was described in Bradshaw (1976) as shy and rare, but most often seen in hazel scrub in upper Teesdale, an unusual habitat and location for this species. Further increases were apparent in 1979, when up to six were noted in the Cox Green/Penshaw Hill area in May, rising to 12 birds in June-July, including fledged young, presumably from the nest found at the same site in May.

The next decade witnessed a sizeable increase in population with flocks well into double-figures, such as the 16 birds at Chopwell Wood in March 1980; while a pair ringed at Beamish Hall in May 1980 included a female with a well-defined brood patch suggesting further evidence of breeding. This measured increase continued with Hamsterley holding a party of five throughout January 1983 and breeding was confirmed in the lower Derwent valley with two young fledged in Thornley Woods. At Ravensworth, ringers caught an unprecedented nine adults on three dates between late May and early June. By the mid-eighties the species appeared to be quite well distributed across the county, although clustered at selected sites, predominantly in the western Gateshead/Lower Derwent valley area and at Hamsterley Forest, Barnard Castle, Finchale and Croxdale. In 1988, a dramatic increase in the reported numbers was recorded, with several parties of two to six being noted at a number of Gateshead sites, Frosterley and Horsley Hop, followed by an exceptional count for Durham of up to 70 birds in the Ravensworth Estate. This flock was present from early March to May, although by 1989 it had declined to a maximum of 40 individuals (Bowey *et al.* 1993).

Numbers at Ravensworth continued to fall to a maximum count of 28 birds in 1990, with small numbers present again in the Derwent valley, although it is possible that these were dispersed birds from the Ravensworth flock. The continuing presence of one to two birds at several Derwent valley sites, such as Victoria Garesfield and Gibside, up to the mid-nineties suggested a small increase in numbers, especially as a small flock numbering up to 14 birds was then present in the Croxdale area of Durham in early 1994. At Ravensworth, there was a maximum of 45-50 birds in March 1992, with numbers declining to between 10-20 birds by April and May. Elsewhere in the county new sites to host the species included Hetton Bogs, in 1992, and in 1993, Lumley Castle. An accelerating decline however, was apparent from 1995 onwards, as whilst the Ravensworth flock reached a maximum count of 25 in February 1995 it had dipped to just 15 by March 1996, although 19 were present in December 1997. Fewer efforts were made to study birds at this site at this time and it is possible that more may have been present. Elsewhere in the county, only a count of seven at Gibside Hall in January 1996 and six at Croxdale in February 1997 were noteworthy; the remainder of sightings concerned one to two birds in the Derwent valley and at Hamsterley, although birds at 'new sites' - Witton-le-Wear, Langley Moor and Page Bank were notable and suggestive of expansions in range. Symptomatic of possible small fluctuations in breeding success, counts during 1998 were slightly higher than normal, with three to five present at Victoria Garesfield, Gibside Chapel, Croxdale Woods and Hamsterley at various times of the year.

A party of 10 birds flying over Victoria Garesfield in January 2000 briefly suggested an upturn in numbers, but in fact, this remains the largest count of the species to the end of 2010. Through the 2000s the species continued its frustratingly elusive status. It followed a pattern of sightings, with singles or pairs of birds and occasional small gatherings being noted on single dates, largely in the first quarter or last few months of the year, most often at favoured sites with occasional sightings elsewhere. Singles or pairs of Hawfinches were consistently recorded at both Wolsingham and then Croxdale Hall (from 2004 onwards), with four in January to March 2004 at the former sites and nine in January 2006 at the latter location being exceptional counts. Away from these locations, Hawfinches were also recorded at Lambton/Chester-le-Street, a number of sites in the Derwent valley, mainly Victoria Garesfield and Chopwell, as well as at Ryton Willows Hetton Bogs, Durham City and Castle Eden.

Ravensworth has been undoubtedly the main site for Hawfinches during the last 30 years in terms of sheer numbers and breeding population, but the last record received from this site was of a single male trapped in 2000. Although a reduction in observer visits contributed to this decline in records, there had been a reduction in numbers there since the peak of up to 70 birds were recorded in 1988. As Langston *et al.* (2002) point out, Hawfinches are noted for their sporadic occurrence at different sites in winter. Hence, the early months of the year generally produce more records and these may indicate dispersals of birds from winter-feeding areas to traditional spring haunts such as the 15 over Lockhaugh Farm on 15 February 1989. By the mid-2000s, the only site where the species was seen with any regularity was Croxdale Hall. In 2006, birds were present there throughout January and up mid-February, with a maximum count of nine individuals, but none were recorded there in 2010 or 2011. It is possible that other sites may hold similarly-sized winter flocks, but as Frankis (1995) suggested, with respect to Northumberland, such locations may be private and there is reluctance from observers to submit records from such sites. The provision of artificial foodstuffs occasionally tempts wintering birds and single Hawfinches were noted at the Low Barns feeding station in 2008 and 2009.

As a migrant in Durham, Temperley noted no evidence, but the occasional appearance of birds in the 'modern' era at coastal stations reinforces the theory of small scale immigration. Records of possible autumnal migrants in the county are rare and concern individuals at Marsden in October and November 1997, Fulwell in November 2000 and then a passage bird at Hartlepool in October 2005. At the same site, a migrant was seen flying inland on 10 October 2010, with another migrant at Hawthorn a day later. The autumn of 2011 produced at least four birds that could be classed as coastal migrants, including three in the Whitburn/Marsden area. There are even fewer spring records. An adult female was ringed at Hartlepool on 19 April 1962, although it was later found dead at the same site the next day (Bell 1962). Birds were seen at Sunderland and Boldon Flats in March and April in 1976, with an individual also found at the latter site on 9th to 10 February 2006, and then in late April of that year, a bird was at Whitburn. A migrant also visited a garden on the Shearwate estate, Whitburn on 3 April 2011.

The sudden appearance of the large flock at Ravensworth in the late 1980s is indicative of possible longer distance movements within Britain. Such large flocks and dramatic fluctuations in numbers are not without precedent elsewhere. At East Wreatham Heath in Norfolk, during 1973 to 1985 up to 183 birds were present, and a regular gathering at Blenheim Palace in Oxfordshire saw Hawfinch 'bird-days' drop by 90% over the period 1984 to 1990. Explanations for this have included fluctuations in seed crop of certain tree species, such as hornbeam *Carpinus betula* (Wernham *et al.* 2002). This tree species is absent at Ravensworth, and another explanation that might have more credence is the suggestion that the high storms of 1987, which caused massive loss of mature timber in southern England, may have contributed to the decrease in Hawfinch numbers triggering some dispersal of birds away from traditional sites; hence the sudden appearance of birds at Ravensworth (Langston *et al.* 2002).

Distribution & Movements

This species is a local breeder over much of Europe, eastwards to Siberia and eastern Asia. Northern populations of continental Hawfinches are known to make southwest movements, wintering in northern Italy and southern France, with the whole of the southern Russian population deserting its breeding grounds for warmer climes (Hagemeijer & Blair 1997). In Britain, it appears to be largely sedentary, though small numbers of autumn and spring migrants, with occasional winter influxes, occur along the east coast to the northern isles (Wernham *et al.* 2002).

The regular presence of Hawfinches at Ravensworth prompted concerted ringing efforts, especially during 1988-1991, to ascertain possible origins and movements. A total of 32 adult birds were ringed at this site between 1983 and 2000. An analysis of the sex ratios (Milton 2005) suggests a definite bias towards males (20) as opposed to females (12), although the reason for this is difficult to ascertain and might be linked to males being perhaps more active as they indulge in flight-chase courtship behaviour in early spring (Snow *et al.* 1998). One ringed in 1985 was a very young bird, only recently fledged. From this ringing effort, the only recovery was of a male ringed on 3 May 1992 and reported dead at Chase Park, Whickham (3km north of the ringing site) in April 1993, having flown into a window. At the time there were only 22 recoveries of this species ever from British ringed birds. This short scale movement is typical, as ringing data analysed by Marchant and Simons (Wernham *et al.* 2002) suggests a median movement of just 12.5km. In Durham, the occasional appearance of birds away from the usual localities, such as the bird flying west over Shibdon Pond in September 1987 and an individual with Yellowhammers *Emberiza citrinella* at Clara Vale in November 1988, suggests that such limited local movements within Durham may be usual.

Snow Bunting
Plectrophenax nivalis

**An uncommon winter visitor and passage migrant, primarily to coastal
localities, but also occurs on high moorlands in small numbers.
It was formerly much more abundant in parts of the county's coastal strip.**

Historical review

The 'snowflake' of the past, this charming species' regular winter habitat is along the tide-line of sandy beaches and dunes. When bad weather prevails however it is forced up onto to coastal fields and onto the paths and more open areas of nearby sparsely vegetated ground, such as Seaton Common and the Marsden/South Shields Leas. It was described by George Temperley (1951), as a *"regular winter visitor in varying numbers"*. He said that it was usually coastally distributed though, also less regularly noted on the moorlands in the west of the county. Temperley said the *"status of this species does not appear to have altered in a century"*.

Hogg (1845) reported it as arriving in flocks at the coast around the end of October in the south east of Durham, frequenting as it did sea drifts and sea banks at the mouth of the Tees. Nelson recorded its regular arrival and presence in flocks at Teesmouth (Nelson 1907). Inland, Robson mentioned that the species was *"only a winter visitor"* to the valley (Robson, 1896) but these probably referred to sightings of birds higher up the Derwent valley, in moorland or moorland-fringing habitats.

Temperley (1951) reported that flocks were present throughout the winter at Teesmouth; their numbers peaking in the period November to January, with birds leaving during February and March. He documented his personal observations of three flocks numbering 25 to 50 birds between North Gare and Greatham Creek on 13 December 1928, indicating the numbers that were regularly present at Teesmouth at that time. George Temperley believed that it was less regular inland, though he highlighted Backhouse's (1885) description of it as *"a winter visitant of regular occurrence"* in upper Teesdale. In 1885, Backhouse commented that it appeared to stay later in that year, not leaving until the end of April or the beginning of May. In 1948/1949, they were exceptionally numerous both at the coast and inland (Temperley 1951).

Its pattern of occurrence in the county seemed to change in the 1950s, but this may have in some part been related to an increase in the number of observers. In the latter part of that decade large numbers of birds were recorded in the north of the county and a period followed that saw the largest numbers ever documented in Durham. For instance, up to 350 birds were in a flock in the South Shields to Marsden area during the winter of 1956/1957; some of these were noted a little inland at a slum demolition zone in the heart of South Shields.

During the winter of 1961/1962 at Teesmouth, a large flock of 270 was at Graythorp on 19 October and this rose to an estimated 1,000 birds by 25 October, with numbers stable at this level until 25 November. This remains the largest number recorded locally. This trend for the occurrence of large flocks persisted for a number of years with birds not always restricted to the coastal zone. In the winter of 1963/1964, up to 400 were in the coastal strip between South Shields and Whitburn, while a long way inland one was on Muggleswick Common on 1 March 1964. The following winter witnessed another large flock of up to 400 birds in the Whitburn area during the early winter months. In 1966, what may have been the Whitburn flock appeared on stubble fields near Horden in late November, when up to 300 birds were present. In 1967, a maximum count of 97 was at Teesmouth on 4 January, with 125 at Seaton Carew on 30 December. In the same year a male was well inland on Newbiggin Common on 12 November.

In his review of fifty-years of records from the Shotton Colliery area from the middle of the 20th century, David Simpson, commented upon this trend, saying *"this species used to be a regular visitor to Edder Acres Farm, near Shotton Colliery, mainly to the stubble fields in hard weather, but only in small numbers, with parties of perhaps up to five birds being counted. Then, in 1967 and 1968, large numbers visited an area of rough ground between Shotton Colliery and the Shotton Brickworks Pond. The flock grew to 370 on 17th February 1968"*. This remains the largest truly inland flock seen in County Durham. In other years flocks of up to 25 birds were often seen flying over Shotton Colliery in the late autumn and winter months.

The late 1960s and early 1970s were the last period in which large numbers frequented the northern part of the county, with peak counts between South Shields and Whitburn regularly in the hundreds. For example: 300 were at Whitburn on 16 January 1969; 200 there in January/February 1970 and 650 were present by the end of that year, 600 remaining into 1971, but rapidly dispersing through the month. Numbers thereafter never reached such heights with no more than 40 recorded in 1972/1973, and peak counts of c.100 in 1974, 200 at South Shields in December 1975 and 100 in 1976. The count of 150 in early 1977 was the last time a flock exceeded 100 in this part of the county.

The other main wintering area for the species during this period was in the North Gare and neighbouring Seaton Common area of Teesside. Flocks here were regularly in three-figures in the first two and last two months of the year with the largest recorded flocks being 500 at Seal Sands on 16 December 1971, 300 at Teesmouth in January 1972 and c.450 there in December of that year, with 350 remaining into January 1973. Flocks of more than 300 were also present in December 1975 and January 1977 but a winter peak in the range of 100-150 was more typical in this area during the latter half of the 1970s.

No other locations held numbers comparable to these two main sites, although the coastal fields between Ryhope and Seaham occasionally attracted flocks in excess of three-figures. Inland occurrences are regular for this species with birds not infrequently recorded in bleak, windswept locations in the west of the county. Numbers here tend to be much lower than at the coast but up to 35 were noted near Stanhope in February 1972. In 1975, notable inland reports were 30 at Cow Green Reservoir on 18 January, flocks of 15 and 32 on moorland above Eastgate and, around the same time, 20 at Middlehope Moor on 2 March 1975. Also on high ground in the west, 13 were at Hisehope Moor on 14 January 1979. Further reports from inland, albeit not as far west, were flocks of up to 45 at Gateshead from 1 December 1970 until 16 February 1971, and 25 at Ryton in February 1973.

Recent status

During the last three decades numbers in the county have not come close to the counts recorded over the 20 year period from the late 1950s, through the 1960s and into the mid-1970s. While varying in number according to weather conditions and national occurrence, individual maximum flock size was lower and the number of flocks was also, in most instances, fewer.

Although the South Shields to Whitburn stretch of coast became less productive in terms of numbers, the early part of the 1980s still featured some high counts around the favoured location of North Gare. Up to 300 were present there at the end of December 1981 and in January 1982, with a similar peak in the winter of 1982/1983. Since then however, numbers have also fallen here with peak winter counts of 100-140 during the late 1980s. During the 1990s three-figure counts became the exception rather than the rule anywhere in County Durham. For example, 150 at Easington Colliery on 10 November 1991 was the last gathering of over 100 in the county until 6 November 2004, when North Gare attracted a flock of at least 120. A flock of 137 at North Gare, on 12 November 2008, was reminiscent of days gone by for some older observers and a novel experience for younger eyes.

Despite much reduced numbers, the coastal strip still remains the place where most birds occur with favoured locations being as in years gone by. That said, small numbers of birds can be encountered almost anywhere along the more open areas of coast, particularly during the peak migration months of October-November and February-March.

Small numbers of birds probably still regularly winter inland on the county's western uplands, but this propensity is in all likelihood very much under-recorded, despite commentary on its occurrence from the nineteenth century, courtesy of Backhouse (1885). In such situations, the presence of birds may only to come to light when they are found feeding on bare ground close to road edges and the margins of paths in what is a vast, open landscape. One observer noted small numbers of birds in such situations on at least five occasions over a period of 12-15 years, examples being at Horseshoe Hill in the Derwent valley in winter 1995, on Hunstanworth Common in November 2002 and in Teesdale in November 2007. One of the more regular locations in the uplands is Harthope Moor and quarries between Teesdale and Weardale, where small numbers of birds were recorded in 1988, 1994 and 2001 with up to 25 noted there over the winter of 1998/1999, while Swinhopeburn held 20-30 birds during February 2001. More concentrated effort at such inhospitable winter locations would no doubt produce more sightings.

Away from upland locations and the coastal strip, modern inland records are much fewer than they once were, though occasionally numbers of birds are noted in such situations; for example, about 60 birds flew inland over Wynyard Woodland Park on 19 November 1987 (Blick 2009). A review of such records in the Gateshead area, revealed just five records over a twelve year period from the early 1980s to the mid-1990s. On 4 January 1984, two were seen flying west over Maryside Hill, near Clara Vale and on 30 March 1989, a female was flushed from the bare ground at the side of some small ponds near Wardley. Two flew north over Heavygate Farm, Chopwell on 5 December 1990 and a female was at the MetroCentre Pools on 26 November 1994 with a party of six birds noted at Pelaw Quarries on 27 November 1994.

The first of the autumn's birds usually appear in early October, although early birds are not infrequently seen during the second half of September; the earliest recorded arrival concerned an individual at Whitburn on 13 September 2009. The number of birds builds up over several weeks, with the peak autumn numbers usually reached during November. Peak winter counts are generally in December or January with returning migrants occasionally swelling numbers to a winter high during February. In many years the majority of birds have already left before the end of February, but small numbers of passage migrants appear regularly during the whole of March and occasionally into the first part of April. Males, though relatively scarce in number in the county, have occasionally been heard singing during mid- to late March (Blick 2009). May birds are decidedly rare with one at Cleadon Hill over 10th to 12 May 1981 being surpassed as the latest ever in the county by an exceptionally tardy individual at Crimdon on 16 May 1993.

Distribution & movements

Snow Buntings occur as circumpolar breeding birds around the whole of the arctic and sub-arctic zones, with birds penetrating further south in mountainous areas. Some of the breeding areas are tenanted throughout the year but the northerly zones are deserted in winter, when birds move south of the core breeding range. Many of the birds wintering in the British Isles are believed to belong to the *insulae* race that occurs principally on Iceland.

Between 1976 and 2008, almost 2,000 Snow Buntings were ringed at Teesmouth, resulting in a number of interesting recoveries (Blick 2009). Most of these birds were ringed in January or February, near North or at South Gare, on the south side of the estuary. Controls usually came within two to three months of capture (Blick 2009). One first winter female ringed at North Gare on 30 January 1988 was in Glen Shee in Grampian, Scotland, on 4 March 1988, 283km to the north west; perhaps part of the small British breeding population. Another bird ringed as an adult female at Seaton Carew on 6 February 1988 was controlled at Glen Shee, on 10 January 1989 and again on 21 March 1989, when it was colour-ringed. This bird was then seen in the field at Glen Shee later in March and again in February 1990. It was found dead on 20 November 1990 on an oil rig in the North Sea. A number of birds (all adults) were ringed at Seaton Carew during winter 1990 and 1991. All subsequent controls came from sites to the north: one at Cambois, Northumberland; one in Lothian, Scotland, later in the winter and three in Cairngorm and Grampian in spring, presumably being there to breed. One bird that was ringed as an adult female in February 1990 was controlled wintering further south in Norfolk at Hunstanton in January 1991. A bird controlled at Seaton Carew on 18 February 1990, had been ringed on 19 February 1989, 222km south along the east coast, at Titchwell in Norfolk. A study of Snow Buntings ringed further south indicates that birds begin to move north well before the winter is over (Blick, 2009). A colour-ringed individual noted at Whitburn over 8th to 20 March 1991 had been marked earlier that winter in Norfolk. This study of local Snow Buntings indicates that the majority of locally wintering birds are immature females and it is unusual for an individual to return in following winters. Birds ringed at Spurn, Humberside, have been found in Belgium, Norway, Iceland and Orkney in the following springs (Wernham *et al.* 2001).

Lapland Bunting (Lapland Longspur)
Calcarius lapponicus

A scarce passage migrant much more regular in autumn than in spring, and a scarce winter visitor.

Historical review

In 1951, at Temperley's time of review, this species was considered to be a very rare vagrant in Durham, with just a single occurrence in the county, which was noted by Hancock (1874). He recorded that one was shot out of a flock of 'snow flakes' in the neighbourhood of Durham in January 1860. This specimen went to the Durham University Museum. The actual place of shooting, as revealed by William Proctor of the Museum, was near Whitburn, north of Sunderland in January 1860, a location that still attracts the species today.

The species' status in the county, changed dramatically in the winter of 1953/1954 when there was an invasion of this species, particularly into the north western parts of the UK, which no doubt influenced a subsequent series of records at Teesmouth. This commenced in 1953 with one on September 13th and 20th and two on October 18th. The species was then reported on eight days in November, with up to 16 birds on 14th. Five were present on December 5th and 18 on 6th. In January 1954, one or two were noted on three days in January, with one on February 7th and four on 14th. Later that year, a flock of five were counted on March 8th and twelve over the period 13th to 15th of that month. The following winter a flock, of up to 25 birds, was found frequenting a 'rubbish tip' at Teesmouth.

Around this time, three were found on moors, a long way inland, somewhere between Richmond and Barnard Castle on 28 November 1954 (Mather 1986). This would appear to have been somewhere around Barningham Moor, which is the only extant area of moorland between the two named locations, however the exact location can probably never be accurately determined, and the birds may have been just outside of the county. The winter of 1956 brought large numbers to the Marsden area towards the year's end, with up to 40 recorded, but for the rest of the 1950s it remained a scarce and local species. The maximum count was of eight birds at Teesmouth in the winter of 1958-59 with 10 at the same location in November 1959.

Through the first half of the 1960s it was rarely recorded, with no records in 1960, 1963 to 1965 and no more than five birds in 1961/1962. From the mid-1960s however, this species' status in the county began to change, from an intermittently recorded sub-rarity to a bird that was noted on an annual basis, occasionally in larger numbers; no doubt this resulted, in part, from increased observer awareness of the species.

Subsequently, there were records in 1967, 1968, 1969 and every year since, though just one bird was noted in 1981. In 1967 single birds were at Teesmouth on 1 April, 15 November and on two dates in December. In 1968, one was at Teesmouth on 25 February with five at North Gare on 10 November. Two were at this latter site on 8 February 1969, with one at Greatham Creek on 5 November.

Recent status

Today, this species is primarily a winter visitor, which is noted in variable numbers, in most years from the middle of October to mid-March. It is recorded almost exclusively in coastal locations, from South Shields Leas in the north, south to a number of favoured sites around Teesmouth, particularly Seaton Common, the Cowpen Marsh to Brinefields area and the area around Saltholme (Blick 2009).

In the last four decades, the pattern of occurrence of this species has been relatively consistent, with most years seeing single figure counts during the winter periods and a small but well defined autumn passage from mid-September until November. First arrivals tend to be from mid-September and numbers peak in October and November with few seen after March. The earliest recorded were at Ryhope on 3 September 2010 and the latest in spring at Teesmouth on 27 May 1977. Whilst numbers are usually small and fluctuate markedly from year-to-year, there are a few traditional sites that often attract birds, such as the Tees Marshes, the Marsden area of South Shields Leas and the coastal fields around Ryhope, which can hold sizeable concentrations during good years. Unlike Snow Bunting, *Plectrophenax nivalis*, this species avoids barren or rocky terrain with the preferred habitat being short, rough grassland or winter stubble fields, usually at no great distance from the sea. Although occasionally found a little way inland, such as at the Sunderland Academy Pools or around some of the Teesmouth wetland sites, truly inland birds are rare enough to be fully documented. Such records are usually associated with passage periods. In 1972, two birds were found feeding on hay seeds at sheep feeders near Stanhope on 3 February; a single was at Hurworth Burn Reservoir on 28 September 1975; one was at Rainton Meadows on 5 February and 1 March 1998 and in 2006, one was at Pitfield Farm, Little Stainton on 14 April with a summer-plumage male on Ravensworth Fell, Gateshead, on 1 May. The influx of autumn 2010 produced a single bird at Bishop Middleham on 18 September, followed by October sightings at Seaton Pond on 15th, Hetton Lyons on 16th, Rainton Meadows on 18th and Hedworth on 23rd.

The peak winter counts of this species are usually in multiples of tens and in many winters the handful of recorded flocks does not rise above single figures. There was a trend towards smaller numbers through the late 1990s into the first decade of the 21st century; only once in that latter decade did any flock reach or exceed 10 birds. The total number of birds in the county in any autumn or winter probably rarely exceeds 30-40, and in many years it may be as little as a third of those figures.

The number and size of flocks of Lapland Buntings in the county peaked during the first half of the 1970s. During the early 1970s the Tees Marshes regularly held winter concentrations with highest annual counts of: 20 in December 1970; 19 in December 1971; 26 in December 1972 and 14 early in 1973. The late year maximum in 1973 was just six birds during November, and the next few years saw a dearth of records with no double-figure counts anywhere until 19 February 1978, when 10 were again on Seaton Common, Teesside. Other years during the 1970s saw reports of one to four birds with the pattern of recording broadly split between isolated winter sightings and autumn passage migrants.

The 1980s and 1990s followed a similar pattern with Seaton Common again attracting up to 22 birds in January and February 1980. This was however, an isolated occurrence as the next few years were lean ones, with no birds during the autumn of 1980; one at Seaburn on 2 March the only 1981 record, and just a handful of reports during 1982-1985; although these included 10 at Castle Eden Denemouth in December 1984.

During 1986-1987, the Ryhope area was much frequented, with the presence of winter stubble perhaps the main influencing factor in retaining birds in the area for long periods. Early in 1986 up to nine birds were present with at least 30 in December of that year and counts in 1987 included 24 in January and up to 20 in December. Other high counts during this period were 23 on Marsden Leas in March 1986 and 25 at Mere Knolls in November 1987. The latter record was the first at the site for 10 years with birds attracted by autumn stubble fields. In what seemed to be an established pattern, another couple of poor years followed with just two singles noted in 1988 during November; although 1989 was a little better with six at Greenabella Marsh on 8 February and a handful of single bird sightings during the final quarter of the year.

Apart from c.30 on Seaton Common and 15 in rough pasture at nearby Hartlepool Power Station in January 1990, single-figure counts were the norm during the 1990s, but better years occurred during 1993/1994 and in the winter of 1998/1999. In January 1993, up to 12 were regularly seen around Ryhope, while the winter of 1993/1994 was probably the best ever in the county. Seaton Common attracted 16 birds on 28 October 1993 and 40 were counted on Cowpen Marsh in January 1994. In the north of the county, South Shields Leas held birds throughout the first quarter of 1994 with peak monthly counts of 26 on 3 January, 25 on 16 February and 23 on 1 March before numbers declined.

Lapland Bunting, Whitburn, September 2009
(Mark Newsome)

Late 1994 saw a return to normal with single birds recorded on two dates in each of the final three months of the year. The mid-1990s continued in similar fashion with no more than seven at Ryhope in 1995, eight on Seaton Common in 1996, with only one or two being seen anywhere during 1997 and the first half of 1998. Late 1998, however, saw an increase in autumn passage with up to four birds recorded before a party of 11 settled for the winter on South Shields Leas during December. This flock increased to 17 in January 1999 with 12 during February and the last one on 20 March. Late in the year, just four passage birds were recorded in October and November but three were at Ryhope and five on the Leas at Marsden in December.

The 2000s brought a similar pattern of occurrence with no large numbers recorded until the autumn of 2010. Prior to this, the winter of 2002/2003 provided the largest gatherings but no flocks of more than the 11 birds on the Leas in late 2002 were recorded. Since then, the species has been decidedly scarce in Durham; the poorest year being 2008, when there were only four records of autumn passage birds.

In a repeat of 1953, when County Durham recorded the species for only the second time ever, the latter part of 2010 saw a record influx, particularly into northern and north western parts of the UK, with flocks of several hundred noted in the northern and western isles of Scotland. Speculation based on weather conditions and the appearance of unprecedented numbers in south west Iceland, was that these were birds of the race *C. l. calcaratus* originating in Greenland and/or Canada; a movement previously acknowledged in *BWP* and more recently by Garner (2007). The first birds in Durham were also the earliest on record with up to seven noted in the coastal fields south of Ryhope from 3 September. The ploughing of the stubble here meant they only lingered for three or four days. Thereafter there were numerous reports of one to four birds at many coastal locations, as birds moved through the county, with peak counts of 10 at South Shields and 13 at Whitburn. The biggest flock was in the south east, with 45 birds on Seaton Common on 19 October; the largest flock ever recorded in County Durham.

Distribution & movements

The Lapland Bunting has a circumpolar distribution, breeding from the mountainous parts of Scandinavia across Arctic and sub-Arctic Russia and east to Siberia. Birds are also present in Alaska and across Arctic Canada and Greenland. The birds breeding in northern Europe mainly winter in central Asia; American birds wintering in the United States, though the birds wintering around parts of Western Europe are thought to comprise birds originating from both directions (Parkin & Knox 2010).

Pine Bunting
Emberiza leucocephalos

A very rare vagrant from Siberia: one record.

The sole record of Pine Bunting for County Durham is of a female visiting a garden at Langley Moor, Durham with Yellowhammers *Emberiza citrinella* from 22nd until at least 24 March 1998. It was found dead on 28 March, having flown into a window. The specimen was professionally mounted and retained by the finder, Norman Urwin (*British Birds* 96: 604). A brief account appears below:

"On returning from my weekly Sunday morning walk ... I stood in the garden for a while, as I usually do, to see what was about. As I looked around, two birds landed in the hawthorn bush, about ten yards away. To the naked eye both were Yellowhammers, through binoculars one clearly was not! .... The overall impression was one of 'greyness', with no yellow tones visible in the plumage at all... As I puzzled over the bird during the day I became more and more convinced that it was a female Pine Bunting. On my return home ... the bird was still present and was feeding with Yellowhammers on the lawn. I was now able to make a more detailed examination from inside the house ... by 17:30 hours I had decided it was a Pine Bunting" (Urwin 1998).

It was seen briefly on 23 March and during the evenings of both 23rd and 24 March. The bird was found dead below a window on Saturday 28th having suffered a broken neck, presumably resulting from a collision with the glass.

Distribution & movements

The nominate race of Pine Bunting breeds across temperate Russia from the western Urals to the upper Kolyma River southwards to southern Siberia and Mongolia, with an isolated race in central China. Strongly migratory, the species' winter range extends from Iran through northern India and Nepal to southern China, though small numbers have recently been found wintering in Israel and western Italy (Occhiato 2003). It is a rare vagrant to Britain with 49 records by the end of 2010. Almost half of the records have come from Orkney and Shetland, where most have occurred between late October and early November. Elsewhere, the pattern is quite different with just five English records during the peak Scottish period for vagrancy and with most found during January to March, often inland amongst flocks of Yellowhammers *Emberiza citrinella;* the Durham bird neatly fits this pattern of occurrence.

Yellowhammer
Emberiza citrinella

A widespread resident in the lowland areas; much more local in the west of the county.

Historical review

Historically the 'Yellow Bunting' was probably a bird of the widespread scrub habitats of the agricultural landscape; it is still found in areas of hawthorn, bracken and gorse, as well as in more open farmland with hedgerow and patches of scrub. There is little detailed information about this species' status in the county during the 19th century or earlier, but it is presumed that it was common and widespread as noted by Holloway (1996). In relation to the Derwent valley, Robson documented a very late nesting record in the late Victorian period, a nest with three eggs being found near Winlaton on 14 September 1890 (Robson 1896).

Temperley called it "*a very common resident*" (Temperley 1951), though during the first half of the century there was some evidence of a decrease in numbers. This process of decline had been noticed as early as 1905 (Tristram 1905), when it was said to have "*decreased much in numbers of late years*". Temperley highlighted the fact that the decline had been 'marked' in the last decade, i.e. over the period 1939 to 1949. Changes in agricultural land use were suggested as the likely reason. This pattern appears to reflect the situation in the country as a whole, though the national population stabilised from the early 1960s, before further decline commenced in the late 20th century. During the fieldwork for the *Breeding Atlas* (Sharrock 1976), the species was recorded in all of the county's 48 ten-kilometre squares, with the most westerly birds in later years being at Middleton-in-Teesdale (1975) and Cotherstone (1990). Nationally, there has been a notable decline in the numbers of Yellowhammers in Britain, which probably commenced in the 1980s (Parkin & Knox 2010). National data revealed a decline of 56% between 1967 and 2008 (Baillie *et al.* 2010), the greatest reductions occurring through the 1980s and 1990s, but with declines slowing since the Millennium. There have also been notable declines in clutch size, brood size and nest success recorded (Leech & Barimore 2008).

Recent breeding status

This species is one of the most common birds in the county's agricultural landscapes, although it has declined quite dramatically in the last quarter of the twentieth century. Marchant *et al.* (1990) indicated that farmland was very much the optimal habitat for this species, with woodland edge being used, but very much to a lesser degree. Yellowhammers occur in most open country habitat types in the county that have areas of scrubby rough ground with plenty of ground cover for nesting. They are, more or less, absent from built-up areas as a breeding bird, though they do occur in some locations along the edges of housing and industrial estates that abut farmland or open areas such as golf courses.

This trait was well illustrated in the *Atlas* maps (Westerberg & Bowey 2000) and by the gaps which occur in its lowland range; these correspond with the major conurbations of Wearside and South Tyneside. The *Atlas* also showed a number of gaps in distribution, which were probably explained by a lack of observer coverage i.e. in the areas between Washington and Chester-le-Street; Consett and Lanchester; Peterlee and Wheatley Hill, as well as to the north and east of Darlington. In relation to the south east of the county, this species is largely found in the agricultural landscapes away from urban Teesside, there being around 300-350 pairs in the northern Tees area. It is absent from much of the North Tees Marshes and that part of the coastal strip as a breeding bird (Joynt *et al.* 2008).

In general terms, the species is concentrated predominantly in the middle and east of the county with birds thinly spread as far west as Derwent Reservoir. The western edge of the species' breeding range roughly equates to the county's 300m contour. Birds occasionally noted on territory in westerly locations, even as far west as upper Teesdale, are very much less numerous than to the east and somewhat beyond the normal range of this species.

The first singing males are usually reported at the beginning of March but this is more regularly noted from later in that month and into April, with singing birds still on territory in July and August due to the protracted breeding season in which two or even three broods can be raised. Nationally, population densities of up to sixty pairs per square kilometre have been recorded (Gibbons *et al.* 1993), although nothing as high as this is has been recorded locally. In 2007, the largest numbers of breeding birds on farmed land were recorded at Broadwood and Old Hall Farm with 20 pairs each at both of these locations. At Rainton Meadows, in an area dominated by grassland habitats, the documented 14 pairs in 2007 was an increase of four pairs over 2006. On the 2007 Kibblesworth Common Bird Census, 26 pairs represented a decline from 34 and 35 pairs during the 2005 and 2006 breeding seasons. The reason for this local decrease is not known. In 2007, one to five birds held territory at 40 widespread sites, though this is only a fraction of the county's total population.

In common with many other farmland species, one of the major factors thought to have contributed to the decline in numbers both locally and nationally, is the loss of winter stubble as a source of food that supports winter survival and enhances subsequent breeding performance. Indeed, the presence of winter stubble is an indicator of site location for breeding populations in the following spring (Whittingham *et al.* 2005) and improved breeding performance (Gillings *et al.* 2005). By the end of the 1990s, it was suggested that numbers were falling by as much as 10% per annum (Parkin & Knox 2010). Considering local increases and declines, the *Atlas*' population estimate of some 6,400 pairs across the county is probably still a reasonably accurate one, though should perhaps be somewhat reduced to a range of 5,500 to 6,000.

Recent non-breeding status

In winter, the Yellowhammer forms foraging flocks, often in association with other buntings and some finch species. Typically there may be up to 100 birds in such flocks, though counts of over 200 do occur on grain or stubble. The autumn and winter flocking habit in this species is widespread across the county, with up to 50 birds often being found concentrated at locations particularly in the eastern part of the area, such as Bishop Middleham, Cowpen Bewley, Elemore Hall, Hetton Lyons, Newfield, Page Bank, Quarrington Hill, Sherburn and Tursdale. Not infrequently, the highest numbers are recorded in farmland areas that are clustered on or around the magnesian Limestone areas of the county, such as High Sharpley and Seaton Pond. In 2007, the largest count of the first quarter of the year was of 80 birds at Byer's Green, where food was being provided, with flocks of from 40 to 60 being reported from Aykley Heads, Boldon Flats, Dalton Moor, Hilton and Pittington. There appears to have been a decline in the widespread large winter flocks that were once recorded in so many areas; such as the 260 noted on farmland at the edge of Washingwell Woods, near Whickham, on 21 January 1989 and those that were frequently found in fields along the coastal fringe of the county. Although this species was declining in some parts of the UK, at the end of the first decade of the 21st century, it was still doing reasonably well in the eastern half of County Durham. This was evidenced by winter counts made in late 2009 and 2010, when over 300 were at Farnless Farm at Bishop Middleham and 250 at Seaton Pond, both at locations where food was being provided; these were the largest registered flocks in the county for a number of years.

Northern European and Siberian populations are migratory and it may be that some birds reach our area, though there is at present, no ringing or other independent data that supports this assertion. Birds are present in many eastern locations, for example the Coordinated Migrant Count on 4 October 1992 found 89 along the coastal strip (Armstrong 1993). However, as Blick noted very few are seen at the coastal migration points, generally only one or two birds per year (Blick 2009). The presence of obvious migrants is difficult to detect although a single present on 17 October 2005 at South Shields Leas, where this species does not usually occur, might fall into this category.

Distribution & movements

This species breeds across much of Europe, though more sparsely the further south one goes. To the east, birds can be found as far east as western Siberia. Most populations of the species are at least partially migratory, the northerly breeding grounds being deserted during winter. Some populations winter to the south of the breeding range, others in the southerly zones of this range. Most British breeding birds are rather sedentary; some Scandinavian birds may join local breeders in the winter, particularly along the county's eastern fringe.

Local Yellowhammers do not appear to move any great distance during their lives as evidenced by one ringed at Graythorp on 31 December 1960, which was still there on 24 July 1965. There is little evidence of significant movements of local birds, other than of their joining winter flocks, or occasional inward seasonal movement into the county from relatively nearby locations; as demonstrated by two colour-ringed birds that were caught at Clara Vale in February 2008 which had been ringed 6.5km to the west, at Nafferton Farm, near Corbridge, Northumberland as part of a Newcastle University research project. Despite, these observations the absence of birds from the higher areas of the west of the county during the Winter Atlas fieldwork (Lack 1986), contrary to the breeding season presence as per Sharrock (1976), indicated some local altitudinal migration.

Cirl Bunting
Emberiza cirlus

A very rare vagrant from Europe: one record of a returning bird.

Widespread on the Continent, this species has a severely restricted breeding range in southern England and at the time of this sighting, the largely sedentary British population was declining rapidly. The addition of this species to the county list remains one of the most unexpected of ornithological events in the county. The only record of Cirl Bunting for County Durham is of a singing male found by David Sowerbutts by the track-bed of the former Browney Valley Railway, just north west of Langley Park on 24 May 1980. It remained in the area until at least 12 August. Initially located by its unfamiliar song, heard while the observer was walking between Langley Park and Malton; this territorial male was seen and heard by numerous observers during its extended stay. It was frequently seen interacting with the many Yellowhammers *Emberiza citrinella* in the area and a female was 'tentatively identified' nearby on at least one date. Remarkably, what was presumably the same male was found in the same area the following spring; being present from 4 April to 5 June 1981, though sightings were more sporadic. It seems not unreasonable to presume that this bird spent the intervening 1980/1981 winter undetected in one of the local Yellowhammer flocks in the area, though this cannot be proven.

Because of the parlous state of the British breeding population at this time, it is tempting to speculate that this bird may have been of continental origin. There are few records for this species of extra-limital vagrants close to County Durham, though a bird was noted on 3 August 1941, near Gillamoor, Kirby-Moorside, North Yorkshire (*British Birds* 38: 211), and a male was near Redcar, just across the River Tees from Durham, during a large fall of common migrants on 17 September 1960 (Stead 1964). A first-year bird ringed in Sussex on 27 July 1975 was recovered some 638km to the north in Fife, Scotland on 11 June 1976, indicating that, although essentially sedentary, some British birds are capable of long-distance movements (Wernham *et al.* 2002). Parkin & Knox (2010) highlighted that away from its breeding areas in Britain, it remains a very rare bird.

Distribution & movements

Cirl Bunting breeds in north western Africa and throughout much of southern Europe from Iberia to western Turkey and the Balkans. In the west it regularly breeds as far north and west as Brittany in northern France and a relict population survives in south western Britain (Parkin & Knox 2010). The species is largely sedentary but is subject to some localised dispersal during the winter months, usually in response in cold weather. Cirl Buntings formerly bred throughout much of southern England and occasionally further north especially during the 19th century, though a major decline in the population during the 1960s saw the species restricted to south Devon by the end of the 20th century. This decline was largely attributed to agricultural practice, in particular the loss of hedgerows and scrub in which to breed, and the switch from spring to autumn grown cereals resulting in the loss of winter stubble fields on which the species is dependent. The population reached an all time low of 118 pairs in 1989, although targeted conservation measures helped the species recover to an estimated 785 to 975 pairs in England by 2009 (Stanbury *et al.* 2010), the majority of these in south Devon.

Ortolan Bunting
Emberiza hortulana

A rare vagrant from Europe: 16 records, most often from August to October, but one May record.

The first record of Ortolan Bunting for County Durham was of a first-winter bird found by Geoff Iceton and Rob Little at Hartlepool Headland on 14 September 1968; it was later seen by the late Frank Dunn amongst others. Initially located in seafront gardens between the Heugh Breakwater and the Old Pier, it spent much of its time feeding on the beach in the weedy area immediately adjacent to the Old Pier, just above the high-water mark.

All records:

1968	Hartlepool Headland, 14 September
1969	Hartlepool Headland, 14th to 15 August
1970	Hartlepool Headland, male, 7 May
1976	Hartlepool Headland, male, 2 October
1979	Hartlepool Headland, first-winter, 18 August
1983	Hartlepool Headland, first-winter, 20th to 21 August
1986	Hartlepool Headland, first-winter, 27 August
1993	Hartlepool Headland, first-winter, 14th to 16 September
1993	Marsden Quarry, first-winter, 16th to 21 September
1995	Marsden Quarry, first-winter, 8th to 11 September
1995	Hartlepool Headland, first-winter, 9 September
1995	Holy Trinity Churchyard, Seaton Carew, first-winter, 9 September
1995	Spion Kop Cemetery, Hartlepool, first-winter, 10 September
1995	Marsden Quarry, first-winter, 10 September
1999	Mere Knolls Cemetery, Seaburn, first-winter, 21st to 22 September
2005	Zinc Works Bushes, Seaton Snook, Teesmouth, first-winter, 10 September

That the first Ortolan Bunting for County Durham was not recorded until 1968 is difficult to explain as the species was already on the decline in Europe by the late 1960s, and might have been expected to have occurred earlier when the population in northern Europe was much higher. Somewhat ironically, the first Yorkshire specimen was caught on a collier off the north east coast in May 1822 (Mather 1986). At this time, the bird, no longer extant, passed up the Durham coastline before being taken to the Newcastle Museum to be used by Thomas Bewick, when illustrating this species in a later edition of his *Land Birds*.

It remains a very scarce migrant in the county to date, averaging just one to three records per decade, except for an exceptional series of records during the 1990s. The sole spring record is of a male at Hartlepool Headland on 7 May 1970 following a prolonged period of easterly winds. The majority of autumn records refer to first-winter birds in the period from mid-August to late September, with the only exception being an adult male at Hartlepool Headland on 2 October 1976; this remains the only autumn adult to date. The earliest and latest arrivals, both from Hartlepool Headland, are a first-winter bird on 14 August 1969 and an adult on 2 October 1976. The record total of five birds was logged during 8th to 10 September 1995, and included two together in Marsden Quarry on 10 September.

Distribution & movements

Ortolan Bunting breeds across much of Europe, excluding the British Isles, eastwards through western Asia to Mongolia, and winters in sub-Saharan Africa in a belt from Sierra Leone to the Sudan and northern Ethiopia. Formerly more widespread throughout Europe, a serious decline has seen the species become highly fragmented as a breeding species particularly in western and central Europe as well as becoming absent from much of the northern part of its former range. This decline may be linked to the excessive trapping of migrant birds in France, where despite the introduction of legislation to address this in September 2007, the species is still considered a delicacy. The species is a scarce passage migrant to Britain in both spring and autumn, though much more frequent in autumn, chiefly in August and September. The majority of passage birds are now found along the south coast of England, and although the species is frequently recorded along the English east coast as far north as Yorkshire it is becoming progressively scarcer further north. The annual maxima recorded in Britain has declined in line with the decrease of the European breeding population and through the 2000s, the species was averaging around 70 records per annum in Britain. Just one Ortolan Bunting has been ringed in the county, a first-winter bird at Hartlepool Headland on 14 August 1969.

Rustic Bunting
Emberiza rustica

A very rare vagrant from north eastern Europe and Siberia: seven records.

The first Rustic Bunting for County Durham was found by Ken Smith and Alan Vittery at North Gare on 7 September 1958, during a fall of drift migrants (Grey 1959, *British Birds* 53: 173). This was only the ninth record for Britain at the time, and only the third away from Shetland.

All records:

1958	North Gare, Seaton Carew, 7 September	
1976	Hartlepool Headland, 17th to 18 September (*British Birds* 70: 442)	
1991	Sunderland Docks, 12th to 17 October (*British Birds* 86: 534)	
1993	Marsden Quarry, 15th to 16 September (*British Birds* 87: 564)	
1993	Marsden Quarry, 16 September (*British Birds* 87: 564)	
2000	Spion Kop Cemetery, Hartlepool, 25 September (*British Birds* 94: 499)	
2000	Marsden Quarry, 21 October (*British Birds* 95: 521)	

All of the records in County Durham have occurred in September and October, which is regarded as the classic time for this species on the English east coast. Typically brief in their appearance, the only bird to have made a protracted stay in the county was the often elusive bird that frequented a small sycamore copse at Sunderland South Dock for six days in October 1991. More unusually, two were present at Marsden Quarry in mid-September 1993 following a period of north easterly gales and rain; with 48 records, this was a record year for Rustic Bunting in Britain. Despite the eight spring records from Northumberland there are no spring records for County Durham.

Distribution & movements

Rustic Bunting breeds in forested bogs from northern Scandinavia eastwards through northern Russia and Siberia to Kamchatka. A noted long distance migrant, the majority of the population including Eurasian birds spends the winter in eastern China, Japan and Korea, though small numbers also winter in southern Kazakhstan and the Xingjian region of China. It is a regular vagrant to Britain in both spring and autumn and had amassed a total of over 450 records by the end of 2005, by which time the species was no longer considered by BBRC. Almost half of these records have come from the Shetland Isles, which also accounts for the majority of spring records principally in May and June, with the species being particularly scarce in spring outside of Scotland. Autumn records are more widespread along the entire east coast of Britain, and to a lesser extent along the south coast of England as far west as the Isles of Scilly, with most occurring in September and October, though occasionally as late as November; there are also three winter records.

Little Bunting
Emberiza pusilla

A very rare vagrant from north eastern Europe and Siberia: seven records.

The first Little Bunting for County Durham was seen and subsequently shot on the sea walls at Seaton Snook, Teesmouth by the late C. Braithwaite and the late C. E Milburn on 11 October 1902. Writing to George Bolam, Milburn described the event: *"It happened to be my good fortune to be the first to detect the Little Bunting on the Durham side of Teesmouth. It was whilst the late C. Braithwaite and I were ranging the bents and sea-walls in search of migrants. After watching the bird and flushing it once or twice from the rough growth along the inner wall-foot, it made over the main wall, where it would have been difficult to keep it in view, so I told my companion to secure it, which he did. This was on 11th October, 1902."* The specimen was preserved and identified as being a female, though is perhaps more likely to have been a first winter bird. It was later sent to W. R. Ogilvie-Grant who exhibited it at a meeting of the British Ornithologists Club on 22 October 1902, whereupon it was accepted as being only the second record of this species for Britain (*Bulletin. B.O.C.* XIII: 14). The first British Bird was caught near Brighton during the autumn of 1864. A long period passed before Durham's second record at Marsden Hall, on 17 October 1972. Since then, there have been just five further records.

All records:
1902	Seaton Snook, Teesmouth, 11 October, shot (*Zoologist* 1902: 466)
1972	Marsden Hall, 17th to 18 October (*British Birds* 66: 354)
1976	Branksome, Darlington, 13 November (*British Birds* 70: 443)
1983	Hartlepool Headland, 30 September to 1 October (*British Birds* 77: 559)
2005	Hartlepool Headland, 10th to 11 September
2005	Marsden Quarry, 16th to 20 October
2008	Blackhall Rocks, 20 November

There have been just seven records for the county; Marsden Hall and Hartlepool Headland have attracted two birds each. Overall, this is a rather low total compared to the equivalent pattern of occurrence for neighbouring east coast counties; for example, there have been over 65 for Northumberland and eight from the area around Redcar, immediately south of Durham. The county's only inland bird, at Branksome near Darlington in November 1976, indicates that birds may be missed if they are part of inland finch and bunting flocks.

The records in County Durham are largely typical in their date and location, with a few exceptions. The bird at Hartlepool Headland in September 2005 is notable for its early date, though the species has been recorded on Shetland as early as 1 September (Pennington *et al.* 2004). The bird present at Blackhall Rocks on 20 November 2008 is noteworthy for the late date at a coastal location.

Distribution & movements

Little Bunting breeds in damp scrub in the taiga zone from Finland eastwards through Russia to eastern Siberia, though is absent from Kamchatka. It has expanded its range in that direction considerably in recent decades, with a consequent increase in British records; its national status changing from a very rare vagrant to a scarce passage migrant. It winters well to the south of its breeding range, largely in southern China.

Its breeding range has been spreading slowly westwards since the first recorded breeding in Finland in the 1930s, and following a marked increase in the population during the 1980s, it now breeds regularly in small numbers in northern Norway and north eastern Sweden. A long distance migrant, it winters well to the south of its breeding range in eastern Nepal, Burma, southern China, and much of northern Thailand and Vietnam. Best regarded as a rare passage migrant to Britain, records have risen in line with the westward spread of the European breeding population. It was removed from the list of species considered by BBRC in 1993, by which time there had been over 600 records, though over half of these have been from the Shetland Isles. Although recorded in all months, including several recent instances of successful wintering, most are recorded in late September and October. The ten-year average during 1990-1999 was of 29 birds per annum, though 59 were recorded in 2000 in what was to be a record year for the species in Britain.

Yellow-breasted Bunting
Emberiza aureola

A very rare vagrant from Siberia: one record.

The sole record of Yellow-breasted Bunting is of a first-winter bird found by Stuart Ling in stubble fields just north of Ryhope Dene on 22 September 1991 (*British Birds* 86: 535). An account of the find appears in *Birds in Durham 1991* (Ling 1992). This was a long-expected addition to Durham's list of birds; 11 had been recorded in nearby Northumberland. The bird at Ryhope in 1991 was on an entirely typical date and was one of seven birds recorded in Britain that year although the other six sightings were all on the Shetland Isles (*British Birds* 85: 551).

Distribution & movements

Yellow-breasted Bunting breeds widely across Russia and Siberia to Kamchatka, southwards to north east China and Japan, and winters throughout much of South East Asia, including the foothills of the eastern Himalayas. The species formerly bred in central Finland and this decline has been mirrored in the species' Russian stronghold where numbers have dropped by 20-29% during 1990-2000 (Birdlife International 2004). This decline may be linked to excessive trapping of this species in its winter range. It is a regular vagrant to Britain with over 230 occurrences by the end of 2010, although it is a speciality of the Northern Isles of Scotland, with almost two thirds of records coming from Shetland, in particular Fair Isle. Most are found during the period from late August to mid-September, though there are occasional records until early October. The recent contraction of the species' breeding range has seen a corresponding drop in the number of records in Britain, with just 30 since 2000.

Reed Bunting (Common Reed Bunting)
Emberiza schoeniclus

A common but local resident breeder; also a passage migrant and winter visitor.

Historical review

In the latter part of the Nineteenth century, Robson thought it "*somewhat common in marshy or ill-drained meadows*" (Robson, 1896) but today they are increasingly found in drier habitats as well. For Temperley (1951), the Reed Bunting was a resident, a passage migrant and a winter visitor, which was "*fairly well distributed as a breeding species wherever there are suitable marshy places in the low country*". It was one of the most plentiful species noted in the marshes around Darlington, which probably included the once extensive wetland habitats in the area of Bradbury and Mordon Carrs, to the north west of the town. It was said to be common '*about South Shields*'; more widespread in winter, also occurring in drier habitats at this time of the year. Migrant birds were noted in small numbers at Darlington Sewage Farm in the final week of April.

Indicative of the species' decline as a result of the loss of its favoured wetland habitats, was the article that noted that this species used to breed 'freely' along the River Team but that numbers there were reduced when the river was straightened or otherwise interfered with. It was commented that in some years, it was never seen. On 3 April 1951 however, a number of males were observed at Rowletch Burn between Birtley and Brown's Buildings. A single had also been seen near Birtley Station the previous week (*The Vasculum* Vol. XXXVI 1951).

In the last four decades, a larger proportion of the population has started using a wider range of habitats; a reflection of a national trend that was first noted in Hertfordshire in the 1930s, but which was more widely documented from the 1960s (Gibbons *et al.* 1993). The *Breeding Atlas* (Sharrock 1976) showed the species to be well distributed throughout County Durham, but the results of the fieldwork for the *New Atlas* (Gibbon *et al.* 1993) indicated that 'low country' well described what continues to be the species' preferred altitudinal range.

Recent breeding status

Typically, this bunting has been considered, a largely lowland species and one can appreciate the altitudinal constraints placed upon it in Durham by realizing that less than 3% of occupied tetrads in the *Atlas* were located above the 130m contour line (Westerberg & Bowey 2000). Gibbons *et al.* (1993) also highlighted the reduced breeding densities to be found at higher altitudes. The relatively small numbers of occupied areas in the upland parts of the west of the county are largely concentrated in the river valleys. That said, the run of mild winters through the 1990s and the first few years of the 21st century, has no doubt facilitated the species' move into the fringes of some of the county's western reservoirs, with singing males noted not only in damp *Juncus*-dominated damp pastures but also in dry bracken-covered slopes at altitudes of 350-370m above sea level at the Balderhead, Derwent, Selset and Waskerley Reservoirs.

In Durham, the species is quite widely distributed and in addition to lowland wetland sites it is also found in a number of drier locations such as field headlands and in some instances, crops. This spread into different habitats may have started back in the 1930s, accelerating through the 1960s (Bell 1969), and the species is now often found alongside Sedge Warbler *Acrocephalus schoenobaenus* in oilseed rape (K. Bowey pers. obs.). There is data available showing breeding densities are four times greater in oilseed rape than on other farmland such as cereal crops or even set aside, and it appears that the presence of rape is crucial in reducing this species' dependency on wetlands (Gruar *et al.* 2006). In 2007, five singing males were in a single rape field between Ryhope and Seaham Hall. As observed in the *Atlas*, this habit may be widespread in lowland County Durham, although not widely reported. At Ryton Willows, Reed Buntings have for many years been nesting in gorse scrub alongside Yellowhammers *Emberiza citrinella* (Bowey *et al.* 1993).

As represented in the *Atlas*, some of the largest concentrations of occupied tetrads for this species are located in the Wear lowlands, around the centre of the county, with other significant concentrations at the mouth of the Tees and along the River Don in the northeast (Westerberg & Bowey 2000). In 1995, a total of 37 pairs were counted in the agricultural belt running from Moorsley to Dawdon, giving some indication of breeding densities in such habitats; while in 2008, breeding birds were reported from no fewer than 92 locations, from coastal rough grasslands and the Tees Marshes to as far west as Balderhead Reservoir, not that far from the county's Cumbrian border (Newsome 2009).

The densest breeding congregations of Reed Buntings tend to be in marshland habitats, which are bordered or merge into scrub or willow/birch *Salix* and *Betula* woodland edge. Ideal habitats can be found at locations such as Shibdon Pond and Rainton Meadows. Breeding season records for 2007, regarded as indicative of the broader situation in the county, included 15 singing males at Rainton Meadows (no change from 2006) with 20 territories at Brancepeth Beck, where active management of the site was considered to have brought about an increase from 12 pairs in 2006 (Newsome 2008). There were also population increases in the Hetton area with the two main areas there holding a combined total of 18 pairs; more than double the eight pairs present during the 2006 breeding season - perhaps an indication of decreased winter mortality. A three-kilometre coastal strip around Noses Point held eight singing males (Newsome 2008).

As highlighted in the *Atlas*, recent studies have suggested that there has been a long-term decline in this species (Westerberg & Bowey 2000), and like many wetland birds, it has been adversely affected by wetland drainage (Prys-Jones 1984). The declines were particularly prevalent between 1975 and 1983 (Marchant *et al.* 1990), and most severe in the north and west of Britain (Gibbons *et al.* 1993). The increased use of herbicides, which reduces the volume of available weed seeds to this and other granivorous birds, such as Linnet *Carduelis cannabina* and Tree Sparrow *Passer montanus*, along with other aspects of agricultural intensification, have also been implicated in the species' decline (O'Connor & Shrubb 1986). Whilst the reduction in numbers does appear to have been manifested in the county, it does not appear to have occurred to the same degree as in some parts of the country. Sites that still have high densities of breeding territories include: Rainton Meadows (22), Hetton Lyons (18), Hetton Bogs, Shibdon Pond, Butterby and Blackhall Rocks at the coast, with survey data from County Durham in 2008 suggesting populations in excess of 50 birds in the best tetrads (Newsome 2009).

The *Atlas* made a population estimate for the breeding population of between 510 and 850 pairs. This was based upon a calculation that between four and six territories were found per occupied tetrad in the borough of Gateshead and extrapolating this to the whole county, working on the assumption that this was typical for the county in all occupied tetrads (Westerberg & Bowey 2000). In the species' core range in the county, there is no evidence to suggest that there has been any significant change since the publication of the *Atlas*. Much additional and detailed information for the species' status in the Teesmouth area was gleaned through the survey work for *The Breeding Birds of Cleveland*. The breeding distribution of the species in that area is concentrated around Teesmouth and the North Tees Marshes in particular, with census work estimating as many as 300 pairs (Joynt *et al.* 2008). Most of these birds can be found in and around the marshes at Teesmouth, with good numbers also along the river corridor and smaller numbers of pairs, scattered across inland agricultural landscapes and locations, away from the developed areas (Joynt *et al.* 2008).

Although in common with many other seed-eating passerines there has been an overall decline in Reed Buntings populations, with a 17% national decline between 1967 and 2008 (Baillie *et al.* 2010), this does not give the full picture. There was a sharp increase in numbers in the second half of the 1960s but this was reversed in the 1970s and it is tempting to speculate that these changes may have been associated with a move towards drier habitats, followed by the adverse effects of agricultural intensification thereafter. Populations have remained relatively stable since the 1980s and there may even be some evidence of small scale local increases since 2000.

Recent non-breeding status

Birds flock together in winter and there is also some evidence of altitudinal migration, for birds occur in moorland habitat round Smiddyshaw Reservoir in summer, but are absent from this area after October.

During the 1970s, large winter roosts gathered at Shibdon Pond, where there were regularly over 100 together in winter from 1972 to 1976; the largest count being 300 on 15 February 1975. More recently, typical roost sizes at that site have numbered just 30 to 40 birds. There have been some reports of large feeding flocks elsewhere in the county during the winter months: in 2005, over 100 were gathered at Murton in March with 200 at nearby Dawdon in December of that year. In 2006, 125 were attracted to a feeding station at Dalton Moor, where at least 50 gathered during January and February 2007. Large areas of set-aside at High Sharpley and Seaton Pond attracted up to 140 at the former site on 26 November and 120 at the latter five days later.

An individual that visited a bird table at Heighington on 29 March 1979 is illustrative of the increased range of habitats now used by this species; such behaviour has become more widespread, though not common, in recent decades and is being increasingly noted in County Durham. Indeed, in suburban gardens bordering suitable agricultural land, Reed Bunting can be amongst the commonest species at feeding stations, with counts of up to 12 birds at one such locality near Jarrow in the winter of 2010/11 (M. Newsome pers. obs.). Prolonged periods of snow-cover can impact severely upon Reed Bunting populations, as it is largely a ground-feeding species (Lack 1986).

Distribution & movements

Reed buntings breed over much of Europe, much more locally in the south. Birds breed east through Russia as far as Siberia and through central Asia as far as Japan. The northerly breeding birds, which penetrate into the boreal zones of northern Norway, tend to be migratory, wintering to the south but within the breeding range. Much of the British population is largely sedentary or at least makes much less extensive movements (Parkin & Knox 2010).

Increased numbers are seen in coastal areas particularly during the autumn migration period in most years at well-watched sites such as Marsden and Hartlepool, which indicates that autumn passage birds occur, albeit in relatively small numbers but with very few in spring. Nine at Hartlepool Headland on 7 October 1977 is the largest count of apparent migrants; these birds arrived alongside large numbers of thrushes and other migrants. One ringed at Vest-Agder in Norway on 14 October 2001, which was at the Long Drag on 13 October 2002, demonstrates that some individuals of this species cross the North Sea to winter in Durham.

Judging by the few ringing recoveries, it seems likely that the flocks, when present, are made up of mostly local birds, with a few from elsewhere in northern Britain and the occasional continental bird. Ringing has shown that there is some movement of birds within Britain. Many thousands of Reed Buntings have been ringed in Durham, in particular around Teesmouth, but in relation to the numbers ringed there have been few birds that have demonstrated significant movements, suggesting that many local birds have a rather sedentary nature. This is borne out by the local movements of birds ringed in the north of the county in the early 1990s at Shibdon Pond, from where a bird travelled to Big Waters at Seaton Burn in Northumberland, some 12km to the north with an almost reverse movement of a bird from Big Waters to Lockhaugh in the lower Derwent valley, a distance of 16km. A similar distance was covered by a bird ringed a few decades earlier at Gosforth Park, Newcastle on 27 October 1962, which was recovered at Birtley on 3 April 1963.

One of the longest distances travelled by a Durham-ringed bird was by the female ringed on 16 October 1977 at Hargreaves Quarry, Teesmouth that was controlled at South Milton Ley, Thurlestone in Devon on 26 November 1977; a journey of 522km to the south west and perhaps an example of a female moving further than males to winter. Another ringed in this area, on 23 July 1994, was even further afield on Jersey on 22 February 1997 (Blick 2009). Meanwhile, one ringed near Hexham, Northumberland, on 26 July 1997 was controlled in Hargreaves Quarry on 18 October 1997. In the other direction, one ringed at Pocklington, Humberside, on 29 December 2000 was at Billingham Bottoms on 9 June 2001; perhaps a Teesside-breeding bird that had been wintering in southerly climes.

Possibly the severity of the weather influences how far birds move during the winter - a number of Teesmouth ringed Reed buntings have travelled south and west, perhaps to avoid harsh weather such as: one ringed at Haverton Hole on 13 August 1983 was near Rotherham, South Yorkshire, on 26 January 1984; the bird ringed at Saltholme Pools on 23 July 1988 which was at Walcot, Shropshire, on 1 December 1988; a bird ringed at Haverton Hole on 30 August 1997 that was at Wing, Buckinghamshire, on 22 January 1998 and one ringed by the Long Drag on 21 September 2002 was at West Felton, Shropshire, on 24 November 2002 (Blick 2000).

Corn Bunting
Emberiza calandra

A once common resident, which now scarce and becoming increasingly rare, largely confined to the centre and east of the county.

Historical review
The Corn Bunting is closely associated with arable crops and has declined nationally with changes in agricultural practice; a decline that has also been very much evident in County Durham over recent decades.

In 1840, Hutchinson gave its status in Durham as '*common*' and some 30 years later, Hancock (1874) called it a '*common resident*'. An extract from *Notes on the Avi-fauna of Upper Teesdale* named this species as Common Bunting, rather than Corn Bunting, which was indicative of its status at the time as being "*exceedingly common throughout the summer in the meadows*" (Backhouse 1884). "*A resident of local occurrence*", is how Temperley (1951) described it, but he remarked on its being "*less numerous than formerly*". Commenting on its decline, Donald, Wilson & Shepherd (1994), highlighted that "*information on changes in the status of the Corn Bunting between North Yorkshire and the Scottish border is scant, although the species was described as a 'common resident' in Northumberland and Durham by Hancock in 1874 (Hancock 1874)*". Nelson (1907) said that it was scarce in the more wild and moorland tracts but not uncommon in the cultivated district of the north west of Yorkshire (such as the upper Teesdale area). Bolam (1912) described it as a "*common resident*" in neighbouring Northumberland, most common perhaps along the coast. The same author wrote 20 years later that the species "*has decreased in some places since so much land went out of cultivation*" (Bolam 1932).

The Rev. George Courtenay, commenting of his time when living in Sunderland, said, "*The outskirts of Sunderland are strong in Corn Buntings, alike on the Fulwell side to the north and the Hylton side to the west. You might almost depend on finding them in these parts at any time of the year. I have counted (in December 1918) as many as six together on the telegraph wires in Hylton Road. And I have a note for May 13th of the same year 'Corn Buntings all the way from Ford Hall to Offerton bridle path, singing from hedge and tree, from post and telegraph wire, and even from the ground'*" (*The Vasculum* Vol. XIX 1934).

Further declines in Durham were noted by Temperley (1951), who found that the species was more coastal in its distribution than in previous years and less common throughout. He recorded a reduction in both range and numbers, noting winter flocks along the coast, which he believed to be of 'local stock'. He reported that the species was still common in Teesdale, between Barnard Castle and Middleton, and also in Weardale, between Stanhope and St. John's Chapel. It was noted in a 1940 article by C.J. Gent, about the Corn Bunting, that in the Tees Valley it was "*... much less common than formerly, decreases being reported from all parts of the area*". Gent however recorded it at a number of sites in Durham and that it was fairly plentiful at Usworth and Washington (*The Vasculum* Vol. XXVI 1940).

In the eastern portion of the county, a long-term set of observations by David Simpson in the Shotton Colliery area, commencing in the 1950s, neatly summarises the species' decline locally. As he stated, from it being "*an everyday bird in this area to there being none whatsoever, all within my bird watching lifetime*" (Simpson 2011). There used to be winter roosts of up to 70 birds at Shotton Brickworks Pond in the 1950s and 1960s. "*On 2 July 1956, I did a 'survey' around Shotton Colliery, walking west, south of the village to Dixon's Estate, north to the Pemberton Arms Pond then east along the B1283 and on towards the A19, then south along the A19. This 'transect' was of approx three and a half miles and in this distance I counted seven pairs* (or singing males)". The last local singing male was one he saw at Shotton Pond in spring and summer 1997, "*today there are none in this same area. Like the time, I ask myself where they have all gone"* (Simpson 2011). Around this time, further north, the species was said to be 'everywhere' along the coastal strip at South Shields Leas, which was in the 1950s a rich mosaic of arable crops and limestone grassland (J. Coulson pers. comm.).

During the 1960s, despite decreases in the decades since the end of the Second World War, numbers were still high enough in the south of the county to produce significantly-sized, winter congregations; for example at Teesmouth *c.*130 birds were at Graythorp on 10 December 1966 with 62, probably some of the same birds, at nearby Cowpen Marsh on 9 Feb 1967, which were the largest respective annual counts in the county at that time. One of the highest counts in the south east of the county was of 180 birds between Hart and Hartlepool on 5 February 1970 (Blick 2009).

The *Breeding Atlas* (Sharrock 1976) recorded its presence in all of the county's ten-kilometre squares west of Crook and Lanchester, with a more sporadic distribution further up the dales and with possible breeding recorded on the south side of the Tees, high in upper Teesdale (Mather 1986). The species was also well represented as a breeding bird along the mid-Tees river corridor up to the early 1980s, but since then, the only truly western reports were from Middleton-in-Teesdale in 1974 and the Barnard Castle area in May 1986. Despite the very sharp national decline in the species, particularly from the mid-1970s, there were still good numbers in the lower Tees valley during that decade, with 64 at Teesside in January 1976, 80 at Cowpen Bewley on 1 January 1978 and a similar number at Teesmouth late in 1981. Also in the south of the county, a large flock of 138 was gathered at Darlington on 14 November 1976.

The 1980s saw an accelerating decline with large winter flocks becoming more and more unusual. There were, however, counts of 80 at Houghton and 60 at Longnewton in 1986 and *c.*80 were concentrated into coastal stubble at Ryhope in 1991 by heavy snow inland. The 1990s produced winter flocks of 72 at Philadelphia near Houghton in 1994 and 70 in set-aside fields at Shotton Colliery in January 1997 but there have been no gatherings of such magnitude since then.

A summary of the species, from the long studied 'Gateshead area', which includes the lower Derwent valley, neatly demonstrates the decline of this species in the context of Durham County. Writing of the Derwent valley, Robson (1896) described the species simply as *"resident in the valley"*. Some 55 years later, Temperley commented that it was becoming much scarcer in the *"Vale of Derwent"* although at his time of writing, he recorded that it had been recently noted around both Winlaton and Greenside. Further marked decreases occurred in this area from the early 1970s. It was once spread sparsely across the whole of the Gateshead borough, being found breeding at Greenside, Kibblesworth, Ryton and in the Team valley, with regular winter flocks noted at one or two localities. Ryton was a favourite wintering spot for the species, with flocks of 50-60 birds being present there in February 1973 and February 1976 (Bowey *et al.* 1993). As recently as January 1984, a flock of 15 birds was found near Chopwell, but by the middle of the 1990s there were only increasingly sporadic records of singing males across the area. Prior to the building of the Ryton bypass in the late 1980s three to five males could still be found annually in the agricultural land to the east of Greenside but these birds were lost as their habitat was destroyed (Bowey *et al.* 1993). Direct habitat destruction however, was not the principal cause of this species' decline in this area. To the mid- to late 1990s, areas still occasionally holding birds included the farmland around Kibblesworth and Lamesley, and in the far east of the borough at Follingsby. But by this time the average season would probably see no more than three to five territorial males in the whole of the borough. In 1995, there was a bird in territory at Kibblesworth and one was present the following summer a little distance away, at Birtley STW; a juvenile there indicated successful local breeding. Two singing birds at Lamesley during April 1999 are, to date, the last recorded for Gateshead and as the decline continued apace, none were left by the early part of the 21st century.

Recent breeding status

The Corn Bunting, like a number of other previously common farmland breeding species, has suffered a major decline in its numbers over the past 100 years (Marchant *et al.* 1990, Mead 2000). This was exacerbated by the adoption of intensive farming practices after the Second World War. Such agricultural intensification led to a very steep decline in the county's small and increasingly fragmented population of what was a formerly common and widespread species. Data published in 2010 (Baille *et al.* 2010) showed an overall national decline in the population of 86% between 1967 and 2008, with particularly sharp decreases noted from the mid-1970s. While this decline is generally ascribed to agricultural change, it is more specifically due to a combination of poor winter survival resulting from a lack of winter stubble offering a reliable food supply, coupled with low productivity.

Although breeding data suggests that there has been an increase in survival rate in terms of the number of chicks reared per nest (Crick 1997, Donald 1997), it appears that fewer birds now produce second broods (Brickle & Harper 2002).

During the period between 1977 and 1988, Common Bird Census indices recorded a decline of around 60% nationally (Marchant *et al*. 1990), but while there has been some local stabilisation of the population, possibly because of the adoption of more environmentally-aware agricultural policies (Newsome 2006-2010), the long-term downward trend continues to the present. Declines of 80% have been recorded across England between 1970 and 2001 with further national declines of 39% noted by the BBS index over the period 1994 to 2006 (Raven *et al*. 2007).

At the time of the production of the *Atlas*, the mapped distribution of Corn Bunting was believed to be an accurate depiction of the species' breeding distribution in the county (Westerberg & Bowey 2000), but there have been further reductions since then. The declines occurred earliest, further west in Durham, and the Corn Bunting is now entirely absent from the county's uplands and probably has been for a generation. The species was still present in the mid-Tees valley, as late as the early 1980s, for it was noted in territory at Hutton Magna in the summer of 1982 and at Wycliffe-on-Tees in the summers of both 1982 and 1983 (K. Bowey pers. comm.). It had not been recorded west of the A68 for more than two decades until 2010, when six birds were found close to Staindrop. In February 2010, four were noted to the north of Crook, just east of the A68, and in 2008 a single was noted at West Auckland.

The *Atlas* noted the species' breeding distribution as being delineated by a line running south, in a rough demarcation, from Gateshead, through Durham City, down to Bishop Auckland. In 2000 (Westerberg & Bowey 2000), it occurred in the north no further than 12km from the coast; while in the south, the breeding range extended inland for some 30-35km. In the last 10 years the species' distribution has become ever more patchy across Durham and it has been lost from a number of areas in the south east, the north east and centrally. In the south east of the county, to the north of the Tees estuary, there were estimated to be around 20 pairs between Billingham and Hartlepool in 1983, but numbers crashed with none being known to be present by 2007 (Joynt *et al*. 2008). During *The Breeding Birds of Cleveland* survey work, between 1999 and 2006, the farmland around Elwick held a concentration of up to seven singing males, but survey work in that area in 2007 revealed just two 'pairs' (Joynt *et al*. 2008). Birds are also present, in tiny numbers in the Hart and Longnewton areas, but by 2010, this species' hold in these areas was considered 'tenuous' at best.

The agricultural corridor between Sunderland and South Shields was well populated to at least the start of the 1980s. A healthy population of breeding birds was located on Cleadon Hills, where a CBC study area from 1976 to 1980, held 12 pairs and was recognised at that time by the BTO as having the highest density of territories of any CBC plot in England. (T. I. Mills pers. obs.). By the early 2000s the numbers were greatly diminished and that decade saw them disappear completely. A pocket of birds to the north east of Bishop Auckland was present into the early 2000s but were lost over the next ten years. Today, the species is most widespread in the belt of arable land running south, down the coast, from Sunderland to Easington and extending inland to Houghton-le-Spring and Pittington; occurring in central pockets of the county around Bishop Middleham, over the Magnesian Limestone grasslands and the arable farmland that flourishes over some parts of that strata.

For the *Atlas*, targeted observer coverage of the species in the breeding seasons of 1993 and 1994, indicated that around 80-100 males were holding territory in the county (Armstrong 1993-1994). The complex mating strategies of the species (Gibbons *et al*. 1993) make simple estimations of numbers based upon the number of singing males, problematic. It seems fair to say however, that during the late 1990s the county's population of Corn Bunting was considered to be not much in excess of one hundred 'pairs'. The 1994 targeting of this species by the Durham Bird Club produced records for a minimum of 84 singing males at 23 locations, with two birds as far west as Hedleyhope Fell at 285m above sea level. By 2000 the future of this species in Durham looked bleak, with less than 30 singing males reported from just 14 locations, none of which held more than four singing birds. The rapidity of the decline of this species in the north east of England was illustrated by Blick's statement about the population of Corn Bunting in Cleveland, where the breeding population was estimated at about 500 pairs in 1985, 25 pairs ten years later in 1995 and just two singing males in 2009 (Blick 2009).

The increase in the area of winter stubble and set-aside brought about by government subsidies however, may have saved this species from extinction in the county. Since the low around the Millennium, there have been signs that the population decline has halted, and with the introduction of land management strategies in areas where birds remain, there are even signs of some localised recovery. In 2006, the majority of records concerned singing males from April to early July, with almost 50 of these noted 'across' the county, virtually all of which could be found in the lowlands east of Durham; the most westerly record came from Binchester, west of Spennymoor. Since then, the population of singing males was more or less stable with 40 in 2007, 35-45 in 2008 and a maximum of 48 in 2009. The current stronghold for the species is around Bishop Middleham, on the Magnesian Limestone area of the county, with a minimum of 25 singing males in this area alone in 2006, and 14 singing males at Farnless Farm in 2009. In 2010, this was the primary site for the local application of subsidised interventions on behalf of this species, including successive strip-planting of seed-bearing plants. An early sign of the positive effects of these efforts came on 21 August 2009 from Bishop Middleham, where a post-breeding gathering of 34 birds included at least 12 juveniles. Also in 2009, the area between Mordon and Preston-le-Skerne held eight to ten singing males with eight other sites containing singing males in small numbers. Breeding data for 2010, suggested that the county population was reasonably stable with approximately 38 singing males at 14 locations, a situation mirrored in 2011.

Marchant *et al.* (1990) noted the Corn Bunting's apparent link to barley and the benefits it derived from the availability of winter stubbles. Across its breeding range in Britain the species shows a close affinity for arable land with cereal crops, in particular barley *Hordeum vulgare* (Gibbons *et al.* 1993), hence its popular old-fashioned folk name 'fat bird of the barley' (Cocker & Mabey 2005). Prominent farmland features such as fence posts and power lines provide an important function as song posts (Gibbons *et al.* 1993). Clearly, a species so sensitive to agricultural change has the potential to be severely affected by any intensification of local farming practices. Whilst even at the turn of the 21st century the prognosis was grim, there is now some hope that the species can be saved, provided that targeted interventions can be sustained.

Recent non-breeding status

Typically, there are now few reports of this species outside of the breeding season. Previous winter concentrations would appear to have been largely influenced by the availability of food, primarily from winter stubble or more recently set-aside, with a secondary factor of proximity to the coast perhaps having greater influence in periods of inland snow cover.

Since 2000, winter maxima have been much smaller than historically. In the winter of 2002, there were 30-40 at the three main sites of Bishop Middleham, Hart and Sedgefield, with a post-breeding gathering of 50 birds at Bishop Middleham in 2003. A single flock of 46 was noted near Murton on 1 March 2005, with 32 reported at Hart in the first winter period of the same year. In 2006 and 2007, very small numbers were recorded, with 17 at Dalton Piercy being by far the best count; while 2008 produced two flocks totalling 55 birds on 5 March. The period immediately before the breeding season would appear to be a good time for larger flocks to be found, perhaps the result of reduced winter feeding concentrating birds in favoured areas. In 2010, an impressive flock of 50 was at Farnless Farm in January, amongst a combined 650-strong mixed bunting flock, attracted by specific management for farmland birds. A post-breeding flock of 26 was noted at Sherburn in August, with 24 at Pittington in November and 18 at Mordon in December. During 2011, a similar situation was evident with the peak post-breeding flock being 28 at Sherburn on 4 August.

Distribution & movements

Corn Buntings breed over most of Europe, and into Northern Africa, though its northerly limit is south of Scandinavia and most of the Baltic. The species in Britain is a sedentary one, though in times past when numbers were larger, it gathered in large wintering flocks, so some breeding areas appeared deserted in winter (BWP). The species has experienced very considerable declines over much of its range, which has been particularly marked in Britain. The only known control of a ringed bird in the county, concerns one that was ringed at Newton Bewley on 14 February 1984 and was near Seal Sands on 18 January 1985 (Blick 2009).

Brown-headed Cowbird
Molothrus ater

A very rare vagrant from North America: one record.

The sole record of Brown-headed Cowbird for County Durham is of a male photographed in a Seaburn garden by local resident John Pillans on 10 May 2010. John first noticed the bird at around 11:00hrs at the feeding station in his garden and realising that it was something unusual he took some photographs and video images to assist him in identifying it (Pillans 2010). News of this incredible occurrence came to light on 3 June, when the photographs were circulated to ex-Durham bird watcher David Parnaby in Scotland. He had been told of this unusual bird from his father, a resident of Seaburn. David confirmed the identification as a Brown-headed Cowbird. In a direct parallel to this occurrence, the first for Northumberland was photographed visiting a garden at Belford from 25 April until 2 May 2009.

Distribution & movements
Brown-headed Cowbird breeds throughout much of North America from British Columbia and Newfoundland southwards to northern Mexico. Southern populations are largely sedentary though those from the northern part of the range are known to be partially migratory and head south during the winter (Dolbeer 1982). It is an extremely rare vagrant to Britain with just five records including the bird at Seaburn in 2010. The first British record was on Islay, Strathclyde on 21 April 1988, though there were no further records until three were seen in 2009. With the exception of a bird in Pembrokeshire in July 2009, all of the British records have been in the period from late April until early May, which corresponds with the northward migration of the species in North America. There were also single birds in France in May 2010 and Norway in June 2010.

Appendix 1

For reference, the following are the definitions of Categories A, B, C, D and E of the British List as described by the British Ornithologists' Union (BOU).

A Species that have been recorded in an apparently natural state at least once since 1 January 1950.

B Species that were recorded in an apparently natural state at least once between 1 January 1800 and 31 December 1949, but have not been recorded subsequently.

C Species that, although introduced, now derive from the resulting self-sustaining populations.

C1 *Naturalized introduced species* – species that have occurred only as a result of introduction, e.g. Egyptian Goose *Alopochen aegyptiacus*

C2 *Naturalized established species* - species with established populations resulting from introduction by Man, but which also occur in an apparently natural state, e.g. Greylag Goose *Anser anser*

C3 *Naturalized re-established species* - species with populations successfully re-established by Man in areas of former occurrence, e.g. Red Kite *Milvus milvus*

C4 *Naturalized feral species* - domesticated species with populations established in the wild, e.g. Rock Pigeon (Dove)/Feral Pigeon *Columba livia* .

C5 *Vagrant naturalized species* - species from established naturalized populations abroad, e.g. possibly some Ruddy Shelducks *Tadorna ferruginea* occurring in Britain. There are currently no species in category C5.

C6 *Former naturalized species* – species formerly placed in C1 whose naturalized populations are either no longer self-sustaining or are considered extinct, e.g. Lady Amherst's Pheasant *Chrysolophus amherstiae* .

D Species that would otherwise appear in Category A except that there is reasonable doubt that they have ever occurred in a natural state. Species placed in Category D only form no part of the British List, and are not included in the species totals.

E Species that have been recorded as introductions, human-assisted transportees or escapees from captivity, and whose breeding populations (if any) are thought not to be self-sustaining. Species in Category E that have bred in the wild in Britain are designated as E*. Category E species form no part of the British List (unless already included within Categories A, B or C).

Category D species

All of the following species have been placed in Category D of the British List by the BOU. These species do not form part of the main list, and are essentially held in a holding category, awaiting further information or records to allow a decision to be made as to their provenance. The category is described by the BOU as "Species that would otherwise appear in Category A except that there is reasonable doubt that they have ever occurred in a natural state". Great White Pelican *Pelecanus onocrotalus* is also currently held in Category D on the basis of a wide ranging bird in 1975, though is not included here as the bird in County Durham was specifically rejected as an escape (BOU 2009).

Ross's Goose
Anser rossi

A very rare vagrant from arctic Canada, or escape from captivity: one record.

The only County Durham record of this potential vagrant from North America is of an adult found by Martin Blick at Saltholme Pools, Teesmouth on 5 October 2007. Initially seen in flight around midday amongst a flock of around 80 Pink-footed Geese *Anser brachyrhynchus*, fortunately the flock settled on Back Saltholme for four and a half hours, before heading off south east over Middlesbrough. This bird had been seen previously at various sites in Northumberland in late September, and was presumed to be one of up to three birds which have regularly wintered amongst 'Pink-feet' in Norfolk since November 2001.

Ross's Goose breeds in the Canadian arctic, and winters in the southern United States and occasionally northern Mexico. Historically, the species wintered solely in the Sacramento Valley in California, though a large increase in population (from 3,000 to over 180,000 birds) has led to it now winter regularly in both Texas and Louisiana, with small numbers along the eastern seaboard (Scott 1995). There are numerous records of Ross's Goose in Europe, including Britain, though these have primarily been deemed escapes from captivity. Only in the Netherlands, where at least two birds have wintered amongst Barnacle Geese *Branta leucopsis* since 1985 (*Dutch Birding* 26: 100-106), is it accepted as a wild vagrant. Recent British records accompanying flocks of Pink-footed and Barnacle Geese are thought to relate to genuine vagrants from North America, and the species was placed in Category D by the BOU in 2005.

White-headed Duck
Oxyjura leucocephala

A very rare vagrant from Spain and western and central Asia, or escape from captivity: one record.

The sole record of this species for County Durham is of an adult male found by John Dunnett on the Reclamation Pond, Teesmouth on 29 March 2004. It remained on the North Tees Marshes until 24 October, often to be found on Saltholme Pools, and was often seen to be very aggressive towards nearby Ruddy Ducks *Oxyjura jamaicensis*. Remarkably, it re-appeared on the same date in 2005, though for a much shorter stay, being present from 29 March to 7 April. This bird was considered to be the same individual that spent the winters of 2003/2004, 2004/2005, and 2005/2006 at Broadwater Gravel Pits in Essex, being last seen there on the 22 January 2006.

White-headed Duck breeds in Spain, North Africa and western Asia. The current population is estimated at around 7,900-13,100 individuals, though just 2,500 are to be found in Spain (Birdlife International 2011). The Spanish population has increased dramatically from just 22 birds in 1977 following targeted conservation measures and is thought to be the origin of vagrants to northern Europe including Britain. There have been several records in Britain since the first in 1978, though it is not included in Category A of the British List as it widely found in captivity and there are many records of known escapes. Nonetheless, there are historical records from the Netherlands and elsewhere in Europe that suggest that it may occur as a genuine vagrant to Britain.

White-headed Duck, North Tees Marshes, summer 2004
(B.Hanson)

Red-headed Bunting
Emberiza bruniceps

A very rare vagrant from central Asia, or escape from captivity: four records.

All records:

1965	Hartlepool Headland, male, 11 September
1971	Hartlepool Headland, male, 3rd to 4 June
1976	Dorman's Pool, Teesmouth, male, Teesmouth, 5 June
1989	Haverton Hole, Teesmouth, male, 2 September (trapped)

The first three records in County Durham are all from when the species was still very common in captivity, with up to 2,000 birds being imported in Britain each year during the 1960s and 1970s (Evans 1994). The bird at Haverton Hole was dismissed by the observers as an escape, and although subsequent to the export ban, the species was still on sale in a Hartlepool pet shop until at least the mid-1980s (G Joynt *pers. comm.*). All of the records in Durham are thought to refer to escapes from captivity.

Red-headed Bunting breeds in central Asia and winters mainly in the Indian subcontinent. This species has had a chequered history as vagrant to Britain being originally included on the British List before removal in 1968 due to the large numbers held in captivity at the time casting doubts as to the origin of British records. There had been almost 300 birds recorded in Britain by 1990 (Evans 1994), though there was a sharp decline in records since an export ban on wild birds was imposed by the Indian Government in 1982, with just three spring records from 1999 to 2006 (*BB* 100: 540-551); broadly comparable with the vagrancy pattern of other Asian overshooting species.

Unconfirmed and Unacceptable Records

Records of the following species from County Durham have all been published with some degree of certainty in relation to their identification, though for a variety of reasons, they do not fully satisfy the criteria for being included in the list of birds of Durham. In some instances, the identification was proven and the species was believed to have occurred within the Durham recording area, but this latter criterion could not be stated to have been unequivocally met. In one instance, it seems that there may be a good case to include the species on the Durham list, and as a potential first record for the British Isles.

Red-breasted Goose
Branta ruficollis

George Temperley did not officially include this species in his list of birds of Durham, but he mentioned it in the text of *A History of the Birds of Durham*, on the strength of information from Hancock (1874), which documented two Durham records. Tunstall had a specimen in his collection which had been shot near London in 1776, but he also mentioned a bird that was taken alive in the neighbourhood of Wycliffe-on-Tees, where he lived and was kept on a nearby pond until its death in 1785. It is not known exactly where this individual was captured, though it may have been in Yorkshire as Tunstall lived on the south bank of the Tees. Its remains were not thought to have been preserved and no specimen exists. The second of Hancock's records came from John Hogg (1845), who said that two specimens were seen at the mouth of the Tees 'of late years'. One of these was said to have been shot at Cowpen Marsh, though the exact details are not known, and Temperley expressed his doubts about the veracity of this record. None of these early records were accepted by BBRC and as such, the species is not included in the Durham list.

A subsequent record of Red-breasted Goose relates to an adult bird seen with Greylag Geese *Anser anser* on Dorman's Pool, Teesmouth on 7 January 1998, which was later seen in Yorkshire at both Tophill Low and Nosterfield from 17 January to 13 March 1998. Although considered of dubious origin by the original observer, this bird was initially accepted as a wild bird by BBRC (*British Birds* 92: 563), though it was reviewed in 2000 and considered to be of captive origin (*British Birds* 94: 461). As such, the record is unacceptable as a county first.

Red-breasted Goose breeds in Siberia and winters principally around the Black and Caspian Seas. There had been a total of 77 British records by the end of 2010.

Swinhoe's Petrel (Swinhoe's Storm Petrel)
Oceanodroma monorhis

A female trapped and ringed at Tynemouth, Northumberland during the early hours of 28 July 1993 (*British Birds* 87:509) was released from Tynemouth pier and watched flying out to sea in the moonlight. This bird was judged to have crossed the centre of the River Tyne into County Durham by those present at the time; a distance of around 200 metres this was never officially ratified by BBRC and as such, the species cannot be officially added to the county list.

Swinhoe's Petrel breeds on islands in the Pacific Ocean off China, Japan and Korea, and spends the rest of the year at sea ranging as far as the Indian Ocean and the Arabian Sea. Five birds have been recorded in Britain, three of these during nocturnal trapping sessions for Storm Petrels *Hydrobates pelagicus* at Tynemouth between 1989 and 1994, including the bird above which was trapped each July between 1990 and 1994. Elsewhere in Western Europe birds have occurred in Ireland, Spain, France, Norway and Madeira, most of which have been trapped and ringed.

Little Crake
Porzana parva

This species was mentioned by Lofthouse (1887) in his *The River Tees: its marshes and their fauna*. A bird was said to have been obtained in "*the Cowpen Marsh area*" some time prior to 1887. There is no mention of this record in Temperley (1951), though Stead (1964) in his *"The Birds of Tees-side"* stated that although there was a specimen of Little Crake held in the Middlesbrough Museum, it was not possible to determine if it had been shot locally; he did not accept the record. A subsequent record is of a singing male at Shibdon Pond on 25 June 1996 (Armstrong 1997). Although published in *Birds in Durham 1996*, with the caveat that full details were awaited, this record was never submitted to BBRC and as such remains unacceptable.

Little Crake breeds in France and across much of Eastern Europe into central Asia. Its wintering range is poorly known but is thought to include both Sub-Saharan Africa and the Indian Subcontinent. There had been a total of 100 British records by the end of 2010, most of which are from the southern half of England.

Sooty Tern
Sterna fuscata

Clayton (1899) made reference to this species in his *Ornithological notes for 1896-7*. A bird was said to have been shot at Teesmouth in the autumn of 1896. This record was not accepted by Stead (1964) and is not listed by Temperley (1951) in his review of the county's avifauna. This record does not form one of the 26 British records accepted by BBRC to date.

Sooty Tern is essentially oceanic, though is an abundant breeder on islands in the tropics and can be found in the Caribbean, the Atlantic, Pacific and Indian Oceans, as well as on islands in the Red Sea. It ranges widely around the tropics outside the breeding season.

Blue-tailed Bee-eater
Meriops philippinus

A bird of this species was recorded by Hancock (1874) as having been shot near Seaton Carew in 1862, but Temperley (1951) said that it was not admitted to the British list and he believed that the specimen was shot at

Branch End on the Yorkshire side of the Tees. Mather (1986) pointed out that the record of this species in August 1862, documented as being taken at Seaton Snook, on north side of Tees (as reported to John Hancock by Thomas Hann of Byers Green), must have been caught on the south side of the Tees estuary (the Yorkshire side), based on the fact that it was said the bird was shot from a perch on a heap of slag, and that no slag-heaps existed at Seaton Snook at the time the specimen was procured.

Blue-tailed Bee-eater is native to South East Asia and is an extremely unlikely vagrant to the British Isles. Along with Madagascar Bee-eater *Merops superciliosus*, it was historically classed as a race of Blue-cheeked Bee-eater *Merops persicus* and as such is included here. Blue-cheeked Bee-eater is a very rare vagrant to Britain with ten records since the first in 1921. A specimen of this species, possibly the bird from Teesmouth, remains in the Hancock collection at the Great North Museum, Newcastle-upon-Tyne.

Cliff Swallow (American Cliff Swallow)
Petrochelidon pyrrhonota

A juvenile was watched flying around the Pilot Station at South Gare, North Yorkshire during the afternoon of 23 October 1988 (*British Birds* 82: 537-538) before being seen to fly west across the river mouth. This bird almost certainly crossed the county boundary, though unfortunately the foggy conditions prevented the observers from confirming this; consequently, this species cannot be included in the county avifauna.

Cliff Swallow breeds across much of North America and winters in western South America from Venezuela southwards to north eastern Argentina. It is a very rare vagrant to Western Europe with just nine British records.

Cedar Waxwing
Bombycilla cedrorum

Two were reported by Newton (1871-1872) as having been taken at Stockton-on-Tees, early in 1850, though he later retracted the record (Alexander & Fitter 1955).

Cedar Waxwing breeds across much of Canada and the southern United States and winters as winters as far south as Central America, and occasionally northern South America. There have been two British records.

Yellow Warbler (American Yellow Warbler)
Setophaga petechia

Yellow Warbler, Axwell Park, May 1904
(K. Bowey, courtesy NHSN)

One record: One was killed against telegraph wires at Axwell Park in the Derwent valley, near Blaydon upon Tyne in the second or third week of May 1904; though this was not recognised as a bird of wild origin by the BOU. The specimen was displayed by a Mr E. Bidwell at a meeting of the British Ornithologists Club (Snow 1992). In discussion, Dr. Sclater said "*that he saw no reason why the bird should not reach England as an accidental autumn visitor*"; it would appear that it was not added to the British List on the basis that it was thought unlikely to "*cross more than three thousand miles of ocean without food or human aid*" (Saunders 1907). The specimen itself was considered to be of the eastern form *aestiva*, which is the most likely race to reach Britain as a vagrant and its occurrence was given credence by Alexander & Fitter (1955), though this record does not appear to have been subsequently reviewed in the light of further occurrences in Europe. At the time of occurrence, this would have been the first Nearctic wood warbler to have occurred in Britain and it was perhaps too lightly dismissed as ship-assisted or an escape.

Yellow Warbler is an abundant breeder throughout much of North America and winters from Central America to northern South America. A noted early migrant, most birds head southward during mid- to late August at a time when there are few Atlantic storms; a reason often cited to account for the relatively few British records of this abundant North American species. There have been five British records, including the first at Bardsey Island, Gwynedd in August 1964, and a further four in Ireland; all have fallen between 24 August and 4 November.

Category E species: Escapes and Failed Introductions

The identification of most of the species in this section is not in doubt, though none are thought to have occurred in County Durham in a wild state. Some have been released, as in the case of Chukar *Alectoris chukar*, whilst others have escaped from captivity. Some of the species involved are also known to occur as vagrants to Britain, though the origin of the birds recorded in County Durham is not thought to be a wild one, though in some instances, it may have been.

Helmeted Guineafowl *Numida meleagris*

Two records: a single was in roadside Hawthorns at Philadelphia, Houghton-le-Spring on the 10 August 2008, and one at Sherburn village on 13 April 2009. This species is native to Sub-Saharan Africa and has been widely domesticated; neither of these individuals is likely to have wandered far from 'home'.

Wild Turkey *Meleagris gallopavo*

One record: a bird was found dead near Shotton Colliery in 2004. A native of North and Central America, it has been domestically reared in Europe since the 16th century.

Chukar Partridge *Alectoris chukar*

At least six records thought to involve pure birds, though the exact picture is complicated by the widespread release of hybrids with Red-legged Partridge *Alectoris rufa* for hunting purposes. Sightings thought to relate to non-hybrids are: three in Baldersdale, on 16 April 1989; at Barmston Pond, on 12 March 1990; several in the Lamesley area during the summer of 1990; three at Salter's Gate, on 19 April 1992 with at least one still present in May; two at Lartington, in January 1994; and, near Thwaite, on 29 May 2009. In addition to these records, birds showing hybrid characteristics were noted at: Barmston Pond on 12 March 1990; two at Whickham Golf Course, in early April 1991; and, three at Muggleswick Common, on 11 January 1994. Native to western and southern Asia, this species was released in large numbers during the 1980s for shooting, though more frequently as a hybrid with Red-legged Partridge. These proved unsuccessful in the wild and their presence led to a steep decline in the numbers of pure Red-legged Partridge, prompting the Nature Conservancy Council (now Natural England) to ban releases of non-native or non-naturalised species in the United Kingdom, in an attempt to halt the decline. From January 1992, it became illegal to release Chukars and hybrids. The bird at Thwaite in 2009 however, demonstrates that illegal releases, or escapes, of this species still take place occasionally.

Silver Pheasant *Lophura nycthermera*

One record: a male was watched walking along the road towards Balder Grange, Cotherstone on 20 June 1992. This sedentary species is native to South East Asia.

Reeve's Pheasant *Syrmaticus reevesii*

Small numbers of birds were known to have been released in the Gibside Estate in lower Derwent valley in the latter part of 1991 and once again in 1992, many subsequent reports are thought to be related to these releases.

An adult male was noted at the Far Pasture Wetland on 14 October 1992 and two females were there five days later (Bowey *et al.* 1993). Birds were in that area seen on a number of occasions during 1993 and the full-length tail feathers of a male, which may have been predated, were found near Lockhaugh Bank Woods in mid-February of that year. Birds were noted again in and around Gibside in 1994, occasionally visiting Far Pasture, in the Derwent Walk Country Park. The species was also noted in the Ravensworth Estate in 1994, with one bird being seen at the nearby Birtley Sewage Treatment works during April. A male was again at Far Pasture on 1 January 1996. In 2001 and 2002, a male, which had escaped locally, was in Edders Acre Dene, near Shotton Colliery. On 10 September 2002, a male was found dead by the roadside of the A19 near Billingham Bottoms. This species is endemic to China, though was introduced unsuccessfully to Bedfordshire and Inverness-shire in the 19th century. Small scale releases for shooting still occur, though the species seems unlikely to become established in Britain in the near future.

Golden Pheasant *Chrysolophus pictus*

Seven records: a male at Ravensworth Grange, Kibblesworth on 13 April 1980; a male at Whickham, on 24 April 1980; a bird extracted from a mist-net at Crimdon Dene, on 31 December 1993; a male at Great Lumley on 21 April 1995; a male seen at both Whitburn and South Shields during 5th to 6 May 1995; a male seen at Shotton Pond, Shotton Colliery on 17 March and 16 November 1994, and again in 2005, 2006 and last noted on 16 November 2007; and a colour-ringed male at Greatham from 9 December 2009 to February 2010. Originally native to western China, a self-sustaining population exists in East Anglia, having been introduced during the latter part of the 19th century. All Durham sightings relate to escapes from captivity, as the species is essentially sedentary in its introduced range.

Indian Peafowl *Pavo cristatus*

There have been many records of this ornamental species in County Durham since at least the early 1950s. Breeding has occurred or been attempted on at least three occasions. A pair was found living wild in Spen Burn, near Rowlands Gill in the spring of 1991; birds persisted in this area for at least two years. A male at Brackenhill Wood, Shotton Colliery in April 2001, was joined by a female in October and produced five young; the male from this pair was present in the Edder Acres Dene area until 2003. Between 2009 and 2011, a pair on the Shearwater Estate, Whitburn nested, raising four chicks to fledging in 2009, though they were thought to have been predated; breeding was repeated in 2010 and 2011. There are also a number of records from the Sherburn Hill area since 2005, including five males on 30 November 2005, indicating that breeding may occasionally have taken place here. Originally a native of the Indian subcontinent, this species has long been popular in ornamental parks and gardens, though has never become established in Britain.

Fulvous Whistling-duck *Dendrocygna bicolor*

One record: a bird at Belasis Technology Park, Billingham on 16 October 2000 had previously been seen in Albert Park, Middlesbrough, in nearby North Yorkshire on 11 October. This species is found in tropical regions of Central and South America, Sub-Saharan Africa and the Indian Subcontinent

Swan Goose *Anser cygnoides*

One published record of the domesticated form often referred to as 'Chinese Goose': two were on the River Tyne at Clara Vale from 4th until at least the end of April 1996. In a non-domesticated form, this species is native to Mongolia, northern China, and south eastern Russia.

Lesser White-fronted Goose *Anser erythropus*

There have been three records of this potential vagrant to the county, all of which are considered to relate to escapes from captivity. The first at Hurworth Burn Reservoir from 17th to 19 October 1985 was present at the same time as a 'Greenland' White-fronted Goose *Anser albifrons albirostris*, though it was not submitted to BBRC on the basis that it was considered to be of doubtful origin; some observers considered that it showed characteristics consistent with that of a hybrid. The second, an adult at Boldon Flats from 26th to 28 May 2005 accompanied Greylag Geese *Anser anser*, and was later seen at Musselburgh, Lothian on the 29 May, and further north at Loch of Strathbeg, Aberdeenshire for five days from 1 June. The following year, an adult by Saltholme Pools, Teesmouth on the 15 June 2006, may have been the same bird. Native to north-eastern Europe and northern Asia, western birds winter predominantly in south-eastern Europe. This species was formerly a regular vagrant to Britain, though it is now very rare following a severe decline in the European population. Most British records were amongst flocks of White-fronted Geese at WWT Slimbridge in Gloucestershire. A re-introduction programme has existed in Sweden for a number of years, with most of these birds wintering in the Netherlands.

Bar-headed Goose *Anser indicus*

There have been numerous records of this species since the first at Shibdon Pond, through much of April 1984. Often associated with flocks of Greylag Goose *Anser anser* the exact number of birds involved is difficult to assess as birds move freely around the county over a number of years. Notable occurrences include: six heading south west over the Reclamation Pond, on 22 June 1993; three on Seaton Common, on 10 May 1994; and, three at Saltholme Pools, Teesmouth on 11th and 17 June 2002. A bird perched on top of Marsden Rock on 21 May 1988 was noteworthy for its bizarre location, and may have been the bird seen at Teesmouth earlier in the year. Birds showing hybrid characters with Greylag Goose were recorded around the North Tees Marshes and Crookfoot Reservoir in 1993 and in January 1994; and, around Saltholme Pools, Teesmouth in September 2008. This species is a native of central Asia and winters in the Indian subcontinent, though a feral population in the Netherlands is considered self-sustaining. This could be responsible for some of the records in Durham, especially as the majority of sightings of more than one individual are from coastal localities. A large number of feral birds are at large in Britain, including a number of isolated breeding pairs.

Emperor Goose *Anser canagica*

One record: a bird at North Gare, Teesmouth on 26 May 1995, was also seen at nearby South Gare in North Yorkshire later the same day and was found dead there a few days later. This arctic goose is native to Alaska and northern-eastern Siberia.

Black Swan *Cygnus atratus*

There have been many records of this Australian species since the first documented occurrence at Washington Ponds on an unspecified date in 1972 (Unwin 1973). The second was at Saltholme Pools, Teesmouth on 9 June 1979. Single pairs were on the North Tees Marshes from 30 April to 2 May 1980, and from 7 June to 30 September 1998; neither pair showed any signs of breeding activity. Other notable records include two at Seaburn on 19 February 2001; three amongst Mute Swans *Cynus olor* at the Marine Park, South Shields, on 8 March 2003; and, one on the River Tees at Stockton-on-Tees from 29 July 2006 to at least 8 October, joined by a second on 9 September. There have been four records of birds passing the observatory at Whitburn: two on 12 September 1998; one on 18 October 1998; five flew south on 1 May 2007; and a single bird flew north on 29 March 2008, with a party of Whooper Swans *Cygnus cygnus*. Native to Australia, and re-introduced to New Zealand, the species is common in captivity throughout much of Europe; it is considered to be an established feral breeding species in the Netherlands. Whilst some of the records in County Durham might emanate from this source, including the bird past Whitburn in 2008, most are probably from the increasing number of feral birds to be found in Britain, where the species has bred sporadically.

South African Shelduck *Tadorna cana*

Three records, all of which accompanied flocks of Shelduck *Tadorna tadorna*: a male at Seal Sands, Teesmouth from 1 September to 1 November 1967; another male at Seal Sands, Teesmouth from the 23 October to 12 November 1977; and, one at Crookfoot Reservoir and then Saltholme Pools during August and September 2011. This species is native to South Africa, where it is largely resident.

Muscovy Duck *Cairina moschata*

There are numerous reports of this domesticated species in County Durham, though it is often difficult to determine the exact origins of such birds as many are found on ornamental lakes and may form part of established collections. Records date back to at least the 1950s when there were regular records from Shotton Brickworks Pond all of which probably came from nearby allotments where free-flying birds were known to be held; there have been no records from this area since 1980. Up to three free-flying birds were present on the River Tyne at Ryton Willows from November 1994 until early 1995. Many of the documented records in Durham have been since 2002, when the species was thought to be becoming established at Ely, Cambridgeshire and may be added to the British List on the basis of this seemingly self-sustaining population. The species is native to Central and South America, though is very common in its domesticated form and comes in a number of colour morphs, unlike the wild stock which is predominantly black. All of the birds recorded in County Durham doubtless refer to local escapes, as the population in Cambridgeshire is largely sedentary.

Wood Duck *Aix sponsa*

Five records: a female at Ward Jackson Park, Hartlepool from 16 March to 3 April 1993; a male at Shibdon Pond, Gateshead from 23 April to 20 May 1996, was also noted on Axwell Park Lake and at Ryton Willows; a male at Belasis Technology Park, Billingham from 27 August to 6 September 1999; a duck was on the River Tees at Stockton-on-Tees, from 23 December 2004 to 7 May 2005, with the same bird at Bowesfield Marsh, Stockton-on-Tees, on the 25 February and 4 March 2005; and, a male was at Hetton Lyons Country Park, Houghton-le-Spring from the 24 August to 15 September 2007, and again in February 2008. This colourful species is native to North America and might be considered a potential vagrant to Britain as there have been several records from both Iceland and the Azores. The species is commonly kept in captivity and it is unlikely to be admitted to the British List without a ringing recovery. None of the birds recorded in County Durham have been candidates as genuine vagrants.

Maned Duck *Chenonetta jubata*

One record: a female on the Long Drag and Dorman's Pool, Teesmouth from 6th to 14 August 1987. This species is native to Australia.

Ringed Teal *Callonetta leucophrys*

Seven records involving eleven birds: a female near the Teesmouth Field Centre on 9 October 1977; a male at Crookfoot Reservoir, on 20th to 21 October 1993; a female at Billingham Beck Country Park, from mid-February to at least 29 March 1996; a party of three drakes on Greatham Creek Saline Lagoon during the evening of 14 July 2003; a male at Seaton Pond from November until at least 7 December 2004; a female in Ward Jackson Park, Hartlepool, from 15 May 2006 to 19 March 2007; and, a party of three females on the Bottom Tank at RSPB Saltholme on 12 November 2009, with one remaining until the following day. This attractive species is popular in captivity and is native to central South America.

Chiloe Wigeon *Anas sibilatrix*

There have been five records, the first of which was one present intermittently on Seal Sands, Teesmouth from 24 September to 30 October 1976; it also frequented the boating lake at Locke Park, Redcar, North Yorkshire during this period. Subsequent records are: on the Reclamation Pond, Teesmouth, on 22 September 1995; at Fish Lake in Lunedale, on 30 July 1998; at Saltholme Pools, Teesmouth, on 26 November 1999; and, at Castle Lake, Bishop Middleham, on 20 November 2010. This South American species is very popular in captivity and hybrids between it and Wigeon *Anas penelope* are not infrequent, some of which bear a superficial resemblance to American Wigeon *Anas americana*.

Cinnamon Teal *Anas cyanoptera*

Four records probably involving three birds: a male bearing a blue colour-ring at Saltholme Pools, Teesmouth on the 8 May 1994; a female was at WWT Washington from the 9 September until at least the end of November 1998; a male present on newly flooded wetland close to the A66 at Newsham from 15th to 20 April 2003; perhaps the same male at Cowpen Marsh, Teesmouth on 22 April 2003. This species is native to western North America and South America, though it occasionally reaches the eastern seaboard of North America as a vagrant. It has been mooted as a potential vagrant to Britain, though is so common in captivity that it is unlikely to be added to the British List without a ringing recovery.

Chestnut Teal *Anas castanea*

One record: an elusive male at Dorman's Pool, Teesmouth on at least 28 February 2009. This species is native to southern Australia.

White-cheeked Pintail *Anas bahamensis*

Two records doubtless referring to the same bird: at Saltholme Pools, Teesmouth on the 10 September 1995 and by Greatham Creek, Teesmouth on 4 October 1995. Its natural range includes the West Indies, the Galapagos, and South America.

Silver Teal *Anas versicolor*

One record: a male around the North Tees Marshes from 10 July to 2 September 1991. A native of South America which can be found in lowland areas from southern Brazil to the Falkland Islands.

Puna Teal *Anas puna*

One record: a male which frequented both Seal Sands and Dorman's Pool from 17 July to 6 September 1979, with probably the same bird at Locke Park, Redcar, North Yorkshire on the 6 October. A native of South America, it inhabits the Andes of Peru, western Bolivia, and northern Chile.

Rosy-billed Pochard *Netta peposaca*

One record: a female at Hurworth Burn Reservoir on 17 June 2007. This South American species is very popular in captivity.

Hooded Merganser *Lophodytes cucullatus*

There have been two records of this potential vagrant from North America. The first was an eclipse drake at Hurworth Burn Reservoir from 13 June to 18 July 1993. It had a metal ring on its right leg and was widely regarded as an escape; consequently, the record was not submitted to BBRC. The second bird was found at Scaling Dam, North Yorkshire, on 25 August 2009 before relocating to the North Tees Marshes from 1 September to 6 October; subsequently moving back to North Yorkshire. After an absence of seven weeks, presumably the same bird was found at Jackson's Landing, Hartlepool during a northerly gale on 30 November, relocating to the North Tees Marshes, where it remained to 25 April 2010. This bird was aged as a juvenile male and it attracted considerable observer interest. The BBRC felt that the initial late August occurrence date was too early for a genuine vagrant to have crossed the Atlantic and considered it an escape from captivity. Hooded Merganser has had a chequered history in Britain, mainly due to the large number of birds in captivity. It was re-admitted to Category A of the British List on the basis of a bird on North Uist, Western Isles in October 2000, with four subsequent records to the end of 2010, as well as many obvious escapes.

Chilean Flamingo *Phoenicopterus chilensis*

There have been several records of this species, most of which are from the North Tees Marshes. The first documented was from 7th to 15 February 1953, followed by single birds: from 28 November to 17 December 1964; two in October and November 1965, one of these found long dead on 26 February 1966; from 9th to 19 August 1967; from early September 1971 to at least May 1980, with two birds from November 1975 to February 1976; intermittently from 3 July 1982 until January 1998, with two from January 1991 to May 1993. There are observations of a flying south at Hartlepool on 14 May 1993, two south past Whitburn on 14 April and 22 May and south over Seaton Carew on the 20 May 1993. All were thought to relate to the birds present on the North Tees Marshes at the time. The only records not apparently related to the Teesmouth birds are from Derwent Reservoir, where a bird was present from the 28 September to 4 October 1970; and Low Barns from 21st to 30 January 1972. This exotic species is native to South America and is the commonest species of Flamingo to be found in captivity in Europe. A feral breeding population exists in Germany, though this is not yet considered self-sustaining. These birds winter in the Netherlands.

Great White Pelican *Pelecanus onocrotalus*

One record: an adult Saltholme Pools, Teesmouth on 26 August 2006, was present for just over an hour before flying off north-west over Greatham village. This bird was initially seen in north Norfolk on 16 August, before moving to Lancashire from 18th to 24 August. It was found in Northumberland, just two hours after leaving County Durham. It remained at various sites in Northumberland until at least 12 September before doing a 'grand tour' of Britain including a protracted stay on Anglesey in late September, before being last reported back in Northumberland at Berwick-upon-Tweed on 9 October. Prior to its arrival in Britain, this bird had been noted at a number of other sites in Europe, including the Netherlands where it was accepted as being a genuine vagrant (*Dutch Birding* 29: 350). The BOU deemed the bird to be an escape. White Pelican breeds in south-eastern Europe, Asia, and Sub-Saharan Africa; and with European breeders heading to East Africa to winter, must be regarded as a potential vagrant to Britain. All previous records have been classed as escapes. There is one old record of a pelican for Durham, species undetermined, which was found dead beside Castle Eden Dene on 25 August 1856 (Naylor 1996).

Pink-backed Pelican *Pelecanus rufescens*

Two records: The first was of a pelican present at Low Barns throughout July 1960 which was identified by the late Phil Stead as being this species. Coulson (1961) stated that this may have been the same bird reported shortly before on the Isle of May, though this record is dated as the 8 August 1960 by Thom (1986), and is listed as a "*Pelican, not identified as to species*". It seems likely to relate to the same bird heading north after its protracted stay in County Durham. The second was watched flying in off the sea at Lizard Point, Whitburn on 12 September 1999, and was subsequently seen at Budle Bay in Northumberland. This species is native to Sub-Saharan Africa and southern Arabia and is considered an unlikely vagrant to Western Europe, though natural vagrancy is not impossible (Jiguet 2008). Most records in Britain are thought to relate to birds which have escaped from zoos on the near Continent.

Turkey Vulture *Cathartes aura*

One record, a free-flying bird was noted around Chopwell Woods from late July 2005 until at least the beginning of 2006. This bird had escaped from a falconer's exhibition at 2005's Chopwell Festival. A common breeding bird across North and South America, and the most widespread of the New World 'vultures', it is frequently found in captivity in Europe where it is often used in falconry displays.

Harris's Hawk *Parabuteo unicinctus*

At least nine records: a bird with falconer's jesses was watched hunting at Rainton Meadows on 6 November 1998; another with jesses was in allotments near Bowburn on 27 October 2001; one was over Seaton Carew Golf

Course, on 31 October 2001; one in the Jarrow area, from 1 December 2001 into 2002, being noted at both Primrose Lane and the Don valley; one was at Cowpen Bewley Woodland Park, on 13 January 2002, and nearby Greatham Creek, Teesmouth on 27 January 2002; one was near Heighington on 9 February 2009; one almost certainly wearing jesses was present at Edder Acres Farm, Shotton Colliery from 1 to 12 September 2010; one was at Hetton Lyons for a week in April 2011; and, one remained in the Follingsby Lane of Washington through the autumn of 2011 and into 2012. The natural range of this species is from the south western United States southwards into South America. It is a particular favourite amongst falconers and frequently escapes. The few records listed here merely represent published records and are doubtless not representative of the true total.

Red-tailed Hawk *Buteo jamaicensis*

Two records. The first was observed at Joe's Pond, adjacent to Rainton Meadows on 31 August 1993. The second was a female that was hand-captured in April 2006 and housed in the aviaries of the Northern Kites Project, before being given away to a local falconer when its original owner could not be traced. This bird had been present in the Lockhaugh area of Rowlands Gill for a few weeks prior to its capture. This is one of the most widespread North American bird of prey species and although migratory within its range, it is essentially a short distance migrant that is considered an unlikely vagrant to Britain. The species is popular with falconers and is commonly kept in captivity in Britain; the result is that all records are considered to relate to escapes.

Steppe Eagle *Aquila nipalensis*

One record: a bird listed as an escape by Unwin (1972) was at South Shields on 12 July 1971. This species which breeds across eastern Europe and central Asia migrates through the Middle-East in huge numbers to winter in Africa and could be considered a potential vagrant to Britain, having occurred as such in several other European countries. It is however fairly common in captivity in Britain and frequently used in falconry displays. A small eagle, complete with jesses, near Hartlepool on 27th to 28 February and Greatham on 5 March 1966 was referred to as an 'Indian Eagle' may have been this species, considering how common it was in captivity at the time.

Lanner Falcon *Falco biarmicus*

Two records: a first summer, clearly an escaped falconer's bird, with transmitter and jesses, was found at the scrap yards in Blaydon Haughs Industrial Estate, Blaydon, in May 1997 before being taken into captivity; and a large falcon considered by the observers present to relate to this species, was present around the North Tees Marshes on the 19 August 1998, though the many hybrids produced for falconry purposes can be very difficult to differentiate in the field. This species is native to Africa, south eastern Europe and adjacent parts of western Asia, and although partially migratory is considered an unlikely vagrant to Britain.

Purple Swamphen *Porphyrio porphyrio*

One record: one at Barmston Ponds on 7 June 1975. This bird was submitted as the first record of this species for Britain and was accepted in terms of its identification by BBRC (Unwin 1976), though was not accepted by BOU predominantly on the basis that it was not of the nominate west Mediterranean race and most likely an escape from captivity (Unwin 1977). In Europe, Purple Swamphen is restricted to the marshlands of southern Iberia (and adjacent parts of North Africa), France and Sardinia where it is chiefly resident, although another twelve races are found in Africa, tropical Asia and Australasia. There are currently no accepted records of this species in Britain despite a handful of subsequent claims, all of which are presumed to have escaped from captivity.

Demoiselle Crane *Anthropoides virgo*

One record: a party of four adults was in fields at Cowpen Bewley from 6th to 21 July 1967, with two of these remaining on nearby Coatham Marsh until 17 September. At least eight more Demoiselle Cranes were recorded in Britain during 1967, including a party of six at Horsey, Norfolk on 23 April, which later split up into two groups; three being present at Bank Newton, North Yorkshire from 10 July to 5 August 1967; and three around Poole Harbour,

948

Dorset for much of July 1967. None of these birds was ever traced to a collection and, in the case of the Teesmouth birds, the records were not submitted to BBRC as they were considered to be escapes, though with hindsight, perhaps they should have been. The core breeding range of this species is in Central Asia, though small numbers continue to breed in Turkey, with most of the world population wintering in north-western India and Pakistan. Although this species is accepted as a natural vagrant to many countries in Europe, it is also fairly common in captivity, and has not yet been accepted as having occurred as a wild vagrant to Britain despite several records.

'Barbary' Dove *Streptopelia 'risoria'*

Two records: at Crimdon Dene on the 8 June 1977; and in gardens at Shotton Colliery throughout July 2000. This is the domesticated form of African Collared Dove *Streptopelia roseogrisea* which occurs in central Africa south of the Sahara, though it is spreading northwards.

Diamond Dove *Geopelia cuneata*

Three records: at Throston Grange, Hartlepool in September 1997; in gardens on the Shearwater Estate, Whitburn on 21 August 2000; and, in Billingham on 16 September 2003. This Australian species is a popular cage bird.

Croaking Ground-dove *Columbina cruziana*

One record: a single in gardens at Carlton, near Stockton-on-Tees from mid-September 1992 to 27 April 1993. This diminutive South American species is found chiefly in Ecuador and Peru.

Cockatiel *Nymphicus hollandicus*

There have been numerous records of this Australian species. The most interesting records are: one that was watched flying over Billingham being mobbed by Carrion Crows *Corvus corone* on 5 May 2001; one flying over the coastal path near Easington in mid-October 2001; one being pursued by Carrion Crows in gardens at Whitburn Steel on 17 July 2007; one flying north about 0.75km offshore from Whitburn Observatory on 20 August 2007; and, one perched atop the communications mast at Hartlepool Headland on 27 August 2007. Several of these coastal records may have escaped on the near-Continent, as the species has been noted flying in off the sea in Britain on more than one occasion. A widespread breeding bird across Australia, it is one of the most commonly kept cage birds in Britain.

Red-crowned Parakeet *Cyanoromphus novaezelandiae*

One record: a bird was present in a Shotton Colliery garden from 11th to 17 September 2005. Native to New Zealand and once widespread it is now confined to some offshore islands, though is quite widely found in captivity.

Crimson Rosella *Platycercus elegans*

One record: at Hartlepool on 27 June 1998. A widespread breeding bird across Australia, it is a commonly kept in captivity.

Eastern Rosella *Platycercus eximius*

Three records: at Billingham on 25 July 1997; around WWT Washington from 11 September to 20 October 1998; and, at Hartlepool Headland from mid-April 2008 to March 2010, when it is thought to have been taken by a Sparrowhawk *Accipiter nisus*. The species is a native of south eastern Australia, it is widely kept in captivity in Britain.

Western Rosella *Platycercus icterotis*

A bird that had escaped from an aviary at Highfield, Rowland Gill, in late September 2001 was present in the area for at least a fortnight and was seen feeding on elder berries *Sumbucus nigra*. Native to south western Australia, this species is commonly kept in captivity in Britain.

Budgerigar *Melopsittacus undulatus*

There have been numerous reports of this popular Australian cage bird from widely scattered locations within the county. Records date back to at least the 1950's when the species was regularly noted around Shotton Colliery having presumably escaped from local houses in the area, whilst in the north of the county there were at least six records at Shibdon Pond between in the years 1983-1992. There are, however relatively few published records for this species as it such a common cage bird and many free flying escapes have doubtless gone unreported over the years. A widespread breeding bird across Australia, it is probably the most commonly kept cage bird in Britain.

Superb Parrot *Polytelis swainsonii*

One record: a male was watched flying over Cowpen Marsh, Teesmouth, and landed in the small wood on the brinefields adjacent to Greatham Creek on 20 April 1997. The species is a native of south eastern Australia.

Rosy-faced Lovebird *Agapornis roseicollis*

Three records of four birds; two present in Newton Aycliffe from 12th to 22 August 1996, were noted as associating with the local House Sparrow *Passer domesticus* population; a single sat on guttering in Hartlepool on 24 September 1997, and a pale washed-out bird was at Brankin Moor, Darlington on 21 July 2001. Native to arid areas of south western Africa, this species is a very popular cage bird.

Fischer's Lovebird *Agapornis fischeri*

One record: at both Seaton Carew Cemetery and in nearby bushes at North Gare on the 9 August 1975. This species in endemic to northern Tanzania, though a self-sustaining population of escapes exists in southern France.

Grey Parrot *Psittacus erithacus*

Two records: one was reported from the Derwent Walk Country Park during the summer of either 1988 or 1989; and another was along the Derwent Walk near Rowlands Gill in July 1994. This popular cage bird is a native to the rainforests of West and Central Africa.

Senegal Parrot *Poicecephalus senegalus*

Two records, though possibly relating to the same birds. A pair frequented Jarrow Cemetery throughout much of 2008, and although they were seen to use a hole in a tree regularly, there was no indication that any nesting activity had taken place. A bird seen in conifers at Cleadon on 9 September 2009 may have been one of these birds. This popular cage bird is native to West Africa, and is quite widely kept in captivity in Britain.

Blue-crowned Parakeet *Aratinga acuticaudata*

One record: at Ryhope Dene on the 25 November 2002. A native of South America.

Nanday Parakeet *Nandayus nenday*

Two records. A pair was present in the Lockhaugh/Rowlands Gill/Thornley Woods area from July to September 1994, one remained until 16 December when it was last noted at Far Pasture, having often roosted in a nest box at Low Thornley Farm. Another was 'at large' in gardens in Billingham on 4 December 2007, before being re-captured and returned to its owner. A native of southern South America, this species is a popular cage bird.

Monk Parakeet *Myiopsitta monachus*

One record. One was in allotments adjacent to Hetton Lyons Country Park on 20 April 2011. A native of South America, this species is a popular cage bird. There are also large feral populations in Spain, Gibraltar and Belgium, as well as a small but stable population in Middlesex, England.

White-winged Parakeet *Brotogeris versicolurus*

One record: a parakeet at Saltholme Pools on 9 October 2005 was later identified as this species, which is also known as Canary-winged Parakeet. Native to the Amazon basin in South America, this popular cage bird has established several self-sustaining populations in North America.

Eurasian Eagle-Owl *Bubo bubo*

George Temperley (1951) recorded an instance of one being shot on an upland moor of County Durham, from Selby's text of 1831, with the specimen reported to be in Mr Bullock's Museum, though he indicated that John Hancock was doubtful of this record. Gurney (1876) also alluded to this occurrence but likewise expressed his doubts as to its authenticity; though he did report that one had been seen at Seaton Carew but gave no specific date.

There are at least eight recent records: well-substantiated reports exist of a single bird living wild in Chopwell Woods during the autumn of 1991; one at Shotley Bridge from 17 May to 6 June 1996 was aggressive enough to fly at an observer's back on one occasion, though no other signs of territorial behaviour were noted (this bird was later trapped using a chicken carcase as bait in a 'drop-trap' and taken into captivity); a bird was seen on a rooftop in Sunderland on 24 January 1998; an obviously tame bird was in gardens at Hartlepool on 3 October 2001; a male was watched calling from the top of a telegraph pole at Sacriston, Durham on 16 January 2008; and another was found in a dense moorland gulley in Baldersdale on 26 May 2008. In addition, a bird was reported in Chopwell Woods during spring 2007. Finally, one was present in the Cleadon/Boldon Flats area during much of 2011.

This species is native to continental Europe and Asia and is commonly kept in captivity in Britain, frequently escaping. It has bred in a feral state in Britain on a number of occasions, most notably in the context of northern England on Ministry of Defence land north of Richmond, North Yorkshire from at least 1997 to 2005. Over a nine year, the pair at this location raised around 20 young, which were proven by ringing recoveries to have dispersed widely, ringed birds being noted in both Shropshire and southern Scotland. The bird in Baldersdale in May 2008 may have been one of the offspring of this pair, or alternatively, from a pair that bred in Lancashire in 2007 in very similar moorland habitat to that of the Yorkshire pair. Birds have also been reported to have bred in Cumbria recently and may have also bred in south west Northumberland in 2005-2006. Perhaps there are a small number of feral birds at large in the west of the county?

Red-billed Blue Magpie *Urocissa erythrorhyncha*

One record: at North Cemetery, Hartlepool on 16 September 1993. This spectacular species is native to the forests of South East Asia and the western Himalayas.

Azure-winged Magpie *Cyanopica cyana*

One record: a sigle bird was seen at Ryton Willows, Gateshead on 16 May 2011. This species has two widely separated populations, one in the south west of the Iberian Peninsula and the other in eastern Asia.

Greater Necklaced Laughingthrush *Garrulax pectoralis*

One record: a bird photographed in Middle Herrington, Sunderland was present from at least 6th to 8 November 1998, and had probably been seen in nearby Silksworth in the summer of the same year. This species is native to Indochina and the Himalayas.

Elliot's Laughingthrush *Garrulax elliotii*

One record: a single bird was present in gardens and adjacent landscaped areas at Middle Herrington, Sunderland, from 6th to 13 March 1993. That two species of Laughing-thrush should be found in the same suburb of Sunderland in a relatively short period of time suggests a local breeder nearby. This species is endemic to central China.

Blue Whistling Thrush *Myophonus caeruleus*

One record: This species was described as having attempted to breed in Upper Weardale in 1970; it is thought that a pair of birds may have been deliberately released, as documented by Dr. H. M. S. Blair (England 1974). This species, which is sometimes called Violet Whistling Thrush, is native to the Himalayas and South East Asia, and is subject to cheifly altitudinal movements during winter.

Village Weaver *Ploceus cucullatus*

One record: an adult male which regularly visited a garden on the Shearwater Estate, Whitburn from 27th to 31 August 1996. This species is native to Sub-Saharan Africa, and is a very popular cage bird.

Red-billed Quelea *Quelea quelea*

Five records: at North Cemetery, Hartlepool in about 1968 (Blick 2009); one in a garden at Fellside, Whickham in early October 1992; a male was present intermittently at Whitburn Country Park from 1st to 27 September 1993, and was joined by a female from 18th to 20 September, suggesting that they had escaped locally; and, one was on The Leas, South Shields on 20 September 1998. This popular cage bird is native to sub-Saharan Africa, and is considered to be one of the most abundant bird species in the world. Considered an agricultural pest in its native range, the species can form flocks numbering hundreds of thousands of birds.

Black-winged Red Bishop *Euplectes hordeaceus*

One record: a male in breeding plumage was present in gardens at Holywood, Wolsingham from 29 September until at least 7 November 2003. This popular cage bird is native to sub-Saharan Africa.

Zebra Finch *Taeniopygia guttata*

One record: one was present in the Jewish Cemetery, Hartlepool on 13 October 2011. This popular cage bird is native to Australia and Indonesia.

White-rumped Munia *Lonchura striata*

One record: one was photographed at Witton Gilbert on 9 March 2009. This species is native to the Indian Subcontinent and much of South East Asia.

Java Sparrow *Lonchura oryzivora*

One record: at Hartlepool Headland in about 1969 (Blick 2009). This attractive species is native to Java and Bali, Indonesia, and is a popular cage bird, though both trapping and habitat loss have contributed to a major decline in its native range.

Atlantic Canary *Serinus canaria*

Records of this species date back to at least the 1950's when one observer in Shotton Colliery recalls seeing numerous escapes, a pattern which continued through the 1960's, at a time when aviculture was much more popular than it is today. There have been at least nine recent sightings, plus a breeding record, though many more must go unrecorded. Records include:: a yellow singing male at Thornley Woodlands Centre, Rowlands Gill on at least 21 March 1998; on cliffs north of Whitburn Observatory on 20 September 1998; at Saltholme Pools, Teesmouth on 16 June 1999; a bird was at Shibdon Pond in the late summer of 2000; one was with Chaffinches *Fringilla coelebs* at Romaldkirk on 25 March 2002; and a male at Sherburn on 16 April 2003; one at Shotton Colliery on 25 July 2006; a male was noted coming to feeders in a garden in Blackhall Mill, in spring 2009; and, one was noted as surviving for at least six months at Easington Lane during the summer and autumn of 2011. The breeding record concerns a pair in Blaydon in spring 2005. A nest was located in late May and by early June, it contained five eggs. However, the outcome is unknown and, as there were no subsequent records of a family party, was assumed to have been unsuccessful. This species is a very popular cage bird and comes in a number of different captive-bred colour varieties. In a wild state it is endemic to the Atlantic Islands of The Azores, The Canary Islands, and Madeira.

Chestnut Bunting *Emberiza rutila*

One record: a male was in gardens at Wheatall Drive, Whitburn from 17th to 20 May 2000. This species breeds in south-central and south-eastern Siberia, though it is a relatively short distance migrant heading south westwards to winter in Burma, southern China and South-East Asia. Nonetheless, it has been recorded as a vagrant to several European countries, with most records being of juvenile or first-winter birds in the period from late September to November. There have been at least seven other occurrences of Chestnut Bunting in Britain, though unlike the vagrancy pattern in the rest of Europe, most of these birds have been in June, with just two in autumn, both adults in early September. In October 2008, the species was moved from Category D of the British list, all records being considered to be escapes from captivity. The species is fairly common in captivity and all British records occurred prior to the EU ban on wild bird imports imposed in 2007.

Appendix 2

Systematic List Authors

The naming of an author for many of the species texts in the systematic list of *Birds of Durham* is not a simple exercise. Many, if not most, of the species texts were written and developed very much as a collaborative process, though each was steered by a lead author. In some instances, texts benefitted from the inputs of many people and in some situations some species texts might rightly have been said to have been co-authored. The authors list below highlights the lead author's main inputs. Almost universally, the texts were better for this cooperative approach. In some fewer cases, the lead authors were largely responsible for 'their' specialist species. Space limitations mean that not all species can be mentioned for the main species authors, Chris Bell and Keith Bowey, who prepared the greatest number of texts. Where an author is not named for a particular species, it should be inferred that it was prepared by one or the other, though often with much support and input from other authors.

The panel of authors comprised: Paul Anderson, Tony Armstrong, Chris Bell, Peter Bell, Keith Bowey, Ian Mills, Fred Milton, Mark Newsome, Dave Raw and John Strowger. Additional significant inputs to some species texts came from a number of people outside of this group, including: Derek Charlton (Redstart, Pied Flycatcher and some common woodland species); Mike Harbinson (who contributed to all texts for the common wildfowl); John Olley (Hobby and Stonechat); David Sowerbutts (Goosander, Red-breasted Merganser, Rook, Magpie, Whinchat, Swift and a number of other species); Phil Warren (gamebirds); and, Steve Westerberg (Kingfisher, Dipper, titmice and other woodland birds) – with apologies for any accidental omissions. The main species texts developed by each lead author are listed below.

Lead Author	Main Species
Paul Anderson	Smew, Ruddy Duck, Hobby, Avocet, Black-headed Gull, Black Tern, Waxwing, Swallow, House Martin, Sand Martin, Tree Pipit, Meadow Pipit, Linnet, Bullfinch, Crossbill, Snow Bunting, Lapland Bunting, Reed Bunting, Yellowhammer and Corn Bunting
Tony Armstrong	Goshawk, Sparrowhawk, Buzzard, Kestrel, Peregrine, Ruff, Jack Snipe, Oystercatcher, Little Ringed Plover, Ringed Plover, Golden Plover, Grey Plover, Lapwing, Turnstone, Green Woodpecker, Great Spotted Woodpecker, Lesser Spotted Woodpecker, Carrion Crow, Hooded Crow, Raven, Black Redstart and Water Pipit
Chris Bell	Most of the scarce and rare species plus Ring-necked Parakeet, Knot, Sanderling, Little Stint, Curlew Sandpiper, Purple Sandpiper, Black-tailed Godwit, Bar-tailed Godwit and Common Sandpiper
Peter Bell	Eider, Jack Snipe, Snipe, Woodcock, Whimbrel, Curlew, Green Sandpiper, Spotted Redshank, Greenshank, Wood Sandpiper, Redshank, Guillemot, Razorbill, Kittiwake, Sandwich Tern, Common Tern, Roseate Tern, Arctic Tern, Feral Pigeon, Wheatear and Whinchat
Keith Bowey	Mandarin, Shoveler, Goldeneye, Grey Heron, Red Kite, Water Rail, Spotted Crake, Barn Owl, Sedge Warbler and most other species not obviously prepared by other lead authors
Ian Mills	Wood Warbler, Chiffchaff, Willow Warbler, Blackcap, Garden Warbler, Lesser Whitethroat, Whitethroat, Fieldfare, Song Thrush and Redwing
Fred Milton	Hawfinch
Mark Newsome	Most seabird species texts (other than those named for other lead authors) and some scarce and rare species texts in collaboration with Chris Bell, plus swans, geese, Mediterranean Gull, Glaucous Gull, Iceland Gull
David Raw	Little Grebe, Great Crested Grebe, Hen Harrier, Montagu's Harrier, Merlin, Moorhen, Coot, Short-eared Owl, Chaffinch, Brambling, Siskin, Lesser Redpoll, Mealy Redpoll, Greenfinch and Goldfinch
John Strowger	Wigeon, Teal, Dotterel, Dunlin, Ring Ouzel, Pied Flycatcher, Rock Pipit and Twite

Appendix 2a

Selected list of Ornithological Researchers and Research Topics at Durham University

From the 1970s to the early 2000s, the Zoology and then the Biological Sciences department of Durham University was one of the leading ornithological research institutes in the country. Much of this bird-related work was steered by Professor Peter Evans (largely supervising work on wading birds at Teesmouth) and Dr. John Coulson (working mainly on seabirds – Kittiwakes in particular - and upland wading birds). Other research inputs in this area came though the mid-1990s and into the 2000s from by Dr. Chris Thomas and Professor Brian Huntley, of the School of Biological and Biomedical Sciences; the latter specialising in issues relating to climate change.

This list is not an exhaustive one, but many of the key researchers and research subject over the period are outlined below. As with most academic bodies, the research was not restricted to Durham's county boundaries, much work was done around the Farne Islands but also in some more exotic locations. The alphabetically organised list names the researcher and gives a short species and topic summary, where details were available.

Fiona Dixon	Kittiwakes at Marsden, their social behaviour
Neil Duncan	Large gulls on the Isle of May
Euan Dunn	Roseate Terns on Coquet Island
Jackie Fairweather	Kittiwakes at Marsden and North Shields
Jerry Fitzgerald	Gull distribution on the river Tyne
Anne Flowers	Eiders in Northumberland
Sheila Frazer	Shags on Farne Islands
Bob Furness	Great Skuas on Foula
Les Goodyear	Wader ringing Teesmouth
Shirley Goodyear	Distribution of waders at Moor House and upper Teesdale
Murray Grant	Whimbrel on Orkney
Susan Greig	Feeding activity of large gulls on landfill sites in Durham, age and sex effects
Larry Griffin	Rooks, home ranges and colony spacing in Durham
Andy Hodges	Kittiwakes at North Shields
Andrew Hoodless	Woodcock in west Durham and Scotland
Jean Horobin	The return pattern of Fulmars to Marsden; Arctic Terns on Farne Islands
Nigel Langham	Terns on Coquet Island
Rowena Langston	Sanderling at Teesmouth
Ian Marshall	Eiders on Farne Islands
Gabriela McKinnon	Black-headed Gulls in Durham, their feeding and origins
Pat Monaghan	Herring Gulls town nesting in South Shields and Sunderland
Mark O'Connell	Herring and Lesser Black-backed Gulls at Abbeystead, Lancashire
Dave Parish	Lapwings in Upper Teesdale
Tom Pearson	Seabirds feeding around Farne Islands
Kirsti Peck Tree	Nest site selection by small bird species in Hamsterley Forest
Mike Pienkowski	Ringed Plover at Teesmouth
Julie Porter	Kittiwakes at North Shields and Marsden
Dick Potts	Shags on Farne Islands; Starlings in Newcastle
Graham Rankin	Waders on Rockcliffe Marsh, Cumbria
Susan Raven	Town nesting gulls in the north east; distribution of gulls on the Tyne in relation to reduced sewage discharges
James Robinson	Common Terns on Farne Islands
Brian Springett	Terns on Farne Islands
John Strowger	Kittiwakes at Marsden
Callum Thomas	Kittiwakes at North Shields
Patrick Thompson	Lapwing in Upper Teesdale, philopatry and population dynamics
David Townsend	Curlew in Durham

Robin Ward	Wader studies and ringing through much of 1990s and early 2000s at Teesmouth
E. White	Kittiwakes at Marsden
Steve Willis	Riverine birds in west Durham; climate change
Ron Wooller	Kittiwakes at North Shields

Appendix 3

Photographers and illustrators

Photographs were kindly supplied by:

James Alder	Black and white: Desert Wheatear (p 835).
Chris Bell	Black and white: Night-heron (p 246), Barred Warbler (p 727). Colour: Terek Sandpiper (plate 61), Lesser Spotted Woodpecker (plate 89), Red-eyed Vireo (plate 94), Amur Wagtail (plate 126), Little Bunting (plate 134).
Keith Bowey	Black and white: Baillon's Crake (p 330), Stone-curlew (p 350), Long-tailed Skua (p 466), Ivory Gull (p 470), Sabine's Gull (p 471), Great Auk (p 551), Tengmalm's Owl (p 596), Golden Oriole (p 625), Great Reed Warbler (p 752), White's Thrush (p 776), Pine Grosbeak (p 909), Yellow Warbler (p 941). (All photos courtesy NHSN).
John Bridges	Colour: Castle Eden Dene (plate 13), Teesside (plate 14), Purple Sandpiper (plate 54), Curlew (plate 60), Razorbill (plate 80), Skylark (plate 97), Pied Flycatcher (plate 119), Tree Sparrow (plate 125), Yellow Wagtail (plate 124), Common Crossbill (plate 130).
Chris Brown	Colour: Aquatic Warbler (plate 107).
Derek Clayton	Colour: Hooded Crow (plate 93).
Steven Clifton	Black and white: Dusky Warbler (p 709), Citrine Wagtail (p 852). Colour: Water Pipit (plate 128).
Paul Cook	Black and white: Ross's Gull (p 489), Laughing Gull (p 490). Colour: Isabelline Shrike (plate 91).
Dr. John Coulson	Black and white: Kittiwake colony (p 474).
Gary Crowder	Colour: White-throated Robin (plate 115).
Jeff Delve	Colour: Long-toed Stint (plate 53).
Jamie Duffie	Black and white: Bluethroat (p 812).
Richard Dunn	Colour: Glaucous-winged Gull (plate 73).
Steve Egglestone	Colour: Sooty Shearwater (plate 27), Greenish Warbler (plate 100).
Ian Fisher	Colour: Eastern Crowned Warbler (plate 99), Siberian Rubythroat (plate 118).
Ian Forrest	Black and white: Great White Egret (p 250), Semipalmated Sandpiper (p 380), Little Auk (p 555), Mistle Thrush (p 799). Colour: Ring-necked Duck (plate 19), Eider (plate 21), Grey Partridge (plate 24), Black-throated and Great Northern Divers (plate 25), Buzzard (plate 42), Merlin (plate 43), Peregrine Falcon (plate 44), Grey Phalarope (plate 65), White-winged Black Tern (plate 77, Barn Owl (plate 85), Willow Tit (plate 96), Cetti's Warbler (plate 98), Waxwing (plate 110), Black-throated Thrush (plate 113), Wheatear (plate 123), Twite (plate 129).
Tom Francis	Black and white: White-billed Diver (p 214), Terek Sandpiper (p 432), Spotted Sandpiper (p 436), Wilson's Phalarope (p 458), Bonaparte's Gull (p 480), Franklin's Gull (p 491), Nightjar (p 599). Colour: Ross's Gull (plate 71), Paddyfield Warbler (plate 108), White-spotted Bluethroat (plate 121).
Bryan Galloway	Black and white: Ivory Gull (p 471).
Barrie Hanson	Black and white: White-headed Duck (p 938).
Andy Hay (RSPB Images)	Colour: Bee-eater (plate 87).
Dougie Holden	Black and white: Eastern Crowned Warbler (p 699).
Steve Howat	Colour: Red-flanked Bluetail (page 117).
Mike Hunter	Colour: Lesser Yellowlegs (plate 63).
Geoff Iceton	Colour: Cattle Egret (plate 30).
Greg Jack	Colour: Goshawk (plate 41).
Dave Johnson	Colour: Goosander (plate 22), Hen Harrier (plate 38), Pomarine Skua (plate 66), Long-eared Owl (plate 86), Dipper (plate 111), Ring Ouzel (plate 112).
David Kay	Colour: Redstart (plate 58), Corn Bunting (plate 64).
Iain Leach	Black and white: Caspian Tern (p 523). Colour: Arctic Warbler (plate 101).
Paul Macklam	Colour: Buff-breasted Sandpiper (plate 57).

John Malloy	Colour: Black Stork (plate 33), Wilson's Phalarope (plate 64), Sabine's Gull (plate 68), Subalpine Warbler (plate 103), Melodious Warbler (plate 106).
Ian Mills	Colour: Sandhill Crane (plate 46), Laughing Gull (plate 72), Lesser Grey Shrike (plate 92).
Mark Newsome	Black and white: Little Ringed Plover (p 353), Kittiwake Tower (p 477), Lapland Bunting (p 921). Colour: Langdon Beck (plate 2), Balderhead (plate 3), Hamsterley (plate 4), Muggleswick (plate 5), Farnless Farm (plate 6), Low Barns (plate 7), Derwent Valley (plate 9), South Shields (plate 10), Marsden (plate 11), Whitburn (plate 12), Whooper Swan (plate 15), Barnacle Goose (plate 16), Wigeon (plate 17), Garganey (plate 18), Surf Scoter (plate 20), Black Grouse (plate 23), Fulmar (plate 26), Grey Heron (plate 31), Black-necked Grebe (plate 35), Northern Harrier (plate 39), Pallid Harrier (plate 40), Golden Plover (plate 50), Lapwing (plate 51), Snipe (plate 58), Kittiwake (plate 69 & 70), Little Auk (plate 81), Alpine Swift (plate 83), Isabelline Shrike (plate 90), Wood Warbler (plate 102), Grasshopper Warbler (plate 104), Fieldfare (plate 114).
Philip Palmer	Black and white: Arctic Redpoll (p 901) (courtesy NHSN).
Jim Pattinson	Colour: Kentish Plover (plate 47), Pacific Golden Plover (plate 49), Great Knot (plate 52), Spotted Sandpiper (plate 62), Great Spotted Cuckoo (plate 82), Roller (plate 88), Booted Warbler (plate 105), Rose-coloured Starling (plate 109), Thrush Nightingale (plate 116), Ortolan Bunting (plate 133).
Mark Penrose	Colour: Common Nighthawk (plate 84).
John Pillans	Colour: Brown-headed Cowbird (plate 136).
David Raw	Colour: Teesdale (plate 1).
Wayne Richardson	Colour: Short-billed Dowitcher (plate 59), Long-tailed Skua (plate 67), Penduline Tit (plate 95), Richard's Pipit (plate 127).
Ken Sanderson	Black and white: Red Kite (p 280).
Ray Scott	Black and white: Dipper (p 773). Colour: Rainton Meadows (plate 8), Purple Heron (plate 32), Glossy Ibis (plate 34), Red Kite (plate 37), Avocet (plate 48), Little Tern (plate 75).
Glynn Sellors	Colour: Roseate Tern (plate 79).
Martyn Sidwell	Black and white: Quail (p 202), Red-footed Falcon (p 313), Great Spotted Cuckoo (p 575), Short-toed Lark (p 678), White-throated Robin scene (p 807). Colour: Broad-billed Sandpiper (plate 55), Sharp-tailed Sandpiper (plate 56), Glaucous Gull (plate 74), Whiskered Tern (plate 76).
Edward Skinner	Black and white: Dusky Thrush (p 785).
Richard Stephenson	Colour: Siberian Stonechat (plate 122).
Tom Tams	Colour: Storm Petrel (plate 28).
Alan Tate	Colour: Double-crested Cormorant (plate 29).
David Tipling	Black and white: Baillon's Crake (crowd) (p 330).
Norman Urwin	Colour: Pine Bunting (plate 132).
Steve Westerberg	Black and white: Parrot Crossbill (p 906).
Pete Wheeler	Black and white: American Golden Plover (p 360), White-rumped Sandpiper (p 387). Colour: Baillon's Crake (plate 45).
John Wood	Black and white: Gyr Falcon (p 321).
Jeff Youngs	Black and white: Sharp-tailed Sandpiper (p 390). Colour: Lesser Crested Tern (plate 78), Arctic Redpoll (plate 131).
(photographer unknown)	Black and white: Black-winged Stilt (p 347).

All archive images used in Chapter 2 were provided courtesy of the Natural History Society of Northumbria.

The inclusion of illustrations to accompany every species account greatly enhances this publication and the vast majority were provided by Ken Baldridge, with contributions also from James Alder, Chris Gibbins and Michael Watson. Several of other illustrations used were supplied to the Durham Bird Club over the past 30 years by artists now unknown, and thanks are also extended to these.

Appendix 4

Species sponsors and species supporters

Species	Sponsor
Bewick's Swan	Bewick Society
Blackbird	W. M. Harvey Ltd.
Blackcap	W. M. Harvey Ltd.
Curlew	North Pennines AONB Partnership
Eider	Northumberland & Tyneside Bird Club
Great Reed Warbler	Gateshead Council
Hen Harrier	Leengate Welding Supplies
Herring Gull	Hartlepool United FC
Kestrel	Durham County Council (countryside service)
Kingfisher	Go North East Limited
Kittiwake	International Paint
Linnet	Tunstall Hills Protection Group
Little Owl	The Original Bakehouse
Long-eared Owl	The Original Bakehouse
Long-tailed Tit	W. M. Harvey Ltd.
Meadow Pipit	North Pennines AONB Partnership
Merlin	North Pennines AONB Partnership
Osprey	Northumbrian Water Limited
Peregrine Falcon	Chromazone
Pheasant	The Cares Group
Pine Grosbeak	Gateshead Council
Red Kite	Go North East Limited
Redshank	Council for the Preservation of Rural England
Redstart	W. M. Harvey Ltd.
Sand Martin	Banks Group
Sedge Warbler	Glead Ecological & Environmental Services
Short-eared Owl	The Original Bakehouse
Skylark	Northumbrian Water Limited
Spotted Flycatcher	Glaxo Smith Kline
Tawny Owl	The Original Bakehouse
Turnstone	South Tyneside Council
Water Rail	Glead Ecological & Environmental Services
Wigeon	Northumbrian Water Limited
Yellow Wagtail	Northumbrian Water Limited

Species	Supporter
Arctic Redpoll	Gary Woodburn
Barn Owl	Brian Moorhead
Bee-eater	Steve Howat
Black Grouse	Ian Forest
Black-necked Grebe	Colin Wilson
Blue Tit	Robert Luke
Buff-breasted Sandpiper	Stuart McLean
Bullfinch	Richard Cowen
Buzzard	Michael Raven
Chiffchaff	Richard Barnes
Crossbill	Bill Curtis
Dipper	David Taylor
Dunnock	Dave Foster
Dusky Thrush	Russell McAndrew
Eastern Crowned Warbler	Dougie Holden
Golden Plover	Michael Raven
Jackdaw	Paula Charlton
Jay	Ian Forest
Glaucous-winged Gull	Toby Collett
Goldfinch	Stan Coates
Goshawk	Clayton Potts
Grasshopper Warbler	Brian Luke
Great Black-backed Gull	Russell McAndrew
Green Sandpiper	Neil A. Fawcett
Greenshank	Michael Lowes
Green Woodpecker	Wayne O'Reagan
Gull-billed Tern	Peter Bell
Hawfinch	Thorton Kay
Hobby	John Olley
Hoopoe	Allan Rowell
Killdeer	Ian Mills
Lesser Crested Tern	Russell McAndrew
Lesser Spotted Woodpecker	Alan Jones
Little Ringed Plover	Dave Simpson
Mealy Redpoll	Graham Mitchell
Oystercatcher	Susan Grey
Pallas's Warbler	Peter Bell
Pied Flycatcher	Michael L. Passant

Puffin	Joe Sollis
Purple Sandpiper	Russell McAndrew
Quail	Paul Tyndall
Raven	Michael Raven
Red-crested Pochard	Joseph Hughes
Red-necked Phalarope	John Fletcher
Reed Warbler	Russell McAndrew
Ring Ouzel	Leslie Morley
Robin	Mandy Bennett
Rock Dove	Dave Simpson
Roseate Tern	Peter Bell
Shelduck	Russell McAndrew
Short-toed Lark	Peter Bell
Snipe	Leslie Morley
Snow Bunting	Paul Anderson
Sooty Shearwater	Mark Newsome
Stock Dove	Dave Simpson
Swallow	Richard Barnes
Swift	Richard Barnes
Tree Sparrow	James Anderson
Tufted Duck	William Billsborrow
Waxwing	John Lumsden
Wheatear	Leslie Morley
White Wagtail	Elaine Brown
Willow Tit	Derek Charlton
Willow Warbler	Richard Barnes
Woodpigeon	Dave Simpson
Yellowhammer	Brian Chapman

Appendix 5

Map and Gazetteer of birdwatching sites in County Durham

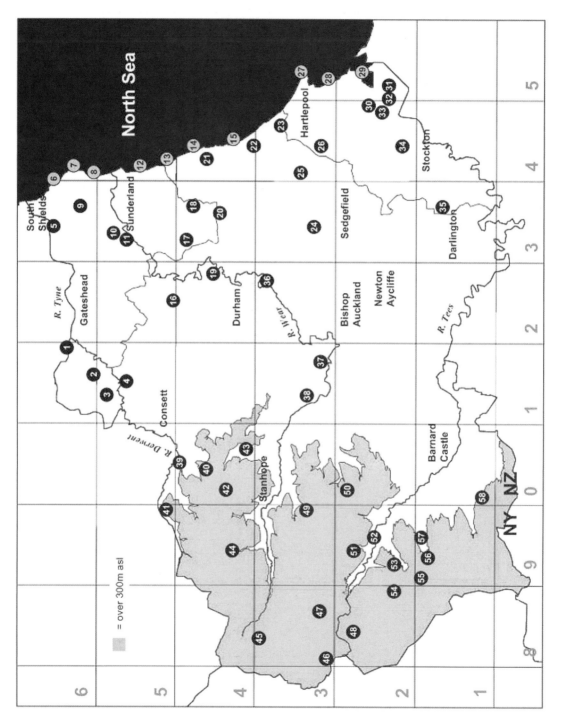

Key to Site References:

By Location Point

Shibdon Pond	1
Derwent Walk Country Park	2
Chopwell Wood	3
Pont Burn Wood	4
Jarrow Slake & R. Don	5
Marsden Quarry & Hall	6
Whiburn Obs & Coastal Park	7
Mere Knolls Cemetery	8
Boldon Flats	9
Barmston Pond	10
Washington WWT	11
Salterfen Rocks	12
Ryhope Dene	13
Dawdon Blast Beach	14
Beacon Hill	15
Waldridge Fell	16
Rainton Meadows & Joe's Pond	17
Hetton Lyons Country Park	18
Brasside Pond	19
Elemore Hall	20
Hawthorn Dene	21
Castle Eden Dene	22
Crimdon Dene	23
Bishop Middleham	24
Hurworth Burn Reservoir	25
Crookfoot Reservoir	26
Hartlepool Headland	27
Seaton Carew	28
Seal Sands	29
Cowpen Marsh	30
Dorman's Pool & Reclamation Pond	31
Saltholme Pools	32
Haverton Hole	33
Billingham Beck	34
Longnewton Reservoir	35
Croxdale	36
Low Barns NR	37
McNeil Bottoms	38
Muggleswick NNR	39
Smiddyshaw Reservoir	40
Derwent Reservoir	41
Waskerley Reservoir	42
Tunstall Reservoir	43
Rookhope	44
Burnhope Reservoir	45
Cow Green Reservoir	46
Langdon Beck	47
Crockley Fell	48
Bollihope Common	49
Hamsterley Forest	50
Bowlees	51
Middleton-in-Teesdale	52
Grassholme Reservoir	53
Selset Reservoir	54
Balderhead Reservoir	55
Blackton Reservoir	56
Hury Reservoir	57
Stang Forest	58

Alphabetical

Balderhead Reservoir	55
Barmston Pond	10
Beacon Hill	15
Billingham Beck	34
Bishop Middleham	24
Blackton Reservoir	56
Boldon Flats	9
Bollihope Common	49
Bowlees	51
Brasside Pond	19
Burnhope Reservoir	45
Castle Eden Dene	22
Chopwell Wood	3
Cow Green Reservoir	46
Cowpen Marsh	30
Crimdon Dene	23
Crockley Fell	48
Crookfoot Reservoir	26
Croxdale	36
Dawdon Blast Beach	14
Derwent Reservoir	41
Derwent Walk Country Park	2
Dorman's Pool & Reclamation Pond	31
Elemore Hall	20
Grassholme Reservoir	53
Hamsterley Forest	50
Hartlepool Headland	27
Haverton Hole	33
Hawthorn Dene	21
Hetton Lyons Country Park	18
Hurworth Burn Reservoir	25
Hury Reservoir	57
Jarrow Slake & R. Don	5
Langdon Beck	47
Longnewton Reservoir	35
Low Barns NR	37
Marsden Quarry & Hall	6
McNeil Bottoms	38
Mere Knolls Cemetery	8
Middleton-in-Teesdale	52
Muggleswick NNR	39
Pont Burn Wood	4
Rainton Meadows & Joe's Pond	17
Rookhope	44
Ryhope Dene	13
Salterfen Rocks	12
Saltholme Pools	32
Seal Sands	29
Seaton Carew	28
Selset Reservoir	54
Shibdon Pond	1
Smiddyshaw Reservoir	40
Stang Forest	58
Tunstall Reservoir	43
Waldridge Fell	16
Washington WWT	11
Waskerley Reservoir	42
Whiburn Obs & Coastal Park	7

The following list of sites and grid references is not intended to be a complete listing of localities mentioned in the book. However, it tries to cover some of the most regularly visited areas, so any reader and prospective visitor can find the location of lesser known haunts.

Durham Wildlife Trust reserves:

Baal Hill Wood	NZ0737	Milkwellburn Wood	NZ1057
Bishop Middleham Quarry	NZ3332	Oakenshaw	NZ2037
Blackhall Rocks	NZ4739	Rabbit Bank Wood	NZ1148
Burnhope Pond	NZ1848	Ragpath Heath	NZ1444
Edmondsley Wood	NZ2349	Rainton Meadows	NZ3248
Hannah's Meadow	NY9318	Raisby Hill Grassland	NZ3335
Hawthorn Dene	NZ4245	Redcar Field	NZ2919
Hesledon Dene	NZ4437	Shibdon Pond	NZ1962
High Wood	NZ1256	Stanley Moss	NZ1538
Joe's Pond	NZ3248	Town Kelloe Bank	NZ3537
Longburn Ford Quarry	NZ0744	Trimdon Grange Quarry	NZ3635
Low Barns	NZ1631	Tudhoe Mill Wood	NZ2535
Malton	NZ1846		

Local Nature Reserves:

Acer Pond	NZ2262	Greatham Beck	NZ5025
Allensford Woods	NZ0850	Green Vale	NZ4119
Barmston Pond	NZ3257	Greencroft Heath	NZ1651
Barwick Pond	NZ4413	Greenvale	NZ4119
Billingham Beck	NZ4522	Hardwick Dene and Elm Tree Wood	NZ4220
Billingham Beck Valley Country	NZ4522	Harelow Heath	NZ1452
Bishop Middleham Wildlife Garden	NZ2331	Harperley and Pea Woods	NZ1853
Black Bobbies Field, Thornaby	NZ4416	Hart to Haswell Walkway	NZ4836
Blackhall Grasslands	NZ4638	Hart Warren Dunes	NZ4936
Bracken Hill Wood	NZ4039	Harton Down Hill	NZ3965
Byerley Park	NZ2526	Hetton Bogs	NZ3448
Castle Eden Walkway	NZ1963	Horden Grasslands	NZ4342
Charlton's Pond	NZ4623	Hylton Dene	NZ3658
Clara Vale	NZ1364	Kyo Bogs	NZ1752
Cleadon Hills	NZ3863	Limekiln Gill	NZ4540
Cong Burn Wood	NZ2550	Little Wood	NZ3437
Cow Plantation	NZ2534	Low Newton Junction	NZ2844
Cowpen Bewley Woodland Country	NZ4825	Marsden Old Quarry	NZ3964
Coxhoe Quarry Wood	NZ3236	Norton Grange Marsh	NZ4420
Cross Lane Meadows	NZ2162	Norwood Nature Park	NZ2461
Crow Trees	NZ3438	Noses Point	NZ4347
Daisy Hill	NZ2548	Pelaw Quarry Pond	NZ3062
Deep Dene, comprising Pontop	NZ1553	Pity Me Carrs	NZ2645
Delight Bank and	NZ1553	Primrose	NZ3264
Dunston Pond	NZ2262	Quarry Wood	NZ4215
Ferryhill Carrs	NZ3032	Raisby Way	NZ3435
Fulwell Quarry	NZ3859	Sacriston Wood	NZ2348

Seaton Dunes and Common	NZ5328	The Moor	NZ2728
Shibdon Pond	NZ1962	The Whinnies	NZ3413
South Stanley Woods	NZ2151	Thorpe Wood	NZ4024
Spion Kop Cemetery, Hartlepool	NZ5134	Tilesheds	NZ3662
Station Burn	NZ3462	Tunstall Hills	NZ3954
Steeley Hill, Cornsay Colliery	NZ1643	Whitburn Point	NZ4163
Stillington Forest Park	NZ3723	Willington North Dene	NZ1935
Summerhill	NZ4831	Windy Nook Nature Park	NZ2761
Tanfield Lea Marsh	NZ1954	Wingate Quarry	NZ3737
The Kittiwake Tower	NZ2763		

National Nature Reserves:

Castle Eden Dene	NZ4339	Cassop Vale	NZ3438
Thrislington	NZ3132	Derwent Gorge & Muggleswick Woods	NZ0548

Other regularly visited bird watching sites:

Arn Gill	NZ0624	Bradbury Carrs	NZ3027
Back Saltholme Pools	NZ5022	Bradley Fell	NZ1262
Balderhead Res.	NY9218	Brancepeth Beck	NZ2436
Barlow Burn	NZ1461	Brasside Ponds	NZ2945
Barlow Fell	NZ1560	Breckon Hill	NZ2950
Barnes Park, Sunderland	NZ3755	Brignall Banks	NZ0611
Barningham Moor	NZ0608	Burdon Moor	NZ2157
Bearpark Hall	NZ2443	Butterknowle	NZ1025
Bedburn	NZ1932	Cassop Moor	NZ3239
Beechburn Farm	NZ1731	Cassop Vale	NZ3338
Belasis Hall Tech Park, Billingham	NZ4723	Cauldron Snout	NY8128
Beldon Burn	NY9449	Causey Woods	NZ2055
Bents Cottages, Whitburn	NZ4061	Chapman's Well	NZ1748
Bent's Park, South Shields	NZ3767	Chilton Moor	NZ3249
Birtley Sewage Works	NZ2556	Chopwell Woods	NZ1358
Bishop Middleham, A1 flashes	NZ3229	Chourdon Point	NZ4446
Bishop Middleham, Castle Lake	NZ3230	Clockburn Lake, Lower Derwent Valley	NZ1860
Bishop Middleham, Farnless Farm	NZ3332	Coatham Wood	NZ3916
Bishop Middleham, Stoney Beck	NZ3331	Cobey Carr, Willington	NZ2034
Bishops Park, Bishop Auckland	NZ2130	Cockfield Fell	NZ1124
Bishopton Sailing Club	NZ3620	Colliery Wood, Boldon Colliery	NZ3562
Black Bank Plantation	NZ1035	Consett	NZ1050
Blackhall Mill	NZ1157	Cornthwaite Park, Whitburn	NZ4061
Blackton Res.	NY9418	Cotherstone	NZ0219
Bollihope Burn	NY9734	Cow Green Res.	NY8130
Bowburn	NZ3138	Cox Green	NZ3255
Bowes	NY9813	Crimdon Denemouth	NZ4936
Bowes Moor	NY9311	Cronkley Fell	NY8427
Bowesfield Marsh	NZ4416	Crookfoot Res.	NZ4331
Bowlees, High Force	NY8828	Croxdale	NZ2737

Dalton Park, Murton	NZ4047	Haverton Hole	NZ4923
Dalton Piercy	NZ4530	Hawthorn Hive	NZ4446
Dawdon Blast Beach	NZ4347	Hawthorn Quarry	NZ4346
Deepdale Wood	NZ0316	Hedleyhope Fell	NZ1541
Derwent Res.	NY9952	Hendon	NZ4155
Dorman's Pool	NZ5122	Herrington CP	NZ3453
Drinkfield Marsh	NZ2817	Hetton Lyons CP	NZ3647
Dryderdale	NZ0833	High Coniscliffe	NZ2215
Dunston Staithes	NZ2362	High Sharpley	NZ3749
Durham City, River Wear	NZ2742	Hill Top	NY9917
Easington Colliery & CP	NZ4343	Hisehope Res.	NZ0246
East Boldon	NZ3661	Holwick Fell	NY8827
Eastgate Quarry	NY9539	Holywell Marsh, Croxdale	NZ2537
Edmundbyers	NZ0049	Horden Dene	NZ4343
Eggleston Common	NY9926	Houghton Dene	NZ3548
Elemore golf course	NZ3545	Houghton Gate	NZ2951
Elemore Hall	NZ3444	Houghton-le-Spring	NZ3349
Elemore Woods	NZ3443	Hudeshope Beck	NY9429
Escombe Lake	NZ1830	Hurworth Burn	NZ4033
Etherley Dene	NZ1929	Hury Res.	NY9619
Etherley Moor	NZ1828	Jackson's Landing, Hartlepool	NZ5133
Far Pasture	NZ1759	James Steel CP	NZ3961
Finchale Woods	NZ2846	Jarrow cem.	NZ3364
Firth Moor CP	NZ3113	Jarrow Slake	NZ3465
Fishburn	NZ3532	Jewish cem., Hartlepool	NZ5134
Flass Vale, Durham	NZ2642	Jock's Bridge	NZ2130
Follingsby Lane	NZ3259	Kelloe Way	NZ3436
Foxcover, Seaham	NZ4348	Kepier Woods, Durham	NZ2743
Framwellgate Moor	NZ2644	Killhope Law	NY8235
Frankland Farm	NZ2844	Kinninvie	NZ0421
Frenchman's Bay, South Shields	NZ3866	Knitsley Fell	NZ0934
Frosterley	NZ0336	Lambton	NZ2952
Garmondsway Moor	NZ3434	Lamesley Pastures	NZ2557
Gaunless Flats	NZ1926	Langdon Beck	NY8531
Gibside	NZ1758	Langleydale Common	NZ0524
Grassholme Res.	NY9422	Lintzgarth Common	NY9242
Great Eppleton	NZ3648	Long Drag	NZ5124
Great Lumley	NZ2849	Longnewton Res.	NZ3616
Greatham Creek	NZ5025	Low Force	NY9027
Greenabella Marsh	NZ5125	Low Middleton	NZ3611
Greta Bridge	NZ0813	Low Willington	NZ2135
Hallgarth	NZ3243	Low Wood	NZ0822
Hamsterley Forest	NZ0329	Lower Derwent Valley CP	NZ1961
Hardwick Hall CP	NZ3429	Lumley Castle	NZ2851
Hargreaves Quarry	NZ5121	Lumley Moor	NZ3148
Hartlepool harbour	NZ5233	Lumley Thicks	NZ3050
Hartlepool Headland	NZ5333	Marine Drive, Hartlepool	NZ5234
Harwood, Upper Teesdale	NY8232	Marsden Bay	NZ4065

Marsden Quarry	NZ3964	Salter's Gate	NZ0742
Marsden Rock	NZ4065	Salterfen Rocks	NZ4153
McNiel Bottoms	NZ1332	Saltholme Pools	NZ5022
Medomsley	NZ1254	Satley	NZ1142
Mere Knolls cem.	NZ3960	Seaburn	NZ4060
Middle End	NY9829	Seaham Hall Dene	NZ4150
Middlestone Moor	NZ2432	Seaham Harbour	NZ4349
Mordon	NZ3226	Seal Sands	NZ5225
Mount Pleasant	NZ3460	Seaton Carew	NZ5229
Mowbray Park	NZ3956	Seaton Common	NZ5328
Muggleswick Woods	NZ0548	Seaton Pond	NZ3849
Murton, Carr House farm	NZ3747	Seaton Snook	NZ5326
Nettlesworth/Plawsworth fields	NZ2548	Sedgeletch	NZ3251
Newbiggin-in-Teesdale	NY9127	Selset Res.	NY9121
Newbottle	NZ3352	Sharpley Woods	NZ3749
Newburn sewage outfall	NZ5131	Sherburn Towers Farm	NZ1659
Newfield	NZ2033	Shotton Pond	NZ4040
Nissan, Washington	NZ3459	Silksworth Lake	NZ3754
No4 Brinefield, Teesmouth	NZ5024	Sled Lane Pond, Crawcrook	NZ1263
North Gare	NZ5428	Smiddyshaw Res.	NZ0446
North Marine Park, South Shields	NZ3767	Souter lighthouse	NZ4064
Observatory Hill, Durham	NZ2641	South Bents	NZ4061
Old Quarrington	NZ3237	South Shields Leas	NZ3866
Pagebank Pond	NZ2234	South Shields, mouth of Tyne	NZ3768
Parton Rocks	NZ5234	St John's Chapel	NY8838
Pawlaw Pike	NY9931	Stadium of Light, Sunderland	NZ3957
Pikeston Fell	NZ0432	Stainton	NZ0618
Pontburn Woods	NZ1455	Stanhope Burn	NY9840
Port Clarence	NZ5021	Stanhope Common	NY9743
Portrack Marsh	NZ4619	Stargate	NZ1663
Pow Hill CP	NZ0151	Sunderland Academy Pools	NZ3960
Quarrington Hill	NZ3437	Sunderland Bridge	NZ2737
Raby Park	NZ1221	Sunderland harbour	NZ4159
Rainton Gate	NZ3245	Sunderland North Dock	NZ4058
Rainton Park Wood	NZ3046	Sunderland South Dock	NZ3958
Ramshaw	NZ1526	Tan Hill	NY8906
Ramside Hall	NZ3144	Tees Barrage	NZ4619
Ravensworth Fell	NZ2258	Temple Park, South Shields	NZ3763
Reclamation Pond	NZ5223	Testo's pool, West Boldon	NZ3461
River Gaunless, Cockfield	NZ1324	The Leas, South Shields	NZ3966
Roker	NZ3959	The Lizard, Marsden	NZ4064
Roker Park	NZ4059	The Stang forest	NZ0208
Rookhope	NY9442	Thornley Woods	NZ1759
Rossmere Park	NZ5029	Timber Beach, River Tyne	NZ2262
Rowlands Gill Viaduct	NZ1759	Timber Beach, River Wear	NZ3758
Ruffside	NY9951	Tow Law	NZ1139
Ryhope	NZ4152	Trimdon	NZ3833
Ryton Willows	NZ1564	Trow Quarry	NZ3866

Trow Rocks	NZ3866	Whitburn CP	NZ4163
Tunstall Res.	NZ0641	Whitburn Golf Club	NZ4064
Victoria Garesfield	NZ1458	Whitburn Range	NZ4162
Waldridge Fell	NZ2449	Whitburn Steel	NZ4161
Ward Jackson Park	NZ4832	Whitesmocks, Durham	NZ2643
Warden Law	NZ3650	Whitworth Hall CP	NZ2334
Waskerley Res.	NZ0244	Willow Pond, Washington	NZ3155
Watergate Park	NZ2260	Witton Park	NZ1730
West Auckland	NZ1725	Witton-le-Wear	NZ1531
West Hall wood, Whitburn	NZ3961	Wolsingham	NZ0736
West Pastures	NZ3360	Woodland Fell	NZ0526
Westgate, Weardale	NY9038	WWT Washington	NZ3356
Whitburn Bird Observatory	NZ4163	Zinc Works road, Teesmouth	NZ5327
Whitburn cem.	NZ4062		

References and Bibliography

General references for the texts contained in the systematic list include all of the annual ornithological reports for Northumberland and Durham published through the 'Natural History Society' in its transactions from 1951 onwards. Similarly used were the *Durham County Bird Report* from 1970 to 1975, the annual reports of the Durham Bird Club, *Birds in Durham*, between 1976 and 2012, and the reports of the Teesmouth Bird Club, the *County Cleveland Bird Report* (later the *Cleveland Bird Report*) over the same period.

To provide a national and broader perspective and context, both breeding atlases (Sharrock 1978 and Gibbons *et al.* 1993) and the winter atlas (Lack 1986) were extensively used; as were key source documents such as *'Population Trends'* (Marchant *et al.* 19903), the *'Historical Atlas'* (Holloway 1996) and the *EEBC Atlas of European Breeding Birds* (Hagemeijer & Blair 1997). As an underpinning source of background information *'BWP'* was routinely referred to. All of the reports of British Birds Rarities Committee (BBRC) were utilised as were those of the Rare Breeding Birds Panel (RBBP), as published annually in British Birds.

Other important sources were: *A Summer Atlas of the Breeding Birds of County Durham* and *The Breeding Birds of Cleveland.* Serial publications often utilised included issues of *The Vasculum* (the North Country Quarterly of Science and Local History) from 1916 onwards, and a range of national publications, in particular *British Birds*, along with a number of older journals, referenced, including The Zoologist and The Naturalist.

Other References

Addinall S. 2005. The Amur Wagtail in County Durham – a new Western Palaearctic bird. *Birding World* 18: 155–158.

Ahmad M. 2005. Franklin's Gulls and Laughing Gulls in Britain and Ireland in November 2005. *Birding World* 18:461-464.

Alexander W. B. and Lack D. 1944. Changes in status among British breeding birds. *British Birds* 38: 62-69.

Alexander W. B. and Fitter R. S. R. 1955. American Landbirds in Western Europe. *British Birds 48:1, 1-13.*

Allan G. 1791-1800. *The Allan Manuscript.*

Allen J. 2007. The Glaucous-winged Gull in Surrey. *Birding World* 20: 150–151.

Almond W. E. 1930. Ruff and Curlew Sandpipers in winter in Durham. *British Birds* 23: 225.

Almond W. E., Nicholson J. B. and Robinson M. G. 1939. Birds of the Tees Valley. *Transactions of the Northern Naturalist's Union* Vol. 1.

Alpin O. V. 1890. On the distribution and period of sojourn in the British Islands of the Spotted Crake. *The Zoologist* XIV (167) 4010-417.
Allport A. M. and Carroll D. 1989. Little Bitterns breeding in South Yorkshire. *British Birds* 82: 442-446.

Alström P., Mild K. and Zetterstrom B. 2003. *Pipits & Wagtails of Europe, Asia and North America.* Christopher Helm/A&C Black, London.

Amar A., Court I. R., Davison M., Downing S., Grimshaw T., Pickford T. and Raw D. 2011. Linking nest histories, remotely sensed land use data and wildlife crime records to explore the impact of grouse moor management on peregrine falcon populations. *Biological Conservation* XXX 2011

Anderson P. 1994. Birding the First Twenty Years. *Birds in Durham 1993.* Durham Bird Club.

Anderson P. 2007. The history of the Avocet in County Durham. *Birds in Durham 2006.* Durham Bird Club.

Appleyard I. 1994. *Ring Ouzels of the Yorkshire Dales* Leeds. W.S. Maney & Sons Ltd.

Arcos J. M. and Oro D. 2004. Pardela Balear *Puffinus mauretanicus.* Pp. 46-50 *in:* Madrono A., Gonzalez C. and Atienza J. C. (eds.) *Libro Rojo de las Aves de Espana.* Madrid, Spain: Direccion General para la Biodiversidad - SEO/BirdLife.

Armstrong A. L. (ed.) 1988. *Birds in Durham 1987.* Durham Bird Club.

Armstrong A. L. (ed.) 1989. *Birds in Durham 1988.* Durham Bird Club.

Armstrong A. L. (ed.) 1990. *Birds in Durham 1989.* Durham Bird Club.

Armstrong A. L. (ed.) 1991. *Birds in Durham 1990.* Durham Bird Club.

Armstrong A. L. (ed.) 1992. *Birds in Durham 1991.* Durham Bird Club.

Armstrong A. L. (ed.) 1993. *Birds in Durham 1992.* Durham Bird Club.

Armstrong A. L. (ed.) 1994. *Birds in Durham 1993.* Durham Bird Club.

Armstrong A. L. (ed.) 1996. *Birds in Durham 1994.* Durham Bird Club.

Armstrong A. L. (ed.) 1997. *Birds in Durham 1995.* Durham Bird Club.

Armstrong A. L. (ed.) 1998. *Birds in Durham 1996.* Durham Bird Club.

Armstrong A. L. (ed.) 1999. *Birds in Durham 1997.* Durham Bird Club.

Armstrong A. L. (ed.) 1999b. *Birds in Durham 1998.* Durham Bird Club.

Armstrong A. L. (ed.) 2003. *Birds in Durham 1999.* Durham Bird Club.

Armstrong A. L. (ed.) 2005. *Birds in Durham 2000.* Durham Bird Club.

Atkinson-Willes G. L. and Matthews G. V. T. 1960. The past status of the Brent Goose *British Birds* 53.

Atkinson-Willes G. L. 1970. Wildfowl situation in England, Scotland and Wales. *Proceeding of the International Regulatory Meeting for the Conservation of Wildfowl Resources,* Leningrad, 1968, 1970

Avery M. I. and Leslie R. 1990. *Birds and Forestry* T&AD Poyser, London.

Backhouse E. 1834. MS. Catalogue of the Birds of the County of Durham *Transactions of the Natural History Society of Northumberland, Durham and Newcastle upon Tyne Vol. II: 182.*

Backhouse J. 1885. Notes on the Avifauna of Upper Teesdale *Naturalist October & November 1885.*

Backhouse J. 1898. *Upper Teesdale Past and Present*, Chapter IX.

Backhouse W. 1846. Additions to Hogg's Catalogue. *The Zoologist* 1846.

Backhouse W. 1854. List of Birds of Darlington. *Longstaffe's History and Antiquities of the Parish of Darlington,* 370-371. Darlington.

Baillie S. R., Marchant J. H., Crick H. Q. P., Noble D. G., Balmer D. E., Coombes R. H., Downie I. S., Freeman S. N., Joys A. C., Leech D. I., Raven M. J., Robinson R. A. and Thewlis R. M. 2006. Breeding Birds in the Wider Countryside: their conservation status 2005. *BTO Research Report No.435.* BTO, Thetford.

Baillie S. R., Marchant J. H., Leech D. I., Renwick A. R., Joys A. C., Noble D. G., Barimore C., Conway G.J., Downie I. S., Risely K. and Robinson R. A. 2010. Breeding Birds in the Wider Countryside: their conservation status 2010. *BTO Research Report No.565.* BTO, Thetford.

Bainbridge I. P. and Minton C. D. T. 1978. The migration and mortality of the Curlew in Britain and Ireland. *Bird Study* 25: 39-50. BTO.

Bainbridge I. P., Evans R. J., Broad R. A., Crooke C. H., Duffy K., Green R. E., Love J. A. and Mudge G. P. 2003. Re-introduction of White-tailed Eagles *Haliaeetus albicilla* to Scotland in Thompson D.B.A., Redpath S.M., Fielding A.H., Marquiss M. and Galbraith C.A. (eds). *Birds of Prey in a Changing Environment.* SNH/The Stationery Office, Edinburgh.

Baines D. 1994. Seasonal differences in habitat selection by Black Grouse *Tetrao tetrix* in the Northern Pennines, England. *Ibis* 136: 39-43.

Baines D. and Hudson P.J. 1995. The decline of Black Grouse in Scotland and northern England. *Bird Study* 42: 122-131.

Baker H., Stroud D. A., Aebischer N. J., Cranswick P. A., Gregory R. D., McSorley C. A., Noble D. G. and Rehfisch M. M. 2006. Population estimates of birds in Great Britain and the United Kingdom. *British Birds* 99:25-44.

Baldridge K. (ed.) 1981. *Birds in Durham 1980.* Durham Bird Club.

Baldridge K. (ed.) 1982. *Birds in Durham 1981.* Durham Bird Club.

Baldridge K. (ed.) 1983. *Birds in Durham 1982.* Durham Bird Club.

Baldridge K. (ed.) 1984. *Birds in Durham 1983.* Durham Bird Club.

Baldridge K. (ed.) 1985. *Birds in Durham 1984.* Durham Bird Club.

Baldridge K. (ed.) 1986. *Birds in Durham 1985.* Durham Bird Club.

Baldridge K. (ed.) 1987. *Birds in Durham 1986.* Durham Bird Club.

Banks A. N., Burton N. H. K., Calladine J.R. and Austin G.E. 2007. Winter Gulls in the UK: Population Estimates from the 2003/04-2005/06 Winter Gull Roost Survey. *BTO Research Report No. 456.* BTO, Thetford.

Bannerman D. A. 1953-1963. *The Birds of the British Isles.* Oliver & Boyd, London & Edinburgh.

Bates B. S. 1991. An Addition to the County Avifauna: Red-eyed Vireo *Vireo olivaceus. Birds in Durham 1990.* Durham Bird Club.

Batten L. A., Bibby C. J., Clement P., Elliot G. D. and Porter R. F. 1990. *Red Data Birds in Britain.* T&AD Poyser for NCC/RSPB, London.

Batten L. A. 2001. European Honey-buzzard Survey 2000 and 2001: preliminary results and request for further surveys. *British Birds* 94: 143-144.

Baxter P. and Gibbins C. 2007. Identification of Kumlien's and American Herring Gull and other large gulls at St John's, Newfoundland. *Birding World* 20: 162-175.

Bayliss M. 1985. Much grunting in the marsh. *BTO News* 138: 8-9.

BBS 2009 & 2010. *The Breeding Bird Survey*: web-based breeding bird data, BTO, JNCC & RSPB.

Beck B. 2000. Short-billed Dowitcher – New to Cleveland. *Cleveland Bird Report for 1999.* Teesmouth Bird Club.

Beekman J. H. 1997. Censuses of NW European Bewick's Swan population January 1990 and 1995. *Swan Specialist Group Newsletter* 6: 7-9.

Bell D. G. 1962-1967. Ornithological Reports for Northumberland & Durham, 1961 to 1966. *Transactions of the Natural History Society of Northumberland, Durham & Newcastle upon Tyne* Vol. XIV No. 4 & 6, XV No. 2 & 4, XVI No. 2. Newcastle upon Tyne

Bell D. G. and Parrack J. D 1968. Ornithological Report for Northumberland & Durham 1967 in *Transactions of the Natural History Society of Northumberland, Durham & Newcastle upon Tyne* Vol. XVI, No. 5. Newcastle upon Tyne.

Bell D. G. and Parrack J. D. 1972. Ornithological Report for Northumberland & Durham 1968. *Transactions of the Natural History Society of Northumberland, Durham & Newcastle upon Tyne.* Newcastle upon Tyne

Bell D. G. 1958-1968. *Teesmouth Bird Reports 1958-1967*, monthly bulletins until 1967, then quarterly. Teesmouth Bird Club.

Bell D. G. (ed.) 1976. *County of Cleveland Bird Report for 1975.* Teesmouth Bird Club.

Bell D. G. (ed.) 1977. *County of Cleveland Bird Report for 1976.* Teesmouth Bird Club.

Bell D. G. (ed.) 1978. *County of Cleveland Bird Report for 1977.* Teesmouth Bird Club.

Bell D. G. (ed.) 1979. *County of Cleveland Bird Report for 1978.* Teesmouth Bird Club.

Bell D. G. (ed.) 1980. *County of Cleveland Bird Report for 1979.* Teesmouth Bird Club.

Bell D. G. (ed.) 1981. *County of Cleveland Bird Report for 1980.* Teesmouth Bird Club.

Bell D. G. (ed.) 1982. *County of Cleveland Bird Report for 1981.* Teesmouth Bird Club.

Bell D. G. (ed.) 1983. *County of Cleveland Bird Report for 1982.* Teesmouth Bird Club.

Bell D. G. (ed.) 1984. *County of Cleveland Bird Report for 1983.* Teesmouth Bird Club.

Bell D. G. (ed.) 1985. *County of Cleveland Bird Report for 1984.* Teesmouth Bird Club.

Bell D. G. (ed.) 1986. *County of Cleveland Bird Report for 1985.* Teesmouth Bird Club.

Bell D. G. (ed.) 1987. *County of Cleveland Bird Report for 1986.* Teesmouth Bird Club.

Bell D. G. (ed.) 1988. *County of Cleveland Bird Report for 1987.* Teesmouth Bird Club.

Bell D. G. (ed.) 1989. *County of Cleveland Bird Report for 1988.* Teesmouth Bird Club.

Bell D. G. (ed.) 1990. *County of Cleveland Bird Report for 1989.* Teesmouth Bird Club.

Bell D. G. (ed.) 1991. *County of Cleveland Bird Report for 1990.* Teesmouth Bird Club.

Bell D. G. (ed.) 1992. *County of Cleveland Bird Report for 1991.* Teesmouth Bird Club.

Bell D. G. (ed.) 1993. *County of Cleveland Bird Report for 1992.* Teesmouth Bird Club.

Bell D. G. (ed.) 1994. *County of Cleveland Bird Report for 1993.* Teesmouth Bird Club.

Bell D. G. (ed.) 1995. *County of Cleveland Bird Report for 1994.* Teesmouth Bird Club.

Bell D. G. (ed.) 1996. *County of Cleveland Bird Report for 1995.* Teesmouth Bird Club.

Bell D. G. (ed.) 1997. *Cleveland Bird Report for 1996.* Teesmouth Bird Club.

Bell D. G. (ed.) 1998. *Cleveland Bird Report for 1997.* Teesmouth Bird Club.

Bell J. 2006. Spring passage of White-billed Divers off southwest Norway. *Birding World* 19 (2); 62-63.

Bell S. C. 2008. Introduced and Exotic birds in County Durham. *Birds in Durham 2007.* Durham Bird Club.

Bell S. C. 2009. Yellow Wagtails in County Durham. *The Lek, Spring 2009.* Durham Bird Club.

Bell S. C. 2010. An Exceptional Influx of Great Grey Shrikes in the County. *The Lek, Winter 2010.* Durham Bird Club.

Benett R. 1844. *The Durham household Book (liber bursarii ecclesiae Dunelmensis); or the accounts of the Bursar of the Monastery of Durham from Pentecost 1530 to Pentecost 1534.* Surtees Society, Durham.

Bengston S. 1970. Breeding behaviour of the Purple Sandpiper *Calidris maritima* in West Spitzbergen. *Ornis Scandinavica,* 1: 17-25.

Berthold P. 1995. Microevolution of migratory behaviour illustrated by the Blackcap *Sylvia atricapilla* 1993 Witherby Lecture. *Bird Study* 42.

Bewick T. 1797. *A History of British Birds. Volume I ... Land Birds.* Newcastle: Beilby & Bewick.

Bewick T. 1804. *A History of British Birds* (Volume 2, 1st edition). Longman & Rees, London.

Bewick T. 1826. *A History of British Birds.* Longman & Co. London

Bibby C. J. and Nattrass M. 1986. Breeding status of the Merlin in Britain. *British Birds* 79: 170-184.

Birdlife International 2004. Birds in Europe; population estimates, trends and conservation status *Birdlife Conservation Series No 12.* Birdlife International, Cambridge.

Birdlife International 2006. *IUCN Red List of Threatened Species.* International Union for Conservation.

Birdlife International 2009. *IUCN Red List of Threatened Species.* International Union for Conservation.

Birdlife International 2011. Species factsheet: *Puffinus mauretanicus.*

Birkhead T. R. 1991. *The Magpies.* T&AD Poyser, London.

Black J. M. and Rees E. C. 1984. The structure and behaviour of the Whooper Swan population wintering at Caerlaverock, Dumfries and Galloway, Scotland: An introductory study. *Wildfowl* 35.

Blair H.M.S. 1944. *The Naturalist.*

Blair H.M.S. 1945. Manx Shearwater breeding on the Durham coast. *British Birds* 38.

Blaker G. B. 1934. *The Barn Owl in England and Wales.* RSPB, London.

Blick M. A. 1978. *The Birds of Tees-side 1968-1973.* Teesmouth Bird Club & Cleveland County Council, Cleveland.

Blick M. A. 1989. Double-crested Cormorant: A New Western Palaearctic Bird. *Birding World* 2.

Blick M. A. 2006. Cetti's Warbler – New to Cleveland. *Cleveland Bird Report for 2005.* Teesmouth Bird Club.

Blick M. A. 2009. *Birds of Cleveland.* Tees Valley Wildlife Trust.

Bolam G. 1912. *The Birds of Northumberland and the Eastern Borders.* Henry Hunter Blair, Alnwick

Bolam G. 1932   A catalogue of the birds of Northumberland. *Transactions of the Natural History Society of Northumberland, Durham & Newcastle upon Tyne* Vol. 8, 1-165.

Bold T. J. 1841. on Honey-buzzard. *The Zoologist.*

Bold T. J. 1850. *The Zoologist* Vol. 8.

Bold T. J. 1855. on Grey Phalarope. *The Zoologist .*

Bolton M., Tyler G., Smith K. and Bamford R. 2007. The impact of predator control on lapwing *Vanellus vanellus* breeding success on wet grassland nature reserves. *Journal of Applied Ecology* Vol. 44, Issue 3.

Bone P., Bowey K. and Coleman J. 1995. Shibdon Pond's Mute Swans, a twelve-year case study. *The Vasculum* 80: 25-49.

BOU 2005. *The British List 2005.* British Ornithologists' Union, Tring.

BOU 2007 - Sangster G., Collinson J. M., Knox A. G., Parkin D. T. and Svensson L. 2007. Taxonomic recommendations for British birds: Fourth report. *Ibis* 149.

Bowey K. 1992. The Mandarin in north-east England, Patterns & Origins. *Birds in Durham 1991.* Durham Bird Club.

Bowey K. 1993. The Barn Owl in County Durham, its history and current status - the results of the 1992 Durham Bird Club/Tyne & Wear Museums Service Survey into the status of the Barn Owl. *The Vasculum*: 78 (3).

Bowey K. 1995. European Storm-petrels without their toes. *British Birds* 88: 2.

Bowey K. 1995. Moorhens feeding on pollen. *British Birds* 88:2, 111-113.

Bowey K. 1997. Grey Heron catching Common Starling in flight. *British Birds* 90:3, 112-114.

Bowey K. 1999. The Spotted Crake in County Durham, a review of its status and distribution. *Birds in Durham 1996.* Durham Bird Club.

Bowey K. 1999b. The distribution of oilseed rape utilization by Sedge Warblers in County Durham (1973-1997). *Birds in Durham 1998.* Durham Bird Club.

Bowey K. 1999c. Birdwatching by Numbers: Five Years of the 'Durham Bird Club Co-ordinated Migrant Count'. *Birds in Durham 1997.* Durham Bird Club.

Bowey K. 2007. The Red Kite *Milvus milvus* - A Review of its Status in County Durham *Birds in Durham 2006*. Durham Bird Club.

Bowey K., Davidson P. and Robson K. 1995. Common Gull in juvenile plumage in late February. *British Birds* 88:1, 47-49.

Bowey K., Donnison A. and Rutherford S. 2007. Wintering Common Goldeneye *Bucephala clangula*, numbers on two adjacent river systems in northeast England in relation to a reduction in the levels of sewage input. *Birds in Durham* 2002/2003. Durham Bird Club.

Bowey K., Rutherford S. and Westerberg S. 1993. *Birds of Gateshead*. Gateshead MBC.

Boyd H. 1954. The 'wreck' of Leach's Petrels in the autumn of 1952. *British Birds* 47:5, 137-163.
Boyle J. R. 1892. *Comprehensive Guide to the County of Durham*. The Walter Scott Publishing Co. Ltd., Durham.

Bradshaw M. E. (ed.) 1976. *The Natural History of Upper Teesdale*. Durham County Conservation Trust.

Bradshaw C., Hodgson M. S., Johnston A. J. and Kerr I. 1990. *Birds in Northumbria 1989*. Northumberland & Tyneside Bird Club.

Branson N. J. B. A. and Minton C.D.T. 1976. Moults, measurements and migrations of the Grey Plover. *Bird Study* 23.

Brickle N. W. and Harper D. G. 2002. Agricultural intensification and the timing of breeding of Corn Buntings *Miliaria calandra*. *Bird Study* 49: 219–228.

Brittingham M. C. and Temple S. A. (ed.) 1983. Have Cowbirds Caused Forest Songbirds to Decline? Bio Science 33: 31-35. *Birds in Europe: Population Estimates, Trends and Conservation Status*. Birdlife International, Cambridge.

Britton D. 1978. Identification of Bonaparte's Gull. *British Birds* 72: 7.

Brown A. and Grice P. 2005. *Birds in England*. T&AD Poyser, London.

Brown R. H. 1925. Late nesting of Moorhen in Durham. *British Birds* 21:3, 61-69.

BTO 2011: *Satellite tagging of Cuckoos* web-based data.

Burton J. F. 1995. *Birds and Climate Change*. Christopher Helm, London.

Burton N. H. K. and Evans P. R. 1997. Survival and winter site-fidelity of Turnstones *Arenaria interpres* and Purple Sandpipers *Calidris maritima* in northeast England. *Bird Study* 44: 35-44.

Burton N. H. K. 2007. Influence of restock age and habitat patchiness on Tree Pipits *Anthus trivialis* breeding in Breckland pine plantations. *Ibis* 149 (supp 2).

Butchart S., Ekstrom J. and Malpas L. 2011. *Spotted Redshank–BirdLife species factsheet*. BirdLife International, web-based data.

Byers J. 1927. Waxwing in Durham. *British Birds* 21:8, 195-207.

Byrkjedal I. 1985. Time-Activity Budget for Breeding Greater Golden-Plovers in Norwegian Mountains. *The Wilson Bulletin* 97: 4.

Byrkjedal I. and Thompson D. 1998. *Tundra Plovers: the Eurasian, Pacific and American Golden Plovers and Grey Plover.* Poyser, London.

Calbrade N., Holt C., Austin G., Mellan H., Hearn R., Stroud D., Wotton S. and Musgrove A. 2010. *Waterbirds in the UK 2008/09: The Wetland Bird Survey.* BTO/WWT/RSPB/JNCC, Slimbridge.

Calvert R. 1884. *Geology & Natural History of the County of Durham.*

Cambell L. H. 1984. The Impact of Changes in Sewage Treatment on Seaducks Wintering in the Firth of Forth, Scotland. *Biol. Conservation* 28.

Carter S .P. 1990. *Studies of Goosander Mergus merganser.* Unpubl. PhD Thesis, University of Durham.

Carter S. P. 1995. *Britain's Birds in 1991/92: the conservation and monitoring review.* British Trust for Ornithology, Thetford/Joint Nature Conservation Committee, Peterborough.

Carter I. 2001. *The Red Kite.* Arlequin Publications, Chelmsford.

Charlton D. 2011. County Summaries for 2010. *The Lek, Spring 2010-Spring 2011.* Durham Bird Club.

Chamberlain D. E. and Crick H.Q.P. 2003. Temporal and spatial associations in aspects of reproductive performance of Lapwings *Vanellus vanellus* in the United Kingdom, 1962–99. *Ardea* 91.

Chandler R. J. 1981. Influxes into Britain and Ireland of Red-necked Grebes and other waterbirds during winter 1978/79. *British Birds* 74:2.

Chapman A. 1889. *Bird Life of the Borders: records of wild sport and natural history on moorland and sea.* Gurney and Jackson, London. (Republished 1990 by The Spredden Press, Stocksfield).

Chislet R. 1952. *Yorkshire Birds.* A. Brown and Sons, London.

Clapham A. R., Tutin T. G. and Warburg E. F. 1985. *Excursion Flora of the British Isles,* 3rd Edition. Cambridge University Press, Cambridge.

Clark F. and McNeil D. A. C. 1980. Cliff-nesting colonies of House Martins, *Delichon urbica*, in Great Britain. *Ibis* 122.

Clark J. A. 2004. Ringing recoveries confirm high wader mortality in severe winters. *Ringing & Migration* 22: 43-50.

Clark J. A., Balmer D. E., Blackburn J. R., Milne L. J., Robinson R. A., Wernham C. V., Adams S. Y. and Griffin B. M. 2002. Bird ringing in Britain and Ireland in 2000. *Ringing & Migration* 21. BTO, Thetford.

Clark J. A., Robinson R. A., Du Feu C., Wright L. J., Conway G. J., Blackburn J. R., Leech D. I., Barber L. J., De Palacio D., Griffin B. M., Moss D. and Schafer S. 2010. Bird Ringing Britain and Ireland in 2009 *Ringing & Migration* 25. BTO, Thetford.

Clark J. A., Dadam D., Robinson R. A., Moss D., Leech D. I , Barber L. J , Blackburn J. R., Conway G. J., De Palacio D., Griffin B. M., and Schafer S. 2011. Bird Ringing Britain and Ireland in 2010 *Ringing & Migration 26.* BTO, Thetford.

Clark J. G. D. C. 1954. *Excavations at Star Carr.* Cambridge.

Clegg G. H. 1931. Nuthatches in Durham. *British Birds* 25: 7.

Clements R. 2000. Range Expansion of the Common Buzzard in Britain. *British Birds* 93, 242-248.

Clements R. 2001. The Hobby in Britain: a new population estimate. *British Birds* 94:9, 402-408.

Clements R. 2002. Common Buzzard in Britain: a new population estimate. *British Birds* 95, 377-383.

Clements R. 2008. The Common Kestrel Population in Britain. *British Birds* 101.

Clifton S. J. and Hedley S. 1995. *Durham Wildlife Audit.* English Nature, Peterbrough/Durham County Council.

Coker M. and Mabey R. 2005. *Birds Britannica.* Chatto & Windus, London.

Coleman J. 1991. The Mute Swan Breeding Survey 1990. *Birds in Northumbria 1990.* Northumberland & Tyneside Bird Club.

Collett T. 2009. The Glaucous-winged Gull in Cleveland – the second British record. *Birding World* 22.

Collinson M., Knox A. G., Parkin D. T. and Sangster G. 2003. Specific status of taxa within the Greenish Warbler complex. *British Birds* 96: 327-331.

Collinson, J. M., Parkin D. T., Knox A. G., Sangster G. and Svensson L. 2008. Species boundaries in the Herring and Lesser Black-backed Gull complex. *British Birds* 101.

Conder P. 1989. *The Wheatear.* Christopher Helm, London.

Coniff R. 1991. Why catfish farmers want to throttle the crow of the sea. *Smithsonian* 22.

Conway G., Wotton S., Henderson I., Langston R., Drewitt A and Currie F. 2007. Status and distribution of European Nightjars *Caprimulgus europaeus* in the UK in 2004. *Bird Study* 54:98-111.

Conway G. J., Burton N. H. K., Handschuh M. and Austin G. E. 2008. UK population estimates from the 2007 Breeding Little Ringed Plover and Ringed Plover Surveys. *Research Report 510.* BTO, Thetford.

Coombs F. 1978. *The Crows.* Batsford, London.

Cooper R. H. W. 1987. Migration strategies of shorebirds during the non-breeding season with particular reference to the Sanderling *Calidris alba. Unpublished PHD thesis,* University of Durham.

Cornwallis R. K. 1961 Four invasions of Waxwings during 1956-60 *British Birds* 54

Corso A. 2001. Head pattern variation in Black-headed Wagtail. *Birding World* 14 (4) 162-167.

Cottier E. J. and Lea D. 1969. Black-tailed godwits, ruffs and black terns breeding on the Ouse Washes. *British Birds* 62: 259-270.

Coulson J. C. 1961. Ornithological Report for Northumberland and Durham for 1960. *Transactions of the Natural History Society of Northumberland, Durham and Newcastle upon Tyne* Vol. XIV, No. 1.

Coulson J. C. 1963. The Status of the Kittiwake in the British Isles. *Bird Study* 10.

Coulson J. C. 1973. *Durham County Bird Report 1972.* Zoology Department, University of Durham.

Coulson J. C. 1974. *Durham County Bird Report 1973.* Zoology Department, University of Durham.

Coulson J. C. 2011. *The Kittiwake.* T&AD Poyser, London.

Coulson J. C., Butterfield J., Duncan N., Thomas C., Monaghan P., Ensor K. and Sheddon C. 1984. The wintering Herring Gulls in northern Britain: the Scandinavian component. *Ornis Scandinavica* 15.

Coulson J. C. and Butterfield J. 1986. Studies on a colony of colour-ringed Herring Gulls *Larus argentatus*: I - Adult survival rates. *Bird Study* 33: 51-54.

Coulson J. C. and Butterfield J. 1986. Studies on a colony of colour-ringed Herring Gulls *Larus argentatus*: II - Colony occupation and feeding outside the breeding season. *Bird Study* 33: 55-59.

Coulson, J. C. and Horobin, J. M. 1972. The annual re-occupation of breeding sites by the Fulmar. *Ibis* 114 (1), 30-42. BOU.

Coulson J. C. and Macdonald A. 1962. Recent changes in the habits of the kittiwake. *British Birds* 55: 171-177.

Coulson J. C. and Strowger J. 1999. The annual mortality rate of Black-legged Kittiwakes in NE England from 1954 to 1998 and a recent exceptionally high mortality. *Waterbirds* 22: 3-13.

Coulson J. C. and White E. 1956. The study of colonies of the Kittiwake *Rissa tridactyla* (L.) *Ibis* 98: 63-79. BOU.

Court I. R., Irving P. V. and Carter I. 2004. Status and Productivity of Peregrine Falcons in the Yorkshire Dales 1978-2002. *British Birds* 97.

Courtenay Rev. G. F. 1931. Birds of Hurworth Burn, County Durham, 1910 to 1929 *Transactions of the Natural History Society of Northumberland, Durham & Newcastle-upon-Tyne*, (New Series) Vol. VII, pt. 2.

Courtenay Rev. G. F. 1933. Birds of Tees-mouth, Durham side 1907-1929. *The Vasculum* Vol. XIX: 81.

Courtenay Rev. G. F. 1934. Some Notes on the Birds of Sunderland. *The Vasculum* Vol. XX.

Coward T. A. 1926. *The Birds of the British Isles and their Eggs*. Frederick Warne & Co, London.

Cowley E. 1979. Sand martin population trends in Britain, 1965-1978. *Bird Study* 26: 113-116.

Cowley E. and Siriwardena G. M. 2005. Long term variation in survival rates of Sand Martins *Riparia riparia*: dependence on breeding and wintering ground weather, age and sex, and their population consequences. *Bird Study* 52:237-251.

Craggs J. D. 1990. Purple Sandpipers. *Hilbre Bird Observatory Report*, 39-50.

Cramp S., Pettet A. and Sharrock J. T. R. 1957. The irruption of tits in autumn 1957. *British Birds* 53:2.

'Cramp et al.' - Cramp S. (ed.). 1977-1994. *The Handbook of the Birds of Europe, the Middle East and North Africa: The Birds of the Western Palaearctic* (BWP), Nine Volumes. Oxford University Press, Oxford.

Cranswick P.A., Bowler J. M., Delany S., Einarsson Ó., Garðarsson A., McElwaine J. G., Rees E. C. and Wells J. H. 1997. Whooper Swans wintering in Britain, Ireland and Iceland: International Census, January 1995. *Wildfowl* 47: 17-30.

Cranswick P. A. , Worden J., Ward R., Rowell H., Hall C., Musgrove A., Hearn R., Holloway S., Banks A., Austin G., Griffin L., Hughes B., Kershaw M., O'Connell M., Pollitt M., Rees E. and Smith L. 2005. *The Wetland Bird Survey 2001-2003*: Wildfowl and Wader Counts.

Crick H. Q. P. 1992. Trends in the breeding performance of Golden Plover in Britain. *Research Report 76*. BTO, Thetford.

Crick H. Q. P. and Sparks T. H. 1999. Climate change related to egg-laying trends. *Nature* 399.

Curtis W. F. 1991. *Yorkshire Bird Report 1989.* Yorkshire Naturalists' Union.
Dale J. E. 1975. Ornithological report for 1975. Yorkshire Naturalists Union.

Davenport D. 1992. The Spring Passage of Long-tailed and Pomarine Skuas in Britain and Ireland. *Birding World* 5: 92-95.

Davidson N. C., Strann K., Crockford N. J., Evans P. R., Richardson J., Standen L. J., Townshend D. J., Uttley J. D., Wilson J. R. and Wood A. R. 1986. The origins of Knots *Calidris canutus* in arctic Norway in spring. *Ornis Scandinavica*, 17: 175-179.

Davis A. and Vinnicombe K. E. 2011. The probable Breeding of Ferruginous Ducks in Avon. *British Birds* 104: 77-83.

Day J. C., Hodgson M. S. and Rossiter B. N. 1995. *The Atlas of Breeding Birds in Northumbria.* Northumberland & Tyneside Bird Club. Newcastle upon Tyne.

Day J. C. 2008. *The First Fifty Years: A Short History of the Northumberland and Tyneside Bird Club* Northumberland & Tyneside Bird Club Bulletin. Newcastle upon Tyne.

Day J. C. and Hodgson M.S. 2003. *The Atlas of Wintering Birds in Northumbria* Northumberland & Tyneside Bird Club. Newcastle upon Tyne.

Dean A. R. 1985. Review of British status and identification of Greenish Warbler. *British Birds* 78: 437-451.

Delany S., Greenwood J. J. D. and Kirby K. 1992. *National Mute Swan Survey 1990.* Unpubl. Report to J.N.C.C., Peterbrough.

del Hoyo J., Elliott A. and Sargatal, J. (eds.) 1996. *Handbook of the Birds of the World* 3: 645.Lynx Edicions.

Dennis R. H. 1987. Osprey Re-colonisation. *RSPB Conservation Review* 1. Sandy.

Dewar J. M. 1934. Northumberland Rook Roosts. *British Birds* 27:4.

Dobinson H. M. and Richards A. J. 1964. The effect of the severe winter of 1962/63 on birds in Britain. *British Birds* 57:10, 373-434.

Dolbeer R. A. 1982. Migration patterns for age and sex classes of blackbirds and starlings. *Journal of Field Ornithology* 53: 28-46.

Donald P. F. 1997. The Ecology and Conservation of Corn Buntings. *UK Natural Conservation No 13* JNCC, Peterborough.

Donald P. F., Wilson J.D. and Shepherd M. 1994. The decline of the Corn Bunting. *British Birds* 87.

Dubois P. 2007. Yellow, Blue-headed, 'Channel' and extralimital Wagtails. *Birding World* 20(3)104-112.

DUBSG - Raw D. (ed.) 1992-2010. *The Annual Reports of the Durham Upland Bird Study Group.*

Duff J. 1850. Crested Tit in Durham. *The Zoologist* Vol. VIII.

Duncan S. 1911. Little Terns Nesting at Teesmouth. *British Birds* 4:2.

Dunn P. J. and Herschfeld E. 1991. Long-tailed Skuas in Britain and Ireland in autumn 1988. *British Birds* 84:121-136.

Dymond J. N., Fraser P. A. and Gantlett S. J. M. 1989. *Rare Birds in Britain and Ireland.* T&AD Poyser, Calton.

Eaton M. A., Ausden M., Burton N., Grice P.V., Hearn R. D., Hewson C. M., Hilton G. M., Noble

D. G., Ratcliffe N. and Rehfisch M. M. 2006. *The state of the UK's birds 2005* RSPB, BTO, WWT, CCW, EN, EHS, and SNH, Sandy, Bedfordshire.

Eaton M. A., Brown A. F., Noble D. G., Musgrove A. J., Hearn R. D., Aebischer, N. J., Gibbons D. W., Evans A. and Gregory R. D. 2009. Birds of Conservation Concern 3 The Population Status of Birds in the United Kingdom, Channel Island and the Isle of Man. *British Birds* 102.

Elliot Sir W. 1876. Memoir of Dr T.C. Jerdon. *History of the Berwickshire Naturalists' Club* 7.

Embleton D. 1884. Note on the Birds seen at Nest House, Felling Shore, in May and June 1884 *Newcastle, Bell, 1884, Medical Tracts* Vol. 54.
England M. D. 1974. A further review of the problem of 'escapes'. *British Birds* 67:5, 177-197.

English Nature 2001. *A Biodiversity Audit of the North East.* North East Biodiversity Forum, 2001.

Etheridge B., Summers R. W. and Green R. E. 1997. The effects of illegal killing and destruction of nests by humans on the population dynamics of the Hen Harrier. *Journal of Applied Ecology* 34.

Evans A. H. (ed.) 1903. *Turner on birds: a short and succinct history of the principal birds noticed by Pliny and Aristotle, first published by Doctor William Turner, 1544.* Cambridge University Press, Cambridge.

Evans J. 1991. *The Red Kite in Wales* Christopher Davies Publishers.

Evans K. L. and Robson R. A. 2004. Barn Swallows and agriculture. *British Birds* 97:218-230.

Evans P. R. 1968. Autumn movements and orientation of waders in north east England and southern Scotland, studied by radar. *Bird Study* 15: 53-64.

Evans P. R. 1978/79. Reclamation of intertidal land: some effects on Shelduck and wader populations in the Tees estuary. *Verhandlung Ornithologische Gesellschaft Bayern* 23: 147-168.

Evans P. R. 1997. Improving the accuracy of predicting the local effects of habitat loss on shorebirds: lessons from the Tees and Orwell Estuary studies. In *Effects of Habitat Loss and Change on Waterbirds*, Goss-Custard J. D., Rufino R. and Luis A. (eds.). The Stationery Office, London.

Evans P. R. 1998. Improving the accuracy of predicting the local effects of habitat loss on shorebirds: lessons from the Tees and Orwell Estuary Studies. *Proceedings of the 10th International Waterfowl Ecology Symposium*

Evans P. R., Herdson D. M., Knights P. J. and Pienkowski M. W. 1979. Short-Term Effects of Reclamation of Part of Seal Sands, Teesmouth, on Wintering Waders and Shelduck: Shorebird Diets, Invertebrate Densities, and the Impact of Predation on the Invertebrates. *Oecologica* 41: 183-206.

Evans P. R. and Pienkowski M. W. 1983. Implications for coastal engineering projects of studies, at the Tees estuary, on the effects of reclamation of intertidal land on shorebird populations. *Waterfowl Science Tech.* 16.

Evans R. 1997. Eurasian Jackdaw Preying on House Martin. *British Birds* 90:5.

Evans S. 2005. Long-Eared Owl Survey Results. *The* Lek, Autumn & Winter 2005. Durham Bird Club.

Evans S. 2006. Owl Study Group - Summer 2006 Sightings Summary. *The Lek, Autumn 2006*. Durham Bird Club.

Ewing S. R., Rebecca G. W., Heavisides A., Court I. R., Lindley P., Ruddock M., Cohen S., Eaton M. A. 2011. Breeding status of Merlins *Falco columbarius* in the UK in 2008. *Bird Study* 58: 379-389.

Fawcett J.W. 1890. *The Birds of Durham* (privately published).

Feare C. J. 1995. Changes in numbers of Common Starling and farming practice in Lincolnshire. *British Birds* 87.

Ferguson-Lees I.J. 1959. Recent reports and news. *British Birds* 52:12, 435.

The Field 1855. Great Snipe shot in Durham.

The Field 1896. Great Snipe shot at Romaldkirk.

Fisher J. 1952. *The Fulmar.* Collins New Naturalist Monograph No. 6, London.

Fisher J. 1966. *The Shell Bird Book (3rd Ed. 1973)*. Ebury and Michael Jospeh, London.

Fisher A. and Flood, B. 2004. A Scopoli's Shearwater off the Isles of Scilly. *Birding World* 17: 334-336.

Fisher I. and Holliday S.T. 2008. *Birds in Northumbria 2007*. Northumberland & Tyneside Bird Club. Newcastle upon Tyne.

Fisher I. and Holliday S.T. 2009. *Birds in Northumbria 2008*. Northumberland & Tyneside Bird Club. Newcastle upon Tyne.

Fitzgerald G. R. and Coulson J. C. 1973. The distribution and feeding ecology of gulls on the tidal beaches of the Rivers Tyne and Wear. *The Vasculum* 58: 29-47.

Flegg J. J. M. 1973. A study of Treecreepers. *Bird Study* 21.

Fletcher J. 2010. *Birdwatchers of Teesmouth 1600–1960, The Events leading up to the formation of the Teesmouth Bird Club.* The Teesmouth Bird Club.

Fletcher K., Aebischer N. J., Baines D., Foster R., Hoodless A. N. 2010. Changes in breeding success and abundance of ground nesting moorland birds in relation to the experimental deployment of legal predator controls. *Journal of Applied Ecology* 47.

Fowler J. A. and Okill J. D. 1988. Recaptures of Storm Petrels tape lured in Shetland. *Ringing and Migration* 9: 49-50.

Fowler, Canon J. T. 1898. *Extracts from the Account Rolls of the Abbey of Durham, from the Original Mss. (Publications of the Surtees Society 100). Volume I.* Durham: Surtees Society, Durham

Fox A. D. 1986. The breeding teal *Anas crecca* of a coastal raised mire in central west Wales. *Bird Study* 33: 18-23.

Fox A. D. 1991. History of the Pochard breeding in Britain. *British Birds* 84:3.

Fox A. D. and Aspinall S. J. 1987. Pomarine Skuas in autumn 1985. *British Birds* 80: 404-42.

Fox A. D., Jarrett N., Gitay H. and Paynter D. 1989. Late summer habitat selection by breeding waterfowl in northern Scotland. *Wildfowl* 40: 106-114.

Fox A. D. and Meek E. R. 1993. History of the Northern Pintail breeding in Britain and Ireland. *British Birds* 86: 151-162.

Fox A. D., Walsh A. J., Norriss D. W., Wilson H. J., Stroud D. A. and Francis I. S. 2006. Twenty-five years of population monitoring - the rise and fall of the Greenland White-fronted Goose *Anser albifrons flavirostris*. *Waterbirds around the world. (*eds.) Boere G.C., Galbraith C.A. and Stroud D. A. The Stationery Office, Edinburgh, UK.

Fox G. T. 1827. *Synopsis of the Newcastle Museum, late the Allan, formerly the Tunstall, or Wycliffe Museum.* T. and J. Hodgson, Newcastle upon Tyne.

Fox G. T. 1831. Notice of some rare Birds recently killed in the counties of Northumberland and Durham. *Transactions of the Natural History Society of Northumberland, Durham & Newcastle-upon-Tyne.*

Frank D. 1992. The influence of feeding conditions on food provisioning of chicks of Common Terns *Sterna hirundo* nesting in the German Wadden Sea. *Ardea* 80: 45-55.

Frankis M. P. 1996. The Status of the Hawfinch in Northumberland. *Birds in Northumbria* 1995 Northumberland & Tyneside Bird Club. Newcastle upon Tyne.

Fraser P. A., Lansdown P. G. and Rogers M. J. 1999. Report on scarce migrant birds in Britain in 1997. *British Birds* 92:12, 618-658.

Fraser P. A. and Rogers M. J. 2006. Report on scarce migrant birds in Britain in 2003 Part 2: Short-toed Lark to Little Bunting. *British Birds* 99:3, 129-147

Fraser P. A, and Rogers M. J. & the Rarities Committee 2007. Report on rare birds in Great Britain in 2005. *British Birds* 100:16–61; 100:72-104.

Fraser P. A and Rogers M. J. & the Rarities Committee 2007b. Report on rare birds in Great Britain in 2006. *British Birds* 100 British Birds 100: 744.

Frazer P. A. and Ryan J. F. 1994. Scarce migrants in Britain and Ireland. Parts 1 & 2. Numbers during 1986-92. *British Birds* 87.

Fuller R. J. 1982. *Bird Habitats in Britain.* T&AD Poyser, Berkhamsted.

Fuller R. J., Gregory R. D., Gibbons D. W., Marchant J. H., Wilson J. D., Baillie S. R. and Carter N. 1995. Population declines and range contractions among lowland farmland birds in Britain. *Conservation Biology* 9: 1425-1441.

Fuller R. J., Noble D. G., Smith K. W. and Vanhinsbergh D. 2005. Recent declines in populations of woodland birds in Britain: a review of possible causes. *British Birds* 98: 116-143.

Fuller E. 1999. The Great Auk. Published privately.

Furness R. W. and Baillie S. R. 1981. Age ratios, wing length and moult as indicators of population structure of Redshank wintering on British estuaries. *Ringing & Migration* 3: 123-132.

Furness R. W. and Ratcliffe N. 2004. Arctic Skua *Stercorarius parasiticus*. In: Mitchell P. I., Newton S.F., Ratcliffe N. and Dunn I. E. (eds.) 2004. Seabird Populations of Britain and Ireland: 160-172. Poyser, London.

Galloway B. and Meek E. R. 1978-1983. *Northumberland's Birds.* Natural History Society of Northumbria 44 (1-3). Newcastle Upon Tyne.

Gantlet S. J. M. (ed.) 1995. Spotted Crake reports. *Birding World* 8.

Gantlett S. J. M. 2002. The Rose-coloured Starling Invasion of summer 2002. *Birding World* 15: 284-286.

Gantlett, S. J. M. 2006. The Leach's Petrels in December 2006. *Birding World* 19 (12): 497-498.

Gardner-Medwin D. 1985. Early Bird Records for Northumberland and Durham, Birds: Ancient and Modern. *Transactions of the Natural History Society of Northumbria* Vol. 54.

Garner M. 2007. Where do our Lapland Buntings come from? *Birding World* 20:5.

Garðarsson A. 1991. Movements of Whooper Swans *Cygnus cygnus* neckbanded in Iceland. In Sears J. and Bacon J. (eds.). Proceedings of the 3rd IWRB International Swan Symposium, 1989. *Wildfowl Special Supplement No. 1*: 189-194.

Gibbons D. W., Reid J. B. and Chapman R. A. 1993. *The New Atlas of Breeding Birds in Britain and Ireland: 1988-1991*. T&AD Poyser, London.

Gilbert R. 1991. Winter movements of Sanderling (*Calidris alba*) between feeding sites. *Acta Oecologica* 12: 281-294.

Giles N. 1992. *Wildlife After Gravel*. Game Conservancy Trust/ARC, Fordingbridge.

Gill F. and Wright M. 2006. *Birds of the World: recommended English Names*. Princeton University Press. Princeton.

Gill R. E., Piersma T., Hufford G., Servranckx R. and Riegen A. 2005. Crossing the ultimate ecological barrier: Evidence for an 11,000km-long non-stop flight from Alaska to New Zealand and eastern Australia by Bar-tailed Godwits. *The Condor*, 107: 1-20.

Gillings S. 2003. Plugging the gaps – winter studies of Eurasian Golden Plovers and Northern Lapwings. *Wader Study Group Bulletin* 100.

Gilroy J. J. and Lees A. C. 2003. Vagrancy theories: are autumn vagrants really reverse migrants? *British Birds* 96: 427-438.

Gladstone H.S. 1924. The distribution of black grouse in Britain. *British Birds* 18: 66-68.

Gladstone H. S. 1943. Comparative prices of game and wildfowl in 1512, 1757, 1807, 1922, 1941 and 1942. *British Birds* 36:7, 122-131.

Glue D. E. and Boswell T. 1994. Comparative nesting ecology of the three British breeding woodpeckers. *British Birds* 87: 253-268.

Glue D. E. and Hammond G. J. 1974. The feeding ecology of the Long-eared Owl in Britain and Ireland. *British Birds* 67: 361-369.

Glue D. E. and Morgan R. 1972. Cuckoo hosts in British habitats. *Bird Study* 19: 187-192.

Glue D. E. and Scott D. 1980. The Breeding Biology of the Little Owl. *British Birds* 73: 167-180

Goddard T. R. 1931. Nightingales in Durham. *British Birds* 25:2.

Goodyer L. R. and Evans P. R. 1980. Movements of shorebirds into and through the Tees estuary. *County of Cleveland Bird Report for 1979*: 45-52.

Gosler A. 1993. *The Great Tit.* Hamlyn, London.

Graham G.G. 1988. *The Flora and Vegetation of County Durham.* The Durham Flora Committee and the Durham County Conservation Trust, Durham.

Grant P. J. 1982 Gulls: a guide to Identification. T&AD Poyser, Calton.

Green R. E. 1988. Effects of environmental factors on timing and success of breeding of Common Snipe *Gallinago gallinago* (Aves: Scolopacidae). *Journal of Applied Ecology* 25.

Green R. E. 1995. The decline of the Corncrake *Crex crex* in Britain continues. *Bird Study* 42: 66-75.

Gregory R. D. and Marchant J. H. 1996. Population Trends of Jays, Magpies, Jackdaws and Carrion Crows in the UK. *Bird Study* 43: 28-37.

Gregory R. D., Wilkinson N. I., Noble D. G., Robinson J. A., Brown A. F., Hughes J., Procter D. and Gibbons D. W. 2002. A priority list for bird conservation in the United Kingdom, Channel Islands and the Isle of Man: Birds of Conservation Concern, 2002-2007. *British Birds* 95:9, 410-448.

Grey F. G. 1958. Ornithological Report for Northumberland and Durham for 1957. *Transactions of the Natural History Society of Northumberland, Durham and Newcastle upon Tyne* Vol. XII, No. 7.

Grey F. G. 1960. Ornithological Report for Northumberland and Durham for 1958. *Transactions of the Natural History Society of Northumberland, Durham and Newcastle upon Tyne* Vol. XIII, No. 5.

Grey F. G. 1960b. Ornithological Report for Northumberland and Durham for 1959. *Transactions of the Natural History Society of Northumberland, Durham and Newcastle upon Tyne* Vol. XIII, No. 8.

Gribble F. C. 1983. Nightjars in Britain and Ireland in 1981. *Bird Study* 23.

Griffin L. R. 1999. Colonization patterns at Rook colonies *Corvus frugilegus*: implications for survey strategies. *Bird Study* 46:170-173.

Griffin L. R. and Thomas C. J. 2000. The spatial distribution and size of rook (*Corvus frugilegus*) breeding colonies is affected by both the distribution of foraging habitat and by inter-colony competition. *Proceedings of the Royal Society of London* B 267: 1463-1467.

Gruar D., Barritt D. and Peach W. J. 2006. Summer utilization of Oilseed Rape by Reed Buntings *Emberiza schoeniclus* and other farmland birds. *Bird Study* 53.

Gurney J. H. 1876. *Rambles of a Naturalist* Additions to the Avifauna of Durham in Rambles of a Naturalist.

Gurney J. H. 1879. On the Claim of the Pine Grosbeak to be regarded as a British Bird. *The Zoologist*, 3rd series, Vol. I.

Gurney J. H. 1890. On the occasional appearance in England of the Crested Tit. The *Zoologist*.

Gurney J. H. 1918. Breeding stations of the Black headed Gull in the British Isles. *Transactions of the Norfolk and Norwich Naturalists Society* Vol. X.

Gurney J. H. 1921. *Some Critical Notes by W. H. Mullens, M.A., L.L.M - From a review of Early Annals of Ornithology* by F.Z.S. in Vol. 8. London (H. F. & G. Witherby).

Hagemeijer E. J. M. and Blair M. J. 1997. *The EBCC Atlas of European Breeding Birds: Their Distribution and Abundance.* T&AD Poyser, London.

Hagemeijer E. J. M. 2002. Atlas van de Nederlandse Broedvogels (*Atlas of the Breeding Birds of the Netherlands*). SOVON 2002.

Hamer K. C, Phillips R. A., Wanless S., Harris M. P, Wood A. G. 2000. Foraging ranges, diets and feeding locations of Gannets *Morus bassanus* in the North Sea: evidence from satellite telemetry. *Marine Ecology Press Series* Vol. 200: 257-264.

Hancock J. *Manuscript notebooks and letters.* Natural History Society of Northumbria.

Hancock J. 1874. A Catalogue of the Birds of Northumberland and Durham. *Transactions of the Natural History Society of Northumberland and Durham & Newcastle upon Tyne* Vol. VI.

Hancock M., Baines D., Gibbons D., Etheridge B. and Shepherd M. 1999. Status of male Black Grouse *Tetrao tetrix* in Britain in 1995-96. *Bird Study* 46: 1-15.

Harradine J. 1985. Duck Shooting in the United Kingdom. *Wildfowl* 36: 81-94.

Harrap S. and Quinn D. 1996. *Tits, Nuthatches & Treecreepers.* Christopher Helm, London.

Harrop A. H. J. 2002. The Ruddy Shelduck in Britain. *British Birds*, 95: 123-128.

Harrop H. and Fray R. 2008. Two-barred Crossbill in the North Isles in July and August 2008. *Birding World* 21: 329-339.

Harrison C. 1988. *The History of the Birds of Britain.* William Collins, London.

Harrison C. J. O. 1980. A Re-examination of British Devensian and earlier Holocene Bird Bones in the

British Museum (Natural History). *Journal of Archaeological Science* 7: 53-68.

Harrison J. W. H. 1928. A Survey of the Lower Tees Marshes and of the re-claimed areas adjoining them. *Transactions of the Natural History Society of Northumberland and Durham & Newcastle upon Tyne* 5: 89-153.

Harrison T. H. and Hollom P. A. D. 1932. The Great Crested Grebe Enquiry 1931. *British Birds* 26.

Hartley C. (ed.) 2009. *Birds and Wildlife in Cumbria 2008.* Cumbria Naturalists' Union.

Hearn, R. D. 2004. Greater White-fronted *Goose Anser albifrons albifrons* (Baltic-North Sea population) in Britain 1960/61 - 1999/2000. *Waterbird Review Series.* The Wildfowl & Wetlands Trust/Joint Nature Conservation Committee, Slimbridge.

Heavisides A. 1987. British and Irish Merlin Recoveries 1911-84. *Ringing & Migration* 8.

Hickling G. 1970. Ornithological Report for the Farne Isles Report for 1968. *Transactions of the Natural History Society of Northumberland and Durham & Newcastle upon Tyne* Vol. XVII, No. 2.

Hill D. and Robertson P. 1988. *The Pheasant: Ecology, management and conservation.* BSP Professional Books, Oxford.

Hirons G. 1980. The significance of roding by Woodcock *Scolopax rusticola*: an alternative explanation based on observation of marked birds. *Ibis* 122.

Hirons G. and Johnson T. H. 1987. A quantitative analysis of habitat preferences of Woodcock *Scolopax rusticola*, in the breeding season. *Ibis* 129.

Hogg J. 1827. Catalogue of the Birds frequenting the Country near Stockton. In *The Natural History of the Vicinity of Stockton.* Stockton.

Hogg J. 1845. A Catalogue of the Birds observed in S.E. Durham and N.W. Cleveland. *The Zoologist*, 1845.

Holden D. and Bilton D. 2009. The Eastern Crowned Warbler in County Durham – a new British Bird. *Birding World* 22: 417-419.

Holliday S. 2011. Some origins of Mediterranean Gulls in Northumberland. Northumberland & Tyneside Bird Club website.

Holling M. and the Rare Breeding Birds Panel 2007. Rare breeding birds in the United Kingdom in 2003 and 2004. *British Birds* 100.

Holling M. and the Rare Breeding Birds Panel 2008. Rare breeding birds in the United Kingdom in 2005. *British Birds* 101.

Holling M. and the Rare Breeding Birds Panel 2009. Rare breeding birds in the United Kingdom in 2006. *British Birds* 102.

Holling M. and the Rare Breeding Birds Panel 2010a. Rare breeding birds in the United Kingdom in 2007. *British Birds* 103.

Holling M. and the Rare Breeding Birds Panel 2010b. Rare breeding birds in the United Kingdom in 2008. *British Birds* 103.

Holling M. and the Rare Breeding Birds Panel 2011. Rare breeding birds in the United Kingdom in 2009. *British Birds* 104.

Holloway S. 1996. *The Historical Atlas of Breeding Birds in Britain and Ireland: 1875-1900.* T&AD Poyser, London.

Holmes J., Walker D., Davies P. and Carter I. 2000. The Illegal Persecution of Raptors in England. *English Nature Research Report 343.* English Nature.

Holmes J., Carter I., Stott M., Hughes J., Davies P. and Walker D. 2003. Raptor persecution in England at the end of the twentieth century. In: *Birds of Prey in a Changing Environment.* Ed. D.B.A. Thompson, S.M. Redpath, A.H. Fielding, M. Marquiss and C.A. Galbraith. Scottish Natural Heritage, Edinburgh, 81–485.

Holt C. A., Austin G. E., Calbrade N. A., Mellan H. J., Mitchell C., Stroud D. A., Wotton S .R. and Musgrove A. J. 2011. *Waterbirds in the UK 2009/10: The Wetland Bird Survey.* BTO/RSPB/JNCC, Thetford.

Holt C. A., Hewson C. M. and Fuller R. J. 2012. The Nightingale in Britain: status, ecology and conservation needs. *British Birds* 105.

Hoodless A. N. and Coulson J. C. 1998. Breeding biology of the Woodcock in Britain. *Bird Study* 45: 195-204.

Hoodless A. N., Inglis J. G., Doucet J. P. and Aebischer N. J. 2008. Vocal individuality in the roding calls of Woodcock and their use to validate a survey method. *Ibis* 150: 80-89.

Hoodless A. N., Lang D., Aebischer N. J., Fuller R. J. and Ewald J. A. 2009. Densities and population estimates of breeding Eurasian Woodcock in Britain in 2003. *Bird Study* 56: 15-25.

Howey D. H. 2008. *Birding in the 1950s and the formation of the Northumberland and Tyneside Bird Club.* Northumberland and Tyneside Bird Club Bulletin.

Howse R. 1878. Preliminary note on the discovery of old sea caves and a raised sea-beach at Whitburn Lizards. *Transactions of Natural History Northumberland, Durham and Newcastle upon Tyne* Vol. VII.

Howse R. 1899. Index Catalogue of the Birds in the Hancock Collection, Newcastle. *Transactions of Natural History Northumberland, Durham and Newcastle upon Tyne*

Hudson N. and the Rarities Committee 2008. Report on rare birds in Britain in 2007. *British Birds* 101.

Hudson N. and the Rarities Committee 2011. Report on Rare Birds in Britain in 2010. *British Birds* 104.

Hudson P. J. 1986. *The Red Grouse: biology and management of a wild gamebird.* Game Conservancy Trust, Fordingbridge.

Hudson P. J. 1992. *Grouse in Space and Time: The population Biology of a Managed Gamebird.* Game Conservancy Trust, Fordingbridge.

Hudson R. 1965. The Spread of the Collared Dove in Britain and Ireland. *British Birds* 58.

Hudson R. 1972. Collared Doves in Britain and Ireland during 1965-70. *British Birds* 65.

Hughes B. and Grussu M. 1994. The Ruddy Duck (*Oxyura jamaicensis*) in the United Kingdom: distribution, monitoring, current research and implications for European colonization. *Oxyura* 7.

Hume R. A. 1980. Identification and ageing of Glaucous and Iceland Gull. In Sharrock J.T.R. (ed.) *Frontiers of Bird Identification*. Macmillan, London.

Hutchinson C. D and Neath B. 1978. Little Gulls in Britain and Ireland. *British Birds* 71.

Hutchinson J. 1840. *Birds of Durham 1840*. MS Catalogue.

Iceton G. (ed.) 1999. *Cleveland Bird Report for 1998.* Teesmouth Bird Club.

Iceton G. (ed.) 2000. *Cleveland Bird Report for 1999.* Teesmouth Bird Club.

Iceton G. (ed.) 2001. *Cleveland Bird Report for 2000.* Teesmouth Bird Club.

Iceton G. (ed.) 2002. *Cleveland Bird Report for 2001.* Teesmouth Bird Club.

INCA 2010. INCA Newsletter, August 2010.

IOC 2010. 25th International Ornithological Congress.

Jackson J. W. 1953. *Archaeology and Palaeontology in British Caving. An Introduction to speleogeology* (ed. Cullingford C. H. D.). Routledge & Keegan Paul Ltd.

Jakobsen J. 1928/1932. *An Etymological Dictionary of The Norn Language in Shetland, Vols 1 and 2.* London and Copenhagen (reprinted Lerwick 1985).

Jenkins R. K. B., Buckton S. T. and Ormerod S. J. 1995. Local movements and population density of Water Rails *Rallus aquaticus* in a small inland reedbed. *Bird Study* 42.

Jessop L. 1999. Bird Specimens Figured by Thomas Bewick Surviving in the Hancock Museum. *Newcastle upon Tyne Transactions of the Natural History Society of Northumbria* Vol. 59, part 3.

Jessop L. and Gordon D. Unpubl. 2007. *The Capercaillie question*. c/o Natural History Society of Northumbria, Newcastle upon Tyne

JNCC 2009/10. Web-based data for seabirds & waders: Eider, Whimbrel, Kittiwake, Sandwich, Common, Roseate & Arctic Tern, Guillemot & Razorbill.

Johnson R. F. and Janiga M. 1995. *Feral Pigeons*. Oxford University Press, New York.

Johnson W. 1905. *Notes on the History, Geology and Entomology of the Vale of Derwent* Vol. V, 87.

Joynt G. J. (ed.) 2004. *Cleveland Bird Report for 2003*. Teesmouth Bird Club.

Joynt G. J. (ed.) 2005. *Cleveland Bird Report for 2004*. Teesmouth Bird Club.

Joynt G. J. (ed.) 2006. *Cleveland Bird Report for 2005*. Teesmouth Bird Club.

Joynt G. J. (ed.) 2007. *Cleveland Bird Report for 2006*. Teesmouth Bird Club.

Joynt G. J. (ed.) 2008. *Cleveland Bird Report for 2007*. Teesmouth Bird Club.

Joynt G. J. (ed.) 2009. *Cleveland Bird Report for 2008*. Teesmouth Bird Club.

Joynt G. J. (ed.) 2010. *Cleveland Bird Report for 2009*. Teesmouth Bird Club.

Joynt G. J. (ed.) 2011. *Cleveland Bird Report for 2010*. Teesmouth Bird Club.

Joynt G. J., Parker E. C. and Fairbrother V. 2008. *The Breeding Birds of Cleveland*. Teesmouth Bird Club, Middlesbrough.

Kaiser E. 1997. Sexual recognition of Common Swift. *British Birds* 90.

Kenward R. E., Marcroström V. and Karlborm M. 1991. The Goshawk *Accipiter gentilis* as a predator and a renewable resource. *Gibier Faune Sauvage* 8.

Kerr I. 1995. *Birds in Northumbria 1994*. Northumberland & Tyneside Bird Club

Kerr I. 2001 *Northumbrian Birds: Their history and status up to the 21st Century*. Northumberland & Tyneside Bird Club. Newcastle upon Tyne.

Kerr I. and Johnston A. J. 1996. *Birds in Northumbria 1995* Northumberland & Tyneside Bird Club. Newcastle upon Tyne

Kirby J. S. 1999. Canada Goose *Branta canadensis,* Introduced: United Kingdom. In: Madsen J., Cracknell G. and Fox T. (eds). *Goose populations of the western Palaearctic: A review of status and distribution*. Wetlands International Publ. No. 48, Wetlands International, Wageningen, The Netherlands. National Environmental Research Institute, Ronde, Denmark.

Kokorev Y. I. and Kuksov, V. A. 2002. Population dynamics of lemmings *Lemmus sibirica* and *Dicrostonyx torquatus,* and Arctic Fox *Alopex lagopus* on the Taimyr peninsula, Siberia, 1960-2001. *Ornis Svecica* 12:139-143.

Koopman K. 2002. Mass, moult Migration and sub-specific status of Red Knot *Calidris canutus* on the Frisian Wadden Sea Coast, The Netherlands. *Wader Study Group Bull.* 97, 30-35.

Lack P. C. 1986. *The Atlas of Wintering Birds in Britain and Ireland*. T&AD Poyser, Calton.

Lack P. C. 1987. The effects of severe hedge cutting on a breeding bird population. *Bird Study* 34: 139-146.

Lack P. C. 1992. *Birds on Lowland Farms.* HMSO, London.

Langston R., Richard G. and Adams R. 2002. The Status of the Hawfinch in the UK 1975-1999. *British Birds* 95: 166-73.

Leech D. and Barimore C. 2008. Is Avian Breeding Success Weathering the Storms? *BTO News* 279.

Lever, C. 1977. *The Naturalized Animals of the British Isles.* Hutchinson, London.

Lewis A. G. J. , Amar A., Chormonond E. C., Stewart F. R. P. 2009. The decline of the Willow Tit in Britain. *British Birds* 102: 7, 386-393.

Lieske D. J., Oliphant L. W., James P. C., Warkentin I. G. and Espier R. H. M. 1997. Age of first breeding in Merlins (*Falco columbarius*). *Auk* 114: 288–290.

Ling S. 1992. An addition to the County Avifauna: Yellow-breasted Bunting *Emberiza aureola*. *Birds in Durham 1991.* Durham Bird Club.

Little R. (ed.) 2003. *Cleveland Bird Report for 2002.* Teesmouth Bird Club.

Little B. and Furness R. W. 1985. Long distance moult migration by British Goosanders. *Ringing and Migration* 6: 77-82.

Lloyd C., Tasker M. L. and Partridge K. 1991. *The Status of Seabirds.* T&AD Poyser, London.

Lock L. and Cook L. 1998. The Little Egret in Britain: a successful colonist. *British Birds*, 91: 273-280.

Lockwood W. B. 1984. *The Oxford Book of British Bird Names.* Oxford University Press, Oxford.

Long J. L. 1981. *Introduced Birds of the World.* David and Charles, London.

Lovat L. 1911. *The Grouse in Health and Disease.* Smith, Elder & Co., London.

Lovegrove R. 1990. *The Kite's Tale: Story of the Red Kite in Wales.* The Royal Society for the Protection of Birds, Sandy.

Lundberg A. and Alatalo R. V. 1992. *The Pied Flycatcher.* T&AD Poyser, London.

Lunn J. and Dale J. E. 1993. Breeding activities of Parrot Crossbills *Loxia pytyopsittacus* in South Yorkshire in 1983. *The Naturalist* 118: 9-12.

Maclean I. M. D., Austin G. E., Rehfisch M. M., Blew J., Crowe O., Delaney S., Devos K., Deceuninck B., Gunther K., Laursen K., Van Roomen M. and Wahl J. 2008. Climate change causes rapid changes in the distribution and site abundance of birds in winter. *Global Change Biology* 14: 2489-2500.

Macpherson H. A. 1901. *The Birds of the County of Cumberland.* Dawsons, London.

Madsen J., Bregnballe T. and Hastrup A. 1992. Impact of the Arctic Fox *Alopex lagopus* on nesting success of geese in Southeast Svalbard. *Polar Research* 11: 35-39.

Madsen J., Cracknell G. and Fox A.D. (eds.) 1999. Goose populations of the Western Palaearctic. A review of status and distribution. *Wetlands International Publication No. 48.* Wetlands International, Wageningen, The Netherlands/National Environmental Research Institute, Rønde, Denmark.

989

Manley G. 1935. Some notes on the climate of North East England. *Q. J. Royal Met. Soc.* 61: 405-410.

Marchant J. H. and Gregory R. D. 1999. Numbers of nesting Rooks *Corvus frugilegus* in the United Kingdom in 1996. *Bird Study* 46: 258–273.

Marchant J. H., Hudson R., Carter S. P. and Whittington P. 1990. *Population Trends in British Breeding Birds.* BTO, Tring/NCC, Peterbrough.

Marquiss M. and Newton I. 1982. The Goshawk in Britain. *British Birds* 75: 243-260.

Marquiss M. and Carss D. 1997. Fish-eating birds and fisheries. *BTO News* 210/211:6-7.

Marshall A. J. and Coombs C. J. F. 1957. The interaction of environmental, internal and behavioural factors in the Rook *Corvus f. frugilegus. Proc. Zool. Soc. London* 128: 545-588.

Martin J. P. 2008. Northern Harrier on Scilly: New to Britain. *British Birds* 101: 8.

Massias A. and Becker P. H. 1990. Nutritive value of food and growth in Common Tern *Sterna hirundo* chicks. *Ornis Scandinavica* 21: 187-194.

Mather J. 1986. *Birds of Yorkshire.* Christopher Helm, London.

Mavor R. A., Parsons M., Heubeck M. and Schmitt S. 2006. *Seabird numbers and breeding success in Britain and Ireland, 2005.* Peterborough, Joint Nature Conservation Committee.

McAndrew R. T. 1972. *Durham County Bird Report 1970.* Zoology Department, University of Durham.

McAndrew R. T. 1972b. *Durham County Bird Report 1971.* Zoology Department, University of Durham.

McAndrew R. T. and Clayton D. 2005. Declining Rooks. *The Spire, Newsletter of Hartlepool Natural History Society* No. 360:1.

McDiarmid H. 1990. The Fishing Industry in North East England. *Seafish Report No. 390.* Sea Fish Indust. Auth. (SFIA).

McElwaine J. G., Wells J. H. and Bowler J. M. 1995. Winter movements of Whooper Swans visiting Ireland: preliminary results. *Irish Birds* 5: 265-78.

Mead C. J. 1979. Colony fidelity and interchange in the Sand Martin. *Bird Study* 26: 99-106.

Mead C. J. 2000. *The State of the Nation's Birds.* BTO, Thetford.

Mearns R. 1983. The Status of the Raven in southern Scotland and Northumbria. *Scottish Birds* 12.

Meek E. R. and Little B. 1977. The spread of the Goosander in Britain and Ireland. *British Birds* 70: 229-237.

Mikkola H. 1985. *The Owls of Europe.* T&AD Poyser, Calton.

Milburn C. E. 1901. Ornithological Notes from Cleveland and Teesmouth 1899. *Cleveland Naturalists' Field Club, Records of Proceedings 1899-1900* Vol. 1, No. 3.

Milburn C.E. 1903. Ornithological Notes from Cleveland and Teesmouth. *Cleveland Naturalists' Field Club, Records of Proceedings 1899-1900.*

Millburn C.E. 1915. The breeding of the Ruff in Durham. *The Vasculum* 1.

Milburn C. E. 1916. The Garganey in south-east Durham. *The Vasculum* 2.

Milburn C. E. 1917. Teesmouth Bird Notes. *The Vasculum* 3.

Milburn C.E. 1932. Ornithological notes in North Yorkshire and South Durham, 1930-31. *Proceedings of Cleveland Naturalists' Field Club 1928-1932,* Vol. 4.

Milburn C.E. 1933. Oystercatcher Breeding in County Durham. *British Birds*, 27: 75.

Miles J. 1992. *Hadrian's Birds.* Miles and Miles of Countryside, Castle Carrock.

Millington R. 2010. Bluethroats in Britain. *Birding World* 23: 156-159.

Milton F. S. 1995. Storm Petrel ringing on the North Durham Coast 1989-1994. *Birds in Durham 1994.* Durham Bird Club.

Milton F. S. 2005. The Status of the Hawfinch (*Coccothraustes coccothraustes*) in County Durham 1970-2000. *Birds in Durham 2000.* Durham Bird Club.

Mitchell P. I., Newton S. F., Ratcliffe N. and Dunn T. E. (eds.) 2004. *Seabird Populations of Britain and Ireland.* Poyser, London.

Mitchell P. I., Scott I. and Evans P. R. 2000. Vulnerability to severe weather and regulation of body mass of Icelandic and British Redshank. *Journal of Avian Biology* 31: 511-521.

Moller A. P. 1983. Breeding habitat selection in the swallow *Hirundo rustica. Bird Study* 30.

Moller A. P. 1989. Population dynamics of a declining Swallow *Hirundo rustica* population. *Journal Animal Ecology* 58.

Moore N. W. 1957. The Past and present status of the Buzzard in British Isles. *British Birds* 50.

Morris A., Burgess D., Fuller R. J., Evans A. D. and Smith K. W. 1994. The status and distribution of Nightjars *Caprimulgus europaeus* in Britain in 1992 - A report to the British Trust for Ornithology. *Bird Study* 41.

Morrison C. A., Robinson R. A., Clark J. A. and Gill J. A. 2010. Spatial and temporal variation in population trends in a long-distance migratory bird. *Diversity & Distributions* 16.

Morrison P. and Gurney M. 2007. Nest boxes for roseate terns *Sterna dougallii* on Coquet Island RSPB reserve, Northumberland, England. *Conservation Evidence* 4: 1-3.

Mullins J. R. 1984. Scarlet Rosefinch breeding in Scotland. *British Birds* 77: 133-135.

Murton R. K. 1968. Breeding, migration and survival of Turtle Doves. *British Birds* 61:5 193-212.

Musgrove A. J., Austin G. E., Hearn R. D., Holt C. A., Stroud D. A. and Wotton S. R. 2011. Overwinter population estimates of British waterbirds. *British Birds,* 104: 364-397.

Nankinov D. N. 2000. Expansion of the Paddyfield Warbler in Europe in the second half of the 20th Century. *Berkut* 9: 102-106.

National Trust 2011. Web-based data for Arctic Tern, the Farne Islands.

Natural England 2008. *A future for the Hen Harrier in England?* Natural England.

The Naturalist on Pectoral Sandpiper 1853: Vol. III, Journal of the Yorkshire Naturalists' Union.

The Naturalist on Pied Flycatcher 1887, Journal of the Yorkshire Naturalists' Union.

The Naturalist James Backhouse on Red-backed shrike 1888, Journal of the Yorkshire Naturalists' Union.

The Naturalist on White Stork 1938, Journal of the Yorkshire Naturalists' Union.

The Naturalist J.R Crawford on Little Auks 1949, Journal of the Yorkshire Naturalists' Union.

Nelson T.H. 1907. *The Birds of Yorkshire*. A Brown & Sons, London.

Nelson T. H. 1911. Arctic and Pomarine Skuas and Sabine's Gull in Yorkshire. *British Birds* 1:170.

Nelson T. H. 1912. Notes on the 1912 'Wreck' of the Little Auk. *British Birds* 5: 282-286.

NERF 2010. *Annual Review 2009.* Report of the North of England Raptor Forum.

Newsome M. (ed.) 2006. *Birds in Durham 2005.* Durham Bird Club.

Newsome M. (ed.) 2007. *Birds in Durham 2002 & 2003.* Durham Bird Club.

Newsome M. 2007. Melodious Warbler at Whitburn – a new bird for the County. *Birds in Durham 2002 & 2003.* Durham Bird Club.

Newsome M. (ed.) 2007b. *Birds in Durham 2006.* Durham Bird Club.

Newsome M. (ed.) 2008. *Birds in Durham 2007.* Durham Bird Club.

Newsome M. (ed.) 2009. *Birds in Durham 2008.* Durham Bird Club.

Newsome M. 2009. The Honey-buzzard passage in autumn 2008. *Birds in Durham 2008.* Durham Bird Club.

Newsome M. (ed.) 2010. *Birds in Durham 2009.* Durham Bird Club.

Newsome M. 2010. Birds New to County Durham – What Next? *The Lek*, Autumn 2010. Durham Bird Club Durham.

Newsome M. (ed.) 2012. *Birds in Durham 2010.* Durham Bird Club.

Newsome M. and Willoughby P. J. 1995. *Flamborough Ornithological Group Bird report for 1995.*

Newton I. 1972. *British Finches* (2nd edition 1985). Collins New Naturalist, London.

Newton I. 1979. *Population Ecology of Raptors.* T&AD Poyser, Calton.

Newton, I. 2006. Movement patterns of Common Crossbills *Loxia curvirostra* in Europe. *Ibis* 148.

Newton I. 2008. Highlights from a long-term study of Sparrowhawks. *British Birds* Vol.10.

Newton I. 2010. *Bird Migration.* Collins New Naturalist 113, London.

Newton I., Meek E. and Little B. 1986. Population and breeding of Northumberland Merlins. *British Birds* 79.

Nethersole-Thompson D. and Nethersole-Thompson M. 1986. Obituary of H. M. S. Blair. *Ibis* 129. BOU.

Nethersole-Thompson D. and Nethersole-Thompson M. 1986. *Waders, their Breeding Haunts and Watchers.* T&AD Poyser, Calton.

Nicholson E. M. 1929. Report on the British Birds Census of Heronries 1928. *British Birds* 22.

Nicholson E. M. and Hollom P. A. D. 1957. The Rarer birds of Prey: Their Present Status in the British Isles. *British Birds* 50:5.

Nightingale B. and Allsopp K. 1994. Invasion of Red-footed Falcons in spring 1992. *British Birds* 87.

Noble-Rollin C. 1928. Fulmar breeding in Durham. *British Birds* 21:3.

Norman R. K. and Saunders D. R. 1969. Status of Little Terns in Great Britain and Ireland in 1967. *British Birds* 62.

Norris C. A. 1945. Summary of a report on the distribution and status of the Corn-Crake *Crex Crex*. *British Birds* 38:9, 162-168.

The Northern Echo '22 February 1890'. Durham.

Northumberland County Records Office EP 57/25-26 *MS Churchwarden's accounts for Corbridge 1676-1745.*

O'Brien M. 2005. Estimating the number of farmland waders breeding in the United Kingdom. *International Wader Studies* 14.

O'Brien S. H., Wilson L. J., Webb A. and Cranswick P. A. 2008. Revised estimate of numbers of wintering Red-throated Divers *Gavia stellata* in Great Britain. *Bird Study* 55(2), 152-160.

O'Connor R.J. and Shrubb M. 1986. *Farming and Birds.* Cambridge University Press, Cambridge.

Ogilvie M. A. 1975. *Ducks in Britain and Europe.* T& AD Poyser, London.

Ogilvie M. and the Rare Breeding Birds Panel 2004. Rare breeding birds in the United Kingdom in 2002. *British Birds* 97: 492-536.

Olley J. 2006. The Stonechat (*Saxicola torquata*): An Association Between Global Warming and its Increasing Population in County Durham. *Birds in Durham 2004.* Durham Bird Club.

Onley D. and Scofield P. 2007. *Albatrosses, Petrels and Shearwaters of the World.* Christopher Helm/A.C. Blacks, London.

Ormerod S. J., Tyler S. J. and Lewis J. M. S. 1985. Is the breeding distribution of Dippers influenced by stream acidity? *Bird Study* 32.

Ormerod S. J. and Tyler S. J. 1991. The influence of stream acidification and land use on the feeding ecology of Grey Wagtails *Motacilla cinerea* in Wales. *Ibis* 133: 53-61.

Owen M., Atkinson-Willes G. L. and Salmon D. G. 1986. *Wildfowl in Great Britain - 2nd ed.* Cambridge University Press, Cambridge.

Owen M., Ogilvie M. A., Fox A. D. and Salmon D. G. 1986. *Numbers and breeding success of Pink-footed and Greylag Geese in Britain, 1976–1985.* Report to NCC & Wildfowl Trust, Slimbridge.

Owen M. and Salmon D.G. 1988. Feral Greylag Gees *Anser anser* in Britain and Ireland 1960-86. *Bird Study* 35.

Palmer P. 2000. *First for Britain and Ireland 1600-1999*. Arlequin Press, Shrewsbury.

Palmer W. J. 1822. *The Tyne and its Tributaries.* George Bell, London.

Parkin D.T., Collinson M., Helbig A. J., Knox A. G. and Sangster G. 2002. The taxonomic status of Carrion and Hooded Crows. *British Birds* 96.

Parkin D. T. and Knox A. G. 2010. *The Status of Birds in Britain & Ireland.* Christopher Helm, London.

Parrack J.D and Bell D.G. 1970. Ornithological Report for Northumberland & Durham 1968. *Transactions of the Natural History Society of Northumberland, Durham & Newcastle upon Tyne* Vol. XVII, No. 2

Parrack J. D., McAndrew R. T., Johnson H. M. and Hicking G. 1972. Ornithological report for Northumberland and Durham for 1969. *Transactions of Natural History Society of Northumberland and Durham and Newcastle upon Tyne* Vol XVII No 5

Parrinder E. R. 1964. Little Ringed Plover in Britain during 1960- 962. *British Birds* 57: 191-198.

Parrinder E. D. 1989. Little Ringed Plovers *Charadrius dubius* in Britain in 1984. *Bird Study* 36.

Parslow J. L. F. 1967. Changes in status among breeding birds in Britain and Ireland. *British Birds* 60:5

Parslow J. L. F. 1973. *The Breeding Birds of Britain and Ireland.* T&AD Poyser, Berkhamsted.

Peach W. J., Baillie S. R. and Underhill L. 1991. Survival of British Sedge Warblers *Acrocephalus schoenobaenus* in relation to West African rainfall. *Ibis* 133.

Peach W. J., Thompson P. S. and Coulson J. C. 1994. Annual and long-term variation in the survival rates of British lapwings *Vanellus vanellus. Journal of Animal Ecology* 63.

PECBMS 2010. *Trends of common birds in Europe, 2010 update.* European Bird Census Council, Prague.

Pennant T. 1776. *British Zoology.* White, London.

Pennington M., Osborn K., Harvey P., Riddington R., Okill D., Ellis P. and Heubeck M. 2004. *The Birds of Shetland.* Christopher Helm, London.

Percival S .M. 1990. Recent Trends in Barn Owl and Tawny Owl Populations in Britain. *BTO Research Report No. 57.*

Perrins C. 1979. *British Tits.* Collins New Naturalist, London.

Petty S. J. 1989. Goshawks: their status, requirements and management. *Forestry Commission Bulletin* 81.

Picozzi N. 1975. A study of the Carrion/ Hooded Crow in N.E. Scotland. *British Birds* 68: 409-419.

Piersma T. 1986. Breeding waders in Europe: A review of population size, productivity and a bibliography of information sources. *Wader Study Group Bulletin* 48, Supplement.

Piersma T., Prokosch P. and Bredin D. 1992. The migration system of Afro-Siberian Knots *Calidris canutus. Wader Study Group Bull.* 64, Suppl., 52-63.

Pillans J. 2010. Brown-headed Cowbird New for Durham. *The Lek, Summer 2010.* Durham Bird Club.

Pithon J. A. and Dytham C. 2001. Determination of the origin of British feral Rose-ringed Parakeets. *British Birds* 94:2, 74-79.

Potts G. R. 1989. The impact of releasing hybrid partridges on wild Red-legged Partridge populations. *Game Conservancy Annual Review* 20.

Potts G. R. 1998. Global dispersion of nesting Hen Harriers *Circus cyaneus*; implications for grouse moors in the UK. *Ibis* 140.

Prater A. J. 1989. Ringed Plover *Charadrius hiaticula* breeding population of the United Kingdom in 1984. *Bird Study* 36.

Prestt I. and Mills D. H. 1966. A Census of the Great Crested Grebe in Britain 1965. *Bird Study* 13.

Proctor W. 1846. List of Birds found in the County in *Sketches of Durham* by Rev. G. Ornsby, as Appendix III, 196-201. G. Andrews, Durham.

Prys-Jones R. P. 1984. Migration patterns of the Reed Bunting, *Emberiza schoeniclus*, and the dependence of wintering distribution on environmental conditions. *Gerfaut* 74.

Pyman G. A. 1960. Report on rare birds in Great Britain and Ireland in 1959 (with 1958 additions). *British Birds* 53.

Rackham J. 1979. Animal Remains, in three Saxo-Norman tenements in Durham City. *Mediaeval Archaeology* 23.

Rackham J. and Allison E. 1981. *An excavation of the Castle Ditch, Newcastle 1974-1976* Archaeol. Aeliana (5th series) 9.

Raine A. F, Sowter D. J., Brown A. F. and Sutherland W. J. 2006. Natal philopatry and local movement patterns of Twite *Carduelis flavirostris*. *Ringing and Migration* 23:89-94.

Ratcliffe D. A. 1963. The status of the Peregrine in Great Britain. *Bird Study* 10.

Ratcliffe D. A. 1976. Observations on the breeding of the Golden Plover in Great Britain. *Bird Study* 19.

Ratcliffe D. A. 1990. *Bird Life of Mountain and Upland.* Cambridge University Press, Cambridge.

Ratcliffe D. A. 1997. *The Raven.* T&AD Poyser, Calton.

Raven J. 1942. *John Ray Naturalist. His Life and Works.* Cambridge University Press, Cambridge.

Raven M. J., Noble D. G. and Baillie S. R. 2007. The Breeding Bird Survey 2006 *BTO Research Report No. 471.* BTO.

Raven S. J. and Coulson J. C. 1997. The distribution and abundance of *Larus* gulls nesting on buildings in Britain and Ireland. *Bird Study* 44.

Raven S. J. and Coulson J. C. 2001. Effects of cleaning a tidal river of sewage on gull numbers: a before and after study of the River Tyne, northeast England. *Bird Study* 48: 48-58.

Raw D. 1985. The Status of the Black-necked Grebe in County Durham. *Birds in Durham 1984.* Durham Bird Club.

Raw D. 1988. Account of Gyrfalcon in County Durham. *Birds in Durham 1987.* Durham Bird Club.

Raw D. 1991. *Alectoris* partridge in Co. Durham. *Birds in Durham 1990.* Durham Bird Club.

Raw D. 1992. *Birds in Durham - A Complete Species Check-list.* Durham Bird Club.

Raw D. 1994. Predicted Additions to The County List *Birds in Durham 1993.* Durham Bird Club

Rebecca G. W. 2001. The contrasting status of the Ring Ouzel in two areas of upper Deeside, north east Scotland between 1991 and 1998. *Scottish Birds* 22: 9-19.

Rebecca G. W. and Bainbridge I. P. 1998. The Breeding Status of the Merlin *Falco columbarius* in Britain 1993-94. *Bird Study* 45.

Redpath S. M. and Thirgood S. J. 1997. Birds of Prey & Red Grouse. *Stationery Office.*

Reed T. 1985. Estimates of British breeding Wader populations. *Wader Study Group Bulletin* 45.

Reeves S. A. and Furness R. W. 2002. Net loss-seabirds gain? Implications of fisheries management for seabirds scavenging discards in the northern North Sea. *Unpublished RSPB Report*, Sandy.

Rewcastle L., Bowey K. and Westerberg S. 1997. Observed winter feeding rate of Common Kingfisher. *British Birds* Vol. 90: 9.

Richmond W. K. 1931. A short account of the present state of bird-life in the Teesmouth. *British Birds* 25:7, 193-197.

Riddington R., Votier S. and Steele J. 2000. The influx of Redpolls into Western Europe, 1995/96. *British Birds*, 93: 59-67.

Risely K., Baillie S. R., Eaton M. A., Joys A. C., Musgrove A. J., Noble D. G., Renwick A. R. and Wright L. J. 2010. The Breeding Bird Survey 2009. *BTO Research Report 559.* British Trust for Ornithology, Thetford.

Roberts S. J., Lewis J. M. S. and Williams I. T. 1999. Breeding European Honey-buzzards in Britain *British Birds* 92.

Robertson I. 1977. Identification and European status of eastern Stonechats. *British Birds* 70:6, 237-245.

Robinson M. G. and Almond W. E. 1931. Late Wheatear in Durham. *British Birds* 24:10.

Robinson R. A. 2005. Bird Facts: profiles of birds occurring in Britain and Ireland. *BTO Research Report 407.* BTO, Thetford.

Robinson R. A., Balmer D. E. and Marchant J. H. 2008. Survival rates of hirundines in relation to British and African rainfall. *Ringing & Migration* 24:1-6.

Robinson R. A. and Clark J. A. 2011. *The Online Ringing Report: Bird Ringing in Britain & Ireland in 2010.* BTO, Thetford.

Robinson R. A., Crick H. Q. P. and Peach W. J. 2003. Population trends in Swallows *Hirundo rustica* breeding in Britain. *Bird Study* 50:1-7.

Robinson R. A., Lawson B., Toms M. P., Peck K. M., Kirkwood J. K., Chantrey J., Clatworthy I. R., Evans A. D., Hughes L. A., Hutchinson O. C., John S. K., Pennycott T. W., Perkins, M. W., Rowley P. S., Simpson V. R., Tyler K. M. and Cunningham A. A. 2010. Emerging infectious disease leads to rapid population declines of common British birds. *PLoS ONE 5(8): e12215.*

Robson T. 1896. *Birds of the Derwent Valley*. Vale of Derwent Naturalists, Consett.

Robson T. 1903. Fauna of the Derwent Valley in *Notes on the History, Geology and Botany of the Vale of Derwent*. Vol. IV. Vale of Derwent Naturalists, Consett.

Rogers M. J. and the Rarities Committee 1984. Report on rare birds in Great Britain in 1983. *British Birds* 77:11.

Rogers M. J. and the Rarities Committee 1991. Report on rare birds in Great Britain in 1990. *British Birds* 84.

Rogers M. J. and the Rarities Committee 1992. Report on rare birds in Great Britain in 1991. *British Birds* 85.

Rogers M. J. and the Rarities Committee 1993. Report on rare birds in Great Britain in 1992. *British Birds* 86.

Rogers M. J. and the Rarities Committee 1995. Report on rare birds in Great Britain in 1994. *British Birds* 88.

Rogers M. J. and the Rarities Committee 1996. Report on rare birds in Great Britain in 1995 *British Birds* 89.

Rogers M. J. and the Rarities Committee 1999. Report on rare birds in Great Britain in 1998. *British Birds* 92.

Rogers M. J. and the Rarities Committee 2000. Report on rare birds in Great Britain in 1999. *British Birds* 93.

Rogers M. J. and the Rarities Committee 2004. Report on rare birds in Great Britain in 2003. *British Birds* 97.

Rogers M. J. and the Rarities Committee 2005. Report on rare birds in Great Britain in 2004. *British Birds* 98.

Rose P. M. and Scott D. A. 1997. Waterfowl Population Estimates - Second Edition. *Wetlands International* Publication 44

Rossiter B. N. 1999. Northumberland's Birds in the 18th and early 19th Centuries: the Contribution of John Wallis (1714-1793). *Transactions of the Natural History Society of Northumbria* 59: 93-136.

Round P. J. 1996. Long-toed Stint in Cornwall: the first record for the Western Palearctic. *British Birds* 89: 12-24.

Routledge's Magazine for Boys, '1 April' 1868.
RSPB 2003. Survey of Yellow Wagtails in Weardale. *Pennine Dales ESA Yellow Wagtail survey* (unpubl. report).

RSPB 2006. *Peak Malpractice*. RSPB Publications, Sandy.

Sage B. L. and Vernon J. D. R. 1978. The 1975 national survey of rookeries. *Bird Study* 25: 64-86.

Sanderson F. J., Donald P. F., Pain D. J., Burfield I. J., Van Bommel F. P. J. 2006. Long term population decline in Afro-Palearctic migrants. *Biological Conservation* 131:93-105.

Sclater J. 1884. On Green woodpecker in *The Zoologist* 1875.

Scott, D.A. and Rose P.M. 1996. Atlas of *Anatidae* Populations in Africa and Western Eurasia. *Wetlands International Publication 41*, Wetlands International, Wageningen, The Netherlands

Scott D. 1997. *The Long-eared Owl*. The Hawk and Owl Trust, London.

Seebohm H. 1883. *A History of British Birds*. R .H Porter, London.

Selby P. J. 1831. A Catalogue of the Birds hitherto met with in the counties of Northumberland and Durham. *Transactions of the Natural History Society of Northumberland, Durham & Newcastle-upon-Tyne* Vol. I.

Sharp C. 1816. List of Birds observed at Hartlepool. In *The History of Hartlepool* (Appendix). Hartlepool.

Sharrock J. T. R. 1976. *The Atlas of Breeding Birds in Britain and Ireland*. T&AD Poyser, Berkhamsted.

Sharrock J. T. R. 1980. *Frontiers of Bird Identification*. Containing Scott, R.E. & Grant P.J. Uncompleted moult in *Sterna* terns and the problem of identification (89-95). Macmillan, London.

Sharrock J. T. R., Ferguson-Lees I. J. and the Rare Breeding Birds Panel 1975. Rare breeding birds in the United Kingdom in 1973. *British Birds* 68: 5-23.

Sharrock J. T. R. and the Rare Breeding Birds Panel 1978. Rare breeding birds in the United Kingdom in 1976. *British Birds* 71.

Shawyer C. 1987. *The Barn Owl in the British Isles - its Past, Present and Future*. Hawk & Owl Trust, London.

Shepherd K. B. 1993. A survey of breeding waders in Baldersdale and Lunedale, County Durham, England in 1993. *RSPB Report*, Sandy.

Shirihai H., Bretagnolle V. and Zino F. 2010. Identification of Fea's, Desertas and Zino's Petrels at sea. *Birding World* 23 (6): 239-275.

Shrubb M. 1990. Effects of agricultural change on nesting Lapwings *Vanellus vanellus* in England and Wales. *Bird Study* 37:115-127.

Shrubb M. and Lack P. C. 1991. The numbers and distribution of Lapwings *Vanellus vanellus* nesting in England and Wales in 1987. *Bird Study* 38: 20-37.

Siggens G. N. 2005. Spring Arrivals. *The Lek*, Winter 2005. Durham Bird Club

Siggens G. N. (ed.) 2005. *Birds in Durham 2001*. Durham Bird Club.

Siggens G. N. (ed.) 2006. *Birds in Durham 2004*. Durham Bird Club

Sim I. M. W., Dillon I. A., Eaton M. A., Etheridge B., Lindley P., Riley H., Saunders R., Sharpe C. and Tickner M. 2007. Status of the Hen Harrier *Circus cyaneus* in the UK and Isle of Man in 2004, and a comparison with the 1988/89 and 1998 surveys. *Bird Study* 54.

Sim I. M. W., Eaton M. A., Setchfield R.P., Warren P. K. and Lindley P. 2008. Abundance of male Black Grouse *Tetrao tetrix* in Britain in 2005, and change since 1995-96. *Bird Study* 55 (3): 304-313.

Simms E. 1978. *British Thrushes*. Collins New Naturalist, London.

Simms E. 1985. *British Warblers*. Collins New Naturalist, London.

Simpson D. 2011. *Simpson's Sightings*. Durham Bird Club (published on-line)

Siriwardena G. M., Baillie S. R., Crick H. Q. P., Wilson J. D. and Gates S. 2000. The demography of lowland farmland birds. In *Proceedings of the 1999 BOU Spring Conference: Ecology and Conservation of Lowland Farmland Birds* (eds. N.J. Aebischer, A.D. Evans, P.V. Grice & J.A. Vickery), pp 117–133. British Ornithologists' Union, Tring.

Slack R. 2009. *Rare Birds, Where and When: An analysis of status and distribution in Britain and Ireland*. Privately published.

Smith F. R. and the Rarities Committee 1972. Report on rare birds in Great Britain in 1971 (with 1967, 1968, 1969 and 1970 additions). *British Birds* 65.

Smith K. W., Reed J. M. and Trevis B. E. 1992. Habitat use and site fidelity of Green Sandpipers *Tringa ochropus* wintering in southern England. *Bird Study* 39: 155-164.

Snow D. (ed.) 1992. *Birds, Discovery and Conservation: 100 Years of the British Ornithologists' Club.* Helm Information Ltd., Robertsbridge.

Soloviev N. Y., Minton C. D. T. and Tomkovich P. S. 2006. Breeding performance of tundra waders in response to rodent abundance and weather from Taimyr to Chukotka, Siberia. *Waterbirds around the World.* The Stationery Office, Edinburgh.

Sowerbutts D. L. 1996. Annual Report of the Project and Surveys Group. *Birds in Durham 1994.* Durham Bird Club

Sowerbutts D. L. 1998. Annual Report of the Project and Surveys Group. *Birds in Durham 1996.* Durham Bird Club.

Spray C., Fraser M. and Coleman J. 1996. *The Swans of Berwick upon Tweed.* Northumbrian Water.

Stanbury A., Davies M., Grice P., Gregory G. and Wotton S. 2010. The status of the Cirl Bunting in the UK in 2009. *British Birds* 103: 702-711.

Starling-Westerberg, A. 2001. The habitat use and diet of Black Grouse *Tetrao tetrix* in the Pennine Hills of northern England. *Bird Study* 48: 76-89.

Stead P. J. 1964. The Birds of Tees-side. *Transactions of Natural History Society of Northumberland and Durham and Newcastle upon Tyne* New Series 15.

Stead P.J. 1969. *The Birds of Tees-side 1962-67.* Teesmouth Bird Club. Middlesbrough.

Stewart B. 1984. Roseate Tern in first-summer plumage. *British Birds* 77: 8, 359-360.

Stewart I. F. 1968-1974. *Teesmouth Bird Reports 1968-1973*, quarterly reports. Teesmouth Bird Club.

Stewart I. F. 1975. *County of Cleveland Bird Report for 1974.* Teesmouth Bird Club.

Stone B. H., Sears J., Cranswick P. A. , Gregory R. D., Gibbons D. W., Rehfisch M. M., Nicholas J. Aebischer N. J. and Reid J. B. 1997. Population estimates of birds in Britain and in the U.K. *British Birds* 90, 1-22.

Stott M. 1998. Hen Harrier breeding success on English grouse moors. *British Birds* 91.

Stott M., Callion J., Kinley I., Raven C. and Roberts J. 2002. *The Breeding Birds of Cumbria.* Cumbria Bird Club

Stroud D. A. and Glue D. 1991. *Britain's Birds in 1989/90: the conservation and monitoring review.* British Trust for Ornithology, Thetford/Nature Conservancy Council, Peterbrough.

Stroud D., Francis I. and Stroud R. 2012. Spotted Crakes breeding in Britain and Ireland: a history and evaluation of current status. *British Birds* 105.

Strowger J., Collins P. and Bowey K. 1997. The Rock Pipit *Anthus petrosus*, in County Durham - including the results of the 1995 Rock Pipit enquiry. *Birds in Durham 1995.* Durham Bird Club.

Strowger J. 1998. The status and breeding biology of the Dotterel *Charadrius morinellus* in northern England during 1972-1995. *Bird Study* 45: 85-91.

Summers-Smith J. D. 1963. *The House Sparrow.* Collins, London.

Summers-Smith J. D. 1980. House Sparrows down coal mines. *British Birds* 73:8.

Summers-Smith J. D. 1988. *The Sparrows.* T&AD Poyser, Calton.

Summers-Smith J. D. 1999. The current status of the House Sparrow in Britain. *British Wildlife* 10.

Summers R.W. 2002. Parrot Crossbills breeding in Abernethy Forest, Highland. *British Birds* 95, 4-11.

Svardson G. 1957. The 'invasion' type of bird migration. *British Birds* 50.

Szep T. 1995. Relationship between west African rainfall and the survival of central European Sand Martins *Riparia riparia. Ibis* 137:162-168.

Tatner P. 1978. A review of House Martin *Delichon urbica* in South Manchester, 1975. *Naturalist* 103.

Taverner J. H. 1959. The Spread of the Eider in Great Britain. *British Birds* 52.

Taverner J. H. 1967. Wintering eiders in England during 1960-65. *British Birds* 60.

Taylor I. 1994. *Barn Owls: predator-prey relationships and conservation.* Cambridge University Press, Cambridge.

Temperley G. W. 1929. On Lapwing in 'Notes & Records' in *The Vasculum* in Vol. XVII.

Temperley G. W. 1931. Immigration of Crossbills in 1930. *British Birds* 22:2.

Temperley G. W. 1932. Waxwings in County Durham. *British Birds* 25:9.

Temperley G. W. 1933. 'Notes & Records' in *The Vasculum* in Vol. XIX.

Temperley G. W. 1934. Some breeding-habits of the British Willow-Tit. *British Birds* 28:6.

Temperley G. W. 1945. Sabine's Gulls in County Durham. *British Birds* 38: 11.

Temperley G. W. 1946. Breeding of the Black Redstart in Britain: a century-old record. *British Birds* 39:4

Temperley G. W. 1949. Ornithological Reports for Northumberland and Durham for 1947 and 1948.

Compiled by George W. Temperley. (Reprinted from *The Naturalist*, July-September, 1948, and July-September, 1949.) *Transactions of the Natural History Society of Northumberland, Durham & Newcastle-upon-Tyne.*

Temperley G. W. 1950. Ornithological Report for Northumberland and Durham for 1949. *Transactions of the Natural History Society of Northumberland, Durham & Newcastle-upon-Tyne.*

Temperley G. W. 1951. A History of the Birds of Durham. *Transactions of the Natural History Society of Northumberland, Durham & Newcastle-upon-Tyne*, Vol. IX.

Temperley G. W. 1953-1957. Ornithological Reports for Northumberland and Durham for 1952-1956. *Transactions of the Natural History Society of Northumberland, Durham & Newcastle-upon-Tyne* Vol. XI, No. 1, 5 & 8, Vol. XII No.2.

Temperley G. W. and Blezzard E. 1951. Status of the Green Woodpecker in northern England. *British Birds* 44.

Thom V. M. 1986. *Birds in Scotland.* T&AD Poyser, London.

Thomsen K. and Hötker H. 2006. The sixth International White Stork Census: 2004-2005. *Waterbirds around the world.* 493-495. The Stationery Office, Edinburgh.

Thorup O. 2006. Breeding Waders in Europe 2000. *International Wader Studies* 14. International Wader Study Group, UK.

Tiainen J., Vickholm I. K., Pakkala T., Piiroinen J. and Yrjola R. 1989. Clutch size, nestling growth and nestling mortality of the Starling, *Sturnus vulgaris* in south Finnish agro-environments. *Ornis Fenn.* 66: 41-48.

Ticehurst N. F. 1923. Some British Birds in the Fourteenth Century. *British Birds* 17: 29-35.

Ticehurst N. F. and Jourdain Rev. F. C. R. 1911. On the distribution of the Nightingale during the breeding season in Great Britain. *British Birds* 5.

Tilman D., Cassman K. G., Matson P. A., Naylor R. and Polasky S. 2002. Agricultural sustainability and intensive production priorities. *Nature* 418:671-677.

Toms P., Crick H. Q. P. and Shawyer C. R. 2001. The Status of breeding barn owls *Tyto alba* in the United Kingdom 1995-97. *Bird Study* 48: 23-37.

Tomkovich P. S. 1992. An analysis of the geographical variability in Knots *Calidris canutus* based on museum skins. *Wader Study Group Bulletin* 64, Supplement, 17-23.

Townshend D. J. 1982. The Lazarus Syndrome in Grey Plovers. *Wader Study Group Bulletin* 54.

Tristram C. H. B. 1905. 'Birds' in *Victoria County History of Durham.* Vol. I, 175-191. London.

Trinder N. 1975. The Feeding Value of Moorland Plants. *ADAS Science Service 1975*: 176-187.

Tsai J. 2003. *Unpublished thesis on Curlew.* Durham University.

Tubbs C. R. 1974. *The Buzzard.* David and Charles, Newton Abbot.

Turner A. 2009. Climate change: a Swallow's eye view. *British Birds* 102, 3-16.

Turner D. 2010. Counts and breeding success of kittiwakes nesting on man-made structures along the River Tyne. *Seabirds* Vol. 23, 111-126.

Tunstall M. 1783-84. Marmaduke Tunstall unpublished manuscript. Wycliffe on Tees, Durham.

Unwin B. (ed.) 1975. *Durham County Bird Report 1974.* Zoology Department, University of Durham.

Unwin B. (Ed.) 1976. *Durham County Bird Report 1975.* Zoology Department, University of Durham.

Unwin B. (Ed.) 1977. *Birds in Durham 1976.* Durham Bird Club.

Unwin B. (Ed.) 1978. *Birds in Durham 1977.* Durham Bird Club.

Unwin B. 1979. Little Gulls on the Co. Durham Coast. *Birds in Durham 1978.* Durham Bird Club.

Unwin B. (ed.) 1979. *Birds in Durham 1978.* Durham Bird Club.

Unwin B. and Sowerbutts D.L. (eds.) 1980. *Birds in Durham 1979.* Durham Bird Club.

Urwin N. 1999. Pine Bunting *Emberiza leucocephala* at Langley Moor, 22nd March 1998: An addition to the avifauna of Vice Co. Durham. *Birds in Durham 1998.* Durham Bird Club.

Van den Berg A. (ed.) 2007. Vier Veldrietzangers op Vlieland. *Dutch Birding* 29: 343-344.

van Ijzendoorn A. L. J. 1944. On Black-necked Grebe. *Limosa* 17.

The Vasculum 1916. 'Notes & Records' in Vol. II.

The Vasculum 1917. 'Notes & Records' in Vol. III.

The Vasculum 1928. 'Notes & Records' in Vol. XIV.

The Vasculum 1929. 'Notes & Records' in Vol. XV.

The Vasculum 1929. 'Notes & Records' in Vol. XVII.

The Vasculum 1930. 'Notes & Records' in Vol. XVI.

The Vasculum 1931. 'Notes & Records' in Vol. XVII.

The Vasculum 1932 . 'Notes & Records' in Vol. XVIII.

The Vasculum 1932. 'Notes & Records' in Vol. XIX.

The Vasculum 1933. 'Notes & Records' in Vol. XIX.

The Vasculum 1934. 'Notes & Records' in Vol. XIX.

The Vasculum 1934. 'Notes & Records' in Vol. XX.

The Vasculum 1935. 'Notes & Records' in Vol. XXI.

The Vasculum 1936. 'Notes & Records' in Vol. XXII.

The Vasculum 1938. 'Notes & Records' in Vol. XXIV.

The Vasculum 1938. 'Notes & Records' in Vol. XXIV.

The Vasculum 1939. 'Notes & Records' in Vol. XXV.

The Vasculum 1940. 'Notes & Records' in Vol. XXVI.

The Vasculum 1942. 'Notes & Records' in Vol. XXVII.

The Vasculum 1944. 'Notes & Records' in Vol. XXIX.

The Vasculum 1945. 'Notes & Records' in Vol. XXX.

The Vasculum 1946. 'Notes & Records' in Vol. XXXI.

The Vasculum 1949. 'Notes & Records' in Vol. XXXXV.

The Vasculum 1949. 'Notes & Records' in Vol. XXXIV.

The Vasculum 1954. 'Notes & Records' in Vol. XXXIX.

The Vasculum 1959. 'Notes & Records' in Vol. XLIV.

The Vasculum 1959. 'Notes & Records' in Vol. XLIV.

The Vasculum 1960. 'Notes & Records' in Vol. XLV.

The Vasculum 1961. 'Notes & Records' in Vol. XLVI.

The Vasculum 1967. 'Notes & Records' in Vol. LII.

The Vasculum 1968. 'Notes & Records' in Vol. LIII.

The Vasculum 1970. 'Notes & Records' in Vol. LV.

Village A. 1987. Numbers, territory size and turnover of Short-eared Owls in relation to vole abundance. *Ornis Scand* 18.

Village A. 1990. *The Kestrel.* Poyser, London.

Vinnicombe K. and Harrop A. H. J. 1999. Ruddy Shelducks in Britain and Ireland 1986-94. *British Birds* 92: 225-255.

Voous K.H. 1960. *Atlas of European Birds.* Nelson, London.

Walker A. F. G. 1970. The moult migration of Yorkshire Canada Geese. *Wildfowl* 21.

Wallace D. I. M. and Bourne W. R. P. 1981. Seabird movements along the east coast of Britain. *British Birds* 74:417-426.

Wallace D. I. M. 1999. History of the Common Rosefinch in Britain and Ireland, 1869-1996. *British Birds* 92: 445-471.

Wallis J. 1769. *Natural history and antiquities of Northumberland: and of so much of the county of Durham as lies between the rivers Tyne and Tweed: commonly called north Bishoprick.* W & W Stahan, London.

Ward R. M. 2000. Migration patterns and moult of Common Terns *Sterna hirundo* and Sandwich Terns *Sterna sandvicensis* using Teesmouth in late summer. *Ringing & Migration* 20: 19-28. BTO.

Ward R. M., Cranswick P. A., Kershaw M., Austin G. E., Brown A. W., Brown L. M., Coleman J. C., Chisholm H. K. and Spray C. J. 2007. Numbers of Mute Swans *Cygnus olor* in Great Britain: Results of the national census in 2002. *Wildfowl* 57: 3-20.

Ward R. M., Evans P. and O'Connell M. 2003. Study of long term changes in bird usage of the Tees Estuary. *WWT Research Department report to English Nature on behalf of the WWT Wetlands Advisory Service.*

Ward R. M. and Joynt G. J. 2009. Long-billed Dowitcher in 2007 – New to Cleveland. *Cleveland Bird Report for 2008.* Teesmouth Bird Club.

Warren P. 2010. Black Grouse Recovery in northern England. *British Wildlife* 21:6, 383-391.

Warren P. and Baines D. 2008. The current status and recent trends in numbers and distribution of Black Grouse *Tetrao tetrix* in northern England. *Bird Study* 55: 94-99.

Watson H. 1944. Crossbill Breeding in Great Britain in 1943. *British Birds* 37:3.

Wernham C. V., Toms M. P., Marchant J. H., Clark J. A., Siriwardena G. M. and Baillie S. R. (eds.) 2002. *The Migration Atlas: movements of the birds of Britain and Ireland* (BTO). T&AD Poyser, London.

Westerberg A. E., Donnison A. and Muirhead L. 1994. *Assessment of the Ornithological Importance of Northumbrian Water Reservoirs.* Northumbrian Water Report, Pity Me, Durham.

Westerberg S. and Bowey K. (eds.) 2000. *A Summer Atlas of the Breeding Birds of County Durham.* The Durham Bird Club, Durham.

Wheeler D. A. 1992. *Wearside Weather.* Jackdaw Press, Sunderland.

White S. J., McCarthy B. and Jones M. (eds.) 2008. *The Birds of Lancashire and North Merseyside.* Hobby Publications.

Whittingham M. J. 1996. *Habitat Requirements of Breeding Golden Plover Pluvialis apricaria.* Unpubl. PhD Thesis, University of Sunderland.

Wiffen T. 2003. *Density of Dippers on the Derwent.* Dissertation for Certificate in Nature Conservation. Newcastle University.

Williamson K. and Ferguson-Lees I. J. 1958. Recent reports and news. *British Birds* 51:7, 277-284.

Wilson A. M., Vickery J. A. and Browne S. J. 2001. Numbers and Distribution of Lapwings *Vanellus vanellus* breeding in England and Wales in 1998. *Bird Study* 48.

Wilson G. R. and Wilson L. P. 1978. Haematology, weight and condition of captive red grouse (*Lagopus lagopus scoticus*) infected with caecal threadworm (*Trichostrongylus tenuis*). *Research in Veterinary Science* 25.

Wilson R. 1992-1997. *Little Tern Project Reports.* Cleveland County Council, INCA & Cleveland Wildlife Trust.

Winkler H., Christie D. A. and Nurney D. 1995. *Woodpeckers - A Guide to the Woodpeckers, Piculets and Wrynecks of the World.* Pica Press, Sussex.

Witherby H. F. 1911. The Crossbill as a British bird. *British Birds* 4:11.

Witherby H. F. 1929. The British Birds Marking Scheme, Progress for 1928. *British Birds* 22:10.

Witherby H. F. and Alexander C. J. 1911. The 1909 irruption of the Crossbill as observed in the British Isles. *British Birds* 4:11, 326-331.

Witherby H. F., Jourdain F. C. R., Ticehurst N. F. and Tucker B. W. 1938-41. *The Handbook of British Birds*, Four Volumes. Witherby, London.

Witherby H. F. and Leach E. P. 1932. Movements of Ringed Birds from Abroad to the British Isles and from the British Islands Abroad. *British Birds* 25:7.

Willughby F. 1678. *The ornithology of Francis Willughby of Middleton in the County of Warwick*, Esq. Fellow of the Royal Society in three books by John Ray, Fellow of the Royal Society. John Martyn, London.

Winstanley D. R., Spencer R. and Williamson K. 1974. Where have all the Whitethroats gone? *Bird Study* 21.

Wotton S. R., Langston R. H. and Gregory R. D. 2002. The breeding status of the Ring Ouzel *Turdus torquatus* in the UK in 1999. *Bird Study* 49: 26-34.

Wynn R. B. and Yesou P. 2007. The changing status of Balearic Shearwater in northwest European waters. *British Birds* 100, 392-406.

Yalden D. W. 1992. The influence of recreational disturbance on Common Sandpipers *Actitis hypoleuca* breeding by an upland reservoir, in England. *Biological Conservation* 61.

Yalden D. W. 2007. The older history of the White-tailed Eagle in Britain. *British Birds* 100:8, 471-480.

Yalden D. W. and Albarella U. 2009. *The History of British Birds*. Oxford University Press, Oxford.

Yarrell W. 1843. *A History of British Birds*. Van Voorst, London.

Yarrell W. 1885. *A History of British Birds*, 4th Edition, (revised and enlarged by Sanders, H.). Van Voorst. London.

The Zoologist on Black-necked Grebe 1845.

The Zoologist Duff J. on Chaffinch 1847.

The Zoologist on Great Shearwater 1846: 4.

The Zoologist on Snowy Owl 1848: 7.

The Zoologist on Honey Buzzard 1848.

The Zoologist Duff J. on Osprey 1849: 5.

The Zoologist on Tengmalm's Owl 1850: 8.

The Zoologist Duff J. on Osprey 1850.

The Zoologist on Red-backed Shrike 1851.

The Zoologist Duff J. on Gyr Falcon 1851.

The Zoologist on Grey Phalarope 1855:13.

The Zoologist on Black Stork 1862.

The Zoologist on Purple Sandpiper 1875.

The Zoologist on Osprey 1884.

The Zoologist on Honey Buzzard 1895.

The Zoologist on Honey Buzzard 1911.

Index of species' English names

A

Auk
  Great 551
  Little 553
Avocet 348

B

Bee-eater 609
  Blue-tailed 940
Bittern 243
  Little 245
Blackbird 780
Blackcap 719
Bluetail
  Red-flanked 808
Bluethroat
  Red-spotted 810
  White-spotted 812
Brambling 877
Bullfinch 910
  Northern 912
Bunting
  Cirl 925
  Corn 932
  Lapland 919
  Little 927
  Ortolan 925
  Pine 922
  Red-headed 939
  Reed 929
  Rustic 927
  Snow 917
  Yellow-breasted 928
Bustard
  Little 343
Buzzard 298
  Rough-legged 301

C

Capercaillie 192
Chaffinch 874
Chiffchaff 712
  Scandinavian 714
  Siberian 715
Chough 635
Coot 337

Cormorant 235
  Continental 238
  Double-crested 240
Corncrake 331
Cowbird
  Brown-headed 936
Crake
  Baillon's 330
  Little 940
  Spotted 327
Crane 340
  Sandhill 342
Crossbill
  Common 902
  Parrot 906
  Two-barred 902
Crow
  Carrion 647
  Hooded 649
Cuckoo 575
  Great Spotted 575
Curlew 426

D

Dipper 772
  Black-bellied 775
Diver
  Black-throated 209
  Great Northern 211
  Red-throated 206
  White-billed 214
Dotterel 358
Dove
  Collared 566
  Rock 558
  Stock 561
  Turtle 569
Dowitcher
  Long-billed 411
  Short-billed 411
Duck
  Ferruginous 156
  Long-tailed 166
  Mandarin 126
  Ring-necked 155
  Ruddy 182
  Tufted 157
  White-headed 938

Dunlin 397
Dunnock 835

E

Eagle
  Golden 304
  White-tailed 282
Egret
  Cattle 247
  Great White 249
  Little 247
Eider 162
  King 165

F

Falcon
  Gyr 320
  Peregrine 321
  Red-footed 312
Fieldfare 786
Firecrest 657
Flycatcher
  Pied 814
  Red-breasted 813
  Spotted 799
Fulmar 215

G

Gadwall 132
Gannet 233
Garganey 144
Godwit
  Bar-tailed 420
  Black-tailed 416
Goldcrest 654
Goldeneye 171
Goldfinch 883
Goosander 179
Goose
  Barnacle 114
  Bean 98
  Brent 116
  Dark-bellied Brent 118
  Egyptian 119
  European White-fronted 103
  Greater Canada 111
  Greenland White-fronted 105

Greylag	106
Pale-bellied Brent	118
Pink-footed	101
Red-breasted	939
Ross's	938
Snow	109
Taiga Bean	100
Tundra Bean	100
White-fronted	103
Goshawk	293

Grebe

Black-necked	269
Great Crested	262
Little	260
Red-necked	265
Slavonian	267
Greenfinch	879
Greenshank	442

Grosbeak

Pine	909

Grouse

Black	188
Red	184
Guillemot	545
Black	552

Gull

Black-headed	481
Bonaparte's	480
Caspian	509
Common	494
Franklin's	491
Glaucous	514
Glaucous-winged	513
Great Black-backed	516
Herring	503
Iceland	511
Ivory	470
Kumlien's	513
Laughing	490
Lesser Black-backed	499
Little	485
Mediterranean	492
Ring-billed	498
Ross's	489
Sabine's	471
Scandinavian Herring	506
Scandinavian Lesser Black-	
backed	502
Yellow-legged	508

H

Harrier

Hen	286
Marsh	283
Montagu's	291
Northern	290
Pallid	290
Hawfinch	913

Heron

Grey	250
Night	246
Purple	254
Hobby	317
Honey-buzzard	273
Hoopoe	611

I

Ibis

Glossy	257

J

Jackdaw	641
Nordic	643
Jay	639

K

Kestrel	308
Killdeer	357
Kingfisher	605

Kite

Black	277, 278
Kittiwake	473
Knot	371
Great	371

L

Lapwing	367

Lark

Shore	682
Short-toed	678
Linnet	890

M

Magpie	636
Mallard	139

Martin

House	691
Sand	684

Merganser

Red-breasted	176
Merlin	313

Moorhen	334

N

Nighthawk	
Common	599
Nightingale	809
Thrush	808
Nightjar	596
Nuthatch	757

O

Oriole

Golden	625
Osprey	305

Ouzel

Ring	777

Owl

Barn	579
Dark-breasted Barn	581
Little	582
Long-eared	588
Short-eared	592
Snowy	581
Tawny	585
Tengmalm's	595
Oystercatcher	343

P

Parakeet

Ring-necked	573

Partridge

Grey	196
Red-legged	193

Petrel

Fea's/Zino's	218
Leach's	231
Storm	228
Swinhoe's	940
Wilson's	227

Phalarope

Grey	459
Red-necked	458
Wilson's	457
Pheasant	203

Pigeon

Feral	558
Pintail	141

Pipit

Meadow	866
Olive-backed	862
Red-throated	869

Richard's	860	**S**		Skylark	679
Rock	869	Sanderling	376	Smew	173
Scandinavian Rock	872	Sandgrouse		Snipe	406
Tawny	861	Pallas's	557	Great	410
Tree	862	Sandpiper		Jack	404
Water	872	Baird's	387	Sparrow	
Plover		Broad-billed	400	House	838
American Golden	360	Buff-breasted	401	Tree	842
Golden	361	Common	432	Sparrowhawk	296
Grey	364	Curlew	390	Spoonbill	258
Kentish	357	Green	437	Starling	766
Little Ringed	351	Marsh	446	Rose-coloured	771
Pacific Golden	361	Pectoral	388	Stilt	
Ringed	354	Purple	393	Black-winged	347
White-tailed	367	Semipalmated	380	Stint	
Pochard	152	Sharp-tailed	390	Little	381
Red-crested	151	Spotted	436	Long-toed	385
Pratincole		Terek	432	Temminck's	384
Collared	351	White-rumped	386	Stonechat	828
Puffin	556	Wood	446	Siberian	827
		Scaup	159	Stone-curlew	350
Q		Lesser	161	Stork	
		Scoter		Black	255
Quail	199	Common	167	White	256
		Surf	169	Swallow	687
R		Velvet	170	Cliff	941
		Serin	879	Red-rumped	693
Rail		Shag	240	Swan	
Water	324	Shearwater		Bewick's	93
Raven	652	Balearic	225	Mute	89
Razorbill	548	Cory's	219	Whooper	96
Redpoll		Great	220	Swift	600
Arctic	900	Macaronesian	227	Alpine	604
Coue's Arctic	901	Manx	223	Pallid	604
Hornemann's	900	Sooty	222		
Lesser	896	Shelduck	122	**T**	
Mealy	899	Ruddy	121		
Redshank	449	Shoveler	147	Teal	135
Spotted	440	Shrike		Blue-winged	146
Redstart	821	Daurian	627	Green-winged	138
Black	818	Great Grey	631	Tern	
Redwing	794	Isabelline	627	Arctic	541
Robin	802	Lesser Grey	630	Black	524
Continental	804	Red-backed	627	Bridled	518
White-throated	806	Turkestan	627	Caspian	523
Roller	610	Woodchat	634	Common	533
Rook	644	Siskin	886	Gull-billed	522
Rosefinch		Skua		Lesser Crested	533
Common	907	Arctic	463	Little	519
Rubythroat		Great	468	Roseate	539
Siberian	806	Long-tailed	466	Sandwich	528
Ruff	402	Pomarine	461	Sooty	940
				Whiskered	524

White-winged Black 527
Thrush
  Black-throated 785
  Dusky 785
  Grey-cheeked 776
  Mistle 797
  Song 790
  White's 775
Tit
  Bearded 676
  Blue 660
  Coal 668
  Continental Coal 670
  Crested 667
  Great 664
  Long-tailed 695
  Marsh 674
  Penduline 660
  Willow 670
Treecreeper 761
Turnstone 455
Twite 893

V

Vireo
  Red-eyed 624

W

Wagtail
  Amur 859
  Black-headed 851
  Blue-headed 850

Citrine 852
Grey 853
Grey-headed 851
Pied 855
White 858
Yellow 845
Warbler
  Aquatic 741
  Arctic 701
  Barred 725
  Blyth's Reed 746
  Booted 738
  Cetti's 694
  Dusky 709
  Eastern Crowned 699
  Garden 723
  Grasshopper 734
  Great Reed 752
  Greenish 699
  Hume's 707
  Icterine 738
  Marsh 746
  Melodious 740
  Northern Willow 718
  Paddyfield 745
  Pallas's 702
  Pallas's Grasshopper 734
  Radde's 708
  Reed 748
  Savi's 737
  Sedge 741
  Subalpine 733
  Willow 716

Wood 710
Yellow 941
Yellow-browed 704
Waxwing 752
  Cedar 941
Wheatear 831
  Desert 835
  Greenland 833
  Pied 834
Whimbrel 423
Whinchat 824
Whitethroat 730
  Lesser 727
Wigeon 128
  American 131
Woodcock 412
Woodlark 678
Woodpecker
  Great Spotted 617
  Green 614
  Lesser Spotted 621
  Northern Great Spotted 620
Woodpigeon 563
Wren 763
Wryneck 612

Y

Yellowhammer 923
Yellowlegs
  Lesser 445

Index of species' scientific names

A

Accipiter
  gentilis   293
  nisus   296
Acrocephalus
  agricola   745
  arundinaceus   752
  dumetorum   746
  paludicola   741
  palustris   746
  schoenobaenus   741
  scirpaceus   748
Actitis
  hypoleucos   432
  macularius   436
Aegithalos
  caudatus   695
Aegolius
  funereus   595
Aix
  galericulata   126
Alauda
  arvensis   679
Alca
  torda   548
Alcedo
  atthis   605
Alectoris
  rufa   193
Alle
  alle   553
Alopochen
  aegyptiaca   119
Anas
  acuta   141
  americana   131
  carolinensis   138
  clypeata   147
  crecca   135
  discors   146
  penelope   128
  platyrhynchos   139
  querquedula   144
  strepera   132
Anser
  albifrons   103
  albifrons albifrons   103
  albifrons flavirostris   105

anser   106
brachyrhynchus   101
caerulescens   109
fabalis   98
fabalis fabalis   100
fabalis rossicus   100
rossi   938
Anthus
  campestris   861
  cervinus   869
  hodgsoni   862
  petrosus   869
  petrosus littoralis   872
  pratensis   866
  richardi   860
  spinoletta   872
  trivialis   862
Apus
  apus   600
  melba   604
  pallidus   604
Aquila
  chrysaetos   304
Ardea
  alba   249
  cinerea   250
  purpurea   254
Arenaria
  interpres   455
Asio
  flammeus   592
  otus   588
Athene
  noctua   582
Aythya
  affinis   161
  collaris   155
  ferina   152
  fuligula   157
  marila   159
  nyroca   156

B

Bombycilla
  cedrorum   941
  garrulus   752
Botaurus
  stellaris   243
Branta

bernicla   116
bernicla bernicla   118
bernicla hrota   118
canadensis   111
leucopsis   114
ruficollis   939
Bubo
  scandiacus   581
Bubulcus
  ibis   247
Bucephala
  clangula   171
Burhinus
  oedicnemus   350
Buteo
  buteo   298
  lagopus   301

C

Calandrella
  brachydactyla   678
Calcarius
  lapponicus   919
Calidris
  acuminata   390
  alba   376
  alpina   397
  bairdii   387
  canutus   371
  ferruginea   390
  fuscicollis   386
  maritima   393
  melanotos   388
  minuta   381
  pusilla   380
  subminuta   385
  temminckii   384
  tenuirostris   371
Calonectris
  diomedea   219
Caprimulgus
  europaeus   596
Carduelis
  cabaret   896
  cannabina   890
  carduelis   883
  flammea   899
  flavirostris   893
  hornemanni   900

hornemanni exilipes 901
hornemanni hornemanni 900
spinus 886
Carpodacus
erythrinus 907
Catharus
minimus 776
Cecropis
daurica 693
Cepphus
grylle 552
Certhia
familiaris 761
Cettia
cetti 694
Charadrius
alexandrinus 357
dubius 351
hiaticula 354
morinellus 358
vociferus 357
Chlidonias
hybrida 524
leucopterus 527
niger 524
Chloris
chloris 879
Chordeiles
minor 599
Chroicocephalus
philadelphia 480
ridibundus 481
Ciconia
ciconia 256
nigra 255
Cinclus
cinclus 772
cinclus cinclus 775
Circus
aeruginosus 283
cyaneus 286
cyaneus hudsonius 290
macrourus 290
pygargus 291
Clamator
glandarius 575
Clangula
hyemalis 166
Coccothraustes
coccothraustes 913
Columba
livia 558

oenas 561
palumbus 563
Coracias
garrulus 610
Corvus
corax 652
cornix 649
corone 647
frugilegus 644
monedula 641
monedula monedula 643
Coturnix
coturnix 199
Crex
crex 331
Cuculus
canorus 575
Cyanistes
caeruleus 660
Cygnus
columbianus 93
cygnus 96
olor 89

D

Delichon
urbicum 691
Dendrocopos
major 617
major major 620
minor 621

E

Egretta
garzetta 247
Emberiza
aureola 928
bruniceps 939
calandra 932
cirlus 925
citrinella 923
hortulana 925
leucocephalos 922
pusilla 927
rustica 927
schoeniclus 929
Eremophila
alpestris 682
Erithacus
rubecula 802
rubecula rubecula 804

F

Falco
columbarius 313
peregrinus 321
rusticolus 320
subbuteo 317
tinnunculus 308
vespertinus 312
Ficedula
hypoleuca 814
parva 813
Fratercula
arctica 556
Fringilla
coelebs 874
montifringilla 877
Fulica
atra 337
Fulmarus
glacialis 215

G

Gallinago
gallinago 406
media 410
Gallinula
chloropus 334
Garrulus
glandarius 639
Gavia
adamsii 214
arctica 209
immer 211
stellata 206
Gelochelidon
nilotica 522
Glareola
pratincola 351
Grus
canadensis 342
grus 340

H

Haematopus
ostralegus 343
Haliaeetus
albicilla 282
Himantopus
himantopus 347
Hippolais
icterina 738

polyglotta	740	
Hirundo		
rustica	687	
Hydrobates		
pelagicus	228	
Hydrocoloeus		
minutus	485	
Hydroprogne		
caspia	523	

I

Iduna	
caligata	738
Irania	
gutturalis	806
Ixobrychus	
minutus	245

J

Jynx	
torquilla	612

L

Lagopus	
lagopus	184
Lanius	
collurio	627
excubitor	631
isabellinus	627
isabellinus isabellinus	627
isabellinus phoenicuroides	
	627
minor	630
senator	634
Larus	
argentatus	503
argentatus argentatus	506
atricilla	490
cachinnans	509
canus	494
delawarensis	498
fuscus	499
fuscus intermedius	502
glaucescens	513
glaucoides	511
glaucoides kumlieni	513
hyperboreus	514
marinus	516
melanocephalus	492
michahellis	508
pipixcan	491

Larvivora	
calliope	806
Limicola	
falcinellus	400
Limnodromus	
griseus	411
scolopaceus	411
Limosa	
lapponica	420
limosa	416
Locustella	
certhiola	734
luscinioides	737
naevia	734
Lophophanes	
cristatus	667
Loxia	
curvirostra	902
leucoptera	902
pytyopsittacus	906
Lullula	
arborea	678
Luscinia	
luscinia	808
megarhynchos	809
svecica	810
svecica cyanecula	812
Lymnocryptes	
minimus	404

M

Melanitta	
fusca	170
nigra	167
perspicillata	169
Mergellus	
albellus	173
Mergus	
merganser	179
serrator	176
Meriops	
philippinus	940
Merops	
apiaster	609
Milvus	
migrans	277
Molothrus	
ater	936
Morus	
bassanus	233
Motacilla	
alba	855

alba alba	858
alba leucopsis	859
cinerea	853
citreola	852
flava	845
flava feldegg	851
flava flava	850
flava thunbergi	851
Muscicapa	
striata	799

N

Netta	
rufina	151
Numenius	
arquata	426
phaeopus	423
Nycticorax	
nycticorax	246

O

Oceanites	
oceanicus	227
Oceanodroma	
leucorhoa	231
monorhis	940
Oenanthe	
deserti	835
oenanthe	831
oenanthe leucorhoa	833
pleschanka	834
Onychoprion	
anaethetus	518
Oriolus	
oriolus	625
Oxyjura	
leucocephala	938
Oxyura	
jamaicensis	182

P

Pagophila	
eburnea	470
Pandion	
haliaetus	305
Panurus	
biarmicus	676
Parus	
major	664
Passer	
domesticus	838

montanus	842	leucorodia	258	Riparia			
Pastor		Plectrophenax		riparia	684		
roseus	771	nivalis	917	Rissa			
Perdix		Plegadis		tridactyla	473		
perdix	196	falcinellus	257				
Periparus		Pluvialis		**S**			
ater	668	apricaria	361				
ater ater	670	dominica	360	Saxicola			
Pernis		fulva	361	maurus	827		
apivorus	273	squatarola	364	rubetra	824		
Petrochelidon		Podiceps		torquatus	828		
pyrrhonota	941	auritus	267	Scolopax			
Phalacrocorax		cristatus	262	rusticola	412		
aristotelis	240	grisegena	265	Serinus			
auritus	240	nigricollis	269	serinus	879		
carbo	235	Poecile		Setophaga			
carbo sinensis	238	montana	670	petechia	941		
Phalaropus		palustris	674	Sitta			
fulicarius	459	Porzana		europaea	757		
lobatus	458	parva	940	Somateria			
tricolor	457	porzana	327	mollissima	162		
Phasianus		pusilla	330	spectabilis	165		
colchicus	203	Prunella		Stercorarius			
Philomachus		modularis	835	longicaudus	466		
pugnax	402	Psittacula		parasiticus	463		
Phoenicurus		krameri	573	pomarinus	461		
ochruros	818	Pterodroma		skua	468		
phoenicurus	821	feae/madeira	218	Sterna			
Phylloscopus		Puffinus		bengalensis	533		
borealis	701	baroli	227	dougallii	539		
collybita	712	gravis	220	fuscata	940		
collybita abietinus	714	griseus	222	hirundo	533		
collybita tristis	715	mauretanicus	225	paradisaea	541		
coronatus	699	puffinus	223	sandvicensis	528		
fuscatus	709	Pyrrhocorax		Sternula			
humei	707	pyrrhocorax	635	albifrons	519		
inornatus	704	Pyrrhula		Streptopelia			
proregulus	702	pyrrhula	910	decaocto	566		
schwarzi	708	pyrrhula pyrrhula	912	turtur	569		
sibilatrix	710			Strix			
trochiloides	699	**R**		aluco	585		
trochilus	716			Sturnus			
trochilus acredula	718	Rallus		vulgaris	766		
Pica		aquaticus	324	Sylvia			
pica	636	Recurvirostra		atricapilla	719		
Picus		avosetta	348	borin	723		
viridis	614	Regulus		cantillans	733		
Pinguinus		ignicapilla	657	communis	730		
impennis	551	regulus	654	curruca	727		
Pinicola		Remiz		nisoria	725		
enucleator	909	pendulinus	660	Syrrhaptes			
Platalea		Rhodostethia		paradoxus	557		
		rosea	489				

T

Tachybaptus	
ruficollis	260
Tadorna	
ferruginea	121
tadorna	122
Tarsiger	
cyanurus	808
Tetrao	
tetrix	188
urogallus	192
Tetrax	
tetrax	343
Tringa	
erythropus	440
flavipes	445
glareola	446
nebularia	442
ochropus	437
stagnatilis	446
totanus	449
Troglodytes	
troglodytes	763
Tryngites	
subruficollis	401
Turdus	
atrogularis	785
eunomus	785
iliacus	794
merula	780
philomelos	790
pilaris	786
torquatus	777
viscivorus	797
Tyto	
alba	579
alba guttata	581

U

Upupa	
epops	611
Uria	
aalge	545

V

Vanellus	
leucurus	367
vanellus	367
Vireo	
olivaceus	624

X

Xema	
sabini	471
Xenus	
cinereus	432

Z

Zoothera	
dauma	775